Beginning & Intermediate Algebra

Sixth Edition

Elayn Martin-Gay

University of New Orleans

PEARSON

Boston   Columbus   Indianapolis   New York   San Francisco
Amsterdam   Cape Town   Dubai   London   Madrid   Milan   Munich   Paris   Montréal   Toronto
Delhi   Mexico City   São Paulo   Sydney   Hong Kong   Seoul   Singapore   Taipei   Tokyo

Editorial Director, Mathematics: *Christine Hoag*
Editor-in-Chief: *Michael Hirsch*
Acquisitions Editor: *Mary Beckwith*
Project Manager Team Lead: *Christina Lepre*
Project Manager: *Lauren Morse*
Sponsoring Editor: *Matt Summers*
Editorial Assistant: *Megan Tripp*
Development Editor: *Dawn Nuttall*
Program Manager Team Lead: *Karen Wernholm*
Program Manager: *Patty Bergin*
Cover and Illustration Design: *Tamara Newnam*
Program Design Lead: *Heather Scott*
Interior Design: *Integra*
Executive Content Manager, MathXL: *Rebecca Williams*
Associate Content Manager, MathXL: *Eric Gregg*
Senior Content Developer, TestGen: *John Flanagan*
Director of Course Production: *Ruth Berry*
Media Producer: *Audra Walsh*
Senior Marketing Manager: *Rachel Ross*
Marketing Manager: *Jennifer Edwards*
Marketing Assistant: *Alexandra Habashi*
Senior Author Support/Technology Specialist: *Joe Vetere*
Procurement Specialist: *Carol Melville*
Production Management and Composition: *Integra Software Services, Pvt. Ltd.*
Text Art: *Scientific Illustrators*
Answer Art: *Integra Software Services, Pvt. Ltd.*

For permission to use copyrighted material, grateful acknowledgment is made to the copyright holders on page P1 which is hereby made an extension of this copyright page.

PEARSON, ALWAYS LEARNING, and MYMATHLAB are exclusive trademarks in the U.S. and/or other countries owned by Pearson Education, Inc. or its affiliates.

Unless otherwise indicated herein, any third-party trademarks that may appear in this work are the property of their respective owners and any references to third-party trademarks, logos or other trade dress are for demonstrative or descriptive purposes only. Such references are not intended to imply any sponsorship, endorsement, authorization, or promotion of Pearson's products by the owners of such marks or any relationship between the owner and Pearson Education, Inc. or its affiliates, authors, licensees or distributors.

Library of Congress Cataloging-in-Publication Data

Martin-Gay, K. Elayn, 1955–
 Beginning & Intermediate Algebra / Elayn Martin-Gay, University of New Orleans.—6th edition.
 pages cm
 ISBN 0-13-419309-1
1. Algebra—Textbooks. I. Title.
 QA152.3.M36 2017
 512.9—dc23   2015010103

1 2 3 4 5 6 7 8 9 10—RRD-W—20 19 18 17 16

www.pearsonhighered.com

ISBN-10: 0-13-419309-1
ISBN-13: 978-0-13-419309-0

Beginning & Intermediate Algebra

This book is dedicated to my sister—Karen Martin Callac Pasch

There's not enough space on this page to write how wonderful she was while walking this earth.

She is in a better place now; and for that, I celebrate.

Contents

Preface   xiii
Applications Index   xxiii

CHAPTER

1

REVIEW OF REAL NUMBERS   1

1.1 Study Skill Tips for Success in Mathematics   2
1.2 Symbols and Sets of Numbers   8
1.3 Fractions and Mixed Numbers   17
1.4 Exponents, Order of Operations, Variable Expressions, and Equations   26
1.5 Adding Real Numbers   36
1.6 Subtracting Real Numbers   44
  Integrated Review—Operations on Real Numbers   51
1.7 Multiplying and Dividing Real Numbers   52
1.8 Properties of Real Numbers   62
  Chapter 1 Vocabulary Check   69
  Chapter 1 Highlights   69
  Chapter 1 Review   73
  Chapter 1 Getting Ready for the Test   76
  Chapter 1 Test   76

CHAPTER

2

EQUATIONS, INEQUALITIES, AND PROBLEM SOLVING   78

2.1 Simplifying Algebraic Expressions   79
2.2 The Addition and Multiplication Properties of Equality   87
2.3 Solving Linear Equations   97
  Integrated Review—Solving Linear Equations   105
2.4 An Introduction to Problem Solving   106
2.5 Formulas and Problem Solving   117
2.6 Percent and Mixture Problem Solving   128
2.7 Further Problem Solving   140
2.8 Solving Linear Inequalities   147
  Chapter 2 Vocabulary Check   159
  Chapter 2 Highlights   159
  Chapter 2 Review   164
  Chapter 2 Getting Ready for the Test   167
  Chapter 2 Test   168
  Chapter 2 Cumulative Review   169

CHAPTER

3

GRAPHING   171

3.1 Reading Graphs and the Rectangular Coordinate System   172
3.2 Graphing Linear Equations   187
3.3 Intercepts   197
3.4 Slope and Rate of Change   205
  Integrated Review—Summary on Slope and Graphing Linear Equations   219
3.5 Equations of Lines   220
3.6 Functions   229
  Chapter 3 Vocabulary Check   241
  Chapter 3 Highlights   241
  Chapter 3 Review   245
  Chapter 3 Getting Ready for the Test   248
  Chapter 3 Test   249
  Chapter 3 Cumulative Review   251

CHAPTER **4**

SOLVING SYSTEMS OF LINEAR EQUATIONS   252

4.1   Solving Systems of Linear Equations by Graphing   253
4.2   Solving Systems of Linear Equations by Substitution   261
4.3   Solving Systems of Linear Equations by Addition   268
     Integrated Review—Solving Systems of Equations   275
4.4   Solving Systems of Linear Equations in Three Variables   276
4.5   Systems of Linear Equations and Problem Solving   283
     Chapter 4 Vocabulary Check   301
     Chapter 4 Highlights   301
     Chapter 4 Review   304
     Chapter 4 Getting Ready for the Test   306
     Chapter 4 Test   307
     Chapter 4 Cumulative Review   308

CHAPTER **5**

EXPONENTS AND POLYNOMIALS   310

5.1   Exponents   311
5.2   Polynomial Functions and Adding and Subtracting Polynomials   322
5.3   Multiplying Polynomials   334
5.4   Special Products   341
     Integrated Review—Exponents and Operations on Polynomials   348
5.5   Negative Exponents and Scientific Notation   348
5.6   Dividing Polynomials   357
5.7   Synthetic Division and the Remainder Theorem   364
     Chapter 5 Vocabulary Check   368
     Chapter 5 Highlights   369
     Chapter 5 Review   371
     Chapter 5 Getting Ready for the Test   374
     Chapter 5 Test   375
     Chapter 5 Cumulative Review   376

CHAPTER **6**

FACTORING POLYNOMIALS   378

6.1   The Greatest Common Factor and Factoring by Grouping   379
6.2   Factoring Trinomials of the Form $x^2 + bx + c$ 387
6.3   Factoring Trinomials of the Form $ax^2 + bx + c$ and Perfect Square Trinomials   394
6.4   Factoring Trinomials of the Form $ax^2 + bx + c$ by Grouping   402
6.5   Factoring Binomials   407
     Integrated Review—Choosing a Factoring Strategy   414
6.6   Solving Quadratic Equations by Factoring   417
6.7   Quadratic Equations and Problem Solving   426
     Chapter 6 Vocabulary Check   435
     Chapter 6 Highlights   436
     Chapter 6 Review   439
     Chapter 6 Getting Ready for the Test   441
     Chapter 6 Test   442
     Chapter 6 Cumulative Review   442

CHAPTER **7**

RATIONAL EXPRESSIONS   444

7.1   Rational Functions and Simplifying Rational Expressions   445
7.2   Multiplying and Dividing Rational Expressions   455
7.3   Adding and Subtracting Rational Expressions with Common Denominators and Least Common Denominator   464
7.4   Adding and Subtracting Rational Expressions with Unlike Denominators   472
7.5   Solving Equations Containing Rational Expressions   478
     Integrated Review—Summary on Rational Expressions   485
7.6   Proportion and Problem Solving with Rational Equations   486
7.7   Simplifying Complex Fractions   499

Chapter 7 Vocabulary Check 506
Chapter 7 Highlights 506
Chapter 7 Review 510
Chapter 7 Getting Ready for the Test 512
Chapter 7 Test 513
Chapter 7 Cumulative Review 514

CHAPTER

8

MORE ON FUNCTIONS AND GRAPHS 516

8.1 Graphing and Writing Linear Functions 517
8.2 Reviewing Function Notation and Graphing Nonlinear Functions 525
   Integrated Review—Summary on Functions and Equations of Lines 533
8.3 Graphing Piecewise-Defined Functions and Shifting and Reflecting Graphs of Functions 534
8.4 Variation and Problem Solving 542
   Chapter 8 Vocabulary Check 551
   Chapter 8 Highlights 552
   Chapter 8 Review 554
   Chapter 8 Getting Ready for the Test 555
   Chapter 8 Test 556
   Chapter 8 Cumulative Review 558

CHAPTER

9

INEQUALITIES AND ABSOLUTE VALUE 559

9.1 Compound Inequalities 560
9.2 Absolute Value Equations 567
9.3 Absolute Value Inequalities 572
   Integrated Review—Solving Compound Inequalities and Absolute Value Equations and Inequalities 578
9.4 Graphing Linear Inequalities in Two Variables and Systems of Linear Inequalities 578
   Chapter 9 Vocabulary Check 587
   Chapter 9 Highlights 588
   Chapter 9 Review 590
   Chapter 9 Getting Ready for the Test 591
   Chapter 9 Test 592
   Chapter 9 Cumulative Review 592

CHAPTER

10

RATIONAL EXPONENTS, RADICALS, AND COMPLEX NUMBERS 595

10.1 Radicals and Radical Functions 596
10.2 Rational Exponents 605
10.3 Simplifying Radical Expressions 612
10.4 Adding, Subtracting, and Multiplying Radical Expressions 620
10.5 Rationalizing Denominators and Numerators of Radical Expressions 626
   Integrated Review—Radicals and Rational Exponents 632
10.6 Radical Equations and Problem Solving 633
10.7 Complex Numbers 643
   Chapter 10 Vocabulary Check 650
   Chapter 10 Highlights 650
   Chapter 10 Review 654
   Chapter 10 Getting Ready for the Test 656
   Chapter 10 Test 657
   Chapter 10 Cumulative Review 658

CHAPTER

11

QUADRATIC EQUATIONS AND FUNCTIONS 660

11.1 Solving Quadratic Equations by Completing the Square 661
11.2 Solving Quadratic Equations by the Quadratic Formula 671
11.3 Solving Equations by Using Quadratic Methods 681
    Integrated Review—Summary on Solving Quadratic Equations 690
11.4 Nonlinear Inequalities in One Variable 691
11.5 Quadratic Functions and Their Graphs 698
11.6 Further Graphing of Quadratic Functions 706
    Chapter 11 Vocabulary Check 714
    Chapter 11 Highlights 714
    Chapter 11 Review 717
    Chapter 11 Getting Ready for the Test 718
    Chapter 11 Test 719
    Chapter 11 Cumulative Review 720

CHAPTER

12

Exponential and Logarithmic Functions 722

12.1 The Algebra of Functions; Composite Functions 723
12.2 Inverse Functions 728
12.3 Exponential Functions 739
12.4 Exponential Growth and Decay Functions 748
12.5 Logarithmic Functions 752
12.6 Properties of Logarithms 760
    Integrated Review—Functions and Properties of Logarithms 766
12.7 Common Logarithms, Natural Logarithms, and Change of Base 767
12.8 Exponential and Logarithmic Equations and Problem Solving 773
    Chapter 12 Vocabulary Check 779
    Chapter 12 Highlights 780
    Chapter 12 Review 783
    Chapter 12 Getting Ready for the Test 785
    Chapter 12 Test 786
    Chapter 12 Cumulative Review 787

CHAPTER

13

Conic Sections 790

13.1 The Parabola and the Circle 791
13.2 The Ellipse and the Hyperbola 800
    Integrated Review—Graphing Conic Sections 807
13.3 Solving Nonlinear Systems of Equations 808
13.4 Nonlinear Inequalities and Systems of Inequalities 813
    Chapter 13 Vocabulary Check 817
    Chapter 13 Highlights 817
    Chapter 13 Review 820
    Chapter 13 Getting Ready for the Test 821
    Chapter 13 Test 821
    Chapter 13 Cumulative Review 822

CHAPTER

14

Sequences, Series, and the Binomial Theorem 824

14.1 Sequences 825
14.2 Arithmetic and Geometric Sequences 829
14.3 Series 837
       Integrated Review—Sequences and Series 842
14.4 Partial Sums of Arithmetic and Geometric Sequences 842
14.5 The Binomial Theorem 849
       Chapter 14 Vocabulary Check 854
       Chapter 14 Highlights 854
       Chapter 14 Review 856
       Chapter 14 Getting Ready for the Test 858
       Chapter 14 Test 858
       Chapter 14 Cumulative Review 859

APPENDICES

A   OPERATIONS ON DECIMALS/TABLE OF PERCENT,
     DECIMAL, AND FRACTION EQUIVALENTS 861
B   REVIEW OF ALGEBRA TOPICS 864
C   AN INTRODUCTION TO USING A GRAPHING UTILITY 889
D   SOLVING SYSTEMS OF EQUATIONS BY MATRICES 894
E   SOLVING SYSTEMS OF EQUATIONS USING DETERMINANTS 899
F   MEAN, MEDIAN, AND MODE 906
G   REVIEW OF ANGLES, LINES, AND SPECIAL TRIANGLES 908
CONTENTS OF STUDENT RESOURCES 915
     STUDENT RESOURCES 916
     STUDY SKILLS BUILDERS 916
     BIGGER PICTURE—STUDY GUIDE OUTLINE 925
     PRACTICE FINAL EXAM 930

Answers to Selected Exercises A1
Index I1
Photo Credits P1

Preface

***Beginning & Intermediate Algebra,* Sixth Edition** was written to provide a solid foundation in algebra for students who might not have previous experience in algebra. Specific care was taken to make sure students have the most up-to-date, relevant text preparation for their next mathematics course or for nonmathematical courses that require an understanding of algebraic fundamentals. I have tried to achieve this by writing a user-friendly text that is keyed to objectives and contains many worked-out examples. As suggested by AMATYC and the NCTM Standards (plus Addenda), real-life and real-data applications, data interpretation, conceptual understanding, problem solving, writing, cooperative learning, appropriate use of technology, number sense, estimation, critical thinking, and geometric concepts are emphasized and integrated throughout the book.

The many factors that contributed to the success of the previous editions have been retained. In preparing the Sixth Edition, I considered comments and suggestions of colleagues, students, and many users of the prior edition throughout the country.

What's New in the Sixth Edition?

- **New Getting Ready for the Test** can be found before each Chapter Test. These exercises help increase student success by helping students prepare for their chapter test. The purpose of these exercises is to check students' conceptual understanding of the topics in the chapter as well as common student errors. It is suggested that students complete and check these exercises before taking a practice Chapter Test. All Getting Ready for the Test exercises are either Multiple Choice or Matching, and all answers can be found in the answer section of this text.

  Video Solutions of all Getting Ready exercises can be found in MyMathLab and on the Interactive DVD Lecture Series. These video solutions contain brief explanations and reminders of material in the chapter. Where applicable, incorrect choices contain explanations.

  Getting Ready for the Test exercise numbers marked in blue indicate that the question is available in **Learning Catalytics**. 1c

- **New Learning Catalytics** is an interactive student response tool that uses students' smartphones, tablets, or laptops to engage them in more sophisticated tasks and thinking. Generate class discussion, guide your lecture, and promote peer-to-peer learning with real-time analytics. Accessible through MyMathLab, instructors can use Learning Catalytics to:

  — Pose a variety of open-ended questions that help your students develop critical thinking skills.

  — Monitor responses to find out where students are struggling.

  — Use real-time data to adjust your instructional strategy and try other ways of engaging your students during class.

  — Manage student interactions by automatically grouping students for discussion, teamwork, and peer-to-peer learning.

  For *Beginning & Intermediate Algebra,* Sixth Edition, new Getting Ready for the Test exercises marked in blue are available in Learning Catalytics. To search for the questions in Learning Catalytics, select **Discipline: Developmental Math,** and **Book: Martin-Gay, Beginning & Intermediate Algebra, 6e;** or search the question library for **MGCOMBO6e Ch** and the chapter number. For example, search **MGCOMBO6e Ch4** for questions from Chapter 4.

- **New Student Success Tips Videos** are 3- to 5-minute video segments designed to be daily reminders to students to continue practicing and maintaining good organizational and study habits. They are organized in three categories and are available in MyMathLab and the Interactive Lecture Series. The categories are:

1. Success Tips that apply to any course in college in general, such as Time Management.

2. Success Tips that apply to any mathematics course. One example is based on understanding that mathematics is a course that requires homework to be completed in a timely fashion.

3. Section- or Content-specific Success Tips to help students avoid common mistakes or to better understand concepts that often prove challenging. One example of this type of tip is how to apply the order of operations to simplify an expression.

- **New Key Concept Activity Lab Workbook** includes Extension Exercises, Exploration Activities, Conceptual Exercises, and Group Activities. These activities are a great way to engage students in conceptual projects and exploration as well as group work.

- **The Martin-Gay MyMathLab** course has been updated and revised to provide more exercise coverage, including assignable video check questions and an expanded video program. There are section lectures videos for every section, which students can also access at the specific objective level; new Getting Ready for the Test video solutions; new Student Success Tips videos; and an increased number of watch clips at the exercise level to help students while doing homework in MathXL.

  Vocabulary, Readiness & Video Check Questions continue to be available in the text and for assignment in MyMathLab. The **Readiness** exercises center on a student's understanding of a concept that is necessary in order to continue to the exercise set. The **video check questions** are included in every section for every learning objective. These exercises are a great way to assess whether students have viewed and understood the key concepts presented in the videos.

- **Exercise Sets Revised and Updated** The text exercise sets have been carefully examined and revised. Special focus was placed on making sure that even- and odd-numbered exercises are paired and that real-life applications are updated.

Key Continuing Resources and Pedagogical Features

- **Interactive DVD Lecture Series**, featuring your text author Elayn Martin-Gay, provides students with active learning at their own pace. The videos offer the following resources and more:

  A complete lecture for each section of the text highlights key examples and exercises from the text. Pop-ups reinforce key terms, definitions, and concepts.

  An interface with menu navigation features allows students to quickly find and focus on the examples and exercises they need to review.

  Interactive Concept Check exercises measure students' understanding of key concepts and common trouble spots.

  New Student Success Tips Videos.

- **The Interactive DVD Lecture Series** also includes the following resources for test prep:

  New Getting Ready for the Chapter Test Videos

  The Chapter Test Prep Videos help students during their most teachable moment—when they are preparing for a test. This innovation provides step-by-step solutions for the exercises found in each Chapter Test. For the Sixth Edition, the chapter test prep videos are also available on YouTube™. The videos are captioned in English and Spanish.

The Practice Final Exam Videos help students prepare for an end-of-course final. Students can watch full video solutions to each exercise in the Practice Final Exam at the end of this text.

● **The Video Organizer** is designed to help students take notes and work practice exercises while watching the Interactive Lecture Series videos (available in MyMathLab and on DVD). All content in the Video Organizer is presented in the same order as it is presented in the videos, making it easy for students to create a course notebook and build good study habits.

— Covers all of the video examples in order.

— Provides ample space for students to write down key definitions and properties.

— Includes Play and Pause button icons to prompt students to follow along with the author for some exercises while they try others on their own.

The Video Organizer is available in a loose-leaf, notebook-ready format. It is also available for download in MyMathLab. Answers to all video questions are available to instructors in MyMathLab and the Instructor's Resource Center.

Key Pedagogical Features

The following key features have been retained and/or updated for the Sixth Edition of the text:

Problem-Solving Process This is formally introduced in Chapter 2 with a four-step process that is integrated throughout the text. The four steps are **Understand, Translate, Solve,** and **Interpret.** The repeated use of these steps in a variety of examples shows their wide applicability. Reinforcing the steps can increase students' comfort level and confidence in tackling problems.

Exercise Sets Revised and Updated The exercise sets have been carefully examined and extensively revised. Special focus was placed on making sure that even- and odd-numbered exercises are paired.

Examples Detailed, step-by-step examples were added, deleted, replaced, or updated as needed. Many examples reflect real life. Additional instructional support is provided in the annotated examples.

Practice Exercises Throughout the text, each worked-out example has a parallel Practice Exercise. These invite students to be actively involved in the learning process. Students should try each Practice Exercise after finishing the corresponding example. Learning by doing will help students grasp ideas before moving on to other concepts. Answers to the Practice Exercises are provided in the back of the text.

Helpful Hints Helpful Hints contain practical advice on applying mathematical concepts. Strategically placed where students are most likely to need immediate reinforcement, Helpful Hints help students avoid common trouble areas and mistakes.

Concept Checks This feature allows students to gauge their grasp of an idea as it is being presented in the text. Concept Checks stress conceptual understanding at the point of use and help suppress misconceived notions before they start. Answers appear at the bottom of the page. Exercises related to Concept Checks are included in the exercise sets.

Mixed Practice Exercises Found in the section exercise sets, these require students to determine the problem type and strategy needed to solve it just as they would need to do on a test.

Integrated Reviews A unique, mid-chapter exercise set that helps students assimilate new skills and concepts that they have learned separately over several sections. These reviews provide yet another opportunity for students to work with mixed exercises as they master the topics.

Vocabulary Check   Provides an opportunity for students to become more familiar with the use of mathematical terms as they strengthen their verbal skills. These appear at the end of each chapter before the Chapter Highlights. Vocabulary, Readiness, and Video Check exercises provide practice at the section level.

Chapter Highlights   Found at the end of every chapter, these contain key definitions and concepts with examples to help students understand and retain what they have learned and help them organize their notes and study for tests.

Chapter Review   The end of every chapter contains a comprehensive review of topics introduced in the chapter. The Chapter Review offers exercises keyed to every section in the chapter, as well as Mixed Review exercises that are not keyed to sections.

Chapter Test and Chapter Test Prep Video   The Chapter Test is structured to include those problems that involve common student errors. The **Chapter Test Prep Videos** give students instant author access to a step-by-step video solution of each exercise in the Chapter Test.

Cumulative Review   Follows every chapter in the text (except Chapter 1). Each odd-numbered exercise contained in the Cumulative Review is an earlier worked example in the text that is referenced in the back of the book along with the answer.

Writing Exercises ╲   These exercises occur in almost every exercise set and require students to provide a written response to explain concepts or justify their thinking.

Applications   Real-world and real-data applications have been thoroughly updated and many new applications are included. These exercises occur in almost every exercise set, show the relevance of mathematics, and help students gradually and continuously develop their problem-solving skills.

Review Exercises   These exercises occur in each exercise set (except in Chapter 1) and are keyed to earlier sections. They review concepts learned earlier in the text that will be needed in the next section or chapter.

Exercise Set Resource Icons   Located at the opening of each exercise set, these icons remind students of the resources available for extra practice and support:

See Student Resource descriptions page xvii for details on the individual resources available.

Exercise Icons   These icons facilitate the assignment of specialized exercises and let students know what resources can support them.

- ▶ Video icon: exercise worked on the Interactive DVD Lecture Series and in MyMathLab.
- △ Triangle icon: identifies exercises involving geometric concepts.
- ╲ Pencil icon: indicates a written response is needed.
- ▦ Calculator icon: optional exercises intended to be solved using a scientific or graphing calculator.

Optional: Calculator Exploration Boxes and Calculator Exercises   The optional Calculator Explorations provide keystrokes and exercises at appropriate points to give an opportunity for students to become familiar with these tools. Section exercises that are best completed by using a calculator are identified by ▦ for ease of assignment.

Student and Instructor Resources

STUDENT RESOURCES

Interactive DVD Lecture Series Videos

Provides students with active learning at their own pace. The videos offer:

- A complete lecture for each text section. The interface allows easy navigation to examples and exercises students need to review.
- Interactive Concept Check exercises
- Student Success Tips Videos
- Practice Final Exam
- Getting Ready for the Chapter Test Videos
- Chapter Test Prep Videos

Video Organizer

Designed to help students take notes and work practice exercises while watching the Interactive Lecture Series videos.

- Covers all of the video examples in order.
- Provides ample space for students to write down key definitions and rules.
- Includes Play and Pause button icons to prompt students to follow along with the author for some exercises while they try others on their own.

Available in loose-leaf, notebook-ready format and in MyMathLab.

Student Solutions Manual

Provides completely worked-out solutions to the odd-numbered section exercises; all exercises in the Integrated Reviews, Chapter Reviews, Chapter Tests, and Cumulative Reviews.

Key Concept Activity Lab Workbook includes Extension Exercises, Exploration Activities, Conceptual Exercises, and Group Activities.

INSTRUCTOR RESOURCES

Annotated Instructor's Edition

Contains all the content found in the student edition, plus the following:

- Classroom example paired to each example
- Answers to exercises on the same text page
- Teaching Tips throughout the text, placed at key points
- Video Answer Section

Instructor's Resource Manual with Tests and Mini-Lectures

- Mini-lectures for each text section
- Additional Practice worksheets for each section
- Several forms of test per chapter—free response and multiple choice
- Answers to all items

Instructor's Solutions Manual
TestGen® (Available for download from the IRC)

Instructor-to-Instructor Videos—available in the Instructor Resources section of the MyMathLab course.

Online Resources
MyMathLab® (access code required)

MathXL® (access code required)

Get the most out of
MyMathLab®

MyMathLab is the world's leading online resource for teaching and learning mathematics. MyMathLab helps students and instructors improve results and provides engaging experiences and personalized learning for each student so learning can happen in any environment. Plus, it offers flexible and time-saving course-management features to allow instructors to easily manage their classes while remaining in complete control, regardless of course format.

Personalized Support for Students

- MyMathLab comes with many learning resources—eText, animations, videos, and more—all designed to support your students as they progress through their course.

- The Adaptive Study Plan acts as a personal tutor, updating in real time based on student performance to provide personalized recommendations on what to work on next. With the new Companion Study Plan assignments, instructors can now assign the Study Plan as a prerequisite to a test or quiz, helping to guide students through concepts they need to master.

- Personalized Homework allows instructors to create homework assignments tailored to each student's specific needs by focusing on just the topics they have not yet mastered.

Used by nearly 4 million students each year, the MyMathLab and MyStatLab family of products delivers consistent, measurable gains in student learning outcomes, retention, and subsequent course success.

Acknowledgments

Many people helped me develop this text, and I will attempt to thank some of them here. Cindy Trimble was *invaluable* for contributing to the overall accuracy of the text. Dawn Nuttall, Emily Keaton, and Suellen Robinson were *invaluable* for their many suggestions and contributions during the development and writing of this Sixth Edition. Courtney Slade, Chakira Lane, Patty Bergin, and Lauren Morse provided guidance throughout the production process.

A very special thank you goes to my editor, Mary Beckwith, for being there 24/7/365, as my students say. Last, my thanks to the staff at Pearson for all their support: Michael Hirsch, Rachel Ross, Heather Scott, Michelle Renda, Chris Hoag, and Paul Corey.

I would like to thank the following reviewers for their input and suggestions:

Rosalie Abraham, *Florida Community College—Jacksonville*
Ana Bacica, *Brazosport College*
Nelson Collins, *Joliet Junior College*
Nancy Desilet, *Carroll Community College*
Elizabeth Eagle, *University of North Carolina—Charlotte*
Dorothy French, *Community College of Philadelphia*
Sharda Gudehithla, *Wilbur Wright College*
Pauline Hall, *Iowa State University*
Debra R. Hill, *University of North Carolina—Charlotte*
Glenn Jablonski, *Triton College*
Sue Kellicut, *Seminole State College*
Jean McArthur, *Joliet Junior College*
Mary T. McMahon, *North Central College*
Owen Mertens, *Missouri State University*
Jeri Rogers, *Seminole State College*
William Stammerman, *Des Moines Area Community College*
Patrick Stevens, *Joliet Junior College*
Arnavaz Taraporevala, *New York City College of Technology*

I would also like to thank the following dedicated group of instructors who participated in our focus groups, Martin-Gay Summits, and our design review for the series. Their feedback and insights have helped to strengthen this edition of the text. These instructors include:

Billie Anderson, *Tyler Junior College*
Cedric Atkins, *Mott Community College*
Lois Beardon, *Schoolcraft College*
Laurel Berry, *Bryant & Stratton College*
John Beyers, *University of Maryland*
Bob Brown, *Community College of Baltimore County—Essex*
Lisa Brown, *Community College of Baltimore County—Essex*
NeKeith Brown, *Richland College*
Gail Burkett, *Palm Beach State College*
Cheryl Cantwell, *Seminole State College*
Ivette Chuca, *El Paso Community College*
Jackie Cohen, *Augusta State College*
Julie Dewan, *Mohawk Valley Community College*
Monette Elizalde, *Palo Alto College*
Kiel Ellis, *Delgado Community College*
Janice Ervin, *Central Piedmont Community College*
Richard Fielding, *Southwestern College*
Dena Frickey, *Delgado Community College*
Cindy Gaddis, *Tyler Junior College*
Gary Garland, *Tarrant County Community College*
Kim Ghiselin, *State College of Florida*
Nita Graham, *St. Louis Community College*

Kim Granger, *St. Louis Community College*
Pauline Hall, *Iowa State University*
Pat Hussey, *Triton College*
Dorothy Johnson, *Lorain County Community College*
Sonya Johnson, *Central Piedmont Community College*
Ann Jones, *Spartanburg Community College*
Irene Jones, *Fullerton College*
Paul Jones, *University of Cincinnati*
Mike Kirby, *Tidewater Community College*
Kathy Kopelousous, *Lewis and Clark Community College*
Tara LaFrance, *Delgado Community College*
John LaMaster, *Indiana Purdue University Fort Wayne*
Nancy Lange, *Inver Hills Community College*
Judy Langer, *Westchester Community College*
Kathy Lavelle, *Westchester Community College*
Lisa Lindloff, *McLennan Community College*
Sandy Lofstock, *St. Petersburg College*
Nicole Mabine, *North Lake College*
Jean McArthur, *Joliet Junior College*
Kevin McCandless, *Evergreen Valley College*
Ena Michael, *State College of Florida*
Armando Perez, *Laredo Community College*
Davidson Pierre, *State College of Florida*
Marilyn Platt, *Gaston College*
Chris Riola, *Moraine Valley Community College*
Carole Shapero, *Oakton Community College*
Janet Sibol, *Hillsborough Community College*
Anne Smallen, *Mohawk Valley Community College*
Barbara Stoner, *Reading Area Community College*
Jennifer Strehler, *Oakton Community College*
Ellen Stutes, *Louisiana State University Eunice*
Tanomo Taguchi, *Fullerton College*
Robyn Toman, *Anne Arundel Community College*
MaryAnn Tuerk, *Elgin Community College*
Walter Wang, *Baruch College*
Leigh Ann Wheeler, *Greenville Technical Community College*
Darlene Williams, *Delgado Community College*
Valerie Wright, *Central Piedmont Community College*

A special thank you to those students who participated in our design review: Katherine Browne, Mike Bulfin, Nancy Canipe, Ashley Carpenter, Jeff Chojnachi, Roxanne Davis, Mike Dieter, Amy Dombrowski, Kay Herring, Todd Jaycox, Kaleena Levan, Matt Montgomery, Tony Plese, Abigail Polkinghorn, Harley Price, Eli Robinson, Avery Rosen, Robyn Schott, Cynthia Thomas, and Sherry Ward.

Elayn Martin-Gay

About the Author

Elayn Martin-Gay has taught mathematics at the University of New Orleans for more than 25 years. Her numerous teaching awards include the local University Alumni Association's Award for Excellence in Teaching, and Outstanding Developmental Educator at University of New Orleans, presented by the Louisiana Association of Developmental Educators.

Prior to writing textbooks, Elayn Martin-Gay developed an acclaimed series of lecture videos to support developmental mathematics students in their quest for success. These highly successful videos originally served as the foundation material for her texts. Today, the videos are specific to each book in the Martin-Gay series. The author has also created Chapter Test Prep videos to help students during their most "teachable moment"—as they prepare for a test—along with Instructor-to-Instructor videos that provide teaching tips, hints, and suggestions for each developmental mathematics course, including basic mathematics, prealgebra, beginning algebra & intermediate algebra. Her most recent innovations are the Algebra Prep Apps for the iPhone and iPod Touch. These Apps embrace the different learning styles, schedules, and paces of students and provide them with quality math tutoring.

Elayn is the author of 12 published textbooks as well as multimedia interactive mathematics, all specializing in developmental mathematics courses. She has participated as an author across the broadest range of educational materials: textbooks, videos, tutorial software, and courseware. This offers an opportunity of various combinations for an integrated teaching and learning package offering great consistency for the student.

Applications Index

A

Academics. *See* Education
Agriculture
    bug spray mixtures, 497, 750
    combine rental fees, 857
    cranberry-producing states, 16, 137
    DDT pesticides, 750
    farm sizes in U.S., 184, 680
    farmland prices, 219
    farms, number of, 138, 307
    weed killer mixtures, 497
Animals & Insects
    bear populations, 784
    beetle species, 114
    bison populations, 751
    bug sprays, 497, 750, 848, 857
    cheetah running speeds, 461
    condor populations, 785
    crane births, 857
    cricket chirps, 116, 126, 127
    DDT pesticides, 750
    dog medicine dosages, 240, 532
    dog run width, 119
    fish tank dividers, 638
    flying fish speeds, 128
    goldfish numbers in tanks, 126
    gorilla births, 839
    grasshopper species, 114
    hyenas overtaking giraffes, 499
    insecticides, 848, 857
    mosquitoes, 747, 767, 857
    opossum deaths, 841
    otter births, 841
    owl populations, 841
    pen dimensions, 127, 679, 813
    pet types owned in U.S., 130
    pet-related expenditures, 183
    pine beetle infestations, 856
    piranha fish tank dimensions, 126
    prairie dog populations, 787, 932
    puppy weight gain, 827
    rat populations, 751
    sparrow populations, 828
    wolf populations, 778
    wood duck populations, 787
Astronomy & Space
    alignment of planets, 471
    comet distance from Earth, 355
    gamma ray conversion by Sun, 356
    Jupiter, 373
    light travel time/distance, 127–128, 356
    magnitude of stars, 16–17
    meteorite weights, 96, 114
    Milky Way, 373
    moon's light reaching Earth, 357
    moon's surface area, 640
    orbit of planets and comets, 806–807
    planet temperatures, 61
    Sun's light reaching Earth, 357
    telescope elevation above sea level, 355
    weight of objects in relation to Earth's center, 549
    weights on Earth *vs.* other planets, 495
Automobiles
    age of, 218
    bus speeds, 145, 494, 497, 498
    car speeds, 145, 493–494, 496, 497, 498, 511, 549, 640, 658
    compact cars, cost of operating, 218
    dealership discounts, 136
    driver's licenses, 195
    fatalities, 298
    fuel economy, 218
    motorcycle speeds, 498
    registered vehicles on road, 138
    sales, 228, 848
    traffic tickets, 146, 497
    used car values, 138, 181, 554
Aviation
    airplane seats, 876
    airplane speed in still air, 296, 497, 498
    airport elevations, 50
    airport traffic, 718, 870, 877
    hang glider flight rate, 128
    hypersonic flight time around Earth, 128
    jet *vs.* car distances, 497
    jet *vs.* propeller plane speeds, 145, 497
    runway length, 127
    SpaceShipOne rocket plane speed, 463
    vertical elevation changes, 50
    wind speeds, 296, 496, 497, 498

B

Business & Industry
    advertising, 220, 848
    balancing company books, 498
    book store closures, 228
    break-even point, 147, 291–292, 299
    car rental fees, 296, 587
    charity donations, 845
    Coca-Cola production, 137
    Coca-Cola sign dimensions, 124
    consulting fees, 511
    Cyber Monday, 746
    defective products, 514
    delivery service daily operating costs, 642
    depreciation of copiers, 827
    diamond production, 114, 532

Business & Industry (*continued*)

discounts, 131, 136, 166

downsizing, 138, 165, 168

Dunkin' Donuts stores, 228

employee age, 274

employee production numbers and hourly wages, 185

employment decline, 524, 751, 875, 876

employment growth, 167, 228, 298, 524, 875

faxes and fax machines, 848, 876

food manufacturing plants, 137

gross profit margin, 454

group/bulk pricing, 287–288, 297

Home Depot revenue, 195

home prices, 524

hourly minimum wage, 238–239, 533

labor estimates, 491–492, 494, 496, 497, 498, 499, 511, 550, 684–685, 689, 691, 717, 720

laundromat prices, 213

manufacturing costs, 245, 299, 434, 450–451, 453, 510, 511, 513, 549, 713, 728, 788

manufacturing volumes, 204, 355

markup and new price, 166

NASDAQ sign dimensions, 124

net income, 43, 77

net sales, 176

occupations predicted to increase, 275

online shopping, 706, 874

original price after discount, 166

percent increase/decrease, 136, 166, 167

postage for large envelopes, 239

price and demand, 670, 813

price decrease and new price, 138, 872

price per items purchased, 294, 296, 659

price to sales ratio, 524

pricing and sales relationship, 228–229, 297

profits, 228, 454, 524, 713, 728

proofreading rates, 497

quantity pricing, 184, 245

restaurant employees, 874

restaurant sales, 213, 680

restaurants in U.S., 228, 524

retail sales, 706

revenue, 195, 299, 372, 453, 728, 857

salary after pay raise, 136

salary growth, 832, 836, 841, 848, 856, 857

sale prices, 137–138, 224, 752

sales tax, 859

sales volume, predicting, 224–225

volume of items sold at original *vs.* reduced prices, 297

Walmart stores, 186

word processing, 587, 684–685

work rates, 491–492, 494, 496, 497, 498, 499, 511, 550, 684–685, 689, 691, 717, 720

years on market and profit relationship, 228

C

Cars. *See* Automobiles

Chemistry

Avogadro's number, 356

eyewash stations, 134

freezing and boiling points of water, 15

gas pressure and Boyle's law, 545, 555

greenhouse gases, 746

lotion mixtures, 139

methane gas emissions, 713–714

nickel, half-life of, 752

nuclear waste, 746

pH of liquids, 760

radioactive material, 744, 746, 751, 760, 836, 841, 857

solution mixtures, 133–134, 136, 138, 162, 166, 251, 290–291, 296, 297, 299, 300, 305, 307, 497, 658, 930

sulfur dioxide emissions, 516, 523

uranium, half-life of, 752

Communications & Technology

area codes, 111, 168, 930

cell phone discounts, 131

cell phone use, 78, 166, 611

computer assembly, 848

computer discounts, 872

computer rentals, 848

computer values, 180–181

country codes, 114

digital media use, 298

Dish Network subscribers, 250

email, 874

engineers, 193

faxes and fax machines, 848, 876

Google searches, 373

households with computers, 195–196, 217

Internet advertising, 220

Internet crime complaints, 136

Internet usage, 166, 172, 310, 333, 680, 877

light bulbs, 877

mobile devices, time spent on, 137

music streaming, 387

newspaper circulation figures, 228

radio stations in U.S., 268

security keypads, 813

smart televisions, 787

social media, 402, 559, 577

software revenue, 372

switchboard connections, 434

television assembly, 857

Wi-Fi enabled cell phones, 713, 864, 875, 878

ZIP codes, 875

Construction & Home Improvement

balsa wood stick lengths, 443, 658

baseboard and carpeting measurements, 124

beam lengths, 113

beams, 113, 333, 550

blueprint measurements, 495

board lengths, 92, 95, 104, 108, 113, 115, 165, 477

board pricing, 184

building values, 554

carpet rolls, 843

column weight, 547–548, 550

computer desk length, 95

dams, 660

deck dimensions, 168, 442, 497
doors, 679
fencing, 125, 251, 300, 873
fertilizer needs, 126
gardens, 116, 119, 125, 251, 300, 435, 492, 840, 859
golden ratio, 679
grass seed, 125
housing starts *vs.* housing completions, 566
ladders, 433
lawn care, 125, 126
measurement conversions, 460–461, 463
molding lengths, 75, 333
painting houses, 511
picture frames, 125, 876
pipe length, 656
roofing pitch, 212, 217, 218
roofing time, 721
rope lengths, 93, 112
sewer pipe slope, 217
siding section lengths, 115
spotlight placement, 640
sprinklers, 689
stained glass windows, 679
steel section lengths, 112
string/wire lengths, 93, 95, 114, 115, 167, 433, 604, 640
swimming pools, 165, 321, 363, 435, 497
trees planted, 840
wall border, 125
washer circumference, 158
wire placement, 637–638, 640

D

Demographics
age groups predicted to increase on workforce, 274
bill collectors, 298
birth rate in U.S., 138
child care centers, 76
driver's licenses, 195
engineers, 193
Internet usage, 166, 172, 310, 333
joggers, 195
metropolitan populations, 869, 876
occupations predicted to increase, 275
octuplet birth weights, 74
pet types owned in U.S., 130
population growth, 748–749, 751, 784, 785, 787, 828, 834, 859, 876
population per square mile of land, 228
population size, 775, 778, 779, 787
postal carriers, 298
registered nurses, 192
water use per person, 250, 549
world population, 356
istance. *See* Time & Distance

E

Economics & Finance. *See also* Personal Finances
coin/bill denominations, 142–143, 145, 146, 166, 294, 296, 305, 306

compound interest, 666–667, 669, 743, 747, 770, 772–773, 776, 778, 783, 784, 785, 787, 859, 860
interest rates, 36, 434, 666–667, 669, 670, 717, 743, 747, 770, 772–773, 776, 778, 783, 784, 785, 787, 859, 860
investment amounts, 143–144, 145, 146, 166, 168, 295, 514
loans, money needed to pay off, 321
money problems, 142–143
national debts, 356
shares of stock owned, 296
simple interest, 145, 146
stamp denominations, 296, 306
stock market gains and losses, 61, 73, 75, 77
stock prices, 296
Education
ACT Assessment scores, 166, 300
admission rates, 15
alumni donations, 844
associate degrees, 246, 378
bachelor's degrees, 267–268, 378
book page numbers, 114
classrooms, 96, 114, 496
college budgeting, 155
combination lock codes, 114
desired employment benefits, 138
graduate and undergraduate student enrollment, 15, 96, 527–528, 828
high school graduates, 387
hours spent studying, 184
Internet access in classrooms, 138
IQ scores, 642
learning curves, 778
president salaries, 876
students per teacher, 183
study abroad students, 746
summer school students, 751
test scores, 158, 567, 875
textbook prices, 876
tuition and fees, 132, 247
Entertainment & Recreation
allowances, 828
auditorium seats, 836, 856
card game scores, 50
casino gaming, 461
deep-sea diving, 15
diving, 15, 61
DVD sale prices, 166
Easter eggs, 158
exercise bikes, 836
Ferris wheels, 799
fund-raiser attendance, 297
gambling, 848
group rate admissions to events, 287–288
hang gliders, 128, 429
ice sculpting, 843
iTunes expenditures, 186
jogging, 195, 305, 496, 688
movie admission prices, 185, 204
movie industry revenue, 183
movie patron ages, 877

Entertainment & Recreation (*continued*)
movie theater screens, 26, 113, 133, 204
movie theater seats, 828, 857
movie ticket sales, 250
museums and art galleries, 73
music CDs, 136, 450–451
music streaming, 387
national park visits, 245, 331–332, 434
Netflix growth, 722, 742–743, 746
ping-pong tables, 363
pool, 848
poster contests, 679
pyramids formed by surfers, 841
Redbox rentals, 488–489
sail dimensions, 126, 428–429, 440, 497, 515
smart televisions, 787
snowboarding, 872
summer camp tournaments, 784
swimming, 165
tickets sold by type, 145, 287–288, 305
tourism expenditures, 217
tourist destinations, 171, 182
video games, 116
zorbing, 595, 620

F
Finance. *See* Economics & Finance; Personal Finances
Food & Nutrition
barbecues, 471
breakfast item prices, 305
calories burned while walking/bicycling, 157
calories in food items, 495, 497
candy mixtures, 300, 305
cheese consumption and production, 298, 572, 746
coffee blends, 137, 297
cook preparation time, 498
dinner cost with tip, 136
drink machines, coin denominations in, 86
fishery products, domestic and imported, 252, 260
frozen yogurt store revenue, 857
fruit companies, 228
grocery store displays, 836
liter-bottles of Pepsi, 489
nut mixtures, 137, 297, 497
nutrition labels, 139
pepper hotness (Scoville units), 139
percent decrease/increase of consumption, 138
pizza sizes, 126
rabbit food mixtures, 299
red meat and poultry consumption, 283–284
restaurant sales, 213
trail mix ingredients, 139
vitamin A and body weight, 681
yogurt production, 248

G
Geography
continent/regional percentage of Earth's land, 136
desert areas, 96, 114

earthquake magnitudes, 768–769, 772, 874
elevation, 10, 15, 42, 47, 50, 61
federally owned land, 874
Newgrange tomb, 790, 799
ponds, 494, 511, 656, 688
river length, 96
river lenth, 96
rope needed to wrap around Earth, 126
Sarsen Circle of Stonehenge, 798–799
state counties, 115
tallest buildings in U.S., 907
tornado classification, 168
volcano heights, 161
volcano surface area, 620
wildfires, 177
Geology
diamond production, 114, 532
glacier flow rates, 117–118, 128
lava flow rates, 118, 127
mixtures, 138
stalactites and stalagmites, 128
Geometry
angle measurements, 15, 50, 74, 95, 96, 109–110, 113, 114, 115, 116, 293–294, 297, 299, 300, 308, 478, 485, 649, 875, 876
area, 24, 35–36, 74, 127, 136, 138, 320, 332, 339, 340, 346, 347, 356, 363, 368, 373, 374, 376, 387, 402, 432, 433, 440, 464, 477, 510, 532, 546, 626, 641
billboard dimensions, 127, 165
boxes/cubes, 36, 122, 127, 320, 321, 339, 356, 368, 372, 532, 689
circles, 24–25, 74, 158, 320, 432, 532, 550–551
circumference, 158, 550–551
complementary angle measurements, 50, 93, 95, 115, 297, 478, 485
cones, 550, 620, 632
cylinders, 320, 546, 551
Fibonacci sequence, 824, 829
flag dimensions, 113
fraction representations in, 24–25, 74
geodesic dome measurements, 115
golden rectangles, 116
hang glider dimensions, 429
Hoberman Sphere volume, 127
parallelograms, 15, 113, 127, 138, 320, 363, 368, 432, 515
Pentagon floor space dimensions, 115, 463
pentagons, 105, 126
percent decrease/increase problems, 136, 138
perimeter, 25, 35–36, 74, 86, 104, 105, 122–123, 126–127, 157, 165, 196, 294, 299, 305, 306, 333, 363, 374, 393, 402, 407, 432, 440, 441, 471, 477, 510, 546, 604, 625–626, 812, 873, 877
polygons, 546
Pythagorean theorem, 430–431, 636–638, 914
quadrilaterals, 96, 114, 299, 300, 432, 440
radius, 432, 532, 632
rectangles, 24, 35–36, 86, 116, 122–123, 136, 157, 165, 196, 294, 320, 339, 340, 346, 373, 374, 393, 432, 433, 434, 440, 441, 464, 477, 625, 679, 873
sail dimensions, 126, 428–429, 440, 497, 515
sign dimensions, 120–121, 124, 125, 298, 877

spheres, 549, 555, 632

squares, 136, 320, 339, 346, 363, 373, 402, 432, 433, 440, 471, 670, 873

supplementary angle measurements, 50, 93, 95, 115, 297, 478, 485

surface area, 321, 334, 372, 546, 555, 620, 640

trapezoids, 432, 471, 625, 626

triangles, 15, 24, 36, 86, 96, 104, 105, 113, 114, 115, 116, 127, 138, 157, 293, 294, 298, 299, 300, 305, 306, 308, 339, 356, 374, 430–431, 432, 434, 440, 441, 442, 464, 490, 495, 498, 511, 512, 514, 604, 625, 639, 641, 649, 670, 679, 717, 824, 875, 876, 877, 913–914

Vietnam Veterans Memorial angle measurements, 109–110

volume, 36, 122, 127, 320, 321, 339, 356, 363, 368, 532, 550, 551, 632

Washington Monument height and base, 165

Government. *See* Politics & Government

H

Health & Medicine

bacterial cultures, 828, 834, 841

basal metabolic rate, 611

blinking rate of human eye, 116

body mass index, 454

body surface area of humans, 604

breast cancer pink ribbons, 127

cephalic index, 454

dog medicine dosages, 240, 532

flu epidemics, 778

fungal cultures, 841

hospital heights, 877

infectious diseases, 828

kidney transplants, 246

medication administration, 97, 453, 477

octuplet birth weights, 74

organ transplants, 219, 246

pediatric dosages, 453, 477

radiation, 784

registered nurses, 192

smoking and pulse rate, 173

treadmills, 131

virus cultures, 836

woman's height given femur bone length, 240, 532

yeast cultures, 856, 857

Home Improvement. *See* Construction & Home Improvement

I

Industry. *See* Business & Industry

Insects. *See* Animals & Insects

M

Medicine. *See* Health & Medicine

Nutrition. *See* Food & Nutrition

P

Personal Finances

bank account balances, 47, 295, 649–650

bankruptcy, 514

charge account balances, 50

donations, 844–845

interest rates, 36, 434, 666–667, 669, 670, 717, 743, 747, 770, 772–773, 776, 778, 783, 784, 785, 787, 859, 860

loans, money needed to pay off, 321

money problems, 142–143

retirement party budgeting, 157

salary after pay raise, 136

salary growth, 832, 836, 841, 848, 856, 857

sales needed to ensure monthly salary, 166

savings accounts, 15, 295, 848

wedding budget, 155, 157, 587

Physics

angstroms, 373

angular frequency of oscillations, 612

currents and resistance, 549

Doppler effect, 505

Earth's interior temperature, 355

force exerted by tractors, 641

Hoberman Sphere volume, 127

horsepower, 550, 551

pendulum arc, 836, 841, 846, 856, 859

pendulum period, 641

speed of waves traveling over stretched string, 612

springs stretching and Hooke's law, 543–544

velocity, 604, 658

weight of objects in relation to Earth's center, 549

wind power generated, 498

Politics & Government

Democrats *vs.* Republicans, 109

governors, 109

mayoral elections, 95

national debts, 356

representatives, 109, 251

Supreme Court decisions, 138

R

Real Estate

condominium sales and price relationships, 225

depreciation, 229

plot perimeter, 104

property values, 836

Recreation. *See* Entertainment & Recreation

S

Safety. *See* Transportation & Safety

School. *See* Education

Space. *See* Astronomy & Space

Sports

baseball earned run average, 505

baseball game admissions, 288

baseball game attendance, 260

baseball Hall of Fame admittance, 16

baseball payroll and team wins, 557

baseball runs batted in, 295

baseball slugging percentage, 454

baseball team wins, 877

basketball player heights, 157

Sports (*continued*)
  basketball points scored, 295, 299–300
  bowling average, 157
  disc throwing records, 139
  football stadiums, 876
  football yards lost/gained, 61, 77
  golf flags, 440
  golf scores, 43, 58, 167
  golf tournament participants, 749–750
  hockey payrolls, 876
  ice hockey penalty killing percentage, 477
  NASCAR grandstand seats, 876
  NASCAR speeds, 690
  Olympics, 114, 461, 877
  quarterback rating, 454
  racquetball, 856
  stock cars, 463
  Super Bowl attendance, 182
  Tour de France, 166

T

Technology. *See* Communications & Technology
Temperature & Weather
  atmospheric pressure, 747, 778
  average temperatures, 43, 51, 127, 234, 250
  changes in, 40, 42, 50, 61, 77
  Earth's interior temperature, 355
  highest and lowest temperatures, 40, 42, 50, 127, 166, 680
  inequality statements regarding, 15
  of planets, 61
  rainfall data, 300
  snowfall at distances from Equator, 184
  sunrise times, 233
  sunset times, 238
  temperature conversions, 119–120, 121, 123, 125, 127, 166, 567, 724
  thermometer readings, 38
  tornado classification, 168
  tornadoes, 168, 874
Time & Distance
  airplane speed in still air, 296, 497, 498
  bicycling speeds, 496, 688
  bicycling travel time, 140, 296
  boat speed in still water, 305, 496, 497, 511, 514
  boats traveling apart at right angles, 435
  bus speeds, 145, 494, 497, 498
  car speeds, 145, 493–494, 496, 497, 498, 511, 549, 640, 658
  catamaran auto ferry speed, 125
  comet distance from Earth, 355
  conveyor belt speeds, 496
  current speeds, 296, 305
  Daytona 500 speeds, 690
  distance saved, 675–676, 678–679, 720, 931
  distance traveled over time, 166, 717
  driving distance, 145
  driving speeds, 36, 146, 493–494, 496, 497, 498, 685–686, 688
  driving time, 125, 127

  dropped/falling objects, 35, 228, 325, 331, 372, 376, 413–414, 433, 434, 440, 441, 442, 524, 557, 641, 669–670, 676–677, 679–680, 828, 836, 846, 848, 856, 857
  free-fall time/distance, 427, 848, 859
  hiking trails, 25, 141, 308
  hyenas overtaking giraffes, 499
  hypersonic flight time around Earth, 128
  jet *vs.* car distances, 497
  lakes/ponds, distance across, 656, 800
  light intensity by distance from source, 549, 550
  light travel time/distance, 127–128, 356
  moon's light to reach Earth, 357
  motorcycle speeds, 498
  objects traveling in opposite directions, 146, 168, 288–290, 297, 308, 496, 822, 930
  of images and objects to focal length, 444
  pendulum swings, 836, 841, 846, 856, 859
  rate and, 117–118
  rope needed to wrap around Earth, 126
  rowing against current, 496
  rowing distance, 146
  rowing rate in still water, 296
  sight distance from a height, 549, 641
  Sun's light to reach Earth, 357
  thrown/launched objects, 393, 426, 433, 440, 697, 712–713, 717, 718, 720, 859, 931
  traffic tickets, 146, 497
  train travel speeds, 115, 128, 141–142, 166, 168, 496, 930
  travel time, 140–141
  walking/running speeds, 305, 496, 688, 691
  walking/running time, 166, 296, 305
  wind speeds, 296, 496, 497, 498, 550
Transportation & Safety
  bridge lengths, 95
  bridges, 220, 800, 822
  bus speeds, 145, 494, 497, 498
  car speeds, 145, 493–494, 496, 497, 498, 511, 549, 640, 658
  catamaran auto ferry speed, 125
  cell phone use while driving, 166
  cloverleaf exits, 658
  grade of roads/railroad tracks, 213, 217, 377
  interstate highway length, 96
  motorcycle speeds, 498
  parking lot dimensions, 125
  railroad tracks, 213, 217
  road sign dimensions, 120–121, 125, 298, 377, 877
  taxi cab fares, 586
  traffic tickets, 146, 497
  train fares for children and adults, 295
  wheelchair ramps, 217
  yield signs, 125

V

Vehicles. *See* Automobiles

W

Weather. *See* Temperature & Weather

CHAPTER 1

Review of Real Numbers

1.1 Study Skill Tips for Success in Mathematics

1.2 Symbols and Sets of Numbers

1.3 Fractions and Mixed Numbers

1.4 Exponents, Order of Operations, Variable Expressions, and Equations

1.5 Adding Real Numbers

1.6 Subtracting Real Numbers

Integrated Review–Operations on Real Numbers

1.7 Multiplying and Dividing Real Numbers

1.8 Properties of Real Numbers

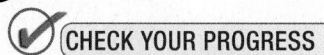

CHECK YOUR PROGRESS

Vocabulary Check

Chapter Highlights

Chapter Review

Getting Ready for the Test

Chapter Test

In this chapter, we review the basic symbols and words—the language—of arithmetic and introduce using variables in place of numbers. This is our starting place in the study of algebra.

A Selection of Resources for Success in this Mathematics Course

Textbook

Instructor

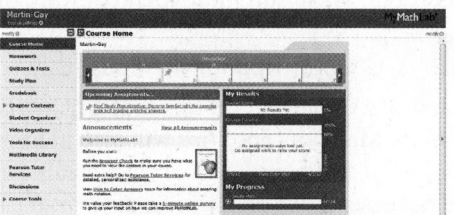

MyMathLab and MathXL

Video Organizer

Video Organizer

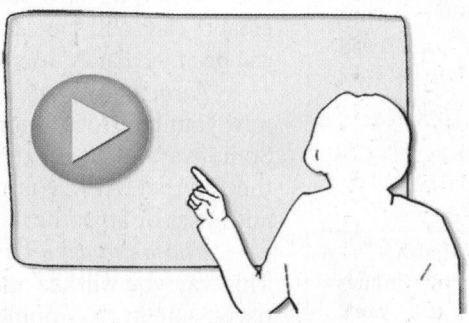

Interactive Lecture Series

For more information about the resources illustrated above, read Section 1.1.

1.1 Study Skill Tips for Success in Mathematics

OBJECTIVES

1 Get Ready for This Course.

2 Understand Some General Tips for Success.

3 Know How to Use This Text.

4 Know How to Use Text Resources.

5 Get Help as Soon as You Need It.

6 Learn How to Prepare for and Take an Exam.

7 Develop Good Time Management.

Before reading Section 1.1, you might want to ask yourself a few questions.

1. When you took your last math course, were you organized? Were your notes and materials from that course easy to find, or were they disorganized and hard to find—if you saved them at all?

2. Were you satisfied—really satisfied—with your performance in that course? In other words, do you feel that your outcome represented your best effort?

If the answer is "no" to these questions, then it is time to make a change. Changing to or resuming good study skill habits is not a process you can start and stop as you please. It is something that you must remember and practice each and every day. To begin, continue reading this section.

OBJECTIVE

1 Getting Ready for This Course

Now that you have decided to take this course, remember that a *positive attitude* will make all the difference in the world. Your belief that you can succeed is just as important as your commitment to this course. Make sure you are ready for this course by having the time and positive attitude that it takes to succeed.

Make sure that you are familiar with the way that this course is being taught. Is it a traditional course, in which you have a printed textbook and meet with an instructor? Is it taught totally online, and your textbook is electronic and you e-mail your instructor? Or is your course structured somewhere in between these two methods? (Not all of the tips that follow will apply to all forms of instruction.)

Also make sure that you have scheduled your math course for a time that will give you the best chance for success. For example, if you are also working, you may want to check with your employer to make sure that your work hours will not conflict with your course schedule.

On the day of your first class period, double-check your schedule and allow yourself extra time to arrive on time in case of traffic problems or difficulty locating your classroom. Make sure that you are aware of and bring all necessary class materials.

OBJECTIVE

2 General Tips for Success

Below are some general tips that will increase your chance for success in a mathematics class. Many of these tips will also help you in other courses you may be taking.

Most important! Organize your class materials. In the next couple pages, many ideas will be presented to help you organize your class materials—notes, any handouts, completed homework, previous tests, etc. In general, you MUST have these materials organized. All of them will be valuable references throughout your course and when studying for upcoming tests and the final exam. One way to make sure you can locate these materials when you need them is to use a three-ring binder. This binder should be used solely for your mathematics class and should be brought to each and every class or lab. This way, any material can be immediately inserted in a section of this binder and will be there when you need it.

Form study groups and/or exchange names and e-mail addresses. Depending on how your course is taught, you may want to keep in contact with your fellow students. Some ways of doing this are to form a study group—whether in person or through the Internet. Also, you may want to ask if anyone is interested in exchanging e-mail addresses or any other form of contact.

Choose to attend all class periods. If possible, sit near the front of the classroom. This way, you will see and hear the presentation better. It may also be easier for you to participate in classroom activities.

Do your homework. You've probably heard the phrase "practice makes perfect" in relation to music and sports. It also applies to mathematics. You will find that the more time you spend solving mathematics exercises, the easier the process becomes. Be sure to schedule enough time to complete your assignments before the due date assigned by your instructor.

Helpful Hint

MyMathLab® and MathXL®
When assignments are turned in online, keep a hard copy of your complete written work. You will need to refer to your written work to be able to ask questions and to study for tests later.

Helpful Hint

MyMathLab® and MathXL®
If you are doing your homework online, you can work and re-work those exercises that you struggle with until you master them. Try working through all the assigned exercises twice before the due date.

Helpful Hint

MyMathLab® and MathXL®
If you are completing your homework online, it's important to work each exercise on paper before submitting the answer. That way, you can check your work and follow your steps to find and correct any mistakes.

Helpful Hint

MyMathLab® and MathXL®
Be aware of assignments and due dates set by your instructor. Don't wait until the last minute to submit work online.

Check your work. Review the steps you took while working a problem. Learn to check your answers in the original exercises. You may also compare your answers with the "Answers to Selected Exercises" section in the back of the book. If you have made a mistake, try to figure out what went wrong. Then correct your mistake. If you can't find what went wrong, **don't** erase your work or throw it away. Show your work to your instructor, a tutor in a math lab, or a classmate. It is easier for someone to find where you had trouble if he or she looks at your original work.

Learn from your mistakes and be patient with yourself. Everyone, even your instructor, makes mistakes. (That definitely includes me—Elayn Martin-Gay.) Use your errors to learn and to become a better math student. The key is finding and understanding your errors.

Was your mistake a careless one, or did you make it because you can't read your own math writing? If so, try to work more slowly or write more neatly and make a conscious effort to carefully check your work.

Did you make a mistake because you don't understand a concept? Take the time to review the concept or ask questions to better understand it.

Did you skip too many steps? Skipping steps or trying to do too many steps mentally may lead to preventable mistakes.

Know how to get help if you need it. It's all right to ask for help. In fact, it's a good idea to ask for help whenever there is something that you don't understand. Make sure you know when your instructor has office hours and how to find his or her office. Find out whether math tutoring services are available on your campus. Check on the hours, location, and requirements of the tutoring service.

Don't be afraid to ask questions. You are not the only person in class with questions. Other students are normally grateful that someone has spoken up.

Turn in assignments on time. This way, you can be sure that you will not lose points for being late. Show every step of a problem and be neat and organized. Also be sure that you understand which problems are assigned for homework. If allowed, you can always double-check the assignment with another student in your class.

OBJECTIVE

3   **Knowing and Using Your Text**

Flip through the pages of this text or view the e-text pages on a computer screen. Start noticing examples, exercise sets, end-of-chapter material, and so on. Every text is organized in some manner. Learn the way this text is organized by reading about and then finding an example in your text of each type of resource listed below. Finding and using these resources throughout your course will increase your chance of success.

- *Practice Exercises.* Each example in every section has a parallel Practice exercise. As you read a section, try each Practice exercise after you've finished the corresponding example. This "learn-by-doing" approach will help you grasp ideas before you move on to other concepts. Answers are at the back of the text.

- *Symbols at the Beginning of an Exercise Set.* If you need help with a particular section, the symbols listed at the beginning of each exercise set will remind you of the numerous resources available.

- *Objectives.* The main section of exercises in each exercise set is referenced by an example(s). There is also often a section of exercises entitled "Mixed Practice," which is referenced by two or more examples or sections. These are mixed exercises written to prepare you for your next exam. Use all of this referencing if you have trouble completing an assignment from the exercise set.

- *Icons (Symbols).* Make sure that you understand the meaning of the icons that are beside many exercises. ▶ tells you that the corresponding exercise may be viewed on the video segment that corresponds to that section. ＼ tells you that this exercise is a writing exercise in which you should answer in complete sentences. △ tells you that the exercise involves geometry. ▦ tells you that this exercise is worked more efficiently with the aid of a calculator. Also, a feature called Graphing Calculator Explorations may be found before select exercise sets.

- *Integrated Reviews.* Found in the middle of each chapter, these reviews offer you a chance to practice—in one place—the many concepts that you have learned separately over several sections.
- *End-of-Chapter Opportunities.* There are many opportunities at the end of each chapter to help you understand the concepts of the chapter.

  Vocabulary Checks contain key vocabulary terms introduced in the chapter.

  Chapter Highlights contain chapter summaries and examples.

  Chapter Reviews contain review exercises. The first part is organized section by section and the second part contains a set of mixed exercises.

  Getting Ready for the Tests contain conceptual exercises written to prepare students for chapter test directions as well as mixed sections of exercises.

  Chapter Tests are sample tests to help you prepare for an exam. The Chapter Test Prep Videos found in the Interactive Lecture Series, MyMathLab, and YouTube provide the video solution to each question on each Chapter Test.

  Cumulative Reviews start at Chapter 2 and are reviews consisting of material from the beginning of the book to the end of that particular chapter.
- *Student Resources in Your Textbook.* You will find a **Student Resources** section at the back of this textbook. It contains the following to help you study and prepare for tests:

  Study Skills Builders contain study skills advice. To increase your chance for success in the course, read these study tips and answer the questions.

  Bigger Picture—Study Guide Outline provides you with a study guide outline of the course, with examples.

  Practice Final provides you with a Practice Final Exam to help you prepare for a final. The video solutions to each question are provided in the Interactive DVD Lecture Series and within MyMathLab®.
- *Resources to Check Your Work.* The **Answers to Selected Exercises** section provides answers to all odd-numbered section exercises and all integrated review and chapter test exercises.

OBJECTIVE

4 Knowing and Using Video and Notebook Organizer Resources

Video Resources

Below is a list of video resources that are all made by me—the author of your text, Elayn Martin-Gay. By making these videos, I can be sure that the methods presented are consistent with those in the text.

- *Interactive DVD Lecture Series.* Exercises marked with a ▶ are fully worked out by the author on the DVDs and within MyMathLab. The lecture series provides approximately 20 minutes of instruction per section and is organized by Objective.
- *Chapter Test Prep Videos.* These videos provide solutions to all of the Chapter Test exercises worked out by the author. They can be found in MyMathLab, the Interactive Lecture series, and YouTube. This supplement is very helpful before a test or exam.
- *Student Success Tips.* These video segments are about 3 minutes long and are daily reminders to help you continue practicing and maintaining good organizational and study habits.
- *Final Exam Videos.* These video segments provide solutions to each question. These videos can be found within MyMathLab and the Interactive Lecture Series.

Notebook Organizer Resource

This resource is in three-ring notebook ready form. It is to be inserted in a three-ring binder and completed. This resource is numbered according to the sections in your text to which they refer.

- *Video Organizer.* This organizer is closely tied to the Interactive Lecture (Video) Series. Each section should be completed while watching a lecture video on the same section. Once completed, you will have a set of notes to accompany the Lecture (Video) Series section by section.

OBJECTIVE

5   Getting Help

If you have trouble completing assignments or understanding the mathematics, get help as soon as you need it! This tip is presented as an objective on its own because it is so important. In mathematics, usually the material presented in one section builds on your understanding of the previous section. This means that if you don't understand the concepts covered during a class period, there is a good chance that you will not understand the concepts covered during the next class period. If this happens to you, get help as soon as you can.

Where can you get help? Many suggestions have been made in this section on where to get help, and now it is up to you to get it. Try your instructor, a tutoring center, or a math lab, or you may want to form a study group with fellow classmates. If you do decide to see your instructor or go to a tutoring center, make sure that you have a neat notebook and are ready with your questions.

OBJECTIVE

6   Preparing for and Taking an Exam

Make sure that you allow yourself plenty of time to prepare for a test. If you think that you are a little "math anxious," it may be that you are not preparing for a test in a way that will ensure success. The way that you prepare for a test in mathematics is important. To prepare for a test:

1. Review your previous homework assignments.
2. Review any notes from class and section-level quizzes you have taken. (If this is a final exam, also review chapter tests you have taken.)
3. Review concepts and definitions by reading the Chapter Highlights at the end of each chapter.
4. Practice working out exercises by completing the Chapter Review found at the end of each chapter. (If this is a final exam, go through a Cumulative Review. There is one found at the end of each chapter except Chapter 1. Choose the review found at the end of the latest chapter that you have covered in your course.) *Don't stop here!*
5. It is important that you place yourself in conditions similar to test conditions to find out how you will perform. In other words, as soon as you feel that you know the material, get a few blank sheets of paper and take a sample test. There is a Chapter Test available at the end of each chapter, or you can work selected problems from the Chapter Review. Your instructor may also provide you with a review sheet. During this sample test, do not use your notes or your textbook. Then check your sample test. If your sample test is the Chapter Test in the text, don't forget that the video solutions are in MyMathLab, the Interactive Lecture Series, and YouTube. If you are not satisfied with the results, study the areas that you are weak in and try again.
6. On the day of the test, allow yourself plenty of time to arrive where you will be taking your exam.

When taking your test:

1. Read the directions on the test carefully.
2. Read each problem carefully as you take the test. Make sure that you answer the question asked.
3. Watch your time and pace yourself so that you can attempt each problem on your test.
4. If you have time, check your work and answers.
5. Do not turn your test in early. If you have extra time, spend it double-checking your work.

7 Managing Your Time

As a college student, you know the demands that classes, homework, work, and family place on your time. Some days you probably wonder how you'll ever get everything done. One key to managing your time is developing a schedule. Here are some hints for making a schedule:

1. Make a list of all your weekly commitments for the term. Include classes, work, regular meetings, extracurricular activities, etc. You may also find it helpful to list such things as laundry, regular workouts, grocery shopping, etc.

2. Next, estimate the time needed for each item on the list. Also make a note of how often you will need to do each item. Don't forget to include time estimates for the reading, studying, and homework you do outside of your classes. You may want to ask your instructor for help estimating the time needed.

3. In the exercise set that follows, you are asked to block out a typical week on the schedule grid given. Start with items with fixed time slots like classes and work.

4. Next, include the items on your list with flexible time slots. Think carefully about how best to schedule items such as study time.

5. Don't fill up every time slot on the schedule. Remember that you need to allow time for eating, sleeping, and relaxing! You should also allow a little extra time in case some items take longer than planned.

6. If you find that your weekly schedule is too full for you to handle, you may need to make some changes in your workload, classload, or other areas of your life. You may want to talk to your advisor, manager or supervisor at work, or someone in your college's academic counseling center for help with such decisions.

1.1 Exercise Set MyMathLab

1. What is your instructor's name?

2. What are your instructor's office location and office hours?

3. What is the best way to contact your instructor?

4. Do you have the name and contact information of at least one other student in class?

5. Will your instructor allow you to use a calculator in this class?

6. Why is it important that you write step-by-step solutions to homework exercises and keep a hard copy of all work submitted?

7. Is there a tutoring service available on campus? If so, what are its hours? What services are available?

8. Have you attempted this course before? If so, write down ways that you might improve your chances of success during this next attempt.

9. List some steps that you can take if you begin having trouble understanding the material or completing an assignment. If you are completing your homework in MyMathLab® and MathXL®, list the resources you can use for help.

10. How many hours of studying does your instructor advise for each hour of instruction?

11. What does the ＼ icon in this text mean?

12. What does the △ icon in this text mean?

13. What does the ▶ icon in this text mean?

14. What are Practice exercises?

15. When might be the best time to work a Practice exercise?

16. Where are the answers to Practice exercises?

17. What answers are contained in this text and where are they?

18. What are Study Skills Builders and where are they?

19. What and where are Integrated Reviews?

20. How many times is it suggested that you work through the homework exercises in MathXL® before the submission deadline?

21. How far in advance of the assigned due date is it suggested that homework be submitted online? Why?

22. Chapter Highlights are found at the end of each chapter. Find the Chapter 1 Highlights and explain how you might use it and how it might be helpful.

23. Chapter Reviews are found at the end of each chapter. Find the Chapter 1 Review and explain how you might use it and how it might be useful.

24. Chapter Tests are at the end of each chapter. Find the Chapter 1 Test and explain how you might use it and how it might be helpful when preparing for an exam on Chapter 1. Include how the Chapter Test Prep Videos may help. If you are working in MyMathLab® and MathXL®, how can you use previous homework assignments to study?

25. What is the Video Organizer? Explain the contents and how it might be used.

26. Explain how the Video Organizer can help you when watching a lecture video.

27. Read or reread Objective 7 and fill out the schedule grid below.

	Monday	*Tuesday*	*Wednesday*	*Thursday*	*Friday*	*Saturday*	*Sunday*
1:00 a.m.							
2:00 a.m.							
3:00 a.m.							
4:00 a.m.							
5:00 a.m.							
6:00 a.m.							
7:00 a.m.							
8:00 a.m.							
9:00 a.m.							
10:00 a.m.							
11:00 a.m.							
Noon							
1:00 p.m.							
2:00 p.m.							
3:00 p.m.							
4:00 p.m.							
5:00 p.m.							
6:00 p.m.							
7:00 p.m.							
8:00 p.m.							
9:00 p.m.							
10:00 p.m.							
11:00 p.m.							
Midnight							

1.2 Symbols and Sets of Numbers

OBJECTIVES

1 Use a Number Line to Order Numbers.

2 Translate Sentences into Mathematical Statements.

3 Identify Natural Numbers, Whole Numbers, Integers, Rational Numbers, Irrational Numbers, and Real Numbers.

4 Find the Absolute Value of a Real Number.

OBJECTIVE

1 Using a Number Line to Order Numbers

We begin with a review of the set of natural numbers and the set of whole numbers and how we use symbols to compare these numbers. A **set** is a collection of objects, each of which is called a **member** or **element** of the set. A pair of brace symbols { } encloses the list of elements and is translated as "the set of" or "the set containing."

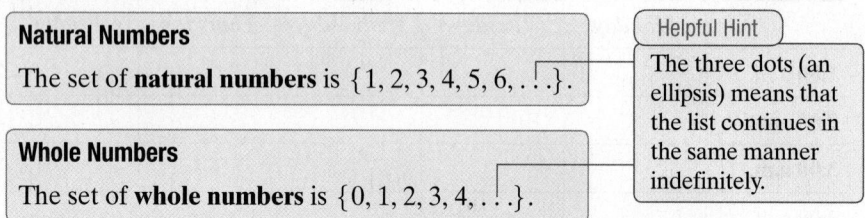

Natural Numbers

The set of **natural numbers** is $\{1, 2, 3, 4, 5, 6, \ldots\}$.

Whole Numbers

The set of **whole numbers** is $\{0, 1, 2, 3, 4, \ldots\}$.

> **Helpful Hint**
> The three dots (an ellipsis) means that the list continues in the same manner indefinitely.

These numbers can be pictured on a **number line.** We will use number lines often to help us visualize distance and relationships between numbers.

To draw a number line, first draw a line. Choose a point on the line and label it 0. To the right of 0, label any other point 1. Being careful to use the same distance as from 0 to 1, mark off equally spaced distances. Label these points 2, 3, 4, 5, and so on. Since the whole numbers continue indefinitely, it is not possible to show every whole number on this number line. The arrow at the right end of the line indicates that the pattern continues indefinitely.

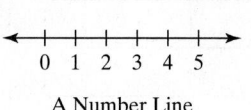

A Number Line

Picturing whole numbers on a number line helps us see the order of the numbers. Symbols can be used to describe concisely in writing the order that we see.

The **equal symbol** $=$ means "is equal to."

The symbol $\neq$ means "is not equal to."

These symbols may be used to form a **mathematical statement.** The statement might be true or it might be false. The two statements below are both true.

$2 = 2$ states that "two is equal to two."

$2 \neq 6$ states that "two is not equal to six."

If two numbers are not equal, one number is larger than the other.
The symbol $>$ means "is greater than."
The symbol $<$ means "is less than." For example,

$3 < 5$ states that "three is less than five."

$2 > 0$ states that "two is greater than zero."

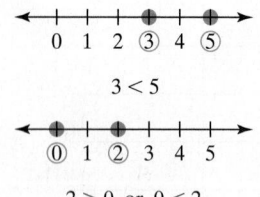

$3 < 5$

$2 > 0$ or $0 < 2$

On a number line, we see that a number **to the right of** another number is **larger.** Similarly, a number **to the left of** another number is smaller. For example, 3 is to the left of 5 on a number line, which means that 3 is less than 5, or $3 < 5$. Similarly, 2 is to the right of 0 on a number line, which means 2 is greater than 0, or $2 > 0$. Since 0 is to the left of 2, we can also say that 0 is less than 2, or $0 < 2$.

The symbols $\neq$, $<$, and $>$ are called **inequality symbols.**

> **Helpful Hint**
>
> Notice that $2 > 0$ has exactly the same meaning as $0 < 2$. Switching the order of the numbers and reversing the direction of the inequality symbol does not change the meaning of the statement.
>
> $3 < 5$ has the same meaning as $5 > 3$.
>
> Also notice that, when the statement is true, the inequality arrow points to the smaller number.

EXAMPLE 1 Insert $<$, $>$, or $=$ in the space between each pair of numbers to make each statement true

   a. 2 3   **b.** 7 4   **c.** 72 27

Solution

   a. $2 < 3$ since 2 is to the left of 3 on a number line.

   b. $7 > 4$ since 7 is to the right of 4 on a number line.

   c. $72 > 27$ since 72 is to the right of 27 on a number line.   ☐

PRACTICE
1   Insert $<$, $>$, or $=$ in the space between each pair of numbers to make each statement true.

   a. 5   8   **b.** 6   4   **c.** 16   82   ■

Two other symbols are used to compare numbers.
The symbol $\leq$ means "is less than or equal to."
The symbol $\geq$ means "is greater than or equal to." For example,

      $7 \leq 10$ states that "seven is less than or equal to ten."

This statement is true since $7 < 10$ is true. If either $7 < 10$ or $7 = 10$ is true, then $7 \leq 10$ is true.

      $3 \geq 3$ states that "three is greater than or equal to three."

This statement is true since $3 = 3$ is true. If either $3 > 3$ or $3 = 3$ is true, then $3 \geq 3$ is true.

     The statement $6 \geq 10$ is false since neither $6 > 10$ nor $6 = 10$ is true. The symbols $\leq$ and $\geq$ are also called **inequality symbols.**

EXAMPLE 2 Tell whether each statement is true or false.

   a. $8 \geq 8$ **b.** $8 \leq 8$ **c.** $23 \leq 0$ **d.** $23 \geq 0$

Solution

   a. True. Since $8 = 8$ is true, then $8 \geq 8$ is true.

   b. True. Since $8 = 8$ is true, then $8 \leq 8$ is true.

   c. False. Since neither $23 < 0$ nor $23 = 0$ is true, then $23 \leq 0$ is false.

   d. True. Since $23 > 0$ is true, then $23 \geq 0$ is true.   ☐

PRACTICE
2   Tell whether each statement is true or false.

   a. $9 \geq 3$ **b.** $3 \geq 8$ **c.** $25 \leq 25$ **d.** $4 \leq 14$ ■

OBJECTIVE
2 **Translating Sentences** ▶

Now, let's use the symbols discussed to translate sentences into mathematical statements.

EXAMPLE 3 Translate each sentence into a mathematical statement.

   a. Nine is less than or equal to eleven.

   b. Eight is greater than one.

   c. Three is not equal to four.

Solution

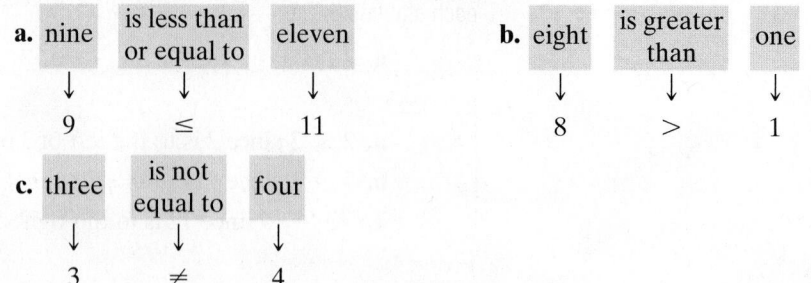

a. nine │ is less than or equal to │ eleven

9 $\leq$ 11

b. eight │ is greater than │ one

8 $>$ 1

c. three │ is not equal to │ four

3 $\neq$ 4

PRACTICE
3 Translate each sentence into a mathematical statement.

a. Three is less than eight.

b. Fifteen is greater than or equal to nine.

c. Six is not equal to seven.

OBJECTIVE
3 Identifying Common Sets of Numbers

Whole numbers are not sufficient to describe many situations in the real world. For example, quantities less than zero must sometimes be represented, such as temperatures less than 0 degrees.

Numbers Less Than Zero on a Number Line

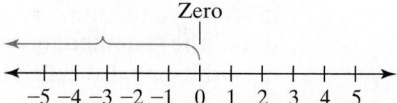

Zero

Numbers less than 0 are to the left of 0 and are labeled −1, −2, −3, and so on. A − sign, such as the one in −1, tells us that the number is to the left of 0 on a number line. In words, −1 is read "negative one." A + sign or no sign tells us that a number lies to the right of 0 on a number line. For example, 3 and +3 both mean positive three.

The numbers we have pictured are called the set of **integers.** Integers to the left of 0 are called **negative integers;** integers to the right of 0 are called **positive integers.** The integer **0 is neither positive nor negative.**

negative integers │ positive integers

−5 −4 −3 −2 −1 0 1 2 3 4 5

Integers

The set of **integers** is { . . . , −3, −2, −1, 0, 1, 2, 3, . . .}.

The ellipses (three dots) to the left and to the right indicate that the positive integers and the negative integers continue indefinitely.

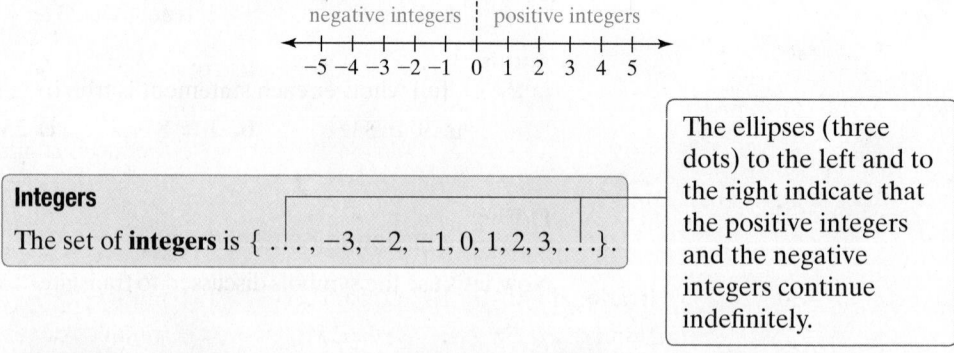

EXAMPLE 4 Use an integer to express the number in the following. "Pole of Inaccessibility, Antarctica, is the coldest location in the world, with an average annual temperature of 72 degrees below zero." (_Source: The Guinness Book of Records_)

Solution The integer −72 represents 72 degrees below zero.

PRACTICE
4 Use an integer to express the number in the following. The elevation of Laguna Salada in Mexico is 10 meters below sea level. (_Source: The World Almanac_)

A problem with integers in real-life settings arises when quantities are smaller than some integer but greater than the next smallest integer. On a number line, these quantities may be visualized by points between integers. Some of these quantities between integers can be represented as a quotient of integers. For example,

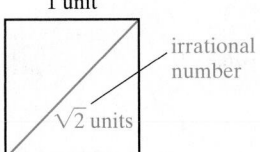

The point on a number line halfway between 0 and 1 can be represented by $\frac{1}{2}$, a quotient of integers.

The point on a number line halfway between 0 and -1 can be represented by $-\frac{1}{2}$. Other quotients of integers and their graphs are shown to the left.

These numbers, each of which can be represented as a quotient of integers, are examples of **rational numbers.** It's not possible to list the set of rational numbers using the notation that we have been using. For this reason, we will use a different notation.

Rational Numbers

$$\left\{ \frac{a}{b} \,\middle|\, a \text{ and } b \text{ are integers and } b \neq 0 \right\}$$

We read this set as "the set of all numbers $\frac{a}{b}$ such that a and b are integers and **b is not equal to 0.**" Notice that every integer is also a rational number since each integer can be expressed as a quotient of integers. For example, the integer 5 is also a rational number since $5 = \frac{5}{1}$.

The number line also contains points that cannot be expressed as quotients of integers. These numbers are called **irrational numbers** because they cannot be represented by rational numbers. For example, $\sqrt{2}$ and π are irrational numbers.

1 unit

$\sqrt{2}$ units

irrational number

Irrational Numbers

The set of **irrational numbers** is

{Nonrational numbers that correspond to points on a number line}.

That is, an irrational number is a number that cannot be expressed as a quotient of integers.

Both rational numbers and irrational numbers can be written as decimal numbers. The decimal equivalent of a rational number will either terminate or repeat in a pattern. For example, upon dividing we find that

Rational Numbers
$$\begin{cases} \dfrac{3}{4} = 0.75 \text{ (decimal number terminates or ends)} \\[2mm] \dfrac{2}{3} = 0.66666\ldots \text{ (decimal number repeats in a pattern)} \end{cases}$$

The decimal representation of an irrational number will neither terminate nor repeat. For example, the decimal representations of irrational numbers $\sqrt{2}$ and π are

Irrational Numbers
$$\begin{cases} \sqrt{2} = 1.414213562\ldots \text{ (decimal number does not terminate or repeat in a pattern)} \\[2mm] \pi = 3.141592653\ldots \text{ (decimal number does not terminate or repeat in a pattern)} \end{cases}$$

(For further review of decimals, see the Appendix.)

Combining the rational numbers with the irrational numbers gives the set of **real numbers.** One and only one point on a number line corresponds to each real number.

Real Numbers

The set of **real numbers** is

{All numbers that correspond to points on a number line}

Helpful Hint

From our previous definitions, we have that

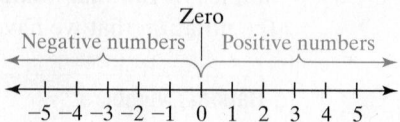

On the following number line, we see that real numbers can be positive, negative, or 0. Numbers to the left of 0 are called **negative numbers;** numbers to the right of 0 are called **positive numbers.** Positive and negative numbers are also called **signed numbers.**

Zero

Negative numbers | Positive numbers

$\overset{\longleftarrow}{\underset{-5\ -4\ -3\ -2\ -1\quad 0\quad 1\quad 2\quad 3\quad 4\quad 5}{\rule{6cm}{0.4pt}}}$

Several different sets of numbers have been discussed in this section. The following diagram shows the relationships among these sets of real numbers.

Common Sets of Numbers

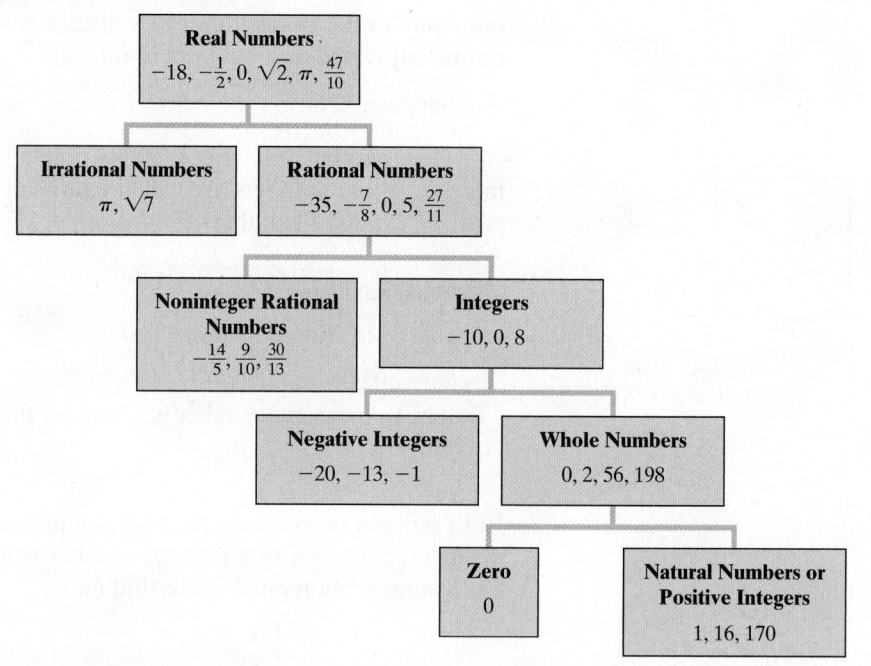

EXAMPLE 5 Given the set $\left\{-2, 0, \dfrac{1}{4}, -1.5, 112, -3, 11, \sqrt{2}\right\}$, list the numbers in this set that belong to the set of:

    a. Natural numbers   **b.** Whole numbers

    c. Integers   **d.** Rational numbers

    e. Irrational numbers   **f.** Real numbers

Solution

    a. The natural numbers are 11 and 112.

    b. The whole numbers are 0, 11, and 112.

    c. The integers are $-3, -2, 0, 11,$ and 112.

    d. Recall that integers are rational numbers also. The rational numbers are $-3, -2, -1.5, 0, \dfrac{1}{4}, 11,$ and 112.

    e. The irrational number is $\sqrt{2}$.

    f. The real numbers are all numbers in the given set.

PRACTICE
5   Given the set $\left\{25, \frac{7}{3}, -15, -\frac{3}{4}, \sqrt{5}, -3.7, 8.8, -99\right\}$, list the numbers in this set that belong to the set of:

a. Natural numbers   **b.** Whole numbers

c. Integers   **d.** Rational numbers

e. Irrational numbers   **f.** Real numbers

We now extend the meaning and use of inequality symbols such as $<$ and $>$ to all real numbers.

Order Property for Real Numbers

For any two real numbers a and b, a is less than b if a is to the left of b on a number line.

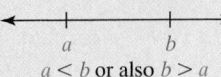

$a < b$ or also $b > a$

EXAMPLE 6   Insert $<$, $>$, or $=$ in the appropriate space to make each statement true.

a. $-1 \quad\quad 0$ **b.** $7 \quad\quad \frac{14}{2}$ **c.** $-5 \quad\quad -6$

Solution

a. $-1 < 0$ since -1 is to the left of 0 on a number line.

$$-2 \ -1 \ \ 0 \ \ 1 \ \ 2$$

$$-1 < 0$$

b. $7 = \frac{14}{2}$ since $\frac{14}{2}$ simplifies to 7.

c. $-5 > -6$ since -5 is to the right of -6 on a number line.

$$-7 \ -6 \ -5 \ -4 \ -3$$

$$-5 > -6$$

PRACTICE
6   Insert $<$, $>$, or $=$ in the appropriate space to make each statement true.

a. $0 \quad\quad 3$ **b.** $15 \quad\quad -5$ **c.** $3 \quad\quad \frac{12}{4}$

OBJECTIVE
4   Finding the Absolute Value of a Real Number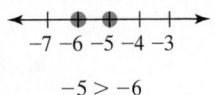

A number line also helps us visualize the distance between numbers. The distance between a real number a and 0 is given a special name called the **absolute value** of a. "The absolute value of a" is written in symbols as $|a|$.

Absolute Value

The absolute value of a real number a, denoted by $|a|$, is the distance between a and 0 on a number line.

For example, $|3| = 3$ and $|-3| = 3$ since both 3 and -3 are a distance of 3 units from 0 on a number line.

Helpful Hint

Since $|a|$ is a distance, $|a|$ is always either positive or 0, never negative. That is, **for any real number a, $|a| \geq 0$.**

EXAMPLE 7   Find the absolute value of each number.

    a. $|4|$ **b.** $|-5|$ **c.** $|0|$ **d.** $\left|-\dfrac{1}{2}\right|$ **e.** $|5.6|$

Solution

    a. $|4| = 4$ since 4 is 4 units from 0 on a number line.

    b. $|-5| = 5$ since -5 is 5 units from 0 on a number line.

    c. $|0| = 0$ since 0 is 0 units from 0 on a number line.

    d. $\left|-\dfrac{1}{2}\right| = \dfrac{1}{2}$ since $-\dfrac{1}{2}$ is $\dfrac{1}{2}$ unit from 0 on a number line.

    e. $|5.6| = 5.6$ since 5.6 is 5.6 units from 0 on a number line.

PRACTICE
7   Find the absolute value of each number.

    a. $|-8|$ **b.** $|9|$ **c.** $|-2.5|$ **d.** $\left|\dfrac{5}{11}\right|$ **e.** $|\sqrt{3}|$

EXAMPLE 8   Insert $<$, $>$, or $=$ in the appropriate space to make each statement true.

    a. $|0|$ 2   **b.** $|-5|$ 5   **c.** $|-3|$ $|-2|$ **d.** $|5|$ $|6|$ **e.** $|-7|$ $|6|$

Solution

    a. $|0| < 2$ since $|0| = 0$ and $0 < 2$.   **b.** $|-5| = 5$ since $5 = 5$.

    c. $|-3| > |-2|$ since $3 > 2$.   **d.** $|5| < |6|$ since $5 < 6$.

    e. $|-7| > |6|$ since $7 > 6$.

PRACTICE
8   Insert $<$, $>$, or $=$ in the appropriate space to make each statement true.

    a. $|8|$ $|-8|$ **b.** $|-3|$ 0   **c.** $|-7|$ $|-11|$ **d.** $|3|$ $|2|$ **e.** $|0|$ $|-4|$

✔ Vocabulary, Readiness & Video Check

Use the choices below to fill in each blank.

real	natural	whole	irrational
$|b|$	inequality	integers	rational

1. The _____ numbers are $\{0, 1, 2, 3, 4, \ldots\}$.

2. The _____ numbers are $\{1, 2, 3, 4, 5, \ldots\}$.

3. The symbols $\neq$, $\leq$, and $>$ are called _____ symbols.

4. The _____ are $\{\ldots, -3, -2, -1, 0, 1, 2, 3, \ldots\}$.

5. The _____ numbers are {all numbers that correspond to points on a number line}.

6. The _____ numbers are $\left\{\dfrac{a}{b} \,\middle|\, a \text{ and } b \text{ are integers, } b \neq 0\right\}$.

7. The _____ numbers are {nonrational numbers that correspond to points on a number line}.

8. The distance between a number b and 0 on a number line is _____.

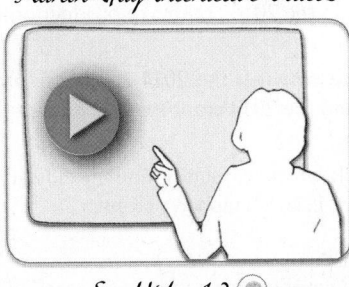

Martin-Gay Interactive Videos

See Video 1.2 ◉

Watch the section lecture video and answer the following questions.

OBJECTIVE 1
9. In Example 2, why is the symbol < inserted between the two numbers?

OBJECTIVE 2
10. Write the sentence given in ▦ Example 4 and translate it to a mathematical statement, using symbols.

OBJECTIVE 3
11. Which sets of numbers does the number in ▦ Example 6 belong to? Why is this number not an irrational number?

OBJECTIVE 4
12. Complete this statement based on the lecture given before ▦ Example 8. The _____ of a real number a, denoted by $|a|$, is the distance between a and 0 on a number line.

1.2 Exercise Set MyMathLab® ▶

Insert <, >, or = in the appropriate space to make the statement true. See Example 1.

1. 7   3
2. 9   15
3. 6.26   6.26
4. 2.13   1.13
5. 0   7
6. 20   0
7. −2   2
8. −4   −6

9. The freezing point of water is 32° Fahrenheit. The boiling point of water is 212° Fahrenheit. Write an inequality statement using < or > comparing the numbers 32 and 212.

10. The freezing point of water is 0° Celsius. The boiling point of water is 100° Celsius. Write an inequality statement using < or > comparing the numbers 0 and 100.

△ 11. An angle measuring 30° is shown and an angle measuring 45° is shown. Use the inequality symbol ≤ or ≥ to write a statement comparing the numbers 30 and 45.

△ 12. The sum of the measures of the angles of a triangle is 180°. The sum of the measures of the angles of a parallelogram is 360°. Use the inequality symbol ≤ or ≥ to write a statement comparing the numbers 360 and 180.

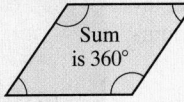

Are the following statements true or false? See Examples 2 and 6.

13. 11 ≤ 11
14. 4 ≥ 7
15. 10 > 11
16. 17 > 16
17. 3 + 8 ≥ 3(8)
18. 8 · 8 ≤ 8 · 7
19. 9 > 0
20. 4 < 7
21. −6 > −2
22. 0 < −15

TRANSLATING

Write each sentence as a mathematical statement. See Example 3.

23. Eight is less than twelve.
24. Fifteen is greater than five.
25. Five is greater than or equal to four.
26. Negative ten is less than or equal to thirty-seven.
27. Fifteen is not equal to negative two.
28. Negative seven is not equal to seven.

Use integers to represent the values in each statement. See Example 4.

29. The highest elevation in California is Mt. Whitney, with an altitude of 14,494 feet. The lowest elevation in California is Death Valley, with an altitude of 282 feet below sea level. (*Source:* U.S. Geological Survey)

30. Driskill Mountain, in Louisiana, has an altitude of 535 feet. New Orleans, Louisiana, lies 8 feet below sea level. (*Source:* U.S. Geological Survey)

31. The number of graduate students at the University of Texas at Austin is 28,000 fewer than the number of undergraduate students. (*Source:* University of Texas at Austin)

32. The number of students admitted to the class of 2014 at UCLA is 80,784 fewer students than the number that had applied. (*Source:* UCLA)

33. Aaron Miller deposited $350 in his savings account. He later withdrew $126.

34. Aris Peña was deep-sea diving. During her dive, she ascended 30 feet and later descended 50 feet.

Tell which set or sets each number belongs to: natural numbers, whole numbers, integers, rational numbers, irrational numbers, and real numbers. See Example 5.

35. 0
36. $\frac{1}{4}$
37. −2
38. $-\frac{1}{2}$

39. 6

40. 5

41. $\frac{2}{3}$

42. $\sqrt{3}$

43. $-\sqrt{5}$

44. $-1\frac{5}{9}$

Tell whether each statement is true or false.

45. Every rational number is also an integer.

46. Every negative number is also a rational number.

47. Every natural number is positive.

48. Every rational number is also a real number.

49. 0 is a real number.

50. Every real number is also a rational number.

51. Every whole number is an integer.

52. $\frac{1}{2}$ is an integer.

53. A number can be both rational and irrational.

54. Every whole number is positive.

Insert $<$, $>$, or $=$ in the appropriate space to make a true statement. See Examples 6 through 8.

55. -10 -100

56. -200 -20

57. 32 5.2

58. 7.1 -7

59. $\frac{18}{3}$ $\frac{24}{3}$

60. $\frac{8}{2}$ $\frac{12}{3}$

61. -51 -50

62. $|-20|$ -200

63. $|-5|$ -4

64. 0 $|0|$

65. $|-1|$ $|1|$

66. $\left|\frac{2}{5}\right|$ $\left|-\frac{2}{5}\right|$

67. $|-2|$ $|-3|$

68. -500 $|-50|$

69. $|0|$ $|-8|$

70. $|-12|$ $\frac{24}{2}$

CONCEPT EXTENSIONS

The graph below is called a bar graph. This particular bar graph shows cranberry production from the top five cranberry-producing states.

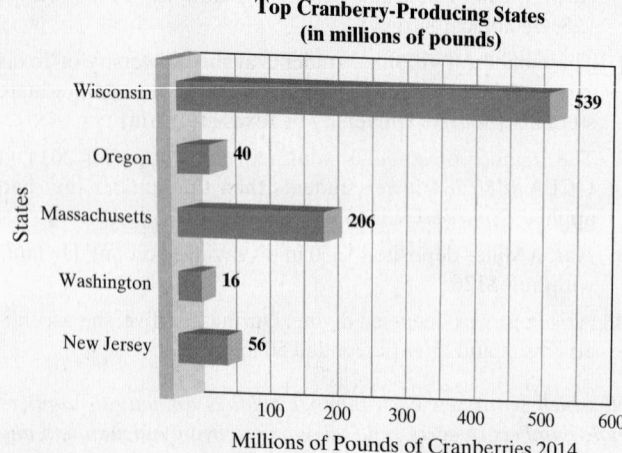

Top Cranberry-Producing States
(in millions of pounds)

Wisconsin — 539
Oregon — 40
Massachusetts — 206
Washington — 16
New Jersey — 56

Millions of Pounds of Cranberries 2014

States

Data from National Agricultural Statistics Service

71. Write an inequality comparing the 2014 cranberry production in Oregon with the 2014 cranberry production in Washington.

72. Write an inequality comparing the 2014 cranberry production in Massachusetts with the 2014 cranberry production in Wisconsin.

73. Determine the difference between the 2014 cranberry production in Washington and the 2014 cranberry production in New Jersey.

74. According to the bar graph, which two states were the closest in terms of millions of pounds in 2014 cranberry crops?

This bar graph shows the number of people admitted into the Baseball Hall of Fame since its founding. Each bar represents a decade, and the height of the bar represents the number of Hall of Famers admitted in that decade.

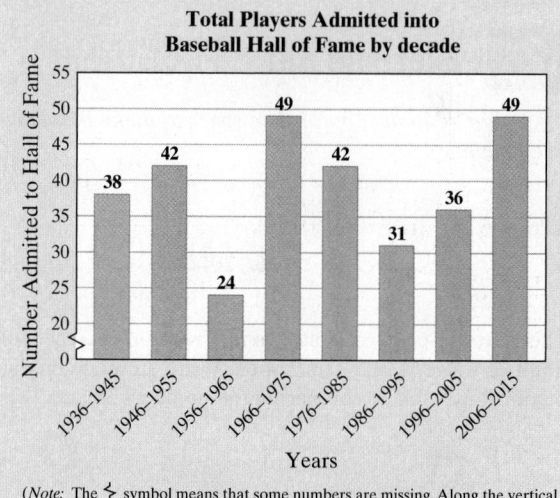

Total Players Admitted into
Baseball Hall of Fame by decade

Number Admitted to Hall of Fame

1936–1945: 38
1946–1955: 42
1956–1965: 24
1966–1975: 49
1976–1985: 42
1986–1995: 31
1996–2005: 36
2006–2015: 49

Years

(*Note:* The ⚡ symbol means that some numbers are missing. Along the vertical data line, notice the numbers between 0 and 20 are missing or not shown.)

Source: BASEBALL-Reference.com

75. In which decade(s) was the number of players admitted the greatest?

76. What was the greatest number of players admitted shown?

77. In which decade(s) was the number of players admitted greater than 40?

78. In which decade(s) was the number of players admitted fewer than 30?

79. Write an inequality statement comparing the number of players admitted in 1936–1945 and in 1966–1975.

80. Do you notice any trends shown by this bar graph?

The apparent magnitude of a star is the measure of its brightness as seen by someone on Earth. The smaller the apparent magnitude, the brighter the star. Use the apparent magnitudes in the table on page 17 to answer Exercises 81 through 86.

81. The apparent magnitude of the Sun is -26.7. The apparent magnitude of the star Arcturus is -0.04. Write an inequality statement comparing the numbers -0.04 and -26.7.

Star	Apparent Magnitude	Star	Apparent Magnitude
Arcturus	−0.04	Spica	0.98
Sirius	−1.46	Rigel	0.12
Vega	0.03	Regulus	1.35
Antares	0.96	Canopus	−0.72
Sun	−26.7	Hadar	0.61

(Data from *Norton's Star Atlas and Reference Handbook*, 20th Edition, edited by Ian Ridpath. © 2004 Pearson Education, Inc.)

82. The apparent magnitude of Antares is 0.96. The apparent magnitude of Spica is 0.98. Write an inequality statement comparing the numbers 0.96 and 0.98.

83. Which is brighter, the Sun or Arcturus?

84. Which is dimmer, Antares or Spica?

85. Which star listed is the brightest?

86. Which star listed is the dimmest?

Rewrite the following inequalities so that the inequality symbol points in the opposite direction and the resulting statement has the same meaning as the given one.

87. $25 \geq 20$

88. $-13 \leq 13$

89. $0 < 6$

90. $75 > 73$

91. $-10 > -12$

92. $-4 < -2$

93. In your own words, explain how to find the absolute value of a number.

94. Give an example of a real-life situation that can be described with integers but not with whole numbers.

1.3   Fractions and Mixed Numbers

OBJECTIVES

1. Write Fractions in Simplest Form.
2. Multiply and Divide Fractions.
3. Add and Subtract Fractions.
4. Perform Operations on Mixed Numbers.

OBJECTIVE

1   Writing Fractions in Simplest Form

A quotient of two numbers such as $\frac{2}{9}$ is called a **fraction.** The parts of a fraction are:

Fraction bar → $\dfrac{2}{9}$ ← Numerator
← Denominator

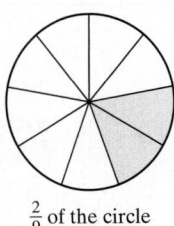

$\frac{2}{9}$ of the circle is shaded.

A fraction may be used to refer to part of a whole. For example, $\frac{2}{9}$ of the circle above is shaded. The denominator 9 tells us how many equal parts the whole circle is divided into, and the numerator 2 tells us how many equal parts are shaded.

To simplify fractions, we can factor the numerator and the denominator. In the statement $3 \cdot 5 = 15$, 3 and 5 are called **factors** and 15 is the **product.** (The raised dot symbol indicates multiplication.)

$$3 \quad \cdot \quad 5 \quad = \quad 15$$
$$\uparrow \qquad \uparrow \qquad \qquad \uparrow$$
$$\text{factor} \quad \text{factor} \qquad \text{product}$$

To **factor** 15 means to write it as a product. The number 15 can be factored as $3 \cdot 5$ or as $1 \cdot 15$.

A fraction is said to be **simplified** or in **lowest terms** when the numerator and the denominator have no factors in common other than 1. For example, the fraction $\frac{5}{11}$ is in lowest terms since 5 and 11 have no common factors other than 1.

To help us simplify fractions, we write the numerator and the denominator as products of **prime numbers.**

> **Prime Number and Composite Number**
>
> A **prime number** is a natural number, other than 1, whose only factors are 1 and itself. The first few prime numbers are
>
> $$2, 3, 5, 7, 11, 13, 17, 19, 23, 29, \text{ and so on.}$$
>
> A natural number, other than 1, that is not a prime number is called a **composite number.**

Helpful Hint

The natural number 1 is neither prime nor composite.

Every composite number can be written as a product of prime numbers. We call this product of prime numbers the **prime factorization** of the composite number.

EXAMPLE 1   Write each of the following numbers as a product of primes.

a. 40   **b.** 63

Solution

a. First, write 40 as the product of any two whole numbers other than 1.

$$40 = 4 \cdot 10$$

Next, factor each of these numbers. Continue this process until all of the factors are prime numbers.

$$40 = 4 \quad \cdot \quad 10$$
$$\quad\quad \downarrow \searrow \quad \downarrow \searrow$$
$$= 2 \cdot 2 \cdot 2 \cdot 5$$

All the factors are now prime numbers. Then 40 written as a product of primes is

$$40 = 2 \cdot 2 \cdot 2 \cdot 5$$

b. $63 = 9 \quad \cdot \quad 7$
$$\quad\quad \downarrow \searrow \quad \downarrow$$
$$= 3 \cdot 3 \cdot 7$$

PRACTICE

1   Write each of the following numbers as a product of primes.

a. 36   **b.** 200

To use prime factors to write a fraction in lowest terms (or simplified form), apply the fundamental principle of fractions.

> **Fundamental Principle of Fractions**
>
> If $\dfrac{a}{b}$ is a fraction and c is a nonzero real number, then
>
> $$\frac{a \cdot c}{b \cdot c} = \frac{a}{b}$$

To understand why this is true, we use the fact that since c is not zero, $\dfrac{c}{c} = 1$.

$$\frac{a \cdot c}{b \cdot c} = \frac{a}{b} \cdot \frac{c}{c} = \frac{a}{b} \cdot 1 = \frac{a}{b}$$

We will call this process dividing out the common factor of c.

EXAMPLE 2   Simplify each fraction (write it in lowest terms).

a. $\dfrac{42}{49}$ **b.** $\dfrac{11}{27}$ **c.** $\dfrac{88}{20}$

<u>Solution</u>

a. Write the numerator and the denominator as products of primes; then apply the fundamental principle to the common factor 7.

$$\frac{42}{49} = \frac{2 \cdot 3 \cdot 7}{7 \cdot 7} = \frac{2 \cdot 3}{7} \cdot \frac{7}{7} = \frac{2 \cdot 3}{7} = \frac{6}{7}$$

b. $\dfrac{11}{27} = \dfrac{11}{3 \cdot 3 \cdot 3}$

There are no common factors other than 1, so $\dfrac{11}{27}$ is already in simplest form.

c. $\dfrac{88}{20} = \dfrac{2 \cdot 2 \cdot 2 \cdot 11}{2 \cdot 2 \cdot 5} = \dfrac{2}{2} \cdot \dfrac{2}{2} \cdot \dfrac{2 \cdot 11}{5} = \dfrac{22}{5}$

PRACTICE
2   Write each fraction in lowest terms.

a. $\dfrac{63}{72}$ **b.** $\dfrac{64}{12}$ **c.** $\dfrac{7}{25}$

✔ **CONCEPT CHECK**

Explain the error in the following steps.

a. $\dfrac{15}{55} = \dfrac{\cancel{1}\,5}{\cancel{5}\,5} = \dfrac{1}{5}$ **b.** $\dfrac{6}{7} = \dfrac{5 + 1}{5 + 2} = \dfrac{1}{2}$

OBJECTIVE
2   **Multiplying and Dividing Fractions**

To multiply two fractions, multiply numerator times numerator to obtain the numerator of the product; multiply denominator times denominator to obtain the denominator of the product.

> **Multiplying Fractions**
>
> $$\frac{a}{b} \cdot \frac{c}{d} = \frac{a \cdot c}{b \cdot d} \quad \text{if } b \neq 0 \text{ and } d \neq 0$$

EXAMPLE 3   Multiply $\dfrac{2}{15}$ and $\dfrac{5}{13}$. Simplify the product if possible.

<u>Solution</u> $\dfrac{2}{15} \cdot \dfrac{5}{13} = \dfrac{2 \cdot 5}{15 \cdot 13}$ Multiply numerators.
Multiply denominators.

Next, simplify the product by dividing the numerator and the denominator by any common factors.

$$= \frac{2 \cdot \overset{1}{\cancel{5}}}{3 \cdot \underset{1}{\cancel{5}} \cdot 13}$$

$$= \frac{2}{39}$$

PRACTICE
3   Multiply $\dfrac{3}{8}$ and $\dfrac{7}{9}$. Simplify the product if possible.

Answers to Concept Check:
answers may vary

Before dividing fractions, we first define **reciprocals.** Two fractions are reciprocals of each other if their product is 1.

For example:

The reciprocal of $\frac{2}{3}$ is $\frac{3}{2}$ because $\frac{2}{3} \cdot \frac{3}{2} = \frac{6}{6} = 1$.

The reciprocal of 5 is $\frac{1}{5}$ because $5 \cdot \frac{1}{5} = \frac{5}{1} \cdot \frac{1}{5} = \frac{5}{5} = 1$.

To divide fractions, multiply the first fraction by the reciprocal of the second fraction.

Dividing Fractions

$$\frac{a}{b} \div \frac{c}{d} = \frac{a}{b} \cdot \frac{d}{c}, \qquad \text{if } b \neq 0, d \neq 0, \text{ and } c \neq 0$$

EXAMPLE 4   Divide. Simplify all quotients if possible.

a. $\dfrac{4}{5} \div \dfrac{5}{16}$ 　　　　**b.** $\dfrac{7}{10} \div 14$ 　　　　**c.** $\dfrac{3}{8} \div \dfrac{3}{10}$

Solution

a. $\dfrac{4}{5} \div \dfrac{5}{16} = \dfrac{4}{5} \cdot \dfrac{16}{5} = \dfrac{4 \cdot 16}{5 \cdot 5} = \dfrac{64}{25}$ ⎫ The numerator and denominator have no common factors.

b. $\dfrac{7}{10} \div 14 = \dfrac{7}{10} \div \dfrac{14}{1} = \dfrac{7}{10} \cdot \dfrac{1}{14} = \dfrac{\overset{1}{\cancel{7}} \cdot 1}{2 \cdot 5 \cdot 2 \cdot \underset{1}{\cancel{7}}} = \dfrac{1}{20}$

c. $\dfrac{3}{8} \div \dfrac{3}{10} = \dfrac{3}{8} \cdot \dfrac{10}{3} = \dfrac{\overset{1}{\cancel{3}} \cdot \overset{1}{\cancel{2}} \cdot 5}{2 \cdot 2 \cdot 2 \cdot \underset{1}{\cancel{3}}} = \dfrac{5}{4}$

PRACTICE
4   Divide. Simplify all quotients if possible.

a. $\dfrac{3}{4} \div \dfrac{4}{9}$ 　　　　**b.** $\dfrac{5}{12} \div 15$ 　　　　**c.** $\dfrac{7}{6} \div \dfrac{7}{15}$ ∎

OBJECTIVE
3   Adding and Subtracting Fractions

To add or subtract fractions with the same denominator, combine numerators and place the sum or difference over the common denominator.

Adding and Subtracting Fractions with the Same Denominator

$$\frac{a}{b} + \frac{c}{b} = \frac{a + c}{b}, \qquad \text{if } b \neq 0$$

$$\frac{a}{b} - \frac{c}{b} = \frac{a - c}{b}, \qquad \text{if } b \neq 0$$

EXAMPLE 5   Add or subtract as indicated. Simplify each result if possible.

a. $\dfrac{2}{7} + \dfrac{4}{7}$ 　　**b.** $\dfrac{3}{10} + \dfrac{2}{10}$ 　　**c.** $\dfrac{9}{7} - \dfrac{2}{7}$ 　　**d.** $\dfrac{5}{3} - \dfrac{1}{3}$

Solution

a. $\dfrac{2}{7} + \dfrac{4}{7} = \dfrac{2 + 4}{7} = \dfrac{6}{7}$ 　　**b.** $\dfrac{3}{10} + \dfrac{2}{10} = \dfrac{3 + 2}{10} = \dfrac{5}{10} = \dfrac{\overset{1}{\cancel{5}}}{2 \cdot \underset{1}{\cancel{5}}} = \dfrac{1}{2}$

c. $\dfrac{9}{7} - \dfrac{2}{7} = \dfrac{9 - 2}{7} = \dfrac{7}{7} = 1$ 　　**d.** $\dfrac{5}{3} - \dfrac{1}{3} = \dfrac{5 - 1}{3} = \dfrac{4}{3}$ □

　Add or subtract as indicated. Simplify each result if possible.

　　a. $\dfrac{8}{5} - \dfrac{3}{5}$　　　**b.** $\dfrac{8}{5} - \dfrac{2}{5}$　　　**c.** $\dfrac{3}{5} + \dfrac{1}{5}$　　　**d.** $\dfrac{5}{12} + \dfrac{1}{12}$　　■

To add or subtract fractions without the same denominator, first write the fractions as equivalent fractions with a common denominator. **Equivalent fractions** are fractions that represent the same quantity. For example,

$$\frac{3}{4} \text{ and } \frac{12}{16} \text{ are equivalent fractions}$$

since they represent the same portion of a whole, as the diagram shows. Count the larger squares, and the shaded portion is $\dfrac{3}{4}$. Count the smaller squares, and the shaded portion is $\dfrac{12}{16}$. Thus, $\dfrac{3}{4} = \dfrac{12}{16}$.

We can write equivalent fractions by multiplying a given fraction by 1, as shown in the next example. Multiplying a fraction by 1 does not change the value of the fraction.

Whole

$$\frac{3}{4} = \frac{12}{16}$$

EXAMPLE 6　Write $\dfrac{2}{5}$ as an equivalent fraction with a denominator of 20.

Solution　Since $5 \cdot 4 = 20$, multiply the fraction by $\dfrac{4}{4}$. Multiplying by $\dfrac{4}{4} = 1$ does not change the value of the fraction.

—— Multiply by $\dfrac{4}{4}$ or 1.

$$\frac{2}{5} = \frac{2}{5} \cdot \frac{4}{4} = \frac{2 \cdot 4}{2 \cdot 4} = \frac{8}{20}$$

Thus, $\dfrac{2}{5} = \dfrac{8}{20}$.　　□

　Write $\dfrac{2}{3}$ as an equivalent fraction with a denominator of 21.　　■

To add or subtract with different denominators, we first write the fractions as **equivalent fractions** with the same denominator. We use the smallest or **least common denominator,** or **LCD.** (The LCD is the same as the least common multiple of the denominators.)

EXAMPLE 7　Add or subtract as indicated. Write each answer in simplest form.

　　a. $\dfrac{2}{5} + \dfrac{1}{4}$　　　**b.** $\dfrac{19}{6} - \dfrac{23}{12}$　　　**c.** $\dfrac{1}{2} + \dfrac{17}{22} - \dfrac{2}{11}$

Solution

　a. Fractions must have a common denominator before they can be added or subtracted. Since 20 is the smallest number that both 5 and 4 divide into evenly, 20 is the **least common denominator** (LCD). Write both fractions as equivalent fractions with denominators of 20. Since

$$\frac{2}{5} \cdot \frac{4}{4} = \frac{2 \cdot 4}{5 \cdot 4} = \frac{8}{20} \qquad \text{and} \qquad \frac{1}{4} \cdot \frac{5}{5} = \frac{1 \cdot 5}{4 \cdot 5} = \frac{5}{20}$$

　　　then

$$\frac{2}{5} + \frac{1}{4} = \frac{8}{20} + \frac{5}{20} = \frac{13}{20}$$

b. The LCD is 12. We write both fractions as equivalent fractions with denominators of 12.

$$\frac{19}{6} - \frac{23}{12} = \frac{38}{12} - \frac{23}{12}$$

$$= \frac{15}{12} = \frac{\overset{1}{\cancel{3}} \cdot 5}{2 \cdot 2 \cdot \underset{1}{\cancel{3}}} = \frac{5}{4}$$

c. The LCD for denominators 2, 22, and 11 is 22. First, write each fraction as an equivalent fraction with a denominator of 22. Then add or subtract from left to right.

$$\frac{1}{2} = \frac{1}{2} \cdot \frac{11}{11} = \frac{11}{22}, \qquad \frac{17}{22} = \frac{17}{22}, \qquad \text{and} \qquad \frac{2}{11} = \frac{2}{11} \cdot \frac{2}{2} = \frac{4}{22}$$

Then

$$\frac{1}{2} + \frac{17}{22} - \frac{2}{11} = \frac{11}{22} + \frac{17}{22} - \frac{4}{22} = \frac{24}{22} = \frac{12}{11}$$

PRACTICE

7 Add or subtract as indicated. Write answers in simplest form.

a. $\dfrac{5}{11} + \dfrac{1}{7}$ **b.** $\dfrac{5}{21} - \dfrac{1}{6}$ **c.** $\dfrac{1}{3} + \dfrac{29}{30} - \dfrac{4}{5}$

OBJECTIVE

4 **Performing Operations on Mixed Numbers** ▶

To multiply or divide mixed numbers, first write each mixed number as an improper fraction. To recall how this is done, let's write $3\dfrac{1}{5}$ as an improper fraction.

$$3\frac{1}{5} = 3 + \frac{1}{5} = \frac{15}{5} + \frac{1}{5} = \frac{16}{5}$$

Because of the steps above, notice that we can use a shortcut process for writing a mixed number as an improper fraction.

$$3\frac{1}{5} = \frac{5 \cdot 3 + 1}{5} = \frac{16}{5}$$

EXAMPLE 8 Divide: $2\dfrac{1}{8} \div 1\dfrac{2}{3}$

Solution First write each mixed number as an improper fraction.

$$2\frac{1}{8} = \frac{8 \cdot 2 + 1}{8} = \frac{17}{8}; \qquad 1\frac{2}{3} = \frac{3 \cdot 1 + 2}{3} = \frac{5}{3}$$

Now divide as usual.

$$2\frac{1}{8} \div 1\frac{2}{3} = \frac{17}{8} \div \frac{5}{3} = \frac{17}{8} \cdot \frac{3}{5} = \frac{51}{40}$$

The fraction $\dfrac{51}{40}$ is improper. To write it as an equivalent mixed number, remember that the fraction bar means division and divide.

$$\begin{array}{r} 1\frac{11}{40} \\ 40\overline{)51} \\ \underline{-40} \\ 11 \end{array}$$

Thus, the quotient is $\dfrac{51}{40}$ or $1\dfrac{11}{40}$.

PRACTICE

8 Multiply: $5\dfrac{1}{6} \cdot 4\dfrac{2}{5}$

As a general rule, if the original exercise contains mixed numbers, write the result as a mixed number if possible.

When adding or subtracting mixed numbers, you might want to use the following method.

EXAMPLE 9   Subtract: $50\frac{1}{6} - 38\frac{1}{3}$

Solution

$$50\frac{1}{6} = \quad 50\frac{1}{6} = \quad 49\frac{7}{6} \qquad 50\frac{1}{6} = 49 + 1 + \frac{1}{6} = 49\frac{7}{6}$$

$$\underline{-38\frac{1}{3}} = \underline{-38\frac{2}{6}} = \underline{-38\frac{2}{6}}$$

$$\qquad\qquad\qquad\qquad 11\frac{5}{6}$$

PRACTICE
9   Subtract: $76\frac{1}{12} - 35\frac{1}{4}$

✔ Vocabulary, Readiness & Video Check

Use the choices below to fill in each blank. Some choices may be used more than once.

simplified	reciprocals	equivalent	denominator
product	factors	fraction	numerator

1. A quotient of two numbers, such as $\frac{5}{8}$, is called a(n) _____.

2. In the fraction $\frac{3}{11}$, the number 3 is called the _____ and the number 11 is called the _____.

3. To factor a number means to write it as a(n) _____.

4. A fraction is said to be _____ when the numerator and the denominator have no common factors other than 1.

5. In $7 \cdot 3 = 21$, the numbers 7 and 3 are called _____ and the number 21 is called the _____.

6. The fractions $\frac{2}{9}$ and $\frac{9}{2}$ are called _____.

7. Fractions that represent the same quantity are called _____ fractions.

Martin-Gay Interactive Videos

See Video 1.3 ◉

Watch the section lecture video and answer the following questions.

OBJECTIVE 1
8. What is the common factor in the numerator and denominator of ▦ Example 1? What principle is used to simplify this fraction?

OBJECTIVE 2
9. During the solving of ▦ Example 3, what two things change in the first step?

OBJECTIVE 3
10. What is the first step needed in order to subtract the fractions in ▦ Example 6 and why?

OBJECTIVE 4
11. For ▦ Example 7, why is the sum not left as $4\frac{7}{6}$?

1.3 Exercise Set MyMathLab®

Represent the shaded part of each geometric figure by a fraction.

1.
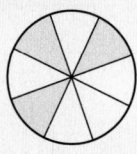

2.
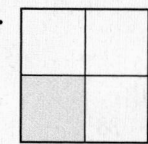

3.
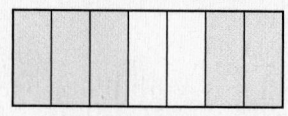

4.
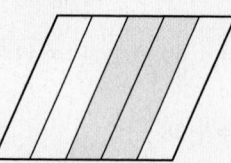

Write each number as a product of primes. See Example 1.

5. 33 **6.** 60 **7.** 98

8. 27 **9.** 20 **10.** 56

11. 75 **12.** 32 **13.** 45

14. 24

Write the fraction in lowest terms. See Example 2.

15. $\frac{2}{4}$ **16.** $\frac{3}{6}$ **17.** $\frac{10}{15}$

18. $\frac{15}{20}$ **19.** $\frac{3}{7}$ **20.** $\frac{5}{9}$

21. $\frac{18}{30}$ **22.** $\frac{42}{45}$ **23.** $\frac{120}{244}$

24. $\frac{360}{700}$

Multiply or divide as indicated. Simplify the answer if possible. See Examples 3 and 4.

25. $\frac{1}{2} \cdot \frac{3}{4}$ **26.** $\frac{7}{11} \cdot \frac{3}{5}$ **27.** $\frac{2}{3} \cdot \frac{3}{4}$

28. $\frac{7}{8} \cdot \frac{3}{21}$ **29.** $\frac{1}{2} \div \frac{7}{12}$ **30.** $\frac{7}{12} \div \frac{1}{2}$

31. $\frac{3}{4} \div \frac{1}{20}$ **32.** $\frac{3}{5} \div \frac{9}{10}$ **33.** $\frac{7}{10} \cdot \frac{5}{21}$

34. $\frac{3}{35} \cdot \frac{10}{63}$ **35.** $\frac{25}{9} \cdot \frac{1}{3}$ **36.** $\frac{1}{4} \cdot \frac{19}{6}$

The area of a plane figure is a measure of the amount of surface of the figure. Find the area of each figure below. (The area of a rectangle is the product of its length and width. The area of a triangle is $\frac{1}{2}$ the product of its base and height.)

37.
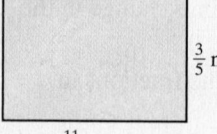

38.
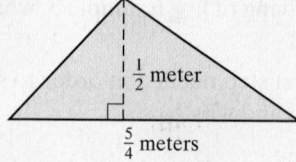

39.
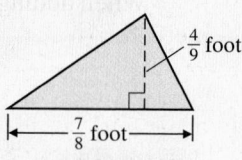

40.
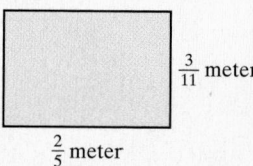

Add or subtract as indicated. Write the answer in lowest terms. See Example 5.

41. $\frac{4}{5} - \frac{1}{5}$ **42.** $\frac{6}{7} - \frac{1}{7}$

43. $\frac{4}{5} + \frac{1}{5}$ **44.** $\frac{6}{7} + \frac{1}{7}$

45. $\frac{17}{21} - \frac{10}{21}$ **46.** $\frac{18}{35} - \frac{11}{35}$

47. $\frac{23}{105} + \frac{4}{105}$ **48.** $\frac{13}{132} + \frac{35}{132}$

Write each fraction as an equivalent fraction with the given denominator. See Example 6.

49. $\frac{7}{10}$ with a denominator of 30

50. $\frac{2}{3}$ with a denominator of 9

51. $\frac{2}{9}$ with a denominator of 18

52. $\frac{8}{7}$ with a denominator of 56

53. $\frac{4}{5}$ with a denominator of 20

54. $\frac{4}{5}$ with a denominator of 25

Add or subtract as indicated. Write the answer in simplest form. See Example 7.

55. $\frac{2}{3} + \frac{3}{7}$ **56.** $\frac{3}{4} + \frac{1}{6}$

57. $\frac{4}{15} - \frac{1}{12}$ **58.** $\frac{11}{12} - \frac{1}{16}$

59. $\frac{5}{22} - \frac{5}{33}$ **60.** $\frac{7}{10} - \frac{8}{15}$

61. $\frac{12}{5} - 1$ **62.** $2 - \frac{3}{8}$

Each circle in Exercises 63–68 represents a whole, or 1. Use subtraction to determine the unknown part of the circle.

63.
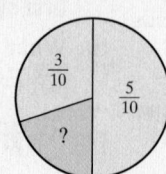

64.
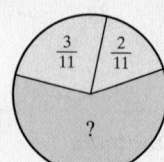

65.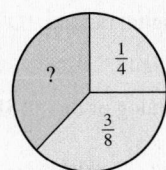

66.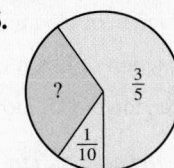

67.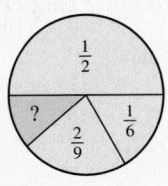

68.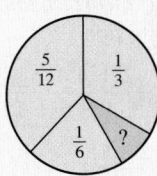

Perform the indicated operations. See Examples 8 and 9.

69. $5\frac{1}{9} \cdot 3\frac{2}{3}$

70. $2\frac{3}{4} \cdot 1\frac{7}{8}$

71. $8\frac{3}{5} \div 2\frac{9}{10}$

72. $1\frac{7}{8} \div 3\frac{8}{9}$

73. $17\frac{2}{5} + 30\frac{2}{3}$

74. $26\frac{11}{20} + 40\frac{7}{10}$

75. $8\frac{11}{12} - 1\frac{5}{6}$

76. $4\frac{7}{8} - 2\frac{3}{16}$

MIXED PRACTICE

Perform the following operations. Write answers in simplest form.

77. $\frac{10}{21} + \frac{5}{21}$

78. $\frac{11}{35} + \frac{3}{35}$

79. $\frac{10}{3} - \frac{5}{21}$

80. $\frac{11}{7} - \frac{3}{35}$

81. $\frac{2}{3} \cdot \frac{3}{5}$

82. $\frac{3}{4} \cdot \frac{7}{12}$

83. $\frac{2}{3} \div \frac{3}{5}$

84. $\frac{3}{4} \div \frac{7}{12}$

85. $5 + \frac{2}{3}$

86. $7 + \frac{1}{10}$

87. $7\frac{2}{5} \div \frac{1}{5}$

88. $9\frac{5}{6} \div \frac{1}{6}$

89. $\frac{1}{2} - \frac{14}{33}$

90. $\frac{7}{15} - \frac{7}{25}$

91. $\frac{23}{105} - \frac{2}{105}$

92. $\frac{57}{132} - \frac{13}{132}$

93. $1\frac{1}{2} + 3\frac{2}{3}$

94. $2\frac{3}{5} + 4\frac{7}{10}$

95. $\frac{2}{3} - \frac{5}{9} + \frac{5}{6}$

96. $\frac{8}{11} - \frac{1}{4} + \frac{1}{2}$

The perimeter of a plane figure is the total distance around the figure. Find the perimeter of each figure in Exercises 97 and 98.

△ **97.**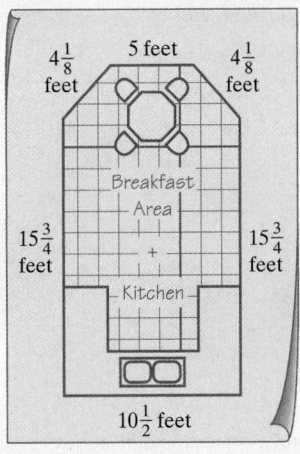

△ **98.**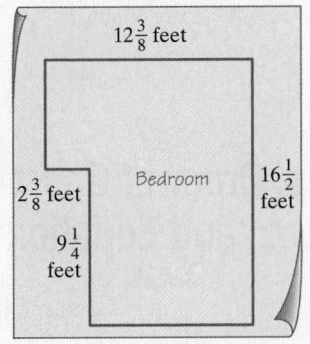

99. In your own words, explain how to add two fractions with different denominators.

100. In your own words, explain how to multiply two fractions.

The following trail chart is given to visitors at the Lakeview Forest Preserve.

Trail Name	Distance (miles)
Robin Path	$3\frac{1}{2}$
Red Falls	$5\frac{1}{2}$
Green Way	$2\frac{1}{8}$
Autumn Walk	$1\frac{3}{4}$

101. How much longer is Red Falls Trail than Green Way Trail?

102. Find the total distance traveled by someone who hiked along all four trails.

CONCEPT EXTENSIONS

The graph shown is called a circle graph or a pie chart. Use the graph to answer Exercises 103 through 106.

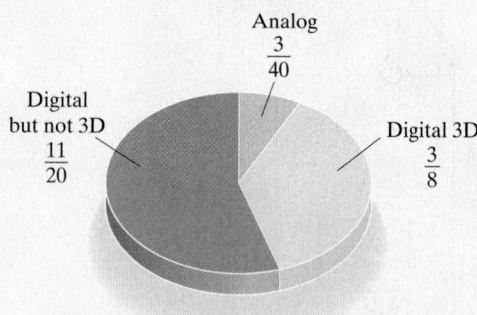

Fraction of U.S. Screens by Type

Analog
$\frac{3}{40}$

Digital but not 3D
$\frac{11}{20}$

Digital 3D
$\frac{3}{8}$

Data from Motion Picture Association of America

103. What fraction of U.S. movie screens are analog?

104. What fraction of U.S. movie screens are digital but not 3D?

105. What fraction of U.S. movie screens are digital?

106. What fraction of U.S. movie screens are analog or digital 3D?

For Exercises 107 through 110, determine whether the work is correct or incorrect. If incorrect, find the error and correct. See the Concept Check in this section.

107. $\frac{12}{24} \overset{?}{=} \frac{2 + 4 + 6}{2 + 4 + 6 + 12} = \frac{1}{12}$

108. $\frac{30}{60} \overset{?}{=} \frac{2 \cdot 3 \cdot 5}{2 \cdot 2 \cdot 3 \cdot 5} = \frac{1}{2}$

109. $\frac{2}{7} + \frac{9}{7} \overset{?}{=} \frac{11}{14}$

110. $\frac{16}{28} \overset{?}{=} \frac{2 \cdot 5 + 6 \cdot 1}{2 \cdot 5 + 6 \cdot 3} = \frac{1}{3}$

1.4 Exponents, Order of Operations, Variable Expressions, and Equations

OBJECTIVES

1 Define and Use Exponents and the Order of Operations.

2 Evaluate Algebraic Expressions, Given Replacement Values for Variables.

3 Determine Whether a Number Is a Solution of a Given Equation.

4 Translate Phrases into Expressions and Sentences into Statements.

OBJECTIVE

1 Using Exponents and the Order of Operations

Frequently in algebra, products occur that contain repeated multiplication of the same factor. For example, the volume of a cube whose sides each measure 2 centimeters is $(2 \cdot 2 \cdot 2)$ cubic centimeters. We may use **exponential notation** to write such products in a more compact form. For example,

$$2 \cdot 2 \cdot 2 \quad \text{may be written as} \quad 2^3.$$

The 2 in 2^3 is called the **base**; it is the repeated factor. The 3 in 2^3 is called the **exponent** and is the number of times the base is used as a factor. The expression 2^3 is called an **exponential expression**.

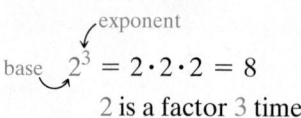

base $2^3 = 2 \cdot 2 \cdot 2 = 8$
exponent
2 is a factor 3 times

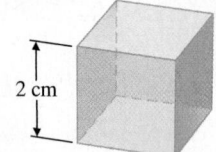

2 cm

Volume is $(2 \cdot 2 \cdot 2)$ cubic centimeters.

EXAMPLE 1 Evaluate the following.

a. 3^2 [read as "3 squared" or as "3 to the second power"]

b. 5^3 [read as "5 cubed" or as "5 to the third power"]

c. 2^4 [read as "2 to the fourth power"]

d. 7^1 **e.** $\left(\frac{3}{7}\right)^2$

Solution

a. $3^2 = 3 \cdot 3 = 9$ **b.** $5^3 = 5 \cdot 5 \cdot 5 = 125$

c. $2^4 = 2 \cdot 2 \cdot 2 \cdot 2 = 16$ **d.** $7^1 = 7$

e. $\left(\frac{3}{7}\right)^2 = \left(\frac{3}{7}\right)\left(\frac{3}{7}\right) = \frac{9}{49}$

PRACTICE

1   Evaluate.

    a. 1^3 **b.** 5^2 **c.** $\left(\dfrac{1}{10}\right)^2$ **d.** 9^1 **e.** $\left(\dfrac{2}{5}\right)^3$

> **Helpful Hint**
>
> $2^3 \neq 2 \cdot 3$ since 2^3 indicates repeated **multiplication** of the same factor.
>
> $$2^3 = 2 \cdot 2 \cdot 2 = 8, \text{ whereas } 2 \cdot 3 = 6.$$

Using symbols for mathematical operations is a great convenience. However, the more operation symbols present in an expression, the more careful we must be when performing the indicated operation. For example, in the expression $2 + 3 \cdot 7$, do we add first or multiply first? To eliminate confusion, **grouping symbols** are used. Examples of grouping symbols are parentheses ( ), brackets [ ], braces { }, and the fraction bar. If we wish $2 + 3 \cdot 7$ to be simplified by adding first, we enclose $2 + 3$ in parentheses.

$$(2 + 3) \cdot 7 = 5 \cdot 7 = 35$$

If we wish to multiply first, $3 \cdot 7$ may be enclosed in parentheses.

$$2 + (3 \cdot 7) = 2 + 21 = 23$$

    To eliminate confusion when no grouping symbols are present, use the following agreed-upon order of operations.

> **Order of Operations**
>
> Simplify expressions using the order below. If grouping symbols such as parentheses are present, simplify expressions within those first, starting with the innermost set. If fraction bars are present, simplify the numerator and the denominator separately.
>
> **1.** Evaluate exponential expressions.
>
> **2.** Perform multiplications or divisions in order from left to right.
>
> **3.** Perform additions or subtractions in order from left to right.

Now simplify $2 + 3 \cdot 7$. There are no grouping symbols and no exponents, so we multiply and then add.

$$2 + 3 \cdot 7 = 2 + 21 \quad \text{Multiply.}$$
$$= 23 \qquad \text{Add.}$$

EXAMPLE 2   Simplify each expression.

    a. $6 \div 3 + 5^2$ **b.** $20 \div 5 \cdot 4$ **c.** $\dfrac{2(12 + 3)}{|-15|}$ **d.** $3 \cdot 4^2$ **e.** $\dfrac{3}{2} \cdot \dfrac{1}{2} - \dfrac{1}{2}$

**Solution**

    a. Evaluate 5^2 first.

$$6 \div 3 + 5^2 = 6 \div 3 + 25$$

Next divide, then add.

$$= 2 + 25 \quad \text{Divide.}$$
$$= 27 \qquad \text{Add.}$$

    b. $20 \div 5 \cdot 4 = 4 \cdot 4$
$$= 16$$

> **Helpful Hint**
>
> Remember to multiply or divide in order from left to right.

c. First, simplify the numerator and the denominator separately.

$$\frac{2(12 + 3)}{|-15|} = \frac{2(15)}{15} \quad \text{Simplify numerator and denominator separately.}$$

$$= \frac{30}{15}$$

$$= 2 \quad \text{Simplify.}$$

d. In this example, only the 4 is squared. The factor of 3 is not part of the base because no grouping symbol includes it as part of the base.

$$3 \cdot 4^2 = 3 \cdot 16 \quad \text{Evaluate the exponential expression.}$$

$$= 48 \quad \text{Multiply.}$$

e. The order of operations applies to operations with fractions in exactly the same way as it applies to operations with whole numbers.

$$\frac{3}{2} \cdot \frac{1}{2} - \frac{1}{2} = \frac{3}{4} - \frac{1}{2} \quad \text{Multiply.}$$

$$= \frac{3}{4} - \frac{2}{4} \quad \text{The least common denominator is 4.}$$

$$= \frac{1}{4} \quad \text{Subtract.}$$

PRACTICE
2 Simplify each expression.

a. $6 + 3 \cdot 9$ 　　　**b.** $4^3 \div 8 + 3$ 　　　**c.** $\left(\dfrac{2}{3}\right)^2 \cdot |-8|$

d. $\dfrac{9(14 - 6)}{|-2|}$ 　　　**e.** $\dfrac{7}{4} \cdot \dfrac{1}{4} - \dfrac{1}{4}$

> **Helpful Hint**
>
> Be careful when evaluating an exponential expression. In $3 \cdot 4^2$, the exponent 2 applies only to the base 4. In $(3 \cdot 4)^2$, the parentheses are a grouping symbol, so the exponent 2 applies to the product $3 \cdot 4$. Thus, we multiply first.
>
> $$3 \cdot 4^2 = 3 \cdot 16 = 48 \qquad (3 \cdot 4)^2 = (12)^2 = 144$$

Expressions that include many grouping symbols can be confusing. When simplifying these expressions, keep in mind that grouping symbols separate the expression into distinct parts. Each is then simplified separately.

EXAMPLE 3 Simplify: $\dfrac{3 + |4 - 3| + 2^2}{6 - 3}$

Solution The fraction bar serves as a grouping symbol and separates the numerator and denominator. Simplify each separately. Also, the absolute value bars here serve as a grouping symbol. We begin in the numerator by simplifying within the absolute value bars.

$$\frac{3 + |4 - 3| + 2^2}{6 - 3} = \frac{3 + |1| + 2^2}{6 - 3} \quad \text{Simplify the expression inside the absolute value bars.}$$

$$= \frac{3 + 1 + 2^2}{3} \quad \text{Find the absolute value and simplify the denominator.}$$

$$= \frac{3 + 1 + 4}{3} \quad \text{Evaluate the exponential expression.}$$

$$= \frac{8}{3} \quad \text{Simplify the numerator.}$$

PRACTICE
3 Simplify: $\dfrac{6^2 - 5}{3 + |6 - 5| \cdot 8}$

EXAMPLE 4   Simplify: $3[4 + 2(10 - 1)]$

Solution   Notice that both parentheses and brackets are used as grouping symbols. Start with the innermost set of grouping symbols.

$$
\begin{aligned}
3[4 + 2(10 - 1)] &= 3[4 + 2(9)] && \text{Simplify the expression in parentheses.} \\
&= 3[4 + 18] && \text{Multiply.} \\
&= 3[22] && \text{Add.} \\
&= 66 && \text{Multiply.}
\end{aligned}
$$

> **Helpful Hint**
> Be sure to follow order of operations and resist the temptation to incorrectly add 4 and 2 first.

PRACTICE
4   Simplify: $4[25 - 3(5 + 3)]$

EXAMPLE 5   Simplify: $\dfrac{8 + 2 \cdot 3}{2^2 - 1}$

Solution

$$
\frac{8 + 2 \cdot 3}{2^2 - 1} = \frac{8 + 6}{4 - 1} = \frac{14}{3}
$$

PRACTICE
5   Simplify: $\dfrac{36 \div 9 + 5}{5^2 - 3}$

OBJECTIVE
2   Evaluating Algebraic Expressions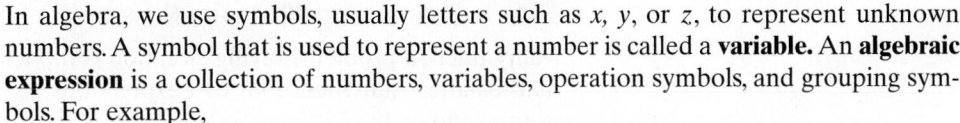

In algebra, we use symbols, usually letters such as *x, y,* or *z,* to represent unknown numbers. A symbol that is used to represent a number is called a **variable.** An **algebraic expression** is a collection of numbers, variables, operation symbols, and grouping symbols. For example,

$$
2x, \quad -3, \quad 2x + 10, \quad 5(p^2 + 1), \quad \text{and} \quad \frac{3y^2 - 6y + 1}{5}
$$

are algebraic expressions.

Expression	Meaning
$2x$	$2 \cdot x$
$5(p^2 + 1)$	$5 \cdot (p^2 + 1)$
$3y^2$	$3 \cdot y^2$
xy	$x \cdot y$

If we give a specific value to a variable, we can **evaluate an algebraic expression.** To evaluate an algebraic expression means to find its numerical value once we know the values of the variables.

Algebraic expressions are often used in problem solving. For example, the expression

$$
16t^2
$$

gives the distance in feet (neglecting air resistance) that an object will fall in *t* seconds.

EXAMPLE 6 Evaluate each expression if $x = 3$ and $y = 2$.

a. $2x - y$ **b.** $\dfrac{3x}{2y}$ **c.** $\dfrac{x}{y} + \dfrac{y}{2}$ **d.** $x^2 - y^2$

Solution

a. Replace x with 3 and y with 2.

$$2x - y = 2(3) - 2 \quad \text{Let } x = 3 \text{ and } y = 2.$$
$$= 6 - 2 \quad \text{Multiply.}$$
$$= 4 \quad \text{Subtract.}$$

b. $\dfrac{3x}{2y} = \dfrac{3 \cdot 3}{2 \cdot 2} = \dfrac{9}{4}$ Let $x = 3$ and $y = 2$.

c. Replace x with 3 and y with 2. Then simplify.

$$\frac{x}{y} + \frac{y}{2} = \frac{3}{2} + \frac{2}{2} = \frac{5}{2}$$

d. Replace x with 3 and y with 2.

$$x^2 - y^2 = 3^2 - 2^2 = 9 - 4 = 5$$

PRACTICE
6 Evaluate each expression if $x = 2$ and $y = 5$.

a. $2x + y$ **b.** $\dfrac{4x}{3y}$ **c.** $\dfrac{3}{x} + \dfrac{x}{y}$ **d.** $x^3 + y^2$

OBJECTIVE
3 Determining Whether a Number Is a Solution of an Equation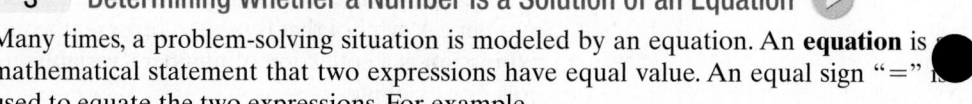

Many times, a problem-solving situation is modeled by an equation. An **equation** is a mathematical statement that two expressions have equal value. An equal sign "=" is used to equate the two expressions. For example,

$$3 + 2 = 5, \quad 7x = 35, \quad \frac{2(x - 1)}{3} = 0, \text{ and } I = PRT \text{ are all equations.}$$

> **Helpful Hint**
> An equation contains an equal sign "=". An algebraic expression does not.

> ✔ **CONCEPT CHECK**
> Which of the following are equations? Which are expressions?
> **a.** $5x = 8$ **b.** $5x - 8$ **c.** $12y + 3x$ **d.** $12y = 3x$

When an equation contains a variable, deciding which values of the variable make the equation a true statement is called **solving** the equation for the variable. A **solution** of an equation is a value for the variable that makes the equation true. For example, 3 is a solution of the equation $x + 4 = 7$ because if x is replaced with 3, the statement is true.

$$x + 4 = 7$$
$$\downarrow$$
$$3 + 4 = 7 \quad \text{Replace } x \text{ with 3.}$$
$$7 = 7 \quad \text{True}$$

Similarly, 1 is not a solution of the equation $x + 4 = 7$ because $1 + 4 = 7$ is **not** a true statement.

EXAMPLE 7   Decide whether 2 is a solution of $3x + 10 = 8x$.

Solution   Replace x with 2 and see if a true statement results.

$$3x + 10 = 8x \qquad \text{Original equation}$$
$$3(2) + 10 \stackrel{?}{=} 8(2) \qquad \text{Replace } x \text{ with 2.}$$
$$6 + 10 \stackrel{?}{=} 16 \qquad \text{Simplify each side.}$$
$$16 = 16 \qquad \text{True}$$

Since we arrived at a true statement after replacing x with 2 and simplifying both sides of the equation, 2 is a solution of the equation. □

PRACTICE
7   Decide whether 4 is a solution of $9x - 6 = 7x$. ■

OBJECTIVE
4   Translating Phrases to Expressions and Sentences to Statements ▶

Now that we know how to represent an unknown number by a variable, let's practice translating phrases into algebraic expressions and sentences into statements. Oftentimes, solving problems requires the ability to translate word phrases and sentences into symbols. Below is a list of some key words and phrases to help us translate.

> **Helpful Hint**
> Order matters when subtracting and dividing, so be especially careful with these translations.

Addition (+)	Subtraction (−)	Multiplication (·)	Division (÷)	Equality (=)
Sum	Difference of	Product	Quotient	Equals
Plus	Minus	Times	Divide	Gives
Added to	Subtracted from	Multiply	Into	Is/Was/Should be
More than	Less than	Twice	Ratio	Yields
Increased by	Decreased by	Of	Divided by	Amounts to
Total	Less			Represents/ Is the same as

EXAMPLE 8   Write an algebraic expression that represents each phrase. Let the variable x represent the unknown number.

   a. The sum of a number and 3   **b.** The product of 3 and a number
   c. Twice a number   **d.** 10 decreased by a number
   e. 5 times a number, increased by 7

Solution

   a. $x + 3$ since "sum" means to add
   b. $3 \cdot x$ and $3x$ are both ways to denote the product of 3 and x
   c. $2 \cdot x$ or $2x$
   d. $10 - x$ because "decreased by" means to subtract
   e. $\underbrace{5x}_{5 \text{ times a number}} + 7$ □

PRACTICE
8   Write an algebraic expression that represents each phrase. Let the variable x represent the unknown number.

   a. Six times a number   **b.** A number decreased by 8
   c. The product of a number and 9   **d.** Two times a number, plus 3
   e. The sum of 7 and a number ■

> **Helpful Hint**
>
> Make sure you understand the difference when translating phrases containing "decreased by," "subtracted from," and "less than."
>
> | Phrase | Translation | |
> |--------|-------------|---|
> | A number decreased by 10 | $x - 10$ | |
> | A number subtracted from 10 | $10 - x$ | Notice the order. |
> | 10 less than a number | $x - 10$ | |
> | A number less 10 | $x - 10$ | |

Now let's practice translating sentences into equations.

EXAMPLE 9 Write each sentence as an equation or inequality. Let x represent the unknown number.

    a. The quotient of 15 and a number is 4.

    b. Three subtracted from 12 is a number.

    c. Four times a number, added to 17, is not equal to 21.

    d. Triple a number is less than 48.

Solution

a. In words:

the quotient of 15 and a number	is	4
↓	↓	↓

Translate: $\dfrac{15}{x} = 4$

b. In words:

three subtracted **from** 12	is	a number
↓	↓	↓

Translate: $12 - 3 = x$

Care must be taken when the operation is subtraction. The expression $3 - 12$ would be incorrect. Notice that $3 - 12 \neq 12 - 3$.

c. In words:

four times a number	added to	17	is not equal to	21
↓	↓	↓	↓	↓

Translate: $4x + 17 \neq 21$

d. In words:

triple a number	is less than	48
↓	↓	↓

Translate: $3x < 48$

PRACTICE

9 Write each sentence as an equation or inequality. Let x represent the unknown number.

    a. A number increased by 7 is equal to 13.

    b. Two less than a number is 11.

    c. Double a number, added to 9, is not equal to 25.

    d. Five times 11 is greater than or equal to an unknown number.

Graphing Calculator Explorations

Exponents

To evaluate exponential expressions on a scientific calculator, find the key marked $\boxed{y^x}$ or $\boxed{\wedge}$. To evaluate, for example, 3^5, press the following keys: $\boxed{3}\ \boxed{y^x}\ \boxed{5}\ \boxed{=}$ or $\boxed{3}\ \boxed{\wedge}\ \boxed{5}\ \boxed{=}$.

$$\updownarrow \text{ or}$$
$$\boxed{\text{ENTER}}$$

The display should read $\boxed{\quad 243}$ or $\boxed{\begin{array}{l} 3\wedge 5 \\ \qquad 243 \end{array}}$

Order of Operations

Some calculators follow the order of operations, and others do not. To see whether your calculator has the order of operations built in, use your calculator to find $2 + 3 \cdot 4$. To do this, press the following sequence of keys:

$$\boxed{2}\ \boxed{+}\ \boxed{3}\ \boxed{\times}\ \boxed{4}\ \boxed{=}.$$
$$\updownarrow \text{ or}$$
$$\boxed{\text{ENTER}}$$

The correct answer is 14 because the order of operations is to multiply before we add. If the calculator displays $\boxed{\quad 14}$, then it has the order of operations built in.

Even if the order of operations is built in, parentheses must sometimes be inserted. For example, to simplify $\dfrac{5}{12 - 7}$, press the keys

$$\boxed{5}\ \boxed{\div}\ \boxed{(}\ \boxed{1}\ \boxed{2}\ \boxed{-}\ \boxed{7}\ \boxed{)}\ \boxed{=}.$$
$$\updownarrow \text{ or}$$
$$\boxed{\text{ENTER}}$$

The display should read $\boxed{\quad 1}$ or $\boxed{\begin{array}{l} 5/(12 - 7) \\ \qquad\qquad 1 \end{array}}$

Use a calculator to evaluate each expression.

1. 5^4 **2.** 7^4

3. 9^5 **4.** 8^6

5. $2(20 - 5)$ **6.** $3(14 - 7) + 21$

7. $24(862 - 455) + 89$ **8.** $99 + (401 + 962)$

9. $\dfrac{4623 + 129}{36 - 34}$ **10.** $\dfrac{956 - 452}{89 - 86}$

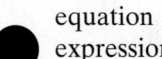

Vocabulary, Readiness & Video Check

Use the choices below to fill in each blank.

equation	variable	base	grouping
expression	solution	solving	exponent

1. In the expression 5^2, the 5 is called the _____ and the 2 is called the _____.

2. The symbols ( ), [ ], and { } are examples of _____ symbols.

3. A symbol that is used to represent a number is called a(n) _____.

4. A collection of numbers, variables, operation symbols, and grouping symbols is called a(n) _____.

5. A mathematical statement that two expressions are equal is called a(n) _____.

6. A value for the variable that makes an equation a true statement is called a(n) _____.

7. Deciding what values of a variable make an equation a true statement is called _____ the equation.

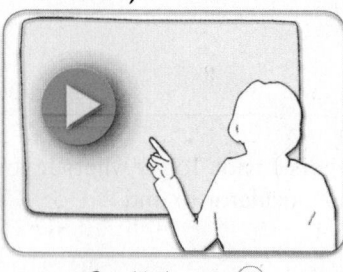

Martin-Gay Interactive Videos

Watch the section lecture video and answer the following questions.

OBJECTIVE 1

8. In ▦ Example 3 and the lecture before, what is the main point made about the order of operations?

OBJECTIVE 2

9. What happens with the replacement value for z in ▦ Example 6 and why?

OBJECTIVE 3

10. Is the value 0 a solution of the equation given in ▦ Example 9? How is this determined?

OBJECTIVE 4

11. Earlier in this video the point was made that equations have =, while expressions do not. In the lecture before ▦ Example 10, translating from English to math is discussed and another difference between expressions and equations is explained. What is it?

See Video 1.4 ◉

1.4   Exercise Set   MyMathLab® ▶

Evaluate. See Example 1.

1. 3^5

2. 2^5

▶ 3. 3^3

4. 4^4

5. 1^5

6. 1^8

7. 5^1

8. 8^1

9. 7^2

10. 9^2

▶ 11. $\left(\dfrac{2}{3}\right)^4$

12. $\left(\dfrac{6}{11}\right)^2$

13. $\left(\dfrac{1}{5}\right)^3$

14. $\left(\dfrac{1}{2}\right)^5$

15. $(1.2)^2$

16. $(1.5)^2$

17. $(0.04)^3$

18. $(0.03)^3$

MIXED PRACTICE

Simplify each expression. See Examples 2 through 5.

▶ 19. $5 + 6 \cdot 2$

20. $8 + 5 \cdot 3$

21. $4 \cdot 8 - 6 \cdot 2$

22. $12 \cdot 5 - 3 \cdot 6$

23. $2(8 - 3)$

24. $5(6 - 2)$

25. $2 + (5 - 2) + 4^2$

26. $6 - 2 \cdot 2 + 2^5$

27. $5 \cdot 3^2$

28. $2 \cdot 5^2$

29. $\dfrac{1}{4} \cdot \dfrac{2}{3} - \dfrac{1}{6}$

30. $\dfrac{3}{4} \cdot \dfrac{1}{2} + \dfrac{2}{3}$

▶ 31. $2[5 + 2(8 - 3)]$

32. $3[4 + 3(6 - 4)]$

33. $\dfrac{19 - 3 \cdot 5}{6 - 4}$

34. $\dfrac{4 \cdot 3 + 2}{4 + 3 \cdot 2}$

▶ 35. $\dfrac{|6 - 2| + 3}{8 + 2 \cdot 5}$

36. $\dfrac{15 - |3 - 1|}{12 - 3 \cdot 2}$

37. $\dfrac{3 + 3(5 + 3)}{3^2 + 1}$

38. $\dfrac{3 + 6(8 - 5)}{4^2 + 2}$

39. $\dfrac{6 + |8 - 2| + 3^2}{18 - 3}$

40. $\dfrac{16 + |13 - 5| + 4^2}{17 - 5}$

41. $2 + 3[10(4 \cdot 5 - 16) - 30]$

42. $3 + 4[8(5 \cdot 5 - 20) - 39]$

43. $\left(\dfrac{2}{3}\right)^3 + \dfrac{1}{9} + \dfrac{1}{3} \cdot \dfrac{4}{3}$

44. $\left(\dfrac{3}{8}\right)^2 + \dfrac{1}{4} + \dfrac{1}{8} \cdot \dfrac{3}{2}$

For Exercises 45 and 46, match each expression in the first column with its value in the second column.

45.
a. $(6 + 2) \cdot (5 + 3)$	19
b. $(6 + 2) \cdot 5 + 3$	22
c. $6 + 2 \cdot 5 + 3$	64
d. $6 + 2 \cdot (5 + 3)$	43

46.
a. $(1 + 4) \cdot 6 - 3$	15
b. $1 + 4 \cdot (6 - 3)$	13
c. $1 + 4 \cdot 6 - 3$	27
d. $(1 + 4) \cdot (6 - 3)$	22

Evaluate each expression when $x = 1$, $y = 3$, and $z = 5$. See Example 6.

47. $3y$

48. $4x$

49. $\dfrac{z}{5x}$

50. $\dfrac{y}{2z}$

51. $3x - 2$

52. $6y - 8$

53. $|2x + 3y|$

54. $|5z - 2y|$

55. $xy + z$

56. $yz - x$

57. $5y^2$

58. $2z^2$

Evaluate each expression if $x = 12$, $y = 8$, and $z = 4$. See Example 6.

59. $\dfrac{x}{z} + 3y$

60. $\dfrac{y}{z} + 8x$

61. $x^2 - 3y + x$

62. $y^2 - 3x + y$

63. $\dfrac{x^2 + z}{y^2 + 2z}$

64. $\dfrac{y^2 + x}{x^2 + 3y}$

Neglecting air resistance, the expression $16t^2$ gives the distance in feet an object will fall in t seconds.

65. Complete the chart below. To evaluate $16t^2$, remember to first find t^2, then multiply by 16.

Time t (in seconds)	Distance $16t^2$ (in feet)
1	
2	
3	
4	

66. Does an object fall the same distance *during* each second? Why or why not? (See Exercise 65.)

Decide whether the given number is a solution of the given equation. See Example 7.

67. Is 5 a solution of $3x + 30 = 9x$?

68. Is 6 a solution of $2x + 7 = 3x$?

69. Is 0 a solution of $2x + 6 = 5x - 1$?

70. Is 2 a solution of $4x + 2 = x + 8$?

71. Is 8 a solution of $2x - 5 = 5$?

72. Is 6 a solution of $3x - 10 = 8$?

73. Is 2 a solution of $x + 6 = x + 6$?

74. Is 10 a solution of $x + 6 = x + 6$?

75. Is 0 a solution of $x = 5x + 15$?

76. Is 1 a solution of $4 = 1 - x$?

TRANSLATING

Write each phrase as an algebraic expression. Let x represent the unknown number. See Example 8.

77. Fifteen more than a number

78. A number increased by 9

79. Five subtracted from a number

80. Five decreased by a number

81. The ratio of a number and 4

82. The quotient of a number and 9

83. Three times a number, increased by 22

84. Twice a number, decreased by 72

TRANSLATING

Write each sentence as an equation or inequality. Use x to represent any unknown number. See Example 9.

85. One increased by two equals the quotient of nine and three.

86. Four subtracted from eight is equal to two squared.

87. Three is not equal to four divided by two.

88. The difference of sixteen and four is greater than ten.

89. The sum of 5 and a number is 20.

90. Seven subtracted from a number is 0.

91. The product of 7.6 and a number is 17.

92. 9.1 times a number equals 4.

93. Thirteen minus three times a number is 13.

94. Eight added to twice a number is 42.

CONCEPT EXTENSIONS

Fill in each blank with one of the following:

add   subtract   multiply   divide

95. To simplify the expression $1 + 3 \cdot 6$, first _____.

96. To simplify the expression $(1 + 3) \cdot 6$, first _____.

97. To simplify the expression $(20 - 4) \cdot 2$, first _____.

98. To simplify the expression $20 - 4 \div 2$, first _____.

99. Are parentheses necessary in the expression $2 + (3 \cdot 5)$? Explain your answer.

100. Are parentheses necessary in the expression $(2 + 3) \cdot 5$? Explain your answer.

△ *Recall that perimeter measures the distance around a plane figure and area measures the amount of surface of a plane figure. The expression $2l + 2w$ gives the perimeter of the rectangle below (measured in units), and the expression lw gives its area (measured in square units). Complete the chart below for the given lengths and widths. Be sure to include units.*

	Length: l	Width: w	Perimeter of Rectangle: $2l + 2w$	Area of Rectangle: lw
101.	4 in.	3 in.		
102.	6 in.	1 in.		
103.	5.3 in.	1.7 in.		
104.	4.6 in.	2.4 in.		

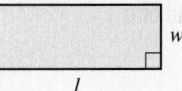

105. Study the perimeters and areas found in the chart on the previous page. Do you notice any trends?

106. In your own words, explain the difference between an expression and an equation.

107. Insert one set of parentheses so that the following expression simplifies to 32.

$$20 - 4 \cdot 4 \div 2$$

108. Insert one set of parentheses so that the following expression simplifies to 28.

$$2 \cdot 5 + 3^2$$

Determine whether each is an expression or an equation. See the Concept Check in this section.

109. a. $5x + 6$

  b. $2a = 7$

  c. $3a + 2 = 9$

  d. $4x + 3y - 8z$

  e. $5^2 - 2(6 - 2)$

110. a. $3x^2 - 26$

  b. $3x^2 - 26 = 1$

  c. $2x - 5 = 7x - 5$

  d. $9y + x - 8$

  e. $3^2 - 4(5 - 3)$

111. Why is 4^3 usually read as "four cubed"? (*Hint:* What is the volume of the **cube** below?)

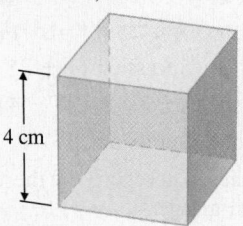

4 cm

112. Why is 8^2 usually read as "eight squared"? (*Hint:* What is the area of the **square** below?)

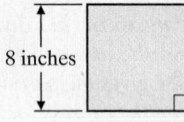

8 inches

113. Write any expression, using 3 or more numbers, that simplifies to 11.

114. Write any expression, using 4 or more numbers, that simplifies to 7.

115. The area of a figure is the total enclosed surface of the figure. Area is measured in square units. The expression lw represents the area of a rectangle when l is its length and w is its width. Find the area of the following rectangular shaped lot.

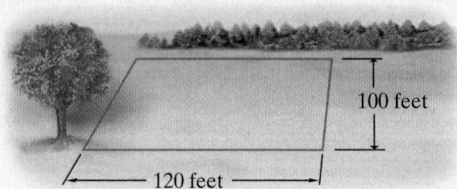

100 feet

120 feet

116. A trapezoid is a four-sided figure with exactly one pair of parallel sides. The expression $\frac{1}{2}h(B + b)$ represents its area, when B and b are the lengths of the two parallel sides and h is the height between these sides. Find the area if $B = 15$ inches, $b = 7$ inches, and $h = 5$ inches.

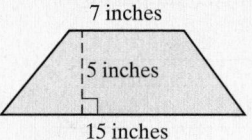

7 inches

5 inches

15 inches

117. The expression $\frac{d}{t}$ represents the average speed in miles per hour if a distance of d miles is traveled in t hours. Find the speed to the nearest whole number if the distance between Dallas, Texas, and Kaw City, Oklahoma, is 432 miles, and it takes Peter Callac 8.5 hours to drive the distance.

118. The expression $\frac{I}{PT}$ represents the rate of interest being charged if a loan of P dollars for T years required I dollars in interest to be paid. Find the interest rate if a $650 loan for 3 years to buy a used IBM personal computer requires $126.75 in interest to be paid.

1.5 Adding Real Numbers

OBJECTIVES

1 Add Real Numbers.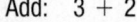

2 Solve Applications That Involve Addition of Real Numbers.

3 Find the Opposite of a Number.

OBJECTIVE

1 Adding Real Numbers

Real numbers can be added, subtracted, multiplied, divided, and raised to powers, just as whole numbers can. We use a number line to help picture the addition of real numbers. We begin by adding numbers with the same sign.

EXAMPLE 1   Add: $3 + 2$

<u>Solution</u> Recall that 3 and 2 are called **addends.** We start at 0 on a number line and draw an arrow representing the addend 3. This arrow is three units long and points to

the right since 3 is positive. From the tip of this arrow, we draw another arrow, representing the addend 2. The number below the tip of this arrow is the sum, 5.

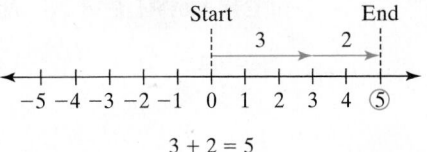

$$3 + 2 = 5$$

PRACTICE

1 Add using a number line: $2 + 4$

EXAMPLE 2 Add: $-1 + (-2)$

Solution Here, -1 and -2 are addends. We start at 0 on a number line and draw an arrow representing -1. This arrow is one unit long and points to the left since -1 is negative. From the tip of this arrow, we draw another arrow, representing -2. The number below the tip of this arrow is the sum, -3.

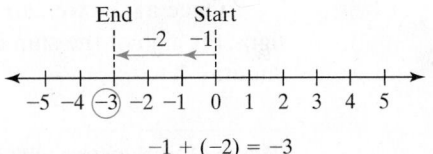

$$-1 + (-2) = -3$$

PRACTICE

2 Add using a number line: $-2 + (-3)$

Thinking of signed numbers as money earned or lost might help make addition more meaningful. Earnings can be thought of as positive numbers. If \$1 is earned and later another \$3 is earned, the total amount earned is \$4. In other words, $1 + 3 = 4$.

On the other hand, losses can be thought of as negative numbers. If \$1 is lost and later another \$3 is lost, a total of \$4 is lost. In other words,

$$(-1) + (-3) = -4.$$

Using a number line each time we add two numbers can be time consuming. Instead, we can notice patterns in the previous examples and write rules for adding signed numbers. When adding two numbers with the same sign, notice that the sign of the sum is the same as the sign of the addends.

Adding Two Numbers with the Same Sign

Add their absolute values. Use their common sign as the sign of the sum.

EXAMPLE 3 Add.

a. $-3 + (-7)$ **b.** $-1 + (-20)$ **c.** $-2 + (-10)$

Solution Notice that each time, we are adding numbers with the same sign.

a. $-3 + (-7) = -10$ ← Add their absolute values: $3 + 7 = 10$.
   └── Use their common sign.

b. $-1 + (-20) = -21$ ← Add their absolute values: $1 + 20 = 21$.
   └── Common sign.

c. $-2 + (-10) = -12$ ← Add their absolute values.
   └── Common sign.

PRACTICE

3 Add. **a.** $-5 + (-8)$ **b.** $-31 + (-1)$

Adding numbers whose signs are not the same can also be pictured on a number line.

EXAMPLE 4 Add: $-4 + 6$

Solution

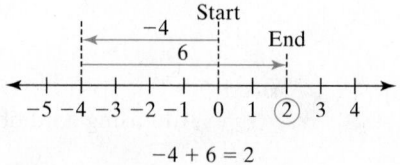

$$-4 + 6 = 2$$

PRACTICE
4 Add using a number line: $-3 + 8$

Using temperature as an example, if a thermometer registers 4 degrees below 0 degrees and then rises 6 degrees, the new temperature is 2 degrees above 0 degrees. Thus, it is reasonable that $-4 + 6 = 2$.

Once again, we can observe a pattern: when adding two numbers with different signs, the sign of the sum is the same as the sign of the addend whose absolute value is larger.

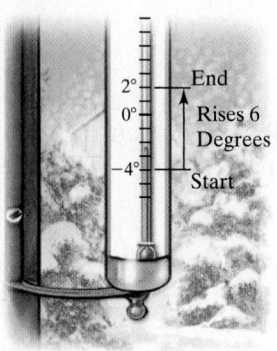

Adding Two Numbers with Different Signs

Subtract the smaller absolute value from the larger absolute value. Use the sign of the number whose absolute value is larger as the sign of the sum.

EXAMPLE 5 Add.

   a. $3 + (-7)$ **b.** $-2 + 10$ **c.** $0.2 + (-0.5)$

Solution Notice that each time, we are adding numbers with different signs.

   a. $3 + (-7) = -4$ ← Subtract their absolute values: $7 - 3 = 4$.
                         The negative number, -7, has the larger absolute value so the sum is negative.

   b. $-2 + 10 = 8$ ← Subtract their absolute values: $10 - 2 = 8$.
                         The positive number, 10, has the larger absolute value so the sum is positive.

   c. $0.2 + (-0.5) = -0.3$ ← Subtract their absolute values: $0.5 - 0.2 = 0.3$.
                         The negative number, -0.5, has the larger absolute value so the sum is negative.

PRACTICE
5 Add.

   a. $15 + (-18)$ **b.** $-19 + 20$ **c.** $-0.6 + 0.4$

In general, we have the following:

Adding Real Numbers

To add two real numbers
1. with the *same sign*, add their absolute values. Use their common sign as the sign of the answer.
2. with *different signs*, subtract their absolute values. Give the answer the same sign as the number with the larger absolute value.

EXAMPLE 6   Add.

    a. $-8 + (-11)$ **b.** $-5 + 35$ **c.** $0.6 + (-1.1)$

    d. $-\dfrac{7}{10} + \left(-\dfrac{1}{10}\right)$ **e.** $11.4 + (-4.7)$ **f.** $-\dfrac{3}{8} + \dfrac{2}{5}$

Solution

    a. $-8 + (-11) = -19$ Same sign. Add absolute values and use the common sign.

    b. $-5 + 35 = 30$ Different signs. Subtract absolute values and use the sign of the number with the larger absolute value.

    c. $0.6 + (-1.1) = -0.5$ Different signs.

    d. $-\dfrac{7}{10} + \left(-\dfrac{1}{10}\right) = -\dfrac{8}{10} = -\dfrac{4}{5}$ Same sign.

    e. $11.4 + (-4.7) = 6.7$

    f. $-\dfrac{3}{8} + \dfrac{2}{5} = -\dfrac{15}{40} + \dfrac{16}{40} = \dfrac{1}{40}$ □

> **Helpful Hint**
>
> Don't forget that a common denominator is needed when adding or subtracting fractions. The common denominator here is 40.

PRACTICE
6   Add.

    a. $-\dfrac{3}{5} + \left(-\dfrac{2}{5}\right)$ **b.** $3 + (-9)$

    c. $2.2 + (-1.7)$ **d.** $-\dfrac{2}{7} + \dfrac{3}{10}$ ■

EXAMPLE 7   Add.

    a. $3 + (-7) + (-8)$ **b.** $[7 + (-10)] + [-2 + |-4|]$

Solution

    a. Perform the additions from left to right.

$$3 + (-7) + (-8) = -4 + (-8) \quad \text{Adding numbers with different signs.}$$
$$= -12 \quad \text{Adding numbers with like signs.}$$

    b. Simplify inside brackets first.

$$[7 + (-10)] + [-2 + |-4|] = [-3] + [-2 + 4]$$
$$= [-3] + [2]$$
$$= -1 \quad \text{Add.} \quad □$$

> **Helpful Hint**
>
> Don't forget that the brackets are a grouping symbol. We simplify within them first.

PRACTICE
7   Add.

    a. $8 + (-5) + (-9)$ **b.** $[-8 + 5] + [-5 + |-2|]$ ■

✔ **CONCEPT CHECK**

What is wrong with the following calculation?

$5 + (-22) = 17$ (crossed out)

Answer to Concept Check:
$5 + (-22) = -17$

OBJECTIVE

2   Solving Applications by Adding Real Numbers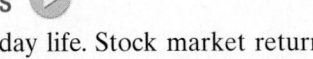

Positive and negative numbers are often used in everyday life. Stock market returns show gains and losses as positive and negative numbers. Temperatures in cold climates often dip into the negative range, commonly referred to as "below zero" temperatures. Bank statements report deposits and withdrawals as positive and negative numbers.

EXAMPLE 8   Calculating Temperature

In Philadelphia, Pennsylvania, the record extreme high temperature is 104°F. Decrease this temperature by 115 degrees, and the result is the record extreme low temperature. Find this temperature. (*Source:* National Climatic Data Center)

Solution:

In words:	extreme low temperature	=	extreme high temperature	+	decrease of 115°
	↓		↓		↓
Translate:	extreme low temperature	=	104	+	(−115)

$$= -11$$

The record extreme low temperature in Philadelphia, Pennsylvania, is −11°F. ☐

PRACTICE
8   If the temperature was −7° Fahrenheit at 6 a.m., and it rose 4 degrees by 7 a.m. and then rose another 7 degrees in the hour from 7 a.m. to 8 a.m., what was the temperature at 8 a.m.? ■

OBJECTIVE
3   Finding the Opposite of a Number

To help us subtract real numbers in the next section, we first review the concept of opposites. The graphs of 4 and −4 are shown on a number line below.

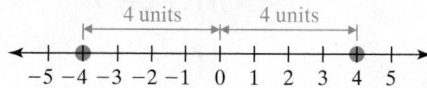

Notice that 4 and −4 lie on opposite sides of 0, and each is 4 units away from 0.

This relationship between −4 and +4 is an important one. Such numbers are known as **opposites** or **additive inverses** of each other.

> **Opposites or Additive Inverses**
>
> Two numbers that are the same distance from 0 but lie on opposite sides of 0 are called **opposites** or **additive inverses** of each other.

EXAMPLE 9   Find the opposite or additive inverse of each number.

   a. 5   **b.** −6   **c.** $\frac{1}{2}$ **d.** −4.5

Solution

   a. The opposite of 5 is −5. Notice that 5 and −5 are on opposite sides of 0 when plotted on a number line and are equal distances away.

   b. The opposite of −6 is 6.

   c. The opposite of $\frac{1}{2}$ is $-\frac{1}{2}$.

   d. The opposite of −4.5 is 4.5. ☐

PRACTICE
9   Find the opposite or additive inverse of each number.

   a. $-\frac{5}{9}$ **b.** 8   **c.** 6.2   **d.** −3 ■

We use the symbol "−" to represent the phrase "the opposite of" or "the additive inverse of." In general, if a is a number, we write the opposite or additive inverse of a as $-a$. We know that the opposite of -3 is 3. Notice that this translates as

the opposite of -3 is 3

$$- \quad (-3) \quad = \quad 3$$

This is true in general.

If a is a number, then $-(-a) = a$.

EXAMPLE 10   Simplify each expression.

a. $-(-10)$ **b.** $-\left(-\dfrac{1}{2}\right)$ **c.** $-(-2x)$ **d.** $-|-6|$

Solution

a. $-(-10) = 10$ **b.** $-\left(-\dfrac{1}{2}\right) = \dfrac{1}{2}$ **c.** $-(-2x) = 2x$

d. Since $|-6| = 6$, then $-|-6| = -6$.

PRACTICE
10   Simplify each expression.

a. $-|-15|$ **b.** $-\left(-\dfrac{3}{5}\right)$ **c.** $-(-5y)$ **d.** $-(-8)$

Let's discover another characteristic about opposites. Notice that the sum of a number and its opposite is 0.

$$10 + (-10) = 0$$
$$-3 + 3 = 0$$
$$\dfrac{1}{2} + \left(-\dfrac{1}{2}\right) = 0$$

In general, we can write the following:

The sum of a number a and its opposite $-a$ is 0.

$$a + (-a) = 0$$

This is why opposites are also called additive inverses. Notice that this also means that the opposite of 0 is then 0 since $0 + 0 = 0$.

✔ Vocabulary, Readiness & Video Check

Use the choices below to fill in each blank. Not all choices will be used.

a positive number n opposites
a negative number 0 $-n$

1. Two numbers that are the same distance from 0 but lie on opposite sides of 0 are called _____.

2. If n is a number, then $n + (-n) = $ _____.

3. If n is a number, then $-(-n) = $ _____.

4. The sum of two negative numbers is always _____.

Martin-Gay Interactive Videos

See Video 1.5

Watch the section lecture video and answer the following questions.

OBJECTIVE 1
5. Complete this statement based on the lecture given before Example 1. To add two numbers with the same sign, add their _____ and use their common sign as the sign of the sum.

OBJECTIVE 1
6. What is the sign of the sum in Example 6 and why?

OBJECTIVE 2
7. What is the real life application of negative numbers used in Example 9? The answer to Example 9 is −6. What does this number mean in the context of the problem?

OBJECTIVE 3
8. Example 12 illustrates the idea that if a is a real number, the opposite of $-a$ is a. Example 13 looks similar to Example 12, but it's actually quite different. Explain the difference.

1.5   Exercise Set   MyMathLab®

MIXED PRACTICE

Add. See Examples 1 through 7.

1. $-6 + 3$
2. $9 + (-12)$
3. $-6 + (-8)$
4. $-6 + (-14)$
5. $8 + (-7)$
6. $6 + (-4)$
7. $-14 + 2$
8. $-10 + 5$
9. $-2 + (-3)$
10. $-7 + (-4)$
11. $-9 + (-3)$
12. $7 + (-5)$
13. $-7 + 3$
14. $-5 + 9$
15. $10 + (-3)$
16. $8 + (-6)$
17. $5 + (-7)$
18. $3 + (-6)$
19. $-16 + 16$
20. $23 + (-23)$
21. $27 + (-46)$
22. $53 + (-37)$
23. $-18 + 49$
24. $-26 + 14$
25. $-33 + (-14)$
26. $-18 + (-26)$
27. $6.3 + (-8.4)$
28. $9.2 + (-11.4)$
29. $|-8| + (-16)$
30. $|-6| + (-61)$
31. $117 + (-79)$
32. $144 + (-88)$
33. $-9.6 + (-3.5)$
34. $-6.7 + (-7.6)$
35. $-\dfrac{3}{8} + \dfrac{5}{8}$
36. $-\dfrac{5}{12} + \dfrac{7}{12}$
37. $-\dfrac{7}{16} + \dfrac{1}{4}$
38. $-\dfrac{5}{9} + \dfrac{1}{3}$
39. $-\dfrac{7}{10} + \left(-\dfrac{3}{5}\right)$
40. $-\dfrac{5}{6} + \left(-\dfrac{2}{3}\right)$
41. $-15 + 9 + (-2)$
42. $-9 + 15 + (-5)$
43. $-21 + (-16) + (-22)$
44. $-18 + (-6) + (-40)$
45. $-23 + 16 + (-2)$
46. $-14 + (-3) + 11$
47. $|5 + (-10)|$
48. $|7 + (-17)|$
49. $6 + (-4) + 9$
50. $8 + (-2) + 7$
51. $[-17 + (-4)] + [-12 + 15]$
52. $[-2 + (-7)] + [-11 + 22]$
53. $|9 + (-12)| + |-16|$
54. $|43 + (-73)| + |-20|$
55. $-1.3 + [0.5 + (-0.3) + 0.4]$
56. $-3.7 + [0.1 + (-0.6) + 8.1]$

Solve. See Example 8.

57. The low temperature in Anoka, Minnesota, was −15° last night. During the day, it rose only 9°. Find the high temperature for the day.

58. On January 2, 1943, the temperature was −4° at 7:30 a.m. in Spearfish, South Dakota. Incredibly, it got 49° warmer in the next 2 minutes. To what temperature did it rise by 7:32?

59. The lowest point in Africa is −512 feet at Lake Assal in Djibouti. If you are standing at a point 658 feet above Lake Assal, what is your elevation? (*Source:* Microsoft Encarta)

60. The lowest elevation in Australia is −52 feet at Lake Eyre. If you are standing at a point 439 feet above Lake Eyre, what is your elevation? (*Source:* National Geographic Society)

A negative net income results when a company spends more money than it brings in.

61. Johnson Outdoors Inc. had the following quarterly net incomes during its 2014 fiscal year. (*Source:* Johnsonoutdoors.com)

Quarter of Fiscal 2014	Net Income (in millions)
First	$7.4
Second	$4.7
Third	−$0.8
Fourth	−$2.2

What was the total net income for fiscal year 2014?

62. LeapFrog Enterprises Inc. had the following quarterly net incomes during its 2013 fiscal year. (*Source:* Leapfroginvestors.com)

Quarter of Fiscal 2013	Net Income (in millions)
First	−$3.0
Second	−$3.2
Third	$26.4
Fourth	$63.9

What was the total net income for fiscal year 2013?

In golf, scores that are under par for the entire round are shown as negative scores; positive scores are shown for scores that are over par, and 0 is par.

63. Austin Ernst was the winner of the 2014 Portland Classic in Oregon. Her scores were −3, −3, −3, and −5. What was her overall score? (*Source:* Ladies Professional Golf Association)

64. During the 2015 Hyundai Tournament of champions in Maui, Hawaii, Patrick Reed won with scores of −6, −4, −5, and −6. What was his overall score? (*Source:* Professional Golfers' Association of America)

Find each additive inverse or opposite. See Example 9.

65. 6   **66.** 4   **67.** −2

68. −8   **69.** 0   **70.** $-\dfrac{1}{4}$

71. $|-6|$ **72.** $|-11|$

Simplify each of the following. See Example 10.

73. $-|-2|$ **74.** $-(-3)$ **75.** $-|0|$

76. $\left|-\dfrac{2}{3}\right|$ **77.** $-\left|-\dfrac{2}{3}\right|$ **78.** $-(-7)$

Decide whether the given number is a solution of the given equation.

79. Is −4 a solution of $x + 9 = 5$?

80. Is 10 a solution of $7 = -x + 3$?

81. Is −1 a solution of $y + (-3) = -7$?

82. Is −6 a solution of $1 = y + 7$?

CONCEPT EXTENSIONS

The following bar graph shows each month's average daily low temperature in degrees Fahrenheit for Barrow, Alaska. Use this graph to answer Exercises 83 through 88.

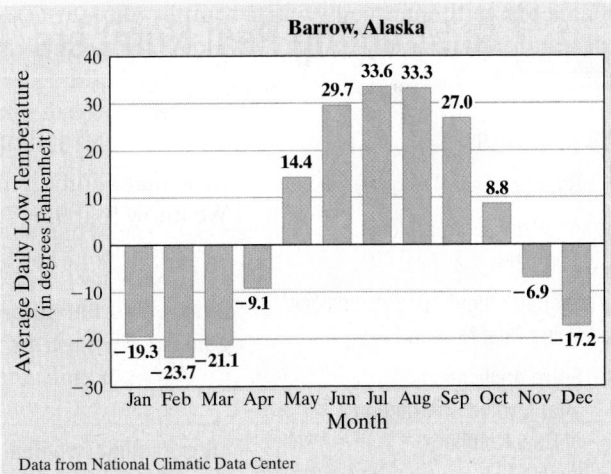

Data from National Climatic Data Center

83. For what month is the graphed temperature the highest?

84. For what month is the graphed temperature the lowest?

85. For what month is the graphed temperature positive *and* closest to 0°?

86. For what month is the graphed temperature negative *and* closest to 0°?

87. Find the average of the temperatures shown for the months of April, May, and October. (To find the average of three temperatures, find their sum and divide by 3.)

88. Find the average of the temperatures shown for the months of January, September, and October.

Each calculation below is incorrect. Find the error and correct it. See the Concept Check in this section.

89. $7 + (-10) \overset{?}{=} 17$ **90.** $-4 + 14 \overset{?}{=} -18$

91. $-10 + (-12) \overset{?}{=} -120$ **92.** $-15 + (-17) \overset{?}{=} 32$

If a is a positive number and b is a negative number, fill in the blanks with the words positive or negative.

93. $-a$ is _____.   **94.** $-b$ is _____.

95. $a + a$ is _____.   **96.** $b + b$ is _____.

For Exercises 97 through 100, determine whether each statement is true or false.

97. The sum of two negative numbers is always a negative number.

98. The sum of two positive numbers is always a positive number.

99. The sum of a positive number and a negative number is always a negative number.

100. The sum of zero and a negative number is always a negative number.

101. In your own words, explain how to find the opposite of a number.

102. In your own words, explain why 0 is the only number that is its own opposite.

103. Explain why adding a negative number to another negative number always gives a negative sum.

104. When a positive and a negative number are added, sometimes the sum is positive, sometimes it is zero, and sometimes it is negative. Explain why and when this happens.

1.6 Subtracting Real Numbers

OBJECTIVES

1 Subtract Real Numbers.

2 Add and Subtract Real Numbers.

3 Evaluate Algebraic Expressions Using Real Numbers.

4 Solve Applications That Involve Subtraction of Real Numbers.

5 Find Complementary and Supplementary Angles.

OBJECTIVE

1 Subtracting Real Numbers

Now that addition of signed numbers has been discussed, we can explore subtraction. We know that $9 - 7 = 2$. Notice that $9 + (-7) = 2$ also. This means that

$$9 - 7 = 9 + (-7)$$

Notice that the difference of 9 and 7 is the same as the sum of 9 and the opposite of 7. In general, we have the following.

Subtracting Two Real Numbers

If a and b are real numbers, then $a - b = a + (-b)$.

In other words, to find the difference of two numbers, add the first number to the opposite of the second number.

EXAMPLE I Subtract.

a. $-13 - 4$ **b.** $5 - (-6)$ **c.** $3 - 6$ **d.** $-1 - (-7)$

Solution

a. $\underset{\text{opposite}}{-13 \overset{\text{add}}{-} 4} = -13 + (-4)$ Add -13 to the opposite of $+4$, which is -4.

$ = -17$

b. $\underset{\text{opposite}}{5 \overset{\text{add}}{-} (-6)} = 5 + (6)$ Add 5 to the opposite of -6, which is 6.

$ = 11$

c. $3 - 6 = 3 + (-6)$ Add 3 to the opposite of 6, which is -6.

$ = -3$

d. $-1 - (-7) = -1 + (7) = 6$ □

PRACTICE

1 Subtract.

a. $-7 - 6$ **b.** $-8 - (-1)$ **c.** $9 - (-3)$ **d.** $5 - 7$ ■

Helpful Hint

Study the patterns indicated.

No change ⌐   Change to addition.
                     Change to opposite.

$$5 - 11 = 5 + (-11) = -6$$
$$-3 - 4 = -3 + (-4) = -7$$
$$7 - (-1) = 7 + (1) = 8$$

EXAMPLE 2   Subtract.

a. $5.3 - (-4.6)$

b. $-\dfrac{3}{10} - \dfrac{5}{10}$

c. $-\dfrac{2}{3} - \left(-\dfrac{4}{5}\right)$

Solution

a. $5.3 - (-4.6) = 5.3 + (4.6) = 9.9$

b. $-\dfrac{3}{10} - \dfrac{5}{10} = -\dfrac{3}{10} + \left(-\dfrac{5}{10}\right) = -\dfrac{8}{10} = -\dfrac{4}{5}$

c. $-\dfrac{2}{3} - \left(-\dfrac{4}{5}\right) = -\dfrac{2}{3} + \left(\dfrac{4}{5}\right) = -\dfrac{10}{15} + \dfrac{12}{15} = \dfrac{2}{15}$ The common denominator is 15. ☐

PRACTICE
2   Subtract.

a. $8.4 - (-2.5)$

b. $-\dfrac{5}{8} - \left(-\dfrac{1}{8}\right)$

c. $-\dfrac{3}{4} - \dfrac{1}{5}$ ■

EXAMPLE 3   Subtract 8 from -4.

Solution Be careful when interpreting this: The order of numbers in subtraction is important. 8 is to be subtracted **from** -4.

$$-4 - 8 = -4 + (-8) = -12$$ ☐

PRACTICE
3   Subtract 5 from -2.   ■

OBJECTIVE
2   Adding and Subtracting Real Numbers

If an expression contains additions and subtractions, just write the subtractions as equivalent additions. Then simplify from left to right.

EXAMPLE 4   Simplify each expression.

a. $-14 - 8 + 10 - (-6)$

b. $1.6 - (-10.3) + (-5.6)$

Solution

a. $-14 - 8 + 10 - (-6) = -14 + (-8) + 10 + 6$
$$= -6$$

b. $1.6 - (-10.3) + (-5.6) = 1.6 + 10.3 + (-5.6)$
$$= 6.3$$ ☐

PRACTICE
4   Simplify each expression.

a. $-15 - 2 - (-4) + 7$

b. $3.5 + (-4.1) - (-6.7)$ ■

When an expression contains parentheses and brackets, remember the order of operations. Start with the innermost set of parentheses or brackets and work your way outward.

EXAMPLE 5 Simplify each expression.

    a. $-3 + [(-2 - 5) - 2]$ **b.** $2^3 - |10| + [-6 - (-5)]$

Solution

a. Start with the innermost set of parentheses. Rewrite $-2 - 5$ as a sum.

$$
\begin{aligned}
-3 + [(-2 - 5) - 2] &= -3 + [(-2 + (-5)) - 2] \\
&= -3 + [(-7) - 2] &&\text{Add: } -2 + (-5). \\
&= -3 + [-7 + (-2)] &&\text{Write } -7 - 2 \text{ as a sum.} \\
&= -3 + [-9] &&\text{Add.} \\
&= -12 &&\text{Add.}
\end{aligned}
$$

b. Start simplifying the expression inside the brackets by writing $-6 - (-5)$ as a sum.

$$
\begin{aligned}
2^3 - |10| + [-6 - (-5)] &= 2^3 - |10| + [-6 + 5] \\
&= 2^3 - |10| + [-1] &&\text{Add.} \\
&= 8 - 10 + (-1) &&\text{Evaluate } 2^3 \text{ and } |10|. \\
&= 8 + (-10) + (-1) &&\text{Write } 8 - 10 \text{ as a sum.} \\
&= -2 + (-1) &&\text{Add.} \\
&= -3 &&\text{Add.} \quad \square
\end{aligned}
$$

PRACTICE

5 Simplify each expression.

    a. $-4 + [(-8 - 3) - 5]$ **b.** $|-13| - 3^2 + [2 - (-7)]$ ■

OBJECTIVE

3 Evaluating Algebraic Expressions

Knowing how to evaluate expressions for given replacement values is helpful when checking solutions of equations and when solving problems whose unknowns satisfy given expressions. The next example illustrates this.

EXAMPLE 6 Find the value of each expression when $x = 2$ and $y = -5$.

    a. $\dfrac{x - y}{12 + x}$ **b.** $x^2 - 3y$

Solution

a. Replace x with 2 and y with -5. Be sure to put parentheses around -5 to separate signs. Then simplify the resulting expression.

$$
\begin{aligned}
\frac{x - y}{12 + x} &= \frac{2 - (-5)}{12 + 2} \\
&= \frac{2 + 5}{14} \\
&= \frac{7}{14} = \frac{1}{2}
\end{aligned}
$$

b. Replace x with 2 and y with -5 and simplify.

$$
\begin{aligned}
x^2 - 3y &= 2^2 - 3(-5) \\
&= 4 - 3(-5) \\
&= 4 - (-15) \\
&= 4 + 15 \\
&= 19
\end{aligned}
$$
$\square$

PRACTICE
6   Find the value of each expression when $x = -3$ and $y = 4$.

a. $\dfrac{7 - x}{2y + x}$

b. $y^2 + x$

Helpful Hint

For additional help when replacing variables with replacement values, first place parentheses around any variables.
For Example 6b on the previous page, we have

$$x^2 - 3y = \underbrace{(x)^2 - 3(y)}_{\substack{\text{Place parentheses} \\ \text{around variables}}} = \underbrace{(2)^2 - 3(-5)}_{\substack{\text{Replace variables} \\ \text{with values}}} = 4 - 3(-5) = 4 - (-15) = 4 + 15 = 19$$

OBJECTIVE

4   **Solving Applications by Subtracting Real Numbers**

One use of positive and negative numbers is in recording altitudes above and below sea level, as shown in the next example.

EXAMPLE 7   Finding a Change in Elevation

The highest point in the United States is the top of Mount McKinley, at a height of 20,320 feet above sea level. The lowest point is Death Valley, California, which is 282 feet below sea level. How much higher is Mount McKinley than Death Valley? (*Source:* U.S. Geological Survey)

Solution: To find "how much higher," we subtract. Don't forget that since Death Valley is 282 feet *below* sea level, we represent its height by −282. Draw a diagram to help visualize the problem.

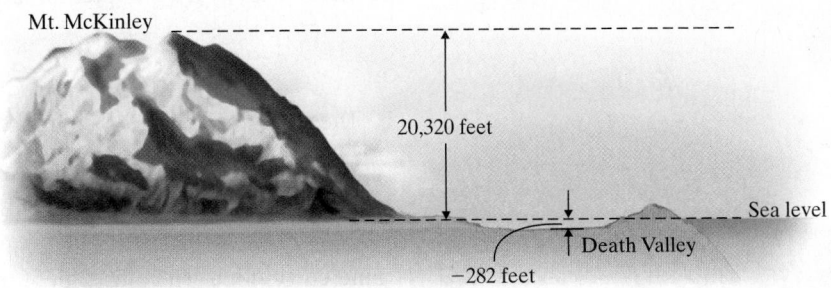

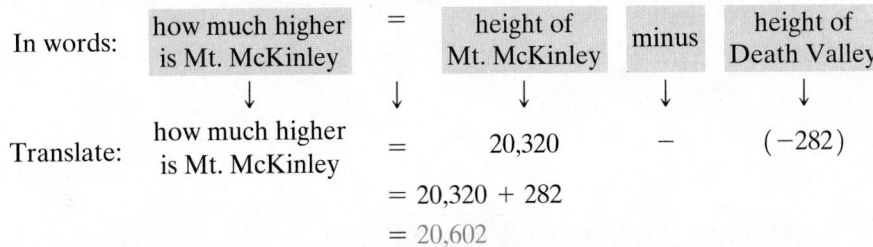

In words: how much higher is Mt. McKinley = height of Mt. McKinley minus height of Death Valley

Translate: how much higher is Mt. McKinley = 20,320 − (−282)

= 20,320 + 282

= 20,602

Thus, Mount McKinley is 20,602 feet higher than Death Valley.   □

PRACTICE
7   On Tuesday morning, a bank account balance was $282. On Thursday, the account balance had dropped to −$75. Find the overall change in this account balance.

OBJECTIVE

> **5** Finding Complementary and Supplementary Angles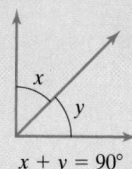

A knowledge of geometric concepts is needed by many professionals, such as doctors, carpenters, electronic technicians, gardeners, machinists, and pilots, just to name a few. With this in mind, we review the geometric concepts of **complementary** and **supplementary angles.**

Complementary and Supplementary Angles

Two angles are **complementary** if their sum is 90°.

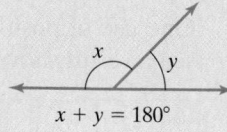

$$x + y = 90°$$

Two angles are **supplementary** if their sum is 180°.

$$x + y = 180°$$

△ **EXAMPLE 8** Find each unknown complementary or supplementary angle.

a.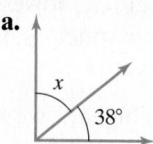

b.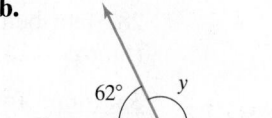

Solution

a. These angles are complementary, so their sum is 90°. This means that x is 90° − 38°.

$$x = 90° − 38° = 52°$$

b. These angles are supplementary, so their sum is 180°. This means that y is 180° − 62°.

$$y = 180° − 62° = 118°$$

PRACTICE

8 Find each unknown complementary or supplementary angle.

a.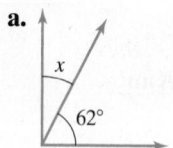

b.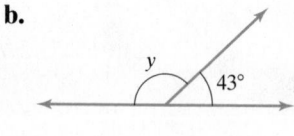

✓ | **Vocabulary, Readiness & Video Check**

Translate each phrase. Let x represent "a number." Use the choices below to fill in each blank.

$$7 − x \qquad x − 7$$

1. 7 minus a number _____

2. 7 subtracted from a number _____

3. A number decreased by 7 _____

4. 7 less a number _____

5. A number less than 7 _____

6. A number subtracted from 7 _____

Multiple choice: Select the correct lettered response following each exercise.

7. To evaluate $x - y$ for $x = -10$ and $y = -14$, we replace x with -10 and y with -14 and evaluate _____.

   a. $10 - 14$ **b.** $-10 - 14$ **c.** $-14 - 10$ **d.** $-10 - (-14)$

8. The expression $-5 - 10$ equals _____.

   a. $5 - 10$ **b.** $5 + 10$ **c.** $-5 + (-10)$ **d.** $10 - 5$

Martin-Gay Interactive Videos

See Video 1.6

Watch the section lecture video and answer the following questions.

OBJECTIVE 1

9. Complete this statement based on the lecture given before Example 1. To subtract two real numbers, change the operation to _____ and take the _____ of the second number.

OBJECTIVE 2

10. When simplifying Example 5, what is the result of the first step and why is the expression rewritten in this way?

OBJECTIVE 3

11. In Example 7, why are you told to be especially careful when working with the replacement value in the numerator?

OBJECTIVE 4

12. For Example 8, why is the overall vertical change represented as a negative number?

OBJECTIVE 5

13. The definition of supplementary angles is given just before Example 9. Explain how this definition is used to solve Example 9.

1.6   Exercise Set   MyMathLab®

MIXED PRACTICE

Subtract. See Examples 1 and 2.

1. $-6 - 4$ **2.** $-12 - 8$

3. $4 - 9$ **4.** $8 - 11$

5. $16 - (-3)$ **6.** $12 - (-5)$

7. $\dfrac{1}{2} - \dfrac{1}{3}$ **8.** $\dfrac{3}{4} - \dfrac{7}{8}$

9. $-16 - (-18)$ **10.** $-20 - (-48)$

11. $-6 - 5$ **12.** $-8 - 4$

13. $7 - (-4)$ **14.** $3 - (-6)$

15. $-6 - (-11)$ **16.** $-4 - (-16)$

17. $16 - (-21)$ **18.** $15 - (-33)$

19. $9.7 - 16.1$ **20.** $8.3 - 11.2$

21. $-44 - 27$ **22.** $-36 - 51$

23. $-21 - (-21)$ **24.** $-17 - (-17)$

25. $-2.6 - (-6.7)$ **26.** $-6.1 - (-5.3)$

27. $-\dfrac{3}{11} - \left(-\dfrac{5}{11}\right)$ **28.** $-\dfrac{4}{7} - \left(-\dfrac{1}{7}\right)$

29. $-\dfrac{1}{6} - \dfrac{3}{4}$ **30.** $-\dfrac{1}{10} - \dfrac{7}{8}$

31. $8.3 - (-0.62)$ **32.** $4.3 - (-0.87)$

TRANSLATING

Translate each phrase to an expression and simplify. See Example 3.

33. Subtract -5 from 8.   **34.** Subtract 3 from -2.

35. Subtract -1 from -6.   **36.** Subtract 17 from 1.

37. Subtract 8 from 7.   **38.** Subtract 9 from -4.

39. Decrease -8 by 15.   **40.** Decrease 11 by -14.

Simplify each expression. (Remember the order of operations.) See Examples 4 and 5.

41. $-10 - (-8) + (-4) - 20$

42. $-16 - (-3) + (-11) - 14$

43. $5 - 9 + (-4) - 8 - 8$

44. $7 - 12 + (-5) - 2 + (-2)$

45. $-6 - (2 - 11)$ **46.** $-9 - (3 - 8)$

47. $3^3 - 8 \cdot 9$ **48.** $2^3 - 6 \cdot 3$

49. $2 - 3(8 - 6)$ **50.** $4 - 6(7 - 3)$

51. $(3 - 6) + 4^2$ **52.** $(2 - 3) + 5^2$

53. $-2 + [(8 - 11) - (-2 - 9)]$

54. $-5 + [(4 - 15) - (-6) - 8]$

55. $|-3| + 2^2 + [-4 - (-6)]$

56. $|-2| + 6^2 + (-3 - 8)$

Evaluate each expression when $x = -5$, $y = 4$, *and* $t = 10$.
See Example 6.

57. $x - y$

58. $y - x$

59. $|x| + 2t - 8y$

60. $|x + t - 7y|$

61. $\dfrac{9 - x}{y + 6}$

62. $\dfrac{15 - x}{y + 2}$

63. $y^2 - x$

64. $t^2 - x$

65. $\dfrac{|x - (-10)|}{2t}$

66. $\dfrac{|5y - x|}{6t}$

Solve. See Example 7.

67. Within 24 hours in 1916, the temperature in Browning, Montana, fell from 44°F to −56°F. How large a drop in temperature was this?

68. Much of New Orleans is below sea level. If George descends 12 feet from an elevation of 5 feet above sea level, what is his new elevation?

69. The coldest temperature ever recorded on Earth was −129°F in Antarctica. The warmest temperature ever recorded was 134°F in Death Valley, California. How many degrees warmer is 134°F than −129°F? (*Source:* World Meteorological Organization)

70. The coldest temperature ever recorded in the United States was −80°F in Alaska. The warmest temperature ever recorded was 134°F in California. How many degrees warmer is 134°F than −80°F? (*Source: The World Almanac*)

71. Mauna Kea in Hawaii has an elevation of 13,796 feet above sea level. The Mid-America Trench in the Pacific Ocean has an elevation of 21,857 feet below sea level. Find the difference in elevation between those two points. (*Source:* National Geographic Society and Defense Mapping Agency)

72. A woman received a statement of her charge account at Old Navy. She spent $93 on purchases last month. She returned an $18 top because she didn't like the color. She also returned a $26 nightshirt because it was damaged. What does she actually owe on her account?

73. A commercial jetliner hits an air pocket and drops 250 feet. After climbing 120 feet, it drops another 178 feet. What is its overall vertical change?

74. In some card games, it is possible to have a negative score. Lavonne Schultz currently has a score of 15 points. She then loses 24 points. What is her new score?

75. The highest point in Africa is Mt. Kilimanjaro, Tanzania, at an elevation of 19,340 feet. The lowest point is Lake Assal, Djibouti, at 512 feet below sea level. How much higher is

Mt. Kilimanjaro than Lake Assal? (*Source:* National Geographic Society)

76. The airport in Bishop, California, is at an elevation of 4101 feet above sea level. The nearby Furnace Creek Airport in Death Valley, California, is at an elevation of 226 feet below sea level. How much higher in elevation is the Bishop Airport than the Furnace Creek Airport? (*Source:* National Climatic Data Center)

Find each unknown complementary or supplementary angle. See Example 8.

77.

78.

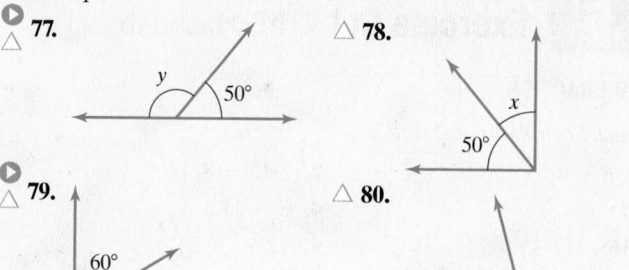

79.

80.

Decide whether the given number is a solution of the given equation.

81. Is −4 a solution of $x - 9 = 5$?

82. Is 3 a solution of $x - 10 = -7$?

83. Is −2 a solution of $-x + 6 = -x - 1$?

84. Is −10 a solution of $-x - 6 = -x - 1$?

85. Is 2 a solution of $-x - 13 = -15$?

86. Is 5 a solution of $4 = 1 - x$?

MIXED PRACTICE—TRANSLATING (*SECTIONS 1.5, 1.6*)

Translate each phrase to an algebraic expression. Use "x" to represent "a number."

87. The sum of −5 and a number.

88. The difference of −3 and a number.

89. Subtract a number from −20.

90. Add a number and −36.

CONCEPT EXTENSIONS

Recall the bar graph from Section 1.5. It shows each month's average daily low temperature in degrees Fahrenheit for Barrow, Alaska. Use this graph to answer Exercises 91 through 94.

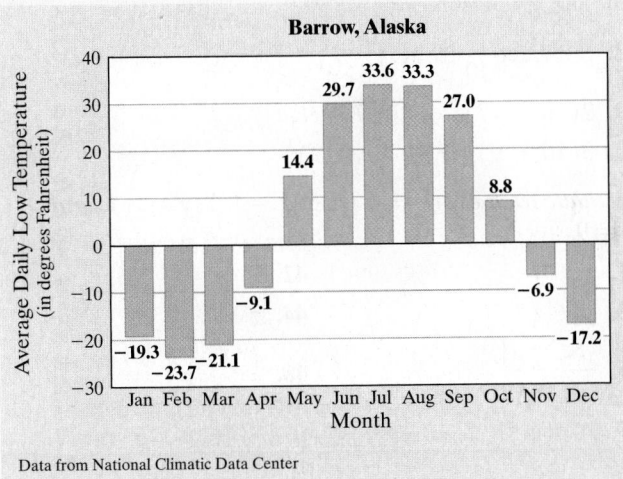

Barrow, Alaska

Data from National Climatic Data Center

91. Record the monthly increases and decreases in the low temperature from the previous month.

Month	Monthly Increase or Decrease (from the previous month)
February	
March	
April	
May	
June	

92. Record the monthly increases and decreases in the low temperature from the previous month.

Month	Monthly Increase or Decrease (from the previous month)
July	
August	
September	
October	
November	
December	

93. Which month had the greatest increase in temperature?

94. Which month had the greatest decrease in temperature?

95. Find two numbers whose difference is −5.

96. Find two numbers whose difference is −9.

*Each calculation below is **incorrect**. Find the error and correct it.*

97. $9 - (-7) \stackrel{?}{=} 2$ **98.** $-4 - 8 \stackrel{?}{=} 4$

99. $10 - 30 \stackrel{?}{=} 20$ **100.** $-3 - (-10) \stackrel{?}{=} -13$

If p is a positive number and n is a negative number, determine whether each statement is true or false. Explain your answer.

101. $p - n$ is always a positive number.

102. $n - p$ is always a negative number.

103. $|n| - |p|$ is always a positive number.

104. $|n - p|$ is always a positive number.

Without calculating, determine whether each answer is positive or negative. Then use a calculator to find the exact difference.

105. $56{,}875 - 87{,}262$

106. $4.362 - 7.0086$

Integrated Review   Operations on Real Numbers

Sections 1.1–1.6

Answer the following with positive, negative, or 0.

1. The opposite of a positive number is a _____ number.

2. The sum of two negative numbers is a _____ number.

3. The absolute value of a negative number is a _____ number.

4. The absolute value of zero is _____.

5. The reciprocal of a positive number is a _____ number.

6. The sum of a number and its opposite is _____.

7. The absolute value of a positive number is a _____ number.

8. The opposite of a negative number is a _____ number.

Fill in the chart:

Number	Opposite	Absolute Value
9. $\frac{1}{7}$		
10. $-\frac{12}{5}$		
11.	-3	
12.		$\frac{9}{11}$

Perform each indicated operation and simplify.

13. $-19 + (-23)$ **14.** $7 - (-3)$

15. $-15 + 17$ **16.** $-8 - 10$

17. $18 + (-25)$ **18.** $-2 + (-37)$

19. $-14 - (-12)$ **20.** $5 - 14$

21. $4.5 - 7.9$ **22.** $-8.6 - 1.2$

23. $-\dfrac{3}{4} - \dfrac{1}{7}$ **24.** $\dfrac{2}{3} - \dfrac{7}{8}$

25. $-9 - (-7) + 4 - 6$

26. $11 - 20 + (-3) - 12$

27. $24 - 6(14 - 11)$

28. $30 - 5(10 - 8)$

29. $(7 - 17) + 4^2$

30. $9^2 + (10 - 30)$

31. $|-9| + 3^2 + (-4 - 20)$

32. $|-4 - 5| + 5^2 + (-50)$

33. $-7 + [(1 - 2) + (-2 - 9)]$

34. $-6 + [(-3 + 7) + (4 - 15)]$

35. Subtract 5 from 1.

36. Subtract -2 from -3.

37. Subtract $-\dfrac{2}{5}$ from $\dfrac{1}{4}$.

38. Subtract $\dfrac{1}{10}$ from $-\dfrac{5}{8}$.

39. $2(19 - 17)^3 - 3(-7 + 9)^2$

40. $3(10 - 9)^2 + 6(20 - 19)^3$

Evaluate each expression when $x = -2, y = -1,$ and $z = 9$.

41. $x - y$ **42.** $x + y$

43. $y + z$ **44.** $z - y$

45. $\dfrac{|5z - x|}{y - x}$ **46.** $\dfrac{|-x - y + z|}{2z}$

1.7 Multiplying and Dividing Real Numbers

OBJECTIVES

1 Multiply Real Numbers.

2 Find the Reciprocal of a Real Number.

3 Divide Real Numbers.

4 Evaluate Expressions Using Real Numbers.

5 Solve Applications That Involve Multiplication or Division of Real Numbers.

OBJECTIVE

1 Multiplying Real Numbers

In this section, we discover patterns for multiplying and dividing real numbers. To discover sign rules for multiplication, recall that multiplication is repeated addition. Thus $3 \cdot 2$ means that 2 is an addend 3 times. That is,

$$2 + 2 + 2 = 3 \cdot 2$$

which equals 6. Similarly, $3 \cdot (-2)$ means -2 is an addend 3 times. That is,

$$(-2) + (-2) + (-2) = 3 \cdot (-2)$$

Since $(-2) + (-2) + (-2) = -6, 3 \cdot (-2) = -6$. This suggests that the product of a positive number and a negative number is a negative number.

What about the product of two negative numbers? To find out, consider the following pattern.

Factor decreases by 1 each time

$$\begin{aligned}-3 \cdot 2 &= -6 \\ -3 \cdot 1 &= -3 \\ -3 \cdot 0 &= 0 \end{aligned}\Big\} \text{ Product increases by 3 each time.}$$

This pattern continues as

Factor decreases by 1 each time

$$\begin{aligned}-3 \cdot -1 &= 3 \\ -3 \cdot -2 &= 6 \end{aligned}\Big\} \text{ Product increases by 3 each time.}$$

This suggests that the product of two negative numbers is a positive number.

Multiplying Real Numbers

1. The product of two numbers with the *same* sign is a positive number.

2. The product of two numbers with *different* signs is a negative number.

EXAMPLE 1   Multiply.

  a. $(-8)(4)$ **b.** $14(-1)$ **c.** $-9(-10)$

Solution

  a. $-8(4) = -32$ **b.** $14(-1) = -14$ **c.** $-9(-10) = 90$ ☐

PRACTICE

1   Multiply.

  a. $8(-5)$ **b.** $(-3)(-4)$ **c.** $(-6)(9)$ ■

We know that any whole number multiplied by zero equals zero. This remains true for real numbers.

> **Zero as a Factor**
>
> If b is a real number, then $b \cdot 0 = 0$. Also, $0 \cdot b = 0$.

EXAMPLE 2   Perform the indicated operations.

  a. $(7)(0)(-6)$ **b.** $(-2)(-3)(-4)$
  c. $(-1)(5)(-9)$ **d.** $(-4)(-11) - (5)(-2)$

Solution

  a. By the order of operations, we multiply from left to right. Notice that, because one of the factors is 0, the product is 0.

$$(7)(0)(-6) = 0(-6) = 0$$

  b. Multiply two factors at a time, from left to right.

$$(-2)(-3)(-4) = (6)(-4) \quad \text{Multiply } (-2)(-3).$$
$$= -24$$

  c. Multiply from left to right.

$$(-1)(5)(-9) = (-5)(-9) \quad \text{Multiply } (-1)(5).$$
$$= 45$$

  d. Follow the rules for order of operations.

$$(-4)(-11) - (5)(-2) = 44 - (-10) \quad \text{Find each product.}$$
$$= 44 + 10 \qquad \text{Add 44 to the opposite of } -10.$$
$$= 54 \qquad \text{Add.} \qquad ☐$$

PRACTICE

2   Perform the indicated operations.

  a. $(-1)(-5)(-6)$ **b.** $(-3)(-2)(4)$
  c. $(-4)(0)(5)$ **d.** $(-2)(-3) - (-4)(5)$ ■

> **Helpful Hint**
>
> You may have noticed from the example that if we multiply:
> - an _even_ number of negative numbers, the product is _positive_.
> - an _odd_ number of negative numbers, the product is _negative_.

✔ **CONCEPT CHECK**

What is the sign of the product of five negative numbers? Explain.

Answer to Concept Check:
negative

Multiplying signed decimals or fractions is carried out exactly the same way as multiplying integers.

EXAMPLE 3 Multiply.

a. $(-1.2)(0.05)$ 　　　　b. $\dfrac{2}{3} \cdot \left(-\dfrac{7}{10}\right)$ 　　　　c. $\left(-\dfrac{4}{5}\right)(-20)$

Solution

a. The product of two numbers with different signs is negative.

$$(-1.2)(0.05) = -[(1.2)(0.05)]$$
$$= -0.06$$

b. $\dfrac{2}{3} \cdot \left(-\dfrac{7}{10}\right) = -\dfrac{2 \cdot 7}{3 \cdot 10} = -\dfrac{2 \cdot 7}{3 \cdot 2 \cdot 5} = -\dfrac{7}{15}$

c. $\left(-\dfrac{4}{5}\right)(-20) = \dfrac{4 \cdot 20}{5 \cdot 1} = \dfrac{4 \cdot 4 \cdot 5}{5 \cdot 1} = \dfrac{16}{1}$ or 16

PRACTICE
3 Multiply.

a. $(0.23)(-0.2)$ 　　　　b. $\left(-\dfrac{3}{5}\right) \cdot \left(\dfrac{4}{9}\right)$ 　　　　c. $\left(-\dfrac{7}{12}\right)(-24)$

Now that we know how to multiply positive and negative numbers, let's see how we find the values of $(-4)^2$ and -4^2, for example. Although these two expressions look similar, the difference between the two is the parentheses. In $(-4)^2$, the parentheses tell us that the base, or repeated factor, is -4. In -4^2, only 4 is the base. Thus,

$$(-4)^2 = (-4)(-4) = 16 \quad \text{The base is } -4.$$
$$-4^2 = -(4 \cdot 4) = -16 \quad \text{The base is } 4.$$

EXAMPLE 4 Evaluate.

a. $(-2)^3$ 　　　　b. -2^3 　　　　c. $(-3)^2$ 　　　　d. -3^2

Solution

a. $(-2)^3 = (-2)(-2)(-2) = -8$ 　　The base is -2.
b. $-2^3 = -(2 \cdot 2 \cdot 2) = -8$ 　　　　The base is 2.
c. $(-3)^2 = (-3)(-3) = 9$ 　　　　The base is -3.
d. $-3^2 = -(3 \cdot 3) = -9$ 　　　　The base is 3.

PRACTICE
4 Evaluate.

a. $(-6)^2$ 　　b. -6^2 　　c. $(-4)^3$ 　　d. -4^3

| Helpful Hint |

Be careful when identifying the base of an exponential expression.

$$(-3)^2 \qquad\qquad -3^2$$
$$\text{Base is } -3 \qquad\qquad \text{Base is } 3$$
$$(-3)^2 = (-3)(-3) = 9 \qquad -3^2 = -(3 \cdot 3) = -9$$

OBJECTIVE

2   Finding Reciprocals

Just as every difference of two numbers $a - b$ can be written as the sum $a + (-b)$, so too every quotient of two numbers can be written as a product. For example, the quotient $6 \div 3$ can be written as $6 \cdot \dfrac{1}{3}$. Recall that the pair of numbers 3 and $\dfrac{1}{3}$ has a special relationship. Their product is 1 and they are called reciprocals or **multiplicative inverses** of each other.

> **Reciprocals or Multiplicative Inverses**
>
> Two numbers whose product is 1 are called reciprocals or multiplicative inverses of each other.

Notice that **0 has no multiplicative inverse** since 0 multiplied by any number is never 1 but always 0.

EXAMPLE 5   Find the reciprocal of each number.

    a. 22   **b.** $\dfrac{3}{16}$ **c.** -10 **d.** $-\dfrac{9}{13}$

Solution

    a. The reciprocal of 22 is $\dfrac{1}{22}$ since $22 \cdot \dfrac{1}{22} = 1$.

    b. The reciprocal of $\dfrac{3}{16}$ is $\dfrac{16}{3}$ since $\dfrac{3}{16} \cdot \dfrac{16}{3} = 1$.

    c. The reciprocal of -10 is $-\dfrac{1}{10}$.

    d. The reciprocal of $-\dfrac{9}{13}$ is $-\dfrac{13}{9}$.   □

PRACTICE

5   Find the reciprocal of each number.

    a. $\dfrac{8}{3}$ **b.** 15   **c.** $-\dfrac{2}{7}$ **d.** -5 ■

OBJECTIVE

3   Dividing Real Numbers

We may now write a quotient as an equivalent product.

> **Quotient of Two Real Numbers**
>
> If a and b are real numbers and b is not 0, then
>
> $$a \div b = \frac{a}{b} = a \cdot \frac{1}{b}$$

In other words, the quotient of two real numbers is the product of the first number and the multiplicative inverse or reciprocal of the second number.

EXAMPLE 6   Use the definition of the quotient of two numbers to divide.

    a. $-18 \div 3$ **b.** $\dfrac{-14}{-2}$ **c.** $\dfrac{20}{-4}$

Solution

    a. $-18 \div 3 = -18 \cdot \dfrac{1}{3} = -6$ **b.** $\dfrac{-14}{-2} = -14 \cdot -\dfrac{1}{2} = 7$

    c. $\dfrac{20}{-4} = 20 \cdot -\dfrac{1}{4} = -5$ □

(Continued on next page)

PRACTICE

6 Use the definition of the quotient of two numbers to divide.

a. $\dfrac{16}{-2}$ **b.** $24 \div (-6)$ **c.** $\dfrac{-35}{-7}$

Since the quotient $a \div b$ can be written as the product $a \cdot \dfrac{1}{b}$, it follows that sign patterns for dividing two real numbers are the same as sign patterns for multiplying two real numbers.

> **Multiplying and Dividing Real Numbers**
>
> 1. The product or quotient of two numbers with the *same* sign is a positive number.
> 2. The product or quotient of two numbers with *different* signs is a negative number.

EXAMPLE 7 Divide.

a. $\dfrac{-24}{-4}$ **b.** $\dfrac{-36}{3}$ **c.** $\dfrac{2}{3} \div \left(-\dfrac{5}{4}\right)$ **d.** $-\dfrac{3}{2} \div 9$

Solution

a. $\dfrac{-24}{-4} = 6$ **b.** $\dfrac{-36}{3} = -12$

c. $\dfrac{2}{3} \div \left(-\dfrac{5}{4}\right) = \dfrac{2}{3} \cdot \left(-\dfrac{4}{5}\right) = -\dfrac{8}{15}$

d. $-\dfrac{3}{2} \div 9 = -\dfrac{3}{2} \cdot \dfrac{1}{9} = -\dfrac{3 \cdot 1}{2 \cdot 9} = -\dfrac{3 \cdot 1}{2 \cdot 3 \cdot 3} = -\dfrac{1}{6}$

PRACTICE

7 Divide.

a. $\dfrac{-18}{-6}$ **b.** $\dfrac{-48}{3}$ **c.** $\dfrac{3}{5} \div \left(-\dfrac{1}{2}\right)$ **d.** $-\dfrac{4}{9} \div 8$

> ✔ **CONCEPT CHECK**
>
> What is wrong with the following calculation?
>
> $\dfrac{-36}{-9} \cancel{=} -4$

The definition of the quotient of two real numbers does not allow for division by 0 because 0 does not have a multiplicative inverse. There is no number we can multiply 0 by to get 1. How then do we interpret $\dfrac{3}{0}$? We say that division by 0 is not allowed or not defined and that $\dfrac{3}{0}$ does not represent a real number. The denominator of a fraction can never be 0.

Can the numerator of a fraction be 0? Can we divide 0 by a number? Yes. For example,

$$\dfrac{0}{3} = 0 \cdot \dfrac{1}{3} = 0$$

In general, the quotient of 0 and any nonzero number is 0.

> **Zero as a Divisor or Dividend**
>
> 1. The quotient of any nonzero real number and 0 is undefined. In symbols, if $a \neq 0$, $\dfrac{a}{0}$ is **undefined.**
> 2. The quotient of 0 and any real number except 0 is 0. In symbols, if $a \neq 0$, $\dfrac{0}{a} = 0$.

EXAMPLE 8   Perform the indicated operations.

a. $\dfrac{1}{0}$ b. $\dfrac{0}{-3}$ c. $\dfrac{0(-8)}{2}$

Solution

a. $\dfrac{1}{0}$ is undefined   b. $\dfrac{0}{-3} = 0$ c. $\dfrac{0(-8)}{2} = \dfrac{0}{2} = 0$ □

PRACTICE
8   Perform the indicated operations.

a. $\dfrac{0}{-2}$ b. $\dfrac{-4}{0}$ c. $\dfrac{-5}{6(0)}$ ■

Notice that $\dfrac{12}{-2} = -6, -\dfrac{12}{2} = -6$, and $\dfrac{-12}{2} = -6$. This means that

$$\frac{12}{-2} = -\frac{12}{2} = \frac{-12}{2}$$

In words, a single negative sign in a fraction can be written in the denominator, in the numerator, or in front of the fraction without changing the value of the fraction. Thus,

$$\frac{1}{-7} = \frac{-1}{7} = -\frac{1}{7}$$

In general, if a and b are real numbers, $b \neq 0$, then $\dfrac{a}{-b} = \dfrac{-a}{b} = -\dfrac{a}{b}$.

OBJECTIVE
4   Evaluating Expressions

Examples combining basic arithmetic operations along with the principles of order of operations help us review these concepts.

EXAMPLE 9   Simplify each expression.

a. $\dfrac{(-12)(-3) + 3}{-7 - (-2)}$ b. $\dfrac{2(-3)^2 - 20}{-5 + 4}$

Solution

a. First, simplify the numerator and denominator separately, then divide.

$$\frac{(-12)(-3) + 3}{-7 - (-2)} = \frac{36 + 3}{-7 + 2}$$

$$= \frac{39}{-5} \text{ or } -\frac{39}{5}$$

b. Simplify the numerator and denominator separately, then divide.

$$\frac{2(-3)^2 - 20}{-5 + 4} = \frac{2 \cdot 9 - 20}{-5 + 4} = \frac{18 - 20}{-5 + 4} = \frac{-2}{-1} = 2$$ □

PRACTICE
9   Simplify each expression.

a. $\dfrac{(-8)(-11) - 4}{-9 - (-4)}$ b. $\dfrac{3(-2)^3 - 9}{-6 + 3}$ ■

Using what we have learned about multiplying and dividing real numbers, we continue to practice evaluating algebraic expressions.

EXAMPLE 10 If $x = -2$ and $y = -4$, evaluate each expression.

   a. $5x - y$ **b.** $x^4 - y^2$ **c.** $\dfrac{3x}{2y}$

Solution

   a. Replace x with -2 and y with -4 and simplify.

$$5x - y = 5(-2) - (-4) = -10 - (-4) = -10 + 4 = -6$$

   b. Replace x with -2 and y with -4.

$$
\begin{aligned}
x^4 - y^2 &= (-2)^4 - (-4)^2 &&\text{Substitute the given values for the variables.} \\
&= 16 - (16) &&\text{Evaluate exponential expressions.} \\
&= 0 &&\text{Subtract.}
\end{aligned}
$$

   c. Replace x with -2 and y with -4 and simplify.

$$\frac{3x}{2y} = \frac{3(-2)}{2(-4)} = \frac{-6}{-8} = \frac{3}{4}$$

PRACTICE
10 If $x = -5$ and $y = -2$, evaluate each expression.

   a. $7y - x$ **b.** $x^2 - y^3$ **c.** $\dfrac{2x}{3y}$

OBJECTIVE
5 Solving Applications That Involve Multiplying or Dividing Numbers ▶

Many real-life problems involve multiplication and division of numbers.

EXAMPLE 11 Calculating a Total Golf Score

A professional golfer finished seven strokes under par (-7) for each of three days of a tournament. What was her total score for the tournament?

Solution Although the key word is "total," since this is repeated addition of the same number, we multiply.

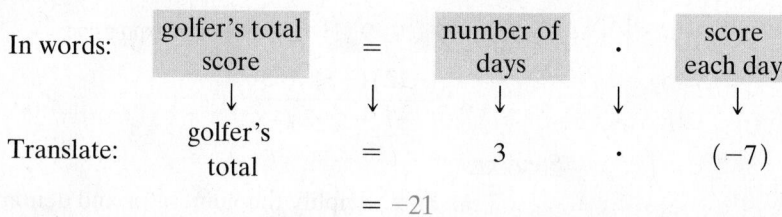

In words: | golfer's total score | = | number of days | · | score each day |

Translate: golfer's total = 3 · (-7)

$$= -21$$

Thus, the golfer's total score was -21, or 21 strokes under par.

PRACTICE
11 A card player had a score of -13 for each of four games. Find the total score.

Graphing Calculator Explorations

Entering Negative Numbers on a Scientific Calculator

To enter a negative number on a scientific calculator, find a key marked $\boxed{+/-}$. (On some calculators, this key is marked $\boxed{\text{CHS}}$ for "change sign.") To enter -8, for example, press the keys $\boxed{8}$ $\boxed{+/-}$. The display will read $\boxed{-8}$.

Entering Negative Numbers on a Graphing Calculator

To enter a negative number on a graphing calculator, find a key marked $\boxed{(-)}$. Do not confuse this key with the key $\boxed{-}$, which is used for subtraction. To enter -8, for example, press the keys $\boxed{(-)}\,\boxed{8}$. The display will read $\boxed{-8}$.

Operations with Real Numbers

To evaluate $-2(7 - 9) - 20$ on a calculator, press the keys

$\boxed{2}\,\boxed{+/-}\,\boxed{\times}\,\boxed{(}\,\boxed{(}\,\boxed{7}\,\boxed{-}\,\boxed{9}\,\boxed{)}\,\boxed{-}\,\boxed{2}\,\boxed{0}\,\boxed{=}$ or

$\boxed{(-)}\,\boxed{2}\,\boxed{(}\,\boxed{(}\,\boxed{7}\,\boxed{-}\,\boxed{9}\,\boxed{)}\,\boxed{-}\,\boxed{2}\,\boxed{0}\,\boxed{\text{ENTER}}$

The display will read $\boxed{-16}$ or $\boxed{\begin{array}{l}-2(7 - 9) - 20 \\ \hspace{3.5cm} -16\end{array}}$.

Use a calculator to simplify each expression.

1. $-38(26 - 27)$

2. $-59(-8) + 1726$

3. $134 + 25(68 - 91)$

4. $45(32) - 8(218)$

5. $\dfrac{-50(294)}{175 - 265}$

6. $\dfrac{-444 - 444.8}{-181 - 324}$

7. $9^5 - 4550$

8. $5^8 - 6259$

9. $(-125)^2$ (Be careful.)

10. -125^2 (Be careful.)

✓ Vocabulary, Readiness & Video Check

Use the choices below to fill in each blank. Some choices may be used more than once.

positive   0   negative   undefined

1. If n is a real number, then $n \cdot 0 = $ _____ and $0 \cdot n = $ _____.

2. If n is a real number, but not 0, then $\dfrac{0}{n} = $ _____ and we say $\dfrac{n}{0}$ is _____.

3. The product of two negative numbers is a _____ number.

4. The quotient of two negative numbers is a _____ number.

5. The quotient of a positive number and a negative number is a _____ number.

6. The product of a positive number and a negative number is a _____ number.

7. The reciprocal of a positive number is a _____ number.

8. The opposite of a positive number is a _____ number.

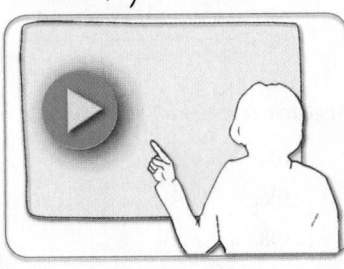

Martin-Gay Interactive Videos

See Video 1.7 ◉

Watch the section lecture video and answer the following questions.

OBJECTIVE 1

9. Explain the significance of the use of parentheses when comparing 🎞 Examples 6 and 7.

OBJECTIVE 2

10. In 🎞 Example 9, why is the reciprocal equal to $\dfrac{3}{2}$ and not $-\dfrac{3}{2}$?

OBJECTIVE 3

11. Before 🎞 Example 11, the sign rules for division of real numbers are discussed. Are the sign rules for division the same as for multiplication? Why or why not?

OBJECTIVE 4

12. In 🎞 Example 17, the importance of placing the replacement values in parentheses when evaluating is emphasized. Why?

OBJECTIVE 5

13. In 🎞 Example 18, explain why each loss of 4 yards is represented by -4 and not 4.

1.7   Exercise Set   MyMathLab® ▶

Multiply. See Examples 1 through 3.

▶ **1.** $-6(4)$

2. $-8(5)$

▶ **3.** $2(-1)$

4. $7(-4)$

▶ **5.** $-5(-10)$

6. $-6(-11)$

7. $-3 \cdot 4$

8. $-2 \cdot 8$

▶ **9.** $-7 \cdot 0$

10. $-6 \cdot 0$

11. $2(-9)$

12. $3(-5)$

13. $-\dfrac{1}{2}\left(-\dfrac{3}{5}\right)$

14. $-\dfrac{1}{8}\left(-\dfrac{1}{3}\right)$

15. $-\dfrac{3}{4}\left(-\dfrac{8}{9}\right)$

16. $-\dfrac{5}{6}\left(-\dfrac{3}{10}\right)$

17. $5(-1.4)$

18. $6(-2.5)$

19. $-0.2(-0.7)$

20. $-0.5(-0.3)$

21. $-10(80)$

22. $-20(60)$

23. $4(-7)$

24. $5(-9)$

25. $(-5)(-5)$

26. $(-7)(-7)$

▶ **27.** $\dfrac{2}{3}\left(-\dfrac{4}{9}\right)$

28. $\dfrac{2}{7}\left(-\dfrac{2}{11}\right)$

29. $-11(11)$

30. $-12(12)$

31. $-\dfrac{20}{25}\left(\dfrac{5}{16}\right)$

32. $-\dfrac{25}{36}\left(\dfrac{6}{15}\right)$

33. $(-1)(2)(-3)(-5)$

34. $(-2)(-3)(-4)(-2)$

Perform the indicated operations. See Example 2.

35. $(-2)(5) - (-11)(3)$

36. $8(-3) - 4(-5)$

37. $(-6)(-1)(-2) - (-5)$

38. $20 - (-4)(3)(-2)$

Decide whether each statement is true or false.

39. The product of three negative integers is negative.

40. The product of three positive integers is positive.

41. The product of four negative integers is negative.

42. The product of four positive integers is positive.

Evaluate. See Example 4.

▶ **43.** $(-2)^4$

▶ **44.** -2^4

45. -1^5

46. $(-1)^5$

47. $(-5)^2$

48. -5^2

49. -7^2

50. $(-7)^2$

Find each reciprocal or multiplicative inverse. See Example 5.

51. 9

52. 100

▶ **53.** $\dfrac{2}{3}$

54. $\dfrac{1}{7}$

▶ **55.** -14

56. -8

57. $-\dfrac{3}{11}$

58. $-\dfrac{6}{13}$

59. 0.2

60. 1.5

61. $\dfrac{1}{-6.3}$

62. $\dfrac{1}{-8.9}$

Divide. See Examples 6 through 8.

▶ **63.** $\dfrac{18}{-2}$

64. $\dfrac{20}{-10}$

65. $\dfrac{-16}{-4}$

66. $\dfrac{-18}{-6}$

67. $\dfrac{-48}{12}$

68. $\dfrac{-60}{5}$

▶ **69.** $\dfrac{0}{-4}$

70. $\dfrac{0}{-9}$

71. $-\dfrac{15}{3}$

72. $-\dfrac{24}{8}$

▶ **73.** $\dfrac{5}{0}$

74. $\dfrac{3}{0}$

▶ **75.** $\dfrac{-12}{-4}$

76. $\dfrac{-45}{-9}$

77. $\dfrac{30}{-2}$

78. $\dfrac{14}{-2}$

79. $\dfrac{6}{7} \div \left(-\dfrac{1}{3}\right)$

80. $\dfrac{4}{5} \div \left(-\dfrac{1}{2}\right)$

▶ **81.** $-\dfrac{5}{9} \div \left(-\dfrac{3}{4}\right)$

82. $-\dfrac{1}{10} \div \left(-\dfrac{8}{11}\right)$

83. $-\dfrac{4}{9} \div \dfrac{4}{9}$

84. $-\dfrac{5}{12} \div \dfrac{5}{12}$

MIXED PRACTICE

Simplify. See Examples 1 through 9.

85. $\dfrac{-9(-3)}{-6}$

86. $\dfrac{-6(-3)}{-4}$

87. $\dfrac{12}{9 - 12}$

88. $\dfrac{-15}{1 - 4}$

89. $\dfrac{-6^2 + 4}{-2}$

90. $\dfrac{3^2 + 4}{5}$

91. $\dfrac{8 + (-4)^2}{4 - 12}$

92. $\dfrac{6 + (-2)^2}{4 - 9}$

93. $\dfrac{22 + (3)(-2)}{-5 - 2}$

94. $\dfrac{-20 + (-4)(3)}{1 - 5}$

95. $\dfrac{-3 - 5^2}{2(-7)}$

96. $\dfrac{-2 - 4^2}{3(-6)}$

▶ **97.** $\dfrac{6 - 2(-3)}{4 - 3(-2)}$

98. $\dfrac{8 - 3(-2)}{2 - 5(-4)}$

99. $\dfrac{-3 - 2(-9)}{-15 - 3(-4)}$

100. $\dfrac{-4 - 8(-2)}{-9 - 2(-3)}$

101. $\dfrac{|5 - 9| + |10 - 15|}{|2(-3)|}$

102. $\dfrac{|-3 + 6| + |-2 + 7|}{|-2 \cdot 2|}$

If $x = -5$ and $y = -3$, evaluate each expression. See Example 10.

103. $3x + 2y$

104. $4x + 5y$

▶ **105.** $2x^2 - y^2$

106. $x^2 - 2y^2$

107. $x^3 + 3y$

108. $y^3 + 3x$

109. $\dfrac{2x - 5}{y - 2}$

110. $\dfrac{2y - 12}{x - 4}$

111. $\dfrac{-3 - y}{x - 4}$

112. $\dfrac{4 - 2x}{y + 3}$

TRANSLATING

Translate each phrase into an expression. Use x to represent "a number." See Example 11.

113. The product of −71 and a number

114. The quotient of −8 and a number

115. Subtract a number from −16.

116. The sum of a number and −12

117. −29 increased by a number

118. The difference of a number and −10

119. Divide a number by −33.

120. Multiply a number by −17.

Solve. See Example 11.

121. A football team lost four yards on each of three consecutive plays. Represent the total loss as a product of signed numbers and find the total loss.

122. Joe Norstrom lost $400 on each of seven consecutive days in the stock market. Represent his total loss as a product of signed numbers and find his total loss.

123. A deep-sea diver must move up or down in the water in short steps to keep from getting a physical condition called the "bends." Suppose a diver moves down from the surface in five steps of 20 feet each. Represent his total movement as a product of signed numbers and find the total depth.

124. A weather forecaster predicts that the temperature will drop five degrees each hour for the next six hours. Represent this drop as a product of signed numbers and find the total drop in temperature.

Decide whether the given number is a solution of the given equation.

125. Is 7 a solution of $-5x = -35$?

126. Is −4 a solution of $2x = x - 1$?

127. Is −20 a solution of $\frac{x}{10} = 2$?

128. Is −3 a solution of $\frac{45}{x} = -15$?

129. Is 5 a solution of $-3x - 5 = -20$?

130. Is −4 a solution of $2x + 4 = x + 8$?

CONCEPT EXTENSIONS

Study the bar graph below showing the average surface temperatures of planets. Use Exercises 131 and 132 to complete the planet temperatures on the graph. (Pluto is now classified as a dwarf planet.)

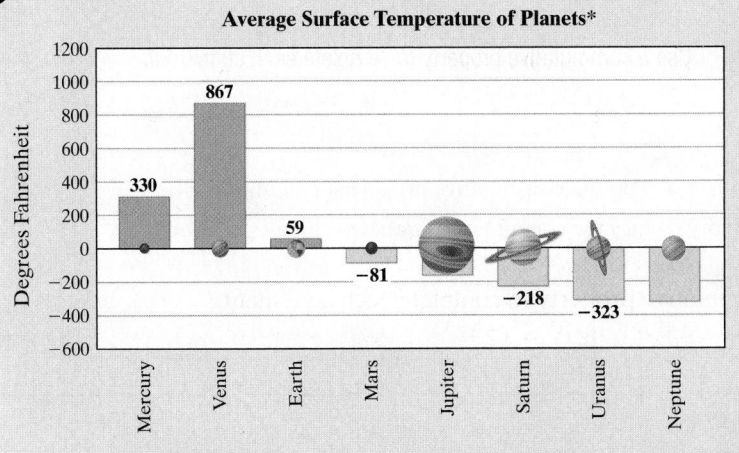

Average Surface Temperature of Planets*

Degrees Fahrenheit

Mercury 330, Venus 867, Earth 59, Mars −81, Saturn −218, Uranus −323

For some planets, the temperature given is the temperature where the atmosphere pressure equals 1 Earth atmosphere; data from The World Almanac

131. The surface temperature of Jupiter is twice the temperature of Mars. Find this temperature.

132. The surface temperature of Neptune is equal to the temperature of Mercury divided by −1. Find this temperature.

133. Explain why the product of an even number of negative numbers is a positive number.

134. If a and b are any real numbers, is the statement $a \cdot b = b \cdot a$ always true? Why or why not?

135. Find any real numbers that are their own reciprocal.

136. Explain why 0 has no reciprocal.

If q is a negative number, r is a negative number, and t is a positive number, determine whether each expression simplifies to a positive or negative number. If it is not possible to determine, state so.

137. $\dfrac{q}{r \cdot t}$

138. $q^2 \cdot r \cdot t$

139. $q + t$

140. $t + r$

141. $t(q + r)$

142. $r(q - t)$

Write each of the following as an expression and evaluate.

143. The sum of −2 and the quotient of −15 and 3

144. The sum of 1 and the product of −8 and −5

145. Twice the sum of −5 and −3

146. 7 subtracted from the quotient of 0 and 5

1.8 Properties of Real Numbers

OBJECTIVES

1 Use the Commutative and Associative Properties.

2 Use the Distributive Property.

3 Use the Identity and Inverse Properties.

OBJECTIVE

1 Using the Commutative and Associative Properties

In this section, we give names to properties of real numbers with which we are already familiar. Throughout this section, the variables *a, b,* and *c* represent real numbers.

We know that order does not matter when adding numbers. For example, we know that $7 + 5$ is the same as $5 + 7$. This property is given a special name—the **commutative property of addition.** We also know that order does not matter when multiplying numbers. For example, we know that $-5(6) = 6(-5)$. This property means that multiplication is commutative also and is called the **commutative property of multiplication.**

Commutative Properties

Addition: $a + b = b + a$

Multiplication: $a \cdot b = b \cdot a$

These properties state that the *order* in which any two real numbers are added or multiplied does not change their sum or product. For example, if we let $a = 3$ and $b = 5$, then the commutative properties guarantee that

$$3 + 5 = 5 + 3 \quad \text{and} \quad 3 \cdot 5 = 5 \cdot 3$$

> **Helpful Hint**
>
> Is subtraction also commutative? Try an example. Does $3 - 2 = 2 - 3$? **No!** The left side of this statement equals 1; the right side equals -1. There is no commutative property of subtraction. Similarly, there is no commutative property for division. For example, $10 \div 2$ does not equal $2 \div 10$.

EXAMPLE 1   Use a commutative property to complete each statement.

a. $x + 5 = $ _____   **b.** $3 \cdot x = $ _____

Solution

a. $x + 5 = 5 + x$ By the commutative property of addition

b. $3 \cdot x = x \cdot 3$ By the commutative property of multiplication

PRACTICE

1   Use a commutative property to complete each statement.

a. $x \cdot 8 = $ ___   **b.** $x + 17 = $ _____

✔ CONCEPT CHECK

Which of the following pairs of actions are commutative?

a. "raking the leaves" and "bagging the leaves"

b. "putting on your left glove" and "putting on your right glove"

c. "putting on your coat" and "putting on your shirt"

d. "reading a novel" and "reading a newspaper"

Let's now discuss grouping numbers. We know that when we add three numbers, the way in which they are grouped or associated does not change their sum. For example, we know that $2 + (3 + 4) = 2 + 7 = 9$. This result is the same if we group the numbers differently. In other words, $(2 + 3) + 4 = 5 + 4 = 9$ also. Thus, $2 + (3 + 4) = (2 + 3) + 4$. This property is called the **associative property of addition.**

We also know that changing the grouping of numbers when multiplying does not change their product. For example, $2 \cdot (3 \cdot 4) = (2 \cdot 3) \cdot 4$ (check it). This is the **associative property of multiplication.**

Associative Properties

Addition: $(a + b) + c = a + (b + c)$

Multiplication: $(a \cdot b) \cdot c = a \cdot (b \cdot c)$

These properties state that the way in which three numbers are *grouped* does not change their sum or their product.

EXAMPLE 2   Use an associative property to complete each statement.

a. $5 + (4 + 6) = $ _____

b. $(-1 \cdot 2) \cdot 5 = $ _____

Solution

a. $5 + (4 + 6) = (5 + 4) + 6$ By the associative property of addition

b. $(-1 \cdot 2) \cdot 5 = -1 \cdot (2 \cdot 5)$ By the associative property of multiplication   □

PRACTICE

2   Use an associative property to complete each statement.

a. $(2 + 9) + 7 = $ _____

b. $-4 \cdot (2 \cdot 7) = $ _____   ■

Helpful Hint

Remember the difference between the commutative properties and the associative properties. The commutative properties have to do with the *order* of numbers, and the associative properties have to do with the *grouping* of numbers.

Let's now illustrate how these properties can help us simplify expressions.

EXAMPLE 3   Simplify each expression.

a. $10 + (x + 12)$ **b.** $-3(7x)$

Solution

a. $10 + (x + 12) = 10 + (12 + x)$ By the commutative property of addition

$= (10 + 12) + x$ By the associative property of addition

$= 22 + x$ Add.

b. $-3(7x) = (-3 \cdot 7)x$ By the associative property of multiplication

$= -21x$ Multiply.   □

PRACTICE

3   Simplify each expression.

a. $(5 + x) + 9$ **b.** $5(-6x)$ ■

OBJECTIVE

2   Using the Distributive Property ▷

The **distributive property of multiplication over addition** is used repeatedly throughout algebra. It is useful because it allows us to write a product as a sum or a sum as a product. We know that $7(2 + 4) = 7(6) = 42$. Compare that with $7(2) + 7(4) = 14 + 28 = 42$. Since both original expressions equal 42, they must equal each other, or

$$7(2 + 4) = 7(2) + 7(4)$$

This is an example of the distributive property. The product on the left side of the equal sign is equal to the sum on the right side. We can think of the 7 as being distributed to each number inside the parentheses.

> **Distributive Property of Multiplication Over Addition**
>
> $$a(b + c) = ab + ac$$

Since multiplication is commutative, this property can also be written as

$$(b + c)a = ba + ca$$

The distributive property can also be extended to more than two numbers inside the parentheses. For example,

$$3(x + y + z) = 3(x) + 3(y) + 3(z)$$
$$= 3x + 3y + 3z$$

Since we define subtraction in terms of addition, the distributive property is also true for subtraction. For example,

$$2(x - y) = 2(x) - 2(y)$$
$$= 2x - 2y$$

EXAMPLE 4 Use the distributive property to write each expression without parentheses. Then simplify if possible.

a. $2(x + y)$ **b.** $-5(-3 + 2z)$ **c.** $5(x + 3y - z)$

d. $-1(2 - y)$ **e.** $-(3 + x - w)$ **f.** $\dfrac{1}{2}(6x + 14) + 10$

Solution

a. $2(x + y) = 2 \cdot x + 2 \cdot y$
            $= 2x + 2y$

b. $-5(-3 + 2z) = -5(-3) + (-5)(2z)$
                 $= 15 - 10z$

c. $5(x + 3y - z) = 5(x) + 5(3y) - 5(z)$
                $= 5x + 15y - 5z$

d. $-1(2 - y) = (-1)(2) - (-1)(y)$
             $= -2 + y$

> **Helpful Hint**
> Notice in part **(e)** that $-(3 + x - w)$ is first rewritten as $-1(3 + x - w)$.

e. $-(3 + x - w) = -1(3 + x - w)$
                 $= (-1)(3) + (-1)(x) - (-1)(w)$
                 $= -3 - x + w$

f. $\dfrac{1}{2}(6x + 14) + 10 = \dfrac{1}{2}(6x) + \dfrac{1}{2}(14) + 10$ Apply the distributive property.
                      $= 3x + 7 + 10$ Multiply.
                      $= 3x + 17$ Add.

PRACTICE

4   Use the distributive property to write each expression without parentheses. Then simplify if possible.

a. $5(x - y)$

b. $-6(4 + 2t)$

c. $2(3x - 4y - z)$

d. $(3 - y) \cdot (-1)$

e. $-(x - 7 + 2s)$

f. $\dfrac{1}{2}(2x + 4) + 9$ ■

We can use the distributive property in reverse to write a sum as a product.

EXAMPLE 5   Use the distributive property to write each sum as a product.

a. $8 \cdot 2 + 8 \cdot x$

b. $7s + 7t$

Solution

a. $8 \cdot 2 + 8 \cdot x = 8(2 + x)$

b. $7s + 7t = 7(s + t)$ □

PRACTICE

5   Use the distributive property to write each sum as a product.

a. $5 \cdot w + 5 \cdot 3$

b. $9w + 9z$ ■

OBJECTIVE

3   Using the Identity and Inverse Properties

Next, we look at the **identity properties.**

The number 0 is called the identity for addition because when 0 is added to any real number, the result is the same real number. In other words, the *identity* of the real number is not changed.

The number 1 is called the identity for multiplication because when a real number is multiplied by 1, the result is the same real number. In other words, the *identity* of the real number is not changed.

Identities for Addition and Multiplication

0 is the identity element for addition.

$$a + 0 = a \quad \text{and} \quad 0 + a = a$$

1 is the identity element for multiplication.

$$a \cdot 1 = a \quad \text{and} \quad 1 \cdot a = a$$

Notice that 0 is the *only* number that can be added to any real number with the result that the sum is the same real number. Also, 1 is the *only* number that can be multiplied by any real number with the result that the product is the same real number.

Additive inverses or **opposites** were introduced in Section 1.5. Two numbers are called additive inverses or opposites if their sum is 0. The additive inverse or opposite of 6 is -6 because $6 + (-6) = 0$. The additive inverse or opposite of -5 is 5 because $-5 + 5 = 0$.

Reciprocals or **multiplicative inverses** were introduced in Section 1.3. Two non-zero numbers are called reciprocals or multiplicative inverses if their product is 1. The reciprocal or multiplicative inverse of $\frac{2}{3}$ is $\frac{3}{2}$ because $\frac{2}{3} \cdot \frac{3}{2} = 1$. Likewise, the reciprocal of -5 is $-\frac{1}{5}$ because $-5\left(-\frac{1}{5}\right) = 1$.

✔ **CONCEPT CHECK**

Which of the following, $1, -\frac{10}{3}, \frac{3}{10}, 0, \frac{10}{3}, -\frac{3}{10}$, is the

a. opposite of $-\frac{3}{10}$? **b.** reciprocal of $-\frac{3}{10}$?

Additive or Multiplicative Inverses

The numbers a and $-a$ are additive inverses or opposites of each other because their sum is 0; that is,

$$a + (-a) = 0$$

The numbers b and $\frac{1}{b}$ (for $b \neq 0$) are reciprocals or multiplicative inverses of each other because their product is 1; that is,

$$b \cdot \frac{1}{b} = 1$$

EXAMPLE 6   Name the property or properties illustrated by each true statement.

Solution

a. $3 \cdot y = y \cdot 3$ Commutative property of multiplication (order changed)

b. $(x + 7) + 9 = x + (7 + 9)$ Associative property of addition (grouping changed)

c. $(b + 0) + 3 = b + 3$ Identity element for addition

d. $0.2 \cdot (z \cdot 5) = 0.2 \cdot (5 \cdot z)$ Commutative property of multiplication (order changed)

e. $-2 \cdot \left(-\frac{1}{2}\right) = 1$ Multiplicative inverse property

f. $-2 + 2 = 0$ Additive inverse property

g. $-6 \cdot (y \cdot 2) = (-6 \cdot 2) \cdot y$ Commutative and associative properties of multiplication (order and grouping changed)

PRACTICE

6   Name the property or properties illustrated by each true statement.

a. $(7 \cdot 3x) \cdot 4 = (3x \cdot 7) \cdot 4$ Commutative property of multiplication

b. $6 + (3 + y) = (6 + 3) + y$ Associative property of addition

c. $8 + (t + 0) = 8 + t$ Identity element for addition

d. $-\frac{3}{4} \cdot \left(-\frac{4}{3}\right) = 1$ Multiplicative inverse property

e. $(2 + x) + 5 = 5 + (2 + x)$ Commutative property of addition

f. $3 + (-3) = 0$ Additive inverse property

g. $(-3b) \cdot 7 = (-3 \cdot 7) \cdot b$ Commutative and associative properties of multiplication

Answers to Concept Check:

a. $\frac{3}{10}$ **b.** $-\frac{10}{3}$

✔ Vocabulary, Readiness & Video Check

Use the choices below to fill in each blank.

distributive property   associative property of multiplication   commutative property of addition
opposites or additive inverses   associative property of addition
reciprocals or multiplicative inverses   commutative property of multiplication

1. $x + 5 = 5 + x$ is a true statement by the _____.

2. $x \cdot 5 = 5 \cdot x$ is a true statement by the _____.

3. $3(y + 6) = 3 \cdot y + 3 \cdot 6$ is true by the _____.

4. $2 \cdot (x \cdot y) = (2 \cdot x) \cdot y$ is a true statement by the _____.

5. $x + (7 + y) = (x + 7) + y$ is a true statement by the _____.

6. The numbers $-\dfrac{2}{3}$ and $-\dfrac{3}{2}$ are called _____.

7. The numbers $-\dfrac{2}{3}$ and $\dfrac{2}{3}$ are called _____.

Martin-Gay Interactive Videos

See Video 1.8 ◉

Watch the section lecture video and answer the following questions.

OBJECTIVE 1

8. The commutative properties are discussed in ▤ Examples 1 and 2 and the associative properties are discussed in ▤ Examples 3–7. What's the one word used again and again to describe the commutative property? The associative property?

OBJECTIVE 2

9. In ▤ Example 10, what point is made about the term 2?

OBJECTIVE 3

10. Complete these statements based on the lecture given before ▤ Example 12.
- The identity element for addition is _____ because if we add _____ to any real number, the result is that real number.
- The identity element for multiplication is _____ because any real number times _____ gives a result of that original real number.

1.8 Exercise Set   MyMathLab® ▷

Use a commutative property to complete each statement. See Example 1.

▷ **1.** $x + 16 =$ _____   **2.** $4 + y =$ _____

3. $-4 \cdot y =$ _____   **4.** $-2 \cdot x =$ _____

▷ **5.** $xy =$ _____   **6.** $ab =$ _____

7. $2x + 13 =$ _____   **8.** $19 + 3y =$ _____

Use an associative property to complete each statement. See Example 2.

▷ **9.** $(xy) \cdot z =$ _____   **10.** $3 \cdot (xy) =$ _____

11. $2 + (a + b) =$ _____   **12.** $(y + 4) + z =$ _____

13. $4 \cdot (ab) =$ _____   **14.** $(-3y) \cdot z =$ _____

15. $(a + b) + c =$ _____

16. $6 + (r + s) =$ _____

Use the commutative and associative properties to simplify each expression. See Example 3.

▷ **17.** $8 + (9 + b)$ **18.** $(r + 3) + 11$

▷ **19.** $4(6y)$ **20.** $2(42x)$

21. $\dfrac{1}{5}(5y)$ **22.** $\dfrac{1}{8}(8z)$

23. $(13 + a) + 13$ **24.** $7 + (x + 4)$

25. $-9(8x)$ **26.** $-3(12y)$

27. $\dfrac{3}{4}\left(\dfrac{4}{3}s\right)$ **28.** $\dfrac{2}{7}\left(\dfrac{7}{2}r\right)$

29. $\dfrac{2}{3} + \left(\dfrac{4}{3} + x\right)$ **30.** $\dfrac{7}{9} + \left(\dfrac{2}{9} + y\right)$

Use the distributive property to write each expression without parentheses. Then simplify the result. See Example 4.

31. $4(x + y)$ **32.** $7(a + b)$

33. $9(x - 6)$ **34.** $11(y - 4)$

35. $2(3x + 5)$ **36.** $5(7 + 8y)$

37. $7(4x - 3)$ **38.** $3(8x - 1)$

▶ 39. $3(6 + x)$ **40.** $2(x + 5)$

41. $-2(y - z)$ **42.** $-3(z - y)$

43. $-7(3y + 5)$

44. $-5(2r + 11)$

45. $5(x + 4m + 2)$

46. $8(3y + z - 6)$

47. $-4(1 - 2m + n)$

48. $-4(4 + 2p + 5q)$

49. $-(5x + 2)$

50. $-(9r + 5)$

▶ 51. $-(r - 3 - 7p)$

52. $-(q - 2 + 6r)$

53. $\dfrac{1}{2}(6x + 8)$

54. $\dfrac{1}{4}(4x - 2)$

55. $-\dfrac{1}{3}(3x - 9y)$

56. $-\dfrac{1}{5}(10a - 25b)$

57. $3(2r + 5) - 7$

58. $10(4s + 6) - 40$

▶ 59. $-9(4x + 8) + 2$

60. $-11(5x + 3) + 10$

61. $-4(4x + 5) - 5$

62. $-6(2x + 1) - 1$

Use the distributive property to write each sum as a product. See Example 5.

63. $4 \cdot 1 + 4 \cdot y$ **64.** $14 \cdot z + 14 \cdot 5$

▶ 65. $11x + 11y$ **66.** $9a + 9b$

67. $(-1) \cdot 5 + (-1) \cdot x$ **68.** $(-3)a + (-3)b$

69. $30a + 30b$ **70.** $25x + 25y$

Name the properties illustrated by each true statement. See Example 6.

71. $3 \cdot 5 = 5 \cdot 3$

72. $4(3 + 8) = 4 \cdot 3 + 4 \cdot 8$

73. $2 + (x + 5) = (2 + x) + 5$

74. $(x + 9) + 3 = (9 + x) + 3$

75. $9(3 + 7) = 9 \cdot 3 + 9 \cdot 7$

▶ 76. $1 \cdot 9 = 9$

77. $(4 \cdot y) \cdot 9 = 4 \cdot (y \cdot 9)$

▶ 78. $6 \cdot \dfrac{1}{6} = 1$

▶ 79. $0 + 6 = 6$

80. $(a + 9) + 6 = a + (9 + 6)$

81. $-4(y + 7) = -4 \cdot y + (-4) \cdot 7$

▶ 82. $(11 + r) + 8 = (r + 11) + 8$

83. $-4 \cdot (8 \cdot 3) = (8 \cdot -4) \cdot 3$

84. $r + 0 = r$

CONCEPT EXTENSIONS

Fill in the table with the opposite (additive inverse) and the reciprocal (multiplicative inverse). Assume that the value of each expression is not 0.

	Expression	Opposite	Reciprocal
85.	8		
86.	$-\dfrac{2}{3}$		
87.	x		
88.	$4y$		
89.			$\dfrac{1}{2x}$
90.		$7x$	

Decide whether each statement is true or false. See the second Concept Check in this section.

91. The opposite of $-\dfrac{a}{2}$ is $-\dfrac{2}{a}$.

92. The reciprocal of $-\dfrac{a}{2}$ is $\dfrac{a}{2}$.

Determine which pairs of actions are commutative. See the first Concept Check in this section.

93. "taking a test" and "studying for the test"

94. "putting on your shoes" and "putting on your socks"

95. "putting on your left shoe" and "putting on your right shoe"

96. "reading the sports section" and "reading the comics section"

97. "mowing the lawn" and "trimming the hedges"

98. "baking a cake" and "eating the cake"

99. "dialing a number" and "turning on the cell phone"

100. "feeding the dog" and "feeding the cat"

Name the property illustrated by each step.

101. a. $\triangle + (\square + \bigcirc) = (\square + \bigcirc) + \triangle$

    b. $= (\bigcirc + \square) + \triangle$

    c. $= \bigcirc + (\square + \triangle)$

102. **a.** $(x + y) + z = x + (y + z)$
     b. $ = (y + z) + x$
     c. $ = (z + y) + x$

103. Explain why 0 is called the identity element for addition.

104. Explain why 1 is called the identity element for multiplication.

105. Write an example that shows that division is not commutative.

106. Write an example that shows that subtraction is not commutative.

Chapter 1 Vocabulary Check

Fill in each blank with one of the words or phrases listed below.

set	inequality symbols	opposites	absolute value	numerator
denominator	grouping symbols	exponent	base	reciprocals
variable	equation	solution		

1. The symbols $\neq$, $<$, and $>$ are called _____.

2. A mathematical statement that two expressions are equal is called a(n) _____.

3. The _____ of a number is the distance between that number and 0 on a number line.

4. A symbol used to represent a number is called a(n) _____.

5. Two numbers that are the same distance from 0 but lie on opposite sides of 0 are called _____.

6. The number in a fraction above the fraction bar is called the _____.

7. A(n) _____ of an equation is a value for the variable that makes the equation a true statement.

8. Two numbers whose product is 1 are called _____.

9. In 2^3, the 2 is called the _____ and the 3 is called the _____.

10. The number in a fraction below the fraction bar is called the _____.

11. Parentheses and brackets are examples of _____.

12. A(n) _____ is a collection of objects.

Chapter 1 Highlights

DEFINITIONS AND CONCEPTS	**EXAMPLES**
Section 1.2 Symbols and Sets of Numbers	
A **set** is a collection of objects, called **elements**, enclosed in braces.	$\{a, c, e\}$
Natural Numbers: $\{1, 2, 3, 4, \dots\}$	Given the set $\left\{-3.4, \sqrt{3}, 0, \dfrac{2}{3}, 5, -4\right\}$, list the numbers that belong to the set of
Whole Numbers: $\{0, 1, 2, 3, 4, \dots\}$	
Integers: $\{\dots, -3, -2, -1, 0, 1, 2, 3, \dots\}$	Natural numbers: 5
Rational Numbers: {real numbers that can be expressed as a quotient of integers}	Whole numbers: $0, 5$
	Integers: $-4, 0, 5$
Irrational Numbers: {real numbers that cannot be expressed as a quotient of integers}	Rational numbers: $-4, -3.4, 0, \dfrac{2}{3}, 5$
	Irrational Numbers: $\sqrt{3}$
Real Numbers: {all numbers that correspond to a point on the number line}	Real numbers: $-4, -3.4, 0, \dfrac{2}{3}, \sqrt{3}, 5$
	(continued)

DEFINITIONS AND CONCEPTS	EXAMPLES

Section 1.2 Symbols and Sets of Numbers (continued)

A line used to picture numbers is called a **number line.**

The **absolute value** of a real number a, denoted by $|a|$, is the distance between a and 0 on a number line.

Symbols: $=$ is equal to

$\neq$ is not equal to

$>$ is greater than

$<$ is less than

$\leq$ is less than or equal to

$\geq$ is greater than or equal to

Order Property for Real Numbers

For any two real numbers a and b, a is less than b if a is to the left of b on a number line.

$$|5| = 5 \qquad |0| = 0 \qquad |-2| = 2$$

$$-7 = -7$$

$$3 \neq -3$$

$$4 > 1$$

$$1 < 4$$

$$6 \leq 6$$

$$18 \geq -\frac{1}{3}$$

$$-3 < 0 \qquad 0 > -3 \qquad 0 < 2.5 \qquad 2.5 > 0$$

Section 1.3 Fractions and Mixed Numbers

A quotient of two integers is called a **fraction.**
The **numerator** of a fraction is the top number.
The **denominator** of a fraction is the bottom number.

If $a \cdot b = c$, then a and b are **factors** and c is the **product.**

$13 \leftarrow$ numerator
$17 \leftarrow$ denominator

$$7 \quad \cdot \quad 9 \quad = \quad 63$$
$$\downarrow \qquad \downarrow \qquad\qquad \downarrow$$
factor   factor   product

A fraction is in **lowest terms** or **simplest form** when the numerator and the denominator have no factors in common other than 1.

$\dfrac{13}{17}$ is in simplest form.

To write a fraction in simplest form, factor the numerator and the denominator; then apply the fundamental principle.

Write in simplest form.

$$\frac{6}{14} = \frac{2 \cdot 3}{2 \cdot 7} = \frac{3}{7}$$

Two fractions are **reciprocals** if their product is 1.
The reciprocal of $\dfrac{a}{b}$ is $\dfrac{b}{a}$.

The reciprocal of $\dfrac{6}{25}$ is $\dfrac{25}{6}$.

To multiply fractions, numerator times numerator is the numerator of the product and denominator times denominator is the denominator of the product.

Perform the indicated operations.

$$\frac{2}{5} \cdot \frac{3}{7} = \frac{6}{35}$$

To divide fractions, multiply the first fraction by the reciprocal of the second fraction.

$$\frac{5}{9} \div \frac{2}{7} = \frac{5}{9} \cdot \frac{7}{2} = \frac{35}{18}$$

To add fractions with the same denominator, add the numerators and place the sum over the common denominator.

$$\frac{5}{11} + \frac{3}{11} = \frac{8}{11}$$

To subtract fractions with the same denominator, subtract the numerators and place the difference over the common denominator.

$$\frac{13}{15} - \frac{3}{15} = \frac{10}{15} = \frac{2}{3}$$

Fractions that represent the same quantity are called **equivalent fractions.**

$$\frac{1}{5} = \frac{1 \cdot 4}{5 \cdot 4} = \frac{4}{20}$$

$\dfrac{1}{5}$ and $\dfrac{4}{20}$ are equivalent fractions.

DEFINITIONS AND CONCEPTS	EXAMPLES

Section 1.4 Exponents, Order of Operations, Variable Expressions, and Equations

The expression a^n is an **exponential expression.** The number a is called the **base;** it is the repeated factor.
The number n is called the **exponent;** it is the number of times that the base is a factor.

Order of Operations

Simplify expressions in the following order. If grouping symbols are present, simplify expressions within those first, starting with the innermost set. Also, simplify the numerator and the denominator of a fraction separately.

1. Simplify exponential expressions.
2. Multiply or divide in order from left to right.
3. Add or subtract in order from left to right.

$$4^3 = 4 \cdot 4 \cdot 4 = 64$$

$$7^2 = 7 \cdot 7 = 49$$

$$\frac{8^2 + 5(7 - 3)}{3 \cdot 7} = \frac{8^2 + 5(4)}{21}$$
$$= \frac{64 + 5(4)}{21}$$
$$= \frac{64 + 20}{21}$$
$$= \frac{84}{21}$$
$$= 4$$

A symbol used to represent a number is called a **variable.**

Examples of variables are:
$$q, x, z$$

An **algebraic expression** is a collection of numbers, variables, operation symbols, and grouping symbols.

Examples of algebraic expressions are:
$$5x, 2(y - 6), \frac{q^2 - 3q + 1}{6}$$

To evaluate an algebraic expression containing a variable, substitute a given number for the variable and simplify.

Evaluate $x^2 - y^2$ if $x = 5$ and $y = 3$.
$$x^2 - y^2 = (5)^2 - (3)^2$$
$$= 25 - 9$$
$$= 16$$

A mathematical statement that two expressions are equal is called an **equation.**

Examples of equations are:
$$3x - 9 = 20$$
$$A = \pi r^2$$

A **solution** of an equation is a value for the variable that makes the equation a true statement.

Determine whether 4 is a solution of $5x + 7 = 27$.
$$5x + 7 = 27$$
$$5(4) + 7 \stackrel{?}{=} 27$$
$$20 + 7 \stackrel{?}{=} 27$$
$$27 = 27 \quad \text{True}$$

4 is a solution.

Section 1.5 Adding Real Numbers

To Add Two Numbers with the Same Sign
1. Add their absolute values.
2. Use their common sign as the sign of the sum.

To Add Two Numbers with Different Signs
1. Subtract their absolute values.
2. Use the sign of the number whose absolute value is larger as the sign of the sum.

Add.

$$10 + 7 = 17$$
$$-3 + (-8) = -11$$

$$-25 + 5 = -20$$
$$14 + (-9) = 5$$

(continued)

DEFINITIONS AND CONCEPTS	EXAMPLES

Section 1.5   Adding Real Numbers (continued)

Two numbers that are the same distance from 0 but lie on opposite sides of 0 are called **opposites** or **additive inverses**. The opposite of a number a is denoted by $-a$.	The opposite of -7 is 7. The opposite of 123 is -123.

The sum of a number a and its opposite, $-a$, is 0.

$$a + (-a) = 0$$

$$-4 + 4 = 0$$
$$12 + (-12) = 0$$

If a is a number, then $-(-a) = a$.

$$-(-8) = 8$$
$$-(-14) = 14$$

Section 1.6   Subtracting Real Numbers

To subtract two numbers a and b, add the first number a to the opposite of the second number b.

$$a - b = a + (-b)$$

Subtract.

$$3 - (-44) = 3 + 44 = 47$$
$$-5 - 22 = -5 + (-22) = -27$$
$$-30 - (-30) = -30 + 30 = 0$$

Section 1.7   Multiplying and Dividing Real Numbers

Quotient of two real numbers

$$\frac{a}{b} = a \cdot \frac{1}{b}$$

Multiplying and Dividing Real Numbers

The product or quotient of two numbers with the same sign is a positive number. The product or quotient of two numbers with different signs is a negative number.

Multiply or divide.

$$\frac{42}{2} = 42 \cdot \frac{1}{2} = 21$$

$$7 \cdot 8 = 56 \quad -7 \cdot (-8) = 56$$
$$-2 \cdot 4 = -8 \quad 2 \cdot (-4) = -8$$
$$\frac{90}{10} = 9 \qquad \frac{-90}{-10} = 9$$
$$\frac{42}{-6} = -7 \qquad \frac{-42}{6} = -7$$

Products and Quotients Involving Zero

The product of 0 and any number is 0.

$$b \cdot 0 = 0 \quad \text{and} \quad 0 \cdot b = 0$$

The quotient of a nonzero number and 0 is undefined.

$$\frac{b}{0} \text{ is undefined.}$$

The quotient of 0 and any nonzero number is 0.

$$\frac{0}{b} = 0$$

$$-4 \cdot 0 = 0 \qquad 0 \cdot \left(-\frac{3}{4}\right) = 0$$

$$\frac{-85}{0} \text{ is undefined.}$$

$$\frac{0}{18} = 0 \qquad \frac{0}{-47} = 0$$

Section 1.8   Properties of Real Numbers

Commutative Properties

Addition: $a + b = b + a$

Multiplication: $a \cdot b = b \cdot a$

$$3 + (-7) = -7 + 3$$
$$-8 \cdot 5 = 5 \cdot (-8)$$

Associative Properties

Addition: $(a + b) + c = a + (b + c)$

Multiplication: $(a \cdot b) \cdot c = a \cdot (b \cdot c)$

$$(5 + 10) + 20 = 5 + (10 + 20)$$
$$(-3 \cdot 2) \cdot 11 = -3 \cdot (2 \cdot 11)$$

DEFINITIONS AND CONCEPTS	EXAMPLES

Section 1.8   Properties of Real Numbers (continued)

Two numbers whose product is 1 are called **multiplicative inverses** or **reciprocals.** The reciprocal of a nonzero number a is $\dfrac{1}{a}$ because $a \cdot \dfrac{1}{a} = 1$.

The reciprocal of 3 is $\dfrac{1}{3}$.

The reciprocal of $-\dfrac{2}{5}$ is $-\dfrac{5}{2}$.

Distributive Property
$$a(b + c) = a \cdot b + a \cdot c$$

$$5(6 + 10) = 5 \cdot 6 + 5 \cdot 10$$
$$-2(3 + x) = -2 \cdot 3 + (-2)(x)$$

Identities
$$a + 0 = a \qquad 0 + a = a$$
$$a \cdot 1 = a \qquad 1 \cdot a = a$$

$$5 + 0 = 5 \qquad 0 + (-2) = -2$$
$$-14 \cdot 1 = -14 \qquad 1 \cdot 27 = 27$$

Inverses

Additive or opposite: $a + (-a) = 0$

$$7 + (-7) = 0$$

Multiplicative or reciprocal: $b \cdot \dfrac{1}{b} = 1$

$$3 \cdot \dfrac{1}{3} = 1$$

Chapter 1 **Review**

(1.2) *Insert* $<, >,$ *or* $=$ *in the appropriate space to make the following statements true.*

1. 8   10

2. 7   2

3. -4 -5

4. $\dfrac{12}{2}$ -8

5. $|-7|$ $|-8|$

6. $|-9|$ -9

7. $-|-1|$ -1

8. $|-14|$ $-(-14)$

9. 1.2   1.02

10. $-\dfrac{3}{2}$ $-\dfrac{3}{4}$

TRANSLATING

Translate each statement into symbols.

11. Four is greater than or equal to negative three.

12. Six is not equal to five.

13. 0.03 is less than 0.3.

14. New York City has 155 museums and 400 art galleries. Write an inequality comparing the numbers 155 and 400. (*Source:* Absolute Trivia.com)

Given the following sets of numbers, list the numbers in each set that also belong to the set of:

a. Natural numbers
b. Whole numbers
c. Integers
d. Rational numbers
e. Irrational numbers
f. Real numbers

15. $\left\{ -6, 0, 1, 1\dfrac{1}{2}, 3, \pi, 9.62 \right\}$

16. $\left\{ -3, -1.6, 2, 5, \dfrac{11}{2}, 15.1, \sqrt{5}, 2\pi \right\}$

The following chart shows the gains and losses in dollars of Density Oil and Gas stock for a particular week.

Day	Gain or Loss in Dollars
Monday	+1
Tuesday	−2
Wednesday	+5
Thursday	+1
Friday	−4

17. Which day showed the greatest loss?

18. Which day showed the greatest gain?

(1.3) *Write the number as a product of prime factors.*

19. 36

20. 120

Perform the indicated operations. Write results in lowest terms.

21. $\dfrac{8}{15} \cdot \dfrac{27}{30}$

22. $\dfrac{7}{8} \div \dfrac{21}{32}$

23. $\dfrac{7}{15} + \dfrac{5}{6}$

24. $\dfrac{3}{4} - \dfrac{3}{20}$

25. $2\dfrac{3}{4} + 6\dfrac{5}{8}$

26. $7\dfrac{1}{6} - 2\dfrac{2}{3}$

27. $5 \div \dfrac{1}{3}$

28. $2 \cdot 8\dfrac{3}{4}$

Each circle represents a whole, or 1. Determine the unknown part of the circle.

29.
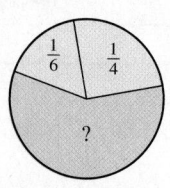

30.
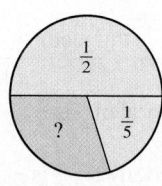

Find the area and the perimeter of each figure.

△ **31.**

△ **32.**

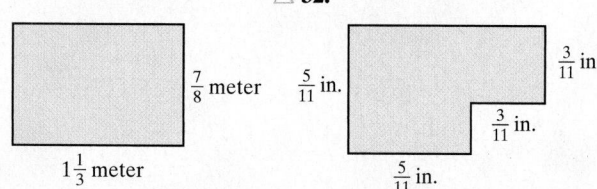

$\frac{7}{8}$ meter

$1\frac{1}{3}$ meter

$\frac{5}{11}$ in.

$\frac{3}{11}$ in.

$\frac{3}{11}$ in.

$\frac{5}{11}$ in.

Octuplets were born in the U.S. in 2009. The following chart gives the octuplets' birthweights. The babies are listed in order of birth.

Baby	Gender	Birthweight (pounds)
Baby A	boy	$2\frac{1}{2}$
Baby B	girl	$2\frac{1}{8}$
Baby C	boy	$3\frac{1}{16}$
Baby D	girl	$2\frac{3}{16}$
Baby E	boy	$1\frac{3}{4}$
Baby F	boy	$2\frac{9}{16}$
Baby G	boy	$1\frac{13}{16}$
Baby H	boy	$2\frac{7}{16}$

33. What was the total weight of the boy octuplets?

34. What was the total weight of the girl octuplets?

35. Find the combined weight of all eight octuplets.

36. Which baby weighed the most?

37. Which baby weighed the least?

38. How much more did the heaviest baby weigh than the lightest baby?

(1.4) Choose the correct answer for each statement.

39. The expression $6 \cdot 3^2 + 2 \cdot 8$ simplifies to
   a. -52 **b.** 448 **c.** 70 **d.** 64

40. The expression $68 - 5 \cdot 2^3$ simplifies to
   a. -232 **b.** 28 **c.** 38 **d.** 504

Simplify each expression.

41. $\left(\frac{2}{7}\right)^2$

42. $\left(\frac{3}{4}\right)^3$

43. $3(1 + 2 \cdot 5) + 4$

44. $8 + 3(2 \cdot 6 - 1)$

45. $\dfrac{4 + |6 - 2| + 8^2}{4 + 6 \cdot 4}$

46. $5[3(2 + 5) - 5]$

TRANSLATING

Translate each word statement into symbols.

47. The difference of twenty and twelve is equal to the product of two and four.

48. The quotient of nine and two is greater than negative five.

Evaluate each expression if $x = 6$, $y = 2$, and $z = 8$.

49. $2x + 3y$

50. $x(y + 2z)$

51. $\dfrac{x}{y} + \dfrac{z}{2y}$

52. $x^2 - 3y^2$

△ **53.** The expression $180 - a - b$ represents the measure of the unknown angle of the given triangle. Replace a with 37 and b with 80 to find the measure of the unknown angle.

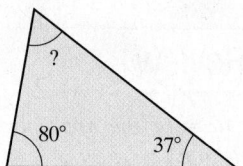

△ **54.** The expression $360 - a - b - c$ represents the measure of the unknown angle of the given quadrilateral. Replace a with 93, b with 80, and c with 82 to find the measure of the unknown angle.

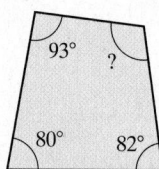

Decide whether the given number is a solution of the given equation.

55. Is $x = 3$ a solution of $7x - 3 = 18$?

56. Is $x = 1$ a solution of $3x^2 + 4 = x - 1$?

(1.5) Find the additive inverse or the opposite.

57. -9

58. $\dfrac{2}{3}$

59. $|-2|$

60. $-|-7|$

Find the following sums.

61. $-15 + 4$

62. $-6 + (-11)$

63. $\dfrac{1}{16} + \left(-\dfrac{1}{4}\right)$

64. $-8 + |-3|$

65. $-4.6 + (-9.3)$

66. $-2.8 + 6.7$

(1.6) *Perform the indicated operations.*

67. $6 - 20$

68. $-3.1 - 8.4$

69. $-6 - (-11)$

70. $4 - 15$

71. $-21 - 16 + 3(8 - 2)$

72. $\dfrac{11 - (-9) + 6(8 - 2)}{2 + 3 \cdot 4}$

Evaluate each expression for $x = 3$, $y = -6$, and $z = -9$. Then choose the correct evaluation.

73. $2x^2 - y + z$

    a. 15   **b.** 3   **c.** 27   **d.** -3

74. $\dfrac{|y - 4x|}{2x}$

    a. 3   **b.** 1   **c.** -1 **d.** -3

75. At the beginning of the week, the price of Density Oil and Gas stock from Exercises 17 and 18 is $50 per share. Find the price of a share of stock at the end of the week.

76. Find the price of a share of stock by the end of the day on Wednesday.

(1.7) *Find the multiplicative inverse or reciprocal.*

77. -6

78. $\dfrac{3}{5}$

Simplify each expression.

79. $6(-8)$

80. $(-2)(-14)$

81. $\dfrac{-18}{-6}$

82. $\dfrac{42}{-3}$

83. $\dfrac{4(-3) + (-8)}{2 + (-2)}$

84. $\dfrac{3(-2)^2 - 5}{-14}$

85. $\dfrac{-6}{0}$

86. $\dfrac{0}{-2}$

87. $-4^2 - (-3 + 5) \div (-1) \cdot 2$

88. $-5^2 - (2 - 20) \div (-3) \cdot 3$

If $x = -5$ and $y = -2$, evaluate each expression.

89. $x^2 - y^4$

90. $x^2 - y^3$

TRANSLATING

Translate each phrase into an expression. Use x to represent a number.

91. The product of -7 and a number

92. The quotient of a number and -13

93. Subtract a number from -20

94. The sum of -1 and a number

(1.8) *Name the property illustrated.*

95. $-6 + 5 = 5 + (-6)$

96. $6 \cdot 1 = 6$

97. $3(8 - 5) = 3 \cdot 8 - 3 \cdot (5)$

98. $4 + (-4) = 0$

99. $2 + (3 + 9) = (2 + 3) + 9$

100. $2 \cdot 8 = 8 \cdot 2$

101. $6(8 + 5) = 6 \cdot 8 + 6 \cdot 5$

102. $(3 \cdot 8) \cdot 4 = 3 \cdot (8 \cdot 4)$

103. $4 \cdot \dfrac{1}{4} = 1$

104. $8 + 0 = 8$

Use the distributive property to write each expression without parentheses.

105. $5(y - 2)$

106. $-3(z + y)$

107. $-(7 - x + 4z)$

108. $\dfrac{1}{2}(6z - 10)$

109. $-4(3x + 5) - 7$

110. $-8(2y + 9) - 1$

MIXED REVIEW

Insert $<$, $>$, or $=$ in the space between each pair of numbers.

111. $-|-11|$ $|11.4|$

112. $-1\dfrac{1}{2}$ $-2\dfrac{1}{2}$

Perform the indicated operations.

113. $-7.2 + (-8.1)$

114. $14 - 20$

115. $4(-20)$

116. $\dfrac{-20}{4}$

117. $-\dfrac{4}{5}\left(\dfrac{5}{16}\right)$

118. $-0.5(-0.3)$

119. $8 \div 2 \cdot 4$

120. $(-2)^4$

121. $\dfrac{-3 - 2(-9)}{-15 - 3(-4)}$

122. $5 + 2[(7 - 5)^2 + (1 - 3)]$

123. $-\dfrac{5}{8} \div \dfrac{3}{4}$

124. $\dfrac{-15 + (-4)^2 + |-9|}{10 - 2 \cdot 5}$

△ **125.** A trim carpenter needs a piece of quarter round molding $6\dfrac{1}{8}$ feet long for a bathroom. She finds a piece $7\dfrac{1}{2}$ feet long. How long a piece does she need to cut from the $7\dfrac{1}{2}$-foot-long molding in order to use it in the bathroom?

Chapter 1 Getting Ready for the Test

|1c|

*All the exercises below are **Multiple Choice**. Choose the correct letter(s). Also, letters may be used more than once.*
Select the given operation between the two numbers.

1. For $-5 + (-3)$, the operation is

   A. addition   **B.** subtraction   **C.** multiplication   **D.** division

2. For $-5(-3)$, the operation is

   A. addition   **B.** subtraction   **C.** multiplication   **D.** division

Identify each as an

   A. equation   or an   **B.** expression

3. $6x + 2 + 4x - 10$ **4.** $6x + 2 = 4x - 10$

5. $-2(x - 1) = 12$ **6.** $-7\left(x + \frac{1}{2}\right) - 22$

For the exercises below, a and b are negative numbers. State whether each expression simplifies to

   A. positive number   **B.** negative number   **C.** 0   **D.** not possible to determine

7. $a + b$ **8.** $a \cdot b$

9. $\frac{a}{b}$ **10.** $a - 0$

11. $0 \cdot b$ **12.** $a - b$

13. $0 + b$ **14.** $\frac{0}{a}$

The exercise statement and the correct answer are given. Select the correct directions.

   A. Find the opposite.   **B.** Find the reciprocal.   **C.** Evaluate or simplify.

15. 5   Answer: $\frac{1}{5}$ **16.** $3 + 2(-8)$ Answer: -13

17. 2^3 Answer: 8   **18.** -7 Answer: 7

Chapter 1 Test   MyMathLab® You Tube™

Translate the statement into symbols.

1. The absolute value of negative seven is greater than five.

2. The sum of nine and five is greater than or equal to four.

Simplify the expression.

3. $-13 + 8$ **4.** $-13 - (-2)$

5. $12 \div 4 \cdot 3 - 6 \cdot 2$ **6.** $(13)(-3)$

7. $(-6)(-2)$ **8.** $\frac{|-16|}{-8}$

9. $\frac{-8}{0}$ **10.** $\frac{|-6| + 2}{5 - 6}$

11. $\frac{1}{2} - \frac{5}{6}$ **12.** $5\frac{3}{4} - 1\frac{1}{8}$

13. $-0.6 + 1.875$ **14.** $3(-4)^2 - 80$

15. $6[5 + 2(3 - 8) - 3]$ **16.** $\frac{-12 + 3 \cdot 8}{4}$

17. $\frac{(-2)(0)(-3)}{-6}$

Insert $<$, $>$, or $=$ in the appropriate space to make each of the following statements true.

18. -3 __ -7 **19.** 4 __ -8

20. 2 __ $|-3|$ **21.** $|-2|$ __ $-1 - (-3)$

22. In the state of Massachusetts, there are 2221 licensed child care centers and 10,993 licensed home-based child care providers. Write an inequality statement comparing the numbers 2221 and 10,993. (*Source:* Children's Foundation)

23. Given $\left\{-5, -1, 0, \frac{1}{4}, 1, 7, 11.6, \sqrt{7}, 3\pi\right\}$, list the numbers in this set that also belong to the set of:

   a. Natural numbers
   b. Whole numbers
   c. Integers
   d. Rational numbers
   e. Irrational numbers
   f. Real numbers

If $x = 6$, $y = -2$, and $z = -3$, evaluate each expression.

24. $x^2 + y^2$ **25.** $x + yz$

26. $2 + 3x - y$ **27.** $\frac{y + z - 1}{x}$

Identify the property illustrated by each expression.

28. $8 + (9 + 3) = (8 + 9) + 3$

29. $6 \cdot 8 = 8 \cdot 6$

30. $-6(2 + 4) = -6 \cdot 2 + (-6) \cdot 4$

31. $\frac{1}{6}(6) = 1$

32. Find the opposite of -9.

33. Find the reciprocal of $-\frac{1}{3}$.

The New Orleans Saints were 22 yards from the goal when the following series of gains and losses occurred.

	Gains and Losses in Yards
First Down	5
Second Down	−10
Third Down	−2
Fourth Down	29

34. During which down did the greatest loss of yardage occur?

35. Was a touchdown scored?

36. The temperature at the Winter Olympics was a frigid 14 degrees below zero in the morning, but by noon it had risen 31 degrees. What was the temperature at noon?

37. A health insurance provider had net incomes of $356 million, $460 million, and −$166 million in 3 consecutive years. What was the health insurance provider's total net income for these three years?

38. A stockbroker decided to sell 280 shares of stock, which decreased in value by $1.50 per share yesterday. How much money did she lose?

Equations, Inequalities, and Problem Solving

2.1 Simplifying Algebraic Expressions

2.2 The Addition and Multiplication Properties of Equality

2.3 Solving Linear Equations

Integrated Review–Solving Linear Equations

2.4 An Introduction to Problem Solving

2.5 Formulas and Problem Solving

2.6 Percent and Mixture Problem Solving

2.7 Further Problem Solving

2.8 Solving Linear Inequalities

 CHECK YOUR PROGRESS

Vocabulary Check
Chapter Highlights
Chapter Review
Getting Ready for the Test
Chapter Test
Cumulative Review

Much of mathematics relates to deciding which statements are true and which are false. For example, the statement $x + 7 = 15$ is an equation stating that the sum $x + 7$ has the same value as 15. Is this statement true or false? It is false for some values of x and true for just one value of x, namely 8. Our purpose in this chapter is to learn ways of deciding which values make an equation or an inequality true.

What Are iOS and Android Systems?

Smartphones, tablets, and iPads are everywhere. What are the operating systems behind these devices? Two such systems are iOS and Android. iOS is a mobile operating system developed by Apple, Inc., for Apple hardware, and Android is a mobile operating system developed by Google, primarily for touchscreen devices such as smartphones and tablet computers. The average American now spends 162 minutes, that is, 2 hours and 42 minutes, daily on one (or more) of these devices. This is an increase of 4 minutes a day from last year, and it seems that this amount of time will only continue to grow.

In Section 2.6, Exercises 33 through 36, we will explore how this time is spent.

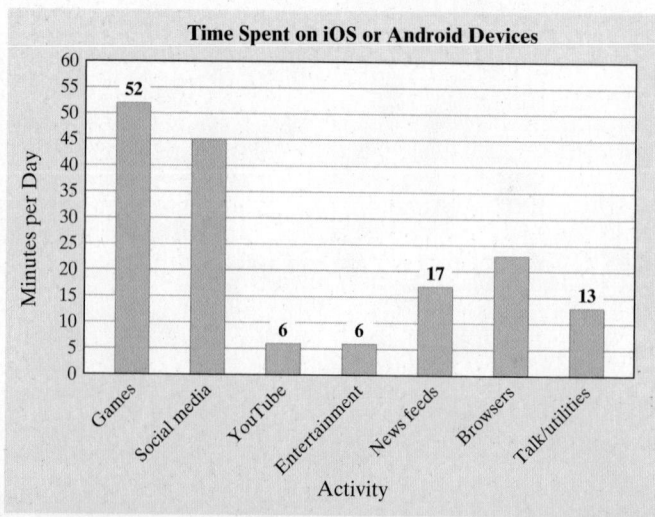

Time Spent on iOS or Android Devices

2.1 Simplifying Algebraic Expressions

OBJECTIVES

1 Identify Terms, Like Terms, and Unlike Terms.

2 Combine Like Terms.

3 Use the Distributive Property to Remove Parentheses.

4 Write Word Phrases as Algebraic Expressions.

As we explore in this section, an expression such as $3x + 2x$ is not as simple as possible because—even without replacing x by a value—we can perform the indicated addition.

OBJECTIVE

1 Identifying Terms, Like Terms, and Unlike Terms

Before we practice simplifying expressions, some new language of algebra is presented. A **term** is a number or the product of a number and variables raised to powers.

Terms

$$-y, \quad 2x^3, \quad -5, \quad 3xz^2, \quad \frac{2}{y}, \quad 0.8z$$

The **numerical coefficient** (sometimes also simply called the **coefficient**) of a term is the numerical factor. The numerical coefficient of $3x$ is 3. Recall that $3x$ means $3 \cdot x$.

Term	Numerical Coefficient
$3x$	3
$\dfrac{y^3}{5}$	$\dfrac{1}{5}$ since $\dfrac{y^3}{5}$ means $\dfrac{1}{5} \cdot y^3$
$0.7ab^3c^5$	0.7
z	1
$-y$	-1
-5	-5

> **Helpful Hint**
>
> The term $-y$ means $-1y$ and thus has a numerical coefficient of -1.
> The term z means $1z$ and thus has a numerical coefficient of 1.

EXAMPLE 1 Identify the numerical coefficient of each term.

   a. $-3y$ **b.** $22z^4$ **c.** y **d.** $-x$ **e.** $\dfrac{x}{7}$

Solution

   a. The numerical coefficient of $-3y$ is -3.

   b. The numerical coefficient of $22z^4$ is 22.

   c. The numerical coefficient of y is 1, since y is $1y$.

   d. The numerical coefficient of $-x$ is -1, since $-x$ is $-1x$.

   e. The numerical coefficient of $\dfrac{x}{7}$ is $\dfrac{1}{7}$, since $\dfrac{x}{7}$ means $\dfrac{1}{7} \cdot x$.

PRACTICE

1 Identify the numerical coefficient of each term.

   a. t **b.** $-7x$ **c.** $-\dfrac{w}{5}$ **d.** $43x^4$ **e.** $-b$

Terms with the same variables raised to exactly the same powers are called **like terms**. Terms that aren't like terms are called **unlike terms**.

Like Terms	Unlike Terms	Reason
$3x, 2x$	$5x, 5x^2$	Why? Same variable x but different powers x and x^2
$-6x^2y, 2x^2y, 4x^2y$	$7y, 3z, 8x^2$	Why? Different variables
$2ab^2c^3, ac^3b^2$	$6abc^3, 6ab^2$	Why? Different variables and different powers

> **Helpful Hint**
>
> In like terms, each variable and its exponent must match exactly, but these factors don't need to be in the same order.
>
> $$2x^2y \text{ and } 3yx^2 \text{ are like terms.}$$

EXAMPLE 2   Determine whether the terms are like or unlike.

     a. $2x, 3x^2$ **b.** $4x^2y, x^2y, -2x^2y$ **c.** $-2yz, -3zy$ **d.** $-x^4, x^4$

Solution

     a. Unlike terms, since the exponents on x are not the same.

     b. Like terms, since each variable and its exponent match.

     c. Like terms, since $zy = yz$ by the commutative property.

     d. Like terms.

PRACTICE
2   Determine whether the terms are like or unlike.

     a. $-4xy, 5yx$ **b.** $5q, -3q^2$

     c. $3ab^2, -2ab^2, 43ab^2$ **d.** $y^5, \dfrac{y^5}{2}$

OBJECTIVE

2   Combining Like Terms

An algebraic expression containing the sum or difference of like terms can be simplified by applying the distributive property. For example, by the distributive property, we rewrite the sum of the like terms $3x + 2x$ as

$$3x + 2x = (3 + 2)x = 5x$$

     Also,

$$-y^2 + 5y^2 = -1y^2 + 5y^2 = (-1 + 5)y^2 = 4y^2$$

Simplifying the sum or difference of like terms is called **combining like terms.**

EXAMPLE 3   Simplify each expression by combining like terms.

     a. $7x - 3x$ **b.** $10y^2 + y^2$

     c. $8x^2 + 2x - 3x$ **d.** $9n^2 - 5n^2 + n^2$

Solution

     a. $7x - 3x = (7 - 3)x = 4x$

     b. $10y^2 + y^2 = 10y^2 + 1y^2 = (10 + 1)y^2 = 11y^2$

     c. $8x^2 + 2x - 3x = 8x^2 + (2 - 3)x = 8x^2 - x$

     d. $9n^2 - 5n^2 + n^2 = (9 - 5 + 1)n^2 = 5n^2$

PRACTICE
3   Simplify each expression by combining like terms.

     a. $-3y + 11y$ **b.** $4x^2 + x^2$

     c. $5x - 3x^2 + 8x^2$ **d.** $20y^2 + 2y^2 - y^2$

The previous example suggests the following:

> **Combining Like Terms**
>
> To **combine like terms,** add the numerical coefficients and multiply the result by the common variable factors.

EXAMPLE 4   Simplify each expression by combining like terms.

    a. $2x + 3x + 5 + 2$ **b.** $-5a - 3 + a + 2$ **c.** $4y - 3y^2$

    d. $2.3x + 5x - 6$ **e.** $-\dfrac{1}{2}b + b$

Solution Use the distributive property to combine like terms.

    a. $2x + 3x + 5 + 2 = (2 + 3)x + (5 + 2)$

                            $= 5x + 7$

    b. $-5a - 3 + a + 2 = -5a + 1a + (-3 + 2)$

                            $= (-5 + 1)a + (-3 + 2)$

                            $= -4a - 1$

    c. $4y - 3y^2$ These two terms cannot be combined because they are unlike terms.

    d. $2.3x + 5x - 6 = (2.3 + 5)x - 6$

                         $= 7.3x - 6$

    e. $-\dfrac{1}{2}b + b = -\dfrac{1}{2}b + 1b = \left(-\dfrac{1}{2} + 1\right)b = \dfrac{1}{2}b$ □

PRACTICE
4   Use the distributive property to combine like terms.

    a. $3y + 8y - 7 + 2$ **b.** $6x - 3 - x - 3$ **c.** $\dfrac{3}{4}t - t$

    d. $9y + 3.2y + 10 + 3$ **e.** $5z - 3z^4$ ■

OBJECTIVE
3   Using the Distributive Property ▶

Simplifying expressions makes frequent use of the distributive property to also remove parentheses.

    It may be helpful to study the examples below.

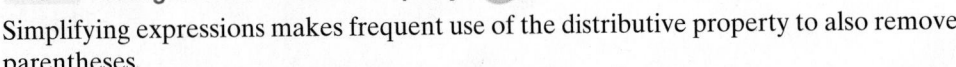

$$+(3a + 2) = +1(3a + 2) = +1(3a) + (+1)(2) = 3a + 2$$
$$\underset{\longrightarrow\text{means}\longrightarrow}{}$$

$$-(3a + 2) = -1(3a + 2) = -1(3a) + (-1)(2) = -3a - 2$$
$$\underset{\longrightarrow\text{means}\longrightarrow}{}$$

EXAMPLE 5   Find each product by using the distributive property to remove parentheses.

    a. $5(3x + 2)$ **b.** $-2(y + 0.3z - 1)$ **c.** $-(9x + y - 2z + 6)$

Solution

    a. $5(3x + 2) = 5 \cdot 3x + 5 \cdot 2$ Apply the distributive property.

                     $= 15x + 10$ Multiply.

(Continued on next page)

b. $-2(y + 0.3z - 1) = -2(y) + (-2)(0.3z) + (-2)(-1)$ Apply the distributive property.
$$= -2y - 0.6z + 2$$ Multiply.

c. $-(9x + y - 2z + 6) = -1(9x + y - 2z + 6)$ Distribute -1 over each term.
$$= -1(9x) - 1(y) - 1(-2z) - 1(6)$$
$$= -9x - y + 2z - 6$$ □

PRACTICE
5   Find each product by using the distributive property to remove parentheses.

a. $3(2x - 7)$ **b.** $-5(x - 0.5z - 5)$
c. $-(2x - y + z - 2)$ ■

> **Helpful Hint**
>
> If a "−" sign precedes parentheses, the sign of each term inside the parentheses is changed when the distributive property is applied to remove parentheses.
>
> **Examples:**
>
> $$-(2x + 1) = -2x - 1 \qquad -(-5x + y - z) = 5x - y + z$$
> $$-(x - 2y) = -x + 2y \qquad -(-3x - 4y - 1) = 3x + 4y + 1$$

When simplifying an expression containing parentheses, we often use the distributive property in both directions—first to remove parentheses and then again to combine any like terms.

EXAMPLE 6   Simplify each expression.

a. $3(2x - 5) + 1$ **b.** $-2(4x + 7) - (3x - 1)$ **c.** $9 + 3(4x - 10)$

Solution

a. $3(2x - 5) + 1 = 6x - 15 + 1$ Apply the distributive property.
$$= 6x - 14$$ Combine like terms.

b. $-2(4x + 7) - (3x - 1) = -8x - 14 - 3x + 1$ Apply the distributive property.
$$= -11x - 13$$ Combine like terms.

c. $9 + 3(4x - 10) = 9 + 12x - 30$ Apply the distributive property.
$$= -21 + 12x$$ Combine like terms.
$$\text{or } 12x - 21$$ □

> **Helpful Hint**
>
> Don't forget to use the distributive property and multiply before adding or subtracting like terms.

PRACTICE
6   Simplify each expression.

a. $4(9x + 1) + 6$ **b.** $-7(2x - 1) - (6 - 3x)$ **c.** $8 - 5(6x + 5)$ ■

EXAMPLE 7   Write the phrase below as an algebraic expression. Then simplify if possible.

"Subtract $4x - 2$ from $2x - 3$."

Solution "Subtract $4x - 2$ **from** $2x - 3$ " translates to $(2x - 3) - (4x - 2)$. Next, simplify the algebraic expression.

$$(2x - 3) - (4x - 2) = 2x - 3 - 4x + 2$$ Apply the distributive property.
$$= -2x - 1$$ Combine like terms.   □

PRACTICE
7   Write the phrase below as an algebraic expression. Then simplify if possible.

"Subtract $7x - 1$ from $2x + 3$."

OBJECTIVE
4   Writing Word Phrases as Algebraic Expressions

Next, we practice writing word phrases as algebraic expressions.

EXAMPLE 8   Write the following phrases as algebraic expressions and simplify if possible. Let x represent the unknown number.

a. Twice a number, plus 6
b. The difference of a number and 4, divided by 7
c. Five added to triple the sum of a number and 1
d. The sum of twice a number, 3 times the number, and 5 times the number

Solution

a. In words: twice a number   plus   6

Translate: $2x$ $+$ 6

b. In words: the difference of a number and 4   divided by   7

Translate: $(x - 4)$ $\div$ 7 or $\dfrac{x - 4}{7}$

c. In words: five   added to   triple   the sum of a number and 1

Translate: 5 $+$ $3 \cdot$ $(x + 1)$

Next, we simplify this expression.

$$5 + 3(x + 1) = 5 + 3x + 3 \quad \text{Use the distributive property.}$$
$$= 8 + 3x \quad \text{Combine like terms.}$$

d. The phrase "the sum of" means that we add.

In words: twice a number   added to   3 times the number   added to   5 times the number

Translate: $2x$ $+$ $3x$ $+$ $5x$

Now let's simplify.

$$2x + 3x + 5x = 10x \quad \text{Combine like terms.}$$

PRACTICE
8   Write the following phrases as algebraic expressions and simplify if possible. Let x represent the unknown number.

a. Three added to double a number
b. Six subtracted from the sum of 5 and a number
c. Two times the sum of 3 and a number, increased by 4
d. The sum of a number, half the number, and 5 times the number

✓ Vocabulary, Readiness & Video Check

Use the choices below to fill in each blank. Some choices may be used more than once.

like   numerical coefficient   term   distributive
unlike   combine like terms   expression

1. $23y^2 + 10y - 6$ is called a(n) _____ while $23y^2$, $10y$, and -6 are each called a(n) _____.
2. To simplify $x + 4x$, we _____.
3. The term y has an understood _____ of 1.
4. The terms $7z$ and $7y$ are _____ terms and the terms $7z$ and $-z$ are _____ terms.
5. For the term $-\frac{1}{2}xy^2$, the number $-\frac{1}{2}$ is the _____.
6. $5(3x - y)$ equals $15x - 5y$ by the _____ property.

Martín-Gay Interactive Videos

See Video 2.1 ◉

Watch the section lecture video and answer the following questions.

OBJECTIVE 1
7. Example 7 shows two terms with exactly the same variables. Why are these terms not considered like terms?

OBJECTIVE 2
8. Example 8 shows us that when combining like terms, we are actually applying what property?

OBJECTIVE 3
9. The expression in Example 11 shows a minus sign before parentheses. When using the distributive property to multiply and remove parentheses, what number are we actually distributing to each term within the parentheses?

OBJECTIVE 4
10. Write the phrase given in Example 14, translate it into an algebraic expression, then simplify it. Why are we able to simplify it?

2.1 Exercise Set   MyMathLab® ▷

Identify the numerical coefficient of each term. See Example 1.

1. $-7y$
2. $3x$
3. x
4. $-y$
5. $17x^2y$
6. $1.2xyz$

Indicate whether the terms in each list are like or unlike. See Example 2.

▷ 7. $5y, -y$
▷ 8. $-2x^2y, 6xy$
9. $2z, 3z^2$
10. $ab^2, -7ab^2$
11. $8wz, \frac{1}{7}zw$
12. $7.4p^3q^2, 6.2p^3q^2r$

Simplify each expression by combining any like terms. See Examples 3 and 4.

13. $7y + 8y$
▷ 14. $3x + 2x$
15. $8w - w + 6w$
16. $c - 7c + 2c$
17. $3b - 5 - 10b - 4$
18. $6g + 5 - 3g - 7$
19. $m - 4m + 2m - 6$
20. $a + 3a - 2 - 7a$
21. $5g - 3 - 5 - 5g$
22. $8p + 4 - 8p - 15$
23. $6.2x - 4 + x - 1.2$
24. $7.9y - 0.7 - y + 0.2$
25. $6x - 5x + x - 3 + 2x$
26. $8h + 13h - 6 + 7h - h$
27. $7x^2 + 8x^2 - 10x^2$
▷ 28. $8x^3 + x^3 - 11x^3$
▷ 29. $6x + 0.5 - 4.3x - 0.4x + 3$
30. $0.4y - 6.7 + y - 0.3 - 2.6y$

Simplify each expression. First use the distributive property to remove any parentheses. See Examples 5 and 6.

31. $5(y - 4)$
32. $7(r - 3)$
33. $-2(x + 2)$
34. $-4(y + 6)$

35. $7(d - 3) + 10$

36. $9(z + 7) - 15$

37. $-5(2x - 3y + 6)$

38. $-2(4x - 3z - 1)$

39. $-(3x - 2y + 1)$

40. $-(y + 5z - 7)$

▶ **41.** $5(x + 2) - (3x - 4)$

42. $4(2x - 3) - 2(x + 1)$

Write each of the following as an algebraic expression. Simplify if possible. See Example 7.

43. Add $6x + 7$ to $4x - 10$.

44. Add $3y - 5$ to $y + 16$.

45. Subtract $7x + 1$ from $3x - 8$.

46. Subtract $4x - 7$ from $12 + x$.

▶ **47.** Subtract $5m - 6$ from $m - 9$.

48. Subtract $m - 3$ from $2m - 6$.

MIXED PRACTICE

Simplify each expression. See Examples 3 through 7.

49. $2k - k - 6$

50. $7c - 8 - c$

51. $-9x + 4x + 18 - 10x$

52. $5y - 14 + 7y - 20y$

53. $-4(3y - 4) + 12y$

54. $-3(2x + 5) - 6x$

55. $3(2x - 5) - 5(x - 4)$

56. $2(6x - 1) - (x - 7)$

57. $-2(3x - 4) + 7x - 6$

58. $8y - 2 - 3(y + 4)$

59. $5k - (3k - 10)$

60. $-11c - (4 - 2c)$

61. Subtract $6x - 1$ from $3x + 4$

62. Subtract $4 + 3y$ from $8 - 5y$

63. $3.4m - 4 - 3.4m - 7$

64. $2.8w - 0.9 - 0.5 - 2.8w$

65. $\frac{1}{3}(7y - 1) + \frac{1}{6}(4y + 7)$

66. $\frac{1}{5}(9y + 2) + \frac{1}{10}(2y - 1)$

67. $2 + 4(6x - 6)$

68. $8 + 4(3x - 4)$

69. $0.5(m + 2) + 0.4m$

70. $0.2(k + 8) - 0.1k$

71. $10 - 3(2x + 3y)$

72. $14 - 11(5m + 3n)$

73. $6(3x - 6) - 2(x + 1) - 17x$

74. $7(2x + 5) - 4(x + 2) - 20x$

75. $\frac{1}{2}(12x - 4) - (x + 5)$

76. $\frac{1}{3}(9x - 6) - (x - 2)$

TRANSLATING

Write each phrase as an algebraic expression and simplify if possible. Let x represent the unknown number. See Examples 7 and 8.

▶ **77.** Twice a number, decreased by four

78. The difference of a number and two, divided by five

79. Seven added to double a number

80. Eight more than triple a number

81. Three-fourths of a number, increased by twelve

82. Eleven, increased by two-thirds of a number

▶ **83.** The sum of 5 times a number and -2, added to 7 times the number

84. The sum of 3 times a number and 10, subtracted from 9 times the number

85. Eight times the sum of a number and six

86. Six times the difference of a number and five

87. Double a number, minus the sum of the number and ten

88. Half a number, minus the product of the number and eight

89. The sum of 2, three times a number, -9, and four times the number

90. The sum of twice a number, -1, five times the number, and -12

REVIEW AND PREVIEW

Evaluate the following expressions for the given values. See Section 1.7.

91. If $x = -1$ and $y = 3$, find $y - x^2$.

92. If $g = 0$ and $h = -4$, find $gh - h^2$.

93. If $a = 2$ and $b = -5$, find $a - b^2$.

94. If $x = -3$, find $x^3 - x^2 + 4$.

95. If $y = -5$ and $z = 0$, find $yz - y^2$.

96. If $x = -2$, find $x^3 - x^2 - x$.

CONCEPT EXTENSIONS

△ **97.** Recall that the perimeter of a figure is the total distance around the figure. Given the following rectangle, express the perimeter as an algebraic expression containing the variable x.

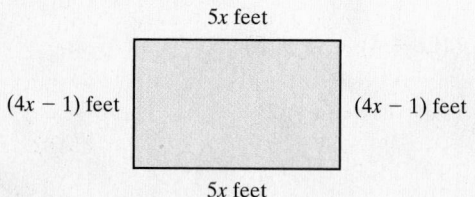

5x feet

(4x − 1) feet   (4x − 1) feet

5x feet

△ **98.** Given the following triangle, express its perimeter as an algebraic expression containing the variable x.

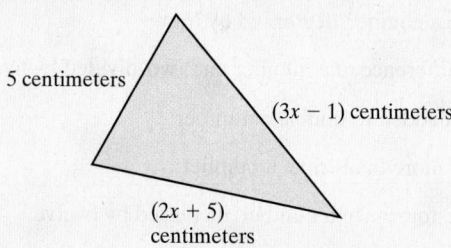

5 centimeters

$(3x − 1)$ centimeters

$(2x + 5)$ centimeters

Given the following two rules, determine whether each scale in Exercises 99 through 102 is balanced.

Rule 1:

1 cone balances 1 cube

Rule 2:

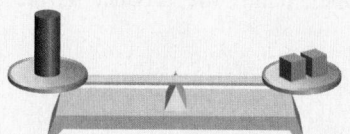

1 cylinder balances 2 cubes

99.

100.

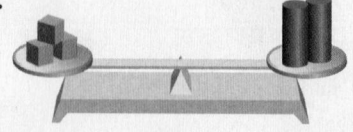

101.

102.

Write each algebraic expression described.

103. Write an expression with 4 terms that simplifies to $3x − 4$.

104. Write an expression of the form _____ (_____ + _____) whose product is $6x + 24$.

105. To convert from feet to inches, we multiply by 12. For example, the number of inches in 2 feet is $12 \cdot 2$ inches. If one board has a length of $(x + 2)$ *feet* and a second board has a length of $(3x − 1)$ *inches*, express their total length in inches as an algebraic expression.

106. The value of 7 nickels is $5 \cdot 7$ cents. Likewise, the value of x nickels is $5x$ cents. If the money box in a drink machine contains x *nickels*, $3x$ *dimes*, and $(30x − 1)$ *quarters*, express their total value in cents as an algebraic expression.

107. In your own words, explain how to combine like terms.

108. Do like terms always contain the same numerical coefficients? Explain your answer.

For Exercises 109 through 114, see the example below.

Example

Simplify: $−3xy + 2x^2y − (2xy − 1)$

Solution

$$−3xy + 2x^2y − (2xy − 1)$$
$$= −3xy + 2x^2y − 2xy + 1 = −5xy + 2x^2y + 1$$

Simplify each expression.

109. $5b^2c^3 + 8b^3c^2 − 7b^3c^2$

110. $4m^4p^2 + m^4p^2 − 5m^2p^4$

111. $3x − (2x^2 − 6x) + 7x^2$

112. $9y^2 − (6xy^2 − 5y^2) − 8xy^2$

113. $−(2x^2y + 3z) + 3z − 5x^2y$

114. $−(7c^3d − 8c) − 5c − 4c^3d$

2.2 | The Addition and Multiplication Properties of Equality

OBJECTIVES

1 Define Linear Equations and Use the Addition Property of Equality to Solve Linear Equations.

2 Use the Multiplication Property of Equality to Solve Linear Equations.

3 Use Both Properties of Equality to Solve Linear Equations.

4 Write Word Phrases as Algebraic Expressions.

OBJECTIVE

1   Defining Linear Equations and Using the Addition Property

Recall from Section 1.4 that an equation is a statement that two expressions have the same value. Also, a value of the variable that makes an equation a true statement is called a solution or root of the equation. The process of finding the solution of an equation is called **solving** the equation for the variable. In this section we concentrate on solving **linear equations** in one variable.

> **Linear Equation in One Variable**
>
> **A linear equation in one variable** can be written in the form
>
> $$ax + b = c$$
>
> where $a, b,$ and c are real numbers and $a \neq 0$.

Evaluating a linear equation for a given value of the variable, as we did in Section 1.4, can tell us whether that value is a solution, but we can't rely on evaluating an equation as our method of solving it.

Instead, to solve a linear equation in x, we write a series of simpler equations, all *equivalent* to the original equation, so that the final equation has the form

$$x = \textbf{number} \quad \text{or} \quad \textbf{number} = x$$

Equivalent equations are equations that have the same solution. This means that the "number" above is the solution to the original equation.

The first property of equality that helps us write simpler equivalent equations is the **addition property of equality.**

> **Addition Property of Equality**
>
> If $a, b,$ and c are real numbers, then
>
> $$a = b \quad \text{and} \quad a + c = b + c$$
>
> are equivalent equations.

This property guarantees that adding the same number to both sides of an equation does not change the solution of the equation. Since subtraction is defined in terms of addition, we may also **subtract the same number from both sides** without changing the solution.

A good way to picture a true equation is as a balanced scale. Since it is balanced, each side of the scale weighs the same amount.

If the same weight is added to or subtracted from each side, the scale remains balanced.

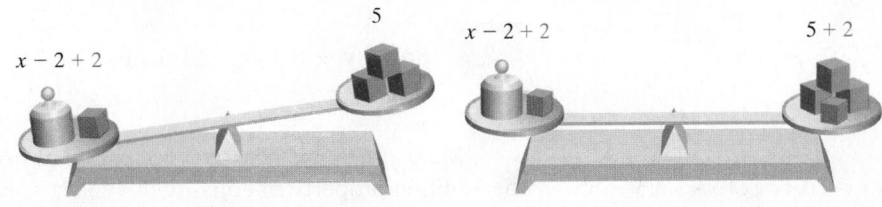

We use the addition property of equality to write equivalent equations until the variable is by itself on one side of the equation, and the equation looks like "$x =$ number" or "number $= x$."

> **EXAMPLE 1**   Solve: $x - 7 = 10$ for x

Solution   To solve for x, we want x alone on one side of the equation. To do this, we add 7 to both sides of the equation.

$$x - 7 = 10$$
$$x - 7 + 7 = 10 + 7 \quad \text{Add 7 to both sides.}$$
$$x = 17 \quad \text{Simplify.}$$

The solution of the equation $x = 17$ is obviously 17. Since we are writing equivalent equations, the solution of the equation $x - 7 = 10$ is also 17.

Check: To check, replace x with 17 in the original equation.

$$x - 7 = 10$$
$$17 - 7 \stackrel{?}{=} 10 \quad \text{Replace } x \text{ with 17 in the original equation.}$$
$$10 = 10 \quad \text{True}$$

Since the statement is true, 17 is the solution.   □

PRACTICE
1   Solve: $x + 3 = -5$ for x ■

✔ **CONCEPT CHECK**

Use the addition property to fill in the blanks so that the middle equation simplifies to the last equation.

$$x - 5 = 3$$
$$x - 5 + \underline{} = 3 + \underline{}$$
$$x = 8$$

> **EXAMPLE 2**   Solve: $y + 0.6 = -1.0$ for y

Solution   To get y alone on one side of the equation, subtract 0.6 from both sides of the equation.

$$y + 0.6 = -1.0$$
$$y + 0.6 - 0.6 = -1.0 - 0.6 \quad \text{Subtract 0.6 from both sides.}$$
$$y = -1.6 \quad \text{Combine like terms.}$$

Check: To check the proposed solution, -1.6, replace y with -1.6 in the original equation.

$$y + 0.6 = -1.0$$
$$-1.6 + 0.6 \stackrel{?}{=} -1.0 \quad \text{Replace } y \text{ with } -1.6 \text{ in the original equation.}$$
$$-1.0 = -1.0 \quad \text{True}$$

The solution is -1.6.   □

PRACTICE
2   Solve: $y - 0.3 = -2.1$ for y ■

Many times, it is best to simplify one or both sides of an equation before applying the addition property of equality.

Answer to Concept Check: 5; 5

EXAMPLE 3 Solve: $2x + 3x - 5 + 7 = 10x + 3 - 6x - 4$

Solution First we simplify both sides of the equation.

$$2x + 3x - 5 + 7 = 10x + 3 - 6x - 4$$
$$5x + 2 = 4x - 1 \qquad \text{Combine like terms on each side of the equation.}$$

Next, we want all terms with a variable on one side of the equation and all numbers on the other side.

$$5x + 2 - 4x = 4x - 1 - 4x \qquad \text{Subtract } 4x \text{ from both sides.}$$
$$x + 2 = -1 \qquad \text{Combine like terms.}$$
$$x + 2 - 2 = -1 - 2 \qquad \text{Subtract 2 from both sides to get } x \text{ alone.}$$
$$x = -3 \qquad \text{Combine like terms.}$$

Check:

$$
\begin{aligned}
2x + 3x - 5 + 7 &= 10x + 3 - 6x - 4 & &\text{Original equation} \\
2(-3) + 3(-3) - 5 + 7 &\overset{?}{=} 10(-3) + 3 - 6(-3) - 4 & &\text{Replace } x \text{ with } -3. \\
-6 - 9 - 5 + 7 &\overset{?}{=} -30 + 3 + 18 - 4 & &\text{Multiply.} \\
-13 &= -13 & &\text{True}
\end{aligned}
$$

The solution is -3. □

PRACTICE
3 Solve: $8x - 5x - 3 + 9 = x + x + 3 - 7$ ∎

If an equation contains parentheses, we use the distributive property to remove them.

EXAMPLE 4 Solve: $7 = -5(2a - 1) - (-11a + 6)$

Solution

$$
\begin{aligned}
7 &= -5(2a - 1) - (-11a + 6) \\
7 &= -10a + 5 + 11a - 6 & &\text{Apply the distributive property.} \\
7 &= a - 1 & &\text{Combine like terms.} \\
7 + 1 &= a - 1 + 1 & &\text{Add 1 to both sides to get } a \text{ alone.} \\
8 &= a & &\text{Combine like terms.}
\end{aligned}
$$

Check to see that 8 is the solution. □

PRACTICE
4 Solve: $2 = 4(2a - 3) - (7a + 4)$ ∎

Helpful Hint
We may solve an equation so that the variable is alone on either side of the equation. For example, $8 = a$ is equivalent to $a = 8$.

When solving equations, we may sometimes encounter an equation such as

$$-x = 5.$$

This equation is not solved for x because x is not isolated. One way to solve this equation for x is to recall that

"$-$" can be read as "the opposite of."

We can read the equation $-x = 5$ then as "the opposite of $x = 5$." If the opposite of x is 5, this means that x is the opposite of 5 or -5.

In summary,

$$-x = 5 \quad \text{and} \quad x = -5$$

are equivalent equations and $x = -5$ is solved for x.

OBJECTIVE

2 Using the Multiplication Property ▶

As useful as the addition property of equality is, it cannot help us solve every type of linear equation in one variable. For example, adding or subtracting a value on both sides of the equation does not help solve

$$\frac{5}{2}x = 15.$$

Instead, we apply another important property of equality, the **multiplication property of equality.**

Multiplication Property of Equality

If $a, b,$ and c are real numbers and $c \neq 0$, then

$$a = b \quad \text{and} \quad ac = bc$$

are equivalent equations.

This property guarantees that multiplying both sides of an equation by the same non-zero number does not change the solution of the equation. Since division is defined in terms of multiplication, we may also **divide both sides of the equation by the same nonzero number** without changing the solution.

EXAMPLE 5 Solve: $\dfrac{5}{2}x = 15$

Solution To get x alone, multiply both sides of the equation by the reciprocal of $\dfrac{5}{2}$, which is $\dfrac{2}{5}$.

$$\frac{5}{2}x = 15$$

> **Helpful Hint**
> Don't forget to multiply *both* sides by $\dfrac{2}{5}$.

$$\frac{2}{5} \cdot \frac{5}{2}x = \frac{2}{5} \cdot 15 \quad \text{Multiply both sides by } \frac{2}{5}.$$

$$\left(\frac{2}{5} \cdot \frac{5}{2}\right)x = \frac{2}{5} \cdot 15 \quad \text{Apply the associative property.}$$

$$1x = 6 \quad \text{Simplify.}$$

or

$$x = 6$$

Check: Replace x with 6 in the original equation.

$$\frac{5}{2}x = 15 \quad \text{Original equation}$$

$$\frac{5}{2}(6) \stackrel{?}{=} 15 \quad \text{Replace } x \text{ with 6.}$$

$$15 = 15 \quad \text{True}$$

The solution is 6.

PRACTICE

5 Solve: $\dfrac{4}{5}x = 16$

In the equation $\frac{5}{2}x = 15$, $\frac{5}{2}$ is the coefficient of x. When the coefficient of x is a *fraction*, we will get x alone by multiplying by the reciprocal. When the coefficient of x is an integer or a decimal, it is usually more convenient to divide both sides by the coefficient. (Dividing by a number is, of course, the same as multiplying by the reciprocal of the number.)

EXAMPLE 6   Solve: $-3x = 33$

Solution   Recall that $-3x$ means $-3 \cdot x$. To get x alone, we divide both sides by the coefficient of x, that is, -3.

$$-3x = 33$$

$$\frac{-3x}{-3} = \frac{33}{-3} \quad \text{Divide both sides by } -3.$$

$$1x = -11 \quad \text{Simplify.}$$

$$x = -11$$

Check:
$$-3x = 33 \quad \text{Original equation}$$

$$-3(-11) \stackrel{?}{=} 33 \quad \text{Replace } x \text{ with } -11.$$

$$33 = 33 \quad \text{True}$$

The solution is -11.   □

PRACTICE
6   Solve: $8x = -96$ ■

EXAMPLE 7   Solve: $\frac{y}{7} = 20$

Solution   Recall that $\frac{y}{7} = \frac{1}{7}y$. To get y alone, we multiply both sides of the equation by 7, the reciprocal of $\frac{1}{7}$.

$$\frac{y}{7} = 20$$

$$\frac{1}{7}y = 20$$

$$7 \cdot \frac{1}{7}y = 7 \cdot 20 \quad \text{Multiply both sides by 7.}$$

$$1y = 140 \quad \text{Simplify.}$$

$$y = 140$$

Check:
$$\frac{y}{7} = 20 \quad \text{Original equation}$$

$$\frac{140}{7} \stackrel{?}{=} 20 \quad \text{Replace } y \text{ with 140.}$$

$$20 = 20 \quad \text{True}$$

The solution is 140.   □

PRACTICE
7   Solve: $\frac{x}{5} = 13$ ■

OBJECTIVE
3   Using Both the Addition and Multiplication Properties ▶

Next, we practice solving equations using both properties.

EXAMPLE 8   Solve: $12a - 8a = 10 + 2a - 13 - 7$

Solution   First, simplify both sides of the equation by combining like terms.

$$12a - 8a = 10 + 2a - 13 - 7$$
$$4a = 2a - 10 \qquad \text{Combine like terms.}$$

To get all terms containing a variable on one side, subtract $2a$ from both sides.

$$4a - 2a = 2a - 10 - 2a \quad \text{Subtract } 2a \text{ from both sides.}$$
$$2a = -10 \qquad\qquad \text{Simplify.}$$
$$\frac{2a}{2} = \frac{-10}{2} \qquad\qquad \text{Divide both sides by 2.}$$
$$a = -5 \qquad\qquad \text{Simplify.}$$

Check: Check by replacing a with -5 in the original equation. The solution is -5. □

PRACTICE
8   Solve: $6b - 11b = 18 + 2b - 6 + 9$ ■

OBJECTIVE
4   Writing Word Phrases as Algebraic Expressions

Next, we practice writing word phrases as algebraic expressions.

EXAMPLE 9

   a. The sum of two numbers is 8. If one number is 3, find the other number.

   b. The sum of two numbers is 8. If one number is x, write an expression representing the other number.

   c. An 8-foot board is cut into two pieces. If one piece is x feet, express the length of the other piece in terms of x.

Solution

   a. If the sum of two numbers is 8 and one number is 3, we find the other number by subtracting 3 from 8. The other number is $8 - 3$ or 5.

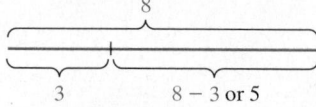

   b. If the sum of two numbers is 8 and one number is x, we find the other number by subtracting x from 8. The other number is represented by $8 - x$.

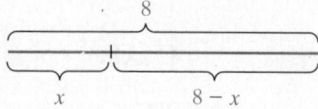

   c. If an 8-foot board is cut into two pieces and one piece is x feet, we find the other length by subtracting x from 8. The other piece is $(8 - x)$ feet.

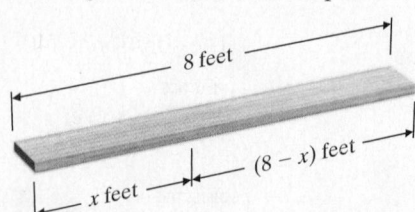

PRACTICE
9

    a. The sum of two numbers is 9. If one number is 2, find the other number.

    b. The sum of two numbers is 9. If one number is x, write an expression representing the other number.

    c. A 9-foot rope is cut into two pieces. If one piece is x feet, express the length of the other piece in terms of x. ∎

EXAMPLE 10   If x is the first of three consecutive integers, express the sum of the three integers in terms of x. Simplify if possible.

<u>Solution</u> An example of three consecutive integers is

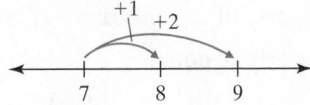

The second consecutive integer is always 1 more than the first, and the third consecutive integer is 2 more than the first. If x is the first of three consecutive integers, the three consecutive integers are

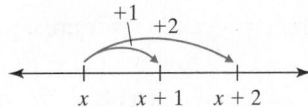

Their sum is

In words: | first integer | + | second integer | + | third integer |

Translate: x + $(x + 1)$ + $(x + 2)$

which simplifies to $3x + 3$. ☐

PRACTICE
10   If x is the first of three consecutive *even* integers, express their sum in terms of x. ∎

Below are examples of consecutive even and odd integers.

    Consecutive Even integers: ***Consecutive Odd integers:***

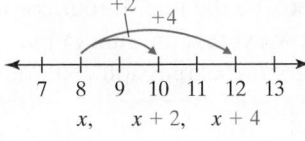   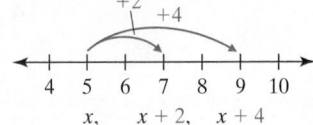

Helpful Hint

If x is an odd integer, then $x + 2$ is the next odd integer. This 2 simply means that odd integers are always 2 units from each other. (The same is true for even integers. They are always 2 units from each other.)

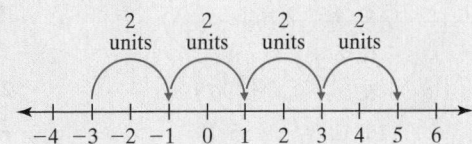

✔ **Vocabulary, Readiness & Video Check**

Use the choices below to fill in each blank. Some choices will be used more than once.

addition	solving	expression	true	multiplication
equivalent	equation	solution	false	

1. The difference between an equation and an expression is that a(n) _____ contains an equal sign, whereas an _____ does not.

2. _____ equations are equations that have the same solution.

3. A value of the variable that makes an equation a true statement is called a(n) _____ of the equation.

4. The process of finding the solution of an equation is called _____ the equation for the variable.

5. By the _____ property of equality, $x = -2$ and $x + 10 = -2 + 10$ are equivalent equations.

6. By the _____ property of equality, $x = -7$ and $x - 5 = -7 - 5$ are equivalent equations.

7. By the _____ property of equality, $y = \frac{1}{2}$ and $5 \cdot y = 5 \cdot \frac{1}{2}$ are equivalent equations.

8. By the _____ property of equality, $9x = -63$ and $\frac{9x}{9} = \frac{-63}{9}$ are equivalent equations.

9. True or false: The equations $x = \frac{1}{2}$ and $\frac{1}{2} = x$ are equivalent equations. _____

10. True or false: The equations $\frac{z}{4} = 10$ and $4 \cdot \frac{z}{4} = 10$ are equivalent equations. _____

Martin-Gay Interactive Videos

See Video 2.2 ◉

Watch the section lecture video and answer the following questions.

OBJECTIVE 1
11. Complete this statement based on the lecture given before ▦ Example 1. The addition property of equality means that if we have an equation, we can add the same real number to _____ of the equation and have an equivalent equation.

OBJECTIVE 2
12. Complete this statement based on the lecture given before ▦ Example 4. We can multiply both sides of an equation by _____ nonzero number and have an equivalent equation.

OBJECTIVE 3
13. Both the addition and multiplication properties of equality are used to solve ▦ Examples 6 and 7. In each of these examples, what property is applied first? What property is applied last? What conclusion, if any, can you make?

OBJECTIVE 4
14. Let x be the first of four consecutive integers, as in ▦ Example 10. Now express the sum of the second integer and the fourth integer as an algebraic expression containing x.

2.2 Exercise Set MyMathLab® ▶

Solve each equation. Check each solution. See Examples 1 and 2.

1. $x + 7 = 10$
2. $x + 14 = 25$

▶ 3. $x - 2 = -4$
4. $y - 9 = 1$

5. $3 + x = -11$
6. $8 + z = -8$

7. $r - 8.6 = -8.1$
8. $t - 9.2 = -6.8$

9. $8x = 7x - 3$
10. $2x = x - 5$

▶ 11. $5b - 0.7 = 6b$
12. $9x + 5.5 = 10x$

13. $7x - 3 = 6x$
14. $18x - 9 = 19x$

Solve each equation. See Examples 3 and 4.

15. $2x + x - 6 = 2x + 5$
16. $7y + 2 = 2y + 4y + 2$

17. $3t - t - 7 = t - 7$
18. $4c + 8 - c = 8 + 2c$

19. $7x + 2x = 8x - 3$
20. $3n + 2n = 7 + 4n$

21. $-2(x + 1) + 3x = 14$

22. $10 = 8(3y - 4) - 23y + 20$

Solve each equation. See Example 6.

▶ 23. $-5x = 20$
24. $-7x = -49$

25. $3x = 0$

26. $-2x = 0$

27. $-x = -12$

28. $-y = 8$

29. $3x + 2x = 50$

30. $-y + 4y = 33$

Solve each equation. See Examples 5 and 7.

31. $\dfrac{2}{3}x = -8$

32. $\dfrac{3}{4}n = -15$

33. $\dfrac{1}{6}d = \dfrac{1}{2}$

34. $\dfrac{1}{8}v = \dfrac{1}{4}$

35. $\dfrac{a}{-2} = 1$

36. $\dfrac{d}{15} = 2$

37. $\dfrac{k}{7} = 0$

38. $\dfrac{f}{-5} = 0$

39. In your own words, explain the addition property of equality.

40. In your own words, explain the multiplication property of equality.

MIXED PRACTICE

Solve each equation. Check each solution. See Examples 1 through 8.

41. $2x - 4 = 16$

42. $3x - 1 = 26$

43. $-x + 2 = 22$

44. $-x + 4 = -24$

45. $6a + 3 = 3$

46. $8t + 5 = 5$

47. $6x + 10 = -20$

48. $-10y + 15 = 5$

49. $5 - 0.3k = 5$

50. $2 + 0.4p = 2$

51. $-2x + \dfrac{1}{2} = \dfrac{7}{2}$

52. $-3n - \dfrac{1}{3} = \dfrac{8}{3}$

53. $\dfrac{x}{3} + 2 = -5$

54. $\dfrac{b}{4} - 1 = -7$

55. $10 = 2x - 1$

56. $12 = 3j - 4$

57. $6z - 8 - z + 3 = 0$

58. $4a + 1 + a - 11 = 0$

59. $10 - 3x - 6 - 9x = 7$

60. $12x + 30 + 8x - 6 = 10$

61. $\dfrac{5}{6}x = 10$

62. $-\dfrac{3}{4}x = 9$

63. $1 = 0.3x - 0.5x - 5$

64. $19 = 0.4x - 0.9x - 6$

65. $z - 5z = 7z - 9 - z$

66. $t - 6t = -13 + t - 3t$

67. $0.4x - 0.6x - 5 = 1$

68. $0.1x - 0.6x - 6 = 19$

69. $6 - 2x + 8 = 10$

70. $-5 - 6y + 6 = 19$

71. $-3a + 6 + 5a = 7a - 8a$

72. $4b - 8 - b = 10b - 3b$

73. $20 = -3(2x + 1) + 7x$

74. $-3 = -5(4x + 3) + 21x$

See Example 9.

75. Two numbers have a sum of 20. If one number is p, express the other number in terms of p.

76. Two numbers have a sum of 13. If one number is y, express the other number in terms of y.

77. A 10-foot board is cut into two pieces. If one piece is x feet long, express the other length in terms of x.

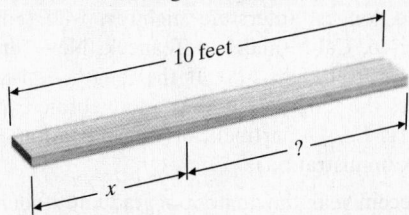

78. A 5-foot piece of string is cut into two pieces. If one piece is x feet long, express the other length in terms of x.

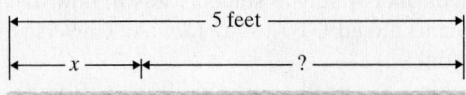

79. Two angles are *supplementary* if their sum is 180°. If one angle measures $x°$, express the measure of its supplement in terms of x.

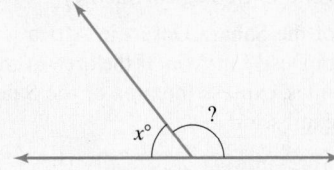

80. Two angles are *complementary* if their sum is 90°. If one angle measures $x°$, express the measure of its complement in terms of x.

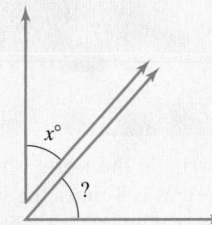

81. In a mayoral election, April Catarella received 284 more votes than Charles Pecot. If Charles received n votes, how many votes did April receive?

82. The length of the top of a computer desk is $1\dfrac{1}{2}$ feet longer than its width. If its width measures m feet, express its length as an algebraic expression in m.

83. The Verrazano-Narrows Bridge in New York City is the longest suspension bridge in North America. The Golden Gate Bridge in San Francisco is 60 feet shorter than the Verrazano-Narrows Bridge. If the length of the Verrazano-Narrows Bridge is m feet, express the length of the Golden Gate Bridge as an algebraic expression in m. (*Source: World Almanac*, 2000).

84. The longest interstate highway in the U.S. is I-90, which connects Seattle, Washington, and Boston, Massachusetts. The second longest interstate highway, I-80 (connecting San Francisco, California, and Teaneck, New Jersey), is 178.5 miles shorter than I-90. If the length of I-80 is m miles, express the length of I-90 as an algebraic expression in m. (*Source:* U.S. Department of Transportation–Federal Highway Administration)

85. In a recent year, the number of graduate students at the University of Texas at Austin was approximately 28,000 fewer than the number of undergraduate students. If the number of undergraduate students was n, how many graduate students attend UT Austin? (*Source:* University of Texas at Austin)

86. The Missouri River is the longest river in the United States. The Mississippi River is 200 miles shorter than the Missouri River. If the length of the Missouri River is r miles, express the length of the Mississippi River as an algebraic expression in r. (*Source:* U.S. Geological Survey)

87. The area of the Sahara Desert in Africa is 7 times the area of the Gobi Desert in Asia. If the area of the Gobi Desert is x square miles, express the area of the Sahara Desert as an algebraic expression in x.

88. The largest meteorite in the world is the Hoba West located in Namibia. Its weight is 3 times the weight of the Armanty meteorite located in Outer Mongolia. If the weight of the Armanty meteorite is y kilograms, express the weight of the Hoba West meteorite as an algebraic expression in y.

Write each algebraic expression described. Simplify if possible. See Example 10.

89. If x represents the first of two consecutive odd integers, express the sum of the two integers in terms of x.

90. If x is the first of four consecutive even integers, write their sum as an algebraic expression in x.

91. If x is the first of four consecutive integers, express the sum of the first integer and the third integer as an algebraic expression containing the variable x.

92. If x is the first of two consecutive integers, express the sum of 20 and the second consecutive integer as an algebraic expression containing the variable x.

93. Classrooms on one side of the science building are all numbered with consecutive even integers. If the first room on this side of the building is numbered x, write an expression in x for the sum of five classroom numbers in a row. Then simplify this expression.

94. Two sides of a quadrilateral have the same length, x, while the other two sides have the same length, both being the next consecutive odd integer. Write the sum of these lengths. Then simplify this expression.

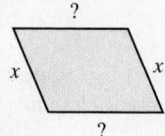

REVIEW AND PREVIEW

Simplify each expression. See Section 2.1.

95. $5x + 2(x - 6)$ **96.** $-7y + 2y - 3(y + 1)$

97. $-(x - 1) + x$ **98.** $-(3a - 3) + 2a - 6$

Insert $<$, $>$, or $=$ in the appropriate space to make each statement true. See Sections 1.2 and 1.7.

99. $(-3)^2$ -3^2 **100.** $(-2)^4$ -2^4

101. $(-2)^3$ -2^3 **102.** $(-4)^3$ -4^3

CONCEPT EXTENSIONS

△ **103.** The sum of the angles of a triangle is 180°. If one angle of a triangle measures $x°$ and a second angle measures $(2x + 7)°$, express the measure of the third angle in terms of x. Simplify the expression.

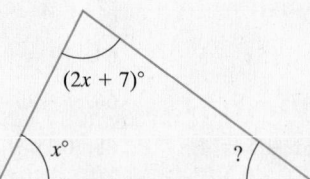

△ **104.** A quadrilateral is a four-sided figure like the one shown below whose angle sum is 360°. If one angle measures $x°$, a second angle measures $3x°$, and a third angle measures $5x°$, express the measure of the fourth angle in terms of x. Simplify the expression.

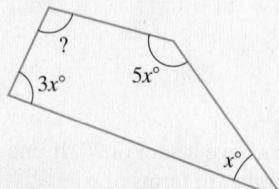

105. Write two terms whose sum is $-3x$.

106. Write four terms whose sum is $2y - 6$.

Use the addition property to fill in numbers between the parentheses so that the middle equation simplifies to the last equation. See the Concept Check in this section.

107. $x - 4 = -9$
      $x - 4 + (\ \ ) = -9 + (\ \ )$
             $x = -5$

108. $a + 9 = 15$
      $a + 9 + (\ \ ) = 15 + (\ \ )$
             $a = 6$

Fill in the blanks with numbers of your choice so that each equation has the given solution. Note: Each blank may be replaced with a different number.

109. _____ $+ x =$ _____; Solution: -3

110. $x -$ _____ $=$ _____; Solution: -10

111. Let $x = 1$ and then $x = 2$ in the equation $x + 5 = x + 6$. Is either number a solution? How many solutions do you think this equation has? Explain your answer.

112. Let $x = 1$ and then $x = 2$ in the equation $x + 3 = x + 3$. Is either number a solution? How many solutions do you think this equation has? Explain your answer.

Fill in the blank with a number so that each equation has the given solution.

113. $6x =$ _____ ; solution: -8

114. _____ $x = 10$; solution: $\dfrac{1}{2}$

115. A licensed nurse practitioner is instructed to give a patient 2100 milligrams of an antibiotic over a period of 36 hours. If the antibiotic is to be given every 4 hours starting immediately, how much antibiotic should be given in each dose? To answer this question, solve the equation $9x = 2100$.

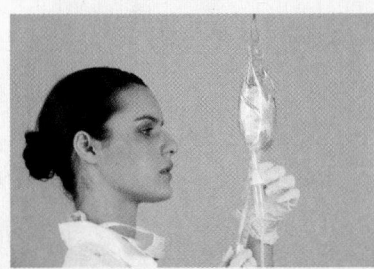

116. Suppose you are a pharmacist and a customer asks you the following question. His child is to receive 13.5 milliliters of a nausea medicine over a period of 54 hours. If the nausea medicine is to be administered every 6 hours starting immediately, how much medicine should be given in each dose?

Use a calculator to determine whether the given value is a solution of the given equation.

117. $8.13 + 5.85y = 20.05y - 8.91$; $y = 1.2$

118. $3(a + 4.6) = 5a + 2.5$; $a = 6.3$

Solve each equation.

119. $-3.6x = 10.62$

120. $4.95y = -31.185$

121. $7x - 5.06 = -4.92$

122. $0.06y + 2.63 = 2.5562$

2.3   Solving Linear Equations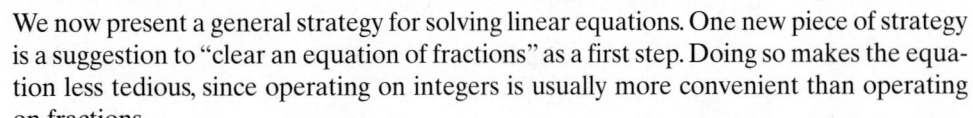

OBJECTIVES

1 Apply a General Strategy for Solving a Linear Equation.

2 Solve Equations Containing Fractions.

3 Solve Equations Containing Decimals.

4 Recognize Identities and Equations with No Solution.

OBJECTIVE

1   Applying a General Strategy for Solving a Linear Equation

We now present a general strategy for solving linear equations. One new piece of strategy is a suggestion to "clear an equation of fractions" as a first step. Doing so makes the equation less tedious, since operating on integers is usually more convenient than operating on fractions.

Solving Linear Equations in One Variable

Step 1. Multiply on both sides by the least common denominator (LCD) to clear the equation of fractions if they occur.

Step 2. Use the distributive property to remove parentheses if they occur.

Step 3. Simplify each side of the equation by combining like terms.

Step 4. Get all variable terms on one side and all numbers on the other side by using the addition property of equality.

Step 5. Get the variable alone by using the multiplication property of equality.

Step 6. Check the solution by substituting it into the original equation.

EXAMPLE 1 Solve: $4(2x - 3) + 7 = 3x + 5$

Solution There are no fractions, so we begin with Step 2.

$$4(2x - 3) + 7 = 3x + 5$$

Step 2. $\quad\quad\quad 8x - 12 + 7 = 3x + 5 \quad\quad$ Apply the distributive property.

Step 3. $\quad\quad\quad\quad 8x - 5 = 3x + 5 \quad\quad$ Combine like terms.

Step 4. Get all variable terms on the same side of the equation by subtracting $3x$ from both sides, then adding 5 to both sides.

$$8x - 5 - 3x = 3x + 5 - 3x \quad\quad \text{Subtract } 3x \text{ from both sides.}$$
$$5x - 5 = 5 \quad\quad \text{Simplify.}$$
$$5x - 5 + 5 = 5 + 5 \quad\quad \text{Add 5 to both sides.}$$
$$5x = 10 \quad\quad \text{Simplify.}$$

Step 5. Use the multiplication property of equality to get x alone.

$$\frac{5x}{5} = \frac{10}{5} \quad\quad \text{Divide both sides by 5.}$$
$$x = 2 \quad\quad \text{Simplify.}$$

Step 6. Check.

	Helpful Hint
When checking solutions, remember to use the original written equation.	

$$4(2x - 3) + 7 = 3x + 5 \quad\quad \text{Original equation}$$
$$4[2(2) - 3] + 7 \stackrel{?}{=} 3(2) + 5 \quad\quad \text{Replace } x \text{ with 2.}$$
$$4(4 - 3) + 7 \stackrel{?}{=} 6 + 5$$
$$4(1) + 7 \stackrel{?}{=} 11$$
$$4 + 7 \stackrel{?}{=} 11$$
$$11 = 11 \quad\quad \text{True}$$

The solution is 2 or the solution set is $\{2\}$.

PRACTICE

1 Solve: $2(4a - 9) + 3 = 5a - 6$

EXAMPLE 2 Solve: $8(2 - t) = -5t$

Solution First, we apply the distributive property.

$$8(2 - t) = -5t$$

Step 2. $\quad\quad 16 - 8t = -5t \quad\quad$ Use the distributive property.

Step 4. $16 - 8t + 8t = -5t + 8t \quad\quad$ To get variable terms on one side, add $8t$ to both sides.

$$16 = 3t \quad\quad \text{Combine like terms.}$$

Step 5. $\quad\quad\quad \dfrac{16}{3} = \dfrac{3t}{3} \quad\quad$ Divide both sides by 3.

$$\frac{16}{3} = t \quad\quad \text{Simplify.}$$

Step 6. Check.

$$8(2 - t) = -5t \quad\quad \text{Original equation}$$
$$8\left(2 - \frac{16}{3}\right) \stackrel{?}{=} -5\left(\frac{16}{3}\right) \quad\quad \text{Replace } t \text{ with } \frac{16}{3}.$$
$$8\left(\frac{6}{3} - \frac{16}{3}\right) \stackrel{?}{=} -\frac{80}{3} \quad\quad \text{The LCD is 3.}$$

$$8\left(-\frac{10}{3}\right) \stackrel{?}{=} -\frac{80}{3} \quad \text{Subtract fractions.}$$

$$-\frac{80}{3} = -\frac{80}{3} \quad \text{True}$$

The solution is $\frac{16}{3}$. □

PRACTICE
2   Solve: $7(x - 3) = -6x$ ■

OBJECTIVE
2   Solving Equations Containing Fractions

If an equation contains fractions, we can clear the equation of fractions by multiplying both sides by the LCD of all denominators. By doing this, we avoid working with time-consuming fractions.

EXAMPLE 3   Solve: $\frac{x}{2} - 1 = \frac{2}{3}x - 3$

Solution   We begin by clearing fractions. To do this, we multiply both sides of the equation by the LCD of 2 and 3, which is 6.

$$\frac{x}{2} - 1 = \frac{2}{3}x - 3$$

Step 1. $6\left(\frac{x}{2} - 1\right) = 6\left(\frac{2}{3}x - 3\right)$ Multiply both sides by the LCD, 6.

Step 2. $6\left(\frac{x}{2}\right) - 6(1) = 6\left(\frac{2}{3}x\right) - 6(3)$ Apply the distributive property.

$$3x - 6 = 4x - 18 \quad \text{Simplify.}$$

> **Helpful Hint**
> Don't forget to multiply *each* term by the LCD.

There are no longer grouping symbols and no like terms on either side of the equation, so we continue with Step 4.

$$3x - 6 = 4x - 18$$

Step 4. $3x - 6 - 3x = 4x - 18 - 3x$ To get variable terms on one side, subtract $3x$ from both sides.

$$-6 = x - 18 \quad \text{Simplify.}$$

$$-6 + 18 = x - 18 + 18 \quad \text{Add 18 to both sides.}$$

$$12 = x \quad \text{Simplify.}$$

Step 5.   The variable is now alone, so there is no need to apply the multiplication property of equality.

Step 6.   Check.

$$\frac{x}{2} - 1 = \frac{2}{3}x - 3 \quad \text{Original equation}$$

$$\frac{12}{2} - 1 \stackrel{?}{=} \frac{2}{3} \cdot 12 - 3 \quad \text{Replace } x \text{ with 12.}$$

$$6 - 1 \stackrel{?}{=} 8 - 3 \quad \text{Simplify.}$$

$$5 = 5 \quad \text{True}$$

The solution is 12. □

PRACTICE
3   Solve: $\frac{3}{5}x - 2 = \frac{2}{3}x - 1$ ■

EXAMPLE 4   Solve: $\dfrac{2(a + 3)}{3} = 6a + 2$

Solution We clear the equation of fractions first.

$$\dfrac{2(a + 3)}{3} = 6a + 2$$

Step 1. $3 \cdot \dfrac{2(a + 3)}{3} = 3(6a + 2)$ Clear the fraction by multiplying both sides by the LCD, 3.

$$2(a + 3) = 3(6a + 2)$$

Step 2. Next, we use the distributive property and remove parentheses.

$$2a + 6 = 18a + 6$$ Apply the distributive property.

Step 4. $2a + 6 - 6 = 18a + 6 - 6$ Subtract 6 from both sides.

$$2a = 18a$$

$$2a - 18a = 18a - 18a$$ Subtract 18a from both sides.

$$-16a = 0$$

Step 5. $\dfrac{-16a}{-16} = \dfrac{0}{-16}$ Divide both sides by -16.

$$a = 0$$ Write the fraction in simplest form.

Step 6. To check, replace a with 0 in the original equation. The solution is 0.   □

PRACTICE
4   Solve: $\dfrac{4(y + 3)}{3} = 5y - 7$

> **Helpful Hint**
>
> Remember: When solving an equation, it makes no difference on which side of the equation variable terms lie. Just make sure that constant terms (number terms) lie on the other side.

OBJECTIVE
3   Solving Equations Containing Decimals

When solving a problem about money, you may need to solve an equation containing decimals. If you choose, you may multiply to clear the equation of decimals.

EXAMPLE 5   Solve: $0.25x + 0.10(x - 3) = 0.05(22)$

Solution First we clear this equation of decimals by multiplying both sides of the equation by 100. Recall that multiplying a decimal number by 100 has the effect of moving the decimal point 2 places to the right.

$$0.25x + 0.10(x - 3) = 0.05(22)$$

> **Helpful Hint**
>
> By the distributive property, 0.10 is multiplied by x and -3. Thus to multiply each term here by 100, we only need to multiply 0.10 by 100.

Step 1. $0.25x + 0.10(x - 3) = 0.05(22)$ Multiply both sides by 100.

$$25x + 10(x - 3) = 5(22)$$

Step 2. $25x + 10x - 30 = 110$ Apply the distributive property.

Step 3. $35x - 30 = 110$ Combine like terms.

Step 4. $35x - 30 + 30 = 110 + 30$ Add 30 to both sides.

$$35x = 140$$ Combine like terms.

Step 5.
$$\frac{35x}{35} = \frac{140}{35} \quad \text{Divide both sides by 35.}$$
$$x = 4$$

Step 6. To check, replace x with 4 in the original equation. The solution is 4.   □

PRACTICE
5   Solve: $0.35x + 0.09(x + 4) = 0.03(12)$ ■

OBJECTIVE
4   Recognizing Identities and Equations with No Solution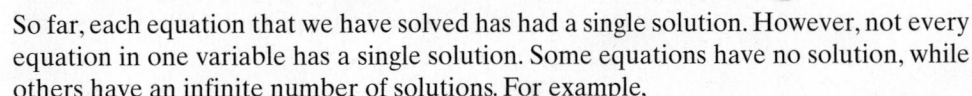

So far, each equation that we have solved has had a single solution. However, not every equation in one variable has a single solution. Some equations have no solution, while others have an infinite number of solutions. For example,

$$x + 5 = x + 7$$

has no solution since no matter which **real number** we replace x with, the equation is false.

real number $+ 5 =$ same **real number** $+ 7$ **FALSE**

On the other hand,

$$x + 6 = x + 6$$

has infinitely many solutions since x can be replaced by any real number and the equation is always true.

real number $+ 6 =$ same **real number** $+ 6$ **TRUE**

The equation $x + 6 = x + 6$ is called an **identity.** The next two examples illustrate special equations like these.

EXAMPLE 6   Solve: $-2(x - 5) + 10 = -3(x + 2) + x$

Solution

$$
\begin{array}{ll}
-2(x - 5) + 10 = -3(x + 2) + x & \\
-2x + 10 + 10 = -3x - 6 + x & \text{Apply the distributive property on both sides.} \\
-2x + 20 = -2x - 6 & \text{Combine like terms.} \\
-2x + 20 + 2x = -2x - 6 + 2x & \text{Add } 2x \text{ to both sides.} \\
20 = -6 & \text{Combine like terms.}
\end{array}
$$

The final equation contains no variable terms, and there is no value for x that makes $20 = -6$ a true equation. We conclude that there is **no solution** to this equation. In set notation, we can indicate that there is no solution with the empty set, $\{\ \}$, or use the empty set or null set symbol, $\varnothing$. In this chapter, we will simply write *no solution*.   □

PRACTICE
6   Solve: $4(x + 4) - x = 2(x + 11) + x$ ■

EXAMPLE 7   Solve: $3(x - 4) = 3x - 12$

Solution

$$
\begin{array}{ll}
3(x - 4) = 3x - 12 & \\
3x - 12 = 3x - 12 & \text{Apply the distributive property.}
\end{array}
$$

(Continued on next page)

The left side of the equation is now identical to the right side. Every real number may be substituted for x and a true statement will result. We arrive at the same conclusion if we continue.

$$3x - 12 = 3x - 12$$
$$3x - 12 + 12 = 3x - 12 + 12 \quad \text{Add 12 to both sides.}$$
$$3x = 3x \quad \text{Combine like terms.}$$
$$3x - 3x = 3x - 3x \quad \text{Subtract } 3x \text{ from both sides.}$$
$$0 = 0$$

Again, one side of the equation is identical to the other side. Thus, $3(x - 4) = 3x - 12$ is an **identity** and **all real numbers** are solutions. In set notation, this is $\{$ all real numbers $\}$. ☐

PRACTICE
7 Solve: $12x - 18 = 9(x - 2) + 3x$ ∎

 CONCEPT CHECK

Suppose you have simplified several equations and obtain the following results. What can you conclude about the solutions of the original equation?

a. $7 = 7$ **b.** $x = 0$ **c.** $7 = -4$

 Graphing Calculator Explorations

Checking Equations

We can use a calculator to check possible solutions of equations. To do this, replace the variable by the possible solution and evaluate both sides of the equation separately.

Equation: $3x - 4 = 2(x + 6)$ *Solution: $x = 16$*

$$3x - 4 = 2(x + 6) \quad \text{Original equation}$$
$$3(16) - 4 \stackrel{?}{=} 2(16 + 6) \quad \text{Replace } x \text{ with 16.}$$

Now evaluate each side with your calculator.

Evaluate left side:

$\boxed{3}$ $\boxed{\times}$ $\boxed{16}$ $\boxed{-}$ $\boxed{4}$ then $\boxed{=}$ or $\boxed{\text{ENTER}}$ Display: $\boxed{44}$ or $\boxed{\begin{array}{l} 3*16 - 4 \\ \hspace{2.5em} 44 \end{array}}$

Evaluate right side:

$\boxed{2}$ $\boxed{(}$ $\boxed{16}$ $\boxed{+}$ $\boxed{6}$ $\boxed{)}$ then $\boxed{=}$ or $\boxed{\text{ENTER}}$ Display: $\boxed{44}$ or $\boxed{\begin{array}{l} 2(16 + 6) \\ \hspace{3em} 44 \end{array}}$

Since the left side equals the right side, the solution checks.

Use a calculator to check the possible solutions to each equation.

1. $2x = 48 + 6x$; $x = -12$ **2.** $-3x - 7 = 3x - 1$; $x = -1$

3. $5x - 2.6 = 2(x + 0.8)$; $x = 4.4$ **4.** $-1.6x - 3.9 = -6.9x - 25.6$; $x = 5$

5. $\dfrac{564x}{4} = 200x - 11(649)$; $x = 121$ **6.** $20(x - 39) = 5x - 432$; $x = 23.2$

Answers to Concept Check:
a. Every real number is a solution.
b. The solution is 0.
c. There is no solution.

✔ Vocabulary, Readiness & Video Check

Throughout algebra, it is important to be able to identify equations and expressions.

Remember,
- an equation contains an equal sign and
- an expression does not.

Also,
- we solve equations and
- we simplify or perform operations on expressions.

Identify each as an equation or an expression.

1. $x = -7$ _____

2. $x - 7$ _____

3. $4y - 6 + 9y + 1$ _____

4. $4y - 6 = 9y + 1$ _____

5. $\dfrac{1}{x} - \dfrac{x-1}{8}$ _____

6. $\dfrac{1}{x} - \dfrac{x-1}{8} = 6$ _____

7. $0.1x + 9 = 0.2x$ _____

8. $0.1x^2 + 9y - 0.2x^2$ _____

Martin-Gay Interactive Videos

See Video 2.3 ◉

Watch the section lecture video and answer the following questions.

OBJECTIVE 1

9. The general strategy for solving linear equations in one variable is discussed after ▦ Example 1. How many properties are mentioned in this strategy and what are they?

OBJECTIVE 2

10. In the first step for solving ▦ Example 2, both sides of the equation are being multiplied by the LCD. Why is the distributive property mentioned?

OBJECTIVE 3

11. In ▦ Example 3, why is the number of decimal places in each term of the equation important?

OBJECTIVE 4

12. Complete each statement based on ▦ Examples 4 and 5.

When solving an equation and all variable terms subtract out:

(a) If you have a true statement, then the equation has ____ solution(s).

(b) If you have a false statement, then the equation has ____ solution(s).

2.3   Exercise Set   MyMathLab® ▶

Solve each equation. See Examples 1 and 2.

1. $-4y + 10 = -2(3y + 1)$

2. $-3x + 1 = -2(4x + 2)$

3. $15x - 8 = 10 + 9x$

4. $15x - 5 = 7 + 12x$

5. $-2(3x - 4) = 2x$

6. $-(5x - 10) = 5x$

▶ **7.** $5(2x - 1) - 2(3x) = 1$

8. $3(2 - 5x) + 4(6x) = 12$

9. $-6(x - 3) - 26 = -8$

10. $-4(n - 4) - 23 = -7$

11. $8 - 2(a + 1) = 9 + a$

12. $5 - 6(2 + b) = b - 14$

13. $4x + 3 = -3 + 2x + 14$

14. $6y - 8 = -6 + 3y + 13$

15. $-2y - 10 = 5y + 18$

16. $-7n + 5 = 8n - 10$

Solve each equation. See Examples 3 through 5.

17. $\dfrac{2}{3}x + \dfrac{4}{3} = -\dfrac{2}{3}$

18. $\dfrac{4}{5}x - \dfrac{8}{5} = -\dfrac{16}{5}$

19. $\dfrac{3}{4}x - \dfrac{1}{2} = 1$

20. $\dfrac{2}{9}x - \dfrac{1}{3} = 1$

▶ **21.** $0.50x + 0.15(70) = 35.5$

22. $0.40x + 0.06(30) = 9.8$

23. $\dfrac{2(x + 1)}{4} = 3x - 2$

24. $\dfrac{3(y + 3)}{5} = 2y + 6$

25. $x + \dfrac{7}{6} = 2x - \dfrac{7}{6}$

26. $\dfrac{5}{2}x - 1 = x + \dfrac{1}{4}$

27. $0.12(y - 6) + 0.06y = 0.08y - 0.70$

28. $0.60(z - 300) + 0.05z = 0.70z - 205$

Solve each equation. See Examples 6 and 7.

29. $4(3x + 2) = 12x + 8$

30. $14x + 7 = 7(2x + 1)$

31. $\dfrac{x}{4} + 1 = \dfrac{x}{4}$

32. $\dfrac{x}{3} - 2 = \dfrac{x}{3}$

33. $3x - 7 = 3(x + 1)$

34. $2(x - 5) = 2x + 10$

35. $-2(6x - 5) + 4 = -12x + 14$

36. $-5(4y - 3) + 2 = -20y + 17$

MIXED PRACTICE

Solve. See Examples 1 through 7.

37. $\dfrac{6(3 - z)}{5} = -z$

38. $\dfrac{4(5 - w)}{3} = -w$

39. $-3(2t - 5) + 2t = 5t - 4$

40. $-(4a - 7) - 5a = 10 + a$

41. $5y + 2(y - 6) = 4(y + 1) - 2$

42. $9x + 3(x - 4) = 10(x - 5) + 7$

43. $\dfrac{3(x - 5)}{2} = \dfrac{2(x + 5)}{3}$

44. $\dfrac{5(x - 1)}{4} = \dfrac{3(x + 1)}{2}$

45. $0.7x - 2.3 = 0.5$

46. $0.9x - 4.1 = 0.4$

▶ **47.** $5x - 5 = 2(x + 1) + 3x - 7$

48. $3(2x - 1) + 5 = 6x + 2$

49. $4(2n + 1) = 3(6n + 3) + 1$

50. $4(4y + 2) = 2(1 + 6y) + 8$

51. $x + \dfrac{5}{4} = \dfrac{3}{4}x$

52. $\dfrac{7}{8}x + \dfrac{1}{4} = \dfrac{3}{4}x$

▶ **53.** $\dfrac{x}{2} - 1 = \dfrac{x}{5} + 2$

54. $\dfrac{x}{5} - 7 = \dfrac{x}{3} - 5$

▶ **55.** $2(x + 3) - 5 = 5x - 3(1 + x)$

56. $4(2 + x) + 1 = 7x - 3(x - 2)$

57. $0.06 - 0.01(x + 1) = -0.02(2 - x)$

58. $-0.01(5x + 4) = 0.04 - 0.01(x + 4)$

59. $\dfrac{9}{2} + \dfrac{5}{2}y = 2y - 4$

60. $3 - \dfrac{1}{2}x = 5x - 8$

61. $-2y - 10 = 5y + 18$

62. $7n + 5 = 10n - 10$

63. $0.6x - 0.1 = 0.5x + 0.2$

64. $0.2x - 0.1 = 0.6x - 2.1$

65. $0.02(6t - 3) = 0.12(t - 2) + 0.18$

66. $0.03(2m + 7) = 0.06(5 + m) - 0.09$

TRANSLATING

Write each phrase as an algebraic expression. Use x for the unknown number. See Section 2.1.

67. A number subtracted from -8

68. Three times a number

69. The sum of -3 and twice a number

70. The difference of 8 and twice a number

71. The product of 9 and the sum of a number and 20

72. The quotient of -12 and the difference of a number and 3

See Section 2.2.

73. A plot of land is in the shape of a triangle. If one side is x meters, a second side is $(2x - 3)$ meters, and a third side is $(3x - 5)$ meters, express the perimeter of the lot as a simplified expression in x.

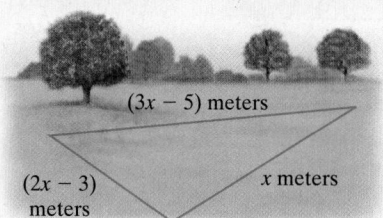

$(3x - 5)$ meters

$(2x - 3)$ meters

x meters

74. A portion of a board has length x feet. The other part has length $(7x - 9)$ feet. Express the total length of the board as a simplified expression in x.

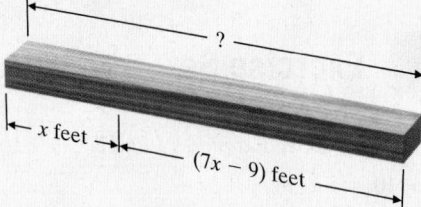

?

x feet

$(7x - 9)$ feet

CONCEPT EXTENSIONS

See the Concept Check in this section.

75. a. Solve: $x + 3 = x + 3$ for x

   b. If you simplify an equation and get $0 = 0$, what can you conclude about the solution(s) of the original equation?

   c. On your own, construct an equation for which every real number is a solution.

76. a. Solve: $x + 3 = x + 5$ for x

   b. If you simplify an equation and get $3 = 5$, what can you conclude about the solution(s) of the original equation?

   c. On your own, construct an equation that has no solution.

For Exercises 77 through 82, match each equation in the first column with its solution in the second column. Items in the second column may be used more than once.

77. $5x + 1 = 5x + 1$

78. $3x + 1 = 3x + 2$

79. $2x - 6x - 10 = -4x + 3 - 10$

80. $x - 11x - 3 = -10x - 1 - 2$

81. $9x - 20 = 8x - 20$

82. $-x + 15 = x + 15$

A. all real numbers

B. no solution

C. 0

83. Explain the difference between simplifying an expression and solving an equation.

84. On your own, write an expression and then an equation. Label each.

For Exercises 85 and 86, **a.** *Write an equation for perimeter.* **b.** *Solve the equation in part (a).* **c.** *Find the length of each side.*

△ **85.** The perimeter of a geometric figure is the sum of the lengths of its sides. The perimeter of the following pentagon (five-sided figure) is 28 centimeters.

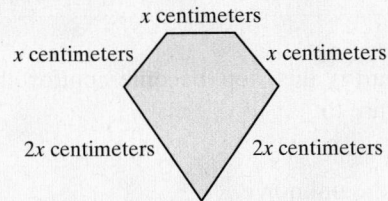

x centimeters

x centimeters x centimeters

$2x$ centimeters $2x$ centimeters

△ **86.** The perimeter of the following triangle is 35 meters.

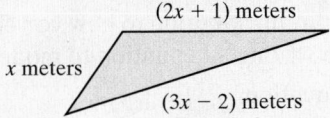

$(2x + 1)$ meters

x meters

$(3x - 2)$ meters

Fill in the blanks with numbers of your choice so that each equation has the given solution. Note: Each blank may be replaced by a different number.

87. $x + \underline{\quad} = 2x - \underline{\quad}$; solution: 9

88. $-5x - \underline{\quad} = \underline{\quad}$; solution: 2

Solve.

89. $1000(7x - 10) = 50(412 + 100x)$

90. $1000(x + 40) = 100(16 + 7x)$

91. $0.035x + 5.112 = 0.010x + 5.107$

92. $0.127x - 2.685 = 0.027x - 2.38$

For Exercises 93 through 96, see the example below.

Example

Solve: $t(t + 4) = t^2 - 2t + 6$

Solution $t(t + 4) = t^2 - 2t + 6$

$t^2 + 4t = t^2 - 2t + 6$

$t^2 + 4t - t^2 = t^2 - 2t + 6 - t^2$

$4t = -2t + 6$

$4t + 2t = -2t + 6 + 2t$

$6t = 6$

$t = 1$

Solve each equation.

93. $x(x - 3) = x^2 + 5x + 7$

94. $t^2 - 6t = t(8 + t)$

95. $2z(z + 6) = 2z^2 + 12z - 8$

96. $y^2 - 4y + 10 = y(y - 5)$

| **Integrated Review** | Solving Linear Equations

Sections 2.1–2.3

Solve. Feel free to use the steps given in Section 2.3.

1. $x - 10 = -4$

2. $y + 14 = -3$

3. $9y = 108$

4. $-3x = 78$

5. $-6x + 7 = 25$

6. $5y - 42 = -47$

7. $\dfrac{2}{3}x = 9$

8. $\dfrac{4}{5}z = 10$

9. $\dfrac{r}{-4} = -2$

10. $\dfrac{y}{-8} = 8$

11. $6 - 2x + 8 = 10$

12. $-5 - 6y + 6 = 19$

13. $2x - 7 = 2x - 27$

14. $3 + 8y = 8y - 2$

15. $-3a + 6 + 5a = 7a - 8a$

16. $4b - 8 - b = 10b - 3b$

17. $-\dfrac{2}{3}x = \dfrac{5}{9}$

18. $-\dfrac{3}{8}y = -\dfrac{1}{16}$

19. $10 = -6n + 16$

20. $-5 = -2m + 7$

21. $3(5c - 1) - 2 = 13c + 3$

22. $4(3t + 4) - 20 = 3 + 5t$

23. $\dfrac{2(z + 3)}{3} = 5 - z$

24. $\dfrac{3(w + 2)}{4} = 2w + 3$

25. $-2(2x - 5) = -3x + 7 - x + 3$

26. $-4(5x - 2) = -12x + 4 - 8x + 4$

27. $0.02(6t - 3) = 0.04(t - 2) + 0.02$

28. $0.03(m + 7) = 0.02(5 - m) + 0.03$

29. $-3y = \dfrac{4(y - 1)}{5}$

30. $-4x = \dfrac{5(1 - x)}{6}$

31. $\dfrac{5}{3}x - \dfrac{7}{3} = x$

32. $\dfrac{7}{5}n + \dfrac{3}{5} = -n$

33. $9(3x - 1) = -4 + 49$

34. $12(2x + 1) = -6 + 66$

35. $\dfrac{1}{10}(3x - 7) = \dfrac{3}{10}x + 5$

36. $\dfrac{1}{7}(2x - 5) = \dfrac{2}{7}x + 1$

37. $5 + 2(3x - 6) = -4(6x - 7)$

38. $3 + 5(2x - 4) = -7(5x + 2)$

2.4 An Introduction to Problem Solving

OBJECTIVES

1 Solve Problems Involving Direct Translations.

2 Solve Problems Involving Relationships Among Unknown Quantities.

3 Solve Problems Involving Consecutive Integers.

OBJECTIVE

1 Solving Direct Translation Problems

In previous sections, we practiced writing word phrases and sentences as algebraic expressions and equations to help prepare for problem solving. We now use these translations to help write equations that model a problem. The problem-solving steps given next may be helpful.

General Strategy for Problem Solving

1. **UNDERSTAND** the problem. During this step, become comfortable with the problem. Some ways of doing this are to:

   Read and reread the problem.

   Choose a variable to represent the unknown.

   Construct a drawing whenever possible.

   Propose a solution and check. Pay careful attention to how you check your proposed solution. This will help when writing an equation to model the problem.

2. **TRANSLATE** the problem into an equation.

3. **SOLVE** the equation.

4. **INTERPRET** the results: *Check* the proposed solution in the stated problem and state your conclusion.

Much of problem solving involves a direct translation from a sentence to an equation.

EXAMPLE 1 Finding an Unknown Number

Twice a number, added to seven, is the same as three subtracted from the number. Find the number.

Solution Translate the sentence into an equation and solve.

In words:	twice a number	added to	seven	is the same as	three subtracted from the number
	↓	↓	↓	↓	↓
Translate:	$2x$	$+$	7	$=$	$x - 3$

> **Helpful Hint**
>
> Order matters when subtracting (and dividing), so be especially careful with these translations.

To solve, begin by subtracting x from both sides to get all variable terms on one side.

$$2x + 7 = x - 3$$
$$2x + 7 - x = x - 3 - x \quad \text{Subtract } x \text{ from both sides.}$$
$$x + 7 = -3 \quad \text{Combine like terms.}$$
$$x + 7 - 7 = -3 - 7 \quad \text{Subtract 7 from both sides.}$$
$$x = -10 \quad \text{Combine like terms.}$$

Check the solution in the problem as it was originally stated. To do so, replace "number" in the sentence with -10. Twice "-10" added to 7 is the same as 3 subtracted from "-10."

$$2(-10) + 7 = -10 - 3$$
$$-13 = -13$$

The unknown number is -10.   □

PRACTICE

1   Three times a number, minus 6, is the same as two times a number, plus 3. Find the number.   ∎

┌─ **Helpful Hint** ───┐
When checking solutions, go back to the original stated problem, rather than to your equation, in case errors have been made in translating into an equation.
└──┘

EXAMPLE 2   Finding an Unknown Number

Twice the sum of a number and 4 is the same as four times the number, decreased by 12. Find the number.

Solution

1. UNDERSTAND. Read and reread the problem. If we let

$$x = \text{the unknown number, then}$$

"the sum of a number and 4" translates to "$x + 4$" and "four times the number" translates to "$4x$."

2. TRANSLATE.

twice	the sum of a number and 4	is the same as	four times the number	decreased by	12
↓	↓	↓	↓	↓	↓
2	$(x + 4)$	$=$	$4x$	$-$	12

3. SOLVE.

$$2(x + 4) = 4x - 12$$
$$2x + 8 = 4x - 12 \qquad \text{Apply the distributive property.}$$
$$2x + 8 - 4x = 4x - 12 - 4x \qquad \text{Subtract } 4x \text{ from both sides.}$$
$$-2x + 8 = -12$$
$$-2x + 8 - 8 = -12 - 8 \qquad \text{Subtract 8 from both sides.}$$
$$-2x = -20$$
$$\frac{-2x}{-2} = \frac{-20}{-2} \qquad \text{Divide both sides by } -2.$$
$$x = 10$$

4. INTERPRET.

Check: Check this solution in the problem as it was originally stated. To do so, replace "number" with 10. Twice the sum of "10" and 4 is 28, which is the same as 4 times "10" decreased by 12.

State: The number is 10.   □

PRACTICE

2   Three times a number, decreased by 4, is the same as double the difference of the number and 1.   ∎

OBJECTIVE

2 Solving Problems Involving Relationships Among Unknown Quantities ▷

The next three examples have to do with relationships among unknown quantities.

EXAMPLE 3 Finding the Length of a Board

Balsa wood sticks are commonly used for building models (for example, bridge models). A 48-inch balsa wood stick is to be cut into two pieces so that the longer piece is 3 times the shorter. Find the length of each piece.

Solution

1. **UNDERSTAND the problem.** To do so, read and reread the problem. You may also want to propose a solution. For example, if 10 inches represents the length of the shorter piece, then $3(10) = 30$ inches is the length of the longer piece, since it is 3 times the length of the shorter piece. This guess gives a total stick length of 10 inches + 30 inches = 40 inches, too short. However, the purpose of proposing a solution is not to guess correctly but to help understand the problem better and how to model it.

    Since the length of the longer piece is given in terms of the length of the shorter piece, let's let

$$x = \text{length of shorter piece; then}$$
$$3x = \text{length of longer piece.}$$

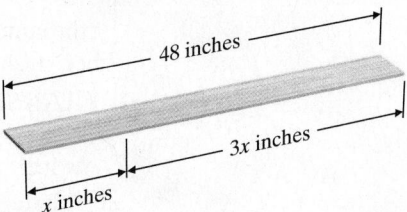

2. **TRANSLATE the problem.** First, we write the equation in words.

length of shorter piece	added to	length of longer piece	equals	total length of stick
↓	↓	↓	↓	↓
x	$+$	$3x$	$=$	48

3. **SOLVE.**

$$x + 3x = 48$$
$$4x = 48 \quad \text{Combine like terms.}$$
$$\frac{4x}{4} = \frac{48}{4} \quad \text{Divide both sides by 4.}$$
$$x = 12$$

4. **INTERPRET.**

Check: Check the solution in the stated problem. If the shorter piece of stick is 12 inches, the longer piece is $3 \cdot (12 \text{ inches}) = 36$ inches, and the sum of the two pieces is 12 inches + 36 inches = 48 inches.

State: The shorter piece of balsa wood is 12 inches, and the longer piece of balsa wood is 36 inches.

Helpful Hint

Make sure that units are included in your answer, if appropriate.

PRACTICE

3 A 45-inch board is to be cut into two pieces so that the longer piece is 4 times the shorter. Find the length of each piece.

EXAMPLE 4   Finding the Number of Republican and Democratic Representatives

The 114th Congress began on January 3, 2015, and had a total of 435 Democratic and Republican representatives. There were 59 fewer Democratic representatives than Republican. Find the number of representatives from each party. (*Source:* Congress.gov)

Solution

1. UNDERSTAND the problem. Read and reread the problem. Let's suppose that there were 200 Republican representatives. Since there were 59 fewer Democrats than Republicans, there must have been 200 − 59 = 141 Democrats. The total number of Republicans and Democrats was then 200 + 141 = 341. This is incorrect since the total should be 435, but we now have a better understanding of the problem.

In general, if we let

$$x = \text{number of Republicans, then}$$
$$x - 59 = \text{number of Democrats.}$$

2. TRANSLATE the problem. First, we write the equation in words.

number of Republicans	added to	number of Democrats	equals	435
↓	↓	↓	↓	↓
x	$+$	$(x - 59)$	$=$	435

3. SOLVE.

$$x + (x - 59) = 435$$
$$2x - 59 = 435 \qquad \text{Combine like terms.}$$
$$2x - 59 + 59 = 435 + 59 \quad \text{Add 59 to both sides.}$$
$$2x = 494$$
$$\frac{2x}{2} = \frac{494}{2} \qquad \text{Divide both sides by 2.}$$
$$x = 247$$

4. INTERPRET.

Check: If there were 247 Republican representatives, then there were 247 − 59 = 188 Democratic representatives. The total number of representatives was then 247 + 188 = 435. The results check.

State: There were 247 Republican and 188 Democratic representatives at the beginning of the 114th Congress. □

PRACTICE
4   In 2015, there were 7 fewer Democratic State Governors than Republican State Governors. Find the number of State Governors from each party. Alaska had an independent governor, so use a total of 49, representing the other 49 states. (*Source:* National Conference of State Legislatures). ▪

△ **EXAMPLE 5**   Finding Angle Measures

If the two walls of the Vietnam Veterans Memorial in Washington, D.C., were connected, an isosceles triangle would be formed. The measure of the third angle is 97.5° more than the measure of either of the other two equal angles. Find the measure of the third angle. (*Source:* National Park Service)

(Continued on next page)

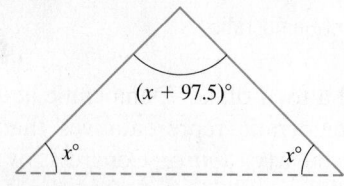

<u>Solution</u>

1. UNDERSTAND. Read and reread the problem. We then draw a diagram (recall that an isosceles triangle has two angles with the same measure) and let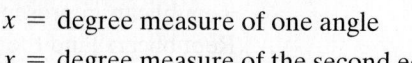

$$x = \text{degree measure of one angle}$$
$$x = \text{degree measure of the second equal angle}$$
$$x + 97.5 = \text{degree measure of the third angle}$$

2. TRANSLATE. Recall that the sum of the measures of the angles of a triangle equals 180.

measure of first angle	measure of second angle	measure of third angle	equals	180
↓	↓	↓	↓	↓
x +	x	$+\ (x + 97.5)$	$=$	180

3. SOLVE.

$$x + x + (x + 97.5) = 180$$
$$3x + 97.5 = 180 \qquad \text{Combine like terms.}$$
$$3x + 97.5 - 97.5 = 180 - 97.5 \qquad \text{Subtract 97.5 from both sides.}$$
$$3x = 82.5$$
$$\frac{3x}{3} = \frac{82.5}{3} \qquad \text{Divide both sides by 3.}$$
$$x = 27.5$$

4. INTERPRET.

Check: If $x = 27.5$, then the measure of the third angle is $x + 97.5 = 125$. The sum of the angles is then $27.5 + 27.5 + 125 = 180$, the correct sum.

State: The third angle measures 125°.*

PRACTICE

5 The second angle of a triangle measures three times as large as the first. If the third angle measures 55° more than the first, find the measures of all three angles. ■

OBJECTIVE

3 Solving Consecutive Integer Problems

The next example has to do with consecutive integers. Recall what we have learned thus far about these integers.

	Example	*General Representation*
Consecutive Integers	11, 12, 13 ↳+1 ↳+1	Let x be an integer. $x,\ x + 1, x + 2$ ↳+1 ↳+1
Consecutive Even Integers	38, 40, 42 ↳+2 ↳+2	Let x be an **even** integer. $x,\ x + 2, x + 4$ ↳+2 ↳+2
Consecutive Odd Integers	57, 59, 61 ↳+2 ↳+2	Let x be an **odd** integer. $x,\ x + 2, x + 4$ ↳+2 ↳+2

* The two walls actually meet at an angle of 125 degrees 12 minutes. The measurement of 97.5° given in the problem is an approximation.

EXAMPLE 6   Some states have a single area code for the entire state. Two such states have area codes that are consecutive odd integers. If the sum of these integers is 1208, find the two area codes. (*Source:* North American Numbering Plan Administration)

Solution

1. UNDERSTAND. Read and reread the problem. If we let

$$x = \text{the first odd integer, then}$$
$$x + 2 = \text{the next odd integer}$$

> **Helpful Hint**
> Remember, the 2 here means that odd integers are 2 units apart, for example, the odd integers 13 and 13 + 2 = 15.

2. TRANSLATE.

first odd integer	the sum of	next odd integer	is	1208
↓	↓	↓	↓	↓
x	$+$	$(x + 2)$	$=$	1208

3. SOLVE.

$$x + x + 2 = 1208$$
$$2x + 2 = 1208$$
$$2x + 2 - 2 = 1208 - 2$$
$$2x = 1206$$
$$\frac{2x}{2} = \frac{1206}{2}$$
$$x = 603$$

4. INTERPRET.

Check: If $x = 603$, then the next odd integer $x + 2 = 603 + 2 = 605$. Notice their sum, $603 + 605 = 1208$, as needed.

State: The area codes are 603 and 605.

Note: New Hampshire's area code is 603 and South Dakota's area code is 605.   □

PRACTICE
6   The sum of three consecutive even integers is 144. Find the integers.   ■

✔ Vocabulary, Readiness & Video Check

Fill in the table.

	A number:			
1.	x	→ Double the number:	→ Double the number, decreased by 31:	
2.	A number: x	→ Three times the number:	→ Three times the number, increased by 17:	
3.	A number: x	→ The sum of the number and 5:	→ Twice the sum of the number and 5:	
4.	A number: x	→ The difference of the number and 11:	→ Seven times the difference of the number and 11:	
5.	A number: y	→ The difference of 20 and the number:	→ The difference of 20 and the number, divided by 3:	
6.	A number: y	→ The sum of −10 and the number:	→ The sum of −10 and the number, divided by 9:	

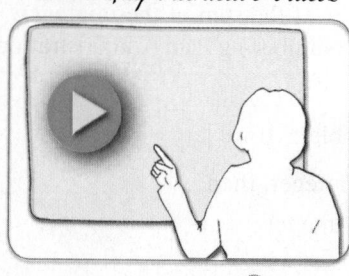

Martin-Gay Interactive Videos

See Video 2.4

Watch the section lecture video and answer the following questions.

OBJECTIVE 1
7. At the end of Example 1, where are you told is the best place to check the solution of an application problem?

OBJECTIVE 2
8. The solution of the equation for ▦ Example 3 is $x = 43$. Why is this not the solution of the application?

OBJECTIVE 3
9. What are two things that should be checked to make sure the solution of ▦ Example 4 is correct?

2.4 Exercise Set MyMathLab®

TRANSLATING

Write each of the following as an equation. Then solve. See Examples 1 and 2.

1. The sum of six times a number, and 1, is equal to five times the number. Find the number.

2. The difference of three times a number, and 1, is the same as twice the number. Find the number.

3. Three times a number, minus 6, is equal to two times the number, plus 8. Find the number.

4. The sum of 4 times a number, and -2, is equal to the sum of 5 times the number, and -2. Find the number.

▶ **5.** Twice the difference of a number and 8 is equal to three times the sum of the number and 3. Find the number.

6. Five times the sum of a number and -1 is the same as 6 times the difference of the number and 5. Find the number.

7. Twice the sum of -2 and a number is the same as the number decreased by $\frac{1}{2}$. Find the number.

8. If the difference of a number and four is doubled, the result is $\frac{1}{4}$ less than the number. Find the number.

Solve. For Exercises 9 and 10, the solutions have been started for you. See Examples 3 through 5.

9. A 25-inch piece of steel is cut into three pieces so that the second piece is twice as long as the first piece, and the third piece is one inch more than five times the length of the first piece. Find the lengths of the pieces.

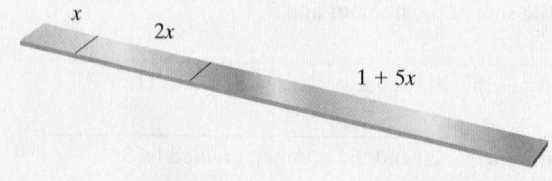

Start the solution:

1. UNDERSTAND the problem. Reread it as many times as needed.

2. TRANSLATE into an equation. (Fill in the blanks below.)

total length of steel	equals	length of first piece	plus	length of second piece	plus	length of third piece
↓	↓	↓	↓	↓	↓	↓
25	=	___	+	___	+	___

Finish with:

3. SOLVE and **4.** INTERPRET

10. A 46-foot piece of rope is cut into three pieces so that the second piece is three times as long as the first piece, and the third piece is two feet more than seven times the length of the first piece. Find the lengths of the pieces.

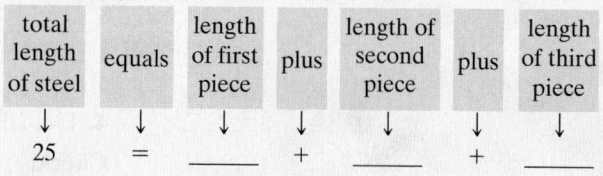

Start the solution:

1. UNDERSTAND the problem. Reread it as many times as needed.

2. TRANSLATE into an equation. (Fill in the blanks below.)

total length of rope	equals	length of first piece	plus	length of second piece	plus	length of third piece
↓	↓	↓	↓	↓	↓	↓
46	=	___	+	___	+	___

Finish with:

3. SOLVE and **4.** INTERPRET

11. A 40-inch board is to be cut into three pieces so that the second piece is twice as long as the first piece, and the third piece is 5 times as long as the first piece. If x represents the length of the first piece, find the lengths of all three pieces.

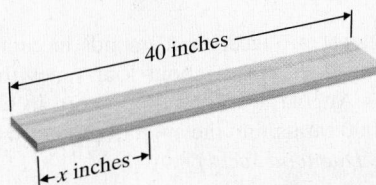

12. A 21-foot beam is to be cut into three pieces so that the second and the third piece are each 3 times the length of the first piece. If x represents the length of the shorter piece, find the lengths of all three pieces.

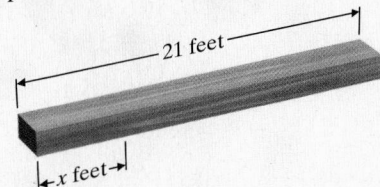

For Exercises 13 and 14, use each table to find the value of x. Then write a sentence to explain, in words, the meaning of the value of x. (Source: IHS Screen Digest)

13. The fastest growing type of movie screen in 2014 was 3D.

Type of Screen	Number of Screens
3D	x
Non-3D	10,973 more than x
Total screens	43,265

14. The vast majority of movie screens in the United States in 2014 were digital.

Type of Screen	Number of Screens
Analog	x
Digital	39,771 more than x
Total screens	43,265

△ **15.** The flag of Equatorial Guinea contains an isosceles triangle. (Recall that an isosceles triangle contains two angles with the same measure.) If the measure of the third angle of the triangle is 30° more than twice the measure of either of the other two angles, find the measure of each angle of the triangle. (*Hint:* Recall that the sum of the measures of the angles of a triangle is 180°.)

△ **16.** The flag of Brazil contains a parallelogram. One angle of the parallelogram is 15° less than twice the measure of the angle next to it. Find the measure of each angle. (*Hint:* Recall that opposite angles of a parallelogram have the same measure and that the sum of the measures of the angles is 360°.)

Solve. See Example 6. For Exercises 17 through 24, fill in the table. Most of the first row has been completed for you.

	First Integer →	Next Integers →			Indicated Sum
17. Three consecutive integers:	Integer: x	$x + 1$	$x + 2$		Sum of the three consecutive integers, simplified:
18. Three consecutive integers:	Integer: x				Sum of the second and third consecutive integers, simplified:
19. Three consecutive even integers:	Even integer: x				Sum of the first and third even consecutive integers, simplified:
20. Three consecutive odd integers:	Odd integer: x				Sum of the three consecutive odd integers, simplified:
21. Four consecutive integers:	Integer: x				Sum of the four consecutive integers, simplified:
22. Four consecutive integers:	Integer: x				Sum of the first and fourth consecutive integers, simplified:
23. Three consecutive odd integers:	Odd integer: x				Sum of the second and third consecutive odd integers, simplified:
24. Three consecutive even integers:	Even integer: x				Sum of the three consecutive even integers, simplified:

25. The left and right page numbers of an open book are two consecutive integers whose sum is 469. Find these page numbers.

26. The room numbers of two adjacent classrooms are two consecutive even numbers. If their sum is 654, find the classroom numbers.

27. To make an international telephone call, you need the code for the country you are calling. The codes for Belgium, France, and Spain are three consecutive integers whose sum is 99. Find the code for each country. (*Source: The World Almanac and Book of Facts*, 2007)

28. To make an international telephone call, you need the code for the country you are calling. The codes for Mali Republic, Côte d'Ivoire, and Niger are three consecutive odd integers whose sum is 675. Find the code for each country.

MIXED PRACTICE

Solve. See Examples 1 through 6.

29. The area of the Sahara Desert is 7 times the area of the Gobi Desert. If the sum of their areas is 4,000,000 square miles, find the area of each desert.

30. The largest meteorite in the world is the Hoba West, located in Namibia. Its weight is 3 times the weight of the Armanty meteorite, located in Outer Mongolia. If the sum of their weights is 88 tons, find the weight of each.

31. A 17-foot piece of string is cut into two pieces so that the longer piece is 2 feet longer than twice the length of the shorter piece. Find the lengths of both pieces.

32. A 25-foot wire is to be cut so that the longer piece is one foot longer than 5 times the length of the shorter piece. Find the length of each piece.

33. Five times a number, subtracted from ten, is triple the number. Find the number.

34. Nine is equal to ten subtracted from double a number. Find the number.

35. The greatest producer of diamonds in carats is Botswana. This country produces about four times the amount produced in Angola. If the total produced in both countries is 40,000,000 carats, find the amount produced in each country. (*Source: Diamond Facts.*)

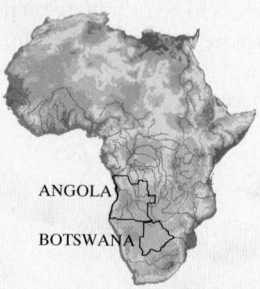

36. Beetles have the greatest number of different species. There are twenty times the number of beetle species as grasshopper species, and the total number of species for both is 420,000. Find the number of species for each type of insect.

37. The measures of the angles of a triangle are 3 consecutive even integers. Find the measure of each angle.

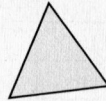

38. A quadrilateral is a polygon with 4 sides. The sum of the measures of the 4 angles in a quadrilateral is 360°. If the measures of the angles of a quadrilateral are consecutive odd integers, find the measures.

39. For the 2014 Winter Olympics, the total number of medals won by athletes from each of the countries of Netherlands, Canada, and Norway are three consecutive integers whose sum is 75. Find the number of medals for each country.

40. The code to unlock a student's combination lock happens to be three consecutive odd integers whose sum is 51. Find the integers.

41. If the sum of a number and five is tripled, the result is one less than twice the number. Find the number.

42. Twice the sum of a number and six equals three times the sum of the number and four. Find the number.

43. Two angles are supplementary if their sum is 180°. A larger angle measures eight degrees more than three times the measure of a smaller angle. If x represents the measure of the smaller angle and these two angles are supplementary, find the measure of each angle.

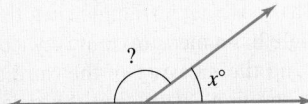

44. Two angles are complementary if their sum is 90°. A larger angle measures three degrees less than twice the measure of a smaller angle. If x represents the measure of the smaller angle and these two angles are complementary, find the measure of each angle.

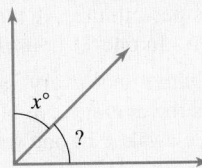

45. If the quotient of a number and 4 is added to $\frac{1}{2}$, the result is $\frac{3}{4}$. Find the number.

46. The sum of $\frac{1}{5}$ and twice a number is equal to $\frac{4}{5}$ subtracted from three times the number. Find the number.

47. The sum of $\frac{2}{3}$ and four times a number is equal to $\frac{5}{6}$ subtracted from five times the number. Find the number.

48. If $\frac{3}{4}$ is added to three times a number, the result is $\frac{1}{2}$ subtracted from twice the number. Find the number.

49. Currently, the two fastest trains in the world are both Chinese, the Shanghai Maglev and the Harmony CRH. The sum of their fastest speeds is 503 miles per hour. If the maximum speed of the Maglev is 31 miles per hour faster than the speed of the Harmony, find the speeds of each. (*Source:* Railway-technology.com)

50. The Pentagon is the world's largest office building in terms of floor space. It has three times the amount of floor space as the Empire State Building. If the total floor space for these two buildings is approximately 8700 thousand square feet, find the floor space of each building.

51. One-third of a number is five-sixths. Find the number.

52. Seven-eighths of a number is one-half. Find the number.

53. The number of counties in California and the number of counties in Montana are consecutive even integers whose sum is 114. If California has more counties than Montana, how many counties does each state have? (*Source: The World Almanac and Book of Facts*)

54. A student is building a bookcase with stepped shelves for her dorm room. She buys a 48-inch board and wants to cut the board into three pieces with lengths equal to three consecutive even integers. Find the three board lengths.

55. A geodesic dome, based on the design by Buckminster Fuller, is composed of two types of triangular panels. One of these is an isosceles triangle. In one geodesic dome, the measure of the third angle is 76.5° more than the measure of either of the two equal angles. Find the measure of the third angle. (*Source: Buckminster Fuller Institute*)

56. The measures of the angles of a particular triangle are such that the second and third angles are each four times larger than the smallest angle. Find the measures of the angles of this triangle.

57. A 30-foot piece of siding is cut into three pieces so that the second piece is four times as long as the first piece and the third piece is five times as long as the first piece. If x represents the length of the first piece, find the lengths of all three pieces.

58. A 48-foot-long piece of cable wire is to be cut into three pieces so that the second piece is five times as long as the first piece and the third piece is six times as long as the first piece. If x represents the length of the first piece, find the lengths of all three pieces.

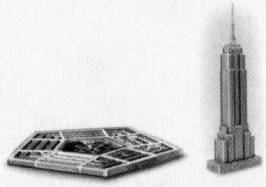

The graph below shows the best-selling video games for 2014. Use this graph for Exercises 59 through 64.

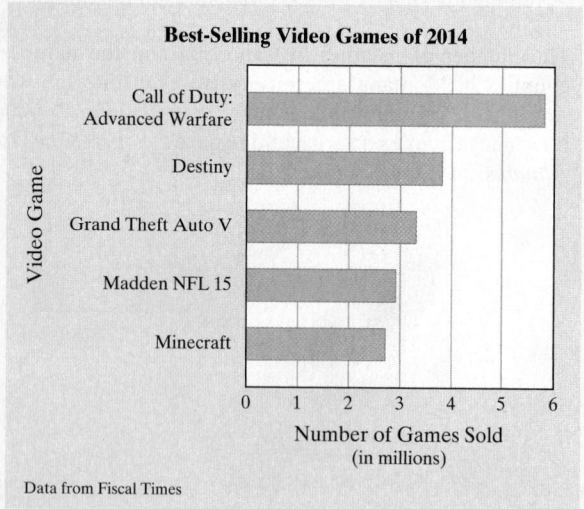

Best-Selling Video Games of 2014

Data from Fiscal Times

59. Which video game is the best-selling game of 2014?

60. Which video games sold between 3 and 5 million copies in 2014?

61. Madden NFL 15 and Destiny together sold 6.7 million copies in 2014. Destiny sold 0.9 million copies more than Madden NFL 15. Find the number of sales of each video game.

62. Grand Theft Auto V and Minecraft together sold 6 million copies in 2014. Grand Theft Auto V sold 0.6 million more copies than Minecraft. Find the number of sales of each video game.

Compare the lengths of the bars in the graph with your results for the exercises below. Are your answers reasonable?

63. Exercise 61   **64.** Exercise 62

REVIEW AND PREVIEW

Evaluate each expression for the given values. See Section 1.4.

65. $2W + 2L$; $W = 7$ and $L = 10$

66. $\frac{1}{2}Bh$; $B = 14$ and $h = 22$

67. πr^2; $r = 15$ **68.** $r \cdot t$; $r = 15$ and $t = 2$

CONCEPT EXTENSIONS

69. A golden rectangle is a rectangle whose length is approximately 1.6 times its width. The early Greeks thought that a rectangle with these dimensions was the most pleasing to the eye, and examples of the golden rectangle are found in many early works of art. For example, the Parthenon in Athens contains many examples of golden rectangles.

Mike Hallahan would like to plant a rectangular garden in the shape of a golden rectangle. If he has 78 feet of fencing available, find the dimensions of the garden.

70. Dr. Dorothy Smith gave the students in her geometry class at the University of New Orleans the following question. Is it possible to construct a triangle such that the second angle of the triangle has a measure that is twice the measure of the first angle and the measure of the third angle is 5 times the measure of the first? If so, find the measure of each angle. (*Hint:* Recall that the sum of the measures of the angles of a triangle is 180°.)

71. Only male crickets chirp. They chirp at different rates depending on their species and the temperature of their environment. Suppose a certain species is currently chirping at a rate of 90 chirps per minute. At this rate, how many chirps occur in one hour? In one 24-hour day? In one year?

72. The human eye blinks once every 5 seconds on average. How many times does the average eye blink in one hour? In one 16-hour day while awake? In one year while awake?

73. In your own words, explain why a solution of a word problem should be checked using the original wording of the problem and not the equation written from the wording.

74. Give an example of how you recently solved a problem using mathematics.

Recall from Exercise 69 that a golden rectangle is a rectangle whose length is approximately 1.6 times its width.

75. It is thought that for about 75% of adults, a rectangle in the shape of the golden rectangle is the most pleasing to the eye. Draw three rectangles, one in the shape of the golden rectangle, and poll your class. Do the results agree with the percentage given above?

76. Examples of golden rectangles can be found today in architecture and manufacturing packaging. Find an example of a golden rectangle in your home. A few suggestions: the front face of a book, the floor of a room, the front of a box of food.

For Exercises 77 and 78, measure the dimensions of each rectangle and decide which one best approximates the shape of a golden rectangle.

77.

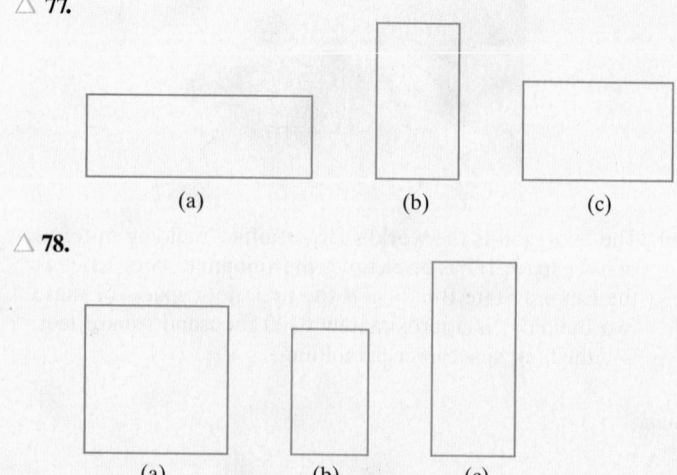

(a)   (b)   (c)

78.

(a)   (b)   (c)

2.5   Formulas and Problem Solving

OBJECTIVES

1   Use Formulas to Solve Problems.

2   Solve a Formula or Equation for One of Its Variables.

OBJECTIVE

1   Using Formulas to Solve Problems

An equation that describes a known relationship among quantities, such as distance, time, volume, weight, and money, is called a **formula.** These quantities are represented by letters and are thus variables of the formula. Here are some common formulas and their meanings.

Formulas	Their Meanings
$A = lw$	Area of a rectangle = length · width
$I = PRT$	Simple interest = principal · rate · time
$P = a + b + c$	Perimeter of a triangle = side a + side b + side c
$d = rt$	Distance = rate · time
$V = lwh$	Volume of a rectangular solid = length · width · height
$F = \left(\dfrac{9}{5}\right)C + 32$ or $F = 1.8C + 32$	Degrees Fahrenheit = $\left(\dfrac{9}{5}\right)$ · degrees Celsius + 32

Formulas are valuable tools because they allow us to calculate measurements as long as we know certain other measurements. For example, if we know we traveled a distance of 100 miles at a rate of 40 miles per hour, we can replace the variables d and r in the formula $d = rt$ and find our time, t.

$$d = rt \quad \text{Formula.}$$
$$100 = 40t \quad \text{Replace } d \text{ with 100 and } r \text{ with 40.}$$

This is a linear equation in one variable, t. To solve for t, divide both sides of the equation by 40.

$$\frac{100}{40} = \frac{40t}{40} \quad \text{Divide both sides by 40.}$$
$$\frac{5}{2} = t \quad \text{Simplify.}$$

The time traveled is $\dfrac{5}{2}$ hours, or $2\dfrac{1}{2}$ hours, or 2.5 hours.

In this section, we solve problems that can be modeled by known formulas. We use the same problem-solving steps that were introduced in the previous section. These steps have been slightly revised to include formulas.

EXAMPLE 1   Finding Time Given Rate and Distance

A glacier is a giant mass of rocks and ice that flows downhill like a river. Portage Glacier in Alaska is about 6 miles, or 31,680 *feet,* long and moves 400 *feet* per year. Icebergs are created when the front end of the glacier flows into Portage Lake. How long does it take for ice at the head (beginning) of the glacier to reach the lake?

(Continued on next page)

Solution

1. **UNDERSTAND.** Read and reread the problem. The appropriate formula needed to solve this problem is the distance formula, $d = rt$. To become familiar with this formula, let's find the distance that ice traveling at a rate of 400 feet per year travels in 100 years. To do so, we let time t be 100 years and rate r be the given 400 feet per year and substitute these values into the formula $d = rt$. We then have that distance $d = 400(100) = 40,000$ feet. Since we are interested in finding how long it takes ice to travel 31,680 feet, we now know that it is less than 100 years.

   Since we are using the formula $d = rt$, we let

   $t =$ the time in years for ice to reach the lake

   $r =$ rate or speed of ice

   $d =$ distance from beginning of glacier to lake.

2. **TRANSLATE.** To translate into an equation, we use the formula $d = rt$ and let distance $d = 31,680$ feet and rate $r = 400$ feet per year.

$$d = r \cdot t$$
$$31{,}680 = 400 \cdot t \quad \text{Let } d = 31{,}680 \text{ and } r = 400.$$

3. **SOLVE.** Solve the equation for t. To solve for t, divide both sides by 400.

$$\frac{31{,}680}{400} = \frac{400 \cdot t}{400} \quad \text{Divide both sides by 400.}$$
$$79.2 = t \quad \text{Simplify.}$$

4. **INTERPRET.**

Check: To check, substitute 79.2 for t and 400 for r in the distance formula and check to see that the distance is 31,680 feet.

State: It takes 79.2 years for the ice at the head of Portage Glacier to reach the lake. □

> **Helpful Hint**
> Don't forget to include units if appropriate.

PRACTICE

1 The Stromboli Volcano, in Italy, began erupting in 2002 and continues to be active after a dormant period of over 17 years. In 2007, a volcanologist measured the lava flow to be moving at 5 meters/second. If the path the lava follows to the sea is 580 meters long, how long does it take the lava to reach the sea? (_Source:_ Thorsten Boeckel and CNN)

△ **EXAMPLE 2**   Calculating the Length of a Garden

Charles Pecot can afford enough fencing to enclose a rectangular garden with a perimeter of 140 feet. If the width of his garden must be 30 feet, find the length.

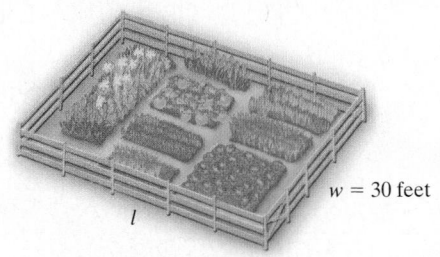

$w = 30$ feet

Solution

1. UNDERSTAND. Read and reread the problem. The formula needed to solve this problem is the formula for the perimeter of a rectangle, $P = 2l + 2w$. Before continuing, let's become familiar with this formula.

    l = the length of the rectangular garden

    w = the width of the rectangular garden

    P = perimeter of the garden

2. TRANSLATE. To translate into an equation, we use the formula $P = 2l + 2w$ and let perimeter $P = 140$ feet and width $w = 30$ feet.

    $$P = 2l + 2w$$
    $$140 = 2l + 2(30) \quad \text{Let } P = 140 \text{ and } w = 30.$$

3. SOLVE.

    $$140 = 2l + 2(30)$$
    $$140 = 2l + 60 \qquad \text{Multiply } 2(30).$$
    $$140 - 60 = 2l + 60 - 60 \quad \text{Subtract 60 from both sides.}$$
    $$80 = 2l \qquad\qquad \text{Combine like terms.}$$
    $$40 = l \qquad\qquad \text{Divide both sides by 2.}$$

4. INTERPRET.

    Check: Substitute 40 for l and 30 for w in the perimeter formula and check to see that the perimeter is 140 feet.

    State: The length of the rectangular garden is 40 feet.   □

PRACTICE
2   Evelyn Gryk fenced in part of her backyard for a dog run. The dog run was 40 feet in length and used 98 feet of fencing. Find the width of the dog run.   ■

EXAMPLE 3   Finding an Equivalent Temperature

The average minimum temperature for July in Shanghai, China, is 77° Fahrenheit. Find the equivalent temperature in degrees Celsius.

Solution

1. UNDERSTAND. Read and reread the problem. A formula that can be used to solve this problem is the formula for converting degrees Celsius to degrees Fahrenheit, $F = \dfrac{9}{5}C + 32$. Before continuing, become familiar with this formula. Using this formula, we let

    C = temperature in degrees Celsius, and

    F = temperature in degrees Fahrenheit.

(Continued on next page)

2. TRANSLATE. To translate into an equation, we use the formula $F = \frac{9}{5}C + 32$ and let degrees Fahrenheit $F = 77$.

Formula: $F = \frac{9}{5}C + 32$

Substitute: $77 = \frac{9}{5}C + 32$ Let $F = 77$.

3. SOLVE.

$$77 = \frac{9}{5}C + 32$$

$$77 - 32 = \frac{9}{5}C + 32 - 32 \quad \text{Subtract 32 from both sides.}$$

$$45 = \frac{9}{5}C \quad \text{Combine like terms.}$$

$$\frac{5}{9} \cdot 45 = \frac{5}{9} \cdot \frac{9}{5}C \quad \text{Multiply both sides by } \frac{5}{9}.$$

$$25 = C \quad \text{Simplify.}$$

4. INTERPRET.

Check: To check, replace C with 25 and F with 77 in the formula and see that a true statement results.

State: Thus, 77° Fahrenheit is equivalent to 25° Celsius.

Note: There is a formula for directly converting degrees Fahrenheit to degrees Celsius. It is $C = \frac{5}{9}(F - 32)$, as we shall see in Example 8.   □

PRACTICE

3   The average minimum temperature for July in Sydney, Australia, is 8° Celsius. Find the equivalent temperature in degrees Fahrenheit.   ■

In the next example, we again use the formula for perimeter of a rectangle as in Example 2. In Example 2, we knew the width of the rectangle. In this example, both the length and width are unknown.

△   **EXAMPLE 4**   Finding Road Sign Dimensions

The length of a rectangular road sign is 2 feet less than three times its width. Find the dimensions if the perimeter is 28 feet.

Solution

1. UNDERSTAND. Read and reread the problem. Recall that the formula for the perimeter of a rectangle is $P = 2l + 2w$. Draw a rectangle and guess the solution. If the width of the rectangular sign is 5 feet, its length is 2 feet less than 3 times the width or $3(5 \text{ feet}) - 2 \text{ feet} = 13 \text{ feet}$. The perimeter P of the rectangle is then $2(13 \text{ feet}) + 2(5 \text{ feet}) = 36 \text{ feet}$, too much. We now know that the width is less than 5 feet.

Proposed rectangle:

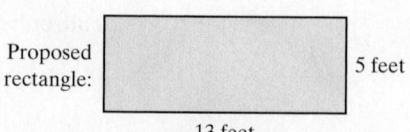

5 feet

13 feet

Let

w = the width of the rectangular sign; then
$3w - 2$ = the length of the sign.

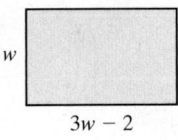

$3w - 2$

Draw a rectangle and label it with the assigned variables.

2. TRANSLATE.

Formula: $P = 2l + 2w$ or

Substitute: $28 = 2(3w - 2) + 2w.$

3. SOLVE.

$$28 = 2(3w - 2) + 2w$$
$$28 = 6w - 4 + 2w \qquad \text{Apply the distributive property.}$$
$$28 = 8w - 4$$
$$28 + 4 = 8w - 4 + 4 \qquad \text{Add 4 to both sides.}$$
$$32 = 8w$$
$$\frac{32}{8} = \frac{8w}{8} \qquad \text{Divide both sides by 8.}$$
$$4 = w$$

4. INTERPRET.

Check: If the width of the sign is 4 feet, the length of the sign is $3(4 \text{ feet}) - 2 \text{ feet} = 10 \text{ feet}.$ This gives a perimeter of $P = 2(4 \text{ feet}) + 2(10 \text{ feet}) = 28 \text{ feet},$ the correct perimeter.

State: The width of the sign is 4 feet, and the length of the sign is 10 feet.   □

PRACTICE

4   The new street signs along Route 114 have a length that is 3 inches more than 5 times the width. Find the dimensions of the signs if the perimeter of the signs is 66 inches.   ■

OBJECTIVE

2   Solving a Formula for One of Its Variables ▶

We say that the formula

$$F = \frac{9}{5}C + 32$$

is solved for F because F is alone on one side of the equation, and the other side of the equation contains no F's. Suppose that we need to convert many Fahrenheit temperatures to equivalent degrees Celsius. In this case, it is easier to perform this task by solving the formula $F = \frac{9}{5}C + 32$ for C. (See Example 8.) For this reason, it is important to be able to solve an equation for any one of its specified variables. For example, the formula $d = rt$ is solved for d in terms of r and t. We can also solve $d = rt$ for t in terms of d and r. To solve for t, divide both sides of the equation by r.

$$d = rt$$
$$\frac{d}{r} = \frac{rt}{r} \qquad \text{Divide both sides by } r.$$
$$\frac{d}{r} = t \qquad \text{Simplify.}$$

To solve a formula or an equation for a specified variable, we use the same steps as for solving a linear equation. These steps are listed next.

> **Solving Equations for a Specified Variable**
>
> **Step 1.**   Multiply on both sides to clear the equation of fractions if they occur.
>
> **Step 2.**   Use the distributive property to remove parentheses if they occur.
>
> **Step 3.**   Simplify each side of the equation by combining like terms.
>
> **Step 4.**   Get all terms containing the specified variable on one side and all other terms on the other side by using the addition property of equality.
>
> **Step 5.**   Get the specified variable alone by using the multiplication property of equality.

EXAMPLE 5   Solve: $V = lwh$ for l

Solution   This formula is used to find the volume of a box. To solve for l, divide both sides by wh.

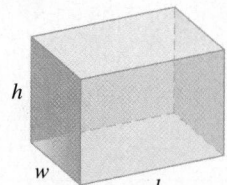

$$V = lwh$$

$$\frac{V}{wh} = \frac{lwh}{wh} \qquad \text{Divide both sides by } wh.$$

$$\frac{V}{wh} = l \qquad \text{Simplify.}$$

Since we have l alone on one side of the equation, we have solved for l in terms of V, w, and h. Remember that it does not matter on which side of the equation we isolate the variable.   □

PRACTICE
5   Solve: $I = PRT$ for R

EXAMPLE 6   Solve: $y = mx + b$ for x

Solution   The term containing the variable we are solving for, mx, is on the right side of the equation. Get mx alone by subtracting b from both sides.

$$y = mx + b$$

$$y - b = mx + b - b \qquad \text{Subtract } b \text{ from both sides.}$$

$$y - b = mx \qquad \text{Combine like terms.}$$

Next, solve for x by dividing both sides by m.

$$\frac{y - b}{m} = \frac{mx}{m}$$

$$\frac{y - b}{m} = x \qquad \text{Simplify.}$$ □

PRACTICE
6   Solve: $H = 5as + 10a$ for s

✔ **CONCEPT CHECK**

Solve:

a.    for ▢

b.    for ▢

 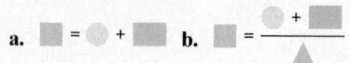
EXAMPLE 7   Solve: $P = 2l + 2w$ for w

Solution   This formula relates the perimeter of a rectangle to its length and width. Find the term containing the variable w. Get this term, $2w$, alone by subtracting $2l$ from both sides.

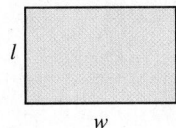

l

w

Helpful Hint

The 2's may *not* be divided out here. Although 2 is a factor of the denominator, 2 is *not* a factor of the numerator since it is not a factor of both terms in the numerator.

$$P = 2l + 2w$$

$P - 2l = 2l + 2w - 2l$ Subtract $2l$ from both sides.

$P - 2l = 2w$ Combine like terms.

$\dfrac{P - 2l}{2} = \dfrac{2w}{2}$ Divide both sides by 2.

$\dfrac{P - 2l}{2} = w$ Simplify. □

PRACTICE

7   Solve: $N = F + d(n - 1)$ for d ■

The next example has an equation containing a fraction. We will first clear the equation of fractions and then solve for the specified variable.

EXAMPLE 8   Solve: $F = \dfrac{9}{5}C + 32$ for C

Solution

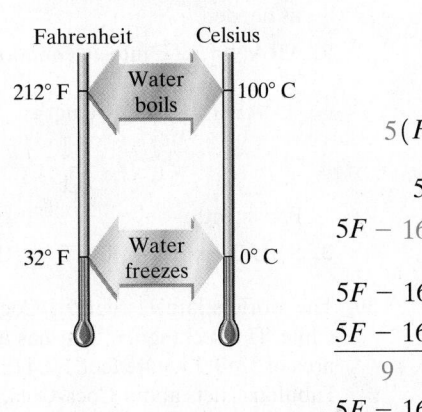

$$F = \frac{9}{5}C + 32$$

$5(F) = 5\left(\dfrac{9}{5}C + 32\right)$ Clear the fraction by multiplying both sides by the LCD.

$5F = 9C + 160$ Distribute the 5.

$5F - 160 = 9C + 160 - 160$ To get the term containing the variable C alone, subtract 160 from both sides.

$5F - 160 = 9C$ Combine like terms.

$\dfrac{5F - 160}{9} = \dfrac{9C}{9}$ Divide both sides by 9.

$\dfrac{5F - 160}{9} = C$ Simplify.

Note: An equivalent way to write this formula is $C = \dfrac{5}{9}(F - 32)$. □

PRACTICE

8   Solve: $A = \dfrac{1}{2}a(b + B)$ for B ■

✓ Vocabulary, Readiness & Video Check

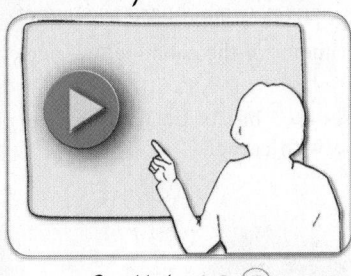

Martin-Gay Interactive Videos

See Video 2.5 ◉

Watch the section lecture video and answer the following questions.

OBJECTIVE 1

1. Complete this statement based on the lecture given before 🎬 Example 1. A formula is an equation that describes known _____ among quantities.

OBJECTIVE 1

2. In 🎬 Example 2, how are the units for the solution determined?

OBJECTIVE 2

3. During 🎬 Example 4, why is the equation $5x = 30$ shown?

2.5 Exercise Set MyMathLab® ▶

Substitute the given values into each given formula and solve for the unknown variable. If necessary, round to one decimal place. See Examples 1 through 4.

△ **1.** $A = bh$; $A = 45, b = 15$ (Area of a parallelogram)

2. $d = rt$; $d = 195, t = 3$ (Distance formula)

△ **3.** $S = 4lw + 2wh$; $S = 102, l = 7, w = 3$ (Surface area of a special rectangular box)

△ **4.** $V = lwh$; $l = 14, w = 8, h = 3$ (Volume of a rectangular box)

△ **5.** $A = \frac{1}{2}h(B + b)$; $A = 180, B = 11, b = 7$ (Area of a trapezoid)

△ **6.** $A = \frac{1}{2}h(B + b)$; $A = 60, B = 7, b = 3$ (Area of a trapezoid)

△ **7.** $P = a + b + c$; $P = 30, a = 8, b = 10$ (Perimeter of a triangle)

△ **8.** $V = \frac{1}{3}Ah$; $V = 45, h = 5$ (Volume of a pyramid)

▦△ **9.** $C = 2\pi r$; $C = 15.7$ (use the approximation 3.14 or a calculator approximation for π) (Circumference of a circle)

▦△ **10.** $A = \pi r^2$; $r = 4.5$ (use the approximation 3.14 or a calculator approximation for π) (Area of a circle)

▦△ **11.** $I = PRT$; $I = 3750, P = 25,000, R = 0.05$ (Simple interest formula)

▦△ **12.** $I = PRT$; $I = 1,056,000, R = 0.055, T = 6$ (Simple interest formula)

▦△ **13.** $V = \frac{1}{3}\pi r^2 h$; $V = 565.2, r = 6$ (use a calculator approximation for π) (Volume of a cone)

▦△ **14.** $V = \frac{4}{3}\pi r^3$; $r = 3$ (use a calculator approximation for π) (Volume of a sphere)

Solve each formula for the specified variable. See Examples 5 through 8.

15. $f = 5gh$ for h

△ **16.** $A = \pi ab$ for b

▶ **17.** $V = lwh$ for w

18. $T = mnr$ for n

19. $3x + y = 7$ for y

20. $-x + y = 13$ for y

21. $A = P + PRT$ for R

22. $A = P + PRT$ for T

23. $V = \frac{1}{3}Ah$ for A

24. $D = \frac{1}{4}fk$ for k

25. $P = a + b + c$ for a

26. $PR = x + y + z + w$ for z

▶△ **27.** $S = 2\pi rh + 2\pi r^2$ for h

△ **28.** $S = 4lw + 2wh$ for h

Solve. For Exercises 29 and 30, the solutions have been started for you. See Examples 1 through 4.

△ **29.** The iconic NASDAQ sign in New York's Times Square has a width of 84 feet and an area of 10,080 square feet. Find the height (or length) of the sign. (*Source:* livedesignonline.com)

Start the solution:

1. **UNDERSTAND** the problem. Reread it as many times as needed.

2. **TRANSLATE** into an equation. (Fill in the blanks below.)

Area	=	length	times	width
↓	↓	↓	↓	↓
___	=	x	·	___

Finish with:

3. **SOLVE** and   4. **INTERPRET**

△ **30.** The world's largest sign for Coca-Cola is located in Arica, Chile. The rectangular sign has a length of 400 feet and an area of 52,400 square feet. Find the width of the sign. (*Source:* Fabulous Facts about Coca-Cola, Atlanta, GA)

Start the solution:

1. **UNDERSTAND** the problem. Reread it as many times as needed.

2. **TRANSLATE** into an equation. (Fill in the blanks below.)

Area	=	length	times	width
↓	↓	↓	↓	↓
___	=	___	·	x

Finish with:

3. **SOLVE** and   4. **INTERPRET**

31. For the purpose of purchasing new baseboard and carpet,

   a. Find the area and perimeter of the room below (neglecting doors).

   b. Identify whether baseboard has to do with area or perimeter and the same with carpet.

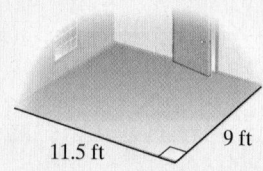

11.5 ft   9 ft

32. For the purpose of purchasing lumber for a new fence and seed to plant grass,

a. Find the area and perimeter of the yard below.

b. Identify whether a fence has to do with area or perimeter and the same with grass seed.

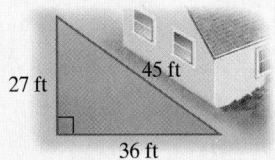

33. A frame shop charges according to both the amount of framing needed to surround the picture and the amount of glass needed to cover the picture.

a. Find the area and perimeter of the trapezoid-shaped framed picture below.

b. Identify whether the amount of framing has to do with perimeter or area and the same with the amount of glass.

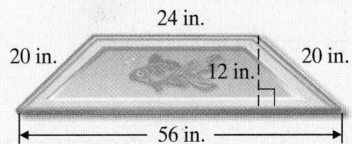

34. A decorator is painting and placing a border completely around the parallelogram-shaped wall.

a. Find the area and perimeter of the wall below.

b. Identify whether the border has to do with perimeter or area and the same with paint.

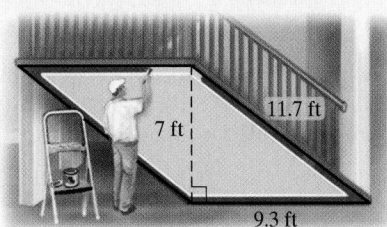

35. Convert Nome, Alaska's 14°F high temperature to Celsius.

36. Convert Paris, France's low temperature of −5°C to Fahrenheit.

37. An architect designs a rectangular flower garden such that the width is exactly two-thirds of the length. If 260 feet of antique picket fencing are to be used to enclose the garden, find the dimensions of the garden.

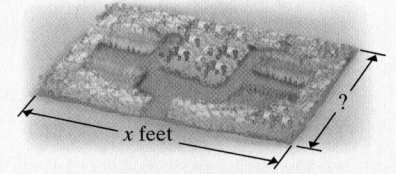

38. If the length of a rectangular parking lot is 10 meters less than twice its width, and the perimeter is 400 meters, find the length of the parking lot.

39. A flower bed is in the shape of a triangle with one side twice the length of the shortest side, and the third side is 30 feet more than the length of the shortest side. Find the dimensions if the perimeter is 102 feet.

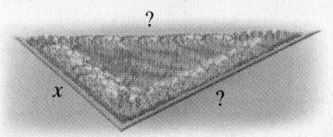

40. The perimeter of a yield sign in the shape of an isosceles triangle is 22 feet. If the shortest side is 2 feet less than the other two sides, find the length of the shortest side. (*Hint:* An isosceles triangle has two sides the same length.)

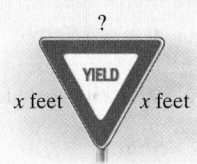

41. The Cat is a high-speed catamaran auto ferry that operates between Bar Harbor, Maine, and Yarmouth, Nova Scotia. The Cat can make the 138-mile trip in about $2\frac{1}{2}$ hours. Find the catamaran speed for this trip. (*Source:* Bay Ferries)

42. A family is planning their vacation to Disney World. They will drive from a small town outside New Orleans, Louisiana, to Orlando, Florida, a distance of 700 miles. They plan to average a rate of 55 mph. How long will this trip take?

△ **43.** Piranha fish require 1.5 cubic feet of water per fish to maintain a healthy environment. Find the maximum number of piranhas you could put in a tank measuring 8 feet by 3 feet by 6 feet.

6 feet

3 feet   8 feet

△ **44.** Find the maximum number of goldfish you can put in a cylindrical tank whose diameter is 8 meters and whose height is 3 meters if each goldfish needs 2 cubic meters water.

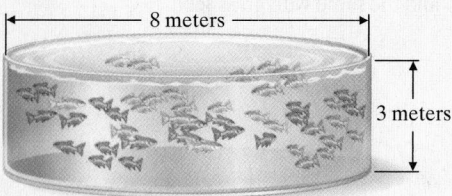

8 meters

3 meters

Dolbear's Law states the relationship between the rate at which Snowy Tree crickets chirp and the air temperature of their environment. The formula is

$$T = 50 + \frac{N - 40}{4}, \text{ where}$$

T = temperature in degrees Fahrenheit and
N = number of chirps per minute

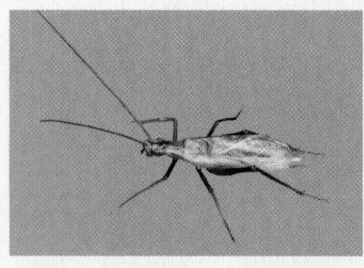

45. If $N = 86$, find the temperature in degrees Fahrenheit, T.

46. If $N = 94$, find the temperature in degrees Fahrenheit, T.

47. If $T = 55°F$, find the number of chirps per minute.

48. If $T = 65°F$, find the number of chirps per minute.

Use the results of Exercises 45–48 to complete each sentence with "increases" or "decreases."

49. As the number of cricket chirps per minute increases, the air temperature of their environment _____.

50. As the air temperature of their environment decreases, the number of cricket chirps per minute _____.

△ **51.** A lawn is in the shape of a trapezoid with a height of 60 feet and bases of 70 feet and 130 feet. How many whole bags of fertilizer must be purchased to cover the lawn if each bag covers 4000 square feet?

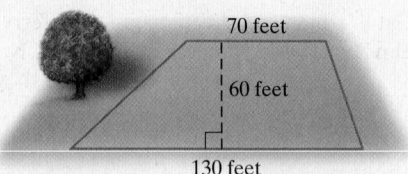

70 feet

60 feet

130 feet

△ **52.** If the area of a right-triangularly shaped sail is 20 square feet and its base is 5 feet, find the height of the sail.

?

5 feet

△ **53.** Maria's Pizza sells one 16-inch cheese pizza or two 10-inch cheese pizzas for $9.99. Determine which size gives more pizza.

16 inches   10 inches   10 inches

△ **54.** Find how much rope is needed to wrap around Earth at the equator if the radius of Earth is 4000 miles. (*Hint:* Use 3.14 for π and the formula for circumference.)

△ **55.** The perimeter of a geometric figure is the sum of the lengths of its sides. If the perimeter of the following pentagon (five-sided figure) is 48 meters, find the length of each side.

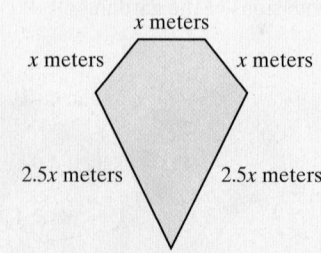

x meters

x meters x meters

2.5x meters   2.5x meters

△ **56.** The perimeter of the following triangle is 82 feet. Find the length of each side.

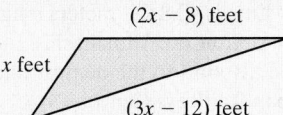

$(2x - 8)$ feet

x feet

$(3x - 12)$ feet

57. The Hawaiian volcano Kilauea is one of the world's most active volcanoes and has had continuous eruptive activity since 1983. Erupting lava flows through a tube system about 11 kilometers to the sea. Assume a lava flow speed of 0.5 kilometer per hour and calculate how long it takes to reach the sea.

58. The world's largest pink ribbon, the sign of the fight against breast cancer, was erected out of pink Post-it® notes on a billboard in New York City in October 2004. If the area of the rectangular billboard covered by the ribbon was approximately 3990 square feet, and the width of the billboard was approximately 57 feet, what was the height of this billboard?

△ **59.** The perimeter of an equilateral triangle is 7 inches more than the perimeter of a square, and the side of the triangle is 5 inches longer than the side of the square. Find the side of the triangle. (*Hint:* An equilateral triangle has three sides the same length.)

△ **60.** A square animal pen and a pen shaped like an equilateral triangle have equal perimeters. Find the length of the sides of each pen if the sides of the triangular pen are fifteen less than twice the sides of the square pen.

61. Find how long it takes a person to drive 135 miles on I-10 if she merges onto I-10 at 10 a.m. and drives nonstop with her cruise control set on 60 mph.

62. Beaumont, Texas, is about 150 miles from Toledo Bend. If Leo Miller leaves Beaumont at 4 a.m. and averages 45 mph, when should he arrive at Toledo Bend?

△ **63.** The longest runway at Los Angeles International Airport has the shape of a rectangle and an area of 1,813,500 square feet. This runway is 150 feet wide. How long is the runway? (*Source:* Los Angeles World Airports)

64. Normal room temperature is about 78°F. Convert this temperature to Celsius.

65. The highest temperature ever recorded in Europe was 122°F in Seville, Spain, in August 1881. Convert this record high temperature to Celsius. (*Source:* National Climatic Data Center)

66. The lowest temperature ever recorded in Oceania was −10°C at the Haleakala Summit in Maui, Hawaii, in January 1961. Convert this record low temperature to Fahrenheit. (*Source:* National Climatic Data Center)

67. The average temperature on the planet Mercury is 167°C. Convert this temperature to degrees Fahrenheit. (*Source:* National Space Science Data Center)

68. The average temperature on the planet Jupiter is −227°F. Convert this temperature to degrees Celsius. Round to the nearest degree. (*Source:* National Space Science Data Center)

△ **69.** The Hoberman Sphere is a toy ball that expands and contracts. When it is completely closed, it has a diameter of 9.5 inches. Find the volume of the Hoberman Sphere when

it is completely closed. Use 3.14 for π. Round to the nearest whole cubic inch. (*Hint:* Volume of a sphere $= \frac{4}{3}\pi r^3$. *Source:* Hoberman Designs, Inc.)

70. When the Hoberman Sphere (see Exercise 69) is completely expanded, its diameter is 30 inches. Find the volume of the Hoberman Sphere when it is completely expanded. Use 3.14 for π. (*Source:* Hoberman Designs, Inc.)

REVIEW AND PREVIEW

Write each percent as a decimal.

71. 32%   **72.** 8%

73. 200%   **74.** 0.5%

Write each decimal as a percent.

75. 0.17   **76.** 0.03

77. 7.2   **78.** 5

CONCEPT EXTENSIONS

△ **79.** The formula $V = lwh$ is used to find the volume of a box. If the length of a box is doubled, the width is doubled, and the height is doubled, how does this affect the volume? Explain your answer.

△ **80.** The formula $A = bh$ is used to find the area of a parallelogram. If the base of a parallelogram is doubled and its height is doubled, how does this affect the area? Explain your answer.

81. Use the Dolbear's Law formula for Exercises 45–48 and calculate when the number of cricket chirps per minute is the same as the temperature in degrees Fahrenheit. (*Hint:* Replace T with N and solve for N or replace N with T and solve for T.)

82. Find the temperature at which the Celsius measurement and the Fahrenheit measurement are the same number.

Solve.

83. $N = R + \dfrac{V}{G}$ for V (Urban forestry: tree plantings per year)

84. $B = \dfrac{F}{P - V}$ for V (Business: break-even point)

Solve. See the Concept Check in this section.

85. ■ − ● · ▮ = ▲   for ●

86. ◣ · ▮ + △ = ●   for ■

87. The distance from the Sun to Earth is approximately 93,000,000 miles. If light travels at a rate of 186,000 miles per second, how long does it take light from the Sun to reach us?

88. Light travels at a rate of 186,000 miles per second. If our moon is 238,860 miles from Earth, how long does it take light from the moon to reach us? (Round to the nearest tenth of a second.)

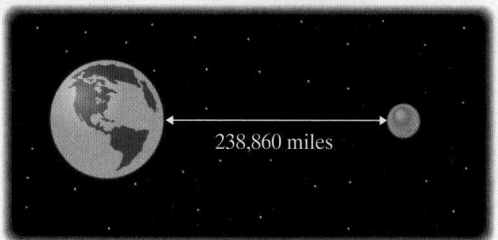

238,860 miles

89. A glacier is a giant mass of rocks and ice that flows downhill like a river. Exit Glacier, near Seward, Alaska, moves at a rate of 20 inches a day. Find the distance in feet the glacier moves in a year. (Assume 365 days a year. Round to 2 decimal places.)

90. Flying fish do not *actually* fly, but glide. They have been known to travel a distance of 1300 feet at a rate of 20 miles per hour. How many seconds does it take to travel this distance? (*Hint:* First convert miles per hour to feet per second. Recall that 1 mile = 5280 feet. Round to the nearest tenth of a second.)

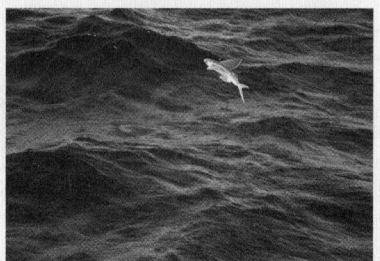

91. A Japanese "bullet" train set a new world record for train speed at 581 kilometers per hour during a manned test run on the Yamanashi Maglev Test Line in 2003. The Yamanashi Maglev Test Line is 42.8 kilometers long. How many *minutes* would a test run on the Yamanashi Line last at this record-setting speed? Round to the nearest hundredth of a minute. (*Source:* Japan Railways Central Co.)

92. The Boeing X-51 is an unmanned demonstration aircraft for hypersonic flight testing. In May 2010, it successfully completed a free flight at about 3800 mph. Neglecting altitude, if the circumference of Earth is approximately 25,000 miles, how long would it take for the X-51 to travel around Earth? Give your answer in hours and minutes rounded to the nearest whole minute.

93. In the United States, a notable hang glider flight was a 303-mile, $8\frac{1}{2}$-hour flight from New Mexico to Kansas. What was the average rate during this flight?

94. Stalactites join stalagmites to form columns. A column found at Natural Bridge Caverns near San Antonio, Texas, rises 15 feet and has a *diameter* of only 2 inches. Find the volume of this column in cubic inches. Round to the nearest tenth of a cubic inch. (*Hint:* Use the formula for volume of a cylinder and use a calculator approximation for π.)

2.6 Percent and Mixture Problem Solving

OBJECTIVES

1 Solve Percent Equations.

2 Solve Discount and Mark-up Problems.

3 Solve Percent of Increase and Percent of Decrease Problems.

4 Solve Mixture Problems.

This section is devoted to solving problems in the categories listed. The same problem-solving steps used in previous sections are also followed in this section. They are listed below for review.

General Strategy for Problem Solving

1. **UNDERSTAND** the problem. During this step, become comfortable with the problem. Some ways of doing this are as follows:

   Read and reread the problem.

   Choose a variable to represent the unknown.

   Construct a drawing whenever possible.

   Propose a solution and check. Pay careful attention to how you check your proposed solution. This will help when writing an equation to model the problem.

2. **TRANSLATE** the problem into an equation.

3. **SOLVE** the equation.

4. **INTERPRET** the results: *Check* the proposed solution in the stated problem and *state* your conclusion.

^{OBJECTIVE}
1 Solving Percent Equations

Many of today's statistics are given in terms of percent: a basketball player's free throw percent, current interest rates, stock market trends, and nutrition labeling, just to name a few. In this section, we first explore percent, percent equations, and applications involving percents. See an appendix if a further review of percents is needed.

EXAMPLE 1   The number 63 is what percent of 72?

Solution

1. UNDERSTAND. Read and reread the problem. Next, let's suppose that the percent is 80%. To check, we find 80% of 72.

$$80\% \text{ of } 72 = 0.80(72) = 57.6$$

This is close but not 63. At this point, though, we have a better understanding of the problem, we know the correct answer is close to and greater than 80%, and we know how to check our proposed solution later.

Let x = the unknown percent.

2. TRANSLATE. Recall that "is" means "equals" and "of" signifies multiplying. Let's translate the sentence directly.

the number 63	is	what percent	of	72
↓	↓	↓	↓	↓
63	=	x	·	72

3. SOLVE.

$$63 = 72x$$
$$0.875 = x \qquad \text{Divide both sides by 72.}$$
$$87.5\% = x \qquad \text{Write as a percent.}$$

4. INTERPRET.

Check: Verify that 87.5% of 72 is 63.

State: The number 63 is 87.5% of 72.

^{PRACTICE}
1   The number 35 is what percent of 56?

EXAMPLE 2   The number 120 is 15% of what number?

Solution

1. UNDERSTAND. Read and reread the problem.

Let x = the unknown number.

2. TRANSLATE.

the number 120	is	15%	of	what number
↓	↓	↓	↓	↓
120	=	15%	·	x

3. SOLVE.

$$120 = 0.15x \quad \text{Write 15\% as 0.15.}$$
$$800 = x \quad \text{Divide both sides by 0.15.}$$

(Continued on next page)

4. INTERPRET.

Check: Check the proposed solution by finding 15% of 800 and verifying that the result is 120.

State: Thus, 120 is 15% of 800.

PRACTICE
2 The number 198 is 55% of what number?

The next example contains a circle graph. This particular circle graph shows percents of pets owned in the United States. Since the circle graph represents all pets owned in the United States, the percents should add to 100%.

> **Helpful Hint**
>
> The percents in a circle graph should have a sum of 100%.

EXAMPLE 3   The circle graph below shows the breakdown of total pets owned in the United States. Use this graph to answer the questions.

Pets Owned in the United States

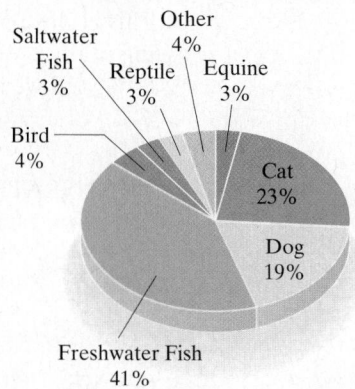

Data from American Pet Products Association's Industry Statistics and Trends, 2014

a. What percent of pets owned in the United States are cats or dogs?

b. What percent of pets owned in the United States are not birds?

c. In 2014, 396.12 million pets were owned in the United States. How many of these were cats? (Round to the nearest tenth of a million.)

Solution

a. From the circle graph, we see that 23% of pets owned are cats and 19% are dogs; thus,

$$23\% + 19\% = 42\%.$$

42% of pets owned are cats or dogs.

b. The circle graph percents have a sum of 100%; thus, the percent of pets owned in the United States that are not birds is

$$100\% - 4\% = 96\%.$$

c. To find the number of cats owned, we find

$$23\% \text{ of } 396.12 = 0.23(396.12)$$
$$= 91.1076$$
$$\approx 91.1 \quad \text{Rounded to the nearest tenth of a million.}$$

Thus, about 91.1 million cats were owned in the United States.

PRACTICE
3 Use the Example 3 circle graph to answer each question.

a. What percent of pets owned in the United States are freshwater fish or saltwater fish?

b. What percent of pets owned in the United States are not equines (horses, ponies, etc.)?

c. In 2014, 396.12 million pets were owned in the United States. How many of these were dogs? (Round to the nearest tenth of a million.)

OBJECTIVE

2 Solving Discount and Mark-up Problems

The next example has to do with discounting the price of a cell phone.

> **EXAMPLE 4**   Cell Phones Unlimited recently reduced the price of a $140 phone by 20%. What are the discount and the new price?

Solution

1. UNDERSTAND. Read and reread the problem. Make sure you understand the meaning of the word *discount*. Discount is the amount of money by which the cost of an item has been decreased. To find the discount, we simply find 20% of $140. In other words, we have the formulas,

$$\text{discount} = \text{percent} \cdot \text{original price} \quad \text{Then}$$

$$\text{new price} = \text{original price} - \text{discount}$$

2, 3. TRANSLATE and **SOLVE.**

$$\begin{aligned}
\text{discount} \quad &= \quad \text{percent} \quad \cdot \quad \text{original price} \\
&= \quad 20\% \quad \cdot \quad \$140 \\
&= \quad 0.20 \quad \cdot \quad \$140 \\
&= \$28
\end{aligned}$$

Thus, the discount in price is $28.

$$\begin{aligned}
\text{new price} \quad &= \quad \text{original price} \quad - \quad \text{discount} \\
&= \quad \$140 \quad - \quad \$28 \\
&= \$112
\end{aligned}$$

4. INTERPRET.

Check: Check your calculations in the formulas; see also whether our results are reasonable. They are.

State: The discount in price is $28, and the new price is $112.   □

PRACTICE

4   A used treadmill, originally purchased for $480, was sold at a garage sale at a discount of 85% of the original price. What were the discount and the new price?   ■

A concept similar to discount is mark-up. What is the difference between the two? A discount is subtracted from the original price while a mark-up is added to the original price. For mark-ups,

$$\text{mark-up} = \text{percent} \cdot \text{original price}$$

$$\text{new price} = \text{original price} + \text{mark-up}$$

Helpful Hint

Discounts are subtracted from the original price, while mark-ups are added.

Mark-up exercises can be found in Exercise Set 2.6 in the form of calculating tips.

OBJECTIVE

3 Solving Percent of Increase and Percent of Decrease Problems

Percent of increase or percent of decrease is a common way to describe how some measurement has increased or decreased. For example, crime increased by 8%, teachers received a 5.5% increase in salary, or a company decreased its number of employees by 10%. The next example is a review of percent of increase.

EXAMPLE 5 Calculating the Percent Increase of Attending College

The tuition and fees cost of attending a public four-year college rose from $4020 in 1996 to $7610 in 2011. Find the percent of increase. Round to the nearest tenth of a percent. (*Source:* The College Board)

Solution

1. UNDERSTAND. Read and reread the problem. Notice that the new tuition, $7610, is almost double the old tuition of $4020. Because of that, we know that the percent of increase is close to 100%. To see this, let's guess that the percent of increase is 100%. To check, we find 100% of $4020 to find the *increase* in cost. Then we add this increase to $4020 to find the *new cost*. In other words, $100\%(\$4020) = 1.00(\$4020) = \$4020$, the *increase* in cost. The *new cost* would be old cost + increase = $4020 + $4020 = $8040, close to the actual new cost of $7610. We now know that the increase is close to, but less than, 100% and we know how to check our proposed solution.

   Let x = the percent of increase.

2. TRANSLATE. First, find the **increase** and then the **percent of increase.** The increase in cost is found by:

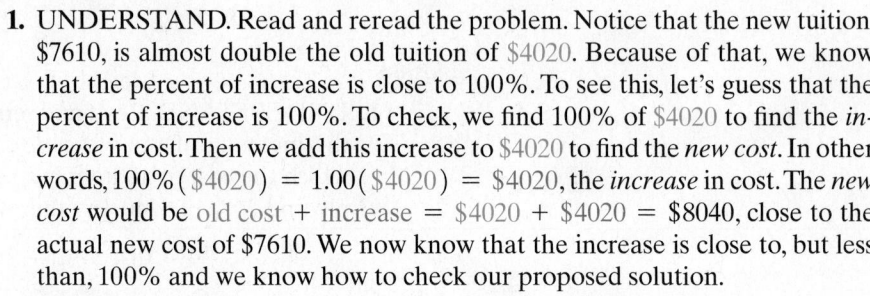

In words: increase = new cost − old cost

Translate: increase = $7610 − $4020

                        = $3590

   Next, find the percent of increase. The percent of increase or percent of decrease is always a percent of the original number or, in this case, the old cost.

In words: increase is what percent of old cost

Translate: $3590 = x · $4020

3. SOLVE.

   $3590 = 4020x$

   $0.893 \approx x$ Divide both sides by 4020 and round to 3 decimal places.

   $89.3\% \approx x$ Write as a percent.

4. INTERPRET.

   Check: Check the proposed solution

   State: The percent of increase in cost is approximately 89.3%. ☐

PRACTICE

5 The tuition and fees cost of attending a public two-year college rose from $1900 in 1996 to $2710 in 2011. Find the percent of increase. Round to the nearest tenth of a percent. (*Source:* The College Board) ■

Percent of decrease is found using a method similar to that in Example 5. First find the decrease, then determine what percent of the original or first amount is that decrease.

Read the next example carefully. For Example 5, we were asked to find percent of increase. In Example 6, we are given the percent of increase and asked to find the number before the increase.

EXAMPLE 6   Most of the movie screens globally project analog film, but the number of cinemas using digital is increasing. Find the number of digital screens worldwide in 2012 if, after a 25% increase, the number in 2013 was 111,809. Round to the nearest whole number. (*Source: Motion Picture Association of America*)

Solution

1. UNDERSTAND. Read and reread the problem. Let's guess a solution and see how we would check our guess. If the number of digital screens worldwide in 2012 was 80,000, we would see if 80,000 plus the increase is 111,809; that is,

$$80{,}000 + 25\%(80{,}000) = 80{,}000 + 0.25(80{,}000) = 80{,}000 + 20{,}000 = 100{,}000$$

Since 100,000 is too small, we know that our guess of 80,000 is too small. We also have a better understanding of the problem. Let

$$x = \text{number of digital screens in 2012.}$$

2. TRANSLATE. To translate into an equation, we remember that

In words:	number of digital screens in 2012	plus	increase	equals	number of digital screens in 2013
Translate:	x	$+$	$0.25x$	$=$	$111{,}809$

3. SOLVE.

$$1.25x = 111{,}809 \quad \text{Add like terms.}$$
$$x = \frac{111{,}809}{1.25}$$
$$x \approx 89{,}447$$

4. INTERPRET.

Check: Recall that x represents the number of digital screens worldwide in 2012. If this number is approximately 89,447, let's see if 89,447 plus the increase is close to 111,809. (We use the word "close" because it is rounded.)

$$89{,}447 + 25\%(89{,}447) = 89{,}447 + 0.25(89{,}447) = 1.25(89{,}447) = 111{,}808.75$$

which is close to 111,809.

State: There were approximately 89,447 digital screens worldwide in 2012.   □

PRACTICE
6   The fastest-growing sector of digital theater screens is 3D. Find the number of digital 3D screens in the United States and Canada in 2012 if, after a 7% increase, the number in 2013 was 15,782. Round to the nearest whole. (*Source:* MPAA)

OBJECTIVE
4   Solving Mixture Problems

Mixture problems involve two or more quantities being combined to form a new mixture. These applications range from Dow Chemical's need to form a chemical mixture of a required strength to Planter's Peanut Company's need to find the correct mixture of peanuts and cashews, given taste and price constraints.

EXAMPLE 7   Calculating Percent for a Lab Experiment

A chemist working on his doctoral degree at Massachusetts Institute of Technology needs 12 liters of a 50% acid solution for a lab experiment. The stockroom has only 40% and 70% solutions. How much of each solution should be mixed together to form 12 liters of a 50% solution?

(Continued on next page)

Solution

1. **UNDERSTAND.** First, read and reread the problem a few times. Next, guess a solution. Suppose that we start with 7 liters of the 40% solution. Then we need $12 - 7 = 5$ liters of the 70% solution. To see if this is indeed the solution, find the amount of pure acid in 7 liters of the 40% solution, in 5 liters of the 70% solution, and in 12 liters of a 50% solution, the required amount and strength.

number of liters	×	acid strength	=	amount of pure acid
7 liters	×	40%		7(0.40) or 2.8 liters
5 liters	×	70%		5(0.70) or 3.5 liters
12 liters	×	50%		12(0.50) or 6 liters

Since 2.8 liters $+ 3.5$ liters $= 6.3$ liters and not 6, our guess is incorrect, but we have gained some valuable insight into how to model and check this problem.

Let

$$x = \text{number of liters of 40\% solution; then}$$
$$12 - x = \text{number of liters of 70\% solution.}$$

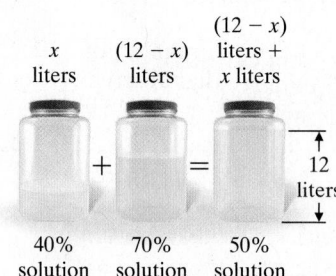

x liters $(12 - x)$ liters $= $ $(12 - x)$ liters $+$ x liters $= 12$ liters

40% solution $+$ 70% solution $=$ 50% solution

2. **TRANSLATE.** To help us translate into an equation, the following table summarizes the information given. Recall that the amount of acid in each solution is found by multiplying the acid strength of each solution by the number of liters.

	No. of Liters · Acid Strength = Amount of Acid		
40% Solution	x	40%	$0.40x$
70% Solution	$12 - x$	70%	$0.70(12 - x)$
50% Solution Needed	12	50%	$0.50(12)$

The amount of acid in the final solution is the sum of the amounts of acid in the two beginning solutions.

In words: acid in 40% solution $+$ acid in 70% solution $=$ acid in 50% mixture

Translate: $0.40x + 0.70(12 - x) = 0.50(12)$

3. **SOLVE.**

$$0.40x + 0.70(12 - x) = 0.50(12)$$
$$0.4x + 8.4 - 0.7x = 6 \qquad \text{Apply the distributive property.}$$
$$-0.3x + 8.4 = 6 \qquad \text{Combine like terms.}$$
$$-0.3x = -2.4 \qquad \text{Subtract 8.4 from both sides.}$$
$$x = 8 \qquad \text{Divide both sides by } -0.3.$$

4. **INTERPRET.**

Check: To check, recall how we checked our guess.

State: If 8 liters of the 40% solution are mixed with $12 - 8$ or 4 liters of the 70% solution, the result is 12 liters of a 50% solution. □

PRACTICE
7 Hamida Barash was responsible for refilling the eyewash stations in the chemistry lab. She needed 6 liters of 3% strength eyewash to refill the dispensers. The supply room only had 2% and 5% eyewash in stock. How much of each solution should she mix to produce the needed 3% strength eyewash?

Vocabulary, Readiness & Video Check

Tell whether the percent labels in the circle graphs are correct.

1.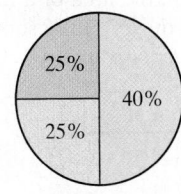
2.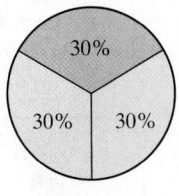
3.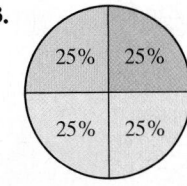
4.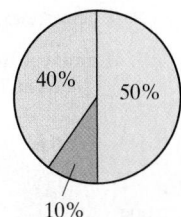

Martin-Gay Interactive Videos

See Video 2.6

Watch the section lecture video and answer the following questions.

OBJECTIVE 1
5. Answer these questions based on how Example 2 was translated into an equation.
  a. What does "is" translate to?
  b. What does "of" translate to?
  c. How do you write a percent as an equivalent decimal?

OBJECTIVE 2
6. At the end of Example 3 you are told that the process for finding discount is *almost* the same as finding mark-up.
  a. How is discount similar?
  b. How does discount differ?

OBJECTIVE 3
7. According to Example 4, what amount must you find before you can find a percent of increase in price? How do you find this amount?

OBJECTIVE 4
8. The following problem is worded like Example 6 in the video, but using different quantities.

How much of an alloy that is 10% copper should be mixed with 400 ounces of an alloy that is 30% copper in order to get an alloy that is 20% copper? Fill in the table and set up an equation that could be used to solve for the unknowns (do not actually solve). Use Example 6 in the video as a model for your work.

Alloy	Ounces	Copper Strength	Amount of Copper

2.6   Exercise Set   MyMathLab

Find each number described. For Exercises 1 and 2, the solutions have been started for you. See Examples 1 and 2.

1. What number is 16% of 70?

Start the solution:

1. UNDERSTAND the problem. Reread it as many times as needed.
2. TRANSLATE into an equation. (Fill in the blanks below.)

What number	is	16%	of	70
↓	↓	↓	↓	↓
x	___	0.16	___	70

Finish with:
3. SOLVE and **4.** INTERPRET

2. What number is 88% of 1000?

Start the solution:

1. UNDERSTAND the problem. Reread it as many times as needed.
2. TRANSLATE into an equation. (Fill in the blanks below.)

What number	is	88%	of	1000
↓	↓	↓	↓	↓
x	___	0.88	___	1000

Finish with:
3. SOLVE and **4.** INTERPRET

3. The number 28.6 is what percent of 52?

4. The number 87.2 is what percent of 436?

▶ **5.** The number 45 is 25% of what number?

6. The number 126 is 35% of what number?

The circle graph below shows each continent/region's percent of Earth's land. Use this graph for Exercises 7 through 10. See Example 3.

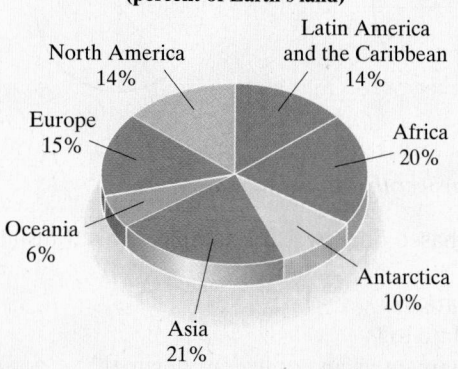

Continent/Region Land Area
(percent of Earth's land)

7. What percent of land is not included in North America?

8. What percent of land is in Asia, Antarctica, or Africa?

9. If Earth's land area is 56.4 million square miles, find the land area of Europe.

10. If Earth's land area is 56.4 million square miles, find the land area of Africa.

Solve. If needed, round answers to the nearest cent. See Example 4.

11. A used automobile dealership recently reduced the price of a used compact car by 8%. If the price of the car before discount was $18,500, find the discount and the new price.

12. A music store is advertising a 25%-off sale on all new releases. Find the discount and the sale price of a newly released CD that regularly sells for $12.50.

13. A birthday celebration meal is $40.50 including tax. Find the total cost if a 15% tip is added to the given cost.

▶ **14.** A retirement dinner for two is $65.40 including tax. Find the total cost if a 20% tip is added to the given cost.

Solve. See Example 5.

Use the graph below for Exercises 15 and 16.

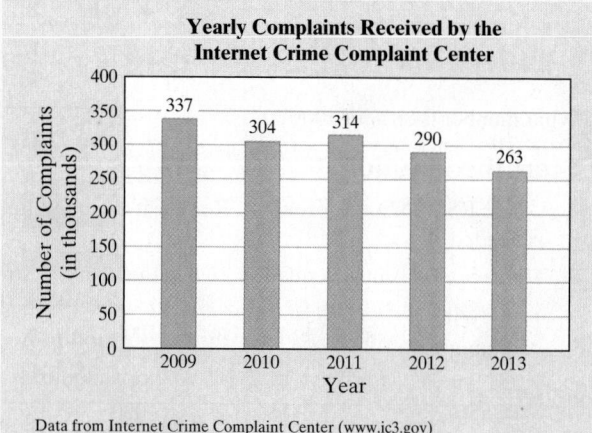

Yearly Complaints Received by the
Internet Crime Complaint Center

Data from Internet Crime Complaint Center (www.ic3.gov)

15. The number of Internet-crime complaints decreased from 2012 to 2013. Find the percent of decrease. Round to the nearest tenth of a percent.

16. The number of Internet-crime complaints decreased from 2011 to 2012. Find the percent of decrease. Round to the nearest tenth of a percent.

17. By decreasing each dimension by 1 unit, the area of a rectangle decreased from 40 square feet (on the left) to 28 square feet (on the right). Find the percent of decrease in area.

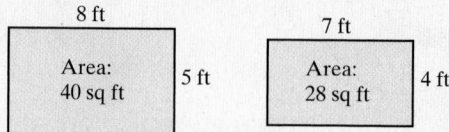

18. By decreasing the length of the side by one unit, the area of a square decreased from 100 square meters to 81 square meters. Find the percent of decrease in area.

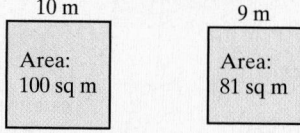

Solve. See Example 6.

19. Find the original price of a pair of shoes if the sale price is $78 after a 25% discount.

20. Find the original price of a popular pair of shoes if the increased price is $80 after a 25% increase.

▶ **21.** Find last year's salary if, after a 4% pay raise, this year's salary is $44,200.

22. Find last year's salary if, after a 3% pay raise, this year's salary is $55,620.

Solve. For each exercise, a table is given for you to complete and use to write an equation that models the situation. See Example 7.

23. How much pure acid should be mixed with 2 gallons of a 40% acid solution in order to get a 70% acid solution?

	Number of Gallons	·	Acid Strength	=	Amount of Acid
Pure Acid			100%		
40% Acid Solution					
70% Acid Solution Needed					

24. How many cubic centimeters (cc) of a 25% antibiotic solution should be added to 10 cubic centimeters of a 60% antibiotic solution to get a 30% antibiotic solution?

	Number of Cubic cm	·	Antibiotic Strength	=	Amount of Antibiotic
25% Antibiotic Solution					
60% Antibiotic Solution					
30% Antibiotic Solution Needed					

25. Community Coffee Company wants a new flavor of Cajun coffee. How many pounds of coffee worth $7 a pound should be added to 14 pounds of coffee worth $4 a pound to get a mixture worth $5 a pound?

	Number of Pounds	·	Cost per Pound	=	Value
$7 per lb Coffee					
$4 per lb Coffee					
$5 per lb Coffee Wanted					

26. Planter's Peanut Company wants to mix 20 pounds of peanuts worth $3 a pound with cashews worth $5 a pound in order to make an experimental mix worth $3.50 a pound. How many pounds of cashews should be added to the peanuts?

	Number of Pounds	·	Cost per Pound	=	Value
$3 per lb Peanuts					
$5 per lb Cashews					
$3.50 per lb Mixture Wanted					

MIXED PRACTICE

Solve. If needed, round money amounts to two decimal places and all other amounts to one decimal place. See Examples 1 through 7.

27. Find 23% of 20.

28. Find 140% of 86.

29. The number 40 is 80% of what number?

30. The number 56.25 is 45% of what number?

31. The number 144 is what percent of 480?

32. The number 42 is what percent of 35?

The average American spends 162 minutes per day on his or her mobile device. The graph below is a result of mobile measurement research on how Americans spend their time daily on their connected iOS or Android devices. Use this graph for Exercises 33 through 36.

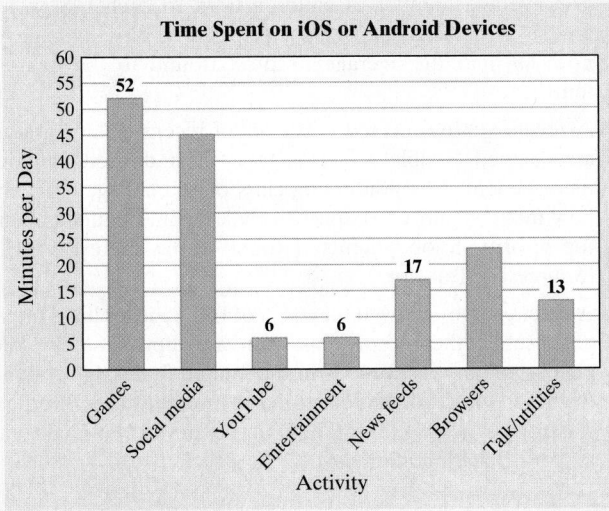

Time Spent on iOS or Android Devices

Data from Flurry Analytics

33. Estimate the amount of time the average American spends on social media daily.

34. Estimate the amount of time spent on Internet browsers.

35. What percent of the time the average American spends on connected devices is devoted to games? (Round to the nearest tenth of a percent.)

36. As newspaper circulation has decreased, more and more people are reading news items online by checking Internet news sites, television station sites, and other news-collecting sites. What percent of the average American's online time is spent following news? (Round to the nearest tenth of a percent.)

For Exercises 37 and 38, fill in the percent column in each table. Each table contains a worked-out example.

37.

Top Cranberry-Producing States in 2014
(in millions of pounds)

State	Millions of Pounds	Percent of Total (rounded to nearest percent)
Wisconsin	539	
Oregon	40	
Massachusetts	206	
New Jersey	56	Example: $\frac{56}{857} \approx 7\%$
Washington	16	
Total	857	

Data from National Agricultural Statistics Service

38.

2013 Worldwide Unit Case Volume for Coca-Cola
(in billions of cases)

World Region	Case Volume	Percent of Total (rounded to nearest percent)
North America	5.9	
Latin America	8.2	
Europe	3.9	Example: $\frac{3.9}{28.2} \approx 14\%$
Eurasia and Africa	4.3	
Pacific	5.9	
Total	28.2	

Data from Coca-Cola Company

39. Nordstrom advertised a 25%-off sale. If a London Fog coat originally sold for $256, find the decrease in price and the sale price.

40. A gasoline station decreased the price of a $0.95 cola by 15%. Find the decrease in price and the new price.

41. In 2010, Americans consumed approximately 11.1 pounds of romaine and leaf lettuce per capita. By 2013, consumption had dropped to 10.3 pounds per capita. Find the percent of decrease. Round to the nearest tenth of a percent. (*Source:* USDA)

42. Due to arid conditions in lettuce-growing areas at the end of 2014, the cost of growing and shipping iceberg lettuce to stores was about $1.19. In certain areas, consumers purchased lettuce for about $1.49 a head. Find the percent of increase. Round to the nearest tenth of a percent. (*Source:* USDA)

43. The number of registered vehicles on the road in the United States is constantly increasing. In 2013, there were approximately 256 million registered vehicles. This represents an increase of 2% over 2006. How many registered vehicles were there in the United States in 2006? Round to the nearest million. (*Source:* Federal Highway Safety Administration)

44. A student at the University of New Orleans makes money by buying and selling used cars. Charles bought a used car and later sold it for a 20% profit. If he sold it for $4680, how much did Charles pay for the car?

45. By doubling each dimension, the area of a parallelogram increased from 36 square centimeters to 144 square centimeters. Find the percent of increase in area.

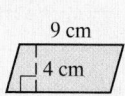 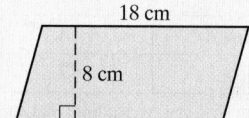

9 cm   4 cm   18 cm   8 cm

46. By doubling each dimension, the area of a triangle increased from 6 square miles to 24 square miles. Find the percent of increase in area.

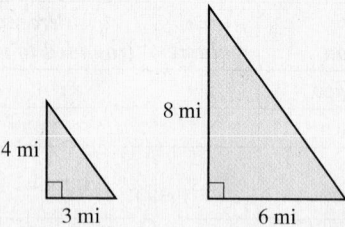

4 mi   3 mi   8 mi   6 mi

47. How much of an alloy that is 20% copper should be mixed with 200 ounces of an alloy that is 50% copper to get an alloy that is 30% copper?

48. How much water should be added to 30 gallons of a solution that is 70% antifreeze to get a mixture that is 60% antifreeze?

49. The number of farms in the United States is decreasing. In 1982 there were approximately 2.5 million farms, while in 2014, there were only approximately 2.1 million farms. Find the percent of decrease in the number of farms. (*Source:* USDA agricultural census)

50. During the 2007–2008 term, the Supreme Court made 72 decisions, while during the 2013–2014 term they made 88 decisions. Find the percent of increase in the number of decisions. Round to the nearest tenth of a percent. (*Source:* supremecourt.gov)

51. A company recently downsized its number of employees by 35%. If there are still 78 employees, how many employees were there prior to the layoffs?

52. The average number of children born to each U.S. woman has decreased by 44% since 1920. If this average is now 1.9, find the average in 1920. Round to the nearest tenth.

53. A recent survey showed that 42% of recent college graduates named flexible hours as their most desired employment benefit. In a graduating class of 860 college students, how many would you expect to rank flexible hours as their top priority in job benefits? (Round to the nearest whole.) (*Source:* JobTrak.com)

54. A recent survey showed that 64% of U.S. colleges have Internet access in their classrooms. There are approximately 9800 post-secondary institutions in the United States. How many of these would you expect to have Internet access in their classrooms? (*Source:* Market Data Retrieval, National Center for Education Statistics)

55. A new self-tanning lotion for everyday use is to be sold. First, an experimental lotion mixture is made by mixing 800 ounces of everyday moisturizing lotion worth $0.30 an ounce with self-tanning lotion worth $3 per ounce. If the experimental lotion is to cost $1.20 per ounce, how many ounces of the self-tanning lotion should be in the mixture?

56. The owner of a local chocolate shop wants to develop a new trail mix. How many pounds of chocolate-covered peanuts worth $5 a pound should be mixed with 10 pounds of granola bites worth $2 a pound to get a mixture worth $3 per pound?

57. Scoville units are used to measure the hotness of a pepper. Measuring 577 thousand Scoville units, the "Red Savina" habañero pepper was known as the hottest chili pepper. That changed with the discovery of the Naga Jolokia pepper from India. It measures 48% hotter than the habañero. Find the measure of the Naga Jolokia pepper. Round to the nearest thousand units.

58. As of this writing, the women's record for throwing a disc (like a heavy Frisbee) was set by Valarie Jenkins of the United States in 2008. Her throw was 148.00 meters. The men's world record was set by Christian Sandstrom of Sweden in 2002. His throw was 68.9% farther than Valarie's. Find the distance of his throw. Round to the nearest meter. (*Source:* World Flying Disc Federation)

REVIEW AND PREVIEW

Place $<, >,$ *or* $=$ *in the appropriate space to make each a true statement. See Sections 1.2, 1.3, 1.5, and 1.7.*

59. $-5 \quad -7$

60. $\dfrac{12}{3} \quad 2^2$

61. $|-5| \quad -(-5)$

62. $-3^3 \quad (-3)^3$

63. $(-3)^2 \quad -3^2$

64. $|-2| \quad -|-2|$

CONCEPT EXTENSIONS

65. Is it possible to mix a 10% acid solution and a 40% acid solution to obtain a 60% acid solution? Why or why not?

66. Must the percents in a circle graph have a sum of 100%? Why or why not?

67. A trail mix is made by combining peanuts worth $3 a pound, raisins worth $2 a pound, and M&M's worth $4 a pound. Would it make good business sense to sell the trail mix for $1.98 a pound? Why or why not?

68. a. Can an item be marked up by more than 100%? Why or why not?

   b. Can an item be discounted by more than 100%? Why or why not?

Standardized nutrition labels like the one below have been displayed on food items since 1994. The percent column on the right shows the percent of daily values (based on a 2000-calorie diet) shown at the bottom of the label. For example, a serving of this food contains 4 grams of total fat when the recommended daily fat based on a 2000-calorie diet is less than 65 grams of fat. This means that $\dfrac{4}{65}$ or approximately 6% (as shown) of your daily recommended fat is taken in by eating a serving of this food. Use this nutrition label to answer Exercises 69 through 71.

Nutrition Facts

Serving Size 18 Crackers (31g)
Servings Per Container About 9

Amount Per Serving

Calories 130   Calories from Fat 35

% Daily Value*

Total Fat 4g	**6%**
Saturated Fat 0.5g	**3%**
Polyunsaturated Fat 0g	
Monounsaturated Fat 1.5g	
Cholesterol 0mg	**0%**
Sodium 230mg	*x*
Total Carbohydrate 23g	*y*
Dietary Fiber 2g	**8%**
Sugars 3g	
Protein 2g	

Vitamin A 0%   •   Vitamin C 0%
Calcium 2%   •   Iron 6%

* Percent Daily Values are based on a 2,000 calorie diet. Your daily values may be higher or lower depending on your calorie needs.

	Calories	2,000	2,500
Total Fat	Less than	65g	80g
Sat. Fat	Less than	20g	25g
Cholesterol	Less than	300mg	300mg
Sodium	Less than	2400mg	2400mg
Total Carbohydrate		300g	375g
Dietary Fiber		25g	30g

69. Based on a 2000-calorie diet, what percent of daily value of sodium is contained in a serving of this food? In other words, find *x* in the label. (Round to the nearest tenth of a percent.)

70. Based on a 2000-calorie diet, what percent of daily value of total carbohydrate is contained in a serving of this food? In other words, find *y* in the label. (Round to the nearest tenth of a percent.)

71. Notice on the nutrition label that one serving of this food contains 130 calories and 35 of these calories are from fat. Find the percent of calories from fat. (Round to the nearest tenth of a percent.) It is recommended that no more than 30% of calorie intake come from fat. Does this food satisfy this recommendation?

Use the nutrition label below to answer Exercises 72 through 74.

NUTRITIONAL INFORMATION PER SERVING

Serving Size: 9.8 oz   Servings Per Container: 1

Calories 280	Polyunsaturated Fat 1g
Protein 12g	Saturated Fat 3g
Carbohydrate 45g	Cholesterol 20mg
Fat 6g	Sodium 520mg
Percent of Calories from Fat....?	Potassium 220mg

72. If fat contains approximately 9 calories per gram, find the percent of calories from fat in one serving of this food. (Round to the nearest tenth of a percent.)

73. If protein contains approximately 4 calories per gram, find the percent of calories from protein from one serving of this food. (Round to the nearest tenth of a percent.)

74. Find a food that contains more than 30% of its calories per serving from fat. Analyze the nutrition label and verify that the percents shown are correct.

2.7 Further Problem Solving

OBJECTIVES

1 Solve Problems Involving Distance. ▷

2 Solve Problems Involving Money. ▷

3 Solve Problems Involving Interest. ▷

This section is devoted to solving problems in the categories listed. The same problem-solving steps used in previous sections are also followed in this section. They are listed below for review.

General Strategy for Problem Solving

1. **UNDERSTAND** the problem. During this step, become comfortable with the problem. Some ways of doing this are to:

   Read and reread the problem.

   Choose a variable to represent the unknown.

   Construct a drawing whenever possible.

   Propose a solution and check. Pay careful attention to how you check your proposed solution. This will help when writing an equation to model the problem.

2. **TRANSLATE** the problem into an equation.

3. **SOLVE** the equation.

4. **INTERPRET** the results: *Check* the proposed solution in the stated problem and *state* your conclusion.

OBJECTIVE

1 Solving Distance Problems

Our first example involves distance. For a review of the distance formula, $d = r \cdot t$, see Section 2.5, Example 1, and the table before the example.

EXAMPLE 1   Finding Time Given Rate and Distance

Marie Antonio, a bicycling enthusiast, rode her 21-speed at an average speed of 18 miles per hour on level roads and then slowed down to an average of 10 mph on the hilly roads of the trip. If she covered a distance of 98 miles, how long did the entire trip take if traveling the level roads took the same time as traveling the hilly roads?

Solution

1. **UNDERSTAND** the problem. To do so, read and reread the problem. The formula $d = r \cdot t$ is needed. At this time, let's guess a solution. Suppose that she spent 2 hours traveling on the level roads. This means that she also spent 2 hours traveling on the hilly roads, since the times spent were the same. What is her total distance? Her distance on the level road is rate · time $= 18(2) = 36$ miles. Her distance on the hilly roads is rate · time $= 10(2) = 20$ miles. This gives a total distance of 36 miles + 20 miles = 56 miles, not the correct distance of 98 miles. Remember that the purpose of guessing a solution is not to guess correctly (although this may happen) but to help understand the problem better and how to model it with an equation. We are looking for the length of the entire trip, so we begin by letting

   $$x = \text{the time spent on level roads.}$$

   Because the same amount of time is spent on hilly roads, then also

   $$x = \text{the time spent on hilly roads.}$$

2. **TRANSLATE.** To help us translate into an equation, we now summarize the information from the problem on the following chart. Fill in the rates given and the variables used to represent the times and use the formula $d = r \cdot t$ to fill in the distance column.

	Rate ·	Time =	Distance
Level	18	x	$18x$
Hilly	10	x	$10x$

Since the entire trip covered 98 miles, we have that

In words: | total distance | = | level distance | + | hilly distance |

Translate: 98   =   18x   +   10x

3. SOLVE.

$$98 = 28x \quad \text{Add like terms.}$$

$$\frac{98}{28} = \frac{28x}{28} \quad \text{Divide both sides by 28.}$$

$$3.5 = x$$

4. INTERPRET the results.

Check: Recall that x represents the time spent on the level portion of the trip and the time spent on the hilly portion. If Marie rides for 3.5 hours at 18 mph, her distance is $18(3.5) = 63$ miles. If Marie rides for 3.5 hours at 10 mph, her distance is $10(3.5) = 35$ miles. The total distance is 63 miles + 35 miles = 98 miles, the required distance.

State: The time of the entire trip is then 3.5 hours + 3.5 hours or 7 hours. □

PRACTICE
1   Sat Tranh took a short hike with his friends up Mt. Wachusett. They hiked uphill at a steady pace of 1.5 miles per hour and downhill at a rate of 4 miles per hour. If the time to climb the mountain took an hour more than the time to hike down, how long was the entire hike? ■

EXAMPLE 2   **Finding Train Speeds**

The Kansas City Southern Railway operates in 10 states and Mexico. Suppose two trains leave Neosho, Missouri, at the same time. One travels north and the other travels south at a speed that is 15 miles per hour faster. In 2 hours, the trains are 230 miles apart. Find the speed of each train.

Kansas City Southern Railway

Solution

1. **UNDERSTAND** the problem. Read and reread the problem. Guess a solution and check. Let's let

$$x = \text{speed of train traveling north.}$$

Because the train traveling south is 15 mph faster, we have

$$x + 15 = \text{speed of train traveling south.}$$

2. **TRANSLATE.** Just as for Example 1, let's summarize our information on a chart. Use the formula $d = r \cdot t$ to fill in the distance column.

	r	$\cdot$ $t =$	d
North Train	x	2	$2x$
South Train	$x + 15$	2	$2(x + 15)$

Since the total distance between the trains is 230 miles, we have

In words: | north train distance | + | south train distance | = | total distance |

Translate: $2x$ + $2(x + 15)$ =   230

(Continued on next page)

3. SOLVE. $2x + 2x + 30 = 230$ Use the distributive property.

$4x + 30 = 230$ Combine like terms.

$4x = 200$ Subtract 30 from both sides.

$\dfrac{4x}{4} = \dfrac{200}{4}$ Divide both sides by 4.

$x = 50$ Simplify.

4. INTERPRET the results.

Check: Recall that x is the speed of the train traveling north, or 50 mph. In 2 hours, this train travels a distance of $2(50) = 100$ miles. The speed of the train traveling south is $x + 15$ or $50 + 15 = 65$ mph. In 2 hours, this train travels $2(65) = 130$ miles. The total distance of the trains is 100 miles $+ 130$ miles $= 230$ miles, the required distance.

State: The northbound train's speed is 50 mph and the southbound train's speed is 65 mph. □

PRACTICE

2 The Kansas City Southern Railway has a station in Mexico City, Mexico. Suppose two trains leave Mexico City at the same time. One travels east and the other west at a speed that is 10 mph slower. In 1.5 hours, the trains are 171 miles apart. Find the speed of each train. ■

OBJECTIVE

2 Solving Money Problems

The next example has to do with finding an unknown number of a certain denomination of coin or bill. These problems are extremely useful in that they help you understand the difference between the number of coins or bills and the total value of the money.

For example, suppose there are seven $5 bills. The *number* of $5 bills is 7 and the *total value* of the money is $5(7) = $35.

Study the table below for more examples.

Denomination of Coin or Bill	Number of Coins or Bills	Value of Coins or Bills
20-dollar bills		$20(17) = $340
nickels		$0.05(31) = $1.55
quarters		$0.25(x) = $0.25x

EXAMPLE 3   **Finding Numbers of Denominations**

Part of the proceeds from a local talent show was $2420 worth of $10 and $20 bills. If there were 37 more $20 bills than $10 bills, find the number of each denomination.

Solution

1. UNDERSTAND the problem. To do so, read and reread the problem. If you'd like, let's guess a solution. Suppose that there are 25 $10 bills. Since there are 37 more $20 bills, we have $25 + 37 = 62$ $20 bills. The total amount of money is $10(25) + $20(62) = 1490, below the given amount of $2420. Remember that our purpose for guessing is to help us understand the problem better.

We are looking for the number of each denomination, so we let

$$x = \text{number of } \$10 \text{ bills.}$$

There are 37 more $20 bills, so

$$x + 37 = \text{number of } \$20 \text{ bills.}$$

2. TRANSLATE. To help us translate into an equation, study the table below.

Denomination	Number of Bills	Value of Bills (in dollars)
$10 bills	x	$10x$
$20 bills	$x + 37$	$20(x + 37)$

Since the total value of these bills is $2420, we have

In words:

value of $10 bills   plus   value of $20 bills   is   2420

Translate: $10x$ $+$ $20(x + 37)$ $=$ 2420

3. SOLVE:

$$10x + 20x + 740 = 2420 \quad \text{Use the distributive property.}$$
$$30x + 740 = 2420 \quad \text{Add like terms.}$$
$$30x = 1680 \quad \text{Subtract 740 from both sides.}$$
$$\frac{30x}{30} = \frac{1680}{30} \quad \text{Divide both sides by 30.}$$
$$x = 56$$

4. INTERPRET the results.

Check: Since x represents the number of $10 bills, we have 56 $10 bills and $56 + 37$, or 93, $20 bills. The total amount of these bills is $\$10(56) + \$20(93) = \$2420$, the correct total.

State: There are 56 $10 bills and 93 $20 bills.   □

PRACTICE

3 A stack of $5 and $20 bills was counted by the treasurer of an organization. The total value of the money was $1710 and there were 47 more $5 bills than $20 bills. Find the number of each type of bill.   ■

OBJECTIVE

3 Solving Interest Problems

The next example is an investment problem. For a review of the simple interest formula, $I = PRT$, see the table at the beginning of Section 2.5 and Exercises 11 and 12 in that exercise set.

EXAMPLE 4   Finding the Investment Amount

Rajiv Puri invested part of his $20,000 inheritance in a mutual fund account that pays 7% simple interest yearly and the rest in a certificate of deposit that pays 9% simple interest yearly. At the end of one year, Rajiv's investments earned $1550. Find the amount he invested at each rate.

Solution

1. UNDERSTAND. Read and reread the problem. Next, guess a solution. Suppose that Rajiv invested $8000 in the 7% fund and the rest, $12,000, in the fund paying 9%. To check, find his interest after one year. Recall the formula $I = PRT$, so the interest from the 7% fund $= \$8000(0.07)(1) = \$560$. The interest from the 9% fund $= \$12,000(0.09)(1) = \$1080$. The sum of the interests is $\$560 + \$1080 = \$1640$. Our guess is incorrect, since the sum of the interests is not $1550, but we now have a better understanding of the problem.

Let

$$x = \text{amount of money in the account paying 7\%.}$$

The rest of the money is $20,000 less x or

$$20,000 - x = \text{amount of money in the account paying 9\%.}$$

(Continued on next page)

2. TRANSLATE. We apply the simple interest formula $I = PRT$ and organize our information in the following chart. Since there are two rates of interest and two amounts invested, we apply the formula twice.

	Principal ·	Rate ·	Time =	Interest
7% Fund	x	0.07	1	$x(0.07)(1)$ or $0.07x$
9% Fund	$20,000 - x$	0.09	1	$(20,000 - x)(0.09)(1)$ or $0.09(20,000 - x)$
Total	20,000			1550

The total interest earned, $1550, is the sum of the interest earned at 7% and the interest earned at 9%.

In words: interest at 7% + interest at 9% = total interest

Translate: $0.07x$ + $0.09(20,000 - x)$ = 1550

3. SOLVE.

$$0.07x + 0.09(20,000 - x) = 1550$$

$0.07x + 1800 - 0.09x = 1550$	Apply the distributive property.
$1800 - 0.02x = 1550$	Combine like terms.
$-0.02x = -250$	Subtract 1800 from both sides.
$x = 12,500$	Divide both sides by -0.02.

4. INTERPRET.

Check: If $x = 12,500$, then $20,000 - x = 20,000 - 12,500$ or 7500. These solutions are reasonable, since their sum is $20,000 as required. The annual interest on $12,500 at 7% is $875; the annual interest on $7500 at 9% is $675, and $875 + $675 = $1550.

State: The amount invested at 7% was $12,500. The amount invested at 9% was $7500.   □

PRACTICE
4   Suzanne Scarpulla invested $30,000, part of it in a high-risk venture that yielded 11.5% per year and the rest in a secure mutual fund paying interest of 6% per year. At the end of one year, Suzanne's investments earned $2790. Find the amount she invested at each rate.   ■

✓ Vocabulary, Readiness & Video Check

Martin-Gay Interactive Videos

See Video 2.7 ◉

Watch the section lecture video and answer the following questions.

OBJECTIVE
1

1. The following problem is worded like 🎞 Example 1 but using different quantities.

How long will it take a bus traveling at 55 miles per hour to overtake a car traveling at 50 mph if the car had a 3 hour head start? Fill in the table and set up an equation that could be used to solve for the unknown (do not actually solve). Use 🎞 Example 1 in the video as a model for your work.

	r ·	t =	d
bus			
car			

OBJECTIVE
2

2. In the lecture before 🎞 Example 3, what important point are you told to remember when working with applications that have to do with money?

OBJECTIVE
3

3. The following problem is worded like Example 4 in the video, but using different quantities.

How can $36,000 be invested, part at 6% annual simple interest and the remainder at 4% annual simple interest, so that the annual simple interest earned by the two accounts is equal? Fill in the table and set up an equation that could be used to solve for the unknowns (do not actually solve). Use Example 4 in the video as a model for your work.

P	$\cdot$	$R \cdot$	$T =$	I

2.7 Exercise Set MyMathLab®

Solve. See Examples 1 and 2.

1. A jet plane traveling at 500 mph overtakes a propeller plane traveling at 200 mph that had a 2-hour head start. How far from the starting point are the planes?

2. How long will it take a bus traveling at 60 miles per hour to overtake a car traveling at 40 mph if the car had a 1.5-hour head start?

3. A bus traveled on a level road for 3 hours at an average speed 20 miles per hour faster than it traveled on a winding road. The time spent on the winding road was 4 hours. Find the average speed on the level road if the entire trip was 305 miles.

4. The Jones family drove to Disneyland at 50 miles per hour and returned on the same route at 40 mph. Find the distance to Disneyland if the total driving time was 7.2 hours.

Complete the table. The first and sixth rows have been completed for you. See Example 3.

	Number of Coins or Bills	Value of Coins or Bills (in dollars)
pennies	x	$0.01x$
5. *dimes*	y	
6. *quarters*	z	
7. *nickels*	$(x + 7)$	
8. *half-dollars*	$(20 - z)$	
$5 bills	$9x$	$5(9x)$
9. *$20 bills*	$4y$	
10. *$100 bills*	$97z$	
11. *$50 bills*	$(35 - x)$	
12. *$10 bills*	$(15 - y)$	

13. Part of the proceeds from a garage sale was $280 worth of $5 and $10 bills. If there were 20 more $5 bills than $10 bills, find the number of each denomination.

	Number of Bills	Value of Bills
$5 bills		
$10 bills		
Total		

14. A bank teller is counting $20 and $50 bills. If there are six times as many $20 bills as $50 bills and the total amount of money is $3910, find the number of each denomination.

	Number of Bills	Value of Bills
$20 bills		
$50 bills		
Total		

Solve. See Example 4.

15. Zoya Lon invested part of her $25,000 advance at 8% annual simple interest and the rest at 9% annual simple interest. If her total yearly interest from both accounts was $2135, find the amount invested at each rate.

16. Karen Waugtal invested some money at 9% annual simple interest and $250 more than that amount at 10% annual simple interest. If her total yearly interest was $101, how much was invested at each rate?

17. Sam Mathius invested part of his $10,000 bonus in a fund that paid an 11% profit and invested the rest in stock that suffered a 4% loss. Find the amount of each investment if his overall net profit was $650.

18. Bruce Blossum invested a sum of money at 10% annual simple interest and invested twice that amount at 12% annual simple interest. If his total yearly income from both investments was $2890, how much was invested at each rate?

19. The Concordia Theatre contains 500 seats, and the ticket prices for a recent play were $43 for adults and $28 for children. For one sold-out matinee, if the total proceeds were $16,805, how many of each type of ticket were sold?

20. A zoo in Oklahoma charged $22 for adults and $15 for children. During a summer day, 732 zoo tickets were sold, and the total receipts were $12,912. How many child and how many adult tickets were sold?

MIXED PRACTICE

21. Two cars leave Richmond, Virginia, at the same time after visiting the nearby Richmond International Speedway. The cars travel in opposite directions, one traveling north at 56 mph and one traveling south at 47 mph. When will the two cars be 206 miles apart?

22. Two cars leave Las Vegas, Nevada, at the same time after visiting the Las Vegas Motor Speedway. The cars travel in opposite directions, one traveling northeast at 65 mph and one traveling southwest at 41 mph. When will the two cars be 530 miles apart?

▶ 23. How can $54,000 be invested, part at 8% annual simple interest and the remainder at 10% annual simple interest, so that the interest earned by the two accounts will be equal?

24. Ms. Mills invested her $20,000 bonus in two accounts. She took a 4% loss on one investment and made a 12% profit on another investment but ended up breaking even. How much was invested in each account?

25. Alan and Dave Schaferkötter leave from the same point driving in opposite directions, Alan driving at 55 miles per hour and Dave at 65 mph. Alan has a one-hour head start. How long will they be able to talk on their car phones if the phones have a 250-mile range?

26. Kathleen and Cade Williams leave simultaneously from the same point, hiking in opposite directions, Kathleen walking at 4 miles per hour and Cade at 5 mph. How long can they talk on their walkie-talkies if the walkie-talkies have a 20-mile radius?

27. Suppose two trains leave Corpus Christi, Texas, at the same time, traveling in opposite directions. One train travels 10 mph faster than the other. In 2.5 hours, the trains are 205 miles apart. Find the speed of each train.

28. Suppose two trains leave Edmonton, Canada, at the same time, traveling in opposite directions. One train travels 8 mph faster than the other. In 1.5 hours, the trains are 162 miles apart. Find the speed of each train.

29. A youth organization collected nickels and dimes for a charity drive. By the end of the 1-day drive, the youths had collected $56.35. If there were three times as many dimes as nickels, how many of each type of coin was collected?

30. A collection of dimes and quarters is retrieved from a soft drink machine. There are five times as many dimes as quarters and the total value of the coins is $27.75. Find the number of dimes and the number of quarters.

31. A truck and a van leave the same location at the same time and travel in opposite directions. The truck's speed is 52 mph and the van's speed is 63 mph. When will the truck and the van be 460 miles apart?

32. Two cars leave the same location at the same time and travel in opposite directions. One car's speed is 65 mph and the other car's speed is 45 mph. When will the two cars be 330 miles apart?

33. Two cars leave Pecos, Texas, at the same time and both travel east on Interstate 20. The first car's speed is 70 mph and the second car's speed is 58 mph. When will the cars be 30 miles apart?

34. Two cars leave Savannah, Georgia, at the same time and both travel north on Interstate 95. The first car's speed is 40 mph and the second car's speed is 50 mph. When will the cars be 20 miles apart?

35. If $3000 is invested at 6% annual simple interest, how much should be invested at 9% annual simple interest so that the total yearly income from both investments is $585?

36. Trudy Waterbury, a financial planner, invested a certain amount of money at 9% annual simple interest, twice that amount at 10% annual simple interest, and three times that amount at 11% annual simple interest. Find the amount invested at each rate if her total yearly income from the investments was $2790.

▶ 37. Two hikers are 11 miles apart and walking toward each other. They meet in 2 hours. Find the rate of each hiker if one hiker walks 1.1 mph faster than the other.

38. Nedra and Latonya Dominguez are 12 miles apart hiking toward each other. How long will it take them to meet if Nedra walks at 3 mph and Latonya walks 1 mph faster?

39. Mark Martin can row upstream at 5 mph and downstream at 11 mph. If Mark starts rowing upstream until he gets tired and then rows downstream to his starting point, how far did Mark row if the entire trip took 4 hours?

40. On a 255-mile trip, Gary Alessandrini traveled at an average speed of 70 mph, got a speeding ticket, and then traveled at 60 mph for the remainder of the trip. If the entire trip took 4.5 hours and the speeding ticket stop took 30 minutes, how long did Gary speed before getting stopped?

REVIEW AND PREVIEW

Perform the indicated operations. See Sections 1.5 and 1.6.

41. $3 + (-7)$

42. $(-2) + (-8)$

43. $\dfrac{3}{4} - \dfrac{3}{16}$

44. $-11 + 2.9$

45. $-5 - (-1)$

46. $-12 - 3$

CONCEPT EXTENSIONS

47. A stack of $20, $50, and $100 bills was retrieved as part of an FBI investigation. There were 46 more $50 bills than $100 bills. Also, the number of $20 bills was 7 times the number of $100 bills. If the total value of the money was $9550, find the number of each type of bill.

48. A man places his pocket change in a jar every day. The jar is full and his children have counted the change. The total value is $44.86. Let x represent the number of quarters and use the information below to find the number of each type of coin.

There are: 136 more dimes than quarters

8 times as many nickels as quarters

32 more than 16 times as many pennies as quarters

To "break even" in a manufacturing business, revenue R (income)
must equal the cost C of production, or R = C.

49. The cost C to produce x number of skateboards is given by $C = 100 + 20x$. The skateboards are sold wholesale for $24 each, so revenue R is given by $R = 24x$. Find how many skateboards the manufacturer needs to produce and sell to break even. (*Hint:* Set the expression for R equal to the expression for C, then solve for x.)

50. The revenue R from selling x number of computer boards is given by $R = 60x$, and the cost C of producing them is given by $C = 50x + 5000$. Find how many boards must be sold to break even. Find how much money is needed to produce the break-even number of boards.

51. The cost C of producing x number of paperback books is given by $C = 4.50x + 2400$. Income R from these books is given by $R = 7.50x$. Find how many books should be produced and sold to break even.

52. Find the break-even quantity for a company that makes x number of computer monitors at a cost C given by $C = 875 + 70x$ and receives revenue R given by $R = 105x$.

53. Exercises 49 through 52 involve finding the break-even point for manufacturing. Discuss what happens if a company makes and sells fewer products than the break-even point. Discuss what happens if more products than the break-even point are made and sold.

2.8   Solving Linear Inequalities

OBJECTIVES

1 Define Linear Inequality in One Variable, Graph Solution Sets on a Number Line, and Use Interval Notation.

2 Solve Linear Inequalities.

3 Solve Compound Inequalities.

4 Solve Inequality Applications.

OBJECTIVE

1   Graphing Solution Sets to Linear Inequalities and Using Interval Notation

In Chapter 1, we reviewed these inequality symbols and their meanings:

< means "is less than"   ≤ means "is less than or equal to"

> means "is greater than"   ≥ means "is greater than or equal to"

Equations	Inequalities
$x = 3$	$x \leq 3$
$5n - 6 = 14$	$5n - 6 < 14$
$12 = 7 - 3y$	$12 \geq 7 - 3y$
$\dfrac{x}{4} - 6 = 1$	$\dfrac{x}{4} - 6 > 1$

A linear inequality is similar to a linear equation except that the equality symbol is replaced with an inequality symbol.

Linear Inequality in One Variable

A **linear inequality in one variable** is an inequality that can be written in the form

$$ax + b < c$$

where a, b, and c are real numbers and a is not 0.

This definition and all other definitions, properties, and steps in this section also hold true for the inequality symbols >, ≥, and ≤.

A **solution of an inequality** is a value of the variable that makes the inequality a true statement. The solution set is the set of all solutions. For the inequality $x < 3$, replacing x with any number less than 3, that is, to the left of 3 on a number line, makes the resulting inequality true. This means that any number less than 3 is a solution of the inequality $x < 3$.

Since there are infinitely many such numbers, we cannot list all the solutions of the inequality. We *can* use set notation and write

$$\{x \mid x < 3\}.$$

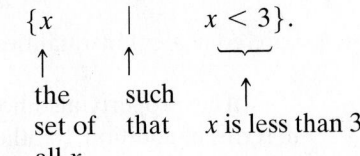

Recall that this is read   the set of all x such that x is less than 3.

We can also picture the solutions on a number line. If we use open/closed-circle notation, the graph of $\{x \mid x < 3\}$ looks like the following.

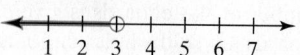

In this text, a convenient notation, called **interval notation,** will be used to write solution sets of inequalities. To help us understand this notation, a different graphing notation will be used. Instead of an open circle, we use a parenthesis; instead of a closed circle, we use a bracket. With this new notation, the graph of $\{x \mid x < 3\}$ now looks like

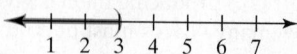

and can be represented in interval notation as $(-\infty, 3)$. The symbol $-\infty$, read as "negative infinity," does not indicate a number but does indicate that the shaded arrow to the left never ends. In other words, the interval $(-\infty, 3)$ includes *all* numbers less than 3.

Picturing the solutions of an inequality on a number line is called **graphing** the solutions or graphing the inequality, and the picture is called the **graph** of the inequality.

To graph $\{x \mid x \le 3\}$ or simply $x \le 3$, shade the numbers to the left of 3 and place a bracket at 3 on the number line as shown in the margin. The bracket indicates that 3 **is** a solution: 3 **is** less than or equal to 3. In interval notation, we write $(-\infty, 3]$.

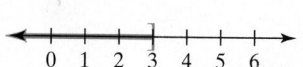

<table>
<tr><td>Helpful Hint</td></tr>
</table>

When writing an inequality in interval notation, it may be easier to graph the inequality first, then write it in interval notation. To help, think of the number line as approaching $-\infty$ to the left and $+\infty$ or ∞ to the right. Then simply write the interval notation by following your shading from left to right.

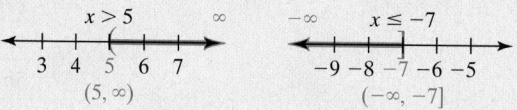

EXAMPLE 1   Graph $x \ge -1$. Then write the solutions in interval notation.

Solution We place a bracket at -1 since the inequality symbol is $\ge$ and -1 is greater than or equal to -1. Then we shade to the right of -1.

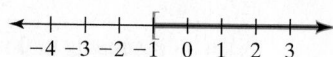

In interval notation, this is $[-1, \infty)$.

PRACTICE

1   Graph $x < 5$. Then write the solutions in interval notation.

OBJECTIVE

2   **Solving Linear Inequalities** ▶

When solutions of a linear inequality are not immediately obvious, they are found through a process similar to the one used to solve a linear equation. Our goal is to get the variable alone, and we use properties of inequality similar to properties of equality.

Addition Property of Inequality

If a, b, and c are real numbers, then

$$a < b \quad \text{and} \quad a + c < b + c$$

are equivalent inequalities.

This property also holds true for subtracting values, since subtraction is defined in terms of addition. In other words, adding or subtracting the same quantity from both sides of an inequality does not change the solution of the inequality.

EXAMPLE 2   Solve $x + 4 \leq -6$ for x. Graph the solution set and write it in interval notation.

**Solution** To solve for x, subtract 4 from both sides of the inequality.

$$x + 4 \leq -6 \qquad \text{Original inequality}$$
$$x + 4 - 4 \leq -6 - 4 \qquad \text{Subtract 4 from both sides.}$$
$$x \leq -10 \qquad \text{Simplify.}$$

(number line graphed from -12 to -6 with solid dot at -10 and arrow to the left)

$$-12 \;-11 \;-10 \;-9 \;-8 \;-7 \;-6$$

The solution set is $(-\infty, -10]$.   □

PRACTICE
2   Solve $x + 11 \geq 6$. Graph the solution set and write it in interval notation.   ■

> **Helpful Hint**
> Notice that any number less than or equal to -10 is a solution of $x \leq -10$. For example, solutions include
>
> $$-10, \; -200, \; -11\frac{1}{2}, \; -7\pi, \; -\sqrt{130}, \; -50.3$$

An important difference between linear equations and linear inequalities is shown when we multiply or divide both sides of an inequality by a nonzero real number. For example, start with the true statement $6 < 8$ and multiply both sides by 2. As we see below, the resulting inequality is also true.

$$6 < 8 \qquad \text{True}$$
$$2(6) < 2(8) \qquad \text{Multiply both sides by 2.}$$
$$12 < 16 \qquad \text{True}$$

But if we start with the same true statement $6 < 8$ and multiply both sides by -2, the resulting inequality is not a true statement.

$$6 < 8 \qquad \text{True}$$
$$-2(6) < -2(8) \qquad \text{Multiply both sides by } -2.$$
$$-12 < -16 \qquad \text{False}$$

Notice, however, that if we reverse the direction of the inequality symbol, the resulting inequality is true.

$$-12 < -16 \quad \text{False}$$
$$-12 > -16 \quad \text{True}$$

This demonstrates the multiplication property of inequality.

> **Multiplication Property of Inequality**
> **1.** If a, b, and c are real numbers, and c is **positive,** then
> $$a < b \qquad \text{and} \qquad ac < bc$$
> are equivalent inequalities.
> **2.** If a, b, and c are real numbers, and c is **negative,** then
> $$a < b \qquad \text{and} \qquad ac > bc$$
> are equivalent inequalities.

Because division is defined in terms of multiplication, this property also holds true when dividing both sides of an inequality by a nonzero number. If we multiply or divide both sides of an inequality by a negative number, **the direction of the inequality symbol must be reversed for the inequalities to remain equivalent.**

> **Helpful Hint**
> Whenever both sides of an inequality are multiplied or divided by a negative number, the direction of the inequality symbol **must be** reversed to form an equivalent inequality.

EXAMPLE 3 Solve $-2x \leq -4$. Graph the solution set and write it in interval notation.

Solution Remember to reverse the direction of the inequality symbol when dividing by a negative number.

$$-2x \leq -4$$
$$\frac{-2x}{-2} \geq \frac{-4}{-2} \quad \text{Divide both sides by } -2 \text{ and reverse the direction of the inequality symbol.}$$
$$x \geq 2 \quad \text{Simplify.}$$

> **Helpful Hint**
> Don't forget to reverse the direction of the inequality symbol.

The solution set $[2, \infty)$ is graphed as shown.

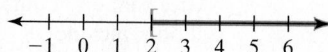

PRACTICE

3 Solve $-5x \geq -15$. Graph the solution set and write it in interval notation.

EXAMPLE 4 Solve $2x < -4$. Graph the solution set and write it in interval notation.

Solution

> **Helpful Hint**
> Do not reverse the inequality symbol.

$$2x < -4$$
$$\frac{2x}{2} < \frac{-4}{2} \quad \text{Divide both sides by 2.}$$
$$\quad \text{Do not reverse the direction of the inequality symbol.}$$
$$x < -2 \quad \text{Simplify.}$$

The solution set $(-\infty, -2)$ is graphed as shown.

PRACTICE

4 Solve $3x > -9$. Graph the solution set and write it in interval notation.

✔ **CONCEPT CHECK**

Fill in the blank with $<$, $>$, $\leq$, or $\geq$.

a. Since $-8 < -4$, then $3(-8)$_____$3(-4)$.

b. Since $5 \geq -2$, then $\dfrac{5}{-7}$_____$\dfrac{-2}{-7}$.

c. If $a < b$, then $2a$_____$2b$.

d. If $a \geq b$, then $\dfrac{a}{-3}$_____$\dfrac{b}{-3}$.

Answers to Concept Check:
a. $<$ **b.** $\leq$ **c.** $<$ **d.** $\leq$

The following steps may be helpful when solving inequalities. Notice that these steps are similar to the ones given in Section 2.3 for solving equations.

Solving Linear Inequalities in One Variable

Step 1. Clear the inequality of fractions by multiplying both sides of the inequality by the least common denominator (LCD) of all fractions in the inequality.

Step 2. Remove grouping symbols such as parentheses by using the distributive property.

Step 3. Simplify each side of the inequality by combining like terms.

Step 4. Write the inequality with variable terms on one side and numbers on the other side by using the addition property of inequality.

Step 5. Get the variable alone by using the multiplication property of inequality.

Helpful Hint

Don't forget that if both sides of an inequality are multiplied or divided by a negative number, the direction of the inequality symbol must be reversed.

EXAMPLE 5   Solve $-4x + 7 \geq -9$. Graph the solution set and write it in interval notation.

Solution
$$-4x + 7 \geq -9$$
$$-4x + 7 - 7 \geq -9 - 7 \quad \text{Subtract 7 from both sides.}$$
$$-4x \geq -16 \quad \text{Simplify.}$$
$$\frac{-4x}{-4} \leq \frac{-16}{-4} \quad \begin{array}{l}\text{Divide both sides by } -4 \text{ and reverse} \\ \text{the direction of the inequality symbol.}\end{array}$$
$$x \leq 4 \quad \text{Simplify.}$$

The solution set $(-\infty, 4]$ is graphed as shown.

$$\xleftarrow{\quad} \overset{\qquad\quad 3\ \ 4\ \ 5\ \ 6\ \ 7\ \ 8}{+\ \ +\ \ +\ \ +\ \ +\ \ +} \xrightarrow{\quad}$$

$\square$

PRACTICE

5   Solve $45 - 7x \leq -4$. Graph the solution set and write it in interval notation.   ■

EXAMPLE 6   Solve $2x + 7 \leq x - 11$. Graph the solution set and write it in interval notation.

Solution
$$2x + 7 \leq x - 11$$
$$2x + 7 - x \leq x - 11 - x \quad \text{Subtract } x \text{ from both sides.}$$
$$x + 7 \leq -11 \quad \text{Combine like terms.}$$
$$x + 7 - 7 \leq -11 - 7 \quad \text{Subtract 7 from both sides.}$$
$$x \leq -18 \quad \text{Combine like terms.}$$

The graph of the solution set $(-\infty, -18]$ is shown.

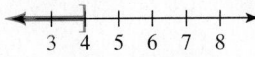

$\square$

PRACTICE

6   Solve $3x + 20 \leq 2x + 13$. Graph the solution set and write it in interval notation.   ■

EXAMPLE 7 Solve $-5x + 7 < 2(x - 3)$. Graph the solution set and write it in interval notation.

Solution

$$-5x + 7 < 2(x - 3)$$

$-5x + 7 < 2x - 6$	Apply the distributive property.
$-5x + 7 - 2x < 2x - 6 - 2x$	Subtract $2x$ from both sides.
$-7x + 7 < -6$	Combine like terms.
$-7x + 7 - 7 < -6 - 7$	Subtract 7 from both sides.
$-7x < -13$	Combine like terms.
$\dfrac{-7x}{-7} > \dfrac{-13}{-7}$	Divide both sides by -7 and reverse the direction of the inequality symbol.
$x > \dfrac{13}{7}$	Simplify.

The graph of the solution set $\left(\dfrac{13}{7}, \infty \right)$ is shown.

PRACTICE

7 Solve $6 - 5x > 3(x - 4)$. Graph the solution set and write it in interval notation.

EXAMPLE 8 Solve $2(x - 3) - 5 \le 3(x + 2) - 18$. Graph the solution set and write it in interval notation.

Solution

$$2(x - 3) - 5 \le 3(x + 2) - 18$$

$2x - 6 - 5 \le 3x + 6 - 18$	Apply the distributive property.
$2x - 11 \le 3x - 12$	Combine like terms.
$-x - 11 \le -12$	Subtract $3x$ from both sides.
$-x \le -1$	Add 11 to both sides.
$\dfrac{-x}{-1} \ge \dfrac{-1}{-1}$	Divide both sides by -1 and reverse the direction of the inequality symbol.
$x \ge 1$	Simplify.

The graph of the solution set $[1, \infty)$ is shown.

PRACTICE

8 Solve $3(x - 4) - 5 \le 5(x - 1) - 12$. Graph the solution set and write it in interval notation.

OBJECTIVE

3 Solving Compound Inequalities ▶

Inequalities containing one inequality symbol are called **simple inequalities,** while inequalities containing two inequality symbols are called **compound inequalities.** A compound inequality is really two simple inequalities in one. The compound inequality

$$3 < x < 5 \quad \text{means} \quad 3 < x \text{ and } x < 5$$

This can be read "x is greater than 3 and less than 5."

A solution of a compound inequality is a value that is a solution of both of the simple inequalities that make up the compound inequality. For example,

$$4\frac{1}{2} \text{ is a solution of } 3 < x < 5 \text{ since } 3 < 4\frac{1}{2} \text{ and } 4\frac{1}{2} < 5.$$

To graph $3 < x < 5$, place parentheses at both 3 and 5 and shade between.

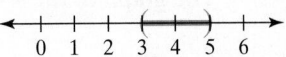

EXAMPLE 9 Graph $2 < x \le 4$. Write the solutions in interval notation.

Solution Graph all numbers greater than 2 and less than or equal to 4. Place a parenthesis at 2, a bracket at 4, and shade between.

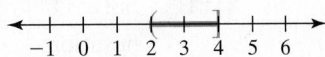

In interval notation, this is $(2, 4]$.

PRACTICE
9 Graph $-3 \le x < 1$. Write the solutions in interval notation.

When we solve a simple inequality, we isolate the variable on one side of the inequality. When we solve a compound inequality, we isolate the variable in the middle part of the inequality. Also, when solving a compound inequality, we must perform the same operation to all **three** parts of the inequality: left, middle, and right.

EXAMPLE 10 Solve $-1 \le 2x - 3 < 5$. Graph the solution set and write it in interval notation.

Solution
$$-1 \le 2x - 3 < 5$$
$$-1 + 3 \le 2x - 3 + 3 < 5 + 3 \quad \text{Add 3 to all three parts.}$$
$$2 \le 2x < 8 \quad \text{Combine like terms.}$$
$$\frac{2}{2} \le \frac{2x}{2} < \frac{8}{2} \quad \text{Divide all three parts by 2.}$$
$$1 \le x < 4 \quad \text{Simplify.}$$

The graph of the solution set $[1, 4)$ is shown.

PRACTICE
10 Solve $-4 < 3x + 2 \le 8$. Graph the solution set and write it in interval notation.

EXAMPLE 11 Solve $3 \le \dfrac{3x}{2} + 4 \le 5$. Graph the solution set and write it in interval notation.

Solution
$$3 \le \frac{3x}{2} + 4 \le 5$$

$$2(3) \le 2\left(\frac{3x}{2} + 4\right) \le 2(5) \quad \begin{array}{l}\text{Multiply all three parts by 2 to clear}\\\text{the fraction.}\end{array}$$
$$6 \le 3x + 8 \le 10 \quad \text{Distribute.}$$
$$-2 \le 3x \le 2 \quad \text{Subtract 8 from all three parts.}$$
$$\frac{-2}{3} \le \frac{3x}{3} \le \frac{2}{3} \quad \text{Divide all three parts by 3.}$$
$$-\frac{2}{3} \le x \le \frac{2}{3} \quad \text{Simplify.}$$

(Continued on next page)

The graph of the solution set $\left[-\dfrac{2}{3}, \dfrac{2}{3} \right]$ is shown.

$$\begin{array}{c} \quad -\tfrac{2}{3} \quad \tfrac{2}{3} \\ \xleftarrow{\;\;\;\;\;\; \underset{-2\;\;-1\;\;\;0\;\;\;1\;\;\;2\;\;\;3}{+\;\;\;+\;\;[\;\;]\;\;+\;\;+\;\;+} \;\;\;\;\;\;}\rightarrow \end{array}$$

PRACTICE

11 Solve $1 < \dfrac{3}{4}x + 5 < 6$. Graph the solution set and write it in interval notation.

OBJECTIVE

4 Solving Inequality Applications

Problems containing words such as "at least," "at most," "between," "no more than," and "no less than" usually indicate that an inequality should be solved instead of an equation. In solving applications involving linear inequalities, use the same procedure you use to solve applications involving linear equations.

Some Inequality Translations			
≥	≤	<	>
at least	at most	is less than	is greater than
no less than	no more than		

EXAMPLE 12 12 subtracted from 3 times a number is less than 21. Find all numbers that make this statement true.

Solution

1. **UNDERSTAND.** Read and reread the problem. This is a direct translation problem, and let's let

$$x = \text{the unknown number.}$$

2. **TRANSLATE.**

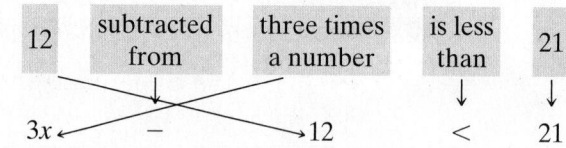

3. **SOLVE.** $3x - 12 < 21$

$\qquad\qquad 3x < 33$ Add 12 to both sides.

$\qquad\qquad \dfrac{3x}{3} < \dfrac{33}{3}$ Divide both sides by 3 and do not reverse the direction of the inequality symbol.

$\qquad\qquad x < 11$ Simplify.

4. **INTERPRET.**

Check: Check the translation; then let's choose a number less than 11 to see if it checks. For example, let's check 10. 12 subtracted from 3 times 10 is 12 subtracted from 30, or 18. Since 18 is less than 21, the number 10 checks.

State: All numbers less than 11 make the original statement true.

PRACTICE

12 Twice a number, subtracted from 35, is greater than 15. Find all numbers that make this true.

EXAMPLE 13   **Staying within Budget**

Marie Chase and Jonathan Edwards are having their wedding reception at the Gallery Reception Hall. They may spend at most $2000 for the reception. If the reception hall charges a $100 cleanup fee plus $36 per person, find the greatest number of people that they can invite and still stay within their budget.

Solution

1. **UNDERSTAND.** Read and reread the problem. Next, guess a solution. If 40 people attend the reception, the cost is
$100 + $36(40) = $100 + $1440 = $1540. Let

$$x = \text{the number of people who attend the reception.}$$

2. **TRANSLATE.**

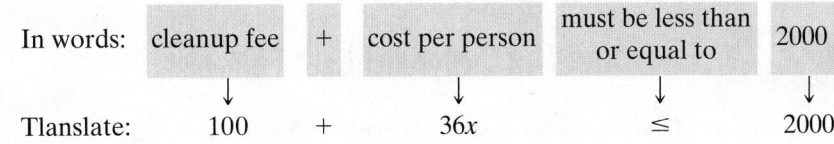

In words:	cleanup fee	+	cost per person	must be less than or equal to	2000
Tlanslate:	100	+	$36x$	$\leq$	2000

3. **SOLVE.**

$$100 + 36x \leq 2000$$
$$36x \leq 1900 \quad \text{Subtract 100 from both sides.}$$
$$x \leq 52\frac{7}{9} \quad \text{Divide both sides by 36.}$$

4. **INTERPRET.**

Check: Since x represents the number of people, we round down to the nearest whole, or 52. Notice that if 52 people attend, the cost is

$$\$100 + \$36(52) = \$1972. \text{ If 53 people attend, the cost is}$$
$$\$100 + \$36(53) = \$2008, \text{ which is more than the given 2000.}$$

State: Marie Chase and Jonathan Edwards can invite at most 52 people to the reception.   □

PRACTICE
13   Kasonga is eager to begin his education at his local community college. He has budgeted $1500 for college this semester. His local college charges a $300 matriculation fee and costs an average of $375 for tuition, fees, and books for each three-credit course. Find the greatest number of classes Kasonga can afford to take this semester.   ■

✔ **Vocabulary, Readiness & Video Check**

Use the choices below to fill in each blank. Choices may be used more than once.

expression   inequality   equation

1. $6x - 7(x + 9)$ _____

2. $6x = 7(x + 9)$ _____

3. $6x < 7(x + 9)$ _____

4. $5y - 2 \geq -38$ _____

Decide which number listed is not a solution of each given inequality.

5. $x \geq -3;\ -3, 0, -5, \pi$ _____

6. $x < 6;\ -6, |-6|, 0, -3.2$ _____

Martin-Gay Interactive Videos

See Video 2.8 ⊙

Watch the section lecture video and answer the following questions.

OBJECTIVE 1

7. Using ▦ Example 1 from the video as a reference, explain the connection between the graph of an inequality and interval notation.

OBJECTIVE 2

8. The steps for solving a linear inequality in one variable are discussed in the lecture before ▦ Example 6. Why are you told to be very careful when you use Step 5?

OBJECTIVE 3

9. For ▦ Example 8, explain how the solving would change if the compound inequality simplified to $0 < -3x < 14$ instead of $0 < 3x < 14$.

OBJECTIVE 4

10. What is the phrase in ▦ Example 9 that tells you to translate to an *inequality*? What does this phrase translate to?

2.8 Exercise Set   MyMathLab® ▷

Graph each set of numbers given in interval notation. Then write an inequality statement in x describing the numbers graphed.

1. $[2, \infty)$

2. $(-3, \infty)$

3. $(-\infty, -5)$

4. $(-\infty, 4]$

Graph each inequality on a number line. Then write the solutions in interval notation. See Example 1.

▷ **5.** $x \leq -1$

6. $y < 0$

7. $x < \dfrac{1}{2}$

8. $z < -\dfrac{2}{3}$

9. $y \geq 5$

▷ **10.** $x > 3$

Solve each inequality. Graph the solution set and write it in interval notation. See Examples 2 through 4.

11. $2x < -6$

12. $3x > -9$

▷ **13.** $x - 2 \geq -7$

14. $x + 4 \leq 1$

▷ **15.** $-8x \leq 16$

16. $-5x < 20$

Solve each inequality. Graph the solution set and write it in interval notation. See Examples 5 and 6.

17. $3x - 5 > 2x - 8$

18. $3 - 7x \geq 10 - 8x$

19. $4x - 1 \leq 5x - 2x$

20. $7x + 3 < 9x - 3x$

Solve each inequality. Graph the solution set and write it in interval notation. See Examples 7 and 8.

21. $x - 7 < 3(x + 1)$

22. $3x + 9 \leq 5(x - 1)$

23. $-6x + 2 \geq 2(5 - x)$

24. $-7x + 4 > 3(4 - x)$

25. $4(3x - 1) \leq 5(2x - 4)$

26. $3(5x - 4) \leq 4(3x - 2)$

27. $3(x + 2) - 6 > -2(x - 3) + 14$

28. $7(x - 2) + x \leq -4(5 - x) - 12$

MIXED PRACTICE

Solve the following inequalities. Graph each solution set and write it in interval notation.

29. $-2x \leq -40$

30. $-7x > 21$

31. $-9 + x > 7$

32. $y - 4 \leq 1$

33. $3x - 7 < 6x + 2$

34. $2x - 1 \geq 4x - 5$

35. $5x - 7x \geq x + 2$

36. $4 - x < 8x + 2x$

37. $\dfrac{3}{4}x > 2$

38. $\dfrac{5}{6}x \geq -8$

39. $3(x - 5) < 2(2x - 1)$

40. $5(x + 4) < 4(2x + 3)$

41. $4(2x + 1) < 4$

42. $6(2 - x) \geq 12$

43. $-5x + 4 \geq -4(x - 1)$

44. $-6x + 2 < -3(x + 4)$

▷ **45.** $-2(x - 4) - 3x < -(4x + 1) + 2x$

46. $-5(1 - x) + x \leq -(6 - 2x) + 6$

47. $\dfrac{1}{4}(x + 4) < \dfrac{1}{5}(2x + 3)$

48. $\dfrac{1}{3}(3x - 1) < \dfrac{1}{2}(x + 4)$

Graph each inequality. Then write the solutions in interval notation. See Example 9.

49. $-1 < x < 3$

50. $2 \le y \le 3$

51. $0 \le y < 2$

52. $-1 \le x \le 4$

Solve each inequality. Graph the solution set and write it in interval notation. See Examples 10 and 11.

53. $-3 < 3x < 6$

54. $-5 < 2x < -2$

55. $2 \le 3x - 10 \le 5$

56. $4 \le 5x - 6 \le 19$

57. $-4 < 2(x - 3) \le 4$

58. $0 < 4(x + 5) \le 8$

59. $-2 < 3x - 5 < 7$

60. $1 < 4 + 2x \le 7$

61. $-6 < 3(x - 2) \le 8$

62. $-5 \le 2(x + 4) < 8$

Solve the following. For Exercises 65 and 66, the solutions have been started for you. See Examples 12 and 13.

63. Six more than twice a number is greater than negative fourteen. Find all numbers that make this statement true.

64. One more than five times a number is less than or equal to ten. Find all such numbers.

65. The perimeter of a rectangle is to be no greater than 100 centimeters and the width must be 15 centimeters. Find the maximum length of the rectangle.

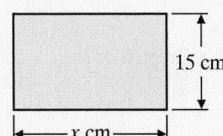

15 cm

x cm

Start the solution:

1. UNDERSTAND the problem. Reread it as many times as needed.

2. TRANSLATE into an equation. (Fill in the blanks below.)

the perimeter of the rectangle	is no greater than	100
↓	↓	↓
$x + 15 + x + 15$	_____	100

Finish with:

3. SOLVE and 4. INTERPRET

66. One side of a triangle is three times as long as another side, and the third side is 12 inches long. If the perimeter can be no longer than 32 inches, find the maximum lengths of the other two sides.

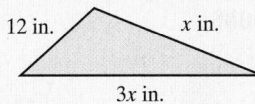

12 in. x in.

$3x$ in.

Start the solution:

1. UNDERSTAND the problem. Reread it as many times as needed.

2. TRANSLATE into an equation. (Fill in the blanks below.)

the perimeter of the triangle	is no greater than	32
↓	↓	↓
$12 + 3x + x$	_____	32

Finish with:

3. SOLVE and 4. INTERPRET

67. Ben Holladay bowled 146 and 201 in his first two games. What must he bowl in his third game to have an average of at least 180? (*Hint:* The average of a list of numbers is their sum divided by the number of numbers in the list.)

68. On an NBA team, the two forwards measure 6′8″ and 6′6″ tall and the two guards measure 6′0″ and 5′9″ tall. How tall should the center be if they wish to have a starting team average height of at least 6′5″?

69. Dennis and Nancy Wood are celebrating their 30th wedding anniversary by having a reception at Tiffany Oaks reception hall. They have budgeted $3000 for their reception. If the reception hall charges a $50.00 cleanup fee plus $34 per person, find the greatest number of people that they may invite and still stay within their budget.

70. A surprise retirement party is being planned for Pratap Puri. A total of $860 has been collected for the event, which is to be held at a local reception hall. This reception hall charges a cleanup fee of $40 and $15 per person for drinks and light snacks. Find the greatest number of people that may be invited and still stay within the $860 budget.

71. A 150-pound person uses 5.8 calories per minute when walking at a speed of 4 mph. How long must a person walk at this speed to use at least 200 calories? Round up to the next minute. (*Source:* Home & Garden Bulletin No. 72)

72. A 170-pound person uses 5.3 calories per minute when bicycling at a speed of 5.5 mph. How long must a person ride a bike at this speed to use at least 200 calories? Round up to the next minute. (*Source:* Same as Exercise 71)

73. Twice a number, increased by one, is between negative five and seven. Find all such numbers.

74. Half a number, decreased by four, is between two and three. Find all such numbers.

REVIEW AND PREVIEW

Evaluate the following. See Section 1.4.

75. 2^3

76. 3^3

77. 1^{12}

78. 0^5

79. $\left(\dfrac{4}{7}\right)^2$

80. $\left(\dfrac{2}{3}\right)^3$

CONCEPT EXTENSIONS

Fill in the box with $<, >, \leq,$ or $\geq$. See the Concept Check in this section.

81. Since $3 < 5$, then $3(-4) \,\square\, 5(-4)$.

82. If $m \leq n$, then $2m \,\square\, 2n$.

83. If $m \leq n$, then $-2m \,\square\, -2n$.

84. If $-x < y$, then $x \,\square\, -y$.

85. When solving an inequality, when must you reverse the direction of the inequality symbol?

86. If both sides of the inequality $-3x < 30$ are divided by -3, do you reverse the direction of the inequality symbol? Why or why not?

Solve.

87. Eric Daly has scores of 75, 83, and 85 on his history tests. Use an inequality to find the scores he can make on his final exam to receive a B in the class. The final exam counts as **two** tests, and a B is received if the final course average is greater than or equal to 80.

88. Maria Lipco has scores of 85, 95, and 92 on her algebra tests. Use an inequality to find the scores she can make on her final exam to receive an A in the course. The final exam counts as **three** tests, and an A is received if the final course average is greater than or equal to 90. Round to one decimal place.

89. Explain how solving a linear inequality is similar to solving a linear equation.

90. Explain how solving a linear inequality is different from solving a linear equation.

91. Explain how solving a linear inequality is different from solving a compound inequality.

92. Explain how solving a linear inequality is similar to solving a compound inequality.

93. The formula $C = 3.14d$ can be used to approximate the circumference of a circle given its diameter. Waldo Manufacturing manufactures and sells a certain washer with an outside circumference of 3 centimeters. The company has decided that a washer whose actual circumference is in the interval $2.9 \leq C \leq 3.1$ centimeters is acceptable. Use a compound inequality and find the corresponding interval for diameters of these washers. (Round to three decimal places.)

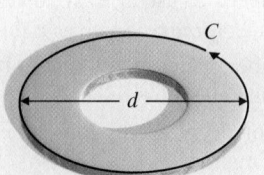

94. A company manufactures plastic Easter eggs that open. The company has determined that if the circumference of the opening of each part of the egg is in the interval $118 \leq C \leq 122$ millimeters, the eggs will open and close comfortably. Use a compound inequality and find the corresponding interval for diameters of these openings. (Round to two decimal places.)

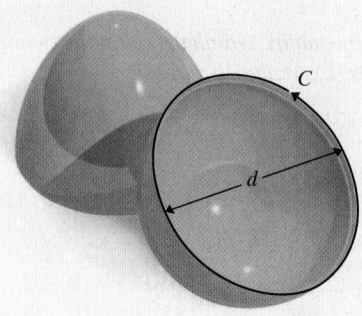

For Exercises 95 through 98, see the example below.

Solve $x(x - 6) > x^2 - 5x + 6$. Graph the solution set and write it in interval notation.

Solution

$$x(x - 6) > x^2 - 5x + 6$$
$$x^2 - 6x > x^2 - 5x + 6$$
$$x^2 - 6x - x^2 > x^2 - 5x + 6 - x^2$$
$$-6x > -5x + 6$$
$$-x > 6$$
$$\frac{-x}{-1} < \frac{6}{-1}$$
$$x < -6$$

The solution set $(-\infty, -6)$ is graphed as shown.

$$\xleftarrow{\quad\quad})\;{+}\;{+}\;{+}\;{+}\;{+}\;{\rightarrow}$$
$$-7\;-6\;-5\;-4\;-3\;-2\;-1$$

Solve each inequality. Graph the solution set and write it in interval notation.

95. $x(x + 4) > x^2 - 2x + 6$

96. $x(x - 3) \geq x^2 - 5x - 8$

97. $x^2 + 6x - 10 < x(x - 10)$

98. $x^2 - 4x + 8 < x(x + 8)$

Chapter 2   Vocabulary Check

Fill in each blank with one of the words or phrases listed below.

like terms	numerical coefficient	linear inequality in one variable	no solution
equivalent equations	formula	compound inequalities	reversed
linear equation in one variable	unlike terms	all real numbers	the same

1. Terms with the same variables raised to exactly the same powers are called _____.

2. If terms are not like terms, they are _____.

3. A(n) _____ can be written in the form $ax + b = c$.

4. A(n) _____ can be written in the form $ax + b < c$ (or $>, \leq, \geq$).

5. Inequalities containing two inequality symbols are called _____.

6. An equation that describes a known relationship among quantities is called a(n) _____.

7. The _____ of a term is its numerical factor.

8. Equations that have the same solution are called _____.

9. The solution(s) of the equation $x + 5 = x + 5$ is/are _____.

10. The solution(s) of the equation $x + 5 = x + 4$ is/are _____.

11. If both sides of an inequality are multiplied or divided by the same positive number, the direction of the inequality symbol is _____.

12. If both sides of an inequality are multiplied or divided by the same negative number, the direction of the inequality symbol is _____.

Chapter 2   Highlights

DEFINITIONS AND CONCEPTS   **EXAMPLES**

Section 2.1   Simplifying Algebraic Expressions

The **numerical coefficient** of a **term** is its numerical factor.

Term	Numerical Coefficient
$-7y$	-7
x	1
$\frac{1}{5}a^2b$	$\frac{1}{5}$

Terms with the same variables raised to exactly the same powers are **like terms.**

Like Terms	Unlike Terms
$12x, -x$	$3y, 3y^2$
$-2xy, 5yx$	$7a^2b, -2ab^2$

To combine like terms, add the numerical coefficients and multiply the result by the common variable factor.

$$9y + 3y = 12y$$

$$-4z^2 + 5z^2 - 6z^2 = -5z^2$$

To remove parentheses, apply the distributive property.

$$-4(x + 7) + 10(3x - 1)$$

$$= -4x - 28 + 30x - 10$$

$$= 26x - 38$$

Section 2.2   The Addition and Multiplication Properties of Equality

A **linear equation in one variable** can be written in the form $ax + b = c$ where a, b, and c are real numbers and $a \neq 0$.

Equivalent equations are equations that have the same solution.

Linear Equations

$$-3x + 7 = 2$$
$$3(x - 1) = -8(x + 5) + 4$$

$x - 7 = 10$ and $x = 17$
are equivalent equations.

Addition Property of Equality

Adding the same number to or subtracting the same number from both sides of an equation does not change its solution.

$$y + 9 = 3$$
$$y + 9 - 9 = 3 - 9$$
$$y = -6$$

Multiplication Property of Equality

Multiplying both sides or dividing both sides of an equation by the same nonzero number does not change its solution.

$$\frac{2}{3}a = 18$$
$$\frac{3}{2}\left(\frac{2}{3}a\right) = \frac{3}{2}(18)$$
$$a = 27$$

Section 2.3   Solving Linear Equations

To Solve Linear Equations

Solve: $\dfrac{5(-2x + 9)}{6} + 3 = \dfrac{1}{2}$

1. Clear the equation of fractions.

  1. $6 \cdot \dfrac{5(-2x + 9)}{6} + 6 \cdot 3 = 6 \cdot \dfrac{1}{2}$

$$5(-2x + 9) + 18 = 3$$

2. Remove any grouping symbols such as parentheses.

  2. $-10x + 45 + 18 = 3$ Distributive property

3. Simplify each side by combining like terms.

  3. $-10x + 63 = 3$ Combine like terms.

4. Write variable terms on one side and numbers on the other side using the addition property of equality.

  4. $-10x + 63 - 63 = 3 - 63$ Subtract 63.

$$-10x = -60$$

5. Get the variable alone using the multiplication property of equality.

  5. $\dfrac{-10x}{-10} = \dfrac{-60}{-10}$ Divide by -10.

$$x = 6$$

6. Check by substituting in the original equation.

  6. $\dfrac{5(-2x + 9)}{6} + 3 = \dfrac{1}{2}$

$$\dfrac{5(-2 \cdot 6 + 9)}{6} + 3 \stackrel{?}{=} \dfrac{1}{2}$$

$$\dfrac{5(-3)}{6} + 3 \stackrel{?}{=} \dfrac{1}{2}$$

$$-\dfrac{5}{2} + \dfrac{6}{2} \stackrel{?}{=} \dfrac{1}{2}$$

$$\dfrac{1}{2} = \dfrac{1}{2} \quad \text{True}$$

DEFINITIONS AND CONCEPTS	EXAMPLES

Section 2.4   An Introduction to Problem Solving

Problem-Solving Steps

The height of the Hudson volcano in Chile is twice the height of the Kiska volcano in the Aleutian Islands. If the sum of their heights is 12,870 feet, find the height of each.

1. UNDERSTAND the problem.

1. Read and reread the problem. Guess a solution and check your guess.

Let x be the height of the Kiska volcano. Then $2x$ is the height of the Hudson volcano.

x⌐ $2x$⌐
Kiska   Hudson

2. TRANSLATE the problem.

2. In words:

height of Kiska	added to	height of Hudson	is	12,870
↓	↓	↓	↓	↓
Translate: x	$+$	$2x$	$=$	12,870

3. SOLVE.

3.
$$x + 2x = 12{,}870$$
$$3x = 12{,}870$$
$$x = 4290$$

4. INTERPRET the results.

4. *Check:* If x is 4290, then $2x$ is 2(4290) or 8580. Their sum is 4290 + 8580 or 12,870, the required amount.

State: The Kiska volcano is 4290 feet high and the Hudson volcano is 8580 feet high.

Section 2.5   Formulas and Problem Solving

Formulas

$$A = lw \text{ (area of a rectangle)}$$
$$I = PRT \text{ (simple interest)}$$

An equation that describes a known relationship among quantities is called a **formula.**

If all values for the variables in a formula are known except for one, this unknown value may be found by substituting in the known values and solving.

If $d = 182$ miles and $r = 52$ miles per hour in the formula $d = r \cdot t$, find t.

$$d = r \cdot t$$
$$182 = 52 \cdot t \quad \text{Let } d = 182 \text{ and } r = 52$$
$$3.5 = t$$

The time is 3.5 hours.

To solve a formula for a specified variable, use the same steps as for solving a linear equation. Treat the specified variable as the only variable of the equation.

Solve $P = 2l + 2w$ for l.
$$P = 2l + 2w$$
$$P - 2w = 2l + 2w - 2w \quad \text{Subtract } 2w.$$
$$P - 2w = 2l$$
$$\frac{P - 2w}{2} = \frac{2l}{2} \qquad \text{Divide by 2.}$$
$$\frac{P - 2w}{2} = l \qquad \text{Simplify.}$$

DEFINITIONS AND CONCEPTS	EXAMPLES

Section 2.6 Percent and Mixture Problem Solving

Use the same problem-solving steps to solve a problem containing percents.	32% of what number is 36.8?
1. UNDERSTAND.	**1.** Read and reread. Propose a solution and check. Let x = the unknown number.

2. TRANSLATE.	**2.** 32% of what number is 36.8
	    ↓ ↓ ↓ ↓ ↓
	   32% · x = 36.8

3. SOLVE:	**3.** *Solve:* $32\% \cdot x = 36.8$
	   $0.32x = 36.8$
	   $\dfrac{0.32x}{0.32} = \dfrac{36.8}{0.32}$ Divide by 0.32.
	   $x = 115$ Simplify.

4. INTERPRET.	**4.** *Check, then state:* 32% of 115 is 36.8.
Use the same problem-solving steps to solve a problem about mixtures.	How many liters of a 20% acid solution must be mixed with a 50% acid solution to obtain 12 liters of a 30% solution?
1. UNDERSTAND.	**1.** Read and reread. Guess a solution and check. Let x = number of liters of 20% solution. Then $12 - x$ = number of liters of 50% solution.

2. TRANSLATE.

2.

	No. of Liters	·	Acid Strength	=	Amount of Acid
20% Solution	x		20%		$0.20x$
50% Solution	$12 - x$		50%		$0.50(12 - x)$
30% Solution Needed	12		30%		$0.30(12)$

In words:

acid in 20% solution	+	acid in 50% solution	=	acid in 30% solution
↓		↓		↓

Translate: $0.20x$ + $0.50(12 - x)$ = $0.30(12)$

3. SOLVE.

3. Solve: $0.20x + 0.50(12 - x) = 0.30(12)$

$0.20x + 6 - 0.50x = 3.6$ Apply the distributive

$-0.30x + 6 = 3.6$ property.

$-0.30x = -2.4$ Subtract 6.

$x = 8$ Divide by -0.30.

4. INTERPRET.

4. *Check, then state:*

If 8 liters of a 20% acid solution are mixed with $12 - 8$ or 4 liters of a 50% acid solution, the result is 12 liters of a 30% solution.

DEFINITIONS AND CONCEPTS	EXAMPLES

Section 2.7   Further Problem Solving

Problem-Solving Steps

A collection of dimes and quarters has a total value of $19.55. If there are three times as many quarters as dimes, find the number of quarters.

1. UNDERSTAND.

1. Read and reread. Propose a solution and check.

Let x = number of dimes, then

$3x$ = number of quarters.

2. TRANSLATE.

2. In words:

value of dimes	+	value of quarters	=	19.55
↓		↓		↓

Translate: $0.10x$ + $0.25(3x)$ =   19.55

3. SOLVE.

3. Solve: $0.10x + 0.75x = 19.55$ Multiply.

$0.85x = 19.55$ Add like terms.

$x = 23$ Divide by 0.85.

4. INTERPRET.

4. *Check, then state:*

The number of dimes is 23 and the number of quarters is 3(23) or 69. The total value of this money is

$0.10(23) + 0.25(69) = 19.55$, so our result checks.

The number of quarters is 69.

Section 2.8   Solving Linear Inequalities

A **linear inequality in one variable** is an inequality that can be written in one of the forms:

$$ax + b < c \qquad ax + b \le c$$
$$ax + b > c \qquad ax + b \ge c$$

where a, b, and c are real numbers and a is not 0.

Linear Inequalities

$$2x + 3 < 6 \qquad 5(x - 6) \ge 10$$

$$\frac{x - 2}{5} > \frac{5x + 7}{2} \qquad \frac{-(x + 8)}{9} \le \frac{-2x}{11}$$

Addition Property of Inequality

Adding the same number to or subtracting the same number from both sides of an inequality does not change the solutions.

$$y + 4 \le -1$$
$$y + 4 - 4 \le -1 - 4 \quad \text{Subtract 4.}$$
$$y \le -5$$

$(-\infty, -5]$ ← | | | | | | | | →
$\qquad\qquad -6\ -5\ -4\ -3\ -2\ -1\ \ 0\ \ 1\ \ 2$

Multiplication Property of Inequality

Multiplying or dividing both sides of an inequality by the same positive number does not change its solutions.

$$\frac{1}{3}x > -2$$
$$3\left(\frac{1}{3}x\right) > 3 \cdot -2 \quad \text{Multiply by 3.}$$
$$x > -6 \quad (-6, \infty) \quad ← (| | | | | →$$
$\qquad\qquad\qquad\qquad\qquad -6\ -4\ -2\ \ 0\ \ 2$

Multiplying or dividing both sides of an inequality by the same **negative number and reversing the direction of the inequality symbol** does not change its solutions.

$$-2x \le 4$$
$$\frac{-2x}{-2} \ge \frac{4}{-2} \quad \text{Divide by } -2, \text{ reverse inequality symbol.}$$
$$x \ge -2 \quad [-2, \infty) \quad ← | [| | | | →$$
$\qquad\qquad\qquad\qquad\qquad -3\ -2\ -1\ \ 0\ \ 1\ \ 2$

(continued)

DEFINITIONS AND CONCEPTS	EXAMPLES

Section 2.8   Solving Linear Inequalities (continued)

To Solve Linear Inequalities

1. Clear the equation of fractions.

2. Remove grouping symbols.

3. Simplify each side by combining like terms.

4. Write variable terms on one side and numbers on the other side, using the addition property of inequality.

5. Get the variable alone, using the multiplication property of inequality.

Solve: $3(x + 2) \le -2 + 8$

1. No fractions to clear. $3(x + 2) \le -2 + 8$

2. $\quad\quad 3x + 6 \le -2 + 8$ Distributive property

3. $\quad\quad 3x + 6 \le 6$ Combine like terms.

4. $\quad 3x + 6 - 6 \le 6 - 6$ Subtract 6.

$\quad\quad 3x \le 0$

5. $\quad\quad \dfrac{3x}{3} \le \dfrac{0}{3}$ Divide by 3.

$\quad\quad x \le 0 \quad (-\infty, 0]$

Inequalities containing two inequality symbols are called **compound inequalities**.

Compound Inequalities

$$-2 < x < 6$$

$$5 \le 3(x - 6) < \frac{20}{3}$$

To solve a compound inequality, isolate the variable in the middle part of the inequality. Perform the same operation to all three parts of the inequality: left, middle, right.

Solve: $\quad -2 < 3x + 1 < 7$

$-2 - 1 < 3x + 1 - 1 < 7 - 1$ Subtract 1.

$\quad\quad -3 < 3x < 6$

$\quad\quad \dfrac{-3}{3} < \dfrac{3x}{3} < \dfrac{6}{3}$ Divide by 3.

$\quad\quad -1 < x < 2 \quad (-1, 2)$

Chapter 2 Review

(2.1) Simplify the following expressions.

1. $5x - x + 2x$

2. $0.2z - 4.6x - 7.4z$

3. $\dfrac{1}{2}x + 3 + \dfrac{7}{2}x - 5$

4. $\dfrac{4}{5}y + 1 + \dfrac{6}{5}y + 2$

5. $2(n - 4) + n - 10$

6. $3(w + 2) - (12 - w)$

7. Subtract $7x - 2$ from $x + 5$.

8. Subtract $1.4y - 3$ from $y - 0.7$.

Write each of the following as algebraic expressions.

9. Three times a number decreased by 7

10. Twice the sum of a number and 2.8 added to 3 times the number

(2.2) Solve each equation.

11. $8x + 4 = 9x$

12. $5y - 3 = 6y$

13. $\dfrac{2}{7}x + \dfrac{5}{7}x = 6$

14. $3x - 5 = 4x + 1$

15. $2x - 6 = x - 6$

16. $4(x + 3) = 3(1 + x)$

17. $6(3 + n) = 5(n - 1)$

18. $5(2 + x) - 3(3x + 2) = -5(x - 6) + 2$

Use the addition property to fill in the blanks so that the middle equation simplifies to the last equation.

19. $\quad\quad x - 5 = 3$

$x - 5 + \underline{\quad} = 3 + \underline{\quad}$

$\quad\quad x = 8$

20. $\quad\quad x + 9 = -2$

$x + 9 - \underline{\quad} = -2 - \underline{\quad}$

$\quad\quad x = -11$

Choose the correct algebraic expression.

21. The sum of two numbers is 10. If one number is x, express the other number in terms of x.

  a. $x - 10$ b. $10 - x$

  c. $10 + x$ d. $10x$

22. Mandy is 5 inches taller than Melissa. If x inches represents the height of Mandy, express Melissa's height in terms of x.

  a. $x - 5$ b. $5 - x$

  c. $5 + x$ d. $5x$

△ 23. If one angle measures $x°$, express the measure of its complement in terms of x.

  a. $(180 - x)°$ b. $(90 - x)°$

  c. $(x - 180)°$ d. $(x - 90)°$

△ **24.** If one angle measures $(x + 5)°$, express the measure of its supplement in terms of x.

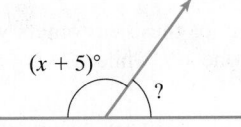

   a. $(185 + x)°$

   b. $(95 + x)°$

   c. $(175 - x)°$

   d. $(x - 170)°$

Solve each equation.

25. $\frac{3}{4}x = -9$ **26.** $\frac{x}{6} = \frac{2}{3}$

27. $-5x = 0$ **28.** $-y = 7$

29. $0.2x = 0.15$ **30.** $\frac{-x}{3} = 1$

31. $-3x + 1 = 19$ **32.** $5x + 25 = 20$

33. $7(x - 1) + 9 = 5x$ **34.** $7x - 6 = 5x - 3$

35. $-5x + \frac{3}{7} = \frac{10}{7}$ **36.** $5x + x = 9 + 4x - 1 + 6$

37. Write the sum of three consecutive integers as an expression in x. Let x be the first integer.

38. Write the sum of the first and fourth of four consecutive even integers. Let x be the first even integer.

(2.3) *Solve each equation.*

39. $\frac{5}{3}x + 4 = \frac{2}{3}x$ **40.** $\frac{7}{8}x + 1 = \frac{5}{8}x$

41. $-(5x + 1) = -7x + 3$ **42.** $-4(2x + 1) = -5x + 5$

43. $-6(2x - 5) = -3(9 + 4x)$

44. $3(8y - 1) = 6(5 + 4y)$

45. $\frac{3(2 - z)}{5} = z$ **46.** $\frac{4(n + 2)}{5} = -n$

47. $0.5(2n - 3) - 0.1 = 0.4(6 + 2n)$

48. $-9 - 5a = 3(6a - 1)$ **49.** $\frac{5(c + 1)}{6} = 2c - 3$

50. $\frac{2(8 - a)}{3} = 4 - 4a$

51. $200(70x - 3560) = -179(150x - 19{,}300)$

52. $1.72y - 0.04y = 0.42$

(2.4) *Solve each of the following.*

53. The height of the Washington Monument is 50.5 inches more than 10 times the length of a side of its square base. If the sum of these two dimensions is 7327 inches, find the height of the Washington Monument. (*Source:* National Park Service)

54. A 12-foot board is to be divided into two pieces so that one piece is twice as long as the other. If x represents the length of the shorter piece, find the length of each piece.

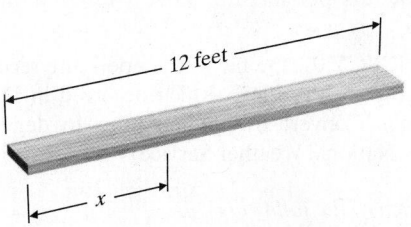

55. In 2015, Target made the decision to close all their Canadian retail stores. Before this decision, Target operated 1926 stores. The number of Target US stores was 69 less than 14 times the number of Target Canada stores. How many Target stores were located in each country?

56. Find three consecutive integers whose sum is -114.

57. The quotient of a number and 3 is the same as the difference of the number and two. Find the number.

58. Double the sum of a number and 6 is the opposite of the number. Find the number.

(2.5) *Substitute the given values into the given formulas and solve for the unknown variable.*

59. $P = 2l + 2w$; $P = 46, l = 14$

60. $V = lwh$; $V = 192, l = 8, w = 6$

Solve each equation as indicated.

61. $y = mx + b$ for m

62. $r = vst - 5$ for s

63. $2y - 5x = 7$ for x

64. $3x - 6y = -2$ for y

△ **65.** $C = \pi D$ for π

△ **66.** $C = 2\pi r$ for π

△ **67.** A swimming pool holds 900 cubic meters of water. If its length is 20 meters and its height is 3 meters, find its width.

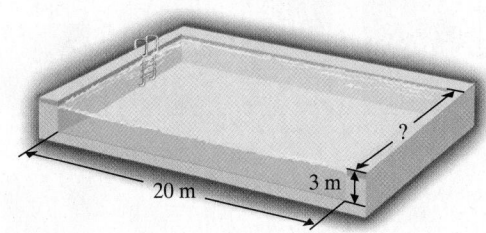

68. The perimeter of a rectangular billboard is 60 feet and has a length 6 feet longer than its width. Find the dimensions of the billboard.

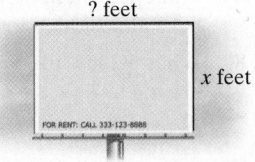

69. A charity 10K race is given annually to benefit a local hospice organization. How long will it take to run/walk a 10K race (10 kilometers or 10,000 meters) if your average pace is 125 **meters** per minute? Give your time in hours and minutes.

70. On April 28, 2001, the highest temperature recorded in the United States was 104°F, which occurred in Death Valley, California. Convert this temperature to degrees Celsius. (*Source:* National Weather Service)

(2.6) *Find each of the following.*

71. The number 9 is what percent of 45?

72. The number 59.5 is what percent of 85?

73. The number 137.5 is 125% of what number?

74. The number 768 is 60% of what number?

75. The price of a small diamond ring was recently increased by 11%. If the ring originally cost $1900, find the mark-up and the new price of the ring.

76. A recent survey found that 79% of Americans use the Internet. If a city has a population of 76,000, how many people in that city would you expect to use the Internet? (*Source:* PEW)

77. Thirty gallons of a 20% acid solution is needed for an experiment. Only 40% and 10% acid solutions are available. How much of each should be mixed to form the needed solution?

78. The ACT Assessment is a college entrance exam taken by about 57% of college-bound students. The national average score was 20.7 in 1993 and rose to 21.0 in 2014. Find the percent of increase. (Round to the nearest hundredth of a percent.)

The graph below shows the percents of cell phone users who have engaged in various behaviors while driving and talking on their cell phones. Use this graph to answer Exercises 79 through 82.

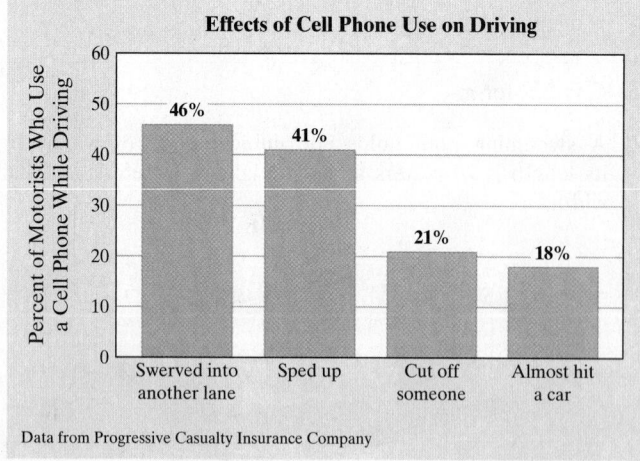

Effects of Cell Phone Use on Driving

Data from Progressive Casualty Insurance Company

79. What percent of motorists who use a cell phone while driving have almost hit another car?

80. What is the most common effect of cell phone use on driving?

Suppose that a cell-phone service has an estimated 4600 customers who use their cell phones while driving. Use this information for Exercises 81 and 82.

81. How many of these customers would you expect to have cut someone off while driving and talking on their cell phones?

82. How many of these customers would you expect to have sped up while driving and talking on their cell phones?

83. If a price decreases from $250 to $170, what is the percent of decrease?

84. Find the original price of a DVD if the sale price is $19.20 after a 20% discount.

85. Vincenzo Nibali of Italy won the 2014 Tour de France. Suppose he rides a bicycle up a category 2 climb at 10 km/hr and rides down the same distance at a speed of 50 km/hr. Find the distance traveled if the total time on the mountain was 3 hours.

(2.7) *Solve.*

86. A $50,000 retirement pension is to be invested into two accounts: a money market fund that pays 8.5% and a certificate of deposit that pays 10.5%. How much should be invested at each rate to provide a yearly interest income of $4550?

87. A pay phone is holding its maximum number of 500 coins consisting of nickels, dimes, and quarters. The number of quarters is twice the number of dimes. If the value of all the coins is $88.00, how many nickels were in the pay phone?

88. How long will it take an Amtrak passenger train to catch up to a freight train if their speeds are 60 and 45 mph and the freight train had an hour and a half head start?

(2.8) *Solve and graph the solutions of each of the following inequalities.*

89. $x > 0$

90. $x \leq -2$

91. $0.5 \leq y < 1.5$

92. $-1 < x < 1$

93. $-3x > 12$

94. $-2x \geq -20$

95. $x + 4 \geq 6x - 16$

96. $5x - 7 > 8x + 5$

97. $-3 < 4x - 1 < 2$

98. $2 \leq 3x - 4 < 6$

99. $4(2x - 5) \leq 5x - 1$

100. $-2(x - 5) > 2(3x - 2)$

101. Tina earns $175 per week plus a 5% commission on all her sales. Find the minimum amount of sales to ensure that she earns at least $300 per week.

102. Ellen Catarella shot rounds of 76, 82, and 79 golfing. What must she shoot on her next round so that her average will be below 80?

MIXED REVIEW

Solve each equation.

103. $6x + 2x - 1 = 5x + 11$

104. $2(3y - 4) = 6 + 7y$

105. $4(3 - a) - (6a + 9) = -12a$

106. $\dfrac{x}{3} - 2 = 5$

107. $2(y + 5) = 2y + 10$

108. $7x - 3x + 2 = 2(2x - 1)$

Solve.

109. The sum of six and twice a number is equal to seven less than the number. Find the number.

110. A 23-inch piece of string is to be cut into two pieces so that the length of the longer piece is three more than four times the shorter piece. If x represents the length of the shorter piece, find the lengths of both pieces.

Solve for the specified variable.

111. $V = \dfrac{1}{3} Ah$ for h

112. What number is 26% of 85?

113. The number 72 is 45% of what number?

114. A company recently increased its number of employees from 235 to 282. Find the percent of increase.

Solve each inequality. Graph the solution set.

115. $4x - 7 > 3x + 2$

116. $-5x < 20$

117. $-3(1 + 2x) + x \geq -(3 - x)$

Chapter 2 Getting Ready for the Test

MULTIPLE CHOICE *Exercises 1–4 below are given. Choose the best directions (choice A, B, C, or D) below for each exercise.*

   A. Solve for x.   **B.** Simplify.   **C.** Identify the numerical coefficient.   **D.** Are these like or unlike terms?

1. Given: $-3x^2$

2. Given: $4x - 5 = 2x + 3$

3. Given: $5x^2$ and $4x$

4. Given: $4x - 5 + 2x + 3$

MULTIPLE CHOICE

5. Subtracting $100z$ from $8m$ translates to

   A. $100z - 8m$ **B.** $8m - 100z$ **C.** $-800zm$ **D.** $92zm$

6. Subtracting $7x - 1$ from $9y$ translates to:

   A. $7x - 1 - 9y$ **B.** $9y - 7x - 1$ **C.** $9y - (7x - 1)$ **D.** $7x - 1 - (9y)$

MATCHING *Match each equation in the first column with its solution in the second column. Items in the second column may be used more than once.*

7. $7x + 6 = 7x + 9$ **A.** all real numbers

8. $2y - 5 = 2y - 5$ **B.** no solution

9. $11x - 13 = 10x - 13$ **C.** the solution is 0

10. $x + 15 = -x + 15$

MULTIPLE CHOICE

11. To solve $5(3x - 2) = -(x + 20)$, we first use the distributive property and remove parentheses by multiplying. Once this is done, the equation is

   A. $15x - 2 = -x + 20$ **B.** $15x - 10 = -x - 20$ **C.** $15x - 10 = -x + 20$ **D.** $15x - 7 = -x - 20$

12. To solve $\dfrac{8x}{3} + 1 = \dfrac{x - 2}{10}$ we multiply through by the LCD, 30. Once this is done, the simplified equation is

   A. $80x + 1 = 3x - 6$ **B.** $80x + 6 = 3x - 6$ **C.** $8x + 1 = x - 2$ **D.** $80x + 30 = 3x - 6$

Chapter 2 Test MyMathLab® You Tube

Simplify each of the following expressions.

1. $2y - 6 - y - 4$

2. $2.7x + 6.1 + 3.2x - 4.9$

3. $4(x - 2) - 3(2x - 6)$

4. $7 + 2(5y - 3)$

Solve each of the following equations.

5. $-\dfrac{4}{5}x = 4$

6. $4(n - 5) = -(4 - 2n)$

7. $5y - 7 + y = -(y + 3y)$

8. $4z + 1 - z = 1 + z$

9. $\dfrac{2(x + 6)}{3} = x - 5$

10. $\dfrac{1}{2} - x + \dfrac{3}{2} = x - 4$

11. $-0.3(x - 4) + x = 0.5(3 - x)$

12. $-4(a + 1) - 3a = -7(2a - 3)$

13. $-2(x - 3) = x + 5 - 3x$

14. Find the value of x if $y = -14$, $m = -2$, and $b = -2$ in the formula $y = mx + b$.

Solve each of the following equations for the indicated variable.

15. $V = \pi r^2 h$ for h

16. $3x - 4y = 10$ for y

Solve each of the following inequalities. Graph each solution set and write it in interval notation.

17. $3x - 5 \geq 7x + 3$

18. $x + 6 > 4x - 6$

19. $-2 < 3x + 1 < 8$

20. $\dfrac{2(5x + 1)}{3} > 2$

Solve each of the following applications.

21. A number increased by two-thirds of the number is 35. Find the number.

22. A rectangular deck is to be built so that the width and length are two consecutive even integers and the perimeter is 252 feet. Find the dimensions of the deck.

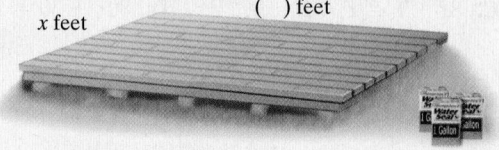

x feet   ( ) feet

23. Some states have a single area code for the entire state. Two such states have area codes where one is double the other. If the sum of these integers is 1203, find the two area codes. *(Source:* North American Numbering Plan Administration*)*

24. Sedric Angell invested an amount of money in Amoxil stock that earned an annual 10% return, and then he invested twice that amount in IBM stock that earned an annual 12% return. If his total return from both investments was $2890, find how much he invested in each stock.

25. Two trains leave Los Angeles simultaneously traveling on the same track in opposite directions at speeds of 50 and 64 mph. How long will it take before they are 285 miles apart?

The following graph shows the breakdown of tornadoes occurring in the United States by strength. The corresponding Fujita Tornado Scale categories are shown in parentheses. Use this graph to answer Exercise 26.

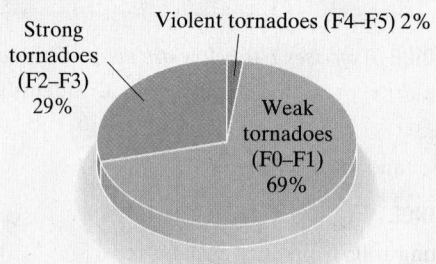

Strong tornadoes (F2–F3) 29%

Violent tornadoes (F4–F5) 2%

Weak tornadoes (F0–F1) 69%

Data from National Climatic Data Center

26. According to the National Climatic Data Center, in an average year, about 800 tornadoes are reported in the United States. How many of these would you expect to be classified as "weak" tornadoes?

27. The number 72 is what percent of 180?

28. The number of employees of a company decreased from 225 to 189. Find this percent of decrease.

1. Given the set $\left\{-2, 0, \frac{1}{4}, -1.5, 112, -3, 11, \sqrt{2}\right\}$, list the numbers in this set that belong to the set of:
   a. Natural numbers
   b. Whole numbers
   c. Integers
   d. Rational numbers
   e. Irrational numbers
   f. Real numbers

2. Given the set $\left\{7, 2, -\frac{1}{5}, 0, \sqrt{3}, -185, 8\right\}$, list the numbers in this set that belong to the set of:
   a. Natural numbers
   b. Whole numbers
   c. Integers
   d. Rational numbers
   e. Irrational numbers
   f. Real numbers

3. Find the absolute value of each number.
   a. $|4|$
   b. $|-5|$
   c. $|0|$
   d. $\left|-\frac{1}{2}\right|$
   e. $|5.6|$

4. Find the absolute value of each number.
   a. $|5|$
   b. $|-8|$
   c. $\left|-\frac{2}{3}\right|$

5. Write each of the following numbers as a product of primes.
   a. 40
   b. 63

6. Write each number as a product of primes.
   a. 44
   b. 90

7. Write $\frac{2}{5}$ as an equivalent fraction with a denominator of 20.

8. Write $\frac{2}{3}$ as an equivalent fraction with a denominator of 24.

9. Simplify: $3[4 + 2(10 - 1)]$

10. Simplify: $5[16 - 4(2 + 1)]$

11. Decide whether 2 is a solution of $3x + 10 = 8x$.

12. Decide whether 3 is a solution of $5x - 2 = 4x$.

Add.

13. $-1 + (-2)$

14. $(-2) + (-8)$

15. $-4 + 6$

16. $-3 + 10$

17. Simplify each expression.
   a. $-(-10)$
   b. $-\left(-\frac{1}{2}\right)$
   c. $-(-2x)$
   d. $-|-6|$

18. Simplify each expression.
   a. $-(-5)$
   b. $-\left(-\frac{2}{3}\right)$
   c. $-(-a)$
   d. $-|-3|$

19. Subtract.
   a. $5.3 - (-4.6)$
   b. $-\frac{3}{10} - \frac{5}{10}$
   c. $-\frac{2}{3} - \left(-\frac{4}{5}\right)$

20. Subtract
   a. $-2.7 - 8.4$
   b. $-\frac{4}{5} - \left(-\frac{3}{5}\right)$
   c. $\frac{1}{4} - \left(-\frac{1}{2}\right)$

21. Find each unknown complementary or supplementary angle.
   a.   b.

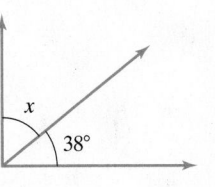 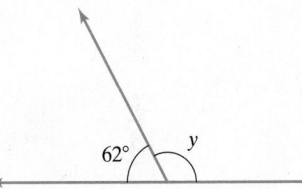

22. Find each unknown complementary or supplementary angle.
   a.   b.

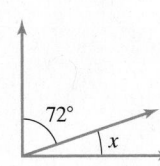 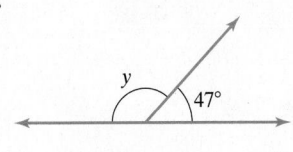

23. Multiply.
   a. $(-1.2)(0.05)$
   b. $\frac{2}{3} \cdot \left(-\frac{7}{10}\right)$
   c. $\left(-\frac{4}{5}\right)(-20)$

24. Find each product.
   a. $(4.5)(-0.08)$
   b. $-\frac{3}{4} \cdot \left(-\frac{8}{17}\right)$

25. Divide.
   a. $\frac{-24}{-4}$
   b. $\frac{-36}{3}$
   c. $\frac{2}{3} \div \left(-\frac{5}{4}\right)$
   d. $-\frac{3}{2} \div 9$

26. Divide.
   a. $\frac{-32}{8}$
   b. $\frac{-108}{-12}$
   c. $\frac{-5}{7} \div \left(\frac{-9}{2}\right)$

27. Use a commutative property to complete each statement.
   a. $x + 5 = $ _____
   b. $3 \cdot x = $ _____

28. Use a commutative property to complete each statement.
   a. $y + 1 = $ _____
   b. $y \cdot 4 = $ _____

29. Use the distributive property to write each sum as a product.
 a. $8 \cdot 2 + 8 \cdot x$
 b. $7s + 7t$

30. Use the distributive property to write each sum as a product.
 a. $4 \cdot y + 4 \cdot \dfrac{1}{3}$
 b. $0.10x + 0.10y$

31. Subtract $4x - 2$ from $2x - 3$.

32. Subtract $10x + 3$ from $-5x + 1$.

Solve.

33. $y + 0.6 = -1.0$

34. $\dfrac{5}{6} + x = \dfrac{2}{3}$

35. $7 = -5(2a - 1) - (-11a + 6)$

36. $-3x + 1 - (-4x - 6) = 10$

37. $\dfrac{y}{7} = 20$

38. $\dfrac{x}{4} = 18$

39. $4(2x - 3) + 7 = 3x + 5$

40. $6x + 5 = 4(x + 4) - 1$

41. Twice the sum of a number and 4 is the same as four times the number, decreased by 12. Find the number.

42. A number increased by 4 is the same as 3 times the number decreased by 8. Find the number.

43. Solve: $V = lwh$ for l.

44. Solve: $C = 2\pi r$ for r.

45. Solve $x + 4 \le -6$ for x. Graph the solution set and write it in interval notation.

46. Solve $x - 3 > 2$ for x. Graph the solution set and write it in interval notation.

Graphing

3.1 Reading Graphs and the Rectangular Coordinate System

3.2 Graphing Linear Equations

3.3 Intercepts

3.4 Slope and Rate of Change

Integrated Review—Summary on Slope and Graphing Linear Equations

3.5 Equations of Lines

3.6 Functions

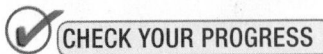 CHECK YOUR PROGRESS

Vocabulary Check
Chapter Highlights
Chapter Review
Getting Ready for the Test
Chapter Test
Cumulative Review

In the previous chapter, we learned to solve and graph the solutions of linear equations and inequalities in one variable. Now we define and present techniques for solving and graphing linear equations and inequalities in two variables.

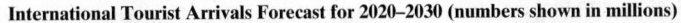
International Tourist Arrivals Forecast for 2020–2030 (numbers shown in millions)

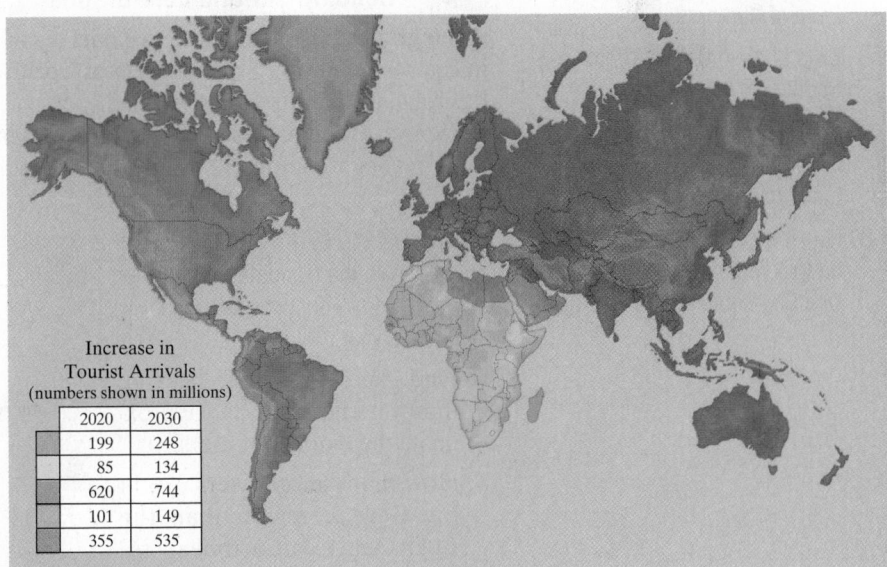

Increase in Tourist Arrivals (numbers shown in millions)

2020	2030
199	248
85	134
620	744
101	149
355	535

What Is Tourism Toward 2030?

Tourism 2020 Vision is the World Tourism Organization's long-term forecast of world tourism through 2020. *Tourism Towards 2030* is its new program title for longer-term forecasts to 2030. The broken-line graph below shows the forecast for number of tourists, which is extremely important as these numbers greatly affect a country's economy. In Section 3.1, Exercises 1 through 6, we read a bar graph showing the top tourist destinations by country.

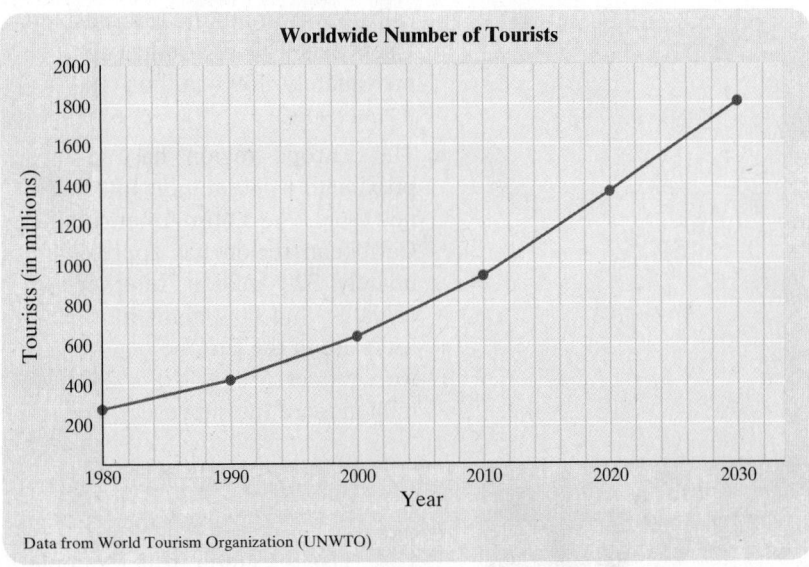

Data from World Tourism Organization (UNWTO)

3.1 Reading Graphs and the Rectangular Coordinate System

OBJECTIVES

1 Read Bar and Line Graphs. ▶

2 Define the Rectangular Coordinate System and Plot Ordered Pairs of Numbers. ▶

3 Graph Paired Data to Create a Scatter Diagram. ▶

4 Determine Whether an Ordered Pair Is a Solution of an Equation in Two Variables. ▶

5 Find the Missing Coordinate of an Ordered Pair Solution, Given One Coordinate of the Pair. ▶

In today's world, where the exchange of information must be fast and entertaining, graphs are becoming increasingly popular. They provide a quick way of making comparisons, drawing conclusions, and approximating quantities.

OBJECTIVE

1   Reading Bar and Line Graphs

A **bar graph** consists of a series of bars arranged vertically or horizontally. The bar graph in Example 1 shows a comparison of worldwide Internet users by region. The names of the regions are listed vertically and a bar is shown for each region. Corresponding to the length of the bar for each region is a number along a horizontal axis. These horizontal numbers are numbers of Internet users in millions.

EXAMPLE I   The bar graph shows the estimated number of Internet users worldwide by region as of a recent year.

a. Find the region that has the most Internet users and approximate the number of users.

b. How many more users are in the Europe region than the Latin America/Caribbean region?

Solution

a. Since these bars are arranged horizontally, we look for the longest bar, which is the bar representing Asia. To approximate the number associated with this region, we move from the right edge of this bar vertically downward to the Internet Users axis. This region has approximately 1390 million Internet users.

b. The Europe region has approximately 580 million Internet users. The Latin America/Caribbean region has approximately 320 million Internet users. To find how many more users are in the Europe region, we subtract $580 - 320 = 260$ million more Internet users.

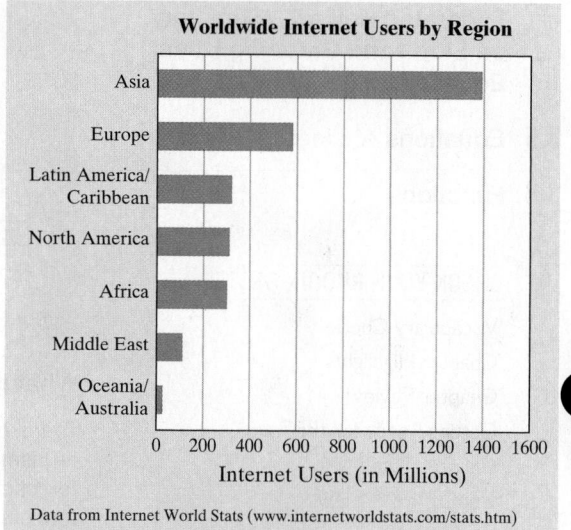

Data from Internet World Stats (www.internetworldstats.com/stats.htm)

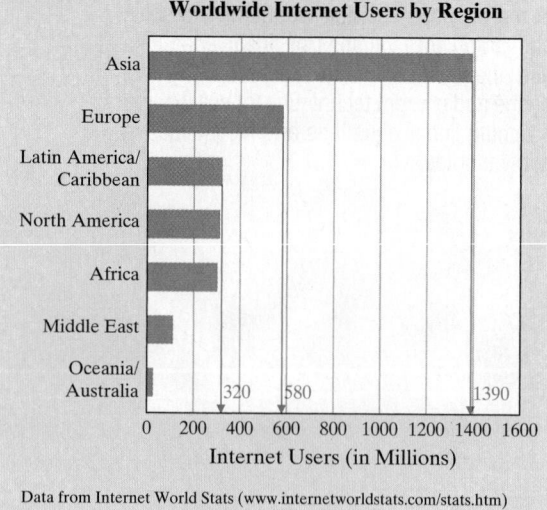

Data from Internet World Stats (www.internetworldstats.com/stats.htm)

PRACTICE

1   Use the graph from Example 1 to answer the following.

a. Find the region with the fewest Internet users and approximate the number of users.

b. How many more users are in the Middle East region than in the Oceania/Australia region?

A **line graph** consists of a series of points connected by a line. The next graph is an example of a line graph. It is also sometimes called a **broken-line graph.**

EXAMPLE 2   The line graph shows the relationship between time spent smoking a cigarette and pulse rate. Time is recorded along the horizontal axis in minutes, with 0 minutes being the moment a smoker lights a cigarette. Pulse is recorded along the vertical axis in heartbeats per minute.

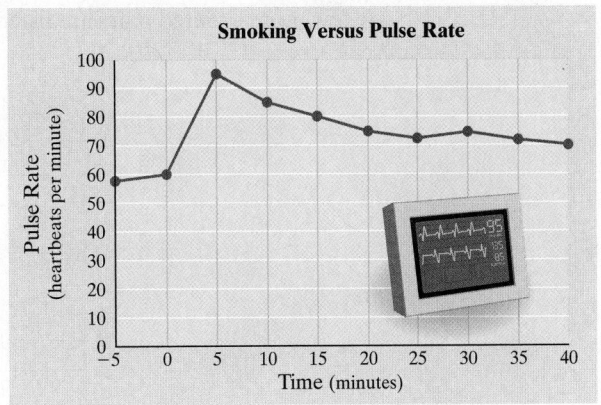

a. What is the pulse rate 15 minutes after a cigarette is lit?

b. When is the pulse rate the lowest?

c. When does the pulse rate show the greatest change?

Solution

a. We locate the number 15 along the time axis and move vertically upward until the line is reached. From this point on the line, we move horizontally to the left until the pulse rate axis is reached. Reading the number of beats per minute, we find that the pulse rate is 80 beats per minute 15 minutes after a cigarette is lit.

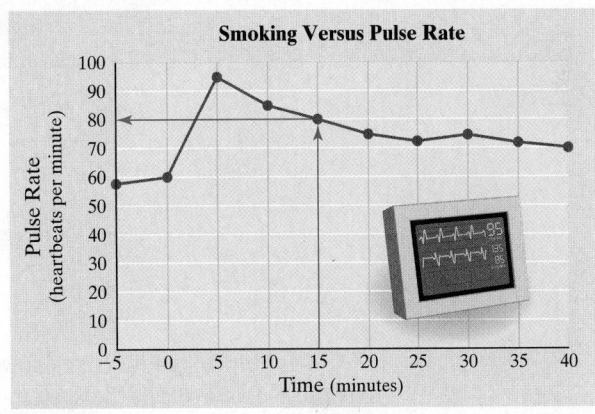

b. We find the lowest point of the line graph, which represents the lowest pulse rate. From this point, we move vertically downward to the time axis. We find that the pulse rate is the lowest at −5 minutes, which means 5 minutes *before* lighting a cigarette.

c. The pulse rate shows the greatest change during the 5 minutes between 0 and 5. Notice that the line graph is *steepest* between 0 and 5 minutes.   □

PRACTICE

2   Use the graph from Example 2 to answer the following.

a. What is the pulse rate 40 minutes after lighting a cigarette?

b. What is the pulse rate when the cigarette is being lit?

c. When is the pulse rate the highest?

OBJECTIVE

2 Defining the Rectangular Coordinate System and Plotting Ordered Pairs of Numbers ▶

Notice in the previous graph that two numbers are associated with each point of the graph. For example, we discussed earlier that 15 minutes after lighting a cigarette, the pulse rate is 80 beats per minute. If we agree to write the time first and the pulse rate second, we can say there is a point on the graph corresponding to the **ordered pair** of numbers (15, 80). A few more ordered pairs are listed alongside their corresponding points.

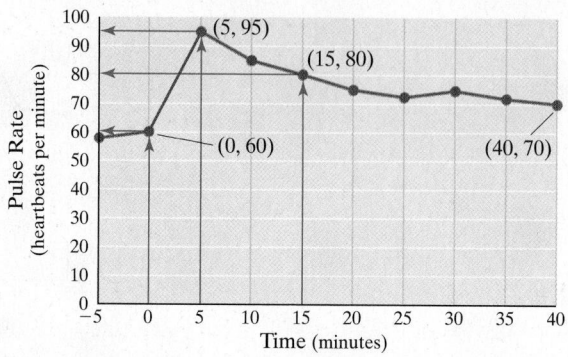

In general, we use this same ordered pair idea to describe the location of a point in a plane (such as a piece of paper). We start with a horizontal and a vertical axis. Each axis is a number line and, for the sake of consistency, we construct our axes to intersect at the 0 coordinate of both. This point of intersection is called the **origin**. Notice that these two number lines or axes divide the plane into four regions called **quadrants**. The quadrants are usually numbered with Roman numerals as shown. The axes are not considered to be in any quadrant.

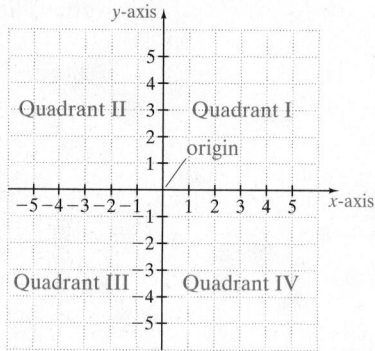

It is helpful to label axes, so we label the horizontal axis the **x-axis** and the vertical axis the **y-axis.** We call the system described above the **rectangular coordinate system.**

Just as with the pulse rate graph, we can then describe the locations of points by ordered pairs of numbers. We list the horizontal **x-axis** measurement first and the vertical **y-axis** measurement second.

To plot or graph the point corresponding to the ordered pair

$$(a, b)$$

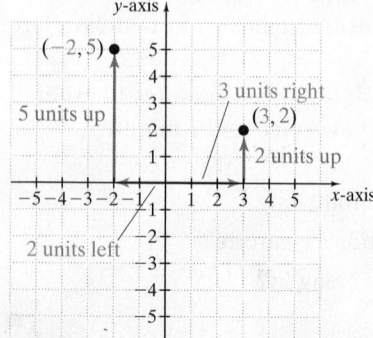

we start at the origin. We then move a units left or right (right if a is positive, left if a is negative). From there, we move b units up or down (up if b is positive, down if b is negative). For example, to plot the point corresponding to the ordered pair (3, 2), we start at the origin, move 3 units right and from there move 2 units up. (See the figure to the left.) The x-value, 3, is called the **x-coordinate** and the y-value, 2, is called the **y-coordinate.** From now on, we will call the point with coordinates (3, 2) simply the point (3, 2). The point (−2, 5) is graphed to the left also.

Does the order in which the coordinates are listed matter? Yes! Notice below that the point corresponding to the ordered pair $(2, 3)$ is in a different location than the point corresponding to $(3, 2)$. These two ordered pairs of numbers describe two different points of the plane.

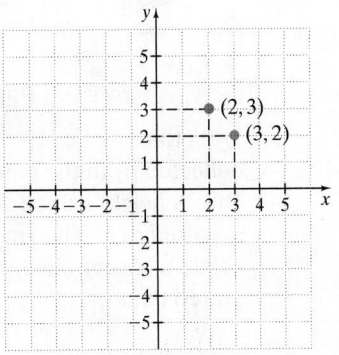

✔ **CONCEPT CHECK**

Is the graph of the point $(-5, 1)$ in the same location as the graph of the point $(1, -5)$? Explain.

Helpful Hint

Don't forget that **each ordered pair corresponds to exactly one point in the plane and that each point in the plane corresponds to exactly one ordered pair.**

EXAMPLE 3   On a single coordinate system, plot each ordered pair. State in which quadrant, if any, each point lies.

a. $(5, 3)$ **b.** $(-5, 3)$ **c.** $(-2, -4)$ **d.** $(1, -2)$ **e.** $(0, 0)$

f. $(0, 2)$ **g.** $(-5, 0)$ **h.** $\left(0, -5\frac{1}{2}\right)$ **i.** $\left(4\frac{2}{3}, -3\right)$

Solution

a. Point $(5, 3)$ lies in quadrant I.

b. Point $(-5, 3)$ lies in quadrant II.

c. Point $(-2, -4)$ lies in quadrant III.

d. Point $(1, -2)$ lies in quadrant IV.

e.–h. Points $(0, 0)$, $(0, 2)$, $(-5, 0)$, and $\left(0, -5\frac{1}{2}\right)$ lie on axes, so they are not in any quadrant.

i. Point $\left(4\frac{2}{3}, -3\right)$ lies in quadrant IV.

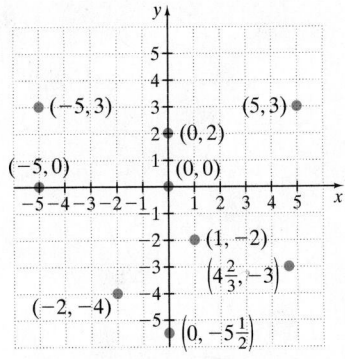

PRACTICE

3   On a single coordinate system, plot each ordered pair. State in which quadrant, if any, each point lies.

a. $(4, -3)$ **b.** $(-3, 5)$ **c.** $(0, 4)$ **d.** $(-6, 1)$

e. $(-2, 0)$ **f.** $(5, 5)$ **g.** $\left(3\frac{1}{2}, 1\frac{1}{2}\right)$ **h.** $(-4, -5)$

Answer to Concept Check:
The graph of point $(-5, 1)$ lies in quadrant II and the graph of point $(1, -5)$ lies in quadrant IV. They are *not* in the same location.

From Example 3, notice that the *y*-coordinate of any point on the *x*-axis is 0. For example, the point $(-5, 0)$ lies on the *x*-axis. Also, the *x*-coordinate of any point on the *y*-axis is 0. For example, the point $(0, 2)$ lies on the *y*-axis.

✔ **CONCEPT CHECK**

For each description of a point in the rectangular coordinate system, write an ordered pair that represents it.

a. Point *A* is located three units to the left of the *y*-axis and five units above the *x*-axis.
b. Point *B* is located six units below the origin.

OBJECTIVE

3 Graphing Paired Data

Data that can be represented as an ordered pair is called **paired data.** Many types of data collected from the real world are paired data. For instance, the annual measurement of a child's height can be written as an ordered pair of the form (year, height in inches) and is paired data. The graph of paired data as points in the rectangular coordinate system is called a **scatter diagram.** Scatter diagrams can be used to look for patterns and trends in paired data.

EXAMPLE 4 The table gives the annual net sales (in billions of dollars) for Target stores for the years shown. (*Source:* Corporate.target.com)

Year	Target Net Sales (in billions of dollars)
2009	65
2010	67
2011	70
2012	73
2013	73
2014	73

a. Write this paired data as a set of ordered pairs of the form (year, sales in billions of dollars).

b. Create a scatter diagram of the paired data.

c. What trend in the paired data does the scatter diagram show?

Solution

a. The ordered pairs are (2009, 65), (2010, 67), (2011, 70), (2012, 73), (2013, 73), (2014, 73).

b. We begin by plotting the ordered pairs. Because the *x*-coordinate in each ordered pair is a year, we label the *x*-axis "Year" and mark the horizontal axis with the years given. Then we label the *y*-axis or vertical axis "Net Sales (in billions of dollars)." In this case we can mark the vertical axis in multiples of 2. Since no net sale is less than 64, we use the notation ⌇ to skip to 64, then proceed by multiples of 2.

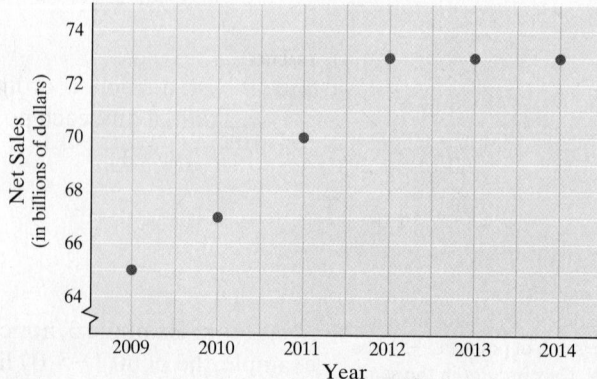

c. The scatter diagram shows that Target net sales were constant or increasing over the years 2009–2014.   □

PRACTICE
4   The table gives the approximate annual number of wildfires (in thousands) that have occurred in the United States for the years shown. (*Source:* National Interagency Fire Center)

Year	Wildfires (in thousands)
2008	79
2009	79
2010	72
2011	73
2012	56
2013	48
2014	46

a. Write this paired data as a set of ordered pairs of the form (year, number of wildfires in thousands).

b. Create a scatter diagram of the paired data.

OBJECTIVE

4   Determining Whether an Ordered Pair Is a Solution ▶

Let's see how we can use ordered pairs to record solutions of equations containing two variables. An equation in one variable such as $x + 1 = 5$ has one solution, which is 4: The number 4 is the value of the variable x that makes the equation true.

An equation in two variables, such as $2x + y = 8$, has solutions consisting of two values, one for x and one for y. For example, $x = 3$ and $y = 2$ is a solution of $2x + y = 8$ because, if x is replaced with 3 and y with 2, we get a true statement.

$$2x + y = 8$$
$$2(3) + 2 = 8$$
$$8 = 8 \quad \text{True}$$

The solution $x = 3$ and $y = 2$ can be written as $(3, 2)$, an **ordered pair** of numbers. The first number, 3, is the x-value and the second number, 2, is the y-value.

In general, an ordered pair is a **solution** of an equation in two variables if replacing the variables by the values of the ordered pair results in a true statement.

EXAMPLE 5   Determine whether each ordered pair is a solution of the equation $x - 2y = 6$.

    a. $(6, 0)$ **b.** $(0, 3)$ **c.** $\left(1, -\dfrac{5}{2}\right)$

Solution

    a. Let $x = 6$ and $y = 0$ in the equation $x - 2y = 6$.

$$x - 2y = 6$$
$$6 - 2(0) = 6 \quad \text{Replace } x \text{ with 6 and } y \text{ with 0.}$$
$$6 - 0 = 6 \quad \text{Simplify.}$$
$$6 = 6 \quad \text{True}$$

$(6, 0)$ is a solution, since $6 = 6$ is a true statement.

(Continued on next page)

b. Let $x = 0$ and $y = 3$.

$$x - 2y = 6$$
$$0 - 2(3) = 6 \quad \text{Replace } x \text{ with 0 and } y \text{ with 3.}$$
$$0 - 6 = 6$$
$$-6 = 6 \quad \text{False}$$

$(0, 3)$ is *not* a solution, since $-6 = 6$ is a false statement.

c. Let $x = 1$ and $y = -\dfrac{5}{2}$ in the equation.

$$x - 2y = 6$$
$$1 - 2\left(-\frac{5}{2}\right) = 6 \quad \text{Replace } x \text{ with 1 and } y \text{ with } -\frac{5}{2}.$$
$$1 + 5 = 6$$
$$6 = 6 \quad \text{True}$$

$\left(1, -\dfrac{5}{2}\right)$ is a solution, since $6 = 6$ is a true statement.   □

PRACTICE

5   Determine whether each ordered pair is a solution of the equation $x + 3y = 6$.

a. $(3, 1)$ **b.** $(6, 0)$ **c.** $\left(-2, \dfrac{2}{3}\right)$ ■

OBJECTIVE

5   Completing Ordered Pair Solutions ▷

If one value of an ordered pair solution of an equation is known, the other value can be determined. To find the unknown value, replace one variable in the equation by its known value. Doing so results in an equation with just one variable that can be solved for the variable using the methods of Chapter 2.

EXAMPLE 6   Complete the following ordered pair solutions of the equation $3x + y = 12$.

a. $(0, \ )$ **b.** $(, 6)$ **c.** $(-1, \ )$

Solution

a. In the ordered pair $(0, \ )$, the x-value is 0. Let $x = 0$ in the equation and solve for y.

$$3x + y = 12$$
$$3(0) + y = 12 \quad \text{Replace } x \text{ with 0.}$$
$$0 + y = 12$$
$$y = 12$$

The completed ordered pair is $(0, 12)$.

b. In the ordered pair $(, 6)$, the y-value is 6. Let $y = 6$ in the equation and solve for x.

$$3x + y = 12$$
$$3x + 6 = 12 \quad \text{Replace } y \text{ with 6.}$$
$$3x = 6 \quad \text{Subtract 6 from both sides.}$$
$$x = 2 \quad \text{Divide both sides by 3.}$$

The ordered pair is $(2, 6)$.

c. In the ordered pair $(-1, \quad)$, the x-value is -1. Let $x = -1$ in the equation and solve for y.

$$3x + y = 12$$
$$3(-1) + y = 12 \quad \text{Replace } x \text{ with } -1.$$
$$-3 + y = 12$$
$$y = 15 \quad \text{Add 3 to both sides.}$$

The ordered pair is $(-1, 15)$. ☐

PRACTICE
6   Complete the following ordered pair solutions of the equation $2x - y = 8$.

a. $(0, \quad)$ **b.** $(\quad, 4)$ **c.** $(-3, \quad)$ ■

Solutions of equations in two variables can also be recorded in a **table of values,** as shown in the next example.

EXAMPLE 7   Complete the table for the equation $y = 3x$.

	x	y
a.	-1	
b.		0
c.		-9

Solution

a. Replace x with -1 in the equation and solve for y.

$$y = 3x$$
$$y = 3(-1) \quad \text{Let } x = -1.$$
$$y = -3$$

The ordered pair is $(-1, -3)$.

b. Replace y with 0 in the equation and solve for x.

$$y = 3x$$
$$0 = 3x \quad \text{Let } y = 0.$$
$$0 = x \quad \text{Divide both sides by 3.}$$

The ordered pair is $(0, 0)$.

c. Replace y with -9 in the equation and solve for x.

$$y = 3x$$
$$-9 = 3x \quad \text{Let } y = -9.$$
$$-3 = x \quad \text{Divide both sides by 3.}$$

The ordered pair is $(-3, -9)$. The completed table is shown to the left. ☐

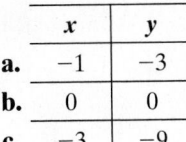

	x	y
a.	-1	-3
b.	0	0
c.	-3	-9

PRACTICE
7   Complete the table for the equation $y = -4x$.

	x	y
a.	-2	
b.		-12
c.	0	

■

EXAMPLE 8 Complete the table for the equation

$$y = \frac{1}{2}x - 5.$$

	x	y
a.	-2	
b.	0	
c.		0

Solution

a. Let $x = -2$.

$$y = \frac{1}{2}x - 5$$

$$y = \frac{1}{2}(-2) - 5$$

$$y = -1 - 5$$

$$y = -6$$

b. Let $x = 0$.

$$y = \frac{1}{2}x - 5$$

$$y = \frac{1}{2}(0) - 5$$

$$y = 0 - 5$$

$$y = -5$$

c. Let $y = 0$.

$$y = \frac{1}{2}x - 5$$

$$0 = \frac{1}{2}x - 5 \quad \text{Now, solve for } x.$$

$$5 = \frac{1}{2}x \quad \text{Add 5.}$$

$$10 = x \quad \text{Multiply by 2.}$$

Ordered pairs: **a.** $(-2, -6)$, **b.** $(0, -5)$, **c.** $(10, 0)$

The completed table is

	x	y
a.	-2	-6
b.	0	-5
c.	10	0

PRACTICE
8 Complete the table for the equation $y = \frac{1}{5}x - 2$.

	x	y
a.	-10	
b.	0	
c.		0

EXAMPLE 9 Finding the Value of a Computer

A computer was recently purchased for a small business for $2000. The business manager predicts that the computer will be used for 5 years and the value in dollars y of the computer in x years is $y = -300x + 2000$. Complete the table.

x	0	1	2	3	4	5
y						

Solution To find the value of y when x is 0, replace x with 0 in the equation. We use this same procedure to find y when x is 1 and when x is 2.

When $x = 0$,

$$y = -300x + 2000$$

$$y = -300 \cdot 0 + 2000$$

$$y = 0 + 2000$$

$$y = 2000$$

When $x = 1$,

$$y = -300x + 2000$$

$$y = -300 \cdot 1 + 2000$$

$$y = -300 + 2000$$

$$y = 1700$$

When $x = 2$,

$$y = -300x + 2000$$

$$y = -300 \cdot 2 + 2000$$

$$y = -600 + 2000$$

$$y = 1400$$

We have the ordered pairs $(0, 2000)$, $(1, 1700)$, and $(2, 1400)$. This means that in 0 years, the value of the computer is $2000, in 1 year the value of the computer is $1700, and in

2 years the value is $1400. To complete the table of values, we continue the procedure for $x = 3$, $x = 4$, and $x = 5$.

When x = 3,

$y = -300x + 2000$

$y = -300 \cdot 3 + 2000$

$y = -900 + 2000$

$y = 1100$

When x = 4,

$y = -300x + 2000$

$y = -300 \cdot 4 + 2000$

$y = -1200 + 2000$

$y = 800$

When x = 5,

$y = -300x + 2000$

$y = -300 \cdot 5 + 2000$

$y = -1500 + 2000$

$y = 500$

The completed table is

x	0	1	2	3	4	5
y	2000	1700	1400	1100	800	500

PRACTICE

9   A college student purchased a used car for $12,000. The student predicted that she would need to use the car for four years and the value in dollars y of the car in x years is $y = -1800x + 12,000$. Complete this table.

x	0	1	2	3	4
y	12,000	10,200	8400	6600	4800

The ordered pair solutions recorded in the completed table for the example above are graphed below. Notice that the graph gives a visual picture of the decrease in value of the computer.

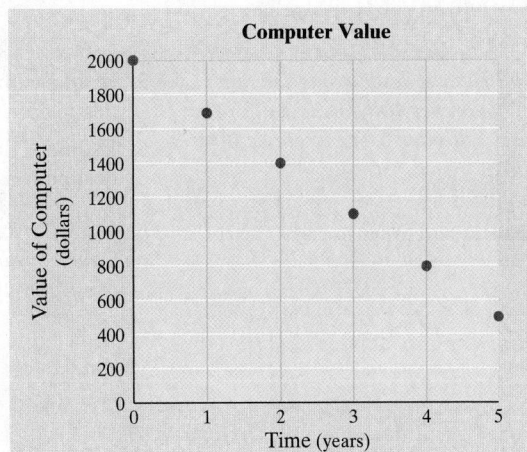

x	y
0	2000
1	1700
2	1400
3	1100
4	800
5	500

✔ Vocabulary, Readiness & Video Check

Use the choices below to fill in each blank. The exercises below all have to do with the rectangular coordinate system.

origin	x-coordinate	x-axis	one	four
quadrants	y-coordinate	y-axis	solution	

1. The horizontal axis is called the _____ and the vertical axis is called the _____.

2. The intersection of the horizontal axis and the vertical axis is a point called the _____.

3. The axes divide the plane into regions called _____. There are _____ of these regions.

4. In the ordered pair of numbers $(-2, 5)$, the number -2 is called the _____ and the number 5 is called the _____.

5. Each ordered pair of numbers corresponds to _____ point in the plane.

6. An ordered pair is a(n) _____ of an equation in two variables if replacing the variables by the coordinates of the ordered pair results in a true statement.

Martin-Gay Interactive Videos

See Video 3.1 ○

Watch the section lecture video and answer the following questions.

OBJECTIVE 1

7. Examples 1–3 ask you to answer questions about a bar graph. What information is provided on the horizontal axis of this bar graph? On the vertical axis?

OBJECTIVE 2

8. Several points are plotted in ▦ Examples 4–11. Where do you always start when plotting a point? How does the 1st coordinate tell you to move? How does the 2nd coordinate tell you to move?

OBJECTIVE 3

9. In the lecture before ▦ Example 12, what connection is made between data and graphing?

OBJECTIVE 4

10. An ordered pair is a solution of an equation if, when the variables are replaced with their values, a true statement results. In ▦ Example 13, three ordered pairs are tested. What are the last two points to be tested? What lesson can be learned by the results of testing these two points and why?

OBJECTIVE 5

11. In ▦ Example 14, when one variable of a linear equation in two variables is replaced by a replacement value, what type of equation results?

3.1 Exercise Set MyMathLab® ▷

The following bar graph shows the top 10 tourist destinations and the number of tourists that visit each destination per year forecasted for 2020. Use this graph to answer Exercises 1 through 6. See Example 1.

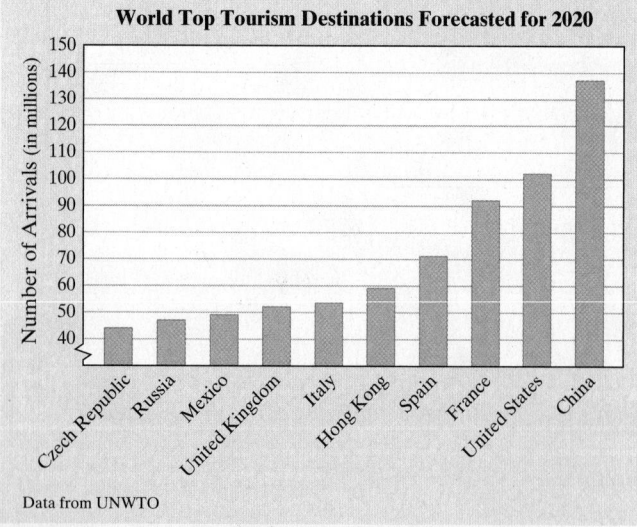

World Top Tourism Destinations Forecasted for 2020

Data from UNWTO

▷ **1.** Which location shown is predicted to be the most popular tourist destination?

2. Which location shown is predicted to be the least popular tourist destination?

▷ **3.** Which locations shown are predicted to have more than 70 million tourists per year?

4. Which locations shown are predicted to have more than 100 million tourists per year?

▷ **5.** Estimate the predicted number of tourists per year whose destination is Italy.

6. Estimate the predicted number of tourists per year whose destination is Mexico.

The following line graph shows the paid attendance at each Super Bowl game from 2008 through 2014. Use this graph to answer Exercises 7 through 10. See Example 2.

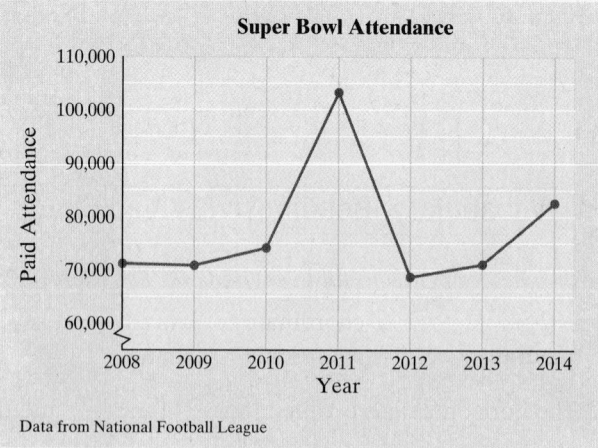

Super Bowl Attendance

Data from National Football League

7. Estimate the Super Bowl attendance in 2014.

8. Estimate the Super Bowl attendance in 2010.

9. Find the year on the graph with the greatest Super Bowl attendance and approximate that attendance.

10. Find the year on the graph with the least Super Bowl attendance and approximate that attendance.

The line graph below shows the number of students per teacher in U.S. public elementary and secondary schools. Use this graph for Exercises 11 through 16. See Example 2.

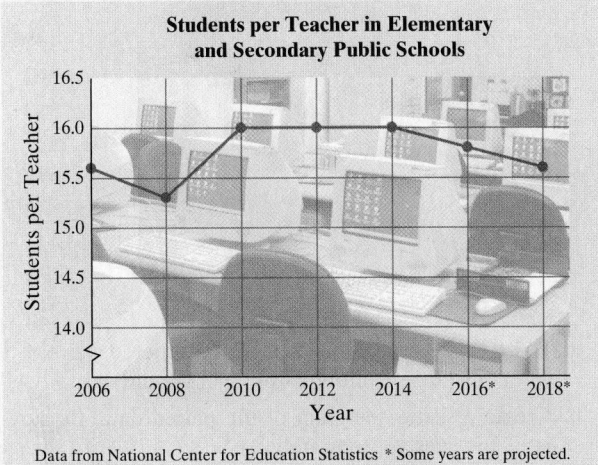

Students per Teacher in Elementary and Secondary Public Schools

Data from National Center for Education Statistics * Some years are projected.

11. Approximate the number of students per teacher predicted in 2016.

12. Approximate the number of students per teacher predicted in 2018.

13. Between what years shown did the greatest increase in number of students per teacher occur?

14. What was the first year shown that the number of students per teacher fell below 15.5?

15. During what period was the student per teacher number at 16?

16. Discuss any trends shown by this line graph.

Plot each ordered pair. State in which quadrant or on which axis each point lies. See Example 3.

17. a. $(1, 5)$ **b.** $(-5, -2)$
    c. $(-3, 0)$ **d.** $(0, -1)$
    e. $(2, -4)$ **f.** $\left(-1, 4\frac{1}{2}\right)$
    g. $(3.7, 2.2)$ **h.** $\left(\frac{1}{2}, -3\right)$

18. a. $(2, 4)$ **b.** $(0, 2)$
    c. $(-2, 1)$ **d.** $(-3, -3)$
    e. $\left(3\frac{3}{4}, 0\right)$ **f.** $(5, -4)$
    g. $(-3.4, 4.8)$ **h.** $\left(\frac{1}{3}, -5\right)$

Find the x- and y-coordinates of each labeled point. See Example 3.

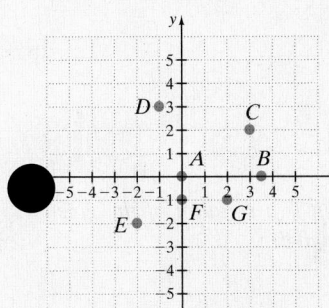

19. *A*
20. *B*
21. *C*
22. *D*
23. *E*
24. *F*
25. *G*

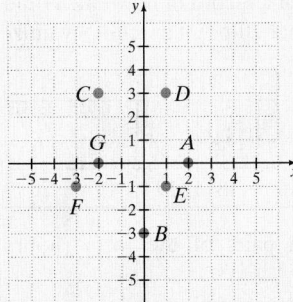

26. *A*
27. *B*
28. *C*
29. *D*
30. *E*
31. *F*
32. *G*

Solve. See Example 4.

33. The table shows the worldwide box office (in billions of dollars) for the movie industry during the years shown. (*Source: Motion Picture Association of America*)

Year	Box Office (in billions of dollars)
2010	21.0
2011	22.4
2012	23.9
2013	25.0
2014	28.0

a. Write this paired data as a set of ordered pairs of the form (year, box office).

b. In your own words, write the meaning of the ordered pair (2014, 28.0).

c. Create a scatter diagram of the paired data. Be sure to label the axes appropriately.

d. What trend in the paired data does the scatter diagram show?

34. The table shows the amount of money (in billions of dollars) Americans spent on their pets for the years shown. (*Source: American Pet Products Manufacturers Association*)

Year	Pet-Related Expenditures (in billions of dollars)
2011	51.0
2012	53.3
2013	55.7
2014	58.5

a. Write this paired data as a set of ordered pairs of the form (year, pet-related expenditures).

b. In your own words, write the meaning of the ordered pair (2014, 58.5).

c. Create a scatter diagram of the paired data. Be sure to label the axes appropriately.

d. What trend in the paired data does the scatter diagram show?

35. Minh, a psychology student, kept a record of how much time she spent studying for each of her 20-point psychology quizzes and her score on each quiz.

Hours Spent Studying	0.50	0.75	1.00	1.25	1.50	1.50	1.75	2.00
Quiz Score	10	12	15	16	18	19	19	20

   a. Write the data as ordered pairs of the form (hours spent studying, quiz score).

   b. In your own words, write the meaning of the ordered pair (1.25, 16).

   c. Create a scatter diagram of the paired data. Be sure to label the axes appropriately.

   d. What might Minh conclude from the scatter diagram?

36. A local lumberyard uses quantity pricing. The table shows the price per board for different amounts of lumber purchased.

Price per Board (in dollars)	Number of Boards Purchased
8.00	1
7.50	10
6.50	25
5.00	50
2.00	100

   a. Write the data as ordered pairs of the form (price per board, number of boards purchased).

   b. In your own words, write the meaning of the ordered pair (2.00, 100).

   c. Create a scatter diagram of the paired data. Be sure to label the axes appropriately.

   d. What trend in the paired data does the scatter diagram show?

37. The table shows the distance from the equator (in miles) and the average annual snowfall (in inches) for each of eight selected U.S. cities. (*Source:* National Climatic Data Center, Wake Forest University Albatross Project)

City	Distance from Equator (in miles)	Average Annual Snowfall (in inches)
1. Atlanta, GA	2313	2
2. Austin, TX	2085	1
3. Baltimore, MD	2711	21
4. Chicago, IL	2869	39
5. Detroit, MI	2920	42
6. Juneau, AK	4038	99
7. Miami, FL	1783	0
8. Winston-Salem, NC	2493	9

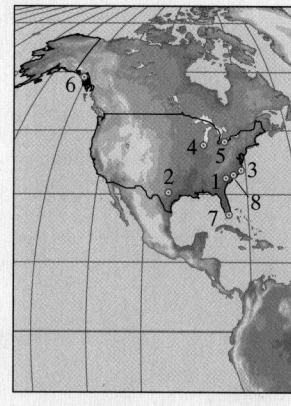

   a. Write this paired data as a set of ordered pairs of the form (distance from equator, average annual snowfall).

   b. Create a scatter diagram of the paired data. Be sure to label the axes appropriately.

   c. What trend in the paired data does the scatter diagram show?

38. The table shows the average farm size (in acres) in the United States during the years shown. (*Source:* National Agricultural Statistics Service)

Year	Average Farm Size (in acres)
2009	418
2010	418
2011	429
2012	433
2013	435
2014	437

   a. Write this paired data as a set of ordered pairs of the form (year, average farm size).

   b. Create a scatter diagram of the paired data. Be sure to label the axes appropriately.

Determine whether each ordered pair is a solution of the given linear equation. See Example 5.

39. $2x + y = 7$; $(3, 1), (7, 0), (0, 7)$

40. $3x + y = 8$; $(2, 3), (0, 8), (8, 0)$

41. $x = -\frac{1}{3}y$; $(0, 0), (3, -9)$

42. $y = -\frac{1}{2}x$; $(0, 0), (4, 2)$

43. $x = 5$; $(4, 5), (5, 4), (5, 0)$

44. $y = -2$; $(-2, 2), (2, -2), (0, -2)$

Complete each ordered pair so that it is a solution of the given linear equation. See Examples 6 through 8.

45. $x - 4y = 4$; $(\quad, -2), (4, \quad)$

46. $x - 5y = -1$; $(\quad, -2), (4, \quad)$

47. $y = \dfrac{1}{4}x - 3$; $(-8, \quad)$, $(\quad, 1)$

48. $y = \dfrac{1}{5}x - 2$; $(-10, \quad)$, $(\quad, 1)$

Complete the table of ordered pairs for each linear equation. See Examples 6 through 8.

49. $y = -7x$

x	y
0	
−1	
	2

50. $y = -9x$

x	y
	0
−3	
	2

51. $y = -x + 2$

x	y
0	
	0
−3	

52. $x = -y + 4$

x	y
	0
0	
	−3

53. $y = \dfrac{1}{2}x$

x	y
0	
−6	
	1

54. $y = \dfrac{1}{3}x$

x	y
0	
−6	
	1

55. $x + 3y = 6$

x	y
0	
	0
	1

56. $2x + y = 4$

x	y
	4
2	
	2

57. $y = 2x - 12$

x	y
0	
	−2
3	

58. $y = 5x + 10$

x	y
	0
	5
0	

59. $2x + 7y = 5$

x	y
0	
	0
	1

60. $x - 6y = 3$

x	y
0	
	1
	−1

MIXED PRACTICE

Complete the table of ordered pairs for each equation. Then plot the ordered pair solutions. See Examples 1 through 7.

61. $x = -5y$

x	y
	0
	1
10	

62. $y = -3x$

x	y
0	
	−2
	9

63. $y = \dfrac{1}{3}x + 2$

x	y
0	
−3	
	0

64. $y = \dfrac{1}{2}x + 3$

x	y
0	
−4	
	0

Solve. See Example 9.

65. The cost in dollars y of producing x computer desks is given by $y = 80x + 5000$.

a. Complete the table.

x	100	200	300
y			

b. Find the number of computer desks that can be produced for $8600. (*Hint:* Find x when $y = 8600$.)

66. The hourly wage y of an employee at a certain production company is given by $y = 0.25x + 9$ where x is the number of units produced by the employee in an hour.

a. Complete the table.

x	0	1	5	10
y				

b. Find the number of units that an employee must produce each hour to earn an hourly wage of $12.25. (*Hint:* Find x when $y = 12.25$.)

67. The average annual cinema admission price y (in dollars) from 2010 through 2014 is given by $y = 0.09x + 7.85$. In this equation, x represents the number of years after 2010. (*Source:* Motion Picture Association of America)

a. Complete the table.

x	0	2	4
y			

b. Find the year in which the average cinema admission price was approximately $8.10.

(*Hint:* Find x when $y = 8.10$ and round to the nearest whole number.)

c. Use the given equation to predict when the cinema admission price might be $10.00. (Use the hint for part (b).)

d. In your own words, write the meaning of the ordered pair (1, 7.94).

68. The average amount of money y spent per person per year on music from iTunes from 2008 to 2013 can be approximated by $y = -5.74x + 38.05$. In this equation, x represents the number of years since 2008. (*Source:* billboard.com)

a. Complete the table.

x	1	3	5
y			

b. Find the year in which the yearly average amount spent on music in iTunes was approximately $26.00. (*Hint:* Find x when $y = 26$ and round to the nearest whole number.)

The graph below shows the number of U.S. Walmart stores for each year. Use this graph to answer Exercises 69 through 72.

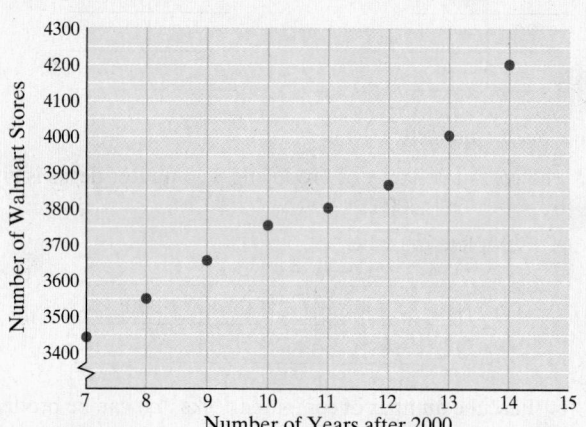

Data from Walmart

69. The ordered pair (14, 4203) is a point of the graph. Write a sentence describing the meaning of this ordered pair.

70. The ordered pair (12, 3868) is a point of the graph. Write a sentence describing the meaning of this ordered pair.

71. Estimate the increase in Walmart stores for years 8, 9, and 10.

72. Use a straightedge or ruler and this graph to predict the number of Walmart stores in the year 2018.

73. Describe what is similar about the coordinates of points whose graph lies on the x-axis.

74. Describe what is similar about the coordinates of points whose graph lies on the y-axis.

REVIEW AND PREVIEW

Solve each equation for y. See Section 2.5.

75. $x + y = 5$

76. $x - y = 3$

77. $2x + 4y = 5$

78. $5x + 2y = 7$

79. $10x = -5y$

80. $4y = -8x$

81. $x - 3y = 6$

82. $2x - 9y = -20$

CONCEPT EXTENSIONS

Answer each exercise with true or false.

83. Point $(-1, 5)$ lies in quadrant IV.

84. Point $(3, 0)$ lies on the y-axis.

85. For the point $\left(-\frac{1}{2}, 1.5\right)$, the first value, $-\frac{1}{2}$, is the x-coordinate and the second value, 1.5, is the y-coordinate.

86. The ordered pair $\left(2, \frac{2}{3}\right)$ is a solution of $2x - 3y = 6$.

For Exercises 87 through 91, fill in each blank with "0," "positive," or "negative." For Exercises 92 and 93, fill in each blank with "x" or "y."

Point	Location
87. (_____ , _____)	quadrant III
88. (_____ , _____)	quadrant I
89. (_____ , _____)	quadrant IV
90. (_____ , _____)	quadrant II
91. (_____ , _____)	origin
92. (number, 0)	__ -axis
93. (0, number)	__ -axis

94. Give an example of an ordered pair whose location is in (or on)

a. quadrant I   b. quadrant II

c. quadrant III   d. quadrant IV

e. x-axis   f. y-axis

Solve. See the first Concept Check in this section.

95. Is the graph of $(3, 0)$ in the same location as the graph of $(0, 3)$? Explain why or why not.

96. Give the coordinates of a point such that if the coordinates are reversed, their location is the same.

97. In general, what points can have coordinates reversed and still have the same location?

98. In your own words, describe how to plot or graph an ordered pair of numbers.

Write an ordered pair for each point described. See the second Concept Check in this section.

99. Point C is four units to the right of the y-axis and seven units below the x-axis.

100. Point D is three units to the left of the origin.

Solve.

△ **101.** Find the perimeter of the rectangle whose vertices are the points with coordinates $(-1, 5)$, $(3, 5)$, $(3, -4)$, and $(-1, -4)$.

△ **102.** Find the area of the rectangle whose vertices are the points with coordinates $(5, 2)$, $(5, -6)$, $(0, -6)$, and $(0, 2)$.

103. Three vertices of a rectangle are $(-2, -3)$, $(-7, -3)$, and $(-7, 6)$.

a. Find the coordinates of the fourth vertex of the rectangle.

b. Find the perimeter of the rectangle.

c. Find the area of the rectangle.

104. Three vertices of a square are $(-4, -1)$, $(-4, 8)$, and $(5, 8)$.

a. Find the coordinates of the fourth vertex of the square.

b. Find the perimeter of the square.

c. Find the area of the square.

3.2 Graphing Linear Equations

OBJECTIVES

1 Identify Linear Equations.

2 Graph a Linear Equation by Finding and Plotting Ordered Pair Solutions.

OBJECTIVE

1 Identifying Linear Equations

In the previous section, we found that equations in two variables may have more than one solution. For example, both $(6, 0)$ and $(2, -2)$ are solutions of the equation $x - 2y = 6$. In fact, this equation has an infinite number of solutions. Other solutions include $(0, -3)$, $(4, -1)$, and $(-2, -4)$. If we graph these solutions, notice that a pattern appears.

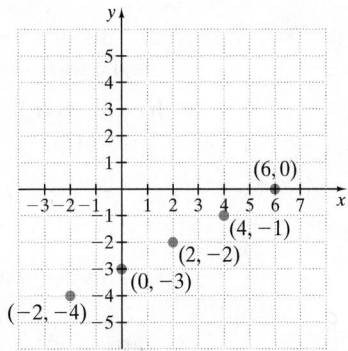

These solutions all appear to lie on the same line, which has been filled in below. It can be shown that every ordered pair solution of the equation corresponds to a point on this line, and every point on this line corresponds to an ordered pair solution. Thus, we say that this line is the **graph of the equation** $x - 2y = 6$.

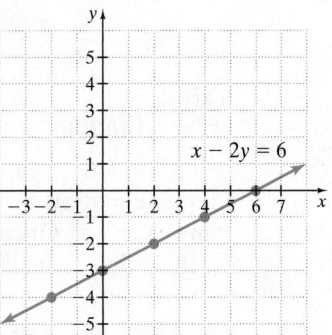

> **Helpful Hint**
>
> Notice that we can only show a part of a line on a graph. The arrowheads on each end of the line remind us that the line actually extends indefinitely in both directions.

The equation $x - 2y = 6$ is called a **linear equation in two variables** and **the graph of every linear equation in two variables is a line.**

> **Linear Equation in Two Variables**
>
> A linear equation in two variables is an equation that can be written in the form
>
> $$Ax + By = C$$
>
> where A, B, and C are real numbers and A and B are not both 0. **The graph of a linear equation in two variables is a straight line.**

The form $Ax + By = C$ is called **standard form**.

> **Helpful Hint**
>
> Notice in the form $Ax + By = C$, the understood exponent on both x and y is 1.

Examples of Linear Equations in Two Variables

$$2x + y = 8 \qquad -2x = 7y \qquad y = \frac{1}{3}x + 2 \qquad y = 7$$
(Standard Form)

Before we graph linear equations in two variables, let's practice identifying these equations.

EXAMPLE 1 Determine whether each equation is a linear equation in two variables.

a. $x - 1.5y = -1.6$ **b.** $y = -2x$ **c.** $x + y^2 = 9$ **d.** $x = 5$

Solution

a. This is a linear equation in two variables because it is written in the form $Ax + By = C$ with $A = 1$, $B = -1.5$, and $C = -1.6$.

b. This is a linear equation in two variables because it can be written in the form $Ax + By = C$.

$$y = -2x$$
$$2x + y = 0 \qquad \text{Add } 2x \text{ to both sides.}$$

c. This is *not* a linear equation in two variables because y is squared.

d. This is a linear equation in two variables because it can be written in the form $Ax + By = C$.

$$x = 5$$
$$x + 0y = 5 \qquad \text{Add } 0 \cdot y. \qquad \square$$

PRACTICE

1 Determine whether each equation is a linear equation in two variables.

a. $3x + 2.7y = -5.3$ **b.** $x^2 + y = 8$

c. $y = 12$ **d.** $5x = -3y$

OBJECTIVE

2 Graphing Linear Equations by Plotting Ordered Pair Solutions

From geometry, we know that a straight line is determined by just two points. Graphing a linear equation in two variables, then, requires that we find just two of its infinitely many solutions. Once we do so, we plot the solution points and draw the line connecting the points. Usually, we find a third solution as well, as a check.

EXAMPLE 2 Graph the linear equation $2x + y = 5$.

Solution Find three ordered pair solutions of $2x + y = 5$. To do this, choose a value for one variable, x or y, and solve for the other variable. For example, let $x = 1$. Then $2x + y = 5$ becomes

$$2x + y = 5$$
$$2(1) + y = 5 \qquad \text{Replace } x \text{ with 1.}$$
$$2 + y = 5 \qquad \text{Multiply.}$$
$$y = 3 \qquad \text{Subtract 2 from both sides.}$$

Since $y = 3$ when $x = 1$, the ordered pair $(1, 3)$ is a solution of $2x + y = 5$. Next, let $x = 0$.

$$2x + y = 5$$
$$2(0) + y = 5 \qquad \text{Replace } x \text{ with 0.}$$
$$0 + y = 5$$
$$y = 5$$

The ordered pair $(0, 5)$ is a second solution.

The two solutions found so far allow us to draw the straight line that is the graph of all solutions of $2x + y = 5$. However, we find a third ordered pair as a check. Let $y = -1$.

$$2x + y = 5$$
$$2x + (-1) = 5 \quad \text{Replace } y \text{ with } -1.$$
$$2x - 1 = 5$$
$$2x = 6 \quad \text{Add 1 to both sides.}$$
$$x = 3 \quad \text{Divide both sides by 2.}$$

The third solution is $(3, -1)$. These three ordered pair solutions are listed in table form as shown. The graph of $2x + y = 5$ is the line through the three points.

x	y
1	3
0	5
3	-1

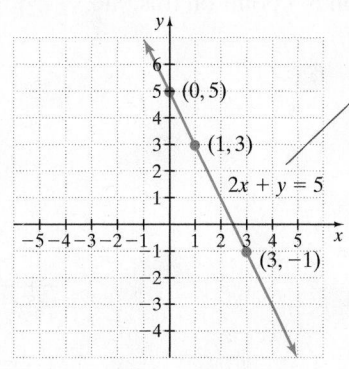

> **Helpful Hint**
>
> All three points should fall on the same straight line. If not, check your ordered pair solutions for a mistake.

PRACTICE
2 Graph the linear equation $x + 3y = 9$.

EXAMPLE 3 Graph the linear equation $-5x + 3y = 15$.

Solution Find three ordered pair solutions of $-5x + 3y = 15$.

Let $x = 0$.	**Let $y = 0$.**	**Let $x = -2$.**
$-5x + 3y = 15$	$-5x + 3y = 15$	$-5x + 3y = 15$
$-5 \cdot 0 + 3y = 15$	$-5x + 3 \cdot 0 = 15$	$-5(-2) + 3y = 15$
$0 + 3y = 15$	$-5x + 0 = 15$	$10 + 3y = 15$
$3y = 15$	$-5x = 15$	$3y = 5$
$y = 5$	$x = -3$	$y = \dfrac{5}{3}$

The ordered pairs are $(0, 5)$, $(-3, 0)$, and $\left(-2, \dfrac{5}{3}\right)$. The graph of $-5x + 3y = 15$ is the line through the three points.

x	y
0	5
-3	0
-2	$\dfrac{5}{3} = 1\dfrac{2}{3}$

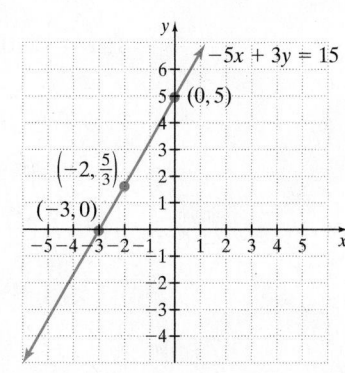

PRACTICE
3 Graph the linear equation $3x - 4y = 12$.

EXAMPLE 4   Graph the linear equation $y = 3x$.

Solution To graph this linear equation, we find three ordered pair solutions. Since the equation is solved for y, choose three x-values.

If $x = 2, y = 3 \cdot 2 = 6$.
If $x = 0, y = 3 \cdot 0 = 0$.
If $x = -1, y = 3 \cdot -1 = -3$.

x	y
2	6
0	0
−1	−3

Next, graph the ordered pair solutions listed in the table above and draw a line through the plotted points as shown below. The line is the graph of $y = 3x$. Every point on the graph represents an ordered pair solution of the equation and every ordered pair solution is a point on this line.

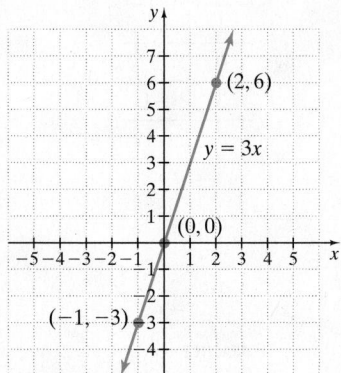

PRACTICE
4   Graph the linear equation $y = -2x$.

Helpful Hint

When graphing a linear equation in two variables, if it is
- solved for y, it may be easier to find ordered pair solutions by choosing x-values. If it is
- solved for x, it may be easier to find ordered pair solutions by choosing y-values.

EXAMPLE 5   Graph the linear equation $y = -\dfrac{1}{3}x + 2$.

Solution Find three ordered pair solutions, graph the solutions, and draw a line through the plotted solutions. To avoid fractions, choose x-values that are multiples of 3 to substitute in the equation. When a multiple of 3 is multiplied by $-\dfrac{1}{3}$, the result is an integer. See the calculations below used to fill in the table.

If $x = 6$, then $y = -\dfrac{1}{3} \cdot 6 + 2 = -2 + 2 = 0$

If $x = 0$, then $y = -\dfrac{1}{3} \cdot 0 + 2 = 0 + 2 = 2$

If $x = -3$, then $y = -\dfrac{1}{3} \cdot -3 + 2 = 1 + 2 = 3$

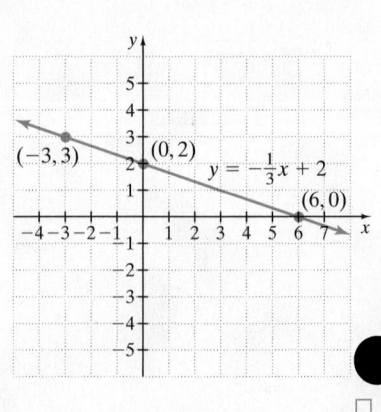

x	y
6	0
0	2
−3	3

PRACTICE
5   Graph the linear equation $y = \dfrac{1}{2}x + 3$.

Let's compare the graphs in Examples 4 and 5. The graph of $y = 3x$ tilts upward (as we follow the line from left to right) and the graph of $y = -\dfrac{1}{3}x + 2$ tilts downward (as we follow the line from left to right). We will learn more about the tilt, or slope, of a line in Section 3.4.

EXAMPLE 6   Graph the linear equation $y = -2$.

Solution The equation $y = -2$ can be written in standard form as $0x + y = -2$. No matter what value we replace x with, y is always -2.

x	y
0	−2
3	−2
−2	−2

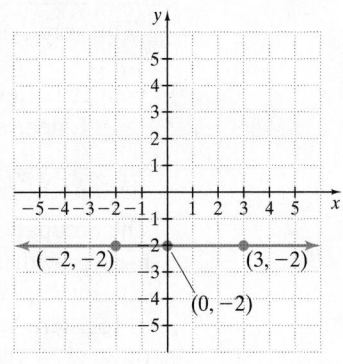

Notice that the graph of $y = -2$ is a horizontal line.   □

PRACTICE
6   Graph the linear equation $x = -2$.   ■

EXAMPLE 7   Graph the linear equation $y = 3x + 6$ and compare this graph with the graph of $y = 3x$ in Example 4.

Solution Find ordered pair solutions, graph the solutions, and draw a line through the plotted solutions. We choose x-values and substitute in the equation $y = 3x + 6$.

If $x = -3$, then $y = 3(-3) + 6 = -3$.
If $x = 0$, then $y = 3(0) + 6 = 6$.
If $x = 1$, then $y = 3(1) + 6 = 9$.

x	y
−3	−3
0	6
1	9

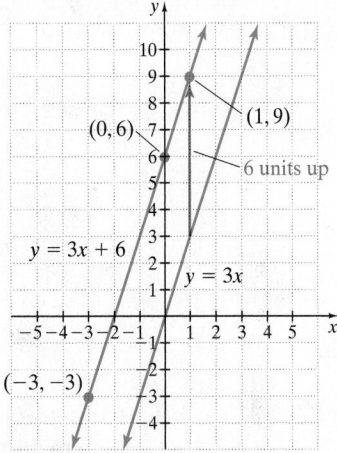

The most startling similarity is that both graphs appear to have the same upward tilt as we move from left to right. Also, the graph of $y = 3x$ crosses the y-axis at the origin, while the graph of $y = 3x + 6$ crosses the y-axis at 6. In fact, the graph of $y = 3x + 6$ is the same as the graph of $y = 3x$ moved vertically upward 6 units.   □

(Continued on next page)

PRACTICE
7 Graph the linear equation $y = -2x + 3$ and compare this graph with the graph of $y = -2x$ in Practice 4.

Notice that the graph of $y = 3x + 6$ crosses the y-axis at 6. This happens because when $x = 0$, $y = 3x + 6$ becomes $y = 3 \cdot 0 + 6 = 6$. The graph contains the point $(0, 6)$, which is on the y-axis.

In general, if a linear equation in two variables is solved for y, we say that it is written in the form $y = mx + b$. The graph of this equation contains the point $(0, b)$ because when $x = 0$, $y = mx + b$ is $y = m \cdot 0 + b = b$.

> The graph of $y = mx + b$ crosses the y-axis at $(0, b)$.

We will review this again in Section 3.5.

Linear equations are often used to model real data as seen in the next example.

EXAMPLE 8 **Estimating the Number of Registered Nurses**

The occupation expected to have the most employment growth in the next few years is registered nurse. The number of people y (in thousands) employed as registered nurses in the United States can be estimated by the linear equation $y = 58.1x + 2619$, where x is the number of years after the year 2008. (*Source:* U.S. Bureau of Labor Statistics)

a. Graph the equation.

b. Use the graph to predict the number of registered nurses in the year 2018.

Solution

a. To graph $y = 58.1x + 2619$, choose x-values and substitute in the equation.

If $x = 0$, then $y = 58.1(0) + 2619 = 2619$.

If $x = 2$, then $y = 58.1(2) + 2619 = 2735.2$.

If $x = 5$, then $y = 58.1(5) + 2619 = 2909.5$.

x	y
0	2619
2	2735.2
5	2909.5

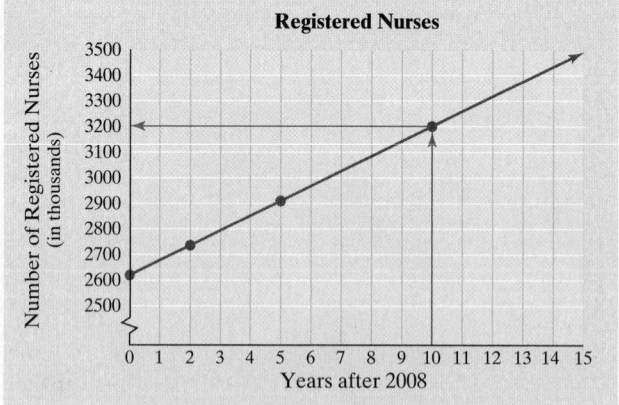

b. To use the graph to *predict* the number of registered nurses in the year 2018, we need to find the y-coordinate that corresponds to $x = 10$. (10 years after 2008 is the year 2018.) To do so, find 10 on the x-axis. Move vertically upward to the graphed line and then horizontally to the left. We approximate the number on the y-axis to be 3200. Thus, in the year 2018, we predict that there will be 3200 thousand registered nurses. (The actual value, using 10 for x, is 3200.) □

PRACTICE

8   One of the occupations expected to have a large growth in employment in the next few years is computer software application engineers. The number of people y (in thousands) employed as computer software application engineers in the United States can be estimated by the linear equation $y = 17.5x + 515$, where x is the number of years after 2008. (*Source:* Based on data from the Bureau of Labor Statistics)

a. Graph the equation.

b. Use the graph to predict the number of computer software application engineers in the year 2020.   ■

> **Helpful Hint**
>
> Make sure you understand that models are mathematical approximations of the data for the known years. (For example, see the model in Example 8.) Any number of unknown factors can affect future years, so be cautious when using models to predict.

Graphing Calculator Explorations

In this section, we begin an optional study of graphing calculators and graphing software packages for computers. These graphers use the same point plotting technique that was introduced in this section. The advantage of this graphing technology is, of course, that graphing calculators and computers can find and plot ordered pair solutions much faster than we can. Note, however, that the features described in these boxes may not be available on all graphing calculators.

The rectangular screen where a portion of the rectangular coordinate system is displayed is called a **window.** We call it a **standard window** for graphing when both the x- and y-axes show coordinates between -10 and 10. This information is often displayed in the window menu on a graphing calculator as

$$\text{Xmin} = -10$$
$$\text{Xmax} = 10$$
$$\text{Xscl} = 1 \qquad \text{The scale on the } x\text{-axis is one unit per tick mark.}$$
$$\text{Ymin} = -10$$
$$\text{Ymax} = 10$$
$$\text{Yscl} = 1 \qquad \text{The scale on the } y\text{-axis is one unit per tick mark.}$$

To use a graphing calculator to graph the equation $y = 2x + 3$, press the $\boxed{Y =}$ key and enter the keystrokes $\boxed{2}\ \boxed{x}\ \boxed{+}\ \boxed{3}$. The top row should now read $Y_1 = 2x + 3$. Next press the $\boxed{\text{GRAPH}}$ key, and the display should look like this:

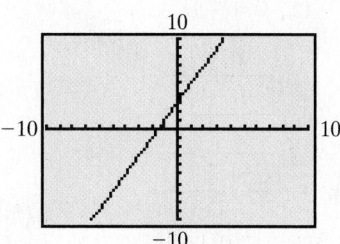

Use a standard window and graph the following linear equations. (Unless otherwise stated, use a standard window when graphing.)

1. $y = -3x + 7$ **2.** $y = -x + 5$ **3.** $y = 2.5x - 7.9$

4. $y = -1.3x + 5.2$ **5.** $y = -\dfrac{3}{10}x + \dfrac{32}{5}$ **6.** $y = \dfrac{2}{9}x - \dfrac{22}{3}$

Vocabulary, Readiness & Video Check

Martín-Gay Interactive Videos

See Video 3.2 ⦿

Watch the section lecture video and answer the following questions.

OBJECTIVE
1

1. Exponents aren't mentioned in the definition of a linear equation in two variables. However, in determining whether Example 3 is a linear equation in two variables, the exponents or powers on the variables are discussed. Explain.

OBJECTIVE
2

2. In the lecture before ▦ Example 5, it's mentioned that you need only two points to determine a line. Why then are three ordered pair solutions found in ▦ Examples 5–7?

OBJECTIVE
2

3. What does a graphed line represent as discussed at the end of ▦ Examples 5 and 7?

3.2 Exercise Set MyMathLab® ▸

Determine whether each equation is a linear equation in two variables. See Example 1.

1. $-x = 3y + 10$

2. $y = x - 15$

3. $x = y$

4. $x = y^3$

5. $x^2 + 2y = 0$

6. $0.01x - 0.2y = 8.8$

7. $y = -1$

8. $x = 25$

For each equation, find three ordered pair solutions by completing the table. Then use the ordered pairs to graph the equation. See Examples 2 through 7.

9. $x - y = 6$

x	y
	0
4	
	−1

10. $x - y = 4$

x	y
	0
	2
	−1

11. $y = -4x$

x	y
1	
0	
−1	

12. $y = -5x$

x	y
1	
0	
−1	

13. $y = \dfrac{1}{3}x$

x	y
0	
6	
−3	

14. $y = \dfrac{1}{2}x$

x	y
0	
−4	
2	

15. $y = -4x + 3$

x	y
0	
1	
2	

16. $y = -5x + 2$

x	y
0	
1	
2	

MIXED PRACTICE

Graph each linear equation. See Examples 2 through 7.

17. $x + y = 1$

18. $x + y = 7$

19. $x - y = -2$

20. $-x + y = 6$

▸ **21.** $x - 2y = 6$

22. $-x + 5y = 5$

23. $y = 6x + 3$

24. $y = -2x + 7$

25. $x = -4$

26. $y = 5$

27. $y = 3$

28. $x = -1$

29. $y = x$

30. $y = -x$

▸ **31.** $x = -3y$

32. $x = -5y$

33. $x + 3y = 9$

34. $2x + y = 2$

▸ **35.** $y = \dfrac{1}{2}x + 2$

36. $y = \dfrac{1}{4}x + 3$

37. $3x - 2y = 12$

38. $2x - 7y = 14$

39. $y = -3.5x + 4$

40. $y = -1.5x - 3$

Graph each pair of linear equations on the same set of axes. Discuss how the graphs are similar and how they are different. See Example 7.

41. $y = 5x; y = 5x + 4$

42. $y = 2x; y = 2x + 5$

43. $y = -2x; y = -2x - 3$

44. $y = x; y = x - 7$

45. $y = \dfrac{1}{2}x; y = \dfrac{1}{2}x + 2$

46. $y = -\dfrac{1}{4}x; y = -\dfrac{1}{4}x + 3$

The graph of $y = 5x$ is given below as well as Figures A–D. For Exercises 47 through 50, match each equation with its graph. Hint: Recall that if an equation is written in the form $y = mx + b$, its graph crosses the y-axis at $(0, b)$.

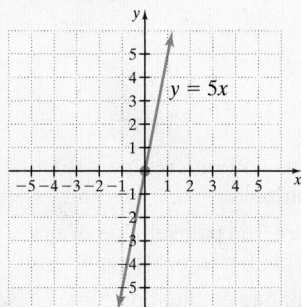

47. $y = 5x + 5$ **48.** $y = 5x - 4$

49. $y = 5x - 1$ **50.** $y = 5x + 2$

A.

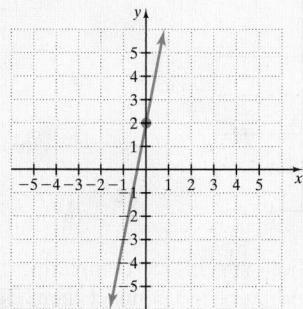

B.

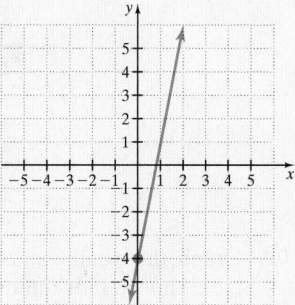

C.

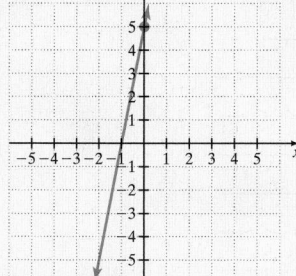

D.

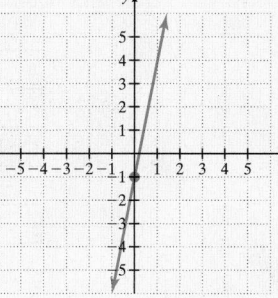

Solve. See Example 8.

51. Jogging is one of the few sports that has been consistently increasing over the past few years. The number of people jogging (in millions) from the years 2000 to 2009 is given by the equation $y = x + 23$, where x is the number of years after 2000. (*Source:* Based on data from the National Sporting Goods Association)

a. Use this equation or a graph of it to complete the ordered pair (8,   ).

b. Write a sentence explaining the meaning of the answer to part (a).

c. If this trend continues, how many joggers will there be in 2017?

52. The revenue y (in billions of dollars) for Home Depot stores during the years 2010 through 2014 is given by the equation $y = 3.2x + 65.2$, where x is the number of years after 2010. (*Source:* Based on data from Home Depot stores)

a. Use this equation or a graph of it to complete the ordered pair (3,   ).

b. Write a sentence explaining the meaning of the answer to part (a).

c. If this trend continues, predict the revenue for Home Depot stores for the year 2018.

53. One American rite of passage is a driver's license. The number of people y (in millions) who have a driver's license can be estimated by the linear equation $y = 2.2x + 190$, where x is the number of years after 2000. (*Source:* Federal Highway Administration)

a. Use this equation to complete the ordered pair (12,   ).

b. Write a sentence explaining the meaning of the ordered pair in part (a).

c. If this trend continues, predict the number of people with driver's licenses in 2020.

54. The percent of U.S. households y with at least one computer can be approximated by the linear equation $y = 2.4x + 51$, where x is the number of years since 2000. (*Source:* Pew Research)

a. Use the equation to complete the ordered pair (10,   ).

b. Write a sentence explaining the meaning of the ordered pair found in part (a).

c. If this trend continues, predict the percent of U.S. households that have at least one computer in 2018.

d. Explain any issues with your answer to part (c).

REVIEW AND PREVIEW

Solve. See Section 3.1.

△ **55.** The coordinates of three vertices of a rectangle are $(-2, 5)$, $(4, 5)$, and $(-2, -1)$. Find the coordinates of the fourth vertex.

△ **56.** The coordinates of two vertices of a square are $(-3, -1)$ and $(2, -1)$. Find the coordinates of two pairs of points possible for the third and fourth vertices.

Complete each table.

57. $x - y = -3$

x	y
0	
	0

58. $y - x = 5$

x	y
0	
	0

59. $y = 2x$

x	y
0	
	0

60. $x = -3y$

x	y
0	
	0

CONCEPT EXTENSIONS

Write each statement as an equation in two variables. Then graph the equation.

61. The y-value is 5 more than the x-value.

62. The y-value is twice the x-value.

63. Two times the x-value, added to three times the y-value is 6.

64. Five times the x-value, added to twice the y-value is -10.

Solve.

△ **65.** The perimeter of the trapezoid below is 22 centimeters. Write a linear equation in two variables for the perimeter. Find y if x is 3 cm.

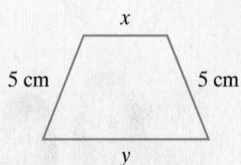

△ **66.** The perimeter of the rectangle below is 50 miles. Write a linear equation in two variables for this perimeter. Use this equation to find x when y is 20.

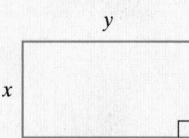

67. Explain how to find ordered pair solutions of linear equations in two variables.

68. If (a, b) is an ordered pair solution of $x + y = 5$, is (b, a) also a solution? Explain why or why not.

69. Graph the nonlinear equation $y = x^2$ by completing the table shown. Plot the ordered pairs and connect them with a smooth curve.

x	y
0	
1	
-1	
2	
-2	

70. Graph the nonlinear equation $y = |x|$ by completing the table shown. Plot the ordered pairs and connect them. This curve is "V" shaped.

x	y
0	
1	
-1	
2	
-2	

3.3 Intercepts

OBJECTIVES

1 Identify Intercepts of a Graph.

2 Graph a Linear Equation by Finding and Plotting Intercepts.

3 Identify and Graph Vertical and Horizontal Lines.

OBJECTIVE

1   Identifying Intercepts

In this section, we graph linear equations in two variables by identifying intercepts. For example, the graph of $y = 4x - 8$ is shown on the right. Notice that this graph crosses the y-axis at the point $(0, -8)$. This point is called the **y-intercept.** Likewise, the graph crosses the x-axis at $(2, 0)$, and this point is called the **x-intercept.**

The intercepts are $(2, 0)$ and $(0, -8)$.

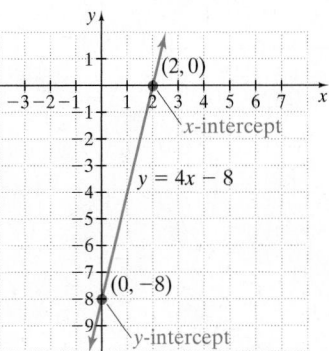

> **Helpful Hint**
>
> If a graph crosses the x-axis at $(-3, 0)$ and the y-axis at $(0, 7)$, then
>
> $$\underset{\substack{\uparrow \\ x\text{-intercept}}}{(-3, 0)} \qquad \underset{\substack{\uparrow \\ y\text{-intercept}}}{(0, 7)}$$
>
> Notice that for the y-intercept, the x-value is 0 and for the x-intercept, the y-value is 0.
> **Note:** Sometimes in mathematics, you may see just the number -3 stated as the x-intercept, and 7 stated as the y-intercept.

EXAMPLES   Identify the x- and y-intercepts.

1.

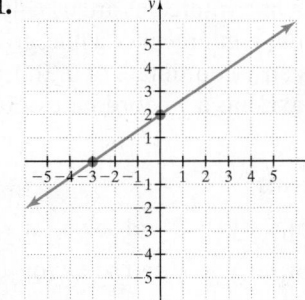

Solution

x-intercept: $(-3, 0)$
y-intercept: $(0, 2)$

2.

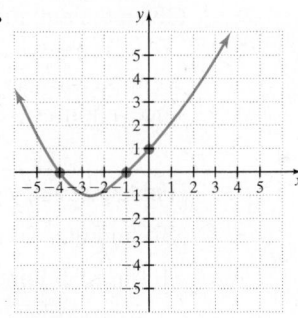

Solution

x-intercepts: $(-4, 0), (-1, 0)$
y-intercept: $(0, 1)$

> **Helpful Hint**
>
> Notice that any time $(0, 0)$ is a point of a graph, then it is an x-intercept and a y-intercept. Why? It is the only point that lies on both axes.

3.

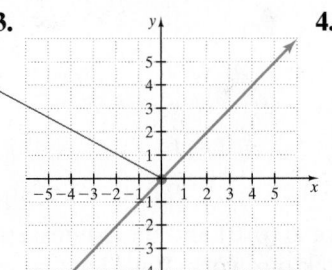

Solution

x-intercept: $(0, 0)$
y-intercept: $(0, 0)$

4.

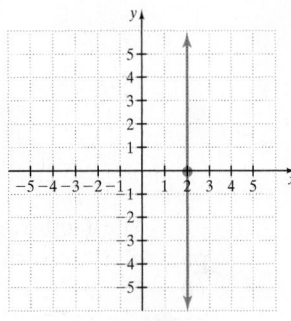

Solution

x-intercept: $(2, 0)$
y-intercept: none

5.

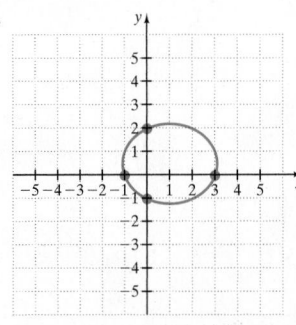

Solution

x-intercepts:
$(-1, 0), (3, 0)$
y-intercepts:
$(0, 2), (0, -1)$

(Continued on next page)

PRACTICE

1–5 Identify the *x*- and *y*-intercepts.

1.

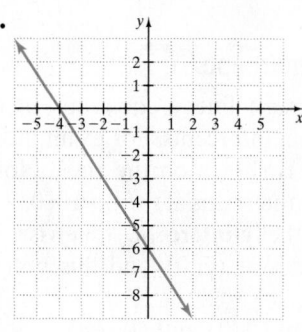

2.

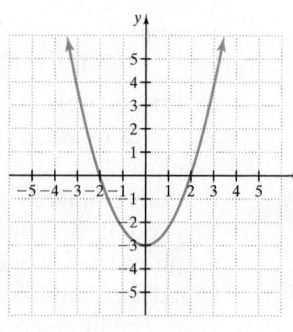

3.

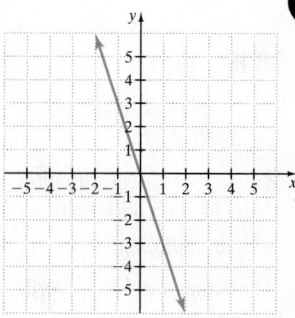

4.

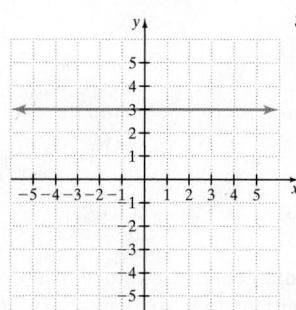

5.

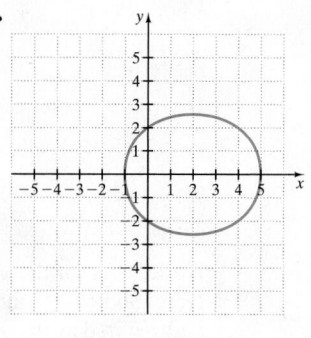

OBJECTIVE

2 Using Intercepts to Graph a Linear Equation ▶

Given the equation of a line, intercepts are usually easy to find since one coordinate is 0.

One way to find the *y*-intercept of a line, given its equation, is to let $x = 0$, since a point on the *y*-axis has an *x*-coordinate of 0. To find the *x*-intercept of a line, let $y = 0$, since a point on the *x*-axis has a *y*-coordinate of 0.

Finding *x*- and *y*-intercepts

To find the *x*-intercept, let $y = 0$ and solve for *x*.

To find the *y*-intercept, let $x = 0$ and solve for *y*.

EXAMPLE 6 Graph $x - 3y = 6$ by finding and plotting intercepts.

Solution Let $y = 0$ to find the *x*-intercept and let $x = 0$ to find the *y*-intercept.

$$\text{Let } y = 0 \qquad\qquad \text{Let } x = 0$$
$$x - 3y = 6 \qquad\qquad x - 3y = 6$$
$$x - 3(0) = 6 \qquad\qquad 0 - 3y = 6$$
$$x - 0 = 6 \qquad\qquad -3y = 6$$
$$x = 6 \qquad\qquad y = -2$$

The *x*-intercept is $(6, 0)$ and the *y*-intercept is $(0, -2)$. We find a third ordered pair solution to check our work. If we let $y = -1$, then $x = 3$. Plot the points $(6, 0)$, $(0, -2)$, and $(3, -1)$. The graph of $x - 3y = 6$ is the line drawn through these points, as shown.

x	y
6	0
0	-2
3	-1

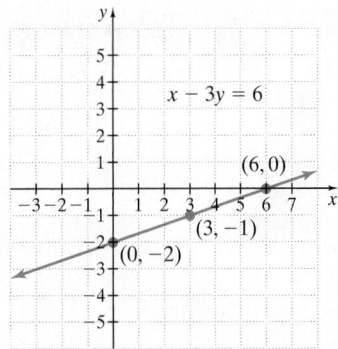

PRACTICE

6 Graph $x + 2y = -4$ by finding and plotting intercepts.

EXAMPLE 7 Graph $x = -2y$ by plotting intercepts.

Solution Let $y = 0$ to find the x-intercept and $x = 0$ to find the y-intercept.

$$\text{Let } y = 0 \qquad\qquad \text{Let } x = 0$$
$$x = -2y \qquad\qquad\quad x = -2y$$
$$x = -2(0) \qquad\qquad 0 = -2y$$
$$x = 0 \qquad\qquad\quad\ \ 0 = y$$

Both the x-intercept and y-intercept are $(0, 0)$. In other words, when $x = 0$, then $y = 0$, which gives the ordered pair $(0, 0)$. Also, when $y = 0$, then $x = 0$, which gives the same ordered pair $(0, 0)$. This happens when the graph passes through the origin. Since two points are needed to determine a line, we must find at least one more ordered pair that satisfies $x = -2y$. We will let $y = -1$ to find a second ordered pair solution and let $y = 1$ as a checkpoint.

$$\text{Let } y = -1 \qquad\qquad \text{Let } y = 1$$
$$x = -2(-1) \qquad\qquad\ x = -2(1)$$
$$x = 2 \qquad\qquad\qquad\ x = -2$$

The ordered pairs are $(0, 0)$, $(2, -1)$, and $(-2, 1)$. Plot these points to graph $x = -2y$.

x	y
0	0
2	-1
-2	1

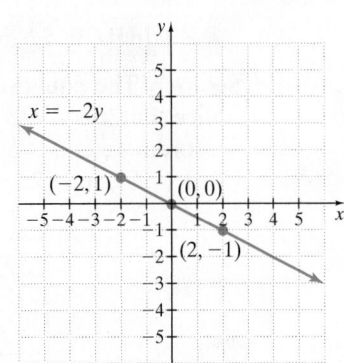

PRACTICE

7 Graph $x = 3y$ by plotting intercepts.

EXAMPLE 8   Graph: $4x = 3y - 9$

Solution Find the x- and y-intercepts, and then choose $x = 2$ to find a checkpoint.

Let $y = 0$	Let $x = 0$	Let $x = 2$
$4x = 3(0) - 9$	$4 \cdot 0 = 3y - 9$	$4(2) = 3y - 9$
$4x = -9$	$9 = 3y$	$8 = 3y - 9$
Solve for x.	Solve for y.	Solve for y.
$x = -\dfrac{9}{4}$ or $-2\dfrac{1}{4}$	$3 = y$	$17 = 3y$
		$\dfrac{17}{3} = y$ or $y = 5\dfrac{2}{3}$

The ordered pairs are $\left(-2\dfrac{1}{4}, 0\right)$, $(0, 3)$, and $\left(2, 5\dfrac{2}{3}\right)$. The equation $4x = 3y - 9$ is graphed as follows.

x	y
$-2\dfrac{1}{4}$	0
0	3
2	$5\dfrac{2}{3}$

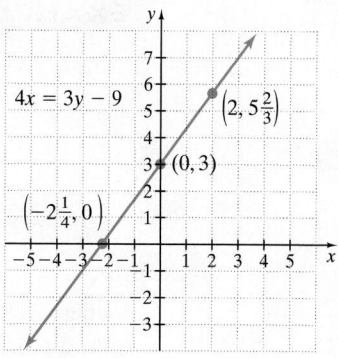

PRACTICE
8   Graph: $3x = 2y + 4$

OBJECTIVE
3   Graphing Vertical and Horizontal Lines

The equation $x = c$, where c is a real number constant, is a linear equation in two variables because it can be written in the form $x + 0y = c$. The graph of this equation is a vertical line as shown in the next example.

EXAMPLE 9   Graph: $x = 2$

Solution The equation $x = 2$ can be written as $x + 0y = 2$. For any y-value chosen, notice that x is 2. No other value for x satisfies $x + 0y = 2$. Any ordered pair whose x-coordinate is 2 is a solution of $x + 0y = 2$. We will use the ordered pair solutions $(2, 3)$, $(2, 0)$, and $(2, -3)$ to graph $x = 2$.

x	y
2	3
2	0
2	-3

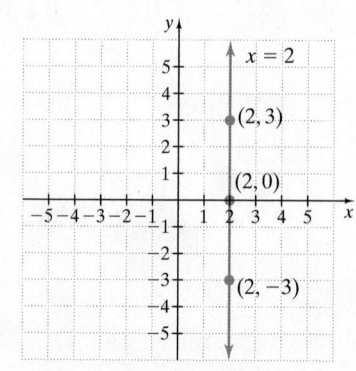

The graph is a vertical line with x-intercept $(2, 0)$. Note that this graph has no y-intercept because x is never 0.   □

PRACTICE
9   Graph: $x = -2$ ■

Vertical Lines

The graph of $x = c$, where c is a real number, is a vertical line with x-intercept $(c, 0)$.

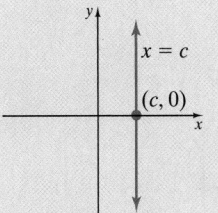

EXAMPLE 10   Graph: $y = -3$

Solution The equation $y = -3$ can be written as $0x + y = -3$. For any x-value chosen, y is -3. If we choose $4, 1,$ and -2 as x-values, the ordered pair solutions are $(4, -3)$, $(1, -3)$, and $(-2, -3)$. Use these ordered pairs to graph $y = -3$. The graph is a horizontal line with y-intercept $(0, -3)$ and no x-intercept.

x	y
4	−3
1	−3
−2	−3

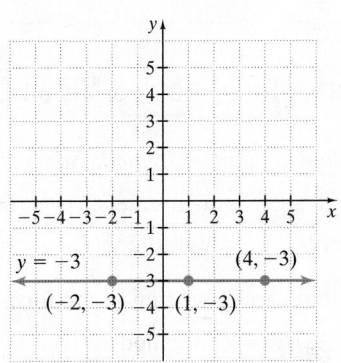

□

PRACTICE
10   Graph: $y = 2$ ■

Horizontal Lines

The graph of $y = c$, where c is a real number, is a horizontal line with y-intercept $(0, c)$.

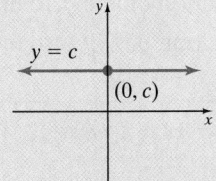

Graphing Calculator Explorations

You may have noticed that to use the $\boxed{Y =}$ key on a grapher to graph an equation, the equation must be solved for y. For example, to graph $2x + 3y = 7$, we solve this equation for y.

$$2x + 3y = 7$$

$$3y = -2x + 7 \qquad \text{Subtract } 2x \text{ from both sides.}$$

$$\frac{3y}{3} = -\frac{2x}{3} + \frac{7}{3} \qquad \text{Divide both sides by 3.}$$

$$y = -\frac{2}{3}x + \frac{7}{3} \qquad \text{Simplify.}$$

To graph $2x + 3y = 7$ or $y = -\frac{2}{3}x + \frac{7}{3}$, press the $\boxed{Y =}$ key and enter

$$Y_1 = -\frac{2}{3}x + \frac{7}{3}$$

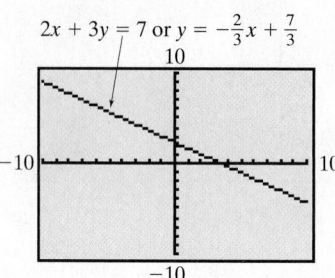

$2x + 3y = 7$ or $y = -\frac{2}{3}x + \frac{7}{3}$

Graph each linear equation.

1. $x = 3.78y$ **2.** $-2.61y = x$ **3.** $3x + 7y = 21$

4. $-4x + 6y = 21$ **5.** $-2.2x + 6.8y = 15.5$ **6.** $5.9x - 0.8y = -10.4$

✔ **Vocabulary, Readiness & Video Check**

Use the choices below to fill in each blank. Some choices may be used more than once. Exercises 1 and 2 come from Section 3.2.

x	vertical	x-intercept	linear
y	horizontal	y-intercept	standard

1. An equation that can be written in the form $Ax + By = C$ is called a(n) _____ equation in two variables.

2. The form $Ax + By = C$ is called _____ form.

3. The graph of the equation $y = -1$ is a _____ line.

4. The graph of the equation $x = 5$ is a _____ line.

5. A point where a graph crosses the y-axis is called a(n) _____.

6. A point where a graph crosses the x-axis is called a(n) _____.

7. Given an equation of a line, to find the x-intercept (if there is one), let _____ = 0 and solve for _____.

8. Given an equation of a line, to find the y-intercept (if there is one), let _____ = 0 and solve for _____.

Martin-Gay Interactive Videos

See Video 3.3 ◉

Watch the section lecture video and answer the following questions.

OBJECTIVE 1

9. At the end of ▦ Example 2, patterns are discussed. What reason is given for why x-intercepts have y-values of 0? For why y-intercepts have x-values of 0?

OBJECTIVE 2

10. In ▦ Example 3, the goal is to use the x- and y-intercepts to graph a line. Yet once the two intercepts are found, a third point is also found before the line is graphed. Why do you think this practice of finding a third point is continued?

OBJECTIVE 3

11. From ▦ Examples 5 and 6, what can you say about the coefficient of x when the equation of a horizontal line is written as $Ax + By = C$? What about the coefficient of y when the equation of a vertical line is written as $Ax + By = C$?

3.3   Exercise Set   MyMathLab

Identify the intercepts. See Examples 1 through 5.

1.

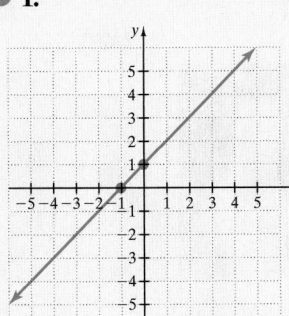

2.

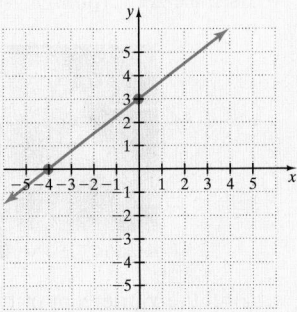

3.

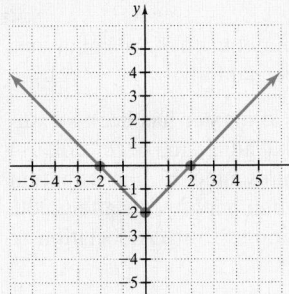

4.

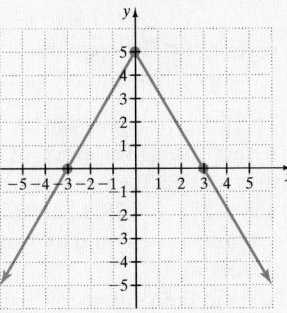

5.

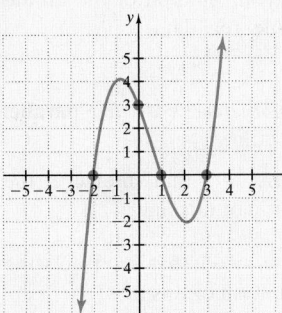

6.

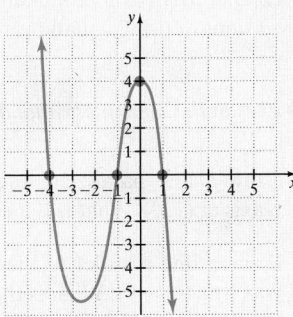

7.

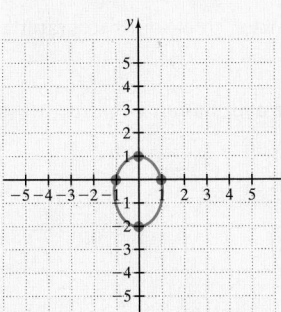

8.

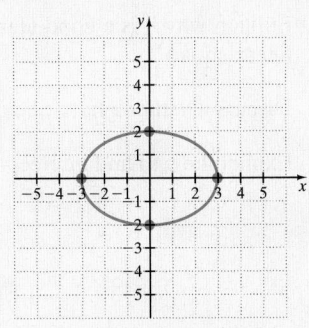

Solve. See Examples 1 through 5.

9. What is the greatest number of intercepts for a line?

10. What is the least number of intercepts for a line?

11. What is the least number of intercepts for a circle?

12. What is the greatest number of intercepts for a circle?

Graph each linear equation by finding and plotting its intercepts. See Examples 6 through 8.

13. $x - y = 3$ **14.** $x - y = -4$ **15.** $x = 5y$

16. $x = 2y$ **17.** $-x + 2y = 6$ **18.** $x - 2y = -8$

19. $2x - 4y = 8$ **20.** $2x + 3y = 6$ **21.** $y = 2x$

22. $y = -2x$ **23.** $y = 3x + 6$ **24.** $y = 2x + 10$

Graph each linear equation. See Examples 9 and 10.

25. $x = -1$ **26.** $y = 5$ **27.** $y = 0$

28. $x = 0$ **29.** $y + 7 = 0$ **30.** $x - 2 = 0$

31. $x + 3 = 0$ **32.** $y - 6 = 0$

MIXED PRACTICE

Graph each linear equation. See Examples 6 through 10.

33. $x = y$ **34.** $x = -y$

35. $x + 8y = 8$ **36.** $x + 3y = 9$

37. $5 = 6x - y$ **38.** $4 = x - 3y$

39. $-x + 10y = 11$ **40.** $-x + 9y = 10$

41. $x = -4\frac{1}{2}$ **42.** $x = -1\frac{3}{4}$

43. $y = 3\frac{1}{4}$ **44.** $y = 2\frac{1}{2}$

45. $y = -\frac{2}{3}x + 1$ **46.** $y = -\frac{3}{5}x + 3$

47. $4x - 6y + 2 = 0$ **48.** $9x - 6y + 3 = 0$

For Exercises 49 through 54, match each equation with its graph. See graphs A–F below and on the next page.

49. $y = 3$ **50.** $y = 2x + 2$

51. $x = -1$ **52.** $x = 3$

53. $y = 2x + 3$ **54.** $y = -2x$

A.

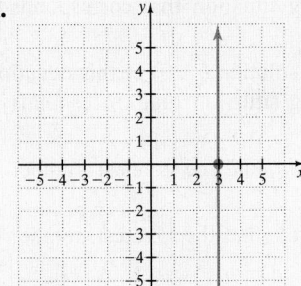

B.

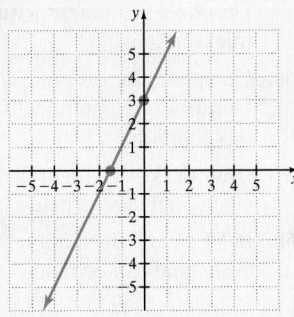

C.

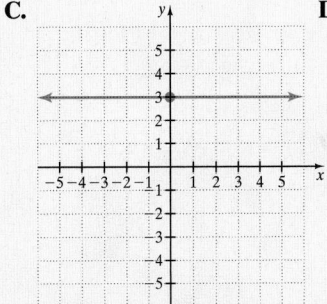

D.

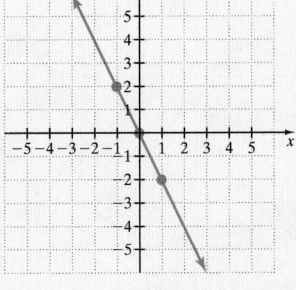

E.

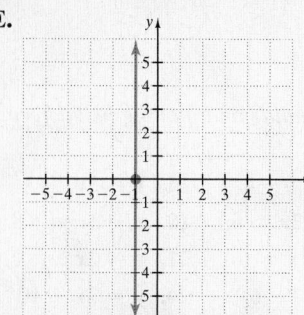

F.

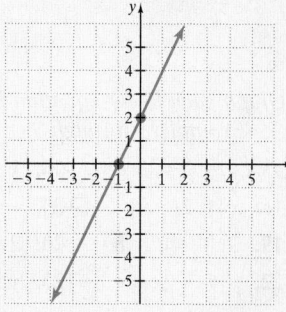

REVIEW AND PREVIEW

Simplify. See Sections 1.4 through 1.7.

55. $\dfrac{-6 - 3}{2 - 8}$

56. $\dfrac{4 - 5}{-1 - 0}$

57. $\dfrac{-8 - (-2)}{-3 - (-2)}$

58. $\dfrac{12 - 3}{10 - 9}$

59. $\dfrac{0 - 6}{5 - 0}$

60. $\dfrac{2 - 2}{3 - 5}$

CONCEPT EXTENSIONS

Answer the following true or false.

61. All lines have an x-intercept *and* a y-intercept.

62. The graph of $y = 4x$ contains the point $(0, 0)$.

63. The graph of $x + y = 5$ has an x-intercept of $(5, 0)$ and a y-intercept of $(0, 5)$.

64. The graph of $y = 5x$ contains the point $(5, 1)$.

The production supervisor at Alexandra's Office Products finds that it takes 3 hours to manufacture a particular office chair and 6 hours to manufacture an office desk. A total of 1200 hours is available to produce office chairs and desks of this style. The linear equation that models this situation is $3x + 6y = 1200$, where x represents the number of chairs produced and y the number of desks manufactured. Use this information for Exercises 65 through 68.

65. Complete the ordered pair solution $(0, \ )$ of this equation. Describe the manufacturing situation that corresponds to this solution.

66. Complete the ordered pair solution $(\ , 0)$ of this equation. Describe the manufacturing situation that corresponds to this solution.

67. If 50 desks are manufactured, find the greatest number of chairs that can be made.

68. If 50 chairs are manufactured, find the greatest number of desks that can be made.

Solve.

69. Since 2009, the number of analog theater screens has been on the decline in the U.S. The number of analog movie screens y each year can be estimated by the equation $y = -7000x + 31{,}800$, where x represents the number of years since 2009. (*Source:* MPAA)

a. Find the x-intercept of this equation. Round to the nearest tenth.

b. What does this x-intercept mean?

c. Use part (b) to comment on your opinion of the limitations of using equations to model real data.

70. The price of admission to a movie theater has been steadily increasing. The price of regular admission y (in dollars) to a movie theater may be represented by the equation $y = 0.24x + 5.28$, where x is the number of years after 2000. (*Source:* Based on data from Motion Picture Association of America)

a. Find the x-intercept of this equation.

b. What does this x-intercept mean?

c. Use part (b) to comment on your opinion of the limitations of using equations to model real data.

Two lines in the same plane that do not intersect are called **parallel lines.**

△ **71.** Draw a line parallel to the line $x = 5$ that intersects the x-axis at $(1, 0)$. What is the equation of this line?

△ **72.** Draw a line parallel to the line $y = -1$ that intersects the y-axis at $(0, -4)$. What is the equation of this line?

73. Discuss whether a vertical line ever has a y-intercept.

74. Explain why it is a good idea to use three points to graph a linear equation.

75. Discuss whether a horizontal line ever has an x-intercept.

76. Explain how to find intercepts.

3.4 Slope and Rate of Change

OBJECTIVES

1 Find the Slope of a Line Given Two Points of the Line.

2 Find the Slope of a Line Given Its Equation.

3 Find the Slopes of Horizontal and Vertical Lines.

4 Compare the Slopes of Parallel and Perpendicular Lines.

5 Interpret Slope as a Rate of Change.

OBJECTIVE

1 Finding the Slope of a Line Given Two Points of the Line

Thus far, much of this chapter has been devoted to graphing lines. You have probably noticed by now that a key feature of a line is its slant or steepness. In mathematics, the slant or steepness of a line is formally known as its **slope.** We measure the slope of a line by the ratio of vertical change to the corresponding horizontal change as we move along the line.

On the line below, for example, suppose that we begin at the point $(1, 2)$ and move to the point $(4, 6)$. The vertical change is the change in y-coordinates: $6 - 2$ or 4 units. The corresponding horizontal change is the change in x-coordinates: $4 - 1 = 3$ units. The ratio of these changes is

$$\text{slope} = \frac{\text{change in } y \text{ (vertical change)}}{\text{change in } x \text{ (horizontal change)}} = \frac{4}{3}$$

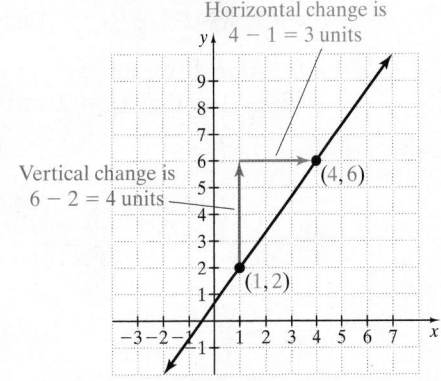

The slope of this line, then, is $\frac{4}{3}$. This means that for every 4 units of change in y-coordinates, there is a corresponding change of 3 units in x-coordinates.

> **Helpful Hint**
>
> It makes no difference what two points of a line are chosen to find its slope. The slope of a line is the same everywhere on the line.
>
>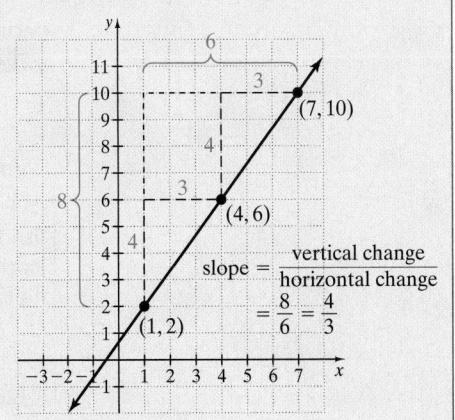

To find the slope of a line, then, choose two points of the line. Label the x-coordinates of the two points x_1 and x_2 (read "x sub one" and "x sub two"), and label the corresponding y-coordinates y_1 and y_2.

The vertical change or **rise** between these points is the difference in the y-coordinates: $y_2 - y_1$. The horizontal change or **run** between the points is the difference of the x-coordinates: $x_2 - x_1$. The slope of the line is the ratio of $y_2 - y_1$ to $x_2 - x_1$, and we traditionally use the letter m to denote slope: $m = \dfrac{y_2 - y_1}{x_2 - x_1}$.

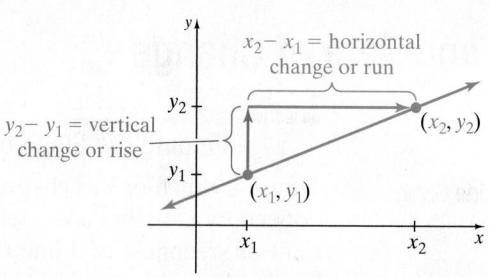

Slope of a Line

The slope m of the line containing the points (x_1, y_1) and (x_2, y_2) is given by

$$m = \frac{\text{rise}}{\text{run}} = \frac{\text{change in } y}{\text{change in } x} = \frac{y_2 - y_1}{x_2 - x_1}, \qquad \text{as long as } x_2 \neq x_1$$

EXAMPLE 1 Find the slope of the line through $(-1, 5)$ and $(2, -3)$. Graph the line.

Solution If we let (x_1, y_1) be $(-1, 5)$, then $x_1 = -1$ and $y_1 = 5$. Also, let (x_2, y_2) be $(2, -3)$ so that $x_2 = 2$ and $y_2 = -3$. Then, by the definition of slope,

$$m = \frac{y_2 - y_1}{x_2 - x_1}$$

$$= \frac{-3 - 5}{2 - (-1)}$$

$$= \frac{-8}{3} = -\frac{8}{3}$$

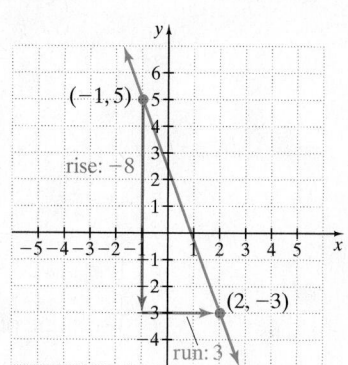

The slope of the line is $-\dfrac{8}{3}$.

PRACTICE
1 Find the slope of the line through $(-4, 11)$ and $(2, 5)$.

Helpful Hint

When finding slope, it makes no difference which point is identified as (x_1, y_1) and which is identified as (x_2, y_2). Just remember that whatever y-value is first in the numerator, its corresponding x-value must be first in the denominator. Another way to calculate the slope in Example 1 is:

$$m = \frac{y_2 - y_1}{x_2 - x_1} = \frac{5 - (-3)}{-1 - 2} = \frac{8}{-3} \quad \text{or} \quad -\frac{8}{3} \quad \leftarrow \text{Same slope as found in Example 1.}$$

✔ **CONCEPT CHECK**
The points $(-2, -5)$, $(0, -2)$, $(4, 4)$, and $(10, 13)$ all lie on the same line. Work with a partner and verify that the slope is the same no matter which points are used to find slope.

Answer to Concept Check:
$$m = \frac{3}{2}$$

EXAMPLE 2   Find the slope of the line through $(-1, -2)$ and $(2, 4)$. Graph the line.

<u>Solution</u>   Let (x_1, y_1) be $(2, 4)$ and (x_2, y_2) be $(-1, -2)$.

$$m = \frac{y_2 - y_1}{x_2 - x_1}$$

$$= \frac{-2 - 4}{-1 - 2} \quad \begin{matrix} y\text{-value} \\ \text{corresponding } x\text{-value} \end{matrix}$$

$$= \frac{-6}{-3} = 2$$

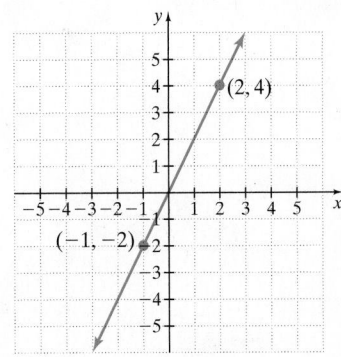

Helpful Hint

The slope for Example 2 is the same if we let (x_1, y_1) be $(-1, -2)$ and (x_2, y_2) be $(2, 4)$.

$$m = \frac{\overset{y\text{-value}}{\overbrace{4 - (-2)}}}{\underset{\text{corresponding } x\text{-value}}{\underbrace{2 - (-1)}}} = \frac{6}{3} = 2$$

PRACTICE

2   Find the slope of the line through $(-3, -1)$ and $(3, 1)$.

✔ **CONCEPT CHECK**

What is wrong with the following slope calculation for the points $(3, 5)$ and $(-2, 6)$?

$$m = \frac{5 - 6}{-2 - 3} = \frac{-1}{-5} = \frac{1}{5}$$

Notice that the slope of the line in Example 1 is negative, whereas the slope of the line in Example 2 is positive. Let your eye follow the line with negative slope from left to right and notice that the line "goes down." Following the line with positive slope from left to right, notice that the line "goes up." This is true in general.

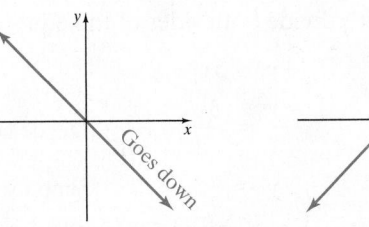

Negative slope   Positive slope

Helpful Hint

To decide whether a line "goes up" or "goes down," always follow the line from left to right.

OBJECTIVE

2   Finding the Slope of a Line Given Its Equation

As we have seen, the slope of a line is defined by two points on the line. Thus, if we know the equation of a line, we can find its slope by finding two of its points. For example, let's find the slope of the line

$$y = 3x + 2$$

To find two points, we can choose two values for x and substitute to find corresponding y-values. If $x = 0$, for example, $y = 3 \cdot 0 + 2$ or $y = 2$. If $x = 1$, $y = 3 \cdot 1 + 2$ or $y = 5$. This gives the ordered pairs $(0, 2)$ and $(1, 5)$. Using the definition for slope, we have

$$m = \frac{5 - 2}{1 - 0} = \frac{3}{1} = 3 \quad \text{The slope is 3.}$$

Notice that the slope, 3, is the same as the coefficient of x in the equation $y = 3x + 2$.

Also, recall from Section 3.2 that the graph of an equation of the form $y = mx + b$ has y-intercept $(0, b)$.

Answer to Concept Check:
The order in which the x- and y-values are used must be the same.

$$m = \frac{5 - 6}{3 - (-2)} = \frac{-1}{5} = -\frac{1}{5}$$

This means that the *y*-intercept of the graph of $y = 3x + 2$ is $(0, 2)$. This is true in general and the form $y = mx + b$ is appropriately called the **slope-intercept form.**

$$y = mx + b$$
slope   *y*-intercept
$(0, b)$

> **Slope-Intercept Form**
>
> When a linear equation in two variables is written in slope-intercept form,
>
> $$y = mx + b$$
>
> *m* is the slope of the line and $(0, b)$ is the *y*-intercept of the line.

EXAMPLE 3 Find the slope and *y*-intercept of the line whose equation is $y = \dfrac{3}{4}x + 6$

Solution The equation is in slope-intercept form, $y = mx + b$.

$$y = \frac{3}{4}x + 6$$

The coefficient of x, $\dfrac{3}{4}$, is the slope and the constant term, 6, is the *y*-value of the *y*-intercept, $(0, 6)$. ☐

PRACTICE
3 Find the slope and *y*-intercept of the line whose equation is $y = \dfrac{2}{3}x - 2$. ■

EXAMPLE 4 Find the slope and the *y*-intercept of the line whose equation is $-y = 5x - 2$.

Solution Remember, the equation must be solved for *y* (not $-y$) in order for it to be written in slope-intercept form.

To solve for *y*, let's divide both sides of the equation by -1.

$$-y = 5x - 2$$
$$\frac{-y}{-1} = \frac{5x}{-1} - \frac{2}{-1} \qquad \text{Divide both sides by } -1.$$
$$y = -5x + 2 \qquad \text{Simplify.}$$

The coefficient of x, -5, is the slope and the constant term, 2, is the *y*-value of the *y*-intercept, $(0, 2)$. ☐

PRACTICE
4 Find the slope and *y*-intercept of the line whose equation is $-y = -6x + 5$. ■

EXAMPLE 5 Find the slope and the *y*-intercept of the line whose equation is $3x - 4y = 4$.

Solution Write the equation in slope-intercept form by solving for *y*.

$$3x - 4y = 4$$
$$-4y = -3x + 4 \qquad \text{Subtract } 3x \text{ from both sides.}$$
$$\frac{-4y}{-4} = \frac{-3x}{-4} + \frac{4}{-4} \qquad \text{Divide both sides by } -4.$$
$$y = \frac{3}{4}x - 1 \qquad \text{Simplify.}$$

The coefficient of x, $\dfrac{3}{4}$, is the slope, and the *y*-intercept is $(0, -1)$. ☐

PRACTICE

5   Find the slope and the *y*-intercept of the line whose equation is $5x + 2y = 8$.   ■

OBJECTIVE

3   Finding Slopes of Horizontal and Vertical Lines

Recall that if a line tilts upward from left to right, its slope is positive. If a line tilts downward from left to right, its slope is negative. Let's now find the slopes of two special lines, horizontal and vertical lines.

EXAMPLE 6   Find the slope of the line $y = -1$.

<u>Solution</u>   Recall that $y = -1$ is a horizontal line with *y*-intercept $(0, -1)$. To find the slope, find two ordered pair solutions of $y = -1$. Solutions of $y = -1$ must have a *y*-value of -1. Let's use points $(2, -1)$ and $(-3, -1)$, which are on the line.

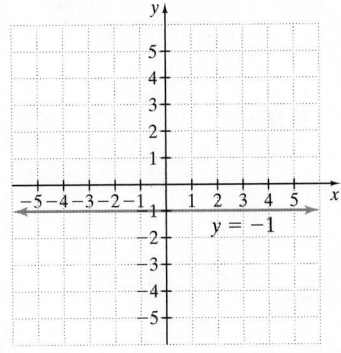

$$m = \frac{y_2 - y_1}{x_2 - x_1} = \frac{-1 - (-1)}{-3 - 2} = \frac{0}{-5} = 0$$

The slope of the line $y = -1$ is 0 and its graph is shown.   □

PRACTICE

6   Find the slope of the line $y = 3$.   ■

Any two points of a horizontal line will have the same *y*-values. This means that the *y*-values will always have a difference of 0 for all horizontal lines. Thus, **all horizontal lines have a slope of 0.**

EXAMPLE 7   Find the slope of the line $x = 5$.

<u>Solution</u>   Recall that the graph of $x = 5$ is a vertical line with *x*-intercept $(5, 0)$.

To find the slope, find two ordered pair solutions of $x = 5$. Solutions of $x = 5$ must have an *x*-value of 5. Let's use points $(5, 0)$ and $(5, 4)$, which are on the line.

$$m = \frac{y_2 - y_1}{x_2 - x_1} = \frac{4 - 0}{5 - 5} = \frac{4}{0}$$

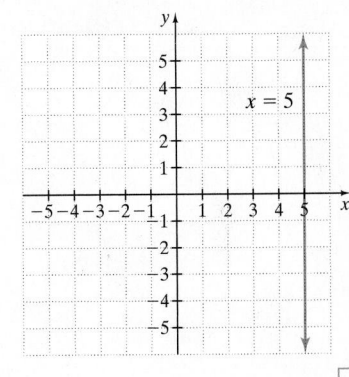

Since $\dfrac{4}{0}$ is undefined, we say the slope of the vertical line $x = 5$ is undefined, and its graph is shown.   □

PRACTICE

7   Find the slope of the line $x = -4$.   ■

Any two points of a vertical line will have the same *x*-values. This means that the *x*-values will always have a difference of 0 for all vertical lines. Thus **all vertical lines have undefined slope.**

> **Helpful Hint**
>
> Slope of 0 and undefined slope are not the same. Vertical lines have undefined slope or no slope, while horizontal lines have a slope of 0.

Here is a general review of slope.

Summary of Slope

Slope m of the line through (x_1, y_1) and (x_2, y_2) is given by the equation $m = \dfrac{y_2 - y_1}{x_2 - x_1}$.

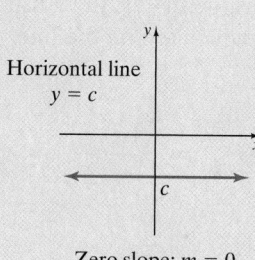

Upward
line

Positive slope: $m > 0$

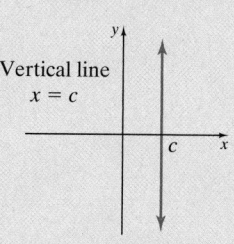

Downward
line

Negative slope: $m < 0$

Horizontal line
$y = c$

Zero slope: $m = 0$

Vertical line
$x = c$

Undefined slope or no slope

OBJECTIVE

4 Slopes of Parallel and Perpendicular Lines

Two lines in the same plane are **parallel** if they do not intersect. Slopes of lines can help us determine whether lines are parallel. Parallel lines have the same steepness, so it follows that they have the same slope.

For example, the graphs of

$$y = -2x + 4$$

and

$$y = -2x - 3$$

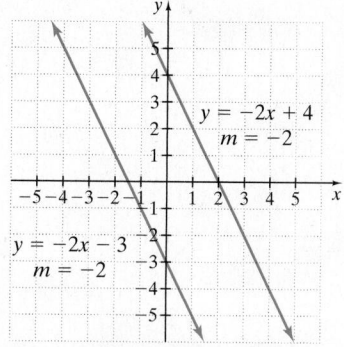

$y = -2x + 4$
$m = -2$

$y = -2x - 3$
$m = -2$

are shown. These lines have the same slope, -2. They also have different y-intercepts, so the lines are distinct and parallel. (If the y-intercepts were the same also, the lines would be the same.)

Parallel Lines

Nonvertical parallel lines have the same slope and different y-intercepts.

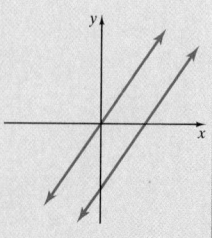

Two lines are **perpendicular** if they lie in the same plane and meet at a 90° (right) angle. How do the slopes of perpendicular lines compare? The product of the slopes of two perpendicular lines is -1.

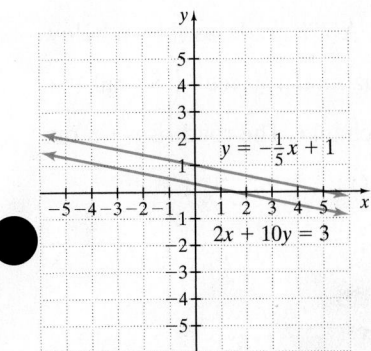

For example, the graphs of

$$y = 4x + 1$$

and

$$y = -\frac{1}{4}x - 3$$

are shown. The slopes of the lines are 4 and $-\frac{1}{4}$. Their product is $4\left(-\frac{1}{4}\right) = -1$, so the lines are perpendicular.

Perpendicular Lines

If the product of the slopes of two lines is -1, then the lines are perpendicular.

    (Two nonvertical lines are perpendicular if the slope of one is the negative reciprocal of the slope of the other.)

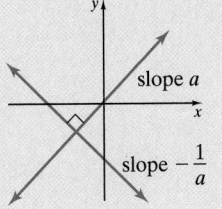

Helpful Hint

Here are examples of numbers that are negative (opposite) reciprocals.

Number	Negative Reciprocal	Their Product Is -1.
$\frac{2}{3}$	$-\frac{3}{2}$	$\frac{2}{3} \cdot -\frac{3}{2} = -\frac{6}{6} = -1$
-5 or $-\frac{5}{1}$	$\frac{1}{5}$	$-5 \cdot \frac{1}{5} = -\frac{5}{5} = -1$

Helpful Hint

Here are a few important facts about vertical and horizontal lines.

- Two distinct vertical lines are parallel.
- Two distinct horizontal lines are parallel.
- A horizontal line and a vertical line are always perpendicular.

EXAMPLE 8 Determine whether each pair of lines is parallel, perpendicular, or neither.

a. $y = -\frac{1}{5}x + 1$ **b.** $x + y = 3$ **c.** $3x + y = 5$

    $2x + 10y = 3$ $-x + y = 4$ $2x + 3y = 6$

Solution

a. The slope of the line $y = -\frac{1}{5}x + 1$ is $-\frac{1}{5}$. We find the slope of the second line by solving its equation for y.

$$2x + 10y = 3$$
$$10y = -2x + 3 \qquad \text{Subtract } 2x \text{ from both sides.}$$
$$y = \frac{-2}{10}x + \frac{3}{10} \qquad \text{Divide both sides by 10.}$$
$$y = -\frac{1}{5}x + \frac{3}{10} \qquad \text{Simplify.}$$

The slope of this line is $-\frac{1}{5}$ also. Since the lines have the same slope and different y-intercepts, they are parallel, as shown in the margin.

(Continued on next page)

b. To find each slope, we solve each equation for y.

$$x + y = 3 \qquad\qquad -x + y = 4$$
$$y = -x + 3 \qquad\qquad y = x + 4$$

↑ The slope is -1.   ↑ The slope is 1.

The slopes are not the same, so the lines are not parallel. Next we check the product of the slopes: $(-1)(1) = -1$. Since the product is -1, the lines are perpendicular, as shown in the figure.

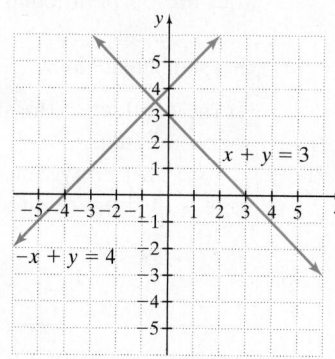

c. We solve each equation for y to find each slope. The slopes are -3 and $-\dfrac{2}{3}$.

The slopes are not the same and their product is not -1. Thus, the lines are neither parallel nor perpendicular. ☐

PRACTICE

8   Determine whether each pair of lines is parallel, perpendicular, or neither.

a. $y = -5x + 1$ **b.** $x + y = 11$ **c.** $2x + 3y = 21$
      $x - 5y = 10$ $2x + y = 11$ $6y = -4x - 2$

✔ **CONCEPT CHECK**

Consider the line $-6x + 2y = 1$.

a. Write the equations of two lines parallel to this line.
b. Write the equations of two lines perpendicular to this line.

OBJECTIVE

5   **Slope as a Rate of Change** ▶

Slope can also be interpreted as a rate of change. In other words, slope tells us how fast y is changing with respect to x. To see this, let's look at a few of the many real-world applications of slope. For example, the pitch of a roof, used by builders and architects, is its slope. The pitch of the roof on the left is $\dfrac{7}{10}\left(\dfrac{\text{rise}}{\text{run}}\right)$. This means that the roof rises vertically 7 feet for every horizontal 10 feet. The rate of change for the roof is 7 vertical feet (y) per 10 horizontal feet (x).

The grade of a road is its slope written as a percent. A 7% grade, as shown below, means that the road rises (or falls) 7 feet for every horizontal 100 feet. $\Big($Recall that $7\% = \dfrac{7}{100}.\Big)$ Here, the slope of $\dfrac{7}{100}$ gives us the rate of change. The road rises (in our diagram) 7 vertical feet (y) for every 100 horizontal feet (x).

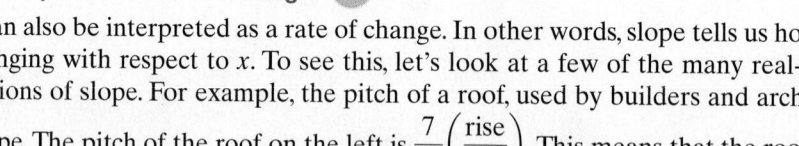

$\dfrac{7}{10}$ pitch

7 feet

10 feet

Answers to Concept Check:
a. any two lines with $m = 3$ and

y-intercept not $\left(0, \dfrac{1}{2}\right)$

b. any two lines with $m = -\dfrac{1}{3}$

$\dfrac{7}{100} = 7\%$ grade

7 feet

100 feet

EXAMPLE 9   **Finding the Grade of a Road**

At one part of the road to the summit of Pikes Peak, the road rises at a rate of 15 vertical feet for a horizontal distance of 250 feet. Find the grade of the road.

Solution Recall that the grade of a road is its slope written as a percent.

$$\text{grade} = \frac{\text{rise}}{\text{run}} = \frac{15}{250} = 0.06 = 6\%$$

15 feet

250 feet

The grade is 6%.

PRACTICE
9   One part of the Mt. Washington (New Hampshire) cog railway rises about 1794 feet over a horizontal distance of 7176 feet. Find the grade of this part of the railway.

EXAMPLE 10   **Finding the Slope of a Line**

The following graph shows annual food and drink sales y (in billions of dollars) for year x.

a. Find the slope of the line and attach the proper units for the rate of change.

b. Then write a sentence explaining the meaning of the slope for this application.

Solution

a. Use (2000, 377) and (2014, 685) to calculate slope.

$$m = \frac{685 - 377}{2014 - 2000} = \frac{308}{14} = \frac{22 \text{ billion dollars}}{1 \text{ year}}$$

b. This means that the rate of change of restaurant food and drink sales increases by 22 billion dollars every 1 year, or \$22 billion per year.

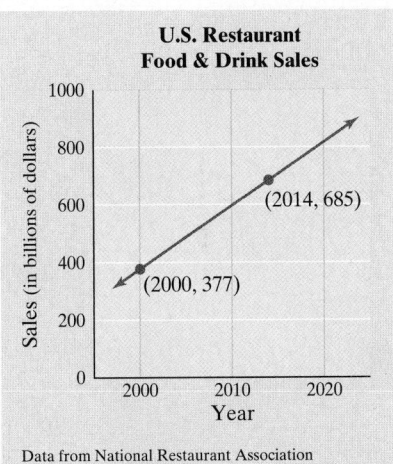

U.S. Restaurant Food & Drink Sales

(2014, 685)

(2000, 377)

Data from National Restaurant Association

PRACTICE
10   The following graph shows the cost y (in dollars) of having laundry done at the Wash-n-Fold, where x is the number of pounds of laundry.

a. Find the slope of the line, and attach the proper units for the rate of change.

b. Then write a sentence explaining the meaning of the slope for this application.

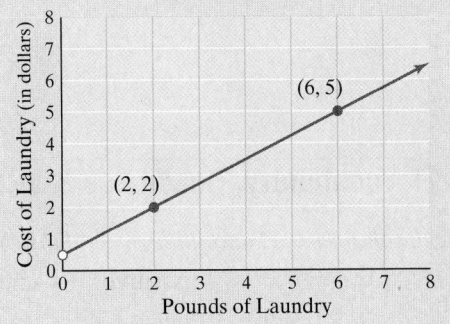

Cost of Laundry

(6, 5)

(2, 2)

Pounds of Laundry

Graphing Calculator Explorations

It is possible to use a grapher to sketch the graph of more than one equation on the same set of axes. This feature can be used to confirm our findings from Section 3.2 when we learned that the graph of an equation written in the form $y = mx + b$ has a y-intercept of b. For example, graph the equations $y = \frac{2}{5}x$, $y = \frac{2}{5}x + 7$, and $y = \frac{2}{5}x - 4$

(Continued on next page)

on the same set of axes. To do so, press the ⎡Y =⎤ key and enter the equations on the first three lines.

$$Y_1 = \left(\frac{2}{5}\right)x$$

$$Y_2 = \left(\frac{2}{5}\right)x + 7$$

$$Y_3 = \left(\frac{2}{5}\right)x - 4$$

The screen should look like:

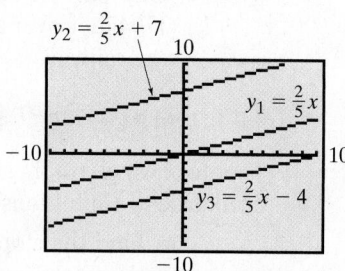

Notice that all three graphs appear to have the same positive slope. The graph of $y = \frac{2}{5}x + 7$ is the graph of $y = \frac{2}{5}x$ moved 7 units upward with a y-intercept of 7.

Also, the graph of $y = \frac{2}{5}x - 4$ is the graph of $y = \frac{2}{5}x$ moved 4 units downward with a y-intercept of -4.

Graph the equations on the same set of axes. Describe the similarities and differences in their graphs. Use the standard window setting or any other convenient window setting.

1. $y = 3.8x, y = 3.8x - 3, y = 3.8x + 7$

2. $y = -4.9x, y = -4.9x + 1, y = -4.9x + 8$

3. $y = \frac{1}{4}x; y = \frac{1}{4}x + 5, y = \frac{1}{4}x - 8$

4. $y = -\frac{3}{4}x, y = -\frac{3}{4}x - 5, y = -\frac{3}{4}x + 6$

✓ Vocabulary, Readiness & Video Check

Use the choices below to fill in each blank. Not all choices will be used.

m	x	0	positive	undefined
b	y	slope	negative	

1. The measure of the steepness or tilt of a line is called _____.

2. If an equation is written in the form $y = mx + b$, the value of the letter _____ is the value of the slope of the graph.

3. The slope of a horizontal line is _____.

4. The slope of a vertical line is _____.

5. If the graph of a line moves upward from left to right, the line has _____ slope.

6. If the graph of a line moves downward from left to right, the line has _____ slope.

7. Given two points of a line, slope $= \dfrac{\text{change in } \_\_}{\text{change in } \_\_}$.

Martin-Gay Interactive Videos

See Video 3.4 ⊙

Watch the section lecture video and answer the following questions.

OBJECTIVE 1

8. What important point is made during ▤ Example 1 having to do with the order of the points in the slope formula?

OBJECTIVE 2

9. From ▤ Example 5, how do you write an equation in "slope-intercept form"? Once the equation is in slope-intercept form, how do you identify the slope?

OBJECTIVE 3

10. In the lecture after ▤ Example 8, different slopes are summarized. What is the difference between zero slope and undefined slope? What does "no slope" mean?

OBJECTIVE 4

11. From ▤ Example 10, what form of the equations is best to determine if two lines are parallel or perpendicular? Why?

OBJECTIVE 5

12. Writing the slope as a rate of change in ▤ Example 11 gave real-life meaning to the slope. What step in the general strategy for problem solving does this correspond to?

3.4   Exercise Set   MyMathLab® ▷

Find the slope of the line that passes through the given points. See Examples 1 and 2.

▷ **1.** $(-1, 5)$ and $(6, -2)$

2. $(3, 1)$ and $(2, 6)$

▷ **3.** $(-4, 3)$ and $(-4, 5)$

4. $(6, -6)$ and $(6, 2)$

5. $(-2, 8)$ and $(1, 6)$

6. $(4, -3)$ and $(2, 2)$

7. $(5, 1)$ and $(-2, 1)$

8. $(0, 13)$ and $(-4, 13)$

Find the slope of each line if it exists. See Examples 1 and 2.

9.

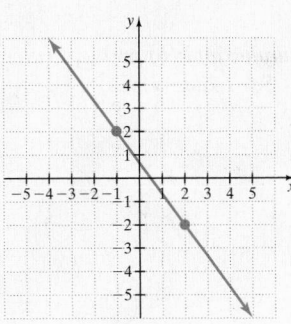

10.

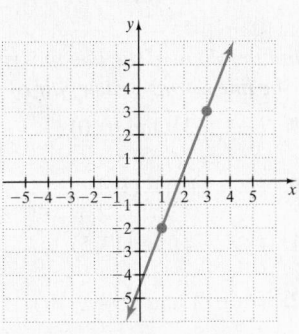

11.

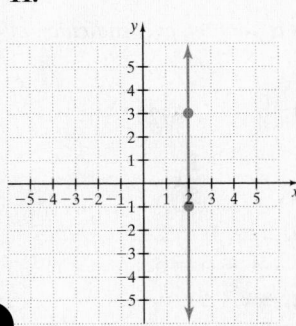

12.

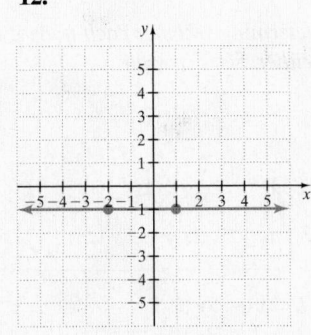

▷ **13.**

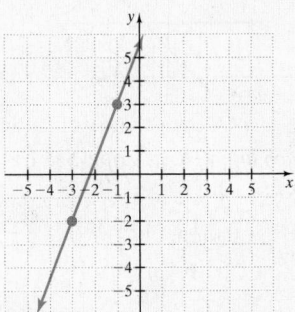

14.

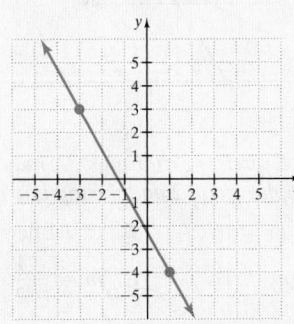

State whether the slope of the line is positive, negative, 0, or is undefined. See the top box on p. 210.

15.

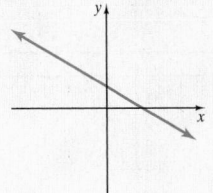

16.

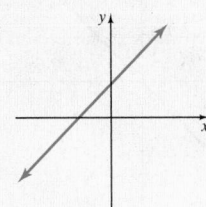

17.

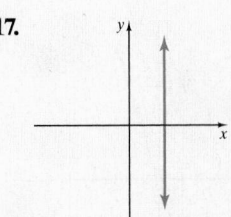

18.

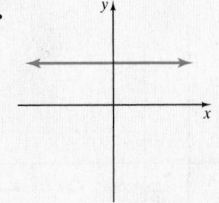

Decide whether a line with the given slope is upward, downward, horizontal, or vertical. See the top box on p. 210.

19. $m = \dfrac{7}{6}$ _____

20. $m = -3$ _____

21. $m = 0$ _____

22. m is undefined. _____

For each graph, determine which line has the greater slope. See the top box on p. 210.

23.

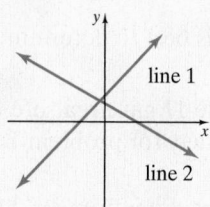

24.

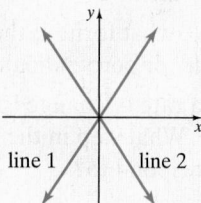

25.

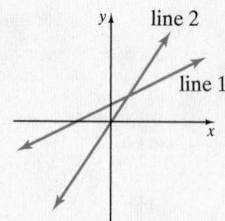

26.

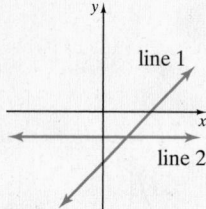

In Exercises 27 through 32, match each line with its slope. See Examples 1 and 2 and the top box on p. 210.

A. $m = 0$
B. undefined slope
C. $m = 3$
D. $m = 1$
E. $m = -\dfrac{1}{2}$
F. $m = -\dfrac{3}{4}$

27.

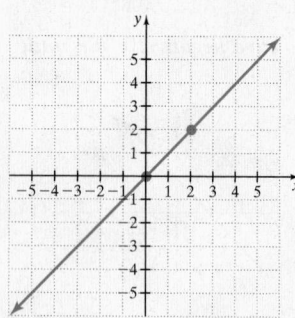

28.

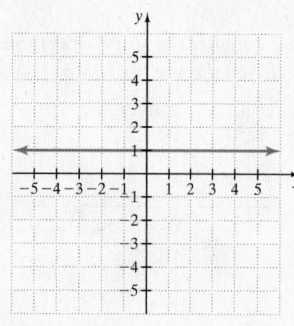

29.

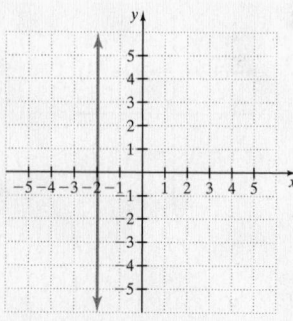

30.

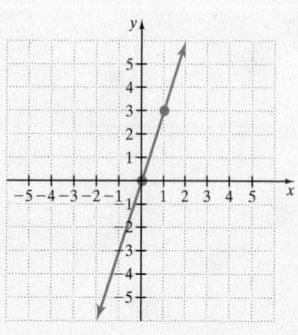

31.

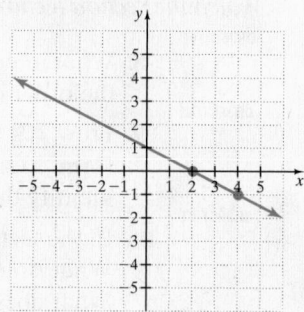

32.

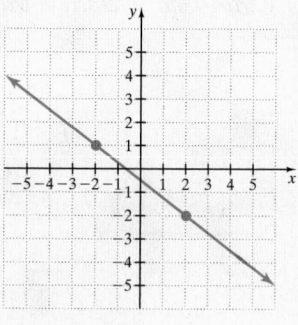

Find the slope of each line. See Examples 6 and 7.

33. $x = 6$
34. $y = 4$
35. $y = -4$
36. $x = 2$
37. $x = -3$
38. $y = -11$
39. $y = 0$
40. $x = 0$

MIXED PRACTICE

Find the slope of each line. See Examples 3 through 7.

41. $y = 5x - 2$
42. $y = -2x + 6$
43. $y = -0.3x + 2.5$
44. $y = -7.6x - 0.1$
▶ **45.** $2x + y = 7$
46. $-5x + y = 10$
▶ **47.** $2x - 3y = 10$
48. $3x - 5y = 1$
▶ **49.** $x = 1$
50. $y = -2$
51. $x = 2y$
52. $x = -4y$
▶ **53.** $y = -3$
54. $x = 5$
55. $-3x - 4y = 6$
56. $-4x - 7y = 9$
57. $20x - 5y = 1.2$
58. $24x - 3y = 5.7$

△ *Find the slope of the line that is (a) parallel and (b) perpendicular to the line through each pair of points. See Example 8.*

59. $(-3, -3)$ and $(0, 0)$

60. $(6, -2)$ and $(1, 4)$

61. $(-8, -4)$ and $(3, 5)$

62. $(6, -1)$ and $(-4, -10)$

△ *Determine whether each pair of lines is parallel, perpendicular, or neither. See Example 8.*

▶ **63.** $y = \dfrac{2}{9}x + 3$

$y = -\dfrac{2}{9}x$

64. $y = \dfrac{1}{5}x + 20$

$y = -\dfrac{1}{5}x$

65. $x - 3y = -6$
$y = 3x - 9$

66. $y = 4x - 2$
$4x + y = 5$

67. $6x = 5y + 1$
$-12x + 10y = 1$

68. $-x + 2y = -2$
$2x = 4y + 3$

69. $6 + 4x = 3y$
$3x + 4y = 8$

▶ **70.** $10 + 3x = 5y$
$5x + 3y = 1$

The pitch of a roof is its slope. Find the pitch of each roof shown. See Example 9. (Note: Pitch of a roof is a positive value.)

71.

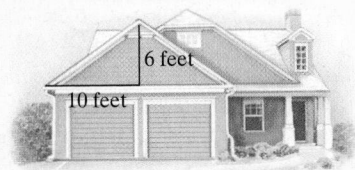

72.

The grade of a road is its slope written as a percent. Find the grade of each road shown. See Example 9. (Note: Grade of a road is a positive value.)

73.

74.

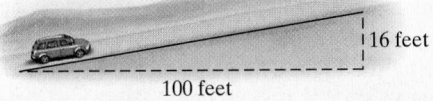

75. One of Japan's superconducting "bullet" trains is researched and tested at the Yamanashi Maglev Test Line near Otsuki City. The steepest section of the track has a rise of 2580 meters for a horizontal distance of 6450 meters. What is the grade of this section of track? (*Source:* Japan Railways Central Co.)

76. Professional plumbers suggest that a sewer pipe should rise 0.25 inch for every horizontal foot. Find the recommended slope for a sewer pipe. Round to the nearest hundredth.

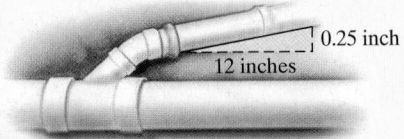

77. There has been controversy over the past few years about the world's steepest street. The *Guinness Book of Records*

actually listed Baldwin Street, in Dunedin, New Zealand, as the world's steepest street, but Canton Avenue in the Pittsburgh neighborhood of Beechview may be steeper. Calculate each grade to the nearest percent.

		Grade (%)
Canton Avenue	for every 30 meters of horizontal distance, the vertical change is 11 meters	
Baldwin Street	for every 2.86 meters of horizontal distance, the vertical change is 1 meter	

78. According to federal regulations, a wheelchair ramp should rise no more than 1 foot for a horizontal distance of 12 feet. Write the slope as a grade. Round to the nearest tenth of a percent.

For Exercises 79 through 82, find the slope of each line and write a sentence explaining the meaning of the slope as a rate of change. Don't forget to attach the proper units. See Example 10.

79. This graph approximates the number of U.S. households that have computers *y* (in millions) for year *x*. (*Source:* U.S. census and statistics)

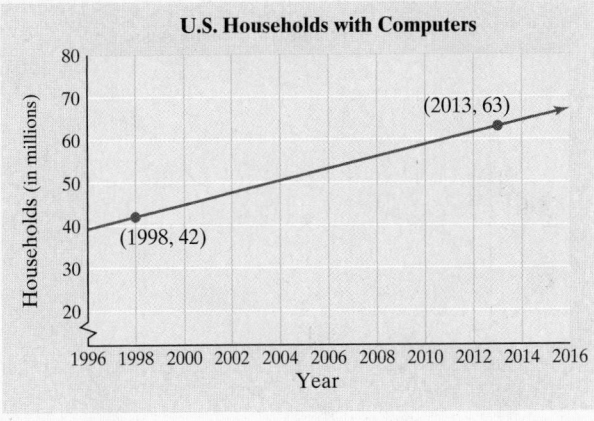

80. The graph approximates the amount of money *y* (in billions of dollars) spent worldwide on leisure travel and tourism for year *x*. (*Source:* World Travel and Tourism Council)

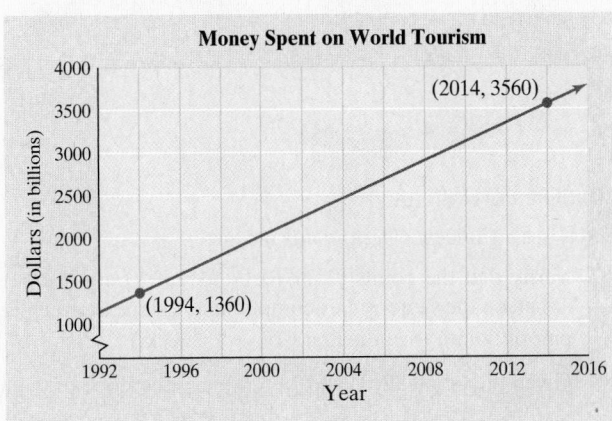

For Exercises 81 and 82, write the slope as a decimal.

▶ **81.** The graph below shows the total cost *y* (in dollars) of owning and operating a compact car where *x* is the number of miles driven.

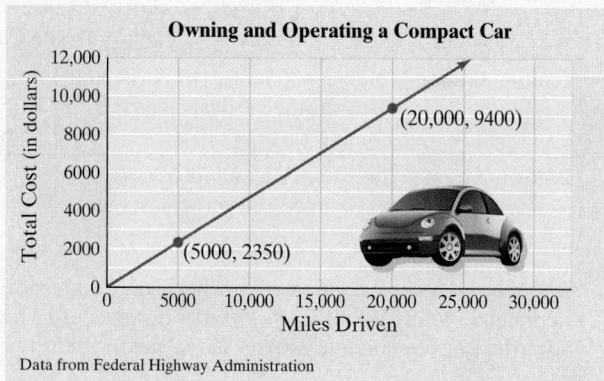

Owning and Operating a Compact Car

Total Cost (in dollars) vs. Miles Driven

(20,000, 9400)

(5000, 2350)

Data from Federal Highway Administration

82. Americans are keeping their cars longer. The graph below shows the median age *y* (in years) of automobiles in the United States for the years shown. (*Source:* Bureau of Transportation Statistics)

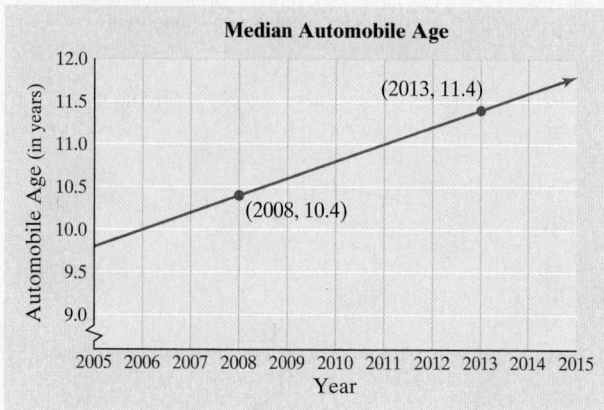

Median Automobile Age

Automobile Age (in years) vs. Year

(2013, 11.4)

(2008, 10.4)

REVIEW AND PREVIEW

Solve each equation for y. See Section 2.5.

83. $y - (-6) = 2(x - 4)$

84. $y - 7 = -9(x - 6)$

85. $y - 1 = -6(x - (-2))$

86. $y - (-3) = 4(x - (-5))$

CONCEPT EXTENSIONS

Solve. See a Concept Check in this section.

87. Verify that the points $(2, 1), (0, 0), (-2, -1)$ and $(-4, -2)$ are all on the same line by computing the slope between each pair of points. (See the first Concept Check.)

88. Given the points $(2, 3)$ and $(-5, 1)$, can the slope of the line through these points be calculated by $\dfrac{1 - 3}{2 - (-5)}$? Why or why not? (See the second Concept Check.)

89. Write the equations of three lines parallel to $10x - 5y = -7$. (See the third Concept Check.)

90. Write the equations of two lines perpendicular to $10x - 5y = -7$. (See the third Concept Check.)

The following line graph shows the average fuel economy (in miles per gallon) of passenger automobiles produced during each of the model years shown. Use this graph to answer Exercises 91 through 96.

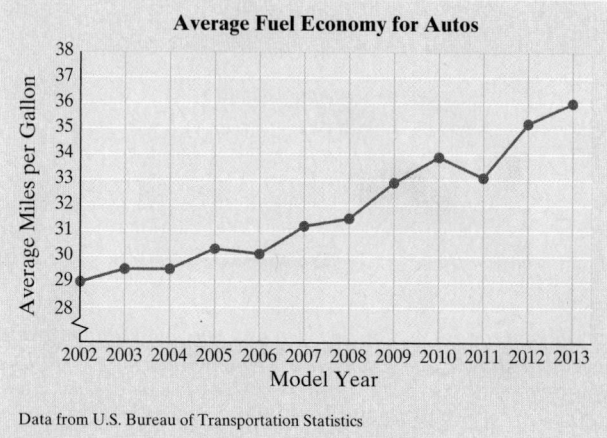

Average Fuel Economy for Autos

Average Miles per Gallon vs. Model Year

Data from U.S. Bureau of Transportation Statistics

91. Between what two years shown was there a decrease in average fuel economy for automobiles?

92. What was the average fuel economy (in miles per gallon) for automobiles produced during 2008?

93. During which of the model years shown was average fuel economy the lowest? What was the average fuel economy that year?

94. During which of the model years shown was average fuel economy the highest? What was the average fuel economy for that year?

95. Of the following line segments, which has the greatest slope: from 2008 to 2009, 2009 to 2010, or 2011 to 2012?

96. Which line segment has a slope of 0?

Solve.

97. Find *x* so that the pitch of the roof is $\dfrac{1}{3}$.

18 feet

x

98. Find *x* so that the pitch of the roof is $\dfrac{2}{5}$.

4 feet

x

99. Approximately 27,000 organ transplants were performed in the United States in 2004. In 2014, the number rose to approximately 29,500. (*Source:* Organ Procurement and Transplantation Network)

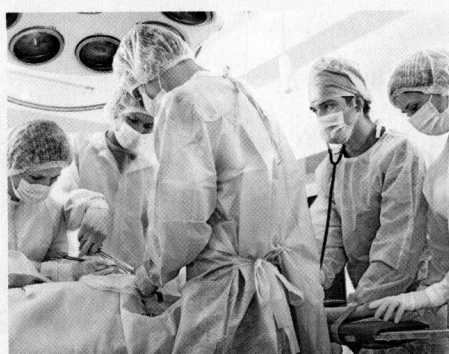

   a. Write two ordered pairs of the form (year, number of organ transplants).

   b. Find the slope of the line between the two points.

   c. Write a sentence explaining the meaning of the slope as a rate of change.

100. The average price of an acre of midgrade Iowa farmland in 2008 was $2300. In 2012, the average price for midgrade farmland was $8300 per acre. (*Source:* National Agricultural Statistics Service)

   a. Write two ordered pairs of the form (year, price of an acre).

   b. Find the slope of the line through the two points.

   c. Write a sentence explaining the meaning of the slope as a rate of change.

101. Show that a triangle with vertices at the points $(1, 1)$, $(-4, 4)$, and $(-3, 0)$ is a right triangle.

102. Show that the quadrilateral with vertices $(1, 3)$, $(2, 1)$, $(-4, 0)$, and $(-3, -2)$ is a parallelogram.

Find the slope of the line through the given points.

103. $(2.1, 6.7)$ and $(-8.3, 9.3)$

104. $(-3.8, 1.2)$ and $(-2.2, 4.5)$

105. $(2.3, 0.2)$ and $(7.9, 5.1)$

106. $(14.3, -10.1)$ and $(9.8, -2.9)$

107. The graph of $y = -\frac{1}{3}x + 2$ has a slope of $-\frac{1}{3}$. The graph of $y = -2x + 2$ has a slope of -2. The graph of $y = -4x + 2$ has a slope of -4. Graph all three equations on a single coordinate system. As the absolute value of the slope becomes larger, how does the steepness of the line change?

108. The graph of $y = \frac{1}{2}x$ has a slope of $\frac{1}{2}$. The graph of $y = 3x$ has a slope of 3. The graph of $y = 5x$ has a slope of 5. Graph all three equations on a single coordinate system. As slope becomes larger, how does the steepness of the line change?

Integrated Review — Summary on Slope and Graphing Linear Equations

Sections 3.1–3.4

Find the slope of each line.

1.

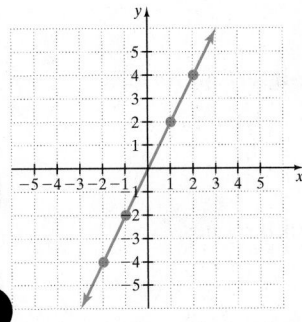

2.

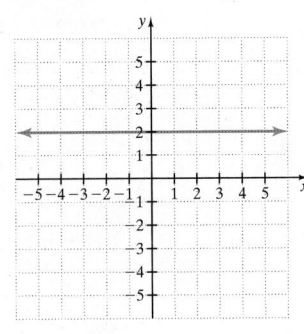

3.

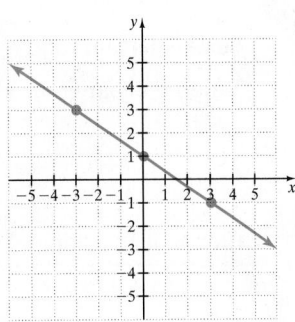

4.

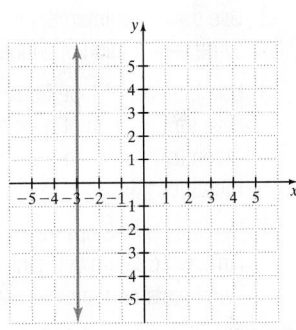

Graph each linear equation.

5. $y = -2x$

6. $x + y = 3$

7. $x = -1$

8. $y = 4$

9. $x - 2y = 6$

10. $y = 3x + 2$

11. $5x + 3y = 15$

12. $2x - 4y = 8$

Determine whether the lines through the points are parallel, perpendicular, or neither.

13. $y = -\dfrac{1}{5}x + \dfrac{1}{3}$

     $3x = -15y$

14. $x - y = \dfrac{1}{2}$

     $3x - y = \dfrac{1}{2}$

15. In the years 2002 through 2013 the number of bridges on public roads (in thousands) in the United States can be modeled by the linear equation $y = 4.09x + 490$, where x is the number of years after 2002 and y is the number of bridges (in thousands). (*Source:* U.S. Dept. of Transportation)

   a. Find the y-intercept of the line.

   b. Write a sentence explaining the meaning of this intercept.

   c. Find the slope of this line.

   d. Write a sentence explaining the meaning of the slope as a rate of change.

16. Online advertising is a means of promoting products and services using the Internet. The revenue y (in billions of dollars) for online advertising for the years 2009 through 2014 is given by $y = 5.3x + 22.7$, where x is the number of years after 2009.

   a. Use this equation to complete the ordered pair $(4, \quad)$.

   b. Write a sentence explaining the meaning of the answer to part (a).

3.5 Equations of Lines

OBJECTIVES

1 Use the Slope-Intercept Form to Graph a Linear Equation.

2 Use the Slope-Intercept Form to Write an Equation of a Line.

3 Use the Point-Slope Form to Find an Equation of a Line Given Its Slope and a Point on the Line.

4 Use the Point-Slope Form to Find an Equation of a Line Given Two Points on the Line.

5 Find Equations of Vertical and Horizontal Lines.

6 Use the Point-Slope Form to Solve Problems.

Recall that the form $y = mx + b$ is appropriately called the *slope-intercept form* of a linear equation.

       slope y-intercept is $(0, b)$

> ### Slope-Intercept Form
>
> When a linear equation in two variables is written in **slope-intercept form,**
>
> $$y = mx + b$$
>
>   slope $(0, b), y$-intercept
>
> then m is the slope of the line and $(0, b)$ is the y-intercept of the line.

OBJECTIVE

1 Using the Slope-Intercept Form to Graph an Equation

We can use the slope-intercept form of the equation of a line to graph a linear equation.

EXAMPLE 1   Use the slope-intercept form to graph the equation

$$y = \frac{3}{5}x - 2.$$

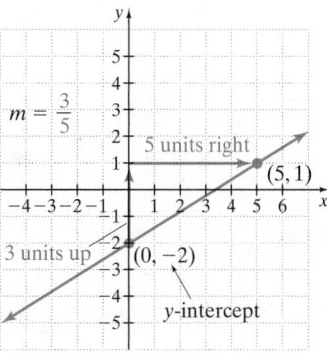

Solution   Since the equation $y = \frac{3}{5}x - 2$ is written in slope-intercept form $y = mx + b$, the slope of its graph is $\frac{3}{5}$ and the y-intercept is $(0, -2)$. To graph this equation, we begin by plotting the point $(0, -2)$. From this point, we can find another point of the graph by using the slope $\frac{3}{5}$ and recalling that slope is $\frac{\text{rise}}{\text{run}}$. We start at the y-intercept and move 3 units up since the numerator of the slope is 3; then we move 5 units to the right since the denominator of the slope is 5. We stop at the point $(5, 1)$. The line through $(0, -2)$ and $(5, 1)$ is the graph of $y = \frac{3}{5}x - 2$.   □

PRACTICE

1   Graph: $y = \frac{2}{3}x - 5$ ■

EXAMPLE 2   Use the slope-intercept form to graph the equation $4x + y = 1$.

Solution   First we write the given equation in slope-intercept form.

$$4x + y = 1$$
$$y = -4x + 1$$

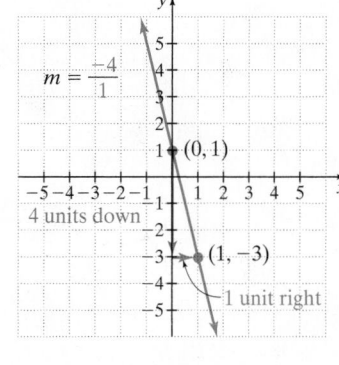

The graph of this equation will have slope -4 and y-intercept $(0, 1)$. To graph this line, we first plot the point $(0, 1)$. To find another point of the graph, we use the slope -4, which can be written as $\frac{-4}{1}$ $\left(\frac{4}{-1} \text{ could also be used} \right)$. We start at the point $(0, 1)$ and move 4 units down (since the numerator of the slope is -4) and then 1 unit to the right (since the denominator of the slope is 1).

We arrive at the point $(1, -3)$. The line through $(0, 1)$ and $(1, -3)$ is the graph of $4x + y = 1$.   □

> **Helpful Hint**
>
> In Example 2, if we interpret the slope of -4 as $\frac{4}{-1}$, we arrive at $(-1, 5)$ for a second point. Notice that this point is also on the line.

PRACTICE

2   Use the slope-intercept form to graph the equation $3x - y = 2$.   ■

OBJECTIVE

2   Using the Slope-Intercept Form to Write an Equation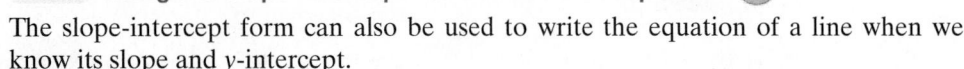

The slope-intercept form can also be used to write the equation of a line when we know its slope and y-intercept.

EXAMPLE 3   Find an equation of the line with y-intercept $(0, -3)$ and slope of $\frac{1}{4}$.

Solution   We are given the slope and the y-intercept. We let $m = \frac{1}{4}$ and $b = -3$ and write the equation in slope-intercept form, $y = mx + b$.

$$y = mx + b$$
$$y = \frac{1}{4}x + (-3) \quad \text{Let } m = \frac{1}{4} \text{ and } b = -3.$$
$$y = \frac{1}{4}x - 3 \quad \text{Simplify.}$$
                                        □

(Continued on next page)

3 Find an equation of the line with y-intercept $(0, 7)$ and slope of $\dfrac{1}{2}$. ■

OBJECTIVE

3 Writing an Equation Given Slope and a Point

Thus far, we have seen that we can write an equation of a line if we know its slope and y-intercept. We can also write an equation of a line if we know its slope and any point on the line. To see how we do this, let m represent slope and (x_1, y_1) represent the point on the line. Then if (x, y) is any other point on the line, we have that

$$\frac{y - y_1}{x - x_1} = m$$

$$y - y_1 = m(x - x_1) \quad \text{Multiply both sides by } (x - x_1).$$

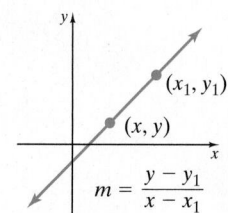

This is the *point-slope form* of the equation of a line.

Point-Slope Form of the Equation of a Line

The **point-slope form** of the equation of a line is

$$y - y_1 = m(x - x_1)$$

where the up-arrow under m indicates slope and the up-arrow under (x_1, y_1) indicates point on the line.

where m is the slope of the line and (x_1, y_1) is a point on the line.

EXAMPLE 4 Find an equation of the line with slope -2 that passes through $(-1, 5)$. Write the equation in slope-intercept form, $y = mx + b$, and in standard form, $Ax + By = C$.

Solution Since the slope and a point on the line are given, we use point-slope form $y - y_1 = m(x - x_1)$ to write the equation. Let $m = -2$ and $(-1, 5) = (x_1, y_1)$.

$$y - y_1 = m(x - x_1)$$
$$y - 5 = -2[x - (-1)] \quad \text{Let } m = -2 \text{ and } (x_1, y_1) = (-1, 5).$$
$$y - 5 = -2(x + 1) \quad \text{Simplify.}$$
$$y - 5 = -2x - 2 \quad \text{Use the distributive property.}$$

To write the equation in slope-intercept form, $y = mx + b$, we simply solve the equation for y. To do this, we add 5 to both sides.

$$y - 5 = -2x - 2$$
$$y = -2x + 3 \quad \text{Slope-intercept form.}$$
$$2x + y = 3 \quad \text{Add } 2x \text{ to both sides and we have standard form.} \qquad \square$$

4 Find an equation of the line passing through $(2, 3)$ with slope 4. Write the equation in standard form: $Ax + By = C$. ■

OBJECTIVE

4 Writing an Equation Given Two Points

We can also find an equation of a line when we are given any two points on the line.

EXAMPLE 5   Find an equation of the line through $(2, 5)$ and $(-3, 4)$. Write the equation in standard form.

**Solution**   First, use the two given points to find the slope of the line.

$$m = \frac{4 - 5}{-3 - 2} = \frac{-1}{-5} = \frac{1}{5}$$

Next we use the slope $\frac{1}{5}$ and either one of the given points to write the equation in point-slope form. We use $(2, 5)$. Let $x_1 = 2$, $y_1 = 5$, and $m = \frac{1}{5}$.

$$y - y_1 = m(x - x_1) \qquad \text{Use point-slope form.}$$

$$y - 5 = \frac{1}{5}(x - 2) \qquad \text{Let } x_1 = 2, y_1 = 5, \text{ and } m = \frac{1}{5}.$$

$$5(y - 5) = 5 \cdot \frac{1}{5}(x - 2) \qquad \text{Multiply both sides by 5 to clear fractions.}$$

$$5y - 25 = x - 2 \qquad \text{Use the distributive property and simplify.}$$

$$-x + 5y - 25 = -2 \qquad \text{Subtract } x \text{ from both sides.}$$

$$-x + 5y = 23 \qquad \text{Add 25 to both sides.} \qquad \square$$

PRACTICE
5   Find an equation of the line through $(-1, 6)$ and $(3, 1)$. Write the equation in standard form.   ■

> **Helpful Hint**
>
> Multiply both sides of the equation $-x + 5y = 23$ by -1, and it becomes $x - 5y = -23$. Both $-x + 5y = 23$ and $x - 5y = -23$ are in standard form, and they are equations of the same line.

OBJECTIVE
5   Finding Equations of Vertical and Horizontal Lines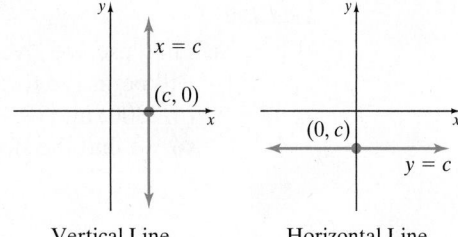

Recall from Section 3.3 that:

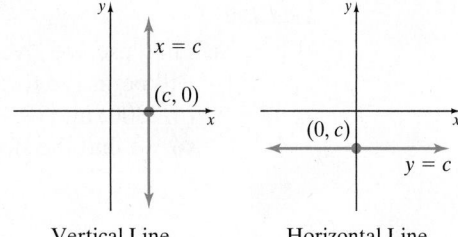

Vertical Line   Horizontal Line

EXAMPLE 6   Find an equation of the vertical line through $(-1, 5)$.

**Solution** The equation of a vertical line can be written in the form $x = c$, so an equation for the vertical line passing through $(-1, 5)$ is $x = -1$.

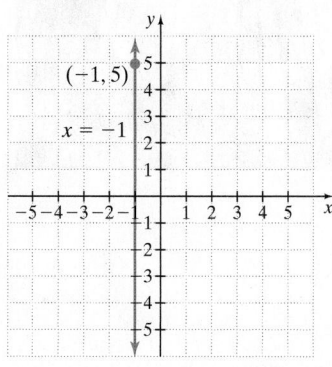

PRACTICE
6   Find an equation of the vertical line through $(3, -2)$.   ■

⚠ **EXAMPLE 7**   Find an equation of the line parallel to the line $y = 5$ and passing through $(-2, -3)$.

Solution Since the graph of $y = 5$ is a horizontal line, any line parallel to it is also horizontal. The equation of a horizontal line can be written in the form $y = c$. An equation for the horizontal line passing through

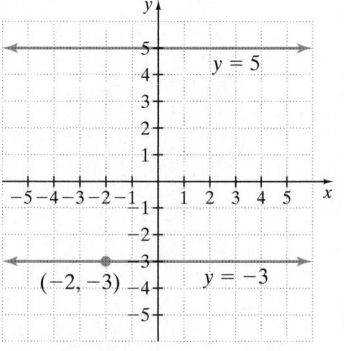

$$(-2, -3) \text{ is } y = -3.$$

PRACTICE

7   Find an equation of the line parallel to the line $y = -2$ and passing through $(4, 3)$.

OBJECTIVE

6   Using the Point-Slope Form to Solve Problems

Problems occurring in many fields can be modeled by linear equations in two variables. The next example is from the field of marketing and shows how consumer demand of a product depends on the price of the product.

EXAMPLE 8   **Predicting the Sales of T-Shirts**

A web-based T-shirt company has learned that by pricing a clearance-sale T-shirt at $6, sales will reach 2000 T-shirts per day. Raising the price to $8 will cause the sales to fall to 1500 T-shirts per day.

   a. Assume that the relationship between sales price and number of T-shirts sold is linear and write an equation describing this relationship. Write the equation in slope-intercept form.

   b. Predict the daily sales of T-shirts if the price is $7.50.

Solution

   a. First, use the given information and write two ordered pairs. Ordered pairs will be in the form (sales price, number sold) so that our ordered pairs are $(6, 2000)$ and $(8, 1500)$. Use the point-slope form to write an equation. To do so, we find the slope of the line that contains these points.

$$m = \frac{2000 - 1500}{6 - 8} = \frac{500}{-2} = -250$$

Next, use the slope and either one of the points to write the equation in point-slope form. We use $(6, 2000)$.

$$
\begin{array}{ll}
y - y_1 = m(x - x_1) & \text{Use point-slope form.} \\
y - 2000 = -250(x - 6) & \text{Let } x_1 = 6, y_1 = 2000, \text{ and } m = -250. \\
y - 2000 = -250x + 1500 & \text{Use the distributive property.} \\
y = -250x + 3500 & \text{Write in slope-intercept form.}
\end{array}
$$

   b. To predict the sales if the price is $7.50, we find y when $x = 7.50$.

$$
\begin{array}{ll}
y = -250x + 3500 & \\
y = -250(7.50) + 3500 & \text{Let } x = 7.50. \\
y = -1875 + 3500 & \\
y = 1625 &
\end{array}
$$

If the price is $7.50, sales will reach 1625 T-shirts per day.

PRACTICE
8 The new *Camelot* condos were selling at a rate of 30 per month when they were priced at $150,000 each. Lowering the price to $120,000 caused the sales to rise to 50 condos per month.

a. Assume that the relationship between number of condos sold and price is linear, and write an equation describing this relationship. Write the equation in slope-intercept form. Use ordered pairs of the form (number sold, sales price).

b. How should the condos be priced if the developer wishes to sell 60 condos per month? ■

The preceding example may also be solved by using ordered pairs of the form (sales price, number sold).

Forms of Linear Equations

$Ax + By = C$	**Standard form** of a linear equation. A and B are not both 0.
$y = mx + b$	**Slope-intercept form** of a linear equation. The slope is m and the y-intercept is $(0, b)$.
$y - y_1 = m(x - x_1)$	**Point-slope form** of a linear equation. The slope is m and (x_1, y_1) is a point on the line.
$y = c$	**Horizontal line** The slope is 0 and the y-intercept is $(0, c)$.
$x = c$	**Vertical line** The slope is undefined and the x-intercept is $(c, 0)$.

Parallel and Perpendicular Lines

Nonvertical parallel lines have the same slope.

The product of the slopes of two nonvertical perpendicular lines is -1.

Graphing Calculator Explorations

A grapher is a very useful tool for discovering patterns. To discover the change in the graph of a linear equation caused by a change in slope, try the following. Use a standard window and graph a linear equation in the form $y = mx + b$. Recall that the graph of such an equation will have slope m and y-intercept b.

First, graph $y = x + 3$. To do so, press the $\boxed{Y=}$ key and enter $Y_1 = x + 3$. Notice that this graph has slope 1 and that the y-intercept is 3. Next, on the same set of axes, graph $y = 2x + 3$ and $y = 3x + 3$ by pressing $\boxed{Y=}$ and entering $Y_2 = 2x + 3$ and $Y_3 = 3x + 3$.

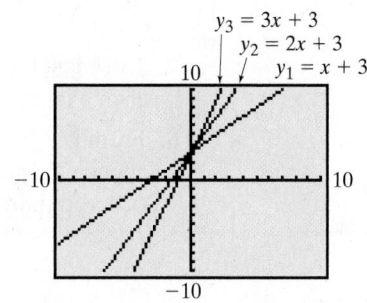

(Continued on next page)

Notice the difference in the graph of each equation as the slope changes from 1 to 2 to 3. How would the graph of $y = 5x + 3$ appear? To see the change in the graph caused by a change in negative slope, try graphing $y = -x + 3$, $y = -2x + 3$, and $y = -3x + 3$ on the same set of axes.

Use a grapher to graph the following equations. For each exercise, graph the first equation and use its graph to predict the appearance of the other equations. Then graph the other equations on the same set of axes and check your prediction.

1. $y = x$; $y = 6x$, $y = -6x$

2. $y = -x$; $y = -5x$, $y = -10x$

3. $y = \frac{1}{2}x + 2$; $y = \frac{3}{4}x + 2$, $y = x + 2$

4. $y = x + 1$; $y = \frac{5}{4}x + 1$, $y = \frac{5}{2}x + 1$

5. $y = -7x + 5$; $y = 7x + 5$

6. $y = 3x - 1$; $y = -3x - 1$

✔ Vocabulary, Readiness & Video Check

Use the choices below to fill in each blank. Some choices may be used more than once and some not at all.

b	(y_1, x_1)	point-slope	vertical	standard
m	(x_1, y_1)	slope-intercept	horizontal	

1. The form $y = mx + b$ is called _____ form. When a linear equation in two variables is written in this form, _____ is the slope of its graph and $(0, \underline{\hspace{1cm}})$ is its y-intercept.

2. The form $y - y_1 = m(x - x_1)$ is called _____ form. When a linear equation in two variables is written in this form, _____ is the slope of its graph and _____ is a point on the graph.

Martin-Gay Interactive Videos

See Video 3.5 ⊙

Watch the section lecture video and answer the following questions.

OBJECTIVE 1

3. We can use the slope-intercept form to graph a line. Complete these statements based on ▦ Example 1. Start by graphing the _____.
Find another point by applying the slope to this point—rewrite the slope as a _____ if necessary.

OBJECTIVE 2

4. In ▦ Example 3, what is the y-intercept?

OBJECTIVE 3

5. In ▦ Example 4 we use the point-slope form to find the equation of a line given the slope and a point. How do we then write this equation in standard form?

OBJECTIVE 4

6. The lecture before ▦ Example 5 discusses how to find the equation of a line given two points. Is there any circumstance when you might want to use the slope-intercept form to find the equation of a line given two points? If so, when?

OBJECTIVE 5

7. Solve ▦ Examples 6 and 7 again, this time using the point $(-2, -3)$ in each exercise.

OBJECTIVE 6

8. From ▦ Example 8, we are told to use ordered pairs of the form (time, speed). Explain why it is important to keep track of how you define your ordered pairs and/or define your variables.

3.5 Exercise Set MyMathLab®

Use the slope-intercept form to graph each equation. See Examples 1 and 2.

1. $y = 2x + 1$

2. $y = -4x - 1$

3. $y = \frac{2}{3}x + 5$

4. $y = \frac{1}{4}x - 3$

5. $y = -5x$

6. $y = -6x$

7. $4x + y = 6$

8. $-3x + y = 2$

9. $4x - 7y = -14$

10. $3x - 4y = 4$

11. $x = \frac{5}{4}y$

12. $x = \frac{3}{2}y$

Write an equation of the line with each given slope, m, and y-intercept, (0, b). See Example 3.

13. $m = 5, b = 3$

14. $m = -3, b = -3$

15. $m = -4, b = -\frac{1}{6}$

16. $m = 2, b = \frac{3}{4}$

17. $m = \frac{2}{3}, b = 0$

18. $m = -\frac{4}{5}, b = 0$

19. $m = 0, b = -8$

20. $m = 0, b = -2$

21. $m = -\frac{1}{5}, b = \frac{1}{9}$

22. $m = \frac{1}{2}, b = -\frac{1}{3}$

Find an equation of each line with the given slope that passes through the given point. Write the equation in the form Ax + By = C. See Example 4.

23. $m = 6;\ \ (2, 2)$

24. $m = 4;\ \ (1, 3)$

25. $m = -8;\ \ (-1, -5)$

26. $m = -2;\ \ (-11, -12)$

27. $m = \frac{3}{2};\ \ (5, -6)$

28. $m = \frac{2}{3};\ \ (-8, 9)$

29. $m = -\frac{1}{2};\ \ (-3, 0)$

30. $m = -\frac{1}{5};\ \ (4, 0)$

Find an equation of the line passing through each pair of points. Write the equation in the form Ax + By = C. See Example 5.

31. $(3, 2)$ and $(5, 6)$

32. $(6, 2)$ and $(8, 8)$

33. $(-1, 3)$ and $(-2, -5)$

34. $(-4, 0)$ and $(6, -1)$

35. $(2, 3)$ and $(-1, -1)$

36. $(7, 10)$ and $(-1, -1)$

37. $(0, 0)$ and $\left(-\frac{1}{8}, \frac{1}{13}\right)$

38. $(0, 0)$ and $\left(-\frac{1}{2}, \frac{1}{3}\right)$

Find an equation of each line. See Example 6.

39. Vertical line through $(0, 2)$

40. Horizontal line through $(1, 4)$

41. Horizontal line through $(-1, 3)$

42. Vertical line through $(-1, 3)$

43. Vertical line through $\left(-\frac{7}{3}, -\frac{2}{5}\right)$

44. Horizontal line through $\left(\frac{2}{7}, 0\right)$

Find an equation of each line. See Example 7.

45. Parallel to $y = 5$, through $(1, 2)$

46. Perpendicular to $y = 5$, through $(1, 2)$

47. Perpendicular to $x = -3$, through $(-2, 5)$

48. Parallel to $y = -4$, through $(0, -3)$

49. Parallel to $x = 0$, through $(6, -8)$

50. Perpendicular to $x = 7$, through $(-5, 0)$

MIXED PRACTICE

See Examples 1 through 7. Find an equation of each line described. Write each equation in slope-intercept form (solved for y), when possible.

51. With slope $-\frac{1}{2}$, through $\left(0, \frac{5}{3}\right)$

52. With slope $\frac{5}{7}$, through $(0, -3)$

53. Through $(10, 7)$ and $(7, 10)$

54. Through $(5, -6)$ and $(-6, 5)$

55. With undefined slope, through $\left(-\frac{3}{4}, 1\right)$

56. With slope 0, through $(6.7, 12.1)$

57. Slope 1, through $(-7, 9)$

58. Slope 5, through $(6, -8)$

59. Slope -5, y-intercept $(0, 7)$

60. Slope -2; y-intercept $(0, -4)$

61. Through $(6, 7)$, parallel to the x-axis

62. Through $(1, -5)$, parallel to the y-axis

63. Through $(2, 3)$ and $(0, 0)$

64. Through $(4, 7)$ and $(0, 0)$

65. Through $(-2, -3)$, perpendicular to the y-axis

66. Through $(0, 12)$, perpendicular to the x-axis

67. Slope $-\frac{4}{7}$, through $(-1, -2)$

68. Slope $-\frac{3}{5}$, through $(4, 4)$

Solve. Assume each exercise describes a linear relationship. Write the equations in slope-intercept form. See Example 8.

69. A rock is dropped from the top of a 400-foot cliff. After 1 second, the rock is traveling 32 feet per second. After 3 seconds, the rock is traveling 96 feet per second.

400 feet

a. Assume that the relationship between time and speed is linear and write an equation describing this relationship. Use ordered pairs of the form (time, speed).

b. Use this equation to determine the speed of the rock 4 seconds after it was dropped.

70. A Hawaiian fruit company is studying the sales of a pineapple sauce to see if this product is to be continued. At the end of its first year, profits on this product amounted to $30,000. At the end of the fourth year, profits were $66,000.

a. Assume that the relationship between years on the market and profit is linear and write an equation describing this relationship. Use ordered pairs of the form (years on the market, profit).

b. Use this equation to predict the profit at the end of 7 years.

71. Sales of automobiles worldwide continues to grow as emerging economies provide their citizens more purchasing power. In 2011, there were 66 thousand new car sales worldwide, and in 2014, this number increased to 72 thousand cars. (*Source:* statista)

a. Write an equation describing the relationship between time and the number of new automobiles sold worldwide. Use ordered pairs of the form (years past 2011, number of automobiles sold).

b. Use this equation to predict the number of new car sales worldwide in the year 2018.

72. In 2004, the restaurant industry in the United States employed about 11.9 million people. In 2014, this number increased to 13.5 million employees. (*Source:* National Restaurant Association)

a. Write an equation describing the relationship between time and the number of restaurant industry employees. Use ordered pairs of the form (years past 2004, number of restaurant employees in millions).

b. Use this equation to predict the number of restaurant industry employees in 2018. Round to the nearest tenth of a million.

73. In 2006, the U.S. population (or persons) per square mile of land area was 85. In 2014, the persons per square mile was 89.48.

a. Write an equation describing the relationship between year and persons per square mile. Use ordered pairs of the form (years past 2006, persons per square mile).

b. Use this equation to predict the persons per square mile in 2018. Round to the nearest tenth of a person.

74. Dunkin' Donuts started as a New England brand coffee and donut shop, but has expanded recently throughout the world. In 2011, there were 10,083 stores worldwide. By 2014, this had grown to 10,884. (*Source:* Dunkin' Donuts)

a. Write an equation describing the relationship between year and number of Dunkin' Donuts stores. Use ordered pairs of the form (years past 2011, number of Dunkin' Donuts stores).

b. Use this equation to predict the number of Dunkin' Donuts stores in 2018.

75. It has been said that newspapers are disappearing, replaced by various electronic media. In 2003, newspaper circulation (the number of copies distributed in a day) was about 55 million. In 2013, this had dropped to 44 million. (*Source:* Newspaper Association of America).

a. Write an equation describing the relationship between time and circulation. Use ordered pairs of the form (years past 2003, number of newspapers circulated in millions).

b. Use this equation to predict newspaper circulation in 2018.

76. A certain chain of book stores is slowly closing down stores. Suppose that in 2006 there were 3991 stores and in 2010 there were 3200 stores.

a. Write an equation describing the relationship between time and number of store locations. Use ordered pairs of the form (years past 2006, number of stores).

b. Use this equation to predict the number of stores in 2018.

77. The Pool Fun Company has learned that, by pricing a newly released Fun Noodle at $3, sales will reach 10,000 Fun Noodles per day during the summer. Raising the price to $5 will cause sales to fall to 8000 Fun Noodles per day.

a. Assume that the relationship between price and number of Fun Noodles sold is linear and write an equation describing this relationship. Use ordered pairs of the form (price, number sold).

b. Predict the daily sales of Fun Noodles if the price is $3.50.

78. The value of a building bought in 2000 may be depreciated (or decreased) as time passes for income tax purposes. Seven years after the building was bought, this value was $225,000 and 12 years after it was bought, this value was $195,000.

    a. If the relationship between number of years past 2000 and the depreciated value of the building is linear, write an equation describing this relationship. Use ordered pairs of the form (years past 2000, value of building).

    b. Use this equation to estimate the depreciated value of the building in 2018.

REVIEW AND PREVIEW

Find the value of $x^2 - 3x + 1$ for each given value of x. See Section 1.7.

79. 2   **80.** 5

81. −1   **82.** −3

For each graph, determine whether any x-values correspond to two or more y-values. See Section 3.1.

83.

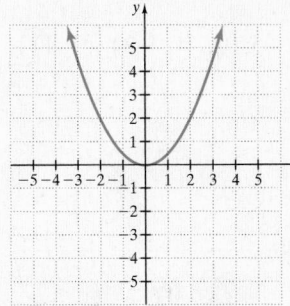

84.

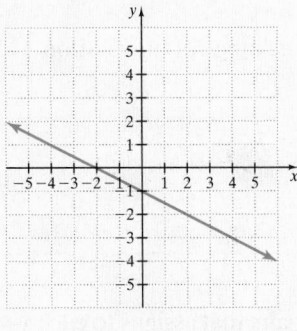

85.

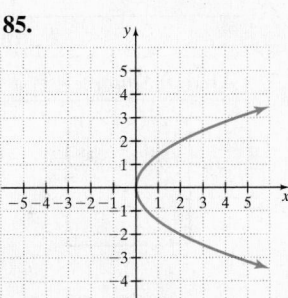

86.

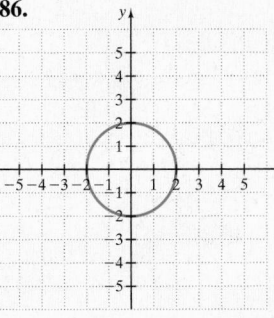

CONCEPT EXTENSIONS

For Exercises 87 through 90, identify the form that the linear equation in two variables is written in. For Exercises 91 and 92, identify the appearance of the graph of the equation.

87. $y - 7 = 4(x + 3)$; _____ form

88. $5x - 9y = 11$; _____ form

89. $y = \frac{3}{4}x - \frac{1}{3}$; _____ form

90. $y + 2 = -\frac{1}{3}(x - 2)$; _____ form

91. $y = \frac{1}{2}$; _____ line **92.** $x = -17$; _____ line

93. Given the equation of a nonvertical line, explain how to find the slope without finding two points on the line.

94. Given two points on a nonvertical line, explain how to use the point-slope form to find the equation of the line.

95. Write an equation in standard form of the line that contains the point $(-1, 2)$ and is

    a. parallel to the line $y = 3x - 1$.

    b. perpendicular to the line $y = 3x - 1$.

96. Write an equation in standard form of the line that contains the point $(4, 0)$ and is

    a. parallel to the line $y = -2x + 3$.

    b. perpendicular to the line $y = -2x + 3$.

97. Write an equation in standard form of the line that contains the point $(3, -5)$ and is

    a. parallel to the line $3x + 2y = 7$.

    b. perpendicular to the line $3x + 2y = 7$.

98. Write an equation in standard form of the line that contains the point $(-2, 4)$ and is

    a. parallel to the line $x + 3y = 6$.

    b. perpendicular to the line $x + 3y = 6$.

3.6   Functions

OBJECTIVES

1. Identify Relations, Domains, and Ranges.
2. Identify Functions.
3. Use the Vertical Line Test.
4. Use Function Notation.

OBJECTIVE

1   Identifying Relations, Domains, and Ranges

In previous sections, we have discussed the relationships between two quantities. For example, the relationship between the length of the side of a square x and its area y is described by the equation $y = x^2$. Ordered pairs can be used to write down solutions of this equation. For example, $(2, 4)$ is a solution of $y = x^2$, and this notation tells us that the x-value 2 is related to the y-value 4 for this equation. In other words, when the length of the side of a square is 2 units, its area is 4 square units.

Examples of Relationships Between Two Quantities

Area of Square: $y = x^2$	Equation of Line: $y = x + 2$	Internet Advertising Revenue
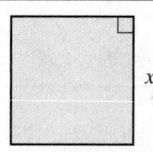	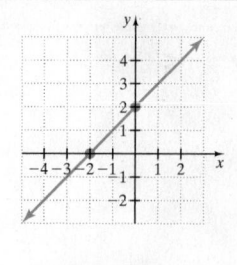	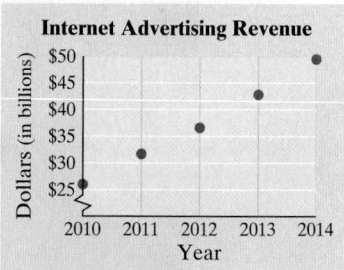

Some Ordered Pairs		Some Ordered Pairs		Ordered Pairs	
x	**y**	**x**	**y**	**Year**	**Billions of Dollars**
2	4	−3	−1	2010	26
5	25	0	2	2011	31.7
7	49	2	4	2012	36.6
12	144	9	11	2013	42.8
				2014	49.5

(*Source:* PricewaterhouseCoopers LLP)

A set of ordered pairs is called a **relation.** The set of all x-coordinates is called the **domain** of a relation, and the set of all y-coordinates is called the **range** of a relation. Equations such as $y = x^2$ are also called relations since equations in two variables define a set of ordered pair solutions.

EXAMPLE 1 Find the domain and the range of the relation $\{(0, 2), (3, 3), (-1, 0), (3, -2)\}$.

Solution The domain is the set of all x-values or $\{-1, 0, 3\}$, and the range is the set of all y-values, or $\{-2, 0, 2, 3\}$. □

PRACTICE
1 Find the domain and the range of the relation $\{(1, 3), (5, 0), (0, -2), (5, 4)\}$.

■

OBJECTIVE
2 Identifying Functions
Some relations are also functions.

> **Function**
>
> A function is a set of ordered pairs that assigns to each x-value exactly one y-value.

In other words, a function cannot have two ordered pairs with the same x-coordinate but different y-coordinates.

EXAMPLE 2 Determine whether each relation is also a function.

    a. $\{(-1, 1), (2, 3), (7, 3), (8, 6)\}$ **b.** $\{(0, -2), (1, 5), (0, 3), (7, 7)\}$

Solution

    a. Although the ordered pairs (2, 3) and (7, 3) have the same y-value, each x-value is assigned to only one y-value, so this set of ordered pairs is a function.

    b. The x-value 0 is assigned to two y-values, −2 and 3, so this set of ordered pairs is not a function. □

PRACTICE
2   Determine whether each relation is also a function.

  a. $\{(4, 1), (3, -2), (8, 5), (-5, 3)\}$ **b.** $\{(1, 2), (-4, 3), (0, 8), (1, 4)\}$ ■

Relations and functions can be described by a graph of their ordered pairs.

EXAMPLE 3   Determine whether each graph is the graph of a function.

a.

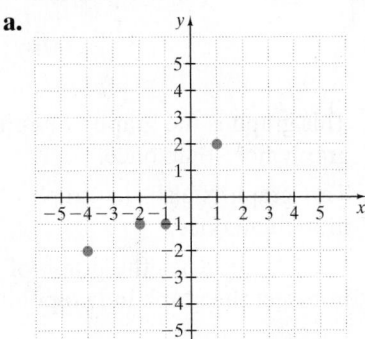

b.

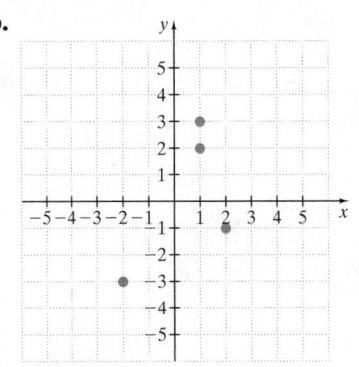

Solution

a. This is the graph of the relation $\{(-4, -2), (-2, -1), (-1, -1), (1, 2)\}$. Each x-coordinate has exactly one y-coordinate, so this is the graph of a function.

b. This is the graph of the relation $\{(-2, -3), (1, 2), (1, 3), (2, -1)\}$. The x-coordinate 1 is paired with two y-coordinates, 2 and 3, so this is not the graph of a function. □

PRACTICE
3   Determine whether each graph is the graph of a function.

a.

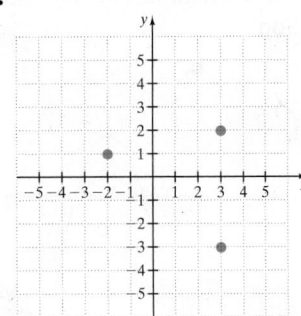

b.

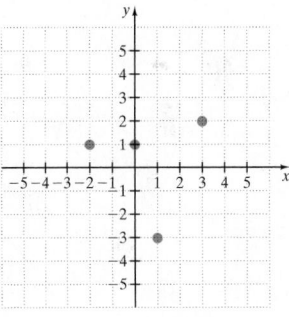

■

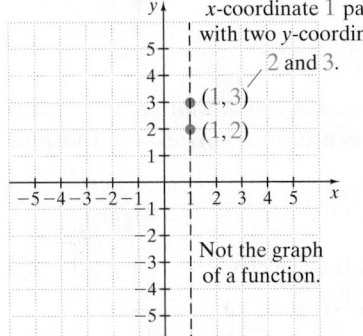

x-coordinate 1 paired with two *y*-coordinates, 2 and 3.

(1, 3)
(1, 2)

Not the graph of a function.

OBJECTIVE
3   Using the Vertical Line Test ▶

The graph in Example 3(b) was not the graph of a function because the x-coordinate 1 was paired with two y-coordinates, 2 and 3. Notice that when an x-coordinate is paired with more than one y-coordinate, a vertical line can be drawn that will intersect the graph at more than one point. We can use this fact to determine whether a relation is also a function. We call this the **vertical line test.**

> **Vertical Line Test**
>
> If a vertical line can be drawn so that it intersects a graph more than once, the graph is not the graph of a function.

This vertical line test works for all types of graphs on the rectangular coordinate system.

EXAMPLE 4 Use the vertical line test to determine whether each graph is the graph of a function.

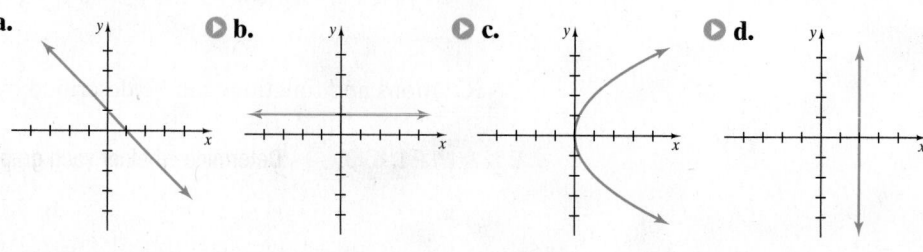

a. **b.** **c.** **d.**

Solution

a. This graph is the graph of a function since no vertical line will intersect this graph more than once.

b. This graph is also the graph of a function; no vertical line will intersect it more than once.

c. This graph is not the graph of a function. Vertical lines can be drawn that intersect the graph in two points. An example of one is shown.

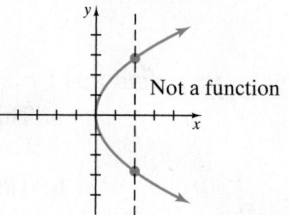

Not a function

d. This graph is not the graph of a function. A vertical line can be drawn that intersects this line at every point.

PRACTICE

4 Use the vertical line test to determine whether each graph is the graph of a function.

a. **b.** **c.** **d.**

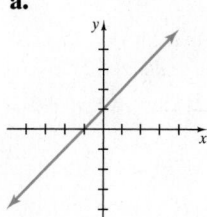 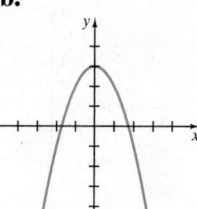 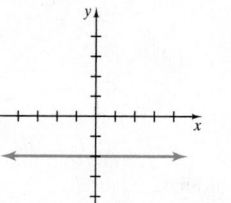 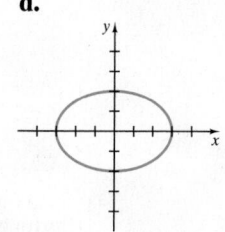

Recall that the graph of a linear equation is a line, and a line that is not vertical will pass the vertical line test. **Thus, all linear equations are functions except those of the form $x = c$, which are vertical lines.**

EXAMPLE 5 Decide whether the equation describes a function.

    a. $y = x$ **b.** $y = 2x + 1$ **c.** $y = 5$ **d.** $x = -1$

Solution **a, b,** and **c** are functions because their graphs are nonvertical lines. **d** is not a function because its graph is a vertical line.

PRACTICE

5 Decide whether the equation describes a function.

    a. $y = 2x$ **b.** $y = -3x - 1$ **c.** $y = 8$ **d.** $x = 2$

Examples of functions can often be found in magazines, newspapers, books, and other printed material in the form of tables or graphs such as that in Example 6.

EXAMPLE 6   The graph shows the sunrise time for Indianapolis, Indiana, for the year. Use this graph to answer the questions.

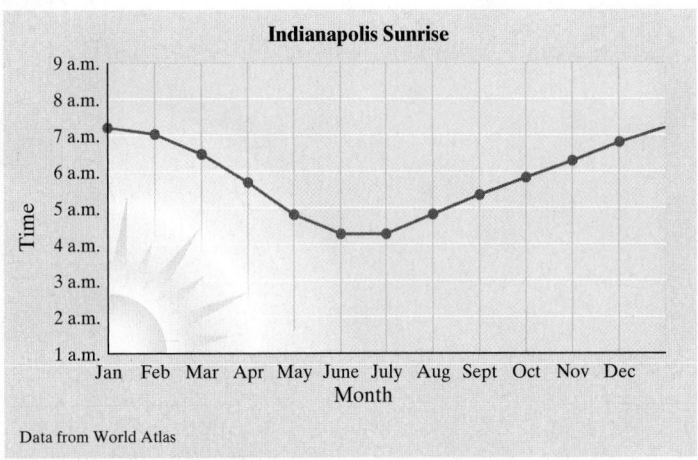

Data from World Atlas

a. Approximate the time of sunrise on February 1.

b. Approximately when does the sun rise at 5 a.m.?

c. Is this the graph of a function?

Solution

a. To approximate the time of sunrise on February 1, we find the mark on the horizontal axis that corresponds to February 1. From this mark, we move vertically upward until the graph is reached. From that point on the graph, we move horizontally to the left until the vertical axis is reached. The vertical axis there reads 7 a.m.

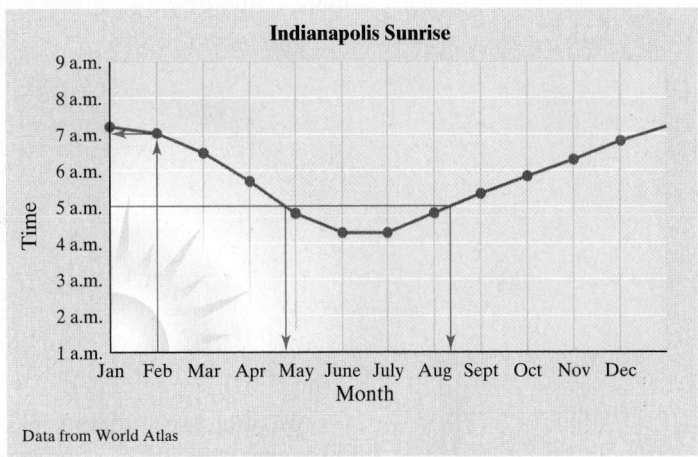

Data from World Atlas

b. To approximate when the sun rises at 5 a.m., we find 5 a.m. on the time axis and move horizontally to the right. Notice that we will reach the graph twice, corresponding to two dates for which the sun rises at 5 a.m. We follow both points on the graph vertically downward until the horizontal axis is reached. The sun rises at 5 a.m. at approximately the end of the month of April and near the middle of the month of August.

c. The graph is the graph of a function since it passes the vertical line test. In other words, for every day of the year in Indianapolis, there is exactly one sunrise time. □

(Continued on next page)

PRACTICE
6 The graph shows the average monthly temperature for Chicago, Illinois, for the year. Use this graph to answer the questions.

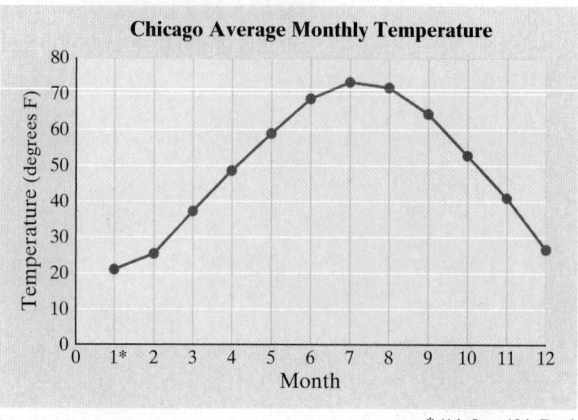

Chicago Average Monthly Temperature

* (1 is Jan.; 12 is Dec.)

a. Approximate the average monthly temperature for June.

b. For what month is the average monthly temperature 25°?

c. Is this the graph of a function?

OBJECTIVE
4   Using Function Notation

The graph of the linear equation $y = 2x + 1$ passes the vertical line test, so we say that $y = 2x + 1$ is a function. In other words, $y = 2x + 1$ gives us a rule for writing ordered pairs where every x-coordinate is paired with one and only one y-coordinate.

The variable y is a function of the variable x. For each value of x, there is only one value of y. Thus, we say the variable x is the **independent variable** because any value in the domain can be assigned to x. The variable y is the **dependent variable** because its value depends on x.

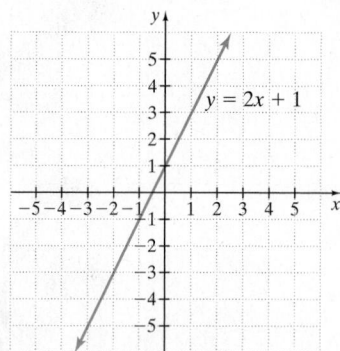

We often use letters such as f, g, and h to name functions. For example, the symbol $f(x)$ means *function of x* and is read "f of x." This notation is called **function notation.** The equation $y = 2x + 1$ can be written as $f(x) = 2x + 1$ using function notation, and these equations mean the same thing. In other words, $y = f(x)$.

The notation $f(1)$ means to replace x with 1 and find the resulting y or function value. Since

$$f(x) = 2x + 1$$

then

$$f(1) = 2(1) + 1$$
$$= 3$$

This means that, when $x = 1$, y or $f(x) = 3$, and we have the ordered pair $(1, 3)$. Now let's find $f(2)$, $f(0)$, and $f(-1)$.

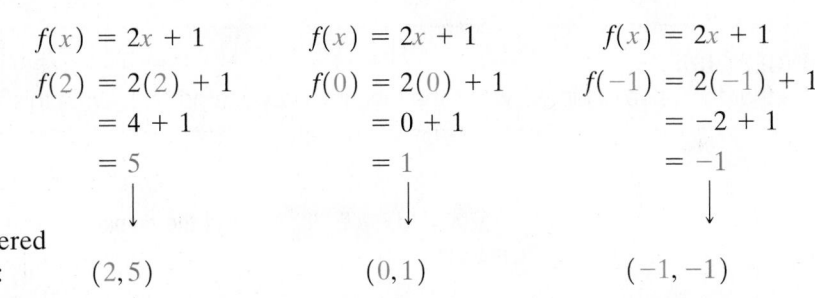

$$f(x) = 2x + 1 \qquad\qquad f(x) = 2x + 1 \qquad\qquad f(x) = 2x + 1$$
$$f(2) = 2(2) + 1 \qquad f(0) = 2(0) + 1 \qquad f(-1) = 2(-1) + 1$$
$$= 4 + 1 \qquad\qquad = 0 + 1 \qquad\qquad = -2 + 1$$
$$= 5 \qquad\qquad\qquad = 1 \qquad\qquad\qquad = -1$$

Helpful Hint
Note that, for example, if $f(2) = 5$, the corresponding ordered pair is $(2, 5)$.

Ordered
Pair: $(2, 5)$ $(0, 1)$ $(-1, -1)$

Helpful Hint
Note that $f(x)$ is a special symbol in mathematics used to denote a function. The symbol $f(x)$ is read "f of x." It does **not** mean $f \cdot x$ (f times x).

EXAMPLE 7   Given $g(x) = x^2 - 3$, find the following. Then write the corresponding ordered pairs generated.

   a. $g(2)$ **b.** $g(-2)$ **c.** $g(0)$

Solution

a. $g(x) = x^2 - 3$ **b.** $g(x) = x^2 - 3$ **c.** $g(x) = x^2 - 3$
$g(2) = 2^2 - 3$ $g(-2) = (-2)^2 - 3$ $g(0) = 0^2 - 3$
$= 4 - 3$ $= 4 - 3$ $= 0 - 3$
$= 1$ $= 1$ $= -3$

Ordered Pairs:	$g(2) = 1$ gives $(2, 1)$	$g(-2) = 1$ gives $(-2, 1)$	$g(0) = -3$ gives $(0, -3)$

PRACTICE
7   Given $h(x) = x^2 + 5$, find the following. Then write the corresponding ordered pairs generated.

   a. $h(2)$ **b.** $h(-5)$ **c.** $h(0)$

We now practice finding the domain and the range of a function. The domains of our functions will be the set of all possible real numbers that x can be replaced by. The range is the set of corresponding y-values.

EXAMPLE 8   Find the domain of each function.

   a. $g(x) = \dfrac{1}{x}$ **b.** $f(x) = 2x + 1$

Solution

   a. Recall that we cannot divide by 0, so the domain of $g(x)$ is the set of all real numbers except 0. In interval notation, we can write $(-\infty, 0) \cup (0, \infty)$.

   b. In this function, x can be any real number. The domain of $f(x)$ is the set of all real numbers, or $(-\infty, \infty)$ in interval notation.

PRACTICE
8   Find the domain of each function.

   a. $h(x) = 6x + 3$ **b.** $f(x) = \dfrac{1}{x^2}$

✔ CONCEPT CHECK

Suppose that the value of *f* is −7 when the function is evaluated at 2. Write this situation in function notation.

EXAMPLE 9 Find the domain and the range of each function graphed. Use interval notation.

a.

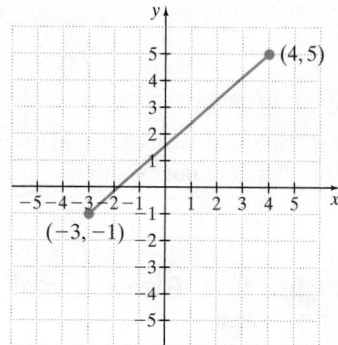

b.

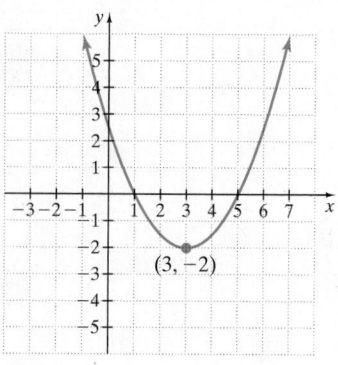

Solution

a.

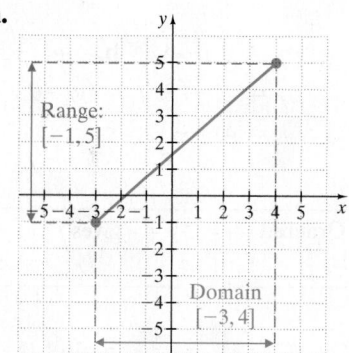

b.

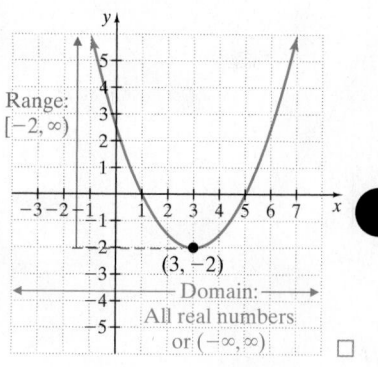

PRACTICE
9 Find the domain and the range of each function graphed. Use interval notation.

a.

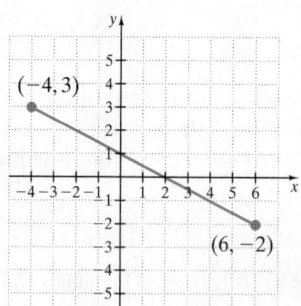

b.

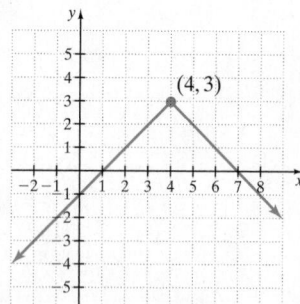

Answer to Concept Check:
$f(2) = -7$

Vocabulary, Readiness & Video Check

Use the choices below to fill in each blank. Some choices may not be used.

$x = c$	horizontal	domain	relation
$y = c$	vertical	range	function

1. A set of ordered pairs is called a(n) _____ .

2. A set of ordered pairs that assigns to each *x*-value exactly one *y*-value is called a(n) _____ .

3. The set of all *y*-coordinates of a relation is called the _____ .

4. The set of all *x*-coordinates of a relation is called the _____ .

5. All linear equations are functions except those whose graphs are _____ lines.

6. All linear equations are functions except those whose equations are of the form _____ .

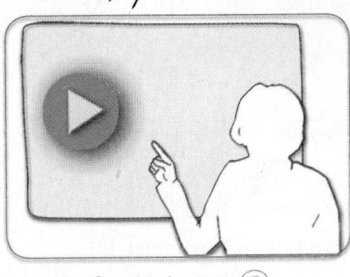

Martin-Gay Interactive Videos

See Video 3.6

Watch the section lecture video and answer the following questions.

OBJECTIVE 1

7. In the lecture before Example 1, relations are discussed. Why can an equation in two variables define a relation?

OBJECTIVE 2

8. Based on Examples 2 and 3, can a set of ordered pairs with no repeated *x*-values, but with repeated *y*-values be a function? For example: $\{(0, 4), (-3, 4), (2, 4)\}$.

OBJECTIVE 3

9. After reviewing Example 8, explain why the vertical line test works.

OBJECTIVE 4

10. In Example 10, three function values were found and their corresponding ordered pairs were written. For example, $f(0) = 2$ corresponds to $(0, 2)$. Write the other two function values found and their corresponding ordered pairs.

3.6 Exercise Set MyMathLab

Find the domain and the range of each relation. See Example 1.

1. $\{(2, 4), (0, 0), (-7, 10), (10, -7)\}$

2. $\{(3, -6), (1, 4), (-2, -2)\}$

3. $\{(0, -2), (1, -2), (5, -2)\}$

4. $\{(5, 0), (5, -3), (5, 4), (5, 3)\}$

Determine whether each relation is also a function. See Example 2.

5. $\{(1, 1), (2, 2), (-3, -3), (0, 0)\}$

6. $\{(11, 6), (-1, -2), (0, 0), (3, -2)\}$

7. $\{(-1, 0), (-1, 6), (-1, 8)\}$

8. $\{(1, 2), (3, 2), (1, 4)\}$

MIXED PRACTICE

Determine whether each graph is the graph of a function. See Examples 3 and 4.

9.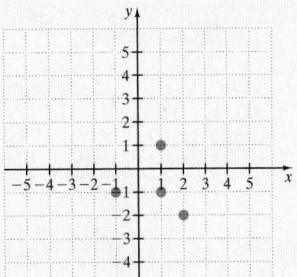

10.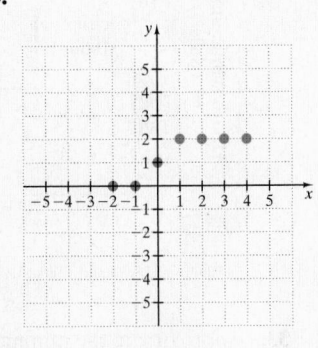

11.

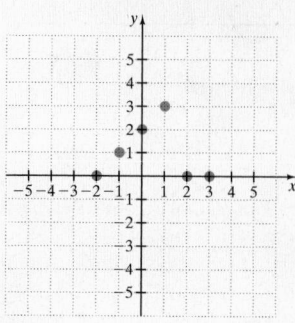

12.

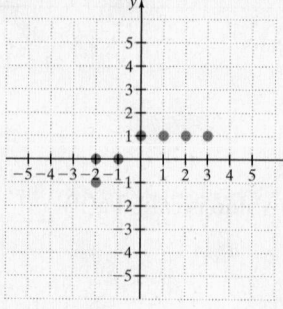

13.

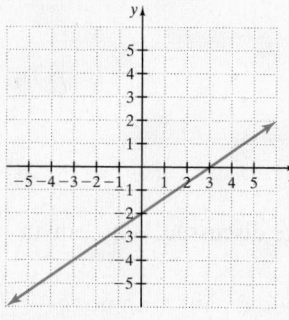

14.

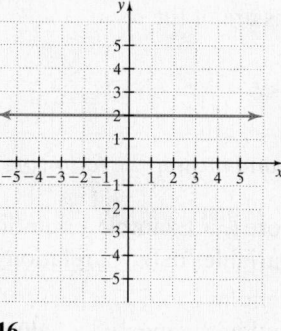

15.

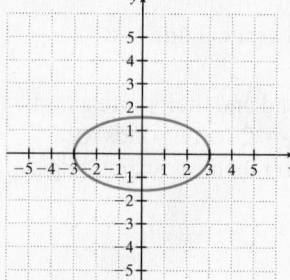

16.

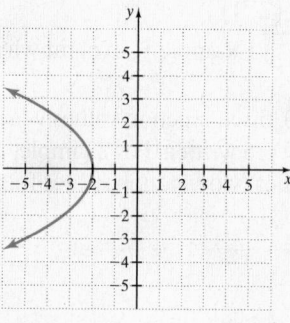

Decide whether the equation describes a function. See Example 5.

17. $y = x + 1$

18. $y = x - 1$

19. $y - x = 7$

20. $2x - 3y = 9$

21. $y = 6$

22. $x = 3$

23. $x = -2$

24. $y = -9$

25. $x = y^2$

26. $y = x^2 - 3$

The graph shows the sunset times for Seward, Alaska. Use this graph to answer Exercises 27 through 32. See Example 6.

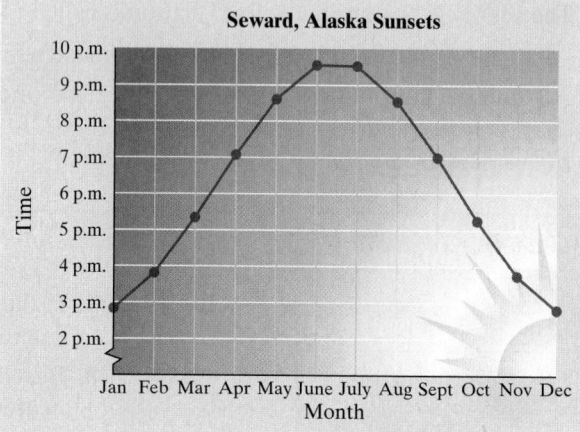

27. Approximate the time of sunset on June 1.

28. Approximate the time of sunset on November 1.

29. Approximate when the sunset is at 3 p.m.

30. Approximate when the sunset is at 9 p.m.

31. Is this graph the graph of a function? Why or why not?

32. Do you think a graph of sunset times for any location will always be a function? Why or why not?

This graph shows the U.S. hourly minimum wage for each year shown. Use this graph to answer Exercises 33 through 38. See Example 6.

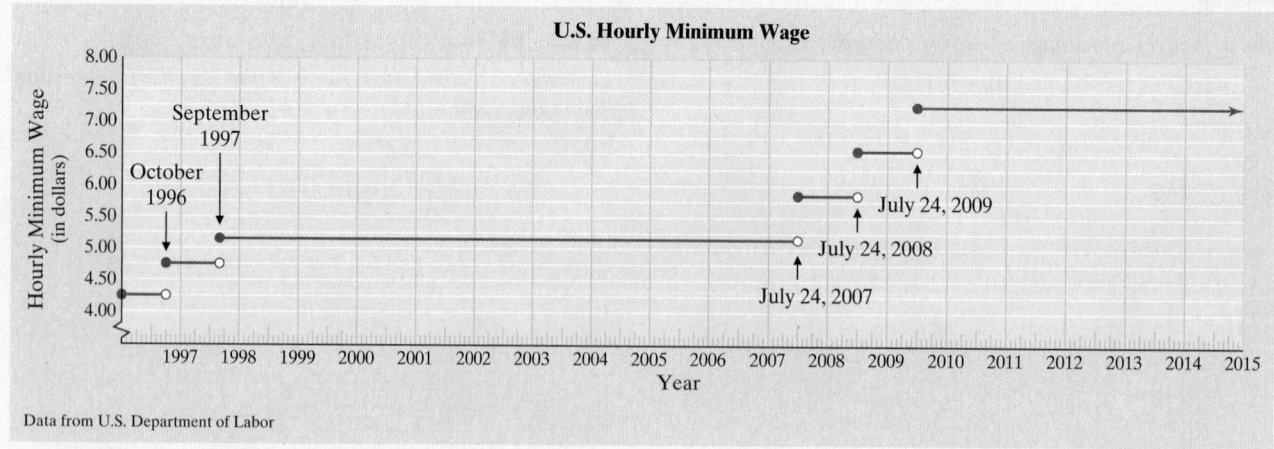

33. Approximate the minimum wage before October 1996.

34. Approximate the minimum wage in 2006.

35. Approximate the year when the minimum wage increased to over $7.00 per hour.

36. According to the graph, what hourly wage was in effect for the greatest number of years?

37. Is this graph the graph of a function? Why or why not?

38. Do you think that a similar graph of your hourly wage on January 1 of every year (whether you are working or not) will be the graph of a function? Why or why not?

This graph shows the cost of mailing a large envelope through the U.S. Postal Service by weight at this time. Use this graph to answer Exercises 39 through 44.

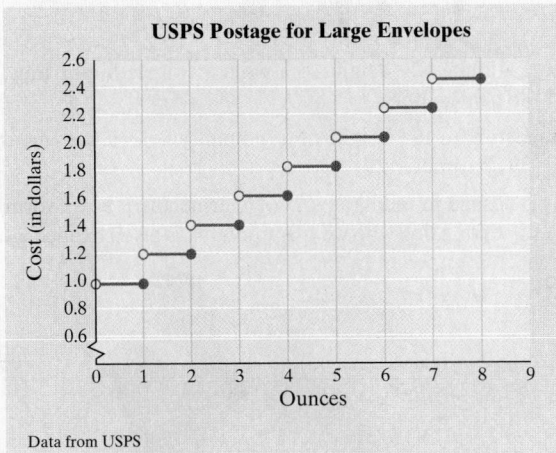

Data from USPS

39. Approximate the postage to mail a large envelope weighing more than 4 ounces but not more than 5 ounces.

40. Approximate the postage to mail a large envelope weighing more than 7 ounces but not more than 8 ounces.

41. Give the weight of a large envelope that costs about $2.00 to mail.

42. If you have $1.50, what is the weight of the largest envelope you can mail for that money?

43. Is this graph a function? Why or why not?

44. Do you think that a similar graph of postage to mail a first-class letter will be the graph of a function? Why or why not?

Find $f(-2), f(0),$ and $f(3)$ for each function. See Example 7.

45. $f(x) = 2x - 5$ **46.** $f(x) = 3 - 7x$

47. $f(x) = x^2 + 2$ **48.** $f(x) = x^2 - 4$

49. $f(x) = 3x$ **50.** $f(x) = -3x$

51. $f(x) = |x|$ **52.** $f(x) = |2 - x|$

Find $h(-1), h(0),$ and $h(4)$ for each function. See Example 7.

53. $h(x) = -5x$ **54.** $h(x) = -3x$

55. $h(x) = 2x^2 + 3$ **56.** $h(x) = 3x^2$

For each given function value, write a corresponding ordered pair.

57. $f(3) = 6$ **58.** $f(7) = -2$

59. $g(0) = -\dfrac{1}{2}$ **60.** $g(0) = -\dfrac{7}{8}$

61. $h(-2) = 9$ **62.** $h(-10) = 1$

Find the domain of each function. See Example 8.

63. $f(x) = 3x - 7$ **64.** $g(x) = 5 - 2x$

65. $h(x) = \dfrac{1}{x + 5}$ **66.** $f(x) = \dfrac{1}{x - 6}$

67. $g(x) = |x + 1|$ **68.** $h(x) = |2x|$

Find the domain and the range of each relation graphed. See Example 9.

69.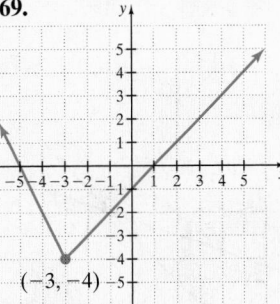

70.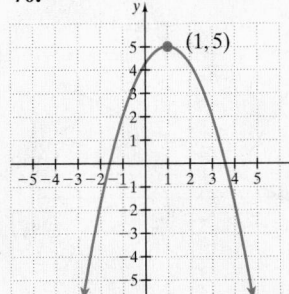

71.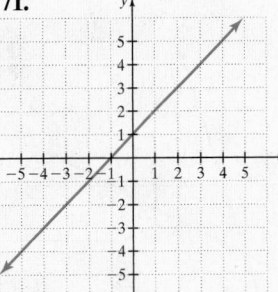

72.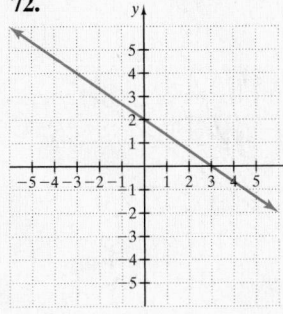

73.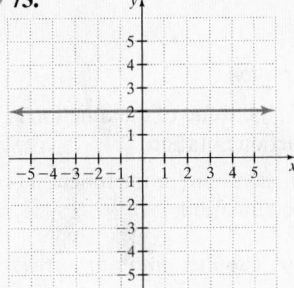

74.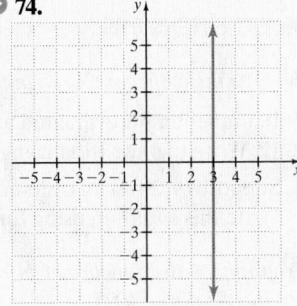

Use the graph of f below to answer Exercises 75 through 80.

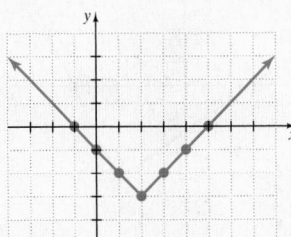

75. Complete the ordered pair solution for f. $(0, \;\;)$

76. Complete the ordered pair solution for f. $(3, \;\;)$

77. $f(0) = $ _____?

78. $f(3) = $ _____?

79. If $f(x) = 0$, find the value(s) of x.

80. If $f(x) = -1$, find the value(s) of x.

REVIEW AND PREVIEW

Find the coordinates of the point of intersection. See Section 3.1.

81.

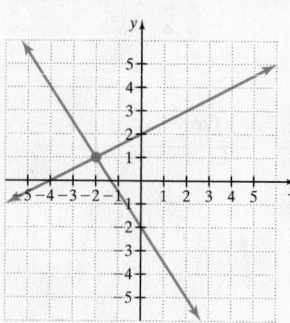

82.

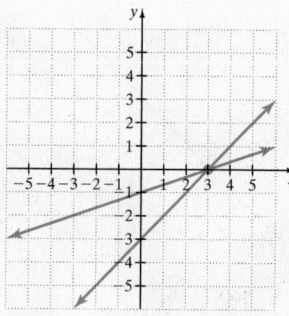

83.

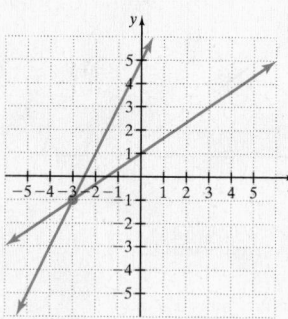

84.

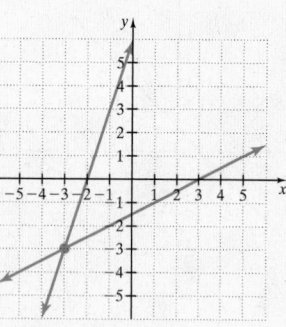

CONCEPT EXTENSIONS

Solve. See the Concept Check in this section.

85. If a function f is evaluated at -5, the value of the function is 12. Write this situation using function notation.

86. Suppose $(9, 20)$ is an ordered-pair solution for the function g. Write this situation using function notation.

The graph of the function f is below. Use this graph to answer Exercises 87 through 90.

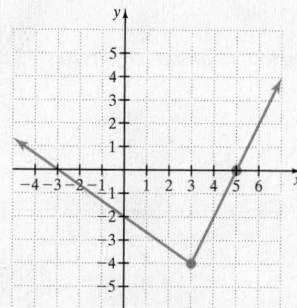

87. Write the coordinates of the lowest point of the graph.

88. Write the answer to Exercise 87 in function notation.

89. An x-intercept of this graph is $(5, 0)$. Write this using function notation.

90. Write the other x-intercept of this graph (see Exercise 89) using function notation.

91. Forensic scientists use the function

$$H(x) = 2.59x + 47.24$$

to estimate the height of a woman in centimeters given the length x of her femur bone.

a. Estimate the height of a woman whose femur measures 46 centimeters.

b. Estimate the height of a woman whose femur measures 39 centimeters.

92. The dosage in milligrams D of Ivermectin, a heartworm preventive for a dog who weighs x pounds, is given by the function

$$D(x) = \frac{136}{25}x$$

a. Find the proper dosage for a dog that weighs 35 pounds.

b. Find the proper dosage for a dog that weighs 70 pounds.

93. In your own words, define **(a)** function; **(b)** domain; **(c)** range.

94. Explain the vertical line test and how it is used.

95. Since $y = x + 7$ is a function, rewrite the equation using function notation.

96. Since $y = 3$ is a function, rewrite this equation using function notation.

See the example below for Exercises 97 through 100.

Example

If $f(x) = x^2 + 2x + 1$, find $f(\pi)$.

Solution:

$$f(x) = x^2 + 2x + 1$$
$$f(\pi) = \pi^2 + 2\pi + 1$$

Given the following functions, find the indicated values.

97. $f(x) = 2x + 7$

    a. $f(2)$ **b.** $f(a)$

98. $g(x) = -3x + 12$

    a. $g(5)$ **b.** $g(r)$

99. $h(x) = x^2 + 7$

    a. $h(3)$ **b.** $h(a)$

100. $f(x) = x^2 - 12$

    a. $f(12)$ **b.** $f(a)$

Chapter 3 Vocabulary Check

Fill in each blank with one of the words listed below.

relation	function	domain	range	standard	slope-intercept
y-axis	x-axis	solution	linear	slope	point-slope
x-intercept	y-intercept	y	x		

1. An ordered pair is a(n) _____ of an equation in two variables if replacing the variables by the coordinates of the ordered pair results in a true statement.
2. The vertical number line in the rectangular coordinate system is called the _____.
3. A(n) _____ equation can be written in the form $Ax + By = C$.
4. A(n) _____ is a point of the graph where the graph crosses the x-axis.
5. The form $Ax + By = C$ is called _____ form.
6. A(n) _____ is a point of the graph where the graph crosses the y-axis.
7. The equation $y = 7x - 5$ is written in _____ form.
8. The equation $y + 1 = 7(x - 2)$ is written in _____ form.
9. To find an x-intercept of a graph, let _____ $= 0$.
10. The horizontal number line in the rectangular coordinate system is called the _____.
11. To find a y-intercept of a graph, let _____ $= 0$.
12. The _____ of a line measures the steepness or tilt of a line.
13. A set of ordered pairs that assigns to each x-value exactly one y-value is called a(n) _____.
14. The set of all x-coordinates of a relation is called the _____ of the relation.
15. The set of all y-coordinates of a relation is called the _____ of the relation.
16. A set of ordered pairs is called a(n) _____.

Chapter 3 Highlights

DEFINITIONS AND CONCEPTS	EXAMPLES

Section 3.1   Reading Graphs and the Rectangular Coordinate System

The **rectangular coordinate system** consists of a plane and a vertical and a horizontal number line intersecting at their 0 coordinates. The vertical number line is called the **y-axis** and the horizontal number line is called the **x-axis.** The point of intersection of the axes is called the **origin**.

To **plot** or **graph** an ordered pair means to find its corresponding point on a rectangular coordinate system.

To plot or graph an ordered pair such as $(3, -2)$, start at the origin. Move 3 units to the right and, from there, 2 units down.

To plot or graph $(-3, 4)$, start at the origin. Move 3 units to the left and, from there, 4 units up.

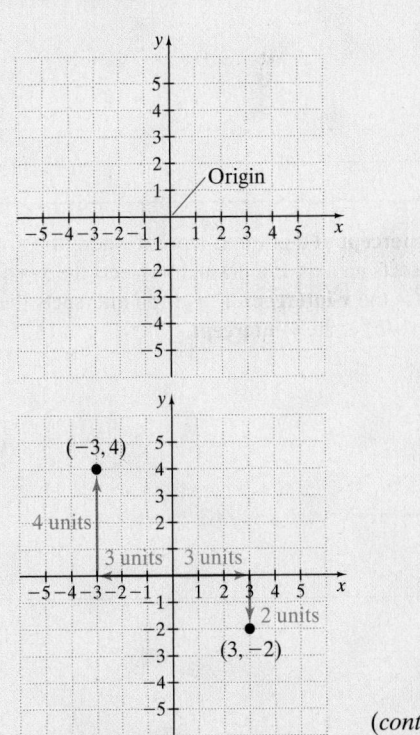

(continued)

DEFINITIONS AND CONCEPTS	EXAMPLES

Section 3.1 Reading Graphs and the Rectangular Coordinate System (continued)

An ordered pair is a **solution** of an equation in two variables if replacing the variables by the coordinates of the ordered pair results in a true statement.

Determine whether $(-1, 5)$ is a solution of $2x + 3y = 13$.

$$2x + 3y = 13$$
$$2(-1) + 3 \cdot 5 = 13 \quad \text{Let } x = -1, y = 5$$
$$-2 + 15 = 13$$
$$13 = 13 \quad \text{True}$$

If one coordinate of an ordered pair solution is known, the other value can be determined by substitution.

Complete the ordered pair solution $(0,\ )$ for the equation $x - 6y = 12$.

$$x - 6y = 12$$
$$0 - 6y = 12 \quad \text{Let } x = 0.$$
$$\frac{-6y}{-6} = \frac{12}{-6} \quad \text{Divide by } -6.$$
$$y = -2$$

The ordered pair solution is $(0, -2)$.

Section 3.2 Graphing Linear Equations

A **linear equation in two variables** is an equation that can be written in the form $Ax + By = C$ where A and B are not both 0. The form $Ax + By = C$ is called **standard form.**

Linear Equations

$$3x + 2y = -6 \qquad x = -5$$
$$y = 3 \qquad y = -x + 10$$

$x + y = 10$ is in standard form.

To graph a linear equation in two variables, find three ordered pair solutions. Plot the solution points and draw the line connecting the points.

Graph $x - 2y = 5$.

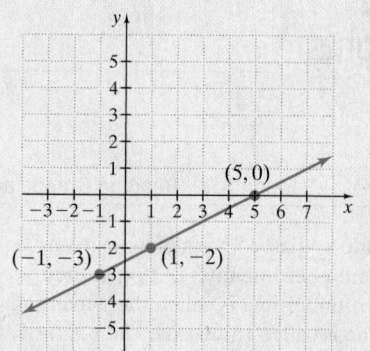

x	y
5	0
1	−2
−1	−3

Section 3.3 Intercepts

An **intercept** of a graph is a point where the graph intersects an axis. If a graph intersects the x-axis at a, then $(a, 0)$ is the **x-intercept.** If a graph intersects the y-axis at b, then $(0, b)$ is the **y-intercept.**

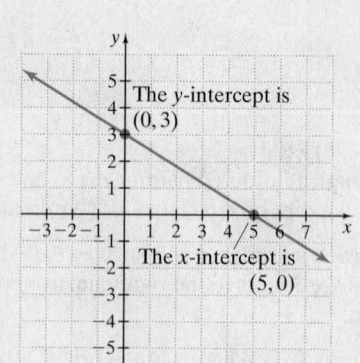

DEFINITIONS AND CONCEPTS	EXAMPLES

Section 3.3 Intercepts (continued)

To find the x-intercept, let $y = 0$ and solve for x.

To find the y-intercept, let $x = 0$ and solve for y.

Graph $2x - 5y = -10$ by finding intercepts.

$$\text{If } y = 0, \text{ then} \qquad \text{If } x = 0, \text{ then}$$
$$2x - 5 \cdot 0 = -10 \qquad 2 \cdot 0 - 5y = -10$$
$$2x = -10 \qquad -5y = -10$$
$$\frac{2x}{2} = \frac{-10}{2} \qquad \frac{-5y}{-5} = \frac{-10}{-5}$$
$$x = -5 \qquad y = 2$$

The x-intercept is $(-5, 0)$. The y-intercept is $(0, 2)$.

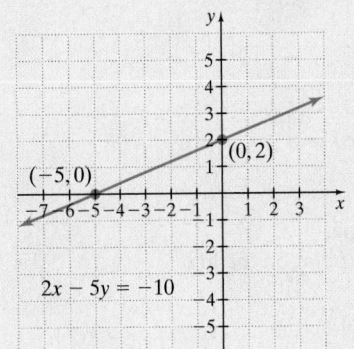

The graph of $x = c$ is a vertical line with x-intercept $(c, 0)$.

The graph of $y = c$ is a horizontal line with y-intercept $(0, c)$.

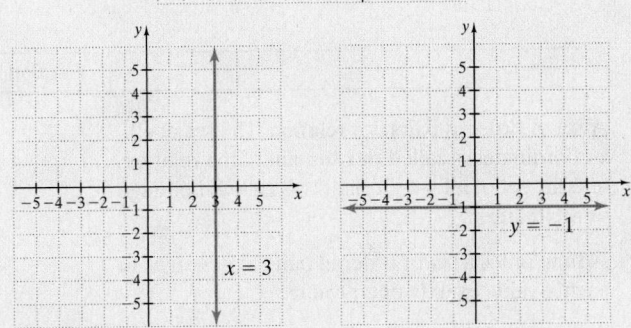

Section 3.4 Slope and Rate of Change

The **slope m** of the line through points (x_1, y_1) and (x_2, y_2) is given by

$$m = \frac{y_2 - y_1}{x_2 - x_1} \qquad \text{as long as } x_2 \neq x_1$$

A horizontal line has slope 0.
The slope of a vertical line is undefined.

The slope of the line through points $(-1, 6)$ and $(-5, 8)$ is

$$m = \frac{y_2 - y_1}{x_2 - x_1} = \frac{8 - 6}{-5 - (-1)} = \frac{2}{-4} = -\frac{1}{2}$$

The slope of the line $y = -5$ is 0.
The line $x = 3$ has undefined slope.

Nonvertical parallel lines have the same slope.

Two nonvertical lines are perpendicular if the slope of one is the negative reciprocal of the slope of the other.

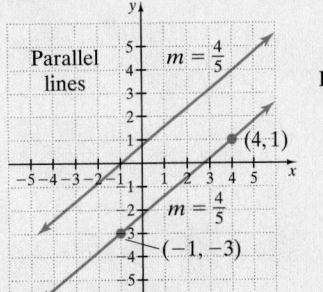

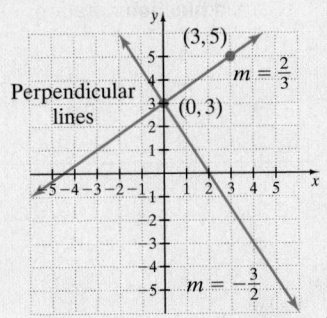

DEFINITIONS AND CONCEPTS	EXAMPLES

Section 3.5   Equations of Lines

Slope-Intercept Form

$$y = mx + b$$

m is the slope of the line.
$(0, b)$ is the y-intercept.

Find the slope and the y-intercept of the line whose equation is $2x + 3y = 6$.

Solve for y:

$$2x + 3y = 6$$
$$3y = -2x + 6 \quad \text{Subtract } 2x.$$
$$y = -\frac{2}{3}x + 2 \quad \text{Divide by 3.}$$

The slope of the line is $-\frac{2}{3}$ and the y-intercept is $(0, 2)$.

Find an equation of the line with slope 3 and y-intercept $(0, -1)$.

The equation is $y = 3x - 1$.

Point-Slope Form

$$y - y_1 = m(x - x_1)$$

m is the slope.
(x_1, y_1) is a point on the line.

Find an equation of the line with slope $\frac{3}{4}$ that contains the point $(-1, 5)$.

$$y - 5 = \frac{3}{4}[x - (-1)]$$

$$4(y - 5) = 3(x + 1) \quad \text{Multiply by 4.}$$
$$4y - 20 = 3x + 3 \quad \text{Distribute.}$$
$$-3x + 4y = 23 \quad \text{Subtract } 3x \text{ and add 20.}$$

Section 3.6   Functions

A set of ordered pairs is a **relation.** The set of all x-coordinates is called the **domain** of the relation, and the set of all y-coordinates is called the **range** of the relation.

The domain of the relation $\{(0, 5), (2, 5), (4, 5), (5, -2)\}$ is $\{0, 2, 4, 5\}$. The range is $\{-2, 5\}$.

A **function** is a set of ordered pairs that assigns to each x-value exactly one y-value.

Which are graphs of functions?

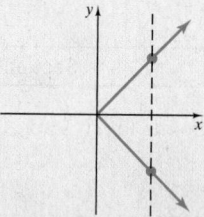

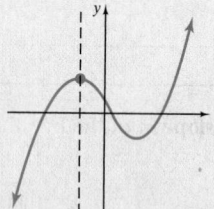

Vertical Line Test

If a vertical line can be drawn so that it intersects a graph more than once, the graph is not the graph of a function.

This graph is not the graph of a function.

This graph is the graph of a function.

The symbol $f(x)$ means **function of x.** This notation is called **function** notation.

If $f(x) = 2x^2 + 6x - 1$, find $f(3)$.

$$f(3) = 2(3)^2 + 6 \cdot 3 - 1$$
$$= 2 \cdot 9 + 18 - 1$$
$$= 18 + 18 - 1$$
$$= 35$$

Chapter 3 Review

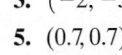

(3.1) *Plot the following ordered pairs on a Cartesian coordinate system.*

1. $(-7, 0)$

2. $\left(0, 4\frac{4}{5}\right)$

3. $(-2, -5)$

4. $(1, -3)$

5. $(0.7, 0.7)$

6. $(-6, 4)$

7. A local office supply store uses quantity pricing. The table shows the price per box of #10 security envelopes for different numbers of envelopes in a box purchased.

Price per Box of Envelopes (in dollars)	Number of Envelopes in Box
5.00	50
8.50	100
20.00	250
27.00	500

   a. Write each paired data as an ordered pair of the form (price per box of envelopes, number of envelopes in box).

   b. Create a scatter diagram of the paired data. Be sure to label the axes appropriately.

8. The table shows the annual overnight stays in national parks. (*Source:* National Park Service)

Year	Overnight Stays in National Parks (in millions)
2008	13.9
2009	14.6
2010	14.6
2011	14.0
2012	14.3
2013	13.5

   a. Write each paired data as an ordered pair of the form (year, number of overnight stays).

   b. Create a scatter diagram of the paired data. Be sure to label the axes properly.

Determine whether each ordered pair is a solution of the given equation.

9. $7x - 8y = 56$; $(0, 56), (8, 0)$

10. $-2x + 5y = 10$; $(-5, 0), (1, 1)$

11. $x = 13$; $(13, 5), (13, 13)$

12. $y = 2$; $(7, 2), (2, 7)$

Complete the ordered pairs so that each is a solution of the given equation.

13. $-2 + y = 6x$; $(7, \quad)$

14. $y = 3x + 5$; $(\quad, -8)$

Complete the table of values for each given equation; then plot the ordered pairs. Use a single coordinate system for each exercise.

15. $9 = -3x + 4y$

x	y
	0
	3
9	

16. $x = 2y$

x	y
	0
	5
	-5

The cost in dollars of producing x compact disc holders is given by $y = 5x + 2000$. Use this equation for Exercises 17 and 18.

17. Complete the following table.

x	y
1	
100	
1000	

18. Find the number of compact disc holders that can be produced for $6430.

(3.2) *Graph each linear equation.*

19. $x - y = 1$

20. $x + y = 6$

21. $x - 3y = 12$

22. $5x - y = -8$

23. $x = 3y$

24. $y = -2x$

25. $2x - 3y = 6$

26. $4x - 3y = 12$

(3.3) *Identify the intercepts.*

27.

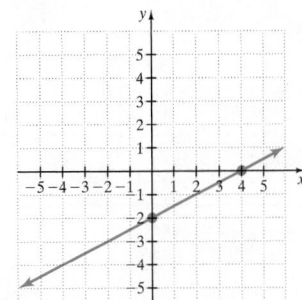

28.

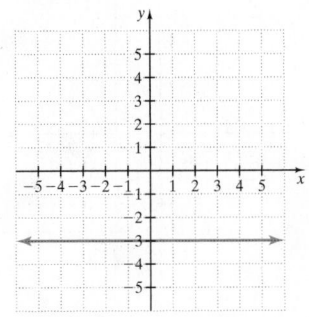

29.

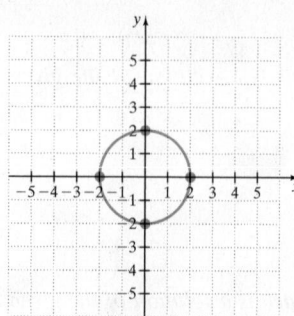

30.

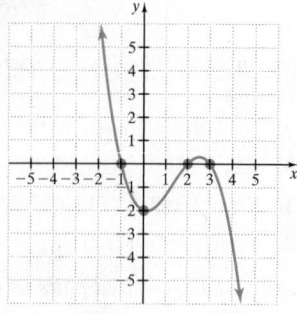

Graph each linear equation by finding its intercepts.

31. $x - 3y = 12$

32. $-4x + y = 8$

33. $y = -3$

34. $x = 5$

35. $y = -3x$

36. $x = 5y$

37. $x - 2 = 0$

38. $y + 6 = 0$

(3.4) Find the slope of each line.

39.

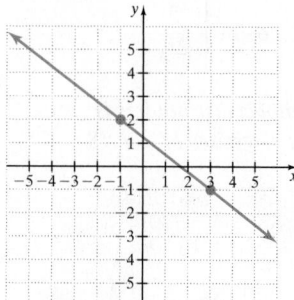

40.

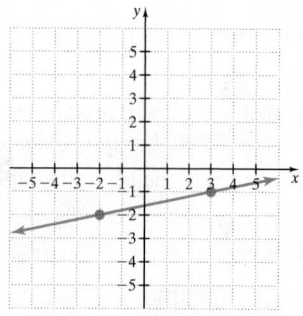

For Exercises 41 through 44, match each slope with its line.

A.

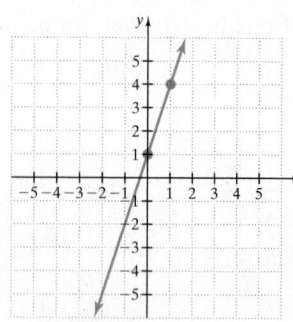

B.

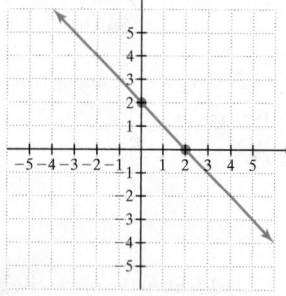

C.

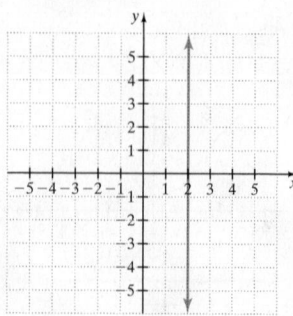

D.

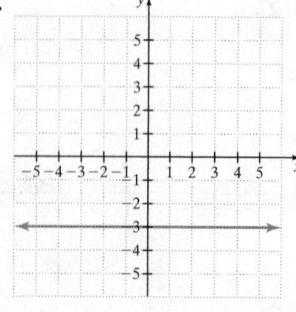

41. $m = 0$

42. $m = -1$

43. undefined slope

44. $m = 3$

Find the slope of the line that goes through the given points.

45. $(2, 5)$ and $(6, 8)$

46. $(4, 7)$ and $(1, 2)$

47. $(1, 3)$ and $(-2, -9)$

48. $(-4, 1)$ and $(3, -6)$

Find the slope of each line.

49. $y = 3x + 7$

50. $x - 2y = 4$

51. $y = -2$

52. $x = 0$

Determine whether each pair of lines is parallel, perpendicular, or neither.

53. $x - y = -6$
$x + y = 3$

54. $3x + y = 7$
$-3x - y = 10$

55. $y = 4x + \dfrac{1}{2}$
$4x + 2y = 1$

56. $x = 4$
$y = -2$

Find the slope of each line. Then write a sentence explaining the meaning of the slope as a rate of change. Don't forget to attach the proper units.

57. The graph below approximates the number of U.S. college students (in thousands) earning an associate's degree for each year x.

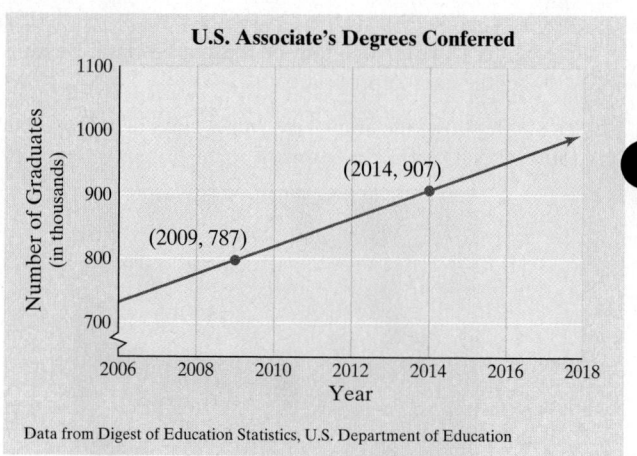

Data from Digest of Education Statistics, U.S. Department of Education

58. The graph below approximates the number of kidney transplants y in the United States for year x.

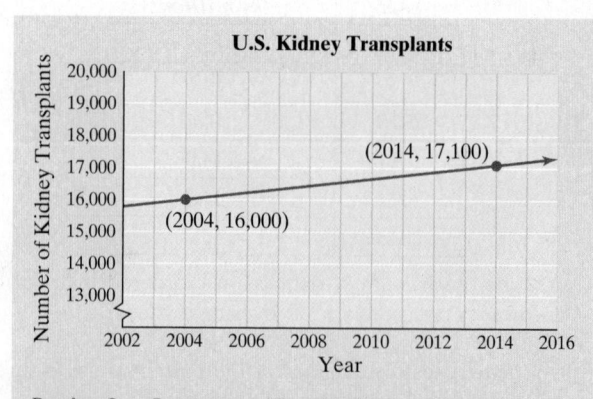

Data from Organ Procurement and Transplantation Network, U.S. Department of Health and Human Services

(3.5) *Determine the slope and the y-intercept of the graph of each equation.*

59. $3x + y = 7$

60. $x - 6y = -1$

61. $y = 2$

62. $x = -5$

Use the slope-intercept form to graph each equation.

63. $y = 3x - 1$

64. $y = -3x$

65. $5x - 3y = 15$

66. $-x + 2y = 8$

Write an equation of each line in slope-intercept form.

67. slope -5; y-intercept $\left(0, \dfrac{1}{2}\right)$

68. slope $\dfrac{2}{3}$; y-intercept $(0, 6)$

For exercises 69 through 72, match each equation with its graph.

A.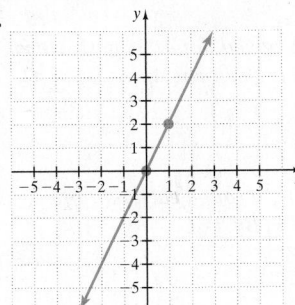

B.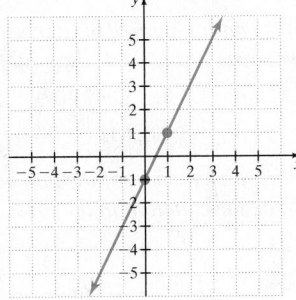

C.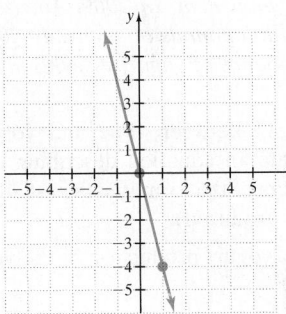

D.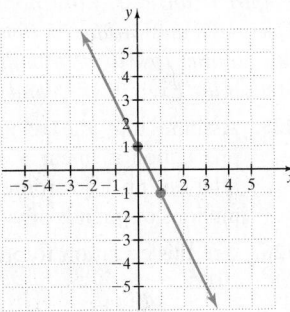

69. $y = -4x$

70. $y = 2x$

71. $y = 2x - 1$

72. $y = -2x + 1$

College is getting more expensive every year. The average cost for tuition and fees at a public two-year college y from 1995 through 2011 can be approximated by the linear equation $y = 56x + 1859$, where x is the number of years after 1995. Use this information for Exercises 73 and 74. (Source: The College Board: Trends in College Pricing 2010)

73. Find the y-intercept of this equation.

74. What does the y-intercept mean?

Write an equation of each line. Write each equation in standard form, $Ax + By = C$, or $x = c$ or $y = c$ form.

75. With slope -3, through $(0, -5)$

76. With slope $\dfrac{1}{2}$, through $\left(0, -\dfrac{7}{2}\right)$

77. With slope 0, through $(-2, -3)$

78. With slope 0, through the origin

79. With slope -6, through $(2, -1)$

80. With slope 12, through $\left(\dfrac{1}{2}, 5\right)$

81. Through $(0, 6)$ and $(6, 0)$

82. Through $(0, -4)$ and $(-8, 0)$

83. Vertical line, through $(5, 7)$

84. Horizontal line, through $(-6, 8)$

85. Through $(6, 0)$, perpendicular to $y = 8$

86. Through $(10, 12)$, perpendicular to $x = -2$

(3.6) *Determine whether each of the following is a function.*

87. $\{(7, 1), (7, 5), (2, 6)\}$

88. $\{(0, -1), (5, -1), (2, 2)\}$

89. $7x - 6y = 1$

90. $y = 7$

91. $x = 2$

92. $y = x^3$

93.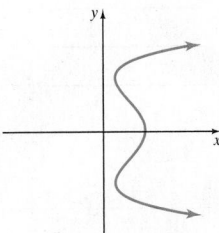

94.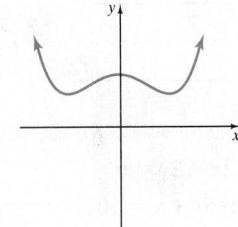

Given the following functions, find the indicated function values.

95. Given $f(x) = -2x + 6$, find

    a. $f(0)$ **b.** $f(-2)$ **c.** $f\left(\dfrac{1}{2}\right)$

96. Given $h(x) = -5 - 3x$, find

    a. $h(2)$ **b.** $h(-3)$ **c.** $h(0)$

97. Given $g(x) = x^2 + 12x$, find

    a. $g(3)$ **b.** $g(-5)$ **c.** $g(0)$

98. Given $h(x) = 6 - |x|$, find

    a. $h(-1)$ **b.** $h(1)$ **c.** $h(-4)$

Find the domain of each function.

99. $f(x) = 2x + 7$

100. $g(x) = \dfrac{7}{x - 2}$

Find the domain and the range of each function graphed.

101.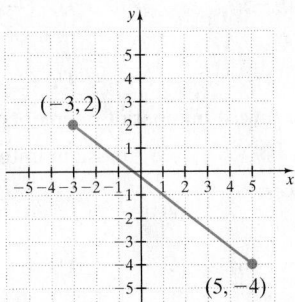

102.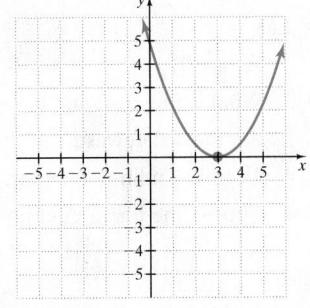

103.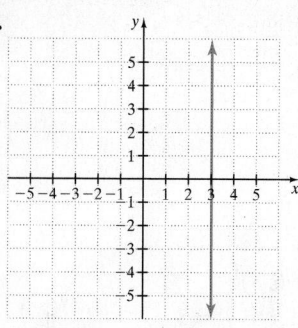

104.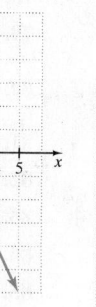

Find the slope of each line.

117.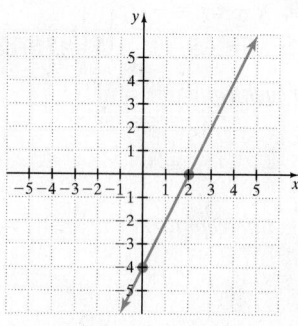

118.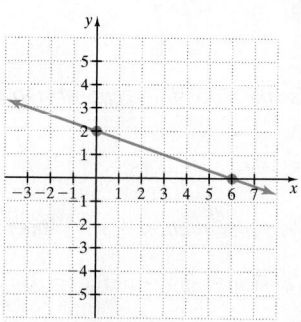

MIXED REVIEW

Complete the table of values for each given equation.

105. $2x - 5y = 9$

x	y
	1
2	
	-3

106. $x = -3y$

x	y
0	
	1
6	

Find the intercepts for each equation.

107. $2x - 3y = 6$

108. $-5x + y = 10$

Graph each linear equation.

109. $x - 5y = 10$

110. $x + y = 4$

111. $y = -4x$

112. $2x + 3y = -6$

113. $x = 3$

114. $y = -2$

Find the slope of the line that passes through each pair of points.

115. $(3, -5)$ and $(-4, 2)$

116. $(1, 3)$ and $(-6, -8)$

Determine the slope and y-intercept of the graph of each equation.

119. $-2x + 3y = -15$

120. $6x + y - 2 = 0$

Write an equation of the line with the given slope that passes through the given point. Write the equation in the form $Ax + By = C$.

121. $m = -5; (3, -7)$

122. $m = 3; (0, 6)$

Write an equation of the line passing through each pair of points. Write the equation in the form $Ax + By = C$.

123. $(-3, 9)$ and $(-2, 5)$

124. $(3, 1)$ and $(5, -9)$

Yogurt is an ever more popular food item. In 2009, American Dairy affiliates produced 3800 million pounds of yogurt. In 2012, this number rose to 4400 million pounds of yogurt. Use this information for Exercises 125 and 126.

125. Assume that the relationship between time and yogurt production is linear and write an equation describing this relationship. Use ordered pairs of the form (years past 2009, millions of pounds of yogurt produced).

126. Use this equation to predict yogurt production in the year 2018.

Chapter 3 Getting Ready for the Test

MULTIPLE CHOICE For Exercises 1 and 2, choose the ordered pair that is NOT a solution of the linear equation.

▸ **1.** $x - y = 5$

    A. $(7, 2)$ **B.** $(0, -5)$ **C.** $(-2, 3)$ **D.** $(-2, -7)$

▸ **2.** $y = 4$

    A. $(4, 0)$ **B.** $(0, 4)$ **C.** $(2, 4)$ **D.** $(100, 4)$

▸ **3.** What is the most and then the fewest number of intercepts a line may have?

    A. most: 2; fewest: 1   **B.** most: infinite number; fewest: 1   **C.** most: 2; fewest: 0

    D. most: infinite number; fewest: 0

▸ **4.** Choose the linear equation:

    A. $\sqrt{x} - 3y = 7$ **B.** $2x = 6^2$ **C.** $4x^3 + 6y^3 = 5^3$ **D.** $y = |x|$

MATCHING *Match each graph in the rectangular system with its slope to the right. Each slope may be used only once.*

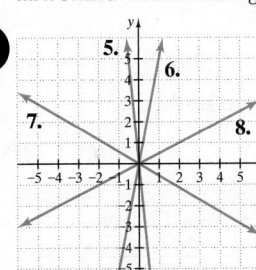

▶ **5.**   **A.** $m = 5$

▶ **6.**   **B.** $m = -10$

▶ **7.**   **C.** $m = \dfrac{1}{2}$

▶ **8.**   **D.** $m = -\dfrac{4}{7}$

MULTIPLE CHOICE *For Exercises 9 and 10, choose the best answer.*

▶ **9.** An ordered pair solution for the function $f(x)$ is $(0, 5)$. This solution using function notation is:

   A. $f(5) = 0$ **B.** $f(5) = f(0)$ **C.** $f(0) = 5$ **D.** $0 = 5$

▶ **10.** Given: $(2, 3)$ and $(0, 9)$. Final Answer: $y = -3x + 9$. Select the correct instructions:

   A. Find the slope of the line through the two points.

   B. Find an equation of the line through the two points. Write the equation in standard form.

   C. Find an equation of the line through the two points. Write the equation in slope-intercept form.

MULTIPLE CHOICE *For Exercises 11–14, use the graph to fill in each blank using the choices below.*

   A. -2 **B.** 2 **C.** 4 **D.** 0 **E.** 3

▶ **11.** $f(0) = $ _____.

▶ **12.** $f(4) = $ _____.

▶ **13.** If $f(x) = 0$, then $x = $ _____ or $x = $ _____.

▶ **14.** $f(1) = $ _____.

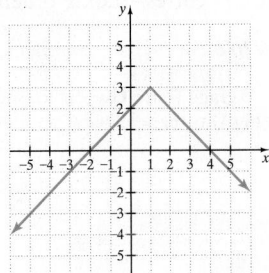

Chapter 3 Test   MyMathLab®   You Tube™

Graph the following.

▶ **1.** $y = \dfrac{1}{2}x$

▶ **2.** $2x + y = 8$

▶ **3.** $5x - 7y = 10$

▶ **4.** $y = -1$

▶ **5.** $x - 3 = 0$

For Exercises 6 through 10, find the slopes of the lines.

▶ **6.**   ▶ **7.**

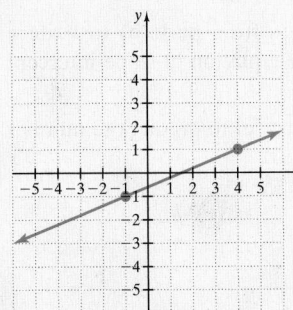   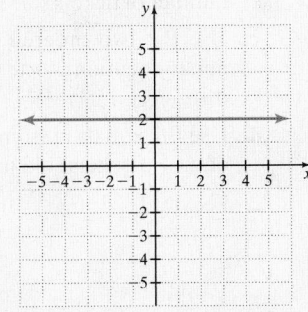

▶ **8.** Through $(6, -5)$ and $(-1, 2)$

▶ **9.** $-3x + y = 5$

▶ **10.** $x = 6$

▶ **11.** Determine the slope and the y-intercept of the graph of $7x - 3y = 2$.

▶ **12.** Determine whether the graphs of $y = 2x - 6$ and $-4x = 2y$ are parallel lines, perpendicular lines, or neither.

Find equations of the following lines. Write the equation in standard form.

▶ **13.** With slope of $-\dfrac{1}{4}$, through $(2, 2)$

▶ **14.** Through the origin and $(6, -7)$

▶ **15.** Through $(2, -5)$ and $(1, 3)$

▶ **16.** Through $(-5, -1)$ and parallel to $x = 7$

▶ **17.** With slope $\dfrac{1}{8}$ and y-intercept $(0, 12)$

Determine whether each graph is the graph of a function.

▶ **18.**   ▶ **19.**

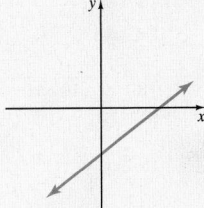   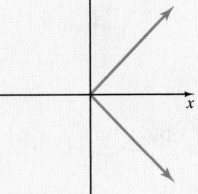

Given the following function, find the indicated function values.

20. $h(x) = x^3 - x$

    a. $h(-1)$ **b.** $h(0)$ **c.** $h(4)$

21. Find the domain of $y = \dfrac{1}{x + 1}$.

For Exercises 22 and 23, **a.** *Identify the x- and y-intercepts.* **b.** *Find the domain and the range of each function graphed.*

22.

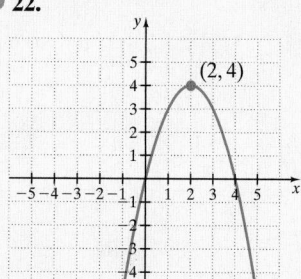

23.

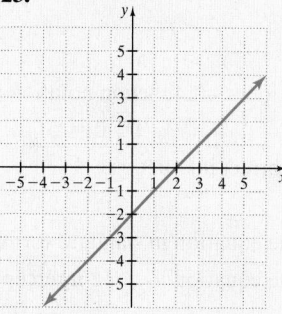

24. If $f(7) = 20$, write the corresponding ordered pair.

Use the bar graph below to answer Exercises 25 and 26.

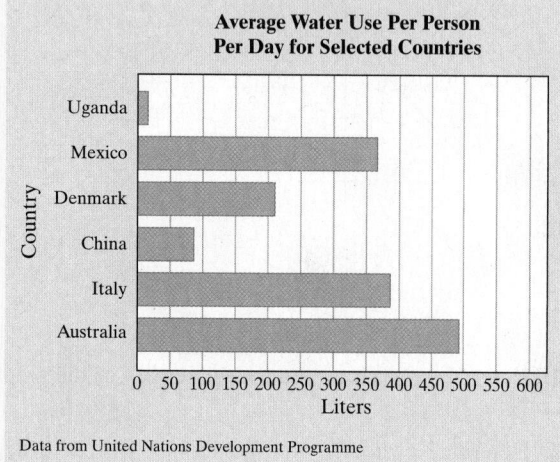

Average Water Use Per Person Per Day for Selected Countries

Data from United Nations Development Programme

25. Estimate the average water use per person per day in Denmark.

26. Estimate the average water use per person per day in Australia.

Use this graph to answer Exercises 27 through 29.

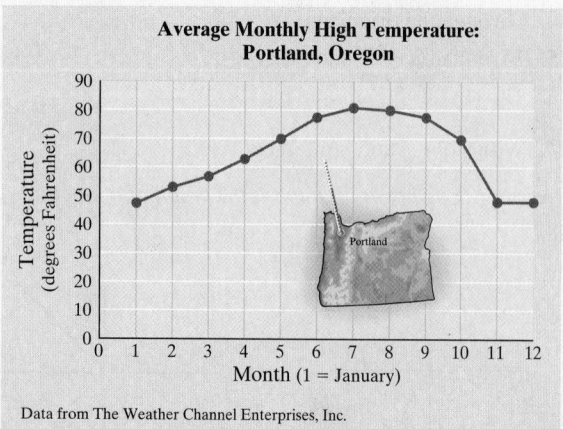

Average Monthly High Temperature: Portland, Oregon

Data from The Weather Channel Enterprises, Inc.

27. During what month is the average high temperature the greatest?

28. Approximate the average high temperature for the month of April.

29. During what month(s) is the average high temperature below 60°F?

30. The table gives the number of Dish Network subscribers (in millions) for the years shown. (*Source:* Dish Network)

Year	Dish Network Subscribers (in millions)
2008	13.68
2009	14.10
2010	14.13
2011	13.97
2012	14.06
2013	14.06
2014	13.98

    a. Write this data as a set of ordered pairs of the form (year, number of Dish Network subscribers in millions).

    b. Create a scatter diagram of the data. Be sure to label the axes properly.

31. This graph approximates the number of movie ticket sales y (in millions) for the year x.

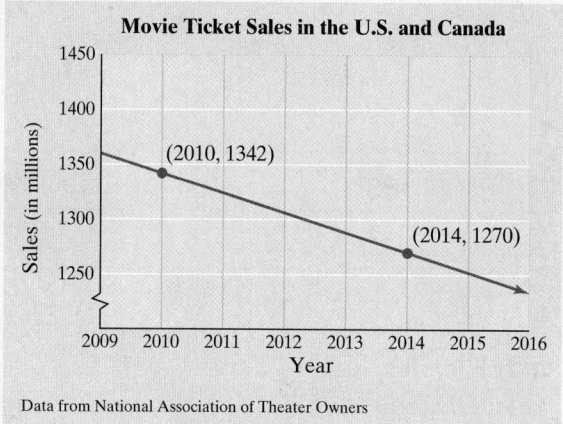

Movie Ticket Sales in the U.S. and Canada

Data from National Association of Theater Owners

    a. Find the slope of the line. Then write a sentence explaining the meaning of the slope as a rate of change. Don't forget to attach the proper units.

    b. Write two ordered pairs of the form (years past 2000, number of tickets sold in millions).

    c. Use the two ordered pairs from part (b) to write a linear equation. Write the equation in slope-intercept form.

    d. Use the equation from part (c) to predict the number of movie tickets sold in 2020.

Chapter 3 Cumulative Review

1. Insert $<$, $>$, or $=$ in the space between each pair of numbers to make each statement true.
   a. 2   3
   b. 7   4
   c. 72   27

2. Write the fraction $\dfrac{56}{64}$ in lowest terms.

3. Multiply $\dfrac{2}{15}$ and $\dfrac{5}{13}$. Simplify the product if possible.

4. Add: $\dfrac{10}{3} + \dfrac{5}{21}$

5. Simplify: $\dfrac{3 + |4 - 3| + 2^2}{6 - 3}$

6. Simplify: $16 - 3 \cdot 3 + 2^4$

7. Add.
   a. $-8 + (-11)$
   b. $-5 + 35$
   c. $0.6 + (-1.1)$
   d. $-\dfrac{7}{10} + \left(-\dfrac{1}{10}\right)$
   e. $11.4 + (-4.7)$
   f. $-\dfrac{3}{8} + \dfrac{2}{5}$

8. Simplify: $|9 + (-20)| + |-10|$

9. Simplify each expression.
   a. $-14 - 8 + 10 - (-6)$
   b. $1.6 - (-10.3) + (-5.6)$

10. Simplify: $-9 - (3 - 8)$

11. If $x = -2$ and $y = -4$, evaluate each expression.
    a. $5x - y$
    b. $x^4 - y^2$
    c. $\dfrac{3x}{2y}$

12. Is -20 a solution of $\dfrac{x}{-10} = 2$?

13. Simplify each expression.
    a. $10 + (x + 12)$
    b. $-3(7x)$

14. Simplify: $(12 + x) - (4x - 7)$

15. Identify the numerical coefficient of each term.
    a. $-3y$
    b. $22z^4$
    c. y
    d. $-x$
    e. $\dfrac{x}{7}$

16. Multiply: $-5(x - 7)$

17. Solve $x - 7 = 10$ for x.

18. Solve: $5(3 + z) - (8z + 9) = -4$

19. Solve: $\dfrac{5}{2}x = 15$

20. Solve: $\dfrac{x}{4} - 1 = -7$

21. If x is the first of three consecutive integers, express the sum of the three integers in terms of x. Simplify if possible.

22. Solve: $\dfrac{x}{3} - 2 = \dfrac{x}{3}$

23. Solve: $\dfrac{2(a + 3)}{3} = 6a + 2$

24. Solve: $x + 2y = 6$ for y

25. The 114th Congress began on January 3, 2015, and had a total of 435 Democratic and Republican representatives. There were 59 fewer Democratic representatives than Republican. Find the number of representatives from each party. (*Source:* congress.gov)

26. Solve $5(x + 4) \geq 4(2x + 3)$. Write the solution set in interval notation.

27. Charles Pecot can afford enough fencing to enclose a rectangular garden with a perimeter of 140 feet. If the width of his garden must be 30 feet, find the length.

28. Solve $-3 < 4x - 1 \leq 2$. Write the solution set in interval notation.

29. Solve: $y = mx + b$ for x

30. Complete the table for $y = -5x$.

x	y
0	
-1	
	-10

31. A chemist working on his doctoral degree at Massachusetts Institute of Technology needs 12 liters of a 50% acid solution for a lab experiment. The stockroom has only 40% and 70% solutions. How much of each solution should be mixed together to form 12 liters of a 50% solution?

32. Graph: $y = -3x + 5$

33. Graph: $x \geq -1$. Then write the solutions in interval notation.

34. Find the x- and y-intercepts of $2x + 4y = -8$.

35. Solve $-1 \leq 2x - 3 < 5$. Graph the solution set and write it in interval notation.

36. Graph $x = 2$ on a rectangular coordinate system.

37. Determine whether each ordered pair is a solution of the equation $x - 2y = 6$.
    a. $(6, 0)$
    b. $(0, 3)$
    c. $\left(1, -\dfrac{5}{2}\right)$

38. Find the slope of the line through $(0, 5)$ and $(-5, 4)$.

39. Determine whether each equation is a linear equation in two variables.
    a. $x - 1.5y = -1.6$
    b. $y = -2x$
    c. $x + y^2 = 9$
    d. $x = 5$

40. Find the slope of the line $x = -10$.

41. Find the slope of the line $y = -1$.

42. Find the slope and y-intercept of the line whose equation is $2x - 5y = 10$.

43. Find an equation of the line with y-intercept $(0, -3)$ and slope of $\dfrac{1}{4}$.

44. Write an equation of the line through $(2, 3)$ and $(0, 0)$. Write the equation in standard form.

Solving Systems of Linear Equations

4.1 Solving Systems of Linear Equations by Graphing

4.2 Solving Systems of Linear Equations by Substitution

4.3 Solving Systems of Linear Equations by Addition

Integrated Review—Solving Systems of Equations

4.4 Solving Systems of Linear Equations in Three Variables

4.5 Systems of Linear Equations and Problem Solving

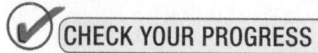

CHECK YOUR PROGRESS

Vocabulary Check

Chapter Highlights

Chapter Review

Getting Ready for the Test

Chapter Test

Cumulative Review

In Chapter 3, we graphed equations containing two variables. Equations like these are often needed to represent relationships between two different values. There are also many real-life opportunities to compare and contrast two such equations, called a system of equations. This chapter presents linear systems and ways we solve these systems and apply them to real-life situations.

Do You Know That We Often Import and Export the Same Type of Item?

The economic importance of the United States' fishing industry extends well beyond the coastal communities, for which it is a vital industry. Commercial fishing operations, including seafood wholesalers, processors, and retailers, contribute billions of dollars annually to the United States economy. Recreational fishing is also a major factor in the economy of the Gulf coast and the South Atlantic states.

Depending on the supply and demand of our country's needs, we also import and export fish of the same type. In Section 4.1, Exercises 69 and 70, we will explore the exportation and importation of Pacific salmon over the years.

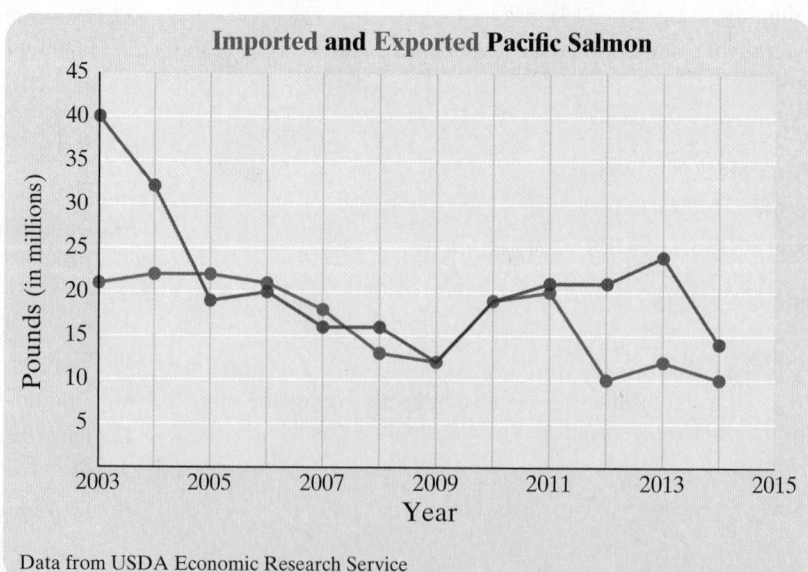

4.1 Solving Systems of Linear Equations by Graphing

OBJECTIVES

1 Determine if an Ordered Pair Is a Solution of a System of Equations in Two Variables.

2 Solve a System of Linear Equations by Graphing.

3 Without Graphing, Determine the Number of Solutions of a System.

A **system of linear equations** consists of two or more linear equations. In this section, we focus on solving systems of linear equations containing two equations in two variables. Examples of such linear systems are

$$\begin{cases} 3x - 3y = 0 \\ x = 2y \end{cases} \quad \begin{cases} x - y = 0 \\ 2x + y = 10 \end{cases} \quad \begin{cases} y = 7x - 1 \\ y = 4 \end{cases}$$

OBJECTIVE

1 Deciding Whether an Ordered Pair Is a Solution

A **solution** of a system of two equations in two variables is an ordered pair of numbers that is a solution of both equations in the system.

EXAMPLE 1 Determine whether $(12, 6)$ is a solution of the system.

$$\begin{cases} 2x - 3y = 6 \\ x = 2y \end{cases}$$

Solution To determine whether $(12, 6)$ is a solution of the system, we replace x with 12 and y with 6 in both equations.

$2x - 3y = 6$	First equation	$x = 2y$	Second equation
$2(12) - 3(6) \stackrel{?}{=} 6$	Let $x = 12$ and $y = 6$.	$12 \stackrel{?}{=} 2(6)$	Let $x = 12$ and $y = 6$.
$24 - 18 \stackrel{?}{=} 6$	Simplify.	$12 = 12$	True
$6 = 6$	True		

Since $(12, 6)$ is a solution of both equations, it is a solution of the system. ☐

PRACTICE

1 Determine whether $(4, 12)$ is a solution of the system.

$$\begin{cases} 4x - y = 2 \\ y = 3x \end{cases}$$

EXAMPLE 2 Determine whether $(-1, 2)$ is a solution of the system.

$$\begin{cases} x + 2y = 3 \\ 4x - y = 6 \end{cases}$$

Solution We replace x with -1 and y with 2 in both equations.

$x + 2y = 3$	First equation	$4x - y = 6$	Second equation
$-1 + 2(2) \stackrel{?}{=} 3$	Let $x = -1$ and $y = 2$.	$4(-1) - 2 \stackrel{?}{=} 6$	Let $x = -1$ and $y = 2$.
$-1 + 4 \stackrel{?}{=} 3$	Simplify.	$-4 - 2 \stackrel{?}{=} 6$	Simplify.
$3 = 3$	True	$-6 = 6$	False

$(-1, 2)$ is not a solution of the second equation, $4x - y = 6$, so it is not a solution of the system. ☐

PRACTICE

2 Determine whether $(-4, 1)$ is a solution of the system.

$$\begin{cases} x - 3y = -7 \\ 2x + 9y = 1 \end{cases}$$

OBJECTIVE

2 Solving Systems of Equations by Graphing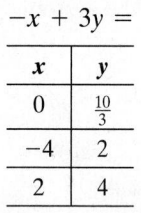

Since a solution of a system of two equations in two variables is a solution common to both equations, it is also a point common to the graphs of both equations. Let's practice finding solutions of both equations in a system—that is, solutions of a system—by graphing and identifying points of intersection.

EXAMPLE 3 Solve the system of equations by graphing.

$$\begin{cases} -x + 3y = 10 \\ x + y = 2 \end{cases}$$

Solution On a single set of axes, graph each linear equation.

$-x + 3y = 10$

x	y
0	$\frac{10}{3}$
-4	2
2	4

$x + y = 2$

x	y
0	2
2	0
1	1

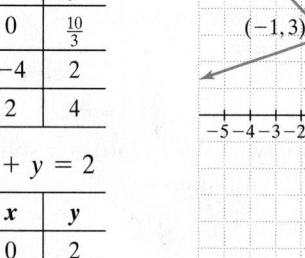

Helpful Hint
The point of intersection gives the solution of the system.

The two lines appear to intersect at the point $(-1, 3)$. To check, we replace x with -1 and y with 3 in both equations.

$$-x + 3y = 10 \quad \text{First equation} \qquad\qquad x + y = 2 \quad \text{Second equation}$$
$$-(-1) + 3(3) \overset{?}{=} 10 \quad \text{Let } x = -1 \text{ and } y = 3. \qquad -1 + 3 \overset{?}{=} 2 \quad \text{Let } x = -1 \text{ and } y = 3.$$
$$1 + 9 \overset{?}{=} 10 \quad \text{Simplify.} \qquad\qquad\qquad 2 = 2 \quad \text{True}$$
$$10 = 10 \quad \text{True}$$

$(-1, 3)$ checks, so it is the solution of the system. □

PRACTICE

3 Solve the system of equations by graphing.

$$\begin{cases} x - y = 3 \\ x + 2y = 18 \end{cases}$$

Helpful Hint
Neatly drawn graphs can help when you are estimating the solution of a system of linear equations by graphing.

In the example above, notice that the two lines intersected in a point. This means that the system has 1 solution.

EXAMPLE 4   Solve the system of equations by graphing.

$$\begin{cases} 2x + 3y = -2 \\ x = 2 \end{cases}$$

Solution We graph each linear equation on a single set of axes.

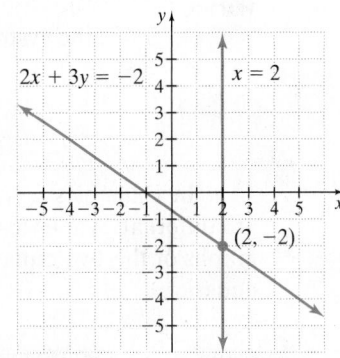

The two lines appear to intersect at the point $(2, -2)$. To determine whether $(2, -2)$ is the solution, we replace x with 2 and y with -2 in both equations.

$2x + 3y = -2$	First equation	$x = 2$	Second equation
$2(2) + 3(-2) \stackrel{?}{=} -2$	Let $x = 2$ and $y = -2$.	$2 \stackrel{?}{=} 2$	Let $x = 2$.
$4 + (-6) \stackrel{?}{=} -2$	Simplify.	$2 = 2$	True
$-2 = -2$	True		

Since a true statement results in both equations, $(2, -2)$ is the solution of the system. □

PRACTICE
4   Solve the system of equations by graphing.

$$\begin{cases} -4x + 3y = -3 \\ y = -5 \end{cases}$$

A system of equations that has at least one solution as in Examples 3 and 4 is said to be a **consistent system.** A system that has no solution is said to be an **inconsistent system.**

EXAMPLE 5   Solve the following system of equations by graphing.

$$\begin{cases} 2x + y = 7 \\ 2y = -4x \end{cases}$$

Solution Graph the two lines in the system.

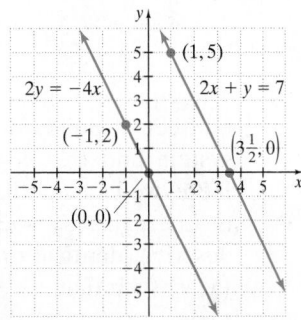

The lines **appear** to be parallel. To confirm this, write both equations in slope-intercept form by solving each equation for y.

$2x + y = 7$	First equation	$2y = -4x$ Second equation
$y = -2x + 7$ Subtract $2x$ from both sides.		$\dfrac{2y}{2} = \dfrac{-4x}{2}$ Divide both sides by 2.
		$y = -2x$

Recall that when an equation is written in slope-intercept form, the coefficient of x is the slope. Since both equations have the same slope, -2, but different y-intercepts, the lines are parallel and have no points in common. Thus, there is no solution of the system and the system is inconsistent. To indicate this, we can say the system has no solution or the solution set is $\{\,\}$ or $\varnothing$.

PRACTICE
5 Solve the system of equations by graphing.

$$\begin{cases} 3y = 9x \\ 6x - 2y = 12 \end{cases}$$

In Examples 3, 4, and 5, the graphs of the two linear equations of each system are different. When this happens, we call these equations **independent equations.** If the graphs of the two equations in a system are identical, we call the equations **dependent equations.**

EXAMPLE 6   Solve the system of equations by graphing.

$$\begin{cases} x - y = 3 \\ -x + y = -3 \end{cases}$$

Solution Graph each line.

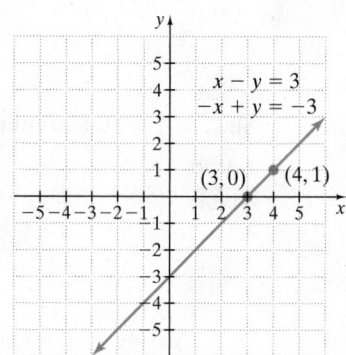

These graphs **appear** to be identical. To confirm this, write each equation in slope-intercept form.

$x - y = 3$ First equation $-x + y = -3$ Second equation

$\quad -y = -x + 3$ Subtract x from both sides. $\quad\ \ \boldsymbol{y = x - 3}$ Add x to both sides.

$\dfrac{-y}{-1} = \dfrac{-x}{-1} + \dfrac{3}{-1}$ Divide both sides by -1.

$\quad \boldsymbol{y = x - 3}$

The equations are identical and so must be their graphs. The lines have an infinite number of points in common. Thus, there is an infinite number of solutions of the system and this is a consistent system. The equations are dependent equations. Here, we can say that there are an infinite number of solutions or the solution set is $\{(x, y)\,|\,x - y = 3\}$ or equivalently $\{(x, y)\,|\,-x + y = -3\}$ since the equations describe identical ordered pairs. The second set is read "the set of all ordered pairs (x, y) such that $-x + y = -3$".

PRACTICE
6 Solve the system of equations by graphing.

$$\begin{cases} x - y = 4 \\ -2x + 2y = -8 \end{cases}$$

As we have seen, three different situations can occur when graphing the two lines associated with the two equations in a linear system:

One point of intersection: one solution

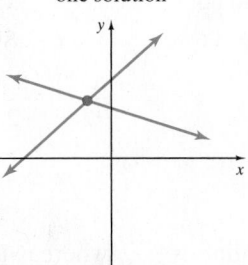

Consistent system
(at least one solution)
Independent equations
(graphs of equations differ)

Parallel lines: no solution

Inconsistent system
(no solution)
Independent equations
(graphs of equations differ)

Same line: infinite number of solutions

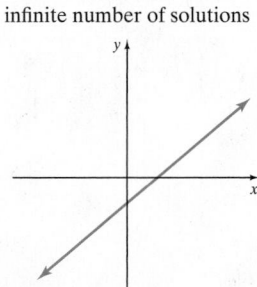

Consistent system
(at least one solution)
Dependent equations
(graphs of equations identical)

OBJECTIVE

3 Finding the Number of Solutions of a System without Graphing ▶

You may have suspected by now that graphing alone is not an accurate way to solve a system of linear equations. For example, a solution of $\left(\frac{1}{2}, \frac{2}{9}\right)$ is unlikely to be read correctly from a graph. The next two sections present two accurate methods of solving these systems. In the meantime, we can decide how many solutions a system has by writing each equation in the slope-intercept form.

EXAMPLE 7 Without graphing, determine the number of solutions of the system.

$$\begin{cases} \frac{1}{2}x - y = 2 \\ x = 2y + 5 \end{cases}$$

Solution First write each equation in slope-intercept form.

$\frac{1}{2}x - y = 2$ First equation

$\frac{1}{2}x = y + 2$ Add y to both sides.

$\frac{1}{2}x - 2 = y$ Subtract 2 from both sides.

$x = 2y + 5$ Second equation

$x - 5 = 2y$ Subtract 5 from both sides.

$\frac{x}{2} - \frac{5}{2} = \frac{2y}{2}$ Divide both sides by 2.

$\frac{1}{2}x - \frac{5}{2} = y$ Simplify.

The slope of each line is $\frac{1}{2}$, but they have different y-intercepts. This tells us that the lines representing these equations are parallel. Since the lines are parallel, the system has no solution and is inconsistent. □

PRACTICE

7 Without graphing, determine the number of solutions of the system.

$$\begin{cases} 5x + 4y = 6 \\ x - y = 3 \end{cases}$$

EXAMPLE 8 Without graphing, determine the number of solutions of the system.

$$\begin{cases} 3x - y = 4 \\ x + 2y = 8 \end{cases}$$

(Continued on next page)

Solution Once again, the slope-intercept form helps determine how many solutions this system has.

$3x - y = 4$	First equation	$x + 2y = 8$	Second equation
$3x = y + 4$	Add y to both sides.	$x = -2y + 8$	Subtract $2y$ from both sides.
$3x - 4 = y$	Subtract 4 from both sides.	$x - 8 = -2y$	Subtract 8 from both sides.
		$\dfrac{x}{-2} - \dfrac{8}{-2} = \dfrac{-2y}{-2}$	Divide both sides by -2.
		$-\dfrac{1}{2}x + 4 = y$	Simplify.

The slope of the second line is $-\dfrac{1}{2}$, whereas the slope of the first line is 3. Since the slopes are not equal, the two lines are neither parallel nor identical and must intersect. Therefore, this system has one solution and is consistent. □

PRACTICE

8 Without graphing, determine the number of solutions of the system.

$$\begin{cases} -\dfrac{2}{3}x + y = 6 \\ 3y = 2x + 5 \end{cases}$$

Graphing Calculator Explorations

A graphing calculator may be used to approximate solutions of systems of equations. For example, to approximate the solution of the system

$$\begin{cases} y = -3.14x - 1.35 \\ y = 4.88x + 5.25, \end{cases}$$

first graph each equation on the same set of axes. Then use the intersect feature of your calculator to approximate the point of intersection.

The approximate point of intersection is $(-0.82, 1.23)$.

Solve each system of equations. Approximate the solutions to two decimal places.

1. $\begin{cases} y = -2.68x + 1.21 \\ y = 5.22x - 1.68 \end{cases}$
2. $\begin{cases} y = 4.25x + 3.89 \\ y = -1.88x + 3.21 \end{cases}$

3. $\begin{cases} 4.3x - 2.9y = 5.6 \\ 8.1x + 7.6y = -14.1 \end{cases}$
4. $\begin{cases} -3.6x - 8.6y = 10 \\ -4.5x + 9.6y = -7.7 \end{cases}$

 ## Vocabulary, Readiness & Video Check

Fill in each blank with one of the words or phrases listed below.

system of linear equations   solution   consistent
dependent   inconsistent   independent

1. In a system of linear equations in two variables, if the graphs of the equations are the same, the equations are _____ equations.

2. Two or more linear equations are called a(n) _____.

3. A system of equations that has at least one solution is called a(n) _____ system.

4. A(n) _____ of a system of two equations in two variables is an ordered pair of numbers that is a solution of both equations in the system.

5. A system of equations that has no solution is called a(n) _____ system.

6. In a system of linear equations in two variables, if the graphs of the equations are different, the equations are _____ equations.

Martin-Gay Interactive Videos

See Video 4.1

Watch the section lecture video and answer the following questions.

OBJECTIVE 1

7. In Example 1, the first ordered pair is a solution of the first equation of the system. Why is this not enough to determine whether the first ordered pair is a solution of the system?

OBJECTIVE 2

8. From Examples 2 and 3, why is finding the solution of a system of equations from a graph considered "guessing" and this proposed solution checked algebraically?

OBJECTIVE 3

9. From Examples 5–7, explain how the slope-intercept form tells us how many solutions a system of equations has.

4.1   Exercise Set   MyMathLab®

Each rectangular coordinate system shows the graph of the equations in a system of equations. Use each graph to determine the number of solutions for each associated system. If the system has only one solution, give its coordinates.

1. 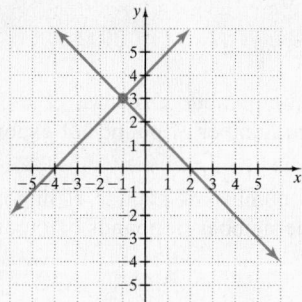   **2.** 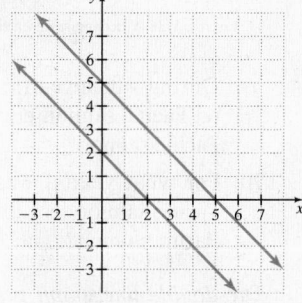   **3.** 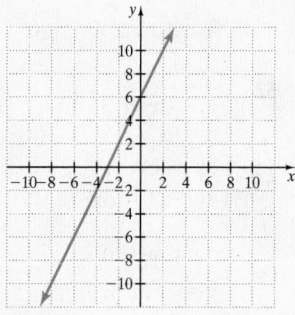   **4.**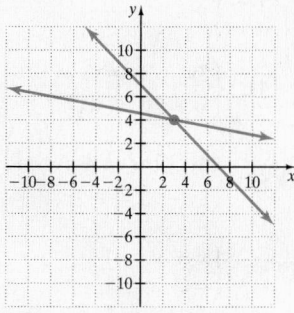

Determine whether each ordered pair is a solution of the system of linear equations. See Examples 1 and 2.

5. $\begin{cases} x + y = 8 \\ 3x + 2y = 21 \end{cases}$
   a. $(2, 4)$
   b. $(5, 3)$

6. $\begin{cases} 2x + y = 5 \\ x + 3y = 5 \end{cases}$
   a. $(5, 0)$
   b. $(2, 1)$

7. $\begin{cases} 3x - y = 5 \\ x + 2y = 11 \end{cases}$
   a. $(3, 4)$
   b. $(0, -5)$

8. $\begin{cases} 2x - 3y = 8 \\ x - 2y = 6 \end{cases}$
   a. $(-2, -4)$
   b. $(7, 2)$

9. $\begin{cases} 2y = 4x + 6 \\ 2x - y = -3 \end{cases}$
   a. $(-3, -3)$
   b. $(0, 3)$

10. $\begin{cases} x + 5y = -4 \\ -2x = 10y + 8 \end{cases}$
   a. $(-4, 0)$
   b. $(6, -2)$

11. $\begin{cases} -2 = x - 7y \\ 6x - y = 13 \end{cases}$
   a. $(-2, 0)$
   b. $\left(\dfrac{1}{2}, \dfrac{5}{14} \right)$

12. $\begin{cases} 4x = 1 - y \\ x - 3y = -8 \end{cases}$
   a. $(0, 1)$
   b. $\left(\dfrac{1}{6}, \dfrac{1}{3} \right)$

MIXED PRACTICE

Solve each system of linear equations by graphing. See Examples 3 through 6.

13. $\begin{cases} x + y = 4 \\ x - y = 2 \end{cases}$

14. $\begin{cases} x + y = 3 \\ x - y = 5 \end{cases}$

15. $\begin{cases} x + y = 6 \\ -x + y = -6 \end{cases}$

16. $\begin{cases} x + y = 1 \\ -x + y = -3 \end{cases}$

17. $\begin{cases} y = 2x \\ 3x - y = -2 \end{cases}$

18. $\begin{cases} y = -3x \\ 2x - y = -5 \end{cases}$

19. $\begin{cases} y = x + 1 \\ y = 2x - 1 \end{cases}$

20. $\begin{cases} y = 3x - 4 \\ y = x + 2 \end{cases}$

21. $\begin{cases} 2x + y = 0 \\ 3x + y = 1 \end{cases}$

22. $\begin{cases} 2x + y = 1 \\ 3x + y = 0 \end{cases}$

23. $\begin{cases} y = -x - 1 \\ y = 2x + 5 \end{cases}$

24. $\begin{cases} y = x - 1 \\ y = -3x - 5 \end{cases}$

25. $\begin{cases} x + y = 5 \\ x + y = 6 \end{cases}$

26. $\begin{cases} x - y = 4 \\ x - y = 1 \end{cases}$

27. $\begin{cases} 2x - y = 6 \\ y = 2 \end{cases}$

28. $\begin{cases} x + y = 5 \\ x = 4 \end{cases}$

29. $\begin{cases} x - 2y = 2 \\ 3x + 2y = -2 \end{cases}$

30. $\begin{cases} x + 3y = 7 \\ 2x - 3y = -4 \end{cases}$

31. $\begin{cases} 2x + y = 4 \\ 6x = -3y + 6 \end{cases}$

32. $\begin{cases} y + 2x = 3 \\ 4x = 2 - 2y \end{cases}$

33. $\begin{cases} y - 3x = -2 \\ 6x - 2y = 4 \end{cases}$

34. $\begin{cases} x - 2y = -6 \\ -2x + 4y = 12 \end{cases}$

35. $\begin{cases} x = 3 \\ y = -1 \end{cases}$

36. $\begin{cases} x = -5 \\ y = 3 \end{cases}$

37. $\begin{cases} y = x - 2 \\ y = 2x + 3 \end{cases}$

38. $\begin{cases} y = x + 5 \\ y = -2x - 4 \end{cases}$

39. $\begin{cases} 2x - 3y = -2 \\ -3x + 5y = 5 \end{cases}$

40. $\begin{cases} 4x - y = 7 \\ 2x - 3y = -9 \end{cases}$

41. $\begin{cases} 6x - y = 4 \\ \dfrac{1}{2}y = -2 + 3x \end{cases}$ **42.** $\begin{cases} 3x - y = 6 \\ \dfrac{1}{3}y = -2 + x \end{cases}$

Without graphing, decide.

  a. Are the graphs of the equations identical lines, parallel lines, or lines intersecting at a single point?

  b. How many solutions does the system have? See Examples 7 and 8.

43. $\begin{cases} 4x + y = 24 \\ x + 2y = 2 \end{cases}$ **44.** $\begin{cases} 3x + y = 1 \\ 3x + 2y = 6 \end{cases}$

45. $\begin{cases} 2x + y = 0 \\ 2y = 6 - 4x \end{cases}$ **46.** $\begin{cases} 3x + y = 0 \\ 2y = -6x \end{cases}$

47. $\begin{cases} 6x - y = 4 \\ \dfrac{1}{2}y = -2 + 3x \end{cases}$ **48.** $\begin{cases} 3x - y = 2 \\ \dfrac{1}{3}y = -2 + 3x \end{cases}$

49. $\begin{cases} x = 5 \\ y = -2 \end{cases}$ **50.** $\begin{cases} y = 3 \\ x = -4 \end{cases}$

51. $\begin{cases} 3y - 2x = 3 \\ x + 2y = 9 \end{cases}$ **52.** $\begin{cases} 2y = x + 2 \\ y + 2x = 3 \end{cases}$

53. $\begin{cases} 6y + 4x = 6 \\ 3y - 3 = -2x \end{cases}$ **54.** $\begin{cases} 8y + 6x = 4 \\ 4y - 2 = 3x \end{cases}$

55. $\begin{cases} x + y = 4 \\ x + y = 3 \end{cases}$ **56.** $\begin{cases} 2x + y = 0 \\ y = -2x + 1 \end{cases}$

REVIEW AND PREVIEW

Solve each equation. See Section 2.3.

57. $5(x - 3) + 3x = 1$

58. $-2x + 3(x + 6) = 17$

59. $4\left(\dfrac{y + 1}{2}\right) + 3y = 0$

60. $-y + 12\left(\dfrac{y - 1}{4}\right) = 3$

61. $8a - 2(3a - 1) = 6$

62. $3z - (4z - 2) = 9$

CONCEPT EXTENSIONS

63. Draw a graph of two linear equations whose associated system has the solution $(-1, 4)$.

64. Draw a graph of two linear equations whose associated system has the solution $(3, -2)$.

65. Draw a graph of two linear equations whose associated system has no solution.

66. Draw a graph of two linear equations whose associated system has an infinite number of solutions.

67. The ordered pair $(-2, 3)$ is a solution of all three independent equations:

$$x + y = 1$$
$$2x - y = -7$$
$$x + 3y = 7$$

Describe the graph of all three equations on the same axes.

68. Explain how to use a graph to determine the number of solutions of a system.

The double line graph below shows the number of pounds of fresh Pacific salmon imported to or exported from the United States during the given years. Use this graph to answer Exercises 69 and 70.

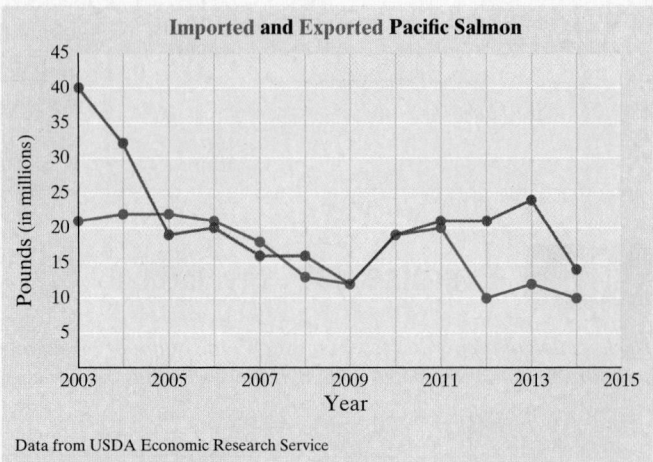

Data from USDA Economic Research Service

69. During which year(s) did the number of pounds of imported Pacific salmon equal the number of pounds of exported Pacific salmon?

70. For what year(s) was the number of pounds of imported Pacific salmon less than the number of pounds of exported Pacific salmon?

The double line graph below shows the average attendance per game for the years shown for the Minnesota Twins and the Texas Rangers baseball teams. Use this for Exercises 71 and 72.

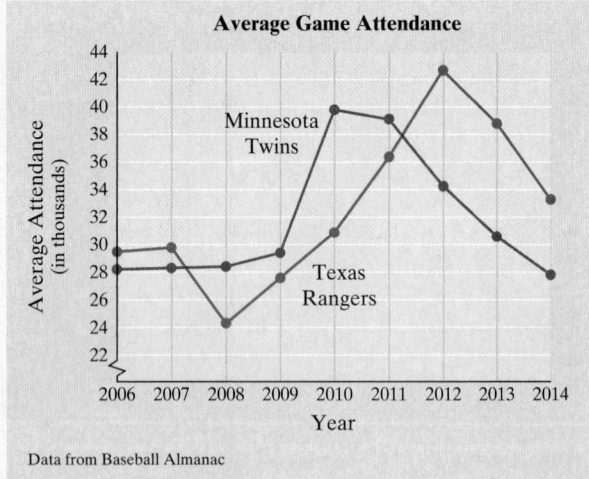

Data from Baseball Almanac

71. In what year(s) was the average attendance per game for the Texas Rangers greater than the average attendance per game for the Minnesota Twins?

72. In what year was the average attendance per game for the Texas Rangers closest to the average attendance per game for the Minnesota Twins, 2007 or 2012?

73. Construct a system of two linear equations that has $(1, 3)$ as a solution.

74. Construct a system of two linear equations that has $(0, 7)$ as a solution.

75. Below are tables of values for two linear equations.
  a. Find a solution of the corresponding system.
  b. Graph several ordered pairs from each table and sketch the two lines.

x	y
1	3
2	5
3	7
4	9
5	11

x	y
1	6
2	7
3	8
4	9
5	10

  c. Does your graph confirm the solution from part (a)?

76. Tables of values for two linear equations are shown.
  a. Find a solution of the corresponding system.
  b. Graph several ordered pairs from each table and sketch the two lines.

x	y
-3	5
-1	1
0	-1
1	-3
2	-5

x	y
-3	7
-1	1
0	-2
1	-5
2	-8

  c. Does your graph confirm the solution from part (a)?

77. Explain how writing each equation in a linear system in slope-intercept form helps determine the number of solutions of a system.

78. Is it possible for a system of two linear equations in two variables to be inconsistent but with dependent equations? Why or why not?

4.2 | Solving Systems of Linear Equations by Substitution

OBJECTIVE

1 Use the Substitution Method to Solve a System of Linear Equations.

OBJECTIVE

1 Using the Substitution Method

As we stated in the preceding section, graphing alone is not an accurate way to solve a system of linear equations. In this section, we discuss a second, more accurate method for solving systems of equations. This method is called the **substitution method** and is introduced in the next example.

EXAMPLE 1 Solve the system:

$$\begin{cases} 2x + y = 10 & \text{First equation} \\ x = y + 2 & \text{Second equation} \end{cases}$$

Solution The second equation in this system is $x = y + 2$. This tells us that x and $y + 2$ have the same value. This means that we may substitute $y + 2$ for x in the first equation.

$$2x + y = 10 \quad \text{First equation}$$

$$2(y + 2) + y = 10 \quad \text{Substitute } y + 2 \text{ for } x \text{ since } x = y + 2.$$

Notice that this equation now has one variable, y. Let's now solve this equation for y.

Helpful Hint
Don't forget the distributive property.

$$2(y + 2) + y = 10$$
$$2y + 4 + y = 10 \quad \text{Use the distributive property.}$$
$$3y + 4 = 10 \quad \text{Combine like terms.}$$
$$3y = 6 \quad \text{Subtract 4 from both sides.}$$
$$y = 2 \quad \text{Divide both sides by 3.}$$

Now we know that the y-value of the ordered pair solution of the system is 2. To find the corresponding x-value, we replace y with 2 in the equation $x = y + 2$ and solve for x.

$$x = y + 2$$
$$x = 2 + 2 \quad \text{Let } y = 2.$$
$$x = 4$$

(Continued on next page)

The solution of the system is the ordered pair $(4, 2)$. Since an ordered pair solution must satisfy both linear equations in the system, we could have chosen the equation $2x + y = 10$ to find the corresponding x-value. The resulting x-value is the same.

Check: We check to see that $(4, 2)$ satisfies both equations of the original system.

First Equation	**Second Equation**
$2x + y = 10$	$x = y + 2$
$2(4) + 2 \overset{?}{=} 10$	$4 \overset{?}{=} 2 + 2$ Let $x = 4$ and $y = 2$.
$10 = 10$ True	$4 = 4$ True

The solution of the system is $(4, 2)$.

A graph of the two equations shows the two lines intersecting at the point $(4, 2)$.

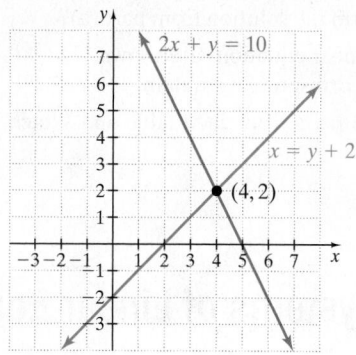

PRACTICE

1   Solve the system:

$$\begin{cases} 2x - y = 9 \\ x = y + 1 \end{cases}$$

EXAMPLE 2   Solve the system:

$$\begin{cases} 5x - y = -2 \\ y = 3x \end{cases}$$

__Solution__ The second equation is solved for y in terms of x. We substitute $3x$ for y in the first equation.

$$5x - y = -2 \quad \text{First equation}$$

$$5x - (3x) = -2 \quad \text{Substitute } 3x \text{ for } y.$$

Now we solve for x.

$$5x - 3x = -2$$
$$2x = -2 \quad \text{Combine like terms.}$$
$$x = -1 \quad \text{Divide both sides by 2.}$$

The x-value of the ordered pair solution is -1. To find the corresponding y-value, we replace x with -1 in the second equation $y = 3x$.

$$y = 3x \quad \text{Second equation}$$
$$y = 3(-1) \quad \text{Let } x = -1.$$
$$y = -3$$

Check to see that the solution of the system is $(-1, -3)$.

PRACTICE

2 Solve the system:

$$\begin{cases} 7x - y = -15 \\ y = 2x \end{cases}$$

To solve a system of equations by substitution, we first need an equation solved for one of its variables, as in Examples 1 and 2. If neither equation in a system is solved for x or y, this will be our first step.

EXAMPLE 3 Solve the system:

$$\begin{cases} x + 2y = 7 \\ 2x + 2y = 13 \end{cases}$$

Solution We choose one of the equations and solve for x or y. We will solve the first equation for x by subtracting $2y$ from both sides.

$$x + 2y = 7 \qquad \text{First equation}$$
$$x = 7 - 2y \quad \text{Subtract } 2y \text{ from both sides.}$$

Since $x = 7 - 2y$, we now substitute $7 - 2y$ for x in the second equation and solve for y.

$$2x + 2y = 13 \quad \text{Second equation}$$
$$2(7 - 2y) + 2y = 13 \quad \text{Let } x = 7 - 2y.$$
$$14 - 4y + 2y = 13 \quad \text{Use the distributive property.}$$
$$14 - 2y = 13 \quad \text{Simplify.}$$
$$-2y = -1 \quad \text{Subtract 14 from both sides.}$$
$$y = \frac{1}{2} \quad \text{Divide both sides by } -2.$$

> **Helpful Hint**
> Don't forget to insert parentheses when substituting $7 - 2y$ for x.

To find x, we let $y = \frac{1}{2}$ in the equation $x = 7 - 2y$.

$$x = 7 - 2y$$
$$x = 7 - 2\left(\frac{1}{2}\right) \quad \text{Let } y = \frac{1}{2}.$$
$$x = 7 - 1$$
$$x = 6$$

> **Helpful Hint**
> To find x, any equation in two variables equivalent to the original equations of the system may be used. We used this equation since it is solved for x.

The solution is $\left(6, \frac{1}{2}\right)$. Check the solution in both equations of the original system.

PRACTICE

3 Solve the system:

$$\begin{cases} x + 3y = 6 \\ 2x + 3y = 10 \end{cases}$$

The following steps may be used to solve a system of equations by the substitution method.

Solving a System of Two Linear Equations by the Substitution Method

Step 1. Solve one of the equations for one of its variables.

Step 2. Substitute the expression for the variable found in Step 1 into the other equation.

Step 3. Solve the equation from Step 2 to find the value of one variable.

Step 4. Substitute the value found in Step 3 in any equation containing both variables to find the value of the other variable.

Step 5. Check the proposed solution in the original system.

✔ **CONCEPT CHECK**

As you solve the system $\begin{cases} 2x + y = -5 \\ x - y = 5 \end{cases}$ you find that $y = -5$. Is this the solution of the system?

EXAMPLE 4 Solve the system:

$$\begin{cases} 7x - 3y = -14 \\ -3x + y = 6 \end{cases}$$

Solution To avoid introducing fractions, we will solve the second equation for y.

$$-3x + y = 6 \qquad \text{Second equation}$$
$$y = 3x + 6$$

Next, substitute $3x + 6$ for y in the first equation.

$$7x - 3y = -14 \quad \text{First equation}$$
$$7x - 3(\overbrace{3x + 6}) = -14 \quad \text{Let } y = 3x + 6.$$
$$7x - 9x - 18 = -14 \quad \text{Use the distributive property.}$$
$$-2x - 18 = -14 \quad \text{Simplify.}$$
$$-2x = 4 \quad \text{Add 18 to both sides.}$$
$$x = -2 \quad \text{Divide both sides by } -2.$$

To find the corresponding y-value, substitute -2 for x in the equation $y = 3x + 6$. Then $y = 3(-2) + 6$ or $y = 0$. The solution of the system is $(-2, 0)$. Check this solution in both equations of the system.

PRACTICE

4 Solve the system:

$$\begin{cases} 5x + 3y = -9 \\ -2x + y = 8 \end{cases}$$

Helpful Hint

When solving a system of equations by the substitution method, begin by solving an equation for one of its variables. If possible, solve for a variable that has a coefficient of 1 or -1. This way, we avoid working with time-consuming fractions.

✔ **CONCEPT CHECK**

To avoid fractions, which of the equations below would you use to solve for x?
a. $3x - 4y = 15$ **b.** $14 - 3y = 8x$ **c.** $7y + x = 12$

Answer to Concept Check:
No, the solution will be an ordered pair.

Answer to Concept Check:
c

EXAMPLE 5 Solve the system:

$$\begin{cases} \dfrac{1}{2}x - y = 3 \\ x = 6 + 2y \end{cases}$$

Solution The second equation is already solved for x in terms of y. Thus we substitute $6 + 2y$ for x in the first equation and solve for y.

$$\frac{1}{2}x - y = 3 \quad \text{First equation}$$

$$\frac{1}{2}(6 + 2y) - y = 3 \quad \text{Let } x = 6 + 2y.$$

$$3 + y - y = 3 \quad \text{Use the distributive property.}$$

$$3 = 3 \quad \text{Simplify.}$$

Arriving at a true statement such as $3 = 3$ indicates that the two linear equations in the original system are equivalent. This means that their graphs are identical and there is an infinite number of solutions of the system. Any solution of one equation is also a solution of the other. For the solution, we can write "infinite number of solutions" or, in set notation, $\left\{ (x, y) \,\middle|\, \frac{1}{2}x - y = 3 \right\}$ or $\{(x, y) \,|\, x = 6 + 2y\}$.

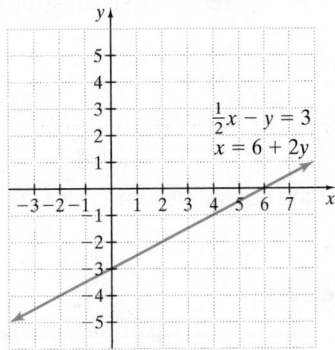

PRACTICE
5   Solve the system:

$$\begin{cases} \dfrac{1}{4}x - y = 2 \\ x = 4y + 8 \end{cases}$$

EXAMPLE 6   Use substitution to solve the system.

$$\begin{cases} 6x + 12y = 5 \\ -4x - 8y = 0 \end{cases}$$

Solution Choose the second equation and solve for y.

$$-4x - 8y = 0 \quad \text{Second equation}$$

$$-8y = 4x \quad \text{Add } 4x \text{ to both sides.}$$

$$\frac{-8y}{-8} = \frac{4x}{-8} \quad \text{Divide both sides by } -8.$$

$$y = -\frac{1}{2}x \quad \text{Simplify.}$$

Now replace y with $-\dfrac{1}{2}x$ in the first equation.

$$6x + 12y = 5 \quad \text{First equation}$$

$$6x + 12\left(-\frac{1}{2}x\right) = 5 \quad \text{Let } y = -\frac{1}{2}x.$$

$$6x + (-6x) = 5 \quad \text{Simplify.}$$

$$0 = 5 \quad \text{Combine like terms.}$$

(Continued on next page)

The false statement $0 = 5$ indicates that this system has no solution and is inconsistent. The graph of the linear equations in the system is a pair of parallel lines. For the solution, we can write "no solution" or, in set notation, write { } or $\varnothing$.

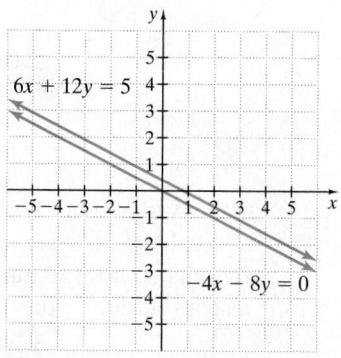

PRACTICE

6   Use substitution to solve the system.

$$\begin{cases} 4x - 3y = 12 \\ -8x + 6y = -30 \end{cases}$$

 CONCEPT CHECK

Describe how the graphs of the equations in a system appear if the system has

a. no solution

b. one solution

c. an infinite number of solutions

✔ **Vocabulary, Readiness & Video Check**

Give the solution of each system. If the system has no solution or an infinite number of solutions, say so. If the system has one solution, find it.

1. $\begin{cases} y = 4x \\ -3x + y = 1 \end{cases}$
When solving, you obtain $x = 1$

2. $\begin{cases} 4x - y = 17 \\ -8x + 2y = 0 \end{cases}$
When solving, you obtain $0 = 34$

3. $\begin{cases} 4x - y = 17 \\ -8x + 2y = -34 \end{cases}$
When solving, you obtain $0 = 0$

4. $\begin{cases} 5x + 2y = 25 \\ x = y + 5 \end{cases}$
When solving, you obtain $y = 0$

5. $\begin{cases} x + y = 0 \\ 7x - 7y = 0 \end{cases}$
When solving, you obtain $x = 0$

6. $\begin{cases} y = -2x + 5 \\ 4x + 2y = 10 \end{cases}$
When solving, you obtain $0 = 0$

Martin-Gay Interactive Videos

See Video 4.2 ◉

Watch the section lecture video and answer the following question.

OBJECTIVE

1   **7.** The systems in ▦ Examples 2–4 all need one of their equations solved for a variable as a first step. What important part of the substitution method is emphasized in each example?

Answers to Concept Check:

a. parallel lines

b. intersect at one point

c. identical graphs

4.2 Exercise Set MyMathLab®

Solve each system of equations by the substitution method. See Examples 1 and 2.

1. $\begin{cases} x + y = 3 \\ x = 2y \end{cases}$

2. $\begin{cases} x + y = 20 \\ x = 3y \end{cases}$

3. $\begin{cases} x + y = 6 \\ y = -3x \end{cases}$

4. $\begin{cases} x + y = 6 \\ y = -4x \end{cases}$

5. $\begin{cases} y = 3x + 1 \\ 4y - 8x = 12 \end{cases}$

6. $\begin{cases} y = 2x + 3 \\ 5y - 7x = 18 \end{cases}$

7. $\begin{cases} y = 2x + 9 \\ y = 7x + 10 \end{cases}$

8. $\begin{cases} y = 5x - 3 \\ y = 8x + 4 \end{cases}$

MIXED PRACTICE

Solve each system of equations by the substitution method. See Examples 1 through 6.

9. $\begin{cases} 3x - 4y = 10 \\ y = x - 3 \end{cases}$

10. $\begin{cases} 4x - 3y = 10 \\ y = x - 5 \end{cases}$

11. $\begin{cases} x + 2y = 6 \\ 2x + 3y = 8 \end{cases}$

12. $\begin{cases} x + 3y = -5 \\ 2x + 2y = 6 \end{cases}$

13. $\begin{cases} 3x + 2y = 16 \\ x = 3y - 2 \end{cases}$

14. $\begin{cases} 2x + 3y = 18 \\ x = 2y - 5 \end{cases}$

15. $\begin{cases} 2x - 5y = 1 \\ 3x + y = -7 \end{cases}$

16. $\begin{cases} 3y - x = 6 \\ 4x + 12y = 0 \end{cases}$

17. $\begin{cases} 4x + 2y = 5 \\ -2x = y + 4 \end{cases}$

18. $\begin{cases} 2y = x + 2 \\ 6x - 12y = 0 \end{cases}$

19. $\begin{cases} 4x + y = 11 \\ 2x + 5y = 1 \end{cases}$

20. $\begin{cases} 3x + y = -14 \\ 4x + 3y = -22 \end{cases}$

21. $\begin{cases} x + 2y + 5 = -4 + 5y - x \\ \quad\quad 2x + x = y + 4 \end{cases}$
(*Hint:* First simplify each equation.)

22. $\begin{cases} 5x + 4y - 2 = -6 + 7y - 3x \\ \quad\quad 3x + 4x = y + 3 \end{cases}$
(*Hint:* See Exercise 21.)

23. $\begin{cases} 6x - 3y = 5 \\ x + 2y = 0 \end{cases}$

24. $\begin{cases} 10x - 5y = -21 \\ x + 3y = 0 \end{cases}$

25. $\begin{cases} 3x - y = 1 \\ 2x - 3y = 10 \end{cases}$

26. $\begin{cases} 2x - y = -7 \\ 4x - 3y = -11 \end{cases}$

27. $\begin{cases} -x + 2y = 10 \\ -2x + 3y = 18 \end{cases}$

28. $\begin{cases} -x + 3y = 18 \\ -3x + 2y = 19 \end{cases}$

29. $\begin{cases} 5x + 10y = 20 \\ 2x + 6y = 10 \end{cases}$

30. $\begin{cases} 6x + 3y = 12 \\ 9x + 6y = 15 \end{cases}$

31. $\begin{cases} 3x + 6y = 9 \\ 4x + 8y = 16 \end{cases}$

32. $\begin{cases} 2x + 4y = 6 \\ 5x + 10y = 16 \end{cases}$

33. $\begin{cases} \dfrac{1}{3}x - y = 2 \\ x - 3y = 6 \end{cases}$

34. $\begin{cases} \dfrac{1}{4}x - 2y = 1 \\ x - 8y = 4 \end{cases}$

35. $\begin{cases} x = \dfrac{3}{4}y - 1 \\ 8x - 5y = -6 \end{cases}$

36. $\begin{cases} x = \dfrac{5}{6}y - 2 \\ 12x - 5y = -9 \end{cases}$

Solve each system by the substitution method. First, simplify each equation by combining like terms.

37. $\begin{cases} -5y + 6y = 3x + 2(x - 5) - 3x + 5 \\ 4(x + y) - x + y = -12 \end{cases}$

38. $\begin{cases} 5x + 2y - 4x - 2y = 2(2y + 6) - 7 \\ 3(2x - y) - 4x = 1 + 9 \end{cases}$

REVIEW AND PREVIEW

Write equivalent equations by multiplying both sides of the given equation by the given nonzero number. See Section 2.2.

39. $3x + 2y = 6$ by -2

40. $-x + y = 10$ by 5

41. $-4x + y = 3$ by 3

42. $5a - 7b = -4$ by -4

Add the expressions by combining any like terms. See Section 2.1.

43. $\begin{array}{r} 3n + 6m \\ \underline{2n - 6m} \end{array}$

44. $\begin{array}{r} -2x + 5y \\ \underline{2x + 11y} \end{array}$

45. $\begin{array}{r} -5a - 7b \\ \underline{5a - 8b} \end{array}$

46. $\begin{array}{r} 9q + p \\ \underline{-9q - p} \end{array}$

CONCEPT EXTENSIONS

47. Explain how to identify a system with no solution when using the substitution method.

48. Occasionally, when using the substitution method, we obtain the equation $0 = 0$. Explain how this result indicates that the graphs of the equations in the system are identical.

Solve. See a Concept Check in this section.

49. As you solve the system $\begin{cases} 3x - y = -6 \\ -3x + 2y = 7 \end{cases}$, you find that $y = 1$. Is this the solution of the system?

50. As you solve the system $\begin{cases} x = 5y \\ y = 2x \end{cases}$, you find that $x = 0$ and $y = 0$. What is the solution of this system?

51. To avoid fractions, which of the equations below would you use if solving for y? Explain why.

    a. $\dfrac{1}{2}x - 4y = \dfrac{3}{4}$ **b.** $8x - 5y = 13$

    c. $7x - y = 19$

52. Give the number of solutions of a system if the graphs of the equations in the system are

    a. lines intersecting in one point

    b. parallel lines

    c. same line

53. The number of men and women receiving bachelor's degrees each year has been steadily increasing. For the years 1970 through 2014, the number of men receiving degrees (in thousands) is given by the equation $y = 3.9x + 443$, and for women the equation is $y = 14.2x + 314$, where x is the number of years after 1970. (*Source:* National Center for Education Statistics)

a. Use the substitution method to solve this system of equations. (Round your final results to the nearest whole numbers.)

b. Explain the meaning of your answer to part (a).

c. Sketch a graph of the system of equations. Write a sentence describing the trends for men and women receiving bachelor's degrees.

54. The number of Adult Contemporary Music radio stations in the United States from 2000 to 2014 is given by the equation $y = 6x + 734$, where x is the number of years after 2000. The number of Spanish radio stations is given by $y = 13x + 542$ for the same time period. (*Source:* M Street Corporation)

a. Use the substitution method to solve this system of equations. (Round your numbers to the nearest tenth.)

b. Explain the meaning of your answer to part (a).

c. Sketch a graph of the system of equations. Write a sentence describing the trends in the popularity of these two types of music formats.

Solve each system by substitution. When necessary, round answers to the nearest hundredth.

55. $\begin{cases} y = 5.1x + 14.56 \\ y = -2x - 3.9 \end{cases}$

56. $\begin{cases} y = 3.1x - 16.35 \\ y = -9.7x + 28.45 \end{cases}$

57. $\begin{cases} 3x + 2y = 14.05 \\ 5x + y = 18.5 \end{cases}$

58. $\begin{cases} x + y = -15.2 \\ -2x + 5y = -19.3 \end{cases}$

4.3   Solving Systems of Linear Equations by Addition ▶

OBJECTIVE

1 Use the Addition Method to Solve a System of Linear Equations. ▶

OBJECTIVE

1   Using the Addition Method ▶

We have seen that substitution is an accurate way to solve a linear system. Another method for solving a system of equations accurately is the **addition** or **elimination method.** The addition method is based on the addition property of equality: adding equal quantities to both sides of an equation does not change the solution of the equation. In symbols,

$$\text{if } A = B \text{ and } C = D, \text{ then } A + C = B + D.$$

EXAMPLE 1 Solve the system:

$$\begin{cases} x + y = 7 \\ x - y = 5 \end{cases}$$

Solution Since the left side of each equation is equal to the right side, we add equal quantities by adding the left sides of the equations together and the right sides of the equations together. This adding eliminates the variable y and gives us an equation in one variable, x. We can then solve for x.

> **Helpful Hint**
> Notice in Example 1 that our goal when solving a system of equations by the addition method is to eliminate a variable when adding the equations.

$x + y = 7$	First equation
$\underline{x - y = 5}$	Second equation
$2x \quad\quad = 12$	Add the equations.
$x = 6$	Divide both sides by 2.

The x-value of the solution is 6. To find the corresponding y-value, let $x = 6$ in either equation of the system. We will use the first equation.

$x + y = 7$	First equation
$6 + y = 7$	Let $x = 6$.
$y = 7 - 6$	Solve for y.
$y = 1$	Simplify.

Check: The solution is $(6, 1)$. Check this in both equations.

First Equation	*Second Equation*
$x + y = 7$	$x - y = 5$
$6 + 1 \overset{?}{=} 7$	$6 - 1 \overset{?}{=} 5$ Let $x = 6$ and $y = 1$.
$7 = 7$ True	$5 = 5$ True

Thus, the solution of the system is $(6, 1)$ and the graphs of the two equations intersect at the point $(6, 1)$ as shown next.

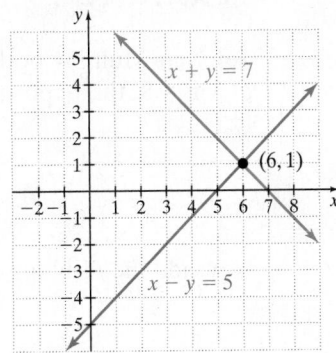

PRACTICE
1   Solve the system: $\begin{cases} x - y = 2 \\ x + y = 8 \end{cases}$

EXAMPLE 2   Solve the system: $\begin{cases} -2x + y = 2 \\ -x + 3y = -4 \end{cases}$

Solution If we simply add the two equations, the result is still an equation in two variables. However, remember from Example 1 that our goal is to eliminate one of the variables. Notice what happens if we multiply *both sides* of the first equation by -3, which we are allowed to do by the multiplication property of equality. The system

$$\begin{cases} -3(-2x + y) = -3(2) \\ -x + 3y = -4 \end{cases} \quad \text{simplifies to} \quad \begin{cases} 6x - 3y = -6 \\ -x + 3y = -4 \end{cases}$$

Now add the resulting equations and the y-variable is eliminated.

$$
\begin{array}{rl}
6x - 3y = -6 & \\
\underline{-x + 3y = -4} & \\
5x \quad\quad = -10 & \text{Add.} \\
x = -2 & \text{Divide both sides by 5.}
\end{array}
$$

To find the corresponding y-value, let $x = -2$ in any of the preceding equations containing both variables. We use the first equation of the original system.

$$
\begin{array}{rl}
-2x + y = 2 & \text{First equation} \\
-2(-2) + y = 2 & \text{Let } x = -2. \\
4 + y = 2 & \\
y = -2 & \text{Subtract 4 from both sides.}
\end{array}
$$

The solution is $(-2, -2)$. Check this ordered pair in both equations of the original system.

PRACTICE
2   Solve the system: $\begin{cases} x - 2y = 11 \\ 3x - y = 13 \end{cases}$

> **Helpful Hint**
>
> When finding the second value of an ordered pair solution, any equation equivalent to one of the original equations in the system may be used.

In Example 2, the decision to multiply the first equation by -3 was no accident. **To eliminate a variable** when adding two equations, **the coefficient of the variable in one equation must be the opposite of its coefficient in the other equation.**

> **Helpful Hint**
>
> Be sure to multiply *both sides* of an equation by a chosen number when solving by the addition method. A common mistake is to multiply only the side containing the variables.

EXAMPLE 3 Solve the system: $\begin{cases} 2x - y = 7 \\ 8x - 4y = 1 \end{cases}$

Solution Multiply both sides of the first equation by -4 and the resulting coefficient of x is -8, the opposite of 8, the coefficient of x in the second equation. The system becomes

> **Helpful Hint**
>
> Don't forget to multiply **both** sides by -4.

$$\begin{cases} -4(2x - y) = -4(7) \\ 8x - 4y = 1 \end{cases} \quad \text{simplifies to} \quad \begin{cases} -8x + 4y = -28 \\ 8x - 4y = 1 \end{cases}$$

Now add the resulting equations.

$$\begin{array}{r} -8x + 4y = -28 \\ \underline{8x - 4y = 1} \quad \text{Add the equations.} \\ 0 = -27 \quad \text{False} \end{array}$$

When we add the equations, both variables are eliminated and we have $0 = -27$, a false statement. This means that the system has no solution. The graphs of these equations are parallel lines. For the solution, we can write "no solution" or, in set notation, $\{\ \}$ or $\varnothing$ □

PRACTICE
3 Solve the system: $\begin{cases} x - 3y = 5 \\ 2x - 6y = -3 \end{cases}$

EXAMPLE 4 Solve the system: $\begin{cases} 3x - 2y = 2 \\ -9x + 6y = -6 \end{cases}$

Solution First we multiply both sides of the first equation by 3, then we add the resulting equations.

$$\begin{cases} 3(3x - 2y) = 3(2) \\ -9x + 6y = -6 \end{cases} \quad \text{simplifies to} \quad \begin{cases} 9x - 6y = 6 \\ \underline{-9x + 6y = -6} \quad \text{Add the equations.} \\ 0 = 0 \quad \text{True} \end{cases}$$

Both variables are eliminated and we have $0 = 0$, a true statement. Whenever you eliminate a variable and get the equation $0 = 0$, the system has an infinite number of solutions. The graphs of these equations are identical. The solution is "infinite number of solutions" or, in set notation, $\{(x, y) \mid 3x - 2y = 2\}$ or $\{(x, y) \mid -9x + 6y = -6\}$ □

PRACTICE
4 Solve the system: $\begin{cases} 4x - 3y = 5 \\ -8x + 6y = -10 \end{cases}$

✔ **CONCEPT CHECK**

Suppose you are solving the system

$$\begin{cases} 3x + 8y = -5 \\ 2x - 4y = 3. \end{cases}$$

You decide to use the addition method and begin by multiplying both sides of the first equation by -2. In which of the following was the multiplication performed correctly? Explain.

a. $-6x - 16y = -5$ **b.** $-6x - 16y = 10$

EXAMPLE 5   Solve the system: $\begin{cases} 3x + 4y = 13 \\ 5x - 9y = 6 \end{cases}$

Solution We can eliminate the variable y by multiplying the first equation by 9 and the second equation by 4.

$$\begin{cases} 9(3x + 4y) = 9(13) \\ 4(5x - 9y) = 4(6) \end{cases} \text{ simplifies to } \begin{cases} 27x + 36y = 117 \\ \underline{20x - 36y = 24} \\ 47x = 141 \\ x = 3 \end{cases}$$

47x = 141 Add the equations.

 x = 3 Divide both sides by 47.

To find the corresponding y-value, we let $x = 3$ in any equation in this example containing two variables. Doing so in any of these equations will give $y = 1$. The solution of this system is $(3, 1)$. Check to see that $(3, 1)$ satisfies each equation in the original system. $\qquad\square$

PRACTICE
5 Solve the system: $\begin{cases} 4x + 3y = 14 \\ 3x - 2y = 2 \end{cases}$

If we had decided to eliminate x instead of y in Example 5, the first equation could have been multiplied by 5 and the second by -3. Try solving the original system this way to check that the solution is $(3, 1)$.

The following steps summarize how to solve a system of linear equations by the addition method.

Solving a System of Two Linear Equations by the Addition Method

Step 1. Rewrite each equation in standard form $Ax + By = C$.

Step 2. If necessary, multiply one or both equations by a nonzero number so that the coefficients of a chosen variable in the system are opposites.

Step 3. Add the equations.

Step 4. Find the value of one variable by solving the resulting equation from Step 3.

Step 5. Find the value of the second variable by substituting the value found in Step 4 into either of the original equations.

Step 6. Check the proposed solution in the original system.

✔ **CONCEPT CHECK**

Suppose you are solving the system

$$\begin{cases} -4x + 7y = 6 \\ x + 2y = 5 \end{cases}$$

by the addition method.
a. What step(s) should you take if you wish to eliminate x when adding the equations?
b. What step(s) should you take if you wish to eliminate y when adding the equations?

EXAMPLE 6   Solve the system: $\begin{cases} -x - \dfrac{y}{2} = \dfrac{5}{2} \\ \dfrac{x}{6} - \dfrac{y}{2} = 0 \end{cases}$

Answers to Concept Check:
a. multiply the second equation by 4
b. possible answer: multiply the first equation by -2 and the second equation by 7

Solution: We begin by clearing each equation of fractions. To do so, we multiply both sides of the first equation by the LCD, 2, and both sides of the second equation by the LCD, 6. Then the system

(Continued on next page)

$$\begin{cases} 2\left(-x - \dfrac{y}{2}\right) = 2\left(\dfrac{5}{2}\right) \\ 6\left(\dfrac{x}{6} - \dfrac{y}{2}\right) = 6(0) \end{cases} \quad \text{simplifies to} \quad \begin{cases} -2x - y = 5 \\ x - 3y = 0 \end{cases}.$$

We can now eliminate the variable x by multiplying the second equation by 2.

$$\begin{cases} -2x - y = 5 \\ 2(x - 3y) = 2(0) \end{cases} \quad \text{simplifies to} \quad \begin{cases} -2x - y = 5 \\ \underline{2x - 6y = 0} \end{cases}$$

$$-7y = 5 \qquad \text{Add the equations.}$$

$$y = -\dfrac{5}{7} \qquad \text{Solve for } y.$$

To find x, we could replace y with $-\dfrac{5}{7}$ in one of the equations with two variables.

Instead, let's go back to the simplified system and multiply by appropriate factors to eliminate the variable y and solve for x. To do this, we multiply the first equation by -3. Then the system

$$\begin{cases} -3(-2x - y) = -3(5) \\ x - 3y = 0 \end{cases} \quad \text{simplifies to} \quad \begin{cases} 6x + 3y = -15 \\ \underline{x - 3y = \quad 0} \end{cases}$$

$$7x \qquad = -15 \qquad \text{Add the equations.}$$

$$x = -\dfrac{15}{7} \qquad \text{Solve for } x.$$

Check the ordered pair $\left(-\dfrac{15}{7}, -\dfrac{5}{7}\right)$ in both equations of the original system. The solution is $\left(-\dfrac{15}{7}, -\dfrac{5}{7}\right)$.

PRACTICE
6   Solve the system: $\begin{cases} -2x + \dfrac{3y}{2} = 5 \\ -\dfrac{x}{2} - \dfrac{y}{4} = \dfrac{1}{2} \end{cases}$

✓ Vocabulary, Readiness & Video Check

Given the system $\begin{cases} 3x - 2y = -9 \\ x + 5y = 14 \end{cases}$ *read each row (Step 1, Step 2, and Result). Then answer whether the result is true or false.*

	Step 1	Step 2	Result	True or False?
1.	Multiply 2nd equation through by −3.	Add the resulting equation to the 1st equation.	The y's are eliminated.	
2.	Multiply 2nd equation through by −3.	Add the resulting equation to the 1st equation.	The x's are eliminated.	
3.	Multiply 1st equation by 5 and 2nd equation by 2.	Add the two new equations.	The y's are eliminated.	
4.	Multiply 1st equation by 5 and 2nd equation by −2.	Add the two new equations.	The y's are eliminated.	

Martin-Gay Interactive Videos

See Video 4.3 ⚫

Watch the section lecture video and answer the following question.

OBJECTIVE 1

5. For the addition/elimination methods, sometimes you need to multiply an equation through by a nonzero number so that the coefficients of a variable are opposites, as is shown in Example 2. What property allows us to do this? What important reminder is made at this step?

4.3 Exercise Set MyMathLab® ▶

Solve each system of equations by the addition method. See Example 1.

1. $\begin{cases} 3x + y = 5 \\ 6x - y = 4 \end{cases}$

2. $\begin{cases} 4x + y = 13 \\ 2x - y = 5 \end{cases}$

3. $\begin{cases} x - 2y = 8 \\ -x + 5y = -17 \end{cases}$

4. $\begin{cases} x - 2y = -11 \\ -x + 5y = 23 \end{cases}$

Solve each system of equations by the addition method. If a system contains fractions or decimals, you may want to clear each equation of fractions or decimals first. See Examples 2 through 6.

5. $\begin{cases} 3x + y = -11 \\ 6x - 2y = -2 \end{cases}$

6. $\begin{cases} 4x + y = -13 \\ 6x - 3y = -15 \end{cases}$

7. $\begin{cases} 3x + 2y = 11 \\ 5x - 2y = 29 \end{cases}$

8. $\begin{cases} 4x + 2y = 2 \\ 3x - 2y = 12 \end{cases}$

9. $\begin{cases} x + 5y = 18 \\ 3x + 2y = -11 \end{cases}$

10. $\begin{cases} x + 4y = 14 \\ 5x + 3y = 2 \end{cases}$

11. $\begin{cases} x + y = 6 \\ x - y = 6 \end{cases}$

12. $\begin{cases} x - y = 1 \\ -x + 2y = 0 \end{cases}$

13. $\begin{cases} 2x + 3y = 0 \\ 4x + 6y = 3 \end{cases}$

14. $\begin{cases} 3x + y = 4 \\ 9x + 3y = 6 \end{cases}$

15. $\begin{cases} -x + 5y = -1 \\ 3x - 15y = 3 \end{cases}$

16. $\begin{cases} 2x + y = 6 \\ 4x + 2y = 12 \end{cases}$

17. $\begin{cases} 3x - 2y = 7 \\ 5x + 4y = 8 \end{cases}$

18. $\begin{cases} 6x - 5y = 25 \\ 4x + 15y = 13 \end{cases}$

19. $\begin{cases} 8x = -11y - 16 \\ 2x + 3y = -4 \end{cases}$

20. $\begin{cases} 10x + 3y = -12 \\ 5x = -4y - 16 \end{cases}$

21. $\begin{cases} 4x - 3y = 7 \\ 7x + 5y = 2 \end{cases}$

22. $\begin{cases} -2x + 3y = 10 \\ 3x + 4y = 2 \end{cases}$

23. $\begin{cases} 4x - 6y = 8 \\ 6x - 9y = 16 \end{cases}$

24. $\begin{cases} 9x - 3y = 12 \\ 12x - 4y = 18 \end{cases}$

25. $\begin{cases} 2x - 5y = 4 \\ 3x - 2y = 4 \end{cases}$

26. $\begin{cases} 6x - 5y = 7 \\ 4x - 6y = 7 \end{cases}$

27. $\begin{cases} \dfrac{x}{3} + \dfrac{y}{6} = 1 \\ \dfrac{x}{2} - \dfrac{y}{4} = 0 \end{cases}$

28. $\begin{cases} \dfrac{x}{2} + \dfrac{y}{8} = 3 \\ x - \dfrac{y}{4} = 0 \end{cases}$

29. $\begin{cases} \dfrac{10}{3}x + 4y = -4 \\ 5x + 6y = -6 \end{cases}$

30. $\begin{cases} \dfrac{3}{2}x + 4y = 1 \\ 9x + 24y = 5 \end{cases}$

31. $\begin{cases} x - \dfrac{y}{3} = -1 \\ -\dfrac{x}{2} + \dfrac{y}{8} = \dfrac{1}{4} \end{cases}$

32. $\begin{cases} 2x - \dfrac{3y}{4} = -3 \\ x + \dfrac{y}{9} = \dfrac{13}{3} \end{cases}$

33. $\begin{cases} -4(x + 2) = 3y \\ 2x - 2y = 3 \end{cases}$

34. $\begin{cases} -9(x + 3) = 8y \\ 3x - 3y = 8 \end{cases}$

35. $\begin{cases} \dfrac{x}{3} - y = 2 \\ -\dfrac{x}{2} + \dfrac{3y}{2} = -3 \end{cases}$

36. $\begin{cases} \dfrac{x}{2} + \dfrac{y}{4} = 1 \\ -\dfrac{x}{4} - \dfrac{y}{8} = 1 \end{cases}$

37. $\begin{cases} \dfrac{3}{5}x - y = -\dfrac{4}{5} \\ 3x + \dfrac{y}{2} = -\dfrac{9}{5} \end{cases}$

38. $\begin{cases} 3x + \dfrac{7}{2}y = \dfrac{3}{4} \\ -\dfrac{x}{2} + \dfrac{5}{3}y = -\dfrac{5}{4} \end{cases}$

39. $\begin{cases} 3.5x + 2.5y = 17 \\ -1.5x - 7.5y = -33 \end{cases}$

40. $\begin{cases} -2.5x - 6.5y = 47 \\ 0.5x - 4.5y = 37 \end{cases}$

41. $\begin{cases} 0.02x + 0.04y = 0.09 \\ -0.1x + 0.3y = 0.8 \end{cases}$

42. $\begin{cases} 0.04x - 0.05y = 0.105 \\ 0.2x - 0.6y = 1.05 \end{cases}$

MIXED PRACTICE

Solve each system by either the addition method or the substitution method.

43. $\begin{cases} 2x - 3y = -11 \\ y = 4x - 3 \end{cases}$

44. $\begin{cases} 4x - 5y = 6 \\ y = 3x - 10 \end{cases}$

45. $\begin{cases} x + 2y = 1 \\ 3x + 4y = -1 \end{cases}$

46. $\begin{cases} x + 3y = 5 \\ 5x + 6y = -2 \end{cases}$

47. $\begin{cases} 2y = x + 6 \\ 3x - 2y = -6 \end{cases}$

48. $\begin{cases} 3y = x + 14 \\ 2x - 3y = -16 \end{cases}$

49. $\begin{cases} y = 2x - 3 \\ y = 5x - 18 \end{cases}$

50. $\begin{cases} y = 6x - 5 \\ y = 4x - 11 \end{cases}$

51. $\begin{cases} x + \dfrac{1}{6}y = \dfrac{1}{2} \\ 3x + 2y = 3 \end{cases}$

52. $\begin{cases} x + \dfrac{1}{3}y = \dfrac{5}{12} \\ 8x + 3y = 4 \end{cases}$

53. $\begin{cases} \dfrac{x+2}{2} = \dfrac{y+11}{3} \\ \dfrac{x}{2} = \dfrac{2y+16}{6} \end{cases}$

54. $\begin{cases} \dfrac{x+5}{2} = \dfrac{y+14}{4} \\ \dfrac{x}{3} = \dfrac{2y+2}{6} \end{cases}$

55. $\begin{cases} 2x + 3y = 14 \\ 3x - 4y = -69.1 \end{cases}$

56. $\begin{cases} 5x - 2y = -19.8 \\ -3x + 5y = -3.7 \end{cases}$

REVIEW AND PREVIEW

Translating *Rewrite the following sentences using mathematical symbols. Do not solve the equations. See Sections 2.3 and 2.4.*

57. Twice a number, added to 6, is 3 less than the number.

58. The sum of three consecutive integers is 66.

59. Three times a number, subtracted from 20, is 2.

60. Twice the sum of 8 and a number is the difference of the number and 20.

61. The product of 4 and the sum of a number and 6 is twice the number.

62. If the quotient of twice a number and 7 is subtracted from the reciprocal of the number, the result is 2.

CONCEPT EXTENSIONS

Solve. See a Concept Check in this section.

63. To solve this system by the addition method and eliminate the variable y,

$$\begin{cases} 4x + 2y = -7 \\ 3x - y = -12 \end{cases}$$

by what value would you multiply the second equation? What do you get when you complete the multiplication?

64. Given the system of linear equations $\begin{cases} 3x - y = -8 \\ 5x + 3y = 2 \end{cases}$, use the addition method and

   a. Solve the system by eliminating x.

   b. Solve the system by eliminating y.

65. Suppose you are solving the system

$$\begin{cases} 3x + 8y = -5 \\ 2x - 4y = 3. \end{cases}$$

You decide to use the addition method by multiplying both sides of the second equation by 2. In which of the following was the multiplication performed correctly? Explain.
   a. $4x - 8y = 3$
   b. $4x - 8y = 6$

66. Suppose you are solving the system

$$\begin{cases} -2x - y = 0 \\ -2x + 3y = 6. \end{cases}$$

You decide to use the addition method by multiplying both sides of the first equation by 3, then adding the resulting equation to the second equation. Which of the following is the correct sum? Explain.

   a. $-8x = 6$

   b. $-8x = 9$

67. When solving a system of equations by the addition method, how do we know when the system has no solution?

68. Explain why the addition method might be preferred over the substitution method for solving the system $\begin{cases} 2x - 3y = 5 \\ 5x + 2y = 6. \end{cases}$

69. Use the system of linear equations below to answer the questions.

$$\begin{cases} x + y = 5 \\ 3x + 3y = b \end{cases}$$

   a. Find the value of b so that the system has an infinite number of solutions.

   b. Find a value of b so that there are no solutions to the system.

70. Use the system of linear equations below to answer the questions.

$$\begin{cases} x + y = 4 \\ 2x + by = 8 \end{cases}$$

   a. Find the value of b so that the system has an infinite number of solutions.

   b. Find a value of b so that the system has a single solution.

Solve each system by the addition method.

71. $\begin{cases} 1.2x + 3.4y = 27.6 \\ 7.2x - 1.7y = -46.56 \end{cases}$

72. $\begin{cases} 5.1x - 2.4y = 3.15 \\ -15.3x + 1.2y = 27.75 \end{cases}$

Solve.

73. According to the Bureau of Labor Statistics, the percent of workers age 20 to 24 predicted for 2012 to 2022 can be approximated by $0.36x + y = 70.9$. The percent of workers age 55 to 64 years for the same years can be approximated by $0.3x - y = -64.5$. For both equations, x is the number of years since 2012, and y is the percent of the workforce. (*Source:* Bureau of Labor Statistics)

   a. Use the addition method to solve the system: $\begin{cases} 0.36x + y = 70.9 \\ 0.30x - y = -64.5 \end{cases}$

   (Eliminate y first and solve for x. Round this result to the nearest whole. Then find y and round to the nearest whole.)

   b. Use your result from part (a) and estimate the year in which both percents are the same.

   c. Use your results from parts (a) and (b) to estimate the percent of workers age 20–24 and those age 55–64 in the year that we found in part (b).

74. Two occupations predicted to greatly increase in the number of jobs are construction workers and personal care aides. The number of construction workers predicted for 2012 through 2022 can be approximated by $32.5x - y = -1284$. The number of personal care aides predicted for 2012 through 2022 can be approximated by $58x - y = -1191$. For both equations, x is the number of years since 2012, and y is the number of jobs (in thousands). (*Source:* Bureau of Labor Statistics)

a. Use the addition method to solve this system of equations:

$$\begin{cases} 32.5x - y = -1284 \\ 58x - y = -1191 \end{cases}$$

(Eliminate y first and solve for x. Round this result to the nearest whole.)

b. Interpret the solution from part (a).

c. Using the year in your answer to part (b), estimate the number of construction workers or personal care aides in that year.

Integrated Review   Solving Systems of Equations

Sections 4.1–4.3

Solve each system by either the addition method or the substitution method.

1. $\begin{cases} 2x - 3y = -11 \\ y = 4x - 3 \end{cases}$

2. $\begin{cases} 4x - 5y = 6 \\ y = 3x - 10 \end{cases}$

3. $\begin{cases} x + y = 3 \\ x - y = 7 \end{cases}$

4. $\begin{cases} x - y = 20 \\ x + y = -8 \end{cases}$

5. $\begin{cases} x + 2y = 1 \\ 3x + 4y = -1 \end{cases}$

6. $\begin{cases} x + 3y = 5 \\ 5x + 6y = -2 \end{cases}$

7. $\begin{cases} y = x + 3 \\ 3x - 2y = -6 \end{cases}$

8. $\begin{cases} y = -2x \\ 2x - 3y = -16 \end{cases}$

9. $\begin{cases} y = 2x - 3 \\ y = 5x - 18 \end{cases}$

10. $\begin{cases} y = 6x - 5 \\ y = 4x - 11 \end{cases}$

11. $\begin{cases} x + \dfrac{1}{6}y = \dfrac{1}{2} \\ 3x + 2y = 3 \end{cases}$

12. $\begin{cases} x + \dfrac{1}{3}y = \dfrac{5}{12} \\ 8x + 3y = 4 \end{cases}$

13. $\begin{cases} x - 5y = 1 \\ -2x + 10y = 3 \end{cases}$

14. $\begin{cases} -x + 2y = 3 \\ 3x - 6y = -9 \end{cases}$

15. $\begin{cases} 0.2x - 0.3y = -0.95 \\ 0.4x + 0.1y = 0.55 \end{cases}$

16. $\begin{cases} 0.08x - 0.04y = -0.11 \\ 0.02x - 0.06y = -0.09 \end{cases}$

17. $\begin{cases} x = 3y - 7 \\ 2x - 6y = -14 \end{cases}$

18. $\begin{cases} y = \dfrac{x}{2} - 3 \\ 2x - 4y = 0 \end{cases}$

19. $\begin{cases} 2x + 5y = -1 \\ 3x - 4y = 33 \end{cases}$

20. $\begin{cases} 7x - 3y = 2 \\ 6x + 5y = -21 \end{cases}$

21. Which method, substitution or addition, would you prefer to use to solve the system below? Explain your reasoning.

$$\begin{cases} 3x + 2y = -2 \\ y = -2x \end{cases}$$

22. Which method, substitution or addition, would you prefer to use to solve the system below? Explain your reasoning.

$$\begin{cases} 3x - 2y = -3 \\ 6x + 2y = 12 \end{cases}$$

4.4 Solving Systems of Linear Equations in Three Variables

OBJECTIVE

1 Solve a System of Three Linear Equations in Three Variables.

In this section, the algebraic methods of solving systems of two linear equations in two variables are extended to systems of three linear equations in three variables. We call the equation $3x - y + z = -15$, for example, a **linear equation in three variables** since there are three variables and each variable is raised only to the power 1. A solution of this equation is an **ordered triple (x, y, z)** that makes the equation a true statement. For example, the ordered triple $(2, 0, -21)$ is a solution of $3x - y + z = -15$ since replacing x with 2, y with 0, and z with -21 yields the true statement $3(2) - 0 + (-21) = -15$. The graph of this equation is a plane in three-dimensional space, just as the graph of a linear equation in two variables is a line in two-dimensional space.

Although we will not discuss the techniques for graphing equations in three variables, visualizing the possible patterns of intersecting planes gives us insight into the possible patterns of solutions of a system of three three-variable linear equations. There are four possible patterns.

1. Three planes have a single point in common. This point represents the single solution of the system. This system is **consistent.**

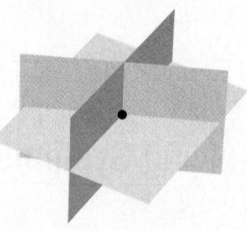

2. Three planes intersect at no point common to all three. This system has no solution. A few ways that this can occur are shown. This system is **inconsistent.**

3. Three planes intersect at all the points of a single line. The system has infinitely many solutions. This system is **consistent.**

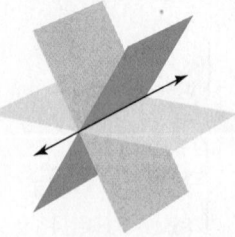

4. Three planes coincide at all points on the plane. The system is consistent, and the equations are **dependent.**

OBJECTIVE

1 Solving a System of Three Linear Equations in Three Variables

Just as with systems of two equations in two variables, we can use the elimination or substitution method to solve a system of three equations in three variables. To use the elimination method, we eliminate a variable and obtain a system of two equations in two variables. Then we use the methods we learned in the previous section to solve the system of two equations.

EXAMPLE 1   Solve the system:

$$\begin{cases} 3x - y + z = -15 & \text{Equation (1)} \\ x + 2y - z = 1 & \text{Equation (2)} \\ 2x + 3y - 2z = 0 & \text{Equation (3)} \end{cases}$$

Solution Add equations (1) and (2) to eliminate z.

$$\begin{array}{r} 3x - y + z = -15 \\ x + 2y - z = 1 \\ \hline 4x + y \quad\quad = -14 \end{array} \quad \text{Equation (4)}$$

Next, add two *other* equations and *eliminate z again*. To do so, multiply both sides of equation (1) by 2 and add this resulting equation to equation (3). Then

$$\begin{cases} 2(3x - y + z) = 2(-15) \\ 2x + 3y - 2z = 0 \end{cases} \text{ simplifies to } \begin{cases} 6x - 2y + 2z = -30 \\ 2x + 3y - 2z = 0 \\ \hline 8x + y \quad\quad = -30 \end{cases} \text{ Equation (5)}$$

> **Helpful Hint**
> Don't forget to add two other equations besides equations (1) and (2) *and* to **eliminate the same variable.**

Now solve equations (4) and (5) for x and y. To solve by elimination, multiply both sides of equation (4) by -1 and add this resulting equation to equation (5). Then

$$\begin{cases} -1(4x + y) = -1(-14) \\ 8x + y = -30 \end{cases} \text{ simplifies to } \begin{cases} -4x - y = 14 \\ 8x + y = -30 \\ \hline 4x \quad\quad = -16 \quad \text{Add the equations.} \\ x = -4 \quad \text{Solve for } x. \end{cases}$$

Replace x with -4 in equation (4) or (5).

$$\begin{array}{ll} 4x + y = -14 & \text{Equation (4)} \\ 4(-4) + y = -14 & \text{Let } x = -4. \\ y = 2 & \text{Solve for } y. \end{array}$$

Finally, replace x with -4 and y with 2 in equation (1), (2), or (3).

$$\begin{array}{ll} x + 2y - z = 1 & \text{Equation (2)} \\ -4 + 2(2) - z = 1 & \text{Let } x = -4 \text{ and } y = 2. \\ -4 + 4 - z = 1 & \\ -z = 1 & \\ z = -1 & \end{array}$$

The solution is $(-4, 2, -1)$. To check, let $x = -4$, $y = 2$, and $z = -1$ in all three original equations of the system.

Equation (1)	**Equation (2)**	**Equation (3)**
$3x - y + z = -15$	$x + 2y - z = 1$	$2x + 3y - 2z = 0$
$3(-4) - 2 + (-1) \stackrel{?}{=} -15$	$-4 + 2(2) - (-1) \stackrel{?}{=} 1$	$2(-4) + 3(2) - 2(-1) \stackrel{?}{=} 0$
$-12 - 2 - 1 \stackrel{?}{=} -15$	$-4 + 4 + 1 \stackrel{?}{=} 1$	$-8 + 6 + 2 \stackrel{?}{=} 0$
$-15 = -15$ True	$1 = 1$ True	$0 = 0$ True

(Continued on next page)

All three statements are true, so the solution is $(-4, 2, -1)$.

PRACTICE
1 Solve the system: $\begin{cases} 3x + 2y - z = 0 \\ x - y + 5z = 2 \\ 2x + 3y + 3z = 7 \end{cases}$

EXAMPLE 2 Solve the system:

$$\begin{cases} 2x - 4y + 8z = 2 & (1) \\ -x - 3y + z = 11 & (2) \\ x - 2y + 4z = 0 & (3) \end{cases}$$

Solution Add equations (2) and (3) to eliminate x, and the new equation is

$$-5y + 5z = 11 \quad (4)$$

To eliminate x again, multiply both sides of equation (2) by 2 and add the resulting equation to equation (1). Then

$$\begin{cases} 2x - 4y + 8z = 2 \\ 2(-x - 3y + z) = 2(11) \end{cases} \quad \begin{array}{l} \text{simplifies} \\ \text{to} \end{array} \quad \begin{cases} 2x - 4y + 8z = 2 \\ \underline{-2x - 6y + 2z = 22} \\ \qquad -10y + 10z = 24 \quad (5) \end{cases}$$

Next, solve for y and z using equations (4) and (5). Multiply both sides of equation (4) by -2 and add the resulting equation to equation (5).

$$\begin{cases} -2(-5y + 5z) = -2(11) \\ -10y + 10z = 24 \end{cases} \quad \begin{array}{l} \text{simplifies} \\ \text{to} \end{array} \quad \begin{cases} 10y - 10z = -22 \\ \underline{-10y + 10z = 24} \\ \qquad\qquad 0 = 2 \quad \text{False} \end{cases}$$

Since the statement is false, this system is inconsistent and has no solution. The solution set is the empty set $\{\ \}$ or $\varnothing$.

PRACTICE
2 Solve the system: $\begin{cases} 6x - 3y + 12z = 4 \\ -6x + 4y - 2z = 7 \\ -2x + y - 4z = 3 \end{cases}$

The elimination method is summarized next.

Solving a System of Three Linear Equations by the Elimination Method

Step 1. Write each equation in standard form $Ax + By + Cz = D$.

Step 2. Choose a pair of equations and use the equations to eliminate a variable.

Step 3. Choose any **other** pair of equations and eliminate the **same variable** as in Step 2.

Step 4. Two equations in two variables should be obtained from Step 2 and Step 3. Use methods from Sections 4.1 through 4.3 to solve this system for both variables.

Helpful Hint
Make sure you read closely and follow Step 3.

Step 5. To solve for the third variable, substitute the values of the variables found in Step 4 into any of the original equations containing the third variable.

Step 6. Check the ordered triple solution in _all three_ original equations.

✔ **CONCEPT CHECK**

In the system

$$\begin{cases} x + y + z = 6 & \text{Equation (1)} \\ 2x - y + z = 3 & \text{Equation (2)} \\ x + 2y + 3z = 14 & \text{Equation (3)} \end{cases}$$

equations (1) and (2) are used to eliminate y. Which action could best be used to finish solving? Why?

a. Use (1) and (2) to eliminate z. **b.** Use (2) and (3) to eliminate y.

c. Use (1) and (3) to eliminate x.

EXAMPLE 3 Solve the system:

$$\begin{cases} 2x + 4y = 1 & (1) \\ 4x - 4z = -1 & (2) \\ y - 4z = -3 & (3) \end{cases}$$

Solution Notice that equation (2) has no term containing the variable y. Let us eliminate y using equations (1) and (3). Multiply both sides of equation (3) by -4 and add the resulting equation to equation (1). Then

$$\begin{cases} 2x + 4y = 1 \\ -4(y - 4z) = -4(-3) \end{cases} \quad \text{simplifies to} \quad \begin{cases} 2x + 4y = 1 \\ \underline{-4y + 16z = 12} \\ 2x + 16z = 13 \quad (4) \end{cases}$$

Next, solve for z using equations (4) and (2). Multiply both sides of equation (4) by -2 and add the resulting equation to equation (2).

$$\begin{cases} -2(2x + 16z) = -2(13) \\ 4x - 4z = -1 \end{cases} \quad \text{simplifies to} \quad \begin{cases} -4x - 32z = -26 \\ \underline{4x - 4z = -1} \\ -36z = -27 \\ z = \dfrac{3}{4} \end{cases}$$

Replace z with $\dfrac{3}{4}$ in equation (3) and solve for y.

$$y - 4\left(\frac{3}{4}\right) = -3 \quad \text{Let } z = \frac{3}{4} \text{ in equation (3).}$$

$$y - 3 = -3$$

$$y = 0$$

Replace y with 0 in equation (1) and solve for x.

$$2x + 4(0) = 1$$

$$2x = 1$$

$$x = \frac{1}{2}$$

The solution is $\left(\dfrac{1}{2}, 0, \dfrac{3}{4}\right)$. Check to see that this solution satisfies all three equations of the system. □

PRACTICE

3 Solve the system: $\begin{cases} 3x + 4y = 0 \\ 9x - 4z = 6 \\ -2y + 7z = 1 \end{cases}$

EXAMPLE 4 Solve the system:

$$\begin{cases} x - 5y - 2z = 6 & (1) \\ -2x + 10y + 4z = -12 & (2) \\ \dfrac{1}{2}x - \dfrac{5}{2}y - z = 3 & (3) \end{cases}$$

Solution Multiply both sides of equation (3) by 2 to eliminate fractions and multiply both sides of equation (2) by $-\dfrac{1}{2}$ so that the coefficient of x is 1. The resulting system is then

$$\begin{cases} x - 5y - 2z = 6 & (1) \\ x - 5y - 2z = 6 & \text{Multiply (2) by } -\dfrac{1}{2}. \\ x - 5y - 2z = 6 & \text{Multiply (3) by 2.} \end{cases}$$

All three equations are identical, and therefore equations (1), (2), and (3) are all equivalent. There are infinitely many solutions of this system. The equations are dependent. The solution set can be written as $\{(x, y, z) \mid x - 5y - 2z = 6\}$. □

PRACTICE
4 Solve the system:
$$\begin{cases} 2x + y - 3z = 6 \\ x + \dfrac{1}{2}y - \dfrac{3}{2}z = 3 \\ -4x - 2y + 6z = -12 \end{cases}$$

As mentioned earlier, we can also use the substitution method to solve a system of linear equations in three variables.

EXAMPLE 5 Solve the system:

$$\begin{cases} x - 4y - 5z = 35 & (1) \\ x - 3y = 0 & (2) \\ -y + z = -55 & (3) \end{cases}$$

Solution Notice in equations (2) and (3) that a variable is missing. Also notice that both equations contain the variable y. Let's use the substitution method by solving equation (2) for x and equation (3) for z and substituting the results in equation (1).

$$x - 3y = 0 \qquad (2)$$
$$x = 3y \qquad \text{Solve equation (2) for } x.$$
$$-y + z = -55 \qquad (3)$$
$$z = y - 55 \qquad \text{Solve equation (3) for } z.$$

Now substitute $3y$ for x and $y - 55$ for z in equation (1).

Helpful Hint
Do not forget to distribute.

$$x - 4y - 5z = 35 \qquad (1)$$
$$3y - 4y - 5(y - 55) = 35 \qquad \text{Let } x = 3y \text{ and } z = y - 55.$$
$$3y - 4y - 5y + 275 = 35 \qquad \text{Use the distributive law and multiply.}$$
$$-6y + 275 = 35 \qquad \text{Combine like terms.}$$
$$-6y = -240 \qquad \text{Subtract 275 from both sides.}$$
$$y = 40 \qquad \text{Solve.}$$

To find x, recall that $x = 3y$ and substitute 40 for y. Then $x = 3y$ becomes $x = 3 \cdot 40 = 120$. To find z, recall that $z = y - 55$ and substitute 40 for y, also. Then $z = y - 55$ becomes $z = 40 - 55 = -15$. The solution is $(120, 40, -15)$. □

PRACTICE
5 Solve the system:
$$\begin{cases} x + 2y + 4z = 16 \\ x + 2z = -4 \\ y - 3z = 30 \end{cases}$$

Vocabulary, Readiness & Video Check

Solve.

1. Choose the equation(s) that has $(-1, 3, 1)$ as a solution.
   a. $x + y + z = 3$ **b.** $-x + y + z = 5$ **c.** $-x + y + 2z = 0$ **d.** $x + 2y - 3z = 2$

2. Choose the equation(s) that has $(2, 1, -4)$ as a solution.
   a. $x + y + z = -1$ **b.** $x - y - z = -3$ **c.** $2x - y + z = -1$ **d.** $-x - 3y - z = -1$

3. Use the result of Exercise 1 to determine whether $(-1, 3, 1)$ is a solution of the system below. Explain your answer.
$$\begin{cases} x + y + z = 3 \\ -x + y + z = 5 \\ x + 2y - 3z = 2 \end{cases}$$

4. Use the result of Exercise 2 to determine whether $(2, 1, -4)$ is a solution of the system below. Explain your answer.
$$\begin{cases} x + y + z = -1 \\ x - y - z = -3 \\ 2x - y + z = -1 \end{cases}$$

Martin-Gay Interactive Videos

Watch the section lecture video and answer the following question.

OBJECTIVE
1

5. From ▦ Example 1 and the lecture before, why does Step 3 stress that the same variable be eliminated from two other equations?

See Video 4.4 ◉

4.4   **Exercise Set**   MyMathLab® ▸

Solve each system. See Examples 1 through 5.

1. $\begin{cases} x - y + z = -4 \\ 3x + 2y - z = 5 \\ -2x + 3y - z = 15 \end{cases}$

2. $\begin{cases} x + y - z = -1 \\ -4x - y + 2z = -7 \\ 2x - 2y - 5z = 7 \end{cases}$

9. $\begin{cases} 4x - y + 2z = 5 \\ 2y + z = 4 \\ 4x + y + 3z = 10 \end{cases}$

10. $\begin{cases} 5y - 7z = 14 \\ 2x + y + 4z = 10 \\ 2x + 6y - 3z = 30 \end{cases}$

3. $\begin{cases} x + y = 3 \\ 2y = 10 \\ 3x + 2y - 3z = 1 \end{cases}$

4. $\begin{cases} 5x = 5 \\ 2x + y = 4 \\ 3x + y - 4z = -15 \end{cases}$

11. $\begin{cases} x + 5z = 0 \\ 5x + y = 0 \\ y - 3z = 0 \end{cases}$

12. $\begin{cases} x - 5y = 0 \\ x - z = 0 \\ -x + 5z = 0 \end{cases}$

5. $\begin{cases} 2x + 2y + z = 1 \\ -x + y + 2z = 3 \\ x + 2y + 4z = 0 \end{cases}$

6. $\begin{cases} 2x - 3y + z = 5 \\ x + y + z = 0 \\ 4x + 2y + 4z = 4 \end{cases}$

13. $\begin{cases} 6x - 5z = 17 \\ 5x - y + 3z = -1 \\ 2x + y = -41 \end{cases}$

14. $\begin{cases} x + 2y = 6 \\ 7x + 3y + z = -33 \\ x - z = 16 \end{cases}$

7. $\begin{cases} x - 2y + z = -5 \\ -3x + 6y - 3z = 15 \\ 2x - 4y + 2z = -10 \end{cases}$

8. $\begin{cases} 3x + y - 2z = 2 \\ -6x - 2y + 4z = -4 \\ 9x + 3y - 6z = 6 \end{cases}$

15. $\begin{cases} x + y + z = 8 \\ 2x - y - z = 10 \\ x - 2y - 3z = 22 \end{cases}$

16. $\begin{cases} 5x + y + 3z = 1 \\ x - y + 3z = -7 \\ -x + y = 1 \end{cases}$

17. $\begin{cases} x + 2y - z = 5 \\ 6x + y + z = 7 \\ 2x + 4y - 2z = 5 \end{cases}$
18. $\begin{cases} 4x - y + 3z = 10 \\ x + y - z = 5 \\ 8x - 2y + 6z = 10 \end{cases}$

19. $\begin{cases} 2x - 3y + z = 2 \\ x - 5y + 5z = 3 \\ 3x + y - 3z = 5 \end{cases}$
20. $\begin{cases} 4x + y - z = 8 \\ x - y + 2z = 3 \\ 3x - y + z = 6 \end{cases}$

21. $\begin{cases} -2x - 4y + 6z = -8 \\ x + 2y - 3z = 4 \\ 4x + 8y - 12z = 16 \end{cases}$
22. $\begin{cases} -6x + 12y + 3z = -6 \\ 2x - 4y - z = 2 \\ -x + 2y + \frac{z}{2} = -1 \end{cases}$

23. $\begin{cases} 2x + 2y - 3z = 1 \\ y + 2z = -14 \\ 3x - 2y = -1 \end{cases}$
24. $\begin{cases} 7x + 4y = 10 \\ x - 4y + 2z = 6 \\ y - 2z = -1 \end{cases}$

25. $\begin{cases} x + 2y - z = 5 \\ -3x - 2y - 3z = 11 \\ 4x + 4y + 5z = -18 \end{cases}$
26. $\begin{cases} 3x - 3y + z = -1 \\ 3x - y - z = 3 \\ -6x + 4y + 3z = -8 \end{cases}$

27. $\begin{cases} \frac{3}{4}x - \frac{1}{3}y + \frac{1}{2}z = 9 \\ \frac{1}{6}x + \frac{1}{3}y - \frac{1}{2}z = 2 \\ \frac{1}{2}x - y + \frac{1}{2}z = 2 \end{cases}$
28. $\begin{cases} \frac{1}{3}x - \frac{1}{4}y + z = -9 \\ \frac{1}{2}x - \frac{1}{3}y - \frac{1}{4}z = -6 \\ x - \frac{1}{2}y - z = -8 \end{cases}$

REVIEW AND PREVIEW

Translating Solve. See Section 2.4.

29. The sum of two numbers is 45 and one number is twice the other. Find the numbers.

30. The difference of two numbers is 5. Twice the smaller number added to five times the larger number is 53. Find the numbers.

Solve. See Section 2.3.

31. $2(x - 1) - 3x = x - 12$

32. $7(2x - 1) + 4 = 11(3x - 2)$

33. $-y - 5(y + 5) = 3y - 10$

34. $z - 3(z + 7) = 6(2z + 1)$

CONCEPT EXTENSIONS

35. Write a single linear equation in three variables that has $(-1, 2, -4)$ as a solution. (There are many possibilities.) Explain the process you used to write an equation.

36. Write a system of three linear equations in three variables that has $(2, 1, 5)$ as a solution. (There are many possibilities.) Explain the process you used to write an equation.

37. Write a system of linear equations in three variables that has the solution $(-1, 2, -4)$. Explain the process you used to write your system.

38. When solving a system of three equations in three unknowns, explain how to determine that a system has no solution.

39. The fraction $\frac{1}{24}$ can be written as the following sum:

$$\frac{1}{24} = \frac{x}{8} + \frac{y}{4} + \frac{z}{3}$$

where the numbers x, y, and z are solutions of

$$\begin{cases} x + y + z = 1 \\ 2x - y + z = 0 \\ -x + 2y + 2z = -1 \end{cases}$$

Solve the system and see that the sum of the fractions is $\frac{1}{24}$.

40. The fraction $\frac{1}{18}$ can be written as the following sum:

$$\frac{1}{18} = \frac{x}{2} + \frac{y}{3} + \frac{z}{9}$$

where the numbers x, y, and z are solutions of

$$\begin{cases} x + 3y + z = -3 \\ -x + y + 2z = -14 \\ 3x + 2y - z = 12 \end{cases}$$

Solve the system and see that the sum of the fractions is $\frac{1}{18}$.

Solving systems involving more than three variables can be accomplished with methods similar to those encountered in this section. Apply what you already know to solve each system of equations in four variables.

41. $\begin{cases} x + y - w = 0 \\ y + 2z + w = 3 \\ x - z = 1 \\ 2x - y - w = -1 \end{cases}$

42. $\begin{cases} 5x + 4y = 29 \\ y + z - w = -2 \\ 5x + z = 23 \\ y - z + w = 4 \end{cases}$

43. $\begin{cases} x + y + z + w = 5 \\ 2x + y + z + w = 6 \\ x + y + z = 2 \\ x + y = 0 \end{cases}$

44. $\begin{cases} 2x - z = -1 \\ y + z + w = 9 \\ y - 2w = -6 \\ x + y = 3 \end{cases}$

45. Write a system of three linear equations in three variables that are dependent equations.

46. How many solutions are there to the system in Exercise 45?

4.5 | Systems of Linear Equations and Problem Solving

OBJECTIVES

1 Solve Problems That Can Be Modeled by a System of Two Linear Equations.

2 Solve Problems with Cost and Revenue Functions.

3 Solve Problems That Can Be Modeled by a System of Three Linear Equations.

OBJECTIVE

1   Solving Problems Modeled by Systems of Two Equations

Thus far, we have solved problems by writing one-variable equations and solving for the variable. Some of these problems can be solved, perhaps more easily, by writing a system of equations, as illustrated in this section.

> **EXAMPLE 1**   **Predicting Equal Consumption of Red Meat and Poultry**
>
> America's consumption of red meat has decreased most years since 2005, while consumption of poultry has decreased also, but by a much smaller amount. The function $y = -2.03x + 119.05$ approximates the annual pounds of red meat consumed per capita, where x is the number of years since 2005. The function $y = -0.41x + 103.21$ approximates the annual pounds of poultry consumed per capita, where x is also the number of years since 2005. Based on this trend, determine the year when the annual consumption of red meat and poultry is equal. (*Source:* USDA: Economic Research Service)

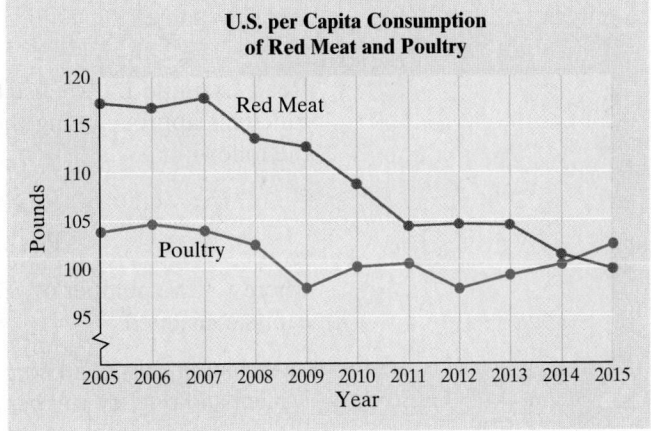

Solution

1. UNDERSTAND. Read and reread the problem and guess a year. Without using the graph for assistance, guess the year 2025. This year is 20 years since 2005, so $x = 20$. Now let $x = 20$ in each given function.

   Red meat: $y = -2.03x + 119.05 = -2.03(20) + 119.05 = 78.45$ pounds

   Poultry: $y = -0.41x + 103.21 = -0.41(20) + 103.21 = 95.01$ pounds

   Since the projected pounds in 2025 for red meat and poultry are not the same, we guessed incorrectly, but we do have a better understanding of the problem. We know that the year is earlier than 2025. Why? Because consumption of red meat for 2025 is less than consumption of poultry.

2. TRANSLATE. We are already given the system of equations.

3. SOLVE. We want to know the year x in which pounds y are the same, so we solve the system

$$\begin{cases} y = -2.03x + 119.05 \\ y = -0.41x + 103.21 \end{cases}$$

   Since both equations are solved for y, one way to solve is to use the substitution method.

(Continued on next page)

$$y = -0.41x + 103.21 \quad \text{Second equation}$$

$$-2.03x + 119.05 = -0.41x + 103.21 \quad \text{Let } y = -2.03x + 119.05.$$

$$-1.62x = -15.84$$

$$x = \frac{-15.84}{-1.62} \approx 9.78$$

4. INTERPRET. Since we are only asked to find the year, we need only solve for x.

Check: To check, see whether $x \approx 9.78$ gives approximately the same number of pounds of red meat and poultry.

Red meat: $y = -2.03x + 119.05 = -2.03(9.78) + 119.05 = 99.1966$ pounds
Poultry: $y = -0.41x + 103.21 = -0.41(9.78) + 103.21 = 99.2002$ pounds

Since we rounded the number of years, the numbers of pounds do differ slightly. They differ only by 0.0036, so we can assume we solved correctly.

State: The consumption of red meat and poultry will be the same about 9.78 years after 2005, or 2014.78. Thus, in the year 2014, we calculate the consumption is the same. □

PRACTICE

1 Read Example 1. If we use the years 2012, 2013, 2014, and 2015 only to write functions approximating the consumption of red meat and poultry, we have the following:

$$\text{Red meat: } y = -0.91x + 110.45$$
$$\text{Poultry: } \quad y = 0.11x + 98.6$$

where x is the number of years since 2005 and y is the pounds per year per capita consumed.

a. Assuming this trend continues, predict the year when consumption of red meat and poultry will be the same. Round to the nearest year.

b. Does your answer differ from the answer to Example 1? Why or why not?

For Example 1, the equations in the system were given to us. Let's now practice writing our own system of equations that we will use to solve an application.

Many of the applications solved earlier using one-variable equations can also be solved using two equations in **two** variables. We use the same problem-solving steps that have been used throughout this text. The only difference is that two variables are assigned to represent the two unknown quantities and that the stated problem is translated into **two** equations.

Problem-Solving Steps

Step 1. UNDERSTAND the problem. During this step, become comfortable with the problem. Some ways of doing this are to

> Read and reread the problem.
>
> Choose two variables to represent the two unknowns.
>
> Construct a drawing if possible.
>
> Propose a solution and check. Pay careful attention to how you check your proposed solution. This will help when writing equations to model the problem.

Step 2. TRANSLATE the problem into two equations.

Step 3. SOLVE the system of equations.

Step 4. INTERPRET the results: **Check** the proposed solution in the stated problem and **state** your conclusion.

EXAMPLE 2   Finding Unknown Numbers

Find two numbers whose sum is 37 and whose difference is 21.

Solution

1. UNDERSTAND. Read and reread the problem. Suppose that one number is 20. If their sum is 37, the other number is 17 because $20 + 17 = 37$. Is their difference 21? No; $20 - 17 = 3$. Our proposed solution is incorrect, but we now have a better understanding of the problem.

    Since we are looking for two numbers, we let

$$x = \text{first number}$$
$$y = \text{second number}$$

2. TRANSLATE. Since we have assigned two variables to this problem, we translate our problem into two equations.

In words: two numbers whose sum   is   37
                       ↓   ↓   ↓
Translate: $x + y$ $=$ 37

In words: two numbers whose difference   is   21
                       ↓   ↓   ↓
Translate: $x - y$ $=$ 21

3. SOLVE. Now we solve the system

$$\begin{cases} x + y = 37 \\ x - y = 21 \end{cases}$$

    Notice that the coefficients of the variable y are opposites. Let's then solve by the addition method and begin by adding the equations.

$$\begin{array}{rl} x + y = 37 & \\ \underline{x - y = 21} & \\ 2x \phantom{{}+ y} = 58 & \text{Add the equations.} \\ x = \dfrac{58}{2} = 29 & \text{Divide both sides by 2.} \end{array}$$

    Now we let $x = 29$ in the first equation to find y.

$$\begin{array}{rl} x + y = 37 & \text{First equation} \\ 29 + y = 37 & \\ y = 8 & \text{Subtract 29 from both sides.} \end{array}$$

4. INTERPRET. The solution of the system is $(29, 8)$.

Check: Notice that the sum of 29 and 8 is $29 + 8 = 37$, the required sum. Their difference is $29 - 8 = 21$, the required difference.

State: The numbers are 29 and 8.   □

PRACTICE
2   Find two numbers whose sum is 30 and whose difference is 6.   ■

EXAMPLE 3   Finding Unknown Numbers

A first number is 4 less than a second number. Four times the first number is 6 more than twice the second. Find the numbers.

(Continued on next page)

Solution

1. UNDERSTAND. Read and reread the problem and guess a solution. If a first number is 10 and this is 4 less than a second number, the second number is 14. Four times the first number is 4(10), or 40. This is not equal to 6 more than twice the second number, which is 2(14) + 6 or 34. Although we guessed incorrectly, we now have a better understanding of the problem.

    Since we are looking for two numbers, we will let

    $$x = \text{first number}$$
    $$y = \text{second number}$$

2. TRANSLATE. Since we have assigned two variables to this problem, we will translate the given facts into two equations. For the first statement we have

    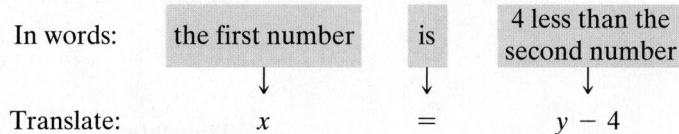

In words:	the first number	is	4 less than the second number
Translate:	x	$=$	$y - 4$

    Next we translate the second statement into an equation.

    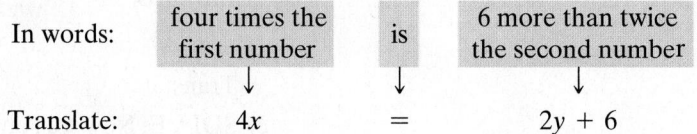

In words:	four times the first number	is	6 more than twice the second number
Translate:	$4x$	$=$	$2y + 6$

3. SOLVE. Here we solve the system

    $$\begin{cases} x = y - 4 \\ 4x = 2y + 6 \end{cases}$$

    Since the first equation expresses x in terms of y, we will use substitution. We substitute $y - 4$ for x in the second equation and solve for y.

    $$4x = 2y + 6 \quad \text{Second equation}$$
    $$4(y - 4) = 2y + 6 \quad \text{Let } x = y - 4.$$
    $$4y - 16 = 2y + 6$$
    $$2y = 22$$
    $$y = 11$$

    Now we replace y with 11 in the equation $x = y - 4$ and solve for x. Then $x = y - 4$ becomes $x = 11 - 4 = 7$. The ordered pair solution of the system is (7, 11).

4. INTERPRET. Since the solution of the system is (7, 11), then the first number we are looking for is 7 and the second number is 11.

    Check: Notice that 7 _is_ 4 less than 11, and 4 times 7 _is_ 6 more than twice 11. The proposed numbers, 7 and 11, are correct.

    State: The numbers are 7 and 11. □

PRACTICE

3 A first number is 5 more than a second number. Twice the first number is 2 less than 3 times the second number. Find the numbers.

EXAMPLE 4   Solving a Problem about Prices

The Cirque du Soleil show Varekai is performing locally. Matinee admission for 4 adults and 2 children is $374, while admission for 2 adults and 3 children is $285.

a. What is the price of an adult's ticket?

b. What is the price of a child's ticket?

c. Suppose that a special rate of $1000 is offered for groups of 20 persons. Should a group of 4 adults and 16 children use the group rate? Why or why not?

Solution

1. UNDERSTAND. Read and reread the problem and guess a solution. Let's suppose that the price of an adult's ticket is $50 and the price of a child's ticket is $40. To check our proposed solution, let's see if admission for 4 adults and 2 children is $374. Admission for 4 adults is 4($50) or $200 and admission for 2 children is 2($40) or $80. This gives a total admission of $200 + $80 = $280, not the required $374. Again, though, we have accomplished the purpose of this process: We have a better understanding of the problem. To continue, we let

$$A = \text{the price of an adult's ticket}$$
$$C = \text{the price of a child's ticket}$$

2. TRANSLATE. We translate the problem into two equations using both variables.

In words:	admission for 4 adults	and	admission for 2 children	is	$374
	↓	↓	↓	↓	↓
Translate:	$4A$	$+$	$2C$	$=$	374

In words:	admission for 2 adults	and	admission for 3 children	is	$285
	↓	↓	↓	↓	↓
Translate:	$2A$	$+$	$3C$	$=$	285

3. SOLVE. We solve the system.

$$\begin{cases} 4A + 2C = 374 \\ 2A + 3C = 285 \end{cases}$$

Since both equations are written in standard form, we solve by the addition method. First we multiply the second equation by -2 so that when we add the equations, we eliminate the variable A. Then the system

$$\begin{cases} 4A + 2C = 374 \\ -2(2A + 3C) = -2(285) \end{cases} \quad \text{simplifies to} \quad \begin{cases} 4A + 2C = 374 \\ -4A - 6C = -570 \end{cases}$$

Add the equations.

$$ -4C = -196$$
$$ C = 49 \text{ or } \$49, \text{ the child's ticket price}$$

(Continued on next page)

To find A, we replace C with 49 in the first equation.

$$4A + 2C = 374 \quad \text{First equation}$$
$$4A + 2(49) = 374 \quad \text{Let } C = 49$$
$$4A + 98 = 374$$
$$4A = 276$$
$$A = 69 \text{ or } \$69, \text{ the adult's ticket price}$$

4. INTERPRET.

Check: Notice that 4 adults and 2 children will pay

$4(\$69) + 2(\$49) = \$276 + \$98 = \$374$, the required amount. Also, the price for 2 adults and 3 children is $2(\$69) + 3(\$49) = \$138 + \$147 = \$285$, the required amount.

State: Answer the three original questions.

a. Since $A = 69$, the price of an adult's ticket is $69.

b. Since $C = 49$, the price of a child's ticket is $49.

c. The regular admission price for 4 adults and 16 children is

$$4(\$69) + 16(\$49) = \$276 + \$784$$
$$= \$1060$$

This is $60 more than the special group rate of $1000, so they should request the group rate. □

PRACTICE
4 It is considered a premium game when the Red Sox or the Yankees come to Texas to play the Rangers. Admission for one of these games for three adults and three children under 14 is $75, while admission for two adults and four children is $62. (*Source:* MLB.com, Texas Rangers)

a. What is the price of an adult's admission?

b. What is the price of a child's admission?

c. Suppose that a special rate of $200 is offered for groups of 20 persons. Should a group of 5 adults and 15 children use the group rate? Why or why not? ■

EXAMPLE 5 Finding the Rate of Speed

Two cars leave Indianapolis, one traveling east and the other west. After 3 hours, they are 297 miles apart. If one car is traveling 5 mph faster than the other, what is the speed of each?

Solution

1. UNDERSTAND. Read and reread the problem. Let's guess a solution and use the formula $d = rt$ (distance = rate · time) to check. Suppose that one car is traveling at a rate of 55 miles per hour. This means that the other car is traveling at a rate of 50 miles per hour since we are told that one car is traveling 5 mph faster than the other. To find the distance apart after 3 hours, we will first find the distance traveled by each car. One car's distance is rate · time = 55(3) = 165 miles. The other car's distance is rate · time = 50(3) = 150 miles. Since one car is traveling east and the other west, their distance apart is the sum of their distances, or 165 miles + 150 miles = 315 miles. Although this distance apart is not the required distance of 297 miles, we now have a better understanding of the problem.

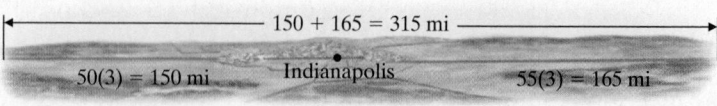
150 + 165 = 315 mi

50(3) = 150 mi   Indianapolis   55(3) = 165 mi

Let's model the problem with a system of equations. We will let

$$x = \text{speed of one car}$$
$$y = \text{speed of the other car}$$

We summarize the information on the following chart. Both cars have traveled 3 hours. Since distance = rate · time, their distances are $3x$ and $3y$ miles, respectively.

	Rate	· Time	= Distance
One Car	x	3	$3x$
Other Car	y	3	$3y$

2. TRANSLATE. We can now translate the stated conditions into two equations.

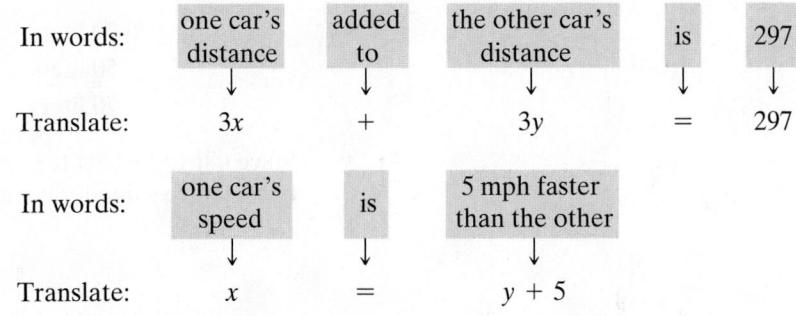

In words:

one car's distance	added to	the other car's distance	is	297
↓	↓	↓	↓	↓

Translate: $3x$ $+$ $3y$ $=$ 297

In words:

one car's speed	is	5 mph faster than the other
↓	↓	↓

Translate: x $=$ $y + 5$

3. SOLVE. Here we solve the system

$$\begin{cases} 3x + 3y = 297 \\ x \qquad = y + 5 \end{cases}$$

Again, the substitution method is appropriate. We replace x with $y + 5$ in the first equation and solve for y.

$$3x + 3y = 297 \quad \text{First equation}$$

$$3\overbrace{(y + 5)} + 3y = 297 \quad \text{Let } x = y + 5.$$
$$3y + 15 + 3y = 297$$
$$6y = 282$$
$$y = 47$$

To find x, we replace y with 47 in the equation $x = y + 5$. Then $x = 47 + 5 = 52$. The ordered pair solution of the system is $(52, 47)$.

4. INTERPRET. The solution $(52, 47)$ means that the cars are traveling at 52 mph and 47 mph, respectively.

Check: Notice that one car is traveling 5 mph faster than the other. Also, if one car travels 52 mph for 3 hours, the distance is $3(52) = 156$ miles. The other car traveling for 3 hours at 47 mph travels a distance of $3(47) = 141$ miles. The sum of the distances $156 + 141$ is 297 miles, the required distance.

> **Helpful Hint**
> Don't forget to attach units if appropriate.

State: The cars are traveling at 52 mph and 47 mph.   □

PRACTICE
5

In 2007, the French train TGV V150 became the fastest conventional rail train in the world. It broke the 1990 record of the next fastest conventional rail train, the French TGV Atlantique. Assume the V150 and the Atlantique left the same station in Paris, with one heading west and one heading east. After 2 hours, they were 2150 kilometers apart. If the V150 is 75 kph faster than the Atlantique, what is the speed of each?

EXAMPLE 6 **Mixing Solutions**

Lynn Pike, a pharmacist, needs 70 liters of a 50% alcohol solution. She has available a 30% alcohol solution and an 80% alcohol solution. How many liters of each solution should she mix to obtain 70 liters of a 50% alcohol solution?

Solution

1. UNDERSTAND. Read and reread the problem. Next, guess the solution. Suppose that we need 20 liters of the 30% solution. Then we need $70 - 20 = 50$ liters of the 80% solution. To see if this gives us 70 liters of a 50% alcohol solution, let's find the amount of pure alcohol in each solution.

number of liters	×	alcohol strength	=	amount of pure alcohol
20 liters	×	0.30	=	6 liters
50 liters	×	0.80	=	40 liters
70 liters	×	0.50	=	35 liters

Since 6 liters + 40 liters = 46 liters and not 35 liters, our guess is incorrect, but we have gained some insight as to how to model and check this problem.

We will let

x = amount of 30% solution, in liters

y = amount of 80% solution, in liters

and use a table to organize the given data.

	Number of Liters	Alcohol Strength	Amount of Pure Alcohol
30% Solution	x	30%	$0.30x$
80% Solution	y	80%	$0.80y$
50% Solution Needed	70	50%	$(0.50)(70)$

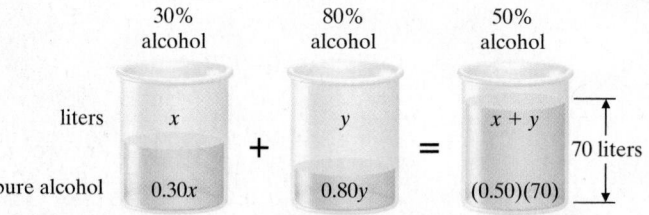

2. TRANSLATE. We translate the stated conditions into two equations.

In words:	amount of 30% solution	+	amount of 80% solution	=	70
Translate:	x	+	y	=	70

In words:	amount of pure alcohol in 30% solution	+	amount of pure alcohol in 80% solution	=	amount of pure alcohol in 50% solution
Translate:	$0.30x$	+	$0.80y$	=	$(0.50)(70)$

3. SOLVE. Here we solve the system

$$\begin{cases} x + y = 70 \\ 0.30x + 0.80y = (0.50)(70) \end{cases}$$

To solve this system, we use the elimination method. We multiply both sides of the first equation by -3 and both sides of the second equation by 10. Then

$$\begin{cases} -3(x + y) = -3(70) \\ 10(0.30x + 0.80y) = 10(0.50)(70) \end{cases} \quad \begin{matrix} \text{simplifies} \\ \text{to} \end{matrix} \quad \begin{cases} -3x - 3y = -210 \\ \underline{3x + 8y = 350} \\ 5y = 140 \\ y = 28 \end{cases}$$

Now we replace y with 28 in the equation $x + y = 70$ and find that $x + 28 = 70$, or $x = 42$.

The ordered pair solution of the system is $(42, 28)$.

4. INTERPRET.

Check: Check the solution in the same way that we checked our guess.

State: The pharmacist needs to mix 42 liters of 30% solution and 28 liters of 80% solution to obtain 70 liters of 50% solution. ☐

PRACTICE
6 Keith Robinson is a chemistry teacher who needs 1 liter of a solution of 5% hydrochloric acid (HCl) to carry out an experiment. If he only has a stock solution of 99% HCl, how much water (0% acid) and how much stock solution (99%) of HCl must he mix to get 1 liter of 5% solution? Round answers to the nearest hundredth of a liter. ◼

 **CONCEPT CHECK**

Suppose you mix an amount of 25% acid solution with an amount of 60% acid solution. You then calculate the acid strength of the resulting acid mixture. For which of the following results should you suspect an error in your calculation? Why?

a. 14% **b.** 32% **c.** 55%

OBJECTIVE
2 Solving Problems with Cost and Revenue Functions

Recall that businesses are often computing cost and revenue functions or equations to predict sales, to determine whether prices need to be adjusted, and to see whether the company is making or losing money. Recall also that the value at which revenue equals cost is called the break-even point. When revenue is less than cost, the company is losing money; when revenue is greater than cost, the company is making money.

EXAMPLE 7 **Finding a Break-Even Point**

A manufacturing company recently purchased $3000 worth of new equipment to offer new personalized stationery to its customers. The cost of producing a package of personalized stationery is $3.00, and it is sold for $5.50. Find the number of packages that must be sold for the company to break even.

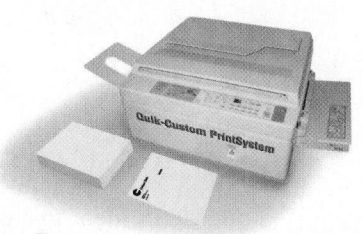

Answer to Concept Check:
a; answers may vary

(Continued on next page)

Solution

1. **UNDERSTAND.** Read and reread the problem. Notice that the cost to the company will include a one-time cost of $3000 for the equipment and then $3.00 per package produced. The revenue will be $5.50 per package sold. To model this problem, we will let

$$x = \text{number of packages of personalized stationery}$$
$$C(x) = \text{total cost of producing } x \text{ packages of stationery}$$
$$R(x) = \text{total revenue from selling } x \text{ packages of stationery}$$

2. **TRANSLATE.** The revenue equation is

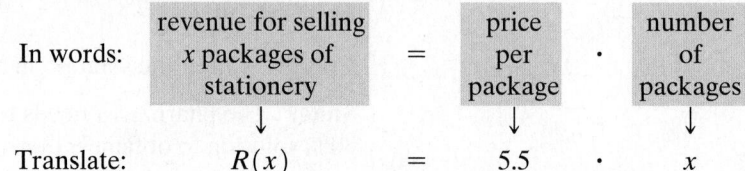

In words:	revenue for selling x packages of stationery	=	price per package	·	number of packages
	↓		↓		↓
Translate:	$R(x)$	=	5.5	·	x

The cost equation is

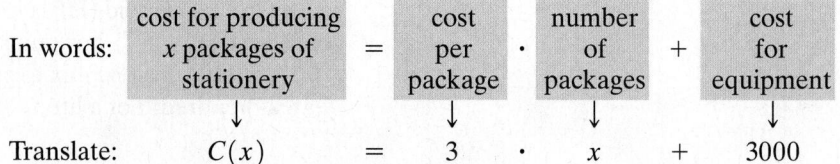

In words:	cost for producing x packages of stationery	=	cost per package	·	number of packages	+	cost for equipment
	↓		↓		↓		↓
Translate:	$C(x)$	=	3	·	x	+	3000

Since the break-even point is when $R(x) = C(x)$, we solve the equation

$$5.5x = 3x + 3000$$

3. **SOLVE.**

$$5.5x = 3x + 3000$$
$$2.5x = 3000 \qquad \text{Subtract } 3x \text{ from both sides.}$$
$$x = 1200 \qquad \text{Divide both sides by 2.5.}$$

4. **INTERPRET.**

Check: To see whether the break-even point occurs when 1200 packages are produced and sold, see if revenue equals cost when $x = 1200$. When $x = 1200, R(x) = 5.5x = 5.5(1200) = 6600$ and $C(x) = 3x + 3000 = 3(1200) + 3000 = 6600$. Since $R(1200) = C(1200) = 6600$, the break-even point is 1200.

State: The company must sell 1200 packages of stationery to break even. The graph of this system is shown.

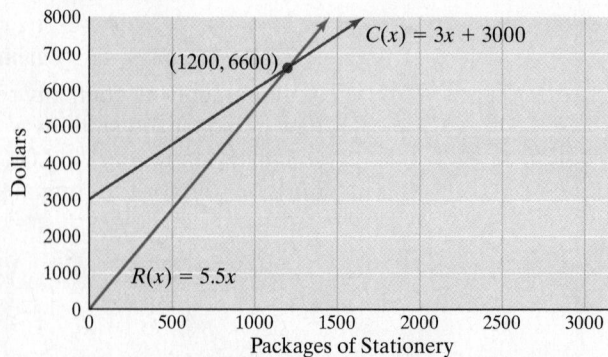

PRACTICE
7 An online-only electronics firm recently purchased $3000 worth of new equipment to create shock-proof packaging for its products. The cost of producing one shock-proof package is $2.50, and the firm charges the customer $4.50 for the packaging. Find the number of packages that must be sold for the company to break even.

OBJECTIVE

3   Solving Problems Modeled by Systems of Three Equations

To introduce problem solving by writing a system of three linear equations in three variables, we solve a problem about triangles.

EXAMPLE 8   **Finding Angle Measures**

The measure of the largest angle of a triangle is 80° more than the measure of the smallest angle, and the measure of the remaining angle is 10° more than the measure of the smallest angle. Find the measure of each angle.

Solution

1. UNDERSTAND. Read and reread the problem. Recall that the sum of the measures of the angles of a triangle is 180°. Then guess a solution. If the smallest angle measures 20°, the measure of the largest angle is 80° more, or 20° + 80° = 100°. The measure of the remaining angle is 10° more than the measure of the smallest angle, or 20° + 10° = 30°. The sum of these three angles is 20° + 100° + 30° = 150°, not the required 180°. We now know that the measure of the smallest angle is greater than 20°.

   To model this problem, we will let

   x = degree measure of the smallest angle

   y = degree measure of the largest angle

   z = degree measure of the remaining angle

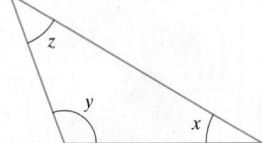

2. TRANSLATE. We translate the given information into three equations.

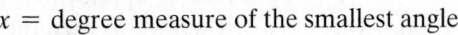

In words:	the sum of the measures	=	180
Translate:	$x + y + z$	=	180

In words:	the largest angle	is	80 more than the smallest angle
Translate:	y	=	$x + 80$

In words:	the remaining angle	is	10 more than the smallest angle
Translate:	z	=	$x + 10$

3. SOLVE. We solve the system

$$\begin{cases} x + y + z = 180 \\ y = x + 80 \\ z = x + 10 \end{cases}$$

Since y and z are both expressed in terms of x, we will solve using the substitution method. We substitute $y = x + 80$ and $z = x + 10$ in the first equation. Then

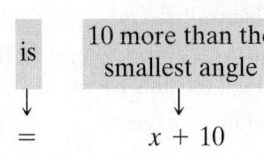

$$x + y + z = 180 \quad \text{First equation}$$

$$x + (x + 80) + (x + 10) = 180 \quad \text{Let } y = x + 80 \text{ and } z = x + 10.$$

$$3x + 90 = 180$$

$$3x = 90$$

$$x = 30$$

Then $y = x + 80 = 30 + 80 = 110$, and $z = x + 10 = 30 + 10 = 40$. The ordered triple solution is (30, 110, 40).

(Continued on next page)

4. INTERPRET.

Check: Notice that $30° + 40° + 110° = 180°$. Also, the measure of the largest angle, $110°$, is $80°$ more than the measure of the smallest angle, $30°$. The measure of the remaining angle, $40°$, is $10°$ more than the measure of the smallest angle, $30°$. □

PRACTICE

8 The measure of the largest angle of a triangle is $40°$ more than the measure of the smallest angle, and the measure of the remaining angle is $20°$ more than the measure of the smallest angle. Find the measure of each angle. ■

✔ Vocabulary, Readiness & Video Check

Martin-Gay Interactive Videos

See Video 4.5 ◉

Watch the section lecture video and answer the following questions.

OBJECTIVE 1

1. In Example 1 and the lecture before, the problem-solving steps for solving applications are mentioned. What is the difference here from when we've used these steps in the past?

OBJECTIVE 2

2. Based on Example 6, explain the meaning of a break-even point. How do you find the break-even point algebraically?

OBJECTIVE 3

3. In Example 7, why is the ordered triple not the final stated solution to the application?

4.5 Exercise Set MyMathLab ▷

Without actually solving each problem, choose each correct solution by deciding which choice satisfies the given conditions.

△ **1.** The length of a rectangle is 3 feet longer than the width. The perimeter is 30 feet. Find the dimensions of the rectangle.

  a. length = 8 feet; width = 5 feet

  b. length = 8 feet; width = 7 feet

  c. length = 9 feet; width = 6 feet

△ **2.** An isosceles triangle, a triangle with two sides of equal length, has a perimeter of 20 inches. Each of the equal sides is one inch longer than the third side. Find the lengths of the three sides.

  a. 6 inches, 6 inches, and 7 inches

  b. 7 inches, 7 inches, and 6 inches

  c. 6 inches, 7 inches, and 8 inches

3. Two computer disks and three notebooks cost $17. However, five computer disks and four notebooks cost $32. Find the price of each.

  a. notebook = $4; computer disk = $3

  b. notebook = $3; computer disk = $4

  c. notebook = $5; computer disk = $2

4. Two music CDs and four music cassette tapes cost a total of $40. However, three music CDs and five cassette tapes cost $55. Find the price of each.

  a. CD = $12; cassette = $4

  b. CD = $15; cassette = $2

  c. CD = $10; cassette = $5

5. Kesha has a total of 100 coins, all of which are either dimes or quarters. The total value of the coins is $13.00. Find the number of each type of coin.

  a. 80 dimes; 20 quarters

  b. 20 dimes; 44 quarters

  c. 60 dimes; 40 quarters

6. Samuel has 28 gallons of saline solution available in two large containers at his pharmacy. One container holds three times as much as the other container. Find the capacity of each container.

  a. 15 gallons; 5 gallons

  b. 20 gallons; 8 gallons

  c. 21 gallons; 7 gallons

TRANSLATING

Write a system of equations in x and y describing each situation. Do not solve the system. See Example 2.

7. A smaller number and a larger number add up to 15 and have a difference of 7. (Let x be the larger number.)

8. The total of two numbers is 16. The first number plus 2 more than 3 times the second equals 18. (Let x be the first number.)

9. Keiko has a total of $6500, which she has invested in two accounts. The larger account is $800 greater than the smaller account. (Let x be the amount of money in the larger account.)

10. Dominique has four times as much money in his savings account as in his checking account. The total amount is $2300. (Let x be the amount of money in his checking account.)

MIXED PRACTICE

Solve. See Examples 1 through 6. For Exercises 13 and 14, the solutions have been started for you.

11. Two numbers total 83 and have a difference of 17. Find the two numbers.

12. The sum of two numbers is 76 and their difference is 52. Find the two numbers.

13. One number is two more than a second number. Twice the first is 4 less than 3 times the second. Find the numbers.

Start the solution:

1. UNDERSTAND the problem. Since we are looking for two numbers, let

$$x = \text{one number}$$
$$y = \text{second number}$$

2. TRANSLATE. Since we have assigned two variables, we will translate the facts into two equations. (Fill in the blanks.)

First equation:

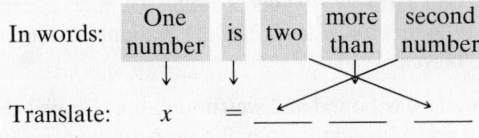

Translate: x =   ___ ___ ___

Second equation:

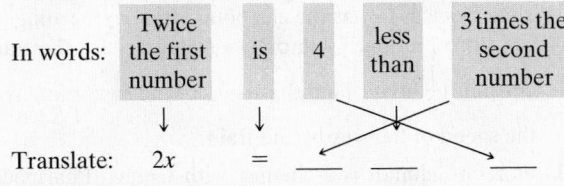

Translate: $2x$ =   ___ ___ ___

3. SOLVE the system and

4. INTERPRET the results.

14. Three times one number minus a second is 8, and the sum of the numbers is 12. Find the numbers.

Start the solution:

1. UNDERSTAND the problem. Since we are looking for two numbers, let

$$x = \text{one number}$$
$$y = \text{second number}$$

2. TRANSLATE. Since we have assigned two variables, we will translate the facts into two equations. (Fill in the blanks.)

First equation:

In words: | Three times one number | minus | a second number | is | 8

Translate: $3x$ ___ ___   = 8

Second equation:

In words: | The sum of the numbers | is | 12

Translate: x +   ___ ___   12

Finish with:

3. SOLVE the system and

4. INTERPRET the results.

15. A first number plus twice a second number is 8. Twice the first number, plus the second totals 25. Find the numbers.

16. One number is 4 more than twice the second number. Their total is 25. Find the numbers.

17. Adrian Gonzalez of the Los Angeles Dodgers led Major League Baseball in runs batted in for the 2014 regular season. Mike Trout of the Los Angeles Angels of Anaheim, who came in second to Gonzalez, had 5 fewer runs batted in for the 2014 regular season. Together, these players brought home 227 runs during the regular 2014 season. How many runs batted in was accounted for by each player? (*Source:* MLB)

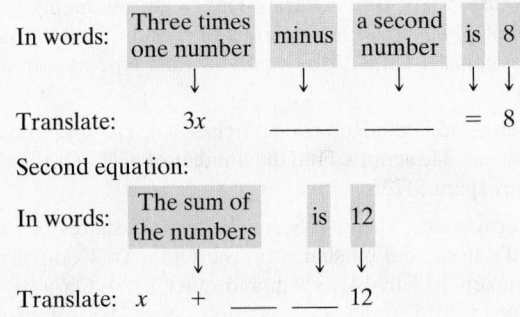

18. The highest scorer during the 2013 WNBA regular season was Angel McCoughtry of the Atlanta Dream. Over the season, McCoughtry scored 60 more points than the second-highest scorer, Diana Taurasi of the Phoenix Mercury. Together, McCoughtry and Taurasi scored 1362 points during the 2013 regular season. How many points did each player score over the course of the season? (*Source:* WNBA)

19. Ann Marie Jones has been pricing Amtrak train fares for a group trip to New York. Three adults and four children must pay $159. Two adults and three children must pay $112. Find the price of an adult's ticket and find the price of a child's ticket.

20. Last month, Jerry Papa purchased two DVDs and five CDs at Wall-to-Wall Sound for $65. This month, he bought four DVDs and three CDs for $81. Find the price of each DVD and find the price of each CD.

21. Johnston and Betsy Waring have a jar containing 80 coins, all of which are either quarters or nickels. The total value of the coins is $14.60. How many of each type of coin do they have?

22. Keith and Sarah Robinson purchased 40 stamps, a mixture of 49¢ and 34¢ stamps. Find the number of each type of stamp if they spent $17.35.

23. Norman and Suzanne Scarpulla own 35 shares of McDonald's stock and 69 shares of The Ohio Art Company stock (makers of Etch A Sketch and other toys). On a particular day in 2015, their stock portfolio consisting of these two stocks was worth $4000. The McDonald's stock was $92 more per share than The Ohio Art Company stock. What was the price of each stock on that day? (*Source:* Yahoo Finance)

24. Katy Biagini has investments in Google and Nintendo stock. During a particular day in 2015, Google stock was at $528 per share, and Nintendo stock was at $13 per share. Katy's portfolio made up of these two stocks was worth $19,809 at that time. If Katy owns 16 more shares of Google stock than she owns of Nintendo stock, how many shares of each type of stock does she own? (*Source:* Yahoo Finance)

25. Twice last month, Judy Carter rented a car from Enterprise in Fresno, California, and traveled around the Southwest on business. Enterprise rents this car for a daily fee plus an additional charge per mile driven. Judy recalls that her first trip lasted 4 days, she drove 450 miles, and the rental cost her $240.50. On her second business trip, she drove the same model of car a distance of 200 miles in 3 days and paid $146.00 for the rental. Find the daily fee and the mileage charge.

26. Joan Gundersen rented the same car model twice from Hertz, which rents this car model for a daily fee plus an additional charge per mile driven. Joan recalls that the car rented for 5 days and driven for 300 miles cost her $178, while the same model car rented for 4 days and driven for 500 miles cost $197. Find the daily fee and find the mileage charge.

27. Pratap Puri rowed 18 miles down the Delaware River in 2 hours, but the return trip took him $4\frac{1}{2}$ hours. Find the rate Pratap can row in still water and find the rate of the current.

Let x = rate Pratap can row in still water
y = rate of the current

d	=	r	$\cdot$	t
Downstream		$x + y$		
Upstream		$x - y$		

28. The Jonathan Schultz family took a canoe 10 miles down the Allegheny River in $1\frac{1}{4}$ hours. After lunch, it took them 4 hours to return. Find the rate of the current.

Let x = rate the family can row in still water
y = rate of the current

d	=	r	$\cdot$	t
Downstream		$x + y$		
Upstream		$x - y$		

29. Dave and Sandy Hartranft are frequent flyers with Delta Airlines. They often fly from Philadelphia to Chicago, a distance of 780 miles. On one particular trip, they fly into the wind, and the flight takes 2 hours. The return trip, with the wind behind them, only takes $1\frac{1}{2}$ hours. If the wind speed is the same on each trip, find the speed of the wind and find the speed of the plane in still air.

30. With a strong wind behind it, a United Airlines jet flies 2400 miles from Los Angeles to Orlando in $4\frac{3}{4}$ hours. The return trip takes 6 hours because the plane flies into the wind. If the wind speed is the same on each trip, find the speed of the plane in still air and find the wind speed to the nearest tenth of a mile per hour.

31. Kevin Briley began a 114-mile bicycle trip to build up stamina for a triathlete competition. Unfortunately, his bicycle chain broke, so he finished the trip walking. The whole trip took 6 hours. If Kevin walks at a rate of 4 miles per hour and rides at 24 miles per hour, find the amount of time he spent on the bicycle.

32. In Canada, eastbound and westbound trains travel along the same track, with sidings to pull onto to avoid accidents. Two trains are now 150 miles apart, with the westbound train traveling twice as fast as the eastbound train. A warning must be issued to pull one train onto a siding or else the trains will crash in $1\frac{1}{4}$ hours. Find the speed of the eastbound train and the speed of the westbound train.

33. Doreen Schmidt is a chemist with Gemco Pharmaceutical. She needs to prepare 12 liters of a 9% hydrochloric acid solution. Find the amount of a 4% solution and the amount of a 12% solution she should mix to get this solution.

Concentration Rate	Liters of Solution	Liters of Pure Acid
0.04	x	$0.04x$
0.12	y	?
0.09	12	?

34. Elise Everly is preparing 15 liters of a 25% saline solution. Elise has two other saline solutions, with strengths of 40% and 10%. Find the amount of 40% solution and the amount of 10% solution she should mix to get 15 liters of a 25% solution.

Concentration Rate	Liters of Solution	Liters of Pure Salt
0.40	x	0.40x
0.10	y	?
0.25	15	?

35. Wayne Osby blends coffee for a local coffee café. He needs to prepare 200 pounds of blended coffee beans selling for $3.95 per pound. He intends to do this by blending together a high-quality bean costing $4.95 per pound and a cheaper bean costing $2.65 per pound. To the nearest pound, find how much high-quality coffee bean and how much cheaper coffee bean he should blend.

36. Macadamia nuts cost an astounding $16.50 per pound, but research by an independent firm says that mixed nuts sell better if macadamias are included. The standard mix costs $9.25 per pound. Find how many pounds of macadamias and how many pounds of the standard mix should be combined to produce 40 pounds that will cost $10 per pound. Find the amounts to the nearest tenth of a pound.

37. Recall that two angles are complementary if the sum of their measures is 90°. Find the measures of two complementary angles if one angle is twice the other.

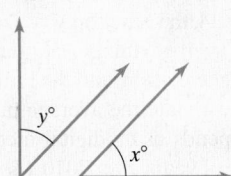

38. Recall that two angles are supplementary if the sum of their measures is 180°. Find the measures of two supplementary angles if one angle is 20° more than four times the other. (See art at top of next column.)

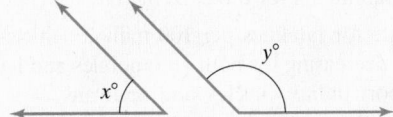

39. Find the measures of two complementary angles if one angle is 10° more than three times the other.

40. Find the measures of two supplementary angles if one angle is 18° more than twice the other.

41. Kathi and Robert Hawn had a pottery stand at the annual Skippack Craft Fair. They sold some of their pottery at the original price of $9.50 each but later decreased the price of each by $2. If they sold all 90 pieces and took in $721, find how many they sold at the original price and how many they sold at the reduced price.

42. A charity fund-raiser consisted of a spaghetti supper where a total of 387 people were fed. They charged $6.80 for adults and half price for children. If they took in $2444.60, find how many adults and how many children attended the supper.

43. The length of the Santa Fe National Historic Trail is approximately 1200 miles between Old Franklin, Missouri, and Santa Fe, New Mexico. Suppose that two groups of hikers, one from each town, start walking the trail toward each other. They meet after a total hiking time of 240 hours. If one group travels $\frac{1}{2}$ mile per hour slower than the other group, find the rate of each group. (*Source:* National Park Service)

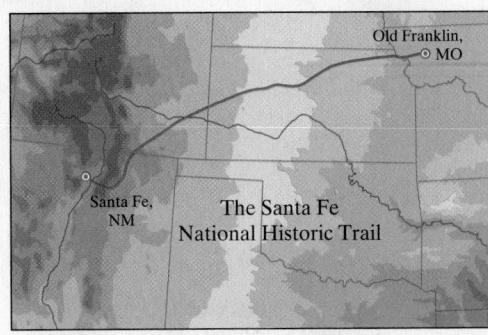

44. California 1 is a historic highway that stretches 123 miles along the coast from Monterey to Morro Bay. Suppose that two antique cars start driving this highway, one from each town. They meet after 3 hours. Find the rate of each car if one car travels 1 mile per hour faster than the other car. (*Source:* National Geographic)

45. A 30% solution of fertilizer is to be mixed with a 60% solution of fertilizer to get 150 gallons of a 50% solution. How many gallons of the 30% solution and 60% solution should be mixed?

46. A 10% acid solution is to be mixed with a 50% acid solution to get 120 ounces of a 20% acid solution. How many ounces of the 10% solution and 50% solution should be mixed?

47. Traffic signs are regulated by the *Manual on Uniform Traffic Control Devices* (MUTCD). According to this manual, if the sign below is placed on a freeway, its perimeter must be 144 inches. Also, its length is 12 inches longer than its width. Find the dimensions of this sign.

48. According to the MUTCD (see Exercise 47), this sign must have a perimeter of 60 inches. Also, its length must be 6 inches longer than its width. Find the dimensions of this sign.

49. The annual U.S. consumption of cheddar cheese has been decreasing in recent years, while the consumption of mozzarella cheese has been increasing. For the years 2008 through 2013, the function $y = -0.24x + 10.6$ approximates the annual U.S. per capita consumption of cheddar cheese in pounds, and the function $y = 0.21x + 10.5$ approximates the annual U.S. per capita consumption of mozzarella cheese in pounds. For both functions, x is the number of years after 2008. (*Source:* Based on data from the U.S. Department of Agriculture)

a. Explain how the given function for cheddar cheese verifies that the consumption of cheddar cheese has decreased, while the given function for mozzarella verifies that the consumption of mozzarella cheese has increased.

b. Based on this information, determine the year in which the pounds of cheddar cheese consumed equaled the pounds of mozzarella cheese consumed.

50. Two of the major jobs defined by the U.S. Department of Labor are postal carrier and bill collector. The number of postal carriers is predicted to decline over the next ten years, while the number of bill collectors is predicted to increase. For the years 2012 to 2022, the function $y = -13,910x + 491,600$ approximates the number of postal carriers in the United States, while the function $y = 5820x + 397,400$ approximates the number of bill collectors. For both functions, x is the number of years after 2012. (*Source:* Based on data from the Bureau of Labor Statistics)

a. Explain how the decrease in postal carrier jobs can be verified by the given function, and the increase in bill collector jobs can be verified by the given function.

b. Based on this information, determine the year in which the number of postal carriers and the number of bill collectors in the United States is the same.

51. The average American adult is spending ever more time on digital media (online or mobile device) than watching television. From 2010 to 2014, the function $y = 0.04x + 4.27$ can be used to estimate the average number of hours per day that an adult spends watching television, and the function $y = 0.67x + 3.01$ can be used to estimate the average number of hours per day that an adult spends using digital media. For both functions, x is the number of years after 2010. (*Source:* eMarketer)

a. Based on this information, determine the year in which the hours of television viewing is the same as the hours of digital media use.

b. Use these functions to predict how many hours the average American adult spends on television viewing and on digital media for the current year.

52. The rate for fatalities per 100 million vehicle-miles has been slowly decreasing for both automobiles and light trucks (pickups, sport utility vehicles, and minivans). For the years 2007 to 2012, the function $y = -0.08x + 1.08$ can be used to estimate the rate of fatalities per 100 million vehicle-miles for light trucks during this period, and the function $y = -0.03x + 1.01$ can be used to estimate the rate of fatalities per 100 million vehicle-miles for automobiles during this period. For both functions, x is the number of years since 2007. (*Source:* Bureau of Transportation Statistics, U.S. Department of Transportation)

a. Based on this data, estimate the year in which the fatality rate for light trucks equaled the fatality rate for automobiles.

b. Use these equations to predict the fatality rate per 100 million vehicle-miles for light trucks and automobiles in the current year.

△ **53.** In the figure, line *l* and line *m* are parallel lines cut by transversal *t*. Find the values of *x* and *y*.

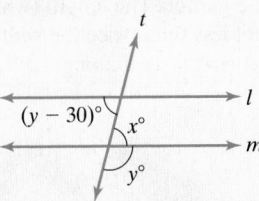

△ **54.** Find the values of *x* and *y* in the following isosceles triangle.

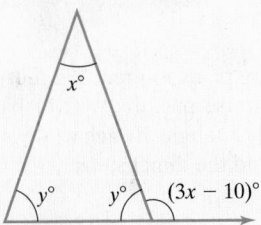

Given the cost function C(x) and the revenue function R(x), find the number of units x that must be sold to break even. See Example 7.

55. $C(x) = 30x + 10,000\ R(x) = 46x$

56. $C(x) = 12x + 15,000\ R(x) = 32x$

57. $C(x) = 1.2x + 1500\ R(x) = 1.7x$

58. $C(x) = 0.8x + 900\ R(x) = 2x$

59. $C(x) = 75x + 16,000\ R(x) = 200x$

60. $C(x) = 105x + 70,000\ R(x) = 245x$

61. The planning department of Abstract Office Supplies has been asked to determine whether the company should introduce a new computer desk next year. The department estimates that $6000 of new manufacturing equipment will need to be purchased and that the cost of constructing each desk will be $200. The department also estimates that the revenue from each desk will be $450.

   a. Determine the revenue function $R(x)$ from the sale of *x* desks.

   b. Determine the cost function $C(x)$ for manufacturing *x* desks.

   c. Find the break-even point.

62. Baskets, Inc., is planning to introduce a new woven basket. The company estimates that $500 worth of new equipment will be needed to manufacture this new type of basket and that it will cost $15 per basket to manufacture. The company also estimates that the revenue from each basket will be $31.

   a. Determine the revenue function $R(x)$ from the sale of *x* baskets.

   b. Determine the cost function $C(x)$ for manufacturing *x* baskets.

   c. Find the break-even point. Round up to the next whole basket.

Solve. See Example 6.

63. Rabbits in a lab are to be kept on a strict daily diet that includes 30 grams of protein, 16 grams of fat, and 24 grams of carbohydrates. The scientist has only three food mixes available with the following grams of nutrients per unit.

	Protein	*Fat*	*Carbohydrate*
Mix A	4	6	3
Mix B	6	1	2
Mix C	4	1	12

Find how many units of each mix are needed daily to meet each rabbit's dietary need.

64. Gerry Gundersen mixes different solutions with concentrations of 25%, 40%, and 50% to get 200 liters of a 32% solution. If he uses twice as much of the 25% solution as of the 40% solution, find how many liters of each kind he uses.

△ **65.** The perimeter of a quadrilateral (four-sided polygon) is 29 inches. The longest side is twice as long as the shortest side. The other two sides are equally long and are 2 inches longer than the shortest side. Find the lengths of all four sides.

△ **66.** The measure of the largest angle of a triangle is 90° more than the measure of the smallest angle, and the measure of the remaining angle is 30° more than the measure of the smallest angle. Find the measure of each angle.

67. The sum of three numbers is 40. The first number is five more than the second number. It is also twice the third. Find the numbers.

68. The sum of the digits of a three-digit number is 15. The tens-place digit is twice the hundreds-place digit, and the ones-place digit is 1 less than the hundreds-place digit. Find the three-digit number.

69. During the 2014 regular NBA season, the top-scoring player was Kevin Durant of the Oklahoma City Thunder. Durant scored a total of 2593 points during the regular season. The number of free throws (each worth one point) he made was 127 more than 3 times the number of three-point field goals he made. The number of two-point field goals he made was 46 less than the number of free throws he made. How many free throws, two-point field goals, and three-point field goals did Kevin Durant make during the 2013–2014 NBA season? (*Source:* National Basketball Association)

70. For 2014, the WNBA's top scorer was Maya Moore of the Minnesota Lynx. She scored a total of 812 points during the regular season. The number of two-point field goals that Moore made was 15 less than 4 times the number of three-point field goals she made. The number of free throws (each worth one point) was 73 fewer than the number of two-point field goals she made. Find how many free throws, two-point field goals, and three-point field goals Maya Moore made during the 2014 regular season. (*Source:* Women's National Basketball Association)

△ **71.** Find the values of x, y, and z in the following triangle.

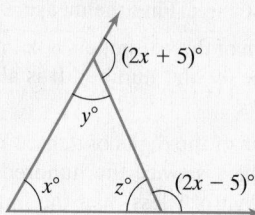

$(2x + 5)°$
$y°$
$x°$ $z°$ $(2x - 5)°$

△ **72.** The sum of the measures of the angles of a quadrilateral is 360°. Find the values of x, y, and z in the following quadrilateral.

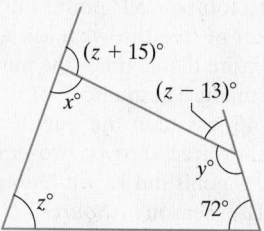

$(z + 15)°$
$x°$
$(z - 13)°$
$y°$
$z°$ $72°$

REVIEW AND PREVIEW

Solve each linear inequality. Write your solution in interval notation. See Section 2.8.

73. $-3x < -9$

74. $2x - 7 \le 5x + 11$

75. $4(2x - 1) \ge 0$

76. $\frac{2}{3}x < \frac{1}{3}$

CONCEPT EXTENSIONS

Solve. See the Concept Check in this section.

77. Suppose you mix an amount of candy costing $0.49 a pound with candy costing $0.65 a pound. Which of the following costs per pound could result?

a. $0.58   **b.** $0.72   **c.** $0.29

78. Suppose you mix a 50% acid solution with pure acid (100%). Which of the following acid strengths are possible for the resulting acid mixture?

a. 25%   **b.** 150%

c. 62%   **d.** 90%

79. Dale and Sharon Mahnke have decided to fence off a garden plot behind their house, using their house as the "fence" along one side of the garden. The length (which runs parallel to the house) is 3 feet less than twice the width. Find the dimensions if 33 feet of fencing is used along the three sides requiring it.

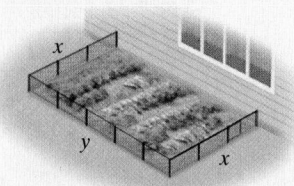

x
y
x

80. Judy McElroy plans to erect 152 feet of fencing around her rectangular horse pasture. A river bank serves as one side length of the rectangle. If each width is 4 feet longer than half the length, find the dimensions.

x
x
y

81. Find the values of a, b, and c such that the equation $y = ax^2 + bx + c$ has ordered pair solutions $(1, 6)$, $(-1, -2)$, and $(0, -1)$. To do so, substitute each ordered pair solution into the equation. Each time, the result is an equation in three unknowns: a, b, and c. Then solve the resulting system of three linear equations in three unknowns, a, b, and c.

82. Find the values of a, b, and c such that the equation $y = ax^2 + bx + c$ has ordered pair solutions $(1, 2)$, $(2, 3)$, and $(-1, 6)$. (*Hint:* See Exercise 81.)

83. Data (x, y) for the total number (in thousands) of college-bound students who took the ACT assessment in the year x are approximately $(3, 927)$, $(11, 1179)$, and $(19, 1495)$, where $x = 3$ represents 1993 and $x = 11$ represents 2001. Find the values a, b, and c such that the equation $y = ax^2 + bx + c$ models these data. According to your model, how many students will take the ACT in 2015? (*Source:* ACT, Inc.)

84. Monthly normal rainfall data (x, y) for Portland, Oregon, are $(4, 2.47)$, $(7, 0.58)$, and $(8, 1.07)$, where x represents the month (with $x = 1$ representing January) and y represents rainfall in inches. Find the values of a, b, and c rounded to 2 decimal places such that the equation $y = ax^2 + bx + c$ models this data. According to your model, how much rain should Portland expect during September? (*Source:* National Climatic Data Center)

The function $f(x) = 0.42x + 8.4$ represents the U.S. annual bottled water consumption (in billions of gallons) and the function $f(x) = -0.17x + 14.2$ represents the U.S. annual soda consumption (in billions of gallons). For both functions, x is the number of years since 2009, and these functions are good for the years 2009–2013.

85. Solve the system formed by these functions. Round each coordinate to the nearest whole number.

86. Use your answer to Exercise 85 to predict the year in which the bottled water and soda consumption will be the same.

Chapter 4 Vocabulary Check

Fill in each blank with one of the words or phrases listed below.

system of linear equations   solution   consistent   independent

dependent   inconsistent   substitution   addition

1. In a system of linear equations in two variables, if the graphs of the equations are the same, the equations are _____ equations.

2. Two or more linear equations are called a(n) _____.

3. A system of equations that has at least one solution is called a(n) _____ system.

4. A(n) _____ of a system of two equations in two variables is an ordered pair of numbers that is a solution of both equations in the system.

5. Two algebraic methods for solving systems of equations are _____ and _____.

6. A system of equations that has no solution is called a(n) _____ system.

7. In a system of linear equations in two variables, if the graphs of the equations are different, the equations are _____ equations.

Chapter 4 Highlights

DEFINITIONS AND CONCEPTS	EXAMPLES

Section 4.1 Solving Systems of Linear Equations by Graphing

A **solution** of a system of two equations in two variables is an ordered pair of numbers that is a solution of both equations in the system.

Determine whether $(-1, 3)$ is a solution of the system:

$$\begin{cases} 2x - y = -5 \\ x = 3y - 10 \end{cases}$$

Replace x with -1 and y with 3 in both equations.

$$2x - y = -5 \qquad\qquad x = 3y - 10$$
$$2(-1) - 3 \overset{?}{=} -5 \qquad -1 \overset{?}{=} 3 \cdot 3 - 10$$
$$-5 = -5 \ \text{True} \qquad -1 = -1 \ \text{True}$$

$(-1, 3)$ is a solution of the system.

Graphically, a solution of a system is a point common to the graphs of both equations.

Solve by graphing. $\begin{cases} 3x - 2y = -3 \\ x + y = 4 \end{cases}$

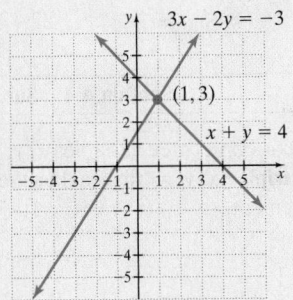

A system of equations with at least one solution is a **consistent system**. A system that has no solution is an **inconsistent system**.

If the graphs of two linear equations are identical, the equations are **dependent**. If their graphs are different, the equations are **independent**.

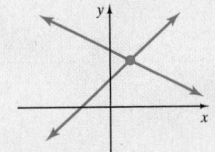

Consistent
and independent

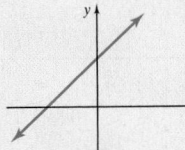

Consistent
and dependent

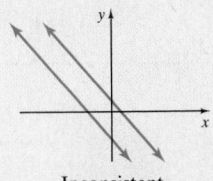

Inconsistent
and independent

DEFINITIONS AND CONCEPTS	EXAMPLES

Section 4.2   Solving Systems of Linear Equations by Substitution

To solve a system of linear equations by the substitution method:

Step 1.   Solve one equation for a variable.

Step 2.   Substitute the expression for the variable into the other equation.

Step 3.   Solve the equation from Step 2 to find the value of one variable.

Step 4.   Substitute the value from Step 3 in either original equation to find the value of the other variable.

Step 5.   Check the solution in both equations.

Solve by substitution.

$$\begin{cases} 3x + 2y = 1 \\ x = y - 3 \end{cases}$$

Substitute $y - 3$ for x in the first equation.

$$3x + 2y = 1$$
$$3(y - 3) + 2y = 1$$
$$3y - 9 + 2y = 1$$
$$5y = 10$$
$$y = 2 \quad \text{Divide by 5.}$$

To find x, substitute 2 for y in $x = y - 3$ so that $x = 2 - 3$ or -1. The solution $(-1, 2)$ checks.

Section 4.3   Solving Systems of Linear Equations by Addition

To solve a system of linear equations by the addition method:

Step 1.   Rewrite each equation in standard form $Ax + By = C$.

Step 2.   If necessary, multiply one or both equations by a nonzero number so that the coefficients of a variable are opposites.

Step 3.   Add the equations.

Step 4.   Find the value of one variable by solving the resulting equation.

Step 5.   Substitute the value from Step 4 into either original equation to find the value of the other variable.

Step 6.   Check the solution in both equations.

If solving a system of linear equations by substitution or addition yields a true statement such as $-2 = -2$, then the graphs of the equations in the system are identical and there is an infinite number of solutions of the system.

Solve by addition.

$$\begin{cases} x - 2y = 8 \\ 3x + y = -4 \end{cases}$$

Multiply both sides of the first equation by -3.

$$\begin{cases} -3x + 6y = -24 \\ \underline{3x + y = -4} \\ 7y = -28 \quad \text{Add.} \\ y = -4 \quad \text{Divide by 7.} \end{cases}$$

To find x, let $y = -4$ in an original equation.

$$x - 2(-4) = 8 \quad \text{First equation}$$
$$x + 8 = 8$$
$$x = 0$$

The solution $(0, -4)$ checks.

Solve: $\begin{cases} 2x - 6y = -2 \\ x = 3y - 1 \end{cases}$

Substitute $3y - 1$ for x in the first equation.

$$2(3y - 1) - 6y = -2$$
$$6y - 2 - 6y = -2$$
$$-2 = -2 \quad \text{True}$$

The system has an infinite number of solutions. In set notation, we write $\{(x, y) \mid 2x - 6y = -2\}$ or $\{(x, y) \mid x = 3y - 1\}$.

Section 4.4   Solving Systems of Linear Equations in Three Variables

A **solution** of an equation in three variables x, y, and z is an **ordered triple** (x, y, z) that makes the equation a true statement.

Verify that $(-2, 1, 3)$ is a solution of $2x + 3y - 2z = -7$. Replace x with -2, y with 1, and z with 3.

$$2(-2) + 3(1) - 2(3) \overset{?}{=} -7$$
$$-4 + 3 - 6 \overset{?}{=} -7$$
$$-7 = -7 \quad \text{True}$$

$(-2, 1, 3)$ is a solution.

DEFINITIONS AND CONCEPTS	EXAMPLES

Section 4.4   Solving Systems of Linear Equations in Three Variables (Continued)

Solving a System of Three Linear Equations by the Elimination Method

Step 1. Write each equation in standard form, $Ax + By + Cz = D$.

Step 2. Choose a pair of equations and use them to eliminate a variable.

Step 3. Choose any other pair of equations and eliminate the same variable.

Step 4. Solve the system of two equations in two variables from Steps 2 and 3.

Step 5. Solve for the third variable by substituting the values of the variables from Step 4 into any of the original equations.

Step 6. Check the solution in all three original equations.

Solve.

$$\begin{cases} 2x + y - z = 0 & (1) \\ x - y - 2z = -6 & (2) \\ -3x - 2y + 3z = -22 & (3) \end{cases}$$

1. Each equation is written in standard form.

2.
$$\begin{array}{l} 2x + y - z = 0 \quad (1) \\ \underline{x - y - 2z = -6} \quad (2) \\ 3x \quad - 3z = -6 \quad (4) \quad \text{Add.} \end{array}$$

3. Eliminate y from equations (1) and (3) also.

$$\begin{array}{ll} 4x + 2y - 2z = 0 & \text{Multiply equation} \\ \underline{-3x - 2y + 3z = -22} \quad (3) & \text{(1) by 2} \\ x \quad + z = -22 \quad (5) & \text{Add.} \end{array}$$

4. Solve:
$$\begin{cases} 3x - 3z = -6 & (4) \\ x + z = -22 & (5) \end{cases}$$

$$\begin{array}{l} x - z = -2 \qquad \text{Divide equation (4) by 3.} \\ \underline{x + z = -22} \\ 2x \quad = -24 \qquad \text{Add.} \\ x \quad = -12 \end{array}$$

To find z, use equation (5).

$$\begin{array}{l} x + z = -22 \\ -12 + z = -22 \\ z = -10 \end{array}$$

5. To find y, use equation (1).

$$\begin{array}{l} 2x + y - z = 0 \\ 2(-12) + y - (-10) = 0 \\ -24 + y + 10 = 0 \\ y = 14 \end{array}$$

6. The solution $(-12, 14, -10)$ checks.

Section 4.5   Systems of Linear Equations and Problem Solving

Problem-solving steps

1. UNDERSTAND. Read and reread the problem.

Two angles are supplementary if their sum is $180°$.

The larger of two supplementary angles is three times the smaller, decreased by twelve. Find the measure of each angle. Let

$$x = \text{measure of smaller angle}$$
$$y = \text{measure of larger angle}$$

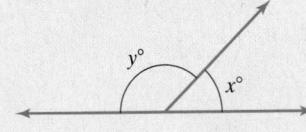

(continued)

DEFINITIONS AND CONCEPTS	EXAMPLES

Section 4.5   Systems of Linear Equations and Problem Solving (Continued)

2. TRANSLATE.

In words:

the sum of supplementary angles	is	180°

Translate: $x + y$ $=$ 180

In words:

larger angle	is	3 times smaller	decreased by	12

Translate: y $=$ $3x$ $-$ 12

3. SOLVE.

Solve the system:

$$\begin{cases} x + y = 180 \\ y = 3x - 12 \end{cases}$$

Use the substitution method and replace y with $3x - 12$ in the first equation.

$$x + y = 180$$
$$x + (3x - 12) = 180$$
$$4x = 192$$
$$x = 48$$

Since $y = 3x - 12$, $y = 3 \cdot 48 - 12$ or 132.

4. INTERPRET.

The solution checks. The smaller angle measures 48° and the larger angle measures 132°.

Chapter 4 Review

(4.1) Determine whether each of the following ordered pairs satisfies the system of linear equations.

1. $\begin{cases} 2x - 3y = 12 \\ 3x + 4y = 1 \end{cases}$

   a. $(12, 4)$ **b.** $(3, -2)$ **c.** $(-3, 6)$

2. $\begin{cases} 4x + y = 0 \\ -8x - 5y = 9 \end{cases}$

   a. $\left(\dfrac{3}{4}, -3\right)$ **b.** $(-2, 8)$ **c.** $\left(\dfrac{1}{2}, -2\right)$

3. $\begin{cases} 5x - 6y = 18 \\ 2y - x = -4 \end{cases}$

   a. $(-6, -8)$ **b.** $\left(3, \dfrac{5}{2}\right)$ **c.** $\left(3, -\dfrac{1}{2}\right)$

4. $\begin{cases} 2x + 3y = 1 \\ 3y - x = 4 \end{cases}$

   a. $(2, 2)$ **b.** $(-1, 1)$ **c.** $(2, -1)$

Solve each system of equations by graphing.

5. $\begin{cases} x + y = 5 \\ x - y = 1 \end{cases}$

6. $\begin{cases} x + y = 3 \\ x - y = -1 \end{cases}$

7. $\begin{cases} x = 5 \\ y = -1 \end{cases}$

8. $\begin{cases} x = -3 \\ y = 2 \end{cases}$

9. $\begin{cases} 2x + y = 5 \\ x = -3y \end{cases}$

10. $\begin{cases} 3x + y = -2 \\ y = -5x \end{cases}$

11. $\begin{cases} y = 3x \\ -6x + 2y = 6 \end{cases}$

12. $\begin{cases} x - 2y = 2 \\ -2x + 4y = -4 \end{cases}$

(4.2) Solve each system of equations by the substitution method.

13. $\begin{cases} y = 2x + 6 \\ 3x - 2y = -11 \end{cases}$

14. $\begin{cases} y = 3x - 7 \\ 2x - 3y = 7 \end{cases}$

15. $\begin{cases} x + 3y = -3 \\ 2x + y = 4 \end{cases}$

16. $\begin{cases} 3x + y = 11 \\ x + 2y = 12 \end{cases}$

17. $\begin{cases} 4y = 2x + 6 \\ x - 2y = -3 \end{cases}$

18. $\begin{cases} 9x = 6y + 3 \\ 6x - 4y = 2 \end{cases}$

19. $\begin{cases} x + y = 6 \\ y = -x - 4 \end{cases}$

20. $\begin{cases} -3x + y = 6 \\ y = 3x + 2 \end{cases}$

(4.3) *Solve each system of equations by the addition method.*

21. $\begin{cases} 2x + 3y = -6 \\ x - 3y = -12 \end{cases}$ **22.** $\begin{cases} 4x + y = 15 \\ -4x + 3y = -19 \end{cases}$

23. $\begin{cases} 2x - 3y = -15 \\ x + 4y = 31 \end{cases}$ **24.** $\begin{cases} x - 5y = -22 \\ 4x + 3y = 4 \end{cases}$

25. $\begin{cases} 2x - 6y = -1 \\ -x + 3y = \dfrac{1}{2} \end{cases}$ **26.** $\begin{cases} 0.6x - 0.3y = -1.5 \\ 0.04x - 0.02y = -0.1 \end{cases}$

27. $\begin{cases} \dfrac{3}{4}x + \dfrac{2}{3}y = 2 \\ x + \dfrac{y}{3} = 6 \end{cases}$ **28.** $\begin{cases} 10x + 2y = 0 \\ 3x + 5y = 33 \end{cases}$

(4.4) *Solve each system of equations in three variables.*

29. $\begin{cases} x \quad\;\; + z = 4 \\ 2x - y \quad\;\; = 4 \\ x + y - z = 0 \end{cases}$

30. $\begin{cases} 2x + 5y \quad\;\; = 4 \\ x - 5y + z = -1 \\ 4x \quad\quad - z = 11 \end{cases}$

31. $\begin{cases} \quad\;\; 4y + 2z = 5 \\ 2x + 8y \quad\;\; = 5 \\ 6x + \quad\;\; 4z = 1 \end{cases}$

32. $\begin{cases} 5x + 7y \quad\;\; = 9 \\ \quad\;\; 14y - z = 28 \\ 4x \quad\;\; + 2z = -4 \end{cases}$

33. $\begin{cases} 3x - 2y + 2z = 5 \\ -x + 6y + z = 4 \\ 3x + 14y + 7z = 20 \end{cases}$

34. $\begin{cases} x + 2y + 3z = 11 \\ \quad\;\; y + 2z = 3 \\ 2x \quad\quad + 2z = 10 \end{cases}$

35. $\begin{cases} 7x - 3y + 2z = 0 \\ 4x - 4y - z = 2 \\ 5x + 2y + 3z = 1 \end{cases}$

36. $\begin{cases} x - 3y - 5z = -5 \\ 4x - 2y + 3z = 13 \\ 5x + 3y + 4z = 22 \end{cases}$

(4.5) *Solve each problem by writing and solving a system of linear equations.*

37. The sum of two numbers is 16. Three times the larger number decreased by the smaller number is 72. Find the two numbers.

38. The Forrest Theater can seat a total of 360 people. They take in $15,150 when every seat is sold. If orchestra section tickets cost $45 and balcony tickets cost $35, find the number of seats in the orchestra section and the number of seats in the balcony.

39. A riverboat can head 340 miles upriver in 19 hours, but the return trip takes only 14 hours. Find the current of the river and

find the speed of the riverboat in still water to the nearest tenth of a mile.

	d	$=$	r	$\cdot$	t
Upriver	340		$x - y$		19
Downriver	340		$x + y$		14

40. Find the amount of a 6% acid solution and the amount of a 14% acid solution Pat Mayfield should combine to prepare 50 cc (cubic centimeters) of a 12% solution.

41. A deli charges $3.80 for a breakfast of three eggs and four strips of bacon. The charge is $2.75 for two eggs and three strips of bacon. Find the cost of each egg and the cost of each strip of bacon.

42. An exercise enthusiast alternates between jogging and walking. He traveled 15 miles during the past 3 hours. He jogs at a rate of 7.5 miles per hour and walks at a rate of 4 miles per hour. Find how much time, to the nearest hundredth of an hour, he actually spent jogging and how much time he spent walking.

43. Chris Kringler has $2.77 in her coin jar—all in pennies, nickels, and dimes. If she has 53 coins in all and four more nickels than dimes, find how many of each type of coin she has.

44. An employee at See's Candy Store needs a special mixture of candy. She has creme-filled chocolates that sell for $3.00 per pound, chocolate-covered nuts that sell for $2.70 per pound, and chocolate-covered raisins that sell for $2.25 per pound. She wants to have twice as many pounds of raisins as pounds of nuts in the mixture. Find how many pounds of each she should use to make 45 pounds worth $2.80 per pound.

45. The perimeter of an isosceles (two sides equal) triangle is 73 centimeters. If the unequal side is 7 centimeters longer than the two equal sides, find the lengths of the three sides.

46. The sum of three numbers is 295. One number is five more than a second and twice the third. Find the numbers.

MIXED REVIEW

Solve each system of equations by graphing.

47. $\begin{cases} x - 2y = 1 \\ 2x + 3y = -12 \end{cases}$ **48.** $\begin{cases} 3x - y = -4 \\ 6x - 2y = -8 \end{cases}$

Solve each system of equations.

49. $\begin{cases} x + 4y = 11 \\ 5x - 9y = -3 \end{cases}$ **50.** $\begin{cases} x + 9y = 16 \\ 3x - 8y = 13 \end{cases}$

51. $\begin{cases} y = -2x \\ 4x + 7y = -15 \end{cases}$

52. $\begin{cases} 3y = 2x + 15 \\ -2x + 3y = 21 \end{cases}$

53. $\begin{cases} 3x - y = 4 \\ 4y = 12x - 16 \end{cases}$

54. $\begin{cases} x + y = 19 \\ x - y = -3 \end{cases}$

55. $\begin{cases} x - 3y = -11 \\ 4x + 5y = -10 \end{cases}$ **56.** $\begin{cases} -x - 15y = 44 \\ 2x + 3y = 20 \end{cases}$

57. $\begin{cases} x - 3y + 2z = 0 \\ 9y - z = 22 \\ 5x + 3z = 10 \end{cases}$

58. $\begin{cases} x - 4y = 4 \\ \dfrac{1}{8}x - \dfrac{1}{2}y = 3 \end{cases}$

Solve each problem by writing and solving a system of linear equations.

59. The sum of two numbers is 12. Three times the smaller number increased by the larger number is 20. Find the numbers.

60. The difference of two numbers is -18 Twice the smaller decreased by the larger is -23 Find the two numbers.

61. Emma Hodges has a jar containing 65 coins, all of which are either nickels or dimes. The total value of the coins is $5.30. How many of each type does she have?

62. Sarah and Owen Hebert purchased 26 stamps, a mixture of 49¢ and 34¢ stamps. Find the number of each type of stamp if they spent $11.39.

63. The perimeter of a triangle is 126 units. The length of one side is twice the length of the shortest side. The length of the third side is fourteen more than the length of the shortest side. Find the length of the sides of the triangle.

Chapter 4 Getting Ready for the Test

1. MULTIPLE CHOICE *The ordered pair* $(-1, 2)$ *is a solution of which system?*

A. $\begin{cases} 5x - y = -7 \\ x - y = 3 \end{cases}$ **B.** $\begin{cases} 3x - y = -5 \\ x + y = 1 \end{cases}$

C. $\begin{cases} x = 2 \\ x + y = 1 \end{cases}$ **D.** $\begin{cases} y = -1 \\ x + y = -3 \end{cases}$

2. MULTIPLE CHOICE *When solving a system of two linear equations in two variables, all variables subtract out and the resulting equation is* $0 = 5$. *What does this mean?*

A. the solution is $(0, 5)$ **B.** the system has an infinite number of solutions   **C.** the system has no solution

MATCHING *Match each system with its solution. Letter choices may be used more than once or not at all.*

3. $\begin{cases} y = 5x + 2 \\ y = -5x + 2 \end{cases}$ **4.** $\begin{cases} y = \dfrac{1}{2}x - 3 \\ y = \dfrac{1}{2}x + 7 \end{cases}$

A. no solution
B. one solution
C. two solutions
D. an infinite number of solutions

5. $\begin{cases} y = 4x + 2 \\ 8x - 2y = -4 \end{cases}$ **6.** $\begin{cases} y = 6x \\ y = -\dfrac{1}{6}x \end{cases}$

MULTIPLE CHOICE *Choose the correct choice for Exercises 7 and 8. The system for these exercises is:*

$$\begin{cases} 5x - y = -8 \\ 2x + 3y = 1 \end{cases}$$

7. When solving, if we decide to multiply the first equation above by 3, the result of the first equation is:

A. $15x - 3y = -8$ **B.** $6x + 9y = 1$ **C.** $6x + 9y = 3$ **D.** $15x - 3y = -24$

8. When solving, if we decide to multiply the second equation above by -5, the result of the second equation is:

A. $-10x - 15y = 1$ **B.** $-25x + 5y = 40$ **C.** $-10x - 15y = -5$ **D.** $-25x + 5y = -8$

9. The ordered triple $(1, 0, -2)$ is a solution of which system?

A. $\begin{cases} 5x - y - z = -7 \\ x + y + z = -1 \\ y + z = -1 \end{cases}$ **B.** $\begin{cases} 3x - y - 2z = 7 \\ x + y + z = -1 \\ 5x + z = 3 \end{cases}$ **C.** $\begin{cases} x = -2 \\ x + y + z = -1 \\ x + z = 1 \end{cases}$ **D.** $\begin{cases} y = 1 \\ x + y + z = -1 \\ x + y = -1 \end{cases}$

MULTIPLE CHOICE *Select the correct choice. The system for Exercises 10–12 is:*

$$\begin{cases} 3x - y + 2z = 3 & \text{Equation (1)} \\ 4x + y - z = 5 & \text{Equation (2)} \\ -x + 5y + 3z = 12 & \text{Equation (3)} \end{cases}$$

The choices for Exercises 10–12 are:

A. $[\text{Equation}(1)] + [\text{Equation}(2)]$

C. $[\text{Equation}(1)] + 2 \cdot [\text{Equation}(2)]$

B. $[\text{Equation}(1)] + 3 \cdot [\text{Equation}(3)]$

D. $[\text{Equation}(1)] + [\text{Equation}(3)]$

▶ 10. Which choice eliminates the variable x?

▶ 11. Which choice eliminates the variable y?

▶ 12. Which choice eliminates the variable z?

Chapter 4 Test   MyMathLab® You Tube

Answer each question true or false.

▶ 1. A system of two linear equations in two variables can have exactly two solutions.

▶ 2. Although $(1, 4)$ is not a solution of $x + 2y = 6$, it can still be a solution of the system $\begin{cases} x + 2y = 6 \\ x + y = 5 \end{cases}$.

▶ 3. If the two equations in a system of linear equations are added and the result is $3 = 0$, the system has no solution.

▶ 4. If the two equations in a system of linear equations are added and the result is $3x = 0$, the system has no solution.

Is the ordered pair a solution of the given linear system?

▶ 5. $\begin{cases} 2x - 3y = 5 \\ 6x + y = 1 \end{cases}$; $(1, -1)$

▶ 6. $\begin{cases} 4x - 3y = 24 \\ 4x + 5y = -8 \end{cases}$; $(3, -4)$

▶ 7. Use graphing to find the solutions of the system $\begin{cases} y - x = 6 \\ y + 2x = -6 \end{cases}$

▶ 8. Use the substitution method to solve the system $\begin{cases} 3x - 2y = -14 \\ x + 3y = -1 \end{cases}$

▶ 9. Use the substitution method to solve the system $\begin{cases} \dfrac{1}{2}x + 2y = -\dfrac{15}{4} \\ 4x = -y \end{cases}$

▶ 10. Use the addition method to solve the system $\begin{cases} 3x + 5y = 2 \\ 2x - 3y = 14 \end{cases}$

▶ 11. Use the addition method to solve the system $\begin{cases} 4x - 6y = 7 \\ -2x + 3y = 0 \end{cases}$

Solve each system using the substitution method or the addition method.

▶ 12. $\begin{cases} 3x + y = 7 \\ 4x + 3y = 1 \end{cases}$

▶ 13. $\begin{cases} 3(2x + y) = 4x + 20 \\ x - 2y = 3 \end{cases}$

▶ 14. $\begin{cases} \dfrac{x - 3}{2} = \dfrac{2 - y}{4} \\ \dfrac{7 - 2x}{3} = \dfrac{y}{2} \end{cases}$

Solve each problem by writing and using a system of linear equations.

▶ 15. Two numbers have a sum of 124 and a difference of 32. Find the numbers.

▶ 16. Find the amount of a 12% saline solution a lab assistant should add to 80 cc (cubic centimeters) of a 22% saline solution to have a 16% solution.

▶ 17. Texas and Missouri are the states with the most farms. Texas has 140 thousand more farms than Missouri and the total number of farms for these two states is 356 thousand. Find the number of farms for each state.

▶ **18.** Two hikers start at opposite ends of the St. Tammany Trail and walk toward each other. The trail is 36 miles long and they meet in 4 hours. If one hiker walks twice as fast as the other, find both hiking speeds.

▶ **19.** $\begin{cases} 2x - 3y & = 4 \\ 3y + 2z = 2 \\ x \quad - z = -5 \end{cases}$

20. $\begin{cases} 3x - 2y - z = -1 \\ 2x - 2y \quad = 4 \\ 2x \quad - 2z = -12 \end{cases}$

▶ **21.** The measure of the largest angle of a triangle is three less than 5 times the measure of the smallest angle. The measure of the remaining angle is 1 less than twice the measure of the smallest angle. Find the measure of each angle.

Chapter 4 Cumulative Review

1. Insert $<, >$, or $=$ in the appropriate space between the paired numbers to make each statement true.

   a. -1 0 **b.** 7 $\dfrac{14}{2}$ **c.** -5 -6

2. Evaluate.
   a. 5^2 **b.** 2^5

3. Name the property or properties illustrated by each true statement.
   a. $3 \cdot y = y \cdot 3$
   b. $(x + 7) + 9 = x + (7 + 9)$
   c. $(b + 0) + 3 = b + 3$
   d. $0.2 \cdot (z \cdot 5) = 0.2 \cdot (5 \cdot z)$
   e. $-2 \cdot \left(-\dfrac{1}{2}\right) = 1$
   f. $-2 + 2 = 0$
   g. $-6 \cdot (y \cdot 2) = (-6 \cdot 2) \cdot y$

4. Evaluate $y^2 - 3x$ for $x = 8$ and $y = 5$.

5. Write the phrase as an algebraic expression, then simplify if possible: Subtract $4x - 2$ from $2x - 3$.

6. Simplify: $7 - 12 + (-5) - 2 + (-2)$

7. Solve: $7 = -5(2a - 1) - (-11a + 6)$

8. Evaluate $2y^2 - x^2$ for $x = -7$ and $y = -3$.

9. Solve: $\dfrac{5}{2}x = 15$

10. Simplify: $0.4y - 6.7 + y - 0.3 - 2.6y$

11. Solve: $\dfrac{x}{2} - 1 = \dfrac{2}{3}x - 3$

12. Solve: $7(x - 2) - 6(x + 1) = 20$

13. Twice the sum of a number and 4 is the same as four times the number, decreased by 12. Find the number.

14. Solve: $5(y - 5) = 5y + 10$

15. Solve $y = mx + b$ for x.

16. Five times the sum of a number and -1 is the same as 6 times the number. Find the number.

17. Solve $-2x \leq -4$ Write the solution set in interval notation.

18. Solve $P = a + b + c$ for b.

19. Graph $x = -2y$ by finding and plotting intercepts.

20. Solve $3x + 7 \geq x - 9$. Write the solution set in interval notation.

21. Find the slope of the line through $(-1, 5)$ and $(2, -3)$.

22. Complete the table of values for $x - 3y = 3$.

x	y
	-1
3	
	2

23. Find the slope and y-intercept of the line whose equation is $y = \dfrac{3}{4}x + 6$.

24. Find the slope of a line parallel to the line passing through $(-1, 3)$ and $(2, -8)$.

25. Find the slope and the y-intercept of the line whose equation is $3x - 4y = 4$.

26. Find the slope and y-intercept of the line whose equation is $y = 7x$.

27. Find an equation of the line passing through $(-1, 5)$ with slope -2. Write the equation in slope-intercept form, $y = mx + b$, and in standard form, $Ax + By = C$.

28. Determine whether the lines are parallel, perpendicular, or neither.

$$y = 4x - 5$$
$$-4x + y = 7$$

29. Find an equation of the vertical line through $(-1, 5)$.

30. Write an equation of the line with slope -5, through $(-2, 3)$.

31. Find the domain and the range of the relation $\{(0, 2), (3, 3), (-1, 0), (3, -2)\}$.

32. If $f(x) = 5x^2 - 6$, find $f(0)$ and $f(-2)$.

33. Determine whether each relation is also a function.
   a. $\{(-1, 1), (2, 3), (7, 3), (8, 6)\}$
   b. $\{(0, -2), (1, 5), (0, 3), (7, 7)\}$

34. Determine whether each graph is also the graph of a function.

a.   **b.**   **c.**

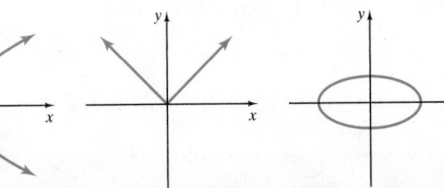

35. Determine the number of solutions of the system.
$$\begin{cases} 3x - y = 4 \\ x + 2y = 8 \end{cases}$$

36. Determine whether each ordered pair is a solution of the given system.
$$\begin{cases} 2x - y = 6 \\ 3x + 2y = -5 \end{cases}$$

  a. $(1, -4)$ **b.** $(0, 6)$ **c.** $(3, 0)$

Solve each system.

37. $\begin{cases} x + 2y = 7 \\ 2x + 2y = 13 \end{cases}$

38. $\begin{cases} 3x - 4y = 10 \\ \qquad y = 2x \end{cases}$

39. $\begin{cases} x + y = 7 \\ x - y = 5 \end{cases}$

40. $\begin{cases} x = 5y - 3 \\ x = 8y + 4 \end{cases}$

41. Solve the system.
$$\begin{cases} 3x - y + z = -15 \\ x + 2y - z = 1 \\ 2x + 3y - 2z = 0 \end{cases}$$

42. Solve the system.
$$\begin{cases} x - 2y + z = 0 \\ 3x - y - 2z = -15 \\ 2x - 3y + 3z = 7 \end{cases}$$

43. A first number is 4 less than a second number. Four times the first number is 6 more than twice the second. Find the numbers.

44. Find two numbers whose sum is 37 and whose difference is 21.

CHAPTER 5

Exponents and Polynomials

5.1 Exponents

5.2 Polynomial Functions and Adding and Subtracting Polynomials

5.3 Multiplying Polynomials

5.4 Special Products

Integrated Review—Exponents and Operations on Polynomials

5.5 Negative Exponents and Scientific Notation

5.6 Dividing Polynomials

5.7 Synthetic Division and the Remainder Theorem

Vocabulary Check

Chapter Highlights

Chapter Review

Getting Ready for the Test

Chapter Test

Cumulative Review

Recall from Chapter 1 that an exponent is a shorthand notation for repeated factors. This chapter explores additional concepts about exponents and exponential expressions. An especially useful type of exponential expression is a polynomial. Polynomials model many real-world phenomena. In this chapter, we focus on polynomials and operations on polynomials.

Can You Imagine a World Without the Internet?

In 1995, less than 1% of the world population was connected to the Internet. By 2015, that number had increased to 40%. Technology changes so fast that, if this trend continues, by the time you read this, far more than 40% of the world population will be connected to the Internet. The circle graph below shows Internet users by region of the world in 2015. In Section 5.2, Exercises 99 and 100, we explore more about the growth of Internet users.

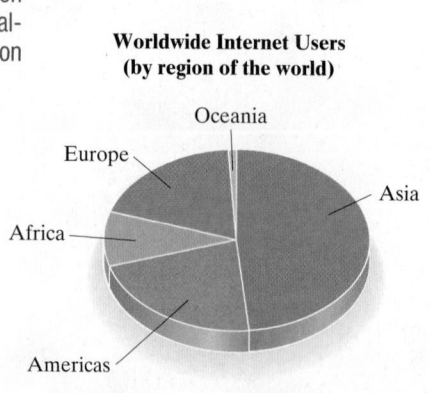

Worldwide Internet Users (by region of the world)

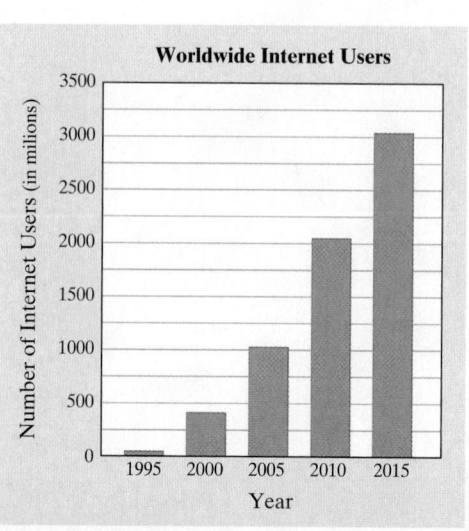

Data from International Telecommunication Union and United Nations Population Division

5.1 Exponents

OBJECTIVES

1 Evaluate Exponential Expressions.

2 Use the Product Rule for Exponents.

3 Use the Power Rule for Exponents.

4 Use the Power Rules for Products and Quotients.

5 Use the Quotient Rule for Exponents, and Define a Number Raised to the 0 Power.

6 Decide Which Rule(s) to Use to Simplify an Expression.

OBJECTIVE

1 Evaluating Exponential Expressions

As we reviewed in Section 1.4, an exponent is a shorthand notation for repeated factors. For example, $2 \cdot 2 \cdot 2 \cdot 2 \cdot 2$ can be written as 2^5. The expression 2^5 is called an **exponential expression**. It is also called the fifth **power** of 2, or we say that 2 is **raised** to the fifth power.

$$5^6 = \underbrace{5 \cdot 5 \cdot 5 \cdot 5 \cdot 5 \cdot 5}_{6 \text{ factors; each factor is } 5} \qquad \text{and} \qquad (-3)^4 = \underbrace{(-3) \cdot (-3) \cdot (-3) \cdot (-3)}_{4 \text{ factors; each factor is } -3}$$

The **base** of an exponential expression is the repeated factor. The **exponent** is the number of times that the base is used as a factor.

$$5^6 \overset{\curvearrowleft \text{exponent}}{\underset{\curvearrowleft \text{base}}{}} \qquad (-3)^4 \overset{\curvearrowleft \text{exponent}}{\underset{\curvearrowleft \text{base}}{}}$$

EXAMPLE 1 Evaluate each expression.

a. 2^3 **b.** 3^1 **c.** $(-4)^2$ **d.** -4^2 **e.** $\left(\dfrac{1}{2}\right)^4$ **f.** $(0.5)^3$ **g.** $4 \cdot 3^2$

Solution

a. $2^3 = 2 \cdot 2 \cdot 2 = 8$

b. To raise 3 to the first power means to use 3 as a factor only once. Therefore, $3^1 = 3$. Also, when no exponent is shown, the exponent is assumed to be 1.

c. $(-4)^2 = (-4)(-4) = 16$ **d.** $-4^2 = -(4 \cdot 4) = -16$

e. $\left(\dfrac{1}{2}\right)^4 = \dfrac{1}{2} \cdot \dfrac{1}{2} \cdot \dfrac{1}{2} \cdot \dfrac{1}{2} = \dfrac{1}{16}$ **f.** $(0.5)^3 = (0.5)(0.5)(0.5) = 0.125$

g. $4 \cdot 3^2 = 4 \cdot 9 = 36$ □

PRACTICE

1 Evaluate each expression.

a. 3^3 **b.** 4^1 **c.** $(-8)^2$ **d.** -8^2

e. $\left(\dfrac{3}{4}\right)^3$ **f.** $(0.3)^4$ **g.** $3 \cdot 5^2$ ▪

Notice how similar -4^2 is to $(-4)^2$ in the example above. The difference between the two is the parentheses. In $(-4)^2$, the parentheses tell us that the base, or repeated factor, is -4. In -4^2, only 4 is the base.

> **Helpful Hint**
>
> Be careful when identifying the base of an exponential expression. Pay close attention to the use of parentheses.
>
> $\qquad (-3)^2 \qquad\qquad\qquad -3^2 \qquad\qquad\qquad 2 \cdot 3^2$
> $\quad$ The base is -3. $\qquad$ The base is 3. $\qquad$ The base is 3.
> $(-3)^2 = (-3)(-3) = 9 \quad -3^2 = -(3 \cdot 3) = -9 \quad 2 \cdot 3^2 = 2 \cdot 3 \cdot 3 = 18$

An exponent has the same meaning whether the base is a number or a variable. If x is a real number and n is a positive integer, then x^n is the product of n factors, each of which is x.

$$x^n = \underbrace{x \cdot x \cdot x \cdot x \cdot x \cdot \ldots \cdot x}_{n \text{ factors of } x}$$

EXAMPLE 2 Evaluate each expression for the given value of x.

    a. $2x^3$; x is 5   **b.** $\dfrac{9}{x^2}$; x is -3

Solution   **a.** If x is 5, $2x^3 = 2 \cdot (5)^3$ **b.** If x is -3, $\dfrac{9}{x^2} = \dfrac{9}{(-3)^2}$

$$= 2 \cdot (5 \cdot 5 \cdot 5)$$
$$= 2 \cdot 125$$
$$= 250$$

$$= \dfrac{9}{(-3)(-3)}$$
$$= \dfrac{9}{9}$$
$$= 1 \qquad \square$$

PRACTICE

2   Evaluate each expression for the given value of x.

    a. $3x^4$; x is 3   **b.** $\dfrac{6}{x^2}$; x is -4

OBJECTIVE

2   **Using the Product Rule** ▶

Exponential expressions can be multiplied, divided, added, subtracted, and themselves raised to powers. By our definition of an exponent,

$$5^4 \cdot 5^3 = \underbrace{(5 \cdot 5 \cdot 5 \cdot 5)}_{4 \text{ factors of } 5} \cdot \underbrace{(5 \cdot 5 \cdot 5)}_{3 \text{ factors of } 5}$$

$$= \underbrace{5 \cdot 5 \cdot 5 \cdot 5 \cdot 5 \cdot 5 \cdot 5}_{7 \text{ factors of } 5}$$

$$= 5^7$$

Also,

$$x^2 \cdot x^3 = (x \cdot x) \cdot (x \cdot x \cdot x)$$
$$= x \cdot x \cdot x \cdot x \cdot x$$
$$= x^5$$

In both cases, notice that the result is exactly the same if the exponents are added.

$$5^4 \cdot 5^3 = 5^{4+3} = 5^7 \qquad \text{and} \qquad x^2 \cdot x^3 = x^{2+3} = x^5$$

This suggests the following rule.

Product Rule for Exponents

If m and n are positive integers and a is a real number, then

$$a^m \cdot a^n = a^{m+n} \leftarrow \text{Add exponents.}$$
                ↑_____ Keep common base.

For example, $3^5 \cdot 3^7 = 3^{5+7} = 3^{12} \leftarrow$ Add exponents.
                      ↑_____ Keep common base.

Helpful Hint

Don't forget that

$$3^5 \cdot 3^7 \neq 9^{12} \leftarrow \text{Add exponents.}$$
                ↑_____ **Common base *not* kept.**

$$3^5 \cdot 3^7 = \underbrace{3 \cdot 3 \cdot 3 \cdot 3 \cdot 3}_{5 \text{ factors of } 3} \cdot \underbrace{3 \cdot 3 \cdot 3 \cdot 3 \cdot 3 \cdot 3 \cdot 3}_{7 \text{ factors of } 3}$$

$$= 3^{12} \quad 12 \text{ factors of } 3, \textit{not } 9$$

In other words, to multiply two exponential expressions with the **same base,** we keep the base and add the exponents. We call this **simplifying** the exponential expression.

EXAMPLE 3   Use the product rule to simplify.

a. $4^2 \cdot 4^5$ b. $x^4 \cdot x^6$ c. $y^3 \cdot y$

▶ d. $y^3 \cdot y^2 \cdot y^7$ ▶ e. $(-5)^7 \cdot (-5)^8$ f. $a^2 \cdot b^2$

Solution

a. $4^2 \cdot 4^5 = 4^{2+5} = 4^7 \leftarrow$ Add exponents.
      └── **Keep** common base.

b. $x^4 \cdot x^6 = x^{4+6} = x^{10}$

c. $y^3 \cdot y = y^3 \cdot y^1$
   $= y^{3+1}$
   $= y^4$

> **Helpful Hint**
> Don't forget that if no exponent is written, it is assumed to be 1.

d. $y^3 \cdot y^2 \cdot y^7 = y^{3+2+7} = y^{12}$

e. $(-5)^7 \cdot (-5)^8 = (-5)^{7+8} = (-5)^{15}$

f. $a^2 \cdot b^2$ Cannot be simplified because a and b are different bases.   □

PRACTICE
3   Use the product rule to simplify.

a. $3^4 \cdot 3^6$ b. $y^3 \cdot y^2$

c. $z \cdot z^4$ d. $x^3 \cdot x^2 \cdot x^6$

e. $(-2)^5 \cdot (-2)^3$ f. $b^3 \cdot t^5$ ■

✔ **CONCEPT CHECK**
Where possible, use the product rule to simplify the expression.
a. $z^2 \cdot z^{14}$ b. $x^2 \cdot y^{14}$ c. $9^8 \cdot 9^3$ d. $9^8 \cdot 2^7$

EXAMPLE 4   Use the product rule to simplify $(2x^2)(-3x^5)$.

Solution Recall that $2x^2$ means $2 \cdot x^2$ and $-3x^5$ means $-3 \cdot x^5$.

$(2x^2)(-3x^5) = 2 \cdot x^2 \cdot -3 \cdot x^5$ Remove parentheses.

$= 2 \cdot -3 \cdot x^2 \cdot x^5$ Group factors with common bases.

$= -6x^7$ Simplify.   □

PRACTICE
4   Use the product rule to simplify $(-5y^3)(-3y^4)$.   ■

EXAMPLE 5   Simplify.

a. $(x^2y)(x^3y^2)$ b. $(-a^7b^4)(3ab^9)$

Solution

a. $(x^2y)(x^3y^2) = (x^2 \cdot x^3) \cdot (y^1 \cdot y^2)$ Group like bases and write y as y^1.
   $= x^5 \cdot y^3$ or x^5y^3 Multiply.

b. $(-a^7b^4)(3ab^9) = (-1 \cdot 3) \cdot (a^7 \cdot a^1) \cdot (b^4 \cdot b^9)$
   $= -3a^8b^{13}$ □

PRACTICE
5   Simplify.

a. $(y^7z^3)(y^5z)$ b. $(-m^4n^4)(7mn^{10})$ ■

> **Helpful Hint**
>
> These examples will remind you of the difference between adding and multiplying terms.
> **Addition**
>
> $$5x^3 + 3x^3 = (5 + 3)x^3 = 8x^3 \quad \text{By the distributive property.}$$
> $$7x + 4x^2 = 7x + 4x^2 \qquad\qquad \text{Cannot be combined.}$$
>
> **Multiplication**
>
> $$(5x^3)(3x^3) = 5 \cdot 3 \cdot x^3 \cdot x^3 = 15x^{3+3} = 15x^6 \quad \text{By the product rule.}$$
> $$(7x)(4x^2) = 7 \cdot 4 \cdot x \cdot x^2 = 28x^{1+2} = 28x^3 \quad \text{By the product rule.}$$

OBJECTIVE

3 Using the Power Rule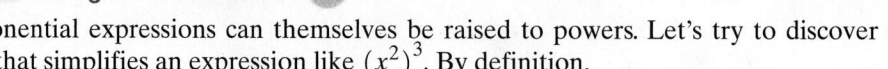

Exponential expressions can themselves be raised to powers. Let's try to discover a rule that simplifies an expression like $(x^2)^3$. By definition,

$$(x^2)^3 = \underbrace{(x^2)(x^2)(x^2)}_{3 \text{ factors of } x^2}$$

which can be simplified by the product rule for exponents.

$$(x^2)^3 = (x^2)(x^2)(x^2) = x^{2+2+2} = x^6$$

Notice that the result is exactly the same if we multiply the exponents.

$$(x^2)^3 = x^{2 \cdot 3} = x^6$$

The following property states this result.

> **Power Rule for Exponents**
>
> If m and n are positive integers and a is a real number, then
>
> $$(a^m)^n = a^{mn} \leftarrow \text{Multiply exponents.}$$
> $$\phantom{(a^m)^n = a^{mn}} \text{Keep common base.}$$

For example, $(7^2)^5 = 7^{2 \cdot 5} = 7^{10} \leftarrow$ Multiply exponents.
Keep common base.

To raise a power to a power, keep the base and multiply the exponents.

> **EXAMPLE 6**   Use the power rule to simplify.
>
> **a.** $(y^8)^2$ **b.** $(8^4)^5$ ▷ **c.** $[(-5)^3]^7$
>
> **Solution**
>
> **a.** $(y^8)^2 = y^{8 \cdot 2} = y^{16}$ **b.** $(8^4)^5 = 8^{4 \cdot 5} = 8^{20}$ **c.** $[(-5)^3]^7 = (-5)^{21}$ □

PRACTICE

6   Use the power rule to simplify.

   a. $(z^3)^7$ **b.** $(4^9)^2$ **c.** $[(-2)^3]^5$ ∎

> **Helpful Hint**
>
> Take a moment to make sure that you understand when to apply the product rule and when to apply the power rule.
>
> | *Product Rule* → *Add Exponents* | *Power Rule* → *Multiply Exponents* |
> | --- | --- |
> | $x^5 \cdot x^7 = x^{5+7} = x^{12}$ | $(x^5)^7 = x^{5 \cdot 7} = x^{35}$ |
> | $y^6 \cdot y^2 = y^{6+2} = y^8$ | $(y^6)^2 = y^{6 \cdot 2} = y^{12}$ |

OBJECTIVE

4 Using the Power Rules for Products and Quotients

When the base of an exponential expression is a product, the definition of x^n still applies. To simplify $(xy)^3$, for example,

$$(xy)^3 = (xy)(xy)(xy) \quad (xy)^3 \text{ means 3 factors of } (xy).$$
$$= x \cdot x \cdot x \cdot y \cdot y \cdot y \quad \text{Group factors with common bases.}$$
$$= x^3 y^3 \quad \text{Simplify.}$$

Notice that to simplify the expression $(xy)^3$, we raise each factor within the parentheses to a power of 3.

$$(xy)^3 = x^3 y^3$$

In general, we have the following rule.

Power of a Product Rule

If n is a positive integer and a and b are real numbers, then

$$(ab)^n = a^n b^n$$

For example, $(3x)^5 = 3^5 x^5$.

In other words, to raise a product to a power, we raise each factor to the power.

EXAMPLE 7   Simplify each expression.

a. $(st)^4$ **b.** $(2a)^3$ **c.** $\left(\dfrac{1}{3}mn^3\right)^2$ **d.** $(-5x^2 y^3 z)^2$

Solution

a. $(st)^4 = s^4 \cdot t^4 = s^4 t^4$ — Use the power of a product rule.

b. $(2a)^3 = 2^3 \cdot a^3 = 8a^3$ — Use the power of a product rule.

c. $\left(\dfrac{1}{3}mn^3\right)^2 = \left(\dfrac{1}{3}\right)^2 \cdot (m)^2 \cdot (n^3)^2 = \dfrac{1}{9}m^2 n^6$ — Use the power of a product rule.

d. $(-5x^2 y^3 z)^2 = (-5)^2 \cdot (x^2)^2 \cdot (y^3)^2 \cdot (z^1)^2$ — Use the power of a product rule.
$$= 25x^4 y^6 z^2 \quad \text{Use the power rule for exponents. } \square$$

PRACTICE

7   Simplify each expression.

a. $(pr)^5$ **b.** $(6b)^2$ **c.** $\left(\dfrac{1}{4}x^2 y\right)^3$ **d.** $(-3a^3 b^4 c)^4$ ■

Let's see what happens when we raise a quotient to a power. To simplify $\left(\dfrac{x}{y}\right)^3$, for example,

$$\left(\frac{x}{y}\right)^3 = \left(\frac{x}{y}\right)\left(\frac{x}{y}\right)\left(\frac{x}{y}\right) \quad \left(\frac{x}{y}\right)^3 \text{ means 3 factors of } \left(\frac{x}{y}\right)$$
$$= \frac{x \cdot x \cdot x}{y \cdot y \cdot y} \quad \text{Multiply fractions.}$$
$$= \frac{x^3}{y^3} \quad \text{Simplify.}$$

Notice that to simplify the expression $\left(\dfrac{x}{y}\right)^3$, we raise both the numerator and the denominator to a power of 3.

$$\left(\frac{x}{y}\right)^3 = \frac{x^3}{y^3}$$

In general, we have the following.

> **Power of a Quotient Rule**
>
> If n is a positive integer and a and c are real numbers, then
>
> $$\left(\frac{a}{c}\right)^n = \frac{a^n}{c^n}, \quad c \neq 0$$

For example, $\left(\dfrac{y}{7}\right)^4 = \dfrac{y^4}{7^4}$.

In other words, to raise a quotient to a power, we raise both the numerator and the denominator to the power.

EXAMPLE 8 Simplify each expression.

a. $\left(\dfrac{m}{n}\right)^7$ **b.** $\left(\dfrac{x^3}{3y^5}\right)^4$

Solution

a. $\left(\dfrac{m}{n}\right)^7 = \dfrac{m^7}{n^7}, n \neq 0$ Use the power of a quotient rule.

b. $\left(\dfrac{x^3}{3y^5}\right)^4 = \dfrac{(x^3)^4}{3^4 \cdot (y^5)^4}, y \neq 0$ Use the power of a product or quotient rule.

$\qquad\qquad = \dfrac{x^{12}}{81y^{20}}$ Use the power rule for exponents.   ☐

PRACTICE
8 Simplify each expression.

a. $\left(\dfrac{x}{y^2}\right)^5$ **b.** $\left(\dfrac{2a^4}{b^3}\right)^5$ ■

OBJECTIVE

5 Using the Quotient Rule and Defining the Zero Exponent

Another pattern for simplifying exponential expressions involves quotients.

To simplify an expression like $\dfrac{x^5}{x^3}$, in which the numerator and the denominator have a common base, we can apply the fundamental principle of fractions and divide the numerator and the denominator by the common base factors. Assume for the remainder of this section that denominators are not 0.

$$\frac{x^5}{x^3} = \frac{x \cdot x \cdot x \cdot x \cdot x}{x \cdot x \cdot x}$$

$$= \frac{x \cdot x \cdot x \cdot x \cdot x}{x \cdot x \cdot x}$$

$$= x \cdot x$$

$$= x^2$$

Notice that the result is exactly the same if we subtract exponents of the common bases.

$$\frac{x^5}{x^3} = x^{5-3} = x^2$$

The quotient rule for exponents states this result in a general way.

> **Quotient Rule for Exponents**
>
> If m and n are positive integers and a is a real number, then
>
> $$\frac{a^m}{a^n} = a^{m-n}$$
>
> as long as a is not 0.

For example, $\dfrac{x^6}{x^2} = x^{6-2} = x^4$.

In other words, to divide one exponential expression by another with a common base, keep the base and subtract exponents.

EXAMPLE 9 Simplify each quotient.

▶ **a.** $\dfrac{x^5}{x^2}$ **b.** $\dfrac{4^7}{4^3}$ **c.** $\dfrac{(-3)^5}{(-3)^2}$ **d.** $\dfrac{s^2}{t^3}$ **e.** $\dfrac{2x^5y^2}{xy}$

Solution

a. $\dfrac{x^5}{x^2} = x^{5-2} = x^3$ Use the quotient rule.

b. $\dfrac{4^7}{4^3} = 4^{7-3} = 4^4 = 256$ Use the quotient rule.

c. $\dfrac{(-3)^5}{(-3)^2} = (-3)^3 = -27$ Use the quotient rule.

d. $\dfrac{s^2}{t^3}$ Cannot be simplified because s and t are different bases.

e. Begin by grouping common bases.

$$\frac{2x^5y^2}{xy} = 2 \cdot \frac{x^5}{x^1} \cdot \frac{y^2}{y^1}$$

$$= 2 \cdot (x^{5-1}) \cdot (y^{2-1})$$ Use the quotient rule.

$$= 2x^4y^1 \quad \text{or} \quad 2x^4y$$ ☐

PRACTICE

9 Simplify each quotient.

a. $\dfrac{z^8}{z^4}$ **b.** $\dfrac{(-5)^5}{(-5)^3}$ **c.** $\dfrac{8^8}{8^6}$ **d.** $\dfrac{q^5}{t^2}$ **e.** $\dfrac{6x^3y^7}{xy^5}$ ■

✔ **CONCEPT CHECK**

Suppose you are simplifying each expression. Tell whether you would *add* the exponents, *subtract* the exponents, *multiply* the exponents, *divide* the exponents, or *none of these*.

a. $(x^{63})^{21}$ **b.** $\dfrac{y^{15}}{y^3}$ **c.** $z^{16} + z^8$ **d.** $w^{45} \cdot w^9$

Let's now give meaning to an expression such as x^0. To do so, we will simplify $\dfrac{x^3}{x^3}$ in two ways and compare the results.

$$\frac{x^3}{x^3} = x^{3-3} = x^0$$ Apply the quotient rule.

$$\frac{x^3}{x^3} = \frac{x \cdot x \cdot x}{x \cdot x \cdot x} = 1$$ Apply the fundamental principle for fractions.

Answers to Concept Check:
a. multiply   **b.** subtract
c. none of these   **d.** add

Since $\dfrac{x^3}{x^3} = x^0$ and $\dfrac{x^3}{x^3} = 1$, we define that $x^0 = 1$ as long as x is not 0.

> **Zero Exponent**
>
> $a^0 = 1$, as long as a is not 0.

In other words, any base raised to the 0 power is 1 as long as the base is not 0.

EXAMPLE 10   Simplify each expression.

    a. 3^0 **b.** $(5x^3y^2)^0$ **c.** $(-5)^0$ **d.** -5^0 **e.** $\left(\dfrac{3}{100}\right)^0$ **f.** $4x^0$

Solution

    a. $3^0 = 1$

    b. Assume that neither x nor y is zero.
$$(5x^3y^2)^0 = 1$$

    c. $(-5)^0 = 1$

    d. $-5^0 = -1 \cdot 5^0 = -1 \cdot 1 = -1$

    e. $\left(\dfrac{3}{100}\right)^0 = 1$

    f. $4x^0 = 4 \cdot x^0 = 4 \cdot 1 = 4$

PRACTICE
10   Simplify the following expressions.

    a. -3^0 **b.** $(-3)^0$ **c.** 8^0 **d.** $(0.2)^0$

    e. $(7a^2y^4)^0$ **f.** $7y^0$

OBJECTIVE

6   Deciding Which Rule to Use ▶

Let's practice deciding which rule(s) to use to simplify. We will continue this discussion with more examples in Section 5.5.

EXAMPLE 11   Simplify each expression.

    a. $x^7 \cdot x^4$ **b.** $\left(\dfrac{t}{2}\right)^4$ **c.** $(9y^5)^2$

Solution

    a. Here we have a product, so we use the product rule to simplify.
$$x^7 \cdot x^4 = x^{7+4} = x^{11}$$

    b. This is a quotient raised to a power, so we use the power of a quotient rule.
$$\left(\dfrac{t}{2}\right)^4 = \dfrac{t^4}{2^4} = \dfrac{t^4}{16}$$

    c. This is a product raised to a power, so we use the power of a product rule.
$$(9y^5)^2 = 9^2(y^5)^2 = 81y^{10}$$

PRACTICE
11   **a.** $\left(\dfrac{z}{12}\right)^2$ **b.** $(4x^6)^3$ **c.** $y^{10} \cdot y^3$

EXAMPLE 12   Simplify each expression.

    a. $4^2 - 4^0$ **b.** $(x^0)^3 + (2^0)^5$ **c.** $\left(\dfrac{3y^7}{6x^5}\right)^2$ **d.** $\dfrac{(2a^3b^4)^3}{-8a^9b^2}$

Solution

    a. $4^2 - 4^0 = 16 - 1 = 15$ Remember that $4^0 = 1$.

    b. $(x^0)^3 + (2^0)^5 = 1^3 + 1^5 = 1 + 1 = 2$

    c. $\left(\dfrac{3y^7}{6x^5}\right)^2 = \dfrac{3^2(y^7)^2}{6^2(x^5)^2} = \dfrac{9 \cdot y^{14}}{36 \cdot x^{10}} = \dfrac{y^{14}}{4x^{10}}$

    d. $\dfrac{(2a^3b^4)^3}{-8a^9b^2} = \dfrac{2^3(a^3)^3(b^4)^3}{-8a^9b^2} = \dfrac{8a^9b^{12}}{-8a^9b^2} = -1 \cdot (a^{9-9}) \cdot (b^{12-2})$

                               $= -1 \cdot a^0 \cdot b^{10} = -1 \cdot 1 \cdot b^{10} = -b^{10}$

PRACTICE
12   Simplify each expression.

    a. $8^2 - 8^0$ **b.** $(z^0)^6 + (4^0)^5$ **c.** $\left(\dfrac{5x^3}{15y^4}\right)^2$ **d.** $\dfrac{(2z^8x^5)^4}{-16z^2x^{20}}$

✔ ┃ **Vocabulary, Readiness & Video Check** ┃

Use the choices below to fill in each blank. Some choices may be used more than once.

    0   base   add

    1   exponent   multiply

1. Repeated multiplication of the same factor can be written using a(n) _____.

2. In 5^2, the 2 is called the _____ and the 5 is called the _____.

3. To simplify $x^2 \cdot x^7$, keep the base and _____ the exponents.

4. To simplify $(x^3)^6$, keep the base and _____ the exponents.

5. The understood exponent on the term y is _____.

6. If $x^\square = 1$, the exponent is _____.

Martin-Gay Interactive Videos

See Video 5.1 ◉

Watch the section lecture video and answer the following questions.

OBJECTIVE
1

7. ▦ Examples 3 and 4 illustrate how to find the base of an exponential expression both with and without parentheses. Explain how identifying the base of ▦ Example 7 is similar to identifying the base of ▦ Example 4.

OBJECTIVE
2

8. Why were the commutative and associative properties applied in ▦ Example 12? Were these properties used in another example?

OBJECTIVE
3

9. What point is made at the end of ▦ Example 15?

OBJECTIVE
4

10. Although it's not especially emphasized in ▦ Example 20, what is helpful to remind yourself about the -2 in the problem?

OBJECTIVE
5

11. In ▦ Example 24, which exponent rule is used to show that any nonzero base raised to zero is 1?

OBJECTIVE
6

12. When simplifying an exponential expression that's a fraction, will you always use the quotient rule? Refer to ▦ Example 30 for this objective to support your answer.

5.1 Exercise Set MyMathLab®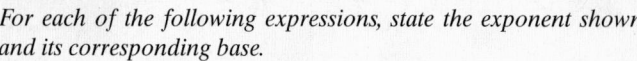

For each of the following expressions, state the exponent shown and its corresponding base.

1. 3^2

2. $(-3)^6$

3. -4^2

4. $5 \cdot 3^4$

5. $5x^2$

6. $(5x)^2$

Evaluate each expression. See Example 1.

7. 7^2

8. -3^2

9. $(-5)^1$

10. $(-3)^2$

11. -2^4

12. -4^3

13. $(-2)^4$

14. $(-4)^3$

15. $(0.1)^5$

16. $(0.2)^5$

17. $\left(\dfrac{1}{3}\right)^4$

18. $\left(-\dfrac{1}{9}\right)^2$

19. $7 \cdot 2^5$

20. $9 \cdot 1^7$

21. $-2 \cdot 5^3$

22. $-4 \cdot 3^3$

Evaluate each expression for the replacement values given. See Example 2.

23. $x^2; x = -2$

24. $x^3; x = -2$

25. $5x^3; x = 3$

26. $4x^2; x = -1$

27. $2xy^2; x = 3$ and $y = 5$

28. $-4x^2y^3; x = 2$ and $y = -1$

29. $\dfrac{2z^4}{5}; z = -2$

30. $\dfrac{10}{3y^3}; y = 5$

Use the product rule to simplify each expression. Write the results using exponents. See Examples 3 through 5.

31. $x^2 \cdot x^5$

32. $y^2 \cdot y$

33. $(-3)^3 \cdot (-3)^9$

34. $(-5)^7 \cdot (-5)^6$

35. $(5y^4)(3y)$

36. $(-2z^3)(-2z^2)$

37. $(x^9y)(x^{10}y^5)$

38. $(a^2b)(a^{13}b^{17})$

39. $(-8mn^6)(9m^2n^2)$

40. $(-7a^3b^3)(7a^{19}b)$

41. $(4z^{10})(-6z^7)(z^3)$

42. $(12x^5)(-x^6)(x^4)$

43. The rectangle below has width $4x^2$ feet and length $5x^3$ feet. Find its area as an expression in x. $(A = l \cdot w)$

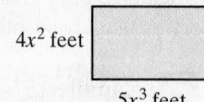

$4x^2$ feet

$5x^3$ feet

44. The parallelogram below has base length $9y^7$ meters and height $2y^{10}$ meters. Find its area as an expression in y. $(A = b \cdot h)$

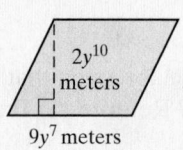

$2y^{10}$ meters

$9y^7$ meters

MIXED PRACTICE

Use the power rule and the power of a product or quotient rule to simplify each expression. See Examples 6 through 8.

45. $(x^9)^4$

46. $(y^7)^5$

47. $(pq)^8$

48. $(ab)^6$

49. $(2a^5)^3$

50. $(4x^6)^2$

51. $(x^2y^3)^5$

52. $(a^4b)^7$

53. $(-7a^2b^5c)^2$

54. $(-3x^7yz^2)^3$

55. $\left(\dfrac{r}{s}\right)^9$

56. $\left(\dfrac{q}{t}\right)^{11}$

57. $\left(\dfrac{mp}{n}\right)^5$

58. $\left(\dfrac{xy}{7}\right)^2$

59. $\left(\dfrac{-2xz}{y^5}\right)^2$

60. $\left(\dfrac{xy^4}{-3z^3}\right)^3$

61. The square shown has sides of length $8z^5$ decimeters. Find its area. $(A = s^2)$

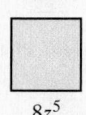

$8z^5$ decimeters

62. Given the circle below with radius $5y$ centimeters, find its area. Do not approximate π. $(A = \pi r^2)$

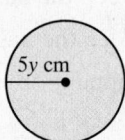

$5y$ cm

63. The vault below is in the shape of a cube. If each side is $3y^4$ feet, find its volume. $(V = s^3)$

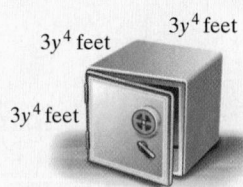

$3y^4$ feet

$3y^4$ feet

$3y^4$ feet

64. The silo shown is in the shape of a cylinder. If its radius is $4x$ meters and its height is $5x^3$ meters, find its volume. Do not approximate π. $(V = \pi r^2h)$

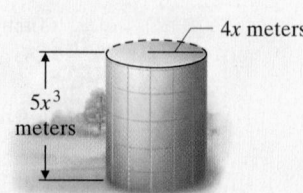

$4x$ meters

$5x^3$ meters

Use the quotient rule and simplify each expression. See Example 9.

65. $\dfrac{x^3}{x}$

66. $\dfrac{y^{10}}{y^9}$

67. $\dfrac{(-4)^6}{(-4)^3}$

68. $\dfrac{(-6)^{13}}{(-6)^{11}}$

69. $\dfrac{p^7 q^{20}}{pq^{15}}$

70. $\dfrac{x^8 y^6}{xy^5}$

71. $\dfrac{7x^2 y^6}{14x^2 y^3}$

72. $\dfrac{9a^4 b^7}{27ab^2}$

Simplify each expression. See Example 10.

73. 7^0

74. 23^0

75. $(2x)^0$

76. $(4y)^0$

77. $-7x^0$

78. $-2x^0$

79. $5^0 + y^0$

80. $-3^0 + 4^0$

MIXED PRACTICE

Simplify each expression. See Examples 1 through 12.

81. -9^2

82. $(-9)^2$

83. $\left(\dfrac{1}{4}\right)^3$

84. $\left(\dfrac{2}{3}\right)^3$

85. $b^4 b^2$

86. $y^4 y$

87. $a^2 a^3 a^4$

88. $x^2 x^{15} x^9$

89. $(2x^3)(-8x^4)$

90. $(3y^4)(-5y)$

91. $(a^7 b^{12})(a^4 b^8)$

92. $(y^2 z^2)(y^{15} z^{13})$

93. $(-2mn^6)(-13m^8 n)$

94. $(-3s^5 t)(-7st^{10})$

95. $(z^4)^{10}$

96. $(t^5)^{11}$

97. $(4ab)^3$

98. $(2ab)^4$

99. $(-6xyz^3)^2$

100. $(-3xy^2 a^3)^3$

101. $\dfrac{3x^5}{x^4}$

102. $\dfrac{5x^9}{x^3}$

103. $(9xy)^2$

104. $(2ab)^5$

105. $2^3 + 2^0$

106. $7^2 - 7^0$

107. $\left(\dfrac{3y^5}{6x^4}\right)^3$

108. $\left(\dfrac{2ab}{6yz}\right)^4$

109. $\dfrac{2x^3 y^2 z}{xyz}$

110. $\dfrac{x^{12} y^{13}}{x^5 y^7}$

111. $(5^0)^3 + (y^0)^7$

112. $(9^0)^4 + (z^0)^5$

113. $\left(\dfrac{5x^9}{10y^{11}}\right)^2$

114. $\left(\dfrac{3a^4}{9b^5}\right)^2$

115. $\dfrac{(2a^5 b^3)^4}{-16a^{20} b^7}$

116. $\dfrac{(2x^6 y^2)^5}{-32x^{20} y^{10}}$

REVIEW AND PREVIEW

Simplify each expression by combining any like terms. Use the distributive property to remove any parentheses. See Section 2.1.

117. $y - 10 + y$

118. $-6z + 20 - 3z$

119. $7x + 2 - 8x - 6$

120. $10y - 14 - y - 14$

121. $2(x - 5) + 3(5 - x)$

122. $-3(w + 7) + 5(w + 1)$

CONCEPT EXTENSIONS

Solve. See the Concept Checks in this section. For Exercises 123 through 126, match the expression with the operation needed to simplify each. A letter may be used more than once and a letter may not be used at all.

123. $(x^{14})^{23}$

124. $x^{14} \cdot x^{23}$

125. $x^{14} + x^{23}$

126. $\dfrac{x^{35}}{x^{17}}$

A. Add the exponents

B. Subtract the exponents

C. Multiply the exponents

D. Divide the exponents

E. None of these

Fill in the boxes so that each statement is true. (More than one answer is possible for each exercise.)

127. $x^\square \cdot x^\square = x^{12}$

128. $(x^\square)^\square = x^{20}$

129. $\dfrac{y^\square}{y^\square} = y^7$

130. $(y^\square)^\square \cdot (y^\square)^\square = y^{30}$

131. The formula $V = x^3$ can be used to find the volume V of a cube with side length x. Find the volume of a cube with side length 7 meters. (Volume is measured in cubic units.)

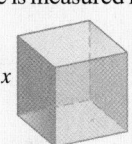

132. The formula $S = 6x^2$ can be used to find the surface area S of a cube with side length x. Find the surface area of a cube with side length 5 meters. (Surface area is measured in square units.)

133. To find the amount of water that a swimming pool in the shape of a cube can hold, do we use the formula for volume of the cube or surface area of the cube? (See Exercises 131 and 132.)

134. To find the amount of material needed to cover an ottoman in the shape of a cube, do we use the formula for volume of the cube or surface area of the cube? (See Exercises 131 and 132.)

135. Explain why $(-5)^4 = 625$, while $-5^4 = -625$.

136. Explain why $5 \cdot 4^2 = 80$, while $(5 \cdot 4)^2 = 400$.

137. In your own words, explain why $5^0 = 1$.

138. In your own words, explain when $(-3)^n$ is positive and when it is negative.

Simplify each expression. Assume that variables represent positive integers.

139. $x^{5a} x^{4a}$

140. $b^{9a} b^{4a}$

141. $(a^b)^5$

142. $(2a^{4b})^4$

143. $\dfrac{x^{9a}}{x^{4a}}$

144. $\dfrac{y^{15b}}{y^{6b}}$

Solve. Round money amounts to 2 decimal places.

145. Suppose you borrow money for 6 months. If the interest is compounded monthly, the formula $A = P\left(1 + \dfrac{r}{12}\right)^6$ gives the total amount A to be repaid at the end of 6 months. For a loan of $P = \$1000$ and interest rate of 9% ($r = 0.09$), how much money is needed to pay off the loan?

146. Suppose you borrow money for 3 years. If the interest is compounded quarterly, the formula $A = P\left(1 + \dfrac{r}{4}\right)^{12}$ gives the total amount A to be repaid at the end of 3 years. For a loan of $10,000 and interest rate of 8% ($r = 0.08$), how much money is needed to pay off the loan in 3 years?

5.2 Polynomial Functions and Adding and Subtracting Polynomials

OBJECTIVES

1 Define Polynomial, Monomial, Binomial, Trinomial, and Degree.

2 Define Polynomial Functions.

3 Simplify a Polynomial by Combining Like Terms.

4 Add and Subtract Polynomials.

OBJECTIVE

1 Defining Polynomial, Monomial, Binomial, Trinomial, and Degree

In this section, we introduce a special algebraic expression called a polynomial. Let's first review some definitions presented in Section 2.1.

Recall that a **term** is a number or the product of a number and variables raised to powers. The terms of the expression $4x^2 + 3x$ are $4x^2$ and $3x$.

The terms of the expression $9x^4 - 7x - 1$ are $9x^4$, $-7x$, and -1.

Expression	*Terms*
$4x^2 + 3x$	$4x^2, 3x$
$9x^4 - 7x - 1$	$9x^4, -7x, -1$
$7y^3$	$7y^3$
5	5

The **numerical coefficient** of a term, or simply the **coefficient**, is the numerical factor of each term. If no numerical factor appears in the term, then the coefficient is understood to be 1. If the term is a number only, it is called a **constant term** or simply a **constant**.

Term	*Coefficient*
x^5	1
$3x^2$	3
$-4x$	-4
$-x^2y$	-1
3 (constant)	3

Now we are ready to define a polynomial.

Polynomial

A **polynomial in x** is a finite sum of terms of the form ax^n, where a is a real number and n is a whole number.

For example,

$$x^5 - 3x^3 + 2x^2 - 5x + 1$$

is a polynomial. Notice that this polynomial is written in **descending powers** of x because the powers of x decrease from left to right. (Recall that the term 1 can be thought of as $1x^0$.)

On the other hand,

$$x^{-5} + 2x - 3$$

is **not** a polynomial because it contains an exponent, -5, that is not a whole number. (We study negative exponents in Section 5.5 of this chapter.)

Some polynomials are given special names.

Types of Polynomials

A **monomial** is a polynomial with exactly one term.

A **binomial** is a polynomial with exactly two terms.

A **trinomial** is a polynomial with exactly three terms.

The following are examples of monomials, binomials, and trinomials. Each of these examples is also a polynomial.

POLYNOMIALS

Monomials	*Binomials*	*Trinomials*	*More than Three Terms*
ax^2	$x + y$	$x^2 + 4xy + y^2$	$5x^3 - 6x^2 + 3x - 6$
$-3z$	$3p + 2$	$x^5 + 7x^2 - x$	$-y^5 + y^4 - 3y^3 - y^2 + y$
4	$4x^2 - 7$	$-q^4 + q^3 - 2q$	$x^6 + x^4 - x^3 + 1$

Each term of a polynomial has a **degree.**

> ### Degree of a Term
> The degree of a term is the sum of the exponents on the variables contained in the term.

EXAMPLE 1   Find the degree of each term.

    a. $3x^2$ **b.** -2^3x^5 **c.** y **d.** $12x^2yz^3$ **e.** 5

Solution

    a. The exponent on x is 2, so the degree of the term is 2.

    b. The exponent on x is 5, so the degree of the term is 5. (Recall that the degree is the sum of the exponents on only the *variables*.)

    c. The degree of y, or y^1, is 1.

    d. The degree is the sum of the exponents on the variables, or $2 + 1 + 3 = 6$.

    e. The degree of 5, which can be written as $5x^0$, is 0.   □

PRACTICE
1   Find the degree of each term.

    a. $5y^3$ **b.** $10xy$ **c.** z **d.** $-3a^2b^5c$ **e.** 8   ■

From the preceding, we can say that **the degree of a constant is 0.**
    Each polynomial also has a degree.

> ### Degree of a Polynomial
> The degree of a polynomial is the greatest degree of any term of the polynomial.

EXAMPLE 2   Find the degree of each polynomial and tell whether the polynomial is a monomial, binomial, trinomial, or none of these.

    ▶ **a.** $-2t^2 + 3t + 6$ **b.** $15x - 10$ **c.** $7x + 3x^3 + 2x^2 - 1$

Solution

    a. The degree of the trinomial $-2t^2 + 3t + 6$ is 2, the greatest degree of any of its terms.

    b. The degree of the binomial $15x - 10$ or $15x^1 - 10$ is 1.

    c. The degree of the polynomial $7x + 3x^3 + 2x^2 - 1$ is 3.   □

PRACTICE
2   Find the degree of each polynomial and tell whether the polynomial is a monomial, binomial, trinomial, or none of these.

    a. $5b^2 - 3b + 7$ **b.** $7t + 3$

    c. $5x^2 + 3x - 6x^3 + 4$ ■

EXAMPLE 3 Complete the table for the polynomial

$$7x^2y - 6xy + x^2 - 3y + 7$$

Use the table to give the degree of the polynomial.

Solution

Term	Numerical Coefficient	Degree of Term
$7x^2y$	7	3
$-6xy$	-6	2
x^2	1	2
$-3y$	-3	1
7	7	0

The degree of the polynomial is 3.

PRACTICE

3 Complete the table for the polynomial $-3x^3y^2 + 4xy^2 - y^2 + 3x - 2$.

Term	Numerical Coefficient	Degree of Term
$-3x^3y^2$		
$4xy^2$		
$-y^2$		
$3x$		
-2		

OBJECTIVE

2 Defining Polynomial Functions

At times, it is convenient to use function notation to represent polynomials. For example, we may write $P(x)$ to represent the polynomial $3x^2 - 2x - 5$. In symbols, this is

$$P(x) = 3x^2 - 2x - 5$$

This function is called a **polynomial function** because the expression $3x^2 - 2x - 5$ is a polynomial.

> **Helpful Hint**
>
> Recall that the symbol $P(x)$ **does not mean** P times x. It is a special symbol used to denote a function.

EXAMPLE 4 If $P(x) = 3x^2 - 2x - 5$, find the following.

   a. $P(1)$ **b.** $P(-2)$

Solution

   a. Substitute 1 for x in $P(x) = 3x^2 - 2x - 5$ and simplify.

$$P(x) = 3x^2 - 2x - 5$$
$$P(1) = 3(1)^2 - 2(1) - 5 = -4$$

b. Substitute -2 for x in $P(x) = 3x^2 - 2x - 5$ and simplify.

$$P(x) = 3x^2 - 2x - 5$$
$$P(-2) = 3(-2)^2 - 2(-2) - 5 = 11$$

PRACTICE
4   If $P(x) = -2x^2 - x + 7$, find
  a. $P(1)$ **b.** $P(-4)$

Many real-world phenomena are modeled by polynomial functions. If the polynomial function model is given, we can often find the solution of a problem by evaluating the function at a certain value.

EXAMPLE 5   **Finding the Height of a Dropped Object**

The Swiss Re Building, in London, is a unique building. Londoners often refer to it as the "pickle building." The building is 592.1 feet tall. An object is dropped from the highest point of this building. Neglecting air resistance, the height in feet of the object above ground at time t seconds is given by the polynomial function $P(t) = -16t^2 + 592.1$. Find the height of the object when $t = 1$ second, and when $t = 6$ seconds.

<u>Solution</u>   To find each height, we find $P(1)$ and $P(6)$.

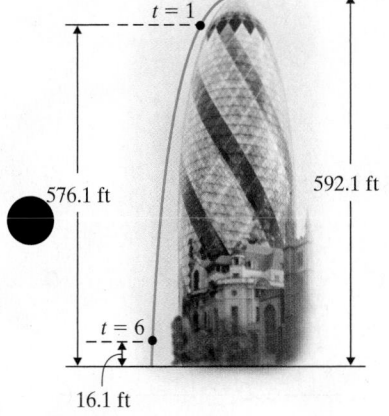

$$P(t) = -16t^2 + 592.1$$
$$P(1) = -16(1)^2 + 592.1 \quad \text{Replace } t \text{ with 1.}$$
$$P(1) = -16 + 592.1$$
$$P(1) = 576.1$$

The height of the object at 1 second is 576.1 feet.

$$P(t) = -16t^2 + 592.1$$
$$P(6) = -16(6)^2 + 592.1 \quad \text{Replace } t \text{ with 6.}$$
$$P(6) = -576 + 592.1$$
$$P(6) = 16.1$$

The height of the object at 6 seconds is 16.1 feet.

PRACTICE
5   The cliff divers of Acapulco dive 130 feet into La Quebrada several times a day for the entertainment of the tourists. If a tourist is standing near the diving platform and drops his camera off the cliff, the height of the camera above the water at time t seconds is given by the polynomial function $P(t) = -16t^2 + 130$. Find the height of the camera when $t = 1$ second and when $t = 2$ seconds.

OBJECTIVE
3   Simplifying Polynomials by Combining Like Terms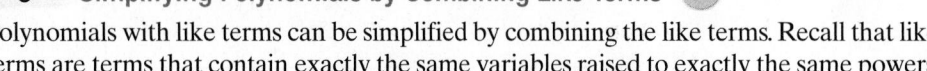

Polynomials with like terms can be simplified by combining the like terms. Recall that like terms are terms that contain exactly the same variables raised to exactly the same powers.

Like Terms	*Unlike Terms*
$5x^2, -7x^2$	$3x, 3y$
$y, 2y$	$-2x^2, -5x$
$\frac{1}{2}a^2b, -a^2b$	$6st^2, 4s^2t$

Only like terms can be combined. We combine like terms by applying the distributive property.

EXAMPLE 6 Simplify each polynomial by combining any like terms.

a. $-3x + 7x$

b. $x + 3x^2$

c. $9x^3 + x^3$

d. $11x^2 + 5 + 2x^2 - 7$

e. $\frac{2}{5}x^4 + \frac{2}{3}x^3 - x^2 + \frac{1}{10}x^4 - \frac{1}{6}x^3$

Solution

a. $-3x + 7x = (-3 + 7)x = 4x$

b. $x + 3x^2$ These terms cannot be combined because x and $3x^2$ are not like terms.

c. $9x^3 + x^3 = 9x^3 + 1x^3 = 10x^3$

d. $11x^2 + 5 + 2x^2 - 7 = 11x^2 + 2x^2 + 5 - 7$

$\qquad = 13x^2 - 2$ Combine like terms.

e. $\frac{2}{5}x^4 + \frac{2}{3}x^3 - x^2 + \frac{1}{10}x^4 - \frac{1}{6}x^3$

$\qquad = \left(\frac{2}{5} + \frac{1}{10}\right)x^4 + \left(\frac{2}{3} - \frac{1}{6}\right)x^3 - x^2$

$\qquad = \left(\frac{4}{10} + \frac{1}{10}\right)x^4 + \left(\frac{4}{6} - \frac{1}{6}\right)x^3 - x^2$

$\qquad = \frac{5}{10}x^4 + \frac{3}{6}x^3 - x^2$

$\qquad = \frac{1}{2}x^4 + \frac{1}{2}x^3 - x^2$ □

PRACTICE
6 Simplify each polynomial by combining any like terms.

a. $-4y + 2y$

b. $z + 5z^3$

c. $15x^3 - x^3$

d. $7a^2 - 5 - 3a^2 - 7$

e. $\frac{3}{8}x^3 - x^2 + \frac{5}{6}x^4 + \frac{1}{12}x^3 - \frac{1}{2}x^4$

✔ **CONCEPT CHECK**

When combining like terms in the expression $5x - 8x^2 - 8x$, which of the following is the proper result?

a. $-11x^2$ b. $-8x^2 - 3x$ c. $-11x$ d. $-11x^4$

EXAMPLE 7 Combine like terms to simplify.

$$-9x^2 + 3xy - 5y^2 + 7yx$$

Solution

$$-9x^2 + 3xy - 5y^2 + 7yx = -9x^2 + (3 + 7)xy - 5y^2$$
$$= -9x^2 + 10xy - 5y^2$$ □

Helpful Hint

This term can be written as $7yx$ or $7xy$.

PRACTICE
7 Combine like terms to simplify: $9xy - 3x^2 - 4yx + 5y^2$.

EXAMPLE 8 Write a polynomial that describes the total area of the squares and rectangles shown on page 327. Then simplify the polynomial.

Solution

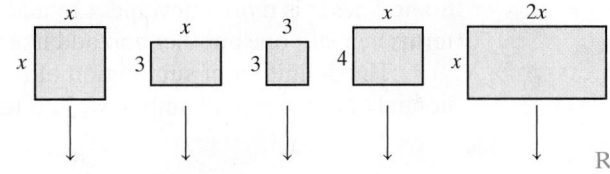

Area: $\quad x \cdot x \quad + \quad 3 \cdot x \quad + \quad 3 \cdot 3 \quad + \quad 4 \cdot x \quad + \quad x \cdot 2x$ Recall that the area of a rectangle is length times width.

$$= x^2 + 3x + 9 + 4x + 2x^2$$
$$= 3x^2 + 7x + 9 \qquad \text{Combine like terms.} \quad \square$$

PRACTICE

8   Write a polynomial that describes the total area of the squares and rectangles shown below. Then simplify the polynomial.

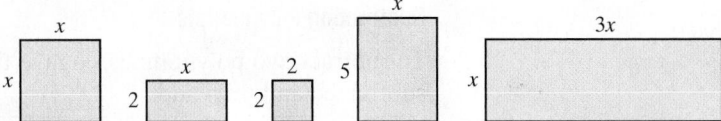

■

OBJECTIVE

4   **Adding and Subtracting Polynomials**

We now practice adding and subtracting polynomials.

> **Adding Polynomials**
>
> To add polynomials, combine all like terms.

EXAMPLE 9   Add.

    a. $(7x^3y - xy^3 + 11) + (6x^3y - 4)$
    b. $(3a^3 - b + 2a - 5) + (a + b + 5)$

Solution

    a. To add, remove the parentheses and group like terms.

$$(7x^3y - xy^3 + 11) + (6x^3y - 4)$$
$$= 7x^3y - xy^3 + 11 + 6x^3y - 4$$
$$= 7x^3y + 6x^3y - xy^3 + 11 - 4 \qquad \text{Group like terms.}$$
$$= 13x^2y - xy^3 + 7 \qquad \text{Combine like terms.}$$

    b. $\quad (3a^3 - b + 2a - 5) + (a + b + 5)$

$$= 3a^3 - b + 2a - 5 + a + b + 5$$
$$= 3a^3 - b + b + 2a + a - 5 + 5 \qquad \text{Group like terms.}$$
$$= 3a^3 + 3a \qquad \text{Combine like terms.} \quad \square$$

PRACTICE

9   Add.

    a. $(4y^2 + x - 3y - 7) + (x + y^2 - 2)$
    b. $(-8a^2b - ab^2 + 10) + (-2ab^2 - 10)$ ■

EXAMPLE 10   Add $(11x^3 - 12x^2 + x - 3)$ and $(x^3 - 10x + 5)$.

Solution

$$(11x^3 - 12x^2 + x - 3) + (x^3 - 10x + 5)$$
$$= 11x^3 + x^3 - 12x^2 + x - 10x - 3 + 5 \qquad \text{Group like terms.}$$
$$= 12x^3 - 12x^2 - 9x + 2 \qquad \text{Combine like terms.} \quad \square$$

PRACTICE

10   Add $(3x^2 - 9x + 11)$ and $(-3x^2 + 7x^3 + 3x - 4)$ ■

Sometimes it is more convenient to add polynomials vertically. To do this, line up like terms beneath one another and add like terms.

The definition of subtraction of real numbers can be extended to apply to polynomials. To subtract a number, we add its opposite.

$$a - b = a + (-b)$$

Likewise, to subtract a polynomial, we add its opposite. In other words, if P and Q are polynomials, then

$$P - Q = P + (-Q)$$

The polynomial $-Q$ is the **opposite,** or **additive inverse,** of the polynomial Q. We can find $-Q$ by writing the opposite of each term of Q.

Subtracting Polynomials

To subtract two polynomials, change the signs of the terms of the polynomial being subtracted and then add.

✔ **CONCEPT CHECK**

Which polynomial is the opposite of $16x^3 - 5x + 7$?
a. $-16x^3 - 5x + 7$ **b.** $-16x^3 + 5x - 7$ **c.** $16x^3 + 5x + 7$ **d.** $-16x^3 + 5x + 7$

EXAMPLE 11   Subtract: $(2x^3 + 8x^2 - 6x) - (2x^3 - x^2 + 1)$

Solution   First, change the sign of each term of the second polynomial and then add.

> **Helpful Hint**
> Notice the sign of each term is changed.

$$(2x^3 + 8x^2 - 6x) - (2x^3 - x^2 + 1) = (2x^3 + 8x^2 - 6x) + (-2x^3 + x^2 - 1)$$
$$= 2x^3 - 2x^3 + 8x^2 + x^2 - 6x - 1$$
$$= 9x^2 - 6x - 1 \quad \text{Combine like terms.} \quad \square$$

PRACTICE
11   Subtract: $(3x^3 - 5x^2 + 4x) - (x^3 - x^2 + 6)$ $2x^3 - 4x^2 + 4x - 6$ ■

EXAMPLE 12   Subtract $(5z - 7)$ from the sum of $(8z + 11)$ and $(9z - 2)$.

Solution   Notice that $(5z - 7)$ is to be subtracted **from** a sum. The translation is

$$[(8z + 11) + (9z - 2)] - (5z - 7)$$
$$= 8z + 11 + 9z - 2 - 5z + 7 \quad \text{Remove grouping symbols.}$$
$$= 8z + 9z - 5z + 11 - 2 + 7 \quad \text{Group like terms.}$$
$$= 12z + 16 \quad\quad\quad\quad\quad \text{Combine like terms.} \quad \square$$

PRACTICE
12   Subtract $(3x + 5)$ from the sum of $(8x - 11)$ and $(2x + 5)$.   ■

EXAMPLE 13   Add or subtract as indicated.

    a. $(3x^2 - 6xy + 5y^2) + (-2x^2 + 8xy - y^2)$
    b. $(9a^2b^2 + 6ab - 3ab^2) - (5b^2a + 2ab - 3 - 9b^2)$

Solution

    a. $(3x^2 - 6xy + 5y^2) + (-2x^2 + 8xy - y^2)$
$$= 3x^2 - 6xy + 5y^2 - 2x^2 + 8xy - y^2$$
$$= x^2 + 2xy + 4y^2 \quad\quad\quad \text{Combine like terms.}$$

b. $(9a^2b^2 + 6ab - 3ab^2) - (5b^2a + 2ab - 3 - 9b^2)$

$= 9a^2b^2 + 6ab - 3ab^2 - 5b^2a - 2ab + 3 + 9b^2$ Change the sign of each term of the polynomial being subtracted.

$= 9a^2b^2 + 4ab - 8ab^2 + 3 + 9b^2$ Combine like terms.   □

PRACTICE
13   Add or subtract as indicated.

a. $(3a^2 - 4ab + 7b^2) + (-8a^2 + 3ab - b^2)$

b. $(5x^2y^2 - 6xy - 4xy^2) - (2x^2y^2 + 4xy - 5 + 6y^2)$ ■

✔ **CONCEPT CHECK**

If possible, simplify each expression by performing the indicated operation.

a. $2y + y$ **b.** $2y \cdot y$ **c.** $-2y - y$ **d.** $(-2y)(-y)$ **e.** $2x + y$

To add or subtract polynomials vertically, just remember to line up like terms. For example, perform the subtraction $(10x^3y^2 - 7x^2y^2) - (4x^3y^2 - 3x^2y^2 + 2y^2)$ vertically.

Add the opposite of the second polynomial.

$$\begin{array}{l} 10x^3y^2 - 7x^2y^2 \\ \underline{-(4x^3y^2 - 3x^2y^2 + 2y^2)} \end{array} \text{ is equivalent to } \begin{array}{l} 10x^3y^2 - 7x^2y^2 \\ \underline{-4x^3y^2 + 3x^2y^2 - 2y^2} \\ 6x^3y^2 - 4x^2y^2 - 2y^2 \end{array}$$

✔ **CONCEPT CHECK**

Why is the following subtraction incorrect?

$$(7z - 5) - (3z - 4)$$
$$= 7z - 5 - 3z - 4$$
$$= 4z - 9$$

Polynomial functions, like polynomials, can be added, subtracted, multiplied, and divided. For example, if

$$P(x) = x^2 + x + 1$$

then

$$2P(x) = 2(x^2 + x + 1) = 2x^2 + 2x + 2 \quad \text{Use the distributive property.}$$

Also, if $Q(x) = 5x^2 - 1$, then $P(x) + Q(x) = (x^2 + x + 1) + (5x^2 - 1)$
$$= 6x^2 + x.$$

A useful business and economics application of subtracting polynomial functions is finding the profit function $P(x)$ when given a revenue function $R(x)$ and a cost function $C(x)$. In business, it is true that

$$\text{profit} = \text{revenue} - \text{cost, or}$$
$$P(x) = R(x) - C(x)$$

For example, if the revenue function is $R(x) = 7x$ and the cost function is $C(x) = 2x + 5000$, then the profit function is

$$P(x) = R(x) - C(x)$$

or

$$P(x) = 7x - (2x + 5000) \quad \text{Substitute } R(x) = 7x$$
$$P(x) = 5x - 5000 \quad \text{and } C(x) = 2x + 5000.$$

Graphing Calculator Explorations

A graphing calculator may be used to visualize addition and subtraction of polynomials in one variable. For example, to visualize the following polynomial subtraction statement

$$(3x^2 - 6x + 9) - (x^2 - 5x + 6) = 2x^2 - x + 3$$

graph both

$$Y_1 = (3x^2 - 6x + 9) - (x^2 - 5x + 6) \quad \text{Left side of equation}$$

and

$$Y_2 = 2x^2 - x + 3 \quad \text{Right side of equation}$$

on the same screen and see that their graphs coincide. (*Note:* If the graphs do not coincide, we can be sure that a mistake has been made in combining polynomials or in calculator keystrokes. If the graphs appear to coincide, we cannot be sure that our work is correct. This is because it is possible for the graphs to differ so slightly that we do not notice it.)

The graphs of Y_1 and Y_2 are shown. The graphs appear to coincide, so the subtraction statement

$$(3x^2 - 6x + 9) - (x^2 - 5x + 6) = 2x^2 - x + 3$$

appears to be correct.

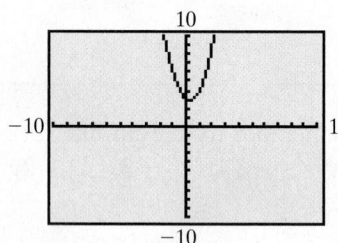

Perform the indicated operations. Then visualize by using the procedure described above.

1. $(2x^2 + 7x + 6) + (x^3 - 6x^2 - 14)$
2. $(-14x^3 - x + 2) + (-x^3 + 3x^2 + 4x)$
3. $(1.8x^2 - 6.8x - 1.7) - (3.9x^2 - 3.6x)$
4. $(-4.8x^2 + 12.5x - 7.8) - (3.1x^2 - 7.8x)$
5. $(1.29x - 5.68) + (7.69x^2 - 2.55x + 10.98)$
6. $(-0.98x^2 - 1.56x + 5.57) + (4.36x - 3.71)$

✓ Vocabulary, Readiness & Video Check

Use the choices below to fill in each blank. Not all choices will be used.

least	monomial	trinomial	coefficient
greatest	binomial	constant	

1. A _____ is a polynomial with exactly 2 terms.
2. A _____ is a polynomial with exactly one term.
3. A _____ is a polynomial with exactly three terms.
4. The numerical factor of a term is called the _____ .
5. A number term is also called a _____ .
6. The degree of a polynomial is the _____ degree of any term of the polynomial.

Martin-Gay Interactive Videos

See Video 5.2

Watch the section lecture video and answer the following questions.

OBJECTIVE 1
7. For Example 2, why is the degree of each **term** found when the example asks for the degree of the **polynomial** only?

OBJECTIVE 2
8. From Example 3, how do you find the value of a polynomial function $Q(x)$ given a value for x?

OBJECTIVE 3
9. When combining any like terms in a polynomial, as in Examples 4–6, what are we doing to the polynomial?

OBJECTIVE 4
10. From Example 7, when we simply remove parentheses and combine the like terms of two polynomials, what operation do we perform? Is this true of Examples 9–11?

5.2   Exercise Set   MyMathLab®

Find the degree of each of the following polynomials and determine whether it is a monomial, binomial, trinomial, or none of these. See Examples 1 through 3.

1. $x + 2$

2. $-6y^2 + 4$

3. $9m^3 - 5m^2 + 4m - 8$

4. $a + 5a^2 + 3a^3 - 4a^4$

5. $12x^4y - x^2y^2 - 12x^2y^4$

6. $7r^2s^2 + 2rs - 3rs^5$

7. $3 - 5x^8$

8. $5y^7 + 2$

In the second column, write the degree of the polynomial in the first column. See Examples 1 through 3.

Polynomial	Degree
9. $3xy^2 - 4$	
10. $8x^2y^2$	
11. $5a^2 - 2a + 1$	
12. $4z^6 + 3z^2$	

if $P(x) = x^2 + x + 1$ and $Q(x) = 5x^2 - 1$, find the following. See Examples 4 and 5.

13. $P(7)$

14. $Q(4)$

15. $Q(-10)$

16. $P(-4)$

17. $P(0)$

18. $Q(0)$

19. $Q\left(\dfrac{1}{4}\right)$

20. $P\left(\dfrac{1}{2}\right)$

The CN Tower in Toronto, Ontario, is 1821 feet tall and is the world's tallest self-supporting structure. An object is dropped from the Skypod of the Tower, which is at 1150 feet. Neglecting air resistance, the height of the object at time t seconds is given by the polynomial function $P(t) = -16t^2 + 1150$. Find the height of the object at the given times.

	Time, t (in seconds)	Height $P(t) = -16t^2 + 1150$
21.	1	
22.	7	
23.	3	
24.	6	

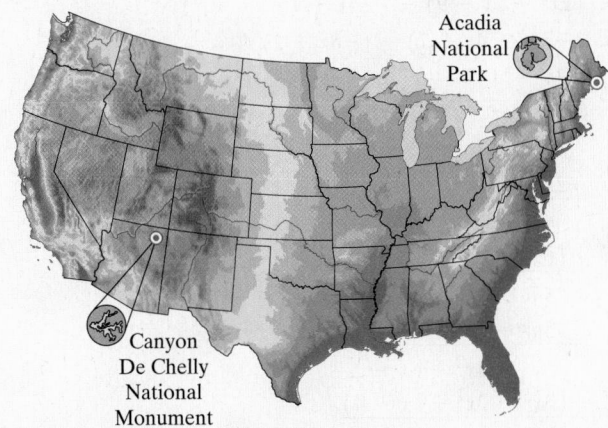

Acadia National Park

Canyon De Chelly National Monument

25. The polynomial $-7.5x^2 + 103x + 2000$ models the yearly number of visitors (in thousands) x years after 2006 at Acadia National Park in Maine. Use this polynomial to estimate the number of visitors to the park in 2016.

26. The polynomial $-0.13x^2 + x + 827$ models the yearly number of visitors (in thousands) x years after 2006 at Canyon De Chelly National Monument in Arizona. Use this polynomial to estimate the number of visitors to the park in 2010.

Simplify each of the following by combining like terms. See Examples 6 and 7.

▶ **27.** $14x^2 + 9x^2$

28. $18x^3 - 4x^3$

▶ **29.** $15x^2 - 3x^2 - y$

30. $12k^3 - 9k^3 + 11$

31. $8s - 5s + 4s$

32. $5y + 7y - 6y$

▶ **33.** $0.1y^2 - 1.2y^2 + 6.7 - 1.9$

34. $7.6y + 3.2y^2 - 8y - 2.5y^2$

35. $\dfrac{2}{5}x^2 - \dfrac{1}{3}x^3 + x^2 - \dfrac{1}{4}x^3 + 6$

36. $\dfrac{1}{6}x^4 - \dfrac{1}{7}x^2 + 5 - \dfrac{1}{2}x^4 - \dfrac{3}{7}x^2 + \dfrac{1}{3}$

37. $6a^2 - 4ab + 7b^2 - a^2 - 5ab + 9b^2$

38. $x^2y + xy - y + 10x^2y - 2y + xy$

Perform the indicated operations. See Examples 9 through 13.

▶ **39.** $(-7x + 5) + (-3x^2 + 7x + 5)$

40. $(3x - 8) + (4x^2 - 3x + 3)$

▶ **41.** $(2x^2 + 5) - (3x^2 - 9)$

42. $(5x^2 + 4) - (-2y^2 + 4)$

43. $3x - (5x - 9)$

44. $4 - (-y - 4)$

45. $(2x^2 + 3x - 9) - (-4x + 7)$

46. $(-7x^2 + 4x + 7) - (-8x + 2)$

▶ **47.** $\begin{array}{r} 3t^2 + 4 \\ +5t^2 - 8 \\ \hline \end{array}$

48. $\begin{array}{r} 7x^3 + 3 \\ +2x^3 - 1 \\ \hline \end{array}$

49. $\begin{array}{r} 4z^2 - 8z + 3 \\ -(6z^2 + 8z - 3) \\ \hline \end{array}$

50. $\begin{array}{r} 5u^5 - 4u^2 + 3u - 7 \\ -(3u^5 + 6u^2 - 8u + 2) \\ \hline \end{array}$

▶ **51.** $\begin{array}{r} 5x^3 - 4x^2 + 6x - 2 \\ -(3x^3 - 2x^2 - x - 4) \\ \hline \end{array}$

52. $\begin{array}{r} 7a^2 - 9a + 6 \\ -(11a^2 - 4a + 2) \\ \hline \end{array}$

MIXED PRACTICE

Perform the indicated operations.

53. $(-3y^2 - 4y) + (2y^2 + y - 1)$

54. $(7x^2 + 2x - 9) + (-3x^2 + 5)$

55. $(5x + 8) - (-2x^2 - 6x + 8)$

56. $(-6y^2 + 3y - 4) - (9y^2 - 3y)$

57. $(-8x^4 + 7x) + (-8x^4 + x + 9)$

58. $(6y^5 - 6y^3 + 4) + (-2y^5 - 8y^3 - 7)$

59. $(3x^2 + 5x - 8) + (5x^2 + 9x + 12) - (x^2 - 14)$

60. $(-a^2 + 1) - (a^2 - 3) + (5a^2 - 6a + 7)$

TRANSLATING

Perform each indicated operation.

61. Subtract $4x$ from $(7x - 3)$.

62. Subtract y from $(y^2 - 4y + 1)$.

63. Add $(4x^2 - 6x + 1)$ and $(3x^2 + 2x + 1)$.

64. Add $(-3x^2 - 5x + 2)$ and $(x^2 - 6x + 9)$.

▶ **65.** Subtract $(19x^2 + 5)$ from $(81x^2 + 10)$.

66. Subtract $(2x + xy)$ from $(3x - 9xy)$.

67. Subtract $(5x + 7)$ from $(7x^2 + 3x + 9)$.

68. Subtract $(5y^2 + 8y + 2)$ from $(7y^2 + 9x - 8)$.

69. Subtract $(2x + 2)$ from the sum of $(8x + 1)$ and $(6x + 3)$.

70. Subtract $(-12x - 3)$ from the sum of $(-5x - 7)$ and $(12x + 3)$.

71. Subtract $(4y^2 - 6y - 3)$ from the sum of $(8y^2 + 7)$ and $(6y + 9)$.

72. Subtract $(4x^2 - 2x + 2)$ from the sum of $(x^2 + 7x + 1)$ and $(7x + 5)$.

Find the area of each figure. Write a polynomial that describes the total area of the rectangles and squares shown in Exercises 73–74. Then simplify the polynomial. See Example 8.

△ **73.**

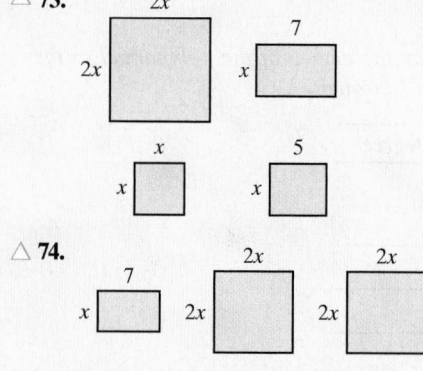

△ **74.**

Add or subtract as indicated. See Examples 9 through 13.

75. $(9a + 6b - 5) + (-11a - 7b + 6)$

76. $(3x - 2 + 6y) + (7x - 2 - y)$

77. $(4x^2 + y^2 + 3) - (x^2 + y^2 - 2)$

78. $(7a^2 - 3b^2 + 10) - (-2a^2 + b^2 - 12)$

79. $(x^2 + 2xy - y^2) + (5x^2 - 4xy + 20y^2)$

80. $(a^2 - ab + 4b^2) + (6a^2 + 8ab - b^2)$

81. $(11r^2s + 16rs - 3 - 2r^2s^2) - (3sr^2 + 5 - 9r^2s^2)$

82. $(3x^2y - 6xy + x^2y^2 - 5) - (11x^2y^2 - 1 + 5yx^2)$

Simplify each polynomial by combining like terms.

83. $7.75x + 9.16x^2 - 1.27 - 14.58x^2 - 18.34$

84. $1.85x^2 - 3.76x + 9.25x^2 + 10.76 - 4.21x$

Perform each indicated operation.

85. $[(7.9y^4 - 6.8y^3 + 3.3y) + (6.1y^3 - 5)]$
$- (4.2y^4 + 1.1y - 1)$

86. $[(1.2x^2 - 3x + 9.1) - (7.8x^2 - 3.1 + 8)] + (1.2x - 6)$

REVIEW AND PREVIEW

Multiply. See Section 5.1.

87. $3x(2x)$ **88.** $-7x(x)$

89. $(12x^3)(-x^5)$ **90.** $6r^3(7r^{10})$

91. $10x^2(20xy^2)$ **92.** $-z^2y(11zy)$

CONCEPT EXTENSIONS

Recall that the perimeter of a figure is the sum of the lengths of its sides. For Exercises 93 through 96, find the perimeter of each polynomial.

△ **93.**

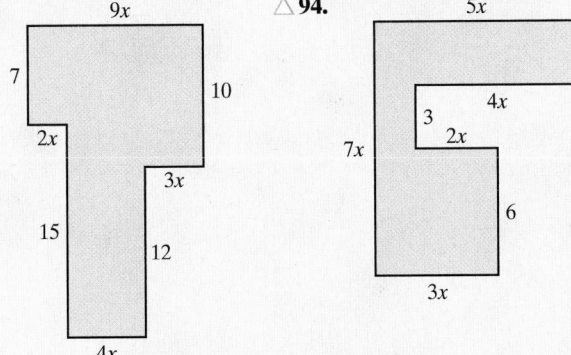

△ **94.**

△ **95.**

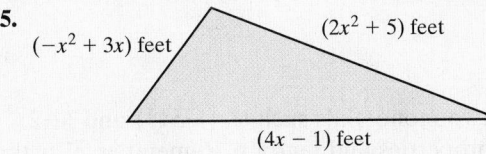

△ **96.**

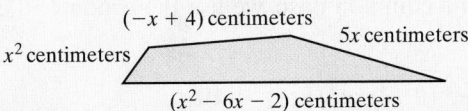

△ **97.** A wooden beam is $(4y^2 + 4y + 1)$ meters long. If a piece $(y^2 - 10)$ meters is cut, express the length of the remaining piece of beam as a polynomial in y.

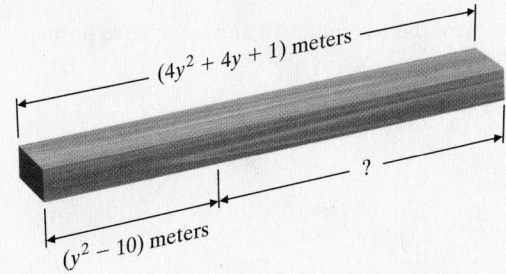

△ **98.** A piece of quarter-round molding is $(13x - 7)$ inches long. If a piece $(2x + 2)$ inches is removed, express the length of the remaining piece of molding as a polynomial in x.

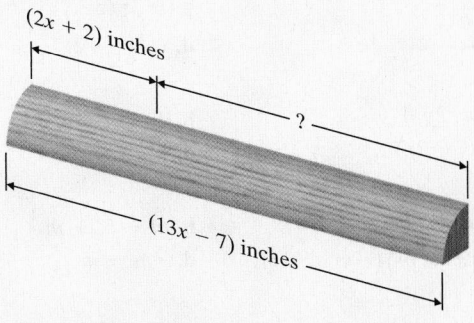

The number of worldwide Internet users (in millions) x years after the year 2000 is given by the polynomial $4.8x^2 + 104x + 431$ for the years 1995 through 2015. Use this polynomial for Exercises 99 and 100.

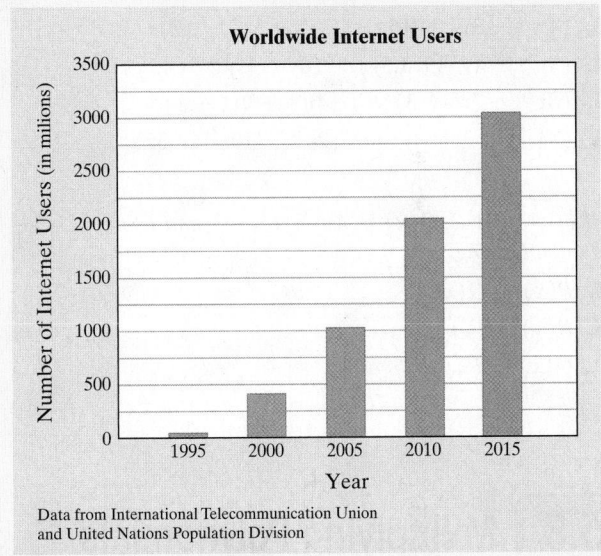

Data from International Telecommunication Union and United Nations Population Division

99. Estimate the number of Internet users in the world in 2015.

100. Use the given polynomial to predict the number of Internet users in the world in 2020.

101. Describe how to find the degree of a term.

102. Describe how to find the degree of a polynomial.

103. Explain why xyz is a monomial while $x + y + z$ is a trinomial.

104. Explain why the degree of the term $5y^3$ is 3 and the degree of the polynomial $2y + y + 2y$ is 1.

Match each expression on the left with its simplification on the right. Not all letters on the right must be used, and a letter may be used more than once.

105. $10y - 6y^2 - y$ **A.** $3y$

106. $5x + 5x$ **B.** $9y - 6y^2$

107. $(5x - 3) + (5x - 3)$ **C.** $10x$

108. $(15x - 3) - (5x - 3)$ **D.** $25x^2$

                                  E. $10x - 6$

                                  F. none of these

Simplify each expression by performing the indicated operation. Explain how you arrived at each answer. See the third Concept Check in this section.

109. a. $z + 3z$ **b.** $z \cdot 3z$
   c. $-z - 3z$ **d.** $(-z)(-3z)$

110. a. $2y + y$ **b.** $2y \cdot y$
   c. $-2y - y$ **d.** $(-2y)(-y)$

111. a. $m \cdot m \cdot m$ **b.** $m + m + m$
   c. $(-m)(-m)(-m)$ **d.** $-m - m - m$

112. a. $x + x$ **b.** $x \cdot x$
   c. $-x - x$ **d.** $(-x)(-x)$

Perform the indicated operations.

113. $(4x^{2a} - 3x^a + 0.5) - (x^{2a} - 5x^a - 0.2)$
114. $(9y^{5a} - 4y^{3a} + 1.5y) - (6y^{5a} - y^{3a} + 4.7y)$
115. $(8x^{2y} - 7x^y + 3) + (-4x^{2y} + 9x^y - 14)$
116. $(14z^{5x} + 3z^{2x} + z) - (2z^{5x} - 10z^{2x} + 3z)$

If $P(x) = 3x + 3$, $Q(x) = 4x^2 - 6x + 3$, and $R(x) = 5x^2 - 7$, find the following.

117. $P(x) + Q(x)$
118. $R(x) + P(x)$
119. $Q(x) - R(x)$

120. $P(x) - Q(x)$
121. $2[Q(x)] - R(x)$
122. $-5[P(x)] - Q(x)$

*If $P(x)$ is the polynomial given, find **a.** $P(a)$, **b.** $P(-x)$, and **c.** $P(x + h)$.*

123. $P(x) = 2x - 3$
124. $P(x) = 8x + 3$
125. $P(x) = 4x$
126. $P(x) = -4x$

Fill in the squares so that each is a true statement.

127. $3x^\square + 4x^2 = 7x^\square$
128. $9y^7 + 3y^\square = 12y^7$
129. $2x^\square + 3x^\square - 5x^\square + 4x^\square = 6x^4 - 2x^3$
130. $3y^\square + 7y^\square - 2y^\square - y^\square = 10y^5 - 3y^2$

Write a polynomial that describes the surface area of each figure. (Recall that the surface area of a solid is the sum of the areas of the faces or sides of the solid.)

△ **131.**

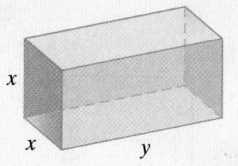

△ **132.**

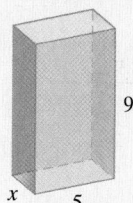

5.3 Multiplying Polynomials

OBJECTIVES

1 Multiply Monomials.
2 Use the Distributive Property to Multiply Polynomials.
3 Multiply Polynomials Vertically.

OBJECTIVE

1 Multiplying Monomials

Recall from Section 5.1 that to multiply two monomials such as $(-5x^3)$ and $(-2x^4)$, we use the associative and commutative properties and regroup. Remember, also, that to multiply exponential expressions with a common base, we use the product rule for exponents and add exponents.

$$(-5x^3)(-2x^4) = (-5)(-2)(x^3)(x^4) = 10x^7$$

EXAMPLES Multiply.

1. $6x \cdot 4x = (6 \cdot 4)(x \cdot x)$ Use the commutative and associative properties.

$= 24x^2$ Multiply.

2. $-7x^2 \cdot 0.2x^5 = (-7 \cdot 0.2)(x^2 \cdot x^5)$

$= -1.4x^7$

3. $\left(-\dfrac{1}{3}x^5\right)\left(-\dfrac{2}{9}x\right) = \left(-\dfrac{1}{3} \cdot -\dfrac{2}{9}\right) \cdot (x^5 \cdot x)$

$= \dfrac{2}{27}x^6$

1–3   Multiply.

1. $5y \cdot 2y$ 2. $(5z^3) \cdot (-0.4z^5)$ 3. $\left(-\dfrac{1}{9}b^6\right)\left(-\dfrac{7}{8}b^3\right)$ ■

✔ **CONCEPT CHECK**
Simplify.
a. $3x \cdot 2x$ b. $3x + 2x$

OBJECTIVE
2   **Using the Distributive Property to Multiply Polynomials** ▶

To multiply polynomials that are not monomials, use the distributive property.

EXAMPLE 4   Use the distributive property to find each product.

a. $5x(2x^3 + 6)$ b. $-3x^2(5x^2 + 6x - 1)$

Solution

a. $5x(2x^3 + 6) = 5x(2x^3) + 5x(6)$ Use the distributive property.

$\qquad = 10x^4 + 30x$ Multiply.

b. $-3x^2(5x^2 + 6x - 1)$

$\qquad = (-3x^2)(5x^2) + (-3x^2)(6x) + (-3x^2)(-1)$ Use the distributive property.

$\qquad = -15x^4 - 18x^3 + 3x^2$ Multiply.   □

PRACTICE
4   Use the distributive property to find each product.

a. $3x(9x^5 + 11)$ b. $-6x^3(2x^2 - 9x + 2)$ ■

We also use the distributive property to multiply two binomials. To multiply $(x + 3)$ by $(x + 1)$, distribute the factor $(x + 3)$ first.

$(x + 3)(x + 1) = x(x + 1) + 3(x + 1)$ Distribute $(x + 3)$.

$\qquad = x(x) + x(1) + 3(x) + 3(1)$ Apply the distributive property a second time.

$\qquad = x^2 + x + 3x + 3$ Multiply.

$\qquad = x^2 + 4x + 3$ Combine like terms.

This idea can be expanded so that we can multiply any two polynomials.

To Multiply Two Polynomials

Multiply each term of the first polynomial by each term of the second polynomial and then combine like terms.

Answers to Concept Check:
a. $6x^2$ b. $5x$

EXAMPLE 5 Multiply: $(3x + 2)(2x - 5)$

Solution Multiply each term of the first binomial by each term of the second.

$$(3x + 2)(2x - 5) = 3x(2x) + 3x(-5) + 2(2x) + 2(-5)$$
$$= 6x^2 - 15x + 4x - 10 \qquad \text{Multiply.}$$
$$= 6x^2 - 11x - 10 \qquad \text{Combine like terms.} \ \square$$

PRACTICE
5 Multiply: $(5x - 2)(2x + 3)$ ∎

EXAMPLE 6 Multiply: $(2x - y)^2$

Solution Recall that $a^2 = a \cdot a$, so $(2x - y)^2 = (2x - y)(2x - y)$. Multiply each term of the first polynomial by each term of the second.

$$(2x - y)(2x - y) = 2x(2x) + 2x(-y) + (-y)(2x) + (-y)(-y)$$
$$= 4x^2 - 2xy - 2xy + y^2 \qquad \text{Multiply.}$$
$$= 4x^2 - 4xy + y^2 \qquad \text{Combine like terms.} \ \square$$

PRACTICE
6 Multiply: $(5x - 3y)^2$ ∎

✔ **CONCEPT CHECK**
Square where indicated. Simplify if possible.
a. $(4a)^2 + (3b)^2$ **b.** $(4a + 3b)^2$

EXAMPLE 7 Multiply $(t + 2)$ by $(3t^2 - 4t + 2)$.

Solution Multiply each term of the first polynomial by each term of the second.

$$(t + 2)(3t^2 - 4t + 2) = t(3t^2) + t(-4t) + t(2) + 2(3t^2) + 2(-4t) + 2(2)$$
$$= 3t^3 - 4t^2 + 2t + 6t^2 - 8t + 4$$
$$= 3t^3 + 2t^2 - 6t + 4 \quad \text{Combine like terms.} \ \square$$

PRACTICE
7 Multiply $(y + 4)$ by $(2y^2 - 3y + 5)$. ∎

EXAMPLE 8 Multiply: $(3a + b)^3$

Solution Write $(3a + b)^3$ as $(3a + b)(3a + b)(3a + b)$.

$$(3a + b)(3a + b)(3a + b) = (9a^2 + 3ab + 3ab + b^2)(3a + b)$$
$$= (9a^2 + 6ab + b^2)(3a + b)$$
$$= 9a^2(3a + b) + 6ab(3a + b) + b^2(3a + b)$$
$$= 27a^3 + 9a^2b + 18a^2b + 6ab^2 + 3ab^2 + b^3$$
$$= 27a^3 + 27a^2b + 9ab^2 + b^3 \qquad \square$$

Answers to Concept Check:
a. $16a^2 + 9b^2$
b. $16a^2 + 24ab + 9b^2$

PRACTICE
8 Multiply: $(s + 2t)^3$ ∎

OBJECTIVE

3 Multiplying Polynomials Vertically

Another convenient method for multiplying polynomials is to use a vertical format similar to the format used to multiply real numbers. We demonstrate this method by multiplying $(3y^2 - 4y + 1)$ by $(y + 2)$.

EXAMPLE 9 Multiply $(3y^2 - 4y + 1)$ by $(y + 2)$. Use a vertical format.

Solution

$$
\begin{array}{r}
3y^2 - 4y + 1 \\
\times \quad\quad y + 2 \\
\hline
6y^2 - 8y + 2 \\
3y^3 - 4y^2 + \quad y \\
\hline
3y^3 + 2y^2 - 7y + 2
\end{array}
$$

1st, multiply $3y^2 - 4y + 1$ by 2.
2nd, multiply $3y^2 - 4y + 1$ by y.
Line up like terms.
3rd, combine like terms.

> **Helpful Hint**
> Make sure like terms are lined up.

Thus, $(3y^2 - 4y + 1)(y + 2) = 3y^3 + 2y^2 - 7y + 2$. ☐

PRACTICE

9 Multiply $(5x^2 - 3x + 5)$ by $(x - 4)$. Use a vertical format. ∎

When multiplying vertically, be careful if a power is missing; you may want to leave space in the partial products and take care that like terms are lined up.

EXAMPLE 10 Multiply $(2x^3 - 3x + 4)$ by $(x^2 + 1)$. Use a vertical format.

Solution

$$
\begin{array}{r}
2x^3 - 3x + 4 \\
\times \quad\quad x^2 + 1 \\
\hline
2x^3 \quad\quad - 3x + 4 \\
2x^5 - 3x^3 + 4x^2 \\
\hline
2x^5 - \quad x^3 + 4x^2 - 3x + 4
\end{array}
$$

Leave space for missing powers of x.
← Line up like terms.
Combine like terms. ☐

PRACTICE

10 Multiply $(x^3 - 2x^2 + 1)$ by $(x^2 + 2)$ using a vertical format. ∎

EXAMPLE 11 Find the product of $(2x^2 - 3x + 4)$ and $(x^2 + 5x - 2)$ using a vertical format.

Solution First, we arrange the polynomials in a vertical format. Then we multiply each term of the second polynomial by each term of the first polynomial.

$$
\begin{array}{r}
2x^2 - \quad 3x + 4 \\
\times \quad\quad x^2 + \quad 5x - 2 \\
\hline
-4x^2 + \quad 6x - 8 \\
10x^3 - 15x^2 + 20x \\
2x^4 - \quad 3x^3 + \quad 4x^2 \\
\hline
2x^4 + \quad 7x^3 - 15x^2 + 26x - 8
\end{array}
$$

Multiply $2x^2 - 3x + 4$ by -2.
Multiply $2x^2 - 3x + 4$ by $5x$.
Multiply $2x^2 - 3x + 4$ by x^2.
Combine like terms. ☐

PRACTICE

11 Find the product of $(5x^2 + 2x - 2)$ and $(x^2 - x + 3)$ using a vertical format. ∎

✓ **Vocabulary, Readiness & Video Check**

Fill in each blank with the correct choice.

1. The expression $5x(3x + 2)$ equals $5x \cdot 3x + 5x \cdot 2$ by the _____ property.

   a. commutative   **b.** associative   **c.** distributive

2. The expression $(x + 4)(7x - 1)$ equals $x(7x - 1) + 4(7x - 1)$ by the _____ property.

   a. commutative   **b.** associative   **c.** distributive

3. The expression $(5y - 1)^2$ equals _____.

   a. $2(5y - 1)$ **b.** $(5y - 1)(5y + 1)$ **c.** $(5y - 1)(5y - 1)$

4. The expression $9x \cdot 3x$ equals _____.

   a. $27x$ **b.** $27x^2$ **c.** $12x$ **d.** $12x^2$

Martin-Gay Interactive Videos

See Video 5.3 ⬤

Watch the section lecture video and answer the following questions.

OBJECTIVE 1   **5.** For Example 1, we use the product property to multiply the monomials. Is it possible to add the same two monomials? Why or why not?

OBJECTIVE 2   **6.** What property and what exponent rule is used in ▦ Examples 2–6?

OBJECTIVE 3   **7.** Would you say the vertical format used in ▦ Example 7 also applies the distributive property? Explain.

5.3 Exercise Set   MyMathLab® ▶

Multiply. See Examples 1 through 3.

1. $-4n^3 \cdot 7n^7$

2. $9t^6(-3t^5)$

3. $(-3.1x^3)(4x^9)$

4. $(-5.2x^4)(3x^4)$

▶ **5.** $\left(-\dfrac{1}{3}y^2\right)\left(\dfrac{2}{5}y\right)$

6. $\left(-\dfrac{3}{4}y^7\right)\left(\dfrac{1}{7}y^4\right)$

7. $(2x)(-3x^2)(4x^5)$

8. $(x)(5x^4)(-6x^7)$

Multiply. See Example 4.

▶ **9.** $3x(2x + 5)$

10. $2x(6x + 3)$

11. $-2a(a + 4)$

12. $-3a(2a + 7)$

13. $3x(2x^2 - 3x + 4)$

14. $4x(5x^2 - 6x - 10)$

15. $-2a^2(3a^2 - 2a + 3)$

16. $-4b^2(3b^3 - 12b^2 - 6)$

▶ **17.** $-y(4x^3 - 7x^2y + xy^2 + 3y^3)$

18. $-x(6y^3 - 5xy^2 + x^2y - 5x^3)$

19. $\dfrac{1}{2}x^2(8x^2 - 6x + 1)$

20. $\dfrac{1}{3}y^2(9y^2 - 6y + 1)$

Multiply. See Examples 5 and 6.

21. $(x + 4)(x + 3)$

22. $(x + 2)(x + 9)$

▶ **23.** $(a + 7)(a - 2)$

24. $(y - 10)(y + 11)$

25. $\left(x + \dfrac{2}{3}\right)\left(x - \dfrac{1}{3}\right)$

26. $\left(x + \dfrac{3}{5}\right)\left(x - \dfrac{2}{5}\right)$

27. $(3x^2 + 1)(4x^2 + 7)$

28. $(5x^2 + 2)(6x^2 + 2)$

29. $(2y - 4)^2$

30. $(6x - 7)^2$

31. $(4x - 3)(3x - 5)$

32. $(8x - 3)(2x - 4)$

▶ **33.** $(3x^2 + 1)^2$

34. $(x^2 + 4)^2$

35. Perform the indicated operations.

   a. $4y^2(-y^2)$

   b. $4y^2 - y^2$

   c. Explain the difference between the two expressions.

36. Perform the indicated operations.

   a. $9x^2(-10x^2)$

   b. $9x^2 - 10x^2$

   c. Explain the difference between the two expressions.

Multiply. See Example 7.

37. $(x - 2)(x^2 - 3x + 7)$

38. $(x + 3)(x^2 + 5x - 8)$

39. $(x + 5)(x^3 - 3x + 4)$

40. $(a + 2)(a^3 - 3a^2 + 7)$

41. $(2a - 3)(5a^2 - 6a + 4)$

42. $(3 + b)(2 - 5b - 3b^2)$

Multiply. See Example 8.

43. $(x + 2)^3$

44. $(y - 1)^3$

45. $(2y - 3)^3$

46. $(3x + 4)^3$

Multiply vertically. See Examples 9 through 11.

47. $(2x - 11)(6x + 1)$

48. $(4x - 7)(5x + 1)$

49. $(5x + 1)(2x^2 + 4x - 1)$

50. $(4x - 5)(8x^2 + 2x - 4)$

51. $(x^2 + 5x - 7)(2x^2 - 7x - 9)$

52. $(3x^2 - x + 2)(x^2 + 2x + 1)$

MIXED PRACTICE

Multiply. See Examples 1 through 11.

53. $-1.2y(-7y^6)$

54. $-4.2x(-2x^5)$

55. $-3x(x^2 + 2x - 8)$

56. $-5x(x^2 - 3x + 10)$

57. $(x + 19)(2x + 1)$

58. $(3y + 4)(y + 11)$

59. $\left(x + \dfrac{1}{7}\right)\left(x - \dfrac{3}{7}\right)$

60. $\left(m + \dfrac{2}{9}\right)\left(m - \dfrac{1}{9}\right)$

61. $(3y + 5)^2$

62. $(7y + 2)^2$

63. $(a + 4)(a^2 - 6a + 6)$

64. $(t + 3)(t^2 - 5t + 5)$

65. $(2x - 5)^3$

66. $(3y - 1)^3$

67. $(4x + 5)(8x^2 + 2x - 4)$

68. $(5x + 4)(x^2 - x + 4)$

69. $(3x^2 + 2x - 4)(2x^2 - 4x + 3)$

70. $(a^2 + 3a - 2)(2a^2 - 5a - 1)$

Express as the product of polynomials. Then multiply.

71. Find the area of the rectangle.

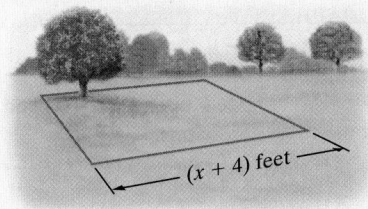

$(2x + 5)$ yards

$(2x - 5)$ yards

72. Find the area of the square field.

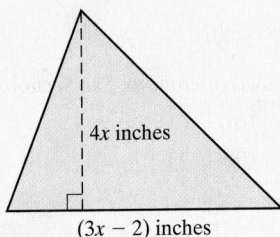

$(x + 4)$ feet

73. Find the area of the triangle.

$\left(\text{Triangle Area} = \dfrac{1}{2} \cdot \text{base} \cdot \text{height}\right)$

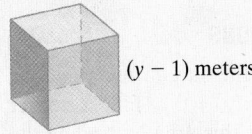

$4x$ inches

$(3x - 2)$ inches

74. Find the volume of the cube.

$(\text{Volume} = \text{length} \cdot \text{width} \cdot \text{height})$

$(y - 1)$ meters

REVIEW AND PREVIEW

In this section, we review operations on monomials. Study the table below, then proceed. See Sections 2.1, 5.1, and 5.2. (Continued on next page)

Operations on Monomials	
Multiply	Review the product rule for exponents.
Divide	Review the quotient rule for exponents.
Add or Subtract	Remember, we may only combine like terms.

Perform the operations on the monomials if possible. The first two rows have been completed for you.

Monomials	Add	Subtract	Multiply	Divide
$6x, 3x$	$6x + 3x = 9x$	$6x - 3x = 3x$	$6x \cdot 3x = 18x^2$	$\dfrac{6x}{3x} = 2$
$-12x^2, 2x$	$-12x^2 + 2x$, can't be simplified	$-12x^2 - 2x$, can't be simplified	$-12x^2 \cdot 2x = -24x^3$	$\dfrac{-12x^2}{2x} = -6x$
75. $5a, 15a$				
76. $4y^7, 4y^3$				
77. $-3y^5, 9y^4$				
78. $-14x^2, 2x^2$				

CONCEPT EXTENSIONS

79. Perform each indicated operation. Explain the difference between the two expressions.

   a. $(3x + 5) + (3x + 7)$

   b. $(3x + 5)(3x + 7)$

80. Perform each indicated operation. Explain the difference between the two expressions.

   a. $(8x - 3) - (5x - 2)$

   b. $(8x - 3)(5x - 2)$

MIXED PRACTICE

Perform the indicated operations. See Sections 5.2 and 5.3.

81. $(3x - 1) + (10x - 6)$

82. $(2x - 1) + (10x - 7)$

83. $(3x - 1)(10x - 6)$

84. $(2x - 1)(10x - 7)$

85. $(3x - 1) - (10x - 6)$

86. $(2x - 1) - (10x - 7)$

CONCEPT EXTENSIONS

△ **87.** The area of the larger rectangle on the right is $x(x + 3)$. Find another expression for this area by finding the sum of the areas of the two smaller rectangles.

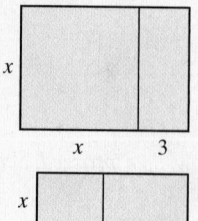

△ **88.** Write an expression for the area of the larger rectangle on the right in two different ways.

△ **89.** The area of the figure on the right is $(x + 2)(x + 3)$. Find another expression for this area by finding the sum of the areas of the four smaller rectangles.

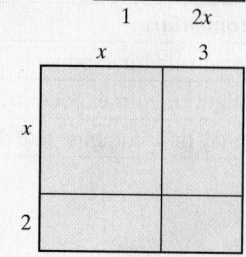

△ **90.** Write an expression for the area of the figure in two different ways.

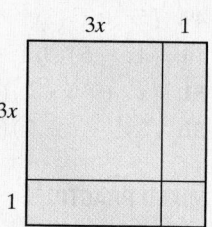

Simplify. See the Concept Checks in this section.

91. $5a + 6a$ **92.** $5a \cdot 6a$

Square where indicated. Simplify if possible.

93. $(5x)^2 + (2y)^2$ **94.** $(5x + 2y)^2$

95. Multiply each of the following polynomials.

   a. $(a + b)(a - b)$

   b. $(2x + 3y)(2x - 3y)$

   c. $(4x + 7)(4x - 7)$

   d. Can you make a general statement about all products of the form $(x + y)(x - y)$?

96. Evaluate each of the following.

   a. $(2 + 3)^2; 2^2 + 3^2$

   b. $(8 + 10)^2; 8^2 + 10^2$

   c. Does $(a + b)^2 = a^2 + b^2$ no matter what the values of a and b are? Why or why not?

△ **97.** Write a polynomial that describes the area of the shaded region. (Find the area of the larger square minus the area of the smaller square.)

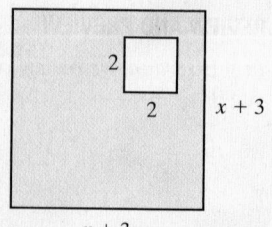

△ **98.** Write a polynomial that describes the area of the shaded region. (See Exercise 97.)

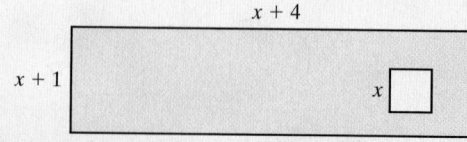

5.4 Special Products

OBJECTIVES

1 Multiply Two Binomials Using the FOIL Method. ▷

2 Square a Binomial. ▷

3 Multiply the Sum and Difference of Two Terms. ▷

4 Use Special Products to Multiply Binomials. ▷

OBJECTIVE

1 Using the FOIL Method ▷

In this section, we multiply binomials using special products. First, a special order for multiplying binomials called the FOIL order or method is introduced. This method is demonstrated by multiplying $(3x + 1)$ by $(2x + 5)$ as shown below.

> **The FOIL Method**
>
> **F** stands for the product of the **First** terms. $(3x + 1)(2x + 5)$
>
> $$(3x)(2x) = 6x^2 \quad \textbf{F}$$
>
> **O** stands for the product of the **Outer** terms. $(3x + 1)(2x + 5)$
>
> $$(3x)(5) = 15x \quad \textbf{O}$$
>
> **I** stands for the product of the **Inner** terms. $(3x + 1)(2x + 5)$
>
> $$(1)(2x) = 2x \quad \textbf{I}$$
>
> **L** stands for the product of the **Last** terms. $(3x + 1)(2x + 5)$
>
> $$(1)(5) = 5 \quad \textbf{L}$$

$$
\begin{array}{cccc}
\text{F} & \text{O} & \text{I} & \text{L}
\end{array}
$$
$$(3x + 1)(2x + 5) = 6x^2 + 15x + 2x + 5$$
$$= 6x^2 + 17x + 5 \qquad \text{Combine like terms.}$$

✔ **CONCEPT CHECK**

Multiply $(3x + 1)(2x + 5)$ using methods from the last section. Show that the product is still $6x^2 + 17x + 5$.

EXAMPLE 1   Multiply $(x - 3)(x + 4)$ by the FOIL method.

Solution

$$
(x - 3)(x + 4) = \overset{\text{F}}{(x)(x)} + \overset{\text{O}}{(x)(4)} + \overset{\text{I}}{(-3)(x)} + \overset{\text{L}}{(-3)(4)}
$$

$$= x^2 + 4x - 3x - 12$$
$$= x^2 + x - 12 \qquad \text{Combine like terms.} \quad \square$$

Helpful Hint

Remember that the FOIL order for multiplying can be used only for the product of two binomials.

PRACTICE

1   Multiply $(x + 2)(x - 5)$ by the FOIL method. ■

EXAMPLE 2   Multiply $(5x - 7)(x - 2)$ by the FOIL method.

Solution

$$
(5x - 7)(x - 2) = \overset{\text{F}}{5x(x)} + \overset{\text{O}}{5x(-2)} + \overset{\text{I}}{(-7)(x)} + \overset{\text{L}}{(-7)(-2)}
$$

$$= 5x^2 - 10x - 7x + 14$$
$$= 5x^2 - 17x + 14 \qquad \text{Combine like terms.} \quad \square$$

Answer to Concept Check:
Multiply and simplify.
$3x(2x + 5) + 1(2x + 5)$

PRACTICE

2   Multiply $(4x - 9)(x - 1)$ by the FOIL method. ■

EXAMPLE 3 Multiply: $2(y + 6)(2y - 1)$

$$
\begin{array}{l}
 \text{F} \quad \text{O} \quad \text{I} \quad \text{L}\\
\underline{\text{Solution}} \; 2(y + 6)(2y - 1) = 2(2y^2 - 1y + 12y - 6)\\
 = 2(2y^2 + 11y - 6) \quad \text{Simplify inside parentheses.}\\
 = 4y^2 + 22y - 12 \quad \text{Now use the distributive property.} \quad \square
\end{array}
$$

PRACTICE
3 Multiply: $3(x + 5)(3x - 1)$

OBJECTIVE
2 Squaring Binomials ▶

Now, try squaring a binomial using the FOIL method.

EXAMPLE 4 Multiply: $(3y + 1)^2$

$\underline{\text{Solution}}$ $(3y + 1)^2 = (3y + 1)(3y + 1)$

$$
\begin{array}{l}
 \text{F} \text{O} \text{I} \text{L}\\
= (3y)(3y) + (3y)(1) + 1(3y) + 1(1)\\
= 9y^2 + 3y + 3y + 1\\
= 9y^2 + 6y + 1 \square
\end{array}
$$

PRACTICE
4 Multiply: $(4x - 1)^2$

Notice the pattern that appears in Example 4.

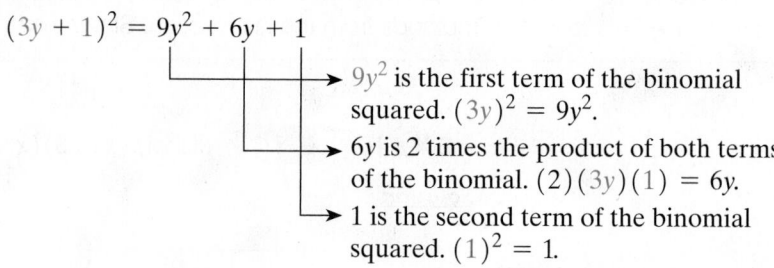

$(3y + 1)^2 = 9y^2 + 6y + 1$

$9y^2$ is the first term of the binomial squared. $(3y)^2 = 9y^2$.

$6y$ is 2 times the product of both terms of the binomial. $(2)(3y)(1) = 6y$.

1 is the second term of the binomial squared. $(1)^2 = 1$.

This pattern leads to the following, which can be used when squaring a binomial. We call these **special products.**

Squaring a Binomial

A binomial squared is equal to the square of the first term plus or minus twice the product of both terms plus the square of the second term.

$$(a + b)^2 = a^2 + 2ab + b^2$$
$$(a - b)^2 = a^2 - 2ab + b^2$$

This product can be visualized geometrically.

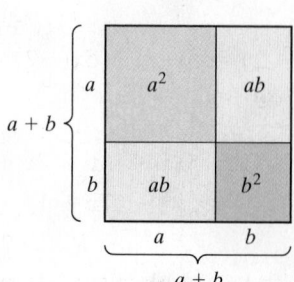

The area of the large square is side · side.

$$\text{Area} = (a + b)(a + b) = (a + b)^2$$

The area of the large square is also the sum of the areas of the smaller rectangles.

$$\text{Area} = a^2 + ab + ab + b^2 = a^2 + 2ab + b^2$$

Thus, $(a + b)^2 = a^2 + 2ab + b^2$.

EXAMPLE 5   Use a special product to square each binomial.

a. $(t + 2)^2$ **b.** $(p - q)^2$ **c.** $(2x + 5)^2$ **d.** $(x^2 - 7y)^2$

Solution

first term squared	plus or minus	twice the product of the terms	plus	second term squared

a. $(t + 2)^2$ $=$ t^2 $+$ $2(t)(2)$ $+$ $2^2 = t^2 + 4t + 4$

b. $(p - q)^2$ $=$ p^2 $-$ $2(p)(q)$ $+$ $q^2 = p^2 - 2pq + q^2$

c. $(2x + 5)^2$ $=$ $(2x)^2$ $+$ $2(2x)(5)$ $+$ $5^2 = 4x^2 + 20x + 25$

d. $(x^2 - 7y)^2$ $=$ $(x^2)^2$ $-$ $2(x^2)(7y)$ $+$ $(7y^2) = x^4 - 14x^2y + 49y^2$

☐

PRACTICE

5   Use a special product to square each binomial.

a. $(b + 3)^2$ **b.** $(x - y)^2$

c. $(3y + 2)^2$ **d.** $(a^2 - 5b)^2$ ■

Helpful Hint

Notice that

$$(a + b)^2 \neq a^2 + b^2 \quad \text{The middle term } 2ab \text{ is missing.}$$
$$(a + b)^2 = (a + b)(a + b) = a^2 + 2ab + b^2$$

Likewise,

$$(a - b)^2 \neq a^2 - b^2$$
$$(a - b)^2 = (a - b)(a - b) = a^2 - 2ab + b^2$$

OBJECTIVE

3   Multiplying the Sum and Difference of Two Terms

Another special product is the product of the sum and difference of the same two terms, such as $(x + y)(x - y)$. Finding this product by the FOIL method, we see a pattern emerge.

$$(x + y)(x - y) = x^2 - xy + xy - y^2$$
$$= x^2 - y^2$$

Notice that the middle two terms subtract out. This is because the **O**uter product is the opposite of the **I**nner product. Only the **difference of squares** remains.

Multiplying the Sum and Difference of Two Terms

The product of the sum and difference of two terms is the square of the first term minus the square of the second term.

$$(a + b)(a - b) = a^2 - b^2$$

EXAMPLE 6 Use a special product to multiply.

a. $4(x + 4)(x - 4)$ **b.** $(6t + 7)(6t - 7)$ **c.** $\left(x - \frac{1}{4}\right)\left(x + \frac{1}{4}\right)$

d. $(2p - q)(2p + q)$ **e.** $(3x^2 - 5y)(3x^2 + 5y)$

Solution

first term squared	minus	second term squared
↓	↓	↙

a. $4(x + 4)(x - 4) = 4(x^2 - 4^2) = 4(x^2 - 16) = 4x^2 - 64$

b. $(6t + 7)(6t - 7) = (6t)^2 - 7^2 = 36t^2 - 49$

c. $\left(x - \frac{1}{4}\right)\left(x + \frac{1}{4}\right) = x^2 - \left(\frac{1}{4}\right)^2 = x^2 - \frac{1}{16}$

d. $(2p - q)(2p + q) = (2p)^2 - q^2 = 4p^2 - q^2$

e. $(3x^2 - 5y)(3x^2 + 5y) = (3x^2)^2 - (5y)^2 = 9x^4 - 25y^2$

PRACTICE
6 Use a special product to multiply.

a. $3(x + 5)(x - 5)$ **b.** $(4b - 3)(4b + 3)$

c. $\left(x + \frac{2}{3}\right)\left(x - \frac{2}{3}\right)$ **d.** $(5s + t)(5s - t)$

e. $(2y - 3z^2)(2y + 3z^2)$

✔ **CONCEPT CHECK**

Match expression number 1 and number 2 to the equivalent expression or expressions in the list below.

1. $(a + b)^2$ **2.** $(a + b)(a - b)$

A. $(a + b)(a + b)$ **B.** $a^2 - b^2$ **C.** $a^2 + b^2$ **D.** $a^2 - 2ab + b^2$ **E.** $a^2 + 2ab + b^2$

OBJECTIVE
4 Using Special Products ▶

Let's now practice multiplying polynomials in general. If possible, use a special product.

EXAMPLE 7 Use a special product to multiply, if possible.

a. $(x - 5)(3x + 4)$ **b.** $(7x + 4)^2$ **c.** $(y - 0.6)(y + 0.6)$

d. $\left(y^4 + \frac{2}{5}\right)\left(3y^2 - \frac{1}{5}\right)$ **e.** $(a - 3)(a^2 + 2a - 1)$

Solution

a. $(x - 5)(3x + 4) = 3x^2 + 4x - 15x - 20$ FOIL.

$= 3x^2 - 11x - 20$

b. $(7x + 4)^2 = (7x)^2 + 2(7x)(4) + 4^2$ Squaring a binomial.

$= 49x^2 + 56x + 16$

c. $(y - 0.6)(y + 0.6) = y^2 - (0.6)^2 = y^2 - 0.36$ Multiplying the sum and difference of 2 terms.

d. $\left(y^4 + \frac{2}{5}\right)\left(3y^2 - \frac{1}{5}\right) = 3y^6 - \frac{1}{5}y^4 + \frac{6}{5}y^2 - \frac{2}{25}$ FOIL.

e. I've inserted this product as a reminder that since it is not a binomial times a binomial, the FOIL order may not be used.

Answers to Concept Check:
1. A and E **2.** B

$$(a - 3)(a^2 + 2a - 1) = a(a^2 + 2a - 1) - 3(a^2 + 2a - 1)$$

Multiplying each term of the binomial by each term of the trinomial.

$$= a^3 + 2a^2 - a - 3a^2 - 6a + 3$$

$$= a^3 - a^2 - 7a + 3$$

□

PRACTICE
7 Use a special product to multiply, if possible.

a. $(4x + 3)(x - 6)$

b. $(7b - 2)^2$

c. $(x + 0.4)(x - 0.4)$

d. $\left(x^2 - \dfrac{3}{7}\right)\left(3x^4 + \dfrac{2}{7}\right)$

e. $(x + 1)(x^2 + 5x - 2)$

■

> **Helpful Hint**
> • When multiplying two binomials, you may always use the FOIL order or method.
> • When multiplying any two polynomials, you may always use the distributive property to find the product.

✓ Vocabulary, Readiness & Video Check

Answer each exercise true or false.

1. $(x + 4)^2 = x^2 + 16$

2. For $(x + 6)(2x - 1)$ the product of the first terms is $2x^2$.

3. $(x + 4)(x - 4) = x^2 + 16$

4. The product $(x - 1)(x^3 + 3x - 1)$ is a polynomial of degree 5.

Martin-Gay Interactive Videos

See Video 5.4 ⊙

Watch the section lecture video and answer the following questions.

OBJECTIVE 1
5. From ⊞ Examples 1–3, for what type of multiplication problem is the FOIL order of multiplication used?

OBJECTIVE 2
6. Name at least one other method you can use to multiply ⊞ Example 4.

OBJECTIVE 3
7. From ⊞ Example 5, why does multiplying the sum and difference of the same two terms always give you a binomial answer?

OBJECTIVE 4
8. Why was the FOIL method not used for ⊞ Example 10?

5.4 Exercise Set MyMathLab® ▶

Multiply using the FOIL method. See Examples 1 through 3.

▶ **1.** $(x + 3)(x + 4)$

2. $(x + 5)(x - 1)$

3. $(x - 5)(x + 10)$

4. $(y - 12)(y + 4)$

5. $(5x - 6)(x + 2)$

6. $(3y - 5)(2y - 7)$

7. $5(y - 6)(4y - 1)$

8. $2(x - 11)(2x - 9)$

9. $(2x + 5)(3x - 1)$

10. $(6x + 2)(x - 2)$

11. $\left(x - \dfrac{1}{3}\right)\left(x + \dfrac{2}{3}\right)$

12. $\left(x - \dfrac{2}{5}\right)\left(x + \dfrac{1}{5}\right)$

Multiply. See Examples 4 and 5.

13. $(x + 2)^2$

14. $(x + 7)^2$

▶ **15.** $(2x - 1)^2$

16. $(7x - 3)^2$

17. $(3a - 5)^2$

18. $(5a + 2)^2$

▶ **19.** $(5x + 9)^2$

20. $(6s - 2)^2$

Multiply. See Example 6.

21. $(a - 7)(a + 7)$

22. $(b + 3)(b - 3)$

23. $(3x - 1)(3x + 1)$

24. $(4x - 5)(4x + 5)$

25. $\left(3x - \dfrac{1}{2}\right)\left(3x + \dfrac{1}{2}\right)$

26. $\left(10x + \dfrac{2}{7}\right)\left(10x - \dfrac{2}{7}\right)$

27. $(9x + y)(9x - y)$

28. $(2x - y)(2x + y)$

29. $(2x + 0.1)(2x - 0.1)$

30. $(5x - 1.3)(5x + 1.3)$

MIXED PRACTICE

Multiply. See Example 7.

31. $(a + 5)(a + 4)$

32. $(a - 5)(a - 7)$

33. $(a + 7)^2$

34. $(b - 2)^2$

35. $(4a + 1)(3a - 1)$

36. $(6a + 7)(6a + 5)$

37. $(x + 2)(x - 2)$

38. $(x - 10)(x + 10)$

39. $(3a + 1)^2$

40. $(4a - 2)^2$

41. $(x^2 + y)(4x - y^4)$

42. $(x^3 - 2)(5x + y)$

43. $(x + 3)(x^2 - 6x + 1)$

44. $(x - 2)(x^2 - 4x + 2)$

45. $(2a - 3)^2$

46. $(5b - 4x)^2$

47. $(5x - 6z)(5x + 6z)$

48. $(11x - 7y)(11x + 7y)$

49. $(x^5 - 3)(x^5 - 5)$

50. $(a^4 + 5)(a^4 + 6)$

51. $(x + 0.8)(x - 0.8)$

52. $(y - 0.9)(y + 0.9)$

53. $(a^3 + 11)(a^4 - 3)$

54. $(x^5 + 5)(x^2 - 8)$

55. $3(x - 2)^2$

56. $2(3b + 7)^2$

57. $(3b + 7)(2b - 5)$

58. $(3y - 13)(y - 3)$

59. $(7p - 8)(7p + 8)$

60. $(3s - 4)(3s + 4)$

61. $\left(\dfrac{1}{3}a^2 - 7\right)\left(\dfrac{1}{3}a^2 + 7\right)$

62. $\left(\dfrac{2}{3}a - b^2\right)\left(\dfrac{2}{3}a + b^2\right)$

63. $5x^2(3x^2 - x + 2)$

64. $4x^3(2x^2 + 5x - 1)$

65. $(2r - 3s)(2r + 3s)$

66. $(6r - 2x)(6r + 2x)$

67. $(3x - 7y)^2$

68. $(4s - 2y)^2$

69. $(4x + 5)(4x - 5)$

70. $(3x + 5)(3x - 5)$

71. $(8x + 4)^2$

72. $(3x + 2)^2$

73. $\left(a - \dfrac{1}{2}y\right)\left(a + \dfrac{1}{2}y\right)$

74. $\left(\dfrac{a}{2} + 4y\right)\left(\dfrac{a}{2} - 4y\right)$

75. $\left(\dfrac{1}{5}x - y\right)\left(\dfrac{1}{5}x + y\right)$

76. $\left(\dfrac{y}{6} - 8\right)\left(\dfrac{y}{6} + 8\right)$

77. $(a + 1)(3a^2 - a + 1)$

78. $(b + 3)(2b^2 + b - 3)$

Express each as a product of polynomials in x. Then multiply and simplify.

△ **79.** Find the area of the square rug shown if its side is $(2x + 1)$ feet.

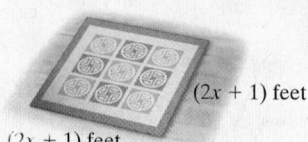

$(2x + 1)$ feet

$(2x + 1)$ feet

△ **80.** Find the area of the rectangular canvas if its length is $(3x - 2)$ inches and its width is $(x - 4)$ inches.

$(x - 4)$ inches

$(3x - 2)$ inches

REVIEW AND PREVIEW

Simplify each expression. See Section 5.1.

81. $\dfrac{50b^{10}}{70b^5}$

82. $\dfrac{x^3 y^6}{xy^2}$

83. $\dfrac{8a^{17}b^{15}}{-4a^7b^{10}}$

84. $\dfrac{-6a^8 y}{3a^4 y}$

85. $\dfrac{2x^4 y^{12}}{3x^4 y^4}$

86. $\dfrac{-48ab^6}{32ab^3}$

Find the slope of each line. See Section 3.4.

87.

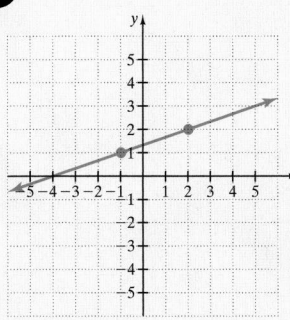

88.

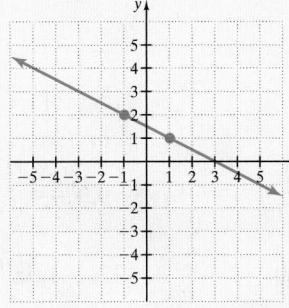

89.

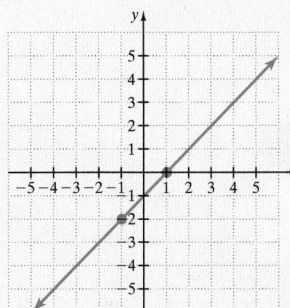

90.

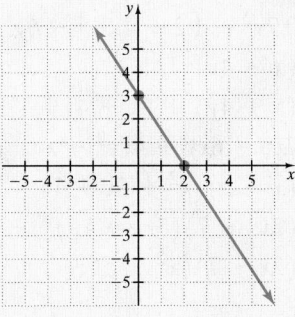

CONCEPT EXTENSIONS

Match each expression on the left to the equivalent expression on the right. See the second Concept Check in this section.

91. $(a - b)^2$ **A.** $a^2 - b^2$

92. $(a - b)(a + b)$ **B.** $a^2 + b^2$

93. $(a + b)^2$ **C.** $a^2 - 2ab + b^2$

94. $(a + b)^2(a - b)^2$ **D.** $a^2 + 2ab + b^2$

          E. none of these

Fill in the squares so that a true statement forms.

95. $(x^\square + 7)(x^\square + 3) = x^4 + 10x^2 + 21$

96. $(5x^\square - 2)^2 = 25x^6 - 20x^3 + 4$

Find the area of the shaded figure. To do so, subtract the area of the smaller square(s) from the area of the larger geometric figure.

△ **97.**

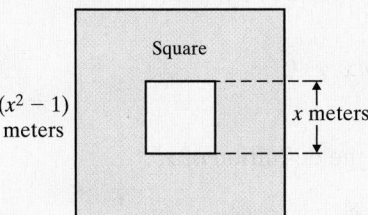

△ **98.**

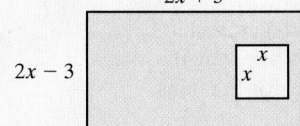

99.

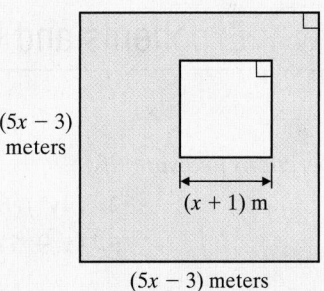

△ **100.**

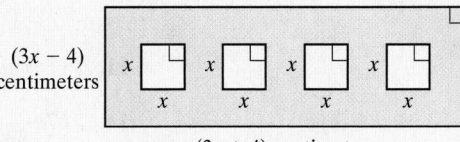

For Exercises 101 and 102, find the area of the shaded figure.

△ **101.**

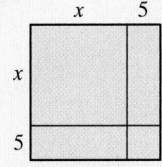

△ **102.**

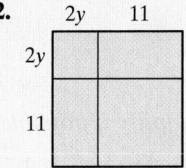

103. In your own words, describe the different methods that can be used to find the product $(2x - 5)(3x + 1)$.

104. In your own words, describe the different methods that can be used to find the product $(5x + 1)^2$.

105. Suppose that a classmate asked you why $(2x + 1)^2$ is **not** $(4x^2 + 1)$. Write down your response to this classmate.

106. Suppose that a classmate asked you why $(2x + 1)^2$ **is** $(4x^2 + 4x + 1)$. Write down your response to this classmate.

107. Using your own words, explain how to square a binomial such as $(a + b)^2$.

108. Explain how to find the product of two binomials using the FOIL method.

Find each product. For example,

$$[(a + b) - 2][(a + b) + 2] = (a + b)^2 - 2^2$$
$$= a^2 + 2ab + b^2 - 4$$

109. $[(x + y) - 3][(x + y) + 3]$

110. $[(a + c) - 5][(a + c) + 5]$

111. $[(a - 3) + b][(a - 3) - b]$

112. $[(x - 2) + y][(x - 2) - y]$

Integrated Review | **Exponents and Operations on Polynomials**

Sections 5.1–5.4

Perform the indicated operations and simplify.

1. $(5x^2)(7x^3)$

2. $(4y^2)(8y^7)$

3. -4^2

4. $(-4)^2$

5. $(x-5)(2x+1)$

6. $(3x-2)(x+5)$

7. $(x-5)+(2x+1)$

8. $(3x-2)+(x+5)$

9. $\dfrac{7x^9y^{12}}{x^3y^{10}}$

10. $\dfrac{20a^2b^8}{14a^2b^2}$

11. $(12m^7n^6)^2$

12. $(4y^9z^{10})^3$

13. $3(4y-3)(4y+3)$

14. $2(7x-1)(7x+1)$

15. $(x^7y^5)^9$

16. $(3^1x^9)^3$

17. $(7x^2-2x+3)-(5x^2+9)$

18. $(10x^2+7x-9)-(4x^2-6x+2)$

19. $0.7y^2-1.2+1.8y^2-6y+1$

20. $7.8x^2-6.8x+3.3+0.6x^2-9$

21. $(x+4y)^2$

22. $(y-9z)^2$

23. $(x+4y)+(x+4y)$

24. $(y-9z)+(y-9z)$

25. $7x^2-6xy+4(y^2-xy)$

26. $5a^2-3ab+6(b^2-a^2)$

27. $(x-3)(x^2+5x-1)$

28. $(x+1)(x^2-3x-2)$

29. $(2x^3-7)(3x^2+10)$

30. $(5x^3-1)(4x^4+5)$

31. $(2x-7)(x^2-6x+1)$

32. $(5x-1)(x^2+2x-3)$

Perform the indicated operations and simplify if possible.

33. $5x^3+5y^3$

34. $(5x^3)(5y^3)$

35. $(5x^3)^3$

36. $\dfrac{5x^3}{5y^3}$

37. $x+x$

38. $x\cdot x$

5.5 Negative Exponents and Scientific Notation

OBJECTIVES

1 Simplify Expressions Containing Negative Exponents.

2 Use All the Rules and Definitions for Exponents to Simplify Exponential Expressions.

3 Write Numbers in Scientific Notation.

4 Convert Numbers from Scientific Notation to Standard Form.

5 Perform Operations on Numbers Written in Scientific Notation.

OBJECTIVE

1 Simplifying Expressions Containing Negative Exponents

Our work with exponential expressions so far has been limited to exponents that are positive integers or 0. Here we expand to give meaning to an expression like x^{-3}.

Suppose that we wish to simplify the expression $\dfrac{x^2}{x^5}$. If we use the quotient rule for exponents, we subtract exponents:

$$\frac{x^2}{x^5} = x^{2-5} = x^{-3}, \quad x \neq 0$$

But what does x^{-3} mean? Let's simplify $\dfrac{x^2}{x^5}$ using the definition of x^n.

$$\frac{x^2}{x^5} = \frac{x \cdot x}{x \cdot x \cdot x \cdot x \cdot x}$$

$$= \frac{\overset{1}{\cancel{x}} \cdot \overset{1}{\cancel{x}}}{\underset{1}{\cancel{x}} \cdot \underset{1}{\cancel{x}} \cdot x \cdot x \cdot x}$$ Divide numerator and denominator by common factors by applying the fundamental principle for fractions.

$$= \frac{1}{x^3}$$

If the quotient rule is to hold true for negative exponents, then x^{-3} must equal $\dfrac{1}{x^3}$. From this example, we state the definition for negative exponents.

Negative Exponents

If a is a real number other than 0 and n is an integer, then

$$a^{-n} = \frac{1}{a^n}$$

For example, $x^{-3} = \dfrac{1}{x^3}$.

In other words, another way to write a^{-n} is to take its reciprocal and change the sign of its exponent.

EXAMPLE 1   Simplify by writing each expression with positive exponents only.

a. 3^{-2} **b.** $2x^{-3}$ **c.** $2^{-1} + 4^{-1}$ **d.** $(-2)^{-4}$ **e.** y^{-4}

Solution

a. $3^{-2} = \dfrac{1}{3^2} = \dfrac{1}{9}$ Use the definition of negative exponents.

b. $2x^{-3} = 2 \cdot \dfrac{1}{x^3} = \dfrac{2}{x^3}$ Use the definition of negative exponents.

c. $2^{-1} + 4^{-1} = \dfrac{1}{2} + \dfrac{1}{4} = \dfrac{2}{4} + \dfrac{1}{4} = \dfrac{3}{4}$

d. $(-2)^{-4} = \dfrac{1}{(-2)^4} = \dfrac{1}{(-2)(-2)(-2)(-2)} = \dfrac{1}{16}$

e. $y^{-4} = \dfrac{1}{y^4}$

> **Helpful Hint**
> Don't forget that since there are no parentheses, only x is the base for the exponent -3.

PRACTICE

1   Simplify by writing each expression with positive exponents only.

a. 5^{-3} **b.** $3y^{-4}$ **c.** $3^{-1} + 2^{-1}$ **d.** $(-5)^{-2}$ **e.** x^{-5}

> **Helpful Hint**
> A negative exponent *does not affect* the sign of its base.
> Remember: Another way to write a^{-n} is to take its reciprocal and change the sign of its exponent: $a^{-n} = \dfrac{1}{a^n}$. For example,
>
> $$x^{-2} = \frac{1}{x^2}, \qquad 2^{-3} = \frac{1}{2^3} \ \text{ or } \ \frac{1}{8}$$
>
> $$\frac{1}{y^{-4}} = \frac{1}{\frac{1}{y^4}} = y^4, \qquad \frac{1}{5^{-2}} = 5^2 \ \text{ or } \ 25$$

From the preceding Helpful Hint, we know that $x^{-2} = \dfrac{1}{x^2}$ and $\dfrac{1}{y^{-4}} = y^4$. We can use this to include another statement in our definition of negative exponents.

Negative Exponents

If a is a real number other than 0 and n is an integer, then

$$a^{-n} = \frac{1}{a^n} \quad \text{and} \quad \frac{1}{a^{-n}} = a^n$$

EXAMPLE 2 Simplify each expression. Write results using positive exponents only.

a. $\dfrac{1}{x^{-3}}$ b. $\dfrac{1}{3^{-4}}$ c. $\dfrac{p^{-4}}{q^{-9}}$ d. $\dfrac{5^{-3}}{2^{-5}}$

Solution

a. $\dfrac{1}{x^{-3}} = \dfrac{x^3}{1} = x^3$ b. $\dfrac{1}{3^{-4}} = \dfrac{3^4}{1} = 81$

c. $\dfrac{p^{-4}}{q^{-9}} = \dfrac{q^9}{p^4}$ d. $\dfrac{5^{-3}}{2^{-5}} = \dfrac{2^5}{5^3} = \dfrac{32}{125}$

PRACTICE
2 Simplify each expression. Write results using positive exponents only.

a. $\dfrac{1}{s^{-5}}$ b. $\dfrac{1}{2^{-3}}$ c. $\dfrac{x^{-7}}{y^{-5}}$ d. $\dfrac{4^{-3}}{3^{-2}}$

EXAMPLE 3 Simplify each expression. Write answers with positive exponents.

a. $\dfrac{y}{y^{-2}}$ b. $\dfrac{3}{x^{-4}}$ c. $\dfrac{x^{-5}}{x^7}$ d. $\left(\dfrac{2}{3}\right)^{-3}$

Solution

a. $\dfrac{y}{y^{-2}} = \dfrac{y^1}{y^{-2}} = y^{1-(-2)} = y^3$ Remember that $\dfrac{a^m}{a^n} = a^{m-n}$.

b. $\dfrac{3}{x^{-4}} = 3 \cdot \dfrac{1}{x^{-4}} = 3 \cdot x^4$ or $3x^4$

c. $\dfrac{x^{-5}}{x^7} = x^{-5-7} = x^{-12} = \dfrac{1}{x^{12}}$

d. $\left(\dfrac{2}{3}\right)^{-3} = \dfrac{2^{-3}}{3^{-3}} = \dfrac{3^3}{2^3} = \dfrac{27}{8}$

PRACTICE
3 Simplify each expression. Write answers with positive exponents.

a. $\dfrac{x^{-3}}{x^2}$ b. $\dfrac{5}{y^{-7}}$ c. $\dfrac{z}{z^{-4}}$ d. $\left(\dfrac{5}{9}\right)^{-2}$

OBJECTIVE
2 Simplifying Exponential Expressions

All the previously stated rules for exponents apply for negative exponents also. Here is a summary of the rules and definitions for exponents.

Summary of Exponent Rules

If m and n are integers and a, b, and c are real numbers, then:

Product rule for exponents: $a^m \cdot a^n = a^{m+n}$

Power rule for exponents: $(a^m)^n = a^{m \cdot n}$

Power of a product: $(ab)^n = a^n b^n$

Power of a quotient: $\left(\dfrac{a}{c}\right)^n = \dfrac{a^n}{c^n}$, $c \neq 0$

Quotient rule for exponents: $\dfrac{a^m}{a^n} = a^{m-n}$, $a \neq 0$

Zero exponent: $a^0 = 1$, $a \neq 0$

Negative exponent: $a^{-n} = \dfrac{1}{a^n}$, $a \neq 0$

EXAMPLE 4   Simplify the following expressions. Write each result using positive exponents only.

a. $(y^{-3}z^6)^{-6}$ **b.** $\dfrac{(2x^3)^4x}{x^7}$ **c.** $\left(\dfrac{3a^2}{b}\right)^{-3}$ **d.** $\dfrac{4^{-1}x^{-3}y}{4^{-3}x^2y^{-6}}$ **e.** $\left(\dfrac{-2x^3y}{xy^{-1}}\right)^3$

Solution

a. $(y^{-3}z^6)^{-6} = y^{18}\cdot z^{-36} = \dfrac{y^{18}}{z^{36}}$

b. $\dfrac{(2x^3)^4x}{x^7} = \dfrac{2^4\cdot x^{12}\cdot x}{x^7} = \dfrac{16\cdot x^{12+1}}{x^7} = \dfrac{16x^{13}}{x^7} = 16x^{13-7} = 16x^6$ Use the power rule.

c. $\left(\dfrac{3a^2}{b}\right)^{-3} = \dfrac{3^{-3}(a^2)^{-3}}{b^{-3}}$ Raise each factor in the numerator and the denominator to the -3 power.

$= \dfrac{3^{-3}a^{-6}}{b^{-3}}$ Use the power rule.

$= \dfrac{b^3}{3^3a^6}$ Use the negative exponent rule.

$= \dfrac{b^3}{27a^6}$ Write 3^3 as 27.

d. $\dfrac{4^{-1}x^{-3}y}{4^{-3}x^2y^{-6}} = 4^{-1-(-3)}x^{-3-2}y^{1-(-6)} = 4^2x^{-5}y^7 = \dfrac{4^2y^7}{x^5} = \dfrac{16y^7}{x^5}$

e. $\left(\dfrac{-2x^3y}{xy^{-1}}\right)^3 = \dfrac{(-2)^3x^9y^3}{x^3y^{-3}} = \dfrac{-8x^9y^3}{x^3y^{-3}} = -8x^{9-3}y^{3-(-3)} = -8x^6y^6$ ☐

PRACTICE
4   Simplify the following expressions. Write each result using positive exponents only.

a. $(a^4b^{-3})^{-5}$ **b.** $\dfrac{x^2(x^5)^3}{x^7}$ **c.** $\left(\dfrac{5p^8}{q}\right)^{-2}$

d. $\dfrac{6^{-2}x^{-4}y^{-7}}{6^{-3}x^3y^{-9}}$ **e.** $\left(\dfrac{-3x^4y}{x^2y^{-2}}\right)^3$ ■

OBJECTIVE
3   **Writing Numbers in Scientific Notation**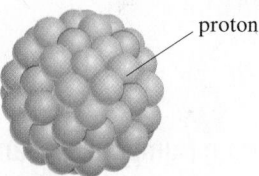

Both very large and very small numbers frequently occur in many fields of science. For example, the distance between the Sun and the dwarf planet Pluto is approximately 5,906,000,000 kilometers, and the mass of a proton is approximately 0.00000000000000000000000165 gram. It can be tedious to write these numbers in this standard decimal notation, so **scientific notation** is used as a convenient shorthand for expressing very large and very small numbers.

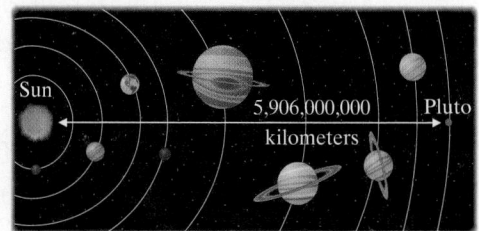

proton

Mass of proton is approximately
0.000 000 000 000 000 000 000 001 65 gram

Sun   5,906,000,000   Pluto
kilometers

Scientific Notation

A positive number is written in scientific notation if it is written as the product of a number a, where $1 \le a < 10$, and an integer power r of 10:

$$a \times 10^r$$

The numbers below are written in scientific notation. The $\times$ sign for multiplication is used as part of the notation.

$2.03 \times 10^2 \qquad 7.362 \times 10^7 \qquad \overbrace{5.906 \times 10^9}$ (Distance between the Sun and Pluto)
$1 \times 10^{-3} \qquad 8.1 \times 10^{-5} \qquad \underbrace{1.65 \times 10^{-24}}$ (Mass of a proton)

The following steps are useful when writing numbers in scientific notation.

> **To Write a Number in Scientific Notation**
>
> **Step 1.** Move the decimal point in the original number to the left or right so that the new number has a value between 1 and 10 (including 1).
>
> **Step 2.** Count the number of decimal places the decimal point is moved in Step 1. If the original number is 10 or greater, the count is positive. If the original number is less than 1, the count is negative.
>
> **Step 3.** Multiply the new number in Step 1 by 10 raised to an exponent equal to the count found in Step 2.

EXAMPLE 5 Write each number in scientific notation.

    a. 367,000,000 **b.** 0.000003 **c.** 20,520,000,000 **d.** 0.00085

Solution

  a. Step 1. Move the decimal point until the number is between 1 and 10.

367,000,000.
        8 places

  Step 2. The decimal point is moved 8 places, and the original number is 10 or greater, so the count is positive 8.

  Step 3. $367{,}000{,}000 = 3.67 \times 10^8$.

  b. Step 1. Move the decimal point until the number is between 1 and 10.

0.000003
    6 places

  Step 2. The decimal point is moved 6 places, and the original number is less than 1, so the count is -6.

  Step 3. $0.000003 = 3.0 \times 10^{-6}$

  c. $20{,}520{,}000{,}000 = 2.052 \times 10^{10}$

  d. $0.00085 = 8.5 \times 10^{-4}$

PRACTICE
5 Write each number in scientific notation.

    a. 0.000007   **b.** 20,700,000

    c. 0.0043   **d.** 812,000,000

OBJECTIVE
4 Converting Numbers to Standard Form

A number written in scientific notation can be rewritten in standard form. For example, to write 8.63×10^3 in standard form, recall that $10^3 = 1000$.

$$8.63 \times 10^3 = 8.63(1000) = 8630$$

Notice that the exponent on the 10 is positive 3, and we moved the decimal point 3 places to the right.

To write 7.29×10^{-3} in standard form, recall that $10^{-3} = \dfrac{1}{10^3} = \dfrac{1}{1000}$.

$$7.29 \times 10^{-3} = 7.29\left(\frac{1}{1000}\right) = \frac{7.29}{1000} = 0.00729$$

The exponent on the 10 is negative 3, and we moved the decimal to the left 3 places.

In general, **to write a scientific notation number in standard form,** move the decimal point the same number of places as the exponent on 10. If the exponent is positive, move the decimal point to the right; if the exponent is negative, move the decimal point to the left.

EXAMPLE 6   Write each number in standard notation, without exponents.

   a. 1.02×10^5 **b.** 7.358×10^{-3} **c.** 8.4×10^7 **d.** 3.007×10^{-5}

Solution

   a. Move the decimal point 5 places to the right.
   $$1.02 \times 10^5 = 102{,}000.$$

   b. Move the decimal point 3 places to the left.
   $$7.358 \times 10^{-3} = 0.007358$$

   c. $8.4 \times 10^7 = 84{,}000{,}000.$ 7 places to the right

   d. $3.007 \times 10^{-5} = 0.00003007$ 5 places to the left   □

PRACTICE
6   Write each number in standard notation, without exponents.
   a. 3.67×10^{-4} **b.** 8.954×10^6
   c. 2.009×10^{-5} **d.** 4.054×10^3 ■

✔ **CONCEPT CHECK**
Which number in each pair is larger?
a. 7.8×10^3 or 2.1×10^5 **b.** 9.2×10^{-2} or 2.7×10^4 **c.** 5.6×10^{-4} or 6.3×10^{-5}

OBJECTIVE
5   **Performing Operations with Scientific Notation**

Performing operations on numbers written in scientific notation uses the rules and definitions for exponents.

EXAMPLE 7   Perform each indicated operation. Write each result in standard decimal notation.

   a. $(8 \times 10^{-6})(7 \times 10^3)$ **b.** $\dfrac{12 \times 10^2}{6 \times 10^{-3}}$

Solution

   a. $(8 \times 10^{-6})(7 \times 10^3) = (8 \cdot 7) \times (10^{-6} \cdot 10^3)$
   $$= 56 \times 10^{-3}$$
   $$= 0.056$$

   b. $\dfrac{12 \times 10^2}{6 \times 10^{-3}} = \dfrac{12}{6} \times 10^{2-(-3)} = 2 \times 10^5 = 200{,}000$ □

PRACTICE
7   Perform each indicated operation. Write each result in standard decimal notation.

   a. $(5 \times 10^{-4})(8 \times 10^6)$ **b.** $\dfrac{64 \times 10^3}{32 \times 10^{-7}}$ ■

Calculator Explorations

Scientific Notation

To enter a number written in scientific notation on a scientific calculator, locate the scientific notation key, which may be marked $\boxed{EE}$ or $\boxed{EXP}$. To enter 3.1×10^7, press $\boxed{3.1}$ $\boxed{EE}$ $\boxed{7}$. The display should read $\boxed{3.1 \quad 07}$.

Enter each number written in scientific notation on your calculator.

1. 5.31×10^3 **2.** -4.8×10^{14}

3. 6.6×10^{-9} **4.** -9.9811×10^{-2}

Multiply each of the following on your calculator. Notice the form of the result.

5. $3,000,000 \times 5,000,000$

6. $230,000 \times 1000$

Multiply each of the following on your calculator. Write the product in scientific notation.

7. $(3.26 \times 10^6)(2.5 \times 10^{13})$

8. $(8.76 \times 10^{-4})(1.237 \times 10^9)$

Vocabulary, Readiness & Video Check

Fill in each blank with the correct choice.

1. The expression x^{-3} equals _____.

   a. $-x^3$ **b.** $\dfrac{1}{x^3}$ **c.** $\dfrac{-1}{x^3}$ **d.** $\dfrac{1}{x^{-3}}$

2. The expression 5^{-4} equals _____.

   a. -20 **b.** -625 **c.** $\dfrac{1}{20}$ **d.** $\dfrac{1}{625}$

3. The number 3.021×10^{-3} is written in _____.

   a. standard form   **b.** expanded form

   c. scientific notation

4. The number 0.0261 is written in _____.

   a. standard form   **b.** expanded form

   c. scientific notation

Martin-Gay Interactive Videos

See Video 5.5

Watch the section lecture video and answer the following questions.

OBJECTIVE 1
5. What important reminder is made at the end of ▤ Example 1?

OBJECTIVE 2
6. Name all the rules and definitions used to simplify ▤ Example 8.

OBJECTIVE 3
7. From ▤ Examples 9 and 10, explain how the movement of the decimal point in step 1 suggests the sign of the exponent on the number 10.

OBJECTIVE 4
8. From ▤ Example 11, what part of a number written in scientific notation is key in telling you how to write the number in standard form?

OBJECTIVE 5
9. For ▤ Example 13, what exponent rules were needed to evaluate?

5.5 Exercise Set   MyMathLab® ▶

Simplify each expression. Write each result using positive exponents only. See Examples 1 through 3.

1. 4^{-3} **2.** 6^{-2} **3.** $(-3)^{-4}$

4. $(-3)^{-5}$ **5.** $7x^{-3}$ **6.** $(7x)^{-3}$

7. $\left(\dfrac{1}{2}\right)^{-5}$ **8.** $\left(\dfrac{1}{8}\right)^{-2}$ **9.** $\left(-\dfrac{1}{4}\right)^{-3}$

10. $\left(-\dfrac{1}{8}\right)^{-2}$ **11.** $3^{-1} + 5^{-1}$ **12.** $4^{-1} + 4^{-2}$

13. $\dfrac{1}{p^{-3}}$

14. $\dfrac{1}{q^{-5}}$

15. $\dfrac{p^{-5}}{q^{-4}}$

16. $\dfrac{r^{-5}}{s^{-2}}$

17. $\dfrac{x^{-2}}{x}$

18. $\dfrac{y}{y^{-3}}$

19. $\dfrac{z^{-4}}{z^{-7}}$

20. $\dfrac{x^{-4}}{x^{-1}}$

21. $3^{-2} + 3^{-1}$

22. $4^{-2} - 4^{-3}$

23. $\dfrac{-1}{p^{-4}}$

24. $\dfrac{-1}{y^{-6}}$

25. $-2^0 - 3^0$

26. $5^0 + (-5)^0$

MIXED PRACTICE

Simplify each expression. Write each result using positive exponents only. See Examples 1 through 4.

27. $\dfrac{x^2 x^5}{x^3}$

28. $\dfrac{y^4 y^5}{y^6}$

29. $\dfrac{p^2 p}{p^{-1}}$

30. $\dfrac{y^3 y}{y^{-2}}$

31. $\dfrac{(m^5)^4 m}{m^{10}}$

32. $\dfrac{(x^2)^8 x}{x^9}$

33. $\dfrac{r}{r^{-3} r^{-2}}$

34. $\dfrac{p}{p^{-3} q^{-5}}$

35. $(x^5 y^3)^{-3}$

36. $(z^5 x^5)^{-3}$

37. $\dfrac{(x^2)^3}{x^{10}}$

38. $\dfrac{(y^4)^2}{y^{12}}$

39. $\dfrac{(a^5)^2}{(a^3)^4}$

40. $\dfrac{(x^2)^5}{(x^4)^3}$

41. $\dfrac{8k^4}{2k}$

42. $\dfrac{27r^4}{3r^6}$

43. $\dfrac{-6m^4}{-2m^3}$

44. $\dfrac{15a^4}{-15a^5}$

45. $\dfrac{-24a^6 b}{6ab^2}$

46. $\dfrac{-5x^4 y^5}{15x^4 y^2}$

47. $(-2x^3 y^{-4})(3x^{-1} y)$

48. $(-5a^4 b^{-7})(-a^{-4} b^3)$

49. $(a^{-5} b^2)^{-6}$

50. $(4^{-1} x^5)^{-2}$

51. $\left(\dfrac{x^{-2} y^4}{x^3 y^7}\right)^2$

52. $\left(\dfrac{a^5 b}{a^7 b^{-2}}\right)^{-3}$

53. $\dfrac{4^2 z^{-3}}{4^3 z^{-5}}$

54. $\dfrac{3^{-1} x^4}{3^3 x^{-7}}$

55. $\dfrac{2^{-3} x^{-4}}{2^2 x}$

56. $\dfrac{5^{-1} z^7}{5^{-2} z^9}$

57. $\dfrac{7ab^{-4}}{7^{-1} a^{-3} b^2}$

58. $\dfrac{6^{-5} x^{-1} y^2}{6^{-2} x^{-4} y^4}$

59. $\left(\dfrac{a^{-5} b}{ab^3}\right)^{-4}$

60. $\left(\dfrac{r^{-2} s^{-3}}{r^{-4} s^{-3}}\right)^{-3}$

61. $\dfrac{(xy^3)^5}{(xy)^{-4}}$

62. $\dfrac{(rs)^{-3}}{(r^2 s^3)^2}$

63. $\dfrac{(-2xy^{-3})^{-3}}{(xy^{-1})^{-1}}$

64. $\dfrac{(-3x^2 y^2)^{-2}}{(xyz)^{-2}}$

65. $\dfrac{6x^2 y^3}{-7xy^5}$

66. $\dfrac{-8xa^2 b}{-5xa^5 b}$

67. $\dfrac{(a^4 b^{-7})^{-5}}{(5a^2 b^{-1})^{-2}}$

68. $\dfrac{(a^6 b^{-2})^4}{(4a^{-3} b^{-3})^3}$

Write each number in scientific notation. See Example 5.

69. 78,000

70. 9,300,000,000

71. 0.00000167

72. 0.00000017

73. 0.00635

74. 0.00194

75. 1,160,000

76. 700,000

77. More than 2,000,000,000 pencils are manufactured in the United States annually. Write this number in scientific notation. (*Source:* AbsoluteTrivia.com)

78. The temperature at the interior of the Earth is 20,000,000 degrees Celsius. Write 20,000,000 in scientific notation.

79. As of this writing, the world's largest optical telescope is the Gran Telescopio Canaris, located in La Palma, Canary Islands, Spain. The elevation of this telescope is 2400 meters above sea level. Write 2400 in scientific notation.

80. In March 2004, the European Space Agency launched the Rosetta spacecraft, whose mission was to deliver the Philae lander to explore comet 67P/Churyumov-Gerasimenko. The lander finally arrived on the comet in late 2014. This comet is currently more than 320,000,000 miles from Earth. Write 320,000,000 in scientific notation. (*Source:* European Space Agency)

Write each number in standard notation. See Example 6.

81. 8.673×10^{-10}

82. 9.056×10^{-4}

83. 3.3×10^{-2}

84. 4.8×10^{-6}

85. 2.032×10^4

86. 9.07×10^{10}

87. Each second, the Sun converts 7.0×10^8 tons of hydrogen into helium and energy in the form of gamma rays. Write this number in standard notation. (*Source:* Students for the Exploration and Development of Space)

88. In chemistry, Avogadro's number is the number of atoms in one mole of an element. Avogadro's number is $6.02214199 \times 10^{23}$. Write this number in standard notation. (*Source:* National Institute of Standards and Technology)

89. The distance light travels in 1 year is 9.46×10^{12} kilometers. Write this number in standard notation.

90. The population of the world is 7.3×10^9. Write this number in standard notation. (*Source:* UN World Population Clock)

MIXED PRACTICE

See Examples 5 and 6. Below are some interesting facts about selected countries' external debts at a certain time. These are public and private debts owed to nonresidents of that country. If a number is written in standard form, write it in scientific notation. If a number is written in scientific notation, write it in standard form. (Source: CIA World Factbook)

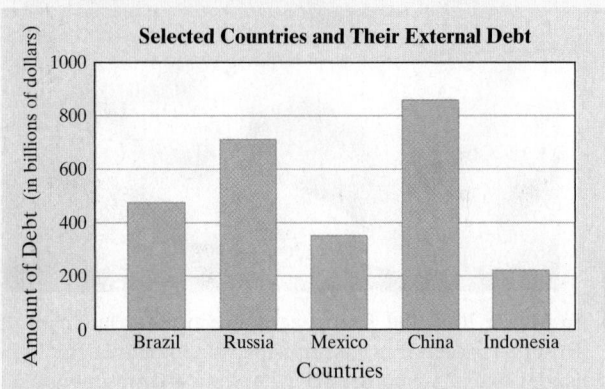

91. The external debt of Russia at a certain time was $714,000,000,000.

92. The amount by which Russia's debt was greater than Mexico's debt was $359,000,000,000.

93. At a certain time, China's external debt was 8.63×10^{11}.

94. At a certain time, the external debt of the United States was 1.5×10^{13}.

95. At a certain time, the estimated per person share of the United States external debt was 4.7×10^4.

96. The bar graph shows the external debt of five countries. Estimate the height of the tallest bar and the shortest bar in standard notation. Then write each number in scientific notation.

Evaluate each expression using exponential rules. Write each result in standard notation. See Example 7.

97. $(1.2 \times 10^{-3})(3 \times 10^{-2})$

98. $(2.5 \times 10^6)(2 \times 10^{-6})$

99. $(4 \times 10^{-10})(7 \times 10^{-9})$

100. $(5 \times 10^6)(4 \times 10^{-8})$

101. $\dfrac{8 \times 10^{-1}}{16 \times 10^5}$

102. $\dfrac{25 \times 10^{-4}}{5 \times 10^{-9}}$

103. $\dfrac{1.4 \times 10^{-2}}{7 \times 10^{-8}}$

104. $\dfrac{0.4 \times 10^5}{0.2 \times 10^{11}}$

REVIEW AND PREVIEW

Simplify the following. See Section 5.1.

105. $\dfrac{5x^7}{3x^4}$

106. $\dfrac{27y^{14}}{3y^7}$

107. $\dfrac{15z^4 y^3}{21zy}$

108. $\dfrac{18a^7 b^{17}}{30a^7 b}$

Use the distributive property and multiply. See Sections 5.3 and 5.5.

109. $\dfrac{1}{y}(5y^2 - 6y + 5)$

110. $\dfrac{2}{x}(3x^5 + x^4 - 2)$

CONCEPT EXTENSIONS

△ 111. Find the volume of the cube.

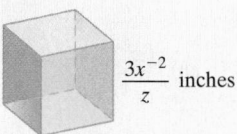

$\dfrac{3x^{-2}}{z}$ inches

△ 112. Find the area of the triangle.

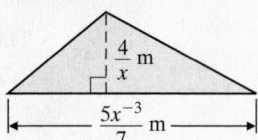

$\dfrac{4}{x}$ m

$\dfrac{5x^{-3}}{7}$ m

Simplify.

113. $(2a^3)^3 a^4 + a^5 a^8$

114. $(2a^3)^3 a^{-3} + a^{11} a^{-5}$

Fill in the boxes so that each statement is true. (More than one answer may be possible for these exercises.)

115. $x^{\square} = \dfrac{1}{x^5}$

116. $7^{\square} = \dfrac{1}{49}$

117. $z^{\square} \cdot z^{\square} = z^{-10}$

118. $(x^{\square})^{\square} = x^{-15}$

119. Which is larger? See the Concept Check in this section.
    a. 9.7×10^{-2} or 1.3×10^1
    b. 8.6×10^5 or 4.4×10^7
    c. 6.1×10^{-2} or 5.6×10^{-4}

120. Determine whether each statement is true or false.
    a. $5^{-1} < 5^{-2}$
    b. $\left(\dfrac{1}{5}\right)^{-1} < \left(\dfrac{1}{5}\right)^{-2}$
    c. $a^{-1} < a^{-2}$ for all nonzero numbers.

121. It was stated earlier that for an integer n,

$$x^{-n} = \frac{1}{x^n}, \quad x \neq 0$$

Explain why x may not equal 0.

122. The quotient rule states that

$$\frac{a^m}{a^n} = a^{m-n}, a \neq 0.$$

Explain why a may not equal 0.

Simplify each expression. Assume that variables represent positive integers.

123. $a^{-4m} \cdot a^{5m}$ **124.** $(x^{-3s})^3$

125. $(3y^{2z})^3$ **126.** $a^{4m+1} \cdot a^4$

Simplify each expression. Write each result in standard notation.

127. $(2.63 \times 10^{12})(-1.5 \times 10^{-10})$

128. $(6.785 \times 10^{-4})(4.68 \times 10^{10})$

Light travels at a rate of 1.86×10^5 miles per second. Use this information and the distance formula $d = r \cdot t$ to answer Exercises 129 and 130.

129. If the distance from the moon to Earth is 238,857 miles, find how long it takes the reflected light of the moon to reach Earth. (Round to the nearest tenth of a second.)

130. If the distance from the Sun to Earth is 93,000,000 miles, find how long it takes the light of the Sun to reach Earth.

5.6 Dividing Polynomials

OBJECTIVES

1 Divide a Polynomial by a Monomial.

2 Use Long Division to Divide a Polynomial by Another Polynomial.

OBJECTIVE

1 Dividing by a Monomial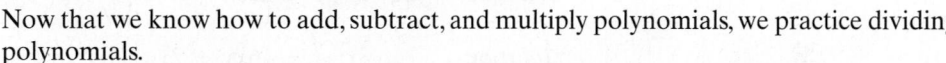

Now that we know how to add, subtract, and multiply polynomials, we practice dividing polynomials.

To divide a polynomial by a monomial, recall addition of fractions. Fractions that have a common denominator are added by adding the numerators:

$$\frac{a}{c} + \frac{b}{c} = \frac{a + b}{c}$$

If we read this equation from right to left and let a, b, and c be monomials, $c \neq 0$, we have the following:

Dividing a Polynomial by a Monomial

Divide each term of the polynomial by the monomial.

$$\frac{a + b}{c} = \frac{a}{c} + \frac{b}{c}, \quad c \neq 0$$

Throughout this section, we assume that denominators are not 0.

EXAMPLE 1 Divide $6m^2 + 2m$ by $2m$.

Solution We begin by writing the quotient in fraction form. Then we divide each term of the polynomial $6m^2 + 2m$ by the monomial $2m$.

$$\frac{6m^2 + 2m}{2m} = \frac{6m^2}{2m} + \frac{2m}{2m}$$

$$= 3m + 1 \quad \text{Simplify.}$$

Check: We know that if $\dfrac{6m^2 + 2m}{2m} = 3m + 1$, then $2m \cdot (3m + 1)$ must equal $6m^2 + 2m$. Thus, to check, we multiply.

$$2m(3m + 1) = 2m(3m) + 2m(1) = 6m^2 + 2m$$

The quotient $3m + 1$ checks. ☐

PRACTICE

1 Divide $8t^3 + 4t^2$ by $4t^2$

EXAMPLE 2 Divide: $\dfrac{9x^5 - 12x^2 + 3x}{3x^2}$

Solution

$$\frac{9x^5 - 12x^2 + 3x}{3x^2} = \frac{9x^5}{3x^2} - \frac{12x^2}{3x^2} + \frac{3x}{3x^2} \qquad \text{Divide each term by } 3x^2.$$

$$= 3x^3 - 4 + \frac{1}{x} \qquad \text{Simplify.}$$

Notice that the quotient is not a polynomial because of the term $\dfrac{1}{x}$. This expression is called a rational expression—we will study rational expressions further in Chapter 7. Although the quotient of two polynomials is not always a polynomial, we may still check by multiplying.

Check: $3x^2\left(3x^3 - 4 + \dfrac{1}{x}\right) = 3x^2(3x^3) - 3x^2(4) + 3x^2\left(\dfrac{1}{x}\right)$

$$= 9x^5 - 12x^2 + 3x \qquad \square$$

PRACTICE
2 Divide: $\dfrac{16x^6 + 20x^3 - 12x}{4x^2}$

EXAMPLE 3 Divide: $\dfrac{8x^2y^2 - 16xy + 2x}{4xy}$

Solution

$$\frac{8x^2y^2 - 16xy + 2x}{4xy} = \frac{8x^2y^2}{4xy} - \frac{16xy}{4xy} + \frac{2x}{4xy} \qquad \text{Divide each term by } 4xy.$$

$$= 2xy - 4 + \frac{1}{2y} \qquad \text{Simplify.}$$

Check: $4xy\left(2xy - 4 + \dfrac{1}{2y}\right) = 4xy(2xy) - 4xy(4) + 4xy\left(\dfrac{1}{2y}\right)$

$$= 8x^2y^2 - 16xy + 2x \qquad \square$$

PRACTICE
3 Divide: $\dfrac{15x^4y^4 - 10xy + y}{5xy}$

✔ **CONCEPT CHECK**

In which of the following is $\dfrac{x + 5}{5}$ simplified correctly?

a. $\dfrac{x}{5} + 1$ **b.** x **c.** $x + 1$

OBJECTIVE
2 Using Long Division to Divide by a Polynomial ▶

To divide a polynomial by a polynomial other than a monomial, we use a process known as long division. Polynomial long division is similar to number long division, so we review long division by dividing 13 into 3660.

$$
\begin{array}{r}
281 \\
13\overline{)3660} \\
\underline{26} \\
106 \\
\underline{104} \\
20 \\
\underline{13} \\
7
\end{array}
$$

> **Helpful Hint**
> Recall that 3660 is called the dividend.

$2 \cdot 13 = 26$

Subtract and bring down the next digit in the dividend.

$8 \cdot 13 = 104$

Subtract and bring down the next digit in the dividend.

$1 \cdot 13 = 13$

Subtract. There are no more digits to bring down, so the remainder is 7.

The quotient is $281\,R7$, which can be written as $281\dfrac{7}{13}.$ ← remainder
← divisor

Recall that division can be checked by multiplication. To check a division problem such as this one, we see that

$$13 \cdot 281 + 7 = 3660$$

Now we demonstrate long division of polynomials.

EXAMPLE 4   Divide $x^2 + 7x + 12$ by $x + 3$ using long division.

Solution

To subtract, change the signs of these terms and add.

$$
\begin{array}{r}
x \phantom{{}+7x+12} \\
x + 3\overline{)x^2 + 7x + 12} \\
x^2 \mp 3x \\
\hline
4x + 12
\end{array}
$$

How many times does x divide x^2? $\dfrac{x^2}{x} = x.$
Multiply: $x(x + 3)$
Subtract and bring down the next term.

Now we repeat this process.

$$
\begin{array}{r}
x + 4 \\
x + 3\overline{)x^2 + 7x + 12} \\
x^2 \mp 3x \\
\hline
4x + 12 \\
4x \mp 12 \\
\hline
0
\end{array}
$$

How many times does x divide $4x$? $\dfrac{4x}{x} = 4.$

To subtract, change the signs of these terms and add.

Multiply: $4(x + 3)$
Subtract. The remainder is 0.

The quotient is $x + 4$.

Check: We check by multiplying.

divisor	·	quotient	+	remainder	=	dividend
$(x + 3)$	·	$(x + 4)$	+	0	=	$x^2 + 7x + 12$

The quotient checks.   □

PRACTICE
4   Divide $x^2 + 5x + 6$ by $x + 2$ using long division.   ■

EXAMPLE 5   Divide $6x^2 + 10x - 5$ by $3x - 1$ using long division.

Solution

$$
\begin{array}{r}
2x + 4 \\
3x - 1\overline{)6x^2 + 10x - 5} \\
6x^2 \mp 2x \\
\hline
12x - 5 \\
12x \mp 4 \\
\hline
-1
\end{array}
$$

$\dfrac{6x^2}{3x} = 2x$, so $2x$ is a term of the quotient.
Multiply $2x(3x - 1)$.
Subtract and bring down the next term.
$\dfrac{12x}{3x} = 4$, multiply $4(3x - 1)$
Subtract. The remainder is -1.

Thus $(6x^2 + 10x - 5)$ divided by $(3x - 1)$ is $(2x + 4)$ with a remainder of -1. This can be written as

$$\frac{6x^2 + 10x - 5}{3x - 1} = 2x + 4 + \frac{-1}{3x - 1} \quad \begin{array}{l} \leftarrow \text{remainder} \\ \leftarrow \text{divisor} \end{array}$$

Check: To check, we multiply $(3x - 1)(2x + 4)$. Then we add the remainder, -1, to this product.

$$(3x - 1)(2x + 4) + (-1) = (6x^2 + 12x - 2x - 4) - 1$$
$$= 6x^2 + 10x - 5$$

The quotient checks.   □

PRACTICE
5   Divide $4x^2 + 8x - 7$ by $2x + 1$ using long division.   ■

In Example 5, the degree of the divisor, $3x - 1$, is 1 and the degree of the remainder, -1, is 0. The division process is continued until the degree of the remainder polynomial is less than the degree of the divisor polynomial.

Writing the dividend and divisor in a form with descending order of powers and with no missing terms is helpful when dividing polynomials.

EXAMPLE 6 Divide: $\dfrac{4x^2 + 7 + 8x^3}{2x + 3}$

Solution Before we begin the division process, we rewrite

$$4x^2 + 7 + 8x^3 \quad \text{as} \quad 8x^3 + 4x^2 + 0x + 7$$

Notice that we have written the polynomial in descending order and have represented the missing x^1-term by $0x$.

$$
\begin{array}{r}
4x^2 - 4x + 6 \\
2x + 3 \overline{)\,8x^3 + 4x^2 + 0x + 7} \\
\underline{8x^3 \mp 12x^2} \\
-8x^2 + 0x \\
\underline{\pm 8x^2 \pm 12x} \\
12x + 7 \\
\underline{12x \mp 18} \\
-11 \quad \text{Remainder}
\end{array}
$$

Thus, $\dfrac{4x^2 + 7 + 8x^3}{2x + 3} = 4x^2 - 4x + 6 + \dfrac{-11}{2x + 3}$.

PRACTICE
6 Divide: $\dfrac{11x - 3 + 9x^3}{3x + 2}$

EXAMPLE 7 Divide: $\dfrac{2x^4 - x^3 + 3x^2 + x - 1}{x^2 + 1}$

Solution Before dividing, rewrite the divisor polynomial

$$x^2 + 1 \quad \text{as} \quad x^2 + 0x + 1$$

The $0x$ term represents the missing x^1-term in the divisor.

$$
\begin{array}{r}
2x^2 - x + 1 \\
x^2 + 0x + 1 \overline{)\,2x^4 - x^3 + 3x^2 + x - 1} \\
\underline{2x^4 \mp 0x^3 \mp 2x^2} \\
-x^3 + x^2 + x \\
\underline{\pm x^3 \mp 0x^2 \mp x} \\
x^2 + 2x - 1 \\
\underline{x^2 \mp 0x \mp 1} \\
2x - 2 \quad \text{Remainder}
\end{array}
$$

Thus, $\dfrac{2x^4 - x^3 + 3x^2 + x - 1}{x^2 + 1} = 2x^2 - x + 1 + \dfrac{2x - 2}{x^2 + 1}$.

PRACTICE
7 Divide: $\dfrac{3x^4 - 2x^3 - 3x^2 + x + 4}{x^2 + 2}$

EXAMPLE 8 Divide $x^3 - 8$ by $x - 2$.

Solution: Notice that the polynomial $x^3 - 8$ is missing an x^2-term and an x-term. We'll represent these terms by inserting $0x^2$ and $0x$.

$$
\begin{array}{r}
x^2 + 2x + 4 \\
x - 2 \overline{\smash{\big)}\ x^3 + 0x^2 + 0x - 8} \\
\underline{-x^3 + 2x^2} \\
2x^2 + 0x \\
\underline{-2x^2 + 4x} \\
4x - 8 \\
\underline{-4x + 8} \\
0
\end{array}
$$

Thus, $\dfrac{x^3 - 8}{x - 2} = x^2 + 2x + 4$.

Check: To check, see that $(x^2 + 2x + 4)(x - 2) = x^3 - 8$. □

PRACTICE
8 Divide $x^3 + 27$ by $x + 3$. ■

✔ **Vocabulary, Readiness & Video Check**

Use the choices below to fill in each blank. Choices may be used more than once.

dividend   divisor   quotient

1. In $6\overline{)18}^{\,3}$, the 18 is the _____, the 3 is the _____, and the 6 is the _____.

2. In $x + 1 \overline{)x^2 + 3x + 2}^{\,x + 2}$, the $x + 1$ is the _____, the $x^2 + 3x + 2$ is the _____, and the $x + 2$ is the _____.

Simplify each expression mentally.

3. $\dfrac{a^6}{a^4}$

4. $\dfrac{p^8}{p^3}$

5. $\dfrac{y^2}{y}$

6. $\dfrac{a^3}{a}$

Martín-Gay Interactive Videos

See Video 5.6 ◉

Watch the section lecture video and answer the following questions.

OBJECTIVE 1
7. The lecture before ▦ Example 1 begins with adding two fractions with the same denominator. From there, the lecture continues to a method for dividing a polynomial by a monomial. What role does the monomial play in the fraction example?

OBJECTIVE 2
8. In ▦ Example 5, you're told that although you don't have to fill in missing powers in the divisor and the dividend, it really is a good idea to do so. Why?

5.6 **Exercise Set** MyMathLab® ▶

Perform each division. See Examples 1 through 3.

1. $\dfrac{12x^4 + 3x^2}{x}$

2. $\dfrac{15x^2 - 9x^5}{x}$

3. $\dfrac{20x^3 - 30x^2 + 5x + 5}{5}$

4. $\dfrac{8x^3 - 4x^2 + 6x + 2}{2}$

5. $\dfrac{15p^3 + 18p^2}{3p}$

6. $\dfrac{14m^2 - 27m^3}{7m}$

7. $\dfrac{-9x^4 + 18x^5}{6x^5}$

8. $\dfrac{6x^5 + 3x^4}{3x^4}$

▶ 9. $\dfrac{-9x^5 + 3x^4 - 12}{3x^3}$

10. $\dfrac{6a^2 - 4a + 12}{-2a^2}$

11. $\dfrac{4x^4 - 6x^3 + 7}{-4x^4}$

12. $\dfrac{-12a^3 + 36a - 15}{3a}$

Find each quotient using long division. See Examples 4 and 5.

▶ 13. $\dfrac{x^2 + 4x + 3}{x + 3}$

14. $\dfrac{x^2 + 7x + 10}{x + 5}$

15. $\dfrac{2x^2 + 13x + 15}{x + 5}$

16. $\dfrac{3x^2 + 8x + 4}{x + 2}$

17. $\dfrac{2x^2 - 7x + 3}{x - 4}$

18. $\dfrac{3x^2 - x - 4}{x - 1}$

19. $\dfrac{9a^3 - 3a^2 - 3a + 4}{3a + 2}$

20. $\dfrac{4x^3 + 12x^2 + x - 14}{2x + 3}$

21. $\dfrac{8x^2 + 10x + 1}{2x + 1}$

22. $\dfrac{3x^2 + 17x + 7}{3x + 2}$

23. $\dfrac{2x^3 + 2x^2 - 17x + 8}{x - 2}$

24. $\dfrac{4x^3 + 11x^2 - 8x - 10}{x + 3}$

Find each quotient using long division. Don't forget to write the polynomials in descending order and fill in any missing terms. See Examples 6 through 8.

25. $\dfrac{x^2 - 36}{x - 6}$

26. $\dfrac{a^2 - 49}{a - 7}$

▶ 27. $\dfrac{x^3 - 27}{x - 3}$

28. $\dfrac{x^3 + 64}{x + 4}$

29. $\dfrac{1 - 3x^2}{x + 2}$

30. $\dfrac{7 - 5x^2}{x + 3}$

31. $\dfrac{-4b + 4b^2 - 5}{2b - 1}$

32. $\dfrac{-3y + 2y^2 - 15}{2y + 5}$

MIXED PRACTICE

Divide. If the divisor contains 2 or more terms, use long division. See Examples 1 through 8.

33. $\dfrac{a^2b^2 - ab^3}{ab}$

34. $\dfrac{m^3n^2 - mn^4}{mn}$

35. $\dfrac{8x^2 + 6x - 27}{2x - 3}$

36. $\dfrac{18w^2 + 18w - 8}{3w + 4}$

37. $\dfrac{2x^2y + 8x^2y^2 - xy^2}{2xy}$

38. $\dfrac{11x^3y^3 - 33xy + x^2y^2}{11xy}$

▶ 39. $\dfrac{2b^3 + 9b^2 + 6b - 4}{b + 4}$

40. $\dfrac{2x^3 + 3x^2 - 3x + 4}{x + 2}$

41. $\dfrac{5x^2 + 28x - 10}{x + 6}$

42. $\dfrac{2x^2 + x - 15}{x + 3}$

43. $\dfrac{10x^3 - 24x^2 - 10x}{10x}$

44. $\dfrac{2x^3 + 12x^2 + 16}{4x^2}$

45. $\dfrac{6x^2 + 17x - 4}{x + 3}$

46. $\dfrac{2x^2 - 9x + 15}{x - 6}$

47. $\dfrac{30x^2 - 17x + 2}{5x - 2}$

48. $\dfrac{4x^2 - 13x - 12}{4x + 3}$

49. $\dfrac{3x^4 - 9x^3 + 12}{-3x}$

50. $\dfrac{8y^6 - 3y^2 - 4y}{4y}$

51. $\dfrac{x^3 + 6x^2 + 18x + 27}{x + 3}$

52. $\dfrac{x^3 - 8x^2 + 32x - 64}{x - 4}$

53. $\dfrac{y^3 + 3y^2 + 4}{y - 2}$

54. $\dfrac{3x^3 + 11x + 12}{x + 4}$

55. $\dfrac{5 - 6x^2}{x - 2}$

56. $\dfrac{3 - 7x^2}{x - 3}$

Divide.

57. $\dfrac{x^5 + x^2}{x^2 + x}$

58. $\dfrac{x^6 - x^4}{x^3 + 1}$

REVIEW AND PREVIEW

Multiply each expression. See Section 5.3.

59. $2a(a^2 + 1)$

60. $-4a(3a^2 - 4)$

61. $2x(x^2 + 7x - 5)$

62. $4y(y^2 - 8y - 4)$

63. $-3xy(xy^2 + 7x^2y + 8)$

64. $-9xy(4xyz + 7xy^2z + 2)$

65. $9ab(ab^2c + 4bc - 8)$

66. $-7sr(6s^2r + 9sr^2 + 9rs + 8)$

CONCEPT EXTENSIONS

67. The perimeter of a square is $(12x^3 + 4x - 16)$ feet. Find the length of its side.

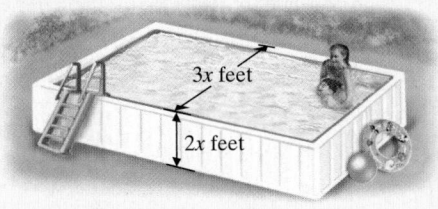

Perimeter is
$(12x^3 + 4x - 16)$ feet

△ **68.** The volume of the swimming pool shown is $(36x^5 - 12x^3 + 6x^2)$ cubic feet. If its height is $2x$ feet and its width is $3x$ feet, find its length.

3x feet

2x feet

69. In which of the following is $\dfrac{a + 7}{7}$ simplified correctly? See the Concept Check in this section.

  a. $a + 1$ **b.** a **c.** $\dfrac{a}{7} + 1$

70. In which of the following is $\dfrac{5x + 15}{5}$ simplified correctly? See the Concept Check in this section.

  a. $x + 15$ **b.** $x + 3$ **c.** $x + 1$

71. Explain how to check a polynomial long division result when the remainder is 0.

72. Explain how to check a polynomial long division result when the remainder is not 0.

△ **73.** The area of the following parallelogram is $(10x^2 + 31x + 15)$ square meters. If its base is $(5x + 3)$ meters, find its height.

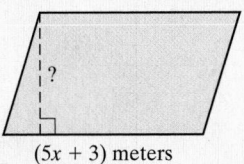

$(5x + 3)$ meters

△ **74.** The area of the top of the Ping-Pong table is $(49x^2 + 70x - 200)$ square inches. If its length is $(7x + 20)$ inches, find its width.

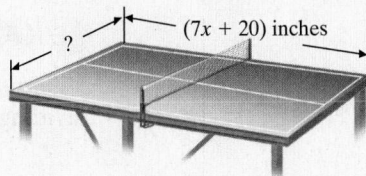

$(7x + 20)$ inches

?

75. $(18x^{10a} - 12x^{8a} + 14x^{5a} - 2x^{3a}) \div 2x^{3a}$

76. $(25y^{11b} + 5y^{6b} - 20y^{3b} + 100y^b) \div 5y^b$

5.7 | Synthetic Division and the Remainder Theorem

OBJECTIVES

1 Use Synthetic Division to Divide a Polynomial by a Binomial.

2 Use the Remainder Theorem to Evaluate Polynomials.

OBJECTIVE

1 Using Synthetic Division

When a polynomial is to be divided by a binomial of the form $x - c$, a shortcut process called **synthetic division** may be used. On the left is an example of long division, and on the right, the same example showing the coefficients of the variables only.

$$
\begin{array}{r}
2x^2 + 5x + 2 \\
x - 3{\overline{\smash{\big)}\,2x^3 - x^2 - 13x + 1}} \\
\underline{2x^3 - 6x^2} \\
5x^2 - 13x \\
\underline{5x^2 - 15x} \\
2x + 1 \\
\underline{2x - 6} \\
7
\end{array}
\qquad
\begin{array}{r}
2 \quad\; 5 \quad\; 2 \\
1 - 3{\overline{\smash{\big)}\,2 - 1 - 13 + 1}} \\
\underline{2 - 6} \\
5 - 13 \\
\underline{5 - 15} \\
2 + 1 \\
\underline{2 - 6} \\
7
\end{array}
$$

Notice that as long as we keep coefficients of powers of x in the same column, we can perform division of polynomials by performing algebraic operations on the coefficients only. This shortcut process of dividing with coefficients only in a special format is called synthetic division. To find $(2x^3 - x^2 - 13x + 1) \div (x - 3)$ by synthetic division, follow the next example.

EXAMPLE 1 Use synthetic division to divide $2x^3 - x^2 - 13x + 1$ by $x - 3$.

<u>**Solution**</u> To use synthetic division, the divisor must be in the form $x - c$. Since we are dividing by $x - 3$, c is 3. Write down 3 and the coefficients of the dividend.

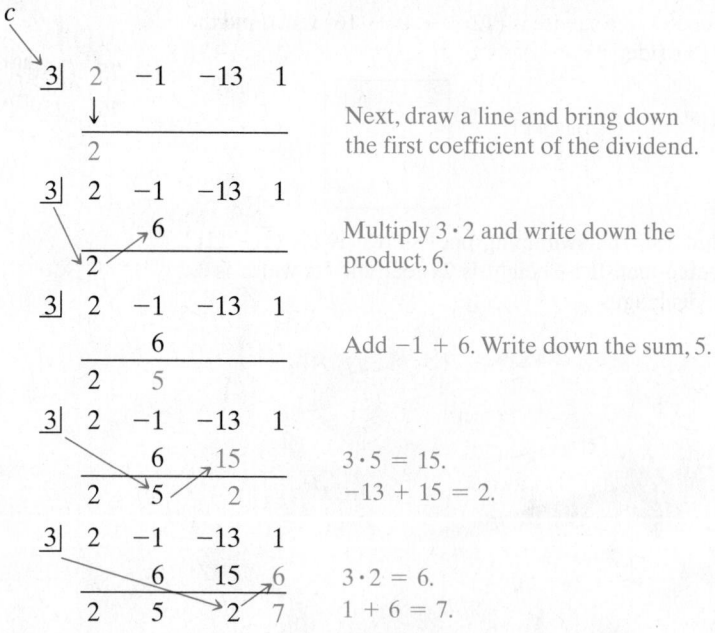

The quotient is found in the bottom row. The numbers 2, 5, and 2 are the coefficients of the quotient polynomial, and the number 7 is the remainder. The degree of the quotient polynomial is one less than the degree of the dividend. In our example, the degree of the dividend is 3, so the degree of the quotient polynomial is 2. As we found when we performed the long division, the quotient is

$$2x^2 + 5x + 2, \quad \text{remainder } 7$$

or

$$2x^2 + 5x + 2 + \frac{7}{x - 3}. \qquad \square$$

PRACTICE

1 Use synthetic division to divide $4x^3 - 3x^2 + 6x + 5$ by $x - 1$ ■

EXAMPLE 2   Use synthetic division to divide $x^4 - 2x^3 - 11x^2 + 5x + 34$ by $x + 2$.

Solution The divisor is $x + 2$, which we write in the form $x - c$ as $x - (-2)$. Thus, c is -2. The dividend coefficients are $1, -2, -11, 5,$ and 34.

$$
\begin{array}{r}
c \\
\end{array}
$$

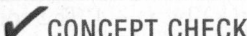

$$
\begin{array}{r|rrrrr}
-2 & 1 & -2 & -11 & 5 & 34 \\
   &   & -2 & 8 & 6 & -22 \\
\hline
   & 1 & -4 & -3 & 11 & 12 \\
\end{array}
$$

The dividend is a fourth-degree polynomial, so the quotient polynomial is a third-degree polynomial. The quotient is $x^3 - 4x^2 - 3x + 11$ with a remainder of 12. Thus,

$$\frac{x^4 - 2x^3 - 11x^2 + 5x + 34}{x + 2} = x^3 - 4x^2 - 3x + 11 + \frac{12}{x + 2}. \qquad \square$$

PRACTICE

2 Use synthetic division to divide $x^4 + 3x^3 - 5x^2 + 6x + 12$ by $x + 3$.   ■

✔ **CONCEPT CHECK**

Which division problems are candidates for the synthetic division process?

a. $(3x^2 + 5) \div (x + 4)$ **b.** $(x^3 - x^2 + 2) \div (3x^3 - 2)$ **c.** $(y^4 + y - 3) \div (x^2 + 1)$ **d.** $x^5 \div (x - 5)$

Helpful Hint
Before dividing by synthetic division, write the dividend in descending order of variable exponents. Any "missing powers" of the variable should be represented by 0 times the variable raised to the missing power.

EXAMPLE 3   If $P(x) = 2x^3 - 4x^2 + 5$

a. Find $P(2)$ by substitution.

b. Use synthetic division to find the remainder when $P(x)$ is divided by $x - 2$.

Solution

a. $P(x) = 2x^3 - 4x^2 + 5$

$P(2) = 2(2)^3 - 4(2)^2 + 5$

$= 2(8) - 4(4) + 5 = 16 - 16 + 5 = 5$

Thus, $P(2) = 5$.

Answer to Concept Check:
a and d

(Continued on next page)

b. The coefficients of $P(x)$ are 2, -4, 0, and 5. The number 0 is the coefficient of the missing power of x^1. The divisor is $x - 2$, so c is 2.

$$
\begin{array}{r|rrrr}
\overset{c}{\searrow} & & & & \\
2 & 2 & -4 & 0 & 5 \\
& & 4 & 0 & 0 \\
\hline
& 2 & 0 & 0 & 5 \text{ remainder}
\end{array}
$$

The remainder when $P(x)$ is divided by $x - 2$ is 5.

PRACTICE

3 If $P(x) = x^3 - 5x - 2$,

a. Find $P(2)$ by substitution.

b. Use synthetic division to find the remainder when $P(x)$ is divided by $x - 2$.

OBJECTIVE

2 Using the Remainder Theorem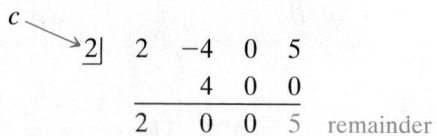

Notice in the preceding example that $P(2) = 5$ and that the remainder when $P(x)$ is divided by $x - 2$ is 5. This is no accident. This illustrates the **remainder theorem.**

Remainder Theorem

If a polynomial $P(x)$ is divided by $x - c$, then the remainder is $P(c)$.

EXAMPLE 4 Use the remainder theorem and synthetic division to find $P(4)$ if

$$P(x) = 4x^6 - 25x^5 + 35x^4 + 17x^2.$$

Solution To find $P(4)$ by the remainder theorem, we divide $P(x)$ by $x - 4$. The coefficients of $P(x)$ are 4, -25, 35, 0, 17, 0, and 0. Also, c is 4.

$$
\begin{array}{r|rrrrrrr}
\overset{c}{\searrow} & & & & & & & \\
4 & 4 & -25 & 35 & 0 & 17 & 0 & 0 \\
& & 16 & -36 & -4 & -16 & 4 & 16 \\
\hline
& 4 & -9 & -1 & -4 & 1 & 4 & 16 \text{ remainder}
\end{array}
$$

Thus, $P(4) = 16$, the remainder.

PRACTICE

4 Use the remainder theorem and synthetic division to find $P(3)$ if $P(x) = 2x^5 - 18x^4 + 90x^2 + 59x$.

✓ Vocabulary, Readiness & Video Check

Martin-Gay Interactive Videos

See Video 5.7 ◉

Watch the section lecture video and answer the following questions.

OBJECTIVE
1

1. From ▦ Example 1, once you've completed the synthetic division, what does the bottom row of numbers mean? What is the degree of the quotient?

OBJECTIVE
2

2. From ▦ Example 4, given a polynomial function $P(x)$, under what circumstances might it be easier/faster to use the remainder theorem to find $P(c)$ rather than substituting the value c for x and then simplifying?

5.7 Exercise Set MyMathLab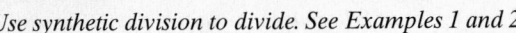

Use synthetic division to divide. See Examples 1 and 2.

1. $(x^2 + 3x - 40) \div (x - 5)$

2. $(x^2 - 14x + 24) \div (x - 2)$

3. $(x^2 + 5x - 6) \div (x + 6)$

4. $(x^2 + 12x + 32) \div (x + 4)$

5. $(x^3 - 7x^2 - 13x + 5) \div (x - 2)$

6. $(x^3 + 6x^2 + 4x - 7) \div (x + 5)$

7. $(4x^2 - 9) \div (x - 2)$

8. $(3x^2 - 4) \div (x - 1)$

*For the given polynomial P(x) and the given c, find P(c) by **(a)** direct substitution and **(b)** the remainder theorem. See Examples 3 and 4.*

9. $P(x) = 3x^2 - 4x - 1; P(2)$

10. $P(x) = x^2 - x + 3; P(5)$

11. $P(x) = 4x^4 + 7x^2 + 9x - 1; P(-2)$

12. $P(x) = 8x^5 + 7x + 4; P(-3)$

13. $P(x) = x^5 + 3x^4 + 3x - 7; P(-1)$

14. $P(x) = 5x^4 - 4x^3 + 2x - 1; P(-1)$

MIXED PRACTICE

Use synthetic division to divide.

15. $(x^3 - 3x^2 + 2) \div (x - 3)$

16. $(x^2 + 12) \div (x + 2)$

17. $(6x^2 + 13x + 8) \div (x + 1)$

18. $(x^3 - 5x^2 + 7x - 4) \div (x - 3)$

19. $(2x^4 - 13x^3 + 16x^2 - 9x + 20) \div (x - 5)$

20. $(3x^4 + 5x^3 - x^2 + x - 2) \div (x + 2)$

21. $(3x^2 - 15) \div (x + 3)$

22. $(3x^2 + 7x - 6) \div (x + 4)$

23. $(3x^3 - 6x^2 + 4x + 5) \div \left(x - \dfrac{1}{2}\right)$

24. $(8x^3 - 6x^2 - 5x + 3) \div \left(x + \dfrac{3}{4}\right)$

25. $(3x^3 + 2x^2 - 4x + 1) \div \left(x - \dfrac{1}{3}\right)$

26. $(9y^3 + 9y^2 - y + 2) \div \left(y + \dfrac{2}{3}\right)$

27. $(7x^2 - 4x + 12 + 3x^3) \div (x + 1)$

28. $(x^4 + 4x^3 - x^2 - 16x - 4) \div (x - 2)$

29. $(x^3 - 1) \div (x - 1)$

30. $(y^3 - 8) \div (y - 2)$

31. $(2x^3 + 12x^2 - 3x - 20) \div (x + 6)$

32. $(4x^3 + 12x^2 + x - 12) \div (x + 3)$

For the given polynomial P(x) and the given c, use the remainder theorem to find P(c).

33. $P(x) = x^3 + 3x^2 - 7x + 4; 1$

34. $P(x) = x^3 + 5x^2 - 4x - 6; 2$

35. $P(x) = 3x^3 - 7x^2 - 2x + 5; -3$

36. $P(x) = 4x^3 + 5x^2 - 6x - 4; -2$

37. $P(x) = 4x^4 + x^2 - 2; -1$

38. $P(x) = x^4 - 3x^2 - 2x + 5; -2$

39. $P(x) = 2x^4 - 3x^2 - 2; \dfrac{1}{3}$

40. $P(x) = 4x^4 - 2x^3 + x^2 - x - 4; \dfrac{1}{2}$

41. $P(x) = x^5 + x^4 - x^3 + 3; \dfrac{1}{2}$

42. $P(x) = x^5 - 2x^3 + 4x^2 - 5x + 6; \dfrac{2}{3}$

43. Explain an advantage of using the remainder theorem instead of direct substitution.

44. Explain an advantage of using synthetic division instead of long division.

REVIEW AND PREVIEW

Solve each equation for x. See Section 2.3.

45. $7x + 2 = x - 3$

46. $4 - 2x = 17 - 5x$

47. $\dfrac{x}{3} - 5 = 13$

48. $\dfrac{2x}{9} + 1 = \dfrac{7}{9}$

Evaluate. See Section 5.1.

49. 2^3

50. 3^4

51. $(-2)^5$

52. -2^5

53. $3 \cdot 4^2$

54. $4 \cdot 3^3$

Evaluate each expression for the given replacement value. See Section 5.1.

55. x^2 if x is -5

56. x^3 if x is -5

57. $2x^3$ if x is -1

58. $3x^2$ if x is -1

CONCEPT EXTENSIONS

Determine whether each division problem is a candidate for the synthetic division process. See the Concept Check in this section.

59. $(5x^2 - 3x + 2) \div (x + 2)$

60. $(x^4 - 6) \div (x^3 + 3x - 1)$

61. $(x^7 - 2) \div (x^5 + 1)$

62. $(3x^2 + 7x - 1) \div \left(x - \dfrac{1}{3}\right)$

△ **63.** If the area of a parallelogram is $(x^4 - 23x^2 + 9x - 5)$ square centimeters and its base is $(x + 5)$ centimeters, find its height.

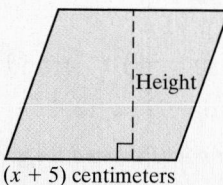

$(x + 5)$ centimeters

△ **64.** If the volume of a box is $(x^4 + 6x^3 - 7x^2)$ cubic meters, its height is x^2 meters, and its length is $(x + 7)$ meters, find its width.

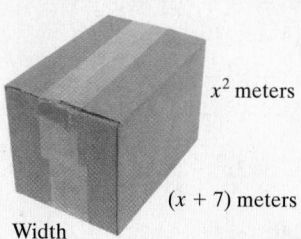

x^2 meters

$(x + 7)$ meters

Width

Divide.

65. $\left(x^4 + \dfrac{2}{3}x^3 + x\right) \div (x - 1)$

66. $\left(2x^3 + \dfrac{9}{2}x^2 - 4x - 10\right) \div (x + 2)$

We say that 2 is a factor of 8 because 2 divides 8 evenly, or with a remainder of 0. In the same manner, the polynomial $x - 2$ is a factor of the polynomial $x^3 - 14x^2 + 24x$ because the remainder is 0 when $x^3 - 14x^2 + 24x$ is divided by $x - 2$. Use this information for Exercises 67 through 69.

67. Use synthetic division to show that $x + 3$ is a factor of $x^3 + 3x^2 + 4x + 12$.

68. Use synthetic division to show that $x - 2$ is a factor of $x^3 - 2x^2 - 3x + 6$.

69. From the remainder theorem, the polynomial $x - c$ is a factor of a polynomial function $P(x)$ if $P(c)$ is what value?

70. If a polynomial is divided by $x - 5$, the quotient is $2x^2 + 5x - 6$ and the remainder is 3. Find the original polynomial.

71. If a polynomial is divided by $x + 3$, the quotient is $x^2 - x + 10$ and the remainder is -2. Find the original polynomial.

Chapter 5 Vocabulary Check

Fill in each blank with one of the words or phrases listed below.

| term | coefficient | monomial | binomial | trinomial |
| polynomials | degree of a term | distributive | FOIL | degree of a polynomial |

1. A _____ is a number or the product of numbers and variables raised to powers.

2. The _____ method may be used when multiplying two binomials.

3. A polynomial with exactly three terms is called a _____.

4. The _____ is the greatest degree of any term of the polynomial.

5. A polynomial with exactly two terms is called a _____.

6. The _____ of a term is its numerical factor.

7. The _____ is the sum of the exponents on the variables in the term.

8. A polynomial with exactly one term is called a _____.

9. Monomials, binomials, and trinomials are all examples of _____.

10. The _____ property is used to multiply $2x(x - 4)$.

Chapter 5   Highlights

DEFINITIONS AND CONCEPTS	EXAMPLES

Section 5.1   Exponents

a^n means the product of n factors, each of which is a.

$$3^2 = 3 \cdot 3 = 9$$
$$(-5)^3 = (-5)(-5)(-5) = -125$$
$$\left(\frac{1}{2}\right)^4 = \frac{1}{2} \cdot \frac{1}{2} \cdot \frac{1}{2} \cdot \frac{1}{2} = \frac{1}{16}$$

If m and n are integers and no denominators are 0,

Product Rule: $a^m \cdot a^n = a^{m+n}$

$$x^2 \cdot x^7 = x^{2+7} = x^9$$

Power Rule: $(a^m)^n = a^{mn}$

$$(5^3)^8 = 5^{3 \cdot 8} = 5^{24}$$

Power of a Product Rule: $(ab)^n = a^n b^n$

$$(7y)^4 = 7^4 y^4$$

Power of a Quotient Rule: $\left(\dfrac{a}{b}\right)^n = \dfrac{a^n}{b^n}$

$$\left(\frac{x}{8}\right)^3 = \frac{x^3}{8^3}$$

Quotient Rule: $\dfrac{a^m}{a^n} = a^{m-n}$

$$\frac{x^9}{x^4} = x^{9-4} = x^5$$

Zero Exponent: $a^0 = 1$, $a \neq 0$.

$$5^0 = 1, \quad x^0 = 1, x \neq 0$$

Section 5.2   Polynomial Functions and Adding and Subtracting Polynomials

Terms

A **term** is a number or the product of numbers and variables raised to powers.

$$-5x, 7a^2b, \frac{1}{4}y^4, 0.2$$

The **numerical coefficient** or **coefficient** of a term is its numerical factor.

Term	Coefficient
$7x^2$	7
y	1
$-a^2b$	-1

A **polynomial** is a finite sum of terms in which all variables have exponents that are nonnegative integers and no variables appear in the denominator.

Polynomials

$$1.3x^2 \quad \text{(monomial)}$$
$$-\frac{1}{3}y + 5 \quad \text{(binomial)}$$
$$6z^2 - 5z + 7 \quad \text{(trinomial)}$$

A function P is a **polynomial function** if $P(x)$ is a polynomial.

For the polynomial function

$$P(x) = -x^2 + 6x - 12, \text{find } P(-2)$$
$$P(-2) = -(-2)^2 + 6(-2) - 12 = -28.$$

The **degree of a term** is the sum of the exponents on the variables in the term.

Term	Degree
$-5x^3$	3
3 (or $3x^0$)	0
$2a^2b^2c$	$2 + 2 + 1 = 5$

The **degree of a polynomial** is the greatest degree of any term of the polynomial.

Polynomial	Degree
$5x^2 - 3x + 2$	2
$7y + 8y^2z^3 - 12$	$2 + 3 = 5$

(continued)

DEFINITIONS AND CONCEPTS	EXAMPLES

Section 5.2 Polynomial Functions and Adding and Subtracting Polynomials (continued)

To add polynomials, add or combine like terms.

Add:

$$(7x^2 - 3x + 2) + (-5x - 6) = 7x^2 - 3x + 2 - 5x - 6$$
$$= 7x^2 - 8x - 4$$

To subtract two polynomials, change the signs of the terms of the second polynomial, then add.

Subtract:

$$(17y^2 - 2y + 1) - (-3y^3 + 5y - 6)$$
$$= (17y^2 - 2y + 1) + (3y^3 - 5y + 6)$$
$$= 17y^2 - 2y + 1 + 3y^3 - 5y + 6$$
$$= 3y^3 + 17y^2 - 7y + 7$$

Section 5.3 Multiplying Polynomials

To multiply two polynomials, multiply each term of one polynomial by each term of the other polynomial and then combine like terms.

Multiply:

$$(2x + 1)(5x^2 - 6x + 2)$$

$$= 2x(5x^2 - 6x + 2) + 1(5x^2 - 6x + 2)$$
$$= 10x^3 - 12x^2 + 4x + 5x^2 - 6x + 2$$
$$= 10x^3 - 7x^2 - 2x + 2$$

Section 5.4 Special Products

The **FOIL method** may be used when multiplying two binomials.

Multiply: $(5x - 3)(2x + 3)$

$$(5x - 3)(2x + 3) = (5x)(2x) + (5x)(3) + (-3)(2x) + (-3)(3)$$
$$= 10x^2 + 15x - 6x - 9$$
$$= 10x^2 + 9x - 9$$

Squaring a Binomial

$$(a + b)^2 = a^2 + 2ab + b^2$$

$$(a - b)^2 = a^2 - 2ab + b^2$$

Square each binomial.

$$(x + 5)^2 = x^2 + 2(x)(5) + 5^2$$
$$= x^2 + 10x + 25$$
$$(3x - 2y)^2 = (3x)^2 - 2(3x)(2y) + (2y)^2$$
$$= 9x^2 - 12xy + 4y^2$$

Multiplying the Sum and Difference of Two Terms

$$(a + b)(a - b) = a^2 - b^2$$

Multiply:

$$(6y + 5)(6y - 5) = (6y)^2 - 5^2$$
$$= 36y^2 - 25$$

Section 5.5 Negative Exponents and Scientific Notation

If $a \neq 0$ and n is an integer,

$$a^{-n} = \frac{1}{a^n}$$

Rules for exponents are true for positive and negative integers.

$$3^{-2} = \frac{1}{3^2} = \frac{1}{9}; 5x^{-2} = \frac{5}{x^2}$$

Simplify: $\left(\dfrac{x^{-2}y}{x^5}\right)^{-2} = \dfrac{x^4 y^{-2}}{x^{-10}}$

$$= x^{4-(-10)}y^{-2}$$
$$= \frac{x^{14}}{y^2}$$

DEFINITIONS AND CONCEPTS	EXAMPLES

Section 5.5 Negative Exponents and Scientific Notation (continued)

A positive number is written in scientific notation if it is written as the product of a number a, $1 \le a < 10$, and an integer power r of 10. $$a \times 10^r$$	Write each number in scientific notation. $12{,}000 = 1.2 \times 10^4$ $0.00000568 = 5.68 \times 10^{-6}$

Section 5.6 Dividing Polynomials

To divide a polynomial by a monomial: $$\frac{a+b}{c} = \frac{a}{c} + \frac{b}{c}$$	Divide: $$\frac{15x^5 - 10x^3 + 5x^2 - 2x}{5x^2} = \frac{15x^5}{5x^2} - \frac{10x^3}{5x^2} + \frac{5x^2}{5x^2} - \frac{2x}{5x^2}$$ $$= 3x^3 - 2x + 1 - \frac{2}{5x}$$
To divide a polynomial by a polynomial other than a monomial, use long division.	$$5x - 1 + \frac{-4}{2x+3}$$ $2x+3\overline{)10x^2 + 13x - 7}$ $\underline{10x^2 \not\!+ 15x}$ $-2x - 7$ $\underline{\not\!- 2x \not\!- 3}$ -4

Section 5.7 Synthetic Division and the Remainder Theorem

A shortcut method called **synthetic division** may be used to divide a polynomial by a binomial of the form $x - c$.	Use synthetic division to divide $2x^3 - x^2 - 8x - 1$ by $x - 2$. $\begin{array}{r|rrrr} 2 & 2 & -1 & -8 & -1 \\ & \downarrow & 4 & 6 & -4 \\ \hline & 2 & 3 & -2 & -5 \end{array}$ The quotient is $2x^2 + 3x - 2 - \dfrac{5}{x-2}$.

Chapter 5 Review

(5.1) *State the base and the exponent for each expression.*

1. 7^9
2. $(-5)^4$
3. -5^4
4. x^6

Evaluate each expression.

5. 8^3
6. $(-6)^2$
7. -6^2
8. $-4^3 - 4^0$
9. $(3b)^0$
10. $\dfrac{8b}{8b}$

Simplify each expression.

11. $y^2 \cdot y^7$
12. $x^9 \cdot x^5$
13. $(2x^5)(-3x^6)$
14. $(-5y^3)(4y^4)$
15. $(x^4)^2$
16. $(y^3)^5$
17. $(3y^6)^4$
18. $(2x^3)^3$

19. $\dfrac{x^9}{x^4}$
20. $\dfrac{z^{12}}{z^5}$
21. $\dfrac{a^5 b^4}{ab}$
22. $\dfrac{x^4 y^6}{xy}$
23. $\dfrac{3x^4 y^{10}}{12xy^6}$
24. $\dfrac{2x^7 y^8}{8xy^2}$
25. $5a^7(2a^4)^3$
26. $(2x)^2(9x)$
27. $(-5a)^0 + 7^0 + 8^0$
28. $8x^0 + 9^0$

Simplify the given expression and choose the correct result.

29. $\left(\dfrac{3x^4}{4y}\right)^3$

a. $\dfrac{27x^{64}}{64y^3}$
b. $\dfrac{27x^{12}}{64y^3}$
c. $\dfrac{9x^{12}}{12y^3}$
d. $\dfrac{3x^{12}}{4y^3}$

30. $\left(\dfrac{5a^6}{b^3}\right)^2$

  a. $\dfrac{10a^{12}}{b^6}$ **b.** $\dfrac{25a^{36}}{b^9}$ **c.** $\dfrac{25a^{12}}{b^6}$ **d.** $25a^{12}b^6$

(5.2) Find the degree of each term.

31. $-5x^4y^3$ **32.** $10x^3y^2z$

33. $35a^5bc^2$ **34.** $95xyz$

Find the degree of each polynomial.

35. $y^5 + 7x - 8x^4$

36. $9y^2 + 30y + 25$

37. $-14x^2y - 28x^2y^3 - 42x^2y^2$

38. $6x^2y^2z^2 + 5x^2y^3 - 12xyz$

39. The Glass Bridge Skywalk is suspended 4000 feet over the Colorado River at the very edge of the Grand Canyon. Neglecting air resistance, the height of an object dropped from the Skywalk at time t seconds is given by the polynomial function $P(t) = -16t^2 + 4000$. Find the height of the object at the given times.

t	0 seconds	1 second	3 seconds	5 seconds
$P(t) = -16t^2 + 4000$				

△ **40.** The surface area of a box with a square base and a height of 5 units is given by the polynomial function $P(x) = 2x^2 + 20x$. Fill in the table below by evaluating the given values of x.

x	1	3	5.1	10
$P(x) = 2x^2 + 20x$				

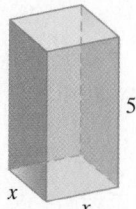

Combine like terms in each expression.

41. $6a^2 + 4a + 9a^2$

42. $21x^2 + 3x + x^2 + 6$

43. $4a^2b - 3b^2 - 8q^2 - 10a^2b + 7q^2$

44. $2s^{14} + 3s^{13} + 12s^{12} - s^{10}$

Add or subtract as indicated.

45. $(3x^2 + 2x + 6) + (5x^2 + x)$

46. $(2x^5 + 3x^4 + 4x^3 + 5x^2) + (4x^2 + 7x + 6)$

47. $(-5y^2 + 3) - (2y^2 + 4)$

48. $(3x^2 - 7xy + 7y^2) - (4x^2 - xy + 9y^2)$

TRANSLATING

Perform the indicated operations.

49. Subtract $(3x - y)$ from $(7x - 14y)$.

50. Subtract $(4x^2 + 8x - 7)$ from the sum of $(x^2 + 7x + 9)$ and $(x^2 + 4)$.

If $P(x) = 9x^2 - 7x + 8$, find the following.

51. $P(6)$ **52.** $P(-2)$

53. Find the perimeter of the rectangle.

$x^2y + 5$ cm

$2x^2y - 6x + 1$ cm

54. With the ownership of computers growing rapidly, the market for new software is also increasing. The revenue for software publishers (in millions of dollars) in the United States from 2001 to 2006 can be represented by the polynomial function $f(x) = 754x^2 - 228x + 80{,}134$ where x is the number of years since 2001. Use this model to find the revenues from software sales in 2009. (*Source:* Software & Information Industry Association)

(5.3) Multiply each expression.

55. $4(2a + 7)$

56. $9(6a - 3)$

57. $-7x(x^2 + 5)$

58. $-8y(4y^2 - 6)$

59. $(3a^3 - 4a + 1)(-2a)$

60. $(6b^3 - 4b + 2)(7b)$

61. $(2x + 2)(x - 7)$

62. $(2x - 5)(3x + 2)$

63. $(x - 9)^2$

64. $(x - 12)^2$

65. $(4a - 1)(a + 7)$

66. $(6a - 1)(7a + 3)$

67. $(5x + 2)^2$

68. $(3x + 5)^2$

69. $(x + 7)(x^3 + 4x - 5)$

70. $(x + 2)(x^5 + x + 1)$

71. $(x^2 + 2x + 4)(x^2 + 2x - 4)$

72. $(x^3 + 4x + 4)(x^3 + 4x - 4)$

73. $(x + 7)^3$

74. $(2x - 5)^3$

(5.4) Use special products to multiply each of the following.

75. $(x + 7)^2$

76. $(x - 5)^2$

77. $(3x - 7)^2$

78. $(4x + 2)^2$

79. $(5x - 9)^2$

80. $(5x + 1)(5x - 1)$

81. $(7x + 4)(7x - 4)$

82. $(a + 2b)(a - 2b)$

83. $(2x - 6)(2x + 6)$

84. $(4a^2 - 2b)(4a^2 + 2b)$

Express each as a product of polynomials in x. Then multiply and simplify.

△ **85.** Find the area of the square if its side is $(3x - 1)$ meters.

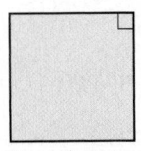

$(3x - 1)$ meters

△ **86.** Find the area of the rectangle.

$(x - 1)$ miles
$(5x + 2)$ miles

(5.5) Simplify each expression.

87. 7^{-2}

88. -7^{-2}

89. $2x^{-4}$

90. $(2x)^{-4}$

91. $\left(\dfrac{1}{5}\right)^{-3}$

92. $\left(\dfrac{-2}{3}\right)^{-2}$

93. $2^0 + 2^{-4}$

94. $6^{-1} - 7^{-1}$

Simplify each expression. Write each answer using positive exponents only.

95. $\dfrac{x^5}{x^{-3}}$

96. $\dfrac{z^4}{z^{-4}}$

97. $\dfrac{r^{-3}}{r^{-4}}$

98. $\dfrac{y^{-2}}{y^{-5}}$

99. $\left(\dfrac{bc^{-2}}{bc^{-3}}\right)^4$

100. $\left(\dfrac{x^{-3}y^{-4}}{x^{-2}y^{-5}}\right)^{-3}$

101. $\dfrac{x^{-4}y^{-6}}{x^2y^7}$

102. $\dfrac{a^5b^{-5}}{a^{-5}b^5}$

103. $a^{6m}a^{5m}$

104. $\dfrac{(x^{5+h})^3}{x^5}$

105. $(3xy^{2z})^3$

106. $a^{m+2}a^{m+3}$

Write each number in scientific notation.

107. 0.00027

108. 0.8868

109. 80,800,000

110. 868,000

111. Google.com is an Internet search engine that handles 2,500,000,000 searches every day. Write 2,500,000,000 in scientific notation. (*Source:* Google, Inc.)

112. The approximate diameter of the Milky Way galaxy is 150,000 light years. Write this number in scientific notation. (*Source:* NASA IMAGE/POETRY Education and Public Outreach Program)

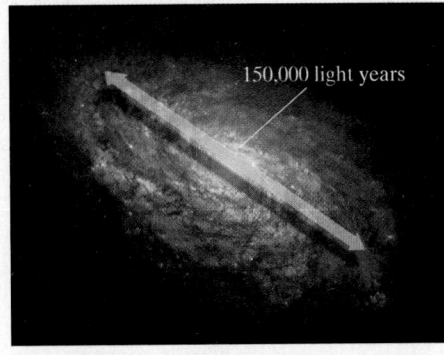

150,000 light years

Write each number in standard notation.

113. 8.67×10^5

114. 3.86×10^{-3}

115. 8.6×10^{-4}

116. 8.936×10^5

117. The volume of the planet Jupiter is 1.43128×10^{15} cubic kilometers. Write this number in standard notation. (*Source:* National Space Science Data Center)

118. An angstrom is a unit of measure, equal to 1×10^{-10} meter, used for measuring wavelengths or the diameters of atoms. Write this number in standard notation. (*Source:* National Institute of Standards and Technology)

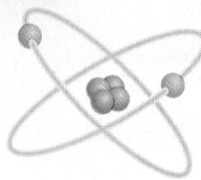

Simplify. Express each result in standard form.

119. $(8 \times 10^4)(2 \times 10^{-7})$

120. $\dfrac{8 \times 10^4}{2 \times 10^{-7}}$

(5.6) Divide.

121. $\dfrac{x^2 + 21x + 49}{7x^2}$

122. $\dfrac{5a^3b - 15ab^2 + 20ab}{-5ab}$

123. $(a^2 - a + 4) \div (a - 2)$

124. $(4x^2 + 20x + 7) \div (x + 5)$

125. $\dfrac{a^3 + a^2 + 2a + 6}{a - 2}$

126. $\dfrac{9b^3 - 18b^2 + 8b - 1}{3b - 2}$

127. $\dfrac{4x^4 - 4x^3 + x^2 + 4x - 3}{2x - 1}$

128. $\dfrac{-10x^2 - x^3 - 21x + 18}{x - 6}$

△ **129.** The area of the rectangle below is $(15x^3 - 3x^2 + 60)$ square feet. If its length is $3x^2$ feet, find its width.

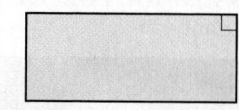

Area is $(15x^3 - 3x^2 + 60)$ sq feet

△ **130.** The perimeter of the equilateral triangle below is $(21a^3b^6 + 3a - 3)$ units. Find the length of a side.

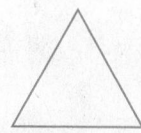

Perimeter is
$(21a^3b^6 + 3a - 3)$ units

(5.7) *Use synthetic division to find each quotient.*

131. $(3x^3 + 12x - 4) \div (x - 2)$

132. $(3x^3 + 2x^2 - 4x - 1) \div \left(x + \dfrac{3}{2}\right)$

133. $(x^5 - 1) \div (x + 1)$

134. $(x^3 - 81) \div (x - 3)$

135. $(x^3 - x^2 + 3x^4 - 2) \div (x - 4)$

136. $(3x^4 - 2x^2 + 10) \div (x + 2)$

If $P(x) = 3x^5 - 9x + 7$, use the remainder theorem to find the following.

137. $P(4)$

138. $P(-5)$

MIXED REVIEW

Evaluate.

139. $\left(-\dfrac{1}{2}\right)^3$

Simplify each expression. Write each answer using positive exponents only.

140. $(4xy^2)(x^3y^5)$

141. $\dfrac{18x^9}{27x^3}$

142. $\left(\dfrac{3a^4}{b^2}\right)^3$

143. $(2x^{-4}y^3)^{-4}$

144. $\dfrac{a^{-3}b^6}{9^{-1}a^{-5}b^{-2}}$

Perform the indicated operations and simplify.

145. $(6x + 2) + (5x - 7)$

146. $(-y^2 - 4) + (3y^2 - 6)$

147. $(8y^2 - 3y + 1) - (3y^2 + 2)$

148. $(5x^2 + 2x - 6) - (-x - 4)$

149. $4x(7x^2 + 3)$

150. $(2x + 5)(3x - 2)$

151. $(x - 3)(x^2 + 4x - 6)$

152. $(7x - 2)(4x - 9)$

Use special products to multiply.

153. $(5x + 4)^2$

154. $(6x + 3)(6x - 3)$

Divide.

155. $\dfrac{8a^4 - 2a^3 + 4a - 5}{2a^3}$

156. $\dfrac{x^2 + 2x + 10}{x + 5}$

157. $\dfrac{4x^3 + 8x^2 - 11x + 4}{2x - 3}$

Chapter 5 Getting Ready for the Test

MATCHING *Match the expression with the exponent operation needed to simplify. Letters may be used more than once or not at all.*

▶ **1.** $x^2 \cdot x^5$

▶ **2.** $(x^2)^5$

▶ **3.** $x^2 + x^5$

▶ **4.** $\dfrac{x^5}{x^2}$

    A. multiply the exponents

    B. divide the exponents

    C. add the exponents

    D. subtract the exponents

    E. this expression will not simplify

MATCHING *Match the operation with the result when the operation is performed on the given terms. Letters may be used more than once or not at all.*

Given Terms: 20y and 4y

▶ 5. Add the terms
▶ 6. Subtract the terms
▶ 7. Multiply the terms
▶ 8. Divide the terms.

A. $80y$
B. $24y^2$
C. $16y$
D. 16

E. $80y^2$
F. $24y$
G. $16y^2$
H. $5y$
I. 5

MULTIPLE CHOICE *The expression 5^{-1} is equivalent to*

▶ 9. **A.** -5 **B.** 4 **C.** $\dfrac{1}{5}$ **D.** $-\dfrac{1}{5}$

MULTIPLE CHOICE *The expression 2^{-3} is equivalent to*

▶ 10. **A.** -6 **B.** -1 **C.** $-\dfrac{1}{6}$ **D.** $\dfrac{1}{8}$

MATCHING *Match each expression with its simplified form. Letters may be used more than once or not at all.*

▶ 11. $y + y + y$
▶ 12. $y \cdot y \cdot y$
▶ 13. $(-y)(-y)(-y)$
▶ 14. $-y - y - y$

A. $3y^3$
B. y^3
C. $3y$
D. $-3y$

E. $-3y^3$
F. $-y^3$

MULTIPLE CHOICE *Choose the division exercise that can be performed using the synthetic division process.*

▶ 15. **A.** $(x^3 - 5x + 15) \div (x^2 - 5)$

   C. $(2x^3 - 5x^2 + 7) \div \left(x - \dfrac{1}{2}\right)$

   B. $(y^4 - y^3 + y^2 - 2) \div (y^3 + 1)$

   D. $(z^5 - 4) \div (z^4 + 2z - 2)$

Chapter 5 Test   MyMathLab® You Tube™

Evaluate each expression.

▶ 1. 2^5
▶ 2. $(-3)^4$
▶ 3. -3^4
▶ 4. 4^{-3}

Simplify each exponential expression. Write the result using only positive exponents.

▶ 5. $(3x^2)(-5x^9)$
▶ 6. $\dfrac{y^7}{y^2}$
▶ 7. $\dfrac{r^{-8}}{r^{-3}}$
▶ 8. $\left(\dfrac{x^2 y^3}{x^3 y^{-4}}\right)^2$
▶ 9. $\dfrac{6^2 x^{-4} y^{-1}}{6^3 x^{-3} y^7}$

Express each number in scientific notation.

▶ 10. 563,000
▶ 11. 0.0000863

Write each number in standard notation.

▶ 12. 1.5×10^{-3}
▶ 13. 6.23×10^4

▶ 14. Simplify. Write the answer in standard notation.

$$(1.2 \times 10^5)(3 \times 10^{-7})$$

▶ 15. **a.** Complete the table for the polynomial $4xy^2 + 7xyz + x^3y - 2$.

Term	Numerical Coefficient	Degree of Term
$4xy^2$		
$7xyz$		
x^3y		
-2		

   b. What is the degree of the polynomial?

▶ 16. Simplify by combining like terms.

$$5x^2 + 4xy - 7x^2 + 11 + 8xy$$

Perform each indicated operation.

17. $(8x^3 + 7x^2 + 4x - 7) + (8x^3 - 7x - 6)$

18. $5x^3 + x^2 + 5x - 2 - (8x^3 - 4x^2 + x - 7)$

19. Subtract $(4x + 2)$ from the sum of $(8x^2 + 7x + 5)$ and $(x^3 - 8)$.

Multiply.

20. $(3x + 7)(x^2 + 5x + 2)$

21. $3x^2(2x^2 - 3x + 7)$

22. $(x + 7)(3x - 5)$

23. $\left(3x - \dfrac{1}{5}\right)\left(3x + \dfrac{1}{5}\right)$

24. $(4x - 2)^2$

25. $(8x + 3)^2$

26. $(x^2 - 9b)(x^2 + 9b)$

Solve.

27. The height of the Bank of China in Hong Kong is 1001 feet. Neglecting air resistance, the height of an object dropped from this building at time t seconds is given by the polynomial function $P(t) = -16t^2 + 1001$. Find the height of the object at the given times below.

t	**0** seconds	**1** second	**3** seconds	**5** seconds
$P(t) = -16t^2 + 1001$				

28. Find the area of the top of the table. Express the area as a product, then multiply and simplify.

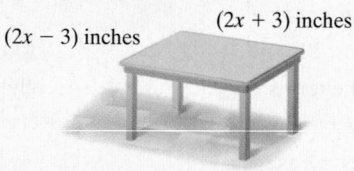

$(2x - 3)$ inches $(2x + 3)$ inches

Divide.

29. $\dfrac{4x^2 + 24xy - 7x}{8xy}$

30. $(x^2 + 7x + 10) \div (x + 5)$

31. $\dfrac{27x^3 - 8}{3x + 2}$

32. A pebble is hurled upward from the top of the Canada Trust Tower, which is 880 feet tall, with an initial velocity of 96 feet per second. Neglecting air resistance, the height $h(t)$ of the pebble after t seconds is given by the polynomial function

$$h(t) = -16t^2 + 96t + 880$$

  a. Find the height of the pebble when $t = 1$.

  b. Find the height of the pebble when $t = 5.1$.

33. Use synthetic division to divide $(4x^4 - 3x^3 - x - 1)$ by $(x + 3)$.

34. If $P(x) = 4x^4 + 7x^2 - 2x - 5$, use the remainder theorem to find $P(-2)$.

Chapter 5   Cumulative Review

1. Tell whether each statement is true or false.

  a. $8 \geq 8$ **b.** $8 \leq 8$

  c. $23 \leq 0$ **d.** $23 \geq 0$

2. Find the absolute value of each number.

  a. $|-7.2|$ **b.** $|0|$ **c.** $\left|-\dfrac{1}{2}\right|$

3. Divide. Simplify all quotients if possible.

  a. $\dfrac{4}{5} \div \dfrac{5}{16}$ **b.** $\dfrac{7}{10} \div 14$ **c.** $\dfrac{3}{8} \div \dfrac{3}{10}$

4. Multiply. Write products in lowest terms.

  a. $\dfrac{3}{4} \cdot \dfrac{7}{21}$ **b.** $\dfrac{1}{2} \cdot 4\dfrac{5}{6}$

5. Evaluate the following.

  a. 3^2 **b.** 5^3 **c.** 2^4

  d. 7^1 **e.** $\left(\dfrac{3}{7}\right)^2$

6. Evaluate $\dfrac{2x - 7y}{x^2}$ for $x = 5$ and $y = 1$.

7. Add.

  a. $-3 + (-7)$ **b.** $-1 + (-20)$

  c. $-2 + (-10)$

8. Simplify: $8 + 3(2 \cdot 6 - 1)$

9. Subtract 8 from -4.

10. Is $x = 1$ a solution of $5x^2 + 2 = x - 8$?

11. Find the reciprocal of each number.

  a. 22 **b.** $\dfrac{3}{16}$

  c. -10 **d.** $-\dfrac{9}{13}$

12. Subtract.

  a. $7 - 40$ **b.** $-5 - (-10)$

13. Use an associative property to complete each statement.

  a. $5 + (4 + 6) =$ _____

  b. $(-1 \cdot 2) \cdot 5 =$ _____

14. Simplify: $\dfrac{4(-3) + (-8)}{5 + (-5)}$

15. Simplify each expression.

  a. $10 + (x + 12)$

  b. $-3(7x)$

16. Use the distributive property to write $-2(x + 3y - z)$ without parentheses.

17. Find each product by using the distributive property to remove parentheses.

  a. $5(3x + 2)$

  b. $-2(y + 0.3z - 1)$

  c. $-(9x + y - 2z + 6)$

18. Simplify: $2(6x - 1) - (x - 7)$

19. Solve $x - 7 = 10$ for x.

20. Write the phrase as an algebraic expression: double a number, subtracted from the sum of the number and seven

21. Solve: $\dfrac{5}{2}x = 15$

22. Solve: $2x + \dfrac{1}{8} = x - \dfrac{3}{8}$

23. Twice a number, added to seven, is the same as three subtracted from the number. Find the number.

24. Solve: $10 = 5j - 2$

25. Twice the sum of a number and 4 is the same as four times the number, decreased by 12. Find the number.

26. Solve: $\dfrac{7x + 5}{3} = x + 3$

27. The length of a rectangular road sign is 2 feet less than three times its width. Find the dimensions if the perimeter is 28 feet.

28. Graph $x < 5$ and write the solutions in interval notation.

29. Solve $F = \dfrac{9}{5}C + 32$ for C.

30. Find the slope of each line.
    a. $x = -1$
    b. $y = 7$

31. Graph: $2 < x \le 4$

32. Recall that the grade of a road is its slope written as a percent. Find the grade of the road shown.

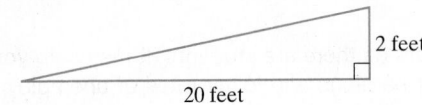

2 feet

20 feet

33. Complete the following ordered pair solutions for the equation $3x + y = 12$.
    a. $(0, \ )$
    b. $(\ , 6)$
    c. $(-1, \ )$

34. Solve the system: $\begin{cases} 3x + 2y = -8 \\ 2x - 6y = -9 \end{cases}$

35. Graph the linear equation: $2x + y = 5$

36. Solve the system: $\begin{cases} x = -3y + 3 \\ 2x + 9y = 5 \end{cases}$

37. Graph: $x = 2$

38. Evaluate.
    a. $(-5)^2$ **b.** -5^2
    c. $2 \cdot 5^2$

39. Find the slope of the line $x = 5$.

40. Simplify: $\dfrac{(z^2)^3 \cdot z^7}{z^9}$

41. Subtract: $(2x^3 + 8x^2 - 6x) - (2x^3 - x^2 + 1)$

42. Subtract: $(5y^2 - 6) - (y^2 + 2)$

43. Use the product rule to simplify $(2x^2)(-3x^5)$.

44. Find the value of $-x^2$ when
    a. $x = 2$
    b. $x = -2$

45. Add $(11x^3 - 12x^2 + x - 3)$ and $(x^3 - 10x + 5)$.

46. Multiply: $(10x^2 - 3)(10x^2 + 3)$

47. Multiply: $(2x - y)^2$

48. Multiply: $(10x^2 + 3)^2$

49. Divide $6m^2 + 2m$ by $2m$.

50. Evaluate.
    a. 5^{-1} **b.** 7^{-2}

Factoring Polynomials

6.1 The Greatest Common Factor and Factoring by Grouping

6.2 Factoring Trinomials of the Form $x^2 + bx + c$

6.3 Factoring Trinomials of the Form $ax^2 + bx + c$ and Perfect Square Trinomials

6.4 Factoring Trinomials of the Form $ax^2 + bx + c$ by Grouping

6.5 Factoring Binomials

Integrated Review—Choosing a Factoring Strategy

6.6 Solving Quadratic Equations by Factoring

6.7 Quadratic Equations and Problem Solving

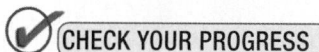

Vocabulary Check

Chapter Highlights

Chapter Review

Getting Ready for the Test

Chapter Test

Cumulative Review

In Chapter 5, we learned how to multiply polynomials. This chapter deals with an operation that is the reverse process of multiplying, called *factoring*. Factoring is an important algebraic skill because this process allows us to write a sum as a product.

At the end of this chapter, we use factoring to help us solve equations other than linear equations, and in Chapter 7, we use factoring to simplify and perform arithmetic operations on rational expressions.

Why Are You in College?

There are probably as many answers as there are students. It may help you to know that college graduates have higher earnings and lower rates of unemployment. The double line graph below shows the increasing number of associate and bachelor degrees awarded over the years. It is also enlightening to know that an increasing number of high school graduates are looking to higher education.

In Exercise 110 of Section 6.1, we will explore how many students graduate from U.S. high schools each year, and how many of those may expect to go to college.

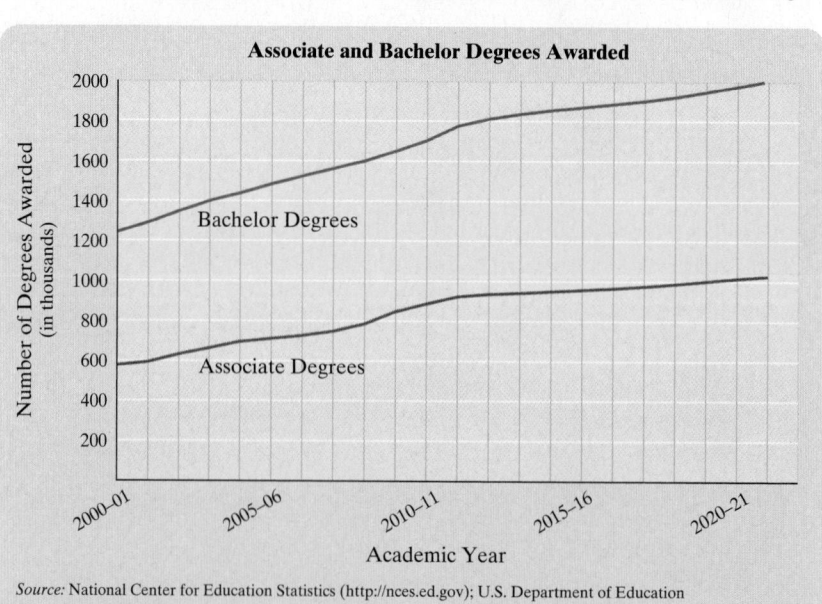

Source: National Center for Education Statistics (http://nces.ed.gov); U.S. Department of Education

Note: Some years are projected.

6.1 The Greatest Common Factor and Factoring by Grouping

OBJECTIVES

1 Find the Greatest Common Factor of a List of Integers.

2 Find the Greatest Common Factor of a List of Terms.

3 Factor Out the Greatest Common Factor from a Polynomial.

4 Factor a Polynomial by Grouping.

In the product $2 \cdot 3 = 6$, the numbers 2 and 3 are called **factors** of 6 and $2 \cdot 3$ is a **factored form** of 6. This is true of polynomials also. Since $(x + 2)(x + 3) = x^2 + 5x + 6$, $(x + 2)$ and $(x + 3)$ are factors of $x^2 + 5x + 6$, and $(x + 2)(x + 3)$ is a factored form of the polynomial.

$$
\underbrace{2 \cdot 3}_{\text{a factored form of 6}} = 6
$$

$$
\underbrace{x^2 \cdot x^3}_{\text{a factored form of } x^5} = x^5
$$

factor factor product   factor factor product

$$
\underbrace{(x + 2)(x + 3)}_{\text{a factored form of } x^2 + 5x + 6} = x^2 + 5x + 6
$$

factor factor   product

The process of writing a polynomial as a product is called **factoring** the polynomial.

Do you see that factoring is the reverse process of multiplying?

$$
x^2 + 5x + 6 = (x + 2)(x + 3)
$$

factoring ⟶

⟵ multiplying

✔ **CONCEPT CHECK**

Multiply: $2(x - 4)$

What do you think the result of factoring $2x - 8$ would be? Why?

OBJECTIVE

1 Finding the Greatest Common Factor of a List of Integers

The first step in factoring a polynomial is to see whether the terms of the polynomial have a common factor. If there is one, we can write the polynomial as a product by **factoring out** the common factor. We will usually factor out the **greatest common factor (GCF)**.

The GCF of a list of integers is the largest integer that is a factor of all the integers in the list. For example, the GCF of 12 and 20 is 4 because 4 is the largest integer that is a factor of both 12 and 20. With large integers, the GCF may not be found easily by inspection. When this happens, use the following steps.

Finding the GCF of a List of Integers

Step 1. Write each number as a product of prime numbers.

Step 2. Identify the common prime factors.

Step 3. The product of all common prime factors found in Step 2 is the greatest common factor. If there are no common prime factors, the greatest common factor is 1.

Recall from Section 1.3 that a prime number is a whole number other than 1 whose only factors are 1 and itself.

EXAMPLE 1 Find the GCF of each list of numbers.

   a. 28 and 40   **b.** 55 and 21   **c.** 15, 18, and 66

Solution

   a. Write each number as a product of primes.

$$28 = 2 \cdot 2 \cdot 7 = 2^2 \cdot 7$$
$$40 = 2 \cdot 2 \cdot 2 \cdot 5 = 2^3 \cdot 5$$

There are two common factors, each of which is 2, so the GCF is

$$\text{GCF} = 2 \cdot 2 = 4$$

   b. $55 = 5 \cdot 11$
       $21 = 3 \cdot 7$

There are no common prime factors; thus, the GCF is 1.

   c. $15 = 3 \cdot 5$
       $18 = 2 \cdot 3 \cdot 3 = 2 \cdot 3^2$
       $66 = 2 \cdot 3 \cdot 11$

The only prime factor common to all three numbers is 3, so the GCF is

$$\text{GCF} = 3$$

PRACTICE

1   Find the GCF of each list of numbers.

   a. 36 and 42   **b.** 35 and 44   **c.** 12, 16, and 40

OBJECTIVE

2   **Finding the Greatest Common Factor of a List of Terms**

The greatest common factor of a list of variables raised to powers is found in a similar way. For example, the GCF of x^2, x^3, and x^5 is x^2 because each term contains a factor of x^2 and no higher power of x is a factor of each term.

$$x^2 = x \cdot x$$
$$x^3 = x \cdot x \cdot x$$
$$x^5 = x \cdot x \cdot x \cdot x \cdot x$$

There are two common factors, each of which is x, so the GCF $= x \cdot x$ or x^2.

   From this example, we see that **the GCF of a list of common variables raised to powers is the variable raised to the smallest exponent in the list.**

EXAMPLE 2 Find the GCF of each list of terms.

   a. x^3, x^7, and x^5 **b.** y, y^4, and y^7

Solution

   a. The GCF is x^3, since 3 is the smallest exponent to which x is raised.

   b. The GCF is y^1 or y, since 1 is the smallest exponent on y.

PRACTICE

2   Find the GCF of each list of terms.

   a. y^7, y^4, and y^6 **b.** x, x^4, and x^2

   In general, the **greatest common factor (GCF) of a list of terms** is the product of the GCF of the numerical coefficients and the GCF of the variable factors.

$$20x^2y^2 = 2 \cdot 2 \cdot 5 \cdot x \cdot x \cdot y \cdot y$$
$$6xy^3 = 2 \cdot 3 \cdot x \cdot y \cdot y \cdot y$$
$$\text{GCF} = 2 \cdot x \cdot y \cdot y = 2xy^2$$

> **Helpful Hint**
>
> Remember that the GCF of a list of terms contains the smallest exponent on each common variable.
>
> The GCF of x^5y^6, x^2y^7, and x^3y^4 is x^2y^4. ⎯⎯⎯ Smallest exponent on x
> ⎯⎯⎯ Smallest exponent on y

EXAMPLE 3   Find the GCF of each list of terms.

  a. $6x^2$, $10x^3$, and $-8x$ **b.** $-18y^2$, $-63y^3$, and $27y^4$ **c.** a^3b^2, a^5b, and a^6b^2

Solution

  a. $6x^2 = 2 \cdot 3 \cdot x^2$
  $10x^3 = 2 \cdot 5 \cdot x^3$ } → The GCF of x^2, x^3, and x^1 is x^1 or x.
  $-8x = -1 \cdot 2 \cdot 2 \cdot 2 \cdot x^1$
  GCF $= 2 \cdot x^1$ or $2x$

  b. $-18y^2 = -1 \cdot 2 \cdot 3 \cdot 3 \cdot y^2$
  $-63y^3 = -1 \cdot 3 \cdot 3 \cdot 7 \cdot y^3$ } → The GCF of y^2, y^3, and y^4 is y^2.
  $27y^4 = 3 \cdot 3 \cdot 3 \cdot y^4$
  GCF $= 3 \cdot 3 \cdot y^2$ or $9y^2$

  c. The GCF of a^3, a^5, and a^6 is a^3.
  The GCF of b^2, b, and b^2 is b. Thus,
  the GCF of a^3b^2, a^5b, and a^6b^2 is a^3b. ☐

PRACTICE
3 Find the GCF of each list of terms.
  a. $5y^4$, $15y^2$, and $-20y^3$ **b.** $4x^2$, x^3, and $3x^8$ **c.** a^4b^2, a^3b^5, and a^2b^3 ■

OBJECTIVE
3 Factoring Out the Greatest Common Factor ▶

The first step in factoring a polynomial is to find the GCF of its terms. Once we do so, we can write the polynomial as a product by **factoring out** the GCF.

The polynomial $8x + 14$, for example, contains two terms: $8x$ and 14. The GCF of these terms is 2. We factor out 2 from each term by writing each term as a product of 2 and the term's remaining factors.

$$8x + 14 = 2 \cdot 4x + 2 \cdot 7$$

Using the distributive property, we can write

$$8x + 14 = 2 \cdot 4x + 2 \cdot 7$$
$$= 2(4x + 7)$$

Thus, a factored form of $8x + 14$ is $2(4x + 7)$. We can check by multiplying:

$$2(4x + 7) = 2 \cdot 4x + 2 \cdot 7 = 8x + 14.$$

> **Helpful Hint**
>
> A factored form of $8x + 14$ is _not_
>
> $$2 \cdot 4x + 2 \cdot 7$$
>
> Although the _terms_ have been factored (written as products), the _polynomial_ $8x + 14$ has not been factored (written as a product).
> A factored form of $8x + 14$ is the _product_ $2(4x + 7)$.

✔ **CONCEPT CHECK**

Which of the following is/are factored form(s) of $7t + 21$?

a. 7   **b.** $7 \cdot t + 7 \cdot 3$ **c.** $7(t + 3)$ **d.** $7(t + 21)$

EXAMPLE 4   Factor each polynomial by factoring out the GCF.

a. $6t + 18$ **b.** $y^5 - y^7$

Solution

a. The GCF of terms $6t$ and 18 is 6.

$$6t + 18 = 6 \cdot t + 6 \cdot 3$$
$$= 6(t + 3) \quad \text{Apply the distributive property.}$$

Our work can be checked by multiplying 6 and $(t + 3)$.

$$6(t + 3) = 6 \cdot t + 6 \cdot 3 = 6t + 18, \text{ the original polynomial.}$$

b. The GCF of y^5 and y^7 is y^5. Thus,

$$y^5 - y^7 = y^5(1) - y^5(y^2)$$
$$= y^5(1 - y^2)$$

> **Helpful Hint**
> Don't forget the 1.

PRACTICE

4   Factor each polynomial by factoring out the GCF.

a. $4t + 12$ **b.** $y^8 + y^4$

EXAMPLE 5   Factor: $-9a^5 + 18a^2 - 3a$

Solution

$$-9a^5 + 18a^2 - 3a = (3a)(-3a^4) + (3a)(6a) + (3a)(-1)$$
$$= 3a(-3a^4 + 6a - 1)$$

> **Helpful Hint**
> Don't forget the -1.

PRACTICE

5   Factor: $-8b^6 + 16b^4 - 8b^2$

In Example 5, we could have chosen to factor out a $-3a$ instead of $3a$. If we factor out a $-3a$, we have

$$-9a^5 + 18a^2 - 3a = (-3a)(3a^4) + (-3a)(-6a) + (-3a)(1)$$
$$= -3a(3a^4 - 6a + 1)$$

> **Helpful Hint**
> Notice the changes in signs when factoring out $-3a$.

EXAMPLES   Factor.

6. $6a^4 - 12a = 6a(a^3 - 2)$

7. $\dfrac{3}{7}x^4 + \dfrac{1}{7}x^3 - \dfrac{5}{7}x^2 = \dfrac{1}{7}x^2(3x^2 + x - 5)$

8. $15p^2q^4 + 20p^3q^5 + 5p^3q^3 = 5p^2q^3(3q + 4pq^2 + p)$

PRACTICE

6–8   Factor.

6. $5x^4 - 20x$ **7.** $\dfrac{5}{9}z^5 + \dfrac{1}{9}z^4 - \dfrac{2}{9}z^3$ **8.** $8a^2b^4 - 20a^3b^3 + 12ab^3$

Answer to Concept Check: c

EXAMPLE 9   Factor: $5(x + 3) + y(x + 3)$

Solution The binomial $(x + 3)$ is the greatest common factor. Use the distributive property to factor out $(x + 3)$.

$$5(x + 3) + y(x + 3) = (x + 3)(5 + y)$$ □

PRACTICE
9   Factor: $8(y - 2) + x(y - 2)$ ■

EXAMPLE 10   Factor: $3m^2n(a + b) - (a + b)$

Solution The greatest common factor is $(a + b)$.

$$3m^2n(a + b) - 1(a + b) = (a + b)(3m^2n - 1)$$ □

PRACTICE
10   Factor: $7xy^3(p + q) - (p + q)$ ■

OBJECTIVE

4   Factoring by Grouping ▶

Once the GCF is factored out, we can often continue to factor the polynomial, using a variety of techniques. We discuss here a technique for factoring polynomials called **factoring by grouping.**

EXAMPLE 11   Factor $xy + 2x + 3y + 6$ by grouping. Check by multiplying.

Solution The GCF of the first two terms is x, and the GCF of the last two terms is 3.

$$xy + 2x + 3y + 6 = (xy + 2x) + (3y + 6) \quad \text{Group terms.}$$
$$= \underbrace{x(y + 2) + 3(y + 2)} \qquad \text{Factor out GCF from each grouping.}$$

> **Helpful Hint**
> Notice that this form, $x(y + 2) + 3(y + 2)$, is *not* a factored form of the original polynomial. It is a sum, not a product.

Next we factor out the common binomial factor, $(y + 2)$.

$$x(y + 2) + 3(y + 2) = (y + 2)(x + 3)$$

Now the result is a factored form because it is a product. We were able to write the polynomial as a product because of the common binomial factor, $(y + 2)$, that appeared. If this does not happen, try rearranging the terms of the original polynomial.

Check: Multiply: $(y + 2)$ by $(x + 3)$

$$(y + 2)(x + 3) = xy + 2x + 3y + 6,$$

the original polynomial.
Thus, a factored form of $xy + 2x + 3y + 6$ is the product $(y + 2)(x + 3)$.   □

PRACTICE
11   Factor $xy + 3y + 4x + 12$ by grouping. Check by multiplying.   ■

You may want to try these steps when factoring by grouping.

To Factor a Four-Term Polynomial by Grouping

Step 1. Group the terms in two groups of two terms so that each group has a common factor.

Step 2. Factor out the GCF from each group.

Step 3. If there is now a common binomial factor in the groups, factor it out.

Step 4. If not, rearrange the terms and try these steps again.

EXAMPLES Factor by grouping.

12. $15x^3 - 10x^2 + 6x - 4$

$= (15x^3 - 10x^2) + (6x - 4)$ Group the terms.

$= 5x^2(3x - 2) + 2(3x - 2)$ Factor each group.

$= (3x - 2)(5x^2 + 2)$ Factor out the common factor, $(3x - 2)$.

13. $3x^2 + 4xy - 3x - 4y$

$= (3x^2 + 4xy) + (-3x - 4y)$

$= x(3x + 4y) - 1(3x + 4y)$ Factor each group. A -1 is factored from the second pair of terms so that there is a common factor, $(3x + 4y)$.

$= (3x + 4y)(x - 1)$ Factor out the common factor, $(3x + 4y)$.

14. $2a^2 + 5ab + 2a + 5b$

$= (2a^2 + 5ab) + (2a + 5b)$

$= a(2a + 5b) + 1(2a + 5b)$ Factor each group. An understood 1 is written before $(2a + 5b)$ to help remember that $(2a + 5b)$ is $1(2a + 5b)$.

$= (2a + 5b)(a + 1)$ Factor out the common factor, $(2a + 5b)$. □

> **Helpful Hint**
> Notice the factor of 1 is written when $(2a + 5b)$ is factored out.

PRACTICE
12–14

12. Factor $40x^3 - 24x^2 + 15x - 9$ by grouping.

13. Factor $2xy + 3y^2 - 2x - 3y$ by grouping.

14. Factor $7a^3 + 5a^2 + 7a + 5$ by grouping.

EXAMPLES Factor by grouping.

15. $3xy + 2 - 3x - 2y$

Notice that the first two terms have no common factor other than 1. However, if we rearrange these terms, a grouping emerges that does lead to a common factor.

$3xy + 2 - 3x - 2y$

$= (3xy - 3x) + (-2y + 2)$

$= 3x(y - 1) - 2(y - 1)$ Factor -2 from the second group so that there is a common factor, $(y - 1)$.

$= (y - 1)(3x - 2)$ Factor out the common factor, $(y - 1)$.

16. $5x - 10 + x^3 - x^2 = 5(x - 2) + x^2(x - 1)$

There is no common binomial factor that can now be factored out. No matter how we rearrange the terms, no grouping will lead to a common factor. Thus, this polynomial is not factorable by grouping. □

PRACTICE
15–16

15. Factor $4xy + 15 - 12x - 5y$ by grouping.

16. Factor $9y - 18 + y^3 - 4y^2$ by grouping.

> **Helpful Hint**
>
> One more reminder: When **factoring** a polynomial, make sure the polynomial is written as a **product.** For example, it is true that
>
> $$3x^2 + 4xy - 3x - 4y = \underbrace{x(3x + 4y) - 1(3x + 4y)}_{\text{but is not a factored form}},$$
>
> since it is a **sum (difference)**, not a **product.** A factored form of $3x^2 + 4xy - 3x - 4y$ is the product $(3x + 4y)(x - 1)$.

Factoring out a greatest common factor first makes factoring by any method easier, as we see in the next example.

EXAMPLE 17 Factor: $4ax - 4ab - 2bx + 2b^2$

Solution First, factor out the common factor 2 from all four terms.

$$4ax - 4ab - 2bx + 2b^2$$

$$= 2(2ax - 2ab - bx + b^2) \quad \text{Factor out 2 from all four terms.}$$

$$= 2[2a(x - b) - b(x - b)] \quad \text{Factor each pair of terms. A "}-b\text{" is factored from the second pair so that there is a common factor, } x - b.$$

$$= 2(x - b)(2a - b) \quad \text{Factor out the common binomial.} \qquad \square$$

PRACTICE
17 Factor: $3xy - 3ay - 6ax + 6a^2$ ■

> **Helpful Hint**
>
> Throughout this chapter, we will be factoring polynomials. Even when the instructions do not so state, it is always a good idea to check your answers by multiplying.

✓ Vocabulary, Readiness & Video Check

Use the choices below to fill in each blank. Some choices may be used more than once and some may not be used at all.

> greatest common factor   factors   factoring   true   false   least   greatest

1. Since $5 \cdot 4 = 20$, the numbers 5 and 4 are called _____ of 20.

2. The _____ of a list of integers is the largest integer that is a factor of all the integers in the list.

3. The greatest common factor of a list of common variables raised to powers is the variable raised to the _____ exponent in the list.

4. The process of writing a polynomial as a product is called _____.

5. True or false: A factored form of $7x + 21 + xy + 3y$ is $7(x + 3) + y(x + 3)$. _____

6. True or false: A factored form of $3x^3 + 6x + x^2 + 2$ is $3x(x^2 + 2)$. _____

Martin-Gay Interactive Videos

See Video 6.1 ◉

Watch the section lecture video and answer the following questions.

OBJECTIVE 1
7. Based on ▦ Example 1, give a general definition for the greatest common factor (GCF) of a list of numbers.

OBJECTIVE 2
8. In ▦ Example 3, why are the numbers factored out, but not the variables?

OBJECTIVE 3
9. From ▦ Example 5, how can the number of terms in the other factor once you factor out the GCF help you determine if your factorization is correct?

OBJECTIVE 4
10. In ▦ Examples 7 and 8, what are you reminded to always do first when factoring a polynomial? Also, explain how a polynomial looks that suggests it might be factored by grouping.

6.1 Exercise Set MyMathLab® ▶

Find the GCF for each list. See Examples 1 through 3.

1. $32, 36$
2. $36, 90$
3. $18, 42, 84$
4. $30, 75, 135$
5. $24, 14, 21$
6. $15, 25, 27$
7. y^2, y^4, y^7
8. x^3, x^2, x^5
9. z^7, z^9, z^{11}
10. y^8, y^{10}, y^{12}
11. $x^{10}y^2, xy^2, x^3y^3$
12. p^7q, p^8q^2, p^9q^3
13. $14x, 21$
14. $20y, 15$
15. $12y^4, 20y^3$
16. $32x^5, 18x^2$
17. $-10x^2, 15x^3$
18. $-21x^3, 14x$
19. $12x^3, -6x^4, 3x^5$
20. $15y^2, 5y^7, -20y^3$

21. $-18x^2y, 9x^3y^3, 36x^3y$ **22.** $7x^3y^3, -21x^2y^2, 14xy^4$

23. $20a^6b^2c^8, 50a^7b$ **24.** $40x^7y^2z, 64x^9y$

Factor out the GCF from each polynomial. See Examples 4 through 10.

25. $3a + 6$ **26.** $18a + 12$

▷ **27.** $30x - 15$ **28.** $42x - 7$

29. $x^3 + 5x^2$ **30.** $y^5 + 6y^4$

31. $6y^4 + 2y^3$ **32.** $5x^2 + 10x^6$

33. $4x - 8y + 4$

34. $7x + 21y - 7$

35. $6x^3 - 9x^2 + 12x$

36. $12x^3 + 16x^2 - 8x$

37. $a^7b^6 - a^3b^2 + a^2b^5 - a^2b^2$

38. $x^9y^6 + x^3y^5 - x^4y^3 + x^3y^3$

39. $8x^5 + 16x^4 - 20x^3 + 12$

40. $9y^6 - 27y^4 + 18y^2 + 6$

41. $\frac{1}{3}x^4 + \frac{2}{3}x^3 - \frac{4}{3}x^5 + \frac{1}{3}x$

42. $\frac{2}{5}y^7 - \frac{4}{5}y^5 + \frac{3}{5}y^2 - \frac{2}{5}y$

▷ **43.** $y(x^2 + 2) + 3(x^2 + 2)$

44. $x(y^2 + 1) - 3(y^2 + 1)$

45. $z(y + 4) - 3(y + 4)$

46. $8(x + 2) - y(x + 2)$

47. $r(z^2 - 6) + (z^2 - 6)$

48. $q(b^3 - 5) + (b^3 - 5)$

Factor a negative number or a GCF with a negative coefficient from each polynomial. See Example 5.

49. $-2x - 14$ **50.** $-7y - 21$

51. $-2x^5 + x^7$ **52.** $-5y^3 + y^6$

53. $-3a^4 + 9a^3 - 3a^2$

54. $-5m^6 + 10m^5 - 5m^3$

Factor each four-term polynomial by grouping. If this is not possible, write "not factorable by grouping." See Examples 11 through 17.

55. $x^3 + 2x^2 + 5x + 10$

56. $x^3 + 4x^2 + 3x + 12$

57. $5x + 15 + xy + 3y$

58. $xy + y + 2x + 2$

59. $6x^3 - 4x^2 + 15x - 10$

60. $16x^3 - 28x^2 + 12x - 21$

61. $5m^3 + 6mn + 5m^2 + 6n$

62. $8w^2 + 7wv + 8w + 7v$

63. $2y - 8 + xy - 4x$

64. $6x - 42 + xy - 7y$

65. $2x^3 - x^2 + 8x - 4$

66. $2x^3 - x^2 - 10x + 5$

67. $3x - 3 + x^3 - 4x^2$

68. $7x - 21 + x^3 - 2x^2$

69. $4x^2 - 8xy - 3x + 6y$

▷ **70.** $5xy - 15x - 6y + 18$

71. $5q^2 - 4pq - 5q + 4p$

72. $6m^2 - 5mn - 6m + 5n$

73. $2x^4 + 5x^3 + 2x^2 + 5x$

74. $4y^4 + y^2 + 20y^3 + 5y$

75. $12x^2y - 42x^2 - 4y + 14$

76. $90 + 15y^2 - 18x - 3xy^2$

MIXED PRACTICE

Factor. See Examples 4 through 17.

77. $32xy - 18x^2$

78. $10xy - 15x^2$

79. $y(x + 2) - 3(x + 2)$

80. $z(y - 4) + 3(y - 4)$

▷ **81.** $14x^3y + 7x^2y - 7xy$

82. $5x^3y - 15x^2y + 10xy$

83. $28x^3 - 7x^2 + 12x - 3$

84. $15x^3 + 5x^2 - 6x - 2$

85. $-40x^8y^6 - 16x^9y^5$

86. $-21x^3y - 49x^2y^2$

▷ **87.** $6a^2 + 9ab^2 + 6ab + 9b^3$

88. $16x^2 + 4xy^2 + 8xy + 2y^3$

REVIEW AND PREVIEW

Multiply. See Sections 5.3 and 5.4.

89. $(x + 2)(x + 5)$

90. $(y + 3)(y + 6)$

91. $(b + 1)(b - 4)$

92. $(x - 5)(x + 10)$

Fill in the chart by finding two numbers that have the given product and sum. The first column is filled in for you.

	93.	**94.**	**95.**	**96.**	**97.**	**98.**
Two Numbers	4, 7					
Their Product	28	12	20	8	16	−10	−24
Their Sum	11	8	9	−9	−10	3	−5

CONCEPT EXTENSIONS

See the Concept Checks in this section.

99. Which of the following is/are factored form(s) of $8a - 24$?

    a. $8 \cdot a - 24$ **b.** $8(a - 3)$

    c. $4(2a - 12)$ **d.** $8 \cdot a - 2 \cdot 12$

100. Which of the following is/are factored form(s) of $-2x + 14$?

    a. $-2(x + 7)$ **b.** $-2 \cdot x + 14$

    c. $-2(x - 14)$ **d.** $-2(x - 7)$

Determine whether the following expressions are factored.

101. $(a + 6)(a + 2)$

102. $(x + 5)(x + y)$

103. $5(2y + z) - b(2y + z)$

104. $3x(a + 2b) + 2(a + 2b)$

105. Construct a binomial whose greatest common factor is $5a^3$. (*Hint:* Multiply $5a^3$ by a binomial whose terms contain no common factor other than 1. $5a^3(\square + \square)$.)

106. Construct a trinomial whose greatest common factor is $2x^2$. See the hint for Exercise 105.

107. Explain how you can tell whether a polynomial is written in factored form.

108. Construct a four-term polynomial that can be factored by grouping.

109. The percent of total music industry revenues from streaming in the United States each year during 2007 through 2014 can be modeled by the polynomial $0.6x^2 - 0.6x + 3.6$, where x is the number of years since 2007. (*Source:* Recording Industry Association of America)

   a. Find the percent of music industry revenue derived from streaming in 2013. To do so, let $x = 6$ and evaluate $0.6x^2 - 0.6x + 3.6$. Round to the nearest percent.

   b. Use this expression to predict the percent revenue derived from streaming in 2018. Round to the nearest percent.

   c. Factor the polynomial $0.6x^2 - 0.6x + 3.6$ by factoring 0.6 from each term.

110. The number (in thousands) of students who graduated from U.S. high schools, both public and private, each year during 2000 through 2013 can be modeled by $-3x^2 + 78x + 2904$, where x is the number of years since 2000. (*Source:* National Center for Educational Statistics)

   a. Find the number of students who graduated from U.S. high schools in 2010. To do so, let $x = 10$ and evaluate $-3x^2 + 78x + 2904$.

   b. Use this expression to predict the number of students who will graduate from U.S. high schools in 2018.

   c. Factor the polynomial $-3x^2 + 78x + 2904$ by factoring -3 from each term.

d. For the year 2010, the National Center for Higher Education determined that 62.5% of U.S. high school graduates went on to higher education. Using your answer from part **a**, determine how many of those graduating in 2010 pursued higher education.

Write an expression for the area of each shaded region. Then write the expression as a factored polynomial.

△ **111.**

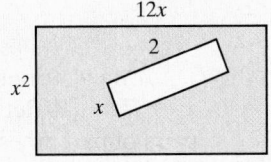

△ **112.**

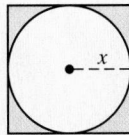

*Write an expression for the length of each rectangle. (**Hint:** Factor the area binomial and recall that Area = width · length.)*

△ **113.**

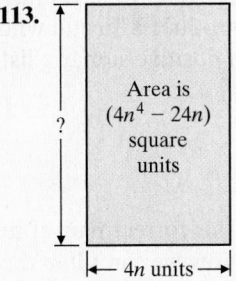

△ **114.**

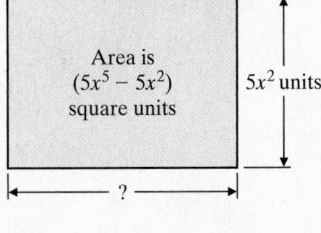

Factor each polynomial by grouping.

115. $x^{2n} + 2x^n + 3x^n + 6$
   (*Hint:* Don't forget that $x^{2n} = x^n \cdot x^n$.)

116. $x^{2n} + 6x^n + 10x^n + 60$

117. $3x^{2n} + 21x^n - 5x^n - 35$

118. $12x^{2n} - 10x^n - 30x^n + 25$

6.2   Factoring Trinomials of the Form $x^2 + bx + c$ ▷

OBJECTIVES

1 Factor Trinomials of the Form $x^2 + bx + c$. ▷

2 Factor Out the Greatest Common Factor and Then Factor a Trinomial of the Form $x^2 + bx + c$. ▷

OBJECTIVE

1   Factoring Trinomials of the Form $x^2 + bx + c$ ▷

In this section, we factor trinomials of the form $x^2 + bx + c$, such as

$$x^2 + 4x + 3, \quad x^2 - 8x + 15, \quad x^2 + 4x - 12, \quad r^2 - r - 42$$

Notice that for these trinomials, the coefficient of the squared variable is 1.

Recall that factoring means to write as a product and that factoring and multiplying are reverse processes. Using the FOIL method of multiplying binomials, we have that

$$\overset{\text{F \quad O \quad I \quad L}}{(x + 3)(x + 1) = x^2 + 1x + 3x + 3}$$
$$= x^2 + 4x + 3$$

Thus, a factored form of $x^2 + 4x + 3$ is $(x + 3)(x + 1)$.

Notice that the product of the first terms of the binomials is $x \cdot x = x^2$, the first term of the trinomial. Also, the product of the last two terms of the binomials is $3 \cdot 1 = 3$, the third term of the trinomial. The sum of these same terms is $3 + 1 = 4$, the coefficient of the middle term, x, of the trinomial.

The product of these numbers is 3.

$$x^2 + 4x + 3 = (x + 3)(x + 1)$$

The sum of these numbers is 4.

Many trinomials, such as the one above, factor into two binomials. To factor $x^2 + 7x + 10$, let's assume that it factors into two binomials and begin by writing two pairs of parentheses. The first term of the trinomial is x^2, so we use x and x as the first terms of the binomial factors.

$$x^2 + 7x + 10 = (x + \square)(x + \square)$$

To determine the last term of each binomial factor, we look for two integers whose product is 10 and whose sum is 7. Since our numbers must have a positive product and a positive sum, we list pairs of positive integer factors of 10 only.

Positive Factors of 10	Sum of Factors
1, 10	$1 + 10 = 11$
2, 5	$2 + 5 = 7$

The correct pair of numbers is 2 and 5 because their product is 10 and their sum is 7. Now we can fill in the last terms of the binomial factors.

$$x^2 + 7x + 10 = (x + 2)(x + 5)$$

Check: To see if we have factored correctly, multiply.

$$(x + 2)(x + 5) = x^2 + 5x + 2x + 10$$
$$= x^2 + 7x + 10 \qquad \text{Combine like terms.}$$

Helpful Hint

Since multiplication is commutative, the factored form of $x^2 + 7x + 10$ can be written as either $(x + 2)(x + 5)$ or $(x + 5)(x + 2)$.

Factoring a Trinomial of the Form $x^2 + bx + c$

The factored form of $x^2 + bx + c$ is

The product of these numbers is c.

$$x^2 + bx + c = (x + \square)(x + \square)$$

The sum of these numbers is b.

EXAMPLE 1 Factor: $x^2 + 7x + 12$

Solution We begin by writing the first terms of the binomial factors.

$$(x + \square)(x + \square)$$

Next we look for two numbers whose product is 12 and whose sum is 7. Since our numbers must have a positive product and a positive sum, we look at pairs of positive factors of 12 only.

Positive Factors of 12	Sum of Factors
1, 12	13
2, 6	8
3, 4	7

Correct sum, so the numbers are 3 and 4.

Thus, $x^2 + 7x + 12 = (x + 3)(x + 4)$

Check: $(x + 3)(x + 4) = x^2 + 4x + 3x + 12 = x^2 + 7x + 12.$ □

PRACTICE
1 Factor: $x^2 + 5x + 6$ ■

EXAMPLE 2 Factor: $x^2 - 12x + 35$

Solution Again, we begin by writing the first terms of the binomials.

$$(x + \square)(x + \square)$$

Now we look for two numbers whose product is 35 and whose sum is -12. Since our numbers must have a positive product and a negative sum, we look at pairs of negative factors of 35 only.

Negative Factors of 35	Sum of Factors
$-1, -35$	-36
$-5, -7$	-12

Correct sum, so the numbers are -5 and -7.

Thus, $x^2 - 12x + 35 = (x - 5)(x - 7)$

Check: To check, multiply $(x - 5)(x - 7)$. □

PRACTICE
2 Factor: $x^2 - 17x + 70$ ■

EXAMPLE 3 Factor: $x^2 + 4x - 12$

Solution $x^2 + 4x - 12 = (x + \square)(x + \square)$

We look for two numbers whose product is -12 and whose sum is 4. Since our numbers must have a negative product, we look at pairs of factors with opposite signs.

Factors of -12	Sum of Factors
$-1, 12$	11
$1, -12$	-11
$-2, 6$	4
$2, -6$	-4
$-3, 4$	1
$3, -4$	-1

Correct sum, so the numbers are -2 and 6.

Thus, $x^2 + 4x - 12 = (x - 2)(x + 6)$ □

PRACTICE
3 Factor: $x^2 + 5x - 14$ ■

EXAMPLE 4   Factor: $r^2 - r - 42$

Solution Because the variable in this trinomial is r, the first term of each binomial factor is r.

$$r^2 - r - 42 = (r + \square)(r + \square)$$

Now we look for two numbers whose product is -42 and whose sum is -1, the numerical coefficient of r. The numbers are 6 and -7. Therefore,

$$r^2 - r - 42 = (r + 6)(r - 7)$$ □

PRACTICE
  4   Factor: $p^2 - 2p - 63$ ■

EXAMPLE 5   Factor: $a^2 + 2a + 10$

Solution Look for two numbers whose product is 10 and whose sum is 2. Neither 1 and 10 nor 2 and 5 give the required sum, 2. We conclude that $a^2 + 2a + 10$ is not factorable with integers. A polynomial such as $a^2 + 2a + 10$ is called a **prime polynomial.**   □

PRACTICE
  5   Factor: $b^2 + 5b + 1$ ■

EXAMPLE 6   Factor: $x^2 + 7xy + 6y^2$

Solution

$$x^2 + 7xy + 6y^2 = (x + \square)(x + \square)$$

Recall that the middle term $7xy$ is the same as $7yx$. Thus, we can see that $7y$ is the "coefficient" of x. We then look for two terms whose product is $6y^2$ and whose sum is $7y$. The terms are $6y$ and $1y$ or $6y$ and y because $6y \cdot y = 6y^2$ and $6y + y = 7y$. Therefore,

$$x^2 + 7xy + 6y^2 = (x + 6y)(x + y)$$ □

PRACTICE
  6   Factor: $x^2 + 7xy + 12y^2$ ■

EXAMPLE 7   Factor: $x^4 + 5x^2 + 6$

Solution As usual, we begin by writing the first terms of the binomials. Since the greatest power of x in this polynomial is x^4, we write

$$(x^2 + \square)(x^2 + \square) \quad \text{since } x^2 \cdot x^2 = x^4$$

Now we look for two factors of 6 whose sum is 5. The numbers are 2 and 3. Thus,

$$x^4 + 5x^2 + 6 = (x^2 + 2)(x^2 + 3)$$ □

PRACTICE
  7   Factor: $x^4 + 13x^2 + 12$ ■

If the terms of a polynomial are not written in descending powers of the variable, you may want to do so before factoring.

EXAMPLE 8   Factor: $40 - 13t + t^2$

Solution First, we rearrange terms so that the trinomial is written in descending powers of t.

$$40 - 13t + t^2 = t^2 - 13t + 40$$

Next, try to factor.

$$t^2 - 13t + 40 = (t + \Box)(t + \Box)$$

Now we look for two factors of 40 whose sum is -13. The numbers are -8 and -5. Thus,

$$t^2 - 13t + 40 = (t - 8)(t - 5)$$ □

PRACTICE

8   Factor: $48 - 14x + x^2$ ■

The following sign patterns may be useful when factoring trinomials.

> **Helpful Hint**
>
> A positive constant in a trinomial tells us to look for two numbers with the same sign. The sign of the coefficient of the middle term tells us whether the signs are both positive or both negative.
>
> | both | same | | both | same |
> | positive | sign | | negative | sign |
>
> $$x^2 + 10x + 16 = (x + 2)(x + 8)   \quad   x^2 - 10x + 16 = (x - 2)(x - 8)$$
>
> A negative constant in a trinomial tells us to look for two numbers with opposite signs.
>
> | opposite | opposite |
> | signs | signs |
>
> $$x^2 + 6x - 16 = (x + 8)(x - 2)   \quad   x^2 - 6x - 16 = (x - 8)(x + 2)$$

OBJECTIVE

2   Factoring Out the Greatest Common Factor

Remember that the first step in factoring any polynomial is to factor out the greatest common factor (if there is one other than 1 or -1).

EXAMPLE 9   Factor: $3m^2 - 24m - 60$

Solution First we factor out the greatest common factor, 3, from each term.

$$3m^2 - 24m - 60 = 3(m^2 - 8m - 20)$$

Now we factor $m^2 - 8m - 20$ by looking for two factors of -20 whose sum is -8. The factors are -10 and 2. Therefore, the complete factored form is

$$3m^2 - 24m - 60 = 3(m + 2)(m - 10)$$ □

> **Helpful Hint**
>
> Remember to write the common factor 3 as part of the factored form.

PRACTICE

9   Factor: $4x^2 - 24x + 36$ ■

EXAMPLE 10   Factor: $2x^4 - 26x^3 + 84x^2$

Solution

$$2x^4 - 26x^3 + 84x^2 = 2x^2(x^2 - 13x + 42)   \quad \text{Factor out common factor, } 2x^2.$$
$$= 2x^2(x - 6)(x - 7)   \quad \text{Factor } x^2 - 13x + 42.$$ □

PRACTICE

10   Factor: $3y^4 - 18y^3 - 21y^2$ ■

Vocabulary, Readiness & Video Check

Fill in each blank with "true" or "false."

1. To factor $x^2 + 7x + 6$, we look for two numbers whose product is 6 and whose sum is 7. _____

2. We can write the factorization $(y + 2)(y + 4)$ also as $(y + 4)(y + 2)$. _____

3. The factorization $(4x - 12)(x - 5)$ is completely factored. _____

4. The factorization $(x + 2y)(x + y)$ may also be written as $(x + 2y)^2$. _____

Complete each factored form.

5. $x^2 + 9x + 20 = (x + 4)(x \quad)$

6. $x^2 + 12x + 35 = (x + 5)(x \quad)$

7. $x^2 - 7x + 12 = (x - 4)(x \quad)$

8. $x^2 - 13x + 22 = (x - 2)(x \quad)$

9. $x^2 + 4x + 4 = (x + 2)(x \quad)$

10. $x^2 + 10x + 24 = (x + 6)(x \quad)$

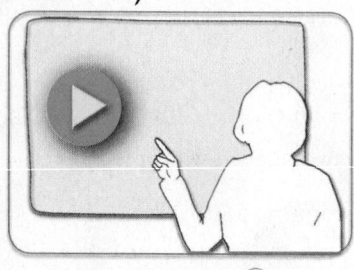

Martin-Gay Interactive Videos

See Video 6.2

Watch the section lecture video and answer the following questions.

OBJECTIVE 1

11. In Example 2, why are only negative factors of 15 considered?

OBJECTIVE 2

12. In Example 5, we know we need a positive and a negative factor of −10. How do we determine which factor is negative?

6.2 Exercise Set MyMathLab®

Factor each trinomial completely. If a polynomial can't be factored, write "prime." See Examples 1 through 8.

1. $x^2 + 7x + 6$

2. $x^2 + 6x + 8$

3. $y^2 - 10y + 9$

4. $y^2 - 12y + 11$

5. $x^2 - 6x + 9$

6. $x^2 - 10x + 25$

7. $x^2 - 3x - 18$

8. $x^2 - x - 30$

9. $x^2 + 3x - 70$

10. $x^2 + 4x - 32$

11. $x^2 + 5x + 2$

12. $x^2 - 7x + 5$

13. $x^2 + 8xy + 15y^2$

14. $x^2 + 6xy + 8y^2$

15. $a^4 - 2a^2 - 15$

16. $y^4 - 3y^2 - 70$

17. $13 + 14m + m^2$

18. $17 + 18n + n^2$

19. $10t - 24 + t^2$

20. $6q - 27 + q^2$

21. $a^2 - 10ab + 16b^2$

22. $a^2 - 9ab + 18b^2$

MIXED PRACTICE

Factor each trinomial completely. Some of these trinomials contain a greatest common factor (other than 1). Don't forget to factor out the GCF first. See Examples 1 through 10.

23. $2z^2 + 20z + 32$

24. $3x^2 + 30x + 63$

25. $2x^3 - 18x^2 + 40x$

26. $3x^3 - 12x^2 - 36x$

27. $x^2 - 3xy - 4y^2$

28. $x^2 - 4xy - 77y^2$

29. $x^2 + 15x + 36$

30. $x^2 + 19x + 60$

31. $x^2 - x - 2$

32. $x^2 - 5x - 14$

33. $r^2 - 16r + 48$

34. $r^2 - 10r + 21$

35. $x^2 + xy - 2y^2$

36. $x^2 - xy - 6y^2$

37. $3x^2 + 9x - 30$

38. $4x^2 - 4x - 48$

39. $3x^2 - 60x + 108$

40. $2x^2 - 24x + 70$

41. $x^2 - 18x - 144$

42. $x^2 + x - 42$

43. $r^2 - 3r + 6$

44. $x^2 + 4x - 10$

45. $x^2 - 8x + 15$

46. $x^2 - 9x + 14$

47. $6x^3 + 54x^2 + 120x$

48. $3x^3 + 3x^2 - 126x$

49. $4x^2y + 4xy - 12y$

50. $3x^2y - 9xy + 45y$

51. $x^2 - 4x - 21$

52. $x^2 - 4x - 32$

53. $x^2 + 7xy + 10y^2$

54. $x^2 - 2xy - 15y^2$

55. $64 + 24t + 2t^2$

56. $50 + 20t + 2t^2$

57. $x^3 - 2x^2 - 24x$

58. $x^3 - 3x^2 - 28x$

59. $2t^5 - 14t^4 + 24t^3$

60. $3x^6 + 30x^5 + 72x^4$

61. $5x^3y - 25x^2y^2 - 120xy^3$

62. $7a^3b - 35a^2b^2 + 42ab^3$

63. $162 - 45m + 3m^2$

64. $48 - 20n + 2n^2$

65. $-x^2 + 12x - 11$ (Factor out -1 first.)

66. $-x^2 + 8x - 7$ (Factor out -1 first.)

67. $\frac{1}{2}y^2 - \frac{9}{2}y - 11$ (Factor out $\frac{1}{2}$ first.)

68. $\frac{1}{3}y^2 - \frac{5}{3}y - 8$ (Factor out $\frac{1}{3}$ first.)

69. $x^3y^2 + x^2y - 20x$

70. $a^2b^3 + ab^2 - 30b$

REVIEW AND PREVIEW

Multiply. See Sections 5.3 and 5.4.

71. $(2x + 1)(x + 5)$

72. $(3x + 2)(x + 4)$

73. $(5y - 4)(3y - 1)$

74. $(4z - 7)(7z - 1)$

75. $(a + 3b)(9a - 4b)$

76. $(y - 5x)(6y + 5x)$

CONCEPT EXTENSIONS

77. Write a polynomial that factors as $(x - 3)(x + 8)$.

78. To factor $x^2 + 13x + 42$, think of two numbers whose _____ is 42 and whose _____ is 13.

Complete each sentence in your own words.

79. If $x^2 + bx + c$ is factorable and c is negative, then the signs of the last-term factors of the binomials are opposite because…

80. If $x^2 + bx + c$ is factorable and c is positive, then the signs of the last-term factors of the binomials are the same because…

Remember that perimeter means distance around. Write the perimeter of each rectangle as a simplified polynomial. Then factor the polynomial.

81.

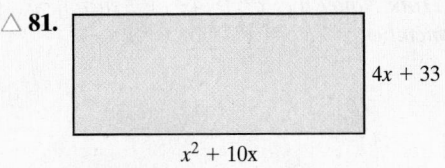

$4x + 33$

$x^2 + 10x$

82.

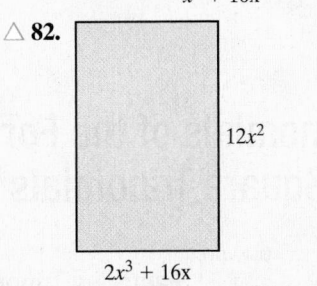

$12x^2$

$2x^3 + 16x$

83. An object is thrown upward from the top of an 80-foot building with an initial velocity of 64 feet per second. Neglecting air resistance, the height of the object after t seconds is given by $-16t^2 + 64t + 80$. Factor this polynomial.

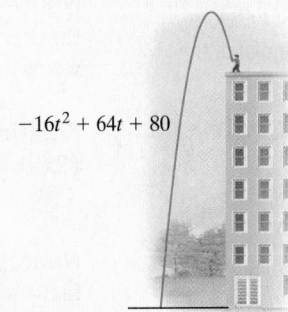

$-16t^2 + 64t + 80$

84. An object is thrown upward from the top of a 112-foot building with an initial velocity of 96 feet per second. Neglecting air resistance, the height of the object after t seconds is given by $-16t^2 + 96t + 112$. Factor this polynomial.

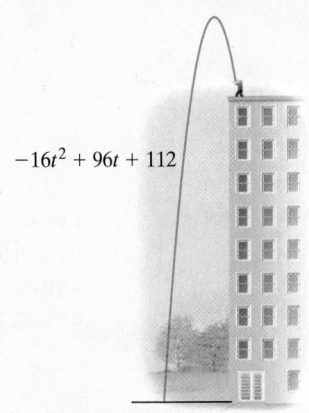

$-16t^2 + 96t + 112$

Factor each trinomial completely.

85. $x^2 + \dfrac{1}{2}x + \dfrac{1}{16}$

86. $x^2 + x + \dfrac{1}{4}$

87. $z^2(x + 1) - 3z(x + 1) - 70(x + 1)$

88. $y^2(x + 1) - 2y(x + 1) - 15(x + 1)$

Factor each trinomial. (**Hint:** *Notice that* $x^{2n} + 4x^n + 3$ *factors as* $(x^n + 1)(x^n + 3)$. ***Remember:*** $x^n \cdot x^n = x^{n+n}$ *or* x^{2n}.)

89. $x^{2n} + 8x^n - 20$

90. $x^{2n} + 5x^n + 6$

Find a positive value of c so that each trinomial is factorable.

91. $x^2 + 6x + c$

92. $t^2 + 8t + c$

93. $y^2 - 4y + c$

94. $n^2 - 16n + c$

Find a positive value of b so that each trinomial is factorable.

95. $x^2 + bx + 15$

96. $y^2 + by + 20$

97. $m^2 + bm - 27$

98. $x^2 + bx - 14$

6.3 | Factoring Trinomials of the Form $ax^2 + bx + c$ and Perfect Square Trinomials ▶

OBJECTIVES

1. Factor Trinomials of the Form $ax^2 + bx + c$, Where $a \neq 1$. ▶
2. Factor Out the GCF Before Factoring a Trinomial of the Form $ax^2 + bx + c$. ▶
3. Factor Perfect Square Trinomials. ▶

OBJECTIVE

1 Factoring Trinomials of the Form $ax^2 + bx + c$ ▶

In this section, we factor trinomials of the form $ax^2 + bx + c$, such as

$$3x^2 + 11x + 6, \qquad 8x^2 - 22x + 5, \qquad \text{and} \qquad 2x^2 + 13x - 7$$

Notice that the coefficient of the squared variable in these trinomials is a number other than 1. We will factor these trinomials using a trial-and-check method based on our work in the last section.

To begin, let's review the relationship between the numerical coefficients of the trinomial and the numerical coefficients of its factored form. For example, since $(2x + 1)(x + 6) = 2x^2 + 13x + 6$,

a factored form of $2x^2 + 13x + 6$ is $(2x + 1)(x + 6)$

Notice that $2x$ and x are factors of $2x^2$, the first term of the trinomial. Also, 6 and 1 are factors of 6, the last term of the trinomial, as shown:

$$\overset{\overbrace{\qquad 2x \cdot x \qquad}}{2x^2 + 13x + 6 = (2x + 1)(x + 6)}_{\underbrace{\qquad\qquad\qquad 1 \cdot 6 \qquad}}$$

Also notice that $13x$, the middle term, is the sum of the following products:

$$2x^2 + 13x + 6 = (2x + 1)(x + 6)$$

$$\begin{array}{c} 1x \\ +12x \\ \hline 13x \end{array} \qquad \text{Middle term}$$

Let's use this pattern to factor $5x^2 + 7x + 2$. First, we find factors of $5x^2$. Since all numerical coefficients in this trinomial are positive, we will use factors with positive numerical coefficients only. Thus, the factors of $5x^2$ are $5x$ and x. Let's try these factors as first terms of the binomials. Thus far, we have

$$5x^2 + 7x + 2 = (5x + \square)(x + \square)$$

Next, we need to find positive factors of 2. Positive factors of 2 are 1 and 2. Now we try possible combinations of these factors as second terms of the binomials until we obtain a middle term of $7x$.

$$(5x + 1)(x + 2) = 5x^2 + 11x + 2$$

$1x$
$+10x$
$11x$ ⟶ Incorrect middle term

Let's try switching factors 2 and 1.

$$(5x + 2)(x + 1) = 5x^2 + 7x + 2$$

$2x$
$+5x$
$7x$ ⟶ Correct middle term

Thus the factored form of $5x^2 + 7x + 2$ is $(5x + 2)(x + 1)$. To check, we multiply $(5x + 2)$ and $(x + 1)$. The product is $5x^2 + 7x + 2$.

EXAMPLE 1   Factor: $3x^2 + 11x + 6$

Solution Since all numerical coefficients are positive, we use factors with positive numerical coefficients. We first find factors of $3x^2$.

$$\text{Factors of } 3x^2: \quad 3x^2 = 3x \cdot x$$

If factorable, the trinomial will be of the form

$$3x^2 + 11x + 6 = (3x + \Box)(x + \Box)$$

Next we factor 6.

$$\text{Factors of 6:} \quad 6 = 1 \cdot 6, \quad 6 = 2 \cdot 3$$

Now we try combinations of factors of 6 until a middle term of $11x$ is obtained. Let's try 1 and 6 first.

$$(3x + 1)(x + 6) = 3x^2 + 19x + 6$$

$1x$
$+18x$
$19x$ ⟶ **Incorrect** middle term

Now let's next try 6 and 1.

$$(3x + 6)(x + 1)$$

Before multiplying, notice that the terms of the factor $3x + 6$ have a common factor of 3. The terms of the original trinomial $3x^2 + 11x + 6$ have no common factor other than 1, so the terms of its factors will also contain no common factor other than 1. This means that $(3x + 6)(x + 1)$ is not a factored form.

Next let's try 2 and 3 as last terms.

$$(3x + 2)(x + 3) = 3x^2 + 11x + 6$$

$2x$
$+9x$
$11x$ ⟶ **Correct** middle term

Thus a factored form of $3x^2 + 11x + 6$ is $(3x + 2)(x + 3)$. □

PRACTICE
1   Factor: $2x^2 + 11x + 15$

> **Helpful Hint**
>
> If the terms of a trinomial have no common factor (other than 1), then the terms of neither of its binomial factors will contain a common factor (other than 1).

✔ **CONCEPT CHECK**

Do the terms of $3x^2 + 29x + 18$ have a common factor? Without multiplying, decide which of the following factored forms could not be a factored form of $3x^2 + 29x + 18$.

a. $(3x + 18)(x + 1)$ **b.** $(3x + 2)(x + 9)$ **c.** $(3x + 6)(x + 3)$ **d.** $(3x + 9)(x + 2)$

EXAMPLE 2   Factor: $8x^2 - 22x + 5$

<u>Solution</u> Factors of $8x^2$: $8x^2 = 8x \cdot x$, $8x^2 = 4x \cdot 2x$

We'll try $8x$ and x.

$$8x^2 - 22x + 5 = (8x + \square)(x + \square)$$

Since the middle term, $-22x$, has a negative numerical coefficient, we factor 5 into negative factors.

$$\text{Factors of 5:   } 5 = -1 \cdot -5$$

Let's try -1 and -5.

$$(8x - 1)(x - 5) = 8x^2 - 41x + 5$$
$$-1x$$
$$+(-40x)$$
$$-41x \longrightarrow \text{Incorrect middle term}$$

Now let's try -5 and -1.

$$(8x - 5)(x - 1) = 8x^2 - 13x + 5$$
$$-5x$$
$$+(-8x)$$
$$-13x \longrightarrow \text{Incorrect middle term}$$

Don't give up yet! We can still try other factors of $8x^2$. Let's try $4x$ and $2x$ with -1 and -5.

$$(4x - 1)(2x - 5) = 8x^2 - 22x + 5$$
$$-2x$$
$$+(-20x)$$
$$-22x \longrightarrow \text{Correct middle term}$$

A factored form of $8x^2 - 22x + 5$ is $(4x - 1)(2x - 5)$.

PRACTICE

2   Factor: $15x^2 - 22x + 8$

EXAMPLE 3   Factor: $2x^2 + 13x - 7$

<u>Solution</u> Factors of $2x^2$: $2x^2 = 2x \cdot x$

Factors of -7: $-7 = -1 \cdot 7$, $-7 = 1 \cdot -7$

Answers to Concept Check:
no; a, c, d

We try possible combinations of these factors:

$$(2x + 1)(x - 7) = 2x^2 - 13x - 7 \quad \text{Incorrect middle term}$$
$$(2x - 1)(x + 7) = 2x^2 + 13x - 7 \quad \text{Correct middle term}$$

A factored form of $2x^2 + 13x - 7$ is $(2x - 1)(x + 7)$.   □

PRACTICE
3   Factor: $4x^2 + 11x - 3$ ■

EXAMPLE 4   Factor: $10x^2 - 13xy - 3y^2$

Solution Factors of $10x^2$: $10x^2 = 10x \cdot x, \quad 10x^2 = 2x \cdot 5x$

Factors of $-3y^2$: $-3y^2 = -3y \cdot y, \quad -3y^2 = 3y \cdot -y$

We try some combinations of these factors:

$$\underset{\downarrow}{\overset{\text{Correct}}{}} \qquad \underset{\downarrow}{\overset{\text{Correct}}{}}$$

$$(10x - 3y)(x + y) = 10x^2 + 7xy - 3y^2$$
$$(x + 3y)(10x - y) = 10x^2 + 29xy - 3y^2$$
$$(5x + 3y)(2x - y) = 10x^2 + xy - 3y^2$$
$$(2x - 3y)(5x + y) = 10x^2 - 13xy - 3y^2 \quad \text{Correct middle term}$$

A factored form of $10x^2 - 13xy - 3y^2$ is $(2x - 3y)(5x + y)$.   □

PRACTICE
4   Factor: $21x^2 + 11xy - 2y^2$ ■

EXAMPLE 5   Factor: $3x^4 - 5x^2 - 8$

Solution Factors of $3x^4$: $3x^4 = 3x^2 \cdot x^2$

Factors of -8: $-8 = -2 \cdot 4, \ -8 = 2 \cdot -4, \ -8 = -1 \cdot 8, \ -8 = 1 \cdot -8$

Try combinations of these factors:

$$\underset{\downarrow}{\overset{\text{Correct}}{}} \qquad \underset{\downarrow}{\overset{\text{Correct}}{}}$$

$$(3x^2 - 2)(x^2 + 4) = 3x^4 + 10x^2 - 8$$
$$(3x^2 + 4)(x^2 - 2) = 3x^4 - 2x^2 - 8$$
$$(3x^2 + 8)(x^2 - 1) = 3x^4 + 5x^2 - 8 \quad \text{Incorrect sign on middle term, so switch signs in binomial factors.}$$
$$(3x^2 - 8)(x^2 + 1) = 3x^4 - 5x^2 - 8 \quad \text{Correct middle term.}$$

A factored form of $3x^4 - 5x^2 - 8$ is $(3x^2 - 8)(x^2 + 1)$.   □

PRACTICE
5   Factor: $2x^4 - 5x^2 - 7$ ■

> **Helpful Hint**
>
> Study the last two lines of Example 5. If a factoring attempt gives you a middle term whose numerical coefficient is the opposite of the desired numerical coefficient, try switching the signs of the last terms in the binomials.
>
> Switched signs $\begin{cases} (3x^2 + 8)(x^2 - 1) = 3x^4 + 5x^2 - 8 \quad \text{Middle term: } +5x^2 \\ (3x^2 - 8)(x^2 + 1) = 3x^4 - 5x^2 - 8 \quad \text{Middle term: } -5x^2 \end{cases}$

2 Factoring out the Greatest Common Factor

Don't forget that the first step in factoring any polynomial is to look for a greatest common factor to factor out.

EXAMPLE 6   Factor: $24x^4 + 40x^3 + 6x^2$

Solution Notice that all three terms have a greatest common factor of $2x^2$. Thus we factor out $2x^2$ from all three terms first.

$$24x^4 + 40x^3 + 6x^2 = 2x^2(12x^2 + 20x + 3)$$

Next we factor $12x^2 + 20x + 3$.

Factors of $12x^2$: $12x^2 = 4x \cdot 3x$, $12x^2 = 12x \cdot x$, $12x^2 = 6x \cdot 2x$

Since all terms in the trinomial have positive numerical coefficients, we factor 3 using positive factors only.

Factors of 3: $3 = 1 \cdot 3$

We try some combinations of the factors.

$$2x^2(4x + 3)(3x + 1) = 2x^2(12x^2 + 13x + 3)$$
$$2x^2(12x + 1)(x + 3) = 2x^2(12x^2 + 37x + 3)$$
$$2x^2(2x + 3)(6x + 1) = 2x^2(12x^2 + 20x + 3) \quad \text{Correct middle term}$$

A factored form of $24x^4 + 40x^3 + 6x^2$ is $2x^2(2x + 3)(6x + 1)$.   □

> **Helpful Hint**
>
> Don't forget to include the greatest common factor in the factored form.

PRACTICE
6   Factor: $3x^3 + 17x^2 + 10x$

When the term containing the squared variable has a negative coefficient, you may want to first factor out a common factor of -1.

EXAMPLE 7   Factor: $-6x^2 - 13x + 5$

Solution We begin by factoring out a common factor of -1.

$$-6x^2 - 13x + 5 = -1(6x^2 + 13x - 5) \quad \text{Factor out } -1.$$
$$= -1(3x - 1)(2x + 5) \quad \text{Factor } 6x^2 + 13x - 5.$$ □

PRACTICE
7   Factor: $-8x^2 + 2x + 3$

■

3 Factoring Perfect Square Trinomials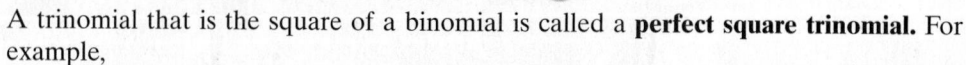

A trinomial that is the square of a binomial is called a **perfect square trinomial**. For example,

$$(x + 3)^2 = (x + 3)(x + 3)$$
$$= x^2 + 6x + 9$$

Thus $x^2 + 6x + 9$ is a perfect square trinomial.

In Chapter 5, we discovered special product formulas for squaring binomials.

$$(a + b)^2 = a^2 + 2ab + b^2 \quad \text{and} \quad (a - b)^2 = a^2 - 2ab + b^2$$

Because multiplication and factoring are reverse processes, we can now use these special products to help us factor perfect square trinomials. If we reverse these equations, we have the following.

Factoring Perfect Square Trinomials

$$a^2 + 2ab + b^2 = (a + b)^2$$
$$a^2 - 2ab + b^2 = (a - b)^2$$

Helpful Hint

Notice that for both given forms of a perfect square trinomial, the last term is positive. This is because the last term is a square.

To use these equations to help us factor, we must first be able to recognize a perfect square trinomial. A trinomial is a perfect square when

1. two terms, a^2 and b^2, are squares and
2. the remaining term is $2 \cdot a \cdot b$ or $-2 \cdot a \cdot b$. That is, this term is twice the product of a and b, or its opposite.

When a trinomial fits this description, its factored form is $(a + b)^2$ or $(a - b)^2$.

EXAMPLE 8   Factor: $x^2 + 12x + 36$

Solution   First, is this a perfect square trinomial?

$$x^2 + 12x + 36$$

1. $x^2 = (x)^2$ and $36 = 6^2$.
2. Is the middle term $2 \cdot x \cdot 6$? Yes, $2 \cdot x \cdot 6 = 12x$, the middle term.

Thus, $x^2 + 12x + 36$ factors as $(x + 6)^2$.   □

PRACTICE
8   Factor: $x^2 + 14x + 49$ ▪

EXAMPLE 9   Factor: $25x^2 + 25xy + 4y^2$

Solution   Is this a perfect square trinomial?

$$25x^2 + 25xy + 4y^2$$

1. $25x^2 = (5x)^2$ and $4y^2 = (2y)^2$.
2. Is the middle term $2 \cdot 5x \cdot 2y$? **No**, $2 \cdot 5x \cdot 2y = 20xy$, **not the middle term** $25xy$.

Helpful Hint

A perfect square trinomial can also be factored by other methods.

Therefore, $25x^2 + 25xy + 4y^2$ is not a perfect square trinomial. It is factorable, though. Using earlier techniques, we find that $25x^2 + 25xy + 4y^2$ factors as $(5x + 4y)(5x + y)$.   □

PRACTICE
9   Factor: $4x^2 + 20xy + 9y^2$ ▪

EXAMPLE 10   Factor: $4m^4 - 4m^2 + 1$

Solution   Is this a perfect square trinomial?

$$4m^4 - 4m^2 + 1$$

1. $4m^4 = (2m^2)^2$ and $1 = 1^2$.
2. Is the middle term $2 \cdot 2m^2 \cdot 1$ or $-2 \cdot 2m^2 \cdot 1$? Yes, $-2 \cdot 2m^2 \cdot 1 = -4m^2$, the middle term.

Thus, $4m^4 - 4m^2 + 1$ factors as $(2m^2 - 1)^2$.   □

PRACTICE
10   Factor: $36n^4 - 12n^2 + 1$ ▪

EXAMPLE 11 Factor: $162x^3 - 144x^2 + 32x$

<u>Solution</u> Don't forget to look first for a common factor. There is a greatest common factor of $2x$ in this trinomial.

$$162x^3 - 144x^2 + 32x = 2x(81x^2 - 72x + 16)$$
$$= 2x[(9x)^2 - 2 \cdot 9x \cdot 4 + 4^2]$$
$$= 2x(9x - 4)^2$$

PRACTICE
11 Factor: $12x^3 - 84x^2 + 147x$

✓ Vocabulary, Readiness & Video Check

Use the choices below to fill in each blank. Some choices will be used more than once and some not used at all.

$5y^2$	$(x + 5y)^2$	perfect square trinomial
$(5y)^2$	$(x - 5y)^2$	perfect square binomial

1. A _____ is a trinomial that is the square of a binomial.
2. The term $25y^2$ written as a square is _____.
3. The expression $x^2 + 10xy + 25y^2$ is called a _____.
4. The factorization $(x + 5y)(x + 5y)$ may also be written as _____.

Complete each factorization.

5. $2x^2 + 5x + 3$ factors as $(2x + 3)(\ ? \ )$.
   a. $(x + 3)$ **b.** $(2x + 1)$ **c.** $(3x + 4)$ **d.** $(x + 1)$

6. $7x^2 + 9x + 2$ factors as $(7x + 2)(\ ? \ )$.
   a. $(3x + 1)$ **b.** $(x + 1)$ **c.** $(x + 2)$ **d.** $(7x + 1)$

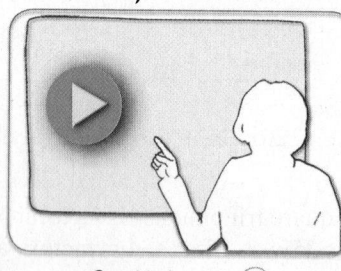

Martin-Gay Interactive Videos

See Video 6.3

Watch the section lecture video and answer the following questions.

OBJECTIVE 1
7. From Example 1, explain in general terms how you would go about factoring a trinomial with a first-term coefficient $\neq 1$.

OBJECTIVE 2
8. From Examples 3 and 5, how can factoring the GCF from a trinomial help you save time when trying to factor the remaining trinomial?

OBJECTIVE 3
9. Describe in words the special patterns that the trinomials in Examples 7 and 8 have that identify them as perfect square trinomials.

6.3 Exercise Set MyMathLab®

Complete each factored form. See Examples 1 through 5, and 8 through 10.

1. $5x^2 + 22x + 8 = (5x + 2)$
2. $2y^2 + 27y + 25 = (2y + 25)$
3. $50x^2 + 15x - 2 = (5x + 2)$
4. $6y^2 + 11y - 10 = (2y + 5)$
5. $25x^2 - 20x + 4 = (5x - 2)$
6. $4y^2 - 20y + 25 = (2y - 5)$

Factor completely. See Examples 1 through 5.

7. $2x^2 + 13x + 15$
8. $3x^2 + 8x + 4$
9. $8y^2 - 17y + 9$
10. $21x^2 - 31x + 10$
11. $2x^2 - 9x - 5$
12. $36r^2 - 5r - 24$
13. $20r^2 + 27r - 8$

14. $3x^2 + 20x - 63$

15. $10x^2 + 31x + 3$

16. $12x^2 + 17x + 5$

17. $2m^2 + 17m + 10$

18. $3n^2 + 20n + 5$

19. $6x^2 - 13xy + 5y^2$

20. $8x^2 - 14xy + 3y^2$

21. $15m^2 - 16m - 15$

22. $25n^2 - 5n - 6$

Factor completely. See Examples 1 through 7.

23. $12x^3 + 11x^2 + 2x$

24. $8a^3 + 14a^2 + 3a$

25. $21b^2 - 48b - 45$

26. $12x^2 - 14x - 10$

27. $7z + 12z^2 - 12$

28. $16t + 15t^2 - 15$

29. $6x^2y^2 - 2xy^2 - 60y^2$

30. $8x^2y + 34xy - 84y$

31. $4x^2 - 8x - 21$

32. $6x^2 - 11x - 10$

33. $-x^2 + 2x + 24$

34. $-x^2 + 4x + 21$

35. $4x^3 - 9x^2 - 9x$

36. $6x^3 - 31x^2 + 5x$

37. $24x^2 - 58x + 9$

38. $36x^2 + 55x - 14$

Factor each perfect square trinomial completely. See Examples 8 through 11.

39. $x^2 + 22x + 121$

40. $x^2 + 18x + 81$

41. $x^2 - 16x + 64$

42. $x^2 - 12x + 36$

43. $16a^2 - 24a + 9$

44. $25x^2 - 20x + 4$

45. $x^4 + 4x^2 + 4$

46. $m^4 + 10m^2 + 25$

47. $2n^2 - 28n + 98$

48. $3y^2 - 6y + 3$

49. $16y^2 + 40y + 25$

50. $9y^2 + 48y + 64$

MIXED PRACTICE

Factor each trinomial completely. See Examples 1 through 11 and Section 6.2.

51. $2x^2 - 7x - 99$

52. $2x^2 + 7x - 72$

53. $24x^2 + 41x + 12$

54. $24x^2 - 49x + 15$

55. $3a^2 + 10ab + 3b^2$

56. $2a^2 + 11ab + 5b^2$

57. $-9x + 20 + x^2$

58. $-7x + 12 + x^2$

59. $p^2 + 12pq + 36q^2$

60. $m^2 + 20mn + 100n^2$

61. $x^2y^2 - 10xy + 25$

62. $x^2y^2 - 14xy + 49$

63. $40a^2b + 9ab - 9b$

64. $24y^2x + 7yx - 5x$

65. $30x^3 + 38x^2 + 12x$

66. $6x^3 - 28x^2 + 16x$

67. $6y^3 - 8y^2 - 30y$

68. $12x^3 - 34x^2 + 24x$

69. $10x^4 + 25x^3y - 15x^2y^2$

70. $42x^4 - 99x^3y - 15x^2y^2$

71. $-14x^2 + 39x - 10$

72. $-15x^2 + 26x - 8$

73. $16p^4 - 40p^3 + 25p^2$

74. $9q^4 - 42q^3 + 49q^2$

75. $x + 3x^2 - 2$

76. $y + 8y^2 - 9$

77. $8x^2 + 6xy - 27y^2$

78. $54a^2 + 39ab - 8b^2$

79. $1 + 6x^2 + x^4$

80. $1 + 16x^2 + x^4$

81. $9x^2 - 24xy + 16y^2$

82. $25x^2 - 60xy + 36y^2$

83. $18x^2 - 9x - 14$

84. $42a^2 - 43a + 6$

85. $-27t + 7t^2 - 4$

86. $-3t + 4t^2 - 7$

87. $49p^2 - 7p - 2$

88. $3r^2 + 10r - 8$

89. $m^3 + 18m^2 + 81m$

90. $y^3 + 12y^2 + 36y$

91. $5x^2y^2 + 20xy + 1$

92. $3a^2b^2 + 12ab + 1$

93. $6a^5 + 37a^3b^2 + 6ab^4$

94. $5m^5 + 26m^3h^2 + 5mh^4$

REVIEW AND PREVIEW

Multiply the following. See Sections 5.3 and 5.4.

95. $(x - 2)(x + 2)$

96. $(y^2 + 3)(y^2 - 3)$

97. $(a + 3)(a^2 - 3a + 9)$

98. $(z - 2)(z^2 + 2z + 4)$

The following graph shows the percent of online adults who participate in various social media sites. See Section 3.1.

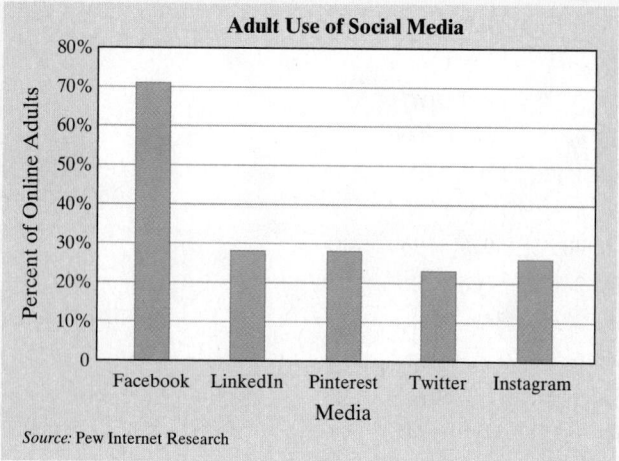

Adult Use of Social Media

Source: Pew Internet Research

99. Which social medium has the highest percent of online adult users?

100. Which social medium has the lowest percent of online adult users?

101. Describe any trend you see.

102. Why don't the percents shown in the graph add up to 100%?

CONCEPT EXTENSIONS

See the Concept Check in this section.

103. Do the terms of $4x^2 + 19x + 12$ have a common factor (other than 1)?

104. Without multiplying, decide which of the following factored forms is not a factored form of $4x^2 + 19x + 12$.

   a. $(2x + 4)(2x + 3)$ **b.** $(4x + 4)(x + 3)$

   c. $(4x + 3)(x + 4)$ **d.** $(2x + 2)(2x + 6)$

105. Describe a perfect square trinomial.

106. Write the perfect square trinomial that factors as $(x + 3y)^2$.

Write the perimeter of each figure as a simplified polynomial. Then factor the polynomial.

107.

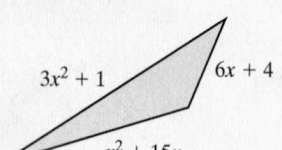

$3x^2 + 1$ $6x + 4$

$x^2 + 15x$

108.

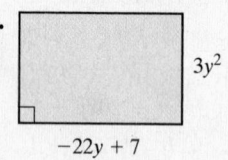

$3y^2$

$-22y + 7$

Factor each trinomial completely.

109. $4x^2 + 2x + \dfrac{1}{4}$

110. $27x^2 + 2x - \dfrac{1}{9}$

111. $4x^2(y - 1)^2 + 10x(y - 1)^2 + 25(y - 1)^2$

112. $3x^2(a + 3)^3 - 10x(a + 3)^3 + 25(a + 3)^3$

113. Fill in the blank so that $x^2 +$ _____ $x + 16$ is a perfect square trinomial.

114. Fill in the blank so that $9x^2 +$ _____ $x + 25$ is a perfect square trinomial.

The area of the largest square in the figure is $(a + b)^2$. Use this figure to answer Exercises 115 and 116.

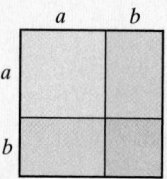

a b

a

b

△ **115.** Write the area of the largest square as the sum of the areas of the smaller squares and rectangles.

△ **116.** What factoring formula from this section is visually represented by this square?

Find a positive value of b so that each trinomial is factorable.

117. $3x^2 + bx - 5$ **118.** $2y^2 + by + 3$

Find a positive value of c so that each trinomial is factorable.

119. $5x^2 + 7x + c$ **120.** $11y^2 - 40y + c$

Factor completely. Don't forget to first factor out the greatest common factor.

121. $-12x^3y^2 + 3x^2y^2 + 15xy^2$

122. $-12r^3x^2 + 38r^2x^2 + 14rx^2$

123. $4x^2(y - 1)^2 + 20x(y - 1)^2 + 25(y - 1)^2$

124. $3x^2(a + 3)^3 - 28x(a + 3)^3 + 25(a + 3)^3$

Factor.

125. $3x^{2n} + 17x^n + 10$

126. $2x^{2n} + 5x^n - 12$

127. In your own words, describe the steps you will use to factor a trinomial.

6.4   Factoring Trinomials of the Form $ax^2 + bx + c$ by Grouping

OBJECTIVE

1 Use the Grouping Method to Factor Trinomials of the Form $ax^2 + bx + c$.

OBJECTIVE

1   Using the Grouping Method

There is an alternative method that can be used to factor trinomials of the form $ax^2 + bx + c, a \neq 1$. This method is called the **grouping method** because it uses factoring by grouping as we learned in Section 6.1.

To see how this method works, recall from Section 6.2 that to factor a trinomial such as $x^2 + 11x + 30$, we find two numbers such that

$$\text{Product is 30}$$
$$\downarrow$$
$$x^2 + 11x + 30$$
$$\downarrow$$
$$\text{Sum is 11.}$$

To factor a trinomial such as $2x^2 + 11x + 12$ by grouping, we use an extension of the method in Section 6.2. Here we look for two numbers such that

$$\text{Product is } 2 \cdot 12 = 24$$
$$2x^2 + 11x + 12$$
$$\downarrow$$
$$\text{Sum is 11.}$$

This time, we use the two numbers to write

$$2x^2 + 11x + 12 \text{ as}$$
$$= 2x^2 + \square x + \square x + 12$$

Then we factor by grouping. Since we want a positive product, 24, and a positive sum, 11, we consider pairs of positive factors of 24 only.

Factors of 24	Sum of Factors	
1, 24	25	
2, 12	14	
3, 8	11	Correct sum

The factors are 3 and 8. Now we use these factors to write the middle term $11x$ as $3x + 8x$ (or $8x + 3x$). We replace $11x$ with $3x + 8x$ in the original trinomial and then we can factor by grouping.

$$
\begin{aligned}
2x^2 + 11x + 12 &= 2x^2 + 3x + 8x + 12 \\
&= (2x^2 + 3x) + (8x + 12) \quad \text{Group the terms.} \\
&= x(2x + 3) + 4(2x + 3) \quad \text{Factor each group.} \\
&= (2x + 3)(x + 4) \quad \text{Factor out } (2x + 3).
\end{aligned}
$$

In general, we have the following procedure.

> **To Factor Trinomials by Grouping**
>
> **Step 1.** Factor out the greatest common factor if there is one other than 1.
>
> **Step 2.** For the resulting trinomial $ax^2 + bx + c$, find two numbers whose product is $a \cdot c$ and whose sum is b.
>
> **Step 3.** Write the middle term, bx, using the factors found in Step 2.
>
> **Step 4.** Factor by grouping.

EXAMPLE 1   Factor $3x^2 + 31x + 10$ by grouping.

Solution

Step 1. The terms of this trinomial contain no greatest common factor other than 1 (or −1).

Step 2. In $3x^2 + 31x + 10$, $a = 3$, $b = 31$, and $c = 10$.

(Continued on the next page)

Let's find two numbers whose product is $a \cdot c$ or $3(10) = 30$ and whose sum is b or 31. The numbers are 1 and 30, as shown in the table below.

Factors of 30	Sum of Factors	
5, 6	11	
3, 10	13	
2, 15	17	
1, 30	31	Correct sum

Step 3. Write $31x$ as $1x + 30x$ so that $3x^2 + 31x + 10 = 3x^2 + 1x + 30x + 10$.

Step 4. Factor by grouping.
$$3x^2 + 1x + 30x + 10 = x(3x + 1) + 10(3x + 1)$$
$$= (3x + 1)(x + 10) \qquad \square$$

PRACTICE

1 Factor $5x^2 + 61x + 12$ by grouping. ∎

EXAMPLE 2 Factor $8x^2 - 14x + 5$ by grouping.

Solution

Step 1. The terms of this trinomial contain no greatest common factor other than 1.

Step 2. This trinomial is of the form $ax^2 + bx + c$ with $a = 8$, $b = -14$, and $c = 5$. Find two numbers whose product is $a \cdot c$ or $8 \cdot 5 = 40$, and whose sum is b or -14.

The numbers are -4 and -10, as shown in the table below.

Factors of 40	Sum of Factors	
$-40, -1$	-41	
$-20, -2$	-22	
$-10, -4$	-14	Correct sum

Step 3. Write $-14x$ as $-4x - 10x$ so that
$$8x^2 - 14x + 5 = 8x^2 - 4x - 10x + 5$$

Step 4. Factor by grouping.
$$8x^2 - 4x - 10x + 5 = 4x(2x - 1) - 5(2x - 1)$$
$$= (2x - 1)(4x - 5) \qquad \square$$

PRACTICE

2 Factor $12x^2 - 19x + 5$ by grouping. ∎

EXAMPLE 3 Factor $6x^2 - 2x - 20$ by grouping.

Solution

Step 1. First factor out the greatest common factor, 2.
$$6x^2 - 2x - 20 = 2(3x^2 - x - 10)$$

Step 2. Next, notice that $a = 3$, $b = -1$, and $c = -10$ in the resulting trinomial. Find two numbers whose product is $a \cdot c$ or $3(-10) = -30$ and whose sum is $b, -1$. The numbers are -6 and 5.

Step 3. $3x^2 - x - 10 = 3x^2 - 6x + 5x - 10$

Step 4. $3x^2 - 6x + 5x - 10 = 3x(x - 2) + 5(x - 2)$
$$= (x - 2)(3x + 5)$$

The factored form of $6x^2 - 2x - 20 = 2(x - 2)(3x + 5)$.

$\llcorner$ Don't forget to include the GCF, 2.   □

PRACTICE
3   Factor $30x^2 - 14x - 4$ by grouping.   ■

EXAMPLE 4   Factor $18y^4 + 21y^3 - 60y^2$ by grouping.

Solution

Step 1. First factor out the greatest common factor, $3y^2$.
$$18y^4 + 21y^3 - 60y^2 = 3y^2(6y^2 + 7y - 20)$$

Step 2. Notice that $a = 6$, $b = 7$, and $c = -20$ in the resulting trinomial. Find two numbers whose product is $a \cdot c$ or $6(-20) = -120$ and whose sum is 7. It may help to factor -120 as a product of primes and -1.
$$-120 = 2 \cdot 2 \cdot 2 \cdot 3 \cdot 5 \cdot (-1)$$

Then choose pairings of factors until you have a pairing whose sum is 7.

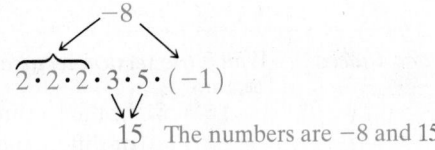

15   The numbers are -8 and 15.

Step 3. $6y^2 + 7y - 20 = 6y^2 - 8y + 15y - 20$
Step 4. $6y^2 - 8y + 15y - 20 = 2y(3y - 4) + 5(3y - 4)$
$$= (3y - 4)(2y + 5)$$

The factored form of $18y^4 + 21y^3 - 60y^2$ is $3y^2(3y - 4)(2y + 5)$.

$\llcorner$ Don't forget to include the GCF, $3y^2$
from **Step 1**.   □

PRACTICE
4   Factor $40m^4 + 5m^3 - 35m^2$ by grouping.   ■

EXAMPLE 5   Factor $4x^2 + 20x + 25$ by grouping.

Solution

Step 1. The terms of this trinomial contain no greatest common factor other than 1 (or -1).

Step 2. In $4x^2 + 20x + 25$, $a = 4$, $b = 20$, and $c = 25$. Find two numbers whose product is $a \cdot c$ or $4 \cdot 25 = 100$ and whose sum is 20. The numbers are 10 and 10.

Step 3. Write $20x$ as $10x + 10x$ so that
$$4x^2 + 20x + 25 = 4x^2 + 10x + 10x + 25$$

Step 4. Factor by grouping.
$$4x^2 + 10x + 10x + 25 = 2x(2x + 5) + 5(2x + 5)$$
$$= (2x + 5)(2x + 5)$$

The factored form of $4x^2 + 20x + 25$ is $(2x + 5)(2x + 5)$ or $(2x + 5)^2$.   □

PRACTICE
5   Factor $16x^2 + 24x + 9$ by grouping.   ■

A trinomial that is the square of a binomial, such as the trinomial in Example 5, is called a **perfect square trinomial.** From Chapter 5, there are special product formulas we can use to help us recognize and factor these trinomials. To study these formulas further, see Section 6.3, Objective 3.

> **Helpful Hint**
>
> **Remember:** A perfect square trinomial, such as the one in Example 5, may be factored by special product formulas or by other methods of factoring trinomials, such as by grouping.

✓ Vocabulary, Readiness & Video Check

For each trinomial $ax^2 + bx + c$, choose two numbers whose product is $a \cdot c$ and whose sum is b.

1. $x^2 + 6x + 8$
  a. 4, 2   **b.** 7, 1   **c.** 6, 2   **d.** 6, 8

2. $x^2 + 11x + 24$
  a. 6, 4   **b.** 24, 1   **c.** 8, 3   **d.** 2, 12

3. $2x^2 + 13x + 6$
  a. 2, 6   **b.** 12, 1   **c.** 13, 1   **d.** 3, 4

4. $4x^2 + 8x + 3$
  a. 4, 3   **b.** 4, 4   **c.** 12, 1   **d.** 2, 6

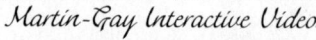

Martin-Gay Interactive Videos

See Video 6.4

Watch the section lecture video and answer the following question.

OBJECTIVE 1

5. In the lecture following 🎞 Example 1, why does writing a term as the sum or difference of two terms suggest we'd then try to factor by grouping?

6.4 Exercise Set MyMathLab®

Factor each polynomial by grouping. Notice that Step 3 has already been done in these exercises. See Examples 1 through 5.

1. $x^2 + 3x + 2x + 6$

2. $x^2 + 5x + 3x + 15$

3. $y^2 + 8y - 2y - 16$

4. $z^2 + 10z - 7z - 70$

5. $8x^2 - 5x - 24x + 15$

6. $4x^2 - 9x - 32x + 72$

7. $5x^4 - 3x^2 + 25x^2 - 15$

8. $2y^4 - 10y^2 + 7y^2 - 35$

MIXED PRACTICE

Factor each trinomial by grouping. Exercises 9–12 are broken into parts to help you get started. See Examples 1 through 5.

9. $6x^2 + 11x + 3$

  a. Find two numbers whose product is $6 \cdot 3 = 18$ and whose sum is 11.

  b. Write $11x$ using the factors from part **a.**

  c. Factor by grouping.

10. $8x^2 + 14x + 3$

  a. Find two numbers whose product is $8 \cdot 3 = 24$ and whose sum is 14.

  b. Write $14x$ using the factors from part **a.**

  c. Factor by grouping.

11. $15x^2 - 23x + 4$

  a. Find two numbers whose product is $15 \cdot 4 = 60$ and whose sum is -23.

  b. Write $-23x$ using the factors from part **a.**

  c. Factor by grouping.

12. $6x^2 - 13x + 5$

  a. Find two numbers whose product is $6 \cdot 5 = 30$ and whose sum is -13.

  b. Write $-13x$ using the factors from part **a.**

  c. Factor by grouping.

13. $21y^2 + 17y + 2$

14. $15x^2 + 11x + 2$

15. $7x^2 - 4x - 11$

16. $8x^2 - x - 9$

17. $10x^2 - 9x + 2$

18. $30x^2 - 23x + 3$

19. $2x^2 - 7x + 5$

20. $2x^2 - 7x + 3$

21. $12x + 4x^2 + 9$

22. $20x + 25x^2 + 4$

23. $4x^2 - 8x - 21$

24. $6x^2 - 11x - 10$

25. $10x^2 - 23x + 12$

26. $21x^2 - 13x + 2$

27. $2x^3 + 13x^2 + 15x$

28. $3x^3 + 8x^2 + 4x$

29. $16y^2 - 34y + 18$

30. $4y^2 - 2y - 12$

31. $-13x + 6 + 6x^2$

32. $-25x + 12 + 12x^2$

33. $54a^2 - 9a - 30$

34. $30a^2 + 38a - 20$

35. $20a^3 + 37a^2 + 8a$

36. $10a^3 + 17a^2 + 3a$

37. $12x^3 - 27x^2 - 27x$

38. $30x^3 - 155x^2 + 25x$

39. $3x^2y + 4xy^2 + y^3$

40. $6r^2t + 7rt^2 + t^3$

41. $20z^2 + 7z + 1$

42. $36z^2 + 6z + 1$

43. $5x^2 + 50xy + 125y^2$

44. $3x^2 + 42xy + 147y^2$

45. $24a^2 - 6ab - 30b^2$

46. $30a^2 + 5ab - 25b^2$

47. $15p^4 + 31p^3q + 2p^2q^2$

48. $20s^4 + 61s^3t + 3s^2t^2$

49. $162a^4 - 72a^2 + 8$

50. $32n^4 - 112n^2 + 98$

51. $35 + 12x + x^2$

52. $33 + 14x + x^2$

53. $6 - 11x + 5x^2$

54. $5 - 12x + 7x^2$

REVIEW AND PREVIEW

Multiply. See Sections 5.3 and 5.4.

55. $(x - 2)(x + 2)$

56. $(y - 5)(y + 5)$

57. $(y + 4)(y + 4)$

58. $(x + 7)(x + 7)$

59. $(9z + 5)(9z - 5)$

60. $(8y + 9)(8y - 9)$

61. $(x - 3)(x^2 + 3x + 9)$

62. $(2z - 1)(4z^2 + 2z + 1)$

CONCEPT EXTENSIONS

Write the perimeter of each figure as a simplified polynomial. Then factor the polynomial.

63.

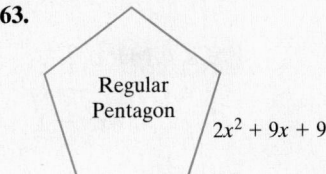

Regular Pentagon $2x^2 + 9x + 9$

64.

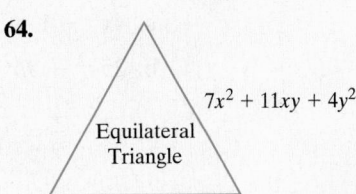

Equilateral Triangle $7x^2 + 11xy + 4y^2$

Factor each polynomial by grouping.

65. $x^{2n} + 2x^n + 3x^n + 6$

(*Hint:* Don't forget that $x^{2n} = x^n \cdot x^n$.)

66. $x^{2n} + 6x^n + 10x^n + 60$

67. $3x^{2n} + 16x^n - 35$

68. $12x^{2n} - 40x^n + 25$

69. In your own words, explain how to factor a trinomial by grouping.

6.5   Factoring Binomials

OBJECTIVES

1. Factor the Difference of Two Squares.

2. Factor the Sum or Difference of Two Cubes.

OBJECTIVE

1   Factoring the Difference of Two Squares

When learning to multiply binomials in Chapter 5, we studied a special product, the product of the sum and difference of two terms, a and b:

$$(a + b)(a - b) = a^2 - b^2$$

For example, the product of $x + 3$ and $x - 3$ is

$$(x + 3)(x - 3) = x^2 - 9$$

The binomial $x^2 - 9$ is called a **difference of squares**. In this section, we reverse the pattern for the product of a sum and difference to factor the binomial difference of squares.

Factoring the Difference of Two Squares

$$a^2 - b^2 = (a + b)(a - b)$$

Helpful Hint

Since multiplication is commutative, remember that the order of factors does not matter. In other words,

$$a^2 - b^2 = (a + b)(a - b) \text{ or } (a - b)(a + b)$$

EXAMPLE 1 Factor: $x^2 - 25$

Solution $x^2 - 25$ is the difference of two squares since $x^2 - 25 = x^2 - 5^2$. Therefore,

$$x^2 - 25 = x^2 - 5^2 = (x + 5)(x - 5)$$

Multiply to check.

PRACTICE
1 Factor: $x^2 - 81$

EXAMPLE 2 Factor each difference of squares.

  a. $4x^2 - 1$ **b.** $25a^2 - 9b^2$ **c.** $y^2 - \dfrac{4}{9}$

Solution

  a. $4x^2 - 1 = (2x)^2 - 1^2 = (2x + 1)(2x - 1)$

  b. $25a^2 - 9b^2 = (5a)^2 - (3b)^2 = (5a + 3b)(5a - 3b)$

  c. $y^2 - \dfrac{4}{9} = y^2 - \left(\dfrac{2}{3}\right)^2 = \left(y + \dfrac{2}{3}\right)\left(y - \dfrac{2}{3}\right)$

PRACTICE
2 Factor each difference of squares.

  a. $9x^2 - 1$ **b.** $36a^2 - 49b^2$ **c.** $p^2 - \dfrac{25}{36}$

EXAMPLE 3 Factor: $x^4 - y^6$

Solution This is a difference of squares since $x^4 = (x^2)^2$ and $y^6 = (y^3)^2$. Thus,

$$x^4 - y^6 = (x^2)^2 - (y^3)^2 = (x^2 + y^3)(x^2 - y^3)$$

PRACTICE
3 Factor: $p^4 - q^{10}$

EXAMPLE 4 Factor each binomial.

  a. $y^4 - 16$ **b.** $x^2 + 4$

Solution

  a. $y^4 - 16 = (y^2)^2 - 4^2$

  $= (y^2 + 4) \underbrace{(y^2 - 4)}$ Factor the difference of two squares.

  This binomial can be factored further since it is the difference of two squares.

  $= (y^2 + 4)(y + 2)(y - 2)$ Factor the difference of two squares.

b. $x^2 + 4$

Note that the binomial $x^2 + 4$ is the *sum* of two squares since we can write $x^2 + 4$ as $x^2 + 2^2$. We might try to factor using $(x + 2)(x + 2)$ or $(x - 2)(x - 2)$. But when we multiply to check, we find that neither factoring is correct.

$$(x + 2)(x + 2) = x^2 + 4x + 4$$

$$(x - 2)(x - 2) = x^2 - 4x + 4$$

In both cases, the product is a trinomial, not the required binomial. In fact, $x^2 + 4$ is a prime polynomial. □

PRACTICE
4   Factor each binomial.

  a. $z^4 - 81$ **b.** $m^2 + 49$ ■

Helpful Hint

When factoring, don't forget:

- See whether the terms have a greatest common factor (GCF) (other than 1) that can be factored out.
- Other than the GCF, the **sum** of two squares cannot be factored using real numbers.
- Factor completely. Always check to see whether any factors can be factored further.

EXAMPLES   Factor each binomial.

5. $4x^3 - 49x = x(4x^2 - 49)$ Factor out the GCF, x.

$\qquad\qquad\quad = x[(2x)^2 - 7^2]$

$\qquad\qquad\quad = x(2x + 7)(2x - 7)$ Factor the difference of two squares.

6. $162x^4 - 2 = 2(81x^4 - 1)$ Factor out the GCF, 2.

$\qquad\qquad\quad = 2(9x^2 + 1)(9x^2 - 1)$ Factor the difference of two squares.

$\qquad\qquad\quad = 2(9x^2 + 1)(3x + 1)(3x - 1)$ Factor the difference of two squares. □

PRACTICE
5–6   Factor each binomial.

  5. $36y^3 - 25y$ **6.** $80y^4 - 5$ ■

EXAMPLE 7   Factor: $-49x^2 + 16$

Solution Factor as is, or, if you like, rearrange terms.

Factor as is: $-49x^2 + 16 = -1(49x^2 - 16)$ Factor out -1.

$\qquad\qquad\qquad\qquad = -1(7x + 4)(7x - 4)$ Factor the difference of two squares.

Rewrite binomial: $-49x^2 + 16 = 16 - 49x^2 = 4^2 - (7x)^2$

$\qquad\qquad\qquad\qquad\qquad = (4 + 7x)(4 - 7x)$

Helpful Hint

When rearranging terms, keep in mind that the sign of a term is in front of the term.

Both factorizations are correct and are equal. To see this, factor -1 from $(4 - 7x)$ in the second factorization. □

PRACTICE
7   Factor: $-9x^2 + 100$ ■

OBJECTIVE

2 **Factoring the Sum or Difference of Two Cubes**

Although the sum of two squares usually does not factor, the sum or difference of two cubes can be factored and reveals factoring patterns. The pattern for the sum of cubes is illustrated by multiplying the binomial $x + y$ and the trinomial $x^2 - xy + y^2$.

$$
\begin{array}{r}
x^2 - xy + y^2 \\
x + y \\
\hline
x^2y - xy^2 + y^3 \\
x^3 - x^2y + xy^2 \\
\hline
x^3 \qquad\qquad + y^3
\end{array}
$$

Thus, $(x + y)(x^2 - xy + y^2) = x^3 + y^3$ Sum of cubes

The pattern for the difference of two cubes is illustrated by multiplying the binomial $x - y$ by the trinomial $x^2 + xy + y^2$. The result is

$$(x - y)(x^2 + xy + y^2) = x^3 - y^3 \quad \text{Difference of cubes}$$

Factoring the Sum or Difference of Two Cubes

$$a^3 + b^3 = (a + b)(a^2 - ab + b^2)$$
$$a^3 - b^3 = (a - b)(a^2 + ab + b^2)$$

Recall that "factor" means "to write as a product." Above are patterns for writing sums and differences as products.

EXAMPLE 8   Factor: $x^3 + 8$

Solution First, write the binomial in the form $a^3 + b^3$.

$$x^3 + 8 = x^3 + 2^3 \quad \text{Write in the form } a^3 + b^3.$$

If we replace a with x and b with 2 in the formula above, we have

$$
\begin{aligned}
x^3 + 2^3 &= (x + 2)[x^2 - (x)(2) + 2^2] \\
&= (x + 2)(x^2 - 2x + 4)
\end{aligned}
$$

PRACTICE

8   Factor: $x^3 + 64$

Helpful Hint

When factoring sums or differences of cubes, notice the sign patterns.

same sign

$$x^3 + y^3 = (x + y)(x^2 - xy + y^2)$$

opposite signs   always positive

same sign

$$x^3 - y^3 = (x - y)(x^2 + xy + y^2)$$

opposite signs   always positive

EXAMPLE 9   Factor: $y^3 - 27$

Solution $y^3 - 27 = y^3 - 3^3$ Write in the form $a^3 - b^3$.

$$
\begin{aligned}
&= (y - 3)[y^2 + (y)(3) + 3^2] \\
&= (y - 3)(y^2 + 3y + 9)
\end{aligned}
$$

PRACTICE

9   Factor: $x^3 - 125$

EXAMPLE 10   Factor: $64x^3 + 1$

Solution
$$64x^3 + 1 = (4x)^3 + 1^3$$
$$= (4x + 1)[(4x)^2 - (4x)(1) + 1^2]$$
$$= (4x + 1)(16x^2 - 4x + 1) \qquad \square$$

PRACTICE
10   Factor: $27y^3 + 1$ ∎

EXAMPLE 11   Factor: $54a^3 - 16b^3$

Solution Remember to factor out the greatest common factor first before using other factoring methods.

$$54a^3 - 16b^3 = 2(27a^3 - 8b^3) \qquad \text{Factor out the GCF, 2.}$$
$$= 2[(3a)^3 - (2b)^3] \quad \text{Difference of two cubes}$$
$$= 2(3a - 2b)[(3a)^2 + (3a)(2b) + (2b)^2]$$
$$= 2(3a - 2b)(9a^2 + 6ab + 4b^2) \qquad \square$$

PRACTICE
11   Factor: $32x^3 - 500y^3$ ∎

Graphing Calculator Explorations

Graphing

A graphing calculator is a convenient tool for evaluating an expression at a given replacement value. For example, let's evaluate $x^2 - 6x$ when $x = 2$. To do so, store the value 2 in the variable x and then enter and evaluate the algebraic expression.

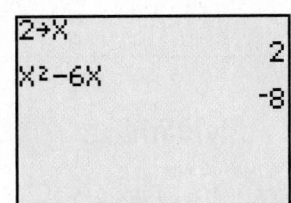

The value of $x^2 - 6x$ when $x = 2$ is -8. You may want to use this method for evaluating expressions as you explore the following.

We can use a graphing calculator to explore factoring patterns numerically. Use your calculator to evaluate $x^2 - 2x + 1$, $x^2 - 2x - 1$, and $(x - 1)^2$ for each value of x given in the table. What do you observe?

	$x^2 - 2x + 1$	$x^2 - 2x - 1$	$(x - 1)^2$
$x = 5$			
$x = -3$			
$x = 2.7$			
$x = -12.1$			
$x = 0$			

Notice in each case that $x^2 - 2x - 1 \neq (x - 1)^2$. Because for each x in the table the value of $x^2 - 2x + 1$ and the value of $(x - 1)^2$ are the same, we might guess that $x^2 - 2x + 1 = (x - 1)^2$. We can verify our guess algebraically with multiplication:

$$(x - 1)(x - 1) = x^2 - x - x + 1 = x^2 - 2x + 1$$

Vocabulary, Readiness & Video Check

Use the choices below to fill in each blank. Some choices may be used more than once and some choices may not be used at all.

true	difference of two squares	sum of two cubes
false	difference of two cubes	

1. The expression $x^3 - 27$ is called a _____.
2. The expression $x^2 - 49$ is called a _____.
3. The expression $z^3 + 1$ is called a _____.
4. True or false: The binomial $y^2 + 9$ factors as $(y + 3)^2$. _____

Write each term as a square.

5. $49x^2$ **6.** $25y^4$

Write each term as a cube.

7. $8y^3$ **8.** x^6

Martin-Gay Interactive Videos

See Video 6.5 ⦾

Watch the section lecture video and answer the following questions.

OBJECTIVE 1
9. In ▦ Examples 1 and 2, what are two reasons the original binomial is rewritten so that each term is a square?

OBJECTIVE 1
10. From ▦ Example 3, what is a prime polynomial?

OBJECTIVE 2
11. In ▦ Examples 6–8, what tips are given to remember how to factor the sum or difference of two cubes rather than memorizing the formulas?

6.5 Exercise Set MyMathLab® ▶

Factor each binomial completely. See Examples 1 through 7.

▶ **1.** $x^2 - 4$

2. $x^2 - 36$

3. $81p^2 - 1$

4. $49m^2 - 1$

5. $25y^2 - 9$

6. $49a^2 - 16$

▶ **7.** $121m^2 - 100n^2$

8. $169a^2 - 49b^2$

9. $x^2y^2 - 1$

10. $a^2b^2 - 16$

11. $x^2 - \dfrac{1}{4}$

12. $y^2 - \dfrac{1}{16}$

13. $-4r^2 + 1$

14. $-9t^2 + 1$

▶ **15.** $16r^2 + 1$

16. $49y^2 + 1$

17. $-36 + x^2$

18. $-1 + y^2$

19. $m^4 - 1$

20. $n^4 - 16$

21. $m^4 - n^{18}$

22. $n^4 - r^6$

Factor the sum or difference of two cubes. See Examples 8 through 11.

▶ **23.** $x^3 + 125$

24. $p^3 + 1$

25. $8a^3 - 1$

26. $27y^3 - 1$

27. $m^3 + 27n^3$

28. $y^3 + 64z^3$

29. $5k^3 + 40$

30. $6r^3 + 162$

▶ **31.** $x^3y^3 - 64$

32. $a^3b^3 - 8$

33. $250r^3 - 128t^3$

34. $24x^3 - 81y^3$

MIXED PRACTICE

Factor each binomial completely. See Examples 1 through 11.

35. $r^2 - 64$

36. $q^2 - 121$

37. $x^2 - 169y^2$

38. $x^2 - 225y^2$

39. $27 - t^3$

40. $125 - r^3$

41. $18r^2 - 8$

42. $32t^2 - 50$

43. $9xy^2 - 4x$

44. $36x^2y - 25y$

45. $8m^3 + 64$

46. $2x^3 + 54$

47. $xy^3 - 9xyz^2$

48. $x^3y - 4xy^3$

49. $36x^2 - 64y^2$

50. $225a^2 - 81b^2$

51. $144 - 81x^2$

52. $12x^2 - 27$

53. $x^3y^3 - z^6$

54. $a^3b^3 - c^9$

55. $49 - \dfrac{9}{25}m^2$

56. $100 - \dfrac{4}{81}n^2$

57. $t^3 + 343$

58. $s^3 + 216$

59. $n^3 + 49n$

60. $y^3 + 64y$

61. $x^6 - 81x^2$

62. $n^9 - n^5$

63. $64p^3q - 81pq^3$

64. $100x^3y - 49xy^3$

65. $27x^2y^3 + xy^2$

66. $8x^3y^3 + x^3y$

67. $125a^4 - 64ab^3$

68. $64m^4 - 27mn^3$

69. $16x^4 - 64x^2$

70. $25y^4 - 100y^2$

REVIEW AND PREVIEW

Solve each equation. See Section 2.2.

71. $x - 6 = 0$

72. $y + 5 = 0$

73. $2m + 4 = 0$

74. $3x - 9 = 0$

75. $5z - 1 = 0$

76. $4a + 2 = 0$

CONCEPT EXTENSIONS

Factor each expression completely.

77. $(x + 2)^2 - y^2$

78. $(y - 6)^2 - z^2$

79. $a^2(b - 4) - 16(b - 4)$

80. $m^2(n + 8) - 9(n + 8)$

81. $(x^2 + 6x + 9) - 4y^2$ (*Hint:* Factor the trinomial in parentheses first.)

82. $(x^2 + 2x + 1) - 36y^2$

83. $x^{2n} - 100$

84. $x^{2n} - 81$

85. What binomial multiplied by $(x - 6)$ gives the difference of two squares?

86. What binomial multiplied by $(5 + y)$ gives the difference of two squares?

87. In your own words, explain how to tell whether a binomial is a difference of squares. Then explain how to factor a difference of squares.

88. In your own words, explain how to tell whether a binomial is a sum of cubes. Then explain how to factor a sum of cubes.

89. The Toroweap Overlook, on the North Rim of the Grand Canyon, lies 3000 vertical feet above the Colorado River. The view is spectacular, and the sheer drop is dramatic. A film crew creating a documentary about the Grand Canyon has built a camera platform 136 feet above the Overlook. A camera filter comes loose and falls to the river below. The height of the filter above the river after t seconds is given by the expression $3136 - 16t^2$.

   a. Find the height of the filter above the river after 3 seconds.

   b. Find the height of the filter above the river after 10 seconds.

   c. To the nearest whole second, estimate when the filter lands in the river.

   d. Factor $3136 - 16t^2$.

90. An object is dropped from the top of Pittsburgh's USX Tower, which is 841 feet tall. (*Source: World Almanac research*) The height of the object after t seconds is given by the expression $841 - 16t^2$.

   a. Find the height of the object after 2 seconds.

   b. Find the height of the object after 5 seconds.

   c. To the nearest whole second, estimate when the object hits the ground.

   d. Factor $841 - 16t^2$.

841 feet

91. At this writing, the tallest completed building in the world is the Burj Khalifa, in Dubai, measuring a height of 2717 feet. (*Source:* Council on Tall Buildings and Urban Habitat) Suppose an action picture is being filmed there and a stunt man is making his way to the top of the spire. He sways in the wind and drops a clip from the height of 2704 feet. The height of the clip after t seconds is given by the expression $2704 - 16t^2$.

  a. Find the height of the clip after 3 seconds.

  b. Find the height of the clip after 7 seconds.

  c. To the nearest whole second, estimate when the clip will hit the ground.

  d. Factor $2704 - 16t^2$.

92. A performer with the Moscow Circus is planning a stunt involving a free fall from the top of the Moscow State University building, which is 784 feet tall. (*Source:* Council of Tall Buildings and Urban Habitat) Neglecting air resistance, the performer's height above gigantic cushions positioned at ground level after t seconds is given by the expression $784 - 16t^2$.

  a. Find the performer's height after 2 seconds.

  b. Find the performer's height after 5 seconds.

  c. To the nearest whole second, estimate when the performer reaches the cushions positioned at ground level.

  d. Factor $784 - 16t^2$.

Integrated Review — Choosing a Factoring Strategy

Sections 6.1–6.5

The following steps may be helpful when factoring polynomials.

> **Factoring a Polynomial**
>
> **Step 1.** Are there any common factors? If so, factor out the GCF.
>
> **Step 2.** How many terms are in the polynomial?
>
>   **a.** If there are **two** terms, decide if one of the following can be applied.
>
>   **i.** Difference of two squares: $a^2 - b^2 = (a + b)(a - b)$.
>
>   **ii.** Difference of two cubes: $a^3 - b^3 = (a - b)(a^2 + ab + b^2)$.
>
>   **iii.** Sum of two cubes: $a^3 + b^3 = (a + b)(a^2 - ab + b^2)$.
>
>   **b.** If there are **three** terms, try one of the following.
>
>   **i.** Perfect square trinomial: $a^2 + 2ab + b^2 = (a + b)^2$
> $a^2 - 2ab + b^2 = (a - b)^2$.
>
>   **ii.** If not a perfect square trinomial, factor using the methods presented in Sections 6.2 through 6.4.
>
>   **c.** If there are **four** or more terms, try factoring by grouping.
>
> **Step 3.** See if any factors in the factored polynomial can be factored further.
>
> **Step 4.** Check by multiplying.

Study the next five examples to help you use the steps above.

> **EXAMPLE 1** Factor: $10t^2 - 17t + 3$

Solution

Step 1. The terms of this polynomial have no common factor (other than 1).

Step 2. There are three terms, so this polynomial is a trinomial. This trinomial is not a perfect square trinomial, so factor using methods from earlier sections.

$$\text{Factors of } 10t^2: \quad 10t^2 = 2t \cdot 5t, \qquad 10t^2 = t \cdot 10t$$

Since the middle term, $-17t$, has a negative numerical coefficient, find negative factors of 3.

$$\text{Factors of 3:} \quad 3 = -1 \cdot -3$$

Try different combinations of these factors. The correct combination is

$$(2t - 3)(5t - 1) = 10t^2 - 17t + 3$$
$$-15t$$
$$-2t$$
$$\overline{-17t} \quad \text{Correct middle term}$$

Step 3. No factor can be factored further, so we have factored completely.

Step 4. To check, multiply $2t - 3$ and $5t - 1$.

$$(2t - 3)(5t - 1) = 10t^2 - 2t - 15t + 3 = 10t^2 - 17t + 3$$

The factored form of $10t^2 - 17t + 3$ is $(2t - 3)(5t - 1)$. □

PRACTICE
1   Factor: $6x^2 - 11x + 3$ ■

EXAMPLE 2   Factor: $2x^3 + 3x^2 - 2x - 3$

Solution

Step 1. There are no factors common to all terms.

Step 2. Try factoring by grouping since this polynomial has four terms.

$$2x^3 + 3x^2 - 2x - 3 = x^2(2x + 3) - 1(2x + 3) \quad \text{Factor out the greatest common factor for each pair of terms.}$$

$$= (2x + 3)(x^2 - 1) \quad \text{Factor out } 2x + 3.$$

Step 3. The binomial $x^2 - 1$ can be factored further. It is the difference of two squares.

$$= (2x + 3)(x + 1)(x - 1) \quad \text{Factor } x^2 - 1 \text{ as a difference of squares.}$$

Step 4. Check by finding the product of the three binomials. The polynomial factored completely is $(2x + 3)(x + 1)(x - 1)$. □

PRACTICE
2   Factor: $3x^3 + x^2 - 12x - 4$ ■

EXAMPLE 3   Factor: $12m^2 - 3n^2$

Solution

Step 1. The terms of this binomial contain a greatest common factor of 3.

$$12m^2 - 3n^2 = 3(4m^2 - n^2) \quad \text{Factor out the greatest common factor.}$$

Step 2. The binomial $4m^2 - n^2$ is a difference of squares.

$$= 3(2m + n)(2m - n) \quad \text{Factor the difference of squares.}$$

Step 3. No factor can be factored further.

Step 4. We check by multiplying.

$$3(2m + n)(2m - n) = 3(4m^2 - n^2) = 12m^2 - 3n^2$$

The factored form of $12m^2 - 3n^2$ is $3(2m + n)(2m - n)$. □

PRACTICE
3   Factor: $27x^2 - 3y^2$ ■

EXAMPLE 4 Factor: $x^3 + 27y^3$

Solution

Step 1. The terms of this binomial contain no common factor (other than 1).

Step 2. This binomial is the sum of two cubes.

$$\begin{aligned}
x^3 + 27y^3 &= (x)^3 + (3y)^3 \\
&= (x + 3y)[x^2 - x(3y) + (3y)^2] \\
&= (x + 3y)(x^2 - 3xy + 9y^2)
\end{aligned}$$

Step 3. No factor can be factored further.

Step 4. We check by multiplying.

$$\begin{aligned}
(x + 3y)(x^2 - 3xy + 9y^2) &= x(x^2 - 3xy + 9y^2) + 3y(x^2 - 3xy + 9y^2) \\
&= x^3 - 3x^2y + 9xy^2 + 3x^2y - 9xy^2 + 27y^3 \\
&= x^3 + 27y^3
\end{aligned}$$

Thus, $x^3 + 27y^3$ factored completely is $(x + 3y)(x^2 - 3xy + 9y^2)$.

PRACTICE
4 Factor: $8a^3 + b^3$

EXAMPLE 5 Factor: $30a^2b^3 + 55a^2b^2 - 35a^2b$

Solution

Step 1. $30a^2b^3 + 55a^2b^2 - 35a^2b = 5a^2b(6b^2 + 11b - 7)$ Factor out the GCF.

Step 2. $= 5a^2b(2b - 1)(3b + 7)$ Factor the resulting trinomial.

Step 3. No factor can be factored further.

Step 4. Check by multiplying.

The trinomial factored completely is $5a^2b(2b - 1)(3b + 7)$.

PRACTICE
5 Factor: $60x^3y^2 - 66x^2y^2 - 36xy^2$

Factor the following completely.

1. $x^2 + 2xy + y^2$

2. $x^2 - 2xy + y^2$

3. $a^2 + 11a - 12$

4. $a^2 - 11a + 10$

5. $a^2 - a - 6$

6. $a^2 - 2a + 1$

7. $x^2 + 2x + 1$

8. $x^2 + x - 2$

9. $x^2 + 4x + 3$

10. $x^2 + x - 6$

11. $x^2 + 7x + 12$

12. $x^2 + x - 12$

13. $x^2 + 3x - 4$

14. $x^2 - 7x + 10$

15. $x^2 + 2x - 15$

16. $x^2 + 11x + 30$

17. $x^2 - x - 30$

18. $x^2 + 11x + 24$

19. $2x^2 - 98$

20. $3x^2 - 75$

21. $x^2 + 3x + xy + 3y$

22. $3y - 21 + xy - 7x$

23. $x^2 + 6x - 16$

24. $x^2 - 3x - 28$

▶ **25.** $4x^3 + 20x^2 - 56x$

26. $6x^3 - 6x^2 - 120x$

27. $12x^2 + 34x + 24$

28. $8a^2 + 6ab - 5b^2$

29. $4a^2 - b^2$

30. $28 - 13x - 6x^2$

31. $20 - 3x - 2x^2$

32. $x^2 - 2x + 4$

33. $a^2 + a - 3$

34. $6y^2 + y - 15$

35. $4x^2 - x - 5$

36. $x^2y - y^3$

37. $4t^2 + 36$

38. $x^2 + x + xy + y$

39. $ax + 2x + a + 2$

40. $18x^3 - 63x^2 + 9x$

41. $12a^3 - 24a^2 + 4a$

42. $x^2 + 14x - 32$

43. $x^2 - 14x - 48$

44. $16a^2 - 56ab + 49b^2$

45. $25p^2 - 70pq + 49q^2$

46. $7x^2 + 24xy + 9y^2$

▶ **47.** $125 - 8y^3$

48. $64x^3 + 27$

49. $-x^2 - x + 30$

50. $-x^2 + 6x - 8$

51. $14 + 5x - x^2$

52. $3 - 2x - x^2$

53. $3x^4y + 6x^3y - 72x^2y$

54. $2x^3y + 8x^2y^2 - 10xy^3$

55. $5x^3y^2 - 40x^2y^3 + 35xy^4$

56. $4x^4y - 8x^3y - 60x^2y$

57. $12x^3y + 243xy$

58. $6x^3y^2 + 8xy^2$

59. $4 - x^2$

60. $9 - y^2$

61. $3rs - s + 12r - 4$

62. $x^3 - 2x^2 + 3x - 6$

63. $4x^2 - 8xy - 3x + 6y$

64. $4x^2 - 2xy - 7yz + 14xz$

65. $6x^2 + 18xy + 12y^2$

66. $12x^2 + 46xy - 8y^2$

67. $xy^2 - 4x + 3y^2 - 12$

68. $x^2y^2 - 9x^2 + 3y^2 - 27$

69. $5(x + y) + x(x + y)$

70. $7(x - y) + y(x - y)$

71. $14t^2 - 9t + 1$

72. $3t^2 - 5t + 1$

73. $3x^2 + 2x - 5$

74. $7x^2 + 19x - 6$

75. $x^2 + 9xy - 36y^2$

76. $3x^2 + 10xy - 8y^2$

77. $1 - 8ab - 20a^2b^2$

78. $1 - 7ab - 60a^2b^2$

79. $9 - 10x^2 + x^4$

80. $36 - 13x^2 + x^4$

81. $x^4 - 14x^2 - 32$

82. $x^4 - 22x^2 - 75$

83. $x^2 - 23x + 120$

84. $y^2 + 22y + 96$

85. $6x^3 - 28x^2 + 16x$

86. $6y^3 - 8y^2 - 30y$

87. $27x^3 - 125y^3$

88. $216y^3 - z^3$

89. $x^3y^3 + 8z^3$

90. $27a^3b^3 + 8$

91. $2xy - 72x^3y$

92. $2x^3 - 18x$

93. $x^3 + 6x^2 - 4x - 24$

94. $x^3 - 2x^2 - 36x + 72$

95. $6a^3 + 10a^2$

96. $4n^2 - 6n$

97. $a^2(a + 2) + 2(a + 2)$

98. $a - b + x(a - b)$

99. $x^3 - 28 + 7x^2 - 4x$

100. $a^3 - 45 - 9a + 5a^2$

CONCEPT EXTENSIONS

Factor.

101. $(x - y)^2 - z^2$

102. $(x + 2y)^2 - 9$

103. $81 - (5x + 1)^2$

104. $b^2 - (4a + c)^2$

105. Explain why it makes good sense to factor out the GCF first, before using other methods of factoring.

106. The sum of two squares usually does not factor. Is the sum of two squares $9x^2 + 81y^2$ factorable?

107. Which of the following are equivalent to $(x + 10)(x - 7)$?

   a. $(x - 7)(x + 10)$ **b.** $-1(x + 10)(x - 7)$

   c. $-1(x + 10)(7 - x)$ **d.** $-1(-x - 10)(7 - x)$

108. Which of the following are equivalent to $(x - 2)(x - 5)$?

   a. $-1(x + 2)(x + 5)$ **b.** $(x - 5)(x - 2)$

   c. $(5 - x)(2 - x)$ **d.** $-1(x + 2)(x - 5)$

6.6 Solving Quadratic Equations by Factoring

OBJECTIVES

1 Solve Quadratic Equations by Factoring.

2 Solve Equations with Degree Greater than 2 by Factoring.

3 Find the *x*-Intercepts of the Graph of a Quadratic Equation in Two Variables.

In this section, we introduce a new type of equation—the **quadratic equation**.

> **Quadratic Equation**
>
> A quadratic equation is one that can be written in the form
>
> $$ax^2 + bx + c = 0$$
>
> where $a, b,$ and c are real numbers and $a \neq 0$.

Some examples of quadratic equations are shown below.

$$x^2 - 9x - 22 = 0 \qquad 4x^2 - 28 = -49 \qquad x(2x - 7) = 4$$

The form $ax^2 + bx + c = 0$ is called the **standard form** of a quadratic equation. The quadratic equation $x^2 - 9x - 22 = 0$ is the only equation above that is in standard form.

144 feet

Quadratic equations model many real-life situations. For example, let's suppose we want to know how long before a person diving from a 144-foot cliff reaches the ocean. The answer to this question is found by solving the quadratic equation $-16t^2 + 144 = 0$. (See Example 1 in Section 6.7.)

OBJECTIVE

1 Solving Quadratic Equations by Factoring

Some quadratic equations can be solved by making use of factoring and the **zero factor property**.

> **Zero Factor Property**
>
> If a and b are real numbers and if $ab = 0$, then $a = 0$ or $b = 0$.

This property states that if the product of two numbers is 0 then at least one of the numbers must be 0.

EXAMPLE 1 Solve: $(x - 3)(x + 1) = 0$

Solution If this equation is to be a true statement, then either the factor $x - 3$ must be 0 or the factor $x + 1$ must be 0. In other words, either

$$x - 3 = 0 \quad \text{or} \quad x + 1 = 0$$

If we solve these two linear equations, we have

$$x = 3 \quad \text{or} \quad x = -1$$

Thus, 3 and -1 are both solutions of the equation $(x - 3)(x + 1) = 0$. To check, we replace x with 3 in the original equation. Then we replace x with -1 in the original equation.

Check: Let $x = 3$.

$(x - 3)(x + 1) = 0$

$(3 - 3)(3 + 1) \stackrel{?}{=} 0$ Replace x with 3.

$0(4) = 0$ True

Let $x = -1$.

$(x - 3)(x + 1) = 0$

$(-1 - 3)(-1 + 1) \stackrel{?}{=} 0$ Replace x with -1.

$(-4)(0) = 0$ True

The solutions are 3 and -1, or we say that the solution set is $\{-1, 3\}$.   □

PRACTICE

1 Solve: $(x + 4)(x - 5) = 0$

> **Helpful Hint**
>
> The zero factor property says that *if a product is 0, then a factor is 0*.
>
> If $a \cdot b = 0$, then $a = 0$ or $b = 0$.
>
> If $x(x + 5) = 0$, then $x = 0$ or $x + 5 = 0$.
>
> If $(x + 7)(2x - 3) = 0$, then $x + 7 = 0$ or $2x - 3 = 0$.
>
> Use this property only when the product is 0.
> For example, if $a \cdot b = 8$, we do not know the value of a or b. The values may be $a = 2, b = 4$ or $a = 8, b = 1$, or any other two numbers whose product is 8.

EXAMPLE 2 Solve: $(x - 5)(2x + 7) = 0$

Solution: The product is 0. By the zero factor property, this is true only when a factor is 0. To solve, we set each factor equal to 0 and solve the resulting linear equations.

$$(x - 5)(2x + 7) = 0$$

$$x - 5 = 0 \quad \text{or} \quad 2x + 7 = 0$$

$$x = 5 \quad \text{or} \quad 2x = -7$$

$$x = -\frac{7}{2}$$

Check: Let $x = 5$.

$$(x - 5)(2x + 7) = 0$$
$$(5 - 5)(2 \cdot 5 + 7) \stackrel{?}{=} 0 \quad \text{Replace } x \text{ with 5.}$$
$$0 \cdot 17 \stackrel{?}{=} 0$$
$$0 = 0 \quad \text{True}$$

Let $x = -\dfrac{7}{2}$.

$$(x - 5)(2x + 7) = 0$$
$$\left(-\frac{7}{2} - 5\right)\left(2\left(-\frac{7}{2}\right) + 7\right) \stackrel{?}{=} 0 \quad \text{Replace } x \text{ with } -\frac{7}{2}.$$
$$\left(-\frac{17}{2}\right)(-7 + 7) \stackrel{?}{=} 0$$
$$\left(-\frac{17}{2}\right) \cdot 0 \stackrel{?}{=} 0$$
$$0 = 0 \quad \text{True}$$

The solutions are 5 and $-\dfrac{7}{2}$.

PRACTICE
2 Solve: $(x - 12)(4x + 3) = 0$

EXAMPLE 3 Solve: $x(5x - 2) = 0$

Solution
$$x(5x - 2) = 0$$
$$x = 0 \quad \text{or} \quad 5x - 2 = 0 \quad \text{Use the zero factor property.}$$
$$5x = 2$$
$$x = \frac{2}{5}$$

Check: Let $x = 0$.

$$x(5x - 2) = 0$$
$$0(5 \cdot 0 - 2) \stackrel{?}{=} 0 \quad \text{Replace } x \text{ with 0.}$$
$$0(-2) \stackrel{?}{=} 0$$
$$0 = 0 \quad \text{True}$$

Let $x = \dfrac{2}{5}$.

$$x(5x - 2) = 0$$
$$\frac{2}{5}\left(5 \cdot \frac{2}{5} - 2\right) \stackrel{?}{=} 0 \quad \text{Replace } x \text{ with } \frac{2}{5}.$$
$$\frac{2}{5}(2 - 2) \stackrel{?}{=} 0$$
$$\frac{2}{5}(0) \stackrel{?}{=} 0$$
$$0 = 0 \quad \text{True}$$

The solutions are 0 and $\dfrac{2}{5}$.

PRACTICE
3 Solve: $x(7x - 6) = 0$

EXAMPLE 4 Solve: $x^2 - 9x - 22 = 0$

Solution One side of the equation is 0. However, to use the zero factor property, one side of the equation must be 0 *and* the other side must be written as a product (must be factored). Thus, we must first factor this polynomial.

$$x^2 - 9x - 22 = 0$$
$$(x - 11)(x + 2) = 0 \quad \text{Factor.}$$

(Continued on the next page)

Now we can apply the zero factor property.

$$x - 11 = 0 \quad \text{or} \quad x + 2 = 0$$
$$x = 11 \text{ or} \qquad x = -2$$

Check: Let $x = 11$.
$$x^2 - 9x - 22 = 0$$
$$11^2 - 9 \cdot 11 - 22 \stackrel{?}{=} 0$$
$$121 - 99 - 22 \stackrel{?}{=} 0$$
$$22 - 22 \stackrel{?}{=} 0$$
$$0 = 0 \quad \text{True}$$

Let $x = -2$.
$$x^2 - 9x - 22 = 0$$
$$(-2)^2 - 9(-2) - 22 \stackrel{?}{=} 0$$
$$4 + 18 - 22 \stackrel{?}{=} 0$$
$$22 - 22 \stackrel{?}{=} 0$$
$$0 = 0 \quad \text{True}$$

The solutions are 11 and -2.

PRACTICE
4   Solve: $x^2 - 8x - 48 = 0$

EXAMPLE 5   Solve: $4x^2 - 28x = -49$

Solution First we rewrite the equation in standard form so that one side is 0. Then we factor the polynomial.

$$4x^2 - 28x = -49$$
$$4x^2 - 28x + 49 = 0 \qquad \text{Write in standard form by adding 49 to both sides.}$$
$$(2x - 7)(2x - 7) = 0 \qquad \text{Factor.}$$

Next we use the zero factor property and set each factor equal to 0. Since the factors are the same, the related equations will give the same solution.

$$2x - 7 = 0 \quad \text{or} \quad 2x - 7 = 0 \quad \text{Set each factor equal to 0.}$$
$$2x = 7 \quad \text{or} \qquad 2x = 7 \quad \text{Solve.}$$
$$x = \frac{7}{2} \quad \text{or} \qquad x = \frac{7}{2}$$

Check: Although $\frac{7}{2}$ occurs twice, there is a single solution. Check this solution in the original equation. The solution is $\frac{7}{2}$.

PRACTICE
5   Solve: $9x^2 - 24x = -16$

The following steps may be used to solve a quadratic equation by factoring.

> **To Solve Quadratic Equations by Factoring**
>
> **Step 1.** Write the equation in standard form so that one side of the equation is 0.
>
> **Step 2.** Factor the quadratic expression completely.
>
> **Step 3.** Set each factor containing a variable equal to 0.
>
> **Step 4.** Solve the resulting equations.
>
> **Step 5.** Check each solution in the original equation.

Since it is not always possible to factor a quadratic polynomial, not all quadratic equations can be solved by factoring. Other methods of solving quadratic equations are presented in Chapter 9.

EXAMPLE 6   Solve: $x(2x - 7) = 4$

Solution   First we write the equation in standard form; then we factor.

$$x(2x - 7) = 4$$
$$2x^2 - 7x = 4 \qquad \text{Multiply.}$$
$$2x^2 - 7x - 4 = 0 \qquad \text{Write in standard form.}$$
$$(2x + 1)(x - 4) = 0 \qquad \text{Factor.}$$
$$2x + 1 = 0 \quad \text{or} \quad x - 4 = 0 \qquad \text{Set each factor equal to zero.}$$
$$2x = -1 \text{ or} \qquad x = 4 \qquad \text{Solve.}$$
$$x = -\frac{1}{2}$$

Check the solutions in the original equation. The solutions are $-\dfrac{1}{2}$ and 4.   □

PRACTICE
6   Solve: $x(3x + 7) = 6$ ■

Helpful Hint

To solve the equation $x(2x - 7) = 4$, do **not** set each factor equal to 4. Remember that to apply the zero factor property, one side of the equation must be 0 and the other side of the equation must be in factored form.

✔ **CONCEPT CHECK**

Explain the error and solve the equation correctly.

$$(x - 3)(x + 1) = 5$$
$$x - 3 = 5 \quad \text{or} \quad x + 1 = 5$$
$$x = 8 \quad \text{or} \qquad x = 4$$

EXAMPLE 7   Solve: $-2x^2 - 4x + 30 = 0$

Solution   The equation is in standard form, so we begin by factoring out the greatest common factor, −2.

$$-2x^2 - 4x + 30 = 0$$
$$-2(x^2 + 2x - 15) = 0 \qquad \text{Factor out } -2.$$
$$-2(x + 5)(x - 3) = 0 \qquad \text{Factor the quadratic.}$$

Next, set each factor **containing a variable** equal to 0.

$$x + 5 = 0 \qquad \text{or} \qquad x - 3 = 0 \qquad \text{Set each factor containing a variable equal to 0.}$$
$$x = -5 \qquad \text{or} \qquad x = 3 \qquad \text{Solve.}$$

Note: The factor −2 is a constant term containing no variables and can never equal 0. The solutions are −5 and 3.   □

PRACTICE
7   Solve: $-3x^2 - 6x + 72 = 0$ ■

Answer to Concept Check:
To use the zero factor property, one side of the equation must be 0, not 5. Correctly, $(x - 3)(x + 1) = 5$, $x^2 - 2x - 3 = 5, x^2 - 2x - 8 = 0,$ $(x - 4)(x + 2) = 0, x - 4 = 0$ or $x + 2 = 0, x = 4$ or $x = -2$.

OBJECTIVE

2   Solving Equations with Degree Greater than Two by Factoring ▶

Some equations involving polynomials of degree higher than 2 may also be solved by factoring and then applying the zero factor property.

EXAMPLE 8 Solve: $3x^3 - 12x = 0$

Solution Factor the left side of the equation. Begin by factoring out the greatest common factor, $3x$.

$$3x^3 - 12x = 0$$
$$3x(x^2 - 4) = 0 \quad \text{Factor out the GCF, } 3x.$$
$$3x(x + 2)(x - 2) = 0 \quad \text{Factor } x^2 - 4, \text{ a difference of squares.}$$

$3x = 0$ or $x + 2 = 0$ or $x - 2 = 0$ Set each factor equal to 0.

$x = 0$ or $x = -2$ or $x = 2$ Solve.

Thus, the equation $3x^3 - 12x = 0$ has three solutions: 0, -2, and 2. To check, replace x with each solution in the original equation.

Let x = 0.

$$3(0)^3 - 12(0) \overset{?}{=} 0$$
$$0 = 0$$

Let x = -2.

$$3(-2)^3 - 12(-2) \overset{?}{=} 0$$
$$3(-8) + 24 \overset{?}{=} 0$$
$$0 = 0$$

Let x = 2.

$$3(2)^3 - 12(2) \overset{?}{=} 0$$
$$3(8) - 24 \overset{?}{=} 0$$
$$0 = 0$$

Substituting 0, -2, or 2 into the original equation results each time in a true equation. The solutions are 0, -2, and 2. ☐

PRACTICE

8 Solve: $7x^3 - 63x = 0$ ∎

EXAMPLE 9 Solve: $(5x - 1)(2x^2 + 15x + 18) = 0$

Solution

$$(5x - 1)(2x^2 + 15x + 18) = 0$$
$$(5x - 1)(2x + 3)(x + 6) = 0 \quad \text{Factor the trinomial.}$$

$5x - 1 = 0$ or $2x + 3 = 0$ or $x + 6 = 0$ Set each factor equal to 0.

$5x = 1$ or $2x = -3$ or $x = -6$ Solve.

$$x = \frac{1}{5} \text{ or } \qquad x = -\frac{3}{2}$$

The solutions are $\frac{1}{5}$, $-\frac{3}{2}$, and -6. Check by replacing x with each solution in the original equation. The solutions are -6, $-\frac{3}{2}$, and $\frac{1}{5}$. ☐

PRACTICE

9 Solve: $(3x - 2)(2x^2 - 13x + 15) = 0$ ∎

EXAMPLE 10 Solve: $2x^3 - 4x^2 - 30x = 0$

Solution Begin by factoring out the GCF, $2x$.

$$2x^3 - 4x^2 - 30x = 0$$
$$2x(x^2 - 2x - 15) = 0 \quad \text{Factor out the GCF, } 2x.$$
$$2x(x - 5)(x + 3) = 0 \quad \text{Factor the quadratic.}$$

$2x = 0$ or $x - 5 = 0$ or $x + 3 = 0$ Set each factor containing a variable equal to 0.

$x = 0$ or $x = 5$ or $x = -3$ Solve.

Check by replacing x with each solution in the cubic equation. The solutions are -3, 0, and 5. ☐

PRACTICE

10 Solve: $5x^3 + 5x^2 - 30x = 0$ ∎

OBJECTIVE

3   Finding *x*-Intercepts of the Graph of a Quadratic Equation

In Chapter 3, we graphed linear equations in two variables, such as $y = 5x - 6$. Recall that to find the *x*-intercept of the graph of a linear equation, let $y = 0$ and solve for *x*. This is also how to find the *x*-intercepts of the graph of a **quadratic equation in two variables,** such as $y = x^2 - 5x + 4$.

> **EXAMPLE II**   Find the *x*-intercepts of the graph of $y = x^2 - 5x + 4$.

Solution Let $y = 0$ and solve for *x*.

$$y = x^2 - 5x + 4$$
$$0 = x^2 - 5x + 4 \qquad \text{Let } y = 0.$$
$$0 = (x - 1)(x - 4) \quad \text{Factor.}$$
$$x - 1 = 0 \quad \text{or} \quad x - 4 = 0 \quad \text{Set each factor equal to 0.}$$
$$x = 1 \quad \text{or} \quad x = 4 \quad \text{Solve.}$$

The *x*-intercepts of the graph of $y = x^2 - 5x + 4$ are $(1, 0)$ and $(4, 0)$.

The graph of $y = x^2 - 5x + 4$ is shown in the margin.

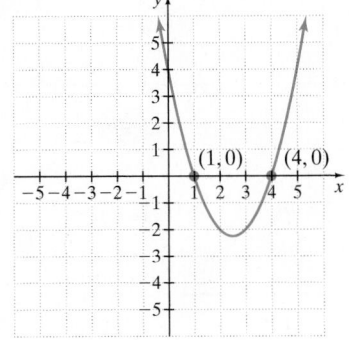

PRACTICE

11   Find the *x*-intercepts of the graph of $y = x^2 - 6x + 8$.

In general, a quadratic equation in two variables is one that can be written in the form $y = ax^2 + bx + c$ where $a \neq 0$. The graph of such an equation is called a **parabola** and will open up or down depending on the sign of *a*.

Notice that the *x*-intercepts of the graph of $y = ax^2 + bx + c$ are the real number solutions of $0 = ax^2 + bx + c$. Also, the real number solutions of $0 = ax^2 + bx + c$ are the *x*-intercepts of the graph of $y = ax^2 + bx + c$. We study more about graphs of quadratic equations in two variables in Chapter 9.

Graph of $y = ax^2 + bx + c$
x-intercepts are solutions of $0 = ax^2 + bx + c$

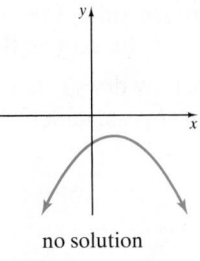

no solution

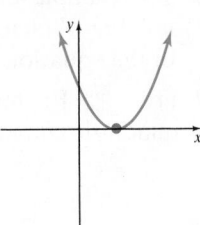

1 solution

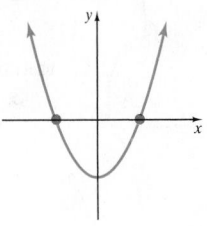

2 solutions

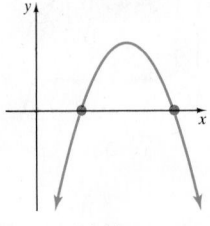

2 solutions

Graphing Calculator Explorations

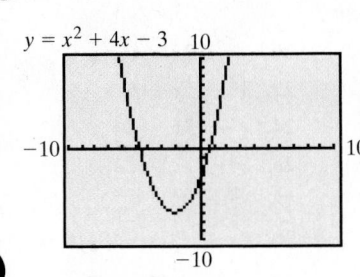

A grapher may be used to find solutions of a quadratic equation whether the related quadratic polynomial is factorable or not. For example, let's use a grapher to approximate the solutions of $0 = x^2 + 4x - 3$. To do so, graph $y_1 = x^2 + 4x - 3$. Recall that the *x*-intercepts of this graph are the solutions of $0 = x^2 + 4x - 3$.

Notice that the graph appears to have an *x*-intercept between -5 and -4 and one between 0 and 1. Many graphers contain a TRACE feature. This feature activates a graph cursor that can be used to *trace* along a graph while the corresponding *x*- and *y*-coordinates are shown on the screen. Use the TRACE feature to confirm that *x*-intercepts lie between -5 and -4 and between 0 and 1. To approximate the *x*-intercepts to the nearest tenth, use a ROOT or a ZOOM feature on your grapher or redefine the viewing window. (A ROOT feature calculates the *x*-intercept. A ZOOM feature magnifies the viewing window around a specific location such as

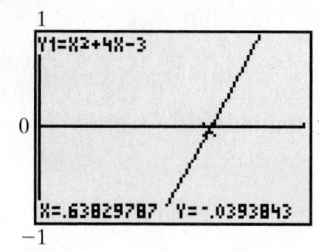

the graph cursor.) If we redefine the window to $[0, 1]$ on the x-axis and $[-1, 1]$ on the y-axis, the graph to the left is generated.

By using the TRACE feature, we can conclude that one x-intercept is approximately 0.6 to the nearest tenth. By repeating these steps for the other x-intercept, we find that it is approximately -4.6.

Use a grapher to approximate the real number solutions to the nearest tenth. If an equation has no real number solution, state so.

1. $3x^2 - 4x - 6 = 0$ **2.** $x^2 - x - 9 = 0$

3. $2x^2 + x + 2 = 0$ **4.** $-4x^2 - 5x - 4 = 0$

5. $-x^2 + x + 5 = 0$ **6.** $10x^2 + 6x - 3 = 0$

✔ Vocabulary, Readiness & Video Check

Use the choices below to fill in each blank. Not all choices will be used.

$-3, 5$ $a = 0$ or $b = 0$ 0 linear

$3, -5$ quadratic 1

1. An equation that can be written in the form $ax^2 + bx + c = 0$, with $a \neq 0$, is called a _____ equation.

2. If the product of two numbers is 0, then at least one of the numbers must be _____.

3. The solutions to $(x - 3)(x + 5) = 0$ are _____.

4. If $a \cdot b = 0$, then _____.

Martin-Gay Interactive Videos

See Video 6.6 ◉

Watch the section lecture video and answer the following questions.

OBJECTIVE
1
5. As shown in Examples 1–3, what two things have to be true in order to use the zero factor property?

OBJECTIVE
2
6. Example 4 implies that the zero factor property can be used with any number of factors on one side of the equation so long as the other side of the equation is zero. Why do you think this is true?

OBJECTIVE
3
7. From 🎞 Example 5, how does finding the x-intercepts of the graph of a quadratic equation in two variables lead to solving a quadratic equation?

6.6 Exercise Set   MyMathLab® ▶

Solve each equation. See Examples 1 through 3.

1. $(x - 6)(x - 7) = 0$
2. $(x - 10)(x - 5) = 0$
3. $(x - 2)(x + 1) = 0$
4. $(x + 4)(x - 10) = 0$
5. $(x + 9)(x + 17) = 0$
6. $(x + 11)(x + 1) = 0$
7. $x(x + 6) = 0$
8. $x(x - 7) = 0$
9. $3x(x - 8) = 0$
10. $2x(x + 12) = 0$
▶ **11.** $(2x + 3)(4x - 5) = 0$
12. $(3x - 2)(5x + 1) = 0$
13. $(2x - 7)(7x + 2) = 0$
14. $(9x + 1)(4x - 3) = 0$
15. $\left(x - \dfrac{1}{2}\right)\left(x + \dfrac{1}{3}\right) = 0$
16. $\left(x + \dfrac{2}{9}\right)\left(x - \dfrac{1}{4}\right) = 0$
17. $(x + 0.2)(x + 1.5) = 0$
18. $(x + 1.7)(x + 2.3) = 0$

Solve. See Examples 4 through 7.

19. $x^2 - 13x + 36 = 0$
20. $x^2 + 2x - 63 = 0$
▶ **21.** $x^2 + 2x - 8 = 0$
22. $x^2 - 5x + 6 = 0$
23. $x^2 - 7x = 0$
24. $x^2 - 3x = 0$
25. $x^2 - 4x = 32$
26. $x^2 - 5x = 24$
27. $x^2 = 16$
28. $x^2 = 9$
29. $(x + 4)(x - 9) = 4x$
30. $(x + 3)(x + 8) = x$
▶ **31.** $x(3x - 1) = 14$
32. $x(4x - 11) = 3$
33. $-3x^2 + 75 = 0$
34. $-2y^2 + 72 = 0$
35. $24x^2 + 44x = 8$
36. $6x^2 + 57x = 30$

Solve each equation. See Examples 8 through 10.

37. $x^3 - 12x^2 + 32x = 0$

38. $x^3 - 14x^2 + 49x = 0$

39. $(4x - 3)(16x^2 - 24x + 9) = 0$

40. $(2x + 5)(4x^2 + 20x + 25) = 0$

41. $4x^3 - x = 0$ **42.** $4y^3 - 36y = 0$

43. $32x^3 - 4x^2 - 6x = 0$ **44.** $15x^3 + 24x^2 - 63x = 0$

MIXED PRACTICE

Solve each equation. See Examples 1 through 10. (A few exercises are linear equations.)

45. $(x + 3)(x - 2) = 0$ **46.** $(x - 6)(x + 7) = 0$

47. $x^2 + 20x = 0$ **48.** $x^2 + 15x = 0$

49. $4(x - 7) = 6$ **50.** $5(3 - 4x) = 9$

51. $4y^2 - 1 = 0$ **52.** $4y^2 - 81 = 0$

53. $(2x + 3)(2x^2 - 5x - 3) = 0$

54. $(2x - 9)(x^2 + 5x - 36) = 0$

55. $x^2 - 15 = -2x$ **56.** $x^2 - 26 = -11x$

57. $30x^2 - 11x - 30 = 0$ **58.** $12x^2 + 7x - 12 = 0$

59. $5x^2 - 6x - 8 = 0$ **60.** $9x^2 + 7x = 2$

61. $6y^2 - 22y - 40 = 0$ **62.** $3x^2 - 6x - 9 = 0$

63. $(y - 2)(y + 3) = 6$ **64.** $(y - 5)(y - 2) = 28$

65. $3x^3 + 19x^2 - 72x = 0$

66. $36x^3 + x^2 - 21x = 0$

67. $x^2 + 14x + 49 = 0$

68. $x^2 + 22x + 121 = 0$

69. $12y = 8y^2$

70. $9y = 6y^2$

71. $7x^3 - 7x = 0$

72. $3x^3 - 27x = 0$

73. $3x^2 + 8x - 11 = 13 - 6x$

74. $2x^2 + 12x - 1 = 4 + 3x$

75. $3x^2 - 20x = -4x^2 - 7x - 6$

76. $4x^2 - 20x = -5x^2 - 6x - 5$

Find the x-intercepts of the graph of each equation. See Example 11.

77. $y = (3x + 4)(x - 1)$

78. $y = (5x - 3)(x - 4)$

79. $y = x^2 - 3x - 10$

80. $y = x^2 + 7x + 6$

81. $y = 2x^2 + 11x - 6$

82. $y = 4x^2 + 11x + 6$

For Exercises 83 through 88, match each equation with its graph. See Example 11.

83. $y = (x + 2)(x - 1)$ **84.** $y = (x - 5)(x + 2)$

85. $y = x(x + 3)$ **86.** $y = x(x - 4)$

87. $y = 2x^2 - 8$ **88.** $y = 2x^2 - 2$

A. 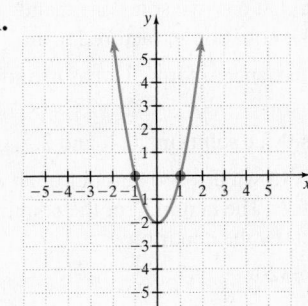   **B.**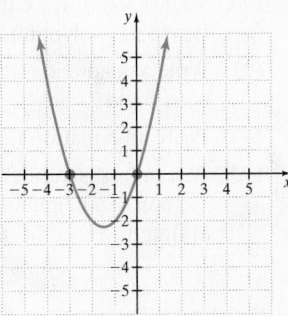

C. 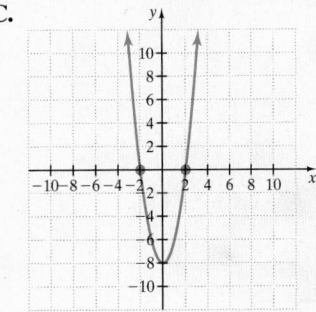   **D.**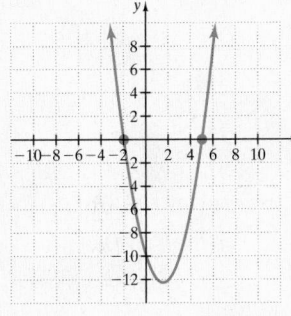

E. 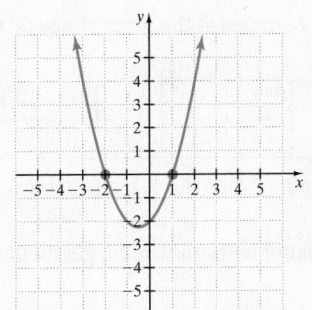   **F.**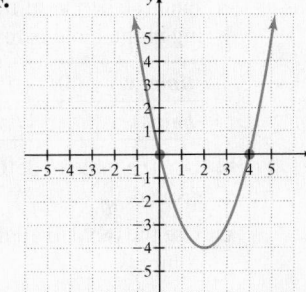

REVIEW AND PREVIEW

Perform the following operations. Write all results in lowest terms. See Section 1.3.

89. $\dfrac{3}{5} + \dfrac{4}{9}$ **90.** $\dfrac{2}{3} + \dfrac{3}{7}$

91. $\dfrac{7}{10} - \dfrac{5}{12}$ **92.** $\dfrac{5}{9} - \dfrac{5}{12}$

93. $\dfrac{7}{8} \div \dfrac{7}{15}$ **94.** $\dfrac{5}{12} - \dfrac{3}{10}$

95. $\dfrac{4}{5} \cdot \dfrac{7}{8}$ **96.** $\dfrac{3}{7} \cdot \dfrac{12}{17}$

CONCEPT EXTENSIONS

For Exercises 97 and 98, see the Concept Check in this section.

97. Explain the error and solve correctly:

$$x(x - 2) = 8$$
$$x = 8 \quad \text{or} \quad x - 2 = 8$$
$$x = 10$$

98. Explain the error and solve correctly:
$$(x - 4)(x + 2) = 0$$
$$x = -4 \quad \text{or} \quad x = 2$$

99. Write a quadratic equation that has two solutions, 6 and -1. Leave the polynomial in the equation in factored form.

100. Write a quadratic equation that has two solutions, 0 and -2. Leave the polynomial in the equation in factored form.

101. Write a quadratic equation in standard form that has two solutions, 5 and 7.

102. Write an equation that has three solutions, 0, 1, and 2.

103. A compass is accidentally thrown upward and out of an air balloon at a height of 300 feet. The height, y, of the compass at time x in seconds is given by the equation
$$y = -16x^2 + 20x + 300$$

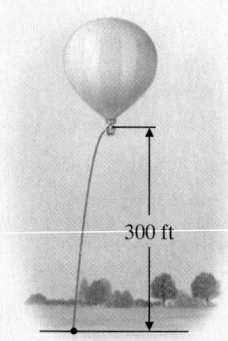
300 ft

a. Find the height of the compass at the given times by filling in the table below.

time, x	0	1	2	3	4	5	6
height, y							

b. Use the table to determine when the compass strikes the ground.

c. Use the table to approximate the maximum height of the compass.

d. Plot the points (x, y) on a rectangular coordinate system and connect them with a smooth curve. Explain your results.

104. A rocket is fired upward from the ground with an initial velocity of 100 feet per second. The height, y, of the rocket at any time x is given by the equation
$$y = -16x^2 + 100x$$

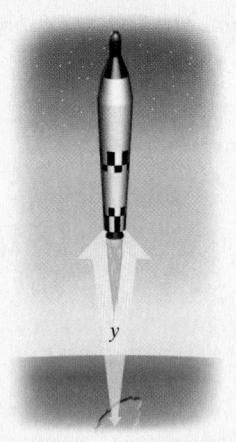
y

a. Find the height of the rocket at the given times by filling in the table below.

time, x	0	1	2	3	4	5	6	7
height, y								

b. Use the table to approximate when the rocket strikes the ground to the nearest second.

c. Use the table to approximate the maximum height of the rocket.

d. Plot the points (x, y) on a rectangular coordinate system and connect them with a smooth curve. Explain your results.

Solve each equation. First, multiply the binomials.

To solve $(x - 6)(2x - 3) = (x + 2)(x + 9)$, see below.
$$(x - 6)(2x - 3) = (x + 2)(x + 9)$$
$$2x^2 - 15x + 18 = x^2 + 11x + 18$$
$$x^2 - 26x = 0$$
$$x(x - 26) = 0$$
$$x = 0 \quad \text{or} \quad x - 26 = 0$$
$$x = 26$$

105. $(x - 3)(3x + 4) = (x + 2)(x - 6)$

106. $(2x - 3)(x + 6) = (x - 9)(x + 2)$

107. $(2x - 3)(x + 8) = (x - 6)(x + 4)$

108. $(x + 6)(x - 6) = (2x - 9)(x + 4)$

6.7 Quadratic Equations and Problem Solving

OBJECTIVE

1 Solve Problems That Can Be Modeled by Quadratic Equations.

OBJECTIVE

1 Solving Problems Modeled by Quadratic Equations

Some problems may be modeled by quadratic equations. To solve these problems, we use the same problem-solving steps that were introduced in Section 2.4. When solving these problems, keep in mind that a solution of an equation that models a problem may not be a solution of the problem. For example, a person's age or the length of a rectangle is always a positive number. Discard solutions that do not make sense as solutions of the problem.

EXAMPLE 1   **Finding Free-Fall Time**

Since the 1940s, one of the top tourist attractions in Acapulco, Mexico, is watching the La Quebrada cliff divers. The divers' platform is about 144 feet above the sea. These divers must time their descent just right, since they land in the crashing Pacific in an inlet that is at most $9\frac{1}{2}$ feet deep. Neglecting air resistance, the height h in feet of a cliff diver above the ocean after t seconds is given by the quadratic equation $h = -16t^2 + 144$.
   Find how long it takes the diver to reach the ocean.

Solution

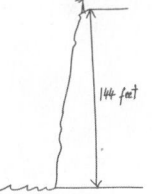

1. UNDERSTAND. Read and reread the problem. Then draw a picture of the problem.

   The equation $h = -16t^2 + 144$ models the height of the falling diver at time t. Familiarize yourself with this equation by finding the height of the diver at time $t = 1$ second and $t = 2$ seconds.

   When $t = 1$ second, the height of the diver is $h = -16(1)^2 + 144 = 128$ feet.

   When $t = 2$ seconds, the height of the diver is $h = -16(2)^2 + 144 = 80$ feet.

2. TRANSLATE. To find how long it takes the diver to reach the ocean, we want to know the value of t for which $h = 0$.

3. SOLVE.

   $$0 = -16t^2 + 144$$
   $$0 = -16(t^2 - 9) \qquad \text{Factor out } -16.$$
   $$0 = -16(t - 3)(t + 3) \quad \text{Factor completely.}$$
   $$t - 3 = 0 \quad \text{or} \quad t + 3 = 0 \quad \text{Set each factor containing a variable equal to 0.}$$
   $$t = 3 \quad \text{or} \qquad t = -3 \quad \text{Solve.}$$

4. INTERPRET.   Since the time t cannot be negative, the proposed solution is 3 seconds.

   Check: Verify that the height of the diver when t is 3 seconds is 0.

   When $t = 3$ seconds, $h = -16(3)^2 + 144 = -144 + 144 = 0.$

   State: It takes the diver 3 seconds to reach the ocean.   □

PRACTICE

1   Cliff divers also frequent the falls at Waimea Falls Park in Oahu, Hawaii. One of the popular diving spots is 64 feet high. Neglecting air resistance, the height of a diver above the pool after t seconds is $h = -16t^2 + 64$. Find how long it takes a diver to reach the pool.   ■

EXAMPLE 2   **Finding an Unknown Number**

The square of a number plus three times the number is 70. Find the number.

Solution

1. UNDERSTAND. Read and reread the problem. Suppose that the number is 5. The square of 5 is 5^2 or 25. Three times 5 is 15. Then $25 + 15 = 40$, not 70, so the number is not 5. Remember, the purpose of proposing a number, such as 5, is to understand the problem better. Now that we do, we will let $x =$ the number.

(Continued on the next page)

2. TRANSLATE.

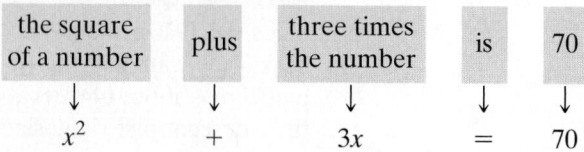

the square of a number	plus	three times the number	is	70
↓	↓	↓	↓	↓
x^2	$+$	$3x$	$=$	70

3. SOLVE.

$$x^2 + 3x = 70$$
$$x^2 + 3x - 70 = 0 \qquad \text{Subtract 70 from both sides.}$$
$$(x + 10)(x - 7) = 0 \qquad \text{Factor.}$$
$$x + 10 = 0 \quad \text{or} \quad x - 7 = 0 \qquad \text{Set each factor equal to 0.}$$
$$x = -10 \qquad\qquad x = 7 \qquad \text{Solve.}$$

4. INTERPRET.

Check: The square of -10 is $(-10)^2$, or 100. Three times -10 is $3(-10)$ or -30. Then $100 + (-30) = 70$, the correct sum, so -10 checks.

The square of 7 is 7^2 or 49. Three times 7 is $3(7)$, or 21. Then $49 + 21 = 70$, the correct sum, so 7 checks.

State: There are two numbers. They are -10 and 7.   □

PRACTICE
2   The square of a number minus eight times the number is equal to forty-eight. Find the number. ■

△ **EXAMPLE 3**   **Finding the Dimensions of a Sail**

The height of a triangular sail is 2 meters less than twice the length of the base. If the sail has an area of 30 square meters, find the length of its base and the height.

Solution

1. UNDERSTAND. Read and reread the problem. Since we are finding the length of the base and the height, we let

$$x = \text{the length of the base}$$

and since the height is 2 meters less than twice the base,

$$2x - 2 = \text{the height.}$$

An illustration is shown to the right.

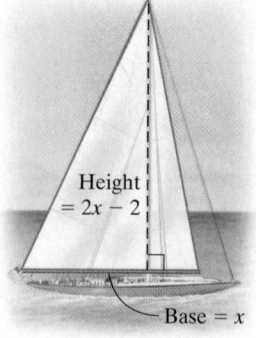

Height = $2x - 2$

Base = x

2. TRANSLATE. We are given that the area of the triangle is 30 square meters, so we use the formula for area of a triangle.

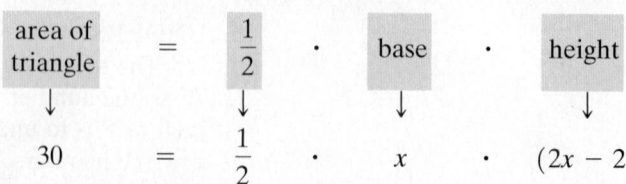

area of triangle	$=$	$\frac{1}{2}$	$\cdot$	base	$\cdot$	height
↓		↓		↓		↓
30	$=$	$\frac{1}{2}$	$\cdot$	x	$\cdot$	$(2x - 2)$

3. SOLVE. Now we solve the quadratic equation.

$$30 = \frac{1}{2}x(2x - 2)$$

$30 = x^2 - x$	Multiply.
$x^2 - x - 30 = 0$	Write in standard form.
$(x - 6)(x + 5) = 0$	Factor.
$x - 6 = 0 \quad \text{or} \quad x + 5 = 0$	Set each factor equal to 0.
$x = 6 \qquad\qquad x = -5$	

4. INTERPRET. Since x represents the length of the base, we discard the solution -5. The base of a triangle cannot be negative. The base is then 6 meters and the height is $2(6) - 2 = 10$ meters.

Check: To check this problem, we recall that $\frac{1}{2}$ base $\cdot$ height $=$ area, or

$$\frac{1}{2}(6)(10) = 30. \quad \text{The required area}$$

State: The base of the triangular sail is 6 meters and the height is 10 meters. ☐

PRACTICE
3 An engineering team from Georgia Tech earned second place in a flight competition, with their triangular shaped paper hang glider. The base of their prize-winning entry was 1 foot less than three times the height. If the area of the triangular glider wing was 210 square feet, find the dimensions of the wing. (*Source: The Technique* [Georgia Tech's newspaper], April 18, 2003)

Study the following diagrams for a review of consecutive integers.

Examples

If x is the first integer, then consecutive integers are
$x, x + 1, x + 2, \ldots$

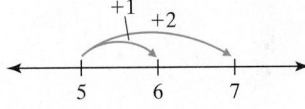

If x is the first even integer, then consecutive even integers are
$x, x + 2, x + 4, \ldots$

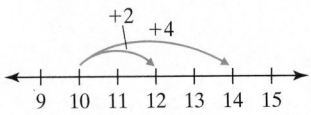

If x is the first odd integer, then consecutive odd integers are
$x, x + 2, x + 4, \ldots$

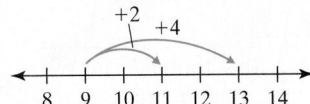

EXAMPLE 4 Finding Consecutive Even Integers

Find two consecutive even integers whose product is 34 more than their sum.

Solution

1. UNDERSTAND. Read and reread the problem. Let's just choose two consecutive even integers to help us better understand the problem. Let's choose 10 and 12. Their product is $10(12) = 120$ and their sum is $10 + 12 = 22$. The product is $120 - 22$, or 98, greater than the sum. Thus our guess is incorrect, but we have a better understanding of this example.

Let's let x and $x + 2$ be the consecutive even integers.

(Continued on the next page)

2. TRANSLATE.

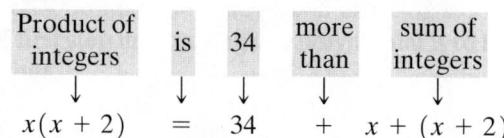

$$x(x + 2) = 34 + x + (x + 2)$$

3. SOLVE. Now we solve the equation.

$$x(x + 2) = 34 + x + (x + 2)$$

$x^2 + 2x = 34 + x + x + 2$	Multiply.
$x^2 + 2x = 2x + 36$	Combine like terms.
$x^2 - 36 = 0$	Write in standard form.
$(x + 6)(x - 6) = 0$	Factor.
$x + 6 = 0$ or $x - 6 = 0$	Set each factor equal to 0.
$x = -6$ $x = 6$	Solve.

4. INTERPRET. If $x = -6$, then $x + 2 = -6 + 2$, or -4.
If $x = 6$, then $x + 2 = 6 + 2$, or 8.

Check: $-6, -4$

$$-6(-4) \overset{?}{=} 34 + (-6) + (-4)$$

$$24 \overset{?}{=} 34 + (-10)$$

$$24 = 24 \qquad \text{True}$$

$6, 8$

$$6(8) \overset{?}{=} 34 + 6 + 8$$

$$48 \overset{?}{=} 34 + 14$$

$$48 = 48 \qquad \text{True}$$

State: The two consecutive even integers are -6 and -4 or 6 and 8.   □

PRACTICE

4   Find two consecutive integers whose product is 41 more than their sum.

The next example uses the **Pythagorean theorem** and consecutive integers. Before we review this theorem, recall that a **right triangle** is a triangle that contains a 90° or right angle. The **hypotenuse** of a right triangle is the side opposite the right angle and is the longest side of the triangle. The **legs** of a right triangle are the other sides of the triangle.

Pythagorean Theorem

In a right triangle, the sum of the squares of the lengths of the two legs is equal to the square of the length of the hypotenuse.

$$(\text{leg})^2 + (\text{leg})^2 = (\text{hypotenuse})^2 \qquad \text{or} \qquad a^2 + b^2 = c^2$$

> **Helpful Hint**
> If you use this formula, don't forget that c represents the length of the hypotenuse.

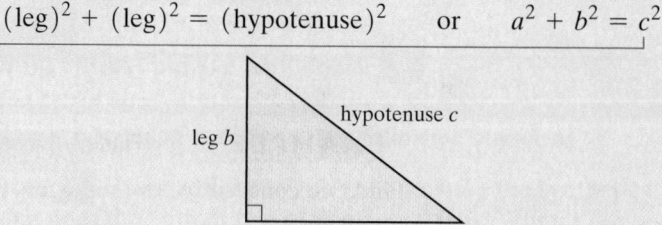

△ **EXAMPLE 5**   **Finding the Dimensions of a Triangle**

Find the lengths of the sides of a right triangle if the lengths can be expressed as three consecutive even integers.

Solution

1. **UNDERSTAND.** Read and reread the problem. Let's suppose that the length of one leg of the right triangle is 4 units. Then the other leg is the next even integer, or 6 units, and the hypotenuse of the triangle is the next even integer, or 8 units. Remember that the hypotenuse is the longest side. Let's see if a triangle with sides of these lengths forms a right triangle. To do this, we check to see whether the Pythagorean theorem holds true.

$$4^2 + 6^2 \overset{?}{=} 8^2$$
$$16 + 36 \overset{?}{=} 64$$
$$52 = 64 \quad \text{False}$$

Our proposed numbers do not check, but we now have a better understanding of the problem.

We let x, $x + 2$, and $x + 4$ be three consecutive even integers. Since these integers represent lengths of the sides of a right triangle, we have the following.

$$x = \text{one leg}$$
$$x + 2 = \text{other leg}$$
$$x + 4 = \text{hypotenuse (longest side)}$$

2. **TRANSLATE.** By the Pythagorean theorem, we have that

$$(\text{leg})^2 + (\text{leg})^2 = (\text{hypotenuse})^2$$
$$(x)^2 + (x + 2)^2 = (x + 4)^2$$

3. **SOLVE.** Now we solve the equation.

$$x^2 + (x + 2)^2 = (x + 4)^2$$
$$x^2 + x^2 + 4x + 4 = x^2 + 8x + 16 \qquad \text{Multiply.}$$
$$2x^2 + 4x + 4 = x^2 + 8x + 16 \qquad \text{Combine like terms.}$$
$$x^2 - 4x - 12 = 0 \qquad \text{Write in standard form.}$$
$$(x - 6)(x + 2) = 0 \qquad \text{Factor.}$$
$$x - 6 = 0 \quad \text{or} \quad x + 2 = 0 \qquad \text{Set each factor equal to 0.}$$
$$x = 6 \qquad\qquad x = -2$$

4. **INTERPRET.** We discard $x = -2$ since length cannot be negative. If $x = 6$, then $x + 2 = 8$ and $x + 4 = 10$.

Check: Verify that

$$(\text{leg})^2 + (\text{leg})^2 = (\text{hypotenuse})^2$$
$$6^2 + 8^2 \overset{?}{=} 10^2$$
$$36 + 64 \overset{?}{=} 100$$
$$100 = 100 \qquad\qquad \text{True}$$

State: The sides of the right triangle have lengths 6 units, 8 units, and 10 units.

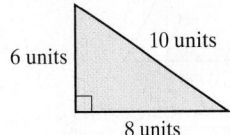

PRACTICE

5 Find the dimensions of a right triangle where the second leg is 1 unit less than double the first leg, and the hypotenuse is 1 unit more than double the length of the first leg.

✔ Vocabulary, Readiness & Video Check

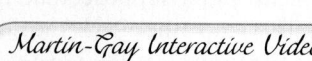

Martin-Gay Interactive Videos

See Video 6.7 ⊙

Watch the section lecture video and answer the following question.

OBJECTIVE
1

1. In each of Examples 1–3, why aren't both solutions of the translated equation accepted as solutions of the application?

6.7 Exercise Set MyMathLab® ▶

MIXED PRACTICE
See Examples 1 through 5 for all exercises.

TRANSLATING

For Exercises 1 through 6, represent each given condition using a single variable, x.

△ **1.** The length and width of a rectangle whose length is 4 centimeters more than its width

△ **2.** The length and width of a rectangle whose length is twice its width

3. Two consecutive odd integers

4. Two consecutive even integers

△ **5.** The base and height of a triangle whose height is one more than four times its base

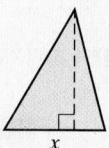

△ **6.** The base and height of a trapezoid whose base is three less than five times its height

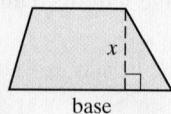

Use the information given to find the dimensions of each figure.

△ **7.** The *area* of the square is 121 square units. Find the length of its sides.

△ **8.** The *area* of the rectangle is 84 square inches. Find its length and width.

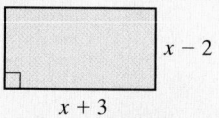

△ **9.** The *perimeter* of the quadrilateral is 120 centimeters. Find the lengths of the sides.

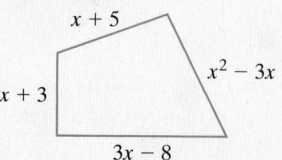

△ **10.** The *perimeter* of the triangle is 85 feet. Find the lengths of its sides.

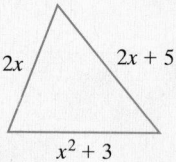

△ **11.** The *area* of the parallelogram is 96 square miles. Find its base and height.

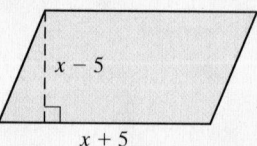

△ **12.** The *area* of the circle is 25π square kilometers. Find its radius.

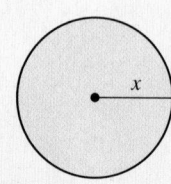

Solve.

13. An object is thrown upward from the top of an 80-foot building with an initial velocity of 64 feet per second. The height h of the object after t seconds is given by the quadratic equation $h = -16t^2 + 64t + 80$. When will the object hit the ground?

14. A hang glider pilot accidentally drops her compass from the top of a 400-foot cliff. The height h of the compass after t seconds is given by the quadratic equation $h = -16t^2 + 400$. When will the compass hit the ground?

15. The width of a rectangle is 7 centimeters less than twice its length. Its area is 30 square centimeters. Find the dimensions of the rectangle.

△ **16.** The length of a rectangle is 9 inches more than its width. Its area is 112 square inches. Find the dimensions of the rectangle.

The equation $D = \frac{1}{2}n(n - 3)$ gives the number of diagonals D for a polygon with n sides. For example, a polygon with 6 sides has $D = \frac{1}{2} \cdot 6(6 - 3)$ or $D = 9$ diagonals. (See if you can count all 9 diagonals. Some are shown in the figure.) Use this equation, $D = \frac{1}{2}n(n - 3)$, for Exercises 17 through 20.

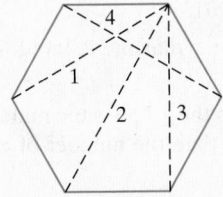

△ **17.** Find the number of diagonals for a polygon that has 12 sides.

△ **18.** Find the number of diagonals for a polygon that has 15 sides.

△ **19.** Find the number of sides n for a polygon that has 35 diagonals.

△ **20.** Find the number of sides n for a polygon that has 14 diagonals.

Solve.

21. The sum of a number and its square is 132. Find the number(s).

▶ **22.** The sum of a number and its square is 182. Find the number(s).

23. The product of two consecutive room numbers is 210. Find the room numbers.

▶ **24.** The product of two consecutive page numbers is 420. Find the page numbers.

25. A ladder is leaning against a building so that the distance from the ground to the top of the ladder is one foot less than the length of the ladder. Find the length of the ladder if the distance from the bottom of the ladder to the building is 5 feet.

26. Use the given figure to find the length of the guy wire.

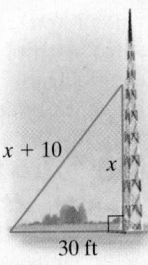

△ **27.** If the sides of a square are increased by 3 inches, the area becomes 64 square inches. Find the length of the sides of the original square.

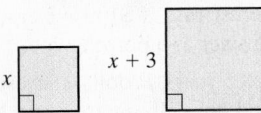

△ **28.** If the sides of a square are increased by 5 meters, the area becomes 100 square meters. Find the length of the sides of the original square.

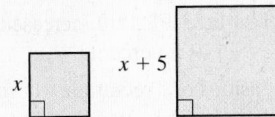

△ **29.** One leg of a right triangle is 4 millimeters longer than the smaller leg and the hypotenuse is 8 millimeters longer than the smaller leg. Find the lengths of the sides of the triangle.

△ **30.** One leg of a right triangle is 9 centimeters longer than the other leg and the hypotenuse is 45 centimeters. Find the lengths of the legs of the triangle.

△ **31.** The length of the base of a triangle is twice its height. If the area of the triangle is 100 square kilometers, find the height.

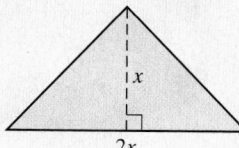

△ **32.** The height of a triangle is 2 millimeters less than the base. If the area is 60 square millimeters, find the base.

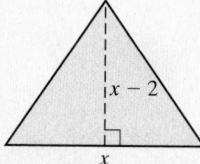

△ **33.** Find the length of the shorter leg of a right triangle if the longer leg is 12 feet more than the shorter leg and the hypotenuse is 12 feet less than twice the shorter leg.

△ **34.** Find the length of the shorter leg of a right triangle if the longer leg is 10 miles more than the shorter leg and the hypotenuse is 10 miles less than twice the shorter leg.

35. An object is dropped from 39 feet below the tip of the pinnacle atop one of the 1483-foot-tall Petronas Twin Towers in Kuala Lumpur, Malaysia. (*Source:* Council on Tall Buildings and Urban Habitat) The height h of the object after t seconds is given by the equation $h = -16t^2 + 1444$. Find how many seconds pass before the object reaches the ground.

36. An object is dropped from the top of 311 South Wacker Drive, a 961-foot-tall office building in Chicago. (*Source:* Council on Tall Buildings and Urban Habitat) The height h of the object after t seconds is given by the equation $h = -16t^2 + 961$. Find how many seconds pass before the object reaches the ground.

37. At the end of 2 years, P dollars invested at an interest rate r compounded annually increases to an amount, A dollars, given by

$$A = P(1 + r)^2$$

Find the interest rate if $100 increased to $144 in 2 years. Write your answer as a percent.

38. At the end of 2 years, P dollars invested at an interest rate r compounded annually increases to an amount, A dollars, given by

$$A = P(1 + r)^2$$

Find the interest rate if $2000 increased to $2420 in 2 years. Write your answer as a percent.

△ **39.** Find the dimensions of a rectangle whose width is 7 miles less than its length and whose area is 120 square miles.

△ **40.** Find the dimensions of a rectangle whose width is 2 inches less than half its length and whose area is 160 square inches.

41. If the cost, C, for manufacturing x units of a certain product is given by $C = x^2 - 15x + 50$, find the number of units manufactured at a cost of $9500.

42. If a switchboard handles n telephones, the number C of telephone connections it can make simultaneously is given by the equation $C = \dfrac{n(n - 1)}{2}$. Find how many telephones are handled by a switchboard making 120 telephone connections simultaneously.

REVIEW AND PREVIEW

The following double-line graph shows a comparison of the number of annual visitors (in millions) to Glacier National Park and Gettysburg National Military Park for the years shown. Use this graph to answer Exercises 43 through 50. See Section 3.1.

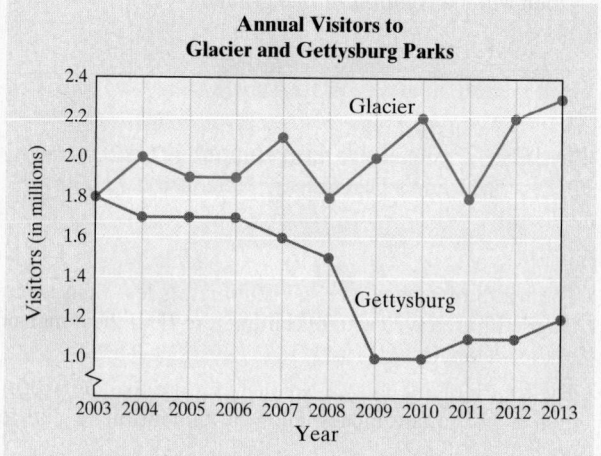

43. Approximate the number of visitors to Glacier National Park in 2011.

44. Approximate the number of visitors to Gettysburg National Military Park in 2011.

45. Approximate the number of visitors to Glacier National Park in 2013.

46. Approximate the number of visitors to Gettysburg National Military Park in 2013.

47. Determine the year that the colored lines in this graph intersect.

48. For what years on the graph is the number of visitors to Glacier Park greater than the number of visitors to Gettysburg Park?

49. In your own words, explain the meaning of the point of intersection in the graph.

50. Describe the trends shown in this graph and speculate as to why these trends have occurred.

Write each fraction in simplest form. See Section 1.3.

51. $\dfrac{20}{35}$ **52.** $\dfrac{24}{32}$ **53.** $\dfrac{27}{18}$

54. $\dfrac{15}{27}$ **55.** $\dfrac{14}{42}$ **56.** $\dfrac{45}{50}$

CONCEPT EXTENSIONS

57. Two boats travel at right angles to each other after leaving the same dock at the same time. One hour later, the boats are 17 miles apart. If one boat travels 7 miles per hour faster than the other boat, find the rate of each boat.

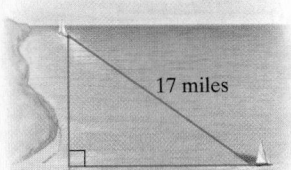

17 miles

58. The side of a square equals the width of a rectangle. The length of the rectangle is 6 meters longer than its width. The sum of the areas of the square and the rectangle is 176 square meters. Find the side of the square.

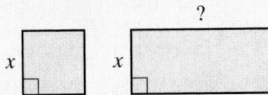

59. The sum of two numbers is 20, and the sum of their squares is 218. Find the numbers.

60. The sum of two numbers is 25, and the sum of their squares is 325. Find the numbers.

61. A rectangular pool is surrounded by a walk 4 meters wide. The pool is 6 meters longer than its width. If the total area of the pool and walk is 576 square meters more than the area of the pool, find the dimensions of the pool.

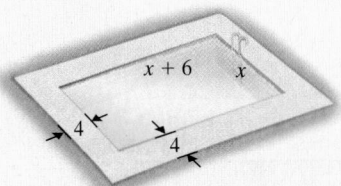

62. A rectangular garden is surrounded by a walk of uniform width. The area of the garden is 180 square yards. If the dimensions of the garden plus the walk are 16 yards by 24 yards, find the width of the walk.

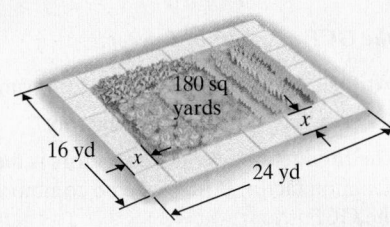

63. Write down two numbers whose sum is 10. Square each number and find the sum of the squares. Use this work to write a word problem like Exercise 59. Then give the word problem to a classmate to solve.

64. Write down two numbers whose sum is 12. Square each number and find the sum of the squares. Use this work to write a word problem like Exercise 60. Then give the word problem to a classmate to solve.

Chapter 6 **Vocabulary Check**

Fill in each blank with one of the words or phrases listed below. Not all choices will be used and some choices may be used more than once.

factoring	quadratic equation	perfect square trinomial	0
greatest common factor	hypotenuse	sum of two cubes	1
difference of two cubes	difference of two squares	triangle	leg

1. An equation that can be written in the form $ax^2 + bx + c = 0$ (with a not 0) is called a(n) _____.

2. _____ is the process of writing an expression as a product.

3. The _____ of a list of terms is the product of all common factors.

4. A trinomial that is the square of some binomial is called a(n) _____.

5. The expression $a^2 - b^2$ is called a(n) _____.

6. The expression $a^3 - b^3$ is called a(n) _____.

7. The expression $a^3 + b^3$ is called a(n) _____.

8. By the zero factor property, if the product of two numbers is 0, then at least one of the numbers must be _____.

9. In a right triangle, the side opposite the right angle is called the _____.

10. In a right triangle, each side adjacent to the right angle is called a _____.

11. The Pythagorean theorem states that $(\text{leg})^2 + (\text{leg})^2 = ($ _____ $)^2$.

Chapter 6 Highlights

DEFINITIONS AND CONCEPTS	EXAMPLES

Section 6.1 The Greatest Common Factor and Factoring by Grouping

Factoring is the process of writing an expression as a product.

Factor:
$$6 = 2 \cdot 3$$
$$x^2 + 5x + 6 = (x + 2)(x + 3)$$

To Find the GCF of a List of Integers

Step 1. Write each number as a product of primes.

Step 2. Identify the common prime factors.

Step 3. The product of all common factors is the greatest common factor. If there are no common prime factors, the GCF is 1.

Find the GCF of 12, 36, and 48.

$$12 = 2 \cdot 2 \cdot 3$$
$$36 = 2 \cdot 2 \cdot 3 \cdot 3$$
$$48 = 2 \cdot 2 \cdot 2 \cdot 2 \cdot 3$$
$$\text{GCF} = 2 \cdot 2 \cdot 3 = 12$$

The GCF of a list of common variables raised to powers is the variable raised to the smallest exponent in the list.

The GCF of z^5, z^3, and z^{10} is z^3.

The GCF of a list of terms is the product of all common factors.

Find the GCF of $8x^2y$, $10x^3y^2$, and $26x^2y^3$.

The GCF of 8, 10, and 26 is 2.

The GCF of x^2, x^3, and x^2 is x^2.

The GCF of y, y^2, and y^3 is y.

The GCF of the terms is $2x^2y$.

To Factor by Grouping

Step 1. Arrange the terms so that the first two terms have a common factor and the last two have a common factor.

Step 2. For each pair of terms, factor out the pair's GCF.

Step 3. If there is now a common binomial factor, factor it out.

Step 4. If there is no common binomial factor, begin again, rearranging the terms differently. If no rearrangement leads to a common binomial factor, the polynomial cannot be factored by grouping.

Factor: $10ax + 15a - 6xy - 9y$

Step 1. $10ax + 15a - 6xy - 9y$

Step 2. $5a(2x + 3) - 3y(2x + 3)$

Step 3. $(2x + 3)(5a - 3y)$

Section 6.2 Factoring Trinomials of the Form $x^2 + bx + c$

The product of these numbers is c.

$$x^2 + bx + c = (x + \square)(x + \square)$$

The sum of these numbers is b.

Factor: $x^2 + 7x + 12$

$$3 + 4 = 7 \qquad 3 \cdot 4 = 12$$

$$x^2 + 7x + 12 = (x + 3)(x + 4)$$

DEFINITIONS AND CONCEPTS	EXAMPLES

Section 6.3   Factoring Trinomials of the Form $ax^2 + bx + c$ and Perfect Square Trinomials

To factor $ax^2 + bx + c$, try various combinations of factors of ax^2 and c until a middle term of bx is obtained when checking.

Factor: $3x^2 + 14x - 5$

Factors of $3x^2$: $3x, x$

Factors of -5: $-1, 5$ and $1, -5$.

$$(3x - 1)(x + 5)$$
$$-1x$$
$$+ 15x$$
$$\overline{\quad 14x \quad} \text{ **Correct** middle term}$$

A **perfect square trinomial** is a trinomial that is the square of some binomial.

Perfect square trinomial = square of binomial
$$x^2 + 4x + 4 = (x + 2)^2$$
$$25x^2 - 10x + 1 = (5x - 1)^2$$

Factoring Perfect Square Trinomials

$$a^2 + 2ab + b^2 = (a + b)^2$$
$$a^2 - 2ab + b^2 = (a - b)^2$$

Factor.

$$x^2 + 6x + 9 = x^2 + 2 \cdot x \cdot 3 + 3^2 = (x + 3)^2$$
$$4x^2 - 12x + 9 = (2x)^2 - 2 \cdot 2x \cdot 3 + 3^2 = (2x - 3)^2$$

Section 6.4   Factoring Trinomials of the Form $ax^2 + bx + c$ by Grouping

To Factor $ax^2 + bx + c$ by Grouping

Step 1. Find two numbers whose product is $a \cdot c$ and whose sum is b.

Step 2. Rewrite bx, using the factors found in Step 1.

Step 3. Factor by grouping.

Factor: $3x^2 + 14x - 5$

Step 1. Find two numbers whose product is $3 \cdot (-5)$ or -15 and whose sum is 14. They are 15 and -1.

Step 2. $3x^2 + 14x - 5$
$= 3x^2 + 15x - 1x - 5$

Step 3. $= 3x(x + 5) - 1(x + 5)$
$= (x + 5)(3x - 1)$

Section 6.5   Factoring Binomials

Difference of Squares

$$a^2 - b^2 = (a + b)(a - b)$$

Sum or Difference of Cubes

$$a^3 + b^3 = (a + b)(a^2 - ab + b^2)$$
$$a^3 - b^3 = (a - b)(a^2 + ab + b^2)$$

Factor.

$$x^2 - 9 = x^2 - 3^2 = (x + 3)(x - 3)$$

$$y^3 + 8 = y^3 + 2^3 = (y + 2)(y^2 - 2y + 4)$$
$$125z^3 - 1 = (5z)^3 - 1^3 = (5z - 1)(25z^2 + 5z + 1)$$

Integrated Review—Choosing a Factoring Strategy

To Factor a Polynomial

Step 1. Factor out the GCF.

Step 2. **a.** If two terms
  i. $a^2 - b^2 = (a + b)(a - b)$
  ii. $a^3 - b^3 = (a - b)(a^2 + ab + b^2)$
  iii. $a^3 + b^3 = (a + b)(a^2 - ab + b^2)$
 b. If three terms
  i. $a^2 + 2ab + b^2 = (a + b)^2$
  $a^2 - 2ab + b^2 = (a - b)^2$
  ii. Methods in Sections 6.2 through 6.4
 c. If four or more terms, try factoring by grouping.

Factor: $2x^4 - 6x^2 - 8$

Step 1. $2x^4 - 6x^2 - 8 = 2(x^4 - 3x^2 - 4)$

Step 2. b. ii. $= 2(x^2 + 1)(x^2 - 4)$

(continued)

DEFINITIONS AND CONCEPTS	EXAMPLES

Integrated Review—Choosing a Factoring Strategy (continued)

Step 3. See if any factors can be factored further.

Step 4. Check by multiplying.

Step 3. $= 2(x^2 + 1)(x + 2)(x - 2)$

Step 4. Check by multiplying.

$$2(x^2 + 1)(x + 2)(x - 2) = 2(x^2 + 1)(x^2 - 4)$$
$$= 2(x^4 - 3x^2 - 4)$$
$$= 2x^4 - 6x^2 - 8$$

Section 6.6   Solving Quadratic Equations by Factoring

A **quadratic equation** is an equation that can be written in the form $ax^2 + bx + c = 0$ with a not 0.

The form $ax^2 + bx + c = 0$ is called the **standard form** of a quadratic equation.

Quadratic Equation	*Standard Form*
$x^2 = 16$	$x^2 - 16 = 0$
$y = -2y^2 + 5$	$2y^2 + y - 5 = 0$

Zero Factor Property

If a and b are real numbers and if $ab = 0$, then $a = 0$ or $b = 0$.

If $(x + 3)(x - 1) = 0$, then $x + 3 = 0$ or $x - 1 = 0$

To Solve Quadratic Equations by Factoring

Step 1. Write the equation in standard form: $ax^2 + bx + c = 0$.

Step 2. Factor the quadratic.

Step 3. Set each factor containing a variable equal to 0.

Step 4. Solve the equations.

Solve: $3x^2 = 13x - 4$

Step 1. $3x^2 - 13x + 4 = 0$

Step 2. $(3x - 1)(x - 4) = 0$

Step 3. $3x - 1 = 0$ or $x - 4 = 0$

Step 4. $3x = 1$ or $x = 4$

$$x = \frac{1}{3}$$

Step 5. Check in the original equation.

Step 5. Check both $\frac{1}{3}$ and 4 in the original equation.

Section 6.7   Quadratic Equations and Problem Solving

Problem-Solving Steps

A garden is in the shape of a rectangle whose length is two feet more than its width. If the area of the garden is 35 square feet, find its dimensions.

1. UNDERSTAND the problem.

1. Read and reread the problem. Guess a solution and check your guess.

Let x be the width of the rectangular garden. Then $x + 2$ is the length.

2. TRANSLATE.

2. In words: length $\cdot$ width $=$ area

Translate: $(x + 2)$ $\cdot$ x $=$ 35

DEFINITIONS AND CONCEPTS	EXAMPLES

Section 6.7   Quadratic Equations and Problem Solving (continued)

3. SOLVE.

4. INTERPRET.

3.
$$(x + 2)x = 35$$
$$x^2 + 2x - 35 = 0$$
$$(x - 5)(x + 7) = 0$$
$$x - 5 = 0 \quad \text{or} \quad x + 7 = 0$$
$$x = 5 \quad \text{or} \quad x = -7$$

4. Discard the solution of -7 since x represents width.

Check: If x is 5 feet then $x + 2 = 5 + 2 = 7$ feet. The area of a rectangle whose width is 5 feet and whose length is 7 feet is (5 feet)(7 feet) or 35 square feet.

State: The garden is 5 feet by 7 feet.

Chapter 6 Review

(6.1) Complete the factoring.

1. $6x^2 - 15x = 3x(\quad)$
2. $2x^3y + 6x^2y^2 + 8xy^3 = 2xy(\quad)$

Factor the GCF from each polynomial.

3. $20x^2 + 12x$
4. $6x^2y^2 - 3xy^3$
5. $3x(2x + 3) - 5(2x + 3)$
6. $5x(x + 1) - (x + 1)$

Factor each polynomial by grouping.

7. $3x^2 - 3x + 2x - 2$
8. $3a^2 + 9ab + 3b^2 + ab$
9. $10a^2 + 5ab + 7b^2 + 14ab$
10. $6x^2 + 10x - 3x - 5$

(6.2) Factor each trinomial.

11. $x^2 + 6x + 8$
12. $x^2 - 11x + 24$
13. $x^2 + x + 2$
14. $x^2 - x + 2$
15. $x^2 + 4xy - 12y^2$
16. $x^2 + 8xy + 15y^2$
17. $72 - 18x - 2x^2$
18. $32 + 12x - 4x^2$
19. $10a^3 - 110a^2 + 100a$
20. $5y^3 - 50y^2 + 120y$
21. To factor $x^2 + 2x - 48$, think of two numbers whose product is ____ and whose sum is ____.
22. What is the first step in factoring $3x^2 + 15x + 30$?

(6.3) or (6.4) Factor each trinomial.

23. $2x^2 + 13x + 6$
24. $4x^2 + 4x - 3$
25. $6x^2 + 5xy - 4y^2$
26. $18x^2 - 9xy - 20y^2$
27. $10y^3 + 25y^2 - 60y$
28. $60y^3 - 39y^2 + 6y$
29. $18x^2 - 60x + 50$
30. $4x^2 - 28xy + 49y^2$

(6.5) Factor each binomial.

31. $4x^2 - 9$
32. $9t^2 - 25s^2$
33. $16x^2 + y^2$
34. $x^3 - 8y^3$
35. $8x^3 + 27$
36. $2x^3 + 8x$
37. $54 - 2x^3y^3$
38. $9x^2 - 4y^2$
39. $16x^4 - 1$
40. $x^4 + 16$

(6.6) Solve the following equations.

41. $(x + 6)(x - 2) = 0$
42. $3x(x + 1)(7x - 2) = 0$
43. $4(5x + 1)(x + 3) = 0$
44. $x^2 + 8x + 7 = 0$
45. $x^2 - 2x - 24 = 0$
46. $x^2 + 10x = -25$

47. $x(x - 10) = -16$

48. $(3x - 1)(9x^2 - 6x + 1) = 0$

49. $56x^2 - 5x - 6 = 0$

50. $20x^2 - 7x - 6 = 0$

51. $5(3x + 2) = 4$

52. $6x^2 - 3x + 8 = 0$

53. $12 - 5t = -3$

54. $5x^3 + 20x^2 + 20x = 0$

55. $4t^3 - 5t^2 - 21t = 0$

56. Write a quadratic equation that has the two solutions 4 and 5.

(6.7) *Use the given information to choose the correct dimensions.*

△ 57. The perimeter of a rectangle is 24 inches. The length is twice the width. Find the dimensions of the rectangle.

   a. 5 inches by 7 inches

   b. 5 inches by 10 inches

   c. 4 inches by 8 inches

   d. 2 inches by 10 inches

△ 58. The area of a rectangle is 80 square meters. The length is one more than three times the width. Find the dimensions of the rectangle.

   a. 8 meters by 10 meters

   b. 4 meters by 13 meters

   c. 4 meters by 20 meters

   d. 5 meters by 16 meters

Use the given information to find the dimensions of each figure.

△ 59. The *area* of the square is 81 square units. Find the length of a side.

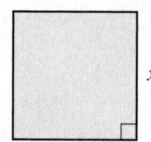

△ 60. The *perimeter* of the quadrilateral is 47 units. Find the lengths of the sides.

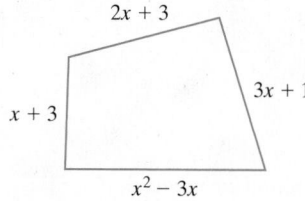

△ 61. A flag for a local organization is in the shape of a rectangle whose length is 15 inches less than twice its width. If the area of the flag is 500 square inches, find its dimensions.

△ 62. The base of a triangular sail is four times its height. If the area of the triangle is 162 square yards, find the base.

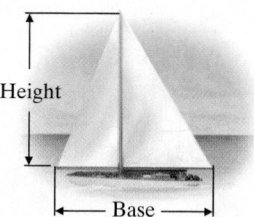

63. Find two consecutive positive integers whose product is 380.

64. Find two consecutive positive even integers whose product is 440.

65. A rocket is fired from the ground with an initial velocity of 440 feet per second. Its height h after t seconds is given by the equation

$$h = -16t^2 + 440t$$

2800 ft

   a. Find how many seconds pass before the rocket reaches a height of 2800 feet. Explain why two answers are obtained.

   b. Find how many seconds pass before the rocket reaches the ground again.

△ 66. An architect's squaring instrument is in the shape of a right triangle. Find the length of the longer leg of the right triangle if the hypotenuse is 8 centimeters longer than the longer leg and the shorter leg is 8 centimeters shorter than the longer leg.

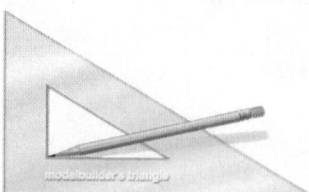

MIXED REVIEW

Factor completely.

67. $7x - 63$

68. $11x(4x - 3) - 6(4x - 3)$

69. $m^2 - \dfrac{4}{25}$

70. $3x^3 - 4x^2 + 6x - 8$

71. $xy + 2x - y - 2$

72. $2x^2 + 2x - 24$

73. $3x^3 - 30x^2 + 27x$

74. $4x^2 - 81$

75. $2x^2 - 18$

76. $16x^2 - 24x + 9$

77. $5x^2 + 20x + 20$

78. $2x^2 + 5x - 12$

79. $4x^2y - 6xy^2$

80. $125x^3 + 27$

81. $24x^2 - 3x - 18$

82. $(x + 7)^2 - y^2$

83. $x^2(x + 3) - 4(x + 3)$

84. $54a^3b - 2b$

Write the perimeter of each figure as a simplified polynomial. Then factor each polynomial.

△ **85.**

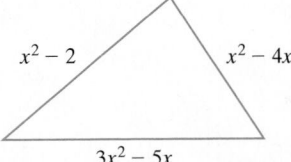

△ **86.**

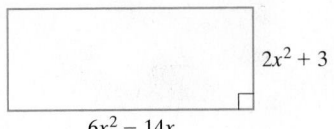

Solve.

87. $2x^2 - x - 28 = 0$

88. $x^2 - 2x = 15$

89. $2x(x + 7)(x + 4) = 0$

90. $x(x - 5) = -6$

91. $x^2 = 16x$

Solve.

△ **92.** The perimeter of the following triangle is 48 inches. Find the lengths of its sides.

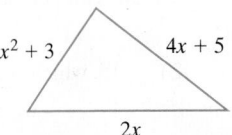

93. The width of a rectangle is 4 inches less than its length. Its area is 12 square inches. Find the dimensions of the rectangle.

94. A 6-foot-tall person drops an object from the top of the Westin Peachtree Plaza in Atlanta, Georgia. The Westin building is 723 feet tall. (*Source: World Almanac* research) The height h of the object after t seconds is given by the equation $h = -16t^2 + 729$. Find how many seconds pass before the object reaches the ground.

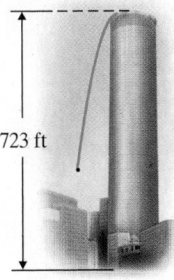

723 ft

Write an expression for the area of the shaded region. Then write the expression as a factored polynomial.

△ **95.**

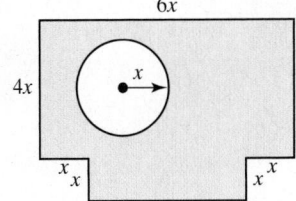

1c

Chapter 6 Getting Ready for the Test

*All the exercises below are **Multiple Choice**. Choose the correct letter. Also, letters may be used more than once.*

▶ **1.** The greatest common factor of the terms of $10x^4 - 70x^3 + 2x^2 - 14x$ is

   A. $2x^2$ **B.** $2x$ **C.** $7x^2$ **D.** $7x$

▶ **2.** Choose the expression that is NOT a factored form of $9y^3 - 18y^2$.

   A. $9(y^3 - 2y^2)$ **B.** $9y(y^2 - 2y)$ **C.** $9y^2(y - 2)$ **D.** $9 \cdot y^3 - 18 \cdot y^2$

Identify each expression as:

   A. A factored expression or   **B.** Not a factored expression

▶ **3.** $(x - 1)(x + 5)$

▶ **4.** $z(z + 12)(z - 12)$

▶ **5.** $y(x - 6) + 1(x - 6)$

▶ **6.** $m \cdot m - 5 \cdot 5$

▶ 7. Choose the correct factored form for $4x^2 + 16$ or select "can't be factored."

    A. can't be factored   **B.** $4(x^2 + 4)$ **C.** $4(x + 2)^2$ **D.** $4(x + 2)(x - 2)$

▶ 8. Which of the binomials can't be factored using real numbers?

    A. $x^2 + 64$ **B.** $x^2 - 64$ **C.** $x^3 + 64$ **D.** $x^3 - 64$

▶ 9. To solve $x(x + 2) = 15$, which is an incorrect next step?

    A. $x^2 + 2x = 15$ **B.** $x(x + 2) - 15 = 0$ **C.** $x = 15$ and $x + 2 = 15$

Chapter 6 Test   MyMathLab® You Tube

Factor each polynomial completely. If a polynomial cannot be factored, write "prime."

▶ 1. $x^2 + 11x + 28$

▶ 2. $49 - m^2$

▶ 3. $y^2 + 22y + 121$

▶ 4. $4(a + 3) - y(a + 3)$

▶ 5. $x^2 + 4$

▶ 6. $y^2 - 8y - 48$

▶ 7. $x^2 + x - 10$

▶ 8. $9x^3 + 39x^2 + 12x$

▶ 9. $3a^2 + 3ab - 7a - 7b$

▶ 10. $3x^2 - 5x + 2$

▶ 11. $x^2 + 14xy + 24y^2$

▶ 12. $180 - 5x^2$

▶ 13. $6t^2 - t - 5$

▶ 14. $xy^2 - 7y^2 - 4x + 28$

▶ 15. $x - x^5$

▶ 16. $-xy^3 - x^3y$

▶ 17. $64x^3 - 1$

▶ 18. $8y^3 - 64$

Solve each equation.

▶ 19. $(x - 3)(x + 9) = 0$

▶ 20. $x^2 + 5x = 14$

▶ 21. $x(x + 6) = 7$

▶ 22. $3x(2x - 3)(3x + 4) = 0$

▶ 23. $5t^3 - 45t = 0$

▶ 24. $t^2 - 2t - 15 = 0$

▶ 25. $6x^2 = 15x$

Solve each problem.

▶ △ 26. A deck for a home is in the shape of a triangle. The length of the base of the triangle is 9 feet longer than its altitude. If the area of the triangle is 68 square feet, find the length of the base.

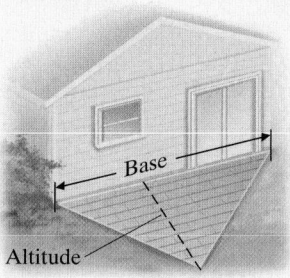

▶ 27. The sum of two numbers is 17 and the sum of their squares is 145. Find the numbers.

▶ 28. An object is dropped from the top of the Woolworth Building on Broadway in New York City. The height h of the object after t seconds is given by the equation

$$h = -16t^2 + 784$$

Find how many seconds pass before the object reaches the ground.

▶ △ 29. Find the lengths of the sides of a right triangle if the hypotenuse is 10 centimeters longer than the shorter leg and 5 centimeters longer than the longer leg.

Chapter 6 Cumulative Review

1. Translate each sentence into a mathematical statement.

    a. Nine is less than or equal to eleven.

    b. Eight is greater than one.

    c. Three is not equal to four.

2. Insert $<$ or $>$ in the space to make each statement true.

    a. $|-5|$ $|-3|$ **b.** $|0|$ $|-2|$

3. Simplify each fraction (write it in lowest terms).

    a. $\dfrac{42}{49}$ **b.** $\dfrac{11}{27}$ **c.** $\dfrac{88}{20}$

4. Evaluate $\dfrac{x}{y} + 5x$ if $x = 20$ and $y = 10$.

5. Simplify: $\dfrac{8 + 2 \cdot 3}{2^2 - 1}$

6. Evaluate $\dfrac{x}{y} + 5x$ if $x = -20$ and $y = 10$.

7. Add.

    a. $3 + (-7) + (-8)$

    b. $[7 + (-10)] + [-2 + |-4|]$

8. Evaluate $\dfrac{x}{y} + 5x$ if $x = -20$ and $y = -10$.

9. Multiply.
   a. $(-8)(4)$
   b. $14(-1)$
   c. $(-9)(-10)$

10. Simplify: $5 - 2(3x - 7)$

11. Simplify each expression by combining like terms.
    a. $7x - 3x$
    b. $10y^2 + y^2$
    c. $8x^2 + 2x - 3x$
    d. $9n^2 - 5n^2 + n^2$

12. Solve: $0.8y + 0.2(y - 1) = 1.8$

Solve.

13. $\dfrac{y}{7} = 20$

14. $\dfrac{x}{-7} = -4$

15. $-3x = 33$

16. $-\dfrac{2}{3}x = -22$

17. $8(2 - t) = -5t$

18. $-z = \dfrac{7z + 3}{5}$

19. Balsa wood sticks are commonly used to build models (for example, bridge models). A 48-inch balsa wood stick is to be cut into two pieces so that the longer piece is 3 times the shorter. Find the length of each piece.

20. Solve $3x + 9 \le 5(x - 1)$. Write the solution set using interval notation.

21. Graph the linear equation: $y = -\dfrac{1}{3}x + 2$

22. Is the ordered pair $(-1, 2)$ a solution of $-7x - 8y = -9$?

23. Find the slope and the y-intercept of the line whose equation is $3x - 4y = 4$.

24. Find the slope of the line through $(5, -6)$ and $(5, 2)$.

25. Evaluate each expression for the given value of x.
    a. $2x^3$; x is 5
    b. $\dfrac{9}{x^2}$; x is -3

26. Find the slope and y-intercept of the line whose equation is $7x - 3y = 2$.

27. Find the degree of each term.
    a. $3x^2$
    b. $-2^3 x^5$
    c. y
    d. $12x^2 y z^3$
    e. 5

28. Find an equation of the vertical line through $(0, 7)$.

29. Subtract: $(2x^3 + 8x^2 - 6x) - (2x^3 - x^2 + 1)$

30. Find an equation of the line with slope 4 and y-intercept $\left(0, \dfrac{1}{2}\right)$. Write the equation in standard form.

31. Multiply: $(3x + 2)(2x - 5)$

32. Write an equation of the line through $(-4, 0)$ and $(6, -1)$. Write the equation in standard form.

33. Multiply: $(3y + 1)^2$

34. Solve the system: $\begin{cases} -x + 3y = 18 \\ -3x + 2y = 19 \end{cases}$

35. Simplify by writing each expression with positive exponents only.
    a. 3^{-2}
    b. $2x^{-3}$
    c. $2^{-1} + 4^{-1}$
    d. $(-2)^{-4}$
    e. y^{-4}

36. Simplify: $\dfrac{(5a^7)^2}{a^5}$

37. Write each number in scientific notation.
    a. $367{,}000{,}000$
    b. 0.000003
    c. $20{,}520{,}000{,}000$
    d. 0.00085

38. Multiply: $(3x - 7y)^2$

39. Divide $x^2 + 7x + 12$ by $x + 3$ using long division.

40. Simplify: $\dfrac{(xy)^{-3}}{(x^5 y^6)^3}$

41. Find the GCF of each list of terms.
    a. x^3, x^7, and x^5
    b. y, y^4, and y^7

Factor.

42. $z^3 + 7z + z^2 + 7$

43. $x^2 + 7x + 12$

44. $2x^3 + 2x^2 - 84x$

45. $8x^2 - 22x + 5$

46. $-4x^2 - 23x + 6$

47. $25a^2 - 9b^2$

48. $9xy^2 - 16x$

49. Solve: $(x - 3)(x + 1) = 0$

50. Solve: $x^2 - 13x = -36$

CHAPTER

7

Rational Expressions

7.1 Rational Functions and Simplifying Rational Expressions

7.2 Multiplying and Dividing Rational Expressions

7.3 Adding and Subtracting Rational Expressions with Common Denominators and Least Common Denominator

7.4 Adding and Subtracting Rational Expressions with Unlike Denominators

7.5 Solving Equations Containing Rational Expressions

Integrated Review—Summary on Rational Expressions

7.6 Proportion and Problem Solving with Rational Equations

7.7 Simplifying Complex Fractions

CHECK YOUR PROGRESS

Vocabulary Check
Chapter Highlights
Chapter Review
Getting Ready for the Test
Chapter Test
Cumulative Review

In this chapter, we expand our knowledge of algebraic expressions to include algebraic fractions, called *rational expressions*. We explore the operations of addition, subtraction, multiplication, and division using principles similar to the principles for numerical fractions.

Side Rear-View Mirror

Telescope

Magnifying Glass

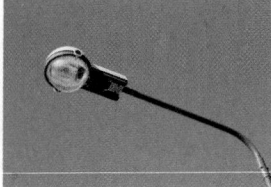

Street Light Reflector

Sunglasses

Camera

What Do the Above Have in Common?

All the useful objects above contain convex mirrors or lenses or were made with convex mirrors or lenses. Basically, all of these objects were made using the rational equation below, called the Gaussian Mirror/Lens Formula. This equation or formula relates an object distance and image distance to the focal length. In general, the focal length is a measure of how strongly a lens converges or diverges light.

Of course, this is just one equation containing rational expressions. There are uses of rational expressions everywhere from health to sports statistics to driving safety. For some applications, see Section 7.1, Exercises 59 through 66, and Section 7.5, Exercises 43 through 52.

Gaussian Mirror/Lens Formula

$$\frac{1}{o} + \frac{1}{i} = \frac{1}{f}$$

$$\frac{1}{\text{object distance}} + \frac{1}{\text{image distance}} = \frac{1}{\text{focal length}}$$

Image distance (negative because of location of image)
$i = -15$ cm

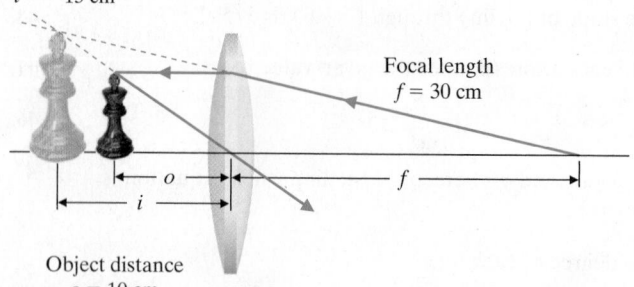

Focal length
$f = 30$ cm

Object distance
$o = 10$ cm

7.1 Rational Functions and Simplifying Rational Expressions

OBJECTIVES

1 Find the Domain of a Rational Function. ▶

2 Simplify or Write Rational Expressions in Lowest Terms. ▶

3 Write Equivalent Rational Expressions of the Form $-\dfrac{a}{b} = \dfrac{-a}{b} = \dfrac{a}{-b}$. ▶

4 Use Rational Functions in Applications. ▶

As we reviewed in Chapter 1, a rational number is a number that can be written as a quotient of integers. A **rational expression** is also a quotient; it is a quotient of polynomials.

> **Rational Expression**
>
> A rational expression is an expression that can be written in the form
> $$\frac{P}{Q},$$
> where P and Q are polynomials and $Q \neq 0$.

Rational Expressions

$$-\frac{2}{7} \qquad \frac{3y^3}{8} \qquad \frac{-4p}{p^3 + 2p + 1} \qquad \frac{5x^2 - 3x + 2}{3x + 7}$$

The first rational expression (or fraction) above is $-\dfrac{2}{7}$. For a negative fraction such as $-\dfrac{2}{7}$, recall from Section 1.7 that

$$-\frac{2}{7} = \frac{2}{-7} = \frac{-2}{7}$$

In general, for any fraction,

$$\frac{-a}{b} = \frac{a}{-b} = -\frac{a}{b}, \qquad b \neq 0$$

This is also true for rational expressions. For example,

$$\underbrace{\frac{-(x + 2)}{x}}_{} = \frac{x + 2}{-x} = -\frac{x + 2}{x}$$

↑
Notice the parentheses.

Rational expressions are sometimes used to describe functions. For example, we call the function $f(x) = \dfrac{x^2 + 2}{x - 3}$ a **rational function** since $\dfrac{x^2 + 2}{x - 3}$ is a rational expression.

OBJECTIVE

1   Finding the Domain of a Rational Function ▶

As with fractions, a rational expression is **undefined** if the denominator is 0. If a variable in a rational expression is replaced with a number that makes the denominator 0, we say that the rational expression is **undefined** for this value of the variable. For example, the rational expression $\dfrac{x^2 + 2}{x - 3}$ is undefined when x is 3, because replacing x with 3 results in a denominator of 0. For this reason, we must exclude 3 from the domain of the function $f(x) = \dfrac{x^2 + 2}{x - 3}$.

The domain of f is then

$$\{x \mid x \text{ is a real number and } x \neq 3\}$$

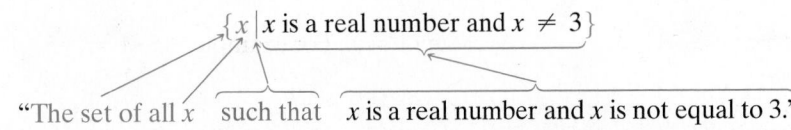

"The set of all x such that x is a real number and x is not equal to 3."

In this section, we will use this set builder notation to write domains. Unless told otherwise, we assume that the domain of a function described by an equation is the set of all real numbers for which the equation is defined.

EXAMPLE 1 Find the domain of each rational function.

a. $f(x) = \dfrac{8x^3 + 7x^2 + 20}{2}$ **b.** $g(x) = \dfrac{5x^2 - 3}{x - 1}$ **c.** $f(x) = \dfrac{7x - 2}{x^2 - 2x - 15}$

Solution The domain of each function will contain all real numbers except those values that make the denominator 0.

a. No matter what the value of x, the denominator of $f(x) = \dfrac{8x^3 + 7x^2 + 20}{2}$ is never 0, so the domain of f is $\{x \mid x \text{ is a real number}\}$.

b. To find the values of x that make the denominator of $g(x)$ equal to 0, we solve the equation "denominator = 0":

$$x - 1 = 0, \quad \text{or} \quad x = 1$$

The domain must exclude 1 since the rational expression is undefined when x is 1. The domain of g is $\{x \mid x \text{ is a real number and } x \neq 1\}$.

c. We find the domain by setting the denominator equal to 0.

$$x^2 - 2x - 15 = 0 \quad \text{Set the denominator equal to 0 and solve.}$$
$$(x - 5)(x + 3) = 0$$
$$x - 5 = 0 \quad \text{or} \quad x + 3 = 0$$
$$x = 5 \quad \text{or} \quad x = -3$$

If x is replaced with 5 or with -3, the rational expression is undefined. The domain of f is $\{x \mid x \text{ is a real number and } x \neq 5, x \neq -3\}$.

PRACTICE
1 Find the domain of each rational function.

a. $f(x) = \dfrac{4x^5 - 3x^2 + 2}{-6}$ **b.** $g(x) = \dfrac{6x^2 + 1}{x + 3}$ **c.** $h(x) = \dfrac{8x - 3}{x^2 - 5x + 6}$

✔ **CONCEPT CHECK**

For which of these values (if any) is the rational expression $\dfrac{x - 3}{x^2 + 2}$ undefined?

a. 2 **b.** 3 **c.** -2 **d.** 0 **e.** None of these

OBJECTIVE
2 Simplifying Rational Expressions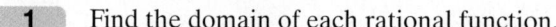

Recall that a fraction is in lowest terms or simplest form if the numerator and denominator have no common factors other than 1 (or -1). For example, $\dfrac{3}{13}$ is in lowest terms since 3 and 13 have no common factors other than 1 (or -1).

To **simplify** a rational expression, or to write it in lowest terms, we use a method similar to simplifying a fraction.

Recall that to simplify a fraction, we essentially "remove factors of 1." Our ability to do this comes from these facts:

• If $c \neq 0$, then $\dfrac{c}{c} = 1$. For example, $\dfrac{7}{7} = 1$ and $\dfrac{-8.65}{-8.65} = 1$.

Answer to Concept Check: e

• $n \cdot 1 = n$. For example, $-5 \cdot 1 = -5$, $126.8 \cdot 1 = 126.8$, and $\dfrac{a}{b} \cdot 1 = \dfrac{a}{b}, b \neq 0$.

In other words, we have the following:

$$\frac{a \cdot c}{b \cdot c} = \frac{a}{b} \cdot \frac{c}{c} = \frac{a}{b}$$

Since $\dfrac{a}{b} \cdot 1 = \dfrac{a}{b}$

Let's practice simplifying a fraction by simplifying $\dfrac{15}{65}$.

$$\frac{15}{65} = \frac{3 \cdot 5}{13 \cdot 5} = \frac{3}{13} \cdot \frac{5}{5} = \frac{3}{13} \cdot 1 = \frac{3}{13}$$

Let's use the same technique and simplify the rational expression $\dfrac{x^2 - 9}{x^2 + x - 6}$.

$$\frac{x^2 - 9}{x^2 + x - 6} = \frac{(x - 3)(x + 3)}{(x - 2)(x + 3)} \qquad \text{Factor the numerator and the denominator.}$$

$$= \frac{(x - 3)\,(x + 3)}{(x - 2)\,(x + 3)} \qquad \text{Look for common factors.}$$

$$= \frac{x - 3}{x - 2} \cdot \frac{x + 3}{x + 3}$$

$$= \frac{x - 3}{x - 2} \cdot 1 \qquad \text{Write } \frac{x + 3}{x + 3} \text{ as 1.}$$

$$= \frac{x - 3}{x - 2} \qquad \text{Multiply to remove a factor of 1.}$$

This "removing a factor of 1" is stated in the principle below:

Fundamental Principle of Rational Expressions

For any rational expression $\dfrac{P}{Q}$ and any polynomial R, where $R \neq 0$,

$$\frac{PR}{QR} = \frac{P}{Q} \cdot \frac{R}{R} = \frac{P}{Q} \cdot 1 = \frac{P}{Q}$$

or, simply,

$$\frac{PR}{QR} = \frac{P}{Q}$$

In general, the following steps may be used to simplify rational expressions or to write a rational expression in lowest terms.

Simplifying or Writing a Rational Expression in Lowest Terms

Step 1. Completely factor the numerator and denominator of the rational expression.

Step 2. Divide out factors common to the numerator and denominator. (This is the same as "removing a factor of 1.")

For now, we assume that variables in a rational expression do not represent values that make the denominator 0.

EXAMPLE 2   Simplify each rational expression.

a. $\dfrac{2x^2}{10x^3 - 2x^2}$

b. $\dfrac{9x^2 + 13x + 4}{8x^2 + x - 7}$

(Continued on next page)

Solution

a. $\dfrac{2x^2}{10x^3 - 2x^2} = \dfrac{2x^2 \cdot 1}{2x^2\,(5x - 1)} = 1 \cdot \dfrac{1}{5x - 1} = \dfrac{1}{5x - 1}$

b. $\dfrac{9x^2 + 13x + 4}{8x^2 + x - 7} = \dfrac{(9x + 4)\,(x + 1)}{(8x - 7)\,(x + 1)}$ Factor the numerator and denominator.

$\qquad\qquad\qquad\;\; = \dfrac{9x + 4}{8x - 7} \cdot 1$ Since $\dfrac{x + 1}{x + 1} = 1$

$\qquad\qquad\qquad\;\; = \dfrac{9x + 4}{8x - 7}$ Simplest form   □

PRACTICE

2   Simplify each rational expression.

a. $\dfrac{5z^4}{10z^5 - 5z^4}$ **b.** $\dfrac{5x^2 + 13x + 6}{6x^2 + 7x - 10}$ ■

Just as for numerical fractions, we can use a shortcut notation. Remember that as long as exact factors in both the numerator and denominator are divided out, we are "removing a factor of 1." We will use the following notation to show this:

$$\dfrac{x^2 - 9}{x^2 + x - 6} = \dfrac{(x - 3)\,(x + 3)}{(x - 2)\,(x + 3)}$$ A factor of 1 is identified by the shading.

$$\qquad\qquad\;\; = \dfrac{x - 3}{x - 2}$$ Remove a factor of 1.

Thus, the rational expression $\dfrac{x^2 - 9}{x^2 + x - 6}$ has the same value as the rational expression $\dfrac{x - 3}{x - 2}$ for all values of x except 2 and -3. (Remember that when x is 2, the denominator of both rational expressions is 0 and when x is -3, the original rational expression has a denominator of 0.)

As we simplify rational expressions, we will assume that the simplified rational expression is equal to the original rational expression for all real numbers except those for which the original denominator is 0.

EXAMPLE 3   Simplify each rational expression.

a. $\dfrac{2 + x}{x + 2}$ **b.** $\dfrac{2 - x}{x - 2}$

Solution

a. $\dfrac{2 + x}{x + 2} = \dfrac{x + 2}{x + 2} = 1$ By the commutative property of addition, $2 + x = x + 2$.   **b.** $\dfrac{2 - x}{x - 2}$

The terms in the numerator of $\dfrac{2 - x}{x - 2}$ differ by sign from the terms of the denominator, so the polynomials are opposites of each other and the expression simplifies to -1. To see this, we factor out -1 from the numerator or the denominator. If -1 is factored from the numerator, then

$$\dfrac{2 - x}{x - 2} = \dfrac{-1(-2 + x)}{x - 2} = \dfrac{-1\,(x - 2)}{x - 2} = \dfrac{-1}{1} = -1$$

If -1 is factored from the denominator, the result is the same.

Helpful Hint

When the numerator and the denominator of a rational expression are opposites of each other, the expression simplifies to -1.

$$\dfrac{2 - x}{x - 2} = \dfrac{2 - x}{-1(-x + 2)} = \dfrac{2 - x}{-1\,(2 - x)} = \dfrac{1}{-1} = -1$$ □

PRACTICE
3 Simplify each rational expression.

a. $\dfrac{x + 3}{3 + x}$ b. $\dfrac{3 - x}{x - 3}$

EXAMPLE 4 Simplify: $\dfrac{18 - 2x^2}{x^2 - 2x - 3}$

Solution

$$\dfrac{18 - 2x^2}{x^2 - 2x - 3} = \dfrac{2(9 - x^2)}{(x + 1)(x - 3)}$$ Factor.

$$= \dfrac{2(3 + x)(3 - x)}{(x + 1)(x - 3)}$$ Factor completely.

$$= \dfrac{2(3 + x) \cdot -1\,(x - 3)}{(x + 1)\,(x - 3)}$$ Notice the opposites $3 - x$ and $x - 3$. Write $3 - x$ as $-1(x - 3)$ and simplify.

$$= -\dfrac{2(3 + x)}{x + 1}$$

PRACTICE
4 Simplify: $\dfrac{20 - 5x^2}{x^2 + x - 6}$

Helpful Hint

When simplifying a rational expression, we look for **common *factors*, not common *terms*.**

$\dfrac{x \cdot (x + 2)}{x \cdot x} = \dfrac{x + 2}{x}$
Common factors. These can be divided out.

$\dfrac{x + 2}{x}$
Common terms. There is no factor of 1 that can be generated.

✔ **CONCEPT CHECK**

Recall that we can only remove *factors* of 1. Which of the following are *not* true? Explain why.

a. $\dfrac{3 - 1}{3 + 5}$ simplifies to $-\dfrac{1}{5}$.   b. $\dfrac{2x + 10}{2}$ simplifies to $x + 5$.

c. $\dfrac{37}{72}$ simplifies to $\dfrac{3}{2}$.   d. $\dfrac{2x + 3}{2}$ simplifies to $x + 3$.

EXAMPLE 5 Simplify each rational expression.

a. $\dfrac{x^3 + 8}{2 + x}$ b. $\dfrac{2y^2 + 2}{y^3 - 5y^2 + y - 5}$

Solution

a. $\dfrac{x^3 + 8}{2 + x} = \dfrac{(x + 2)(x^2 - 2x + 4)}{x + 2}$ Factor the sum of the two cubes.

$$= x^2 - 2x + 4$$ Divide out common factors.

b. $\dfrac{2y^2 + 2}{y^3 - 5y^2 + y - 5} = \dfrac{2(y^2 + 1)}{(y^3 - 5y^2) + (y - 5)}$ Factor the numerator; group the denominator.

$$= \dfrac{2(y^2 + 1)}{y^2(y - 5) + 1(y - 5)}$$ Factor the denominator by grouping.

$$= \dfrac{2(y^2 + 1)}{(y - 5)(y^2 + 1)}$$

$$= \dfrac{2}{y - 5}$$ Divide out common factors.

(Continued on the next page)

PRACTICE
5 Simplify each rational expression.

a. $\dfrac{x^3 + 64}{4 + x}$

b. $\dfrac{5z^2 + 10}{z^3 - 3z^2 + 2z - 6}$

✔ **CONCEPT CHECK**

Does $\dfrac{n}{n + 2}$ simplify to $\dfrac{1}{2}$? Why or why not?

OBJECTIVE

3 Writing Equivalent Forms of Rational Expressions

From Example 3, we have

$$\frac{2 + x}{x + 2} = \frac{x + 2}{x + 2} = 1 \quad \text{and} \quad \frac{2 - x}{x - 2} = \frac{2 - x}{-1\,(2 - x)} = \frac{1}{-1} = -1.$$

When performing operations on rational expressions, equivalent forms of answers often result. For this reason, it is very important to be able to recognize equivalent answers.

EXAMPLE 6 List some equivalent forms of $-\dfrac{5x - 1}{x + 9}$.

Solution To do so, recall that $-\dfrac{a}{b} = \dfrac{-a}{b} = \dfrac{a}{-b}$. Thus

$$-\frac{5x - 1}{x + 9} = \frac{-(5x - 1)}{x + 9} = \frac{-5x + 1}{x + 9} \quad \text{or} \quad \frac{1 - 5x}{x + 9}$$

Also,

$$-\frac{5x - 1}{x + 9} = \frac{5x - 1}{-(x + 9)} = \frac{5x - 1}{-x - 9} \quad \text{or} \quad \frac{5x - 1}{-9 - x}$$

Thus $-\dfrac{5x - 1}{x + 9} = \dfrac{-(5x - 1)}{x + 9} = \dfrac{-5x + 1}{x + 9} = \dfrac{5x - 1}{-(x + 9)} = \dfrac{5x - 1}{-x - 9}$

Helpful Hint

Remember, a negative sign in front of a fraction or rational expression may be moved to the numerator or the denominator, but *not* both.

PRACTICE
6 List some equivalent forms of $-\dfrac{x + 3}{6x - 11}$.

Keep in mind that many rational expressions may look different, but in fact be equivalent.

OBJECTIVE

4 Using Rational Functions in Applications

Rational functions occur often in real-life situations.

EXAMPLE 7 Cost for Pressing Compact Discs

For the ICL Production Company, the rational function $C(x) = \dfrac{2.6x + 10{,}000}{x}$ describes the company's cost per disc of pressing x compact discs. Find the cost per disc for pressing:

a. 100 compact discs   b. 1000 compact discs

<u>*Solution*</u>

a. $C(100) = \dfrac{2.6(100) + 10{,}000}{100} = \dfrac{10{,}260}{100} = 102.6$

The cost per disc for pressing 100 compact discs is \$102.60.

b. $C(1000) = \dfrac{2.6(1000) + 10{,}000}{1000} = \dfrac{12{,}600}{1000} = 12.6$

The cost per disc for pressing 1000 compact discs is \$12.60. Notice that as more compact discs are produced, the cost per disc decreases. □

PRACTICE
7 A company's cost per tee shirt for silk screening x tee shirts is given by the rational function $C(x) = \dfrac{3.2x + 400}{x}$. Find the cost per tee shirt for printing:

a. 100 tee shirts   b. 1000 tee shirts ■

Graphing Calculator Explorations

(Note: The information below about *connected* mode and *dot* mode may not apply to your graphing calculator.)

Recall that since the rational expression $\dfrac{7x - 2}{(x - 2)(x + 5)}$ is not defined when $x = 2$ or when $x = -5$, we say that the domain of the rational function $f(x) = \dfrac{7x - 2}{(x - 2)(x + 5)}$ is all real numbers except 2 and -5. This domain can be written as $\{x \mid x$ is a real number and $x \neq 2, x \neq -5\}$. This means that the graph of $f(x)$ should not cross the vertical lines $x = 2$ and $x = -5$. The graph of $f(x)$ in *connected* mode is to the left. In connected mode the graphing calculator tries to connect all dots of the graph so that the result is a smooth curve. This is what has happened in the graph. Notice that the graph appears to contain vertical lines at $x = 2$ and at $x = -5$. We know that this cannot happen because the function is not defined at $x = 2$ and at $x = -5$. We also know that this cannot happen because the graph of this function would not pass the vertical line test.

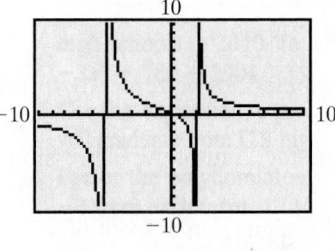

The graph of $f(x)$ in *dot* mode is to the left. In dot mode the graphing calculator will not connect dots with a smooth curve. Notice that the vertical lines have disappeared, and we have a better picture of the graph. The graph, however, actually appears more like the hand-drawn graph below. By using a Table feature, a Calculate Value feature, or by tracing, we can see that the function is not defined at $x = 2$ and at $x = -5$.

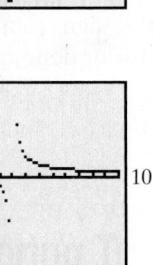

Find the domain of each rational function. Then graph each rational function and use the graph to confirm the domain.

1. $f(x) = \dfrac{x + 1}{x^2 - 4}$

2. $g(x) = \dfrac{5x}{x^2 - 9}$

3. $h(x) = \dfrac{x^2}{2x^2 + 7x - 4}$

4. $f(x) = \dfrac{3x + 2}{4x^2 - 19x - 5}$

✔ Vocabulary, Readiness & Video Check

Use the choices below to fill in each blank. Some choices may not be used.

1	true	rational	simplified	$\dfrac{-a}{-b}$	$\dfrac{-a}{b}$	$\dfrac{a}{-b}$
-1	false	domain	0			

1. A _____ expression is an expression that can be written as the quotient $\dfrac{P}{Q}$ of two polynomials P and Q as long as $Q \neq 0$.

2. A rational expression is undefined if the denominator is _____.

3. The _____ of the rational function $f(x) = \dfrac{2}{x}$ is $\{x \mid x \text{ is a real number and } x \neq 0\}$.

4. A rational expression is _____ if the numerator and denominator have no common factors other than 1 or -1.

5. The expression $\dfrac{x^2 + 2}{2 + x^2}$ simplifies to _____.

6. The expression $\dfrac{y - z}{z - y}$ simplifies to _____.

7. For a rational expression, $-\dfrac{a}{b} = $ _____ $=$ _____.

8. True or false: $\dfrac{a - 6}{a + 2} = \dfrac{-(a - 6)}{-(a + 2)} = \dfrac{-a + 6}{-a - 2}$. _____

Martin-Gay Interactive Videos

Watch the section lecture video and answer the following questions.

See Video 7.1 ●

OBJECTIVE 1
9. Why can't the denominators of rational expressions be zero? How can we find the domain of a rational function?

OBJECTIVE 2
10. In ▦ Example 6, why isn't a factor of x divided out of the expression at the end?

OBJECTIVE 3
11. From ▦ Example 8, if we move a negative sign from in front of a rational expression to either the numerator or denominator, when do we insert parentheses and why?

OBJECTIVE 4
12. From ▦ Example 9, why do we subtract parts a. and b. to find the answer to part c.?

7.1 Exercise Set MyMathLab® ▶

Find the domain of each rational expression. See Example 1.

1. $f(x) = \dfrac{5x - 7}{4}$

2. $g(x) = \dfrac{4 - 3x}{2}$

3. $s(t) = \dfrac{t^2 + 1}{2t}$

4. $v(t) = -\dfrac{5t + t^2}{3t}$

▶ 5. $f(x) = \dfrac{3x}{7 - x}$

6. $f(x) = \dfrac{-4x}{-2 + x}$

7. $f(x) = \dfrac{x}{3x - 1}$

8. $g(x) = \dfrac{-2}{2x + 5}$

9. $R(x) = \dfrac{3 + 2x}{x^3 + x^2 - 2x}$

10. $h(x) = \dfrac{5 - 3x}{2x^2 - 14x + 20}$

▶ 11. $C(x) = \dfrac{x + 3}{x^2 - 4}$

12. $R(x) = \dfrac{5}{x^2 - 7x}$

Study Example 6. Then list four equivalent forms for each rational expression.

13. $-\dfrac{x - 10}{x + 8}$

14. $-\dfrac{x + 11}{x - 4}$

15. $-\dfrac{5y - 3}{y - 12}$

16. $-\dfrac{8y - 1}{y - 15}$

MIXED PRACTICE

Simplify each expression. See Examples 2 through 5.

17. $\dfrac{x + 7}{7 + x}$

18. $\dfrac{y + 9}{9 + y}$

19. $\dfrac{x - 7}{7 - x}$

20. $\dfrac{y - 9}{9 - y}$

21. $\dfrac{2}{8x + 16}$

22. $\dfrac{3}{9x + 6}$

23. $\dfrac{-5a - 5b}{a + b}$

24. $\dfrac{-4x - 4y}{x + y}$

25. $\dfrac{7x + 35}{x^2 + 5x}$

26. $\dfrac{9x + 99}{x^2 + 11x}$

27. $\dfrac{x + 5}{x^2 - 4x - 45}$

28. $\dfrac{x - 3}{x^2 - 6x + 9}$

29. $\dfrac{5x^2 + 11x + 2}{x + 2}$

30. $\dfrac{12x^2 + 4x - 1}{2x + 1}$

31. $\dfrac{x^3 + 7x^2}{x^2 + 5x - 14}$

32. $\dfrac{x^4 - 10x^3}{x^2 - 17x + 70}$

33. $\dfrac{2x^2 - 8}{4x - 8}$

34. $\dfrac{5x^2 - 500}{35x + 350}$

35. $\dfrac{4 - x^2}{x - 2}$

36. $\dfrac{49 - y^2}{y - 7}$

37. $\dfrac{11x^2 - 22x^3}{6x - 12x^2}$

38. $\dfrac{24y^2 - 8y^3}{15y - 5y^2}$

39. $\dfrac{x^2 + xy + 2x + 2y}{x + 2}$

40. $\dfrac{ab + ac + b^2 + bc}{b + c}$

41. $\dfrac{x^3 + 8}{x + 2}$

42. $\dfrac{x^3 + 64}{x + 4}$

43. $\dfrac{x^3 - 1}{1 - x}$

44. $\dfrac{3 - x}{x^3 - 27}$

45. $\dfrac{2xy + 5x - 2y - 5}{3xy + 4x - 3y - 4}$

46. $\dfrac{2xy + 2x - 3y - 3}{2xy + 4x - 3y - 6}$

47. $\dfrac{3x^2 - 5x - 2}{6x^3 + 2x^2 + 3x + 1}$

48. $\dfrac{2x^2 - x - 3}{2x^3 - 3x^2 + 2x - 3}$

49. $\dfrac{9x^2 - 15x + 25}{27x^3 + 125}$

50. $\dfrac{8x^3 - 27}{4x^2 + 6x + 9}$

MIXED PRACTICE

Simplify each expression. Then determine whether the given answer is correct. See Examples 3 through 6.

51. $\dfrac{9 - x^2}{x - 3}$; Answer: $-3 - x$

52. $\dfrac{100 - x^2}{x - 10}$; Answer: $-10 - x$

53. $\dfrac{7 - 34x - 5x^2}{25x^2 - 1}$; Answer: $\dfrac{x + 7}{-5x - 1}$

54. $\dfrac{2 - 15x - 8x^2}{64x^2 - 1}$; Answer: $\dfrac{x + 2}{-8x - 1}$

Find each function value. See Example 7.

55. If $f(x) = \dfrac{x + 8}{2x - 1}$, find $f(2), f(0),$ and $f(-1)$.

56. If $f(x) = \dfrac{x - 2}{-5 + x}$, find $f(-5), f(0),$ and $f(10)$.

57. If $g(x) = \dfrac{x^2 + 8}{x^3 - 25x}$, find $g(3), g(-2),$ and $g(1)$.

58. If $s(t) = \dfrac{t^3 + 1}{t^2 + 1}$, find $s(-1), s(1),$ and $s(2)$.

Solve. See Example 7.

59. The total revenue from the sale of a popular book is approximated by the rational function $R(x) = \dfrac{1000x^2}{x^2 + 4}$, where x is the number of years since publication and $R(x)$ is the total revenue in millions of dollars.

 a. Find the total revenue at the end of the first year.

 b. Find the total revenue at the end of the second year.

 c. Find the revenue during the second year only.

 d. Find the domain of function R.

60. The function $f(x) = \dfrac{100,000x}{100 - x}$ models the cost in dollars for removing x percent of the pollutants from a bayou in which a nearby company dumped creosol.

 a. Find the cost of removing 20% of the pollutants from the bayou. [*Hint:* Find $f(20)$.]

 b. Find the cost of removing 60% of the pollutants and then 80% of the pollutants.

 c. Find $f(90)$, then $f(95)$, and then $f(99)$. What happens to the cost as x approaches 100%?

 d. Find the domain of function f.

61. The dose of medicine prescribed for a child depends on the child's age A in years and the adult dose D for the medication. Young's Rule is a formula used by pediatricians that gives a child's dose C as

$$C = \dfrac{DA}{A + 12}$$

Suppose that an 8-year-old child needs medication, and the normal adult dose is 1000 mg. What size dose should the child receive?

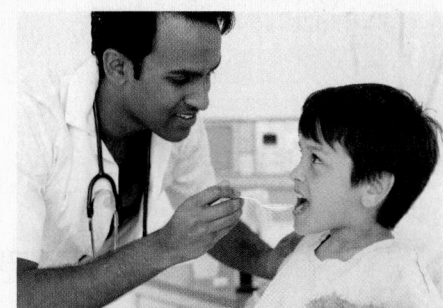

62. Calculating body-mass index is a way to gauge whether a person should lose weight. Doctors recommend that body-mass index values fall between 18.5 and 25. The formula for body-mass index B is

$$B = \frac{703w}{h^2}$$

where w is weight in pounds and h is height in inches. Should a 148-pound person who is 5 feet 6 inches tall lose weight?

63. Anthropologists and forensic scientists use a measure called the cephalic index to help classify skulls. The cephalic index of a skull with width W and length L from front to back is given by the formula

$$C = \frac{100W}{L}$$

A long skull has an index value less than 75, a medium skull has an index value between 75 and 85, and a broad skull has an index value over 85. Find the cephalic index of a skull that is 5 inches wide and 6.4 inches long. Classify the skull.

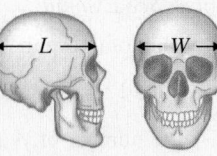

64. A company's gross profit margin P can be computed with the formula $P = \dfrac{R - C}{R}$, where R = the company's revenue and C = cost of goods sold. During a recent fiscal year, computer company Apple had revenues of $32.5 billion and cost of goods sold $21.3 billion. (*Source:* Apple, Inc.) What was Apple's gross profit margin in this year? Express the answer as a percent, rounded to the nearest tenth of a percent.

65. A baseball player's slugging average S can be calculated with the following formula:

$$S = \frac{h + d + 2t + 3r}{b}, \text{ where } h = \text{number of hits,}$$

d = number of doubles, t = number of triples, r = number of home runs, and b = number of at bats. In 2014, Jose Abreu of the Chicago White Sox led Major League Baseball in slugging average. During the 2014 season, Abreu had 556 at bats, 176 hits, 35 doubles, 2 triples, and 36 home runs. (*Source:* Major League Baseball) Calculate Abreu's slugging average. Round to three decimal places.

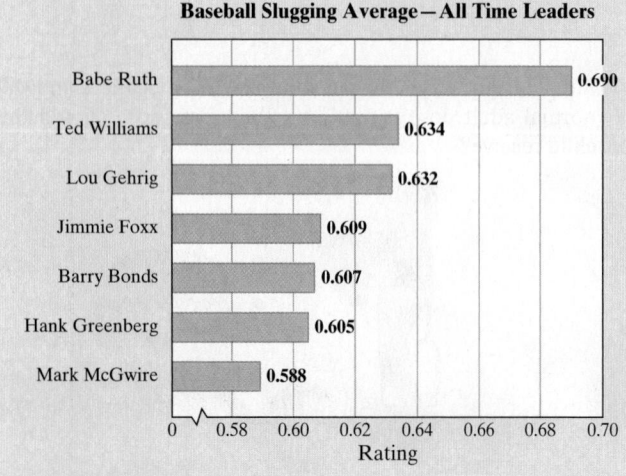

Baseball Slugging Average — All Time Leaders

	Rating
Babe Ruth	0.690
Ted Williams	0.634
Lou Gehrig	0.632
Jimmie Foxx	0.609
Barry Bonds	0.607
Hank Greenberg	0.605
Mark McGwire	0.588

Source: Baseball Almanac

66. To calculate a quarterback's rating in NCAA football, you may use the formula $\dfrac{100C + 330T - 200I + 8.4Y}{A}$, where C = the number of completed passes, A = the number of attempted passes, T = the number of touchdown passes, Y = the number of yards in the completed passes, and I = the number of interceptions. Marcus Mariota of the Oregon Ducks was selected as the 2014 winner of the Heisman Memorial Trophy as the Most Outstanding Football Player. Mariota, a junior quarterback, ended the season with 445 attempts, 304 completions, 4454 yards, 42 touchdowns, and only 4 interceptions. Calculate Mariota's quarterback rating for the 2014 season. (*Source:* NCAA) Round the answer to the nearest tenth.

REVIEW AND PREVIEW

Perform each indicated operation. See Section 1.3.

67. $\dfrac{1}{3} \cdot \dfrac{9}{11}$ **68.** $\dfrac{5}{27} \cdot \dfrac{2}{5}$ **69.** $\dfrac{1}{3} \div \dfrac{1}{4}$

70. $\dfrac{7}{8} \div \dfrac{1}{2}$ **71.** $\dfrac{13}{20} \div \dfrac{2}{9}$ **72.** $\dfrac{8}{15} \div \dfrac{5}{8}$

CONCEPT EXTENSIONS

Which of the following are incorrect and why? See the second Concept Check in this section.

73. $\dfrac{5a - 15}{5}$ simplifies to $a - 3$?

74. $\dfrac{7m - 9}{7}$ simplifies to $m - 9$?

75. $\dfrac{1 + 2}{1 + 3}$ simplifies to $\dfrac{2}{3}$?

76. $\dfrac{46}{54}$ simplifies to $\dfrac{6}{5}$?

Determine whether each rational expression can be simplified. If yes, does it simplify to $1, -1$, or neither? (Do not actually simplify.)

77. $\dfrac{x}{x + 7}$ **78.** $\dfrac{x + 9}{x - 9}$ **79.** $\dfrac{3 + x}{x + 3}$

80. $\dfrac{8 + x}{x + 8}$ **81.** $\dfrac{5 - x}{x - 5}$ **82.** $\dfrac{x - 7}{-x + 7}$

83. Does $\dfrac{x}{x + 5}$ simplify to $\dfrac{1}{5}$? Why or why not?

84. Does $\dfrac{x + 7}{x}$ simplify to 7? Why or why not?

85. In your own words explain how to simplify a rational expression.

86. In your own words, explain how to find the domain of a rational function.

87. Graph a portion of the function $f(x) = \dfrac{20x}{100 - x}$. To do so, complete the given table, plot the points, and then connect the plotted points with a smooth curve. (Note: The domain of this function is all real numbers except 100. We are graphing just a portion of this function.)

x	0	10	30	50	70	90	95	99
y or $f(x)$								

88. The domain of the function $f(x) = \dfrac{1}{x}$ is all real numbers except 0. This means that the graph of this function will be in two pieces: one piece corresponding to x values less than 0 and one piece corresponding to x values greater than 0. Graph the function by completing the following tables, separately plotting the points, and connecting each set of plotted points with a smooth curve.

x	$\dfrac{1}{4}$	$\dfrac{1}{2}$	1	2	4
y or $f(x)$					

x	-4	-2	-1	$-\dfrac{1}{2}$	$-\dfrac{1}{4}$
y or $f(x)$					

How does the graph of $y = \dfrac{x^2 - 9}{x - 3}$ compare to the graph of $y = x + 3$? Recall that $\dfrac{x^2 - 9}{x - 3} = \dfrac{(x + 3)(x - 3)}{x - 3} = x + 3$ as

long as x is not 3. This means that the graph of $y = \dfrac{x^2 - 9}{x - 3}$ is the same as the graph of $y = x + 3$ with $x \neq 3$. To graph $y = \dfrac{x^2 - 9}{x - 3}$, then, graph the linear equation $y = x + 3$ and place an open dot on the graph at 3. This open dot or interruption of the line at 3 means $x \neq 3$.

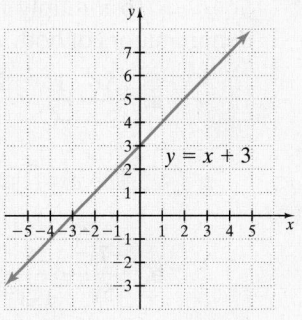

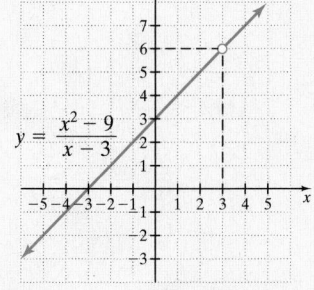

89. Graph: $y = \dfrac{x^2 - 16}{x - 4}$

90. Graph: $y = \dfrac{x^2 - 25}{x + 5}$

91. Graph: $y = \dfrac{x^2 - 6x + 8}{x - 2}$

92. Graph: $y = \dfrac{x^2 + x - 12}{x + 4}$

7.2   Multiplying and Dividing Rational Expressions

OBJECTIVES

1. Multiply Rational Expressions.
2. Divide Rational Expressions.
3. Multiply or Divide Rational Expressions.
4. Convert between Units of Measure.

OBJECTIVE

1   Multiplying Rational Expressions

Just as simplifying rational expressions is similar to simplifying number fractions, multiplying and dividing rational expressions is similar to multiplying and dividing number fractions.

Fractions	*Rational Expressions*
Multiply: $\dfrac{3}{5} \cdot \dfrac{10}{11}$	Multiply: $\dfrac{x - 3}{x + 5} \cdot \dfrac{2x + 10}{x^2 - 9}$

Multiply numerators and multiply denominators.

$$\dfrac{3}{5} \cdot \dfrac{10}{11} = \dfrac{3 \cdot 10}{5 \cdot 11} \qquad\qquad \dfrac{x - 3}{x + 5} \cdot \dfrac{2x + 10}{x^2 - 9} = \dfrac{(x - 3) \cdot (2x + 10)}{(x + 5) \cdot (x^2 - 9)}$$

Simplify by factoring numerators and denominators.

$$= \dfrac{3 \cdot 2 \cdot 5}{5 \cdot 11} \qquad\qquad\qquad = \dfrac{(x - 3) \cdot 2\,(x + 5)}{(x + 5)\,(x + 3)\,(x - 3)}$$

Apply the fundamental principle.

$$= \dfrac{3 \cdot 2}{11} \quad \text{or} \quad \dfrac{6}{11} \qquad\qquad = \dfrac{2}{x + 3}$$

Multiplying Rational Expressions

If $\dfrac{P}{Q}$ and $\dfrac{R}{S}$ are rational expressions, then

$$\dfrac{P}{Q} \cdot \dfrac{R}{S} = \dfrac{PR}{QS}, \qquad Q \neq 0, S \neq 0$$

To multiply rational expressions, multiply the numerators and multiply the denominators.

Note: Recall that for Sections 7.1 through 7.4, we assume variables in rational expressions have only those replacement values for which the expressions are defined.

EXAMPLE 1   Multiply.

a. $\dfrac{25x}{2} \cdot \dfrac{1}{y^3}$

b. $\dfrac{-7x^2}{5y} \cdot \dfrac{3y^5}{14x^2}$

Solution To multiply rational expressions, multiply the numerators and multiply the denominators of both expressions. Then simplify if possible.

a. $\dfrac{25x}{2} \cdot \dfrac{1}{y^3} = \dfrac{25x \cdot 1}{2 \cdot y^3} = \dfrac{25x}{2y^3}$

The expression $\dfrac{25x}{2y^3}$ is in simplest form.

b. $\dfrac{-7x^2}{5y} \cdot \dfrac{3y^5}{14x^2} = \dfrac{-7x^2 \cdot 3y^5}{5y \cdot 14x^2}$ Multiply.

The expression $\dfrac{-7x^2 \cdot 3y^5}{5y \cdot 14x^2}$ is not in simplest form, so we factor the numerator and the denominator and divide out common factors.

$$= \dfrac{-1 \cdot 7 \cdot 3 \cdot x^2 \cdot y \cdot y^4}{5 \cdot 2 \cdot 7 \cdot x^2 \cdot y}$$

$$= -\dfrac{3y^4}{10}$$

> **Helpful Hint**
>
> It is the Fundamental Principle of Fractions that allows us to simplify.

PRACTICE
1   Multiply.

a. $\dfrac{4a}{5} \cdot \dfrac{3}{b^2}$

b. $\dfrac{-3p^4}{q^2} \cdot \dfrac{2q^3}{9p^4}$

When multiplying rational expressions, it is usually best to factor each numerator and denominator before multiplying. This will help us when we divide out common factors to write the product in lowest terms.

EXAMPLE 2   Multiply: $\dfrac{x^2 + x}{3x} \cdot \dfrac{6}{5x + 5}$

Solution $\dfrac{x^2 + x}{3x} \cdot \dfrac{6}{5x + 5} = \dfrac{x(x + 1)}{3x} \cdot \dfrac{2 \cdot 3}{5(x + 1)}$ Factor numerators and denominators.

$$= \dfrac{x(x + 1) \cdot 2 \cdot 3}{3x \cdot 5(x + 1)}$$ Multiply.

$$= \dfrac{2}{5}$$ Simplify by dividing out common factors.

PRACTICE
2   Multiply: $\dfrac{x^2 - x}{5x} \cdot \dfrac{15}{x^2 - 1}$

The following steps may be used to multiply rational expressions.

Multiplying Rational Expressions

Step 1. Completely factor numerators and denominators.

Step 2. Multiply numerators and multiply denominators.

Step 3. Simplify or write the product in lowest terms by dividing out common factors.

 **CONCEPT CHECK**

Which of the following is a true statement?

a. $\dfrac{1}{3} \cdot \dfrac{1}{2} = \dfrac{1}{5}$ b. $\dfrac{2}{x} \cdot \dfrac{5}{x} = \dfrac{10}{x}$ c. $\dfrac{3}{x} \cdot \dfrac{1}{2} = \dfrac{3}{2x}$ d. $\dfrac{x}{7} \cdot \dfrac{x+5}{4} = \dfrac{2x+5}{28}$

EXAMPLE 3   Multiply: $\dfrac{3x+3}{5x-5x^2} \cdot \dfrac{2x^2+x-3}{4x^2-9}$

Solution

$$\dfrac{3x+3}{5x-5x^2} \cdot \dfrac{2x^2+x-3}{4x^2-9} = \dfrac{3(x+1)}{5x(1-x)} \cdot \dfrac{(2x+3)(x-1)}{(2x-3)(2x+3)} \quad \text{Factor.}$$

$$= \dfrac{3(x+1)(2x+3)(x-1)}{5x(1-x)(2x-3)(2x+3)} \quad \text{Multiply.}$$

$$= \dfrac{3(x+1)(x-1)}{5x(1-x)(2x-3)} \quad \begin{array}{l}\text{Divide out common}\\\text{factors.}\end{array}$$

Next, recall that $x-1$ and $1-x$ are opposites so that $x-1 = -1(1-x)$.

$$= \dfrac{3(x+1)(-1)(1-x)}{5x(1-x)(2x-3)} \quad \begin{array}{l}\text{Write } x-1 \text{ as}\\ -1(1-x).\end{array}$$

$$= \dfrac{-3(x+1)}{5x(2x-3)} \quad \text{or} \quad -\dfrac{3(x+1)}{5x(2x-3)} \quad \begin{array}{l}\text{Divide out common}\\\text{factors.} \quad \square\end{array}$$

PRACTICE
3   Multiply: $\dfrac{6-3x}{6x+6x^2} \cdot \dfrac{3x^2-2x-5}{x^2-4}$

OBJECTIVE
2   **Dividing Rational Expressions**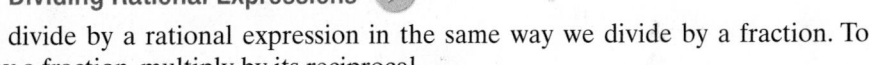

We can divide by a rational expression in the same way we divide by a fraction. To divide by a fraction, multiply by its reciprocal.

> **Helpful Hint**
>
> Don't forget how to find reciprocals. The reciprocal of $\dfrac{a}{b}$ is $\dfrac{b}{a}, a \neq 0, b \neq 0$.

For example, to divide $\dfrac{3}{2}$ by $\dfrac{7}{8}$, multiply $\dfrac{3}{2}$ by $\dfrac{8}{7}$.

$$\dfrac{3}{2} \div \dfrac{7}{8} = \dfrac{3}{2} \cdot \dfrac{8}{7} = \dfrac{3 \cdot 4 \cdot 2}{2 \cdot 7} = \dfrac{12}{7}$$

Dividing Rational Expressions

If $\dfrac{P}{Q}$ and $\dfrac{R}{S}$ are rational expressions and $\dfrac{R}{S}$ is not 0, then

$$\dfrac{P}{Q} \div \dfrac{R}{S} = \dfrac{P}{Q} \cdot \dfrac{S}{R} = \dfrac{PS}{QR}$$

To divide two rational expressions, multiply the first rational expression by the reciprocal of the second rational expression.

EXAMPLE 4 Divide: $\dfrac{3x^3y^7}{40} \div \dfrac{4x^3}{y^2}$

Solution

$$\dfrac{3x^3y^7}{40} \div \dfrac{4x^3}{y^2} = \dfrac{3x^3y^7}{40} \cdot \dfrac{y^2}{4x^3} \quad \text{Multiply by the reciprocal of } \dfrac{4x^3}{y^2}.$$

$$= \dfrac{3x^3y^9}{160x^3}$$

$$= \dfrac{3y^9}{160} \quad \text{Simplify.} \qquad \square$$

PRACTICE

4 Divide: $\dfrac{5a^3b^2}{24} \div \dfrac{10a^5}{6}$

EXAMPLE 5 Divide: $\dfrac{(x+2)^2}{10} \div \dfrac{2x+4}{5}$

Solution

$$\dfrac{(x+2)^2}{10} \div \dfrac{2x+4}{5} = \dfrac{(x+2)^2}{10} \cdot \dfrac{5}{2x+4} \quad \text{Multiply by the reciprocal of } \dfrac{2x+4}{5}.$$

$$= \dfrac{(x+2)(x+2) \cdot 5}{5 \cdot 2 \cdot 2 \cdot (x+2)} \quad \text{Factor and multiply.}$$

$$= \dfrac{x+2}{4} \quad \text{Simplify.} \qquad \square$$

> **Helpful Hint**
>
> Remember, **to Divide by a Rational Expression**, multiply by its reciprocal.

PRACTICE

5 Divide: $\dfrac{(x-5)^2}{3} \div \dfrac{4x-20}{9}$

The following may be used to divide by a rational expression.

> **Dividing by a Rational Expression**
>
> Multiply by its reciprocal.

EXAMPLE 6 Divide: $\dfrac{6x+2}{x^2-1} \div \dfrac{3x^2+x}{x-1}$

Solution

$$\dfrac{6x+2}{x^2-1} \div \dfrac{3x^2+x}{x-1} = \dfrac{6x+2}{x^2-1} \cdot \dfrac{x-1}{3x^2+x} \quad \text{Multiply by the reciprocal.}$$

$$= \dfrac{2(3x+1)(x-1)}{(x+1)(x-1) \cdot x(3x+1)} \quad \text{Factor and multiply.}$$

$$= \dfrac{2}{x(x+1)} \quad \text{Simplify.} \qquad \square$$

PRACTICE

6 Divide: $\dfrac{10x-2}{x^2-9} \div \dfrac{5x^2-x}{x+3}$

EXAMPLE 7 Divide: $\dfrac{2x^2 - 11x + 5}{5x - 25} \div \dfrac{4x - 2}{10}$

Solution

$$\dfrac{2x^2 - 11x + 5}{5x - 25} \div \dfrac{4x - 2}{10} = \dfrac{2x^2 - 11x + 5}{5x - 25} \cdot \dfrac{10}{4x - 2} \quad \text{Multiply by the reciprocal.}$$

$$= \dfrac{(2x - 1)(x - 5) \cdot 2 \cdot 5}{5(x - 5) \cdot 2(2x - 1)} \quad \text{Factor and multiply.}$$

$$= \dfrac{1}{1} \quad \text{or} \quad 1 \quad \text{Simplify.} \qquad \square$$

PRACTICE
7 Divide: $\dfrac{3x^2 - 11x - 4}{2x - 8} \div \dfrac{9x + 3}{6}$ ∎

OBJECTIVE
3 Multiplying or Dividing Rational Expressions ▶

Let's make sure that we understand the difference between multiplying and dividing rational expressions.

Rational Expressions	
Multiplication	Multiply the numerators and multiply the denominators.
Division	Multiply by the reciprocal of the divisor.

EXAMPLE 8 Multiply or divide as indicated.

a. $\dfrac{x - 4}{5} \cdot \dfrac{x}{x - 4}$ **b.** $\dfrac{x - 4}{5} \div \dfrac{x}{x - 4}$ **c.** $\dfrac{x^2 - 4}{2x + 6} \cdot \dfrac{x^2 + 4x + 3}{2 - x}$

Solution

a. $\dfrac{x - 4}{5} \cdot \dfrac{x}{x - 4} = \dfrac{(x - 4) \cdot x}{5 \cdot (x - 4)} = \dfrac{x}{5}$

b. $\dfrac{x - 4}{5} \div \dfrac{x}{x - 4} = \dfrac{x - 4}{5} \cdot \dfrac{x - 4}{x} = \dfrac{(x - 4)^2}{5x}$

c. $\dfrac{x^2 - 4}{2x + 6} \cdot \dfrac{x^2 + 4x + 3}{2 - x} = \dfrac{(x - 2)(x + 2) \cdot (x + 1)(x + 3)}{2(x + 3) \cdot (2 - x)} \quad \begin{array}{l}\text{Factor and}\\\text{multiply.}\end{array}$

$$= \dfrac{(x - 2)(x + 2) \cdot (x + 1)(x + 3)}{2(x + 3) \cdot (2 - x)}$$

$$= \dfrac{-1(x + 2)(x + 1)}{2} \quad \begin{array}{l}\text{Divide out com-}\\\text{mon factors. Recall}\\\text{that } \dfrac{x - 2}{2 - x} = -1\end{array}$$

$$= -\dfrac{(x + 2)(x + 1)}{2} \qquad \square$$

PRACTICE
8 Multiply or divide as indicated.

a. $\dfrac{y + 9}{8x} \cdot \dfrac{y + 9}{2x}$ **b.** $\dfrac{y + 9}{8x} \div \dfrac{y + 9}{2}$ **c.** $\dfrac{35x - 7x^2}{x^2 - 25} \cdot \dfrac{x^2 + 3x - 10}{x^2 + 4x}$ ∎

_{OBJECTIVE}

4 Converting Between Units of Measure

How many square inches are in 1 square foot?

How many cubic feet are in a cubic yard?

If you have trouble answering these questions, this section will be helpful to you.

Now that we know how to multiply fractions and rational expressions, we can use this knowledge to help us convert between units of measure. To do so, we will use **unit fractions**. A unit fraction is a fraction that equals 1. For example, since 12 in. = 1 ft, we have the unit fractions

$$\frac{12 \text{ in.}}{1 \text{ ft}} = 1 \quad \text{and} \quad \frac{1 \text{ ft}}{12 \text{ in.}} = 1$$

EXAMPLE 9 18 square feet = _____ square yards

Solution Let's multiply 18 square feet by a unit fraction that has square feet in the denominator and square yards in the numerator. From the diagram, you can see that

1 square yard = 9 square feet

Thus,

$$18 \text{ sq ft} = \frac{18 \text{ sq ft}}{1} \cdot 1 = \frac{\overset{2}{\cancel{18} \text{ sq ft}}}{1} \cdot \frac{1 \text{ sq yd}}{\underset{1}{\cancel{9} \text{ sq ft}}}$$

$$= \frac{2 \cdot 1}{1 \cdot 1} \text{ sq yd} = 2 \text{ sq yd}$$

1 yd = 3 ft

1 yd = 3 ft

Area: 1 sq yd or 9 sq ft

Thus, 18 sq ft = 2 sq yd.

Draw a diagram of 18 sq ft to help you see that this is reasonable.

_{PRACTICE}

9 288 square inches = _____ square feet

EXAMPLE 10 5.2 square yards = _____ square feet

Solution

$$5.2 \text{ sq yd} = \frac{5.2 \text{ sq yd}}{1} \cdot 1 = \frac{5.2 \text{ sq yd}}{1} \cdot \frac{9 \text{ sq ft}}{1 \text{ sq yd}} \quad \begin{matrix} \leftarrow \text{Units converting to} \\ \leftarrow \text{Units given} \end{matrix}$$

$$= \frac{5.2 \cdot 9}{1 \cdot 1} \text{ sq ft}$$

$$= 46.8 \text{ sq ft}$$

Thus, 5.2 sq yd = 46.8 sq ft.

Draw a diagram to see that this is reasonable.

_{PRACTICE}

10 3.5 square feet = _____ square inches

EXAMPLE 11 Converting from Cubic Feet to Cubic Yards

The largest building in the world by volume is The Boeing Company's Everett, Washington, factory complex, where Boeing's wide-body jetliners, the 747, 767, and 777, are built. The volume of this factory complex is 472,370,319 cubic feet. Find the volume of this Boeing facility in cubic yards. (_Source:_ The Boeing Company)

Solution There are 27 cubic feet in 1 cubic yard. (See the diagram.)

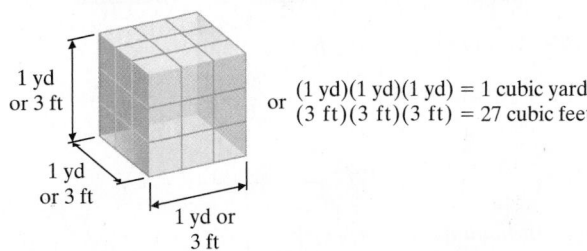

$$1 \text{ yd} \atop \text{or } 3 \text{ ft}$$ or $\begin{matrix}(1 \text{ yd})(1 \text{ yd})(1 \text{ yd}) = 1 \text{ cubic yard} \\ (3 \text{ ft})(3 \text{ ft})(3 \text{ ft}) = 27 \text{ cubic feet}\end{matrix}$

$$472{,}370{,}319 \text{ cu ft} = 472{,}370{,}319 \text{ cu ft} \cdot \frac{1 \text{ cu yd}}{27 \text{ cu ft}}$$

$$= \frac{472{,}370{,}319}{27} \text{ cu yd}$$

$$= 17{,}495{,}197 \text{ cu yd}$$

PRACTICE
11   The largest casino in the world is the Venetian, in Macau, on the southern tip of China. The gaming area for this casino is approximately 61,000 _square yards_. Find the size of the gaming area in _square feet_. (_Source: USA Today_)

Helpful Hint

When converting among units of measurement, if possible write the unit fraction so that **the numerator contains the units you are converting to** and **the denominator contains the original units**.

Unit fraction

$$48 \text{ in.} = \frac{48 \text{ in.}}{1} \cdot \overbrace{\frac{1 \text{ ft}}{12 \text{ in.}}}^{} \quad \begin{matrix}\leftarrow \text{Units converting to} \\ \leftarrow \text{Original units}\end{matrix}$$

$$= \frac{48}{12} \text{ ft} = 4 \text{ ft}$$

EXAMPLE 12   At the 2012 Summer Olympics, Jamaican athlete Usain Bolt won the gold medal in the men's 100-meter track event. He ran the distance at an average speed of 34.1 feet per second. Convert this speed to miles per hour. (_Source:_ International Olympic Committee)

Solution Recall that 1 mile = 5280 feet and 1 hour = 3600 seconds $(60 \cdot 60)$.

Unit fractions

$$34.1 \text{ feet/second} = \frac{34.1 \text{ feet}}{1 \text{ second}} \cdot \overbrace{\frac{3600 \text{ seconds}}{1 \text{ hour}}}^{} \cdot \overbrace{\frac{1 \text{ mile}}{5280 \text{ feet}}}^{}$$

$$= \frac{34.1 \cdot 3600}{5280} \text{ miles/hour}$$

$$\approx 23.3 \text{ miles/hour (rounded to the nearest tenth)}$$

PRACTICE
12   The cheetah is the fastest land animal, being clocked at about 102.7 feet per second. Convert this to miles per hour. Round to the nearest tenth. (_Source: World Almanac and Book of Facts_)

✔ **Vocabulary, Readiness & Video Check**

Use one of the choices below to fill in the blank.

opposites   reciprocals

1. The expressions $\dfrac{x}{2y}$ and $\dfrac{2y}{x}$ are called _____.

Multiply or divide as indicated.

2. $\dfrac{a}{b} \cdot \dfrac{c}{d} =$ _____

3. $\dfrac{a}{b} \div \dfrac{c}{d} =$ _____

4. $\dfrac{x}{7} \cdot \dfrac{x}{6} =$ _____

5. $\dfrac{x}{7} \div \dfrac{x}{6} =$ _____

Martin-Gay Interactive Videos

See Video 7.2 ⦿

Watch the section lecture video and answer the following questions.

OBJECTIVE 1

6. Would you say a person needs to be quite comfortable with factoring polynomials in order to be successful with multiplying rational expressions? Explain, referencing ▦ Example 2 in your answer.

OBJECTIVE 2

7. Based on the lecture before ▦ Example 3, complete the following statements. Dividing rational expressions is exactly like dividing _____. Therefore, to divide by a rational expression, multiply by its _____.

OBJECTIVE 3

8. In ▦ Examples 4 and 5, determining the operation is the first step in deciding how to simplify. Why do you think this is so?

OBJECTIVE 4

9. When converting between units of measurement, a unit fraction may be used. What units are used in the numerator and what units are used in the denominator of your unit fraction?

7.2 Exercise Set MyMathLab® ▸

Find each product and simplify if possible. See Examples 1 through 3.

1. $\dfrac{3x}{y^2} \cdot \dfrac{7y}{4x}$

2. $\dfrac{9x^2}{y} \cdot \dfrac{4y}{3x^3}$

▸ **3.** $\dfrac{8x}{2} \cdot \dfrac{x^5}{4x^2}$

4. $\dfrac{6x^2}{10x^3} \cdot \dfrac{5x}{12}$

5. $-\dfrac{5a^2b}{30a^2b^2} \cdot b^3$

6. $-\dfrac{9x^3y^2}{18xy^5} \cdot y^3$

7. $\dfrac{x}{2x - 14} \cdot \dfrac{x^2 - 7x}{5}$

8. $\dfrac{4x - 24}{20x} \cdot \dfrac{5}{x - 6}$

9. $\dfrac{6x + 6}{5} \cdot \dfrac{10}{36x + 36}$

10. $\dfrac{x^2 + x}{8} \cdot \dfrac{16}{x + 1}$

11. $\dfrac{(m + n)^2}{m - n} \cdot \dfrac{m}{m^2 + mn}$

12. $\dfrac{(m - n)^2}{m + n} \cdot \dfrac{m}{m^2 - mn}$

13. $\dfrac{x^2 - 25}{x^2 - 3x - 10} \cdot \dfrac{x + 2}{x}$

14. $\dfrac{a^2 - 4a + 4}{a^2 - 4} \cdot \dfrac{a + 3}{a - 2}$

15. $\dfrac{x^2 + 6x + 8}{x^2 + x - 20} \cdot \dfrac{x^2 + 2x - 15}{x^2 + 8x + 16}$

16. $\dfrac{x^2 + 9x + 20}{x^2 - 15x + 44} \cdot \dfrac{x^2 - 11x + 28}{x^2 + 12x + 35}$

Find each quotient and simplify. See Examples 4 through 7.

17. $\dfrac{5x^7}{2x^5} \div \dfrac{15x}{4x^3}$

18. $\dfrac{9y^4}{6y} \div \dfrac{y^2}{3}$

19. $\dfrac{8x^2}{y^3} \div \dfrac{4x^2y^3}{6}$

20. $\dfrac{7a^2b}{3ab^2} \div \dfrac{21a^2b^2}{14ab}$

21. $\dfrac{(x - 6)(x + 4)}{4x} \div \dfrac{2x - 12}{8x^2}$

22. $\dfrac{(x + 3)^2}{5} \div \dfrac{5x + 15}{25}$

23. $\dfrac{3x^2}{x^2 - 1} \div \dfrac{x^5}{(x + 1)^2}$

24. $\dfrac{9x^5}{a^2 - b^2} \div \dfrac{27x^2}{3b - 3a}$

25. $\dfrac{m^2 - n^2}{m + n} \div \dfrac{m}{m^2 + nm}$

26. $\dfrac{(m - n)^2}{m + n} \div \dfrac{m^2 - mn}{m}$

▸ **27.** $\dfrac{x + 2}{7 - x} \div \dfrac{x^2 - 5x + 6}{x^2 - 9x + 14}$

28. $\dfrac{x - 3}{2 - x} \div \dfrac{x^2 + 3x - 18}{x^2 + 2x - 8}$

29. $\dfrac{x^2 + 7x + 10}{x - 1} \div \dfrac{x^2 + 2x - 15}{x - 1}$

30. $\dfrac{x + 1}{(x + 1)(2x + 3)} \div \dfrac{20x + 100}{2x + 3}$

MIXED PRACTICE

Multiply or divide as indicated. See Examples 1 through 8.

31. $\dfrac{5x - 10}{12} \div \dfrac{4x - 8}{8}$

32. $\dfrac{6x + 6}{5} \div \dfrac{9x + 9}{10}$

33. $\dfrac{x^2 + 5x}{8} \cdot \dfrac{9}{3x + 15}$

34. $\dfrac{3x^2 + 12x}{6} \cdot \dfrac{9}{2x + 8}$

35. $\dfrac{7}{6p^2 + q} \div \dfrac{14}{18p^2 + 3q}$

36. $\dfrac{3x + 6}{20} \div \dfrac{4x + 8}{8}$

37. $\dfrac{3x + 4y}{x^2 + 4xy + 4y^2} \cdot \dfrac{x + 2y}{2}$

38. $\dfrac{x^2 - y^2}{3x^2 + 3xy} \cdot \dfrac{3x^2 + 6x}{3x^2 - 2xy - y^2}$

39. $\dfrac{(x + 2)^2}{x - 2} \div \dfrac{x^2 - 4}{2x - 4}$

40. $\dfrac{x + 3}{x^2 - 9} \div \dfrac{5x + 15}{(x - 3)^2}$

41. $\dfrac{x^2 - 4}{24x} \div \dfrac{2 - x}{6xy}$

42. $\dfrac{3y}{3 - x} \div \dfrac{12xy}{x^2 - 9}$

43. $\dfrac{a^2 + 7a + 12}{a^2 + 5a + 6} \cdot \dfrac{a^2 + 8a + 15}{a^2 + 5a + 4}$

44. $\dfrac{b^2 + 2b - 3}{b^2 + b - 2} \cdot \dfrac{b^2 - 4}{b^2 + 6b + 8}$

45. $\dfrac{5x - 20}{3x^2 + x} \cdot \dfrac{3x^2 + 13x + 4}{x^2 - 16}$

46. $\dfrac{9x + 18}{4x^2 - 3x} \cdot \dfrac{4x^2 - 11x + 6}{x^2 - 4}$

47. $\dfrac{8n^2 - 18}{2n^2 - 5n + 3} \div \dfrac{6n^2 + 7n - 3}{n^2 - 9n + 8}$

48. $\dfrac{36n^2 - 64}{3n^2 + 10n + 8} \div \dfrac{3n^2 - 13n + 12}{n^2 - 5n - 14}$

49. Find the quotient of $\dfrac{x^2 - 9}{2x}$ and $\dfrac{x + 3}{8x^4}$.

50. Find the quotient of $\dfrac{4x^2 + 4x + 1}{4x + 2}$ and $\dfrac{4x + 2}{16}$.

Multiply or divide as indicated. Some of these expressions contain 4-term polynomials and sums and differences of cubes. See Examples 1 through 8.

51. $\dfrac{a^2 + ac + ba + bc}{a - b} \div \dfrac{a + c}{a + b}$

52. $\dfrac{x^2 + 2x - xy - 2y}{x^2 - y^2} \div \dfrac{2x + 4}{x + y}$

53. $\dfrac{3x^2 + 8x + 5}{x^2 + 8x + 7} \cdot \dfrac{x + 7}{x^2 + 4}$

54. $\dfrac{16x^2 + 2x}{16x^2 + 10x + 1} \cdot \dfrac{1}{4x^2 + 2x}$

55. $\dfrac{x^3 + 8}{x^2 - 2x + 4} \cdot \dfrac{4}{x^2 - 4}$

56. $\dfrac{9y}{3y - 3} \cdot \dfrac{y^3 - 1}{y^3 + y^2 + y}$

57. $\dfrac{a^2 - ab}{6a^2 + 6ab} \div \dfrac{a^3 - b^3}{a^2 - b^2}$

58. $\dfrac{x^3 + 27y^3}{6x} \div \dfrac{x^2 - 9y^2}{x^2 - 3xy}$

Convert as indicated. See Examples 9 through 12.

59. 10 square feet = _____ square inches.

60. 1008 square inches = _____ square feet.

61. 45 square feet = _____ square yards.

62. 2 square yards = _____ square inches.

63. 3 cubic yards = _____ cubic feet.

64. 2 cubic yards = _____ cubic inches.

65. 50 miles per hour = _____ feet per second (round to the nearest whole).

66. 10 feet per second = _____ miles per hour (round to the nearest tenth).

67. 6.3 square yards = _____ square feet.

68. 3.6 square yards = _____ square feet.

69. In January 2010, the Burj Khalifa Tower officially became the tallest building in the world. This tower has a curtain wall (the exterior skin of the building) that is approximately 133,500 square yards. Convert this to square feet. (*Source:* Burj Khalifa)

70. The Pentagon, headquarters for the Department of Defense, contains 3,705,793 square feet of office and storage space. Convert this to square yards. Round to the nearest square yard. (*Source:* U.S. Department of Defense)

71. On February 14, 2014, Brian Smith set a new stock car world speed record of 396.7 feet per second on the Space Shuttle landing runway at The John F. Kennedy Space Center. Convert this speed to miles per hour. Round to the nearest tenth. (*Source:* Vox Media)

72. On October 4, 2004, the rocket plane *SpaceShipOne* shot to an altitude of more than 100 km for the second time inside a week to claim the $10 million Ansari X-Prize. At one point in its flight, *SpaceShipOne* was traveling past Mach 1, about 930 miles per hour. Find this speed in feet per second. (*Source:* Space.com)

REVIEW AND PREVIEW

Perform each indicated operation. See Section 1.3.

73. $\dfrac{1}{5} + \dfrac{4}{5}$

74. $\dfrac{3}{15} + \dfrac{6}{15}$

75. $\dfrac{9}{9} - \dfrac{19}{9}$

76. $\dfrac{4}{3} - \dfrac{8}{3}$

77. $\dfrac{6}{5} + \left(\dfrac{1}{5} - \dfrac{8}{5}\right)$

78. $-\dfrac{3}{2} + \left(\dfrac{1}{2} - \dfrac{3}{2}\right)$

Graph each linear equation. See Section 3.2.

79. $x - 2y = 6$

80. $5x - y = 10$

CONCEPT EXTENSIONS

Identify each statement as true or false. If false, correct the multiplication. See the Concept Check in this section.

81. $\dfrac{4}{a} \cdot \dfrac{1}{b} = \dfrac{4}{ab}$

82. $\dfrac{2}{3} \cdot \dfrac{2}{4} = \dfrac{2}{7}$

83. $\dfrac{x}{5} \cdot \dfrac{x+3}{4} = \dfrac{2x+3}{20}$

84. $\dfrac{7}{a} \cdot \dfrac{3}{a} = \dfrac{21}{a}$

85. Find the area of the rectangle.

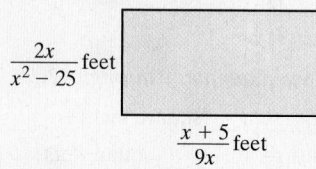

$\dfrac{2x}{x^2-25}$ feet

$\dfrac{x+5}{9x}$ feet

86. Find the area of the square.

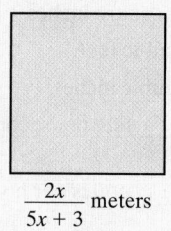

$\dfrac{2x}{5x+3}$ meters

Multiply or divide as indicated.

87. $\left(\dfrac{x^2 - y^2}{x^2 + y^2} \div \dfrac{x^2 - y^2}{3x} \right) \cdot \dfrac{x^2 + y^2}{6}$

88. $\left(\dfrac{x^2 - 9}{x^2 - 1} \cdot \dfrac{x^2 + 2x + 1}{2x^2 + 9x + 9} \right) \div \dfrac{2x + 3}{1 - x}$

89. $\left(\dfrac{2a + b}{b^2} \cdot \dfrac{3a^2 - 2ab}{ab + 2b^2} \right) \div \dfrac{a^2 - 3ab + 2b^2}{5ab - 10b^2}$

90. $\left(\dfrac{x^2 y^2 - xy}{4x - 4y} \div \dfrac{3y - 3x}{8x - 8y} \right) \cdot \dfrac{y - x}{8}$

91. In your own words, explain how you multiply rational expressions.

92. Explain how dividing rational expressions is similar to dividing rational numbers.

7.3 Adding and Subtracting Rational Expressions with Common Denominators and Least Common Denominator

OBJECTIVES

1 Add and Subtract Rational Expressions with the Same Denominator.

2 Find the Least Common Denominator of a List of Rational Expressions.

3 Write a Rational Expression as an Equivalent Expression Whose Denominator Is Given.

OBJECTIVE

1 Adding and Subtracting Rational Expressions with the Same Denominator

Like multiplication and division, addition and subtraction of rational expressions is similar to addition and subtraction of rational numbers. In this section, we add and subtract rational expressions with a common (or the same) denominator.

Add: $\dfrac{6}{5} + \dfrac{2}{5}$ ⎪ Add: $\dfrac{9}{x+2} + \dfrac{3}{x+2}$

Add the numerators and place the sum over the common denominator.

$\dfrac{6}{5} + \dfrac{2}{5} = \dfrac{6+2}{5}$ ⎪ $\dfrac{9}{x+2} + \dfrac{3}{x+2} = \dfrac{9+3}{x+2}$

$= \dfrac{8}{5}$ Simplify. ⎪ $= \dfrac{12}{x+2}$ Simplify.

Adding and Subtracting Rational Expressions with Common Denominators

If $\dfrac{P}{R}$ and $\dfrac{Q}{R}$ are rational expressions, then

$$\dfrac{P}{R} + \dfrac{Q}{R} = \dfrac{P + Q}{R} \qquad \text{and} \qquad \dfrac{P}{R} - \dfrac{Q}{R} = \dfrac{P - Q}{R}$$

To add or subtract rational expressions, add or subtract the numerators and place the sum or difference over the common denominator.

EXAMPLE 1 Add: $\dfrac{5m}{2n} + \dfrac{m}{2n}$

Solution $\dfrac{5m}{2n} + \dfrac{m}{2n} = \dfrac{5m + m}{2n}$ Add the numerators.

$= \dfrac{6m}{2n}$ Simplify the numerator by combining like terms.

$= \dfrac{3m}{n}$ Simplify by applying the fundamental principle. □

PRACTICE 1 Add: $\dfrac{7a}{4b} + \dfrac{a}{4b}$

EXAMPLE 2 Subtract: $\dfrac{2y}{2y - 7} - \dfrac{7}{2y - 7}$

Solution $\dfrac{2y}{2y - 7} - \dfrac{7}{2y - 7} = \dfrac{2y - 7}{2y - 7}$ Subtract the numerators.

$= \dfrac{1}{1}$ or 1   Simplify. □

PRACTICE 2 Subtract: $\dfrac{3x}{3x - 2} - \dfrac{2}{3x - 2}$

EXAMPLE 3 Subtract: $\dfrac{3x^2 + 2x}{x - 1} - \dfrac{10x - 5}{x - 1}$

Solution $\dfrac{3x^2 + 2x}{x - 1} - \dfrac{10x - 5}{x - 1} = \dfrac{(3x^2 + 2x) - (10x - 5)}{x - 1}$ Subtract the numerators. Notice the parentheses.

$= \dfrac{3x^2 + 2x - 10x + 5}{x - 1}$ Use the distributive property.

$= \dfrac{3x^2 - 8x + 5}{x - 1}$ Combine like terms.

$= \dfrac{(x - 1)(3x - 5)}{x - 1}$ Factor.

$= 3x - 5$ Simplify. □

> **Helpful Hint**
> Parentheses are inserted so that the entire numerator, $10x - 5$, is subtracted.

PRACTICE 3 Subtract: $\dfrac{4x^2 + 15x}{x + 3} - \dfrac{8x + 15}{x + 3}$

> **Helpful Hint**
> Notice how the numerator $10x - 5$ has been subtracted in Example 3.
> This $-$ sign applies to the entire numerator, $10x - 5$.   So parentheses are inserted here to indicate this.
>
> $$\dfrac{3x^2 + 2x}{x - 1} - \dfrac{10x - 5}{x - 1} = \dfrac{3x^2 + 2x - (10x - 5)}{x - 1}$$

OBJECTIVE

2 Finding the Least Common Denominator

To add and subtract fractions with **unlike** denominators, first find the least common denominator (LCD) and then write all fractions as equivalent fractions with the LCD.

For example, suppose we add $\frac{8}{3}$ and $\frac{2}{5}$. The LCD of denominators 3 and 5 is 15, since 15 is the least common multiple (LCM) of 3 and 5. That is, 15 is the smallest number that both 3 and 5 divide into evenly.

Next, rewrite each fraction so that its denominator is 15.

$$\frac{8}{3} + \frac{2}{5} = \frac{8(5)}{3(5)} + \frac{2(3)}{5(3)} = \frac{40}{15} + \frac{6}{15} = \frac{40 + 6}{15} = \frac{46}{15}$$

We are multiplying by 1.

To add or subtract rational expressions with unlike denominators, we also first find the LCD and then write all rational expressions as equivalent expressions with the LCD. The **least common denominator (LCD) of a list of rational expressions** is a polynomial of least degree whose factors include all the factors of the denominators in the list.

Finding the Least Common Denominator (LCD)

Step 1. Factor each denominator completely.

Step 2. The least common denominator (LCD) is the product of all unique factors found in Step 1, each raised to a power equal to the greatest number of times that the factor appears in any one factored denominator.

EXAMPLE 4 Find the LCD for each pair.

a. $\frac{1}{8}, \frac{3}{22}$

b. $\frac{7}{5x}, \frac{6}{15x^2}$

Solution

a. Start by finding the prime factorization of each denominator.

$$8 = 2 \cdot 2 \cdot 2 = 2^3 \quad \text{and}$$
$$22 = 2 \cdot 11$$

Next, write the product of all the unique factors, each raised to a power equal to the greatest number of times that the factor appears in any denominator.

The greatest number of times that the factor 2 appears is 3.

The greatest number of times that the factor 11 appears is 1.

$$\text{LCD} = 2^3 \cdot 11^1 = 8 \cdot 11 = 88$$

b. Factor each denominator.

$$5x = 5 \cdot x \quad \text{and}$$
$$15x^2 = 3 \cdot 5 \cdot x^2$$

The greatest number of times that the factor 5 appears is 1.

The greatest number of times that the factor 3 appears is 1.

The greatest number of times that the factor x appears is 2.

$$\text{LCD} = 3^1 \cdot 5^1 \cdot x^2 = 15x^2$$

PRACTICE

4 Find the LCD for each pair.

a. $\frac{3}{14}, \frac{5}{21}$

b. $\frac{4}{9y}, \frac{11}{15y^3}$

EXAMPLE 5   Find the LCD for each pair.

a. $\dfrac{7x}{x + 2}$ and $\dfrac{5x^2}{x - 2}$ **b.** $\dfrac{3}{x}$ and $\dfrac{6}{x + 4}$

Solution

a. The denominators $x + 2$ and $x - 2$ are completely factored already. The factor $x + 2$ appears once and the factor $x - 2$ appears once.

$$\text{LCD} = (x + 2)(x - 2)$$

b. The denominators x and $x + 4$ cannot be factored further. The factor x appears once and the factor $x + 4$ appears once.

$$\text{LCD} = x(x + 4)$$

PRACTICE
5   Find the LCD for each pair.

a. $\dfrac{16}{y - 5}$ and $\dfrac{3y^3}{y - 4}$ **b.** $\dfrac{8}{a}$ and $\dfrac{5}{a + 2}$

EXAMPLE 6   Find the LCD of $\dfrac{6m^2}{3m + 15}$ and $\dfrac{2}{(m + 5)^2}$.

Solution   We factor each denominator.

$$3m + 15 = 3(m + 5)$$
$$(m + 5)^2 = (m + 5)^2 \quad \text{This denominator is already factored.}$$

The greatest number of times that the factor 3 appears is 1.

The greatest number of times that the factor $m + 5$ appears *in any one denominator* is 2.

$$\text{LCD} = 3(m + 5)^2$$

PRACTICE
6   Find the LCD of $\dfrac{2x^3}{(2x - 1)^2}$ and $\dfrac{5x}{6x - 3}$.

✔ **CONCEPT CHECK**

Choose the correct LCD of $\dfrac{x}{(x + 1)^2}$ and $\dfrac{5}{x + 1}$.

a. $x + 1$ **b.** $(x + 1)^2$ **c.** $(x + 1)^3$ **d.** $5x(x + 1)^2$

EXAMPLE 7   Find the LCD of $\dfrac{t - 10}{t^2 - t - 6}$ and $\dfrac{t + 5}{t^2 + 3t + 2}$.

Solution   Start by factoring each denominator.

$$t^2 - t - 6 = (t - 3)(t + 2)$$
$$t^2 + 3t + 2 = (t + 1)(t + 2)$$
$$\text{LCD} = (t - 3)(t + 2)(t + 1)$$

PRACTICE
7   Find the LCD of $\dfrac{x - 5}{x^2 + 5x + 4}$ and $\dfrac{x + 8}{x^2 - 16}$.

Answer to Concept Check: b

EXAMPLE 8 Find the LCD of $\dfrac{2}{x-2}$ and $\dfrac{10}{2-x}$.

Solution The denominators $x-2$ and $2-x$ are opposites. That is, $2-x = -1(x-2)$. Use $x-2$ or $2-x$ as the LCD.

$$LCD = x - 2 \qquad \text{or} \qquad LCD = 2 - x$$

PRACTICE
8 Find the LCD of $\dfrac{5}{3-x}$ and $\dfrac{4}{x-3}$.

OBJECTIVE
3 Writing Equivalent Rational Expressions ▶

Next we practice writing a rational expression as an equivalent rational expression with a given denominator. To do this, we multiply by a form of 1. Recall that multiplying an expression by 1 produces an equivalent expression. In other words,

$$\frac{P}{Q} = \frac{P}{Q} \cdot 1 = \frac{P}{Q} \cdot \frac{R}{R} = \frac{PR}{QR}.$$

EXAMPLE 9 Write each rational expression as an equivalent rational expression with the given denominator.

a. $\dfrac{4b}{9a} = \dfrac{}{27a^2b}$ **b.** $\dfrac{7x}{2x+5} = \dfrac{}{6x+15}$

Solution

a. We can ask ourselves: "What do we multiply $9a$ by to get $27a^2b$?" The answer is $3ab$, since $9a(3ab) = 27a^2b$. So we multiply by 1 in the form of $\dfrac{3ab}{3ab}$.

$$\frac{4b}{9a} = \frac{4b}{9a} \cdot 1$$

$$= \frac{4b}{9a} \cdot \frac{3ab}{3ab}$$

$$= \frac{4b(3ab)}{9a(3ab)} = \frac{12ab^2}{27a^2b}$$

b. First, factor the denominator on the right.

$$\frac{7x}{2x+5} = \frac{}{3(2x+5)}$$

To obtain the denominator on the right from the denominator on the left, we multiply $\dfrac{7x}{2x+5}$ by 1 in the form of $\dfrac{3}{3}$.

$$\frac{7x}{2x+5} = \frac{7x}{2x+5} \cdot \frac{3}{3} = \frac{7x \cdot 3}{(2x+5) \cdot 3} = \frac{21x}{3(2x+5)} \text{ or } \frac{21x}{6x+15}$$

PRACTICE
9 Write each rational expression as an equivalent fraction with the given denominator.

a. $\dfrac{3x}{5y} = \dfrac{}{35xy^2}$ **b.** $\dfrac{9x}{4x+7} = \dfrac{}{8x+14}$

EXAMPLE 10   Write the rational expression as an equivalent rational expression with the given denominator.

$$\frac{5}{x^2 - 4} = \frac{}{(x - 2)(x + 2)(x - 4)}$$

Solution   First, factor the denominator $x^2 - 4$ as $(x - 2)(x + 2)$.

If we multiply the original denominator $(x - 2)(x + 2)$ by $x - 4$, the result is the new denominator $(x - 2)(x + 2)(x - 4)$. Thus, we multiply by 1 in the form of $\dfrac{x - 4}{x - 4}$.

$$\frac{5}{x^2 - 4} = \underbrace{\frac{5}{(x - 2)(x + 2)}}_{\substack{\text{Factored} \\ \text{denominator}}} = \frac{5}{(x - 2)(x + 2)} \cdot \frac{x - 4}{x - 4}$$

$$= \frac{5(x - 4)}{(x - 2)(x + 2)(x - 4)}$$

$$= \frac{5x - 20}{(x - 2)(x + 2)(x - 4)}$$ □

PRACTICE
10   Write the rational expression as an equivalent rational expression with the given denominator.

$$\frac{3}{x^2 - 2x - 15} = \frac{}{(x - 2)(x + 3)(x - 5)}$$ ■

✓ Vocabulary, Readiness & Video Check

Use the choices below to fill in each blank. Not all choices will be used.

$$\frac{9}{22} \qquad \frac{5}{22} \qquad \frac{9}{11} \qquad \frac{5}{11} \qquad \frac{ac}{b} \qquad \frac{a - c}{b} \qquad \frac{a + c}{b} \qquad \frac{5 - 6 + x}{x} \qquad \frac{5 - (6 + x)}{x}$$

1. $\dfrac{7}{11} + \dfrac{2}{11} = $ _____

2. $\dfrac{7}{11} - \dfrac{2}{11} = $ _____

3. $\dfrac{a}{b} + \dfrac{c}{b} = $ _____

4. $\dfrac{a}{b} - \dfrac{c}{b} = $ _____

5. $\dfrac{5}{x} - \dfrac{6 + x}{x} = $ _____

Martin-Gay Interactive Videos

See Video 7.3 ◉

Watch the section lecture video and answer the following questions.

OBJECTIVE 1
6. In ▦ Example 3, why is it important to place parentheses around the second numerator when writing as one expression?

OBJECTIVE 2
7. In ▦ Examples 4 and 5, we factor the denominators completely. How does this help determine the LCD?

OBJECTIVE 3
8. Based on ▦ Example 6, complete the following statements. To write an equivalent rational expression, you multiply the _____ of a rational expression by the same expression as the denominator. This means you're multiplying the original rational expression by a factor of _____ and therefore not changing the _____ of the original expression.

7.3 Exercise Set MyMathLab®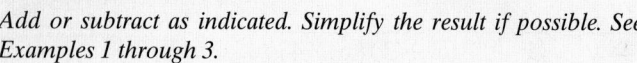

Add or subtract as indicated. Simplify the result if possible. See Examples 1 through 3.

1. $\dfrac{a+1}{13} + \dfrac{8}{13}$

2. $\dfrac{x+1}{7} + \dfrac{6}{7}$

3. $\dfrac{4m}{3n} + \dfrac{5m}{3n}$

4. $\dfrac{3p}{2q} + \dfrac{11p}{2q}$

5. $\dfrac{4m}{m-6} - \dfrac{24}{m-6}$

6. $\dfrac{8y}{y-2} - \dfrac{16}{y-2}$

7. $\dfrac{9}{3+y} + \dfrac{y+1}{3+y}$

8. $\dfrac{9}{y+9} + \dfrac{y-5}{y+9}$

9. $\dfrac{5x^2+4x}{x-1} - \dfrac{6x+3}{x-1}$

10. $\dfrac{x^2+9x}{x+7} - \dfrac{4x+14}{x+7}$

11. $\dfrac{4a}{a^2+2a-15} - \dfrac{12}{a^2+2a-15}$

12. $\dfrac{3y}{y^2+3y-10} - \dfrac{6}{y^2+3y-10}$

13. $\dfrac{2x+3}{x^2-x-30} - \dfrac{x-2}{x^2-x-30}$

14. $\dfrac{3x-1}{x^2+5x-6} - \dfrac{2x-7}{x^2+5x-6}$

15. $\dfrac{2x+1}{x-3} + \dfrac{3x+6}{x-3}$

16. $\dfrac{4p-3}{2p+7} + \dfrac{3p+8}{2p+7}$

17. $\dfrac{2x^2}{x-5} - \dfrac{25+x^2}{x-5}$

18. $\dfrac{6x^2}{2x-5} - \dfrac{25+2x^2}{2x-5}$

19. $\dfrac{5x+4}{x-1} - \dfrac{2x+7}{x-1}$

20. $\dfrac{7x+1}{x-4} - \dfrac{2x+21}{x-4}$

Find the LCD for each list of rational expressions. See Examples 4 through 8.

21. $\dfrac{19}{2x}, \dfrac{5}{4x^3}$

22. $\dfrac{17x}{4y^5}, \dfrac{2}{8y}$

23. $\dfrac{9}{8x}, \dfrac{3}{2x+4}$

24. $\dfrac{1}{6y}, \dfrac{3x}{4y+12}$

25. $\dfrac{2}{x+3}, \dfrac{5}{x-2}$

26. $\dfrac{-6}{x-1}, \dfrac{4}{x+5}$

27. $\dfrac{x}{x+6}, \dfrac{10}{3x+18}$

28. $\dfrac{12}{x+5}, \dfrac{x}{4x+20}$

29. $\dfrac{8x^2}{(x-6)^2}, \dfrac{13x}{5x-30}$

30. $\dfrac{9x^2}{7x-14}, \dfrac{6x}{(x-2)^2}$

31. $\dfrac{1}{3x+3}, \dfrac{7}{2x^2+4x+2}$

32. $\dfrac{19x+5}{4x-12}, \dfrac{3}{2x^2-12x+18}$

33. $\dfrac{5}{x-8}, \dfrac{3}{8-x}$

34. $\dfrac{2x+5}{3x-7}, \dfrac{5}{7-3x}$

35. $\dfrac{5x+1}{x^2+3x-4}, \dfrac{3x}{x^2+2x-3}$

36. $\dfrac{4}{x^2+4x+3}, \dfrac{4x-2}{x^2+10x+21}$

37. $\dfrac{2x}{3x^2+4x+1}, \dfrac{7}{2x^2-x-1}$

38. $\dfrac{3x}{4x^2+5x+1}, \dfrac{5}{3x^2-2x-1}$

39. $\dfrac{1}{x^2-16}, \dfrac{x+6}{2x^3-8x^2}$

40. $\dfrac{5}{x^2-25}, \dfrac{x+9}{3x^3-15x^2}$

Rewrite each rational expression as an equivalent rational expression with the given denominator. See Examples 9 and 10.

41. $\dfrac{3}{2x} = \dfrac{}{4x^2}$

42. $\dfrac{3}{9y^5} = \dfrac{}{72y^9}$

43. $\dfrac{6}{3a} = \dfrac{}{12ab^2}$

44. $\dfrac{5}{4y^2x} = \dfrac{}{32y^3x^2}$

45. $\dfrac{9}{2x+6} = \dfrac{}{2y(x+3)}$

46. $\dfrac{4x+1}{3x+6} = \dfrac{}{3y(x+2)}$

47. $\dfrac{9a+2}{5a+10} = \dfrac{}{5b(a+2)}$

48. $\dfrac{5+y}{2x^2+10} = \dfrac{}{4(x^2+5)}$

49. $\dfrac{x}{x^3+6x^2+8x} = \dfrac{}{x(x+4)(x+2)(x+1)}$

50. $\dfrac{5x}{x^3+2x^2-3x} = \dfrac{}{x(x-1)(x-5)(x+3)}$

51. $\dfrac{9y-1}{15x^2-30} = \dfrac{}{30x^2-60}$

52. $\dfrac{6m-5}{3x^2-9} = \dfrac{}{12x^2-36}$

MIXED PRACTICE SECTIONS 7.2 AND 7.3

Perform the indicated operations.

53. $\dfrac{5x}{7} + \dfrac{9x}{7}$

54. $\dfrac{5x}{7} \cdot \dfrac{9x}{7}$

55. $\dfrac{x+3}{4} \div \dfrac{2x-1}{4}$

56. $\dfrac{x+3}{4} - \dfrac{2x-1}{4}$

57. $\dfrac{x^2}{x-6} - \dfrac{5x+6}{x-6}$

58. $\dfrac{x^2+5x}{x^2-25} \cdot \dfrac{3x-15}{x^2}$

59. $\dfrac{-2x}{x^3 - 8x} + \dfrac{3x}{x^3 - 8x}$

60. $\dfrac{-2x}{x^3 - 8x} \div \dfrac{3x}{x^3 - 8x}$

61. $\dfrac{12x - 6}{x^2 + 3x} \cdot \dfrac{4x^2 + 13x + 3}{4x^2 - 1}$

62. $\dfrac{x^3 + 7x^2}{3x^3 - x^2} \div \dfrac{5x^2 + 36x + 7}{9x^2 - 1}$

REVIEW AND PREVIEW

Perform each indicated operation. See Section 1.3.

63. $\dfrac{2}{3} + \dfrac{5}{7}$

64. $\dfrac{9}{10} - \dfrac{3}{5}$

65. $\dfrac{1}{6} - \dfrac{3}{4}$

66. $\dfrac{11}{15} + \dfrac{5}{9}$

67. $\dfrac{1}{12} + \dfrac{3}{20}$

68. $\dfrac{7}{30} + \dfrac{3}{18}$

CONCEPT EXTENSIONS

For Exercises 69 and 70, see the Concept Check in this section.

69. Choose the correct LCD of $\dfrac{11a^3}{4a - 20}$ and $\dfrac{15a^3}{(a - 5)^2}$.

    a. $4a(a - 5)(a + 5)$ **b.** $a - 5$ **c.** $(a - 5)^2$

    d. $4(a - 5)^2$ **e.** $(4a - 20)(a - 5)^2$

70. Choose the correct LCD of $\dfrac{5}{14x^2}$ and $\dfrac{y}{6x^3}$.

    a. $84x^5$ **b.** $84x^3$

    c. $42x^3$ **d.** $42x^5$

For Exercises 71 and 72, an algebra student approaches you with each incorrect solution. Find the error and correct the work shown below.

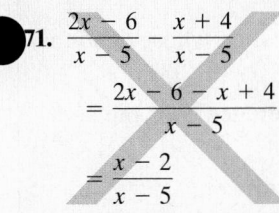

71. $\dfrac{2x - 6}{x - 5} - \dfrac{x + 4}{x - 5}$

$= \dfrac{2x - 6 - x + 4}{x - 5}$

$= \dfrac{x - 2}{x - 5}$

72. $\dfrac{x}{x + 3} + \dfrac{2}{x + 3}$

$= \dfrac{x + 2}{x + 3}$

$= \dfrac{2}{3}$

Multiple choice. Select the correct result.

73. $\dfrac{3}{x} + \dfrac{y}{x} =$

    a. $\dfrac{3 + y}{x^2}$ **b.** $\dfrac{3 + y}{2x}$ **c.** $\dfrac{3 + y}{x}$

74. $\dfrac{3}{x} - \dfrac{y}{x} =$

    a. $\dfrac{3 - y}{x^2}$ **b.** $\dfrac{3 - y}{2x}$ **c.** $\dfrac{3 - y}{x}$

75. $\dfrac{3}{x} \cdot \dfrac{y}{x} =$

    a. $\dfrac{3y}{x}$ **b.** $\dfrac{3y}{x^2}$ **c.** $3y$

76. $\dfrac{3}{x} \div \dfrac{y}{x} =$

    a. $\dfrac{3}{y}$ **b.** $\dfrac{y}{3}$ **c.** $\dfrac{3}{x^2 y}$

Write each rational expression as an equivalent expression with a denominator of $x - 2$.

77. $\dfrac{5}{2 - x}$

78. $\dfrac{8y}{2 - x}$

79. $-\dfrac{7 + x}{2 - x}$

80. $\dfrac{x - 3}{-(x - 2)}$

△ 81. A square has a side of length $\dfrac{5}{x - 2}$ meters. Express its perimeter as a rational expression.

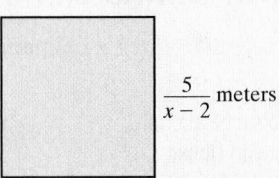

$\dfrac{5}{x - 2}$ meters

△ 82. A trapezoid has sides of the indicated lengths. Find its perimeter.

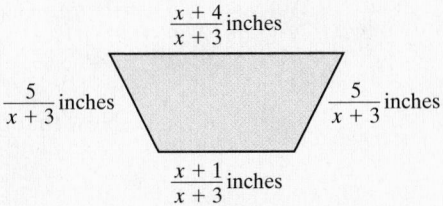

$\dfrac{x + 4}{x + 3}$ inches

$\dfrac{5}{x + 3}$ inches $\dfrac{5}{x + 3}$ inches

$\dfrac{x + 1}{x + 3}$ inches

83. Write two rational expressions with the same denominator whose sum is $\dfrac{5}{3x - 1}$.

84. Write two rational expressions with the same denominator whose difference is $\dfrac{x - 7}{x^2 + 1}$.

85. The planet Mercury revolves around the Sun in 88 Earth days. It takes Jupiter 4332 Earth days to make one revolution around the Sun. (*Source:* National Space Science Data Center) If the two planets are aligned as shown in the figure, how long will it take for them to align again?

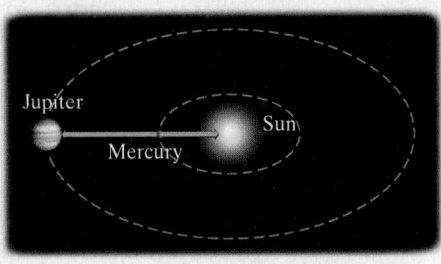

86. You are throwing a barbecue and you want to make sure that you purchase the same number of hot dogs as hot dog buns. Hot dogs come 8 to a package and hot dog buns come 12 to a package. What is the least number of each type of package you should buy?

87. Write some instructions to help a friend who is having difficulty finding the LCD of two rational expressions.

88. Explain why the LCD of the rational expressions $\dfrac{7}{x + 1}$ and $\dfrac{9x}{(x + 1)^2}$ is $(x + 1)^2$ and not $(x + 1)^3$.

89. In your own words, describe how to add or subtract two rational expressions with the same denominators.

90. Explain the similarities between subtracting $\dfrac{3}{8}$ from $\dfrac{7}{8}$ and subtracting $\dfrac{6}{x + 3}$ from $\dfrac{9}{x + 3}$.

7.4 Adding and Subtracting Rational Expressions with Unlike Denominators ▶

OBJECTIVE

1 Add and Subtract Rational Expressions with Unlike Denominators. ▶

OBJECTIVE

1 Adding and Subtracting Rational Expressions with Unlike Denominators ▶

Let's add $\dfrac{3}{8}$ and $\dfrac{1}{6}$. From the previous section, the LCD of 8 and 6 is 24. Now let's write equivalent fractions with denominator 24 by multiplying by different forms of 1.

$$\frac{3}{8} = \frac{3}{8} \cdot 1 = \frac{3}{8} \cdot \frac{3}{3} = \frac{3 \cdot 3}{8 \cdot 3} = \frac{9}{24}$$

$$\frac{1}{6} = \frac{1}{6} \cdot 1 = \frac{1}{6} \cdot \frac{4}{4} = \frac{1 \cdot 4}{6 \cdot 4} = \frac{4}{24}$$

Now that the denominators are the same, we may add.

$$\frac{3}{8} + \frac{1}{6} = \frac{9}{24} + \frac{4}{24} = \frac{9 + 4}{24} = \frac{13}{24}$$

We add or subtract rational expressions the same way. You may want to use the steps below.

Adding or Subtracting Rational Expressions with Unlike Denominators

Step 1. Find the LCD of the rational expressions.

Step 2. Rewrite each rational expression as an equivalent expression whose denominator is the LCD found in Step 1.

Step 3. Add or subtract numerators and write the sum or difference over the common denominator.

Step 4. Simplify or write the rational expression in simplest form.

EXAMPLE 1 Perform each indicated operation.

a. $\dfrac{a}{4} - \dfrac{2a}{8}$ **b.** $\dfrac{3}{10x^2} + \dfrac{7}{25x}$

Solution

a. First, we must find the LCD. Since $4 = 2^2$ and $8 = 2^3$, the LCD $= 2^3 = 8$. Next we write each fraction as an equivalent fraction with the denominator 8, then we subtract.

$$\frac{a}{4} - \frac{2a}{8} = \frac{a(2)}{4(2)} - \frac{2a}{8} = \frac{2a}{8} - \frac{2a}{8} = \frac{2a - 2a}{8} = \frac{0}{8} = 0$$

Multiplying the numerator and denominator by 2 is the same as multiplying by $\dfrac{2}{2}$ or 1.

b. Since $10x^2 = 2 \cdot 5 \cdot x \cdot x$ and $25x = 5 \cdot 5 \cdot x$, the LCD $= 2 \cdot 5^2 \cdot x^2 = 50x^2$. We write each fraction as an equivalent fraction with a denominator of $50x^2$.

$$\frac{3}{10x^2} + \frac{7}{25x} = \frac{3(5)}{10x^2(5)} + \frac{7(2x)}{25x(2x)}$$

$$= \frac{15}{50x^2} + \frac{14x}{50x^2}$$

$$= \frac{15 + 14x}{50x^2} \qquad \text{Add numerators. Write the sum over the common denominator.} \quad \square$$

PRACTICE
1   Perform each indicated operation.

a. $\dfrac{2x}{5} - \dfrac{6x}{15}$

b. $\dfrac{7}{8a} + \dfrac{5}{12a^2}$

EXAMPLE 2   Subtract: $\dfrac{6x}{x^2 - 4} - \dfrac{3}{x + 2}$

Solution Since $x^2 - 4 = (x + 2)(x - 2)$, the LCD $= (x - 2)(x + 2)$. We write equivalent expressions with the LCD as denominators.

$$\frac{6x}{x^2 - 4} - \frac{3}{x + 2} = \frac{6x}{(x - 2)(x + 2)} - \frac{3(x - 2)}{(x + 2)(x - 2)}$$

$$= \frac{6x - 3(x - 2)}{(x + 2)(x - 2)} \quad \text{Subtract numerators. Write the difference over the common denominator.}$$

$$= \frac{6x - 3x + 6}{(x + 2)(x - 2)} \quad \text{Apply the distributive property in the numerator.}$$

$$= \frac{3x + 6}{(x + 2)(x - 2)} \quad \text{Combine like terms in the numerator.}$$

Next we factor the numerator to see if this rational expression can be simplified.

$$= \frac{3(x + 2)}{(x + 2)(x - 2)} \quad \text{Factor.}$$

$$= \frac{3}{x - 2} \quad \text{Divide out common factors to simplify.} \quad \square$$

PRACTICE
2   Subtract: $\dfrac{12x}{x^2 - 25} - \dfrac{6}{x + 5}$

EXAMPLE 3   Add: $\dfrac{2}{3t} + \dfrac{5}{t + 1}$

Solution The LCD is $3t(t + 1)$. We write each rational expression as an equivalent rational expression with a denominator of $3t(t + 1)$.

$$\frac{2}{3t} + \frac{5}{t + 1} = \frac{2(t + 1)}{3t(t + 1)} + \frac{5(3t)}{(t + 1)(3t)}$$

$$= \frac{2(t + 1) + 5(3t)}{3t(t + 1)} \quad \text{Add numerators. Write the sum over the common denominator.}$$

$$= \frac{2t + 2 + 15t}{3t(t + 1)} \quad \text{Apply the distributive property in the numerator.}$$

$$= \frac{17t + 2}{3t(t + 1)} \quad \text{Combine like terms in the numerator.} \quad \square$$

PRACTICE
3   Add: $\dfrac{3}{5y} + \dfrac{2}{y + 1}$

EXAMPLE 4   Subtract: $\dfrac{7}{x - 3} - \dfrac{9}{3 - x}$

Solution To find a common denominator, we notice that $x - 3$ and $3 - x$ are opposites. That is, $3 - x = -(x - 3)$. We write the denominator $3 - x$ as $-(x - 3)$ and simplify.

(Continued on next page)

$$\frac{7}{x-3} - \frac{9}{3-x} = \frac{7}{x-3} - \frac{9}{-(x-3)}$$

$$= \frac{7}{x-3} - \frac{-9}{x-3} \qquad \text{Apply } \frac{a}{-b} = \frac{-a}{b}.$$

$$= \frac{7-(-9)}{x-3} \qquad \text{Subtract numerators. Write the difference over the common denominator.}$$

$$= \frac{16}{x-3}$$

PRACTICE
4 Subtract: $\dfrac{6}{x-5} - \dfrac{7}{5-x}$

EXAMPLE 5 Add: $1 + \dfrac{m}{m+1}$

Solution Recall that 1 is the same as $\dfrac{1}{1}$. The LCD of $\dfrac{1}{1}$ and $\dfrac{m}{m+1}$ is $m+1$.

$$1 + \frac{m}{m+1} = \frac{1}{1} + \frac{m}{m+1} \qquad \text{Write 1 as } \frac{1}{1}.$$

$$= \frac{1(m+1)}{1(m+1)} + \frac{m}{m+1} \qquad \text{Multiply both the numerator and the denominator of } \frac{1}{1} \text{ by } m+1.$$

$$= \frac{m+1+m}{m+1} \qquad \text{Add numerators. Write the sum over the common denominator.}$$

$$= \frac{2m+1}{m+1} \qquad \text{Combine like terms in the numerator.}$$

PRACTICE
5 Add: $2 + \dfrac{b}{b+3}$

EXAMPLE 6 Subtract: $\dfrac{3}{2x^2 + x} - \dfrac{2x}{6x+3}$

Solution First, we factor the denominators.

$$\frac{3}{2x^2+x} - \frac{2x}{6x+3} = \frac{3}{x(2x+1)} - \frac{2x}{3(2x+1)}$$

The LCD is $3x(2x+1)$. We write equivalent expressions with denominators of $3x(2x+1)$.

$$= \frac{3(3)}{x(2x+1)(3)} - \frac{2x(x)}{3(2x+1)(x)}$$

$$= \frac{9-2x^2}{3x(2x+1)} \qquad \text{Subtract numerators. Write the difference over the common denominator.}$$

PRACTICE
6 Subtract: $\dfrac{5}{2x^2+3x} - \dfrac{3x}{4x+6}$

EXAMPLE 7 Add: $\dfrac{2x}{x^2 + 2x + 1} + \dfrac{x}{x^2 - 1}$

Solution First we factor the denominators.

$$\frac{2x}{x^2 + 2x + 1} + \frac{x}{x^2 - 1} = \frac{2x}{(x + 1)(x + 1)} + \frac{x}{(x + 1)(x - 1)}$$

Now we write the rational expressions as equivalent expressions with denominators of $(x + 1)(x + 1)(x - 1)$, the LCD.

$$= \frac{2x(x - 1)}{(x + 1)(x + 1)(x - 1)} + \frac{x(x + 1)}{(x + 1)(x - 1)(x + 1)}$$

$$= \frac{2x(x - 1) + x(x + 1)}{(x + 1)^2(x - 1)} \quad \text{Add numerators. Write the sum over the common denominator.}$$

$$= \frac{2x^2 - 2x + x^2 + x}{(x + 1)^2(x - 1)} \quad \text{Apply the distributive property in the numerator.}$$

$$= \frac{3x^2 - x}{(x + 1)^2(x - 1)} \quad \text{or} \quad \frac{x(3x - 1)}{(x + 1)^2(x - 1)}$$

The numerator was factored as a last step to see if the rational expression could be simplified further. Since there are no factors common to the numerator and the denominator, we can't simplify further. □

PRACTICE
7 Add: $\dfrac{2x}{x^2 + 7x + 12} + \dfrac{3x}{x^2 - 9}$ ■

✓ Vocabulary, Readiness & Video Check

Match each exercise with the first step needed to perform the operation. Do not actually perform the operation.

1. $\dfrac{3}{4} - \dfrac{y}{4}$ **2.** $\dfrac{2}{a} \cdot \dfrac{3}{(a + 6)}$ **3.** $\dfrac{x + 1}{x} \div \dfrac{x - 1}{x}$ **4.** $\dfrac{9}{x - 2} - \dfrac{x}{x + 2}$

A. Multiply the first rational expression by the reciprocal of the second rational expression.

B. Find the LCD. Write each expression as an equivalent expression with the LCD as denominator.

C. Multiply numerators and multiply denominators.

D. Subtract numerators. Place the difference over a common denominator.

Martin-Gay Interactive Videos

See Video 7.4 ◉

Watch the section lecture video and answer the following question.

OBJECTIVE
1   **5.** What special case is shown in ▦ Example 2 and what's the purpose of presenting it?

7.4 Exercise Set MyMathLab®

MIXED PRACTICE

Perform each indicated operation. Simplify if possible. See Examples 1 through 7.

1. $\dfrac{4}{2x} + \dfrac{9}{3x}$

2. $\dfrac{15}{7a} + \dfrac{8}{6a}$

3. $\dfrac{15a}{b} - \dfrac{6b}{5}$

4. $\dfrac{4c}{d} - \dfrac{8d}{5}$

▶ 5. $\dfrac{3}{x} + \dfrac{5}{2x^2}$

6. $\dfrac{14}{3x^2} + \dfrac{6}{x}$

7. $\dfrac{6}{x+1} + \dfrac{10}{2x+2}$

8. $\dfrac{8}{x+4} - \dfrac{3}{3x+12}$

9. $\dfrac{3}{x+2} - \dfrac{2x}{x^2-4}$

10. $\dfrac{5}{x-4} + \dfrac{4x}{x^2-16}$

11. $\dfrac{3}{4x} + \dfrac{8}{x-2}$

12. $\dfrac{5}{y^2} - \dfrac{y}{2y+1}$

▶ 13. $\dfrac{6}{x-3} + \dfrac{8}{3-x}$

14. $\dfrac{15}{y-4} + \dfrac{20}{4-y}$

15. $\dfrac{9}{x-3} + \dfrac{9}{3-x}$

16. $\dfrac{5}{a-7} + \dfrac{5}{7-a}$

17. $\dfrac{-8}{x^2-1} - \dfrac{7}{1-x^2}$

18. $\dfrac{-9}{25x^2-1} + \dfrac{7}{1-25x^2}$

19. $\dfrac{5}{x} + 2$

20. $\dfrac{7}{x^2} - 5x$

21. $\dfrac{5}{x-2} + 6$

22. $\dfrac{6y}{y+5} + 1$

▶ 23. $\dfrac{y+2}{y+3} - 2$

24. $\dfrac{7}{2x-3} - 3$

25. $\dfrac{-x+2}{x} - \dfrac{x-6}{4x}$

26. $\dfrac{-y+1}{y} - \dfrac{2y-5}{3y}$

27. $\dfrac{5x}{x+2} - \dfrac{3x-4}{x+2}$

28. $\dfrac{7x}{x-3} - \dfrac{4x+9}{x-3}$

29. $\dfrac{3x^4}{7} - \dfrac{4x^2}{21}$

30. $\dfrac{5x}{6} + \dfrac{11x^2}{2}$

31. $\dfrac{1}{x+3} - \dfrac{1}{(x+3)^2}$

32. $\dfrac{5x}{(x-2)^2} - \dfrac{3}{x-2}$

33. $\dfrac{4}{5b} + \dfrac{1}{b-1}$

34. $\dfrac{1}{y+5} + \dfrac{2}{3y}$

35. $\dfrac{2}{m} + 1$

36. $\dfrac{6}{x} - 1$

37. $\dfrac{2x}{x-7} - \dfrac{x}{x-2}$

38. $\dfrac{9x}{x-10} - \dfrac{x}{x-3}$

39. $\dfrac{6}{1-2x} - \dfrac{4}{2x-1}$

40. $\dfrac{10}{3n-4} - \dfrac{5}{4-3n}$

41. $\dfrac{7}{(x+1)(x-1)} + \dfrac{8}{(x+1)^2}$

42. $\dfrac{5}{(x+1)(x+5)} - \dfrac{2}{(x+5)^2}$

43. $\dfrac{x}{x^2-1} - \dfrac{2}{x^2-2x+1}$

44. $\dfrac{x}{x^2-4} - \dfrac{5}{x^2-4x+4}$

▶ 45. $\dfrac{3a}{2a+6} - \dfrac{a-1}{a+3}$

46. $\dfrac{1}{x+y} - \dfrac{y}{x^2-y^2}$

47. $\dfrac{y-1}{2y+3} + \dfrac{3}{(2y+3)^2}$

48. $\dfrac{x-6}{5x+1} + \dfrac{6}{(5x+1)^2}$

49. $\dfrac{5}{2-x} + \dfrac{x}{2x-4}$

50. $\dfrac{-1}{a-2} + \dfrac{4}{4-2a}$

51. $\dfrac{15}{x^2+6x+9} + \dfrac{2}{x+3}$

52. $\dfrac{2}{x^2+4x+4} + \dfrac{1}{x+2}$

53. $\dfrac{13}{x^2-5x+6} - \dfrac{5}{x-3}$

54. $\dfrac{-7}{y^2 - 3y + 2} - \dfrac{2}{y - 1}$

55. $\dfrac{70}{m^2 - 100} + \dfrac{7}{2(m + 10)}$

56. $\dfrac{27}{y^2 - 81} + \dfrac{3}{2(y + 9)}$

57. $\dfrac{x + 8}{x^2 - 5x - 6} + \dfrac{x + 1}{x^2 - 4x - 5}$

58. $\dfrac{x + 4}{x^2 + 12x + 20} + \dfrac{x + 1}{x^2 + 8x - 20}$

59. $\dfrac{5}{4n^2 - 12n + 8} - \dfrac{3}{3n^2 - 6n}$

60. $\dfrac{6}{5y^2 - 25y + 30} - \dfrac{2}{4y^2 - 8y}$

MIXED PRACTICE

Perform the indicated operations. Addition, subtraction, multiplication, and division of rational expressions are included here.

61. $\dfrac{15x}{x + 8} \cdot \dfrac{2x + 16}{3x}$

62. $\dfrac{9z + 5}{15} \cdot \dfrac{5z}{81z^2 - 25}$

63. $\dfrac{8x + 7}{3x + 5} - \dfrac{2x - 3}{3x + 5}$

64. $\dfrac{2z^2}{4z - 1} - \dfrac{z - 2z^2}{4z - 1}$

65. $\dfrac{5a + 10}{18} \div \dfrac{a^2 - 4}{10a}$

66. $\dfrac{9}{x^2 - 1} \div \dfrac{12}{3x + 3}$

67. $\dfrac{5}{x^2 - 3x + 2} + \dfrac{1}{x - 2}$

68. $\dfrac{4}{2x^2 + 5x - 3} + \dfrac{2}{x + 3}$

REVIEW AND PREVIEW

Solve the following linear and quadratic equations. See Sections 2.3 and 6.6.

69. $3x + 5 = 7$

70. $5x - 1 = 8$

71. $2x^2 - x - 1 = 0$

72. $4x^2 - 9 = 0$

73. $4(x + 6) + 3 = -3$

74. $2(3x + 1) + 15 = -7$

CONCEPT EXTENSIONS

Perform each indicated operation.

75. $\dfrac{3}{x} - \dfrac{2x}{x^2 - 1} + \dfrac{5}{x + 1}$

76. $\dfrac{5}{x - 2} + \dfrac{7x}{x^2 - 4} - \dfrac{11}{x}$

77. $\dfrac{5}{x^2 - 4} + \dfrac{2}{x^2 - 4x + 4} - \dfrac{3}{x^2 - x - 6}$

78. $\dfrac{8}{x^2 + 6x + 5} - \dfrac{3x}{x^2 + 4x - 5} + \dfrac{2}{x^2 - 1}$

79. $\dfrac{9}{x^2 + 9x + 14} - \dfrac{3x}{x^2 + 10x + 21} + \dfrac{x + 4}{x^2 + 5x + 6}$

80. $\dfrac{x + 10}{x^2 - 3x - 4} - \dfrac{8}{x^2 + 6x + 5} - \dfrac{9}{x^2 + x - 20}$

81. A board of length $\dfrac{3}{x + 4}$ inches was cut into two pieces. If one piece is $\dfrac{1}{x - 4}$ inches, express the length of the other piece as a rational expression.

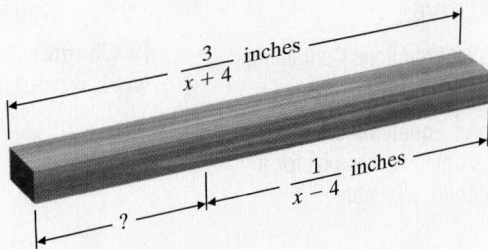

82. The length of a rectangle is $\dfrac{3}{y - 5}$ feet, while its width is $\dfrac{2}{y}$ feet. Find its perimeter and then find its area.

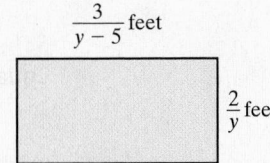

83. In ice hockey, penalty killing percentage is a statistic calculated as $1 - \dfrac{G}{P}$, where G = opponent's power play goals and P = opponent's power play opportunities. Simplify this expression.

84. The dose of medicine prescribed for a child depends on the child's age A in years and the adult dose D for the medication. Two expressions that give a child's dose are Young's Rule, $\dfrac{DA}{A + 12}$, and Cowling's Rule, $\dfrac{D(A + 1)}{24}$. Find an expression for the difference in the doses given by these expressions.

85. Explain when the LCD of the rational expressions in a sum is the product of the denominators.

86. Explain when the LCD is the same as one of the denominators of a rational expression to be added or subtracted.

△ **87.** Two angles are said to be complementary if the sum of their measures is 90°. If one angle measures $\dfrac{40}{x}$ degrees, find the measure of its complement.

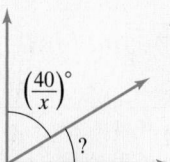

△ **88.** Two angles are said to be supplementary if the sum of their measures is 180°. If one angle measures $\dfrac{x+2}{x}$ degrees, find the measure of its supplement.

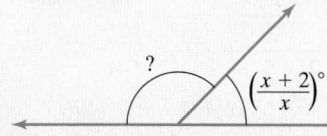

89. In your own words, explain how to add two rational expressions with different denominators.

90. In your own words, explain how to subtract two rational expressions with different denominators.

7.5 Solving Equations Containing Rational Expressions

OBJECTIVES

1. Solve Equations Containing Rational Expressions.
2. Solve Equations Containing Rational Expressions for a Specified Variable.

OBJECTIVE

1 Solving Equations Containing Rational Expressions

In Chapter 2, we solved equations containing fractions. In this section, we continue the work we began in Chapter 2 by solving equations containing rational expressions.

Examples of Equations Containing Rational Expressions

$$\frac{x}{2} + \frac{8}{3} = \frac{1}{6} \quad \text{and} \quad \frac{4x}{x^2 + x - 30} + \frac{2}{x - 5} = \frac{1}{x + 6}$$

To solve equations such as these, use the multiplication property of equality to clear the equation of fractions by multiplying both sides of the equation by the LCD.

EXAMPLE I Solve: $\dfrac{x}{2} + \dfrac{8}{3} = \dfrac{1}{6}$

Solution The LCD of denominators 2, 3, and 6 is 6, so we multiply both sides of the equation by 6.

$$6\left(\frac{x}{2} + \frac{8}{3}\right) = 6\left(\frac{1}{6}\right)$$

> **Helpful Hint**
> Make sure that *each* term is multiplied by the LCD, 6.

$$6\left(\frac{x}{2}\right) + 6\left(\frac{8}{3}\right) = 6\left(\frac{1}{6}\right) \quad \text{Use the distributive property.}$$

$$3 \cdot x + 16 = 1 \quad \text{Multiply and simplify.}$$

$$3x = -15 \quad \text{Subtract 16 from both sides.}$$

$$x = -5 \quad \text{Divide both sides by 3.}$$

Check: To check, we replace x with -5 in the original equation.

$$\frac{x}{2} + \frac{8}{3} = \frac{1}{6}$$

$$\frac{-5}{2} + \frac{8}{3} \stackrel{?}{=} \frac{1}{6} \quad \text{Replace } x \text{ with } -5$$

$$\frac{1}{6} = \frac{1}{6} \quad \text{True}$$

This number checks, so the solution is -5.

PRACTICE
1 Solve: $\dfrac{x}{3} + \dfrac{4}{5} = \dfrac{2}{15}$

EXAMPLE 2   Solve: $\dfrac{t-4}{2} - \dfrac{t-3}{9} = \dfrac{5}{18}$

Solution   The LCD of denominators 2, 9, and 18 is 18, so we multiply both sides of the equation by 18.

$$18\left(\frac{t-4}{2} - \frac{t-3}{9}\right) = 18\left(\frac{5}{18}\right)$$

Helpful Hint	
Multiply *each* term by 18.	

$$18\left(\frac{t-4}{2}\right) - 18\left(\frac{t-3}{9}\right) = 18\left(\frac{5}{18}\right) \quad \text{Use the distributive property.}$$

$$9(t-4) - 2(t-3) = 5 \qquad\qquad \text{Simplify.}$$

$$9t - 36 - 2t + 6 = 5 \qquad\qquad \text{Use the distributive property.}$$

$$7t - 30 = 5 \qquad\qquad\qquad \text{Combine like terms.}$$

$$7t = 35$$

$$t = 5 \qquad\qquad\qquad\quad \text{Solve for } t.$$

Check:

$$\frac{t-4}{2} - \frac{t-3}{9} = \frac{5}{18}$$

$$\frac{5-4}{2} - \frac{5-3}{9} \stackrel{?}{=} \frac{5}{18} \quad \text{Replace } t \text{ with 5.}$$

$$\frac{1}{2} - \frac{2}{9} \stackrel{?}{=} \frac{5}{18} \quad \text{Simplify.}$$

$$\frac{5}{18} = \frac{5}{18} \quad \text{True}$$

The solution is 5.   □

PRACTICE
2   Solve: $\dfrac{x+4}{4} - \dfrac{x-3}{3} = \dfrac{11}{12}$

Recall from Section 7.1 that a rational expression is defined for all real numbers except those that make the denominator of the expression 0. This means that if an equation contains *rational expressions with variables in the denominator*, we must be certain that a proposed solution does not make any denominator 0. If replacing the variable with the proposed solution makes any denominator 0, the rational expression is undefined and this proposed solution must be rejected.

EXAMPLE 3   Solve: $3 - \dfrac{6}{x} = x + 8$

Solution   In this equation, 0 cannot be a solution because if x is 0, the rational expression $\dfrac{6}{x}$ is undefined. The LCD is x, so we multiply both sides of the equation by x.

$$x\left(3 - \frac{6}{x}\right) = x(x + 8)$$

Helpful Hint	
Multiply *each* term by x.	

$$x(3) - x\left(\frac{6}{x}\right) = x \cdot x + x \cdot 8 \quad \text{Use the distributive property.}$$

$$3x - 6 = x^2 + 8x \quad \text{Simplify.}$$

Now we write the quadratic equation in standard form and solve for x.

$$0 = x^2 + 5x + 6$$

$$0 = (x + 3)(x + 2) \qquad\qquad \text{Factor.}$$

$$x + 3 = 0 \quad \text{or} \quad x + 2 = 0 \quad \text{Set each factor equal to 0 and solve.}$$

$$x = -3 \qquad\qquad x = -2$$

(Continued on next page)

Notice that neither -3 nor -2 makes the denominator in the original equation equal to 0.

Check: To check these solutions, we replace x in the original equation by -3, and then by -2.

If $x = -3$:

$$3 - \frac{6}{x} = x + 8$$

$$3 - \frac{6}{-3} \stackrel{?}{=} -3 + 8$$

$$3 - (-2) \stackrel{?}{=} 5$$

$$5 = 5 \qquad \text{True}$$

If $x = -2$:

$$3 - \frac{6}{x} = x + 8$$

$$3 - \frac{6}{-2} \stackrel{?}{=} -2 + 8$$

$$3 - (-3) \stackrel{?}{=} 6$$

$$6 = 6 \qquad \text{True}$$

Both -3 and -2 are solutions. ☐

PRACTICE
3 Solve: $8 + \dfrac{7}{x} = x + 2$

The following steps may be used to solve an equation containing rational expressions.

> **Solving an Equation Containing Rational Expressions**
>
> **Step 1.** Multiply both sides of the equation by the LCD of all rational expressions in the equation.
>
> **Step 2.** Remove any grouping symbols and solve the resulting equation.
>
> **Step 3.** Check the solution in the original equation.

EXAMPLE 4 Solve: $\dfrac{4x}{x^2 + x - 30} + \dfrac{2}{x - 5} = \dfrac{1}{x + 6}$

Solution The denominator $x^2 + x - 30$ factors as $(x + 6)(x - 5)$. The LCD is then $(x + 6)(x - 5)$, so we multiply both sides of the equation by this LCD.

$$(x + 6)(x - 5)\left(\frac{4x}{x^2 + x - 30} + \frac{2}{x - 5}\right) = (x + 6)(x - 5)\left(\frac{1}{x + 6}\right) \quad \text{Multiply by the LCD.}$$

$$(x + 6)(x - 5) \cdot \frac{4x}{x^2 + x - 30} + (x + 6)(x - 5) \cdot \frac{2}{x - 5} \quad \text{Apply the distributive property.}$$

$$= (x + 6)(x - 5) \cdot \frac{1}{x + 6}$$

$$4x + 2(x + 6) = x - 5 \qquad \text{Simplify.}$$

$$4x + 2x + 12 = x - 5 \qquad \text{Apply the distributive property.}$$

$$6x + 12 = x - 5 \qquad \text{Combine like terms.}$$

$$5x = -17$$

$$x = -\frac{17}{5} \qquad \text{Divide both sides by 5.}$$

Check: Check by replacing x with $-\dfrac{17}{5}$ in the original equation. The solution is $-\dfrac{17}{5}$.

PRACTICE
4 Solve: $\dfrac{6x}{x^2 - 5x - 14} - \dfrac{3}{x + 2} = \dfrac{1}{x - 7}$

EXAMPLE 5   Solve: $\dfrac{2x}{x-4} = \dfrac{8}{x-4} + 1$

Solution Multiply both sides by the LCD, $x - 4$.

$$(x-4)\left(\frac{2x}{x-4}\right) = (x-4)\left(\frac{8}{x-4}+1\right)$$ Multiply by the LCD. Notice that 4 cannot be a solution.

$$(x-4)\cdot\frac{2x}{x-4} = (x-4)\cdot\frac{8}{x-4} + (x-4)\cdot 1$$ Use the distributive property.

$$2x = 8 + (x-4)$$ Simplify.

$$2x = 4 + x$$

$$x = 4$$

Notice that 4 makes the denominators 0 in the original equation. Therefore, 4 is *not* a solution.

This equation has *no solution*. □

PRACTICE
5   Solve: $\dfrac{7x}{x-2} = \dfrac{14}{x-2} + 4$ ■

> **Helpful Hint**
>
> As we can see from Example 5, it is important to check the proposed solution(s) in the *original* equation.

✔ **CONCEPT CHECK**

When can we clear fractions by multiplying through by the LCD?

a. When adding or subtracting rational expressions
b. When solving an equation containing rational expressions
c. Both of these
d. Neither of these

EXAMPLE 6   Solve: $x + \dfrac{14}{x-2} = \dfrac{7x}{x-2} + 1$

Solution Notice the denominators in this equation. We can see that 2 can't be a solution. The LCD is $x - 2$, so we multiply both sides of the equation by $x - 2$.

$$(x-2)\left(x + \frac{14}{x-2}\right) = (x-2)\left(\frac{7x}{x-2}+1\right)$$

$$(x-2)(x) + (x-2)\left(\frac{14}{x-2}\right) = (x-2)\left(\frac{7x}{x-2}\right) + (x-2)(1)$$

$$x^2 - 2x + 14 = 7x + x - 2$$ Simplify.

$$x^2 - 2x + 14 = 8x - 2$$ Combine like terms.

$$x^2 - 10x + 16 = 0$$ Write the quadratic equation in standard form.

$$(x-8)(x-2) = 0$$ Factor.

$$x - 8 = 0 \quad \text{or} \quad x - 2 = 0$$ Set each factor equal to 0.

$$x = 8 \qquad\qquad x = 2$$ Solve.

(Continued on next page)

As we have already noted, 2 can't be a solution of the original equation. So we need only replace x with 8 in the original equation. We find that 8 is a solution; the only solution is 8.

PRACTICE
6   Solve: $x + \dfrac{x}{x-5} = \dfrac{5}{x-5} - 7$

OBJECTIVE
2   Solving Equations for a Specified Variable

The last example in this section is an equation containing several variables, and we are directed to solve for one of the variables. The steps used in the preceding examples can be applied to solve equations for a specified variable as well.

EXAMPLE 7   Solve $\dfrac{1}{a} + \dfrac{1}{b} = \dfrac{1}{x}$ for x.

Solution This type of equation often models a work problem, as we shall see in Section 7.6. The LCD is abx, so we multiply both sides by abx.

$$abx\left(\frac{1}{a} + \frac{1}{b}\right) = abx\left(\frac{1}{x}\right)$$

$$abx\left(\frac{1}{a}\right) + abx\left(\frac{1}{b}\right) = abx \cdot \frac{1}{x}$$

$bx + ax = ab$	Simplify.
$x(b + a) = ab$	Factor out x from each term on the left side.
$\dfrac{x(b + a)}{b + a} = \dfrac{ab}{b + a}$	Divide both sides by $b + a$.
$x = \dfrac{ab}{b + a}$	Simplify.

This equation is now solved for x.

PRACTICE
7   Solve $\dfrac{1}{a} + \dfrac{1}{b} = \dfrac{1}{x}$ for b.

Graphing Calculator Explorations

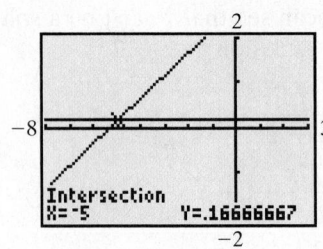

A graphing calculator may be used to check solutions of equations containing rational expressions. For example, to check the solution of Example 1, $\dfrac{x}{2} + \dfrac{8}{3} = \dfrac{1}{6}$, graph $Y_1 = \dfrac{x}{2} + \dfrac{8}{3}$ and $Y_2 = \dfrac{1}{6}$.

Use TRACE and ZOOM, or use INTERSECT, to find the point of intersection. The point of intersection has an x-value of -5, so the solution of the equation is -5.

Use a graphing calculator to check the examples of this section.

1. Example 2   2. Example 3

3. Example 5   4. Example 6

Vocabulary, Readiness & Video Check

Multiple choice. Choose the correct response.

1. Multiply both sides of the equation $\dfrac{3x}{2} + 5 = \dfrac{1}{4}$ by 4. The result is:

   a. $3x + 5 = 1$ **b.** $6x + 5 = 1$ **c.** $6x + 20 = 1$ **d.** $6x + 9 = 1$

2. Multiply both sides of the equation $\dfrac{1}{x} - \dfrac{3}{5x} = 2$ by $5x$. The result is:

   a. $1 - 3 = 10x$ **b.** $5 - 3 = 10x$ **c.** $1 - 3 = 7x$ **d.** $5 - 3 = 7x$

Choose the correct LCD for the fractions in each equation.

3. Equation: $\dfrac{9}{x} + \dfrac{3}{4} = \dfrac{1}{12}$; LCD: _____

   a. $4x$ **b.** $12x$ **c.** $48x$ **d.** x

4. Equation: $\dfrac{8}{3x} - \dfrac{1}{x} = \dfrac{7}{9}$; LCD: _____

   a. x **b.** $3x$ **c.** $27x$ **d.** $9x$

5. Equation: $\dfrac{9}{x - 1} = \dfrac{7}{(x - 1)^2}$; LCD: _____

   a. $(x - 1)^2$ **b.** $x - 1$ **c.** $(x - 1)^3$ **d.** 63

6. Equation: $\dfrac{1}{x - 2} - \dfrac{3}{x^2 - 4} = 8$; LCD: _____

   a. $(x - 2)$ **b.** $x + 2$ **c.** $x^2 - 4$ **d.** $(x - 2)(x^2 - 4)$

Martin-Gay Interactive Videos

See Video 7.5 ◉

Watch the section lecture video and answer the following questions.

OBJECTIVE 1
7. After multiplying through by the LCD and then simplifying, why is it important to take a moment and determine whether you have a linear or a quadratic equation before you finish solving the problem?

OBJECTIVE 1
8. From ▦ Examples 2–5, what extra step is needed when checking solutions of an equation containing rational expressions?

OBJECTIVE 2
9. The steps for solving ▦ Example 6 for a specified variable are the same as what other steps? How do you treat this specified variable?

7.5 Exercise Set   MyMathLab®

Solve each equation and check each solution. See Examples 1 through 3.

1. $\dfrac{x}{5} + 3 = 9$ **2.** $\dfrac{x}{5} - 2 = 9$

3. $\dfrac{x}{2} + \dfrac{5x}{4} = \dfrac{x}{12}$ **4.** $\dfrac{x}{6} + \dfrac{4x}{3} = \dfrac{x}{18}$

5. $2 - \dfrac{8}{x} = 6$ **6.** $5 + \dfrac{4}{x} = 1$

7. $2 + \dfrac{10}{x} = x + 5$ **8.** $6 + \dfrac{5}{y} = y - \dfrac{2}{y}$

9. $\dfrac{a}{5} = \dfrac{a - 3}{2}$ **10.** $\dfrac{b}{5} = \dfrac{b + 2}{6}$

▶ **11.** $\dfrac{x - 3}{5} + \dfrac{x - 2}{2} = \dfrac{1}{2}$ **12.** $\dfrac{a + 5}{4} + \dfrac{a + 5}{2} = \dfrac{a}{8}$

Solve each equation and check each proposed solution. See Examples 4 through 6.

13. $\dfrac{3}{2a - 5} = -1$ **14.** $\dfrac{6}{4 - 3x} = -3$

15. $\dfrac{4y}{y - 4} + 5 = \dfrac{5y}{y - 4}$ **16.** $\dfrac{2a}{a + 2} - 5 = \dfrac{7a}{a + 2}$

17. $2 + \dfrac{3}{a - 3} = \dfrac{a}{a - 3}$

18. $\dfrac{2y}{y - 2} - \dfrac{4}{y - 2} = 4$

19. $\dfrac{1}{x + 3} + \dfrac{6}{x^2 - 9} = 1$

20. $\dfrac{1}{x + 2} + \dfrac{4}{x^2 - 4} = 1$

21. $\dfrac{2y}{y + 4} + \dfrac{4}{y + 4} = 3$

22. $\dfrac{5y}{y + 1} - \dfrac{3}{y + 1} = 4$

23. $\dfrac{2x}{x + 2} - 2 = \dfrac{x - 8}{x - 2}$

24. $\dfrac{4y}{y - 3} - 3 = \dfrac{3y - 1}{y + 3}$

MIXED PRACTICE

Solve each equation. See Examples 1 through 6.

25. $\dfrac{2}{y} + \dfrac{1}{2} = \dfrac{5}{2y}$

26. $\dfrac{6}{3y} + \dfrac{3}{y} = 1$

27. $\dfrac{a}{a - 6} = \dfrac{-2}{a - 1}$

28. $\dfrac{5}{x - 6} = \dfrac{x}{x - 2}$

29. $\dfrac{11}{2x} + \dfrac{2}{3} = \dfrac{7}{2x}$

30. $\dfrac{5}{3} - \dfrac{3}{2x} = \dfrac{3}{2}$

31. $\dfrac{2}{x - 2} + 1 = \dfrac{x}{x + 2}$

32. $1 + \dfrac{3}{x + 1} = \dfrac{x}{x - 1}$

33. $\dfrac{x + 1}{3} - \dfrac{x - 1}{6} = \dfrac{1}{6}$

34. $\dfrac{3x}{5} - \dfrac{x - 6}{3} = -\dfrac{2}{5}$

35. $\dfrac{t}{t - 4} = \dfrac{t + 4}{6}$

36. $\dfrac{15}{x + 4} = \dfrac{x - 4}{x}$

37. $\dfrac{y}{2y + 2} + \dfrac{2y - 16}{4y + 4} = \dfrac{2y - 3}{y + 1}$

38. $\dfrac{1}{x + 2} = \dfrac{4}{x^2 - 4} - \dfrac{1}{x - 2}$

39. $\dfrac{4r - 4}{r^2 + 5r - 14} + \dfrac{2}{r + 7} = \dfrac{1}{r - 2}$

40. $\dfrac{3}{x + 3} = \dfrac{12x + 19}{x^2 + 7x + 12} - \dfrac{5}{x + 4}$

41. $\dfrac{x + 1}{x + 3} = \dfrac{x^2 - 11x}{x^2 + x - 6} - \dfrac{x - 3}{x - 2}$

42. $\dfrac{2t + 3}{t - 1} - \dfrac{2}{t + 3} = \dfrac{5 - 6t}{t^2 + 2t - 3}$

Solve each equation for the indicated variable. See Example 7.

43. $R = \dfrac{E}{I}$ for I (Electronics: resistance of a circuit)

44. $T = \dfrac{V}{Q}$ for Q (Water purification: settling time)

45. $T = \dfrac{2U}{B + E}$ for B (Merchandising: stock turnover rate)

46. $i = \dfrac{A}{t + B}$ for t (Hydrology: rainfall intensity)

47. $B = \dfrac{705w}{h^2}$ for w (Health: body-mass index)

48. $\dfrac{A}{W} = L$ for W (Geometry: area of a rectangle)

49. $N = R + \dfrac{V}{G}$ for G (Urban forestry: tree plantings per year)

50. $C = \dfrac{D(A + 1)}{24}$ for A (Medicine: Cowling's Rule for child's dose)

51. $\dfrac{C}{\pi r} = 2$ for r (Geometry: circumference of a circle)

52. $W = \dfrac{CE^2}{2}$ for C (Electronics: energy stored in a capacitor)

53. $\dfrac{1}{y} + \dfrac{1}{3} = \dfrac{1}{x}$ for x

54. $\dfrac{1}{5} + \dfrac{2}{y} = \dfrac{1}{x}$ for x

REVIEW AND PREVIEW

TRANSLATING

Write each phrase as an expression.

55. The reciprocal of x

56. The reciprocal of $x + 1$

57. The reciprocal of x, added to the reciprocal of 2

58. The reciprocal of x, subtracted from the reciprocal of 5

Answer each question.

59. If a tank is filled in 3 hours, what fractional part of the tank is filled in 1 hour?

60. If a strip of beach is cleaned in 4 hours, what fractional part of the beach is cleaned in 1 hour?

Identify the x- and y-intercepts. See Section 3.3.

61.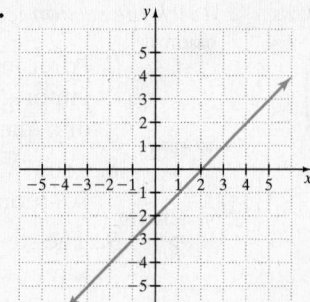

62.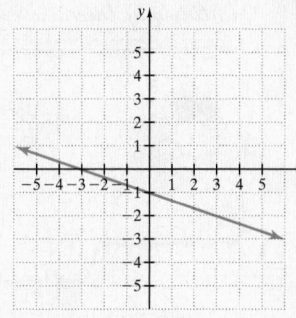

63.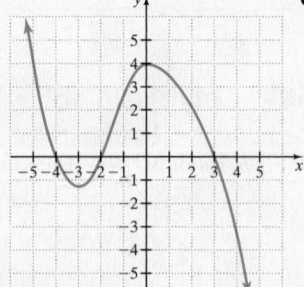

64.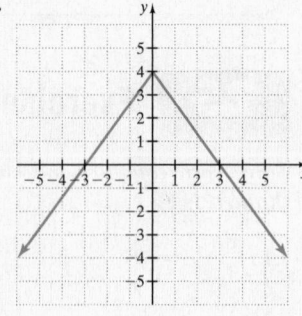

CONCEPT EXTENSIONS

65. Explain the difference between solving an equation such as $\dfrac{x}{2} + \dfrac{3}{4} = \dfrac{x}{4}$ for x and performing an operation such as adding $\dfrac{x}{2} + \dfrac{3}{4}$.

66. When solving an equation such as $\dfrac{y}{4} = \dfrac{y}{2} - \dfrac{1}{4}$, we may multiply all terms by 4. When subtracting two rational expressions such as $\dfrac{y}{2} - \dfrac{1}{4}$, we may not. Explain why.

Determine whether each of the following is an equation or an expression. If it is an equation, then solve it for its variable. If it is an expression, perform the indicated operation.

67. $\dfrac{1}{x} + \dfrac{5}{9}$

68. $\dfrac{1}{x} + \dfrac{5}{9} = \dfrac{2}{3}$

69. $\dfrac{5}{x-1} - \dfrac{2}{x} = \dfrac{5}{x(x-1)}$

70. $\dfrac{5}{x-1} - \dfrac{2}{x}$

Recall that two angles are supplementary if the sum of their measures is 180°. Find the measures of the following supplementary angles.

△ **71.**

△ **72.**

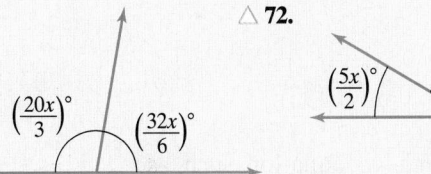

Recall that two angles are complementary if the sum of their measures is 90°. Find the measures of the following complementary angles.

△ **73.**

△ **74.**

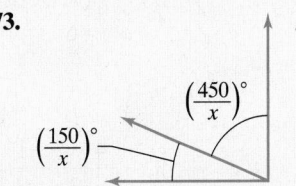
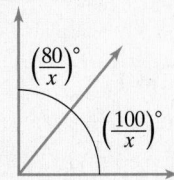

Solve each equation.

75. $\dfrac{5}{a^2 + 4a + 3} + \dfrac{2}{a^2 + a - 6} - \dfrac{3}{a^2 - a - 2} = 0$

76. $\dfrac{-2}{a^2 + 2a - 8} + \dfrac{1}{a^2 + 9a + 20} = \dfrac{-4}{a^2 + 3a - 10}$

Integrated Review | Summary on Rational Expressions

Sections 7.1–7.5

It is important to know the difference between performing operations with rational expressions and solving an equation containing rational expressions. Study the examples below.

Performing Operations with Rational Expressions

Adding: $\dfrac{1}{x} + \dfrac{1}{x+5} = \dfrac{1 \cdot (x+5)}{x(x+5)} + \dfrac{1 \cdot x}{x(x+5)} = \dfrac{x+5+x}{x(x+5)} = \dfrac{2x+5}{x(x+5)}$

Subtracting: $\dfrac{3}{x} - \dfrac{5}{x^2 y} = \dfrac{3 \cdot xy}{x \cdot xy} - \dfrac{5}{x^2 y} = \dfrac{3xy - 5}{x^2 y}$

Multiplying: $\dfrac{2}{x} \cdot \dfrac{5}{x-1} = \dfrac{2 \cdot 5}{x(x-1)} = \dfrac{10}{x(x-1)}$

Dividing: $\dfrac{4}{2x+1} \div \dfrac{x-3}{x} = \dfrac{4}{2x+1} \cdot \dfrac{x}{x-3} = \dfrac{4x}{(2x+1)(x-3)}$

Solving an Equation Containing Rational Expressions

To solve an equation containing rational expressions, we clear the equation of fractions by multiplying both sides by the LCD.

$$\dfrac{3}{x} - \dfrac{5}{x-1} = \dfrac{1}{x(x-1)} \qquad \text{Note that } x \text{ can't be 0 or 1.}$$

$$x(x-1)\left(\dfrac{3}{x}\right) - x(x-1)\left(\dfrac{5}{x-1}\right) = x(x-1) \cdot \dfrac{1}{x(x-1)} \qquad \text{Multiply both sides by the LCD.}$$

$$3(x-1) - 5x = 1 \qquad \text{Simplify.}$$

$$3x - 3 - 5x = 1 \qquad \text{Use the distributive property.}$$

$$-2x - 3 = 1 \qquad \text{Combine like terms.}$$

$$-2x = 4 \qquad \text{Add 3 to both sides.}$$

$$x = -2 \qquad \text{Divide both sides by } -2.$$

Determine whether each of the following is an equation or an expression. If it is an equation, solve it for its variable. If it is an expression, perform the indicated operation.

1. $\dfrac{1}{x} + \dfrac{2}{3}$

2. $\dfrac{3}{a} + \dfrac{5}{6}$

3. $\dfrac{1}{x} + \dfrac{2}{3} = \dfrac{3}{x}$

4. $\dfrac{3}{a} + \dfrac{5}{6} = 1$

5. $\dfrac{2}{x-1} - \dfrac{1}{x}$

6. $\dfrac{4}{x-3} - \dfrac{1}{x}$

7. $\dfrac{2}{x+1} - \dfrac{1}{x} = 1$

8. $\dfrac{4}{x-3} - \dfrac{1}{x} = \dfrac{6}{x(x-3)}$

9. $\dfrac{15x}{x+8} \cdot \dfrac{2x+16}{3x}$

10. $\dfrac{9z+5}{15} \cdot \dfrac{5z}{81z^2 - 25}$

11. $\dfrac{2x+1}{x-3} + \dfrac{3x+6}{x-3}$

12. $\dfrac{4p-3}{2p+7} + \dfrac{3p+8}{2p+7}$

13. $\dfrac{x+5}{7} = \dfrac{8}{2}$

14. $\dfrac{1}{2} = \dfrac{x-1}{8}$

15. $\dfrac{5a+10}{18} \div \dfrac{a^2 - 4}{10a}$

16. $\dfrac{9}{x^2 - 1} + \dfrac{12}{3x+3}$

17. $\dfrac{x+2}{3x-1} + \dfrac{5}{(3x-1)^2}$

18. $\dfrac{4}{(2x-5)^2} + \dfrac{x+1}{2x-5}$

19. $\dfrac{x-7}{x} - \dfrac{x+2}{5x}$

20. $\dfrac{10x-9}{x} - \dfrac{x-4}{3x}$

21. $\dfrac{3}{x+3} = \dfrac{5}{x^2 - 9} - \dfrac{2}{x-3}$

22. $\dfrac{9}{x^2 - 4} + \dfrac{2}{x+2} = \dfrac{-1}{x-2}$

23. Explain the difference between solving an equation, such as $\dfrac{x}{5} + \dfrac{3}{10} = \dfrac{x}{10}$, for x and performing an operation such as adding $\dfrac{x}{5} + \dfrac{3}{10}$.

24. When solving an equation such as $\dfrac{y}{10} = \dfrac{y}{5} - \dfrac{1}{10}$, we may multiply all terms by 10. When subtracting two rational expressions such as $\dfrac{y}{5} - \dfrac{1}{10}$, we may not. Explain why.

7.6 Proportion and Problem Solving with Rational Equations

OBJECTIVES

1 Solve Proportions.

2 Use Proportions to Solve Problems.

3 Solve Problems about Numbers.

4 Solve Problems about Work.

5 Solve Problems about Distance.

OBJECTIVE

1 Solving Proportions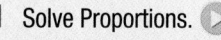

A **ratio** is the quotient of two numbers or two quantities. For example, the ratio of 2 to 5 can be written as $\dfrac{2}{5}$, the quotient of 2 and 5.

If two ratios are equal, we say the ratios are **in proportion** to each other. A **proportion** is a mathematical statement that two ratios are equal.

For example, the equation $\dfrac{1}{2} = \dfrac{4}{8}$ is a proportion, as is $\dfrac{x}{5} = \dfrac{8}{10}$, because both sides of the equations are ratios. When we want to emphasize the equation as a proportion, we

read the proportion $\dfrac{1}{2} = \dfrac{4}{8}$ as "one is to two as four is to eight"

In a proportion, cross products are equal. To understand cross products, let's start with the proportion

$$\dfrac{a}{b} = \dfrac{c}{d}$$

and multiply both sides by the LCD, bd.

$$bd\left(\frac{a}{b}\right) = bd\left(\frac{c}{d}\right)$$ Multiply both sides by the LCD, bd.

$$\underbrace{ad}_{} = \underbrace{bc}_{}$$ Simplify.

Cross product   Cross product

Notice why ad and bc are called cross products.

$$ad \nwarrow \qquad \nearrow bc$$
$$\frac{a}{b} \overset{\times}{=} \frac{c}{d}$$

Cross Products

If $\dfrac{a}{b} = \dfrac{c}{d}$, then $ad = bc$.

For example, if

$$\frac{1}{2} = \frac{4}{8}, \text{ then } 1 \cdot 8 = 2 \cdot 4 \text{ or}$$

$$8 = 8$$

Notice that a proportion contains four numbers (or expressions). If any three numbers are known, we can solve and find the fourth number.

EXAMPLE 1   Solve for x: $\dfrac{45}{x} = \dfrac{5}{7}$

Solution   This is an equation with rational expressions as well as a proportion. Below are two ways to solve.

Since this is a rational equation, we can use the methods of the previous section.	Since this is also a proportion, we may set cross products equal.
$$\frac{45}{x} = \frac{5}{7}$$	$$\frac{45}{x} \overset{\times}{=} \frac{5}{7}$$
$7x \cdot \dfrac{45}{x} = 7x \cdot \dfrac{5}{7}$ Multiply both sides by the LCD, $7x$.	
$7 \cdot 45 = x \cdot 5$ Divide out common factors.	$45 \cdot 7 = x \cdot 5$ Set cross products equal.
$315 = 5x$ Multiply.	$315 = 5x$ Multiply.
$\dfrac{315}{5} = \dfrac{5x}{5}$ Divide both sides by 5.	$\dfrac{315}{5} = \dfrac{5x}{5}$ Divide both sides by 5.
$63 = x$ Simplify.	$63 = x$ Simplify.

Check: Both methods give us a solution of 63. To check, substitute 63 for x in the original proportion. The solution is 63.   □

PRACTICE
1   Solve for x: $\dfrac{36}{x} = \dfrac{4}{11}$ ∎

In this section, if the rational equation is a proportion, we will use cross products to solve.

> **EXAMPLE 2**   Solve for x: $\dfrac{x-5}{3} = \dfrac{x+2}{5}$

Solution

$$\dfrac{x-5}{3} \bowtie \dfrac{x+2}{5}$$

$5(x-5) = 3(x+2)$ Set cross products equal.

$5x - 25 = 3x + 6$ Multiply.

$5x = 3x + 31$ Add 25 to both sides.

$2x = 31$ Subtract $3x$ from both sides.

$\dfrac{2x}{2} = \dfrac{31}{2}$ Divide both sides by 2.

$x = \dfrac{31}{2}$

Check: Verify that $\dfrac{31}{2}$ is the solution.

PRACTICE
2   Solve for x: $\dfrac{3x+2}{9} = \dfrac{x-1}{2}$

OBJECTIVE
2 Using Proportions to Solve Problems

Proportions can be used to model and solve many real-life problems. When using proportions in this way, it is important to judge whether the solution is reasonable. Doing so helps us decide if the proportion has been formed correctly. We use the same problem-solving steps that were introduced in Section 2.4.

> **EXAMPLE 3**   **Calculating the Cost of a Redbox Rental**

Not everyone is streaming movies. Many people still rent movies and video games from Redbox. If renting 3 movies costs $4.50, how much should 5 movies cost?

Solution

1. **UNDERSTAND. Read and reread the problem.** We know that the cost of renting 5 movies is more than the cost of renting three movies, or $4.50, and less than the cost of 6 rentals, which is double the cost of 3 rentals, or $2(\$4.50) = \$9.00$. Let's suppose that 5 rentals cost $7.00. To check, we see if 3 rentals is to 5 rentals as the *price* of three rentals is to the *price* of 5 rentals. In other words, we see if

$$\dfrac{3 \text{ rentals}}{5 \text{ rentals}} = \dfrac{price \text{ of 3 rentals}}{price \text{ of 5 rentals}}$$

or

$$\dfrac{3}{5} \bowtie \dfrac{4.50}{7}$$

or

$3(7) = 5(4.50)$ Set cross products equal.

$21 = 22.5$ Not a true statement.

Thus, $7 is not correct, but we now have a better understanding of the problem. Let x = price of renting 5 videos.

2. TRANSLATE.

$$\frac{3 \text{ rentals}}{5 \text{ rentals}} = \frac{price \text{ of } 3 \text{ rentals}}{price \text{ of } 5 \text{ rentals}}$$

$$\frac{3}{5} = \frac{4.50}{x}$$

3. SOLVE.

$$\frac{3}{5} = \frac{4.50}{x}$$

$3x = 5(4.50)$ Set cross products equal.

$3x = 22.50$

$x = 7.50$ Divide both sides by **3**.

4. INTERPRET.

Check: Verify that 3 rentals is to 5 rentals as $4.50 is to $7.50. Also, notice that our solution is a reasonable one as discussed in Step 1.

State: Five rentals from Redbox cost $7.50.   □

PRACTICE
3 Four 2-liter bottles of Diet Pepsi cost $5.16. How much will seven 2-liter bottles cost?   ■

> **Helpful Hint**
>
> The proportion $\dfrac{5 \text{ rentals}}{3 \text{ rentals}} = \dfrac{price \text{ of } 5 \text{ rentals}}{price \text{ of } 3 \text{ rentals}}$ could also have been used to solve Example 3. Notice that the cross products are the same.

Similar triangles have the same shape but not necessarily the same size. In similar triangles, the measures of corresponding angles are equal, and corresponding sides are in proportion.

If triangle ABC and triangle XYZ shown are similar, then we know that the measure of angle A = the measure of angle X, the measure of angle B = the measure of angle Y, and the measure of angle C = the measure of angle Z. We also know that corresponding sides are in proportion: $\dfrac{a}{x} = \dfrac{b}{y} = \dfrac{c}{z}$.

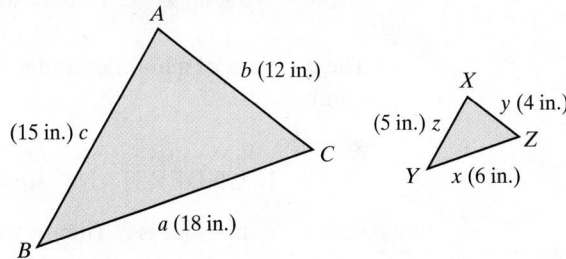

In this section, we will position similar triangles so that they have the same orientation.

To show that corresponding sides are in proportion for the triangles above, we write the ratios of the corresponding sides.

$$\frac{a}{x} = \frac{18}{6} = 3 \qquad \frac{b}{y} = \frac{12}{4} = 3 \qquad \frac{c}{z} = \frac{15}{5} = 3$$

EXAMPLE 4 Finding the Length of a Side of a Triangle

If the following two triangles are similar, find the unknown length x.

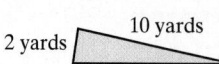

2 yards   10 yards   3 yards x yards

Solution

1. UNDERSTAND. Read the problem and study the figure.

2. TRANSLATE. Since the triangles are similar, their corresponding sides are in proportion and we have

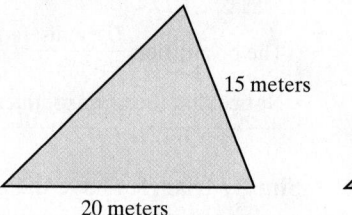

$$\frac{2}{3} = \frac{10}{x}$$

3. SOLVE. To solve, we multiply both sides by the LCD, $3x$, or cross multiply.

$$2x = 30$$
$$x = 15 \quad \text{Divide both sides by 2.}$$

4. INTERPRET.

Check: To check, replace x with 15 in the original proportion and see that a true statement results.

State: The unknown length is 15 yards. □

PRACTICE
4   If the following two triangles are similar, find x.

15 meters

x meters

20 meters   8 meters

OBJECTIVE
3 Solving Problems about Numbers

Let's continue to solve problems. The remaining problems are all modeled by rational equations.

EXAMPLE 5 Finding an Unknown Number

The quotient of a number and 6, minus $\frac{5}{3}$, is the quotient of the number and 2. Find the number.

Solution

1. UNDERSTAND. Read and reread the problem. Suppose that the unknown number is 2. Then we see if the quotient of 2 and 6, or $\frac{2}{6}$, minus $\frac{5}{3}$ is equal to the quotient of 2 and 2, or $\frac{2}{2}$.

$$\frac{2}{6} - \frac{5}{3} = \frac{1}{3} - \frac{5}{3} = -\frac{4}{3}, \text{ not } \frac{2}{2}$$

Don't forget that the purpose of a proposed solution is to better understand the problem.

Let $x =$ the unknown number.

2. TRANSLATE.

In words:

the quotient of x and 6	minus	$\frac{5}{3}$	is	the quotient of x and 2
↓	↓	↓	↓	↓

Translate: $\dfrac{x}{6}$ $-$ $\dfrac{5}{3}$ $=$ $\dfrac{x}{2}$

3. SOLVE. Here, we solve the equation $\dfrac{x}{6} - \dfrac{5}{3} = \dfrac{x}{2}$. We begin by multiplying both sides of the equation by the LCD, 6.

$$6\left(\frac{x}{6} - \frac{5}{3}\right) = 6\left(\frac{x}{2}\right)$$

$$6\left(\frac{x}{6}\right) - 6\left(\frac{5}{3}\right) = 6\left(\frac{x}{2}\right) \qquad \text{Apply the distributive property.}$$

$$x - 10 = 3x \qquad \text{Simplify.}$$

$$-10 = 2x \qquad \text{Subtract } x \text{ from both sides.}$$

$$\frac{-10}{2} = \frac{2x}{2} \qquad \text{Divide both sides by 2.}$$

$$-5 = x \qquad \text{Simplify.}$$

4. INTERPRET.

Check: To check, we verify that "the quotient of -5 and 6 minus $\dfrac{5}{3}$ is the quotient of -5 and 2," or $-\dfrac{5}{6} - \dfrac{5}{3} = -\dfrac{5}{2}$.

State: The unknown number is -5.   □

PRACTICE

5   The quotient of a number and 5, minus $\dfrac{3}{2}$, is the quotient of the number and 10. Find the number.   ∎

OBJECTIVE

4   Solving Problems about Work

The next example is often called a work problem. Work problems usually involve people or machines doing a certain task.

EXAMPLE 6   Finding Work Rates

Sam Waterton and Frank Schaffer work in a plant that manufactures automobiles. Sam can complete a quality control tour of the plant in 3 hours, while his assistant, Frank, needs 7 hours to complete the same job. The regional manager is coming to inspect the plant facilities, so both Sam and Frank are directed to complete a quality control tour at the same time. How long will this take?

Solution

1. **UNDERSTAND.** Read and reread the problem. The key idea here is the relationship between the **time** (hours) it takes to complete the job and the **part of the job** completed in 1 unit of time (hour). For example, if the **time** it takes Sam to complete the job is 3 hours, the **part of the job** he can complete in 1 hour is $\dfrac{1}{3}$. Similarly, Frank can complete $\dfrac{1}{7}$ of the job in 1 hour.

   Let $x =$ the **time** in hours it takes Sam and Frank to complete the job together. Then $\dfrac{1}{x} =$ the **part of the job** they complete in 1 hour.

(Continued on next page)

	Hours to Complete Total Job	Part of Job Completed in 1 Hour
Sam	3	$\dfrac{1}{3}$
Frank	7	$\dfrac{1}{7}$
Together	x	$\dfrac{1}{x}$

2. TRANSLATE.

In words:	part of job Sam completes in 1 hour	added to	part of job Frank completes in 1 hour	is equal to	part of job they complete together in 1 hour
	↓	↓	↓	↓	↓
Translate:	$\dfrac{1}{3}$	$+$	$\dfrac{1}{7}$	$=$	$\dfrac{1}{x}$

3. SOLVE. Here, we solve the equation $\dfrac{1}{3} + \dfrac{1}{7} = \dfrac{1}{x}$. We begin by multiplying both sides of the equation by the LCD, $21x$.

$$21x\left(\dfrac{1}{3}\right) + 21x\left(\dfrac{1}{7}\right) = 21x\left(\dfrac{1}{x}\right)$$

$$7x + 3x = 21 \qquad\qquad \text{Simplify.}$$

$$10x = 21$$

$$x = \dfrac{21}{10} \quad \text{or} \quad 2\dfrac{1}{10} \text{ hours}$$

4. INTERPRET.

Check: Our proposed solution is $2\dfrac{1}{10}$ hours. This proposed solution is reasonable since $2\dfrac{1}{10}$ hours is more than half of Sam's time and less than half of Frank's time.

Check this solution in the originally *stated* problem.

State: Sam and Frank can complete the quality control tour in $2\dfrac{1}{10}$ hours.   □

PRACTICE
6 Cindy Liu and Mary Beckwith own a landscaping company. Cindy can complete a certain garden planting in 3 hours, while Mary takes 4 hours to complete the same job. If both of them work together, how long will it take to plant the garden?   ■

✔ **CONCEPT CHECK**
Solve $E = mc^2$
a. for m.   **b.** for c^2.

Answers to Concept Check:
a. $m = \dfrac{E}{c^2}$ **b.** $c^2 = \dfrac{E}{m}$

OBJECTIVE
5 Solving Problems About Distance

Next we look at a problem solved by the distance formula,

$$d = r \cdot t$$

EXAMPLE 7   **Finding Speeds of Vehicles**

A car travels 180 miles in the same time that a truck travels 120 miles. If the car's speed is 20 miles per hour faster than the truck's, find the car's speed and the truck's speed.

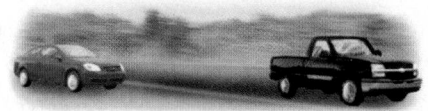

Solution

1. UNDERSTAND. Read and reread the problem. Suppose that the truck's speed is 45 miles per hour. Then the car's speed is 20 miles per hour more, or 65 miles per hour.

   We are given that the car travels 180 miles in the same time that the truck travels 120 miles. To find the time it takes the car to travel 180 miles, remember that since $d = rt$, we know that $\dfrac{d}{r} = t$.

Car's Time	**Truck's Time**
$t = \dfrac{d}{r} = \dfrac{180}{65} = 2\dfrac{50}{65} = 2\dfrac{10}{13}$ hours	$t = \dfrac{d}{r} = \dfrac{120}{45} = 2\dfrac{30}{45} = 2\dfrac{2}{3}$ hours

   Since the times are not the same, our proposed solution is not correct. But we have a better understanding of the problem.

   Let x = the speed of the truck.

   Since the car's speed is 20 miles per hour faster than the truck's, then

   $$x + 20 = \text{the speed of the car.}$$

   Use the formula $d = r \cdot t$ or **d**istance = **r**ate · **t**ime. Prepare a chart to organize the information in the problem.

   > **Helpful Hint**
   > If $d = r \cdot t$,
   > then $t = \dfrac{d}{r}$
   > or *time* = $\dfrac{distance}{rate}$.

	Distance	=	Rate	·	Time
Truck	120		x		$\begin{cases} 120 \leftarrow \text{distance} \\ x \leftarrow \text{rate} \end{cases}$
Car	180		$x + 20$		$\begin{cases} 180 \leftarrow \text{distance} \\ x + 20 \leftarrow \text{rate} \end{cases}$

2. TRANSLATE. Since the car and the truck travel the same amount of time, we have that

   In words: car's time   =   truck's time

   Translate: $\dfrac{180}{x + 20}$ = $\dfrac{120}{x}$

3. SOLVE. We begin by multiplying both sides of the equation by the LCD, $x(x + 20)$, or cross multiplying.

   $$\dfrac{180}{x + 20} = \dfrac{120}{x}$$

   $$180x = 120(x + 20)$$
   $$180x = 120x + 2400 \quad \text{Use the distributive property.}$$
   $$60x = 2400 \quad \text{Subtract } 120x \text{ from both sides.}$$
   $$x = 40 \quad \text{Divide both sides by 60.}$$

(Continued on next page)

4. INTERPRET. The speed of the truck is 40 miles per hour. The speed of the car must then be $x + 20$ or 60 miles per hour.

Check: Find the time it takes the car to travel 180 miles and the time it takes the truck to travel 120 miles.

Car's Time	Truck's Time
$t = \dfrac{d}{r} = \dfrac{180}{60} = 3$ hours	$t = \dfrac{d}{r} = \dfrac{120}{40} = 3$ hours

Since both travel the same amount of time, the proposed solution is correct.

State: The car's speed is 60 miles per hour and the truck's speed is 40 miles per hour. ☐

PRACTICE
7 A bus travels 180 miles in the same time that a car travels 240 miles. If the car's speed is 15 miles per hour faster than the speed of the bus, find the speed of the car and the speed of the bus. ■

✓ Vocabulary, Readiness & Video Check

Without solving algebraically, select the best choice for each exercise.

1. One person can complete a job in 7 hours. A second person can complete the same job in 5 hours. How long will it take them to complete the job if they work together?

a. more than 7 hours

b. between 5 and 7 hours

c. less than 5 hours

2. One inlet pipe can fill a pond in 30 hours. A second inlet pipe can fill the same pond in 25 hours. How long before the pond is filled if both inlet pipes are on?

a. less than 25 hours

b. between 25 and 30 hours

c. more than 30 hours

TRANSLATING

Given the variable in the first column, use the phrase in the second column to translate into an expression and then continue to the phrase in the third column to translate into another expression.

3.	A number: x	The reciprocal of the number:	The reciprocal of the number, decreased by 3:
4.	A number: y	The reciprocal of the number:	The reciprocal of the number, increased by 2:
5.	A number: z	The sum of the number and 5:	The reciprocal of the sum of the number and 5:
6.	A number: x	The difference of the number and 1:	The reciprocal of the difference of the number and 1:
7.	A number: y	Twice the number:	Eleven divided by twice the number:
8.	A number: z	Triple the number:	Negative ten divided by triple the number:

Martin-Gay Interactive Videos

See Video 7.6 ◉

Watch the section lecture video and answer the following questions.

OBJECTIVE 1

9. Based on ▣ Examples 1 and 2, can proportions only be solved by using cross products? Explain.

OBJECTIVE 2

10. In ▣ Example 3 we are told there are many ways to set up a correct proportion. Why does this fact make it even more important to check that your solution is reasonable?

OBJECTIVE 3

11. What words or phrases in ▣ Example 5 told you to translate into an equation containing rational expressions?

OBJECTIVE 4

12. From ▣ Example 6, how can you determine a somewhat reasonable answer to a work problem before you even begin to solve it?

OBJECTIVE 5

13. The following problem is worded like ▣ Example 7 in the video, but using different quantities.

A car travels 325 miles in the same time that a motorcycle travels 290 miles. If the car's speed is 7 miles per hour more than the motorcycle's, find the speed of the car and the speed of the motorcycle. Fill in the table and set up an equation based on this problem (do not solve). Use ▣ Example 7 in the video as a model for your work.

	d =	r ·	t
car			
motorcycle			

7.6 Exercise Set MyMathLab® ▸

Solve each proportion. See Examples 1 and 2. For additional exercises on proportion and proportion applications, see Appendix B.

1. $\dfrac{2}{3} = \dfrac{x}{6}$

2. $\dfrac{x}{2} = \dfrac{16}{6}$

▸ 3. $\dfrac{x}{10} = \dfrac{5}{9}$

4. $\dfrac{9}{4x} = \dfrac{6}{2}$

▸ 5. $\dfrac{x+1}{2x+3} = \dfrac{2}{3}$

6. $\dfrac{x+1}{x+2} = \dfrac{5}{3}$

7. $\dfrac{9}{5} = \dfrac{12}{3x+2}$

8. $\dfrac{6}{11} = \dfrac{27}{3x-2}$

Solve. See Example 3.

9. The ratio of the weight of an object on Earth to the weight of the same object on Pluto is 100 to 3. If an elephant weighs 4100 pounds on Earth, find the elephant's weight on Pluto.

10. If a 170-pound person weighs approximately 65 pounds on Mars, about how much does a 9000-pound satellite weigh? Round your answer to the nearest pound.

▸ 11. There are 110 calories per 177.4 grams of Frosted Flakes cereal. Find how many calories are in 212.5 grams of this cereal. Round to the nearest whole calorie.

12. On an architect's blueprint, 1 inch corresponds to 4 feet. Find the length of a wall represented by a line that is $3\dfrac{7}{8}$ inches long on the blueprint.

Find the unknown length x or y in the following pairs of similar triangles. See Example 4.

△ 13.

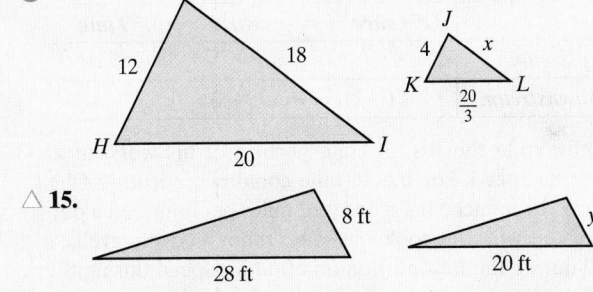

△ 14.

△ 15.

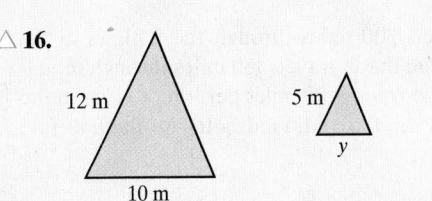

△ 16.

Solve the following. See Example 5.

17. Three times the reciprocal of a number equals 9 times the reciprocal of 6. Find the number.

18. Twelve divided by the sum of x and 2 equals the quotient of 4 and the difference of x and 2. Find x.

19. If twice a number added to 3 is divided by the number plus 1, the result is three halves. Find the number.

20. A number added to the product of 6 and the reciprocal of the number equals -5. Find the number.

See Example 6.

21. Smith Engineering found that an experienced surveyor surveys a roadbed in 4 hours. An apprentice surveyor needs 5 hours to survey the same stretch of road. If the two work together, find how long it takes them to complete the job.

22. An experienced bricklayer constructs a small wall in 3 hours. The apprentice completes the job in 6 hours. Find how long it takes if they work together.

23. In 2 minutes, a conveyor belt moves 300 pounds of recyclable aluminum from the delivery truck to a storage area. A smaller belt moves the same quantity of cans the same distance in 6 minutes. If both belts are used, find how long it takes to move the cans to the storage area.

24. Find how long it takes the conveyor belts described in Exercise 23 to move 1200 pounds of cans. (*Hint:* Think of 1200 pounds as four 300-pound jobs.)

See Example 7.

25. A jogger begins her workout by jogging to the park, a distance of 12 miles. She then jogs home at the same speed but along a different route. This return trip is 18 miles and her time is one hour longer. Complete the accompanying chart and use it to find her jogging speed.

	Distance	=	Rate	·	Time
Trip to Park	12				
Return Trip	18				

26. A boat can travel 9 miles upstream in the same amount of time it takes to travel 11 miles downstream. If the current of the river is 3 miles per hour, complete the chart below and use it to find the speed of the boat in still water.

	Distance	=	Rate	·	Time
Upstream	9		$r - 3$		
Downstream	11		$r + 3$		

27. A cyclist rode the first 20-mile portion of his workout at a constant speed. For the 16-mile cooldown portion of his workout, he reduced his speed by 2 miles per hour. Each portion of the workout took the same time. Find the cyclist's speed during the first portion and find his speed during the cooldown portion.

28. A semi-truck travels 300 miles through the flatland in the same amount of time that it travels 180 miles through mountains. The rate of the truck is 20 miles per hour slower in the mountains than in the flatland. Find both the flatland rate and mountain rate.

MIXED PRACTICE

Solve the following. See Examples 1 through 7. (Note: Some exercises can be modeled by equations without rational expressions.)

29. A human factors expert recommends that there be at least 9 square feet of floor space in a college classroom for every student in the class. Find the minimum floor space that 40 students need.

30. Due to space problems at a local university, a 20-foot by 12-foot conference room is converted into a classroom. Find the maximum number of students the room can accommodate. (See Exercise 29.)

31. One-fourth equals the quotient of a number and 8. Find the number.

32. Four times a number added to 5 is divided by 6. The result is $\frac{7}{2}$. Find the number.

33. Marcus and Tony work for Lombardo's Pipe and Concrete. Mr. Lombardo is preparing an estimate for a customer. He knows that Marcus lays a slab of concrete in 6 hours. Tony lays the same size slab in 4 hours. If both work on the job and the cost of labor is $45.00 per hour, decide what the labor estimate should be.

34. Mr. Dodson can paint his house by himself in 4 days. His son needs an additional day to complete the job if he works by himself. If they work together, find how long it takes to paint the house.

35. A pilot can travel 400 miles with the wind in the same amount of time as 336 miles against the wind. Find the speed of the wind if the pilot's speed in still air is 230 miles per hour.

36. A fisherman on Pearl River rows 9 miles downstream in the same amount of time he rows 3 miles upstream. If the current is 6 miles per hour, find how long it takes him to cover the 12 miles.

37. Find the unknown length y.

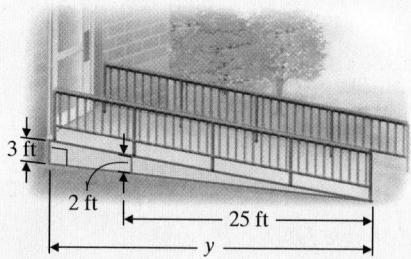

38. Find the unknown length y.

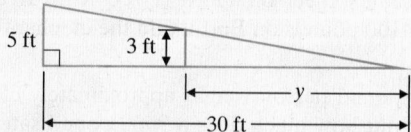

39. Suppose two trains leave Holbrook, Arizona, at the same time, traveling in opposite directions. One train travels 10 mph faster than the other. In 3.5 hours, the trains are 322 miles apart. Find the speed of each train.

40. Suppose two cars leave Brinkley, Arkansas, at the same time, traveling in opposite directions. One car travels 8 mph faster than the other car. In 2.5 hours, the cars are 280 miles apart. Find the speed of each car.

41. Two divided by the difference of a number and 3 minus 4 divided by the number plus 3, equals 8 times the reciprocal of the difference of the number squared and 9. What is the number?

42. If 15 times the reciprocal of a number is added to the ratio of 9 times the number minus 7 and the number plus 2, the result is 9. What is the number?

43. A pilot flies 630 miles with a tailwind of 35 miles per hour. Against the wind, he flies only 455 miles in the same amount of time. Find the rate of the plane in still air.

44. A marketing manager travels 1080 miles in a corporate jet and then an additional 240 miles by car. If the car ride takes one hour longer than the jet ride takes, and if the rate of the jet is 6 times the rate of the car, find the time the manager travels by jet and find the time the manager travels by car.

45. To mix weed killer with water correctly, it is necessary to mix 8 teaspoons of weed killer with 2 gallons of water. Find how many gallons of water are needed to mix with the entire box if it contains 36 teaspoons of weed killer.

46. The directions for a certain bug spray concentrate is to mix 3 ounces of concentrate with 2 gallons of water. How many ounces of concentrate are needed to mix with 5 gallons of water?

47. A boater travels 16 miles per hour on the water on a still day. During one particular windy day, he finds that he travels 48 miles with the wind behind him in the same amount of time that he travels 16 miles into the wind. Find the rate of the wind.

Let x be the rate of the wind.

	r	$\times$	t	$=$	d
with wind	$16 + x$				48
into wind	$16 - x$				16

48. The current on a portion of the Mississippi River is 3 miles per hour. A barge can go 6 miles upstream in the same amount of time it takes to go 10 miles downstream. Find the speed of the boat in still water.

Let x be the speed of the boat in still water.

	r	$\times$	t	$=$	d
upstream	$x - 3$				6
downstream	$x + 3$				10

49. Two hikers are 11 miles apart and walking toward each other. They meet in 2 hours. Find the rate of each hiker if one hiker walks 1.1 mph faster than the other.

50. On a 255-mile trip, Gary Alessandrini traveled at an average speed of 70 mph, got a speeding ticket, and then traveled at 60 mph for the remainder of the trip. If the entire trip took 4.5 hours and the speeding ticket stop took 30 minutes, how long did Gary speed before getting stopped?

51. One custodian cleans a suite of offices in 3 hours. When a second worker is asked to join the regular custodian, the job takes only $1\frac{1}{2}$ hours. How long does it take the second worker to do the same job alone?

52. One person proofreads copy for a small newspaper in 4 hours. If a second proofreader is also employed, the job can be done in $2\frac{1}{2}$ hours. How long does it take the second proofreader to do the same job alone?

△ **53.** An architect is completing the plans for a triangular deck. Use the diagram below to find the unknown dimension.

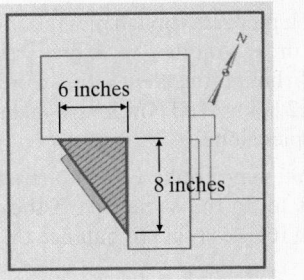
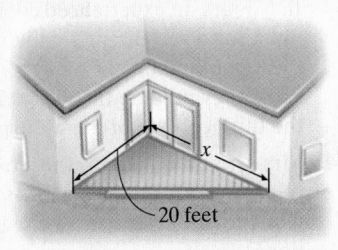

△ **54.** A student wishes to make a small model of a triangular mainsail to study the effects of wind on the sail. The smaller model will be the same shape as a regular-size sailboat's mainsail. Use the following diagram to find the unknown dimensions.

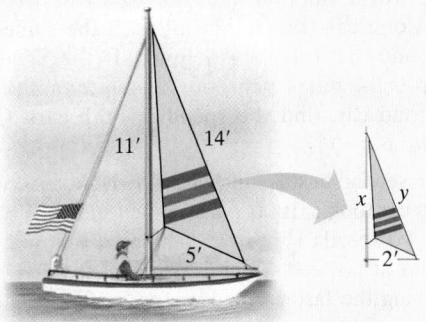

55. The manufacturers of cans of salted mixed nuts state that the ratio of peanuts to other nuts is 3 to 2. If 324 peanuts are in a can, find how many other nuts should also be in the can.

56. There are 1280 calories in a 14-ounce portion of Eagle Brand Milk. Find how many calories are in 2 ounces of Eagle Brand Milk.

57. A jet plane traveling at 500 mph overtakes a propeller plane traveling at 200 mph that had a 2-hour head start. How far from the starting point are the planes?

58. How long will it take a bus traveling at 60 miles per hour to overtake a car traveling at 40 miles per hour if the car had a 1.5-hour head start?

59. One pipe fills a storage pool in 20 hours. A second pipe fills the same pool in 15 hours. When a third pipe is added and all three are used to fill the pool, it takes only 6 hours. Find how long it takes the third pipe to do the job.

60. One pump fills a tank in 9 hours. A second pump fills the same tank in 6 hours. When a third pump is added and all three are used to fill the tank, it takes only 3 hours. Find how long it takes the third pump to fill the tank.

61. A car travels 280 miles in the same time that a motorcycle travels 240 miles. If the car's speed is 10 miles per hour more than the motorcycle's, find the speed of the car and the speed of the motorcycle.

62. A bus traveled on a straight road for 3 hours at an average speed 20 miles per hour faster than it traveled on a winding road. The time spent on the winding road was 4 hours. Find the average speed on the straight road if the entire trip was 305 miles.

63. In 6 hours, an experienced cook prepares enough pies to supply a local restaurant's daily order. Another cook prepares the same number of pies in 7 hours. Together with a third cook, they prepare the pies in 2 hours. Find how long it takes the third cook to prepare the pies alone.

64. Mrs. Smith balances the company books in 8 hours. It takes her assistant 12 hours to do the same job. If they work together, find how long it takes them to balance the books.

65. The quotient of a number and 3, minus 1, equals $\frac{5}{3}$. Find the number.

66. The quotient of a number and 5, minus 1, equals $\frac{7}{5}$. Find the number.

67. Currently, the Toyota Camry is the best-selling car in the world. Suppose that during a test drive of two Camrys, one car travels 224 miles in the same time that the second car travels 175 miles. If the speed of the first car is 14 miles per hour faster than the speed of the second car, find the speed of both cars. (*Source: Kelley Blue Book*)

68. The second best-selling car is the Honda Accord. A driver of this car took a day trip along the California coastline driving at two speeds. He drove 70 miles at a slower speed and 300 miles at a speed 40 miles per hour faster. If the time spent driving the faster speed was twice that spent at the slower speed, find the two speeds during the trip. (*Source: Kelley Blue Book*)

69. A pilot can fly an MD-11 2160 miles with the wind in the same time she can fly 1920 miles against the wind. If the speed of the wind is 30 mph, find the speed of the plane in still air. (*Source*: Air Transport Association of America)

70. A pilot can fly a DC-10 1365 miles against the wind in the same time he can fly 1575 miles with the wind. If the speed of the plane in still air is 490 miles per hour, find the speed of the wind. (*Source*: Air Transport Association of America)

Given that the following pairs of triangles are similar, find each unknown length.

71.

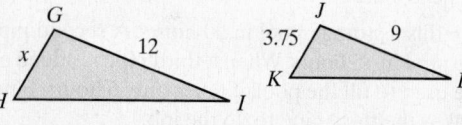

72.

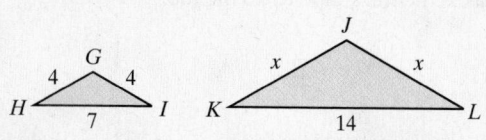

73.

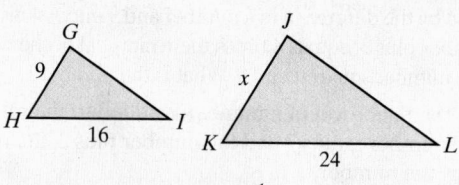

74.

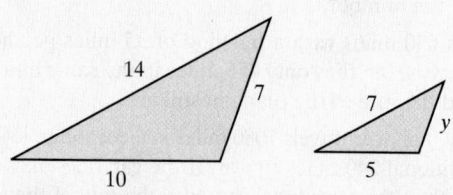

REVIEW AND PREVIEW

Find the slope of the line through each pair of points. Use the slope to determine whether the line is vertical, horizontal, or moves upward or downward from left to right. See Section 3.4.

75. $(-2, 5), (4, -3)$

76. $(0, 4), (2, 10)$

77. $(-3, -6), (1, 5)$

78. $(-2, 7), (3, -2)$

79. $(3, 7), (3, -2)$

80. $(0, -4), (2, -4)$

CONCEPT EXTENSIONS

The following bar graph shows the capacity of the United States to generate electricity from the wind in the years shown. Use this graph for Exercises 81 and 82.

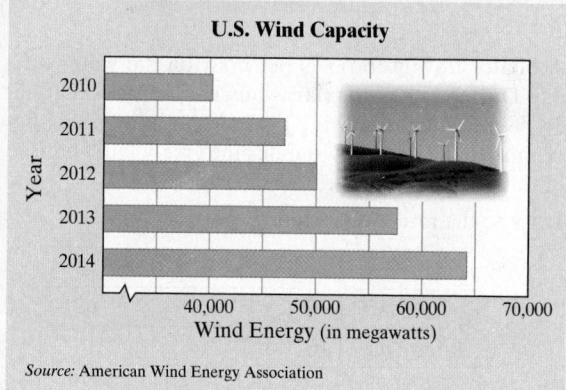

Source: American Wind Energy Association

81. Find the approximate megawatt capacity in 2014.

82. Find the approximate megawatt capacity in 2013.

In general, 1000 megawatts will serve the average electricity needs of 560,000 people. Use this fact and the preceding graph to answer Exercises 83 and 84.

83. In 2014, the number of megawatts that were generated from wind would serve the electricity needs of how many people?

84. How many megawatts of electricity are needed to serve the city or town in which you live?

For Exercises 85 and 86 decide whether we can immediately use cross products to solve for x. Do not actually solve.

85. $\frac{2-x}{5} = \frac{1+x}{3}$

86. $\frac{2}{5} - x = \frac{1+x}{3}$

Solve.

87. One pump fills a tank 3 times as fast as another pump. If the pumps work together, they fill the tank in 21 minutes. How long does it take each pump to fill the tank?

88. It takes 9 hours for pump A to fill a tank alone. Pump B takes 15 hours to fill the same tank alone. If pumps A, B, and C are used, the tank fills in 5 hours. How long does it take pump C to fill the tank alone?

89. For what value of x is $\dfrac{x}{x-1}$ in proportion to $\dfrac{x+1}{x}$? Explain your result.

90. If x is 10, is $\dfrac{2}{x}$ in proportion to $\dfrac{x}{50}$? Explain why or why not.

91. Person A can complete a job in 5 hours, and person B can complete the same job in 3 hours. Without solving algebraically, discuss reasonable and unreasonable answers for how long it would take them to complete the job together.

92. A hyena spots a giraffe 0.5 mile away and begins running toward it. The giraffe starts running away from the hyena just as the hyena begins running toward it. A hyena can run at a speed of 40 mph and a giraffe can run at 32 mph. How long will it take for the hyena to overtake the giraffe? (*Source: World Almanac and Book of Facts*)

Solve. See the Concept Check in this section.

Solve D = RT

93. for R **94.** for T

7.7   Simplifying Complex Fractions

OBJECTIVES

1 Simplify Complex Fractions by Simplifying the Numerator and Denominator and Then Dividing.

2 Simplify Complex Fractions by Multiplying by a Common Denominator.

3 Simplify Expressions with Negative Exponents.

A rational expression whose numerator, denominator, or both contain one or more rational expressions is called a **complex rational expression** or a **complex fraction.**

Complex Fractions

$$\dfrac{\dfrac{1}{a}}{\dfrac{b}{2}} \qquad \dfrac{\dfrac{x}{2y^2}}{\dfrac{6x-2}{9y}} \qquad \dfrac{x+\dfrac{1}{y}}{y+1}$$

The parts of a complex fraction are

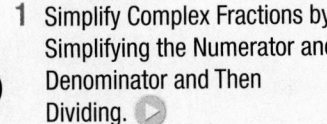 ← Numerator of complex fraction

                      ← Main fraction bar

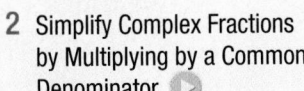 ← Denominator of complex fraction

Our goal in this section is to simplify complex fractions. A complex fraction is simplified when it is in the form $\dfrac{P}{Q}$, where P and Q are polynomials that have no common factors. Two methods of simplifying complex fractions are introduced. The first method evolves from the definition of a fraction as a quotient.

OBJECTIVE

1 Simplifying Complex Fractions: Method 1

Simplifying a Complex Fraction: Method I

Step 1. Simplify the numerator and the denominator of the complex fraction so that each is a single fraction.

Step 2. Perform the indicated division by multiplying the numerator of the complex fraction by the reciprocal of the denominator of the complex fraction.

Step 3. Simplify if possible.

EXAMPLE 1 Simplify each complex fraction.

a. $\dfrac{\dfrac{2x}{27y^2}}{\dfrac{6x^2}{9}}$

b. $\dfrac{\dfrac{5x}{x+2}}{\dfrac{10}{x-2}}$

c. $\dfrac{\dfrac{x}{y^2}+\dfrac{1}{y}}{\dfrac{y}{x^2}+\dfrac{1}{x}}$

Solution

a. The numerator of the complex fraction is already a single fraction, and so is the denominator. Perform the indicated division by multiplying the numerator, $\dfrac{2x}{27y^2}$, by the reciprocal of the denominator, $\dfrac{6x^2}{9}$. Then simplify.

$$\dfrac{\dfrac{2x}{27y^2}}{\dfrac{6x^2}{9}}=\dfrac{2x}{27y^2}\div\dfrac{6x^2}{9}$$

$$=\dfrac{2x}{27y^2}\cdot\dfrac{9}{6x^2} \qquad \text{Multiply by the reciprocal of } \dfrac{6x^2}{9}.$$

$$=\dfrac{2x\cdot 9}{27y^2\cdot 6x^2}$$

$$=\dfrac{1}{9xy^2}$$

Helpful Hint
Both the numerator and denominator are single fractions, so we perform the indicated division.

b. $\dfrac{\left\{\dfrac{5x}{x+2}\right.}{\left\{\dfrac{10}{x-2}\right.}=\dfrac{5x}{x+2}\div\dfrac{10}{x-2}=\dfrac{5x}{x+2}\cdot\dfrac{x-2}{10}$ $\quad$ Multiply by the reciprocal of $\dfrac{10}{x-2}$.

$$=\dfrac{5x(x-2)}{2\cdot 5(x+2)}$$

$$=\dfrac{x(x-2)}{2(x+2)} \qquad \text{Simplify.}$$

c. First simplify the numerator and the denominator of the complex fraction separately so that each is a single fraction. Then perform the indicated division.

$$\dfrac{\dfrac{x}{y^2}+\dfrac{1}{y}}{\dfrac{y}{x^2}+\dfrac{1}{x}}=\dfrac{\dfrac{x}{y^2}+\dfrac{1\cdot y}{y\cdot y}}{\dfrac{y}{x^2}+\dfrac{1\cdot x}{x\cdot x}} \qquad \begin{array}{l}\text{Simplify the numerator. The LCD is } y^2.\\[4pt]\text{Simplify the denominator. The LCD is } x^2.\end{array}$$

$$=\dfrac{\dfrac{x+y}{y^2}}{\dfrac{y+x}{x^2}} \qquad \text{Add.}$$

$$=\dfrac{x+y}{y^2}\cdot\dfrac{x^2}{y+x} \qquad \text{Multiply by the reciprocal of } \dfrac{y+x}{x^2}.$$

$$=\dfrac{x^2\,(x+y)}{y^2\,(y+x)}$$

$$=\dfrac{x^2}{y^2} \qquad \text{Simplify.}$$

□

PRACTICE
1   Simplify each complex fraction.

a. $\dfrac{\dfrac{5k}{36m}}{\dfrac{15k}{9}}$

b. $\dfrac{\dfrac{8x}{x-4}}{\dfrac{3}{x+4}}$

c. $\dfrac{\dfrac{5}{a}+\dfrac{b}{a^2}}{\dfrac{5a}{b^2}+\dfrac{1}{b}}$ ■

✔ **CONCEPT CHECK**

Which of the following are equivalent to $\dfrac{\dfrac{5}{y}}{\dfrac{2}{z}}$?

a. $\dfrac{5}{y} \div \dfrac{2}{z}$ **b.** $\dfrac{5}{y} \cdot \dfrac{z}{2}$ **c.** $\dfrac{5}{y} \div \dfrac{z}{2}$

OBJECTIVE
2   Simplifying Complex Fractions: Method 2

Next we look at another method of simplifying complex fractions. With this method, we multiply the numerator and the denominator of the complex fraction by the LCD of all fractions in the complex fraction.

> **Simplifying a Complex Fraction: Method 2**
>
> **Step 1.** Multiply the numerator and the denominator of the complex fraction by the LCD of the fractions in both the numerator and the denominator.
>
> **Step 2.** Simplify.

EXAMPLE 2   Simplify each complex fraction.

a. $\dfrac{\dfrac{5x}{x+2}}{\dfrac{10}{x-2}}$

b. $\dfrac{\dfrac{x}{y^2}+\dfrac{1}{y}}{\dfrac{y}{x^2}+\dfrac{1}{x}}$

Solution

a. Notice we are reworking Example 1(b) using method 2. The least common denominator of $\dfrac{5x}{x+2}$ and $\dfrac{10}{x-2}$ is $(x+2)(x-2)$. Multiply both the numerator, $\dfrac{5x}{x+2}$, and the denominator, $\dfrac{10}{x-2}$, by the LCD.

$$\dfrac{\dfrac{5x}{x+2}}{\dfrac{10}{x-2}} = \dfrac{\left(\dfrac{5x}{x+2}\right)\cdot (x+2)(x-2)}{\left(\dfrac{10}{x-2}\right)\cdot (x+2)(x-2)} \quad \text{Multiply numerator and denominator by the LCD.}$$

$$= \dfrac{5x\cdot(x-2)}{2\cdot 5\cdot(x+2)} \quad \text{Simplify.}$$

$$= \dfrac{x(x-2)}{2(x+2)} \quad \text{Simplify.}$$

b. Here, we are reworking Example 1(c) using method 2. The least common denominator of $\dfrac{x}{y^2}, \dfrac{1}{y}, \dfrac{y}{x^2}$, and $\dfrac{1}{x}$ is x^2y^2.

(Continued on next page)

$$\frac{\frac{x}{y^2} + \frac{1}{y}}{\frac{y}{x^2} + \frac{1}{x}} = \frac{\left(\frac{x}{y^2} + \frac{1}{y}\right) \cdot x^2 y^2}{\left(\frac{y}{x^2} + \frac{1}{x}\right) \cdot x^2 y^2}$$

Multiply the numerator and denominator by the LCD.

$$= \frac{\frac{x}{y^2} \cdot x^2 y^2 + \frac{1}{y} \cdot x^2 y^2}{\frac{y}{x^2} \cdot x^2 y^2 + \frac{1}{x} \cdot x^2 y^2}$$

Use the distributive property.

$$= \frac{x^3 + x^2 y}{y^3 + xy^2}$$

Simplify.

$$= \frac{x^2(x + y)}{y^2(y + x)}$$

Factor.

$$= \frac{x^2}{y^2}$$

Simplify. □

PRACTICE

2 Use method 2 to simplify.

a. $\dfrac{\dfrac{8x}{x - 4}}{\dfrac{3}{x + 4}}$

b. $\dfrac{\dfrac{b}{a^2} + \dfrac{1}{a}}{\dfrac{a}{b^2} + \dfrac{1}{b}}$

OBJECTIVE

3 Simplifying Expressions with Negative Exponents

If an expression contains negative exponents, write the expression as an equivalent expression with positive exponents.

EXAMPLE 3 Simplify:

$$\frac{x^{-1} + 2xy^{-1}}{x^{-2} - x^{-2}y^{-1}}$$

<u>Solution</u> This fraction does not appear to be a complex fraction. If we write it by using only positive exponents, however, we see that it is a complex fraction.

$$\frac{x^{-1} + 2xy^{-1}}{x^{-2} - x^{-2}y^{-1}} = \frac{\dfrac{1}{x} + \dfrac{2x}{y}}{\dfrac{1}{x^2} - \dfrac{1}{x^2 y}}$$

The LCD of $\dfrac{1}{x}, \dfrac{2x}{y}, \dfrac{1}{x^2}$, and $\dfrac{1}{x^2 y}$ is $x^2 y$. Multiply both the numerator and denominator by $x^2 y$.

$$= \frac{\left(\dfrac{1}{x} + \dfrac{2x}{y}\right) \cdot x^2 y}{\left(\dfrac{1}{x^2} - \dfrac{1}{x^2 y}\right) \cdot x^2 y}$$

$$= \frac{\dfrac{1}{x} \cdot x^2 y + \dfrac{2x}{y} \cdot x^2 y}{\dfrac{1}{x^2} \cdot x^2 y - \dfrac{1}{x^2 y} \cdot x^2 y}$$

Apply the distributive property.

$$= \frac{xy + 2x^3}{y - 1} \quad \text{or} \quad \frac{x(y + 2x^2)}{y - 1}$$

Simplify. □

PRACTICE
3   Simplify: $\dfrac{3x^{-1} + x^{-2}y^{-1}}{y^{-2} + xy^{-1}}$

EXAMPLE 4   Simplify: $\dfrac{(2x)^{-1} + 1}{2x^{-1} - 1}$

Solution $\dfrac{(2x)^{-1} + 1}{2x^{-1} - 1} = \dfrac{\dfrac{1}{2x} + 1}{\dfrac{2}{x} - 1}$ Write using positive exponents.

$= \dfrac{\left(\dfrac{1}{2x} + 1\right) \cdot 2x}{\left(\dfrac{2}{x} - 1\right) \cdot 2x}$ The LCD of $\dfrac{1}{2x}$ and $\dfrac{2}{x}$ is $2x$.

$= \dfrac{\dfrac{1}{2x} \cdot 2x + 1 \cdot 2x}{\dfrac{2}{x} \cdot 2x - 1 \cdot 2x}$ Use distributive property.

$= \dfrac{1 + 2x}{4 - 2x}$ or $\dfrac{1 + 2x}{2(2 - x)}$ Simplify.

Helpful Hint

Don't forget that $(2x)^{-1} = \dfrac{1}{2x}$,

but $2x^{-1} = 2 \cdot \dfrac{1}{x} = \dfrac{2}{x}$.

PRACTICE
4   Simplify: $\dfrac{(3x)^{-1} - 2}{5x^{-1} + 2}$

Vocabulary, Readiness & Video Check

Complete the steps by writing the simplified complex fraction.

1. $\dfrac{\dfrac{7}{x}}{\dfrac{1}{x} + \dfrac{z}{x}} = \dfrac{x\left(\dfrac{7}{x}\right)}{x\left(\dfrac{1}{x}\right) + x\left(\dfrac{z}{x}\right)} = $ _____

2. $\dfrac{\dfrac{x}{4}}{\dfrac{x^2}{2} + \dfrac{1}{4}} = \dfrac{4\left(\dfrac{x}{4}\right)}{4\left(\dfrac{x^2}{2}\right) + 4\left(\dfrac{1}{4}\right)} = $ _____

Write with positive exponents.

3. $x^{-2} = $ _____

4. $y^{-3} = $ _____

5. $2x^{-1} = $ _____

6. $(2x)^{-1} = $ _____

7. $(9y)^{-1} = $ _____

8. $9y^{-2} = $ _____

Martin-Gay Interactive Videos

See Video 7.7

Watch the section lecture video and answer the following questions.

OBJECTIVE 1

9. From ▦ Example 2, before you can rewrite the complex fraction as division, describe how it must appear.

OBJECTIVE 2

10. How does finding an LCD in method 2, as in ▦ Example 3, differ from finding an LCD in method 1? In your answer, mention the purpose of the LCD in each method.

OBJECTIVE 3

11. Based on ▦ Example 4, what connection is there between negative exponents and complex fractions?

7.7 Exercise Set MyMathLab®

Simplify each complex fraction. See Examples 1 and 2.

1. $\dfrac{\dfrac{10}{3x}}{\dfrac{5}{6x}}$

2. $\dfrac{\dfrac{15}{2x}}{\dfrac{5}{6x}}$

3. $\dfrac{1+\dfrac{2}{5}}{2+\dfrac{3}{5}}$

4. $\dfrac{2+\dfrac{1}{7}}{3-\dfrac{4}{7}}$

5. $\dfrac{\dfrac{4}{x-1}}{\dfrac{x}{x-1}}$

6. $\dfrac{\dfrac{x}{x+2}}{\dfrac{2}{x+2}}$

7. $\dfrac{1-\dfrac{2}{x}}{x+\dfrac{4}{9x}}$

8. $\dfrac{5-\dfrac{3}{x}}{x+\dfrac{2}{3x}}$

9. $\dfrac{\dfrac{4x^2-y^2}{xy}}{\dfrac{2}{y}-\dfrac{1}{x}}$

10. $\dfrac{\dfrac{x^2-9y^2}{xy}}{\dfrac{1}{y}-\dfrac{3}{x}}$

11. $\dfrac{\dfrac{x+1}{3}}{\dfrac{2x-1}{6}}$

12. $\dfrac{\dfrac{x+3}{12}}{\dfrac{4x-5}{15}}$

13. $\dfrac{\dfrac{2}{x}+\dfrac{3}{x^2}}{\dfrac{4}{x^2}-\dfrac{9}{x}}$

14. $\dfrac{\dfrac{2}{x^2}+\dfrac{1}{x}}{\dfrac{4}{x^2}-\dfrac{1}{x}}$

15. $\dfrac{\dfrac{1}{x}+\dfrac{2}{x^2}}{x+\dfrac{8}{x^2}}$

16. $\dfrac{\dfrac{1}{y}+\dfrac{3}{y^2}}{y+\dfrac{27}{y^2}}$

17. $\dfrac{\dfrac{4}{5-x}+\dfrac{5}{x-5}}{\dfrac{2}{x}+\dfrac{3}{x-5}}$

18. $\dfrac{\dfrac{3}{x-4}-\dfrac{2}{4-x}}{\dfrac{2}{x-4}-\dfrac{2}{x}}$

19. $\dfrac{\dfrac{x+2}{x}-\dfrac{2}{x-1}}{\dfrac{x+1}{x}+\dfrac{x+1}{x-1}}$

20. $\dfrac{\dfrac{5}{a+2}-\dfrac{1}{a-2}}{\dfrac{3}{2+a}+\dfrac{6}{2-a}}$

21. $\dfrac{\dfrac{2}{x}+3}{\dfrac{4}{x^2}-9}$

22. $\dfrac{2+\dfrac{1}{x}}{4x-\dfrac{1}{x}}$

23. $\dfrac{1-\dfrac{x}{y}}{\dfrac{x^2}{y^2}-1}$

24. $\dfrac{1-\dfrac{2}{x}}{x-\dfrac{4}{x}}$

25. $\dfrac{\dfrac{-2x}{x-y}}{\dfrac{y}{x^2}}$

26. $\dfrac{\dfrac{7y}{x^2+xy}}{\dfrac{y^2}{x^2}}$

27. $\dfrac{\dfrac{2}{x}+\dfrac{1}{x^2}}{\dfrac{y}{x^2}}$

28. $\dfrac{\dfrac{5}{x^2}-\dfrac{2}{x}}{\dfrac{1}{x}+2}$

29. $\dfrac{\dfrac{x}{9}-\dfrac{1}{x}}{1+\dfrac{3}{x}}$

30. $\dfrac{\dfrac{x}{4}-\dfrac{4}{x}}{1-\dfrac{4}{x}}$

31. $\dfrac{\dfrac{x-1}{x^2-4}}{1+\dfrac{1}{x-2}}$

32. $\dfrac{\dfrac{x+3}{x^2-9}}{1+\dfrac{1}{x-3}}$

33. $\dfrac{\dfrac{2}{x+5}+\dfrac{4}{x+3}}{\dfrac{3x+13}{x^2+8x+15}}$

34. $\dfrac{\dfrac{2}{x+2}+\dfrac{6}{x+7}}{\dfrac{4x+13}{x^2+9x+14}}$

Simplify. See Examples 3 and 4.

35. $\dfrac{x^{-1}}{x^{-2}+y^{-2}}$

36. $\dfrac{a^{-3}+b^{-1}}{a^{-2}}$

37. $\dfrac{2a^{-1}+3b^{-2}}{a^{-1}-b^{-1}}$

38. $\dfrac{x^{-1}+y^{-1}}{3x^{-2}+5y^{-2}}$

39. $\dfrac{1}{x-x^{-1}}$

40. $\dfrac{x^{-2}}{x+3x^{-1}}$

41. $\dfrac{a^{-1}+1}{a^{-1}-1}$

42. $\dfrac{a^{-1}-4}{4+a^{-1}}$

43. $\dfrac{3x^{-1}+(2y)^{-1}}{x^{-2}}$

44. $\dfrac{5x^{-2}-3y^{-1}}{x^{-1}+y^{-1}}$

45. $\dfrac{2a^{-1}+(2a)^{-1}}{a^{-1}+2a^{-2}}$

46. $\dfrac{a^{-1}+2a^{-2}}{2a^{-1}+(2a)^{-1}}$

47. $\dfrac{5x^{-1}+2y^{-1}}{x^{-2}y^{-2}}$

48. $\dfrac{x^{-2}y^{-2}}{5x^{-1}+2y^{-1}}$

49. $\dfrac{5x^{-1}-2y^{-1}}{25x^{-2}-4y^{-2}}$

50. $\dfrac{3x^{-1}+3y^{-1}}{4x^{-2}-9y^{-2}}$

REVIEW AND PREVIEW

Simplify. See Section 5.1.

51. $\dfrac{3x^3y^2}{12x}$

52. $\dfrac{-36xb^3}{9xb^2}$

53. $\dfrac{144x^5y^5}{-16x^2y}$

54. $\dfrac{48x^3y^2}{-4xy}$

Solve the following. See Section 3.6.

55. If $P(x)=-x^2$, find $P(-3)$.

56. If $f(x)=x^2-6$, find $f(-1)$.

CONCEPT EXTENSIONS

Solve. See the Concept Check in this section.

57. Which of the following are equivalent to $\dfrac{\frac{x+1}{9}}{\frac{y-2}{5}}$?

a. $\dfrac{x+1}{9} \div \dfrac{y-2}{5}$ **b.** $\dfrac{x+1}{9} \cdot \dfrac{y-2}{5}$ **c.** $\dfrac{x+1}{9} \cdot \dfrac{5}{y-2}$

58. Which of the following are equivalent to $\dfrac{\frac{a}{7}}{\frac{b}{13}}$?

a. $\dfrac{a}{7} \cdot \dfrac{b}{13}$ **b.** $\dfrac{a}{7} \div \dfrac{b}{13}$ **c.** $\dfrac{a}{7} \div \dfrac{13}{b}$ **d.** $\dfrac{a}{7} \cdot \dfrac{13}{b}$

59. When the source of a sound is traveling toward a listener, the pitch that the listener hears due to the Doppler effect is given by the complex rational expression $\dfrac{a}{1 - \frac{s}{770}}$, where a is the actual pitch of the sound and s is the speed of the sound source. Simplify this expression.

60. In baseball, the earned run average (ERA) statistic gives the average number of earned runs scored on a pitcher per game. It is computed with the following expression: $\dfrac{E}{\frac{I}{9}}$, where E is the number of earned runs scored on a pitcher and I is the total number of innings pitched by the pitcher. Simplify this expression.

61. Which of the following are equivalent to $\dfrac{\frac{1}{x}}{\frac{3}{y}}$?

a. $\dfrac{1}{x} \div \dfrac{3}{y}$ **b.** $\dfrac{1}{x} \cdot \dfrac{y}{3}$ **c.** $\dfrac{1}{x} \div \dfrac{y}{3}$

62. Which of the following are equivalent to $\dfrac{\frac{5}{2}}{a}$?

a. $\dfrac{5}{1} \div \dfrac{2}{a}$ **b.** $\dfrac{1}{5} \div \dfrac{2}{a}$ **c.** $\dfrac{5}{1} \cdot \dfrac{2}{a}$

63. In your own words, explain one method for simplifying a complex fraction.

64. Explain your favorite method for simplifying a complex fraction and why.

Simplify.

65. $\dfrac{1}{1 + (1+x)^{-1}}$

66. $\dfrac{(x+2)^{-1} + (x-2)^{-1}}{(x^2 - 4)^{-1}}$

67. $\dfrac{x}{1 - \dfrac{1}{1 + \frac{1}{x}}}$

68. $\dfrac{x}{1 - \dfrac{1}{1 - \frac{1}{x}}}$

69. $\dfrac{\dfrac{2}{y^2} - \dfrac{5}{xy} - \dfrac{3}{x^2}}{\dfrac{2}{y^2} + \dfrac{7}{xy} + \dfrac{3}{x^2}}$

70. $\dfrac{\dfrac{2}{x^2} - \dfrac{1}{xy} - \dfrac{1}{y^2}}{\dfrac{1}{x^2} - \dfrac{3}{xy} + \dfrac{2}{y^2}}$

71. $\dfrac{3(a+1)^{-1} + 4a^{-2}}{(a^3 + a^2)^{-1}}$

72. $\dfrac{9x^{-1} - 5(x-y)^{-1}}{4(x-y)^{-1}}$

*In the study of calculus, the difference quotient $\dfrac{f(a+h) - f(a)}{h}$ is often found and simplified. Find and simplify this quotient for each function $f(x)$ by following steps **a** through **d**.*

a. Find $(a + h)$. **b.** Find $f(a)$.

c. Use steps **a** and **b** to find $\dfrac{f(a+h) - f(a)}{h}$

d. Simplify the result of step **c**.

73. $f(x) = \dfrac{1}{x}$

74. $f(x) = \dfrac{5}{x}$

75. $\dfrac{3}{x+1}$

76. $\dfrac{2}{x^2}$

Chapter 7 Vocabulary Check

Fill in each blank with one of the words or phrases listed below. Not all choices will be used.

least common denominator   simplifying   reciprocals   numerator $\dfrac{-a}{b}$ $\dfrac{a}{-b}$

cross products   ratio   proportion

rational expression   domain   complex fraction   denominator $\dfrac{-a}{-b}$

1. A(n) _____ is an expression that can be written in the form $\dfrac{P}{Q}$, where P and Q are polynomials and Q is not 0.

2. In a(n) _____, the numerator or denominator or both may contain fractions.

3. For a rational expression, $-\dfrac{a}{b} =$ _____ = _____.

4. A rational expression is undefined when the _____ is 0.

5. The process of writing a rational expression in lowest terms is called _____.

6. The expressions $\dfrac{2x}{7}$ and $\dfrac{7}{2x}$ are called _____.

7. The _____ of a list of rational expressions is a polynomial of least degree whose factors include all factors of the denominators in the list.

8. A(n) _____ is the quotient of two numbers.

9. $\dfrac{x}{2} = \dfrac{7}{16}$ is an example of a(n) _____.

10. If $\dfrac{a}{b} = \dfrac{c}{d}$, then ad and bc are called _____.

11. The _____ of the rational function $f(x) = \dfrac{1}{x - 3}$ is $\{x \mid x \text{ is a real number}, x \neq 3\}$.

Chapter 7 Highlights

DEFINITIONS AND CONCEPTS	EXAMPLES

Section 7.1   Rational Functions and Simplifying Rational Expressions

A **rational expression** is an expression that can be written in the form $\dfrac{P}{Q}$, where P and Q are polynomials and Q does not equal 0.

$$\frac{7y^3}{4}, \frac{x^2 + 6x + 1}{x - 3}, \frac{-5}{s^3 + 8}$$

To find values for which a rational expression is undefined, find values for which the denominator is 0.

Find any values for which the expression $\dfrac{5y}{y^2 - 4y + 3}$ is undefined.

$$y^2 - 4y + 3 = 0 \quad \text{Set the denominator equal to 0.}$$
$$(y - 3)(y - 1) = 0 \quad \text{Factor.}$$
$$y - 3 = 0 \quad \text{or} \quad y - 1 = 0 \quad \text{Set each factor equal to 0.}$$
$$y = 3 \qquad\qquad y = 1 \quad \text{Solve.}$$

The expression is undefined when y is 3 and when y is 1.

To Simplify a Rational Expression

Step 1. Factor the numerator and denominator.

Step 2. Divide out factors common to the numerator and denominator. (This is the same as removing a factor of 1.)

A **rational function** is a function described by a rational expression.

Simplify: $\dfrac{4x + 20}{x^2 - 25}$

$$\frac{4x + 20}{x^2 - 25} = \frac{4(x + 5)}{(x + 5)(x - 5)} = \frac{4}{x - 5}$$

$$f(x) = \frac{2x - 6}{7}, h(t) = \frac{t^2 - 3t + 5}{t - 1}$$

DEFINITIONS AND CONCEPTS	EXAMPLES

Section 7.2 Multiplying and Dividing Rational Expressions

To Multiply Rational Expressions

Step 1. Factor numerators and denominators.

Step 2. Multiply numerators and multiply denominators.

Step 3. Write the product in simplest form.

$$\frac{P}{Q} \cdot \frac{R}{S} = \frac{PR}{QS}$$

Multiply: $\dfrac{4x + 4}{2x - 3} \cdot \dfrac{2x^2 + x - 6}{x^2 - 1}$

$$\frac{4x + 4}{2x - 3} \cdot \frac{2x^2 + x - 6}{x^2 - 1} = \frac{4(x + 1)}{2x - 3} \cdot \frac{(2x - 3)(x + 2)}{(x + 1)(x - 1)}$$

$$= \frac{4(x + 1)(2x - 3)(x + 2)}{(2x - 3)(x + 1)(x - 1)}$$

$$= \frac{4(x + 2)}{x - 1}$$

To Divide by a Rational Expression

To divide by a rational expression, multiply by the reciprocal.

$$\frac{P}{Q} \div \frac{R}{S} = \frac{P}{Q} \cdot \frac{S}{R} = \frac{PS}{QR}$$

Divide: $\dfrac{15x + 5}{3x^2 - 14x - 5} \div \dfrac{15}{3x - 12}$

$$\frac{15x + 5}{3x^2 - 14x - 5} \div \frac{15}{3x - 12} = \frac{5(3x + 1)}{(3x + 1)(x - 5)} \cdot \frac{3(x - 4)}{3 \cdot 5}$$

$$= \frac{x - 4}{x - 5}$$

Section 7.3 Adding and Subtracting Rational Expressions with Common Denominators and Least Common Denominator

To Add or Subtract Rational Expressions with the Same Denominator

To add or subtract rational expressions with the same denominator, add or subtract numerators and place the sum or difference over the common denominator.

$$\frac{P}{R} + \frac{Q}{R} = \frac{P + Q}{R}$$

$$\frac{P}{R} - \frac{Q}{R} = \frac{P - Q}{R}$$

Perform indicated operations.

$$\frac{5}{x + 1} + \frac{x}{x + 1} = \frac{5 + x}{x + 1}$$

$$\frac{2y + 7}{y^2 - 9} - \frac{y + 4}{y^2 - 9} = \frac{(2y + 7) - (y + 4)}{y^2 - 9}$$

$$= \frac{2y + 7 - y - 4}{y^2 - 9}$$

$$= \frac{y + 3}{(y + 3)(y - 3)}$$

$$= \frac{1}{y - 3}$$

To Find the Least Common Denominator (LCD)

Step 1. Factor the denominators.

Step 2. The LCD is the product of all unique factors, each raised to a power equal to the greatest number of times that it appears in any one factored denominator.

Find the LCD for

$$\frac{7x}{x^2 + 10x + 25} \text{ and } \frac{11}{3x^2 + 15x}$$

$$x^2 + 10x + 25 = (x + 5)(x + 5)$$

$$3x^2 + 15x = 3x(x + 5)$$

LCD is $3x(x + 5)(x + 5)$ or $3x(x + 5)^2$

Section 7.4 Adding and Subtracting Rational Expressions with Unlike Denominators

To Add or Subtract Rational Expressions with Unlike Denominators

Step 1. Find the LCD.

Step 2. Rewrite each rational expression as an equivalent expression whose denominator is the LCD.

Perform the indicated operation.

$$\frac{9x + 3}{x^2 - 9} - \frac{5}{x - 3}$$

$$= \frac{9x + 3}{(x + 3)(x - 3)} - \frac{5}{x - 3}$$

(continued)

DEFINITIONS AND CONCEPTS	EXAMPLES

Section 7.4 Adding and Subtracting Rational Expressions with Unlike Denominators (continued)

Step 3. Add or subtract numerators and place the sum or difference over the common denominator.

Step 4. Write the result in simplest form.

LCD is $(x + 3)(x - 3)$.

$$= \frac{9x + 3}{(x + 3)(x - 3)} - \frac{5(x + 3)}{(x - 3)(x + 3)}$$

$$= \frac{9x + 3 - 5(x + 3)}{(x + 3)(x - 3)}$$

$$= \frac{9x + 3 - 5x - 15}{(x + 3)(x - 3)}$$

$$= \frac{4x - 12}{(x + 3)(x - 3)}$$

$$= \frac{4(x - 3)}{(x + 3)(x - 3)} = \frac{4}{x + 3}$$

Section 7.5 Solving Equations Containing Rational Expressions

To Solve an Equation Containing Rational Expressions

Step 1. Multiply both sides of the equation by the LCD of all rational expressions in the equation.

Step 2. Remove any grouping symbols and solve the resulting equation.

Step 3. Check the solution in the original equation.

Solve: $\dfrac{5x}{x + 2} + 3 = \dfrac{4x - 6}{x + 2}$

$$(x + 2)\left(\frac{5x}{x + 2} + 3\right) = (x + 2)\left(\frac{4x - 6}{x + 2}\right)$$

$$(x + 2)\left(\frac{5x}{x + 2}\right) + (x + 2)(3) = (x + 2)\left(\frac{4x - 6}{x + 2}\right)$$

$$5x + 3x + 6 = 4x - 6$$

$$4x = -12$$

$$x = -3$$

The solution checks and the solution is -3.

Section 7.6 Proportion and Problem Solving with Rational Equations

A **ratio** is the quotient of two numbers or two quantities. A **proportion** is a mathematical statement that two ratios are equal.

Cross products

$$\text{If } \frac{a}{b} = \frac{c}{d}, \text{ then } ad = bc.$$

Proportions

$$\frac{2}{3} = \frac{8}{12} \qquad \frac{x}{7} = \frac{15}{35}$$

Cross Products

$2 \cdot 12$ or 24 $\qquad\qquad\qquad\qquad$ $3 \cdot 8$ or 24

$$\frac{2}{3} = \frac{8}{12}$$

Solve: $\dfrac{3}{4} = \dfrac{x}{x - 1}$

$$\frac{3}{4} = \frac{x}{x - 1}$$

$$3(x - 1) = 4x \qquad \text{Set cross products equal.}$$

$$3x - 3 = 4x$$

$$-3 = x$$

DEFINITIONS AND CONCEPTS	EXAMPLES

Section 7.6 Proportion and Problem Solving with Rational Equations (continued)

Problem-Solving Steps

1. UNDERSTAND. Read and reread the problem.

A small plane and a car leave Kansas City, Missouri, and head for Minneapolis, Minnesota, a distance of 450 miles. The speed of the plane is 3 times the speed of the car, and the plane arrives 6 hours ahead of the car. Find the speed of the car.

Let $x =$ the speed of the car.

Then $3x =$ the speed of the plane.

	Distance	= Rate	· Time
Car	450	x	$\dfrac{450}{x} \left(\dfrac{\text{distance}}{\text{rate}}\right)$
Plane	450	$3x$	$\dfrac{450}{3x} \left(\dfrac{\text{distance}}{\text{rate}}\right)$

2. TRANSLATE.

In words: plane's time + 6 hours = car's time

Translate: $\dfrac{450}{3x}$ + 6 + $\dfrac{450}{x}$

3. SOLVE.

$$\frac{450}{3x} + 6 = \frac{450}{x}$$

$$3x\left(\frac{450}{3x}\right) + 3x(6) = 3x\left(\frac{450}{x}\right)$$

$$450 + 18x = 1350$$

$$18x = 900$$

$$x = 50$$

4. INTERPRET.

Check this solution in the originally stated problem. **State** the conclusion: The speed of the car is 50 miles per hour.

Section 7.7 Simplifying Complex Fractions

Method 1: Simplify the numerator and the denominator so that each is a single fraction. Then perform the indicated division and simplify if possible.

Simplify: $\dfrac{\dfrac{x+2}{x}}{x - \dfrac{4}{x}}$

Method 1: $\dfrac{\dfrac{x+2}{x}}{\dfrac{x \cdot x}{1 \cdot x} - \dfrac{4}{x}} = \dfrac{\dfrac{x+2}{x}}{\dfrac{x^2 - 4}{x}}$

$$= \frac{x+2}{x} \cdot \frac{x}{(x+2)(x-2)} = \frac{1}{x-2}$$

Method 2: Multiply the numerator and the denominator of the complex fraction by the LCD of the fractions in both the numerator and the denominator. Then simplify if possible.

Method 2: $\dfrac{\left(\dfrac{x+2}{x}\right) \cdot x}{\left(x - \dfrac{4}{x}\right) \cdot x} = \dfrac{\dfrac{x+2}{x} \cdot x}{x \cdot x - \dfrac{4}{x} \cdot x}$

$$= \frac{x+2}{x^2 - 4} = \frac{x+2}{(x+2)(x-2)} = \frac{1}{x-2}$$

Chapter 7 Review

(7.1) *Find the domain for each rational function.*

1. $f(x) = \dfrac{3 - 5x}{7}$

2. $g(x) = \dfrac{2x + 4}{11}$

3. $f(x) = \dfrac{-3x^2}{x - 5}$

4. $h(x) = \dfrac{4x}{3x - 12}$

5. $f(x) = \dfrac{x^3 + 2}{x^2 + 8x}$

6. $G(x) = \dfrac{20}{3x^2 - 48}$

Simplify each rational expression.

7. $\dfrac{x + 12}{12 - x}$

8. $\dfrac{2x}{2x^2 - 2x}$

9. $\dfrac{x + 7}{x^2 - 49}$

10. $\dfrac{2x^2 + 4x - 30}{x^2 + x - 20}$

Simplify each expression. This section contains four-term polynomials and sums and differences of two cubes.

11. $\dfrac{x^2 + xa + xb + ab}{x^2 - xc + bx - bc}$

12. $\dfrac{x^2 + 5x - 2x - 10}{x^2 - 3x - 2x + 6}$

13. $\dfrac{4 - x}{x^3 - 64}$

14. $\dfrac{x^2 - 4}{x^3 + 8}$

The average cost (per bookcase) of manufacturing x bookcases is given by the rational function

$$C(x) = \dfrac{35x + 4200}{x}$$

15. Find the average cost per bookcase of manufacturing 50 bookcases.

16. Find the average cost per bookcase of manufacturing 100 bookcases.

(7.2) *Perform each indicated operation and simplify.*

17. $\dfrac{15x^3y^2}{z} \cdot \dfrac{z}{5xy^3}$

18. $\dfrac{-y^3}{8} \cdot \dfrac{9x^2}{y^3}$

19. $\dfrac{x^2 - 9}{x^2 - 4} \cdot \dfrac{x - 2}{x + 3}$

20. $\dfrac{2x + 5}{x - 6} \cdot \dfrac{2x}{-x + 6}$

21. $\dfrac{x^2 - 5x - 24}{x^2 - x - 12} \div \dfrac{x^2 - 10x + 16}{x^2 + x - 6}$

22. $\dfrac{4x + 4y}{xy^2} \div \dfrac{3x + 3y}{x^2y}$

23. $\dfrac{x^2 + x - 42}{x - 3} \cdot \dfrac{(x - 3)^2}{x + 7}$

24. $\dfrac{2a + 2b}{3} \cdot \dfrac{a - b}{a^2 - b^2}$

25. $\dfrac{2x^2 - 9x + 9}{8x - 12} \div \dfrac{x^2 - 3x}{2x}$

26. $\dfrac{x^2 - y^2}{x^2 + xy} \div \dfrac{3x^2 - 2xy - y^2}{3x^2 + 6x}$

27. $\dfrac{x - y}{4} \div \dfrac{y^2 - 2y - xy + 2x}{16x + 24}$

28. $\dfrac{5 + x}{7} \div \dfrac{xy + 5y - 3x - 15}{7y - 35}$

(7.3) *Perform each indicated operation and simplify.*

29. $\dfrac{x}{x^2 + 9x + 14} + \dfrac{7}{x^2 + 9x + 14}$

30. $\dfrac{x}{x^2 + 2x - 15} + \dfrac{5}{x^2 + 2x - 15}$

31. $\dfrac{4x - 5}{3x^2} - \dfrac{2x + 5}{3x^2}$

32. $\dfrac{9x + 7}{6x^2} - \dfrac{3x + 4}{6x^2}$

Find the LCD of each pair of rational expressions.

33. $\dfrac{x + 4}{2x}, \dfrac{3}{7x}$

34. $\dfrac{x - 2}{x^2 - 5x - 24}, \dfrac{3}{x^2 + 11x + 24}$

Rewrite each rational expression as an equivalent expression whose denominator is the given polynomial.

35. $\dfrac{5}{7x} = \dfrac{}{14x^3y}$

36. $\dfrac{9}{4y} = \dfrac{}{16y^3x}$

37. $\dfrac{x + 2}{x^2 + 11x + 18} = \dfrac{}{(x + 2)(x - 5)(x + 9)}$

38. $\dfrac{3x - 5}{x^2 + 4x + 4} = \dfrac{}{(x + 2)^2(x + 3)}$

(7.4) *Perform each indicated operation and simplify.*

39. $\dfrac{4}{5x^2} - \dfrac{6}{y}$

40. $\dfrac{2}{x - 3} - \dfrac{4}{x - 1}$

41. $\dfrac{4}{x + 3} - 2$

42. $\dfrac{3}{x^2 + 2x - 8} + \dfrac{2}{x^2 - 3x + 2}$

43. $\dfrac{2x - 5}{6x + 9} - \dfrac{4}{2x^2 + 3x}$

44. $\dfrac{x - 1}{x^2 - 2x + 1} - \dfrac{x + 1}{x - 1}$

Find the perimeter and the area of each figure.

△ 45.

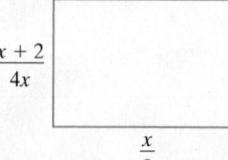

46.

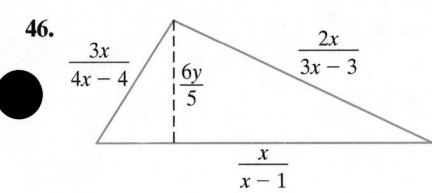

(7.5) *Solve each equation.*

47. $\dfrac{n}{10} = 9 - \dfrac{n}{5}$

48. $\dfrac{2}{x+1} - \dfrac{1}{x-2} = -\dfrac{1}{2}$

49. $\dfrac{y}{2y+2} + \dfrac{2y-16}{4y+4} = \dfrac{y-3}{y+1}$

50. $\dfrac{2}{x-3} - \dfrac{4}{x+3} = \dfrac{8}{x^2-9}$

51. $\dfrac{x-3}{x+1} - \dfrac{x-6}{x+5} = 0$

52. $x + 5 = \dfrac{6}{x}$

Solve the equation for the indicated variable.

53. $\dfrac{4A}{5b} = x^2$, for b

54. $\dfrac{x}{7} + \dfrac{y}{8} = 10$, for y

(7.6) *Solve each proportion.*

55. $\dfrac{x}{2} = \dfrac{12}{4}$

56. $\dfrac{20}{1} = \dfrac{x}{25}$

57. $\dfrac{2}{x-1} = \dfrac{3}{x+3}$

58. $\dfrac{4}{y-3} = \dfrac{2}{y-3}$

olve.

59. A machine can process 300 parts in 20 minutes. Find how many parts can be processed in 45 minutes.

60. As his consulting fee, Mr. Visconti charges $90.00 per day. Find how much he charges for 3 hours of consulting. Assume an 8-hour work day.

61. Five times the reciprocal of a number equals the sum of $\dfrac{3}{2}$ the reciprocal of the number and $\dfrac{7}{6}$. What is the number?

62. The reciprocal of a number equals the reciprocal of the difference of 4 and the number. Find the number.

63. A car travels 90 miles in the same time that a car traveling 10 miles per hour slower travels 60 miles. Find the speed of each car.

64. The current in a bayou near Lafayette, Louisiana, is 4 miles per hour. A paddle boat travels 48 miles upstream in the same amount of time it takes to travel 72 miles downstream. Find the speed of the boat in still water.

65. When Mark and Maria manicure Mr. Stergeon's lawn, it takes them 5 hours. If Mark works alone, it takes 7 hours. Find how long it takes Maria alone.

66. It takes pipe A 20 days to fill a fish pond. Pipe B takes 15 days. Find how long it takes both pipes together to fill the pond.

Given that the pairs of triangles are similar, find each unknown ngth x.

△ **67.**

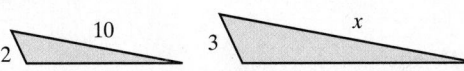

△ **68.**

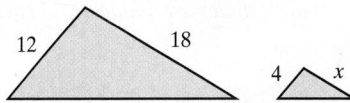

(7.7) *Simplify each complex fraction.*

69. $\dfrac{\dfrac{5x}{27}}{-\dfrac{10xy}{21}}$

70. $\dfrac{\dfrac{3}{5} + \dfrac{2}{7}}{\dfrac{1}{5} + \dfrac{5}{6}}$

71. $\dfrac{3 - \dfrac{1}{y}}{2 - \dfrac{1}{y}}$

72. $\dfrac{\dfrac{6}{x+2} + 4}{\dfrac{8}{x+2} - 4}$

73. $\dfrac{\dfrac{x-3}{x+3} + \dfrac{x+3}{x-3}}{\dfrac{x-3}{x+3} - \dfrac{x+3}{x-3}}$

74. $\dfrac{\dfrac{3}{x-1} - \dfrac{2}{1-x}}{\dfrac{2}{x-1} - \dfrac{2}{x}}$

75. $\dfrac{x + y^{-1}}{\dfrac{x}{y}}$

76. $\dfrac{x - xy^{-1}}{\dfrac{1+x}{y}}$

MIXED REVIEW

Simplify each rational expression.

77. $\dfrac{4x+12}{8x^2+24x}$

78. $\dfrac{x^3 - 6x^2 + 9x}{x^2 + 4x - 21}$

Perform the indicated operations and simplify.

79. $\dfrac{x^2+9x+20}{x^2-25} \cdot \dfrac{x^2-9x+20}{x^2+8x+16}$

80. $\dfrac{x^2-x-72}{x^2-x-30} \div \dfrac{x^2+6x-27}{x^2-9x+18}$

81. $\dfrac{x}{x^2-36} + \dfrac{6}{x^2-36}$

82. $\dfrac{5x-1}{4x} - \dfrac{3x-2}{4x}$

83. $\dfrac{4}{3x^2+8x-3} + \dfrac{2}{3x^2-7x+2}$

84. $\dfrac{3x}{x^2+9x+14} - \dfrac{6x}{x^2+4x-21}$

Solve.

85. $\dfrac{4}{a-1} + 2 = \dfrac{3}{a-1}$

86. $\dfrac{x}{x+3} + 4 = \dfrac{x}{x+3}$

Solve.

87. The quotient of twice a number and three, minus one-sixth, is the quotient of the number and two. Find the number.

88. Mr. Crocker can paint his house by himself in three days. His son will need an additional day to complete the job if he works alone. If they work together, find how long it takes to paint the house.

Given that the following pairs of triangles are similar, find each unknown length.

89.

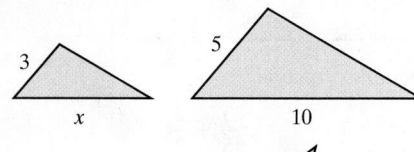

90.

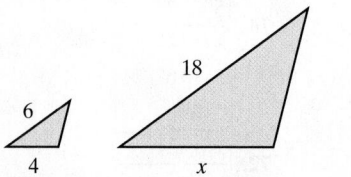

Simplify each complex fraction.

91. $\dfrac{\dfrac{1}{4}}{\dfrac{1}{3} + \dfrac{1}{2}}$

92. $\dfrac{4 + \dfrac{2}{x}}{6 + \dfrac{3}{x}}$

93. $\dfrac{y^{-2}}{1 - y^{-2}}$

94. $\dfrac{4 + x^{-1}}{3 + x^{-1}}$

1c

Chapter 7 Getting Ready for the Test

MULTIPLE CHOICE *Select the correct choice.*

▷ **1.** Choose the expression that is equivalent to $\dfrac{-x}{4 - x}$.

   A. $\dfrac{1}{4}$ **B.** $-\dfrac{1}{4}$ **C.** $\dfrac{x}{4 - x}$ **D.** $\dfrac{x}{x - 4}$

▷ **2.** For which of these values (if any) is the expression $\dfrac{x + 3}{x^2 + 9}$ undefined?

   A. -3 **B.** 3 **C.** -3 and 3 **D.** 0 **E.** none of these

MATCHING *Match each rational expression to its simplified form. Letters may be used more than once.*

▷ **3.** $\dfrac{y - 6}{6 - y}$ **4.** $\dfrac{y + 3}{3 + y}$ **A.** 1

▷ **5.** $\dfrac{x - 2}{-2 + x}$ **6.** $\dfrac{m - 4}{m + 4}$ **B.** -1

                                           C. neither 1 nor -1

MULTIPLE CHOICE *Select the correct choice.*

▷ **7.** $\dfrac{8}{x^2} \cdot \dfrac{4}{x^2} =$

   A. $\dfrac{32}{x^2}$ **B.** $\dfrac{2}{x^2}$ **C.** $\dfrac{32}{x^4}$ **D.** 2 **E.** $\dfrac{1}{2}$

▷ **8.** $\dfrac{8}{x^2} \div \dfrac{4}{x^2} =$

   A. $\dfrac{32}{x^2}$ **B.** $\dfrac{2}{x^2}$ **C.** $\dfrac{32}{x^4}$ **D.** 2 **E.** $\dfrac{1}{2}$

▷ **9.** $\dfrac{8}{x^2} + \dfrac{4}{x^2} =$

   A. $\dfrac{32}{x^2}$ **B.** $\dfrac{2}{x^2}$ **C.** $\dfrac{12}{x^4}$ **D.** $\dfrac{12}{x^2}$

▷ **10.** $\dfrac{7x}{x - 1} - \dfrac{5 + 2x}{x - 1} =$

   A. 5 **B.** $\dfrac{9x - 5}{x - 1}$ **C.** $\dfrac{5}{x - 1}$ **D.** $\dfrac{14}{x - 1}$

▷ **11.** The LCD of $\dfrac{9}{25x}$ and $\dfrac{z}{10x^3}$ is

   A. $250x^4$ **B.** $250x$ **C.** $50x^4$ **D.** $50x^3$

12. The LCD of $\dfrac{5}{4x + 8}$ and $\dfrac{9}{8x - 8}$ is

    A. $(4x + 8)(8x - 8)$ **B.** $32(x + 2)(x - 1)$ **C.** $4(x + 2)(x - 1)$ **D.** $8(x + 2)(x - 1)$

MULTIPLE CHOICE *Identify each as an*   **A.** *expression or* **B.** *equation.*
Letters may be used more than once or not at all.

13. $\dfrac{5}{x} + \dfrac{1}{3}$ **14.** $\dfrac{5}{x} + \dfrac{1}{3} = \dfrac{2}{x}$ **15.** $\dfrac{a + 5}{11} = 9$ **16.** $\dfrac{a + 5}{11} \cdot 9$

MULTIPLE CHOICE *Select the correct choice.*

17. Multiply the given equation through by the LCD of its terms. Choose the correct equivalent equation once this is done.
    Given Equation: $\dfrac{x + 3}{4} + \dfrac{5}{6} = 3$
    A. $(x + 3) + 5 = 3$ **B.** $3(x + 3) + 2 \cdot 5 = 3$ **C.** $3(x + 3) + 2 \cdot 5 = 12 \cdot 3$ **D.** $6(x + 3) + 4 \cdot 5 = 3$

18. Multiply the given equation through by the LCD of its terms. Choose the correct equivalent equation once this is done.
    Given equation: $3 - \dfrac{10x}{4(x + 1)} = \dfrac{5}{6(x + 1)}$
    A. $3 - 10x = 5$ **B.** $3 - 3 \cdot 10x = 2 \cdot 5$ **C.** $3 \cdot 12(x + 1) - 3 \cdot 10x = 2 \cdot 5$ **D.** $4(x + 1) - 3 \cdot 10x = 2 \cdot 5$

19. Translate into an equation. Let x be the unknown number. "The quotient of a number and 5 equals the sum of that number and 12."

    A. $\dfrac{x}{5} = x + 12$ **B.** $\dfrac{5}{x} = x + 12$ **C.** $\dfrac{x}{5} = x \cdot 12$ **D.** $\dfrac{x}{5} \cdot (x + 12)$

20. Write $\dfrac{2x^{-1}}{y^{-2} + (5x)^{-1}}$ without negative exponents.

    A. $\dfrac{\dfrac{2}{x}}{\dfrac{1}{y^2} + \dfrac{1}{5x}}$ **B.** $\dfrac{\dfrac{1}{2x}}{\dfrac{1}{y^2} + \dfrac{1}{5x}}$ **C.** $\dfrac{y^2 + 5x}{2x}$ **D.** $\dfrac{\dfrac{2}{x}}{\dfrac{1}{y^2} + \dfrac{5}{x}}$

Chapter 7 Test   MyMathLab®   You Tube

1. Find the domain of the rational function

$$g(x) = \frac{9x^2 - 9}{x^2 + 4x + 3}$$

2. For a certain computer desk, the average cost C (in dollars) per desk manufactured is

$$C = \frac{100x + 3000}{x}$$

where x is the number of desks manufactured.

  a. Find the average cost per desk when manufacturing 200 computer desks.

  b. Find the average cost per desk when manufacturing 1000 computer desks.

Simplify each rational expression.

3. $\dfrac{3x - 6}{5x - 10}$

4. $\dfrac{x + 6}{x^2 + 12x + 36}$

5. $\dfrac{x + 3}{x^3 + 27}$

6. $\dfrac{2m^3 - 2m^2 - 12m}{m^2 - 5m + 6}$

7. $\dfrac{ay + 3a + 2y + 6}{ay + 3a + 5y + 15}$

8. $\dfrac{y - x}{x^2 - y^2}$

Perform the indicated operation and simplify if possible.

9. $\dfrac{3}{x - 1} \cdot (5x - 5)$

10. $\dfrac{y^2 - 5y + 6}{2y + 4} \cdot \dfrac{y + 2}{2y - 6}$

11. $\dfrac{15x}{2x + 5} - \dfrac{6 - 4x}{2x + 5}$

12. $\dfrac{5a}{a^2 - a - 6} - \dfrac{2}{a - 3}$

13. $\dfrac{6}{x^2 - 1} + \dfrac{3}{x + 1}$

14. $\dfrac{x^2 - 9}{x^2 - 3x} \div \dfrac{xy + 5x + 3y + 15}{2x + 10}$

▶ **15.** $\dfrac{x + 2}{x^2 + 11x + 18} + \dfrac{5}{x^2 - 3x - 10}$

Solve each equation.

▶ **16.** $\dfrac{4}{y} - \dfrac{5}{3} = \dfrac{-1}{5}$ **17.** $\dfrac{5}{y + 1} = \dfrac{4}{y + 2}$

▶ **18.** $\dfrac{a}{a - 3} = \dfrac{3}{a - 3} - \dfrac{3}{2}$

▶ **19.** $x - \dfrac{14}{x - 1} = 4 - \dfrac{2x}{x - 1}$

▶ **20.** $\dfrac{10}{x^2 - 25} = \dfrac{3}{x + 5} + \dfrac{1}{x - 5}$

Simplify each complex fraction.

▶ **21.** $\dfrac{\dfrac{5x^2}{yz^2}}{\dfrac{10x}{z^3}}$ ▶ **22.** $\dfrac{5 - \dfrac{1}{y^2}}{\dfrac{1}{y} + \dfrac{2}{y^2}}$ ▶ **23.** $\dfrac{\dfrac{b}{a} - \dfrac{a}{b}}{\dfrac{1}{b} + \dfrac{1}{a}}$

▶ **24.** In a sample of 85 fluorescent bulbs, 3 were found to be defective. At this rate, how many defective bulbs should be found in 510 bulbs?

▶ **25.** One number plus five times its reciprocal is equal to six. Find the number.

▶ **26.** A pleasure boat traveling down the Red River takes the same time to go 14 miles upstream as it takes to go 16 miles downstream. If the current of the river is 2 miles per hour, find the speed of the boat in still water.

▶ **27.** An inlet pipe can fill a tank in 12 hours. A second pipe can fill the tank in 15 hours. If both pipes are used, find how long it takes to fill the tank.

▶ **28.** Given that the two triangles are similar, find x.

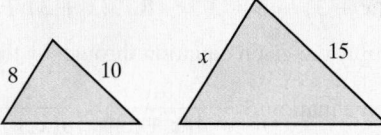

Chapter 7 Cumulative Review

TRANSLATING

1. Write each sentence as an equation or inequality. Let x represent the unknown number.

  a. The quotient of 15 and a number is 4.

  b. Three subtracted from 12 is a number.

  c. Four times a number, added to 17, is not equal to 21.

  d. Triple a number is less than 48.

2. Write each sentence as an equation. Let x represent the unknown number.

  a. The difference of 12 and a number is -45.

  b. The product of 12 and a number is -45.

  c. A number less 10 is twice the number.

3. Rajiv Puri invested part of his $20,000 inheritance in a mutual fund account that pays 7% simple interest yearly and the rest in a certificate of deposit that pays 9% simple interest yearly. At the end of one year, Rajiv's investments earned $1550. Find the amount he invested at each rate.

4. The number of non-business bankruptcies has increased over the years. In 2002, the number of non-business bankruptcies was 80,000 less than twice the number in 1994. If the total of non-business bankruptcies for these two years is 2,290,000 find the number of non-business bankruptcies for each year. (*Source:* American Bankruptcy Institute)

5. Graph $x - 3y = 6$ by finding and plotting intercepts.

6. Find the slope of the line whose equation is $7x + 2y = 9$.

7. Use the product rule to simplify each expression.

  a. $4^2 \cdot 4^5$ **b.** $x^4 \cdot x^6$

  c. $y^3 \cdot y$ **d.** $y^3 \cdot y^2 \cdot y^7$

  e. $(-5)^7 \cdot (-5)^8$ **f.** $a^2 \cdot b^2$

8. Simplify.

  a. $\dfrac{x^9}{x^7}$ **b.** $\dfrac{x^{19}y^5}{xy}$

  c. $(x^5y^2)^3$ **d.** $(-3a^2b)(5a^3b)$

9. Subtract $(5z - 7)$ from the sum of $(8z + 11)$ and $(9z - 2)$.

10. Subtract $(9x^2 - 6x + 2)$ from $(x + 1)$.

11. Multiply: $(3a + b)^3$

12. Multiply: $(2x + 1)(5x^2 - x + 2)$

13. Use a special product to square each binomial.

  a. $(t + 2)^2$ **b.** $(p - q)^2$

  c. $(2x + 5)^2$

  d. $(x^2 - 7y)^2$

14. Multiply.

  a. $(x + 9)^2$

  b. $(2x + 1)(2x - 1)$

  c. $8x(x^2 + 1)(x^2 - 1)$

15. Simplify each expression. Write results using positive exponents only.

  a. $\dfrac{1}{x^{-3}}$ **b.** $\dfrac{1}{3^{-4}}$

  c. $\dfrac{p^{-4}}{q^{-9}}$ **d.** $\dfrac{5^{-3}}{2^{-5}}$

16. Simplify. Write results with positive exponents only.

  a. 5^{-3} **b.** $\dfrac{9}{x^{-7}}$ **c.** $\dfrac{11^{-1}}{7^{-2}}$

17. Divide: $\dfrac{4x^2 + 7 + 8x^3}{2x + 3}$

18. Divide $(4x^3 - 9x + 2)$ by $(x - 4)$.

19. Find the GCF of each list of numbers.
  a. 28 and 40
  b. 55 and 21
  c. 15, 18, and 66

20. Find the GCF of $9x^2$, $6x^3$, and $21x^5$.

Factor.

21. $-9a^5 + 18a^2 - 3a$

22. $7x^6 - 7x^5 + 7x^4$

23. $3m^2 - 24m - 60$

24. $-2a^2 + 10a + 12$

25. $3x^2 + 11x + 6$

26. $10m^2 - 7m + 1$

27. $x^2 + 12x + 36$

28. $4x^2 + 12x + 9$

29. $x^2 + 4$

30. $x^2 - 4$

31. $x^3 + 8$

32. $27y^3 - 1$

33. $2x^3 + 3x^2 - 2x - 3$

34. $3x^3 + 5x^2 - 12x - 20$

35. $12m^2 - 3n^2$

36. $x^5 - x$

37. Solve: $x(2x - 7) = 4$

38. Solve: $3x^2 + 5x = 2$

39. Find the x-intercepts of the graph of $y = x^2 - 5x + 4$.

40. Find the x-intercepts of the graph of $y = x^2 - x - 6$.

41. The height of a triangular sail is 2 meters less than twice the length of the base. If the sail has an area of 30 square meters, find the length of its base and the height.

42. The height of a parallelogram is 5 feet more than three times its base. If the area of the parallelogram is 182 square feet, find the length of its base and height.

43. Simplify: $\dfrac{18 - 2x^2}{x^2 - 2x - 3}$

44. Simplify: $\dfrac{2x^2 - 50}{4x^4 - 20x^3}$

45. Divide: $\dfrac{6x + 2}{x^2 - 1} \div \dfrac{3x^2 + x}{x - 1}$

46. Multiply: $\dfrac{6x^2 - 18x}{3x^2 - 2x} \cdot \dfrac{15x - 10}{x^2 - 9}$

47. Simplify: $\dfrac{(2x)^{-1} + 1}{2x^{-1} - 1}$

48. Simplify: $\dfrac{\dfrac{m}{3} + \dfrac{n}{6}}{\dfrac{m + n}{12}}$

More on Functions and Graphs

8.1 Graphing and Writing Linear Functions

8.2 Reviewing Function Notation and Graphing Nonlinear Functions

Integrated Review—Summary on Functions and Equations of Lines

8.3 Graphing Piecewise-Defined Functions and Shifting and Reflecting Graphs of Functions

8.4 Variation and Problem Solving

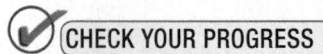
CHECK YOUR PROGRESS

Vocabulary Check

Chapter Highlights

Chapter Review

Getting Ready for the Test

Chapter Test

Cumulative Review

In Section 3.6, we introduced the notion of relation and the notion of function, perhaps the single most important and useful concepts in all of mathematics. In this chapter, we explore the concept of functions further.

Sulfur dioxide in the environment contributes to air pollution and acid rain. Under the Clean Air Act, the EPA sets standards for sulfur dioxide concentration. Nationally, average sulfur dioxide concentrations have decreased substantially over the years.

The bar graph below shows the decrease in sulfur dioxide emissions in the United States. Notice that the two functions, $f(x)$ and $g(x)$, both approximate the amount of sulfur dioxide emissions (in millions of tons) in the United States. Also, for both functions, x is the number of years since 1970. In Section 8.1, Exercises 69–74, we use these functions to predict the further reduction of sulfur dioxide emissions.

$$f(x) = -0.51x + 31.04$$
$$g(x) = 0.0004x^2 - 0.53x + 31$$

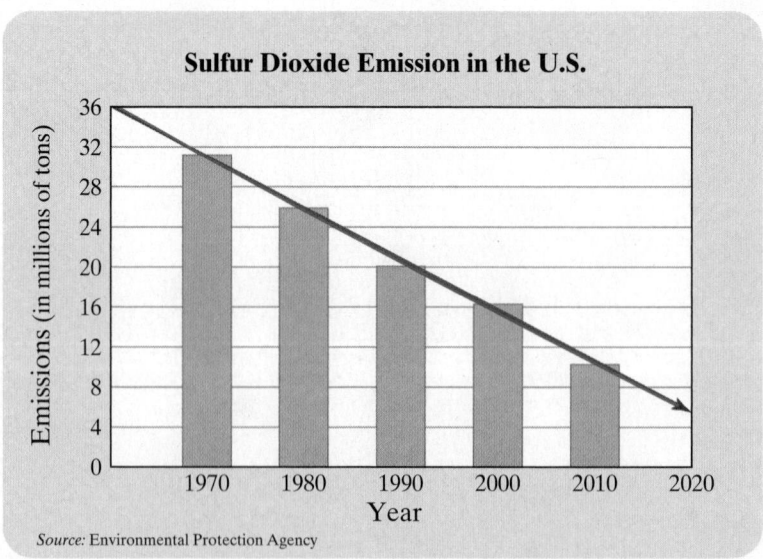

Sulfur Dioxide Emission in the U.S.

Source: Environmental Protection Agency

8.1 Graphing and Writing Linear Functions

OBJECTIVES

1 Graph Linear Functions.

2 Write an Equation of a Line Using Function Notation.

3 Find Equations of Parallel and Perpendicular Lines.

OBJECTIVE

1 Graphing Linear Functions

In this section, we identify and graph linear functions. By the vertical line test, Section 3.6, we know that all linear equations except those whose graphs are vertical lines are functions. Thus, all linear equations except those of the form $x = c$ (vertical lines) are linear functions. For example, we know from Section 3.2 that $y = 2x$ is a linear equation in two variables. Its graph is shown.

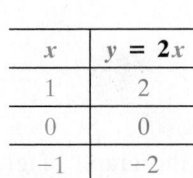

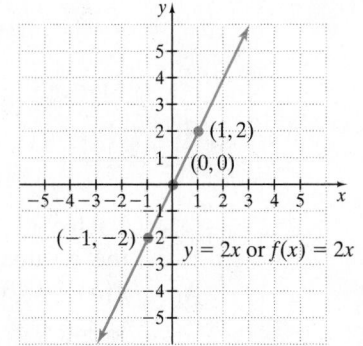

Because this graph passes the vertical line test, we know that $y = 2x$ is a function. If we want to emphasize that this equation describes a function, we may write $y = 2x$ as $f(x) = 2x$.

EXAMPLE 1 Graph $g(x) = 2x + 1$. Compare this graph with the graph of $f(x) = 2x$.

Solution To graph $g(x) = 2x + 1$, find three ordered pair solutions.

x	$f(x) = 2x$	$g(x) = 2x + 1$
0	0	1
−1	−2	−1
1	2	3

add 1

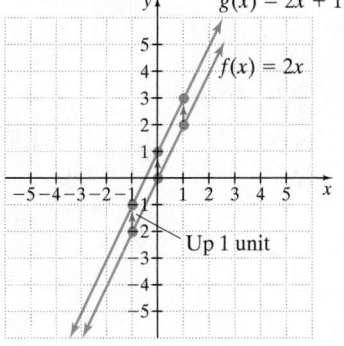

Notice that y-values for the graph of $g(x) = 2x + 1$ are obtained by adding 1 to each y-value of each corresponding point of the graph of $f(x) = 2x$. The graph of $g(x) = 2x + 1$ is the same as the graph of $f(x) = 2x$ shifted upward 1 unit. □

PRACTICE

1 Graph $g(x) = 4x - 3$ and $f(x) = 4x$ on the same axes. ■

If a linear function is solved for y, we can easily use function notation to describe it by replacing y with $f(x)$. Recall the slope-intercept form of a linear equation, $y = mx + b$, where m is the slope of the line and $(0, b)$ is the y-intercept. Since this form is solved for y, we use it to define a linear function.

In general, a **linear function** is a function that can be written in the form $f(x) = mx + b$. For example, $g(x) = 2x + 1$ is in this form, with $m = 2$ and $b = 1$. Thus, the slope of the linear function $g(x)$ is 2 and the y-intercept is $(0, 1)$.

EXAMPLE 2 Graph the linear functions $f(x) = -3x$ and $g(x) = -3x - 6$ on the same set of axes.

Solution To graph $f(x)$ and $g(x)$, find ordered pair solutions.

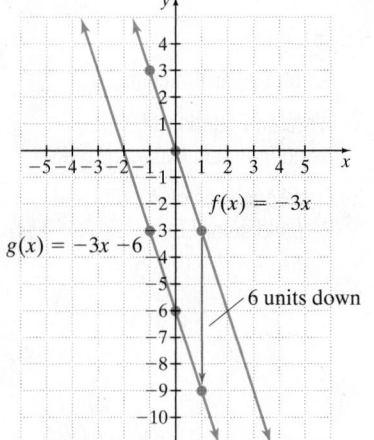

x	$f(x) = -3x$	$g(x) = -3x - 6$
0	0	−6
1	−3	−9
−1	3	−3
−2	6	0

(subtract 6)
(subtract 6)

$g(x) = -3x - 6$
$f(x) = -3x$
6 units down

Each y-value for the graph of $g(x) = -3x - 6$ is obtained by subtracting 6 from the y-value of the corresponding point of the graph of $f(x) = -3x$. The graph of $g(x) = -3x - 6$ is the same as the graph of $f(x) = -3x$ shifted down 6 units. □

PRACTICE

2 Graph the linear functions $f(x) = -2x$ and $g(x) = -2x + 5$ on the same set of axes. ■

OBJECTIVE

2 Writing Equations of Lines Using Function Notation

We now practice writing linear functions.

↑ ↖

This means the graph is a line that passes the vertical line test.

Below is a review of some tools we can use.

$y = mx + b$	**Slope-intercept form** of a linear equation. The slope is m, and the y-intercept is $(0, b)$.
$y - y_1 = m(x - x_1)$	**Point-slope form** of a linear equation. The slope is m, and (x_1, y_1) is a point on the line.
$y = c$	**Horizontal line** The slope is 0, and the y-intercept is $(0, c)$.

Note: $x = c$, whose graph is a vertical line, is not included above, as these equations do **not** define functions.

EXAMPLE 3 Find an equation of the line with slope -3 and y-intercept $(0, -5)$. Write the equation using function notation.

Solution Because we know the slope and the y-intercept, we use the slope-intercept form with $m = -3$ and $b = -5$.

$$y = mx + b \qquad \text{Slope-intercept form}$$
$$y = -3 \cdot x + (-5) \quad \text{Let } m = -3 \text{ and } b = -5.$$
$$y = -3x - 5 \qquad \text{Simplify.}$$

This equation is solved for y. To write using function notation, we replace y with $f(x)$.

$$f(x) = -3x - 5$$ □

PRACTICE

3 Find an equation of the line with slope -4 and y-intercept $(0, -3)$. Write the equation using function notation. ■

EXAMPLE 4   Find an equation of the line through points $(4, 0)$ and $(-4, -5)$. Write the equation using function notation.

Solution First, find the slope of the line.

$$m = \frac{-5 - 0}{-4 - 4} = \frac{-5}{-8} = \frac{5}{8}$$

Next, make use of the point-slope form. Replace (x_1, y_1) by either $(4, 0)$ or $(-4, -5)$ in the point-slope equation. We will choose the point $(4, 0)$. The line through $(4, 0)$ with slope $\frac{5}{8}$ is

$$y - y_1 = m(x - x_1) \quad \text{Point-slope form.}$$

$$y - 0 = \frac{5}{8}(x - 4) \quad \text{Let } m = \frac{5}{8} \text{ and } (x_1, y_1) = (4, 0).$$

$$8y = 5(x - 4) \quad \text{Multiply both sides by 8.}$$

$$8y = 5x - 20 \quad \text{Apply the distributive property.}$$

To write the equation using function notation, we solve for y, then replace y with $f(x)$.

$$8y = 5x - 20$$

$$y = \frac{5}{8}x - \frac{20}{8} \quad \text{Divide both sides by 8.}$$

$$f(x) = \frac{5}{8}x - \frac{5}{2} \quad \text{Write using function notation.} \qquad \square$$

PRACTICE

4   Find an equation of the line through points $(-1, 2)$ and $(2, 0)$. Write the equation using function notation. ∎

Helpful Hint

If two points of a line are given, either one may be used with the point-slope form to write an equation of the line.

EXAMPLE 5   Find an equation of the horizontal line containing the point $(2, 3)$. Write the equation using function notation.

Solution A horizontal line has an equation of the form $y = c$. Since the line contains the point $(2, 3)$, the equation is $y = 3$, as shown to the right.

Using function notation, the equation is

$$f(x) = 3.$$

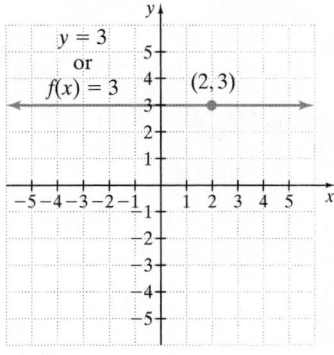

PRACTICE

5   Find the equation of the horizontal line containing the point $(6, -2)$. Use function notation. ∎

OBJECTIVE

3 Finding Equations of Parallel and Perpendicular Lines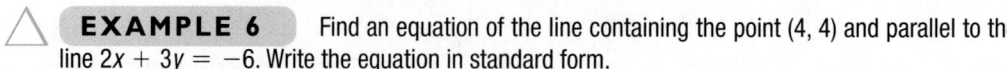

Next, we find equations of parallel and perpendicular lines.

> ⚠ **EXAMPLE 6** Find an equation of the line containing the point (4, 4) and parallel to the line $2x + 3y = -6$. Write the equation in standard form.

Solution Because the line we want to find is *parallel* to the line $2x + 3y = -6$, the two lines must have equal slopes. Find the slope of $2x + 3y = -6$ by writing it in the form $y = mx + b$. In other words, solve the equation for y.

$$2x + 3y = -6$$
$$3y = -2x - 6 \quad \text{Subtract } 2x \text{ from both sides.}$$
$$y = \frac{-2x}{3} - \frac{6}{3} \quad \text{Divide by 3.}$$
$$y = -\frac{2}{3}x - 2 \quad \text{Write in slope-intercept form.}$$

The slope of this line is $-\frac{2}{3}$. Thus, a line parallel to this line will also have a slope of $-\frac{2}{3}$.

The equation we are asked to find describes a line containing the point (4, 4) with a slope of $-\frac{2}{3}$. We use the point-slope form.

$$y - y_1 = m(x - x_1)$$
$$y - 4 = -\frac{2}{3}(x - 4) \quad \text{Let } m = -\frac{2}{3}, x_1 = 4, \text{ and } y_1 = 4.$$
$$3(y - 4) = -2(x - 4) \quad \text{Multiply both sides by 3.}$$
$$3y - 12 = -2x + 8 \quad \text{Apply the distributive property.}$$
$$2x + 3y = 20 \quad \text{Write in standard form.} \qquad \square$$

> **Helpful Hint**
>
> Multiply both sides of the equation $2x + 3y = 20$ by -1 and it becomes $-2x - 3y = -20$. Both equations are in standard form, and their graphs are the same line.

PRACTICE

6 Find an equation of the line containing the point $(8, -3)$ and parallel to the line $3x + 4y = 1$. Write the equation in standard form. ▪

> **EXAMPLE 7** Write a function that describes the line containing the point (4, 4) and perpendicular to the line $2x + 3y = -6$.

Solution In the previous example, we found that the slope of the line $2x + 3y = -6$ is $-\frac{2}{3}$. A line perpendicular to this line will have a slope that is the negative reciprocal of $-\frac{2}{3}$, or $\frac{3}{2}$. From the point-slope equation, we have

$$y - y_1 = m(x - x_1)$$
$$y - 4 = \frac{3}{2}(x - 4) \quad \text{Let } x_1 = 4, y_1 = 4 \text{ and } m = \frac{3}{2}.$$
$$2(y - 4) = 3(x - 4) \quad \text{Multiply both sides by 2.}$$
$$2y - 8 = 3x - 12 \quad \text{Apply the distributive property.}$$
$$2y = 3x - 4 \quad \text{Add 8 to both sides.}$$
$$y = \frac{3}{2}x - 2 \quad \text{Divide both sides by 2.}$$
$$f(x) = \frac{3}{2}x - 2 \quad \text{Write using function notation.} \qquad \square$$

PRACTICE

7   Write a function that describes the line containing the point $(8, -3)$ and perpendicular to the line $3x + 4y = 1$.

Graphing Calculator Explorations

You may have noticed by now that to use the $\boxed{Y =}$ key on a graphing calculator to graph an equation, the equation must be solved for y.

Graph each function by first solving the function for y.

1. $x = 3.5y$

2. $-2.7y = x$

3. $5.78x + 2.31y = 10.98$

4. $-7.22x + 3.89y = 12.57$

5. $y - |x| = 3.78$

6. $3y - 5x^2 = 6x - 4$

7. $y - 5.6x^2 = 7.7x + 1.5$

8. $y + 2.6|x| = -3.2$

 ### Vocabulary, Readiness & Video Check

Use the choices given to fill in each blank. Some choices may not be used.

linear	$(0, b)$	m
quadratic	$(b, 0)$	mx

1. A _____ function can be written in the form $f(x) = mx + b$.
2. In the form $f(x) = mx + b$, the y-intercept is _____ and the slope is _____.

State the slope and the y-intercept of the graph of each function.

3. $f(x) = -4x + 12$

4. $g(x) = \dfrac{2}{3}x - \dfrac{7}{2}$

5. $g(x) = 5x$

6. $f(x) = -x$

Decide whether the lines are parallel, perpendicular, or neither.

7. $y = 12x + 6$
   $y = 12x - 2$

8. $y = -5x + 8$
   $y = -5x - 8$

9. $y = -9x + 3$
   $y = \dfrac{3}{2}x - 7$

10. $y = 2x - 12$
    $y = \dfrac{1}{2}x - 6$

Martin-Gay Interactive Videos

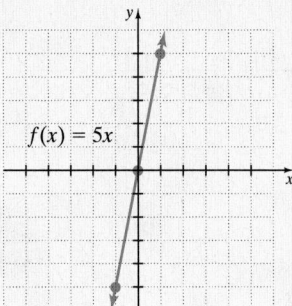

See Video 8.1 ◉

Watch the section lecture video and answer the following questions.

OBJECTIVE
1 **11.** Based on the lecture before ▦ Example 1, in what form can a linear function be written?

OBJECTIVE
2 **12.** From ▦ Example 2, given a *y*-intercept, how do you know which value to use for *b* in the slope-intercept form?

OBJECTIVE
2 **13.** ▦ Example 4 discusses how to find an equation of a line given two points. Under what circumstances might the slope-intercept form be chosen over the point-slope form to find an equation?

OBJECTIVE
3 **14.** Solve ▦ Example 6 again, this time writing the equation of the line in function notation *parallel* to the given line through the given point.

8.1 Exercise Set MyMathLab® ▶

Graph each linear function. See Examples 1 and 2.

1. $f(x) = -2x$
2. $f(x) = 2x$

▶ **3.** $f(x) = -2x + 3$
4. $f(x) = 2x + 6$

5. $f(x) = \frac{1}{2}x$
6. $f(x) = \frac{1}{3}x$

7. $f(x) = \frac{1}{2}x - 4$
8. $f(x) = \frac{1}{3}x - 2$

The graph of $f(x) = 5x$ follows. Use this graph to match each linear function with its graph. See Examples 1 and 2.

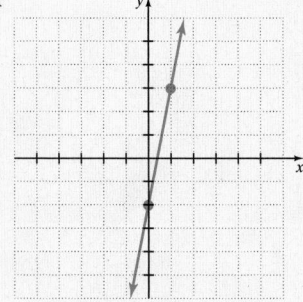

A

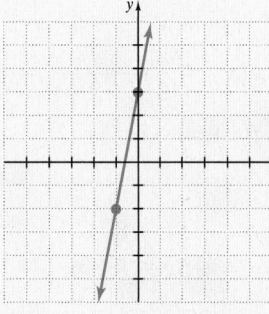

B

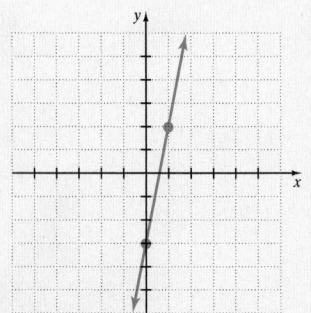

C

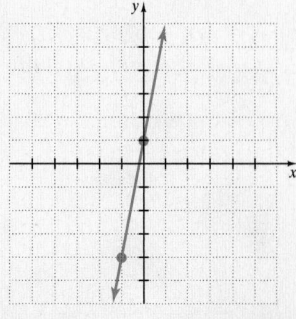

D

9. $f(x) = 5x - 3$
10. $f(x) = 5x - 2$

11. $f(x) = 5x + 1$
12. $f(x) = 5x + 3$

Use function notation to write the equation of each line with the given slope and y-intercept. See Example 3.

▶ **13.** Slope -1; *y*-intercept $(0, 1)$

14. Slope $\frac{1}{2}$; *y*-intercept $(0, -6)$

15. Slope 2; *y*-intercept $\left(0, \frac{3}{4}\right)$

16. Slope -3; *y*-intercept $\left(0, -\frac{1}{5}\right)$

17. Slope $\frac{2}{7}$; *y*-intercept $(0, 0)$

18. Slope $-\frac{4}{5}$; *y*-intercept $(0, 0)$

Find an equation of the line with the given slope and containing the given point. Write the equation using function notation. See Example 3.

▶ **19.** Slope 3; through $(1, 2)$

20. Slope 4; through $(5, 1)$

21. Slope -2; through $(1, -3)$

22. Slope -4; through $(2, -4)$

23. Slope $\frac{1}{2}$; through $(-6, 2)$

24. Slope $\frac{2}{3}$; through $(-9, 4)$

25. Slope $-\frac{9}{10}$; through $(-3, 0)$

26. Slope $-\frac{1}{5}$; through $(4, -6)$

Find an equation of the line passing through the given points. Use function notation to write the equation. See Example 4.

27. $(2, 0), (4, 6)$

28. $(3, 0), (7, 8)$

29. $(-2, 5), (-6, 13)$

30. $(7, -4), (2, 6)$

31. $(-2, -4), (-4, -3)$

32. $(-9, -2), (-3, 10)$

33. $(-3, -8), (-6, -9)$

34. $(8, -3), (4, -8)$

35. $\left(\dfrac{3}{5}, \dfrac{2}{5}\right)$ and $\left(-\dfrac{1}{5}, \dfrac{7}{10}\right)$

36. $\left(\dfrac{1}{2}, -\dfrac{1}{4}\right)$ and $\left(\dfrac{3}{2}, \dfrac{3}{4}\right)$

Write an equation of each line using function notation. See Example 5.

37. Slope 0; through $(-2, -4)$

38. Horizontal; through $(-3, 1)$

39. Horizontal; through $(0, 5)$

40. Slope 0; through $(-10, 23)$

Find an equation of each line. Write the equation using function notation. See Examples 6 and 7.

41. Through $(3, 8)$; parallel to $f(x) = 4x - 2$

42. Through $(1, 5)$; parallel to $f(x) = 3x - 4$

43. Through $(2, -5)$; perpendicular to $3y = x - 6$

44. Through $(-4, 8)$; perpendicular to $2x - 3y = 1$

45. Through $(-2, -3)$; parallel to $3x + 2y = 5$

46. Through $(-2, -3)$; perpendicular to $3x + 2y = 5$

MIXED PRACTICE

Find an equation of each line. Write the equation in standard form unless indicated otherwise. See Examples 3 through 7.

47. Slope 2; through $(-2, 3)$

48. Slope 3; through $(-4, 2)$

49. Through $(1, 6)$ and $(5, 2)$; use function notation

50. Through $(2, 9)$ and $(8, 6)$; use function notation

51. With slope $-\dfrac{1}{2}$; y-intercept 11; use function notation

52. With slope -4; y-intercept $\dfrac{2}{9}$; use function notation

53. Through $(-7, -4)$ and $(0, -6)$

54. Through $(2, -8)$ and $(-4, -3)$

55. Slope $-\dfrac{4}{3}$; through $(-5, 0)$

56. Slope $-\dfrac{3}{5}$; through $(4, -1)$

57. Horizontal line; through $(-2, -10)$; use function notation

58. Horizontal line; through $(1, 0)$; use function notation

59. Through $(6, -2)$; parallel to the line $2x + 4y = 9$

60. Through $(8, -3)$; parallel to the line $6x + 2y = 5$

61. Slope 0; through $(-9, 12)$; use function notation

62. Slope 0; through $(10, -8)$; use function notation

63. Through $(6, 1)$; parallel to the line $8x - y = 9$

64. Through $(3, 5)$; perpendicular to the line $2x - y = 8$

65. Through $(5, -6)$; perpendicular to $y = 9$

66. Through $(-3, -5)$; parallel to $y = 9$

67. Through $(2, -8)$ and $(-6, -5)$; use function notation

68. Through $(-4, -2)$ and $(-6, 5)$; use function notation

REVIEW AND PREVIEW

From the Chapter 8 opener, we have two functions to describe the amount of sulfur dioxide emissions in the United States. For both functions, x is the number of years since 1970 and y (or f(x) or g(x)) is the amount of emissions in millions of tons.

$$f(x) = -0.51x + 31.04 \quad \text{or} \quad g(x) = 0.0004x^2 - 0.53x + 31$$

Use this for Exercises 69–74. See Section 3.6.

69. Find $f(20)$ and describe in words what this means.

70. Find $g(20)$ and describe in words what this means.

71. Assume the trend of $g(x)$ continues. Find $g(50)$ and describe in words what this means.

72. Assume the trend of $f(x)$ continues. Find $f(50)$ and describe in words what this means.

73. Use Exercises 69–72 and compare $f(20)$ and $g(20)$, then $f(50)$ and $g(50)$. As x increases, are the function values staying about the same or not? Explain your answer.

74. Use the Chapter 8 opener graph and study the graphs of $f(x)$ and $g(x)$. Use these graphs to answer Exercise 73. Explain your answer.

CONCEPT EXTENSIONS

Find an equation of each line graphed. Write the equation using function notation. (Hint: Use each graph to write 2 ordered pair solutions. Find the slope of each line, then refer to Example 3 or 4 to complete.)

75.

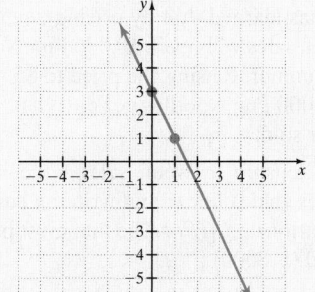

76.

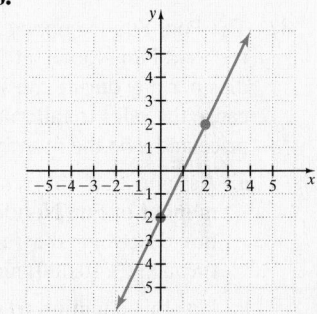

77. **78.**

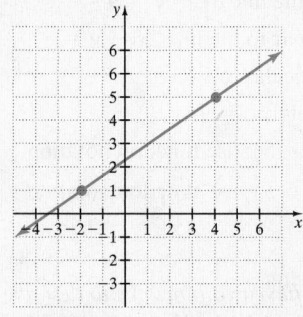
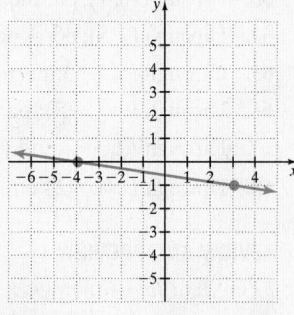

Solve.

79. A rock is dropped from the top of a 400-foot building. After 1 second, the rock is traveling 32 feet per second. After 3 seconds, the rock is traveling 96 feet per second. Let y be the rate of descent and x be the number of seconds since the rock was dropped.

a. Write a linear equation that relates time x to rate y. [*Hint:* Use the ordered pairs $(1, 32)$ and $(3, 96)$.]

b. Use this equation to determine the rate of travel of the rock 4 seconds after it was dropped.

80. A fruit company recently released a new applesauce. By the end of its first year, profits on this product amounted to $30,000. The anticipated profit for the end of the fourth year is $66,000. The ratio of change in time to change in profit is constant. Let x be years and y be profit.

a. Write a linear equation that relates profit and time. [*Hint:* Use the ordered pairs $(1, 30,000)$ and $(4, 66,000)$.]

b. Use this equation to predict the company's profit at the end of the seventh year.

c. Predict when the profit should reach $126,000.

81. The Whammo Company has learned that by pricing a newly released Frisbee at $6, sales will reach 2000 per day. Raising the price to $8 will cause the sales to fall to 1500 per day. Assume that the ratio of change in price to change in daily sales is constant and let x be the price of the Frisbee and y be number of sales.

a. Find the linear equation that models the price–sales relationship for this Frisbee. [*Hint:* The line must pass through $(6, 2000)$ and $(8, 1500)$.]

b. Use this equation to predict the daily sales of Frisbees if the price is set at $7.50.

82. The Pool Fun Company has learned that by pricing a newly released Fun Noodle at $3, sales will reach 10,000 Fun Noodles per day during the summer. Raising the price to $5 will cause the sales to fall to 8000 Fun Noodles per day. Let x be price and y be the number sold.

a. Assume that the relationship between sales price and number of Fun Noodles sold is linear and write an equation describing this relationship. [*Hint:* The line must pass through $(3, 10,000)$ and $(5, 8000)$.]

b. Use this equation to predict the daily sales of Fun Noodles if the price is $3.50.

83. The number of people employed in the United States as nurse practitioners was 110 thousand in 2012. By 2022, this number is expected to rise to 147 thousand. Let y be the number of nurse practitioners (in thousands) employed in the United States in the year x, where $x = 0$ represents 2012. (*Source:* U.S. Bureau of Labor Statistics)

a. Write a linear equation that models the number of people (in thousands) employed as nurse practitioners in year x.

b. Use this equation to estimate the number of people employed as nurse practitioners in 2018.

84. In 2010, Target had 355,000 employees. By 2015, due to store closures, the number of employees was reduced to 347,000. Let y be the number of Target employees in the year x, where $x = 0$ represents 2010. (*Source:* Target Corporation)

a. Write a linear equation that models the decrease in the number of Target employees, in terms of the year x.

b. Use this equation to predict the number of Target employees in 2018.

85. In 2014, the average price of a new home sold in the United States was $343,200. In 2004, the average price of a new home in the United States was $271,500. Let y be the average price of a new home in the year x, where $x = 0$ represents the year 2004. (*Source:* Based on data from U.S. census)

a. Write a linear equation that models the average price of a new home in terms of the year x. [*Hint:* The line must pass through the points $(0, 271,500)$ and $(10, 343,200)$.]

b. Use this equation to predict the average price of a new home in 2016.

86. The number of McDonald's restaurants worldwide in 2014 was 36,258. In 2009, there were 32,737 McDonald's restaurants worldwide. Let y be the number of McDonald's restaurants in the year x, where $x = 0$ represents the year 2009. (*Source:* McDonald's Corporation)

a. Write a linear equation that models the growth in the number of McDonald's restaurants worldwide in terms of the year x. [*Hint:* The line must pass through the points $(0, 32,737)$ and $(5, 36,258)$.]

b. Use this equation to predict the number of McDonald's restaurants worldwide in 2016.

Example:

Find an equation of the perpendicular bisector of the line segment whose endpoints are $(2, 6)$ and $(0, -2)$.

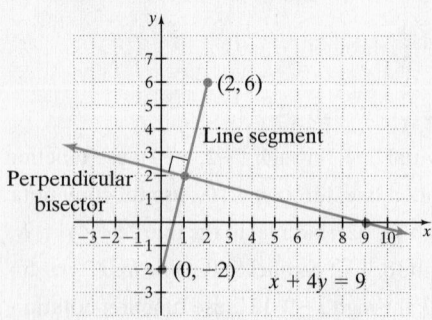

Solution:

A perpendicular bisector is a line that contains the midpoint of the given segment and is perpendicular to the segment.

Step 1   The midpoint of the segment with endpoints $(2, 6)$ and $(0, -2)$ is $(1, 2)$.

Step 2   The slope of the segment containing points $(2, 6)$ and $(0, -2)$ is 4.

Step 3   A line perpendicular to this line segment will have slope of $-\dfrac{1}{4}$.

Step 4   The equation of the line through the midpoint $(1, 2)$ with a slope of $-\dfrac{1}{4}$ will be the equation of the perpendicular bisector. This equation in standard form is $x + 4y = 9$.

Find an equation of the perpendicular bisector of the line segment whose endpoints are given. See the previous example.

△ **87.** $(3, -1); (-5, 1)$

△ **88.** $(-6, -3); (-8, -1)$

△ **89.** $(-2, 6); (-22, -4)$

△ **90.** $(5, 8); (7, 2)$

△ **91.** $(2, 3); (-4, 7)$

△ **92.** $(-6, 8); (-4, -2)$

↘ **93.** Describe how to check to see if the graph of $2x - 4y = 7$ passes through the points $(1.4, -1.05)$ and $(0, -1.75)$. Then follow your directions and check these points.

8.2   Reviewing Function Notation and Graphing Nonlinear Functions

OBJECTIVES

1   Review Function Notation.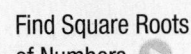

2   Find Square Roots of Numbers.

3   Graph Nonlinear Functions.

In the previous section, we studied linear equations that described functions. Not all equations in two variables are linear equations, and not all graphs of equations in two variables are lines. In Chapter 3, we saw graphs of nonlinear equations, some of which were functions since they passed the vertical line test. In this section, we study the functions whose graphs may not be lines. First, let's review function notation.

OBJECTIVE

1   Reviewing Function Notation

Suppose we have a function f such that $f(2) = -1$. Recall this means that when $x = 2$, $f(x)$, or $y, = -1$. Thus, the graph of f passes through $(2, -1)$.

> **Helpful Hint**
>
> Remember that $f(x)$ is a special symbol in mathematics used to denote a function. The symbol $f(x)$ is read "f of x." It does *not* mean $f \cdot x$ (f times x).

> ✔ **CONCEPT CHECK**
>
> Suppose $y = f(x)$ and we are told that $f(3) = 9$. Which is not true?
> **a.** When $x = 3$, $y = 9$.
> **b.** A possible function is $f(x) = x^2$.
> **c.** A point on the graph of the function is $(3, 9)$.
> **d.** A possible function is $f(x) = 2x + 4$.

If it helps, think of a function, f, as a machine that has been programmed with a certain correspondence or rule. An input value (a member of the domain) is then fed into the machine, the machine does the correspondence or rule, and the result is the output (a member of the range).

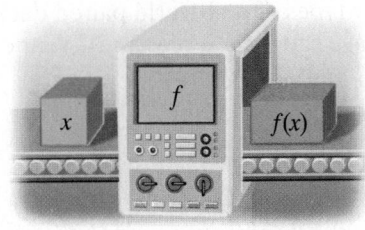

EXAMPLE 1 Given the graphs of the functions f and g, find each function value by inspecting the graphs.

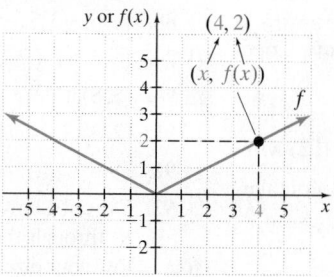 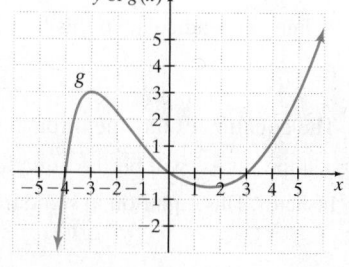

a. $f(4)$ **b.** $f(-2)$ **c.** $g(5)$ **d.** $g(0)$

e. Find all x-values such that $f(x) = 1$.

f. Find all x-values such that $g(x) = 0$.

Solution

a. To find $f(4)$, find the y-value when $x = 4$. We see from the graph that when $x = 4$, y or $f(x) = 2$. Thus, $f(4) = 2$.

b. $f(-2) = 1$ from the ordered pair $(-2, 1)$.

c. $g(5) = 3$ from the ordered pair $(5, 3)$.

d. $g(0) = 0$ from the ordered pair $(0, 0)$.

e. To find x-values such that $f(x) = 1$, we are looking for any ordered pairs on the graph of f whose $f(x)$ or y-value is 1. They are $(2, 1)$ and $(-2, 1)$. Thus $f(2) = 1$ and $f(-2) = 1$. The x-values are 2 and -2.

f. Find ordered pairs on the graph of g whose $g(x)$ or y-value is 0. They are $(3, 0)$, $(0, 0)$, and $(-4, 0)$. Thus $g(3) = 0$, $g(0) = 0$, and $g(-4) = 0$. The x-values are 3, 0, and -4.   □

PRACTICE

1 Given the graphs of the functions f and g, find each function value by inspecting the graphs.

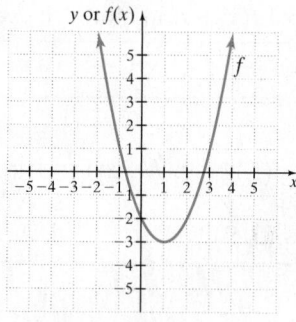 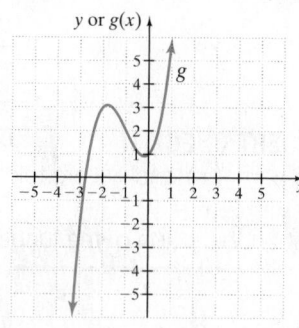

a. $f(1)$ **b.** $f(0)$ **c.** $g(-2)$ **d.** $g(0)$

e. Find all x-values such that $f(x) = 1$.

f. Find all x-values such that $g(x) = -2$.

Many types of real-world paired data form functions. The broken-line graphs on the next page show the 2-year and 4-year enrollments in post-secondary institutions.

EXAMPLE 2   The following graph shows the 2-year and 4-year enrollments in postsecondary institutions as functions of time.

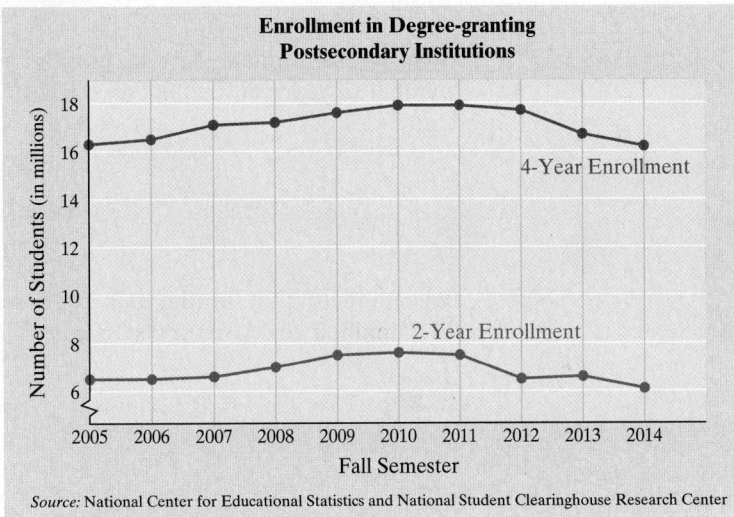

Source: National Center for Educational Statistics and National Student Clearinghouse Research Center

a. Approximate the 2-year enrollment in 2013.

b. In Fall 2007, the enrollment in 4-year colleges was 17.1 million students. Find the increase in 4-year college enrollment from Fall 2007 to Fall 2011.

Solution

a. Find the semester Fall 2013 and move upward until you reach the bottom broken-line graph. From this point on the graph, move horizontally to the left until the vertical axis is reached. In Fall 2013, approximately 6.6 million students, or 6,600,000 students, were enrolled in 2-year degree-granting institutions.

b. The increase in enrollment in 4-year postsecondary institutions from Fall 2007 to Fall 2011 is 17.9 million − 17.1 million = 0.8 million or 800,000 students.   □

PRACTICE
2   Use the graph in Example 2 and approximate the 2-year enrollment in Fall 2010. ■

Notice that each graph separately in Example 2 is the graph of a function since for each year there is only one 4-year institution enrollment figure and only one 2-year enrollment figure. Often, businesses depend on equations that closely fit data-defined functions like these to model the data and predict future trends. For example, the function $f(x) = -0.08x^2 + 1.5x + 10.3$ approximates the data for the red graph and the function $g(x) = -0.06x^2 + 1.1x + 2.2$ approximates the data for the blue graph. For each function, x is the number of years since 2000, and $f(x)$ is the number of students (in millions). The graphs and the data functions are shown next.

Helpful Hint

Each function graphed is the graph of a function and passes the vertical line test.

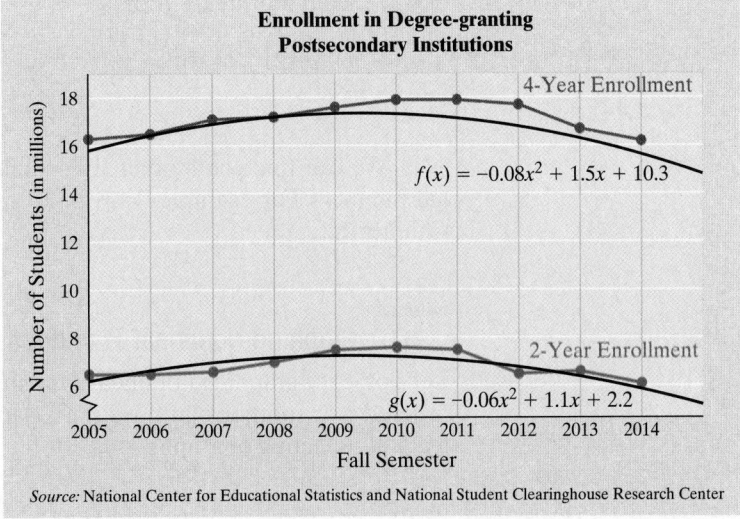

Source: National Center for Educational Statistics and National Student Clearinghouse Research Center

EXAMPLE 3 Use the function $f(x) = 0.19x + 10.5$ and the discussion following Example 2 to estimate the total full-time enrollment in degree-granting postsecondary institutions for fall 2015.

Solution To estimate the total enrollment in fall 2015, remember that x represents the number of years since 2000, so $x = 2015 - 2000 = 15$. Use $f(x) = 0.19x + 10.5$ and find $f(15)$.

$$f(x) = 0.19x + 10.5$$
$$f(15) = 0.19(15) + 10.5$$
$$= 13.35$$

We estimate that in the fall 2015 semester, the total full-time enrollment was 13.35 million, or 13,350,000 students. □

PRACTICE

3 Use $f(x) = 0.12x + 6.5$ to approximate the total part-time enrollment in fall 2015. ∎

OBJECTIVE

2 Finding Square Roots of Numbers ▷

Later in this section, we graph the square root function, $f(x) = \sqrt{x}$. To prepare for this graph, let's review finding square roots of numbers.

The opposite of squaring a number is taking the **square root** of a number. For example, since the square of 4, or 4^2, is 16, we say that a square root of 16 is 4. The notation $\sqrt{a}$ is used to denote the **positive, or principal, square root** of a nonnegative number a. We then have in symbols that $\sqrt{16} = 4$. The negative square root of 16 is written $-\sqrt{16} = -4$. The square root of a negative number, such as $\sqrt{-16}$, is not a real number. Why? There is no real number that, when squared, gives a negative number.

EXAMPLE 4 Find the square roots.

    a. $\sqrt{9}$ **b.** $\sqrt{25}$ **c.** $\sqrt{\dfrac{1}{4}}$ **d.** $-\sqrt{36}$ **e.** $\sqrt{-36}$ **f.** $\sqrt{0}$

Solution

    a. $\sqrt{9} = 3$ since 3 is positive and $3^2 = 9$.   **b.** $\sqrt{25} = 5$ since $5^2 = 25$.

    c. $\sqrt{\dfrac{1}{4}} = \dfrac{1}{2}$ since $\left(\dfrac{1}{2}\right)^2 = \dfrac{1}{4}$.   **d.** $-\sqrt{36} = -6$

    e. $\sqrt{-36}$ is not a real number.   **f.** $\sqrt{0} = 0$ since $0^2 = 0$. □

PRACTICE

4 Find the square roots.

    a. $\sqrt{121}$ **b.** $\sqrt{\dfrac{1}{16}}$ **c.** $-\sqrt{64}$ **d.** $\sqrt{-64}$ **e.** $\sqrt{100}$ ∎

We can find roots other than square roots. Also, not all roots simplify to rational numbers. For example, $\sqrt{3} \approx 1.7$ using a calculator. We study radicals further in Chapter 10.

OBJECTIVE

3 Graphing Nonlinear Functions ▷

Let's practice graphing nonlinear functions. In this section, we graph by plotting enough ordered pair solutions until we see a pattern. In the next section, we learn about shifting and reflecting of graphs.

EXAMPLE 5   Graph: $f(x) = x^2$

__Solution__ This equation is not linear because the x^2-term does not allow us to write it in the form $Ax + By = C$. Its graph is not a line. We begin by finding ordered pair solutions. Because $f(x) = y$, feel free to think of this equation as $f(x) = x^2$ or $y = x^2$. This graph is solved for y, so we choose x-values and find corresponding y-values.

If $x = -3$, then $f(-3) = (-3)^2$, or 9.

If $x = -2$, then $f(-2) = (-2)^2$, or 4.

If $x = -1$, then $f(-1) = (-1)^2$, or 1.

If $x = 0$, then $f(0) = 0^2$, or 0.

If $x = 1$, then $f(1) = 1^2$, or 1.

If $x = 2$, then $f(2) = 2^2$, or 4.

If $x = 3$, then $f(3) = 3^2$, or 9.

x	y or $f(x)$
-3	9
-2	4
-1	1
0	0
1	1
2	4
3	9

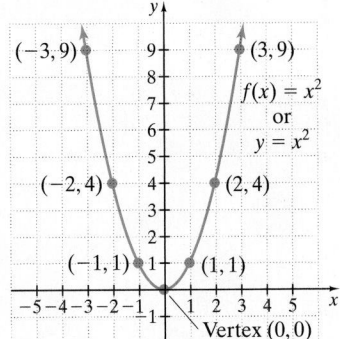

Study the table a moment and look for patterns. Notice that the ordered pair solution $(0, 0)$ contains the smallest y-value because any other x-value squared will give a positive result. This means that the point $(0, 0)$ will be the lowest point on the graph. Also notice that all other y-values correspond to two different x-values. For example, $3^2 = 9$ and also $(-3)^2 = 9$. This means that the graph will be a mirror image of itself across the y-axis. Connect the plotted points with a smooth curve to sketch the graph.

This curve is given a special name, a **parabola.** We will study more about parabolas in later chapters.

PRACTICE
5   Graph: $f(x) = 3x^2$

EXAMPLE 6   Graph the nonlinear function $f(x) = |x|$.

__Solution__ This is not a linear equation since it cannot be written in the form $Ax + By = C$. Its graph is not a line. Because we do not know the shape of this graph, we find many ordered pair solutions. We will choose x-values and substitute to find corresponding y-values.

If $x = -3$, then $f(-3) = |-3|$, or 3.

If $x = -2$, then $f(-2) = |-2|$, or 2.

If $x = -1$, then $f(-1) = |-1|$, or 1.

If $x = 0$, then $f(0) = |0|$, or 0.

If $x = 1$, then $f(1) = |1|$, or 1.

If $x = 2$, then $f(2) = |2|$, or 2.

If $x = 3$, then $f(3) = |3|$, or 3.

x	y or $f(x)$
-3	3
-2	2
-1	1
0	0
1	1
2	2
3	3

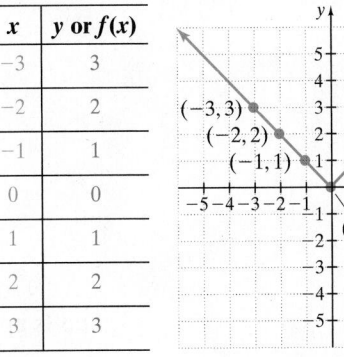

Again, study the table of values for a moment and notice any patterns.

From the plotted ordered pairs, we see that the graph of this absolute value equation is V-shaped.

PRACTICE
6   Graph: $f(x) = -|x|$

EXAMPLE 7 Graph the nonlinear function $f(x) = \sqrt{x}$.

Solution To graph this square root function, we identify the domain, evaluate the function for several values of x, plot the resulting points, and connect the points with a smooth curve. Since $\sqrt{x}$ represents the nonnegative square root of x, the domain of this function is the set of all nonnegative numbers, $\{x \mid x \geq 0\}$, or $[0, \infty)$. We have approximated $\sqrt{3}$ below to help us locate the point corresponding to $(3, \sqrt{3})$.

If $x = 0$, then $f(0) = \sqrt{0}$, or 0.

If $x = 1$, then $f(1) = \sqrt{1}$, or 1.

If $x = 3$, then $f(3) = \sqrt{3}$, or 1.7.

If $x = 4$, then $f(4) = \sqrt{4}$, or 2.

If $x = 9$, then $f(9) = \sqrt{9}$, or 3.

x	y or $f(x)$
0	0
1	1
3	$\sqrt{3} \approx 1.7$
4	2
9	3

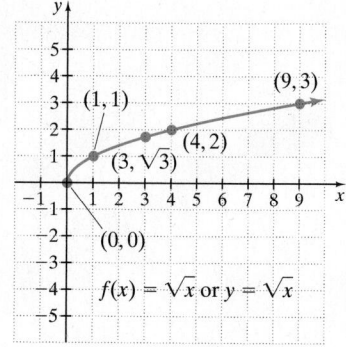

PRACTICE

7 Graph: $f(x) = \sqrt{x} + 1$

Graphing Calculator Explorations

It is possible to use a graphing calculator to sketch the graph of more than one equation on the same set of axes. For example, graph the functions $f(x) = x^2$ and $g(x) = x^2 + 4$ on the same set of axes.

To graph on the same set of axes, press the $\boxed{Y =}$ key and enter the equations on the first two lines.

$$Y_1 = x^2$$
$$Y_2 = x^2 + 4$$

Then press the $\boxed{\text{GRAPH}}$ key as usual. The screen should look like this.

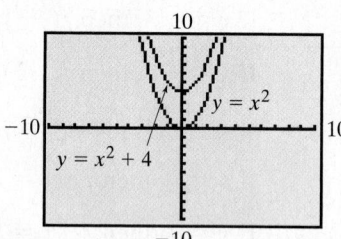

Notice that the graph of y or $g(x) = x^2 + 4$ is the graph of $y = x^2$ moved 4 units upward.

Graph each pair of functions on the same set of axes.

1. $f(x) = |x|$
   $g(x) = |x| + 1$

2. $f(x) = x^2$
   $h(x) = x^2 - 5$

3. $f(x) = x$
   $H(x) = x - 6$

4. $f(x) = |x|$
   $G(x) = |x| + 3$

5. $f(x) = -x^2$
   $F(x) = -x^2 + 7$

6. $f(x) = x$
   $F(x) = x + 2$

✓ Vocabulary, Readiness & Video Check

Use the choices below to fill in each blank. Some choices may not be used.

$(1.7, -2)$	line	parabola	-6	-9
$(-2, 1.7)$	V-shaped	6		9

1. The graph of $y = |x|$ looks _____.
2. The graph of $y = x^2$ is a _____.
3. If $f(-2) = 1.7$, the corresponding ordered pair is _____.
4. If $f(x) = x^2$, then $f(-3) = $ _____.

Martin-Gay Interactive Videos

See Video 8.2 ◉

Watch the section lecture video and answer the following questions.

OBJECTIVE 1

5. From 🎞 Examples 1 and 2, what is the connection between function notation to evaluate a function at certain values and ordered pair solutions of the function?

OBJECTIVE 2

6. Explain why 🎞 Example 6 does not simplify to a real number.

OBJECTIVE 3

7. Based on 🎞 Examples 7, 8, and 9, complete the following statements. When graphing a nonlinear equation, first recognize it as a nonlinear equation and know that the graph is _____ a line. If you don't know the _____ of the graph, plot enough points until you see a pattern.

8.2   Exercise Set   MyMathLab ▶

Use the graph of the following function f(x) to find each value. See Examples 1 and 2.

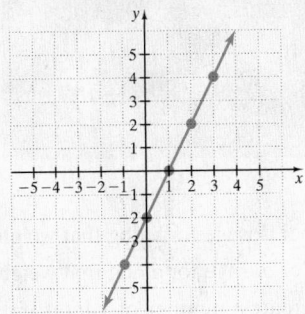

▶ **1.** $f(1)$
 2. $f(0)$
▶ **3.** $f(-1)$
 4. $f(2)$
▶ **5.** Find x such that $f(x) = 4$.
 6. Find x such that $f(x) = -6$.

Use the graph of the functions below to answer Exercises 7 through 18. See Examples 1 and 2.

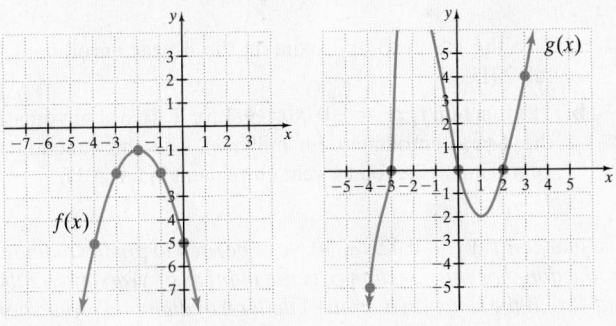

7. If $f(1) = -10$, write the corresponding ordered pair.
8. If $f(-5) = -10$, write the corresponding ordered pair.
9. If $g(4) = 56$, write the corresponding ordered pair.
10. If $g(-2) = 8$, write the corresponding ordered pair.
11. Find $f(-1)$.
12. Find $f(-2)$.

13. Find $g(2)$.

14. Find $g(-4)$.

15. Find all values of x such that $f(x) = -5$.

16. Find all values of x such that $f(x) = -2$.

17. Find all positive values of x such that $g(x) = 4$.

18. Find all values of x such that $g(x) = 0$.

Find the following roots. See Example 4.

▶ 19. $\sqrt{49}$

20. $\sqrt{144}$

21. $-\sqrt{\dfrac{4}{9}}$

22. $-\sqrt{\dfrac{4}{25}}$

23. $\sqrt{64}$

24. $\sqrt{4}$

▶ 25. $\sqrt{81}$

26. $\sqrt{1}$

▶ 27. $\sqrt{-100}$

28. $\sqrt{-25}$

MIXED PRACTICE

Graph each function by finding and plotting ordered pair solutions. See Examples 5 through 7.

29. $f(x) = x^2 + 3$

30. $g(x) = (x + 2)^2$

▶ 31. $h(x) = |x| - 2$

32. $f(x) = |x - 2|$

▶ 33. $g(x) = 2x^2$

34. $h(x) = 5x^2$

35. $f(x) = 5x - 1$

36. $g(x) = -3x + 2$

▶ 37. $f(x) = \sqrt{x} + 1$

38. $f(x) = \sqrt{x} - 1$

39. $g(x) = -2|x|$

40. $g(x) = -3|x|$

41. $h(x) = \sqrt{x} + 2$

42. $h(x) = \sqrt{x + 2}$

Use the graph on page 527 to answer the following. Also see Example 3.

43. **a.** Use the graph to approximate the 2-year enrollment in fall 2011.

b. The function $f(x) = -0.06x^2 + 1.1x + 2.2$ approximates the 2-year enrollment (in millions). Use this function to approximate the 2-year enrollment in fall 2011.

44. **a.** Use the graph to approximate the 4-year enrollment in fall 2011.

b. The function $f(x) = -0.08x^2 + 1.5x + 10.3$ approximates the 4-year enrollment (in millions). Use this function to approximate the total 4-year enrollment in fall 2011.

The function $f(x) = 0.42x + 10.5$ can be used to predict diamond production. For this function, x is the number of years after 2000, and $f(x)$ is the value (in billions of dollars) of diamond production.

45. Use the function in the directions above to estimate diamond production in 2012.

46. Use the function in the directions above to predict diamond production in 2015.

The function $A(r) = \pi r^2$ may be used to find the area of a circle if we are given its radius.

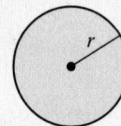

△ 47. Find the area of a circle whose radius is 5 centimeters. (Do not approximate π.)

△ 48. Find the area of a circular garden whose radius is 8 feet. (Do not approximate π.)

The function $V(x) = x^3$ may be used to find the volume of a cube if we are given the length x of a side.

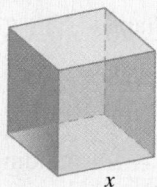

49. Find the volume of a cube whose side is 14 inches.

50. Find the volume of a die whose side is 1.7 centimeters.

Forensic scientists use the following functions to find the height of a woman if they are given the height of her femur bone f or her tibia bone t in centimeters.

$$H(f) = 2.59f + 47.24$$
$$H(t) = 2.72t + 61.28$$

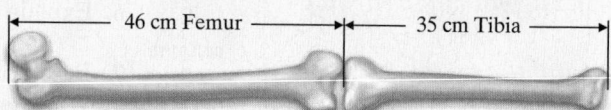

46 cm Femur — 35 cm Tibia

51. Find the height of a woman whose femur measures 46 centimeters.

52. Find the height of a woman whose tibia measures 35 centimeters.

The dosage in milligrams D of Ivermectin, a heartworm preventive, for a dog who weighs x pounds is given by

$$D(x) = \frac{136}{25}x$$

53. Find the proper dosage for a dog that weighs 30 pounds.

54. Find the proper dosage for a dog that weighs 50 pounds.

╲ 55. What is the greatest number of x-intercepts that a function may have? Explain your answer.

╲ 56. What is the greatest number of y-intercepts that a function may have? Explain your answer.

REVIEW AND PREVIEW

Solve the following equations. See Section 2.3.

57. $3(x - 2) + 5x = 6x - 16$

58. $5 + 7(x + 1) = 12 + 10x$

59. $3x + \dfrac{2}{5} = \dfrac{1}{10}$

60. $\dfrac{1}{6} + 2x = \dfrac{2}{3}$

CONCEPT EXTENSIONS

For Exercises 61 through 64, match each description with the graph that best illustrates it.

61. Moe worked 40 hours per week until the fall semester started. He quit and didn't work again until he worked 60 hours a week during the holiday season starting mid-December.

62. Kawana worked 40 hours a week for her father during the summer. She slowly cut back her hours to not working at all during the fall semester. During the holiday season in December, she started working again and increased her hours to 60 hours per week.

63. Wendy worked from July through February, never quitting. She worked between 10 and 30 hours per week.

64. Bartholomew worked from July through February. During the holiday season between mid-November and the beginning of January, he worked 40 hours per week. The rest of the time, he worked between 10 and 40 hours per week.

a.
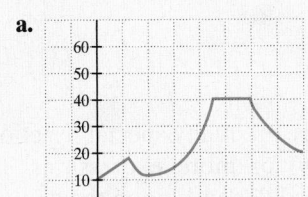

b.
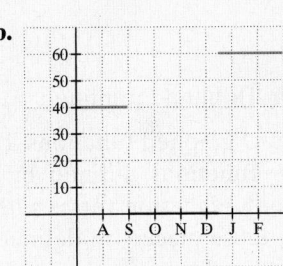

c.
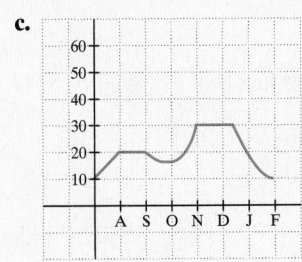

d.
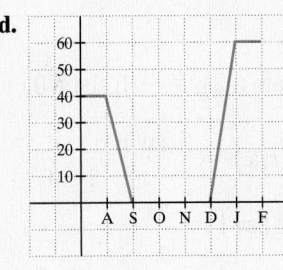

This broken-line graph shows the hourly minimum wage and the years it increased. Use this graph for Exercises 65 through 68.

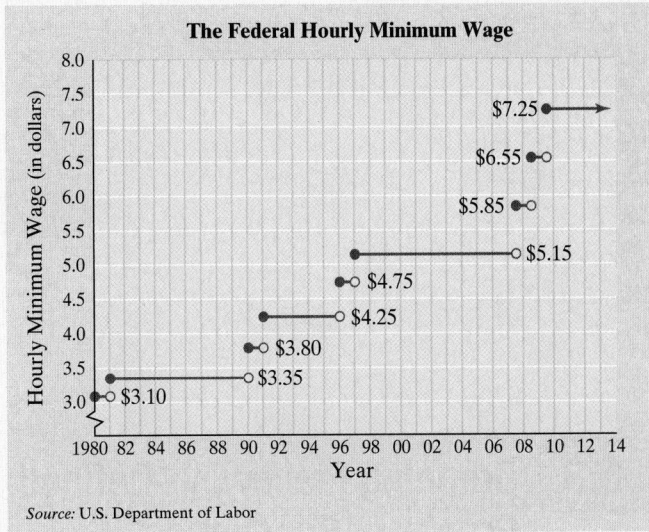

Source: U.S. Department of Labor

65. What was the first year that the minimum hourly wage rose above $5.00?

66. What was the first year that the minimum hourly wage rose above $6.00?

67. Why do you think that this graph is shaped the way it is?

68. The federal hourly minimum wage started in 1938 at $0.25. How much has it increased by 2011?

69. Graph $y = x^2 - 4x + 7$. Let $x = 0, 1, 2, 3, 4$ to generate ordered pair solutions.

70. Graph $y = x^2 + 2x + 3$. Let $x = -3, -2, -1, 0, 1$ to generate ordered pair solutions.

71. The function $f(x) = [x]$ is called the greatest integer function and its graph is similar to the graph above. The value of $[x]$ is the greatest integer less than or equal to x. For example, $f(1.5) = [1.5] = 1$ since the greatest integer ≤ 1.5 is 1. Sketch the graph of $f(x) = [x]$.

Integrated Review Summary on Functions and Equations of Lines

Sections 8.1–8.2

Find the slope and y-intercept of the graph of each function.

1. $f(x) = 3x - 5$

2. $f(x) = \dfrac{5}{2}x - \dfrac{7}{2}$

Determine whether each pair of lines is parallel, perpendicular, or neither.

3. $f(x) = 8x - 6$
   $g(x) = 8x + 6$

4. $f(x) = \dfrac{2}{3}x + 1$
   $2y + 3x = 1$

Find the equation of each line. Write the equation using function notation.

5. Through $(1, 6)$ and $(5, 2)$

6. Through $(2, -8)$ and $(-6, -5)$

7. Through $(-1, -5)$; parallel to $3x - y = 5$

8. Through $(0, 4)$; perpendicular to $4x - 5y = 10$

9. Through $(2, -3)$; perpendicular to $4x + y = \dfrac{2}{3}$

10. Through $(-1, 0)$; parallel to $5x + 2y = 2$

Determine whether each function is linear or not. Then graph the function.

11. $f(x) = 4x - 2$

12. $f(x) = 6x - 5$

13. $g(x) = |x| + 3$

14. $h(x) = |x| + 2$

15. $f(x) = 2x^2$

16. $F(x) = 3x^2$

17. $h(x) = x^2 - 3$

18. $G(x) = x^2 + 3$

19. $F(x) = -2x$

20. $H(x) = -3x$

21. $G(x) = |x + 2|$

22. $g(x) = |x - 1|$

23. $f(x) = \dfrac{1}{3}x - 1$

24. $f(x) = \dfrac{1}{2}x - 3$

25. $g(x) = -\dfrac{3}{2}x + 1$

26. $G(x) = -\dfrac{2}{3}x + 1$

8.3 Graphing Piecewise-Defined Functions and Shifting and Reflecting Graphs of Functions ▷

OBJECTIVES

1 Graph Piecewise-Defined Functions. ▷

2 Vertical and Horizontal Shifts. ▷

3 Reflect Graphs. ▷

OBJECTIVE

1 Graphing Piecewise-Defined Functions ▷

Thus far in Chapter 8, we have graphed functions. There are many special functions. In this objective, we study functions defined by two or more expressions. The expression used to complete the function varies with and depends upon the value of *x*. Before we actually graph these piecewise-defined functions, let's practice finding function values.

EXAMPLE 1   Evaluate $f(2)$, $f(-6)$, and $f(0)$ for the function

$$f(x) = \begin{cases} 2x + 3 & \text{if } x \le 0 \\ -x - 1 & \text{if } x > 0 \end{cases}$$

Then write your results in ordered pair form.

<u>*Solution*</u> Take a moment and study this function. It is a single function defined by two expressions depending on the value of *x*. From above, if $x \le 0$, use $f(x) = 2x + 3$. If $x > 0$, use $f(x) = -x - 1$. Thus

$f(2) = -(2) - 1$	$f(-6) = 2(-6) + 3$	$f(0) = 2(0) + 3$
$\quad = -3 \;$ since $2 > 0$	$\quad = -9 \;$ since $-6 \le 0$	$\quad = 3 \;$ since $0 \le 0$
$f(2) = -3$	$f(-6) = -9$	$f(0) = 3$
Ordered pairs: $(2, -3)$	$(-6, -9)$	$(0, 3)$

PRACTICE

1   Evaluate $f(4)$, $f(-2)$, and $f(0)$ for the function

$$f(x) = \begin{cases} -4x - 2 & \text{if } x \le 0 \\ x + 1 & \text{if } x > 0 \end{cases}$$

Now, let's graph a piecewise-defined function.

EXAMPLE 2   Graph: $f(x) = \begin{cases} 2x + 3 & \text{if } x \le 0 \\ -x - 1 & \text{if } x > 0 \end{cases}$

Solution Let's graph each piece.

If $x \leq 0$,

$$f(x) = 2x + 3$$

Values ≤ 0

x	$f(x) = 2x + 3$
0	3 Closed circle
-1	1
-2	-1

If $x > 0$,

$$f(x) = -x - 1$$

Values > 0

x	$f(x) = -x - 1$
1	-2
2	-3
3	-4

The graph of the first part of $f(x)$ listed will look like a ray with a closed-circle end point at $(0, 3)$. The graph of the second part of $f(x)$ listed will look like a ray with an open-circle end point. To find the exact location of the open-circle end point, use $f(x) = -x - 1$ and find $f(0)$. Since $f(0) = -0 - 1 = -1$, we graph the values from the second table and place an open circle at $(0, -1)$.

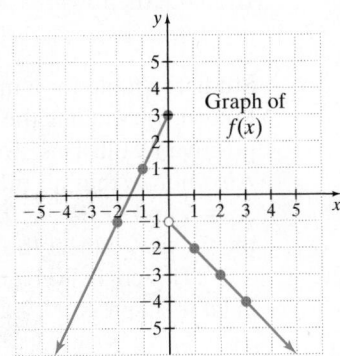

Notice that this graph is the graph of a function because it passes the vertical line test. The domain of this function is $(-\infty, \infty)$ and the range is $(-\infty, 3]$.   □

PRACTICE

2 Graph:

$$f(x) = \begin{cases} -4x - 2 & \text{if } x \leq 0 \\ x + 1 & \text{if } x > 0 \end{cases}$$

■

OBJECTIVE

2 **Vertical and Horizontal Shifting**

Review of Common Graphs

We now take common graphs and learn how more complicated graphs are actually formed by shifting and reflecting these common graphs. These shifts and reflections are called transformations, and it is possible to combine transformations. A knowledge of these transformations will help you simplify future graphs.

Let's begin with a review of the graphs of four common functions. Many of these functions we graphed in earlier sections.

First, **let's graph the linear function $f(x) = x$ or $y = x$**. Ordered pair solutions of this graph consist of ordered pairs whose x- and y-values are the same.

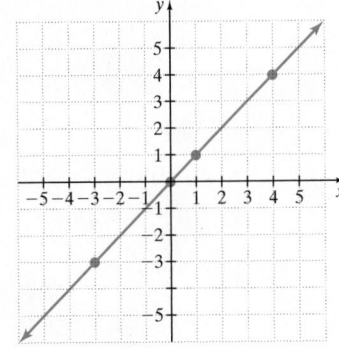

x	y or $f(x) = x$
-3	-3
0	0
1	1
4	4

These common nonlinear functions; $f(x) = x^2$, $f(x) = |x|$, and $f(x) = \sqrt{x}$, were graphed in Section 8.2. If you are familiar with these functions and their graphs, pass this material and continue with the Common Graphs box on the next page.

Next, **let's graph the nonlinear function $f(x) = x^2$ or $y = x^2$.**

This equation is not linear because the x^2 term does not allow us to write it in the form $Ax + By = C$. Its graph is not a line. We begin by finding ordered pair solutions. Because this graph is solved for $f(x)$, or y, we choose x-values and find corresponding $f(x)$, or y-values.

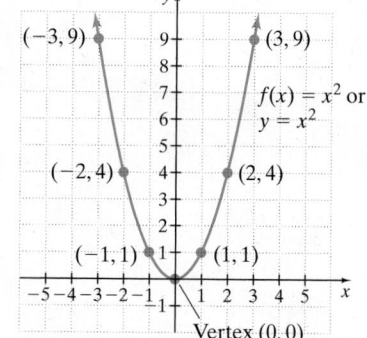

If $x = -3$, then $y = (-3)^2$, or 9.

If $x = -2$, then $y = (-2)^2$, or 4.

If $x = -1$, then $y = (-1)^2$, or 1.

If $x = 0$, then $y = 0^2$, or 0.

If $x = 1$, then $y = 1^2$, or 1.

If $x = 2$, then $y = 2^2$, or 4.

If $x = 3$, then $y = 3^2$, or 9.

x	$f(x)$ or y
-3	9
-2	4
-1	1
0	0
1	1
2	4
3	9

Study the table for a moment and look for patterns. Notice that the ordered pair solution $(0, 0)$ contains the smallest y-value because any other x-value squared will give a positive result. This means that the point $(0, 0)$ will be the lowest point on the graph. Also notice that all other y-values correspond to two different x-values, for example, $3^2 = 9$ and $(-3)^2 = 9$. This means that the graph will be a mirror image of itself across the y-axis. Connect the plotted points with a smooth curve to sketch its graph.

This curve is given a special name, a **parabola.** We will study more about parabolas in later chapters.

Next, **let's graph another nonlinear function, $f(x) = |x|$ or $y = |x|$.**

This is not a linear equation since it cannot be written in the form $Ax + By = C$. Its graph is not a line. Because we do not know the shape of this graph, we find many ordered pair solutions. We will choose x-values and substitute to find corresponding y-values.

If $x = -3$, then $y = |-3|$, or 3.

If $x = -2$, then $y = |-2|$, or 2.

If $x = -1$, then $y = |-1|$, or 1.

If $x = 0$, then $y = |0|$, or 0.

If $x = 1$, then $y = |1|$, or 1.

If $x = 2$, then $y = |2|$, or 2.

If $x = 3$, then $y = |3|$, or 3.

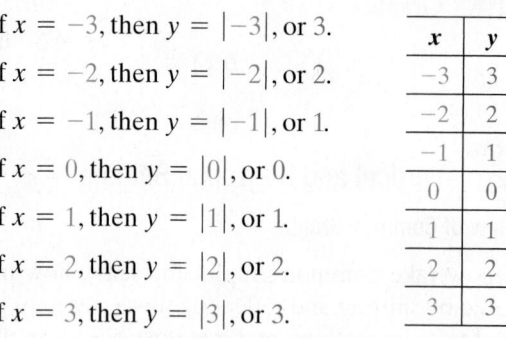

x	y
-3	3
-2	2
-1	1
0	0
1	1
2	2
3	3

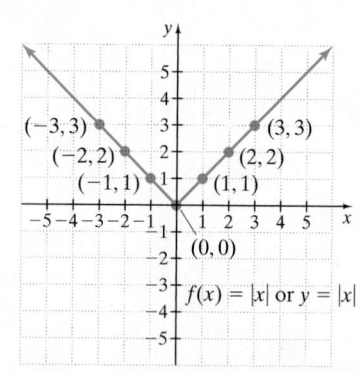

Again, study the table of values for a moment and notice any patterns.

From the plotted ordered pairs, we see that the graph of this absolute value equation is V-shaped.

Finally, consider a fourth common function, $f(x) = \sqrt{x}$ or $y = \sqrt{x}$. For this graph, you need to recall basic facts about square roots and use your calculator to approximate some square roots to help locate points. Recall also that the square root of a negative number is not a real number, so be careful when finding your domain.

Now **let's graph the square root function $f(x) = \sqrt{x}$ or $y = \sqrt{x}$.**

To graph, we identify the domain, evaluate the function for several values of x, plot the resulting points, and connect the points with a smooth curve. Since $\sqrt{x}$ represents the nonnegative square root of x, the domain of this function is the set of all nonnegative numbers, $\{x \mid x \geq 0\}$, or $[0, \infty)$. We have approximated $\sqrt{3}$ on the next page to help us locate the point corresponding to $(3, \sqrt{3})$.

If $x = 0$, then $y = \sqrt{0}$, or 0.

If $x = 1$, then $y = \sqrt{1}$, or 1.

If $x = 3$, then $y = \sqrt{3}$, or 1.7.

If $x = 4$, then $y = \sqrt{4}$, or 2.

If $x = 9$, then $y = \sqrt{9}$, or 3.

x	$f(x) = \sqrt{x}$
0	0
1	1
3	$\sqrt{3} \approx 1.7$
4	2
9	3

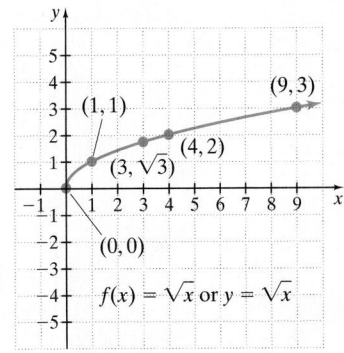

Notice that the graph of this function passes the vertical line test, as expected.

Below is a summary of our four common graphs. Take a moment and study these graphs. Your success in the rest of this section depends on your knowledge of these graphs.

Common Graphs

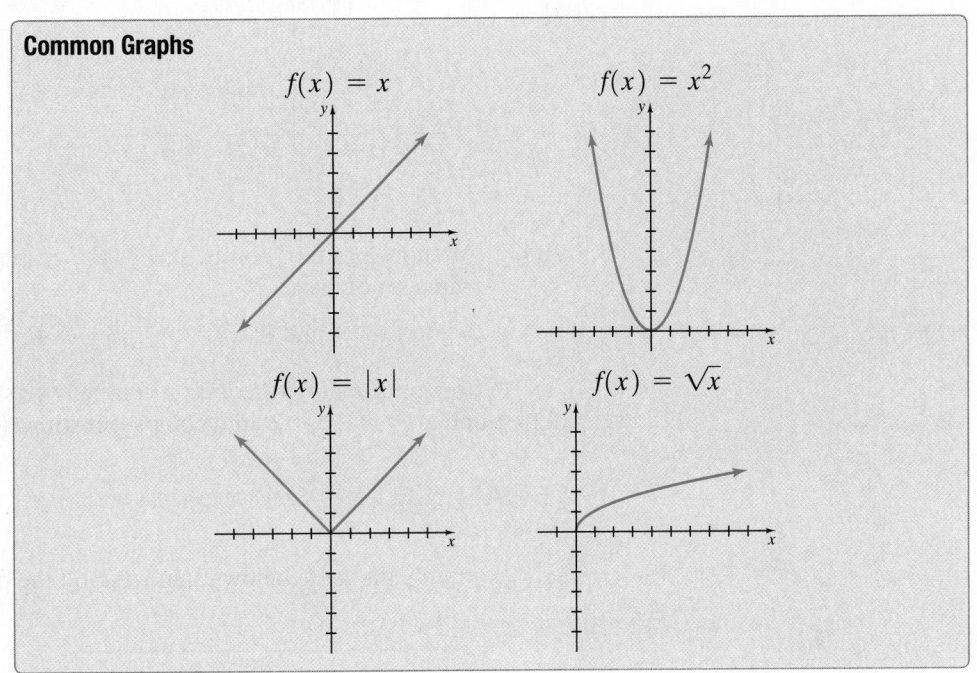

$f(x) = x$

$f(x) = x^2$

$f(x) = |x|$

$f(x) = \sqrt{x}$

Your knowledge of the slope–intercept form, $f(x) = mx + b$, will help you understand simple shifting or transformations such as vertical shifts. For example, what is the difference between the graphs of $f(x) = x$ and $g(x) = x + 3$?

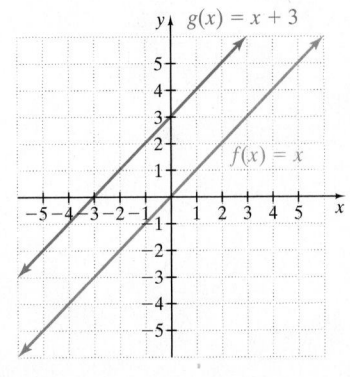

$f(x) = x$

slope, $m = 1$

y-intercept is $(0, 0)$

$g(x) = x + 3$

slope, $m = 1$

y-intercept is $(0, 3)$

Notice that the graph of $g(x) = x + 3$ is the same as the graph of $f(x) = x$, but moved upward 3 units. This is an example of a **vertical shift** and is true for graphs in general.

Vertical Shifts (Upward and Downward)
Let *k* be a Positive Number

Graph of	Same As	Moved
$g(x) = f(x) + k$	$f(x)$	k units upward
$g(x) = f(x) - k$	$f(x)$	k units downward

EXAMPLES Without plotting points, sketch the graph of each pair of functions on the same set of axes.

3. $f(x) = x^2$ and $g(x) = x^2 + 2$ **4.** $f(x) = \sqrt{x}$ and $g(x) = \sqrt{x} - 3$

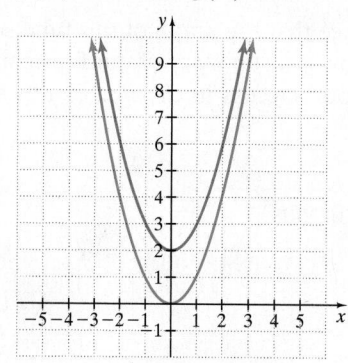

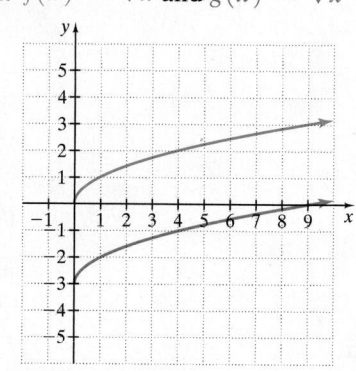

PRACTICE
3-4 Without plotting points, sketch the graphs of each pair of functions on the same set of axes.

3. $f(x) = x^2$ and $g(x) = x^2 - 3$ **4.** $f(x) = \sqrt{x}$ and $g(x) = \sqrt{x} + 1$

A horizontal shift to the left or right may be slightly more difficult to understand. Let's graph $g(x) = |x - 2|$ and compare it with $f(x) = |x|$.

EXAMPLE 5 Sketch the graphs of $f(x) = |x|$ and $g(x) = |x - 2|$ on the same set of axes.

Solution Study the table below to understand the placement of both graphs.

x	$f(x) = |x|$	$g(x) = |x - 2|$
-3	3	5
-2	2	4
-1	1	3
0	0	2
1	1	1
2	2	0
3	3	1

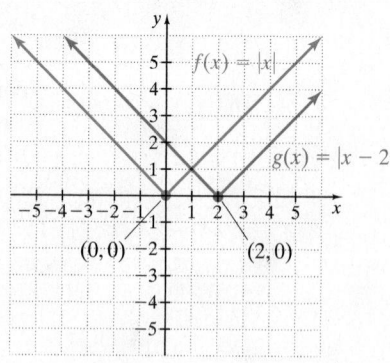

PRACTICE
5 Sketch the graphs of $f(x) = |x|$ and $g(x) = |x - 3|$ on the same set of axes.

The graph of $g(x) = |x - 2|$ is the same as the graph of $f(x) = |x|$, but moved 2 units to the right. This is an example of a **horizontal shift** and is true for graphs in general.

Horizontal Shift (To the Left or Right)
Let *h* be a Positive Number

Graph of	Same as	Moved
$g(x) = f(x - h)$	$f(x)$	*h* units to the right
$g(x) = f(x + h)$	$f(x)$	*h* units to the left

> **Helpful Hint**
> Notice that $f(x - h)$ corresponds to a shift to the right and $f(x + h)$ corresponds to a shift to the left.

Vertical and horizontal shifts can be combined.

EXAMPLE 6   Sketch the graphs of $f(x) = x^2$ and $g(x) = (x - 2)^2 + 1$ on the same set of axes.

Solution The graph of $g(x)$ is the same as the graph of $f(x)$ shifted 2 units to the right and 1 unit up.

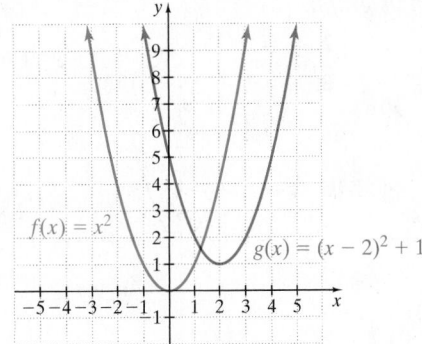

PRACTICE
6   Sketch the graphs of $f(x) = |x|$ and $g(x) = |x - 2| + 3$ on the same set of axes.

OBJECTIVE
3   Reflecting Graphs ▶

Another type of transformation is called a **reflection.** In this section, we will study reflections (mirror images) about the *x*-axis only. For example, take a moment and study these two graphs. The graph of $g(x) = -x^2$ can be verified, as usual, by plotting points.

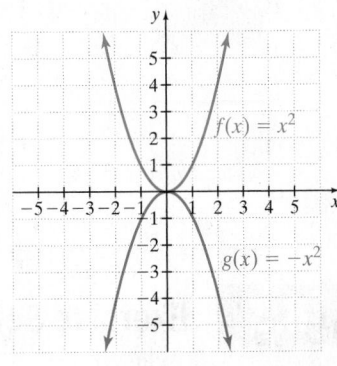

Reflection about the *x*-Axis

The graph of $g(x) = -f(x)$ is the graph of $f(x)$ reflected about the *x*-axis.

EXAMPLE 7 Sketch the graph of $h(x) = -|x - 3| + 2$.

Solution The graph of $h(x) = -|x - 3| + 2$ is the same as the graph of $f(x) = |x|$ reflected about the x-axis, then moved three units to the right and two units upward.

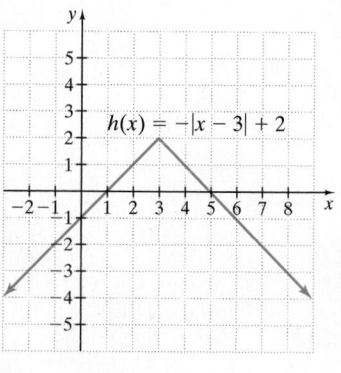

PRACTICE
7 Sketch the graph of $h(x) = -(x + 2)^2 - 1$.

There are other transformations, such as stretching, that won't be covered in this section. For a review of this transformation, see the Appendix.

✔ Vocabulary, Readiness & Video Check

Match each equation with its graph.

1. $y = \sqrt{x}$ **2.** $y = x^2$ **3.** $y = x$ **4.** $y = |x|$

A.
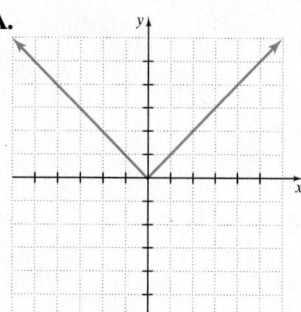

B.
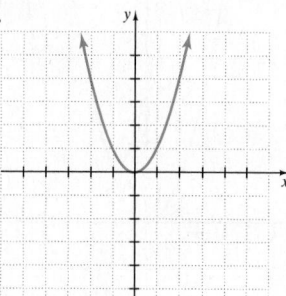

C.
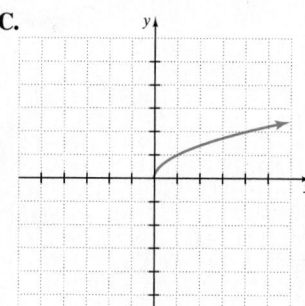

D.
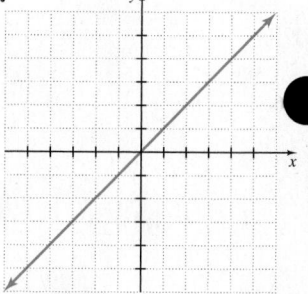

Martin-Gay Interactive Videos

See Video 8.3 ◉

Watch the section lecture video and answer the following questions.

OBJECTIVE 1
5. In ▦ Example 1, only one piece of the function is defined for the value $x = -1$. Why do we find $f(-1)$ for $f(x) = x + 3$?

OBJECTIVE 2
6. For ▦ Examples 2–8, why is it helpful to be familiar with common graphs and their basic shapes?

OBJECTIVE 3
7. Based on the lecture before ▦ Example 9, complete the following statement. The graph of $f(x) = -\sqrt{x} + 6$ has the same shape as the graph of $f(x) = \sqrt{x} + 6$ but it is reflected about the _____.

8.3 Exercise Set MyMathLab®

Graph each piecewise-defined function. See Examples 1 and 2.

1. $f(x) = \begin{cases} 2x & \text{if } x < 0 \\ x + 1 & \text{if } x \geq 0 \end{cases}$

2. $f(x) = \begin{cases} 3x & \text{if } x < 0 \\ x + 2 & \text{if } x \geq 0 \end{cases}$

3. $f(x) = \begin{cases} 4x + 5 & \text{if } x \leq 0 \\ \dfrac{1}{4}x + 2 & \text{if } x > 0 \end{cases}$

4. $f(x) = \begin{cases} 5x + 4 & \text{if } x \leq 0 \\ \dfrac{1}{3}x - 1 & \text{if } x > 0 \end{cases}$

5. $g(x) = \begin{cases} -x & \text{if } x \le 1 \\ 2x + 1 & \text{if } x > 1 \end{cases}$

6. $g(x) = \begin{cases} 3x - 1 & \text{if } x \le 2 \\ -x & \text{if } x > 2 \end{cases}$

7. $f(x) = \begin{cases} 5 & \text{if } x < -2 \\ 3 & \text{if } x \ge -2 \end{cases}$ **8.** $f(x) = \begin{cases} 4 & \text{if } x < -3 \\ -2 & \text{if } x \ge -3 \end{cases}$

MIXED PRACTICE

(Sections 3.2, 3.6) Graph each piecewise-defined function. Use the graph to determine the domain and range of the function. See Examples 1 and 2.

9. $f(x) = \begin{cases} -2x & \text{if } x \le 0 \\ 2x + 1 & \text{if } x > 0 \end{cases}$

10. $g(x) = \begin{cases} -3x & \text{if } x \le 0 \\ 3x + 2 & \text{if } x > 0 \end{cases}$

11. $h(x) = \begin{cases} 5x - 5 & \text{if } x < 2 \\ -x + 3 & \text{if } x \ge 2 \end{cases}$

12. $f(x) = \begin{cases} 4x - 4 & \text{if } x < 2 \\ -x + 1 & \text{if } x \ge 2 \end{cases}$

13. $f(x) = \begin{cases} x + 3 & \text{if } x < -1 \\ -2x + 4 & \text{if } x \ge -1 \end{cases}$

14. $h(x) = \begin{cases} x + 2 & \text{if } x < 1 \\ 2x + 1 & \text{if } x \ge 1 \end{cases}$

15. $g(x) = \begin{cases} -2 & \text{if } x \le 0 \\ -4 & \text{if } x \ge 1 \end{cases}$

16. $f(x) = \begin{cases} -1 & \text{if } x \le 0 \\ -3 & \text{if } x \ge 2 \end{cases}$

MIXED PRACTICE

Sketch the graph of each function. See Examples 3 through 6.

17. $f(x) = |x| + 3$ **18.** $f(x) = |x| - 2$

19. $f(x) = \sqrt{x} - 2$ **20.** $f(x) = \sqrt{x} + 3$

21. $f(x) = |x - 4|$ **22.** $f(x) = |x + 3|$

23. $f(x) = \sqrt{x + 2}$ **24.** $f(x) = \sqrt{x - 2}$

25. $y = (x - 4)^2$ **26.** $y = (x + 4)^2$

27. $f(x) = x^2 + 4$ **28.** $f(x) = x^2 - 4$

29. $f(x) = \sqrt{x - 2} + 3$ **30.** $f(x) = \sqrt{x - 1} + 3$

31. $f(x) = |x - 1| + 5$ **32.** $f(x) = |x - 3| + 2$

33. $f(x) = \sqrt{x + 1} + 1$ **34.** $f(x) = \sqrt{x + 3} + 2$

35. $f(x) = |x + 3| - 1$ **36.** $f(x) = |x + 1| - 4$

37. $g(x) = (x - 1)^2 - 1$ **38.** $h(x) = (x + 2)^2 + 2$

39. $f(x) = (x + 3)^2 - 2$ **40.** $f(x) = (x + 2)^2 + 4$

Sketch the graph of each function. See Examples 3 through 7.

41. $f(x) = -(x - 1)^2$ **42.** $g(x) = -(x + 2)^2$

43. $h(x) = -\sqrt{x} + 3$ **44.** $f(x) = -\sqrt{x + 3}$

45. $h(x) = -|x + 2| + 3$ **46.** $g(x) = -|x + 1| + 1$

47. $f(x) = (x - 3) + 2$ **48.** $f(x) = (x - 1) + 4$

REVIEW AND PREVIEW

Match each equation with its graph.

49. $y = -1$ **50.** $x = -1$

51. $x = 3$ **52.** $y = 3$

A.
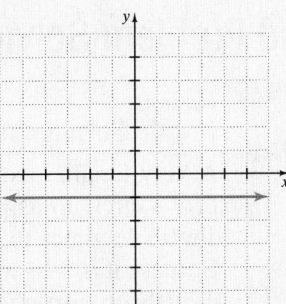

B.
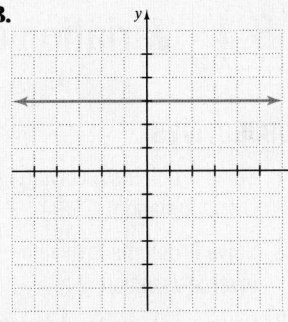

C.
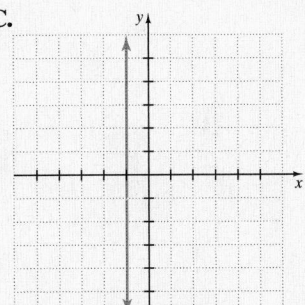

D.
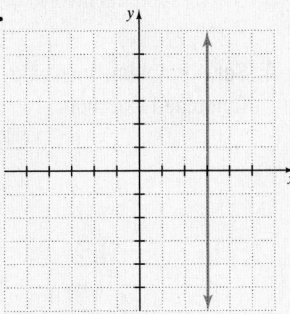

CONCEPT EXTENSIONS

53. Draw a graph whose domain is $(-\infty, 5]$ and whose range is $[2, \infty)$.

54. In your own words, describe how to graph a piecewise-defined function.

55. Graph: $f(x) = \begin{cases} -\dfrac{1}{2}x & \text{if } x \le 0 \\ x + 1 & \text{if } 0 < x \le 2 \\ 2x - 1 & \text{if } x > 2 \end{cases}$

56. Graph: $f(x) = \begin{cases} -\dfrac{1}{3}x & \text{if } x \le 0 \\ x + 2 & \text{if } 0 < x \le 4 \\ 3x - 4 & \text{if } x > 4 \end{cases}$

Write the domain and range of the following exercises.

57. Exercise 29   **58.** Exercise 30

59. Exercise 45   **60.** Exercise 46

Without graphing, find the domain of each function.

61. $f(x) = 5\sqrt{x - 20} + 1$

62. $g(x) = -3\sqrt{x + 5}$

63. $h(x) = 5|x - 20| + 1$

64. $f(x) = -3|x + 5.7|$

65. $g(x) = 9 - \sqrt{x + 103}$

66. $h(x) = \sqrt{x - 17} - 3$

Sketch the graph of each piecewise-defined function. Write the domain and range of each function.

67. $f(x) = \begin{cases} |x| & \text{if } x \le 0 \\ x^2 & \text{if } x > 0 \end{cases}$ **68.** $f(x) = \begin{cases} x^2 & \text{if } x < 0 \\ \sqrt{x} & \text{if } x \ge 0 \end{cases}$

69. $g(x) = \begin{cases} |x - 2| & \text{if } x < 0 \\ -x^2 & \text{if } x \ge 0 \end{cases}$

70. $g(x) = \begin{cases} -|x + 1| - 1 & \text{if } x < -2 \\ \sqrt{x + 2} - 4 & \text{if } x \ge -2 \end{cases}$

8.4 Variation and Problem Solving

OBJECTIVES

1 Solve Problems Involving Direct Variation.

2 Solve Problems Involving Inverse Variation.

3 Solve Problems Involving Joint Variation.

4 Solve Problems Involving Combined Variation.

OBJECTIVE

1 Solving Problems Involving Direct Variation

A very familiar example of direct variation is the relationship of the circumference C of a circle to its radius r. The formula $C = 2\pi r$ expresses that the circumference is always 2π times the radius. In other words, C is always a constant multiple (2π) of r. Because it is, we say that **C varies directly as r,** that **C varies directly with r,** or that **C is directly proportional to r.**

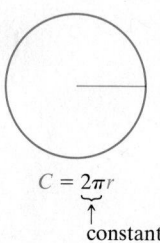

$$C = 2\pi r$$
constant

> **Direct Variation**
>
> **y varies directly as x,** or **y is directly proportional to x,** if there is a nonzero constant k such that
>
> $$y = kx$$
>
> The number k is called the **constant of variation** or the **constant of proportionality.**

In the above definition, the relationship described between x and y is a linear one. In other words, the graph of $y = kx$ is a line. The slope of the line is k, and the line passes through the origin.

For example, the graph of the direct variation equation $C = 2\pi r$ is shown. The horizontal axis represents the radius r, and the vertical axis is the circumference C. From the graph, we can read that when the radius is 6 units, the circumference is approximately 38 units. Also, when the circumference is 45 units, the radius is between 7 and 8 units. Notice that as the radius increases, the circumference increases.

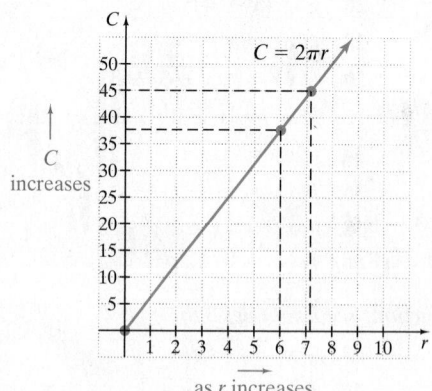

EXAMPLE 1 Suppose that y varies directly as x. If y is 5 when x is 30, find the constant of variation and the direct variation equation.

Solution Since *y* varies directly as *x*, we write $y = kx$. If $y = 5$ when $x = 30$, we have that

$$y = kx$$
$$5 = k(30) \quad \text{Replace } y \text{ with 5 and } x \text{ with 30.}$$
$$\frac{1}{6} = k \quad \text{Solve for } k.$$

The constant of variation is $\frac{1}{6}$.

After finding the constant of variation *k*, the direct variation equation can be written as $y = \frac{1}{6}x$. ☐

PRACTICE

1 Suppose that *y* varies directly as *x*. If *y* is 20 when *x* is 15, find the constant of variation and the direct variation equation. ■

EXAMPLE 2 **Using Direct Variation and Hooke's Law**

Hooke's law states that the distance a spring stretches is directly proportional to the weight attached to the spring. If a 40-pound weight attached to the spring stretches the spring 5 inches, find the distance that a 65-pound weight attached to the spring stretches the spring.

Solution

1. UNDERSTAND. Read and reread the problem. Notice that we are given that the distance a spring stretches is **directly proportional** to the weight attached. We let

   d = the distance stretched and

   w = the weight attached.

   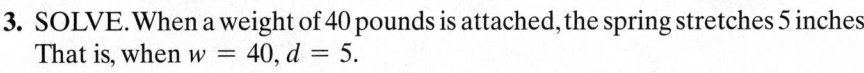

   The constant of variation is represented by *k*.

2. TRANSLATE. Because *d* is directly proportional to *w*, we write

   $$d = kw$$

3. SOLVE. When a weight of 40 pounds is attached, the spring stretches 5 inches. That is, when $w = 40$, $d = 5$.

   $$d = kw$$
   $$5 = k(40) \quad \text{Replace } d \text{ with 5 and } w \text{ with 40.}$$
   $$\frac{1}{8} = k \quad \text{Solve for } k.$$

Now when we replace *k* with $\frac{1}{8}$ in the equation $d = kw$, we have

$$d = \frac{1}{8}w$$

To find the stretch when a weight of 65 pounds is attached, we replace *w* with 65 to find *d*.

$$d = \frac{1}{8}(65)$$
$$= \frac{65}{8} = 8\frac{1}{8} \quad \text{or} \quad 8.125$$

(Continued on next page)

4. INTERPRET.

Check: Check the proposed solution of 8.125 inches in the original problem

State: The spring stetches 8.125 inches when a 65-pound weight is attached.   □

PRACTICE
2   Use Hooke's law as stated in Example 2. If a 36-pound weight attached to a spring stretches the spring 9 inches, find the distance that a 75-pound weight attached to the spring stretches the spring.   ■

OBJECTIVE
2   Solving Problems Involving Inverse Variation

When y is proportional to the **reciprocal** of another variable x, we say that **y varies inversely as x**, or that **y is inversely proportional to x**. An example of the inverse variation relationship is the relationship between the pressure that a gas exerts and the volume of its container. As the volume of a container decreases, the pressure of the gas it contains increases.

Inverse Variation

y varies inversely as x, or **y is inversely proportional to x,** if there is a nonzero constant k such that

$$y = \frac{k}{x}$$

The number k is called the **constant of variation** or the **constant of proportionality.**

Notice that $y = \dfrac{k}{x}$ is a rational equation. Its graph for $k > 0$ and $x > 0$ is shown. From the graph, we can see that as x increases, y decreases.

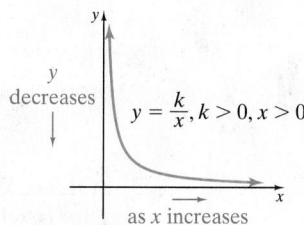

EXAMPLE 3   Suppose that u varies inversely as w. If u is 3 when w is 5, find the constant of variation and the inverse variation equation.

Solution Since u varies inversely as w, we have $u = \dfrac{k}{w}$. We let $u = 3$ and $w = 5$, and we solve for k.

$$u = \frac{k}{w}$$

$$3 = \frac{k}{5} \quad \text{Let } u = 3 \text{ and } w = 5.$$

$$15 = k \quad \text{Multiply both sides by 5.}$$

The constant of variation k is 15. This gives the inverse variation equation

$$u = \frac{15}{w}$$

□

PRACTICE
3 Suppose that b varies inversely as a. If b is 5 when a is 9, find the constant of variation and the inverse variation equation. ■

> **EXAMPLE 4** **Using Inverse Variation and Boyle's Law**

Boyle's law says that if the temperature stays the same, the pressure P of a gas is inversely proportional to the volume V. If a cylinder in a steam engine has a pressure of 960 kilopascals when the volume is 1.4 cubic meters, find the pressure when the volume increases to 2.5 cubic meters.

Solution

1. UNDERSTAND. Read and reread the problem. Notice that we are given that the pressure of a gas is *inversely proportional* to the volume. We will let P = the pressure and V = the volume. The constant of variation is represented by k.

2. TRANSLATE. Because P is inversely proportional to V, we write

$$P = \frac{k}{V}$$

When P = 960 kilopascals, the volume V = 1.4 cubic meters. We use this information to find k.

$$960 = \frac{k}{1.4} \quad \text{Let } P = 960 \text{ and } V = 1.4.$$

$$1344 = k \quad \text{Multiply both sides by 1.4.}$$

Thus, the value of k is 1344. Replacing k with 1344 in the variation equation, we have

$$P = \frac{1344}{V}$$

Next we find P when V is 2.5 cubic meters.

3. SOLVE.

$$P = \frac{1344}{2.5} \quad \text{Let } V = 2.5.$$

$$= 537.6$$

4. INTERPRET.

Check: Check the proposed solution in the original problem.

State: When the volume is 2.5 cubic meters, the pressure is 537.6 kilopascals. □

PRACTICE
4 Use Boyle's law as stated in Example 4. If $P = 350$ kilopascals when $V = 2.8$ cubic meters, find the pressure when the volume decreases to 1.5 cubic meters. ■

OBJECTIVE
3 Solving Problems Involving Joint Variation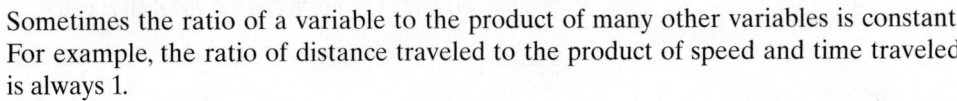

Sometimes the ratio of a variable to the product of many other variables is constant. For example, the ratio of distance traveled to the product of speed and time traveled is always 1.

$$\frac{d}{rt} = 1 \quad \text{or} \quad d = rt$$

Such a relationship is called **joint variation.**

> **Joint Variation**
>
> If the ratio of a variable y to the product of two or more variables is constant, then y **varies jointly as,** or **is jointly proportional to,** the other variables. If
>
> $$y = kxz$$
>
> then the number k is the **constant of variation** or the **constant of proportionality.**

✔ **CONCEPT CHECK**

Which type of variation is represented by the equation $xy = 8$? Explain.

a. Direct variation   b. Inverse variation   c. Joint variation

△ **EXAMPLE 5**   **Expressing Surface Area**

The lateral surface area of a cylinder varies jointly as its radius and height. Express this surface area S in terms of radius r and height h.

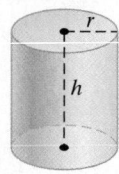

Solution Because the surface area varies jointly as the radius r and the height h, we equate S to a constant multiple of r and h.

$$S = krh$$

In the equation, $S = krh$, it can be determined that the constant k is 2π, and we then have the formula $S = 2\pi rh$. (The lateral surface area formula does not include the areas of the two circular bases.) □

PRACTICE

5   The area of a regular polygon varies jointly as its apothem and its perimeter. Express the area in terms of the apothem a and the perimeter p. ■

OBJECTIVE

4   Solving Problems Involving Combined Variation

Some examples of variation involve combinations of direct, inverse, and joint variation. We will call these variations **combined variation.**

EXAMPLE 6   Suppose that y varies directly as the square of x. If y is 24 when x is 2, find the constant of variation and the variation equation.

Solution Since y varies directly as the square of x, we have

$$y = kx^2$$

Answer to Concept Check:
b; answers may vary

Now let $y = 24$ and $x = 2$ and solve for k.

$$y = kx^2$$
$$24 = k \cdot 2^2$$
$$24 = 4k$$
$$6 = k$$

The constant of variation is 6, so the variation equation is

$$y = 6x^2 \qquad \square$$

PRACTICE
6   Suppose that y varies inversely as the cube of x. If y is $\dfrac{1}{2}$ when x is 2, find the constant of variation and the variation equation. ∎

△   **EXAMPLE 7**   **Finding Column Weight**

The maximum weight that a circular column can support is directly proportional to the fourth power of its diameter and is inversely proportional to the square of its height. A 2-meter-diameter column that is 8 meters in height can support 1 ton. Find the weight that a 1-meter-diameter column that is 4 meters in height can support.

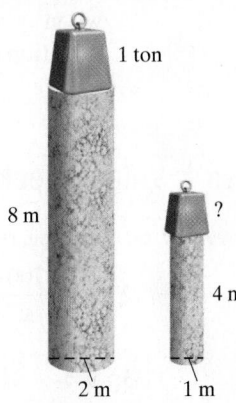

Solution

1. UNDERSTAND. Read and reread the problem. Let w = weight, d = diameter, h = height, and k = the constant of variation.

2. TRANSLATE. Since w is directly proportional to d^4 and inversely proportional to h^2, we have

$$w = \frac{kd^4}{h^2}$$

3. SOLVE. To find k, we are given that a 2-meter-diameter column that is 8 meters in height can support 1 ton. That is, $w = 1$ when $d = 2$ and $h = 8$, or

$$1 = \frac{k \cdot 2^4}{8^2} \qquad \text{Let } w = 1, d = 2, \text{ and } h = 8.$$

$$1 = \frac{k \cdot 16}{64}$$

$$4 = k \qquad \text{Solve for } k.$$

(Continued on next page)

Now replace k with 4 in the equation $w = \dfrac{kd^4}{h^2}$ and we have

$$w = \frac{4d^4}{h^2}$$

To find weight w for a 1-meter-diameter column that is 4 meters in height, let $d = 1$ and $h = 4$.

$$w = \frac{4 \cdot 1^4}{4^2}$$

$$w = \frac{4}{16} = \frac{1}{4}$$

4. INTERPRET.

Check: Check the proposed solution in the original problem.

State: The 1-meter-diameter column that is 4 meters in height can support $\frac{1}{4}$ ton of weight.

PRACTICE

7 Suppose that y varies directly as z and inversely as the cube of x. If y is 15 when $z = 5$ and $x = 3$, find the constant of variation and the variation equation.

 Vocabulary, Readiness & Video Check

State whether each equation represents direct, inverse, or joint variation.

1. $y = 5x$

2. $y = \dfrac{700}{x}$

3. $y = 5xz$

4. $y = \dfrac{1}{2}abc$

5. $y = \dfrac{9.1}{x}$

6. $y = 2.3x$

7. $y = \dfrac{2}{3}x$

8. $y = 3.1st$

Martin-Gay Interactive Videos

See Video 8.4

Watch the section lecture video and answer the following questions.

OBJECTIVE 1

9. Based on the lecture before ▦ Example 1, what kind of equation is a direct variation equation? What does k, the constant of variation, represent in this equation?

OBJECTIVE 2

10. In ▦ Example 3, why is it not necessary to replace the given values of x and y in the inverse variation equation in order to find k?

OBJECTIVE 3

11. Based on ▦ Example 5 and the lecture before, what is the variation equation for "y varies jointly as the square of a and the fifth power of b"?

OBJECTIVE 4

12. From ▦ Example 6, what kind of variation does a combined variation application involve?

8.4 Exercise Set MyMathLab

If y varies directly as x, find the constant of variation and the direct variation equation for each situation. See Example 1.

1. $y = 4$ when $x = 20$
2. $y = 5$ when $x = 40$
3. $y = 6$ when $x = 4$
4. $y = 12$ when $x = 8$
5. $y = 7$ when $x = \frac{1}{2}$
6. $y = 11$ when $x = \frac{1}{3}$
7. $y = 0.2$ when $x = 0.8$
8. $y = 0.4$ when $x = 2.5$

Solve. See Example 2.

9. The weight of a synthetic ball varies directly with the cube of its radius. A ball with a radius of 2 inches weighs 1.20 pounds. Find the weight of a ball of the same material with a 3-inch radius.

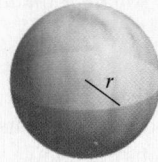

10. At sea, the distance to the horizon is directly proportional to the square root of the elevation of the observer. If a person who is 36 feet above the water can see 7.4 miles, find how far a person 64 feet above the water can see. Round to the nearest tenth of a mile.

11. The amount W of water used varies directly with the population N of people. Kansas City has a population of 467,000 people and used 36.5 billion gallons of water in fiscal year 2013. Find how many billion gallons of water we would expect Springfield to use if we know that its population is 160,000. Round to the nearest tenth of a billion. (*Source: Kansas City Water Board*)

12. Charles's law states that if the pressure P stays the same, the volume V of a gas is directly proportional to its temperature T. If a balloon is filled with 20 cubic meters of a gas at a temperature of 300 K, find the new volume if the temperature rises to 360 K while the pressure stays the same.

If y varies inversely as x, find the constant of variation and the inverse variation equation for each situation. See Example 3.

13. $y = 6$ when $x = 5$
14. $y = 20$ when $x = 9$
15. $y = 100$ when $x = 7$
16. $y = 63$ when $x = 3$
17. $y = \frac{1}{8}$ when $x = 16$
18. $y = \frac{1}{10}$ when $x = 40$
19. $y = 0.2$ when $x = 0.7$
20. $y = 0.6$ when $x = 0.3$

Solve. See Example 4.

21. Pairs of markings a set distance apart are made on highways so that police can detect drivers exceeding the speed limit. Over a fixed distance, the speed R varies inversely with the time T. In one particular pair of markings, R is 45 mph when T is 6 seconds. Find the speed of a car that travels the given distance in 5 seconds.

22. The weight of an object on or above the surface of Earth varies inversely as the square of the distance between the object and Earth's center. If a person weighs 160 pounds on Earth's surface, find the individual's weight if he moves 200 miles above Earth. Round to the nearest whole pound. (Assume that Earth's radius is 4000 miles.)

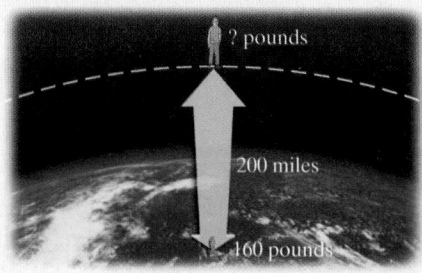

23. If the voltage V in an electric circuit is held constant, the current I is inversely proportional to the resistance R. If the current is 40 amperes when the resistance is 270 ohms, find the current when the resistance is 150 ohms.

24. Because it is more efficient to produce larger numbers of items, the cost of producing a certain computer DVD is inversely proportional to the number produced. If 4000 can be produced at a cost of $1.20 each, find the cost per DVD when 6000 are produced.

25. The intensity I of light varies inversely as the square of the distance d from the light source. If the distance from the light source is doubled (see the figure), determine what happens to the intensity of light at the new location.

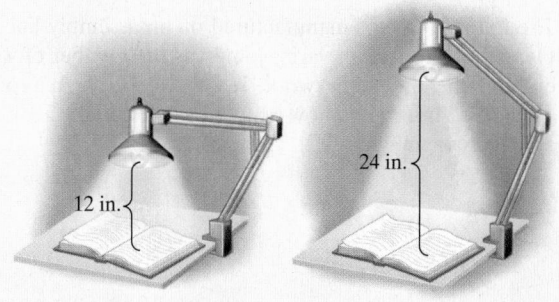

△ **26.** The maximum weight that a circular column can hold is inversely proportional to the square of its height. If an 8-foot column can hold 2 tons, find how much weight a 10-foot column can hold.

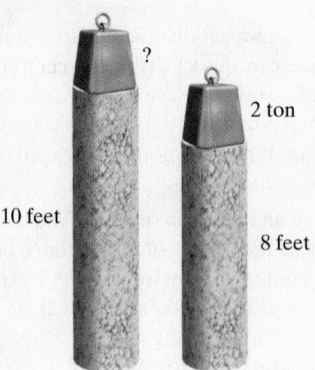

?

2 ton

10 feet

8 feet

MIXED PRACTICE

Write each statement as an equation. Use k as the constant of variation. See Example 5.

27. x varies jointly as y and z.

28. P varies jointly as R and the square of S.

29. r varies jointly as s and the cube of t.

30. a varies jointly as b and c.

For each statement, find the constant of variation and the variation equation. See Examples 5 and 6.

31. y varies directly as the cube of x; $y = 9$ when $x = 3$

32. y varies directly as the cube of x; $y = 32$ when $x = 4$

33. y varies directly as the square root of x; $y = 0.4$ when $x = 4$

34. y varies directly as the square root of x; $y = 2.1$ when $x = 9$

35. y varies inversely as the square of x; $y = 0.052$ when $x = 5$

36. y varies inversely as the square of x; $y = 0.011$ when $x = 10$

▶ **37.** y varies jointly as x and the cube of z; $y = 120$ when $x = 5$ and $z = 2$

38. y varies jointly as x and the square of z; $y = 360$ when $x = 4$ and $z = 3$

Solve. See Example 7.

△ **39.** The maximum weight that a rectangular beam can support varies jointly as its width and the square of its height and inversely as its length. If a beam $\frac{1}{2}$ foot wide, $\frac{1}{3}$ foot high, and 10 feet long can support 12 tons, find how much a similar beam can support if the beam is $\frac{2}{3}$ foot wide, $\frac{1}{2}$ foot high, and 16 feet long.

40. The number of cars manufactured on an assembly line at a General Motors plant varies jointly as the number of workers and the time they work. If 200 workers can produce 60 cars in 2 hours, find how many cars 240 workers should be able to make in 3 hours.

△ **41.** The volume of a cone varies jointly as its height and the square of its radius. If the volume of a cone is 32π cubic inches when the radius is 4 inches and the height is 6 inches, find the volume of a cone when the radius is 3 inches and the height is 5 inches.

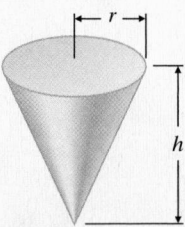

△ **42.** When a wind blows perpendicularly against a flat surface, its force is jointly proportional to the surface area and the speed of the wind. A sail whose surface area is 12 square feet experiences a 20-pound force when the wind speed is 10 miles per hour. Find the force on an 8-square-foot sail if the wind speed is 12 miles per hour.

43. The intensity of light (in foot-candles) varies inversely as the square of x, the distance in feet from the light source. The intensity of light 2 feet from the source is 80 foot-candles. How far away is the source if the intensity of light is 5 foot-candles?

44. The horsepower that can be safely transmitted to a shaft varies jointly as the shaft's angular speed of rotation (in revolutions per minute) and the cube of its diameter. A 2-inch shaft making 120 revolutions per minute safely transmits 40 horsepower. Find how much horsepower can be safely transmitted by a 3-inch shaft making 80 revolutions per minute.

MIXED PRACTICE

Write an equation to describe each variation. Use k for the constant of proportionality. See Examples 1 through 7.

45. y varies directly as x

46. p varies directly as q

47. a varies inversely as b

48. y varies inversely as x

49. y varies jointly as x and z

50. y varies jointly as q, r, and t

51. y varies inversely as x^3

52. y varies inversely as a^4

53. y varies directly as x and inversely as p^2

54. y varies directly as a^5 and inversely as b

REVIEW AND PREVIEW

Find the exact circumference and area of each circle. See Section 2.3.

△ **55.**

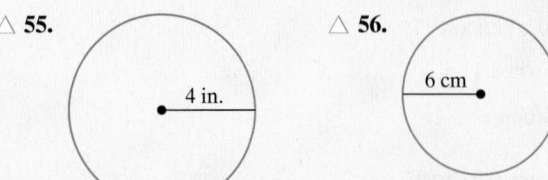

4 in.

△ **56.**

6 cm

△ **57.**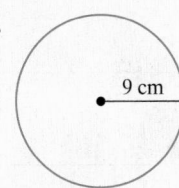
9 cm

△ **58.**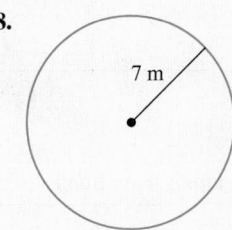
7 m

Simplify. See Sections 1.2, 1.4, and 1.5.

59. $|-1.2|$

60. $|-3|$

61. $-|7|$

62. $|0|$

63. $-\left|-\dfrac{1}{2}\right|$

64. $-\left|\dfrac{1}{5}\right|$

65. $\left(\dfrac{2}{3}\right)^3$

66. $\left(\dfrac{5}{11}\right)^2$

CONCEPT EXTENSIONS

Solve. See the Concept Check in this section. Choose the type of variation that each equation represents. **a.** *Direct variation* **b.** *Inverse variation* **c.** *Joint variation*

67. $y = \dfrac{2}{3}x$

68. $y = \dfrac{0.6}{x}$

69. $y = 9ab$

70. $xy = \dfrac{2}{11}$

71. The horsepower to drive a boat varies directly as the cube of the speed of the boat. If the speed of the boat is to double, determine the corresponding increase in horsepower required.

72. The volume of a cylinder varies jointly as the height and the square of the radius. If the height is halved and the radius is doubled, determine what happens to the volume.

73. Suppose that y varies directly as x. If x is doubled, what is the effect on y?

74. Suppose that y varies directly as x^2. If x is doubled, what is the effect on y?

Complete the following table for the inverse variation $y = \dfrac{k}{x}$ for each given value of k. Plot the points on a rectangular coordinate system.

x	$\dfrac{1}{4}$	$\dfrac{1}{2}$	1	2	4
$y = \dfrac{k}{x}$					

75. $k = 3$ **76.** $k = 1$ **77.** $k = \dfrac{1}{2}$ **78.** $k = 5$

Chapter 8 Vocabulary Check

Fill in each blank with one of the words or phrases listed below.

slope-intercept	directly	slope
jointly	parallel	perpendicular
function	inversely	linear function

1. _____ lines have the same slope and different y-intercepts.

2. _____ form of a linear equation in two variables is $y = mx + b$.

3. A(n) _____ is a relation in which each first component in the ordered pairs corresponds to exactly one second component.

4. In the equation $y = 4x - 2$, the coefficient of x is the _____ of its corresponding graph.

5. Two lines are _____ if the product of their slopes is -1.

6. A(n) _____ is a function that can be written in the form $f(x) = mx + b$.

7. In the equation $y = kx$, y varies _____ as x.

8. In the equation $y = \dfrac{k}{x}$, y varies _____ as x.

9. In the equation $y = kxz$, y varies _____ as x and z.

Chapter 8 Highlights

DEFINITIONS AND CONCEPTS	EXAMPLES

Section 8.1 Graphing and Writing Linear Functions

A **linear function** is a function that can be written in the form $f(x) = mx + b$.

Linear functions

$$f(x) = -3, g(x) = 5x, h(x) = -\frac{1}{3}x - 7$$

To graph a linear function, find three ordered pair solutions. Graph the solutions and draw a line through the plotted points.

Graph: $f(x) = -2x$

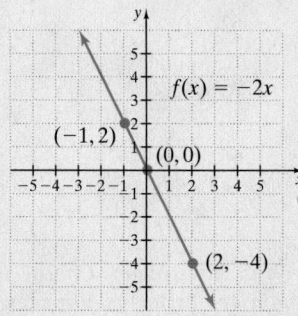

x	y or $f(x)$
-1	2
0	0
2	-4

The point-slope form of the equation of a line is $y - y_1 = m(x - x_1)$, where m is the slope of the line and (x_1, y_1) is a point on the line.

Find an equation of the line parallel to $g(x) = 2x - 1$ and containing the point $(1, -4)$. Write the equation using function notation.

Since we want a parallel line, use the same slope as $g(x)$, which is 2.

$$y - y_1 = m(x - x_1)$$
$$y - (-4) = 2(x - 1)$$
$$y + 4 = 2x - 2$$
$$y = 2x - 6 \qquad \text{Solve for } y.$$
$$f(x) = 2x - 6 \qquad \text{Let } y = f(x)$$

The graph of $y = mx + b$ is the same as the graph of $y = mx$, but shifted $|b|$ units up if b is positive and $|b|$ units down if b is negative.

Graph: $g(x) = -2x + 3$

This is the same as the graph of $f(x) = -2x$ shifted 3 units up.

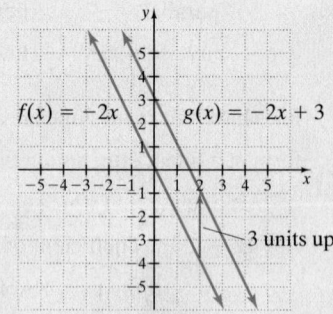

DEFINITIONS AND CONCEPTS	EXAMPLES

Section 8.2   Reviewing Function Notation and Graphing Nonlinear Functions

To graph a function that is not linear, find a sufficient number of ordered pair solutions so that a pattern may be discovered.

Graph: $f(x) = x^2 + 2$

x	y or $f(x)$
-2	6
-1	3
0	2
1	3
2	6

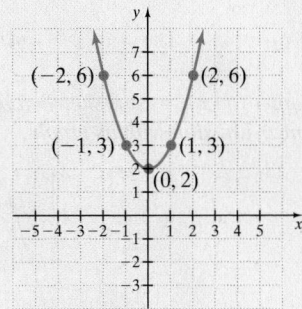

Section 8.3   Graphing Piecewise-Defined Functions and Shifting and Reflecting Graphs of Functions

Vertical shifts (upward and downward):
Let k be a positive number.

Graph of	*Same as*	*Moved*
$g(x) = f(x) + k$	$f(x)$	
$g(x) = f(x) - k$	$f(x)$	

Horizontal shift (to the left or right):
Let h be a positive number.

Graph of	*Same as*	*Moved*
$g(x) = f(x - h)$	$f(x)$	h units to the right
$g(x) = f(x + h)$	$f(x)$	h units to the left

Reflection about the x-axis

The graph of $g(x) = -f(x)$ is the graph of $f(x)$ reflected about the x-axis.

The graph of $h(x) = -|x - 3| + 1$ is the same as the graph of $f(x) = |x|$, reflected about the x-axis, shifted 3 units right, then 1 unit up.

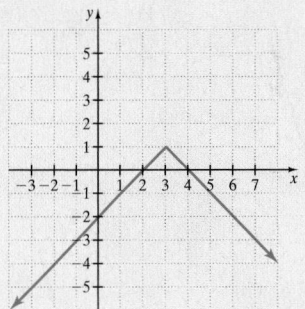

Section 8.4   Variation and Problem Solving

y **varies directly as** x, or y is **directly proportional to** x, if there is a nonzero constant k such that

$$y = kx$$

y **varies inversely as** x, or y is **inversely proportional to** x, if there is a nonzero constant k such that

$$y = \frac{k}{x}$$

y **varies jointly as** x and z or y is **jointly proportional to** x and z if there is a nonzero constant k such that

$$y = kxz$$

The circumference of a circle C varies directly as its radius r.

$$C = \underset{k}{2\pi} r$$

Pressure P varies inversely with volume V.

$$P = \frac{k}{V}$$

The lateral surface area S of a cylinder varies jointly as its radius r and height h.

$$S = \underset{k}{2\pi} rh$$

Chapter 8 Review

(8.1) *Graph each linear function.*

1. $f(x) = x$　　　　**2.** $f(x) = -\dfrac{1}{3}x$

3. $g(x) = 4x - 1$　　　**4.** $F(x) = -\dfrac{2}{3}x + 2$

The graph of $f(x) = 3x$ is sketched below. Use this graph to match each linear function with its graph.

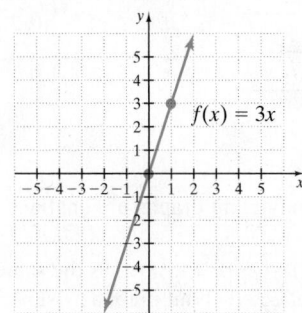

$f(x) = 3x$

A

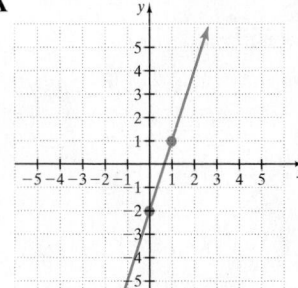

B

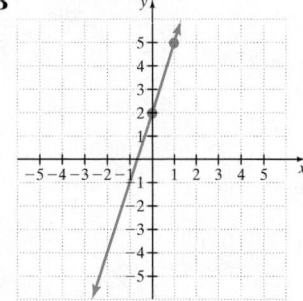

C

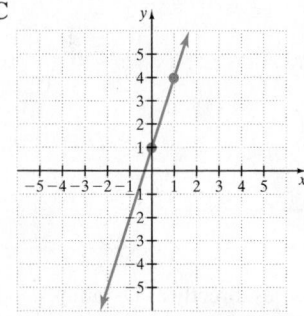

D

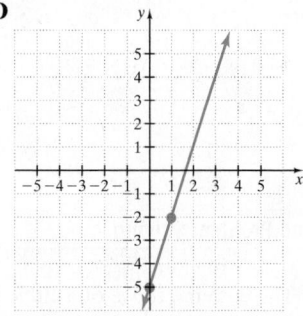

5. $f(x) = 3x + 1$　　　**6.** $f(x) = 3x - 2$

7. $f(x) = 3x + 2$　　　**8.** $f(x) = 3x - 5$

Find the slope and y-intercept of each function.

9. $f(x) = \dfrac{2}{5}x - \dfrac{4}{3}$　　**10.** $f(x) = -\dfrac{2}{7}x + \dfrac{3}{2}$

Find the standard form equation of each line satisfying the given conditions.

11. Slope 2; through $(5, -2)$

12. Through $(-3, 5)$; slope 3

13. Through $(-5, 3)$ and $(-4, -8)$

14. Through $(-6, -1)$ and $(-4, -2)$

15. Through $(-2, -5)$; parallel to $y = 8$

16. Through $(-2, 3)$; perpendicular to $x = 4$

Find an equation of each line satisfying the given conditions. Write each equation using function notation.

17. Horizontal; through $(3, -1)$

18. Slope $-\dfrac{2}{3}$; y-intercept $(0, 4)$

19. Slope -1; y-intercept $(0, -2)$

△ **20.** Through $(2, -6)$; parallel to $6x + 3y = 5$

△ **21.** Through $(-4, -2)$; parallel to $3x + 2y = 8$

△ **22.** Through $(-6, -1)$; perpendicular to $4x + 3y = 5$

23. Through $(-4, 5)$; perpendicular to $2x - 3y = 6$

24. The value of an automobile bought in 2010 continues to decrease as time passes. Two years after the car was bought, it was worth $21,500; four years after it was bought, it was worth $18,300.

   a. Assuming that this relationship between the number of years past 2010 and the value of the car is linear, write an equation describing this relationship. [*Hint*: Use ordered pairs of the form (years past 2010, value of the automobile).]

   b. Use this equation to estimate the value of the automobile in 2016.

△ **25.** The value of a building bought in 2000 continues to increase as time passes. Seven years after the building was bought, it was worth $210,000; 12 years after it was bought, it was worth $270,000.

   a. Assuming that this relationship between the number of years past 2000 and the value of the building is linear, write an equation describing this relationship. [*Hint*: Use ordered pairs of the form (years past 2000, value of the building).]

   b. Use this equation to estimate the value of the building in 2018.

26. Decide whether the lines are parallel, perpendicular, or neither.

$$-x + 3y = 2$$
$$6x - 18y = 3$$

(8.2) *Use the graph of the function on the next page to answer Exercises 27 through 30.*

27. Find $f(-1)$.　　　　**28.** Find $f(1)$.

29. Find all values of x such that $f(x) = 1$.

30. Find all values of x such that $f(x) = -1$.

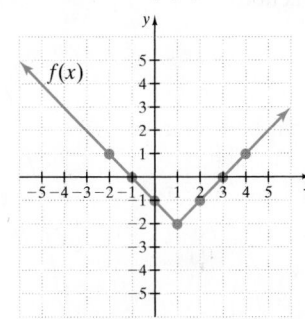

Determine whether each function is linear or not. Then graph the function.

31. $f(x) = 3x$

32. $f(x) = 5x$

33. $g(x) = |x| + 4$

34. $h(x) = x^2 + 4$

35. $F(x) = -\dfrac{1}{2}x + 2$

36. $G(x) = -x + 5$

37. $y = -1.36x$

38. $y = 2.1x + 5.9$

39. $H(x) = (x - 2)^2$

40. $f(x) = -|x - 3|$

(8.3) *Graph each function.*

41. $g(x) = \begin{cases} -\dfrac{1}{5}x & \text{if } x \le -1 \\ -4x + 2 & \text{if } x > -1 \end{cases}$

42. $f(x) = \begin{cases} -3x & \text{if } x < 0 \\ x - 3 & \text{if } x \ge 0 \end{cases}$

Graph each function.

43. $f(x) = \sqrt{x - 4}$

44. $y = \sqrt{x} - 4$

45. $h(x) = -(x + 3)^2 - 1$

46. $g(x) = |x - 2| - 2$

(8.4) *Solve each variation problem.*

47. A is directly proportional to B. If $A = 6$ when $B = 14$, find A when $B = 21$.

48. C is inversely proportional to D. If $C = 12$ when $D = 8$, find C when $D = 24$.

49. According to Boyle's law, the pressure exerted by a gas is inversely proportional to the volume, as long as the temperature stays the same. If a gas exerts a pressure of 1250 pounds per square inch when the volume is 2 cubic feet, find the volume when the pressure is 800 pounds per square inch.

△ **50.** The surface area of a sphere varies directly as the square of its radius. If the surface area is 36π square inches when the radius is 3 inches, find the surface area when the radius is 4 inches.

MIXED REVIEW

Write an equation of the line satisfying each set of conditions. Write the equation in the form $f(x) = mx + b$.

51. Slope 0; through $\left(-4, \dfrac{9}{2}\right)$

52. Slope $\dfrac{3}{4}$; through $(-8, -4)$

53. Through $(-3, 8)$ and $(-2, 3)$

54. Through $(-6, 1)$; parallel to $y = -\dfrac{3}{2}x + 11$

55. Through $(-5, 7)$; perpendicular to $5x - 4y = 10$

Graph each piecewise-defined function.

56. $g(x) = \begin{cases} 4x - 3 & \text{if } x \le 1 \\ 2x & \text{if } x > 1 \end{cases}$

57. $f(x) = \begin{cases} x - 2 & \text{if } x \le 0 \\ -\dfrac{x}{3} & \text{if } x \ge 3 \end{cases}$

Graph each function.

58. $f(x) = |x + 1| - 3$

59. $f(x) = \sqrt{x - 2}$

60. y is inversely proportional to x. If $y = 14$ when $x = 6$, find y when $x = 21$.

1c

Chapter 8 Getting Ready for the Test

MULTIPLE CHOICE *For Exercises 1 through 5, choose the best answer.*

⊳ **1.** If $f(x) = -x^2$, find the value of $f(-3)$.

    A. 6   **B.** -6 **C.** 9   **D.** -9

⊳ **2.** If $f(x) = 2x - 3$, find the value of $f(a + h)$.

    A. $2a + h - 3$ **B.** $2x(a + h) - 3$ **C.** $2a + 2h - 3$ **D.** $2a + 2h$

⊳ **3.** An ordered pair solution for the function $g(x)$ is $(-6, 0)$. This solution written using function notation is:

    A. $g(0) = -6$ **B.** $g(-6) = 0$ **C.** $g(-6) = g(0)$ **D.** $-6 = 0$

⊳ **4.** Suppose $y = f(x)$ and we are given that $f(2) = 8$. Which is not true?

    A. When $x = 2$, $y = 8$.

    B. A possible function is $f(x) = x^3$.

    C. A possible function is $f(x) = x - 6$.

    D. A point on the graph of the function is $(2, 8)$.

● 5. Given: $(0, 4)$ and $(5, 0)$. Final answer: $f(x) = -\dfrac{4}{5}x + 4$. Select the correct instructions.

   A. Find the slope of the line through the two points.

   B. Find an equation of the line through the two points. Write the equation in standard form.

   C. Find an equation of the line through the two points. Write the equation using function notation.

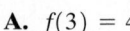

MULTIPLE CHOICE *For Exercises 6 through 11 use the given graph to fill in each blank using the choices below. Letters may be used more than once or not at all.*

A. $f(3) = 4$ **B.** $f(4) = 3$ **C.** 6   **D.** -9 **E.** 2

F. 0   **G.** $(-\infty, \infty)$ **H.** $(-\infty, 4]$ **I.** $(-\infty, 3]$

● 6. The vertex written in function notation is _____.

● 7. $f(0) =$ _____   ● 8. If $f(x) = 0$, then $x =$ _____ or $x =$ _____.

● 9. $f(8) =$ _____   ● 10. The domain of $f(x)$ is _____.   ● 11. The range of $f(x)$ is _____.

MULTIPLE CHOICE

● 12. The graph of $f(x) = (x - 2)^2$ is:

A. 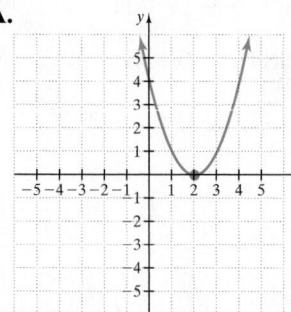   **B.** 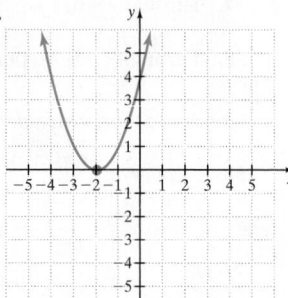   **C.** 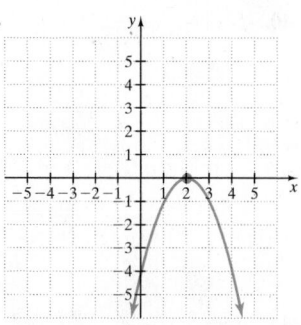   **D.**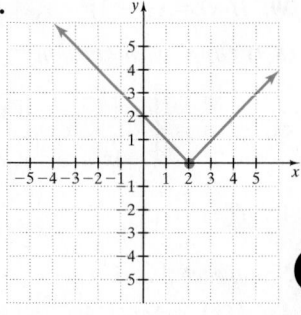

MATCHING *Match each statement with its translation. Use each letter once.*

● 13. y varies inversely as x

● 14. y varies directly as x

● 15. y varies jointly as x and z

● 16. y varies directly as x^2

● 17. y varies inversely as x^2

A. $y = kx^2$

B. $y = kx$

C. $y = kxz$

D. $y = \dfrac{k}{x}$

E. $y = \dfrac{k}{x^2}$

Chapter 8 Test   MyMathLab® You Tube

Use the graph of the function f to find each value.

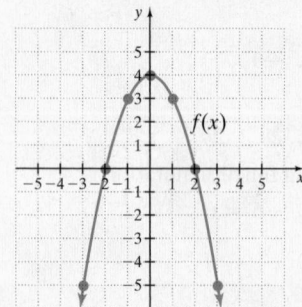

● 1. Find $f(1)$.

● 2. Find $f(-3)$.

● 3. Find all values of x such that $f(x) = 0$.

● 4. Find all values of x such that $f(x) = 4$.

Graph each line.

● 5. $2x - 3y = -6$ ● 6. $f(x) = \dfrac{2}{3}x$

Find an equation of each line satisfying the given conditions. Write Exercises 7–9 in standard form. Write Exercises 10–12 using function notation.

● 7. Horizontal; through $(2, -8)$

● 8. Through $(4, -1)$; slope -3

● 9. Through $(0, -2)$; slope 5

● 10. Through $(4, -2)$ and $(6, -3)$

△ 11. Through $(-1, 2)$; perpendicular to $3x - y = 4$

△ 12. Parallel to $2y + x = 3$; through $(3, -2)$

13. Line L_1 has the equation $2x - 5y = 8$. Line L_2 passes through the points $(1, 4)$ and $(-1, -1)$. Determine whether these lines are parallel lines, perpendicular lines, or neither.

Find the domain and range of each relation. Also determine whether the relation is a function.

14.

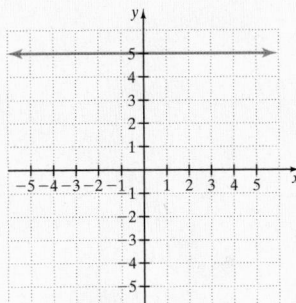

15.

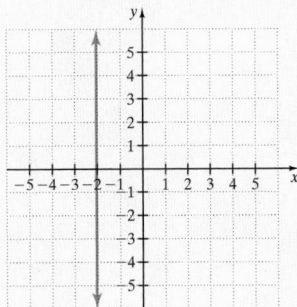

16.

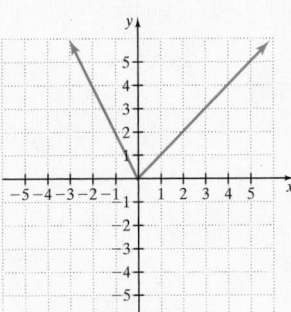

17.

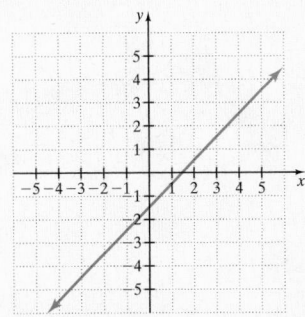

18. For the 2014 Major League Baseball season, the following linear function describes the relationship between a team's payroll x (in millions of dollars) and the number of games y that team won during the regular season.

$$f(x) = 0.024x + 79.44$$

Round to the nearest whole. (*Source:* Major League Baseball)

a. According to this equation, how many games would have been won during the 2014 season by a team with a payroll of $90 million?

b. The New York Yankees had a payroll of $204 million in 2014. According to this equation, how many games would they have won during the season?

c. According to this equation, what payroll would have been necessary in 2014 to have won 95 games during the season?

d. Find and interpret the slope of the equation.

Graph each function. For Exercises 19 and 21, state the domain and the range of the function.

19. $f(x) = \begin{cases} -\dfrac{1}{2}x & \text{if } x \le 0 \\ 2x - 3 & \text{if } x > 0 \end{cases}$

20. $f(x) = (x - 4)^2$

21. $g(x) = -|x + 2| - 1$

22. $h(x) = \sqrt{x} - 1$

23. Suppose that W is inversely proportional to V. If $W = 20$ when $V = 12$, find W when $V = 15$.

24. Suppose that Q is jointly proportional to R and the square of S. If $Q = 24$ when $R = 3$ and $S = 4$, find Q when $R = 2$ and $S = 3$.

25. When an anvil is dropped into a gorge, the speed with which it strikes the ground is directly proportional to the square root of the distance it falls. An anvil that falls 400 feet hits the ground at a speed of 160 feet per second. Find the height of a cliff over the gorge if a dropped anvil hits the ground at a speed of 128 feet per second.

Chapter 8 Cumulative Review

1. Simplify: $3[4 + 2(10 - 1)]$
2. Simplify: $5[3 + 6(8 - 5)]$

Find the value of each expression when $x = 2$ and $y = -5$.

3. **a.** $\dfrac{x - y}{12 + x}$ **b.** $x^2 - 3y$

4. **a.** $\dfrac{x + y}{3y}$ **b.** $y^2 - x$

Solve.

5. $-3x = 33$

6. $\dfrac{2}{3}y = 7$

7. $8(2 - t) = -5t$

8. $5x - 9 = 5x - 29$

9. Solve $y = mx + b$ for x.

10. Solve $y = 7x - 2$ for x.

11. $-4x + 7 \geq -9$ Write the solution using interval notation.

12. $-5x - 6 < 3x + 1$ Write the solution using interval notation.

13. Find the slope of the line $y = -1$.

14. Find the slope of a line parallel to the line passing through the points $(0, 7)$ and $(-1, 0)$.

15. Given $g(x) = x^2 - 3$, find the following. Then write the corresponding ordered pairs generated.
    a. $g(2)$ **b.** $g(-2)$
    c. $g(0)$

16. Given $f(x) = 3 - x^2$, find the following. Then write the corresponding ordered pairs generated.
    a. $f(2)$ **b.** $f(-2)$
    c. $f(0)$

17. Solve the system: $\begin{cases} 2x + y = 10 \\ x = y + 2 \end{cases}$

18. Solve the system: $\begin{cases} 3y = x + 10 \\ 2x + 5y = 24 \end{cases}$

19. Solve the system: $\begin{cases} -x - \dfrac{y}{2} = \dfrac{5}{2} \\ \dfrac{x}{6} - \dfrac{y}{2} = 0 \end{cases}$

20. Solve the system: $\begin{cases} \dfrac{x}{2} + y = \dfrac{5}{6} \\ 2x - y = \dfrac{5}{6} \end{cases}$

21. Divide $x^2 + 7x + 12$ by $x + 3$ using long division.

22. Divide: $\dfrac{5x^2y - 6xy + 2}{6xy}$

Factor out the GCF (greatest common factor).

23. **a.** $6t + 18$
    b. $y^5 - y^7$

24. **a.** $5y - 20$
    b. $z^{10} - z^3$

Factor completely.

25. $x^2 + 4x - 12$
26. $x^2 - 10x + 21$
27. $10x^2 - 13xy - 3y^2$
28. $12a^2 + 5ab - 2b^2$
29. $x^3 + 8$
30. $y^3 - 27$

Solve.

31. $x^2 - 9x - 22 = 0$
32. $y^2 - 5y = -6$

Simplify.

33. **a.** $\dfrac{2x^2}{10x^3 - 2x^2}$

    b. $\dfrac{9x^2 + 13x + 4}{8x^2 + x - 7}$

34. **a.** $\dfrac{33x^4y^2}{3xy}$

    b. $\dfrac{9y}{90y^2 + 9y}$

Perform the indicated operations.

35. $\dfrac{3x + 3}{5x - 5x^2} \cdot \dfrac{2x^2 + x - 3}{4x^2 - 9}$

36. $\dfrac{2x}{x - 6} - \dfrac{x + 6}{x - 6}$

37. $\dfrac{3x^2 + 2x}{x - 1} - \dfrac{10x - 5}{x - 1}$

38. $\dfrac{9}{y^2} - 4y$

Solve.

39. $3 - \dfrac{6}{x} = x + 8$

40. $\dfrac{x}{2} + \dfrac{x}{5} = \dfrac{x - 7}{20}$

Find an equation of the line through the given points. Write the equation using function notation.

41. $(4, 0)$ and $(-4, -5)$

42. $(-1, 3)$ and $(-2, 7)$

CHAPTER 9

Inequalities and Absolute Value

9.1 Compound Inequalities

9.2 Absolute Value Equations

9.3 Absolute Value Inequalities

Integrated Review—
Solving Compound
Inequalities and Absolute
Value Equations and
Inequalities

9.4 Graphing Linear
Inequalities in Two
Variables and
Systems of Linear
Inequalities

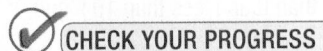

CHECK YOUR PROGRESS

Vocabulary Check

Chapter Highlights

Chapter Review

Getting Ready for the Test

Chapter Test

Cumulative Review

Mathematics is a tool for solving problems in such diverse fields as transportation, engineering, economics, medicine, business, and biology. We solve problems using mathematics by modeling real-world phenomena with mathematical equations or inequalities. Our ability to solve problems using mathematics, then, depends in part on our ability to solve different types of equations and inequalities. This chapter includes solving absolute value equations and inequalities and other types of inequalities.

Growth of Social Network Users

Whatever your interests, there are many social media choices that fit your preferences. For example, we may check on family in Facebook, tweet latest activities, look for new ideas in Pinterest, or go to Instagram or WhatsApp or Tumblr. In a recent year, U.S. users spent more than 121.8 billion minutes monthly on social media. This is consistent over much of the world. The following graph demonstrates the current and projected growth in social network users worldwide.

In Section 9.3, Exercises 83 and 84, we will find the projected growth in the number of social network users worldwide as well as the percent of increase.

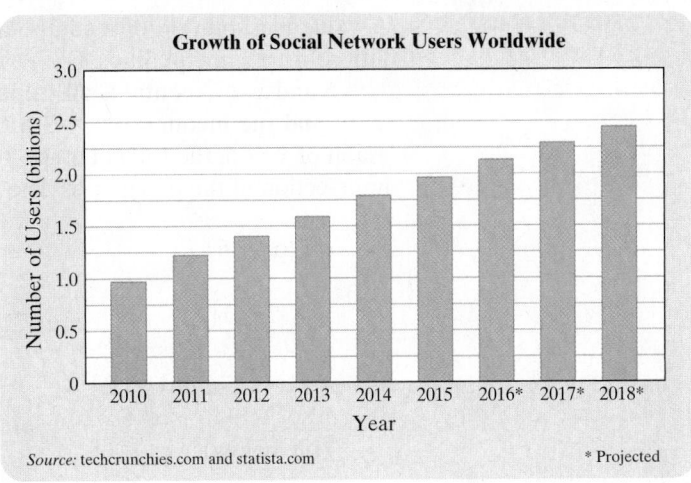

Growth of Social Network Users Worldwide

Source: techcrunchies.com and statista.com   * Projected

9.1 | Compound Inequalities

OBJECTIVES

1 Find the Intersection of Two Sets.

2 Solve Compound Inequalities Containing **and**.

3 Find the Union of Two Sets.

4 Solve Compound Inequalities Containing **or**.

Two inequalities joined by the word **and** or **or** are called **compound inequalities.**

Compound Inequalities

$$x + 3 < 8 \quad and \quad x > 2$$

$$\frac{2x}{3} \geq 5 \quad or \quad -x + 10 < 7$$

OBJECTIVE

1   Finding the Intersection of Two Sets

The solution set of a compound inequality formed by the word **and** is the **intersection** of the solution sets of the two inequalities. We use the symbol ∩ to represent "intersection."

Intersection of Two Sets

The intersection of two sets, A and B, is the set of all elements common to both sets. A intersect B is denoted by $A \cap B$.

$A \cap B$

A B

EXAMPLE 1   If $A = \{x \,|\, x$ is an even number greater than 0 and less than 10$\}$ and $B = \{3, 4, 5, 6\}$, find $A \cap B$.

Solution   Let's list the elements in set A.

$$A = \{2, 4, 6, 8\}$$

The numbers 4 and 6 are in sets A and B. The intersection is $\{4, 6\}$.   □

PRACTICE

1   If $A = \{x \,|\, x$ is an odd number greater than 0 and less than 10$\}$ and $B = \{1, 2, 3, 4\}$, find $A \cap B$.   ■

OBJECTIVE

2   Solving Compound Inequalities Containing "and"

A value is a solution of a compound inequality formed by the word **and** if it is a solution of _both_ inequalities. For example, the solution set of the compound inequality $x \leq 5$ and $x \geq 3$ contains all values of x that make the inequality $x \leq 5$ a true statement **and** the inequality $x \geq 3$ a true statement. The first graph shown below is the graph of $x \leq 5$, the second graph is the graph of $x \geq 3$, and the third graph shows the intersection of the two graphs. The third graph is the graph of $x \leq 5$ **and** $x \geq 3$.

$\{x \,|\, x \leq 5\}$ $(-\infty, 5]$

$\{x \,|\, x \geq 3\}$ $[3, \infty)$

$\{x \,|\, x \leq 5 \ and \ x \geq 3\}$
also $\{x \,|\, 3 \leq x \leq 5\}$
(see below) $[3, 5]$

Since $x \geq 3$ is the same as $3 \leq x$, the compound inequality $3 \leq x$ and $x \leq 5$ can be written in a more compact form as $3 \leq x \leq 5$. The solution set $\{x \,|\, 3 \leq x \leq 5\}$ includes all numbers that are greater than or equal to 3 and at the same time less than or equal to 5.

In interval notation, the set $\{x \,|\, x \leq 5 \ and \ x \geq 3\}$ or the set $\{x \,|\, 3 \leq x \leq 5\}$ is written as $[3, 5]$.

Don't forget that some compound inequalities containing "and" can be written in a more compact form.

Compound Inequality	Compact Form	Interval Notation
$2 \leq x$ *and* $x \leq 6$	$2 \leq x \leq 6$	$[2, 6]$

Graph:

EXAMPLE 2   Solve: $x - 7 < 2$ *and* $2x + 1 < 9$

<u>Solution</u> First we solve each inequality separately.

$$
\begin{array}{ccc}
x - 7 < 2 & and & 2x + 1 < 9 \\
x < 9 & and & 2x < 8 \\
x < 9 & and & x < 4
\end{array}
$$

Now we can graph the two intervals on two number lines and find their intersection. Their intersection is shown on the third number line.

$\{x \mid x < 9\}$ $(-\infty, 9)$

$\{x \mid x < 4\}$ $(-\infty, 4)$

$\{x \mid x < 9 \text{ and } x < 4\} = \{x \mid x < 4\}$ $(-\infty, 4)$

The solution set is $(-\infty, 4)$.

PRACTICE

2   Solve $x + 3 < 8$ *and* $2x - 1 < 3$. Write the solution set in interval notation.

EXAMPLE 3   Solve: $2x \geq 0$ *and* $4x - 1 \leq -9$

<u>Solution</u> First we solve each inequality separately.

$$
\begin{array}{ccc}
2x \geq 0 & and & 4x - 1 \leq -9 \\
x \geq 0 & and & 4x \leq -8 \\
x \geq 0 & and & x \leq -2
\end{array}
$$

Now we can graph the two intervals and find their intersection.

$\{x \mid x \geq 0\}$ $[0, \infty)$

$\{x \mid x \leq -2\}$ $(-\infty, -2]$

$\{x \mid x \geq 0 \text{ and } x \leq -2\} = \varnothing$ $\varnothing$

There is no number that is greater than or equal to 0 *and* less than or equal to -2. The solution set can be written as $\{\ \}$ or $\varnothing$.

PRACTICE

3   Solve $4x \leq 0$ *and* $3x + 2 > 8$. Write the solution set in interval notation.

Example 3 shows that some compound inequalities have no solution. Also, some have all real numbers as solutions.

To solve a compound inequality written in a compact form, such as $2 < 4 - x < 7$, we get x alone in the "middle part." Since a compound inequality is really two inequalities in one statement, we must perform the same operations on all three parts of the inequality. For example:

$$2 < 4 - x < 7 \text{ means } 2 < 4 - x \quad \textit{and} \quad 4 - x < 7,$$

EXAMPLE 4   Solve: $2 < 4 - x < 7$

Solution To get x alone, we first subtract 4 from all three parts.

$$2 < 4 - x < 7$$
$$2 - 4 < 4 - x - 4 < 7 - 4 \quad \text{Subtract 4 from all three parts.}$$
$$-2 < -x < 3 \quad \text{Simplify.}$$
$$\frac{-2}{-1} > \frac{-x}{-1} > \frac{3}{-1} \quad \text{Divide all three parts by } -1 \text{ and reverse the inequality symbols.}$$
$$2 > x > -3$$

> **Helpful Hint**
> Don't forget to reverse both inequality symbols.

This is equivalent to $-3 < x < 2$.

The solution set in interval notation is $(-3, 2)$, and its graph is shown.

$$\xleftarrow{\quad} \overset{(}{\underset{-4}{\;}} \; \overset{}{\underset{-3}{|}} \; \overset{}{\underset{-2}{|}} \; \overset{}{\underset{-1}{|}} \; \overset{}{\underset{0}{|}} \; \overset{}{\underset{1}{|}} \; \overset{)}{\underset{2}{\;}} \; \overset{}{\underset{3}{|}} \xrightarrow{\quad}$$

PRACTICE

4   Solve $3 < 5 - x < 9$. Write the solution set in interval notation.

EXAMPLE 5   Solve: $-1 \leq \dfrac{2x}{3} + 5 \leq 2$

Solution First, clear the inequality of fractions by multiplying all three parts by the LCD, 3.

$$-1 \leq \frac{2x}{3} + 5 \leq 2$$

$$3(-1) \leq 3\left(\frac{2x}{3} + 5\right) \leq 3(2) \quad \text{Multiply all three parts by the LCD, 3.}$$

$$-3 \leq 2x + 15 \leq 6 \quad \text{Use the distributive property and multiply.}$$
$$-3 - 15 \leq 2x + 15 - 15 \leq 6 - 15 \quad \text{Subtract 15 from all three parts.}$$
$$-18 \leq 2x \leq -9 \quad \text{Simplify.}$$
$$\frac{-18}{2} \leq \frac{2x}{2} \leq \frac{-9}{2} \quad \text{Divide all three parts by 2.}$$

$$-9 \leq x \leq -\frac{9}{2} \quad \text{Simplify.}$$

The graph of the solution is shown.

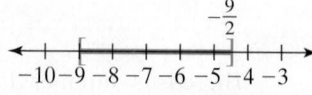

The solution set in interval notation is $\left[-9, -\dfrac{9}{2}\right]$.

PRACTICE

5   Solve $-4 \leq \dfrac{x}{2} - 1 \leq 3$. Write the solution set in interval notation.

OBJECTIVE

3 Finding the Union of Two Sets

The solution set of a compound inequality formed by the word **or** is the **union** of the solution sets of the two inequalities. We use the symbol $\cup$ to denote "union."

> **Helpful Hint**
> The word *either* in this definition means "one or the other or both."

> **Union of Two Sets**
>
> The **union** of two sets, A and B, is the set of elements that belong to *either* of the sets. A union B is denoted by $A \cup B$.
>
> $A \cup B$

EXAMPLE 6 If $A = \{x \mid x$ is an even number greater than 0 and less than 10$\}$ and $B = \{3, 4, 5, 6\}$, find $A \cup B$.

Solution Recall from Example 1 that $A = \{2, 4, 6, 8\}$. The numbers that are in either set or both sets are $\{2, 3, 4, 5, 6, 8\}$. This set is the union. $\square$

PRACTICE

6 If $A = \{x \mid x$ is an odd number greater than 0 and less than 10$\}$ and $B = \{2, 3, 4, 5, 6\}$, find $A \cup B$. ∎

OBJECTIVE

4 Solving Compound Inequalities Containing "or"

A value is a solution of a compound inequality formed by the word **or** if it is a solution of **either** inequality. For example, the solution set of the compound inequality $x \leq 1$ **or** $x \geq 3$ contains all numbers that make the inequality $x \leq 1$ a true statement **or** the inequality $x \geq 3$ a true statement.

$\{x \mid x \leq 1\}$ ⟷ -1 0 1 2 3 4 5 6 $(-\infty, 1]$

$\{x \mid x \geq 3\}$ ⟷ -1 0 1 2 3 4 5 6 $[3, \infty)$

$\{x \mid x \leq 1 \text{ or } x \geq 3\}$ ⟷ -1 0 1 2 3 4 5 6 $(-\infty, 1] \cup [3, \infty)$

In interval notation, the set $\{x \mid x \leq 1 \text{ or } x \geq 3\}$ is written as $(-\infty, 1] \cup [3, \infty)$.

EXAMPLE 7 Solve: $5x - 3 \leq 10 \text{ or } x + 1 \geq 5$

Solution First we solve each inequality separately.

$$5x - 3 \leq 10 \quad or \quad x + 1 \geq 5$$
$$5x \leq 13 \quad or \quad x \geq 4$$
$$x \leq \frac{13}{5} \quad or \quad x \geq 4$$

Now we can graph each interval and find their union.

$\left\{x \mid x \leq \dfrac{13}{5}\right\}$ ⟷ $\frac{13}{5}$ -1 0 1 2 3 4 5 6 $\left(-\infty, \dfrac{13}{5}\right]$

$\{x \mid x \geq 4\}$ ⟷ -1 0 1 2 3 4 5 6 $[4, \infty)$

$\left\{x \mid x \leq \dfrac{13}{5} \text{ or } x \geq 4\right\}$ ⟷ $\frac{13}{5}$ -1 0 1 2 3 4 5 6 $\left(-\infty, \dfrac{13}{5}\right] \cup [4, \infty)$

Continued on next page

The solution set is $\left(-\infty, \dfrac{13}{5}\right] \cup [4, \infty)$.

PRACTICE
7 Solve $8x + 5 \leq 8$ or $x - 1 \geq 2$. Write the solution set in interval notation. ■

EXAMPLE 8   Solve: $-2x - 5 < -3$ *or* $6x < 0$

Solution First we solve each inequality separately.

$$-2x - 5 < -3 \quad or \quad 6x < 0$$
$$-2x < 2 \quad or \quad x < 0$$
$$x > -1 \quad or \quad x < 0$$

Now we can graph each interval and find their union.

$\{x \mid x > -1\}$ $(-1, \infty)$

$\{x \mid x < 0\}$ $(-\infty, 0)$

$\{x \mid x > -1 \ or \ x < 0\}$ $(-\infty, \infty)$

$=$ all real numbers

The solution set is $(-\infty, \infty)$.

PRACTICE
8 Solve $-3x - 2 > -8$ *or* $5x > 0$. Write the solution set in interval notation. ■

✔ **CONCEPT CHECK**
Which of the following is *not* a correct way to represent the set of all numbers between -3 and 5?
a. $\{x \mid -3 < x < 5\}$ **b.** $-3 < x \ or \ x < 5$
c. $(-3, 5)$ **d.** $x > -3 \ and \ x < 5$

Answer to Concept Check:
b is not correct

✔ **Vocabulary, Readiness & Video Check**

Use the choices below to fill in each blank.

or   ∪   ∅
and   ∩   compound

1. Two inequalities joined by the word "and" or "or" are called _____ inequalities.

2. The word _____ means intersection.

3. The word _____ means union.

4. The symbol _____ represents intersection.

5. The symbol _____ represents union.

6. The symbol _____ is the empty set.

Watch the section lecture video and answer the following questions.

OBJECTIVE 1

7. Based on ▦ Example 1 and the lecture before, complete the following statement. For an element to be in the intersection of sets A and B, the element must be in set A _____ in set B.

OBJECTIVE 2

8. In ▦ Example 2, how can using three number lines help us find the solution to this "and" compound inequality?

OBJECTIVE 3

9. Based on ▦ Example 4 and the lecture before, complete the following statement. For an element to be in the union of sets A and B, the element must be in set A _____ in set B.

OBJECTIVE 4

10. In ▦ Example 5, how can using three number lines help us find the solution to this "or" compound inequality?

See Video 9.1

9.1   Exercise Set   MyMathLab®

MIXED PRACTICE

If $A = \{x \mid x$ is an even integer$\}$, $B = \{x \mid x$ is an odd integer$\}$, $C = \{2, 3, 4, 5\}$, and $D = \{4, 5, 6, 7\}$, list the elements of each set. See Examples 1 and 6.

1. $C \cup D$
2. $C \cap D$
3. $A \cap D$
4. $A \cup D$
5. $A \cup B$
6. $A \cap B$
7. $B \cap D$
8. $B \cup D$
9. $B \cup C$
10. $B \cap C$
11. $A \cap C$
12. $A \cup C$

Solve each compound inequality. Graph the solution set and write it in interval notation. See Examples 2 and 3.

13. $x < 1$ and $x > -3$
14. $x \leq 0$ and $x \geq -2$
15. $x \leq -3$ and $x \geq -2$
16. $x < 2$ and $x > 4$
17. $x < -1$ and $x < 1$
18. $x \geq -4$ and $x > 1$

Solve each compound inequality. Write solutions in interval notation. See Examples 2 and 3.

19. $x + 1 \geq 7$ and $3x - 1 \geq 5$
20. $x + 2 \geq 3$ and $5x - 1 \geq 9$
21. $4x + 2 \leq -10$ and $2x \leq 0$
22. $2x + 4 > 0$ and $4x > 0$
23. $-2x < -8$ and $x - 5 < 5$
24. $-7x \leq -21$ and $x - 20 \leq -15$

Solve each compound inequality. See Examples 4 and 5.

25. $5 < x - 6 < 11$
26. $-2 \leq x + 3 \leq 0$
27. $-2 \leq 3x - 5 \leq 7$
28. $1 < 4 + 2x < 7$
29. $1 \leq \frac{2}{3}x + 3 \leq 4$
30. $-2 < \frac{1}{2}x - 5 < 1$
31. $-5 \leq \frac{-3x + 1}{4} \leq 2$
32. $-4 \leq \frac{-2x + 5}{3} \leq 1$

Solve each compound inequality. Graph the solution set and write it in interval notation. See Examples 7 and 8.

33. $x < 4$ or $x < 5$
34. $x \geq -2$ or $x \leq 2$
35. $x \leq -4$ or $x \geq 1$
36. $x < 0$ or $x < 1$
37. $x > 0$ or $x < 3$
38. $x \geq -3$ or $x \leq -4$

Solve each compound inequality. Write solutions in interval notation. See Examples 7 and 8.

39. $-2x \leq -4$ or $5x - 20 \geq 5$
40. $-5x \leq 10$ or $3x - 5 \geq 1$
41. $x + 4 < 0$ or $6x > -12$
42. $x + 9 < 0$ or $4x > -12$
43. $3(x - 1) < 12$ or $x + 7 > 10$
44. $5(x - 1) \geq -5$ or $5 + x \leq 11$

MIXED PRACTICE

Solve each compound inequality. Write solutions in interval notation. See Examples 1 through 8.

45. $x < \frac{2}{3}$ and $x > -\frac{1}{2}$
46. $x < \frac{5}{7}$ and $x < 1$
47. $x < \frac{2}{3}$ or $x > -\frac{1}{2}$
48. $x < \frac{5}{7}$ or $x < 1$
49. $0 \leq 2x - 3 \leq 9$
50. $3 < 5x + 1 < 11$
51. $\frac{1}{2} < x - \frac{3}{4} < 2$

52. $\frac{2}{3} < x + \frac{1}{2} < 4$

53. $x + 3 \geq 3 \text{ and } x + 3 \leq 2$

54. $2x - 1 \geq 3 \text{ and } -x > 2$

55. $3x \geq 5 \text{ or } -\frac{5}{8}x - 6 > 1$

56. $\frac{3}{8}x + 1 \leq 0 \text{ or } -2x < -4$

57. $0 < \frac{5 - 2x}{3} < 5$

58. $-2 < \frac{-2x - 1}{3} < 2$

59. $-6 < 3(x - 2) \leq 8$

60. $-5 < 2(x + 4) < 8$

61. $-x + 5 > 6 \text{ and } 1 + 2x \leq -5$

62. $5x \leq 0 \text{ and } -x + 5 < 8$

◐ 63. $3x + 2 \leq 5 \text{ or } 7x > 29$

64. $-x < 7 \text{ or } 3x + 1 < -20$

65. $5 - x > 7 \text{ and } 2x + 3 \geq 13$

66. $-2x < -6 \text{ or } 1 - x > -2$

67. $-\frac{1}{2} \leq \frac{4x - 1}{6} < \frac{5}{6}$

68. $-\frac{1}{2} \leq \frac{3x - 1}{10} < \frac{1}{2}$

69. $\frac{1}{15} < \frac{8 - 3x}{15} < \frac{4}{5}$

70. $-\frac{1}{4} < \frac{6 - x}{12} < -\frac{1}{6}$

71. $0.3 < 0.2x - 0.9 < 1.5$

72. $-0.7 \leq 0.4x + 0.8 < 0.5$

REVIEW AND PREVIEW

Evaluate the following. See Sections 1.5 and 1.6.

73. $|-7| - |19|$

74. $|-7 - 19|$

75. $-(-6) - |-10|$

76. $|-4| - (-4) + |-20|$

Find by inspection all values for x that make each equation true.

77. $|x| = 7$　　　　　**78.** $|x| = 5$

79. $|x| = 0$

80. $|x| = -2$

CONCEPT EXTENSIONS

Use the graph to answer Exercises 81 and 82.

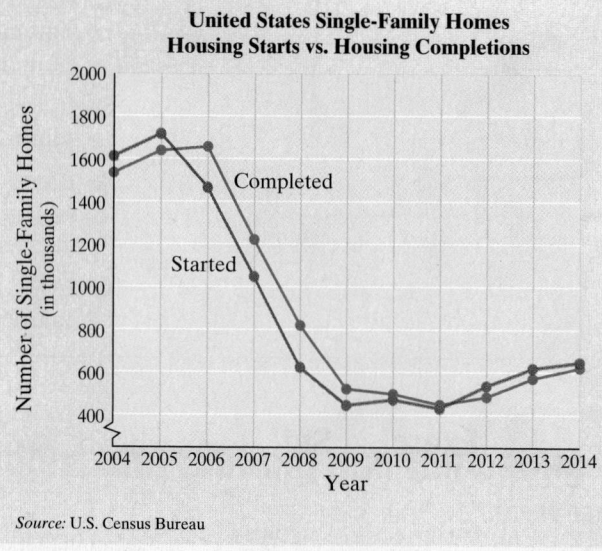

United States Single-Family Homes Housing Starts vs. Housing Completions

Source: U.S. Census Bureau

81. For which years were the number of single-family housing starts greater than 1500 and the number of single-family home completions greater than 1500?

82. For which years were the number of single-family housing starts less than 500 or the number of single-family housing completions greater than 1500?

83. In your own words, describe how to find the union of two sets.

84. In your own words, describe how to find the intersection of two sets.

*Solve each compound inequality for x. See the example below. To solve $x - 6 < 3x < 2x + 5$, notice that this inequality contains a variable not only in the middle but also on the left and the right. When this occurs, we solve by rewriting the inequality using the word **and**.*

$$x - 6 < 3x \quad \text{and} \quad 3x < 2x + 5$$
$$-6 < 2x \quad \text{and} \quad x < 5$$
$$-3 < x$$
$$x > -3 \quad \text{and} \quad x < 5$$

$x > -3$

$x < 5$

$-3 < x < 5 \text{ or } (-3, 5)$

85. $2x - 3 < 3x + 1 < 4x - 5$

86. $x + 3 < 2x + 1 < 4x + 6$

87. $-3(x - 2) \leq 3 - 2x \leq 10 - 3x$

88. $7x - 1 \leq 7 + 5x \leq 3(1 + 2x)$

89. $5x - 8 < 2(2 + x) < -2(1 + 2x)$

90. $1 + 2x < 3(2 + x) < 1 + 4x$

The formula for converting Fahrenheit temperatures to Celsius temperatures is $C = \dfrac{5}{9}(F - 32)$. Use this formula for Exercises 91 and 92.

91. During a recent year, the temperatures in Chicago ranged from $-29°C$ to $35°C$. Use a compound inequality to convert these temperatures to Fahrenheit temperatures.

92. In Oslo, the average temperature ranges from $-10°C$ to $18°C$. Use a compound inequality to convert these temperatures to the Fahrenheit scale.

Solve.

93. Christian D'Angelo has scores of 68, 65, 75, and 78 on his algebra tests. Use a compound inequality to find the scores he can make on his final exam to receive a C in the course. The final exam counts as two tests, and a C is received if the final course average is from 70 to 79.

94. Wendy Wood has scores of 80, 90, 82, and 75 on her chemistry tests. Use a compound inequality to find the range of scores she can make on her final exam to receive a B in the course. The final exam counts as two tests, and a B is received if the final course average is from 80 to 89.

9.2   Absolute Value Equations

OBJECTIVE

1 Solve Absolute Value Equations.

OBJECTIVE

1   Solving Absolute Value Equations

In Chapter 1, we defined the absolute value of a number as its distance from 0 on a number line.

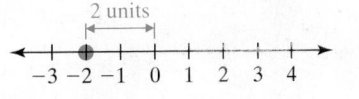

$$|-2| = 2 \text{ and } |3| = 3$$

In this section, we concentrate on solving equations containing the absolute value of a variable or a variable expression. Examples of absolute value equations are

$$|x| = 3 \qquad -5 = |2y + 7| \qquad |z - 6.7| = |3z + 1.2|$$

Since distance and absolute value are so closely related, absolute value equations and inequalities (see Section 9.3) are extremely useful in solving distance-type problems such as calculating the possible error in a measurement.

For the absolute value equation $|x| = 3$, its solution set will contain all numbers whose distance from 0 is 3 units. Two numbers are 3 units away from 0 on the number line: 3 and -3.

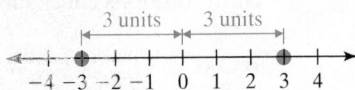

Thus, the solution set of the equation $|x| = 3$ is $\{3, -3\}$. This suggests the following:

> **Solving Equations of the Form $|X| = a$**
>
> If a is a positive number, then $|X| = a$ is equivalent to $X = a$ or $X = -a$.

EXAMPLE 1   Solve: $|p| = 2$

Solution Since 2 is positive, $|p| = 2$ is equivalent to $p = 2$ or $p = -2$.

To check, let $p = 2$ and then $p = -2$ in the original equation.

$|p| = 2$ Original equation		$|p| = 2$ Original equation
$|2| \stackrel{?}{=} 2$ Let $p = 2$.		$|-2| \stackrel{?}{=} 2$ Let $p = -2$.
$2 = 2$ True		$2 = 2$ True

The solutions are 2 and -2 or the solution set is $\{2, -2\}$. □

PRACTICE

1   Solve: $|q| = 13$

If the expression inside the absolute value bars is more complicated than a single variable, we can still apply the absolute value property.

> **Helpful Hint**
>
> For the equation $|X| = a$ in the box on the previous page, X can be a single variable or a variable expression.

EXAMPLE 2 Solve: $|5w + 3| = 7$

Solution Here the expression inside the absolute value bars is $5w + 3$. If we think of the expression $5w + 3$ as X in the absolute value property, we see that $|X| = 7$ is equivalent to

$$X = 7 \quad \text{or} \quad X = -7$$

Then substitute $5w + 3$ for X, and we have

$$5w + 3 = 7 \quad \text{or} \quad 5w + 3 = -7$$

Solve these two equations for w.

$$
\begin{aligned}
5w + 3 &= 7 \quad &\text{or} \quad 5w + 3 &= -7 \\
5w &= 4 \quad &\text{or} \quad 5w &= -10 \\
w &= \frac{4}{5} \quad &\text{or} \quad w &= -2
\end{aligned}
$$

Check: To check, let $w = -2$ and then $w = \frac{4}{5}$ in the original equation.

Let $w = -2$

$$|5(-2) + 3| \stackrel{?}{=} 7$$

$$|-10 + 3| \stackrel{?}{=} 7$$

$$|-7| \stackrel{?}{=} 7$$

$$7 = 7 \quad \text{True}$$

Let $w = \frac{4}{5}$

$$\left|5\left(\frac{4}{5}\right) + 3\right| \stackrel{?}{=} 7$$

$$|4 + 3| \stackrel{?}{=} 7$$

$$|7| \stackrel{?}{=} 7$$

$$7 = 7 \quad \text{True}$$

Both solutions check, and the solutions are -2 and $\frac{4}{5}$ or the solution set is $\left\{-2, \frac{4}{5}\right\}$. □

PRACTICE
2 Solve: $|2x - 3| = 5$

EXAMPLE 3 Solve: $\left|\dfrac{x}{2} - 1\right| = 11$

Solution $\left|\dfrac{x}{2} - 1\right| = 11$ is equivalent to

$$\frac{x}{2} - 1 = 11 \qquad \text{or} \qquad \frac{x}{2} - 1 = -11$$

$$2\left(\frac{x}{2} - 1\right) = 2(11) \quad \text{or} \quad 2\left(\frac{x}{2} - 1\right) = 2(-11) \qquad \text{Clear fractions.}$$

$$x - 2 = 22 \qquad \text{or} \qquad x - 2 = -22 \qquad \text{Apply the distributive property.}$$

$$x = 24 \qquad \text{or} \qquad x = -20$$

The solutions are 24 and -20.

PRACTICE
3 Solve: $\left|\dfrac{x}{5} + 1\right| = 15$

To apply the absolute value property, first make sure that the absolute value expression is isolated.

> **Helpful Hint**
>
> If the equation has a single absolute value expression containing variables, isolate the absolute value expression first.

EXAMPLE 4   Solve: $|2x| + 5 = 7$

Solution We want the absolute value expression alone on one side of the equation, so begin by subtracting 5 from both sides. Then apply the absolute value property.

$$|2x| + 5 = 7$$
$$|2x| = 2 \qquad\qquad \text{Subtract 5 from both sides.}$$
$$2x = 2 \quad \text{or} \quad 2x = -2$$
$$x = 1 \quad \text{or} \quad x = -1$$

The solutions are -1 and 1. □

PRACTICE
4   Solve: $|3x| + 8 = 14$ ■

EXAMPLE 5   Solve: $|y| = 0$

Solution We are looking for all numbers whose distance from 0 is zero units. The only number is 0. The solution is 0. □

PRACTICE
5   Solve: $|z| = 0$ ■

The next two examples illustrate a special case for absolute value equations. This special case occurs when an isolated absolute value is equal to a negative number.

EXAMPLE 6   Solve: $2|x| + 25 = 23$

Solution First, isolate the absolute value.

$$2|x| + 25 = 23$$
$$2|x| = -2 \quad \text{Subtract 25 from both sides.}$$
$$|x| = -1 \quad \text{Divide both sides by 2.}$$

The absolute value of a number is never negative, so this equation has no solution. The solution set is $\{\ \}$ or $\varnothing$. □

PRACTICE
6   Solve: $3|z| + 9 = 7$ ■

EXAMPLE 7   Solve: $\left|\dfrac{3x + 1}{2}\right| = -2$

Solution Again, the absolute value of any expression is never negative, so no solution exists. The solution set is $\{\ \}$ or $\varnothing$. □

PRACTICE
7   Solve: $\left|\dfrac{5x + 3}{4}\right| = -8$ ■

Given two absolute value expressions, we might ask, when are the absolute values of two expressions equal? To see the answer, notice that

$$|2| = |2|, \quad |-2| = |-2|, \quad |-2| = |2|, \quad \text{and} \quad |2| = |-2|$$

same   same   opposites   opposites

Two absolute value expressions are equal when the expressions inside the absolute value bars are equal to or are opposites of each other.

EXAMPLE 8 Solve: $|3x + 2| = |5x - 8|$

Solution This equation is true if the expressions inside the absolute value bars are equal to or are opposites of each other.

$$3x + 2 = 5x - 8 \quad \text{or} \quad 3x + 2 = -(5x - 8)$$

Next, solve each equation.

$$\begin{aligned}
3x + 2 &= 5x - 8 & \text{or} \quad 3x + 2 &= -5x + 8 \\
-2x + 2 &= -8 & \text{or} \quad 8x + 2 &= 8 \\
-2x &= -10 & \text{or} \quad 8x &= 6 \\
x &= 5 & \text{or} \quad x &= \frac{3}{4}
\end{aligned}$$

The solutions are $\frac{3}{4}$ and 5.   □

PRACTICE
8 Solve: $|2x + 4| = |3x - 1|$

EXAMPLE 9 Solve: $|x - 3| = |5 - x|$

Solution

$$\begin{aligned}
x - 3 &= 5 - x & \text{or} & \quad x - 3 = -(5 - x) \\
2x - 3 &= 5 & \text{or} & \quad x - 3 = -5 + x \\
2x &= 8 & \text{or} & \quad x - 3 - x = -5 + x - x \\
x &= 4 & \text{or} & \quad -3 = -5 \quad \text{False}
\end{aligned}$$

Recall from Section 2.3 that when an equation simplifies to a false statement, the equation has no solution. Thus, the only solution for the original absolute value equation is 4.   □

PRACTICE
9 Solve: $|x - 2| = |8 - x|$

✔ **CONCEPT CHECK**
True or false? Absolute value equations always have two solutions. Explain your answer.

The following box summarizes the methods shown for solving absolute value equations.

Absolute Value Equations
$\|X\| = a$ $\begin{cases} \text{If } a \text{ is positive, then solve } X = a \text{ or } X = -a. \\ \text{If } a \text{ is 0, solve } X = 0. \\ \text{If } a \text{ is negative, the equation } \|X\| = a \text{ has no solution.} \end{cases}$
$\|X\| = \|Y\|$ Solve $X = Y$ or $X = -Y$.

✔ Vocabulary, Readiness & Video Check

Match each absolute value equation with the equivalent statement.

1. $|x - 2| = 5$

2. $|x - 2| = 0$

3. $|x - 2| = |x + 3|$

4. $|x + 3| = 5$

5. $|x + 3| = -5$

A. $x - 2 = 0$

B. $x - 2 = x + 3$ or $x - 2 = -(x + 3)$

C. $x - 2 = 5$ or $x - 2 = -5$

D. $\varnothing$

E. $x + 3 = 5$ or $x + 3 = -5$

Martin-Gay Interactive Videos

See Video 9.2

Watch the section lecture videos and answer the following question.

OBJECTIVE 1

6. As explained in Example 3, why is *a* positive in the rule "$|X| = a$ is equivalent to $X = a$ or $X = -a$"?

9.2   Exercise Set   MyMathLab® ▷

Solve each absolute value equation. See Examples 1 through 7.

▷ **1.** $|x| = 7$

2. $|y| = 15$

3. $|3x| = 12.6$

4. $|6n| = 12.6$

5. $|2x - 5| = 9$

6. $|6 + 2n| = 4$

▷ **7.** $\left|\dfrac{x}{2} - 3\right| = 1$

8. $\left|\dfrac{n}{3} + 2\right| = 4$

9. $|z| + 4 = 9$

10. $|x| + 1 = 3$

11. $|3x| + 5 = 14$

12. $|2x| - 6 = 4$

13. $|2x| = 0$

14. $|7z| = 0$

15. $|4n + 1| + 10 = 4$

16. $|3z - 2| + 8 = 1$

17. $|5x - 1| = 0$

18. $|3y + 2| = 0$

Solve. See Examples 8 and 9.

19. $|5x - 7| = |3x + 11|$

20. $|9y + 1| = |6y + 4|$

21. $|z + 8| = |z - 3|$

22. $|2x - 5| = |2x + 5|$

MIXED PRACTICE

Solve each absolute value equation. See Examples 1 through 9.

23. $|x| = 4$

24. $|x| = 1$

25. $|y| = 0$

26. $|y| = 8$

27. $|z| = -2$

28. $|y| = -9$

29. $|7 - 3x| = 7$

30. $|4m + 5| = 5$

31. $|6x| - 1 = 11$

32. $|7z| + 1 = 22$

33. $|4p| = -8$

34. $|5m| = -10$

▷ **35.** $|x - 3| + 3 = 7$

36. $|x + 4| - 4 = 1$

37. $\left|\dfrac{z}{4} + 5\right| = -7$

38. $\left|\dfrac{c}{5} - 1\right| = -2$

39. $|9v - 3| = -8$

40. $|1 - 3b| = -7$

41. $|8n + 1| = 0$

42. $|5x - 2| = 0$

43. $|1 + 6c| - 7 = -3$

44. $|2 + 3m| - 9 = -7$

45. $|5x + 1| = 11$

46. $|8 - 6c| = 1$

47. $|4x - 2| = |-10|$

48. $|3x + 5| = |-4|$

49. $|5x + 1| = |4x - 7|$

50. $|3 + 6n| = |4n + 11|$

51. $|6 + 2x| = -|-7|$

52. $|4 - 5y| = -|-3|$

53. $|2x - 6| = |10 - 2x|$

54. $|4n + 5| = |4n + 3|$

55. $\left|\dfrac{2x - 5}{3}\right| = 7$

56. $\left|\dfrac{1 + 3n}{4}\right| = 4$

57. $2 + |5n| = 17$

58. $8 + |4m| = 24$

59. $\left|\dfrac{2x - 1}{3}\right| = |-5|$

60. $\left|\dfrac{5x + 2}{2}\right| = |-6|$

▷ **61.** $|2y - 3| = |9 - 4y|$

62. $|5z - 1| = |7 - z|$

63. $\left|\dfrac{3n + 2}{8}\right| = |-1|$

64. $\left|\dfrac{2r - 6}{5}\right| = |-2|$

65. $|x + 4| = |7 - x|$

66. $|8 - y| = |y + 2|$

67. $\left|\dfrac{8c - 7}{3}\right| = -|-5|$

68. $\left|\dfrac{5d + 1}{6}\right| = -|-9|$

REVIEW AND PREVIEW

The circle graph shows the types of cheese produced in the United States in 2014. Use this graph to answer Exercises 69 through 72. See Section 2.6.

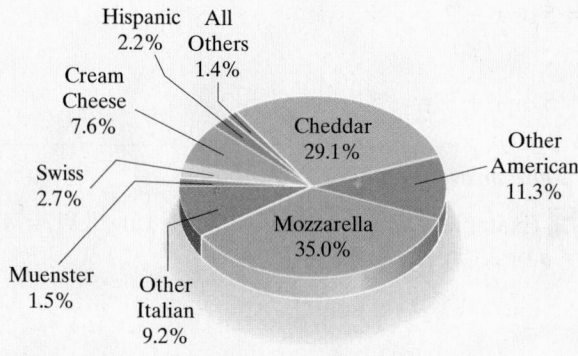

U.S. Cheese[1] Production by Variety, 2014

[1]Excludes Cottage Cheese

Source: USDA, *Dairy Products Annual Survey*

69. In 2014, cheddar cheese made up what percent of U.S. cheese production?

70. Which cheese had the highest U.S. production in 2014?

71. A circle contains 360°. Find the number of degrees found in the 9.2% sector for Other Italian Cheese.

72. In 2014, the total production of cheeses above in the United States was 11,201,000,000 pounds. Find the amount of cream cheese produced during that year.

List five integer solutions of each inequality. See Section 1.2.

73. $|x| \leq 3$

74. $|x| \geq -2$

75. $|y| > -10$

76. $|y| < 0$

CONCEPT EXTENSIONS

Without going through a solution procedure, determine the solution of each absolute value equation or inequality.

77. $|x - 7| = -4$ **78.** $|x - 7| < -4$

79. Write an absolute value equation representing all numbers x whose distance from 0 is 5 units.

80. Write an absolute value equation representing all numbers x whose distance from 0 is 2 units.

81. Explain why some absolute value equations have two solutions.

82. Explain why some absolute value equations have one solution.

83. Write an absolute value equation representing all numbers x whose distance from 1 is 5 units.

84. Write an absolute value equation representing all numbers x whose distance from 7 is 2 units.

85. Describe how solving an absolute value equation such as $|2x - 1| = 3$ is similar to solving an absolute value equation such as $|2x - 1| = |x - 5|$.

86. Describe how solving an absolute value equation such as $|2x - 1| = 3$ is different from solving an absolute value equation such as $|2x - 1| = |x - 5|$.

Write each as an equivalent absolute value equation.

87. $x = 6$ or $x = -6$

88. $2x - 1 = 4$ or $2x - 1 = -4$

89. $x - 2 = 3x - 4$ or $x - 2 = -(3x - 4)$

90. For what value(s) of c will an absolute value equation of the form $|ax + b| = c$ have

a. one solution?

b. no solution?

c. two solutions?

9.3 # Absolute Value Inequalities

OBJECTIVES

1 Solve Absolute Value Inequalities of the Form $|X| < a$.

2 Solve Absolute Value Inequalities of the Form $|X| > a$.

OBJECTIVE

1 Solving Absolute Value Inequalities of the Form $|X| < a$

The solution set of an absolute value inequality such as $|x| < 2$ contains all numbers whose distance from 0 is less than 2 units, as shown below.

Distance from 0: less than 2 units   Distance from 0: less than 2 units

−3 −2 −1 0 1 2 3

The solution set is $\{x | -2 < x < 2\}$, or $(-2, 2)$ in interval notation.

EXAMPLE 1   Solve $|x| \leq 3$ and graph the solution set.

Solution The solution set of this inequality contains all numbers whose distance from 0 is less than or equal to 3. Thus 3, −3, and all numbers between 3 and −3 are in the solution set.

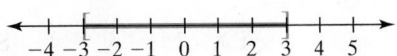

The solution set is $[-3, 3]$.

PRACTICE

1 Solve $|x| < 5$ and graph the solution set.

In general, we have the following.

Solving Absolute Value Inequalities of the Form $|X| < a$

If a is a positive number, then $|X| < a$ is equivalent to $-a < X < a$.

This property also holds true for the inequality symbol $\leq$.

EXAMPLE 2   Solve $|m - 6| < 2$ for m and graph the solution set.

Solution Replace X with $m - 6$ and a with 2 in the preceding property, and we see that

$$|m - 6| < 2 \quad \text{is equivalent to} \quad -2 < m - 6 < 2$$

Solve this compound inequality for m by adding 6 to all three parts.

$$-2 < m - 6 < 2$$
$$-2 + 6 < m - 6 + 6 < 2 + 6 \quad \text{Add 6 to all three parts.}$$
$$4 < m < 8 \quad \text{Simplify.}$$

The solution set is $(4, 8)$, and its graph is shown.

PRACTICE

2 Solve $|b + 1| < 3$ for b and graph the solution set.

Helpful Hint

Before using an absolute value inequality property, isolate the absolute value expression on one side of the inequality.

EXAMPLE 3   Solve $|5x + 1| + 1 \leq 10$ for x and graph the solution set.

Solution First, isolate the absolute value expression by subtracting 1 from both sides.

$$|5x + 1| + 1 \leq 10$$
$$|5x + 1| \leq 10 - 1 \quad \text{Subtract 1 from both sides.}$$
$$|5x + 1| \leq 9 \quad \text{Simplify.}$$

Since 9 is positive, we apply the absolute value property for $|X| \leq a$.

$$-9 \leq 5x + 1 \leq 9$$
$$-9 - 1 \leq 5x + 1 - 1 \leq 9 - 1 \quad \text{Subtract 1 from all three parts.}$$
$$-10 \leq 5x \leq 8 \quad \text{Simplify.}$$
$$-2 \leq x \leq \frac{8}{5} \quad \text{Divide all three parts by 5.}$$

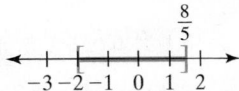

The solution set is $\left[-2, \dfrac{8}{5}\right]$, and the graph is shown above.

PRACTICE

3 Solve $|3x - 2| + 5 \leq 9$ for x and graph the solution set.

EXAMPLE 4 Solve for x: $\left|2x - \dfrac{1}{10}\right| < -13$

Solution The absolute value of a number is always nonnegative and can never be less than -13. Thus this absolute value inequality has no solution. The solution set is $\{\ \}$ or $\varnothing$.

PRACTICE
4 Solve for x: $\left|3x + \dfrac{5}{8}\right| < -4$

EXAMPLE 5 Solve for x: $\left|\dfrac{2(x+1)}{3}\right| \le 0$

Solution Recall that "$\le$" means "is less than or equal to." The absolute value of any expression will never be less than 0, but it may be equal to 0. Thus, to solve $\left|\dfrac{2(x+1)}{3}\right| \le 0$, we solve $\left|\dfrac{2(x+1)}{3}\right| = 0$

$$\dfrac{2(x+1)}{3} = 0$$

$$3\left[\dfrac{2(x+1)}{3}\right] = 3(0) \quad \text{Clear the equation of fractions.}$$

$$2x + 2 = 0 \quad \text{Apply the distributive property.}$$

$$2x = -2 \quad \text{Subtract 2 from both sides.}$$

$$x = -1 \quad \text{Divide both sides by 2.}$$

The solution set is $\{-1\}$.

PRACTICE
5 Solve for x: $\left|\dfrac{3(x-2)}{5}\right| \le 0$

OBJECTIVE
2 Solving Absolute Value Inequalities of the Form $|X| > a$ ▶

Let us now solve an absolute value inequality of the form $|X| > a$, such as $|x| \ge 3$. The solution set contains all numbers whose distance from 0 is 3 or more units. Thus the graph of the solution set contains 3 and all points to the right of 3 on the number line or -3 and all points to the left of -3 on the number line.

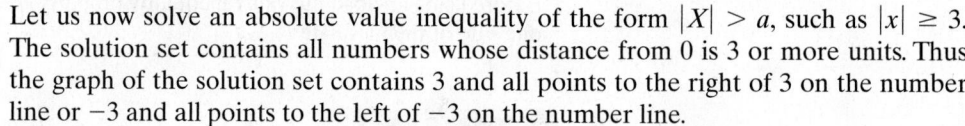

This solution set is written as $\{x \mid x \le -3 \text{ or } x \ge 3\}$. In interval notation, the solution is $(-\infty, -3] \cup [3, \infty)$, since "or" means "union." In general, we have the following.

Solving Absolute Value Inequalities of the Form $|X| > a$

If a is a positive number, then $|X| > a$ is equivalent to $X < -a$ or $X > a$.

This property also holds true for the inequality symbol $\ge$.

EXAMPLE 6 Solve for y: $|y - 3| > 7$

Solution Since 7 is positive, we apply the property for $|X| > a$.

$$|y - 3| > 7 \text{ is equivalent to } y - 3 < -7 \text{ or } y - 3 > 7$$

Next, solve the compound inequality.

$$y - 3 < -7 \qquad \text{or} \qquad y - 3 > 7$$
$$y - 3 + 3 < -7 + 3 \quad \text{or} \quad y - 3 + 3 > 7 + 3 \qquad \text{Add 3 to both sides.}$$
$$y < -4 \qquad \text{or} \qquad y > 10 \qquad \text{Simplify.}$$

The solution set is $(-\infty, -4) \cup (10, \infty)$, and its graph is shown.

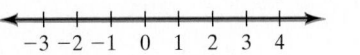

PRACTICE

6 Solve for y: $|y + 4| \geq 6$

Example 7 illustrates another special case of absolute value inequalities when an isolated absolute value expression is less than, less than or equal to, greater than, or greater than or equal to a negative number or 0.

EXAMPLE 7 Solve: $|2x + 9| + 5 > 3$

Solution First isolate the absolute value expression by subtracting 5 from both sides.

$$|2x + 9| + 5 > 3$$
$$|2x + 9| + 5 - 5 > 3 - 5 \qquad \text{Subtract 5 from both sides.}$$
$$|2x + 9| > -2 \qquad \text{Simplify.}$$

The absolute value of any number is always nonnegative and thus is always greater than -2. This inequality and the original inequality are true for all values of x. The solution set is $\{x \mid x \text{ is a real number}\}$ or $(-\infty, \infty)$, and its graph is shown.

PRACTICE

7 Solve $|4x + 3| + 5 > 3$. Graph the solution set.

✔ **CONCEPT CHECK**

Without taking any solution steps, how do you know that the absolute value inequality $|3x - 2| > -9$ has a solution? What is its solution?

EXAMPLE 8 Solve: $\left| \dfrac{x}{3} - 1 \right| - 7 \geq -5$

Solution First, isolate the absolute value expression by adding 7 to both sides.

$$\left| \frac{x}{3} - 1 \right| - 7 \geq -5$$

$$\left| \frac{x}{3} - 1 \right| - 7 + 7 \geq -5 + 7 \qquad \text{Add 7 to both sides.}$$

$$\left| \frac{x}{3} - 1 \right| \geq 2 \qquad \text{Simplify.}$$

Next, write the absolute value inequality as an equivalent compound inequality and solve.

$$\frac{x}{3} - 1 \leq -2 \qquad \text{or} \qquad \frac{x}{3} - 1 \geq 2$$

$$3\left(\frac{x}{3} - 1\right) \leq 3(-2) \qquad \text{or} \qquad 3\left(\frac{x}{3} - 1\right) \geq 3(2) \qquad \text{Clear the inequalities of fractions.}$$

$$x - 3 \leq -6 \qquad \text{or} \qquad x - 3 \geq 6 \qquad \text{Apply the distributive property.}$$

$$x \leq -3 \qquad \text{or} \qquad x \geq 9 \qquad \text{Add 3 to both sides.}$$

Answer to Concept Check:
$(-\infty, \infty)$ since the absolute value is always nonnegative

Continued on next page

The solution set is $(-\infty, -3] \cup [9, \infty)$, and its graph is shown.

$$
\begin{array}{c}
\overset{-3}{} \qquad \overset{9}{} \\
\text{—} \\
-6\,-4\,-2\;\;0\;\;2\;\;4\;\;6\;\;8\;\;10\;\;12
\end{array}
$$

PRACTICE
8 Solve $\left|\dfrac{x}{2} - 3\right| - 5 > -2$. Graph the solution set.

The following box summarizes the types of absolute value equations and inequalities.

Solving Absolute Value Equations and Inequalities with $a > 0$

Algebraic Solution	Solution Graph
$|X| = a$ is equivalent to $X = a$ or $X = -a$.	(graph) $-a \quad a$
$|X| < a$ is equivalent to $-a < X < a$.	(graph) $-a \quad a$
$|X| > a$ is equivalent to $X < -a$ or $X > a$.	(graph) $-a \quad a$

✔ Vocabulary, Readiness & Video Check

Match each absolute value statement with the equivalent statement.

1. $|2x + 1| = 3$

2. $|2x + 1| \le 3$

3. $|2x + 1| < 3$

4. $|2x + 1| \ge 3$

5. $|2x + 1| > 3$

A. $2x + 1 > 3$ or $2x + 1 < -3$

B. $2x + 1 \ge 3$ or $2x + 1 \le -3$

C. $-3 < 2x + 1 < 3$

D. $2x + 1 = 3$ or $2x + 1 = -3$

E. $-3 \le 2x + 1 \le 3$

Martin-Gay Interactive Videos

See Video 9.3 ⬤

Watch the section lecture video and answer the following questions.

OBJECTIVE
1 **6.** In ▦ Example 3, how can you reason that the inequality has no solution even if you don't know the rule?

OBJECTIVE
2 **7.** In ▦ Example 4, why is the union symbol used when the solution is written in interval notation?

9.3 Exercise Set MyMathLab® ▶

Solve each inequality. Then graph the solution set and write it in interval notation. See Examples 1 through 4.

1. $|x| \le 4$

2. $|x| < 6$

3. $|x - 3| < 2$

4. $|y - 7| \le 5$

5. $|x + 3| < 2$

6. $|x + 4| < 6$

7. $|2x + 7| \le 13$

8. $|5x - 3| \le 18$

9. $|x| + 7 \le 12$

10. $|x| + 6 \le 7$

▶ **11.** $|3x - 1| < -5$

12. $|8x - 3| < -2$

13. $|x - 6| - 7 \le -1$

14. $|z + 2| - 7 < -3$

Solve each inequality. Graph the solution set and write it in interval notation. See Examples 6 through 8.

▶ **15.** $|x| > 3$

16. $|y| \ge 4$

17. $|x + 10| \ge 14$

18. $|x - 9| \ge 2$

19. $|x| + 2 > 6$

20. $|x| - 1 > 3$

21. $|5x| > -4$

22. $|4x - 11| > -1$

23. $|6x - 8| + 3 > 7$

24. $|10 + 3x| + 1 > 2$

Solve each inequality. Graph the solution set and write it in interval notation. See Example 5.

25. $|x| \leq 0$

26. $|x| \geq 0$

27. $|8x + 3| > 0$

28. $|5x - 6| < 0$

MIXED PRACTICE

Solve each inequality. Graph the solution set and write it in interval notation. See Examples 1 through 8.

29. $|x| \leq 2$

30. $|z| < 8$

31. $|y| > 1$

32. $|x| \geq 10$

33. $|x - 3| < 8$

34. $|-3 + x| \leq 10$

35. $|0.6x - 3| > 0.6$

36. $|1 + 0.3x| \geq 0.1$

37. $5 + |x| \leq 2$

38. $8 + |x| < 1$

39. $|x| > -4$

40. $|x| \leq -7$

41. $|2x - 7| \leq 11$

42. $|5x + 2| < 8$

43. $|x + 5| + 2 \geq 8$

44. $|-1 + x| - 6 > 2$

45. $|x| > 0$

46. $|x| < 0$

47. $9 + |x| > 7$

48. $5 + |x| \geq 4$

49. $6 + |4x - 1| \leq 9$

50. $-3 + |5x - 2| \leq 4$

51. $\left|\dfrac{2}{3}x + 1\right| > 1$

52. $\left|\dfrac{3}{4}x - 1\right| \geq 2$

53. $|5x + 3| < -6$

54. $|4 + 9x| \geq -6$

55. $\left|\dfrac{8x - 3}{4}\right| \leq 0$

56. $\left|\dfrac{5x + 6}{2}\right| \leq 0$

57. $|1 + 3x| + 4 < 5$

58. $|7x - 3| - 1 \leq 10$

59. $\left|\dfrac{x + 6}{3}\right| > 2$

60. $\left|\dfrac{7 + x}{2}\right| \geq 4$

61. $-15 + |2x - 7| \leq -6$

62. $-9 + |3 + 4x| < -4$

63. $\left|2x + \dfrac{3}{4}\right| - 7 \leq -2$

64. $\left|\dfrac{3}{5} + 4x\right| - 6 < -1$

MIXED PRACTICE

Solve each equation or inequality for x. (Sections 9.2, 9.3)

65. $|2x - 3| < 7$

66. $|2x - 3| > 7$

67. $|2x - 3| = 7$

68. $|5 - 6x| = 29$

69. $|x - 5| \geq 12$

70. $|x + 4| \geq 20$

71. $|9 + 4x| = 0$

72. $|9 + 4x| \geq 0$

73. $|2x + 1| + 4 < 7$

74. $8 + |5x - 3| \geq 11$

75. $|3x - 5| + 4 = 5$

76. $|5x - 3| + 2 = 4$

77. $|x + 11| = -1$

78. $|4x - 4| = -3$

79. $\left|\dfrac{2x - 1}{3}\right| = 6$

80. $\left|\dfrac{6 - x}{4}\right| = 5$

81. $\left|\dfrac{3x - 5}{6}\right| > 5$

82. $\left|\dfrac{4x - 7}{5}\right| < 2$

REVIEW AND PREVIEW

Solve. See Section 2.4.

Many companies predict the growth or decline of various social network use. The following data is from Techcrunchies, a technical information site. Notice that the first table is the predicted increase in the number of people using social network sites worldwide (in millions) and the second is the predicted percent increase in the number of social network site users worldwide.

83. Use the middle column in the table to find the predicted number of social network site users for each year.

Year	Increase in Social Network Users	Predicted Number
2015	$2x$	
2016	$2x + 170$	
2017	$3x - 650$	
Total	6380 million	

84. Use the middle column in the table to find the predicted percent of increase in the number of social network site users for each year.

Year	Percent of Increase in Social Network Users	Predicted Percent of Increase
2015	x	
2016	$3x - 11$	
2017	$2x + 11$	
Total	66%	

Consider the equation $3x - 4y = 12$. For each value of x or y given, find the corresponding value of the other variable that makes the statement true. See Section 2.5.

85. If $x = 2$, find y.

86. If $y = -1$, find x.

87. If $y = -3$, find x.

88. If $x = 4$, find y.

CONCEPT EXTENSIONS

89. Write an absolute value inequality representing all numbers x whose distance from 0 is less than 7 units.

90. Write an absolute value inequality representing all numbers x whose distance from 0 is greater than 4 units.

91. Write $-5 \leq x \leq 5$ as an equivalent inequality containing an absolute value.

92. Write $x > 1$ or $x < -1$ as an equivalent inequality containing an absolute value.

93. Describe how solving $|x - 3| = 5$ is different from solving $|x - 3| < 5$.

94. Describe how solving $|x + 4| = 0$ is similar to solving $|x + 4| \leq 0$.

The expression $|x_T - x|$ is defined to be the absolute error in x,
where x_T is the true value of a quantity and x is the measured value
or value as stored in a computer.

95. If the true value of a quantity is 3.5 and the absolute error must be less than 0.05, find the acceptable measured
values.

96. If the true value of a quantity is 0.2 and the approximate
value stored in a computer is $\dfrac{51}{256}$, find the absolute error.

| Integrated Review | Solving Compound Inequalities and Absolute Value Equations and Inequalities |

Sections 9.1–9.3

Solve each equation or inequality. Write inequality solution sets in interval notation. For inequalities containing "and" or "or",
also graph the solution set notation.

1. $x < 7$ *and* $x > -5$

2. $x < 7$ *or* $x > -5$

3. $|4x - 3| = 1$

4. $|2x + 1| < 5$

5. $|6x| - 9 \geq -3$

6. $|x - 7| = |2x + 11|$

7. $-5 \leq \dfrac{3x - 8}{2} \leq 2$

8. $|9x - 1| = -3$

9. $3x + 2 \leq 5$ *or* $-3x \geq 0$

10. $3x + 2 \leq 5$ *and* $-3x \geq 0$

11. $|3 - x| - 5 \leq -2$

12. $\left|\dfrac{4x + 1}{5}\right| = |-1|$

Match each equation or inequality on the left with the equivalent statement on the right.

13. $|2x + 1| = 5$

14. $|2x + 1| < 5$

15. $|2x + 1| > 5$

16. $x < 3$ *or* $x < 5$

17. $x < 3$ *and* $x < 5$

A. $2x + 1 > 5$ or $2x + 1 < -5$

B. $2x + 1 = 5$ or $2x + 1 = -5$

C. $x < 5$

D. $x < 3$

E. $-5 < 2x + 1 < 5$

9.4   Graphing Linear Inequalities in Two Variables and Systems of Linear Inequalities

OBJECTIVE

1 Graph a Linear Inequality in Two Variables.

2 Solve a System of Linear Inequalities.

In this section, we first learn to graph a single linear inequality in two variables. Then we solve systems of linear inequalities.

Recall that a linear equation in two variables is an equation that can be written in the form $Ax + By = C$ where $A, B,$ and C are real numbers and A and B are not both 0. The definition of a linear inequality is the same except that the equal sign is replaced with an inequality sign.

A **linear inequality in two variables** is an inequality that can be written in one of the forms:

$$Ax + By < C \qquad Ax + By \leq C$$
$$Ax + By > C \qquad Ax + By \geq C$$

where $A, B,$ and C are real numbers and A and B are not both 0. Just as for linear equations in x and y, an ordered pair is a **solution** of an inequality in x and y if replacing the variables by coordinates of the ordered pair results in a true statement.

OBJECTIVE

1 Graphing Linear Inequalities in Two Variables

The linear equation $x - y = 1$ is graphed next. Recall that all points on the line correspond to ordered pairs that satisfy the equation $x - y = 1$.

Notice the line defined by $x - y = 1$ divides the rectangular coordinate system plane into 2 sides. All points on one side of the line satisfy the inequality $x - y < 1$ and all points on the other side satisfy the inequality $x - y > 1$. The graph below shows a few examples of this.

Check	$x - y < 1$
$(1, 3)$	$1 - 3 < 1$ True
$(-2, 1)$	$-2 - 1 < 1$ True
$(-4, -4)$	$-4 - (-4) < 1$ True

Check	$x - y > 1$
$(4, 1)$	$4 - 1 > 1$ True
$(2, -2)$	$2 - (-2) > 1$ True
$(0, -4)$	$0 - (-4) > 1$ True

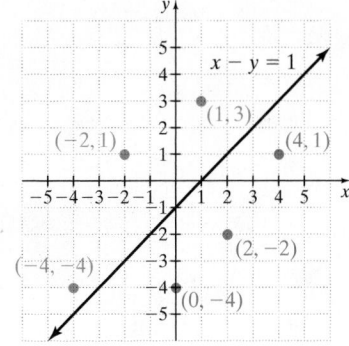

The graph of $x - y < 1$ is the region shaded blue, and the graph of $x - y > 1$ is the region shaded red below.

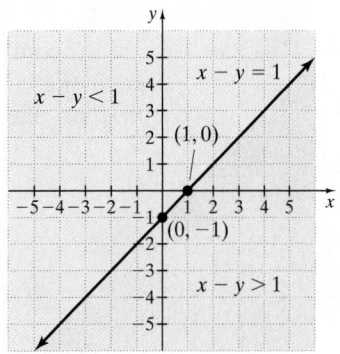

The region to the left of the line and the region to the right of the line are called **half-planes.** Every line divides the plane (similar to a sheet of paper extending indefinitely in all directions) into two half-planes; the line is called the **boundary.**

Recall that the inequality $x - y \leq 1$ means

$$x - y = 1 \quad \text{or} \quad x - y < 1$$

Thus, the graph of $x - y \leq 1$ is the half-plane $x - y < 1$ along with the boundary line $x - y = 1$.

Graphing a Linear Inequality in Two Variables

Step 1. Graph the boundary line found by replacing the inequality sign with an equal sign. If the inequality sign is $>$ or $<$, graph a dashed boundary line (indicating that the points on the line are not solutions of the inequality). If the inequality sign is $\geq$ or $\leq$, graph a solid boundary line (indicating that the points on the line are solutions of the inequality).

Step 2. Choose a point, *not* on the boundary line, as a test point. Substitute the coordinates of this test point into the *original* inequality.

Step 3. If a true statement is obtained in Step 2, shade the half-plane that contains the test point. If a false statement is obtained, shade the half-plane that does not contain the test point.

EXAMPLE 1 Graph: $x + y < 7$

Solution

Step 1. First we graph the boundary line by graphing the equation $x + y = 7$. We graph this boundary as a dashed line because the inequality sign is $<$, and thus the points on the line are not solutions of the inequality $x + y < 7$.

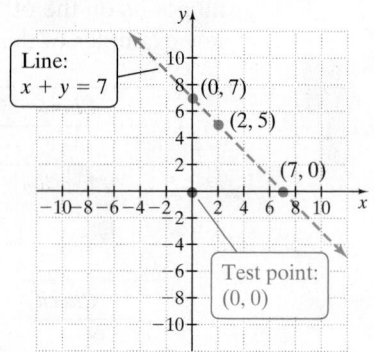

Step 2. Next, we choose a test point, being careful not to choose a point on the boundary line. We choose $(0, 0)$. Substitute the coordinates of $(0, 0)$ into $x + y < 7$.

$$x + y < 7 \quad \text{Original inequality}$$

$$0 + 0 \overset{?}{<} 7 \quad \text{Replace } x \text{ with 0 and } y \text{ with 0.}$$

$$0 < 7 \quad \text{True}$$

Step 3. Since the result is a true statement, $(0, 0)$ is a solution of $x + y < 7$, and every point in the same half-plane as $(0, 0)$ is also a solution. To indicate this, shade the entire half-plane containing $(0, 0)$, as shown.

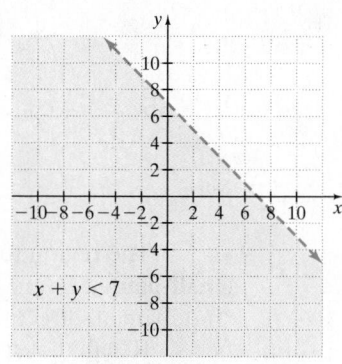

Graph of $x + y < 7$

PRACTICE

1 Graph: $x + y > 5$

✔ **CONCEPT CHECK**

Determine whether $(0, 0)$ is included in the graph of

a. $y \geq 2x + 3$ **b.** $x < 7$ **c.** $2x - 3y < 6$

EXAMPLE 2 Graph: $2x - y \geq 3$

Solution

Step 1. We graph the boundary line by graphing $2x - y = 3$. We draw this line as a solid line because the inequality sign is $\geq$, and thus the points on the line are solutions of $2x - y \geq 3$.

Answers to Concept Check:
a. no **b.** yes **c.** yes

Step 2. Once again, $(0, 0)$ is a convenient test point since it is not on the boundary line. We substitute 0 for x and 0 for y into the original inequality.

$$2x - y \geq 3$$
$$2(0) - 0 \geq 3 \quad \text{Let } x = 0 \text{ and } y = 0.$$
$$0 \geq 3 \quad \text{False}$$

Step 3. Since the statement is false, no point in the half-plane containing $(0, 0)$ is a solution. Therefore, we shade the half-plane that does not contain $(0, 0)$. Every point in the shaded half-plane and every point on the boundary line is a solution of $2x - y \geq 3$.

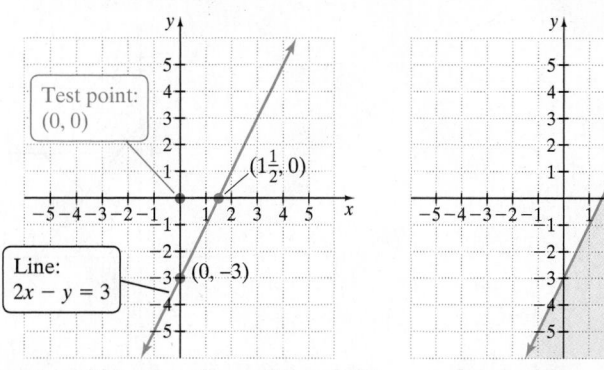

Step 1 and Step 2 on Page 580 and 581   **Graph of $2x - y \geq 3$**

PRACTICE
2 Graph: $3x - y \geq 4$

> **Helpful Hint**
>
> When graphing an inequality, make sure the test point is substituted into the **original inequality.** For Example 2, we substituted the test point $(0, 0)$ into the **original inequality** $2x - y \geq 3$, *not* $2x - y = 3$.

EXAMPLE 3 Graph: $x > 2y$

Solution

Step 1. We find the boundary line by graphing $x = 2y$. The boundary line is a dashed line since the inequality symbol is $>$.

Step 2. We cannot use $(0, 0)$ as a test point because it is a point on the boundary line. We choose instead $(0, 2)$.

$$x > 2y$$
$$0 > 2(2) \quad \text{Let } x = 0 \text{ and } y = 2.$$
$$0 > 4 \quad \text{False}$$

Step 3. Since the statement is false, we shade the half-plane that does not contain the test point $(0, 2)$, as shown.

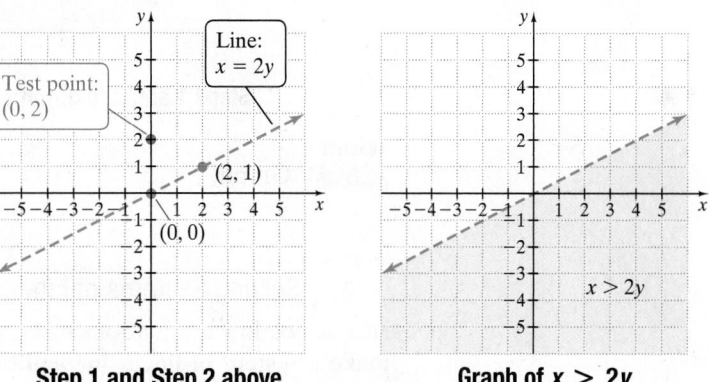

Step 1 and Step 2 above   **Graph of $x > 2y$**

PRACTICE
3 Graph: $x > 3y$

EXAMPLE 4 Graph: $5x + 4y \leq 20$

Solution We graph the solid boundary line $5x + 4y = 20$ and choose $(0, 0)$ as the test point.

$$5x + 4y \leq 20$$
$$5(0) + 4(0) \stackrel{?}{\leq} 20 \quad \text{Let } x = 0 \text{ and } y = 0.$$
$$0 \leq 20 \quad \text{True}$$

We shade the half-plane that contains $(0, 0)$, as shown.

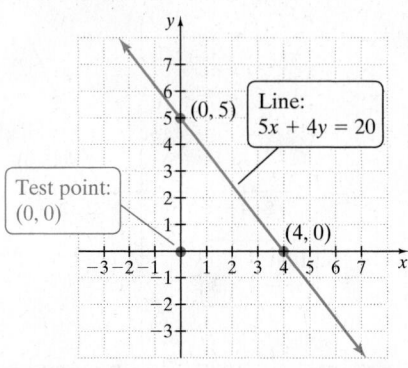 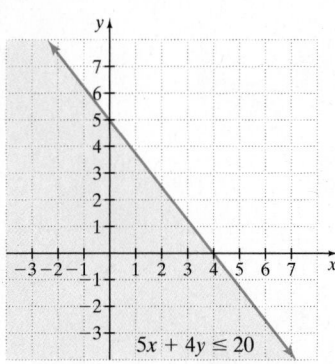

Steps 1 and 2 to graph $5x + 4y \leq 20$ Graph of $5x + 4y \leq 20$

PRACTICE
4 Graph: $3x + 4y \geq 12$

EXAMPLE 5 Graph: $y > 3$

Solution We graph the dashed boundary line $y = 3$ and choose $(0, 0)$ as the test point. (Recall that the graph of $y = 3$ is a horizontal line with y-intercept 3.)

$$y > 3$$
$$0 \stackrel{?}{>} 3 \quad \text{Let } y = 0.$$
$$0 > 3 \quad \text{False}$$

We shade the half-plane that does not contain $(0, 0)$, as shown.

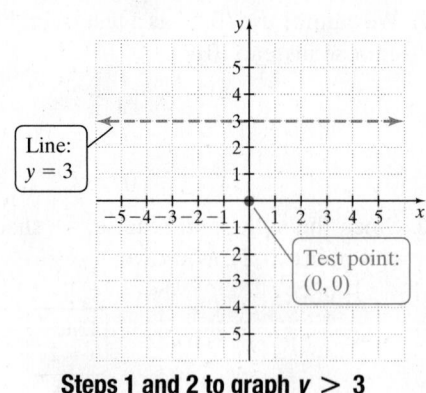

Steps 1 and 2 to graph $y > 3$ Graph of $y > 3$

PRACTICE
5 Graph: $x > 3$

OBJECTIVE
2 Solving Systems of Linear Inequalities

Just as two linear equations make a system of linear equations, two linear inequalities make a **system of linear inequalities.** Systems of inequalities are very important in a process called linear programming. Many businesses use linear programming to find the most profitable way to use limited resources such as employees, machines, or buildings.

A **solution of a system of linear inequalities** is an ordered pair that satisfies each inequality in the system. The set of all such ordered pairs is the solution set of the system. Graphing this set gives us a picture of the solution set. We can graph a system of inequalities by graphing each inequality in the system and identifying the region of overlap.

EXAMPLE 6   Graph the solution of the system: $\begin{cases} 3x \geq y \\ x + 2y \leq 8 \end{cases}$

Solution We begin by graphing each inequality on the same set of axes. The graph of the solution region of the system is the region contained in the graphs of both inequalities. It is their intersection.

First, graph $3x \geq y$. The boundary line is the graph of $3x = y$. Sketch a solid boundary line since the inequality $3x \geq y$ means $3x > y$ or $3x = y$. The test point $(1, 0)$ satisfies the inequality, so shade the half-plane that includes $(1, 0)$.

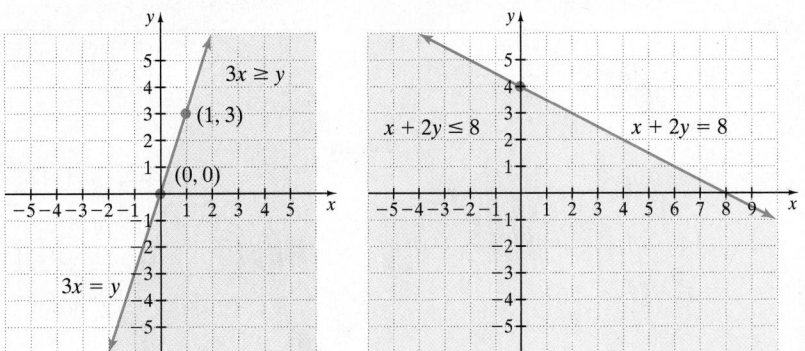

Next, graph $x + 2y \leq 8$ on the same set of axes. (For clarity, the graph of $x + 2y \leq 8$ is shown on a separate set of axes.) Sketch a solid boundary line $x + 2y = 8$. The test point $(0, 0)$ satisfies the inequality $x + 2y \leq 8$, so shade the half-plane that includes $(0, 0)$.

An ordered pair solution of the system must satisfy both inequalities. These solutions are points that lie in both shaded regions. The solution region of the system is the purple shaded region as seen below. This solution region includes parts of both boundary lines.

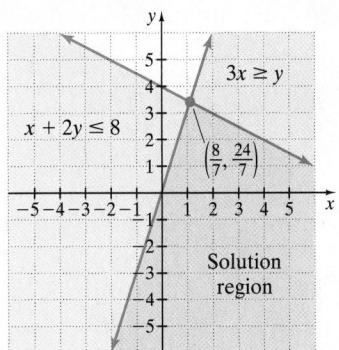

PRACTICE
6   Graph the solution of the system: $\begin{cases} 4x \leq y \\ x + 3y \geq 9 \end{cases}$

In linear programming, it is sometimes necessary to find the coordinates of the **corner point:** the point at which two boundary lines intersect. To find the point of intersection, solve the related linear system

$$\begin{cases} 3x = y \\ x + 2y = 8 \end{cases}$$

by the substitution method or the addition method. The lines intersect at $\left(\dfrac{8}{7}, \dfrac{24}{7}\right)$, the corner point of the graph.

> **Graphing the Solution Region of a System of Linear Inequalities**
>
> **Step 1.** Graph each inequality in the system on the same set of axes.
>
> **Step 2.** The solutions (or solution region) of the system are the points common to the graphs of all the inequalities in the system.

EXAMPLE 7   Graph the solution of the system: $\begin{cases} x - y < 2 \\ x + 2y > -1 \end{cases}$

Solution Graph both inequalities on the same set of axes. Both boundary lines are dashed lines since the inequality symbols are $<$ and $>$. The solution region of the system is the region shown by the purple shading. In this example, the boundary lines are not a part of the solution.

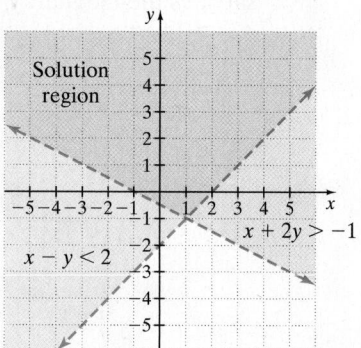

PRACTICE
7   Graph the solution of the system: $\begin{cases} x - y > 4 \\ x + 3y < -4 \end{cases}$

EXAMPLE 8   Graph the solution of the system: $\begin{cases} -3x + 4y < 12 \\ x \geq 2 \end{cases}$

Solution Graph both inequalities on the same set of axes.

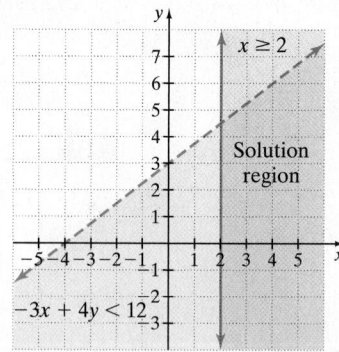

The solution region of the system is the purple shaded region, including a portion of the line $x = 2$.

PRACTICE
8   Graph the solution of the system: $\begin{cases} y \leq 6 \\ -2x + 5y > 10 \end{cases}$

✓ Vocabulary, Readiness & Video Check

Use the choices below to fill in each blank. Some choices may be used more than once and some not at all.

true	$x < 3$	$y < 3$	half-planes	yes
false	$x \leq 3$	$y \leq 3$	linear inequality in two variables	no

1. The statement $5x - 6y < 7$ is an example of a(n) _____.

2. A boundary line divides a plane into two regions called _____.

3. True or false: The graph of $5x - 6y < 7$ includes its corresponding boundary line. _____

4. True or false: When graphing a linear inequality, to determine which side of the boundary line to shade, choose a point *not* on the boundary line. _____

5. True or false: The boundary line for the inequality $5x - 6y < 7$ is the graph of $5x - 6y = 7$. _____

6. The graph of _____ is

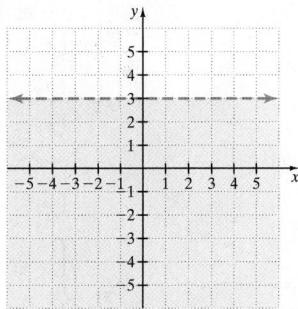

State whether the graph of each inequality includes its corresponding boundary line. Answer yes or no.

7. $y \geq x + 4$ **8.** $x - y > -7$ **9.** $y \geq x$ **10.** $x > 0$

Martin-Gay Interactive Videos

See Video 9.4

Watch the section lecture video and answer the following questions.

OBJECTIVE 1

11. From ▦ Example 1, how do you find the equation of the boundary line? How do you determine if the points on the boundary line are solutions of the inequality?

OBJECTIVE 2

12. In ▦ Example 2, did the graph of the first inequality of the system limit where we could choose the test point for the second inequality? Why or why not?

9.4 Exercise Set MyMathLab ▶

Determine whether the ordered pairs given are solutions of the linear inequality in two variables.

1. $x - y > 3$; $(2, -1)$, $(5, 1)$

2. $y - x < -2$; $(2, 1)$, $(5, -1)$

3. $3x - 5y \leq -4$; $(-1, -1)$, $(4, 0)$

4. $2x + y \geq 10$; $(-1, -4)$, $(5, 0)$

5. $x < -y$; $(0, 2)$, $(-5, 1)$

6. $y > 3x$; $(0, 0)$, $(-1, -4)$

MIXED PRACTICE

Graph each inequality. See Examples 1 through 5.

7. $x + y \leq 1$ **8.** $x + y \geq -2$

9. $2x + y > -4$ **10.** $x + 3y \leq 3$

11. $x + 6y \leq -6$ **12.** $7x + y > -14$

13. $2x + 5y > -10$ **14.** $5x + 2y \leq 10$

15. $x + 2y \leq 3$ **16.** $2x + 3y > -5$

▶ **17.** $2x + 7y > 5$ **18.** $3x + 5y \le -2$

19. $x - 2y \ge 3$ **20.** $4x + y \le 2$

21. $5x + y < 3$ **22.** $x + 2y > -7$

23. $4x + y < 8$ **24.** $9x + 2y \ge -9$

25. $y \ge 2x$ **26.** $x < 5y$ **27.** $x \ge 0$

28. $y \le 0$ **29.** $y \le -3$ **30.** $x > -\dfrac{2}{3}$

31. $2x - 7y > 0$ **32.** $5x + 2y \le 0$ **33.** $3x - 7y \ge 0$

34. $-2x - 9y > 0$ **35.** $x > y$ **36.** $x \le -y$

37. $x - y \le 6$ **38.** $x - y > 10$ **39.** $-\dfrac{1}{4}y + \dfrac{1}{3}x > 1$

40. $\dfrac{1}{2}x - \dfrac{1}{3}y \le -1$ **41.** $-x < 0.4y$ **42.** $0.3x \ge 0.1y$

In Exercises 43–48, match each graph with its inequality.

  a. $x > 2$ **b.** $y < 2$ **c.** $y < 2x$

  d. $y \le -3x$ **e.** $2x + 3y < 6$ **f.** $3x + 2y > 6$

43.
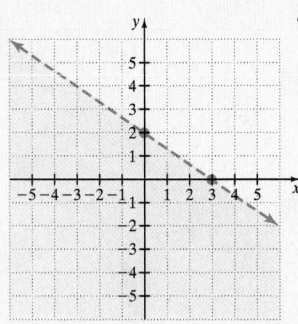

44.
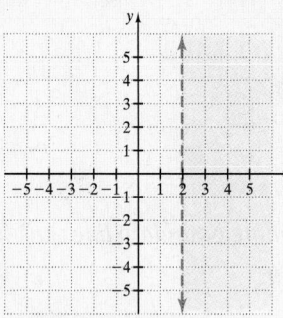

45.
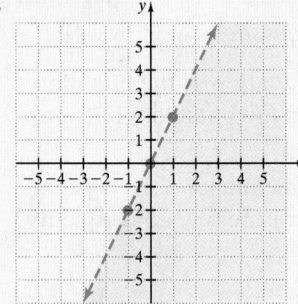

46.
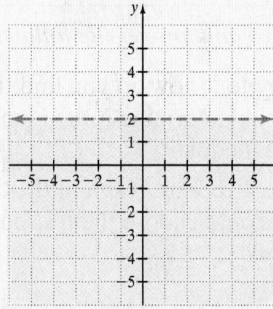

47.
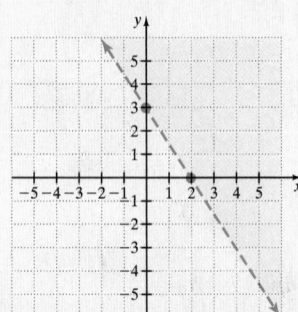

48.
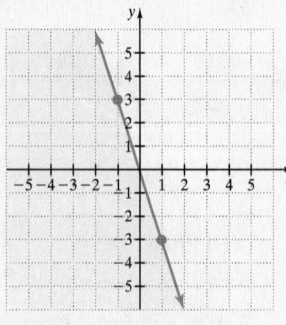

Graph the solution of each system of linear inequalities. See Examples 6 through 8.

49. $\begin{cases} y \ge x + 1 \\ y \ge 3 - x \end{cases}$ **50.** $\begin{cases} y \ge x - 3 \\ y \ge -1 - x \end{cases}$

51. $\begin{cases} y < 3x - 4 \\ y \le x + 2 \end{cases}$ **52.** $\begin{cases} y \le 2x + 1 \\ y > x + 2 \end{cases}$

53. $\begin{cases} y \le -2x - 2 \\ y \ge x + 4 \end{cases}$ **54.** $\begin{cases} y \le 2x + 4 \\ y \ge -x - 5 \end{cases}$

55. $\begin{cases} y \ge -x + 2 \\ y \le 2x + 5 \end{cases}$ **56.** $\begin{cases} y \ge x - 5 \\ y \le -3x + 3 \end{cases}$

▶ **57.** $\begin{cases} x \ge 3y \\ x + 3y \le 6 \end{cases}$ **58.** $\begin{cases} -2x < y \\ x + 2y < 3 \end{cases}$

59. $\begin{cases} y + 2x \ge 0 \\ 5x - 3y \le 12 \end{cases}$ **60.** $\begin{cases} y + 2x \le 0 \\ 5x + 3y \ge -2 \end{cases}$

61. $\begin{cases} 3x - 4y \ge -6 \\ 2x + y \le 7 \end{cases}$ **62.** $\begin{cases} 4x - y \ge -2 \\ 2x + 3y \le -8 \end{cases}$

63. $\begin{cases} x \le 2 \\ y \ge -3 \end{cases}$ **64.** $\begin{cases} x \ge -3 \\ y \ge -2 \end{cases}$

▶ **65.** $\begin{cases} y \ge 1 \\ x < -3 \end{cases}$ **66.** $\begin{cases} y > 2 \\ x \ge -1 \end{cases}$

67. $\begin{cases} 2x + 3y < -8 \\ x \ge -4 \end{cases}$ **68.** $\begin{cases} 3x + 2y \le 6 \\ x < 2 \end{cases}$

69. $\begin{cases} 2x - 5y \le 9 \\ y \le -3 \end{cases}$ **70.** $\begin{cases} 2x + 5y \le -10 \\ y \ge 1 \end{cases}$

71. $\begin{cases} y \ge \dfrac{1}{2}x + 2 \\ y \le \dfrac{1}{2}x - 3 \end{cases}$ **72.** $\begin{cases} y \ge -\dfrac{3}{2}x + 3 \\ y < -\dfrac{3}{2}x + 6 \end{cases}$

REVIEW AND PREVIEW

Evaluate each expression for the given replacement value. See Section 5.1.

73. x^2 if x is -5 **74.** x^3 if x is -5

75. $2x^3$ if x is -1 **76.** $3x^2$ if x is -1

CONCEPT EXTENSIONS

Determine whether $(1, 1)$ is included in each graph. See the Concept Check in this section.

77. $3x + 4y < 8$ **78.** $y > 5x$

79. $y \ge -\dfrac{1}{2}x$ **80.** $x > 3$

81. Write an inequality whose solutions are all pairs of numbers x and y whose sum is at least 13. Graph the inequality.

82. Write an inequality whose solutions are all the pairs of numbers x and y whose sum is at most -4. Graph the inequality.

83. Explain why a point on the boundary line should not be chosen as the test point.

84. Describe the graph of a linear inequality.

85. The price for a taxi cab in a small city is $2.50 per mile, x, while traveling, and $0.25 every minute, y, while waiting. If you have $20 to spend on a cab ride, the inequality

$$2.5x + 0.25y \le 20$$

represents your situation. Graph this inequality in the first quadrant only.

86. A word processor charges \$22 per hour, x, for typing a first draft, and \$15 per hour, y, for making changes and typing a second draft. If you need a document typed and have \$100, the inequality

$$22x + 15y \leq 100$$

represents your situation. Graph the inequality in the first quadrant only.

87. In Exercises 85 and 86, why were you instructed to graph each inequality in the first quadrant only?

88. Scott Sambracci and Sara Thygeson are planning their wedding. They have calculated that they want the cost of their wedding ceremony, x, plus the cost of their reception, y, to be no more than \$5000.

  a. Write an inequality describing this relationship.

  b. Graph this inequality.

  c. Why should we be interested in only quadrant I of this graph?

89. It's the end of the budgeting period for Dennis Fernandes, and he has \$500 left in his budget for car rental expenses. He plans to spend this budget on a sales trip throughout southern Texas. He will rent a car that costs \$30 per day and \$0.15 per mile, and he can spend no more than \$500.

  a. Write an inequality describing this situation. Let x = number of days and let y = number of miles.

  b. Graph this inequality.

  c. Why should we be interested in only quadrant I of this graph?

90. Explain how to decide which region to shade to show the solution region of the following system.

$$\begin{cases} x \geq 3 \\ y \geq -2 \end{cases}$$

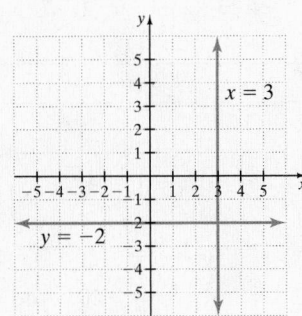

For each system of inequalities, choose the corresponding graph.

91. $\begin{cases} y < 5 \\ x > 3 \end{cases}$ **92.** $\begin{cases} y > 5 \\ x < 3 \end{cases}$ **93.** $\begin{cases} y \leq 5 \\ x < 3 \end{cases}$

94. $\begin{cases} y > 5 \\ x \geq 3 \end{cases}$

A. 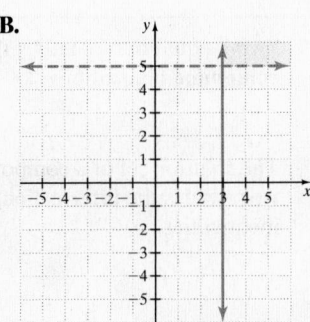   **B.**

C. 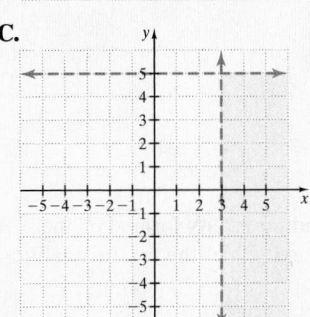   **D.**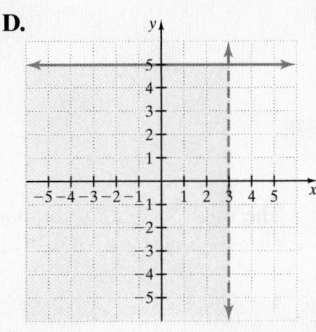

95. Graph the solution: $\begin{cases} 2x - y \leq 6 \\ x \geq 3 \\ y > 2 \end{cases}$

96. Graph the solution: $\begin{cases} x + y < 5 \\ y < 2x \\ x \geq 0 \\ y \geq 0 \end{cases}$

97. Describe the location of the solution region of the system.

$$\begin{cases} x > 0 \\ y > 0 \end{cases}$$

Chapter 9 Vocabulary Check

Fill in each blank with one of the words or phrases listed below.

compound inequality	solution	system of linear inequalities
absolute value	union	intersection

1. The statement "$x < 5$ *or* $x > 7$" is called a(n) _____.

2. The _____ of two sets is the set of all elements common to both sets.

3. The _____ of two sets is the set of all elements that belong to either of the sets.

4. A number's distance from 0 is called its _____.

5. When a variable in an equation is replaced by a number and the resulting equation is true, then that number is called a(n) _____ of the equation.

6. Two or more linear inequalities are called a(n) _____.

Chapter 9 Highlights

DEFINITIONS AND CONCEPTS	EXAMPLES

Section 9.1   Compound Inequalities

Two inequalities joined by the word **and** or **or** are called **compound inequalities.**

Compound Inequalities

$$x - 7 \leq 4 \quad and \quad x \geq -21$$
$$2x + 7 > x - 3 \quad or \quad 5x + 2 > -3$$

The solution set of a compound inequality formed by the word **and** is the **intersection,** $\cap$, of the solution sets of the two inequalities.

Solve for x:

$x < 5$ *and* $x < 3$

$\{x \mid x < 5\}$

$(-\infty, 5)$

$\{x \mid x < 3\}$

$(-\infty, 3)$

$\{x \mid x < 3$
and $x < 5\}$

$(-\infty, 3)$

The solution set of a compound inequality formed by the word **or** is the **union,** $\cup$, of the solution sets of the two inequalities.

Solve for x:

$$x - 2 \geq -3 \quad or \quad 2x \leq -4$$
$$x \geq -1 \quad or \quad x \leq -2$$

$\{x \mid x \geq -1\}$

$[-1, \infty)$

$\{x \mid x \leq -2\}$

$(-\infty, -2]$

$\{x \mid x \leq -2$
or $x \geq -1\}$

$(-\infty, -2]$
$\cup [-1, \infty)$

Section 9.2   Absolute Value Equations

If a is a positive number, then $|x| = a$ is equivalent to $x = a$ or $x = -a$.

Solve for y:

$$|5y - 1| - 7 = 4$$

$$|5y - 1| = 11 \qquad \text{Add 7.}$$
$$5y - 1 = 11 \quad or \quad 5y - 1 = -11$$
$$5y = 12 \quad or \quad 5y = -10 \quad \text{Add 1.}$$
$$y = \frac{12}{5} \quad or \quad y = -2 \quad \text{Divide by 5.}$$

The solutions are -2 and $\frac{12}{5}$.

If a is negative, then $|x| = a$ has no solution.

Solve for x:

$$\left| \frac{x}{2} - 7 \right| = -1$$

The solution set is $\{ \ \}$ or $\varnothing$.

DEFINITIONS AND CONCEPTS	EXAMPLES

Section 9.2 Absolute Value Equations (continued)

If an absolute value equation is of the form $\|x\| = \|y\|$, solve $x = y$ or $x = -y$.	Solve for x: $\|x - 7\| = \|2x + 1\|$ $\quad x - 7 = 2x + 1 \quad$ or $\quad x - 7 = -(2x + 1)$ $\quad\quad x = 2x + 8 \quad$ or $\quad x - 7 = -2x - 1$ $\quad\quad -x = 8 \quad$ or $\quad\quad\quad x = -2x + 6$ $\quad\quad x = -8 \quad$ or $\quad\quad\quad 3x = 6$ $\quad\quad\quad\quad\quad\quad\quad\quad\quad\quad\quad x = 2$ The solutions are -8 and 2.

Section 9.3 Absolute Value Inequalities

If a is a positive number, then $\|x\| < a$ is equivalent to $-a < x < a$.	Solve for y: $\|y - 5\| \le 3$ $\quad -3 \le y - 5 \le 3$ $\quad -3 + 5 \le y - 5 + 5 \le 3 + 5 \quad$ Add 5. $\quad\quad 2 \le y \le 8$ The solution set is $[2, 8]$.
If a is a positive number, then $\|x\| > a$ is equivalent to $x < -a$ or $x > a$.	Solve for x: $\left\|\dfrac{x}{2} - 3\right\| > 7$ $\dfrac{x}{2} - 3 < -7 \quad$ or $\quad \dfrac{x}{2} - 3 > 7$ $x - 6 < -14 \quad$ or $\quad x - 6 > 14 \quad$ Multiply by 2. $\quad\quad x < -8 \quad$ or $\quad\quad x > 20 \quad\quad$ Add 6. The solution set is $(-\infty, -8) \cup (20, \infty)$.

Section 9.4 Graphing Linear Inequalities in Two Variables and Systems of Linear Inequalities

A **linear inequality in two variables** is an inequality that can be written in one of the forms: $\quad Ax + By < C \quad\quad Ax + By \le C$ $\quad Ax + By > C \quad\quad Ax + By \ge C$	Linear Inequalities $\quad\quad 2x - 5y < 6 \quad\quad x \ge -5$ $\quad\quad\quad y > -8x \quad\quad y \le 2$
To graph a linear inequality 1. Graph the boundary line by graphing the related equation. Draw the line solid if the inequality symbol is $\le$ or $\ge$. Draw the line dashed if the inequality symbol is $<$ or $>$. 2. Choose a test point not on the line. Substitute its coordinates into the original inequality. 3. If the resulting inequality is true, shade the half-plane that contains the test point. If the inequality is not true, shade the half-plane that does not contain the test point.	Graph $2x - y \le 4$. 1. Graph $2x - y = 4$. Draw a solid line because the inequality symbol is $\le$. 2. Check the test point $(0, 0)$ in the inequality $2x - y \le 4$. $\quad\quad 2 \cdot 0 - 0 \le 4 \quad$ Let $x = 0$ and $y = 0$. $\quad\quad\quad 0 \le 4 \quad$ True 3. The inequality is true, so we shade the half-plane containing $(0, 0)$. 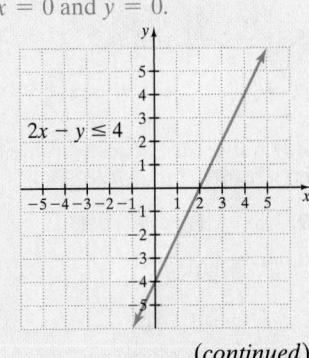

(continued)

DEFINITIONS AND CONCEPTS	EXAMPLES

Section 9.4 Graphing Linear Inequalities in Two Variables and Systems of Linear Inequalities (continued)

A system of linear inequalities consists of two or more linear inequalities.

To graph a system of inequalities, graph each inequality in the system. The overlapping region is the solution of the system.

System of Linear Inequalities

$$\begin{cases} x - y \geq 3 \\ y \leq -2x \end{cases}$$

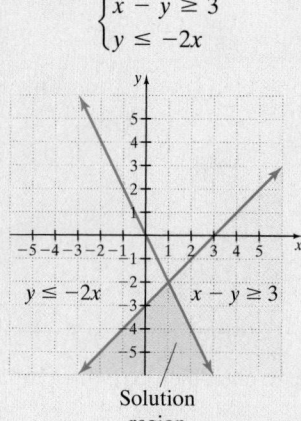

Solution region

Chapter 9 Review

(9.1) Solve each inequality. Write your answers in interval notation.

1. $-3 < 4(2x - 1) < 12$

2. $-2 \leq 8 + 5x < -1$

3. $\dfrac{1}{6} < \dfrac{4x - 3}{3} \leq \dfrac{4}{5}$

4. $-6 < x - (3 - 4x) < -3$

5. $3x - 5 > 6 \ or \ -x < -5$

6. $x \leq 2 \ and \ x > -5$

(9.2) Solve each absolute value equation.

7. $|8 - x| = 3$ **8.** $|x - 7| = 9$

9. $|-3x + 4| = 7$ **10.** $|2x + 9| = 9$

11. $5 + |6x + 1| = 5$ **12.** $|3x - 2| + 6 = 10$

13. $|5 - 6x| + 8 = 3$ **14.** $-5 = |4x - 3|$

15. $\left|\dfrac{3x - 7}{4}\right| = 2$ **16.** $-4 = \left|\dfrac{x - 3}{2}\right| - 5$

17. $|6x + 1| = |15 + 4x|$ **18.** $|x - 3| = |x + 5|$

(9.3) Solve each absolute value inequality. Graph the solution set and write it in interval notation.

19. $|5x - 1| < 9$

20. $|6 + 4x| \geq 10$

21. $|3x| - 8 > 1$

22. $9 + |5x| < 24$

23. $|6x - 5| \leq -1$

24. $|6x - 5| \leq 5$

25. $\left|3x + \dfrac{2}{5}\right| \geq 4$

26. $|5x - 3| > 2$

27. $\left|\dfrac{x}{3} + 6\right| - 8 > -5$

28. $\left|\dfrac{4(x - 1)}{7}\right| + 10 < 2$

(9.4) Graph the following inequalities.

29. $3x - 4y \leq 0$ **30.** $3x - 4y \geq 0$

31. $x + 6y < 6$ **32.** $y \leq -4$

33. $y \geq -7$ **34.** $x \geq -y$

Graph the solution region of the following systems of linear inequalities.

35. $\begin{cases} y \geq 2x - 3 \\ y \leq -2x + 1 \end{cases}$ **36.** $\begin{cases} y \leq -3x - 3 \\ y \leq 2x + 7 \end{cases}$

37. $\begin{cases} x + 2y > 0 \\ x - y \leq 6 \end{cases}$ **38.** $\begin{cases} 4x - y \leq 0 \\ 3x - 2y \geq -5 \end{cases}$

39. $\begin{cases} 3x - 2y \leq 4 \\ 2x + y \geq 5 \end{cases}$ **40.** $\begin{cases} -2x + 3y > -7 \\ x \geq -2 \end{cases}$

MIXED REVIEW

Solve. If an inequality, write your solutions in interval notation.

41. $0 \leq \dfrac{2(3x + 4)}{5} \leq 3$

42. $x \leq 2 \text{ or } x > -5$

43. $-2x \leq 6 \text{ and } -2x + 3 < -7$

44. $|7x| - 26 = -5$

45. $\left|\dfrac{9 - 2x}{5}\right| = -3$

46. $|x - 3| = |7 + 2x|$

47. $|6x - 5| \geq -1$

48. $\left|\dfrac{4x - 3}{5}\right| < 1$

Graph the solutions.

49. $-x \leq y$

50. $x + y > -2$

51. $\begin{cases} -3x + 2y > -1 \\ \qquad\quad y < -2 \end{cases}$

52. $\begin{cases} x - 2y \geq 7 \\ x + y \leq -5 \end{cases}$

Chapter 9 Getting Ready for the Test

MULTIPLE CHOICE *For Exercises 1 and 2, choose the correct interval notation for each set.*

1. $\{x \mid x \leq -11\}$

    A. $(-\infty, -11]$ **B.** $[-11, \infty)$ **C.** $[-11, 11]$ **D.** $(-\infty, -11)$

2. $\{x \mid -5 < x\}$

    A. $(-\infty, -5)$ **B.** $(-5, \infty)$ **C.** $(-5, 5)$ **D.** $(-\infty, -5]$

3. Choose the solution of $|x - 3| = 7$ by checking the given numbers in the original equation.

    A. $\{-7, 7\}$ **B.** $\{-10, 10\}$ **C.** $\{4, 10\}$ **D.** $\{-4, 10\}$

MATCHING *Match each equation or inequality with an equivalent statement. Letters may be used more than once or not at all.*

    A. $5x - 2 = 4 \text{ or } 5x - 2 = -4$ **B.** $5x = 6 \text{ or } 5x = -6$ **C.** $-4 \leq 5x - 2 \leq 4$

    D. $-6 \leq 5x \leq 6$ **E.** $5x - 2 \geq 4 \text{ or } 5x - 2 \leq -4$ **F.** $5x \geq 6 \text{ or } 5x \leq -6$

4. $|5x - 2| \leq 4$ **5.** $|5x - 2| = 4$

6. $|5x - 2| \geq 4$ **7.** $|5x| - 2 = 4$

MATCHING *Match each equation or inequality with its solution. Letters may be used more than once or not at all.*

    A. no solution, or $\varnothing$ **B.** $(-\infty, \infty)$

8. $|x + 3| = -9$ **9.** $|x + 3| < -9$ **10.** $|x + 3| > -9$

MATCHING *Match each system with the graph of its solution region.*

11. $\begin{cases} x \leq -3 \\ y \leq 3 \end{cases}$ **12.** $\begin{cases} x \leq -3 \\ y \geq 3 \end{cases}$ **13.** $\begin{cases} x \geq -3 \\ y \geq 3 \end{cases}$ **14.** $\begin{cases} x \geq -3 \\ y \leq 3 \end{cases}$

A. 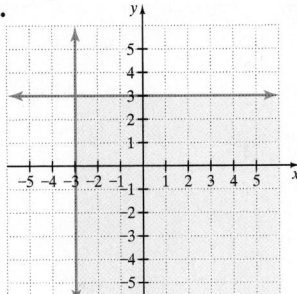   **B.** 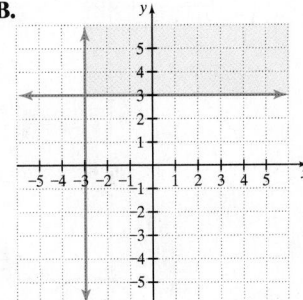   **C.** 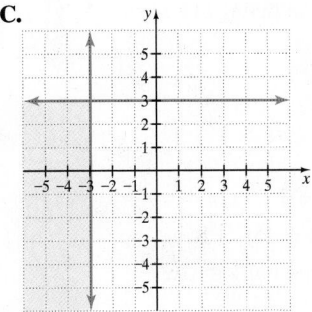   **D.**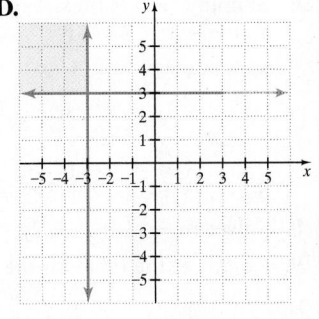

Chapter 9 Test MyMathLab® YouTube

Solve each equation or inequality.

▶ **1.** $|6x - 5| - 3 = -2$

▶ **2.** $|8 - 2t| = -6$

▶ **3.** $|x - 5| = |x + 2|$

▶ **4.** $-3 < 2(x - 3) \le 4$

▶ **5.** $|3x + 1| > 5$

▶ **6.** $|x - 5| - 4 < -2$

▶ **7.** $x \le -2$ and $x \le -5$

▶ **8.** $x \le -2$ or $x \le -5$

▶ **9.** $-x > 1$ and $3x + 3 \ge x - 3$

▶ **10.** $6x + 1 > 5x + 4$ or $1 - x > -4$

▶ **11.** $\left|\dfrac{5x - 7}{2}\right| = 4$

▶ **12.** $\left|17x - \dfrac{1}{5}\right| > -2$

▶ **13.** $-1 \le \dfrac{2x - 5}{3} < 2$

Graph each linear inequality.

▶ **14.** $y > -4x$

▶ **15.** $2x - 3y > -6$

Graph the solutions of the following systems of linear inequalities.

▶ **16.** $\begin{cases} y + 2x \le 4 \\ \qquad y \ge 2 \end{cases}$

▶ **17.** $\begin{cases} 2y - x \ge 1 \\ \ x + y \ge -4 \end{cases}$

Chapter 9 Cumulative Review

1. Find the value of each expression when $x = 2$ and $y = -5$.

a. $\dfrac{x - y}{12 + x}$ **b.** $x^2 - 3y$

2. Find the value of each expression when $x = -4$ and $y = 7$.

a. $\dfrac{x - y}{7 - x}$ **b.** $x^2 + 2y$

3. Simplify each expression.

a. $\dfrac{(-12)(-3) + 3}{-7 - (-2)}$ **b.** $\dfrac{2(-3)^2 - 20}{-5 + 4}$

4. Simplify each expression.

a. $\dfrac{4(-3) - (-6)}{-8 + 4}$ **b.** $\dfrac{3 + (-3)(-2)^3}{-1 - (-4)}$

5. Simplify each expression by combining like terms.

a. $2x + 3x + 5 + 2$

b. $-5a - 3 + a + 2$

c. $4y - 3y^2$

d. $2.3x + 5x - 6$

e. $-\dfrac{1}{2}b + b$

6. Simplify each expression by combining like terms.

a. $4x - 3 + 7 - 5x$

b. $-6y + 3y - 8 + 8y$

c. $2 + 8.1a + a - 6$

d. $2x^2 - 2x$

7. Solve: $2x + 3x - 5 + 7 = 10x + 3 - 6x - 4$

8. Solve: $6y - 11 + 4 + 2y = 8 + 15y - 8y$

9. Complete the table for the equation $y = 3x$.

x	y
-1	
	0
	-9

10. Complete the table for the equation $2x + y = 6$.

x	y
0	
	-2
3	

11. Identify the x- and y-intercepts.

a.
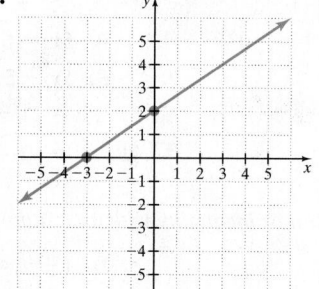

b.
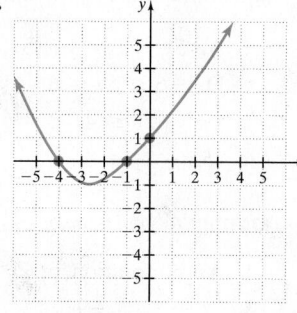

c.
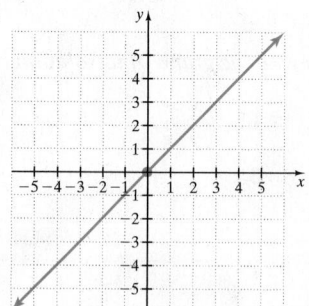

d.
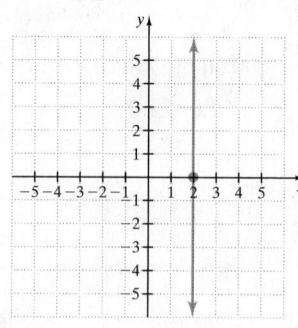

e.
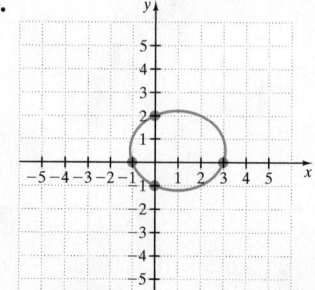

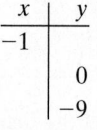

12. Identify the x- and y-intercepts.

a.

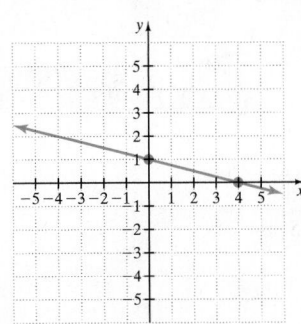

b.

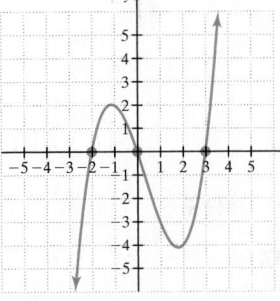

c.

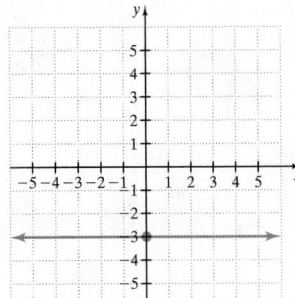

d.

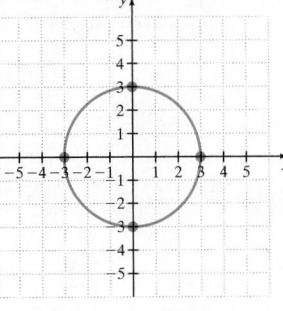

13. Determine whether the graphs of $y = -\frac{1}{5}x + 1$ and $2x + 10y = 3$ are parallel lines, perpendicular lines, or neither.

14. Determine whether the graphs of $y = 3x + 7$ and $x + 3y = -15$ are parallel lines, perpendicular lines, or neither.

15. Find an equation of the line with y-intercept $(0, -3)$ and slope of $\frac{1}{4}$.

16. Find an equation of the line with y-intercept $(0, 4)$ and slope of -2.

17. Find an equation of the line parallel to the line $y = 5$ and passing through $(-2, -3)$.

18. Find an equation of the line perpendicular to $y = 2x + 4$ and passing through $(1, 5)$.

19. Which of the following linear equations are functions?
   a. $y = x$
   b. $y = 2x + 1$
   c. $y = 5$
   d. $x = -1$

20. Which of the following linear equations are functions?
   a. $2x + 3 = y$ **b.** $x + 4 = 0$
   c. $\frac{1}{2}y = 2x$ **d.** $y = 0$

21. Determine whether $(12, 6)$ is a solution of the system.
$$\begin{cases} 2x - 3y = 6 \\ x = 2y \end{cases}$$

22. State whether each of the following ordered pairs is a solution of the system.
$$\begin{cases} 2x + y = 4 \\ x + y = 2 \end{cases}$$
   a. $(1, 1)$
   b. $(2, 0)$

23. Add $(11x^3 - 12x^2 + x - 3)$ and $(x^3 - 10x + 5)$.

24. Combine like terms to simplify.
$4a^2 + 3a - 2a^2 + 7a - 5$.

25. Factor: $x^2 + 7yx + 6y^2$

26. Factor: $3x^2 + 15x + 18$

27. Divide: $\dfrac{3x^3y^7}{40} \div \dfrac{4x^3}{y^2}$

28. Divide: $\dfrac{12x^2y^3}{5} \div \dfrac{3y^2}{x}$

29. Subtract: $\dfrac{2y}{2y - 7} - \dfrac{7}{2y - 7}$

30. Subtract: $\dfrac{-4x^2}{x + 1} - \dfrac{4x}{x + 1}$

31. Add: $\dfrac{2x}{x^2 + 2x + 1} + \dfrac{x}{x^2 - 1}$

32. Add: $\dfrac{3x}{x^2 + 5x + 6} + \dfrac{1}{x^2 + 2x - 3}$

33. Solve: $\dfrac{x}{2} + \dfrac{8}{3} = \dfrac{1}{6}$

34. Solve: $\dfrac{1}{21} + \dfrac{x}{7} = \dfrac{5}{3}$

35. Solve the following system of equations by graphing.
$$\begin{cases} 2x + y = 7 \\ 2y = -4x \end{cases}$$

36. Solve the following system by graphing.
$$\begin{cases} y = x + 2 \\ 2x + y = 5 \end{cases}$$

37. Solve the system.
$$\begin{cases} 7x - 3y = -14 \\ -3x + y = 6 \end{cases}$$

38. Solve the system.
$$\begin{cases} 5x + y = 3 \\ y = -5x \end{cases}$$

39. Solve the system.
$$\begin{cases} 3x - 2y = 2 \\ -9x + 6y = -6 \end{cases}$$

40. Solve the system.
$$\begin{cases} -2x + y = 7 \\ 6x - 3y = -21 \end{cases}$$

41. Graph the solution of the system.
$$\begin{cases} -3x + 4y < 12 \\ x \geq 2 \end{cases}$$

42. Graph the solution of the system.
$$\begin{cases} 2x - y \leq 6 \\ y \geq 2 \end{cases}$$

43. Simplify the following.

a. $x^7 \cdot x^4$

b. $\left(\dfrac{t}{2}\right)^4$

c. $(9y^5)^2$

44. Simplify.

a. $\left(\dfrac{-6x}{y^3}\right)^3$

b. $\dfrac{(2b^2)^5}{a^2 b^7}$

c. $\dfrac{(3y)^2}{y^2}$

d. $\dfrac{(x^2 y^4)^2}{xy^3}$

45. Solve: $(5x - 1)(2x^2 + 15x + 18) = 0$

46. Solve: $(x + 1)(2x^2 - 3x - 5) = 0$

47. Solve: $\dfrac{45}{x} = \dfrac{5}{7}$

48. Solve: $\dfrac{2x + 7}{3} = \dfrac{x - 6}{2}$

Rational Exponents, Radicals, and Complex Numbers

10.1 Radicals and Radical Functions

10.2 Rational Exponents

10.3 Simplifying Radical Expressions

10.4 Adding, Subtracting, and Multiplying Radical Expressions

10.5 Rationalizing Denominators and Numerators of Radical Expressions

Integrated Review—Radicals and Rational Exponents

10.6 Radical Equations and Problem Solving

10.7 Complex Numbers

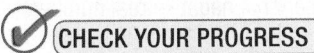

CHECK YOUR PROGRESS

Vocabulary Check
Chapter Highlights
Chapter Review
Getting Ready for the Test
Chapter Test
Cumulative Review

In this chapter, radical notation is reviewed, and then rational exponents are introduced. As the name implies, rational exponents are exponents that are rational numbers. We present an interpretation of rational exponents that is consistent with the meaning and rules already established for integer exponents, and we present two forms of notation for roots: radical and exponent. We conclude this chapter with complex numbers, a natural extension of the real number system.

What Is Zorbing?

New Zealand is the home of some great extreme adventures. For example, bungee jumping, base jumping, and zorbing originated in New Zealand. Invented in 1995, zorbing is climbing inside a large, double-chambered ball and rolling down a hillside track. The creators have also provided zorbs for a show at Sea World, San Diego, created snow zorbing, and produced the zurf, which is zorbing at the beach. Since its inception, zorbing has spread throughout the world, including a site in Amesbury, Massachusetts.

NASA has even investigated using a zorb-like rover to investigate Mars, called the tumbleweed rover.

In Section 10.3, Exercises 125 and 126, we will explore aspects of zorbs and zorbing.

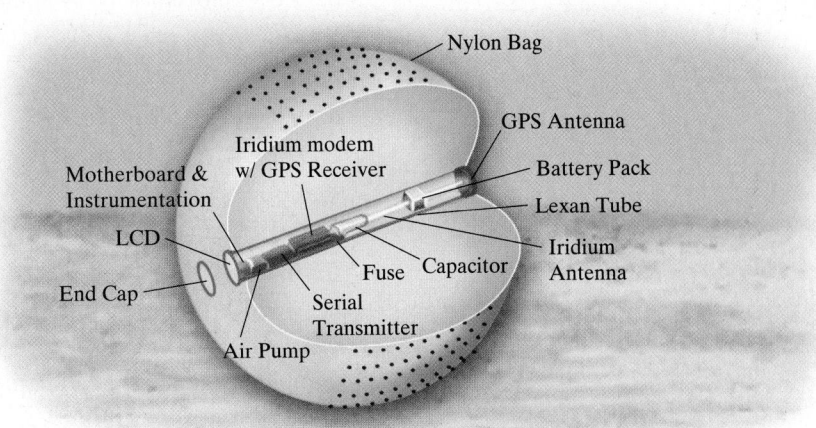

Tumbleweed Rover Architecture

If volume V or surface area A are known, then radius r can be calculated using

$$r = \sqrt[3]{\frac{3V}{4\pi}} \text{ or } r = \sqrt{\frac{A}{4\pi}}$$

10.1 Radicals and Radical Functions

OBJECTIVES

1 Find Square Roots. ▶

2 Approximate Roots. ▶

3 Find Cube Roots. ▶

4 Find nth Roots. ▶

5 Find $\sqrt[n]{a^n}$ Where a Is a Real Number. ▶

6 Graph Square and Cube Root Functions. ▶

OBJECTIVE

1 Finding Square Roots ▶

Recall from Section 8.2 that to find a **square root** of a number a, we find a number that was squared to get a.

Thus, because

$$5^2 = 25 \quad \text{and} \quad (-5)^2 = 25,$$

both 5 and -5 are square roots of 25.

Recall that we denote the **nonnegative**, or **principal, square root** with the **radical sign.**

$$\sqrt{25} = 5$$

We denote the **negative square root** with the **negative radical sign.**

$$-\sqrt{25} = -5$$

An expression containing a radical sign is called a **radical expression.** An expression within, or "under," a radical sign is called a **radicand.**

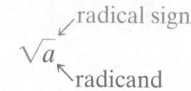

radical expression:

Principal and Negative Square Roots

If a is a nonnegative number, then

$\sqrt{a}$ is the **principal,** or **nonnegative, square root** of a

$-\sqrt{a}$ is the **negative square root** of a

▶ **EXAMPLE 1**   Simplify. Assume that all variables represent nonnegative real numbers.

a. $\sqrt{36}$ **b.** $\sqrt{0}$ **c.** $\sqrt{\dfrac{4}{49}}$ **d.** $\sqrt{0.25}$

e. $\sqrt{x^6}$ **f.** $\sqrt{9x^{12}}$ **g.** $-\sqrt{81}$ **h.** $\sqrt{-81}$

Solution

a. $\sqrt{36} = 6$ because $6^2 = 36$ and 6 is not negative.

b. $\sqrt{0} = 0$ because $0^2 = 0$ and 0 is not negative.

c. $\sqrt{\dfrac{4}{49}} = \dfrac{2}{7}$ because $\left(\dfrac{2}{7}\right)^2 = \dfrac{4}{49}$ and $\dfrac{2}{7}$ is not negative.

d. $\sqrt{0.25} = 0.5$ because $(0.5)^2 = 0.25.$

e. $\sqrt{x^6} = x^3$ because $(x^3)^2 = x^6.$

f. $\sqrt{9x^{12}} = 3x^6$ because $(3x^6)^2 = 9x^{12}.$

g. $-\sqrt{81} = -9.$ The negative in front of the radical indicates the negative square root of 81.

h. $\sqrt{-81}$ is not a real number.

PRACTICE

1   Simplify. Assume that all variables represent nonnegative real numbers.

a. $\sqrt{49}$ **b.** $\sqrt{\dfrac{0}{1}}$ **c.** $\sqrt{\dfrac{16}{81}}$ **d.** $\sqrt{0.64}$

e. $\sqrt{z^8}$ **f.** $\sqrt{16b^4}$ **g.** $-\sqrt{36}$ **h.** $\sqrt{-36}$

Recall from Section 8.2 our discussion of the square root of a negative number. For example, can we simplify $\sqrt{-4}$? That is, can we find a real number whose square is -4? No, there is no real number whose square is -4, and we say that $\sqrt{-4}$ is not a real number. In general:

The square root of a negative number is not a real number.

Helpful Hint

- Remember: $\sqrt{0} = 0$.
- Don't forget that the square root of a negative number is not a real number. For example,

$$\sqrt{-9} \text{ is not a real number}$$

because there is no real number that when multiplied by itself would give a product of -9. In Section 10.7, we will see what kind of a number $\sqrt{-9}$ is.

OBJECTIVE

2 Approximating Roots

Recall that numbers such as 1, 4, 9, and 25 are called **perfect squares,** since $1 = 1^2, 4 = 2^2, 9 = 3^2$, and $25 = 5^2$. Square roots of perfect square radicands simplify to rational numbers. What happens when we try to simplify a root such as $\sqrt{3}$? Since there is no rational number whose square is 3, $\sqrt{3}$ is not a rational number. It is called an **irrational number,** and we can find a decimal **approximation** of it. To find decimal approximations, use a calculator. For example, an approximation for $\sqrt{3}$ is

$$\sqrt{3} \approx 1.732$$

approximation symbol

To see if the approximation is reasonable, notice that since

$$1 < 3 < 4,$$
$$\sqrt{1} < \sqrt{3} < \sqrt{4}, \text{ or}$$
$$1 < \sqrt{3} < 2.$$

1 and 4 are perfect squares closest to but less than and greater than 3, respectively.

We found $\sqrt{3} \approx 1.732$, a number between 1 and 2, so our result is reasonable.

EXAMPLE 2   Use a calculator to approximate $\sqrt{20}$. Round the approximation to 3 decimal places and check to see that your approximation is reasonable.

Solution $\sqrt{20} \approx 4.472$

Is this reasonable? Since $16 < 20 < 25$, $\sqrt{16} < \sqrt{20} < \sqrt{25}$, or $4 < \sqrt{20} < 5$. The approximation is between 4 and 5 and thus is reasonable.   □

PRACTICE

2   Use a calculator to approximate $\sqrt{45}$. Round the approximation to three decimal places and check to see that your approximation is reasonable.   ▪

OBJECTIVE

3 Finding Cube Roots

Finding roots can be extended to other roots such as cube roots. For example, since $2^3 = 8$, we call 2 the **cube root** of 8. In symbols, we write

$$\sqrt[3]{8} = 2$$

Cube Root

The **cube root** of a real number a is written as $\sqrt[3]{a}$, and

$$\sqrt[3]{a} = b \text{ only if } b^3 = a$$

From this definition, we have

$$\sqrt[3]{64} = 4 \text{ since } 4^3 = 64$$
$$\sqrt[3]{-27} = -3 \text{ since } (-3)^3 = -27$$
$$\sqrt[3]{x^3} = x \text{ since } x^3 = x^3$$

Notice that, unlike with square roots, *it is possible to have a negative radicand when finding a cube root.* This is so because the *cube* of a negative number is a negative number. Therefore, the *cube root* of a negative number is a negative number.

EXAMPLE 3 Find the cube roots.

    a. $\sqrt[3]{1}$ **b.** $\sqrt[3]{-64}$ **c.** $\sqrt[3]{\dfrac{8}{125}}$ **d.** $\sqrt[3]{x^6}$ **e.** $\sqrt[3]{-27x^{15}}$

Solution

    a. $\sqrt[3]{1} = 1$ because $1^3 = 1$.

    b. $\sqrt[3]{-64} = -4$ because $(-4)^3 = -64$.

    c. $\sqrt[3]{\dfrac{8}{125}} = \dfrac{2}{5}$ because $\left(\dfrac{2}{5}\right)^3 = \dfrac{8}{125}$.

    d. $\sqrt[3]{x^6} = x^2$ because $(x^2)^3 = x^6$.

    e. $\sqrt[3]{-27x^{15}} = -3x^5$ because $(-3x^5)^3 = -27x^{15}$.

PRACTICE

3 Find the cube roots.

    a. $\sqrt[3]{-1}$ **b.** $\sqrt[3]{27}$ **c.** $\sqrt[3]{\dfrac{27}{64}}$ **d.** $\sqrt[3]{x^{12}}$ **e.** $\sqrt[3]{-8x^3}$

OBJECTIVE

4 Finding *n*th Roots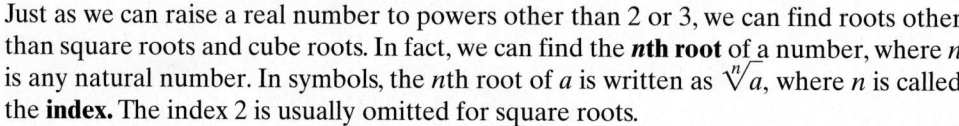

Just as we can raise a real number to powers other than 2 or 3, we can find roots other than square roots and cube roots. In fact, we can find the ***n*th root** of a number, where n is any natural number. In symbols, the *n*th root of a is written as $\sqrt[n]{a}$, where n is called the **index.** The index 2 is usually omitted for square roots.

> **Helpful Hint**
>
> If the index is even, such as $\sqrt{}, \sqrt[4]{}, \sqrt[6]{}$, and so on, the radicand must be nonnegative for the root to be a real number. For example,
>
> $$\sqrt[4]{16} = 2, \text{ but } \sqrt[4]{-16} \text{ is not a real number.}$$
> $$\sqrt[6]{64} = 2, \text{ but } \sqrt[6]{-64} \text{ is not a real number.}$$
>
> If the index is odd, such as $\sqrt[3]{}, \sqrt[5]{}$, and so on, the radicand may be any real number. For example,
>
> $$\sqrt[3]{64} = 4 \quad \text{and} \quad \sqrt[3]{-64} = -4$$
> $$\sqrt[5]{32} = 2 \quad \text{and} \quad \sqrt[5]{-32} = -2$$

> ✔ **CONCEPT CHECK**
> Which one is not a real number?
>
> **a.** $\sqrt[3]{-15}$ **b.** $\sqrt[4]{-15}$ **c.** $\sqrt[5]{-15}$ **d.** $\sqrt{(-15)^2}$

EXAMPLE 4   Simplify the following expressions.

    a. $\sqrt[4]{81}$ **b.** $\sqrt[5]{-243}$ **c.** $-\sqrt{25}$ **d.** $\sqrt[4]{-81}$ **e.** $\sqrt[3]{64x^3}$

Solution

    a. $\sqrt[4]{81} = 3$ because $3^4 = 81$ and 3 is positive.

    b. $\sqrt[5]{-243} = -3$ because $(-3)^5 = -243$.

    c. $-\sqrt{25} = -5$ because -5 is the opposite of $\sqrt{25}$.

    d. $\sqrt[4]{-81}$ is not a real number. There is no real number that, when raised to the fourth power, is -81.

    e. $\sqrt[3]{64x^3} = 4x$ because $(4x)^3 = 64x^3$.   □

PRACTICE
4   Simplify the following expressions.

    a. $\sqrt[4]{10,000}$ **b.** $\sqrt[5]{-1}$ **c.** $-\sqrt{81}$ **d.** $\sqrt[4]{-625}$ **e.** $\sqrt[3]{27x^9}$ ∎

OBJECTIVE
5   Finding $\sqrt[n]{a^n}$ Where a Is a Real Number ▷

Recall that the notation $\sqrt{a^2}$ indicates the positive square root of a^2 only. For example,

$$\sqrt{(-7)^2} = \sqrt{49} = 7$$

    When variables are present in the radicand and it is _unclear whether the variable represents a positive number or a negative number_, absolute value bars are sometimes needed to ensure that the result is a positive number. For example,

$$\sqrt{x^2} = |x|$$

This ensures that the result is positive. This same situation may occur when the index is any _even_ positive integer. When the index is any _odd_ positive integer, absolute value bars are not necessary.

Finding $\sqrt[n]{a^n}$

If n is an _even_ positive integer, then $\sqrt[n]{a^n} = |a|$.

If n is an _odd_ positive integer, then $\sqrt[n]{a^n} = a$.

EXAMPLE 5   Simplify.

    a. $\sqrt{(-3)^2}$ **b.** $\sqrt{x^2}$ **c.** $\sqrt[4]{(x-2)^4}$ **d.** $\sqrt[3]{(-5)^3}$

    e. $\sqrt[5]{(2x-7)^5}$ **f.** $\sqrt{25x^2}$ **g.** $\sqrt{x^2 + 2x + 1}$

Solution

    a. $\sqrt{(-3)^2} = |-3| = 3$ When the index is even, the absolute value bars ensure that our result is not negative.

    b. $\sqrt{x^2} = |x|$

    c. $\sqrt[4]{(x-2)^4} = |x-2|$

    d. $\sqrt[3]{(-5)^3} = -5$

    e. $\sqrt[5]{(2x-7)^5} = 2x - 7$ Absolute value bars are not needed when the index is odd.

    f. $\sqrt{25x^2} = 5|x|$

    g. $\sqrt{x^2 + 2x + 1} = \sqrt{(x+1)^2} = |x+1|$ □

(Continued on next page)

5 Simplify.

a. $\sqrt{(-4)^2}$ b. $\sqrt{x^{14}}$ c. $\sqrt[4]{(x + 7)^4}$ d. $\sqrt[3]{(-7)^3}$

e. $\sqrt[5]{(3x - 5)^5}$ f. $\sqrt{49x^2}$ g. $\sqrt{x^2 + 16x + 64}$

OBJECTIVE

6   Graphing Square and Cube Root Functions

Recall that an equation in x and y describes a function if each x-value is paired with exactly one y-value. With this in mind, does the equation

$$y = \sqrt{x}$$

describe a function? First, notice that replacement values for x must be nonnegative real numbers, since $\sqrt{x}$ is not a real number if $x < 0$. The notation $\sqrt{x}$ denotes the principal square root of x, so for every nonnegative number x, there is exactly one number, $\sqrt{x}$. Therefore, $y = \sqrt{x}$ describes a function, and we may write it as

$$f(x) = \sqrt{x}$$

In general, radical functions are functions of the form

$$f(x) = \sqrt[n]{x}.$$

    Recall that the domain of a function in x is the set of all possible replacement values for x. This means that if n is even, the domain is the set of all nonnegative numbers, or $\{x | x \geq 0\}$ or $[0, \infty)$. If n is odd, the domain is the set of all real numbers, or $(-\infty, \infty)$. Keep this in mind as we find function values.

EXAMPLE 6   If $f(x) = \sqrt{x - 4}$ and $g(x) = \sqrt[3]{x + 2}$, find each function value.

a. $f(8)$ b. $f(6)$ c. $g(-1)$ d. $g(1)$

Solution

a. $f(8) = \sqrt{8 - 4} = \sqrt{4} = 2$ b. $f(6) = \sqrt{6 - 4} = \sqrt{2}$

c. $g(-1) = \sqrt[3]{-1 + 2} = \sqrt[3]{1} = 1$ d. $g(1) = \sqrt[3]{1 + 2} = \sqrt[3]{3}$

6   If $f(x) = \sqrt{x + 5}$ and $g(x) = \sqrt[3]{x - 3}$, find each function value.

a. $f(11)$ b. $f(-1)$ c. $g(11)$ d. $g(-6)$

> **Helpful Hint**
>
> Notice that for the function $f(x) = \sqrt{x - 4}$, the domain includes all real numbers that make the radicand ≥ 0. To see what numbers these are, solve $x - 4 \geq 0$ and find that $x \geq 4$. The domain is $\{x | x \geq 4\}$, or $[4, \infty)$.
>   The domain of the cube root function $g(x) = \sqrt[3]{x + 2}$ is the set of real numbers, or $(-\infty, \infty)$.

EXAMPLE 7   Graph the square root function: $f(x) = \sqrt{x}$

Solution To graph, we identify the domain, evaluate the function for several values of x, plot the resulting points, and connect the points with a smooth curve. Since $\sqrt{x}$ is not a real number for negative values of x, the domain of this function is the set of all nonnegative numbers, $\{x | x \geq 0\}$, or $[0, \infty)$. We have approximated $\sqrt{3}$ in the table on the next page to help us locate the point corresponding to $(3, \sqrt{3})$.

x	$f(x) = \sqrt{x}$
0	0
1	1
3	$\sqrt{3} \approx 1.7$
4	2
9	3

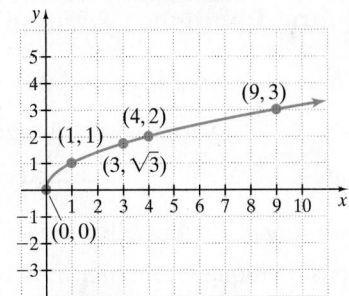

Notice that the graph of this function passes the vertical line test, as expected.   □

PRACTICE
7   Graph the square root function: $h(x) = \sqrt{x + 2}$

The equation $f(x) = \sqrt[3]{x}$ also describes a function. Here, x may be any real number, so the domain of this function is the set of all real numbers, or $(-\infty, \infty)$. A few function values are given next.

$$f(0) = \sqrt[3]{0} = 0$$
$$f(1) = \sqrt[3]{1} = 1$$
$$f(-1) = \sqrt[3]{-1} = -1$$
$$f(6) = \sqrt[3]{6}$$
$$f(-6) = \sqrt[3]{-6}$$
$$f(8) = \sqrt[3]{8} = 2$$
$$f(-8) = \sqrt[3]{-8} = -2$$

Here, there is no rational number whose cube is 6. Thus, the radicals do not simplify to rational numbers.

EXAMPLE 8   Graph the function: $f(x) = \sqrt[3]{x}$

Solution   To graph, we identify the domain, plot points, and connect the points with a smooth curve. The domain of this function is the set of all real numbers. The table comes from the function values obtained earlier. We have approximated $\sqrt[3]{6}$ and $\sqrt[3]{-6}$ for graphing purposes.

x	$f(x) = \sqrt[3]{x}$
0	0
1	1
−1	−1
6	$\sqrt[3]{6} \approx 1.8$
−6	$\sqrt[3]{-6} \approx -1.8$
8	2
−8	−2

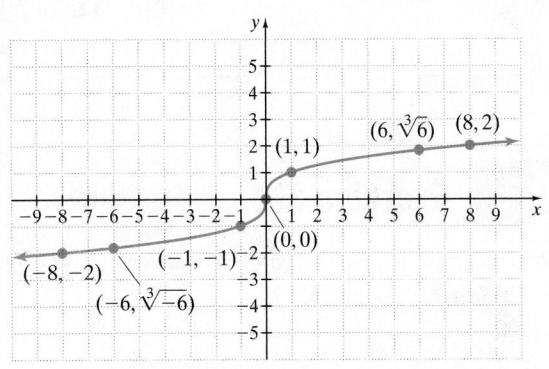

The graph of this function passes the vertical line test, as expected.   □

PRACTICE
8   Graph the function: $f(x) = \sqrt[3]{x} - 4$

✓ Vocabulary, Readiness & Video Check

Use the choices below to fill in each blank. Not all choices will be used.

is	cubes	$-\sqrt{a}$	radical sign	index
is not	squares	$\sqrt{-a}$	radicand	

1. In the expression $\sqrt[n]{a}$, the n is called the _____, the $\sqrt{}$ is called the _____, and a is called the _____.

2. If $\sqrt{a}$ is the positive square root of a, $a \neq 0$, then _____ is the negative square root of a.

3. The square root of a negative number _____ a real number.

4. Numbers such as 1, 4, 9, and 25 are called perfect _____, whereas numbers such as 1, 8, 27, and 125 are called perfect _____.

Fill in the blank.

5. The domain of the function $f(x) = \sqrt{x}$ is _____.

6. The domain of the function $f(x) = \sqrt[3]{x}$ is _____.

7. If $f(16) = 4$, the corresponding ordered pair is _____.

8. If $g(-8) = -2$, the corresponding ordered pair is _____.

Martin-Gay Interactive Videos

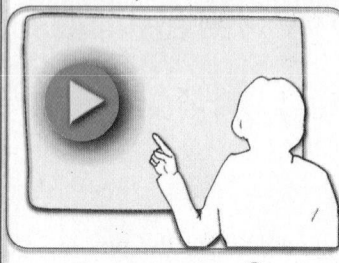

See Video 10.1 ⊙

Watch the section lecture video and answer the following questions.

OBJECTIVE 1
9. From ▦ Examples 5 and 6, when simplifying radicals containing variables with exponents, describe a shortcut you can use.

OBJECTIVE 2
10. From ▦ Example 9, how can you determine a reasonable approximation for a non-perfect square root without using a calculator?

OBJECTIVE 3
11. From ▦ Example 11, what is an important difference between the square root and the cube root of a negative number?

OBJECTIVE 4
12. From ▦ Example 12, what conclusion is made about the even root of a negative number?

OBJECTIVE 5
13. From the lecture before ▦ Example 17, why do you think no absolute value bars are used when n is odd?

OBJECTIVE 6
14. In ▦ Example 19, the domain is found by looking at the graph. How can the domain be found by looking at the function?

10.1 Exercise Set MyMathLab® ▶

Simplify. Assume that variables represent nonnegative real numbers. See Example 1.

1. $\sqrt{100}$ **2.** $\sqrt{400}$

3. $\sqrt{\dfrac{1}{4}}$ **4.** $\sqrt{\dfrac{9}{25}}$

5. $\sqrt{0.0001}$ **6.** $\sqrt{0.04}$

7. $-\sqrt{36}$ **8.** $-\sqrt{9}$

9. $\sqrt{x^{10}}$ **10.** $\sqrt{x^{16}}$

11. $\sqrt{16y^6}$ **12.** $\sqrt{64y^{20}}$

Use a calculator to approximate each square root to 3 decimal places. Check to see that each approximation is reasonable. See Example 2.

13. $\sqrt{7}$ **14.** $\sqrt{11}$

▶ **15.** $\sqrt{38}$ **16.** $\sqrt{56}$

17. $\sqrt{200}$ **18.** $\sqrt{300}$

Find each cube root. See Example 3.

19. $\sqrt[3]{64}$ **20.** $\sqrt[3]{27}$

▶ **21.** $\sqrt[3]{\dfrac{1}{8}}$ **22.** $\sqrt[3]{\dfrac{27}{64}}$

23. $\sqrt[3]{-1}$ **24.** $\sqrt[3]{-125}$

25. $\sqrt[3]{x^{12}}$ **26.** $\sqrt[3]{x^{15}}$

▶ **27.** $\sqrt[3]{-27x^9}$ **28.** $\sqrt[3]{-64x^6}$

Find each root. Assume that all variables represent nonnegative real numbers. See Example 4.

29. $-\sqrt[4]{16}$

30. $\sqrt[5]{-243}$

▶ **31.** $\sqrt[4]{-16}$

32. $\sqrt{-16}$

▶ **33.** $\sqrt[5]{-32}$

34. $\sqrt[5]{-1}$

35. $\sqrt[5]{x^{20}}$

36. $\sqrt[4]{x^{20}}$

▶ **37.** $\sqrt[6]{64x^{12}}$

38. $\sqrt[5]{-32x^{15}}$

39. $\sqrt{81x^4}$

40. $\sqrt[4]{81x^4}$

41. $\sqrt[4]{256x^8}$

42. $\sqrt{256x^8}$

Simplify. Assume that the variables represent any real number. See Example 5.

▶ **43.** $\sqrt{(-8)^2}$

44. $\sqrt{(-7)^2}$

▶ **45.** $\sqrt[3]{(-8)^3}$

46. $\sqrt[5]{(-7)^5}$

47. $\sqrt{4x^2}$

48. $\sqrt[4]{16x^4}$

49. $\sqrt[3]{x^3}$

50. $\sqrt[5]{x^5}$

▶ **51.** $\sqrt{(x-5)^2}$

52. $\sqrt{(y-6)^2}$

53. $\sqrt{x^2+4x+4}$
(*Hint:* Factor the polynomial first.)

54. $\sqrt{x^2-8x+16}$
(*Hint:* Factor the polynomial first.)

MIXED PRACTICE

Simplify each radical. Assume that all variables represent positive real numbers.

55. $-\sqrt{121}$

56. $-\sqrt[3]{125}$

57. $\sqrt[3]{8x^3}$

58. $\sqrt{16x^8}$

59. $\sqrt{y^{12}}$

60. $\sqrt[3]{y^{12}}$

61. $\sqrt{25a^2b^{20}}$

62. $\sqrt{9x^4y^6}$

63. $\sqrt[3]{-27x^{12}y^9}$

64. $\sqrt[3]{-8a^{21}b^6}$

65. $\sqrt[4]{a^{16}b^4}$

66. $\sqrt[4]{x^8y^{12}}$

▶ **67.** $\sqrt[5]{-32x^{10}y^5}$

68. $\sqrt[5]{-243x^5z^{15}}$

69. $\sqrt{\dfrac{25}{49}}$

70. $\sqrt{\dfrac{4}{81}}$

71. $\sqrt{\dfrac{x^{20}}{4y^2}}$

72. $\sqrt{\dfrac{y^{10}}{9x^6}}$

73. $-\sqrt[3]{\dfrac{z^{21}}{27x^3}}$

74. $-\sqrt[3]{\dfrac{64a^3}{b^9}}$

75. $\sqrt[4]{\dfrac{x^4}{16}}$

76. $\sqrt[4]{\dfrac{y^4}{81x^4}}$

If $f(x) = \sqrt{2x+3}$ and $g(x) = \sqrt[3]{x-8}$, find the following function values. See Example 6.

77. $f(0)$

78. $g(0)$

79. $g(7)$

80. $f(-1)$

81. $g(-19)$

82. $f(3)$

83. $f(2)$

84. $g(1)$

Identify the domain and then graph each function. See Example 7.

▶ **85.** $f(x) = \sqrt{x} + 2$

86. $f(x) = \sqrt{x} - 2$

87. $f(x) = \sqrt{x-3}$; use the following table.

x	$f(x)$
3	
4	
7	
12	

88. $f(x) = \sqrt{x+1}$; use the following table.

x	$f(x)$
−1	
0	
3	
8	

Identify the domain and then graph each function. See Example 8.

89. $f(x) = \sqrt[3]{x} + 1$

90. $f(x) = \sqrt[3]{x} - 2$

91. $g(x) = \sqrt[3]{x-1}$; use the following table.

x	$g(x)$
1	
2	
0	
9	
−7	

92. $g(x) = \sqrt[3]{x+1}$; use the following table.

x	$g(x)$
−1	
0	
−2	
7	
−9	

REVIEW AND PREVIEW

Simplify each exponential expression. See Sections 5.1 and 5.5.

93. $(-2x^3y^2)^5$

94. $(4y^6z^7)^3$

95. $(-3x^2y^3z^5)(20x^5y^7)$

96. $(-14a^5bc^2)(2abc^4)$

97. $\dfrac{7x^{-1}y}{14(x^5y^2)^{-2}}$

98. $\dfrac{(2a^{-1}b^2)^3}{(8a^2b)^{-2}}$

CONCEPT EXTENSIONS

Determine whether the following are real numbers. See the Concept Check in this section.

99. $\sqrt{-17}$

100. $\sqrt[3]{-17}$

101. $\sqrt[10]{-17}$

102. $\sqrt[15]{-17}$

Choose the correct letter or letters. No pencil is needed, just think your way through these.

103. Which radical is not a real number?

  a. $\sqrt{3}$ **b.** $-\sqrt{11}$ **c.** $\sqrt[3]{-10}$ **d.** $\sqrt{-10}$

104. Which radical(s) simplify to 3?

  a. $\sqrt{9}$ **b.** $\sqrt{-9}$ **c.** $\sqrt[3]{27}$ **d.** $\sqrt[3]{-27}$

105. Which radical(s) simplify to -3?

  a. $\sqrt{9}$ **b.** $\sqrt{-9}$ **c.** $\sqrt[3]{27}$ **d.** $\sqrt[3]{-27}$

106. Which radical does not simplify to a whole number?

  a. $\sqrt{64}$ **b.** $\sqrt[3]{64}$ **c.** $\sqrt{8}$ **d.** $\sqrt[3]{8}$

For Exercises 107 through 110, do not use a calculator.

107. $\sqrt{160}$ is closest to

  a. 10   **b.** 13   **c.** 20   **d.** 40

108. $\sqrt{1000}$ is closest to

  a. 10   **b.** 30   **c.** 100   **d.** 500

△ **109.** The perimeter of the triangle is closest to

  a. 12   **b.** 18

  c. 66   **d.** 132

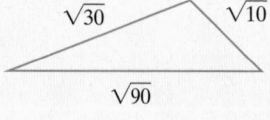

110. The length of the bent wire is closest to

  a. 5   **b.** $\sqrt{28}$

  c. 7   **d.** 14

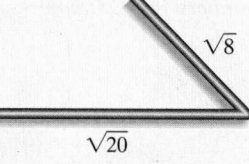

111. Explain why $\sqrt{-64}$ is not a real number.

112. Explain why $\sqrt[3]{-64}$ is a real number.

The Mosteller formula for calculating adult body surface area is $B = \sqrt{\dfrac{hw}{3131}}$, where B is an individual's body surface area in square meters, h is the individual's height in inches, and w is the individual's weight in pounds. Use this information to answer Exercises 113 and 114. Round answers to 2 decimal places.

△ **113.** Find the body surface area of an individual who is 66 inches tall and who weighs 135 pounds.

△ **114.** Find the body surface area of an individual who is 74 inches tall and who weighs 225 pounds.

115. Escape velocity is the minimum speed that an object must reach to escape the pull of a planet's gravity. Escape velocity v is given by the equation $v = \sqrt{\dfrac{2Gm}{r}}$, where m is the mass of the planet, r is its radius, and G is the universal gravitational constant, which has a value of $G = 6.67 \times 10^{-11}$ m³/kg·s². The mass of Earth is 5.97×10^{24} kg, and its radius is 6.37×10^6 m. Use this information to find the escape velocity for Earth in meters per second. Round to the nearest whole number. (*Source*: National Space Science Data Center)

116. Use the formula from Exercise 115 to determine the escape velocity for the moon. The mass of the moon is 7.35×10^{22} kg, and its radius is 1.74×10^6 m. Round to the nearest whole number. (*Source*: National Space Science Data Center)

117. Suppose a classmate tells you that $\sqrt{13} \approx 5.7$. Without a calculator, how can you convince your classmate that he or she must have made an error?

118. Suppose a classmate tells you that $\sqrt[3]{10} \approx 3.2$. Without a calculator, how can you convince your friend that he or she must have made an error?

Use a graphing calculator to verify the domain of each function and its graph.

119. Exercise 85   **120.** Exercise 86

121. Exercise 89   **122.** Exercise 90

10.2 Rational Exponents

OBJECTIVES

1 Understand the Meaning of $a^{1/n}$.

2 Understand the Meaning of $a^{m/n}$.

3 Understand the Meaning of $a^{-m/n}$.

4 Use Rules for Exponents to Simplify Expressions That Contain Rational Exponents.

5 Use Rational Exponents to Simplify Radical Expressions.

OBJECTIVE

1 Understanding the Meaning of $a^{1/n}$

So far in this text, we have not defined expressions with rational exponents such as $3^{1/2}$, $x^{2/3}$, and $-9^{-1/4}$. We will define these expressions so that the rules for exponents will apply to these rational exponents as well.

Suppose that $x = 5^{1/3}$. Then

$$x^3 = (5^{1/3})^3 = 5^{1/3 \cdot 3} = 5^1 \text{ or } 5$$

$$\text{using rules for exponents}$$

Since $x^3 = 5$, x is the number whose cube is 5, or $x = \sqrt[3]{5}$. Notice that we also know that $x = 5^{1/3}$. This means

$$5^{1/3} = \sqrt[3]{5}$$

Definition of $a^{1/n}$

If n is a positive integer greater than 1 and $\sqrt[n]{a}$ is a real number, then

$$a^{1/n} = \sqrt[n]{a}$$

Notice that the denominator of the rational exponent corresponds to the index of the radical.

EXAMPLE 1 Use radical notation to write the following. Simplify if possible.

  a. $4^{1/2}$ **b.** $64^{1/3}$ **c.** $x^{1/4}$ **d.** $0^{1/6}$ **e.** $-9^{1/2}$ **f.** $(81x^8)^{1/4}$ **g.** $5y^{1/3}$

Solution

  a. $4^{1/2} = \sqrt{4} = 2$ **b.** $64^{1/3} = \sqrt[3]{64} = 4$

  c. $x^{1/4} = \sqrt[4]{x}$ **d.** $0^{1/6} = \sqrt[6]{0} = 0$

  e. $-9^{1/2} = -\sqrt{9} = -3$ **f.** $(81x^8)^{1/4} = \sqrt[4]{81x^8} = 3x^2$

  g. $5y^{1/3} = 5\sqrt[3]{y}$

PRACTICE

1 Use radical notation to write the following. Simplify if possible.

  a. $36^{1/2}$ **b.** $1000^{1/3}$ **c.** $x^{1/3}$ **d.** $1^{1/4}$

  e. $-64^{1/2}$ **f.** $(125x^9)^{1/3}$ **g.** $3x^{1/4}$

OBJECTIVE

2 Understanding the Meaning of $a^{m/n}$

As we expand our use of exponents to include $\dfrac{m}{n}$, we define their meaning so that rules for exponents still hold true. For example, by properties of exponents,

$$8^{2/3} = (8^{1/3})^2 = (\sqrt[3]{8})^2 \quad \text{or}$$

$$8^{2/3} = (8^2)^{1/3} = \sqrt[3]{8^2}$$

Definition of $a^{m/n}$

If m and n are positive integers greater than 1 with $\dfrac{m}{n}$ in simplest form, then

$$a^{m/n} = \sqrt[n]{a^m} = (\sqrt[n]{a})^m$$

as long as $\sqrt[n]{a}$ is a real number.

Notice that the denominator n of the rational exponent corresponds to the index of the radical. The numerator m of the rational exponent indicates that the base is to be raised to the mth power. This means

$$8^{2/3} = \sqrt[3]{8^2} = \sqrt[3]{64} = 4 \quad \text{or}$$

$$8^{2/3} = \left(\sqrt[3]{8}\right)^2 = 2^2 = 4$$

From simplifying $8^{2/3}$, can you see that it doesn't matter whether you raise to a power first and then take the nth root or you take the nth root first and then raise to a power?

> **Helpful Hint**
>
> Most of the time, $\left(\sqrt[n]{a}\right)^m$ will be easier to calculate than $\sqrt[n]{a^m}$.

EXAMPLE 2 Use radical notation to write the following. Then simplify if possible.

   a. $4^{3/2}$ **b.** $-16^{3/4}$ **c.** $(-27)^{2/3}$

   d. $\left(\dfrac{1}{9}\right)^{3/2}$ **e.** $(4x - 1)^{3/5}$

Solution

   a. $4^{3/2} = \left(\sqrt{4}\right)^3 = 2^3 = 8$ **b.** $-16^{3/4} = -\left(\sqrt[4]{16}\right)^3 = -(2)^3 = -8$

   c. $(-27)^{2/3} = \left(\sqrt[3]{-27}\right)^2 = (-3)^2 = 9$ **d.** $\left(\dfrac{1}{9}\right)^{3/2} = \left(\sqrt{\dfrac{1}{9}}\right)^3 = \left(\dfrac{1}{3}\right)^3 = \dfrac{1}{27}$

   e. $(4x - 1)^{3/5} = \sqrt[5]{(4x - 1)^3}$ □

PRACTICE
2 Use radical notation to write the following. Simplify if possible.

   a. $16^{3/2}$ **b.** $-1^{3/5}$ **c.** $-(81)^{3/4}$

   d. $\left(\dfrac{1}{25}\right)^{3/2}$ **e.** $(3x + 2)^{5/9}$ ■

> **Helpful Hint**
>
> The _denominator_ of a rational exponent is the index of the corresponding radical. For example, $x^{1/5} = \sqrt[5]{x}$ and $z^{2/3} = \sqrt[3]{z^2}$, or $z^{2/3} = \left(\sqrt[3]{z}\right)^2$.

OBJECTIVE
3 Understanding the Meaning of $a^{-m/n}$

The rational exponents we have given meaning to exclude negative rational numbers. To complete the set of definitions, we define $a^{-m/n}$.

> **Definition of $a^{-m/n}$**
>
> $$a^{-m/n} = \frac{1}{a^{m/n}}$$
>
> as long as $a^{m/n}$ is a nonzero real number.

EXAMPLE 3 Write each expression with a positive exponent, and then simplify.

   a. $16^{-3/4}$ **b.** $(-27)^{-2/3}$

Solution

   a. $16^{-3/4} = \dfrac{1}{16^{3/4}} = \dfrac{1}{\left(\sqrt[4]{16}\right)^3} = \dfrac{1}{2^3} = \dfrac{1}{8}$

   b. $(-27)^{-2/3} = \dfrac{1}{(-27)^{2/3}} = \dfrac{1}{\left(\sqrt[3]{-27}\right)^2} = \dfrac{1}{(-3)^2} = \dfrac{1}{9}$ □

PRACTICE

3   Write each expression with a positive exponent; then simplify.

a. $9^{-3/2}$ **b.** $(-64)^{-2/3}$

Helpful Hint

If an expression contains a negative rational exponent, such as $9^{-3/2}$, you may want to first write the expression with a positive exponent and then interpret the rational exponent. Notice that the sign of the base is not affected by the sign of its exponent. For example,

$$9^{-3/2} = \frac{1}{9^{3/2}} = \frac{1}{(\sqrt{9})^3} = \frac{1}{27}$$

Also,

$$(-27)^{-1/3} = \frac{1}{(-27)^{1/3}} = -\frac{1}{3}$$

✔ CONCEPT CHECK

Which one is correct?

a. $-8^{2/3} = \frac{1}{4}$ **b.** $8^{-2/3} = -\frac{1}{4}$ **c.** $8^{-2/3} = -4$ **d.** $-8^{-2/3} = -\frac{1}{4}$

OBJECTIVE

4   Using Rules for Exponents to Simplify Expressions

It can be shown that the properties of integer exponents hold for rational exponents. By using these properties and definitions, we can now simplify expressions that contain rational exponents.

These rules are repeated here for review.

Note: For the remainder of this chapter, we will assume that variables represent positive real numbers. Since this is so, we need not insert absolute value bars when we simplify even roots.

Summary of Exponent Rules

If m and n are rational numbers, and a, b, and c are numbers for which the expressions below exist, then

Product rule for exponents:	$a^m \cdot a^n = a^{m+n}$
Power rule for exponents:	$(a^m)^n = a^{m \cdot n}$
Power rules for products and quotients:	$(ab)^n = a^n b^n$ and
	$\left(\dfrac{a}{c}\right)^n = \dfrac{a^n}{c^n}, c \neq 0$
Quotient rule for exponents:	$\dfrac{a^m}{a^n} = a^{m-n}, a \neq 0$
Zero exponent:	$a^0 = 1, a \neq 0$
Negative exponent:	$a^{-n} = \dfrac{1}{a^n}, a \neq 0$

EXAMPLE 4   Use properties of exponents to simplify. Write results with only positive exponents.

a. $b^{1/3} \cdot b^{5/3}$ **b.** $x^{1/2} x^{1/3}$ **c.** $\dfrac{7^{1/3}}{7^{4/3}}$

d. $y^{-4/7} \cdot y^{6/7}$ **e.** $\dfrac{(2x^{2/5}y^{-1/3})^5}{x^2 y}$

(Continued on next page)

Solution

a. $b^{1/3} \cdot b^{5/3} = b^{(1/3+5/3)} = b^{6/3} = b^2$ Use the product rule.

b. $x^{1/2}x^{1/3} = x^{(1/2+1/3)} = x^{3/6+2/6} = x^{5/6}$ Use the product rule.

c. $\dfrac{7^{1/3}}{7^{4/3}} = 7^{1/3-4/3} = 7^{-3/3} = 7^{-1} = \dfrac{1}{7}$ Use the quotient rule.

d. $y^{-4/7} \cdot y^{6/7} = y^{-4/7+6/7} = y^{2/7}$ Use the product rule.

e. We begin by using the power rule $(ab)^m = a^m b^m$ to simplify the numerator.

$$\frac{(2x^{2/5}y^{-1/3})^5}{x^2 y} = \frac{2^5(x^{2/5})^5(y^{-1/3})^5}{x^2 y} = \frac{32x^2 y^{-5/3}}{x^2 y}$$ Use the power rule and simplify

$$= 32x^{2-2}y^{-5/3-3/3}$$ Apply the quotient rule.

$$= 32x^0 y^{-8/3}$$

$$= \frac{32}{y^{8/3}}$$ □

PRACTICE
4   Use properties of exponents to simplify.

a. $y^{2/3} \cdot y^{8/3}$ **b.** $x^{3/5} \cdot x^{1/4}$ **c.** $\dfrac{9^{2/7}}{9^{9/7}}$

d. $b^{4/9} \cdot b^{-2/9}$ **e.** $\dfrac{(3x^{1/4}y^{-2/3})^4}{x^4 y}$

EXAMPLE 5   Multiply.

a. $z^{2/3}(z^{1/3} - z^5)$ **b.** $(x^{1/3} - 5)(x^{1/3} + 2)$

Solution

a. $z^{2/3}(z^{1/3} - z^5) = z^{2/3}z^{1/3} - z^{2/3}z^5$ Apply the distributive property.

$$= z^{(2/3+1/3)} - z^{(2/3+5)}$$ Use the product rule.

$$= z^{3/3} - z^{(2/3+15/3)}$$

$$= z - z^{17/3}$$

b. $(x^{1/3} - 5)(x^{1/3} + 2) = x^{2/3} + 2x^{1/3} - 5x^{1/3} - 10$ Think of $(x^{1/3} - 5)$ and

$$= x^{2/3} - 3x^{1/3} - 10$$ $(x^{1/3} + 2)$ as 2 binomials, then multiply using FOIL.   □

PRACTICE
5   Multiply.

a. $x^{3/5}(x^{1/3} - x^2)$ **b.** $(x^{1/2} + 6)(x^{1/2} - 2)$

EXAMPLE 6   Factor $x^{-1/2}$ from the expression $3x^{-1/2} - 7x^{5/2}$. Assume that all variables represent positive numbers.

Solution

$$3x^{-1/2} - 7x^{5/2} = (x^{-1/2})(3) - (x^{-1/2})(7x^{6/2})$$

$$= x^{-1/2}(3 - 7x^3)$$

To check, multiply $x^{-1/2}(3 - 7x^3)$ to see that the product is $3x^{-1/2} - 7x^{5/2}$.   □

PRACTICE
6   Factor $x^{-1/5}$ from the expression $2x^{-1/5} - 7x^{4/5}$.

OBJECTIVE

5   **Using Rational Exponents to Simplify Radical Expressions**

Some radical expressions are easier to simplify when we first write them with rational exponents. Next, use properties of exponents to simplify the expression, and then convert it back to radical notation.

EXAMPLE 7   Use rational exponents to simplify. Assume that variables represent positive numbers.

$$\textbf{a. } \sqrt[8]{x^4} \qquad\qquad \textbf{b. } \sqrt[6]{25} \qquad\qquad \textbf{c. } \sqrt[4]{r^2 s^6}$$

Solution

a. $\sqrt[8]{x^4} = x^{4/8} = x^{1/2} = \sqrt{x}$

b. $\sqrt[6]{25} = 25^{1/6} = (5^2)^{1/6} = 5^{2/6} = 5^{1/3} = \sqrt[3]{5}$

c. $\sqrt[4]{r^2 s^6} = (r^2 s^6)^{1/4} = r^{2/4} s^{6/4} = r^{1/2} s^{3/2} = (rs^3)^{1/2} = \sqrt{rs^3}$

PRACTICE

7   Use rational exponents to simplify. Assume that the variables represent positive numbers.

$$\textbf{a. } \sqrt[9]{x^3} \qquad\qquad \textbf{b. } \sqrt[4]{36} \qquad\qquad \textbf{c. } \sqrt[8]{a^4 b^2}$$

EXAMPLE 8   Use rational exponents to write as a single radical.

$$\textbf{a. } \sqrt{x} \cdot \sqrt[4]{x} \qquad \textbf{b. } \dfrac{\sqrt{x}}{\sqrt[3]{x}} \qquad \textbf{c. } \sqrt[3]{3} \cdot \sqrt{2}$$

Solution

a. $\sqrt{x} \cdot \sqrt[4]{x} = x^{1/2} \cdot x^{1/4} = x^{1/2 + 1/4}$

$\qquad = x^{3/4} = \sqrt[4]{x^3}$

b. $\dfrac{\sqrt{x}}{\sqrt[3]{x}} = \dfrac{x^{1/2}}{x^{1/3}} = x^{1/2 - 1/3} = x^{3/6 - 2/6}$

$\qquad = x^{1/6} = \sqrt[6]{x}$

c. $\sqrt[3]{3} \cdot \sqrt{2} = 3^{1/3} \cdot 2^{1/2}$ Write with rational exponents.

$\qquad = 3^{2/6} \cdot 2^{3/6}$ Write the exponents so that they have the same denominator.

$\qquad = (3^2 \cdot 2^3)^{1/6}$ Use $a^n b^n = (ab)^n$

$\qquad = \sqrt[6]{3^2 \cdot 2^3}$ Write with radical notation.

$\qquad = \sqrt[6]{72}$ Multiply $3^2 \cdot 2^3$.

PRACTICE

8   Use rational exponents to write each of the following as a single radical.

$$\textbf{a. } \sqrt[3]{x} \cdot \sqrt[4]{x} \qquad\qquad \textbf{b. } \dfrac{\sqrt[3]{y}}{\sqrt[5]{y}} \qquad\qquad \textbf{c. } \sqrt[3]{5} \cdot \sqrt{3}$$

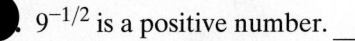

Vocabulary, Readiness & Video Check

Answer each true or false.

1. $9^{-1/2}$ is a positive number. _____

2. $9^{-1/2}$ is a whole number. _____

3. $\dfrac{1}{a^{-m/n}} = a^{m/n}$ (where $a^{m/n}$ is a nonzero real number). _____

Fill in the blank with the correct choice.

4. To simplify $x^{2/3} \cdot x^{1/5}$, _____ the exponents.

    a. add   **b.** subtract   **c.** multiply   **d.** divide

5. To simplify $(x^{2/3})^{1/5}$, _____ the exponents.

    a. add   **b.** subtract   **c.** multiply   **d.** divide

6. To simplify $\dfrac{x^{2/3}}{x^{1/5}}$, _____ the exponents.

    a. add   **b.** subtract   **c.** multiply   **d.** divide

---

Martin-Gay Interactive Videos

See Video 10.2

Watch the section lecture video and answer the following questions.

OBJECTIVE 1

7. After studying Example 2, write $-(3x)^{1/5}$ in radical notation.

OBJECTIVE 2

8. From Examples 3 and 4, in a fractional exponent, what do the numerator and denominator each represent in radical form?

OBJECTIVE 3

9. Based on Example 5, complete the following statements. A negative fractional exponent will move a base from the numerator to the _____ with the fractional exponent becoming _____.

OBJECTIVE 4

10. Based on Examples 7–9, complete the following statements. Assume you have an expression with fractional exponents. If applying the product rule of exponents, you _____ the exponents. If applying the quotient rule of exponents, you _____ the exponents. If applying the power rule of exponents, you _____ the exponents.

OBJECTIVE 5

11. From Example 10, describe a way to simplify a radical of a variable raised to a power if the index and the exponent have a common factor.

10.2 Exercise Set   MyMathLab

Use radical notation to write each expression. Simplify if possible. See Example 1.

1. $49^{1/2}$ **2.** $64^{1/3}$

3. $27^{1/3}$ **4.** $8^{1/3}$

5. $\left(\dfrac{1}{16}\right)^{1/4}$ **6.** $\left(\dfrac{1}{64}\right)^{1/2}$

7. $169^{1/2}$ **8.** $81^{1/4}$

9. $2m^{1/3}$ **10.** $(2m)^{1/3}$

11. $(9x^4)^{1/2}$ **12.** $(16x^8)^{1/2}$

13. $(-27)^{1/3}$ **14.** $-64^{1/2}$

15. $-16^{1/4}$ **16.** $(-32)^{1/5}$

Use radical notation to write each expression. Simplify if possible. See Example 2.

17. $16^{3/4}$ **18.** $4^{5/2}$

19. $(-64)^{2/3}$ **20.** $(-8)^{4/3}$

21. $(-16)^{3/4}$ **22.** $(-9)^{3/2}$

23. $(2x)^{3/5}$ **24.** $2x^{3/5}$

25. $(7x + 2)^{2/3}$ **26.** $(x - 4)^{3/4}$

27. $\left(\dfrac{16}{9}\right)^{3/2}$ **28.** $\left(\dfrac{49}{25}\right)^{3/2}$

Write with positive exponents. Simplify if possible. See Example 3.

29. $8^{-4/3}$ **30.** $64^{-2/3}$

31. $(-64)^{-2/3}$ **32.** $(-8)^{-4/3}$

33. $(-4)^{-3/2}$ **34.** $(-16)^{-5/4}$

35. $x^{-1/4}$ **36.** $y^{-1/6}$

37. $\dfrac{1}{a^{-2/3}}$ **38.** $\dfrac{1}{n^{-8/9}}$

39. $\dfrac{5}{7x^{-3/4}}$ **40.** $\dfrac{2}{3y^{-5/7}}$

Use the properties of exponents to simplify each expression. Write with positive exponents. See Example 4.

41. $a^{2/3}a^{5/3}$ **42.** $b^{9/5}b^{8/5}$

43. $x^{-2/5} \cdot x^{7/5}$ **44.** $y^{4/3} \cdot y^{-1/3}$

45. $3^{1/4} \cdot 3^{3/8}$ **46.** $5^{1/2} \cdot 5^{1/6}$

47. $\dfrac{y^{1/3}}{y^{1/6}}$ **48.** $\dfrac{x^{3/4}}{x^{1/8}}$

49. $(4u^2)^{3/2}$ **50.** $(32^{1/5}x^{2/3})^3$

51. $\dfrac{b^{1/2}b^{3/4}}{-b^{1/4}}$ **52.** $\dfrac{a^{1/4}a^{-1/2}}{a^{2/3}}$

53. $\dfrac{(x^3)^{1/2}}{x^{7/2}}$

54. $\dfrac{y^{11/3}}{(y^5)^{1/3}}$

55. $\dfrac{(3x^{1/4})^3}{x^{1/12}}$

56. $\dfrac{(2x^{1/5})^4}{x^{3/10}}$

57. $\dfrac{(y^3z)^{1/6}}{y^{-1/2}z^{1/3}}$

58. $\dfrac{(m^2n)^{1/4}}{m^{-1/2}n^{5/8}}$

59. $\dfrac{(x^3y^2)^{1/4}}{(x^{-5}y^{-1})^{-1/2}}$

60. $\dfrac{(a^{-2}b^3)^{1/8}}{(a^{-3}b)^{-1/4}}$

Multiply. See Example 5.

61. $y^{1/2}(y^{1/2} - y^{2/3})$

62. $x^{1/2}(x^{1/2} + x^{3/2})$

63. $x^{2/3}(x - 2)$

64. $3x^{1/2}(x + y)$

65. $(2x^{1/3} + 3)(2x^{1/3} - 3)$

66. $(y^{1/2} + 5)(y^{1/2} + 5)$

Factor the given factor from the expression. See Example 6.

67. $x^{8/3}; x^{8/3} + x^{10/3}$

68. $x^{3/2}; x^{5/2} - x^{3/2}$

69. $x^{1/5}; x^{2/5} - 3x^{1/5}$

70. $x^{2/7}; x^{3/7} - 2x^{2/7}$

71. $x^{-1/3}; 5x^{-1/3} + x^{2/3}$

72. $x^{-3/4}; x^{-3/4} + 3x^{1/4}$

Use rational exponents to simplify each radical. Assume that all variables represent positive numbers. See Example 7.

73. $\sqrt[6]{x^3}$

74. $\sqrt[9]{a^3}$

75. $\sqrt[6]{4}$

76. $\sqrt[4]{36}$

77. $\sqrt[4]{16x^2}$

78. $\sqrt[8]{4y^2}$

79. $\sqrt[8]{x^4y^4}$

80. $\sqrt[9]{y^6z^3}$

81. $\sqrt[12]{a^8b^4}$

82. $\sqrt[10]{a^5b^5}$

83. $\sqrt[4]{(x + 3)^2}$

84. $\sqrt[8]{(y + 1)^4}$

Use rational exponents to write as a single radical expression. See Example 8.

85. $\sqrt[3]{y} \cdot \sqrt[5]{y^2}$

86. $\sqrt[3]{y^2} \cdot \sqrt[6]{y}$

87. $\dfrac{\sqrt[3]{b^2}}{\sqrt[4]{b}}$

88. $\dfrac{\sqrt[4]{a}}{\sqrt[5]{a}}$

89. $\sqrt[3]{x} \cdot \sqrt[4]{x} \cdot \sqrt[8]{x^3}$

90. $\sqrt[6]{y} \cdot \sqrt[3]{y} \cdot \sqrt[5]{y^2}$

91. $\dfrac{\sqrt[3]{a^2}}{\sqrt[6]{a}}$

92. $\dfrac{\sqrt[5]{b^2}}{\sqrt[10]{b^3}}$

93. $\sqrt{3} \cdot \sqrt[3]{4}$

94. $\sqrt[3]{5} \cdot \sqrt{2}$

95. $\sqrt[5]{7} \cdot \sqrt[3]{y}$

96. $\sqrt[4]{5} \cdot \sqrt[3]{x}$

97. $\sqrt{5r} \cdot \sqrt[3]{s}$

98. $\sqrt[3]{b} \cdot \sqrt[5]{4a}$

REVIEW AND PREVIEW

Write each integer as a product of two integers such that one of the factors is a perfect square. For example, write 18 as $9 \cdot 2$ because 9 is a perfect square.

99. 75

100. 20

101. 48

102. 45

Write each integer as a product of two integers such that one of the factors is a perfect cube. For example, write 24 as $8 \cdot 3$ because 8 is a perfect cube.

103. 16

104. 56

105. 54

106. 80

CONCEPT EXTENSIONS

Choose the correct letter for each exercise. Letters will be used more than once. No pencil is needed. Just think about the meaning of each expression.

A = 2, B = −2, C = not a real number

107. $4^{1/2}$ _____

108. $-4^{1/2}$ _____

109. $(-4)^{1/2}$ _____

110. $8^{1/3}$ _____

111. $-8^{1/3}$ _____

112. $(-8)^{1/3}$ _____

Basal metabolic rate (BMR) is the number of calories per day a person needs to maintain life. A person's basal metabolic rate $B(w)$ in calories per day can be estimated with the function $B(w) = 70w^{3/4}$, where w is the person's weight in kilograms. Use this information to answer Exercises 113 and 114.

113. Estimate the BMR for a person who weighs 60 kilograms. Round to the nearest calorie. (*Note:* 60 kilograms is approximately 132 pounds.)

114. Estimate the BMR for a person who weighs 90 kilograms. Round to the nearest calorie. (*Note:* 90 kilograms is approximately 198 pounds.)

The number of cellular telephone subscribers in the United States from 2010–2015 can be modeled by $f(x) = 236x^{1/20}$, where $f(x)$ is the number of cellular telephone subscriptions in millions, x years after 2010. (Source: International Telecommunications Union) Use this information to answer Exercises 115 and 116.

115. Use this model to estimate the number of cellular telephone subscriptions in 2015. Round to the nearest tenth of a million.

116. Predict the number of cellular telephone subscriptions in 2020. Round to the nearest tenth of a million.

117. Explain how writing x^{-7} with positive exponents is similar to writing $x^{-1/4}$ with positive exponents.

118. Explain how writing $2x^{-5}$ with positive exponents is similar to writing $2x^{-3/4}$ with positive exponents.

Fill in each box with the correct expression.

119. $\square \cdot a^{2/3} = a^{3/3}$, or a

120. $\square \cdot x^{1/8} = x^{4/8}$, or $x^{1/2}$

121. $\dfrac{\square}{x^{-2/5}} = x^{3/5}$

122. $\dfrac{\square}{y^{-3/4}} = y^{4/4}$, or y

Use a calculator to write a four-decimal-place approximation of each number.

123. $8^{1/4}$

124. $20^{1/5}$

125. $18^{3/5}$

126. $76^{5/7}$

127. In physics, the speed of a wave traveling over a stretched string with tension t and density u is given by the expression $\dfrac{\sqrt{t}}{\sqrt{u}}$. Write this expression with rational exponents.

128. In electronics, the angular frequency of oscillations in a certain type of circuit is given by the expression $(LC)^{-1/2}$. Use radical notation to write this expression.

10.3 Simplifying Radical Expressions

OBJECTIVES

1 Use the Product Rule for Radicals.

2 Use the Quotient Rule for Radicals.

3 Simplify Radicals.

4 Use the Distance and Midpoint Formulas.

OBJECTIVE

1 Using the Product Rule

It is possible to simplify some radicals that do not evaluate to rational numbers. To do so, we use a product rule and a quotient rule for radicals. To discover the product rule, notice the following pattern.

$$\sqrt{9} \cdot \sqrt{4} = 3 \cdot 2 = 6$$
$$\sqrt{9 \cdot 4} = \sqrt{36} = 6$$

Since both expressions simplify to 6, it is true that

$$\sqrt{9} \cdot \sqrt{4} = \sqrt{9 \cdot 4}$$

This pattern suggests the following product rule for radicals.

> **Product Rule for Radicals**
>
> If $\sqrt[n]{a}$ and $\sqrt[n]{b}$ are real numbers, then
> $$\sqrt[n]{a} \cdot \sqrt[n]{b} = \sqrt[n]{ab}$$

Notice that the product rule is the relationship $a^{1/n} \cdot b^{1/n} = (ab)^{1/n}$ stated in radical notation.

EXAMPLE 1 Multiply.

a. $\sqrt{3} \cdot \sqrt{5}$

b. $\sqrt{21} \cdot \sqrt{x}$

c. $\sqrt[3]{4} \cdot \sqrt[3]{2}$

d. $\sqrt[4]{5y^2} \cdot \sqrt[4]{2x^3}$

e. $\sqrt{\dfrac{2}{a}} \cdot \sqrt{\dfrac{b}{3}}$

Solution

a. $\sqrt{3} \cdot \sqrt{5} = \sqrt{3 \cdot 5} = \sqrt{15}$

b. $\sqrt{21} \cdot \sqrt{x} = \sqrt{21x}$

c. $\sqrt[3]{4} \cdot \sqrt[3]{2} = \sqrt[3]{4 \cdot 2} = \sqrt[3]{8} = 2$

d. $\sqrt[4]{5y^2} \cdot \sqrt[4]{2x^3} = \sqrt[4]{5y^2 \cdot 2x^3} = \sqrt[4]{10y^2x^3}$

e. $\sqrt{\dfrac{2}{a}} \cdot \sqrt{\dfrac{b}{3}} = \sqrt{\dfrac{2}{a} \cdot \dfrac{b}{3}} = \sqrt{\dfrac{2b}{3a}}$

PRACTICE

1 Multiply.

a. $\sqrt{5} \cdot \sqrt{7}$

b. $\sqrt{13} \cdot \sqrt{z}$

c. $\sqrt[4]{125} \cdot \sqrt[4]{5}$

d. $\sqrt[3]{5y} \cdot \sqrt[3]{3x^2}$

e. $\sqrt{\dfrac{5}{m}} \cdot \sqrt{\dfrac{t}{2}}$

OBJECTIVE

2   Using the Quotient Rule

To discover a quotient rule for radicals, notice the following pattern.

$$\sqrt{\frac{4}{9}} = \frac{2}{3}$$

$$\frac{\sqrt{4}}{\sqrt{9}} = \frac{2}{3}$$

Since both expressions simplify to $\frac{2}{3}$, it is true that

$$\sqrt{\frac{4}{9}} = \frac{\sqrt{4}}{\sqrt{9}}$$

This pattern suggests the following quotient rule for radicals.

Quotient Rule for Radicals

If $\sqrt[n]{a}$ and $\sqrt[n]{b}$ are real numbers and $\sqrt[n]{b}$ is not zero, then

$$\sqrt[n]{\frac{a}{b}} = \frac{\sqrt[n]{a}}{\sqrt[n]{b}}$$

Notice that the quotient rule is the relationship $\left(\dfrac{a}{b}\right)^{1/n} = \dfrac{a^{1/n}}{b^{1/n}}$ stated in radical notation. We can use the quotient rule to simplify radical expressions by reading the rule from left to right or to divide radicals by reading the rule from right to left.

For example,

$$\sqrt{\frac{x}{16}} = \frac{\sqrt{x}}{\sqrt{16}} = \frac{\sqrt{x}}{4} \qquad \text{Using } \sqrt[n]{\frac{a}{b}} = \frac{\sqrt[n]{a}}{\sqrt[n]{b}}$$

$$\frac{\sqrt{75}}{\sqrt{3}} = \sqrt{\frac{75}{3}} = \sqrt{25} = 5 \quad \text{Using } \frac{\sqrt[n]{a}}{\sqrt[n]{b}} = \sqrt[n]{\frac{a}{b}}$$

Note: *Recall that from Section 10.2 on, we assume that variables represent positive real numbers. Since this is so, we need not insert absolute value bars when we simplify even roots.*

EXAMPLE 2   Use the quotient rule to simplify.

a. $\sqrt{\dfrac{25}{49}}$ **b.** $\sqrt{\dfrac{x}{9}}$ **c.** $\sqrt[3]{\dfrac{8}{27}}$ **d.** $\sqrt[4]{\dfrac{3}{16y^4}}$

Solution

a. $\sqrt{\dfrac{25}{49}} = \dfrac{\sqrt{25}}{\sqrt{49}} = \dfrac{5}{7}$ **b.** $\sqrt{\dfrac{x}{9}} = \dfrac{\sqrt{x}}{\sqrt{9}} = \dfrac{\sqrt{x}}{3}$

c. $\sqrt[3]{\dfrac{8}{27}} = \dfrac{\sqrt[3]{8}}{\sqrt[3]{27}} = \dfrac{2}{3}$ **d.** $\sqrt[4]{\dfrac{3}{16y^4}} = \dfrac{\sqrt[4]{3}}{\sqrt[4]{16y^4}} = \dfrac{\sqrt[4]{3}}{2y}$ □

PRACTICE

2   Use the quotient rule to simplify.

a. $\sqrt{\dfrac{36}{49}}$ **b.** $\sqrt{\dfrac{z}{16}}$ **c.** $\sqrt[3]{\dfrac{125}{8}}$ **d.** $\sqrt[4]{\dfrac{5}{81x^8}}$ ■

OBJECTIVE

3 Simplifying Radicals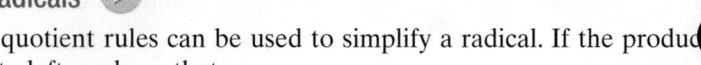

Both the product and quotient rules can be used to simplify a radical. If the product rule is read from right to left, we have that

$$\sqrt[n]{ab} = \sqrt[n]{a} \cdot \sqrt[n]{b}.$$

This is used to simplify the following radicals.

EXAMPLE 3 Simplify the following.

   a. $\sqrt{50}$ **b.** $\sqrt[3]{24}$ **c.** $\sqrt{26}$ **d.** $\sqrt[4]{32}$

Solution

a. Factor 50 such that one factor is the largest perfect square that divides 50. The largest perfect square factor of 50 is 25, so we write 50 as $25 \cdot 2$ and use the product rule for radicals to simplify.

$$\sqrt{50} = \sqrt{25 \cdot 2} = \sqrt{25} \cdot \sqrt{2} = 5\sqrt{2}$$
The largest perfect square factor of 50

> Helpful Hint
>
> Don't forget that, for example, $5\sqrt{2}$ means $5 \cdot \sqrt{2}$.

b. $\sqrt[3]{24} = \sqrt[3]{8 \cdot 3} = \sqrt[3]{8} \cdot \sqrt[3]{3} = 2\sqrt[3]{3}$
The largest perfect cube factor of 24

c. $\sqrt{26}$ The largest perfect square factor of 26 is 1, so $\sqrt{26}$ cannot be simplified further.

d. $\sqrt[4]{32} = \sqrt[4]{16 \cdot 2} = \sqrt[4]{16} \cdot \sqrt[4]{2} = 2\sqrt[4]{2}$
The largest fourth power factor of 32   □

PRACTICE

3 Simplify the following.

   a. $\sqrt{98}$ **b.** $\sqrt[3]{54}$ **c.** $\sqrt{35}$ **d.** $\sqrt[4]{243}$

After simplifying a radical such as a square root, always check the radicand to see that it contains no other perfect square factors. It may, if the largest perfect square factor of the radicand was not originally recognized. For example,

$$\sqrt{200} = \sqrt{4 \cdot 50} = \sqrt{4} \cdot \sqrt{50} = 2\sqrt{50}$$

Notice that the radicand 50 still contains the perfect square factor 25. This is because 4 is not the largest perfect square factor of 200. We continue as follows.

$$2\sqrt{50} = 2\sqrt{25 \cdot 2} = 2 \cdot \sqrt{25} \cdot \sqrt{2} = 2 \cdot 5 \cdot \sqrt{2} = 10\sqrt{2}$$

The radical is now simplified since 2 contains no perfect square factors (other than 1).

> Helpful Hint
>
> To help you recognize largest perfect power factors of a radicand, it will help if you are familiar with some perfect powers. A few are listed below.
>
> | Perfect Squares | 1, | 4, | 9, | 16, | 25, | 36, | 49, | 64, | 81, | 100, | 121, | 144 |
> |---|---|---|---|---|---|---|---|---|---|---|---|---|
> | | 1^2 | 2^2 | 3^2 | 4^2 | 5^2 | 6^2 | 7^2 | 8^2 | 9^2 | 10^2 | 11^2 | 12^2 |
>
> | Perfect Cubes | 1, | 8, | 27, | 64, | 125 |
> |---|---|---|---|---|---|
> | | 1^3 | 2^3 | 3^3 | 4^3 | 5^3 |
>
> | Perfect Fourth Powers | 1, | 16, | 81, | 256 |
> |---|---|---|---|---|
> | | 1^4 | 2^4 | 3^4 | 4^4 |

In general, we say that a radicand of the form $\sqrt[n]{a}$ is simplified when the radicand a contains no factors that are perfect nth powers (other than 1 or -1).

EXAMPLE 4 Use the product rule to simplify.

$$\text{a. } \sqrt{25x^3} \qquad\qquad \text{b. } \sqrt[3]{54x^6y^8} \qquad\qquad \text{c. } \sqrt[4]{81z^{11}}$$

Solution

a. $\sqrt{25x^3} = \sqrt{25x^2 \cdot x}$ Find the largest perfect square factor.

$\qquad\qquad = \sqrt{25x^2} \cdot \sqrt{x}$ Apply the product rule.

$\qquad\qquad = 5x\sqrt{x}$ Simplify.

b. $\sqrt[3]{54x^6y^8} = \sqrt[3]{27 \cdot 2 \cdot x^6 \cdot y^6 \cdot y^2}$ Factor the radicand and identify perfect cube factors.

$\qquad\qquad = \sqrt[3]{27x^6y^6 \cdot 2y^2}$

$\qquad\qquad = \sqrt[3]{27x^6y^6} \cdot \sqrt[3]{2y^2}$ Apply the product rule.

$\qquad\qquad = 3x^2y^2\sqrt[3]{2y^2}$ Simplify.

c. $\sqrt[4]{81z^{11}} = \sqrt[4]{81 \cdot z^8 \cdot z^3}$ Factor the radicand and identify perfect fourth power factors.

$\qquad\qquad = \sqrt[4]{81z^8} \cdot \sqrt[4]{z^3}$ Apply the product rule.

$\qquad\qquad = 3z^2\sqrt[4]{z^3}$ Simplify.   □

PRACTICE
4 Use the product rule to simplify.

$$\text{a. } \sqrt{36z^7} \qquad\qquad \text{b. } \sqrt[3]{32p^4q^7} \qquad\qquad \text{c. } \sqrt[4]{16x^{15}} \qquad\qquad ■$$

EXAMPLE 5 Use the quotient rule to divide, and simplify if possible.

$$\text{a. } \frac{\sqrt{20}}{\sqrt{5}} \qquad \text{b. } \frac{\sqrt{50x}}{2\sqrt{2}} \qquad \text{c. } \frac{7\sqrt[3]{48x^4y^8}}{\sqrt[3]{6y^2}} \qquad \text{d. } \frac{2\sqrt[4]{32a^8b^6}}{\sqrt[4]{a^{-1}b^2}}$$

Solution

a. $\dfrac{\sqrt{20}}{\sqrt{5}} = \sqrt{\dfrac{20}{5}}$ Apply the quotient rule.

$\qquad\quad = \sqrt{4}$ Simplify.

$\qquad\quad = 2$ Simplify.

b. $\dfrac{\sqrt{50x}}{2\sqrt{2}} = \dfrac{1}{2} \cdot \sqrt{\dfrac{50x}{2}}$ Apply the quotient rule.

$\qquad\quad = \dfrac{1}{2} \cdot \sqrt{25x}$ Simplify.

$\qquad\quad = \dfrac{1}{2} \cdot \sqrt{25} \cdot \sqrt{x}$ Factor 25x.

$\qquad\quad = \dfrac{1}{2} \cdot 5 \cdot \sqrt{x}$ Simplify.

$\qquad\quad = \dfrac{5}{2}\sqrt{x}$

c. $\dfrac{7\sqrt[3]{48x^4y^8}}{\sqrt[3]{6y^2}} = 7 \cdot \sqrt[3]{\dfrac{48x^4y^8}{6y^2}}$ Apply the quotient rule.

$\qquad\quad = 7 \cdot \sqrt[3]{8x^4y^6}$ Simplify.

$\qquad\quad = 7\sqrt[3]{8x^3y^6 \cdot x}$ Factor.

$\qquad\quad = 7 \cdot \sqrt[3]{8x^3y^6} \cdot \sqrt[3]{x}$ Apply the product rule.

$\qquad\quad = 7 \cdot 2xy^2 \cdot \sqrt[3]{x}$ Simplify.

$\qquad\quad = 14xy^2\sqrt[3]{x}$

(Continued on next page)

d. $\dfrac{2\sqrt[4]{32a^8b^6}}{\sqrt[4]{a^{-1}b^2}} = 2\sqrt[4]{\dfrac{32a^8b^6}{a^{-1}b^2}} = 2\sqrt[4]{32a^9b^4} = 2\sqrt[4]{16 \cdot a^8 \cdot b^4 \cdot 2 \cdot a}$

$$= 2\sqrt[4]{16a^8b^4} \cdot \sqrt[4]{2a} = 2 \cdot 2a^2b \cdot \sqrt[4]{2a} = 4a^2b\sqrt[4]{2a}$$

PRACTICE

5 Use the quotient rule to divide and simplify.

a. $\dfrac{\sqrt{80}}{\sqrt{5}}$
b. $\dfrac{\sqrt{98z}}{3\sqrt{2}}$
c. $\dfrac{5\sqrt[3]{40x^5y^7}}{\sqrt[3]{5y}}$
d. $\dfrac{3\sqrt[5]{64x^9y^8}}{\sqrt[5]{x^{-1}y^2}}$

 CONCEPT CHECK

Find and correct the error:

$$\dfrac{\sqrt[3]{27}}{\sqrt{9}} = \sqrt[3]{\dfrac{27}{9}} = \sqrt[3]{3}$$

OBJECTIVE

4 Using the Distance and Midpoint Formulas

Now that we know how to simplify radicals, we can derive and use the distance formula. The midpoint formula is often confused with the distance formula, so to clarify both, we will also review the midpoint formula.

The Cartesian coordinate system helps us visualize a distance between points. To find the distance between two points, we use the distance formula, which is derived from the Pythagorean theorem.

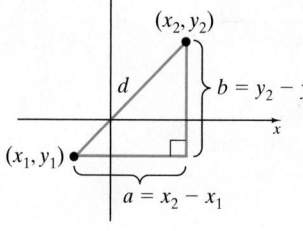

To find the distance d between two points (x_1, y_1) and (x_2, y_2) as shown to the left, notice that the length of leg a is $x_2 - x_1$ and that the length of leg b is $y_2 - y_1$.

Thus, the Pythagorean theorem tells us that

$$d^2 = a^2 + b^2$$

or

$$d^2 = (x_2 - x_1)^2 + (y_2 - y_1)^2$$

or

$$d = \sqrt{(x_2 - x_1)^2 + (y_2 - y_1)^2}$$

This formula gives us the distance between any two points on the real plane.

> **Distance Formula**
>
> The distance d between two points (x_1, y_1) and (x_2, y_2) is given by
> $$d = \sqrt{(x_2 - x_1)^2 + (y_2 - y_1)^2}$$

EXAMPLE 6 Find the distance between $(2, -5)$ and $(1, -4)$. Give the exact distance and a three-decimal-place approximation.

Solution To use the distance formula, it makes no difference which point we call (x_1, y_1) and which point we call (x_2, y_2). We will let $(x_1, y_1) = (2, -5)$ and $(x_2, y_2) = (1, -4)$.

Answer to Concept Check:
$$\dfrac{\sqrt[3]{27}}{\sqrt{9}} = \dfrac{3}{3} = 1$$

$$d = \sqrt{(x_2 - x_1)^2 + (y_2 - y_1)^2}$$
$$= \sqrt{(1 - 2)^2 + [-4 - (-5)]^2}$$
$$= \sqrt{(-1)^2 + (1)^2}$$
$$= \sqrt{1 + 1}$$
$$= \sqrt{2} \approx 1.414$$

The distance between the two points is exactly $\sqrt{2}$ units, or approximately 1.414 units.

PRACTICE

6   Find the distance between $(-3, 7)$ and $(-2, 3)$. Give the exact distance and a three-decimal-place approximation.

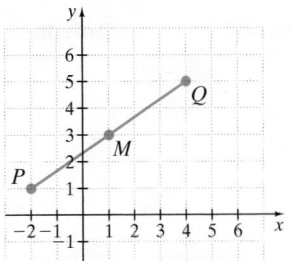

The **midpoint** of a line segment is the **point** located exactly halfway between the two endpoints of the line segment. On the graph to the left, the point M is the midpoint of line segment PQ. Thus, the distance between M and P equals the distance between M and Q.

Note: We usually need no knowledge of roots to calculate the midpoint of a line segment. We review midpoint here only because it is often confused with the distance between two points.

The x-coordinate of M is at half the distance between the x-coordinates of P and Q, and the y-coordinate of M is at half the distance between the y-coordinates of P and Q. That is, the x-coordinate of M is the average of the x-coordinates of P and Q; the y-coordinate of M is the average of the y-coordinates of P and Q.

Midpoint Formula

The midpoint of the line segment whose endpoints are (x_1, y_1) and (x_2, y_2) is the point with coordinates

$$\left(\frac{x_1 + x_2}{2}, \frac{y_1 + y_2}{2} \right)$$

EXAMPLE 7   Find the midpoint of the line segment that joins points $P(-3, 3)$ and $Q(1, 0)$.

Solution Use the midpoint formula. It makes no difference which point we call (x_1, y_1) or which point we call (x_2, y_2). Let $(x_1, y_1) = (-3, 3)$ and $(x_2, y_2) = (1, 0)$.

$$\text{midpoint} = \left(\frac{x_1 + x_2}{2}, \frac{y_1 + y_2}{2} \right)$$
$$= \left(\frac{-3 + 1}{2}, \frac{3 + 0}{2} \right)$$
$$= \left(\frac{-2}{2}, \frac{3}{2} \right)$$
$$= \left(-1, \frac{3}{2} \right)$$

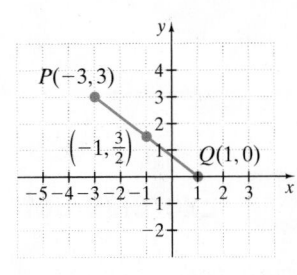

The midpoint of the segment is $\left(-1, \dfrac{3}{2} \right)$.

PRACTICE

7   Find the midpoint of the line segment that joins points $P(5, -2)$ and $Q(8, -6)$.

> **Helpful Hint**
>
> The distance between two points is a distance. The midpoint of a line segment is the point halfway between the endpoints of the segment.

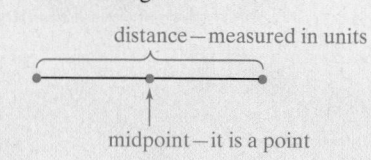

distance—measured in units

midpoint—it is a point

✓ Vocabulary, Readiness & Video Check

Use the choices below to fill in each blank. Some choices may be used more than once.

distance   midpoint   point

1. The _____ of a line segment is the _____ exactly halfway between the two endpoints of the line segment.

2. The _____ between two points is a distance, measured in units.

3. The _____ formula is $d = \sqrt{(x_2 - x_1)^2 + (y_2 - y_1)^2}$.

4. The _____ formula is $\left(\dfrac{x_1 + x_2}{2}, \dfrac{y_1 + y_2}{2} \right)$.

Martin-Gay Interactive Videos

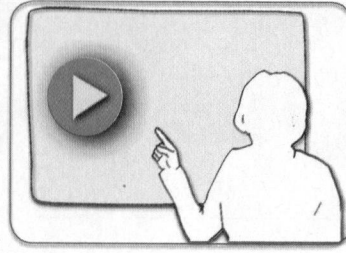

See Video 10.3 ⊙

Watch the section lecture video and answer the following questions.

OBJECTIVE 1

5. From ⊞ Example 1 and the lecture before, in order to apply the product rule for radicals, what must be true about the indexes of the radicals being multiplied?

OBJECTIVE 2

6. From ⊞ Examples 2–6, when might you apply the quotient rule (in either direction) in order to simplify a fractional radical expression?

OBJECTIVE 3

7. From ⊞ Example 8, we know that an even power of a variable is a perfect square factor of the variable, leaving no factor in the radicand once simplified. Therefore, what must be true about the power of any variable left in the radicand of a simplified square root? Explain.

OBJECTIVE 4

8. From ⊞ Example 10, the formula uses the coordinates of two points similar to the slope formula. What caution should you take when replacing values in the formula?

OBJECTIVE 4

9. Based on ⊞ Example 11, complete the following statement. The *x*-value of the midpoint is the _____ of the *x*-values of the endpoints, and the *y*-value of the midpoint is the _____ of the *y*-values of the endpoints.

10.3 Exercise Set   MyMathLab ▸

Use the product rule to multiply. See Example 1.

1. $\sqrt{7} \cdot \sqrt{2}$

2. $\sqrt{11} \cdot \sqrt{10}$

3. $\sqrt[4]{8} \cdot \sqrt[4]{2}$

4. $\sqrt[4]{27} \cdot \sqrt[4]{3}$

5. $\sqrt[3]{4} \cdot \sqrt[3]{9}$

6. $\sqrt[3]{10} \cdot \sqrt[3]{5}$

▸ **7.** $\sqrt{2} \cdot \sqrt{3x}$

8. $\sqrt{3y} \cdot \sqrt{5x}$

9. $\sqrt{\dfrac{7}{x}} \cdot \sqrt{\dfrac{2}{y}}$

10. $\sqrt{\dfrac{6}{m}} \cdot \sqrt{\dfrac{n}{5}}$

11. $\sqrt[4]{4x^3} \cdot \sqrt[4]{5}$

12. $\sqrt[4]{ab^2} \cdot \sqrt[4]{27ab}$

Use the quotient rule to simplify. See Examples 2 and 3.

▸ **13.** $\sqrt{\dfrac{6}{49}}$

14. $\sqrt{\dfrac{8}{81}}$

15. $\sqrt{\dfrac{2}{49}}$

16. $\sqrt{\dfrac{5}{121}}$

▸ **17.** $\sqrt[4]{\dfrac{x^3}{16}}$

18. $\sqrt[4]{\dfrac{y}{81x^4}}$

19. $\sqrt[3]{\dfrac{4}{27}}$

20. $\sqrt[3]{\dfrac{3}{64}}$

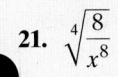

21. $\sqrt[4]{\dfrac{8}{x^8}}$

22. $\sqrt[4]{\dfrac{a^3}{81}}$

23. $\sqrt[3]{\dfrac{2x}{81y^{12}}}$

24. $\sqrt[3]{\dfrac{3}{8x^6}}$

25. $\sqrt{\dfrac{x^2y}{100}}$

26. $\sqrt{\dfrac{y^2z}{36}}$

27. $\sqrt{\dfrac{5x^2}{4y^2}}$

28. $\sqrt{\dfrac{y^{10}}{9x^6}}$

29. $-\sqrt[3]{\dfrac{z^7}{27x^3}}$

30. $-\sqrt[3]{\dfrac{64a}{b^9}}$

Simplify. See Examples 3 and 4.

31. $\sqrt{32}$

32. $\sqrt{27}$

33. $\sqrt[3]{192}$

34. $\sqrt[3]{108}$

35. $5\sqrt{75}$

36. $3\sqrt{8}$

37. $\sqrt{24}$

38. $\sqrt{20}$

39. $\sqrt{100x^5}$

40. $\sqrt{64y^9}$

41. $\sqrt[3]{16y^7}$

42. $\sqrt[3]{64y^9}$

43. $\sqrt[4]{a^8b^7}$

44. $\sqrt[5]{32z^{12}}$

45. $\sqrt{y^5}$

46. $\sqrt[3]{y^5}$

47. $\sqrt{25a^2b^3}$

48. $\sqrt{9x^5y^7}$

49. $\sqrt[5]{-32x^{10}y}$

50. $\sqrt[5]{-243z^9}$

51. $\sqrt[3]{50x^{14}}$

52. $\sqrt[3]{40y^{10}}$

53. $-\sqrt{32a^8b^7}$

54. $-\sqrt{20ab^6}$

55. $\sqrt{9x^7y^9}$

56. $\sqrt{12r^9s^{12}}$

57. $\sqrt[3]{125r^9s^{12}}$

58. $\sqrt[3]{8a^6b^9}$

59. $\sqrt[4]{32x^{12}y^5}$

60. $\sqrt[4]{162x^7y^{20}}$

Use the quotient rule to divide. Then simplify if possible. See Example 5.

61. $\dfrac{\sqrt{14}}{\sqrt{7}}$

62. $\dfrac{\sqrt{45}}{\sqrt{9}}$

63. $\dfrac{\sqrt[3]{24}}{\sqrt[3]{3}}$

64. $\dfrac{\sqrt[3]{10}}{\sqrt[3]{2}}$

65. $\dfrac{5\sqrt[4]{48}}{\sqrt[4]{3}}$

66. $\dfrac{7\sqrt[4]{162}}{\sqrt[4]{2}}$

67. $\dfrac{\sqrt{x^5y^3}}{\sqrt{xy}}$

68. $\dfrac{\sqrt{a^7b^6}}{\sqrt{a^3b^2}}$

69. $\dfrac{8\sqrt[3]{54m^7}}{\sqrt[3]{2m}}$

70. $\dfrac{\sqrt[3]{128x^3}}{-3\sqrt[3]{2x}}$

71. $\dfrac{3\sqrt{100x^2}}{2\sqrt{2x^{-1}}}$

72. $\dfrac{\sqrt{270y^2}}{5\sqrt{3y^{-4}}}$

73. $\dfrac{\sqrt[4]{96a^{10}b^3}}{\sqrt[4]{3a^2b^3}}$

74. $\dfrac{\sqrt[4]{160x^{10}y^5}}{\sqrt[4]{2x^2y^2}}$

75. $\dfrac{\sqrt[5]{64x^{10}y^3}}{\sqrt[5]{2x^3y^{-7}}}$

76. $\dfrac{\sqrt[5]{192x^6y^{12}}}{\sqrt[5]{2x^{-1}y^{-3}}}$

Find the distance between each pair of points. Give the exact distance and a three-decimal-place approximation. See Example 6.

77. $(5,1)$ and $(8,5)$

78. $(2,3)$ and $(14,8)$

79. $(-3,2)$ and $(1,-3)$

80. $(3,-2)$ and $(-4,1)$

81. $(-9,4)$ and $(-8,1)$

82. $(-5,-2)$ and $(-6,-6)$

83. $(0,-\sqrt{2})$ and $(\sqrt{3},0)$

84. $(-\sqrt{5},0)$ and $(0,\sqrt{7})$

85. $(1.7,-3.6)$ and $(-8.6,5.7)$

86. $(9.6,2.5)$ and $(-1.9,-3.7)$

Find the midpoint of the line segment whose endpoints are given. See Example 7.

87. $(6,-8),(2,4)$

88. $(3,9),(7,11)$

89. $(-2,-1),(-8,6)$

90. $(-3,-4),(6,-8)$

91. $(7,3),(-1,-3)$

92. $(-2,5),(-1,6)$

93. $\left(\dfrac{1}{2},\dfrac{3}{8}\right),\left(-\dfrac{3}{2},\dfrac{5}{8}\right)$

94. $\left(-\dfrac{2}{5},\dfrac{7}{15}\right),\left(-\dfrac{2}{5},-\dfrac{4}{15}\right)$

95. $(\sqrt{2},3\sqrt{5}),(\sqrt{2},-2\sqrt{5})$

96. $(\sqrt{8},-\sqrt{12}),(3\sqrt{2},7\sqrt{3})$

97. $(4.6,-3.5),(7.8,-9.8)$

98. $(-4.6,2.1),(-6.7,1.9)$

REVIEW AND PREVIEW

Perform each indicated operation. See Sections 2.1, 5.3, and 5.4.

99. $6x+8x$

100. $(6x)(8x)$

101. $(2x+3)(x-5)$

102. $(2x+3)+(x-5)$

103. $9y^2-8y^2$

104. $(9y^2)(-8y^2)$

105. $-3(x+5)$

106. $-3+x+5$

107. $(x-4)^2$

108. $(2x+1)^2$

CONCEPT EXTENSIONS

Answer true or false. Assume all radicals represent nonzero real numbers.

109. $\sqrt[n]{a}\cdot\sqrt[n]{b}=\sqrt[n]{ab}$, _____

110. $\sqrt[3]{7}\cdot\sqrt[3]{11}=\sqrt[3]{18}$, _____

111. $\sqrt[3]{7}\cdot\sqrt{11}=\sqrt{77}$, _____

112. $\sqrt{x^7y^8}=\sqrt{x^7}\cdot\sqrt{y^8}$, _____

113. $\dfrac{\sqrt[n]{a}}{\sqrt[n]{b}}=\sqrt[n]{\dfrac{a}{b}}$, _____

114. $\dfrac{\sqrt[3]{12}}{\sqrt[3]{4}}=\sqrt[3]{8}$, _____

Find and correct the error. See the Concept Check in this section.

115. $\dfrac{\sqrt[3]{64}}{\sqrt{64}} = \sqrt[3]{\dfrac{64}{64}} = \sqrt[3]{1} = 1$ ✗

116. $\dfrac{\sqrt[4]{16}}{\sqrt{4}} = \sqrt[4]{\dfrac{16}{4}} = \sqrt[4]{4}$ ✗

Simplify. Assume variables represent positive numbers.

117. $\sqrt[5]{x^{35}}$

118. $\sqrt[6]{y^{48}}$

119. $\sqrt[4]{a^{12}b^4c^{20}}$

120. $\sqrt[3]{a^9b^{21}c^3}$

121. $\sqrt[3]{z^{32}}$

122. $\sqrt[5]{x^{49}}$

123. $\sqrt[7]{q^{17}r^{40}s^7}$

124. $\sqrt[4]{p^{11}q^4r^{45}}$

125. The formula for the radius r of a sphere with surface area A is given by $r = \sqrt{\dfrac{A}{4\pi}}$. Calculate the radius of a standard zorb whose outside surface area is 32.17 square meters. Round to the nearest tenth. (A zorb is a large inflated ball within a ball in which a person, strapped inside, may choose to roll down a hill. *Source:* Zorb, Ltd.)

126. NASA has investigated using a zorb-like Mars explorer, called the tumbleweed rover, to explore the Martian plains using wind power alone. Researchers have determined that the optimal rover would need to be efficient at catching any breeze, light enough to move easily, and large enough to roll over any rocks in its path. (*Source:* NASA)

a. If scientists determine that the volume of the sphere-shaped tumbleweed rover should be approximately 4200 cubic feet, find the radius for the rover rounded to the nearest tenth. The formula for the radius r of a sphere with volume V is given by $r = \sqrt[3]{\dfrac{3V}{4\pi}}$.

b. One of the designs being considered for further investigation is the so-called box kite drawn below. Here, in addition to a sphere, there are three disks that look like three great circles at right angles to each other with an exoskeleton to protect the "kites." Find the total area of three great circles each with a diameter of 6 meters. Round to the nearest tenth of a square meter.

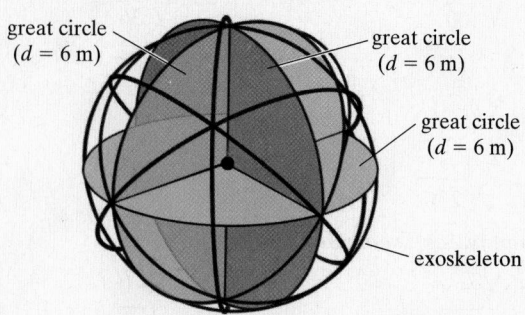

Box Kite design

c. Do you think the box kite design is better than the sphere-alone design? Why or why not?

127. The formula for the lateral surface area A of a cone with height h and radius r is given by

$$A = \pi r \sqrt{r^2 + h^2}$$

a. Find the lateral surface area of a cone whose height is 3 centimeters and whose radius is 4 centimeters.

b. Approximate to two decimal places the lateral surface area of a cone whose height is 7.2 feet and whose radius is 6.8 feet.

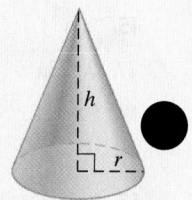

128. Before Mount Vesuvius, a volcano in Italy, erupted violently in 79 C.E., its height was 4190 feet. Vesuvius was roughly cone-shaped, and its base had a radius of approximately 25,200 feet. Use the formula for the lateral surface area of a cone, given in Exercise 127, to approximate the surface area this volcano had before it erupted. (*Source:* Global Volcanism Network)

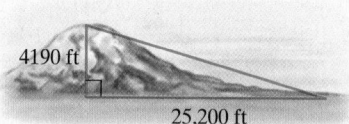

4190 ft

25,200 ft

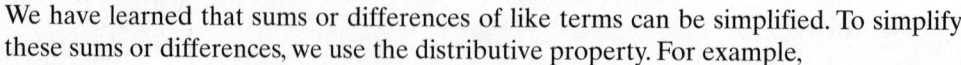

10.4 Adding, Subtracting, and Multiplying Radical Expressions ▶

OBJECTIVES

1 Add or Subtract Radical Expressions. ▶

2 Multiply Radical Expressions. ▶

OBJECTIVE

1 Adding or Subtracting Radical Expressions ▶

We have learned that sums or differences of like terms can be simplified. To simplify these sums or differences, we use the distributive property. For example,

$$2x + 3x = (2 + 3)x = 5x \quad \text{and} \quad 7x^2y - 4x^2y = (7 - 4)x^2y = 3x^2y$$

The distributive property can also be used to add **like radicals.**

Like Radicals

Radicals with the same index and the same radicand are like radicals.

For example, $2\sqrt{7} + 3\sqrt{7} = (2 + 3)\sqrt{7} = 5\sqrt{7}$. Also,

$$\underbrace{2\sqrt{7} + 3\sqrt{7}}_{\text{Like radicals}}$$

$$5\sqrt{3x} - 7\sqrt{3x} = (5 - 7)\sqrt{3x} = -2\sqrt{3x}$$

The expression $2\sqrt{7} + 2\sqrt[3]{7}$ cannot be simplified further since $2\sqrt{7}$ and $2\sqrt[3]{7}$ are not like radicals.

$$\underbrace{2\sqrt{7} + 2\sqrt[3]{7}}_{\text{Unlike radicals}}$$

EXAMPLE 1   Add or subtract as indicated. Assume all variables represent positive real numbers.

a. $4\sqrt{11} + 8\sqrt{11}$ b. $5\sqrt[3]{3x} - 7\sqrt[3]{3x}$ c. $4\sqrt{5} + 4\sqrt[3]{5}$

Solution

a. $4\sqrt{11} + 8\sqrt{11} = (4 + 8)\sqrt{11} = 12\sqrt{11}$
b. $5\sqrt[3]{3x} - 7\sqrt[3]{3x} = (5 - 7)\sqrt[3]{3x} = -2\sqrt[3]{3x}$
c. $4\sqrt{5} + 4\sqrt[3]{5}$
This expression cannot be simplified since $4\sqrt{5}$ and $4\sqrt[3]{5}$ do not contain like radicals.   □

PRACTICE
1   Add or subtract as indicated.
a. $3\sqrt{17} + 5\sqrt{17}$ b. $7\sqrt[3]{5z} - 12\sqrt[3]{5z}$ c. $3\sqrt{2} + 5\sqrt[3]{2}$ ■

When adding or subtracting radicals, always check first to see whether any radicals can be simplified.

✔ CONCEPT CHECK
True or false?

$$\sqrt{a} + \sqrt{b} = \sqrt{a + b}$$

Explain.

EXAMPLE 2   Add or subtract. Assume that variables represent positive real numbers.

a. $\sqrt{20} + 2\sqrt{45}$ b. $\sqrt[3]{54} - 5\sqrt[3]{16} + \sqrt[3]{2}$ c. $\sqrt{27x} - 2\sqrt{9x} + \sqrt{72x}$
d. $\sqrt[3]{98} + \sqrt{98}$ e. $\sqrt[3]{48y^4} + \sqrt[3]{6y^4}$

Solution First, simplify each radical. Then add or subtract any like radicals.

a. $\sqrt{20} + 2\sqrt{45} = \sqrt{4 \cdot 5} + 2\sqrt{9 \cdot 5}$ Factor 20 and 45.

$\qquad\qquad = \sqrt{4} \cdot \sqrt{5} + 2 \cdot \sqrt{9} \cdot \sqrt{5}$ Use the product rule.

$\qquad\qquad = 2 \cdot \sqrt{5} + 2 \cdot 3 \cdot \sqrt{5}$ Simplify $\sqrt{4}$ and $\sqrt{9}$.

$\qquad\qquad = 2\sqrt{5} + 6\sqrt{5}$

$\qquad\qquad = 8\sqrt{5}$ Add like radicals.

b. $\sqrt[3]{54} - 5\sqrt[3]{16} + \sqrt[3]{2}$

$\qquad = \sqrt[3]{27} \cdot \sqrt[3]{2} - 5 \cdot \sqrt[3]{8} \cdot \sqrt[3]{2} + \sqrt[3]{2}$ Factor and use the product rule.

$\qquad = 3 \cdot \sqrt[3]{2} - 5 \cdot 2 \cdot \sqrt[3]{2} + \sqrt[3]{2}$ Simplify $\sqrt[3]{27}$ and $\sqrt[3]{8}$.

$\qquad = 3\sqrt[3]{2} - 10\sqrt[3]{2} + \sqrt[3]{2}$ Write $5 \cdot 2$ as 10.

$\qquad = -6\sqrt[3]{2}$ Combine like radicals.

Answer to Concept Check:
false; answers may vary

(Continued on next page)

c. $\sqrt{27x} - 2\sqrt{9x} + \sqrt{72x}$

$= \sqrt{9} \cdot \sqrt{3x} - 2 \cdot \sqrt{9} \cdot \sqrt{x} + \sqrt{36} \cdot \sqrt{2x}$ Factor and use the product rule.

$= 3 \cdot \sqrt{3x} - 2 \cdot 3 \cdot \sqrt{x} + 6 \cdot \sqrt{2x}$ Simplify $\sqrt{9}$ and $\sqrt{36}$.

$= 3\sqrt{3x} - 6\sqrt{x} + 6\sqrt{2x}$ Write $2 \cdot 3$ as 6.

> **Helpful Hint**
>
> None of these terms contain like radicals. We can simplify no further.

d. $\sqrt[3]{98} + \sqrt{98} = \sqrt[3]{98} + \sqrt{49} \cdot \sqrt{2}$ Factor and use the product rule.

$= \sqrt[3]{98} + 7\sqrt{2}$ No further simplification is possible.

e. $\sqrt[3]{48y^4} + \sqrt[3]{6y^4} = \sqrt[3]{8y^3} \cdot \sqrt[3]{6y} + \sqrt[3]{y^3} \cdot \sqrt[3]{6y}$ Factor and use the product rule.

$= 2y\sqrt[3]{6y} + y\sqrt[3]{6y}$ Simplify $\sqrt[3]{8y^3}$ and $\sqrt[3]{y^3}$.

$= 3y\sqrt[3]{6y}$ Combine like radicals.   ☐

PRACTICE
2 Add or subtract.

a. $\sqrt{24} + 3\sqrt{54}$ **b.** $\sqrt[3]{24} - 4\sqrt[3]{81} + \sqrt[3]{3}$ **c.** $\sqrt{75x} - 3\sqrt{27x} + \sqrt{12x}$
d. $\sqrt{40} + \sqrt[3]{40}$ **e.** $\sqrt[3]{81x^4} + \sqrt[3]{3x^4}$ ■

Let's continue to assume that variables represent positive real numbers.

EXAMPLE 3   Add or subtract as indicated.

a. $\dfrac{\sqrt{45}}{4} - \dfrac{\sqrt{5}}{3}$ **b.** $\sqrt[3]{\dfrac{7x}{8}} + 2\sqrt[3]{7x}$

Solution

a. $\dfrac{\sqrt{45}}{4} - \dfrac{\sqrt{5}}{3} = \dfrac{3\sqrt{5}}{4} - \dfrac{\sqrt{5}}{3}$ To subtract, notice that the LCD is 12.

$= \dfrac{3\sqrt{5} \cdot 3}{4 \cdot 3} - \dfrac{\sqrt{5} \cdot 4}{3 \cdot 4}$ Write each expression as an equivalent expression with a denominator of 12.

$= \dfrac{9\sqrt{5}}{12} - \dfrac{4\sqrt{5}}{12}$ Multiply factors in the numerator and the denominator.

$= \dfrac{5\sqrt{5}}{12}$ Subtract.

b. $\sqrt[3]{\dfrac{7x}{8}} + 2\sqrt[3]{7x} = \dfrac{\sqrt[3]{7x}}{\sqrt[3]{8}} + 2\sqrt[3]{7x}$ Apply the quotient rule for radicals.

$= \dfrac{\sqrt[3]{7x}}{2} + 2\sqrt[3]{7x}$ Simplify.

$= \dfrac{\sqrt[3]{7x}}{2} + \dfrac{2\sqrt[3]{7x} \cdot 2}{2}$ Write each expression as an equivalent expression with a denominator of 2.

$= \dfrac{\sqrt[3]{7x}}{2} + \dfrac{4\sqrt[3]{7x}}{2}$

$= \dfrac{5\sqrt[3]{7x}}{2}$ Add.   ☐

PRACTICE
3 Add or subtract as indicated.

a. $\dfrac{\sqrt{28}}{3} - \dfrac{\sqrt{7}}{4}$ **b.** $\sqrt[3]{\dfrac{6y}{64}} + 3\sqrt[3]{6y}$

OBJECTIVE

2 Multiplying Radical Expressions

We can multiply radical expressions by using many of the same properties used to multiply polynomial expressions. For instance, to multiply $\sqrt{2}(\sqrt{6} - 3\sqrt{2})$, we use the distributive property and multiply $\sqrt{2}$ by each term inside the parentheses.

$$\sqrt{2}(\sqrt{6} - 3\sqrt{2}) = \sqrt{2}(\sqrt{6}) - \sqrt{2}(3\sqrt{2}) \quad \text{Use the distributive property.}$$
$$= \sqrt{2 \cdot 6} - 3\sqrt{2 \cdot 2}$$
$$= \sqrt{2 \cdot 2 \cdot 3} - 3 \cdot 2 \quad \text{Use the product rule for radicals.}$$
$$= 2\sqrt{3} - 6$$

EXAMPLE 4 Multiply.

a. $\sqrt{3}(5 + \sqrt{30})$ **b.** $(\sqrt{5} - \sqrt{6})(\sqrt{7} + 1)$ **c.** $(7\sqrt{x} + 5)(3\sqrt{x} - \sqrt{5})$

d. $(4\sqrt{3} - 1)^2$ **e.** $(\sqrt{2x} - 5)(\sqrt{2x} + 5)$ **f.** $(\sqrt{x-3} + 5)^2$

Solution

a. $\sqrt{3}(5 + \sqrt{30}) = \sqrt{3}(5) + \sqrt{3}(\sqrt{30})$
$$= 5\sqrt{3} + \sqrt{3 \cdot 30}$$
$$= 5\sqrt{3} + \sqrt{3 \cdot 3 \cdot 10}$$
$$= 5\sqrt{3} + 3\sqrt{10}$$

b. To multiply, we can use the FOIL method.

$$\overset{\qquad\quad\text{First}\qquad\text{Outer}\qquad\text{Inner}\qquad\text{Last}}{(\sqrt{5} - \sqrt{6})(\sqrt{7} + 1) = \sqrt{5}\cdot\sqrt{7} + \sqrt{5}\cdot 1 - \sqrt{6}\cdot\sqrt{7} - \sqrt{6}\cdot 1}$$
$$= \sqrt{35} + \sqrt{5} - \sqrt{42} - \sqrt{6}$$

c. $(7\sqrt{x} + 5)(3\sqrt{x} - \sqrt{5}) = 7\sqrt{x}(3\sqrt{x}) - 7\sqrt{x}(\sqrt{5}) + 5(3\sqrt{x}) - 5(\sqrt{5})$
$$= 21x - 7\sqrt{5x} + 15\sqrt{x} - 5\sqrt{5}$$

d. $(4\sqrt{3} - 1)^2 = (4\sqrt{3} - 1)(4\sqrt{3} - 1)$
$$= 4\sqrt{3}(4\sqrt{3}) - 4\sqrt{3}(1) - 1(4\sqrt{3}) - 1(-1)$$
$$= 16 \cdot 3 - 4\sqrt{3} - 4\sqrt{3} + 1$$
$$= 48 - 8\sqrt{3} + 1$$
$$= 49 - 8\sqrt{3}$$

e. $(\sqrt{2x} - 5)(\sqrt{2x} + 5) = \sqrt{2x}\cdot\sqrt{2x} + 5\sqrt{2x} - 5\sqrt{2x} - 5 \cdot 5$
$$= 2x - 25$$

f. $(\sqrt{x-3} + 5)^2 = (\sqrt{x-3})^2 + 2\cdot\sqrt{x-3}\cdot 5 + 5^2$

$$\underset{a \qquad\qquad b \qquad\qquad a^2 \qquad + 2\cdot \qquad a \qquad \cdot b + b^2}{\uparrow \qquad\quad \uparrow \qquad\qquad\quad \uparrow \qquad\quad \uparrow\uparrow \qquad \uparrow \qquad \uparrow\;\;\uparrow}$$

$$= x - 3 + 10\sqrt{x-3} + 25 \quad \text{Simplify.}$$
$$= x + 22 + 10\sqrt{x-3} \quad \text{Combine like terms.} \qquad \square$$

PRACTICE

4 Multiply.

a. $\sqrt{5}(2 + \sqrt{15})$ **b.** $(\sqrt{2} - \sqrt{5})(\sqrt{6} + 2)$

c. $(3\sqrt{z} - 4)(2\sqrt{z} + 3)$ **d.** $(\sqrt{6} - 3)^2$

e. $(\sqrt{5x} + 3)(\sqrt{5x} - 3)$ **f.** $(\sqrt{x+2} + 3)^2$

Vocabulary, Readiness & Video Check

Complete the table with "Like" or "Unlike."

Terms	Like or Unlike Radical Terms?
1. $\sqrt{7}, \sqrt[3]{7}$	
2. $\sqrt[3]{x^2 y}, \sqrt[3]{yx^2}$	
3. $\sqrt[3]{abc}, \sqrt[3]{cba}$	
4. $2x\sqrt{5}, 2x\sqrt{10}$	

Simplify. Assume that all variables represent positive real numbers.

5. $2\sqrt{3} + 4\sqrt{3} =$ _____

6. $5\sqrt{7} + 3\sqrt{7} =$ _____

7. $8\sqrt{x} - \sqrt{x} =$ _____

8. $3\sqrt{y} - \sqrt{y} =$ _____

9. $7\sqrt[3]{x} + \sqrt[3]{x} =$ _____

10. $8\sqrt[3]{z} + \sqrt[3]{z} =$ _____

Martin-Gay Interactive Videos

See Video 10.4

Watch the section lecture video and answer the following questions.

OBJECTIVE 1

11. From ▢ Examples 1 and 2, why should you always check to see if all terms in your expression are simplified before attempting to add or subtract radicals?

OBJECTIVE 2

12. In ▢ Example 4, what are you told to remember about the square root of a positive number?

10.4 Exercise Set MyMathLab®

Add or subtract. See Examples 1 through 3.

1. $\sqrt{8} - \sqrt{32}$

2. $\sqrt{27} - \sqrt{75}$

3. $2\sqrt{2x^3} + 4x\sqrt{8x}$

4. $3\sqrt{45x^3} + x\sqrt{5x}$

5. $2\sqrt{50} - 3\sqrt{125} + \sqrt{98}$

6. $4\sqrt{32} - \sqrt{18} + 2\sqrt{128}$

7. $\sqrt[3]{16x} - \sqrt[3]{54x}$

8. $2\sqrt[3]{3a^4} - 3a\sqrt[3]{81a}$

9. $\sqrt{9b^3} - \sqrt{25b^3} + \sqrt{49b^3}$

10. $\sqrt{4x^7} + 9x^2\sqrt{x^3} - 5x\sqrt{x^5}$

11. $\dfrac{5\sqrt{2}}{3} + \dfrac{2\sqrt{2}}{5}$

12. $\dfrac{\sqrt{3}}{2} + \dfrac{4\sqrt{3}}{3}$

13. $\sqrt[3]{\dfrac{11}{8}} - \dfrac{\sqrt[3]{11}}{6}$

14. $\dfrac{2\sqrt[3]{4}}{7} - \dfrac{\sqrt[3]{4}}{14}$

15. $\dfrac{\sqrt{20x}}{9} + \sqrt{\dfrac{5x}{9}}$

16. $\dfrac{3x\sqrt{7}}{5} + \sqrt{\dfrac{7x^2}{100}}$

17. $7\sqrt{9} - 7 + \sqrt{3}$

18. $\sqrt{16} - 5\sqrt{10} + 7$

19. $2 + 3\sqrt{y^2} - 6\sqrt{y^2} + 5$

20. $3\sqrt{7} - \sqrt[3]{x} + 4\sqrt{7} - 3\sqrt[3]{x}$

21. $3\sqrt{108} - 2\sqrt{18} - 3\sqrt{48}$

22. $-\sqrt{75} + \sqrt{12} - 3\sqrt{3}$

23. $-5\sqrt[3]{625} + \sqrt[3]{40}$

24. $-2\sqrt[3]{108} - \sqrt[3]{32}$

25. $a^3\sqrt{9ab^3} - \sqrt{25a^7b^3} + \sqrt{16a^7b^3}$

26. $\sqrt{4x^7y^5} + 9x^2\sqrt{x^3y^5} - 5xy\sqrt{x^5y^3}$

27. $5y\sqrt{8y} + 2\sqrt{50y^3}$

28. $3\sqrt{8x^2y^3} - 2x\sqrt{32y^3}$

29. $\sqrt[3]{54xy^3} - 5\sqrt[3]{2xy^3} + y\sqrt[3]{128x}$

30. $2\sqrt[3]{24x^3y^4} + 4x\sqrt[3]{81y^4}$

31. $6\sqrt[3]{11} + 8\sqrt{11} - 12\sqrt{11}$

32. $3\sqrt[3]{5} + 4\sqrt{5} - 8\sqrt{5}$

33. $-2\sqrt[4]{x^7} + 3\sqrt[4]{16x^7} - x\sqrt[4]{x^3}$

34. $6\sqrt[3]{24x^3} - 2\sqrt[3]{81x^3} - x\sqrt[3]{3}$

35. $\dfrac{4\sqrt{3}}{3} - \dfrac{\sqrt{12}}{3}$

36. $\dfrac{\sqrt{45}}{10} + \dfrac{7\sqrt{5}}{10}$

37. $\dfrac{\sqrt[3]{8x^4}}{7} + \dfrac{3x\sqrt[3]{x}}{7}$

38. $\dfrac{\sqrt[4]{48}}{5x} - \dfrac{2\sqrt[4]{3}}{10x}$

39. $\sqrt{\dfrac{28}{x^2}} + \sqrt{\dfrac{7}{4x^2}}$

40. $\dfrac{\sqrt{99}}{5x} - \sqrt{\dfrac{44}{x^2}}$

41. $\sqrt[3]{\dfrac{16}{27}} - \dfrac{\sqrt[3]{54}}{6}$

42. $\dfrac{\sqrt[3]{3}}{10} + \sqrt[3]{\dfrac{24}{125}}$

43. $-\dfrac{\sqrt[3]{2x^4}}{9} + \sqrt[3]{\dfrac{250x^4}{27}}$

44. $\dfrac{\sqrt[3]{y^5}}{8} + \dfrac{5y\sqrt[3]{y^2}}{4}$

△ **45.** Find the perimeter of the trapezoid.

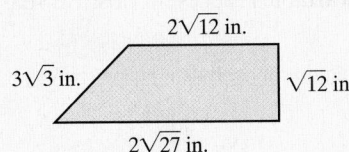

2√12 in.

3√3 in.   √12 in.

2√27 in.

△ **46.** Find the perimeter of the triangle.

√8 m   √32 m

√45 m

Multiply and then simplify if possible. See Example 4.

47. $\sqrt{7}(\sqrt{5} + \sqrt{3})$

48. $\sqrt{5}(\sqrt{15} - \sqrt{35})$

49. $(\sqrt{5} - \sqrt{2})^2$

50. $(3x - \sqrt{2})(3x - \sqrt{2})$

51. $\sqrt{3x}(\sqrt{3} - \sqrt{x})$

52. $\sqrt{5y}(\sqrt{y} + \sqrt{5})$

53. $(2\sqrt{x} - 5)(3\sqrt{x} + 1)$

54. $(8\sqrt{y} + z)(4\sqrt{y} - 1)$

55. $(\sqrt[3]{a} - 4)(\sqrt[3]{a} + 5)$

56. $(\sqrt[3]{a} + 2)(\sqrt[3]{a} + 7)$

57. $6(\sqrt{2} - 2)$

58. $\sqrt{5}(6 - \sqrt{5})$

59. $\sqrt{2}(\sqrt{2} + x\sqrt{6})$

60. $\sqrt{3}(\sqrt{3} - 2\sqrt{5x})$

61. $(2\sqrt{7} + 3\sqrt{5})(\sqrt{7} - 2\sqrt{5})$

62. $(\sqrt{6} - 4\sqrt{2})(3\sqrt{6} + \sqrt{2})$

63. $(\sqrt{x} - y)(\sqrt{x} + y)$

64. $(\sqrt{3x} + 2)(\sqrt{3x} - 2)$

65. $(\sqrt{3} + x)^2$

66. $(\sqrt{y} - 3x)^2$

67. $(\sqrt{5x} - 2\sqrt{3x})(\sqrt{5x} - 3\sqrt{3x})$

68. $(5\sqrt{7x} - \sqrt{2x})(4\sqrt{7x} + 6\sqrt{2x})$

69. $(\sqrt[3]{4} + 2)(\sqrt[3]{2} - 1)$

70. $(\sqrt[3]{3} + \sqrt[3]{2})(\sqrt[3]{9} - \sqrt[3]{4})$

71. $(\sqrt[3]{x} + 1)(\sqrt[3]{x^2} - \sqrt[3]{x} + 1)$

72. $(\sqrt[3]{3x} + 2)(\sqrt[3]{9x^2} - 2\sqrt[3]{3x} + 4)$

73. $(\sqrt{x-1} + 5)^2$

74. $(\sqrt{3x+1} + 2)^2$

75. $(\sqrt{2x+5} - 1)^2$

76. $(\sqrt{x-6} - 7)^2$

REVIEW AND PREVIEW

Factor each numerator and denominator. Then simplify if possible. See Section 7.1.

77. $\dfrac{2x - 14}{2}$

78. $\dfrac{8x - 24y}{4}$

79. $\dfrac{7x - 7y}{x^2 - y^2}$

80. $\dfrac{x^3 - 8}{4x - 8}$

81. $\dfrac{6a^2b - 9ab}{3ab}$

82. $\dfrac{14r - 28r^2s^2}{7rs}$

83. $\dfrac{-4 + 2\sqrt{3}}{6}$

84. $\dfrac{-5 + 10\sqrt{7}}{5}$

CONCEPT EXTENSIONS

△ **85.** Find the perimeter and area of the rectangle.

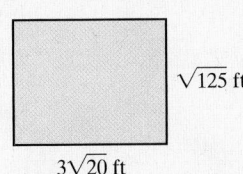

√125 ft

3√20 ft

△ **86.** Find the area and perimeter of the trapezoid. (*Hint:* The area of a trapezoid is the product of half the height $6\sqrt{3}$ meters and the sum of the bases $2\sqrt{63}$ and $7\sqrt{7}$ meters.)

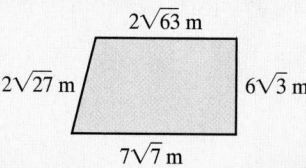

$2\sqrt{63}$ m

$2\sqrt{27}$ m $6\sqrt{3}$ m

$7\sqrt{7}$ m

87. a. Add: $\sqrt{3} + \sqrt{3}$

　　b. Multiply: $\sqrt{3} \cdot \sqrt{3}$

　　c. Describe the differences in parts **a** and **b**.

88. a. Add: $2\sqrt{5} + \sqrt{5}$

　　b. Multiply: $2\sqrt{5} \cdot \sqrt{5}$

　　c. Describe the differences in parts **a** and **b**.

89. Multiply: $(\sqrt{2} + \sqrt{3} - 1)^2$

90. Multiply: $(\sqrt{5} - \sqrt{2} + 1)^2$

91. Explain how simplifying $2x + 3x$ is similar to simplifying $2\sqrt{x} + 3\sqrt{x}$.

92. Explain how multiplying $(x - 2)(x + 3)$ is similar to multiplying $(\sqrt{x} - \sqrt{2})(\sqrt{x} + 3)$.

10.5 Rationalizing Denominators and Numerators of Radical Expressions

OBJECTIVES

1 Rationalize Denominators.

2 Rationalize Denominators Having Two Terms.

3 Rationalize Numerators.

OBJECTIVE

1 Rationalizing Denominators of Radical Expressions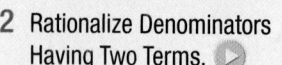

Often in mathematics, it is helpful to write a radical expression such as $\dfrac{\sqrt{3}}{\sqrt{2}}$ either without a radical in the denominator or without a radical in the numerator. The process of writing this expression as an equivalent expression but without a radical in the denominator is called **rationalizing the denominator**. To rationalize the denominator of $\dfrac{\sqrt{3}}{\sqrt{2}}$, we use the fundamental principle of fractions and multiply the numerator and the denominator by $\sqrt{2}$. Recall that this is the same as multiplying by $\dfrac{\sqrt{2}}{\sqrt{2}}$, which simplifies to 1.

$$\frac{\sqrt{3}}{\sqrt{2}} = \frac{\sqrt{3} \cdot \sqrt{2}}{\sqrt{2} \cdot \sqrt{2}} = \frac{\sqrt{6}}{\sqrt{4}} = \frac{\sqrt{6}}{2}$$

In this section, we continue to assume that variables represent positive real numbers.

EXAMPLE 1 Rationalize the denominator of each expression.

a. $\dfrac{2}{\sqrt{5}}$　　**b.** $\dfrac{2\sqrt{16}}{\sqrt{9x}}$　　**c.** $\sqrt[3]{\dfrac{1}{2}}$

Solution

a. To rationalize the denominator, we multiply the numerator and denominator by a factor that makes the radicand in the denominator a perfect square.

$$\frac{2}{\sqrt{5}} = \frac{2 \cdot \sqrt{5}}{\sqrt{5} \cdot \sqrt{5}} = \frac{2\sqrt{5}}{5} \qquad \text{The denominator is now rationalized.}$$

b. First, we simplify the radicals and then rationalize the denominator.

$$\frac{2\sqrt{16}}{\sqrt{9x}} = \frac{2(4)}{3\sqrt{x}} = \frac{8}{3\sqrt{x}}$$

To rationalize the denominator, multiply the numerator and denominator by $\sqrt{x}$. Then

$$\frac{8}{3\sqrt{x}} = \frac{8 \cdot \sqrt{x}}{3\sqrt{x} \cdot \sqrt{x}} = \frac{8\sqrt{x}}{3x}$$

c. $\sqrt[3]{\dfrac{1}{2}} = \dfrac{\sqrt[3]{1}}{\sqrt[3]{2}} = \dfrac{1}{\sqrt[3]{2}}$. Now we rationalize the denominator. Since $\sqrt[3]{2}$ is a cube root, we want to multiply by a value that will make the radicand 2 a perfect cube. If we multiply $\sqrt[3]{2}$ by $\sqrt[3]{2^2}$, we get $\sqrt[3]{2^3} = \sqrt[3]{8} = 2$.

$$\dfrac{1 \cdot \sqrt[3]{2^2}}{\sqrt[3]{2} \cdot \sqrt[3]{2^2}} = \dfrac{\sqrt[3]{4}}{\sqrt[3]{2^3}} = \dfrac{\sqrt[3]{4}}{2}$$ Multiply the numerator and denominator by $\sqrt[3]{2^2}$ and then simplify.   □

PRACTICE
1 Rationalize the denominator of each expression.

a. $\dfrac{5}{\sqrt{3}}$ **b.** $\dfrac{3\sqrt{25}}{\sqrt{4x}}$ **c.** $\sqrt[3]{\dfrac{2}{9}}$ ■

✔ **CONCEPT CHECK**
Determine the smallest number both the numerator and denominator can be multiplied by to rationalize the denominator of the radical expression.

a. $\dfrac{1}{\sqrt[3]{7}}$ **b.** $\dfrac{1}{\sqrt[4]{8}}$

EXAMPLE 2 Rationalize the denominator of $\sqrt{\dfrac{7x}{3y}}$.

Solution $\sqrt{\dfrac{7x}{3y}} = \dfrac{\sqrt{7x}}{\sqrt{3y}}$ Use the quotient rule. No radical may be simplified further.

$= \dfrac{\sqrt{7x} \cdot \sqrt{3y}}{\sqrt{3y} \cdot \sqrt{3y}}$ Multiply numerator and denominator by $\sqrt{3y}$ so that the radicand in the denominator is a perfect square.

$= \dfrac{\sqrt{21xy}}{3y}$ Use the product rule in the numerator and denominator. Remember that $\sqrt{3y} \cdot \sqrt{3y} = 3y$.   □

PRACTICE
2 Rationalize the denominator of $\sqrt{\dfrac{3z}{5y}}$.   ■

EXAMPLE 3 Rationalize the denominator of $\dfrac{\sqrt[4]{x}}{\sqrt[4]{81y^5}}$.

Solution First, simplify each radical if possible.

$\dfrac{\sqrt[4]{x}}{\sqrt[4]{81y^5}} = \dfrac{\sqrt[4]{x}}{\sqrt[4]{81y^4} \cdot \sqrt[4]{y}}$ Use the product rule in the denominator.

$= \dfrac{\sqrt[4]{x}}{3y\sqrt[4]{y}}$ Write $\sqrt[4]{81y^4}$ as $3y$.

$= \dfrac{\sqrt[4]{x} \cdot \sqrt[4]{y^3}}{3y\sqrt[4]{y} \cdot \sqrt[4]{y^3}}$ Multiply numerator and denominator by $\sqrt[4]{y^3}$ so that the radicand in the denominator is a perfect fourth power.

$= \dfrac{\sqrt[4]{xy^3}}{3y\sqrt[4]{y^4}}$ Use the product rule in the numerator and denominator.

$= \dfrac{\sqrt[4]{xy^3}}{3y^2}$ In the denominator, $\sqrt[4]{y^4} = y$ and $3y \cdot y = 3y^2$.   □

PRACTICE
3 Rationalize the denominator of $\dfrac{\sqrt[3]{z^2}}{\sqrt[3]{27x^4}}$.   ■

OBJECTIVE

2 Rationalizing Denominators Having Two Terms ▶

Remember the product of the sum and difference of two terms?

$$(a + b)(a - b) = a^2 - b^2$$

These two expressions are called **conjugates** of each other.

To rationalize a numerator or denominator that is a sum or difference of two terms, we use conjugates. To see how and why this works, let's rationalize the denominator of the expression $\dfrac{5}{\sqrt{3} - 2}$. To do so, we multiply both the numerator and the denominator by $\sqrt{3} + 2$, the **conjugate** of the denominator $\sqrt{3} - 2$, and see what happens.

$$\frac{5}{\sqrt{3} - 2} = \frac{5(\sqrt{3} + 2)}{(\sqrt{3} - 2)(\sqrt{3} + 2)}$$

$$= \frac{5(\sqrt{3} + 2)}{(\sqrt{3})^2 - 2^2} \qquad \text{Multiply the sum and difference}$$
$$\text{of two terms: } (a + b)(a - b) = a^2 - b^2.$$

$$= \frac{5(\sqrt{3} + 2)}{3 - 4}$$

$$= \frac{5(\sqrt{3} + 2)}{-1}$$

$$= -5(\sqrt{3} + 2) \quad \text{or} \quad -5\sqrt{3} - 10$$

Notice in the denominator that the product of $(\sqrt{3} - 2)$ and its conjugate, $(\sqrt{3} + 2)$, is -1. In general, the product of an expression and its conjugate will contain no radical terms. This is why, when rationalizing a denominator or a numerator containing two terms, we multiply by its conjugate. Examples of conjugates are

$$\sqrt{a} - \sqrt{b} \quad \text{and} \quad \sqrt{a} + \sqrt{b}$$
$$x + \sqrt{y} \quad \text{and} \quad x - \sqrt{y}$$

EXAMPLE 4 Rationalize each denominator.

a. $\dfrac{2}{3\sqrt{2} + 4}$ **b.** $\dfrac{\sqrt{6} + 2}{\sqrt{5} - \sqrt{3}}$ **c.** $\dfrac{2\sqrt{m}}{3\sqrt{x} + \sqrt{m}}$

Solution

a. Multiply the numerator and denominator by the conjugate of the denominator, $3\sqrt{2} + 4$.

$$\frac{2}{3\sqrt{2} + 4} = \frac{2(3\sqrt{2} - 4)}{(3\sqrt{2} + 4)(3\sqrt{2} - 4)}$$

$$= \frac{2(3\sqrt{2} - 4)}{(3\sqrt{2})^2 - 4^2}$$

$$= \frac{2(3\sqrt{2} - 4)}{18 - 16}$$

$$= \frac{2(3\sqrt{2} - 4)}{2}, \quad \text{or} \quad 3\sqrt{2} - 4$$

It is often useful to leave a numerator in factored form to help determine whether the expression can be simplified.

b. Multiply the numerator and denominator by the conjugate of $\sqrt{5} - \sqrt{3}$.

$$\frac{\sqrt{6} + 2}{\sqrt{5} - \sqrt{3}} = \frac{(\sqrt{6} + 2)(\sqrt{5} + \sqrt{3})}{(\sqrt{5} - \sqrt{3})(\sqrt{5} + \sqrt{3})}$$

$$= \frac{\sqrt{6}\sqrt{5} + \sqrt{6}\sqrt{3} + 2\sqrt{5} + 2\sqrt{3}}{(\sqrt{5})^2 - (\sqrt{3})^2}$$

$$= \frac{\sqrt{30} + \sqrt{18} + 2\sqrt{5} + 2\sqrt{3}}{5 - 3}$$

$$= \frac{\sqrt{30} + 3\sqrt{2} + 2\sqrt{5} + 2\sqrt{3}}{2}$$

c. Multiply by the conjugate of $3\sqrt{x} + \sqrt{m}$ to eliminate the radicals from the denominator.

$$\frac{2\sqrt{m}}{3\sqrt{x} + \sqrt{m}} = \frac{2\sqrt{m}(3\sqrt{x} - \sqrt{m})}{(3\sqrt{x} + \sqrt{m})(3\sqrt{x} - \sqrt{m})} = \frac{6\sqrt{mx} - 2m}{(3\sqrt{x})^2 - (\sqrt{m})^2}$$

$$= \frac{6\sqrt{mx} - 2m}{9x - m}$$

PRACTICE
4   Rationalize each denominator.

a. $\dfrac{5}{3\sqrt{5} + 2}$ **b.** $\dfrac{\sqrt{2} + 5}{\sqrt{3} - \sqrt{5}}$ **c.** $\dfrac{3\sqrt{x}}{2\sqrt{x} + \sqrt{y}}$

OBJECTIVE
3   **Rationalizing Numerators**

As mentioned earlier, it is also often helpful to write an expression such as $\dfrac{\sqrt{3}}{\sqrt{2}}$ as an equivalent expression without a radical in the numerator. This process is called **rationalizing the numerator.** To rationalize the numerator of $\dfrac{\sqrt{3}}{\sqrt{2}}$, we multiply the numerator and the denominator by $\sqrt{3}$.

$$\frac{\sqrt{3}}{\sqrt{2}} = \frac{\sqrt{3} \cdot \sqrt{3}}{\sqrt{2} \cdot \sqrt{3}} = \frac{\sqrt{9}}{\sqrt{6}} = \frac{3}{\sqrt{6}}$$

EXAMPLE 5   Rationalize the numerator of $\dfrac{\sqrt{7}}{\sqrt{45}}$.

Solution First we simplify $\sqrt{45}$.

$$\frac{\sqrt{7}}{\sqrt{45}} = \frac{\sqrt{7}}{\sqrt{9 \cdot 5}} = \frac{\sqrt{7}}{3\sqrt{5}}$$

Next we rationalize the numerator by multiplying the numerator and the denominator by $\sqrt{7}$.

$$\frac{\sqrt{7}}{3\sqrt{5}} = \frac{\sqrt{7} \cdot \sqrt{7}}{3\sqrt{5} \cdot \sqrt{7}} = \frac{7}{3\sqrt{5 \cdot 7}} = \frac{7}{3\sqrt{35}}$$

PRACTICE
5   Rationalize the numerator of $\dfrac{\sqrt{32}}{\sqrt{80}}$.

EXAMPLE 6 Rationalize the numerator of $\dfrac{\sqrt[3]{2x^2}}{\sqrt[3]{5y}}$.

Solution The numerator and the denominator of this expression are already simplified. To rationalize the numerator, $\sqrt[3]{2x^2}$, we multiply the numerator and denominator by a factor that will make the radicand a perfect cube. If we multiply $\sqrt[3]{2x^2}$ by $\sqrt[3]{4x}$, we get $\sqrt[3]{8x^3} = 2x$.

$$\frac{\sqrt[3]{2x^2}}{\sqrt[3]{5y}} = \frac{\sqrt[3]{2x^2} \cdot \sqrt[3]{4x}}{\sqrt[3]{5y} \cdot \sqrt[3]{4x}} = \frac{\sqrt[3]{8x^3}}{\sqrt[3]{20xy}} = \frac{2x}{\sqrt[3]{20xy}}$$

PRACTICE
6 Rationalize the numerator of $\dfrac{\sqrt[3]{5b}}{\sqrt[3]{2a}}$.

EXAMPLE 7 Rationalize the numerator of $\dfrac{\sqrt{x} + 2}{5}$.

Solution We multiply the numerator and the denominator by the conjugate of the numerator, $\sqrt{x} + 2$.

$$\frac{\sqrt{x} + 2}{5} = \frac{(\sqrt{x} + 2)(\sqrt{x} - 2)}{5(\sqrt{x} - 2)} \qquad \text{Multiply by } \sqrt{x} - 2, \text{ the conjugate of } \sqrt{x} + 2.$$

$$= \frac{(\sqrt{x})^2 - 2^2}{5(\sqrt{x} - 2)} \qquad (a + b)(a - b) = a^2 - b^2$$

$$= \frac{x - 4}{5(\sqrt{x} - 2)}$$

PRACTICE
7 Rationalize the numerator of $\dfrac{\sqrt{x} - 3}{4}$.

✔ Vocabulary, Readiness & Video Check

Use the choices below to fill in each blank. Not all choices will be used.

rationalizing the numerator   conjugate $\dfrac{\sqrt{3}}{\sqrt{3}}$

rationalizing the denominator $\dfrac{5}{5}$

1. The _____ of $a + b$ is $a - b$.

2. The process of writing an equivalent expression, but without a radical in the denominator, is called _____.

3. The process of writing an equivalent expression, but without a radical in the numerator, is called _____.

4. To rationalize the denominator of $\dfrac{5}{\sqrt{3}}$, we multiply by _____.

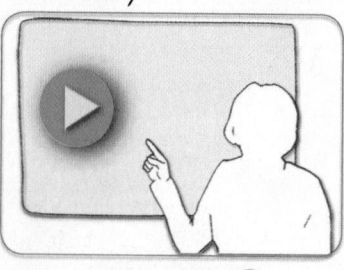
Martin-Gay Interactive Videos

See Video 10.5 ◉

Watch the section lecture video and answer the following questions.

OBJECTIVE 1
5. From ▦ Examples 1–3, what is the goal of rationalizing a denominator?

OBJECTIVE 2
6. From ▦ Example 4, why will multiplying a denominator by its conjugate always rationalize the denominator?

OBJECTIVE 3
7. From ▦ Example 5, is the process of rationalizing a numerator any different from rationalizing a denominator?

10.5   Exercise Set   MyMathLab®

Rationalize each denominator. See Examples 1 through 3.

1. $\dfrac{\sqrt{2}}{\sqrt{7}}$

2. $\dfrac{\sqrt{3}}{\sqrt{2}}$

3. $\sqrt{\dfrac{1}{5}}$

4. $\sqrt{\dfrac{1}{2}}$

5. $\sqrt{\dfrac{4}{x}}$

6. $\sqrt{\dfrac{25}{y}}$

7. $\dfrac{4}{\sqrt[3]{3}}$

8. $\dfrac{6}{\sqrt[3]{9}}$

9. $\dfrac{3}{\sqrt{8x}}$

10. $\dfrac{5}{\sqrt{27a}}$

11. $\dfrac{3}{\sqrt[3]{4x^2}}$

12. $\dfrac{5}{\sqrt[3]{3y}}$

13. $\dfrac{9}{\sqrt{3a}}$

14. $\dfrac{x}{\sqrt{5}}$

15. $\dfrac{3}{\sqrt[3]{2}}$

16. $\dfrac{5}{\sqrt[3]{9}}$

17. $\dfrac{2\sqrt{3}}{\sqrt{7}}$

18. $\dfrac{-5\sqrt{2}}{\sqrt{11}}$

19. $\sqrt{\dfrac{2x}{5y}}$

20. $\sqrt{\dfrac{13a}{2b}}$

21. $\sqrt[3]{\dfrac{3}{5}}$

22. $\sqrt[3]{\dfrac{7}{10}}$

23. $\sqrt{\dfrac{3x}{50}}$

24. $\sqrt{\dfrac{11y}{45}}$

25. $\dfrac{1}{\sqrt{12z}}$

26. $\dfrac{1}{\sqrt{32x}}$

27. $\dfrac{\sqrt[3]{2y^2}}{\sqrt[3]{9x^2}}$

28. $\dfrac{\sqrt[3]{3x}}{\sqrt[3]{4y^4}}$

29. $\sqrt[4]{\dfrac{81}{8}}$

30. $\sqrt[4]{\dfrac{1}{9}}$

31. $\sqrt[4]{\dfrac{16}{9x^7}}$

32. $\sqrt[5]{\dfrac{32}{m^6 n^{13}}}$

33. $\dfrac{5a}{\sqrt[5]{8a^9 b^{11}}}$

34. $\dfrac{9y}{\sqrt[4]{4y^9}}$

Write the conjugate of each expression.

35. $\sqrt{2} + x$

36. $\sqrt{3} + y$

37. $5 - \sqrt{a}$

38. $6 - \sqrt{b}$

39. $-7\sqrt{5} + 8\sqrt{x}$

40. $-9\sqrt{2} - 6\sqrt{y}$

Rationalize each denominator. See Example 4.

41. $\dfrac{6}{2 - \sqrt{7}}$

42. $\dfrac{3}{\sqrt{7} - 4}$

43. $\dfrac{-7}{\sqrt{x} - 3}$

44. $\dfrac{-8}{\sqrt{y} + 4}$

45. $\dfrac{\sqrt{2} - \sqrt{3}}{\sqrt{2} + \sqrt{3}}$

46. $\dfrac{\sqrt{3} + \sqrt{4}}{\sqrt{2} - \sqrt{3}}$

47. $\dfrac{\sqrt{a} + 1}{2\sqrt{a} - \sqrt{b}}$

48. $\dfrac{2\sqrt{a} - 3}{2\sqrt{a} + \sqrt{b}}$

49. $\dfrac{8}{1 + \sqrt{10}}$

50. $\dfrac{-3}{\sqrt{6} - 2}$

51. $\dfrac{\sqrt{x}}{\sqrt{x} + \sqrt{y}}$

52. $\dfrac{2\sqrt{a}}{2\sqrt{x} - \sqrt{y}}$

53. $\dfrac{2\sqrt{3} + \sqrt{6}}{4\sqrt{3} - \sqrt{6}}$

54. $\dfrac{4\sqrt{5} + \sqrt{2}}{2\sqrt{5} - \sqrt{2}}$

Rationalize each numerator. See Examples 5 and 6.

55. $\sqrt{\dfrac{5}{3}}$

56. $\sqrt{\dfrac{3}{2}}$

57. $\sqrt{\dfrac{18}{5}}$

58. $\sqrt{\dfrac{12}{7}}$

59. $\dfrac{\sqrt{4x}}{7}$

60. $\dfrac{\sqrt{3x^5}}{6}$

61. $\dfrac{\sqrt[3]{5y^2}}{\sqrt[3]{4x}}$

62. $\dfrac{\sqrt[3]{4x}}{\sqrt[3]{z^4}}$

63. $\sqrt{\dfrac{2}{5}}$

64. $\sqrt{\dfrac{3}{7}}$

65. $\dfrac{\sqrt{2x}}{11}$

66. $\dfrac{\sqrt{y}}{7}$

67. $\sqrt[3]{\dfrac{7}{8}}$

68. $\sqrt[3]{\dfrac{25}{2}}$

69. $\dfrac{\sqrt[3]{3x^5}}{10}$

70. $\sqrt[3]{\dfrac{9y}{7}}$

71. $\sqrt{\dfrac{18x^4 y^6}{3z}}$

72. $\sqrt{\dfrac{8x^5 y}{2z}}$

Rationalize each numerator. See Example 7.

73. $\dfrac{2 - \sqrt{11}}{6}$

74. $\dfrac{\sqrt{15} + 1}{2}$

75. $\dfrac{2 - \sqrt{7}}{-5}$

76. $\dfrac{\sqrt{5} + 2}{\sqrt{2}}$

77. $\dfrac{\sqrt{x} + 3}{\sqrt{x}}$

78. $\dfrac{5 + \sqrt{2}}{\sqrt{2x}}$

79. $\dfrac{\sqrt{2} - 1}{\sqrt{2} + 1}$

80. $\dfrac{\sqrt{8} - \sqrt{3}}{\sqrt{2} + \sqrt{3}}$

81. $\dfrac{\sqrt{x} + 1}{\sqrt{x} - 1}$

82. $\dfrac{\sqrt{x} + \sqrt{y}}{\sqrt{x} - \sqrt{y}}$

REVIEW AND PREVIEW

Solve each equation. See Sections 2.3 and 6.6.

83. $2x - 7 = 3(x - 4)$
84. $9x - 4 = 7(x - 2)$

85. $(x - 6)(2x + 1) = 0$
86. $(y + 2)(5y + 4) = 0$

87. $x^2 - 8x = -12$
88. $x^3 = x$

CONCEPT EXTENSIONS

△ **89.** The formula of the radius r of a sphere with surface area A is

$$r = \sqrt{\frac{A}{4\pi}}$$

Rationalize the denominator of the radical expression in this formula.

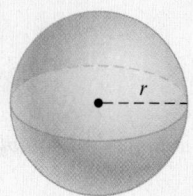

△ **90.** The formula for the radius r of a cone with height 7 centimeters and volume V is

$$r = \sqrt{\frac{3V}{7\pi}}$$

Rationalize the numerator of the radical expression in this formula.

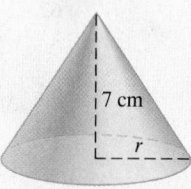

7 cm

r

91. Given $\dfrac{\sqrt{5y^3}}{\sqrt{12x^3}}$, rationalize the denominator by following parts **a** and **b**.

    a. Multiply the numerator and denominator by $\sqrt{12x^3}$.

    b. Multiply the numerator and denominator by $\sqrt{3x}$.

    c. What can you conclude from parts **a** and **b**?

92. Given $\dfrac{\sqrt[3]{5y}}{\sqrt[3]{4}}$, rationalize the denominator by following parts **a** and **b**.

    a. Multiply the numerator and denominator by $\sqrt[3]{16}$.

    b. Multiply the numerator and denominator by $\sqrt[3]{2}$.

    c. What can you conclude from parts **a** and **b**?

Determine the smallest number both the numerator and denominator should be multiplied by to rationalize the denominator of the radical expression. See the Concept Check in this section.

93. $\dfrac{9}{\sqrt[3]{5}}$

94. $\dfrac{5}{\sqrt{27}}$

95. When rationalizing the denominator of $\dfrac{\sqrt{5}}{\sqrt{7}}$, explain why both the numerator and the denominator must be multiplied by $\sqrt{7}$.

96. When rationalizing the numerator of $\dfrac{\sqrt{5}}{\sqrt{7}}$, explain why both the numerator and the denominator must be multiplied by $\sqrt{5}$.

97. Explain why rationalizing the denominator does not change the value of the original expression.

98. Explain why rationalizing the numerator does not change the value of the original expression.

Integrated Review   Radicals and Rational Exponents

Sections 10.1–10.5

Throughout this review, assume that all variables represent positive real numbers.
Find each root.

1. $\sqrt{81}$
    2. $\sqrt[3]{-8}$
    3. $\sqrt[4]{\dfrac{1}{16}}$
    4. $\sqrt{x^6}$

5. $\sqrt[3]{y^9}$
    6. $\sqrt{4y^{10}}$
    7. $\sqrt[5]{-32y^5}$
    8. $\sqrt[4]{81b^{12}}$

Use radical notation to write each expression. Simplify if possible.

9. $36^{1/2}$
    10. $(3y)^{1/4}$
    11. $64^{-2/3}$
    12. $(x + 1)^{3/5}$

Use the properties of exponents to simplify each expression. Write with positive exponents.

13. $y^{-1/6} \cdot y^{7/6}$
    14. $\dfrac{(2x^{1/3})^4}{x^{5/6}}$
    15. $\dfrac{x^{1/4}x^{3/4}}{x^{-1/4}}$
    16. $4^{1/3} \cdot 4^{2/5}$

Use rational exponents to simplify each radical.

17. $\sqrt[3]{8x^6}$
    18. $\sqrt[12]{a^9b^6}$

Use rational exponents to write each as a single radical expression.

19. $\sqrt[4]{x} \cdot \sqrt{x}$
    20. $\sqrt{5} \cdot \sqrt[3]{2}$

Simplify.

21. $\sqrt{40}$

22. $\sqrt[4]{16x^7y^{10}}$

23. $\sqrt[3]{54x^4}$

24. $\sqrt[5]{-64b^{10}}$

Multiply or divide. Then simplify if possible.

25. $\sqrt{5} \cdot \sqrt{x}$

26. $\sqrt[3]{8x} \cdot \sqrt[3]{8x^2}$

27. $\dfrac{\sqrt{98y^6}}{\sqrt{2y}}$

28. $\dfrac{\sqrt[4]{48a^9b^3}}{\sqrt[4]{ab^3}}$

Perform each indicated operation.

29. $\sqrt{20} - \sqrt{75} + 5\sqrt{7}$

30. $\sqrt[3]{54y^4} - y\sqrt[3]{16y}$

31. $\sqrt{3}(\sqrt{5} - \sqrt{2})$

32. $(\sqrt{7} + \sqrt{3})^2$

33. $(2x - \sqrt{5})(2x + \sqrt{5})$

34. $(\sqrt{x+1} - 1)^2$

Rationalize each denominator.

35. $\sqrt{\dfrac{7}{3}}$

36. $\dfrac{5}{\sqrt[3]{2x^2}}$

37. $\dfrac{\sqrt{3} - \sqrt{7}}{2\sqrt{3} + \sqrt{7}}$

Rationalize each numerator.

38. $\sqrt{\dfrac{7}{3}}$

39. $\sqrt[3]{\dfrac{9y}{11}}$

40. $\dfrac{\sqrt{x} - 2}{\sqrt{x}}$

10.6 Radical Equations and Problem Solving

OBJECTIVES

1 Solve Equations That Contain Radical Expressions.

2 Use the Pythagorean Theorem to Model Problems.

OBJECTIVE

1   Solving Equations That Contain Radical Expressions

In this section, we present techniques to solve equations containing radical expressions such as

$$\sqrt{2x - 3} = 9$$

We use the power rule to help us solve these radical equations.

> **Power Rule**
>
> If both sides of an equation are raised to the same power, **all** solutions of the original equation are **among** the solutions of the new equation.

This property *does not* say that raising both sides of an equation to a power yields an equivalent equation. A solution of the new equation *may or may not* be a solution of the original equation. For example, $(-2)^2 = 2^2$, but $-2 \neq 2$. Thus, *each solution of the new equation must be checked* to make sure it is a solution of the original equation. Recall that a proposed solution that is not a solution of the original equation is called an **extraneous solution**.

EXAMPLE 1   Solve: $\sqrt{2x - 3} = 9$

Solution We use the power rule to square both sides of the equation to eliminate the radical.

$$\sqrt{2x - 3} = 9$$
$$(\sqrt{2x - 3})^2 = 9^2$$
$$2x - 3 = 81$$
$$2x = 84$$
$$x = 42$$

Now we check the solution in the original equation.

(Continued on next page)

Check:

$$\sqrt{2x - 3} = 9$$

$$\sqrt{2(42) - 3} \stackrel{?}{=} 9 \quad \text{Let } x = 42.$$

$$\sqrt{84 - 3} \stackrel{?}{=} 9$$

$$\sqrt{81} \stackrel{?}{=} 9$$

$$9 = 9 \quad \text{True}$$

The solution checks, so we conclude that the solution is 42, or the solution set is $\{42\}$. □

PRACTICE

1 Solve: $\sqrt{3x - 5} = 7$ ∎

To solve a radical equation, first isolate a radical on one side of the equation.

EXAMPLE 2 Solve: $\sqrt{-10x - 1} + 3x = 0$

Solution First, isolate the radical on one side of the equation. To do this, we subtract $3x$ from both sides.

$$\sqrt{-10x - 1} + 3x = 0$$

$$\sqrt{-10x - 1} + 3x - 3x = 0 - 3x$$

$$\sqrt{-10x - 1} = -3x$$

Next we use the power rule to eliminate the radical.

$$(\sqrt{-10x - 1})^2 = (-3x)^2$$

$$-10x - 1 = 9x^2$$

Since this is a quadratic equation, we can set the equation equal to 0 and try to solve by factoring.

$$9x^2 + 10x + 1 = 0$$

$$(9x + 1)(x + 1) = 0 \quad \text{Factor.}$$

$$9x + 1 = 0 \quad \text{or} \quad x + 1 = 0 \quad \text{Set each factor equal to 0.}$$

$$x = -\frac{1}{9} \quad \text{or} \quad x = -1$$

Check: Let $x = -\frac{1}{9}$. $\qquad\qquad\qquad\qquad\qquad$ Let $x = -1$.

$$\sqrt{-10x - 1} + 3x = 0 \qquad\qquad\qquad \sqrt{-10x - 1} + 3x = 0$$

$$\sqrt{-10\left(-\frac{1}{9}\right) - 1} + 3\left(-\frac{1}{9}\right) \stackrel{?}{=} 0 \qquad \sqrt{-10(-1) - 1} + 3(-1) \stackrel{?}{=} 0$$

$$\sqrt{\frac{10}{9} - \frac{9}{9}} - \frac{3}{9} \stackrel{?}{=} 0 \qquad\qquad\qquad \sqrt{10 - 1} - 3 \stackrel{?}{=} 0$$

$$\sqrt{\frac{1}{9}} - \frac{1}{3} \stackrel{?}{=} 0 \qquad\qquad\qquad\qquad \sqrt{9} - 3 \stackrel{?}{=} 0$$

$$\frac{1}{3} - \frac{1}{3} = 0 \quad \text{True} \qquad\qquad\qquad 3 - 3 = 0 \quad \text{True}$$

Both solutions check. The solutions are $-\frac{1}{9}$ and -1, or the solution set is $\left\{-\frac{1}{9}, -1\right\}$. □

PRACTICE

2 Solve: $\sqrt{16x - 3} - 4x = 0$ ∎

The following steps may be used to solve a radical equation.

Solving a Radical Equation

Step 1. Isolate one radical on one side of the equation.

Step 2. Raise each side of the equation to a power equal to the index of the radical and simplify.

Step 3. If the equation still contains a radical term, repeat Steps 1 and 2. If not, solve the equation.

Step 4. Check all proposed solutions in the original equation.

EXAMPLE 3   Solve: $\sqrt[3]{x + 1} + 5 = 3$

Solution First we isolate the radical by subtracting 5 from both sides of the equation.

$$\sqrt[3]{x + 1} + 5 = 3$$
$$\sqrt[3]{x + 1} = -2$$

Next we raise both sides of the equation to the third power to eliminate the radical.

$$(\sqrt[3]{x + 1})^3 = (-2)^3$$
$$x + 1 = -8$$
$$x = -9$$

The solution checks in the original equation, so the solution is -9. □

PRACTICE
3   Solve: $\sqrt[3]{x - 2} + 1 = 3$ ■

EXAMPLE 4   Solve: $\sqrt{4 - x} = x - 2$

Solution

$$\sqrt{4 - x} = x - 2$$
$$(\sqrt{4 - x})^2 = (x - 2)^2$$
$$4 - x = x^2 - 4x + 4$$
$$x^2 - 3x = 0 \qquad \text{Write the quadratic equation in standard form.}$$
$$x(x - 3) = 0 \qquad \text{Factor.}$$
$$x = 0 \quad \text{or} \quad x - 3 = 0 \qquad \text{Set each factor equal to 0.}$$
$$x = 3$$

Check:

$\sqrt{4 - x} = x - 2$	$\sqrt{4 - x} = x - 2$
$\sqrt{4 - 0} \stackrel{?}{=} 0 - 2$ Let $x = 0$.	$\sqrt{4 - 3} \stackrel{?}{=} 3 - 2$ Let $x = 3$.
$2 = -2$ False	$1 = 1$ True

The proposed solution 3 checks, but 0 does not. Since 0 is an extraneous solution, the only solution is 3. □

PRACTICE
4   Solve: $\sqrt{16 + x} = x - 4$ ■

Helpful Hint

In Example 4, notice that $(x - 2)^2 = x^2 - 4x + 4$. Make sure binomials are squared correctly.

✔ **CONCEPT CHECK**

How can you immediately tell that the equation $\sqrt{2y + 3} = -4$ has no real solution?

EXAMPLE 5 Solve: $\sqrt{2x + 5} + \sqrt{2x} = 3$

<u>Solution</u> We get one radical alone by subtracting $\sqrt{2x}$ from both sides.

$$\sqrt{2x + 5} + \sqrt{2x} = 3$$
$$\sqrt{2x + 5} = 3 - \sqrt{2x}$$

Now we use the power rule to begin eliminating the radicals. First we square both sides.

$$(\sqrt{2x + 5})^2 = (3 - \sqrt{2x})^2$$
$$2x + 5 = 9 - 6\sqrt{2x} + 2x \quad \text{Multiply } (3 - \sqrt{2x})(3 - \sqrt{2x}).$$

There is still a radical in the equation, so we get a radical alone again. Then we square both sides.

$$2x + 5 = 9 - 6\sqrt{2x} + 2x$$

$6\sqrt{2x} = 4$	Get the radical alone.
$36(2x) = 16$	Square both sides of the equation to eliminate the radical.
$72x = 16$	Multiply.
$x = \dfrac{16}{72}$	Solve.
$x = \dfrac{2}{9}$	Simplify.

The proposed solution, $\dfrac{2}{9}$, checks in the original equation. The solution is $\dfrac{2}{9}$.

PRACTICE
5 Solve: $\sqrt{8x + 1} + \sqrt{3x} = 2$

Helpful Hint

Make sure expressions are squared correctly. In Example 5, we squared $(3 - \sqrt{2x})$ as

$$(3 - \sqrt{2x})^2 = (3 - \sqrt{2x})(3 - \sqrt{2x})$$
$$= 3 \cdot 3 - 3\sqrt{2x} - 3\sqrt{2x} + \sqrt{2x} \cdot \sqrt{2x}$$
$$= 9 - 6\sqrt{2x} + 2x$$

✔ **CONCEPT CHECK**
What is wrong with the following solution?

$$\sqrt{2x + 5} + \sqrt{4 - x} = 8$$
$$(\sqrt{2x + 5} + \sqrt{4 - x})^2 = 8^2$$
$$(2x + 5) + (4 - x) = 64$$
$$x + 9 = 64$$
$$x = 55$$

Answers to Concept Checks:
answers may vary;
$(\sqrt{2x + 5} + \sqrt{4 - x})^2$ is not
$(2x + 5) + (4 - x)$.

OBJECTIVE

2 Using the Pythagorean Theorem

Recall that the Pythagorean theorem states that in a right triangle, the length of the hypotenuse squared equals the sum of the lengths of each of the legs squared.

Pythagorean Theorem

If a and b are the lengths of the legs of a right triangle and c is the length of the hypotenuse, then $a^2 + b^2 = c^2$.

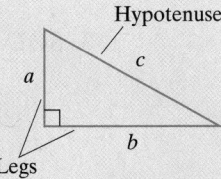

Hypotenuse
c
a
b
Legs

△ **EXAMPLE 6**   Find the length of the unknown leg of the right triangle.

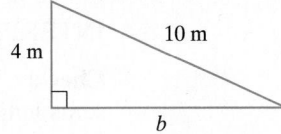

4 m

10 m

b

Solution   In the formula $a^2 + b^2 = c^2$, c is the hypotenuse. Here, $c = 10$, the length of the hypotenuse, and $a = 4$. We solve for b. Then $a^2 + b^2 = c^2$ becomes

$$4^2 + b^2 = 10^2$$
$$16 + b^2 = 100$$
$$b^2 = 84 \qquad \text{Subtract 16 from both sides.}$$
$$b = \pm\sqrt{84} = \pm\sqrt{4 \cdot 21} = \pm 2\sqrt{21}$$

Since b is a length and thus is positive, we will use the positive value only. The unknown leg of the triangle is $2\sqrt{21}$ meters long.   □

PRACTICE

6   Find the length of the unknown leg of the right triangle.

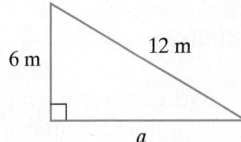

6 m

12 m

a

■

△ **EXAMPLE 7**   **Calculating Placement of a Wire**

A 50-foot supporting wire is to be attached to a 75-foot antenna. Because of surrounding buildings, sidewalks, and roadways, the wire must be anchored exactly 20 feet from the base of the antenna.

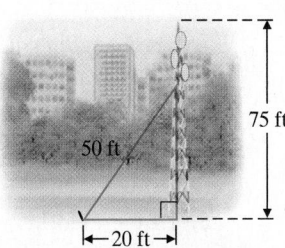

75 ft

50 ft

← 20 ft →

a. How high from the base of the antenna is the wire attached?

b. Local regulations require that a supporting wire be attached at a height no less than $\dfrac{3}{5}$ of the total height of the antenna. From part (a), have local regulations been met?

Solution

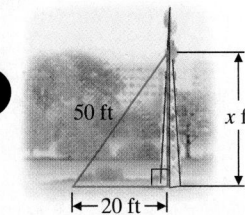

50 ft

x ft

← 20 ft →

1. UNDERSTAND. Read and reread the problem. From the diagram, we notice that a right triangle is formed with hypotenuse 50 feet and one leg 20 feet. Let x be the height from the base of the antenna to the attached wire.

(Continued on next page)

2. TRANSLATE. Use the Pythagorean theorem.

$$a^2 + b^2 = c^2$$

$$20^2 + x^2 = 50^2 \quad a = 20, c = 50$$

3. SOLVE.

$$20^2 + x^2 = 50^2$$

$$400 + x^2 = 2500$$

$$x^2 = 2100 \qquad \text{Subtract 400 from both sides.}$$

$$x = \pm\sqrt{2100}$$

$$= \pm10\sqrt{21}$$

4. INTERPRET. *Check* the work and *state* the solution.

Check: We will use only the positive value, $x = 10\sqrt{21}$, because x represents length. The wire is attached exactly $10\sqrt{21}$ feet from the base of the pole, or approximately 45.8 feet.

State: The supporting wire must be attached at a height no less than $\frac{3}{5}$ of the total height of the antenna. This height is $\frac{3}{5}$ (75 feet), or 45 feet. Since we know from part (a) that the wire is to be attached at a height of approximately 45.8 feet, local regulations have been met. □

PRACTICE
7 Keith Robinson bought two Siamese fighting fish, but when he got home, he found he only had one rectangular tank that was 12 in. long, 7 in. wide, and 5 in. deep. Since the fish must be kept separated, he needed to insert a plastic divider in the diagonal of the tank. He already has a piece that is 5 in. in one dimension, but how long must it be to fit corner to corner in the tank?

Graphing Calculator Explorations

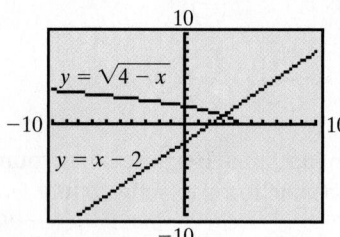

We can use a graphing calculator to solve radical equations. For example, to use a graphing calculator to approximate the solutions of the equation solved in Example 4, we graph the following.

$$Y_1 = \sqrt{4 - x} \quad \text{and} \quad Y_2 = x - 2$$

The x-value of the point of intersection is the solution. Use the Intersect feature or the Zoom and Trace features of your graphing calculator to see that the solution is 3.

Use a graphing calculator to solve each radical equation. Round all solutions to the nearest hundredth.

1. $\sqrt{x + 7} = x$

2. $\sqrt{3x + 5} = 2x$

3. $\sqrt{2x + 1} = \sqrt{2x + 2}$

4. $\sqrt{10x - 1} = \sqrt{-10x + 10} - 1$

5. $1.2x = \sqrt{3.1x + 5}$

6. $\sqrt{1.9x^2 - 2.2} = -0.8x + 3$

Vocabulary, Readiness & Video Check

Use the choices below to fill in each blank. Not all choices will be used.

hypotenuse	right	$x^2 + 25$	$16 - 8\sqrt{7x} + 7x$
extraneous solution	legs	$x^2 - 10x + 25$	$16 + 7x$

1. A proposed solution that is not a solution of the original equation is called a(n) _____.

2. The Pythagorean theorem states that $a^2 + b^2 = c^2$ where a and b are the lengths of the _____ of a(n) _____ triangle and c is the length of the _____.

3. The square of $x - 5$, or $(x - 5)^2 =$ _____.

4. The square of $4 - \sqrt{7x}$, or $(4 - \sqrt{7x})^2 =$ _____.

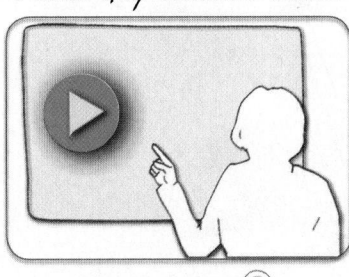

Martin-Gay Interactive Videos

See Video 10.6

Watch the section lecture video and answer the following questions.

OBJECTIVE 1 **5.** From ▦ Examples 1–4, why must you be careful and check your proposed solution(s) in the original equation?

OBJECTIVE 2 **6.** From ▦ Example 5, when solving problems using the Pythagorean theorem, what two things must you remember?

OBJECTIVE 2 **7.** What important reminder is given as the final answer to ▦ Example 5 is being found?

10.6 Exercise Set MyMathLab®

Solve. See Examples 1 and 2.

1. $\sqrt{2x} = 4$
2. $\sqrt{3x} = 3$
3. $\sqrt{x - 3} = 2$
4. $\sqrt{x + 1} = 5$
5. $\sqrt{2x} = -4$
6. $\sqrt{5x} = -5$
7. $\sqrt{4x - 3} - 5 = 0$
8. $\sqrt{x - 3} - 1 = 0$
9. $\sqrt{2x - 3} - 2 = 1$
10. $\sqrt{3x + 3} - 4 = 8$

Solve. See Example 3.

11. $\sqrt[3]{6x} = -3$
12. $\sqrt[3]{4x} = -2$
13. $\sqrt[3]{x - 2} - 3 = 0$
14. $\sqrt[3]{2x - 6} - 4 = 0$

Solve. See Examples 4 and 5.

15. $\sqrt{13 - x} = x - 1$
16. $\sqrt{2x - 3} = 3 - x$
17. $x - \sqrt{4 - 3x} = -8$
18. $2x + \sqrt{x + 1} = 8$
19. $\sqrt{y + 5} = 2 - \sqrt{y - 4}$
20. $\sqrt{x + 3} + \sqrt{x - 5} = 3$
21. $\sqrt{x - 3} + \sqrt{x + 2} = 5$
22. $\sqrt{2x - 4} - \sqrt{3x + 4} = -2$

MIXED PRACTICE

Solve. See Examples 1 through 5.

23. $\sqrt{3x - 2} = 5$
24. $\sqrt{5x - 4} = 9$
25. $-\sqrt{2x} + 4 = -6$
26. $-\sqrt{3x + 9} = -12$
27. $\sqrt{3x + 1} + 2 = 0$
28. $\sqrt{3x + 1} - 2 = 0$
29. $\sqrt[4]{4x + 1} - 2 = 0$
30. $\sqrt[4]{2x - 9} - 3 = 0$
31. $\sqrt{4x - 3} = 7$
32. $\sqrt{3x + 9} = 6$
33. $\sqrt[3]{6x - 3} - 3 = 0$
34. $\sqrt[3]{3x} + 4 = 7$
35. $\sqrt[3]{2x - 3} - 2 = -5$
36. $\sqrt[3]{x - 4} - 5 = -7$
37. $\sqrt{x + 4} = \sqrt{2x - 5}$
38. $\sqrt{3y + 6} = \sqrt{7y - 6}$
39. $x - \sqrt{1 - x} = -5$
40. $x - \sqrt{x - 2} = 4$
41. $\sqrt[3]{-6x - 1} = \sqrt[3]{-2x - 5}$
42. $\sqrt[3]{-4x - 3} = \sqrt[3]{-x - 15}$
43. $\sqrt{5x - 1} - \sqrt{x + 2} = 3$
44. $\sqrt{2x - 1} - 4 = -\sqrt{x - 4}$
45. $\sqrt{2x - 1} = \sqrt{1 - 2x}$
46. $\sqrt{7x - 4} = \sqrt{4 - 7x}$

47. $\sqrt{3x + 4} - 1 = \sqrt{2x + 1}$
48. $\sqrt{x - 2} + 3 = \sqrt{4x + 1}$
49. $\sqrt{y + 3} - \sqrt{y - 3} = 1$
50. $\sqrt{x + 1} - \sqrt{x - 1} = 2$

Find the length of the unknown side of each triangle. See Example 6.

51.
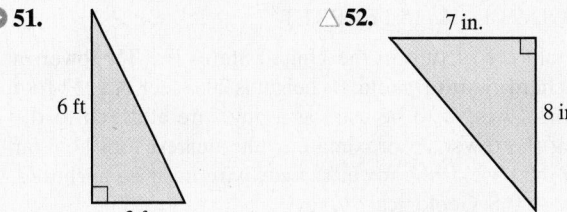
6 ft
3 ft

52.
7 in.
8 in.

53.
3 m
7 m

54.
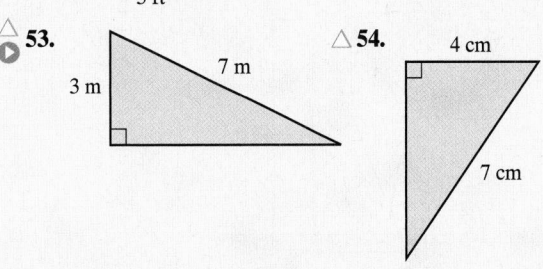
4 cm
7 cm

Find the length of the unknown side of each triangle. Give the exact length and a one-decimal-place approximation. See Example 6.

55.
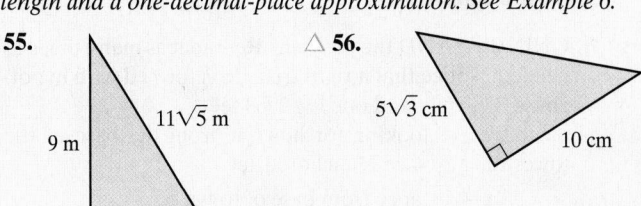
9 m
$11\sqrt{5}$ m

56.
$5\sqrt{3}$ cm
10 cm

57.
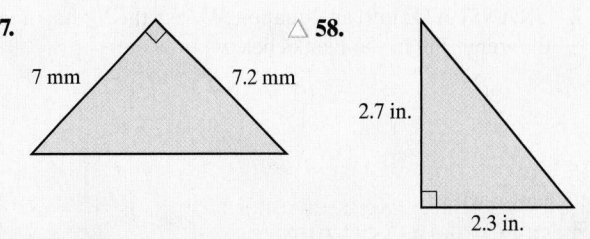
7 mm
7.2 mm

58.
2.7 in.
2.3 in.

Solve. Give exact answers and two-decimal-place approximations where appropriate. For Exercises 59 and 60, the solutions have been started for you. See Example 7.

59. A wire is needed to support a vertical pole 15 feet tall. The cable will be anchored to a stake 8 feet from the base of the pole. How much cable is needed?

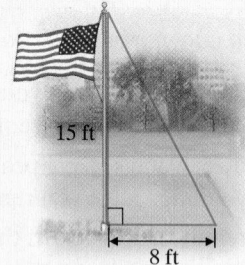

15 ft

8 ft

Start the solution:

1. **UNDERSTAND** the problem. Reread it as many times as needed. Notice that a right triangle is formed with legs of length 8 ft and 15 ft.

   Since we are looking for how much cable is needed, let

   x = amount of cable needed.

2. **TRANSLATE** into an equation. We use the Pythagorean theorem. (Fill in the blanks below.)

$$a^2 \quad + \quad b^2 \quad = \quad c^2$$
$$\downarrow \qquad\qquad \downarrow$$
$$\underline{\quad}^2 \quad + \quad \underline{\quad}^2 \quad = \quad x^2$$

   Finish with:

3. **SOLVE** and **4. INTERPRET**

60. The tallest structure in the United States is a TV tower in Blanchard, North Dakota. Its height is 2063 feet. A 2382-foot length of wire is to be used as a guy wire attached to the top of the tower. Approximate to the nearest foot how far from the base of the tower the guy wire must be anchored. (*Source:* U.S. Geological Survey)

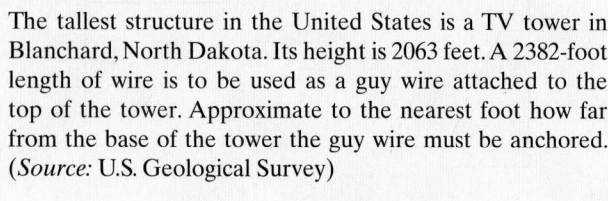

2382 ft   2063 ft

?

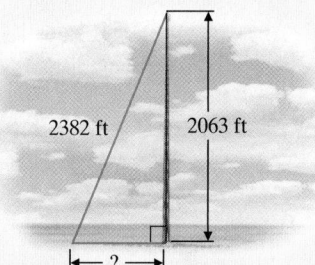

Start the solution:

1. **UNDERSTAND** the problem. Reread it as many times as needed. Notice that a right triangle is formed with hypotenuse 2382 feet and one leg 2063 feet.

   Since we are looking for how far from the base of the tower the guy wire is anchored, let

   x = distance from base of tower to where guy wire is anchored.

2. **TRANSLATE** into an equation. We use the Pythagorean theorem. (Fill in the blanks below.)

$$a^2 \quad + \quad b^2 \quad = \quad c^2$$
$$\downarrow \qquad\qquad \downarrow$$
$$\underline{\quad}^2 \quad + \quad x^2 \quad = \quad \underline{\quad}^2$$

   Finish with:

3. **SOLVE** and **4. INTERPRET**

61. A spotlight is mounted on the eaves of a house 12 feet above the ground. A flower bed runs between the house and the sidewalk, so the closest a ladder can be placed to the house is 5 feet. How long of a ladder is needed so that an electrician can reach the place where the light is mounted?

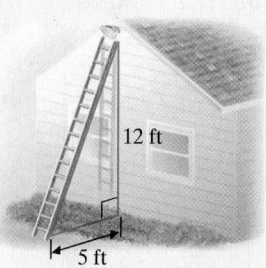

12 ft

5 ft

62. A wire is to be attached to support a telephone pole. Because of surrounding buildings, sidewalks, and roadways, the wire must be anchored exactly 15 feet from the base of the pole. Telephone company workers have only 30 feet of cable, and 2 feet of that must be used to attach the cable to the pole and to the stake on the ground. How high from the base of the pole can the wire be attached?

15 ft

63. The radius of the moon is 1080 miles. Use the formula for the radius r of a sphere given its surface area A,

$$r = \sqrt{\frac{A}{4\pi}}$$

to find the surface area of the moon. Round to the nearest square mile. (*Source:* National Space Science Data Center)

64. Police departments find it very useful to be able to approximate the speed of a car when they are given the distance that the car skidded before it came to a stop. If the road surface is wet concrete, the function $S(x) = \sqrt{10.5x}$ is used, where $S(x)$ is the speed of the car in miles per hour and x is the distance skidded in feet. Find how fast a car was moving if it skidded 280 feet on wet concrete.

65. The formula $v = \sqrt{2gh}$ gives the velocity v, in feet per second, of an object when it falls h feet accelerated by gravity g, in feet per second squared. If g is approximately 32 feet per second squared, find how far an object has fallen if its velocity is 80 feet per second.

66. Two tractors are pulling a tree stump from a field. If two forces A and B pull at right angles (90°) to each other, the size of the resulting force R is given by the formula $R = \sqrt{A^2 + B^2}$. If tractor A is exerting 600 pounds of force and the resulting force is 850 pounds, find how much force tractor B is exerting.

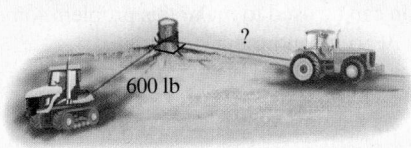
600 lb

In psychology, it has been suggested that the number S of nonsense syllables that a person can repeat consecutively depends on his or her IQ score I according to the equation $S = 2\sqrt{I} - 9$.

67. Use this relationship to estimate the IQ of a person who can repeat 11 nonsense syllables consecutively.

68. Use this relationship to estimate the IQ of a person who can repeat 15 nonsense syllables consecutively.

*The **period** of a pendulum is the time it takes for the pendulum to make one full back-and-forth swing. The period of a pendulum depends on the length of the pendulum. The formula for the period P, in seconds, is $P = 2\pi\sqrt{\dfrac{l}{32}}$, where l is the length of the pendulum in feet. Use this formula for Exercises 69 through 74.*

69. Find the period of a pendulum whose length is 2 feet. Give the exact answer and a two-decimal-place approximation.

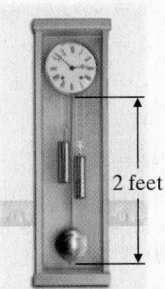
2 feet

70. Klockit sells a 43-inch lyre pendulum. Find the period of this pendulum. Round your answer to two decimal places. (*Hint:* First convert inches to feet.)

71. Find the length of a pendulum whose period is 4 seconds. Round your answer to two decimal places.

72. Find the length of a pendulum whose period is 3 seconds. Round your answer to two decimal places.

73. Study the relationship between period and pendulum length in Exercises 69 through 72 and make a conjecture about this relationship.

74. Galileo experimented with pendulums. He supposedly made conjectures about pendulums of equal length with different bob weights. Try this experiment. Make two pendulums 3 feet long. Attach a heavy weight (lead) to one and a light weight (a cork) to the other. Pull both pendulums back the same angle measure and release. Make a conjecture from your observations.

If the three lengths of the sides of a triangle are known, Heron's formula can be used to find its area. If a, b, and c are the lengths of the three sides, Heron's formula for area is

$$A = \sqrt{s(s - a)(s - b)(s - c)}$$

where s is half the perimeter of the triangle, or $s = \dfrac{1}{2}(a + b + c)$.

Use this formula to find the area of each triangle. Give the exact answer and then a two-decimal-place approximation.

△ **75.**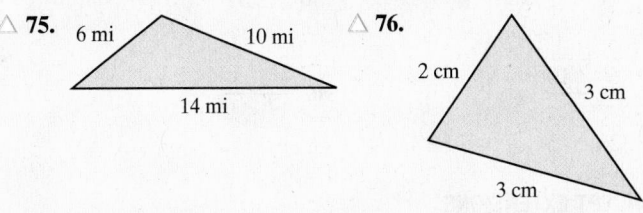
6 mi   10 mi
14 mi

△ **76.**
2 cm   3 cm
3 cm

77. Describe when Heron's formula might be useful.

78. In your own words, explain why you think s in Heron's formula is called the *semiperimeter*.

The maximum distance $D(h)$ in kilometers that a person can see from a height h kilometers above the ground is given by the function $D(h) = 111.7\sqrt{h}$. Use this function for Exercises 79 and 80. Round your answers to two decimal places.

79. Find the height that would allow a person to see 80 kilometers.

80. Find the height that would allow a person to see 40 kilometers.

REVIEW AND PREVIEW

Use the vertical line test to determine whether each graph represents the graph of a function. See Section 3.6.

81.
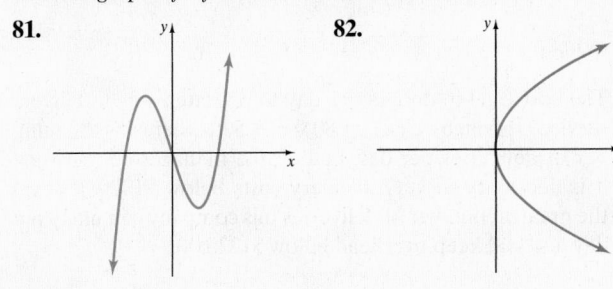

82.

83.

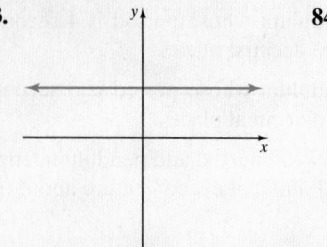

84.

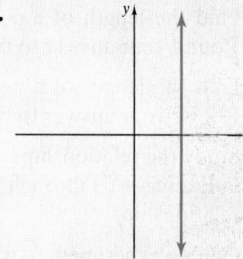

85.

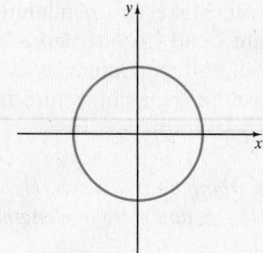

86.

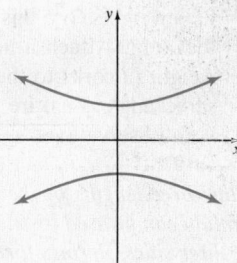

Simplify. See Section 7.7.

87. $\dfrac{\dfrac{x}{6}}{\dfrac{2x}{3} + \dfrac{1}{2}}$

88. $\dfrac{\dfrac{1}{y} + \dfrac{4}{5}}{-\dfrac{3}{20}}$

89. $\dfrac{\dfrac{z}{5} + \dfrac{1}{10}}{\dfrac{z}{20} - \dfrac{z}{5}}$

90. $\dfrac{\dfrac{1}{y} + \dfrac{1}{x}}{\dfrac{1}{y} - \dfrac{1}{x}}$

CONCEPT EXTENSIONS

Find the error in each solution and correct. See the second Concept Check in this section.

91. $\sqrt{5x - 1} + 4 = 7$
$(\sqrt{5x - 1} + 4)^2 = 7^2$
$5x - 1 + 16 = 49$
$5x = 34$
$x = \dfrac{34}{5}$

92. $\sqrt{2x + 3} + 4 = 1$
$\sqrt{2x + 3} = 5$
$(\sqrt{2x + 3})^2 = 5^2$
$2x + 3 = 25$
$2x = 22$
$x = 11$

93. Solve: $\sqrt{\sqrt{x + 3} + \sqrt{x}} = \sqrt{3}$

94. The cost $C(x)$ in dollars per day to operate a small delivery service is given by $C(x) = 80\sqrt[3]{x} + 500$, where x is the number of deliveries per day. In July, the manager decides that it is necessary to keep delivery costs below $1220.00. Find the greatest number of deliveries this company can make per day and still keep overhead below $1220.00.

95. Consider the equations $\sqrt{2x} = 4$ and $\sqrt[3]{2x} = 4$.

   a. Explain the difference in solving these equations.

   b. Explain the similarity in solving these equations.

96. Explain why proposed solutions of radical equations must be checked in the original equation.

For Exercises 97 through 100, see the example below.

Example

Solve $(t^2 - 3t) - 2\sqrt{t^2 - 3t} = 0$.

Solution

Substitution can be used to make this problem somewhat simpler. Since $t^2 - 3t$ occurs more than once, let $x = t^2 - 3t$.

$(t^2 - 3t) - 2\sqrt{t^2 - 3t} = 0$
$x - 2\sqrt{x} = 0$ Let $x = t^2 - 3t$.
$x = 2\sqrt{x}$
$x^2 = (2\sqrt{x})^2$
$x^2 = 4x$
$x^2 - 4x = 0$
$x(x - 4) = 0$
$x = 0$ or $x - 4 = 0$
$x = 4$

Now we "undo" the substitution by replacing x with $t^2 - 3t$.

$x = 0$
$t^2 - 3t = 0$ Replace x with $t^2 - 3t$.
$t(t - 3) = 0$ Factor.
$t = 0$ or $t - 3 = 0$
$t = 3$

Replace x with $t^2 - 3t$ in $x = 4$.

$x = 4$
$t^2 - 3t = 4$ Replace x with $t^2 - 3t$.
$t^2 - 3t - 4 = 0$ Subtract 4.
$(t - 4)(t + 1) = 0$ Factor.
$t - 4 = 0$ or $t + 1 = 0$
$t = 4$ $t = -1$

In this problem, we have four possible solutions for t: 0, 3, 4, and -1. All four solutions check in the original equation, so the solutions are $-1, 0, 3, 4$.

Solve. See the preceding example.

97. $3\sqrt{x^2 - 8x} = x^2 - 8x$

98. $\sqrt{(x^2 - x) + 7} = 2(x^2 - x) - 1$

99. $7 - (x^2 - 3x) = \sqrt{(x^2 - 3x) + 5}$

100. $x^2 + 6x = 4\sqrt{x^2 + 6x}$

10.7 Complex Numbers

OBJECTIVES

1 Write Square Roots of Negative Numbers in the Form *bi*.

2 Add or Subtract Complex Numbers.

3 Multiply Complex Numbers.

4 Divide Complex Numbers.

5 Raise *i* to Powers.

OBJECTIVE

1 Writing Numbers in the Form *bi*

Our work with radical expressions has excluded expressions such as $\sqrt{-16}$ because $\sqrt{-16}$ is not a real number; there is no real number whose square is -16. In this section, we discuss a number system that includes roots of negative numbers. This number system is the **complex number system,** and it includes the set of real numbers as a subset. The complex number system allows us to solve equations such as $x^2 + 1 = 0$ that have no real number solutions. The set of complex numbers includes the **imaginary unit.**

> **Imaginary Unit**
>
> The imaginary unit, written i, is the number whose square is -1. That is,
>
> $$i^2 = -1 \quad \text{and} \quad i = \sqrt{-1}$$

To write the square root of a negative number in terms of i, use the property that if a is a positive number, then

$$\sqrt{-a} = \sqrt{-1} \cdot \sqrt{a}$$
$$= i \cdot \sqrt{a}$$

Using i, we can write $\sqrt{-16}$ as

$$\sqrt{-16} = \sqrt{-1 \cdot 16} = \sqrt{-1} \cdot \sqrt{16} = i \cdot 4, \text{ or } 4i$$

EXAMPLE 1 Write with *i* notation.

a. $\sqrt{-36}$ **b.** $\sqrt{-5}$ **c.** $-\sqrt{-20}$

Solution

a. $\sqrt{-36} = \sqrt{-1 \cdot 36} = \sqrt{-1} \cdot \sqrt{36} = i \cdot 6, \text{ or } 6i$

b. $\sqrt{-5} = \sqrt{-1(5)} = \sqrt{-1} \cdot \sqrt{5} = i\sqrt{5}.$

c. $-\sqrt{-20} = -\sqrt{-1 \cdot 20} = -\sqrt{-1} \cdot \sqrt{4 \cdot 5} = -i \cdot 2\sqrt{5} = -2i\sqrt{5}$ □

> **Helpful Hint**
>
> Since $\sqrt{5}i$ can easily be confused with $\sqrt{5i}$, we write $\sqrt{5}i$ as $i\sqrt{5}$.

PRACTICE

1 Write with *i* notation.

a. $\sqrt{-4}$ **b.** $\sqrt{-7}$ **c.** $-\sqrt{-18}$ ■

The product rule for radicals does not necessarily hold true for imaginary numbers. *To multiply square roots of negative numbers, first we write each number in terms of the imaginary unit i.* For example, to multiply $\sqrt{-4}$ and $\sqrt{-9}$, we first write each number in the form *bi*.

$$\sqrt{-4}\sqrt{-9} = 2i(3i) = 6i^2 = 6(-1) = -6 \quad \text{Correct}$$

We will also use this method to simplify quotients of square roots of negative numbers. Why? The product rule does not work for this example. In other words,

$$\sqrt{-4} \cdot \sqrt{-9} = \sqrt{(-4)(-9)} = \sqrt{36} = 6 \quad \text{Incorrect}$$

EXAMPLE 2 Multiply or divide as indicated.

a. $\sqrt{-3} \cdot \sqrt{-5}$ b. $\sqrt{-36} \cdot \sqrt{-1}$ c. $\sqrt{8} \cdot \sqrt{-2}$ d. $\dfrac{\sqrt{-125}}{\sqrt{5}}$

Solution

a. $\sqrt{-3} \cdot \sqrt{-5} = i\sqrt{3}(i\sqrt{5}) = i^2\sqrt{15} = -1\sqrt{15} = -\sqrt{15}$

b. $\sqrt{-36} \cdot \sqrt{-1} = 6i(i) = 6i^2 = 6(-1) = -6$

c. $\sqrt{8} \cdot \sqrt{-2} = 2\sqrt{2}(i\sqrt{2}) = 2i(\sqrt{2}\sqrt{2}) = 2i(2) = 4i$

d. $\dfrac{\sqrt{-125}}{\sqrt{5}} = \dfrac{i\sqrt{125}}{\sqrt{5}} = i\sqrt{25} = 5i$

PRACTICE
2 Multiply or divide as indicated.

a. $\sqrt{-5} \cdot \sqrt{-6}$ b. $\sqrt{-9} \cdot \sqrt{-1}$ c. $\sqrt{125} \cdot \sqrt{-5}$ d. $\dfrac{\sqrt{-27}}{\sqrt{3}}$

Now that we have practiced working with the imaginary unit, we define complex numbers.

> **Complex Numbers**
>
> A **complex number** is a number that can be written in the form $a + bi$, where a and b are real numbers.

Notice that the set of real numbers is a subset of the complex numbers since any real number can be written in the form of a complex number. For example,

$$16 = 16 + 0i$$

In general, a complex number $a + bi$ is a real number if $b = 0$. Also, a complex number is called a **pure imaginary number** or an imaginary number if $a = 0$ and $b \neq 0$. For example,

$$3i = 0 + 3i \quad \text{and} \quad i\sqrt{7} = 0 + i\sqrt{7}$$

are pure imaginary numbers.

The following diagram shows the relationship between complex numbers and their subsets.

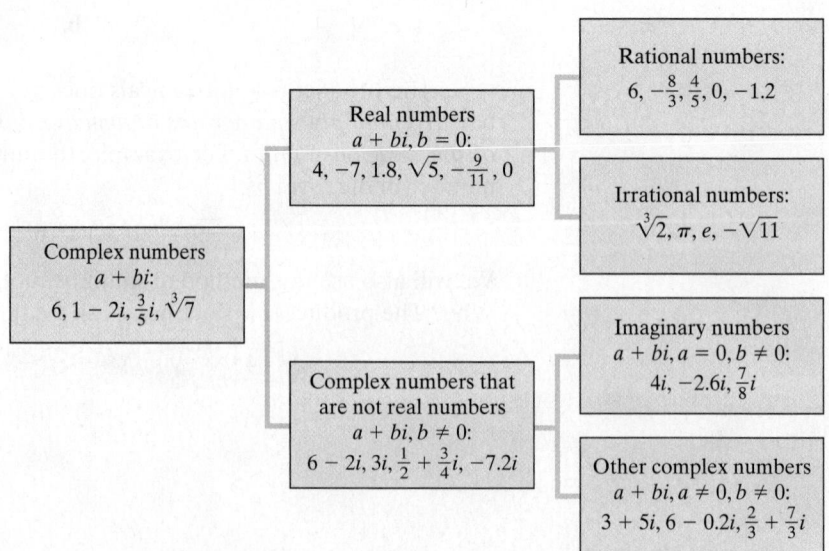

✔ **CONCEPT CHECK**
True or false? Every complex number is also a real number.

OBJECTIVE

2 Adding or Subtracting Complex Numbers

Two complex numbers $a + bi$ and $c + di$ are equal if and only if $a = c$ and $b = d$. Complex numbers can be added or subtracted by adding or subtracting their real parts and then adding or subtracting their imaginary parts.

> **Sum or Difference of Complex Numbers**
>
> If $a + bi$ and $c + di$ are complex numbers, then their sum is
>
> $$(a + bi) + (c + di) = (a + c) + (b + d)i$$
>
> Their difference is
>
> $$(a + bi) - (c + di) = a + bi - c - di = (a - c) + (b - d)i$$

EXAMPLE 3 Add or subtract the complex numbers. Write the sum or difference in the form $a + bi$.

    a. $(2 + 3i) + (-3 + 2i)$ **b.** $5i - (1 - i)$ **c.** $(-3 - 7i) - (-6)$

Solution

    a. $(2 + 3i) + (-3 + 2i) = (2 - 3) + (3 + 2)i = -1 + 5i$

    b. $5i - (1 - i) = 5i - 1 + i$
$$= -1 + (5 + 1)i$$
$$= -1 + 6i$$

    c. $(-3 - 7i) - (-6) = -3 - 7i + 6$
$$= (-3 + 6) - 7i$$
$$= 3 - 7i$$ □

PRACTICE

3 Add or subtract the complex numbers. Write the sum or difference in the form $a + bi$.

    a. $(3 - 5i) + (-4 + i)$ **b.** $4i - (3 - i)$ **c.** $(-5 - 2i) - (-8)$ ■

OBJECTIVE

3 Multiplying Complex Numbers

To multiply two complex numbers of the form $a + bi$, we multiply as though they are binomials. Then we use the relationship $i^2 = -1$ to simplify.

EXAMPLE 4 Multiply the complex numbers. Write the product in the form $a + bi$.

    a. $-7i \cdot 3i$ **b.** $3i(2 - i)$ **c.** $(2 - 5i)(4 + i)$

    d. $(2 - i)^2$ **e.** $(7 + 3i)(7 - 3i)$

Solution

    a. $-7i \cdot 3i = -21i^2$
$$= -21(-1) \quad \text{Replace } i^2 \text{ with } -1.$$
$$= 21 + 0i$$

(Continued on next page)

  b. $3i(2 - i) = 3i \cdot 2 - 3i \cdot i$ Use the distributive property.

$$= 6i - 3i^2 \qquad \text{Multiply.}$$
$$= 6i - 3(-1) \quad \text{Replace } i^2 \text{ with } -1.$$
$$= 6i + 3$$
$$= 3 + 6i \qquad \text{Use the FOIL order below. (First, Outer, Inner, Last)}$$

  c. $(2 - 5i)(4 + i) = 2(4) + 2(i) - 5i(4) - 5i(i)$
$$\qquad\qquad\qquad\qquad\text{F} \qquad \text{O} \qquad \text{I} \qquad \text{L}$$
$$= 8 + 2i - 20i - 5i^2$$
$$= 8 - 18i - 5(-1) \qquad\qquad i^2 = -1$$
$$= 8 - 18i + 5$$
$$= 13 - 18i$$

  d. $(2 - i)^2 = (2 - i)(2 - i)$
$$= 2(2) - 2(i) - 2(i) + i^2$$
$$= 4 - 4i + (-1) \qquad\qquad i^2 = -1$$
$$= 3 - 4i$$

  e. $(7 + 3i)(7 - 3i) = 7(7) - 7(3i) + 3i(7) - 3i(3i)$
$$= 49 - 21i + 21i - 9i^2$$
$$= 49 - 9(-1) \qquad\qquad i^2 = -1$$
$$= 49 + 9$$
$$= 58 + 0i \qquad\qquad\qquad \square$$

PRACTICE

4   Multiply the complex numbers. Write the product in the form $a + bi$.

  a. $-4i \cdot 5i$ **b.** $5i(2 + i)$ **c.** $(2 + 3i)(6 - i)$

  d. $(3 - i)^2$ **e.** $(9 + 2i)(9 - 2i)$

    Notice that if you add, subtract, or multiply two complex numbers, just like real numbers, the result is a complex number.

OBJECTIVE

4 **Dividing Complex Numbers**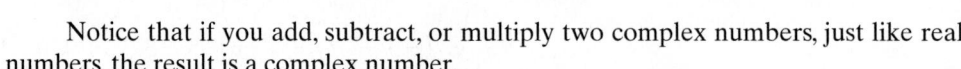

From Example 4(e), notice that the product of $7 + 3i$ and $7 - 3i$ is a real number. These two complex numbers are called **complex conjugates** of one another. In general, we have the following definition.

Complex Conjugates

The complex numbers $(a + bi)$ and $(a - bi)$ are called **complex conjugates** of each other, and

$$(a + bi)(a - bi) = a^2 + b^2.$$

    To see that the product of a complex number $a + bi$ and its conjugate $a - bi$ is the real number $a^2 + b^2$, we multiply.

$$(a + bi)(a - bi) = a^2 - abi + abi - b^2i^2$$
$$= a^2 - b^2(-1)$$
$$= a^2 + b^2$$

We use complex conjugates to divide by a complex number.

EXAMPLE 5   Divide. Write in the form $a + bi$.

    a. $\dfrac{2 + i}{1 - i}$ **b.** $\dfrac{7}{3i}$

Solution

a. Multiply the numerator and denominator by the complex conjugate of $1 - i$ to eliminate the imaginary number in the denominator.

$$\frac{2 + i}{1 - i} = \frac{(2 + i)(1 + i)}{(1 - i)(1 + i)}$$

$$= \frac{2(1) + 2(i) + 1(i) + i^2}{1^2 - i^2}$$

$$= \frac{2 + 3i - 1}{1 + 1} \qquad\qquad i^2 = -1.$$

$$= \frac{1 + 3i}{2} \quad \text{or} \quad \frac{1}{2} + \frac{3}{2}i$$

b. Multiply the numerator and denominator by the conjugate of $3i$. Note that $3i = 0 + 3i$, so its conjugate is $0 - 3i$ or $-3i$.

$$\frac{7}{3i} = \frac{7(-3i)}{(3i)(-3i)} = \frac{-21i}{-9i^2} = \frac{-21i}{-9(-1)} = \frac{-21i}{9} = \frac{-7i}{3} \quad \text{or} \quad 0 - \frac{7}{3}i \quad \square$$

PRACTICE
5   Divide. Write in the form $a + bi$.

    a. $\dfrac{4 - i}{3 + i}$ **b.** $\dfrac{5}{2i}$ ■

> **Helpful Hint**
>
> Recall that division can be checked by multiplication.
>
> To check that $\dfrac{2 + i}{1 - i} = \dfrac{1}{2} + \dfrac{3}{2}i$, in Example 5(a), multiply $\left(\dfrac{1}{2} + \dfrac{3}{2}i\right)(1 - i)$ to verify that the product is $2 + i$.

OBJECTIVE
5   **Finding Powers of i** ▶

We can use the fact that $i^2 = -1$ to find higher powers of i. To find i^3, we rewrite it as the product of i^2 and i.

$$i^3 = i^2 \cdot i = (-1)i = -i$$
$$i^4 = i^2 \cdot i^2 = (-1) \cdot (-1) = 1$$

We continue this process and use the fact that $i^4 = 1$ and $i^2 = -1$ to simplify i^5 and i^6.

$$i^5 = i^4 \cdot i = 1 \cdot i = i$$
$$i^6 = i^4 \cdot i^2 = 1 \cdot (-1) = -1$$

If we continue finding powers of i, we generate the following pattern. Notice that the values i, -1, $-i$, and 1 repeat as i is raised to higher and higher powers.

$i^1 = i$	$i^5 = i$	$i^9 = i$
$i^2 = -1$	$i^6 = -1$	$i^{10} = -1$
$i^3 = -i$	$i^7 = -i$	$i^{11} = -i$
$i^4 = 1$	$i^8 = 1$	$i^{12} = 1$

This pattern allows us to find other powers of i. To do so, we will use the fact that $i^4 = 1$ and rewrite a power of i in terms of i^4. For example,

$$i^{22} = i^{20} \cdot i^2 = (i^4)^5 \cdot i^2 = 1^5 \cdot (-1) = 1 \cdot (-1) = -1.$$

EXAMPLE 6   Find the following powers of i.

a. i^7 **b.** i^{20} **c.** i^{46} **d.** i^{-12}

Solution

a. $i^7 = i^4 \cdot i^3 = 1(-i) = -i$

b. $i^{20} = (i^4)^5 = 1^5 = 1$

c. $i^{46} = i^{44} \cdot i^2 = (i^4)^{11} \cdot i^2 = 1^{11}(-1) = -1$

d. $i^{-12} = \dfrac{1}{i^{12}} = \dfrac{1}{(i^4)^3} = \dfrac{1}{(1)^3} = \dfrac{1}{1} = 1$

PRACTICE
6   Find the following powers of i.

a. i^9 **b.** i^{16} **c.** i^{34} **d.** i^{-24}

 Vocabulary, Readiness & Video Check

Use the choices below to fill in each blank. Not all choices will be used.

-1	$\sqrt{-1}$	real	imaginary unit
1	$\sqrt{1}$	complex	pure imaginary

1. A(n) _____ number is one that can be written in the form $a + bi$, where a and b are real numbers.

2. In the complex number system, i denotes the _____.

3. $i^2 =$ _____

4. $i =$ _____

5. A complex number, $a + bi$, is a(n) _____ number if $b = 0$.

6. A complex number, $a + bi$, is a(n) _____ number if $a = 0$ and $b \neq 0$.

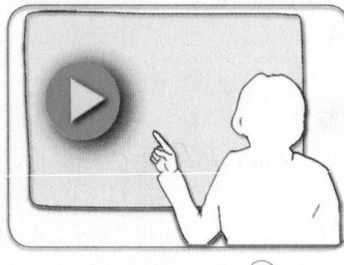

Martin-Gay Interactive Videos

See Video 10.7

Watch the section lecture video and answer the following questions.

OBJECTIVE 1
7. From Example 4, with what rule must you be especially careful when working with imaginary numbers and why?

OBJECTIVE 2
8. In Examples 5 and 6, what is the process of adding and subtracting complex numbers compared to? What important reminder is given about i?

OBJECTIVE 3
9. In Examples 7 and 8, what part of the definition of the imaginary unit i may be used during the multiplication of complex numbers to help simplify products?

OBJECTIVE 4
10. In Example 9, using complex conjugates to divide complex numbers is compared to what process?

OBJECTIVE 5
11. From the lecture before Example 10, what are the first four powers of i whose values keep repeating?

10.7 Exercise Set MyMathLab®

Simplify. See Example 1.

1. $\sqrt{-81}$ **2.** $\sqrt{-49}$ **3.** $\sqrt{-7}$

4. $\sqrt{-3}$ **5.** $-\sqrt{16}$ **6.** $-\sqrt{4}$

7. $\sqrt{-64}$ **8.** $\sqrt{-100}$

Write in terms of i. See Example 1.

9. $\sqrt{-24}$ **10.** $\sqrt{-32}$

11. $-\sqrt{-36}$ **12.** $-\sqrt{-121}$

13. $8\sqrt{-63}$ **14.** $4\sqrt{-20}$

15. $-\sqrt{54}$ **16.** $\sqrt{-63}$

Multiply or divide. See Example 2.

17. $\sqrt{-2} \cdot \sqrt{-7}$ **18.** $\sqrt{-11} \cdot \sqrt{-3}$

19. $\sqrt{-5} \cdot \sqrt{-10}$ **20.** $\sqrt{-2} \cdot \sqrt{-6}$

21. $\sqrt{16} \cdot \sqrt{-1}$ **22.** $\sqrt{3} \cdot \sqrt{-27}$

23. $\dfrac{\sqrt{-9}}{\sqrt{3}}$ **24.** $\dfrac{\sqrt{49}}{\sqrt{-10}}$

25. $\dfrac{\sqrt{-80}}{\sqrt{-10}}$ **26.** $\dfrac{\sqrt{-40}}{\sqrt{-8}}$

Add or subtract. Write the sum or difference in the form a + bi. See Example 3.

27. $(4 - 7i) + (2 + 3i)$ **28.** $(2 - 4i) - (2 - i)$

29. $(6 + 5i) - (8 - i)$ **30.** $(8 - 3i) + (-8 + 3i)$

31. $6 - (8 + 4i)$ **32.** $(9 - 4i) - 9$

Multiply. Write the product in the form a + bi. See Example 4.

33. $-10i \cdot -4i$ **34.** $-2i \cdot -11i$

35. $6i(2 - 3i)$ **36.** $5i(4 - 7i)$

37. $(\sqrt{3} + 2i)(\sqrt{3} - 2i)$ **38.** $(\sqrt{5} - 5i)(\sqrt{5} + 5i)$

39. $(4 - 2i)^2$ **40.** $(6 - 3i)^2$

Write each quotient in the form a + bi. See Example 5.

41. $\dfrac{4}{i}$ **42.** $\dfrac{5}{6i}$

43. $\dfrac{7}{4 + 3i}$ **44.** $\dfrac{9}{1 - 2i}$

45. $\dfrac{3 + 5i}{1 + i}$ **46.** $\dfrac{6 + 2i}{4 - 3i}$

47. $\dfrac{5 - i}{3 - 2i}$ **48.** $\dfrac{6 - i}{2 + i}$

MIXED PRACTICE

Perform each indicated operation. Write the result in the form a + bi.

49. $(7i)(-9i)$ **50.** $(-6i)(-4i)$

51. $(6 - 3i) - (4 - 2i)$ **52.** $(-2 - 4i) - (6 - 8i)$

53. $-3i(-1 + 9i)$ **54.** $-5i(-2 + i)$

55. $\dfrac{4 - 5i}{2i}$ **56.** $\dfrac{6 + 8i}{3i}$

57. $(4 + i)(5 + 2i)$ **58.** $(3 + i)(2 + 4i)$

59. $(6 - 2i)(3 + i)$ **60.** $(2 - 4i)(2 - i)$

61. $(8 - 3i) + (2 + 3i)$ **62.** $(7 + 4i) + (4 - 4i)$

63. $(1 - i)(1 + i)$ **64.** $(6 + 2i)(6 - 2i)$

65. $\dfrac{16 + 15i}{-3i}$ **66.** $\dfrac{2 - 3i}{-7i}$

67. $(9 + 8i)^2$ **68.** $(4 - 7i)^2$

69. $\dfrac{2}{3 + i}$ **70.** $\dfrac{5}{3 - 2i}$

71. $(5 - 6i) - 4i$ **72.** $(6 - 2i) + 7i$

73. $\dfrac{2 - 3i}{2 + i}$ **74.** $\dfrac{6 + 5i}{6 - 5i}$

75. $(2 + 4i) + (6 - 5i)$ **76.** $(5 - 3i) + (7 - 8i)$

77. $(\sqrt{6} + i)(\sqrt{6} - i)$ **78.** $(\sqrt{14} - 4i)(\sqrt{14} + 4i)$

79. $4(2 - i)^2$ **80.** $9(2 - i)^2$

Find each power of i. See Example 6.

81. i^8 **82.** i^{10} **83.** i^{21} **84.** i^{15}

85. i^{11} **86.** i^{40} **87.** i^{-6} **88.** i^{-9}

89. $(2i)^6$ **90.** $(5i)^4$ **91.** $(-3i)^5$ **92.** $(-2i)^7$

REVIEW AND PREVIEW

Recall that the sum of the measures of the angles of a triangle is 180°. Find the unknown angle in each triangle.

△ **93.**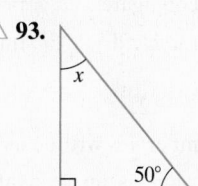

△ **94.**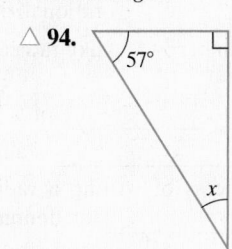

Use synthetic division to divide the following. See Section 5.7.

95. $(x^3 - 6x^2 + 3x - 4) \div (x - 1)$

96. $(5x^4 - 3x^2 + 2) \div (x + 2)$

Thirty people were recently polled about the average monthly balance in their checking accounts. The results of this poll are shown in the following histogram. Use this graph to answer Exercises 97 through 102. See Section 3.1.

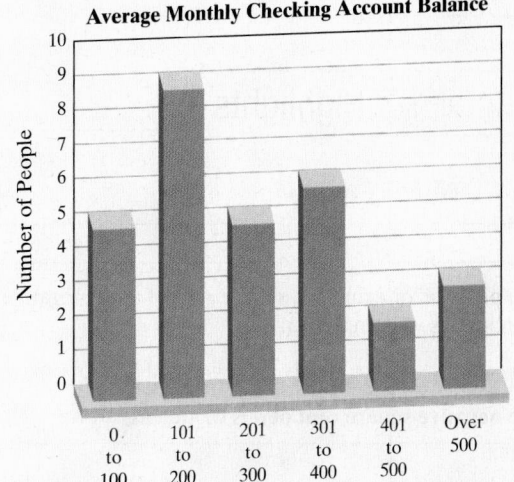

Average Monthly Checking Account Balance

97. How many people polled reported an average checking balance of $201 to $300?

98. How many people polled reported an average checking balance of $0 to $100?

99. How many people polled reported an average checking balance of $200 or less?

100. How many people polled reported an average checking balance of $301 or more?

101. What percent of people polled reported an average checking balance of $201 to $300? Round to the nearest tenth of a percent.

102. What percent of people polled reported an average checking balance of $0 to $100? Round to the nearest tenth of a percent.

CONCEPT EXTENSIONS

Write in the form a + bi.

103. $i^3 - i^4$

104. $i^8 - i^7$

105. $i^6 + i^8$

106. $i^4 + i^{12}$

107. $2 + \sqrt{-9}$

108. $5 - \sqrt{-16}$

109. $\dfrac{6 + \sqrt{-18}}{3}$

110. $\dfrac{4 - \sqrt{-8}}{2}$

111. $\dfrac{5 - \sqrt{-75}}{10}$

112. $\dfrac{7 + \sqrt{-98}}{14}$

113. Describe how to find the conjugate of a complex number.

114. Explain why the product of a complex number and its complex conjugate is a real number.

Simplify.

115. $(8 - \sqrt{-3}) - (2 + \sqrt{-12})$

116. $(8 - \sqrt{-4}) - (2 + \sqrt{-16})$

117. Determine whether $2i$ is a solution of $x^2 + 4 = 0$.

118. Determine whether $-1 + i$ is a solution of $x^2 + 2x = -2$.

Chapter 10 Vocabulary Check

Fill in each blank with one of the words or phrases listed below.

index	rationalizing	conjugate	principal square root	cube root	midpoint
complex number	like radicals	radicand	imaginary unit	distance	

1. The _____ of $\sqrt{3} + 2$ is $\sqrt{3} - 2$.

2. The _____ of a positive number a is written as $\sqrt{a}$.

3. The process of writing a radical expression as an equivalent expression but without a radical in the denominator is called _____ the denominator.

4. The _____, written i, is the number whose square is -1.

5. The _____ of a number is written as $\sqrt[3]{a}$.

6. In the notation $\sqrt[n]{a}$, n is called the _____ and a is called the _____.

7. Radicals with the same index and the same radicand are called _____.

8. A(n) _____ is a number that can be written in the form $a + bi$, where a and b are real numbers.

9. The _____ formula is $d = \sqrt{(x_2 - x_1)^2 + (y_2 - y_1)^2}$.

10. The _____ formula is $\left(\dfrac{x_1 + x_2}{2}, \dfrac{y_1 + y_2}{2}\right)$.

Chapter 10 Highlights

DEFINITIONS AND CONCEPTS	EXAMPLES
Section 10.1 Radicals and Radical Functions	
The **positive**, or **principal**, **square root** of a nonnegative number a is written as $\sqrt{a}$. $$\sqrt{a} = b \text{ only if } b^2 = a \text{ and } b \geq 0$$ The **negative square root** of a is written as $-\sqrt{a}$.	$\sqrt{36} = 6 \qquad \sqrt{\dfrac{9}{100}} = \dfrac{3}{10}$ $-\sqrt{36} = -6 \qquad -\sqrt{0.04} = -0.2$

DEFINITIONS AND CONCEPTS	EXAMPLES

Section 10.1   Radicals and Radical Functions (continued)

The **cube root** of a real number a is written as $\sqrt[3]{a}$.

$$\sqrt[3]{a} = b \text{ only if } b^3 = a$$

If n is an even positive integer, then $\sqrt[n]{a^n} = |a|$.

If n is an odd positive integer, then $\sqrt[n]{a^n} = a$.

A **radical function** in x is a function defined by an expression containing a root of x.

$$\sqrt[3]{27} = 3 \qquad \sqrt[3]{-\frac{1}{8}} = -\frac{1}{2}$$

$$\sqrt[3]{y^6} = y^2 \qquad \sqrt[3]{64x^9} = 4x^3$$

$$\sqrt{(-3)^2} = |-3| = 3$$

$$\sqrt[3]{(-7)^3} = -7$$

If $f(x) = \sqrt{x} + 2$,

$$f(1) = \sqrt{(1)} + 2 = 1 + 2 = 3$$

$$f(3) = \sqrt{(3)} + 2 \approx 3.73$$

Section 10.2   Rational Exponents

$a^{1/n} = \sqrt[n]{a}$ if $\sqrt[n]{a}$ is a real number.

If m and n are positive integers greater than 1 with $\dfrac{m}{n}$ in lowest terms and $\sqrt[n]{a}$ is a real number, then

$$a^{m/n} = \sqrt[n]{a^m} = \left(\sqrt[n]{a}\right)^m$$

$a^{-m/n} = \dfrac{1}{a^{m/n}}$ as long as $a^{m/n}$ is a nonzero number.

Exponent rules are true for rational exponents.

$$81^{1/2} = \sqrt{81} = 9$$

$$(-8x^3)^{1/3} = \sqrt[3]{-8x^3} = -2x$$

$$4^{5/2} = (\sqrt{4})^5 = 2^5 = 32$$

$$27^{2/3} = (\sqrt[3]{27})^2 = 3^2 = 9$$

$$16^{-3/4} = \frac{1}{16^{3/4}} = \frac{1}{(\sqrt[4]{16})^3} = \frac{1}{2^3} = \frac{1}{8}$$

$$x^{2/3} \cdot x^{-5/6} = x^{2/3-5/6} = x^{-1/6} = \frac{1}{x^{1/6}}$$

$$(8^4)^{1/2} = 8^2 = 64$$

$$\frac{a^{4/5}}{a^{-2/5}} = a^{4/5-(-2/5)} = a^{6/5}$$

Section 10.3   Simplifying Radical Expressions

Product and Quotient Rules

If $\sqrt[n]{a}$ and $\sqrt[n]{b}$ are real numbers,

$$\sqrt[n]{a} \cdot \sqrt[n]{b} = \sqrt[n]{a \cdot b}$$

$$\frac{\sqrt[n]{a}}{\sqrt[n]{b}} = \sqrt[n]{\frac{a}{b}}, \text{ provided } \sqrt[n]{b} \neq 0$$

A radical of the form $\sqrt[n]{a}$ is **simplified** when a contains no factors that are perfect nth powers.

Multiply or divide as indicated:

$$\sqrt{11} \cdot \sqrt{3} = \sqrt{33}$$

$$\frac{\sqrt[3]{40x}}{\sqrt[3]{5x}} = \sqrt[3]{8} = 2$$

$$\sqrt{40} = \sqrt{4 \cdot 10} = 2\sqrt{10}$$

$$\sqrt{36x^5} = \sqrt{36x^4 \cdot x} = 6x^2\sqrt{x}$$

$$\sqrt[3]{24x^7y^3} = \sqrt[3]{8x^6y^3 \cdot 3x} = 2x^2y\sqrt[3]{3x}$$

Distance Formula

The distance d between two points (x_1, y_1) and (x_2, y_2) is given by

$$d = \sqrt{(x_2 - x_1)^2 + (y_2 - y_1)^2}$$

Find the distance between points $(-1, 6)$ and $(-2, -4)$.
Let $(x_1, y_1) = (-1, 6)$ and $(x_2, y_2) = (-2, -4)$.

$$d = \sqrt{(x_2 - x_1)^2 + (y_2 - y_1)^2}$$

$$= \sqrt{(-2 - (-1))^2 + (-4 - 6)^2}$$

$$= \sqrt{1 + 100} = \sqrt{101}$$

(continued)

DEFINITIONS AND CONCEPTS	EXAMPLES

Section 10.3 Simplifying Radical Expressions (continued)

Midpoint Formula

The midpoint of the line segment whose endpoints are (x_1, y_1) and (x_2, y_2) is the point with coordinates

$$\left(\frac{x_1 + x_2}{2}, \frac{y_1 + y_2}{2}\right)$$

Find the midpoint of the line segment whose endpoints are $(-1, 6)$ and $(-2, -4)$.

$$\left(\frac{-1 + (-2)}{2}, \frac{6 + (-4)}{2}\right)$$

The midpoint is $\left(-\frac{3}{2}, 1\right)$.

Section 10.4 Adding, Subtracting, and Multiplying Radical Expressions

Radicals with the same index and the same radicand are **like radicals.**

The distributive property can be used to add like radicals.

$$5\sqrt{6} + 2\sqrt{6} = (5 + 2)\sqrt{6} = 7\sqrt{6}$$

$$-\sqrt[3]{3x} - 10\sqrt[3]{3x} + 3\sqrt[3]{10x}$$

$$= (-1 - 10)\sqrt[3]{3x} + 3\sqrt[3]{10x}$$

$$= -11\sqrt[3]{3x} + 3\sqrt[3]{10x}$$

Radical expressions are multiplied by using many of the same properties used to multiply polynomials.

Multiply.

$$(\sqrt{5} - \sqrt{2x})(\sqrt{2} + \sqrt{2x})$$

$$= \sqrt{10} + \sqrt{10x} - \sqrt{4x} - 2x$$

$$= \sqrt{10} + \sqrt{10x} - 2\sqrt{x} - 2x$$

$$(2\sqrt{3} - \sqrt{8x})(2\sqrt{3} + \sqrt{8x})$$

$$= 4(3) - 8x = 12 - 8x$$

Section 10.5 Rationalizing Denominators and Numerators of Radical Expressions

The **conjugate** of $a + b$ is $a - b$.

The conjugate of $\sqrt{7} + \sqrt{3}$ is $\sqrt{7} - \sqrt{3}$.

The process of writing the denominator of a radical expression without a radical is called **rationalizing the denominator.**

Rationalize each denominator.

$$\frac{\sqrt{5}}{\sqrt{3}} = \frac{\sqrt{5} \cdot \sqrt{3}}{\sqrt{3} \cdot \sqrt{3}} = \frac{\sqrt{15}}{3}$$

$$\frac{6}{\sqrt{7} + \sqrt{3}} = \frac{6(\sqrt{7} - \sqrt{3})}{(\sqrt{7} + \sqrt{3})(\sqrt{7} - \sqrt{3})}$$

$$= \frac{6(\sqrt{7} - \sqrt{3})}{7 - 3}$$

$$= \frac{6(\sqrt{7} - \sqrt{3})}{4} = \frac{3(\sqrt{7} - \sqrt{3})}{2}$$

DEFINITIONS AND CONCEPTS	EXAMPLES

Section 10.5   Rationalizing Denominators and Numerators of Radical Expressions (continued)

The process of writing the numerator of a radical expression without a radical is called **rationalizing the numerator.**

Rationalize each numerator.

$$\frac{\sqrt[3]{9}}{\sqrt[3]{5}} = \frac{\sqrt[3]{9} \cdot \sqrt[3]{3}}{\sqrt[3]{5} \cdot \sqrt[3]{3}} = \frac{\sqrt[3]{27}}{\sqrt[3]{15}} = \frac{3}{\sqrt[3]{15}}$$

$$\frac{\sqrt{9} + \sqrt{3x}}{12} = \frac{(\sqrt{9} + \sqrt{3x})(\sqrt{9} - \sqrt{3x})}{12(\sqrt{9} - \sqrt{3x})}$$

$$= \frac{9 - 3x}{12(\sqrt{9} - \sqrt{3x})}$$

$$= \frac{3(3 - x)}{3 \cdot 4(3 - \sqrt{3x})} = \frac{3 - x}{4(3 - \sqrt{3x})}$$

Section 10.6   Radical Equations and Problem Solving

To Solve a Radical Equation

Step 1. Write the equation so that one radical is by itself on one side of the equation.

Step 2. Raise each side of the equation to a power equal to the index of the radical and simplify.

Step 3. If the equation still contains a radical, repeat Steps 1 and 2. If not, solve the equation.

Step 4. Check all proposed solutions in the original equation.

Solve: $x = \sqrt{4x + 9} + 3$

1. $x - 3 = \sqrt{4x + 9}$

2. $(x - 3)^2 = (\sqrt{4x + 9})^2$
       $x^2 - 6x + 9 = 4x + 9$

3. $x^2 - 10x = 0$
       $x(x - 10) = 0$
            $x = 0$ or $x = 10$

4. The proposed solution 10 checks, but 0 does not. The solution is 10.

Section 10.7   Complex Numbers

$i^2 = -1$ and $i = \sqrt{-1}$

A **complex number** is a number that can be written in the form $a + bi$, where a and b are real numbers.

Simplify: $\sqrt{-9}$

$$\sqrt{-9} = \sqrt{-1 \cdot 9} = \sqrt{-1} \cdot \sqrt{9} = i \cdot 3 \text{ or } 3i$$

Complex Numbers	*Written in Form $a + bi$*
12	$12 + 0i$
$-5i$	$0 + (-5)i$
$-2 - 3i$	$-2 + (-3)i$

Multiply,

$$\sqrt{-3} \cdot \sqrt{-7} = i\sqrt{3} \cdot i\sqrt{7}$$
$$= i^2\sqrt{21}$$
$$= -\sqrt{21}$$

To add or subtract complex numbers, add or subtract their real parts and then add or subtract their imaginary parts.

To multiply complex numbers, multiply as though they are binomials.

Perform each indicated operation.

$$(-3 + 2i) - (7 - 4i) = -3 + 2i - 7 + 4i$$
$$= -10 + 6i$$

$$(-7 - 2i)(6 + i) = -42 - 7i - 12i - 2i^2$$
$$= -42 - 19i - 2(-1)$$
$$= -42 - 19i + 2$$
$$= -40 - 19i$$

(continued)

DEFINITIONS AND CONCEPTS	EXAMPLES

Section 10.7   Complex Numbers (continued)

The complex numbers $(a + bi)$ and $(a - bi)$ are called **complex conjugates**.

The complex conjugate of $(3 + 6i)$ is $(3 - 6i)$.

Their product is a real number:

$$(3 - 6i)(3 + 6i) = 9 - 36i^2$$
$$= 9 - 36(-1) = 9 + 36 = 45$$

To divide complex numbers, multiply the numerator and the denominator by the conjugate of the denominator.

Divide.

$$\frac{4}{2 - i} = \frac{4(2 + i)}{(2 - i)(2 + i)}$$
$$= \frac{4(2 + i)}{4 - i^2}$$
$$= \frac{4(2 + i)}{5}$$
$$= \frac{8 + 4i}{5} = \frac{8}{5} + \frac{4}{5}i$$

Chapter 10   Review

(10.1) *Find the root. Assume that all variables represent positive numbers.*

1. $\sqrt{81}$

2. $\sqrt[4]{81}$

3. $\sqrt[3]{-8}$

4. $\sqrt[4]{-16}$

5. $-\sqrt{\dfrac{1}{49}}$

6. $\sqrt{x^{64}}$

7. $-\sqrt{36}$

8. $\sqrt[3]{64}$

9. $\sqrt[3]{-a^6b^9}$

10. $\sqrt{16a^4b^{12}}$

11. $\sqrt[5]{32a^5b^{10}}$

12. $\sqrt[5]{-32x^{15}y^{20}}$

13. $\sqrt{\dfrac{x^{12}}{36y^2}}$

14. $\sqrt[3]{\dfrac{27y^3}{z^{12}}}$

Simplify. Use absolute value bars when necessary.

15. $\sqrt{(-x)^2}$

16. $\sqrt[4]{(x^2 - 4)^4}$

17. $\sqrt[3]{(-27)^3}$

18. $\sqrt[5]{(-5)^5}$

19. $-\sqrt[5]{x^5}$

20. $-\sqrt[3]{x^3}$

21. $\sqrt[4]{16(2y + z)^4}$

22. $\sqrt{25(x - y)^2}$

23. $\sqrt[5]{y^5}$

24. $\sqrt[6]{x^6}$

25. Let $f(x) = \sqrt{x} + 3$.

   a. Find $f(0)$ and $f(9)$.

   b. Find the domain of $f(x)$.

   c. Graph $f(x)$.

26. Let $g(x) = \sqrt[3]{x} - 3$.

   a. Find $g(11)$ and $g(20)$.

   b. Find the domain $g(x)$.

   c. Graph $g(x)$.

(10.2) *Evaluate.*

27. $\left(\dfrac{1}{81}\right)^{1/4}$

28. $\left(-\dfrac{1}{27}\right)^{1/3}$

29. $(-27)^{-1/3}$

30. $(-64)^{-1/3}$

31. $-9^{3/2}$

32. $64^{-1/3}$

33. $(-25)^{5/2}$

34. $\left(\dfrac{25}{49}\right)^{-3/2}$

35. $\left(\dfrac{8}{27}\right)^{-2/3}$

36. $\left(-\dfrac{1}{36}\right)^{-1/4}$

Write with rational exponents.

37. $\sqrt[3]{x^2}$

38. $\sqrt[5]{5x^2y^3}$

Write using radical notation.

39. $y^{4/5}$

40. $5(xy^2z^5)^{1/3}$

41. $(x + 2)^{-1/3}$

42. $(x + 2y)^{-1/2}$

Simplify each expression. Assume that all variables represent positive real numbers. Write with only positive exponents.

43. $a^{1/3}a^{4/3}a^{1/2}$

44. $\dfrac{b^{1/3}}{b^{4/3}}$

45. $(a^{1/2}a^{-2})^3$

46. $(x^{-3}y^6)^{1/3}$

47. $\left(\dfrac{b^{3/4}}{a^{-1/2}}\right)^8$

48. $\dfrac{x^{1/4}x^{-1/2}}{x^{2/3}}$

49. $\left(\dfrac{49c^{5/3}}{a^{-1/4}b^{5/6}}\right)^{-1}$

50. $a^{-1/4}(a^{5/4} - a^{9/4})$

Use a calculator and write a three-decimal-place approximation of each number.

51. $\sqrt{20}$

52. $\sqrt[3]{-39}$

53. $\sqrt[4]{726}$

54. $56^{1/3}$

55. $-78^{3/4}$

56. $105^{-2/3}$

Use rational exponents to write each as a single radical.

57. $\sqrt[3]{2} \cdot \sqrt{7}$

58. $\sqrt[3]{3} \cdot \sqrt[4]{x}$

(10.3) *Perform each indicated operation and then simplify if possible. Assume that all variables represent positive real numbers.*

59. $\sqrt{3} \cdot \sqrt{8}$

60. $\sqrt[3]{7y} \cdot \sqrt[3]{x^2 z}$

61. $\dfrac{\sqrt{44x^3}}{\sqrt{11x}}$

62. $\dfrac{\sqrt[4]{a^6 b^{13}}}{\sqrt[4]{a^2 b}}$

Simplify.

63. $\sqrt{60}$

64. $-\sqrt{75}$

65. $\sqrt[3]{162}$

66. $\sqrt[3]{-32}$

67. $\sqrt{36x^7}$

68. $\sqrt[3]{24a^5 b^7}$

69. $\sqrt{\dfrac{p^{17}}{121}}$

70. $\sqrt[3]{\dfrac{y^5}{27x^6}}$

71. $\sqrt[4]{\dfrac{xy^6}{81}}$

72. $\sqrt{\dfrac{2x^3}{49y^4}}$

The formula for the radius r of a circle of area A is $r = \sqrt{\dfrac{A}{\pi}}$. Use this for Exercises 73 and 74.

73. Find the exact radius of a circle whose area is 25 square meters.

74. Approximate to two decimal places the radius of a circle whose area is 104 square inches.

Find the distance between each pair of points. Give the exact value and a three-decimal-place approximation.

75. $(-6, 3)$ and $(8, 4)$

76. $(-4, -6)$ and $(-1, 5)$

77. $(-1, 5)$ and $(2, -3)$

78. $(-\sqrt{2}, 0)$ and $(0, -4\sqrt{6})$

79. $(-\sqrt{5}, -\sqrt{11})$ and $(-\sqrt{5}, -3\sqrt{11})$

80. $(7.4, -8.6)$ and $(-1.2, 5.6)$

Find the midpoint of each line segment whose endpoints are given.

81. $(2, 6); (-12, 4)$

82. $(-6, -5); (-9, 7)$

83. $(4, -6); (-15, 2)$

84. $\left(0, -\dfrac{3}{8}\right); \left(\dfrac{1}{10}, 0\right)$

85. $\left(\dfrac{3}{4}, -\dfrac{1}{7}\right); \left(-\dfrac{1}{4}, -\dfrac{3}{7}\right)$

86. $(\sqrt{3}, -2\sqrt{6}); (\sqrt{3}, -4\sqrt{6})$

(10.4) *Perform each indicated operation. Assume that all variables represent positive real numbers.*

87. $\sqrt{20} + \sqrt{45} - 7\sqrt{5}$

88. $x\sqrt{75x} - \sqrt{27x^3}$

89. $\sqrt[3]{128} + \sqrt[3]{250}$

90. $3\sqrt[4]{32a^5} - a\sqrt[4]{162a}$

91. $\dfrac{5}{\sqrt{4}} + \dfrac{\sqrt{3}}{3}$

92. $\sqrt{\dfrac{8}{x^2}} - \sqrt{\dfrac{50}{16x^2}}$

93. $2\sqrt{50} - 3\sqrt{125} + \sqrt{98}$

94. $2a\sqrt[4]{32b^5} - 3b\sqrt[4]{162a^4 b} + \sqrt[4]{2a^4 b^5}$

Multiply and then simplify if possible. Assume that all variables represent positive real numbers.

95. $\sqrt{3}(\sqrt{27} - \sqrt{3})$

96. $(\sqrt{x} - 3)^2$

97. $(\sqrt{5} - 5)(2\sqrt{5} + 2)$

98. $(2\sqrt{x} - 3\sqrt{y})(2\sqrt{x} + 3\sqrt{y})$

99. $(\sqrt{a} + 3)(\sqrt{a} - 3)$

100. $(\sqrt[3]{a} + 2)^2$

101. $(\sqrt[3]{5x} + 9)(\sqrt[3]{5x} - 9)$

102. $(\sqrt[3]{a} + 4)(\sqrt[3]{a^2} - 4\sqrt[3]{a} + 16)$

(10.5) *Rationalize each denominator. Assume that all variables represent positive real numbers.*

103. $\dfrac{3}{\sqrt{7}}$

104. $\sqrt{\dfrac{x}{12}}$

105. $\dfrac{5}{\sqrt[3]{4}}$

106. $\sqrt{\dfrac{24x^5}{3y}}$

107. $\sqrt[3]{\dfrac{15x^6 y^7}{z^2}}$

108. $\sqrt[4]{\dfrac{81}{8x^{10}}}$

109. $\dfrac{3}{\sqrt{y} - 2}$

110. $\dfrac{\sqrt{2} - \sqrt{3}}{\sqrt{2} + \sqrt{3}}$

Rationalize each numerator. Assume that all variables represent positive real numbers.

111. $\dfrac{\sqrt{11}}{3}$

112. $\sqrt{\dfrac{18}{y}}$

113. $\dfrac{\sqrt[3]{9}}{7}$

114. $\sqrt{\dfrac{24x^5}{3y^2}}$

115. $\sqrt[3]{\dfrac{xy^2}{10z}}$

116. $\dfrac{\sqrt{x} + 5}{-3}$

(10.6) *Solve each equation.*

117. $\sqrt{y - 7} = 5$

118. $\sqrt{2x} + 10 = 4$

119. $\sqrt[3]{2x - 6} = 4$

120. $\sqrt{x + 6} = \sqrt{x + 2}$

121. $2x - 5\sqrt{x} = 3$

122. $\sqrt{x + 9} = 2 + \sqrt{x - 7}$

Find each unknown length.

△ **123.**

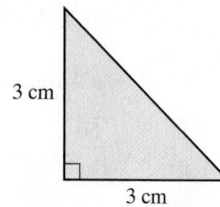

3 cm

3 cm

△ **124.**

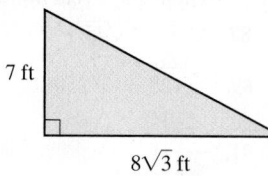

7 ft

8√3 ft

△ **125.** Craig and Daniel Cantwell want to determine the distance *x* across a pond on their property. They are able to measure the distances shown on the following diagram. Find how wide the pond is at the crossing point indicated by the triangle to the nearest tenth of a foot.

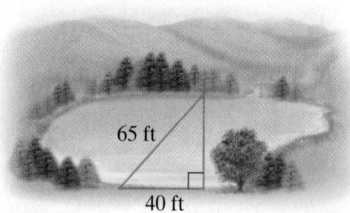

65 ft

40 ft

△ **126.** Andrea Roberts, a pipefitter, needs to connect two underground pipelines that are offset by 3 feet, as pictured in the diagram. Neglecting the joints needed to join the pipes, find the length of the shortest possible connecting pipe rounded to the nearest hundredth of a foot.

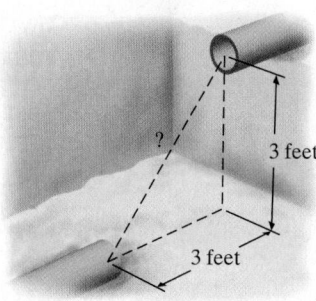

3 feet

?

3 feet

(10.7) *Perform each indicated operation and simplify. Write the results in the form a + bi.*

127. $\sqrt{-8}$

128. $-\sqrt{-6}$

129. $\sqrt{-4} + \sqrt{-16}$

130. $\sqrt{-2} \cdot \sqrt{-5}$

131. $(12 - 6i) + (3 + 2i)$

132. $(-8 - 7i) - (5 - 4i)$

133. $(2i)^6$

134. $(3i)^4$

135. $-3i(6 - 4i)$

136. $(3 + 2i)(1 + i)$

137. $(2 - 3i)^2$

138. $(\sqrt{6} - 9i)(\sqrt{6} + 9i)$

139. $\dfrac{2 + 3i}{2i}$

140. $\dfrac{1 + i}{-3i}$

MIXED REVIEW

Simplify. Use absolute value bars when necessary.

141. $\sqrt[3]{x^3}$

142. $\sqrt{(x + 2)^2}$

Simplify. Assume that all variables represent positive real numbers. If necessary, write answers with positive exponents only.

143. $-\sqrt{100}$

144. $\sqrt[3]{-x^{12}y^3}$

145. $\sqrt[4]{\dfrac{y^{20}}{16x^{12}}}$

146. $9^{1/2}$

147. $64^{-1/2}$

148. $\left(\dfrac{27}{64}\right)^{-2/3}$

149. $\dfrac{(x^{2/3}x^{-3})^3}{x^{-1/2}}$

150. $\sqrt{200x^9}$

151. $\sqrt{\dfrac{3n^3}{121m^{10}}}$

152. $3\sqrt{20} - 7x\sqrt[3]{40} + 3\sqrt[3]{5x^3}$

153. $(2\sqrt{x} - 5)^2$

154. Find the distance between $(-3, 5)$ and $(-8, 9)$.

155. Find the midpoint of the line segment joining $(-3, 8)$ and $(11, 24)$.

Rationalize each denominator.

156. $\dfrac{7}{\sqrt{13}}$

157. $\dfrac{2}{\sqrt{x} + 3}$

Solve.

158. $\sqrt{x} + 2 = x$

159. $\sqrt{2x - 1} + 2 = x$

1c

Chapter 10 Getting Ready for the Test

MULTIPLE CHOICE *Select the correct choice.*

▶ **1.** Which radical simplifies to −4?

   A. $\sqrt{16}$ **B.** $\sqrt{-16}$ **C.** $\sqrt[3]{64}$ **D.** $\sqrt[3]{-64}$

▶ **2.** Which radical does not simplify to a whole number?

   A. $\sqrt{16}$ **B.** $\sqrt[3]{16}$ **C.** $\sqrt{64}$ **D.** $\sqrt[3]{64}$

For Exercises 3–6, identify each as **A.** *a real number* or **B.** *not a real number.*

3. $\sqrt{5}$ **4.** $\sqrt{-11}$ **5.** $\sqrt[3]{-9}$ **6.** $-\sqrt{17}$

MATCHING *Match each expression with its simplified form. Letters may be used more than once or not at all.*

7. $25^{1/2}$ **A.** 5

8. $(-25)^{1/2}$ **B.** -5

9. $(-125)^{1/3}$ **C.** not a real number

10. $-25^{1/2}$

MULTIPLE CHOICE *For Exercises 11–14, identify each statement as* **A.** *true or* **B.** *false.*

11. $\sqrt{5} \cdot \sqrt{10} = \sqrt{15}$ **12.** $\sqrt[3]{4} \cdot \sqrt[3]{9} = \sqrt[3]{36}$ **13.** $\dfrac{\sqrt{12}}{\sqrt{6}} = \sqrt{2}$

14. $\dfrac{\sqrt[3]{10}}{\sqrt[3]{4}} = \sqrt[3]{6}$ **15.** $\sqrt{2} + \sqrt{3} = \sqrt{5}$ **16.** $\sqrt[3]{7} + \sqrt{7} = \sqrt[4]{7}$

MULTIPLE CHOICE *Select the correct choice.*

17. Which expression simplifies to x?

     A. $x^{1/2} + x^{1/2}$ **B.** $x^{1/2} \cdot x^{1/2}$ **C.** $\left(x^{1/2}\right)^2$ **D.** both B and C   **E.** A, B, and C

18. To rationalize the numerator of $\dfrac{\sqrt{x} - 3}{\sqrt{x}}$, we multiply by:

     A. $\dfrac{\sqrt{x}}{\sqrt{x} + 3}$ **B.** $\dfrac{\sqrt{x}}{\sqrt{x}}$ **C.** $\dfrac{\sqrt{x} + 3}{\sqrt{x} + 3}$ **D.** $\dfrac{\sqrt{x} + 3}{\sqrt{x}}$

19. Square both sides of the equation $\sqrt{x + 1} = 3 + \sqrt{x - 1}$. The result is:

     A. $x + 1 = 9 + (x - 1)$ **B.** $x + 1 = 9 + 6\sqrt{x - 1} + (x - 1)$ **C.** $x + 1 = 6 + (x - 1)$

20. $(5 - 2i)^2 =$

     A. $25 + 4i$ **B.** 21   **C.** 29   **D.** $21 - 20i$

21. The expression $\dfrac{3 + \sqrt{-9}}{3}$ simplifies to:

     A. $1 + i$ **B.** $3i$ **C.** $1 + 3i$ **D.** $2i$

Chapter 10 **Test** MyMathLab® You Tube™

Raise to the power or find the root. Assume that all variables represent positive numbers. Write with only positive exponents.

1. $\sqrt{216}$ **2.** $-\sqrt[4]{x^{64}}$

3. $\left(\dfrac{1}{125}\right)^{1/3}$ **4.** $\left(\dfrac{1}{125}\right)^{-1/3}$

5. $\left(\dfrac{8x^3}{27}\right)^{2/3}$ **6.** $\sqrt[3]{-a^{18}b^9}$

7. $\left(\dfrac{64c^{4/3}}{a^{-2/3}b^{5/6}}\right)^{1/2}$ **8.** $a^{-2/3}(a^{5/4} - a^3)$

Find the root. Use absolute value bars when necessary.

9. $\sqrt[4]{(4xy)^4}$ **10.** $\sqrt[3]{(-27)^3}$

Rationalize the denominator. Assume that all variables represent positive numbers.

11. $\sqrt{\dfrac{9}{y}}$ **12.** $\dfrac{4 - \sqrt{x}}{4 + 2\sqrt{x}}$ **13.** $\dfrac{\sqrt[3]{ab}}{\sqrt[3]{ab^2}}$

14. Rationalize the numerator of $\dfrac{\sqrt{6} + x}{8}$ and simplify.

Perform the indicated operations. Assume that all variables represent positive numbers.

15. $\sqrt{125x^3} - 3\sqrt{20x^3}$

16. $\sqrt{3}(\sqrt{16} - \sqrt{2})$

17. $(\sqrt{x} + 1)^2$

18. $(\sqrt{2} - 4)(\sqrt{3} + 1)$

19. $(\sqrt{5} + 5)(\sqrt{5} - 5)$

Use a calculator to approximate each to three decimal places.

20. $\sqrt{561}$ **21.** $386^{-2/3}$

Solve.

22. $x = \sqrt{x - 2} + 2$

23. $\sqrt{x^2 - 7} + 3 = 0$

24. $\sqrt[3]{x + 5} = \sqrt[3]{2x - 1}$

Perform the indicated operation and simplify. Write the result in the form $a + bi$.

25. $\sqrt{-2}$ **26.** $-\sqrt{-8}$

27. $(12 - 6i) - (12 - 3i)$ **28.** $(6 - 2i)(6 + 2i)$

▶ **29.** $(4 + 3i)^2$

▶ **30.** $\dfrac{1 + 4i}{1 - i}$

▶ **31.** Find x.

△

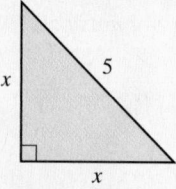

▶ **32.** Identify the domain of $g(x)$. Then complete the accompanying table and graph $g(x)$.

$$g(x) = \sqrt{x + 2}$$

x	-2	-1	2	7
$g(x)$				

▶ **33.** Find the distance between the points $(-6, 3)$ and $(-8, -7)$.

▶ **34.** Find the distance between the points $(-2\sqrt{5}, \sqrt{10})$ and $(-\sqrt{5}, 4\sqrt{10})$.

▶ **35.** Find the midpoint of the line segment whose endpoints are $(-2, -5)$ and $(-6, 12)$.

▶ **36.** Find the midpoint of the line segment whose endpoints are $\left(-\dfrac{2}{3}, -\dfrac{1}{5}\right)$ and $\left(-\dfrac{1}{3}, \dfrac{4}{5}\right)$.

Solve.

▶ **37.** The function $V(r) = \sqrt{2.5r}$ can be used to estimate the maximum safe velocity V in miles per hour at which a car can travel if it is driven along a curved road with a *radius of curvature r* in feet. To the nearest whole number, find the maximum safe speed if a cloverleaf exit on an expressway has a radius of curvature of 300 feet.

▶ **38.** Use the formula from Exercise 37 to find the radius of curvature if the safe velocity is 30 mph.

Chapter 10 Cumulative Review

1. Simplify each expression.
  a. $-3 + [(-2 - 5) - 2]$
  b. $2^3 - |10| + [-6 - (-5)]$

2. Simplify each expression.
  a. $2(x - 3) + (5x + 3)$
  b. $4(3x + 2) - 3(5x - 1)$
  c. $7x + 2(x - 7) - 3x$

3. Solve: $\dfrac{x}{2} - 1 = \dfrac{2}{3}x - 3$

4. Solve: $\dfrac{a - 1}{2} + a = 2 - \dfrac{2a + 7}{8}$

5. A 48-inch balsa wood stick is to be cut into two pieces so that the longer piece is 3 times the shorter. Find the length of each piece.

6. The Smith family owns a lake house 121.5 miles from home. If it takes them $4\dfrac{1}{2}$ hours to drive round-trip from their house to their lake house, find their average speed.

7. Without graphing, determine the number of solutions of the system.
  $$\begin{cases} 3x - y = 4 \\ x + 2y = 8 \end{cases}$$

8. Solve: $|3x - 2| + 5 = 5$

9. Solve the system: $\begin{cases} x + 2y = 7 \\ 2x + 2y = 13 \end{cases}$

10. Solve: $\left|\dfrac{x}{2} - 1\right| \le 0$

11. Solve the system: $\begin{cases} 2x - y = 7 \\ 8x - 4y = 1 \end{cases}$

12. Graph: $y = |x - 2|$

13. Lynn Pike, a pharmacist, needs 70 liters of a 50% alcohol solution. She has available a 30% alcohol solution and an 80% alcohol solution. How many liters of each solution should she mix to obtain 70 liters of a 50% alcohol solution?

14. Find the domain and the range of each relation. Use the vertical line test to determine whether each graph is the graph of a function.

a.

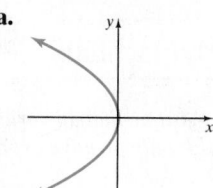

b.

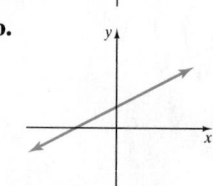

c.

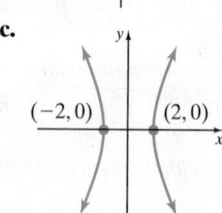

15. If $P(x) = 3x^2 - 2x - 5$, find the following.
  a. $P(1)$ **b.** $P(-2)$

16. Graph: $f(x) = -2$

17. Divide $6m^2 + 2m$ by $2m$.

18. Find the slope of $y = -3$.

19. Use synthetic division to divide $2x^3 - x^2 - 13x + 1$ by $x - 3$.

20. Solve the system.
$$\begin{cases} \dfrac{x}{6} - \dfrac{y}{2} = 1 \\ \dfrac{x}{3} - \dfrac{y}{4} = 2 \end{cases}$$

21. Factor: $40 - 13t + t^2$

22. At a seasonal clearance sale, Nana Long spent $33.75. She paid $3.50 for tee-shirts and $4.25 for shorts. If she bought 9 items, how many of each item did she buy?

23. Simplify each rational expression.

   a. $\dfrac{x^3 + 8}{2 + x}$

   b. $\dfrac{2y^2 + 2}{y^3 - 5y^2 + y - 5}$

24. Use scientific notation to simplify and write the answer in scientific notation. $\dfrac{0.0000035 \times 4000}{0.28}$

25. Solve: $|x - 3| = |5 - x|$

26. Subtract $(2x - 5)$ from the sum of $(5x^2 - 3x + 6)$ and $(4x^2 + 5x - 3)$.

27. Subtract: $\dfrac{3x^2 + 2x}{x - 1} - \dfrac{10x - 5}{x - 1}$

28. Multiply and simplify the product if possible.

   a. $(y - 2)(3y + 4)$

   b. $(3y - 1)(2y^2 + 3y - 1)$

29. Add: $1 + \dfrac{m}{m + 1}$

30. Factor: $x^3 - x^2 + 4x - 4$

31. Simply each complex fraction.

   a. $\dfrac{\dfrac{5x}{x + 2}}{\dfrac{10}{x - 2}}$

   b. $\dfrac{\dfrac{x}{y^2} + \dfrac{1}{y}}{\dfrac{y}{x^2} + \dfrac{1}{x}}$

32. Simplify each rational expression.

   a. $\dfrac{a^3 - 8}{2 - a}$

   b. $\dfrac{3a^2 - 3}{a^3 + 5a^2 - a - 5}$

33. Solve: $|5x + 1| + 1 \le 10$

34. Perform the indicated operations.

   a. $\dfrac{3}{xy^2} - \dfrac{2}{3x^2y}$

   b. $\dfrac{5x}{x + 3} - \dfrac{2x}{x - 3}$

   c. $\dfrac{x}{x - 2} - \dfrac{5}{2 - x}$

35. If the following two triangles are similar, find the unknown length x.

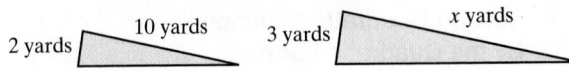

36. Simplify each complex fraction.

   a. $\dfrac{\dfrac{y - 2}{16}}{\dfrac{2y + 3}{12}}$

   b. $\dfrac{\dfrac{x}{16} - \dfrac{1}{x}}{1 - \dfrac{4}{x}}$

37. Find the cube roots.

   a. $\sqrt[3]{1}$ **b.** $\sqrt[3]{-64}$

   c. $\sqrt[3]{\dfrac{8}{125}}$ **d.** $\sqrt[3]{x^6}$

   e. $\sqrt[3]{-27x^{15}}$

38. Divide $x^3 - 2x^2 + 3x - 6$ by $x - 2$.

39. Write each expression with a positive exponent, and then simplify.

   a. $16^{-3/4}$

   b. $(-27)^{-2/3}$

40. Use synthetic division to divide $4y^3 - 12y^2 - y + 12$ by $y - 3$.

41. Rationalize the numerator of $\dfrac{\sqrt{x} + 2}{5}$.

42. Solve: $\dfrac{28}{9 - a^2} = \dfrac{2a}{a - 3} + \dfrac{6}{a + 3}$

43. Suppose that u varies inversely as w. If u is 3 when w is 5, find the constant of variation and the inverse variation equation.

44. Suppose that y varies directly as x. If $y = 0.51$ when $x = 3$, find the constant of variation and the direct variation equation.

Quadratic Equations and Functions

11.1 Solving Quadratic Equations by Completing the Square

11.2 Solving Quadratic Equations by the Quadratic Formula

11.3 Solving Equations by Using Quadratic Methods

Integrated Review—Summary on Solving Quadratic Equations

11.4 Nonlinear Inequalities in One Variable

11.5 Quadratic Functions and Their Graphs

11.6 Further Graphing of Quadratic Functions

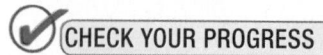

CHECK YOUR PROGRESS

Vocabulary Check

Chapter Highlights

Chapter Review

Getting Ready for the Test

Chapter Test

Cumulative Review

An important part of the study of algebra is learning to model and solve problems. Often, the model of a problem is a quadratic equation or a function containing a second-degree polynomial. In this chapter, we continue the work begun in Chapter 6, when we solved polynomial equations in one variable by factoring. Two additional methods of solving quadratic equations are analyzed as well as methods of solving nonlinear inequalities in one variable.

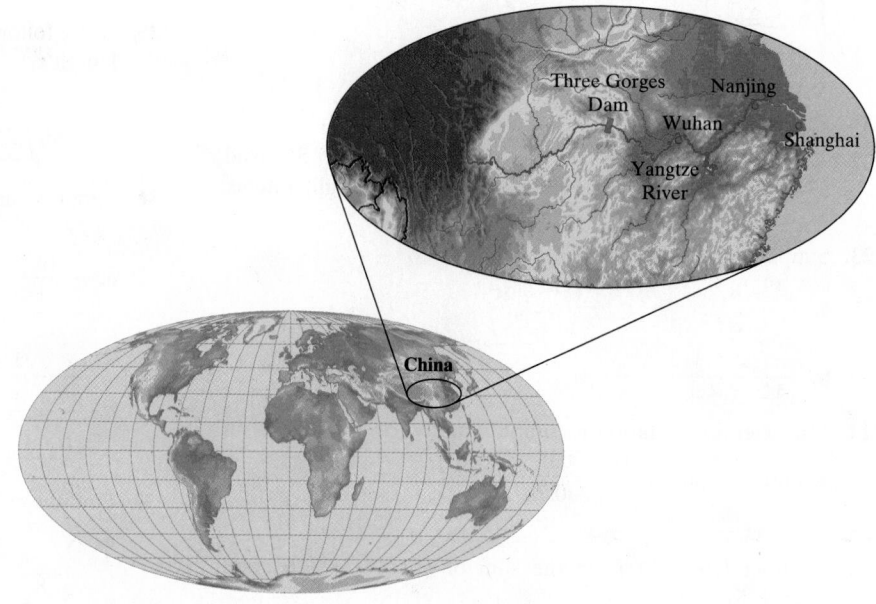

Why Are Dams Built?

One type of a dam is a man-made structure built across a river. Dams are built primarily to control river flow, improve navigation, provide reservoirs for fresh water, or to produce hydroelectric power. The Three Gorges Dam in China is the largest operating hydroelectric facility. While this dam produces the most electricity of any dam in the world, it is only 610 feet in height.

In Section 11.1, Exercises 81 and 82, we will examine the height of some of the great dams.

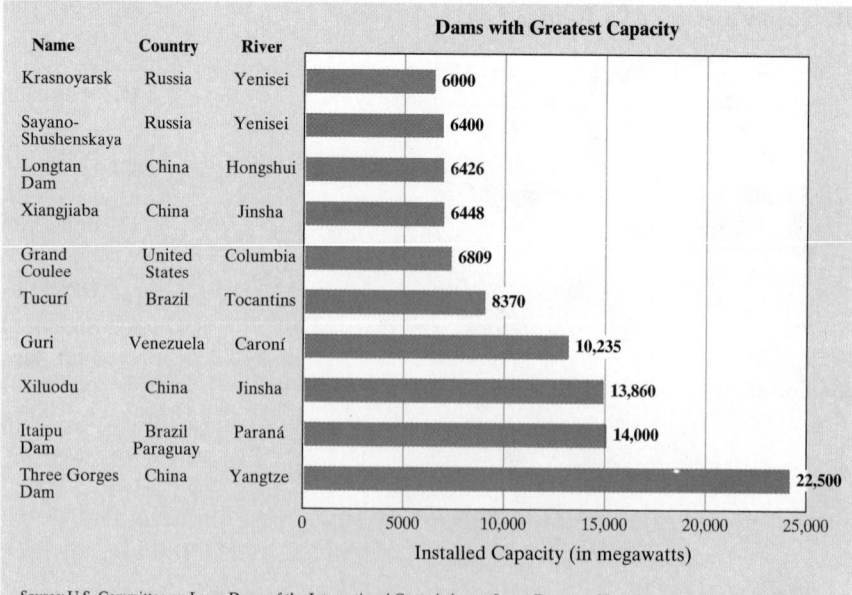

Name	Country	River	Dams with Greatest Capacity (Installed Capacity in megawatts)
Krasnoyarsk	Russia	Yenisei	6000
Sayano-Shushenskaya	Russia	Yenisei	6400
Longtan Dam	China	Hongshui	6426
Xiangjiaba	China	Jinsha	6448
Grand Coulee	United States	Columbia	6809
Tucurí	Brazil	Tocantins	8370
Guri	Venezuela	Caroní	10,235
Xiluodu	China	Jinsha	13,860
Itaipu Dam	Brazil Paraguay	Paraná	14,000
Three Gorges Dam	China	Yangtze	22,500

Source: U.S. Committee on Large Dams of the International Commission on Large Dams   Note: 1 megawatt = 1,000,000,000

11.1 Solving Quadratic Equations by Completing the Square

OBJECTIVES

1 Use the Square Root Property to Solve Quadratic Equations.

2 Solve Quadratic Equations by Completing the Square.

3 Use Quadratic Equations to Solve Problems.

OBJECTIVE

1 Using the Square Root Property

In Chapter 6, we solved quadratic equations by factoring. Recall that a **quadratic,** or **second-degree, equation** is an equation that can be written in the form $ax^2 + bx + c = 0$, where a, b, and c are real numbers and a is not 0. To solve a quadratic equation such as $x^2 = 9$ by factoring, we use the zero factor property. To use the zero factor property, the equation must first be written in standard form, $ax^2 + bx + c = 0$.

$$x^2 = 9$$
$$x^2 - 9 = 0 \qquad \text{Subtract 9 from both sides.}$$
$$(x + 3)(x - 3) = 0 \qquad \text{Factor.}$$
$$x + 3 = 0 \quad \text{or} \quad x - 3 = 0 \qquad \text{Set each factor equal to 0.}$$
$$x = -3 \qquad\qquad x = 3 \qquad \text{Solve.}$$

The solution set is $\{-3, 3\}$, the positive and negative square roots of 9. Not all quadratic equations can be solved by factoring, so we need to explore other methods. Notice that the solutions of the equation $x^2 = 9$ are two numbers whose square is 9.

$$3^2 = 9 \qquad \text{and} \qquad (-3)^2 = 9$$

Thus, we can solve the equation $x^2 = 9$ by taking the square root of both sides. Be sure to include both $\sqrt{9}$ and $-\sqrt{9}$ as solutions since both $\sqrt{9}$ and $-\sqrt{9}$ are numbers whose square is 9.

$$x^2 = 9$$
$$\sqrt{x^2} = \pm\sqrt{9} \qquad \text{The notation } \pm\sqrt{9} \text{ (read as "plus or minus } \sqrt{9}\text{")}$$
$$x = \pm 3 \qquad \text{indicates the pair of numbers } +\sqrt{9} \text{ and } -\sqrt{9}.$$

This illustrates the square root property.

Square Root Property

If b is a real number and if $a^2 = b$, then $a = \pm\sqrt{b}$.

Helpful Hint

The notation ± 3, for example, is read as "plus or minus 3." It is a shorthand notation for the pair of numbers $+3$ and -3.

EXAMPLE 1 Use the square root property to solve $x^2 = 50$.

Solution

$$x^2 = 50$$
$$x = \pm\sqrt{50} \qquad \text{Use the square root property.}$$
$$x = \pm 5\sqrt{2} \qquad \text{Simplify the radical.}$$

Check:

Let $x = 5\sqrt{2}$.

$$x^2 = 50$$
$$(5\sqrt{2})^2 \stackrel{?}{=} 50$$
$$25 \cdot 2 \stackrel{?}{=} 50$$
$$50 = 50 \quad \text{True}$$

Let $x = -5\sqrt{2}$.

$$x^2 = 50$$
$$(-5\sqrt{2})^2 \stackrel{?}{=} 50$$
$$25 \cdot 2 \stackrel{?}{=} 50$$
$$50 = 50 \quad \text{True}$$

The solutions are $5\sqrt{2}$ and $-5\sqrt{2}$, or the solution set is $\{-5\sqrt{2}, 5\sqrt{2}\}$. ☐

PRACTICE

1 Use the square root property to solve $x^2 = 32$.

EXAMPLE 2 Use the square root property to solve $2x^2 - 14 = 0$.

Solution First we get the squared variable alone on one side of the equation.

$$2x^2 - 14 = 0$$
$$2x^2 = 14 \qquad \text{Add 14 to both sides.}$$
$$x^2 = 7 \qquad \text{Divide both sides by 2.}$$
$$x = \pm\sqrt{7} \qquad \text{Use the square root property.}$$

Check to see that the solutions are $\sqrt{7}$ and $-\sqrt{7}$, or the solution set is $\{-\sqrt{7}, \sqrt{7}\}$. □

PRACTICE
2 Use the square root property to solve $5x^2 - 50 = 0$. ■

EXAMPLE 3 Use the square root property to solve $(x + 1)^2 = 12$.

Solution

$$(x + 1)^2 = 12$$
$$x + 1 = \pm\sqrt{12} \qquad \text{Use the square root property.}$$
$$x + 1 = \pm2\sqrt{3} \qquad \text{Simplify the radical.}$$
$$x = -1 \pm 2\sqrt{3} \qquad \text{Subtract 1 from both sides.}$$

> **Helpful Hint**
>
> Don't forget that $-1 \pm 2\sqrt{3}$, for example, means $-1 + 2\sqrt{3}$ and $-1 - 2\sqrt{3}$. In other words, the equation in Example 3 has two solutions.

Check: Below is a check for $-1 + 2\sqrt{3}$. The check for $-1 - 2\sqrt{3}$ is almost the same and is left for you to do on your own.

$$(x + 1)^2 = 12$$
$$(-1 + 2\sqrt{3} + 1)^2 \stackrel{?}{=} 12$$
$$(2\sqrt{3})^2 \stackrel{?}{=} 12$$
$$4 \cdot 3 \stackrel{?}{=} 12$$
$$12 = 12 \qquad \text{True}$$

The solutions are $-1 + 2\sqrt{3}$ and $-1 - 2\sqrt{3}$. □

PRACTICE
3 Use the square root property to solve $(x + 3)^2 = 20$. ■

EXAMPLE 4 Use the square root property to solve $(2x - 5)^2 = -16$.

Solution

$$(2x - 5)^2 = -16$$
$$2x - 5 = \pm\sqrt{-16} \qquad \text{Use the square root property.}$$
$$2x - 5 = \pm4i \qquad \text{Simplify the radical.}$$
$$2x = 5 \pm 4i \qquad \text{Add 5 to both sides.}$$
$$x = \frac{5 \pm 4i}{2} \qquad \text{Divide both sides by 2.}$$

The solutions are $\dfrac{5 + 4i}{2}$ and $\dfrac{5 - 4i}{2}$. □

PRACTICE
4 Use the square root property to solve $(5x - 2)^2 = -9$. ■

✔ CONCEPT CHECK
How do you know just by looking that $(x - 2)^2 = -4$ has complex but not real solutions?

OBJECTIVE

2   **Solving by Completing the Square**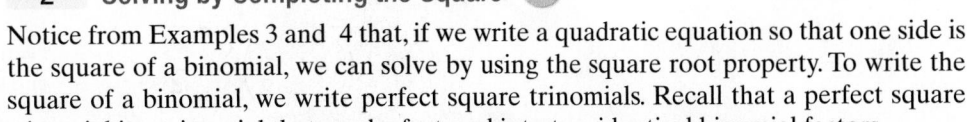

Notice from Examples 3 and 4 that, if we write a quadratic equation so that one side is the square of a binomial, we can solve by using the square root property. To write the square of a binomial, we write perfect square trinomials. Recall that a perfect square trinomial is a trinomial that can be factored into two identical binomial factors.

Perfect Square Trinomials	*Factored Form*
$x^2 + 8x + 16$	$(x + 4)^2$
$x^2 - 6x + 9$	$(x - 3)^2$
$x^2 + 3x + \dfrac{9}{4}$	$\left(x + \dfrac{3}{2}\right)^2$

Notice that for each perfect square trinomial in x, **the constant term of the trinomial is the square of half the coefficient of the x-term.** For example,

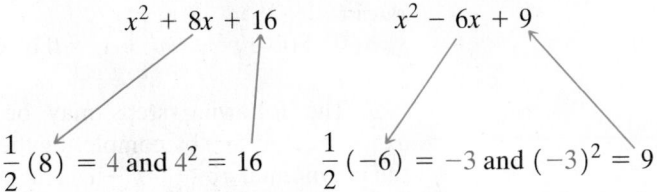

$$x^2 + 8x + 16 \qquad x^2 - 6x + 9$$

$$\frac{1}{2}(8) = 4 \text{ and } 4^2 = 16 \qquad \frac{1}{2}(-6) = -3 \text{ and } (-3)^2 = 9$$

The process of writing a quadratic equation so that one side is a perfect square trinomial is called **completing the square.**

EXAMPLE 5   Solve $p^2 + 2p = 4$ by completing the square.

Solution   First, add the square of half the coefficient of p to both sides so that the resulting trinomial will be a perfect square trinomial. The coefficient of p is 2.

$$\frac{1}{2}(2) = 1 \quad \text{and} \quad 1^2 = 1$$

Add 1 to both sides of the original equation.

$$p^2 + 2p = 4$$
$$p^2 + 2p + 1 = 4 + 1 \quad \text{Add 1 to both sides.}$$
$$(p + 1)^2 = 5 \qquad \text{Factor the trinomial; simplify the right side.}$$

We may now use the square root property and solve for p.

$$p + 1 = \pm\sqrt{5} \qquad \text{Use the square root property.}$$
$$p = -1 \pm \sqrt{5} \quad \text{Subtract 1 from both sides.}$$

Notice that there are two solutions: $-1 + \sqrt{5}$ and $-1 - \sqrt{5}$.   □

PRACTICE

5   Solve $b^2 + 4b = 3$ by completing the square.   ■

EXAMPLE 6   Solve $m^2 - 7m - 1 = 0$ for m by completing the square.

Solution   First, add 1 to both sides of the equation so that the left side has no constant term.

$$m^2 - 7m - 1 = 0$$
$$m^2 - 7m = 1$$

(Continued on next page)

Now find the constant term that makes the left side a perfect square trinomial by squaring half the coefficient of m. Add this constant to both sides of the equation.

$$\frac{1}{2}(-7) = -\frac{7}{2} \quad \text{and} \quad \left(-\frac{7}{2}\right)^2 = \frac{49}{4}$$

$$m^2 - 7m + \frac{49}{4} = 1 + \frac{49}{4} \qquad \text{Add } \frac{49}{4} \text{ to both sides of the equation.}$$

$$\left(m - \frac{7}{2}\right)^2 = \frac{53}{4} \qquad \text{Factor the perfect square trinomial and simplify the right side.}$$

$$m - \frac{7}{2} = \pm\sqrt{\frac{53}{4}} \qquad \text{Apply the square root property.}$$

$$m = \frac{7}{2} \pm \frac{\sqrt{53}}{2} \qquad \text{Add } \frac{7}{2} \text{ to both sides and simplify } \sqrt{\frac{53}{4}}.$$

$$m = \frac{7 \pm \sqrt{53}}{2} \qquad \text{Simplify.}$$

The solutions are $\dfrac{7 + \sqrt{53}}{2}$ and $\dfrac{7 - \sqrt{53}}{2}$.

PRACTICE

6 Solve $p^2 - 3p + 1 = 0$ by completing the square.

The following steps may be used to solve a quadratic equation such as $ax^2 + bx + c = 0$ by completing the square. This method may be used whether or not the polynomial $ax^2 + bx + c$ is factorable.

Solving a Quadratic Equation in x by Completing the Square

Step 1. If the coefficient of x^2 is 1, go to Step 2. Otherwise, divide both sides of the equation by the coefficient of x^2.

Step 2. Isolate all variable terms on one side of the equation.

Step 3. Complete the square for the resulting binomial by adding the square of half of the coefficient of x to both sides of the equation.

Step 4. Factor the resulting perfect square trinomial and write it as the square of a binomial.

Step 5. Use the square root property to solve for x.

EXAMPLE 7 Solve: $2x^2 - 8x + 3 = 0$

Solution Our procedure for finding the constant term to complete the square works only if the coefficient of the squared variable term is 1. Therefore, to solve this equation, the first step is to divide both sides by 2, the coefficient of x^2.

$$2x^2 - 8x + 3 = 0$$

Step 1. $x^2 - 4x + \dfrac{3}{2} = 0$ Divide both sides by 2.

Step 2. $x^2 - 4x = -\dfrac{3}{2}$ Subtract $\dfrac{3}{2}$ from both sides.

Next find the square of half of -4.

$$\frac{1}{2}(-4) = -2 \quad \text{and} \quad (-2)^2 = 4$$

Add 4 to both sides of the equation to complete the square.

Step 3. $x^2 - 4x + 4 = -\dfrac{3}{2} + 4$

Step 4. $(x - 2)^2 = \dfrac{5}{2}$ Factor the perfect square and simplify the right side.

Step 5. $x - 2 = \pm\sqrt{\dfrac{5}{2}}$ Apply the square root property.

$x - 2 = \pm\dfrac{\sqrt{10}}{2}$ Rationalize the denominator.

$x = 2 \pm \dfrac{\sqrt{10}}{2}$ Add 2 to both sides.

$= \dfrac{4}{2} \pm \dfrac{\sqrt{10}}{2}$ Find a common denominator.

$= \dfrac{4 \pm \sqrt{10}}{2}$ Simplify.

The solutions are $\dfrac{4 + \sqrt{10}}{2}$ and $\dfrac{4 - \sqrt{10}}{2}$.

PRACTICE
7   Solve: $3x^2 - 12x + 1 = 0$

EXAMPLE 8   Solve $3x^2 - 9x + 8 = 0$ by completing the square.

Solution $3x^2 - 9x + 8 = 0$

Step 1. $x^2 - 3x + \dfrac{8}{3} = 0$ Divide both sides of the equation by 3.

Step 2. $x^2 - 3x = -\dfrac{8}{3}$ Subtract $\dfrac{8}{3}$ from both sides.

Since $\dfrac{1}{2}(-3) = -\dfrac{3}{2}$ and $\left(-\dfrac{3}{2}\right)^2 = \dfrac{9}{4}$, we add $\dfrac{9}{4}$ to both sides of the equation.

Step 3. $x^2 - 3x + \dfrac{9}{4} = -\dfrac{8}{3} + \dfrac{9}{4}$

Step 4. $\left(x - \dfrac{3}{2}\right)^2 = -\dfrac{5}{12}$ Factor the perfect square trinomial.

Step 5. $x - \dfrac{3}{2} = \pm\sqrt{-\dfrac{5}{12}}$ Apply the square root property.

$x - \dfrac{3}{2} = \pm\dfrac{i\sqrt{5}}{2\sqrt{3}}$ Simplify the radical.

$x - \dfrac{3}{2} = \pm\dfrac{i\sqrt{15}}{6}$ Rationalize the denominator.

$x = \dfrac{3}{2} \pm \dfrac{i\sqrt{15}}{6}$ Add $\dfrac{3}{2}$ to both sides.

$= \dfrac{9}{6} \pm \dfrac{i\sqrt{15}}{6}$ Find a common denominator.

$= \dfrac{9 \pm i\sqrt{15}}{6}$ Simplify.

The solutions are $\dfrac{9 + i\sqrt{15}}{6}$ and $\dfrac{9 - i\sqrt{15}}{6}$.

PRACTICE
8   Solve $2x^2 - 5x + 7 = 0$ by completing the square.

OBJECTIVE

3 Solving Problems Modeled by Quadratic Equations

Recall the **simple interest** formula $I = Prt$, where I is the interest earned, P is the principal, r is the rate of interest, and t is time in years. If $100 is invested at a simple interest rate of 5% annually, at the end of 3 years the total interest I earned is

$$I = P \cdot r \cdot t$$

or

$$I = 100 \cdot 0.05 \cdot 3 = \$15$$

and the new principal is

$$\$100 + \$15 = \$115$$

Most of the time, the interest computed on money borrowed or money deposited is **compound interest.** Compound interest, unlike simple interest, is computed on original principal *and* on interest already earned. To see the difference between simple interest and compound interest, suppose that $100 is invested at a rate of 5% compounded annually. To find the total amount of money at the end of 3 years, we calculate as follows.

$$I = P \cdot r \cdot t$$

First year:	Interest = $100 \cdot 0.05 \cdot 1 = \$5.00$
	New principal = $\$100.00 + \$5.00 = \$105.00$
Second year:	Interest = $\$105.00 \cdot 0.05 \cdot 1 = \$5.25$
	New principal = $\$105.00 + \$5.25 = \$110.25$
Third year:	Interest = $\$110.25 \cdot 0.05 \cdot 1 \approx \$5.51$
	New principal = $\$110.25 + \$5.51 = \$115.76$

At the end of the third year, the total compound interest earned is $15.76, whereas the total simple interest earned is $15.

It is tedious to calculate compound interest as we did above, so we use a compound interest formula. The formula for calculating the total amount of money when interest is compounded annually is

$$A = P(1 + r)^t$$

where P is the original investment, r is the interest rate per compounding period, and t is the number of periods. For example, the amount of money A at the end of 3 years if $100 is invested at 5% compounded annually is

$$A = \$100(1 + 0.05)^3 \approx \$100(1.1576) = \$115.76$$

as we previously calculated.

EXAMPLE 9 Finding Interest Rates

Use the formula $A = P(1 + r)^t$ to find the interest rate r if $2000 compounded annually grows to $2420 in 2 years.

Solution

1. UNDERSTAND the problem. Since the $2000 is compounded annually, we use the compound interest formula. For this example, make sure that you understand the formula for compounding interest annually.

2. TRANSLATE. We substitute the given values into the formula.

$$A = P(1 + r)^t$$

$$2420 = 2000(1 + r)^2 \quad \text{Let } A = 2420, P = 2000, \text{ and } t = 2.$$

3. SOLVE. Solve the equation for r.

$$2420 = 2000(1 + r)^2$$

$$\frac{2420}{2000} = (1 + r)^2 \qquad \text{Divide both sides by 2000.}$$

$$\frac{121}{100} = (1 + r)^2 \qquad \text{Simplify the fraction.}$$

$$\pm\sqrt{\frac{121}{100}} = 1 + r \qquad \text{Use the square root property.}$$

$$\pm\frac{11}{10} = 1 + r \qquad \text{Simplify.}$$

$$-1 \pm \frac{11}{10} = r$$

$$-\frac{10}{10} \pm \frac{11}{10} = r$$

$$\frac{1}{10} = r \quad \text{or} \quad -\frac{21}{10} = r$$

4. INTERPRET. The rate cannot be negative, so we reject $-\dfrac{21}{10}$.

Check: $\dfrac{1}{10} = 0.10 = 10\%$ per year. If we invest \$2000 at 10% compounded annually, in 2 years the amount in the account would be $2000(1 + 0.10)^2 = 2420$ dollars, the desired amount.

State: The interest rate is 10% compounded annually.   ☐

PRACTICE

9   Use the formula from Example 9 to find the interest rate r if \$5000 compounded annually grows to \$5618 in 2 years.   ◼

 Graphing Calculator Explorations

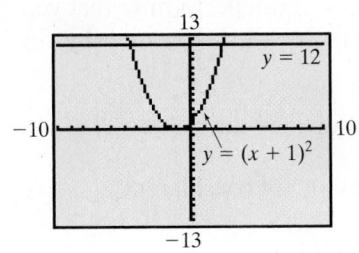

In Section 6.6, we showed how we can use a grapher to approximate real number solutions of a quadratic equation written in standard form. We can also use a grapher to solve a quadratic equation when it is not written in standard form. For example, to solve $(x + 1)^2 = 12$, the quadratic equation in Example 3, we graph the following on the same set of axes. Use Xmin $= -10$, Xmax $= 10$, Ymin $= -13$, and Ymax $= 13$.

$$\text{Y}_1 = (x + 1)^2 \quad \text{and} \quad \text{Y}_2 = 12$$

Use the Intersect feature or the Zoom and Trace features to locate the points of intersection of the graphs. (See your manuals for specific instructions.) The x-values of these points are the solutions of $(x + 1)^2 = 12$. The solutions, rounded to two decimal places, are 2.46 and -4.46.

Check to see that these numbers are approximations of the exact solutions $-1 \pm 2\sqrt{3}$.

Use a graphing calculator to solve each quadratic equation. Round all solutions to the nearest hundredth.

1. $x(x - 5) = 8$ **2.** $x(x + 2) = 5$

3. $x^2 + 0.5x = 0.3x + 1$ **4.** $x^2 - 2.6x = -2.2x + 3$

5. Use a graphing calculator and solve $(2x - 5)^2 = -16$, Example 4 in this section, using the window

$$Xmin = -20$$
$$Xmax = 20$$
$$Xscl = 1$$
$$Ymin = -20$$
$$Ymax = 20$$
$$Yscl = 1$$

Explain the results. Compare your results with the solution found in Example 4.

6. What are the advantages and disadvantages of using a graphing calculator to solve quadratic equations?

✓ Vocabulary, Readiness & Video Check

Use the choices below to fill in each blank. Not all choices will be used.

| binomial | $\sqrt{b}$ | $\pm\sqrt{b}$ | b^2 | 9 | 25 | completing the square |
| quadratic | $-\sqrt{b}$ | $\dfrac{b}{2}$ | $\left(\dfrac{b}{2}\right)^2$ | 3 | 5 | |

1. By the square root property, if b is a real number, and $a^2 = b$, then $a = $ _____.

2. A _____ equation can be written in the form $ax^2 + bx + c = 0, a \neq 0$.

3. The process of writing a quadratic equation so that one side is a perfect square trinomial is called _____.

4. A perfect square trinomial is one that can be factored as a _____ squared.

5. To solve $x^2 + 6x = 10$ by completing the square, add _____ to both sides.

6. To solve $x^2 + bx = c$ by completing the square, add _____ to both sides.

Martin-Gay Interactive Videos

Watch the section lecture video and answer the following questions.

OBJECTIVE 1
7. From ▦ Examples 2 and 3, explain a step you can perform so that you may easily apply the square root property to $2x^2 = 16$. Explain why you perform this step.

OBJECTIVE 2
8. In ▦ Example 5, why is the equation first divided through by 3?

OBJECTIVE 3
9. In ▦ Example 6, why is the negative solution not considered?

See Video 11.1 ◉

11.1 Exercise Set MyMathLab® ▶

Use the square root property to solve each equation. These equations have real number solutions. See Examples 1 through 3.

▶ 1. $x^2 = 16$
2. $x^2 = 49$
3. $x^2 - 7 = 0$
4. $x^2 - 11 = 0$
5. $x^2 = 18$
6. $y^2 = 20$
7. $3z^2 - 30 = 0$
8. $2x^2 - 4 = 0$
9. $(x + 5)^2 = 9$
10. $(y - 3)^2 = 4$

▶ 11. $(z - 6)^2 = 18$
12. $(y + 4)^2 = 27$
13. $(2x - 3)^2 = 8$
14. $(4x + 9)^2 = 6$

Use the square root property to solve each equation. See Examples 1 through 4.

15. $x^2 + 9 = 0$
16. $x^2 + 4 = 0$
▶ 17. $x^2 - 6 = 0$
18. $y^2 - 10 = 0$
19. $2z^2 + 16 = 0$
20. $3p^2 + 36 = 0$

21. $(3x - 1)^2 = -16$

22. $(4y + 2)^2 = -25$

23. $(z + 7)^2 = 5$

24. $(x + 10)^2 = 11$

25. $(x + 3)^2 + 8 = 0$

26. $(y - 4)^2 + 18 = 0$

Add the proper constant to each binomial so that the resulting trinomial is a perfect square trinomial. Then factor the trinomial.

27. $x^2 + 16x +$ _____

28. $y^2 + 2y +$ _____

29. $z^2 - 12z +$ _____

30. $x^2 - 8x +$ _____

31. $p^2 + 9p +$ _____

32. $n^2 + 5n +$ _____

33. $x^2 + x +$ _____

34. $y^2 - y +$ _____

MIXED PRACTICE

Solve each equation by completing the square. These equations have real number solutions. See Examples 5 through 7.

35. $x^2 + 8x = -15$

36. $y^2 + 6y = -8$

37. $x^2 + 6x + 2 = 0$

38. $x^2 - 2x - 2 = 0$

39. $x^2 + x - 1 = 0$

40. $x^2 + 3x - 2 = 0$

41. $x^2 + 2x - 5 = 0$

42. $x^2 - 6x + 3 = 0$

43. $y^2 + y - 7 = 0$

44. $x^2 - 7x - 1 = 0$

45. $3p^2 - 12p + 2 = 0$

46. $2x^2 + 14x - 1 = 0$

47. $4y^2 - 2 = 12y$

48. $6x^2 - 3 = 6x$

49. $2x^2 + 7x = 4$

50. $3x^2 - 4x = 4$

51. $x^2 + 8x + 1 = 0$

52. $x^2 - 10x + 2 = 0$

53. $3y^2 + 6y - 4 = 0$

54. $2y^2 + 12y + 3 = 0$

55. $2x^2 - 3x - 5 = 0$

56. $5x^2 + 3x - 2 = 0$

Solve each equation by completing the square. See Examples 5 through 8.

57. $y^2 + 2y + 2 = 0$

58. $x^2 + 4x + 6 = 0$

59. $y^2 + 6y - 8 = 0$

60. $y^2 + 10y - 26 = 0$

61. $2a^2 + 8a = -12$

62. $3x^2 + 12x = -14$

63. $5x^2 + 15x - 1 = 0$

64. $16y^2 + 16y - 1 = 0$

65. $2x^2 - x + 6 = 0$

66. $4x^2 - 2x + 5 = 0$

67. $x^2 + 10x + 28 = 0$

68. $y^2 + 8y + 18 = 0$

69. $z^2 + 3z - 4 = 0$

70. $y^2 + y - 2 = 0$

71. $2x^2 - 4x = -3$

72. $9x^2 - 36x = -40$

73. $3x^2 + 3x = 5$

74. $10y^2 - 30y = 2$

Use the formula $A = P(1 + r)^t$ to solve Exercises 75 through 78. See Example 9.

75. Find the rate r at which $3000 compounded annually grows to $4320 in 2 years.

76. Find the rate r at which $800 compounded annually grows to $882 in 2 years.

77. Find the rate at which $15,000 compounded annually grows to $16,224 in 2 years.

78. Find the rate at which $2000 compounded annually grows to $2880 in 2 years.

Neglecting air resistance, the distance $s(t)$ in feet traveled by a freely falling object is given by the function $s(t) = 16t^2$, where t is time in seconds. Use this formula to solve Exercises 79 through 82. Round answers to two decimal places.

79. The Petronas Towers in Kuala Lumpur, completed in 1998, are the tallest buildings in Malaysia. Each tower is 1483 feet tall. How long would it take an object to fall to the ground from the top of one of the towers? (*Source:* Council on Tall Buildings and Urban Habitat, Lehigh University)

80. The Burj Khalifa, the tallest building in the world, was completed in 2010 in Dubai. It is estimated to be 2717 feet tall. How long would it take an object to fall to the ground from the top of the building? (*Source:* Council on Tall Buildings and Urban Habitat)

81. The Three Gorges Dam, while producing the most electricity of any dam in the world, is only 610 feet tall. How long would it take an object to fall from the top to the base of the dam? (*Source:* U.S. Committee on Large Dams of the International Commission on Large Dams)

82. The Hoover Dam, located on the Colorado River on the border of Nevada and Arizona near Las Vegas, is 725 feet tall. How long would it take an object to fall from the top to the base of the dam? (*Source:* U.S. Committee on Large Dams of the International Commission on Large Dams)

Solve.

△ **83.** The area of a square room is 225 square feet. Find the dimensions of the room.

△ **84.** The area of a circle is 36π square inches. Find the radius of the circle.

△ **85.** An isosceles right triangle has legs of equal length. If the hypotenuse is 20 centimeters long, find the length of each leg.

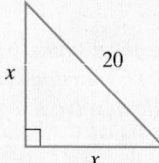

△ **86.** The top of a square coffee table has a diagonal that measures 30 inches. Find the length of each side of the top of the coffee table.

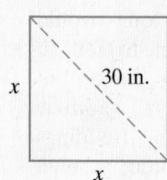

REVIEW AND PREVIEW

Simplify each expression. See Section 10.1.

87. $\dfrac{1}{2} - \sqrt{\dfrac{9}{4}}$

88. $\dfrac{9}{10} - \sqrt{\dfrac{49}{100}}$

Simplify each expression. See Section 10.5.

89. $\dfrac{6 + 4\sqrt{5}}{2}$

90. $\dfrac{10 - 20\sqrt{3}}{2}$

91. $\dfrac{3 - 9\sqrt{2}}{6}$

92. $\dfrac{12 - 8\sqrt{7}}{16}$

Evaluate $\sqrt{b^2 - 4ac}$ for each set of values. See Section 10.3.

93. $a = 2, b = 4, c = -1$

94. $a = 1, b = 6, c = 2$

95. $a = 3, b = -1, c = -2$

96. $a = 1, b = -3, c = -1$

CONCEPT EXTENSIONS

Without solving, determine whether the solutions of each equation are real numbers or complex but not real numbers. See the Concept Check in this section.

97. $(x + 1)^2 = -1$

98. $(y - 5)^2 = -9$

99. $3z^2 = 10$

100. $4x^2 = 17$

101. $(2y - 5)^2 + 7 = 3$

102. $(3m + 2)^2 + 4 = 1$

Find two possible missing terms so that each is a perfect square trinomial.

103. $x^2 + \quad + 16$

104. $y^2 + \quad + 9$

105. $z^2 + \quad + \dfrac{25}{4}$

106. $x^2 + \quad + \dfrac{1}{4}$

107. In your own words, explain how to calculate the number that will complete the square on an expression such as $x^2 - 5x$.

108. In your own words, what is the difference between simple interest and compound interest?

109. If you are depositing money in an account that pays 4%, would you prefer the interest to be simple or compound? Explain your answer.

110. If you are borrowing money at a rate of 10%, would you prefer the interest to be simple or compound? Explain your answer.

A common equation used in business is a demand equation. It expresses the relationship between the unit price of some commodity and the quantity demanded. For Exercises 111 and 112, p represents the unit price and x represents the quantity demanded in thousands.

111. A manufacturing company has found that the demand equation for a certain type of scissors is given by the equation $p = -x^2 + 47$. Find the demand for the scissors if the price is $11 per pair.

112. Acme, Inc., sells desk lamps and has found that the demand equation for a certain style of desk lamp is given by the equation $p = -x^2 + 15$. Find the demand for the desk lamp if the price is $7 per lamp.

11.2 # Solving Quadratic Equations by the Quadratic Formula

OBJECTIVES

1 Solve Quadratic Equations by Using the Quadratic Formula.

2 Determine the Number and Type of Solutions of a Quadratic Equation by Using the Discriminant.

3 Solve Problems Modeled by Quadratic Equations.

OBJECTIVE

1   Solving Quadratic Equations by Using the Quadratic Formula

Any quadratic equation can be solved by completing the square. Since the same sequence of steps is repeated each time we complete the square, let's complete the square for a general quadratic equation, $ax^2 + bx + c = 0, a \neq 0$. By doing so, we find a pattern for the solutions of a quadratic equation known as the **quadratic formula.**

Recall that to complete the square for an equation such as $ax^2 + bx + c = 0$, we first divide both sides by the coefficient of x^2.

$$ax^2 + bx + c = 0$$

$$x^2 + \frac{b}{a}x + \frac{c}{a} = 0 \qquad \text{Divide both sides by } a, \text{ the coefficient of } x^2.$$

$$x^2 + \frac{b}{a}x = -\frac{c}{a} \qquad \text{Subtract the constant } \frac{c}{a} \text{ from both sides.}$$

Next, find the square of half $\frac{b}{a}$, the coefficient of x.

$$\frac{1}{2}\left(\frac{b}{a}\right) = \frac{b}{2a} \quad \text{and} \quad \left(\frac{b}{2a}\right)^2 = \frac{b^2}{4a^2}$$

Add this result to both sides of the equation.

$$x^2 + \frac{b}{a}x + \frac{b^2}{4a^2} = -\frac{c}{a} + \frac{b^2}{4a^2} \qquad \text{Add } \frac{b^2}{4a^2} \text{ to both sides.}$$

$$x^2 + \frac{b}{a}x + \frac{b^2}{4a^2} = \frac{-c \cdot 4a}{a \cdot 4a} + \frac{b^2}{4a^2} \qquad \text{Find a common denominator on the right side.}$$

$$x^2 + \frac{b}{a}x + \frac{b^2}{4a^2} = \frac{b^2 - 4ac}{4a^2} \qquad \text{Simplify the right side.}$$

$$\left(x + \frac{b}{2a}\right)^2 = \frac{b^2 - 4ac}{4a^2} \qquad \text{Factor the perfect square trinomial on the left side.}$$

$$x + \frac{b}{2a} = \pm\sqrt{\frac{b^2 - 4ac}{4a^2}} \qquad \text{Apply the square root property.}$$

$$x + \frac{b}{2a} = \pm\frac{\sqrt{b^2 - 4ac}}{2a} \qquad \text{Simplify the radical.}$$

$$x = -\frac{b}{2a} \pm \frac{\sqrt{b^2 - 4ac}}{2a} \qquad \text{Subtract } \frac{b}{2a} \text{ from both sides.}$$

$$x = \frac{-b \pm \sqrt{b^2 - 4ac}}{2a} \qquad \text{Simplify.}$$

This equation identifies the solutions of the general quadratic equation in standard form and is called the quadratic formula. It can be used to solve any equation written in standard form $ax^2 + bx + c = 0$ as long as a is not 0.

Quadratic Formula

A quadratic equation written in the form $ax^2 + bx + c = 0$ has the solutions

$$x = \frac{-b \pm \sqrt{b^2 - 4ac}}{2a}$$

EXAMPLE 1 Solve $3x^2 + 16x + 5 = 0$ for x.

Solution This equation is in standard form, so $a = 3$, $b = 16$, and $c = 5$. Substitu[te] these values into the quadratic formula.

$$x = \frac{-b \pm \sqrt{b^2 - 4ac}}{2a} \quad \text{Quadratic formula}$$

$$= \frac{-16 \pm \sqrt{16^2 - 4(3)(5)}}{2 \cdot 3} \quad \text{Use } a = 3, b = 16, \text{ and } c = 5.$$

$$= \frac{-16 \pm \sqrt{256 - 60}}{6}$$

$$= \frac{-16 \pm \sqrt{196}}{6} = \frac{-16 \pm 14}{6}$$

$$x = \frac{-16 + 14}{6} = -\frac{1}{3} \quad \text{or} \quad x = \frac{-16 - 14}{6} = -\frac{30}{6} = -5$$

The solutions are $-\dfrac{1}{3}$ and -5, or the solution set is $\left\{-\dfrac{1}{3}, -5\right\}$. □

PRACTICE
1 Solve $3x^2 - 5x - 2 = 0$ for x. ■

> **Helpful Hint**
> To replace a, b, and c correctly in the quadratic formula, write the quadratic equation in standard form $ax^2 + bx + c = 0$.

EXAMPLE 2 Solve: $2x^2 - 4x = 3$

Solution First write the equation in standard form by subtracting 3 from both sides.

$$2x^2 - 4x - 3 = 0$$

Now $a = 2$, $b = -4$, and $c = -3$. Substitute these values into the quadratic formula.

$$x = \frac{-b \pm \sqrt{b^2 - 4ac}}{2a}$$

$$= \frac{-(-4) \pm \sqrt{(-4)^2 - 4(2)(-3)}}{2 \cdot 2}$$

$$= \frac{4 \pm \sqrt{16 + 24}}{4}$$

$$= \frac{4 \pm \sqrt{40}}{4} = \frac{4 \pm 2\sqrt{10}}{4}$$

$$= \frac{2(2 \pm \sqrt{10})}{2 \cdot 2} = \frac{2 \pm \sqrt{10}}{2}$$

The solutions are $\dfrac{2 + \sqrt{10}}{2}$ and $\dfrac{2 - \sqrt{10}}{2}$, or the solution set is $\left\{\dfrac{2 - \sqrt{10}}{2}, \dfrac{2 + \sqrt{10}}{2}\right\}$. □

PRACTICE
2 Solve: $3x^2 - 8x = 2$ ■

> **Helpful Hint**
> To simplify the expression $\dfrac{4 \pm 2\sqrt{10}}{4}$ in the preceding example, note that 2 is factored out of both terms of the numerator *before* simplifying.
>
> $$\frac{4 \pm 2\sqrt{10}}{4} = \frac{2(2 \pm \sqrt{10})}{2 \cdot 2} = \frac{2 \pm \sqrt{10}}{2}$$

✔ **CONCEPT CHECK**

For the quadratic equation $x^2 = 7$, choose the correct substitution for a, b, and c in the standard form $ax^2 + bx + c = 0$.

a. $a = 1, b = 0,$ and $c = -7$
b. $a = 1, b = 0,$ and $c = 7$
c. $a = 0, b = 0,$ and $c = 7$
d. $a = 1, b = 1,$ and $c = -7$

EXAMPLE 3   Solve: $\dfrac{1}{4} m^2 - m + \dfrac{1}{2} = 0$

Solution We could use the quadratic formula with $a = \dfrac{1}{4}$, $b = -1$, and $c = \dfrac{1}{2}$. Instead, we find a simpler, equivalent standard form equation whose coefficients are not fractions.

Multiply both sides of the equation by the LCD 4 to clear fractions.

$$4\left(\frac{1}{4} m^2 - m + \frac{1}{2} \right) = 4 \cdot 0$$

$$m^2 - 4m + 2 = 0 \qquad \text{Simplify.}$$

Substitute $a = 1$, $b = -4$, and $c = 2$ into the quadratic formula and simplify.

$$m = \frac{-(-4) \pm \sqrt{(-4)^2 - 4(1)(2)}}{2 \cdot 1} = \frac{4 \pm \sqrt{16 - 8}}{2}$$

$$= \frac{4 \pm \sqrt{8}}{2} = \frac{4 \pm 2\sqrt{2}}{2} = \frac{2(2 \pm \sqrt{2})}{2}$$

$$= 2 \pm \sqrt{2}$$

The solutions are $2 + \sqrt{2}$ and $2 - \sqrt{2}$.   □

PRACTICE
3   Solve: $\dfrac{1}{8} x^2 - \dfrac{1}{4} x - 2 = 0$ ■

EXAMPLE 4   Solve: $x = -3x^2 - 3$

Solution The equation in standard form is $3x^2 + x + 3 = 0$. Thus, let $a = 3$, $b = 1$, and $c = 3$ in the quadratic formula.

$$x = \frac{-1 \pm \sqrt{1^2 - 4(3)(3)}}{2 \cdot 3} = \frac{-1 \pm \sqrt{1 - 36}}{6} = \frac{-1 \pm \sqrt{-35}}{6} = \frac{-1 \pm i\sqrt{35}}{6}$$

The solutions are $\dfrac{-1 + i\sqrt{35}}{6}$ and $\dfrac{-1 - i\sqrt{35}}{6}$.   □

PRACTICE
4   Solve: $x = -2x^2 - 2$ ■

✔ **CONCEPT CHECK**

What is the first step in solving $-3x^2 = 5x - 4$ using the quadratic formula?

In Example 1, the equation $3x^2 + 16x + 5 = 0$ had two real roots, $-\dfrac{1}{3}$ and -5.

In Example 4, the equation $3x^2 + x + 3 = 0$ (written in standard form) had no real roots. How do their related graphs compare? Recall that the x-intercepts of $f(x) = 3x^2 + 16x + 5$ occur where $f(x) = 0$ or where $3x^2 + 16x + 5 = 0$.

Since this equation has two real roots, the graph has two x-intercepts. Similarly, since the equation $3x^2 + x + 3 = 0$ has no real roots, the graph of $f(x) = 3x^2 + x + 3$ has no x-intercepts.

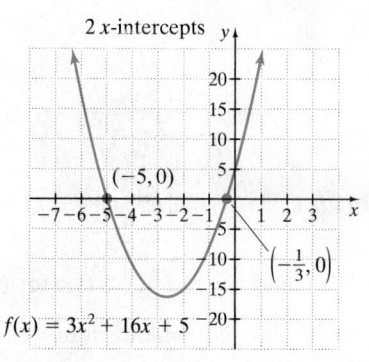

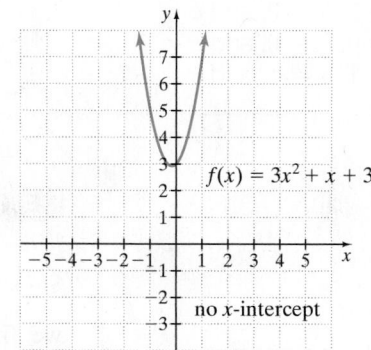

OBJECTIVE

2 Using the Discriminant

In the quadratic formula, $x = \dfrac{-b \pm \sqrt{b^2 - 4ac}}{2a}$, the radicand $b^2 - 4ac$ is called the **discriminant** because, by knowing its value, we can **discriminate** among the possible number and type of solutions of a quadratic equation. Possible values of the discriminant and their meanings are summarized next.

Discriminant

The following table corresponds the discriminant $b^2 - 4ac$ of a quadratic equation of the form $ax^2 + bx + c = 0$ with the number and type of solutions of the equation.

$b^2 - 4ac$	*Number and Type of Solutions*
Positive	Two real solutions
Zero	One real solution
Negative	Two complex but not real solutions

EXAMPLE 5   Use the discriminant to determine the number and type of solutions of each quadratic equation.

    a. $x^2 + 2x + 1 = 0$ **b.** $3x^2 + 2 = 0$ **c.** $2x^2 - 7x - 4 = 0$

Solution

    a. In $x^2 + 2x + 1 = 0$, $a = 1$, $b = 2$, and $c = 1$. Thus,
$$b^2 - 4ac = 2^2 - 4(1)(1) = 0$$
Since $b^2 - 4ac = 0$, this quadratic equation has one real solution.

    b. In this equation, $a = 3$, $b = 0$, $c = 2$. Then $b^2 - 4ac = 0 - 4(3)(2) = -24$. Since $b^2 - 4ac$ is negative, the quadratic equation has two complex but not real solutions.

    c. In this equation, $a = 2$, $b = -7$, and $c = -4$. Then
$$b^2 - 4ac = (-7)^2 - 4(2)(-4) = 81$$
Since $b^2 - 4ac$ is positive, the quadratic equation has two real solutions. ☐

PRACTICE

    5   Use the discriminant to determine the number and type of solutions of each quadratic equation.

        a. $x^2 - 6x + 9 = 0$ **b.** $x^2 - 3x - 1 = 0$ **c.** $7x^2 + 11 = 0$

The discriminant helps us determine the number and type of solutions of a quadratic equation, $ax^2 + bx + c = 0$. Recall that the solutions of this equation are the same as the x-intercepts of its related graph $f(x) = ax^2 + bx + c$. This means that the discriminant of $ax^2 + bx + c = 0$ also tells us the number of x-intercepts for the graph of $f(x) = ax^2 + bx + c$ or, equivalently, $y = ax^2 + bx + c$.

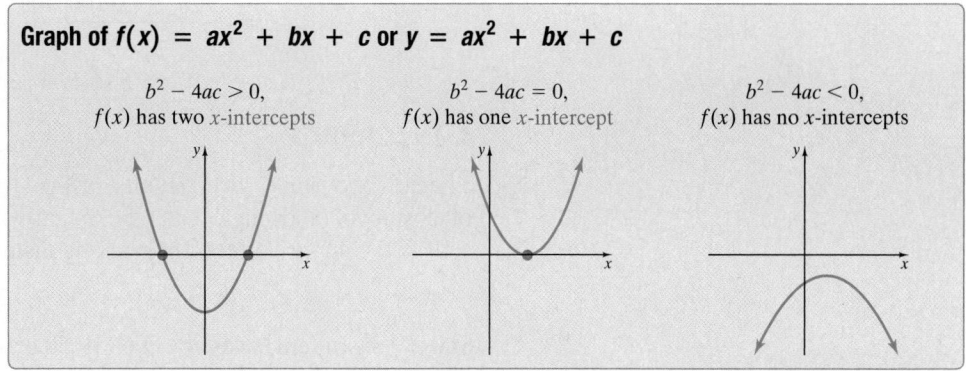

Graph of $f(x) = ax^2 + bx + c$ or $y = ax^2 + bx + c$

| $b^2 - 4ac > 0$, | $b^2 - 4ac = 0$, | $b^2 - 4ac < 0$, |
| $f(x)$ has two x-intercepts | $f(x)$ has one x-intercept | $f(x)$ has no x-intercepts |

OBJECTIVE

3   Solving Problems Modeled by Quadratic Equations

The quadratic formula is useful in solving problems that are modeled by quadratic equations.

△ **EXAMPLE 6**   **Calculating Distance Saved**

At a local university, students often leave the sidewalk and cut across the lawn to save walking distance. Given the diagram below of a favorite place to cut across the lawn, approximate how many feet of walking distance a student saves by cutting across the lawn instead of walking on the sidewalk.

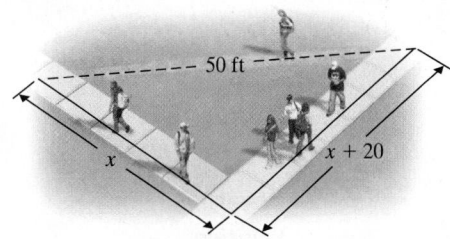

Solution

1. UNDERSTAND. Read and reread the problem. In the diagram, notice that a triangle is formed. Since the corner of the block forms a right angle, we use the Pythagorean theorem for right triangles. You may want to review this theorem.

2. TRANSLATE. By the Pythagorean theorem, we have

$$\text{In words: } (\text{leg})^2 + (\text{leg})^2 = (\text{hypotenuse})^2$$
$$\text{Translate: } x^2 + (x + 20)^2 = 50^2$$

3. SOLVE. Use the quadratic formula to solve.

$$x^2 + x^2 + 40x + 400 = 2500 \quad \text{Square } (x + 20) \text{ and } 50.$$
$$2x^2 + 40x - 2100 = 0 \qquad \text{Set the equation equal to } 0.$$
$$x^2 + 20x - 1050 = 0 \qquad \text{Divide by } 2.$$

(Continued on next page)

Here, $a = 1, b = 20, c = -1050$. By the quadratic formula,

$$x = \frac{-20 \pm \sqrt{20^2 - 4(1)(-1050)}}{2 \cdot 1}$$

$$= \frac{-20 \pm \sqrt{400 + 4200}}{2} = \frac{-20 \pm \sqrt{4600}}{2}$$

$$= \frac{-20 \pm \sqrt{100 \cdot 46}}{2} = \frac{-20 \pm 10\sqrt{46}}{2}$$

$$= -10 \pm 5\sqrt{46} \quad \text{Simplify.}$$

4. INTERPRET.

Check: We have two results using the quadratic formula. The length of a side of a triangle can't be negative, so we reject $-10 - 5\sqrt{46}$. Since $-10 + 5\sqrt{46} \approx 24$ feet, the walking distance along the sidewalk is

$$x + (x + 20) \approx 24 + (24 + 20) = 68 \text{ feet.}$$

State: A student saves about $68 - 50$ or 18 feet of walking distance by cutting across the lawn. □

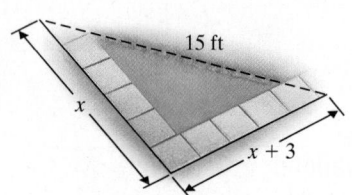

PRACTICE
6 Given the diagram, approximate to the nearest foot how many feet of walking distance a person can save by cutting across the lawn instead of walking on the sidewalk. ▪

EXAMPLE 7 Calculating Landing Time

An object is thrown upward from the top of a 200-foot cliff with a velocity of 12 feet per second. The height h in feet of the object after t seconds is

$$h = -16t^2 + 12t + 200$$

How long after the object is thrown will it strike the ground? Round to the nearest tenth of a second.

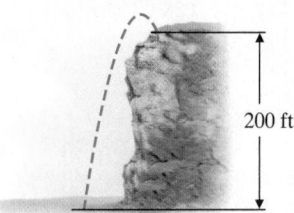

Solution

1. UNDERSTAND. Read and reread the problem.

2. TRANSLATE. Since we want to know when the object strikes the ground, we want to know when the height $h = 0$, or

$$0 = -16t^2 + 12t + 200$$

3. SOLVE. First we divide both sides of the equation by -4.

$$0 = 4t^2 - 3t - 50 \quad \text{Divide both sides by } -4.$$

Here, $a = 4, b = -3$, and $c = -50$. By the quadratic formula,

$$t = \frac{-(-3) \pm \sqrt{(-3)^2 - 4(4)(-50)}}{2 \cdot 4}$$

$$= \frac{3 \pm \sqrt{9 + 800}}{8}$$

$$= \frac{3 \pm \sqrt{809}}{8}$$

4. INTERPRET.

Check: We check our calculations from the quadratic formula. Since the time won't be negative, we reject the proposed solution

$$\frac{3 - \sqrt{809}}{8}.$$

State: The time it takes for the object to strike the ground is exactly

$$\frac{3 + \sqrt{809}}{8} \text{ seconds} \approx 3.9 \text{ seconds.} \qquad \square$$

PRACTICE

7   A toy rocket is shot upward from the top of a 45-foot-tall building, with an initial velocity of 20 feet per second. The height h in feet of the rocket after t seconds is

$$h = -16t^2 + 20t + 45$$

How long after the rocket is launched will it strike the ground? Round to the nearest tenth of a second.   ∎

✔ Vocabulary, Readiness & Video Check

Fill in each blank.

1. The quadratic formula is _____ .

2. For $2x^2 + x + 1 = 0$, if $a = 2$, then $b =$ _____ and $c =$ _____ .

3. For $5x^2 - 5x - 7 = 0$, if $a = 5$, then $b =$ _____ and $c =$ _____ .

4. For $7x^2 - 4 = 0$, if $a = 7$, then $b =$ _____ and $c =$ _____ .

5. For $x^2 + 9 = 0$, if $c = 9$, then $a =$ _____ and $b =$ _____ .

6. The correct simplified form of $\dfrac{5 \pm 10\sqrt{2}}{5}$ is _____ .

   a. $1 \pm 10\sqrt{2}$ **b.** $2\sqrt{2}$ **c.** $1 \pm 2\sqrt{2}$ **d.** $\pm 5\sqrt{2}$

Martin-Gay Interactive Videos

See Video 11.2 ◉

Watch the section lecture video and answer the following questions.

OBJECTIVE 1

7. Based on ▦ Examples 1–3, answer the following.
   a. Must a quadratic equation be written in standard form in order to use the quadratic formula? Why or why not?
   b. Must fractions be cleared from an equation before using the quadratic formula? Why or why not?

OBJECTIVE 2

8. Based on ▦ Example 4 and the lecture before, complete the following statements. The discriminant is the _____ in the quadratic formula and can be used to find the number and type of solutions of a quadratic equation without _____ the equation. To use the discriminant, the quadratic equation needs to be written in _____ form.

OBJECTIVE 3

9. In ▦ Example 5, the value of x is found, which is then used to find the dimensions of the triangle. Yet all this work still does not solve the problem. Explain.

11.2   Exercise Set   MyMathLab®

Use the quadratic formula to solve each equation. These equations have real number solutions only. See Examples 1 through 3.

1. $m^2 + 5m - 6 = 0$

2. $p^2 + 11p - 12 = 0$

3. $2y = 5y^2 - 3$

4. $5x^2 - 3 = 14x$

5. $x^2 - 6x + 9 = 0$

6. $y^2 + 10y + 25 = 0$

▶ 7. $x^2 + 7x + 4 = 0$

8. $y^2 + 5y + 3 = 0$

9. $8m^2 - 2m = 7$

10. $11n^2 - 9n = 1$

11. $3m^2 - 7m = 3$

12. $x^2 - 13 = 5x$

13. $\frac{1}{2}x^2 - x - 1 = 0$

14. $\frac{1}{6}x^2 + x + \frac{1}{3} = 0$

15. $\frac{2}{5}y^2 + \frac{1}{5}y = \frac{3}{5}$

16. $\frac{1}{8}x^2 + x = \frac{5}{2}$

17. $\frac{1}{3}y^2 = y + \frac{1}{6}$

18. $\frac{1}{2}y^2 = y + \frac{1}{2}$

19. $x^2 + 5x = -2$

20. $y^2 - 8 = 4y$

21. $(m + 2)(2m - 6) = 5(m - 1) - 12$

22. $7p(p - 2) + 2(p + 4) = 3$

MIXED PRACTICE

Use the quadratic formula to solve each equation. These equations have real solutions and complex but not real solutions. See Examples 1 through 4.

23. $x^2 + 6x + 13 = 0$

24. $x^2 + 2x + 2 = 0$

▶ 25. $(x + 5)(x - 1) = 2$

26. $x(x + 6) = 2$

27. $6 = -4x^2 + 3x$

28. $2 = -9x^2 - x$

29. $\frac{x^2}{3} - x = \frac{5}{3}$

30. $\frac{x^2}{2} - 3 = -\frac{9}{2}x$

31. $10y^2 + 10y + 3 = 0$

32. $3y^2 + 6y + 5 = 0$

33. $x(6x + 2) = 3$

34. $x(7x + 1) = 2$

▶ 35. $\frac{2}{5}y^2 + \frac{1}{5}y + \frac{3}{5} = 0$

36. $\frac{1}{8}x^2 + x + \frac{5}{2} = 0$

37. $\frac{1}{2}y^2 = y - \frac{1}{2}$

38. $\frac{2}{3}x^2 - \frac{20}{3}x = -\frac{100}{6}$

39. $(n - 2)^2 = 2n$

40. $\left(p - \frac{1}{2}\right)^2 = \frac{p}{2}$

Use the discriminant to determine the number and type of solutions of each equation. See Example 5.

41. $x^2 - 5 = 0$

42. $x^2 - 7 = 0$

43. $4x^2 + 12x = -9$

44. $9x^2 + 1 = 6x$

45. $3x = -2x^2 + 7$

46. $3x^2 = 5 - 7x$

▶ 47. $6 = 4x - 5x^2$

48. $8x = 3 - 9x^2$

49. $9x - 2x^2 + 5 = 0$

50. $5 - 4x + 12x^2 = 0$

Solve. See Examples 6 and 7.

51. Nancy, Thelma, and John Varner live on a corner lot. Often, neighborhood children cut across their lot to save walking distance. Given the diagram below, approximate to the nearest foot how many feet of walking distance is saved by cutting across their property instead of walking around the lot.

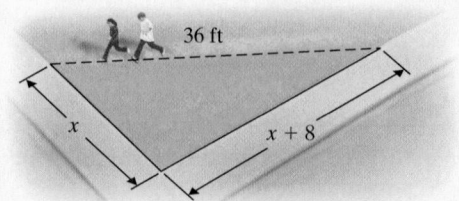

△ **52.** Given the diagram below, approximate to the nearest foot how many feet of walking distance a person saves by cutting across the lawn instead of walking on the sidewalk.

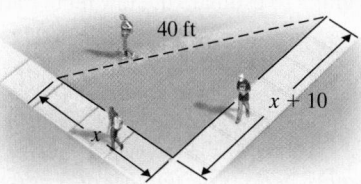

40 ft

x + 10

x

△ **53.** The hypotenuse of an isosceles right triangle is 2 centimeters longer than either of its legs. Find the exact length of each side. (*Hint:* An isosceles right triangle is a right triangle whose legs are the same length.)

△ **54.** The hypotenuse of an isosceles right triangle is one meter longer than either of its legs. Find the length of each side.

△ **55.** Bailey's rectangular dog pen for his Irish setter must have an area of 400 square feet. Also, the length must be 10 feet longer than the width. Find the dimensions of the pen.

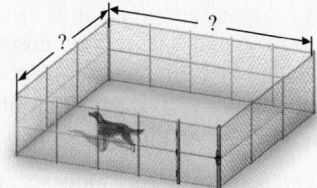

?

?

△ **56.** An entry in the Peach Festival Poster Contest must be rectangular and have an area of 1200 square inches. Furthermore, its length must be 20 inches longer than its width. Find the dimensions each entry must have.

△ **57.** A holding pen for cattle must be square and have a diagonal length of 100 meters.

 a. Find the length of a side of the pen.

 b. Find the area of the pen.

△ **58.** A rectangle is three times longer than it is wide. It has a diagonal of length 50 centimeters.

 a. Find the dimensions of the rectangle.

 b. Find the perimeter of the rectangle.

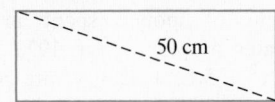

50 cm

△ **59.** The heaviest reported door in the world is the 708.6-ton radiation shield door in the National Institute for Fusion Science at Toki, Japan. If the height of the door is 1.1 feet longer than its width, and its front area (neglecting depth) is 1439.9 square feet, find its width and height [Interesting note: the door is 6.6 feet thick.] (*Source: Guinness World Records*)

 60. Christi and Robbie Wegmann are constructing a rectangular stained glass window whose length is 7.3 inches longer than its width. If the area of the window is 569.9 square inches, find its width and length.

△ **61.** The base of a triangle is four more than twice its height. If the area of the triangle is 42 square centimeters, find its base and height.

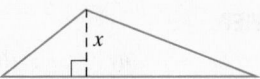

x

 62. If a point *B* divides a line segment such that the smaller portion is to the larger portion as the larger is to the whole, the whole is the length of the *golden ratio*.

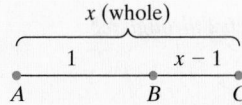

x (whole)

1   x − 1

A   B   C

The golden ratio was thought by the Greeks to be the most pleasing to the eye, and many of their buildings contained numerous examples of the golden ratio. The value of the golden ratio is the positive solution of

$$\begin{array}{l} (\text{smaller}) \\ (\text{larger}) \end{array} \quad \frac{x-1}{1} = \frac{1}{x} \quad \begin{array}{l} (\text{larger}) \\ (\text{whole}) \end{array}$$

Find this value.

The Wollomombi Falls in Australia have a height of 1100 feet. A pebble is thrown upward from the top of the falls with an initial velocity of 20 feet per second. The height of the pebble h after t seconds is given by the equation $h = -16t^2 + 20t + 1100$. Use this equation for Exercises 63 and 64.

 63. How long after the pebble is thrown will it hit the ground? Round to the nearest tenth of a second.

 64. How long after the pebble is thrown will it be 550 feet from the ground? Round to the nearest tenth of a second.

A ball is thrown downward from the top of a 180-foot building with an initial velocity of 20 feet per second. The height of the ball h after t seconds is given by the equation $h = -16t^2 - 20t + 180$. Use this equation to answer Exercises 65 and 66.

 65. How long after the ball is thrown will it strike the ground? Round the result to the nearest tenth of a second.

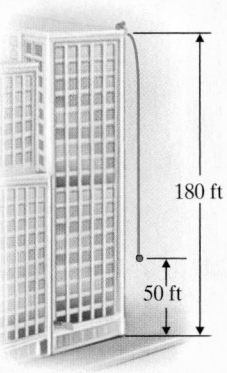

180 ft

50 ft

66. How long after the ball is thrown will it be 50 feet from the ground? Round the result to the nearest tenth of a second.

REVIEW AND PREVIEW

Solve each equation. See Sections 7.5 and 10.6.

67. $\sqrt{5x - 2} = 3$

68. $\sqrt{y + 2} + 7 = 12$

69. $\dfrac{1}{x} + \dfrac{2}{5} = \dfrac{7}{x}$

70. $\dfrac{10}{z} = \dfrac{5}{z} - \dfrac{1}{3}$

Factor. See Sections 6.2 through 6.5.

71. $x^4 + x^2 - 20$

72. $4y^4 + 23y^2 - 6$

73. $z^4 - 13z^2 + 36$

74. $x^4 - 1$

CONCEPT EXTENSIONS

For each quadratic equation, choose the correct substitution for a, b, and c in the standard form $ax^2 + bx + c = 0$.

75. $x^2 = -10$

    a. $a = 1, b = 0, c = -10$

    b. $a = 1, b = 0, c = 10$

    c. $a = 0, b = 1, c = -10$

    d. $a = 1, b = 1, c = 10$

76. $x^2 + 5 = -x$

    a. $a = 1, b = 5, c = -1$

    b. $a = 1, b = -1, c = 5$

    c. $a = 1, b = 5, c = 1$

    d. $a = 1, b = 1, c = 5$

77. Solve Exercise 1 by factoring. Explain the result.

78. Solve Exercise 2 by factoring. Explain the result.

Use the quadratic formula and a calculator to approximate each solution to the nearest tenth.

79. $2x^2 - 6x + 3 = 0$

80. $3.6x^2 + 1.8x - 4.3 = 0$

The accompanying graph shows the daily low temperatures for one week in New Orleans, Louisiana.

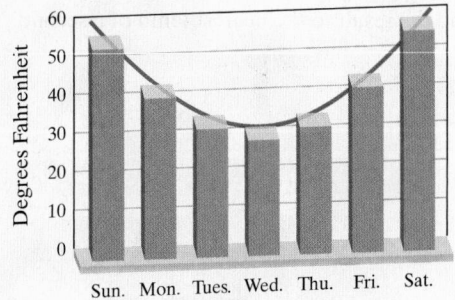

81. Between which days of the week was there the greatest decrease in the low temperature?

82. Between which days of the week was there the greatest increase in the low temperature?

83. Which day of the week had the lowest low temperature?

84. Use the graph to estimate the low temperature on Thursday.

Notice that the shape of the temperature graph is similar to the curve drawn. In fact, this graph can be modeled by the quadratic function $f(x) = 3x^2 - 18x + 56$, where f(x) is the temperature in degrees Fahrenheit and x is the number of days from Sunday. (This graph is shown in blue.) Use this function to answer Exercises 85 and 86.

85. Use the quadratic function given to approximate the temperature on Thursday. Does your answer agree with the graph?

86. Use the function given and the quadratic formula to find when the temperature was 35° F. [*Hint:* Let $f(x) = 35$ and solve for x.] Round your answer to one decimal place and interpret your result. Does your answer agree with the graph?

Solve.

87. The number of Internet users worldwide (in millions) can be modeled by the quadratic equation $f(x) = 5x^2 + 55x + 29$, where $f(x)$ is the number of Internet users worldwide in millions of users, and x is the number of years after 1995. (*Source:* Data from International Telecommunications Union)

    a. Find the number of Internet users worldwide in 2015.

    b. If the trend described in this model continues, find in what year after 1995 the worldwide users of the Internet will reach 4500 million.

88. While the number of farms in the United States was decreasing through the late twentieth century and the early twenty-first century, the average size of farms was increasing. This may be due to consolidation of existing farmlands. The equation $f(x) = -0.5x^2 + 4x + 429$ models the size of the average farm, where $f(x)$ is farm size in acres, and x represents the number of years after 2011. (*Source:* Based on data from the National Agricultural Statistics Service)

    a. Find the size of the average farm in 2012. Round to the nearest acre.

    b. Find the size of the average farm in 2014. Round to the nearest acre.

89. The amount of money spent in restaurants in the United States x years after 1970 can be modeled by $f(x) = 0.2x^2 + 5.1x + 45.3$, where $f(x)$ is the amount of money spent in restaurants (in billions of dollars), and x is the number of years since 1970.

    a. Find the amount the American public spent in restaurants in 2015. Round to the nearest tenth of a billion.

    b. According to this model, in what year after 1970 was the amount of money spent in restaurants $400 billion?

    c. According to this model, in what year after 1970 will the amount of money spent on restaurants equal $1000 billion?

90. The relationship between body weight and the Recommended Dietary Allowance (RDA) for vitamin A in children up to age 10 is modeled by the quadratic equation $y = 0.149x^2 - 4.475x + 406.478$, where y is the RDA for vitamin A in micrograms for a child whose weight is x pounds. (*Source:* Based on data from the Food and Nutrition Board, National Academy of Sciences–Institute of Medicine, 1989)

   a. Determine the vitamin A requirements of a child who weighs 35 pounds.

   b. What is the weight of a child whose RDA of vitamin A is 600 micrograms? Round your answer to the nearest pound.

The solutions of the quadratic equation $ax^2 + bx + c = 0$ *are* $\dfrac{-b + \sqrt{b^2 - 4ac}}{2a}$ *and* $\dfrac{-b - \sqrt{b^2 - 4ac}}{2a}$.

91. Show that the sum of these solutions is $\dfrac{-b}{a}$.

92. Show that the product of these solutions is $\dfrac{c}{a}$.

Use the quadratic formula to solve each quadratic equation.

93. $3x^2 - \sqrt{12}x + 1 = 0$
   (*Hint:* $a = 3, b = -\sqrt{12}, c = 1$)

94. $5x^2 + \sqrt{20}x + 1 = 0$

95. $x^2 + \sqrt{2}x + 1 = 0$

96. $x^2 - \sqrt{2}x + 1 = 0$

97. $2x^2 - \sqrt{3}x - 1 = 0$

98. $7x^2 + \sqrt{7}x - 2 = 0$

99. Use a graphing calculator to solve Exercises 63 and 65.

100. Use a graphing calculator to solve Exercises 64 and 66.

Recall that the discriminant also tells us the number of x-intercepts of the related function.

101. Check the results of Exercise 49 by graphing $y = 9x - 2x^2 + 5$.

102. Check the results of Exercise 50 by graphing $y = 5 - 4x + 12x^2$.

11.3 Solving Equations by Using Quadratic Methods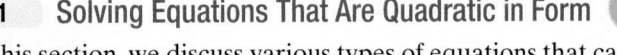

OBJECTIVES

1 Solve Various Equations That Are Quadratic in Form.

2 Solve Problems That Lead to Quadratic Equations.

OBJECTIVE

1 Solving Equations That Are Quadratic in Form

In this section, we discuss various types of equations that can be solved in part by using the methods for solving quadratic equations. We call these equations ones that are quadratic in form.

Once each equation is simplified, you may want to use these steps when deciding which method to use to solve the quadratic equation.

Solving a Quadratic Equation

Step 1. If the equation is in the form $(ax + b)^2 = c$, use the square root property and solve. If not, go to Step 2.

Step 2. Write the equation in standard form: $ax^2 + bx + c = 0$.

Step 3. Try to solve the equation by the factoring method. If not possible, go to Step 4.

Step 4. Solve the equation by the quadratic formula.

The first example is a radical equation that becomes a quadratic equation once we square both sides.

EXAMPLE I   Solve: $x - \sqrt{x} - 6 = 0$

Solution Recall that to solve a radical equation, first get the radical alone on one side of the equation. Then square both sides.

$$x - 6 = \sqrt{x} \qquad \text{Add } \sqrt{x} \text{ to both sides.}$$
$$(x - 6)^2 = (\sqrt{x})^2 \qquad \text{Square both sides.}$$
$$x^2 - 12x + 36 = x$$
$$x^2 - 13x + 36 = 0 \qquad \text{Set the equation equal to 0.}$$
$$(x - 9)(x - 4) = 0$$
$$x - 9 = 0 \quad \text{or} \quad x - 4 = 0$$
$$x = 9 \qquad\qquad x = 4 \qquad\qquad \text{(Continued on next page)}$$

Check:

$$\text{Let } x = 9$$
$$x - \sqrt{x} - 6 = 0$$
$$9 - \sqrt{9} - 6 \stackrel{?}{=} 0$$
$$9 - 3 - 6 \stackrel{?}{=} 0$$
$$0 = 0 \quad \text{True}$$

$$\text{Let } x = 4$$
$$x - \sqrt{x} - 6 = 0$$
$$4 - \sqrt{4} - 6 \stackrel{?}{=} 0$$
$$4 - 2 - 6 \stackrel{?}{=} 0$$
$$-4 = 0 \quad \text{False}$$

The solution is 9 or the solution set is {9}.

PRACTICE
1 Solve: $x - \sqrt{x + 1} - 5 = 0$

EXAMPLE 2 Solve: $\dfrac{3x}{x - 2} - \dfrac{x + 1}{x} = \dfrac{6}{x(x - 2)}$

Solution In this equation, x cannot be either 2 or 0 because these values cause denominators to equal zero. To solve for x, we first multiply both sides of the equation by $x(x - 2)$ to clear the fractions. By the distributive property, this means that we multiply each term by $x(x - 2)$.

$$x(x - 2)\left(\frac{3x}{x - 2}\right) - x(x - 2)\left(\frac{x + 1}{x}\right) = x(x - 2)\left[\frac{6}{x(x - 2)}\right]$$

$$3x^2 - (x - 2)(x + 1) = 6 \quad \text{Simplify.}$$
$$3x^2 - (x^2 - x - 2) = 6 \quad \text{Multiply.}$$
$$3x^2 - x^2 + x + 2 = 6$$
$$2x^2 + x - 4 = 0 \quad \text{Simplify.}$$

This equation cannot be factored using integers, so we solve by the quadratic formula.

$$x = \frac{-1 \pm \sqrt{1^2 - 4(2)(-4)}}{2 \cdot 2} \quad \text{Use } a = 2, b = 1, \text{ and } c = -4 \text{ in the quadratic formula.}$$

$$= \frac{-1 \pm \sqrt{1 + 32}}{4} \quad \text{Simplify.}$$

$$= \frac{-1 \pm \sqrt{33}}{4}$$

Neither proposed solution will make the denominators 0.

The solutions are $\dfrac{-1 + \sqrt{33}}{4}$ and $\dfrac{-1 - \sqrt{33}}{4}$ or the solution set is $\left\{\dfrac{-1 + \sqrt{33}}{4}, \dfrac{-1 - \sqrt{33}}{4}\right\}$.

PRACTICE
2 Solve: $\dfrac{5x}{x + 1} - \dfrac{x + 4}{x} = \dfrac{3}{x(x + 1)}$

EXAMPLE 3 Solve: $p^4 - 3p^2 - 4 = 0$

Solution First we factor the trinomial.

$$p^4 - 3p^2 - 4 = 0$$
$$(p^2 - 4)(p^2 + 1) = 0 \quad \text{Factor.}$$
$$(p - 2)(p + 2)(p^2 + 1) = 0 \quad \text{Factor further.}$$
$$p - 2 = 0 \quad \text{or} \quad p + 2 = 0 \quad \text{or} \quad p^2 + 1 = 0 \quad \text{Set each factor equal to 0 and solve.}$$
$$p = 2 \qquad\qquad p = -2 \qquad\qquad p^2 = -1$$
$$p = \pm\sqrt{-1} = \pm i$$

The solutions are 2, -2, i and $-i$. □

PRACTICE
3 Solve: $p^4 - 7p^2 - 144 = 0$

> **Helpful Hint**
>
> Example 3 can be solved using substitution also. Think of $p^4 - 3p^2 - 4 = 0$ as
>
> $$(p^2)^2 - 3p^2 - 4 = 0 \quad \text{Then let } x = p^2 \text{ and solve and substitute back.}$$
> $$\text{The solutions will be the same.}$$
> $$x^2 - 3x - 4 = 0$$

✔ CONCEPT CHECK

a. True or false? The maximum number of solutions that a quadratic equation can have is 2.
b. True or false? The maximum number of solutions that an equation in quadratic form can have is 2.

EXAMPLE 4 Solve: $(x - 3)^2 - 3(x - 3) - 4 = 0$

Solution Notice that the quantity $(x - 3)$ is repeated in this equation. Sometimes it is helpful to substitute a variable (in this case other than x) for the repeated quantity. We will let $y = x - 3$. Then

$$(x - 3)^2 - 3(x - 3) - 4 = 0$$

becomes

$$y^2 - 3y - 4 = 0 \quad \text{Let } x - 3 = y.$$
$$(y - 4)(y + 1) = 0 \quad \text{Factor.}$$

To solve, we use the zero factor property.

$$y - 4 = 0 \quad \text{or} \quad y + 1 = 0 \quad \text{Set each factor equal to 0.}$$
$$y = 4 \qquad\qquad y = -1 \quad \text{Solve.}$$

> **Helpful Hint**
>
> When using substitution, don't forget to substitute back to the original variable.

To find values of x, we substitute back. That is, we substitute $x - 3$ for y.

$$x - 3 = 4 \quad \text{or} \quad x - 3 = -1$$
$$x = 7 \qquad\qquad x = 2$$

Both 2 and 7 check. The solutions are 2 and 7. □

PRACTICE
4 Solve: $(x + 2)^2 - 2(x + 2) - 3 = 0$

EXAMPLE 5 Solve: $x^{2/3} - 5x^{1/3} + 6 = 0$

Solution The key to solving this equation is recognizing that $x^{2/3} = (x^{1/3})^2$. We replace $x^{1/3}$ with m so that

$$(x^{1/3})^2 - 5x^{1/3} + 6 = 0$$

becomes

$$m^2 - 5m + 6 = 0$$

Now we solve by factoring.

$$m^2 - 5m + 6 = 0$$
$$(m - 3)(m - 2) = 0 \qquad\qquad \text{Factor.}$$
$$m - 3 = 0 \quad \text{or} \quad m - 2 = 0 \quad \text{Set each factor equal to 0.}$$
$$m = 3 \qquad\qquad m = 2$$

(Continued on next page)

Since $m = x^{1/3}$, we have

$$x^{1/3} = 3 \qquad \text{or} \quad x^{1/3} = 2$$
$$x = 3^3 = 27 \quad \text{or} \qquad x = 2^3 = 8$$

Both 8 and 27 check. The solutions are 8 and 27.

PRACTICE
5   Solve: $x^{2/3} - 5x^{1/3} + 4 = 0$

OBJECTIVE
2   Solving Problems That Lead to Quadratic Equations

The next example is a work problem. This problem is modeled by a rational equation that simplifies to a quadratic equation.

EXAMPLE 6   Finding Work Time

Together, an experienced word processor and an apprentice word processor can create a word document in 6 hours. Alone, the experienced word processor can create the document 2 hours faster than the apprentice word processor can. Find the time in which each person can create the word document alone.

Solution

1. UNDERSTAND. Read and reread the problem. The key idea here is the relationship between the *time* (hours) it takes to complete the job and the *part of the job* completed in one unit of time (hour). For example, because they can complete the job together in 6 hours, the *part of the job* they can complete in 1 hour is $\frac{1}{6}$.

Let

$x =$ the *time* in hours it takes the apprentice word processor to complete the job alone, and

$x - 2 =$ the *time* in hours it takes the experienced word processor to complete the job alone.

We can summarize in a chart the information discussed.

	Total Hours to Complete Job	*Part of Job Completed in 1 Hour*
Apprentice Word Processor	x	$\frac{1}{x}$
Experienced Word Processor	$x - 2$	$\frac{1}{x-2}$
Together	6	$\frac{1}{6}$

2. TRANSLATE.

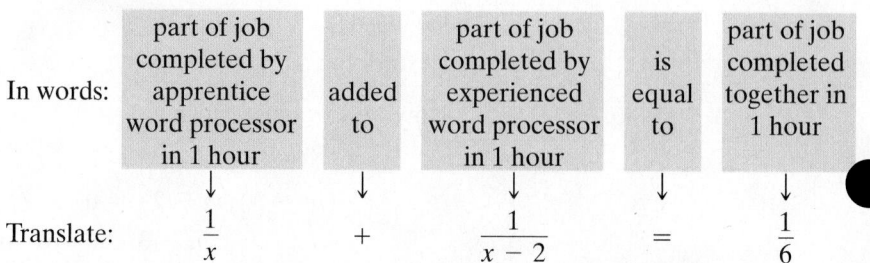

In words:	part of job completed by apprentice word processor in 1 hour	added to	part of job completed by experienced word processor in 1 hour	is equal to	part of job completed together in 1 hour
	↓	↓	↓	↓	↓
Translate:	$\frac{1}{x}$	$+$	$\frac{1}{x-2}$	$=$	$\frac{1}{6}$

3. SOLVE.

$$\frac{1}{x} + \frac{1}{x-2} = \frac{1}{6}$$

$$6x(x-2)\left(\frac{1}{x} + \frac{1}{x-2}\right) = 6x(x-2)\cdot\frac{1}{6}$$ Multiply both sides by the LCD $6x(x-2)$.

$$6x(x-2)\cdot\frac{1}{x} + 6x(x-2)\cdot\frac{1}{x-2} = 6x(x-2)\cdot\frac{1}{6}$$ Use the distributive property.

$$6(x-2) + 6x = x(x-2)$$

$$6x - 12 + 6x = x^2 - 2x$$

$$0 = x^2 - 14x + 12$$

Now we can substitute $a = 1, b = -14,$ and $c = 12$ into the quadratic formula and simplify.

$$x = \frac{-(-14) \pm \sqrt{(-14)^2 - 4(1)(12)}}{2\cdot 1} = \frac{14 \pm \sqrt{148}}{2}^{*}$$

Using a calculator or a square root table, we see that $\sqrt{148} \approx 12.2$ rounded to one decimal place. Thus,

$$x \approx \frac{14 \pm 12.2}{2}$$

$$x \approx \frac{14 + 12.2}{2} = 13.1 \quad \text{or} \quad x \approx \frac{14 - 12.2}{2} = 0.9$$

4. INTERPRET.

Check: If the apprentice word processor completes the job alone in 0.9 hours, the experienced word processor completes the job alone in $x - 2 = 0.9 - 2 = -1.1$ hours. Since this is not possible, we reject the solution of 0.9. The approximate solution thus is 13.1 hours.

State: The apprentice word processor can complete the job alone in approximately 13.1 hours, and the experienced word processor can complete the job alone in approximately

$$x - 2 = 13.1 - 2 = 11.1 \text{ hours}$$ □

PRACTICE

6 Together, Katy and Steve can groom all the dogs at the Barkin' Doggie Day Care in 4 hours. Alone, Katy can groom the dogs 1 hour faster than Steve can groom the dogs alone. Find the time in which each of them can groom the dogs alone. ▪

EXAMPLE 7   **Finding Driving Speeds**

Beach and Fargo are about 400 miles apart. A salesperson travels from Fargo to Beach one day at a certain speed. She returns to Fargo the next day and drives 10 mph faster. Her total travel time was $14\frac{2}{3}$ hours. Find her speed to Beach and the return speed to Fargo.

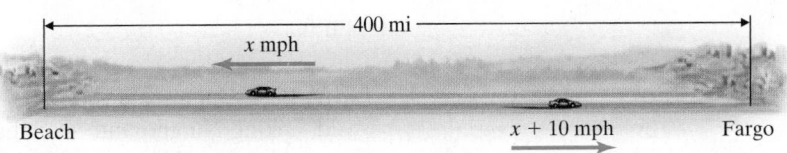

*This expression can be simplified further, but this will suffice because we are approximating.

(Continued on next page)

Solution

1. UNDERSTAND. Read and reread the problem. Let

$$x = \text{the speed to Beach, so}$$

$$x + 10 = \text{the return speed to Fargo.}$$

Then organize the given information in a table.

> **Helpful Hint**
>
> Since $d = rt, t = \dfrac{d}{r}$. The time column was completed using $\dfrac{d}{r}$.

	distance	= rate	· time	
To Beach	400	x	$\dfrac{400}{x}$	← distance ← rate
Return to Fargo	400	$x + 10$	$\dfrac{400}{x + 10}$	← distance ← rate

2. TRANSLATE.

In words: $\boxed{\begin{array}{c}\text{time to}\\\text{Beach}\end{array}}$ $+$ $\boxed{\begin{array}{c}\text{return}\\\text{time to}\\\text{Fargo}\end{array}}$ $=$ $14\dfrac{2}{3}$ hours

Translate: $\dfrac{400}{x} + \dfrac{400}{x + 10} = \dfrac{44}{3}$

3. SOLVE.

$$\frac{400}{x} + \frac{400}{x + 10} = \frac{44}{3}$$

$$\frac{100}{x} + \frac{100}{x + 10} = \frac{11}{3} \qquad \text{Divide both sides by 4.}$$

$$3x(x + 10)\left(\frac{100}{x} + \frac{100}{x + 10}\right) = 3x(x + 10)\cdot\frac{11}{3} \qquad \begin{array}{l}\text{Multiply both}\\\text{sides by the LCD}\\3x(x + 10).\end{array}$$

$$3x(x + 10)\cdot\frac{100}{x} + 3x(x + 10)\cdot\frac{100}{x + 10} = 3x(x + 10)\cdot\frac{11}{3} \qquad \begin{array}{l}\text{Use the distributive}\\\text{property.}\end{array}$$

$$3(x + 10)\cdot 100 + 3x\cdot 100 = x(x + 10)\cdot 11$$

$$300x + 3000 + 300x = 11x^2 + 110x$$

$$0 = 11x^2 - 490x - 3000 \qquad \begin{array}{l}\text{Set equation}\\\text{equal to 0.}\end{array}$$

$$0 = (11x + 60)(x - 50) \qquad \text{Factor.}$$

$$11x + 60 = 0 \quad \text{or} \quad x - 50 = 0 \qquad \begin{array}{l}\text{Set each factor}\\\text{equal to 0.}\end{array}$$

$$x = -\frac{60}{11} \text{ or } -5\frac{5}{11}; \quad x = 50$$

4. INTERPRET.

Check: The speed is not negative, so it's not $-5\dfrac{5}{11}$. The number 50 does check.

State: The speed to Beach was 50 mph, and her return speed to Fargo was 60 mph. □

PRACTICE

7 The 36-km S-shaped Hangzhou Bay Bridge is the longest cross-sea bridge in the world, linking Ningbo and Shanghai, China. A merchant drives over the bridge one morning from Ningbo to Shanghai in very heavy traffic and returns home that night driving 50 km per hour faster. The total travel time was 1.3 hours. Find the speed to Shanghai and the return speed to Ningbo. ▪

✔ Vocabulary, Readiness & Video Check

Martin-Gay Interactive Videos

See Video 11.3 ⊙

Watch the section lecture video and answer the following questions.

OBJECTIVE
1

1. From Examples 1 and 2, what's the main thing to remember when using a substitution in order to solve an equation by quadratic methods?

OBJECTIVE
2

2. In Example 4, the translated equation is actually a rational equation. Explain how we end up solving it using quadratic methods.

11.3 Exercise Set MyMathLab® ▶

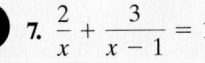

Solve. See Example 1.

1. $2x = \sqrt{10 + 3x}$

2. $3x = \sqrt{8x + 1}$

3. $x - 2\sqrt{x} = 8$

4. $x - \sqrt{2x} = 4$

5. $\sqrt{9x} = x + 2$

6. $\sqrt{16x} = x + 3$

Solve. See Example 2.

7. $\dfrac{2}{x} + \dfrac{3}{x - 1} = 1$

8. $\dfrac{6}{x^2} = \dfrac{3}{x + 1}$

9. $\dfrac{3}{x} + \dfrac{4}{x + 2} = 2$

10. $\dfrac{5}{x - 2} + \dfrac{4}{x + 2} = 1$

11. $\dfrac{7}{x^2 - 5x + 6} = \dfrac{2x}{x - 3} - \dfrac{x}{x - 2}$

12. $\dfrac{11}{2x^2 + x - 15} = \dfrac{5}{2x - 5} - \dfrac{x}{x + 3}$

Solve. See Example 3.

13. $p^4 - 16 = 0$

14. $x^4 + 2x^2 - 3 = 0$

15. $4x^4 + 11x^2 = 3$

16. $z^4 = 81$

17. $z^4 - 13z^2 + 36 = 0$

18. $9x^4 + 5x^2 - 4 = 0$

Solve. See Examples 4 and 5.

19. $x^{2/3} - 3x^{1/3} - 10 = 0$

20. $x^{2/3} + 2x^{1/3} + 1 = 0$

21. $(5n + 1)^2 + 2(5n + 1) - 3 = 0$

22. $(m - 6)^2 + 5(m - 6) + 4 = 0$

23. $2x^{2/3} - 5x^{1/3} = 3$

24. $3x^{2/3} + 11x^{1/3} = 4$

25. $1 + \dfrac{2}{3t - 2} = \dfrac{8}{(3t - 2)^2}$

26. $2 - \dfrac{7}{x + 6} = \dfrac{15}{(x + 6)^2}$

27. $20x^{2/3} - 6x^{1/3} - 2 = 0$

28. $4x^{2/3} + 16x^{1/3} = -15$

MIXED PRACTICE

Solve. See Examples 1 through 5.

29. $a^4 - 5a^2 + 6 = 0$

30. $x^4 - 12x^2 + 11 = 0$

31. $\dfrac{2x}{x - 2} + \dfrac{x}{x + 3} = -\dfrac{5}{x + 3}$

32. $\dfrac{5}{x - 3} + \dfrac{x}{x + 3} = \dfrac{19}{x^2 - 9}$

▶ **33.** $(p + 2)^2 = 9(p + 2) - 20$

34. $2(4m - 3)^2 - 9(4m - 3) = 5$

35. $2x = \sqrt{11x + 3}$

36. $4x = \sqrt{2x + 3}$

37. $x^{2/3} - 8x^{1/3} + 15 = 0$

38. $x^{2/3} - 2x^{1/3} - 8 = 0$

39. $y^3 + 9y - y^2 - 9 = 0$

40. $x^3 + x - 3x^2 - 3 = 0$

41. $2x^{2/3} + 3x^{1/3} - 2 = 0$

42. $6x^{2/3} - 25x^{1/3} - 25 = 0$

43. $x^{-2} - x^{-1} - 6 = 0$

44. $y^{-2} - 8y^{-1} + 7 = 0$

45. $x - \sqrt{x} = 2$

46. $x - \sqrt{3x} = 6$

47. $\dfrac{x}{x-1} + \dfrac{1}{x+1} = \dfrac{2}{x^2-1}$

48. $\dfrac{x}{x-5} + \dfrac{5}{x+5} = -\dfrac{1}{x^2-25}$

49. $p^4 - p^2 - 20 = 0$

50. $x^4 - 10x^2 + 9 = 0$

51. $(x+3)(x^2-3x+9) = 0$

52. $(x-6)(x^2+6x+36) = 0$

53. $1 = \dfrac{4}{x-7} + \dfrac{5}{(x-7)^2}$

54. $3 + \dfrac{1}{2p+4} = \dfrac{10}{(2p+4)^2}$

55. $27y^4 + 15y^2 = 2$

56. $8z^4 + 14z^2 = -5$

57. $x - \sqrt{19 - 2x} - 2 = 0$

58. $x - \sqrt{17 - 4x} - 3 = 0$

Solve. For Exercises 59 and 60, the solutions have been started for you. See Examples 6 and 7.

59. Roma Sherry drove 330 miles from her hometown to Tucson. During her return trip, she was able to increase her speed by 11 miles per hour. If her return trip took 1 hour less time, find her original speed and her speed returning home.

Start the solution:

1. UNDERSTAND the problem. Reread it as many times as needed. Let

$$x = \text{original speed, and}$$
$$x + 11 = \text{return-trip speed.}$$

Organize the information in a table.

	distance =	rate ·	time	
To Tucson	330	x	$\dfrac{330}{x}$	← distance ← rate
Return trip	330	___	$\dfrac{330}{\_\_}$	← distance ← rate

2. TRANSLATE into an equation. (Fill in the blanks below.)

time to Tucson	equals	return trip time	plus	1 hour
↓	↓	↓	↓	↓
___	=	___	+	1

Finish with:

3. SOLVE and **4.** INTERPRET

60. A salesperson drove to Portland, a distance of 300 miles. During the last 80 miles of his trip, heavy rainfall forced him to decrease his speed by 15 miles per hour. If his total driving time was 6 hours, find his original speed and his speed during the rainfall.

Start the solution:

1. UNDERSTAND the problem. Reread it as many times as needed. Let

$$x = \text{original speed, and}$$
$$x - 15 = \text{rainfall speed.}$$

Organize the information in a table.

	distance =	rate ·	time	
First part of trip	300 − 80, or 220	x	$\dfrac{220}{x}$	← distance ← rate
Heavy rainfall part of trip	80	$x - 15$	$\dfrac{80}{\_\_}$	← distance ← rate

2. TRANSLATE into an equation. (Fill in the blanks below.)

time during first part of trip	plus	time during heavy rainfall	equals	6 hr
↓	↓	↓	↓	↓
___	+	___	=	6

Finish with:

3. SOLVE and **4.** INTERPRET

61. A jogger ran 3 miles, decreased her speed by 1 mile per hour, and then ran another 4 miles. If her total time jogging was $1\dfrac{3}{5}$ hours, find her speed for each part of her run.

62. Mark Keaton's workout consists of jogging for 3 miles and then riding his bike for 5 miles at a speed 4 miles per hour faster than he jogs. If his total workout time is 1 hour, find his jogging speed and his biking speed.

63. A Chinese restaurant in Mandeville, Louisiana, has a large goldfish pond around the restaurant. Suppose that an inlet pipe and a hose together can fill the pond in 8 hours. The inlet pipe alone can complete the job in one hour less time than the hose alone. Find the time that the hose can complete the job alone and the time that the inlet pipe can complete the job alone. Round each to the nearest tenth of an hour.

64. A water tank on a farm in Flatonia, Texas, can be filled with a large inlet pipe and a small inlet pipe in 3 hours. The large inlet pipe alone can fill the tank in 2 hours less time than the small inlet pipe alone. Find the time to the nearest tenth of an hour each pipe can fill the tank alone.

65. Bill Shaughnessy and his son Billy can clean the house together in 4 hours. When the son works alone, it takes him an hour longer to clean than it takes his dad alone. Find how long to the nearest tenth of an hour it takes the son to clean alone.

66. Together, Noodles and Freckles eat a 50-pound bag of dog food in 30 days. Noodles by himself eats a 50-pound bag in 2 weeks less time than Freckles does by himself. How many days to the nearest whole day would a 50-pound bag of dog food last Freckles?

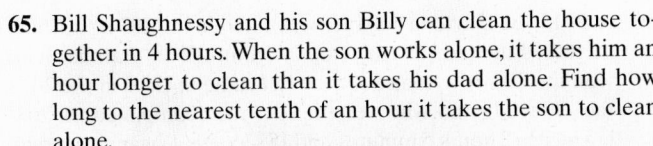

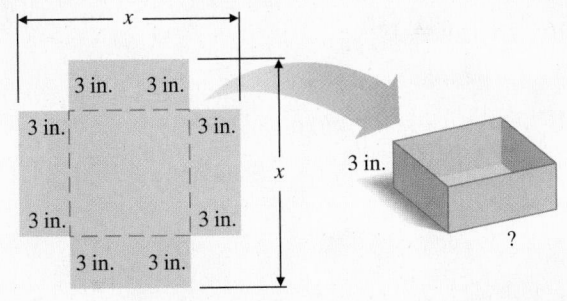

67. The product of a number and 4 less than the number is 96. Find the number.

68. A whole number increased by its square is two more than twice itself. Find the number.

△ **69.** Suppose that an open box is to be made from a square sheet of cardboard by cutting out squares from each corner as shown and then folding along the dotted lines. If the box is to have a volume of 300 cubic inches, find the original dimensions of the sheet of cardboard.

a. The ? in the drawing above will be the length (and the width) of the box as shown. Represent this length in terms of x.

b. Use the formula for volume of a box, $V = l \cdot w \cdot h$, to write an equation in x.

c. Solve the equation for x and give the dimensions of the sheet of cardboard. Check your solution.

70. Suppose that an open box is to be made from a square sheet of cardboard by cutting out squares from each corner as shown and then folding along the dotted lines. If the box is to have

a volume of 128 cubic inches, find the original dimensions of the sheet of cardboard.

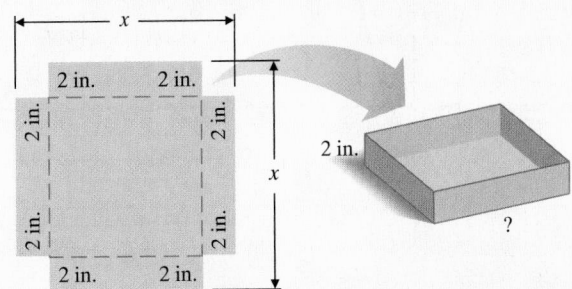

a. The ? in the drawing above will be the length (and the width) of the box as shown. Represent this length in terms of x.

b. Use the formula for volume of a box, $V = l \cdot w \cdot h$, to write an equation in x.

c. Solve the equation for x and give the dimensions of the sheet of cardboard. Check your solution.

△ **71.** A sprinkler that sprays water in a circular pattern is to be used to water a square garden. If the area of the garden is 920 square feet, find the smallest whole number *radius* that the sprinkler can be adjusted to so that the entire garden is watered.

△ **72.** Suppose that a square field has an area of 6270 square feet. See Exercise 71 and find the sprinkler radius.

REVIEW AND PREVIEW

Solve each inequality. See Section 2.8.

73. $\dfrac{5x}{3} + 2 \le 7$

74. $\dfrac{2x}{3} + \dfrac{1}{6} \ge 2$

75. $\dfrac{y-1}{15} > -\dfrac{2}{5}$

76. $\dfrac{z-2}{12} < \dfrac{1}{4}$

Find the domain and range of each graphed relation. Decide which relations are also functions. See Section 3.6.

77.

78.

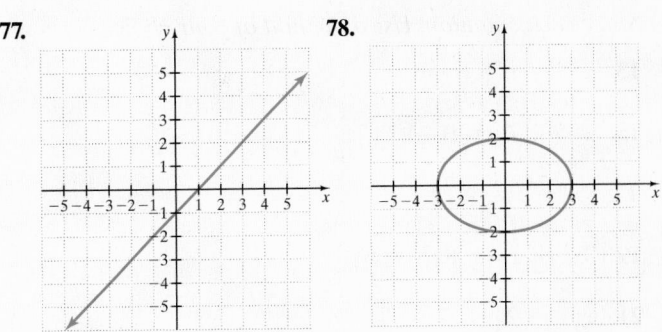

79.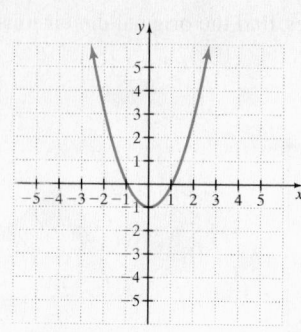

80.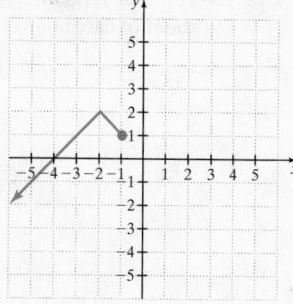

86. $y^3 - 216 = 0$

87. Write a polynomial equation that has three solutions: 2, 5, and -7.

88. Write a polynomial equation that has three solutions: 0, $2i$, and $-2i$.

89. In 2015, the Daytona 500 race was extended to 203 laps, which increased the race distance to 507.5 miles. Dale Earnhardt Jr. won the 2014 Daytona 500 race but came in third in the 2015 race. Joey Logano, who won the 2015 race, completed the 203-lap race in 3 hours, 5 minutes, and 15 seconds. Logano's average speed for the race was 1.193 seconds per lap faster than that of Earnhardt. (Source: Nascar)

  a. Find Joey Logano's average race speed (miles per hour) for the 2015 Daytona 500. Round to the nearest hundredth.

  b. Find Joey Logano's average lap time (seconds per lap) for the 2015 Daytona 500. Round to the nearest thousandth.

  c. Find Dale Earnhart Jr.'s average lap time.

  d. Find Dale Earnhardt Jr.'s average race speed (miles per hour).

90. Use a graphing calculator to solve Exercise 29. Compare the solution with the solution from Exercise 29. Explain any differences.

CONCEPT EXTENSIONS

Solve.

81. $5y^3 + 45y - 5y^2 - 45 = 0$

82. $10x^3 + 10x - 30x^2 - 30 = 0$

83. $3x^{-2} - 3x^{-1} - 18 = 0$

84. $2y^{-2} - 16y^{-1} + 14 = 0$

85. $2x^3 = -54$

Integrated Review Summary on Solving Quadratic Equations

Sections 11.1–11.3

Use the square root property to solve each equation.

1. $x^2 - 10 = 0$

2. $x^2 - 14 = 0$

3. $(x - 1)^2 = 8$

4. $(x + 5)^2 = 12$

Solve each equation by completing the square.

5. $x^2 + 2x - 12 = 0$

6. $x^2 - 12x + 11 = 0$

7. $3x^2 + 3x = 5$

8. $16y^2 + 16y = 1$

Use the quadratic formula to solve each equation.

9. $2x^2 - 4x + 1 = 0$

10. $\frac{1}{2}x^2 + 3x + 2 = 0$

11. $x^2 + 4x = -7$

12. $x^2 + x = -3$

Solve each equation. Use a method of your choice.

13. $x^2 + 3x + 6 = 0$

14. $2x^2 + 18 = 0$

15. $x^2 + 17x = 0$

16. $4x^2 - 2x - 3 = 0$

17. $(x - 2)^2 = 27$

18. $\frac{1}{2}x^2 - 2x + \frac{1}{2} = 0$

19. $3x^2 + 2x = 8$

20. $2x^2 = -5x - 1$

21. $x(x - 2) = 5$

22. $x^2 - 31 = 0$

23. $5x^2 - 55 = 0$

24. $5x^2 + 55 = 0$

25. $x(x + 5) = 66$

26. $5x^2 + 6x - 2 = 0$

27. $2x^2 + 3x = 1$

28. $x - \sqrt{13 - 3x} - 3 = 0$

29. $\dfrac{5x}{x - 2} - \dfrac{x + 1}{x} = \dfrac{3}{x(x - 2)}$

Solve.

△ **30.** The diagonal of a square room measures 20 feet. Find the exact length of a side of the room. Then approximate the length to the nearest tenth of a foot.

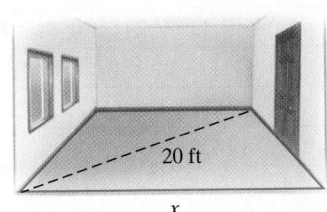

20 ft

x

31. Together, Jack and Lucy Hoag can prepare a crawfish boil for a large party in 4 hours. Lucy alone can complete the job in 2 hours less time than Jack alone. Find the time that each person can prepare the crawfish boil alone. Round each time to the nearest tenth of an hour.

32. Diane Gray exercises at Total Body Gym. On the treadmill, she runs 5 miles, then increases her speed by 1 mile per hour and runs an additional 2 miles. If her total time on the treadmill is $1\dfrac{1}{3}$ hours, find her speed during each part of her run.

11.4 Nonlinear Inequalities in One Variable

OBJECTIVES

1 Solve Polynomial Inequalities of Degree 2 or Greater.

2 Solve Inequalities That Contain Rational Expressions with Variables in the Denominator.

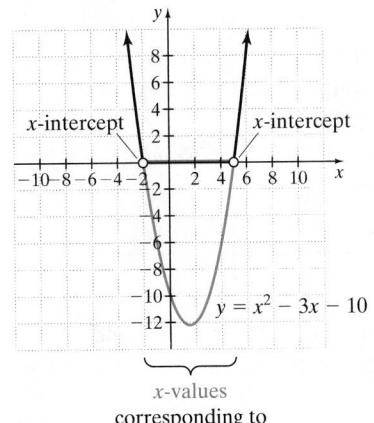

x-intercept   *x*-intercept

$y = x^2 - 3x - 10$

x-values
corresponding to
negative y-values

Helpful Hint

The related equation is found by replacing the inequality symbol with "=".

OBJECTIVE

1   Solving Polynomial Inequalities

Just as we can solve linear inequalities in one variable, so can we also solve quadratic inequalities in one variable. A **quadratic inequality** is an inequality that can be written so that one side is a quadratic expression and the other side is 0. Here are examples of quadratic inequalities in one variable. Each is written in **standard form.**

$$x^2 - 10x + 7 \leq 0 \qquad 3x^2 + 2x - 6 > 0$$
$$2x^2 + 9x - 2 < 0 \qquad x^2 - 3x + 11 \geq 0$$

A solution of a quadratic inequality in one variable is a value of the variable that makes the inequality a true statement.

The value of an expression such as $x^2 - 3x - 10$ will sometimes be positive, sometimes negative, and sometimes 0, depending on the value substituted for x. To solve the inequality $x^2 - 3x - 10 < 0$, we are looking for all values of x that make the expression $x^2 - 3x - 10$ **less than 0,** or **negative.** To understand how we find these values, we'll study the graph of the quadratic function $y = x^2 - 3x - 10$.

Notice that the x-values for which y is positive are separated from the x values for which y is negative by the x-intercepts. (Recall that the x-intercepts correspond to values of x for which $y = 0$.) Thus, the solution set of $x^2 - 3x - 10 < 0$ consists of all real numbers from -2 to 5 or, in interval notation, $(-2, 5)$.

It is not necessary to graph $y = x^2 - 3x - 10$ to solve the related inequality $x^2 - 3x - 10 < 0$. Instead, we can draw a number line representing the x-axis and keep the following in mind: *A region on the number line for which the value of* $x^2 - 3x - 10$ *is positive is separated from a region on the number line for which the value of* $x^2 - 3x - 10$ *is negative by a value for which the expression is 0.*

Let's find these values for which the expression is 0 by solving the related equation:

$$x^2 - 3x - 10 = 0$$
$$(x - 5)(x + 2) = 0 \qquad \text{Factor.}$$
$$x - 5 = 0 \quad \text{or} \quad x + 2 = 0 \qquad \text{Set each factor equal to 0.}$$
$$x = 5 \qquad\qquad x = -2 \qquad \text{Solve.}$$

These two numbers, -2 and 5, divide the number line into three regions. We will call the regions A, B, and C. These regions are important because, if the value of $x^2 - 3x - 10$ is negative when a number from a region is substituted for x, then $x^2 - 3x - 10$ is negative when any number in that region is substituted for x. The same is true if the value of $x^2 - 3x - 10$ is positive for a particular value of x in a region.

To see whether the inequality $x^2 - 3x - 10 < 0$ is true or false in each region, we choose a test point from each region and substitute its value for x in the inequality $x^2 - 3x - 10 < 0$. If the resulting inequality is true, the region containing the test point is a solution region.

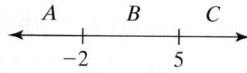

Region	Test Point Value	$(x - 5)(x + 2) < 0$	Result
A	-3	$(-8)(-1) < 0$	False
B	0	$(-5)(2) < 0$	True
C	6	$(1)(8) < 0$	False

The values in region B satisfy the inequality. The numbers -2 and 5 are not included in the solution set since the inequality symbol is $<$. The solution set is $(-2, 5)$, and its graph is shown.

EXAMPLE 1   Solve: $(x + 3)(x - 3) > 0$

__Solution__ First we solve the related equation, $(x + 3)(x - 3) = 0$.

$$(x + 3)(x - 3) = 0$$
$$x + 3 = 0 \quad \text{or} \quad x - 3 = 0$$
$$x = -3 \qquad\qquad x = 3$$

> **Helpful Hint**
>
> Inequality: $(x + 3)(x - 3) > 0$
> Related equation:
> $(x + 3)(x - 3) = 0$

The two numbers -3 and 3 separate the number line into three regions, A, B, and C.

Now we substitute the value of a test point from each region. If the test value satisfies the inequality, every value in the region containing the test value is a solution.

Region	Test Point Value	$(x + 3)(x - 3) > 0$	Result
A	-4	$(-1)(-7) > 0$	True
B	0	$(3)(-3) > 0$	False
C	4	$(7)(1) > 0$	True

The points in regions A and C satisfy the inequality. The numbers -3 and 3 are not included in the solution since the inequality symbol is $>$. The solution set is $(-\infty, -3) \cup (3, \infty)$, and its graph is shown.

PRACTICE

1   Solve: $(x - 4)(x + 3) > 0$

The following steps may be used to solve a polynomial inequality.

Solving a Polynomial Inequality

Step 1. Solve the related equation.

Step 2. Separate the number line into regions with the solutions from Step 1.

Step 3. For each region, choose a test point and determine whether its value satisfies the *original inequality*.

Step 4. The solution set includes the regions whose test point value is a solution. If the inequality symbol is $\le$ or $\ge$, the values from Step 1 are solutions; if $<$ or $>$, they are not.

✔ **CONCEPT CHECK**
When choosing a test point in Step 3, why would the solutions from Step 2 not make good choices for test points?

EXAMPLE 2   Solve: $x^2 - 4x \le 0$

Solution First we solve the related equation, $x^2 - 4x = 0$.

$$x^2 - 4x = 0$$
$$x(x - 4) = 0$$
$$x = 0 \quad \text{or} \quad x = 4$$

The numbers 0 and 4 separate the number line into three regions, A, B, and C.

We check a test value in each region in the original inequality. Values in region B satisfy the inequality. The numbers 0 and 4 are included in the solution since the inequality symbol is $\le$. The solution set is $[0, 4]$, and its graph is shown.

PRACTICE
2   Solve: $x^2 - 8x \le 0$

EXAMPLE 3   Solve: $(x + 2)(x - 1)(x - 5) \le 0$

Solution First we solve $(x + 2)(x - 1)(x - 5) = 0$. By inspection, we see that the solutions are -2, 1, and 5. They separate the number line into four regions, A, B, C, and D. Next we check test points from each region.

Region	Test Point Value	$(x + 2)(x - 1)(x - 5) \le 0$	*Result*
A	-3	$(-1)(-4)(-8) \le 0$	True
B	0	$(2)(-1)(-5) \le 0$	False
C	2	$(4)(1)(-3) \le 0$	True
D	6	$(8)(5)(1) \le 0$	False

(Continued on next page)

Answer to Concept Check:
The solutions found in Step 2 have a value of 0 in the original inequality.

The solution set is $(-\infty, -2] \cup [1, 5]$, and its graph is shown. We include the numbers $-2, 1,$ and 5 because the inequality symbol is $\leq$.

$$
\begin{array}{ccccc}
 & A & B & C & D \\
\end{array}
$$

$$
\xleftarrow{\quad} \underset{\text{T}\ \ -2\ \ \text{F}\ \ 1\ \ \text{T}\ \ 5\ \ \text{F}}{\quad\quad\quad\quad\quad} \xrightarrow{\quad}
$$

PRACTICE
3 Solve: $(x + 3)(x - 2)(x + 1) \leq 0$

OBJECTIVE
2 Solving Rational Inequalities

Inequalities containing rational expressions with variables in the denominator are solved by using a similar procedure.

EXAMPLE 4 Solve: $\dfrac{x + 2}{x - 3} \leq 0$

Solution First we find all values that make the denominator equal to 0. To do this, we solve $x - 3 = 0$ and find that $x = 3$.

Next, we solve the related equation $\dfrac{x + 2}{x - 3} = 0$.

$$\frac{x + 2}{x - 3} = 0$$

$$x + 2 = 0 \qquad \text{Multiply both sides by the LCD, } x - 3.$$

$$x = -2$$

Now we place these numbers on a number line and proceed as before, checking test point values in the original inequality.

$$
\begin{array}{ccc}
A & B & C \\
\end{array}
$$

$$\xleftarrow{\quad} \underset{-2 \qquad\quad 3}{\quad\quad\quad\quad} \xrightarrow{\quad}$$

Choose -3 from region A.	**Choose 0 from region B.**	**Choose 4 from region C.**
$\dfrac{x + 2}{x - 3} \leq 0$	$\dfrac{x + 2}{x - 3} \leq 0$	$\dfrac{x + 2}{x - 3} \leq 0$
$\dfrac{-3 + 2}{-3 - 3} \leq 0$	$\dfrac{0 + 2}{0 - 3} \leq 0$	$\dfrac{4 + 2}{4 - 3} \leq 0$
$\dfrac{-1}{-6} \leq 0$	$-\dfrac{2}{3} \leq 0 \quad \text{True}$	$6 \leq 0 \quad \text{False}$
$\dfrac{1}{6} \leq 0 \quad \text{False}$		

The solution set is $[-2, 3)$. This interval includes -2 because -2 satisfies the original inequality. This interval does not include 3 because 3 would make the denominator 0.

$$
\begin{array}{ccc}
A & B & C \\
\end{array}
$$

$$\xleftarrow{\quad} \underset{\text{F}\ \ -2\ \ \text{T}\ \ 3\ \ \text{F}}{\quad\quad\quad\quad\quad} \xrightarrow{\quad}$$

PRACTICE
4 Solve: $\dfrac{x - 5}{x + 4} \leq 0$

The following steps may be used to solve a rational inequality with variables in the denominator.

Solving a Rational Inequality

Step 1. Solve for values that make all denominators 0.

Step 2. Solve the related equation.

Step 3. Separate the number line into regions with the solutions from Steps 1 and 2.

Step 4. For each region, choose a test point and determine whether its value satisfies the *original inequality*.

Step 5. The solution set includes the regions whose test point value is a solution. Check whether to include values from Step 2. Be sure *not* to include values that make any denominator 0.

EXAMPLE 5   Solve: $\dfrac{5}{x+1} < -2$

Solution First we find values for x that make the denominator equal to 0.

$$x + 1 = 0$$
$$x = -1$$

Next we solve $\dfrac{5}{x+1} = -2$.

$$(x+1) \cdot \dfrac{5}{x+1} = (x+1) \cdot -2 \quad \text{Multiply both sides by the LCD, } x+1.$$
$$5 = -2x - 2 \qquad \text{Simplify.}$$
$$7 = -2x$$
$$-\dfrac{7}{2} = x$$

We use these two solutions to divide a number line into three regions and choose test points. Only a test point value from region B satisfies the *original inequality*. The solution set is $\left(-\dfrac{7}{2}, -1 \right)$, and its graph is shown.

PRACTICE
5   Solve: $\dfrac{7}{x+3} < 5$

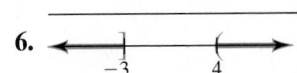

Vocabulary, Readiness & Video Check

Write the graphed solution set in interval notation.

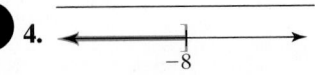

1.
2.
3.

4.
5.
6.

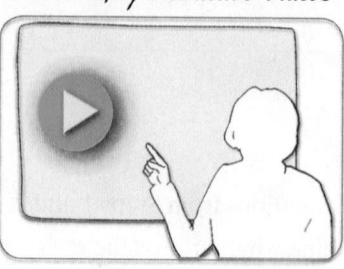

Martin-Gay Interactive Videos

See Video 11.4 ●

OBJECTIVE
1

7. From ▦ Examples 1–3, how does solving a related equation help you solve a polynomial inequality? Are the solutions to the related equation ever solutions to the inequality?

OBJECTIVE
2

8. In ▦ Example 4, one of the values that separates the number line into regions is 4. The inequality is ≥, so why isn't 4 included in the solution set?

11.4 Exercise Set MyMathLab® ▶

Solve each polynomial inequality. Write the solution set in interval notation. See Examples 1 through 3.

1. $(x + 1)(x + 5) > 0$

2. $(x + 1)(x + 5) \le 0$

▶ **3.** $(x - 3)(x + 4) \le 0$

4. $(x + 4)(x - 1) > 0$

5. $x^2 - 7x + 10 \le 0$

6. $x^2 + 8x + 15 \ge 0$

7. $3x^2 + 16x < -5$

8. $2x^2 - 5x < 7$

9. $(x - 6)(x - 4)(x - 2) > 0$

10. $(x - 6)(x - 4)(x - 2) \le 0$

11. $x(x - 1)(x + 4) \le 0$

12. $x(x - 6)(x + 2) > 0$

13. $(x^2 - 9)(x^2 - 4) > 0$

14. $(x^2 - 16)(x^2 - 1) \le 0$

Solve each inequality. Write the solution set in interval notation. See Example 4.

15. $\dfrac{x + 7}{x - 2} < 0$

16. $\dfrac{x - 5}{x - 6} > 0$

17. $\dfrac{5}{x + 1} > 0$

18. $\dfrac{3}{y - 5} < 0$

▶ **19.** $\dfrac{x + 1}{x - 4} \ge 0$

20. $\dfrac{x + 1}{x - 4} \le 0$

Solve each inequality. Write the solution set in interval notation. See Example 5.

21. $\dfrac{3}{x - 2} < 4$

22. $\dfrac{-2}{y + 3} > 2$

23. $\dfrac{x^2 + 6}{5x} \ge 1$

24. $\dfrac{y^2 + 15}{8y} \le 1$

25. $\dfrac{x + 2}{x - 3} < 1$

26. $\dfrac{x - 1}{x + 4} > 2$

MIXED PRACTICE

Solve each inequality. Write the solution set in interval notation.

27. $(2x - 3)(4x + 5) \le 0$

28. $(6x + 7)(7x - 12) > 0$

▶ **29.** $x^2 > x$

30. $x^2 < 25$

31. $(2x - 8)(x + 4)(x - 6) \le 0$

32. $(3x - 12)(x + 5)(2x - 3) \ge 0$

33. $6x^2 - 5x \ge 6$

34. $12x^2 + 11x \le 15$

35. $4x^3 + 16x^2 - 9x - 36 > 0$

36. $x^3 + 2x^2 - 4x - 8 < 0$

▶ **37.** $x^4 - 26x^2 + 25 \ge 0$

38. $16x^4 - 40x^2 + 9 \le 0$

39. $(2x - 7)(3x + 5) > 0$

40. $(4x - 9)(2x + 5) < 0$

41. $\dfrac{x}{x - 10} < 0$

42. $\dfrac{x + 10}{x - 10} > 0$

43. $\dfrac{x - 5}{x + 4} \geq 0$

44. $\dfrac{x - 3}{x + 2} \leq 0$

45. $\dfrac{x(x + 6)}{(x - 7)(x + 1)} \geq 0$

46. $\dfrac{(x - 2)(x + 2)}{(x + 1)(x - 4)} \leq 0$

47. $\dfrac{-1}{x - 1} > -1$

48. $\dfrac{4}{y + 2} < -2$

49. $\dfrac{x}{x + 4} \leq 2$

50. $\dfrac{4x}{x - 3} \geq 5$

51. $\dfrac{z}{z - 5} \geq 2z$

52. $\dfrac{p}{p + 4} \leq 3p$

53. $\dfrac{(x + 1)^2}{5x} > 0$

54. $\dfrac{(2x - 3)^2}{x} < 0$

REVIEW AND PREVIEW

Recall that the graph of $f(x) + K$ is the same as the graph of $f(x)$ shifted K units upward if $K > 0$ and $|K|$ units downward if $K < 0$. Use the graph of $f(x) = |x|$ below to sketch the graph of each function. See Section 8.3.

55. $g(x) = |x| + 2$

56. $H(x) = |x| - 2$

57. $G(x) = |x| - 1$

58. $h(x) = |x| + 5$

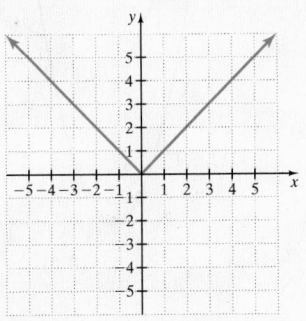

Use the graph of $f(x) = x^2$ below to sketch the graph of each function.

59. $F(x) = x^2 - 3$

60. $h(x) = x^2 - 4$

61. $H(x) = x^2 + 1$

62. $g(x) = x^2 + 3$

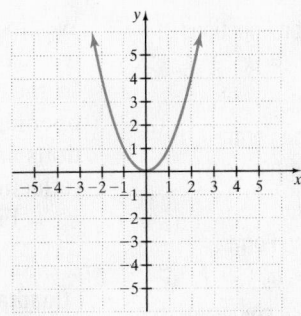

CONCEPT EXTENSIONS

63. Explain why $\dfrac{x + 2}{x - 3} > 0$ and $(x + 2)(x - 3) > 0$ have the same solutions.

64. Explain why $\dfrac{x + 2}{x - 3} \geq 0$ and $(x + 2)(x - 3) \geq 0$ do not have the same solutions.

Find all numbers that satisfy each of the following.

65. A number minus its reciprocal is less than zero. Find the numbers.

66. Twice a number added to its reciprocal is nonnegative. Find the numbers.

67. The total profit function $P(x)$ for a company producing x thousand units is given by

$$P(x) = -2x^2 + 26x - 44$$

Find the values of x for which the company makes a profit. (*Hint:* The company makes a profit when $P(x) > 0$.)

68. A projectile is fired straight up from the ground with an initial velocity of 80 feet per second. Its height $s(t)$ in feet at any time t is given by the function

$$s(t) = -16t^2 + 80t$$

Find the interval of time for which the height of the projectile is greater than 96 feet.

Use a graphing calculator to check each exercise.

69. Exercise 37

70. Exercise 38

71. Exercise 39

72. Exercise 40

11.5 Quadratic Functions and Their Graphs

OBJECTIVES

1 Graph Quadratic Functions of the Form $f(x) = x^2 + k$.

2 Graph Quadratic Functions of the Form $f(x) = (x - h)^2$.

3 Graph Quadratic Functions of the Form $f(x) = (x - h)^2 + k$.

4 Graph Quadratic Functions of the Form $f(x) = ax^2$.

5 Graph Quadratic Functions of the Form $f(x) = a(x - h)^2 + k$.

OBJECTIVE

1 Graphing $f(x) = x^2 + k$

We first graphed the quadratic equation $y = x^2$ in Section 8.2. In Sections 8.2 and 8.3, we learned that this graph defines a function, and we wrote $y = x^2$ as $f(x) = x^2$. In those sections, we discovered that the graph of a quadratic function is a parabola opening upward or downward. In this section, we continue our study of quadratic functions and their graphs. (Much of the contents of this section is a review of shifting and reflecting techniques from Section 8.3, but specific to quadratic functions.)

First, let's recall the definition of a quadratic function.

Quadratic Function

A quadratic function is a function that can be written in the form $f(x) = ax^2 + bx + c$, where $a, b,$ and c are real numbers and $a \neq 0$.

Notice that equations of the form $y = ax^2 + bx + c$, where $a \neq 0$, define quadratic functions, since y is a function of x or $y = f(x)$.

Recall that if $a > 0$, the parabola opens upward and if $a < 0$, the parabola opens downward. Also, the vertex of a parabola is the lowest point if the parabola opens upward and the highest point if the parabola opens downward. The axis of symmetry is the vertical line that passes through the vertex.

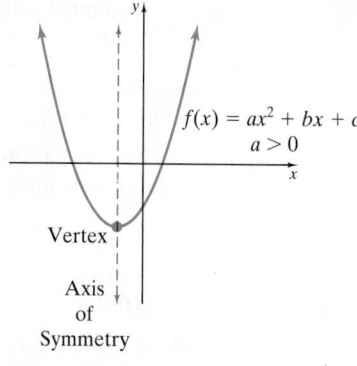
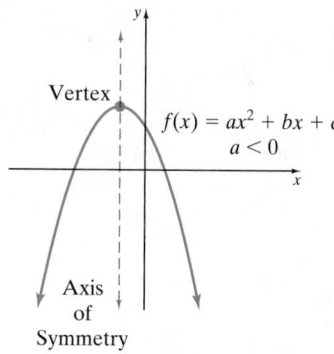

EXAMPLE 1 Graph $f(x) = x^2$ and $g(x) = x^2 + 6$ on the same set of axes.

Solution First we construct a table of values for $f(x)$ and plot the points. Notice that for each x-value, the corresponding value of $g(x)$ must be 6 more than the corresponding value of $f(x)$ since $f(x) = x^2$ and $g(x) = x^2 + 6$. In other words, the graph of $g(x) = x^2 + 6$ is the same as the graph of $f(x) = x^2$ shifted upward 6 units. The axis of symmetry for both graphs is the y-axis.

x	$f(x) = x^2$	$g(x) = x^2 + 6$
-2	4	10
-1	1	7
0	0	6
1	1	7
2	4	10

Each y-value is increased by 6.

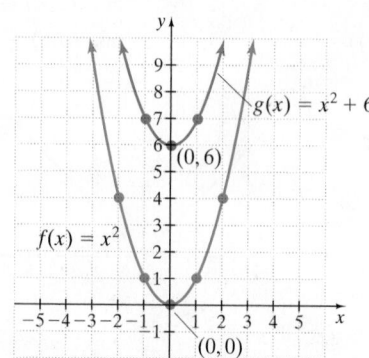

PRACTICE

1 Graph $f(x) = x^2$ and $g(x) = x^2 - 4$ on the same set of axes.

In general, we have the following properties.

> **Graphing the Parabola Defined by $f(x) = x^2 + k$**
>
> If k is positive, the graph of $f(x) = x^2 + k$ is the graph of $y = x^2$ shifted upward k units.
>
> If k is negative, the graph of $f(x) = x^2 + k$ is the graph of $y = x^2$ shifted downward $|k|$ units.
>
> The vertex is $(0, k)$, and the axis of symmetry is the y-axis.

EXAMPLE 2   Graph each function.

    a. $F(x) = x^2 + 2$ **b.** $g(x) = x^2 - 3$

Solution

  a. $F(x) = x^2 + 2$

    The graph of $F(x) = x^2 + 2$ is obtained by shifting the graph of $y = x^2$ upward 2 units.

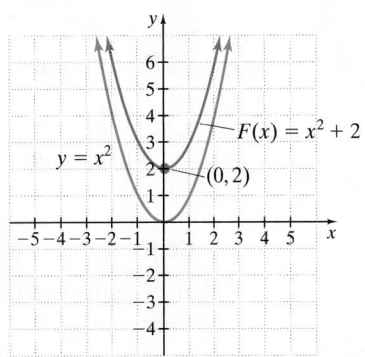

  b. $g(x) = x^2 - 3$

    The graph of $g(x) = x^2 - 3$ is obtained by shifting the graph of $y = x^2$ downward 3 units.

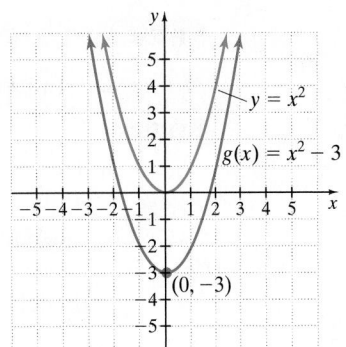

PRACTICE

  2   Graph each function.

    a. $f(x) = x^2 - 5$ **b.** $g(x) = x^2 + 3$

OBJECTIVE

  2   Graphing $f(x) = (x - h)^2$ ▶

Now we will graph functions of the form $f(x) = (x - h)^2$.

EXAMPLE 3 Graph $f(x) = x^2$ and $g(x) = (x - 2)^2$ on the same set of axes.

<u>Solution</u> By plotting points, we see that for each x-value, the corresponding value of $g(x)$ is the same as the value of $f(x)$ when the x-value is increased by 2. Thus, the graph of $g(x) = (x - 2)^2$ is the graph of $f(x) = x^2$ shifted to the right 2 units. The axis of symmetry for the graph of $g(x) = (x - 2)^2$ is also shifted 2 units to the right and is the line $x = 2$.

x	$f(x) = x^2$	x	$g(x) = (x - 2)^2$
-2	4	0	4
-1	1	1	1
0	0	2	0
1	1	3	1
2	4	4	4

Each x-value increased by 2 corresponds to same y-value.

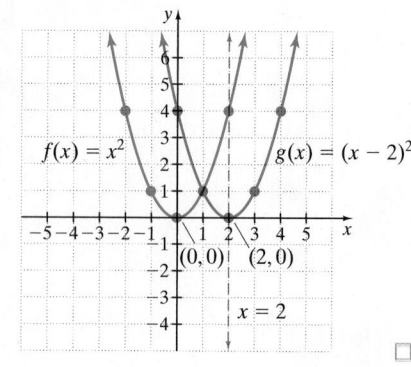

PRACTICE
3 Graph $f(x) = x^2$ and $g(x) = (x + 6)^2$ on the same set of axes.

In general, we have the following properties.

Graphing the Parabola Defined by $f(x) = (x - h)^2$

If h is positive, the graph of $f(x) = (x - h)^2$ is the graph of $y = x^2$ shifted to the right h units.
If h is negative, the graph of $f(x) = (x - h)^2$ is the graph of $y = x^2$ shifted to the left $|h|$ units.
The vertex is $(h, 0)$, and the axis of symmetry is the vertical line $x = h$.

EXAMPLE 4 Graph each function.

    a. $G(x) = (x - 3)^2$ **b.** $F(x) = (x + 1)^2$

<u>Solution</u>

    a. The graph of $G(x) = (x - 3)^2$ is obtained by shifting the graph of $y = x^2$ to the right 3 units. The graph of $G(x)$ is below on the left.

    b. The equation $F(x) = (x + 1)^2$ can be written as $F(x) = [x - (-1)]^2$. The graph of $F(x) = [x - (-1)]^2$ is obtained by shifting the graph of $y = x^2$ to the left 1 unit. The graph of $F(x)$ is below on the right.

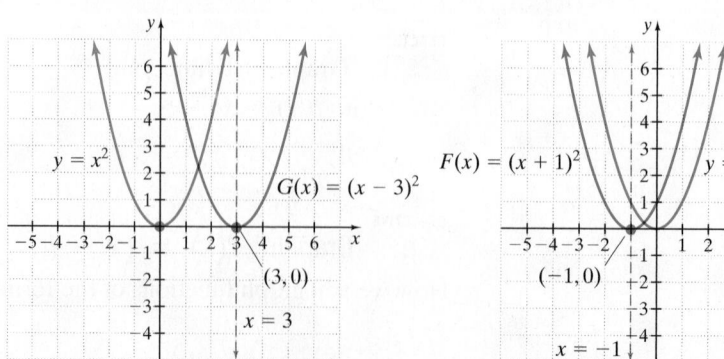

PRACTICE
4   Graph each function.

a. $G(x) = (x + 4)^2$ **b.** $H(x) = (x - 7)^2$ ■

OBJECTIVE
3   Graphing $f(x) = (x - h)^2 + k$ ▶

As we will see in graphing functions of the form $f(x) = (x - h)^2 + k$, it is possible to combine vertical and horizontal shifts.

> **Graphing the Parabola Defined by $f(x) = (x - h)^2 + k$**
>
> The parabola has the same shape as $y = x^2$.
> The vertex is (h, k), and the axis of symmetry is the vertical line $x = h$.

EXAMPLE 5   Graph: $F(x) = (x - 3)^2 + 1$

Solution The graph of $F(x) = (x - 3)^2 + 1$ is the graph of $y = x^2$ shifted 3 units to the right and 1 unit up. The vertex is then $(3, 1)$, and the axis of symmetry is $x = 3$. A few ordered pair solutions are plotted to aid in graphing.

x	$F(x) = (x - 3)^2 + 1$
1	5
2	2
4	2
5	5

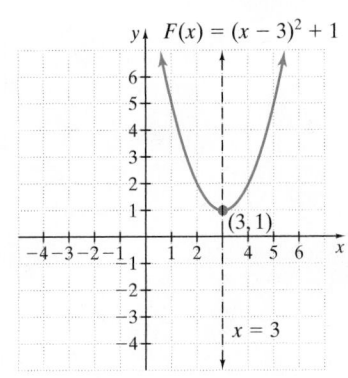

PRACTICE
5   Graph: $f(x) = (x + 2)^2 + 2$ ■

OBJECTIVE
4   Graphing $f(x) = ax^2$ ▶

Next, we discover the change in the shape of the graph when the coefficient of x^2 is not 1.

EXAMPLE 6   Graph $f(x) = x^2$, $g(x) = 3x^2$, and $h(x) = \dfrac{1}{2}x^2$ on the same set of axes.

Solution Comparing the tables of values, we see that for each x-value, the corresponding value of $g(x)$ is triple the corresponding value of $f(x)$. Similarly, the value of $h(x)$ is half the value of $f(x)$.

x	$f(x) = x^2$
−2	4
−1	1
0	0
1	1
2	4

x	$g(x) = 3x^2$
−2	12
−1	3
0	0
1	3
2	12

x	$h(x) = \dfrac{1}{2}x^2$
−2	2
−1	$\dfrac{1}{2}$
0	0
1	$\dfrac{1}{2}$
2	2

(Continued on next page)

The result is that the graph of $g(x) = 3x^2$ is narrower than the graph of $f(x) = x^2$, and the graph of $h(x) = \dfrac{1}{2}x^2$ is wider. The vertex for each graph is $(0, 0)$, and the axis of symmetry is the y-axis.

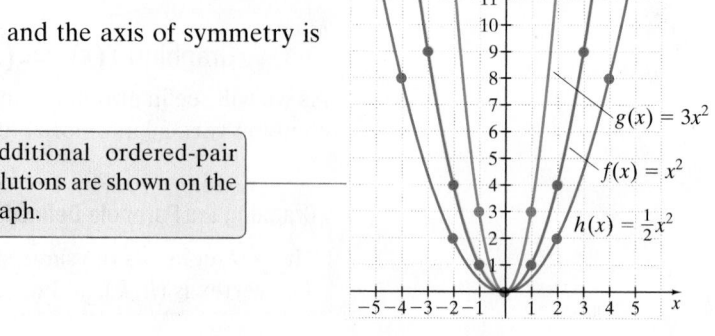

Additional ordered-pair solutions are shown on the graph.

PRACTICE

6 Graph $f(x) = x^2$, $g(x) = 4x^2$, and $h(x) = \dfrac{1}{4}x^2$ on the same set of axes. ■

Graphing the Parabola Defined by $f(x) = ax^2$

If a is positive, the parabola opens upward, and if a is negative, the parabola opens downward.
If $|a| > 1$, the graph of the parabola is narrower than the graph of $y = x^2$.
If $|a| < 1$, the graph of the parabola is wider than the graph of $y = x^2$.

EXAMPLE 7 Graph: $f(x) = -2x^2$

Solution Because $a = -2$, a negative value, this parabola opens downward. Since $|-2| = 2$ and $2 > 1$, the parabola is narrower than the graph of $y = x^2$. The vertex is $(0, 0)$, and the axis of symmetry is the y-axis. We verify this by plotting a few points.

x	$f(x) = -2x^2$
-2	-8
-1	-2
0	0
1	-2
2	-8

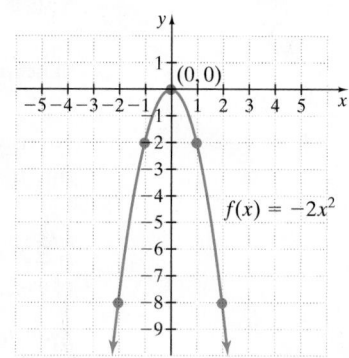

PRACTICE

7 Graph: $f(x) = -\dfrac{1}{2}x^2$ ■

OBJECTIVE

5 Graphing $f(x) = a(x - h)^2 + k$ ▶

Now we will see the shape of the graph of a quadratic function of the form $f(x) = a(x - h)^2 + k$.

EXAMPLE 8   Graph $g(x) = \dfrac{1}{2}(x + 2)^2 + 5$. Find the vertex and the axis of symmetry.

Solution   The function $g(x) = \dfrac{1}{2}(x + 2)^2 + 5$ may be written as $g(x) = \dfrac{1}{2}[x - (-2)]^2 + 5$. Thus, this graph is the same as the graph of $y = x^2$ shifted 2 units to the left and 5 units up, and it is wider because a is $\dfrac{1}{2}$. The vertex is $(-2, 5)$, and the axis of symmetry is $x = -2$. We plot a few points to verify.

x	$g(x) = \dfrac{1}{2}(x + 2)^2 + 5$
-4	7
-3	$5\dfrac{1}{2}$
-2	5
-1	$5\dfrac{1}{2}$
0	7

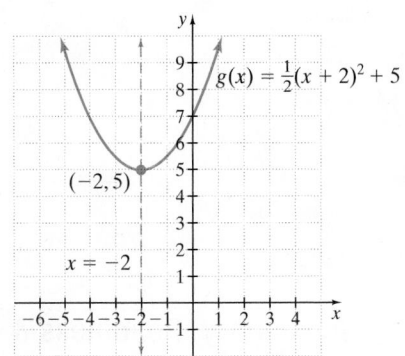

PRACTICE

8   Graph $h(x) = \dfrac{1}{3}(x - 4)^2 - 3$.

In general, the following holds.

Graph of a Quadratic Function

The graph of a quadratic function written in the form $f(x) = a(x - h)^2 + k$ is a parabola with vertex (h, k).

If $a > 0$, the parabola opens upward.
If $a < 0$, the parabola opens downward.

The axis of symmetry is the line whose equation is $x = h$.

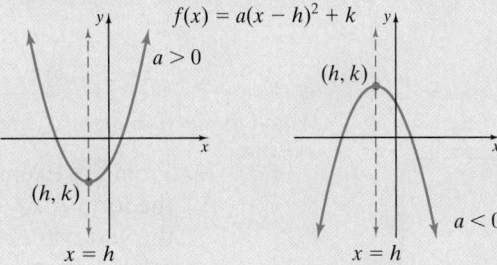

✔ CONCEPT CHECK
Which description of the graph of $f(x) = -0.35(x + 3)^2 - 4$ is correct?
a. The graph opens downward and has its vertex at $(-3, 4)$.
b. The graph opens upward and has its vertex at $(-3, 4)$.
c. The graph opens downward and has its vertex at $(-3, -4)$.
d. The graph is narrower than the graph of $y = x^2$.

Graphing Calculator Explorations

Use a graphing calculator to graph the first function of each pair that follows. Then use its graph to predict the graph of the second function. Check your prediction by graphing both on the same set of axes.

1. $F(x) = \sqrt{x}; G(x) = \sqrt{x} + 1$

2. $g(x) = x^3; H(x) = x^3 - 2$

3. $H(x) = |x|; f(x) = |x - 5|$

4. $h(x) = x^3 + 2; g(x) = (x - 3)^3 + 2$

5. $f(x) = |x + 4|; F(x) = |x + 4| + 3$

6. $G(x) = \sqrt{x} - 2; g(x) = \sqrt{x - 4} - 2$

✔ Vocabulary, Readiness & Video Check

Use the choices below to fill in each blank. Some choices will be used more than once.

upward   highest   parabola   downward   lowest   quadratic

1. A(n) _____ function is one that can be written in the form $f(x) = ax^2 + bx + c, a \neq 0$.

2. The graph of a quadratic function is a(n) _____ opening _____ or _____.

3. If $a > 0$, the graph of the quadratic function opens _____.

4. If $a < 0$, the graph of the quadratic function opens _____.

5. The vertex of a parabola is the _____ point if $a > 0$.

6. The vertex of a parabola is the _____ point if $a < 0$.

State the vertex of the graph of each quadratic function.

7. $f(x) = x^2$

8. $f(x) = -5x^2$

9. $g(x) = (x - 2)^2$

10. $g(x) = (x + 5)^2$

11. $f(x) = 2x^2 + 3$

12. $h(x) = x^2 - 1$

13. $g(x) = (x + 1)^2 + 5$

14. $h(x) = (x - 10)^2 - 7$

Martin-Gay Interactive Videos

See Video 11.5

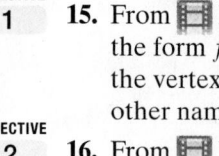

Watch the section lecture video and answer the following questions.

OBJECTIVE 1

15. From ▦ Examples 1 and 2 and the lecture before, how do graphs of the form $f(x) = x^2 + k$ differ from $y = x^2$? Consider the location of the vertex $(0, k)$ on these graphs of the form $f(x) = x^2 + k$—by what other name do we call this point on a graph?

OBJECTIVE 2

16. From ▦ Example 3 and the lecture before, how do graphs of the form $f(x) = (x - h)^2$ differ from $y = x^2$? Consider the location of the vertex $(h, 0)$ on these graphs of the form $f(x) = (x - h)^2$—by what other name do we call this point on a graph?

OBJECTIVE 3

17. From ▦ Example 4 and the lecture before, what general information does the equation $f(x) = (x - h)^2 + k$ tell us about its graph?

OBJECTIVE 4

18. From the lecture before ▦ Example 5, besides the direction a parabola opens, what other graphing information can the value of a tell us?

OBJECTIVE 5

19. In ▦ Examples 6 and 7, what four properties of the graph did we learn from the equation that helped us locate and draw the general shape of the parabola?

11.5 Exercise Set   MyMathLab® ▶

MIXED PRACTICE

Sketch the graph of each quadratic function. Label the vertex and sketch and label the axis of symmetry. See Examples 1 through 5.

▶ **1.** $f(x) = x^2 - 1$

2. $g(x) = x^2 + 3$

▶ **3.** $h(x) = x^2 + 5$

4. $h(x) = x^2 - 4$

5. $g(x) = x^2 + 7$

6. $f(x) = x^2 - 2$

7. $f(x) = (x - 5)^2$

8. $g(x) = (x + 5)^2$

▶ **9.** $h(x) = (x + 2)^2$

10. $H(x) = (x - 1)^2$

11. $G(x) = (x + 3)^2$

12. $f(x) = (x - 6)^2$

▶ **13.** $f(x) = (x - 2)^2 + 5$

14. $g(x) = (x - 6)^2 + 1$

15. $h(x) = (x + 1)^2 + 4$

16. $G(x) = (x + 3)^2 + 3$

17. $g(x) = (x + 2)^2 - 5$

18. $h(x) = (x + 4)^2 - 6$

Sketch the graph of each quadratic function. Label the vertex and sketch and label the axis of symmetry. See Examples 6 and 7.

19. $H(x) = 2x^2$

20. $f(x) = 5x^2$

21. $h(x) = \frac{1}{3}x^2$

22. $f(x) = -\frac{1}{4}x^2$

▶ **23.** $g(x) = -x^2$

24. $g(x) = -3x^2$

Sketch the graph of each quadratic function. Label the vertex and sketch and label the axis of symmetry. See Example 8.

25. $f(x) = 2(x - 1)^2 + 3$

26. $g(x) = 4(x - 4)^2 + 2$

▶ **27.** $h(x) = -3(x + 3)^2 + 1$

28. $f(x) = -(x - 2)^2 - 6$

29. $H(x) = \frac{1}{2}(x - 6)^2 - 3$

30. $G(x) = \frac{1}{5}(x + 4)^2 + 3$

MIXED PRACTICE

Sketch the graph of each quadratic function. Label the vertex and sketch and label the axis of symmetry.

31. $f(x) = -(x - 2)^2$

32. $g(x) = -(x + 6)^2$

33. $F(x) = -x^2 + 4$

34. $H(x) = -x^2 + 10$

35. $F(x) = 2x^2 - 5$

36. $g(x) = \frac{1}{2}x^2 - 2$

37. $h(x) = (x - 6)^2 + 4$

38. $f(x) = (x - 5)^2 + 2$

39. $F(x) = \left(x + \frac{1}{2}\right)^2 - 2$

40. $H(x) = \left(x + \frac{1}{2}\right)^2 - 3$

41. $F(x) = \frac{3}{2}(x + 7)^2 + 1$

42. $g(x) = -\frac{3}{2}(x - 1)^2 - 5$

43. $f(x) = \frac{1}{4}x^2 - 9$

44. $H(x) = \frac{3}{4}x^2 - 2$

45. $G(x) = 5\left(x + \frac{1}{2}\right)^2$

46. $F(x) = 3\left(x - \frac{3}{2}\right)^2$

47. $h(x) = -(x - 1)^2 - 1$

48. $f(x) = -3(x + 2)^2 + 2$

49. $g(x) = \sqrt{3}(x + 5)^2 + \frac{3}{4}$

50. $G(x) = \sqrt{5}(x - 7)^2 - \frac{1}{2}$

▶ **51.** $h(x) = 10(x + 4)^2 - 6$

52. $h(x) = 8(x + 1)^2 + 9$

53. $f(x) = -2(x - 4)^2 + 5$

54. $G(x) = -4(x + 9)^2 - 1$

REVIEW AND PREVIEW

Add the proper constant to each binomial so that the resulting trinomial is a perfect square trinomial. See Section 11.1.

55. $x^2 + 8x$

56. $y^2 + 4y$

57. $z^2 - 16z$

58. $x^2 - 10x$

59. $y^2 + y$

60. $z^2 - 3z$

Solve by completing the square. See Section 11.1.

61. $x^2 + 4x = 12$

62. $y^2 + 6y = -5$

63. $z^2 + 10z - 1 = 0$

64. $x^2 + 14x + 20 = 0$

65. $z^2 - 8z = 2$

66. $y^2 - 10y = 3$

CONCEPT EXTENSIONS

Solve. See the Concept Check in this section.

67. Which description of $f(x) = -213(x - 0.1)^2 + 3.6$ is correct?

Graph Opens	Vertex
a. upward	$(0.1, 3.6)$
b. upward	$(-213, 3.6)$
c. downward	$(0.1, 3.6)$
d. downward	$(-0.1, 3.6)$

68. Which description of $f(x) = 5\left(x + \frac{1}{2}\right)^2 + \frac{1}{2}$ is correct?

Graph Opens	Vertex
a. upward	$\left(\frac{1}{2}, \frac{1}{2}\right)$
b. upward	$\left(-\frac{1}{2}, \frac{1}{2}\right)$
c. downward	$\left(\frac{1}{2}, -\frac{1}{2}\right)$
d. downward	$\left(-\frac{1}{2}, -\frac{1}{2}\right)$

Write the equation of the parabola that has the same shape as $f(x) = 5x^2$ but with the following vertex.

69. $(2, 3)$

70. $(1, 6)$

71. $(-3, 6)$

72. $(4, -1)$

The shifting properties covered in this section apply to the graphs of all functions. Given the graph of $y = f(x)$ below, sketch the graph of each of the following.

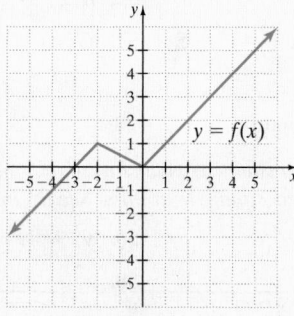

73. $y = f(x) + 1$

74. $y = f(x) - 2$

75. $y = f(x - 3)$

76. $y = f(x + 3)$

77. $y = f(x + 2) + 2$

78. $y = f(x - 1) + 1$

79. The quadratic function $f(x) = 12.5x^2 - 64x + 135$ approximates the annual retail sales from online shopping in the United States from 2005 through 2014, where x is the number of years past 2005 and $f(x)$ is the annual retail sales in billions of dollars. (*Source:* Based on data from the U.S. Bureau of the Census and the Retail Association of America)

  a. Use this function to find the online retail sales in the United States for 2013.

  b. Use this function to predict the online retail sales in the United States for 2020.

80. Use the function in Exercise 79.

  a. Use this function to predict the online retail sales in the United States for 2018.

  b. Look up the annual online retail sales for the United States for the current year.

  c. Based on your answers for parts **a** and **b**, discuss some possible limitations of using this quadratic function to predict data.

11.6 Further Graphing of Quadratic Functions

OBJECTIVES

1 Write Quadratic Functions in the Form $y = a(x - h)^2 + k$.

2 Derive a Formula for Finding the Vertex of a Parabola.

3 Find the Minimum or Maximum Value of a Quadratic Function.

OBJECTIVE

1   Writing Quadratic Functions in the Form $y = a(x - h)^2 + k$

We know that the graph of a quadratic function is a parabola. If a quadratic function is written in the form

$$f(x) = a(x - h)^2 + k$$

we can easily find the vertex (h, k) and graph the parabola. To write a quadratic function in this form, complete the square. (See Section 11.1. for a review of completing the square.)

EXAMPLE 1   Graph $f(x) = x^2 - 4x - 12$. Find the vertex and any intercepts.

<u>Solution</u> The graph of this quadratic function is a parabola. To find the vertex of the parabola, we will write the function in the form $y = (x - h)^2 + k$. To do this, we complete the square on the binomial $x^2 - 4x$. To simplify our work, we let $f(x) = y$.

$$y = x^2 - 4x - 12 \quad \text{Let } f(x) = y.$$

$$y + 12 = x^2 - 4x \qquad \text{Add 12 to both sides to get the } x\text{-variable terms alone.}$$

Now we add the square of half of -4 to both sides.

$$\frac{1}{2}(-4) = -2 \quad \text{and} \quad (-2)^2 = 4$$

$$y + 12 + 4 = x^2 - 4x + 4 \qquad \text{Add 4 to both sides.}$$

$$y + 16 = (x - 2)^2 \qquad \text{Factor the trinomial.}$$

$$y = (x - 2)^2 - 16 \quad \text{Subtract 16 from both sides.}$$

$$f(x) = (x - 2)^2 - 16 \quad \text{Replace } y \text{ with } f(x).$$

From this equation, we can see that the vertex of the parabola is $(2, -16)$, a point in quadrant IV, and the axis of symmetry is the line $x = 2$.

Notice that $a = 1$. Since $a > 0$, the parabola opens upward. This parabola opening upward with vertex $(2, -16)$ will have two x-intercepts and one y-intercept. (See the Helpful Hint after this example.)

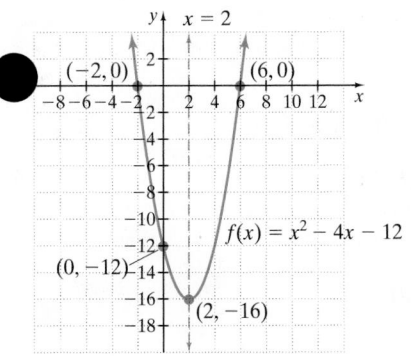

$f(x) = x^2 - 4x - 12$

x-intercepts: let y or $f(x) = 0$

$$f(x) = x^2 - 4x - 12$$
$$0 = x^2 - 4x - 12$$
$$0 = (x - 6)(x + 2)$$
$$0 = x - 6 \quad \text{or} \quad 0 = x + 2$$
$$6 = x \qquad\qquad -2 = x$$

y-intercept: let $x = 0$

$$f(x) = x^2 - 4x - 12$$
$$f(0) = 0^2 - 4 \cdot 0 - 12$$
$$= -12$$

The two x-intercepts are $(6, 0)$ and $(-2, 0)$. The y-intercept is $(0, -12)$. The sketch of $f(x) = x^2 - 4x - 12$ is shown.

Notice that the axis of symmetry is always halfway between the x-intercepts. For this example, halfway between -2 and 6 is $\dfrac{-2 + 6}{2} = 2$, and the axis of symmetry is $x = 2$.   □

PRACTICE

1   Graph $g(x) = x^2 - 2x - 3$. Find the vertex and any intercepts.   ■

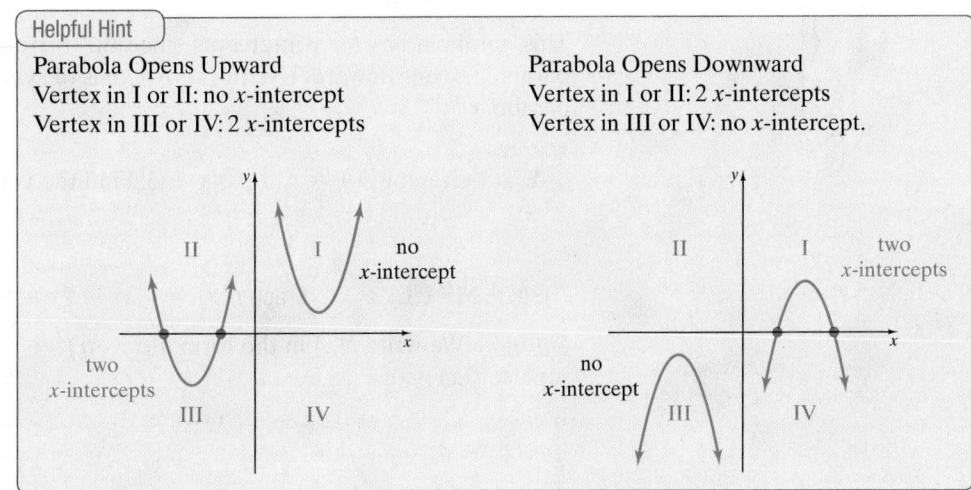

Helpful Hint

Parabola Opens Upward
Vertex in I or II: no x-intercept
Vertex in III or IV: 2 x-intercepts

Parabola Opens Downward
Vertex in I or II: 2 x-intercepts
Vertex in III or IV: no x-intercept.

EXAMPLE 2   Graph $f(x) = 3x^2 + 3x + 1$. Find the vertex and any intercepts.

Solution   Replace $f(x)$ with y and complete the square on x to write the equation in the form $y = a(x - h)^2 + k$.

$$y = 3x^2 + 3x + 1 \quad \text{Replace } f(x) \text{ with } y.$$
$$y - 1 = 3x^2 + 3x \qquad \text{Isolate } x\text{-variable terms.}$$

Factor 3 from the terms $3x^2 + 3x$ so that the coefficient of x^2 is 1.

$$y - 1 = 3(x^2 + x) \quad \text{Factor out 3.}$$

The coefficient of x in the parentheses above is 1. Then $\dfrac{1}{2}(1) = \dfrac{1}{2}$ and $\left(\dfrac{1}{2}\right)^2 = \dfrac{1}{4}$.

Since we are adding $\dfrac{1}{4}$ inside the parentheses, we are really adding $3\left(\dfrac{1}{4}\right)$, so we *must* add $3\left(\dfrac{1}{4}\right)$ to the left side.

(Continued on next page)

$$y - 1 + 3\left(\frac{1}{4}\right) = 3\left(x^2 + x + \frac{1}{4}\right)$$

$$y - \frac{1}{4} = 3\left(x + \frac{1}{2}\right)^2 \qquad \text{Simplify the left side and factor the right side.}$$

$$y = 3\left(x + \frac{1}{2}\right)^2 + \frac{1}{4} \qquad \text{Add } \frac{1}{4} \text{ to both sides.}$$

$$f(x) = 3\left(x + \frac{1}{2}\right)^2 + \frac{1}{4} \qquad \text{Replace } y \text{ with } f(x).$$

Then $a = 3, h = -\frac{1}{2}$, and $k = \frac{1}{4}$. This means that the parabola opens upward with vertex $\left(-\frac{1}{2}, \frac{1}{4}\right)$ and that the axis of symmetry is the line $x = -\frac{1}{2}$.

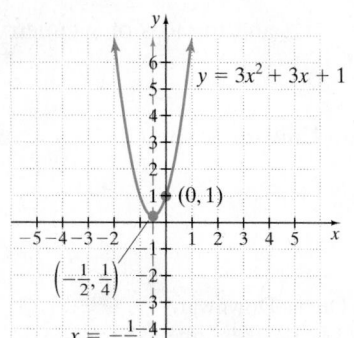

To find the y-intercept, let $x = 0$. Then

$$f(0) = 3(0)^2 + 3(0) + 1 = 1$$

Thus, the y-intercept is $(0, 1)$.

This parabola has no x-intercepts since the vertex is in the second quadrant and the parabola opens upward. Use the vertex, axis of symmetry, and y-intercept to sketch the parabola.   □

PRACTICE
2   Graph $g(x) = 4x^2 + 4x + 3$. Find the vertex and any intercepts.   ■

EXAMPLE 3   Graph $f(x) = -x^2 - 2x + 3$. Find the vertex and any intercepts.

Solution We write $f(x)$ in the form $a(x - h)^2 + k$ by completing the square. First we replace $f(x)$ with y.

$$f(x) = -x^2 - 2x + 3$$

$$y = -x^2 - 2x + 3$$

$$y - 3 = -x^2 - 2x \qquad \begin{array}{l}\text{Subtract 3 from both sides to get}\\ \text{the } x\text{-variable terms alone.}\end{array}$$

$$y - 3 = -1(x^2 + 2x) \qquad \text{Factor } -1 \text{ from the terms } -x^2 - 2x.$$

The coefficient of x is 2. Then $\frac{1}{2}(2) = 1$ and $1^2 = 1$. We add 1 to the right side inside the parentheses and add $-1(1)$ to the left side.

$$y - 3 - 1(1) = -1(x^2 + 2x + 1)$$

$$y - 4 = -1(x + 1)^2 \qquad \begin{array}{l}\text{Simplify the left side and}\\ \text{factor the right side.}\end{array}$$

$$y = -1(x + 1)^2 + 4 \qquad \text{Add 4 to both sides.}$$

$$\underline{f(x) = -1(x + 1)^2 + 4} \qquad \text{Replace } y \text{ with } f(x).$$

> **Helpful Hint**
> This can be written as
> $f(x) = -1[x - (-1)]^2 + 4.$
> Notice that the vertex is $(-1, 4)$.

Since $a = -1$, the parabola opens downward with vertex $(-1, 4)$ and axis of symmetry $x = -1$.

To find the y-intercept, we let $x = 0$. Then

$$f(0) = -0^2 - 2(0) + 3 = 3$$

Thus, $(0, 3)$ is the y-intercept.

To find the x-intercepts, we let y or $f(x) = 0$ and solve for x.

$$f(x) = -x^2 - 2x + 3$$

$$0 = -x^2 - 2x + 3 \qquad \text{Let } f(x) = 0.$$

Now we divide both sides by -1 so that the coefficient of x^2 is 1.

$$\frac{0}{-1} = \frac{-x^2}{-1} - \frac{2x}{-1} + \frac{3}{-1}$$ Divide both sides by -1.

$$0 = x^2 + 2x - 3$$ Simplify.

$$0 = (x + 3)(x - 1)$$ Factor.

$$x + 3 = 0 \quad \text{or} \quad x - 1 = 0$$ Set each factor equal to 0.

$$x = -3 \qquad\qquad x = 1$$ Solve.

The x-intercepts are $(-3, 0)$ and $(1, 0)$. Use these points to sketch the parabola. □

PRACTICE

3 Graph $g(x) = -x^2 + 5x + 6$. Find the vertex and any intercepts. ■

OBJECTIVE

2 Deriving a Formula for Finding the Vertex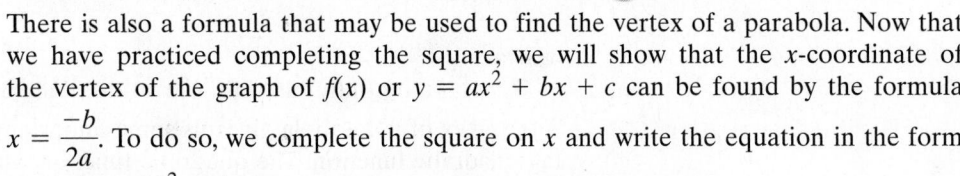

There is also a formula that may be used to find the vertex of a parabola. Now that we have practiced completing the square, we will show that the x-coordinate of the vertex of the graph of $f(x)$ or $y = ax^2 + bx + c$ can be found by the formula $x = \frac{-b}{2a}$. To do so, we complete the square on x and write the equation in the form $y = a(x - h)^2 + k$.

First, isolate the x-variable terms by subtracting c from both sides.

$$y = ax^2 + bx + c$$
$$y - c = ax^2 + bx$$

Next, factor a from the terms $ax^2 + bx$.

$$y - c = a\left(x^2 + \frac{b}{a}x\right)$$

Next, add the square of half of $\frac{b}{a}$, or $\left(\frac{b}{2a}\right)^2 = \frac{b^2}{4a^2}$, to the right side inside the parentheses. Because of the factor a, what we really added was $a\left(\frac{b^2}{4a^2}\right)$, and this must be added to the left side.

$$y - c + a\left(\frac{b^2}{4a^2}\right) = a\left(x^2 + \frac{b}{a}x + \frac{b^2}{4a^2}\right)$$

$$y - c + \frac{b^2}{4a} = a\left(x + \frac{b}{2a}\right)^2$$ Simplify the left side and factor the right side.

$$y = a\left(x + \frac{b}{2a}\right)^2 + c - \frac{b^2}{4a}$$ Add c to both sides and subtract $\frac{b^2}{4a}$ from both sides.

Compare this form with $f(x)$ or $y = a(x - h)^2 + k$ and see that h is $\frac{-b}{2a}$, which means that the x-coordinate of the vertex of the graph of $f(x) = ax^2 + bx + c$ is $\frac{-b}{2a}$.

Vertex Formula

The graph of $f(x) = ax^2 + bx + c$, when $a \neq 0$, is a parabola with vertex

$$\left(\frac{-b}{2a}, f\left(\frac{-b}{2a}\right)\right)$$

Let's use this formula to find the vertex of the parabola we graphed in Example 1.

EXAMPLE 4 Find the vertex of the graph of $f(x) = x^2 - 4x - 12$.

Solution To find the vertex (h, k), notice that for $f(x) = x^2 - 4x - 12$, we have $a = 1, b = -4$, and $c = -12$. Then

$$h = \frac{-b}{2a} = \frac{-(-4)}{2(1)} = 2$$

The x-value of the vertex is 2. To find the corresponding $f(x)$ or y-value, find $f(2)$. Then

$$f(2) = 2^2 - 4(2) - 12 = 4 - 8 - 12 = -16$$

The vertex is $(2, -16)$. These results agree with our findings in Example 1. □

PRACTICE

4 Find the vertex of the graph of $g(x) = x^2 - 2x - 3$. ■

OBJECTIVE

3 Finding Minimum and Maximum Values

The vertex of a parabola gives us some important information about its corresponding quadratic function. The quadratic function whose graph is a parabola that opens upward has a minimum value, and the quadratic function whose graph is a parabola that opens downward has a maximum value. The $f(x)$ or y-value of the vertex is the minimum or maximum value of the function.

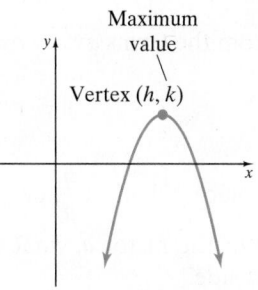
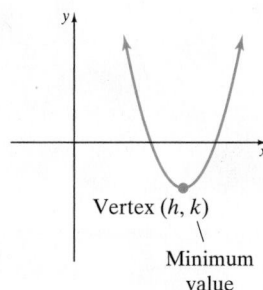

✔ **CONCEPT CHECK**

Without making any calculations, tell whether the graph of $f(x) = 7 - x - 0.3x^2$ has a maximum value or a minimum value. Explain your reasoning.

EXAMPLE 5 Finding Maximum Height

A rock is thrown upward from the ground. Its height in feet above ground after t seconds is given by the function $f(t) = -16t^2 + 20t$. Find the maximum height of the rock and the number of seconds it took for the rock to reach its maximum height.

Solution

1. UNDERSTAND. The maximum height of the rock is the largest value of $f(t)$. Since the function $f(t) = -16t^2 + 20t$ is a quadratic function, its graph is a parabola. It opens downward since $-16 < 0$. Thus, the maximum value of $f(t)$ is the $f(t)$ or y-value of the vertex of its graph.

2. TRANSLATE. To find the vertex (h, k), notice that for $f(t) = -16t^2 + 20t$, $a = -16$, $b = 20$, and $c = 0$. We will use these values and the vertex formula

$$\left(\frac{-b}{2a}, f\left(\frac{-b}{2a}\right)\right)$$

3. SOLVE.

$$h = \frac{-b}{2a} = \frac{-20}{-32} = \frac{5}{8}$$

$$f\left(\frac{5}{8}\right) = -16\left(\frac{5}{8}\right)^2 + 20\left(\frac{5}{8}\right)$$

$$= -16\left(\frac{25}{64}\right) + \frac{25}{2}$$

$$= -\frac{25}{4} + \frac{50}{4} = \frac{25}{4}$$

4. INTERPRET. The graph of $f(t)$ is a parabola opening downward with vertex $\left(\frac{5}{8}, \frac{25}{4}\right)$. This means that the rock's maximum height is $\frac{25}{4}$ feet, or $6\frac{1}{4}$ feet, which was reached in $\frac{5}{8}$ second.   □

PRACTICE
5   A ball is tossed upward from the ground. Its height in feet above ground after t seconds is given by the function $h(t) = -16t^2 + 24t$. Find the maximum height of the ball and the number of seconds it took for the ball to reach the maximum height.   ■

Vocabulary, Readiness & Video Check

Fill in each blank.

1. If a quadratic function is in the form $f(x) = a(x - h)^2 + k$, the vertex of its graph is _____.

2. The graph of $f(x) = ax^2 + bx + c, a \neq 0$, is a parabola whose vertex has x-value _____.

Martin-Gay Interactive Videos

See Video 11.6 ◉

Watch the section lecture video and answer the following questions.

OBJECTIVE 1
3. From Example 1, how does writing a quadratic function in the form $f(x) = a(x - h)^2 + k$ help us graph the function? What procedure can we use to write a quadratic function in this form?

OBJECTIVE 2
4. From Example 2, how can locating the vertex and knowing whether the parabola opens upward or downward potentially help save unnecessary work? Explain.

OBJECTIVE 3
5. From Example 4, when an application involving a quadratic function asks for the maximum or minimum, what part of a parabola should we find?

11.6 Exercise Set MyMathLab®

Fill in each blank.

	Parabola Opens	Vertex Location	Number of x-intercept(s)	Number of y-intercept(s)
1.	up	Q I		
2.	up	Q III		
3.	down	Q II		
4.	down	Q IV		
5.	up	x-axis		
6.	down	x-axis		
7.		Q III	0	
8.		Q I	2	
9.		Q IV	2	
10.		Q II	0	

Find the vertex of the graph of each quadratic function. See Examples 1 through 4.

11. $f(x) = x^2 + 8x + 7$

12. $f(x) = x^2 + 6x + 5$

13. $f(x) = -x^2 + 10x + 5$

14. $f(x) = -x^2 - 8x + 2$

15. $f(x) = 5x^2 - 10x + 3$

16. $f(x) = -3x^2 + 6x + 4$

17. $f(x) = -x^2 + x + 1$

18. $f(x) = x^2 - 9x + 8$

Match each function with its graph. See Examples 1 through 4.

19. $f(x) = x^2 - 4x + 3$

20. $f(x) = x^2 + 2x - 3$

21. $f(x) = x^2 - 2x - 3$

22. $f(x) = x^2 + 4x + 3$

A.

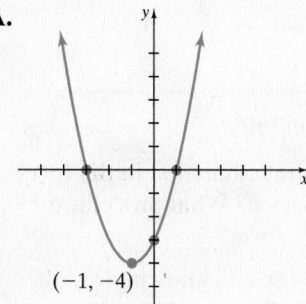

$(-1, -4)$

B.

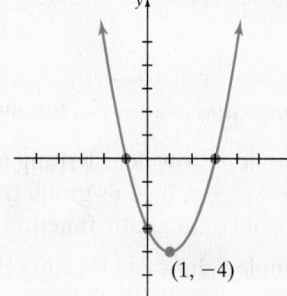

$(1, -4)$

C.

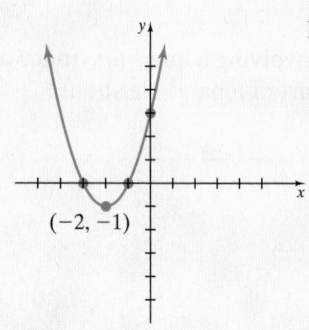

$(-2, -1)$

D.

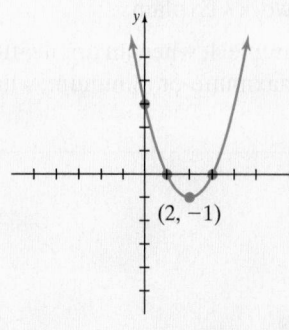

$(2, -1)$

MIXED PRACTICE

Find the vertex of the graph of each quadratic function. Determine whether the graph opens upward or downward, find any intercepts, and sketch the graph. See Examples 1 through 4.

23. $f(x) = x^2 + 4x - 5$

24. $f(x) = x^2 + 2x - 3$

25. $f(x) = -x^2 + 2x - 1$

26. $f(x) = -x^2 + 4x - 4$

27. $f(x) = x^2 - 4$

28. $f(x) = x^2 - 1$

29. $f(x) = 4x^2 + 4x - 3$

30. $f(x) = 2x^2 - x - 3$

31. $f(x) = \frac{1}{2}x^2 + 4x + \frac{15}{2}$

32. $f(x) = \frac{1}{5}x^2 + 2x + \frac{9}{5}$

33. $f(x) = x^2 - 6x + 5$

34. $f(x) = x^2 - 4x + 3$

35. $f(x) = x^2 - 4x + 5$

36. $f(x) = x^2 - 6x + 11$

37. $f(x) = 2x^2 + 4x + 5$

38. $f(x) = 3x^2 + 12x + 16$

39. $f(x) = -2x^2 + 12x$

40. $f(x) = -4x^2 + 8x$

41. $f(x) = x^2 + 1$

42. $f(x) = x^2 + 4$

43. $f(x) = x^2 - 2x - 15$

44. $f(x) = x^2 - x - 12$

45. $f(x) = -5x^2 + 5x$

46. $f(x) = 3x^2 - 12x$

47. $f(x) = -x^2 + 2x - 12$

48. $f(x) = -x^2 + 8x - 17$

49. $f(x) = 3x^2 - 12x + 15$

50. $f(x) = 2x^2 - 8x + 11$

51. $f(x) = x^2 + x - 6$

52. $f(x) = x^2 + 3x - 18$

53. $f(x) = -2x^2 - 3x + 35$

54. $f(x) = 3x^2 - 13x - 10$

Solve. See Example 5.

55. If a projectile is fired straight upward from the ground with an initial speed of 96 feet per second, then its height h in feet after t seconds is given by the equation

$$h(t) = -16t^2 + 96t$$

Find the maximum height of the projectile.

56. If Rheam Gaspar throws a ball upward with an initial speed of 32 feet per second, then its height h in feet after t seconds is given by the equation

$$h(t) = -16t^2 + 32t$$

Find the maximum height of the ball.

57. The cost C in dollars of manufacturing x bicycles at Holladay's Production Plant is given by the function

$$C(x) = 2x^2 - 800x + 92,000.$$

a. Find the number of bicycles that must be manufactured to minimize the cost.

b. Find the minimum cost.

58. The Utah Ski Club sells calendars to raise money. The profit P, in cents, from selling x calendars is given by the equation $P(x) = 360x - x^2$.

a. Find how many calendars must be sold to maximize profit.

b. Find the maximum profit in dollars.

59. Find two numbers whose sum is 60 and whose product is as large as possible. [*Hint:* Let x and $60 - x$ be the two positive numbers. Their product can be described by the function $f(x) = x(60 - x)$.]

60. Find two numbers whose sum is 11 and whose product is as large as possible. (Use the hint for Exercise 59.)

61. Find two numbers whose difference is 10 and whose product is as small as possible. (Use the hint for Exercise 59.)

62. Find two numbers whose difference is 8 and whose product is as small as possible. (Use the hint for Exercise 59.)

△ **63.** The length and width of a rectangle must have a sum of 40. Find the dimensions of the rectangle that will have the maximum area. (Use the hint for Exercise 59.)

△ **64.** The length and width of a rectangle must have a sum of 50. Find the dimensions of the rectangle that will have maximum area. (Use the hint for Exercise 59.)

REVIEW AND PREVIEW

Sketch the graph of each function. See Sections 3.2 and 11.5.

65. $f(x) = x^2 + 2$ **66.** $f(x) = (x - 3)^2$

67. $g(x) = x + 2$ **68.** $h(x) = x - 3$

69. $f(x) = (x + 5)^2 + 2$ **70.** $f(x) = 2(x - 3)^2 + 2$

71. $f(x) = 3(x - 4)^2 + 1$ **72.** $f(x) = (x + 1)^2 + 4$

73. $f(x) = -(x - 4)^2 + \dfrac{3}{2}$ **74.** $f(x) = -2(x + 7)^2 + \dfrac{1}{2}$

CONCEPT EXTENSIONS

Without calculating, tell whether each graph has a minimum value or a maximum value. See the Concept Check in the section.

75. $f(x) = 2x^2 - 5$

76. $g(x) = -7x^2 + x + 1$

77. $f(x) = 3 - \dfrac{1}{2}x^2$

78. $G(x) = 3 - \dfrac{1}{2}x + 0.8x^2$

Find the vertex of the graph of each quadratic function. Determine whether the graph opens upward or downward, find the y-intercept, approximate the x-intercepts to one decimal place, and sketch the graph.

79. $f(x) = x^2 + 10x + 15$ **80.** $f(x) = x^2 - 6x + 4$

81. $f(x) = 3x^2 - 6x + 7$ **82.** $f(x) = 2x^2 + 4x - 1$

Find the maximum or minimum value of each function. Approximate to two decimal places.

83. $f(x) = 2.3x^2 - 6.1x + 3.2$

84. $f(x) = 7.6x^2 + 9.8x - 2.1$

85. $f(x) = -1.9x^2 + 5.6x - 2.7$

86. $f(x) = -5.2x^2 - 3.8x + 5.1$

87. The projected number of Wi-Fi-enabled cell phones in the United States can be modeled by the quadratic function $c(x) = -0.4x^2 + 21x + 35$, where $c(x)$ is the projected number of Wi-Fi-enabled cell phones in millions and x is the number of years after 2009. (*Source:* Techcrunch)

a. Will this function have a maximum or a minimum? How can you tell?

b. According to this model, in what year will the number of Wi-Fi-enabled cell phones in the United States be at its maximum or minimum?

c. What is the maximum/minimum number of Wi-Fi-enabled cell phones predicted? Round to the nearest whole million.

88. Methane is a gas produced by landfills, natural gas systems, and coal mining that contributes to the greenhouse effect and global warming. Projected methane emissions in the United States can be modeled by the quadratic function

$$f(x) = -0.072x^2 + 1.93x + 173.9$$

where $f(x)$ is the amount of methane produced in million metric tons and x is the number of years after 2000. (*Source:* Based on data from the U.S. Environmental Protection Agency, 2000–2020)

a. According to this model, what will U.S. emissions of methane be in 2018? Round to 2 decimal places.

b. Will this function have a maximum or a minimum? How can you tell?

c. In what year will methane emissions in the United States be at their maximum/minimum? Round to the nearest whole year.

d. What is the level of methane emissions for that year? (Use your rounded answer from part **(c)**. Round this answer to 2 decimal places.)

Use a graphing calculator to check each exercise.

89. Exercise 37   **90.** Exercise 38

91. Exercise 47   **92.** Exercise 48

Chapter 11 Vocabulary Check

Fill in each blank with one of the words or phrases listed below.

quadratic formula   quadratic   discriminant $\pm \sqrt{b}$

completing the square   quadratic inequality (h, k) $(0, k)$

$(h, 0)$ $\dfrac{-b}{2a}$

1. The _____ helps us find the number and type of solutions of a quadratic equation.

2. If $a^2 = b$, then $a =$ _____.

3. The graph of $f(x) = ax^2 + bx + c$, where a is not 0, is a parabola whose vertex has x-value _____.

4. A _____ is an inequality that can be written so that one side is a quadratic expression and the other side is 0.

5. The process of writing a quadratic equation so that one side is a perfect square trinomial is called _____.

6. The graph of $f(x) = x^2 + k$ has vertex _____.

7. The graph of $f(x) = (x - h)^2$ has vertex _____.

8. The graph of $f(x) = (x - h)^2 + k$ has vertex _____.

9. The formula $x = \dfrac{-b \pm \sqrt{b^2 - 4ac}}{2a}$ is called the _____.

10. A _____ equation is one that can be written in the form $ax^2 + bx + c = 0$ where $a, b,$ and c are real numbers and a is not 0.

Chapter 11 Highlights

DEFINITIONS AND CONCEPTS	EXAMPLES
Section 11.1 Solving Quadratic Equations by Completing the Square	

Square Root Property	Solve: $(x + 3)^2 = 14$
If b is a real number and if $a^2 = b$, then $a = \pm\sqrt{b}$.	$x + 3 = \pm\sqrt{14}$
	$x = -3 \pm \sqrt{14}$
To Solve a Quadratic Equation in x by Completing the Square	Solve: $3x^2 - 12x - 18 = 0$
Step 1. If the coefficient of x^2 is not 1, divide both sides of the equation by the coefficient of x^2.	**1.** $x^2 - 4x - 6 = 0$
Step 2. Isolate the variable terms.	**2.** $x^2 - 4x = 6$
Step 3. Complete the square by adding the square of half of the coefficient of x to both sides.	**3.** $\dfrac{1}{2}(-4) = -2$ and $(-2)^2 = 4$
	$x^2 - 4x + 4 = 6 + 4$
Step 4. Write the resulting trinomial as the square of a binomial.	**4.** $(x - 2)^2 = 10$
Step 5. Apply the square root property and solve for x.	**5.** $x - 2 = \pm\sqrt{10}$
	$x = 2 \pm \sqrt{10}$

DEFINITIONS AND CONCEPTS	EXAMPLES

Section 11.2   Solving Quadratic Equations by the Quadratic Formula

A quadratic equation written in the form $ax^2 + bx + c = 0$ has solutions

$$x = \frac{-b \pm \sqrt{b^2 - 4ac}}{2a}$$

Solve: $x^2 - x - 3 = 0$

$$a = 1, b = -1, c = -3$$

$$x = \frac{-(-1) \pm \sqrt{(-1)^2 - 4(1)(-3)}}{2 \cdot 1}$$

$$x = \frac{1 \pm \sqrt{13}}{2}$$

Section 11.3   Solving Equations by Using Quadratic Methods

Substitution is often helpful in solving an equation that contains a repeated variable expression.

Solve: $(2x + 1)^2 - 5(2x + 1) + 6 = 0$

Let $m = 2x + 1$. Then

$$m^2 - 5m + 6 = 0 \qquad \text{Let } m = 2x + 1.$$

$$(m - 3)(m - 2) = 0$$

$$m = 3 \quad \text{or} \quad m = 2$$

$$2x + 1 = 3 \quad \text{or} \quad 2x + 1 = 2 \quad \text{Substitute back.}$$

$$x = 1 \quad \text{or} \quad x = \frac{1}{2}$$

Section 11.4   Nonlinear Inequalities in One Variable

To Solve a Polynomial Inequality

Step 1. Write the inequality in standard form.

Step 2. Solve the related equation.

Step 3. Use solutions from Step 2 to separate the number line into regions.

Step 4. Use test points to determine whether values in each region satisfy the original inequality.

Step 5. Write the solution set as the union of regions whose test point value is a solution.

Solve: $x^2 \geq 6x$

1. $x^2 - 6x \geq 0$

2. $x^2 - 6x = 0$

$$x(x - 6) = 0$$

$$x = 0 \quad \text{or} \quad x = 6$$

3.

4.

Region	Test Point Value	$x^2 \geq 6x$	Result
A	-2	$(-2)^2 \geq 6(-2)$	True
B	1	$1^2 \geq 6(1)$	False
C	7	$7^2 \geq 6(7)$	True

5.

The solution set is $(-\infty, 0] \cup [6, \infty)$.

Solve: $\dfrac{6}{x - 1} < -2$

1. $x - 1 = 0$ Set denominator equal to 0.

$$x = 1$$

2. $\dfrac{6}{x - 1} = -2$

$$6 = -2(x - 1) \quad \text{Multiply by } (x - 1).$$

$$6 = -2x + 2$$

$$4 = -2x$$

$$-2 = x$$

To Solve a Rational Inequality

Step 1. Solve for values that make all denominators 0.

Step 2. Solve the related equation.

Step 3. Use solutions from Steps 1 and 2 to separate the number line into regions.

Step 4. Use test points to determine whether values in each region satisfy the original inequality.

Step 5. Write the solution set as the union of regions whose test point value is a solution. Do not include values from step 1.

(continued)

DEFINITIONS AND CONCEPTS	EXAMPLES

Section 11.4   Nonlinear Inequalities in One Variable (continued)

3.

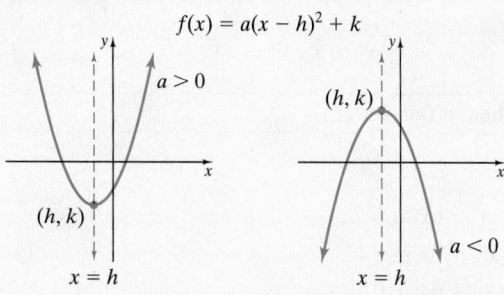

4. Only a test value from region B satisfies the original inequality.

5.

The solution set is $(-2, 1)$.

Section 11.5   Quadratic Functions and Their Graphs

Graph of a Quadratic Function

The graph of a quadratic function written in the form $f(x) = a(x - h)^2 + k$ is a parabola with vertex (h, k). If $a > 0$, the parabola opens upward; if $a < 0$, the parabola opens downward. The axis of symmetry is the line whose equation is $x = h$.

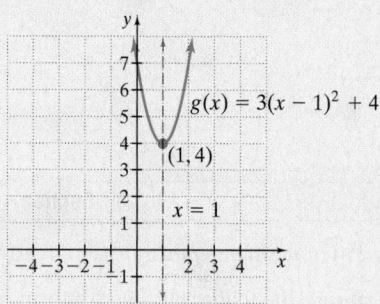

Graph: $g(x) = 3(x - 1)^2 + 4$

The graph is a parabola with vertex $(1, 4)$ and axis of symmetry $x = 1$. Since $a = 3$ is positive, the graph opens upward.

Section 11.6   Further Graphing of Quadratic Functions

The graph of $f(x) = ax^2 + bx + c$, where $a \neq 0$, is a parabola with vertex

$$\left(\frac{-b}{2a}, f\left(\frac{-b}{2a} \right) \right)$$

Graph $f(x) = x^2 - 2x - 8$. Find the vertex and x- and y-intercepts.

$$\frac{-b}{2a} = \frac{-(-2)}{2 \cdot 1} = 1$$

$$f(1) = 1^2 - 2(1) - 8 = -9$$

The vertex is $(1, -9)$.

$$0 = x^2 - 2x - 8$$
$$0 = (x - 4)(x + 2)$$
$$x = 4 \quad \text{or} \quad x = -2$$

The x-intercepts are $(4, 0)$ and $(-2, 0)$.

$$f(0) = 0^2 - 2 \cdot 0 - 8 = -8$$

The y-intercept is $(0, -8)$.

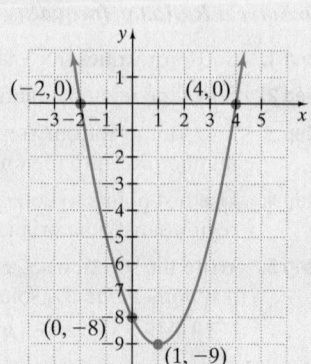

Chapter 11 **Review**

(11.1) *Solve by factoring.*

1. $x^2 - 15x + 14 = 0$
2. $7a^2 = 29a + 30$

Solve by using the square root property.

3. $4m^2 = 196$
4. $(5x - 2)^2 = 2$

Solve by completing the square.

5. $z^2 + 3z + 1 = 0$

6. $(2x + 1)^2 = x$

7. If P dollars are originally invested, the formula $A = P(1 + r)^2$ gives the amount A in an account paying interest rate r compounded annually after 2 years. Find the interest rate r such that \$2500 increases to \$2717 in 2 years. Round the result to the nearest hundredth of a percent.

△ 8. Two ships leave a port at the same time and travel at the same speed. One ship is traveling due north and the other due east. In a few hours, the ships are 150 miles apart. How many miles has each ship traveled? Give the exact answer and a one-decimal-place approximation.

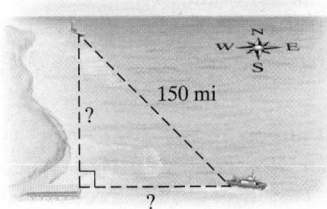

(11.2) *If the discriminant of a quadratic equation has the given value, determine the number and type of solutions of the equation.*

9. -8
10. 48
11. 100
12. 0

Solve by using the quadratic formula.

13. $x^2 - 16x + 64 = 0$
14. $x^2 + 5x = 0$
15. $2x^2 + 3x = 5$
16. $9x^2 + 4 = 2x$
17. $6x^2 + 7 = 5x$
18. $(2x - 3)^2 = x$

19. Cadets graduating from military school usually toss their hats high into the air at the end of the ceremony. One cadet threw his hat so that its distance $d(t)$ in feet above the ground t seconds after it was thrown was $d(t) = -16t^2 + 30t + 6$.
    a. Find the distance above the ground of the hat 1 second after it was thrown.
    b. Find the time it takes the hat to hit the ground. Give the exact time and a one-decimal-place approximation.

20. The hypotenuse of an isosceles right triangle is 6 centimeters longer than either of the legs. Find the length of the legs.

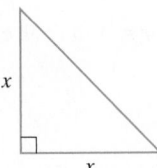

(11.3) *Solve each equation for the variable.*

21. $x^3 = 27$

22. $y^3 = -64$

23. $\dfrac{5}{x} + \dfrac{6}{x - 2} = 3$

24. $x^4 - 21x^2 - 100 = 0$

25. $x^{2/3} - 6x^{1/3} + 5 = 0$

26. $5(x + 3)^2 - 19(x + 3) = 4$

27. $a^6 - a^2 = a^4 - 1$

28. $y^{-2} + y^{-1} = 20$

29. Two postal workers, Jerome Grant and Tim Bozik, can sort a stack of mail in 5 hours. Working alone, Tim can sort the mail in 1 hour less time than Jerome can. Find the time that each postal worker can sort the mail alone. Round the result to one decimal place.

30. A negative number decreased by its reciprocal is $-\dfrac{24}{5}$. Find the number.

(11.4) *Solve each inequality for x. Write each solution set in interval notation.*

31. $2x^2 - 50 \le 0$
32. $\dfrac{1}{4}x^2 < \dfrac{1}{16}$

33. $(x^2 - 4)(x^2 - 25) \le 0$

34. $(x^2 - 16)(x^2 - 1) > 0$

35. $\dfrac{x - 5}{x - 6} < 0$

36. $\dfrac{(4x + 3)(x - 5)}{x(x + 6)} > 0$

37. $(x + 5)(x - 6)(x + 2) \le 0$

38. $x^3 + 3x^2 - 25x - 75 > 0$

39. $\dfrac{x^2 + 4}{3x} \le 1$
40. $\dfrac{3}{x - 2} > 2$

(11.5) *Sketch the graph of each function. Label the vertex and the axis of symmetry.*

41. $f(x) = x^2 - 4$

42. $g(x) = x^2 + 7$

43. $H(x) = 2x^2$

44. $h(x) = -\frac{1}{3}x^2$

45. $F(x) = (x - 1)^2$

46. $G(x) = (x + 5)^2$

47. $f(x) = (x - 4)^2 - 2$

48. $f(x) = -3(x - 1)^2 + 1$

(11.6) *Sketch the graph of each function. Find the vertex and the intercepts.*

49. $f(x) = x^2 + 10x + 25$

50. $f(x) = -x^2 + 6x - 9$

51. $f(x) = 4x^2 - 1$

52. $f(x) = -5x^2 + 5$

53. Find the vertex of the graph of $f(x) = -3x^2 - 5x + 4$. Determine whether the graph opens upward or downward, find the y-intercept, approximate the x-intercepts to one decimal place, and sketch the graph.

54. The function $h(t) = -16t^2 + 120t + 300$ gives the height in feet of a projectile fired from the top of a building after t seconds.
   a. When will the object reach a height of 350 feet? Round your answer to one decimal place.
   b. Explain why part **a** has two answers.

55. Find two numbers whose product is as large as possible, given that their sum is 420.

56. Write an equation of a quadratic function whose graph is a parabola that has vertex $(-3, 7)$. Let the value of a be $-\frac{7}{9}$.

MIXED REVIEW

Solve each equation or inequality.

57. $x^2 - x - 30 = 0$

58. $10x^2 = 3x + 4$

59. $9y^2 = 36$

60. $(9n + 1)^2 = 9$

61. $x^2 + x + 7 = 0$

62. $(3x - 4)^2 = 10x$

63. $x^2 + 11 = 0$

64. $x^2 + 7 = 0$

65. $(5a - 2)^2 - a = 0$

66. $\frac{7}{8} = \frac{8}{x^2}$

67. $x^{2/3} - 6x^{1/3} = -8$

68. $(2x - 3)(4x + 5) \geq 0$

69. $\frac{x(x + 5)}{4x - 3} \geq 0$

70. $\frac{3}{x - 2} > 2$

71. The busiest airport in the world is the Hartsfield-Jackson International Airport in Atlanta, Georgia. The total amount of passenger traffic through Atlanta during the period 2005 through 2014 can be modeled by the equation $y = -111x^2 + 1960x + 85{,}907$, where y is the number of passengers enplaned and deplaned in thousands, and x is the number of years after 2005. (*Source:* Based on data from Airports Council International)
   a. Estimate the passenger traffic at Atlanta's Hartsfield-Jackson International Airport in 2020.
   b. According to this model, will the passenger traffic at Atlanta's Hartsfield-Jackson International Airport continue to grow? Why or why not?

Chapter 11 Getting Ready for the Test

MULTIPLE CHOICE *For each quadratic equation in Exercises 1 and 2, choose the correct substitution for a, b, and c in the standard form* $ax^2 + bx + c = 0$.

1. $x^2 = 8$

    A. $a = 1, b = 0, c = 8$ **B.** $a = 1, b = 8, c = 0$ **C.** $a = 1, b = 0, c = -8$

2. $\frac{1}{9}x^2 + \frac{1}{3} = x$ One correct substitution is $a = \frac{1}{9}$, $b = -1$, $c = \frac{1}{3}$. Find another.

    A. $a = 1, b = -1, c = 3$ **B.** $a = 1, b = -9, c = 3$ **C.** $a = 1, b = 9, c = 3$ **D.** $a = 1, b = -1, c = 1$.

MULTIPLE CHOICE *Select the correct choice.*

3. $\dfrac{-4 \pm \sqrt{-4}}{2}$ simplifies to

   A. $-2 \pm i$ **B.** $-2 \pm 2i$ **C.** $\pm 4i$ **D.** $-2 \pm i\sqrt{2}$

4. $\dfrac{9 \pm \sqrt{27}}{3}$ simplifies to

   A. $3 \pm \sqrt{27}$ **B.** $3 \pm 3\sqrt{3}$ **C.** $9 \pm \sqrt{3}$ **D.** $3 \pm \sqrt{3}$

5. To solve $2x^2 + 16x = 5$ by completing the square, choose the next correct step.

   A. $x^2 + 8x = 5$ **B.** $x^2 + 8x = \dfrac{5}{2}$ **C.** $2x^2 + 16x + 16 = 5$ **D.** $2x^2 + 16x + 64 = 5$

6. To solve $x^2 - 7x = 5$ by completing the square, choose the next correct step.

   A. $x^2 - 7x + 49 = 5$ **B.** $x^2 - 7x + 49 = 5 + 49$ **C.** $x^2 - 7x + \dfrac{49}{4} = 5$ **D.** $x^2 - 7x + \dfrac{49}{4} = 5 + \dfrac{49}{4}$

7. The solution set of $(x + 5)(x - 1) \le 0$ is $[-5, 1]$. Use this to select the solution set of $\dfrac{x + 5}{x - 1} \le 0$.

   A. $[-5, 1]$ **B.** $(-5, 1]$ **C.** $[-5, 1)$ **D.** $(-5, 1)$

8. Choose the correct description of the graph of $f(x) = -103(x - 20)^2 + 5.6$.

   A. opens up; vertex $(20, 5.6)$ **B.** opens up; vertex $(-20, 5.6)$
   C. opens down; vertex $(20, 5.6)$ **D.** opens down; vertex $(20, -5.6)$

9. Choose the correct description of the graph of $f(x) = 0.5(x + 1)^2 - 3$.
   A. opens up; vertex $(-1, -3)$ **B.** opens up; vertex $(1, -3)$
   C. opens down; vertex $(-1, -3)$ **D.** opens down; vertex $(-3, -1)$

10. Select the vertex of the graph of $f(x) = 3x^2 + 12x - 7$.

    A. $(-2, -7)$ **B.** $(-2, -19)$ **C.** $\left(-\dfrac{1}{2}, -12\dfrac{1}{4}\right)$ **D.** $(-12, 281)$

11. Select the x-intercept(s) of the graph of $f(x) = 2x^2 - x - 10$.

    A. $(0, 0)$ **B.** $(5, 0), (-2, 0)$ **C.** $\left(\dfrac{5}{2}, 0\right), (-2, 0)$ **D.** $\left(-\dfrac{5}{2}, 0\right), (2, 0)$

Chapter 11 **Test**   MyMathLab®   You Tube

Solve each equation.

1. $5x^2 - 2x = 7$

2. $(x + 1)^2 = 10$

3. $m^2 - m + 8 = 0$

4. $u^2 - 6u + 2 = 0$

5. $7x^2 + 8x + 1 = 0$

6. $y^2 - 3y = 5$

7. $\dfrac{4}{x + 2} + \dfrac{2x}{x - 2} = \dfrac{6}{x^2 - 4}$

8. $x^5 + 3x^4 = x + 3$

9. $x^6 + 1 = x^4 + x^2$

10. $(x + 1)^2 - 15(x + 1) + 56 = 0$

Solve by completing the square.

11. $x^2 - 6x = -2$

12. $2a^2 + 5 = 4a$

Solve each inequality for x. Write the solution set in interval notation.

13. $2x^2 - 7x > 15$

14. $(x^2 - 16)(x^2 - 25) \ge 0$

15. $\dfrac{5}{x + 3} < 1$

16. $\dfrac{7x - 14}{x^2 - 9} \le 0$

Graph each function. Label the vertex.

17. $f(x) = 3x^2$

18. $G(x) = -2(x - 1)^2 + 5$

Graph each function. Find and label the vertex, y-intercept, and x-intercepts (if any).

19. $h(x) = x^2 - 4x + 4$

20. $F(x) = 2x^2 - 8x + 9$

▶ **21.** Dave and Sandy Hartranft can paint a room together in 4 hours. Working alone, Dave can paint the room in 2 hours less time than Sandy can. Find how long it takes Sandy to paint the room alone.

▶ **22.** A stone is thrown upward from a bridge. The stone's height in feet, $s(t)$, above the water t seconds after the stone is thrown is a function given by the equation $s(t) = -16t^2 + 32t + 256$.

    a. Find the maximum height of the stone.

    b. Find the time it takes the stone to hit the water. Round the answer to two decimal places.

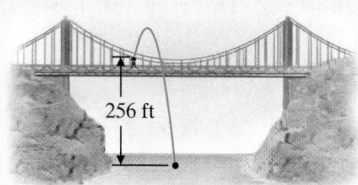

△ **23.** Given the diagram shown, approximate to the nearest tenth of a foot how many feet of walking distance a person saves by cutting across the lawn instead of walking on the sidewalk.

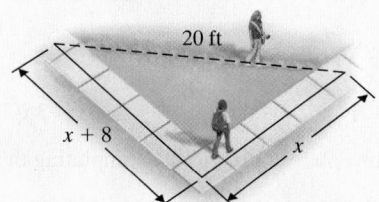

Chapter 11 Cumulative Review

1. Find the value of each expression when $x = 2$ and $y = -5$.

    a. $\dfrac{x - y}{12 + x}$ **b.** $x^2 - 3y$

2. Solve: $|3x - 2| = -5$

3. Simplify each expression by combining like terms.

    a. $2x + 3x + 5 + 2$ **b.** $-5a - 3 + a + 2$

    c. $4y - 3y^2$ **d.** $2.3x + 5x - 6$

    e. $-\dfrac{1}{2}b + b$

4. Use the addition method to solve the system.

$$\begin{cases} -6x + y = 5 \\ 4x - 2y = 6 \end{cases}$$

5. Solve the following system of equations by graphing.

$$\begin{cases} 2x + y = 7 \\ 2y = -4x \end{cases}$$

6. Simplify. Use positive exponents to write each answer.

    a. $(a^{-2}bc^3)^{-3}$ **b.** $\left(\dfrac{a^{-4}b^2}{c^3}\right)^{-2}$

    c. $\left(\dfrac{3a^8b^2}{12a^5b^5}\right)^{-2}$

7. Solve the system: $\begin{cases} 7x - 3y = -14 \\ -3x + y = 6 \end{cases}$

8. Multiply.

    a. $(4a - 3)(7a - 2)$

    b. $(2a + b)(3a - 5b)$

9. Simplify each quotient.

    a. $\dfrac{x^5}{x^2}$ **b.** $\dfrac{4^7}{4^3}$ **c.** $\dfrac{(-3)^5}{(-3)^2}$

    d. $\dfrac{s^2}{t^3}$ **e.** $\dfrac{2x^5y^2}{xy}$

10. Factor.

    a. $9x^3 + 27x^2 - 15x$

    b. $2x(3y - 2) - 5(3y - 2)$

    c. $2xy + 6x - y - 3$

11. If $P(x) = 2x^3 - 4x^2 + 5$

    a. Find $P(2)$ by substitution.

    b. Use synthetic division to find the remainder when $P(x)$ is divided by $x - 2$.

12. Factor: $x^2 - 2x - 48$

13. Solve: $(5x - 1)(2x^2 + 15x + 18) = 0$

14. Factor: $2ax^2 - 12axy + 18ay^2$

15. Simplify the rational expression.

$$\dfrac{2x^2}{10x^3 - 2x^2}$$

16. Solve: $2(a^2 + 2) - 8 = -2a(a - 2) - 5$

17. Simplify: $\dfrac{x^{-1} + 2xy^{-1}}{x^{-2} - x^{-2}y^{-1}}$

18. Find the vertex and any intercepts of $f(x) = x^2 + x - 12$.

19. Factor: $4m^4 - 4m^2 + 1$

20. Simplify: $\dfrac{x^2 - 4x + 4}{2 - x}$

21. The square of a number plus three times the number is 70. Find the number.

22. Subtract: $\dfrac{a + 1}{a^2 - 6a + 8} - \dfrac{3}{16 - a^2}$

23. Use the product rule to simplify.

    a. $\sqrt{25x^3}$ **b.** $\sqrt[3]{54x^6y^8}$

    c. $\sqrt[4]{81z^{11}}$

24. Simplify: $\dfrac{(2a)^{-1} + b^{-1}}{a^{-1} + (2b)^{-1}}$

25. Rationalize the denominator of each expression.

   a. $\dfrac{2}{\sqrt{5}}$ **b.** $\dfrac{2\sqrt{16}}{\sqrt{9x}}$

   c. $\sqrt[3]{\dfrac{1}{2}}$

26. Divide $x^3 - 3x^2 - 10x + 24$ by $x + 3$.

27. Solve: $\sqrt{2x + 5} + \sqrt{2x} = 3$

28. If $P(x) = 4x^3 - 2x^2 + 3$,

   a. Find $P(-2)$ by substitution.

   b. Use synthetic division to find the remainder when $P(x)$ is divided by $x + 2$.

29. Solve: $\dfrac{x}{2} + \dfrac{8}{3} = \dfrac{1}{6}$

30. Solve: $\dfrac{x + 3}{x^2 + 5x + 6} = \dfrac{3}{2x + 4} - \dfrac{1}{x + 3}$

31. The quotient of a number and 6, minus $\dfrac{5}{3}$, is the quotient of the number and 2. Find the number.

32. Mr. Briley can roof his house in 24 hours. His son can roof the same house in 40 hours. If they work together, how long will it take to roof the house?

33. Suppose that y varies directly as x. If y is 5 when x is 30, find the constant of variation and the direct variation equation.

34. Suppose that y varies inversely as x. If y is 8 when x is 24, find the constant of variation and the inverse variation equation.

35. Simplify. Assume that the variables represent any real number.

   a. $\sqrt{(-3)^2}$ **b.** $\sqrt{x^2}$

   c. $\sqrt[4]{(x - 2)^4}$ **d.** $\sqrt[3]{(-5)^3}$

   e. $\sqrt[5]{(2x - 7)^5}$ **f.** $\sqrt{25x^2}$

   g. $\sqrt{x^2 + 2x + 1}$

36. Simplify. Assume that the variables represent any real number.

   a. $\sqrt{(-2)^2}$

   b. $\sqrt{y^2}$

   c. $\sqrt[4]{(a - 3)^4}$

   d. $\sqrt[3]{(-6)^3}$

   e. $\sqrt[5]{(3x - 1)^5}$

37. Use rational exponents to simplify. Assume that variables represent positive numbers.

   a. $\sqrt[8]{x^4}$

   b. $\sqrt[6]{25}$

   c. $\sqrt[4]{r^2 s^6}$

38. Use rational exponents to simplify. Assume that variables represent positive numbers.

   a. $\sqrt[4]{5^2}$

   b. $\sqrt[12]{x^3}$

   c. $\sqrt[6]{x^2 y^4}$

39. Divide. Write in the form $a + bi$.

   a. $\dfrac{2 + i}{1 - i}$

   b. $\dfrac{7}{3i}$

40. Write each product in the form of $a + bi$.

   a. $3i(5 - 2i)$

   b. $(6 - 5i)^2$

   c. $(\sqrt{3} + 2i)(\sqrt{3} - 2i)$

41. Use the square root property to solve $(x + 1)^2 = 12$.

42. Use the square root property to solve $(y - 1)^2 = 24$.

43. Solve: $x - \sqrt{x} - 6 = 0$

44. Use the quadratic formula to solve $m^2 = 4m + 8$.

CHAPTER

12

Exponential and Logarithmic Functions

12.1 The Algebra of Functions; Composite Functions

12.2 Inverse Functions

12.3 Exponential Functions

12.4 Exponential Growth and Decay Functions

12.5 Logarithmic Functions

12.6 Properties of Logarithms

Integrated Review—Functions and Properties of Logarithms

12.7 Common Logarithms, Natural Logarithms, and Change of Base

12.8 Exponential and Logarithmic Equations and Problem Solving

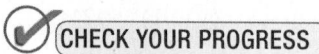

CHECK YOUR PROGRESS

Vocabulary Check

Chapter Highlights

Chapter Review

Getting Ready for the Test

Chapter Test

Cumulative Review

In this chapter, we discuss two closely related functions: exponential and logarithmic functions. These functions are vital to applications in economics, finance, engineering, the sciences, education, and other fields. Models of tumor growth and learning curves are two examples of the uses of exponential and logarithmic functions.

Who Are the Current Players in Streaming Media?

Above, we see some of the streaming media companies trying to get our business in streaming media. Subscriptions cost anywhere from free to about $20 per month. How do we choose? It can depend on what we are looking for—movies, TV shows, live TV, sports shows, or best price.

Interestingly enough, one of the companies is currently owned by Walmart. Also, Google, Inc. owns more than one of these companies. These are just examples that show how competitive and fast-growing is this business of streaming media.

Below, we see the exponential growth in subscriptions of one of the companies above, Netflix.

In Section 12.3, Objective 3 and Exercises 43 and 44, we explore the growth of one streaming media company.

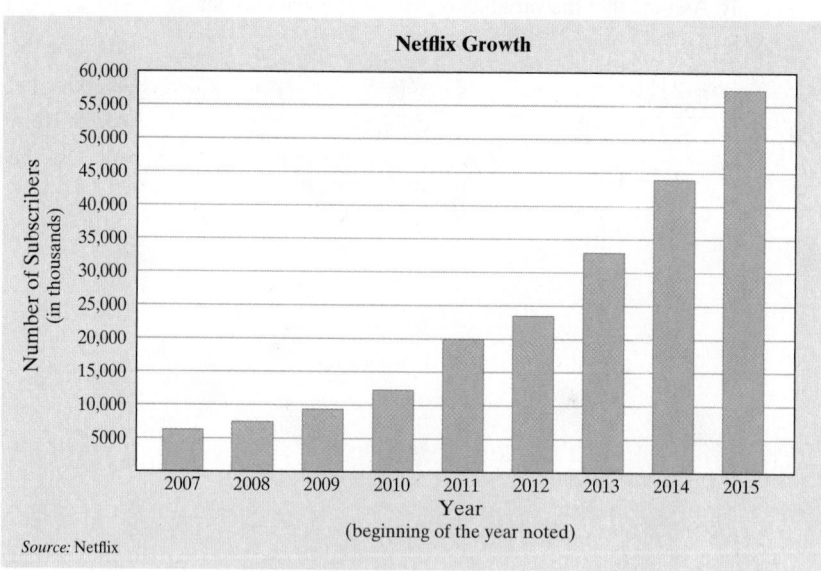

Source: Netflix

12.1 The Algebra of Functions; Composite Functions

OBJECTIVES

1 Add, Subtract, Multiply, and Divide Functions.

2 Construct Composite Functions.

OBJECTIVE

1 Adding, Subtracting, Multiplying, and Dividing Functions

As we have seen in earlier chapters, it is possible to add, subtract, multiply, and divide functions. Although we have not stated it as such, the sums, differences, products, and quotients of functions are themselves functions. For example, if $f(x) = 3x$ and $g(x) = x + 1$, their product, $f(x) \cdot g(x) = 3x(x + 1) = 3x^2 + 3x$, is a new function. We can use the notation $(f \cdot g)(x)$ to denote this new function. Finding the sum, difference, product, and quotient of functions to generate new functions is called the **algebra of functions.**

Algebra of Functions

Let f and g be functions. New functions from f and g are defined as follows.

Sum	$(f + g)(x) = f(x) + g(x)$
Difference	$(f - g)(x) = f(x) - g(x)$
Product	$(f \cdot g)(x) = f(x) \cdot g(x)$
Quotient	$\left(\dfrac{f}{g}\right)(x) = \dfrac{f(x)}{g(x)},\quad g(x) \neq 0$

EXAMPLE 1   If $f(x) = x - 1$ and $g(x) = 2x - 3$, find

a. $(f + g)(x)$ b. $(f - g)(x)$ c. $(f \cdot g)(x)$ d. $\left(\dfrac{f}{g}\right)(x)$

Solution Use the algebra of functions and replace $f(x)$ by $x - 1$ and $g(x)$ by $2x - 3$. Then we simplify.

a. $(f + g)(x) = f(x) + g(x)$
$$= (x - 1) + (2x - 3) = 3x - 4$$

b. $(f - g)(x) = f(x) - g(x)$
$$= (x - 1) - (2x - 3)$$
$$= x - 1 - 2x + 3$$
$$= -x + 2$$

c. $(f \cdot g)(x) = f(x) \cdot g(x)$
$$= (x - 1)(2x - 3)$$
$$= 2x^2 - 5x + 3$$

d. $\left(\dfrac{f}{g}\right)(x) = \dfrac{f(x)}{g(x)} = \dfrac{x - 1}{2x - 3},$ where $x \neq \dfrac{3}{2}$

PRACTICE

1   If $f(x) = x + 2$ and $g(x) = 3x + 5$, find

a. $(f + g)(x)$ b. $(f - g)(x)$ c. $(f \cdot g)(x)$ d. $\left(\dfrac{f}{g}\right)(x)$

There is an interesting but not surprising relationship between the graphs of functions and the graphs of their sum, difference, product, and quotient. For example, the graph of $(f + g)(x)$ can be found by adding the graph of $f(x)$ to the graph of $g(x)$. We add two graphs by adding y-values of corresponding x-values.

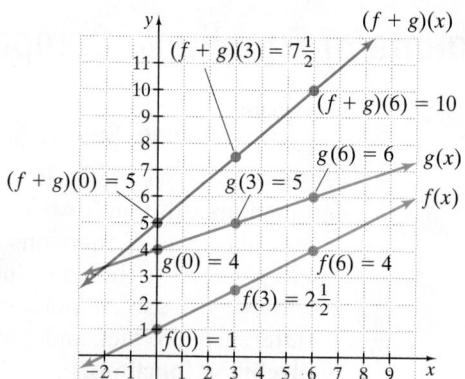

OBJECTIVE

2 Constructing Composite Functions

Another way to combine functions is called **function composition**. To understand this new way of combining functions, study the diagrams below. The right diagram shows an illustration by tables, and the left diagram is the same illustration but by thermometers. In both illustrations, we show degrees Celsius $f(x)$ as a function of degrees Fahrenheit x, then Kelvins $g(x)$ as a function of degrees Celsius x. (The Kelvin scale is a temperature scale devised by Lord Kelvin in 1848.) The first function we will call f, and the second function we will call g.

Table Illustration

x = **Degrees Fahrenheit (Input)**	-13	32	68	212
$f(x)$ = **Degrees Celsius (Output)**	-25	0	20	100

x = **Degrees Celsius (Input)**	-25	0	20	100
$g(x)$ = **Kelvins (Output)**	248.15	273.15	293.15	373.15

Suppose that we want a function that shows a direct conversion from degrees Fahrenheit to Kelvins. In other words, suppose that a function is needed that shows Kelvins as a function of degrees Fahrenheit. This can easily be done because the output of the first function $f(x)$ is the same as the input of the second function. If we use $g(f(x))$ to represent this, then we get the diagrams below.

x = **Degrees Fahrenheit (Input)**	-13	32	68	212
$g(f(x))$ = **Kelvins (Output)**	248.15	273.15	293.15	373.15

For example $g(f(-13)) = 248.15$, and so on.

Since the output of the first function is used as the input of the second function, we write the new function as $g(f(x))$. The new function is formed from the composition of the other two functions. The mathematical symbol for this composition is $(g \circ f)(x)$. Thus, $(g \circ f)(x) = g(f(x))$.

It is possible to find an equation for the composition of the two functions f and g. In other words, we can find a function that converts degrees Fahrenheit directly to Kelvins. The function $f(x) = \dfrac{5}{9}(x - 32)$ converts degrees Fahrenheit to degrees Celsius, and the function $g(x) = x + 273.15$ converts degrees Celsius to Kelvins. Thus,

$$(g \circ f)(x) = g(f(x)) = g\left(\frac{5}{9}(x - 32)\right) = \frac{5}{9}(x - 32) + 273.15$$

In general, the notation $g(f(x))$ means "g composed with f" and can be written as $(g \circ f)(x)$. Also $f(g(x))$, or $(f \circ g)(x)$, means "f composed with g."

Composition of Functions

The composition of functions f and g is

$$(f \circ g)(x) = f(g(x))$$

Helpful Hint

$(f \circ g)(x)$ does not mean the same as $(f \cdot g)(x)$.

$$\underset{\substack{\uparrow \\ \text{Composition of functions}}}{(f \circ g)(x) = f(g(x))} \text{ while } \underset{\substack{\uparrow \\ \text{Multiplication of functions}}}{(f \cdot g)(x) = f(x) \cdot g(x)}$$

EXAMPLE 2   If $f(x) = x^2$ and $g(x) = x + 3$, find each composition.

a. $(f \circ g)(2)$ and $(g \circ f)(2)$ **b.** $(f \circ g)(x)$ and $(g \circ f)(x)$

Solution

a. $(f \circ g)(2) = f(g(2))$

$\qquad\qquad\quad = f(5)$ Replace $g(2)$ with 5. [Since $g(x) = x + 3$, then

$\qquad\qquad\quad = 5^2 = 25$ $g(2) = 2 + 3 = 5$.]

$\quad (g \circ f)(2) = g(f(2))$

$\qquad\qquad\quad = g(4)$ Since $f(x) = x^2$, then $f(2) = 2^2 = 4$.

$\qquad\qquad\quad = 4 + 3 = 7$

b. $(f \circ g)(x) = f(g(x))$

$\qquad\qquad\quad = f(x + 3)$ Replace $g(x)$ with $x + 3$.

$\qquad\qquad\quad = (x + 3)^2$ $f(x + 3) = (x + 3)^2$

$\qquad\qquad\quad = x^2 + 6x + 9$ Square $(x + 3)$.

$\quad (g \circ f)(x) = g(f(x))$

$\qquad\qquad\quad = g(x^2)$ Replace $f(x)$ with x^2.

$\qquad\qquad\quad = x^2 + 3$ $g(x^2) = x^2 + 3$ □

PRACTICE

2   If $f(x) = x^2 + 1$ and $g(x) = 3x - 5$, find

a. $(f \circ g)(4)$ **b.** $(f \circ g)(x)$
$\quad\;\; (g \circ f)(4)$ $(g \circ f)(x)$

EXAMPLE 3   If $f(x) = |x|$ and $g(x) = x - 2$, find each composition.

a. $(f \circ g)(x)$ **b.** $(g \circ f)(x)$

Solution

a. $(f \circ g)(x) = f(g(x)) = f(x - 2) = |x - 2|$

b. $(g \circ f)(x) = g(f(x)) = g(|x|) = |x| - 2$ □

PRACTICE

3   If $f(x) = x^2 + 5$ and $g(x) = x + 3$, find each composition.

a. $(f \circ g)(x)$ **b.** $(g \circ f)(x)$

Helpful Hint

In Examples 2 and 3, notice that $(g \circ f)(x) \neq (f \circ g)(x)$. In general, $(g \circ f)(x)$ *may* or *may not* equal $(f \circ g)(x)$.

EXAMPLE 4 If $f(x) = 5x$, $g(x) = x - 2$, and $h(x) = \sqrt{x}$, write each function as a composition using two of the given functions.

a. $F(x) = \sqrt{x - 2}$ **b.** $G(x) = 5x - 2$

Solution

a. Notice the order in which the function F operates on an input value x. First, 2 is subtracted from x. This is the function $g(x) = x - 2$. Then the square root *of that result* is taken. The square root function is $h(x) = \sqrt{x}$. This means that $F = h \circ g$. To check, we find $h \circ g$.

$$F(x) = (h \circ g)(x) = h(g(x)) = h(x - 2) = \sqrt{x - 2}$$

b. Notice the order in which the function G operates on an input value x. First, x is multiplied by 5, and then 2 is subtracted from the result. This means that $G = g \circ f$. To check, we find $g \circ f$.

$$G(x) = (g \circ f)(x) = g(f(x)) = g(5x) = 5x - 2$$

PRACTICE
4 If $f(x) = 3x$, $g(x) = x - 4$, and $h(x) = |x|$, write each function as a composition using two of the given functions.

a. $F(x) = |x - 4|$ **b.** $G(x) = 3x - 4$

Graphing Calculator Explorations

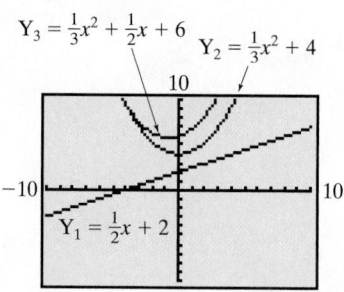

If $f(x) = \dfrac{1}{2}x + 2$ and $g(x) = \dfrac{1}{3}x^2 + 4$, then

$$(f + g)(x) = f(x) + g(x)$$
$$= \left(\frac{1}{2}x + 2\right) + \left(\frac{1}{3}x^2 + 4\right)$$
$$= \frac{1}{3}x^2 + \frac{1}{2}x + 6.$$

To visualize this addition of functions with a graphing calculator, graph

$$Y_1 = \frac{1}{2}x + 2, \qquad Y_2 = \frac{1}{3}x^2 + 4, \qquad Y_3 = \frac{1}{3}x^2 + \frac{1}{2}x + 6$$

Use a TABLE feature to verify that for a given x value, $Y_1 + Y_2 = Y_3$. For example, verify that when $x = 0$, $Y_1 = 2$, $Y_2 = 4$, and $Y_3 = 2 + 4 = 6$.

✓ Vocabulary, Readiness & Video Check

Match each function with its definition.

1. $(f \circ g)(x)$ **4.** $(g \circ f)(x)$ **A.** $g(f(x))$ **D.** $\dfrac{f(x)}{g(x)}, g(x) \neq 0$

2. $(f \cdot g)(x)$ **5.** $\left(\dfrac{f}{g}\right)(x)$ **B.** $f(x) + g(x)$ **E.** $f(x) \cdot g(x)$

3. $(f - g)(x)$ **6.** $(f + g)(x)$ **C.** $f(g(x))$ **F.** $f(x) - g(x)$

Martin-Gay Interactive Videos

See Video 12.1 ⊙

Watch the section lecture video and answer the following questions.

OBJECTIVE
1   **7.** From ▦ Example 1 and the lecture before, we know that $(f + g)(x) = f(x) + g(x)$. Use this fact to explain two ways you can find $(f + g)(2)$.

OBJECTIVE
2   **8.** From ▦ Example 3, given two functions $f(x)$ and $g(x)$, can $f(g(x))$ ever equal $g(f(x))$?

12.1   Exercise Set   MyMathLab® ▶

For the functions f and g, find ***a.*** $(f + g)(x)$, ***b.*** $(f - g)(x)$, ***c.*** $(f \cdot g)(x)$, *and* ***d.*** $\left(\dfrac{f}{g}\right)(x)$. *See Example 1.*

1. $f(x) = x - 7, g(x) = 2x + 1$

2. $f(x) = x + 4, g(x) = 5x - 2$

3. $f(x) = x^2 + 1, g(x) = 5x$

4. $f(x) = x^2 - 2, g(x) = 3x$

5. $f(x) = \sqrt[3]{x}, g(x) = x + 5$

6. $f(x) = \sqrt[3]{x}, g(x) = x - 3$

7. $f(x) = -3x, g(x) = 5x^2$

8. $f(x) = 4x^3, g(x) = -6x$

If $f(x) = x^2 - 6x + 2, g(x) = -2x$, and $h(x) = \sqrt{x}$, find each composition. See Example 2.

9. $(f \circ g)(2)$ **10.** $(h \circ f)(-2)$

11. $(g \circ f)(-1)$ **12.** $(f \circ h)(1)$

13. $(g \circ h)(0)$ **14.** $(h \circ g)(0)$

Find $(f \circ g)(x)$ and $(g \circ f)(x)$. See Examples 2 and 3.

15. $f(x) = x^2 + 1, g(x) = 5x$

16. $f(x) = x - 3, g(x) = x^2$

17. $f(x) = 2x - 3, g(x) = x + 7$

18. $f(x) = x + 10, g(x) = 3x + 1$

19. $f(x) = x^3 + x - 2, g(x) = -2x$

20. $f(x) = -4x, g(x) = x^3 + x^2 - 6$

21. $f(x) = |x|; g(x) = 10x - 3$

22. $f(x) = |x|; g(x) = 14x - 8$

23. $f(x) = \sqrt{x}, g(x) = -5x + 2$

24. $f(x) = 7x - 1, g(x) = \sqrt[3]{x}$

If $f(x) = 3x, g(x) = \sqrt{x}$, and $h(x) = x^2 + 2$, write each function as a composition using two of the given functions. See Example 4.

25. $H(x) = \sqrt{x^2 + 2}$

26. $G(x) = \sqrt{3x}$

27. $F(x) = 9x^2 + 2$

28. $H(x) = 3x^2 + 6$

29. $G(x) = 3\sqrt{x}$

30. $F(x) = x + 2$

Find $f(x)$ and $g(x)$ so that the given function $h(x) = (f \circ g)(x)$.

31. $h(x) = (x + 2)^2$

32. $h(x) = |x - 1|$

33. $h(x) = \sqrt{x + 5} + 2$

34. $h(x) = (3x + 4)^2 + 3$

35. $h(x) = \dfrac{1}{2x - 3}$

36. $h(x) = \dfrac{1}{x + 10}$

REVIEW AND PREVIEW

Solve each equation for y. See Section 2.5.

37. $x = y + 2$

38. $x = y - 5$

39. $x = 3y$

40. $x = -6y$

41. $x = -2y - 7$

42. $x = 4y + 7$

CONCEPT EXTENSIONS

Given that $f(-1) = 4$ $g(-1) = -4$
$$f(0) = 5 \quad g(0) = -3$$
$$f(2) = 7 \quad g(2) = -1$$
$$f(7) = 1 \quad g(7) = 4$$
find each function value.

43. $(f + g)(2)$

44. $(f - g)(7)$

45. $(f \circ g)(2)$

46. $(g \circ f)(2)$

47. $(f \cdot g)(7)$

48. $(f \cdot g)(0)$

49. $\left(\dfrac{f}{g}\right)(-1)$

50. $\left(\dfrac{g}{f}\right)(-1)$

51. If you are given $f(x)$ and $g(x)$, explain in your own words how to find $(f \circ g)(x)$ and then how to find $(g \circ f)(x)$.

52. Given $f(x)$ and $g(x)$, describe in your own words the difference between $(f \circ g)(x)$ and $(f \cdot g)(x)$.

Solve.

53. Business people are concerned with cost functions, revenue functions, and profit functions. Recall that the profit $P(x)$ obtained from x units of a product is equal to the revenue $R(x)$ from selling the x units minus the cost $C(x)$ of manufacturing the x units. Write an equation expressing this relationship among $C(x)$, $R(x)$, and $P(x)$.

54. Suppose the revenue $R(x)$ for x units of a product can be described by $R(x) = 25x$, and the cost $C(x)$ can be described by $C(x) = 50 + x^2 + 4x$. Find the profit $P(x)$ for x units. (See Exercise 53.)

12.2 Inverse Functions

OBJECTIVES

1 Determine Whether a Function Is a One-to-One Function.

2 Use the Horizontal Line Test to Decide Whether a Function Is a One-to-One Function.

3 Find the Inverse of a Function.

4 Find the Equation of the Inverse of a Function.

5 Graph Functions and Their Inverses.

6 Determine Whether Two Functions Are Inverses of Each Other.

OBJECTIVE

1 Determining Whether a Function Is One-To-One

In the next three sections, we begin a study of two new functions: exponential and logarithmic functions. As we learn more about these functions, we will discover that they share a special relation to each other: They are inverses of each other.

Before we study these functions, we need to learn about inverses. We begin by defining one-to-one functions.

Study the following table.

Degrees Fahrenheit (Input)	−31	−13	32	68	149	212
Degrees Celsius (Output)	−35	−25	0	20	65	100

Recall that since each degrees Fahrenheit (input) corresponds to exactly one degrees Celsius (output), this pairing of inputs and outputs does describe a function. Also notice that each output corresponds to exactly one input. This type of function is given a special name—a one-to-one function.

Does the set $f = \{(0, 1), (2, 2), (-3, 5), (7, 6)\}$ describe a one-to-one function? It is a function since each x-value corresponds to a unique y-value. For this particular function f, each y-value also corresponds to a unique x-value. Thus, this function is also a **one-to-one function.**

> **One-to-One Function**
>
> For a **one-to-one function,** each x-value (input) corresponds to only one y-value (output), and each y-value (output) corresponds to only one x-value (input).

EXAMPLE 1 Determine whether each function described is one-to-one.

a. $f = \{(6, 2), (5, 4), (-1, 0), (7, 3)\}$

b. $g = \{(3, 9), (-4, 2), (-3, 9), (0, 0)\}$

c. $h = \{(1, 1), (2, 2), (10, 10), (-5, -5)\}$

d.

Mineral (Input)	Talc	Gypsum	Diamond	Topaz	Stibnite
Hardness on the Mohs Scale (Output)	1	2	10	8	2

e.

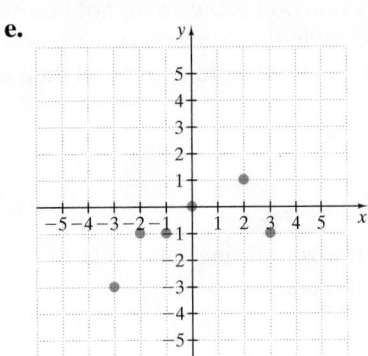

f.

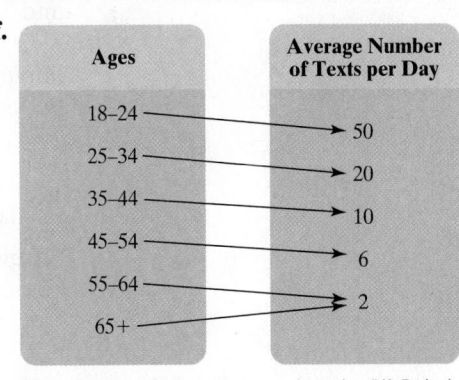

(*Source:* Pew Research Center Internet and American Life Project)

Solution

a. f is one-to-one since each y-value corresponds to only one x-value.

b. g is not one-to-one because the y-value 9 in $(3, 9)$ and $(-3, 9)$ corresponds to different x-values.

c. h is a one-to-one function since each y-value corresponds to only one x-value.

d. This table does not describe a one-to-one function since the output 2 corresponds to two inputs, gypsum and stibnite.

e. This graph does not describe a one-to-one function since the y-value -1 corresponds to three x-values, -2, -1, and 3. (See the graph to the left.)

f. The mapping is not one-to-one since 2 texts per day corresponds to both ages 55–64 and ages 65+. □

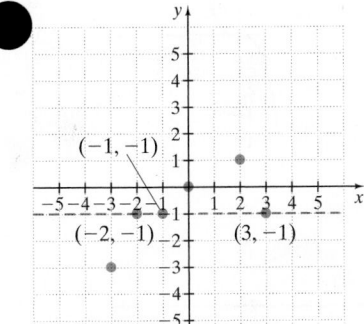

PRACTICE

1 Determine whether each function described is one-to-one.

a. $f = \{(4, -3), (3, -4), (2, 7), (5, 0)\}$

b. $g = \{(8, 4), (-2, 0), (6, 4), (2, 6)\}$

c. $h = \{(2, 4), (1, 3), (4, 6), (-2, 4)\}$

d.

Year	1950	1963	1968	1975	1997	2008
Federal Minimum Wage	$0.75	$1.25	$1.60	$2.10	$5.15	$6.55

e.

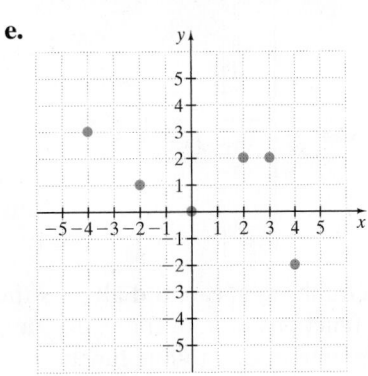

f.

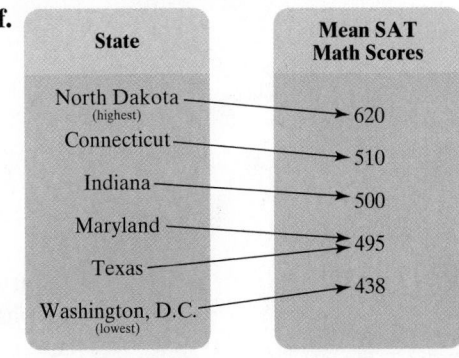

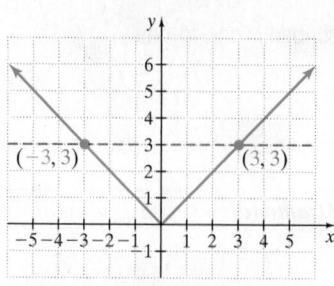

Not a one-to-one function.

2 Using the Horizontal Line Test

Recall that we recognize the graph of a function when it passes the vertical line test. Since every x-value of the function corresponds to exactly one y-value, each vertical line intersects the function's graph at most once. The graph shown (left), for instance, is the graph of a function.

Is this function a *one-to-one* function? The answer is no. To see why not, notice that the y-value of the ordered pair $(-3, 3)$, for example, is the same as the y-value of the ordered pair $(3, 3)$. In other words, the y-value 3 corresponds to two x-values, -3 and 3. This function is therefore not one-to-one.

To test whether a graph is the graph of a one-to-one function, apply the vertical line test to see if it is a function and then apply a similar **horizontal line test** to see if it is a one-to-one function.

Horizontal Line Test

If every horizontal line intersects the graph of a function at most once, then the function is a one-to-one function.

EXAMPLE 2   Determine whether each graph is the graph of a one-to-one function.

a.
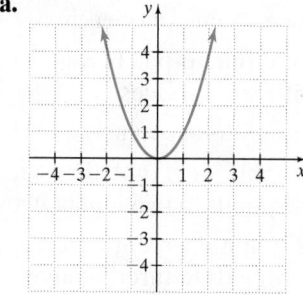

b.
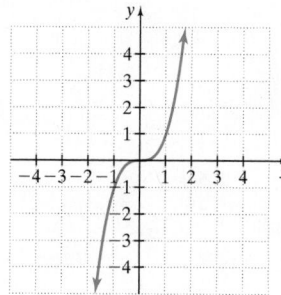

c.
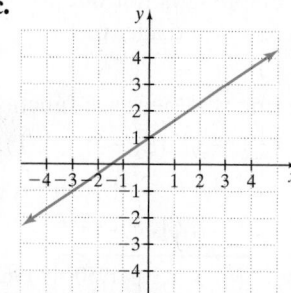

d.
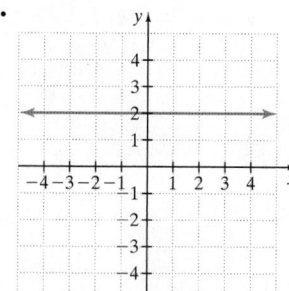

e.
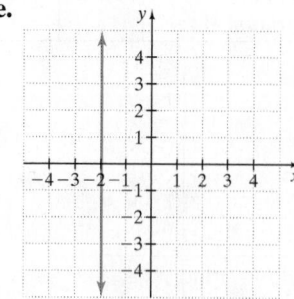

Solution Graphs **a, b, c,** and **d** all pass the vertical line test, so only these graphs are graphs of functions. But, of these, only **b** and **c** pass the horizontal line test, so only **b** and **c** are graphs of one-to-one functions.

PRACTICE

2 Determine whether each graph is the graph of a one-to-one function.

a.
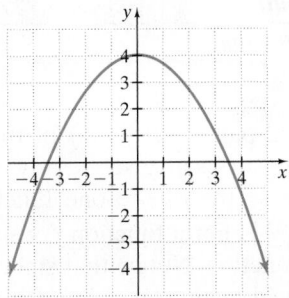

b.
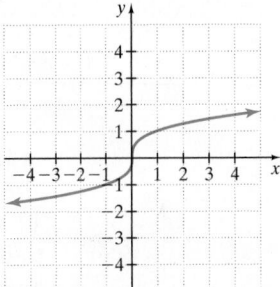

c.
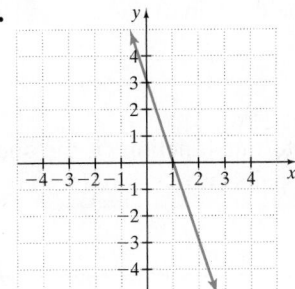

d.
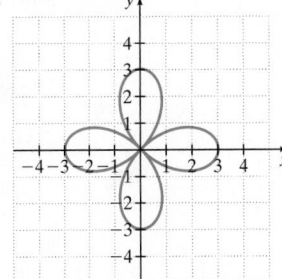

e.
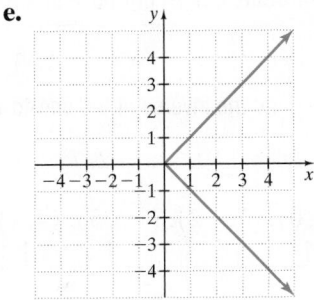

Helpful Hint

All linear equations are one-to-one functions except those whose graphs are horizontal or vertical lines. A vertical line does not pass the vertical line test and hence is not the graph of a function. A horizontal line is the graph of a function but does not pass the horizontal line test and hence is not the graph of a one-to-one function.

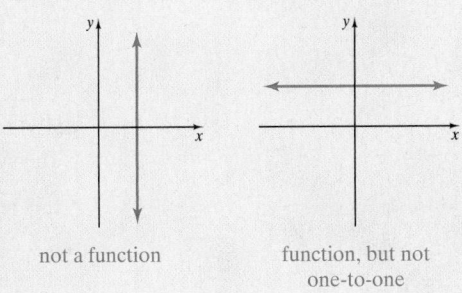

not a function   function, but not one-to-one

OBJECTIVE

3 Finding the Inverse of a Function ▶

One-to-one functions are special in that their graphs pass both the vertical and horizontal line tests. They are special, too, in another sense: For each one-to-one function, we can find its **inverse function** by switching the coordinates of the ordered pairs of the function, or the inputs and the outputs. For example, the inverse of the one-to-one function

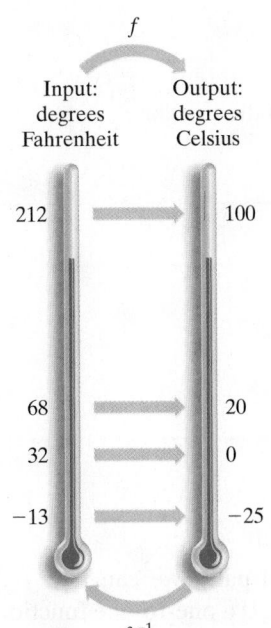

Degrees Fahrenheit (Input)	−31	−13	32	68	149	212
Degrees Celsius (Output)	−35	−25	0	20	65	100

is the function

Degrees Celsius (Input)	−35	−25	0	20	65	100
Degrees Fahrenheit (Output)	−31	−13	32	68	149	212

Notice that the ordered pair $(-31, -35)$ of the function, for example, becomes the ordered pair $(-35, -31)$ of its inverse.

Also, the inverse of the one-to-one function $f = \{(2, -3), (5, 10), (9, 1)\}$ is $\{(-3, 2), (10, 5), (1, 9)\}$. For a function f, we use the notation f^{-1}, read "f inverse," to denote its inverse function. Notice that since the coordinates of each ordered pair have been switched, the domain (set of inputs) of f is the range (set of outputs) of f^{-1}, and the range of f is the domain of f^{-1}.

Inverse Function

The inverse of a one-to-one function f is the one-to-one function f^{-1} that consists of the set of all ordered pairs (y, x) where (x, y) belongs to f.

Helpful Hint

If a function is not one-to-one, it does not have an inverse function.

EXAMPLE 3   Find the inverse of the one-to-one function.

$$f = \{(0, 1), (-2, 7), (3, -6), (4, 4)\}$$

Solution $f^{-1} = \{(1, 0), (7, -2), (-6, 3), (4, 4)\}$
Switch coordinates of each ordered pair.

PRACTICE
3   Find the inverse of the one-to-one function.
$$f = \{(3, 4), (-2, 0), (2, 8), (6, 6)\}$$

Helpful Hint

The symbol f^{-1} is the single symbol that denotes the inverse of the function f.

It is read as "f inverse." This symbol *does not mean* $\dfrac{1}{f}$.

✔ **CONCEPT CHECK**

Suppose that f is a one-to-one function and that $f(1) = 5$.
a. Write the corresponding ordered pair.
b. Write one point that we know must belong to the inverse function f^{-1}.

OBJECTIVE
4   Finding the Equation of the Inverse of a Function ▶

If a one-to-one function f is defined as a set of ordered pairs, we can find f^{-1} by interchanging the x- and y-coordinates of the ordered pairs. If a one-to-one function f is given in the form of an equation, we can find f^{-1} by using a similar procedure.

Finding the Inverse of a One-to-One Function $f(x)$

Step 1. Replace $f(x)$ with y.

Step 2. Interchange x and y.

Step 3. Solve the equation for y.

Step 4. Replace y with the notation $f^{-1}(x)$.

EXAMPLE 4   Find an equation of the inverse of $f(x) = x + 3$.

Solution $f(x) = x + 3$

Step 1. $y = x + 3$ Replace $f(x)$ with y.

Step 2. $x = y + 3$ Interchange x and y.

Step 3. $x - 3 = y$ Solve for y.

Step 4. $f^{-1}(x) = x - 3$ Replace y with $f^{-1}(x)$.

The inverse of $f(x) = x + 3$ is $f^{-1}(x) = x - 3$. Notice that, for example,

$$f(1) = 1 + 3 = 4 \quad \text{and} \quad f^{-1}(4) = 4 - 3 = 1$$

Ordered pair: $(1, 4)$ Ordered pair: $(4, 1)$

The coordinates are switched, as expected.   □

PRACTICE

4   Find the equation of the inverse of $f(x) = 6 - x$.   ■

EXAMPLE 5   Find the equation of the inverse of $f(x) = 3x - 5$. Graph f and f^{-1} on the same set of axes.

Solution $f(x) = 3x - 5$

Step 1. $y = 3x - 5$ Replace $f(x)$ with y.

Step 2. $x = 3y - 5$ Interchange x and y.

Step 3. $3y = x + 5$ Solve for y.

$$y = \frac{x + 5}{3}$$

Step 4. $f^{-1}(x) = \dfrac{x + 5}{3}$ Replace y with $f^{-1}(x)$.

Now we graph $f(x)$ and $f^{-1}(x)$ on the same set of axes. Both $f(x) = 3x - 5$ and $f^{-1}(x) = \dfrac{x + 5}{3}$ are linear functions, so each graph is a line.

$f(x) = 3x - 5$	
x	$y = f(x)$
1	-2
0	-5
$\dfrac{5}{3}$	0

$f^{-1}(x) = \dfrac{x + 5}{3}$	
x	$y = f^{-1}(x)$
-2	1
-5	0
0	$\dfrac{5}{3}$

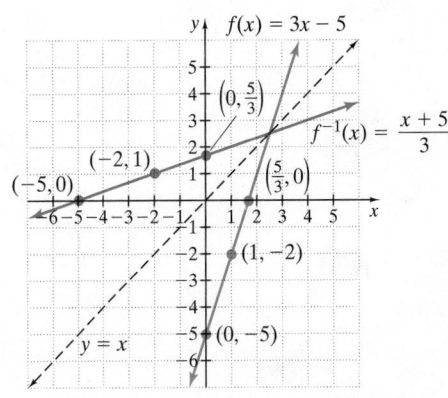

□

(Continued on next page)

PRACTICE
5
Find the equation of the inverse of $f(x) = 5x + 2$. Graph f and f^{-1} on the same set of axes.

OBJECTIVE

5 Graphing Inverse Functions

Notice that the graphs of f and f^{-1} in Example 5 are mirror images of each other, and the "mirror" is the dashed line $y = x$. This is true for every function and its inverse. For this reason, we say that *the graphs of f and f^{-1} are symmetric about the line $y = x$.*

To see why this happens, study the graph of a few ordered pairs and their switched coordinates in the diagram below.

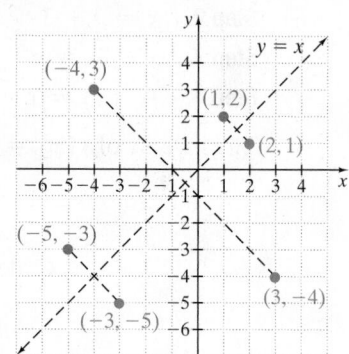

EXAMPLE 6 Graph the inverse of each function.

<u>Solution</u> The function is graphed in blue and the inverse is graphed in red.

a.

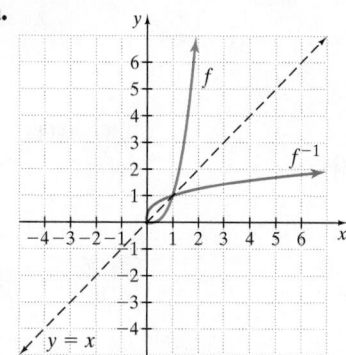

b.

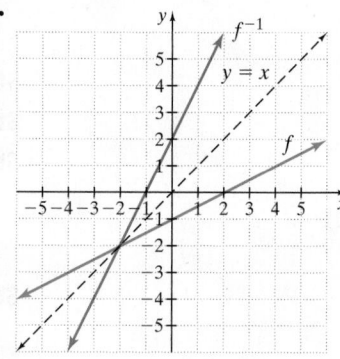

PRACTICE
6
Graph the inverse of each function.

a.

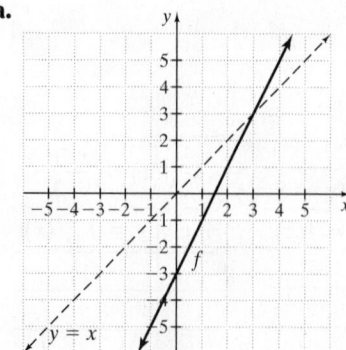

b.

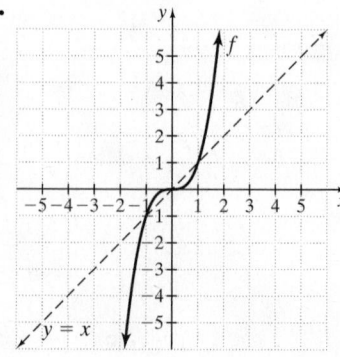

OBJECTIVE

6   Determining Whether Functions Are Inverses of Each Other

Notice in the table of values in Example 5 that $f(0) = -5$ and $f^{-1}(-5) = 0$, as expected. Also, for example, $f(1) = -2$ and $f^{-1}(-2) = 1$. In words, we say that for some input x, the function f^{-1} takes the output of x, called $f(x)$, back to x.

$$x \rightarrow f(x) \quad \text{and} \quad f^{-1}(f(x)) \rightarrow x$$
$$\downarrow \qquad \downarrow \qquad\qquad \downarrow \qquad \downarrow$$
$$f(0) = -5 \quad \text{and} \quad f^{-1}(-5) = 0$$
$$f(1) = -2 \quad \text{and} \quad f^{-1}(-2) = 1$$

In general,

If f is a one-to-one function, then the inverse of f is the function f^{-1} such that
$$(f^{-1} \circ f)(x) = x \quad \text{and} \quad (f \circ f^{-1})(x) = x$$

EXAMPLE 7   Show that if $f(x) = 3x + 2$, then $f^{-1}(x) = \dfrac{x - 2}{3}$.

Solution   See that $(f^{-1} \circ f)(x) = x$ and $(f \circ f^{-1})(x) = x$.

$$(f^{-1} \circ f)(x) = f^{-1}(f(x))$$
$$= f^{-1}(3x + 2) \qquad \text{Replace } f(x) \text{ with } 3x + 2.$$
$$= \frac{3x + 2 - 2}{3}$$
$$= \frac{3x}{3}$$
$$= x$$

$$(f \circ f^{-1})(x) = f(f^{-1}(x))$$
$$= f\left(\frac{x - 2}{3}\right) \qquad \text{Replace } f^{-1}(x) \text{ with } \frac{x - 2}{3}.$$
$$= 3\left(\frac{x - 2}{3}\right) + 2$$
$$= x - 2 + 2$$
$$= x$$

PRACTICE

7   Show that if $f(x) = 4x - 1$, then $f^{-1}(x) = \dfrac{x + 1}{4}$.

Graphing Calculator Explorations

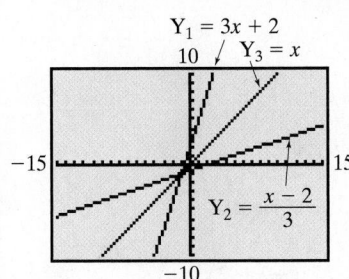

$Y_1 = 3x + 2$
$Y_3 = x$
$Y_2 = \dfrac{x - 2}{3}$

A graphing calculator can be used to visualize the results of Example 7. Recall that the graph of a function f and its inverse f^{-1} are mirror images of each other across the line $y = x$. To see this for the function from Example 7, use a square window and graph

the given function: $Y_1 = 3x + 2$

its inverse: $Y_2 = \dfrac{x - 2}{3}$

and the line: $Y_3 = x$

See Exercises 67–70 in Exercise Set 12.2.

✓ Vocabulary, Readiness & Video Check

Use the choices below to fill in each blank. Some choices will not be used, and some will be used more than once.

vertical	$(3, 7)$	$(11, 2)$	$y = x$	x	true
horizontal	$(7, 3)$	$(2, 11)$	$\dfrac{1}{f}$	the inverse of f	false

1. If $f(2) = 11$, the corresponding ordered pair is _____.

2. If $(7, 3)$ is an ordered pair solution of $f(x)$, and $f(x)$ has an inverse, then an ordered pair solution of $f^{-1}(x)$ is _____.

3. The symbol f^{-1} means _____.

4. True or false: The function notation $f^{-1}(x)$ means $\dfrac{1}{f(x)}$. _____

5. To tell whether a graph is the graph of a function, use the _____ line test.

6. To tell whether the graph of a function is also a one-to-one function, use the _____ line test.

7. The graphs of f and f^{-1} are symmetric about the line _____.

8. Two functions are inverses of each other if $(f \circ f^{-1})(x) =$ _____ and $(f^{-1} \circ f)(x) =$ _____.

Martin-Gay Interactive Videos

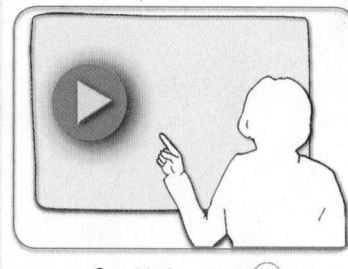

See Video 12.2 ⬤

Watch the section lecture video and answer the following questions.

OBJECTIVE 1
9. From ▦ Example 1 and the definition before, what makes a one-to-one function different from other types of functions?

OBJECTIVE 2
10. From ▦ Examples 2 and 3, if a graph passes the horizontal line test, but not the vertical line test, is it a one-to-one function? Explain.

OBJECTIVE 3
11. From ▦ Example 4 and the lecture before, if you find the inverse of a one-to-one function, is this inverse function also a one-to-one function? How do you know?

OBJECTIVE 4
12. From ▦ Examples 5 and 6, explain why the interchanging of x and y when finding an inverse equation makes sense given the definition of an inverse function.

OBJECTIVE 5
13. From ▦ Example 7, if you have the equation or graph of a one-to-one function, how can you graph its inverse without finding the inverse's equation?

OBJECTIVE 6
14. Based on ▦ Example 8 and the lecture before, what's wrong with the following statement? "If f is a one-to-one function, you can prove that f and f^{-1} are inverses of each other by showing that $f(f^{-1}(x)) = f^{-1}(f(x))$."

12.2 Exercise Set MyMathLab

Determine whether each function is a one-to-one function. If it is one-to-one, list the inverse function by switching coordinates, or inputs and outputs. See Examples 1 and 3.

1. $f = \{(-1, -1), (1, 1), (0, 2), (2, 0)\}$

2. $g = \{(8, 6), (9, 6), (3, 4), (-4, 4)\}$

3. $h = \{(10, 10)\}$

4. $r = \{(1, 2), (3, 4), (5, 6), (6, 7)\}$

▶ 5. $f = \{(11, 12), (4, 3), (3, 4), (6, 6)\}$

6. $g = \{(0, 3), (3, 7), (6, 7), (-2, -2)\}$

7.

Month of 2015 (Input)	January	February	March	April	May	June
Unemployment Rate in Percent (Output)	5.7	5.5	5.5	5.4	5.5	5.3

(*Source:* Bureau of Labor Statistics)

8.

State (Input)	Texas	Massachusetts	Nevada	Idaho	Wisconsin
Number of Two-Year Colleges (Output)	70	22	3	3	31

(*Source:* University of Texas at Austin)

9.

State (Input)	California	Alaska	Arizona	Louisiana	New Mexico	Ohio
Rank in Population (Output)	1	47	16	25	36	7

(*Source:* U.S. Bureau of the Census)

△ **10.**

Shape (Input)	Triangle	Pentagon	Quadrilateral	Hexagon	Decagon
Number of Sides (Output)	3	5	4	6	10

Given the one-to-one function $f(x) = x^3 + 2$, find the following.
[Hint: You do not need to find the equation for $f^{-1}(x)$.]

11. a. $f(1)$
   b. $f^{-1}(3)$

12. a. $f(0)$
   b. $f^{-1}(2)$

13. a. $f(-1)$
   b. $f^{-1}(1)$

14. a. $f(-2)$
   b. $f^{-1}(-6)$

Determine whether the graph of each function is the graph of a one-to-one function. See Example 2.

▶ **15.**

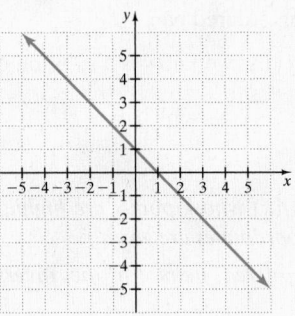

16.

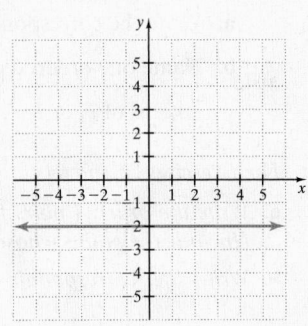

▶ **17.**

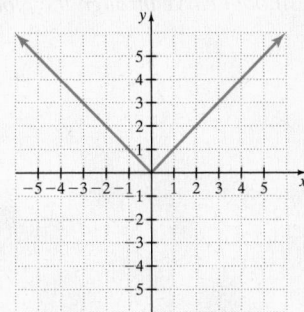

18.

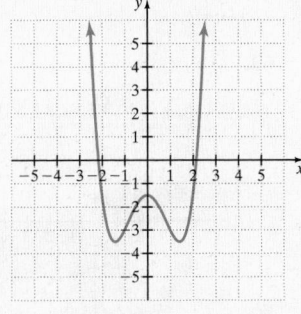

19.

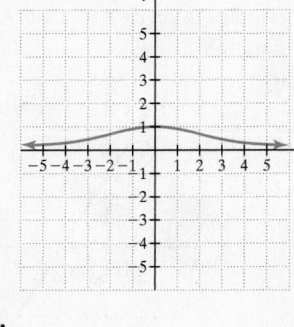

20.

21.

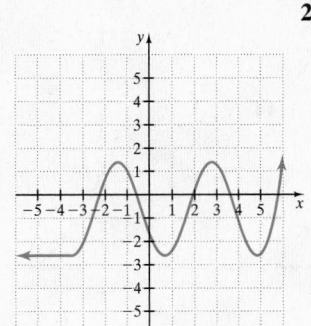

22.

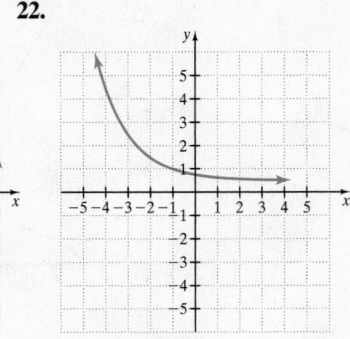

MIXED PRACTICE

Each of the following functions is one-to-one. Find the inverse of each function and graph the function and its inverse on the same set of axes. See Examples 4 and 5.

23. $f(x) = x + 4$

24. $f(x) = x - 5$

▶ **25.** $f(x) = 2x - 3$

26. $f(x) = 4x + 9$

27. $f(x) = \dfrac{1}{2}x - 1$

28. $f(x) = -\dfrac{1}{2}x + 2$

29. $f(x) = x^3$

30. $f(x) = x^3 - 1$

Find the inverse of each one-to-one function. See Examples 4 and 5.

31. $f(x) = 5x + 2$

32. $f(x) = 6x - 1$

33. $f(x) = \dfrac{x - 2}{5}$

34. $f(x) = \dfrac{x - 3}{2}$

35. $f(x) = \sqrt[3]{x}$

36. $f(x) = \sqrt[3]{x + 1}$

▶ **37.** $f(x) = \dfrac{5}{3x + 1}$

38. $f(x) = \dfrac{7}{2x + 4}$

39. $f(x) = (x + 2)^3$

40. $f(x) = (x - 5)^3$

Graph the inverse of each function on the same set of axes. See Example 6.

41.

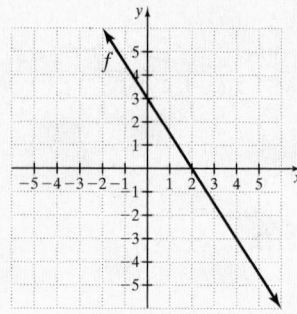

42.

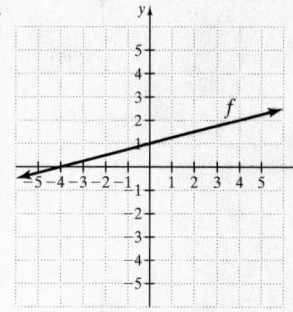

43.

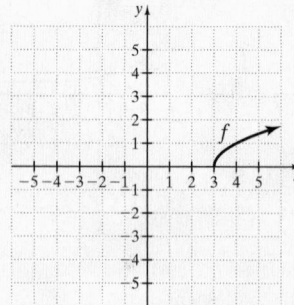

44.

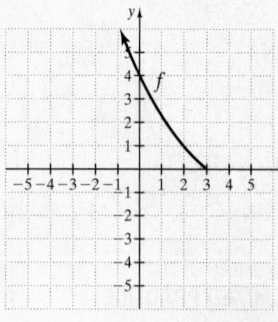

45.

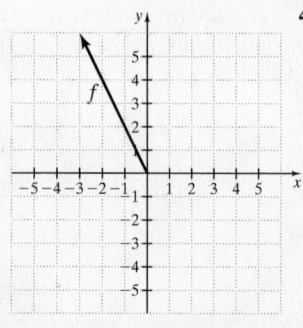

46.

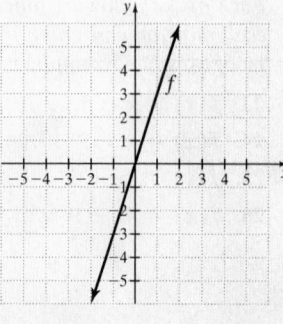

Solve. See Example 7.

▶ **47.** If $f(x) = 2x + 1$, show that $f^{-1}(x) = \dfrac{x - 1}{2}$.

48. If $f(x) = 3x - 10$, show that $f^{-1}(x) = \dfrac{x + 10}{3}$.

49. If $f(x) = x^3 + 6$, show that $f^{-1}(x) = \sqrt[3]{x - 6}$.

50. If $f(x) = x^3 - 5$, show that $f^{-1}(x) = \sqrt[3]{x + 5}$.

REVIEW AND PREVIEW

Evaluate each of the following. See Section 10.2.

51. $25^{1/2}$ **52.** $49^{1/2}$

53. $16^{3/4}$ **54.** $27^{2/3}$

55. $9^{-3/2}$ **56.** $81^{-3/4}$

If $f(x) = 3^x$, find the following. In Exercises 59 and 60, give the exact answer and a two-decimal-place approximation. See Sections 5.1, 8.2, and 10.2.

57. $f(2)$ **58.** $f(0)$

59. $f\left(\dfrac{1}{2}\right)$ **60.** $f\left(\dfrac{2}{3}\right)$

CONCEPT EXTENSIONS

Solve. See the Concept Check in this section.

61. Suppose that f is a one-to-one function and that $f(2) = 9$.

    a. Write the corresponding ordered pair.

    b. Name one ordered pair that we know is a solution of the inverse of f, or f^{-1}.

62. Suppose that F is a one-to-one function and that $F\left(\dfrac{1}{2}\right) = -0.7$.

    a. Write the corresponding ordered pair.

    b. Name one ordered pair that we know is a solution of the inverse of F, or F^{-1}.

For Exercises 63 and 64,

    a. *Write the ordered pairs for $f(x)$ whose points are highlighted. (Include the points whose coordinates are given.)*

    b. *Write the corresponding ordered pairs for the inverse of f, f^{-1}.*

    c. *Graph the ordered pairs for f^{-1} found in part **b**.*

    d. *Graph $f^{-1}(x)$ by drawing a smooth curve through the plotted points.*

63.

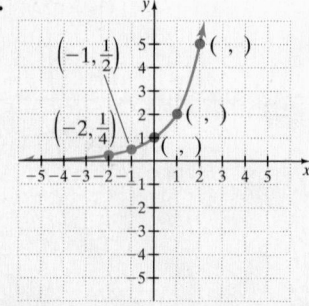

64.

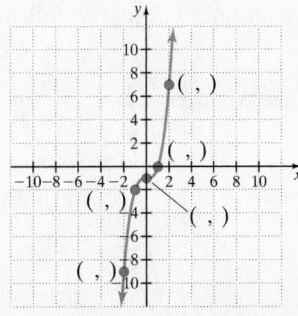

65. If you are given the graph of a function, describe how you can tell from the graph whether a function has an inverse.

66. Describe the appearance of the graphs of a function and its inverse.

Find the inverse of each given one-to-one function. Then use a graphing calculator to graph the function and its inverse on a square window.

67. $f(x) = 3x + 1$

68. $f(x) = -2x - 6$

69. $f(x) = \sqrt[3]{x} + 1$

70. $f(x) = x^3 - 3$

12.3 **Exponential Functions**

OBJECTIVES

1 Graph Exponential Functions.

2 Solve Equations of the Form $b^x = b^y$.

3 Solve Problems Modeled by Exponential Equations.

OBJECTIVE

1 **Graphing Exponential Functions**

In earlier chapters, we gave meaning to exponential expressions such as 2^x, where x is a rational number. For example,

$$2^3 = 2 \cdot 2 \cdot 2 \qquad \text{Three factors; each factor is 2}$$
$$2^{3/2} = (2^{1/2})^3 = \sqrt{2} \cdot \sqrt{2} \cdot \sqrt{2} \quad \text{Three factors; each factor is } \sqrt{2}$$

When x is an irrational number (for example, $\sqrt{3}$), what meaning can we give to $2^{\sqrt{3}}$?

It is beyond the scope of this book to give precise meaning to 2^x if x is irrational. We can confirm your intuition and say that $2^{\sqrt{3}}$ is a real number and, since $1 < \sqrt{3} < 2, 2^1 < 2^{\sqrt{3}} < 2^2$. We can also use a calculator and approximate $2^{\sqrt{3}}: 2^{\sqrt{3}} \approx 3.321997$. In fact, as long as the base b is positive, b^x is a real number for all real numbers x. Finally, the rules of exponents apply whether x is rational or irrational as long as b is positive. In this section, we are interested in functions of the form $f(x) = b^x$, where $b > 0$. A function of this form is called an **exponential function.**

> **Exponential Function**
>
> A function of the form
>
> $$f(x) = b^x$$
>
> is called an **exponential function** if $b > 0$, b is not 1, and x is a real number.

Next, we practice graphing exponential functions.

EXAMPLE 1 Graph the exponential functions defined by $f(x) = 2^x$ and $g(x) = 3^x$ on the same set of axes.

Solution Graph each function by plotting points. Set up a table of values for each of the two functions.

(Continued on next page)

If each set of points is plotted and connected with a smooth curve, the following graphs result.

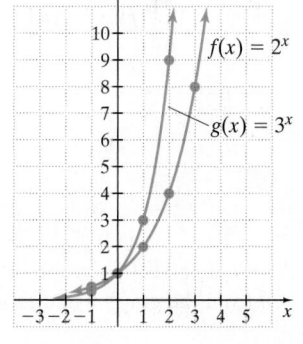

x	0	1	2	3	-1	-2
$f(x)$	1	2	4	8	$\dfrac{1}{2}$	$\dfrac{1}{4}$

$f(x) = 2^x$

x	0	1	2	3	-1	-2
$g(x)$	1	3	9	27	$\dfrac{1}{3}$	$\dfrac{1}{9}$

$g(x) = 3^x$

PRACTICE

1 Graph the exponential functions defined by $f(x) = 2^x$ and $g(x) = 7^x$ on the same set of axes. ∎

A number of things should be noted about the two graphs of exponential functions in Example 1. First, the graphs show that $f(x) = 2^x$ and $g(x) = 3^x$ are one-to-one functions since each graph passes the vertical and horizontal line tests. The y-intercept of each graph is $(0, 1)$, but neither graph has an x-intercept. From the graph, we can also see that the domain of each function is all real numbers and that the range is $(0, \infty)$. We can also see that as x-values are increasing, y-values are increasing also.

EXAMPLE 2 Graph the exponential functions $y = \left(\dfrac{1}{2}\right)^x$ and $y = \left(\dfrac{1}{3}\right)^x$ on the same set of axes.

Solution As before, plot points and connect them with a smooth curve.

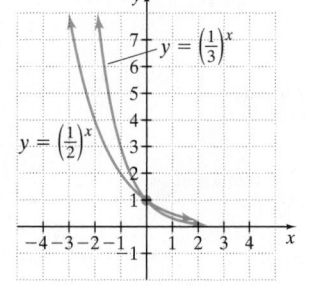

x	0	1	2	3	-1	-2
y	1	$\dfrac{1}{2}$	$\dfrac{1}{4}$	$\dfrac{1}{8}$	2	4

$y = \left(\dfrac{1}{2}\right)^x$

x	0	1	2	3	-1	-2
y	1	$\dfrac{1}{3}$	$\dfrac{1}{9}$	$\dfrac{1}{27}$	3	9

$y = \left(\dfrac{1}{3}\right)^x$

PRACTICE

2 Graph the exponential functions $f(x) = \left(\dfrac{1}{3}\right)^x$ and $g(x) = \left(\dfrac{1}{5}\right)^x$ on the same set of axes. ∎

Each function in Example 2 again is a one-to-one function. The y-intercept of both is $(0, 1)$. The domain is the set of all real numbers, and the range is $(0, \infty)$.

Notice the difference between the graphs of Example 1 and the graphs of Example 2. An exponential function is always increasing if the base is greater than 1.

When the base is between 0 and 1, the graph is always decreasing. The figures on the next page summarize these characteristics of exponential functions.

$$f(x) = b^x, \quad b > 0, \quad b \neq 1$$

- one-to-one function
- y-intercept $(0, 1)$
- no x-intercept

- domain: $(-\infty, \infty)$
- range: $(0, \infty)$

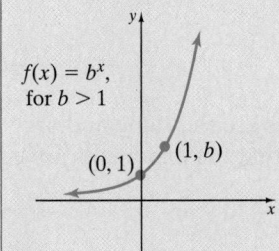

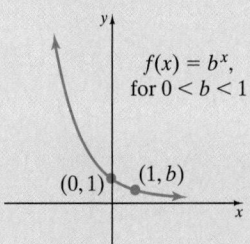

EXAMPLE 3   Graph the exponential function $f(x) = 3^{x+2}$.

Solution   As before, we find and plot a few ordered pair solutions. Then we connect the points with a smooth curve.

$f(x) = 3^{x+2}$	
x	$f(x)$
0	9
−1	3
−2	1
−3	$\dfrac{1}{3}$
−4	$\dfrac{1}{9}$

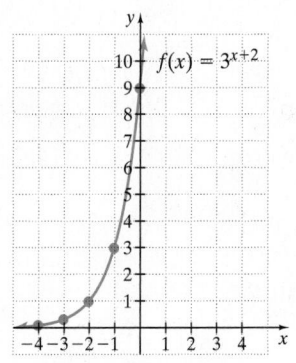

PRACTICE

3   Graph the exponential function $f(x) = 2^{x-3}$.

✔ **CONCEPT CHECK**

Which functions are exponential functions?

a. $f(x) = x^3$ **b.** $g(x) = \left(\dfrac{2}{3}\right)^x$ **c.** $h(x) = 5^{x-2}$ **d.** $w(x) = (2x)^2$

OBJECTIVE

2   **Solving Equations of the Form $b^x = b^y$** ▶

We have seen that an exponential function $y = b^x$ is a one-to-one function. Another way of stating this fact is a property that we can use to solve exponential equations.

Uniqueness of b^x

Let $b > 0$ and $b \neq 1$. Then $b^x = b^y$ is equivalent to $x = y$.

Thus, one way to solve an exponential equation depends on whether it's possible to write each side of the equation with the same base; that is, $b^x = b^y$. We solve by this method first.

EXAMPLE 4   Solve each equation for x.

   a. $2^x = 16$ **b.** $9^x = 27$ **c.** $4^{x+3} = 8^x$

Solution

a. We write 16 as a power of 2 and then use the uniqueness of b^x to solve.

$$2^x = 16$$
$$2^x = 2^4$$

Since the bases are the same and are nonnegative, by the uniqueness of b^x, we then have that the exponents are equal. Thus,

$$x = 4$$

The solution is 4, or the solution set is $\{4\}$.

b. Notice that both 9 and 27 are powers of 3.

$$9^x = 27$$
$$(3^2)^x = 3^3 \qquad \text{Write 9 and 27 as powers of 3.}$$
$$3^{2x} = 3^3$$
$$2x = 3 \qquad \text{Apply the uniqueness of } b^x.$$
$$x = \frac{3}{2} \qquad \text{Divide by 2.}$$

To check, replace x with $\frac{3}{2}$ in the original expression, $9^x = 27$. The solution is $\frac{3}{2}$.

c. Write both 4 and 8 as powers of 2.

$$4^{x+3} = 8^x$$
$$(2^2)^{x+3} = (2^3)^x$$
$$2^{2x+6} = 2^{3x}$$
$$2x + 6 = 3x \qquad \text{Apply the uniqueness of } b^x.$$
$$6 = x \qquad \text{Subtract } 2x \text{ from both sides.}$$

The solution is 6.   □

PRACTICE

4   Solve each equation for x.

   a. $3^x = 9$ **b.** $8^x = 16$ **c.** $125^x = 25^{x-2}$ ■

There is one major problem with the preceding technique. Often the two sides of an equation cannot easily be written as powers of a common base. We explore how to solve an equation such as $4 = 3^x$ with the help of **logarithms** later.

OBJECTIVE

3   Solving Problems Modeled by Exponential Equations

The bar graph on the next page shows the increase in the number of Netflix subscriptions. Notice that the graph of the exponential function $y = 5723(1.33)^x$ approximates the heights of the bars. This is just one example of how the world abounds with patterns that can be modeled by exponential functions. To make these applications realistic, we use numbers that require the use of a calculator.

Netflix Growth

$y = 5723(1.33)^x$
where $x = 0$
corresponds to
the beginning
of 2007

Source: Netflix

Another application of an exponential function has to do with interest rates on loans. The exponential function defined by $A = P\left(1 + \dfrac{r}{n}\right)^{nt}$ models the dollars A accrued (or owed) after P dollars are invested (or loaned) at an annual rate of interest r compounded n times each year for t years. This function is known as the compound interest formula.

EXAMPLE 5 Using the Compound Interest Formula

Find the amount owed at the end of 5 years if \$1600 is loaned at a rate of 9% compounded monthly.

Solution We use the formula $A = P\left(1 + \dfrac{r}{n}\right)^{nt}$, with the following values.

$P = \$1600$ (the amount of the loan)

$r = 9\% = 0.09$ (the annual rate of interest)

$n = 12$ (the number of times interest is compounded each year)

$t = 5$ (the duration of the loan, in years)

$$A = P\left(1 + \frac{r}{n}\right)^{nt} \qquad \text{Compound interest formula}$$

$$= 1600\left(1 + \frac{0.09}{12}\right)^{12(5)} \qquad \text{Substitute known values.}$$

$$= 1600(1.0075)^{60}$$

To approximate A, use the $\boxed{y^x}$ or $\boxed{\wedge}$ key on your calculator.

$$\boxed{2505.0896}$$

Thus, the amount A owed is approximately \$2505.09.

PRACTICE

5 Find the amount owed at the end of 4 years if \$3000 is loaned at a rate of 7% compounded semiannually (twice a year).

EXAMPLE 6   Estimating Percent of Radioactive Material

As a result of a nuclear accident, radioactive debris was carried through the atmosphere. One immediate concern was the impact that the debris had on the milk supply. The percent y of radioactive material in raw milk after t days is estimated by $y = 100(2.7)^{-0.1t}$. Estimate the expected percent of radioactive material in the milk after 30 days.

Solution   Replace t with 30 in the given equation.

$$y = 100(2.7)^{-0.1t}$$
$$= 100(2.7)^{-0.1(30)} \quad \text{Let } t = 30.$$
$$= 100(2.7)^{-3}$$

To approximate the percent y, the following keystrokes may be used on a scientific calculator.

$$\boxed{2.7}\ \boxed{y^x}\ \boxed{3}\ \boxed{+/-}\ \boxed{=}\ \boxed{\times}\ \boxed{100}\ \boxed{=}$$

The display should read

$$\boxed{5.0805263}$$

Thus, approximately 5% of the radioactive material still remained in the milk supply after 30 days.   □

PRACTICE

6   The percent p of light that passes through n successive sheets of a particular glass is given approximately by the function $p(n) = 100(2.7)^{-0.05n}$. Estimate the expected percent of light that will pass through the following numbers of sheets of glass. Round each to the nearest hundredth of a percent.

a. 2 sheets of glass   **b.** 10 sheets of glass

Graphing Calculator Explorations

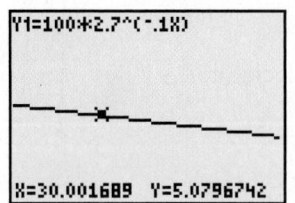

We can use a graphing calculator and its TRACE feature to solve Example 6 graphically.

To estimate the expected percent of radioactive material in the milk after 30 days, enter $Y_1 = 100(2.7)^{-0.1x}$. (The variable t in Example 6 is changed to x here to accommodate our work on the graphing calculator.) The graph does not appear on a standard viewing window, so we need to determine an appropriate viewing window. Because it doesn't make sense to look at radioactivity *before* the nuclear accident, we use Xmin = 0. We are interested in finding the percent of radioactive material in the milk when $x = 30$, so we choose Xmax = 35 to leave enough space to see the graph at $x = 30$. Because the values of y are percents, it seems appropriate that $0 \le y \le 100$. (We also use Xscl = 1 and Yscl = 10.) Now we graph the function.

We can use the TRACE feature to obtain an approximation of the expected percent of radioactive material in the milk when $x = 30$. (A TABLE feature may also be used to approximate the percent.) To obtain a better approximation, let's use the ZOOM feature several times to zoom in near $x = 30$.

The percent of radioactive material in the milk 30 days after the nuclear accident was 5.08%, accurate to two decimal places.

Use a graphing calculator to find each percent. Approximate your solutions so that they are rounded to two decimal places.

1. Estimate the expected percent of radioactive material in the milk 2 days after the nuclear accident.

2. Estimate the expected percent of radioactive material in the milk 10 days after the nuclear accident.

3. Estimate the expected percent of radioactive material in the milk 15 days after the nuclear accident.

4. Estimate the expected percent of radioactive material in the milk 25 days after the nuclear accident.

✔ Vocabulary, Readiness & Video Check

Use the choices to fill in each blank.

1. A function such as $f(x) = 2^x$ is a(n) _____ function.
   A. linear   **B.** quadratic   **C.** exponential

2. If $7^x = 7^y$, then _____.
   A. $x = 7^y$ **B.** $x = y$ **C.** $y = 7^x$ **D.** $7 = 7^y$

Answer the questions about the graph of $y = 2^x$, shown to the right.

3. Is this a function? _____

4. Is this a one-to-one function? _____

5. Is there an x-intercept? _____ If so, name the coordinates. _____

6. Is there a y-intercept? _____ If so, name the coordinates. _____

7. The domain of this function, in interval notation, is _____.

8. The range of this function, in interval notation, is _____.

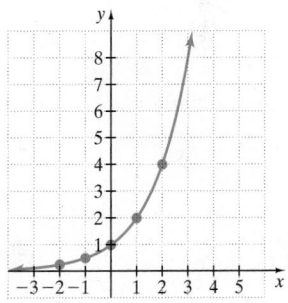

Martin-Gay Interactive Videos

See Video 12.3 ⦿

Watch the section lecture video and answer the following questions.

OBJECTIVE 1
9. From the lecture before ▤ Example 1, what's the main difference between a polynomial function and an exponential function?

OBJECTIVE 2
10. From ▤ Examples 2 and 3, you can only apply the uniqueness of b^x to solve an exponential equation if you're able to do what?

OBJECTIVE 3
11. For ▤ Example 4, write the equation and find how much uranium will remain after 101 days. Round your answer to the nearest tenth.

12.3   Exercise Set   MyMathLab® ▸

Graph each exponential function. See Examples 1 through 3.

1. $y = 5^x$

2. $y = 4^x$

3. $y = 2^x + 1$

4. $y = 3^x - 1$

5. $y = \left(\dfrac{1}{4}\right)^x$

6. $y = \left(\dfrac{1}{5}\right)^x$

7. $y = \left(\dfrac{1}{2}\right)^x - 2$

8. $y = \left(\dfrac{1}{3}\right)^x + 2$

9. $y = -2^x$

10. $y = -3^x$

11. $y = -\left(\dfrac{1}{4}\right)^x$

12. $y = -\left(\dfrac{1}{5}\right)^x$

13. $f(x) = 2^{x+1}$

14. $f(x) = 3^{x-1}$

15. $f(x) = 4^{x-2}$

16. $f(x) = 2^{x+3}$

Match each exponential equation with its graph below. See Examples 1 through 3.

17. $f(x) = \left(\dfrac{1}{2}\right)^x$

18. $f(x) = \left(\dfrac{1}{4}\right)^x$

19. $f(x) = 2^x$

20. $f(x) = 3^x$

A.

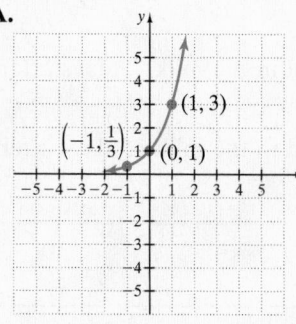

B.

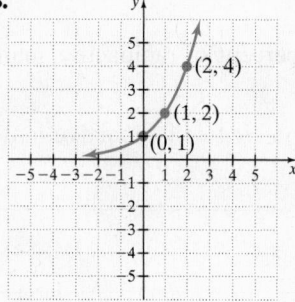

C.

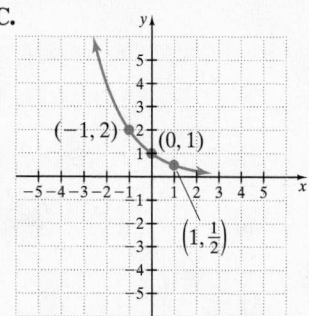

D.

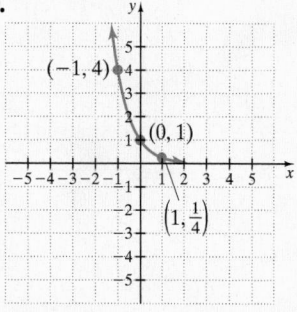

Solve each equation for x. See Example 4.

21. $3^x = 27$

22. $6^x = 36$

▶ **23.** $16^x = 8$

24. $64^x = 16$

25. $32^{2x-3} = 2$

26. $9^{2x+1} = 81$

27. $\dfrac{1}{4} = 2^{3x}$

28. $\dfrac{1}{27} = 3^{2x}$

29. $5^x = 625$

30. $2^x = 64$

31. $4^x = 8$

32. $32^x = 4$

▶ **33.** $27^{x+1} = 9$

34. $125^{x-2} = 25$

35. $81^{x-1} = 27^{2x}$

36. $4^{3x-7} = 32^{2x}$

Solve. Unless otherwise indicated, round results to one decimal place. See Example 6.

▶ **37.** One type of uranium has a radioactive decay rate of 0.4% per day. If 30 pounds of this uranium is available today, how much will still remain after 50 days? Use $y = 30(0.996)^x$ and let x be 50.

38. The nuclear waste from an atomic energy plant decays at a rate of 3% each century. If 150 pounds of nuclear waste is disposed of, how much of it will still remain after 10 centuries? Use $y = 150(0.97)^x$, and let x be 10.

39. Cheese consumption in the United States is currently growing at a rate of 1.5% per year. The equation $y = 29.2(1.015)^x$ models the per capita cheese consumption in the United States from 2005 through 2013. In this equation, y is the per capita cheese consumption (in pounds) and x represents the number of years after 2005. Round answers to the nearest tenth of a pound. (*Source:* National Agricultural Statistics Service)

a. Estimate the per capita cheese consumption in the United States in 2010.

b. Assuming this equation continues to be valid in the future, use the equation to predict the annual per capita consumption of cheese in the United States in 2020.

40. Cyber Monday is the term coined to represent the Monday after Thanksgiving, which is the biggest online spending day of the entire year. The prevailing theory is that after braving the stores on the Thanksgiving weekend, people choose to finish their shopping online once they get back to work. In 2005, a total of $511 million in revenue was collected that day. Answer the following questions using $y = 511(1.16)^x$, where y is Cyber Monday revenue in millions of dollars and x is the number of years after 2005. Round answers to the nearest tenth of a million dollars. (*Source:* ComScore)

a. According to the model, what level of Cyber Monday sales was expected in 2008?

b. If the given model continues to be valid, predict the level of Cyber Monday sales in 2018.

41. The equation $y = 122.1(1.065)^x$ models the number of American college students (in thousands) who studied abroad each year from 1998 through 2013. In this equation, y is the number of American students studying abroad (in thousands) and x represents the number of years after 1998. Round answers to the nearest tenth of a thousand. (*Source:* Based on data from Institute of International Education, Open Doors)

a. Estimate the number of American students studying abroad in 2003.

b. Assuming this equation continues to be valid in the future, use this equation to predict the number of American students studying abroad in 2018.

42. Carbon dioxide (CO_2) is a greenhouse gas that contributes to global warming. Partially due to the combustion of fossil fuel, the amount of CO_2 in Earth's atmosphere has been increasing by 0.6% annually over the past century. In 2005, the concentration of CO_2 in the atmosphere was 379.76 parts per million by volume. To make the following predictions, use $y = 379.76(1.006)^t$, where y is the concentration of CO_2 in parts per million by volume and t is the number of years after 2005. Round to the nearest hundredth. (*Sources:* Based on data from NOAA and Scripps Oceanographic Institute)

a. Predict the concentration of CO_2 in the atmosphere in the year 2020.

b. Predict the concentration of CO_2 in the atmosphere in the year 2030.

The equation $y = 5723(1.33)^x$ models the number of Netflix subscriptions (in thousands) at the beginning of each year from 2007 through 2015. In this equation, $x = 0$ corresponds to the beginning of 2007 and so on. Use this model to solve Exercises 43 and 44. Round answers to the nearest whole number (in thousands). (Source: Netflix)

43. If this growth continues to be modeled by the equation given, predict the number of subscriptions at the beginning of 2020.

44. Predict the number of subscriptions at the beginning of 2018. (See Exercise 43.)

45. An unusually wet spring has caused the size of the Cape Cod mosquito population to increase by 8% each day. If an estimated 200,000 mosquitoes are on Cape Cod on May 12, find how many mosquitoes will inhabit the Cape on May 25. Use $y = 200{,}000(1.08)^x$ where x is the number of days since May 12. Round to the nearest thousand.

46. The atmospheric pressure p, in pascals, on a weather balloon decreases with increasing height. This pressure, measured in millimeters of mercury, is related to the number of kilometers h above sea level by the function $p(h) = 760(2.7)^{-0.145h}$. Round to the nearest tenth of a pascal.

   a. Find the atmospheric pressure at a height of 1 kilometer.

   b. Find the atmospheric pressure at a height of 10 kilometers.

Solve. Use $A = P\left(1 + \dfrac{r}{n}\right)^{nt}$. Round answers to two decimal places. See Example 5.

47. Find the amount Erica owes at the end of 3 years if $6000 is loaned to her at a rate of 8% compounded monthly.

48. Find the amount owed at the end of 5 years if $3000 is loaned at a rate of 10% compounded quarterly.

49. Find the total amount Janina has in a college savings account if $2000 was invested and earned 6% compounded semiannually for 12 years.

50. Find the amount accrued if $500 is invested and earns 7% compounded monthly for 4 years.

REVIEW AND PREVIEW

Solve each equation. See Sections 2.3 and 6.6.

51. $5x - 2 = 18$ **52.** $3x - 7 = 11$

53. $3x - 4 = 3(x + 1)$ **54.** $2 - 6x = 6(1 - x)$

55. $x^2 + 6 = 5x$ **56.** $18 = 11x - x^2$

By inspection, find the value for x that makes each statement true. See Section 5.1.

57. $2^x = 8$ **58.** $3^x = 9$

59. $5^x = \dfrac{1}{5}$ **60.** $4^x = 1$

CONCEPT EXTENSIONS

Is the given function an exponential function? See the Concept Check in this section.

61. $f(x) = 1.5x^2$ **62.** $g(x) = 3^x$

63. $h(x) = \left(\dfrac{1}{2}x\right)^2$ **64.** $F(x) = 0.4^{x+1}$

Match each exponential function with its graph.

65. $f(x) = 2^{-x}$ **66.** $f(x) = \left(\dfrac{1}{2}\right)^{-x}$

67. $f(x) = 4^{-x}$ **68.** $f(x) = \left(\dfrac{1}{3}\right)^{-x}$

A.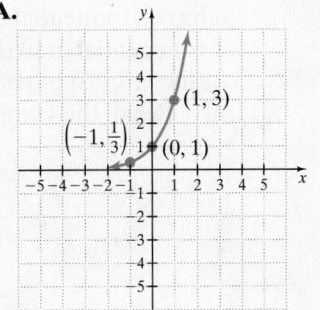

B.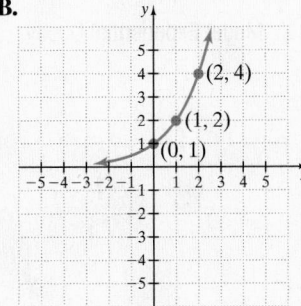

C.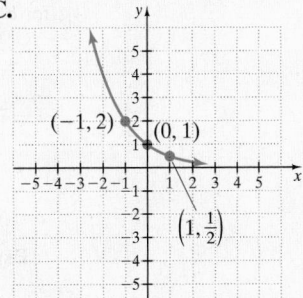

D.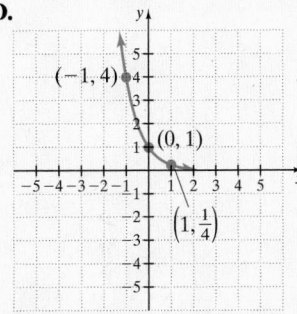

69. Explain why the graph of an exponential function $y = b^x$ contains the point $(1, b)$.

70. Explain why an exponential function $y = b^x$ has a y-intercept of $(0, 1)$.

Graph.

71. $y = |3^x|$ **72.** $y = \left|\left(\dfrac{1}{3}\right)^x\right|$

73. $y = 3^{|x|}$ **74.** $y = \left(\dfrac{1}{3}\right)^{|x|}$

75. Graph $y = 2^x$ and $y = \left(\dfrac{1}{2}\right)^{-x}$ on the same set of axes. Describe what you see and why.

76. Graph $y = 2^x$ and $x = 2^y$ on the same set of axes. Describe what you see.

Use a graphing calculator to solve. Estimate your results to two decimal places.

77. Verify the results of Exercise 37.

78. Verify the results of Exercise 38.

79. From Exercise 37, estimate the number of pounds of uranium that will be available after 100 days.

80. From Exercise 37, estimate the number of pounds of uranium that will be available after 120 days.

12.4 Exponential Growth and Decay Functions

Now that we can graph exponential functions, let's learn about exponential growth and exponential decay.

A quantity that grows or decays by the same percent at regular time periods is said to have **exponential growth** or **exponential decay.** There are many real-life examples of exponential growth and decay, such as population, bacteria, viruses, and radioactive substances, just to name a few.

Recall the graphs of exponential functions.

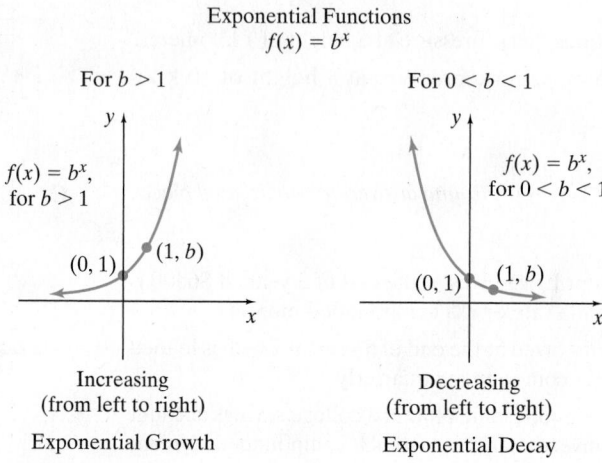

OBJECTIVE

1 Modeling Exponential Growth

We begin with exponential growth, as described below.

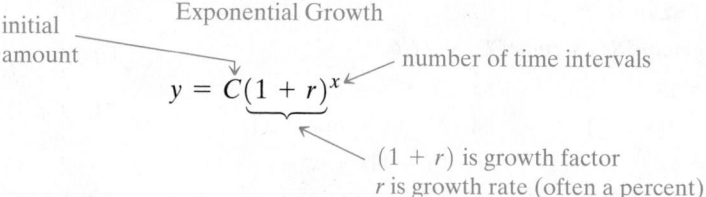

EXAMPLE 1 In 2002, let's suppose a town named Jackson had a population of 15,500 and was consistently increasing by 10% per year. If this yearly increase continues, predict the city's population in 2022. (Round to the nearest whole.)

Solution: Let's begin to understand by calculating the city's population each year:

Time Interval	x = 1	x = 2	3	4	5	and so on …
Year	2003	2004	2005	2006	2007	
Population	17,050	18,755	20,631	22,694	24,963	

$$15,500 + 0.10(15,500) \qquad 17,050 + 0.10(17,050)$$

This is an example of exponential growth, so let's use our formula with

$$C = 15,500; \; r = 0.10; \; x = 2022 - 2002 = 20$$

Then,

$$y = C(1 + r)^x$$
$$= 15,500(1 + 0.10)^{20}$$
$$= 15,500(1.1)^{20}$$
$$\approx 104,276$$

In 2022, we predict the population of Jackson to be 104,276.

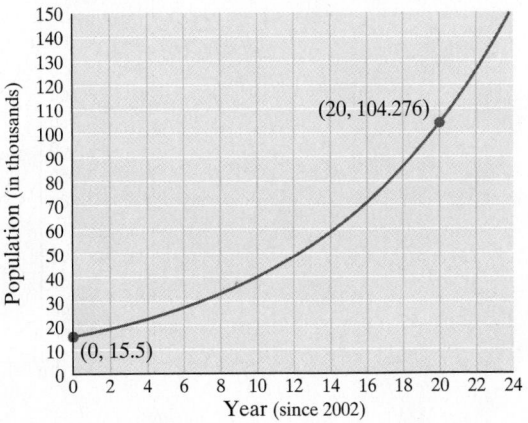

PRACTICE
1   In 2007, the town of Jackson (from Example 1) had a population of 25,000 and started consistently increasing by 12% per year. If this yearly increase continues, predict the city's population in 2022. Round to the nearest whole. ∎

Note: The exponential growth formula, $y = C(1 + r)^x$, should remind you of the compound interest formula from the previous section, $A = P\left(1 + \frac{r}{n}\right)^{nt}$. In fact, if the number of compoundings per year, n, is 1, the interest formula becomes $A = P(1 + r)^t$, which is the exponential growth formula written with different variables.

OBJECTIVE
2   Modeling Exponential Decay

Now let's study exponential decay.

Exponential Decay

initial amount — number of time intervals

$$y = C(1 - r)^x$$

$(1 - r)$ is decay factor
r is decay rate (often a percent)

EXAMPLE 2   A large golf country club holds a singles tournament each year. At the start of the tournament for a particular year, there are 512 players. After each round, half the players are eliminated. How many players remain after 6 rounds?

Solution: This is an example of exponential decay.

Let's begin to understand by calculating the number of players after a few rounds.

Round (same as interval)	1	2	3	4	and so on ...
Players (at end of round)	256	128	64	32	

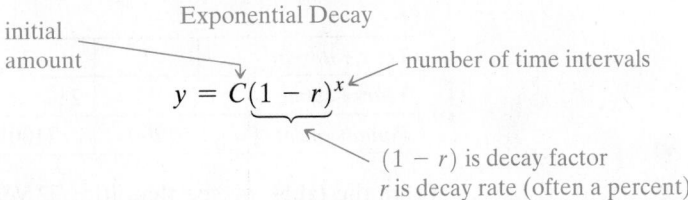

Here, $C = 512; r = \dfrac{1}{2}$ or 50% = 0.50; $x = 6$

Thus,

$$\begin{aligned} y &= 512(1 - 0.50)^6 \\ &= 512(0.50)^6 \\ &= 8 \end{aligned}$$

(Continued on next page)

After 6 rounds, there are 8 players remaining.

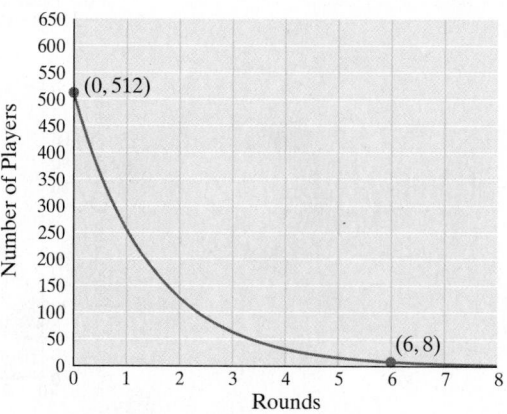

2 A tournament with 800 persons is played so that after each round, the number of players decreases by 30%. Find the number of players after round 9. Round your answer to the nearest whole.

The **half-life** of a substance is the amount of time it takes for half of the substance to decay.

EXAMPLE 3 A form of DDT pesticide (banned in 1972) has a half-life of approximately 15 years. If a storage unit had 400 pounds of DDT, find how much DDT is remaining after 72 years. Round to the nearest tenth of a pound.

Solution: Here, we need to be careful because each time interval is 15 years, the half-life.

Time Interval	1	2	3	4	5	and so on …
Years Passed	15	$2 \cdot 15 = 30$	45	60	75	
Pounds of DDT	200	100	50	25	12.5	

From the table, we see that after 72 years, between 4 and 5 intervals, there should be between 12.5 and 25 pounds of DDT remaining.

Let's calculate x, the number of time intervals.

$$x = \frac{72 \text{ (years)}}{15 \text{ (half-life)}} = 4.8$$

Now, using our exponential decay formula and the definition of half-life, for each time interval x, the decay rate r is $\frac{1}{2}$ or 50% or 0.50.

$$y = 400(1 - 0.50)^{4.8} \text{—time intervals for 72 years}$$

original amount   decay rate

$$y = 400(0.50)^{4.8}$$

$$y \approx 14.4$$

In 72 years, 14.4 pounds of DDT remain.

3 Use the information from Example 3 and calculate how much of a 500-gram sample of DDT will remain after 51 years. Round to the nearest tenth of a gram.

Vocabulary, Readiness & Video Check

Martin-Gay Interactive Videos

See Video 12.4 ⏺

Watch the section lecture video and answer the following questions.

OBJECTIVE
1

1. 🎞 Example 1 reviews exponential growth. Explain how you find the growth rate and the correct number of time intervals.

OBJECTIVE
2

2. Explain how you know that 🎞 Example 2 has to do with exponential decay, and not exponential growth.

OBJECTIVE
2

3. For Example 3, which has to do with half-life, explain how to calculate the number of time intervals. Also, what is the decay rate for half-life and why?

12.4 Exercise Set MyMathLab® ▶

Practice using the exponential growth formula by completing the table below. Round final amounts to the nearest whole. See Example 1.

	Original Amount	Growth Rate per Year	Number of Years, x	Final Amount after x Years of Growth
▶ **1.**	305	5%	8	
2.	402	7%	5	
3.	2000	11%	41	
4.	1000	47%	19	
5.	17	29%	28	
6.	29	61%	12	

Practice using the exponential decay formula by completing the table below. Round final amounts to the nearest whole. See Example 2.

	Original Amount	Decay Rate per Year	Number of Years, x	Final Amount after x Years of Decay
7.	305	5%	8	
8.	402	7%	5	
9.	10,000	12%	15	
10.	15,000	16%	11	
11.	207,000	32%	25	
12.	325,000	29%	31	

MIXED PRACTICE

Solve. Unless noted otherwise, round answers to the nearest whole. See Examples 1 and 2.

13. Suppose a city with population 500,000 has been growing at a rate of 3% per year. If this rate continues, find the population of this city in 12 years.

14. Suppose a city with population 320,000 has been growing at a rate of 4% per year. If this rate continues, find the population of this city in 20 years.

▶ **15.** The number of employees for a certain company has been decreasing each year by 5%. If the company currently has 640 employees and this rate continues, find the number of employees in 10 years.

16. The number of students attending summer school at a local community college has been decreasing each year by 7%. If 984 students currently attend summer school and this rate continues, find the number of students attending summer school in 5 years.

17. National Park Service personnel are trying to increase the size of the bison population of Theodore Roosevelt National Park. If 260 bison currently live in the park, and if the population's rate of growth is 2.5% annually, find how many bison there should be in 10 years.

18. The size of the rat population of a wharf area grows at a rate of 8% monthly. If there are 200 rats in January, find how many rats should be expected by next January.

19. A rare isotope of a nuclear material is very unstable, decaying at a rate of 15% each second. Find how much isotope remains 10 seconds after 5 grams of the isotope is created.

20. An accidental spill of 75 grams of radioactive material in a local stream has led to the presence of radioactive debris decaying at a rate of 4% each day. Find how much debris still remains after 14 days.

Practice using the exponential decay formula with half-lives by completing the table below. The first row has been completed for you. See Example 3.

	Original Amount	Half-Life (in years)	Number of Years	Time Intervals, $x = \left(\dfrac{\text{Years}}{\text{Half-Life}}\right)$ Rounded to Tenths if Needed	Final Amount after x Time Intervals (rounded to tenths)	Is Your Final Amount Reasonable?
	60	8	10	$\dfrac{10}{8} = 1.25$	25.2	yes
21. a.	40	7	14			
b.	40	7	11			
22. a.	200	12	36			
b.	200	12	40			
23.	21	152	500			
24.	35	119	500			

Solve. Round answers to the nearest tenth.

25. A form of nickel has a half-life of 96 years. How much of a 30-gram sample is left after 250 years?

26. A form of uranium has a half-life of 72 years. How much of a 100-gram sample is left after 500 years?

REVIEW AND PREVIEW

By inspection, find the value for x that makes each statement true. See Sections 5.1 and 12.3.

27. $2^x = 8$　　**28.** $3^x = 9$　　**29.** $5^x = \dfrac{1}{5}$　　**30.** $4^x = 1$

CONCEPT EXTENSIONS

31. An item is on sale for 40% off its original price. If it is then marked down an additional 60%, does this mean the item is free? Discuss why or why not.

32. Uranium U-232 has a half-life of 72 years. What eventually happens to a 10 gram sample? Does it ever completely decay and disappear? Discuss why or why not.

12.5 | Logarithmic Functions

OBJECTIVES

1 Write Exponential Equations with Logarithmic Notation and Write Logarithmic Equations with Exponential Notation.

2 Solve Logarithmic Equations by Using Exponential Notation.

3 Identify and Graph Logarithmic Functions.

OBJECTIVE

1　Using Logarithmic Notation

Since the exponential function $f(x) = 2^x$ is a one-to-one function, it has an inverse.

We can create a table of values for f^{-1} by switching the coordinates in the accompanying table of values for $f(x) = 2^x$.

x	$y = f(x)$
-3	$\dfrac{1}{8}$
-2	$\dfrac{1}{4}$
-1	$\dfrac{1}{2}$
0	1
1	2
2	4
3	8

x	$y = f^{-1}(x)$
$\dfrac{1}{8}$	-3
$\dfrac{1}{4}$	-2
$\dfrac{1}{2}$	-1
1	0
2	1
4	2
8	3

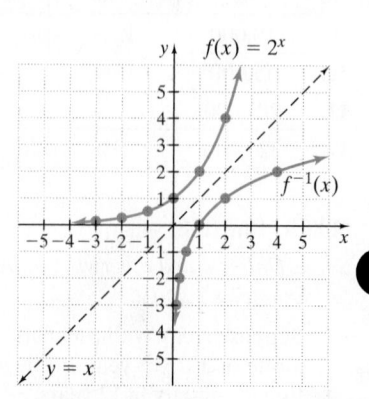

The graphs of $f(x)$ and its inverse are shown on the previous page. Notice that the graphs of f and f^{-1} are symmetric about the line $y = x$, as expected.

Now we would like to be able to write an equation for f^{-1}. To do so, we follow the steps for finding an inverse.

$$f(x) = 2^x$$

Step 1. Replace $f(x)$ by y. $\qquad y = 2^x$

Step 2. Interchange x and y. $\qquad x = 2^y$

Step 3. Solve for y.

At this point, we are stuck. To solve this equation for y, a new notation, the **logarithmic notation,** is needed.

The symbol $\log_b x$ means "the power to which b is raised to produce a result of x." In other words,

$$\log_b x = y \quad \text{means} \quad b^y = x$$

We say that $\log_b x$ is "the logarithm of x to the base b" or "the log of x to the base b."

Logarithmic Definition

If $b > 0$ and $b \neq 1$, then

$$y = \log_b x \text{ means } x = b^y$$

for every $x > 0$ and every real number y.

Helpful Hint

Notice that a *logarithm* is an *exponent*. In other words, $\log_3 9$ is the *power* to which we raise 3 in order to get 9.

Before returning to the function $x = 2^y$ and solving it for y in terms of x, let's practice using the new notation $\log_b x$.

It is important to be able to write exponential equations from logarithmic notation and vice versa. The following table shows examples of both forms.

Logarithmic Equation	*Corresponding Exponential Equation*
$\log_3 9 = 2$	$3^2 = 9$
$\log_6 1 = 0$	$6^0 = 1$
$\log_2 8 = 3$	$2^3 = 8$
$\log_4 \dfrac{1}{16} = -2$	$4^{-2} = \dfrac{1}{16}$
$\log_8 2 = \dfrac{1}{3}$	$8^{1/3} = 2$

EXAMPLE 1 Write each as an exponential equation.

a. $\log_5 25 = 2$ $\qquad$ **b.** $\log_6 \dfrac{1}{6} = -1$ $\qquad$ **c.** $\log_2 \sqrt{2} = \dfrac{1}{2}$ $\qquad$ **d.** $\log_7 x = 5$

Solution

a. $\log_5 25 = 2$ means $5^2 = 25$

b. $\log_6 \dfrac{1}{6} = -1$ means $6^{-1} = \dfrac{1}{6}$

c. $\log_2 \sqrt{2} = \dfrac{1}{2}$ means $2^{1/2} = \sqrt{2}$

d. $\log_7 x = 5$ means $7^5 = x$

(Continued on next page)

1 Write each as an exponential equation.

a. $\log_3 81 = 4$ **b.** $\log_5 \dfrac{1}{5} = -1$ **c.** $\log_7 \sqrt{7} = \dfrac{1}{2}$ **d.** $\log_{13} y = 4$ ■

EXAMPLE 2 Write each as a logarithmic equation.

a. $9^3 = 729$ **b.** $6^{-2} = \dfrac{1}{36}$ **c.** $5^{1/3} = \sqrt[3]{5}$ **d.** $\pi^4 = x$

Solution

a. $9^3 = 729$ means $\log_9 729 = 3$

b. $6^{-2} = \dfrac{1}{36}$ means $\log_6 \dfrac{1}{36} = -2$

c. $5^{1/3} = \sqrt[3]{5}$ means $\log_5 \sqrt[3]{5} = \dfrac{1}{3}$

d. $\pi^4 = x$ means $\log_\pi x = 4$ □

PRACTICE
2 Write each as a logarithmic equation.

a. $4^3 = 64$ **b.** $6^{1/3} = \sqrt[3]{6}$ **c.** $5^{-3} = \dfrac{1}{125}$ **d.** $\pi^7 = z$ ■

EXAMPLE 3 Find the value of each logarithmic expression.

a. $\log_4 16$ **b.** $\log_{10} \dfrac{1}{10}$ **c.** $\log_9 3$

Solution

a. $\log_4 16 = 2$ because $4^2 = 16$

b. $\log_{10} \dfrac{1}{10} = -1$ because $10^{-1} = \dfrac{1}{10}$

c. $\log_9 3 = \dfrac{1}{2}$ because $9^{1/2} = \sqrt{9} = 3$ □

PRACTICE
3 Find the value of each logarithmic expression.

a. $\log_3 9$ **b.** $\log_2 \dfrac{1}{8}$ **c.** $\log_{49} 7$ ■

(Helpful Hint)

Another method for evaluating logarithms such as those in Example 3 is to set the expression equal to x and then write them in exponential form to find x. For example:

a. $\log_4 16 = x$ means $4^x = 16$. Since $4^2 = 16$, $x = 2$ or $\log_4 16 = 2$.

b. $\log_{10} \dfrac{1}{10} = x$ means $10^x = \dfrac{1}{10}$. Since $10^{-1} = \dfrac{1}{10}$, $x = -1$ or $\log_{10} \dfrac{1}{10} = -1$.

c. $\log_9 3 = x$ means $9^x = 3$. Since $9^{1/2} = 3$, $x = \dfrac{1}{2}$ or $\log_9 3 = \dfrac{1}{2}$.

OBJECTIVE

2 Solving Logarithmic Equations

The ability to interchange the logarithmic and exponential forms of a statement is often the key to solving logarithmic equations.

EXAMPLE 4 Solve each equation for x.

a. $\log_4 \dfrac{1}{4} = x$ **b.** $\log_5 x = 3$ **c.** $\log_x 25 = 2$

d. $\log_3 1 = x$ **e.** $\log_b 1 = x$

Solution

a. $\log_4 \dfrac{1}{4} = x$ means $4^x = \dfrac{1}{4}$. Solve $4^x = \dfrac{1}{4}$ for x.

$$4^x = \dfrac{1}{4}$$

$$4^x = 4^{-1}$$

Since the bases are the same, by the uniqueness of b^x, we have that

$$x = -1$$

The solution is -1 or the solution set is $\{-1\}$. To check, see that $\log_4 \dfrac{1}{4} = -1$, since $4^{-1} = \dfrac{1}{4}$.

b. $\log_5 x = 3$

$5^3 = x$ Write as an exponential equation.

$125 = x$

The solution is 125.

c. $\log_x 25 = 2$

$x^2 = 25$ Write as an exponential equation. Here $x > 0$, $x \neq 1$.

$x = 5$

Even though $(-5)^2 = 25$, the base b of a logarithm must be positive. The solution is 5.

d. $\log_3 1 = x$

$3^x = 1$ Write as an exponential equation.

$3^x = 3^0$ Write 1 as 3^0.

$x = 0$ Use the uniqueness of b^x.

The solution is 0.

e. $\log_b 1 = x$

$b^x = 1$ Write as an exponential equation. Here, $b > 0$ and $b \neq 1$.

$b^x = b^0$ Write 1 as b^0.

$x = 0$ Apply the uniqueness of b^x.

The solution is 0. □

PRACTICE

4 Solve each equation for x.

a. $\log_5 \dfrac{1}{25} = x$ **b.** $\log_x 8 = 3$ **c.** $\log_6 x = 2$

d. $\log_{13} 1 = x$ **e.** $\log_h 1 = x$

In Example 4e, we proved an important property of logarithms. That is, $\log_b 1$ is always 0. This property as well as two important others are given next.

Properties of Logarithms

If b is a real number, $b > 0$, and $b \neq 1$, then

1. $\log_b 1 = 0$

2. $\log_b b^x = x$

3. $b^{\log_b x} = x$

To see that **2.** $\log_b b^x = x$, change the logarithmic form to exponential form. Then, $\log_b b^x = x$ means $b^x = b^x$. In exponential form, the statement is true, so in logarithmic form, the statement is also true.

To understand **3.** $b^{\log_b x} = x$, write this exponential equation as an equivalent logarithm.

EXAMPLE 5 Simplify.

a. $\log_3 3^2$ **b.** $\log_7 7^{-1}$ **c.** $5^{\log_5 3}$ **d.** $2^{\log_2 6}$

Solution

a. From Property 2, $\log_3 3^2 = 2$.

b. From Property 2, $\log_7 7^{-1} = -1$.

c. From Property 3, $5^{\log_5 3} = 3$.

d. From Property 3, $2^{\log_2 6} = 6$. □

PRACTICE

5 Simplify.

a. $\log_5 5^4$ **b.** $\log_9 9^{-2}$ **c.** $6^{\log_6 5}$ **d.** $7^{\log_7 4}$

OBJECTIVE

3 Graphing Logarithmic Functions

Let us now return to the function $f(x) = 2^x$ and write an equation for its inverse, $f^{-1}(x)$. Recall our earlier work.

$$f(x) = 2^x$$

Step 1. Replace $f(x)$ by y. $y = 2^x$

Step 2. Interchange x and y. $x = 2^y$

Having gained proficiency with the notation $\log_b x$, we can now complete the steps for writing the inverse equation by writing $x = 2^y$ as an equivalent logarithm.

Step 3. Solve for y. $y = \log_2 x$

Step 4. Replace y with $f^{-1}(x)$. $f^{-1}(x) = \log_2 x$

Thus, $f^{-1}(x) = \log_2 x$ defines a function that is the inverse function of the function $f(x) = 2^x$. The function $f^{-1}(x)$ or $y = \log_2 x$ is called a **logarithmic function.**

Logarithmic Function

If x is a positive real number, b is a constant positive real number, and b is not 1, then a **logarithmic function** is a function that can be defined by

$$f(x) = \log_b x$$

The domain of f is the set of positive real numbers, and the range of f is the set of real numbers.

✔ **CONCEPT CHECK**

Let $f(x) = \log_3 x$ and $g(x) = 3^x$. These two functions are inverses of each other. Since $(2, 9)$ is an ordered pair solution of $g(x)$ or $g(2) = 9$, what ordered pair do we know to be a solution of $f(x)$? Also, find $f(9)$. Explain why.

We can explore logarithmic functions by graphing them.

EXAMPLE 6   Graph the logarithmic function $y = \log_2 x$.

Solution   First we write the equation with exponential notation as $2^y = x$. Then we find some ordered pair solutions that satisfy this equation. Finally, we plot the points and connect them with a smooth curve. The domain of this function is $(0, \infty)$, and the range is all real numbers.

Since $x = 2^y$ is solved for x, we choose y-values and compute corresponding x-values.

If $y = 0, x = 2^0 = 1$

If $y = 1, x = 2^1 = 2$

If $y = 2, x = 2^2 = 4$

If $y = -1, x = 2^{-1} = \dfrac{1}{2}$

$x = 2^y$	y
1	0
2	1
4	2
$\dfrac{1}{2}$	-1

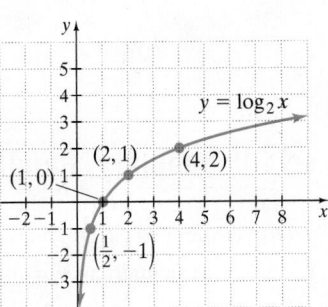

Notice that the x-intercept is $(1, 0)$ and there is no y-intercept.   □

PRACTICE

6   Graph the logarithmic function $y = \log_9 x$.   ▪

EXAMPLE 7   Graph the logarithmic function $f(x) = \log_{1/3} x$.

Solution   Replace $f(x)$ with y and write the result with exponential notation.

$$f(x) = \log_{1/3} x$$
$$y = \log_{1/3} x \quad \text{Replace } f(x) \text{ with } y.$$
$$\left(\frac{1}{3}\right)^y = x \quad \text{Write in exponential form.}$$

Now we can find ordered pair solutions that satisfy $\left(\dfrac{1}{3}\right)^y = x$, plot these points, and connect them with a smooth curve.

If $y = 0, x = \left(\dfrac{1}{3}\right)^0 = 1$

If $y = 1, x = \left(\dfrac{1}{3}\right)^1 = \dfrac{1}{3}$

If $y = -1, x = \left(\dfrac{1}{3}\right)^{-1} = 3$

If $y = -2, x = \left(\dfrac{1}{3}\right)^{-2} = 9$

$x = \left(\dfrac{1}{3}\right)^y$	y
1	0
$\dfrac{1}{3}$	1
3	-1
9	-2

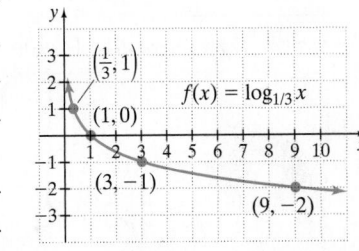

The domain of this function is $(0, \infty)$, and the range is the set of all real numbers. The x-intercept is $(1, 0)$ and there is no y-intercept.   □

PRACTICE

7   Graph the logarithmic function $y = \log_{1/4} x$.   ▪

The following figures summarize characteristics of logarithmic functions.

$$f(x) = \log_b x, b > 0, b \neq 1$$

- one-to-one function
- x-intercept $(1, 0)$
- no y-intercept

- domain: $(0, \infty)$
- range: $(-\infty, \infty)$

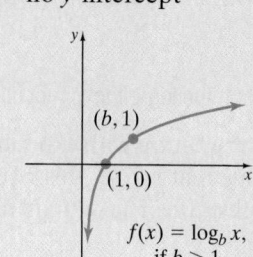

$(b, 1)$

$(1, 0)$

$f(x) = \log_b x,$
if $b > 1$

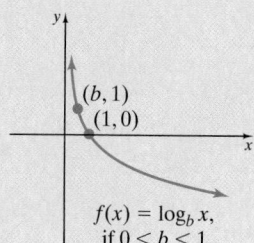

$(b, 1)$
$(1, 0)$

$f(x) = \log_b x,$
if $0 < b < 1$

Vocabulary, Readiness & Video Check

Use the choices to fill in each blank.

1. A function such as $y = \log_2 x$ is a(n) _____ function.

    A. linear   **B.** logarithmic   **C.** quadratic   **D.** exponential

2. If $y = \log_2 x$, then _____.

    A. $x = y$ **B.** $2^x = y$ **C.** $2^y = x$ **D.** $2y = x$

Answer the questions about the graph of $y = \log_2 x$, shown to the left.

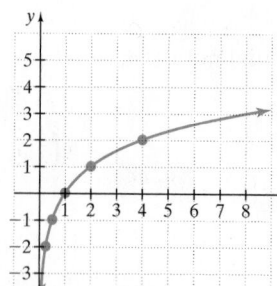

3. Is this a one-to-one function? _____

4. Is there an x-intercept? _____ If so, name the coordinates. _____

5. Is there a y-intercept? _____ If so, name the coordinates. _____

6. The domain of this function, in interval notation, is _____.

7. The range of this function, in interval notation, is _____.

Martin-Gay Interactive Videos

See Video 12.5

Watch the section lecture video and answer the following questions.

OBJECTIVE 1

8. Notice from the definition of a logarithm and from ▦ Examples 1–4 that the logarithm in a logarithmic statement equals the power in the exponent statement, such as $b^y = x$. What conclusion can you make about logarithms and exponents?

OBJECTIVE 2

9. From ▦ Examples 8 and 9, how do you solve a logarithmic equation?

OBJECTIVE 3

10. In ▦ Example 12, why is it easier to choose values for y when finding ordered pairs for the graph?

12.5 Exercise Set MyMathLab ▶

Write each as an exponential equation. See Example 1.

1. $\log_6 36 = 2$

2. $\log_2 32 = 5$

3. $\log_3 \dfrac{1}{27} = -3$

4. $\log_5 \dfrac{1}{25} = -2$

5. $\log_{10} 1000 = 3$

6. $\log_{10} 10 = 1$

7. $\log_9 x = 4$

8. $\log_8 y = 7$

9. $\log_\pi \dfrac{1}{\pi^2} = -2$

10. $\log_e \dfrac{1}{e} = -1$

11. $\log_7 \sqrt{7} = \dfrac{1}{2}$

12. $\log_{11} \sqrt[4]{11} = \dfrac{1}{4}$

13. $\log_{0.7} 0.343 = 3$

14. $\log_{1.2} 1.44 = 2$

15. $\log_3 \dfrac{1}{81} = -4$

16. $\log_{1/4} 16 = -2$

Write each as a logarithmic equation. See Example 2.

17. $2^4 = 16$

18. $5^3 = 125$

19. $10^2 = 100$

20. $10^4 = 10{,}000$

21. $\pi^3 = x$

22. $\pi^5 = y$

23. $10^{-1} = \dfrac{1}{10}$

24. $10^{-2} = \dfrac{1}{100}$

25. $4^{-2} = \dfrac{1}{16}$

26. $3^{-4} = \dfrac{1}{81}$

27. $5^{1/2} = \sqrt{5}$

28. $4^{1/3} = \sqrt[3]{4}$

Find the value of each logarithmic expression. See Examples 3 and 5.

29. $\log_2 8$

30. $\log_7 49$

31. $\log_3 \dfrac{1}{9}$

32. $\log_2 \dfrac{1}{32}$

33. $\log_{25} 5$

34. $\log_8 \dfrac{1}{2}$

35. $\log_{1/2} 2$

36. $\log_{2/3} \dfrac{4}{9}$

37. $\log_6 1$

38. $\log_9 9$

39. $\log_{10} 100$

40. $\log_{10} \dfrac{1}{10}$

41. $\log_3 81$

42. $\log_2 16$

43. $\log_4 \dfrac{1}{64}$

44. $\log_3 \dfrac{1}{9}$

Solve. See Example 4.

45. $\log_3 9 = x$

46. $\log_4 64 = x$

47. $\log_3 x = 4$

48. $\log_2 x = 3$

49. $\log_x 49 = 2$

50. $\log_x 8 = 3$

51. $\log_2 \dfrac{1}{8} = x$

52. $\log_3 \dfrac{1}{81} = x$

53. $\log_3 \dfrac{1}{27} = x$

54. $\log_5 \dfrac{1}{125} = x$

55. $\log_8 x = \dfrac{1}{3}$

56. $\log_9 x = \dfrac{1}{2}$

57. $\log_4 16 = x$

58. $\log_2 16 = x$

59. $\log_{3/4} x = 3$

60. $\log_{2/3} x = 2$

61. $\log_x 100 = 2$

62. $\log_x 27 = 3$

63. $\log_2 2^4 = x$

64. $\log_6 6^{-2} = x$

65. $3^{\log_3 5} = x$

66. $5^{\log_5 7} = x$

67. $\log_x \dfrac{1}{7} = \dfrac{1}{2}$

68. $\log_x 2 = -\dfrac{1}{3}$

Simplify. See Example 5.

69. $\log_5 5^3$

70. $\log_6 6^2$

71. $2^{\log_2 3}$

72. $7^{\log_7 4}$

73. $\log_9 9$

74. $\log_2 2$

75. $\log_8 (8)^{-1}$

76. $\log_{11} (11)^{-1}$

Graph each logarithmic function. Label any intercepts. See Examples 6 and 7.

77. $y = \log_3 x$

78. $y = \log_8 x$

79. $f(x) = \log_{1/4} x$

80. $f(x) = \log_{1/2} x$

81. $f(x) = \log_5 x$

82. $f(x) = \log_6 x$

83. $f(x) = \log_{1/6} x$

84. $f(x) = \log_{1/5} x$

REVIEW AND PREVIEW

Simplify each rational expression. See Section 7.1.

85. $\dfrac{x+3}{3+x}$

86. $\dfrac{x-5}{5-x}$

87. $\dfrac{x^2 - 8x + 16}{2x - 8}$

88. $\dfrac{x^2 - 3x - 10}{2 + x}$

Add or subtract as indicated. See Sections 7.3 and 7.4.

89. $\dfrac{2}{x} + \dfrac{3}{x^2}$

90. $\dfrac{5}{y+1} - \dfrac{4}{y-1}$

91. $\dfrac{3x}{x+3} + \dfrac{9}{x+3}$

92. $\dfrac{m^2}{m+1} - \dfrac{1}{m+1}$

CONCEPT EXTENSIONS

Solve. See the Concept Check in this section.

93. Let $f(x) = \log_5 x$. Then $g(x) = 5^x$ is the inverse of $f(x)$. The ordered pair $(2, 25)$ is a solution of the function $g(x)$.

    a. Write this solution using function notation.

    b. Write an ordered pair that we know to be a solution of $f(x)$.

    c. Use the answer to part **b** and write the solution using function notation.

94. Let $f(x) = \log_{0.3} x$. Then $g(x) = 0.3^x$ is the inverse of $f(x)$. The ordered pair $(3, 0.027)$ is a solution of the function $g(x)$.

    a. Write this solution using function notation.

    b. Write an ordered pair that we know to be a solution of $f(x)$.

    c. Use the answer to part **b** and write the solution using function notation.

95. Explain why negative numbers are not included as logarithmic bases.

96. Explain why 1 is not included as a logarithmic base.

Solve by first writing as an exponent.

97. $\log_7(5x - 2) = 1$ **98.** $\log_3(2x + 4) = 2$

99. Simplify: $\log_3(\log_5 125)$

100. Simplify: $\log_7(\log_4(\log_2 16))$

Graph each function and its given inverse function on the same set of axes. Label any intercepts.

101. $y = 4^x$; $y = \log_4 x$

102. $y = 3^x$; $y = \log_3 x$

103. $y = \left(\dfrac{1}{3}\right)^x$; $y = \log_{1/3} x$

104. $y = \left(\dfrac{1}{2}\right)^x$; $y = \log_{1/2} x$

105. Explain why the graph of the function $y = \log_b x$ contains the point $(1, 0)$ no matter what b is.

106. $\log_3 10$ is between which two integers? Explain your answer.

107. The formula $\log_{10}(1 - k) = \dfrac{-0.3}{H}$ models the relationship between the half-life H of a radioactive material and its rate of decay k. Find the rate of decay of the iodine isotope I-131 if its half-life is 8 days. Round to four decimal places.

108. The formula $\text{pH} = -\log_{10}(\text{H}^+)$ provides the pH for a liquid, where H^+ stands for the concentration of hydronium ions. Find the pH of lemonade, whose concentration of hydronium ions is 0.0050 moles/liter. Round to the nearest tenth.

12.6 | Properties of Logarithms

OBJECTIVES

1 Use the Product Property of Logarithms.

2 Use the Quotient Property of Logarithms.

3 Use the Power Property of Logarithms.

4 Use the Properties of Logarithms Together.

In the previous section, we explored some basic properties of logarithms. We now introduce and explore additional properties. Because a logarithm is an exponent, logarithmic properties are just restatements of exponential properties.

OBJECTIVE

1   Using the Product Property

The first of these properties is called the **product property of logarithms** because it deals with the logarithm of a product.

> **Product Property of Logarithms**
>
> If x, y, and b are positive real numbers and $b \neq 1$, then
> $$\log_b xy = \log_b x + \log_b y$$

To prove this, let $\log_b x = M$ and $\log_b y = N$. Now write each logarithm with exponential notation.

$$\log_b x = M \quad \text{is equivalent to} \quad b^M = x$$

$$\log_b y = N \quad \text{is equivalent to} \quad b^N = y$$

When we multiply the left sides and the right sides of the exponential equations, we have that

$$xy = (b^M)(b^N) = b^{M+N}$$

If we write the equation $xy = b^{M+N}$ in equivalent logarithmic form, we have

$$\log_b xy = M + N$$

But since $M = \log_b x$ and $N = \log_b y$, we can write

$$\log_b xy = \log_b x + \log_b y \quad \text{Let } M = \log_b x \text{ and } N = \log_b y.$$

In other words, the logarithm of a product is the sum of the logarithms of the factors. This property is sometimes used to simplify logarithmic expressions.

In the examples that follow, assume that variables represent positive numbers.

EXAMPLE 1   Write each sum as a single logarithm.

a. $\log_{11} 10 + \log_{11} 3$ **b.** $\log_3 \dfrac{1}{2} + \log_3 12$ **c.** $\log_2(x + 2) + \log_2 x$

Solution   In each case, both terms have a common logarithmic base.

a. $\log_{11}10 + \log_{11}3 = \log_{11}(10 \cdot 3)$ Apply the product property.

$$= \log_{11} 30$$

b. $\log_3 \dfrac{1}{2} + \log_3 12 = \log_3 \left(\dfrac{1}{2} \cdot 12 \right) = \log_3 6$

c. $\log_2(x + 2) + \log_2 x = \log_2[(x + 2) \cdot x] = \log_2(x^2 + 2x)$ ☐

> **Helpful Hint**
>
> Check your logarithm properties. Make sure you understand that $\log_2(x + 2)$ _is not_ $\log_2 x + \log_2 2$.

PRACTICE
1   Write each sum as a single logarithm.

a. $\log_8 5 + \log_8 3$

b. $\log_2 \dfrac{1}{3} + \log_2 18$

c. $\log_5(x - 1) + \log_5(x + 1)$

OBJECTIVE
2   Using the Quotient Property

The second property is the **quotient property of logarithms.**

> **Quotient Property of Logarithms**
>
> If x, y, and b are positive real numbers and $b \neq 1$, then
>
> $$\log_b \frac{x}{y} = \log_b x - \log_b y$$

The proof of the quotient property of logarithms is similar to the proof of the product property. Notice that the quotient property says that the logarithm of a quotient is the difference of the logarithms of the dividend and divisor.

> ✔ **CONCEPT CHECK**
>
> Which of the following is the correct way to rewrite $\log_5 \dfrac{7}{2}$?
>
> **a.** $\log_5 7 - \log_5 2$ **b.** $\log_5(7 - 2)$ **c.** $\dfrac{\log_5 7}{\log_5 2}$ **d.** $\log_5 14$

EXAMPLE 2 Write each difference as a single logarithm.

a. $\log_{10} 27 - \log_{10} 3$ **b.** $\log_5 8 - \log_5 x$ **c.** $\log_3(x^2 + 5) - \log_3(x^2 + 1)$

Solution In each case, both terms have a common logarithmic base.

a. $\log_{10} 27 - \log_{10} 3 = \log_{10} \dfrac{27}{3} = \log_{10} 9$ Apply the quotient property.

b. $\log_5 8 - \log_5 x = \log_5 \dfrac{8}{x}$

c. $\log_3(x^2 + 5) - \log_3(x^2 + 1) = \log_3 \dfrac{x^2 + 5}{x^2 + 1}$ □

PRACTICE
2 Write each difference as a single logarithm.

a. $\log_5 18 - \log_5 6$ **b.** $\log_6 x - \log_6 3$ **c.** $\log_4(x^2 + 1) - \log_4(x^2 + 3)$ ∎

OBJECTIVE
3 Using the Power Property

The third and final property we introduce is the **power property of logarithms.**

> **Power Property of Logarithms**
>
> If x and b are positive real numbers, $b \neq 1$, and r is a real number, then
> $$\log_b x^r = r \log_b x$$

EXAMPLE 3 Use the power property to rewrite each expression.

a. $\log_5 x^3$ **b.** $\log_4 \sqrt{2}$

Solution

a. $\log_5 x^3 = 3 \log_5 x$ **b.** $\log_4 \sqrt{2} = \log_4 2^{1/2} = \dfrac{1}{2} \log_4 2$ □

PRACTICE
3 Use the power property to rewrite each expression.

a. $\log_7 x^8$ **b.** $\log_5 \sqrt[4]{7}$ ∎

OBJECTIVE
4 Using the Properties Together

Many times, we must use more than one property of logarithms to simplify a logarithmic expression.

EXAMPLE 4 Write as a single logarithm.

a. $2 \log_5 3 + 3 \log_5 2$ **b.** $3 \log_9 x - \log_9(x + 1)$
c. $\log_4 25 + \log_4 3 - \log_4 5$

Solution In each case, all terms have a common logarithmic base.

a. $2 \log_5 3 + 3 \log_5 2 = \log_5 3^2 + \log_5 2^3$ Apply the power property.

$= \log_5 9 + \log_5 8$

$= \log_5 (9 \cdot 8)$ Apply the product property.

$= \log_5 72$

b. $3 \log_9 x - \log_9(x + 1) = \log_9 x^3 - \log_9(x + 1)$ Apply the power property.

$= \log_9 \dfrac{x^3}{x + 1}$ Apply the quotient property.

c. Use both the product and quotient properties.

$$\log_4 25 + \log_4 3 - \log_4 5 = \log_4(25 \cdot 3) - \log_4 5 \quad \text{Apply the product property.}$$
$$= \log_4 75 - \log_4 5 \quad \text{Simplify.}$$
$$= \log_4 \frac{75}{5} \quad \text{Apply the quotient property.}$$
$$= \log_4 15 \quad \text{Simplify.} \qquad \square$$

PRACTICE
4 Write as a single logarithm.

a. $2 \log_5 4 + 5 \log_5 2$ b. $2 \log_8 x - \log_8(x + 3)$

c. $\log_7 12 + \log_7 5 - \log_7 4$

EXAMPLE 5   Write each expression as sums or differences of multiples of logarithms.

a. $\log_3 \dfrac{5 \cdot 7}{4}$ b. $\log_2 \dfrac{x^5}{y^2}$

Solution

a. $\log_3 \dfrac{5 \cdot 7}{4} = \log_3(5 \cdot 7) - \log_3 4 \quad \text{Apply the quotient property.}$
$$\qquad\qquad = \log_3 5 + \log_3 7 - \log_3 4 \quad \text{Apply the product property.}$$

b. $\log_2 \dfrac{x^5}{y^2} = \log_2(x^5) - \log_2(y^2) \quad \text{Apply the quotient property.}$
$$\qquad\qquad = 5 \log_2 x - 2 \log_2 y \quad \text{Apply the power property.} \qquad \square$$

PRACTICE
5 Write each expression as sums or differences of multiples of logarithms.

a. $\log_5 \dfrac{4 \cdot 3}{7}$ b. $\log_4 \dfrac{a^2}{b^5}$

> **Helpful Hint**
>
> Notice that we are not able to simplify further a logarithmic expression such as $\log_5(2x - 1)$. None of the basic properties gives a way to write the logarithm of a difference (or sum) in some equivalent form.

✔ **CONCEPT CHECK**

What is wrong with the following?

$$\log_{10}(x^2 + 5) = \log_{10} x^2 + \log_{10} 5$$
$$= 2 \log_{10} x + \log_{10} 5$$

Use a numerical example to demonstrate that the result is incorrect.

EXAMPLE 6   If $\log_b 2 = 0.43$ and $\log_b 3 = 0.68$, use the properties of logarithms to evaluate.

a. $\log_b 6$ b. $\log_b 9$ c. $\log_b \sqrt{2}$

(Continued on next page)

Solution

a. $\log_b 6 = \log_b(2 \cdot 3)$ Write 6 as $2 \cdot 3$.

$= \log_b 2 + \log_b 3$ Apply the product property.

$= 0.43 + 0.68$ Substitute given values.

$= 1.11$ Simplify.

b. $\log_b 9 = \log_b 3^2$ Write 9 as 3^2.

$= 2\log_b 3$ Apply the power property.

$= 2(0.68)$ Substitute 0.68 for $\log_b 3$.

$= 1.36$ Simplify.

c. First, recall that $\sqrt{2} = 2^{1/2}$. Then

$\log_b \sqrt{2} = \log_b 2^{1/2}$ Write $\sqrt{2}$ as $2^{1/2}$.

$= \dfrac{1}{2}\log_b 2$ Apply the power property.

$= \dfrac{1}{2}(0.43)$ Substitute the given value.

$= 0.215$ Simplify.   □

PRACTICE

6   If $\log_b 5 = 0.83$ and $\log_b 3 = 0.56$, use the properties of logarithms to evaluate.

a. $\log_b 15$ **b.** $\log_b 25$ **c.** $\log_b \sqrt{3}$ ■

A summary of the basic properties of logarithms that we have developed so far is given next.

> **Properties of Logarithms**
>
> If x, y, and b are positive real numbers, $b \neq 1$, and r is a real number, then
>
> **1.** $\log_b 1 = 0$ 　　　　　　　　　　　　　**2.** $\log_b b^x = x$
>
> **3.** $b^{\log_b x} = x$ 　　　　　　　　　　　**4.** $\log_b xy = \log_b x + \log_b y$　Product property.
>
> **5.** $\log_b \dfrac{x}{y} = \log_b x - \log_b y$　Quotient property.　**6.** $\log_b x^r = r\log_b x$　Power property.

✔ **Vocabulary, Readiness & Video Check**

Select the correct choice.

1. $\log_b 12 + \log_b 3 = \log_b$ _____
  a. 36　**b.** 15　**c.** 4　**d.** 9

2. $\log_b 12 - \log_b 3 = \log_b$ _____
  a. 36　**b.** 15　**c.** 4　**d.** 9

3. $7\log_b 2 =$ _____
  a. $\log_b 14$　**b.** $\log_b 2^7$　**c.** $\log_b 7^2$　**d.** $(\log_b 2)^7$

4. $\log_b 1 =$ _____
  a. b　**b.** 1　**c.** 0　**d.** no answer

5. $b^{\log_b x} =$ _____
  a. x　**b.** b　**c.** 1　**d.** 0

6. $\log_5 5^2 =$ _____
  a. 25　**b.** 2　**c.** 5^{5^2}　**d.** 32

Martin-Gay Interactive Videos

See Video 12.6 ⊙

Watch the section lecture video and answer the following questions.

OBJECTIVE
1

7. Can the product property of logarithms be used again on the bottom line of ▤ Example 2 to write $\log_{10}(10x^2 + 20)$ as a sum of logarithms, $\log_{10} 10x^2 + \log_{10} 20$? Explain.

OBJECTIVE
2

8. From ▤ Example 3 and the lecture before, what must be true about bases before you can apply the quotient property of logarithms?

OBJECTIVE
3

9. Based on ▤ Example 5, explain why $\log_2 \frac{1}{x} = -\log_2 x$.

OBJECTIVE
4

10. From the lecture before ▤ Example 6, where do the logarithmic properties come from?

12.6 Exercise Set MyMathLab® ▸

Write each sum as a single logarithm. Assume that variables represent positive numbers. See Example 1.

▸ **1.** $\log_5 2 + \log_5 7$

2. $\log_3 8 + \log_3 4$

3. $\log_4 9 + \log_4 x$

4. $\log_2 x + \log_2 y$

5. $\log_6 x + \log_6 (x + 1)$

6. $\log_5 y^3 + \log_5 (y - 7)$

▸ **7.** $\log_{10} 5 + \log_{10} 2 + \log_{10} (x^2 + 2)$

8. $\log_6 3 + \log_6 (x + 4) + \log_6 5$

Write each difference as a single logarithm. Assume that variables represent positive numbers. See Example 2.

9. $\log_5 12 - \log_5 4$

10. $\log_7 20 - \log_7 4$

▸ **11.** $\log_3 8 - \log_3 2$

12. $\log_5 12 - \log_5 3$

13. $\log_2 x - \log_2 y$

14. $\log_3 12 - \log_3 z$

15. $\log_2 (x^2 + 6) - \log_2 (x^2 + 1)$

16. $\log_7 (x + 9) - \log_7 (x^2 + 10)$

Use the power property to rewrite each expression. See Example 3.

▸ **17.** $\log_3 x^2$

18. $\log_2 x^5$

19. $\log_4 5^{-1}$

20. $\log_6 7^{-2}$

▸ **21.** $\log_5 \sqrt{y}$

22. $\log_5 \sqrt[3]{x}$

MIXED PRACTICE

Write each as a single logarithm. Assume that variables represent positive numbers. See Example 4.

23. $\log_2 5 + \log_2 x^3$

24. $\log_5 2 + \log_5 y^2$

25. $3\log_4 2 + \log_4 6$

26. $2\log_3 5 + \log_3 2$

▸ **27.** $3\log_5 x + 6\log_5 z$

28. $2\log_7 y + 6\log_7 z$

29. $\log_4 2 + \log_4 10 - \log_4 5$

30. $\log_6 18 + \log_6 2 - \log_6 9$

31. $\log_7 6 + \log_7 3 - \log_7 4$

32. $\log_8 5 + \log_8 15 - \log_8 20$

33. $\log_{10} x - \log_{10} (x + 1) + \log_{10} (x^2 - 2)$

34. $\log_9 (4x) - \log_9 (x - 3) + \log_9 (x^3 + 1)$

35. $3\log_2 x + \frac{1}{2}\log_2 x - 2\log_2 (x + 1)$

36. $2\log_5 x + \frac{1}{3}\log_5 x - 3\log_5 (x + 5)$

37. $2\log_8 x - \frac{2}{3}\log_8 x + 4\log_8 x$

38. $5\log_6 x - \frac{3}{4}\log_6 x + 3\log_6 x$

MIXED PRACTICE

Write each expression as a sum or difference of multiples of logarithms. Assume that variables represent positive numbers. See Example 5.

39. $\log_3 \frac{4y}{5}$

40. $\log_7 \frac{5x}{4}$

41. $\log_4 \frac{5}{9z}$

42. $\log_9 \frac{7}{8y}$

▸ **43.** $\log_2 \frac{x^3}{y}$

44. $\log_5 \frac{x}{y^4}$

45. $\log_b \sqrt{7x}$

46. $\log_b \sqrt{\frac{3}{y}}$

47. $\log_6 x^4 y^5$

48. $\log_2 y^3 z$

49. $\log_5 x^3 (x + 1)$

50. $\log_3 x^2 (x - 9)$

51. $\log_6 \frac{x^2}{x + 3}$

52. $\log_3 \frac{(x + 5)^2}{x}$

If $\log_b 3 = 0.5$ and $\log_b 5 = 0.7$, evaluate each expression. See Example 6.

53. $\log_b 15$

54. $\log_b 25$

55. $\log_b \frac{5}{3}$

56. $\log_b \frac{3}{5}$

57. $\log_b \sqrt{5}$

58. $\log_b \sqrt[4]{3}$

If $\log_b 2 = 0.43$ and $\log_b 3 = 0.68$, evaluate each expression. See Example 6.

59. $\log_b 8$

60. $\log_b 81$

61. $\log_b \dfrac{3}{9}$

62. $\log_b \dfrac{4}{32}$

63. $\log_b \sqrt{\dfrac{2}{3}}$

64. $\log_b \sqrt{\dfrac{3}{2}}$

REVIEW AND PREVIEW

Graph both functions on the same set of axes. See Sections 12.3 and 12.5.

65. $y = 10^x$

66. $y = \log_{10} x$.

Evaluate each expression. See Section 12.5.

67. $\log_{10} 100$

68. $\log_{10} \dfrac{1}{10}$

69. $\log_7 7^2$

70. $\log_7 \sqrt{7}$

CONCEPT EXTENSIONS

Solve. See the Concept Checks in this section.

71. Which of the following is the correct way to rewrite $\log_3 \dfrac{14}{11}$?

a. $\dfrac{\log_3 14}{\log_3 11}$

b. $\log_3 14 - \log_3 11$

c. $\log_3 (14 - 11)$

d. $\log_3 154$

72. Which of the following is the correct way to rewrite $\log_9 \dfrac{21}{3}$?

a. $\log_9 7$

b. $\log_9 (21 - 3)$

c. $\dfrac{\log_9 21}{\log_9 3}$

d. $\log_9 21 - \log_9 3$

Answer the following true or false. Study your logarithm properties carefully before answering.

73. $\log_2 x^3 = 3 \log_2 x$

74. $\log_3 (x + y) = \log_3 x + \log_3 y$

75. $\dfrac{\log_7 10}{\log_7 5} = \log_7 2$

76. $\log_7 \dfrac{14}{8} = \log_7 14 - \log_7 8$

77. $\dfrac{\log_7 x}{\log_7 y} = (\log_7 x) - (\log_7 y)$

78. $(\log_3 6) \cdot (\log_3 4) = \log_3 24$

79. It is true that $\log_b 8 = \log_b (8 \cdot 1) = \log_b 8 + \log_b 1$. Explain how $\log_b 8$ can equal $\log_b 8 + \log_b 1$.

80. It is true that $\log_b 7 = \log_b \dfrac{7}{1} = \log_b 7 - \log_b 1$. Explain how $\log_b 7$ can equal $\log_b 7 - \log_b 1$.

Integrated Review Functions and Properties of Logarithms

Sections 12.1–12.6

If $f(x) = x - 6$ and $g(x) = x^2 + 1$, find each function.

1. $(f + g)(x)$

2. $(f - g)(x)$

3. $(f \cdot g)(x)$

4. $\left(\dfrac{f}{g}\right)(x)$

If $f(x) = \sqrt{x}$ and $g(x) = 3x - 1$, find each function.

5. $(f \circ g)(x)$

6. $(g \circ f)(x)$

Determine whether each is a one-to-one function. If it is, find its inverse.

7. $f = \{(-2, 6), (4, 8), (2, -6), (3, 3)\}$

8. $g = \{(4, 2), (-1, 3), (5, 3), (7, 1)\}$

Determine from the graph whether each function is one-to-one.

9.

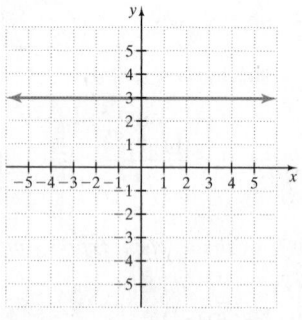

10.

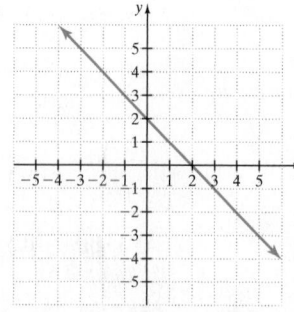

11.

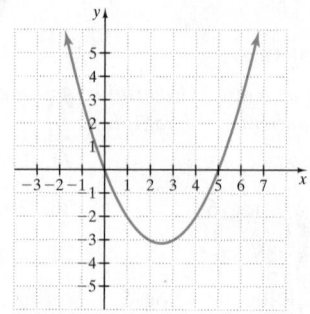

Each function listed is one-to-one. Find the inverse of each function.

12. $f(x) = 3x$

13. $f(x) = x + 4$

14. $f(x) = 5x - 1$

15. $f(x) = 3x + 2$

Graph each function.

16. $y = \left(\dfrac{1}{2}\right)^x$

17. $y = 2^x + 1$

18. $y = \log_3 x$

19. $y = \log_{1/3} x$

Solve.

20. $2^x = 8$

21. $9 = 3^{x-5}$

22. $4^{x-1} = 8^{x+2}$

23. $25^x = 125^{x-1}$

24. $\log_4 16 = x$

25. $\log_{49} 7 = x$

26. $\log_2 x = 5$

27. $\log_x 64 = 3$

28. $\log_x \dfrac{1}{125} = -3$

29. $\log_3 x = -2$

Write each as a single logarithm.

30. $5 \log_2 x$

31. $x \log_2 5$

32. $3 \log_5 x - 5 \log_5 y$

33. $9 \log_5 x + 3 \log_5 y$

34. $\log_2 x + \log_2(x - 3) - \log_2(x^2 + 4)$

35. $\log_3 y - \log_3(y + 2) + \log_3(y^3 + 11)$

Write each expression as sums or differences of multiples of logarithms.

36. $\log_7 \dfrac{9x^2}{y}$

37. $\log_6 \dfrac{5y}{z^2}$

38. An unusually wet spring has caused the size of the mosquito population in a community to increase by 6% each day. If an estimated 100,000 mosquitoes are in the community on April 1, find how many mosquitoes will inhabit the community on April 17. Round to the nearest thousand.

12.7 Common Logarithms, Natural Logarithms, and Change of Base

OBJECTIVES

1 Identify Common Logarithms and Approximate Them by Calculator.

2 Evaluate Common Logarithms of Powers of 10.

3 Identify Natural Logarithms and Approximate Them by Calculator.

4 Evaluate Natural Logarithms of Powers of *e*.

5 Use the Change of Base Formula.

In this section, we look closely at two particular logarithmic bases. These two logarithmic bases are used so frequently that logarithms to their bases are given special names. **Common logarithms** are logarithms to base 10. **Natural logarithms** are logarithms to base *e*, which we introduce in this section. The work in this section is based on the use of a calculator that has both the "common log" ⌊LOG⌋ and the "natural log" ⌊LN⌋ keys.

OBJECTIVE

1 Approximating Common Logarithms

Logarithms to base 10, common logarithms, are used frequently because our number system is a base 10 decimal system. The notation $\log x$ means the same as $\log_{10} x$.

> **Common Logarithms**
>
> $$\log x \text{ means } \log_{10} x$$

EXAMPLE 1   Use a calculator to approximate log 7 to four decimal places.

Solution Press the following sequence of keys.

$$\boxed{7}\ \boxed{\text{LOG}}\ \text{ or }\ \boxed{\text{LOG}}\ \boxed{7}\ \boxed{\text{ENTER}}$$

To four decimal places,

$$\log 7 \approx 0.8451$$

PRACTICE

1   Use a calculator to approximate log 15 to four decimal places.

OBJECTIVE

2   Evaluating Common Logarithms of Powers of 10 ▶

To evaluate the common log of a power of 10, a calculator is not needed. According to the property of logarithms,

$$\log_b b^x = x$$

It follows that if b is replaced with 10, we have

$$\log 10^x = x$$

> **Helpful Hint**
>
> Remember that the understood base here is 10.

EXAMPLE 2   Find the exact value of each logarithm.

a. $\log 10$ **b.** $\log 1000$ **c.** $\log \dfrac{1}{10}$ **d.** $\log \sqrt{10}$

Solution

a. $\log 10 = \log 10^1 = 1$ **b.** $\log 1000 = \log 10^3 = 3$

c. $\log \dfrac{1}{10} = \log 10^{-1} = -1$ **d.** $\log \sqrt{10} = \log 10^{1/2} = \dfrac{1}{2}$ □

PRACTICE

2   Find the exact value of each logarithm.

a. $\log \dfrac{1}{100}$ **b.** $\log 100{,}000$ **c.** $\log \sqrt[5]{10}$ **d.** $\log 0.001$ ■

As we will soon see, equations containing common logarithms are useful models of many natural phenomena.

EXAMPLE 3   Solve $\log x = 1.2$ for x. Give the exact solution and then approximate the solution to four decimal places.

Solution Remember that the base of a common logarithm is understood to be 10.

$$\log x = 1.2$$

> **Helpful Hint**
>
> The understood base is 10.

$$10^{1.2} = x \qquad \text{Write with exponential notation.}$$

The exact solution is $10^{1.2}$. To four decimal places, $x \approx 15.8489$.   □

PRACTICE

3   Solve $\log x = 3.4$ for x. Give the exact solution and then approximate the solution to four decimal places.   ■

The Richter scale measures the intensity, or magnitude, of an earthquake. The formula for the magnitude R of an earthquake is $R = \log\left(\dfrac{a}{T}\right) + B$, where a is the amplitude in micrometers of the vertical motion of the ground at the recording station, T is the number of seconds between successive seismic waves, and B is an adjustment factor that takes into account the weakening of the seismic wave as the distance increases from the epicenter of the earthquake.

EXAMPLE 4 Finding the Magnitude of an Earthquake

Find an earthquake's magnitude on the Richter scale if a recording station measures an amplitude of 300 micrometers and 2.5 seconds between waves. Assume that B is 4.2. Approximate the solution to the nearest tenth.

Solution Substitute the known values into the formula for earthquake intensity.

$$R = \log\left(\frac{a}{T}\right) + B \qquad \text{Richter scale formula}$$

$$= \log\left(\frac{300}{2.5}\right) + 4.2 \quad \text{Let } a = 300, T = 2.5, \text{ and } B = 4.2.$$

$$= \log(120) + 4.2$$

$$\approx 2.1 + 4.2 \qquad \text{Approximate log 120 by 2.1.}$$

$$= 6.3$$

This earthquake had a magnitude of 6.3 on the Richter scale. □

PRACTICE

4 Find an earthquake's magnitude on the Richter scale if a recording station measures an amplitude of 450 micrometers and 4.2 seconds between waves with $B = 3.6$. Approximate the solution to the nearest tenth. ∎

OBJECTIVE

3 Approximating Natural Logarithms

Natural logarithms are also frequently used, especially to describe natural events— hence the label "natural logarithm." Natural logarithms are logarithms to the base e, which is a constant approximately equal to 2.7183. The number e is an irrational number, as is π. The notation $\log_e x$ is usually abbreviated to $\ln x$. (The abbreviation $\ln$ is read "el en.")

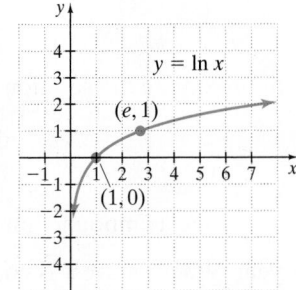
$y = \ln x$
$(e, 1)$
$(1, 0)$

Natural Logarithms

$$\ln x \text{ means } \log_e x$$

The graph of $y = \ln x$ is shown to the left.

EXAMPLE 5 Use a calculator to approximate $\ln 8$ to four decimal places.

Solution Press the following sequence of keys.

$$\boxed{8}\ \boxed{\text{LN}} \quad \text{or} \quad \boxed{\text{LN}}\ \boxed{8}\ \boxed{\text{ENTER}}$$

To four decimal places,

$$\ln 8 \approx 2.0794$$ □

PRACTICE

5 Use a calculator to approximate $\ln 13$ to four decimal places. ∎

OBJECTIVE

4 Evaluating Natural Logarithms of Powers of e.

As a result of the property $\log_b b^x = x$, we know that $\log_e e^x = x$, or **$\ln e^x = x$.**
Since $\ln e^x = x$, $\ln e^5 = 5$, $\ln e^{22} = 22$, and so on. Also,

$$\ln e^1 = 1 \text{ or simply } \ln e = 1.$$

That is why the graph of $y = \ln x$ shown above in the margin passes through $(e, 1)$. If $x = e$, then $y = \ln e = 1$; thus the ordered pair is $(e, 1)$.

EXAMPLE 6 Find the exact value of each natural logarithm.

a. $\ln e^3$ b. $\ln \sqrt[8]{e}$

Solution

a. $\ln e^3 = 3$ b. $\ln \sqrt[8]{e} = \ln e^{1/8} = \dfrac{1}{8}$

PRACTICE
6 Find the exact value of each natural logarithm.

a. $\ln e^4$ b. $\ln \sqrt[3]{e}$

EXAMPLE 7 Solve $\ln 3x = 5$. Give the exact solution and then approximate the solution to four decimal places.

Solution Remember that the base of a natural logarithm is understood to be e.

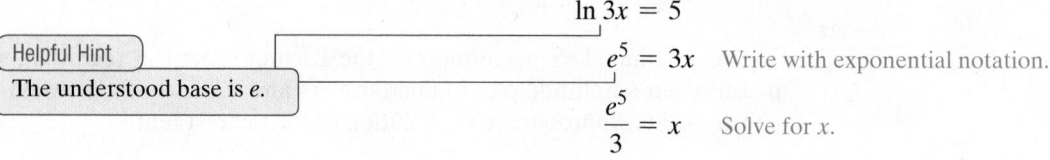

$$\ln 3x = 5$$

Helpful Hint
The understood base is e.

$$e^5 = 3x \quad \text{Write with exponential notation.}$$

$$\frac{e^5}{3} = x \quad \text{Solve for } x.$$

The exact solution is $\dfrac{e^5}{3}$. To four decimal places,

$$x \approx 49.4711.$$

PRACTICE
7 Solve $\ln 5x = 8$. Give the exact solution and then approximate the solution to four decimal places.

Recall from Section 12.3 the formula $A = P\left(1 + \dfrac{r}{n}\right)^{nt}$ for compound interest, where n represents the number of compoundings per year. When interest is compounded continuously, the formula $A = Pe^{rt}$ is used, where r is the annual interest rate, and interest is compounded continuously for t years.

EXAMPLE 8 **Finding Final Loan Payment**

Find the amount owed at the end of 5 years if $1600 is loaned at a rate of 9% compounded continuously.

Solution Use the formula $A = Pe^{rt}$, where

$$P = \$1600 \text{ (the amount of the loan)}$$
$$r = 9\% = 0.09 \text{ (the rate of interest)}$$
$$t = 5 \text{ (the 5-year duration of the loan)}$$
$$A = Pe^{rt}$$
$$= 1600e^{0.09(5)} \quad \text{Substitute in known values.}$$
$$= 1600e^{0.45}$$

Now we can use a calculator to approximate the solution.

$$A \approx 2509.30$$

The total amount of money owed is $2509.30.

PRACTICE
8 Find the amount owed at the end of 4 years if $2400 is borrowed at a rate of 6% compounded continuously.

OBJECTIVE

5 Using the Change of Base Formula

Calculators are handy tools for approximating natural and common logarithms. Unfortunately, some calculators cannot be used to approximate logarithms to bases other than e or 10—at least not directly. In such cases, we use the change of base formula.

Change of Base

If a, b, and c are positive real numbers and neither b nor c is 1, then

$$\log_b a = \frac{\log_c a}{\log_c b}$$

EXAMPLE 9 Approximate $\log_5 3$ to four decimal places.

Solution Use the change of base property to write $\log_5 3$ as a quotient of logarithms to base 10.

$$\log_5 3 = \frac{\log 3}{\log 5} \qquad \text{Use the change of base property. In the change of base property, we let } a = 3, b = 5, \text{ and } c = 10.$$

$$\approx \frac{0.4771213}{0.69897} \qquad \text{Approximate logarithms by calculator.}$$

$$\approx 0.6826062 \qquad \text{Simplify by calculator.}$$

To four decimal places, $\log_5 3 \approx 0.6826$.

PRACTICE

9 Approximate $\log_8 5$ to four decimal places.

✔ **CONCEPT CHECK**

If a graphing calculator cannot directly evaluate logarithms to base 5, describe how you could use the graphing calculator to graph the function $f(x) = \log_5 x$.

✔ **Vocabulary, Readiness & Video Check**

Use the choices to fill in each blank.

1. The base of $\log 7$ is _____.
  a. e **b.** 7   **c.** 10   **d.** no answer

2. The base of $\ln 7$ is _____.
  a. e **b.** 7   **c.** 10   **d.** no answer

3. $\log_{10} 10^7 =$ _____.
  a. e **b.** 7   **c.** 10   **d.** no answer

4. $\log_7 1 =$ _____.
  a. e **b.** 7   **c.** 10   **d.** 0

5. $\log_e e^5 =$ _____.
  a. e **b.** 5   **c.** 0   **d.** 1

6. Study exercise 5 to the left. Then answer: $\ln e^5 =$ _____.
  a. e **b.** 5   **c.** 0   **d.** 1

7. $\log_2 7 =$ _____ (There may be more than one answer.)
  a. $\dfrac{\log 7}{\log 2}$ **b.** $\dfrac{\ln 7}{\ln 2}$ **c.** $\dfrac{\log 2}{\log 7}$ **d.** $\log \dfrac{7}{2}$

Answer to Concept Check:
$$f(x) = \frac{\log x}{\log 5}$$

Martin-Gay Interactive Videos

See Video 12.7

12.7 Exercise Set MyMathLab®

MIXED PRACTICE

Use a calculator to approximate each logarithm to four decimal places. See Examples 1 and 5.

1. $\log 8$ **2.** $\log 6$

3. $\log 2.31$ **4.** $\log 4.86$

5. $\ln 2$ **6.** $\ln 3$

7. $\ln 0.0716$ **8.** $\ln 0.0032$

9. $\log 12.6$ **10.** $\log 25.9$

11. $\ln 5$ **12.** $\ln 7$

13. $\log 41.5$ **14.** $\ln 41.5$

MIXED PRACTICE

Find the exact value. See Examples 2 and 6.

15. $\log 100$ **16.** $\log 10{,}000$

17. $\log \dfrac{1}{1000}$ **18.** $\log \dfrac{1}{100}$

19. $\ln e^2$ **20.** $\ln e^9$

21. $\ln \sqrt[4]{e}$ **22.** $\ln \sqrt[5]{e}$

23. $\log 10^3$ **24.** $\log 10^7$

25. $\ln e^{-7}$ **26.** $\ln e^{-5}$

27. $\log 0.0001$ **28.** $\log 0.001$

29. $\ln \sqrt{e}$ **30.** $\log \sqrt{10}$

Solve each equation for x. Give the exact solution and a four-decimal-place approximation. See Examples 3 and 7.

31. $\ln 2x = 7$ **32.** $\ln 5x = 9$

33. $\log x = 1.3$ **34.** $\log x = 2.1$

35. $\log 2x = 1.1$ **36.** $\log 3x = 1.3$

37. $\ln x = 1.4$ **38.** $\ln x = 2.1$

39. $\ln(3x - 4) = 2.3$

40. $\ln(2x + 5) = 3.4$

41. $\log x = 2.3$

42. $\log x = 3.1$

43. $\ln x = -2.3$

44. $\ln x = -3.7$

45. $\log(2x + 1) = -0.5$

46. $\log(3x - 2) = -0.8$

47. $\ln 4x = 0.18$

48. $\ln 3x = 0.76$

Approximate each logarithm to four decimal places. See Example 9.

49. $\log_2 3$ **50.** $\log_3 2$

51. $\log_{1/2} 5$ **52.** $\log_{1/3} 2$

53. $\log_4 9$ **54.** $\log_9 4$

55. $\log_3 \dfrac{1}{6}$ **56.** $\log_6 \dfrac{2}{3}$

57. $\log_8 6$ **58.** $\log_6 8$

Use the formula $R = \log\left(\dfrac{a}{T}\right) + B$ to find the intensity R on the Richter scale of the earthquakes that fit the descriptions given. Round answers to one decimal place. See Example 4.

59. Amplitude a is 200 micrometers, time T between waves is 1.6 seconds, and B is 2.1.

60. Amplitude a is 150 micrometers, time T between waves is 3.6 seconds, and B is 1.9.

61. Amplitude a is 400 micrometers, time T between waves is 2.6 seconds, and B is 3.1.

62. Amplitude a is 450 micrometers, time T between waves is 4.2 seconds, and B is 2.7.

Use the formula $A = Pe^{rt}$ to solve. See Example 8.

63. Find how much money Dana Jones has after 12 years if $1400 is invested at 8% interest compounded continuously.

64. Determine the amount in an account in which $3500 earns 6% interest compounded continuously for 1 year.

65. Find the amount of money Barbara Mack owes at the end of 4 years if 6% interest is compounded continuously on her $2000 debt.

66. Find the amount of money for which a $2500 certificate of deposit is redeemable if it has been paying 10% interest compounded continuously for 3 years.

REVIEW AND PREVIEW

Solve each equation for x. See Sections 2.3, 2.5, and 6.6.

67. $6x - 3(2 - 5x) = 6$

68. $2x + 3 = 5 - 2(3x - 1)$

69. $2x + 3y = 6x$

70. $4x - 8y = 10x$

71. $x^2 + 7x = -6$

72. $x^2 + 4x = 12$

Solve each system of equations. See Section 4.1.

73. $\begin{cases} x + 2y = -4 \\ 3x - y = 9 \end{cases}$

74. $\begin{cases} 5x + y = 5 \\ -3x - 2y = -10 \end{cases}$

CONCEPT EXTENSIONS

75. Use a calculator to try to approximate log 0. Describe what happens and explain why.

76. Use a calculator to try to approximate ln 0. Describe what happens and explain why.

77. Without using a calculator, explain which of log 50 or ln 50 must be larger and why.

78. Without using a calculator, explain which of $\log 50^{-1}$ or $\ln 50^{-1}$ must be larger and why.

Graph each function by finding ordered pair solutions, plotting the solutions, and then drawing a smooth curve through the plotted points.

79. $f(x) = e^x$

80. $f(x) = e^{2x}$

81. $f(x) = e^{-3x}$

82. $f(x) = e^{-x}$

83. $f(x) = e^x + 2$

84. $f(x) = e^x - 3$

85. $f(x) = e^{x-1}$

86. $f(x) = e^{x+4}$

87. $f(x) = 3e^x$

88. $f(x) = -2e^x$

89. $f(x) = \ln x$

90. $f(x) = \log x$

91. $f(x) = -2 \log x$

92. $f(x) = 3 \ln x$

93. $f(x) = \log(x + 2)$

94. $f(x) = \log(x - 2)$

95. $f(x) = \ln x - 3$

96. $f(x) = \ln x + 3$

97. Graph $f(x) = e^x$ (Exercise 79), $f(x) = e^x + 2$ (Exercise 83), and $f(x) = e^x - 3$ (Exercise 84) on the same screen. Discuss any trends shown on the graphs.

98. Graph $f(x) = \ln x$ (Exercise 89), $f(x) = \ln x - 3$ (Exercise 95), and $f(x) = \ln x + 3$ (Exercise 96) on the same screen. Discuss any trends shown on the graphs.

12.8 Exponential and Logarithmic Equations and Problem Solving

OBJECTIVES

1 Solve Exponential Equations.

2 Solve Logarithmic Equations.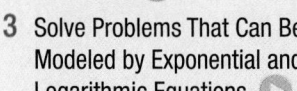

3 Solve Problems That Can Be Modeled by Exponential and Logarithmic Equations.

OBJECTIVE

1 Solving Exponential Equations

In Section 12.3 we solved exponential equations such as $2^x = 16$ by writing 16 as a power of 2 and applying the uniqueness of b^x.

$$2^x = 16$$
$$2^x = 2^4 \quad \text{Write 16 as } 2^4.$$
$$x = 4 \quad \text{Use the uniqueness of } b^x.$$

Solving the equation in this manner is possible since 16 is a power of 2. When solving an equation such as $2^x = a\ number$ and the number is not a power of 2, we use logarithms. For example, to solve an equation such as $3^x = 7$, we use the fact that $f(x) = \log_b x$ is a one-to-one function. Another way of stating this fact is as a property of equality.

> **Logarithm Property of Equality**
>
> Let a, b, and c be real numbers such that $\log_b a$ and $\log_b c$ are real numbers and b is not 1. Then
>
> $$\log_b a = \log_b c \text{ is equivalent to } a = c$$

EXAMPLE 1 Solve: $3^x = 7$

Solution To solve, we use the logarithm property of equality and take the logarithm of both sides. For this example, we use the common logarithm.

$$3^x = 7$$
$$\log 3^x = \log 7 \quad \text{Take the common logarithm of both sides.}$$
$$x \log 3 = \log 7 \quad \text{Apply the power property of logarithms.}$$
$$x = \frac{\log 7}{\log 3} \quad \text{Divide both sides by log 3.}$$

(Continued on next page)

The exact solution is $\dfrac{\log 7}{\log 3}$. If a decimal approximation is preferred,

$$\frac{\log 7}{\log 3} \approx \frac{0.845098}{0.4771213} \approx 1.7712 \text{ to four decimal places.}$$

The solution is $\dfrac{\log 7}{\log 3}$, or *approximately* 1.7712. □

PRACTICE

1 Solve: $5^x = 9$ ■

OBJECTIVE

2 Solving Logarithmic Equations ▶

By applying the appropriate properties of logarithms, we can solve a broad variety of logarithmic equations.

EXAMPLE 2 Solve: $\log_4(x - 2) = 2$

Solution Notice that $x - 2$ must be positive, so x must be greater than 2. With this in mind, we first write the equation with exponential notation.

$$\log_4(x - 2) = 2$$
$$4^2 = x - 2$$
$$16 = x - 2$$
$$18 = x \qquad \text{Add 2 to both sides.}$$

Check: To check, we replace x with 18 in the original equation.

$$\log_4(x - 2) = 2$$
$$\log_4(18 - 2) \overset{?}{=} 2 \qquad \text{Let } x = 18.$$
$$\log_4 16 \overset{?}{=} 2$$
$$4^2 = 16 \qquad \text{True}$$

The solution is 18. □

PRACTICE

2 Solve: $\log_2(x - 1) = 5$ ■

EXAMPLE 3 Solve: $\log_2 x + \log_2(x - 1) = 1$

Solution Notice that $x - 1$ must be positive, so x must be greater than 1. We use the product property on the left side of the equation.

$$\log_2 x + \log_2(x - 1) = 1$$
$$\log_2 x(x - 1) = 1 \qquad \text{Apply the product property.}$$
$$\log_2(x^2 - x) = 1$$

Next we write the equation with exponential notation and solve for x.

$$2^1 = x^2 - x$$
$$0 = x^2 - x - 2 \qquad \text{Subtract 2 from both sides.}$$
$$0 = (x - 2)(x + 1) \qquad \text{Factor.}$$
$$0 = x - 2 \quad \text{or} \quad 0 = x + 1 \qquad \text{Set each factor equal to 0.}$$
$$2 = x \qquad\qquad -1 = x$$

Recall that -1 cannot be a solution because x must be greater than 1. If we forgot this, we would still reject -1 after checking. To see this, we replace x with -1 in the original equation.

$$\log_2 x + \log_2(x - 1) = 1$$
$$\log_2(-1) + \log_2(-1 - 1) \overset{?}{=} 1 \qquad \text{Let } x = -1.$$

Because the logarithm of a negative number is undefined, -1 is rejected. Check to see that the solution is 2. □

PRACTICE
3   Solve: $\log_5 x + \log_5(x + 4) = 1$ ■

EXAMPLE 4   Solve: $\log(x + 2) - \log x = 2$

We use the quotient property of logarithms on the left side of the equation.

Solution $\log(x + 2) - \log x = 2$

$$\log\frac{x + 2}{x} = 2 \qquad \text{Apply the quotient property.}$$

$$10^2 = \frac{x + 2}{x} \qquad \text{Write using exponential notation.}$$

$$100 = \frac{x + 2}{x} \qquad \text{Simplify.}$$

$$100x = x + 2 \qquad \text{Multiply both sides by } x.$$

$$99x = 2 \qquad \text{Subtract } x \text{ from both sides.}$$

$$x = \frac{2}{99} \qquad \text{Divide both sides by 99.}$$

Verify that the solution is $\frac{2}{99}$. □

PRACTICE
4   Solve: $\log(x + 3) - \log x = 1$ ■

OBJECTIVE
3 Solving Problems Modeled by Exponential and Logarithmic Equations ▶

Logarithmic and exponential functions are used in a variety of scientific, technical, and business settings. A few examples follow.

EXAMPLE 5   **Estimating Population Size**

The population size y of a community of lemmings varies according to the relationship $y = y_0 e^{0.15t}$. In this formula, t is time in months, and y_0 is the initial population at time 0. Estimate the population after 6 months if there were originally 5000 lemmings.

Solution   We substitute 5000 for y_0 and 6 for t.

$$y = y_0 e^{0.15t}$$
$$= 5000 e^{0.15(6)} \qquad \text{Let } t = 6 \text{ and } y_0 = 5000.$$
$$= 5000 e^{0.9} \qquad \text{Multiply.}$$

Using a calculator, we find that $y \approx 12{,}298.016$. In 6 months, the population will be approximately 12,300 lemmings. □

PRACTICE
5   The population size y of a group of rabbits varies according to the relationship $y = y_0 e^{0.916t}$. In this formula, t is time in years and y_0 is the initial population at time $t = 0$. Estimate the population in three years if there were originally 60 rabbits. ■

EXAMPLE 6 **Doubling an Investment**

How long does it take an investment of $2000 to double if it is invested at 5% interest compounded quarterly? The necessary formula is $A = P\left(1 + \dfrac{r}{n}\right)^{nt}$, where A is the accrued (or owed) amount, P is the principal invested, r is the annual rate of interest, n is the number of compounding periods per year, and t is the number of years.

<u>Solution</u> We are given that $P = \$2000$ and $r = 5\% = 0.05$. Compounding quarterly means 4 times a year, so $n = 4$. The investment is to double, so A must be $4000. Substitute these values and solve for t.

$$A = P\left(1 + \frac{r}{n}\right)^{nt}$$

$$4000 = 2000\left(1 + \frac{0.05}{4}\right)^{4t} \quad \text{Substitute in known values.}$$

$$4000 = 2000(1.0125)^{4t} \quad \text{Simplify } 1 + \frac{0.05}{4}.$$

$$2 = (1.0125)^{4t} \quad \text{Divide both sides by 2000.}$$

$$\log 2 = \log 1.0125^{4t} \quad \text{Take the logarithm of both sides.}$$

$$\log 2 = 4t(\log 1.0125) \quad \text{Apply the power property.}$$

$$\frac{\log 2}{4 \log 1.0125} = t \quad \text{Divide both sides by } 4 \log 1.0125.$$

$$13.949408 \approx t \quad \text{Approximate by calculator.}$$

Thus, it takes nearly 14 years for the money to double in value. □

PRACTICE
6 How long does it take for an investment of $3000 to double if it is invested at 7% interest compounded monthly? Round to the nearest year. ∎

Graphing Calculator Explorations

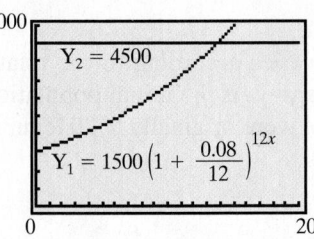

Use a graphing calculator to find how long it takes an investment of $1500 to triple if it is invested at 8% interest compounded monthly.

First, let $P = \$1500$, $r = 0.08$, and $n = 12$ (for 12 months) in the formula

$$A = P\left(1 + \frac{r}{n}\right)^{nt}$$

Notice that when the investment has tripled, the accrued amount A is $4500. Thus,

$$4500 = 1500\left(1 + \frac{0.08}{12}\right)^{12t}$$

Determine an appropriate viewing window and enter and graph the equations

$$Y_1 = 1500\left(1 + \frac{0.08}{12}\right)^{12x}$$

and

$$Y_2 = 4500$$

The point of intersection of the two curves is the solution. The x-coordinate tells how long it takes for the investment to triple.

Use a TRACE feature or an INTERSECT feature to approximate the coordinates of the point of intersection of the two curves. It takes approximately 13.78 years, or 13 years and 9 months, for the investment to triple in value to $4500.

Use this graphical solution method to solve each problem. Round each answer to the nearest hundredth.

1. Find how long it takes an investment of $5000 to grow to $6000 if it is invested at 5% interest compounded quarterly.

2. Find how long it takes an investment of $1000 to double if it is invested at 4.5% interest compounded daily. (Use 365 days in a year.)

3. Find how long it takes an investment of $10,000 to quadruple if it is invested at 6% interest compounded monthly.

4. Find how long it takes $500 to grow to $800 if it is invested at 4% interest compounded semiannually.

✔ Vocabulary, Readiness & Video Check

Martin-Gay Interactive Videos

See Video 12.8

Watch the section lecture video and answer the following questions.

OBJECTIVE 1
1. From the lecture before ▦ Example 1, explain why $\ln(4x - 2) = \ln 3$ is equivalent to $4x - 2 = 3$.

OBJECTIVE 2
2. Why is the possible solution of -8 rejected in ▦ Example 3?

OBJECTIVE 3
3. For ▦ Example 4, write the equation and find the number of years it takes $1000 to double at 7% interest compounded monthly. Explain the similarity to the answer to Example 4. Round your answer to the nearest tenth.

12.8   Exercise Set   MyMathLab® ▶

Solve each equation. Give the exact solution and approximate the solution to four decimal places. See Example 1.

1. $3^x = 6$

2. $4^x = 7$

3. $3^{2x} = 3.8$

4. $5^{3x} = 5.6$

▶ 5. $2^{x-3} = 5$

6. $8^{x-2} = 12$

7. $9^x = 5$

8. $3^x = 11$

9. $4^{x+7} = 3$

10. $6^{x+3} = 2$

MIXED PRACTICE

Solve each equation. See Examples 1 through 4.

▶ 11. $\log_2(x + 5) = 4$

12. $\log_2(x - 5) = 3$

13. $\log_4 2 + \log_4 x = 0$

14. $\log_3 5 + \log_3 x = 1$

15. $\log_2 6 - \log_2 x = 3$

16. $\log_4 10 - \log_4 x = 2$

17. $\log_6(x^2 - x) = 1$

18. $\log_2(x^2 + x) = 1$

▶ 19. $\log_4 x + \log_4(x + 6) = 2$

20. $\log_3 x + \log_3(x + 6) = 3$

21. $\log_5(x + 3) - \log_5 x = 2$

22. $\log_6(x + 2) - \log_6 x = 2$

23. $7^{3x-4} = 11$

24. $5^{2x-6} = 12$

25. $\log_4(x^2 - 3x) = 1$

26. $\log_8(x^2 - 2x) = 1$

27. $e^{6x} = 5$

28. $e^{2x} = 8$

29. $\log_3 x^2 = 4$

30. $\log_2 x^2 = 6$

31. $\ln 5 + \ln x = 0$

32. $\ln 3 + \ln(x - 1) = 0$

33. $3 \log x - \log x^2 = 2$

34. $2 \log x - \log x = 3$

35. $\log_4 x - \log_4(2x - 3) = 3$

36. $\log_2 x - \log_2(3x + 5) = 4$

37. $\log_2 x + \log_2(3x + 1) = 1$

38. $\log_3 x + \log_3(x - 8) = 2$

39. $\log_2 x + \log_2(x + 5) = 1$

40. $\log_4 x + \log_4(x + 7) = 1$

Solve. See Example 5.

41. The size of the wolf population at Isle Royale National Park increases according to the formula $y = y_0 e^{0.043t}$. In this formula, t is time in years and y_0 is the initial population at time 0. If the size of the current population is 83 wolves, find how many there should be in 5 years. Round to the nearest whole number.

42. The number of victims of a flu epidemic is increasing according to the formula $y = y_0 e^{0.075t}$. In this formula, t is time in weeks and y_0 is the given population at time 0. If 20,000 people are currently infected, how many might be infected in 3 weeks? Round to the nearest whole number.

43. The population of the Cook Islands is decreasing according to the formula $y = y_0 e^{-1.044t}$. In this formula, t is the time in years and y_0 is the given population at time 0. If the size of the population in 2014 was 13,700, use the formula to predict the population of the Cook Islands in 2018. Round to the nearest whole number. (*Source: CIA World Factbook*)

44. The population of Aruba is increasing according to the formula $y = y_0 e^{0.307t}$. In this formula, t is the time in years and y_0 is the initial population at time 0. If the size of the population in 2014 was 110,660, use the formula to predict the population of Aruba in 2019. Round to the nearest whole number. (*Source: CIA World Factbook*)

Use the formula $A = P\left(1 + \dfrac{r}{n}\right)^{nt}$ to solve these compound interest problems. Round to the nearest tenth. See Example 6.

▶ 45. Find how long it takes $600 to double if it is invested at 7% interest compounded monthly.

46. Find how long it takes $600 to double if it is invested at 12% interest compounded monthly.

47. Find how long it takes a $1200 investment to earn $200 interest if it is invested at 9% interest compounded quarterly.

48. Find how long it takes a $1500 investment to earn $200 interest if it is invested at 10% compounded semiannually.

49. Find how long it takes $1000 to double if it is invested at 8% interest compounded semiannually.

50. Find how long it takes $1000 to double if it is invested at 8% interest compounded monthly.

The formula $w = 0.00185h^{2.67}$ is used to estimate the normal weight w of a boy h inches tall. Use this formula to solve the height–weight problems. Round to the nearest tenth.

51. Find the expected weight of a boy who is 35 inches tall.

52. Find the expected weight of a boy who is 43 inches tall.

53. Find the expected height of a boy who weighs 85 pounds.

54. Find the expected height of a boy who weighs 140 pounds.

The formula $P = 14.7e^{-0.21x}$ gives the average atmospheric pressure P, in pounds per square inch, at an altitude x, in miles above sea level. Use this formula to solve these pressure problems. Round answers to the nearest tenth.

55. Find the average atmospheric pressure of Denver, which is 1 mile above sea level.

56. Find the average atmospheric pressure of Pikes Peak, which is 2.7 miles above sea level.

57. Find the elevation of a Delta jet if the atmospheric pressure outside the jet is 7.5 lb/sq in.

58. Find the elevation of a remote Himalayan peak if the atmospheric pressure atop the peak is 6.5 lb/sq in.

Psychologists call the graph of the formula $t = \dfrac{1}{c}\ln\left(\dfrac{A}{A - N}\right)$ the learning curve, since the formula relates time t passed, in weeks, to a measure N of learning achieved, to a measure A of maximum learning possible, and to a measure c of an individual's learning style. Round to the nearest week.

59. Norman is learning to type. If he wants to type at a rate of 50 words per minute (N is 50) and his expected maximum rate is 75 words per minute (A is 75), find how many weeks it should take him to achieve his goal. Assume that c is 0.09.

60. An experiment with teaching chimpanzees sign language shows that a typical chimp can master a maximum of 65 signs. Find how many weeks it should take a chimpanzee to master 30 signs if c is 0.03.

61. Janine is working on her dictation skills. She wants to take dictation at a rate of 150 words per minute and believes that the maximum rate she can hope for is 210 words per minute. Find how many weeks it should take her to achieve the 150 words per minute level if c is 0.07.

62. A psychologist is measuring human capability to memorize nonsense syllables. Find how many weeks it should take a subject to learn 15 nonsense syllables if the maximum possible to learn is 24 syllables and c is 0.17.

REVIEW AND PREVIEW

If $x = -2$, $y = 0$, and $z = 3$, find the value of each expression. See Section 1.4.

63. $\dfrac{x^2 - y + 2z}{3x}$

64. $\dfrac{x^3 - 2y + z}{2z}$

65. $\dfrac{3z - 4x + y}{x + 2z}$

66. $\dfrac{4y - 3x + z}{2x + y}$

Find the inverse function of each one-to-one function. See Section 12.2.

67. $f(x) = 5x + 2$

68. $f(x) = \dfrac{x - 3}{4}$

CONCEPT EXTENSIONS

The formula $y = y_0 e^{kt}$ gives the population size y of a population that experiences an annual rate of population growth k (given as a decimal). In this formula, t is time in years and y_0 is the initial population at time 0. Use this formula to solve Exercises 69 and 70.

69. In 2014, the population of Michigan was approximately 9,910,000 and increasing according to the formula $y = y_0 e^{0.0015t}$. Assume that the population continues to increase according to the given formula and predict how many years after 2014 the population of Michigan will be 9,950,000. Round to the nearest tenth of a year. (*Hint:* Let $y_0 = 9,910,000$ and $y = 9,950,000$, and solve for t.) (*Source:* United States Bureau of the Census)

70. The population of Iowa was approximately 3,107,000 in 2014 and increasing according to the formula $y = y_0 e^{0.0198t}$. Assume that the population continues to increase according to the given formula and predict how many years after 2014 the population of Iowa will be 3,500,000. Round to the nearest tenth of a year. (See the hint for Exercise 69.) (*Source:* United States Bureau of the Census)

71. When solving a logarithmic equation, explain why you must check possible solutions in the original equation.

72. Solve $5^x = 9$ by taking the common logarithm of both sides of the equation. Next, solve this equation by taking the natural logarithm of both sides. Compare your solutions. Are they the same? Why or why not?

Use a graphing calculator to solve each equation. For example, to solve Exercise 73, let $Y_1 = e^{0.3x}$ and $Y_2 = 8$ and graph the equations. The x-value of the point of intersection is the solution. Round all solutions to two decimal places.

73. $e^{0.3x} = 8$

74. $10^{0.5x} = 7$

75. $2 \log(-5.6x + 1.3) + x + 1 = 0$

76. $\ln(1.3x - 2.1) + 3.5x - 5 = 0$

77. Check Exercise 23.

78. Check Exercise 24.

79. Check Exercise 31.

80. Check Exercise 32.

Chapter 12 **Vocabulary Check**

Fill in each blank with one of the words or phrases listed below. Some words or phrases may be used more than once.

inverse	common	composition	symmetric	exponential
vertical	logarithmic	natural	half-life	horizontal

1. For a one-to-one function, we can find its _____ function by switching the coordinates of the ordered pairs of the function.

2. The _____ of functions f and g is $(f \circ g)(x) = f(g(x))$.

3. A function of the form $f(x) = b^x$ is called a(n) _____ function if $b > 0$, b is not 1, and x is a real number.

4. The graphs of f and f^{-1} are _____ about the line $y = x$.

5. _____ logarithms are logarithms to base e.

6. _____ logarithms are logarithms to base 10.

7. To see whether a graph is the graph of a one-to-one function, apply the _____ line test to see whether it is a function and then apply the _____ line test to see whether it is a one-to-one function.

8. A(n) _____ function is a function that can be defined by $f(x) = \log_b x$ where x is a positive real number, b is a constant positive real number, and b is not 1.

9. _____ is the amount of time it takes for half of the amount of a substance to decay.

10. A quantity that grows or decays by the same percent at regular time periods is said to have _____ growth or decay.

Chapter 12 Highlights

DEFINITIONS AND CONCEPTS	EXAMPLES

Section 12.1 The Algebra of Functions; Composite Functions

Algebra of Functions

Sum	$(f + g)(x) = f(x) + g(x)$
Difference	$(f - g)(x) = f(x) - g(x)$
Product	$(f \cdot g)(x) = f(x) \cdot g(x)$
Quotient	$\left(\dfrac{f}{g}\right)(x) = \dfrac{f(x)}{g(x)}, g(x) \neq 0$

If $f(x) = 7x$ and $g(x) = x^2 + 1$,

$$(f + g)(x) = f(x) + g(x) = 7x + x^2 + 1$$
$$(f - g)(x) = f(x) - g(x) = 7x - (x^2 + 1)$$
$$= 7x - x^2 - 1$$
$$(f \cdot g)(x) = f(x) \cdot g(x) = 7x(x^2 + 1)$$
$$= 7x^3 + 7x$$
$$\left(\frac{f}{g}\right)(x) = \frac{f(x)}{g(x)} = \frac{7x}{x^2 + 1}$$

Composite Functions

The notation $(f \circ g)(x)$ means "f composed with g."

$$(f \circ g)(x) = f(g(x))$$
$$(g \circ f)(x) = g(f(x))$$

If $f(x) = x^2 + 1$ and $g(x) = x - 5$, find $(f \circ g)(x)$.

$$(f \circ g)(x) = f(g(x))$$
$$= f(x - 5)$$
$$= (x - 5)^2 + 1$$
$$= x^2 - 10x + 26$$

Section 12.2 Inverse Functions

If f is a function, then f is a **one-to-one function** only if each y-value (output) corresponds to only one x-value (input).

Horizontal Line Test

If every horizontal line intersects the graph of a function at most once, then the function is a one-to-one function.

Determine whether each graph is a one-to-one function.

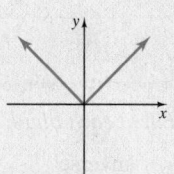

Graphs **A** and **C** pass the vertical line test, so only these are graphs of functions. Of graphs **A** and **C**, only graph **A** passes the horizontal line test, so only graph **A** is the graph of a one-to-one function.

Find the inverse of $f(x) = 2x + 7$.

$$y = 2x + 7 \quad \text{Replace } f(x) \text{ with } y.$$
$$x = 2y + 7 \quad \text{Interchange } x \text{ and } y.$$
$$2y = x - 7 \quad \text{Solve for } y.$$
$$y = \frac{x - 7}{2}$$
$$f^{-1}(x) = \frac{x - 7}{2} \quad \text{Replace } y \text{ with } f^{-1}(x).$$

The **inverse** of a one-to-one function f is the one-to-one function f^{-1} that is the set of all ordered pairs (b, a) such that (a, b) belongs to f.

To Find the Inverse of a One-to-One Function f(x)

Step 1. Replace $f(x)$ with y.

Step 2. Interchange x and y.

Step 3. Solve for y.

Step 4. Replace y with $f^{-1}(x)$.

The inverse of $f(x) = 2x + 7$ is $f^{-1}(x) = \dfrac{x - 7}{2}$.

DEFINITIONS AND CONCEPTS	EXAMPLES

Section 12.3 Exponential Functions

A function of the form $f(x) = b^x$ is an **exponential function,** where $b > 0$, $b \neq 1$, and x is a real number.

Graph the exponential function $y = 4^x$.

x	y
-2	$\dfrac{1}{16}$
-1	$\dfrac{1}{4}$
0	1
1	4
2	16

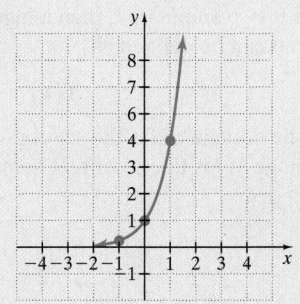

Uniqueness of b^x

If $b > 0$ and $b \neq 1$, then $b^x = b^y$ is equivalent to $x = y$.

Solve $2^{x+5} = 8$.

$2^{x+5} = 2^3$ Write 8 as 2^3.

$x + 5 = 3$ Use the uniqueness of b^x.

$x = -2$ Subtract 5 from both sides.

Section 12.4 Exponential Growth and Decay Functions

A quantity that grows or decays by the same percent at regular time periods is said to have **exponential growth** or **exponential decay**.

Exponential Growth

initial amount   number of time intervals

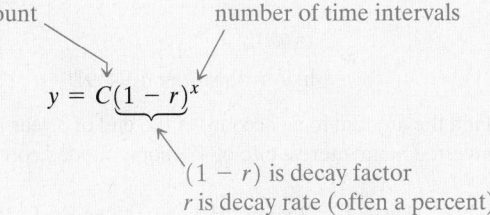

$(1 + r)$ is growth factor
r is growth rate (often a percent)

Exponential Decay

initial amount   number of time intervals

$$y = C\underbrace{(1 - r)}^{x}$$

$(1 - r)$ is decay factor
r is decay rate (often a percent)

A city has a current population of 37,000 that has been increasing at a rate of 3% per year. At this rate, find the city's population in 20 years.

$$y = C(1 + r)^x$$
$$y = 37{,}000(1 + 0.03)^{20}$$
$$y \approx 66{,}826.12$$

In 20 years, the predicted population of the city is 66,826.

A city has a current population of 37,000 that has been decreasing at a rate of 3% per year. At this rate, find the city's population in 20 years.

$$y = C(1 - r)^x$$
$$y = 37{,}000(1 - 0.03)^{20}$$
$$y \approx 20{,}120.39$$

In 20 years, the predicted population of the city is 20,120.

Section 12.5 Logarithmic Functions

Logarithmic Definition

If $b > 0$ and $b \neq 1$, then

$$y = \log_b x \quad \text{means} \quad x = b^y$$

for any positive number x and real number y.

Logarithmic Form	*Corresponding Exponential Statement*
$\log_5 25 = 2$	$5^2 = 25$
$\log_9 3 = \dfrac{1}{2}$	$9^{1/2} = 3$

Properties of Logarithms

If b is a real number, $b > 0$, and $b \neq 1$, then

$$\log_b 1 = 0, \quad \log_b b^x = x, \quad b^{\log_b x} = x$$

$$\log_5 1 = 0, \quad \log_7 7^2 = 2, \quad 3^{\log_3 6} = 6$$

(continued)

DEFINITIONS AND CONCEPTS	EXAMPLES

Section 12.5 Logarithmic Functions (continued)

Logarithmic Function

If $b > 0$ and $b \neq 1$, then a **logarithmic function** is a function that can be defined as

$$f(x) = \log_b x$$

The domain of f is the set of positive real numbers, and the range of f is the set of real numbers.

Graph $y = \log_3 x$.

Write $y = \log_3 x$ as $3^y = x$. Plot the ordered pair solutions listed in the table and connect them with a smooth curve.

x	y
3	1
1	0
$\dfrac{1}{3}$	-1
$\dfrac{1}{9}$	-2

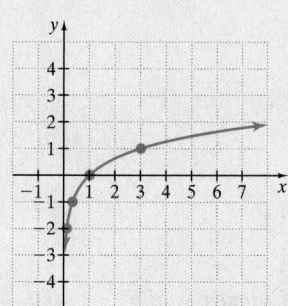

Section 12.6 Properties of Logarithms

Let x, y, and b be positive numbers and $b \neq 1$.

Product Property

$$\log_b xy = \log_b x + \log_b y$$

Quotient Property

$$\log_b \frac{x}{y} = \log_b x - \log_b y$$

Power Property

$$\log_b x^r = r \log_b x$$

Write as a single logarithm.

$2 \log_5 6 + \log_5 x - \log_5(y + 2)$
$= \log_5 6^2 + \log_5 x - \log_5(y + 2)$ Power property
$= \log_5 36 \cdot x - \log_5(y + 2)$ Product property
$= \log_5 \dfrac{36x}{y + 2}$ Quotient property

Section 12.7 Common Logarithms, Natural Logarithms, and Change of Base

Common Logarithms

$$\log x \quad \text{means} \quad \log_{10} x$$

Natural Logarithms

$$\ln x \quad \text{means} \quad \log_e x$$

Continuously Compounded Interest Formula

$$A = Pe^{rt}$$

where r is the annual interest rate for P dollars invested for t years.

$$\log 5 = \log_{10} 5 \approx 0.69897$$

$$\ln 7 = \log_e 7 \approx 1.94591$$

Find the amount in an account at the end of 3 years if $1000 is invested at an interest rate of 4% compounded continuously.

Here, $t = 3$ years, $P = \$1000$, and $r = 0.04$.

$$A = Pe^{rt}$$
$$= 1000e^{0.04(3)}$$
$$\approx \$1127.50$$

Section 12.8 Exponential and Logarithmic Equations and Problem Solving

Logarithm Property of Equality

Let $\log_b a$ and $\log_b c$ be real numbers and $b \neq 1$. Then

$$\log_b a = \log_b c \text{ is equivalent to } a = c$$

Solve $2^x = 5$.

$\log 2^x = \log 5$ Logarithm property of equality

$x \log 2 = \log 5$ Power property

$x = \dfrac{\log 5}{\log 2}$ Divide both sides by log 2.

$x \approx 2.3219$ Use a calculator.

Chapter 12 Review

(12.1) If $f(x) = x - 5$ and $g(x) = 2x + 1$, find

1. $(f + g)(x)$
2. $(f - g)(x)$
3. $(f \cdot g)(x)$
4. $\left(\dfrac{g}{f}\right)(x)$

If $f(x) = x^2 - 2$, $g(x) = x + 1$, and $h(x) = x^3 - x^2$, find each composition.

5. $(f \circ g)(x)$
6. $(g \circ f)(x)$
7. $(h \circ g)(2)$
8. $(f \circ f)(x)$
9. $(f \circ g)(-1)$
10. $(h \circ h)(2)$

(12.2) Determine whether each function is a one-to-one function. If it is one-to-one, list the elements of its inverse.

11. $h = \{(-9, 14), (6, 8), (-11, 12), (15, 15)\}$

12. $f = \{(-5, 5), (0, 4), (13, 5), (11, -6)\}$

13.

U.S. Region (Input)	Northeast	Midwest	South	West
Rank in Housing Starts for 2014 (Output)	4	3	1	2

△ 14.

Shape (Input)	Square	Triangle	Parallelogram	Rectangle
Number of Sides (Output)	4	3	4	4

Given that $f(x) = \sqrt{x + 2}$ is a one-to-one function, find the following.

15. **a.** $f(7)$
    b. $f^{-1}(3)$
16. **a.** $f(-1)$
    b. $f^{-1}(1)$

Determine whether each function is a one-to-one function.

17.
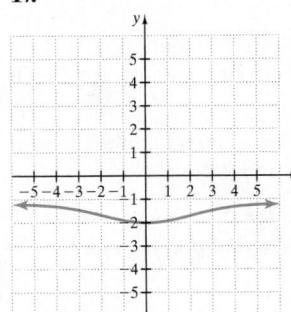

18.
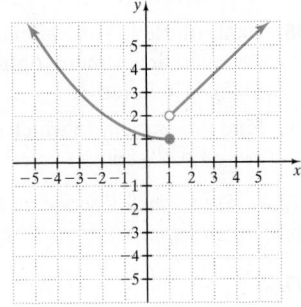

19.
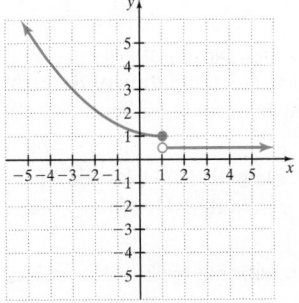

20.
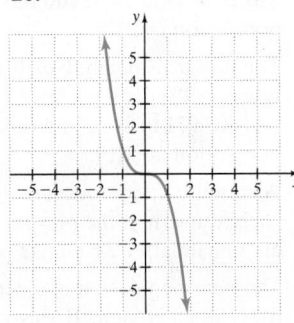

Find an equation defining the inverse function of the given one-to-one function.

21. $f(x) = x - 9$
22. $f(x) = x + 8$
23. $f(x) = 6x + 11$
24. $f(x) = 12x - 1$
25. $f(x) = x^3 - 5$
26. $f(x) = \sqrt[3]{x + 2}$
27. $g(x) = \dfrac{12x - 7}{6}$
28. $r(x) = \dfrac{13x - 5}{2}$

Graph each one-to-one function and its inverse on the same set of axes.

29. $f(x) = -2x + 3$
30. $f(x) = 5x - 5$

(12.3) Solve each equation for x.

31. $4^x = 64$
32. $3^x = \dfrac{1}{9}$
33. $2^{3x} = \dfrac{1}{16}$
34. $5^{2x} = 125$
35. $9^{x+1} = 243$
36. $8^{3x-2} = 4$

Graph each exponential function.

37. $y = 3^x$
38. $y = \left(\dfrac{1}{3}\right)^x$
39. $y = 2^{x-4}$
40. $y = 2^x + 4$

Use the formula $A = P\left(1 + \dfrac{r}{n}\right)^{nt}$ to solve the interest problems. In this formula,

   $A = $ amount accrued (or owed)
   $P = $ principal invested (or loaned)
   $r = $ rate of interest
   $n = $ number of compounding periods per year
   $t = $ time in years

41. Find the amount accrued if $1600 is invested at 9% interest compounded semiannually for 7 years.

42. A total of $800 is invested in a 7% certificate of deposit for which interest is compounded quarterly. Find the value that this certificate will have at the end of 5 years.

(12.4) *Solve. Round each answer to the nearest whole.*

43. The city of Henderson, Nevada, has been growing at a rate of 5.2% per year since 2010. If the population of Henderson was 257,437 in 2010 and this rate continues, predict the city's population in 2020.

44. The city of Raleigh, North Carolina, has been growing at a rate of 6.9% per year since 2010. If the population of Raleigh was 403,971 in 2010 and this rate continues, predict the city's population in 2019.

45. A summer camp tournament starts with 1024 players. After each round, half the players are eliminated. How many players remain after 7 rounds?

46. The bear population in a certain national park is decreasing by 11% each year. If this rate continues, and there is currently an estimated bear population of 1280, find the bear population in 6 years.

(12.5) *Write each equation with logarithmic notation.*

47. $49 = 7^2$

48. $2^{-4} = \dfrac{1}{16}$

Write each logarithmic equation with exponential notation.

49. $\log_{1/2} 16 = -4$

50. $\log_{0.4} 0.064 = 3$

Solve for x.

51. $\log_4 x = -3$

52. $\log_3 x = 2$

53. $\log_3 1 = x$

54. $\log_4 64 = x$

55. $\log_4 4^5 = x$

56. $\log_7 7^{-2} = x$

57. $5^{\log_5 4} = x$

58. $2^{\log_2 9} = x$

59. $\log_2(3x - 1) = 4$

60. $\log_3(2x + 5) = 2$

61. $\log_4(x^2 - 3x) = 1$

62. $\log_8(x^2 + 7x) = 1$

Graph each pair of equations on the same coordinate system.

63. $y = 2^x$ and $y = \log_2 x$

64. $y = \left(\dfrac{1}{2}\right)^x$ and $y = \log_{1/2} x$

(12.6) *Write each of the following as single logarithms.*

65. $\log_3 8 + \log_3 4$

66. $\log_2 6 + \log_2 3$

67. $\log_7 15 - \log_7 20$

68. $\log 18 - \log 12$

69. $\log_{11} 8 + \log_{11} 3 - \log_{11} 6$

70. $\log_5 14 + \log_5 3 - \log_5 21$

71. $2 \log_5 x - 2 \log_5(x + 1) + \log_5 x$

72. $4 \log_3 x - \log_3 x + \log_3(x + 2)$

Use properties of logarithms to write each expression as a sum or difference of multiples of logarithms.

73. $\log_3 \dfrac{x^3}{x + 2}$

74. $\log_4 \dfrac{x + 5}{x^2}$

75. $\log_2 \dfrac{3x^2 y}{z}$

76. $\log_7 \dfrac{yz^3}{x}$

If $\log_b 2 = 0.36$ and $\log_b 5 = 0.83$, find the following.

77. $\log_b 50$

78. $\log_b \dfrac{4}{5}$

(12.7) *Use a calculator to approximate the logarithm to four decimal places.*

79. $\log 3.6$

80. $\log 0.15$

81. $\ln 1.25$

82. $\ln 4.63$

Find the exact value.

83. $\log 1000$

84. $\log \dfrac{1}{10}$

85. $\ln \dfrac{1}{e}$

86. $\ln e^4$

Solve each equation for x.

87. $\ln(2x) = 2$

88. $\ln(3x) = 1.6$

89. $\ln(2x - 3) = -1$

90. $\ln(3x + 1) = 2$

Use the formula $\ln \dfrac{I}{I_0} = -kx$ to solve the radiation problems in Exercises 91 and 92. In this formula,

  x = depth in millimeters

  I = intensity of radiation

  I_0 = initial intensity

  k = a constant measure dependent on the material

Round answers to two decimal places.

91. Find the depth at which the intensity of the radiation passing through a lead shield is reduced to 3% of the original intensity if the value of k is 2.1.

92. If k is 3.2, find the depth at which 2% of the original radiation will penetrate.

Approximate the logarithm to four decimal places.

93. $\log_5 1.6$

94. $\log_3 4$

Use the formula $A = Pe^{rt}$ to solve the interest problems in which interest is compounded continuously. In this formula,

  A = amount accrued (or owed)

  P = principal invested (or loaned)

  r = rate of interest

  t = time in years

95. Bank of New York offers a 5-year, 3% continuously compounded investment option. Find the amount accrued if $1450 is invested.

96. Find the amount to which a $940 investment grows if it is invested at 4% compounded continuously for 3 years.

(12.8) *Solve each exponential equation for x. Give the exact solution and approximate the solution to four decimal places.*

97. $3^{2x} = 7$

98. $6^{3x} = 5$

99. $3^{2x+1} = 6$

100. $4^{3x+2} = 9$

101. $5^{3x-5} = 4$

102. $8^{4x-2} = 3$

103. $5^{x-1} = \dfrac{1}{2}$

104. $4^{x+5} = \dfrac{2}{3}$

Solve the equation for x.

105. $\log_5 2 + \log_5 x = 2$

106. $\log_3 x + \log_3 10 = 2$

107. $\log(5x) - \log(x + 1) = 4$

108. $-\log_6(4x + 7) + \log_6 x = 1$

109. $\log_2 x + \log_2 2x - 3 = 1$

110. $\log_3(x^2 - 8x) = 2$

Use the formula $y = y_0 e^{kt}$ to solve the population growth problems. In this formula,

y = size of population

y_0 = initial count of population

k = rate of growth written as a decimal

t = time

Round each answer to the nearest tenth.

111. In 1987, the population of California condors was only 27 birds. They were all brought in from the wild and an intensive breeding program was instituted. If we assume a yearly growth rate of 11.4%, how long will it take the condor population to reach 425 California condors? (*Source:* U.S. Park Service)

112. France is experiencing an annual growth rate of 0.5%. In 2014, the population of France was approximately 66,030,000. How long will it take for the population to reach 70,000,000? Round to the nearest tenth. (*Source:* CIA World Factbook)

113. In 2014, the population of Australia was approximately 23,130,000. How long will it take Australia to double its population if its growth rate is 1.8% annually? Round to the nearest tenth. (*Source:* CIA World Factbook)

114. Canada's population is increasing in size at a rate of 1.2% per year. How long will it take for the population of 35,160,000 to double in size? Round to the nearest tenth. (*Source:* CIA World Factbook)

Use the compound interest equation $A = P\left(1 + \dfrac{r}{n}\right)^{nt}$ to solve the following. (See the directions for Exercises 41 and 42 for an explanation of this formula.) Round answers to the nearest tenth.

115. Find how long it will take a $5000 investment to grow to $10,000 if it is invested at 8% interest compounded quarterly.

116. An investment of $6000 has grown to $10,000 while the money was invested at 6% interest compounded monthly. Find how long it was invested.

Use a graphing calculator to solve each equation. Round all solutions to two decimal places.

117. $e^x = 2$

118. $10^{0.3x} = 7$

MIXED REVIEW

Solve each equation.

119. $3^x = \dfrac{1}{81}$

120. $7^{4x} = 49$

121. $8^{3x-2} = 32$

122. $9^{x-2} = 27$

123. $\log_4 4 = x$

124. $\log_3 x = 4$

125. $\log_5(x^2 - 4x) = 1$

126. $\log_4(3x - 1) = 2$

127. $\ln x = -3.2$

128. $\log_5 x + \log_5 10 = 2$

129. $\ln x - \ln 2 = 1$

130. $\log_6 x - \log_6(4x + 7) = 1$

1c

Chapter 12 Getting Ready for the Test

MULTIPLE CHOICE *Select the correct choice. For Exercises 1 through 4 we are given:*

$$\begin{cases} f(0) = 4 & g(0) = -2 \\ f(-3) = -5 & g(-3) = 5 \\ f(5) = 9 & g(5) = -3 \\ f(9) = 6 & g(9) = 8 \end{cases}$$

1. $(f + g)(9) =$
   A. 9   **B.** 14   **C.** 48   **D.** 2

2. $(f \circ g)(5) =$
   A. -5   **B.** 6   **C.** -3   **D.** -27

3. $(f \cdot g)(-3) =$
   A. -10   **B.** 9   **C.** 0   **D.** -25

4. $\left(\dfrac{f}{g}\right)(0) =$
   A. $-\dfrac{1}{2}$ **B.** 0   **C.** -2   **D.** undefined

5. Given that f is a one-to-one function and $f(-2) = 7$, choose one ordered pair that must be a solution of the inverse of f, or f^{-1}.
   A. $(-2, 7)$ **B.** $\left(-\dfrac{1}{2}, \dfrac{1}{7}\right)$ **C.** $\left(\dfrac{1}{7}, -\dfrac{1}{2}\right)$ **D.** $(7, -2)$

MATCHING *Match each exponential function with its graph.*

6. $f(x) = 3^{-x}$

7. $f(x) = \left(\dfrac{1}{3}\right)^{-x}$

8. $f(x) = 5^{-x}$

9. $f(x) = \left(\dfrac{1}{5}\right)^{-x}$

A.

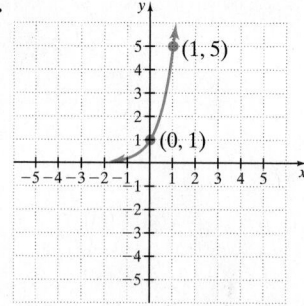

B.

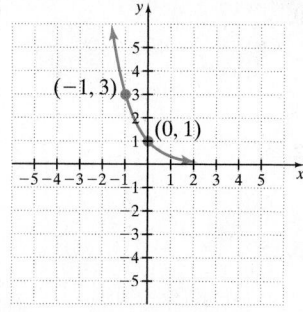

C.

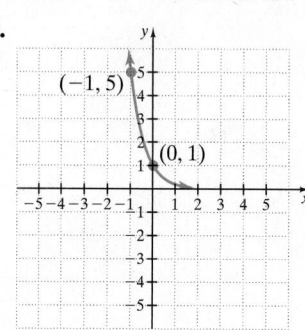

D.

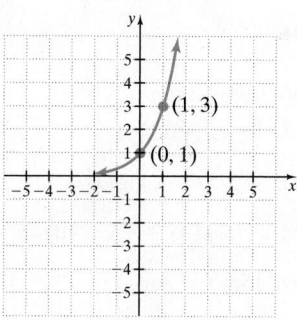

MULTIPLE CHOICE *Select the correct choice.*

10. A substance has a half-life of 32 years. How much of a 50-gram sample is left after 32 years?

   A. 50 g   **B.** 25 g   **C.** 32 g   **D.** 16 g

11. $\log_7 \dfrac{13}{9}$ can be rewritten as:

   A. $\dfrac{\log_7 13}{\log_7 9}$ **B.** $\log_7 13 - \log_7 9$ **C.** $\log_7(13 - 9)$ **D.** $\log_7 117$

12. $\log_3 2$ can be rewritten as:

   A. $\dfrac{\log 2}{\log 3}$ **B.** $\log \dfrac{2}{3}$ **C.** $\dfrac{\ln 2}{\ln 3}$ **D.** both A and C

MULTIPLE CHOICE *Answer each exercise as* **A.** *true or* **B.** *false*

13. $\log_2(a + b) = \log_2 a + \log_2 b$

14. $\log_5 7^2 = 2 \log_5 7$

15. $\dfrac{\log_2 x}{\log_2 y} = (\log_2 x) - (\log_2 y)$

Chapter 12 Test   MyMathLab®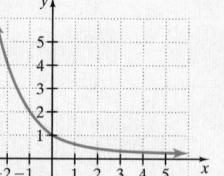

If $f(x) = x$ and $g(x) = 2x - 3$, find the following.

1. $(f \cdot g)(x)$

2. $(f - g)(x)$

If $f(x) = x$, $g(x) = x - 7$, and $h(x) = x^2 - 6x + 5$, find the following.

3. $(f \circ h)(0)$

4. $(g \circ f)(x)$

5. $(g \circ h)(x)$

On the same set of axes, graph the given one-to-one function and its inverse.

6. $f(x) = 7x - 14$

Determine whether the given graph is the graph of a one-to-one function.

7.   8.

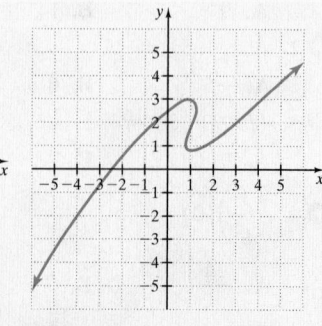

Determine whether each function is one-to-one. If it is one-to-one, find an equation or a set of ordered pairs that defines the inverse function of the given function.

9. $f(x) = 6 - 2x$

10. $f = \{(0, 0), (2, 3), (-1, 5)\}$

11.

Word (Input)	Dog	Cat	House	Desk	Circle
First Letter of Word (Output)	d	c	h	d	c

Use the properties of logarithms to write each expression as a single logarithm.

12. $\log_3 6 + \log_3 4$

13. $\log_5 x + 3 \log_5 x - \log_5(x + 1)$

14. Write the expression $\log_6 \dfrac{2x}{y^3}$ as the sum or difference of multiples of logarithms.

15. If $\log_b 3 = 0.79$ and $\log_b 5 = 1.16$, find the value of $\log_b \dfrac{3}{25}$.

16. Approximate $\log_7 8$ to four decimal places.

17. Solve $8^{x-1} = \dfrac{1}{64}$ for x. Give the exact solution.

18. Solve $3^{2x+5} = 4$ for x. Give the exact solution and approximate the solution to four decimal places.

Solve each logarithmic equation for x. Give the exact solution.

19. $\log_3 x = -2$

20. $\ln \sqrt{e} = x$

21. $\log_8(3x - 2) = 2$

22. $\log_5 x + \log_5 3 = 2$

23. $\log_4(x + 1) - \log_4(x - 2) = 3$

24. Solve $\ln(3x + 7) = 1.31$ rounded to four decimal places.

25. Graph: $y = \left(\dfrac{1}{2}\right)^x + 1$

26. Graph the functions $y = 3^x$ and $y = \log_3 x$ on the same coordinate system.

Use the formula $A = P\left(1 + \dfrac{r}{n}\right)^{nt}$ to solve Exercises 27–29.

27. Find the amount in an account if $4000 is invested for 3 years at 9% interest compounded monthly.

28. Find how long it will take $2000 to grow to $3000 if the money is invested at 7% interest compounded semiannually. Round to the nearest whole year.

29. Suppose you have $3000 to invest. Which investment, rounded to the nearest dollar, yields the greater return over 10 years: 6.5% compounded semiannually or 6% compounded monthly? How much more is yielded by the better investment?

Solve. Round answers to the nearest whole.

30. Suppose a city with population of 150,000 has been decreasing at a rate of 2% per year. If this rate continues, predict the population of the city in 20 years.

31. The prairie dog population of the Grand Forks area now stands at 57,000 animals. If the population is growing at a rate of 2.6% annually, how many prairie dogs will there be in that area 5 years from now?

32. In an attempt to save an endangered species of wood duck, naturalists would like to increase the wood duck population from 400 to 1000 ducks. If the annual population growth rate is 6.2%, how long will it take the naturalists to reach their goal? Round to the nearest whole year.

The reliability of a new model of smart television can be described by the exponential function $R = 1.6^{-(1/3)t}$, where the reliability R is the probability (as a decimal) that the television is still working t years after it is manufactured. Round answers to the nearest hundredth. Then write your answers in percents.

33. What is the probability that the smart television will still work half a year after it is manufactured?

34. What is the probability that the smart television will still work 2 years after it is manufactured?

Chapter 12 Cumulative Review

1. Divide. Simplify all quotients if possible.

   a. $\dfrac{4}{5} \div \dfrac{5}{16}$

   b. $\dfrac{7}{10} \div 14$

   c. $\dfrac{3}{8} \div \dfrac{3}{10}$

2. Solve: $\dfrac{1}{3}(x - 2) = \dfrac{1}{4}(x + 1)$

3. Graph: $f(x) = x^2$

4. Find an equation of the line through $(-2, 6)$ and perpendicular to $f(x) = -3x + 4$. Write the equation using function notation.

5. Solve the system.
$$\begin{cases} x - 5y - 2z = 6 \\ -2x + 10y + 4z = -12 \\ \dfrac{1}{2}x - \dfrac{5}{2}y - z = 3 \end{cases}$$

6. Line l and line m are parallel lines cut by transversal t. Find the values of x and y.

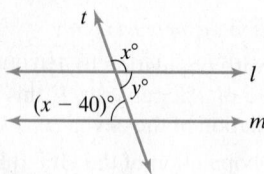

7. Simplify each expression.
a. $-3 + [(-2 - 5) - 2]$
b. $2^3 - |10| + [-6 - (-5)]$

8. Use the power rules to simplify the following. Use positive exponents to write all results.
a. $(4a^3)^2$
b. $\left(-\dfrac{2}{3}\right)^3$
c. $\left(\dfrac{4a^5}{b^3}\right)^3$
d. $\left(\dfrac{3^{-2}}{x}\right)^{-3}$
e. $(a^{-2}b^3c^{-4})^{-2}$

9. For the ICL Production Company, the rational function $C(x) = \dfrac{2.6x + 10,000}{x}$ describes the company's cost per disc of pressing x compact discs. Find the cost per disc for pressing:
a. 100 compact discs
b. 1000 compact discs

10. Multiply.
a. $(3x - 1)^2$
b. $\left(\dfrac{1}{2}x + 3\right)\left(\dfrac{1}{2}x - 3\right)$
c. $(2x - 5)(6x + 7)$

11. Solve: $12a - 8a = 10 + 2a - 13 - 7$

12. Perform the indicated operations and simplify if possible.
$$\dfrac{5}{x - 2} + \dfrac{3}{x^2 + 4x + 4} - \dfrac{6}{x + 2}$$

13. Divide: $\dfrac{8x^2y^2 - 16xy + 2x}{4xy}$

14. Simplify each complex fraction.
a. $\dfrac{\dfrac{a}{5}}{\dfrac{a - 1}{10}}$
b. $\dfrac{\dfrac{3}{2 + a} + \dfrac{6}{2 - a}}{\dfrac{5}{a + 2} - \dfrac{1}{a - 2}}$
c. $\dfrac{x^{-1} + y^{-1}}{xy}$

15. Factor: $3m^2 - 24m - 60$

16. Factor: $5x^2 - 85x + 350$

17. Subtract: $\dfrac{3x^2 + 2x}{x - 1} - \dfrac{10x - 5}{x - 1}$

18. Use synthetic division to divide $(8x^2 - 12x - 7) \div (x - 2)$.

19. Simplify the following expressions.
a. $\sqrt[4]{81}$
b. $\sqrt[5]{-243}$
c. $-\sqrt{25}$
d. $\sqrt[4]{-81}$
e. $\sqrt[3]{64x^3}$

20. Solve: $\dfrac{1}{a + 5} = \dfrac{1}{3a + 6} - \dfrac{a + 2}{a^2 + 7a + 10}$

21. Use rational exponents to write as a single radical.
a. $\sqrt{x} \cdot \sqrt[4]{x}$
b. $\dfrac{\sqrt{x}}{\sqrt[3]{x}}$
c. $\sqrt[3]{3} \cdot \sqrt{2}$

22. Suppose that y varies directly as x. If $y = \dfrac{1}{2}$ when $x = 12$, find the constant of variation and the direct variation equation.

23. Multiply.
a. $\sqrt{3}(5 + \sqrt{30})$
b. $(\sqrt{5} - \sqrt{6})(\sqrt{7} + 1)$
c. $(7\sqrt{x} + 5)(3\sqrt{x} - \sqrt{5})$
d. $(4\sqrt{3} - 1)^2$
e. $(\sqrt{2x} - 5)(\sqrt{2x} + 5)$
f. $(\sqrt{x - 3} + 5)^2$

24. Find each root. Assume that all variables represent nonnegative real numbers.
a. $\sqrt{9}$
b. $\sqrt[3]{-27}$
c. $\sqrt{\dfrac{9}{64}}$
d. $\sqrt[4]{x^{12}}$
b. $\sqrt[3]{-125y^6}$

25. Rationalize the denominator of $\dfrac{\sqrt[4]{x}}{\sqrt[4]{81y^5}}$.

26. Multiply.
a. $a^{1/4}(a^{3/4} - a^8)$
b. $(x^{1/2} - 3)(x^{1/2} + 5)$

27. Solve: $\sqrt{4 - x} = x - 2$

28. Use the quotient rule to divide and simplify if possible.
a. $\dfrac{\sqrt{54}}{\sqrt{6}}$
b. $\dfrac{\sqrt{108a^2}}{3\sqrt{3}}$
c. $\dfrac{3\sqrt[3]{81a^5b^{10}}}{\sqrt[3]{3b^4}}$

29. Solve $3x^2 - 9x + 8 = 0$ by completing the square.

30. Add or subtract as indicated.
a. $\dfrac{\sqrt{20}}{3} + \dfrac{\sqrt{5}}{4}$
b. $\sqrt[3]{\dfrac{24x}{27}} - \dfrac{\sqrt[3]{3x}}{2}$

31. Solve: $\dfrac{3x}{x-2} - \dfrac{x+1}{x} = \dfrac{6}{x(x-2)}$

32. Rationalize the denominator. $\sqrt[3]{\dfrac{27}{m^4 n^8}}$

33. Solve: $x^2 - 4x \le 0$

34. Find the length of the unknown side of the triangle.

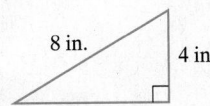

8 in.   4 in.

35. Graph: $F(x) = (x-3)^2 + 1$

36. Find the following powers of i.

  a. i^8 **b.** i^{21} **c.** i^{42}

  d. i^{-13}

37. Solve for x: $\dfrac{45}{x} = \dfrac{5}{7}$

38. Solve $4x^2 + 8x - 1 = 0$ by completing the square.

39. Find an equation of the inverse of $f(x) = x + 3$.

40. Solve by using the quadratic formula.

$$\left(x - \dfrac{1}{2}\right)^2 = \dfrac{x}{2}$$

41. Find the value of each logarithmic expression.

  a. $\log_4 16$ **b.** $\log_{10} \dfrac{1}{10}$

  c. $\log_9 3$

42. Graph: $f(x) = -(x+1)^2 + 1$

CHAPTER 13

Conic Sections

13.1 The Parabola and the Circle

13.2 The Ellipse and the Hyperbola

Integrated Review—Graphing Conic Sections

13.3 Solving Nonlinear Systems of Equations

13.4 Nonlinear Inequalities and Systems of Inequalities

 **CHECK YOUR PROGRESS**

Vocabulary Check

Chapter Highlights

Chapter Review

Getting Ready for the Test

Chapter Test

Cumulative Review

In Chapter 11, we analyzed some of the important connections between a parabola and its equation. Parabolas are interesting in their own right but are more interesting still because they are part of a collection of curves known as conic sections. This chapter is devoted to quadratic equations in two variables and their conic section graphs: the parabola, circle, ellipse, and hyperbola.

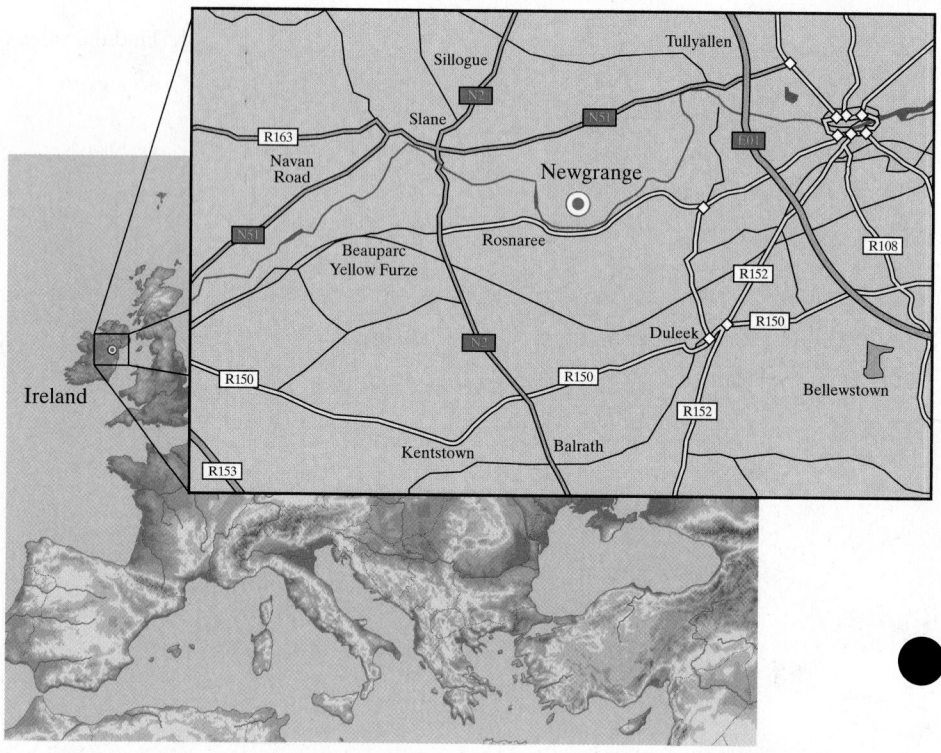

What Is Brú na Bóinne?

Brú na Bóinne means *Palace of the Boyne* and consists of a collection of Neolithic mounds, tombs, standing stones, and other historic enclosures found along the River Boyne in Ireland. This collection of sites is older than Stonehenge and the Great Pyramids.

One of these sites is Newgrange, a passage tomb. The most interesting feature of this tomb is that at dawn on the winter solstice, the shortest day of the year, a shaft of sunlight penetrates the small opening above the entrance and lights the entire passage into the tomb.

Below is a photo of Newgrange along with an overhead drawing. In Section 13.1, Exercise 92, we will explore more information on Newgrange.

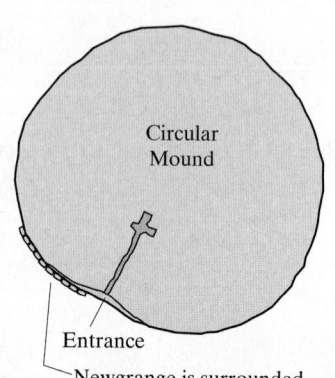

Circular Mound

Entrance

Newgrange is surrounded by 97 large kerbstones (stones used to form a curb)

13.1   The Parabola and the Circle

OBJECTIVES

1 Graph Parabolas of the Form $x = a(y - k)^2 + h$ and $y = a(x - h)^2 + k$.

2 Graph Circles of the Form $(x - h)^2 + (y - k)^2 = r^2$.

3 Find the Center and the Radius of a Circle, Given Its Equation.

4 Write an Equation of a Circle, Given Its Center and Radius.

Conic sections are named so because each conic section is the intersection of a right circular cone and a plane. The circle, parabola, ellipse, and hyperbola are the conic sections.

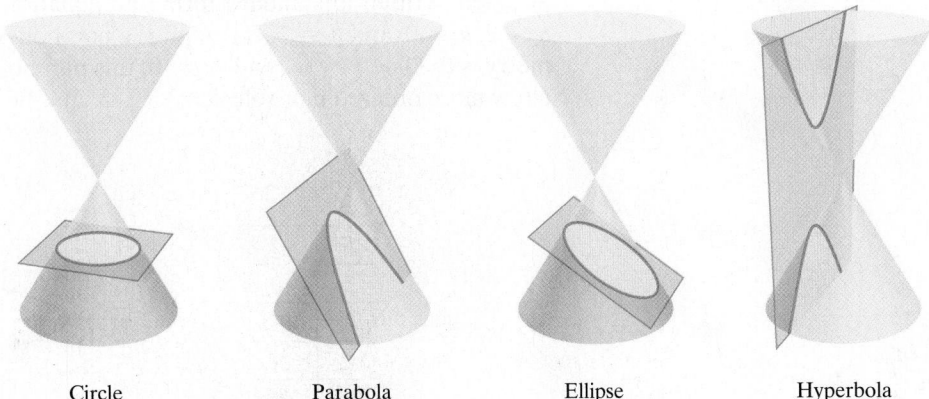

Circle   Parabola   Ellipse   Hyperbola

OBJECTIVE

1   Graphing Parabolas

Thus far, we have seen that $f(x)$ or $y = a(x - h)^2 + k$ is the equation of a parabola that opens upward if $a > 0$ or downward if $a < 0$. Parabolas can also open left or right or even on a slant. Equations of these parabolas are not functions of x, of course, since a parabola opening any way other than upward or downward fails the vertical line test. In this section, we introduce parabolas that open to the left and to the right. Parabolas opening on a slant will not be developed in this book.

Just as $y = a(x - h)^2 + k$ is the equation of a parabola that opens upward or downward, $x = a(y - k)^2 + h$ is the equation of a parabola that opens to the right or to the left. The parabola opens to the right if $a > 0$ and to the left if $a < 0$. The parabola has vertex (h, k), and its axis of symmetry is the line $y = k$.

Parabolas

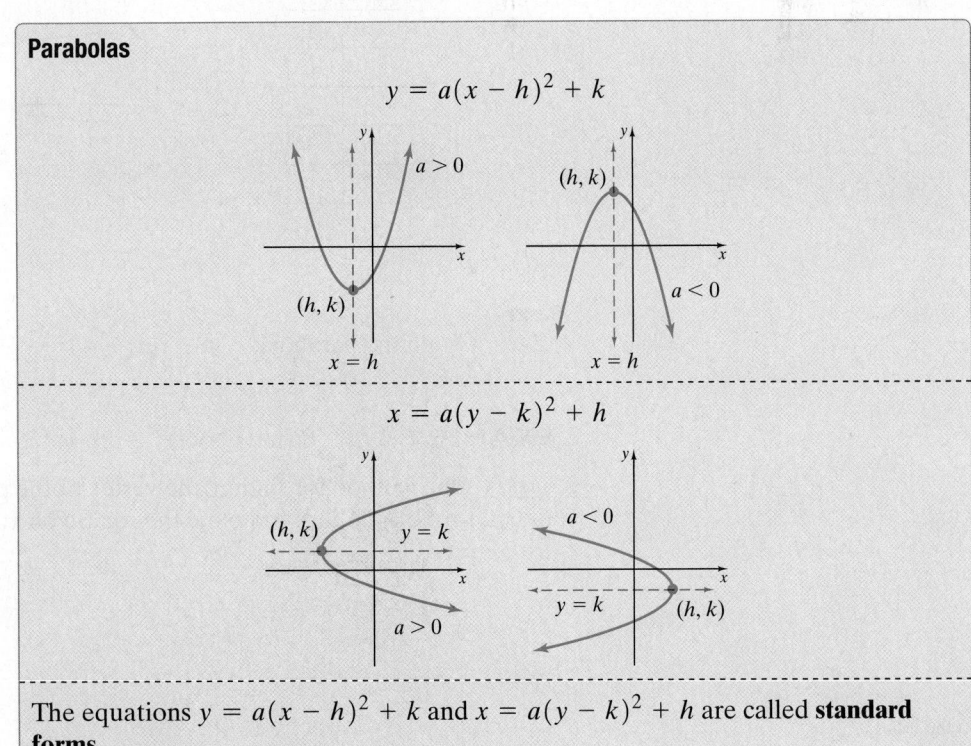

The equations $y = a(x - h)^2 + k$ and $x = a(y - k)^2 + h$ are called **standard forms**.

✔ **CONCEPT CHECK**

Does the graph of the parabola given by the equation $x = -3y^2$ open to the left, to the right, upward, or downward?

EXAMPLE 1   Graph the parabola: $x = 2y^2$

<u>Solution</u> Written in standard form, the equation $x = 2y^2$ is $x = 2(y - 0)^2 + 0$ with $a = 2, k = 0$, and $h = 0$. Its graph is a parabola with vertex $(0, 0)$, and its axis of symmetry is the line $y = 0$. Since $a > 0$, this parabola opens to the right. The table shows a few more ordered pair solutions of $x = 2y^2$. Its graph is also shown.

x	y
8	-2
2	-1
0	0
2	1
8	2

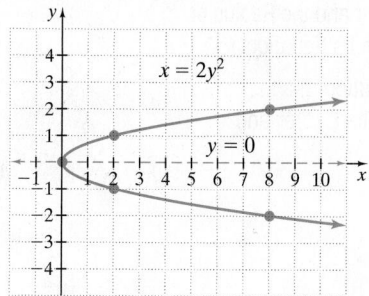

PRACTICE

1   Graph the parabola: $x = \dfrac{1}{2}y^2$

EXAMPLE 2   Graph the parabola: $x = -3(y - 1)^2 + 2$

<u>Solution</u> The equation $x = -3(y - 1)^2 + 2$ is in the form $x = a(y - k)^2 + h$ with $a = -3, k = 1$, and $h = 2$. Since $a < 0$, the parabola opens to the left. The vertex (h, k) is $(2, 1)$, and the axis of symmetry is the line $y = 1$. When $y = 0, x = -1$, so the x-intercept is $(-1, 0)$. Again, we obtain a few ordered pair solutions and then graph the parabola.

x	y
2	1
-1	0
-1	2
-10	3
-10	-1

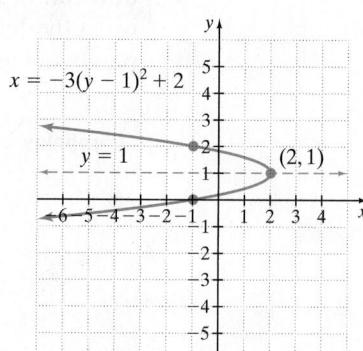

PRACTICE

2   Graph the parabola: $x = -2(y + 4)^2 - 1$

EXAMPLE 3   Graph: $y = -x^2 - 2x + 15$

<u>Solution</u> One method for finding the vertex of the parabola is to write the equation in standard form. To do this, complete the square on x.

$$y - 15 = -x^2 - 2x \qquad \text{Subtract 15 from both sides.}$$

$$y - 15 = -1(x^2 + 2x) \qquad \text{Factor } -1 \text{ from the terms } -x^2 - 2x.$$

The coefficient of x is 2. Find the square of half of 2.

$$\frac{1}{2}(2) = 1 \quad \text{and} \quad 1^2 = 1$$

$$y - 15 - 1(1) = -1(x^2 + 2x + 1) \quad \text{Add } -1(1) \text{ to both sides.}$$
$$y - 16 = -1(x + 1)^2 \qquad \begin{array}{l}\text{Simplify the left side and}\\ \text{factor the right side.}\end{array}$$
$$y = -(x + 1)^2 + 16 \qquad \text{Add 16 to both sides.}$$

The equation is now in standard form $y = a(x - h)^2 + k$ with $a = -1, h = -1$, and $k = 16$.

The vertex is then (h, k), or $(-1, 16)$.

A second method for finding the vertex of the graph of $y = -x^2 - 2x + 15$ is by using the formula $\frac{-b}{2a}$.

Helpful Hint

For $y = -x^2 - 2x + 15$, recall that $a = -1, b = -2$, and $c = 15$.

$$x = \frac{-(-2)}{2(-1)} = \frac{2}{-2} = -1$$

$$y = -(-1)^2 - 2(-1) + 15 = -1 + 2 + 15 = 16$$

Again, we see that the vertex is $(-1, 16)$, and the axis of symmetry is the vertical line $x = -1$. The y-intercept is $(0, 15)$. Now we find a few more ordered pair solutions of $y = -x^2 - 2x + 15$ to graph the parabola.

x	y
-1	16
0	15
-2	15
1	12
-3	12
3	0
-5	0

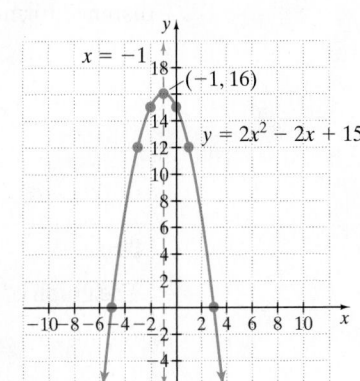

PRACTICE

3 Graph: $y = -x^2 + 4x + 6$

EXAMPLE 4 Graph: $x = 2y^2 + 4y + 5$

Solution Notice that this equation is quadratic in y, so its graph is a parabola that opens to the left or the right. We can complete the square on y, or we can use the formula $\frac{-b}{2a}$ to find the vertex.

Since the equation is quadratic in y, the formula gives us the y-value of the vertex.

$$y = \frac{-4}{2 \cdot 2} = \frac{-4}{4} = -1$$

Now let $y = -1$ in $x = 2y^2 + 4y + 5$ to find the x-value of the vertex.

$$x = 2(-1)^2 + 4(-1) + 5 = 2 \cdot 1 - 4 + 5 = 3$$

(Continued on next page)

The vertex is $(3, -1)$, and the axis of symmetry is the line $y = -1$. The parabola opens to the right since $a > 0$. The x-intercept is $(5, 0)$.

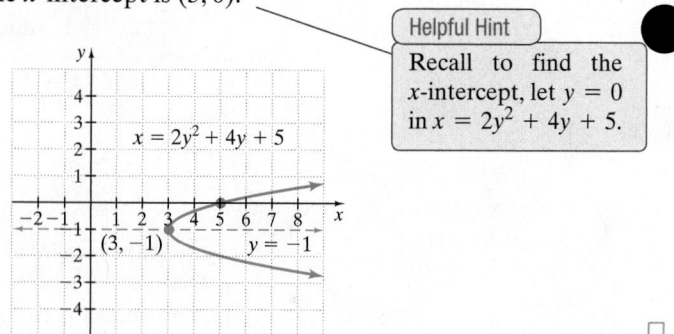

Helpful Hint

Recall to find the x-intercept, let $y = 0$ in $x = 2y^2 + 4y + 5$.

PRACTICE

4 Graph: $x = 3y^2 + 6y + 4$

OBJECTIVE

2 Graphing Circles

Another conic section is the **circle**. A circle is the set of all points in a plane that are the same distance from a fixed point called the **center**. The distance is called the **radius** of the circle. To find a standard equation for a circle, let (h, k) represent the center of the circle and let (x, y) represent any point on the circle. The distance between (h, k) and (x, y) is defined to be the circle's radius, r units. We can find this distance r by using the distance formula.

$$r = \sqrt{(x - h)^2 + (y - k)^2}$$
$$r^2 = (x - h)^2 + (y - k)^2 \qquad \text{Square both sides.}$$

Circle

The graph of $(x - h)^2 + (y - k)^2 = r^2$ is a circle with center (h, k) and radius r.

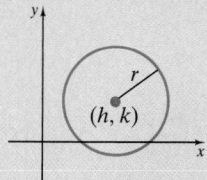

The equation $(x - h)^2 + (y - k)^2 = r^2$ is called **standard form.**

If an equation can be written in the standard form

$$(x - h)^2 + (y - k)^2 = r^2$$

then its graph is a circle, which we can draw by graphing the center (h, k) and using the radius r.

Helpful Hint

Notice that the radius is the *distance* from the center of the circle to any point of the circle. Also notice that the *midpoint* of a diameter of a circle is the center of the circle.

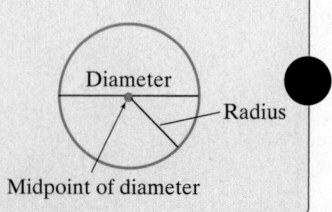

Diameter

Radius

Midpoint of diameter

EXAMPLE 5   Graph: $x^2 + y^2 = 4$

Solution The equation can be written in standard form as

$$(x - 0)^2 + (y - 0)^2 = 2^2$$

The center of the circle is $(0, 0)$, and the radius is 2. Its graph is shown.

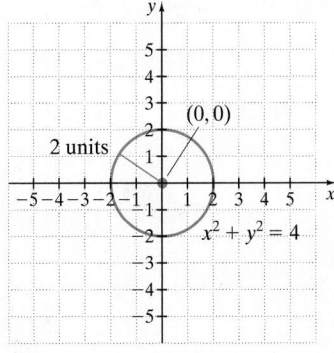

PRACTICE

5   Graph: $x^2 + y^2 = 25$

Helpful Hint

Notice the difference between the equation of a circle and the equation of a parabola.

- The equation of a circle contains both x^2 and y^2 terms on the same side of the equation with equal coefficients.
- The equation of a parabola has either an x^2 term or a y^2 term but not both.

EXAMPLE 6   Graph: $(x + 1)^2 + y^2 = 8$

Solution The equation can be written as $(x + 1)^2 + (y - 0)^2 = 8$ with $h = -1$, $k = 0$, and $r = \sqrt{8}$. The center is $(-1, 0)$, and the radius is $\sqrt{8} = 2\sqrt{2} \approx 2.8$.

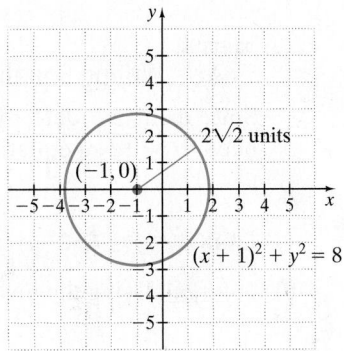

PRACTICE

6   Graph: $(x - 3)^2 + (y + 2)^2 = 4$

✔ **CONCEPT CHECK**

In the graph of the equation $(x - 3)^2 + (y - 2)^2 = 5$, what is the distance between the center of the circle and any point on the circle?

OBJECTIVE

3   Finding the Center and the Radius of a Circle

To find the center and the radius of a circle from its equation, write the equation in standard form. To write the equation of a circle in standard form, we complete the square on both x and y.

Answer to Concept Check:
$\sqrt{5}$ units

EXAMPLE 7   Graph: $x^2 + y^2 + 4x - 8y = 16$

Solution Since this equation contains x^2 and y^2 terms on the same side of the equation with equal coefficients, its graph is a circle. To write the equation in standard form, group the terms involving x and the terms involving y and then complete the square on each variable.

$$(x^2 + 4x) + (y^2 - 8y) = 16$$

Here, $\frac{1}{2}(4) = 2$ and $2^2 = 4$. Also, $\frac{1}{2}(-8) = -4$ and $(-4)^2 = 16$. Add 4 and then 16 to both sides.

$$(x^2 + 4x + 4) + (y^2 - 8y + 16) = 16 + 4 + 16$$
$$(x + 2)^2 + (y - 4)^2 = 36 \qquad \text{Factor.}$$

This circle has center $(-2, 4)$ and radius 6, as shown.

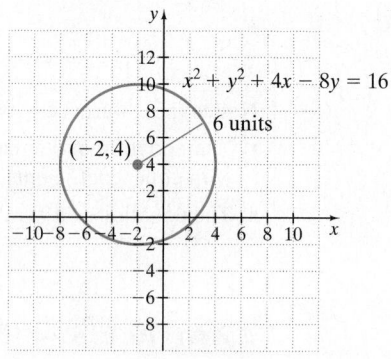

PRACTICE
7   Graph: $x^2 + y^2 + 6x - 2y = 6$

OBJECTIVE

4   **Writing Equations of Circles** ▶

Since a circle is determined entirely by its center and radius, this information is all we need to write an equation of a circle.

EXAMPLE 8   Find an equation of the circle with center $(-7, 3)$ and radius 10.

Solution Using the given values $h = -7, k = 3$, and $r = 10$, we write the equation

$$(x - h)^2 + (y - k)^2 = r^2$$

or

$$[x - (-7)]^2 + (y - 3)^2 = 10^2 \quad \text{Substitute the given values.}$$

or

$$(x + 7)^2 + (y - 3)^2 = 100$$

PRACTICE
8   Find an equation of the circle with center $(-2, -5)$ and radius 9.

Graphing Calculator Explorations

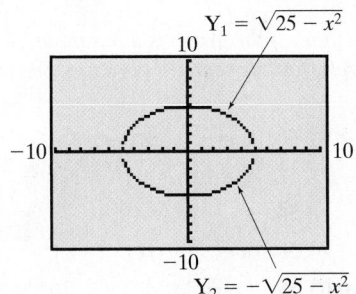

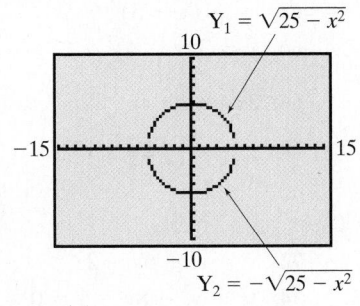

To graph an equation such as $x^2 + y^2 = 25$ with a graphing calculator, we first solve the equation for y.

$$x^2 + y^2 = 25$$
$$y^2 = 25 - x^2$$
$$y = \pm\sqrt{25 - x^2}$$

The graph of $y = \sqrt{25 - x^2}$ will be the top half of the circle, and the graph of $y = -\sqrt{25 - x^2}$ will be the bottom half of the circle.

To graph, press $\boxed{Y=}$ and enter $Y_1 = \sqrt{25 - x^2}$ and $Y_2 = -\sqrt{25 - x^2}$. Insert parentheses around $25 - x^2$ so that $\sqrt{25 - x^2}$ and not $\sqrt{25} - x^2$ is graphed.

The top graph to the left does not appear to be a circle because we are currently using a standard window and the screen is rectangular. This causes the tick marks on the x-axis to be farther apart than the tick marks on the y-axis and, thus, creates the distorted circle. If we want the graph to appear circular, we must define a square window by using a feature of the graphing calculator or by redefining the window to show the x-axis from -15 to 15 and the y-axis from -10 to 10. Using a square window, the graph appears as shown on the bottom to the left.

Use a graphing calculator to graph each circle.

1. $x^2 + y^2 = 55$

2. $x^2 + y^2 = 20$

3. $5x^2 + 5y^2 = 50$

4. $6x^2 + 6y^2 = 105$

5. $2x^2 + 2y^2 - 34 = 0$

6. $4x^2 + 4y^2 - 48 = 0$

7. $7x^2 + 7y^2 - 89 = 0$

8. $3x^2 + 3y^2 - 35 = 0$

Vocabulary, Readiness & Video Check

Use the choices below to fill in each blank. Some choices may be used more than once.

radius	center	vertex
diameter	circle	conic sections

1. The circle, parabola, ellipse, and hyperbola are called the _____ .

2. For a parabola that opens upward, the lowest point is the _____ .

3. A _____ is the set of all points in a plane that are the same distance from a fixed point. The fixed point is called the _____ .

4. The midpoint of a diameter of a circle is the _____ .

5. The distance from the center of a circle to any point of the circle is called the _____ .

6. Twice a circle's radius is its _____ .

Martin-Gay Interactive Videos

See Video 13.1 ◉

Watch the section lecture video and answer the following questions.

OBJECTIVE 1

7. Based on ▦ Example 1 and the lecture before, would you say that parabolas of the form $x = a(y - k)^2 + h$ are functions? Why or why not?

OBJECTIVE 2

8. Based on the lecture before ▦ Example 2, what would be the standard form of a circle with its center at the origin? Simplify your answer.

OBJECTIVE 3

9. From ▦ Example 3, if you know the center and radius of a circle, how can you write that circle's equation?

OBJECTIVE 4

10 From ▦ Example 4, why do we need to complete the square twice when writing this equation of a circle in standard form?

13.1 Exercise Set MyMathLab® ▶

The graph of each equation is a parabola. Determine whether the parabola opens upward, downward, to the left, or to the right. Do not graph. See Examples 1 through 4.

1. $y = x^2 - 7x + 5$

2. $y = -x^2 + 16$

3. $x = -y^2 - y + 2$

4. $x = 3y^2 + 2y - 5$

5. $y = -x^2 + 2x + 1$

6. $x = -y^2 + 2y - 6$

The graph of each equation is a parabola. Find the vertex of the parabola and then graph it. See Examples 1 through 4.

7. $x = 3y^2$ **8.** $x = 5y^2$

9. $x = -2y^2$ **10.** $x = -4y^2$

11. $y = -4x^2$ **12.** $y = -2x^2$

▶ **13.** $x = (y - 2)^2 + 3$ **14.** $x = (y - 4)^2 - 1$

15. $y = -3(x - 1)^2 + 5$ **16.** $y = -4(x - 2)^2 + 2$

17. $x = y^2 + 6y + 8$ **18.** $x = y^2 - 6y + 6$

19. $y = x^2 + 10x + 20$ **20.** $y = x^2 + 4x - 5$

21. $x = -2y^2 + 4y + 6$ **22.** $x = 3y^2 + 6y + 7$

The graph of each equation is a circle. Find the center and the radius and then graph the circle. See Examples 5 through 7.

23. $x^2 + y^2 = 9$ **24.** $x^2 + y^2 = 25$

25. $x^2 + (y - 2)^2 = 1$ **26.** $(x - 3)^2 + y^2 = 9$

▶ **27.** $(x - 5)^2 + (y + 2)^2 = 1$ **28.** $(x + 3)^2 + (y + 3)^2 = 4$

29. $x^2 + y^2 + 6y = 0$

30. $x^2 + 10x + y^2 = 0$

31. $x^2 + y^2 + 2x - 4y = 4$

32. $x^2 + y^2 + 6x - 4y = 3$

33. $(x + 2)^2 + (y - 3)^2 = 7$

34. $(x + 1)^2 + (y - 2)^2 = 5$

35. $x^2 + y^2 - 4x - 8y - 2 = 0$

36. $x^2 + y^2 - 2x - 6y - 5 = 0$

Hint: For Exercises 37 through 42, first divide the equation through by the coefficient of x^2 (or y^2).

37. $3x^2 + 3y^2 = 75$

38. $2x^2 + 2y^2 = 18$

39. $6(x - 4)^2 + 6(y - 1)^2 = 24$

40. $7(x - 1)^2 + 7(y - 3)^2 = 63$

41. $4(x + 1)^2 + 4(y - 3)^2 = 12$

42. $5(x - 2)^2 + 5(y + 1) = 50$

Write an equation of the circle with the given center and radius. See Example 8.

43. $(2, 3); 6$ **44.** $(-7, 6); 2$

45. $(0, 0); \sqrt{3}$ **46.** $(0, -6); \sqrt{2}$

▶ **47.** $(-5, 4); 3\sqrt{5}$ **48.** the origin; $4\sqrt{7}$

MIXED PRACTICE

Sketch the graph of each equation. If the graph is a parabola, find its vertex. If the graph is a circle, find its center and radius. See the Helpful Hint on page 795.

49. $x = y^2 - 3$ **50.** $x = y^2 + 2$

51. $y = (x - 2)^2 - 2$ **52.** $y = (x + 3)^2 + 3$

53. $x^2 + y^2 = 1$ **54.** $x^2 + y^2 = 49$

55. $x = (y + 3)^2 - 1$ **56.** $x = (y - 1)^2 + 4$

57. $(x - 2)^2 + (y - 2)^2 = 16$ **58.** $(x + 3)^2 + (y - 1)^2 = 9$

59. $x = -(y - 1)^2$ **60.** $x = -2(y + 5)^2$

61. $(x - 4)^2 + y^2 = 7$ **62.** $x^2 + (y + 5)^2 = 5$

63. $y = 5(x + 5)^2 + 3$ **64.** $y = 3(x - 4)^2 + 2$

65. $\dfrac{x^2}{8} + \dfrac{y^2}{8} = 2$ **66.** $2x^2 + 2y^2 = \dfrac{1}{2}$

67. $y = x^2 + 7x + 6$ **68.** $y = x^2 - 2x - 15$

▶ **69.** $x^2 + y^2 + 2x + 12y - 12 = 0$

70. $x^2 + y^2 + 6x + 10y - 2 = 0$

71. $x = y^2 + 8y - 4$ **72.** $x = y^2 + 6y + 2$

73. $x^2 - 10y + y^2 + 4 = 0$ **74.** $x^2 + y^2 - 8y + 5 = 0$

75. $x = -3y^2 + 30y$ **76.** $x = -2y^2 - 4y$

77. $5x^2 + 5y^2 = 25$ **78.** $\dfrac{x^2}{3} + \dfrac{y^2}{3} = 2$

79. $y = 5x^2 - 20x + 16$ **80.** $y = 4x^2 - 40x + 105$

REVIEW AND PREVIEW

Graph each equation. See Sections 3.2 and 3.3.

81. $y = 2x + 5$ **82.** $y = -3x + 3$

83. $y = 3$ **84.** $x = -2$

Rationalize each denominator and simplify if possible. See Section 10.5.

85. $\dfrac{1}{\sqrt{3}}$ **86.** $\dfrac{\sqrt{5}}{\sqrt{8}}$

87. $\dfrac{4\sqrt{7}}{\sqrt{6}}$ **88.** $\dfrac{10}{\sqrt{5}}$

CONCEPT EXTENSIONS

For Exercises 89 and 90, explain the error in each statement.

89. The graph of $x = 5(y + 5)^2 + 1$ is a parabola with vertex $(-5, 1)$ and opening to the right.

90. The graph of $x^2 + (y + 3)^2 = 10$ is a circle with center $(0, -3)$ and radius 5.

91. *The Sarsen Circle* The first image that comes to mind when one thinks of Stonehenge is the very large sandstone blocks with sandstone lintels across the top. The Sarsen Circle of Stonehenge is the outer circle of the sandstone blocks, each of which weighs up to 50 tons. There were originally 30 of these monolithic blocks, but only 17 remain upright to this day. The "altar stone" lies at the center of this circle, which has a diameter of 33 meters.

a. What is the radius of the Sarsen Circle?

b. What is the circumference of the Sarsen Circle? Round your result to two decimal places.

c. Since there were originally 30 Sarsen stones located on the circumference, how far apart would the centers of the stones have been? Round to the nearest tenth of a meter.

d. Using the axes in the drawing, what are the coordinates of the center of the circle?

e. Use parts **a** and **d** to write the equation of the Sarsen Circle.

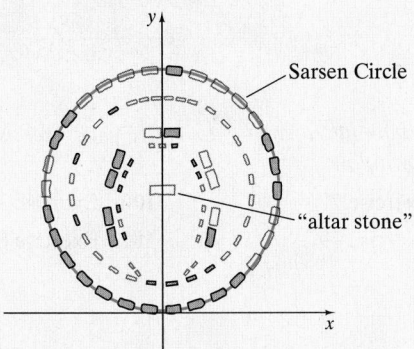

Sarsen Circle

"altar stone"

92. Newgrange is a Stone Age monument in Ireland that is actually a passage tomb. On the outside it looks like a large grass mound, due to the changes in the landscape over the centuries. The tomb is surrounded at its base by a circular curb of 97 stones. The large mound of the tomb is approximately 80 meters in diameter. (See the Chapter 13 Opener.)

a. What is the radius of this structure?

b. What is the circumference of the tomb? Round your result to two decimal places.

c. There were originally 97 kerbstones located on the circumference. How far apart would the centers of the stones have been? Round to the nearest tenth of a meter.

d. Using the axes in the drawing, what are the coordinates of the center of the circle?

e. Use parts **a** and **d** to write the equation of Newgrange.

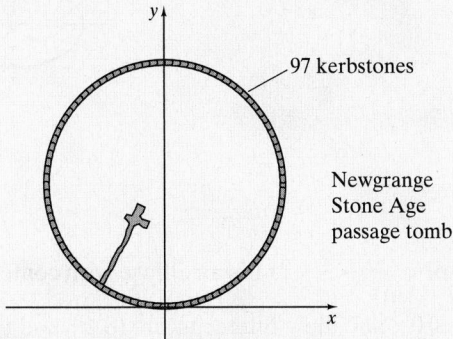

97 kerbstones

Newgrange
Stone Age
passage tomb

93. In 1893, Pittsburgh bridge builder George Ferris designed and built a gigantic revolving steel wheel whose height was 264 feet and diameter was 250 feet. This Ferris wheel opened at the 1893 exposition in Chicago. It had 36 wooden cars, each capable of holding 60 passengers. (*Source: The Handy Science Answer Book*)

a. What was the radius of this Ferris wheel?

b. How close was the wheel to the ground?

c. How high was the center of the wheel from the ground?

d. Using the axes in the drawing, what are the coordinates of the center of the wheel?

e. Use parts **a** and **d** to write an equation of the wheel.

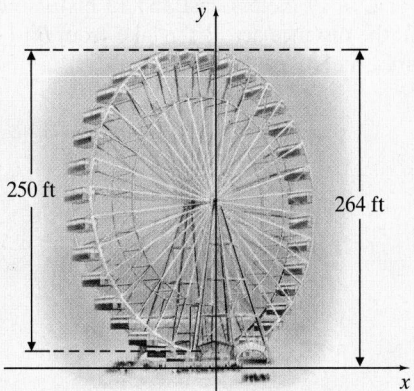

250 ft   264 ft

94. Although there are many larger observation wheels on the horizon, as of this writing the largest observation wheel in the world is the Singapore Flyer. From the Flyer, you can see up to 45 kilometers away. Each of the 28 enclosed capsules holds 28 passengers and completes a full rotation every 32 minutes. Its diameter is 150 meters, and the height of this giant wheel is 165 meters. (*Source:* singaporeflyer.com)

a. What is the radius of the Singapore Flyer?

b. How close is the wheel to the ground?

c. How high is the center of the wheel from the ground?

d. Using the axes in the drawing, what are the coordinates of the center of the wheel?

e. Use parts **a** and **d** to write an equation of the Singapore Flyer.

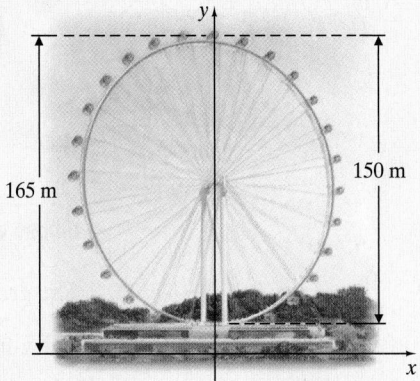

165 m   150 m

95. If you are given a list of equations of circles and parabolas and none are in standard form, explain how you would determine which is an equation of a circle and which is an equation of a parabola. Explain also how you would distinguish the upward or downward parabolas from the left-opening or right-opening parabolas.

△ **96.** Determine whether the triangle with vertices $(2, 6)$, $(0, -2)$, and $(5, 1)$ is an isosceles triangle.

Solve.

97. Two surveyors need to find the distance across a lake. They place a reference pole at point *A* in the diagram. Point *B* is 3 meters east and 1 meter north of the reference point *A*. Point *C* is 19 meters east and 13 meters north of point *A*. Find the distance across the lake, from *B* to *C*. Round to the nearest meter.

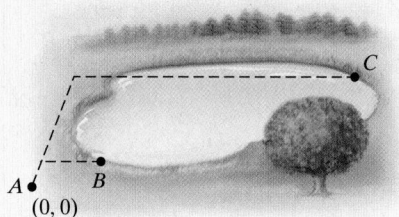

98. A bridge constructed over a bayou has a supporting arch in the shape of a parabola. Find an equation of the parabolic arch if the length of the road over the arch is 100 meters and the maximum height of the arch is 40 meters.

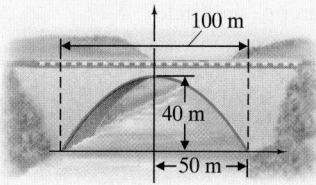

📖 *Use a graphing calculator to verify each exercise. Use a square viewing window.*

99. Exercise 77   **100.** Exercise 78

101. Exercise 79   **102.** Exercise 80

13.2 The Ellipse and the Hyperbola

OBJECTIVES

1 Define and Graph an Ellipse.

2 Define and Graph a Hyperbola.

OBJECTIVE

1 Graphing Ellipses

An **ellipse** can be thought of as the set of points in a plane such that the sum of the distances of those points from two fixed points is constant. Each of the two fixed points is called a **focus.** (The plural of focus is **foci.**) The point midway between the foci is called the **center.**

An ellipse may be drawn by hand by using two push pins, a piece of string, and a pencil. Secure the two push pins in a piece of cardboard, for example, and tie each end of the string to a pin. Use your pencil to pull the string tight and draw the ellipse. The two push pins are the foci of the drawn ellipse.

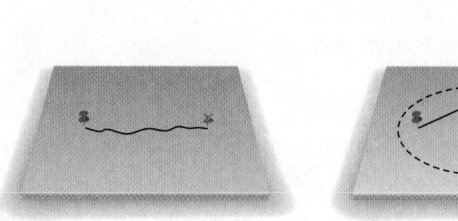 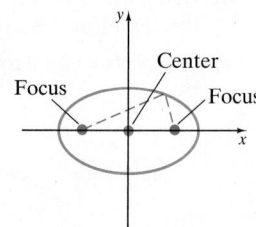

Ellipse with Center (0, 0)

The graph of an equation of the form $\dfrac{x^2}{a^2} + \dfrac{y^2}{b^2} = 1$ is an ellipse with center $(0,0)$.

The *x*-intercepts are $(a,0)$ and $(-a, 0)$, and the *y*-intercepts are $(0,b)$, and $(0, -b)$.

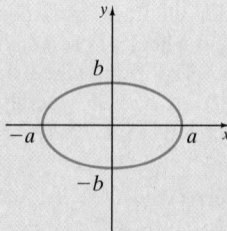

The **standard form** of an ellipse with center $(0,0)$ is $\dfrac{x^2}{a^2} + \dfrac{y^2}{b^2} = 1.$

EXAMPLE 1 Graph: $\dfrac{x^2}{9} + \dfrac{y^2}{16} = 1$

Solution The equation is of the form $\dfrac{x^2}{a^2} + \dfrac{y^2}{b^2} = 1$, with $a = 3$ and $b = 4$, so its graph is an ellipse with center $(0, 0)$, x-intercepts $(3, 0)$ and $(-3, 0)$, and y-intercepts $(0, 4)$ and $(0, -4)$.

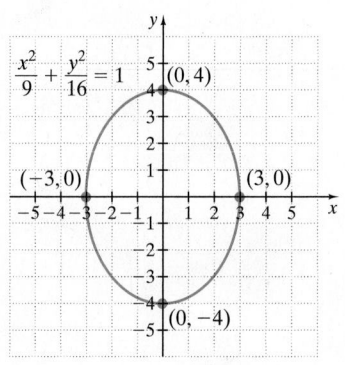

PRACTICE
1 Graph: $\dfrac{x^2}{25} + \dfrac{y^2}{4} = 1$

EXAMPLE 2 Graph: $4x^2 + 16y^2 = 64$

Solution Although this equation contains a sum of squared terms in x and y on the same side of an equation, this is not the equation of a circle since the coefficients of x^2 and y^2 are not the same. The graph of this equation is an ellipse. Since the standard form of the equation of an ellipse has 1 on one side, divide both sides of this equation by 64.

$$4x^2 + 16y^2 = 64$$

$$\dfrac{4x^2}{64} + \dfrac{16y^2}{64} = \dfrac{64}{64} \qquad \text{Divide both sides by 64.}$$

$$\dfrac{x^2}{16} + \dfrac{y^2}{4} = 1 \qquad \text{Simplify.}$$

We now recognize the equation of an ellipse with $a = 4$ and $b = 2$. This ellipse has center $(0, 0)$, x-intercepts $(4, 0)$ and $(-4, 0)$, and y-intercepts $(0, 2)$ and $(0, -2)$.

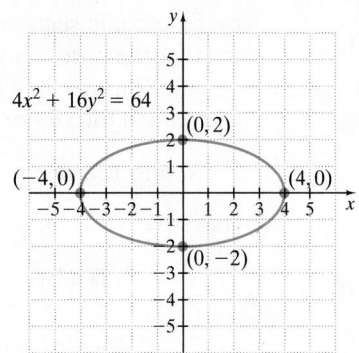

PRACTICE
2 Graph: $9x^2 + 4y^2 = 36$

The center of an ellipse is not always $(0, 0)$, as shown in the next example.

Ellipse with Center (h, k)

The standard form of the equation of an ellipse with center (h, k) is

$$\dfrac{(x - h)^2}{a^2} + \dfrac{(y - k)^2}{b^2} = 1$$

EXAMPLE 3 Graph: $\dfrac{(x+3)^2}{25} + \dfrac{(y-2)^2}{36} = 1$

Solution The center of this ellipse is found in a way that is similar to finding the center of a circle. This ellipse has center $(-3, 2)$. Notice that $a = 5$ and $b = 6$. To find four points on the graph of the ellipse, first graph the center, $(-3, 2)$. Since $a = 5$, count 5 units right and then 5 units left of the point with coordinates $(-3, 2)$. Next, since $b = 6$, start at $(-3, 2)$ and count 6 units up and then 6 units down to find two more points on the ellipse.

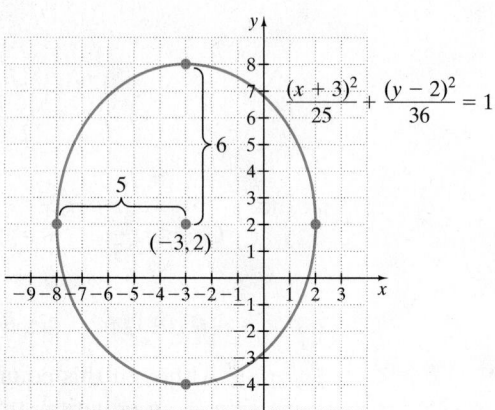

PRACTICE 3 Graph: $\dfrac{(x-4)^2}{49} + \dfrac{(y+1)^2}{81} = 1$

✔ **CONCEPT CHECK**

In the graph of the equation $\dfrac{x^2}{64} + \dfrac{y^2}{36} = 1$, which distance is longer: the distance between the x-intercepts or the distance between the y-intercepts? How much longer? Explain.

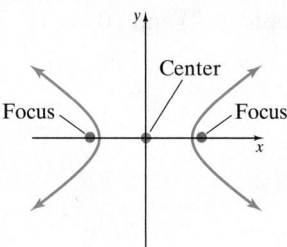

OBJECTIVE

2 Graphing Hyperbolas

The final conic section is the **hyperbola.** A hyperbola is the set of points in a plane such that the absolute value of the difference of the distances from two fixed points is constant. Each of the two fixed points is called a **focus.** The point midway between the foci is called the **center.**

Using the distance formula, we can show that the graph of $\dfrac{x^2}{a^2} - \dfrac{y^2}{b^2} = 1$ is a hyperbola with center $(0, 0)$ and x-intercepts $(a, 0)$ and $(-a, 0)$. Also, the graph of $\dfrac{y^2}{b^2} - \dfrac{x^2}{a^2} = 1$ is a hyperbola with center $(0, 0)$ and y-intercepts $(0, b)$ and $(0, -b)$.

> **Hyperbola with Center (0, 0)**
>
> The graph of an equation of the form $\dfrac{x^2}{a^2} - \dfrac{y^2}{b^2} = 1$ is a hyperbola with center $(0, 0)$ and x-intercepts $(a, 0)$ and $(-a, 0)$.
>
>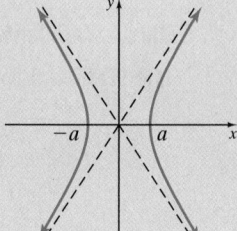

Answer to Concept Check:
x-intercepts, by 4 units

The graph of an equation of the form $\dfrac{y^2}{b^2} - \dfrac{x^2}{a^2} = 1$ is a hyperbola with center $(0, 0)$ and y-intercepts $(0, b)$ and $(0, -b)$.

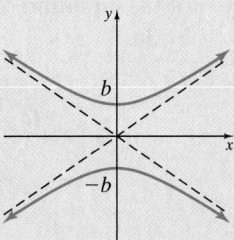

The equations $\dfrac{x^2}{a^2} - \dfrac{y^2}{b^2} = 1$ and $\dfrac{y^2}{b^2} - \dfrac{x^2}{a^2} = 1$ are the **standard forms** for the equation of a hyperbola.

Helpful Hint

Notice the difference between the equation of an ellipse and a hyperbola.

- The equation of the ellipse contains x^2 and y^2 terms on the same side of the equation with same-sign coefficients.
- For a hyperbola, the coefficients of x^2 and y^2 on the same side of the equation have different signs.

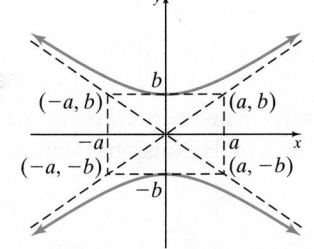

Graphing a hyperbola such as $\dfrac{y^2}{b^2} - \dfrac{x^2}{a^2} = 1$ is made easier by recognizing one of its important characteristics. Examining the figure to the left, notice how the sides of the branches of the hyperbola extend indefinitely and seem to approach the dashed lines in the figure. These dashed lines are called the **asymptotes** of the hyperbola.

To sketch these lines, or asymptotes, draw a rectangle with vertices (a, b), $(-a, b)$, $(a, -b)$, and $(-a, -b)$. The asymptotes of the hyperbola are the extended diagonals of this rectangle.

EXAMPLE 4   Graph: $\dfrac{x^2}{16} - \dfrac{y^2}{25} = 1$

Solution This equation has the form $\dfrac{x^2}{a^2} - \dfrac{y^2}{b^2} = 1$, with $a = 4$ and $b = 5$. Thus, its graph is a hyperbola that opens to the left and right. It has center $(0, 0)$ and x-intercepts $(4, 0)$ and $(-4, 0)$. To aid in graphing the hyperbola, we first sketch its asymptotes. The extended diagonals of the rectangle with corners $(4, 5)$, $(4, -5)$, $(-4, 5)$, and $(-4, -5)$ are the asymptotes of the hyperbola. Then we use the asymptotes to aid in sketching the hyperbola.

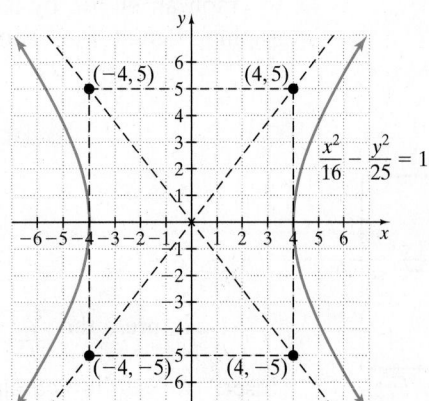

PRACTICE
4   Graph: $\dfrac{x^2}{9} - \dfrac{y^2}{16} = 1$

EXAMPLE 5 Graph: $4y^2 - 9x^2 = 36$

Solution Since this is a difference of squared terms in x and y on the same side of the equation, its graph is a hyperbola as opposed to an ellipse or a circle. The standard form of the equation of a hyperbola has a 1 on one side, so divide both sides of the equation by 36.

$$4y^2 - 9x^2 = 36$$

$$\frac{4y^2}{36} - \frac{9x^2}{36} = \frac{36}{36} \quad \text{Divide both sides by 36.}$$

$$\frac{y^2}{9} - \frac{x^2}{4} = 1 \quad \text{Simplify.}$$

The equation is of the form $\dfrac{y^2}{b^2} - \dfrac{x^2}{a^2} = 1$, with $a = 2$ and $b = 3$, so the hyperbola is centered at $(0, 0)$ with y-intercepts $(0, 3)$ and $(0, -3)$. The sketch of the hyperbola is shown.

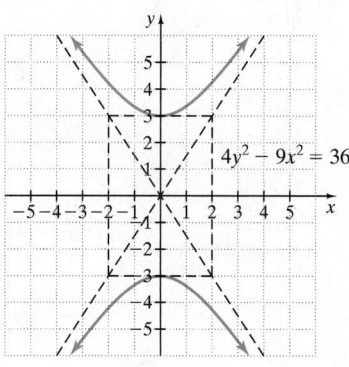

PRACTICE
5 Graph: $9y^2 - 25x^2 = 225$

Although it is beyond the scope of this text, the standard forms of the equations of hyperbolas with center (h, k) are given below. The Concept Extensions section in Exercise Set 13.2 contains some hyperbolas of this form.

> **Hyperbola with Center (h, k)**
>
> Standard forms of the equations of hyperbolas with center (h, k) are:
>
> $$\frac{(x - h)^2}{a^2} - \frac{(y - k)^2}{b^2} = 1 \qquad \frac{(y - k)^2}{b^2} - \frac{(x - h)^2}{a^2} = 1$$

Graphing Calculator Explorations

To graph an ellipse by using a graphing calculator, use the same procedure as for graphing a circle. For example, to graph $x^2 + 3y^2 = 22$, first solve for y.

$$3y^2 = 22 - x^2$$

$$y^2 = \frac{22 - x^2}{3}$$

$$y = \pm\sqrt{\frac{22 - x^2}{3}}$$

Next, press the $\boxed{Y =}$ key and enter $Y_1 = \sqrt{\dfrac{22 - x^2}{3}}$ and $Y_2 = -\sqrt{\dfrac{22 - x^2}{3}}$

(Insert two sets of parentheses in the radicand as $\sqrt{((22 - x^2)/3)}$ so that the desired graph is obtained.) The graph appears as shown to the left.

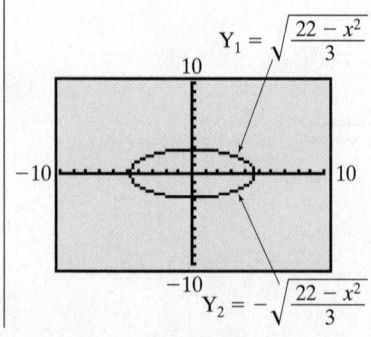

Use a graphing calculator to graph each ellipse in a square window.

1. $10x^2 + y^2 = 32$

2. $x^2 + 6y^2 = 35$

3. $20x^2 + 5y^2 = 100$

4. $4y^2 + 12x^2 = 48$

5. $7.3x^2 + 15.5y^2 = 95.2$

6. $18.8x^2 + 36.1y^2 = 205.8$

✔ Vocabulary, Readiness & Video Check

Use the choices below to fill in each blank. Some choices will be used more than once and some not at all.

ellipse	$(0,0)$	x	$(a,0)$ and $(-a,0)$	$(0,a)$ and $(0,-a)$	focus
hyperbola	center	y	$(b,0)$ and $(-b,0)$	$(0,b)$ and $(0,-b)$	

1. A(n) _____ is the set of points in a plane such that the absolute value of the differences of their distances from two fixed points is constant.

2. A(n) _____ is the set of points in a plane such that the sum of their distances from two fixed points is constant.

For Exercises 1 and 2 above,

3. The two fixed points are each called a _____.

4. The point midway between the foci is called the _____.

5. The graph of $\dfrac{x^2}{a^2} - \dfrac{y^2}{b^2} = 1$ is a(n) _____ with center _____ and _____ -intercepts of _____.

6. The graph of $\dfrac{x^2}{a^2} + \dfrac{y^2}{b^2} = 1$ is a(n) _____ with center _____ and x-intercepts of _____.

Martin-Gay Interactive Videos

Watch the section lecture video and answer the following questions.

OBJECTIVE 1

7. From ▦ Example 1, what information do the values of a and b give us about the graph of an ellipse? Answer this same question for ▦ Example 2.

OBJECTIVE 2

8. From ▦ Example 3, we know the points $(a,b), (a,-b), (-a,b)$, and $(-a,-b)$ are not part of the graph. Explain the role of these points.

See Video 13.2 ◉

13.2 Exercise Set   MyMathLab® ▸

Identify the graph of each equation as an ellipse or a hyperbola. Do not graph. See Examples 1 through 5.

1. $\dfrac{x^2}{16} + \dfrac{y^2}{4} = 1$

2. $\dfrac{x^2}{16} - \dfrac{y^2}{4} = 1$

3. $x^2 - 5y^2 = 3$

4. $-x^2 + 5y^2 = 3$

5. $-\dfrac{y^2}{25} + \dfrac{x^2}{36} = 1$

6. $\dfrac{y^2}{25} + \dfrac{x^2}{36} = 1$

9. $\dfrac{x^2}{9} + y^2 = 1$

10. $x^2 + \dfrac{y^2}{4} = 1$

11. $9x^2 + y^2 = 36$

12. $x^2 + 4y^2 = 16$

13. $4x^2 + 25y^2 = 100$

14. $36x^2 + y^2 = 36$

Sketch the graph of each equation. See Examples 1 and 2.

7. $\dfrac{x^2}{4} + \dfrac{y^2}{25} = 1$

8. $\dfrac{x^2}{16} + \dfrac{y^2}{9} = 1$

Sketch the graph of each equation. See Example 3.

15. $\dfrac{(x+1)^2}{36} + \dfrac{(y-2)^2}{49} = 1$

16. $\dfrac{(x-3)^2}{9} + \dfrac{(y+3)^2}{16} = 1$

▶**17.** $\dfrac{(x-1)^2}{4} + \dfrac{(y-1)^2}{25} = 1$

18. $\dfrac{(x+3)^2}{16} + \dfrac{(y+2)^2}{4} = 1$

Sketch the graph of each equation. See Examples 4 and 5.

19. $\dfrac{x^2}{4} - \dfrac{y^2}{9} = 1$ **20.** $\dfrac{x^2}{36} - \dfrac{y^2}{36} = 1$

21. $\dfrac{y^2}{25} - \dfrac{x^2}{16} = 1$ **22.** $\dfrac{y^2}{25} - \dfrac{x^2}{49} = 1$

23. $x^2 - 4y^2 = 16$ **24.** $4x^2 - y^2 = 36$

25. $16y^2 - x^2 = 16$ **26.** $4y^2 - 25x^2 = 100$

MIXED PRACTICE

Graph each equation. See Examples 1 through 5.

27. $\dfrac{y^2}{36} = 1 - x^2$ **28.** $\dfrac{x^2}{36} = 1 - y^2$

29. $4(x - 1)^2 + 9(y + 2)^2 = 36$

30. $25(x + 3)^2 + 4(y - 3)^2 = 100$

31. $8x^2 + 2y^2 = 32$ **32.** $3x^2 + 12y^2 = 48$

33. $25x^2 - y^2 = 25$ **34.** $x^2 - 9y^2 = 9$

MIXED PRACTICE—SECTIONS 13.1, 13.2

Identify whether each equation, when graphed, will be a parabola, circle, ellipse, or hyperbola. Sketch the graph of each equation.

If a parabola, label the vertex.
If a circle, label the center and note the radius.
If an ellipse, label the center.
If a hyperbola, label the x- or y-intercepts.

35. $(x - 7)^2 + (y - 2)^2 = 4$ **36.** $y = x^2 + 4$

37. $y = x^2 + 12x + 36$ **38.** $\dfrac{x^2}{4} + \dfrac{y^2}{9} = 1$

39. $\dfrac{y^2}{9} - \dfrac{x^2}{9} = 1$ **40.** $\dfrac{x^2}{16} - \dfrac{y^2}{4} = 1$

41. $\dfrac{x^2}{16} + \dfrac{y^2}{4} = 1$ **42.** $x^2 + y^2 = 16$

43. $x = y^2 + 4y - 1$ **44.** $x = -y^2 + 6y$

45. $9x^2 - 4y^2 = 36$ ▶ **46.** $9x^2 + 4y^2 = 36$

47. $\dfrac{(x - 1)^2}{49} + \dfrac{(y + 2)^2}{25} = 1$ ▶ **48.** $y^2 = x^2 + 16$

49. $\left(x + \dfrac{1}{2}\right)^2 + \left(y - \dfrac{1}{2}\right)^2 = 1$ **50.** $y = -2x^2 + 4x - 3$

REVIEW AND PREVIEW

Perform the indicated operations. See Sections 5.1 and 5.2.

51. $(2x^3)(-4x^2)$ **52.** $2x^3 - 4x^3$

53. $-5x^2 + x^2$ **54.** $(-5x^2)(x^2)$

CONCEPT EXTENSIONS

The graph of each equation is an ellipse. Determine which distance is longer, the distance between the x-intercepts or the distance between the y-intercepts. How much longer? See the Concept Check in this section.

55. $\dfrac{x^2}{16} + \dfrac{y^2}{25} = 1$ **56.** $\dfrac{x^2}{100} + \dfrac{y^2}{49} = 1$

57. $4x^2 + y^2 = 16$ **58.** $x^2 + 4y^2 = 36$

59. If you are given a list of equations of circles, parabolas, ellipses, and hyperbolas, explain how you could distinguish the different conic sections from their equations.

60. We know that $x^2 + y^2 = 25$ is the equation of a circle. Rewrite the equation so that the right side is equal to 1. Which type of conic section does this equation form resemble? In fact, the circle is a special case of this type of conic section. Describe the conditions under which this type of conic section is a circle.

The orbits of stars, planets, comets, asteroids, and satellites all have the shape of one of the conic sections. Astronomers use a measure called eccentricity to describe the shape and elongation of an orbital path. For the circle and ellipse, eccentricity e is calculated with the formula $e = \dfrac{c}{d}$, where $c^2 = |a^2 - b^2|$ and d is the larger value of a or b. For a hyperbola, eccentricity e is calculated with the formula $e = \dfrac{c}{d}$, where $c^2 = a^2 + b^2$ and the value of d is equal to a if the hyperbola has x-intercepts or equal to b if the hyperbola has y-intercepts. Use equations A–H to answer Exercises 61–70.

A. $\dfrac{x^2}{36} - \dfrac{y^2}{13} = 1$ **B.** $\dfrac{x^2}{4} + \dfrac{y^2}{4} = 1$ **C.** $\dfrac{x^2}{25} + \dfrac{y^2}{16} = 1$

D. $\dfrac{y^2}{25} - \dfrac{x^2}{39} = 1$ **E.** $\dfrac{x^2}{17} + \dfrac{y^2}{81} = 1$ **F.** $\dfrac{x^2}{36} + \dfrac{y^2}{36} = 1$

G. $\dfrac{x^2}{16} - \dfrac{y^2}{65} = 1$ **H.** $\dfrac{x^2}{144} + \dfrac{y^2}{140} = 1$

61. Identify the type of conic section represented by each of the equations A–H.

62. For each of the equations A–H, identify the values of a^2 and b^2.

63. For each of the equations A–H, calculate the value of c^2 and c.

64. For each of the equations A–H, find the value of d.

65. For each of the equations A–H, calculate the eccentricity e.

66. What do you notice about the values of e for the equations you identified as ellipses?

67. What do you notice about the values of e for the equations you identified as circles?

68. What do you notice about the values of e for the equations you identified as hyperbolas?

69. The eccentricity of a parabola is exactly 1. Use this information and the observations you made in Exercises 66, 67, and 68 to describe a way that could be used to identify the type of conic section based on its eccentricity value.

70. Graph each of the conic sections given in equations A–H. What do you notice about the shape of the ellipses for increasing values of eccentricity? Which is the most elliptical? Which is the least elliptical, that is, the most circular?

71. A planet's orbit about the Sun can be described as an ellipse. Consider the origin of a rectangular coordinate system as the ellipse's center. Suppose that the x-intercepts of the elliptical path of the planet are $\pm 130{,}000{,}000$ and that the y-intercepts are $\pm 125{,}000{,}000$. Write the equation of the elliptical path of the planet.

72. Comets orbit the Sun in elongated ellipses. Consider the Sun as the origin of a rectangular coordinate system. Suppose that the equation of the path of the comet is

$$\frac{(x - 1{,}782{,}000{,}000)^2}{3.42 \times 10^{23}} + \frac{(y - 356{,}400{,}000)^2}{1.368 \times 10^{22}} = 1$$

Find the center of the path of the comet.

73. Use a graphing calculator to verify Exercise 46.

74. Use a graphing calculator to verify Exercise 12.

For Exercises 75 through 80, see the example below.

Example

Sketch the graph of $\dfrac{(x - 2)^2}{25} - \dfrac{(y - 1)^2}{9} = 1$.

Solution

This hyperbola has center $(2, 1)$. Notice that $a = 5$ and $b = 3$.

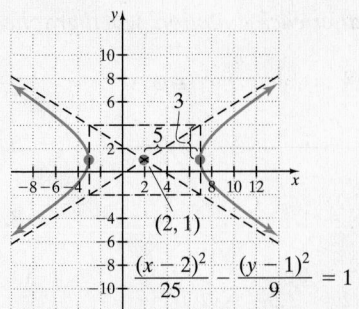

$$\frac{(x - 2)^2}{25} - \frac{(y - 1)^2}{9} = 1$$

Sketch the graph of each equation.

75. $\dfrac{(x - 1)^2}{4} - \dfrac{(y + 1)^2}{25} = 1$ **76.** $\dfrac{(x + 2)^2}{9} - \dfrac{(y - 1)^2}{4} = 1$

77. $\dfrac{y^2}{16} - \dfrac{(x + 3)^2}{9} = 1$ **78.** $\dfrac{(y + 4)^2}{4} - \dfrac{x^2}{25} = 1$

79. $\dfrac{(x + 5)^2}{16} - \dfrac{(y + 2)^2}{25} = 1$ **80.** $\dfrac{(x - 3)^2}{9} - \dfrac{(y - 2)^2}{4} = 1$

| Integrated Review | Graphing Conic Sections |

Sections 13.1–13.2

Following is a summary of conic sections.

Conic Sections

	Standard Form	Graph
Parabola	$y = a(x - h)^2 + k$	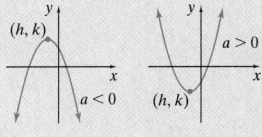
Parabola	$x = a(y - k)^2 + h$	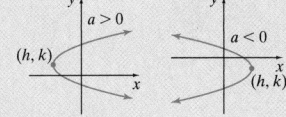
Circle	$(x - h)^2 + (y - k)^2 = r^2$	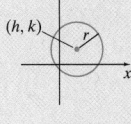
Ellipse center $(0, 0)$	$\dfrac{x^2}{a^2} + \dfrac{y^2}{b^2} = 1$	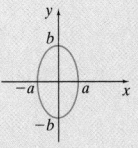
Hyperbola center $(0, 0)$	$\dfrac{x^2}{a^2} - \dfrac{y^2}{b^2} = 1$	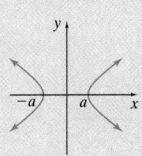
Hyperbola center $(0, 0)$	$\dfrac{y^2}{b^2} - \dfrac{x^2}{a^2} = 1$	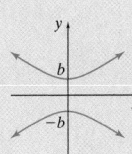

Identify whether each equation, when graphed, will be a parabola, circle, ellipse, or hyperbola. Then graph each equation.

1. $(x - 7)^2 + (y - 2)^2 = 4$

2. $y = x^2 + 4$

3. $y = x^2 + 12x + 36$

4. $\dfrac{x^2}{4} + \dfrac{y^2}{9} = 1$

5. $\dfrac{y^2}{9} - \dfrac{x^2}{9} = 1$

6. $\dfrac{x^2}{16} - \dfrac{y^2}{4} = 1$

7. $\dfrac{x^2}{16} + \dfrac{y^2}{4} = 1$

8. $x^2 + y^2 = 16$

9. $x = y^2 + 4y - 1$

10. $x = -y^2 + 6y$

11. $9x^2 - 4y^2 = 36$

12. $9x^2 + 4y^2 = 36$

13. $\dfrac{(x - 1)^2}{49} + \dfrac{(y + 2)^2}{25} = 1$

14. $y^2 = x^2 + 16$

15. $\left(x + \dfrac{1}{2}\right)^2 + \left(y - \dfrac{1}{2}\right)^2 = 1$

13.3 Solving Nonlinear Systems of Equations ▶

OBJECTIVES

1 Solve a Nonlinear System by Substitution. ▶

2 Solve a Nonlinear System by Elimination. ▶

In Chapter 4, we used graphing, substitution, and elimination methods to find solutions of systems of linear equations in two variables. We now apply these same methods to nonlinear systems of equations in two variables. A **nonlinear system of equations** is a system of equations at least one of which is not linear. Since we will be graphing the equations in each system, we are interested in real number solutions only.

OBJECTIVE

1 Solving Nonlinear Systems by Substitution

First, nonlinear systems are solved by the substitution method.

EXAMPLE 1   Solve the system.

$$\begin{cases} x^2 - 3y = 1 \\ x - y = 1 \end{cases}$$

<u>Solution</u> We can solve this system by substitution if we solve one equation for one of the variables. Solving the first equation for x is not the best choice since doing so introduces a radical. Also, solving for y in the first equation introduces a fraction. We solve the second equation for y.

$$x - y = 1 \quad \text{Second equation}$$
$$x - 1 = y \quad \text{Solve for } y.$$

Replace y with $x - 1$ in the first equation, and then solve for x.

$$x^2 - 3y = 1 \quad \text{First equation}$$
$$x^2 - 3(x - 1) = 1 \quad \text{Replace } y \text{ with } x - 1.$$
$$x^2 - 3x + 3 = 1$$
$$x^2 - 3x + 2 = 0$$
$$(x - 2)(x - 1) = 0$$
$$x = 2 \quad \text{or} \quad x = 1$$

Let $x = 2$ and then let $x = 1$ in the equation $y = x - 1$ to find corresponding y-values.

Let $x = 2$.

$$y = x - 1$$
$$y = 2 - 1 = 1$$

Let $x = 1$.

$$y = x - 1$$
$$y = 1 - 1 = 0$$

The solutions are $(2, 1)$ and $(1, 0)$, or the solution set is $\{(2, 1), (1, 0)\}$. Check both solutions in both equations. Both solutions satisfy both equations, so both are solutions of the system. The graph of each equation in the system is shown next. Intersections of the graphs are at $(2, 1)$ and $(1, 0)$.

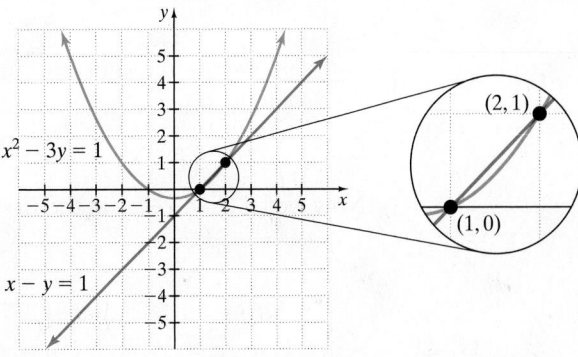

1   Solve the system: $\begin{cases} x^2 - 4y = 4 \\ x + y = -1 \end{cases}$

EXAMPLE 2   Solve the system.

$$\begin{cases} y = \sqrt{x} \\ x^2 + y^2 = 6 \end{cases}$$

Solution This system is ideal for substitution since y is expressed in terms of x in the first equation. Notice that if $y = \sqrt{x}$, then both x and y must be nonnegative if they are real numbers. Substitute $\sqrt{x}$ for y in the second equation, and solve for x.

$$x^2 + y^2 = 6$$
$$x^2 + (\sqrt{x})^2 = 6 \quad \text{Let } y = \sqrt{x}$$
$$x^2 + x = 6$$
$$x^2 + x - 6 = 0$$
$$(x + 3)(x - 2) = 0$$
$$x = -3 \quad \text{or} \quad x = 2$$

The solution -3 is discarded because we have noted that x must be nonnegative. To see this, let $x = -3$ in the first equation. Then let $x = 2$ in the first equation to find the corresponding y-value.

Let $x = -3$.
$$y = \sqrt{x}$$
$$y = \sqrt{-3} \quad \text{Not a real number}$$

Let $x = 2$.
$$y = \sqrt{x}$$
$$y = \sqrt{2}$$

Since we are interested only in real number solutions, the only solution is $(2, \sqrt{2})$. Check to see that this solution satisfies both equations. The graph of each equation in the system is shown to the right.

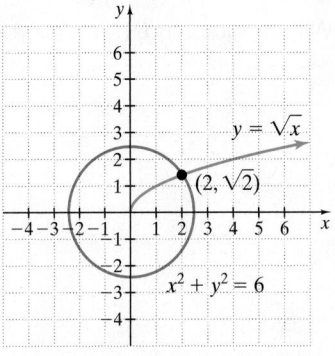

2   Solve the system: $\begin{cases} y = -\sqrt{x} \\ x^2 + y^2 = 20 \end{cases}$

EXAMPLE 3 Solve the system.

$$\begin{cases} x^2 + y^2 = 4 \\ x + y = 3 \end{cases}$$

Solution We use the substitution method and solve the second equation for x.

$$x + y = 3 \qquad \text{Second equation}$$
$$x = 3 - y$$

Now we let $x = 3 - y$ in the first equation.

$$x^2 + y^2 = 4 \quad \text{First equation}$$
$$(3 - y)^2 + y^2 = 4 \quad \text{Let } x = 3 - y.$$
$$9 - 6y + y^2 + y^2 = 4$$
$$2y^2 - 6y + 5 = 0$$

By the quadratic formula, where $a = 2$, $b = -6$, and $c = 5$, we have

$$y = \frac{6 \pm \sqrt{(-6)^2 - 4 \cdot 2 \cdot 5}}{2 \cdot 2} = \frac{6 \pm \sqrt{-4}}{4}$$

Since $\sqrt{-4}$ is not a real number, there is no real solution, or $\varnothing$. Graphically, the circle and the line do not intersect, as shown below.

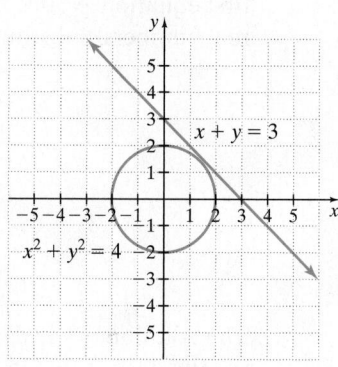

PRACTICE
3 Solve the system: $\begin{cases} x^2 + y^2 = 9 \\ x - y = 5 \end{cases}$

✔ **CONCEPT CHECK**
Without graphing, how can you tell that $x^2 + y^2 = 9$ and $x^2 + y^2 = 16$ do not have any points of intersection?

OBJECTIVE
2 Solving Nonlinear Systems by Elimination ▶

Some nonlinear systems may be solved by the elimination method.

EXAMPLE 4 Solve the system.

$$\begin{cases} x^2 + 2y^2 = 10 \\ x^2 - y^2 = 1 \end{cases}$$

Answer to Concept Check:
$x^2 + y^2 = 9$ is a circle inside the
circle $x^2 + y^2 = 16$; therefore, they
do not have any points of intersection.

Solution We will use the elimination, or addition, method to solve this system. To eliminate x^2 when we add the two equations, multiply both sides of the second equation by -1. Then

$$\begin{cases} x^2 + 2y^2 = 10 \\ (-1)(x^2 - y^2) = -1 \cdot 1 \end{cases}$$ is equivalent to $$\begin{cases} x^2 + 2y^2 = 10 \\ -x^2 + y^2 = -1 \end{cases}$$ Add.

$$3y^2 = 9 \qquad \text{Divide both}$$
$$y^2 = 3 \qquad \text{sides by 3.}$$
$$y = \pm\sqrt{3}$$

To find the corresponding x-values, we let $y = \sqrt{3}$ and $y = -\sqrt{3}$ in either original equation. We choose the second equation.

Let $y = \sqrt{3}$.

$$x^2 - y^2 = 1$$
$$x^2 - (\sqrt{3})^2 = 1$$
$$x^2 - 3 = 1$$
$$x^2 = 4$$
$$x = \pm\sqrt{4} = \pm 2$$

Let $y = -\sqrt{3}$.

$$x^2 - y^2 = 1$$
$$x^2 - (-\sqrt{3})^2 = 1$$
$$x^2 - 3 = 1$$
$$x^2 = 4$$
$$x = \pm\sqrt{4} = \pm 2$$

The solutions are $(2, \sqrt{3})$, $(-2, \sqrt{3})$, $(2, -\sqrt{3})$, and $(-2, -\sqrt{3})$. Check all four ordered pairs in both equations of the system. The graph of each equation in this system is shown.

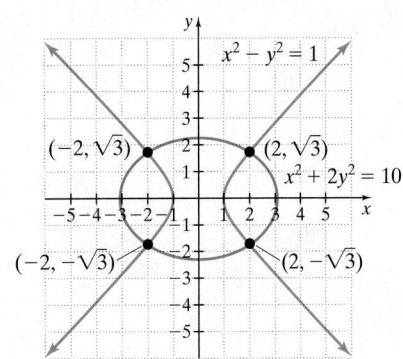

PRACTICE
4 Solve the system: $\begin{cases} x^2 + 4y^2 = 16 \\ x^2 - y^2 = 1 \end{cases}$

✓ Vocabulary, Readiness & Video Check

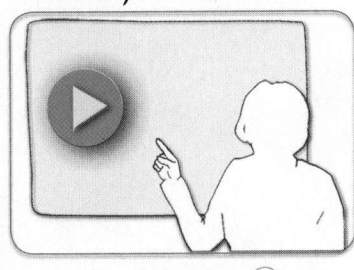

Martin-Gay Interactive Videos

See Video 13.3 ◉

Watch the section lecture video and answer the following questions.

OBJECTIVE 1
1. In ▦ Example 1, why do we choose not to solve either equation for y?

OBJECTIVE 2
2. In ▦ Example 2, what important reminder is made as the second equation is multiplied by a number to get opposite coefficients of x?

13.3 Exercise Set MyMathLab®

MIXED PRACTICE

Solve each nonlinear system of equations for real solutions. See Examples 1 through 4.

1. $\begin{cases} x^2 + y^2 = 25 \\ 4x + 3y = 0 \end{cases}$

2. $\begin{cases} x^2 + y^2 = 25 \\ 3x + 4y = 0 \end{cases}$

3. $\begin{cases} x^2 + 4y^2 = 10 \\ y = x \end{cases}$

4. $\begin{cases} 4x^2 + y^2 = 10 \\ y = x \end{cases}$

▶ 5. $\begin{cases} y^2 = 4 - x \\ x - 2y = 4 \end{cases}$

6. $\begin{cases} x^2 + y^2 = 4 \\ x + y = -2 \end{cases}$

7. $\begin{cases} x^2 + y^2 = 9 \\ 16x^2 - 4y^2 = 64 \end{cases}$

8. $\begin{cases} 4x^2 + 3y^2 = 35 \\ 5x^2 + 2y^2 = 42 \end{cases}$

9. $\begin{cases} x^2 + 2y^2 = 2 \\ x - y = 2 \end{cases}$

10. $\begin{cases} x^2 + 2y^2 = 2 \\ x^2 - 2y^2 = 6 \end{cases}$

11. $\begin{cases} y = x^2 - 3 \\ 4x - y = 6 \end{cases}$

12. $\begin{cases} y = x + 1 \\ x^2 - y^2 = 1 \end{cases}$

13. $\begin{cases} y = x^2 \\ 3x + y = 10 \end{cases}$

14. $\begin{cases} 6x - y = 5 \\ xy = 1 \end{cases}$

15. $\begin{cases} y = 2x^2 + 1 \\ x + y = -1 \end{cases}$

16. $\begin{cases} x^2 + y^2 = 9 \\ x + y = 5 \end{cases}$

17. $\begin{cases} y = x^2 - 4 \\ y = x^2 - 4x \end{cases}$

18. $\begin{cases} x = y^2 - 3 \\ x = y^2 - 3y \end{cases}$

▶ 19. $\begin{cases} 2x^2 + 3y^2 = 14 \\ -x^2 + y^2 = 3 \end{cases}$

20. $\begin{cases} 4x^2 - 2y^2 = 2 \\ -x^2 + y^2 = 2 \end{cases}$

21. $\begin{cases} x^2 + y^2 = 1 \\ x^2 + (y + 3)^2 = 4 \end{cases}$

22. $\begin{cases} x^2 + 2y^2 = 4 \\ x^2 - y^2 = 4 \end{cases}$

23. $\begin{cases} y = x^2 + 2 \\ y = -x^2 + 4 \end{cases}$

24. $\begin{cases} x = -y^2 - 3 \\ x = y^2 - 5 \end{cases}$

25. $\begin{cases} 3x^2 + y^2 = 9 \\ 3x^2 - y^2 = 9 \end{cases}$

26. $\begin{cases} x^2 + y^2 = 25 \\ x = y^2 - 5 \end{cases}$

27. $\begin{cases} x^2 + 3y^2 = 6 \\ x^2 - 3y^2 = 10 \end{cases}$

28. $\begin{cases} x^2 + y^2 = 1 \\ y = x^2 - 9 \end{cases}$

29. $\begin{cases} x^2 + y^2 = 36 \\ y = \dfrac{1}{6}x^2 - 6 \end{cases}$

30. $\begin{cases} x^2 + y^2 = 16 \\ y = -\dfrac{1}{4}x^2 + 4 \end{cases}$

31. $\begin{cases} y = \sqrt{x} \\ x^2 + y^2 = 12 \end{cases}$

32. $\begin{cases} y = \sqrt{x} \\ x^2 + y^2 = 20 \end{cases}$

REVIEW AND PREVIEW

Graph each inequality in two variables. See Section 9.4.

33. $x > -3$ **34.** $y \le 1$

35. $y < 2x - 1$ **36.** $3x - y \le 4$

Find the perimeter of each geometric figure. See Section 5.2.

△ **37.**

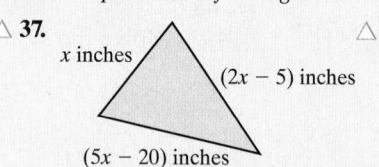

x inches, $(2x - 5)$ inches, $(5x - 20)$ inches

△ **38.**

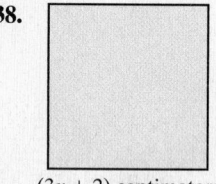

$(3x + 2)$ centimeters

△ **39.** $(x^2 + 3x + 1)$ meters

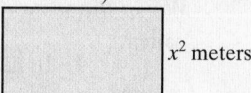

x^2 meters

△ **40.**

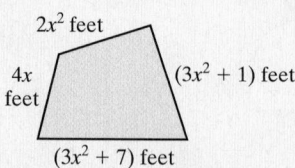

$2x^2$ feet, $4x$ feet, $(3x^2 + 1)$ feet, $(3x^2 + 7)$ feet

CONCEPT EXTENSIONS

For the exercises below, see the Concept Check in this section.

41. Without graphing, how can you tell that the graphs of $x^2 + y^2 = 1$ and $x^2 + y^2 = 4$ do not have any points of intersection?

42. Without solving, how can you tell that the graphs of $y = 2x + 3$ and $y = 2x + 7$ do not have any points of intersection?

43. How many real solutions are possible for a system of equations whose graphs are a circle and a parabola? Draw diagrams to illustrate each possibility.

44. How many real solutions are possible for a system of equations whose graphs are an ellipse and a line? Draw diagrams to illustrate each possibility.

Solve.

45. The sum of the squares of two numbers is 130. The difference of the squares of the two numbers is 32. Find the two numbers.

46. The sum of the squares of two numbers is 20. Their product is 8. Find the two numbers.

47. During the development stage of a new rectangular keypad for a security system, it was decided that the area of the rectangle should be 285 square centimeters and the perimeter should be 68 centimeters. Find the dimensions of the keypad.

48. A rectangular holding pen for cattle is to be designed so that its perimeter is 92 feet and its area is 525 square feet. Find the dimensions of the holding pen.

Recall that in business, a demand function expresses the quantity of a commodity demanded as a function of the commodity's unit price. A supply function expresses the quantity of a commodity supplied as a function of the commodity's unit price. When the quantity produced and supplied is equal to the quantity demanded, then we have what is called **market equilibrium**.

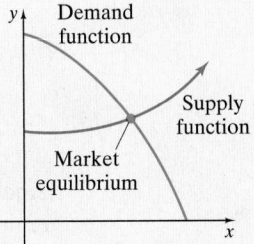

49. The demand function for a certain compact disc is given by the function
$$p = -0.01x^2 - 0.2x + 9$$
and the corresponding supply function is given by
$$p = 0.01x^2 - 0.1x + 3$$
where p is in dollars and x is in thousands of units. Find the equilibrium quantity and the corresponding price by solving the system consisting of the two given equations.

50. The demand function for a certain style of picture frame is given by the function
$$p = -2x^2 + 90$$
and the corresponding supply function is given by
$$p = 9x + 34$$
where p is in dollars and x is in thousands of units. Find the equilibrium quantity and the corresponding price by solving the system consisting of the two given equations.

Use a graphing calculator to verify the results of each exercise.

51. Exercise 3   **52.** Exercise 4

53. Exercise 23   **54.** Exercise 24

13.4 Nonlinear Inequalities and Systems of Inequalities

OBJECTIVES

1 Graph a Nonlinear Inequality.

2 Graph a System of Nonlinear Inequalities. ▷

OBJECTIVE

1 Graphing Nonlinear Inequalities ▷

We can graph a nonlinear inequality in two variables such as $\dfrac{x^2}{9} + \dfrac{y^2}{16} \leq 1$ in a way similar to the way we graphed a linear inequality in two variables in Section 9.4. First, graph the related equation $\dfrac{x^2}{9} + \dfrac{y^2}{16} = 1$. The graph of the equation is our boundary. Then, using test points, we determine and shade the region whose points satisfy the inequality.

EXAMPLE 1   Graph: $\dfrac{x^2}{9} + \dfrac{y^2}{16} \leq 1$

Solution First, graph the equation $\dfrac{x^2}{9} + \dfrac{y^2}{16} = 1$. Sketch a solid curve since the graph of $\dfrac{x^2}{9} + \dfrac{y^2}{16} \leq 1$ includes the graph of $\dfrac{x^2}{9} + \dfrac{y^2}{16} = 1$. The graph is an ellipse, and it

(Continued on next page)

divides the plane into two regions, the "inside" and the "outside" of the ellipse. To determine which region contains the solutions, select a test point in either region and determine whether the coordinates of the point satisfy the inequality. We choose $(0, 0)$ as the test point.

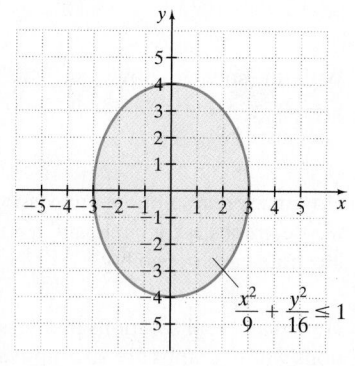

$$\frac{x^2}{9} + \frac{y^2}{16} \le 1$$

$$\frac{0^2}{9} + \frac{0^2}{16} \le 1 \quad \text{Let } x = 0 \text{ and } y = 0.$$

$$0 \le 1 \quad \text{True}$$

Since this statement is true, the solution set is the region containing $(0, 0)$. The graph of the solution set includes the points on and inside the ellipse, as shaded in the figure.

PRACTICE
1 Graph: $\dfrac{x^2}{36} + \dfrac{y^2}{16} \ge 1$

EXAMPLE 2 Graph: $4y^2 > x^2 + 16$

<u>Solution</u> The related equation is $4y^2 = x^2 + 16$. Subtract x^2 from both sides and divide both sides by 16, and we have $\dfrac{y^2}{4} - \dfrac{x^2}{16} = 1$, which is a hyperbola. Graph the hyperbola as a dashed curve since the graph of $4y^2 > x^2 + 16$ does *not* include the graph of $4y^2 = x^2 + 16$. The hyperbola divides the plane into three regions. Select a test point in each region—not on a boundary curve—to determine whether that region contains solutions of the inequality.

Test Region A with $(0, 4)$	*Test Region B with* $(0, 0)$	*Test Region C with* $(0, -4)$
$4y^2 > x^2 + 16$	$4y^2 > x^2 + 16$	$4y^2 > x^2 + 16$
$4(4)^2 > 0^2 + 16$	$4(0)^2 > 0^2 + 16$	$4(-4)^2 > 0^2 + 16$
$64 > 16 \quad$ True	$0 > 16 \quad$ False	$64 > 16 \quad$ True

The graph of the solution set includes the shaded regions A and C only, not the boundary.

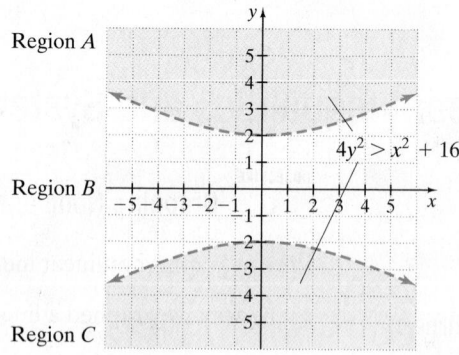

PRACTICE
2 Graph: $16y^2 > 9x^2 + 144$

OBJECTIVE
2 Graphing Systems of Nonlinear Inequalities ▶

In Section 9.4 we graphed systems of linear inequalities. Recall that the graph of a system of inequalities is the intersection of the graphs of the inequalities.

EXAMPLE 3   Graph the system.

$$\begin{cases} x \le 1 - 2y \\ y \le x^2 \end{cases}$$

Solution We graph each inequality on the same set of axes. The intersection is shown in the third graph below. It is the darkest shaded (appears purple) region along with its boundary lines. The coordinates of the points of intersection can be found by solving the related system.

$$\begin{cases} x = 1 - 2y \\ y = x^2 \end{cases}$$

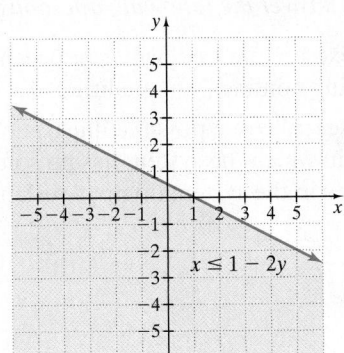 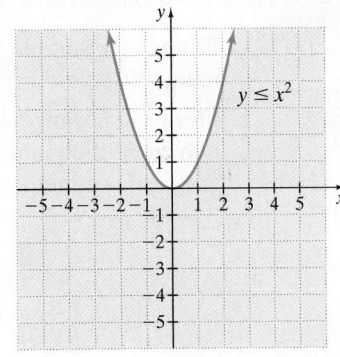 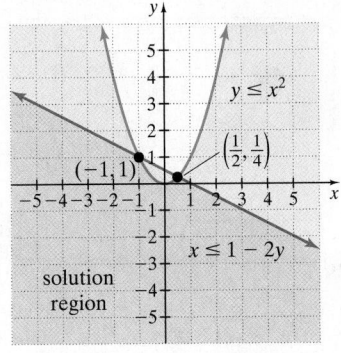

PRACTICE
3   Graph the system: $\begin{cases} y \ge x^2 \\ y \le -3x + 2 \end{cases}$

EXAMPLE 4   Graph the system.

$$\begin{cases} x^2 + y^2 < 25 \\ \dfrac{x^2}{9} - \dfrac{y^2}{25} < 1 \\ y < x + 3 \end{cases}$$

Solution We graph each inequality. The graph of $x^2 + y^2 < 25$ contains points "inside" the circle that has center $(0, 0)$ and radius 5. The graph of $\dfrac{x^2}{9} - \dfrac{y^2}{25} < 1$ is the region between the two branches of the hyperbola with x-intercepts -3 and 3 and center $(0, 0)$. The graph of $y < x + 3$ is the region "below" the line with slope 1 and y-intercept $(0, 3)$. The graph of the solution set of the system is the intersection of all the graphs, the darkest shaded region shown. The boundary of this region is not part of the solution.

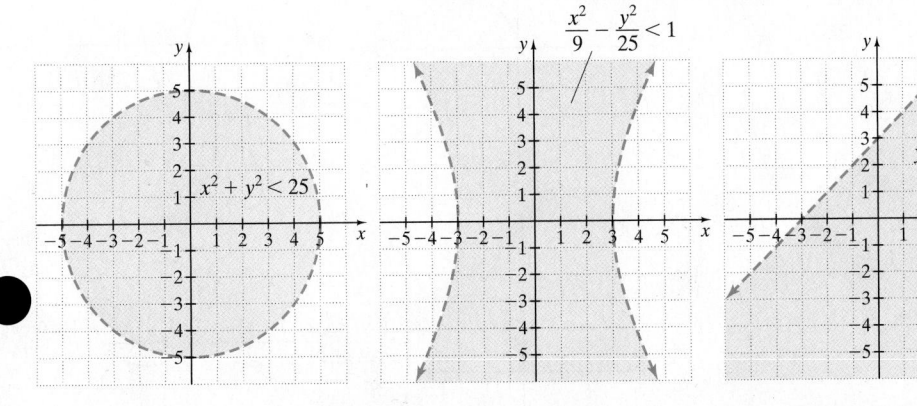 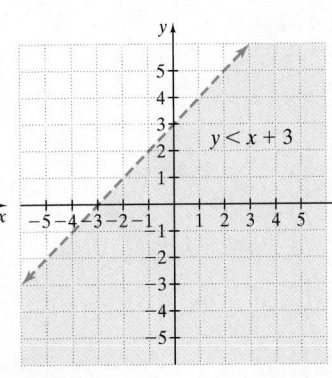 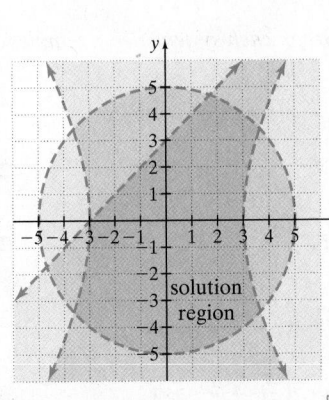

(Continued on next page)

PRACTICE
4 Graph the system: $\begin{cases} x^2 + y^2 < 16 \\ \dfrac{x^2}{4} - \dfrac{y^2}{9} < 1 \\ y < x + 3 \end{cases}$

Vocabulary, Readiness & Video Check

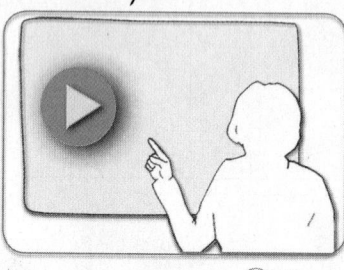

Martin-Gay Interactive Videos

See Video 13.4

Watch the section lecture video and answer the following questions.

OBJECTIVE **1** 1. From Example 1, explain the similarities between graphing linear inequalities and graphing nonlinear inequalities.

OBJECTIVE **2** 2. From Example 2, describe one possible illustration of graphs of two circle inequalities in which the system has no solution—that is, the graph of the inequalities in the system do not overlap.

13.4 Exercise Set MyMathLab®

Graph each inequality. See Examples 1 and 2.

1. $y < x^2$

2. $y < -x^2$

3. $x^2 + y^2 \geq 16$

4. $x^2 + y^2 < 36$

5. $\dfrac{x^2}{4} - y^2 < 1$

6. $x^2 - \dfrac{y^2}{9} \geq 1$

7. $y > (x-1)^2 - 3$

8. $y > (x+3)^2 + 2$

9. $x^2 + y^2 \leq 9$

10. $x^2 + y^2 > 4$

11. $y > -x^2 + 5$

12. $y < -x^2 + 5$

13. $\dfrac{x^2}{4} + \dfrac{y^2}{9} \leq 1$

14. $\dfrac{x^2}{25} + \dfrac{y^2}{4} \geq 1$

15. $\dfrac{y^2}{4} - x^2 \leq 1$

16. $\dfrac{y^2}{16} - \dfrac{x^2}{9} > 1$

17. $y < (x-2)^2 + 1$

18. $y > (x-2)^2 + 1$

19. $y \leq x^2 + x - 2$

20. $y > x^2 + x - 2$

Graph each system. See Examples 3 and 4.

21. $\begin{cases} 4x + 3y \geq 12 \\ x^2 + y^2 < 16 \end{cases}$

22. $\begin{cases} 3x - 4y \leq 12 \\ x^2 + y^2 < 16 \end{cases}$

23. $\begin{cases} x^2 + y^2 \leq 9 \\ x^2 + y^2 \geq 1 \end{cases}$

24. $\begin{cases} x^2 + y^2 \geq 9 \\ x^2 + y^2 \geq 16 \end{cases}$

25. $\begin{cases} y > x^2 \\ y \geq 2x + 1 \end{cases}$

26. $\begin{cases} y \leq -x^2 + 3 \\ y \leq 2x - 1 \end{cases}$

27. $\begin{cases} x^2 + y^2 > 9 \\ y > x^2 \end{cases}$

28. $\begin{cases} x^2 + y^2 \leq 9 \\ y < x^2 \end{cases}$

29. $\begin{cases} \dfrac{x^2}{4} + \dfrac{y^2}{9} \geq 1 \\ x^2 + y^2 \geq 4 \end{cases}$

30. $\begin{cases} x^2 + (y-2)^2 \geq 9 \\ \dfrac{x^2}{4} + \dfrac{y^2}{25} < 1 \end{cases}$

31. $\begin{cases} x^2 - y^2 \geq 1 \\ y \geq 0 \end{cases}$

32. $\begin{cases} x^2 - y^2 \geq 1 \\ x \geq 0 \end{cases}$

33. $\begin{cases} x + y \geq 1 \\ 2x + 3y < 1 \\ x > -3 \end{cases}$

34. $\begin{cases} x - y < -1 \\ 4x - 3y > 0 \\ y > 0 \end{cases}$

35. $\begin{cases} x^2 - y^2 < 1 \\ \dfrac{x^2}{16} + y^2 \leq 1 \\ x \geq -2 \end{cases}$

36. $\begin{cases} x^2 - y^2 \geq 1 \\ \dfrac{x^2}{16} + \dfrac{y^2}{4} \leq 1 \\ y \geq 1 \end{cases}$

REVIEW AND PREVIEW

Determine whether each graph is the graph of a function. See Section 3.6.

37.

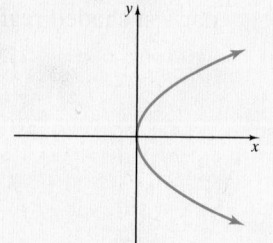

38.

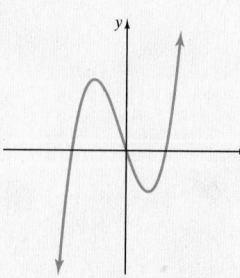

39.

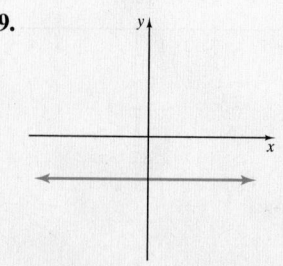

40.

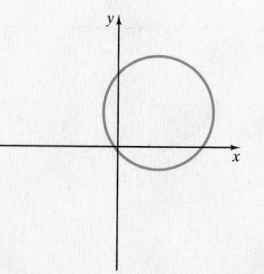

Find each function value if $f(x) = 3x^2 - 2$. See Section 3.2.

41. $f(-1)$

42. $f(-3)$

43. $f(a)$

44. $f(b)$

CONCEPT EXTENSIONS

45. Discuss how graphing a linear inequality such as $x + y < 9$ is similar to graphing a nonlinear inequality such as $x^2 + y^2 < 9$.

46. Discuss how graphing a linear inequality such as $x + y < 9$ is different from graphing a nonlinear inequality such as $x^2 + y^2 < 9$.

47. Graph the system: $\begin{cases} y \le x^2 \\ y \ge x + 2 \\ x \ge 0 \\ y \ge 0 \end{cases}$

48. Graph the system: $\begin{cases} x \ge 0 \\ y \ge 0 \\ y \ge x^2 + 1 \\ y \le 4 - x \end{cases}$

Chapter 13 Vocabulary Check

Fill in each blank with one of the words or phrases listed below.

circle	ellipse	hyperbola
conic sections	vertex	diameter
center	radius	nonlinear system of equations

1. A(n) _____ is the set of all points in a plane that are the same distance from a fixed point, called the _____ .

2. A(n) _____ is a system of equations at least one of which is not linear.

3. A(n) _____ is the set of points in a plane such that the sum of the distances of those points from two fixed points is a constant.

4. In a circle, the distance from the center to a point of the circle is called its _____ .

5. A(n) _____ is the set of points in a plane such that the absolute value of the difference of the distances from two fixed points is constant.

6. The circle, parabola, ellipse, and hyperbola are called the _____ .

7. For a parabola that opens upward, the lowest point is the _____ .

8. Twice a circle's radius is its _____ .

Chapter 13 Highlights

DEFINITIONS AND CONCEPTS	EXAMPLES

Section 13.1 The Parabola and the Circle

Parabolas

$$y = a(x - h)^2 + k$$

Graph: $x = 3y^2 - 12y + 13$

$$x = 3y^2 - 12y + 13$$
$$x - 13 = 3y^2 - 12y$$
$$x - 13 + 3(4) = 3(y^2 - 4y + 4) \quad \text{Add } 3(4) \text{ to both sides.}$$
$$x = 3(y - 2)^2 + 1$$

Since $a = 3$, this parabola opens to the right with vertex $(1, 2)$. Its axis of symmetry is $y = 2$. The x-intercept is $(13, 0)$.

(continued)

DEFINITIONS AND CONCEPTS	EXAMPLES

Section 13.1   The Parabola and the Circle (continued)

$$x = a(y - k)^2 + h$$

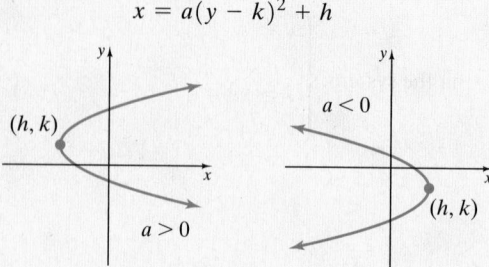

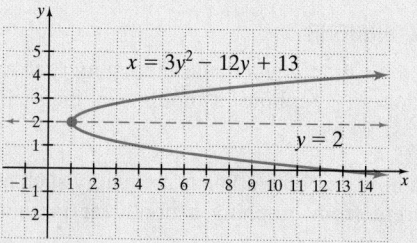

$x = 3y^2 - 12y + 13$

$y = 2$

Circle

The graph of $(x - h)^2 + (y - k)^2 = r^2$ is a circle with center (h, k) and radius r.

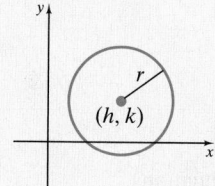

Graph: $x^2 + (y + 3)^2 = 5$

This equation can be written as

$$(x - 0)^2 + (y + 3)^2 = 5 \text{ with } h = 0,$$
$$k = -3, \text{ and } r = \sqrt{5}.$$

The center of this circle is $(0, -3)$, and the radius is $\sqrt{5}$.

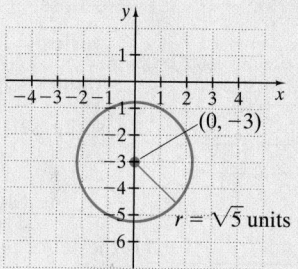

$(0, -3)$

$r = \sqrt{5}$ units

Section 13.2   The Ellipse and the Hyperbola

Ellipse with Center (0, 0)

The graph of an equation of the form $\dfrac{x^2}{a^2} + \dfrac{y^2}{b^2} = 1$ is an ellipse with center $(0, 0)$. The x-intercepts are $(a, 0)$ and $(-a, 0)$, and the y-intercepts are $(0, b)$ and $(0, -b)$.

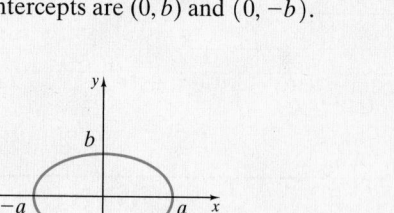

Graph: $4x^2 + 9y^2 = 36$

$$\frac{x^2}{9} + \frac{y^2}{4} = 1 \quad \text{Divide by 36.}$$

$$\frac{x^2}{3^2} + \frac{y^2}{2^2} = 1$$

The ellipse has center $(0, 0)$, x-intercepts $(3, 0)$ and $(-3, 0)$, and y-intercepts $(0, 2)$ and $(0, -2)$.

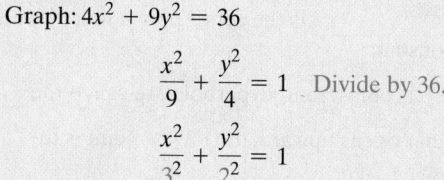

Hyperbola with Center (0, 0)

The graph of an equation of the form $\dfrac{x^2}{a^2} - \dfrac{y^2}{b^2} = 1$ is a hyperbola with center $(0, 0)$ and x-intercepts $(a, 0)$ and $(-a, 0)$.

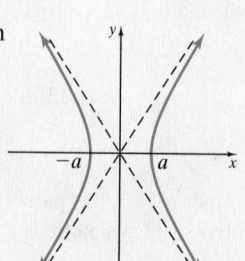

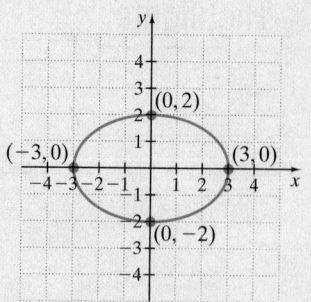

$(0, 2)$

$(-3, 0)$

$(3, 0)$

$(0, -2)$

DEFINITIONS AND CONCEPTS	EXAMPLES

Section 13.2   The Ellipse and the Hyperbola (continued)

The graph of an equation of the form $\dfrac{y^2}{b^2} - \dfrac{x^2}{a^2} = 1$ is a hyperbola with center $(0,0)$ and y-intercepts $(0,b)$ and $(0,-b)$.

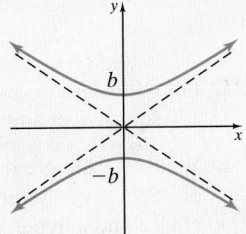

Graph $\dfrac{x^2}{9} - \dfrac{y^2}{4} = 1$. Here $a = 3$ and $b = 2$.

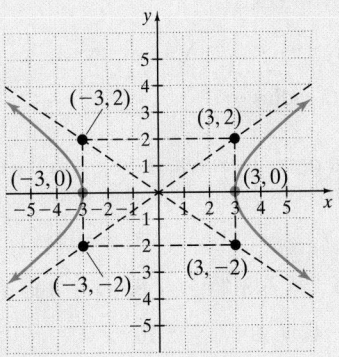

Section 13.3 Solving Nonlinear Systems of Equations

A **nonlinear system of equations** is a system of equations at least one of which is not linear. Both the substitution method and the elimination method may be used to solve a nonlinear system of equations.

Solve the nonlinear system: $\begin{cases} y = x + 2 \\ 2x^2 + y^2 = 3 \end{cases}$

Substitute $x + 2$ for y in the second equation.

$$2x^2 + y^2 = 3$$
$$2x^2 + (x + 2)^2 = 3$$
$$2x^2 + x^2 + 4x + 4 = 3$$
$$3x^2 + 4x + 1 = 0$$
$$(3x + 1)(x + 1) = 0$$
$$x = -\frac{1}{3}, x = -1$$

If $x = -\dfrac{1}{3}, y = x + 2 = -\dfrac{1}{3} + 2 = \dfrac{5}{3}$.

If $x = -1, y = x + 2 = -1 + 2 = 1$.

The solutions are $\left(-\dfrac{1}{3}, \dfrac{5}{3}\right)$ and $(-1, 1)$.

Section 13.4 Nonlinear Inequalities and Systems of Inequalities

The graph of a system of inequalities is the intersection of the graphs of the inequalities.

Graph the system: $\begin{cases} x \geq y^2 \\ x + y \leq 4 \end{cases}$

The graph of the system is the purple-shaded region along with its boundary lines.

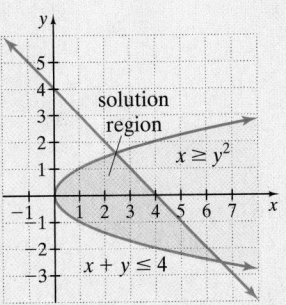

Chapter 13 Review

(13.1) *Write an equation of the circle with the given center and radius.*

1. center $(-4, 4)$, radius 3

2. center $(5, 0)$, radius 5

3. center $(-7, -9)$, radius $\sqrt{11}$

4. center $(0, 0)$, radius $\dfrac{7}{2}$

Sketch the graph of the equation. If the graph is a circle, find its center. If the graph is a parabola, find its vertex.

5. $x^2 + y^2 = 7$

6. $x = 2(y - 5)^2 + 4$

7. $x = -(y + 2)^2 + 3$

8. $(x - 1)^2 + (y - 2)^2 = 4$

9. $y = -x^2 + 4x + 10$

10. $x = -y^2 - 4y + 6$

11. $x = \dfrac{1}{2}y^2 + 2y + 1$

12. $y = -3x^2 + \dfrac{1}{2}x + 4$

13. $x^2 + y^2 + 2x + y = \dfrac{3}{4}$

14. $x^2 + y^2 - 3y = \dfrac{7}{4}$

15. $4x^2 + 4y^2 + 16x + 8y = 1$

16. $3x^2 + 3y^2 + 18x - 12y = -12$

(13.1, 13.2) *Graph each equation.*

17. $x^2 - \dfrac{y^2}{4} = 1$

18. $x^2 + \dfrac{y^2}{4} = 1$

19. $4y^2 + 9x^2 = 36$

20. $-5x^2 + 25y^2 = 125$

21. $x^2 - y^2 = 1$

22. $\dfrac{(x + 3)^2}{9} + \dfrac{(y - 4)^2}{25} = 1$

23. $y = x^2 + 9$

24. $36y^2 - 49x^2 = 1764$

25. $x = 4y^2 - 16$

26. $y = x^2 + 4x + 6$

27. $y^2 + 2(x - 1)^2 = 8$

28. $x - 4y = y^2$

29. $x^2 - 4 = y^2$

30. $x^2 = 4 - y^2$

31. $36y^2 = 576 + 16x^2$

32. $3(x - 7)^2 + 3(y + 4)^2 = 1$

(13.3) *Solve each system of equations.*

33. $\begin{cases} y = 2x - 4 \\ y^2 = 4x \end{cases}$

34. $\begin{cases} x^2 + y^2 = 4 \\ x - y = 4 \end{cases}$

35. $\begin{cases} y = x + 2 \\ y = x^2 \end{cases}$

36. $\begin{cases} 4x - y^2 = 0 \\ 2x^2 + y^2 = 16 \end{cases}$

37. $\begin{cases} x^2 + 4y^2 = 16 \\ x^2 + y^2 = 4 \end{cases}$

38. $\begin{cases} x^2 + 2y = 9 \\ 5x - 2y = 5 \end{cases}$

39. $\begin{cases} y = 3x^2 + 5x - 4 \\ y = 3x^2 - x + 2 \end{cases}$

40. $\begin{cases} x^2 - 3y^2 = 1 \\ 4x^2 + 5y^2 = 21 \end{cases}$

△ 41. Find the length and the width of a room whose area is 150 square feet and whose perimeter is 50 feet.

42. What is the greatest number of real number solutions possible for a system of two equations whose graphs are an ellipse and a hyperbola?

(13.4) *Graph each inequality or system of inequalities.*

43. $y \le -x^2 + 3$

44. $x \le y^2 - 1$

45. $x^2 + y^2 < 9$

46. $\dfrac{x^2}{4} + \dfrac{y^2}{9} \ge 1$

47. $\begin{cases} 3x + 4y \le 12 \\ x - 2y > 6 \end{cases}$

48. $\begin{cases} x^2 + y^2 \le 16 \\ x^2 + y^2 \ge 4 \end{cases}$

49. $\begin{cases} x^2 + y^2 < 4 \\ x^2 - y^2 \le 1 \end{cases}$

50. $\begin{cases} x^2 + y^2 < 4 \\ y \ge x^2 - 1 \\ x \ge 0 \end{cases}$

MIXED REVIEW

51. Write an equation of the circle with center $(-7, 8)$ and radius 5.

Graph each equation.

52. $y = x^2 + 6x + 9$

53. $x = y^2 + 6y + 9$

54. $\dfrac{y^2}{4} - \dfrac{x^2}{16} = 1$

55. $\dfrac{y^2}{4} + \dfrac{x^2}{16} = 1$

56. $\dfrac{(x - 2)^2}{4} + (y - 1)^2 = 1$

57. $y^2 = x^2 + 6$

58. $y^2 + (x - 2)^2 = 10$

59. $3x^2 + 6x + 3y^2 = 9$

60. $x^2 + y^2 - 8y = 0$

61. $6(x - 2)^2 + 9(y + 5)^2 = 36$

62. $\dfrac{x^2}{16} - \dfrac{y^2}{25} = 1$

Solve each system of equations.

63. $\begin{cases} y = x^2 - 5x + 1 \\ y = -x + 6 \end{cases}$

64. $\begin{cases} x^2 + y^2 = 10 \\ 9x^2 + y^2 = 18 \end{cases}$

Graph each inequality or system of inequalities.

65. $x^2 - y^2 < 1$

66. $\begin{cases} y > x^2 \\ x + y \ge 3 \end{cases}$

lc

Chapter 13 Getting Ready for the Test

MATCHING *Match each equation in the first column with the shape of its graph in the second column. Letters may be used more than once or not at all.*

1. $6x^2 + 3y^2 = 24$
2. $6x + 3y = 24$
3. $6x^2 - 3y^2 = 24$
4. $x = 5(y - 2)^2 + 3$
5. $x^2 + y^2 - 4y = 10$
6. $y = x^2 + 2x$

A. parabola
B. circle
C. ellipse
D. hyperbola
E. none of these

MULTIPLE CHOICE *Select the correct choice.*

7. The graph of $(x + 2)^2 + y^2 = 22$ is a circle. Identify its center and radius.
   A. $(2,0); r = 11$ **B.** $(-2,0); r = 11$ **C.** $(2,0); r = \sqrt{22}$ **D.** $(-2,0); r = \sqrt{22}$

8. The graph of $x = 3(y + 2)^2 + 1$ is a parabola. Identify its vertex and direction of opening.
   A. $(-2,1);$ right   **B.** $(1, -2);$ right   **C.** $(2, -1);$ up   **D.** $(1, -2);$ up

9. The graph of $\dfrac{x^2}{100} + \dfrac{y^2}{64} = 1$ is an ellipse. Determine the distance between the y-intercepts.
   A. 10 units   **B.** 20 units   **C.** 8 units   **D.** 16 units

10. The graph of $x^2 - 4y^2 = 36$ is a hyperbola. Determine the distance between the x-intercepts.
    A. 2 units   **B.** 4 units   **C.** 12 units   **D.** 6 units

11. Without graphing, determine the number of points of intersection for the graphs of $x^2 + y^2 = 1$ and $x^2 + y^2 = 9$.
    A. 0   **B.** 1   **C.** 2   **D.** 4

MATCHING *Match the figures in the first column with the letter indicating the maximum number of points of intersection for their graphs. Letters may be used more than once or not at all.*

12. a line and a circle
13. a line and a parabola
14. a circle and an ellipse
15. an ellipse and a hyperbola
16. an ellipse and an ellipse
17. two distinct lines

A. 0
B. 1
C. 2
D. 3
E. 4

Chapter 13 Test MyMathLab® YouTube™

Sketch the graph of each equation.

1. $x^2 + y^2 = 36$
2. $x^2 - y^2 = 36$
3. $16x^2 + 9y^2 = 144$
4. $y = x^2 - 8x + 16$
5. $x^2 + y^2 + 6x = 16$
6. $x = y^2 + 8y - 3$
7. $\dfrac{(x - 4)^2}{16} + \dfrac{(y - 3)^2}{9} = 1$
8. $y^2 - x^2 = 1$

Solve each system.

9. $\begin{cases} x^2 + y^2 = 169 \\ 5x + 12y = 0 \end{cases}$

10. $\begin{cases} x^2 + y^2 = 26 \\ x^2 - 2y^2 = 23 \end{cases}$

11. $\begin{cases} y = x^2 - 5x + 6 \\ y = 2x \end{cases}$

12. $\begin{cases} x^2 + 4y^2 = 5 \\ y = x \end{cases}$

Graph each system.

13. $\begin{cases} 2x + 5y \geq 10 \\ y \geq x^2 + 1 \end{cases}$

14. $\begin{cases} \dfrac{x^2}{4} + y^2 \leq 1 \\ x + y > 1 \end{cases}$

15. $\begin{cases} x^2 + y^2 > 1 \\ \dfrac{x^2}{4} - y^2 \geq 1 \end{cases}$

16. $\begin{cases} x^2 + y^2 \geq 4 \\ x^2 + y^2 < 16 \\ y \geq 0 \end{cases}$

17. Which graph below best resembles the graph of $x = a(y - k)^2 + h$ if $a > 0, h < 0$, and $k > 0$?

A.

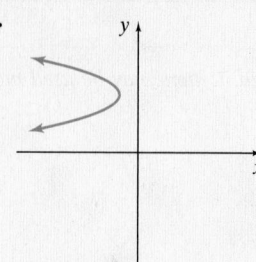

B.

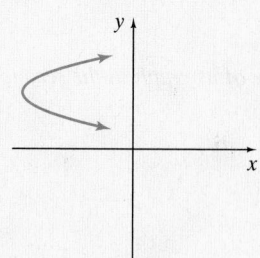

C.

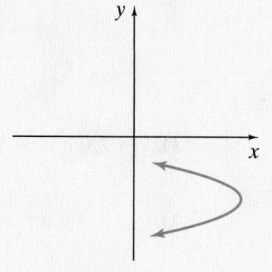

D.

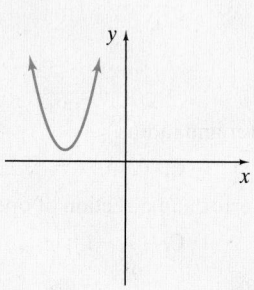

18. A bridge has an arch in the shape of half an ellipse. If the equation of the ellipse, measured in feet, is $100x^2 + 225y^2 = 22{,}500$, find the height of the arch from the road and the width of the arch.

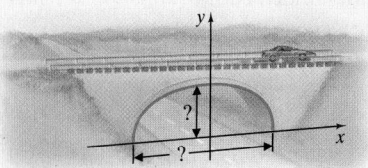

Chapter 13 Cumulative Review

1. Solve $2x \geq 0$ *and* $4x - 1 \leq -9$.

2. Solve $3x + 4 > 1$ *and* $2x - 5 \leq 9$. Write the solution in interval notation.

3. Solve $5x - 3 \leq 10$ *or* $x + 1 \geq 5$.

4. Find the slope of the line that goes through $(3, 2)$ and $(1, -4)$.

Solve.

5. $|5w + 3| = 7$

6. Two planes leave Greensboro, one traveling north and the other south. After 2 hours they are 650 miles apart. If one plane is flying 25 mph faster than the other, what is the speed of each?

7. $\left| \dfrac{x}{2} - 1 \right| = 11$

8. Use the quotient rule to simplify.

a. $\dfrac{4^8}{4^3}$

b. $\dfrac{y^{11}}{y^5}$

c. $\dfrac{32x^7}{4x^6}$

d. $\dfrac{18a^{12}b^6}{12a^8b^6}$

9. Solve: $|3x + 2| = |5x - 8|$

10. Factor.

a. $3y^2 + 14y + 15$

b. $20a^5 + 54a^4 + 10a^3$

c. $(y - 3)^2 - 2(y - 3) - 8$

11. Solve for m: $|m - 6| < 2$

12. Perform the indicated operation and simplify if possible.
$$\dfrac{2}{3a - 15} - \dfrac{a}{25 - a^2}$$

13. Simplify: $\dfrac{x^{-1} + 2xy^{-1}}{x^{-2} - x^{-2}y^{-1}}$

14. Simplify each complex fraction.

a. $(a^{-1} - b^{-1})^{-1}$

b. $\dfrac{2 - \dfrac{1}{x}}{4x - \dfrac{1}{x}}$

15. Solve: $|2x + 9| + 5 > 3$

16. Solve: $\dfrac{2}{x + 3} = \dfrac{1}{x^2 - 9} - \dfrac{1}{x - 3}$

17. Use the remainder theorem and synthetic division to find $P(4)$ if
$$P(x) = 4x^6 - 25x^5 + 35x^4 + 17x^2.$$

18. Suppose that y varies inversely as x. If $y = 3$ when $x = \dfrac{2}{3}$, find the constant of variation and the inverse variation equation.

19. Find the cube roots.

 a. $\sqrt[3]{1}$ **b.** $\sqrt[3]{-64}$ **c.** $\sqrt[3]{\dfrac{8}{125}}$

 d. $\sqrt[3]{x^6}$ **e.** $\sqrt[3]{-27x^{15}}$

20. Multiply and simplify if possible.

 a. $\sqrt{5}\left(2 + \sqrt{15}\right)$

 b. $\left(\sqrt{3} - \sqrt{5}\right)\left(\sqrt{7} - 1\right)$

 c. $\left(2\sqrt{5} - 1\right)^2$

 d. $\left(3\sqrt{2} + 5\right)\left(3\sqrt{2} - 5\right)$

21. Multiply.

 a. $z^{2/3}\left(z^{1/3} - z^5\right)$

 b. $\left(x^{1/3} - 5\right)\left(x^{1/3} + 2\right)$

22. Rationalize the denominator: $\dfrac{-2}{\sqrt{3} + 3}$.

23. Use the quotient rule to divide, and simplify if possible.

 a. $\dfrac{\sqrt{20}}{\sqrt{5}}$ **b.** $\dfrac{\sqrt{50x}}{2\sqrt{2}}$

 c. $\dfrac{7\sqrt[3]{48x^4y^8}}{\sqrt[3]{6y^2}}$ **d.** $\dfrac{2\sqrt[4]{32a^8 b^6}}{\sqrt[4]{a^{-1} b^2}}$

24. Solve: $\sqrt{2x - 3} = x - 3$

25. Add or subtract as indicated.

 a. $\dfrac{\sqrt{45}}{4} - \dfrac{\sqrt{5}}{3}$ **b.** $\sqrt[3]{\dfrac{7x}{8}} + 2\sqrt[3]{7x}$

26. Use the discriminant to determine the number and type of solutions for $9x^2 - 6x = -4$.

27. Rationalize the denominator: $\sqrt{\dfrac{7x}{3y}}$.

28. Solve: $\dfrac{4}{x - 2} - \dfrac{x}{x + 2} = \dfrac{16}{x^2 - 4}$

29. Solve: $\sqrt{2x - 3} = 9$

30. Solve: $x^3 + 2x^2 - 4x \geq 8$

31. Find the following powers of i.

 a. i^7 **b.** i^{20} **c.** i^{46} **d.** i^{-12}

32. Graph: $f(x) = (x + 2)^2 - 1$

33. Solve $p^2 + 2p = 4$ by completing the square.

34. Find the maximum value of $f(x) = -x^2 - 6x + 4$.

35. Solve: $\dfrac{1}{4}m^2 - m + \dfrac{1}{2} = 0$

36. Find the inverse of $f(x) = \dfrac{x + 1}{2}$.

37. Solve: $p^4 - 3p^2 - 4 = 0$

38. Use the quotient rule to simplify.

 a. $\dfrac{\sqrt{32}}{\sqrt{4}}$ **b.** $\dfrac{\sqrt[3]{240y^2}}{5\sqrt[3]{3y^{-4}}}$

 c. $\dfrac{\sqrt[5]{64x^9y^2}}{\sqrt[5]{2x^2y^{-8}}}$

39. Solve: $\dfrac{x + 2}{x - 3} \leq 0$

40. Graph: $4x^2 + 9y^2 = 36$

41. Graph $g(x) = \dfrac{1}{2}(x + 2)^2 + 5$. Find the vertex and the axis of symmetry.

42. Solve each equation for x.

 a. $64^x = 4$ **b.** $125^{x-3} = 25$

 c. $\dfrac{1}{81} = 3^{2x}$

43. Find the vertex of the graph of $f(x) = x^2 - 4x - 12$.

44. Graph the system: $\begin{cases} x + 2y < 8 \\ y \geq x^2 \end{cases}$

45. Find the distance between $(2, -5)$ and $(1, -4)$. Give the exact distance and a three-decimal-place approximation.

46. Solve the system: $\begin{cases} x^2 + y^2 = 36 \\ y = x + 6 \end{cases}$

Sequences, Series, and the Binomial Theorem

14.1 Sequences

14.2 Arithmetic and Geometric Sequences

14.3 Series

Integrated Review—
Sequences and Series

14.4 Partial Sums of Arithmetic and Geometric Sequences

14.5 The Binomial Theorem

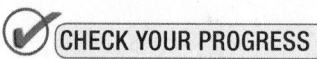

CHECK YOUR PROGRESS

Vocabulary Check
Chapter Highlights
Chapter Review
Getting Ready for the Test
Chapter Test
Cumulative Review

Having explored in some depth the concept of function, we turn now in this final chapter to *sequences*. In one sense, a sequence is simply an ordered list of numbers. In another sense, a sequence is itself a function. Phenomena modeled by such functions are everywhere around us. The starting place for all mathematics is the sequence of natural numbers: 1, 2, 3, 4, and so on.

Sequences lead us to *series*, which are a sum of ordered numbers. Through series, we gain new insight, for example about the expansion of a binomial $(a + b)^n$, the concluding topic of this book.

The Branches of a Tree

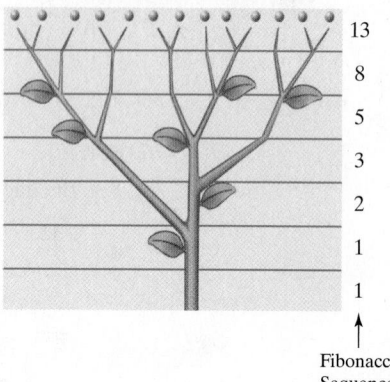

Fibonacci Sequence

The Spirals of a Pineapple
(or a Pine Cone)

Number in each spiral: 5–8–13 or 8–13–21

Ancestry of a Male Bee

Male Bee–comes from female bee only (unfertilized)
Female Bee–comes from female and male bees (fertilized)

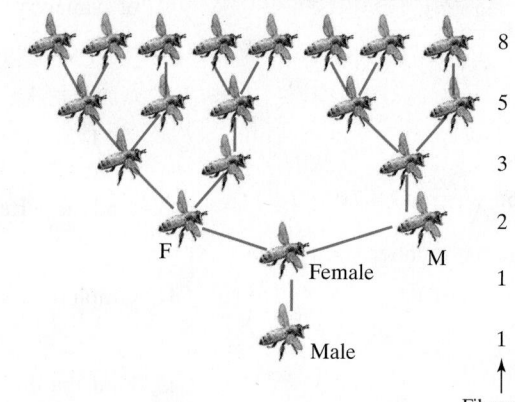

Fibonacci Sequence

Tiling

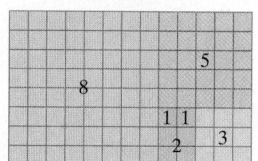

A tiling with squares whose sides are successive Fibonacci numbers in length

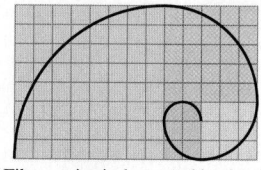

A Fibonacci spiral, created by drawing arcs connecting opposite corners of squares in the Fibonacci tiling

The Fibonacci Sequence and Pascal's Triangle—Is There a Relationship?

The Fibonacci sequence is a special sequence in which the first two terms are 1 and each term thereafter is the sum of the two previous terms:

$$1, 1, 2, 3, 5, 8, 13, 21, \ldots$$

The Fibonacci numbers are named after Leonardo of Pisa, known as Fibonacci, although there is some evidence that these numbers had been described earlier in India.

Study the diagrams above to discover a few interesting facts about this sequence. In Section 14.1, Exercise 46, we have the opportunity to check a formula for this sequence.

In Section 14.5, we introduce Pascal's triangle. There is a relationship between the Fibonacci sequence and Pascal's triangle. To discover this relationship, find the sums of the diagonals of Pascal's triangle.

 **Sequences**

OBJECTIVES

1 Write the Terms of a Sequence Given Its General Term.

2 Find the General Term of a Sequence.

3 Solve Applications That Involve Sequences.

Suppose that a town's present population of 100,000 is growing by 5% each year. After the first year, the town's population will be

$$100,000 + 0.05(100,000) = 105,000$$

After the second year, the town's population will be

$$105,000 + 0.05(105,000) = 110,250$$

After the third year, the town's population will be

$$110,250 + 0.05(110,250) \approx 115,763$$

If we continue to calculate, the town's yearly population can be written as the **infinite sequence** of numbers

$$105,000, 110,250, 115,763, \ldots$$

If we decide to stop calculating after a certain year (say, the fourth year), we obtain the **finite sequence**

$$105,000, \ 110,250, \ 115,763, \ 121,551$$

> **Sequences**
>
> An infinite sequence is a function whose domain is the set of natural numbers $\{1, 2, 3, 4, \ldots\}$.
>
> A finite sequence is a function whose domain is the set of natural numbers $\{1, 2, 3, 4, \ldots, n\}$, where n is some natural number.

OBJECTIVE

1 Writing the Terms of a Sequence

Given the sequence 2, 4, 8, 16, . . . , we say that each number is a **term** of the sequence. Because a sequence is a function, we could describe it by writing $f(n) = 2^n$, where n is a natural number. Instead, we use the notation

$$a_n = 2^n$$

> **Helpful Hint**
>
> When we use the sequence notation, a_n, remember that n is a natural number.

Some function values are

$$a_1 = 2^1 = 2 \qquad \text{First term of the sequence}$$
$$a_2 = 2^2 = 4 \qquad \text{Second term}$$
$$a_3 = 2^3 = 8 \qquad \text{Third term}$$
$$a_4 = 2^4 = 16 \qquad \text{Fourth term}$$
$$a_{10} = 2^{10} = 1024 \qquad \text{Tenth term}$$

The nth term of the sequence a_n is called the **general term.**

> **Helpful Hint**
>
> If it helps, think of a sequence as simply a list of values in which a position is assigned. For the sequence directly above,
>
> Value: 2, 4, 8, 16, . . . , 1024
>
> Position: 1st 2nd 3rd 4th 10th

EXAMPLE 1 Write the first five terms of the sequence whose general term is given by

$$a_n = n^2 - 1$$

Solution Evaluate a_n, where n is 1, 2, 3, 4, and 5.

$$a_n = n^2 - 1$$
$$a_1 = 1^2 - 1 = 0 \qquad \text{Replace } n \text{ with 1}$$

(Continued on next page)

$$a_2 = 2^2 - 1 = 3 \quad \text{Replace } n \text{ with 2.}$$

$$a_3 = 3^2 - 1 = 8 \quad \text{Replace } n \text{ with 3.}$$

$$a_4 = 4^2 - 1 = 15 \quad \text{Replace } n \text{ with 4.}$$

$$a_5 = 5^2 - 1 = 24 \quad \text{Replace } n \text{ with 5.}$$

Thus, the first five terms of the sequence $a_n = n^2 - 1$ are 0, 3, 8, 15, and 24.   □

PRACTICE

1 Write the first five terms of the sequence whose general term is given by $a_n = 5 + n^2$. ∎

EXAMPLE 2 If the general term of a sequence is given by $a_n = \dfrac{(-1)^n}{3n}$, find

a. the first term of the sequence   **b.** a_8

c. the one-hundredth term of the sequence   **d.** a_{15}

Solution

a. $a_1 = \dfrac{(-1)^1}{3(1)} = -\dfrac{1}{3}$ Replace n with 1.

b. $a_8 = \dfrac{(-1)^8}{3(8)} = \dfrac{1}{24}$ Replace n with 8.

c. $a_{100} = \dfrac{(-1)^{100}}{3(100)} = \dfrac{1}{300}$ Replace n with 100.

d. $a_{15} = \dfrac{(-1)^{15}}{3(15)} = -\dfrac{1}{45}$ Replace n with 15.

PRACTICE

2 If the general term of a sequence is given by $a_n = \dfrac{(-1)^n}{5n}$, find

a. the first term of the sequence   **b.** a_4

c. the thirtieth term of the sequence   **d.** a_{19} ∎

OBJECTIVE

2 Finding the General Term of a Sequence ▶

Suppose we know the first few terms of a sequence and want to find a general term that fits the pattern of the first few terms.

EXAMPLE 3 Find a general term a_n of the sequence whose first few terms are given.

a. $1, 4, 9, 16, \ldots$

b. $\dfrac{1}{1}, \dfrac{1}{2}, \dfrac{1}{3}, \dfrac{1}{4}, \dfrac{1}{5}, \ldots$

c. $-3, -6, -9, -12, \ldots$

d. $\dfrac{1}{2}, \dfrac{1}{4}, \dfrac{1}{8}, \dfrac{1}{16}, \ldots$

Solution

a. These numbers are the squares of the first four natural numbers, so a general term might be $a_n = n^2$.

b. These numbers are the reciprocals of the first five natural numbers, so general term might be $a_n = \dfrac{1}{n}$.

c. These numbers are the product of -3 and the first four natural numbers, so a general term might be $a_n = -3n$.

d. Notice that the denominators double each time.

$$\frac{1}{2}, \quad \frac{1}{2\cdot 2}, \quad \frac{1}{2(2\cdot 2)}, \quad \frac{1}{2(2\cdot 2\cdot 2)}$$

or

$$\frac{1}{2^1}, \quad \frac{1}{2^2}, \quad \frac{1}{2^3}, \quad \frac{1}{2^4}$$

We might then suppose that the general term is $a_n = \dfrac{1}{2^n}$.

PRACTICE

3   Find the general term a_n of the sequence whose first few terms are given.

a. $1, 3, 5, 7, \ldots$ **b.** $3, 9, 27, 81, \ldots$

c. $\dfrac{1}{2}, \dfrac{2}{3}, \dfrac{3}{4}, \dfrac{4}{5}, \ldots$ **d.** $-\dfrac{1}{2}, -\dfrac{1}{3}, -\dfrac{1}{4}, -\dfrac{1}{5}, \ldots$

OBJECTIVE

3   Solving Applications Modeled by Sequences

Sequences model many phenomena of the physical world, as illustrated by the following example.

EXAMPLE 4   **Finding a Puppy's Weight Gain**

The amount of weight, in pounds, a puppy gains in each month of its first year is modeled by a sequence whose general term is $a_n = n + 4$, where n is the number of the month. Write the first five terms of the sequence and find how much weight the puppy should gain in its fifth month.

Solution Evaluate $a_n = n + 4$ when n is $1, 2, 3, 4,$ and 5.

$$a_1 = 1 + 4 = 5$$
$$a_2 = 2 + 4 = 6$$
$$a_3 = 3 + 4 = 7$$
$$a_4 = 4 + 4 = 8$$
$$a_5 = 5 + 4 = 9$$

The puppy should gain 9 pounds in its fifth month.

PRACTICE

4   The value v, in dollars, of an office copier depreciates according to the sequence $v_n = 3950(0.8)^n$, where n is the time in years. Find the value of the copier after three years.

✔ Vocabulary, Readiness & Video Check

Use the choices below to fill in each blank.

infinite   finite   general

1. The nth term of the sequence a_n is called the _____ term.

2. A(n) _____ sequence is a function whose domain is $\{1, 2, 3, 4, \ldots, n\}$ where n is some natural number.

3. A(n) _____ sequence is a function whose domain is $\{1, 2, 3, 4, \ldots\}$.

Write the first term of each sequence.

4. $a_n = 7^n; a_1 = $ _____.   **5.** $a_n = \dfrac{(-1)^n}{n}; a_1 = $ _____.   **6.** $a_n = (-1)^n \cdot n^4; a_1 = $ _____.

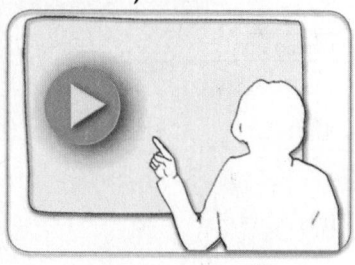

Martin-Gay Interactive Videos

See Video 14.1 🔘

Watch the section lecture video and answer the following questions.

OBJECTIVE 1

7. Based on the lecture before 🎬 Example 1, complete the following statements. A sequence is a_____whose_____is the set of natural numbers. We use_____to mean the general term of a sequence.

OBJECTIVE 2

8. In 🎬 Example 3, why can't the general term be $a_n = (-2)^n$?

OBJECTIVE 3

9. For 🎬 Example 4, write the equation for the specific term and find the allowance amount for day 9 of the vacation.

14.1 Exercise Set   MyMathLab® ▶

Write the first five terms of each sequence, whose general term is given. See Example 1.

1. $a_n = n + 4$

2. $a_n = 5 - n$

3. $a_n = (-1)^n$

4. $a_n = (-2)^n$

▶ **5.** $a_n = \dfrac{1}{n+3}$

6. $a_n = \dfrac{1}{7-n}$

7. $a_n = 2n$

8. $a_n = -6n$

9. $a_n = -n^2$

10. $a_n = n^2 + 2$

11. $a_n = 2^n$

12. $a_n = 3^{n-2}$

13. $a_n = 2n + 5$

14. $a_n = 1 - 3n$

15. $a_n = (-1)^n n^2$

16. $a_n = (-1)^{n+1}(n-1)$

Find the indicated term for each sequence, whose general term is given. See Example 2.

17. $a_n = 3n^2; a_5$

18. $a_n = -n^2; a_{15}$

19. $a_n = 6n - 2; a_{20}$

20. $a_n = 100 - 7n; a_{50}$

21. $a_n = \dfrac{n+3}{n}; a_{15}$

22. $a_n = \dfrac{n}{n+4}; a_{24}$

23. $a_n = (-3)^n; a_6$

24. $a_n = 5^{n+1}; a_3$

25. $a_n = \dfrac{n-2}{n+1}; a_6$

26. $a_n = \dfrac{n+3}{n+4}; a_8$

▶ **27.** $a_n = \dfrac{(-1)^n}{n}; a_8$

28. $a_n = \dfrac{(-1)^n}{2n}; a_{100}$

29. $a_n = -n^2 + 5; a_{10}$

30. $a_n = 8 - n^2; a_{20}$

31. $a_n = \dfrac{(-1)^n}{n+6}; a_{19}$

32. $a_n = \dfrac{n-4}{(-2)^n}; a_6$

Find a general term a_n for each sequence, whose first four terms are given. See Example 3.

33. $3, 7, 11, 15$

34. $2, 7, 12, 17$

▶ **35.** $-2, -4, -8, -16$

36. $-4, 16, -64, 256$

37. $\dfrac{1}{3}, \dfrac{1}{9}, \dfrac{1}{27}, \dfrac{1}{81}$

38. $\dfrac{2}{5}, \dfrac{2}{25}, \dfrac{2}{125}, \dfrac{2}{625}$

Solve. See Example 4.

39. The distance, in feet, that a Thermos dropped from a cliff falls in each consecutive second is modeled by a sequence whose general term is $a_n = 32n - 16$, where n is the number of seconds. Find the distance the Thermos falls in the second, third, and fourth seconds.

40. The population size of a culture of bacteria triples every hour such that its size is modeled by the sequence $a_n = 50(3)^{n-1}$, where n is the number of the hour just beginning. Find the size of the culture at the beginning of the fourth hour and the size of the culture at the beginning of the first hour.

▶ **41.** Mrs. Laser agrees to give her son Mark an allowance of $0.10 on the first day of his 14-day vacation, $0.20 on the second day, $0.40 on the third day, and so on. Write an equation of a sequence whose terms correspond to Mark's allowance. Find the allowance Mark will receive on the last day of his vacation.

42. A small theater has 10 rows with 12 seats in the first row, 15 seats in the second row, 18 seats in the third row, and so on. Write an equation of a sequence whose terms correspond to the seats in each row. Find the number of seats in the eighth row.

43. The number of cases of a new infectious disease is doubling every year such that the number of cases is modeled by a sequence whose general term is $a_n = 75(2)^{n-1}$, where n is the number of the year just beginning. Find how many cases there will be at the beginning of the sixth year. Find how many cases there were at the beginning of the first year.

44. A new college had an initial enrollment of 2700 students in 2000, and each year the enrollment increases by 150 students. Find the enrollment for each of 5 years, beginning with 2000.

45. An endangered species of sparrow had an estimated population of 800 in 2000, and scientists predicted that its population would decrease by half each year. Estimate the population in 2004. Estimate the year the sparrow was extinct.

46. A **Fibonacci sequence** is a special type of sequence in which the first two terms are 1, and each term thereafter is the sum of the two previous terms: 1, 1, 2, 3, 5, 8, etc. The formula for the nth Fibonacci term is $a_n = \dfrac{1}{\sqrt{5}}\left[\left(\dfrac{1 + \sqrt{5}}{2}\right)^n - \left(\dfrac{1 - \sqrt{5}}{2}\right)^n\right]$.
Verify that the first two terms of the Fibonacci sequence are each 1.

REVIEW AND PREVIEW

Sketch the graph of each quadratic function. See Section 11.5.

47. $f(x) = (x - 1)^2 + 3$ **48.** $f(x) = (x - 2)^2 + 1$

49. $f(x) = 2(x + 4)^2 + 2$ **50.** $f(x) = 3(x - 3)^2 + 4$

Find the distance between each pair of points. See Section 10.3.

51. $(-4, -1)$ and $(-7, -3)$

52. $(-2, -1)$ and $(-1, 5)$

53. $(2, -7)$ and $(-3, -3)$

54. $(10, -14)$ and $(5, -11)$

CONCEPT EXTENSIONS

Find the first five terms of each sequence. Round each term after the first to four decimal places.

55. $a_n = \dfrac{1}{\sqrt{n}}$

56. $\dfrac{\sqrt{n}}{\sqrt{n} + 1}$

57. $a_n = \left(1 + \dfrac{1}{n}\right)^n$

58. $a_n = \left(1 + \dfrac{0.05}{n}\right)^n$

14.2 Arithmetic and Geometric Sequences

OBJECTIVES

1 Identify Arithmetic Sequences and Their Common Differences.

2 Identify Geometric Sequences and Their Common Ratios.

OBJECTIVE

1   Identifying Arithmetic Sequences

Find the first four terms of the sequence whose general term is $a_n = 5 + (n - 1)3$.

$$a_1 = 5 + (1 - 1)3 = 5 \qquad \text{Replace } n \text{ with 1.}$$

$$a_2 = 5 + (2 - 1)3 = 8 \qquad \text{Replace } n \text{ with 2.}$$

$$a_3 = 5 + (3 - 1)3 = 11 \qquad \text{Replace } n \text{ with 3.}$$

$$a_4 = 5 + (4 - 1)3 = 14 \qquad \text{Replace } n \text{ with 4.}$$

The first four terms are 5, 8, 11, and 14. Notice that the difference of any two successive terms is 3.

$$8 - 5 = 3$$
$$11 - 8 = 3$$
$$14 - 11 = 3$$
$$\vdots$$
$$a_n - a_{n-1} = 3$$
$$\uparrow \qquad \uparrow$$
$$n\text{th} \qquad \text{previous}$$
$$\text{term} \qquad \text{term}$$

Because the difference of any two successive terms is a constant, we call the sequence an **arithmetic sequence,** or an **arithmetic progression.** The constant difference d in successive terms is called the **common difference.** In this example, d is 3.

> ### Arithmetic Sequence and Common Difference
>
> An **arithmetic sequence** is a sequence in which each term (after the first) differs from the preceding term by a constant amount d. The constant d is called the **common difference** of the sequence.

The sequence 2, 6, 10, 14, 18, . . . is an arithmetic sequence. Its common difference is 4. Given the first term a_1 and the common difference d of an arithmetic sequence, we can find any term of the sequence.

EXAMPLE 1 Write the first five terms of the arithmetic sequence whose first term is 7 and whose common difference is 2.

Solution

$$a_1 = 7$$
$$a_2 = 7 + 2 = 9$$
$$a_3 = 9 + 2 = 11$$
$$a_4 = 11 + 2 = 13$$
$$a_5 = 13 + 2 = 15$$

The first five terms are $7, 9, 11, 13, 15$. ☐

PRACTICE

1 Write the first five terms of the arithmetic sequence whose first term is 4 and whose common difference is 5. ■

Notice the general pattern of the terms in Example 1.

$$a_1 = 7$$
$$a_2 = 7 + 2 = 9 \quad \text{or} \quad a_2 = a_1 + d$$
$$a_3 = 9 + 2 = 11 \quad \text{or} \quad a_3 = a_2 + d = (a_1 + d) + d = a_1 + 2d$$
$$a_4 = 11 + 2 = 13 \quad \text{or} \quad a_4 = a_3 + d = (a_1 + 2d) + d = a_1 + 3d$$
$$a_5 = 13 + 2 = 15 \quad \text{or} \quad a_5 = a_4 + d = (a_1 + 3d) + d = a_1 + 4d$$

(subscript $- 1$) is multiplier

The pattern on the right suggests that the general term a_n of an arithmetic sequence is given by

$$a_n = a_1 + (n - 1)d$$

General Term of an Arithmetic Sequence

The general term a_n of an arithmetic sequence is given by

$$a_n = a_1 + (n - 1)d$$

where a_1 is the first term and d is the common difference.

EXAMPLE 2 Consider the arithmetic sequence whose first term is 3 and whose common difference is -5.

a. Write an expression for the general term a_n.

b. Find the twentieth term of this sequence.

Solution

a. This is an arithmetic sequence whose first term, $a_1 = 3$ and whose common difference, $d = -5$.

$$a_n = a_1 + (n - 1)d \qquad \text{General term of an arithmetic sequence.}$$
$$a_n = 3 + (n - 1)(-5) \qquad \text{Let } a_1 = 3 \text{ and } d = -5.$$
$$= 3 - 5n + 5 \qquad \text{Multiply.}$$
$$= 8 - 5n \qquad \text{Simplify.}$$

The general term of this sequence is $a_n = 8 - 5n$.

b. $\quad a_n = 8 - 5n$
$$a_{20} = 8 - 5 \cdot 20 \qquad \text{Let } n = 20.$$
$$= 8 - 100 = -92$$ ☐

PRACTICE
2   Consider the arithmetic sequence whose first term is 2 and whose common difference is -3.

   a. Write an expression for the general term a_n.

   b. Find the twelfth term of the sequence.   ■

EXAMPLE 3   Find the eleventh term of the arithmetic sequence whose first three terms are 2, 9, and 16.

Solution   Since the sequence is arithmetic, the eleventh term is

$$a_{11} = a_1 + (11 - 1)d = a_1 + 10d$$

We know a_1 is the first term of the sequence, so $a_1 = 2$. Also, d is the constant difference of terms, so $d = a_2 - a_1 = 9 - 2 = 7$. Thus,

$$
\begin{aligned}
a_{11} &= a_1 + 10d \\
&= 2 + 10 \cdot 7 \quad \text{Let } a_1 = 2 \text{ and } d = 7. \\
&= 72
\end{aligned}
$$
□

PRACTICE
3   Find the ninth term of the arithmetic sequence whose first three terms are 3, 9, and 15.   ■

EXAMPLE 4   If the third term of an arithmetic sequence is 12 and the eighth term is 27, find the fifth term.

Solution   We need to find a_1 and d to write the general term, which then enables us to find a_5, the fifth term. The given facts about terms a_3 and a_8 lead to a system of linear equations.

$$
\begin{cases} a_3 = a_1 + (3 - 1)d \\ a_8 = a_1 + (8 - 1)d \end{cases}
\quad \text{or} \quad
\begin{cases} 12 = a_1 + 2d \\ 27 = a_1 + 7d \end{cases}
$$

Next, we solve the system $\begin{cases} 12 = a_1 + 2d \\ 27 = a_1 + 7d \end{cases}$ by elimination. Multiply both sides of the second equation by -1 so that

$$
\begin{cases} 12 = a_1 + 2d \\ -1(27) = -1(a_1 + 7d) \end{cases}
\begin{array}{c} \text{simplifies} \\ \text{to} \end{array}
\begin{cases} 12 = a_1 + 2d \\ -27 = -a_1 - 7d \end{cases}
$$

$$
\begin{aligned}
-15 &= -5d \quad \text{Add the equations.} \\
3 &= d \quad \text{Divide both sides by } -5.
\end{aligned}
$$

To find a_1, let $d = 3$ in $12 = a_1 + 2d$. Then

$$
\begin{aligned}
12 &= a_1 + 2(3) \\
12 &= a_1 + 6 \\
6 &= a_1
\end{aligned}
$$

Thus, $a_1 = 6$ and $d = 3$, so

$$
\begin{aligned}
a_n &= 6 + (n - 1)(3) \\
&= 6 + 3n - 3 \\
&= 3 + 3n
\end{aligned}
$$

and

$$a_5 = 3 + 3 \cdot 5 = 18$$
□

PRACTICE
4   If the third term of an arithmetic sequence is 23 and the eighth term is 63, find the sixth term.   ■

EXAMPLE 5 Finding Salary

Donna Theime has an offer for a job starting at $40,000 per year and guaranteeing her a raise of $1600 per year for the next 5 years. Write the general term for the arithmetic sequence that models Donna's potential annual salaries and find her salary for the fourth year.

Solution The first term, a_1, is $40,000$, and d is 1600. So

$$a_n = a_1 + (n - 1)d \qquad \text{General term of an arithmetic sequence.}$$
$$a_n = 40,000 + (n - 1)(1600)$$
$$= 38,400 + 1600n$$

Thus, $a_4 = 38,400 + 1600 \cdot 4 = 44,800$

Her salary for the fourth year will be $44,800.   □

PRACTICE

5 A starting salary for a consulting company is $57,000 per year with guaranteed annual increases of $2200 for the next 4 years. Write the general term for the arithmetic sequence that models the potential annual salaries and find the salary for the third year.   ■

OBJECTIVE

2 Identifying Geometric Sequences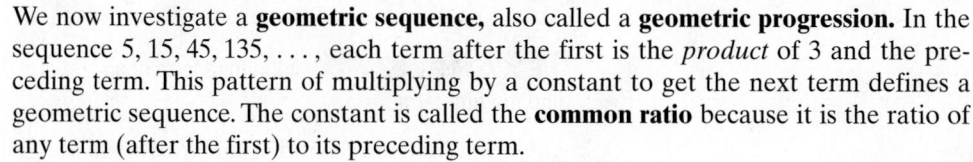

We now investigate a **geometric sequence,** also called a **geometric progression.** In the sequence $5, 15, 45, 135, \ldots$, each term after the first is the *product* of 3 and the preceding term. This pattern of multiplying by a constant to get the next term defines a geometric sequence. The constant is called the **common ratio** because it is the ratio of any term (after the first) to its preceding term.

$$\frac{15}{5} = 3$$
$$\frac{45}{15} = 3$$
$$\frac{135}{45} = 3$$
$$\vdots$$

$$n\text{th term} \longrightarrow \frac{a_n}{a_{n-1}} = 3 \longleftarrow \text{previous term}$$

Geometric Sequence and Common Ratio

A **geometric sequence** is a sequence in which each term (after the first) is obtained by multiplying the preceding term by a constant r. The constant r is called the **common ratio** of the sequence.

The sequence $12, 6, 3, \dfrac{3}{2}, \ldots$ is geometric since each term after the first is the product of the previous term and $\dfrac{1}{2}$.

EXAMPLE 6 Write the first five terms of a geometric sequence whose first term is 7 and whose common ratio is 2.

Solution

$$a_1 = 7$$
$$a_2 = 7(2) = 14$$
$$a_3 = 14(2) = 28$$
$$a_4 = 28(2) = 56$$
$$a_5 = 56(2) = 112$$

The first five terms are $7, 14, 28, 56,$ and 112.   □

PRACTICE

6 Write the first four terms of a geometric sequence whose first term is 8 and whose common ratio is -3 ■

Notice the general pattern of the terms in Example 6.

$$a_1 = 7$$

$$a_2 = 7(2) = 14 \quad \text{or} \quad a_2 = a_1(r)$$

$$a_3 = 14(2) = 28 \quad \text{or} \quad a_3 = a_2(r) = (a_1 \cdot r) \cdot r = a_1 r^2$$

$$a_4 = 28(2) = 56 \quad \text{or} \quad a_4 = a_3(r) = (a_1 \cdot r^2) \cdot r = a_1 r^3$$

$$a_5 = 56(2) = 112 \quad \text{or} \quad a_5 = a_4(r) = (a_1 \cdot r^3) \cdot r = a_1 r^4$$

$$\text{(subscript} - 1) \text{ is power}$$

The pattern on the right above suggests that the general term of a geometric sequence is given by $a_n = a_1 r^{n-1}$.

> **General Term of a Geometric Sequence**
>
> The general term a_n of a geometric sequence is given by
>
> $$a_n = a_1 r^{n-1}$$
>
> where a_1 is the first term and r is the common ratio.

EXAMPLE 7   Find the eighth term of the geometric sequence whose first term is 12 and whose common ratio is $\dfrac{1}{2}$.

Solution Since this is a geometric sequence, the general term a_n is given by

$$a_n = a_1 r^{n-1}$$

Here $a_1 = 12$ and $r = \dfrac{1}{2}$, so $a_n = 12\left(\dfrac{1}{2}\right)^{n-1}$. Evaluate a_n for $n = 8$.

$$a_8 = 12\left(\frac{1}{2}\right)^{8-1} = 12\left(\frac{1}{2}\right)^7 = 12\left(\frac{1}{128}\right) = \frac{3}{32} \qquad \square$$

PRACTICE

7 Find the seventh term of the geometric sequence whose first term is 64 and whose common ratio is $\dfrac{1}{4}$. ■

EXAMPLE 8   Find the fifth term of the geometric sequence whose first three terms are 2, -6, and 18.

Solution Since the sequence is geometric and $a_1 = 2$, the fifth term must be $a_1 r^{5-1}$, or $2r^4$. We know that r is the common ratio of terms, so r must be $\dfrac{-6}{2}$, or -3. Thus,

$$a_5 = 2r^4$$

$$a_5 = 2(-3)^4 = 162 \qquad \square$$

PRACTICE

8 Find the seventh term of the geometric sequence whose first three terms are $-3, 6$, and -12. ■

EXAMPLE 9 If the second term of a geometric sequence is $\frac{5}{4}$ and the third term is $\frac{5}{16}$, find the first term and the common ratio.

Solution To find the common ratio, notice that $\frac{5}{16} \div \frac{5}{4} = \frac{1}{4}$, so $r = \frac{1}{4}$. Then

$$a_2 = a_1\left(\frac{1}{4}\right)^{2-1}$$

$$\frac{5}{4} = a_1\left(\frac{1}{4}\right)^1, \quad \text{or} \quad a_1 = 5 \quad \text{Replace } a_2 \text{ with } \frac{5}{4} \text{ and solve for } a_1.$$

The first term is 5. □

PRACTICE
9 If the second term of a geometric sequence is $\frac{9}{2}$ and the third term is $\frac{27}{4}$, find the first term and the common ratio. ■

EXAMPLE 10 Predicting Population of a Bacterial Culture

The population size of a bacterial culture growing under controlled conditions is doubling each day. Predict how large the culture will be at the beginning of day 7 if it measures 10 units at the beginning of day 1.

Solution Since the culture doubles in size each day, the population sizes are modeled by a geometric sequence. Here $a_1 = 10$ and $r = 2$. Thus,

$$a_n = a_1 r^{n-1} = 10(2)^{n-1} \quad \text{and} \quad a_7 = 10(2)^{7-1} = 640$$

The bacterial culture should measure 640 units at the beginning of day 7. □

PRACTICE
10 After applying a test antibiotic, the population of a bacterial culture is reduced by one-half every day. Predict how large the culture will be at the start of day 7 if it measures 4800 units at the beginning of day 1. ■

✔ Vocabulary, Readiness & Video Check

Use the choices below to fill in each blank. Some choices may be used more than once and some not at all.

| first | arithmetic | difference |
| last | geometric | ratio |

1. A(n) _____ sequence is one in which each term (after the first) is obtained by multiplying the preceding term by a constant r. The constant r is called the common _____.

2. A(n) _____ sequence is one in which each term (after the first) differs from the preceding term by a constant amount d. The constant d is called the common _____.

3. The general term of an arithmetic sequence is $a_n = a_1 + (n - 1)d$ where a_1 is the _____ term and d is the common _____.

4. The general term of a geometric sequence is $a_n = a_1 r^{n-1}$ where a_1 is the _____ term and r is the common _____.

Martin-Gay Interactive Videos

Watch the section lecture video and answer the following questions.

OBJECTIVE
1

5. From the lecture before ▥ Example 1, what makes a sequence an arithmetic sequence?

OBJECTIVE
2

6. From the lecture before ▥ Example 3, what's the difference between an arithmetic and a geometric sequence?

See Video 14.2 ◉

14.2   Exercise Set   MyMathLab® ▸

Write the first five terms of the arithmetic or geometric sequence whose first term, a_1, and common difference, d, or common ratio, r, are given. See Examples 1 and 6.

1. $a_1 = 4; d = 2$

2. $a_1 = 3; d = 10$

▸ **3.** $a_1 = 6; d = -2$

4. $a_1 = -20; d = 3$

5. $a_1 = 1; r = 3$

6. $a_1 = -2; r = 2$

7. $a_1 = 48; r = \dfrac{1}{2}$

8. $a_1 = 1; r = \dfrac{1}{3}$

Find the indicated term of each sequence. See Examples 2 and 7.

9. The eighth term of the arithmetic sequence whose first term is 12 and whose common difference is 3

10. The twelfth term of the arithmetic sequence whose first term is 32 and whose common difference is −4

11. The fourth term of the geometric sequence whose first term is 7 and whose common ratio is −5

12. The fifth term of the geometric sequence whose first term is 3 and whose common ratio is 3

13. The fifteenth term of the arithmetic sequence whose first term is −4 and whose common difference is −4

14. The sixth term of the geometric sequence whose first term is 5 and whose common ratio is −4

Find the indicated term of each sequence. See Examples 3 and 8.

15. The ninth term of the arithmetic sequence 0, 12, 24, . . .

16. The thirteenth term of the arithmetic sequence −3, 0, 3, . . .

▸ **17.** The twenty-fifth term of the arithmetic sequence 20, 18, 16, . . .

18. The ninth term of the geometric sequence 5, 10, 20, . . .

19. The fifth term of the geometric sequence 2, −10, 50, . . .

20. The sixth term of the geometric sequence $\dfrac{1}{2}, \dfrac{3}{2}, \dfrac{9}{2}, \ldots$

Find the indicated term of each sequence. See Examples 4 and 9.

21. The eighth term of the arithmetic sequence whose fourth term is 19 and whose fifteenth term is 52

22. If the second term of an arithmetic sequence is 6 and the tenth term is 30, find the twenty-fifth term.

23. If the second term of an arithmetic progression is −1 and the fourth term is 5, find the ninth term.

24. If the second term of a geometric progression is 15 and the third term is 3, find a_1 and r.

25. If the second term of a geometric progression is $-\dfrac{4}{3}$ and the third term is $\dfrac{8}{3}$, find a_1 and r.

26. If the third term of a geometric sequence is 4 and the fourth term is −12, find a_1 and r.

27. Explain why 14, 10, and 6 may be the first three terms of an arithmetic sequence when it appears we are subtracting instead of adding to get the next term.

28. Explain why 80, 20, and 5 may be the first three terms of a geometric sequence when it appears we are dividing instead of multiplying to get the next term.

MIXED PRACTICE

Given are the first three terms of a sequence that is either arithmetic or geometric. If a sequence is arithmetic, find a_1 and d. If a sequence is geometric, find a_1 and r.

29. 2, 4, 6

30. 8, 16, 24

31. 5, 10, 20

32. 2, 6, 18

33. $\dfrac{1}{2}, \dfrac{1}{10}, \dfrac{1}{50}$

34. $\dfrac{2}{3}, \dfrac{4}{3}, 2$

35. $x, 5x, 25x$

36. $y, -3y, 9y$

37. $p, p + 4, p + 8$

38. $t, t - 1, t - 2$

Find the indicated term of each sequence.

39. The twenty-first term of the arithmetic sequence whose first term is 14 and whose common difference is $\dfrac{1}{4}$

40. The fifth term of the geometric sequence whose first term is 8 and whose common ratio is −3

41. The fourth term of the geometric sequence whose first term is 3 and whose common ratio is $-\dfrac{2}{3}$

42. The fourth term of the arithmetic sequence whose first term is 9 and whose common difference is 5

43. The fifteenth term of the arithmetic sequence $\frac{3}{2}, 2, \frac{5}{2}, \ldots$

44. The eleventh term of the arithmetic sequence $2, \frac{5}{3}, \frac{4}{3}, \ldots$

45. The sixth term of the geometric sequence $24, 8, \frac{8}{3}, \ldots$

46. The eighteenth term of the arithmetic sequence $5, 2, -1, \ldots$

47. If the third term of an arithmetic sequence is 2 and the seventeenth term is -40, find the tenth term.

48. If the third term of a geometric sequence is -28 and the fourth term is -56, find a_1 and r.

Solve. See Examples 5 and 10.

49. An auditorium has 54 seats in the first row, 58 seats in the second row, 62 seats in the third row, and so on. Find the general term of this arithmetic sequence and the number of seats in the twentieth row.

50. A triangular display of cans in a grocery store has 20 cans in the first row, 17 cans in the next row, and so on, in an arithmetic sequence. Find the general term and the number of cans in the fifth row. Find how many rows there are in the display and how many cans are in the top row.

▶ **51.** The initial size of a virus culture is 6 units, and it triples its size every day. Find the general term of the geometric sequence that models the culture's size.

52. A real estate investment broker predicts that a certain property will increase in value 15% each year. Thus, the yearly property values can be modeled by a geometric sequence whose common ratio r is 1.15. If the initial property value was $500,000, write the first four terms of the sequence and predict the value at the end of the third year.

53. A rubber ball is dropped from a height of 486 feet, and it continues to bounce one-third the height from which it last fell. Write out the first five terms of this geometric sequence and find the general term. Find how many bounces it takes for the ball to rebound less than 1 foot.

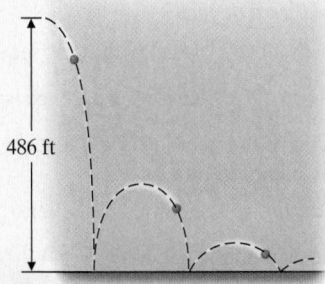

486 ft

54. On the first swing, the length of the arc through which a pendulum swings is 50 inches. The length of each successive swing is 80% of the preceding swing. Determine whether this sequence is arithmetic or geometric. Find the length of the fourth swing.

55. Jose takes a job that offers a monthly starting salary of $4000 and guarantees him a monthly raise of $125 during his first year of training. Find the general term of this arithmetic sequence and his monthly salary at the end of his training.

56. The beginning of an exercise program starts with pedaling 15 minutes on a stationary bike. Each week, this time is increased by 5 minutes. Write the general term of this arithmetic sequence, and find the pedaling time after 7 weeks. Find how many weeks it takes to reach a pedaling time of 1 hour.

57. If a radioactive element has a half-life of 3 hours, then x grams of the element dwindles to $\frac{x}{2}$ grams after 3 hours. If a nuclear reactor has 400 grams of that radioactive element, find the amount of radioactive material after 12 hours.

REVIEW AND PREVIEW

Evaluate. See Section 1.7

58. $5(1) + 5(2) + 5(3) + 5(4)$

59. $\frac{1}{3(1)} + \frac{1}{3(2)} + \frac{1}{3(3)}$

60. $2(2 - 4) + 3(3 - 4) + 4(4 - 4)$

61. $3^0 + 3^1 + 3^2 + 3^3$

62. $\frac{1}{4(1)} + \frac{1}{4(2)} + \frac{1}{4(3)}$

63. $\frac{8 - 1}{8 + 1} + \frac{8 - 2}{8 + 2} + \frac{8 - 3}{8 + 3}$

CONCEPT EXTENSIONS

Write the first four terms of the arithmetic or geometric sequence whose first term, a_1, and common difference, d, or common ratio, r, are given.

64. $a_1 = \$3720, d = -\$268.50$

65. $a_1 = \$11{,}782.40, r = 0.5$

66. $a_1 = 26.8, r = 2.5$

67. $a_1 = 19.652; d = -0.034$

68. Describe a situation in your life that can be modeled by a geometric sequence. Write an equation for the sequence.

69. Describe a situation in your life that can be modeled by an arithmetic sequence. Write an equation for the sequence.

 **14.3** **Series**

OBJECTIVES

1 Identify Finite and Infinite Series and Use Summation Notation.

2 Find Partial Sums.

Month	Money Saved for the Month
1	$100
2	$100 + $10
3	($100 + $10) + $10
4	($100 + $10 + $10) + $10
⋮	⋮

Helpful Hint

Sequence: A listing of terms, a_n, where n is a natural number.
Series: A sum of the terms of a sequence.

OBJECTIVE

1 Identifying Finite and Infinite Series and Using Summation Notation

A person who conscientiously saves money by saving first $100 and then saving $10 more each month than he saved the preceding month (see table in margin) is saving money according to the arithmetic sequence

$$a_n = 100 + 10(n - 1)$$

Following this sequence, he can predict how much money he should save for any particular month. But if he also wants to know how much money *in total* he has saved, say, by the fifth month, he must find the *sum* of the first five terms of the sequence

$$\underbrace{100}_{a_1} + \underbrace{100 + 10}_{a_2} + \underbrace{100 + 20}_{a_3} + \underbrace{100 + 30}_{a_4} + \underbrace{100 + 40}_{a_5}$$

A sum of the terms of a sequence is called a **series** (the plural is also "series"). As our example here suggests, series are frequently used to model financial and natural phenomena.

A series is a **finite series** if it is the sum of a finite number of terms. A series is an **infinite series** if it is the sum of all the terms of an infinite sequence. For example,

Sequence	Series	
$5, 9, 13$	$5 + 9 + 13$	Finite; sum of 3 terms
$5, 9, 13, \ldots$	$5 + 9 + 13 + \cdots$	Infinite
$4, -2, 1, -\frac{1}{2}, \frac{1}{4}$	$4 + (-2) + 1 + \left(-\frac{1}{2}\right) + \left(\frac{1}{4}\right)$	Finite; sum of 5 terms
$4, -2, 1, \ldots$	$4 + (-2) + 1 + \cdots$	Infinite
$3, 6, \ldots, 99$	$3 + 6 + \cdots + 99$	Finite; sum of 33 terms

A shorthand notation for denoting a series when the general term of the sequence is known is called **summation notation.** The Greek uppercase letter **sigma,** Σ, is used to mean "sum." The expression $\sum_{n=1}^{5} (3n + 1)$ is read "the sum of $3n + 1$ as n goes from 1 to 5"; this expression means the sum of the first five terms of the sequence whose general term is $a_n = 3n + 1$. Often, the variable i is used instead of n in summation notation: $\sum_{i=1}^{5} (3i + 1)$. Whether we use $n, i, k,$ or some other variable, the variable is called the **index of summation.** The notation $i = 1$ below the symbol Σ indicates the beginning value of i, and the number 5 above the symbol Σ indicates the ending value of i. Thus, the terms of the sequence are found by successively replacing i with the natural numbers 1, 2, 3, 4, 5. To find the sum, we write out the terms and then add.

$$\sum_{i=1}^{5} (3i + 1) = (3 \cdot 1 + 1) + (3 \cdot 2 + 1) + (3 \cdot 3 + 1)$$
$$+ (3 \cdot 4 + 1) + (3 \cdot 5 + 1)$$
$$= 4 + 7 + 10 + 13 + 16 = 50$$

EXAMPLE 1 Evaluate.

a. $\sum_{i=0}^{6} \frac{i - 2}{2}$ b. $\sum_{i=3}^{5} 2^i$

(Continued on next page)

Solution

a. $\displaystyle\sum_{i=0}^{6}\frac{i-2}{2} = \frac{0-2}{2} + \frac{1-2}{2} + \frac{2-2}{2} + \frac{3-2}{2} + \frac{4-2}{2} + \frac{5-2}{2} + \frac{6-2}{2}$

$= (-1) + \left(-\frac{1}{2}\right) + 0 + \frac{1}{2} + 1 + \frac{3}{2} + 2$

$= \frac{7}{2}$, or $3\frac{1}{2}$

b. $\displaystyle\sum_{i=3}^{5} 2^i = 2^3 + 2^4 + 2^5$

$= 8 + 16 + 32$

$= 56$

PRACTICE
1 Evaluate.

a. $\displaystyle\sum_{i=0}^{4}\frac{i-3}{4}$

b. $\displaystyle\sum_{i=2}^{5} 3^i$

EXAMPLE 2 Write each series with summation notation.

a. $3 + 6 + 9 + 12 + 15$

b. $\dfrac{1}{2} + \dfrac{1}{4} + \dfrac{1}{8} + \dfrac{1}{16}$

Solution

a. Since the _difference_ of each term and the preceding term is 3, the terms correspond to the first five terms of the arithmetic sequence $a_n = a_1 + (n-1)d$ with $a_1 = 3$ and $d = 3$. So $a_n = 3 + (n-1)3 = 3n$ when simplified. Thus in summation notation,

$$3 + 6 + 9 + 12 + 15 = \sum_{i=1}^{5} 3i.$$

b. Since each term is the _product_ of the preceding term and $\dfrac{1}{2}$, these terms correspond to the first four terms of the geometric sequence $a_n = a_1 r^{n-1}$. Here $a_1 = \dfrac{1}{2}$ and $r = \dfrac{1}{2}$, so $a_n = \left(\dfrac{1}{2}\right)\left(\dfrac{1}{2}\right)^{n-1} = \left(\dfrac{1}{2}\right)^{1+(n-1)} = \left(\dfrac{1}{2}\right)^{n}$. In summation notation,

$$\frac{1}{2} + \frac{1}{4} + \frac{1}{8} + \frac{1}{16} = \sum_{i=1}^{4}\left(\frac{1}{2}\right)^{i}$$

PRACTICE
2 Write each series with summation notation.

a. $5 + 10 + 15 + 20 + 25 + 30$

b. $\dfrac{1}{5} + \dfrac{1}{25} + \dfrac{1}{125} + \dfrac{1}{625}$

OBJECTIVE
2 Finding Partial Sums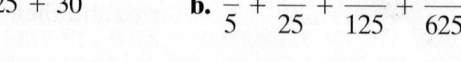

The sum of the first n terms of a sequence is a finite series known as a **partial sum, S_n.** Thus, for the sequence $a_1, a_2, \ldots, a_n$, the first three partial sums are

$$S_1 = a_1$$
$$S_2 = a_1 + a_2$$
$$S_3 = a_1 + a_2 + a_3$$

In general, S_n is the sum of the first n terms of a sequence.

$$S_n = \sum_{i=1}^{n} a_n$$

EXAMPLE 3   Find the sum of the first three terms of the sequence whose general term is $a_n = \dfrac{n + 3}{2n}$.

Solution

$$S_3 = \sum_{i=1}^{3} \frac{i + 3}{2i} = \frac{1 + 3}{2 \cdot 1} + \frac{2 + 3}{2 \cdot 2} + \frac{3 + 3}{2 \cdot 3}$$

$$= 2 + \frac{5}{4} + 1 = 4\frac{1}{4}$$

$\square$

PRACTICE
3   Find the sum of the first four terms of the sequence whose general term is $a_n = \dfrac{2 + 3n}{n^2}$.

$\blacksquare$

The next example illustrates how these sums model real-life phenomena.

EXAMPLE 4   **Number of Baby Gorillas Born**

The number of baby gorillas born at the San Diego Zoo is a sequence defined by $a_n = n(n - 1)$, where n is the number of years the zoo has owned gorillas. Find the total number of baby gorillas born in the _first 4 years_.

Solution   To solve, find the sum

$$S_4 = \sum_{i=1}^{4} i(i - 1)$$

$$= 1(1 - 1) + 2(2 - 1) + 3(3 - 1) + 4(4 - 1)$$

$$= 0 + 2 + 6 + 12 = 20$$

Twenty gorillas were born in the first 4 years.

$\square$

PRACTICE
4   The number of new strawberry plants growing in a garden each year is a sequence defined by $a_n = n(2n - 1)$, where n is the number of years after planting a strawberry plant. Find the total number of strawberry plants after 5 years.

$\blacksquare$

✓ Vocabulary, Readiness & Video Check

Use the choices below to fill in each blank. Not all choices may be used.

index of summation	infinite	sigma	1	7
partial sum	finite	summation	5	

1. A series is a(n) _____ series if it is the sum of all the terms of an infinite sequence.

2. A series is a(n) _____ series if it is the sum of a finite number of terms.

3. A shorthand notation for denoting a series when the general term of the sequence is known is called _____ notation.

4. In the notation $\displaystyle\sum_{i=1}^{7} (5i - 2)$, the Σ is the Greek uppercase letter _____ and the i is called the _____.

5. The sum of the first n terms of a sequence is a finite series known as a _____.

6. For the notation in Exercise 4 above, the beginning value of i is _____ and the ending value of i is _____.

Martin-Gay Interactive Videos

See Video 14.3 ◉

Watch the section lecture video and answer the following questions.

OBJECTIVE
1

7. From the lecture before ▦ Example 1, for the series with the summation notation $\sum_{i=2}^{10} \frac{(-1)^i}{i}$, identify/explain each piece of the notation: $\Sigma, i, 2, 10, \frac{(-1)^i}{i}$.

OBJECTIVE
2

8. From ▦ Example 2 and the lecture before, if you're finding the series S_7 of a sequence, what are you actually finding?

14.3 Exercise Set   MyMathLab® ▷

Evaluate. See Example 1.

1. $\sum_{i=1}^{4} (i - 3)$

2. $\sum_{i=1}^{5} (i + 6)$

▷**3.** $\sum_{i=4}^{7} (2i + 4)$

4. $\sum_{i=2}^{3} (5i - 1)$

5. $\sum_{i=2}^{4} (i^2 - 3)$

6. $\sum_{i=3}^{5} i^3$

7. $\sum_{i=1}^{3} \left(\frac{1}{i + 5} \right)$

8. $\sum_{i=2}^{4} \left(\frac{2}{i + 3} \right)$

9. $\sum_{i=1}^{3} \frac{1}{6i}$

10. $\sum_{i=1}^{3} \frac{1}{3i}$

11. $\sum_{i=2}^{6} 3i$

12. $\sum_{i=3}^{6} -4i$

13. $\sum_{i=3}^{5} i(i + 2)$

14. $\sum_{i=2}^{4} i(i - 3)$

15. $\sum_{i=1}^{5} 2^i$

16. $\sum_{i=1}^{4} 3^{i-1}$

17. $\sum_{i=1}^{4} \frac{4i}{i + 3}$

18. $\sum_{i=2}^{5} \frac{6 - i}{6 + i}$

Write each series with summation notation. See Example 2.

19. $1 + 3 + 5 + 7 + 9$

20. $4 + 7 + 10 + 13$

21. $4 + 12 + 36 + 108$

22. $5 + 10 + 20 + 40 + 80 + 160$

23. $12 + 9 + 6 + 3 + 0 + (-3)$

24. $5 + 1 + (-3) + (-7)$

25. $12 + 4 + \frac{4}{3} + \frac{4}{9}$

26. $80 + 20 + 5 + \frac{5}{4} + \frac{5}{16}$

27. $1 + 4 + 9 + 16 + 25 + 36 + 49$

28. $1 + (-4) + 9 + (-16)$

Find each partial sum. See Example 3.

29. Find the sum of the first two terms of the sequence whose general term is $a_n = (n + 2)(n - 5)$.

▷**30.** Find the sum of the first two terms of the sequence whose general term is $a_n = n(n - 6)$.

31. Find the sum of the first six terms of the sequence whose general term is $a_n = (-1)^n$.

32. Find the sum of the first seven terms of the sequence whose general term is $a_n = (-1)^{n-1}$.

33. Find the sum of the first four terms of the sequence whose general term is $a_n = (n + 3)(n + 1)$.

34. Find the sum of the first five terms of the sequence whose general term is $a_n = \frac{(-1)^n}{2n}$.

35. Find the sum of the first four terms of the sequence whose general term is $a_n = -2n$.

36. Find the sum of the first five terms of the sequence whose general term is $a_n = (n - 1)^2$.

37. Find the sum of the first three terms of the sequence whose general term is $a_n = -\frac{n}{3}$.

38. Find the sum of the first three terms of the sequence whose general term is $a_n = (n + 4)^2$.

Solve. See Example 4.

39. A gardener is making a triangular planting with 1 tree in the first row, 2 trees in the second row, 3 trees in the third row, and so on for 10 rows. Write the sequence that describes the number of trees in each row. Find the total number of trees planted.

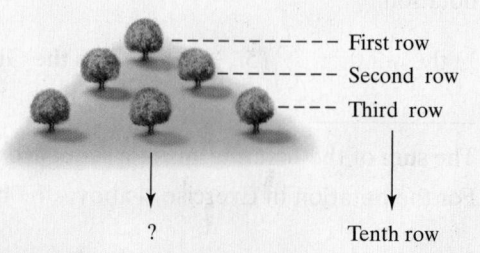

First row
Second row
Third row

?   Tenth row

40. Some surfers at the beach form a human pyramid with 2 surfers in the top row, 3 surfers in the second row, 4 surfers in the third row, and so on. If there are 6 rows in the pyramid, write the sequence that describes the number of surfers in each row of the pyramid. Find the total number of surfers.

Top row

Second row

Third row

Sixth row
?

41. A culture of fungus starts with 6 units and grows according to the sequence defined by $a_n = 6 \cdot 2^{n-1}$, where n is the number of days. Find the total number of fungus units there will be at the end of the fifth day.

42. A bacterial colony begins with 100 bacteria and grows according to the sequence defined by $a_n = 100 \cdot 2^{n-1}$, where n is the number of 6-hour periods. Find the total number of bacteria there will be after 24 hours.

43. The number of species born each year in a new aquarium forms a sequence whose general term is $a_n = (n + 1)(n + 3)$. Find the number of species born in the fourth year, and find the total number born in the first four years.

44. The number of otters born each year in a new aquarium forms a sequence whose general term is $a_n = (n - 1)(n + 3)$. Find the number of otters born in the third year and find the total number of otters born in the first three years.

45. The number of opossums killed each month on a new highway forms the sequence whose general term is $a_n = (n + 1)(n + 2)$, where n is the number of the month. Find the number of opossums killed in the fourth month and find the total number killed in the first four months.

46. In 2014, the population of the Northern Spotted Owl continued to decline, and the owl remained on the endangered species list as old-growth Northwest forests were logged. The size of the decrease in the population in a given year can be estimated by $200 - 6n$ pairs of birds. Find the decrease in the population in 2017 if year 1 is 2014. Find the estimated total decrease in the spotted owl population for the years 2014 through 2017. (*Source:* National Forest Service)

47. The amount of decay in pounds of a radioactive isotope each year is given by the sequence whose general term is $a_n = 100(0.5)^n$, where n is the number of the year. Find the amount of decay in the fourth year and find the total amount of decay in the first four years.

48. A person has a choice between two job offers. Job A has an annual starting salary of $30,000 with guaranteed annual raises of $1200 for the next four years, whereas job B has an annual starting salary of $28,000 with guaranteed annual raises of $2500 for the next four years. Compare the fifth partial sums for each sequence to determine which job would pay more money over the next 5 years.

49. A pendulum swings a length of 40 inches on its first swing. Each successive swing is $\frac{4}{5}$ of the preceding swing. Find the length of the fifth swing and the total length swung during the first five swings. (Round to the nearest tenth of an inch.)

50. Explain the difference between a sequence and a series.

REVIEW AND PREVIEW

Evaluate. See Sections 1.7 and 7.7.

51. $\dfrac{5}{1 - \dfrac{1}{2}}$

52. $\dfrac{-3}{1 - \dfrac{1}{7}}$

53. $\dfrac{\dfrac{1}{3}}{1 - \dfrac{1}{10}}$

54. $\dfrac{\dfrac{6}{11}}{1 - \dfrac{1}{10}}$

55. $\dfrac{3(1 - 2^4)}{1 - 2}$

56. $\dfrac{2(1 - 5^3)}{1 - 5}$

57. $\dfrac{10}{2}(3 + 15)$

58. $\dfrac{12}{2}(2 + 19)$

CONCEPT EXTENSIONS

59. a. Write the sum $\displaystyle\sum_{i=1}^{7}(i + i^2)$ without summation notation.

b. Write the sum $\displaystyle\sum_{i=1}^{7} i + \sum_{i=1}^{7} i^2$ without summation notation.

c. Compare the results of parts **a** and **b**.

d. Do you think the following is true or false? Explain your answer.

$$\sum_{i=1}^{n}(a_n + b_n) = \sum_{i=1}^{n} a_n + \sum_{i=1}^{n} b_n$$

60. a. Write the sum $\displaystyle\sum_{i=1}^{6} 5i^3$ without summation notation.

b. Write the expression $5 \cdot \displaystyle\sum_{i=1}^{6} i^3$ without summation notation.

c. Compare the results of parts **a** and **b**.

d. Do you think the following is true or false? Explain your answer.

$$\sum_{i=1}^{n} c \cdot a_n = c \cdot \sum_{i=1}^{n} a_n, \text{ where } c \text{ is a constant}$$

Integrated Review | Sequences and Series

Sections 14.1–14.3

Write the first five terms of each sequence whose general term is given.

1. $a_n = n - 3$

2. $a_n = \dfrac{7}{1 + n}$

3. $a_n = 3^{n-1}$

4. $a_n = n^2 - 5$

Find the indicated term for each sequence.

5. $(-2)^n; a_6$

6. $-n^2 + 2; a_4$

7. $\dfrac{(-1)^n}{n}; a_{40}$

8. $\dfrac{(-1)^n}{2n}; a_{41}$

Write the first five terms of the arithmetic or geometric sequence whose first term is a_1 and whose common difference, d, or common ratio, r, is given.

9. $a_1 = 7; d = -3$

10. $a_1 = -3; r = 5$

11. $a_1 = 45; r = \dfrac{1}{3}$

12. $a_1 = -12; d = 10$

Find the indicated term of each sequence.

13. The tenth term of the arithmetic sequence whose first term is 20 and whose common difference is 9

14. The sixth term of the geometric sequence whose first term is 64 and whose common ratio is $\dfrac{3}{4}$

15. The seventh term of the geometric sequence $6, -12, 24, \ldots$

16. The twentieth term of the arithmetic sequence $-100, -85, -70, \ldots$

17. The fifth term of the arithmetic sequence whose fourth term is -5 and whose tenth term is -35

18. The fifth term of a geometric sequence whose fourth term is 1 and whose seventh term is $\dfrac{1}{8}$

Evaluate.

19. $\displaystyle\sum_{i=1}^{4} 5i$

20. $\displaystyle\sum_{i=1}^{7} (3i + 2)$

21. $\displaystyle\sum_{i=3}^{7} 2^{i-4}$

22. $\displaystyle\sum_{i=2}^{5} \dfrac{i}{i + 1}$

Find each partial sum.

23. Find the sum of the first three terms of the sequence whose general term is $a_n = n(n - 4)$.

24. Find the sum of the first ten terms of the sequence whose general term is $a_n = (-1)^n(n + 1)$.

14.4 Partial Sums of Arithmetic and Geometric Sequences

OBJECTIVES

1 Find the Partial Sum of an Arithmetic Sequence.

2 Find the Partial Sum of a Geometric Sequence.

3 Find the Sum of the Terms of an Infinite Geometric Sequence.

OBJECTIVE

1 Finding Partial Sums of Arithmetic Sequences

Partial sums S_n are relatively easy to find when n is small—that is, when the number of terms to add is small. But when n is large, finding S_n can be tedious. For a large n, S_n is still relatively easy to find if the addends are terms of an arithmetic sequence or a geometric sequence.

For an arithmetic sequence, $a_n = a_1 + (n - 1)d$ for some first term a_1 and some common difference d. So S_n, the sum of the first n terms, is

$$S_n = a_1 + (a_1 + d) + (a_1 + 2d) + \cdots + [a_1 + (n - 1)d] \quad \text{Equation 1}$$

We might also find S_n by working backward from the nth term a_n, finding the preceding term a_{n-1}, by subtracting d each time.

$$S_n = a_n + (a_n - d) + (a_n - 2d) + \cdots + [a_n - (n - 1)d] \quad \text{Equation 2}$$

Now add the left sides of these two equations and add the right sides.

$$2S_n = (a_1 + a_n) + (a_1 + a_n) + (a_1 + a_n) + \cdots + (a_1 + a_n) \quad \text{Equation 1 + 2}$$

The d terms subtract out, leaving n sums of the first term, a_1, and last term, a_n. Thus, we write

$$2S_n = n(a_1 + a_n)$$

or

$$S_n = \dfrac{n}{2}(a_1 + a_n) \quad \text{Divide both sides by 2.}$$

> **Partial Sum S_n of an Arithmetic Sequence**
>
> The partial sum S_n of the first n terms of an arithmetic sequence is given by
>
> $$S_n = \frac{n}{2}(a_1 + a_n)$$
>
> where a_1 is the first term of the sequence and a_n is the nth term.

EXAMPLE 1   Use the partial sum formula to find the sum of the first six terms of the arithmetic sequence 2, 5, 8, 11, 14, 17, ...

<u>Solution</u>   Use the formula for S_n of an arithmetic sequence, replacing n with 6, a_1 with 2, and a_n with 17.

$$S_n = \frac{n}{2}(a_1 + a_n)$$

$$S_6 = \frac{6}{2}(2 + 17) = 3(19) = 57 \qquad \square$$

PRACTICE

1   Use the partial sum formula to find the sum of the first five terms of the arithmetic sequence 2, 9, 16, 23, 30, ...   ■

EXAMPLE 2   Find the sum of the first 30 positive integers.

<u>Solution</u>   Because 1, 2, 3, . . . , 30 is an arithmetic sequence, use the formula for S_n with $n = 30$, $a_1 = 1$, and $a_n = 30$. Thus,

$$S_n = \frac{n}{2}(a_1 + a_n)$$

$$S_{30} = \frac{30}{2}(1 + 30) = 15(31) = 465 \qquad \square$$

PRACTICE

2   Find the sum of the first 50 positive integers.   ■

EXAMPLE 3   **Stacking Rolls of Carpet**

Rolls of carpet are stacked in 20 rows with 3 rolls in the top row, 4 rolls in the next row, and so on, forming an arithmetic sequence. Find the total number of carpet rolls if there are 22 rolls in the bottom row.

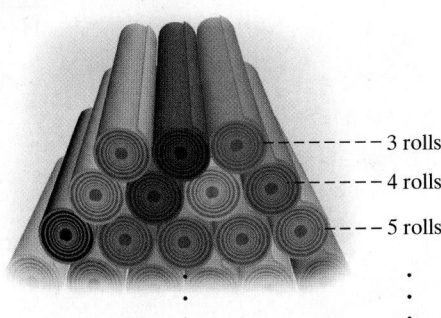

<u>Solution</u>   The list 3, 4, 5, . . . , 22 is the first 20 terms of an arithmetic sequence. Use the formula for S_n with $a_1 = 3$, $a_n = 22$, and $n = 20$ terms. Thus,

$$S_{20} = \frac{20}{2}(3 + 22) = 10(25) = 250$$

There are a total of 250 rolls of carpet. $\square$

PRACTICE

3   An ice sculptor is creating a gigantic castle-facade ice sculpture for First Night festivities in Boston. To get the volume of ice necessary, large blocks of ice were stacked atop each other in 10 rows. The topmost row comprised 6 blocks of ice, the next row 7 blocks of ice, and so on, forming an arithmetic sequence. Find the total number of ice blocks needed if there were 15 blocks in the bottom row.   ■

OBJECTIVE

2 Finding Partial Sums of Geometric Sequences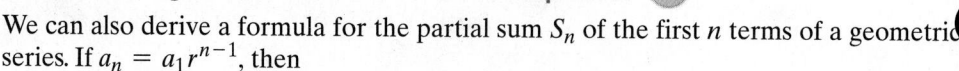

We can also derive a formula for the partial sum S_n of the first n terms of a geometric series. If $a_n = a_1 r^{n-1}$, then

$$S_n = a_1 + a_1 r + a_1 r^2 + \cdots + a_1 r^{n-1}$$

$$\uparrow \qquad \uparrow \qquad \uparrow \qquad\qquad \uparrow$$

1st term   2nd term   3rd term nth term

Multiply each side of the equation by $-r$.

$$-rS_n = -a_1 r - a_1 r^2 - a_1 r^3 - \cdots - a_1 r^n$$

Add the two equations.

$$S_n - rS_n = a_1 + (a_1 r - a_1 r) + (a_1 r^2 - a_1 r^2) + (a_1 r^3 - a_1 r^3) + \cdots - a_1 r^n$$

$$S_n - rS_n = a_1 - a_1 r^n$$

Now factor each side.

$$S_n(1 - r) = a_1(1 - r^n)$$

Solve for S_n by dividing both sides by $1 - r$. Thus,

$$S_n = \frac{a_1(1 - r^n)}{1 - r}$$

as long as r is not 1.

> **Partial Sum S_n of a Geometric Sequence**
>
> The partial sum S_n of the first n terms of a geometric sequence is given by
>
> $$S_n = \frac{a_1(1 - r^n)}{1 - r}$$
>
> where a_1 is the first term of the sequence, r is the common ratio, and $r \neq 1$.

EXAMPLE 4   Find the sum of the first six terms of the geometric sequence 5, 10, 20, 40, 80, 160.

Solution   Use the formula for the partial sum S_n of the terms of a geometric sequence. Here, $n = 6$, the first term $a_1 = 5$, and the common ratio $r = 2$.

$$S_n = \frac{a_1(1 - r^n)}{1 - r}$$

$$S_6 = \frac{5(1 - 2^6)}{1 - 2} = \frac{5(-63)}{-1} = 315$$

PRACTICE

4   Find the sum of the first five terms of the geometric sequence $32, 8, 2, \dfrac{1}{2}, \dfrac{1}{8}$.

EXAMPLE 5   **Finding Amount of Donation**

A grant from an alumnus to a university specified that the university was to receive $800,000 during the first year and 75% of the preceding year's donation during each of the following 5 years. Find the total amount donated during the 6 years.

Solution   The donations are modeled by the first six terms of a geometric sequence. Evaluate S_n when $n = 6$, $a_1 = 800,000$, and $r = 0.75$.

$$S_6 = \frac{800{,}000\left[1 - (0.75)^6\right]}{1 - 0.75}$$

$$= \$2{,}630{,}468.75$$

The total amount donated during the 6 years is $2,630,468.75.

PRACTICE
5 A new youth center is being established in a downtown urban area. A philanthropic charity has agreed to help it get off the ground. The charity has pledged to donate $250,000 in the first year, with 80% of the preceding year's donation for each of the following 6 years. Find the total amount donated during the 7 years. ■

OBJECTIVE

3 Finding Sums of Terms of Infinite Geometric Sequences ▶

Is it possible to find the sum of all the terms of an infinite sequence? Examine the partial sums of the geometric sequence $\frac{1}{2}, \frac{1}{4}, \frac{1}{8}, \ldots$.

$$S_1 = \frac{1}{2}$$

$$S_2 = \frac{1}{2} + \frac{1}{4} = \frac{3}{4}$$

$$S_3 = \frac{1}{2} + \frac{1}{4} + \frac{1}{8} = \frac{7}{8}$$

$$S_4 = \frac{1}{2} + \frac{1}{4} + \frac{1}{8} + \frac{1}{16} = \frac{15}{16}$$

$$S_5 = \frac{1}{2} + \frac{1}{4} + \frac{1}{8} + \frac{1}{16} + \frac{1}{32} = \frac{31}{32}$$

$$\vdots$$

$$S_{10} = \frac{1}{2} + \frac{1}{4} + \frac{1}{8} + \cdots + \frac{1}{2^{10}} = \frac{1023}{1024}$$

Even though each partial sum is larger than the preceding partial sum, we see that each partial sum is closer to 1 than the preceding partial sum. If n gets larger and larger, then S_n gets closer and closer to 1. We say that 1 is the **limit** of S_n and that 1 is the sum of the terms of this infinite sequence. In general, if $|r| < 1$, the following formula gives the sum of the terms of an infinite geometric sequence.

Sum of the Terms of an Infinite Geometric Sequence

The sum S_∞ of the terms of an infinite geometric sequence is given by

$$S_\infty = \frac{a_1}{1 - r}$$

where a_1 is the first term of the sequence, r is the common ratio, and $|r| < 1$. If $|r| \geq 1$, S_∞ does not exist.

What happens for other values of r? For example, in the following geometric sequence, $r = 3$.

$$6, 18, 54, 162, \ldots$$

Here, as n increases, the sum S_n increases also. This time, though, S_n does not get closer and closer to a fixed number but instead increases without bound.

EXAMPLE 6 Find the sum of the terms of the geometric sequence $2, \dfrac{2}{3}, \dfrac{2}{9}, \dfrac{2}{27}, \ldots$

Solution For this geometric sequence, $r = \dfrac{1}{3}$. Since $|r| < 1$, we may use the formula for S_∞ of a geometric sequence with $a_1 = 2$ and $r = \dfrac{1}{3}$.

$$S_\infty = \frac{a_1}{1-r} = \frac{2}{1-\dfrac{1}{3}} = \frac{2}{\dfrac{2}{3}} = 3$$

□

PRACTICE
6 Find the sum of the terms of the geometric sequence $7, \dfrac{7}{4}, \dfrac{7}{16}, \dfrac{7}{64}, \ldots$ ■

The formula for the sum of the terms of an infinite geometric sequence can be used to write a repeating decimal as a fraction. For example,

$$0.33\overline{3} = \frac{3}{10} + \frac{3}{100} + \frac{3}{1000} + \cdots$$

This sum is the sum of the terms of an infinite geometric sequence whose first term a_1 is $\dfrac{3}{10}$ and whose common ratio r is $\dfrac{1}{10}$. Using the formula for S_∞,

$$S_\infty = \frac{a_1}{1-r} = \frac{\dfrac{3}{10}}{1-\dfrac{1}{10}} = \frac{1}{3}$$

So, $0.33\overline{3} = \dfrac{1}{3}$.

EXAMPLE 7 Distance Traveled by a Pendulum

On its first pass, a pendulum swings through an arc whose length is 24 inches. On each pass thereafter, the arc length is 75% of the arc length on the preceding pass. Find the total distance the pendulum travels before it comes to rest.

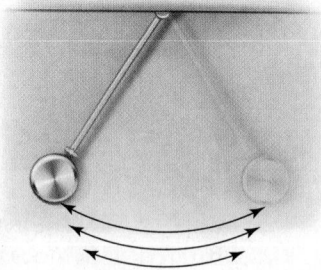

Solution We must find the sum of the terms of an infinite geometric sequence whose first term, a_1, is 24 and whose common ratio, r, is 0.75. Since $|r| < 1$, we may use the formula for S_∞.

$$S_\infty = \frac{a_1}{1-r} = \frac{24}{1-0.75} = \frac{24}{0.25} = 96$$

The pendulum travels a total distance of 96 inches before it comes to rest. □

PRACTICE
7 The manufacturers of the "perpetual bouncing ball" claim that the ball rises to 96% of its dropped height on each bounce of the ball. Find the total distance the ball travels before it comes to rest if it is dropped from a height of 36 inches. ■

Vocabulary, Readiness & Video Check

Decide whether each sequence is geometric or arithmetic.

1. $5, 10, 15, 20, 25, \ldots$ _____

2. $5, 10, 20, 40, 80, \ldots$ _____

3. $-1, 3, -9, 27, -81 \ldots$ _____

4. $-1, 1, 3, 5, 7, \ldots$ _____

5. $-7, 0, 7, 14, 21, \ldots$ _____

6. $-7, 7, -7, 7, -7, \ldots$ _____

Martin-Gay Interactive Videos

See Video 14.4

Watch the section lecture video and answer the following questions.

OBJECTIVE 1

7. From Example 1, suppose you are asked to find the sum of the first 100 terms of an arithmetic sequence in which you were given only the first few terms. You need the 100th term for the partial sum formula—how can you find this term without actually writing down the first 100 terms?

OBJECTIVE 2

8. From the lecture before Example 2, we know $r \neq 1$ in the partial sum formula because it would make the denominator 0. What would a geometric sequence with $r = 1$ look like? How could the partial sum of such a sequence be found (without the given formula)?

OBJECTIVE 3

9. From the lecture before Example 3, why can't you find S_∞ for the geometric sequence $-1, -3, -9, -27, \ldots$?

14.4 Exercise Set MyMathLab®

Use the partial sum formula to find the partial sum of the given arithmetic or geometric sequence. See Examples 1 and 4.

1. Find the sum of the first six terms of the arithmetic sequence $1, 3, 5, 7, \ldots$.

2. Find the sum of the first seven terms of the arithmetic sequence $-7, -11, -15, \ldots$.

3. Find the sum of the first five terms of the geometric sequence $4, 12, 36, \ldots$.

4. Find the sum of the first eight terms of the geometric sequence $-1, 2, -4, \ldots$.

5. Find the sum of the first six terms of the arithmetic sequence $3, 6, 9, \ldots$.

6. Find the sum of the first four terms of the arithmetic sequence $-4, -8, -12, \ldots$.

7. Find the sum of the first four terms of the geometric sequence $2, \dfrac{2}{5}, \dfrac{2}{25}, \ldots$.

8. Find the sum of the first five terms of the geometric sequence $\dfrac{1}{3}, -\dfrac{2}{3}, \dfrac{4}{3}, \ldots$.

Solve. See Example 2.

9. Find the sum of the first ten positive integers.

10. Find the sum of the first eight negative integers.

11. Find the sum of the first four positive odd integers.

12. Find the sum of the first five negative odd integers.

Find the sum of the terms of each infinite geometric sequence. See Example 6.

13. $12, 6, 3, \ldots$

14. $45, 15, 5, \ldots$

15. $\dfrac{1}{10}, \dfrac{1}{100}, \dfrac{1}{1000}, \ldots$

16. $\dfrac{3}{5}, \dfrac{3}{20}, \dfrac{3}{80}, \ldots$

17. $-10, -5, -\dfrac{5}{2}, \ldots$

18. $-16, -4, -1, \ldots$

19. $2, -\dfrac{1}{4}, \dfrac{1}{32}, \ldots$

20. $-3, \dfrac{3}{5}, -\dfrac{3}{25}, \ldots$

21. $\dfrac{2}{3}, -\dfrac{1}{3}, \dfrac{1}{6}, \ldots$

22. $6, -4, \dfrac{8}{3}, \ldots$

MIXED PRACTICE

Solve.

23. Find the sum of the first ten terms of the sequence $-4, 1, 6, \ldots, 41$ where 41 is the tenth term.

24. Find the sum of the first twelve terms of the sequence $-3, -13, -23, \ldots, -113$ where -113 is the twelfth term.

25. Find the sum of the first seven terms of the sequence $3, \dfrac{3}{2}, \dfrac{3}{4}, \ldots$.

26. Find the sum of the first five terms of the sequence $-2, -6, -18, \ldots$.

27. Find the sum of the first five terms of the sequence $-12, 6, -3, \ldots$.

28. Find the sum of the first four terms of the sequence $-\dfrac{1}{4}, -\dfrac{3}{4}, -\dfrac{9}{4}, \ldots$.

29. Find the sum of the first twenty terms of the sequence $\dfrac{1}{2}, \dfrac{1}{4}, 0, \ldots, -\dfrac{17}{4}$ where $-\dfrac{17}{4}$ is the twentieth term.

30. Find the sum of the first fifteen terms of the sequence $-5, -9, -13, \ldots, -61$ where -61 is the fifteenth term.

31. If a_1 is 8 and r is $-\dfrac{2}{3}$, find S_3.

32. If a_1 is 10, a_{18} is $\dfrac{3}{2}$, and d is $-\dfrac{1}{2}$, find S_{18}.

Solve. See Example 3.

33. Modern Car Company has come out with a new car model. Market analysts predict that 4000 cars will be sold in the first month and that sales will drop by 50 cars per month after that during the first year. Write out the first five terms of the sequence and find the number of sold cars predicted for the twelfth month. Find the total predicted number of sold cars for the first year.

34. A company that sends faxes charges $3 for the first page sent and $0.10 less than the preceding page for each additional page sent. The cost per page forms an arithmetic sequence. Write the first five terms of this sequence and use a partial sum to find the cost of sending a nine-page document.

35. Sal has two job offers: Firm *A* starts at $22,000 per year and guarantees raises of $1000 per year, whereas Firm *B* starts at $20,000 and guarantees raises of $1200 per year. Over a 10-year period, determine the more profitable offer.

36. The game of pool uses 15 balls numbered 1 to 15. In the variety called rotation, a player who sinks a ball receives as many points as the number on the ball. Use an arithmetic series to find the score of a player who sinks all 15 balls.

Solve. See Example 5.

37. A woman made $30,000 during the first year she owned her business and made an additional 10% over the previous year in each subsequent year. Find how much she made during her fourth year of business. Find her total earnings during the first four years.

38. In free fall, a parachutist falls 16 feet during the first second, 48 feet during the second second, 80 feet during the third second, and so on. Find how far she falls during the eighth second. Find the total distance she falls during the first 8 seconds.

39. A trainee in a computer company takes 0.9 times as long to assemble each computer as he took to assemble the preceding computer. If it took him 30 minutes to assemble the first computer, find how long it takes him to assemble the fifth computer. Find the total time he takes to assemble the first five computers (round to the nearest minute).

40. On a gambling trip to Reno, Carol doubled her bet each time she lost. If her first losing bet was $5 and she lost six consecutive bets, find how much she lost on the sixth bet. Find the total amount lost on these six bets.

Solve. See Example 7.

41. A ball is dropped from a height of 20 feet and repeatedly rebounds to a height that is $\dfrac{4}{5}$ of its previous height. Find the total distance the ball covers before it comes to rest.

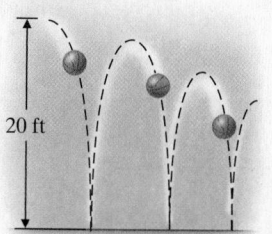

42. A rotating flywheel coming to rest makes 300 revolutions in the first minute and in each minute thereafter makes $\dfrac{2}{5}$ as many revolutions as in the preceding minute. Find how many revolutions the wheel makes before it comes to rest.

MIXED PRACTICE

Solve.

43. In the pool game of rotation, player *A* sinks balls numbered 1 to 9, and player *B* sinks the rest of the balls. Use an arithmetic series to find each player's score (see Exercise 36).

44. A godfather deposited $250 in a savings account on the day his godchild was born. On each subsequent birthday, he deposited $50 more than he deposited the previous year. Find how much money he deposited on his godchild's twenty-first birthday. Find the total amount deposited over the 21 years.

45. During the holiday rush, a business can rent a computer system for $200 the first day, with the rental fee decreasing $5 for each additional day. Find the fee paid for 20 days during the holiday rush.

46. The spraying of a field with insecticide killed 6400 weevils the first day, 1600 the second day, 400 the third day, and so on. Find the total number of weevils killed during the first 5 days.

47. A college student humorously asks his parents to charge him room and board according to this geometric sequence: $0.01 for the first day of the month, $0.02 for the second day, $0.04 for the third day, and so on. Find the total room and board he would pay for 30 days.

48. Following its television advertising campaign, a bank attracted 80 new customers the first day, 120 the second day, 160 the third day, and so on in an arithmetic sequence. Find how many new customers were attracted during the first 5 days following its television campaign.

REVIEW AND PREVIEW

Evaluate. See Section 1.7.

49. $6 \cdot 5 \cdot 4 \cdot 3 \cdot 2 \cdot 1$

50. $8 \cdot 7 \cdot 6 \cdot 5 \cdot 4 \cdot 3 \cdot 2 \cdot 1$

51. $\dfrac{3 \cdot 2 \cdot 1}{2 \cdot 1}$

52. $\dfrac{5 \cdot 4 \cdot 3 \cdot 2 \cdot 1}{3 \cdot 2 \cdot 1}$

Multiply. See Section 5.4.

53. $(x + 5)^2$

54. $(x - 2)^2$

55. $(2x - 1)^3$

56. $(3x + 2)^3$

CONCEPT EXTENSIONS

57. Write $0.88\overline{8}$ as an infinite geometric series and use the formula for S_∞ to write it as a rational number.

58. Write $0.54\overline{54}$ as an infinite geometric series and use the formula for S_∞ to write it as a rational number.

59. Explain whether the sequence $5, 5, 5, \ldots$ is arithmetic, geometric, neither, or both.

60. Describe a situation in everyday life that can be modeled by an infinite geometric series.

14.5 The Binomial Theorem

OBJECTIVES

1. Use Pascal's Triangle to Expand Binomials.
2. Evaluate Factorials.
3. Use the Binomial Theorem to Expand Binomials.
4. Find the *n*th Term in the Expansion of a Binomial Raised to a Positive Power.

In this section, we learn how to **expand** binomials of the form $(a + b)^n$ easily. Expanding a binomial such as $(a + b)^n$ means to write the factored form as a sum. First, we review the patterns in the expansions of $(a + b)^n$.

$$(a + b)^0 = 1 \qquad \text{1 term}$$
$$(a + b)^1 = a + b \qquad \text{2 terms}$$
$$(a + b)^2 = a^2 + 2ab + b^2 \qquad \text{3 terms}$$
$$(a + b)^3 = a^3 + 3a^2b + 3ab^2 + b^3 \qquad \text{4 terms}$$
$$(a + b)^4 = a^4 + 4a^3b + 6a^2b^2 + 4ab^3 + b^4 \qquad \text{5 terms}$$
$$(a + b)^5 = a^5 + 5a^4b + 10a^3b^2 + 10a^2b^3 + 5ab^4 + b^5 \qquad \text{6 terms}$$

Notice the following patterns.

1. The expansion of $(a + b)^n$ contains $n + 1$ terms. For example, for $(a + b)^3$, $n = 3$, and the expansion contains $3 + 1$ terms, or 4 terms.

2. The first term of the expansion of $(a + b)^n$ is a^n, and the last term is b^n.

3. The powers of a decrease by 1 for each term, whereas the powers of b increase by 1 for each term.

4. For each term of the expansion of $(a + b)^n$, the sum of the exponents of a and b is n. (For example, the sum of the exponents of $5a^4b$ is $4 + 1$, or 5, and the sum of the exponents of $10a^3b^2$ is $3 + 2$, or 5.)

OBJECTIVE

1 Using Pascal's Triangle

There are patterns in the coefficients of the terms as well. Written in a triangular array, the coefficients are called **Pascal's triangle.**

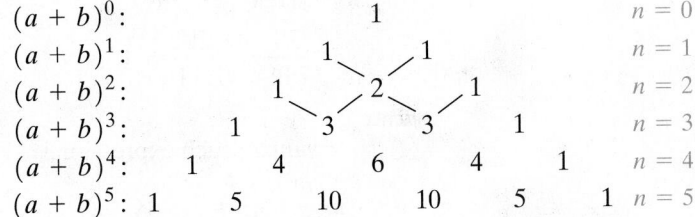

Each row in Pascal's triangle begins and ends with 1. Any other number in a row is the sum of the two closest numbers above it. Using this pattern, we can write the next row for $n = 6$, by first writing the number 1. Then we can add the consecutive numbers in the row for $n = 5$ and write each sum between and below the pair. We complete the row by writing a 1.

$$
\begin{array}{ccccccccccccc}
1 & & 5 & & 10 & & 10 & & 5 & & 1 & & n = 5 \\
& 1 & & 6 & & 15 & & 20 & & 15 & & 6 & & 1 & & n = 6
\end{array}
$$

We can use Pascal's triangle and the patterns noted to expand $(a + b)^n$ without actually multiplying any terms.

EXAMPLE 1   Expand: $(a + b)^6$

Solution Using the $n = 6$ row of Pascal's triangle as the coefficients and following the patterns noted, $(a + b)^6$ can be expanded as

$$(a + b)^6 = a^6 + 6a^5b + 15a^4b^2 + 20a^3b^3 + 15a^2b^4 + 6ab^5 + b^6$$ □

PRACTICE
1   Expand: $(p + r)^7$ ■

OBJECTIVE
2   Evaluating Factorials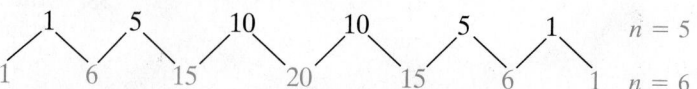

For a large n, the use of Pascal's triangle to find coefficients for $(a + b)^n$ can be tedious. An alternative method for determining these coefficients is based on the concept of a **factorial.**

The **factorial of n,** written $n!$ (read "n factorial"), is the product of the first n consecutive natural numbers.

> **Helpful Hint**
> Remember that $0! = 1$

> **Factorial of n: $n!$**
>
> If n is a natural number, then $n! = n(n - 1)(n - 2)(n - 3) \cdots \cdot 3 \cdot 2 \cdot 1$.
> The factorial of 0, written $0!$, is defined to be 1.

For example, $3! = 3 \cdot 2 \cdot 1 = 6$, $5! = 5 \cdot 4 \cdot 3 \cdot 2 \cdot 1 = 120$, and $0! = 1$.

EXAMPLE 2   Evaluate each expression.

a. $\dfrac{5!}{6!}$ **b.** $\dfrac{10!}{7!3!}$ **c.** $\dfrac{3!}{2!1!}$ **d.** $\dfrac{7!}{7!0!}$

Solution

a. $\dfrac{5!}{6!} = \dfrac{5 \cdot 4 \cdot 3 \cdot 2 \cdot 1}{6 \cdot 5 \cdot 4 \cdot 3 \cdot 2 \cdot 1} = \dfrac{1}{6}$

b. $\dfrac{10!}{7!3!} = \dfrac{10 \cdot 9 \cdot 8 \cdot 7!}{7! \cdot 3 \cdot 2 \cdot 1} = \dfrac{10 \cdot 9 \cdot 8}{3 \cdot 2 \cdot 1} = 10 \cdot 3 \cdot 4 = 120$

c. $\dfrac{3!}{2!1!} = \dfrac{3 \cdot 2 \cdot 1}{2 \cdot 1 \cdot 1} = 3$

d. $\dfrac{7!}{7!0!} = \dfrac{7!}{7! \cdot 1} = 1$ □

PRACTICE
2   Evaluate each expression.

a. $\dfrac{6!}{7!}$ **b.** $\dfrac{8!}{4!2!}$ **c.** $\dfrac{5!}{4!1!}$ **d.** $\dfrac{9!}{9!0!}$ ■

Helpful Hint

We can use a calculator with a factorial key to evaluate a factorial. A calculator uses scientific notation for large results.

OBJECTIVE

3   Using the Binomial Theorem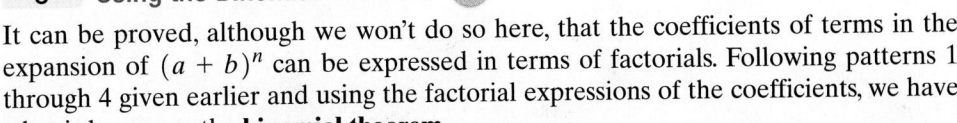

It can be proved, although we won't do so here, that the coefficients of terms in the expansion of $(a + b)^n$ can be expressed in terms of factorials. Following patterns 1 through 4 given earlier and using the factorial expressions of the coefficients, we have what is known as the **binomial theorem.**

Binomial Theorem

If n is a positive integer, then

$$(a + b)^n = a^n + \frac{n}{1!}a^{n-1}b^1 + \frac{n(n-1)}{2!}a^{n-2}b^2$$

$$+ \frac{n(n-1)(n-2)}{3!}a^{n-3}b^3 + \cdots + b^n$$

We call the formula for $(a + b)^n$ given by the binomial theorem the **binomial formula.**

EXAMPLE 3   Use the binomial theorem to expand $(x + y)^{10}$.

Solution   Let $a = x, b = y$, and $n = 10$ in the binomial formula.

$$(x + y)^{10} = x^{10} + \frac{10}{1!}x^9y + \frac{10 \cdot 9}{2!}x^8y^2 + \frac{10 \cdot 9 \cdot 8}{3!}x^7y^3 + \frac{10 \cdot 9 \cdot 8 \cdot 7}{4!}x^6y^4$$

$$+ \frac{10 \cdot 9 \cdot 8 \cdot 7 \cdot 6}{5!}x^5y^5 + \frac{10 \cdot 9 \cdot 8 \cdot 7 \cdot 6 \cdot 5}{6!}x^4y^6$$

$$+ \frac{10 \cdot 9 \cdot 8 \cdot 7 \cdot 6 \cdot 5 \cdot 4}{7!}x^3y^7$$

$$+ \frac{10 \cdot 9 \cdot 8 \cdot 7 \cdot 6 \cdot 5 \cdot 4 \cdot 3}{8!}x^2y^8$$

$$+ \frac{10 \cdot 9 \cdot 8 \cdot 7 \cdot 6 \cdot 5 \cdot 4 \cdot 3 \cdot 2}{9!}xy^9 + y^{10}$$

$$= x^{10} + 10x^9y + 45x^8y^2 + 120x^7y^3 + 210x^6y^4 + 252x^5y^5 + 210x^4y^6$$

$$+ 120x^3y^7 + 45x^2y^8 + 10xy^9 + y^{10}$$

PRACTICE

3   Use the binomial theorem to expand $(a + b)^9$.

EXAMPLE 4   Use the binomial theorem to expand $(x + 2y)^5$.

Solution   Let $a = x$ and $b = 2y$ in the binomial formula.

$$(x + 2y)^5 = x^5 + \frac{5}{1!}x^4(2y) + \frac{5 \cdot 4}{2!}x^3(2y)^2 + \frac{5 \cdot 4 \cdot 3}{3!}x^2(2y)^3$$

$$+ \frac{5 \cdot 4 \cdot 3 \cdot 2}{4!}x(2y)^4 + (2y)^5$$

$$= x^5 + 10x^4y + 40x^3y^2 + 80x^2y^3 + 80xy^4 + 32y^5$$

PRACTICE

4   Use the binomial theorem to expand $(a + 5b)^3$.

EXAMPLE 5　Use the binomial theorem to expand $(3m - n)^4$.

Solution Let $a = 3m$ and $b = -n$ in the binomial formula.

$$(3m - n)^4 = (3m)^4 + \frac{4}{1!}(3m)^3(-n) + \frac{4 \cdot 3}{2!}(3m)^2(-n)^2$$

$$+ \frac{4 \cdot 3 \cdot 2}{3!}(3m)(-n)^3 + (-n)^4$$

$$= 81m^4 - 108m^3n + 54m^2n^2 - 12mn^3 + n^4$$

PRACTICE
5　Use the binomial theorem to expand $(3x - 2y)^3$.

OBJECTIVE
4　Finding the nth Term of a Binomial Expansion ▶

Sometimes it is convenient to find a specific term of a binomial expansion without writing out the entire expansion. By studying the expansion of binomials, a pattern forms for each term. This pattern is most easily stated for the $(r + 1)$st term.

$(r + 1)$st Term in a Binomial Expansion

The $(r + 1)$st term of the expansion of $(a + b)^n$ is $\dfrac{n!}{r!(n - r)!}a^{n-r}b^r$.

EXAMPLE 6　Find the eighth term in the expansion of $(2x - y)^{10}$.

Solution Use the formula with $n = 10$, $a = 2x$, $b = -y$, and $r + 1 = 8$. Notice that since $r + 1 = 8$, $r = 7$.

$$\frac{n!}{r!(n - r)!}a^{n-r}b^r = \frac{10!}{7!3!}(2x)^3(-y)^7$$

$$= 120(8x^3)(-y^7)$$

$$= -960x^3y^7$$

PRACTICE
6　Find the seventh term in the expansion of $(x - 4y)^{11}$.

✓ Vocabulary, Readiness & Video Check

Fill in each blank.

1. $0! =$ _____　**2.** $1! =$ _____　**3.** $4! =$ _____　**4.** $2! =$ _____　**5.** $3!0! =$ _____　**6.** $0!2! =$ _____

Martin-Gay Interactive Videos

See Video 14.5 ◉

Watch the section lecture video and answer the following questions.

OBJECTIVE **1**
7. From ▦ Example 1 and the lecture before, when expanding a binomial such as $(x + y)^7$, what does Pascal's triangle tell you? What does the power on the binomial tell you?

OBJECTIVE **2**
8. From ▦ Example 2 and the lecture before, write the definition of 4! and evaluate it. What is the value of 0!?

OBJECTIVE **3**
9. From ▦ Example 4, what point is made about the terms of a binomial when applying the binomial theorem?

OBJECTIVE **4**
10. In ▦ Example 5, we are looking for the 4th term, so why do we let $r = 3$?

14.5 Exercise Set MyMathLab

Use Pascal's triangle to expand the binomial. See Example 1.

1. $(m + n)^3$
2. $(x + y)^4$
3. $(c + d)^5$
4. $(a + b)^6$
5. $(y - x)^5$
6. $(q - r)^7$
7. Explain how to generate a row of Pascal's triangle.
8. Write the $n = 8$ row of Pascal's triangle.

Evaluate each expression. See Example 2.

9. $\dfrac{8!}{7!}$
10. $\dfrac{6!}{0!}$
11. $\dfrac{7!}{5!}$
12. $\dfrac{8!}{5!}$
13. $\dfrac{10!}{7!2!}$
14. $\dfrac{9!}{5!3!}$
15. $\dfrac{8!}{6!0!}$
16. $\dfrac{10!}{4!6!}$

MIXED PRACTICE

Use the binomial formula to expand each binomial. See Examples 3 through 5.

17. $(a + b)^7$

18. $(x + y)^8$

19. $(a + 2b)^5$

20. $(x + 3y)^6$

21. $(q + r)^9$

22. $(b + c)^6$

23. $(4a + b)^5$

24. $(3m + n)^4$

25. $(5a - 2b)^4$

26. $(m - 4)^6$

27. $(2a + 3b)^3$

28. $(4 - 3x)^5$

29. $(x + 2)^5$
30. $(3 + 2a)^4$

Find the indicated term. See Example 6.

31. The fifth term of the expansion of $(c - d)^5$
32. The fourth term of the expansion of $(x - y)^6$
33. The eighth term of the expansion of $(2c + d)^7$
34. The tenth term of the expansion of $(5x - y)^9$
35. The fourth term of the expansion of $(2r - s)^5$
36. The first term of the expansion of $(3q - 7r)^6$
37. The third term of the expansion of $(x + y)^4$
38. The fourth term of the expansion of $(a + b)^8$
39. The second term of the expansion of $(a + 3b)^{10}$
40. The third term of the expansion of $(m + 5n)^7$

REVIEW AND PREVIEW

Sketch the graph of each function. Decide whether each function is one-to-one. See Sections 3.6 and 12.2.

41. $f(x) = |x|$
42. $g(x) = 3(x - 1)^2$
43. $H(x) = 2x + 3$
44. $F(x) = -2$
45. $f(x) = x^2 + 3$
46. $h(x) = -(x + 1)^2 - 4$

CONCEPT EXTENSIONS

47. Expand the expression $(\sqrt{x} + \sqrt{3})^5$.

48. Find the term containing x^2 in the expansion of $(\sqrt{x} - \sqrt{5})^6$.

Evaluate the following.

The notation $\dbinom{n}{r}$ means $\dfrac{n!}{r!(n - r)!}$. For example,

$$\binom{5}{3} = \frac{5!}{3!(5 - 3)!} = \frac{5!}{3!2!} = \frac{5 \cdot 4 \cdot 3 \cdot 2 \cdot 1}{(3 \cdot 2 \cdot 1) \cdot (2 \cdot 1)} = 10.$$

49. $\dbinom{9}{5}$
50. $\dbinom{4}{3}$
51. $\dbinom{8}{2}$
52. $\dbinom{12}{11}$

53. Show that $\dbinom{n}{n} = 1$ for any whole number n.

Chapter 14 Vocabulary Check

Fill in each blank with one of the words or phrases listed below.

general term	common difference	finite sequence	common ratio	Pascal's triangle
infinite sequence	factorial of n	arithmetic sequence	geometric sequence	series

1. A(n) _____ is a function whose domain is the set of natural numbers $\{1, 2, 3, \ldots, n\}$, where n is some natural number.

2. The _____, written $n!$, is the product of the first n consecutive natural numbers.

3. A(n) _____ is a function whose domain is the set of natural numbers.

4. A(n) _____ is a sequence in which each term (after the first) is obtained by multiplying the preceding term by a constant amount r. The constant r is called the _____ of the sequence.

5. The sum of the terms of a sequence is called a(n) _____.

6. The nth term of the sequence a_n is called the _____.

7. A(n) _____ is a sequence in which each term (after the first) differs from the preceding term by a constant amount d. The constant d is called the _____ of the sequence.

8. A triangular array of the coefficients of the terms of the expansion of $(a + b)^n$ is called _____.

Chapter 14 Highlights

DEFINITIONS AND CONCEPTS	EXAMPLES
Section 14.1 Sequences	
An **infinite sequence** is a function whose domain is the set of natural numbers $\{1, 2, 3, 4, \ldots\}$.	*Infinite Sequence* $$2, 4, 6, 8, 10, \ldots$$
A **finite sequence** is a function whose domain is the set of natural numbers $\{1, 2, 3, 4, \ldots, n\}$, where n is some natural number.	*Finite Sequence* $$1, -2, 3, -4, 5, -6$$
The notation a_n, where n is a natural number, denotes a sequence.	Write the first four terms of the sequence whose general term is $a_n = n^2 + 1$. $$a_1 = 1^2 + 1 = 2$$ $$a_2 = 2^2 + 1 = 5$$ $$a_3 = 3^2 + 1 = 10$$ $$a_4 = 4^2 + 1 = 17$$
Section 14.2 Arithmetic and Geometric Sequences	
An **arithmetic sequence** is a sequence in which each term differs from the preceding term by a constant amount d, called the **common difference.**	*Arithmetic Sequence* $$5, 8, 11, 14, 17, 20, \ldots$$ Here, $a_1 = 5$ and $d = 3$.
The **general term** a_n of an arithmetic sequence is given by $$a_n = a_1 + (n - 1)d$$ where a_1 is the first term and d is the common difference.	The general term is $$a_n = a_1 + (n - 1)d \text{ or }$$ $$a_n = 5 + (n - 1)3$$

DEFINITIONS AND CONCEPTS	EXAMPLES

Section 14.2   Arithmetic and Geometric Sequences (continued)

A **geometric sequence** is a sequence in which each term is obtained by multiplying the preceding term by a constant r, called the **common ratio**.

Geometric Sequence

$$12, -6, 3, -\frac{3}{2}, \ldots$$

Here $a_1 = 12$ and $r = -\frac{1}{2}$.

The **general term** a_n of a geometric sequence is given by

$$a_n = a_1 r^{n-1}$$

where a_1 is the first term and r is the common ratio.

The general term is

$$a_n = a_1 r^{n-1} \text{ or}$$

$$a_n = 12\left(-\frac{1}{2}\right)^{n-1}$$

Section 14.3   Series

A sum of the terms of a sequence is called a **series**.

Sequence	*Series*	
$3, 7, 11, 15$	$3 + 7 + 11 + 15$	finite
$3, 7, 11, 15, \ldots$	$3 + 7 + 11 + 15 + \cdots$	infinite

A shorthand notation for denoting a series is called **summation notation**:

index of summation $\rightarrow \underset{i=1}{\overset{4}{\sum}} \rightarrow$ Greek letter sigma used to mean sum

$$\sum_{i=1}^{4} 3^i = 3^1 + 3^2 + 3^3 + 3^4$$

$$= 3 + 9 + 27 + 81$$

$$= 120$$

Section 14.4   Partial Sums of Arithmetic and Geometric Sequences

Partial sum, S_n, of the first n terms of an arithmetic sequence:

$$S_n = \frac{n}{2}(a_1 + a_n)$$

where a_1 is the first term and a_n is the nth term.

The sum of the first five terms of the arithmetic sequence

$$12, 24, 36, 48, 60, \ldots \text{ is}$$

$$S_5 = \frac{5}{2}(12 + 60) = 180$$

Partial sum, S_n, of the first n terms of a geometric sequence:

$$S_n = \frac{a_1(1 - r^n)}{1 - r}$$

where a_1 is the first term, r is the common ratio, and $r \neq 1$.

The sum of the first five terms of the geometric sequence

$$15, 30, 60, 120, 240, \ldots \text{ is}$$

$$S_5 = \frac{15(1 - 2^5)}{1 - 2} = 465$$

Sum of the terms of an infinite geometric sequence:

$$S_\infty = \frac{a_1}{1 - r}$$

where a_1 is the first term, r is the common ratio, and $|r| < 1$. (If $|r| \geq 1$, S_∞ does not exist.)

The sum of the terms of the infinite geometric sequence

$$1, \frac{1}{3}, \frac{1}{9}, \frac{1}{27}, \ldots \text{ is}$$

$$S_\infty = \frac{1}{1 - \frac{1}{3}} = \frac{3}{2}$$

Section 14.5   The Binomial Theorem

The **factorial of n**, written $n!$, is the product of the first n consecutive natural numbers.

$$5! = 5 \cdot 4 \cdot 3 \cdot 2 \cdot 1 = 120$$

Binomial Theorem

If n is a positive integer, then

$$(a + b)^n = a^n + \frac{n}{1!}a^{n-1}b^1 + \frac{n(n-1)}{2!}a^{n-2}b^2$$

$$+ \frac{n(n-1)(n-2)}{3!}a^{n-3}b^3 + \cdots + b^n$$

Expand: $(3x + y)^4$

$$(3x + y)^4 = (3x)^4 + \frac{4}{1!}(3x)^3(y)^1$$

$$+ \frac{4 \cdot 3}{2!}(3x)^2(y)^2 + \frac{4 \cdot 3 \cdot 2}{3!}(3x)^1y^3 + y^4$$

$$= 81x^4 + 108x^3y + 54x^2y^2 + 12xy^3 + y^4$$

Chapter 14 Review

(14.1) Find the indicated term(s) of the given sequence.

1. The first five terms of the sequence $a_n = -3n^2$

2. The first five terms of the sequence $a_n = n^2 + 2n$

3. The one-hundredth term of the sequence $a_n = \dfrac{(-1)^n}{100}$

4. The fiftieth term of the sequence $a_n = \dfrac{2n}{(-1)^n}$

5. The general term a_n of the sequence $\dfrac{1}{6}, \dfrac{1}{12}, \dfrac{1}{18}, \ldots$

6. The general term a_n of the sequence $-1, 4, -9, 16, \ldots$

(14.2) Solve the following applications.

7. The distance in feet that an olive falling from rest in a vacuum will travel during each second is given by an arithmetic sequence whose general term is $a_n = 32n - 16$, where n is the number of the second. Find the distance the olive will fall during the fifth, sixth, and seventh seconds.

8. A culture of yeast measures 80 and doubles every day in a geometric sequence, where n is the number of the day just ending. Write the measure of the yeast culture at the end of each of the next 5 days. Find how many days it takes the yeast culture to measure at least 10,000.

9. The Colorado Forest Service reported that western pine beetle infestation, which kills trees, affected approximately 97,000 new acres of pine forests in Colorado in 2010. This epidemic seems to be slowing, and the forest service predicted that during the next five years, the beetles would infest 1.5 times the number of new acres per year as the year before. Write out the first five terms of this geometric sequence and find the number of acres of newly infested trees there were in 2014.

10. The first row of an amphitheater contains 50 seats, and each row thereafter contains 8 additional seats. Write the first ten terms of this arithmetic sequence and find the number of seats in the tenth row.

11. Find the first five terms of the geometric sequence whose first term is -2 and whose common ratio is $\dfrac{2}{3}$.

12. Find the first five terms of the arithmetic sequence whose first term is 12 and whose common difference is -1.5.

13. Find the thirtieth term of the arithmetic sequence whose first term is -5 and whose common difference is 4.

14. Find the eleventh term of the arithmetic sequence whose first term is 2 and whose common difference is $\dfrac{3}{4}$.

15. Find the twentieth term of the arithmetic sequence whose first three terms are 12, 7, and 2.

16. Find the sixth term of the geometric sequence whose first three terms are 4, 6, and 9.

17. If the fourth term of an arithmetic sequence is 18 and the twentieth term is 98, find the first term and the common difference.

18. If the third term of a geometric sequence is -48 and the fourth term is 192, find the first term and the common ratio.

19. Find the general term of the sequence $\dfrac{3}{10}, \dfrac{3}{100}, \dfrac{3}{1000}, \ldots$

20. Find a general term that satisfies the terms shown for the sequence $50, 58, 66, \ldots$.

Determine whether each of the following sequences is arithmetic, geometric, or neither. If a sequence is arithmetic, find a_1 and d. If a sequence is geometric, find a_1 and r.

21. $\dfrac{8}{3}, 4, 6, \ldots$

22. $-10.5, -6.1, -1.7$

23. $7x, -14x, 28x$

24. $3x^2, 3x^2 + 5, 3x^2 + 10, \ldots$

Solve the following applications.

25. To test the bounce of a racquetball, the ball is dropped from a height of 8 feet. The ball is judged "good" if it rebounds at least 75% of its previous height with each bounce. Write out the first six terms of this geometric sequence (round to the nearest tenth). Determine if a ball is "good" that rebounds to a height of 2.5 feet after the fifth bounce.

26. A display of oil cans in an auto parts store has 25 cans in the bottom row, 21 cans in the next row, and so on, in an arithmetic progression. Find the general term and the number of cans in the top row.

27. Suppose that you save $1 the first day of a month, $2 the second day, $4 the third day, continuing to double your savings each day. Write the general term of this geometric sequence and find the amount you will save on the tenth day. Estimate the amount you will save on the thirtieth day of the month and check your estimate with a calculator.

28. On the first swing, the length of an arc through which a pendulum swings is 30 inches. The length of the arc for each successive swing is 70% of the preceding swing. Find the length of the arc for the fifth swing.

29. Rosa takes a job that has a monthly starting salary of $900 and guarantees her a monthly raise of $150 during her 6-month training period. Find the general term of this sequence and her salary at the end of her training.

30. A sheet of paper is $\dfrac{1}{512}$ inch thick. By folding the sheet in half, the total thickness will be $\dfrac{1}{256}$ inch. A second fold produces a total thickness of $\dfrac{1}{128}$ inch. Estimate the thickness of the stack after 15 folds and then check your estimate with a calculator.

(14.3) *Write out the terms and find the sum for each of the following.*

31. $\displaystyle\sum_{i=1}^{5}(2i-1)$

32. $\displaystyle\sum_{i=1}^{5}i(i+2)$

33. $\displaystyle\sum_{i=2}^{4}\frac{(-1)^i}{2i}$

34. $\displaystyle\sum_{i=3}^{5}5(-1)^{i-1}$

Write the sum with Σ notation.

35. $1+3+9+27+81+243$

36. $6+2+(-2)+(-6)+(-10)+(-14)+(-18)$

37. $\dfrac{1}{4}+\dfrac{1}{16}+\dfrac{1}{64}+\dfrac{1}{256}$

38. $1+\left(-\dfrac{3}{2}\right)+\dfrac{9}{4}$

Solve.

39. A yeast colony begins with 20 yeast and doubles every 8 hours. Write the sequence that describes the growth of the yeast and find the total yeast after 48 hours.

40. The number of cranes born each year in a new aviary forms a sequence whose general term is $a_n=n^2+2n-1$. Find the number of cranes born in the fourth year and the total number of cranes born in the first four years.

41. Harold has a choice between two job offers. Job A has an annual starting salary of $39,500 with guaranteed annual raises of $2200 for the next four years, whereas job B has an annual starting salary of $41,000 with guaranteed annual raises of $1400 for the next four years. Compare the salaries for the fifth year under each job offer.

42. A sample of radioactive waste is decaying such that the amount decaying in kilograms during year n is $a_n=200(0.5)^n$. Find the amount of decay in the third year and the total amount of decay in the first three years.

(14.4) *Find the partial sum of the given sequence.*

43. S_4 of the sequence $a_n=(n-3)(n+2)$

44. S_6 of the sequence $a_n=n^2$

45. S_5 of the sequence $a_n=-8+(n-1)3$

46. S_3 of the sequence $a_n=5(4)^{n-1}$

47. The sixth partial sum of the sequence $15,19,23,\ldots$

48. The ninth partial sum of the sequence $5,-10,20,\ldots$

49. The sum of the first 30 odd positive integers

50. The sum of the first 20 positive multiples of 7

51. The sum of the first 20 terms of the sequence $8,5,2,\ldots$

52. The sum of the first eight terms of the sequence $\dfrac{3}{4},\dfrac{9}{4},\dfrac{27}{4},\ldots$

53. S_4 if $a_1=6$ and $r=5$

54. S_{100} if $a_1=-3$ and $d=-6$

Find the sum of each infinite geometric sequence.

55. $5,\dfrac{5}{2},\dfrac{5}{4},\ldots$

56. $18,-2,\dfrac{2}{9},\ldots$

57. $-20,-4,-\dfrac{4}{5},\ldots$

58. $0.2,0.02,0.002,\ldots$

Solve.

59. A frozen yogurt store owner cleared $20,000 the first year he owned his business and made an additional 15% over the previous year in each subsequent year. Find how much he made during his fourth year of business. Find his total earnings during the first 4 years (round to the nearest dollar).

60. On his first morning in a television assembly factory, a trainee takes 0.8 times as long to assemble each television as he took to assemble the one before. If it took him 40 minutes to assemble the first television, find how long it takes him to assemble the fourth television. Find the total time he takes to assemble the first four televisions (round to the nearest minute).

61. During the harvest season, a farmer can rent a combine machine for $100 the first day, with the rental fee decreasing $7 for each additional day. Find how much the farmer pays for the rental on the seventh day. Find how much total rent the farmer pays for 7 days.

62. A rubber ball is dropped from a height of 15 feet and rebounds 80% of its previous height after each bounce. Find the total distance the ball travels before it comes to rest.

63. After a pond was sprayed once with insecticide, 1800 mosquitoes were killed the first day, 600 the second day, 200 the third day, and so on. Find the total number of mosquitoes killed during the first 6 days after the spraying (round to the nearest unit).

64. See Exercise 63. Find the day on which the insecticide is no longer effective and find the total number of mosquitoes killed (round to the nearest mosquito).

65. Use the formula S_∞ to write $0.55\overline{5}$ as a fraction.

66. A movie theater has 27 seats in the first row, 30 seats in the second row, 33 seats in the third row, and so on. Find the total number of seats in the theater if there are 20 rows.

(14.5) *Use Pascal's triangle to expand each binomial.*

67. $(x+z)^5$

68. $(y-r)^6$

69. $(2x+y)^4$

70. $(3y-z)^4$

Use the binomial formula to expand the following.

71. $(b + c)^8$

72. $(x - w)^7$

73. $(4m - n)^4$

74. $(p - 2r)^5$

Find the indicated term.

75. The fourth term of the expansion of $(a + b)^7$

76. The eleventh term of the expansion of $(y + 2z)^{10}$

MIXED REVIEW

77. Evaluate: $\displaystyle\sum_{i=1}^{4} i^2(i + 1)$

78. Find the fifteenth term of the arithmetic sequence whose first three terms are $14, 8$, and 2.

79. Find the sum of the infinite geometric sequence $27, 9, 3, 1, \ldots$.

80. Expand: $(2x - 3)^4$

Chapter 14 Getting Ready for the Test

MULTIPLE CHOICE *Select the correct choice.*

1. Find a_3 for the sequence whose general term is $a_n = -n^2$.
    A. 9   **B.** -9 **C.** 6   **D.** -6

MATCHING *Each sequence in the first column is either arithmetic or geometric. If a sequence is arithmetic, match it with the correct a_1 and d in the second column. If geometric, match it with the correct a_1 and r in the second column.*

2. $6, 9, 12$ **A.** $a_1 = 6, r = \dfrac{1}{2}$

3. $6, 9, \dfrac{27}{2}$ **B.** $a_1 = 6, r = \dfrac{3}{2}$

4. $6, 3, 0$ **C.** $a_1 = 6, d = 3$

5. $6, 3, \dfrac{3}{2}$ **D.** $a_1 = 6, d = -3$

MULTIPLE CHOICE *Select the correct choice.*

6. $\displaystyle\sum_{i=3}^{5} (i^2 - 1) =$
    A. 47   **B.** 32   **C.** 8   **D.** 24

7. Find S_{20} of the sequence $a_n = 12 + (n - 1)5$
    A. $\dfrac{119}{2}$ **B.** 1070   **C.** 1190   **D.** 107

8. Find S_{10} of the sequence $a_n = 3(2)^{n-1}$.
    A. 3069   **B.** 1536   **C.** -3069 **D.** 1023

9. Find S_∞ of the sequence $a_n = 3\left(\dfrac{1}{2}\right)^{n-1}$
    A. 3   **B.** $\dfrac{3}{2}$ **C.** $\dfrac{1}{6}$ **D.** 6

10. Evaluate $\dfrac{10 \cdot 9 \cdot 8 \cdot 7 \cdot 6}{5!}$
    A. 3024   **B.** 252   **C.** 30,240   **D.** 6048

Chapter 14 Test   MyMathLab®   YouTube

Find the indicated term(s) of the given sequence.

1. The first five terms of the sequence $a_n = \dfrac{(-1)^n}{n + 4}$

2. The eightieth term of the sequence $a_n = 10 + 3(n - 1)$

3. The general term of the sequence $\dfrac{2}{5}, \dfrac{2}{25}, \dfrac{2}{125}, \ldots$

4. The general term of the sequence $-9, 18, -27, 36, \ldots$

Find the partial sum of the given sequence.

5. S_5 of the sequence $a_n = 5(2)^{n-1}$

6. S_{30} of the sequence $a_n = 18 + (n - 1)(-2)$

7. S_∞ of the sequence $a_1 = 24$ and $r = \dfrac{1}{6}$

8. S_∞ of the sequence $\dfrac{3}{2}, -\dfrac{3}{4}, \dfrac{3}{8}, \ldots$

9. $\displaystyle\sum_{i=1}^{4} i(i-2)$

10. $\displaystyle\sum_{i=2}^{4} 5(2)^i(-1)^{i-1}$

Expand each binomial.

11. $(a-b)^6$

12. $(2x+y)^5$

Solve the following applications.

13. The population of a small town is growing yearly according to the sequence defined by $a_n = 250 + 75(n-1)$, where n is the number of the year just beginning. Predict the population at the beginning of the tenth year. Find the town's initial population.

14. A gardener is making a triangular planting with one shrub in the first row, three shrubs in the second row, five shrubs in the third row, and so on, for eight rows. Write the finite series of this sequence and find the total number of shrubs planted.

15. A pendulum swings through an arc of length 80 centimeters on its first swing. On each successive swing, the length of the arc is $\dfrac{3}{4}$ the length of the arc on the preceding swing. Find the length of the arc on the fourth swing and find the total arc length for the first four swings.

16. See Exercise 15. Find the total arc length before the pendulum comes to rest.

17. A parachutist in free-fall falls 16 feet during the first second, 48 feet during the second second, 80 feet during the third second, and so on. Find how far he falls during the tenth second. Find the total distance he falls during the first 10 seconds.

18. Use the formula S_∞ to write $0.42\overline{42}$ as a fraction.

Chapter 14 Cumulative Review

1. Evaluate.
   a. $(-2)^3$
   b. -2^3
   c. $(-3)^2$
   d. -3^2

2. Simplify each expression.
   a. $3a - (4a+3)$
   b. $(5x-3) + (2x+6)$
   c. $4(2x-5) - 3(5x+1)$

3. Subtract $4x - 2$ from $2x - 3$. Simplify, if possible.

4. Sara bought a digital camera for $344.50 including tax. If the tax rate is 6%, what was the price of the camera before taxes?

5. Find an equation of the line with y-intercept $(0, -3)$ and slope of $\dfrac{1}{4}$.

6. Find an equation of a line through $(3, -2)$ and parallel to $3x - 2y = 6$. Write the equation using function notation.

7. Find an equation of the line through $(2, 5)$ and $(-3, 4)$. Write the equation in standard form.

8. Solve: $y^3 + 5y^2 - y = 5$

9. Use synthetic division to divide $x^4 - 2x^3 - 11x^2 + 5x + 34$ by $x + 2$.

10. Perform the indicated operation and simplify if possible.
$$\frac{5}{3a-6} - \frac{a}{a-2} + \frac{3+2a}{5a-10}$$

11. Simplify the following.
   a. $\sqrt{50}$
   b. $\sqrt[3]{24}$
   c. $\sqrt{26}$
   d. $\sqrt[4]{32}$

12. Solve: $\sqrt{3x+6} - \sqrt{7x-6} = 0$

13. Use the formula $A = P(1+r)^t$ to find the interest rate r if $2000 compounded annually grows to $2420 in 2 years.

14. Rationalize each denominator.
   a. $\sqrt[3]{\dfrac{4}{3x}}$
   b. $\dfrac{\sqrt{2}+1}{\sqrt{2}-1}$

15. Solve: $(x-3)^2 - 3(x-3) - 4 = 0$

16. Solve: $\dfrac{10}{(2x+4)^2} - \dfrac{1}{2x+4} = 3$

17. Solve: $\dfrac{5}{x+1} < -2$

18. Graph $f(x) = (x+2)^2 - 6$. Find the vertex and axis of symmetry.

19. A rock is thrown upward from the ground. Its height in feet above ground after t seconds is given by the function $f(t) = -16t^2 + 20t$. Find the maximum height of the rock and the number of seconds it takes for the rock to reach its maximum height.

20. Find the vertex of $f(x) = x^2 + 3x - 18$.

21. If $f(x) = x^2$ and $g(x) = x + 3$, find each composition.
   a. $(f \circ g)(2)$ and $(g \circ f)(2)$
   b. $(f \circ g)(x)$ and $(g \circ f)(x)$

22. Find the inverse of $f(x) = -2x + 3$.

23. Find the inverse of the one-to-one function.
$f = \{(0,1), (-2,7), (3,-6), (4,4)\}$.

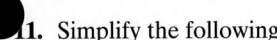

24. If $f(x) = x^2 - 2$ and $g(x) = x + 1$, find each composition.

   a. $(f \circ g)(2)$ and $(g \circ f)(2)$

   b. $(f \circ g)(x)$ and $(g \circ f)(x)$

25. Solve each equation for x.

   a. $2^x = 16$ **b.** $9^x = 27$

   c. $4^{x+3} = 8^x$

26. Solve each equation.

   a. $\log_2 32 = x$ **b.** $\log_4 \dfrac{1}{64} = x$

   c. $\log_{1/2} x = 5$

27. Simplify.

   a. $\log_3 3^2$ **b.** $\log_7 7^{-1}$

   c. $5^{\log_5 3}$ **d.** $2^{\log_2 6}$

28. Solve each equation for x.

   a. $4^x = 64$ **b.** $8^x = 32$

   c. $9^{x+4} = 243^x$

29. Write each sum as a single logarithm.

   a. $\log_{11} 10 + \log_{11} 3$

   b. $\log_3 \dfrac{1}{2} + \log_3 12$

   c. $\log_2(x + 2) + \log_2 x$

30. Find the exact value.

   a. $\log 100{,}000$ **b.** $\log 10^{-3}$

   c. $\ln \sqrt[5]{e}$ **d.** $\ln e^4$

31. Find the amount owed at the end of 5 years if \$1600 is loaned at a rate of 9% compounded continuously.

32. Write each expression as a single logarithm.

   a. $\log_6 5 + \log_6 4$

   b. $\log_8 12 - \log_8 4$

   c. $2 \log_2 x + 3 \log_2 x - 2 \log_2(x - 1)$

33. Solve: $3^x = 7$

34. Using $A = P\left(1 + \dfrac{r}{n}\right)^{nt}$, find how long it takes \$5000 to double if it is invested at 2% interest compounded quarterly. Round to the nearest tenth.

35. Solve: $\log_4(x - 2) = 2$

36. Solve: $\log_4 10 - \log_4 x = 2$

37. Graph: $\dfrac{x^2}{16} - \dfrac{y^2}{25} = 1$

38. Find the distance between $(8, 5)$ and $(-2, 4)$.

39. Solve the system: $\begin{cases} y = \sqrt{x} \\ x^2 + y^2 = 6 \end{cases}$

40. Solve the system: $\begin{cases} x^2 + y^2 = 36 \\ x - y = 6 \end{cases}$

41. Graph: $\dfrac{x^2}{9} + \dfrac{y^2}{16} \le 1$

42. Graph: $\begin{cases} y \ge x^2 \\ y \le 4 \end{cases}$

43. Write the first five terms of the sequence whose general term is given by $a_n = n^2 - 1$.

44. If the general term of a sequence is $a_n = \dfrac{n}{n + 4}$, find a_8.

45. Find the eleventh term of the arithmetic sequence whose first three terms are 2, 9, and 16.

46. Find the sixth term of the geometric sequence $2, 10, 50, \ldots$.

47. Evaluate.

   a. $\displaystyle\sum_{i=0}^{6} \dfrac{i - 2}{2}$ **b.** $\displaystyle\sum_{i=3}^{5} 2^i$

48. Evaluate.

   a. $\displaystyle\sum_{i=0}^{4} i(i + 1)$ **b.** $\displaystyle\sum_{i=0}^{3} 2^i$

49. Find the sum of the first 30 positive integers.

50. Find the third term of the expansion of $(x - y)^6$.

Appendix A

Operations on Decimals/Table of Percent, Decimal, and Fraction Equivalents

A.1 Operations on Decimals

To **add** or **subtract** decimals, write the numbers vertically with decimal points lined up. Add or subtract as with whole numbers and place the decimal point in the answer directly below the decimal points in the problem.

EXAMPLE 1 Add: 5.87 + 23.279 + 0.003

Solution

$$
\begin{array}{r}
5.87 \\
23.279 \\
+\,0.003 \\
\hline
29.152
\end{array}
$$

□

EXAMPLE 2 Subtract: 32.15 − 11.237

Solution

$$
\begin{array}{r}
3\ \overset{1}{2}\ .\ \overset{11}{1}\ \overset{4}{5}\ \overset{10}{0} \\
-\ 1\ 1\ .\ 2\ 3\ 7 \\
\hline
2\ 0\ .\ 9\ 1\ 3
\end{array}
$$

□

To **multiply** decimals, multiply the numbers as if they were whole numbers. The decimal point in the product is placed so that the number of decimal places in the product is the same as the sum of the number of decimal places in the factors.

EXAMPLE 3 Multiply: 0.072 × 3.5

Solution

$$
\begin{array}{rl}
0.072 & \text{3 decimal places} \\
\times\quad 3.5 & \text{1 decimal place} \\
\hline
360 & \\
216\ \ & \\
\hline
0.2520 & \text{4 decimal places}
\end{array}
$$

□

To **divide** decimals, move the decimal point in the divisor to the right of the last digit. Move the decimal point in the dividend the same number of places that the decimal point in the divisor was moved. The decimal point in the quotient lies directly above the decimal point in the dividend.

EXAMPLE 4 Divide: 9.46 ÷ 0.04

Solution

$$
\begin{array}{r}
236.5 \\
04.\overline{)946.0} \\
-8 \\
\hline
14 \\
-12 \\
\hline
26 \\
-24 \\
\hline
20 \\
-20
\end{array}
$$

□

Perform the indicated operations.

1. 9.076 + 8.004

2. 6.3
× 0.05
___

3. 27.004
− 14.2
___

4. 0.0036
7.12
32.502
+ 0.05
___

5. 107.92
+ 3.04
___

6. 7.2 ÷ 4

7. 10 − 7.6

8. 40 ÷ 0.25

9. 126.32 − 97.89

10. 3.62
7.11
12.36
4.15
+ 2.29
___

11. 3.25
× 70
___

12. 26.014
− 7.8
___

13. 8.1 ÷ 3

14. 1.2366
0.005
15.17
+ 0.97
___

15. 55.405 − 6.1711

16. 8.09 + 0.22

17. 60 ÷ 0.75

18. 20 − 12.29

19. 7.612 ÷ 100

20. 8.72
1.12
14.86
3.98
+ 1.99
___

21. 12.312 ÷ 2.7

22. 0.443 ÷ 100

23. 569.2
71.25
+ 8.01
___

24. 3.706 − 2.91

25. 768 − 0.17

26. 63 ÷ 0.28

27. 12 + 0.062

28. 0.42 + 18

29. 76 − 14.52

30. 1.1092 ÷ 0.47

31. 3.311 ÷ 0.43

32. 7.61 + 0.0004

33. 762.12
89.7
+ 11.55
___

34. 444 ÷ 0.6

35. 23.4 − 0.821

36. 3.7 + 5.6

37. 476.12 − 112.97

38. 19.872 ÷ 0.54

39. 0.007 + 7

40. 51.77
+ 3.6
___

A.2 Table of Percent, Decimal, and Fraction Equivalents

Percent, Decimal, and Fraction Equivalents

Percent	Decimal	Fraction
1%	0.01	$\frac{1}{100}$
5%	0.05	$\frac{1}{20}$
10%	0.1	$\frac{1}{10}$
12.5% or $12\frac{1}{2}$%	0.125	$\frac{1}{8}$
$16.\overline{6}$% or $16\frac{2}{3}$%	$0.1\overline{6}$	$\frac{1}{6}$
20%	0.2	$\frac{1}{5}$
25%	0.25	$\frac{1}{4}$
30%	0.3	$\frac{3}{10}$
$33.\overline{3}$% or $33\frac{1}{3}$%	$0.\overline{3}$	$\frac{1}{3}$
37.5% or $37\frac{1}{2}$%	0.375	$\frac{3}{8}$
40%	0.4	$\frac{2}{5}$
50%	0.5	$\frac{1}{2}$
60%	0.6	$\frac{3}{5}$
62.5% or $62\frac{1}{2}$%	0.625	$\frac{5}{8}$
$66.\overline{6}$% or $66\frac{2}{3}$%	$0.\overline{6}$	$\frac{2}{3}$
70%	0.7	$\frac{7}{10}$
75%	0.75	$\frac{3}{4}$
80%	0.8	$\frac{4}{5}$
$83.\overline{3}$% or $83\frac{1}{3}$%	$08.\overline{3}$	$\frac{5}{6}$
87.5% or $87\frac{1}{2}$%	0.875	$\frac{7}{8}$
90%	0.9	$\frac{9}{10}$
100%	1.0	1
110%	1.1	$1\frac{1}{10}$
125%	1.25	$1\frac{1}{4}$
$133.\overline{3}$% or $133\frac{1}{3}$%	$1.\overline{3}$	$1\frac{1}{3}$
150%	1.5	$1\frac{1}{2}$
$166.\overline{6}$% or $166\frac{2}{3}$%	$1.\overline{6}$	$1\frac{2}{3}$
175%	1.75	$1\frac{3}{4}$
200%	2.0	2

Appendix B

Review of Algebra Topics

Recall that equations model many real-life problems. For example, we can use a linear equation to calculate the increase in the number (in billions) of worldwide users of social network sites.

Social network sites allow us to connect with more people, and many sites allow us to project our own image to the world. The number of social network sites is constantly in flux, and today's hot spot is tomorrow's has-been. Current and projected users of social network sites are shown below.

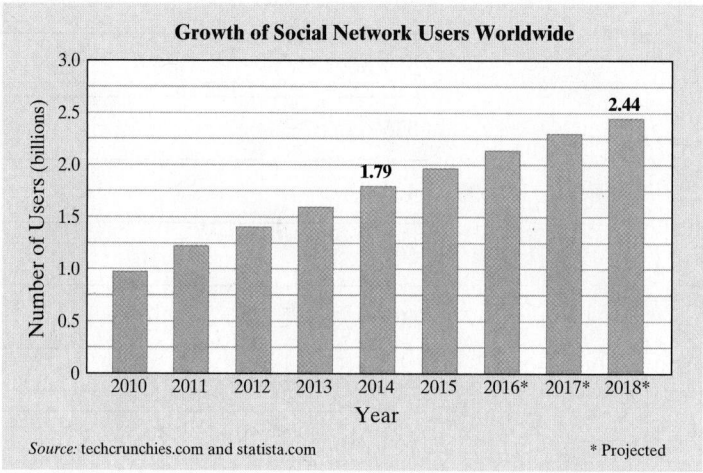

Growth of Social Network Users Worldwide

Source: techcrunchies.com and statista.com   * Projected

To find the projected increase in the number of social network users worldwide from 2014 to 2018, for example, we can use the equation below.

In words:	Increase in social network users	is	users in 2018	minus	users in 2014
Translate:	x	$=$	2.44	$-$	1.79

Since our variable x (increase in social network users) is by itself on one side of the equation, we can find the value of x by simplifying the right side.

$$x = 0.65$$

The projected increase in social network users worldwide from 2014 to 2018 is 0.65 billion people.

The **equation** $x = 2.44 - 1.79$ is a linear equation in one variable. In this section, we review solving linear equations and quadratic equations that can be solved by factoring. We learn other methods for solving quadratic equations in Chapter 11.

B.1   Equations (Linear and Quadratic Solved by Factoring)

OBJECTIVE

1 Solve Linear and Quadratic Equations.

OBJECTIVE

1   Solving Linear and Quadratic Equations

EXAMPLE 1   Solve: $2(x - 3) = 5x - 9$

Solution First, use the distributive property.

$$2(x - 3) = 5x - 9$$
$$2x - 6 = 5x - 9 \quad \text{Use the distributive property.}$$

Next, get variable terms on the same side of the equation by subtracting $5x$ from both sides.

$$2x - 6 - 5x = 5x - 9 - 5x \quad \text{Subtract } 5x \text{ from both sides.}$$
$$-3x - 6 = -9 \quad \text{Simplify.}$$
$$-3x - 6 + 6 = -9 + 6 \quad \text{Add 6 to both sides.}$$
$$-3x = -3 \quad \text{Simplify.}$$
$$\frac{-3x}{-3} = \frac{-3}{-3} \quad \text{Divide both sides by } -3.$$
$$x = 1$$

Let $x = 1$ in the original equation to see that 1 is the solution.   □

PRACTICE

1   Solve: $3(x - 5) = 6x - 3$ ■

Don't forget, if an equation contains fractions, you may want to first clear the equation of fractions by multiplying both sides of the equation by the *least common denominator* (LCD) of all fractions in the equation.

EXAMPLE 2   Solve for y: $\dfrac{y}{3} - \dfrac{y}{4} = \dfrac{1}{6}$

Solution First, clear the equation of fractions by multiplying both sides of the equation by 12, the LCD of denominators 3, 4, and 6.

$$\frac{y}{3} - \frac{y}{4} = \frac{1}{6}$$

$$12\left(\frac{y}{3} - \frac{y}{4}\right) = 12\left(\frac{1}{6}\right) \quad \text{Multiply both sides by the LCD, 12.}$$

$$12\left(\frac{y}{3}\right) - 12\left(\frac{y}{4}\right) = 2 \quad \text{Apply the distributive property.}$$

$$4y - 3y = 2 \quad \text{Simplify.}$$
$$y = 2 \quad \text{Simplify.}$$

Check: To check, let $y = 2$ in the original equation.

$$\frac{y}{3} - \frac{y}{4} = \frac{1}{6} \quad \text{Original equation.}$$

$$\frac{2}{3} - \frac{2}{4} \stackrel{?}{=} \frac{1}{6} \quad \text{Let } y = 2.$$

$$\frac{8}{12} - \frac{6}{12} \stackrel{?}{=} \frac{1}{6} \quad \text{Write fractions with the LCD.}$$

(Continued on next page)

$$\frac{2}{12} \stackrel{?}{=} \frac{1}{6} \quad \text{Subtract.}$$

$$\frac{1}{6} = \frac{1}{6} \quad \text{Simplify.}$$

This is a true statement, so the solution is 2.

PRACTICE
2 Solve for y: $\frac{y}{2} - \frac{y}{5} = \frac{1}{4}$

EXAMPLE 3 Solve: $3(x^2 + 4) + 5 = -6(x^2 + 2x) + 13$

Solution Rewrite the equation so that one side is 0.

$$\begin{aligned}
3(x^2 + 4) + 5 &= -6(x^2 + 2x) + 13 \\
3x^2 + 12 + 5 &= -6x^2 - 12x + 13 \qquad &\text{Apply the distributive property.} \\
9x^2 + 12x + 4 &= 0 &\text{Rewrite the equation so that one side is 0.} \\
(3x + 2)(3x + 2) &= 0 &\text{Factor.} \\
3x + 2 = 0 \quad \text{or} \quad 3x + 2 &= 0 &\text{Set each factor equal to 0.} \\
3x = -2 \quad \text{or} \quad 3x &= -2 \\
x = -\frac{2}{3} \quad \text{or} \quad x &= -\frac{2}{3} &\text{Solve each equation.}
\end{aligned}$$

The solution is $-\frac{2}{3}$. Check by substituting $-\frac{2}{3}$ into the original equation.

PRACTICE
3 Solve: $8(x^2 + 3) + 4 = -8x(x + 3) + 19$

EXAMPLE 4 Solve for x: $\frac{x + 5}{2} + \frac{1}{2} = 2x - \frac{x - 3}{8}$

Solution Multiply both sides of the equation by 8, the LCD of 2 and 8.

$$8\left(\frac{x + 5}{2} + \frac{1}{2}\right) = 8\left(2x - \frac{x - 3}{8}\right) \qquad \text{Multiply both sides by 8.}$$

$$8\left(\frac{x + 5}{2}\right) + 8 \cdot \frac{1}{2} = 8 \cdot 2x - 8\left(\frac{x - 3}{8}\right) \qquad \text{Apply the distributive property.}$$

$$\begin{aligned}
4(x + 5) + 4 &= 16x - (x - 3) &\text{Simplify.} \\
4x + 20 + 4 &= 16x - x + 3 &\text{Use the distributive property to remove parentheses.} \\
4x + 24 &= 15x + 3 &\text{Combine like terms.} \\
-11x + 24 &= 3 &\text{Subtract } 15x \text{ from both sides.} \\
-11x &= -21 &\text{Subtract 24 from both sides.} \\
\frac{-11x}{-11} &= \frac{-21}{-11} &\text{Divide both sides by } -11. \\
x &= \frac{21}{11} &\text{Simplify.}
\end{aligned}$$

> **Helpful Hint**
>
> When we multiply both sides of an equation by a number, the distributive property tells us that each term of the equation is multiplied by the number.

Check: To check, verify that replacing x with $\frac{21}{11}$ makes the original equation true. The solution is $\frac{21}{11}$.

PRACTICE
4 Solve for x: $x - \frac{x - 2}{12} = \frac{x + 3}{4} + \frac{1}{4}$

EXAMPLE 5   Solve: $2x^2 = \dfrac{17}{3}x + 1$

Solution $2x^2 = \dfrac{17}{3}x + 1$

$3(2x^2) = 3\left(\dfrac{17}{3}x + 1\right)$ Clear the equation of fractions.

$6x^2 = 17x + 3$ Apply the distributive property.

$6x^2 - 17x - 3 = 0$ Rewrite the equation in standard form.

$(6x + 1)(x - 3) = 0$ Factor.

$6x + 1 = 0 \quad \text{or} \quad x - 3 = 0$ Set each factor equal to zero.

$6x = -1$

$x = -\dfrac{1}{6} \quad \text{or} \quad x = 3$ Solve each equation.

The solutions are $-\dfrac{1}{6}$ and 3.   □

PRACTICE
5   Solve: $4x^2 = \dfrac{15}{2}x + 1$ ■

B.1   Exercise Set   MyMathLab® ▶

MIXED PRACTICE

Solve each equation. See Examples 1 through 5.

1. $x^2 + 11x + 24 = 0$

2. $y^2 - 10y + 24 = 0$

3. $3x - 4 - 5x = x + 4 + x$

4. $13x - 15x + 8 = 4x + 2 - 24$

5. $12x^2 + 5x - 2 = 0$

6. $3y^2 - y - 14 = 0$

7. $z^2 + 9 = 10z$

8. $n^2 + n = 72$

9. $5(y + 4) = 4(y + 5)$

10. $6(y - 4) = 3(y - 8)$

11. $0.6x - 10 = 1.4x - 14$

12. $0.3x + 2.4 = 0.1x + 4$

13. $x(5x + 2) = 3$

14. $n(2n - 3) = 2$

15. $6x - 2(x - 3) = 4(x + 1) + 4$

16. $10x - 2(x + 4) = 8(x - 2) + 6$

17. $\dfrac{3}{8} + \dfrac{b}{3} = \dfrac{5}{12}$

18. $\dfrac{a}{2} + \dfrac{7}{4} = 5$

19. $x^2 - 6x = x(8 + x)$

20. $n(3 + n) = n^2 + 4n$

21. $\dfrac{z^2}{6} - \dfrac{z}{2} - 3 = 0$

22. $\dfrac{c^2}{20} - \dfrac{c}{4} + \dfrac{1}{5} = 0$

23. $-z + 3(2 + 4z) = 6(z + 1) + 5z$

24. $4(m - 6) - m = 8(m - 3) - 5m$

25. $\dfrac{x^2}{2} + \dfrac{x}{20} = \dfrac{1}{10}$

26. $\dfrac{y^2}{30} = \dfrac{y}{15} + \dfrac{1}{2}$

27. $\dfrac{4t^2}{5} = \dfrac{t}{5} + \dfrac{3}{10}$

28. $\dfrac{5x^2}{6} - \dfrac{7x}{2} + \dfrac{2}{3} = 0$

29. $\dfrac{3t + 1}{8} = \dfrac{5 + 2t}{7} + 2$

30. $4 - \dfrac{2z + 7}{9} = \dfrac{7 - z}{12}$

31. $\dfrac{m - 4}{3} - \dfrac{3m - 1}{5} = 1$

32. $\dfrac{n + 1}{8} - \dfrac{2 - n}{3} = \dfrac{5}{6}$

33. $3x^2 = -x$

34. $y^2 = -5y$

35. $x(x - 3) = x^2 + 5x + 7$

36. $z^2 - 4z + 10 = z(z - 5)$

37. $3(t - 8) + 2t = 7 + t$

38. $7c - 2(3c + 1) = 5(4 - 2c)$

39. $-3(x - 4) + x = 5(3 - x)$

40. $-4(a + 1) - 3a = -7(2a - 3)$

41. $(x - 1)(x + 4) = 24$

42. $(2x - 1)(x + 2) = -3$

43. $\dfrac{x^2}{4} - \dfrac{5}{2}x + 6 = 0$

44. $\dfrac{x^2}{18} + \dfrac{x}{2} + 1 = 0$

45. $y^2 + \dfrac{1}{4} = -y$

46. $\dfrac{x^2}{10} + \dfrac{5}{2} = x$

47. Which solution strategies are incorrect? Why?

    a. Solve $(y - 2)(y + 2) = 4$ by setting each factor equal to 4.

    b. Solve $(x + 1)(x + 3) = 0$ by setting each factor equal to 0.

    c. Solve $z^2 + 5z + 6 = 0$ by factoring $z^2 + 5z + 6$ and setting each factor equal to 0.

    d. Solve $x^2 + 6x + 8 = 10$ by factoring $x^2 + 6x + 8$ and setting each factor equal to 0.

48. Describe two ways a linear equation differs from a quadratic equation.

Find the value of K such that the equations are equivalent.

49. $3.2x + 4 = 5.4x - 7$
$3.2x = 5.4x + K$

50. $-7.6y - 10 = -1.1y + 12$
$-7.6y = -1.1y + K$

51. $\dfrac{x}{6} + 4 = \dfrac{x}{3}$
$x + K = 2x$

52. $\dfrac{5x}{4} + \dfrac{1}{2} = \dfrac{x}{2}$
$5x + K = 2x$

Solve and check.

53. $2.569x = -12.48534$

54. $-9.112y = -47.537304$

55. $2.86z - 8.1258 = -3.75$

56. $1.25x - 20.175 = -8.15$

B.2   Problem Solving

OBJECTIVES

1 Write Algebraic Expressions That Can Be Simplified.

2 Apply the Steps for Problem Solving.

OBJECTIVE

1   Writing and Simplifying Algebraic Expressions

In order to prepare for problem solving, we practice writing algebraic expressions that can be simplified.

    Our first example involves consecutive integers and perimeter. Recall that *consecutive integers* are integers that follow one another in order. Study the examples of consecutive, even, and odd integers and their representations.

Consecutive Integers:

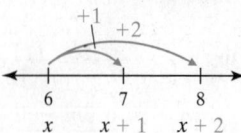

Consecutive Even Integers:

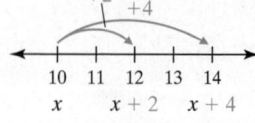

Consecutive Odd Integers:

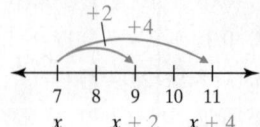

EXAMPLE 1   Write the following as algebraic expressions. Then simplify.

    a. The sum of three consecutive integers, if x is the first consecutive integer.

    △ **b.** The perimeter of a triangle with sides of length x, $5x$, and $6x - 3$.

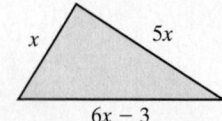

Solution

a. Recall that if x is the first integer, then the next consecutive integer is 1 more, or $x + 1$, and the next consecutive integer is 1 more than $x + 1$, or $x + 2$.

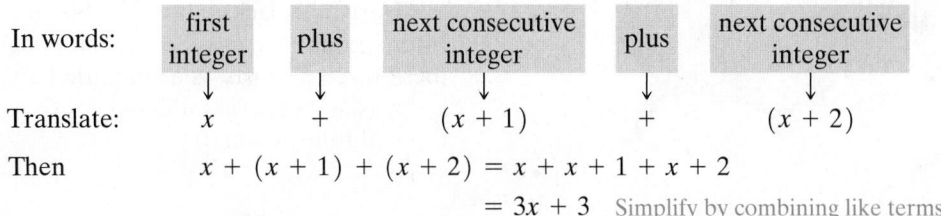

In words: | first integer | plus | next consecutive integer | plus | next consecutive integer |

Translate: $\quad x \qquad + \qquad (x + 1) \qquad + \qquad (x + 2)$

Then $\quad x + (x + 1) + (x + 2) = x + x + 1 + x + 2$

$\qquad\qquad\qquad\qquad\qquad = 3x + 3 \quad$ Simplify by combining like terms.

b. The perimeter of a triangle is the sum of the lengths of the sides.

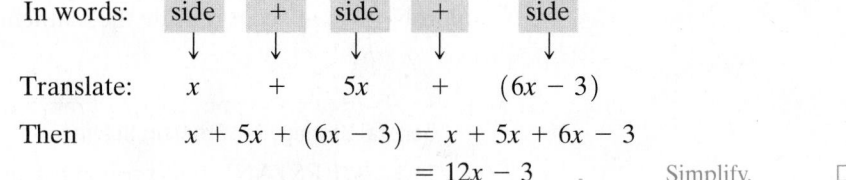

In words: | side | + | side | + | side |

Translate: $\quad x \quad + \quad 5x \quad + \quad (6x - 3)$

Then $\quad x + 5x + (6x - 3) = x + 5x + 6x - 3$

$\qquad\qquad\qquad\qquad\qquad = 12x - 3 \quad$ Simplify. □

PRACTICE
1 Write the following algebraic expressions. Then simplify.

a. The sum of three consecutive odd integers if x is the first consecutive odd integer

b. The perimeter of a trapezoid with bases x and $2x$ and sides of $x + 2$ and $2x - 3$

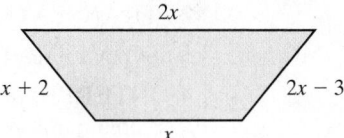

2x

x + 2   2x − 3

x

■

EXAMPLE 2   **Writing Algebraic Expressions Representing Urban Areas**

The most populous urban area in the United States is New York, although it is only the ninth most populous in the world. Tokyo-Yokohama is the most populous urban area. Delhi, India, is third in size. Delhi's urban area population is 4.6 million more than that of New York, and Tokyo-Yokohama's population is twice that of New York, decreased by 3.4 million. Write the sum of the population of these three urban areas as an algebraic expression. Let x be the population of New York (in millions). (*Source:* Demographia World Urban Areas, 2015.01)

Solution

If $x =$ the population of New York (in millions), then

$x + 4.6 =$ the population of Delhi (in millions) and

$2x - 3.4 =$ the population of Tokyo-Yokohama (in millions).

In words: | population of New York | plus | population of Delhi | plus | population of Tokyo-Yokohama |

Translate: $\qquad x \qquad\qquad + \qquad (x + 4.6) \qquad + \qquad (2x - 3.4)$

Then $x + (x + 4.6) + (2x - 3.4) = x + x + 2x + 4.6 - 3.4$

$\qquad\qquad\qquad\qquad\qquad\qquad = 4x + 1.2 \quad$ Combine like terms.

In Exercise 57 we will find the actual populations of these cities. □

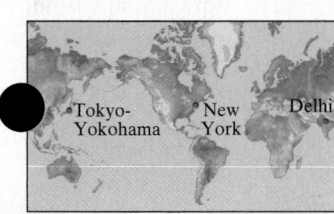

Tokyo-Yokohama   New York   Delhi

Populous Urban Areas

2 The three busiest airports by number of passengers in the United States are in Atlanta, Chicago, and Los Angeles. Chicago's O'Hare airport ha[s] 3.1 million more passengers than Los Angeles International Airport. Atlanta's Hartsfield-Jackson had double the number of passengers as Los Angeles International Airport less 31.9 million. Write the sum of the passengers from these three airports as a simplified algebraic expression. Let x be the number of passengers (in millions) at Los Angeles International. (*Source:* Airport Council International)

OBJECTIVE

2 Applying Steps for Problem Solving

Our main purpose for studying algebra is to solve problems. The following problem-solving strategy will be used throughout this text and may also be used to solve real-life problems that occur outside the mathematics classroom.

General Strategy for Problem Solving

1. **UNDERSTAND** the problem. During this step, become comfortable with the problem. Some ways of doing this are to:

   Read and reread the problem.

   Propose a solution and check. Pay careful attention to how you check your proposed solution. This will help when writing an equation to model the problem.

   Construct a drawing.

   Choose a variable to represent the unknown. (Very important part)

2. **TRANSLATE** the problem into an equation.

3. **SOLVE** the equation.

4. **INTERPRET** the results: *Check* the proposed solution in the stated problem and *state* your conclusion.

Let's review this strategy by solving a problem involving unknown numbers.

EXAMPLE 3 **Finding Unknown Numbers**

Find three numbers such that the second number is 3 more than twice the first number, and the third number is four times the first number. The sum of the three numbers is 164.

Solution

1. UNDERSTAND the problem. First let's read and reread the problem and then propose a solution. For example, if the first number is 25, then the second number is 3 more than twice 25, or 53. The third number is four times 25, or 100. The sum of 25, 53, and 100 is 178, not the required sum, but we have gained some valuable information about the problem. First, we know that the first number is less than 25 since our guess led to a sum greater than the required sum. Also, we have gained some information as to how to model the problem.

   Next let's assign a variable and use this variable to represent any other unknown quantities. If we let

$$x = \text{the first number, then}$$
$$2x + 3 = \text{the second number}$$

3 more than
twice the first number

$$4x = \text{the third number}$$

Helpful Hint

The purpose of guessing a solution is not to guess correctly but to gain confidence and to help understand the problem and how to model it.

2. TRANSLATE the problem into an equation. To do so, we use the fact that the sum of the numbers is 164. First let's write this relationship in words and then translate into an equation.

	first number	added to	second number	added to	third number	is	164
In words:	↓	↓	↓	↓	↓	↓	↓
Translate:	x	$+$	$(2x + 3)$	$+$	$4x$	$=$	164

3. SOLVE the equation.

$$x + (2x + 3) + 4x = 164$$

$x + 2x + 4x + 3 = 164$	Remove parentheses.
$7x + 3 = 164$	Combine like terms.
$7x = 161$	Subtract 3 from both sides.
$x = 23$	Divide both sides by 7.

4. INTERPRET. Here, we *check* our work and *state* the solution. Recall that if the first number $x = 23$, then the second number $2x + 3 = 2 \cdot 23 + 3 = 49$ and the third number $4x = 4 \cdot 23 = 92$.

Check: Is the second number 3 more than twice the first number? Yes, since 3 more than twice 23 is $46 + 3$, or 49. Also, 92 is 4 times 23 and the sum, $23 + 49 + 92 = 164$, is the required sum.

State: The three numbers are 23, 49, and 92.

PRACTICE
3   Find three numbers such that the second number is 8 less than triple the first number, the third number is five times the first number, and the sum of the three numbers is 118.

Many of today's rates and statistics are given as percents. Interest rates, tax rates, nutrition labeling, and percent of households in a given category are just a few examples. Before we practice solving problems containing percents, let's briefly review the meaning of percent and how to find a percent of a number.

The word *percent* means "per hundred," and the symbol % denotes percent.

This means that 23% is 23 per hundred, or $\dfrac{23}{100}$. Also,

$$41\% = \frac{41}{100} = 0.41$$

To find a percent of a number, we multiply.

$$16\% \text{ of } 25 = 16\% \cdot 25 = 0.16 \cdot 25 = 4$$

Thus, 16% of 25 is 4.

Study the table below. It will help you become more familiar with finding percents.

Percent	*Meaning/Shortcut*	*Example*
50%	$\dfrac{1}{2}$ or half of a number	50% of 60 is 30.
25%	$\dfrac{1}{4}$ or a quarter of a number	25% of 60 is 15.
10%	0.1 or $\dfrac{1}{10}$ of a number (move the decimal point 1 place to the left)	10% of 60 is 6.0 or 6.
1%	0.01 or $\dfrac{1}{100}$ of a number (move the decimal point 2 places to the left)	1% of 60 is 0.60 or 0.6.
100%	1 or all of a number	100% of 60 is 60.
200%	2 or double a number	200% of 60 is 120.

✔ **CONCEPT CHECK**

Suppose you are finding 112% of a number x. Which of the following is a correct description of the result? Explain.

a. The result is less than x.   **b.** The result is equal to x.   **c.** The result is greater than x.

Next, we solve a problem containing a percent.

EXAMPLE 4   **Finding the Original Price of a Computer**

Suppose that a computer store just announced an 8% decrease in the price of a particular computer model. If this computer sells for $2162 after the decrease, find the original price of this computer.

Solution

1. UNDERSTAND. Read and reread the problem. Recall that a percent decrease means a percent of the original price. Let's guess that the original price of the computer is $2500. The amount of decrease is then 8% of $2500, or $(0.08)(\$2500) = \$200$. This means that the new price of the computer is the original price minus the decrease, or $2500 − $200 = $2300. Our guess is incorrect, but we now have an idea of how to model this problem. In our model, we will let x = the original price of the computer.

2. TRANSLATE.

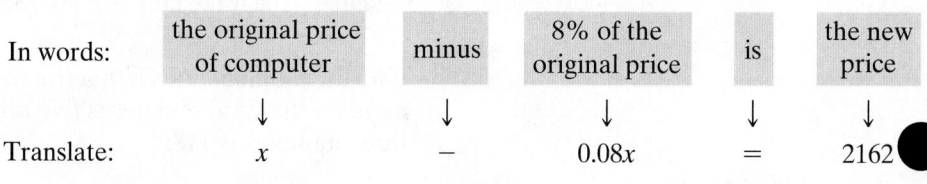

In words:	the original price of computer	minus	8% of the original price	is	the new price
	↓	↓	↓	↓	↓
Translate:	x	−	$0.08x$	=	2162

3. SOLVE the equation.

$$x - 0.08x = 2162$$
$$0.92x = 2162 \qquad \text{Combine like terms.}$$
$$x = \frac{2162}{0.92} = 2350 \quad \text{Divide both sides by 0.92.}$$

4. INTERPRET.

Check: If the original price of the computer was $2350, the new price is

$$\$2350 - (0.08)(\$2350) = \$2350 - \$188$$
$$= \$2162 \qquad \text{The given new price}$$

State: The original price of the computer was $2350. ☐

PRACTICE

4   At the end of the season, the cost of a snowboard was reduced by 40%. If the snowboard sells for $270 after the decrease, find the original price of the board. ∎

Answer to Concept Check:
c; answers may vary

✔ **Vocabulary & Readiness Check**

Fill in each blank with $<$, $>$, or $=$. (Assume that the unknown number is a positive number.)

1. 130% of a number ___ the number.   **2.** 70% of a number ___ the number.

3. 100% of a number ___ the number.   **4.** 200% of a number ___ the number.

Complete the table. The first row has been completed for you.

	First Integer	All Described Integers
Three consecutive integers	18	18, 19, 20
5. Four consecutive integers	31	
6. Three consecutive odd integers	31	
7. Three consecutive even integers	18	
8. Four consecutive even integers	92	
9. Three consecutive integers	y	
10. Three consecutive even integers	z (z is even)	
11. Four consecutive integers	p	
12. Three consecutive odd integers	s (s is odd)	

B.2 Exercise Set MyMathLab®

Write the following as algebraic expressions. Then simplify. See Examples 1 and 2.

△ **1.** The perimeter of a square with side length y.

△ **2.** The perimeter of a rectangle with length x and width $x - 5$.

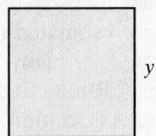

▶ **3.** The sum of three consecutive integers if the first is z.

4. The sum of three consecutive odd integers if the first integer is z.

▶ **5.** The total amount of money (in cents) in x nickels, $(x + 3)$ dimes, and $2x$ quarters. (*Hint:* The value of a nickel is 5 cents, the value of a dime is 10 cents, and the value of a quarter is 25 cents.)

6. The total amount of money (in cents) in y quarters, $7y$ dimes, and $(2y - 1)$ nickels. (Use the hint for Exercise 5.)

△ **7.** A piece of land along Bayou Liberty is to be fenced and subdivided as shown so that each rectangle has the same dimensions. Express the total amount of fencing needed as an algebraic expression in x.

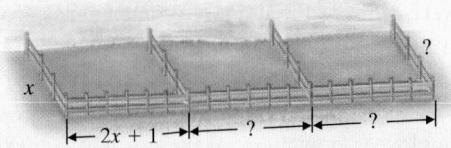

8. A flooded piece of land near the Mississippi River in New Orleans is to be surveyed and divided into 4 rectangles of equal dimension. Express the total amount of fencing needed as an algebraic expression in x.

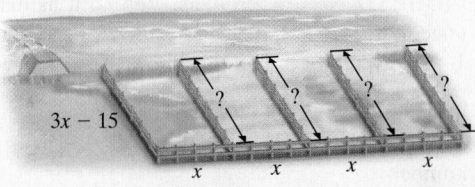

△ **9.** Write the perimeter of the floor plan shown as an algebraic expression in x.

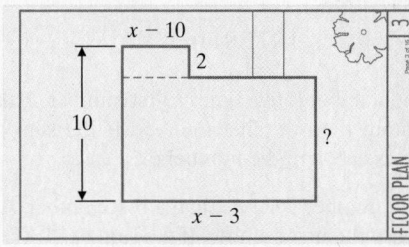

10. Write the perimeter of the floor plan shown as an algebraic expression in x.

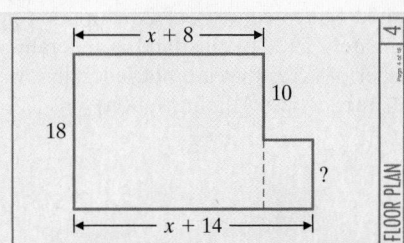

Solve. For Exercises 11 and 12, the solutions have been started for you. See Example 3.

11. Four times the difference of a number and 2 is the same as 2, increased by four times the number, plus twice the number. Find the number.

Start the solution:

1. UNDERSTAND the problem. Reread it as many times as needed.

2. TRANSLATE into an equation. (Fill in the blanks below.)

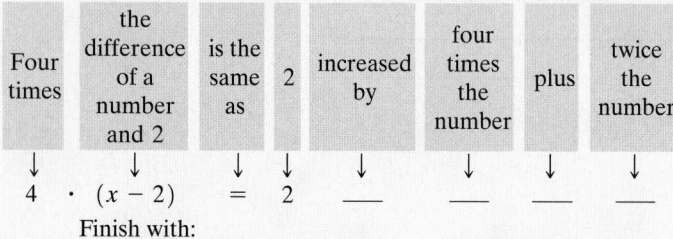

Four times	the difference of a number and 2	is the same as	2	increased by	four times the number	plus	twice the number
↓	↓	↓	↓	↓	↓	↓	↓
4	· (x − 2)	=	2	___	___	___	___

Finish with:

3. SOLVE and **4.** INTERPRET

12. Twice the sum of a number and 3 is the same as five times the number, minus 1, minus four times the number. Find the number.

Start the solution:

1. UNDERSTAND the problem. Reread it as many times as needed.

2. TRANSLATE into an equation. (Fill in the blanks below.)

Twice	the sum of a number and 3	is the same as	five times the number	minus	1	minus	four times the number
↓	↓	↓	↓	↓	↓	↓	↓
2	(x + 3)	=	___	___	1	___	___

Finish with:

3. SOLVE and **4.** INTERPRET

▷ **13.** A second number is five times a first number. A third number is 100 more than the first number. If the sum of the three numbers is 415, find the numbers.

14. A second number is 6 less than a first number. A third number is twice the first number. If the sum of the three numbers is 306, find the numbers.

Solve. See Example 4.

15. The United States consists of 2271 million acres of land. Approximately 29% of this land is federally owned. Find the number of acres that are not federally owned. (*Source:* U.S. General Services Administration)

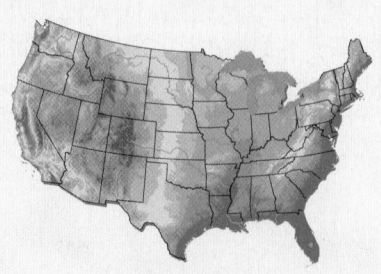

16. The state of Nevada contains the most federally owned acres of land in the United States. If 90% of the state's 70 million acres of land is federally owned, find the number of acres that are not federally owned. (*Source:* U.S. General Services Administration)

17. In 2014, 16,674 earthquakes occurred in the United States. Of these, 89.6% were minor tremors with magnitudes of 4.9 or less on the Richter scale. How many minor earthquakes occurred in the United States in 2014? Round to the nearest whole. (*Source:* U.S. Geological Survey National Earthquake Information Center)

18. Of the 881 tornadoes that occurred in the United States during 2014, 32.2% occurred during the month of June. How many tornadoes occurred in the United States during June 2014? Round to the nearest whole. (*Source:* Storm Prediction Center)

19. In a recent survey, 15% of online shoppers in the United States say that they prefer to do business only with large, well-known retailers. In a group of 1500 online shoppers, how many are willing to do business with any size retailers? (*Source:* Inc.com)

20. In 2014, the restaurant industry employed 10% of the U.S. workforce. If there are estimated to be 147 million Americans in the workforce, how many people are employed by the restaurant industry? Round to the nearest tenth. (*Source:* National Restaurant Association, U.S. Bureau of Labor Statistics)

The following graph is called a circle graph or a pie chart. The circle represents a whole, or in this case, 100%. This particular graph shows the number of minutes per day that people use email at work. Use this graph to answer Exercises 21 through 24.

Time Spent on Email at Work

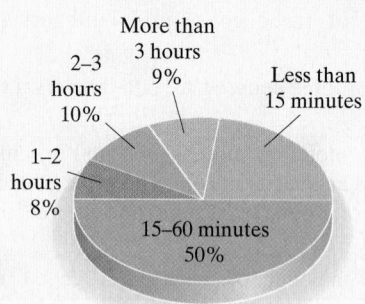

Source: Pew Internet & American Life Project

21. What percent of email users at work spend less than 15 minutes on email per day?

22. Among email users at work, what is the most common time spent on email per day?

23. If it were estimated that a large company has 4633 employees, how many of these would you expect to be using email more than 3 hours per day?

24. If it were estimated that a medium-size company has 250 employees, how many of these would you expect to be using email between 2 and 3 hours per day?

MIXED PRACTICE

Use the diagrams to find the unknown measures of angles or lengths of sides. Recall that the sum of the angle measures of a triangle is 180°.

25.

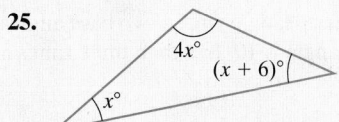

26.

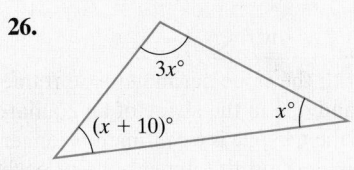

27.

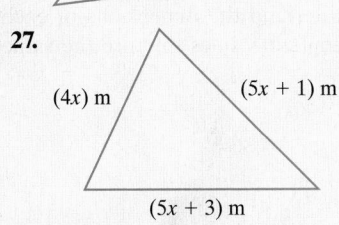

Perimeter is 102 meters.

28.

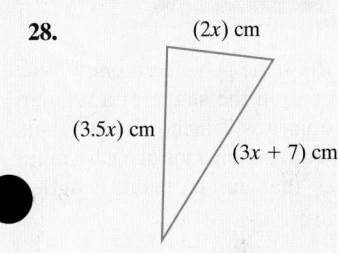

Perimeter is 75 centimeters.

29.

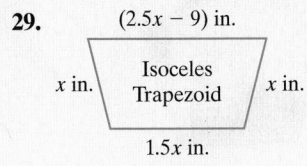

Perimeter is 99 inches.

30.

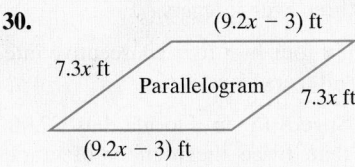

Perimeter is 324 feet.

Solve.

31. The sum of three consecutive integers is 228. Find the integers.

32. The sum of three consecutive odd integers is 327. Find the integers.

33. The ZIP codes of three Nevada locations—Fallon, Fernley, and Gardnerville Ranchos—are three consecutive even integers. If twice the first integer added to the third is 268,222, find each ZIP code.

34. During a recent year, the average SAT scores in math for the states of Alabama, Louisiana, and Michigan were 3 consecutive integers. If the sum of the first integer, second integer, and three times the third integer is 2637, find each score.

Many companies predict the growth or decline of various social network use. The following data is from Techcrunchies, a technical information site. Notice that the first table is the predicted increase in the number of people using social network sites worldwide (in billions) and the second is the predicted percent increase in the number of social network site users worldwide.

35. Use the middle column in the table to find the predicted number of social network site users for each year.

Year	Increase in Social Network Users	Predicted Number
2015	$2x$	
2016	$2x + 170$	
2017	$3x - 650$	
Total	6380 million	

36. Use the middle column in the table to find the predicted percent of increase in the number of social network site users for each year.

Year	Percent of Increase in Social Network Users	Predicted Percent of Increase
2015	x	
2016	$3x - 11$	
2017	$2x + 11$	
Total	66%	

37. The occupations of registered nurse, food service employee, and customer service representative are among the 10 with the largest predicted growth from 2012 to 2022. The number of registered nurses is predicted to grow 317 thousand less than twice the number of food service jobs. The number of customer service jobs is predicted to grow 88 thousand more than half the number of food service jobs. If the total growth of these three jobs is predicted to be 1248 thousand, find the predicted growth of each job. (*Source:* U.S. Department of Labor, Bureau of Labor Statistics)

38. The occupations of reporter, travel agent, and flight attendant are among the 10 jobs with the largest predicted decline from 2012 to 2022. The number of reporter or correspondent jobs is predicted to decline 17 hundred less than the number of travel agent jobs. The number of flight attendant jobs is predicted to decline 21 hundred less than twice the number of travel agent jobs. If the total decline of these three jobs is predicted to be 318 hundred, find the predicted decline of each job. (*Source:* U.S. Department of Labor, Bureau of Labor Statistics)

39. The B767-300ER aircraft has 88 more seats than the B737-200 aircraft. The F-100 has 32 fewer seats than the B737-200 aircraft. If their total number of seats is 413, find the number of seats for each aircraft. (*Source:* Air Transport Association of America)

40. AT&T Stadium, home of the Dallas Cowboys of the NFL, seats approximately 11,200 more fans than does Gillette Stadium, home of the New England Patriots. CenturyLink Field, home of the Seattle Seahawks, seats 3800 fewer than Gillette Stadium. If the total seats in these three stadiums is 213,800, how many seats are in each of the three stadiums?

41. A new fax machine was recently purchased for an office in Hopedale for $464.40 including tax. If the tax rate in Hopedale is 8%, find the price of the fax machine before tax.

42. A premedical student at a local university was complaining that she had just paid $158.60 for her human anatomy book, including tax. Find the price of the book before taxes if the tax rate at this university is 9%.

43. The median compensation for a U.S. university president was $479,000 for the 2013–2014 academic year. Suppose a U.S. university president had a salary of $479,000 for that year. Calculate her salary if she received a 3.6% raise.

44. In 2014, the population of Germany was 80.9 million. This represented a decrease in population of 1.96% from 2004. What was the population of Germany in 2004? Round to the nearest tenth of a million. (*Source:* Population Reference Bureau)

45. In 2014, the population of Angola was 22,400,000 people. From 2014 to 2050, Angola's population is expected to increase by 272%. Find the expected population of Angola in 2050. (*Source:* Population Reference Bureau)

46. Dana, an auto parts supplier headquartered in Toledo, Ohio, recently announced it would be cutting 11,000 jobs worldwide. This is equivalent to 15% of Dana's workforce. Find the size of Dana's workforce prior to this round of job layoffs. Round to the nearest whole. (*Source:* Dana Corporation)

Recall that two angles are complements of each other if their sum is 90°. Two angles are supplements of each other if their sum is 180°. Find the measure of each angle.

47. One angle is three times its supplement increased by 20°. Find the measures of the two supplementary angles.

48. One angle is twice its complement increased by 30°. Find the measure of the two complementary angles.

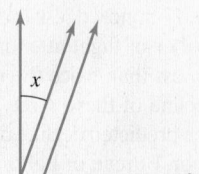

Recall that the sum of the angle measures of a triangle is 180°.

△ **49.** Find the measures of the angles of a triangle if the measure of one angle is twice the measure of a second angle and the third angle measures 3 times the second angle decreased by 12.

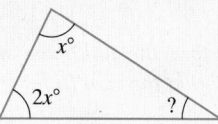

△ **50.** Find the angles of an isoceles triangle whose two base angles are equal and whose third angle is 10° less than three times a base angle.

▶ △ **51.** Two frames are needed with the same perimeter: one frame in the shape of a square and one in the shape of an equilateral triangle. Each side of the triangle is 6 centimeters longer than each side of the square. Find the dimensions of each frame. (An equilateral triangle has sides that are the same length.)

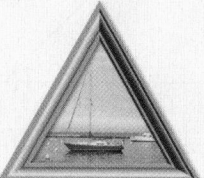

52. Two frames are needed with the same perimeter: one frame in the shape of a square and one in the shape of a regular pentagon. Each side of the square is 7 inches longer than each side of the pentagon. Find the dimensions of each frame. (A regular polygon has sides that are the same length.)

53. The sum of the first and third of three consecutive even integers is 156. Find the three even integers.

54. The sum of the second and fourth of four consecutive integers is 110. Find the four integers.

55. Daytona International Speedway in Florida has 37,000 more grandstand seats than twice the number of grandstand seats at Darlington Motor Raceway in South Carolina. Together, these two race tracks seat 220,000 NASCAR fans. How many seats does each race track have? (*Source:* NASCAR)

56. For the 2014–2015 National Hockey League season, the payroll for the San Jose Sharks was $5,049,585 less than that of the Montreal Canadiens. The total payroll for these two teams was $129,215,719. What were the payrolls for these two teams for the 2014–2015 NHL season?

57. The sum of the populations of the urban areas of New York, Delhi, and Tokyo-Yokohama is 83.6 million. Use this information and Example 2 in this section to find the population of each urban area. Round to the nearest tenth of a million. (*Source:* Demographia World Urban Areas, 2015.01)

58. The airports in Atlanta, Chicago, and Los Angeles have a total of 226 million annual passengers. Use this information and Practice 2 in this section to find the number of passengers at each airport.

59. Suppose the perimeter of the triangle in Example 1(b) in this section is 483 feet. Find the length of each side.

60. Suppose the perimeter of the trapezoid in Practice 1(b) in this section is 110 meters. Find the lengths of its sides and bases.

61. Incandescent, fluorescent, and halogen bulbs are lasting longer today than ever before. On average, the number of bulb hours for a fluorescent bulb is 25 times the number of bulb hours for a halogen bulb. The number of bulb hours for an incandescent bulb is 2500 less than the halogen bulb. If the total number of bulb hours for the three types of bulbs is 105,500, find the number of bulb hours for each type. (*Source: Popular Science* magazine)

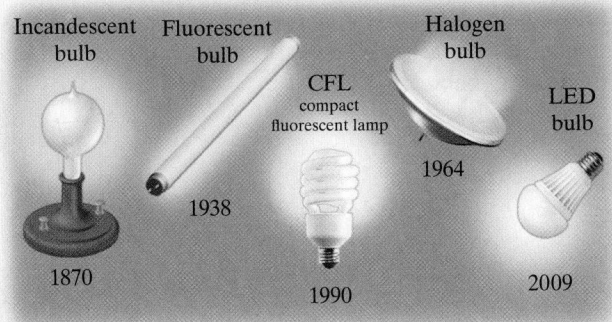

62. LED (light-emitting diode) bulbs and lamps are gaining popularity because of their low energy cost and production of very little heat. Over a 20-year span, the energy cost of a CFL bulb is $26 more than an LED bulb. Also, the energy cost of an incandescent bulb is $18 more than 6 times an LED bulb. If the total energy cost of these three bulb types is $476, find the energy cost of each. (Based on 13cents/kWh; *Source:* Wikipedia)

63. During the 2014 Major League Baseball season, the number of wins for the Toronto Blue Jays, the New York Yankees, and the Milwaukee Brewers were three consecutive integers. Of these three teams, the Yankees had the most wins and the Brewers had the least wins. The total number of wins by these three teams was 249. How many wins did each team have in the 2014 season?

64. In the 2014 Winter Olympics, Norway won more medals than Canada, which won more medals than the Netherlands. If the numbers of medals won by these three countries are consecutive integers whose sum is 75, find the number of medals won by each. (*Source:* Sochi 2014)

65. The three tallest hospitals in the world are Guy's Tower in London, Queen Mary Hospital in Hong Kong, and Galter Pavilion in Chicago. These buildings have a total height of 1320 feet. Guy's Tower is 67 feet taller than Galter Pavilion, and the Queen Mary Hospital is 47 feet taller than Galter Pavilion. Find the heights of the three hospitals.

△ 66. The official manual for traffic signs is the *Manual on Uniform Traffic Control Devices* published by the Government Printing Office. The rectangular sign below has a length 12 inches more than twice its height. If the perimeter of the sign is 312 inches, find its dimensions.

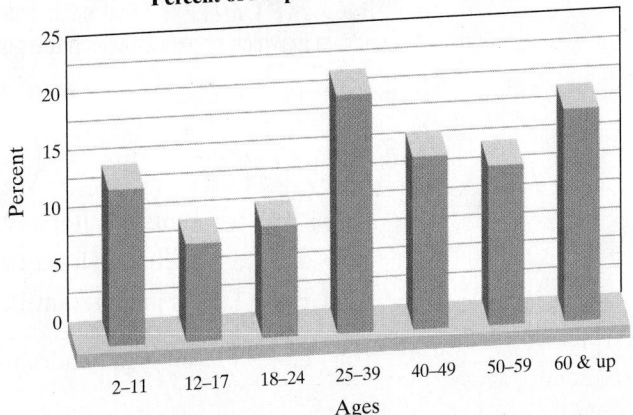

B.3 Graphing

OBJECTIVES

1 Plot Ordered Pairs.
2 Graph Linear Equations.

OBJECTIVE

1 Plotting Ordered Pairs

Graphs are widely used today in newspapers, magazines, and all forms of newsletters. A few examples of graphs are shown here.

Percent of People Who Go to the Movies

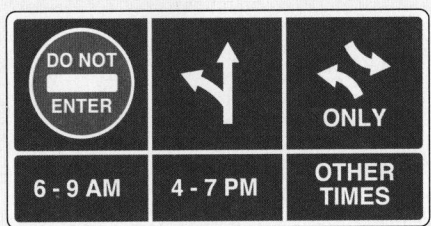

Source: Motion Picture Association of America

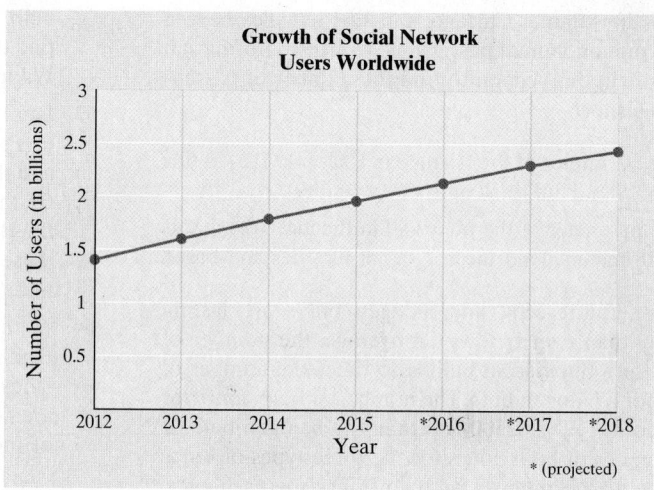

To review how to read these graphs, we review their origin—the rectangular coordinate system. One way to locate points on a plane is by using a **rectangular coordinate system,** which is also called a **Cartesian coordinate system** after its inventor, René Descartes (1596–1650). The diagram on the left below shows the rectangular coordinate system. For further review of this system, see Section 3.1.

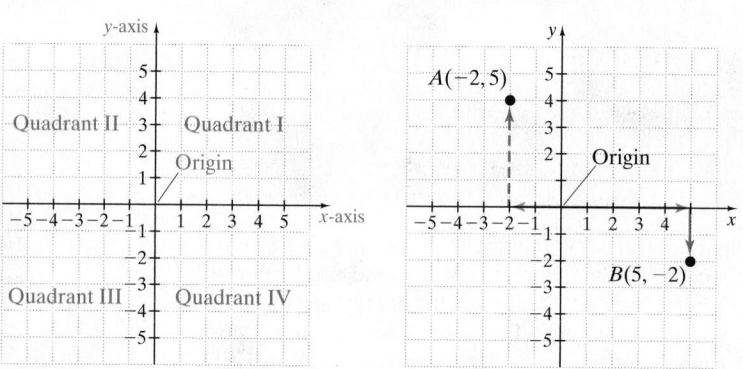

Recall that the location of point A in the figure above is described as 2 units to the left of the origin along the x-axis and 5 units upward parallel to the y-axis. Thus, we identify point A with the ordered pair $(-2, 5)$. Notice that the order of these numbers is *critical.* The x-value -2 is called the **x-coordinate** and is associated with the x-axis. The y-value 5 is called the **y-coordinate** and is associated with the y-axis. Compare the location of point A with the location of point B, which corresponds to the ordered pair $(5, -2)$.

Keep in mind that **each ordered pair corresponds to exactly one point in the real plane and that each point in the plane corresponds to exactly one ordered pair.** Thus, we may refer to the ordered pair (x, y) as the point (x, y).

EXAMPLE 1 Plot each ordered pair on a Cartesian coordinate system and name the quadrant in which or axis on which the point is located.

a. $(2, -1)$ 　　　　　**b.** $(0, 5)$ 　　　　　**c.** $(-3, 5)$

d. $(-2, 0)$ 　　　　　**e.** $\left(-\dfrac{1}{2}, -4\right)$ 　　　　**f.** $(1.5, 1.5)$

Solution The six points are graphed as shown on the next page.

    a. $(2, -1)$ is in quadrant IV. 　　　　**b.** $(0, 5)$ is on the y-axis.

    c. $(-3, 5)$ is in quadrant II. 　　　　**d.** $(-2, 0)$ is on the x-axis.

    e. $\left(-\dfrac{1}{2}, -4\right)$ is in quadrant III. 　　　**f.** $(1.5, 1.5)$ is in quadrant I.

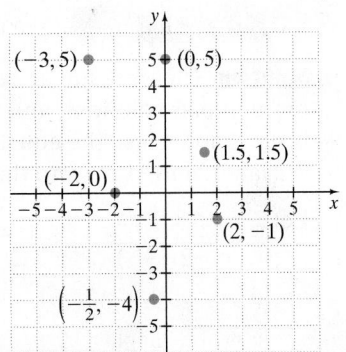

PRACTICE

1 Plot each ordered pair on a Cartesian coordinate system and name the quadrant in which or axis on which the point is located.

a. $(3, -4)$ **b.** $(0, -2)$ **c.** $(-2, 4)$ **d.** $(4, 0)$ **e.** $\left(-1\frac{1}{2}, -2\right)$ **f.** $(2.5, 3.5)$

Notice that the y-coordinate of any point on the x-axis is 0. For example, the point with coordinates $(-2, 0)$ lies on the x-axis. Also, the x-coordinate of any point on the y-axis is 0. For example, the point with coordinates $(0, 5)$ lies on the y-axis. These points that lie on the axes do not lie in any quadrant.

✔ **CONCEPT CHECK**

Which of the following correctly describes the location of the point $(3, -6)$ in a rectangular coordinate system?

a. 3 units to the left of the y-axis and 6 units above the x-axis
b. 3 units above the x-axis and 6 units to the left of the y-axis
c. 3 units to the right of the y-axis and 6 units below the x-axis
d. 3 units below the x-axis and 6 units to the right of the y-axis

---

OBJECTIVE

2 Graphing Linear Equations

Recall that an equation such as $3x - y = 12$ is called a linear equation in two variables, and **the graph of every linear equation in two variables is a line.**

Linear Equation in Two Variables

A linear equation in two variables is an equation that can be written in the form

$$Ax + By = C$$

where A and B are not both 0. This form is called **standard form.**

Some examples of equations in standard form:

$$3x - y = 12$$
$$-2.1x + 5.6y = 0$$

Helpful Hint

Remember: A linear equation is written in standard form when all of the variable terms are on one side of the equation and the constant is on the other side.

Answer to Concept Check:
c

EXAMPLE 2 Graph the equation: $y = -2x + 3$

<u>Solution</u> This is a linear equation. (In standard form it is $2x + y = 3$.) Find three ordered pair solutions, and plot the ordered pairs. The line through the plotted points is the graph. Since the equation is solved for y, let's choose three x-values. We'll choose $0, 2$, and then -1 for x to find our three ordered pair solutions.

Let $x = 0$

$y = -2x + 3$

$y = -2 \cdot 0 + 3$

$y = 3$ Simplify.

Let $x = 2$

$y = -2x + 3$

$y = -2 \cdot 2 + 3$

$y = -1$ Simplify.

Let $x = -1$

$y = -2x + 3$

$y = -2(-1) + 3$

$y = 5$ Simplify.

The three ordered pairs $(0, 3)$, $(2, -1)$, and $(-1, 5)$ are listed in the table and the graph is shown.

x	y
0	3
2	-1
-1	5

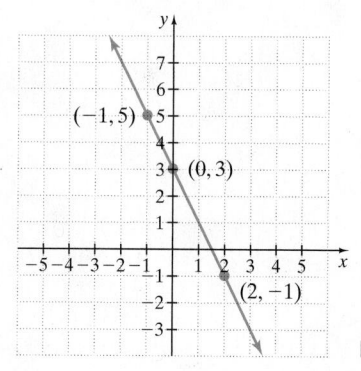

PRACTICE
2 Graph the equation: $y = -3x - 2$

Notice that the graph crosses the y-axis at the point $(0, 3)$. This point is called the **y-intercept.** (You may sometimes see just the number 3 called the y-intercept.) This graph also crosses the x-axis at the point $\left(\frac{3}{2}, 0\right)$. This point is called the **x-intercept.**

(You may also see just the number $\frac{3}{2}$ called the x-intercept.)

Since every point on the y-axis has an x-value of 0, we can find the y-intercept of a graph by letting $x = 0$ and solving for y. Also, every point on the x-axis has a y-value of 0. To find the x-intercept, we let $y = 0$ and solve for x.

Finding x- and y-Intercepts

To find an x-intercept, let $y = 0$ and solve for x.
To find a y-intercept, let $x = 0$ and solve for y.

EXAMPLE 3 Graph the linear equation: $y = \frac{1}{3}x$

<u>Solution</u> To graph, we find ordered pair solutions, plot the ordered pairs, and draw a line through the plotted points. We will choose x-values and substitute in the equation. To avoid fractions, we choose x-values that are multiples of 3. To find the y-intercept, we let $x = 0$.

Helpful Hint

Notice that by using multiples of 3 for x, we avoid fractions.

Helpful Hint

Since the equation $y = \frac{1}{3}x$ is solved for y, we choose x-values for finding points. This way, we simply need to evaluate an expression to find the y-value, as shown.

$y = \frac{1}{3}x$

If $x = 0$, then $y = \frac{1}{3}(0)$, or 0.

If $x = 6$, then $y = \frac{1}{3}(6)$, or 2.

If $x = -3$, then $y = \frac{1}{3}(-3)$, or -1.

x	y
0	0
6	2
-3	-1

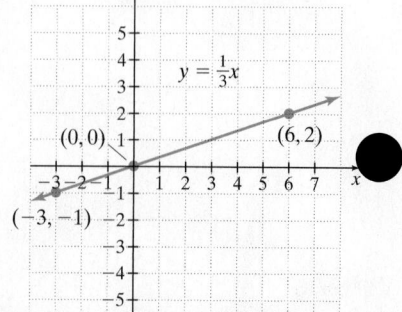

This graph crosses the x-axis at $(0, 0)$ and the y-axis at $(0, 0)$. This means that the x-intercept is $(0, 0)$ and that the y-intercept is $(0, 0)$.   □

PRACTICE
3   Graph the linear equation: $y = -\dfrac{1}{2}x$ ■

B.3   Exercise Set   MyMathLab

Determine the coordinates of each point on the graph.

1. Point A
2. Point B
3. Point C
4. Point D
5. Point E
6. Point F
7. Point G
8. Point H

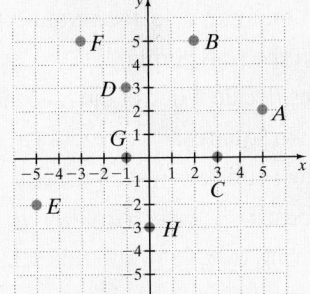

Without graphing, visualize the location of each point. Then give its location by quadrant or x- or y-axis. See Example 1.

9. $(2, 3)$
10. $(0, 5)$
11. $(-2, 7)$
12. $(-3, 0)$
13. $(-1, -4)$
14. $(4, -2)$
15. $(0, -100)$
16. $(10, 30)$
17. $(-10, -30)$
18. $(0, 0)$
19. $(-87, 0)$
20. $(-42, 17)$

Given that x is a positive number and that y is a positive number, determine the quadrant or axis in which each point lies.

21. $(x, -y)$
22. $(-x, y)$
23. $(x, 0)$
24. $(0, -y)$
25. $(-x, -y)$
26. $(0, 0)$

Graph each linear equation. See Examples 2 and 3.

27. $y = -x - 2$
28. $y = -2x + 1$
29. $3x - 4y = 8$
30. $x - 9y = 3$
31. $y = \dfrac{1}{3}x$
32. $y = \dfrac{3}{2}x$
33. $y + 4 = 0$
34. $x = -1.5$

Recall that if $f(2) = 7$, for example, this corresponds to the ordered pair $(2, 7)$ on the graph of f. Use this information and the graphs of f and g below to answer Exercises 35 through 42.

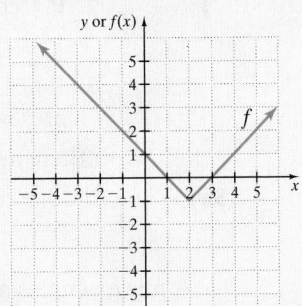
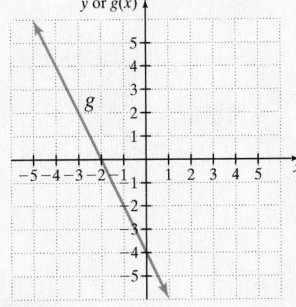

35. $f(4) =$
36. $f(0) =$
37. $g(0) =$
38. $g(-1) =$
39. Find all values for x such that $f(x) = 0$.
40. Find all values for x such that $g(x) = 0$.
41. If $(-1, -2)$ is a point on the graph of g, write this using function notation.
42. If $(-1, 2)$ is a point on the graph of f, write this using function notation.

B.4   Polynomials and Factoring

OBJECTIVES

1 Review Operations on Polynomials.
2 Review Factoring Polynomials.

OBJECTIVE
1   Operations on Polynomials

B.4   Exercise Set, Part 1   MyMathLab

Perform each indicated operation.

1. $(-y^2 + 6y - 1) + (3y^2 - 4y - 10)$
2. $(5z^4 - 6z^2 + z + 1) - (7z^4 - 2z + 1)$
3. Subtract $(x - 5)$ from $(x^2 - 6x + 2)$.
4. $(2x^2 + 6x - 5) + (5x^2 - 10x)$
5. $(5x - 3)^2$
6. $(5x^2 - 14x - 3) \div (5x + 1)$

7. $(2x^4 - 3x^2 + 5x - 2) \div (x + 2)$

8. $(4x - 1)(x^2 - 3x - 2)$

2 Factoring Strategies

The key to proficiency in factoring polynomials is to practice until you are comfortable with each technique. A strategy for factoring polynomials completely is given next.

Factoring a Polynomial

Step 1. Are there any common factors? If so, factor out the greatest common factor.

Step 2. How many terms are in the polynomial?

    a. If there are *two* terms, decide if one of the following formulas may be applied:

      i. Difference of two squares: $a^2 - b^2 = (a - b)(a + b)$

      ii. Difference of two cubes: $a^3 - b^3 = (a - b)(a^2 + ab + b^2)$

      iii. Sum of two cubes: $a^3 + b^3 = (a + b)(a^2 - ab + b^2)$

    b. If there are *three* terms, try one of the following:

      i. Perfect square trinomial: $a^2 + 2ab + b^2 = (a + b)^2$

         $a^2 - 2ab + b^2 = (a - b)^2$

      ii. If not a perfect square trinomial, factor by using the methods presented in Sections 6.2 through 6.4.

    c. If there are *four* or more terms, try factoring by grouping.

Step 3. See whether any factors in the factored polynomial can be factored further.

A few examples are worked for you below.

EXAMPLE 1   Factor each polynomial completely.

    a. $8a^2b - 4ab$ **b.** $36x^2 - 9$ **c.** $2x^2 - 5x - 7$

    d. $5p^2 + 5 + qp^2 + q$ **e.** $9x^2 + 24x + 16$ **f.** $y^2 + 25$

Solution

a. Step 1. The terms have a common factor of $4ab$, which we factor out.

$$8a^2b - 4ab = 4ab(2a - 1)$$

Step 2. There are two terms, but the binomial $2a - 1$ is not the difference of two squares or the sum or difference of two cubes.

Step 3. The factor $2a - 1$ cannot be factored further.

b. Step 1. Factor out a common factor of 9.

$$36x^2 - 9 = 9(4x^2 - 1)$$

Step 2. The factor $4x^2 - 1$ has two terms, and it is the difference of two squares.

$$9(4x^2 - 1) = 9(2x + 1)(2x - 1)$$

Step 3. No factor with more than one term can be factored further.

c. Step 1. The terms of $2x^2 - 5x - 7$ contain no common factor other than 1 or -1.

Step 2. There are three terms. The trinomial is not a perfect square, so we factor by methods from Section 6.3 or 6.4.

$$2x^2 - 5x - 7 = (2x - 7)(x + 1)$$

Step 3. No factor with more than one term can be factored further.

d. Step 1. There is no common factor of all terms of $5p^2 + 5 + qp^2 + q$.

Step 2. The polynomial has four terms, so try factoring by grouping.

$$5p^2 + 5 + qp^2 + q = (5p^2 + 5) + (qp^2 + q) \quad \text{Group the terms.}$$
$$= 5(p^2 + 1) + q(p^2 + 1)$$
$$= (p^2 + 1)(5 + q)$$

Step 3. No factor can be factored further.

e. Step 1. The terms of $9x^2 + 24x + 16$ contain no common factor other than 1 or -1.

Step 2. The trinomial $9x^2 + 24x + 16$ is a perfect square trinomial, and $9x^2 + 24x + 16 = (3x + 4)^2$.

Step 3. No factor can be factored further.

f. Step 1. There is no common factor of $y^2 + 25$ other than 1.

Step 2. This binomial is the sum of two squares and is prime.

Step 3. The binomial $y^2 + 25$ cannot be factored further. □

PRACTICE

1 Factor each polynomial completely.

a. $12x^2y - 3xy$ **b.** $49x^2 - 4$

c. $5x^2 + 2x - 3$ **d.** $3x^2 + 6 + x^3 + 2x$

e. $4x^2 + 20x + 25$ **f.** $b^2 + 100$ ■

EXAMPLE 2 Factor each polynomial completely.

a. $27a^3 - b^3$ **b.** $3n^2m^4 - 48m^6$ **c.** $2x^2 - 12x + 18 - 2z^2$

d. $8x^4y^2 + 125xy^2$ **e.** $(x - 5)^2 - 49y^2$

Solution

a. This binomial is the difference of two cubes.

$$27a^3 - b^3 = (3a)^3 - b^3$$
$$= (3a - b)[(3a)^2 + (3a)(b) + b^2]$$
$$= (3a - b)(9a^2 + 3ab + b^2)$$

b. $3n^2m^4 - 48m^6 = 3m^4(n^2 - 16m^2)$ Factor out the GCF $3m^4$.
$$= 3m^4(n + 4m)(n - 4m) \quad \text{Factor the difference of squares.}$$

c. $2x^2 - 12x + 18 - 2z^2 = 2(x^2 - 6x + 9 - z^2)$ The GCF is 2.
$$= 2[(x^2 - 6x + 9) - z^2] \quad \text{Group the first three terms together.}$$
$$= 2[(x - 3)^2 - z^2] \quad \text{Factor the perfect square trinomial.}$$
$$= 2[(x - 3) + z][(x - 3) - z] \quad \text{Factor the difference of squares.}$$
$$= 2(x - 3 + z)(x - 3 - z)$$

d. $8x^4y^2 + 125xy^2 = xy^2(8x^3 + 125)$ The GCF is xy^2.
$$= xy^2[(2x)^3 + 5^3]$$
$$= xy^2(2x + 5)[(2x)^2 - (2x)(5) + 5^2] \quad \text{Factor the sum of cubes.}$$
$$= xy^2(2x + 5)(4x^2 - 10x + 25)$$

e. This binomial is the difference of squares.

$$(x - 5)^2 - 49y^2 = (x - 5)^2 - (7y)^2$$
$$= [(x - 5) + 7y][(x - 5) - 7y]$$
$$= (x - 5 + 7y)(x - 5 - 7y) \quad □$$

(Continued on next page)

Factor each polynomial completely.

a. $64x^3 + y^3$

b. $7x^2y^2 - 63y^4$

c. $3x^2 + 12x + 12 - 3b^2$

d. $x^5y^4 + 27x^2y$

e. $(x + 7)^2 - 81y^2$

B.4 Exercise Set, Part 2 MyMathLab ▶

Factor completely.

9. $x^2 - 8x + 16 - y^2$

10. $12x^2 - 22x - 20$

11. $x^4 - x$

12. $(2x + 1)^2 - 3(2x + 1) + 2$

13. $14x^2y - 2xy$

14. $24ab^2 - 6ab$

15. $4x^2 - 16$

16. $9x^2 - 81$

17. $3x^2 - 8x - 11$

18. $5x^2 - 2x - 3$

19. $4x^2 + 8x - 12$

20. $6x^2 - 6x - 12$

21. $4x^2 + 36x + 81$

22. $25x^2 + 40x + 16$

23. $8x^3 + 125y^3$

24. $27x^3 - 64y^3$

25. $64x^2y^3 - 8x^2$

26. $27x^5y^4 - 216x^2y$

27. $(x + 5)^3 + y^3$

28. $(y - 1)^3 + 27x^3$

29. $(5a - 3)^2 - 6(5a - 3) + 9$

30. $(4r + 1)^2 + 8(4r + 1) + 16$

31. $7x^2 - 63x$

32. $20x^2 + 23x + 6$

33. $ab - 6a + 7b - 42$

34. $20x^2 - 220x + 600$

35. $x^4 - 1$

36. $15x^2 - 20x$

37. $10x^2 - 7x - 33$

38. $45m^3n^3 - 27m^2n^2$

39. $5a^3b^3 - 50a^3b$

40. $x^4 + x$

41. $16x^2 + 25$

42. $20x^3 + 20y^3$

43. $10x^3 - 210x^2 + 1100x$

44. $9y^2 - 42y + 49$

45. $64a^3b^4 - 27a^3b$

46. $y^4 - 16$

47. $2x^3 - 54$

48. $2sr + 10s - r - 5$

49. $3y^5 - 5y^4 + 6y - 10$

50. $64a^2 + b^2$

51. $100z^3 + 100$

52. $250x^4 - 16x$

53. $4b^2 - 36b + 81$

54. $2a^5 - a^4 + 6a - 3$

55. $(y - 6)^2 + 3(y - 6) + 2$

56. $(c + 2)^2 - 6(c + 2) + 5$

△ **57.** Express the area of the shaded region as a polynomial. Factor the polynomial completely.

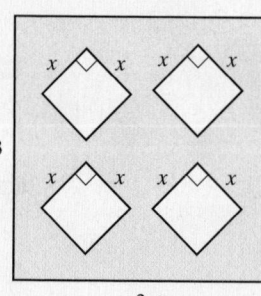

B.5 **Rational Expressions**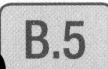

OBJECTIVE

1 Perform Operations on Rational Expressions and Solve Equations Containing Rational Expressions.

OBJECTIVE

1 Performing Operations on Rational Expressions and Solving Equations Containing Rational Expressions

It is very important that you understand the difference between an expression and an equation containing rational expressions. An equation contains an equal sign; an expression does not.

Expression to be Simplified	**Equation to be Solved**

$$\frac{x}{2} + \frac{x}{6}$$

$$\frac{x}{2} + \frac{x}{6} = \frac{2}{3}$$

Write both rational expressions with the LCD, 6, as the denominator.

Multiply both sides by the LCD, 6.

$$\frac{x}{2} + \frac{x}{6} = \frac{x \cdot 3}{2 \cdot 3} + \frac{x}{6}$$

$$6\left(\frac{x}{2} + \frac{x}{6}\right) = 6\left(\frac{2}{3}\right)$$

$$= \frac{3x}{6} + \frac{x}{6}$$

$$3x + x = 4$$

$$4x = 4$$

$$= \frac{4x}{6} = \frac{2x}{3}$$

$$x = 1$$

Check to see that the solution is 1.

> **Helpful Hint**
> Remember: Equations can be cleared of fractions; expressions cannot.

EXAMPLE I   Multiply: $\dfrac{x^3 - 1}{-3x + 3} \cdot \dfrac{15x^2}{x^2 + x + 1}$

Solution

$$\frac{x^3 - 1}{-3x + 3} \cdot \frac{15x^2}{x^2 + x + 1} = \frac{(x - 1)(x^2 + x + 1)}{-3(x - 1)} \cdot \frac{15x^2}{x^2 + x + 1} \quad \text{Factor.}$$

$$= \frac{(x - 1)(x^2 + x + 1) \cdot 3 \cdot 5x^2}{-1 \cdot 3(x - 1)(x^2 + x + 1)} \quad \text{Factor.}$$

$$= \frac{5x^2}{-1} = -5x^2 \quad \text{Simplest form} \quad \square$$

PRACTICE

1 Multiply.

a. $\dfrac{2 + 5n}{3n} \cdot \dfrac{6n + 3}{5n^2 - 3n - 2}$

b. $\dfrac{x^3 - 8}{-6x + 12} \cdot \dfrac{6x^2}{x^2 + 2x + 4}$

EXAMPLE 2   Divide: $\dfrac{8m^2}{3m^2 - 12} \div \dfrac{40}{2 - m}$

Solution

$$\frac{8m^2}{3m^2 - 12} \div \frac{40}{2 - m} = \frac{8m^2}{3m^2 - 12} \cdot \frac{2 - m}{40} \quad \text{Multiply by the reciprocal of the divisor.}$$

$$= \frac{8m^2(2 - m)}{3(m + 2)(m - 2) \cdot 40} \quad \text{Factor and multiply.}$$

$$= \frac{8 \, m^2 \cdot -1 \, (m - 2)}{3(m + 2)(m - 2) \cdot 8 \cdot 5} \quad \text{Write } (2 - m) \text{ as } -1(m - 2).$$

$$= -\frac{m^2}{15(m + 2)} \quad \text{Simplify.} \quad \square$$

(Continued on next page)

PRACTICE
2   Divide.

a. $\dfrac{6y^3}{3y^2 - 27} \div \dfrac{42}{3 - y}$ **b.** $\dfrac{10x^2 + 23x - 5}{5x^2 - 51x + 10} \div \dfrac{2x^2 + 9x + 10}{7x^2 - 68x - 20}$

EXAMPLE 3   Perform the indicated operation.

$$\frac{3}{x + 2} + \frac{2x}{x - 2}$$

Solution   The LCD is the product of the two denominators: $(x + 2)(x - 2)$.

$$\frac{3}{x + 2} + \frac{2x}{x - 2} = \frac{3 \cdot (x - 2)}{(x + 2) \cdot (x - 2)} + \frac{2x \cdot (x + 2)}{(x - 2) \cdot (x + 2)} \qquad \text{Write equivalent rational expressions.}$$

$$= \frac{3x - 6}{(x + 2)(x - 2)} + \frac{2x^2 + 4x}{(x + 2)(x - 2)} \qquad \text{Multiply in the numerators.}$$

$$= \frac{3x - 6 + 2x^2 + 4x}{(x + 2)(x - 2)} \qquad \text{Add the numerators.}$$

$$= \frac{2x^2 + 7x - 6}{(x + 2)(x - 2)} \qquad \text{Simplify the numerator.} \qquad \square$$

PRACTICE
3   Perform the indicated operation.

a. $\dfrac{4}{p^3 q} + \dfrac{3}{5p^4 q}$ **b.** $\dfrac{4}{y + 3} + \dfrac{5y}{y - 3}$

c. $\dfrac{3z - 18}{z - 5} - \dfrac{3}{5 - z}$

EXAMPLE 4   Solve: $\dfrac{2x}{x - 3} + \dfrac{6 - 2x}{x^2 - 9} = \dfrac{x}{x + 3}$

Solution   We factor the second denominator to find that the LCD is $(x + 3)(x - 3)$. We multiply both sides of the equation by $(x + 3)(x - 3)$. By the distributive property, this is the same as multiplying each term by $(x + 3)(x - 3)$.

$$\frac{2x}{x - 3} + \frac{6 - 2x}{x^2 - 9} = \frac{x}{x + 3}$$

$$(x + 3)(x - 3) \cdot \frac{2x}{x - 3} + (x + 3)(x - 3) \cdot \frac{6 - 2x}{(x + 3)(x - 3)}$$

$$= (x + 3)(x - 3)\left(\frac{x}{x + 3}\right)$$

$$2x(x + 3) + (6 - 2x) = x(x - 3) \qquad \text{Simplify.}$$

$$2x^2 + 6x + 6 - 2x = x^2 - 3x \qquad \text{Use the distributive property.}$$

Next we solve this quadratic equation by the factoring method. To do so, we first write the equation so that one side is 0.

$$x^2 + 7x + 6 = 0$$

$$(x + 6)(x + 1) = 0 \qquad \text{Factor.}$$

$$x = -6 \quad \text{or} \quad x = -1 \qquad \text{Set each factor equal to 0.}$$

Neither -6 nor -1 makes any denominator 0, so they are both solutions. The solutions are -6 and -1. $\square$

PRACTICE
4   Solve: $\dfrac{2}{x - 2} - \dfrac{5 + 2x}{x^2 - 4} = \dfrac{x}{x + 2}$

B.5   Exercise Set   MyMathLab® ▶

Perform each indicated operation and simplify, or solve the equation for the variable.

1. $\dfrac{x}{2} = \dfrac{1}{8} + \dfrac{x}{4}$

2. $\dfrac{x}{4} = \dfrac{3}{2} + \dfrac{x}{10}$

3. $\dfrac{1}{8} + \dfrac{x}{4}$

4. $\dfrac{3}{2} + \dfrac{x}{10}$

5. $\dfrac{4}{x + 2} - \dfrac{2}{x - 1}$

6. $\dfrac{5}{x - 2} - \dfrac{10}{x + 4}$

7. $\dfrac{4}{x + 2} = \dfrac{2}{x - 1}$

8. $\dfrac{5}{x - 2} = \dfrac{10}{x + 4}$

9. $\dfrac{2}{x^2 - 4} = \dfrac{1}{x + 2} - \dfrac{3}{x - 2}$

10. $\dfrac{3}{x^2 - 25} = \dfrac{1}{x + 5} + \dfrac{2}{x - 5}$

11. $\dfrac{5}{x^2 - 3x} + \dfrac{4}{2x - 6}$

12. $\dfrac{5}{x^2 - 3x} \div \dfrac{4}{2x - 6}$

13. $\dfrac{x - 1}{x + 1} + \dfrac{x + 7}{x - 1} = \dfrac{4}{x^2 - 1}$

14. $\left(1 - \dfrac{y}{x}\right) \div \left(1 - \dfrac{x}{y}\right)$

15. $\dfrac{a^2 - 9}{a - 6} \cdot \dfrac{a^2 - 5a - 6}{a^2 - a - 6}$

16. $\dfrac{2}{a - 6} + \dfrac{3a}{a^2 - 5a - 6} - \dfrac{a}{5a + 5}$

17. $\dfrac{2x + 3}{3x - 2} = \dfrac{4x + 1}{6x + 1}$

18. $\dfrac{5x - 3}{2x} = \dfrac{10x + 3}{4x + 1}$

19. $\dfrac{a}{9a^2 - 1} + \dfrac{2}{6a - 2}$

20. $\dfrac{3}{4a - 8} - \dfrac{a + 2}{a^2 - 2a}$

21. $-\dfrac{3}{x^2} - \dfrac{1}{x} + 2 = 0$

22. $\dfrac{x}{2x + 6} + \dfrac{5}{x^2 - 9}$

23. $\dfrac{x - 8}{x^2 - x - 2} + \dfrac{2}{x - 2}$

24. $\dfrac{x - 8}{x^2 - x - 2} + \dfrac{2}{x - 2} = \dfrac{3}{x + 1}$

25. $\dfrac{3}{a} - 5 = \dfrac{7}{a} - 1$

26. $\dfrac{7}{3z - 9} + \dfrac{5}{z}$

Use $\dfrac{x}{5} - \dfrac{x}{4} = \dfrac{1}{10}$ and $\dfrac{x}{5} - \dfrac{x}{4} + \dfrac{1}{10}$ for Exercises 27 and 28.

27. a. Which one above is an expression?

   b. Describe the first step to simplify this expression.

   c. Simplify the expression.

28. a. Which one above is an equation?

   b. Describe the first step to solve this equation.

   c. Solve the equation.

For each exercise, choose the correct statement.* Each figure represents a real number, and no denominators are 0.

29. a. $\dfrac{\triangle + \square}{\triangle} = \square$ **b.** $\dfrac{\triangle + \square}{\triangle} = 1 + \dfrac{\square}{\triangle}$

   c. $\dfrac{\triangle + \square}{\triangle} = \dfrac{\square}{\triangle}$ **d.** $\dfrac{\triangle + \square}{\triangle} = 1 + \square$

   e. $\dfrac{\triangle + \square}{\triangle - \square} = -1$

*My thanks to Kelly Champagne for permission to use her Exercises for 29 through 33.

Helpful Hint

Remember: Equations can be cleared of fractions; expressions cannot.

30. **a.** $\dfrac{\triangle}{\square} + \dfrac{\square}{\triangle} = \dfrac{\triangle + \square}{\square + \triangle} = 1$

b. $\dfrac{\triangle}{\square} + \dfrac{\square}{\triangle} = \dfrac{\triangle + \square}{\triangle\square}$

c. $\dfrac{\triangle}{\square} + \dfrac{\square}{\triangle} = \triangle\triangle + \square\square$

d. $\dfrac{\triangle}{\square} + \dfrac{\square}{\triangle} = \dfrac{\triangle\triangle + \square\square}{\square\triangle}$

e. $\dfrac{\triangle}{\square} + \dfrac{\square}{\triangle} = \dfrac{\triangle\square}{\square\triangle} = 1$

31. **a.** $\dfrac{\triangle}{\square} \cdot \dfrac{\bigcirc}{\square} = \dfrac{\triangle\bigcirc}{\square}$

b. $\dfrac{\triangle}{\square} \cdot \dfrac{\bigcirc}{\square} = \triangle\bigcirc$

c. $\dfrac{\triangle}{\square} \cdot \dfrac{\bigcirc}{\square} = \dfrac{\triangle + \bigcirc}{\square + \square}$

d. $\dfrac{\triangle}{\square} \cdot \dfrac{\bigcirc}{\square} = \dfrac{\triangle\bigcirc}{\square\square}$

32. **a.** $\dfrac{\triangle}{\square} \div \dfrac{\bigcirc}{\triangle} = \dfrac{\triangle\triangle}{\square\bigcirc}$

b. $\dfrac{\triangle}{\square} \div \dfrac{\bigcirc}{\triangle} = \dfrac{\bigcirc\square}{\triangle\triangle}$

c. $\dfrac{\triangle}{\square} \div \dfrac{\bigcirc}{\triangle} = \dfrac{\bigcirc}{\square}$

d. $\dfrac{\triangle}{\square} \div \dfrac{\bigcirc}{\triangle} = \dfrac{\triangle + \triangle}{\square + \bigcirc}$

33. **a.** $\dfrac{\frac{\triangle + \square}{\bigcirc}}{\frac{\triangle}{\bigcirc}} = \square$

b. $\dfrac{\frac{\triangle + \square}{\bigcirc}}{\frac{\triangle}{\bigcirc}} = \dfrac{\triangle\triangle + \triangle\square}{\bigcirc\bigcirc}$

c. $\dfrac{\frac{\triangle + \square}{\bigcirc}}{\frac{\triangle}{\bigcirc}} = 1 + \square$

d. $\dfrac{\frac{\triangle + \square}{\bigcirc}}{\frac{\triangle}{\bigcirc}} = \dfrac{\triangle + \square}{\triangle}$

Appendix C

An Introduction to Using a Graphing Utility

The Viewing Window and Interpreting Window Settings

In this appendix, we will use the term **graphing utility** to mean a graphing calculator or a computer software graphing package. All graphing utilities graph equations by plotting points on a screen. While plotting several points can be slow and sometimes tedious for us, a graphing utility can quickly and accurately plot hundreds of points. How does a graphing utility show plotted points? A computer or calculator screen is made up of a grid of small rectangular areas called **pixels.** If a pixel contains a point to be plotted, the pixel is turned "on"; otherwise, the pixel remains "off." The graph of an equation is then a collection of pixels turned "on." The graph of $y = 3x + 1$ from a graphing calculator is shown in Figure A-1. Notice the irregular shape of the line caused by the rectangular pixels.

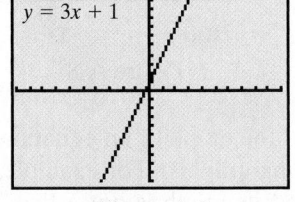

Figure A-1

The portion of the coordinate plane shown on the screen in Figure A-1 is called the **viewing window** or the **viewing rectangle.** Notice the x-axis and the y-axis on the graph. While tick marks are shown on the axes, they are not labeled. This means that from this screen alone, we do not know how many units each tick mark represents. To see what each tick mark represents and the minimum and maximum values on the axes, check the window setting of the graphing utility. It defines the viewing window. The window of the graph of $y = 3x + 1$ shown in Figure A-1 has the following settings (Figure A-2):

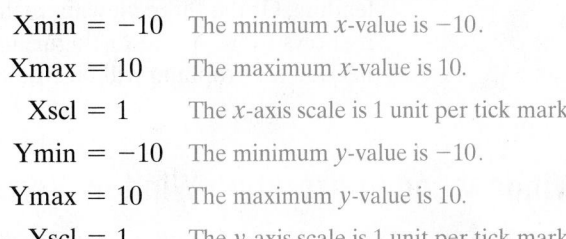

Xmin $= -10$	The minimum x-value is -10.
Xmax $= 10$	The maximum x-value is 10.
Xscl $= 1$	The x-axis scale is 1 unit per tick mark.
Ymin $= -10$	The minimum y-value is -10.
Ymax $= 10$	The maximum y-value is 10.
Yscl $= 1$	The y-axis scale is 1 unit per tick mark.

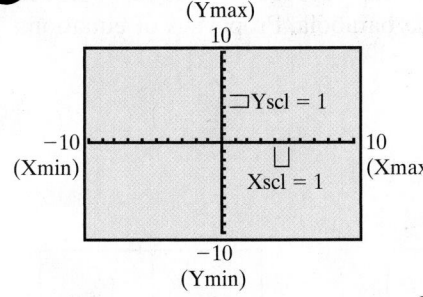

Figure A-2

By knowing the scale, we can find the minimum and the maximum values on the axes simply by counting tick marks. For example, if both the Xscl (x-axis scale) and the Yscl (y-axis scale) are 1 unit per tick mark on the graph in Figure A-3, we can count the tick marks and find that the minimum x-value is -10 and the maximum x-value is 10. Also, the minimum y-value is -10 and the maximum y-value is 10. If the Xscl changes to 2 units per tick mark (shown in Figure A-4), by counting tick marks, we see that the minimum x-value is now -20 and the maximum x-value is now 20.

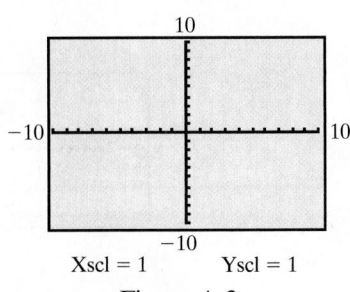

Xscl = 1   Yscl = 1

Figure A-3

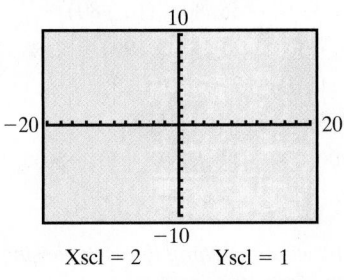

Xscl = 2   Yscl = 1

Figure A-4

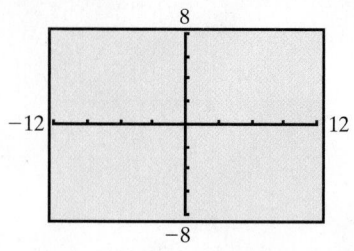

Figure A-5

It is also true that if we know the Xmin and the Xmax values, we can calculate the Xscl by the displayed axes. For example, the Xscl of the graph in Figure A-5 must be 3 units per tick mark for the maximum and minimum x-values to be as shown. Also, the Yscl of that graph must be 2 units per tick mark for the maximum and minimum y-values to be as shown.

We will call the viewing window in Figure A-3 a *standard* viewing window or rectangle. Although a standard viewing window is sufficient for much of this text, special care must be taken to ensure that all key features of a graph are shown. Figures A-6, A-7, and A-8 show the graph of $y = x^2 + 11x - 1$ on three viewing windows. Note that certain viewing windows for this equation are misleading.

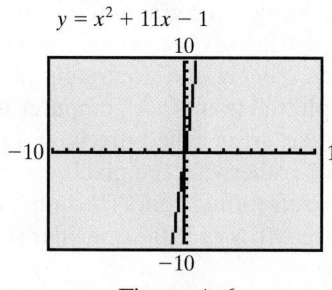

Figure A-6

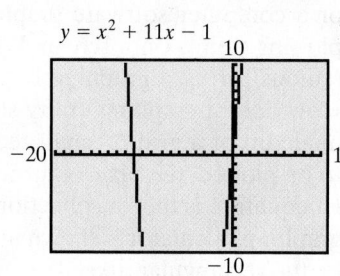

Figure A-7

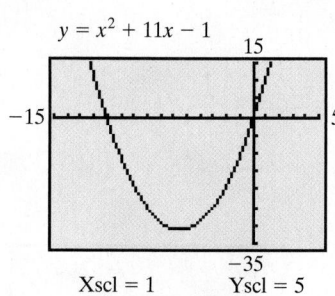

Figure A-8

How do we ensure that all distinguishing features of the graph of an equation are shown? It helps to know about the equation that is being graphed. For example, the equation $y = x^2 + 11x - 1$ is not a linear equation and its graph is not a line. This equation is a quadratic equation and, therefore, its graph is a parabola. By knowing this information, we know that the graph shown in Figure A-6, although correct, is misleading. Of the three viewing rectangles shown, the graph in Figure A-8 is best because it shows more of the distinguishing features of the parabola. Properties of equations needed for graphing will be studied in this text.

The Viewing Window and Interpreting Window Settings Exercise Set

In Exercises 1–4, determine whether all ordered pairs listed will lie within a standard viewing rectangle.

1. $(-9, 0), (5, 8), (1, -8)$

2. $(4, 7), (0, 0), (-8, 9)$

3. $(-11, 0), (2, 2), (7, -5)$

4. $(3, 5), (-3, -5), (15, 0)$

In Exercises 5–10, choose an Xmin, Xmax, Ymin, and Ymax so that all ordered pairs listed will lie within the viewing rectangle.

5. $(-90, 0), (55, 80), (0, -80)$

6. $(4, 70), (20, 20), (-18, 90)$

7. $(-11, 0), (2, 2), (7, -5)$

8. $(3, 5), (-3, -5), (15, 0)$

9. $(200, 200), (50, -50), (70, -50)$

10. $(40, 800), (-30, 500), (15, 0)$

Write the window setting for each viewing window shown. Use the following format:

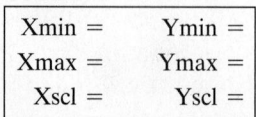

11.

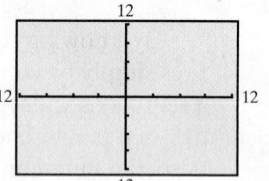

12.

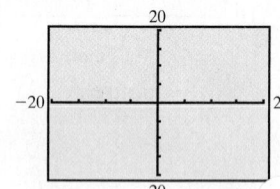

13.

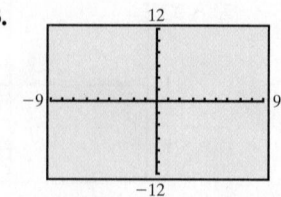

14.

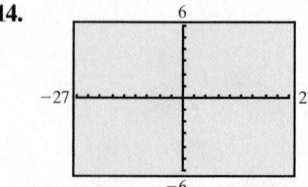

15.

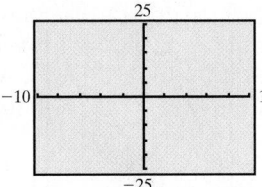

16.

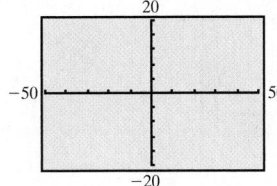

17.

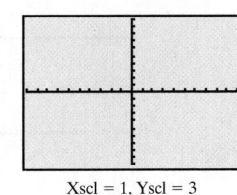

Xscl = 1, Yscl = 3

18.

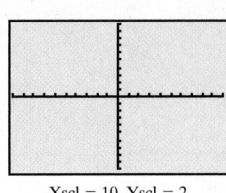

Xscl = 10, Yscl = 2

19.

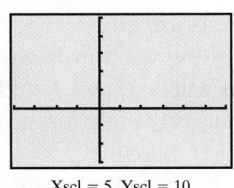

Xscl = 5, Yscl = 10

20.

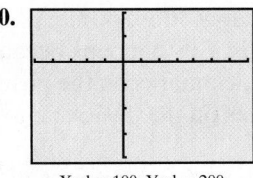

Xscl = 100, Yscl = 200

Graphing Equations and Square Viewing Window

In general, the following steps may be used to graph an equation on a standard viewing window.

> **Graphing an Equation in *x* and *y* with a Graphing Utility on a Standard Viewing Window**
>
> **Step 1:** Solve the equation for *y*.
>
> **Step 2:** Using your graphing utility, enter the equation in the form
> Y = *expression involving x*.
>
> **Step 3:** Activate the graphing utility.

Special care must be taken when entering the *expression involving x* in Step 2. You must be sure that the graphing utility you are using interprets the expression as you want it to. For example, let's graph $3y = 4x$. To do so,

Step 1: Solve the equation for *y*.

$$3y = 4x$$

$$\frac{3y}{3} = \frac{4x}{3}$$

$$y = \frac{4}{3}x$$

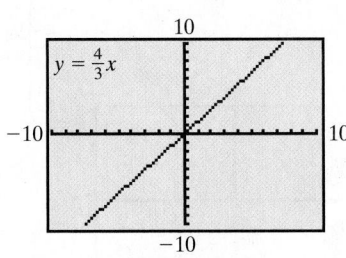

Figure A-9

Step 2: Using your graphing utility, enter the expression $\frac{4}{3}x$ after the Y = prompt.

In order for your graphing utility to correctly interpret the expression, you may need to enter $(4/3)x$ or $(4 \div 3)x$.

Step 3: Activate the graphing utility. The graph should appear as in Figure A-9.

Distinguishing features of the graph of a line include showing all the intercepts of the line. For example, the window of the graph of the line in Figure A-10 does not show both intercepts of the line, but the window of the graph of the same line in Figure A-11 does show both intercepts. Notice the notation below each graph. This is a shorthand notation of the range setting of the graph. This notation means [Xmin, Xmax] by [Ymin, Ymax].

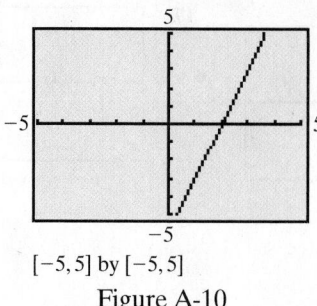

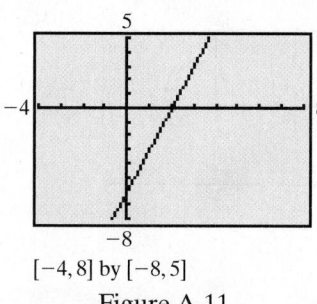

[−5, 5] by [−5, 5]   [−4, 8] by [−8, 5]

Figure A-10   Figure A-11

On a standard viewing window, the tick marks on the *y*-axis are closer together than the tick marks on the *x*-axis. This happens because the viewing window is a rectangle, and so 10 equally spaced tick marks on the positive *y*-axis will be closer together than 10 equally spaced tick marks on the positive *x*-axis. This causes the appearance of graphs to be distorted.

For example, notice the different appearances of the same line graphed using different viewing windows. The line in Figure A-12 is distorted because the tick marks along the *x*-axis are farther apart than the tick marks along the *y*-axis. The graph of the same line in Figure A-13 is not distorted because the viewing rectangle has been selected so that there is equal spacing between tick marks on both axes.

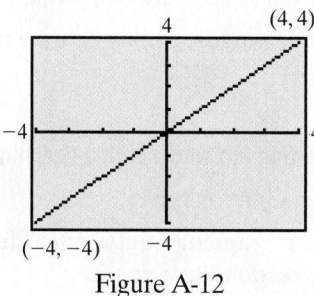

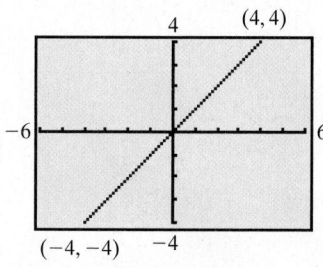

Figure A-12   Figure A-13

We say that the line in Figure A-13 is graphed on a *square* setting. Some graphing utilities have a built-in program that, if activated, will automatically provide a square setting. A square setting is especially helpful when we are graphing perpendicular lines, circles, or when a true geometric perspective is desired. Some examples of square screens are shown in Figures A-14 and A-15.

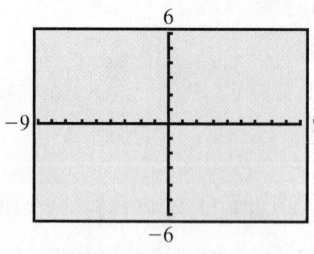

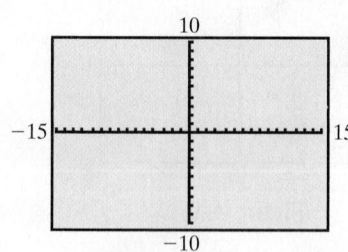

Figure A-14   Figure A-15

Other features of a graphing utility such as Trace, Zoom, Intersect, and Table are discussed in appropriate Graphing Calculator Explorations in this text.

Graphing Equations and Square Viewing Window Exercise Set

Graph each linear equation in two variables using the two different range settings given. Determine which setting shows all intercepts of a line.

1. $y = 2x + 12$
Setting A: $[-10, 10]$ by $[-10, 10]$
Setting B: $[-10, 10]$ by $[-10, 15]$

2. $y = -3x + 25$
Setting A: $[-5, 5]$ by $[-30, 10]$
Setting B: $[-10, 10]$ by $[-10, 30]$

3. $y = -x - 41$
Setting A: $[-50, 10]$ by $[-10, 10]$
Setting B: $[-50, 10]$ by $[-50, 15]$

4. $y = 6x - 18$
Setting A: $[-10, 10]$ by $[-20, 10]$
Setting B: $[-10, 10]$ by $[-10, 10]$

5. $y = \dfrac{1}{2}x - 15$
Setting A: $[-10, 10]$ by $[-20, 10]$
Setting B: $[-10, 35]$ by $[-20, 15]$

6. $y = -\dfrac{2}{3}x - \dfrac{29}{3}$
Setting A: $[-10, 10]$ by $[-10, 10]$
Setting B: $[-15, 5]$ by $[-15, 5]$

The graph of each equation is a line. Use a graphing utility and a standard viewing window to graph each equation.

7. $3x = 5y$ **8.** $7y = -3x$ **9.** $9x - 5y = 30$

10. $4x + 6y = 20$ **11.** $y = -7$ **12.** $y = 2$

13. $x + 10y = -5$ **14.** $x - 5y = 9$

Graph the following equations using the square setting given. Some keystrokes that may be helpful are given.

15. $y = \sqrt{x}$ $[-12, 12]$ by $[-8, 8]$
Suggested keystrokes: $\sqrt{\ } \, x$

16. $y = \sqrt{2x}$ $[-12, 12]$ by $[-8, 8]$
Suggested keystrokes: $\sqrt{\ } \, (2x)$

17. $y = x^2 + 2x + 1$ $[-15, 15]$ by $[-10, 10]$
Suggested keystrokes: $x^2 + 2x + 1$

18. $y = x^2 - 5$ $[-15, 15]$ by $[-10, 10]$
Suggested keystrokes: $x^2 - 5$

19. $y = |x|$ $[-9, 9]$ by $[-6, 6]$
Suggested keystrokes: $ABS(x)$

20. $y = |x - 2|$ $[-9, 9]$ by $[-6, 6]$
Suggested keystrokes: $ABS(x - 2)$

Graph each line. Use a standard viewing window; then, if necessary, change the viewing window so that all intercepts of each line show.

21. $x + 2y = 30$ **22.** $1.5x - 3.7y = 40.3$

Appendix D

Solving Systems of Equations by Matrices

OBJECTIVES

1 Use Matrices to Solve a System of Two Equations. ▶

2 Use Matrices to Solve a System of Three Equations. ▶

By now, you may have noticed that the solution of a system of equations depends on the coefficients of the equations in the system and not on the variables. In this section, we introduce solving a system of equations by a **matrix.**

OBJECTIVE

1 Using Matrices to Solve a System of Two Equations ▶

A matrix (plural: **matrices**) is a rectangular array of numbers. The following are examples of matrices.

$$\begin{bmatrix} 1 & 0 \\ 0 & 1 \end{bmatrix} \qquad \begin{bmatrix} 2 & 1 & 3 & -1 \\ 0 & -1 & 4 & 5 \\ -6 & 2 & 1 & 0 \end{bmatrix} \qquad \begin{bmatrix} a & b & c \\ d & e & f \end{bmatrix}$$

The numbers aligned horizontally in a matrix are in the same **row.** The numbers aligned vertically are in the same **column.**

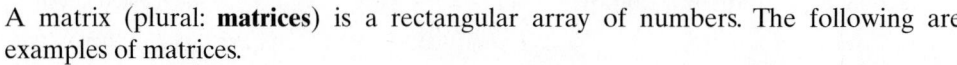

$$\begin{matrix} \text{row 1} \to \\ \text{row 2} \to \end{matrix} \begin{bmatrix} 2 & 1 & 0 \\ -1 & 6 & 2 \end{bmatrix}$$

This matrix has 2 rows and 3 columns. It is called a 2×3 (read "two by three") matrix.

column 1, column 2, column 3

To see the relationship between systems of equations and matrices, study the example below.

> **Helpful Hint**
>
> Before writing the corresponding matrix associated with a system of equations, make sure that the equations are written in standard form.

System of Equations (in standard form)

$$\begin{cases} 2x - 3y = 6 & \text{Equation 1} \\ x + y = 0 & \text{Equation 2} \end{cases}$$

Corresponding Matrix

$$\begin{bmatrix} 2 & -3 & | & 6 \\ 1 & 1 & | & 0 \end{bmatrix} \begin{matrix} \text{Row 1} \\ \text{Row 2} \end{matrix}$$

Notice that the rows of the matrix correspond to the equations in the system. The coefficients of the variables are placed to the left of a vertical dashed line. The constants are placed to the right. Each of the numbers in the matrix is called an **element.**

The method of solving systems by matrices is to write this matrix as an equivalent matrix from which we easily identify the solution. Two matrices are equivalent if they represent systems that have the same solution set. The following **row operations** can be performed on matrices, and the result is an equivalent matrix.

> **Helpful Hint**
>
> Notice that these *row* operations are the same operations that we can perform on *equations* in a system.

> **Elementary Row Operations**
>
> **1.** Any two rows in a matrix may be interchanged.
>
> **2.** The elements of any row may be multiplied (or divided) by the same nonzero number.
>
> **3.** The elements of any row may be multiplied (or divided) by a nonzero number and added to their corresponding elements in any other row.

To solve a system of two equations in x and y by matrices, write the corresponding matrix associated with the system. Then use elementary row operations to write equivalent matrices until you have a matrix of the form

$$\begin{bmatrix} 1 & a & | & b \\ 0 & 1 & | & c \end{bmatrix},$$

where a, b, and c are constants. Why? If a matrix associated with a system of equations is in this form, we can easily solve for x and y. For example,

Matrix		*System of Equations*

$$\begin{bmatrix} 1 & 2 & | & -3 \\ 0 & 1 & | & 5 \end{bmatrix} \quad \text{corresponds to} \quad \begin{cases} 1x + 2y = -3 \\ 0x + 1y = 5 \end{cases} \quad \text{or} \quad \begin{cases} x + 2y = -3 \\ y = 5 \end{cases}$$

In the second equation, we have $y = 5$. Substituting this in the first equation, we have $x + 2(5) = -3$ or $x = -13$. The solution of the system is the ordered pair $(-13, 5)$.

EXAMPLE 1   Use matrices to solve the system.

$$\begin{cases} x + 3y = 5 \\ 2x - y = -4 \end{cases}$$

Solution   The corresponding matrix is $\begin{bmatrix} 1 & 3 & | & 5 \\ 2 & -1 & | & -4 \end{bmatrix}$. We use elementary row operations to write an equivalent matrix that looks like $\begin{bmatrix} 1 & a & | & b \\ 0 & 1 & | & c \end{bmatrix}$.

For the matrix given, the element in the first row, first column is already 1, as desired. Next we write an equivalent matrix with a 0 below the 1. To do this, we multiply row 1 by -2 and add to row 2. *We will change only row 2.*

$$\begin{bmatrix} 1 & 3 & | & 5 \\ -2(1) + 2 & -2(3) + (-1) & | & -2(5) + (-4) \end{bmatrix} \text{ simplifies to } \begin{bmatrix} 1 & 3 & | & 5 \\ 0 & -7 & | & -14 \end{bmatrix}$$

$$\begin{array}{cccccc} \uparrow & \uparrow & \uparrow & \uparrow & \uparrow & \uparrow \\ \text{row 1} & \text{row 2} & \text{row 1} & \text{row 2} & \text{row 1} & \text{row 2} \\ \text{element} & \text{element} & \text{element} & \text{element} & \text{element} & \text{element} \end{array}$$

Now we change the -7 to a 1 by use of an elementary row operation. We divide row 2 by -7, then

$$\begin{bmatrix} 1 & 3 & | & 5 \\ \dfrac{0}{-7} & \dfrac{-7}{-7} & | & \dfrac{-14}{-7} \end{bmatrix} \quad \text{simplifies to} \quad \begin{bmatrix} 1 & 3 & | & 5 \\ 0 & 1 & | & 2 \end{bmatrix}$$

This last matrix corresponds to the system

$$\begin{cases} x + 3y = 5 \\ y = 2 \end{cases}$$

To find x, we let $y = 2$ in the first equation, $x + 3y = 5$.

$$x + 3y = 5 \qquad \text{First equation}$$
$$x + 3(2) = 5 \qquad \text{Let } y = 2.$$
$$x = -1$$

The ordered pair solution is $(-1, 2)$. Check to see that this ordered pair satisfies both equations.   □

PRACTICE

1   Use matrices to solve the system.

$$\begin{cases} x + 4y = -2 \\ 3x - y = 7 \end{cases}$$

EXAMPLE 2 Use matrices to solve the system.

$$\begin{cases} 2x - y = 3 \\ 4x - 2y = 5 \end{cases}$$

Solution The corresponding matrix is $\begin{bmatrix} 2 & -1 & | & 3 \\ 4 & -2 & | & 5 \end{bmatrix}$. To get 1 in the row 1, column 1 position, we divide the elements of row 1 by 2.

$$\begin{bmatrix} \dfrac{2}{2} & -\dfrac{1}{2} & | & \dfrac{3}{2} \\ 4 & -2 & | & 5 \end{bmatrix} \quad \text{simplifies to} \quad \begin{bmatrix} 1 & -\dfrac{1}{2} & | & \dfrac{3}{2} \\ 4 & -2 & | & 5 \end{bmatrix}$$

To get 0 under the 1, we multiply the elements of row 1 by -4 and add the new elements to the elements of row 2.

$$\begin{bmatrix} 1 & -\dfrac{1}{2} & | & \dfrac{3}{2} \\ -4(1) + 4 & -4\left(-\dfrac{1}{2}\right) - 2 & | & -4\left(\dfrac{3}{2}\right) + 5 \end{bmatrix} \quad \text{simplifies to} \quad \begin{bmatrix} 1 & -\dfrac{1}{2} & | & \dfrac{3}{2} \\ 0 & 0 & | & -1 \end{bmatrix}$$

The corresponding system is $\begin{cases} x - \dfrac{1}{2}y = \dfrac{3}{2} \\ 0 = -1 \end{cases}$. The equation $0 = -1$ is false for all y or x values; hence, the system is inconsistent and has no solution. The solution set is $\{ \ \}$ or $\varnothing$. $\square$

PRACTICE

2 Use matrices to solve the system.

$$\begin{cases} x - 3y = 3 \\ -2x + 6y = 4 \end{cases}$$

✔ **CONCEPT CHECK**

Consider the system

$$\begin{cases} 2x - 3y = 8 \\ x + 5y = -3 \end{cases}$$

What is wrong with its corresponding matrix shown below?

$$\begin{bmatrix} 2 & -3 & | & 8 \\ 0 & 5 & | & -3 \end{bmatrix}$$

OBJECTIVE

2 Using Matrices to Solve a System of Three Equations

To solve a system of three equations in three variables using matrices, we will write the corresponding matrix in the form

$$\begin{bmatrix} 1 & a & b & | & d \\ 0 & 1 & c & | & e \\ 0 & 0 & 1 & | & f \end{bmatrix}$$

Answer to Concept Check:

matrix should be $\begin{bmatrix} 2 & -3 & | & 8 \\ 1 & 5 & | & -3 \end{bmatrix}$

EXAMPLE 3   Use matrices to solve the system.

$$\begin{cases} x + 2y + z = 2 \\ -2x - y + 2z = 5 \\ x + 3y - 2z = -8 \end{cases}$$

Solution The corresponding matrix is $\begin{bmatrix} 1 & 2 & 1 & \vdots & 2 \\ -2 & -1 & 2 & \vdots & 5 \\ 1 & 3 & -2 & \vdots & -8 \end{bmatrix}$. Our goal is to write an

equivalent matrix with 1's along the diagonal (see the numbers in red) and 0's below the 1's. The element in row 1, column 1 is already 1. Next we get 0's for each element in the rest of column 1. To do this, first we multiply the elements of row 1 by 2 and add the new elements to row 2. Also, we multiply the elements of row 1 by -1 and add the new elements to the elements of row 3. _We do not change row 1._ Then

$$\begin{bmatrix} 1 & 2 & 1 & \vdots & 2 \\ 2(1) - 2 & 2(2) - 1 & 2(1) + 2 & \vdots & 2(2) + 5 \\ -1(1) + 1 & -1(2) + 3 & -1(1) - 2 & \vdots & -1(2) - 8 \end{bmatrix} \text{ simplifies to } \begin{bmatrix} 1 & 2 & 1 & \vdots & 2 \\ 0 & 3 & 4 & \vdots & 9 \\ 0 & 1 & -3 & \vdots & -10 \end{bmatrix}$$

We continue down the diagonal and use elementary row operations to get 1 where the element 3 is now. To do this, we interchange rows 2 and 3.

$$\begin{bmatrix} 1 & 2 & 1 & \vdots & 2 \\ 0 & 3 & 4 & \vdots & 9 \\ 0 & 1 & -3 & \vdots & -10 \end{bmatrix} \text{ is equivalent to } \begin{bmatrix} 1 & 2 & 1 & \vdots & 2 \\ 0 & 1 & -3 & \vdots & -10 \\ 0 & 3 & 4 & \vdots & 9 \end{bmatrix}$$

Next we want the new row 3, column 2 element to be 0. We multiply the elements of row 2 by -3 and add the result to the elements of row 3.

$$\begin{bmatrix} 1 & 2 & 1 & \vdots & 2 \\ 0 & 1 & -3 & \vdots & -10 \\ -3(0) + 0 & -3(1) + 3 & -3(-3) + 4 & \vdots & -3(-10) + 9 \end{bmatrix} \text{ simplifies to}$$

$$\begin{bmatrix} 1 & 2 & 1 & \vdots & 2 \\ 0 & 1 & -3 & \vdots & -10 \\ 0 & 0 & 13 & \vdots & 39 \end{bmatrix}$$

Finally, we divide the elements of row 3 by 13 so that the final diagonal element is 1.

$$\begin{bmatrix} 1 & 2 & 1 & \vdots & 2 \\ 0 & 1 & -3 & \vdots & -10 \\ \dfrac{0}{13} & \dfrac{0}{13} & \dfrac{13}{13} & \vdots & \dfrac{39}{13} \end{bmatrix} \text{ simplifies to } \begin{bmatrix} 1 & 2 & 1 & \vdots & 2 \\ 0 & 1 & -3 & \vdots & -10 \\ 0 & 0 & 1 & \vdots & 3 \end{bmatrix}$$

This matrix corresponds to the system

$$\begin{cases} x + 2y + z = 2 \\ y - 3z = -10 \\ z = 3 \end{cases}$$

We identify the z-coordinate of the solution as 3. Next, we replace z with 3 in the second equation and solve for y.

$$y - 3z = -10 \quad \text{Second equation}$$
$$y - 3(3) = -10 \quad \text{Let } z = 3.$$
$$y = -1$$

(Continued on next page)

To find x, we let $z = 3$ and $y = -1$ in the first equation.

$$x + 2y + z = 2 \quad \text{First equation}$$
$$x + 2(-1) + 3 = 2 \quad \text{Let } z = 3 \text{ and } y = -1.$$
$$x = 1$$

The ordered triple solution is $(1, -1, 3)$. Check to see that it satisfies all three equations in the original system. □

PRACTICE

3 Use matrices to solve the system.

$$\begin{cases} x + 3y - z = 0 \\ 2x + y + 3z = 5 \\ -x - 2y + 4z = 7 \end{cases}$$

D Exercise Set MyMathLab®

Solve each system of linear equations using matrices. See Example 1.

1. $\begin{cases} x + y = 1 \\ x - 2y = 4 \end{cases}$

2. $\begin{cases} 2x - y = 8 \\ x + 3y = 11 \end{cases}$

3. $\begin{cases} x + 3y = 2 \\ x + 2y = 0 \end{cases}$

4. $\begin{cases} 4x - y = 5 \\ 3x + 3y = 0 \end{cases}$

Solve each system of linear equations using matrices. See Example 2.

5. $\begin{cases} x - 2y = 4 \\ 2x - 4y = 4 \end{cases}$

6. $\begin{cases} -x + 3y = 6 \\ 3x - 9y = 9 \end{cases}$

7. $\begin{cases} 3x - 3y = 9 \\ 2x - 2y = 6 \end{cases}$

8. $\begin{cases} 9x - 3y = 6 \\ -18x + 6y = -12 \end{cases}$

Solve each system of linear equations using matrices. See Example 3.

9. $\begin{cases} x + y = 3 \\ 2y = 10 \\ 3x + 2y - 4z = 12 \end{cases}$

10. $\begin{cases} 5x = 5 \\ 2x + y = 4 \\ 3x + y - 5z = -15 \end{cases}$

11. $\begin{cases} 2y - z = -7 \\ x + 4y + z = -4 \\ 5x - y + 2z = 13 \end{cases}$

12. $\begin{cases} 4y + 3z = -2 \\ 5x - 4y = 1 \\ -5x + 4y + z = -3 \end{cases}$

MIXED PRACTICE

Solve each system of linear equations using matrices. See Examples 1 through 3.

13. $\begin{cases} x - 4 = 0 \\ x + y = 1 \end{cases}$

14. $\begin{cases} 3y = 6 \\ x + y = 7 \end{cases}$

15. $\begin{cases} x + y + z = 2 \\ 2x - z = 5 \\ 3y + z = 2 \end{cases}$

16. $\begin{cases} x + 2y + z = 5 \\ x - y - z = 3 \\ y + z = 2 \end{cases}$

17. $\begin{cases} 5x - 2y = 27 \\ -3x + 5y = 18 \end{cases}$

18. $\begin{cases} 4x - y = 9 \\ 2x + 3y = -27 \end{cases}$

19. $\begin{cases} 4x - 7y = 7 \\ 12x - 21y = 24 \end{cases}$

20. $\begin{cases} 2x - 5y = 12 \\ -4x + 10y = 20 \end{cases}$

21. $\begin{cases} 4x - y + 2z = 5 \\ 2y + z = 4 \\ 4x + y + 3z = 10 \end{cases}$

22. $\begin{cases} 5y - 7z = 14 \\ 2x + y + 4z = 10 \\ 2x + 6y - 3z = 30 \end{cases}$

23. $\begin{cases} 4x + y + z = 3 \\ -x + y - 2z = -11 \\ x + 2y + 2z = -1 \end{cases}$

24. $\begin{cases} x + y + z = 9 \\ 3x - y + z = -1 \\ -2x + 2y - 3z = -2 \end{cases}$

CONCEPT EXTENSIONS

Solve. See the Concept Check in this section.

25. For the system $\begin{cases} x + z = 7 \\ y + 2z = -6, \\ 3x - y = 0 \end{cases}$ which is the correct corresponding matrix?

a. $\begin{bmatrix} 1 & 1 & \vdots & 7 \\ 1 & 2 & \vdots & -6 \\ 3 & -1 & \vdots & 0 \end{bmatrix}$

b. $\begin{bmatrix} 1 & 0 & 1 & \vdots & 7 \\ 1 & 2 & 0 & \vdots & -6 \\ 3 & -1 & 0 & \vdots & 0 \end{bmatrix}$

c. $\begin{bmatrix} 1 & 0 & 1 & \vdots & 7 \\ 0 & 1 & 2 & \vdots & -6 \\ 3 & -1 & 0 & \vdots & 0 \end{bmatrix}$

Appendix E

Solving Systems of Equations Using Determinants

OBJECTIVES

1 Define and Evaluate a 2 × 2 Determinant.

2 Use Cramer's Rule to Solve a System of Two Linear Equations in Two Variables.

3 Define and Evaluate a 3 × 3 Determinant.

4 Use Cramer's Rule to Solve a System of Three Linear Equations in Three Variables.

We have solved systems of two linear equations in two variables in four different ways: graphically, by substitution, by elimination, and by matrices. Now we analyze another method, called **Cramer's rule.**

OBJECTIVE

1 Evaluating 2 × 2 Determinants

Recall that a matrix is a rectangular array of numbers. If a matrix has the same number of rows and columns, it is called a **square matrix.** Examples of square matrices are

$$\begin{bmatrix} 1 & 6 \\ 5 & 2 \end{bmatrix} \qquad \begin{bmatrix} 2 & 4 & 1 \\ 0 & 5 & 2 \\ 3 & 6 & 9 \end{bmatrix}$$

A **determinant** is a real number associated with a square matrix. The determinant of a square matrix is denoted by placing vertical bars around the array of numbers. Thus,

The determinant of the square matrix $\begin{bmatrix} 1 & 6 \\ 5 & 2 \end{bmatrix}$ is $\begin{vmatrix} 1 & 6 \\ 5 & 2 \end{vmatrix}$.

The determinant of the square matrix $\begin{bmatrix} 2 & 4 & 1 \\ 0 & 5 & 2 \\ 3 & 6 & 9 \end{bmatrix}$ is $\begin{vmatrix} 2 & 4 & 1 \\ 0 & 5 & 2 \\ 3 & 6 & 9 \end{vmatrix}$.

We define the determinant of a 2 × 2 matrix first. (Recall that 2 × 2 is read "two by two." It means that the matrix has 2 rows and 2 columns.)

Determinant of a 2 × 2 Matrix

$$\begin{vmatrix} a & b \\ c & d \end{vmatrix} = ad - bc$$

EXAMPLE 1 Evaluate each determinant.

a. $\begin{vmatrix} -1 & 2 \\ 3 & -4 \end{vmatrix}$ **b.** $\begin{vmatrix} 2 & 0 \\ 7 & -5 \end{vmatrix}$

Solution First we identify the values of $a, b, c,$ and d. Then we perform the evaluation.

a. Here $a = -1, b = 2, c = 3,$ and $d = -4.$

$$\begin{vmatrix} -1 & 2 \\ 3 & -4 \end{vmatrix} = ad - bc = (-1)(-4) - (2)(3) = -2$$

b. In this example, $a = 2, b = 0, c = 7,$ and $d = -5.$

$$\begin{vmatrix} 2 & 0 \\ 7 & -5 \end{vmatrix} = ad - bc = 2(-5) - (0)(7) = -10 \qquad \square$$

OBJECTIVE

2 Using Cramer's Rule to Solve a System of Two Linear Equations

To develop Cramer's rule, we solve the system $\begin{cases} ax + by = h \\ cx + dy = k \end{cases}$ using elimination. First, we eliminate y by multiplying both sides of the first equation by d and both sides of the second equation by $-b$ so that the coefficients of y are opposites. The result is that

$$\begin{cases} d(ax + by) = d \cdot h \\ -b(cx + dy) = -b \cdot k \end{cases} \quad \text{simplifies to} \quad \begin{cases} adx + bdy = hd \\ -bcx - bdy = -kb \end{cases}$$

We now add the two equations and solve for x.

$$
\begin{aligned}
adx + bdy &= hd \\
\underline{-bcx - bdy} &= \underline{-kb} \\
adx - bcx &= hd - kb \quad \text{Add the equations.} \\
(ad - bc)x &= hd - kb \\
x &= \frac{hd - kb}{ad - bc} \quad \text{Solve for } x.
\end{aligned}
$$

When we replace x with $\dfrac{hd - kb}{ad - bc}$ in the equation $ax + by = h$ and solve for y, we find that $y = \dfrac{ak - ch}{ad - bc}$.

Notice that the numerator of the value of x is the determinant of

$$\begin{vmatrix} h & b \\ k & d \end{vmatrix} = hd - kb$$

Also, the numerator of the value of y is the determinant of

$$\begin{vmatrix} a & h \\ c & k \end{vmatrix} = ak - hc \ .$$

Finally, the denominators of the values of x and y are the same and are the determinant of

$$\begin{vmatrix} a & b \\ c & d \end{vmatrix} = ad - bc$$

This means that the values of x and y can be written in determinant notation:

$$x = \frac{\begin{vmatrix} h & b \\ k & d \end{vmatrix}}{\begin{vmatrix} a & b \\ c & d \end{vmatrix}} \quad \text{and} \quad y = \frac{\begin{vmatrix} a & h \\ c & k \end{vmatrix}}{\begin{vmatrix} a & b \\ c & d \end{vmatrix}}$$

For convenience, we label the determinants D, D_x, and D_y.

x-coefficients
y-coefficients

$$\begin{vmatrix} a & b \\ c & d \end{vmatrix} = D \qquad \begin{vmatrix} h & b \\ k & d \end{vmatrix} = D_x \qquad \begin{vmatrix} a & h \\ c & k \end{vmatrix} = D_y$$

x-column replaced y-column replaced
by constants   by constants

These determinant formulas for the coordinates of the solution of a system are known as **Cramer's rule.**

Cramer's Rule for Two Linear Equations in Two Variables

The solution of the system $\begin{cases} ax + by = h \\ cx + dy = k \end{cases}$ is given by

$$x = \frac{\begin{vmatrix} h & b \\ k & d \end{vmatrix}}{\begin{vmatrix} a & b \\ c & d \end{vmatrix}} = \frac{D_x}{D} \qquad y = \frac{\begin{vmatrix} a & h \\ c & k \end{vmatrix}}{\begin{vmatrix} a & b \\ c & d \end{vmatrix}} = \frac{D_y}{D}$$

as long as $D = ad - bc$ is not 0.

When $D = 0$, the system is either inconsistent or the equations are dependent. When this happens, we need to use another method to see which is the case.

EXAMPLE 2   Use Cramer's rule to solve the system.

$$\begin{cases} 3x + 4y = -7 \\ x - 2y = -9 \end{cases}$$

Solution   First we find D, D_x, and D_y.

$$\begin{array}{ccc} a & b & h \\ \downarrow & \downarrow & \downarrow \end{array}$$
$$\begin{cases} 3x + 4y = -7 \\ x - 2y = -9 \end{cases}$$
$$\begin{array}{ccc} \uparrow & \uparrow & \uparrow \\ c & d & k \end{array}$$

$$D = \begin{vmatrix} a & b \\ c & d \end{vmatrix} = \begin{vmatrix} 3 & 4 \\ 1 & -2 \end{vmatrix} = 3(-2) - 4(1) = -10$$

$$D_x = \begin{vmatrix} h & b \\ k & d \end{vmatrix} = \begin{vmatrix} -7 & 4 \\ -9 & -2 \end{vmatrix} = (-7)(-2) - 4(-9) = 50$$

$$D_y = \begin{vmatrix} a & h \\ c & d \end{vmatrix} = \begin{vmatrix} 3 & -7 \\ 1 & -9 \end{vmatrix} = 3(-9) - (-7)(1) = -20$$

Then $x = \dfrac{D_x}{D} = \dfrac{50}{-10} = -5$ and $y = \dfrac{D_y}{D} = \dfrac{-20}{-10} = 2$.

The ordered pair solution is $(-5, 2)$.

As always, check the solution in both original equations. $\square$

EXAMPLE 3   Use Cramer's rule to solve the system.

$$\begin{cases} 5x + y = 5 \\ -7x - 2y = -7 \end{cases}$$

Solution   First we find D, D_x, and D_y.

$$D = \begin{vmatrix} 5 & 1 \\ -7 & -2 \end{vmatrix} = 5(-2) - (-7)(1) = -3$$

$$D_x = \begin{vmatrix} 5 & 1 \\ -7 & -2 \end{vmatrix} = 5(-2) - (-7)(1) = -3$$

$$D_y = \begin{vmatrix} 5 & 5 \\ -7 & -7 \end{vmatrix} = 5(-7) - 5(-7) = 0$$

(Continued on next page)

Then

$$x = \frac{D_x}{D} = \frac{-3}{-3} = 1 \qquad y = \frac{D_y}{D} = \frac{0}{-3} = 0$$

The ordered pair solution is $(1, 0)$. $\square$

3 Evaluating 3 × 3 Determinants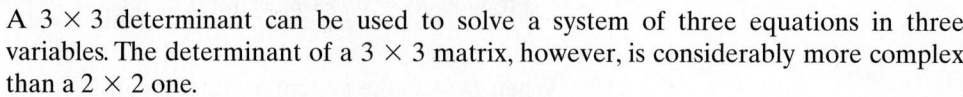

A 3×3 determinant can be used to solve a system of three equations in three variables. The determinant of a 3×3 matrix, however, is considerably more complex than a 2×2 one.

Determinant of a 3 × 3 Matrix

$$\begin{vmatrix} a_1 & b_1 & c_1 \\ a_2 & b_2 & c_2 \\ a_3 & b_3 & c_3 \end{vmatrix} = a_1 \cdot \begin{vmatrix} b_2 & c_2 \\ b_3 & c_3 \end{vmatrix} - a_2 \cdot \begin{vmatrix} b_1 & c_1 \\ b_3 & c_3 \end{vmatrix} + a_3 \cdot \begin{vmatrix} b_1 & c_1 \\ b_2 & c_2 \end{vmatrix}$$

Notice that the determinant of a 3×3 matrix is related to the determinants of three 2×2 matrices. Each determinant of these 2×2 matrices is called a **minor,** and every element of a 3×3 matrix has a minor associated with it. For example, the minor of c_2 is the determinant of the 2×2 matrix found by deleting the row and column containing c_2.

$$\begin{matrix} a_1 & b_1 & c_1 \\ a_2 & b_2 & c_2 \\ a_3 & b_3 & c_3 \end{matrix} \qquad \text{The minor of } c_2 \text{ is} \qquad \begin{vmatrix} a_1 & b_1 \\ a_3 & b_3 \end{vmatrix}$$

Also, the minor of element a_1 is the determinant of the 2×2 matrix that has no row or column containing a_1.

$$\begin{matrix} a_1 & b_1 & c_1 \\ a_2 & b_2 & c_2 \\ a_3 & b_3 & c_3 \end{matrix} \qquad \text{The minor of } a_1 \text{ is} \qquad \begin{vmatrix} b_2 & c_2 \\ b_3 & c_3 \end{vmatrix}$$

So the determinant of a 3×3 matrix can be written as

$$a_1 \cdot (\text{minor of } a_1) - a_2 \cdot (\text{minor of } a_2) + a_3 \cdot (\text{minor of } a_3)$$

Finding the determinant by using minors of elements in the first column is called **expanding** by the minors of the first column. *The value of a determinant can be found by expanding by the minors of any row or column.* The following **array of signs** is helpful in determining whether to add or subtract the product of an element and its minor.

$$\begin{matrix} + & - & + \\ - & + & - \\ + & - & + \end{matrix}$$

If an element is in a position marked $+$, we add. If marked $-$, we subtract.

EXAMPLE 4   Evaluate by expanding by the minors of the given row or column.

$$\begin{vmatrix} 0 & 5 & 1 \\ 1 & 3 & -1 \\ -2 & 2 & 4 \end{vmatrix}$$

a. First column   **b.** Second row

Solution

a. The elements of the first column are $0, 1,$ and $-2.$ The first column of the array of signs is $+, -, +.$

$$\begin{vmatrix} 0 & 5 & 1 \\ 1 & 3 & -1 \\ -2 & 2 & 4 \end{vmatrix} = 0 \cdot \begin{vmatrix} 3 & -1 \\ 2 & 4 \end{vmatrix} - 1 \cdot \begin{vmatrix} 5 & 1 \\ 2 & 4 \end{vmatrix} + (-2) \cdot \begin{vmatrix} 5 & 1 \\ 3 & -1 \end{vmatrix}$$

$$= 0(12 - (-2)) - 1(20 - 2) + (-2)(-5 - 3)$$
$$= 0 - 18 + 16 = -2$$

b. The elements of the second row are $1, 3,$ and $-1.$ This time, the signs begin with $-$ and again alternate.

$$\begin{vmatrix} 0 & 5 & 1 \\ 1 & 3 & -1 \\ -2 & 2 & 4 \end{vmatrix} = -1 \cdot \begin{vmatrix} 5 & 1 \\ 2 & 4 \end{vmatrix} + 3 \cdot \begin{vmatrix} 0 & 1 \\ -2 & 4 \end{vmatrix} - (-1) \cdot \begin{vmatrix} 0 & 5 \\ -2 & 2 \end{vmatrix}$$

$$= -1(20 - 2) + 3(0 - (-2)) - (-1)(0 - (-10))$$
$$= -18 + 6 + 10 = -2$$

Notice that the determinant of the 3×3 matrix is the same regardless of the row or column you select to expand by.   □

✔ **CONCEPT CHECK**

Why would expanding by minors of the second row be a good choice for the determinant $\begin{vmatrix} 3 & 4 & -2 \\ 5 & 0 & 0 \\ 6 & -3 & 7 \end{vmatrix}$?

OBJECTIVE

4   Using Cramer's Rule to Solve a System of Three Linear Equations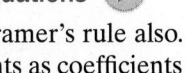

A system of three equations in three variables may be solved with Cramer's rule also. Using the elimination process to solve a system with unknown constants as coefficients leads to the following.

Cramer's Rule for Three Equations in Three Variables

The solution of the system $\begin{cases} a_1x + b_1y + c_1z = k_1 \\ a_2x + b_2y + c_2z = k_2 \\ a_3x + b_3y + c_3z = k_3 \end{cases}$ is given by

$$x = \frac{D_x}{D} \qquad y = \frac{D_y}{D} \qquad \text{and} \qquad z = \frac{D_z}{D}$$

where

$$D = \begin{vmatrix} a_1 & b_1 & c_1 \\ a_2 & b_2 & c_2 \\ a_3 & b_3 & c_3 \end{vmatrix} \qquad D_x = \begin{vmatrix} k_1 & b_1 & c_1 \\ k_2 & b_2 & c_2 \\ k_3 & b_3 & c_3 \end{vmatrix}$$

$$D_y = \begin{vmatrix} a_1 & k_1 & c_1 \\ a_2 & k_2 & c_2 \\ a_3 & k_3 & c_3 \end{vmatrix} \qquad D_z = \begin{vmatrix} a_1 & b_1 & k_1 \\ a_2 & b_2 & k_2 \\ a_3 & b_3 & k_3 \end{vmatrix}$$

as long as D is not 0.

Answer to Concept Check:
Two elements of the second row are 0, which makes calculations easier.

EXAMPLE 5   Use Cramer's rule to solve the system.

$$\begin{cases} x - 2y + z = 4 \\ 3x + y - 2z = 3 \\ 5x + 5y + 3z = -8 \end{cases}$$

Solution   First we find $D, D_x, D_y,$ and D_z. Beginning with D, we expand by the minors of the first column.

$$D = \begin{vmatrix} 1 & -2 & 1 \\ 3 & 1 & -2 \\ 5 & 5 & 3 \end{vmatrix} = 1 \cdot \begin{vmatrix} 1 & -2 \\ 5 & 3 \end{vmatrix} - 3 \cdot \begin{vmatrix} -2 & 1 \\ 5 & 3 \end{vmatrix} + 5 \cdot \begin{vmatrix} -2 & 1 \\ 1 & -2 \end{vmatrix}$$

$$= 1(3 - (-10)) - 3(-6 - 5) + 5(4 - 1)$$

$$= 13 + 33 + 15 = 61$$

$$D_x = \begin{vmatrix} 4 & -2 & 1 \\ 3 & 1 & -2 \\ -8 & 5 & 3 \end{vmatrix} = 4 \cdot \begin{vmatrix} 1 & -2 \\ 5 & 3 \end{vmatrix} - 3 \cdot \begin{vmatrix} -2 & 1 \\ 5 & 3 \end{vmatrix} + (-8) \cdot \begin{vmatrix} -2 & 1 \\ 1 & -2 \end{vmatrix}$$

$$= 4(3 - (-10)) - 3(-6 - 5) + (-8)(4 - 1)$$

$$= 52 + 33 - 24 = 61$$

$$D_y = \begin{vmatrix} 1 & 4 & 1 \\ 3 & 3 & -2 \\ 5 & -8 & 3 \end{vmatrix} = 1 \cdot \begin{vmatrix} 3 & -2 \\ -8 & 3 \end{vmatrix} - 3 \cdot \begin{vmatrix} 4 & 1 \\ -8 & 3 \end{vmatrix} + 5 \cdot \begin{vmatrix} 4 & 1 \\ 3 & -2 \end{vmatrix}$$

$$= 1(9 - 16) - 3(12 - (-8)) + 5(-8 - 3)$$

$$= -7 - 60 - 55 = -122$$

$$D_z = \begin{vmatrix} 1 & -2 & 4 \\ 3 & 1 & 3 \\ 5 & 5 & -8 \end{vmatrix} = 1 \cdot \begin{vmatrix} 1 & 3 \\ 5 & -8 \end{vmatrix} - 3 \cdot \begin{vmatrix} -2 & 4 \\ 5 & -8 \end{vmatrix} + 5 \cdot \begin{vmatrix} -2 & 4 \\ 1 & 3 \end{vmatrix}$$

$$= 1(-8 - 15) - 3(16 - 20) + 5(-6 - 4)$$

$$= -23 + 12 - 50 = -61$$

From these determinants, we calculate the solution:

$$x = \frac{D_x}{D} = \frac{61}{61} = 1 \quad y = \frac{D_y}{D} = \frac{-122}{61} = -2 \quad z = \frac{D_z}{D} = \frac{-61}{61} = -1$$

The ordered triple solution is $(1, -2, -1)$. Check this solution by verifying that it satisfies each equation of the system.   ☐

E   Exercise Set   MyMathLab ▶

Evaluate. See Example 1.

▶ **1.** $\begin{vmatrix} 3 & 5 \\ -1 & 7 \end{vmatrix}$ **2.** $\begin{vmatrix} -5 & 1 \\ 1 & -4 \end{vmatrix}$

3. $\begin{vmatrix} 9 & -2 \\ 4 & -3 \end{vmatrix}$ **4.** $\begin{vmatrix} 4 & 0 \\ 9 & 8 \end{vmatrix}$

5. $\begin{vmatrix} -2 & 9 \\ 4 & -18 \end{vmatrix}$ **6.** $\begin{vmatrix} -40 & 8 \\ 70 & -14 \end{vmatrix}$

Use Cramer's rule, if possible, to solve each system of linear equations. See Examples 2 and 3.

7. $\begin{cases} 2y - 4 = 0 \\ x + 2y = 5 \end{cases}$ **8.** $\begin{cases} 4x - y = 5 \\ 3x - 3 = 0 \end{cases}$

9. $\begin{cases} 3x + y = 1 \\ 2y = 2 - 6x \end{cases}$ **10.** $\begin{cases} y = 2x - 5 \\ 8x - 4y = 20 \end{cases}$

11. $\begin{cases} 5x - 2y = 27 \\ -3x + 5y = 18 \end{cases}$ **12.** $\begin{cases} 4x - y = 9 \\ 2x + 3y = -27 \end{cases}$

Evaluate. See Example 4.

13. $\begin{vmatrix} 2 & 1 & 0 \\ 0 & 5 & -3 \\ 4 & 0 & 2 \end{vmatrix}$ **14.** $\begin{vmatrix} -6 & 4 & 2 \\ 1 & 0 & 5 \\ 0 & 3 & 1 \end{vmatrix}$

15. $\begin{vmatrix} 4 & -6 & 0 \\ -2 & 3 & 0 \\ 4 & -6 & 1 \end{vmatrix}$ **16.** $\begin{vmatrix} 5 & 2 & 1 \\ 3 & -6 & 0 \\ -2 & 8 & 0 \end{vmatrix}$

17. $\begin{vmatrix} 3 & 6 & -3 \\ -1 & -2 & 3 \\ 4 & -1 & 6 \end{vmatrix}$ **18.** $\begin{vmatrix} 2 & -2 & 1 \\ 4 & 1 & 3 \\ 3 & 1 & 2 \end{vmatrix}$

Use Cramer's rule, if possible, to solve each system of linear equations. See Example 5.

19. $\begin{cases} 3x & + z = -1 \\ -x - 3y + z = 7 \\ 3y + z = 5 \end{cases}$ **20.** $\begin{cases} 4y - 3z = -2 \\ 8x - 4y = 4 \\ -8x + 4y + z = -2 \end{cases}$

21. $\begin{cases} x + y + z = 8 \\ 2x - y - z = 10 \\ x - 2y + 3z = 22 \end{cases}$ **22.** $\begin{cases} 5x + y + 3z = 1 \\ x - y - 3z = -7 \\ -x + y = 1 \end{cases}$

MIXED PRACTICE

Evaluate.

23. $\begin{vmatrix} 10 & -1 \\ -4 & 2 \end{vmatrix}$ **24.** $\begin{vmatrix} -6 & 2 \\ 5 & -1 \end{vmatrix}$

25. $\begin{vmatrix} 1 & 0 & 4 \\ 1 & -1 & 2 \\ 3 & 2 & 1 \end{vmatrix}$ **26.** $\begin{vmatrix} 0 & 1 & 2 \\ 3 & -1 & 2 \\ 3 & 2 & -2 \end{vmatrix}$

27. $\begin{vmatrix} \frac{3}{4} & \frac{5}{2} \\ -\frac{1}{6} & \frac{7}{3} \end{vmatrix}$ **28.** $\begin{vmatrix} \frac{5}{7} & \frac{1}{3} \\ \frac{6}{7} & \frac{2}{3} \end{vmatrix}$

29. $\begin{vmatrix} 4 & -2 & 2 \\ 6 & -1 & 3 \\ 2 & 1 & 1 \end{vmatrix}$ **30.** $\begin{vmatrix} 1 & 5 & 0 \\ 7 & 9 & -4 \\ 3 & 2 & -2 \end{vmatrix}$

31. $\begin{vmatrix} -2 & 5 & 4 \\ 5 & -1 & 3 \\ 4 & 1 & 2 \end{vmatrix}$ **32.** $\begin{vmatrix} 5 & -2 & 4 \\ -1 & 5 & 3 \\ 1 & 4 & 2 \end{vmatrix}$

Use Cramer's rule, if possible, to solve each system of linear equations.

33. $\begin{cases} 2x - 5y = 4 \\ x + 2y = -7 \end{cases}$ **34.** $\begin{cases} 3x - y = 2 \\ -5x + 2y = 0 \end{cases}$

35. $\begin{cases} 4x + 2y = 5 \\ 2x + y = -1 \end{cases}$ **36.** $\begin{cases} 3x + 6y = 15 \\ 2x + 4y = 3 \end{cases}$

37. $\begin{cases} 2x + 2y + z = 1 \\ -x + y + 2z = 3 \\ x + 2y + 4z = 0 \end{cases}$ **38.** $\begin{cases} 2x - 3y + z = 5 \\ x + y + z = 0 \\ 4x + 2y + 4z = 4 \end{cases}$

39. $\begin{cases} \frac{2}{3}x - \frac{3}{4}y = -1 \\ -\frac{1}{6}x + \frac{3}{4}y = \frac{5}{2} \end{cases}$ **40.** $\begin{cases} \frac{1}{2}x - \frac{1}{3}y = -3 \\ \frac{1}{8}x + \frac{1}{6}y = 0 \end{cases}$

41. $\begin{cases} 0.7x - 0.2y = -1.6 \\ 0.2x - y = -1.4 \end{cases}$ **42.** $\begin{cases} -0.7x + 0.6y = 1.3 \\ 0.5x - 0.3y = -0.8 \end{cases}$

43. $\begin{cases} -2x + 4y - 2z = 6 \\ x - 2y + z = -3 \\ 3x - 6y + 3z = -9 \end{cases}$ **44.** $\begin{cases} -x - y + 3z = 2 \\ 4x + 4y - 12z = -8 \\ -3x - 3y + 9z = 6 \end{cases}$

45. $\begin{cases} x - 2y + z = -5 \\ 3y + 2z = 4 \\ 3x - y = -2 \end{cases}$ **46.** $\begin{cases} 4x + 5y = 10 \\ 3y + 2z = -6 \\ x + y + z = 3 \end{cases}$

CONCEPT EXTENSIONS

Find the value of x such that each is a true statement.

47. $\begin{vmatrix} 1 & x \\ 2 & 7 \end{vmatrix} = -3$ **48.** $\begin{vmatrix} 6 & 1 \\ -2 & x \end{vmatrix} = 26$

49. If all the elements in a single row of a square matrix are zero, to what does the determinant evaluate? Explain your answer.

50. If all the elements in a single column of a square matrix are 0, to what does the determinant evaluate? Explain your answer.

51. Suppose you are interested in finding the determinant of a 4×4 matrix. Study the pattern shown in the array of signs for a 3×3 matrix. Use the pattern to expand the array of signs for use with a 4×4 matrix.

52. Why would expanding by minors of the third column be a good choice for the determinant $\begin{vmatrix} 3 & 4 & -2 \\ 5 & 7 & 0 \\ 6 & -3 & 0 \end{vmatrix}$?

Find the value of each determinant. To evaluate a 4×4 determinant, select any row or column and expand by the minors. The array of signs for a 4×4 determinant is the same as for a 3×3 determinant except expanded.

53. $\begin{vmatrix} 5 & 0 & 0 & 0 \\ 0 & 4 & 2 & -1 \\ 1 & 3 & -2 & 0 \\ 0 & -3 & 1 & 2 \end{vmatrix}$ **54.** $\begin{vmatrix} 1 & 7 & 0 & -1 \\ 1 & 3 & -2 & 0 \\ 1 & 0 & -1 & 2 \\ 0 & -6 & 2 & 4 \end{vmatrix}$

55. $\begin{vmatrix} 4 & 0 & 2 & 5 \\ 0 & 3 & -1 & 1 \\ 0 & 0 & 2 & 0 \\ 0 & 0 & 0 & 1 \end{vmatrix}$ **56.** $\begin{vmatrix} 2 & 0 & -1 & 4 \\ 6 & 0 & 4 & 1 \\ 2 & 4 & 3 & -1 \\ 4 & 0 & 5 & -4 \end{vmatrix}$

Appendix F

Mean, Median, and Mode

It is sometimes desirable to be able to describe a set of data, or a set of numbers, by a single "middle" number. Three such **measures of central tendency** are the mean, the median, and the mode.

The most common measure of central tendency is the mean (sometimes called the arithmetic mean or the average). The **mean** of a set of data items, denoted by $\bar{x}$, is the sum of the items divided by the number of items.

EXAMPLE 1 Seven students in a psychology class conducted an experiment on mazes. Each student was given a pencil and asked to successfully complete the same maze. The timed results are below.

Student	Ann	Thanh	Carlos	Jesse	Melinda	Ramzi	Dayni
Time (Seconds)	13.2	11.8	10.7	16.2	15.9	13.8	18.5

a. Who completed the maze in the shortest time? Who completed the maze in the longest time?

b. Find the mean.

c. How many students took longer than the mean time? How many students took shorter than the mean time?

Solution

a. Carlos completed the maze in 10.7 seconds, the shortest time. Dayni completed the maze in 18.5 seconds, the longest time.

b. To find the mean, $\bar{x}$, find the sum of the data items and divide by 7, the number of items.

$$\bar{x} = \frac{13.2 + 11.8 + 10.7 + 16.2 + 15.9 + 13.8 + 18.5}{7} = \frac{100.1}{7} = 14.3$$

c. Three students, Jesse, Melinda, and Dayni, had times longer than the mean time. Four students, Ann, Thanh, Carlos, and Ramzi, had times shorter than the mean time. □

Two other measures of central tendency are the median and the mode.

The **median** of an ordered set of numbers is the middle number. If the number of items is even, the median is the mean of the two middle numbers. The **mode** of a set of numbers is the number that occurs most often. It is possible for a data set to have no mode or more than one mode.

EXAMPLE 2 Find the median and the mode of the following list of numbers. These numbers were high temperatures for fourteen consecutive days in a city in Montana.

76, 80, 85, 86, 89, 87, 82, 77, 76, 79, 82, 89, 89, 92

906

Solution

First, write the numbers in order.

76, 76, 77, 79, 80, 82, 82, 85, 86, 87, 89, 89, 89, 92

two middle numbers

mode

Since there is an even number of items, the median is the mean of the two middle numbers.

$$\text{median} = \frac{82 + 85}{2} = 83.5$$

The mode is 89, since 89 occurs most often.

F Exercise Set MyMathLab®

For each of the following data sets, find the mean, the median, and the mode. If necessary, round the mean to one decimal place.

1. 21, 28, 16, 42, 38

2. 42, 35, 36, 40, 50

3. 7.6, 8.2, 8.2, 9.6, 5.7, 9.1

4. 4.9, 7.1, 6.8, 6.8, 5.3, 4.9

5. 0.2, 0.3, 0.5, 0.6, 0.6, 0.9, 0.2, 0.7, 1.1

6. 0.6, 0.6, 0.8, 0.4, 0.5, 0.3, 0.7, 0.8, 0.1

7. 231, 543, 601, 293, 588, 109, 334, 268

8. 451, 356, 478, 776, 892, 500, 467, 780

Eight of the tallest buildings in the United States are listed below. Use this table for Exercises 9 through 12.

Building	Height (feet)
Willis Tower, Chicago, IL	1454
Empire State, New York, NY	1250
Amoco, Chicago, IL	1136
John Hancock Center, Chicago, IL	1127
First Interstate World Center, Los Angeles, CA	1107
Chrysler, New York, NY	1046
NationsBank Tower, Atlanta, GA	1023
Texas Commerce Tower, Houston, TX	1002

9. Find the mean height for the five tallest buildings.

10. Find the median height for the five tallest buildings.

11. Find the median height for the eight tallest buildings.

12. Find the mean height for the eight tallest buildings.

During an experiment, the following times (in seconds) were recorded: 7.8, 6.9, 7.5, 4.7, 6.9, 7.0.

13. Find the mean.

14. Find the median.

15. Find the mode.

In a mathematics class, the following test scores were recorded for a student: 86, 95, 91, 74, 77, 85.

16. Find the mean. Round to the nearest hundredth.

17. Find the median.

18. Find the mode.

The following pulse rates were recorded for a group of fifteen students: 78, 80, 66, 68, 71, 64, 82, 71, 70, 65, 70, 75, 77, 86, 72.

19. Find the mean.

20. Find the median.

21. Find the mode.

22. How many rates were higher than the mean?

23. How many rates were lower than the mean?

24. Have each student in your algebra class take his/her pulse rate. Record the data and find the mean, the median, and the mode.

Find the missing numbers in each list of numbers. (These numbers are not necessarily in numerical order.)

25. __, __, 16, 18, __

The mode is 21.   The mean is 20.

26. __, __, __, __, 40

The mode is 35.   The median is 37.   The mean is 38.

Appendix G

Review of Angles, Lines, and Special Triangles

The word **geometry** is formed from the Greek words **geo,** meaning earth, and **metron,** meaning measure. Geometry literally means to measure the earth.

This section contains a review of some basic geometric ideas. It will be assumed that fundamental ideas of geometry such as point, line, ray, and angle are known. In this appendix, the notation ∠1 is read "angle 1" and the notation $m\angle 1$ is read "the measure of angle 1."

We first review types of angles.

Angles

A **right angle** is an angle whose measure is 90°. A right angle can be indicated by a square drawn at the vertex of the angle, as shown below.

An angle whose measure is more than 0° but less than 90° is called an **acute angle.**

An angle whose measure is greater than 90° but less than 180° is called an **obtuse angle.**

An angle whose measure is 180° is called a **straight angle.**

Two angles are said to be **complementary** if the sum of their measures is 90°. Each angle is called the **complement** of the other.

Two angles are said to be **supplementary** if the sum of their measures is 180°. Each angle is called the **supplement** of the other.

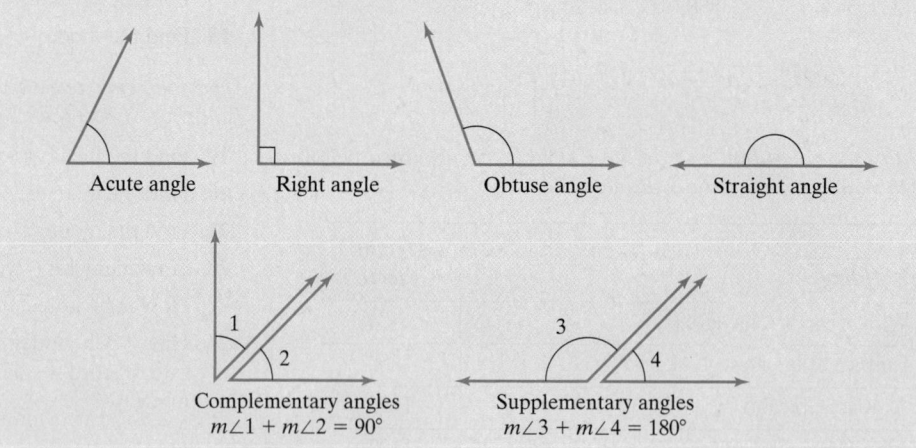

Acute angle   Right angle   Obtuse angle   Straight angle

Complementary angles
$m\angle 1 + m\angle 2 = 90°$

Supplementary angles
$m\angle 3 + m\angle 4 = 180°$

EXAMPLE 1   If an angle measures 28°, find its complement.

Solution Two angles are complementary if the sum of their measures is 90°. The complement of a 28° angle is an angle whose measure is $90° - 28° = 62°$. To check, notice that $28° + 62° = 90°$.   □

Plane is an undefined term that we will describe. A plane can be thought of as a flat surface with infinite length and width, but no thickness. A plane is two dimensional.

908

The arrows in the following diagram indicate that a plane extends indefinitely and has no boundaries.

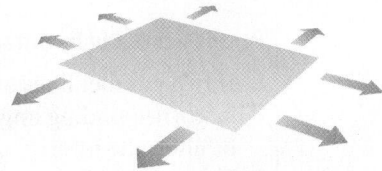

Figures that lie on a plane are called **plane figures.** Lines that lie in the same plane are called **coplanar.**

Lines

Two lines are **parallel** if they lie in the same plane but never meet.

Intersecting lines meet or cross in one point.

Two lines that form right angles when they intersect are said to be **perpendicular.**

Parallel lines   Intersecting
                          lines   Intersecting lines that
                                         are perpendicular

Two intersecting lines form **vertical angles**. Angles 1 and 3 are vertical angles. Also, angles 2 and 4 are vertical angles. It can be shown that **vertical angles have equal measures.**

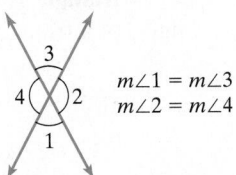

$$m\angle 1 = m\angle 3$$
$$m\angle 2 = m\angle 4$$

Adjacent angles have the same vertex and share a side. Angles 1 and 2 are adjacent angles. Other pairs of adjacent angles are angles 2 and 3, angles 3 and 4, and angles 4 and 1.

A **transversal** is a line that intersects two or more lines in the same plane. Line l is a transversal that intersects lines m and n. The eight angles formed are numbered and certain pairs of these angles are given special names.

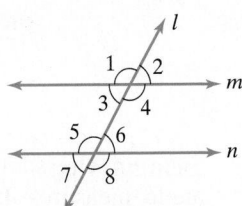

Corresponding angles: $\angle 1$ and $\angle 5$, $\angle 3$ and $\angle 7$, $\angle 2$ and $\angle 6$, and $\angle 4$ and $\angle 8$.

Exterior angles: $\angle 1$, $\angle 2$, $\angle 7$, and $\angle 8$.

Interior angles: $\angle 3$, $\angle 4$, $\angle 5$, and $\angle 6$.

Alternate interior angles: $\angle 3$ and $\angle 6$, $\angle 4$ and $\angle 5$.

These angles and parallel lines are related in the following manner.

Parallel Lines Cut by a Transversal

1. If two parallel lines are cut by a transversal, then
   a. corresponding angles are equal and
   b. alternate interior angles are equal.
2. If corresponding angles formed by two lines and a transversal are equal, then the lines are parallel.
3. If alternate interior angles formed by two lines and a transversal are equal, then the lines are parallel.

EXAMPLE 2 Given that lines *m* and *n* are parallel and that the measure of angle 1 is 100°, find the measures of angles 2, 3, and 4.

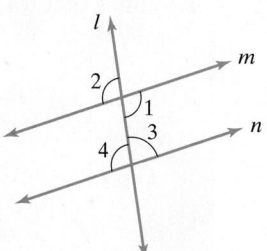

Solution $m\angle 2 = 100°$, since angles 1 and 2 are vertical angles.

$m\angle 4 = 100°$, since angles 1 and 4 are alternate interior angles.

$m\angle 3 = 180° - 100° = 80°$, since angles 4 and 3 are supplementary angles. □

A **polygon** is the union of three or more coplanar line segments that intersect each other only at each endpoint, with each endpoint shared by exactly two segments.

A **triangle** is a polygon with three sides. The sum of the measures of the three angles of a triangle is 180°. In the following figure, $m\angle 1 + m\angle 2 + m\angle 3 = 180°$.

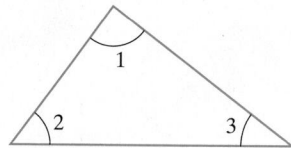

EXAMPLE 3 Find the measure of the third angle of the triangle shown.

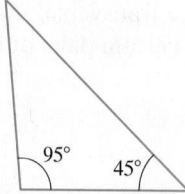

Solution The sum of the measures of the angles of a triangle is 180°. Since one angle measures 45° and the other angle measures 95°, the third angle measures $180° - 45° - 95° = 40°$. □

Two triangles are **congruent** if they have the same size and the same shape. In congruent triangles, the measures of corresponding angles are equal and the lengths of corresponding sides are equal. The following triangles are congruent.

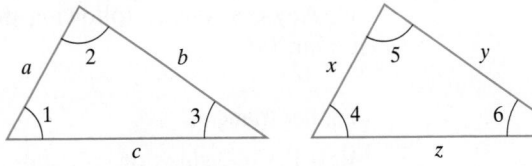

Corresponding angles are equal: $m\angle 1 = m\angle 4, m\angle 2 = m\angle 5$, and $m\angle 3 = m\angle 6$. Also, lengths of corresponding sides are equal: $a = x$, $b = y$, and $c = z$.

Any one of the following may be used to determine whether two triangles are congruent.

Congruent Triangles

1. If the measures of two angles of a triangle equal the measures of two angles of another triangle and the lengths of the sides between each pair of angles are equal, the triangles are congruent.

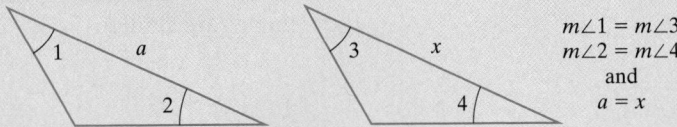

$$m\angle 1 = m\angle 3$$
$$m\angle 2 = m\angle 4$$
and
$$a = x$$

2. If the lengths of the three sides of a triangle equal the lengths of corresponding sides of another triangle, the triangles are congruent.

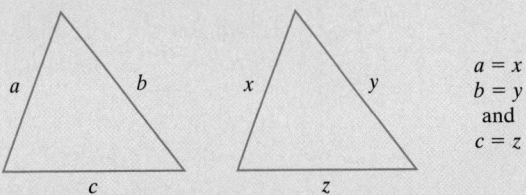

$$a = x$$
$$b = y$$
and
$$c = z$$

3. If the lengths of two sides of a triangle equal the lengths of corresponding sides of another triangle, and the measures of the angles between each pair of sides are equal, the triangles are congruent.

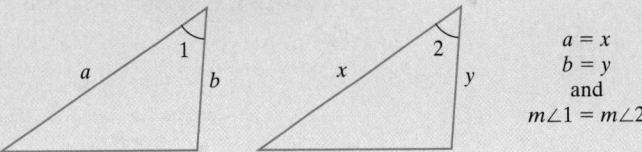

$$a = x$$
$$b = y$$
and
$$m\angle 1 = m\angle 2$$

Two triangles are **similar** if they have the same shape. In similar triangles, the measures of corresponding angles are equal and corresponding sides are in proportion. The following triangles are similar. (All similar triangles drawn in this appendix will be oriented the same.)

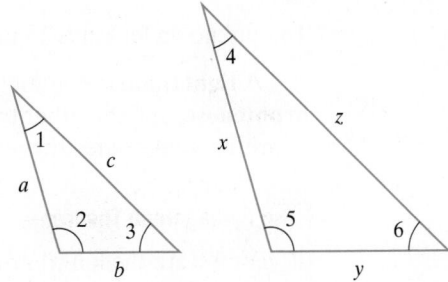

Corresponding angles are equal: $m\angle 1 = m\angle 4, m\angle 2 = m\angle 5$, and $m\angle 3 = m\angle 6$.

Also, corresponding sides are proportional: $\dfrac{a}{x} = \dfrac{b}{y} = \dfrac{c}{z}$.

Any one of the following may be used to determine whether two triangles are similar.

Similar Triangles

1. If the measures of two angles of a triangle equal the measures of two angles of another triangle, the triangles are similar.

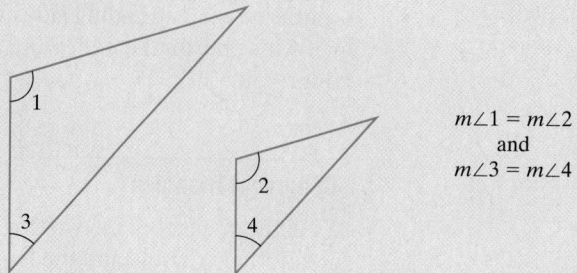

$$m\angle 1 = m\angle 2$$
and
$$m\angle 3 = m\angle 4$$

2. If three sides of one triangle are proportional to three sides of another triangle, the triangles are similar.

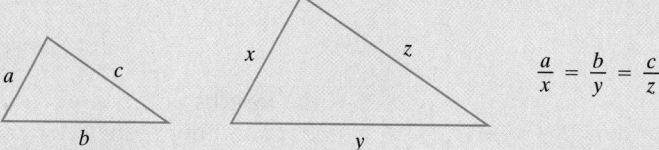

$$\frac{a}{x} = \frac{b}{y} = \frac{c}{z}$$

3. If two sides of a triangle are proportional to two sides of another triangle and the measures of the included angles are equal, the triangles are similar.

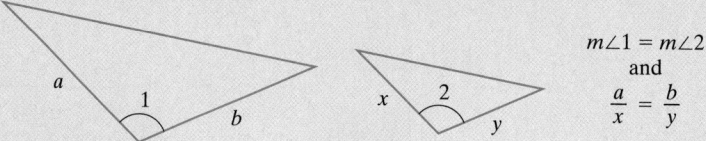

$$m\angle 1 = m\angle 2$$
and
$$\frac{a}{x} = \frac{b}{y}$$

EXAMPLE 4   Given that the following triangles are similar, find the unknown length x.

<u>Solution</u> Since the triangles are similar, corresponding sides are in proportion. Thus, $\frac{2}{3} = \frac{10}{x}$. To solve this equation for x, we multiply both sides by the LCD, $3x$.

$$3x\left(\frac{2}{3}\right) = 3x\left(\frac{10}{x}\right)$$
$$2x = 30$$
$$x = 15$$

The unknown length is 15 units. □

A **right triangle** contains a right angle. The side opposite the right angle is called the **hypotenuse,** and the other two sides are called the **legs.** The **Pythagorean theorem** gives a formula that relates the lengths of the three sides of a right triangle.

The Pythagorean Theorem

If a and b are the lengths of the legs of a right triangle, and c is the length of the hypotenuse, then $a^2 + b^2 = c^2$.

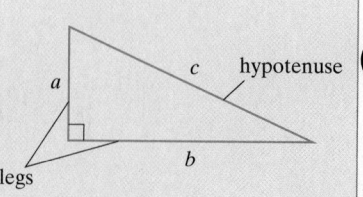

EXAMPLE 5   Find the length of the hypotenuse of a right triangle whose legs have lengths of 3 centimeters and 4 centimeters.

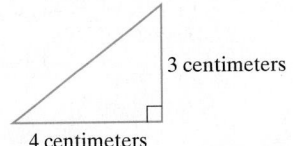

3 centimeters

4 centimeters

<u>Solution</u> Because we have a right triangle, we use the Pythagorean theorem. The legs are 3 centimeters and 4 centimeters, so let $a = 3$ and $b = 4$ in the formula.

$$a^2 + b^2 = c^2$$
$$3^2 + 4^2 = c^2$$
$$9 + 16 = c^2$$
$$25 = c^2$$

Since c represents a length, we assume that c is positive. Thus, if c^2 is 25, c must be 5. The hypotenuse has a length of 5 centimeters.   □

G Exercise Set   MyMathLab® ▶

Find the complement of each angle. See Example 1.

1. $19°$
2. $65°$
3. $70.8°$
4. $45\frac{2}{3}°$
5. $11\frac{1}{4}°$
6. $19.6°$

Find the supplement of each angle.

7. $150°$
8. $90°$
9. $30.2°$
10. $81.9°$
11. $79\frac{1}{2}°$
12. $165\frac{8}{9}°$

13. If lines m and n are parallel, find the measures of angles 1 through 7. See Example 2.

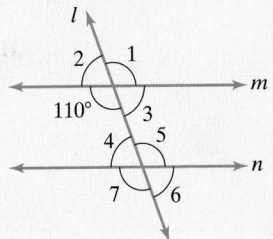

14. If lines m and n are parallel, find the measures of angles 1 through 5. See Example 2.

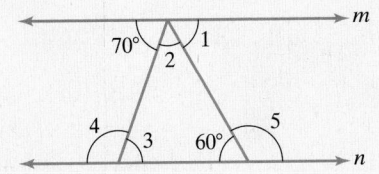

In each of the following, the measures of two angles of a triangle are given. Find the measure of the third angle. See Example 3.

15. $11°, 79°$
16. $8°, 102°$
17. $25°, 65°$
18. $44°, 19°$
19. $30°, 60°$
20. $67°, 23°$

In each of the following, the measure of one angle of a right triangle is given. Find the measures of the other two angles.

21. $45°$
22. $60°$
23. $17°$
24. $30°$
25. $39\frac{3}{4}°$
26. $72.6°$

Given that each of the following pairs of triangles is similar, find the unknown lengths. See Example 4.

27.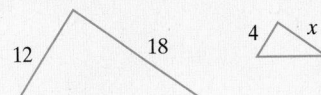

28.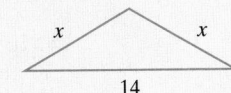

29.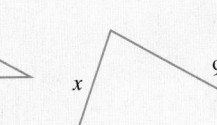

30.

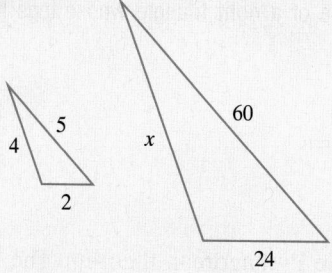

Use the Pythagorean theorem to find the unknown lengths in the right triangles. See Example 5.

31.

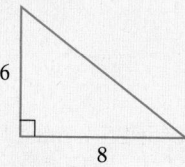

32.

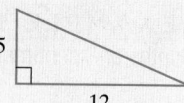

33.

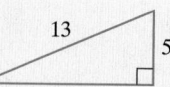

34.

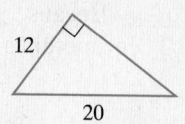

Contents of Student Resources

Study Skills Builders

Attitude and Study Tips:
1. Have You Decided to Complete This Course Successfully?
2. Tips for Studying for an Exam
3. What to Do the Day of an Exam
4. Are You Satisfied with Your Performance on a Particular Quiz or Exam?
5. How Are You Doing?
6. Are You Preparing for Your Final Exam?

Organizing Your Work:
7. Learning New Terms
8. Are You Organized?
9. Organizing a Notebook
10. How Are Your Homework Assignments Going?

MyMathLab and MathXL:
11. Tips for Turning in Your Homework on Time
12. Tips for Doing Your Homework Online
13. Organizing Your Work
14. Getting Help with Your Homework Assignments
15. Tips for Preparing for an Exam
16. How Well Do You Know the Resources Available to You in MyMathLab?

Additional Help Inside and Outside Your Textbook:
17. How Well Do You Know Your Textbook?
18. Are You Familiar with Your Textbook Supplements?
19. Are You Getting All the Mathematics Help That You Need?

The Bigger Picture—Study Guide Outline

Practice Final Exam

Answers to Selected Exercises

Student Resources

Study Skills Builders

Attitude and Study Tips

Study Skills Builder 1

Have You Decided to Complete This Course Successfully?

Ask yourself if one of your current goals is to complete this course successfully.

If it is not a goal of yours, ask yourself why. One common reason is fear of failure. Amazingly enough, fear of failure alone can be strong enough to keep many of us from doing our best in any endeavor.

Another common reason is that you simply haven't taken the time to think about or write down your goals for this course. To help accomplish this, answer the following questions.

Exercises

1. Write down your goal(s) for this course.

2. Now list steps you will take to make sure your goal(s) in Exercise 1 are accomplished.

3. Rate your commitment to this course with a number between 1 and 5. Use the diagram below to help.

High Commitment		Average Commitment		Not Commited at All
5	4	3	2	1

4. If you have rated your personal commitment level (from the exercise above) as a 1, 2, or 3, list the reasons why this is so. Then determine whether it is possible to increase your commitment level to a 4 or 5.

Good luck, and don't forget that a positive attitude will make a big difference.

Study Skills Builder 2

Tips for Studying for an Exam

To prepare for an exam, try the following study techniques:

- Start the study process days before your exam.
- Make sure that you are up to date on your assignments.
- If there is a topic that you are unsure of, use one of the many resources that are available to you. For example,

  See your instructor.

  View a lecture video on the topic.

  Visit a learning resource center on campus.

  Read the textbook material and examples on the topic.

- Reread your notes and carefully review the Chapter Highlights at the end of any chapter.
- Work the review exercises at the end of the chapter.
- Find a quiet place to take the Chapter Test found at the end of the chapter. Do not use any resources when taking this sample test. This way, you will have a clear indication of how prepared you are for your exam. Check your answers and use the Chapter Test Prep Videos to make sure that you correct any missed exercises.

Good luck, and keep a positive attitude.

Exercises

Let's see how you did on your last exam.

1. How many days before your last exam did you start studying for that exam?

2. Were you up to date on your assignments at that time or did you need to catch up on assignments?

3. List the most helpful text supplement (if you used one).

4. List the most helpful campus supplement (if you used one).

5. List your process for preparing for a mathematics test.

6. Was this process helpful? In other words, were you satisfied with your performance on your exam?

7. If not, what changes can you make in your process that will make it more helpful to you?

Study Skills Builder 3

What to Do the Day of an Exam

Your first exam may be soon. On the day of an exam, don't forget to try the following:

- Allow yourself plenty of time to arrive.
- Read the directions on the test carefully.
- Read each problem carefully as you take your test. Make sure that you answer the question asked.
- Watch your time and pace yourself so that you may attempt each problem on your test.
- Check your work and answers.
- *Do not turn your test in early.* If you have extra time, spend it double-checking your work.

Good luck!

Exercises

Answer the following questions based on your most recent mathematics exam, whenever that was.

1. How soon before class did you arrive?
2. Did you read the directions on the test carefully?
3. Did you make sure you answered the question asked for each problem on the exam?
4. Were you able to attempt each problem on your exam?
5. If your answer to Exercise 4 is no, list reasons why.
6. Did you have extra time on your exam?
7. If your answer to Exercise 6 is yes, describe how you spent that extra time.

Study Skills Builder 4

Are You Satisfied with Your Performance on a Particular Quiz or Exam?

If not, don't forget to analyze your quiz or exam and look for common errors. Were most of your errors a result of:

- *Carelessness?* Did you turn in your quiz or exam before the allotted time expired? If so, resolve to use any extra time to check your work.
- *Running out of time?* Answer the questions you are sure of first. Then attempt the questions you are unsure of and delay checking your work until all questions have been answered.
- *Not understanding a concept?* If so, review that concept and correct your work so that you make sure you understand it before the next quiz or the final exam.
- *Test conditions?* When studying for a quiz or exam, make sure you place yourself in conditions similar to test conditions. For example, before your next quiz or exam, take a sample test without the aid of your notes or text.

(For a sample test, see your instructor or use the Chapter Test at the end of each chapter.)

Exercises

1. Have you corrected all your previous quizzes and exams?
2. List any errors you have found common to two or more of your graded papers.
3. Is one of your common errors not understanding a concept? If so, are you making sure you understand all the concepts for the next quiz or exam?
4. Is one of your common errors making careless mistakes? If so, are you now taking all the time allotted to check over your work so that you can minimize the number of careless mistakes?
5. Are you satisfied with your grades thus far on quizzes and tests?
6. If your answer to Exercise 5 is no, are there any more suggestions you can make to your instructor or yourself to help? If so, list them here and share these with your instructor.

Study Skills Builder 5

How Are You Doing?

If you haven't done so yet, take a few moments to think about how you are doing in this course. Are you working toward your goal of successfully completing this course? Is your performance on homework, quizzes, and tests satisfactory? If not, you might want to see your instructor to see whether he/she has any suggestions on how you can improve your performance. Reread Section 1.1 for ideas on places to get help with your mathematics course.

Exercises

Answer the following.

1. List any textbook supplements you are using to help you through this course.
2. List any campus resources you are using to help you through this course.
3. Write a short paragraph describing how you are doing in your mathematics course.
4. If improvement is needed, list ways that you can work toward improving your situation as described in Exercise 3.

Study Skills Builder 6

Are You Preparing for Your Final Exam?

To prepare for your final exam, try the following study techniques:

- Review the material that you will be responsible for on your exam. This includes material from your textbook, your notebook, and any handouts from your instructor.
- Review any formulas that you may need to memorize.
- Check to see if your instructor or mathematics department will be conducting a final exam review.
- Check with your instructor to see whether final exams from previous semesters/quarters are available to students for review.

- Use your previously taken exams as a practice final exam. To do so, rewrite the test questions in mixed order on blank sheets of paper. This will help you prepare for exam conditions.
- If you are unsure of a few concepts, see your instructor or visit a learning lab for assistance. Also, view the video segment of any troublesome sections.
- If you need further exercises to work, try the Cumulative Reviews at the end of the chapters.

Once again, good luck! I hope you are enjoying this textbook and your mathematics course.

Organizing Your Work

Study Skills Builder 7

Learning New Terms

Many of the terms used in this text may be new to you. It will be helpful to make a list of new mathematical terms and symbols as you encounter them and to review them frequently. Placing these new terms (including page references) on 3×5 index cards might help you later when you're preparing for a quiz.

Exercises

1. Name one way you might place a word and its definition on a 3×5 card.
2. How do new terms stand out in this text so that they can be found?

Study Skills Builder 8

Are You Organized?

Have you ever had trouble finding a completed assignment? When it's time to study for a test, are your notes neat and organized? Have you ever had trouble reading your own mathematics handwriting? (Be honest—I have.)

When any of these things happens, it's time to get organized. Here are a few suggestions:

- Write your notes and complete your homework assignments in a notebook with pockets (spiral or ring binder).
- Take class notes in this notebook, and then follow the notes with your completed homework assignment.
- When you receive graded papers or handouts, place them in the notebook pocket so that you will not lose them.
- Mark (possibly with an exclamation point) any notes that seem extra important to you.
- Mark (possibly with a question mark) any notes or homework that you are having trouble with.
- See your instructor or a math tutor to help you with the concepts or exercises that you are having trouble understanding.

- If you are having trouble reading your own handwriting, *slow down* and write your mathematics work clearly!

Exercises

1. Have you been completing your assignments on time?
2. Have you been correcting any exercises you may be having difficulty with?
3. If you are having trouble with a mathematical concept or correcting any homework exercises, have you visited your instructor, a tutor, or your campus math lab?
4. Are you taking lecture notes in your mathematics course? (By the way, these notes should include worked-out examples solved by your instructor.)
5. Is your mathematics course material (handouts, graded papers, lecture notes) organized?
6. If your answer to Exercise 5 is no, take a moment to review your course material. List at least two ways that you might organize it better.

Study Skills Builder 9

Organizing a Notebook

It's never too late to get organized. If you need ideas about organizing a notebook for your mathematics course, try some of these:

- Use a spiral or ring binder notebook with pockets and use it for mathematics only.
- Start each page by writing the book's section number you are working on at the top.
- When your instructor is lecturing, take notes. *Always* include any examples your instructor works for you.
- Place your worked-out homework exercises in your notebook immediately after the lecture notes from that section. This way, a section's worth of material is together.
- Homework exercises: Attempt and check all assigned homework.
- Place graded quizzes in the pockets of your notebook or a special section of your binder.

Exercises

Check your notebook organization by answering the following questions.

1. Do you have a spiral or ring binder notebook for your mathematics course only?
2. Have you ever had to flip through several sheets of notes and work in your mathematics notebook to determine what section's work you are in?
3. Are you now writing the textbook's section number at the top of each notebook page?
4. Have you ever lost or had trouble finding a graded quiz or test?
5. Are you now placing all your graded work in a dedicated place in your notebook?
6. Are you attempting all of your homework and placing all of your work in your notebook?
7. Are you checking and correcting your homework in your notebook? If not, why not?
8. Are you writing in your notebook the examples your instructor works for you in class?

Study Skills Builder 10

How Are Your Homework Assignments Going?

It is very important in mathematics to keep up with homework. Why? Many concepts build on each other. Often your understanding of a day's concepts depends on an understanding of the previous day's material.

Remember that completing your homework assignment involves a lot more than attempting a few of the problems assigned.

To complete a homework assignment, remember these four things:

- Attempt all of it.
- Check it.
- Correct it.
- If needed, ask questions about it.

Exercises

Take a moment to review your completed homework assignments. Answer the questions below based on this review.

1. Approximate the fraction of your homework you have attempted.

2. Approximate the fraction of your homework you have checked (if possible).

3. If you are able to check your homework, have you corrected it when errors have been found?

4. When working homework, if you do not understand a concept, what do you do?

MyMathLab and MathXL

Study Skills Builder 11

Tips for Turning in Your Homework on Time

It is very important to keep up with your mathematics homework assignments. Why? Many concepts in mathematics build upon each other.

Remember these four tips to help ensure that your work is completed on time:

- Know the assignments and due dates set by your instructor.
- Do not wait until the last minute to submit your homework.
- Set a goal to submit your homework 6–8 hours before the scheduled due date in case you have unexpected technology trouble.
- Schedule enough time to complete each assignment.

Following these tips will also help you avoid losing points for late or missed assignments.

Exercises

Take a moment to consider your work on your homework assignments to date and answer the following questions:

1. What percentage of your assignments have you turned in on time?

2. Why might it be a good idea to submit your homework 6–8 hours before the scheduled deadline?

3. If you have missed submitting any homework by the due date, list some of the reasons this occurred.

4. What steps do you plan to take in the future to ensure that your homework is submitted on time?

Study Skills Builder 12

Tips for Doing Your Homework Online

Practice is one of the main keys to success in any mathematics course. Did you know that MyMathLab/MathXL provides you with **immediate feedback** for each exercise? If you are incorrect, you are given hints to work the exercise correctly. You have **unlimited practice opportunities** and can rework any exercises you have trouble with until you master them, and submit homework assignments unlimited times before the deadline.

Remember these success tips when doing your homework online:

- Attempt all assigned exercises.
- Write down (neatly) your step-by-step work for each exercise before entering your answer.
- Use the immediate feedback provided by the program to help you check and correct your work for each exercise.
- Rework any exercises you have trouble with until you master them.

- Work through your homework assignment as many times as necessary until you are satisfied.

Exercises

Take a moment to think about your homework assignments to date and answer the following:

1. Have you attempted all assigned exercises?

2. Of the exercises attempted, have you also written out your work before entering your answer so that you can check it?

3. Are you familiar with how to enter answers using the MathXL player so that you avoid answer entry–type errors?

4. List some ways the immediate feedback and practice supports have helped you with your homework. If you have not used these supports, how do you plan to use them with the given success tips on your next assignment?

Study Skills Builder 13

Organizing Your Work

Have you ever used any readily available paper (such as the back of a flyer, another course assignment, Post-it notes, etc.) to work out homework exercises before entering the answer in MathXL? To save time, have you ever entered answers directly into MathXL without working the exercises on paper? When it's time to study, have you ever been unable to find your completed work or read and follow your own mathematics handwriting?

When any of these things happen, it's time to get organized. Here are some suggestions:

- Write your step-by-step work for each homework exercise (neatly) on lined, loose-leaf paper and keep this in a 3-ring binder.
- Refer to your step-by-step work when you receive feedback that your answer is incorrect in MathXL. Double-check against the steps and hints provided by the program and correct your work accordingly.
- Keep your written homework with your class notes for that section.

- Identify any exercises you are having trouble with and ask questions about them.
- Keep all graded quizzes and tests in this binder as well to study later.

If you follow these suggestions, you and your instructor or tutor will be able to follow your steps and correct any mistakes. You will have a written copy of your work to refer to later to ask questions and study for tests.

Exercises

1. Why is it important to write out your step-by-step work to homework exercises and keep a hard copy of all work submitted online?

2. If you have gotten an incorrect answer, are you able to follow your steps and find your error?

3. If you were asked today to review your previous homework assignments and first test, could you find them? If not, list some ways you might organize your work better.

Study Skills Builder 14

Getting Help with Your Homework Assignments

Many helpful resources are available to you through MathXL to help you work through any homework exercises you may have trouble with. It is important for you to know what these resources are and when and how to use them.

Let's review these features, found in the homework exercises:

- **Help Me Solve This**—provides step-by-step help for the exercise you are working. You must work an additional exercise of the same type (without this help) before you can get credit for having worked it correctly.

- **View an Example**—allows you to view a correctly worked exercise similar to the one you are having trouble with. You can go back to your original exercise and work it on your own.

- **E-Book**—allows you to read examples from your text and find similar exercises.

- **Video****—your text author, Elayn Martin-Gay, works an exercise similar to the one you need help with. **Not all exercises have an accompanying video clip.

- **Ask My Instructor**—allows you to email your instructor for help with an exercise.

Exercises

1. How does the "Help Me Solve This" feature work?

2. If the "View an Example" feature is used, is it necessary to work an additional problem before continuing the assignment?

3. When might be a good time to use the "Video" feature? Do all exercises have an accompanying video clip?

4. Which of the features above have you used? List those you found the most helpful to you.

5. If you haven't used the features discussed, list those you plan to try on your next homework assignment.

Study Skills Builder 15

Tips for Preparing for an Exam

Did you know that you can rework your previous homework assignments in MyMathLab and MathXL? This is a great way to prepare for tests. To do this, open a previous homework assignment and click "similar exercise." This will generate new exercises similar to the homework you have submitted. You can then rework the exercises and assignments until you feel confident that you understand them.

To prepare for an exam, follow these tips:

- Review your written work for your previous homework assignments along with your class notes.

- Identify any exercises or topics that you have questions on or have difficulty understanding.

- Rework your previous assignments in MyMathLab and MathXL until you fully understand them and can do them without help.

- Get help for any topics you feel unsure of or for which you have questions.

Exercises

1. Are your current homework assignments up to date and is your written work for them organized in a binder or notebook? If the answer is no, it's time to get organized. For tips on this, see Study Skills Builder 13—Organizing Your Work.

2. How many days in advance of an exam do you usually start studying?

3. List some ways you think that practicing previous homework assignments can help you prepare for your test.

4. List two or three resources you can use to get help for any topics you are unsure of or have questions on.

Good luck!

Study Skills Builder 16

How Well Do You Know the Resources Available to You in MyMathLab?

Many helpful resources are available to you in MyMathLab. Let's take a moment to locate and explore a few of them now. Go into your MyMathLab course and visit the multimedia library, tools for success, and E-book.

Let's see what you found.

Exercises

1. List the resources available to you in the Multimedia Library.

2. List the resources available to you in the Tools for Success folder.

3. Where did you find the English/Spanish Audio Glossary?

4. Can you view videos from the E-book?

5. Did you find any resources you did not know about? If so, which ones?

6. Which resources have you used most often or found most helpful?

Additional Help Inside and Outside Your Textbook

Study Skills Builder 17

How Well Do You Know Your Textbook?

The following questions will help determine whether you are familiar with your textbook. For additional information, see Section 1.1 in this text.

1. What does the ▶ icon mean?

2. What does the ＼ icon mean?

3. What does the △ icon mean?

4. Where can you find a review for each chapter? What answers to this review can be found in the back of your text?

5. Each chapter contains an overview of the chapter along with examples. What is this feature called?

6. Each chapter contains a review of vocabulary. What is this feature called?

7. Practice exercises are contained in this text. What are they and how can they be used?

8. This text contains a student section in the back entitled Student Resources. List the contents of this section and how they might be helpful.

9. What exercise answers are available in this text? Where are they located?

Study Skills Builder 18

Are You Familiar with Your Textbook Supplements?

Below is a review of some of the student supplements available for additional study. Check to see whether you are using the ones most helpful to you.

- Chapter Test Prep Videos. These videos provide video clip solutions to the Chapter Test exercises in this text. You will find this extremely useful when studying for tests or exams.
- Interactive DVD Lecture Series. These are keyed to each section of the text. The material is presented by me, Elayn Martin-Gay, and I have placed a ▶ by the exercises in the text that I have worked on the video.
- The *Student Solutions Manual.* This contains worked-out solutions to odd-numbered exercises as well as every exercise in the Integrated Reviews, Chapter Reviews, Chapter Tests, and Cumulative Reviews and every Practice exercise.
- Pearson Tutor Center. Mathematics questions may be phoned, faxed, or emailed to this center.

- MyMathLab is a text-specific online course. MathXL is an online homework, tutorial, and assessment system. Take a moment to determine whether these are available to you.

    As usual, your instructor is your best source of information.

Exercises

Let's see how you are doing with textbook supplements.

1. Name one way the Lecture Videos can be helpful to you.
2. Name one way the Chapter Test Prep Video can help you prepare for a chapter test.
3. List any textbook supplements that you have found useful.
4. Have you located and visited a learning resource lab located on your campus?
5. List the textbook supplements that are currently housed in your campus's learning resource lab.

Study Skills Builder 19

Are You Getting All the Mathematics Help That You Need?

Remember that, in addition to your instructor, there are many places to get help with your mathematics course. For example:

- This text has an accompanying video lesson by the author for every section. There are also worked-out video solutions by the author to every Chapter Test exercise.
- The back of the book contains answers to odd-numbered exercises.
- A *Student Solutions Manual* is available that contains worked-out solutions to odd-numbered exercises as well as solutions to every exercise in the Integrated Reviews, Chapter Reviews, Chapter Tests, and Cumulative Reviews and every Practice exercise.

- Don't forget to check with your instructor for other local resources available to you, such as a tutor center.

Exercises

1. List items you find helpful in the text and all student supplements to this text.
2. List all the campus help that is available to you for this course.
3. List any help (besides the textbook) from Exercises 1 and 2 above that you are using.
4. List any help (besides the textbook) that you feel you should try.
5. Write a goal for yourself that includes trying everything you listed in Exercise 4 during the next week.

The Bigger Picture—Study Guide Outline

Simplifying Expressions and Solving Equations and Inequalities

I. Simplifying Expressions

  A. Real Numbers

    1. Add: (Sec. 1.5)

$$-1.7 + (-0.21) = -1.91 \qquad \text{Adding like signs.}$$

Add absolute values. Attach common sign.

$$-7 + 3 = -4 \qquad \text{Adding different signs.}$$

Subtract absolute values. Attach the sign of the number with the larger absolute value.

    2. Subtract: Add the first number to the opposite of the second number. (Sec. 1.6)

$$17 - 25 = 17 + (-25) = -8$$

    3. Multiply or divide: Multiply or divide the two numbers as usual. If the signs are the same, the answer is positive. If the signs are different, the answer is negative. (Sec. 1.7)

$$-10 \cdot 3 = -30, \qquad -81 \div (-3) = 27$$

  B. Exponents (Sec. 5.1 and 5.5)

$$x^7 \cdot x^5 = x^{12}; \ (x^7)^5 = x^{35}; \ \frac{x^7}{x^5} = x^2; \ x^0 = 1; \ 8^{-2} = \frac{1}{8^2} = \frac{1}{64}$$

  C. Polynomials

    1. Add: Combine like terms. (Sec. 5.2)

$$(3y^2 + 6y + 7) + (9y^2 - 11y - 15) = 3y^2 + 6y + 7 + 9y^2 - 11y - 15$$
$$= 12y^2 - 5y - 8$$

    2. Subtract: Change the sign of the terms of the polynomial being subtracted, then add. (Sec. 5.2)

$$(3y^2 + 6y + 7) - (9y^2 - 11y - 15) = 3y^2 + 6y + 7 - 9y^2 + 11y + 15$$
$$= -6y^2 + 17y + 22$$

    3. Multiply: Multiply each term of one polynomial by each term of the other polynomial. (Secs. 5.3 and 5.4)

$$(x + 5)(2x^2 - 3x + 4) = x(2x^2 - 3x + 4) + 5(2x^2 - 3x + 4)$$
$$= 2x^3 - 3x^2 + 4x + 10x^2 - 15x + 20$$
$$= 2x^3 + 7x^2 - 11x + 20$$

    4. Divide: (Sec. 5.6)

      a. To divide by a monomial, divide each term of the polynomial by the monomial.

$$\frac{8x^2 + 2x - 6}{2x} = \frac{8x^2}{2x} + \frac{2x}{2x} - \frac{6}{2x} = 4x + 1 - \frac{3}{x}$$

b. To divide by a polynomial other than a monomial, use long division.

$$2x + 5 \overline{)2x^2 - 7x + 10} \quad x - 6 + \frac{40}{2x + 5}$$

$$\underline{-2x^2 \cancel{\mp} 5x}$$
$$-12x + 10$$
$$\underline{\overset{+}{\cancel{\mp}}12x \overset{+}{\cancel{\mp}} 30}$$
$$40$$

D. Factoring Polynomials

See the Chapter 6 Integrated Review for steps.

$$3x^4 - 78x^2 + 75 = 3(x^4 - 26x^2 + 25) \quad \text{Factor out GCF—always first step.}$$
$$= 3(x^2 - 25)(x^2 - 1) \quad \text{Factor trinomial.}$$
$$= 3(x + 5)(x - 5)(x + 1)(x - 1) \quad \text{Factor further—each}$$
$$\text{difference of squares.}$$

E. Rational Expressions

1. **Simplify:** Factor the numerator and denominator. Then divide out factors of 1 by dividing out common factors in the numerator and denominator. (Sec. 7.1)

$$\frac{x^2 - 9}{7x^2 - 21x} = \frac{(x + 3)(x - 3)}{7x(x - 3)} = \frac{x + 3}{7x}$$

2. **Multiply:** Multiply numerators, then multiply denominators. (Sec. 7.2)

$$\frac{5z}{2z^2 - 9z - 18} \cdot \frac{22z + 33}{10z} = \frac{5 \cdot z}{(2z + 3)(z - 6)} \cdot \frac{11(2z + 3)}{2 \cdot 5 \cdot z} = \frac{11}{2(z - 6)}$$

3. **Divide:** First fraction times the reciprocal of the second fraction. (Sec. 7.2)

$$\frac{14}{x + 5} \div \frac{x + 1}{2} = \frac{14}{x + 5} \cdot \frac{2}{x + 1} = \frac{28}{(x + 5)(x + 1)}$$

4. **Add or subtract:** Must have same denominator. If not, find the LCD and write each fraction as an equivalent fraction with the LCD as denominator. (Sec. 7.4)

$$\frac{9}{10} - \frac{x + 1}{x + 5} = \frac{9(x + 5)}{10(x + 5)} - \frac{10(x + 1)}{10(x + 5)}$$
$$= \frac{9x + 45 - 10x - 10}{10(x + 5)} = \frac{-x + 35}{10(x + 5)}$$

F. Radicals

1. **Simplify square roots:** If possible, factor the radicand so that one factor is a perfect square. Then use the product rule and simplify. (Sec. 10.3)

$$\sqrt{75} = \sqrt{25 \cdot 3} = \sqrt{25} \cdot \sqrt{3} = 5\sqrt{3}$$

2. **Add or subtract:** Only like radicals (same index and radicand) can be added or subtracted. (Sec. 10.4)

$$8\sqrt{10} - \sqrt{40} + \sqrt{5} = 8\sqrt{10} - 2\sqrt{10} + \sqrt{5} = 6\sqrt{10} + \sqrt{5}$$

3. **Multiply or divide:** $\sqrt{a} \cdot \sqrt{b} = \sqrt{ab}$; $\dfrac{\sqrt{a}}{\sqrt{b}} = \sqrt{\dfrac{a}{b}}$. (Sec. 10.4)

$$\sqrt{11} \cdot \sqrt{3} = \sqrt{33}; \quad \frac{\sqrt{140}}{\sqrt{7}} = \sqrt{\frac{140}{7}} = \sqrt{20} = \sqrt{4 \cdot 5} = 2\sqrt{5}$$

4. Rationalizing the denominator: (Sec. 10.5)

  a. If denominator is one term,

$$\frac{5}{\sqrt{11}} = \frac{5 \cdot \sqrt{11}}{\sqrt{11} \cdot \sqrt{11}} = \frac{5\sqrt{11}}{11}$$

  b. If denominator is two terms, multiply by 1 in the form of $\dfrac{\text{conjugate of denominator}}{\text{conjugate of denominator}}$.

$$\frac{13}{3 + \sqrt{2}} = \frac{13}{3 + \sqrt{2}} \cdot \frac{3 - \sqrt{2}}{3 - \sqrt{2}} = \frac{13(3 - \sqrt{2})}{9 - 2} = \frac{13(3 - \sqrt{2})}{7}$$

II. Solving Equations

A. Linear Equations (Sec. 2.3)

$$5(x - 2) = \frac{4(2x + 1)}{3}$$

$$3 \cdot 5(x - 2) = \cancel{3} \cdot \frac{4(2x + 1)}{\cancel{3}}$$

$$15x - 30 = 8x + 4$$

$$7x = 34$$

$$x = \frac{34}{7}$$

B. Quadratic and Higher-Degree Equations (Sec. 6.6, 11.1, 11.2, 11.3)

$$2x^2 - 7x = 9 \qquad\qquad 2x^2 + x - 2 = 0$$

$$2x^2 - 7x - 9 = 0 \qquad\quad a = 2, \ b = 1, \ c = -2$$

$$(2x - 9)(x + 1) = 0$$

$$2x - 9 = 0 \ \text{ or } \ x + 1 = 0 \qquad x = \frac{-1 \pm \sqrt{1^2 - 4(2)(-2)}}{2 \cdot 2}$$

$$x = \frac{9}{2} \ \text{ or } \qquad x = -1 \qquad x = \frac{-1 \pm \sqrt{17}}{4}$$

C. Equations with Rational Expressions (Sec. 7.5)

$$\frac{7}{x - 1} + \frac{3}{x + 1} = \frac{x + 3}{x^2 - 1}$$

$$\cancel{(x - 1)}(x + 1) \cdot \frac{7}{\cancel{x - 1}} + (x - 1)\cancel{(x + 1)} \cdot \frac{3}{\cancel{x + 1}}$$

$$= \cancel{(x - 1)}\cancel{(x + 1)} \cdot \frac{x + 3}{\cancel{(x - 1)}\cancel{(x + 1)}}$$

$$7(x + 1) + 3(x - 1) = x + 3$$

$$7x + 7 + 3x - 3 = x + 3$$

$$9x = -1$$

$$x = -\frac{1}{9}$$

D. Proportions: An equation with two ratios equal. Set cross products equal, then solve. Make sure the proposed solution does not make any denominator 0. (Sec. 7.6)

$$\frac{5}{x} \times \frac{9}{2x - 3}$$

$5(2x - 3) = 9 \cdot x$ Set cross products equal.

$10x - 15 = 9x$ Multiply.

$x = 15$ Write equation with variable terms on one side and constants on the other.

E. Absolute Value Equations (Sec. 9.2)

$|3x - 1| = 8$

$3x - 1 = 8$ or $3x - 1 = -8$

$3x = 9$ or $3x = -7$

$x = 3$ or $x = -\dfrac{7}{3}$

$|x - 5| = |x + 1|$

$x - 5 = x + 1$ or $x - 5 = -(x + 1)$

$\underbrace{-5 = 1}_{\text{No solution}}$ or $x - 5 = -x - 1$

           or $2x = 4$

               $x = 2$

F. Equations with Radicals (Sec. 10.6)

$$\sqrt{5x + 10} - 2 = x$$
$$\sqrt{5x + 10} = x + 2$$
$$(\sqrt{5x + 10})^2 = (x + 2)^2$$
$$5x + 10 = x^2 + 4x + 4$$
$$0 = x^2 - x - 6$$
$$0 = (x - 3)(x + 2)$$
$$x - 3 = 0 \quad \text{or} \quad x + 2 = 0$$
$$x = 3 \quad \text{or} \quad x = -2$$

Both solutions check.

G. Exponential Equations (Secs. 12.3, 12.8)

$$9^x = 27^{x+1}$$
$$(3^2)^x = (3^3)^{x+1}$$
$$3^{2x} = 3^{3x+3}$$
$$2x = 3x + 3$$
$$-3 = x$$

$$5^x = 7$$
$$\log 5^x = \log 7$$
$$x \log 5 = \log 7$$
$$x = \frac{\log 7}{\log 5}$$

H. Logarithmic Equations (Sec. 12.8)

$$\log 7 + \log(x + 3) = \log 5$$
$$\log 7(x + 3) = \log 5$$
$$7(x + 3) = 5$$
$$7x + 21 = 5$$
$$7x = -16$$
$$x = -\frac{16}{7}$$

III. Solving Inequalities

A. Linear Inequalities (Sec. 2.8)

$$-3(x + 2) \geq 6$$
$$-3x - 6 \geq 6$$
$$-3x \geq 12$$
$$\frac{-3x}{-3} \leq \frac{12}{-3}$$
$$x \leq -4 \quad \text{or} \quad (-\infty, -4]$$

B. Compound Inequalities (Sec. 9.1)

$$x \leq 3 \quad \text{and} \quad x < -7$$

$\{x \mid x \leq 3\}$ $(-\infty, 3]$

$\{x \mid x < -7\}$ $(-\infty, -7)$

$\{x \mid x \leq 3 \text{ and } x < -7\}$ $(-\infty, -7)$

$$x \leq 3 \quad \text{or} \quad x < -7$$

$\{x \mid x \leq 3\}$ $(-\infty, 3]$

$\{x \mid x < -7\}$ $(-\infty, -7)$

$\{x \mid x \leq 3 \text{ or } x < -7\}$ $(-\infty, 3]$

C. Absolute Value Inequalities (Sec. 9.3)

$$|x - 5| - 8 < -2$$
$$|x - 5| < 6$$
$$-6 < x - 5 < 6$$
$$-1 < x < 11$$
$$(-1, 11)$$

$$|2x + 1| \geq 17$$
$$2x + 1 \geq 17 \quad \text{or} \quad 2x + 1 \leq -17$$
$$2x \geq 16 \quad \text{or} \quad 2x \leq -18$$
$$x \geq 8 \quad \text{or} \quad x \leq -9$$
$$(-\infty, -9] \cup [8, \infty)$$

D. Nonlinear Inequalities (Sec. 11.4)

$$x^2 - x < 6$$
$$x^2 - x - 6 < 0$$
$$(x - 3)(x + 2) < 0$$

$$(-2, 3)$$

$$\frac{x - 5}{x + 1} \geq 0$$

$$(-\infty, -1) \cup [5, \infty)$$

MyMathLab® You Tube

Evaluate.

1. $6[5 + 2(3 - 8) - 3]$

2. -3^4

3. 4^{-3}

4. $\dfrac{1}{2} - \dfrac{5}{6}$

Perform the indicated operations and simplify if possible.

5. $5x^3 + x^2 + 5x - 2 - (8x^3 - 4x^2 + x - 7)$

6. $(4x - 2)^2$

7. $(3x + 7)(x^2 + 5x + 2)$

Factor.

8. $y^2 - 8y - 48$

9. $9x^3 + 39x^2 + 12x$

10. $180 - 5x^2$

11. $3a^2 + 3ab - 7a - 7b$

12. $8y^3 - 64$

Simplify. Write answer with positive exponents only.

13. $\left(\dfrac{x^2y^3}{x^3y^{-4}}\right)^2$

Solve each equation or inequality. Write inequality answers using interval notation.

14. $-4(a + 1) - 3a = -7(2a - 3)$

15. $3x - 5 \geq 7x + 3$

16. $x(x + 6) = 7$

Graph the following.

17. $5x - 7y = 10$

18. $x - 3 = 0$

Find the slope of each line.

19. Through $(6, -5)$ and $(-1, 2)$

20. $-3x + y = 5$

Write equations of the following lines. Write each equation in standard form.

21. Through $(2, -5)$ and $(1, 3)$

22. Through $(-5, -1)$ and parallel to $x = 7$

Solve each system of equations.

23. $\begin{cases} \dfrac{1}{2}x + 2y = -\dfrac{15}{4} \\ 4x = -y \end{cases}$

24. $\begin{cases} 4x - 6y = 7 \\ -2x + 3y = 0 \end{cases}$

25. Divide by long division: $\dfrac{27x^3 - 8}{3x + 2}$

Answer the questions about functions.

26. If $h(x) = x^3 - x$, find
  a. $h(-1)$
  b. $h(0)$
  c. $h(4)$

27. Identify the *x*- and *y*-intercepts. Then find the domain and range of the function graphed.

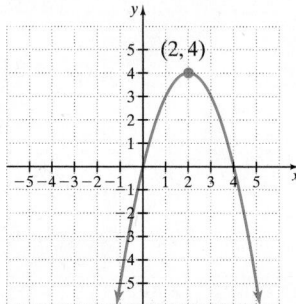

Solve each application.

28. Some states have a single area code for the entire state. Two such states have area codes where one is double the other. If the sum of these integers is 1203, find the two area codes.

29. Two trains leave Los Angeles simultaneously traveling on the same track in opposite directions at speeds of 50 and 64 mph. How long will it take before they are 285 miles apart?

30. Find the amount of a 12% saline solution a lab assistant should add to 80 cc (cubic centimeters) of a 22% saline solution in order to have a 16% solution.

31. Find the domain of the rational function
$$g(x) = \frac{9x^2 - 9}{x^2 + 4x + 3}.$$

Perform the indicated operations and simplify if possible.

32. $\dfrac{15x}{2x + 5} - \dfrac{6 - 4x}{2x + 5}$

33. $\dfrac{x^2 - 9}{x^2 - 3x} \div \dfrac{xy + 5x + 3y + 15}{2x + 10}$

34. $\dfrac{5a}{a^2 - a - 6} - \dfrac{2}{a - 3}$

35. $\dfrac{5 - \dfrac{1}{y^2}}{\dfrac{1}{y} + \dfrac{2}{y^2}}$

Solve each equation.

36. $\dfrac{4}{y} - \dfrac{5}{3} = -\dfrac{1}{5}$

37. $\dfrac{5}{y+1} = \dfrac{4}{y+2}$

38. $\dfrac{a}{a-3} = \dfrac{3}{a-3} - \dfrac{3}{2}$

Solve.

39. One number plus five times its reciprocal is equal to six. Find the number.

Simplify. If needed, write answers with positive exponents only.

40. $\sqrt{216}$

41. $\left(\dfrac{1}{125}\right)^{-1/3}$

42. $\left(\dfrac{64c^{4/3}}{a^{-2/3}b^{5/6}}\right)^{1/2}$

Perform the indicated operations and simplify if possible.

43. $\sqrt{125x^3} - 3\sqrt{20x^3}$

44. $(\sqrt{5}+5)(\sqrt{5}-5)$

Solve each equation or inequality. Write inequality solutions using interval notation.

45. $|6x-5|-3=-2$

46. $-3 < 2(x-3) \le 4$

47. $|3x+1| > 5$

48. $y^2 - 3y = 5$

49. $x = \sqrt{x-2} + 2$

50. $2x^2 - 7x > 15$

Graph the following.

51. $y > -4x$

52. $g(x) = -|x+2| - 1$. Also, find the domain and range of this function.

53. $h(x) = x^2 - 4x + 4$. Label the vertex and any intercepts.

54. $f(x) = \begin{cases} -\dfrac{1}{2}x & \text{if } x \le 0 \\ 2x - 3 & \text{if } x > 0 \end{cases}$. Also, find the domain and range of this function.

Write equations of the following lines. Write each equation using function notation.

55. Through $(4,-2)$ and $(6,-3)$

56. Through $(-1,2)$ and perpendicular to $3x - y = 4$

Find the distance or midpoint.

57. Find the distance between the points $(-6,3)$ and $(-8,-7)$.

58. Find the midpoint of the line segment whose endpoints are $(-2,-5)$ and $(-6,12)$.

Rationalize each denominator. Assume that variables represent positive numbers.

59. $\sqrt{\dfrac{9}{y}}$

60. $\dfrac{4-\sqrt{x}}{4+2\sqrt{x}}$

Solve.

61. Suppose that W is inversely proportional to V. If $W = 20$ when $V = 12$, find W when $V = 15$.

62. Given the diagram shown, approximate to the nearest foot how many feet of walking distance a person saves by cutting across the lawn instead of walking on the sidewalk.

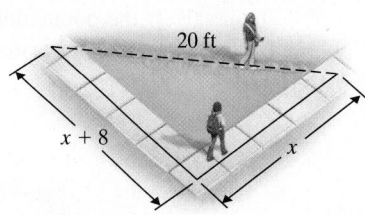

63. A stone is thrown upward from a bridge. The stone's height in feet, $s(t)$, above the water t seconds after the stone is thrown is a function given by the equation

$$s(t) = -16t^2 + 32t + 256$$

a. Find the maximum height of the stone.

b. Find the time it takes the stone to hit the water. Round the answer to two decimal places.

COMPLEX NUMBERS: CHAPTER 10

Perform the indicated operation and simplify. Write the result in the form $a + bi$.

64. $-\sqrt{-8}$

65. $(12 - 6i) - (12 - 3i)$

66. $(4 + 3i)^2$

67. $\dfrac{1+4i}{1-i}$

INVERSE, EXPONENTIAL, AND LOGARITHMIC FUNCTIONS: CHAPTER 12

68. If $g(x) = x - 7$ and $h(x) = x^2 - 6x + 5$, find $(g \circ h)(x)$.

69. Determine whether $f(x) = 6 - 2x$ is a one-to-one function. If it is, find its inverse.

70. Use properties of logarithms to write the expression as a single logarithm.

$$\log_5 x + 3\log_5 x - \log_5(x+1)$$

Solve. Give exact solutions.

71. $8^{x-1} = \dfrac{1}{64}$

72. $3^{2x+5} = 4$ Give the exact solution and a 4-decimal-place approximation.

73. $\log_8(3x - 2) = 2$

74. $\log_4(x + 1) - \log_4(x - 2) = 3$

75. $\ln\sqrt{e} = x$

76. Graph: $y = \left(\dfrac{1}{2}\right)^x + 1$

77. The prairie dog population of the Grand Forks area now stands at 57,000 animals. If the population is growing at a rate of 2.6% annually, how many prairie dogs will there be in that area 5 years from now?

CONIC SECTIONS: CHAPTER 13
Sketch the graph of each equation.

78. $x^2 - y^2 = 36$

79. $16x^2 + 9y^2 = 144$

80. $x^2 + y^2 + 6x = 16$

81. Solve the system: $\begin{cases} x^2 + y^2 = 26 \\ x^2 - 2y^2 = 23 \end{cases}$

SEQUENCES, SERIES, AND THE BINOMIAL THEOREM: CHAPTER 14

82. Find the first five terms of the sequence $a_n = \dfrac{(-1)^n}{n + 4}$.

83. Find the partial sum S_5 of the sequence $a_n = 5(2)^{n-1}$.

84. Find S_∞ of the sequence $\dfrac{3}{2}, -\dfrac{3}{4}, \dfrac{3}{8}, \ldots$.

85. Find: $\displaystyle\sum_{i=1}^{4} i(i - 2)$

86. Expand: $(2x + y)^5$

Answers to Selected Exercises

CHAPTER 1 REVIEW OF REAL NUMBERS

Section 1.2
Practice Exercises

1. a. $<$ **b.** $>$ **c.** $<$ **2. a.** True   **b.** False   **c.** True   **d.** True   **3. a.** $3 < 8$ **b.** $15 \geq 9$ **c.** $6 \neq 7$ **4.** -10 **5. a.** 25
b. 25   **c.** $25, -15, -99$ **d.** $25, \frac{7}{3}, -15, -\frac{3}{4}, -3.7, 8.8, -99$ **e.** $\sqrt{5}$ **f.** $25, \frac{7}{3}, -15, -\frac{3}{4}, \sqrt{5}, -3.7, 8.8, -99$ **6. a.** $<$ **b.** $>$ **c.** $=$
7. a. 8   **b.** 9   **c.** 2.5   **d.** $\frac{5}{11}$ **e.** $\sqrt{3}$ **8. a.** $=$ **b.** $>$ **c.** $<$ **d.** $>$ **e.** $<$

Vocabulary, Readiness & Video Check 1.2

1. whole   **3.** inequality   **5.** real   **7.** irrational   **9.** To form a true statement: $0 < 7$.   **11.** 0 belongs to the whole numbers, the integers, the rational numbers, and the real numbers; because 0 is a rational number, it cannot also be an irrational number

Exercise Set 1.2

1. $>$ **3.** $=$ **5.** $<$ **7.** $<$ **9.** $32 < 212$ **11.** $30 \leq 45$ **13.** true   **15.** false   **17.** false   **19.** true   **21.** false   **23.** $8 < 12$
25. $5 \geq 4$ **27.** $15 \neq -2$ **29.** $14,494; -282$ **31.** $-28,000$ **33.** $350; -126$ **35.** whole, integers, rational, real   **37.** integers, rational, real
39. natural, whole, integers, rational, real   **41.** rational, real   **43.** irrational, real   **45.** false   **47.** true   **49.** true   **51.** true   **53.** false
55. $>$ **57.** $>$ **59.** $<$ **61.** $<$ **63.** $>$ **65.** $=$ **67.** $<$ **69.** $<$ **71.** 40 million $>$ 16 million   **73.** 40 million pounds less, or -40 million   **75.** 1966–1975 and 2006–2015   **77.** 1946–1955, 1966–1975, 1976–1985, 2006–2015   **79.** $38 < 49$ **81.** $-0.04 > -26.7$
83. Sun   **85.** Sun   **87.** $20 \leq 25$ **89.** $6 > 0$ **91.** $-12 < -10$ **93.** answers may vary

Section 1.3
Practice Exercises

1. a. $2 \cdot 2 \cdot 3 \cdot 3$ **b.** $2 \cdot 2 \cdot 2 \cdot 5 \cdot 5$ **2. a.** $\frac{7}{8}$ **b.** $\frac{16}{3}$ **c.** $\frac{7}{25}$ **3.** $\frac{7}{24}$ **4. a.** $\frac{27}{16}$ **b.** $\frac{1}{36}$ **c.** $\frac{5}{2}$ **5. a.** 1   **b.** $\frac{6}{5}$ **c.** $\frac{4}{5}$ **d.** $\frac{1}{2}$
6. $\frac{14}{21}$ **7. a.** $\frac{46}{77}$ **b.** $\frac{1}{14}$ **c.** $\frac{1}{2}$ **8.** $22\frac{11}{15}$ **9.** $40\frac{5}{6}$

Vocabulary, Readiness & Video Check 1.3

1. fraction   **3.** product   **5.** factors, product   **7.** equivalent   **9.** The division operation changes to multiplication and the second fraction $\frac{1}{20}$ changes to its reciprocal $\frac{20}{1}$.   **11.** The number $4\frac{7}{6}$ is not in proper mixed number form as the fraction part, $\frac{7}{6}$, should not be an improper fraction.

Exercise Set 1.3

1. $\frac{3}{8}$ **3.** $\frac{5}{7}$ **5.** $3 \cdot 11$ **7.** $2 \cdot 7 \cdot 7$ **9.** $2 \cdot 2 \cdot 5$ **11.** $3 \cdot 5 \cdot 5$ **13.** $3 \cdot 3 \cdot 5$ **15.** $\frac{1}{2}$ **17.** $\frac{2}{3}$ **19.** $\frac{3}{7}$ **21.** $\frac{3}{5}$ **23.** $\frac{30}{61}$ **25.** $\frac{3}{8}$
27. $\frac{1}{2}$ **29.** $\frac{6}{7}$ **31.** 15   **33.** $\frac{1}{6}$ **35.** $\frac{25}{27}$ **37.** $\frac{11}{20}$ sq mi   **39.** $\frac{7}{36}$ sq ft   **41.** $\frac{3}{5}$ **43.** 1   **45.** $\frac{1}{3}$ **47.** $\frac{9}{35}$ **49.** $\frac{21}{30}$ **51.** $\frac{4}{18}$
53. $\frac{16}{20}$ **55.** $\frac{23}{21}$ **57.** $\frac{11}{60}$ **59.** $\frac{5}{66}$ **61.** $\frac{7}{5}$ **63.** $\frac{1}{5}$ **65.** $\frac{3}{8}$ **67.** $\frac{1}{9}$ **69.** $18\frac{20}{27}$ **71.** $2\frac{28}{29}$ **73.** $48\frac{1}{15}$ **75.** $7\frac{1}{12}$ **77.** $\frac{5}{7}$
79. $\frac{65}{21}$ **81.** $\frac{2}{5}$ **83.** $\frac{10}{9}$ **85.** $\frac{17}{3}$ **87.** 37   **89.** $\frac{5}{66}$ **91.** $\frac{1}{5}$ **93.** $5\frac{1}{6}$ **95.** $\frac{17}{18}$ **97.** $55\frac{1}{4}$ ft   **99.** answers may vary
101. $3\frac{3}{8}$ mi   **103.** $\frac{3}{40}$ **105.** $\frac{37}{40}$ **107.** incorrect; $\frac{12}{24} = \frac{2 \cdot 2 \cdot 3}{2 \cdot 2 \cdot 2 \cdot 3} = \frac{1}{2}$ **109.** incorrect; $\frac{2}{7} + \frac{9}{7} = \frac{11}{7}$

Section 1.4
Practice Exercises

1. a. 1   **b.** 25   **c.** $\frac{1}{100}$ **d.** 9   **e.** $\frac{8}{125}$ **2. a.** 33   **b.** 11   **c.** $\frac{32}{9}$ or $3\frac{5}{9}$ **d.** 36   **e.** $\frac{3}{16}$ **3.** $\frac{31}{11}$ **4.** 4   **5.** $\frac{9}{22}$ **6. a.** 9
b. $\frac{8}{15}$ **c.** $\frac{19}{10}$ **d.** 33   **7.** no   **8. a.** $6x$ **b.** $x - 8$ **c.** $x \cdot 9$ or $9x$ **d.** $2x + 3$ **e.** $7 + x$ **9. a.** $x + 7 = 13$ **b.** $x - 2 = 11$
c. $2x + 9 \neq 25$ **d.** $5(11) \geq x$

Calculator Explorations 1.4

1. 625   **3.** 59,049   **5.** 30   **7.** 9857   **9.** 2376

Vocabulary, Readiness & Video Check 1.4

1. base; exponent   **3.** variable   **5.** equation   **7.** solving   **9.** The replacement value for z is not used because it's not needed—there is no variable z in the given algebraic expression.   **11.** We translate phrases to mathematical expressions and sentences to mathematical equations.

Exercise Set 1.4

1. 243 **3.** 27 **5.** 1 **7.** 5 **9.** 49 **11.** $\frac{16}{81}$ **13.** $\frac{1}{125}$ **15.** 1.44 **17.** 0.000064 **19.** 17 **21.** 20 **23.** 10 **25.** 21

27. 45 **29.** 0 **31.** 30 **33.** 2 **35.** $\frac{7}{18}$ **37.** $\frac{27}{10}$ **39.** $\frac{7}{5}$ **41.** 32 **43.** $\frac{23}{27}$ **45. a.** 64 **b.** 43 **c.** 19 **d.** 22 **47.** 9

49. 1 **51.** 1 **53.** 11 **55.** 8 **57.** 45 **59.** 27 **61.** 132 **63.** $\frac{37}{18}$ **65.** 16; 64; 144; 256 **67.** yes **69.** no **71.** no **73.** yes

75. no **77.** $x + 15$ **79.** $x - 5$ **81.** $\frac{x}{4}$ **83.** $3x + 22$ **85.** $1 + 2 = 9 \div 3$ **87.** $3 \neq 4 \div 2$ **89.** $5 + x = 20$ **91.** $7.6x = 17$

93. $13 - 3x = 13$ **95.** multiply **97.** subtract **99.** no; answers may vary **101.** 14 in., 12 sq in. **103.** 14 in., 9.01 sq in. **105.** Rectangles with the same perimeter can have different areas. **107.** $(20 - 4) \cdot 4 \div 2$ **109. a.** expression **b.** equation **c.** equation **d.** expression **e.** expression **111.** answers may vary **113.** answers may vary, for example, $2(5) + 1$ **115.** 12,000 sq ft **117.** 51 mph

Section 1.5
Practice Exercises

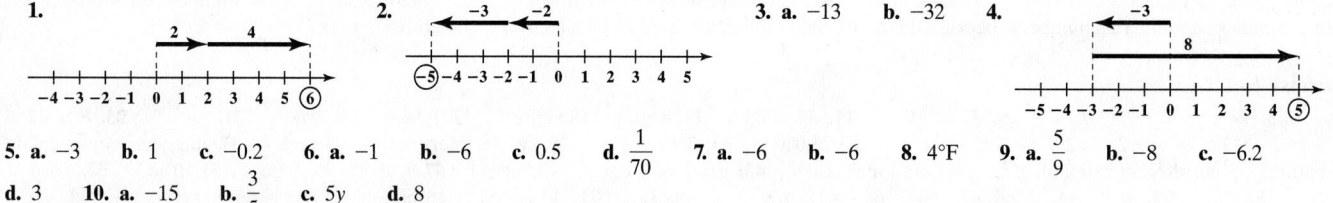

1. **2.** **3. a.** −13 **b.** −32 **4.**

5. a. −3 **b.** 1 **c.** −0.2 **6. a.** −1 **b.** −6 **c.** 0.5 **d.** $\frac{1}{70}$ **7. a.** −6 **b.** −6 **8.** 4°F **9. a.** $\frac{5}{9}$ **b.** −8 **c.** −6.2 **d.** 3 **10. a.** −15 **b.** $\frac{3}{5}$ **c.** $5y$ **d.** 8

Vocabulary, Readiness & Video Check 1.5

1. opposites **3.** n **5.** absolute values **7.** Negative temperatures; the high temperature for the day was −6°F

Exercise Set 1.5

1. −3 **3.** −14 **5.** 1 **7.** −12 **9.** −5 **11.** −12 **13.** −4 **15.** 7 **17.** −2 **19.** 0 **21.** −19 **23.** 31 **25.** −47

27. −2.1 **29.** −8 **31.** 38 **33.** −13.1 **35.** $\frac{2}{8} = \frac{1}{4}$ **37.** $-\frac{3}{16}$ **39.** $-\frac{13}{10}$ **41.** −8 **43.** −59 **45.** −9 **47.** 5 **49.** 11

51. −18 **53.** 19 **55.** −0.7 **57.** −6° **59.** 146 ft **61.** $9.1 million **63.** −14 **65.** −6 **67.** 2 **69.** 0 **71.** −6 **73.** −2

75. 0 **77.** $-\frac{2}{3}$ **79.** yes **81.** no **83.** July **85.** October **87.** 4.7°F **89.** −3 **91.** −22 **93.** negative **95.** positive **97.** true

99. false **101.** answers may vary **103.** answers may vary

Section 1.6
Practice Exercises

1. a. −13 **b.** −7 **c.** 12 **d.** −2 **2. a.** 10.9 **b.** $-\frac{1}{2}$ **c.** $-\frac{19}{20}$ **3.** −7 **4. a.** −6 **b.** 6.1 **5. a.** −20 **b.** 13 **6. a.** 2 **b.** 13 **7.** $357 **8. a.** 28° **b.** 137°

Vocabulary, Readiness & Video Check 1.6

1. $7 - x$ **3.** $x - 7$ **5.** $7 - x$ **7.** $-10 - (-14)$; d **9.** addition; opposite **11.** There's a minus sign in the numerator and the replacement value is negative (notice parentheses are used around the replacement value), and it's always good to be careful when working with negative signs. **13.** In Example 9, you have two supplementary angles and know the measure of one of them. From the definition, you know that two supplementary angles must sum to 180°. Therefore you can subtract the known angle measure from 180° to get the measure of the other angle.

Exercise Set 1.6

1. −10 **3.** −5 **5.** 19 **7.** $\frac{1}{6}$ **9.** 2 **11.** −11 **13.** 11 **15.** 5 **17.** 37 **19.** −6.4 **21.** −71 **23.** 0 **25.** 4.1

27. $\frac{2}{11}$ **29.** $-\frac{11}{12}$ **31.** 8.92 **33.** 13 **35.** −5 **37.** −1 **39.** −23 **41.** −26 **43.** −24 **45.** 3 **47.** −45 **49.** −4

51. 13 **53.** 6 **55.** 9 **57.** −9 **59.** −7 **61.** $\frac{7}{5}$ **63.** 21 **65.** $\frac{1}{4}$ **67.** 100°F **69.** 263°F **71.** 35,653 ft **73.** −308 ft

75. 19,852 ft **77.** 130° **79.** 30° **81.** no **83.** no **85.** yes **87.** $-5 + x$ **89.** $-20 - x$ **91.** −4.4°; 2.6°; 12°; 23.5°; 15.3° **93.** May **95.** answers may vary **97.** 16 **99.** −20 **101.** true; answers may vary **103.** false; answers may vary **105.** negative, −30,387

Integrated Review

1. negative **2.** negative **3.** positive **4.** 0 **5.** positive **6.** 0 **7.** positive **8.** positive **9.** $-\frac{1}{7}; \frac{1}{7}$ **10.** $\frac{12}{5}; \frac{12}{5}$ **11.** 3; 3

12. $-\frac{9}{11}; \frac{9}{11}$ **13.** −42 **14.** 10 **15.** 2 **16.** −18 **17.** −7 **18.** −39 **19.** −2 **20.** −9 **21.** −3.4 **22.** −9.8 **23.** $-\frac{25}{28}$

24. $-\frac{5}{24}$ **25.** −4 **26.** −24 **27.** 6 **28.** 20 **29.** 6 **30.** 61 **31.** −6 **32.** −16 **33.** −19 **34.** −13 **35.** −4 **36.** −1

37. $\frac{13}{20}$ **38.** $-\frac{29}{40}$ **39.** 4 **40.** 9 **41.** −1 **42.** −3 **43.** 8 **44.** 10 **45.** 47 **46.** $\frac{2}{3}$

Section 1.7
Practice Exercises

1. a. −40 **b.** 12 **c.** −54 **2. a.** −30 **b.** 24 **c.** 0 **d.** 26 **3. a.** −0.046 **b.** $-\dfrac{4}{15}$ **c.** 14 **4. a.** 36 **b.** −36 **c.** −64

d. −64 **5. a.** $\dfrac{3}{8}$ **b.** $\dfrac{1}{15}$ **c.** $-\dfrac{7}{2}$ **d.** $-\dfrac{1}{5}$ **6. a.** −8 **b.** −4 **c.** 5 **7. a.** 3 **b.** −16 **c.** $-\dfrac{6}{5}$ **d.** $-\dfrac{1}{18}$ **8. a.** 0

b. undefined **c.** undefined **9. a.** $-\dfrac{84}{5}$ **b.** 11 **10. a.** −9 **b.** 33 **c.** $\dfrac{5}{3}$ **11.** −52

Calculator Explorations 1.7
1. 38 **3.** −441 **5.** 163.$\overline{3}$ **7.** 54,499 **9.** 15,625

Vocabulary, Readiness & Video Check 1.7
1. 0; 0 **3.** positive **5.** negative **7.** positive **9.** The parentheses, or lack of them, determine the base of the expression. In Example 6, $(-2)^4$, the base is −2 and all of −2 is raised to the fourth power. In Example 7, -2^4, the base is 2 and only 2 is raised to the fourth power. **11.** Yes; because division of real numbers is defined in terms of multiplication. **13.** The football team lost 4 yards on each play and a loss of yardage is represented by a negative number.

Exercise Set 1.7

1. −24 **3.** −2 **5.** 50 **7.** −12 **9.** 0 **11.** −18 **13.** $\dfrac{3}{10}$ **15.** $\dfrac{2}{3}$ **17.** −7 **19.** 0.14 **21.** −800 **23.** −28 **25.** 25

27. $-\dfrac{8}{27}$ **29.** −121 **31.** $-\dfrac{1}{4}$ **33.** −30 **35.** 23 **37.** −7 **39.** true **41.** false **43.** 16 **45.** −1 **47.** 25 **49.** −49

51. $\dfrac{1}{9}$ **53.** $\dfrac{3}{2}$ **55.** $-\dfrac{1}{14}$ **57.** $-\dfrac{11}{3}$ **59.** $\dfrac{1}{0.2}$ **61.** −6.3 **63.** −9 **65.** 4 **67.** −4 **69.** 0 **71.** −5 **73.** undefined

75. 3 **77.** −15 **79.** $-\dfrac{18}{7}$ **81.** $\dfrac{20}{27}$ **83.** −1 **85.** $-\dfrac{9}{2}$ **87.** −4 **89.** 16 **91.** −3 **93.** $-\dfrac{16}{7}$ **95.** 2 **97.** $\dfrac{6}{5}$ **99.** −5

101. $\dfrac{3}{2}$ **103.** −21 **105.** 41 **107.** −134 **109.** 3 **111.** 0 **113.** −71 · x or −71x **115.** −16 − x **117.** −29 + x **119.** $\dfrac{x}{-33}$ or

$x \div (-33)$ **121.** 3 · (−4) = −12; a loss of 12 yd **123.** 5(−20) = −100; a depth of 100 ft **125.** yes **127.** no **129.** yes **131.** −162°F

133. answers may vary **135.** 1, −1 **137.** positive **139.** not possible **141.** negative **143.** $-2 + \dfrac{-15}{3}; -7$ **145.** 2[−5 + (−3)]; −16

Section 1.8
Practice Exercises

1. a. 8 · x **b.** 17 + x **2. a.** 2 + (9 + 7) **b.** (−4 · 2) · 7 **3. a.** x + 14 **b.** −30x **4. a.** 5x − 5y **b.** −24 − 12t
c. 6x − 8y − 2z **d.** −3 + y **e.** −x + 7 − 2s **f.** x + 11 **5. a.** 5(w + 3) **b.** 9(w + z) **6. a.** commutative property of multiplication
b. associative property of addition **c.** identity element for addition **d.** multiplicative inverse property **e.** commutative property of addition
f. additive inverse property **g.** commutative and associative properties of multiplication

Vocabulary, Readiness & Video Check 1.8
1. commutative property of addition **3.** distributive property **5.** associative property of addition **7.** opposites or additive inverses
9. 2 is outside the parentheses, so the point is made that you should only distribute the −9 to the terms within the parentheses and not also to the 2.

Exercise Set 1.8

1. 16 + x **3.** y · (−4) **5.** yx **7.** 13 + 2x **9.** x · (yz) **11.** (2 + a) + b **13.** (4a) · b **15.** a + (b + c) **17.** 17 + b
19. 24y **21.** y **23.** 26 + a **25.** −72x **27.** s **29.** 2 + x **31.** 4x + 4y **33.** 9x − 54 **35.** 6x + 10 **37.** 28x − 21
39. 18 + 3x **41.** −2y + 2z **43.** −21y − 35 **45.** 5x + 20m + 10 **47.** −4 + 8m − 4n **49.** −5x − 2 **51.** −r + 3 + 7p
53. 3x + 4 **55.** −x + 3y **57.** 6r + 8 **59.** −36x − 70 **61.** −16x − 25 **63.** 4(1 + y) **65.** 11(x + y) **67.** −1(5 + x)
69. 30(a + b) **71.** commutative property of multiplication **73.** associative property of addition **75.** distributive property
77. associative property of multiplication **79.** identity element for addition **81.** distributive property **83.** commutative and associative

properties of multiplication **85.** −8; $\dfrac{1}{8}$ **87.** −x; $\dfrac{1}{x}$ **89.** 2x; −2x **91.** false **93.** no **95.** yes **97.** yes **99.** no
101. a. commutative property of addition **b.** commutative property of addition **c.** associative property of addition **103.** answers may vary
105. answers may vary

Chapter 1 Vocabulary Check

1. inequality symbols **2.** equation **3.** absolute value **4.** variable **5.** opposites **6.** numerator **7.** solution **8.** reciprocals
9. base; exponent **10.** denominator **11.** grouping symbols **12.** set

Chapter 1 Review

1. < **3.** > **5.** < **7.** = **9.** > **11.** 4 ≥ −3 **13.** 0.03 < 0.3 **15. a.** 1, 3 **b.** 0, 1, 3 **c.** −6, 0, 1, 3

d. −6, 0, 1, $1\dfrac{1}{2}$, 3, 9.62 **e.** π **f.** −6, 0, 1, $1\dfrac{1}{2}$, 3, π, 9.62 **17.** Friday **19.** 2 · 2 · 3 · 3 **21.** $\dfrac{12}{25}$ **23.** $\dfrac{13}{10}$ **25.** $9\dfrac{3}{8}$ **27.** 15

29. $\dfrac{7}{12}$ **31.** $A = 1\dfrac{1}{6}$ sq m; $P = 4\dfrac{5}{12}$ m **33.** $14\dfrac{1}{8}$ lb **35.** $18\dfrac{7}{16}$ lb **37.** Baby E **39.** c **41.** $\dfrac{4}{49}$ **43.** 37 **45.** $\dfrac{18}{7}$

47. 20 − 12 = 2 · 4 **49.** 18 **51.** 5 **53.** 63° **55.** yes **57.** 9 **59.** −2 **61.** −11 **63.** $-\dfrac{3}{16}$ **65.** −13.9 **67.** −14

69. 5 **71.** -19 **73.** a **75.** \$51 **77.** $-\dfrac{1}{6}$ **79.** -48 **81.** 3 **83.** undefined **85.** undefined **87.** -12 **89.** 9
91. $-7 \cdot x$ or $-7x$ **93.** $-20 - x$ **95.** commutative property of addition **97.** distributive property **99.** associative property of addition
101. distributive property **103.** multiplicative inverse property **105.** $5y - 10$ **107.** $-7 + x - 4z$ **109.** $-12z - 27$ **111.** $<$
113. -15.3 **115.** -80 **117.** $-\dfrac{1}{4}$ **119.** 16 **121.** -5 **123.** $-\dfrac{5}{6}$ **125.** $1\dfrac{3}{8}$ ft

Chapter 1 Getting Ready for the Test

1. A **2.** C **3.** B **4.** A **5.** A **6.** B **7.** B **8.** A **9.** A **10.** B **11.** C **12.** D **13.** B **14.** C **15.** B **16.** C
17. C **18.** A

Chapter 1 Test

1. $|-7| > 5$ **2.** $9 + 5 \geq 4$ **3.** -5 **4.** -11 **5.** -3 **6.** -39 **7.** 12 **8.** -2 **9.** undefined **10.** -8 **11.** $-\dfrac{1}{3}$ **12.** $4\dfrac{5}{8}$
13. 1.275 **14.** -32 **15.** -48 **16.** 3 **17.** 0 **18.** $>$ **19.** $>$ **20.** $<$ **21.** $=$ **22.** $2221 < 10,993$ **23. a.** 1, 7 **b.** 0, 1, 7
c. $-5, -1, 0, 1, 7$ **d.** $-5, -1, 0, \dfrac{1}{4}, 1, 7, 11.6$ **e.** $\sqrt{7}, 3\pi$ **f.** $-5, -1, 0, \dfrac{1}{4}, 1, 7, 11.6, \sqrt{7}, 3\pi$ **24.** 40 **25.** 12 **26.** 22 **27.** -1
28. associative property of addition **29.** commutative property of multiplication **30.** distributive property **31.** multiplicative inverse property
32. 9 **33.** -3 **34.** second down **35.** yes **36.** $17°$ **37.** \$650 million **38.** \$420

CHAPTER 2 EQUATIONS, INEQUALITIES, AND PROBLEM SOLVING

Section 2.1
Practice Exercises

1. a. 1 **b.** -7 **c.** $-\dfrac{1}{5}$ **d.** 43 **e.** -1 **2. a.** like terms **b.** unlike terms **c.** like terms **d.** like terms **3. a.** $8y$ **b.** $5x^2$
c. $5x + 5x^2$ **d.** $21y^2$ **4. a.** $11y - 5$ **b.** $5x - 6$ **c.** $-\dfrac{1}{4}t$ **d.** $12.2y + 13$ **e.** $5z - 3z^4$ **5. a.** $6x - 21$ **b.** $-5x + 2.5z + 25$
c. $-2x + y - z + 2$ **6. a.** $36x + 10$ **b.** $-11x + 1$ **c.** $-30x - 17$ **7.** $-5x + 4$ **8. a.** $2x + 3$ **b.** $x - 1$ **c.** $2x + 10$ **d.** $\dfrac{13}{2}x$

Vocabulary, Readiness & Video Check 2.1

1. expression; term **3.** numerical coefficient **5.** numerical coefficient **7.** Although these terms have exactly the same variables, the exponents
on each are not exactly the same — the exponents on x differ in each term. **9.** -1

Exercise Set 2.1

1. -7 **3.** 1 **5.** 17 **7.** like **9.** unlike **11.** like **13.** $15y$ **15.** $13w$ **17.** $-7b - 9$ **19.** $-m - 6$ **21.** -8 **23.** $7.2x - 5.2$
25. $4x - 3$ **27.** $5x^2$ **29.** $1.3x + 3.5$ **31.** $5y - 20$ **33.** $-2x - 4$ **35.** $7d - 11$ **37.** $-10x + 15y - 30$ **39.** $-3x + 2y - 1$
41. $2x + 14$ **43.** $10x - 3$ **45.** $-4x - 9$ **47.** $-4m - 3$ **49.** $k - 6$ **51.** $-15x + 18$ **53.** 16 **55.** $x + 5$ **57.** $x + 2$
59. $2k + 10$ **61.** $-3x + 5$ **63.** -11 **65.** $3y + \dfrac{5}{6}$ **67.** $-22 + 24x$ **69.** $0.9m + 1$ **71.** $10 - 6x - 9y$ **73.** $-x - 38$ **75.** $5x - 7$
77. $2x - 4$ **79.** $2x + 7$ **81.** $\dfrac{3}{4}x + 12$ **83.** $-2 + 12x$ **85.** $8(x + 6)$ or $8x + 48$ **87.** $x - 10$ **89.** $7x - 7$ **91.** 2 **93.** -23
95. -25 **97.** $(18x - 2)$ ft **99.** balanced **101.** balanced **103.** answers may vary **105.** $(15x + 23)$ in. **107.** answers may vary
109. $5b^2c^3 + b^3c^2$ **111.** $5x^2 + 9x$ **113.** $-7x^2y$

Section 2.2
Practice Exercises

1. -8 **2.** -1.8 **3.** -10 **4.** 18 **5.** 20 **6.** -12 **7.** 65 **8.** -3 **9. a.** 7 **b.** $9 - x$ **c.** $(9 - x)$ ft **10.** $3x + 6$

Vocabulary, Readiness & Video Check 2.2

1. equation; expression **3.** solution **5.** addition **7.** multiplication **9.** true **11.** both sides **13.** addition property; multiplication
property; answers may vary

Exercise Set 2.2

1. 3 **3.** -2 **5.** -14 **7.** 0.5 **9.** -3 **11.** -0.7 **13.** 3 **15.** 11 **17.** 0 **19.** -3 **21.** 16 **23.** -4 **25.** 0 **27.** 12
29. 10 **31.** -12 **33.** 3 **35.** -2 **37.** 0 **39.** answers may vary **41.** 10 **43.** -20 **45.** 0 **47.** -5 **49.** 0
51. $-\dfrac{3}{2}$ **53.** -21 **55.** $\dfrac{11}{2}$ **57.** 1 **59.** $-\dfrac{1}{4}$ **61.** 12 **63.** -30 **65.** $\dfrac{9}{10}$ **67.** -30 **69.** 2 **71.** -2 **73.** 23
75. $20 - p$ **77.** $(10 - x)$ ft **79.** $(180 - x)°$ **81.** $(n + 284)$ votes **83.** $(m - 60)$ ft **85.** $(n - 28,000)$ students **87.** $7x$ sq mi
89. $2x + 2$ **91.** $2x + 2$ **93.** $5x + 20$ **95.** $7x - 12$ **97.** 1 **99.** $>$ **101.** $=$ **103.** $(173 - 3x)°$ **105.** answers may vary
107. 4; 4 **109.** answers may vary **111.** answers may vary **113.** -48 **115.** $\dfrac{700}{3}$ mg **117.** solution **119.** -2.95 **121.** 0.02

Section 2.3
Practice Exercises

1. 3   **2.** $\dfrac{21}{13}$ **3.** -15 **4.** 3   **5.** 0   **6.** no solution   **7.** all real numbers

Calculator Explorations 2.3

1. solution   **3.** not a solution   **5.** solution

Vocabulary, Readiness & Video Check 2.3

1. equation   **3.** expression   **5.** expression   **7.** equation   **9.** 3; distributive property, addition property of equality, multiplication property of equality   **11.** The number of decimal places in each number helps you determine the least power of 10 you can multiply through by so you are no longer dealing with decimals.

Exercise Set 2.3

1. -6 **3.** 3   **5.** 1   **7.** $\dfrac{3}{2}$ **9.** 0   **11.** -1 **13.** 4   **15.** -4 **17.** -3 **19.** 2   **21.** 50   **23.** 1   **25.** $\dfrac{7}{3}$ **27.** 0.2

29. all real numbers   **31.** no solution   **33.** no solution   **35.** all real numbers   **37.** 18   **39.** $\dfrac{19}{9}$ **41.** $\dfrac{14}{3}$ **43.** 13   **45.** 4

47. all real numbers   **49.** $-\dfrac{3}{5}$ **51.** -5 **53.** 10   **55.** no solution   **57.** 3   **59.** -17 **61.** -4 **63.** 3   **65.** all real numbers

67. $-8 - x$ **69.** $-3 + 2x$ **71.** $9(x + 20)$ **73.** $(6x - 8)$ m   **75. a.** all real numbers   **b.** answers may vary   **c.** answers may vary
77. A   **79.** B   **81.** C   **83.** answers may vary   **85. a.** $x + x + x + 2x + 2x = 28$ **b.** $x = 4$ **c.** $x = 4$ cm; $2x = 8$ cm

87. answers may vary   **89.** 15.3   **91.** -0.2 **93.** $-\dfrac{7}{8}$ **95.** no solution

Integrated Review

1. 6   **2.** -17 **3.** 12   **4.** -26 **5.** -3 **6.** -1 **7.** $\dfrac{27}{2}$ **8.** $\dfrac{25}{2}$ **9.** 8   **10.** -64 **11.** 2   **12.** -3 **13.** no solution

14. no solution   **15.** -2 **16.** -2 **17.** $-\dfrac{5}{6}$ **18.** $\dfrac{1}{6}$ **19.** 1   **20.** 6   **21.** 4   **22.** 1   **23.** $\dfrac{9}{5}$ **24.** $-\dfrac{6}{5}$ **25.** all real numbers

26. all real numbers   **27.** 0   **28.** -1.6 **29.** $\dfrac{4}{19}$ **30.** $-\dfrac{5}{19}$ **31.** $\dfrac{7}{2}$ **32.** $-\dfrac{1}{4}$ **33.** 2   **34.** 2   **35.** no solution   **36.** no solution

37. $\dfrac{7}{6}$ **38.** $\dfrac{1}{15}$

Section 2.4
Practice Exercises

1. 9   **2.** 2   **3.** 9 in. and 36 in.   **4.** 28 Republican and 21 Democratic governors   **5.** $25°, 75°, 80°$ **6.** $46, 48, 50$

Vocabulary, Readiness & Video Check 2.4

1. $2x; 2x - 31$ **3.** $x + 5; 2(x + 5)$ **5.** $20 - y; \dfrac{20 - y}{3}$ or $(20 - y) \div 3$ **7.** in the statement of the application   **9.** That the 3 angle measures are consecutive even integers and that they sum to $180°$.

Exercise Set 2.4

1. $6x + 1 = 5x; -1$ **3.** $3x - 6 = 2x + 8; 14$ **5.** $2(x - 8) = 3(x + 3); -25$ **7.** $2(-2 + x) = x - \dfrac{1}{2}; \dfrac{7}{2}$ **9.** 3 in.; 6 in.; 16 in.   **11.** 1st piece: 5 in.; 2nd piece: 10 in.; 3rd piece: 25 in.   **13.** 16,146; 27,119; In 2014, 16,146 screens were 3D.   **15.** 1st angle: $37.5°$; 2nd angle: $37.5°$; 3rd angle: $105°$ **17.** $3x + 3$ **19.** $x + 2; x + 4; 2x + 4$ **21.** $x + 1; x + 2; x + 3; 4x + 6$ **23.** $x + 2; x + 4; 2x + 6$ **25.** 234, 235

27. Belgium: 32; France: 33; Spain: 34   **29.** Sahara: 3,500,000 sq mi; Gobi: 500,000 sq mi   **31.** 5 ft, 12 ft   **33.** $\dfrac{5}{4}$ **35.** Botswana: 32,000,000 carats; Angola: 8,000,000 carats   **37.** $58°, 60°, 62°$ **39.** Netherlands: 24; Canada: 25; Norway: 26   **41.** -16 **43.** $43°, 137°$ **45.** 1   **47.** $\dfrac{3}{2}$ **49.** Maglev: 267 mph; Harmony: 236 mph   **51.** $\dfrac{5}{2}$ **53.** California: 58; Montana: 56   **55.** $111°$ **57.** 1st piece: 3 ft; 2nd piece: 12 ft; 3rd piece: 15 ft   **59.** Call of Duty: Advanced Warfare   **61.** Destiny: 3.8 million; Madden NFL 15: 2.9 million   **63.** answers may vary   **65.** 34   **67.** 225π **69.** 15 ft by 24 ft
71. 5400 chirps per hour; 129,600 chirps per day; 47,304,000 chirps per year   **73.** answers may vary   **75.** answers may vary   **77.** c

Section 2.5
Practice Exercises

1. 116 sec or 1 min 56 sec   **2.** 9 ft   **3.** $46.4°$F   **4.** length: 28 in.; width: 5 in.   **5.** $R = \dfrac{I}{PT}$ **6.** $s = \dfrac{H - 10a}{5a}$ **7.** $d = \dfrac{N - F}{n - 1}$

8. $B = \dfrac{2A - ab}{a}$

Vocabulary, Readiness & Video Check 2.5

1. relationships **3.** To show that the process of solving this equation for x—dividing both sides by 5, the coefficient of x—is the same process used to solve a formula for a specific variable. Treat whatever is multiplied by that specific variable as the coefficient—the coefficient is all the factors except that specific variable.

Exercise Set 2.5

1. $h = 3$ **3.** $h = 3$ **5.** $h = 20$ **7.** $c = 12$ **9.** $r \approx 2.5$ **11.** $T = 3$ **13.** $h \approx 15$ **15.** $h = \dfrac{f}{5g}$ **17.** $w = \dfrac{V}{lh}$ **19.** $y = 7 - 3x$
21. $R = \dfrac{A - P}{PT}$ **23.** $A = \dfrac{3V}{h}$ **25.** $a = P - b - c$ **27.** $h = \dfrac{S - 2\pi r^2}{2\pi r}$ **29.** 120 ft **31. a.** area: 103.5 sq ft; perimeter: 41 ft
b. baseboard: perimeter; carpet: area **33. a.** area: 480 sq in.; perimeter: 120 in. **b.** frame: perimeter; glass: area **35.** $-10°$C **37.** length: 78 ft; width: 52 ft **39.** 18 ft, 36 ft, 48 ft **41.** 55.2 mph **43.** 96 piranhas **45.** 61.5°F **47.** 60 chirps per minute **49.** increases **51.** 2 bags
53. one 16-in. pizza **55.** $x = 6$ m, $2.5x = 15$ m **57.** 22 hr **59.** 13 in. **61.** 2.25 hr **63.** 12,090 ft **65.** 50°C **67.** 332.6°F
69. 449 cu in. **71.** 0.32 **73.** 2.00 or 2 **75.** 17% **77.** 720% **79.** multiplies the volume by 8; answers may vary **81.** $53\dfrac{1}{3}$
83. $V = G(N - R)$ **85.** $\bullet = \dfrac{\blacksquare - \blacktriangle}{\blacksquare}$ **87.** 500 sec or $8\dfrac{1}{3}$ min **89.** 608.33 ft **91.** 4.42 min **93.** $35\dfrac{11}{17}$ mph

Section 2.6
Practice Exercises

1. 62.5% **2.** 360 **3. a.** 44% **b.** 97% **c.** 75.3 million dogs **4.** discount: $408; new price: $72 **5.** 42.6% **6.** 14,750 screens
7. 2 liters of 5% eyewash; 4 liters of 2% eyewash

Vocabulary, Readiness & Video Check 2.6

1. no **3.** yes **5. a.** equals; = **b.** multiplication; · **c.** Drop the percent symbol and move the decimal point two places to the left.
7. You must first find the actual amount of increase in price by subtracting the original price from the new price.

Exercise Set 2.6

1. 11.2 **3.** 55% **5.** 180 **7.** 86% **9.** 8.46 million sq mi **11.** discount: $1480; new price: $17,020 **13.** $46.58 **15.** 9.3% **17.** 30%
19. $104 **21.** $42,500 **23.** 2 gal **25.** 7 lb **27.** 4.6 **29.** 50 **31.** 30% **33.** 45 min **35.** 32.1% **37.** 63%; 5%; 24%; 2%; 101%
due to rounding **39.** decrease: $64; sale price: $192 **41.** 7.2% **43.** 251 million **45.** 300% **47.** 400 oz **49.** 16% **51.** 120 employees
53. 361 college students **55.** 400 oz **57.** 854 thousand Scoville units **59.** > **61.** = **63.** > **65.** no; answers may vary
67. no; answers may vary **69.** 9.6% **71.** 26.9%; yes **73.** 17.1%

Section 2.7
Practice Exercises

1. 2.2 hr **2.** eastbound: 62 mph; westbound: 52 mph **3.** 106 $5 bills; 59 $20 bills **4.** $18,000 at 11.5%; $12,000 at 6%

Vocabulary, Readiness & Video Check 2.7

1.

	r	·	t	=	d
bus	55		x		$55x$
car	50		$x + 3$		$50(x + 3)$

; $55x = 50(x + 3)$

3.

	P	·	R	·	T	=	I
	x		0.06		1		$0.06x$
	$36{,}000 - x$		0.04		1		$0.04(36{,}000 - x)$

; $0.06x = 0.04(36{,}000 - x)$

Exercise Set 2.7

1. $666\dfrac{2}{3}$ mi **3.** 55 mph **5.** $0.10y$ **7.** $0.05(x + 7)$ **9.** $20(4y)$ or $80y$ **11.** $50(35 - x)$ **13.** 12 $10 bills; 32 $5 bills **15.** $11,500 at 8%;
$13,500 at 9% **17.** $7000 at 11% profit; $3000 at 4% loss **19.** 187 adult tickets; 313 child tickets **21.** 2 hr **23.** $30,000 at 8%; $24,000 at 10%
25. 2 hr $37\dfrac{1}{2}$ min **27.** 36 mph; 46 mph **29.** 483 dimes; 161 nickels **31.** 4 hr **33.** 2.5 hr **35.** $4500 **37.** 2.2 mph; 3.3 mph **39.** 27.5 mi
41. -4 **43.** $\dfrac{9}{16}$ **45.** -4 **47.** 25 $100 bills; 71 $50 bills; 175 $20 bills **49.** 25 skateboards **51.** 800 books **53.** answers may vary

Section 2.8
Practice Exercises

1. $(-\infty, 5)$ **2.** $[-5, \infty)$ **3.** $(-\infty, 3]$ **4.** $(-3, \infty)$
5. $[7, \infty)$ **6.** $(-\infty, -7]$ **7.** $\left(-\infty, \dfrac{9}{4}\right)$ **8.** $[0, \infty)$
9. $[-3, 1)$ **10.** $(-2, 2]$ **11.** $\left(-\dfrac{16}{3}, \dfrac{4}{3}\right)$ **12.** all numbers less than 10
13. Kasonga can afford at most 3 classes.

Vocabulary, Readiness & Video Check 2.8

1. expression **3.** inequality **5.** -5 **7.** The graph of Example 1 is shaded from $-\infty$ to and including -1, as indicated by a bracket. To write interval notation, you write down what is shaded for the inequality from left to right. A parenthesis is always used with $-\infty$, so from the graph, the interval notation is $(-\infty, -1]$. **9.** You would divide the left, middle, and right by -3 instead of 3, which would reverse the directions of both inequality symbols.

Exercise Set 2.8

1. $x \geq 2$ **3.** $x < -5$ **5.** $(-\infty, -1]$ **7.** $\left(-\infty, \frac{1}{2}\right)$

9. $[5, \infty)$ **11.** $x < -3$, $(-\infty, -3)$ **13.** $x \geq -5$, $[-5, \infty)$

15. $x \geq -2$ $[-2, \infty)$ **17.** $x > -3$, $(-3, \infty)$ **19.** $x \leq 1$, $(-\infty, 1]$

21. $x > -5$, $(-5, \infty)$ **23.** $x \leq -2$, $(-\infty, -2]$ **25.** $x \leq -8$, $(-\infty, -8]$

27. $x > 4$, $(4, \infty)$ **29.** $x \geq 20$, $[20, \infty)$ **31.** $x > 16$, $(16, \infty)$

33. $x > -3$, $(-3, \infty)$ **35.** $x \leq -\frac{2}{3}$, $\left(-\infty, -\frac{2}{3}\right]$ **37.** $x > \frac{8}{3}$, $\left(\frac{8}{3}, \infty\right)$

39. $x > -13$, $(-13, \infty)$ **41.** $x < 0$, $(-\infty, 0)$ **43.** $x \leq 0$, $(-\infty, 0]$

45. $x > 3$, $(3, \infty)$ **47.** $x > \frac{8}{3}$ $\left(\frac{8}{3}, \infty\right)$ **49.** $(-1, 3)$ **51.** $[0, 2)$

53. $(-1, 2)$ **55.** $[4, 5]$ **57.** $(1, 5]$ **59.** $(1, 4)$

61. $\left(0, \frac{14}{3}\right]$ **63.** all numbers greater than -10 **65.** 35 cm **67.** at least 193 **69.** 86 people **71.** at least 35 min

73. $-3 < x < 3$ **75.** 8 **77.** 1 **79.** $\frac{16}{49}$ **81.** $>$ **83.** $\geq$ **85.** when multiplying or dividing by a negative number

87. final exam score ≥ 78.5 **89.** answers may vary **91.** answers may vary **93.** $0.924 \leq d \leq 0.987$ **95.** $(1, \infty)$

97. $\left(-\infty, \frac{5}{8}\right)$

Chapter 2 Vocabulary Check

1. like terms **2.** unlike terms **3.** linear equation in one variable **4.** linear inequality in one variable **5.** compound inequalities **6.** formula **7.** numerical coefficient **8.** equivalent equations **9.** all real numbers **10.** no solution **11.** the same **12.** reversed

Chapter 2 Review

1. $6x$ **3.** $4x - 2$ **5.** $3n - 18$ **7.** $-6x + 7$ **9.** $3x - 7$ **11.** 4 **13.** 6 **15.** 0 **17.** -23 **19.** 5; 5 **21.** b **23.** b **25.** -12

27. 0 **29.** 0.75 **31.** -6 **33.** -1 **35.** $-\frac{1}{5}$ **37.** $3x + 3$ **39.** -4 **41.** 2 **43.** no solution **45.** $\frac{3}{4}$ **47.** 20 **49.** $\frac{23}{7}$

51. 102 **53.** 6665.5 in. **55.** Target Canada: 133; Target US: 1793 **57.** 3 **59.** $w = 9$ **61.** $m = \frac{y - b}{x}$ **63.** $x = \frac{2y - 7}{5}$ **65.** $\pi = \frac{C}{D}$

67. 15 m **69.** 1 hr 20 min **71.** 20% **73.** 110 **75.** mark-up: \$209; new price: \$2109 **77.** 40% solution: 10 gal; 10% solution: 20 gal
79. 18% **81.** 966 customers **83.** 32% **85.** 50 km **87.** 80 nickels **89.** $(0, \infty)$ **91.** $[0.5, 1.5)$

93. $(-\infty, -4)$ **95.** $(-\infty, 4]$ **97.** $\left(-\frac{1}{2}, \frac{3}{4}\right)$ **99.** $\left(-\infty, \frac{19}{3}\right]$

101. \$2500 **103.** 4 **105.** $-\frac{3}{2}$ **107.** all real numbers **109.** -13 **111.** $h = \frac{3V}{A}$ **113.** 160 **115.** $(9, \infty)$

117. $(-\infty, 0]$

Chapter 2 Getting Ready for the Test

1. C **2.** A **3.** D **4.** B **5.** B **6.** C **7.** B **8.** A **9.** C **10.** C **11.** B **12.** D

Chapter 2 Test

1. $y - 10$ **2.** $5.9x + 1.2$ **3.** $-2x + 10$ **4.** $10y + 1$ **5.** -5 **6.** 8 **7.** $\frac{7}{10}$ **8.** 0 **9.** 27 **10.** 3 **11.** 0.25 **12.** $\frac{25}{7}$

13. no solution **14.** $x = 6$ **15.** $h = \frac{V}{\pi r^2}$ **16.** $y = \frac{3x - 10}{4}$ **17.** $(-\infty, -2]$ **18.** $(-\infty, 4)$

19. $\left(-1, \frac{7}{3}\right)$ **20.** $\left(\frac{2}{5}, \infty\right)$ **21.** 21 **22.** 62 ft by 64 ft **23.** 401, 802 **24.** \$8500 at 10%; \$17,000 at 12%

25. $2\frac{1}{2}$ hr **26.** 552 tornadoes **27.** 40% **28.** 16%

Chapter 2 Cumulative Review

1. a. 11, 112   **b.** 0, 11, 112   **c.** $-3, -2, 0, 11, 112$ **d.** $-3, -2, -1.5, 0, \frac{1}{4}, 11, 112$ **e.** $\sqrt{2}$ **f.** all numbers in the given set; Sec. 1.2, Ex. 5
3. a. 4   **b.** 5   **c.** 0   **d.** $\frac{1}{2}$ **e.** 5.6; Sec. 1.2, Ex. 7   **5. a.** $2 \cdot 2 \cdot 2 \cdot 5$ **b.** $3 \cdot 3 \cdot 7$; Sec. 1.3, Ex. 1   **7.** $\frac{8}{20}$; Sec. 1.3, Ex. 6   **9.** 66; Sec. 1.4, Ex. 4
11. 2 is a solution; Sec. 1.4, Ex. 7   **13.** -3; Sec. 1.5, Ex. 2   **15.** 2; Sec. 1.5, Ex. 4   **17. a.** 10   **b.** $\frac{1}{2}$ **c.** $2x$ **d.** -6; Sec. 1.5, Ex. 10
19. a. 9.9   **b.** $-\frac{4}{5}$ **c.** $\frac{2}{15}$; Sec. 1.6, Ex. 2   **21. a.** $52°$ **b.** $118°$; Sec. 1.6, Ex. 8   **23. a.** -0.06 **b.** $-\frac{7}{15}$ **c.** 16; Sec. 1.7, Ex. 3
25. a. 6   **b.** -12 **c.** $-\frac{8}{15}$ **d.** $-\frac{1}{6}$; Sec. 1.7, Ex. 7   **27. a.** $5 + x$ **b.** $x \cdot 3$; Sec. 1.8, Ex. 1   **29. a.** $8(2 + x)$ **b.** $7(s + t)$; Sec. 1.8, Ex. 5
31. $-2x - 1$; Sec. 2.1, Ex. 7   **33.** -1.6; Sec. 2.2, Ex. 2   **35.** 8; Sec. 2.2, Ex. 4   **37.** 140; Sec. 2.2, Ex. 7   **39.** 2; Sec. 2.3, Ex. 1   **41.** 10; Sec. 2.4, Ex. 2
43. $\frac{V}{wh} = l$; Sec. 2.5, Ex. 5   **45.** $(-\infty, -10]$ ⟵———⟍———➤; Sec. 2.8, Ex. 2

CHAPTER 3 GRAPHING

Section 3.1
Practice Exercises

1. a. Oceania/Australia region, 25 million Internet users   **b.** 85 million more Internet users   **2. a.** 70 beats per minute   **b.** 60 beats per minute
c. 5 minutes after lighting   **3.**   **4. a.** $(2008, 79), (2009, 79), (2010, 72), (2011, 73), (2012, 56), (2013, 48), (2014, 46)$

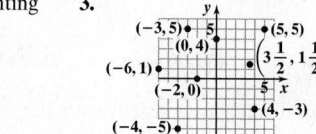

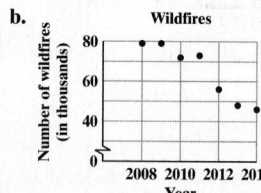

5. a. yes   **b.** yes   **c.** no   **6. a.** $(0, -8)$ **b.** $(6, 4)$ **c.** $(-3, -14)$

7.

x	y
a. -2	8
b. 3	-12
c. 0	0

8.

x	y
a. -10	-4
b. 0	-2
c. 10	0

9.

x	0	1	2	3	4
y	12,000	10,200	8400	6600	4800

Vocabulary, Readiness & Video Check 3.1

1. x-axis; y-axis   **3.** quadrants; four   **5.** one   **7.** horizontal: top tourist destinations; vertical: number of arrivals (in millions) to these destinations
9. Data occurring in pairs of numbers can be written as ordered pairs, called paired data, and then graphed on a coordinate system.
11. a linear equation in one variable

Exercise Set 3.1

1. China   **3.** France, U.S., Spain, and China   **5.** 53 million   **7.** 82,000   **9.** 2011; 103,000   **11.** 15.8   **13.** from 2008 to 2010
15. 2010 to 2014   **17.** $\left(-1, 4\frac{1}{2}\right)$ $(1, 5)$ and $(3.7, 2.2)$ are in quadrant I, $\left(-1, 4\frac{1}{2}\right)$ is in quadrant II, $(-5, -2)$ is in quadrant III, $(2, -4)$ and $\left(\frac{1}{2}, -3\right)$ are in quadrant IV, $(-3, 0)$ lies on the x-axis, $(0, -1)$ lies on the y-axis   **19.** $(0, 0)$
21. $(3, 2)$ **23.** $(-2, -2)$ **25.** $(2, -1)$ **27.** $(0, -3)$ **29.** $(1, 3)$ **31.** $(-3, -1)$

33. a. $(2010, 21.0), (2011, 22.4), (2012, 23.9), (2013, 25.0), (2014, 28.0)$ **b.** In the year 2014, the worldwide box office was $28.0 billion.
c.   **d.** The worldwide box office increased every year.   **35. a.** $(0.50, 10), (0.75, 12), (1.00, 15), (1.25, 16), (1.50, 18),$ $(1.50, 19), (1.75, 19), (2.00, 20)$ **b.** When Minh studied 1.25 hours, her quiz score was 16.   **c.**
d. answers may very

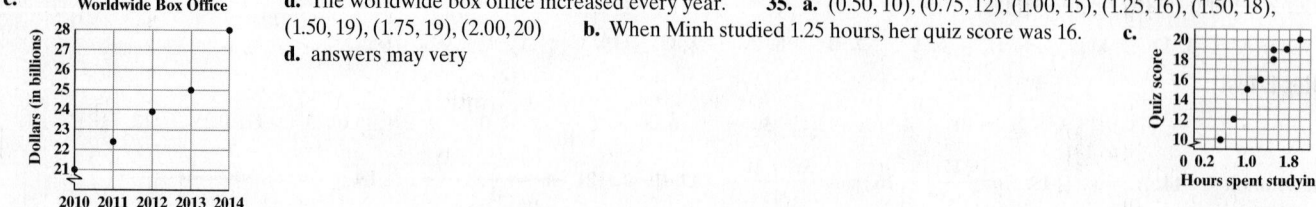

37. a. $(2313, 2), (2085, 1), (2711, 21), (2869, 39), (2920, 42), (4038, 99), (1783, 0), (2493, 9)$ **b.** 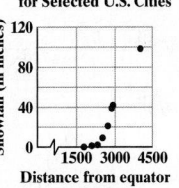 **c.** The farther from the equator, the more snowfall. **39.** yes; no; yes **41.** yes; yes **43.** no; yes; yes

45. $(-4, -2), (4, 0)$ **47.** $(-8, -5), (16, 1)$ **49.** $0; 7; -\dfrac{2}{7}$ **51.** $2; 2; 5$

53. $0; -3; 2$ **55.** $2; 6; 3$ **57.** $-12; 5; -6$ **59.** $\dfrac{5}{7}; \dfrac{5}{2}; -1$ **61.** $0; -5; -2$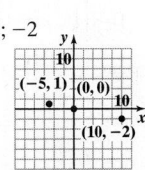

63. $2; 1; -6$ 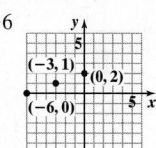 **65. a.** $13{,}000; 21{,}000; 29{,}000$ **b.** 45 desks **67. a.** $7.85; 8.03; 8.21$ **b.** year 3; 2013 **c.** 2034
d. In 2011, the average cinema admission price was $7.94. **69.** In 2014, there were 4203 Walmart stores in the U.S.
71. year 8: 100 stores; year 9: 105 stores; year 10: 100 stores **73.** The y-values are all 0. **75.** $y = 5 - x$

77. $y = -\dfrac{1}{2}x + \dfrac{5}{4}$ **79.** $y = -2x$ **81.** $y = \dfrac{1}{3}x - 2$ **83.** false **85.** true **87.** negative; negative

89. positive; negative **91.** $0; 0$ **93.** y **95.** no; answers may vary **97.** answers may vary **99.** $(4, -7)$ **101.** 26 units **103. a.** $(-2, 6)$
b. 28 units **c.** 45 sq units

Section 3.2
Practice Exercises

1. a. yes **b.** no **c.** yes **d.** yes **2.** 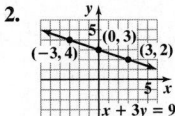 **3.** **4.** **5.**

6. 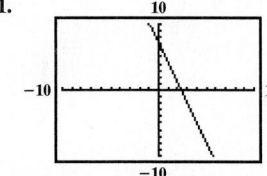 **7.** 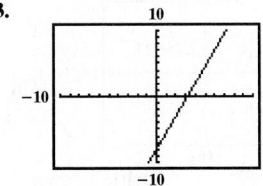 The graph of $y = -2x + 3$ is the same as the graph of $y = -2x$ except that the graph of $y = -2x + 3$ is moved 3 units upward.

8. a. 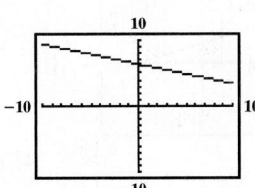 **b.** We predict 725 thousand computer software application engineers in the year 2020.

Calculator Explorations 3.2

1. **3.** **5.**

Vocabulary, Readiness & Video Check 3.2

1. In the definition, x and y both have an understood power of 1. Example 3 shows an equation where y has a power of 2, so it is not a linear equation in two variables. **3.** An infinite number of points make up the line and each point corresponds to an ordered pair that is a solution of the linear equation in two variables.

Exercise Set 3.2

1. yes **3.** yes **5.** no **7.** yes **9.**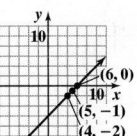

x	y
6	0
4	-2
5	-1

11.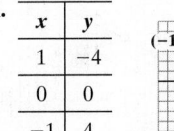

x	y
1	-4
0	0
-1	4

13.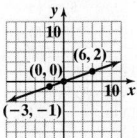

x	y
0	0
6	2
-3	-1

15.

x	y
0	3
1	−1
2	−5

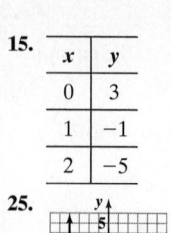

17.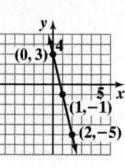

19.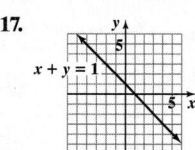

21.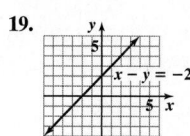

23.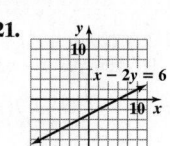

25.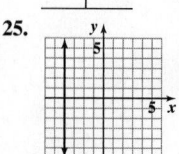

27.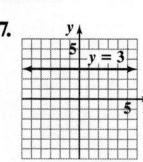

29.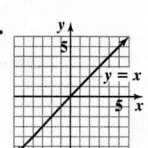

31.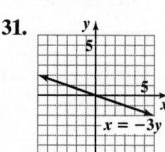

33.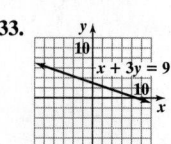

35.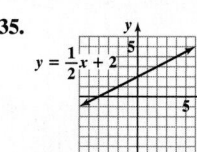

37.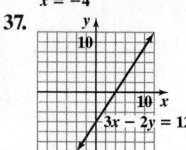

39.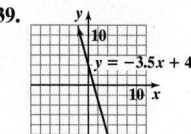

41.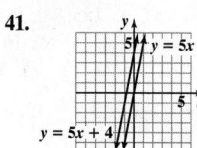

43.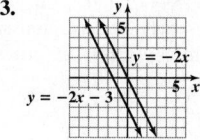

45.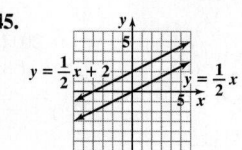

47. C   **49.** D   **51. a.** $(8, 31)$ **b.** In 2008, there were 31 million joggers.   **c.** 40 million joggers   **53. a.** $(12, 216.4)$ **b.** In 2012, there were 216.4 million people with driver's licenses.   **c.** 234 million   **55.** $(4, -1)$ **57.** $3; -3$ **59.** $0; 0$ **61.** $y = x + 5$

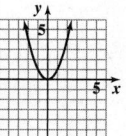

63. $2x + 3y = 6$
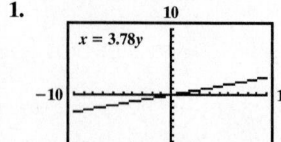

65. $x + y = 12; y = 9\,\text{cm}$ **67.** answers may vary   **69.** $0; 1; 1; 4; 4$

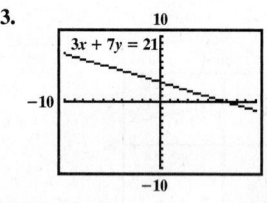

Section 3.3

Practice Exercises

1. x-intercept: $(-4, 0)$
y-intercept: $(0, -6)$ **2.** x-intercepts: $(-2, 0), (2, 0)$
y-intercept: $(0, -3)$ **3.** x-intercept: $(0, 0)$
y-intercept: $(0, 0)$ **4.** x-intercept: none
y-intercept: $(0, 3)$ **5.** x-intercepts: $(-1, 0), (5, 0)$
y-intercepts: $(0, 2), (0, -2)$

6. 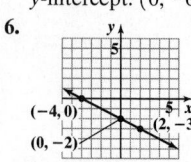   **7.** 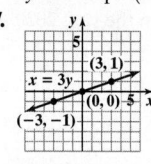   **8.** 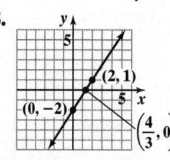   **9.** 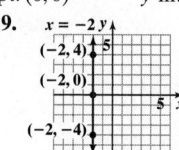   **10.**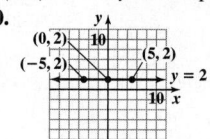

Calculator Explorations 3.3

1. 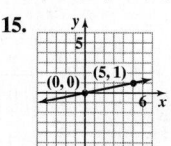   **3.** 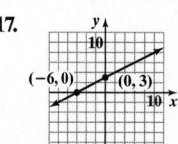   **5.** $-2.2x + 6.8y = 15.5$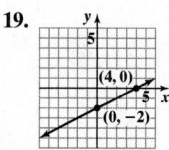

Vocabulary, Readiness & Video Check 3.3

1. linear   **3.** horizontal   **5.** y-intercept   **7.** $y; x$ **9.** Because x-intercepts lie on the x-axis; because y-intercepts lie on the y-axis.
11. For a horizontal line, the coefficient of x will be 0; for a vertical line, the coefficient of y will be 0.

Exercise Set 3.3

1. $(-1, 0); (0, 1)$ **3.** $(-2, 0); (2, 0); (0, -2)$ **5.** $(-2, 0); (1, 0); (3, 0); (0, 3)$ **7.** $(-1, 0); (1, 0); (0, 1); (0, -2)$ **9.** infinite   **11.** 0

13. 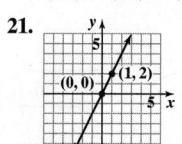   **15.** (additional graphs)   **17.**   **19.**   **21.**   **23.**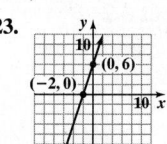

25. 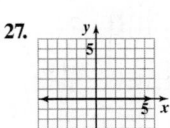 **27.** 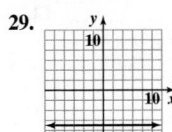 **29.** 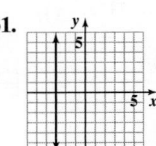 **31.** 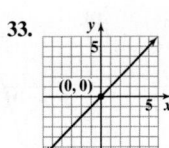 **33.**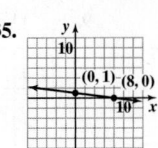

35.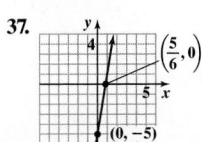

37. 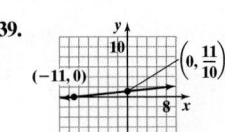 **39.** 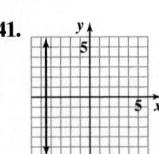 **41.** 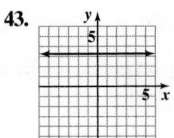 **43.** 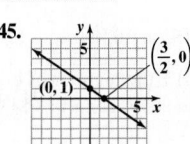 **45.**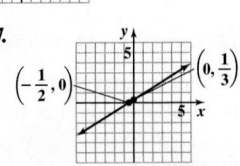

47.

49. C   **51.** E   **53.** B   **55.** $\dfrac{3}{2}$ **57.** 6   **59.** $-\dfrac{6}{5}$ **61.** false   **63.** true   **65.** $(0, 200)$; no chairs and 200 desks are manufactured,

67. 300 chairs   **69. a.** $(4.5, 0)$ **b.** 4.5 years after 2009, there should be no more analog screens.   **c.** answers may vary   **71.** $x = 1$
73. answers may vary   **75.** answers may vary

Section 3.4

Practice Exercises

1. -1 **2.** $\dfrac{1}{3}$ **3.** $m = \dfrac{2}{3}$; y-intercept: $(0, -2)$ **4.** $m = 6$; y-intercept: $(0, -5)$ **5.** $m = -\dfrac{5}{2}$; y-intercept: $(0, 4)$ **6.** $m = 0$

7. slope is undefined   **8. a.** perpendicular   **b.** neither   **c.** parallel   **9.** 25%   **10.** $m = \dfrac{0.75 \text{ dollar}}{1 \text{ pound}}$; The Wash-n-Fold charges $0.75 per pound of laundry.

Calculator Explorations 3.4

1. 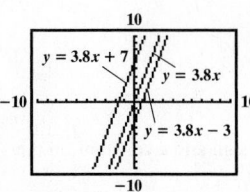   **3.**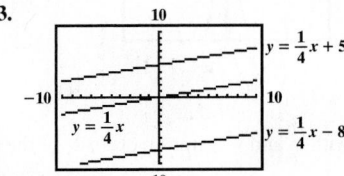

Vocabulary, Readiness & Video Check 3.4

1. slope   **3.** 0   **5.** positive   **7.** y; x **9.** solve the equation for y; the slope is the coefficient of x **11.** slope-intercept form; this form makes the slope easy to see, and you need to compare slopes to determine if two lines are parallel or perpendicular

Exercise Set 3.4

1. -1 **3.** undefined   **5.** $-\dfrac{2}{3}$ **7.** 0   **9.** $m = -\dfrac{4}{3}$ **11.** undefined slope   **13.** $m = \dfrac{5}{2}$ **15.** negative   **17.** undefined   **19.** upward
21. horizontal   **23.** line 1   **25.** line 2   **27.** D   **29.** B   **31.** E   **33.** undefined slope   **35.** $m = 0$ **37.** undefined slope
39. $m = 0$ **41.** $m = 5$ **43.** $m = -0.3$ **45.** $m = -2$ **47.** $m = \dfrac{2}{3}$ **49.** undefined slope   **51.** $m = \dfrac{1}{2}$ **53.** $m = 0$ **55.** $m = -\dfrac{3}{4}$
57. $m = 4$ **59. a.** 1   **b.** -1 **61. a.** $\dfrac{9}{11}$ **b.** $-\dfrac{11}{9}$ **63.** neither   **65.** neither   **67.** parallel   **69.** perpendicular   **71.** $\dfrac{3}{5}$ **73.** 12.5%
75. 40%   **77.** 37%; 35%   **79.** $m = 1.4$ or $\dfrac{1.4}{1}$; every 1 year, there are 1.4 million more U.S. households with computers.   **81.** $m = 0.47$ or $\dfrac{0.47}{1}$; It
costs $0.47 per 1 mile to own and operate a compact car.   **83.** $y = 2x - 14$ **85.** $y = -6x - 11$ **87.** $m = \dfrac{1}{2}$ **89.** answers may vary
91. 2005–2006; 2010–2011   **93.** 2002; 29 mi per gallon   **95.** from 2011 to 2012   **97.** $x = 6$ **99. a.** $(2004, 27,000)$; $(2014, 29,500)$
b. 250 or $\dfrac{250}{1}$ **c.** For the years 2004 through 2014, the number of organ transplants increased at a rate of 250 per 1 year.   **101.** The slope through
$(-3, 0)$ and $(1, 1)$ is $\dfrac{1}{4}$. The slope through $(-3, 0)$ and $(-4, 4)$ is -4. The product of the slopes is -1, so the sides are perpendicular.
103. -0.25 **105.** 0.875   **107.** The line becomes steeper.

Integrated Review

1. $m = 2$ **2.** $m = 0$ **3.** $m = -\dfrac{2}{3}$ **4.** undefined slope   **5.**   **6.**   **7.**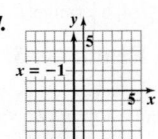

8. 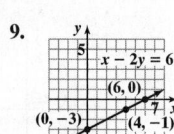 **9.** 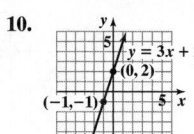 **10.** 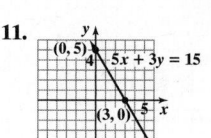 **11.** 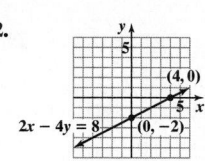 **12.**

13. parallel **14.** neither **15. a.** $(0, 490)$ **b.** In 2002, there were 490 thousand bridges on public roads. **c.** 4.09 or $\dfrac{4.09}{1}$
d. For the years 2002 through 2013, the number of bridges on public roads increased at a rate of 4.09 thousand per 1 year. **16. a.** $(4, 43.9)$
b. In 2013, the revenue for online advertising was $43.9 billion.

Section 3.5
Practice Exercises

1. 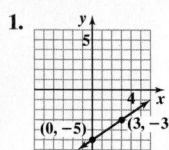 **2.** 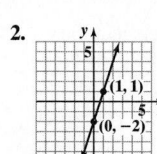 **3.** $y = \dfrac{1}{2}x + 7$ **4.** $4x - y = 5$ **5.** $5x + 4y = 19$ **6.** $x = 3$ **7.** $y = 3$

8. a. $y = -1500x + 195{,}000$ **b.** $105{,}000$

Calculator Explorations 3.5

1. 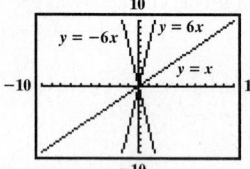 **3.** 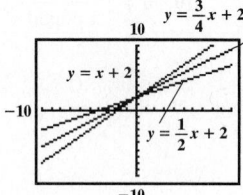 **5.**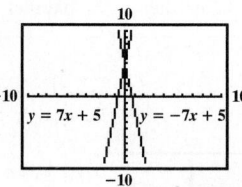

Vocabulary, Readiness & Video Check 3.5

1. slope-intercept; m; b **3.** y-intercept; fraction **5.** Write the equation with x- and y- terms on one side of the equal sign and a constant on the other side. **7.** Example 6: $y = -3$; Example 7: $x = -2$

Exercise Set 3.5

1. 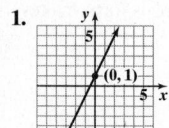 **3.** 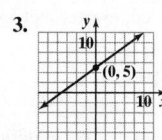 **5.** 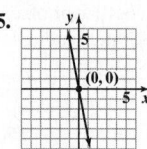 **7.** 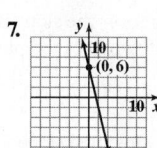 **9.** 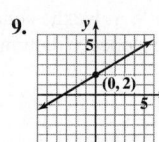 **11.**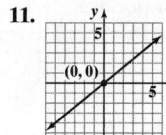

13. $y = 5x + 3$ **15.** $y = -4x - \dfrac{1}{6}$ **17.** $y = \dfrac{2}{3}x$ **19.** $y = -8$ **21.** $y = -\dfrac{1}{5}x + \dfrac{1}{9}$ **23.** $-6x + y = -10$ **25.** $8x + y = -13$
27. $3x - 2y = 27$ **29.** $x + 2y = -3$ **31.** $2x - y = 4$ **33.** $8x - y = -11$ **35.** $4x - 3y = -1$ **37.** $8x + 13y = 0$ **39.** $x = 0$
41. $y = 3$ **43.** $x = -\dfrac{7}{3}$ **45.** $y = 2$ **47.** $y = 5$ **49.** $x = 6$ **51.** $y = -\dfrac{1}{2}x + \dfrac{5}{3}$ **53.** $y = -x + 17$ **55.** $x = -\dfrac{3}{4}$
57. $y = x + 16$ **59.** $y = -5x + 7$ **61.** $y = 7$ **63.** $y = \dfrac{3}{2}x$ **65.** $y = -3$ **67.** $y = -\dfrac{4}{7}x - \dfrac{18}{7}$ **69. a.** $s = 32t$ **b.** 128 ft/sec
71. a. $y = 2x + 66$ **b.** 80 thousand cars **73. a.** $y = 0.56x + 85$ **b.** 91.7 persons per sq mi **75. a.** $y = -1.1x + 55$
b. 38.5 million **77. a.** $S = -1000p + 13{,}000$ **b.** 9500 Fun Noodles **79.** -1 **81.** 5 **83.** no **85.** yes **87.** point-slope
89. slope-intercept **91.** horizontal **93.** answers may vary **95. a.** $3x - y = -5$ **b.** $x + 3y = 5$ **97. a.** $3x + 2y = -1$ **b.** $2x - 3y = 21$

Section 3.6
Practice Exercises

1. domain: $\{0, 1, 5\}$; range: $\{-2, 0, 3, 4\}$ **2. a.** function **b.** not a function **3. a.** not a function **b.** function **4. a.** function
b. function **c.** function **d.** not a function **5. a.** function **b.** function **c.** function **d.** not a function **6. a.** $69°F$ **b.** February
c. yes **7. a.** $h(2) = 9$; $(2, 9)$ **b.** $h(-5) = 30$; $(-5, 30)$ **c.** $h(0) = 5$; $(0, 5)$ **8. a.** domain: $(-\infty, \infty)$ **b.** domain: $(-\infty, 0) \cup (0, \infty)$
9. a. domain: $[-4, 6]$; range: $[-2, 3]$ **b.** domain: $(-\infty, \infty)$; range: $(-\infty, 3]$

Vocabulary, Readiness & Video Check 3.6

1. relation **3.** range **5.** vertical **7.** A relation is a set of ordered pairs and an equation in two variables defines a set of ordered pairs. Therefore, an equation in two variables can also define a relation. **9.** A vertical line represents one x-value paired with many y-values. A function only allows an x-value paired with exactly one y-value, so if a vertical line intersects a graph more than once, there's an x-value paired with more than one y-value, and we don't have a function.

Exercise Set 3.6

1. $\{-7, 0, 2, 10\}$; $\{-7, 0, 4, 10\}$ **3.** $\{0, 1, 5\}$; $\{-2\}$ **5.** yes **7.** no **9.** no **11.** yes **13.** yes **15.** no **17.** yes **19.** yes **21.** yes **23.** no **25.** no **27.** 9:30 p.m. **29.** January 1 and December 1 **31.** yes; it passes the vertical line test **33.** $4.25 per hour **35.** 2009 **37.** yes; answers may vary **39.** $1.80 **41.** more than 5 ounces and less than or equal to 6 ounces **43.** yes; answers may vary

45. $-9, -5, 1$ **47.** $6, 2, 11$ **49.** $-6, 0, 9$ **51.** $2, 0, 3$ **53.** $5, 0, -20$ **55.** $5, 3, 35$ **57.** $(3, 6)$ **59.** $\left(0, -\dfrac{1}{2}\right)$ **61.** $(-2, 9)$

63. $(-\infty, \infty)$ **65.** all real number except -5 or $(-\infty, -5) \cup (-5, \infty)$ **67.** $(-\infty, \infty)$ **69.** domain: $(-\infty, \infty)$; range: $[-4, \infty)$ **71.** domain: $(-\infty, \infty)$; range: $(-\infty, \infty)$ **73.** domain: $(-\infty, \infty)$; range: $\{2\}$ **75.** -1 **77.** -1 **79.** $-1, 5$ **81.** $(-2, 1)$ **83.** $(-3, -1)$ **85.** $f(-5) = 12$ **87.** $(3, -4)$ **89.** $f(5) = 0$ **91. a.** 166.38 cm **b.** 148.25 cm **93.** answers may vary **95.** $f(x) = x + 7$ **97. a.** 11 **b.** $2a + 7$ **99. a.** 16 **b.** $a^2 + 7$

Chapter 3 Vocabulary Check

1. solution **2.** y-axis **3.** linear **4.** x-intercept **5.** standard **6.** y-intercept **7.** slope-intercept **8.** point-slope **9.** y **10.** x-axis **11.** x **12.** slope **13.** function **14.** domain **15.** range **16.** relation

Chapter 3 Review

1. **3.** **5.** **7. a.** $(5.00, 50), (8.50, 100), (20.00, 250), (27.00, 500)$ **b.**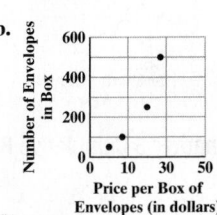

9. no; yes **11.** yes; yes **13.** $(7, 44)$ **15.** $(-3, 0)$; $(1, 3)$; $(9, 9)$ **17.** $2005; 2500; 7000$ **19.**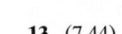

21. **23.** **25.** **27.** $(4, 0), (0, -2)$ **29.** $(-2, 0), (2, 0), (0, 2), (0, -2)$

31. **33.** **35.** **37.** **39.** $m = -\dfrac{3}{4}$ **41.** D **43.** C **45.** $\dfrac{3}{4}$ **47.** 4

49. 3 **51.** 0 **53.** perpendicular **55.** neither **57.** $m = 24$ or $\dfrac{24}{1}$; every 1 year, 24 thousand (24,000) more students graduate with an associate's degree. **59.** $m = -3; (0, 7)$ **61.** $m = 0; (0, 2)$ **63.** **65.** **67.** $y = -5x + \dfrac{1}{2}$ **69.** C

71. B **73.** $(0, 1859)$ **75.** $3x + y = -5$ **77.** $y = -3$ **79.** $6x + y = 11$ **81.** $x + y = 6$ **83.** $x = 5$ **85.** $x = 6$ **87.** no **89.** yes **91.** no **93.** no **95. a.** 6 **b.** 10 **c.** 5 **97. a.** 45 **b.** -35 **c.** 0 **99.** $(-\infty, \infty)$ **101.** domain: $[-3, 5]$ range: $[-4, 2]$ **103.** domain: $\{3\}$; range: $(-\infty, \infty)$ **105.** $7; -1; -3$ **107.** $(3, 0); (0, -2)$ **109.** **111.** **113.**

115. $m = -1$ **117.** $m = 2$ **119.** $m = \dfrac{2}{3}; (0, -5)$ **121.** $5x + y = 8$ **123.** $4x + y = -3$ **125.** $y = 200x + 3800$

Chapter 3 Getting Ready for the Test

1. C **2.** A **3.** B **4.** B **5.** B **6.** A **7.** D **8.** C **9.** C **10.** C **11.** B **12.** D **13.** A or C; A or C **14.** E

Chapter 3 Test

1. 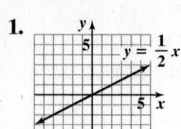 **2.** 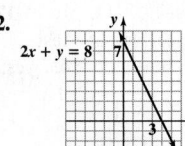 **3.** 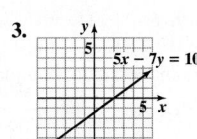 **4.** 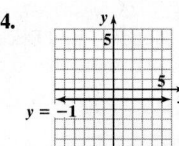 **5.**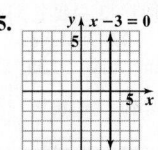

6. $\frac{2}{5}$ **7.** 0 **8.** -1 **9.** 3 **10.** undefined **11.** $m = \frac{7}{3}; \left(0, -\frac{2}{3}\right)$ **12.** neither **13.** $x + 4y = 10$ **14.** $7x + 6y = 0$

15. $8x + y = 11$ **16.** $x = -5$ **17.** $x - 8y = -96$ **18.** yes **19.** no **20. a.** 0 **b.** 0 **c.** 60 **21.** all real numbers except -1 or $(-\infty, -1) \cup (-1, \infty)$ **22. a.** x-intercepts: $(0, 0), (4, 0); y$-intercept: $(0, 0)$ **b.** domain: $(-\infty, \infty);$ range: $(-\infty, 4]$ **23. a.** x-intercept: $(2, 0);$ y-intercept; $(0, -2)$ **b.** domain: $(-\infty, \infty);$ range: $(-\infty, \infty)$ **24.** $(7, 20)$ **25.** 210 liters **26.** 490 liters **27.** July **28.** 63°F **29.** January, February, March, November, December **30. a.** $(2008, 13.68), (2009, 14.1), (2010, 14.13), (2011, 13.97), (2012, 14.06), (2013, 14.06), (2014, 13.98)$

b. 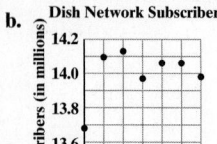 **31. a.** $m = -18$ or $\frac{-18}{1};$ For every 1 year 18 million fewer tickets are sold. **b.** $(10, 1342), (14, 1270)$ **c.** $y = -18x + 1522$

d. In 2020, we predict that 1162 million movie tickets will be sold in the U.S. and Canada.

Chapter 3 Cumulative Review

1. a. $<$ **b.** $>$ **c.** $>;$ Sec. 1.2, Ex. 1 **3.** $\frac{2}{39};$ Sec. 1.3, Ex. 3 **5.** $\frac{8}{3};$ Sec. 1.4, Ex. 3 **7. a.** -19 **b.** 30 **c.** -0.5 **d.** $-\frac{4}{5}$ **e.** 6.7 **f.** $\frac{1}{40};$ Sec. 1.5, Ex. 6 **9. a.** -6 **b.** 6.3; Sec. 1.6, Ex. 4 **11. a.** -6 **b.** 0 **c.** $\frac{3}{4};$ Sec. 1.7, Ex. 10 **13. a.** $22 + x$ **b.** $-21x;$ Sec. 1.8, Ex. 3 **15. a.** -3 **b.** 22 **c.** 1 **d.** -1 **e.** $\frac{1}{7};$ Sec. 2.1, Ex. 1 **17.** 17; Sec. 2.2, Ex. 1 **19.** 6; Sec. 2.2, Ex. 5 **21.** $3x + 3;$ Sec. 2.2, Ex. 10

23. 0; Sec. 2.3, Ex. 4 **25.** 247 Republicans, 188 Democrats; Sec. 2.4, Ex. 4 **27.** 40 ft; Sec. 2.5, Ex. 2 **29.** $\frac{y - b}{m} = x;$ Sec. 2.5, Ex. 6

31. 40% solution: 8 liters; 70% solution: 4 liters; Sec. 2.6, Ex. 7 **33.** $[-1, \infty);$ Sec. 2.8, Ex. 1 **35.** $[1, 4);$ Sec. 2.8, Ex. 10

37. a. solution **b.** not a solution **c.** solution; Sec. 3.1, Ex. 5 **39. a.** yes **b.** yes **c.** no **d.** yes; Sec. 3.2, Ex. 1 **41.** 0; Sec. 3.4, Ex. 6

43. $y = \frac{1}{4}x - 3;$ Sec. 3.5, Ex. 3

CHAPTER 4 SOLVING SYSTEMS OF LINEAR EQUATIONS

Section 4.1

Practice Exercises

1. no **2.** yes **3.** $(8, 5)$ **4.** $(-3, -5)$ **5.** no solution; inconsistent system; { } or $\varnothing$

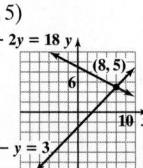 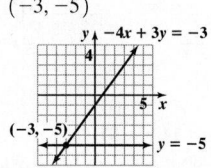 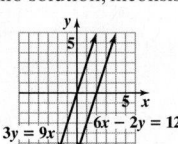

6. infinite number of solutions; consistent, system, $\{(x, y) \mid x - y = 4\}$ or $\{(x, y) \mid -2x + 2y = -8\}$ **7.** one solution **8.** no solution

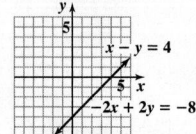

Calculator Explorations 4.1

1. $(0.37, 0.23)$ **3.** $(0.03, -1.89)$

Vocabulary, Readiness & Video Check 4.1

1. dependent **3.** consistent **5.** inconsistent **7.** The ordered pair must satisfy all equations of the system in order to be a solution of the system, so we must check that the ordered pair is a solution of both equations. **9.** Writing the equations of a system in slope-intercept form lets you see their slope and y-intercept. Different slopes mean one solution; same slope with different y-intercepts means no solution; same slope with same y-intercept means infinite number of solutions.

Exercise Set 4.1

1. one solution, $(-1, 3)$ **3.** infinite number of solutions **5. a.** no **b.** yes **7. a.** yes **b.** no **9. a.** yes **b.** yes
11. a. no **b.** no **13.** 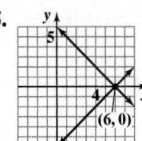 **15.** 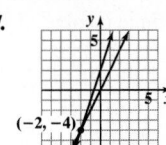 **17.** 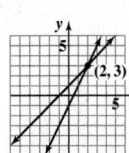 **19.** 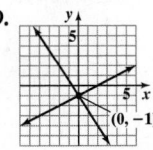 **21.**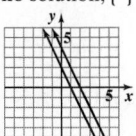

23. 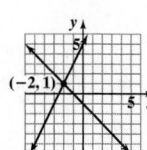 **25.** no solution; { } or $\varnothing$ 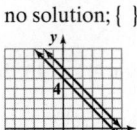 **27.** 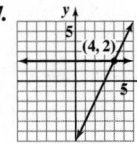 **29.** 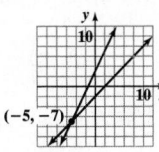 **31.** no solution; { } or $\varnothing$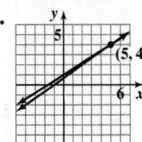

33. infinite number of solutions; 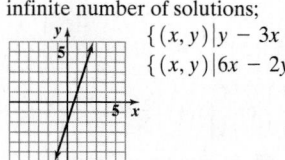 $\{(x, y) \mid y - 3x = -2\}$ or $\{(x, y) \mid 6x - 2y = 4\}$ **35.** 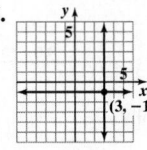 **37.** 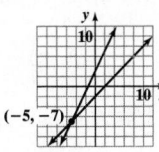 **39.**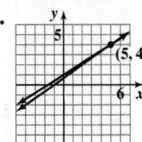

41. infinite number of solutions; 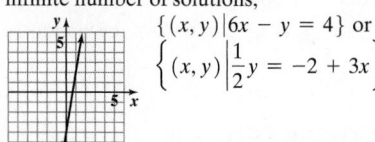 $\{(x, y) \mid 6x - y = 4\}$ or $\left\{(x, y) \mid \frac{1}{2}y = -2 + 3x\right\}$ **43. a.** intersecting; **b.** one solution **45. a.** parallel; **b.** no solution
47. a. identical lines; **b.** infinite number of solutions **49. a.** intersecting; **b.** one solution
51. a. intersecting; **b.** one solution **53. a.** identical lines; **b.** infinite number of solutions
55. a. parallel; **b.** no solution **57.** 2 **59.** $-\dfrac{2}{5}$ **61.** 2

63. answers may vary; possible answer **65.** answers may vary; possible answer **67.** answers may vary **69.** 2009, 2010
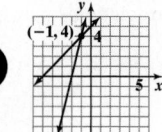 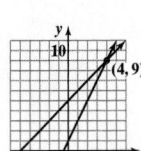

71. 2006, 2007, 2012, 2013, 2014 **73.** answers may vary **75. a.** $(4, 9)$ **b.** [graph] **c.** yes **77.** answers may vary

Section 4.2

Practice Exercises

1. $(8, 7)$ **2.** $(-3, -6)$ **3.** $\left(4, \dfrac{2}{3}\right)$ **4.** $(-3, 2)$ **5.** infinite number of solutions; $\left\{(x, y) \mid \dfrac{1}{4}x - y = 2\right\}$ or $\{(x, y) \mid x = 4y + 8\}$
6. no solution

Vocabulary, Readiness & Video Check 4.2

1. $(1, 4)$ **3.** infinite number of solutions **5.** $(0, 0)$ **7.** You solved one equation for a variable. Now be sure to substitute this expression for the variable into the *other* equation.

Exercise Set 4.2

1. $(2, 1)$ **3.** $(-3, 9)$ **5.** $(2, 7)$ **7.** $\left(-\dfrac{1}{5}, \dfrac{43}{5}\right)$ **9.** $(2, -1)$ **11.** $(-2, 4)$ **13.** $(4, 2)$ **15.** $(-2, -1)$ **17.** no solution; { } or $\varnothing$
19. $(3, -1)$ **21.** $(3, 5)$ **23.** $\left(\dfrac{2}{3}, -\dfrac{1}{3}\right)$ **25.** $(-1, -4)$ **27.** $(-6, 2)$ **29.** $(2, 1)$ **31.** no solution; { } or $\varnothing$ **33.** infinite number of
solutions; $\left\{(x, y) \mid \dfrac{1}{3}x - y = 2\right\}$ or $\{(x, y) \mid x - 3y = 6\}$ **35.** $\left(\dfrac{1}{2}, 2\right)$ **37.** $(1, -3)$ **39.** $-6x - 4y = -12$ **41.** $-12x + 3y = 9$ **43.** $5n$

45. $-15b$ **47.** answers may vary **49.** no; answers may vary **51.** c; answers may vary **53. a.** $(13, 492)$ **b.** In $1970 + 13 = 1983$, the number
of men and women receiving bachelor's degrees was the same. **c.** answers may vary [graph: $y = 14.2x + 314$, $y = 3.9x + 443$] **55.** $(-2.6, 1.3)$ **57.** $(3.28, 2.11)$

Section 4.3

Practice Exercises

1. $(5, 3)$ **2.** $(3, -4)$ **3.** no solution; $\{\ \}$ or $\varnothing$ **4.** infinite number of solutions; $\{(x, y) | 4x - 3y = 5\}$ or $\{(x, y) | -8x + 6y = -10\}$ **5.** $(2, $

6. $\left(-\dfrac{8}{5}, \dfrac{6}{5}\right)$

Vocabulary, Readiness & Video Check 4.3

1. false **3.** true **5.** The multiplication property of equality; be sure to multiply *both* sides of the equation by the number chosen.

Exercise Set 4.3

1. $(1, 2)$ **3.** $(2, -3)$ **5.** $(-2, -5)$ **7.** $(5, -2)$ **9.** $(-7, 5)$ **11.** $(6, 0)$ **13.** no solution; $\{\ \}$ or $\varnothing$ **15.** infinite number of solutions;
$\{(x, y) | -x + 5y = -1\}$ or $\{(x, y) | 3x - 15y = 3\}$ **17.** $\left(2, -\dfrac{1}{2}\right)$ **19.** $(-2, 0)$ **21.** $(1, -1)$ **23.** no solution; $\{\ \}$ or $\varnothing$ **25.** $\left(\dfrac{12}{11}, -\dfrac{4}{11}\right)$

27. $\left(\dfrac{3}{2}, 3\right)$ **29.** infinite number of solutiotns; $\left\{(x, y) \left| \dfrac{10}{3}x + 4y = -4 \right.\right\}$ or $\{(x, y) | 5x + 6y = -6\}$ **31.** $(1, 6)$ **33.** $\left(-\dfrac{1}{2}, -2\right)$ **35.** infinite

number of solutions; $\left\{(x, y) \left| \dfrac{x}{3} - y = 2 \right.\right\}$ or $\left\{(x, y) \left| -\dfrac{x}{2} + \dfrac{3y}{2} = -3 \right.\right\}$ **37.** $\left(-\dfrac{2}{3}, \dfrac{2}{5}\right)$ **39.** $(2, 4)$ **41.** $(-0.5, 2.5)$ **43.** $(2, 5)$

45. $(-3, 2)$ **47.** $(0, 3)$ **49.** $(5, 7)$ **51.** $\left(\dfrac{1}{3}, 1\right)$ **53.** infinite number of solutions; $\left\{(x, y) \left| \dfrac{x + 2}{2} = \dfrac{y + 11}{3} \right.\right\}$ or $\left\{(x, y) \left| \dfrac{x}{2} = \left(\dfrac{2y + 16}{6}\right) \right.\right\}$

55. $(-8.9, 10.6)$ **57.** $2x + 6 = x - 3$ **59.** $20 - 3x = 2$ **61.** $4(n + 6) = 2n$ **63.** $2; 6x - 2y = -24$ **65.** b; answers may vary
67. answers may vary **69. a.** $b = 15$ **b.** any real number except 15 **71.** $(-4.2, 9.6)$ **73. a.** $(10, 67)$ or $(10, 68)$
b. In 2022 $(2012 + 10)$, the percent of workers age 20–24 and the percent of workers age 55–64 will be the same. **c.** 67% or 68% of the workforce
for each of these age groups

Integrated Review

1. $(2, 5)$ **2.** $(4, 2)$ **3.** $(5, -2)$ **4.** $(6, -14)$ **5.** $(-3, 2)$ **6.** $(-4, 3)$ **7.** $(0, 3)$ **8.** $(-2, 4)$ **9.** $(5, 7)$ **10.** $(-3, -23)$
11. $\left(\dfrac{1}{3}, 1\right)$ **12.** $\left(-\dfrac{1}{4}, 2\right)$ **13.** no solution; $\{\ \}$ or $\varnothing$ **14.** infinite number of solutions; $\{(x, y) | -x + 2y = 3\}$ or $\{(x, y) | 3x - 6y = -9\}$
15. $(0.5, 3.5)$ **16.** $(-0.75, 1.25)$ **17.** infinite number of solutions; $\{(x, y) | x = 3y - 7\}$ or $\{(x, y) | 2x - 6y = -14\}$ **18.** no solution; $\{\ \}$ or $\varnothing$
19. $(7, -3)$ **20.** $(-1, -3)$ **21.** answers may vary **22.** answers may vary

Section 4.4

Practice Exercises

1. $(-1, 2, 1)$ **2.** $\{\ \}$ or $\varnothing$ **3.** $\left(\dfrac{2}{3}, -\dfrac{1}{2}, 0\right)$ **4.** $\{(x, y, z) | 2x + y - 3z = 6\}$ **5.** $(6, 15, -5)$

Vocabulary, Readiness & Video Check 4.4

1. a, b, d **3.** yes; answers may vary **5.** Once we have one equation in two variables, we need to get another equation in the *same* two variables,
giving us a system of two equations in two variables. We solve this new system to find the value of two variables. We then substitute these values into an
original equation to find the value of the third.

Exercise Set 4.4

1. $(-1, 5, 2)$ **3.** $(-2, 5, 1)$ **5.** $(-2, 3, -1)$ **7.** $\{(x, y, z) | x - 2y + z = -5\}$ **9.** $\varnothing$ **11.** $(0, 0, 0)$ **13.** $(-3, -35, -7)$
15. $(6, 22, -20)$ **17.** $\varnothing$ **19.** $(3, 2, 2)$ **21.** $\{(x, y, z) | x + 2y - 3z = 4\}$ **23.** $(-3, -4, -5)$ **25.** $\left(0, \dfrac{1}{2}, -4\right)$ **27.** $(12, 6, 4)$
29. 15 and 30 **31.** 5 **33.** $-\dfrac{5}{3}$ **35.** answers may vary **37.** answers may vary **39.** $(1, 1, -1)$ **41.** $(1, 1, 0, 2)$
43. $(1, -1, 2, 3)$ **45.** answers may vary

Section 4.5

Practice Exercises

1. a. 2017 **b.** yes; answers may vary **2.** 18, 12 **3.** 17 and 12 **4. a.** Adult: $19 **b.** Child: $6 **c.** No, the regular rates are less than the
group rate **5.** Atlantique: 500 kph; V150: 575 kph **6.** 0.95 liter of water; 0.05 liter of 99% HCl **7.** 1500 packages **8.** $40°, 60°, 80°$

Vocabulary, Readiness & Video Check 4.5

1. Up to now we've been choosing one variable/unknown and translating to one equation. To solve by a system of equations, we'll choose two variables
to represent two unknowns and translate to two equations. **3.** The ordered triple still needs to be interpreted in the context of the application. Each
value actually represents the angle measure of a triangle, in degrees.

Exercise Set 4.5

1. c **3.** b **5.** a **7.** $\begin{cases} x + y = 15 \\ x - y = 7 \end{cases}$ **9.** $\begin{cases} x + y = 6500 \\ x = y + 800 \end{cases}$ **11.** 33 and 50 **13.** 10 and 8 **15.** 14 and -3 **17.** Gonzalez: 116; Trout: 111
19. child's ticket: $18; adult's ticket: $29 **21.** quarters: 53; nickels: 27 **23.** McDonald's: $99.50; The Ohio Art Company: $7.50 **25.** daily fee: $32;
mileage charge: $0.25 per mi **27.** distance downstream = distance upstream = 18 mi; time downstream: 2 hr; time upstream: $4\dfrac{1}{2}$ hr; still water: 6.5 mph;

current: 2.5 mph   **29.** still air: 455 mph; wind: 65 mph   **31.** $4\frac{1}{2}$ hr   **33.** 12% solution: $7\frac{1}{2}$ liters; 4% solution: $4\frac{1}{2}$ liters   **35.** $4.95 beans: 113 lb;

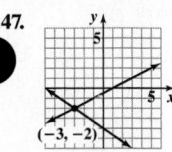 2.65 beans: 87 lb   **37.** $60°, 30°$ **39.** $20°, 70°$ **41.** number sold at $9.50: 23; number sold at $7.50: 67   **43.** $2\frac{1}{4}$ mph and $2\frac{3}{4}$ mph

45. 30%: 50 gal; 60%: 100 gal   **47.** length: 42 in.; width: 30 in.   **49. a.** answers may vary   **b.** 2008   **51. a.** 2012   **b.** answers may vary
53. $x = 75; y = 105$ **55.** 625 units   **57.** 3000 units   **59.** 1280 units   **61. a.** $R(x) = 450x$ **b.** $C(x) = 200x + 6000$ **c.** 24 desks
63. 2 units of Mix A; 3 units of Mix B; 1 unit of Mix C   **65.** 5 in.; 7 in.; 7 in.; 10 in.   **67.** 18, 13, and 9   **69.** free throws; 703 2-pt field goals;

657 3-pt fields goals; 192   **71.** $x = 60; y = 55; z = 65$ **73.** $(3, \infty)$ **75.** $\left[\frac{1}{2}, \infty\right)$ **77.** a   **79.** width: 9 ft; length: 15 ft   **81.** $a = 3, b = 4, c = -1$

83. $a = 0.5, b = 24.5, c = 849$; 2015: 1774 thousand students   **85.** $(10, 13)$

Chapter 4 Vocabulary Check

1. dependent   **2.** system of linear equations   **3.** consistent   **4.** solution   **5.** addition; substitution   **6.** inconsistent   **7.** independent

Chapter 4 Review

1. a. no   **b.** yes   **c.** no   **3. a.** no   **b.** no   **c.** yes   **5.**   **7.**   **9.**

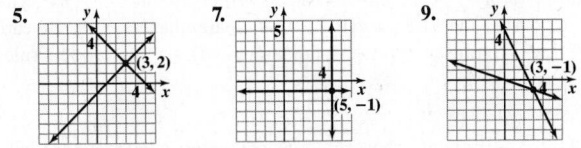

11. no solution; $\{ \}$ or $\varnothing$ **13.** $(-1, 4)$ **15.** $(3, -2)$ **17.** infinite number of solutions; $\{(x, y) | 4y = 2x + 6\}$ or $\{(x, y) | x - 2y = -3\}$
19. no solution; $\{ \}$ or $\varnothing$ **21.** $(-6, 2)$ **23.** $(3, 7)$

25. infinite number of solutions; $\{(x, y) | 2x - 6y = -1\}$ or $\left\{(x, y) \left| -x + 3y = \frac{1}{2}\right.\right\}$ **27.** $(8, -6)$ **29.** $(2, 0, 2)$ **31.** $\left(-\frac{1}{2}, \frac{3}{4}, 1\right)$ **33.** $\varnothing$

35. $(1, 1, -2)$ **37.** -6 and 22 **39.** current of river: 3.2 mph; speed in still water: 21.1 mph   **41.** egg: $0.40; strip of bacon: $0.65
43. 17 pennies; 20 nickels; 16 dimes   **45.** two sides: 22 cm each; third side: 29 cm

47.   **49.** $(3, 2)$ **51.** $\left(1\frac{1}{2}, -3\right)$ **53.** infinite number of solutions; $\{(x, y) | 3x - y = 4\}$ or $\{(x, y) | 4y = 12x - 16\}$ **55.** $(-5, 2)$
57. $(-1, 3, 5)$ **59.** 4 and 8   **61.** 24 nickels and 41 dimes   **63.** 28 units, 42 units, 56 units

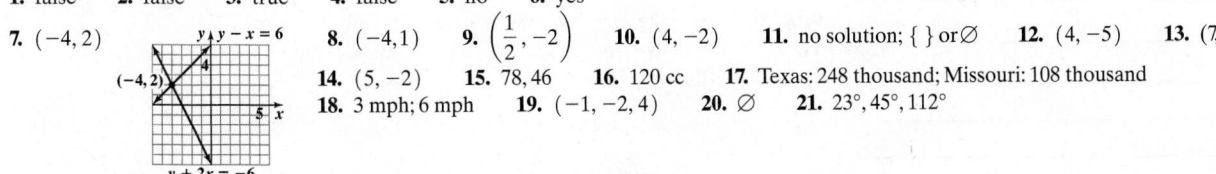

Chapter 4 Getting Ready for the Test

1. B   **2.** C   **3.** B   **4.** A   **5.** D   **6.** B   **7.** D   **8.** C   **9.** B   **10.** B   **11.** A   **12.** C

Chapter 4 Test

1. false   **2.** false   **3.** true   **4.** false   **5.** no   **6.** yes

7. $(-4, 2)$ **8.** $(-4, 1)$ **9.** $\left(\frac{1}{2}, -2\right)$ **10.** $(4, -2)$ **11.** no solution; $\{ \}$ or $\varnothing$ **12.** $(4, -5)$ **13.** $(7, 2)$

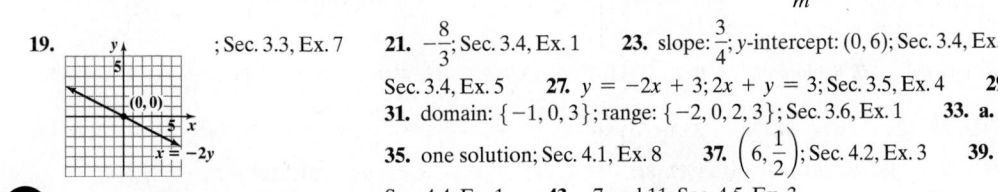

14. $(5, -2)$ **15.** 78, 46   **16.** 120 cc   **17.** Texas: 248 thousand; Missouri: 108 thousand
18. 3 mph; 6 mph   **19.** $(-1, -2, 4)$ **20.** $\varnothing$ **21.** $23°, 45°, 112°$

Chapter 4 Cumulative Review

1. a. $<$ **b.** $=$ **c.** $>$; Sec. 1.2, Ex. 6   **3. a.** commutative property of multiplication   **b.** associative property of addition
c. identity element for addition   **d.** commutative property of multiplication   **e.** multiplicative inverse property   **f.** additive inverse
property   **g.** commutative and associative properties of multiplication; Sec. 1.8, Ex. 6   **5.** $-2x - 1$; Sec. 2.1, Ex. 7   **7.** 8; Sec. 2.2, Ex. 4

9. 6; Sec. 2.2, Ex. 5   **11.** 12; Sec. 2.3, Ex. 3   **13.** 10; Sec. 2.4, Ex. 2   **15.** $x = \dfrac{y - b}{m}$; Sec. 2.5, Ex. 6   **17.** $[2, \infty)$; Sec. 2.8, Ex. 3

19.   ; Sec. 3.3, Ex. 7   **21.** $-\dfrac{8}{3}$; Sec. 3.4, Ex. 1   **23.** slope: $\dfrac{3}{4}$; y-intercept: $(0, 6)$; Sec. 3.4, Ex. 3   **25.** slope: $\dfrac{3}{4}$; y-intercept: $(0, -1)$;

Sec. 3.4, Ex. 5   **27.** $y = -2x + 3$; $2x + y = 3$; Sec. 3.5, Ex. 4   **29.** $x = -1$; Sec. 3.5, Ex. 6
31. domain: $\{-1, 0, 3\}$; range: $\{-2, 0, 2, 3\}$; Sec. 3.6, Ex. 1   **33. a.** function   **b.** not a function; Sec. 3.6, Ex. 2
35. one solution; Sec. 4.1, Ex. 8   **37.** $\left(6, \dfrac{1}{2}\right)$; Sec. 4.2, Ex. 3   **39.** $(6, 1)$; Sec. 4.3, Ex. 1   **41.** $(-4, 2, -1)$;

Sec. 4.4, Ex. 1   **43.** 7 and 11; Sec. 4.5, Ex. 3

CHAPTER 5 EXPONENTS AND POLYNOMIALS

Section 5.1

Practice Exercises

1. a. 27 **b.** 4 **c.** 64 **d.** -64 **e.** $\dfrac{27}{64}$ **f.** 0.0081 **g.** 75 **2. a.** 243 **b.** $\dfrac{3}{8}$ **3. a.** 3^{10} **b.** y^5 **c.** z^5 **d.** x^{11} **e.** $(-2)^8$
f. $b^3 \cdot t^5$ **4.** $15y^7$ **5. a.** $y^{12}z^4$ **b.** $-7m^5n^{14}$ **6. a.** z^{21} **b.** 4^{18} **c.** $(-2)^{15}$ **7. a.** p^5r^5 **b.** $36b^2$ **c.** $\dfrac{1}{64}x^6y^3$ **d.** $81a^{12}b^{16}c^4$
8. a. $\dfrac{x^5}{y^{10}}$ **b.** $\dfrac{32a^{20}}{b^{15}}$ **9. a.** z^4 **b.** 25 **c.** 64 **d.** $\dfrac{q^5}{t^2}$ **e.** $6x^2y^2$ **10. a.** -1 **b.** 1 **c.** 1 **d.** 1 **e.** 1 **f.** 7 **11. a.** $\dfrac{z^2}{144}$
b. $64x^{18}$ **c.** y^{13} **12. a.** 63 **b.** 2 **c.** $\dfrac{x^6}{9y^8}$ **d.** $-z^{30}$

Vocabulary, Readiness & Video Check 5.1

1. exponent **3.** add **5.** 1 **7.** Example 4 can be written as $-4^2 = -1 \cdot 4^2$, which is similar to Example 7, $4 \cdot 3^2$, and shows why the negative sign should not be considered part of the base when there are no parentheses. **9.** Be careful not to confuse the power rule with the product rule. The power rule involves a power raised to a power (exponents are multiplied), and the product rule involves a product (exponents are added). **11.** the quotient rule

Exercise Set 5.1

1. exponent: 2; base: 3 **3.** exponent: 2; base: 4 **5.** exponent: 2; base: x **7.** 49 **9.** -5 **11.** -16 **13.** 16 **15.** 0.00001 **17.** $\dfrac{1}{81}$
19. 224 **21.** -250 **23.** 4 **25.** 135 **27.** 150 **29.** $\dfrac{32}{5}$ **31.** x^7 **33.** $(-3)^{12}$ **35.** $15y^5$ **37.** $x^{19}y^6$ **39.** $-72m^3n^8$ **41.** $-24z^{20}$
43. $20x^5$ sq ft **45.** x^{36} **47.** p^8q^8 **49.** $8a^{15}$ **51.** $x^{10}y^{15}$ **53.** $49a^4b^{10}c^2$ **55.** $\dfrac{r^9}{s^9}$ **57.** $\dfrac{m^5p^5}{n^5}$ **59.** $\dfrac{4x^2z^2}{y^{10}}$ **61.** $64z^{10}$ sq dm
63. $27y^{12}$ cu ft **65.** x^2 **67.** -64 **69.** p^6q^5 **71.** $\dfrac{y^3}{2}$ **73.** 1 **75.** 1 **77.** -7 **79.** 2 **81.** -81 **83.** $\dfrac{1}{64}$ **85.** b^6 **87.** a^9
89. $-16x^7$ **91.** $a^{11}b^{20}$ **93.** $26m^9n^7$ **95.** z^{40} **97.** $64a^3b^3$ **99.** $36x^2y^2z^6$ **101.** $3x$ **103.** $81x^2y^2$ **105.** 9 **107.** $\dfrac{y^{15}}{8x^{12}}$ **109.** $2x^2y$
111. 2 **113.** $\dfrac{x^{18}}{4y^{22}}$ **115.** $-b^5$ **117.** $2y - 10$ **119.** $-x - 4$ **121.** $-x + 5$ **123.** C **125.** E **127.** answers may vary
129. answers may vary **131.** 343 cu m **133.** volume **135.** answers may vary **137.** answers may vary **139.** x^{9a} **141.** a^{5b} **143.** x^5
145. $1045.85

Section 5.2

Practice Exercises

1. a. degree 3 **b.** degree 2 **c.** degree 1 **d.** degree 8 **e.** degree 0 **2. a.** trinomial, degree 2 **b.** binomial, degree 1 **c.** none of these, degree 3
3.

Term	Numerical Coefficient	Degree of Term
$-3x^3y^2$	-3	5
$4xy^2$	4	3
$-y^2$	-1	2
$3x$	3	1
-2	-2	0

4. a. 4 **b.** -21 **5.** 114 ft; 66 ft **6. a.** $-2y$ **b.** $z + 5z^3$ **c.** $14x^3$ **d.** $4a^2 - 12$
e. $\dfrac{1}{3}x^4 + \dfrac{11}{24}x^3 - x^2$ **7.** $-3x^2 + 5xy + 5y^2$ **8.** $x^2 + 2x + 4 + 5x + 3x^2; 4x^2 + 7x + 4$
9. a. $5y^2 - 3y + 2x - 9$ **b.** $-8a^2b - 3ab^2$ **10.** $7x^3 - 6x + 7$
11. $2x^3 - 4x^2 + 4x - 6$ **12.** $7x - 11$ **13. a.** $-5a^2 - ab + 6b^2$
b. $3x^2y^2 - 10xy - 4xy^2 + 5 - 6y^2$

Graphing Calculator Explorations 5.2

1. $x^3 - 4x^2 + 7x - 8$ **3.** $-2.1x^2 - 3.2x - 1.7$ **5.** $7.69x^2 - 1.26x + 5.3$

Vocabulary, Readiness & Video Check 5.2

1. binomial **3.** trinomial **5.** constant **7.** The degree of the polynomial is the greatest degree of any of its terms, so we need to find the degree of each term first. **9.** simplifying it

Exercise Set 5.2

1. 1; binomial **3.** 3; none of these **5.** 6; trinomial **7.** 8; binomial **9.** 3 **11.** 2 **13.** 57 **15.** 499 **17.** 1 **19.** $-\dfrac{11}{16}$ **21.** 1134 ft
23. 1006 ft **25.** 2280 thousand **27.** $23x^2$ **29.** $12x^2 - y$ **31.** $7s$ **33.** $-1.1y^2 + 4.8$ **35.** $-\dfrac{7}{12}x^3 + \dfrac{7}{5}x^2 + 6$ **37.** $5a^2 - 9ab + 16b^2$
39. $-3x^2 + 10$ **41.** $-x^2 + 14$ **43.** $-2x + 9$ **45.** $2x^2 + 7x - 16$ **47.** $8t^2 - 4$ **49.** $-2z^2 - 16z + 6$ **51.** $2x^3 - 2x^2 + 7x + 2$
53. $-y^2 - 3y - 1$ **55.** $2x^2 + 11x$ **57.** $-16x^4 + 8x + 9$ **59.** $7x^2 + 14x + 18$ **61.** $3x - 3$ **63.** $7x^2 - 4x + 2$ **65.** $62x^2 + 5$
67. $7x^2 - 2x + 2$ **69.** $12x + 2$ **71.** $4y^2 + 12y + 19$ **73.** $4x^2 + 7x + x^2 + 5x; 5x^2 + 12x$ **75.** $-2a - b + 1$ **77.** $3x^2 + 5$
79. $6x^2 - 2xy + 19y^2$ **81.** $8r^2s + 16rs - 8 + 7r^2s^2$ **83.** $-5.42x^2 + 7.75x - 19.61$ **85.** $3.7y^4 - 0.7y^3 + 2.2y - 4$ **87.** $6x^2$ **89.** $-12x^8$
91. $200x^3y^2$ **93.** $18x + 44$ **95.** $(x^2 + 7x + 4)$ ft **97.** $(3y^2 + 4y + 11)$ m **99.** 3071 million users **101.** answers may vary
103. answers may vary **105.** B **107.** E **109. a.** $4z$ **b.** $3z^2$ **c.** $-4z$ **d.** $3z^2$; answers may vary **111. a.** m^3 **b.** $3m$ **c.** $-m^3$
d. $-3m$; answers may vary **113.** $3x^{2a} + 2x^a + 0.7$ **115.** $4x^{2y} + 2x^y - 11$ **117.** $4x^2 - 3x + 6$ **119.** $-x^2 - 6x + 10$ **121.** $3x^2 - 12x + 13$
123. a. $2a - 3$ **b.** $-2x - 3$ **c.** $2x + 2h - 3$ **125. a.** $4a$ **b.** $-4x$ **c.** $4x + 4h$ **127.** 2; 2 **129.** 4; 3; 3; 4 **131.** $2x^2 + 4xy$

Section 5.3

Practice Exercises

1. $10y^2$ **2.** $-2z^8$ **3.** $\dfrac{7}{72}b^9$ **4. a.** $27x^6 + 33x$ **b.** $-12x^5 + 54x^4 - 12x^3$ **5.** $10x^2 + 11x - 6$ **6.** $25x^2 - 30xy + 9y^2$
7. $2y^3 + 5y^2 - 7y + 20$ **8.** $s^3 + 6s^2t + 12st^2 + 8t^3$ **9.** $5x^3 - 23x^2 + 17x - 20$ **10.** $x^5 - 2x^4 + 2x^3 - 3x^2 + 2$
11. $5x^4 - 3x^3 + 11x^2 + 8x - 6$

Vocabulary, Readiness & Video Check 5.3

1. c. distributive   **3. c.** $(5y - 1)(5y - 1)$ **5.** No. The monomials are unlike terms.   **7.** Yes. The parentheses have been removed for the vertical format, but every term in the first polynomial is still distributed to every term in the second polynomial.

Exercise Set 5.3

1. $-28n^{10}$ **3.** $-12.4x^{12}$ **5.** $-\dfrac{2}{15}y^3$ **7.** $-24x^8$ **9.** $6x^2 + 15x$ **11.** $-2a^2 - 8a$ **13.** $6x^3 - 9x^2 + 12x$ **15.** $-6a^4 + 4a^3 - 6a^2$

17. $-4x^3y + 7x^2y^2 - xy^3 - 3y^4$ **19.** $4x^4 - 3x^3 + \dfrac{1}{2}x^2$ **21.** $x^2 + 7x + 12$ **23.** $a^2 + 5a - 14$ **25.** $x^2 + \dfrac{1}{3}x - \dfrac{2}{9}$ **27.** $12x^4 + 25x^2 + 7$

29. $4y^2 - 16y + 16$ **31.** $12x^2 - 29x + 15$ **33.** $9x^4 + 6x^2 + 1$ **35. a.** $-4y^4$ **b.** $3y^2$ **c.** answers may vary   **37.** $x^3 - 5x^2 + 13x - 14$
39. $x^4 + 5x^3 - 3x^2 - 11x + 20$ **41.** $10a^3 - 27a^2 + 26a - 12$ **43.** $x^3 + 6x^2 + 12x + 8$ **45.** $8y^3 - 36y^2 + 54y - 27$
47. $12x^2 - 64x - 11$ **49.** $10x^3 + 22x^2 - x - 1$ **51.** $2x^4 + 3x^3 - 58x^2 + 4x + 63$ **53.** $8.4y^7$ **55.** $-3x^3 - 6x^2 + 24x$

57. $2x^2 + 39x + 19$ **59.** $x^2 - \dfrac{2}{7}x - \dfrac{3}{49}$ **61.** $9y^2 + 30y + 25$ **63.** $a^3 - 2a^2 - 18a + 24$ **65.** $8x^3 - 60x^2 + 150x - 125$

67. $32x^3 + 48x^2 - 6x - 20$ **69.** $6x^4 - 8x^3 - 7x^2 + 22x - 12$ **71.** $(4x^2 - 25)$ sq yd   **73.** $(6x^2 - 4x)$ sq in.

75. $5a + 15a = 20a; 5a - 15a = -10a; 5a \cdot 15a = 75a^2; \dfrac{5a}{15a} = \dfrac{1}{3}$ **77.** $-3y^5 + 9y^4$, can't be simplified; $-3y^5 - 9y^4$, can't be simplified;

$-3y^5 \cdot 9y^4 = -27y^9; \dfrac{-3y^5}{9y^4} = -\dfrac{y}{3}$ **79. a.** $6x + 12$ **b.** $9x^2 + 36x + 35$; answers may vary   **81.** $13x - 7$ **83.** $30x^2 - 28x + 6$
85. $-7x + 5$ **87.** $x^2 + 3x$ **89.** $x^2 + 5x + 6$ **91.** $11a$ **93.** $25x^2 + 4y^2$ **95. a.** $a^2 - b^2$ **b.** $4x^2 - 9y^2$ **c.** $16x^2 - 49$
d. answers may vary   **97.** $(x^2 + 6x + 5)$ sq units

Section 5.4

Practice Exercises

1. $x^2 - 3x - 10$ **2.** $4x^2 - 13x + 9$ **3.** $9x^2 + 42x - 15$ **4.** $16x^2 - 8x + 1$ **5. a.** $b^2 + 6b + 9$ **b.** $x^2 - 2xy + y^2$
c. $9y^2 + 12y + 4$ **d.** $a^4 - 10a^2b + 25b^2$ **6. a.** $3x^2 - 75$ **b.** $16b^2 - 9$ **c.** $x^2 - \dfrac{4}{9}$ **d.** $25s^2 - t^2$ **e.** $4y^2 - 9z^4$

7. a. $4x^2 - 21x - 18$ **b.** $49b^2 - 28b + 4$ **c.** $x^2 - 0.16$ **d.** $3x^6 - \dfrac{9}{7}x^4 + \dfrac{2}{7}x^2 - \dfrac{6}{49}$ **e.** $x^3 + 6x^2 + 3x - 2$

Vocabulary, Readiness & Video Check 5.4

1. false   **3.** false   **5.** a binomial times a binomial   **7.** Multiplying gives you four terms, and the two like terms will always subtract out.

Exercise Set 5.4

1. $x^2 + 7x + 12$ **3.** $x^2 + 5x - 50$ **5.** $5x^2 + 4x - 12$ **7.** $20y^2 - 125y + 30$ **9.** $6x^2 + 13x - 5$ **11.** $x^2 + \dfrac{1}{3}x - \dfrac{2}{9}$ **13.** $x^2 + 4x + 4$

15. $4x^2 - 4x + 1$ **17.** $9a^2 - 30a + 25$ **19.** $25x^2 + 90x + 81$ **21.** $a^2 - 49$ **23.** $9x^2 - 1$ **25.** $9x^2 - \dfrac{1}{4}$ **27.** $81x^2 - y^2$
29. $4x^2 - 0.01$ **31.** $a^2 + 9a + 20$ **33.** $a^2 + 14a + 49$ **35.** $12a^2 - a - 1$ **37.** $x^2 - 4$ **39.** $9a^2 + 6a + 1$ **41.** $4x^3 - x^2y^4 + 4xy - y^5$
43. $x^3 - 3x^2 - 17x + 3$ **45.** $4a^2 - 12a + 9$ **47.** $25x^2 - 36z^2$ **49.** $x^{10} - 8x^5 + 15$ **51.** $x^2 - 0.64$ **53.** $a^7 + 11a^4 - 3a^3 - 33$
55. $3x^2 - 12x + 12$ **57.** $6b^2 - b - 35$ **59.** $49p^2 - 64$ **61.** $\dfrac{1}{9}a^4 - 49$ **63.** $15x^4 - 5x^3 + 10x^2$ **65.** $4r^2 - 9s^2$
67. $9x^2 - 42xy + 49y^2$ **69.** $16x^2 - 25$ **71.** $64x^2 + 64x + 16$ **73.** $a^2 - \dfrac{1}{4}y^2$ **75.** $\dfrac{1}{25}x^2 - y^2$ **77.** $3a^3 + 2a^2 + 1$

79. $(2x + 1)(2x + 1)$ sq ft or $(4x^2 + 4x + 1)$ sq ft   **81.** $\dfrac{5b^5}{7}$ **83.** $-2a^{10}b^5$ **85.** $\dfrac{2y^8}{3}$ **87.** $\dfrac{1}{3}$ **89.** 1   **91.** C   **93.** D   **95.** 2; 2
97. $(x^4 - 3x^2 + 1)$ sq m   **99.** $(24x^2 - 32x + 8)$ sq m   **101.** $(x^2 + 10x + 25)$ sq units   **103.** answers may vary   **105.** answers may vary
107. answers may vary   **109.** $x^2 + 2xy + y^2 - 9$ **111.** $a^2 - 6a + 9 - b^2$

Integrated Review

1. $35x^5$ **2.** $32y^9$ **3.** -16 **4.** 16 **5.** $2x^2 - 9x - 5$ **6.** $3x^2 + 13x - 10$ **7.** $3x - 4$ **8.** $4x + 3$ **9.** $7x^6y^2$ **10.** $\dfrac{10b^6}{7}$
11. $144m^{14}n^{12}$ **12.** $64y^{27}z^{30}$ **13.** $48y^2 - 27$ **14.** $98x^2 - 2$ **15.** $x^{63}y^{45}$ **16.** $27x^{27}$ **17.** $2x^2 - 2x - 6$ **18.** $6x^2 + 13x - 11$
19. $2.5y^2 - 6y - 0.2$ **20.** $8.4x^2 - 6.8x - 5.7$ **21.** $x^2 + 8xy + 16y^2$ **22.** $y^2 - 18yz + 81z^2$ **23.** $2x + 8y$ **24.** $2y - 18z$
25. $7x^2 - 10xy + 4y^2$ **26.** $-a^2 - 3ab + 6b^2$ **27.** $x^3 + 2x^2 - 16x + 3$ **28.** $x^3 - 2x^2 - 5x - 2$ **29.** $6x^5 + 20x^3 - 21x^2 - 70$
30. $20x^7 + 25x^3 - 4x^4 - 5$ **31.** $2x^3 - 19x^2 + 44x - 7$ **32.** $5x^3 + 9x^2 - 17x + 3$ **33.** cannot simplify   **34.** $25x^3y^3$ **35.** $125x^9$
36. $\dfrac{x^3}{y^3}$ **37.** $2x$ **38.** x^2

Section 5.5
Practice Exercises

1. a. $\dfrac{1}{125}$ **b.** $\dfrac{3}{y^4}$ **c.** $\dfrac{5}{6}$ **d.** $\dfrac{1}{25}$ **e.** $\dfrac{1}{x^5}$ **2. a.** s^5 **b.** 8 **c.** $\dfrac{y^5}{x^7}$ **d.** $\dfrac{9}{64}$ **3. a.** $\dfrac{1}{x^5}$ **b.** $5y^7$ **c.** z^5 **d.** $\dfrac{81}{25}$ **4. a.** $\dfrac{b^{15}}{a^{20}}$

b. x^{10} **c.** $\dfrac{q^2}{25p^{16}}$ **d.** $\dfrac{6y^2}{x^7}$ **e.** $-27x^6y^9$ **5. a.** 7×10^{-6} **b.** 2.07×10^7 **c.** 4.3×10^{-3} **d.** 8.12×10^8 **6. a.** 0.000367

b. 8,954,000 **c.** 0.00002009 **d.** 4054 **7. a.** 4000 **b.** 20,000,000,000

Calculator Explorations 5.5
1. 5.31 EE 3 **3.** 6.6 EE -9 **5.** 1.5×10^{13} **7.** 8.15×10^{19}

Vocabulary, Readiness & Video Check 5.5

1. b. $\dfrac{1}{x^3}$ **3. c.** scientific notation **5.** A negative exponent has nothing to do with the sign of the simplified result. **7.** When you move the decimal point to the left, the sign of the exponent will be positive; when you move the decimal point to the right, the sign of the exponent will be negative.
9. the quotient rule

Exercise Set 5.5

1. $\dfrac{1}{64}$ **3.** $\dfrac{1}{81}$ **5.** $\dfrac{7}{x^3}$ **7.** 32 **9.** -64 **11.** $\dfrac{8}{15}$ **13.** p^3 **15.** $\dfrac{q^4}{p^5}$ **17.** $\dfrac{1}{x^3}$ **19.** z^3 **21.** $\dfrac{4}{9}$ **23.** $-p^4$ **25.** -2 **27.** x^4

29. p^4 **31.** m^{11} **33.** r^6 **35.** $\dfrac{1}{x^{15}y^9}$ **37.** $\dfrac{1}{x^4}$ **39.** $\dfrac{1}{a^2}$ **41.** $4k^3$ **43.** $3m$ **45.** $-\dfrac{4a^5}{b}$ **47.** $-\dfrac{6x^2}{y^3}$ **49.** $\dfrac{a^{30}}{b^{12}}$ **51.** $\dfrac{1}{x^{10}y^6}$

53. $\dfrac{z^2}{4}$ **55.** $\dfrac{1}{32x^5}$ **57.** $\dfrac{49a^4}{b^6}$ **59.** $a^{24}b^8$ **61.** x^9y^{19} **63.** $-\dfrac{y^8}{8x^2}$ **65.** $-\dfrac{6x}{7y^2}$ **67.** $\dfrac{25b^{33}}{a^{16}}$ **69.** 7.8×10^4 **71.** 1.67×10^{-6}

73. 6.35×10^{-3} **75.** 1.16×10^6 **77.** 2×10^9 **79.** 2.4×10^3 **81.** 0.0000000008673 **83.** 0.033 **85.** 20,320 **87.** 700,000,000
89. 9,460,000,000,000 **91.** $\$7.14 \times 10^{11}$ **93.** \$863,000,000,000 **95.** \$47,000 **97.** 0.000036 **99.** 0.00000000000000000028 **101.** 0.0000005
103. 200,000 **105.** $\dfrac{5x^3}{3}$ **107.** $\dfrac{5z^3y^2}{7}$ **109.** $5y - 6 + \dfrac{5}{y}$ **111.** $\dfrac{27}{x^6z^3}$ cu in. **113.** $9a^{13}$ **115.** -5 **117.** answers may vary
119. a. 1.3×10^1 **b.** 4.4×10^7 **c.** 6.1×10^{-2} **121.** answers may vary **123.** a^m **125.** $27y^{6z}$ **127.** -394.5 **129.** 1.3 sec

Section 5.6
Practice Exercises

1. $2t + 1$ **2.** $4x^4 + 5x - \dfrac{3}{x}$ **3.** $3x^3y^3 - 2 + \dfrac{1}{5x}$ **4.** $x + 3$ **5.** $2x + 3 + \dfrac{-10}{2x + 1}$ or $2x + 3 - \dfrac{10}{2x + 1}$ **6.** $3x^2 - 2x + 5 + \dfrac{-13}{3x + 2}$ or

$3x^2 - 2x + 5 - \dfrac{13}{3x + 2}$ **7.** $3x^2 - 2x - 9 + \dfrac{5x + 22}{x^2 + 2}$ **8.** $x^2 - 3x + 9$

Vocabulary, Readiness & Video Check 5.6
1. dividend, quotient, divisor **3.** a^2 **5.** y **7.** the common denominator

Exercise Set 5.6

1. $12x^3 + 3x$ **3.** $4x^3 - 6x^2 + x + 1$ **5.** $5p^2 + 6p$ **7.** $-\dfrac{3}{2x} + 3$ **9.** $-3x^2 + x - \dfrac{4}{x^3}$ **11.** $-1 + \dfrac{3}{2x} - \dfrac{7}{4x^4}$ **13.** $x + 1$ **15.** $2x + 3$

17. $2x + 1 + \dfrac{7}{x - 4}$ **19.** $3a^2 - 3a + 1 + \dfrac{2}{3a + 2}$ **21.** $4x + 3 - \dfrac{2}{2x + 1}$ **23.** $2x^2 + 6x - 5 - \dfrac{2}{x - 2}$ **25.** $x + 6$ **27.** $x^2 + 3x + 9$

29. $-3x + 6 - \dfrac{11}{x + 2}$ **31.** $2b - 1 - \dfrac{6}{2b - 1}$ **33.** $ab - b^2$ **35.** $4x + 9$ **37.** $x + 4xy - \dfrac{y}{2}$ **39.** $2b^2 + b + 2 - \dfrac{12}{b + 4}$

41. $5x - 2 + \dfrac{2}{x + 6}$ **43.** $x^2 - \dfrac{12x}{5} - 1$ **45.** $6x - 1 - \dfrac{1}{x + 3}$ **47.** $6x - 1$ **49.** $-x^3 + 3x^2 - \dfrac{4}{x}$ **51.** $x^2 + 3x + 9$

53. $y^2 + 5y + 10 + \dfrac{24}{y - 2}$ **55.** $-6x - 12 - \dfrac{19}{x - 2}$ **57.** $x^3 - x^2 + x$ **59.** $2a^3 + 2a$ **61.** $2x^3 + 14x^2 - 10x$ **63.** $-3x^2y^3 - 21x^3y^2 - 24xy$
65. $9a^2b^3c + 36ab^2c - 72ab$ **67.** $(3x^3 + x - 4)$ ft **69.** c **71.** answers may vary **73.** $(2x + 5)$ m **75.** $9x^{7a} - 6x^{5a} + 7x^{2a} - 1$

Section 5.7
Practice Exercises

1. $4x^2 + x + 7 + \dfrac{12}{x - 1}$ **2.** $x^3 - 5x + 21 - \dfrac{51}{x + 3}$ **3. a.** -4 **b.** -4 **4.** 15

Vocabulary, Readiness & Video Check 5.7
1. The last number n is the remainder and the other numbers are the coefficients of the variables in the quotient; the degree of the quotient is one less than the degree of the dividend

Exercise Set 5.7

1. $x + 8$ **3.** $x - 1$ **5.** $x^2 - 5x - 23 - \dfrac{41}{x - 2}$ **7.** $4x + 8 + \dfrac{7}{x - 2}$ **9.** 3 **11.** 73 **13.** -8 **15.** $x^2 + \dfrac{2}{x - 3}$ **17.** $6x + 7 + \dfrac{1}{x + 1}$

19. $2x^3 - 3x^2 + x - 4$ **21.** $3x - 9 + \dfrac{12}{x+3}$ **23.** $3x^2 - \dfrac{9}{2}x + \dfrac{7}{4} + \dfrac{47}{8(x - \frac{1}{2})}$ **25.** $3x^2 + 3x - 3$ **27.** $3x^2 + 4x - 8 + \dfrac{20}{x+1}$

29. $x^2 + x + 1$ **31.** $2x^2 - 3 - \dfrac{2}{x+6}$ **33.** 1   **35.** -133 **37.** 3   **39.** $-\dfrac{187}{81}$ **41.** $\dfrac{95}{32}$ **43.** answers may vary   **45.** $-\dfrac{5}{6}$ **47.** 54

49. 8   **51.** -32 **53.** 48   **55.** 25   **57.** -2 **59.** yes   **61.** no   **63.** $(x^3 - 5x^2 + 2x - 1)$ cm   **65.** $x^3 + \dfrac{5}{3}x^2 + \dfrac{5}{3}x + \dfrac{8}{3} + \dfrac{8}{3(x-1)}$

67. $(x+3)(x^2+4) = x^3 + 3x^2 + 4x + 12$ **69.** 0   **71.** $x^3 + 2x^2 + 7x + 28$

Chapter 5 Vocabulary Check

1. term   **2.** FOIL   **3.** trinomial   **4.** degree of a polynomial   **5.** binomial   **6.** coefficient   **7.** degree of a term   **8.** monomial   **9.** polynomials   **10.** distributive

Chapter 5 Review

1. base: 7; exponent: 9   **3.** base: 5; exponent: 4   **5.** 512   **7.** -36 **9.** 1   **11.** y^9 **13.** $-6x^{11}$ **15.** x^8 **17.** $81y^{24}$ **19.** x^5

21. a^4b^3 **23.** $\dfrac{x^3y^4}{4}$ **25.** $40a^{19}$ **27.** 3   **29.** b   **31.** 7   **33.** 8   **35.** 5   **37.** 5   **39.** 4000 ft; 3984 ft; 3856 ft; 3600 ft   **41.** $15a^2 + 4a$

43. $-6a^2b - 3b^2 - q^2$ **45.** $8x^2 + 3x + 6$ **47.** $-7y^2 - 1$ **49.** $4x - 13y$ **51.** 290   **53.** $(6x^2y - 12x + 12)$ cm   **55.** $8a + 28$

57. $-7x^3 - 35x$ **59.** $-6a^4 + 8a^2 - 2a$ **61.** $2x^2 - 12x - 14$ **63.** $x^2 - 18x + 81$ **65.** $4a^2 + 27a - 7$ **67.** $25x^2 + 20x + 4$

69. $x^4 + 7x^3 + 4x^2 + 23x - 35$ **71.** $x^4 + 4x^3 + 4x^2 - 16$ **73.** $x^3 + 21x^2 + 147x + 343$ **75.** $x^2 + 14x + 49$ **77.** $9x^2 - 42x + 49$

79. $25x^2 - 90x + 81$ **81.** $49x^2 - 16$ **83.** $4x^2 - 36$ **85.** $(9x^2 - 6x + 1)$ sq m   **87.** $\dfrac{1}{49}$ **89.** $\dfrac{2}{x^4}$ **91.** 125   **93.** $\dfrac{17}{16}$ **95.** x^8

97. r **99.** c^4 **101.** $\dfrac{1}{x^6y^{13}}$ **103.** a^{11m} **105.** $27x^3y^{6z}$ **107.** 2.7×10^{-4} **109.** 8.08×10^7 **111.** 2.5×10^9 **113.** 867,000

115. 0.00086   **117.** 1,431,280,000,000,000   **119.** 0.016   **121.** $\dfrac{1}{7} + \dfrac{3}{x} + \dfrac{7}{x^2}$ **123.** $a + 1 + \dfrac{6}{a-2}$ **125.** $a^2 + 3a + 8 + \dfrac{22}{a-2}$

127. $2x^3 - x^2 + 2 - \dfrac{1}{2x-1}$ **129.** $\left(5x - 1 + \dfrac{20}{x^2}\right)$ ft   **131.** $3x^2 + 6x + 24 + \dfrac{44}{x-2}$ **133.** $x^4 - x^3 + x^2 - x + 1 - \dfrac{2}{x+1}$

135. $3x^3 + 13x^2 + 51x + 204 + \dfrac{814}{x-4}$ **137.** 3043   **139.** $-\dfrac{1}{8}$ **141.** $\dfrac{2x^6}{3}$ **143.** $\dfrac{x^{16}}{16y^{12}}$ **145.** $11x - 5$ **147.** $5y^2 - 3y - 1$

149. $28x^3 + 12x$ **151.** $x^3 + x^2 - 18x + 18$ **153.** $25x^2 + 40x + 16$ **155.** $4a - 1 + \dfrac{2}{a^2} - \dfrac{5}{2a^3}$ **157.** $2x^2 + 7x + 5 + \dfrac{19}{2x-3}$

Chapter 5 Getting Ready for the Test

1. C   **2.** A   **3.** E   **4.** D   **5.** F   **6.** C   **7.** E   **8.** I   **9.** C   **10.** D   **11.** C   **12.** B   **13.** F   **14.** D   **15.** C

Chapter 5 Test

1. 32   **2.** 81   **3.** -81 **4.** $\dfrac{1}{64}$ **5.** $-15x^{11}$ **6.** y^5 **7.** $\dfrac{1}{r^5}$ **8.** $\dfrac{y^{14}}{x^2}$ **9.** $\dfrac{1}{6xy^8}$ **10.** 5.63×10^5 **11.** 8.63×10^{-5} **12.** 0.0015

13. 62,300   **14.** 0.036   **15. a.** $4, 3; 7, 3; 1, 4; -2, 0$ **b.** 4   **16.** $-2x^2 + 12xy + 11$ **17.** $16x^3 + 7x^2 - 3x - 13$ **18.** $-3x^3 + 5x^2 + 4x + 5$

19. $x^3 + 8x^2 + 3x - 5$ **20.** $3x^3 + 22x^2 + 41x + 14$ **21.** $6x^4 - 9x^3 + 21x^2$ **22.** $3x^2 + 16x - 35$ **23.** $9x^2 - \dfrac{1}{25}$ **24.** $16x^2 - 16x + 4$

25. $64x^2 + 48x + 9$ **26.** $x^4 - 81b^2$ **27.** 1001 ft; 985 ft; 857 ft; 601 ft   **28.** $(4x^2 - 9)$ sq in.   **29.** $\dfrac{x}{2y} + 3 - \dfrac{7}{8y}$ **30.** $x + 2$

31. $9x^2 - 6x + 4 - \dfrac{16}{3x+2}$ **32. a.** 960 ft   **b.** 953.44 ft   **33.** $4x^3 - 15x^2 + 45x - 136 + \dfrac{407}{x+3}$ **34.** 91

Chapter 5 Cumulative Review

1. a. true   **b.** true   **c.** false   **d.** true; Sec. 1.2, Ex. 2   **3. a.** $\dfrac{64}{25}$ **b.** $\dfrac{1}{20}$ **c.** $\dfrac{5}{4}$; Sec. 1.3, Ex. 4   **5. a.** 9   **b.** 125   **c.** 16   **d.** 7

e. $\dfrac{9}{49}$; Sec. 1.4, Ex. 1   **7. a.** -10 **b.** -21 **c.** -12; Sec. 1.5, Ex. 3   **9.** -12; Sec. 1.6, Ex. 3   **11. a.** $\dfrac{1}{22}$ **b.** $\dfrac{16}{3}$ **c.** $-\dfrac{1}{10}$

d. $-\dfrac{13}{9}$; Sec. 1.7, Ex. 5   **13. a.** $(5+4)+6$ **b.** $-1 \cdot (2 \cdot 5)$; Sec. 1.8, Ex. 2   **15. a.** $22 + x$ **b.** $-21x$; Sec. 1.8, Ex. 3   **17. a.** $15x + 10$

b. $-2y - 0.6z + 2$ **c.** $-9x - y + 2z - 6$; Sec. 2.1, Ex. 5   **19.** 17; Sec. 2.2, Ex. 1   **21.** 6; Sec. 2.3, Ex. 5   **23.** -10; Sec. 2.4, Ex. 1

25. 10; Sec. 2.4, Ex. 2   **27.** width: 4 ft; length: 10 ft; Sec. 2.5, Ex. 4   **29.** $\dfrac{5F - 160}{9} = C$; Sec. 2.5, Ex. 8   **31.** ; Sec. 2.8, Ex. 9

33. a. $(0, 12)$ **b.** $(2, 6)$ **c.** $(-1, 15)$; Sec. 3.1, Ex. 6   **35.** ; Sec. 3.2, Ex. 2   **37.** ; Sec. 3.3, Ex. 9

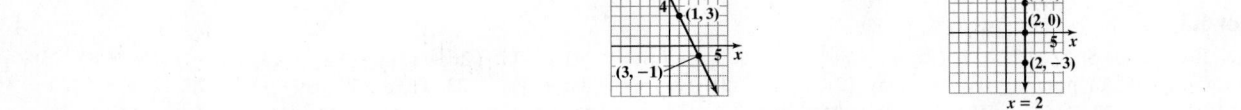

39. undefined slope; Sec. 3.4, Ex. 7   **41.** $9x^2 - 6x - 1$; Sec. 5.2, Ex. 11   **43.** $-6x^7$; Sec. 5.1, Ex. 4   **45.** $12x^3 - 12x^2 - 9x + 2$; Sec. 5.2, Ex. 10

47. $4x^2 - 4xy + y^2$; Sec. 5.3, Ex. 6   **49.** $3m + 1$; Sec. 5.6, Ex. 1

CHAPTER 6 FACTORING POLYNOMIALS

Section 6.1

Practice Exercises

1. a. 6 **b.** 1 **c.** 4 **2. a.** y^4 **b.** x **3. a.** $5y^2$ **b.** x^2 **c.** a^2b^2 **4. a.** $4(t + 3)$ **b.** $y^4(y^4 + 1)$ **5.** $8b^2(-b^4 + 2b^2 - 1)$ or

$-8b^2(b^4 - 2b^2 + 1)$ **6.** $5x(x^3 - 4)$ **7.** $\frac{1}{9}z^3(5z^2 + z - 2)$ **8.** $4ab^3(2ab - 5a^2 + 3)$ **9.** $(y - 2)(8 + x)$ **10.** $(p + q)(7xy^3 - 1)$

11. $(x + 3)(y + 4)$ **12.** $(5x - 3)(8x^2 + 3)$ **13.** $(2x + 3y)(y - 1)$ **14.** $(7a + 5)(a^2 + 1)$ **15.** $(y - 3)(4x - 5)$ **16.** cannot be
factored by grouping **17.** $3(x - a)(y - 2a)$

Vocabulary, Readiness & Video Check 6.1

1. factors **3.** least **5.** false **7.** The GCF of a list of numbers is the largest number that is a factor of all numbers in the list. **9.** When factoring
out a GCF, the number of terms in the other factor should have the same number of terms as your original polynomial.

Exercise Set 6.1

1. 4 **3.** 6 **5.** 1 **7.** y^2 **9.** z^7 **11.** xy^2 **13.** 7 **15.** $4y^3$ **17.** $5x^2$ **19.** $3x^3$ **21.** $9x^2y$ **23.** $10a^6b$ **25.** $3(a + 2)$
27. $15(2x - 1)$ **29.** $x^2(x + 5)$ **31.** $2y^3(3y + 1)$ **33.** $4(x - 2y + 1)$ **35.** $3x(2x^2 - 3x + 4)$ **37.** $a^2b^2(a^5b^4 - a + b^3 - 1)$
39. $4(2x^5 + 4x^4 - 5x^3 + 3)$ **41.** $\frac{1}{3}x(x^3 + 2x^2 - 4x^4 + 1)$ **43.** $(x^2 + 2)(y + 3)$ **45.** $(y + 4)(z - 3)$ **47.** $(z^2 - 6)(r + 1)$
49. $-2(x + 7)$ **51.** $-x^5(2 - x^2)$ **53.** $-3a^2(a^2 - 3a + 1)$ **55.** $(x + 2)(x^2 + 5)$ **57.** $(x + 3)(5 + y)$ **59.** $(3x - 2)(2x^2 + 5)$
61. $(5m^2 + 6n)(m + 1)$ **63.** $(y - 4)(2 + x)$ **65.** $(2x - 1)(x^2 + 4)$ **67.** not factorable by grouping **69.** $(x - 2y)(4x - 3)$
71. $(5q - 4p)(q - 1)$ **73.** $x(x^2 + 1)(2x + 5)$ **75.** $2(2y - 7)(3x^2 - 1)$ **77.** $2x(16y - 9x)$ **79.** $(x + 2)(y - 3)$ **81.** $7xy(2x^2 + x - 1)$
83. $(4x - 1)(7x^2 + 3)$ **85.** $-8x^8y^5(5y + 2x)$ **87.** $3(2a + 3b^2)(a + b)$ **89.** $x^2 + 7x + 10$ **91.** $b^2 - 3b - 4$ **93.** 2, 6 **95.** $-1, -8$
97. $-2, 5$ **99.** b **101.** factored **103.** not factored **105.** answers may vary **107.** answers may vary **109. a.** 22% **b.** 70%
c. $0.6(x^2 - x + 6)$ **111.** $12x^3 - 2x; 2x(6x^2 - 1)$ **113.** $(n^3 - 6)$ units **115.** $(x^n + 2)(x^n + 3)$ **117.** $(3x^n - 5)(x^n + 7)$

Section 6.2

Practice Exercises

1. $(x + 2)(x + 3)$ **2.** $(x - 10)(x - 7)$ **3.** $(x + 7)(x - 2)$ **4.** $(p - 9)(p + 7)$ **5.** prime polynomial **6.** $(x + 3y)(x + 4y)$
7. $(x^2 + 12)(x^2 + 1)$ **8.** $(x - 6)(x - 8)$ **9.** $4(x - 3)(x - 3)$ **10.** $3y^2(y - 7)(y + 1)$

Vocabulary, Readiness & Video Check 6.2

1. true **3.** false **5.** $+5$ **7.** -3 **9.** $+2$ **11.** 15 is positive, so its factors would have to be either both positive or both negative. Since the
factors need to sum to -8, both factors must be negative.

Exercise Set 6.2

1. $(x + 6)(x + 1)$ **3.** $(y - 9)(y - 1)$ **5.** $(x - 3)(x - 3)$ or $(x - 3)^2$ **7.** $(x - 6)(x + 3)$ **9.** $(x + 10)(x - 7)$ **11.** prime
13. $(x + 5y)(x + 3y)$ **15.** $(a^2 - 5)(a^2 + 3)$ **17.** $(m + 13)(m + 1)$ **19.** $(t - 2)(t + 12)$ **21.** $(a - 2b)(a - 8b)$ **23.** $2(z + 8)(z + 2)$
25. $2x(x - 5)(x - 4)$ **27.** $(x - 4y)(x + y)$ **29.** $(x + 12)(x + 3)$ **31.** $(x - 2)(x + 1)$ **33.** $(r - 12)(r - 4)$ **35.** $(x + 2y)(x - y)$
37. $3(x + 5)(x - 2)$ **39.** $3(x - 18)(x - 2)$ **41.** $(x - 24)(x + 6)$ **43.** prime **45.** $(x - 5)(x - 3)$ **47.** $6x(x + 4)(x + 5)$
49. $4y(x^2 + x - 3)$ **51.** $(x - 7)(x + 3)$ **53.** $(x + 5y)(x + 2y)$ **55.** $2(t + 8)(t + 4)$ **57.** $x(x - 6)(x + 4)$ **59.** $2t^3(t - 4)(t - 3)$
61. $5xy(x - 8y)(x + 3y)$ **63.** $3(m - 9)(m - 6)$ **65.** $-1(x - 11)(x - 1)$ **67.** $\frac{1}{2}(y - 11)(y + 2)$ **69.** $x(xy - 4)(xy + 5)$
71. $2x^2 + 11x + 5$ **73.** $15y^2 - 17y + 4$ **75.** $9a^2 + 23ab - 12b^2$ **77.** $x^2 + 5x - 24$ **79.** answers may vary
81. $2x^2 + 28x + 66; 2(x + 3)(x + 11)$ **83.** $-16(t - 5)(t + 1)$ **85.** $\left(x + \frac{1}{4}\right)\left(x + \frac{1}{4}\right)$ or $\left(x + \frac{1}{4}\right)^2$ **87.** $(x + 1)(z - 10)(z + 7)$
89. $(x^n + 10)(x^n - 2)$ **91.** 5; 8; 9 **93.** 3; 4 **95.** 8; 16 **97.** 6; 26

Section 6.3

Practice Exercises

1. $(2x + 5)(x + 3)$ **2.** $(5x - 4)(3x - 2)$ **3.** $(4x - 1)(x + 3)$ **4.** $(7x - y)(3x + 2y)$ **5.** $(2x^2 - 7)(x^2 + 1)$ **6.** $x(3x + 2)(x + 5)$
7. $-1(4x - 3)(2x + 1)$ **8.** $(x + 7)^2$ **9.** $(2x + 9y)(2x + y)$ **10.** $(6n^2 - 1)^2$ **11.** $3x(2x - 7)^2$

Vocabulary, Readiness & Video Check 6.3

1. perfect square trinomial **3.** perfect square trinomial **5.** d **7.** Consider the factors of the first and last terms and the signs of the trinomial.
Continue to check by multiplying until you get the middle term of the trinomial. **9.** The first and last terms are squares, a^2 and b^2, and the middle
term is $2 \cdot a \cdot b$ or $-2 \cdot a \cdot b$.

Exercise Set 6.3

1. $x + 4$ **3.** $10x - 1$ **5.** $5x - 2$ **7.** $(2x + 3)(x + 5)$ **9.** $(y - 1)(8y - 9)$ **11.** $(2x + 1)(x - 5)$ **13.** $(4r - 1)(5r + 8)$
15. $(10x + 1)(x + 3)$ **17.** prime **19.** $(3x - 5y)(2x - y)$ **21.** $(3m - 5)(5m + 3)$ **23.** $x(3x + 2)(4x + 1)$ **25.** $3(7b + 5)(b - 3)$
27. $(3z + 4)(4z - 3)$ **29.** $2y^2(3x - 10)(x + 3)$ **31.** $(2x - 7)(2x + 3)$ **33.** $-1(x - 6)(x + 4)$ **35.** $x(4x + 3)(x - 3)$
37. $(4x - 9)(6x - 1)$ **39.** $(x + 11)^2$ **41.** $(x - 8)^2$ **43.** $(4a - 3)^2$ **45.** $(x^2 + 2)^2$ **47.** $2(n - 7)^2$ **49.** $(4y + 5)^2$

51. $(2x + 11)(x - 9)$ **53.** $(8x + 3)(3x + 4)$ **55.** $(3a + b)(a + 3b)$ **57.** $(x - 4)(x - 5)$ **59.** $(p + 6q)^2$ **61.** $(xy - 5)^2$
63. $b(8a - 3)(5a + 3)$ **65.** $2x(3x + 2)(5x + 3)$ **67.** $2y(3y + 5)(y - 3)$ **69.** $5x^2(2x - y)(x + 3y)$ **71.** $-1(2x - 5)(7x - 2)$
73. $p^2(4p - 5)^2$ **75.** $(3x - 2)(x + 1)$ **77.** $(4x + 9y)(2x - 3y)$ **79.** prime   **81.** $(3x - 4y)^2$ **83.** $(6x - 7)(3x + 2)$
85. $(7t + 1)(t - 4)$ **87.** $(7p + 1)(7p - 2)$ **89.** $m(m + 9)^2$ **91.** prime   **93.** $a(6a^2 + b^2)(a^2 + 6b^2)$ **95.** $x^2 - 4$ **97.** $a^3 + 27$
99. Facebook   **101.** answers may vary   **103.** no   **105.** answers may vary   **107.** $4x^2 + 21x + 5; (4x + 1)(x + 5)$ **109.** $\left(2x + \dfrac{1}{2}\right)^2$
111. $(y - 1)^2(4x^2 + 10x + 25)$ **113.** 8   **115.** $a^2 + 2ab + b^2$ **117.** 2; 14   **119.** 2   **121.** $-3xy^2(4x - 5)(x + 1)$
123. $(y - 1)^2(2x + 5)^2$ **125.** $(3x^n + 2)(x^n + 5)$ **127.** answers may vary

Section 6.4

Practice Exercises

1. $(5x + 1)(x + 12)$ **2.** $(4x - 5)(3x - 1)$ **3.** $2(5x + 1)(3x - 2)$ **4.** $5m^2(8m - 7)(m + 1)$ **5.** $(4x + 3)^2$

Vocabulary, Readiness & Video Check 6.4

1. a   **3.** b   **5.** This gives us a four-term polynomial, which may be factored by grouping.

Exercise Set 6.4

1. $(x + 3)(x + 2)$ **3.** $(y + 8)(y - 2)$ **5.** $(8x - 5)(x - 3)$ **7.** $(5x^2 - 3)(x^2 + 5)$ **9. a.** $9, 2$ **b.** $9x + 2x$ **c.** $(3x + 1)(2x + 3)$
11. a. $-20, -3$ **b.** $-20x - 3x$ **c.** $(3x - 4)(5x - 1)$ **13.** $(3y + 2)(7y + 1)$ **15.** $(7x - 11)(x + 1)$ **17.** $(5x - 2)(2x - 1)$
19. $(2x - 5)(x - 1)$ **21.** $(2x + 3)^2$ **23.** $(2x + 3)(2x - 7)$ **25.** $(5x - 4)(2x - 3)$ **27.** $x(2x + 3)(x + 5)$
29. $2(8y - 9)(y - 1)$ **31.** $(2x - 3)(3x - 2)$ **33.** $3(3a + 2)(6a - 5)$ **35.** $a(4a + 1)(5a + 8)$ **37.** $3x(4x + 3)(x - 3)$
39. $y(3x + y)(x + y)$ **41.** prime   **43.** $5(x + 5y)^2$ **45.** $6(a + b)(4a - 5b)$ **47.** $p^2(15p + q)(p + 2q)$ **49.** $2(9a^2 - 2)^2$
51. $(7 + x)(5 + x)$ or $(x + 7)(x + 5)$ **53.** $(6 - 5x)(1 - x)$ or $(5x - 6)(x - 1)$ **55.** $x^2 - 4$ **57.** $y^2 + 8y + 16$ **59.** $81z^2 - 25$
61. $x^3 - 27$ **63.** $10x^2 + 45x + 45; 5(2x + 3)(x + 3)$ **65.** $(x^n + 2)(x^n + 3)$ **67.** $(3x^n - 5)(x^n + 7)$ **69.** answers may vary

Section 6.5

Practice Exercises

1. $(x + 9)(x - 9)$ **2. a.** $(3x - 1)(3x + 1)$ **b.** $(6a - 7b)(6a + 7b)$ **c.** $\left(p + \dfrac{5}{6}\right)\left(p - \dfrac{5}{6}\right)$ **3.** $(p^2 - q^5)(p^2 + q^5)$
4. a. $(z^2 + 9)(z + 3)(z - 3)$ **b.** prime polynomial   **5.** $y(6y + 5)(6y - 5)$ **6.** $5(4y^2 + 1)(2y + 1)(2y - 1)$
7. $-1(3x + 10)(3x - 10)$ or $(10 + 3x)(10 - 3x)$ **8.** $(x + 4)(x^2 - 4x + 16)$ **9.** $(x - 5)(x^2 + 5x + 25)$ **10.** $(3y + 1)(9y^2 - 3y + 1)$
11. $4(2x - 5y)(4x^2 + 10xy + 25y^2)$

Graphing Calculator Explorations 6.5

	$x^2 - 2x + 1$	$x^2 - 2x - 1$	$(x - 1)^2$
$x = 5$	16	14	16
$x = -3$	16	14	16
$x = 2.7$	2.89	0.89	2.89
$x = -12.1$	171.61	169.61	171.61
$x = 0$	1	-1	1

Vocabulary, Readiness & Video Check 6.5

1. difference of two cubes   **3.** sum of two cubes   **5.** $(7x)^2$ **7.** $(2y)^3$ **9.** In order to recognize the binomial as a difference of squares and also to identify the terms to use in the special factoring formula.   **11.** First rewrite the original binomial with terms writtten as cubes. Answers will then vary depending on your interpretation.

Exercise Set 6.5

1. $(x + 2)(x - 2)$ **3.** $(9p + 1)(9p - 1)$ **5.** $(5y - 3)(5y + 3)$ **7.** $(11m + 10n)(11m - 10n)$ **9.** $(xy - 1)(xy + 1)$
11. $\left(x - \dfrac{1}{2}\right)\left(x + \dfrac{1}{2}\right)$ **13.** $-1(2r + 1)(2r - 1)$ or $(1 - 2r)(1 + 2r)$ **15.** prime   **17.** $(x - 6)(x + 6)$ or $-1(6 + x)(6 - x)$
19. $(m^2 + 1)(m + 1)(m - 1)$ **21.** $(m^2 + n^9)(m^2 - n^9)$ **23.** $(x + 5)(x^2 - 5x + 25)$ **25.** $(2a - 1)(4a^2 + 2a + 1)$
27. $(m + 3n)(m^2 - 3mn + 9n^2)$ **29.** $5(k + 2)(k^2 - 2k + 4)$ **31.** $(xy - 4)(x^2y^2 + 4xy + 16)$ **33.** $2(5r - 4t)(25r^2 + 20rt + 16t^2)$
35. $(r + 8)(r - 8)$ **37.** $(x + 13y)(x - 13y)$ **39.** $(3 - t)(9 + 3t + t^2)$ **41.** $2(3r + 2)(3r - 2)$ **43.** $x(3y + 2)(3y - 2)$
45. $8(m + 2)(m^2 - 2m + 4)$ **47.** $xy(y - 3z)(y + 3z)$ **49.** $4(3x - 4y)(3x + 4y)$ **51.** $9(4 - 3x)(4 + 3x)$ **53.** $(xy - z^2)(x^2y^2 + xyz^2 + z^4)$
55. $\left(7 - \dfrac{3}{5}m\right)\left(7 + \dfrac{3}{5}m\right)$ **57.** $(t + 7)(t^2 - 7t + 49)$ **59.** $n(n^2 + 49)$ **61.** $x^2(x^2 + 9)(x + 3)(x - 3)$ **63.** $pq(8p + 9q)(8p - 9q)$
65. $xy^2(27xy + 1)$ **67.** $a(5a - 4b)(25a^2 + 20ab + 16b^2)$ **69.** $16x^2(x + 2)(x - 2)$ **71.** 6   **73.** -2 **75.** $\dfrac{1}{5}$
77. $(x + 2 + y)(x + 2 - y)$ **79.** $(a + 4)(a - 4)(b - 4)$ **81.** $(x + 3 + 2y)(x + 3 - 2y)$ **83.** $(x^n + 10)(x^n - 10)$
85. $(x + 6)$ **87.** answers may vary   **89. a.** 2992 ft   **b.** 1536 ft   **c.** 14 sec   **d.** $16(14 - t)(14 + t)$ **91. a.** 2560 ft   **b.** 1920 ft
c. 13 sec   **d.** $16(13 + t)(13 - t)$

Integrated Review

Practice Exercises

1. $(3x - 1)(2x - 3)$ **2.** $(3x + 1)(x - 2)(x + 2)$ **3.** $3(3x - y)(3x + y)$ **4.** $(2a + b)(4a^2 - 2ab + b^2)$ **5.** $6xy^2(5x + 2)(2x - 3)$

Exercise Set

1. $(x + y)^2$ **2.** $(x - y)^2$ **3.** $(a + 12)(a - 1)$ **4.** $(a - 10)(a - 1)$ **5.** $(a + 2)(a - 3)$ **6.** $(a - 1)^2$ **7.** $(x + 1)^2$
8. $(x + 2)(x - 1)$ **9.** $(x + 1)(x + 3)$ **10.** $(x + 3)(x - 2)$ **11.** $(x + 3)(x + 4)$ **12.** $(x + 4)(x - 3)$ **13.** $(x + 4)(x - 1)$
14. $(x - 5)(x - 2)$ **15.** $(x + 5)(x - 3)$ **16.** $(x + 6)(x + 5)$ **17.** $(x - 6)(x + 5)$ **18.** $(x + 8)(x + 3)$ **19.** $2(x + 7)(x - 7)$
20. $3(x + 5)(x - 5)$ **21.** $(x + 3)(x + y)$ **22.** $(y - 7)(3 + x)$ **23.** $(x + 8)(x - 2)$ **24.** $(x - 7)(x + 4)$ **25.** $4x(x + 7)(x - 2)$
26. $6x(x - 5)(x + 4)$ **27.** $2(3x + 4)(2x + 3)$ **28.** $(2a - b)(4a + 5b)$ **29.** $(2a + b)(2a - b)$ **30.** $(4 - 3x)(7 + 2x)$
31. $(5 - 2x)(4 + x)$ **32.** prime **33.** prime **34.** $(3y + 5)(2y - 3)$ **35.** $(4x - 5)(x + 1)$ **36.** $y(x + y)(x - y)$ **37.** $4(t^2 + 9)$
38. $(x + 1)(x + y)$ **39.** $(x + 1)(a + 2)$ **40.** $9x(2x^2 - 7x + 1)$ **41.** $4a(3a^2 - 6a + 1)$ **42.** $(x + 16)(x - 2)$ **43.** prime
44. $(4a - 7b)^2$ **45.** $(5p - 7q)^2$ **46.** $(7x + 3y)(x + 3y)$ **47.** $(5 - 2y)(25 + 10y + 4y^2)$ **48.** $(4x + 3)(16x^2 - 12x + 9)$
49. $-(x - 5)(x + 6)$ **50.** $-(x - 2)(x - 4)$ **51.** $(7 - x)(2 + x)$ **52.** $(3 + x)(1 - x)$ **53.** $3x^2y(x + 6)(x - 4)$ **54.** $2xy(x + 5y)(x - y)$
55. $5xy^2(x - 7y)(x - y)$ **56.** $4x^2y(x - 5)(x + 3)$ **57.** $3xy(4x^2 + 81)$ **58.** $2xy^2(3x^2 + 4)$ **59.** $(2 + x)(2 - x)$ **60.** $(3 + y)(3 - y)$
61. $(s + 4)(3r - 1)$ **62.** $(x - 2)(x^2 + 3)$ **63.** $(4x - 3)(x - 2y)$ **64.** $(2x - y)(2x + 7z)$ **65.** $6(x + 2y)(x + y)$
66. $2(x + 4y)(6x - y)$ **67.** $(x + 3)(y + 2)(y - 2)$ **68.** $(y + 3)(y - 3)(x^2 + 3)$ **69.** $(5 + x)(x + y)$ **70.** $(x - y)(7 + y)$
71. $(7t - 1)(2t - 1)$ **72.** prime **73.** $(3x + 5)(x - 1)$ **74.** $(7x - 2)(x + 3)$ **75.** $(x + 12y)(x - 3y)$ **76.** $(3x - 2y)(x + 4y)$
77. $(1 - 10ab)(1 + 2ab)$ **78.** $(1 + 5ab)(1 - 12ab)$ **79.** $(3 + x)(3 - x)(1 + x)(1 - x)$ **80.** $(3 + x)(3 - x)(2 + x)(2 - x)$
81. $(x + 4)(x - 4)(x^2 + 2)$ **82.** $(x + 5)(x - 5)(x^2 + 3)$ **83.** $(x - 15)(x - 8)$ **84.** $(y + 16)(y + 6)$ **85.** $2x(3x - 2)(x - 4)$
86. $2y(3y + 5)(y - 3)$ **87.** $(3x - 5y)(9x^2 + 15xy + 25y^2)$ **88.** $(6y - z)(36y^2 + 6yz + z^2)$ **89.** $(xy + 2z)(x^2y^2 - 2xyz + 4z^2)$
90. $(3ab + 2)(9a^2b^2 - 6ab + 4)$ **91.** $2xy(1 + 6x)(1 - 6x)$ **92.** $2x(x + 3)(x - 3)$ **93.** $(x + 2)(x - 2)(x + 6)$
94. $(x - 2)(x + 6)(x - 6)$ **95.** $2a^2(3a + 5)$ **96.** $2n(2n - 3)$ **97.** $(a^2 + 2)(a + 2)$ **98.** $(a - b)(1 + x)$ **99.** $(x + 2)(x - 2)(x + 7)$
100. $(a + 3)(a - 3)(a + 5)$ **101.** $(x - y + z)(x - y - z)$ **102.** $(x + 2y + 3)(x + 2y - 3)$ **103.** $(9 + 5x + 1)(9 - 5x - 1)$
104. $(b + 4a + c)(b - 4a - c)$ **105.** answers may vary **106.** yes; $9(x^2 + 9y^2)$ **107.** a, c **108.** b, c

Section 6.6

Practice Exercises

1. $-4, 5$ **2.** $-\dfrac{3}{4}, 12$ **3.** $0, \dfrac{6}{7}$ **4.** $-4, 12$ **5.** $\dfrac{4}{3}$ **6.** $-3, \dfrac{2}{3}$ **7.** $-6, 4$ **8.** $-3, 0, 3$ **9.** $\dfrac{2}{3}, \dfrac{3}{2}, 5$ **10.** $-3, 0, 2$
11. The x-intercepts are $(2, 0)$ and $(4, 0)$.

Calculator Explorations 6.6

1. $-0.9, 2.2$ **3.** no real solution **5.** $-1.8, 2.8$

Vocabulary, Readiness & Video Check 6.6

1. quadratic **3.** $3, -5$ **5.** One side of the equation must be a factored polynomial and the other side must be zero. **7.** To find the x-intercepts
of any graph in two variables, we let $y = 0$. Doing this with our quadratic equation gives us an equation $= 0$, which we can try to solve by factoring.

Exercise Set 6.6

1. $6, 7$ **3.** $2, -1$ **5.** $-9, -17$ **7.** $0, -6$ **9.** $0, 8$ **11.** $-\dfrac{3}{2}, \dfrac{5}{4}$ **13.** $\dfrac{7}{2}, -\dfrac{2}{7}$ **15.** $\dfrac{1}{2}, -\dfrac{1}{3}$ **17.** $-0.2, -1.5$ **19.** $9, 4$ **21.** $-4, 2$
23. $0, 7$ **25.** $8, -4$ **27.** $4, -4$ **29.** $-3, 12$ **31.** $\dfrac{7}{3}, -2$ **33.** $-5, 5$ **35.** $-2, \dfrac{1}{6}$ **37.** $0, 4, 8$ **39.** $\dfrac{3}{4}$ **41.** $-\dfrac{1}{2}, 0, \dfrac{1}{2}$ **43.** $-\dfrac{3}{8}, 0, \dfrac{1}{2}$
45. $-3, 2$ **47.** $-20, 0$ **49.** $\dfrac{17}{2}$ **51.** $-\dfrac{1}{2}, \dfrac{1}{2}$ **53.** $-\dfrac{3}{2}, -\dfrac{1}{2}, 3$ **55.** $-5, 3$ **57.** $-\dfrac{5}{6}, \dfrac{6}{5}$ **59.** $2, -\dfrac{4}{5}$ **61.** $-\dfrac{4}{3}, 5$ **63.** $-4, 3$
65. $\dfrac{8}{3}, -9, 0$ **67.** -7 **69.** $0, \dfrac{3}{2}$ **71.** $0, 1, -1$ **73.** $-6, \dfrac{4}{3}$ **75.** $\dfrac{6}{7}, 1$ **77.** $\left(-\dfrac{4}{3}, 0\right), (1, 0)$ **79.** $(-2, 0), (5, 0)$ **81.** $(-6, 0), \left(\dfrac{1}{2}, 0\right)$
83. E **85.** B **87.** C **89.** $\dfrac{47}{45}$ **91.** $\dfrac{17}{60}$ **93.** $\dfrac{15}{8}$ **95.** $\dfrac{7}{10}$ **97.** didn't write equation in standard form; should be $x = 4$ or $x = -2$
99. answers may vary; for example $(x - 6)(x + 1) = 0$ **101.** answers may vary; for example, $x^2 - 12x + 35 = 0$ **103. a.** $300; 304; 276; 216;$
$124; 0; -156$ **b.** 5 sec **c.** 304 ft **d.**
105. $0, \dfrac{1}{2}$ **107.** $0, -15$

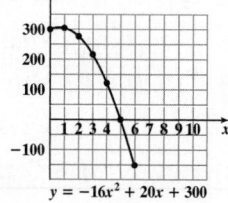

$y = -16x^2 + 20x + 300$

Section 6.7

Practice Exercises

1. 2 sec **2.** There are two numbers. They are -4 and 12. **3.** base: 35 ft; height: 12 ft **4.** 7 and 8 or -6 and -5 **5.** leg: 8 units; leg: 15 units;
hypotenuse: 17 units

Vocabulary, Readiness & Video Check 6.7

1. In applications, the context of the problem needs to be considered. Each exercise resulted in both a positive and a negative solution, and a negative solution is not appropriate for any of the problems.

Exercise Set 6.7

1. width $= x$; length $= x + 4$ **3.** x and $x + 2$ if x is an odd integer   **5.** base $= x$; height $= 4x + 1$ **7.** 11 units   **9.** 15 cm, 13 cm, 70 cm, 22 cm
11. base $= 16$ mi; height $= 6$ mi   **13.** 5 sec   **15.** width $= 5$ cm; length $= 6$ cm   **17.** 54 diagonals   **19.** 10 sides   **21.** -12 or 11   **23.** 14, 15
25. 13 feet   **27.** 5 in.   **29.** 12 mm, 16 mm, 20 mm   **31.** 10 km   **33.** 36 ft   **35.** 9.5 sec   **37.** 20%   **39.** length: 15 mi; width: 8 mi

41. 105 units   **43.** 1.8 million   **45.** 2.3 million   **47.** 2003   **49.** answers may vary   **51.** $\dfrac{4}{7}$ **53.** $\dfrac{3}{2}$ **55.** $\dfrac{1}{3}$

57. slow boat: 8 mph; fast boat: 15 mph   **59.** 13 and 7   **61.** width: 29 m; length: 35 m   **63.** answers may vary

Chapter 6 Vocabulary Check

1. quadratic equation   **2.** Factoring   **3.** greatest common factor   **4.** perfect square trinomial   **5.** difference of two squares   **6.** difference of two cubes   **7.** sum of two cubes   **8.** 0   **9.** hypotenuse   **10.** leg   **11.** hypotenuse

Chapter 6 Review

1. $2x - 5$ **3.** $4x(5x + 3)$ **5.** $(2x + 3)(3x - 5)$ **7.** $(x - 1)(3x + 2)$ **9.** $(2a + b)(5a + 7b)$ **11.** $(x + 4)(x + 2)$ **13.** prime
15. $(x + 6y)(x - 2y)$ **17.** $2(3 - x)(12 + x)$ **19.** $10a(a - 1)(a - 10)$ **21.** $-48, 2$ **23.** $(2x + 1)(x + 6)$ **25.** $(3x + 4y)(2x - y)$
27. $5y(2y - 3)(y + 4)$ **29.** $2(3x - 5)^2$ **31.** $(2x + 3)(2x - 3)$ **33.** prime   **35.** $(2x + 3)(4x^2 - 6x + 9)$ **37.** $2(3 - xy)(9 + 3xy + x^2y^2)$
39. $(4x^2 + 1)(2x + 1)(2x - 1)$ **41.** $-6, 2$ **43.** $-\dfrac{1}{5}, -3$ **45.** $-4, 6$ **47.** $2, 8$ **49.** $-\dfrac{2}{7}, \dfrac{3}{8}$ **51.** $-\dfrac{2}{5}$ **53.** 3   **55.** $0, -\dfrac{7}{4}, 3$
57. c   **59.** 9 units   **61.** width: 20 in.; length: 25 in.   **63.** 19 and 20   **65. a.** 17.5 sec and 10 sec; The rocket reaches a height of 2800 ft on its way up and on its way back down.   **b.** 27.5 sec   **67.** $7(x - 9)$ **69.** $\left(m + \dfrac{2}{5}\right)\left(m - \dfrac{2}{5}\right)$ **71.** $(y + 2)(x - 1)$ **73.** $3x(x - 9)(x - 1)$
75. $2(x + 3)(x - 3)$ **77.** $5(x + 2)^2$ **79.** $2xy(2x - 3y)$ **81.** $3(8x^2 - x - 6)$ **83.** $(x + 3)(x + 2)(x - 2)$
85. $5x^2 - 9x - 2$; $(5x + 1)(x - 2)$ **87.** $-\dfrac{7}{2}, 4$ **89.** $0, -7, -4$ **91.** $0, 16$ **93.** length: 6 in.; width: 2 in.   **95.** $28x^2 - \pi x^2$; $x^2(28 - \pi)$

Chapter 6 Getting Ready for the Test

1. B   **2.** D   **3.** A   **4.** A   **5.** B   **6.** B   **7.** B   **8.** A   **9.** C

Chapter 6 Test

1. $(x + 7)(x + 4)$ **2.** $(7 - m)(7 + m)$ **3.** $(y + 11)^2$ **4.** $(a + 3)(4 - y)$ **5.** prime   **6.** $(y - 12)(y + 4)$ **7.** prime
8. $3x(3x + 1)(x + 4)$ **9.** $(3a - 7)(a + 9)$ **10.** $(3x - 2)(x - 1)$ **11.** $(x + 12y)(x + 2y)$ **12.** $5(6 + x)(6 - x)$
13. $(6t + 5)(t - 1)$ **14.** $(y + 2)(y - 2)(x - 7)$ **15.** $x(1 + x^2)(1 + x)(1 - x)$ **16.** $-xy(y^2 + x^2)$ **17.** $(4x - 1)(16x^2 + 4x + 1)$
18. $8(y - 2)(y^2 + 2y + 4)$ **19.** $-9, 3$ **20.** $-7, 2$ **21.** $-7, 1$ **22.** $0, \dfrac{3}{2}, -\dfrac{4}{3}$ **23.** $0, 3, -3$ **24.** $-3, 5$ **25.** $0, \dfrac{5}{2}$ **26.** 17 ft

27. 8 and 9   **28.** 7 sec   **29.** hypotenuse: 25 cm; legs: 15 cm, 20 cm

Chapter 6 Cumulative Review

1. a. $9 \le 11$ **b.** $8 > 1$ **c.** $3 \ne 4$; Sec. 1.2, Ex. 3   **3. a.** $\dfrac{6}{7}$ **b.** $\dfrac{11}{27}$ **c.** $\dfrac{22}{5}$; Sec. 1.3, Ex. 2   **5.** $\dfrac{14}{3}$; Sec. 1.4, Ex. 5   **7. a.** -12
b. -1; Sec. 1.5, Ex. 7   **9. a.** -32 **b.** -14 **c.** 90; Sec. 1.7, Ex. 1   **11. a.** $4x$ **b.** $11y^2$ **c.** $8x^2 - x$ **d.** $5n^2$; Sec. 2.1, Ex. 3
13. 140; Sec. 2.2, Ex. 7   **15.** -11; Sec. 2.2, Ex. 6   **17.** $\dfrac{16}{3}$; Sec. 2.3, Ex. 2   **19.** shorter: 12 in.; longer: 36 in.; Sec. 2.4, Ex. 3

21.

; Sec. 3.2, Ex. 5   **23.** $m = \dfrac{3}{4}$; y-intercept: $(0, -1)$; Sec. 3.4, Ex. 5   **25. a.** 250   **b.** 1; Sec. 5.1, Ex. 2
27. a. 2   **b.** 5   **c.** 1   **d.** 6   **e.** 0; Sec. 5.2, Ex. 1   **29.** $9x^2 - 6x - 1$; Sec. 5.2, Ex. 11
31. $6x^2 - 11x - 10$; Sec. 5.3, Ex. 5   **33.** $9y^2 + 6y + 1$; Sec. 5.4, Ex. 4   **35. a.** $\dfrac{1}{9}$ **b.** $\dfrac{2}{x^3}$ **c.** $\dfrac{3}{4}$ **d.** $\dfrac{1}{16}$

e. $\dfrac{1}{y^4}$; Sec. 5.5, Ex. 1   **37. a.** 3.67×10^8 **b.** 3.0×10^{-6} **c.** 2.052×10^{10} **d.** 8.5×10^{-4}; Sec. 5.5, Ex. 5   **39.** $x + 4$; Sec. 5.6, Ex. 4
41. a. x^3 **b.** y; Sec. 6.1, Ex. 2   **43.** $(x + 3)(x + 4)$; Sec. 6.2, Ex. 1   **45.** $(4x - 1)(2x - 5)$; Sec. 6.3, Ex. 2   **47.** $(5a + 3b)(5a - 3b)$; Sec. 6.5, Ex. 2b   **49.** $3, -1$; Sec. 6.6, Ex. 1

CHAPTER 7 RATIONAL EXPRESSIONS

Section 7.1
Practice Exercises

1. a. $\{x \mid x \text{ is a real number}\}$ **b.** $\{x \mid x \text{ is a real number and } x \neq -3\}$ **c.** $\{x \mid x \text{ is a real number and } x \neq 2, x \neq 3\}$ **2. a.** $\dfrac{1}{2z - 1}$ **b.** $\dfrac{5x + 3}{6x - 5}$

3. a. 1 **b.** -1 **4.** $-\dfrac{5(2 + x)}{x + 3}$ **5. a.** $x^2 - 4x + 16$ **b.** $\dfrac{5}{z - 3}$ **6.** $\dfrac{-(x + 3)}{6x - 11}; \dfrac{-x - 3}{6x - 11}; \dfrac{x + 3}{-(6x - 11)}; \dfrac{x + 3}{-6x + 11}; \dfrac{x + 3}{11 - 6x}$

7. a. $\$7.20$ **b.** $\$3.60$

Graphing Calculator Explorations 7.1
1. $\{x \mid x \text{ is a real number and } x \neq -2, x \neq 2\}$ 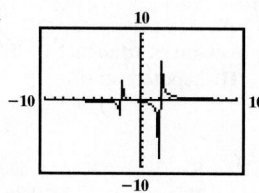 **3.** $\left\{x \mid x \text{ is a real number and } x \neq -4, x \neq \dfrac{1}{2}\right\}$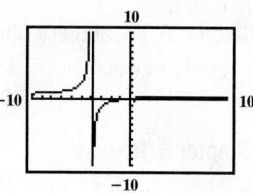

Vocabulary, Readiness & Video Check 7.1

1. rational **3.** domain **5.** 1 **7.** $\dfrac{-a}{b}; \dfrac{a}{-b}$ **9.** Rational expressions are fractions and are therefore undefined if the denominator is zero; the domain of a rational function is all real numbers except those that make the denominator of the related rational expression equal to 0. If a denominator contains variables, set it equal to zero and solve. **11.** We insert parentheses around the numerator or denominator if it has more than one term because the negative sign needs to apply to the entire numerator or denominator.

Exercise Set 7.1

1. $\{x \mid x \text{ is a real number}\}$ **3.** $\{t \mid t \text{ is a real number and } t \neq 0\}$ **5.** $\{x \mid x \text{ is a real number and } x \neq 7\}$ **7.** $\left\{x \mid x \text{ is a real number and } x \neq \dfrac{1}{3}\right\}$

9. $\{x \mid x \text{ is a real number and } x \neq -2, x \neq 0, x \neq 1\}$ **11.** $\{x \mid x \text{ is a real number and } x \neq 2, x \neq -2\}$ **13.** $\dfrac{-(x - 10)}{x + 8}; \dfrac{-x + 10}{x + 8}; \dfrac{x - 10}{-(x + 8)}; \dfrac{x - 10}{-x - 8}$

15. $\dfrac{-(5y - 3)}{y - 12}; \dfrac{-5y + 3}{y - 12}; \dfrac{5y - 3}{-(y - 12)}; \dfrac{5y - 3}{-y + 12}$ **17.** 1 **19.** -1 **21.** $\dfrac{1}{4(x + 2)}$ **23.** -5 **25.** $\dfrac{7}{x}$ **27.** $\dfrac{1}{x - 9}$ **29.** $5x + 1$

31. $\dfrac{x^2}{x - 2}$ **33.** $\dfrac{x + 2}{2}$ **35.** $-(x + 2)$ or $-x - 2$ **37.** $\dfrac{11x}{6}$ **39.** $x + y$ **41.** $x^2 - 2x + 4$ **43.** $-x^2 - x - 1$ **45.** $\dfrac{2y + 5}{3y + 4}$

47. $\dfrac{x - 2}{2x^2 + 1}$ **49.** $\dfrac{1}{3x + 5}$ **51.** correct **53.** correct **55.** $\dfrac{10}{3}, -8, -\dfrac{7}{3}$ **57.** $-\dfrac{17}{48}, \dfrac{2}{7}, -\dfrac{3}{8}$ **59. a.** $\$200$ million **b.** $\$500$ million

c. $\$300$ million **d.** $\{x \mid x \text{ is a real number}\}$ **61.** 400 mg **63.** $C = 78.125$; medium **65.** 0.581 **67.** $\dfrac{3}{11}$ **69.** $\dfrac{4}{3}$ **71.** $\dfrac{117}{40}$

73. correct **75.** incorrect; $\dfrac{1 + 2}{1 + 3} = \dfrac{3}{4}$ **77.** no **79.** yes; 1 **81.** yes; -1 **83.** no; answers may vary **85.** answers may vary

87. $0, \dfrac{20}{9}, \dfrac{60}{7}, 20, \dfrac{140}{3}, 180, 380, 1980;$ 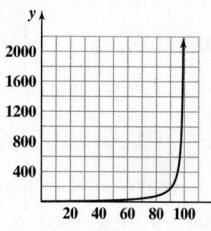 **89.** 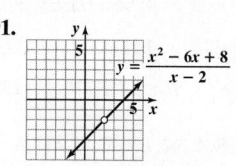 $y = \dfrac{x^2 - 16}{x - 4}$ **91.** $y = \dfrac{x^2 - 6x + 8}{x - 2}$

Section 7.2
Practice Exercises

1. a. $\dfrac{12a}{5b^2}$ **b.** $-\dfrac{2q}{3}$ **2.** $\dfrac{3}{x + 1}$ **3.** $-\dfrac{3x - 5}{2x(x + 2)}$ **4.** $\dfrac{b^2}{8a^2}$ **5.** $\dfrac{3(x - 5)}{4}$ **6.** $\dfrac{2}{x(x - 3)}$ **7.** 1 **8. a.** $\dfrac{(y + 9)^2}{16x^2}$ **b.** $\dfrac{1}{4x}$

c. $-\dfrac{7(x - 2)}{x + 4}$ **9.** 2 sq ft **10.** 504 sq in. **11.** 549,000 sq ft **12.** 70.0 miles per hour

Vocabulary, Readiness & Video Check 7.2

1. reciprocals **3.** $\dfrac{a \cdot d}{b \cdot c}$ or $\dfrac{ad}{bc}$ **5.** $\dfrac{6}{7}$ **7.** fractions; reciprocal **9.** The units in the unit fraction consist of $\dfrac{\text{units converting to}}{\text{original units}}$.

Exercise Set 7.2

1. $\dfrac{21}{4y}$ **3.** x^4 **5.** $-\dfrac{b^2}{6}$ **7.** $\dfrac{x^2}{10}$ **9.** $\dfrac{1}{3}$ **11.** $\dfrac{m+n}{m-n}$ **13.** $\dfrac{x+5}{x}$ **15.** $\dfrac{(x+2)(x-3)}{(x-4)(x+4)}$ **17.** $\dfrac{2x^4}{3}$ **19.** $\dfrac{12}{y^6}$ **21.** $x(x+4)$

23. $\dfrac{3(x+1)}{x^3(x-1)}$ **25.** m^2-n^2 **27.** $-\dfrac{x+2}{x-3}$ **29.** $\dfrac{x+2}{x-3}$ **31.** $\dfrac{5}{6}$ **33.** $\dfrac{3x}{8}$ **35.** $\dfrac{3}{2}$ **37.** $\dfrac{3x+4y}{2(x+2y)}$ **39.** $\dfrac{2(x+2)}{x-2}$ **41.** $-\dfrac{y(x+2)}{4}$

43. $\dfrac{(a+5)(a+3)}{(a+2)(a+1)}$ **45.** $\dfrac{5}{x}$ **47.** $\dfrac{2(n-8)}{3n-1}$ **49.** $4x^3(x-3)$ **51.** $\dfrac{(a+b)^2}{a-b}$ **53.** $\dfrac{3x+5}{x^2+4}$ **55.** $\dfrac{4}{x-2}$ **57.** $\dfrac{a-b}{6(a^2+ab+b^2)}$

59. 1440 **61.** 5 **63.** 81 **65.** 73 **67.** 56.7 **69.** 1,201,500 sq ft **71.** 270.5 miles/hour **73.** 1 **75.** $-\dfrac{10}{9}$ **77.** $-\dfrac{1}{5}$

79.
81. true **83.** false; $\dfrac{x^2+3x}{20}$ **85.** $\dfrac{2}{9(x-5)}$ sq ft **87.** $\dfrac{x}{2}$ **89.** $\dfrac{5a(2a+b)(3a-2b)}{b^2(a-b)(a+2b)}$ **91.** answers may vary

Section 7.3
Practice Exercises

1. $\dfrac{2a}{b}$ **2.** 1 **3.** $4x-5$ **4. a.** 42 **b.** $45y^3$ **5. a.** $(y-5)(y-4)$ **b.** $a(a+2)$ **6.** $3(2x-1)^2$ **7.** $(x+4)(x+1)(x-4)$

8. $3-x$ or $x-3$ **9. a.** $\dfrac{21x^2y}{35xy^2}$ **b.** $\dfrac{18x}{8x+14}$ **10.** $\dfrac{3x-6}{(x-2)(x+3)(x-5)}$

Vocabulary, Readiness & Video Check 7.3

1. $\dfrac{9}{11}$ **3.** $\dfrac{a+c}{b}$ **5.** $\dfrac{5-(6+x)}{x}$ **7.** We factor denominators into the smallest factors—including coefficients—so we can determine the most number of times each unique factor occurs in any one denominator for the LCD.

Exercise Set 7.3

1. $\dfrac{a+9}{13}$ **3.** $\dfrac{3m}{n}$ **5.** 4 **7.** $\dfrac{y+10}{3+y}$ **9.** $5x+3$ **11.** $\dfrac{4}{a+5}$ **13.** $\dfrac{1}{x-6}$ **15.** $\dfrac{5x+7}{x-3}$ **17.** $x+5$ **19.** 3 **21.** $4x^3$
23. $8x(x+2)$ **25.** $(x+3)(x-2)$ **27.** $3(x+6)$ **29.** $5(x-6)^2$ **31.** $6(x+1)^2$ **33.** $x-8$ or $8-x$ **35.** $(x-1)(x+4)(x+3)$
37. $(3x+1)(x+1)(x-1)(2x+1)$ **39.** $2x^2(x+4)(x-4)$ **41.** $\dfrac{6x}{4x^2}$ **43.** $\dfrac{24b^2}{12ab^2}$ **45.** $\dfrac{9y}{2y(x+3)}$ **47.** $\dfrac{9ab+2b}{5b(a+2)}$
49. $\dfrac{x^2+x}{x(x+4)(x+2)(x+1)}$ **51.** $\dfrac{18y-2}{30x^2-60}$ **53.** $2x$ **55.** $\dfrac{x+3}{2x-1}$ **57.** $x+1$ **59.** $\dfrac{1}{x^2-8}$ **61.** $\dfrac{6(4x+1)}{x(2x+1)}$ **63.** $\dfrac{29}{21}$
65. $-\dfrac{7}{12}$ **67.** $\dfrac{7}{30}$ **69.** d **71.** answers may vary **73.** c **75.** b **77.** $-\dfrac{5}{x-2}$ **79.** $\dfrac{7+x}{x-2}$ **81.** $\dfrac{20}{x-2}$ m **83.** answers may vary
85. 95,304 Earth days **87.** answers may vary **89.** answers may vary

Section 7.4
Practice Exercises

1. a. 0 **b.** $\dfrac{21a+10}{24a^2}$ **2.** $\dfrac{6}{x-5}$ **3.** $\dfrac{13y+3}{5y(y+1)}$ **4.** $\dfrac{13}{x-5}$ **5.** $\dfrac{3b+6}{b+3}$ or $\dfrac{3(b+2)}{b+3}$ **6.** $\dfrac{10-3x^2}{2x(2x+3)}$ **7.** $\dfrac{x(5x+6)}{(x+4)(x+3)(x-3)}$

Vocabulary, Readiness & Video Check 7.4

1. D **3.** A **5.** The problem adds two rational expressions with denominators that are opposites of each other. Recognizing this special case can save you time and effort. If you recognize that one denominator is -1 times the other denominator, you may save time.

Exercise Set 7.4

1. $\dfrac{5}{x}$ **3.** $\dfrac{75a-6b^2}{5b}$ **5.** $\dfrac{6x+5}{2x^2}$ **7.** $\dfrac{11}{x+1}$ **9.** $\dfrac{x-6}{(x-2)(x+2)}$ **11.** $\dfrac{35x-6}{4x(x-2)}$ **13.** $-\dfrac{2}{x-3}$ **15.** 0 **17.** $-\dfrac{1}{x^2-1}$
19. $\dfrac{5+2x}{x}$ **21.** $\dfrac{6x-7}{x-2}$ **23.** $-\dfrac{y+4}{y+3}$ **25.** $\dfrac{-5x+14}{4x}$ or $-\dfrac{5x-14}{4x}$ **27.** 2 **29.** $\dfrac{9x^4-4x^2}{21}$ **31.** $\dfrac{x+2}{(x+3)^2}$ **33.** $\dfrac{9b-4}{5b(b-1)}$
35. $\dfrac{2+m}{m}$ **37.** $\dfrac{x^2+3x}{(x-7)(x-2)}$ or $\dfrac{x(x+3)}{(x-7)(x-2)}$ **39.** $\dfrac{10}{1-2x}$ **41.** $\dfrac{15x-1}{(x+1)^2(x-1)}$ **43.** $\dfrac{x^2-3x-2}{(x-1)^2(x+1)}$
45. $\dfrac{a+2}{2(a+3)}$ **47.** $\dfrac{y(2y+1)}{(2y+3)^2}$ **49.** $\dfrac{x-10}{2(x-2)}$ **51.** $\dfrac{2x+21}{(x+3)^2}$ **53.** $\dfrac{-5x+23}{(x-2)(x-3)}$ **55.** $\dfrac{7}{2(m-10)}$
57. $\dfrac{2x^2-2x-46}{(x+1)(x-6)(x-5)}$ or $\dfrac{2(x^2-x-23)}{(x+1)(x-6)(x-5)}$ **59.** $\dfrac{n+4}{4n(n-1)(n-2)}$ **61.** 10 **63.** 2 **65.** $\dfrac{25a}{9(a-2)}$

67. $\dfrac{x+4}{(x-2)(x-1)}$ **69.** $x=\dfrac{2}{3}$ **71.** $x=-\dfrac{1}{2}, x=1$ **73.** $x=-\dfrac{15}{2}$ **75.** $\dfrac{6x^2-5x-3}{x(x+1)(x-1)}$ **77.** $\dfrac{4x^2-15x+6}{(x-2)^2(x+2)(x-3)}$

79. $\dfrac{-2x^2+14x+55}{(x+2)(x+7)(x+3)}$ **81.** $\dfrac{2x-16}{(x+4)(x-4)}$ in. **83.** $\dfrac{P-G}{P}$ **85.** answers may vary **87.** $\left(\dfrac{90x-40}{x}\right)^{\circ}$ **89.** answers may vary

Section 7.5
Practice Exercises

1. -2 **2.** 13 **3.** $-1, 7$ **4.** $-\dfrac{19}{2}$ **5.** No solution **6.** -8 **7.** $b=\dfrac{ax}{a-x}$

Graphing Calculator Explorations 7.5

1. **3.**

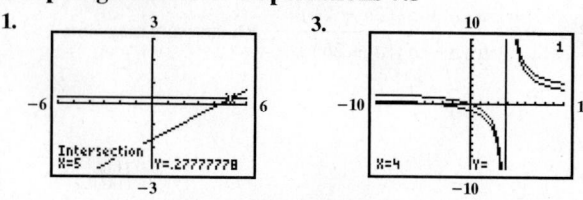

Vocabulary, Readiness & Video Check 7.5

1. c **3.** b **5.** a **7.** These equations are solved in very different ways, so you need to determine the next correct move to make. For a linear equation, you first "move" variable terms to one side and numbers to the other; for a quadratic equation, you first set the equation equal to 0. **9.** the steps for solving an equation containing rational expressions; as if it's the only variable in the equation

Exercise Set 7.5

1. 30 **3.** 0 **5.** -2 **7.** $-5, 2$ **9.** 5 **11.** 3 **13.** 1 **15.** 5 **17.** no solution **19.** 4 **21.** -8 **23.** 6, -4 **25.** 1

27. 3, -4 **29.** -3 **31.** 0 **33.** -2 **35.** 8, -2 **37.** no solution **39.** 3 **41.** $-11, 1$ **43.** $I=\dfrac{E}{R}$ **45.** $B=\dfrac{2U-TE}{T}$

47. $w=\dfrac{Bh^2}{705}$ **49.** $G=\dfrac{V}{N-R}$ **51.** $r=\dfrac{C}{2\pi}$ **53.** $x=\dfrac{3y}{3+y}$ **55.** $\dfrac{1}{x}$ **57.** $\dfrac{1}{x}+\dfrac{1}{2}$ **59.** $\dfrac{1}{3}$ **61.** $(2,0), (0,-2)$

63. $(-4,0), (-2,0), (3,0), (0,4)$ **65.** answers may vary **67.** $\dfrac{5x+9}{9x}$ **69.** no solution **71.** $100°, 80°$ **73.** $22.5°, 67.5°$ **75.** $\dfrac{17}{4}$

Integrated Review

1. expression; $\dfrac{3+2x}{3x}$ **2.** expression; $\dfrac{18+5a}{6a}$ **3.** equation; 3 **4.** equation; 18 **5.** expression; $\dfrac{x+1}{x(x-1)}$ **6.** expression; $\dfrac{3(x+1)}{x(x-3)}$

7. equation; no solution **8.** equation; 1 **9.** expression; 10 **10.** expression; $\dfrac{z}{3(9z-5)}$ **11.** expression; $\dfrac{5x+7}{x-3}$ **12.** expression; $\dfrac{7p+5}{2p+7}$

13. equation; 23 **14.** equation; 5 **15.** expression; $\dfrac{25a}{9(a-2)}$ **16.** expression; $\dfrac{4x+5}{(x+1)(x-1)}$ **17.** expression; $\dfrac{3x^2+5x+3}{(3x-1)^2}$

18. expression; $\dfrac{2x^2-3x-1}{(2x-5)^2}$ **19.** expression; $\dfrac{4x-37}{5x}$ **20.** expression; $\dfrac{29x-23}{3x}$ **21.** equation; $\dfrac{8}{5}$ **22.** equation; $-\dfrac{7}{3}$

23. answers may vary **24.** answers may vary

Section 7.6
Practice Exercises

1. 99 **2.** $\dfrac{13}{3}$ **3.** \$9.03 **4.** 6 **5.** 15 **6.** $1\dfrac{5}{7}$ hr **7.** bus: 45 mph; car: 60 mph

Vocabulary, Readiness & Video Check 7.6

1. c **3.** $\dfrac{1}{x}; \dfrac{1}{x}-3$ **5.** $z+5; \dfrac{1}{z+5}$ **7.** $2y; \dfrac{11}{2y}$ **9.** No. Proportions are actually equations containing rational expressions, so they can also be solved by using the steps to solve those equations. **11.** divided by, quotient **13.** $\dfrac{325}{x+7}=\dfrac{290}{x}$

Exercise Set 7.6

1. 4 **3.** $\dfrac{50}{9}$ **5.** -3 **7.** $\dfrac{14}{9}$ **9.** 123 lb **11.** 132 cal **13.** $y=21.25$ **15.** $y=5\dfrac{5}{7}$ ft **17.** 2 **19.** -3 **21.** $2\dfrac{2}{9}$ hr **23.** $1\dfrac{1}{2}$ min

25. trip to park rate: r; to park time: $\dfrac{12}{r}$; return trip rate: r; return time: $\dfrac{18}{r}=\dfrac{12}{r}+1$; $r=6$ mph **27.** 1st portion: 10 mph; cooldown: 8 mph

29. 360 sq ft **31.** 2 **33.** \$108.00 **35.** 20 mph **37.** $y=37\dfrac{1}{2}$ ft **39.** 41 mph; 51 mph **41.** 5 **43.** 217 mph **45.** 9 gal

47. 8 mph **49.** 2.2 mph; 3.3 mph **51.** 3 hr **53.** $26\dfrac{2}{3}$ ft **55.** 216 nuts **57.** $666\dfrac{2}{3}$ mi **59.** 20 hr **61.** car: 70 mph; motorcycle: 60 mph

63. $5\dfrac{1}{4}$ hr **65.** 8 **67.** first car: 64 mph; second car: 50 mph **69.** 510 mph **71.** $x=5$ **73.** $x=13.5$ **75.** $-\dfrac{4}{3}$; downward

77. $\frac{11}{4}$; upward   **79.** undefined slope; vertical   **81.** 64,000 megawatts   **83.** 35,840,000 people   **85.** yes   **87.** first pump: 28 min; second pump: 84 min   **89.** none; answers may vary   **91.** answers may vary   **93.** $R = \frac{D}{T}$

Section 7.7

Practice Exercises

1. a. $\frac{1}{12m}$ **b.** $\frac{8x(x+4)}{3(x-4)}$ **c.** $\frac{b^2}{a^2}$ **2. a.** $\frac{8x(x+4)}{3(x-4)}$ **b.** $\frac{b^2}{a^2}$ **3.** $\frac{y(3xy+1)}{x^2(1+xy)}$ **4.** $\frac{1-6x}{15+6x}$ or $\frac{1-6x}{3(5+2x)}$

Vocabulary, Readiness & Video Check 7.7

1. $\frac{7}{1+z}$ **3.** $\frac{1}{x^2}$ **5.** $\frac{2}{x}$ **7.** $\frac{1}{9y}$ **9.** a single fraction in the numerator and in the denominator   **11.** Since a negative exponent moves its base from a numerator to a denominator of the expression, a rational expression containing negative exponents can become a complex fraction when rewritten with positive exponents.

Exercise Set 7.7

1. 4   **3.** $\frac{7}{13}$ **5.** $\frac{4}{x}$ **7.** $\frac{9(x-2)}{9x^2+4}$ **9.** $2x+y$ **11.** $\frac{2(x+1)}{2x-1}$ **13.** $\frac{2x+3}{4-9x}$ **15.** $\frac{1}{x^2-2x+4}$ **17.** $\frac{x}{5(x-2)}$ **19.** $\frac{x-2}{2x-1}$

21. $\frac{x}{2-3x}$ **23.** $-\frac{y}{x+y}$ **25.** $-\frac{2x^3}{y(x-y)}$ **27.** $\frac{2x+1}{y}$ **29.** $\frac{x-3}{9}$ **31.** $\frac{1}{x+2}$ **33.** 2   **35.** $\frac{xy^2}{x^2+y^2}$ **37.** $\frac{2b^2+3a}{b(b-a)}$

39. $\frac{x}{(x+1)(x-1)}$ **41.** $\frac{1+a}{1-a}$ **43.** $\frac{x(x+6y)}{2y}$ **45.** $\frac{5a}{2(a+2)}$ **47.** $xy(5y+2x)$ **49.** $\frac{xy}{2x+5y}$ **51.** $\frac{x^2y^2}{4}$ **53.** $-9x^3y^4$

55. -9 **57.** a and c   **59.** $\frac{770a}{770-s}$ **61.** a, b   **63.** answers may vary   **65.** $\frac{1+x}{2+x}$ **67.** $x(x+1)$ **69.** $\frac{x-3y}{x+3y}$ **71.** $3a^2+4a+4$

73. a. $\frac{1}{a+h}$ **b.** $\frac{1}{a}$ **c.** $\frac{\frac{1}{a+h}-\frac{1}{a}}{h}$ **d.** $-\frac{1}{a(a+h)}$ **75. a.** $\frac{3}{a+h+1}$ **b.** $\frac{3}{a+1}$ **c.** $\frac{\frac{3}{a+h+1}-\frac{3}{a+1}}{h}$ **d.** $\frac{-3}{(a+h+1)(a+1)}$

Chapter 7 Vocabulary Check

1. rational expression   **2.** complex fraction   **3.** $\frac{-a}{b}; \frac{a}{-b}$ **4.** denominator   **5.** simplifying   **6.** reciprocals   **7.** least common denominator
8. ratio   **9.** proportion   **10.** cross products   **11.** domain

Chapter 7 Review

1. $\{x \mid x \text{ is a real number}\}$ **3.** $\{x \mid x \text{ is a real number and } x \neq 5\}$ **5.** $\{x \mid x \text{ is a real number and } x \neq 0, x \neq 8\}$ **7.** -1 **9.** $\frac{1}{x-7}$

11. $\frac{x+a}{x-c}$ **13.** $-\frac{1}{x^2+4x+16}$ **15.** \$119   **17.** $\frac{3x^2}{y}$ **19.** $\frac{x-3}{x+2}$ **21.** $\frac{x+3}{x-4}$ **23.** $(x-6)(x-3)$ **25.** $\frac{1}{2}$ **27.** $-\frac{2(2x+3)}{y-2}$

29. $\frac{1}{x+2}$ **31.** $\frac{2x-10}{3x^2}$ **33.** $14x$ **35.** $\frac{10x^2y}{14x^3y}$ **37.** $\frac{x^2-3x-10}{(x+2)(x-5)(x+9)}$ **39.** $\frac{4y-30x^2}{5x^2y}$ **41.** $\frac{-2x-2}{x+3}$ **43.** $\frac{x-4}{3x}$

45. $\frac{x^2+2x+4}{4x}; \frac{x+2}{32}$ **47.** 30   **49.** no solution   **51.** $\frac{9}{7}$ **53.** $b = \frac{4A}{5x^2}$ **55.** 6   **57.** 9   **59.** 675 parts   **61.** 3

63. fast car speed: 30 mph; slow car speed: 20 mph   **65.** $17\frac{1}{2}$ hr   **67.** $x = 15$ **69.** $-\frac{7}{18y}$ **71.** $\frac{3y-1}{2y-1}$ **73.** $-\frac{x^2+9}{6x}$ **75.** $\frac{xy+1}{x}$

77. $\frac{1}{2x}$ **79.** $\frac{x-4}{x+4}$ **81.** $\frac{1}{x-6}$ **83.** $\frac{2}{(x+3)(x-2)}$ **85.** $\frac{1}{2}$ **87.** 1   **89.** $x = 6$ **91.** $\frac{3}{10}$ **93.** $\frac{1}{y^2-1}$

Chapter 7 Getting Ready for the Test

1. D   **2.** E   **3.** B   **4.** A   **5.** A   **6.** C   **7.** C   **8.** D   **9.** D   **10.** A   **11.** D   **12.** D   **13.** A   **14.** B   **15.** B
16. A   **17.** C   **18.** C   **19.** A   **20.** A

Chapter 7 Test

1. $\{x \mid x \text{ is a real number}, x \neq -1, x \neq -3\}$ **2. a.** \$115   **b.** \$103   **3.** $\frac{3}{5}$ **4.** $\frac{1}{x+6}$ **5.** $\frac{1}{x^2-3x+9}$ **6.** $\frac{2m(m+2)}{m-2}$ **7.** $\frac{a+2}{a+5}$

8. $-\frac{1}{x+y}$ **9.** 15   **10.** $\frac{y-2}{4}$ **11.** $\frac{19x-6}{2x+5}$ **12.** $\frac{3a-4}{(a-3)(a+2)}$ **13.** $\frac{3}{x-1}$ **14.** $\frac{2(x+5)}{x(y+5)}$ **15.** $\frac{x^2+2x+35}{(x+9)(x+2)(x-5)}$

16. $\frac{30}{11}$ **17.** -6 **18.** no solution   **19.** $-2, 5$ **20.** no solution   **21.** $\frac{xz}{2y}$ **22.** $\frac{5y^2-1}{y+2}$ **23.** $b-a$ **24.** 18 bulbs   **25.** 5 or 1

26. 30 mph   **27.** $6\frac{2}{3}$ hr   **28.** $x = 12$

Chapter 7 Cumulative Review

1. a. $\dfrac{15}{x} = 4$ **b.** $12 - 3 = x$ **c.** $4x + 17 \neq 21$ **d.** $3x < 48$; Sec. 1.4, Ex. 9 **3.** amount at 7%: \$12,500; amount at 9%: \$7500; Sec. 2.7, Ex. 4

5. 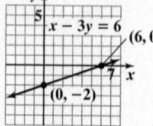 ; Sec. 3.3, Ex. 6 **7. a.** 4^7 **b.** x^{10} **c.** y^4 **d.** y^{12} **e.** $(-5)^{15}$ **f.** a^2b^2; Sec. 5.1, Ex. 3 **9.** $12z + 16$; Sec. 5.2, Ex. 12

11. $27a^3 + 27a^2b + 9ab^2 + b^3$; Sec. 5.3, Ex. 8 **13. a.** $t^2 + 4t + 4$ **b.** $p^2 - 2pq + q^2$ **c.** $4x^2 + 20x + 25$ **d.** $x^4 - 14x^2y + 49y^2$; Sec. 5.4, Ex. 5

15. a. x^3 **b.** 81 **c.** $\dfrac{q^9}{p^4}$ **d.** $\dfrac{32}{125}$; Sec. 5.5, Ex. 2 **17.** $4x^2 - 4x + 6 + \dfrac{-11}{2x + 3}$; Sec. 5.6, Ex. 6 **19. a.** 4 **b.** 1 **c.** 3; Sec. 6.1, Ex. 1

21. $-3a(3a^4 - 6a + 1)$; Sec. 6.1, Ex. 5 **23.** $3(m + 2)(m - 10)$; Sec. 6.2, Ex. 9 **25.** $(3x + 2)(x + 3)$; Sec. 6.3, Ex. 1 **27.** $(x + 6)^2$; Sec. 6.3, Ex. 8

29. prime polynomial; Sec. 6.5, Ex. 4b **31.** $(x + 2)(x^2 - 2x + 4)$; Sec. 6.5, Ex. 8 **33.** $(2x + 3)(x + 1)(x - 1)$; Ch. 6 Int. Rev., Ex. 2

35. $3(2m + n)(2m - n)$; Ch. 6 Int. Rev., Ex. 3 **37.** $-\dfrac{1}{2}, 4$; Sec. 6.6, Ex. 6 **39.** $(1, 0), (4, 0)$; Sec. 6.6, Ex. 11 **41.** base: 6 m; height: 10 m; Sec. 6.7, Ex. 3

43. $-\dfrac{2(3 + x)}{x + 1}$; Sec. 7.1, Ex. 4 **45.** $\dfrac{2}{x(x + 1)}$; Sec. 7.2, Ex. 6 **47.** $\dfrac{1 + 2x}{2(2 - x)}$; Sec. 7.7, Ex. 4

CHAPTER 8 MORE ON FUNCTIONS AND GRAPHS

Section 8.1
Practice Exercises

1. 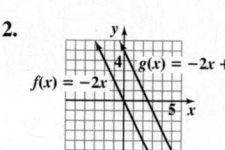 **2.** **3.** $f(x) = -4x - 3$ **4.** $f(x) = -\dfrac{2}{3}x + \dfrac{4}{3}$ **5.** $f(x) = -2$ **6.** $3x + 4y = 12$ **7.** $f(x) = \dfrac{4}{3}x - \dfrac{41}{3}$

Graphing Calculator Explorations 8.1

1. $y = \dfrac{x}{3.5}$ 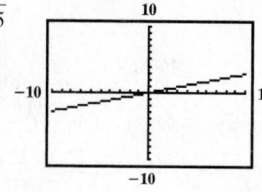 **3.** $y = -\dfrac{5.78}{2.31}x + \dfrac{10.98}{2.31}$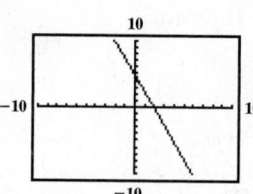

5. $y = |x| + 3.78$ 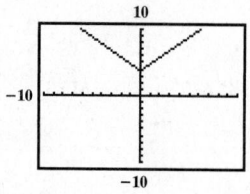 **7.** $y = 5.6x^2 + 7.7x + 1.5$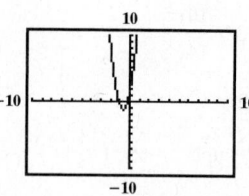

Vocabulary, Readiness & Video Check 8.1

1. linear **3.** $m = -4$, y-intercept: $(0, 12)$ **5.** $m = 5$, y-intercept: $(0, 0)$ **7.** parallel **9.** neither **11.** $f(x) = mx + b$, or slope-intercept form **13.** if one of the two points given is the y-intercept

Exercise Set 8.1

1. 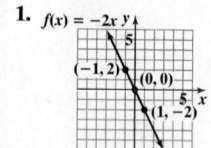 **3.** 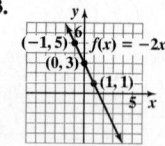 **5.** 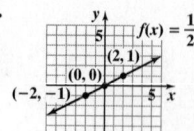 **7.** 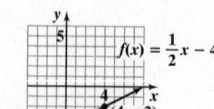 **9.** C **11.** D **13.** $f(x) = -x + 1$

15. $f(x) = 2x + \dfrac{3}{4}$ **17.** $f(x) = \dfrac{2}{7}x$ **19.** $f(x) = 3x - 1$ **21.** $f(x) = -2x - 1$ **23.** $f(x) = \dfrac{1}{2}x + 5$ **25.** $f(x) = -\dfrac{9}{10}x - \dfrac{27}{10}$

27. $f(x) = 3x - 6$ **29.** $f(x) = -2x + 1$ **31.** $f(x) = -\dfrac{1}{2}x - 5$ **33.** $f(x) = \dfrac{1}{3}x - 7$ **35.** $f(x) = -\dfrac{3}{8}x + \dfrac{5}{8}$ **37.** $f(x) = -4$

39. $f(x) = 5$ **41.** $f(x) = 4x - 4$ **43.** $f(x) = -3x + 1$ **45.** $f(x) = -\dfrac{3}{2}x - 6$ **47.** $2x - y = -7$ **49.** $f(x) = -x + 7$

51. $f(x) = -\frac{1}{2}x + 11$ **53.** $2x + 7y = -42$ **55.** $4x + 3y = -20$ **57.** $f(x) = -10$ **59.** $x + 2y = 2$ **61.** $f(x) = 12$

63. $8x - y = 47$ **65.** $x = 5$ **67.** $f(x) = -\frac{3}{8}x - \frac{29}{4}$ **69.** 20.84; in 1990 there were about 20.84 million tons of sulfur dioxide emissions in the U.S.

71. 5.5; In 2020, we predict there will be about 5.5 million tons of sulfur dioxide emissions in the U.S.   **73.** answers may vary   **75.** $f(x) = -2x + 3$

77. $f(x) = \frac{2}{3}x + \frac{7}{3}$ **79. a.** $y = 32x$ **b.** 128 ft per sec   **81. a.** $y = -250x + 3500$ **b.** 1625 Frisbees   **83. a.** $y = 3.7x + 110$

b. 132.2 thousand   **85. a.** $y = 7170x + 271,500$ **b.** $357,540   **87.** $-4x + y = 4$ **89.** $2x + y = -23$ **91.** $3x - 2y = -13$

93. answers may vary

Section 8.2

Practice Exercises

1. a. -3 **b.** -2 **c.** 3   **d.** 1   **e.** -1 and 3   **f.** -3 **2.** 13 million or 13,000,000 students   **3.** 8.3 million or 8,300,000 students

4. a. 11   **b.** $\frac{1}{4}$ **c.** -8 **d.** not a real number   **e.** 10   **5.** 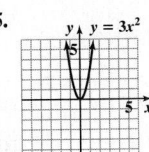   **6.** 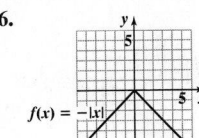   **7.**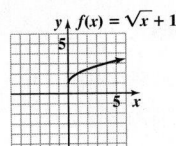

Graphing Calculator Explorations 8.2

1. 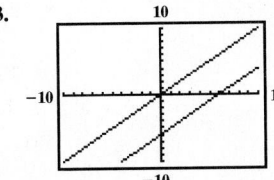   **3.** 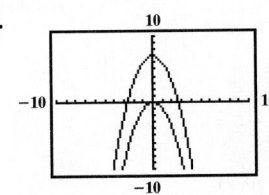   **5.**

Vocabulary, Readiness & Video Check 8.2

1. V-shaped   **3.** $(-2, 1.7)$ **5.** Using function notation, the replacement value for x and the resulting $f(x)$ or y-value corresponds to an ordered pair (x, y) solution of the function   **7.** not; shape

Exercise Set 8.2

1. 0   **3.** -4 **5.** 3   **7.** $(1, -10)$ **9.** $(4, 56)$ **11.** $f(-1) = -2$ **13.** $g(2) = 0$ **15.** $-4, 0$ **17.** 3   **19.** 7   **21.** $-\frac{2}{3}$

23. 8   **25.** 9   **27.** not a real number   **29.**   **31.** 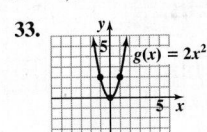   **33.**   **35.**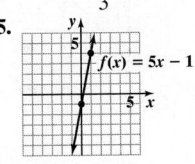

37. 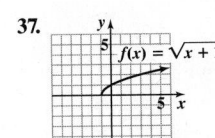   **39.** 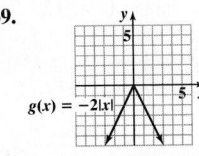   **41.** 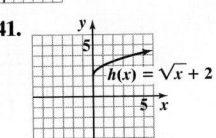   **43. a.** 7.6 million or 7,600,000   **b.** 7.04 million or 7,040,000

45. $15.54 billion   **47.** 25π sq cm   **49.** 2744 cu in.

51. 166.38 cm   **53.** 163.2 mg   **55.** infinite number

57. -5 **59.** $-\frac{1}{10}$ **61.** b   **63.** c   **65.** 1997

67. answers may vary

69.   **71.**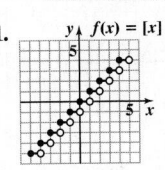

Integrated Review

1. $m = 3$; $(0, -5)$ **2.** $m = \frac{5}{2}$; $\left(0, -\frac{7}{2}\right)$ **3.** parallel   **4.** perpendicular   **5.** $f(x) = -x + 7$ **6.** $f(x) = -\frac{3}{8}x - \frac{29}{4}$ **7.** $f(x) = 3x - 2$

8. $f(x) = -\frac{5}{4}x + 4$ **9.** $f(x) = \frac{1}{4}x - \frac{7}{2}$ **10.** $f(x) = -\frac{5}{2}x - \frac{5}{2}$

11. linear   **12.** linear   **13.** not linear   **14.** not linear   **15.** not linear   **16.** not linear

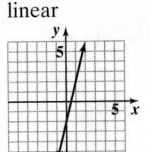   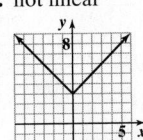   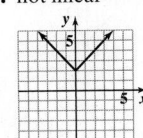   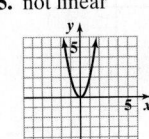   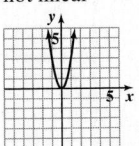

17. not linear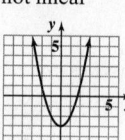

18. not linear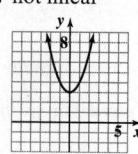

19. linear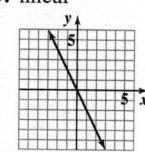

20. linear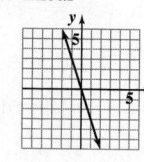

21. not linear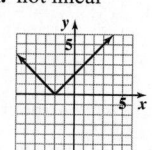

22. not linear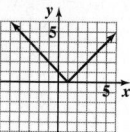

23. linear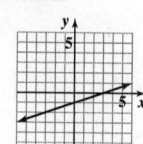

24. linear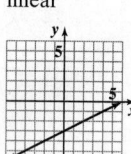

25. linear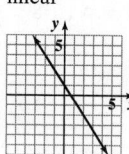

26. linear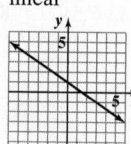

Section 8.3

Practice Exercises

1. $f(4) = 5; f(-2) = 6; f(0) = -2$

2.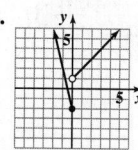

3.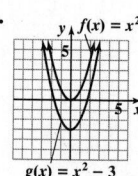

4.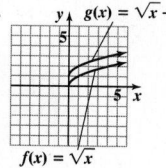

5.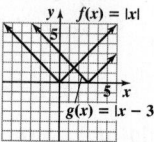

6.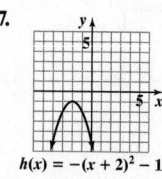

7.

Vocabulary, Readiness & Video Check 8.3

1. C   **3.** D   **5.** Although $f(x) = x + 3$ isn't defined for $x = -1$, we need to clearly indicate the point where this piece of the graph ends. Therefore, we find this point and graph it as an open circle.   **7.** x-axis

Exercise Set 8.3

1.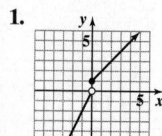

3.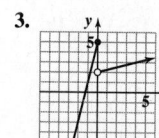

5.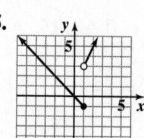

7.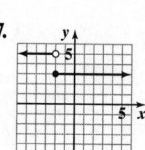

9. 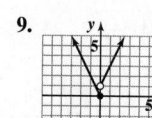 domain: $(-\infty, \infty)$; range: $[0, \infty)$

11. 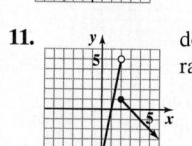 domain: $(-\infty, \infty)$; range: $(-\infty, 5)$

13. 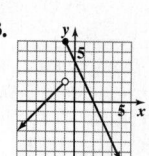 domain: $(-\infty, \infty)$; range: $(-\infty, 6]$

15. 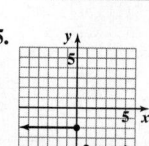 domain: $(-\infty, 0] \cup [1, \infty)$; range: $\{-4, -2\}$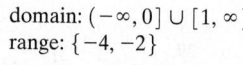

17.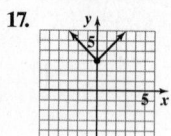

19.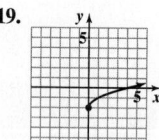

21.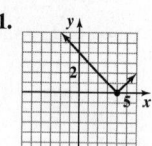

23.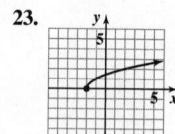

25.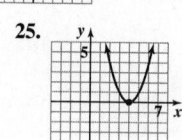

27.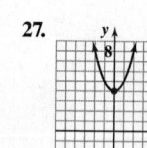

29.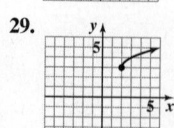

31.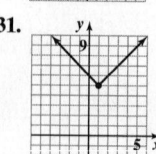

33.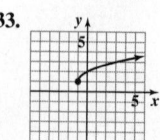

35.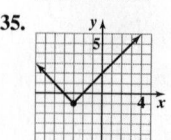

37.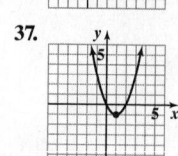

39.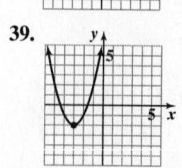

41.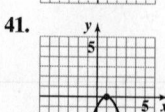

43.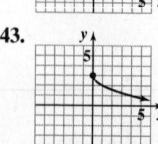

45.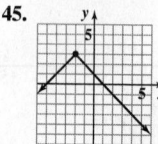

47.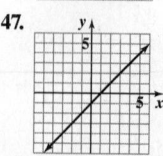

49. A   **51.** D   **53.** answers may vary

55.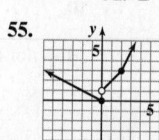

57. domain: $[2, \infty)$; range: $[3, \infty)$

59. domain: $(-\infty, \infty)$; range: $(-\infty, 3]$ **61.** $[20, \infty)$ **63.** $(-\infty, \infty)$ **65.** $[-103, \infty)$

67. 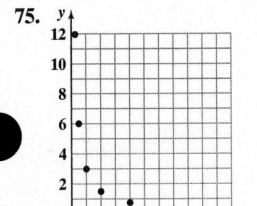   domain: $(-\infty, \infty)$;   **69.** 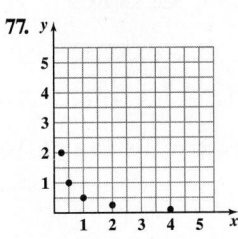   domain: $(-\infty, \infty)$;
range: $[0, \infty)$ range: $(-\infty, 0] \cup (2, \infty)$

Section 8.4
Practice Exercises

1. $k = \dfrac{4}{3}; y = \dfrac{4}{3}x$ **2.** $18\dfrac{3}{4}$ in.   **3.** $k = 45; b = \dfrac{45}{a}$ **4.** $653\dfrac{1}{3}$ kilopascals   **5.** $A = kpa$ **6.** $k = 4; y = \dfrac{4}{x^3}$ **7.** $k = 81; y = \dfrac{81z}{x^3}$

Vocabulary, Readiness & Video Check 8.4

1. direct   **3.** joint   **5.** inverse   **7.** direct   **9.** linear; slope   **11.** $y = ka^2b^5$

Exercise Set 8.4

1. $k = \dfrac{1}{5}; y = \dfrac{1}{5}x$ **3.** $k = \dfrac{3}{2}; y = \dfrac{3}{2}x$ **5.** $k = 14; y = 14x$ **7.** $k = 0.25; y = 0.25x$ **9.** 4.05 lb   **11.** 12.5 billion gal   **13.** $k = 30; y = \dfrac{30}{x}$

15. $k = 700; y = \dfrac{700}{x}$ **17.** $k = 2; y = \dfrac{2}{x}$ **19.** $k = 0.14; y = \dfrac{0.14}{x}$ **21.** 54 mph   **23.** 72 amps   **25.** divided by 4   **27.** $x = kyz$

29. $r = kst^3$ **31.** $k = \dfrac{1}{3}; y = \dfrac{1}{3}x^3$ **33.** $k = 0.2; y = 0.2\sqrt{x}$ **35.** $k = 1.3; y = \dfrac{1.3}{x^2}$ **37.** $k = 3; y = 3xz^3$ **39.** 22.5 tons   **41.** 15π cu in.

43. 8 ft   **45.** $y = kx$ **47.** $a = \dfrac{k}{b}$ **49.** $y = kxz$ **51.** $y = \dfrac{k}{x^3}$ **53.** $y = \dfrac{kx}{p^2}$ **55.** $C = 8\pi$ in.; $A = 16\pi$ sq in.

57. $C = 18\pi$ cm; $A = 81\pi$ sq cm   **59.** 1.2   **61.** -7 **63.** $\dfrac{-1}{2}$ **65.** $\dfrac{8}{27}$ **67.** a   **69.** c   **71.** multiplied by 8   **73.** multiplied by 2

75. 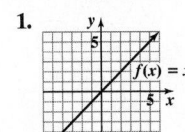   **77.**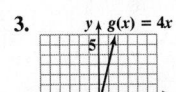

Chapter 8 Vocabulary Check

1. Parallel   **2.** Slope-intercept   **3.** function   **4.** slope   **5.** perpendicular   **6.** linear function   **7.** directly
8. inversely   **9.** jointly

Chapter 8 Review

1. 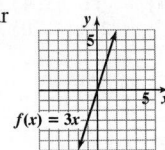   **3.** 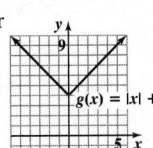   **5.** C   **7.** B   **9.** $m = \dfrac{2}{5}$; y-intercept $\left(0, -\dfrac{4}{3}\right)$ **11.** $2x - y = 12$ **13.** $11x + y = -52$

15. $y = -5$ **17.** $f(x) = -1$ **19.** $f(x) = -x - 2$ **21.** $f(x) = -\dfrac{3}{2}x - 8$

23. $f(x) = -\dfrac{3}{2}x - 1$ **25. a.** $y = 12{,}000x + 126{,}000$ **b.** \$342,000   **27.** 0   **29.** $-2, 4$

31. linear 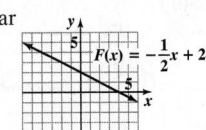   **33.** not linear 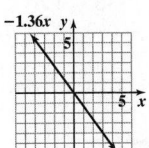   **35.** linear 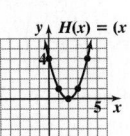   **37.** linear $y = -1.36x$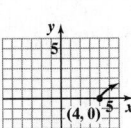

39. not linear 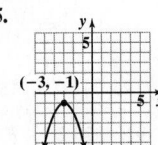   **41.** 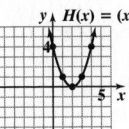   **43.** 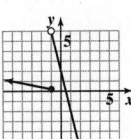   **45.** 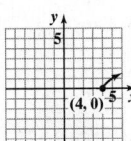   **47.** 9   **49.** 3.125 cu ft   **51.** $f(x) = \dfrac{9}{2}$

53. $f(x) = -5x - 7$ **55.** $f(x) = -\dfrac{4}{5}x + 3$ **57.** 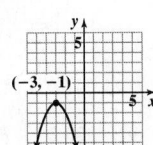   **59.**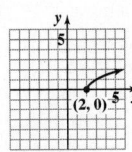

Chapter 8 Getting Ready for the Test

1. D **2.** C **3.** B **4.** C **5.** C **6.** B **7.** D **8.** C or E, C or E **9.** D **10.** G **11.** I **12.** A **13.** D **14.** B **15.** C **16.** A **17.** E

Chapter 8 Test

1. 3 **2.** −5 **3.** 2, −2 **4.** 0 **5.** 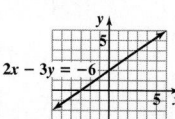 **6.** 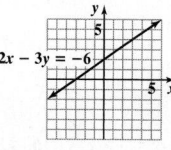 **7.** $y = -8$ **8.** $3x + y = 11$ **9.** $5x - y = 2$

10. $f(x) = -\dfrac{1}{2}x$ **11.** $f(x) = -\dfrac{1}{3}x + \dfrac{5}{3}$ **12.** $f(x) = -\dfrac{1}{2}x - \dfrac{1}{2}$ **13.** neither **14.** domain: $(-\infty, \infty)$; range: $\{5\}$; function
15. domain: $\{-2\}$; range: $(-\infty, \infty)$; not a function **16.** domain: $(-\infty, \infty)$; range: $[0, \infty)$; function **17.** domain: $(-\infty, \infty)$; range: $(-\infty, \infty)$;
function **18. a.** 82 games **b.** 84 games **c.** \$648 million **d.** $m = 0.024$; Every million dollars spent on payroll increases winnings by 0.024 game.
19. 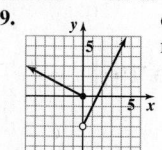 domain: $(-\infty, \infty)$; **20.** 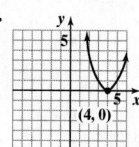 **21.** 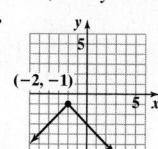 domain: $(-\infty, \infty)$; **22.**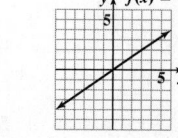
range: $(-3, \infty)$ range: $(-\infty, -1]$

23. 16 **24.** 9 **25.** 256 ft

Chapter 8 Cumulative Review

1. 66; Sec. 1.4, Ex. 4 **3. a.** $\dfrac{1}{2}$ **b.** 19; Sec. 1.6, Ex. 6 **5.** −11; Sec. 2.2, Ex. 6 **7.** $\dfrac{16}{3}$; Sec. 2.3, Ex. 2 **9.** $x = \dfrac{y - b}{m}$; Sec. 2.5, Ex. 6
11. $(-\infty, 4]$; Sec. 2.8, Ex. 5 **13.** 0; Sec. 3.4, Ex. 6 **15. a.** 1; (2, 1) **b.** 1; (−2, 1) **c.** −3; (0, −3); Sec. 3.6, Ex. 7 **17.** (4, 2); Sec. 4.2, Ex. 1
19. $\left(-\dfrac{15}{7}, -\dfrac{5}{7}\right)$; Sec. 4.3, Ex. 6 **21.** $x + 4$; Sec. 5.6, Ex. 4 **23. a.** $6(t + 3)$ **b.** $y^5(1 - y^2)$; Sec. 6.1, Ex. 4
25. $(x - 2)(x + 6)$; Sec. 6.2, Ex. 3 **27.** $(2x - 3y)(5x + y)$; Sec. 6.3, Ex. 4 **29.** $(x + 2)(x^2 - 2x + 4)$; Sec. 6.5, Ex. 8
31. 11, −2; Sec. 6.6, Ex. 4 **33. a.** $\dfrac{1}{5x - 1}$ **b.** $\dfrac{9x + 4}{8x - 7}$; Sec. 7.1, Ex. 2 **35.** $-\dfrac{3(x + 1)}{5x(2x - 3)}$; Sec. 7.2, Ex. 3 **37.** $3x - 5$; Sec. 7.3, Ex. 3
39. −3, −2; Sec. 7.5, Ex. 3 **41.** $f(x) = \dfrac{5}{8}x - \dfrac{5}{2}$; Sec. 8.1, Ex. 4

CHAPTER 9 INEQUALITIES AND ABSOLUTE VALUE

Section 9.1
Practice Exercises

1. $\{1, 3\}$ **2.** $(-\infty, 2)$ **3.** $\{\ \}$ or $\varnothing$ **4.** $(-4, 2)$ **5.** $[-6, 8]$ **6.** $\{1, 2, 3, 4, 5, 6, 7, 9\}$ **7.** $\left(-\infty, \dfrac{3}{8}\right] \cup [3, \infty)$ **8.** $(-\infty, \infty)$

Vocabulary, Readiness & Video Check 9.1
1. compound **3.** or **5.** $\cup$ **7.** and **9.** or

Exercise Set 9.1
1. $\{2, 3, 4, 5, 6, 7\}$ **3.** $\{4, 6\}$ **5.** $\{\ldots, -2, -1, 0, 1, \ldots\}$ **7.** $\{5, 7\}$ **9.** $\{x \mid x$ is an odd integer or $x = 2$ or $x = 4\}$ **11.** $\{2, 4\}$
13. ⟵─(───)──⟶ $(-3, 1)$ **15.** ⟵──┼──⟶ $\varnothing$ **17.** ⟵────⟶ $(-\infty, -1)$ **19.** $[6, \infty)$ **21.** $(-\infty, -3]$ **23.** $(4, 10)$
$\qquad\; -3 \quad\; 1 \qquad\qquad\qquad\quad 0 \qquad\qquad\qquad\quad -1$
25. $(11, 17)$ **27.** $[1, 4]$ **29.** $\left[-3, \dfrac{3}{2}\right]$ **31.** $\left[-\dfrac{7}{3}, 7\right]$ **33.** ⟵──⟶ $(-\infty, 5)$ **35.** ⟵─┼──┼─⟶ $(-\infty, -4] \cup [1, \infty)$
$\qquad\qquad\qquad\qquad\qquad\qquad\qquad\qquad\qquad\qquad\qquad\qquad\; 5 \qquad\qquad\qquad\qquad\quad -4 \quad 1$
37. ⟵──┼──⟶ $(-\infty, \infty)$ **39.** $[2, \infty)$ **41.** $(-\infty, -4) \cup (-2, \infty)$ **43.** $(-\infty, \infty)$ **45.** $\left(-\dfrac{1}{2}, \dfrac{2}{3}\right)$ **47.** $(-\infty, \infty)$ **49.** $\left[\dfrac{3}{2}, 6\right]$
$\qquad\quad 0$
51. $\left(\dfrac{5}{4}, \dfrac{11}{4}\right)$ **53.** $\varnothing$ **55.** $\left(-\infty, -\dfrac{56}{5}\right) \cup \left(\dfrac{5}{3}, \infty\right)$ **57.** $\left(-5, \dfrac{5}{2}\right)$ **59.** $\left(0, \dfrac{14}{3}\right]$ **61.** $(-\infty, -3]$ **63.** $(-\infty, 1] \cup \left(\dfrac{29}{7}, \infty\right)$ **65.** $\varnothing$
67. $\left[-\dfrac{1}{2}, \dfrac{3}{2}\right)$ **69.** $\left(-\dfrac{4}{3}, \dfrac{7}{3}\right)$ **71.** $(6, 12)$ **73.** −12 **75.** −4 **77.** −7, 7 **79.** 0 **81.** 2004, 2005 **83.** answers may vary
85. $(6, \infty)$ **87.** $[3, 7]$ **89.** $(-\infty, -1)$ **91.** $-20.2° \le F \le 95°$ **93.** $67 \le$ final score ≤ 94

Section 9.2
Practice Exercises

1. −13, 13 **2.** −1, 4 **3.** −80, 70 **4.** −2, 2 **5.** 0 **6.** $\{\ \}$ or $\varnothing$ **7.** $\{\ \}$ or $\varnothing$ **8.** $-\dfrac{3}{5}, 5$ **9.** 5

Vocabulary, Readiness & Video Check 9.2

1. C   **3.** B   **5.** D

Exercise Set 9.2

1. $-7, 7$ **3.** $4.2, -4.2$ **5.** $7, -2$ **7.** $8, 4$ **9.** $5, -5$ **11.** $3, -3$ **13.** 0 **15.** $\varnothing$ **17.** $\dfrac{1}{5}$ **19.** $9, -\dfrac{1}{2}$ **21.** $-\dfrac{5}{2}$ **23.** $4, -4$

25. 0 **27.** $\varnothing$ **29.** $0, \dfrac{14}{3}$ **31.** $2, -2$ **33.** $\varnothing$ **35.** $7, -1$ **37.** $\varnothing$ **39.** $\varnothing$ **41.** $-\dfrac{1}{8}$ **43.** $\dfrac{1}{2}, -\dfrac{5}{6}$ **45.** $2, -\dfrac{12}{5}$ **47.** $3, -2$

49. $-8, \dfrac{2}{3}$ **51.** $\varnothing$ **53.** 4 **55.** $13, -8$ **57.** $3, -3$ **59.** $8, -7$ **61.** $2, 3$ **63.** $2, -\dfrac{10}{3}$ **65.** $\dfrac{3}{2}$ **67.** $\varnothing$ **69.** 29.1%

71. $33.12°$ **73.** answers may vary   **75.** answers may vary   **77.** $\varnothing$ **79.** $|x| = 5$ **81.** answers may vary   **83.** $|x-1| = 5$ **85.** answers may vary   **87.** $|x| = 6$ **89.** $|x - 2| = |3x - 4|$

Section 9.3
Practice Exercises

1. $(-5, 5)$ **2.** $(-4, 2)$ **3.** $\left[-\dfrac{2}{3}, 2\right]$ **4.** { } or $\varnothing$ **5.** $\{2\}$

6. $(-\infty, -10] \cup [2, \infty)$ **7.** $(-\infty, \infty)$ **8.** $(-\infty, 0) \cup (12, \infty)$

Vocabulary, Readiness & Video Check 9.3

1. D   **3.** C   **5.** A   **7.** The solution set involves "or" and "or" means "union."

Exercise Set 9.3

1. $[-4, 4]$ **3.** $(1, 5)$ **5.** $(-5, -1)$ **7.** $[-10, 3]$

9. $[-5, 5]$ **11.** $\varnothing$ **13.** $[0, 12]$ **15.** $(-\infty, -3) \cup (3, \infty)$

17. $(-\infty, -24] \cup [4, \infty)$ **19.** $(-\infty, -4) \cup (4, \infty)$ **21.** $(-\infty, \infty)$

23. $\left(-\infty, \dfrac{2}{3}\right) \cup (2, \infty)$ **25.** $\{0\}$ **27.** $\left(-\infty, -\dfrac{3}{8}\right) \cup \left(-\dfrac{3}{8}, \infty\right)$ **29.** $[-2, 2]$

31. $(-\infty, -1) \cup (1, \infty)$ **33.** $(-5, 11)$ **35.** $(-\infty, 4) \cup (6, \infty)$ **37.** $\varnothing$

39. $(-\infty, \infty)$ **41.** $[-2, 9]$ **43.** $(-\infty, -11] \cup [1, \infty)$ **45.** $(-\infty, 0) \cup (0, \infty)$

47. $(-\infty, \infty)$ **49.** $\left[-\dfrac{1}{2}, 1\right]$ **51.** $(-\infty, -3) \cup (0, \infty)$ **53.** $\varnothing$

55. $\left\{\dfrac{3}{8}\right\}$ **57.** $\left(-\dfrac{2}{3}, 0\right)$ **59.** $(-\infty, -12) \cup (0, \infty)$ **61.** $[-1, 8]$

63. $\left[-\dfrac{23}{8}, \dfrac{17}{8}\right]$ **65.** $(-2, 5)$ **67.** $5, -2$ **69.** $(-\infty, -7] \cup [17, \infty)$ **71.** $-\dfrac{9}{4}$ **73.** $(-2, 1)$ **75.** $2, \dfrac{4}{3}$ **77.** $\varnothing$

79. $\dfrac{19}{2}, -\dfrac{17}{2}$ **81.** $\left(-\infty, -\dfrac{25}{3}\right) \cup \left(\dfrac{35}{3}, \infty\right)$ **83.** 1960 million; 2130 million; 2290 million   **85.** -1.5 **87.** 0 **89.** $|x| < 7$

91. $|x| \le 5$ **93.** answers may vary   **95.** $3.45 < x < 3.55$

Integrated Review

1. $(-5, 7)$ **2.** $(-\infty, \infty)$ **3.** $1, \dfrac{1}{2}$ **4.** $(-3, 2)$ **5.** $(-\infty, -1] \cup [1, \infty)$ **6.** $-18, -\dfrac{4}{3}$ **7.** $\left[-\dfrac{2}{3}, 4\right]$

8. $\varnothing$ **9.** $(-\infty, 1]$ **10.** $(-\infty, 0]$ **11.** $[0, 6]$ **12.** $-\dfrac{3}{2}, 1$ **13.** B   **14.** E   **15.** A   **16.** C   **17.** D

Section 9.4
Practice Exercises

1.   **2.**   **3.**   **4.**   **5.**   **6.**

7. 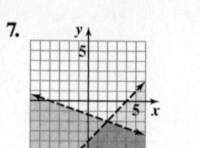 **8.**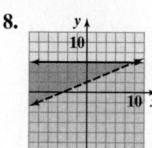

Vocabulary, Readiness & Video Check 9.4

1. linear inequality in two variables **3.** false **5.** true **7.** yes **9.** yes **11.** We find the boundary line equation by replacing the inequality symbol with $=$. The points on this line are solutions (line is solid) if the inequality is $\geq$ or $\leq$; they are not solutions (line is dashed) if the inequality is $>$ or $<$.

Exercise Set 9.4

1. no; yes **3.** no; no **5.** no; yes **7.** 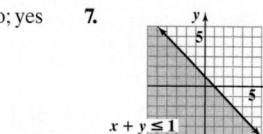 **9.** 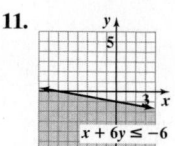 **11.** **13.**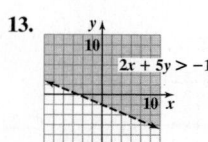

15. 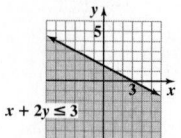 **17.** **19.** 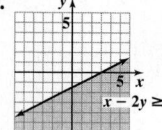 **21.** **23.**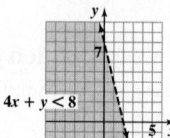

25. 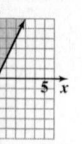 **27.** **29.** **31.** **33.** **35.**

37. **39.** **41.** **43.** e **45.** c **47.** f **49.**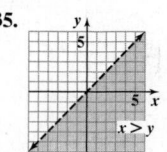

51. 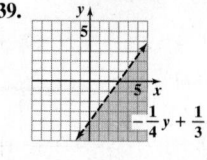 **53.** 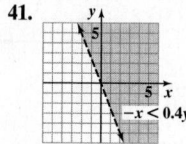 **55.** 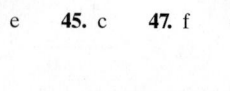 **57.** 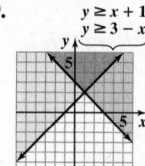 **59.**

61. 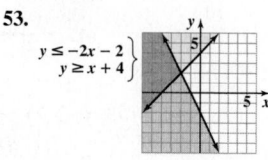 **63.** 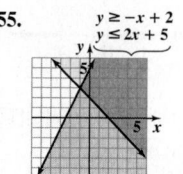 **65.** 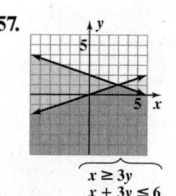 **67.** 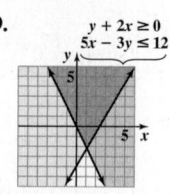 **69.**

71. no solution; { } or $\varnothing$ **73.** 25 **75.** -2 **77.** yes **79.** yes **81.** $x + y \geq 13$ 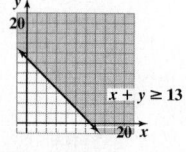 **83.** answers may vary

85. 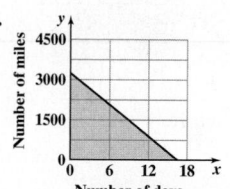 **87.** answers may vary **89. a.** $30x + 0.15y \leq 500$ **b.** **c.** answers may vary **91.** C **93.** D

95. 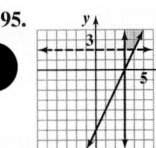 **97.** answers may vary

Chapter 9 Vocabulary Check

1. compound inequality **2.** intersection **3.** union **4.** absolute value **5.** solution **6.** system of linear inequalities

Chapter 9 Review

1. $\left(\dfrac{1}{8}, 2\right)$ **3.** $\left(\dfrac{7}{8}, \dfrac{27}{20}\right]$ **5.** $\left(\dfrac{11}{3}, \infty\right)$ **7.** $5, 11$ **9.** $-1, \dfrac{11}{3}$ **11.** $-\dfrac{1}{6}$ **13.** $\varnothing$ **15.** $5, -\dfrac{1}{3}$ **17.** $7, -\dfrac{8}{5}$ **19.** $\left(-\dfrac{8}{5}, 2\right)$

21. 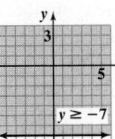 $(-\infty, -3) \cup (3, \infty)$ **23.** $\varnothing$ **25.** $\left(\infty, -\dfrac{22}{15}\right] \cup \left[\dfrac{6}{5}, \infty\right)$

27. $(-\infty, -27) \cup (-9, \infty)$ **29.** 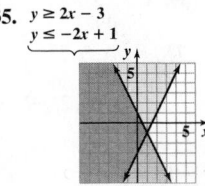 $3x - 4y \le 0$ **31.** 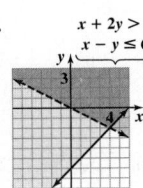 $x + 6y < 6$ **33.** 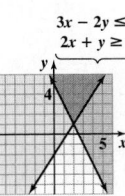 $y \ge -7$

35. $y \ge 2x - 3$
 $y \le -2x + 1$ 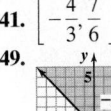 **37.** $x + 2y > 0$
 $x - y \le 6$ **39.** $3x - 2y \le 4$
 $2x + y \ge 5$ **41.** $\left[-\dfrac{4}{3}, \dfrac{7}{6}\right]$ **43.** $(5, \infty)$ **45.** $\varnothing$ **47.** $(-\infty, \infty)$

49. $-x \le y$ **51.** 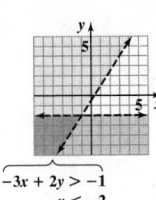 $-3x + 2y > -1$
 $y < -2$

Chapter 9 Getting Ready for the Test

1. A **2.** B **3.** D **4.** C **5.** A **6.** E **7.** B **8.** A **9.** A **10.** B **11.** C **12.** D **13.** B **14.** A

Chapter 9 Test

1. $1, \dfrac{2}{3}$ **2.** $\varnothing$ **3.** $\dfrac{3}{2}$ **4.** $\left(\dfrac{3}{2}, 5\right]$ **5.** $(-\infty, -2) \cup \left(\dfrac{4}{3}, \infty\right)$ **6.** $(3, 7)$ **7.** $(-\infty, -5]$ **8.** $(-\infty, -2]$ **9.** $[-3, -1)$ **10.** $(-\infty, \infty)$

11. $3, -\dfrac{1}{5}$ **12.** $(-\infty, \infty)$ **13.** $\left[1, \dfrac{11}{2}\right)$ **14.** $y > -4x$ **15.** $2x - 3y > -6$ **16.** $y + 2x \le 4$
 $y \ge 2$ **17.** $2y - x \ge 1$
 $x + y \ge -4$

Chapter 9 Cumulative Review

1. a. $\dfrac{1}{2}$ **b.** 19; Sec. 1.6, Ex. 6 **3. a.** $-\dfrac{39}{5}$ **b.** 2; Sec. 1.7, Ex. 9 **5. a.** $5x + 7$ **b.** $-4a - 1$ **c.** $4y - 3y^2$ **d.** $7.3x - 6$ **e.** $\dfrac{1}{2}b$; Sec. 2.1, Ex. 4

7. -3; Sec. 2.2, Ex. 3 **9.**

x	y
-1	-3
0	0
-3	-9

; Sec. 3.1, Ex. 7 **11. a.** x-int: $(-3, 0)$; y-int: $(0, 2)$ **b.** x-int: $(-4, 0), (-1, 0)$; y-int: $(0, 1)$ **c.** x-int and y-int: $(0, 0)$
 d. x-int: $(2, 0)$; y-int: none **e.** x-int: $(-1, 0), (3, 0)$; y-int: $(0, -1), (0, 2)$; Sec. 3.3, Ex. 1–5
 13. parallel; Sec. 3.4, Ex. 8a **15.** $y = \dfrac{1}{4}x - 3$; Sec. 3.5, Ex. 3 **17.** $y = -3$; Sec. 3.5, Ex. 7

19. a, b, c; Sec. 3.6, Ex. 5 **21.** solution; Sec. 4.1, Ex. 1 **23.** $12x^3 - 12x^2 - 9x + 2$; Sec. 5.2, Ex. 10

25. $(x + 6y)(x + y)$; Sec. 6.2, Ex. 6 **27.** $\dfrac{3y^9}{160}$; Sec. 7.2, Ex. 4 **29.** 1; Sec. 7.3, Ex. 2 **31.** $\dfrac{x(3x - 1)}{(x + 1)^2(x - 1)}$; Sec. 7.4, Ex. 7 **33.** -5; Sec. 7.5, Ex. 1

35. no solution; $\{\ \}$ or $\varnothing$; Sec. 4.1, Ex. 5 **37.** $(-2, 0)$; Sec. 4.2, Ex. 4 **39.** $\{(x, y) \mid 3x - 2y = 2\}$ or $\{(x, y) \mid -9x + 6y = -6\}$; Sec. 4.3, Ex. 4

41. 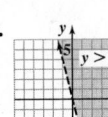 ; Sec. 9.4, Ex. 8
 $-3x + 4y < 12$
 $x \ge 2$ **43. a.** x^{11} **b.** $\dfrac{t^4}{16}$ **c.** $81y^{10}$; Sec. 5.1, Ex. 11 **45.** $-6, -\dfrac{3}{2}, \dfrac{1}{5}$; Sec. 6.6, Ex. 9 **47.** 63; Sec. 7.6, Ex. 1

CHAPTER 10 RATIONAL EXPONENTS, RADICALS, AND COMPLEX NUMBERS

Section 10.1
Practice Exercises

1. a. 7 **b.** 0 **c.** $\frac{4}{9}$ **d.** 0.8 **e.** z^4 **f.** $4b^2$ **g.** -6 **h.** not a real number **2.** 6.708 **3. a.** -1 **b.** 3 **c.** $\frac{3}{4}$ **d.** x^4
e. $-2x$ **4. a.** 10 **b.** -1 **c.** -9 **d.** not a real number **e.** $3x^3$ **5. a.** 4 **b.** $|x^7|$ **c.** $|x+7|$ **d.** -7 **e.** $3x-5$
f. $7|x|$ **g.** $|x+8|$ **6. a.** 4 **b.** 2 **c.** 2 **d.** $\sqrt[3]{-9}$ **7.** **8.**

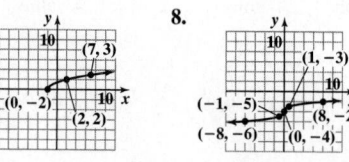

Vocabulary, Readiness & Video Check 10.1

1. index; radical sign; radicand **3.** is not **5.** $[0,\infty)$ **7.** $(16, 4)$ **9.** Divide the index into each exponent in the radicand. **11.** The square root of a negative number is not a real number, but the cube root of a negative number is a real number. **13.** For odd roots, there's only one root/answer whether the radicand is positive or negative, so absolute value bars aren't needed.

Exercise Set 10.1

1. 10 **3.** $\frac{1}{2}$ **5.** 0.01 **7.** -6 **9.** x^5 **11.** $4y^3$ **13.** 2.646 **15.** 6.164 **17.** 14.142 **19.** 4 **21.** $\frac{1}{2}$ **23.** -1 **25.** x^4 **27.** $-3x^3$
29. -2 **31.** not a real number **33.** -2 **35.** x^4 **37.** $2x^2$ **39.** $9x^2$ **41.** $4x^2$ **43.** 8 **45.** -8 **47.** $2|x|$ **49.** x **51.** $|x-5|$
53. $|x+2|$ **55.** -11 **57.** $2x$ **59.** y^6 **61.** $5ab^{10}$ **63.** $-3x^4y^3$ **65.** a^4b **67.** $-2x^2y$ **69.** $\frac{5}{7}$ **71.** $\frac{x^{10}}{2y}$ **73.** $-\frac{z^7}{3x}$ **75.** $\frac{x}{2}$
77. $\sqrt{3}$ **79.** -1 **81.** -3 **83.** $\sqrt{7}$ **85.** $[0,\infty)$; **87.** $[3,\infty)$; 0, 1, 2, 3 **89.** $(-\infty, \infty)$; **91.** $(-\infty, \infty)$; 0, 1, -1, 2, -2

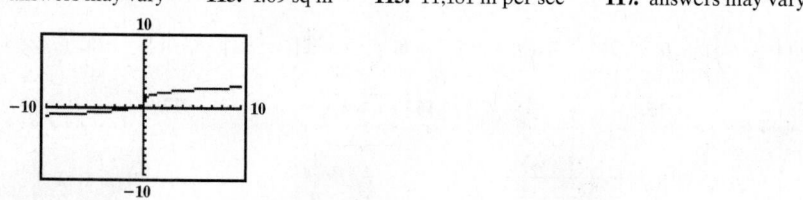 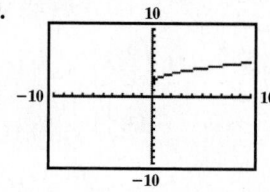

93. $-32x^{15}y^{10}$ **95.** $-60x^7y^{10}z^5$ **97.** $\frac{x^9y^5}{2}$ **99.** not a real number **101.** not a real number **103.** d **105.** d **107.** b **109.** b
111. answers may vary **113.** 1.69 sq m **115.** 11,181 m per sec **117.** answers may vary **119.**
121.

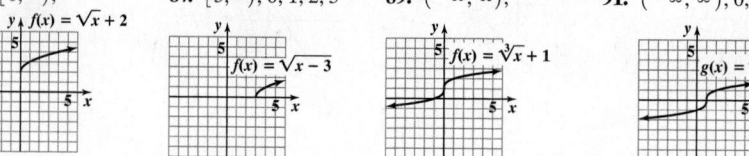

Section 10.2
Practice Exercises

1. a. 6 **b.** 10 **c.** $\sqrt[3]{x}$ **d.** 1 **e.** -8 **f.** $5x^3$ **g.** $3\sqrt[4]{x}$ **2. a.** 64 **b.** -1 **c.** -27 **d.** $\frac{1}{125}$ **e.** $\sqrt[9]{(3x+2)^5}$
3. a. $\frac{1}{27}$ **b.** $\frac{1}{16}$ **4. a.** $y^{10/3}$ **b.** $x^{17/20}$ **c.** $\frac{1}{9}$ **d.** $b^{2/9}$ **e.** $\frac{81}{x^3y^{11/3}}$ **5. a.** $x^{14/15} - x^{13/5}$ **b.** $x + 4x^{1/2} - 12$ **6.** $x^{-1/5}(2 - 7x)$
7. a. $\sqrt[3]{x}$ **b.** $\sqrt{6}$ **c.** $\sqrt[4]{a^2b}$ **8. a.** $\sqrt[12]{x^7}$ **b.** $\sqrt[15]{y^2}$ **c.** $\sqrt[6]{675}$

Vocabulary, Readiness & Video Check 10.2

1. true **3.** true **5.** multiply, c **7.** $-\sqrt[5]{3x}$ **9.** denominator; positive **11.** Write the radical using an equivalent fractional exponent form, simplify the fraction, then write as a radical again.

Exercise Set 10.2

1. 7 **3.** 3 **5.** $\frac{1}{2}$ **7.** 13 **9.** $2\sqrt[3]{m}$ **11.** $3x^2$ **13.** -3 **15.** -2 **17.** 8 **19.** 16 **21.** not a real number **23.** $\sqrt[5]{(2x)^3}$
25. $\sqrt[3]{(7x+2)^2}$ **27.** $\frac{64}{27}$ **29.** $\frac{1}{16}$ **31.** $\frac{1}{16}$ **33.** not a real number **35.** $\frac{1}{x^{1/4}}$ **37.** $a^{2/3}$ **39.** $\frac{5x^{3/4}}{7}$ **41.** $a^{7/3}$ **43.** x **45.** $3^{5/8}$
47. $y^{1/6}$ **49.** $8u^3$ **51.** $-b$ **53.** $\frac{1}{x^2}$ **55.** $27x^{2/3}$ **57.** $\frac{y}{z^{1/6}}$ **59.** $\frac{1}{x^{7/4}}$ **61.** $y - y^{7/6}$ **63.** $x^{5/3} - 2x^{2/3}$ **65.** $4x^{2/3} - 9$
67. $x^{8/3}(1 + x^{2/3})$ **69.** $x^{1/5}(x^{1/5} - 3)$ **71.** $x^{-1/3}(5 + x)$ **73.** $\sqrt{x}$ **75.** $\sqrt[3]{2}$ **77.** $2\sqrt{x}$ **79.** $\sqrt{xy}$ **81.** $\sqrt[3]{a^2b}$ **83.** $\sqrt{x+3}$
85. $\sqrt[15]{y^{11}}$ **87.** $\sqrt[12]{b^5}$ **89.** $\sqrt[24]{x^{23}}$ **91.** $\sqrt{a}$ **93.** $\sqrt[6]{432}$ **95.** $\sqrt[15]{343y^5}$ **97.** $\sqrt[6]{125r^3s^2}$ **99.** $25 \cdot 3$ **101.** $16 \cdot 3$ or $4 \cdot 12$

103. $8 \cdot 2$ **105.** $27 \cdot 2$ **107.** A   **109.** C   **111.** B   **113.** 1509 calories   **115.** 255.8 million   **117.** answers may vary   **119.** $a^{1/3}$

121. $x^{1/5}$ **123.** 1.6818   **125.** 5.6645   **127.** $\dfrac{t^{1/2}}{u^{1/2}}$

Section 10.3
Practice Exercises

1. a. $\sqrt{35}$ **b.** $\sqrt{13z}$ **c.** 5   **d.** $\sqrt[3]{15x^2y}$ **e.** $\sqrt{\dfrac{5t}{2m}}$ **2. a.** $\dfrac{6}{7}$ **b.** $\dfrac{\sqrt{z}}{4}$ **c.** $\dfrac{5}{2}$ **d.** $\dfrac{\sqrt[4]{5}}{3x^2}$ **3. a.** $7\sqrt{2}$ **b.** $3\sqrt[3]{2}$

c. $\sqrt{35}$ **d.** $3\sqrt[4]{3}$ **4. a.** $6z^3\sqrt{z}$ **b.** $2pq^2\sqrt[3]{4pq}$ **c.** $2x^3\sqrt[4]{x^3}$ **5. a.** 4   **b.** $\dfrac{7}{3}\sqrt{z}$ **c.** $10xy^2\sqrt[3]{x^2}$ **d.** $6x^2y\sqrt[5]{2y}$

6. $\sqrt{17}$ units ≈ 4.123 units   **7.** $\left(\dfrac{13}{2}, -4\right)$

Vocabulary, Readiness & Video Check 10.3

1. midpoint; point   **3.** distance   **5.** the indexes must be the same   **7.** The power must be 1. Any even power is a perfect square and will leave no factor in the radicand; any higher odd power can have an even power factored from it, leaving one factor remaining in the radicand.   **9.** average; average

Exercise Set 10.3

1. $\sqrt{14}$ **3.** 2   **5.** $\sqrt[3]{36}$ **7.** $\sqrt{6x}$ **9.** $\sqrt{\dfrac{14}{xy}}$ **11.** $\sqrt[4]{20x^3}$ **13.** $\dfrac{\sqrt{6}}{7}$ **15.** $\dfrac{\sqrt{2}}{7}$ **17.** $\dfrac{\sqrt[4]{x^3}}{2}$ **19.** $\dfrac{\sqrt[3]{4}}{3}$ **21.** $\dfrac{\sqrt[4]{8}}{x^2}$

23. $\dfrac{\sqrt[3]{2x}}{3y^4\sqrt[3]{3}}$ **25.** $\dfrac{x\sqrt{y}}{10}$ **27.** $\dfrac{x\sqrt{5}}{2y}$ **29.** $-\dfrac{z^2\sqrt[3]{z}}{3x}$ **31.** $4\sqrt{2}$ **33.** $4\sqrt[3]{3}$ **35.** $25\sqrt{3}$ **37.** $2\sqrt{6}$ **39.** $10x^2\sqrt{x}$ **41.** $2y^2\sqrt[3]{2y}$

43. $a^2b\sqrt[4]{b^3}$ **45.** $y^2\sqrt{y}$ **47.** $5ab\sqrt{b}$ **49.** $-2x^2\sqrt[5]{y}$ **51.** $x^4\sqrt[3]{50x^2}$ **53.** $-4a^4b^3\sqrt{2b}$ **55.** $3x^3y^4\sqrt{xy}$ **57.** $5r^3s^4$

59. $2x^3y\sqrt[4]{2y}$ **61.** $\sqrt{2}$ **63.** 2   **65.** 10   **67.** x^2y **69.** $24m^2$ **71.** $\dfrac{15x\sqrt{2x}}{2}$ or $\dfrac{15x}{2}\sqrt{2x}$ **73.** $2a^2\sqrt[4]{2}$ **75.** $2xy^2\sqrt[5]{x^2}$ **77.** 5 units

79. $\sqrt{41}$ units ≈ 6.403 units   **81.** $\sqrt{10}$ units ≈ 3.162 units   **83.** $\sqrt{5}$ units ≈ 2.236 units   **85.** $\sqrt{192.58}$ units ≈ 13.877 units   **87.** $(4, -2)$

89. $\left(-5, \dfrac{5}{2}\right)$ **91.** $(3,0)$ **93.** $\left(-\dfrac{1}{2}, \dfrac{1}{2}\right)$ **95.** $\left(\sqrt{2}, \dfrac{\sqrt{5}}{2}\right)$ **97.** $(6.2, -6.65)$ **99.** $14x$ **101.** $2x^2 - 7x - 15$ **103.** y^2

105. $-3x - 15$ **107.** $x^2 - 8x + 16$ **109.** true   **111.** false   **113.** true   **115.** $\dfrac{\sqrt[3]{64}}{\sqrt{64}} = \dfrac{4}{8} = \dfrac{1}{2}$ **117.** x^7 **119.** a^3bc^5 **121.** $z^{10}\sqrt[3]{z^2}$

123. $q^2r^5s\sqrt[4]{q^3r^5}$ **125.** 1.6 m   **127. a.** 20π sq cm   **b.** 211.57 sq ft

Section 10.4
Practice Exercises

1. a. $8\sqrt{17}$ **b.** $-5\sqrt[3]{5z}$ **c.** $3\sqrt{2} + 5\sqrt[3]{2}$ **2. a.** $11\sqrt{6}$ **b.** $-9\sqrt[3]{3}$ **c.** $-2\sqrt{3x}$ **d.** $2\sqrt{10} + 2\sqrt[3]{5}$ **e.** $4x\sqrt[3]{3x}$

3. a. $\dfrac{5\sqrt{7}}{12}$ **b.** $\dfrac{13\sqrt[3]{6y}}{4}$ **4. a.** $2\sqrt{5} + 5\sqrt{3}$ **b.** $2\sqrt{3} + 2\sqrt{2} - \sqrt{30} - 2\sqrt{5}$ **c.** $6z + \sqrt{z} - 12$ **d.** $-6\sqrt{6} + 15$

e. $5x - 9$ **f.** $6\sqrt{x + 2} + x + 11$

Vocabulary, Readiness & Video Check 10.4

1. Unlike   **3.** Like   **5.** $6\sqrt{3}$ **7.** $7\sqrt{x}$ **9.** $8\sqrt[3]{x}$ **11.** Sometimes you can't see that there are like radicals until you simplify, so you may incorrectly think you cannot add or subtract if you don't simplify first.

Exercise Set 10.4

1. $-2\sqrt{2}$ **3.** $10x\sqrt{2x}$ **5.** $17\sqrt{2} - 15\sqrt{5}$ **7.** $-\sqrt[3]{2x}$ **9.** $5b\sqrt{b}$ **11.** $\dfrac{31\sqrt{2}}{15}$ **13.** $\dfrac{\sqrt[3]{11}}{3}$ **15.** $\dfrac{5\sqrt{5x}}{9}$ **17.** $14 + \sqrt{3}$

19. $7 - 3y$ **21.** $6\sqrt{3} - 6\sqrt{2}$ **23.** $-23\sqrt[3]{5}$ **25.** $2a^3b\sqrt{ab}$ **27.** $20y\sqrt{2y}$ **29.** $2y\sqrt[3]{2x}$ **31.** $6\sqrt[3]{11} - 4\sqrt{11}$ **33.** $3x\sqrt[4]{x^3}$ **35.** $\dfrac{2\sqrt{3}}{3}$

37. $\dfrac{5x\sqrt[3]{x}}{7}$ **39.** $\dfrac{5\sqrt{7}}{2x}$ **41.** $\dfrac{\sqrt[3]{2}}{6}$ **43.** $\dfrac{14x\sqrt[3]{2x}}{9}$ **45.** $15\sqrt{3}$ in.   **47.** $\sqrt{35} + \sqrt{21}$ **49.** $7 - 2\sqrt{10}$ **51.** $3\sqrt{x} - x\sqrt{3}$

53. $6x - 13\sqrt{x} - 5$ **55.** $\sqrt[3]{a^2} + \sqrt[3]{a} - 20$ **57.** $6\sqrt{2} - 12$ **59.** $2 + 2x\sqrt{3}$ **61.** $-16 - \sqrt{35}$ **63.** $x - y^2$ **65.** $3 + 2x\sqrt{3} + x^2$

67. $23x - 5x\sqrt{15}$ **69.** $2\sqrt[3]{2} - \sqrt[3]{4}$ **71.** $x + 1$ **73.** $x + 24 + 10\sqrt{x - 1}$ **75.** $2x + 6 - 2\sqrt{2x + 5}$ **77.** $x - 7$ **79.** $\dfrac{7}{x + y}$

81. $2a - 3$ **83.** $\dfrac{-2 + \sqrt{3}}{3}$ **85.** $22\sqrt{5}$ ft; 150 sq ft   **87. a.** $2\sqrt{3}$ **b.** 3   **c.** answers may vary   **89.** $2\sqrt{6} - 2\sqrt{2} - 2\sqrt{3} + 6$

91. answers may vary

Section 10.5
Practice Exercises

1. a. $\dfrac{5\sqrt{3}}{3}$ **b.** $\dfrac{15\sqrt{x}}{2x}$ **c.** $\dfrac{\sqrt[3]{6}}{3}$ **2.** $\dfrac{\sqrt{15yz}}{5y}$ **3.** $\dfrac{\sqrt[3]{z^2x^2}}{3x^2}$ **4. a.** $\dfrac{5(3\sqrt{5}-2)}{41}$ **b.** $\dfrac{\sqrt{6}+\sqrt{10}+5\sqrt{3}+5\sqrt{5}}{-2}$ **c.** $\dfrac{6x-3\sqrt{xy}}{4x-y}$

5. $\dfrac{2}{\sqrt{10}}$ **6.** $\dfrac{5b}{\sqrt[3]{50ab^2}}$ **7.** $\dfrac{x-9}{4(\sqrt{x}+3)}$

Vocabulary, Readiness & Video Check 10.5

1. conjugate **3.** rationalizing the numerator **5.** To write an equivalent expression without a radical in the denominator. **7.** No, except for the fact you're working with numerators, the process is the same.

Exercise Set 10.5

1. $\dfrac{\sqrt{14}}{7}$ **3.** $\dfrac{\sqrt{5}}{5}$ **5.** $\dfrac{2\sqrt{x}}{x}$ **7.** $\dfrac{4\sqrt[3]{9}}{3}$ **9.** $\dfrac{3\sqrt{2x}}{4x}$ **11.** $\dfrac{3\sqrt[3]{2x}}{2x}$ **13.** $\dfrac{3\sqrt{3a}}{a}$ **15.** $\dfrac{3\sqrt[3]{4}}{2}$ **17.** $\dfrac{2\sqrt{21}}{7}$ **19.** $\dfrac{\sqrt{10xy}}{5y}$ **21.** $\dfrac{\sqrt[3]{75}}{5}$

23. $\dfrac{\sqrt{6x}}{10}$ **25.** $\dfrac{\sqrt{3z}}{6z}$ **27.** $\dfrac{\sqrt[4]{6xy^2}}{3x}$ **29.** $\dfrac{3\sqrt[4]{2}}{2}$ **31.** $\dfrac{2\sqrt[4]{9x}}{3x^2}$ **33.** $\dfrac{5\sqrt[5]{4ab^4}}{2ab^3}$ **35.** $\sqrt{2}-x$ **37.** $5+\sqrt{a}$ **39.** $-7\sqrt{5}-8\sqrt{x}$

41. $-2(2+\sqrt{7})$ **43.** $\dfrac{7(3+\sqrt{x})}{9-x}$ **45.** $-5+2\sqrt{6}$ **47.** $\dfrac{2a+2\sqrt{a}+\sqrt{ab}+\sqrt{b}}{4a-b}$ **49.** $-\dfrac{8(1-\sqrt{10})}{9}$ **51.** $\dfrac{x-\sqrt{xy}}{x-y}$

53. $\dfrac{5+3\sqrt{2}}{7}$ **55.** $\dfrac{5}{\sqrt{15}}$ **57.** $\dfrac{6}{\sqrt{10}}$ **59.** $\dfrac{2x}{7\sqrt{x}}$ **61.** $\dfrac{5y}{\sqrt[3]{100xy}}$ **63.** $\dfrac{2}{\sqrt{10}}$ **65.** $\dfrac{2x}{11\sqrt{2x}}$ **67.** $\dfrac{7}{2\sqrt[3]{49}}$ **69.** $\dfrac{3x^2}{10\sqrt[3]{9x}}$ **71.** $\dfrac{6x^2y^3}{\sqrt{6z}}$

73. $\dfrac{-7}{12+6\sqrt{11}}$ **75.** $\dfrac{3}{10+5\sqrt{7}}$ **77.** $\dfrac{x-9}{x-3\sqrt{x}}$ **79.** $\dfrac{1}{3+2\sqrt{2}}$ **81.** $\dfrac{x-1}{x-2\sqrt{x}+1}$ **83.** 5 **85.** $-\dfrac{1}{2}$, 6 **87.** 2, 6 **89.** $r=\dfrac{\sqrt{A\pi}}{2\pi}$

91. a. $\dfrac{y\sqrt{15xy}}{6x^2}$ **b.** $\dfrac{y\sqrt{15xy}}{6x^2}$ **c.** answers may vary **93.** $\sqrt[3]{25}$ **95.** answers may vary **97.** answers may vary

Integrated Review

1. 9 **2.** -2 **3.** $\dfrac{1}{2}$ **4.** x^3 **5.** y^3 **6.** $2y^5$ **7.** $-2y$ **8.** $3b^3$ **9.** 6 **10.** $\sqrt[4]{3y}$ **11.** $\dfrac{1}{16}$ **12.** $\sqrt[5]{(x+1)^3}$ **13.** y

14. $16x^{1/2}$ **15.** $x^{5/4}$ **16.** $4^{11/15}$ **17.** $2x^2$ **18.** $\sqrt[4]{a^3b^2}$ **19.** $\sqrt[4]{x^3}$ **20.** $\sqrt[6]{500}$ **21.** $2\sqrt{10}$ **22.** $2xy^2\sqrt[4]{x^3y^2}$ **23.** $3x\sqrt[3]{2x}$

24. $-2b^2\sqrt[5]{2}$ **25.** $\sqrt{5x}$ **26.** $4x$ **27.** $7y^2\sqrt{y}$ **28.** $2a^2\sqrt[4]{3}$ **29.** $2\sqrt{5}-5\sqrt{3}+5\sqrt{7}$ **30.** $y\sqrt[3]{2y}$ **31.** $\sqrt{15}-\sqrt{6}$ **32.** $10+2\sqrt{21}$

33. $4x^2-5$ **34.** $x+2-2\sqrt{x+1}$ **35.** $\dfrac{\sqrt{21}}{3}$ **36.** $\dfrac{5\sqrt[3]{4x}}{2x}$ **37.** $\dfrac{13-3\sqrt{21}}{5}$ **38.** $\dfrac{7}{\sqrt{21}}$ **39.** $\dfrac{3y}{\sqrt[3]{33y^2}}$ **40.** $\dfrac{x-4}{x+2\sqrt{x}}$

Section 10.6
Practice Exercises

1. 18 **2.** $\dfrac{1}{4},\dfrac{3}{4}$ **3.** 10 **4.** 9 **5.** $\dfrac{3}{25}$ **6.** $6\sqrt{3}$ m **7.** $\sqrt{193}$ in. ≈ 13.89 in.

Graphing Calculator Explorations 10.6

1. 3.19 **3.** $\varnothing$ **5.** 3.23

Vocabulary, Readiness & Video Check 10.6

1. extraneous solution **3.** $x^2-10x+25$ **5.** Applying the power rule can result in an equation with more solutions than the original equation, so you need to check all proposed solutions in the original equation. **7.** Our answer is either a positive square root of a value or a negative square root of a value. We're looking for a length, which must be positive, so our answer must be the positive square root.

Exercise Set 10.6

1. 8 **3.** 7 **5.** $\varnothing$ **7.** 7 **9.** 6 **11.** $-\dfrac{9}{2}$ **13.** 29 **15.** 4 **17.** -4 **19.** $\varnothing$ **21.** 7 **23.** 9 **25.** 50 **27.** $\varnothing$

29. $\dfrac{15}{4}$ **31.** 13 **33.** 5 **35.** -12 **37.** 9 **39.** -3 **41.** 1 **43.** 1 **45.** $\dfrac{1}{2}$ **47.** 0, 4 **49.** $\dfrac{37}{4}$ **51.** $3\sqrt{5}$ ft

53. $2\sqrt{10}$ m **55.** $2\sqrt{131}$ m ≈ 22.9 m **57.** $\sqrt{100.84}$ mm ≈ 10.0 mm **59.** 17 ft **61.** 13 ft **63.** 14,657,415 sq mi **65.** 100 ft

67. 100 **69.** $\dfrac{\pi}{2}$ sec ≈ 1.57 sec **71.** 12.97 ft **73.** answers may vary **75.** $15\sqrt{3}$ sq mi ≈ 25.98 sq mi **77.** answers may vary

79. 0.51 km **81.** function **83.** function **85.** not a function **87.** $\dfrac{x}{4x+3}$ **89.** $-\dfrac{4z+2}{3z}$

91. $\sqrt{5x-1}+4=7$ **93.** 1 **95. a.–b.** answers may vary **97.** $-1,0,8,9$ **99.** $-1,4$
$\sqrt{5x-1}=3$
$(\sqrt{5x-1})^2=3^2$
$5x-1=9$
$5x=10$
$x=2$

Section 10.7

Practice Exercises

1. a. $2i$ **b.** $i\sqrt{7}$ **c.** $-3i\sqrt{2}$ **2. a.** $-\sqrt{30}$ **b.** -3 **c.** $25i$ **d.** $3i$ **3. a.** $-1 - 4i$ **b.** $-3 + 5i$ **c.** $3 - 2i$

4. a. $20 + 0i$ **b.** $-5 + 10i$ **c.** $15 + 16i$ **d.** $8 - 6i$ **e.** $85 + 0i$ **5. a.** $\dfrac{11}{10} - \dfrac{7}{10}i$ **b.** $0 - \dfrac{5}{2}i$ **6. a.** i **b.** 1 **c.** -1 **d.** 1

Vocabulary, Readiness & Video Check 10.7

1. complex   **3.** -1 **5.** real   **7.** The product rule for radicals; you need to first simplify each separate radical and have nonnegative radicands before applying the product rule.   **9.** The fact that $i^2 = -1$.   **11.** $i, i^2 = -1, i^3 = -i, i^4 = 1$

Exercise Set 10.7

1. $9i$ **3.** $i\sqrt{7}$ **5.** -4 **7.** $8i$ **9.** $2i\sqrt{6}$ **11.** $-6i$ **13.** $24i\sqrt{7}$ **15.** $-3\sqrt{6}$ **17.** $-\sqrt{14}$ **19.** $-5\sqrt{2}$ **21.** $4i$ **23.** $i\sqrt{3}$

25. $2\sqrt{2}$ **27.** $6 - 4i$ **29.** $-2 + 6i$ **31.** $-2 - 4i$ **33.** $-40 + 0i$ **35.** $18 + 12i$ **37.** $7 + 0i$ **39.** $12 - 16i$ **41.** $0 - 4i$

43. $\dfrac{28}{25} - \dfrac{21}{25}i$ **45.** $4 + i$ **47.** $\dfrac{17}{13} + \dfrac{7}{13}i$ **49.** $63 + 0i$ **51.** $2 - i$ **53.** $27 + 3i$ **55.** $-\dfrac{5}{2} - 2i$ **57.** $18 + 13i$ **59.** $20 + 0i$

61. $10 + 0i$ **63.** $2 + 0i$ **65.** $-5 + \dfrac{16}{3}i$ **67.** $17 + 144i$ **69.** $\dfrac{3}{5} - \dfrac{1}{5}i$ **71.** $5 - 10i$ **73.** $\dfrac{1}{5} - \dfrac{8}{5}i$ **75.** $8 - i$ **77.** $7 + 0i$

79. $12 - 16i$ **81.** 1 **83.** i **85.** $-i$ **87.** -1 **89.** -64 **91.** $-243i$ **93.** $40°$ **95.** $x^2 - 5x - 2 - \dfrac{6}{x - 1}$ **97.** 5 people

99. 14 people   **101.** 16.7%   **103.** $-1 - i$ **105.** $0 + 0i$ **107.** $2 + 3i$ **109.** $2 + i\sqrt{2}$ **111.** $\dfrac{1}{2} - \dfrac{\sqrt{3}}{2}i$ **113.** answers may vary

115. $6 - 3i\sqrt{3}$ **117.** yes

Chapter 10 Vocabulary Check

1. conjugate   **2.** principal square root   **3.** rationalizing   **4.** imaginary unit   **5.** cube root   **6.** index; radicand   **7.** like radicals
8. complex number   **9.** distance   **10.** midpoint

Chapter 10 Review

1. 9 **3.** -2 **5.** $-\dfrac{1}{7}$ **7.** -6 **9.** $-a^2b^3$ **11.** $2ab^2$ **13.** $\dfrac{x^6}{6y}$ **15.** $|-x|$ **17.** -27 **19.** $-x$ **21.** $2|2y + z|$ **23.** y **25. a.** $3, 6$

b. $[0, \infty)$ **c.** 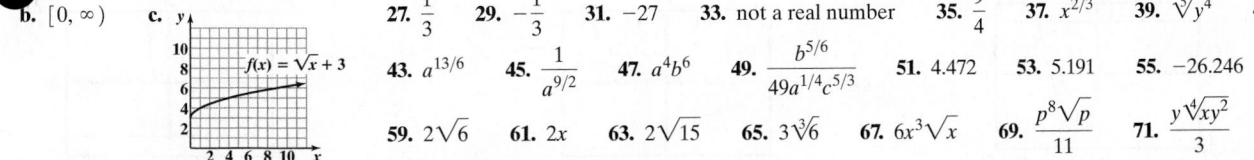   **27.** $\dfrac{1}{3}$ **29.** $-\dfrac{1}{3}$ **31.** -27 **33.** not a real number   **35.** $\dfrac{9}{4}$ **37.** $x^{2/3}$ **39.** $\sqrt[5]{y^4}$ **41.** $\dfrac{1}{\sqrt[3]{x + 2}}$

43. $a^{13/6}$ **45.** $\dfrac{1}{a^{9/2}}$ **47.** a^4b^6 **49.** $\dfrac{b^{5/6}}{49a^{1/4}c^{5/3}}$ **51.** 4.472 **53.** 5.191 **55.** -26.246 **57.** $\sqrt[6]{1372}$

59. $2\sqrt{6}$ **61.** $2x$ **63.** $2\sqrt{15}$ **65.** $3\sqrt[3]{6}$ **67.** $6x^3\sqrt{x}$ **69.** $\dfrac{p^8\sqrt{p}}{11}$ **71.** $\dfrac{y\sqrt[4]{xy^2}}{3}$

73. $\dfrac{5}{\sqrt{\pi}}$ m or $\dfrac{5\sqrt{\pi}}{\pi}$ m   **75.** $\sqrt{197}$ units ≈ 14.036 units   **77.** $\sqrt{73}$ units ≈ 8.544 units   **79.** $2\sqrt{11}$ units ≈ 6.633 units   **81.** $(-5, 5)$

83. $\left(-\dfrac{11}{2}, -2\right)$ **85.** $\left(\dfrac{1}{4}, -\dfrac{2}{7}\right)$ **87.** $-2\sqrt{5}$ **89.** $9\sqrt[3]{2}$ **91.** $\dfrac{15 + 2\sqrt{3}}{6}$ **93.** $17\sqrt{2} - 15\sqrt{5}$ **95.** 6 **97.** $-8\sqrt{5}$

99. $a - 9$ **101.** $\sqrt[3]{25x^2} - 81$ **103.** $\dfrac{3\sqrt{7}}{7}$ **105.** $\dfrac{5\sqrt[3]{2}}{2}$ **107.** $\dfrac{x^2y^2\sqrt[3]{15yz}}{z}$ **109.** $\dfrac{3\sqrt{y} + 6}{y - 4}$ **111.** $\dfrac{11}{3\sqrt{11}}$ **113.** $\dfrac{3}{7\sqrt[3]{3}}$ **115.** $\dfrac{xy}{\sqrt[3]{10x^2yz}}$

117. 32 **119.** 35 **121.** 9 **123.** $3\sqrt{2}$ cm   **125.** 51.2 ft   **127.** $0 + 2i\sqrt{2}$ **129.** $0 + 6i$ **131.** $15 - 4i$ **133.** -64

135. $-12 - 18i$ **137.** $-5 - 12i$ **139.** $\dfrac{3}{2} - i$ **141.** x **143.** -10 **145.** $\dfrac{y^5}{2x^3}$ **147.** $\dfrac{1}{8}$ **149.** $\dfrac{1}{x^{13/2}}$ **151.** $\dfrac{n\sqrt{3n}}{11m^5}$

153. $4x - 20\sqrt{x} + 25$ **155.** $(4, 16)$ **157.** $\dfrac{2\sqrt{x} - 6}{x - 9}$ **159.** 5

Chapter 10 Getting Ready for the Test

1. D   **2.** B   **3.** A   **4.** B   **5.** A   **6.** A   **7.** A   **8.** C   **9.** B   **10.** B   **11.** B   **12.** A   **13.** A   **14.** B   **15.** B
16. B   **17.** D   **18.** C   **19.** B   **20.** D   **21.** A

Chapter 10 Test

1. $6\sqrt{6}$ **2.** $-x^{16}$ **3.** $\dfrac{1}{5}$ **4.** 5 **5.** $\dfrac{4x^2}{9}$ **6.** $-a^6b^3$ **7.** $\dfrac{8a^{1/3}c^{2/3}}{b^{5/12}}$ **8.** $a^{7/12} - a^{7/3}$ **9.** $|4xy|$ or $4|xy|$ **10.** -27 **11.** $\dfrac{3\sqrt{y}}{y}$

12. $\dfrac{8 - 6\sqrt{x} + x}{8 - 2x}$ **13.** $\dfrac{\sqrt[3]{b^2}}{b}$ **14.** $\dfrac{6 - x^2}{8(\sqrt{6} - x)}$ **15.** $-x\sqrt{5x}$ **16.** $4\sqrt{3} - \sqrt{6}$ **17.** $x + 2\sqrt{x} + 1$ **18.** $\sqrt{6} + \sqrt{2} - 4\sqrt{3} - 4$

19. -20 **20.** 23.685 **21.** 0.019 **22.** $2, 3$ **23.** $\varnothing$ **24.** 6 **25.** $0 + i\sqrt{2}$ **26.** $0 - 2i\sqrt{2}$ **27.** $0 - 3i$ **28.** $40 + 0i$ **29.** $7 + 24i$

30. $-\dfrac{3}{2} + \dfrac{5}{2}i$ **31.** $\dfrac{5\sqrt{2}}{2}$ **32.** $[-2, \infty)$; $0, 1, 2, 3$;

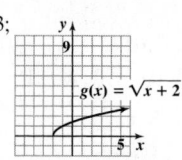

33. $2\sqrt{26}$ units   **34.** $\sqrt{95}$ units   **35.** $\left(-4, \dfrac{7}{2}\right)$ **36.** $\left(-\dfrac{1}{2}, \dfrac{3}{10}\right)$

37. 27 mph   **38.** 360 ft

Chapter 10 Cumulative Review

1. a. -12 **b.** -3; Sec. 1.6, Ex. 5   **3.** 12; Sec. 2.3, Ex. 3   **5.** 12 in., 36 in.; Sec. 2.4, Ex. 3   **7.** one; Sec. 4.1, Ex. 8   **9.** $\left(6, \dfrac{1}{2}\right)$; Sec. 4.2, Ex. 3

11. no solution; $\{\ \}$ or $\varnothing$; Sec. 4.3, Ex. 3   **13.** 30% solution: 42 L; 80% solution: 28 L; Sec. 4.5, Ex. 6   **15. a.** -4 **b.** 11; Sec. 5.2, Ex. 4

17. $3m + 1$; Sec. 5.6, Ex. 1   **19.** $2x^2 + 5x + 2 + \dfrac{7}{x-3}$; Sec. 5.7, Ex. 1   **21.** $(t-8)(t-5)$; Sec. 6.2, Ex. 8   **23. a.** $x^2 - 2x + 4$

b. $\dfrac{2}{y-5}$; Sec. 7.1, Ex. 5   **25.** 4; Sec. 9.2, Ex. 9   **27.** $3x - 5$; Sec. 7.3, Ex. 3   **29.** $\dfrac{2m+1}{m+1}$; Sec. 7.4, Ex. 5   **31. a.** $\dfrac{x(x-2)}{2(x+2)}$ **b.** $\dfrac{x^2}{y^2}$; Sec. 7.7, Ex. 2

33. $\left[-2, \dfrac{8}{5}\right]$; Sec. 9.3, Ex. 3   **35.** 15 yd; Sec. 7.6, Ex. 4   **37. a.** 1 **b.** -4 **c.** $\dfrac{2}{5}$ **d.** x^2 **e.** $-3x^5$; Sec. 10.1, Ex. 3   **39. a.** $\dfrac{1}{8}$ **b.** $\dfrac{1}{9}$; Sec. 10.2, Ex. 3

41. $\dfrac{x-4}{5(\sqrt{x}-2)}$; Sec. 10.5, Ex. 7   **43.** constant of variation: 15, $u = \dfrac{15}{w}$; Sec. 8.4, Ex. 3

CHAPTER 11 QUADRATIC EQUATIONS AND FUNCTIONS

Section 11.1

Practice Exercises

1. $-4\sqrt{2}, 4\sqrt{2}$ **2.** $-\sqrt{10}, \sqrt{10}$ **3.** $-3 - 2\sqrt{5}, -3 + 2\sqrt{5}$ **4.** $\dfrac{2+3i}{5}, \dfrac{2-3i}{5}$ **5.** $-2 - \sqrt{7}, -2 + \sqrt{7}$ **6.** $\dfrac{3-\sqrt{5}}{2}, \dfrac{3+\sqrt{5}}{2}$

7. $\dfrac{6-\sqrt{33}}{3}, \dfrac{6+\sqrt{33}}{3}$ **8.** $\dfrac{5-i\sqrt{31}}{4}, \dfrac{5+i\sqrt{31}}{4}$ **9.** 6%

Graphing Calculator Explorations 11.1

1. $-1.27, 6.27$

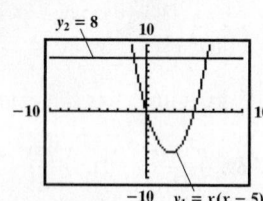

3. $-1.10, 0.90$

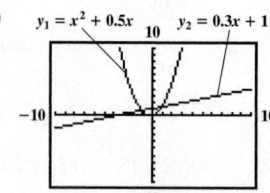

5. no real solutions

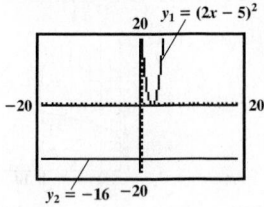

Vocabulary, Readiness & Video Check 11.1

1. $\pm\sqrt{b}$ **3.** completing the square   **5.** 9   **7.** We need a quantity squared by itself on one side of the equation. The only quantity squared is x, so we need to divide both sides by 2 before applying the square root property.   **9.** We're looking for an interest rate, so a negative value does not make sense.

Exercise Set 11.1

1. $-4, 4$ **3.** $-\sqrt{7}, \sqrt{7}$ **5.** $-3\sqrt{2}, 3\sqrt{2}$ **7.** $-\sqrt{10}, \sqrt{10}$ **9.** $-8, -2$ **11.** $6 - 3\sqrt{2}, 6 + 3\sqrt{2}$ **13.** $\dfrac{3-2\sqrt{2}}{2}, \dfrac{3+2\sqrt{2}}{2}$

15. $-3i, 3i$ **17.** $-\sqrt{6}, \sqrt{6}$ **19.** $-2i\sqrt{2}, 2i\sqrt{2}$ **21.** $\dfrac{1-4i}{3}, \dfrac{1+4i}{3}$ **23.** $-7 - \sqrt{5}, -7 + \sqrt{5}$ **25.** $-3 - 2i\sqrt{2}, -3 + 2i\sqrt{2}$

27. $x^2 + 16x + 64 = (x+8)^2$ **29.** $z^2 - 12z + 36 = (z-6)^2$ **31.** $p^2 + 9p + \dfrac{81}{4} = \left(p + \dfrac{9}{2}\right)^2$ **33.** $x^2 + x + \dfrac{1}{4} = \left(x + \dfrac{1}{2}\right)^2$

35. $-5, -3$ **37.** $-3 - \sqrt{7}, -3 + \sqrt{7}$ **39.** $\dfrac{-1-\sqrt{5}}{2}, \dfrac{-1+\sqrt{5}}{2}$ **41.** $-1 - \sqrt{6}, -1 + \sqrt{6}$ **43.** $\dfrac{-1-\sqrt{29}}{2}, \dfrac{-1+\sqrt{29}}{2}$

45. $\dfrac{6-\sqrt{30}}{3}, \dfrac{6+\sqrt{30}}{3}$ **47.** $\dfrac{3-\sqrt{11}}{2}, \dfrac{3+\sqrt{11}}{2}$ **49.** $-4, \dfrac{1}{2}$ **51.** $-4 - \sqrt{15}, -4 + \sqrt{15}$ **53.** $\dfrac{-3-\sqrt{21}}{3}, \dfrac{-3+\sqrt{21}}{3}$ **55.** $-1, \dfrac{5}{2}$

57. $-1 - i, -1 + i$ **59.** $3 - \sqrt{17}, 3 + \sqrt{17}$ **61.** $-2 - i\sqrt{2}, -2 + i\sqrt{2}$ **63.** $\dfrac{-15-7\sqrt{5}}{10}, \dfrac{-15+7\sqrt{5}}{10}$ **65.** $\dfrac{1-i\sqrt{47}}{4}, \dfrac{1+i\sqrt{47}}{4}$

67. $-5 - i\sqrt{3}, -5 + i\sqrt{3}$ **69.** $-4, 1$ **71.** $\dfrac{2-i\sqrt{2}}{2}, \dfrac{2+i\sqrt{2}}{2}$ **73.** $\dfrac{-3-\sqrt{69}}{6}, \dfrac{-3+\sqrt{69}}{6}$ **75.** 20%   **77.** 4%   **79.** 9.63 sec

81. 6.17 sec   **83.** 15 ft by 15 ft   **85.** $10\sqrt{2}$ cm   **87.** -1 **89.** $3 + 2\sqrt{5}$ **91.** $\dfrac{1-3\sqrt{2}}{2}$ **93.** $2\sqrt{6}$ **95.** 5   **97.** complex, but not

real numbers   **99.** real numbers   **101.** complex, but not real numbers   **103.** $-8x, 8x$ **105.** $-5z, 5z$ **107.** answers may vary
109. compound; answers may vary   **111.** 6 thousand scissors

Section 11.2

Practice Exercises

1. $2, -\dfrac{1}{3}$ **2.** $\dfrac{4 - \sqrt{22}}{3}, \dfrac{4 + \sqrt{22}}{3}$ **3.** $1 - \sqrt{17}, 1 + \sqrt{17}$ **4.** $\dfrac{-1 - i\sqrt{15}}{4}, \dfrac{1 + i\sqrt{15}}{4}$ **5. a.** one real solution **b.** two real solutions

c. two complex but not real solutions **6.** 6 ft **7.** 2.4 sec

Vocabulary, Readiness & Video Check 11.2

1. $x = \dfrac{-b \pm \sqrt{b^2 - 4ac}}{2a}$ **3.** $-5; -7$ **5.** $1; 0$ **7. a.** Yes, in order to make sure we have correct values for a, b, and c. **b.** No; clearing

fractions makes the work less tedious, but it's not a necessary step. **9.** With applications, we need to make sure we answer the question(s) asked. Here we're asked how much distance is saved, so once the dimensions of the triangle are known, further calculations are needed to answer this question and solve the problem.

Exercise Set 11.2

1. $-6, 1$ **3.** $-\dfrac{3}{5}, 1$ **5.** 3 **7.** $\dfrac{-7 - \sqrt{33}}{2}, \dfrac{-7 + \sqrt{33}}{2}$ **9.** $\dfrac{1 - \sqrt{57}}{8}, \dfrac{1 + \sqrt{57}}{8}$ **11.** $\dfrac{7 - \sqrt{85}}{6}, \dfrac{7 + \sqrt{85}}{6}$ **13.** $1 - \sqrt{3}, 1 + \sqrt{3}$

15. $-\dfrac{3}{2}, 1$ **17.** $\dfrac{3 - \sqrt{11}}{2}, \dfrac{3 + \sqrt{11}}{2}$ **19.** $\dfrac{-5 - \sqrt{17}}{2}, \dfrac{-5 + \sqrt{17}}{2}$ **21.** $\dfrac{5}{2}, 1$ **23.** $-3 - 2i, -3 + 2i$ **25.** $-2 - \sqrt{11}, -2 + \sqrt{11}$

27. $\dfrac{3 - i\sqrt{87}}{8}, \dfrac{3 + i\sqrt{87}}{8}$ **29.** $\dfrac{3 - \sqrt{29}}{2}, \dfrac{3 + \sqrt{29}}{2}$ **31.** $\dfrac{-5 - i\sqrt{5}}{10}, \dfrac{-5 + i\sqrt{5}}{10}$ **33.** $\dfrac{-1 - \sqrt{19}}{6}, \dfrac{-1 + \sqrt{19}}{6}$

35. $\dfrac{-1 - i\sqrt{23}}{4}, \dfrac{-1 + i\sqrt{23}}{4}$ **37.** 1 **39.** $3 - \sqrt{5}, 3 + \sqrt{5}$ **41.** two real solutions **43.** one real solution **45.** two real solutions

47. two complex but not real solutions **49.** two real solutions **51.** 14 ft **53.** $(2 + 2\sqrt{2})$ cm, $(2 + 2\sqrt{2})$ cm, $(4 + 2\sqrt{2})$ cm

55. width: $(-5 + 5\sqrt{17})$ ft; length: $(5 + 5\sqrt{17})$ ft **57. a.** $50\sqrt{2}$ m **b.** 5000 sq m **59.** 37.4 ft by 38.5 ft **61.** base: $(2 + 2\sqrt{43})$ cm; height:

$(-1 + \sqrt{43})$ cm **63.** 8.9 sec **65.** 2.8 sec **67.** $\dfrac{11}{5}$ **69.** 15 **71.** $(x^2 + 5)(x + 2)(x - 2)$ **73.** $(z + 3)(z - 3)(z + 2)(z - 2)$

75. b **77.** answers may vary **79.** 0.6, 2.4 **81.** Sunday to Monday **83.** Wednesday **85.** $f(4) = 32$; yes **87. a.** 3129 million

b. 2019 **89. a.** \$679.8 billion **b.** 2001 **c.** 2027 **91.** answers may vary **93.** $\dfrac{\sqrt{3}}{3}$ **95.** $\dfrac{-\sqrt{2} - i\sqrt{2}}{2}, \dfrac{-\sqrt{2} + i\sqrt{2}}{2}$

97. $\dfrac{\sqrt{3} - \sqrt{11}}{4}, \dfrac{\sqrt{3} + \sqrt{11}}{4}$ **99.** 8.9 sec: 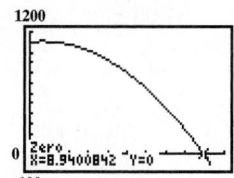 2.8 sec: 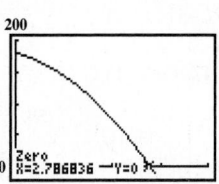 **101.** two real solutions

Section 11.3

Practice Exercises

1. 8 **2.** $\dfrac{5 \pm \sqrt{137}}{8}$ **3.** $4, -4, 3i, -3i$ **4.** $1, -3$ **5.** $1, 64$ **6.** Katy: $\dfrac{7 + \sqrt{65}}{2} \approx 7.5$ hr; Steve: $\dfrac{9 + \sqrt{65}}{2} \approx 8.5$ hr

7. to Shanghai: 40 km/hr; to Ningbo: 90 km/hr

Vocabulary, Readiness & Video Check 11.3

1. The values we get for the substituted variable are *not* our final answers. Remember to always substitute back to the original variable and solve for it if necessary.

Exercise Set 11.3

1. 2 **3.** 16 **5.** 1, 4 **7.** $3 - \sqrt{7}, 3 + \sqrt{7}$ **9.** $\dfrac{3 - \sqrt{57}}{4}, \dfrac{3 + \sqrt{57}}{4}$ **11.** $\dfrac{1 - \sqrt{29}}{2}, \dfrac{1 + \sqrt{29}}{2}$ **13.** $-2, 2, -2i, 2i$

15. $-\dfrac{1}{2}, \dfrac{1}{2}, -i\sqrt{3}, i\sqrt{3}$ **17.** $-3, 3, -2, 2$ **19.** $125, -8$ **21.** $-\dfrac{4}{5}, 0$ **23.** $-\dfrac{1}{8}, 27$ **25.** $-\dfrac{2}{3}, \dfrac{4}{3}$ **27.** $-\dfrac{1}{125}, \dfrac{1}{8}$ **29.** $-\sqrt{2}, \sqrt{2}, -\sqrt{3}, \sqrt{3}$

31. $\dfrac{-9 - \sqrt{201}}{6}, \dfrac{-9 + \sqrt{201}}{6}$ **33.** 2, 3 **35.** 3 **37.** 27, 125 **39.** $1, -3i, 3i$ **41.** $\dfrac{1}{8}, -8$ **43.** $-\dfrac{1}{2}, \dfrac{1}{3}$ **45.** 4 **47.** -3

49. $-\sqrt{5}, \sqrt{5}, -2i, 2i$ **51.** $-3, \dfrac{3 - 3i\sqrt{3}}{2}, \dfrac{3 + 3i\sqrt{3}}{2}$ **53.** 6, 12 **55.** $-\dfrac{1}{3}, \dfrac{1}{3}, -\dfrac{i\sqrt{6}}{3}, \dfrac{i\sqrt{6}}{3}$ **57.** 5 **59.** 55 mph: 66 mph

61. 5 mph, then 4 mph **63.** inlet pipe: 15.5 hr; hose: 16.5 hr **65.** 8.5 hr **67.** 12 or -8 **69. a.** $(x - 6)$ in. **b.** $300 = (x - 6) \cdot (x - 6) \cdot 3$ **c.** 16 in. by 16 in. **71.** 22 feet **73.** $(-\infty, 3]$ **75.** $(-5, \infty)$ **77.** domain: $(-\infty, \infty)$; range: $(-\infty, \infty)$; function **79.** domain: $(-\infty, \infty)$;

range: $[-1, \infty)$; function **81.** $1, -3i, 3i$ **83.** $-\dfrac{1}{2}, \dfrac{1}{3}$ **85.** $-3, \dfrac{3 - 3i\sqrt{3}}{2}, \dfrac{3 + 3i\sqrt{3}}{2}$ **87.** answers may vary **89. a.** approximately 164.37 mph

b. 54.754 sec/lap **c.** 55.947 sec/lap **d.** 160.87 mph

Integrated Review

1. $-\sqrt{10}, \sqrt{10}$ **2.** $-\sqrt{14}, \sqrt{14}$ **3.** $1 - 2\sqrt{2}, 1 + 2\sqrt{2}$ **4.** $-5 - 2\sqrt{3}, -5 + 2\sqrt{3}$ **5.** $-1 - \sqrt{13}, -1 + \sqrt{13}$ **6.** $1, 11$

7. $\dfrac{-3 - \sqrt{69}}{6}, \dfrac{-3 + \sqrt{69}}{6}$ **8.** $\dfrac{-2 - \sqrt{5}}{4}, \dfrac{-2 + \sqrt{5}}{4}$ **9.** $\dfrac{2 - \sqrt{2}}{2}, \dfrac{2 + \sqrt{2}}{2}$ **10.** $-3 - \sqrt{5}, -3 + \sqrt{5}$ **11.** $-2 - i\sqrt{3}, -2 + i\sqrt{3}$

12. $\dfrac{-1 - i\sqrt{11}}{2}, \dfrac{-1 + i\sqrt{11}}{2}$ **13.** $\dfrac{-3 - i\sqrt{15}}{2}, \dfrac{-3 + i\sqrt{15}}{2}$ **14.** $-3i, 3i$ **15.** $-17, 0$ **16.** $\dfrac{1 - \sqrt{13}}{4}, \dfrac{1 + \sqrt{13}}{4}$ **17.** $2 - 3\sqrt{3}, 2 + 3\sqrt{3}$

18. $2 - \sqrt{3}, 2 + \sqrt{3}$ **19.** $-2, \dfrac{4}{3}$ **20.** $\dfrac{-5 - \sqrt{17}}{4}, \dfrac{-5 + \sqrt{17}}{4}$ **21.** $1 - \sqrt{6}, 1 + \sqrt{6}$ **22.** $-\sqrt{31}, \sqrt{31}$ **23.** $-\sqrt{11}, \sqrt{11}$

24. $-i\sqrt{11}, i\sqrt{11}$ **25.** $-11, 6$ **26.** $\dfrac{-3 - \sqrt{19}}{5}, \dfrac{-3 + \sqrt{19}}{5}$ **27.** $\dfrac{-3 - \sqrt{17}}{4}, \dfrac{-3 + \sqrt{17}}{4}$ **28.** 4 **29.** $\dfrac{-1 - \sqrt{17}}{8}, \dfrac{-1 + \sqrt{17}}{8}$

30. $10\sqrt{2}\,\text{ft} \approx 14.1\,\text{ft}$ **31.** Jack: 9.1 hr; Lucy: 7.1 hr   **32.** 5 mph during the first part, then 6 mph

Section 11.4
Practice Exercises

1. $(-\infty, -3) \cup (4, \infty)$ **2.** $[0, 8]$ **3.** $(-\infty, -3] \cup [-1, 2]$ **4.** $(-4, 5]$ **5.** $(-\infty, -3) \cup \left(-\dfrac{8}{5}, \infty\right)$

Vocabulary, Readiness & Video Check 11.4

1. $[-7, 3)$ **3.** $(-\infty, 0]$ **5.** $(-\infty, -12) \cup [-10, \infty)$ **7.** We use the solutions of the related equation to divide the number line into regions that either entirely are or entirely are not solution regions; the solutions of the related equation are solutions of the inequality only if the inequality symbol is $\leq$ or $\geq$.

Exercise Set 11.4

1. $(-\infty, -5) \cup (-1, \infty)$ **3.** $[-4, 3]$ **5.** $[2, 5]$ **7.** $\left(-5, -\dfrac{1}{3}\right)$ **9.** $(2, 4) \cup (6, \infty)$ **11.** $(-\infty, -4] \cup [0, 1]$

13. $(-\infty, -3) \cup (-2, 2) \cup (3, \infty)$ **15.** $(-7, 2)$ **17.** $(-1, \infty)$ **19.** $(-\infty, -1] \cup (4, \infty)$ **21.** $(-\infty, 2) \cup \left(\dfrac{11}{4}, \infty\right)$ **23.** $(0, 2] \cup [3, \infty)$

25. $(-\infty, 3)$ **27.** $\left[-\dfrac{5}{4}, \dfrac{3}{2}\right]$ **29.** $(-\infty, 0) \cup (1, \infty)$ **31.** $(-\infty, -4] \cup [4, 6]$ **33.** $\left(-\infty, -\dfrac{2}{3}\right] \cup \left[\dfrac{3}{2}, \infty\right)$

35. $\left(-4, -\dfrac{3}{2}\right) \cup \left(\dfrac{3}{2}, \infty\right)$ **37.** $(-\infty, -5] \cup [-1, 1] \cup [5, \infty)$ **39.** $\left(-\infty, -\dfrac{5}{3}\right) \cup \left(\dfrac{7}{2}, \infty\right)$ **41.** $(0, 10)$ **43.** $(-\infty, -4) \cup [5, \infty)$

45. $(-\infty, -6] \cup (-1, 0] \cup (7, \infty)$ **47.** $(-\infty, 1) \cup (2, \infty)$ **49.** $(-\infty, -8] \cup (-4, \infty)$ **51.** $(-\infty, 0] \cup \left(5, \dfrac{11}{2}\right]$ **53.** $(0, \infty)$

55. 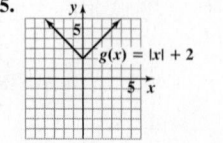   **57.** 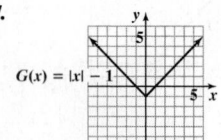   **59.** 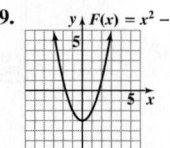   **61.** 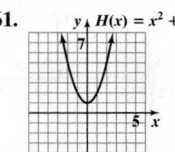   **63.** answers may vary   **65.** $(-\infty, -1) \cup (0, 1)$, or any number less than -1 or between 0 and 1

67. x is between 2 and 11   **69.** 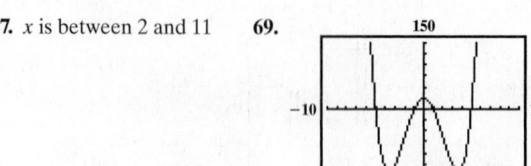   **71.**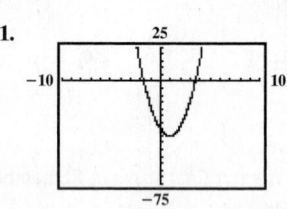

Section 11.5
Practice Exercises

1. 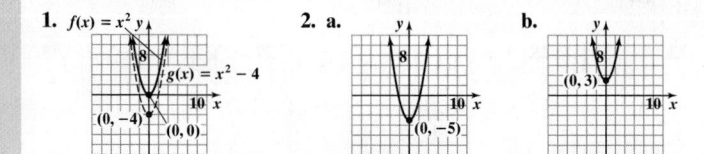   **2. a.** 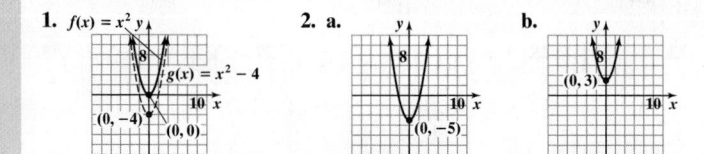 **b.**   **3.**   **4. a.** **b.**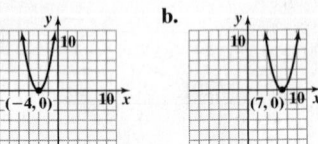

5. **6.** 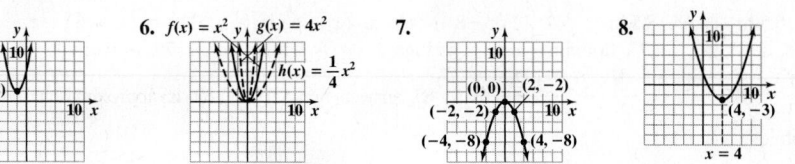 **7.**   **8.**

Section 11.2

Practice Exercises

1. $2, -\dfrac{1}{3}$ **2.** $\dfrac{4 - \sqrt{22}}{3}, \dfrac{4 + \sqrt{22}}{3}$ **3.** $1 - \sqrt{17}, 1 + \sqrt{17}$ **4.** $\dfrac{-1 - i\sqrt{15}}{4}, \dfrac{1 + i\sqrt{15}}{4}$ **5. a.** one real solution   **b.** two real solutions
c. two complex but not real solutions   **6.** 6 ft   **7.** 2.4 sec

Vocabulary, Readiness & Video Check 11.2

1. $x = \dfrac{-b \pm \sqrt{b^2 - 4ac}}{2a}$ **3.** $-5; -7$ **5.** $1; 0$ **7. a.** Yes, in order to make sure we have correct values for $a, b,$ and c.   **b.** No; clearing
fractions makes the work less tedious, but it's not a necessary step.   **9.** With applications, we need to make sure we answer the question(s) asked.
Here we're asked how much distance is saved, so once the dimensions of the triangle are known, further calculations are needed to answer this question
and solve the problem.

Exercise Set 11.2

1. $-6, 1$ **3.** $-\dfrac{3}{5}, 1$ **5.** 3   **7.** $\dfrac{-7 - \sqrt{33}}{2}, \dfrac{-7 + \sqrt{33}}{2}$ **9.** $\dfrac{1 - \sqrt{57}}{8}, \dfrac{1 + \sqrt{57}}{8}$ **11.** $\dfrac{7 - \sqrt{85}}{6}, \dfrac{7 + \sqrt{85}}{6}$ **13.** $1 - \sqrt{3}, 1 + \sqrt{3}$

15. $-\dfrac{3}{2}, 1$ **17.** $\dfrac{3 - \sqrt{11}}{2}, \dfrac{3 + \sqrt{11}}{2}$ **19.** $\dfrac{-5 - \sqrt{17}}{2}, \dfrac{-5 + \sqrt{17}}{2}$ **21.** $\dfrac{5}{2}, 1$ **23.** $-3 - 2i, -3 + 2i$ **25.** $-2 - \sqrt{11}, -2 + \sqrt{11}$

27. $\dfrac{3 - i\sqrt{87}}{8}, \dfrac{3 + i\sqrt{87}}{8}$ **29.** $\dfrac{3 - \sqrt{29}}{2}, \dfrac{3 + \sqrt{29}}{2}$ **31.** $\dfrac{-5 - i\sqrt{5}}{10}, \dfrac{-5 + i\sqrt{5}}{10}$ **33.** $\dfrac{-1 - \sqrt{19}}{6}, \dfrac{-1 + \sqrt{19}}{6}$

35. $\dfrac{-1 - i\sqrt{23}}{4}, \dfrac{-1 + i\sqrt{23}}{4}$ **37.** 1   **39.** $3 - \sqrt{5}, 3 + \sqrt{5}$ **41.** two real solutions   **43.** one real solution   **45.** two real solutions
47. two complex but not real solutions   **49.** two real solutions   **51.** 14 ft   **53.** $(2 + 2\sqrt{2})$ cm, $(2 + 2\sqrt{2})$ cm, $(4 + 2\sqrt{2})$ cm
55. width: $(-5 + 5\sqrt{17})$ ft; length: $(5 + 5\sqrt{17})$ ft   **57. a.** $50\sqrt{2}$ m   **b.** 5000 sq m   **59.** 37.4 ft by 38.5 ft   **61.** base: $(2 + 2\sqrt{43})$ cm; height:
$(-1 + \sqrt{43})$ cm   **63.** 8.9 sec   **65.** 2.8 sec   **67.** $\dfrac{11}{5}$ **69.** 15   **71.** $(x^2 + 5)(x + 2)(x - 2)$ **73.** $(z + 3)(z - 3)(z + 2)(z - 2)$

75. b   **77.** answers may vary   **79.** 0.6, 2.4   **81.** Sunday to Monday   **83.** Wednesday   **85.** $f(4) = 32$; yes   **87. a.** 3129 million
b. 2019   **89. a.** \$679.8 billion   **b.** 2001   **c.** 2027   **91.** answers may vary   **93.** $\dfrac{\sqrt{3}}{3}$ **95.** $\dfrac{-\sqrt{2} - i\sqrt{2}}{2}, \dfrac{-\sqrt{2} + i\sqrt{2}}{2}$

$\dfrac{\sqrt{3} - \sqrt{11}}{4}, \dfrac{\sqrt{3} + \sqrt{11}}{4}$ **99.** 8.9 sec: 1200   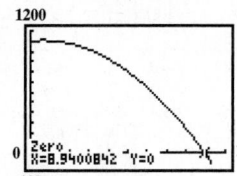   2.8 sec: 200   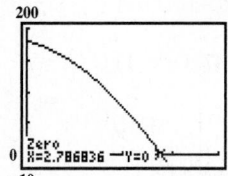   **101.** two real solutions

Section 11.3

Practice Exercises

1. 8   **2.** $\dfrac{5 \pm \sqrt{137}}{8}$ **3.** $4, -4, 3i, -3i$ **4.** $1, -3$ **5.** $1, 64$ **6.** Katy: $\dfrac{7 + \sqrt{65}}{2} \approx 7.5$ hr; Steve: $\dfrac{9 + \sqrt{65}}{2} \approx 8.5$ hr
7. to Shanghai: 40 km/hr; to Ningbo: 90 km/hr

Vocabulary, Readiness & Video Check 11.3

1. The values we get for the substituted variable are *not* our final answers. Remember to always substitute back to the original variable and solve for it if
necessary.

Exercise Set 11.3

1. 2   **3.** 16   **5.** 1, 4   **7.** $3 - \sqrt{7}, 3 + \sqrt{7}$ **9.** $\dfrac{3 - \sqrt{57}}{4}, \dfrac{3 + \sqrt{57}}{4}$ **11.** $\dfrac{1 - \sqrt{29}}{2}, \dfrac{1 + \sqrt{29}}{2}$ **13.** $-2, 2, -2i, 2i$

15. $-\dfrac{1}{2}, \dfrac{1}{2}, -i\sqrt{3}, i\sqrt{3}$ **17.** $-3, 3, -2, 2$ **19.** $125, -8$ **21.** $-\dfrac{4}{5}, 0$ **23.** $-\dfrac{1}{8}, 27$ **25.** $-\dfrac{2}{3}, \dfrac{4}{3}$ **27.** $-\dfrac{1}{125}, \dfrac{1}{8}$ **29.** $-\sqrt{2}, \sqrt{2}, -\sqrt{3}, \sqrt{3}$

31. $\dfrac{-9 - \sqrt{201}}{6}, \dfrac{-9 + \sqrt{201}}{6}$ **33.** 2, 3   **35.** 3   **37.** 27, 125   **39.** $1, -3i, 3i$ **41.** $\dfrac{1}{8}, -8$ **43.** $-\dfrac{1}{2}, \dfrac{1}{3}$ **45.** 4   **47.** -3

49. $-\sqrt{5}, \sqrt{5}, -2i, 2i$ **51.** $-3, \dfrac{3 - 3i\sqrt{3}}{2}, \dfrac{3 + 3i\sqrt{3}}{2}$ **53.** 6, 12   **55.** $-\dfrac{1}{3}, \dfrac{1}{3}, -\dfrac{i\sqrt{6}}{3}, \dfrac{i\sqrt{6}}{3}$ **57.** 5   **59.** 55 mph: 66 mph
61. 5 mph, then 4 mph   **63.** inlet pipe: 15.5 hr; hose: 16.5 hr   **65.** 8.5 hr   **67.** 12 or -8 **69. a.** $(x - 6)$ in.   **b.** $300 = (x - 6) \cdot (x - 6) \cdot 3$
c. 16 in. by 16 in.   **71.** 22 feet   **73.** $(-\infty, 3]$ **75.** $(-5, \infty)$ **77.** domain: $(-\infty, \infty)$; range: $(-\infty, \infty)$; function   **79.** domain: $(-\infty, \infty)$;
range: $[-1, \infty)$; function   **81.** $1, -3i, 3i$ **83.** $-\dfrac{1}{2}, \dfrac{1}{3}$ **85.** $-3, \dfrac{3 - 3i\sqrt{3}}{2}, \dfrac{3 + 3i\sqrt{3}}{2}$ **87.** answers may vary   **89. a.** approximately 164.37 mph
b. 54.754 sec/lap   **c.** 55.947 sec/lap   **d.** 160.87 mph

Integrated Review

1. $-\sqrt{10}, \sqrt{10}$ **2.** $-\sqrt{14}, \sqrt{14}$ **3.** $1 - 2\sqrt{2}, 1 + 2\sqrt{2}$ **4.** $-5 - 2\sqrt{3}, -5 + 2\sqrt{3}$ **5.** $-1 - \sqrt{13}, -1 + \sqrt{13}$ **6.** $1, 11$

7. $\dfrac{-3 - \sqrt{69}}{6}, \dfrac{-3 + \sqrt{69}}{6}$ **8.** $\dfrac{-2 - \sqrt{5}}{4}, \dfrac{-2 + \sqrt{5}}{4}$ **9.** $\dfrac{2 - \sqrt{2}}{2}, \dfrac{2 + \sqrt{2}}{2}$ **10.** $-3 - \sqrt{5}, -3 + \sqrt{5}$ **11.** $-2 - i\sqrt{3}, -2 + i\sqrt{3}$

12. $\dfrac{-1 - i\sqrt{11}}{2}, \dfrac{-1 + i\sqrt{11}}{2}$ **13.** $\dfrac{-3 - i\sqrt{15}}{2}, \dfrac{-3 + i\sqrt{15}}{2}$ **14.** $-3i, 3i$ **15.** $-17, 0$ **16.** $\dfrac{1 - \sqrt{13}}{4}, \dfrac{1 + \sqrt{13}}{4}$ **17.** $2 - 3\sqrt{3}, 2 + 3\sqrt{3}$

18. $2 - \sqrt{3}, 2 + \sqrt{3}$ **19.** $-2, \dfrac{4}{3}$ **20.** $\dfrac{-5 - \sqrt{17}}{4}, \dfrac{-5 + \sqrt{17}}{4}$ **21.** $1 - \sqrt{6}, 1 + \sqrt{6}$ **22.** $-\sqrt{31}, \sqrt{31}$ **23.** $-\sqrt{11}, \sqrt{11}$

24. $-i\sqrt{11}, i\sqrt{11}$ **25.** $-11, 6$ **26.** $\dfrac{-3 - \sqrt{19}}{5}, \dfrac{-3 + \sqrt{19}}{5}$ **27.** $\dfrac{-3 - \sqrt{17}}{4}, \dfrac{-3 + \sqrt{17}}{4}$ **28.** 4 **29.** $\dfrac{-1 - \sqrt{17}}{8}, \dfrac{-1 + \sqrt{17}}{8}$

30. $10\sqrt{2}\,\text{ft} \approx 14.1\,\text{ft}$ **31.** Jack: 9.1 hr; Lucy: 7.1 hr   **32.** 5 mph during the first part, then 6 mph

Section 11.4
Practice Exercises

1. $(-\infty, -3) \cup (4, \infty)$ **2.** $[0, 8]$ **3.** $(-\infty, -3] \cup [-1, 2]$ **4.** $(-4, 5]$ **5.** $(-\infty, -3) \cup \left(-\dfrac{8}{5}, \infty\right)$

Vocabulary, Readiness & Video Check 11.4

1. $[-7, 3)$ **3.** $(-\infty, 0]$ **5.** $(-\infty, -12) \cup [-10, \infty)$ **7.** We use the solutions of the related equation to divide the number line into regions that either entirely are or entirely are not solution regions; the solutions of the related equation are solutions of the inequality only if the inequality symbol is $\le$ or $\ge$.

Exercise Set 11.4

1. $(-\infty, -5) \cup (-1, \infty)$ **3.** $[-4, 3]$ **5.** $[2, 5]$ **7.** $\left(-5, -\dfrac{1}{3}\right)$ **9.** $(2, 4) \cup (6, \infty)$ **11.** $(-\infty, -4] \cup [0, 1]$

13. $(-\infty, -3) \cup (-2, 2) \cup (3, \infty)$ **15.** $(-7, 2)$ **17.** $(-1, \infty)$ **19.** $(-\infty, -1] \cup (4, \infty)$ **21.** $(-\infty, 2) \cup \left(\dfrac{11}{4}, \infty\right)$ **23.** $(0, 2] \cup [3, \infty)$

25. $(-\infty, 3)$ **27.** $\left[-\dfrac{5}{4}, \dfrac{3}{2}\right]$ **29.** $(-\infty, 0) \cup (1, \infty)$ **31.** $(-\infty, -4] \cup [4, 6]$ **33.** $\left(-\infty, -\dfrac{2}{3}\right] \cup \left[\dfrac{3}{2}, \infty\right)$

35. $\left(-4, -\dfrac{3}{2}\right) \cup \left(\dfrac{3}{2}, \infty\right)$ **37.** $(-\infty, -5] \cup [-1, 1] \cup [5, \infty)$ **39.** $\left(-\infty, -\dfrac{5}{3}\right) \cup \left(\dfrac{7}{2}, \infty\right)$ **41.** $(0, 10)$ **43.** $(-\infty, -4) \cup [5, \infty)$

45. $(-\infty, -6] \cup (-1, 0] \cup (7, \infty)$ **47.** $(-\infty, 1) \cup (2, \infty)$ **49.** $(-\infty, -8] \cup (-4, \infty)$ **51.** $(-\infty, 0] \cup \left(5, \dfrac{11}{2}\right]$ **53.** $(0, \infty)$

55. 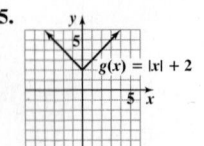   **57.** 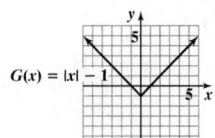   **59.** 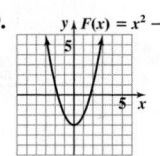   **61.** 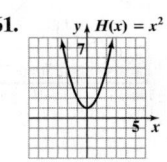   **63.** answers may vary   **65.** $(-\infty, -1) \cup (0, 1)$, or any number less than -1 or between 0 and 1

67. x is between 2 and 11   **69.** 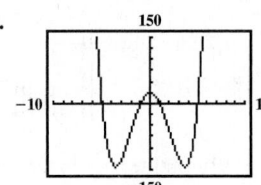   **71.**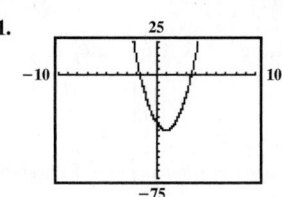

Section 11.5
Practice Exercises

1. 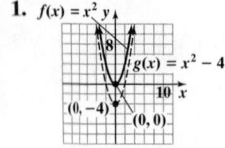   **2. a.** 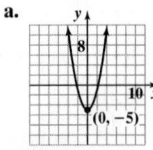   **b.** 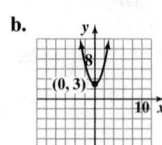   **3.** 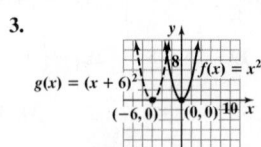   **4. a.** 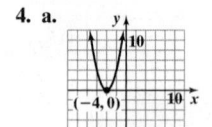   **b.**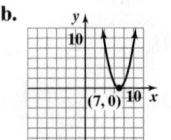

5. 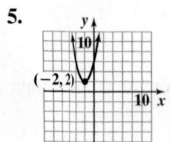   **6.** 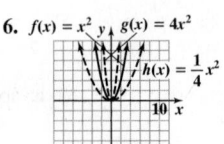   **7.** 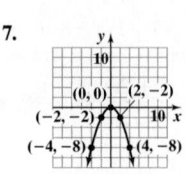   **8.**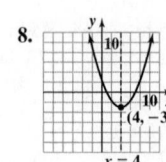

Graphing Calculator Explorations 11.5

1. 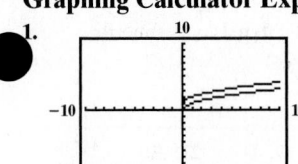   **3.** 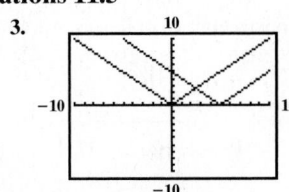   **5.**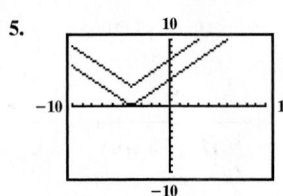

Vocabulary, Readiness & Video Check 11.5

1. quadratic   **3.** upward   **5.** lowest   **7.** $(0, 0)$ **9.** $(2, 0)$ **11.** $(0, 3)$ **13.** $(-1, 5)$ **15.** Graphs of the form $f(x) = x^2 + k$ shift up or down the y-axis k units from $y = x^2$; the y-intercept.   **17.** The vertex, (h, k), and the axis of symmetry, $x = h$; the basic shape of $y = x^2$ does not change.   **19.** the coordinates of the vertex, whether the graph opens upward or downward, whether the graph is narrower or wider than $y = x^2$, and the graph's axis of symmetry

Exercise Set 11.5

1. 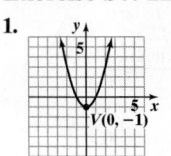   **3.** 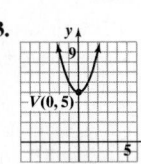   **5.** 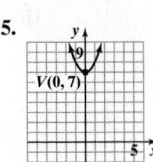   **7.** 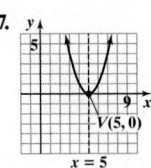   **9.** 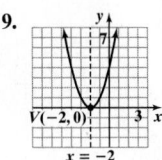   **11.**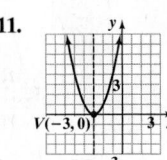

13. 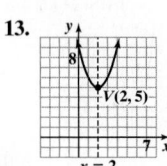   **15.** 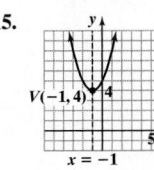   **17.** 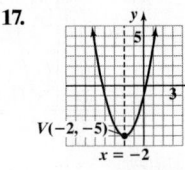   **19.** 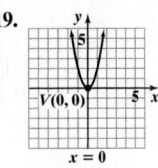   **21.** 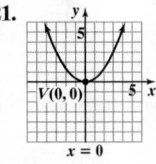   **23.**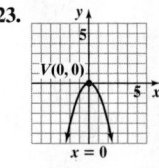

25. 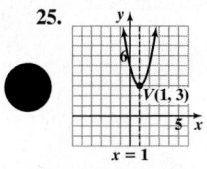   **27.** 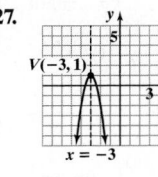   **29.** 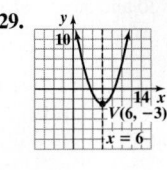   **31.** 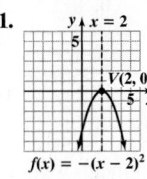   **33.** 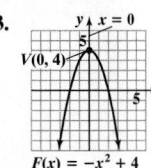   **35.**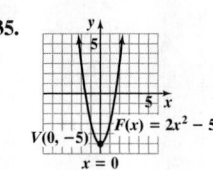

37. 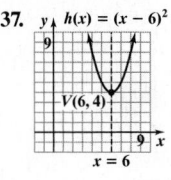   **39.** 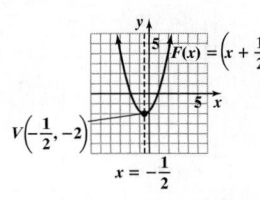   **41.** 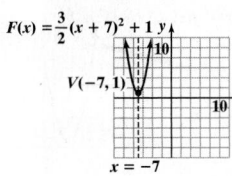   **43.**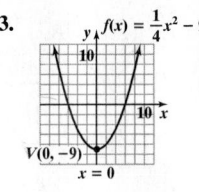

45. 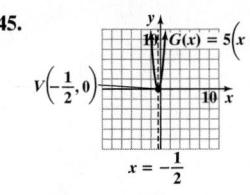   **47.** 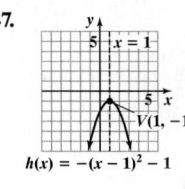   **49.** $g(x) = \sqrt{3}(x + 5)^2 + \frac{3}{4}$ 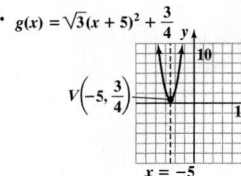   **51.**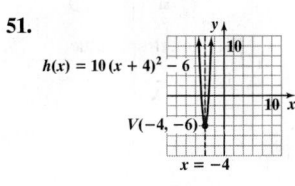

53. 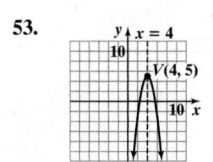   **55.** $x^2 + 8x + 16$ **57.** $z^2 - 16z + 64$ **59.** $y^2 + y + \frac{1}{4}$ **61.** $-6, 2$ **63.** $-5 - \sqrt{26}, -5 + \sqrt{26}$ **65.** $4 - 3\sqrt{2}, 4 + 3\sqrt{2}$ **67.** c   **69.** $f(x) = 5(x - 2)^2 + 3$ **71.** $f(x) = 5(x + 3)^2 + 6$

73. 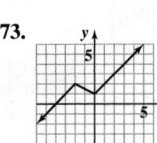   **75.** 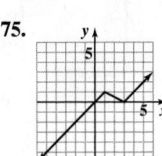   **77.** 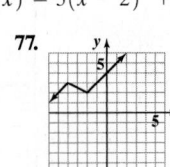   **79. a.** \$423 billion   **b.** \$1987.5 billion

Section 11.6

Practice Exercises

1. 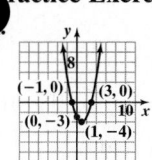   **2.** $\left(-\frac{1}{2}, 2\right)$ 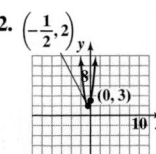   **3.** 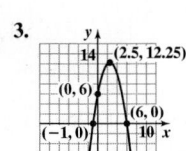   **4.** $(1, -4)$ **5.** Maximum height 9 feet in $\frac{3}{4}$ second

Vocabulary, Readiness & Video Check 11.6

1. (h, k) **3.** We can immediately identify the vertex (h, k), whether the parabola opens upward or downward, and know its axis of symmetry; completing the square.   **5.** the vertex

Exercise Set 11.6

1. $0; 1$ **3.** $2; 1$ **5.** $1; 1$ **7.** down   **9.** up   **11.** $(-4, -9)$ **13.** $(5, 30)$ **15.** $(1, -2)$ **17.** $\left(\frac{1}{2}, \frac{5}{4}\right)$ **19.** D   **21.** B

23. 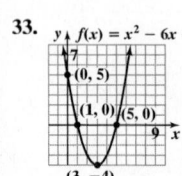   **25.** 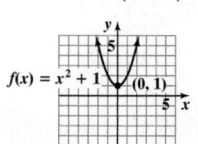   **27.** 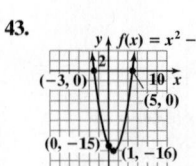   **29.** 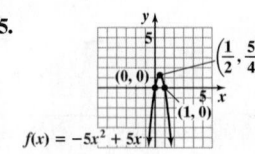   **31.**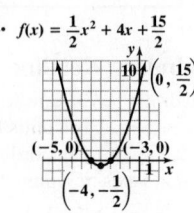

33. 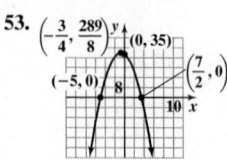   **35.** 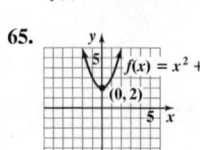   **37.** 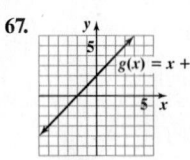   **39.** 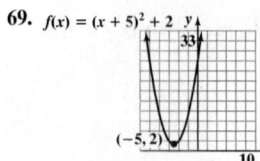   **41.**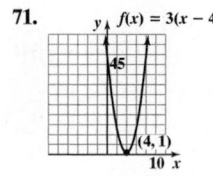

43. 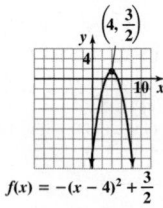   **45.** 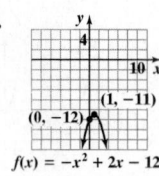   **47.** 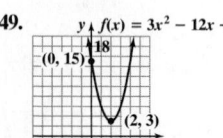   **49.** 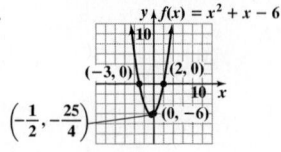   **51.**

53. 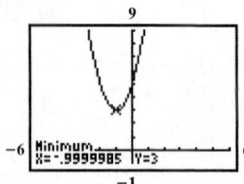   **55.** 144 ft   **57. a.** 200 bicycles   **b.** $12,000   **59.** 30 and 30   **61.** 5, -5 **63.** length: 20 units; width: 20 units

65.   **67.**   **69.** $f(x) = (x + 5)^2 + 2$ **71.** $f(x) = 3(x - 4)^2 + 1$ **73.** $f(x) = -(x - 4)^2 + \frac{3}{2}$

75. minimum value   **77.** maximum value   **79.** $f(x) = x^2 + 10x + 15$ **81.**   **83.** -0.84 **85.** 1.43

87. a. maximum; answers may vary   **b.** 2035   **c.** 311 million   **89.** 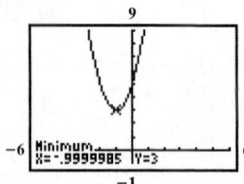   **91.**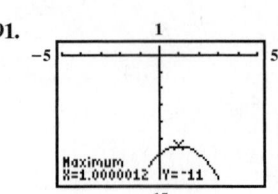

Chapter 11 Vocabulary Check

1. discriminant   **2.** $\pm\sqrt{b}$ **3.** $\frac{-b}{2a}$ **4.** quadratic inequality   **5.** completing the square   **6.** $(0, k)$ **7.** $(h, 0)$ **8.** (h, k)
9. quadratic formula   **10.** quadratic

Chapter 11 Review

1. $14, 1$ **3.** $-7, 7$ **5.** $\dfrac{-3 - \sqrt{5}}{2}, \dfrac{-3 + \sqrt{5}}{2}$ **7.** 4.25%   **9.** two complex but not real solutions   **11.** two real solutions   **13.** 8

15. $-\dfrac{5}{2}, 1$ **17.** $\dfrac{5 - i\sqrt{143}}{12}, \dfrac{5 + i\sqrt{143}}{12}$ **19. a.** 20 ft **b.** $\dfrac{15 + \sqrt{321}}{16}$ sec; 2.1 sec **21.** $3, \dfrac{-3 - 3i\sqrt{3}}{2}, \dfrac{-3 + 3i\sqrt{3}}{2}$ **23.** $\dfrac{2}{3}, 5$ **25.** 1, 125

27. $-1, 1, -i, i$ **29.** Jerome: 10.5 hr; Tim: 9.5 hr **31.** $[-5, 5]$ **33.** $[-5, -2] \cup [2, 5]$ **35.** $(5, 6)$ **37.** $(-\infty, -5] \cup [-2, 6]$ **39.** $(-\infty, 0)$

41. 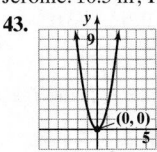 **43.** 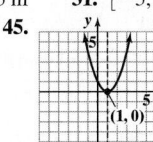 **45.** **47.** 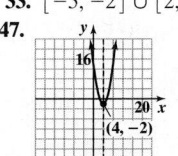 **49.** 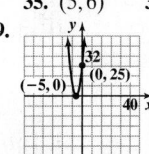 **51.**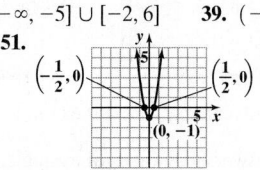

53. $\left(-\dfrac{5}{6}, \dfrac{73}{12}\right)$ 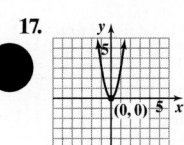 **55.** 210 and 210 **57.** $-5, 6$ **59.** $-2, 2$ **61.** $\dfrac{-1 - 3i\sqrt{3}}{2}, \dfrac{-1 + 3i\sqrt{3}}{2}$ **63.** $-i\sqrt{11}, i\sqrt{11}$

65. $\dfrac{21 - \sqrt{41}}{50}, \dfrac{21 + \sqrt{41}}{50}$ **67.** 8, 64 **69.** $[-5, 0] \cup \left(\dfrac{3}{4}, \infty\right)$ **71. a.** 90,332 thousand passengers **b.** no; answers may vary

Chapter 11 Getting Ready for the Test

1. C **2.** B **3.** A **4.** D **5.** B **6.** D **7.** C **8.** C **9.** A **10.** B **11.** C

Chapter 11 Test

1. $\dfrac{7}{5}, -1$ **2.** $-1 - \sqrt{10}, -1 + \sqrt{10}$ **3.** $\dfrac{1 - i\sqrt{31}}{2}, \dfrac{1 + i\sqrt{31}}{2}$ **4.** $3 - \sqrt{7}, 3 + \sqrt{7}$ **5.** $-\dfrac{1}{7}, -1$ **6.** $\dfrac{3 - \sqrt{29}}{2}, \dfrac{3 + \sqrt{29}}{2}$

7. $-2 - \sqrt{11}, -2 + \sqrt{11}$ **8.** $-1, 1, -i, i, -3$ **9.** $-1, 1, -i, i$ **10.** 6, 7 **11.** $3 - \sqrt{7}, 3 + \sqrt{7}$ **12.** $\dfrac{2 - i\sqrt{6}}{2}, \dfrac{2 + i\sqrt{6}}{2}$

13. $\left(-\infty, -\dfrac{3}{2}\right) \cup (5, \infty)$ **14.** $(-\infty, -5] \cup [-4, 4] \cup [5, \infty)$ **15.** $(-\infty, -3) \cup (2, \infty)$ **16.** $(-\infty, -3) \cup [2, 3)$

17. 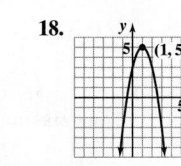 **18.** 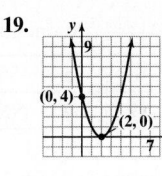 **19.** 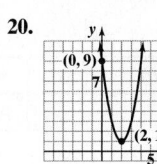 **20.** **21.** $(5 + \sqrt{17})$ hr ≈ 9.12 hr **22. a.** 272 ft **b.** 5.12 sec **23.** 7.2 ft

Chapter 11 Cumulative Review

1. a. $\dfrac{1}{2}$ **b.** 19; Sec. 1.6, Ex. 6 **3. a.** $5x + 7$ **b.** $-4a - 1$ **c.** $4y - 3y^2$ **d.** $7.3x - 6$ **e.** $\dfrac{1}{2}b$; Sec. 2.1, Ex. 4 **5.** no solution; $\{\ \}$ or $\varnothing$; Sec. 4.1, Ex. 5

7. $(-2, 0)$; Sec. 4.2, Ex. 4 **9. a.** x^3 **b.** 256 **c.** -27 **d.** cannot be simplified **e.** $2x^4y$; Sec. 5.1, Ex. 9 **11. a.** 5 **b.** 5; Sec. 5.7, Ex. 3

13. $-6, -\dfrac{3}{2}, \dfrac{1}{5}$; Sec. 6.6, Ex. 9 **15.** $\dfrac{1}{5x - 1}$; Sec. 7.1, Ex. 2a **17.** $\dfrac{xy + 2x^3}{y - 1}$; Sec. 7.7, Ex. 3 **19.** $(2m^2 - 1)^2$; Sec. 6.3, Ex. 10 **21.** -10 and 7; Sec.

6.7, Ex. 2 **23. a.** $5x\sqrt{x}$ **b.** $3x^2y^2\sqrt[3]{2y^2}$ **c.** $3z^2\sqrt[4]{z^3}$; Sec. 10.3, Ex. 4 **25. a.** $\dfrac{2\sqrt{5}}{5}$ **b.** $\dfrac{8\sqrt{x}}{3x}$ **c.** $\dfrac{\sqrt[3]{4}}{2}$; Sec. 10.5, Ex. 1 **27.** $\dfrac{2}{9}$; Sec. 10.6, Ex. 5

29. -5; Sec. 7.5, Ex. 1 **31.** -5; Sec. 7.6, Ex. 5 **33.** $k = \dfrac{1}{6}; y = \dfrac{1}{6}x$; Sec. 8.4, Ex. 1 **35. a.** 3 **b.** $|x|$ **c.** $|x - 2|$ **d.** -5 **e.** $2x - 7$

f. $5|x|$ **g.** $|x + 1|$; Sec. 10.1, Ex. 5 **37. a.** $\sqrt{x}$ **b.** $\sqrt[3]{5}$ **c.** $\sqrt{rs^3}$; Sec. 10.2, Ex. 7 **39. a.** $\dfrac{1}{2} + \dfrac{3}{2}i$ **b.** $0 - \dfrac{7}{3}i$; Sec. 10.7, Ex. 5

41. $-1 + 2\sqrt{3}, -1 - 2\sqrt{3}$; Sec. 11.1, Ex. 3 **43.** 9; Sec. 11.3, Ex. 1

CHAPTER 12 EXPONENTIAL AND LOGARITHMIC FUNCTIONS

Section 12.1

Practice Exercises

1. a. $4x + 7$ **b.** $-2x - 3$ **c.** $3x^2 + 11x + 10$ **d.** $\dfrac{x + 2}{3x + 5}$, where $x \neq -\dfrac{5}{3}$ **2. a.** 50; 46 **b.** $9x^2 - 30x + 26; 3x^2 - 2$

3. a. $x^2 + 6x + 14$ **b.** $x^2 + 8$ **4. a.** $(h \circ g)(x)$ **b.** $(g \circ f)(x)$

Vocabulary, Readiness & Video Check 12.1

C **3.** F **5.** D **7.** You can find $(f + g)(x)$ and then find $(f + g)(2)$ or you can find $f(2)$ and $g(2)$ and then add those results.

Exercise Set 12.1

1. a. $3x - 6$ **b.** $-x - 8$ **c.** $2x^2 - 13x - 7$ **d.** $\dfrac{x - 7}{2x + 1}$, where $x \neq -\dfrac{1}{2}$ **3. a.** $x^2 + 5x + 1$ **b.** $x^2 - 5x + 1$ **c.** $5x^3 + 5x$

d. $\dfrac{x^2 + 1}{5x}$, where $x \neq 0$ **5. a.** $\sqrt[3]{x} + x + 5$ **b.** $\sqrt[3]{x} - x - 5$ **c.** $x\sqrt[3]{x} + 5\sqrt[3]{x}$ **d.** $\dfrac{\sqrt[3]{x}}{x + 5}$, where $x \neq -5$

7. a. $5x^2 - 3x$ **b.** $-5x^2 - 3x$ **c.** $-15x^3$ **d.** $-\dfrac{3}{5x}$, where $x \neq 0$ **9.** 42 **11.** -18 **13.** 0

15. $(f \circ g)(x) = 25x^2 + 1; (g \circ f)(x) = 5x^2 + 5$ **17.** $(f \circ g)(x) = 2x + 11; (g \circ f)(x) = 2x + 4$

19. $(f \circ g)(x) = -8x^3 - 2x - 2; (g \circ f)(x) = -2x^3 - 2x + 4$ **21.** $(f \circ g)(x) = |10x - 3|; (g \circ f)(x) = 10|x| - 3$

23. $(f \circ g)(x) = \sqrt{-5x + 2}; (g \circ f)(x) = -5\sqrt{x} + 2$ **25.** $H(x) = (g \circ h)(x)$ **27.** $F(x) = (h \circ f)(x)$ **29.** $G(x) = (f \circ g)(x)$

31. answers may vary; for example, $g(x) = x + 2$ and $f(x) = x^2$ **33.** answers may vary; for example, $g(x) = x + 5$ and $f(x) = \sqrt{x} + 2$

35. answers may vary; for example, $g(x) = 2x - 3$ and $f(x) = \dfrac{1}{x}$ **37.** $y = x - 2$ **39.** $y = \dfrac{x}{3}$ **41.** $y = -\dfrac{x + 7}{2}$ **43.** 6 **45.** 4

47. 4 **49.** -1 **51.** answers may vary **53.** $P(x) = R(x) - C(x)$

Section 12.2

Practice Exercises

1. a. one-to-one **b.** not one-to-one **c.** not one-to-one **d.** one-to-one **e.** not one-to-one **f.** not one-to-one **2. a.** no, not one-to-one
b. yes **c.** yes **d.** no, not a function **e.** no, not a function **3.** $f^{-1} = \{(4, 3), (0, -2), (8, 2), (6, 6)\}$ **4.** $f^{-1}(x) = 6 - x$

5. 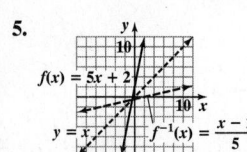 **6. a.** 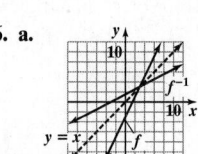 **b.** 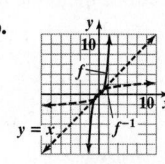 **7.** $f(f^{-1}(x)) = f\left(\dfrac{x + 1}{4}\right) = 4\left(\dfrac{x + 1}{4}\right) - 1 = x + 1 - 1 = x$

$f^{-1}(f(x)) = f^{-1}(4x - 1) = \dfrac{(4x - 1) + 1}{4} = \dfrac{4x}{4} = x$

Vocabulary, Readiness & Video Check 12.2

1. $(2, 11)$ **3.** the inverse of f **5.** vertical **7.** $y = x$ **9.** Every function must have each x-value correspond to only one y-value. A one-to-one function must also have each y-value correspond to only one x-value. **11.** Yes; by the definition of an inverse function. **13.** Once you know some points of the original equation or graph, you can switch the x's and y's of these points to find points that satisfy the inverse and then graph it. You can also check that the two graphs (the original and the inverse) are symmetric about the line $y = x$.

Exercise Set 12.2

1. one-to-one; $f^{-1} = \{(-1, -1), (1, 1), (2, 0), (0, 2)\}$ **3.** one-to-one; $h^{-1} = \{(10, 10)\}$ **5.** one-to-one; $f^{-1} = \{(12, 11), (3, 4), (4, 3), (6, 6)\}$
7. not one-to-one **9.** one-to-one;

Rank in Population (Input)	1	47	16	25	36	7
State (Output)	CA	AK	AZ	LA	NM	OH

11. a. 3 **b.** 1 **13. a.** 1
b. -1 **15.** one-to-one
17. not one-to-one

19. one-to-one **21.** not one-to-one

23. $f^{-1}(x) = x - 4$ 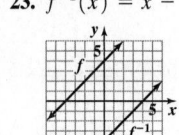 **25.** $f^{-1}(x) = \dfrac{x + 3}{2}$ 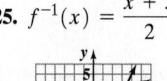 **27.** $f^{-1}(x) = 2x + 2$ 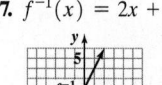 **29.** $f^{-1}(x) = \sqrt[3]{x}$ 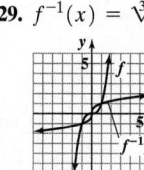 **31.** $f^{-1}(x) = \dfrac{x - 2}{5}$ **33.** $f^{-1}(x) = 5x + 2$

35. $f^{-1}(x) = x^3$ **37.** $f^{-1}(x) = \dfrac{5 - x}{3x}$
39. $f^{-1}(x) = \sqrt[3]{x} - 2$

41. 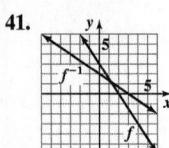 **43.** 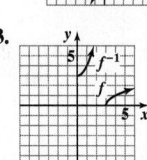 **45.** 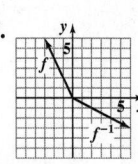 **47.** $(f \circ f^{-1})(x) = x; (f^{-1} \circ f)(x) = x$ **49.** $(f \circ f^{-1})(x) = x; (f^{-1} \circ f)(x) = x$

51. 5 **53.** 8 **55.** $\dfrac{1}{27}$ **57.** 9 **59.** $3^{1/2} \approx 1.73$ **61. a.** $(2, 9)$ **b.** $(9, 2)$

63. a. $\left(-2, \dfrac{1}{4}\right), \left(-1, \dfrac{1}{2}\right), (0, 1), (1, 2), (2, 5)$ **b.** $\left(\dfrac{1}{4}, -2\right), \left(\dfrac{1}{2}, -1\right), (1, 0), (2, 1), (5, 2)$ **c.** 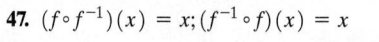 **d.**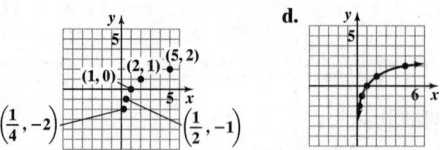

65. answers may vary **67.** $f^{-1}(x) = \dfrac{x - 1}{3}$; 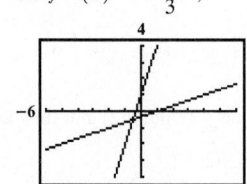 **69.** $f^{-1}(x) = x^3 - 1$;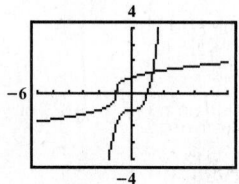

Section 12.3

Practice Exercises

1. 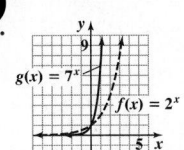 **2.** 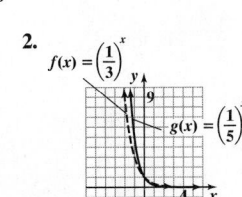 **3.** 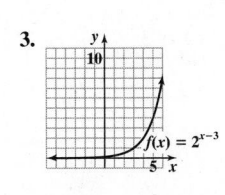 **4. a.** 2 **b.** $\frac{4}{3}$ **c.** -4 **5.** \$3950.43 **6. a.** 90.54% **b.** 60.86%

Graphing Calculator Explorations 12.3

1. 81.98%; 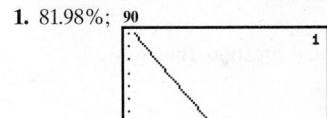 **3.** 22.54%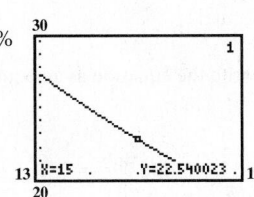

Vocabulary, Readiness & Video Check 12.3

1. c. exponential **3.** yes **5.** no; none **7.** $(-\infty, \infty)$ **9.** In a polynomial function, the base is the variable and the exponent is the constant; in an exponential function, the base is the constant and the exponent is the variable. **11.** $y = 30(0.996)^{101} \approx 20.0$ lb

Exercise Set 12.3

1. 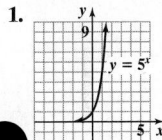 **3.** 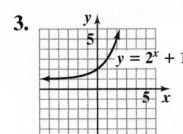 **5.** 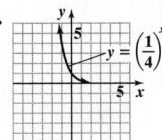 **7.** 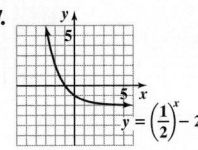 **9.** 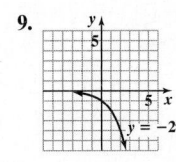 **11.**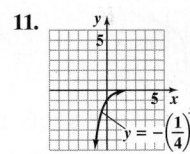

13. 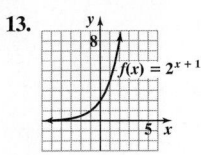 **15.** 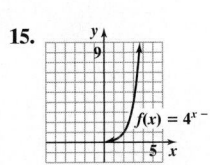 **17.** C **19.** B **21.** 3 **23.** $\frac{3}{4}$ **25.** $\frac{8}{5}$ **27.** $-\frac{2}{3}$ **29.** 4 **31.** $\frac{3}{2}$

33. $-\frac{1}{3}$ **35.** -2 **37.** 24.6 lb **39. a.** 31.5 lb per person **b.** 36.5 lb per person **41. a.** 167.3 thousand students **b.** 430.2 thousand students **43.** 233,182 thousand **45.** 544,000 mosquitoes **47.** \$7621.42 **49.** \$4065.59 **51.** 4 **53.** $\varnothing$ **55.** 2, 3 **57.** 3 **59.** -1 **61.** no

63. no **65.** C **67.** D **69.** answers may vary **71.** **73.** **75.** The graphs are the same since $\left(\frac{1}{2}\right)^{-x} = 2^x$. **77.** 24.55 lb **79.** 20.09 lb

Section 12.4

Practice Exercises

1. 136,839 **2.** 32 **3.** 47.4 g

Vocabulary, Readiness & Video Check 12.4

1. For Example 1, the growth rate is given as 5% per year. Since this is "per year," the number of time intervals is the "number of years," or 8.

3. time intervals = years/half-life; the decay rate is 50% or $\frac{1}{2}$ because half-life is the amount of time it takes half of a substance to decay

Exercise Set 12.4

1. 451 **3.** 144,302 **5.** 21,231 **7.** 202 **9.** 1470 **11.** 13 **13.** 712,880 **15.** 383 **17.** 333 bison **19.** 1 g **21. a.** $\frac{14}{7} = 2$; 10; yes

b. $\frac{11}{7} \approx 1.6$; 13.2; yes **23.** $\frac{500}{152} \approx 3.3$; 2.1; yes **25.** 4.9 g **27.** 3 **29.** -1 **31.** no; answers may vary

Section 12.5
Practice Exercises

1. a. $3^4 = 81$ **b.** $5^{-1} = \dfrac{1}{5}$ **c.** $7^{1/2} = \sqrt{7}$ **d.** $13^4 = y$ **2. a.** $\log_4 64 = 3$ **b.** $\log_6 \sqrt[3]{6} = \dfrac{1}{3}$ **c.** $\log_5 \dfrac{1}{125} = -3$ **d.** $\log_\pi z = 7$

3. a. 2   **b.** -3 **c.** $\dfrac{1}{2}$ **4. a.** -2 **b.** 2   **c.** 36   **d.** 0   **e.** 0   **5. a.** 4   **b.** -2 **c.** 5   **d.** 4

6. 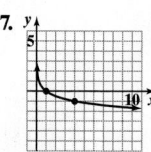   **7.**

Vocabulary, Readiness & Video Check 12.5

1. B. logarithmic   **3.** yes   **5.** no; none   **7.** $(-\infty, \infty)$ **9.** First write the equation as an equivalent exponential equation. Then solve.

Exercise Set 12.5

1. $6^2 = 36$ **3.** $3^{-3} = \dfrac{1}{27}$ **5.** $10^3 = 1000$ **7.** $9^4 = x$ **9.** $\pi^{-2} = \dfrac{1}{\pi^2}$ **11.** $7^{1/2} = \sqrt{7}$ **13.** $0.7^3 = 0.343$ **15.** $3^{-4} = \dfrac{1}{81}$

17. $\log_2 16 = 4$ **19.** $\log_{10} 100 = 2$ **21.** $\log_\pi x = 3$ **23.** $\log_{10} \dfrac{1}{10} = -1$ **25.** $\log_4 \dfrac{1}{16} = -2$ **27.** $\log_5 \sqrt{5} = \dfrac{1}{2}$

29. 3   **31.** -2 **33.** $\dfrac{1}{2}$ **35.** -1 **37.** 0   **39.** 2   **41.** 4   **43.** -3 **45.** 2   **47.** 81   **49.** 7   **51.** -3

53. -3 **55.** 2   **57.** 2   **59.** $\dfrac{27}{64}$ **61.** 10   **63.** 4   **65.** 5   **67.** $\dfrac{1}{49}$ **69.** 3   **71.** 3   **73.** 1   **75.** -1

77.   **79.**   **81.**   **83.**   **85.** 1   **87.** $\dfrac{x-4}{2}$ **89.** $\dfrac{2x+3}{x^2}$ **91.** 3

93. a. $g(2) = 25$ **b.** $(25, 2)$ **c.** $f(25) = 2$ **95.** answers may vary   **97.** $\dfrac{9}{5}$ **99.** 1

101. 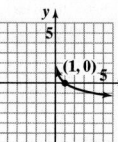   **103.** 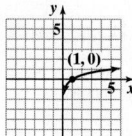   **105.** answers may vary   **107.** 0.0827

Section 12.6
Practice Exercises

1. a. $\log_8 15$ **b.** $\log_2 6$ **c.** $\log_5(x^2 - 1)$ **2. a.** $\log_5 3$ **b.** $\log_6 \dfrac{x}{3}$ **c.** $\log_4 \dfrac{x^2+1}{x^2+3}$ **3. a.** $8 \log_7 x$ **b.** $\dfrac{1}{4}\log_5 7$

4. a. $\log_5 512$ **b.** $\log_8 \dfrac{x^2}{x+3}$ **c.** $\log_7 15$ **5. a.** $\log_5 4 + \log_5 3 - \log_5 7$ **b.** $2 \log_4 a - 5 \log_4 b$ **6. a.** 1.39   **b.** 1.66   **c.** 0.28

Vocabulary, Readiness & Video Check 12.6

1. a. 36   **3. b.** $\log_b 2^7$ **5. a.** x **7.** No, the product property says the logarithm of a product can be written as a sum of logarithms—the expression in Example 2 is a logarithm of a sum.   **9.** Since $\dfrac{1}{x} = x^{-1}$, this gives us $\log_2 x^{-1}$. Using the power property, we get $-1 \log_2 x$ or $-\log_2 x$.

Exercise Set 12.6

1. $\log_5 14$ **3.** $\log_4 9x$ **5.** $\log_6(x^2 + x)$ **7.** $\log_{10}(10x^2 + 20)$ **9.** $\log_5 3$ **11.** $\log_3 4$ **13.** $\log_2 \dfrac{x}{y}$ **15.** $\log_2 \dfrac{x^2+6}{x^2+1}$ **17.** $2 \log_3 x$

19. $-1 \log_4 5 = -\log_4 5$ **21.** $\dfrac{1}{2}\log_5 y$ **23.** $\log_2 5x^3$ **25.** $\log_4 48$ **27.** $\log_5 x^3 z^6$ **29.** $\log_4 4$, or 1   **31.** $\log_7 \dfrac{9}{2}$ **33.** $\log_{10} \dfrac{x^3 - 2x}{x+1}$

35. $\log_2 \dfrac{x^{7/2}}{(x+1)^2}$ **37.** $\log_8 x^{16/3}$ **39.** $\log_3 4 + \log_3 y - \log_3 5$ **41.** $\log_4 5 - \log_4 9 - \log_4 z$ **43.** $3 \log_2 x - \log_2 y$ **45.** $\dfrac{1}{2}\log_b 7 + \dfrac{1}{2}\log_b x$

47. $4 \log_6 x + 5 \log_6 y$ **49.** $3 \log_5 x + \log_5(x+1)$ **51.** $2 \log_6 x - \log_6(x+3)$ **53.** 1.2   **55.** 0.2   **57.** 0.35   **59.** 1.29   **61.** -0.68

63. -0.125 **65–66.**   **67.** 2   **69.** 2   **71.** b   **73.** true   **75.** false   **77.** false   **79.** because $\log_b 1 = 0$

Integrated Review

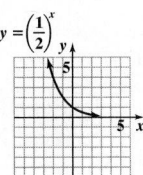 **1.** $x^2 + x - 5$ **2.** $-x^2 + x - 7$ **3.** $x^3 - 6x^2 + x - 6$ **4.** $\dfrac{x - 6}{x^2 + 1}$ **5.** $\sqrt{3x - 1}$ **6.** $3\sqrt{x} - 1$

7. one-to-one; $\{(6, -2), (8, 4), (-6, 2), (3, 3)\}$ **8.** not one-to-one   **9.** not one-to-one   **10.** one-to-one   **11.** not one-to-one

12. $f^{-1}(x) = \dfrac{x}{3}$ **13.** $f^{-1}(x) = x - 4$ **14.** $f^{-1}(x) = \dfrac{x + 1}{5}$ **15.** $f^{-1}(x) = \dfrac{x - 2}{3}$

16. $y = \left(\dfrac{1}{2}\right)^x$ **17.** $y = 2^x + 1$ **18.** $y = \log_3 x$ **19.**   **20.** 3

$y = \log_{1/3} x$

21. 7   **22.** -8 **23.** 3   **24.** 2   **25.** $\dfrac{1}{2}$ **26.** 32   **27.** 4   **28.** 5   **29.** $\dfrac{1}{9}$ **30.** $\log_2 x^5$ **31.** $\log_2 5^x$ **32.** $\log_5 \dfrac{x^3}{y^5}$ **33.** $\log_5 x^9 y^3$

34. $\log_2 \dfrac{x^2 - 3x}{x^2 + 4}$ **35.** $\log_3 \dfrac{y^4 + 11y}{y + 2}$ **36.** $\log_7 9 + 2\log_7 x - \log_7 y$ **37.** $\log_6 5 + \log_6 y - 2\log_6 z$ **38.** 254,000 mosquitoes

Section 12.7

Practice Exercises

1. 1.1761   **2. a.** -2 **b.** 5   **c.** $\dfrac{1}{5}$ **d.** -3 **3.** $10^{3.4} \approx 2511.8864$ **4.** 5.6   **5.** 2.5649   **6. a.** 4   **b.** $\dfrac{1}{3}$ **7.** $\dfrac{e^8}{5} \approx 596.1916$

8. \$3051.00   **9.** 0.7740

Vocabulary, Readiness & Video Check 12.7

1. c. 10   **3. b.** 7   **5. b.** 5   **7. a.** $\dfrac{\log 7}{\log 2}$ or **b.** $\dfrac{\ln 7}{\ln 2}$ **9.** The understood base of a common logarithm is 10. If you're finding the common logarithm of a known power of 10, then the common logarithm is the known power of 10.   **11.** $\log_b b^x = x$

Exercise Set 12.7

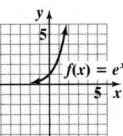 **1.** 0.9031   **3.** 0.3636   **5.** 0.6931   **7.** -2.6367 **9.** 1.1004   **11.** 1.6094   **13.** 1.6180   **15.** 2   **17.** -3 **19.** 2   **21.** $\dfrac{1}{4}$

23. 3   **25.** -7 **27.** -4 **29.** $\dfrac{1}{2}$ **31.** $\dfrac{e^7}{2} \approx 548.3166$ **33.** $10^{1.3} \approx 19.9526$ **35.** $\dfrac{10^{1.1}}{2} \approx 6.2946$ **37.** $e^{1.4} \approx 4.0552$

39. $\dfrac{4 + e^{2.3}}{3} \approx 4.6581$ **41.** $10^{2.3} \approx 199.5262$ **43.** $e^{-2.3} \approx 0.1003$ **45.** $\dfrac{10^{-0.5} - 1}{2} \approx -0.3419$ **47.** $\dfrac{e^{0.18}}{4} \approx 0.2993$

49. 1.5850   **51.** -2.3219 **53.** 1.5850   **55.** -1.6309 **57.** 0.8617   **59.** 4.2   **61.** 5.3   **63.** \$3656.38   **65.** \$2542.50

67. $\dfrac{4}{7}$ **69.** $x = \dfrac{3y}{4}$ **71.** $-6, -1$ **73.** $(2, -3)$ **75.** answers may vary   **77.** ln 50; answers may vary

79. $f(x) = e^x$ **81.** 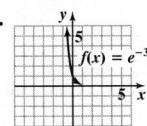 $f(x) = e^{-3x}$ **83.** 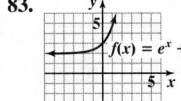 $f(x) = e^x + 2$ **85.** 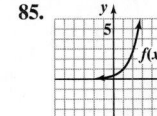 $f(x) = e^{x-1}$ **87.** 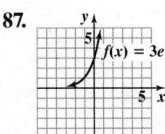 $f(x) = 3e^x$ **89.** 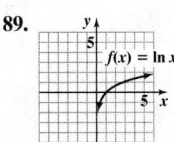 $f(x) = \ln x$

91. 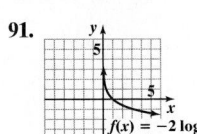 $f(x) = -2\log x$ **93.** 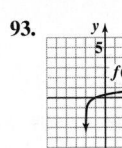 $f(x) = \log(x + 2)$ **95.** 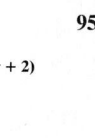 $f(x) = \ln x - 3$ **97.** answers may vary;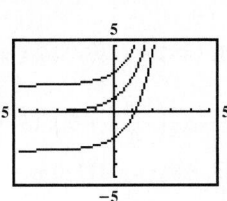

Section 12.8

Practice Exercises

1. $\dfrac{\log 9}{\log 5} \approx 1.3652$ **2.** 33   **3.** 1   **4.** $\dfrac{1}{3}$ **5.** 937 rabbits   **6.** 10 yr

Graphing Calculator Explorations 12.8

1. 3.67 years, or 3 years and 8 months   **3.** 23.16 years, or 23 years and 2 months

Vocabulary, Readiness & Video Check 12.8

1. $\ln(4x - 2) = \ln 3$ is the same as $\log_e(4x - 2) = \log_e 3$. Therefore, from the logarithm property of equality, we know that $4x - 2 = 3$.

3. $2000 = 1000\left(1 + \dfrac{0.07}{12}\right)^{12 \cdot t} \approx 9.9$ yr; As long as the interest rate and compounding are the same, it takes any amount of money the same time to double.

Exercise Set 12.8

1. $\dfrac{\log 6}{\log 3}$; 1.6309 **3.** $\dfrac{\log 3.8}{2 \log 3}$; 0.6076 **5.** $\dfrac{3 \log 2 + \log 5}{\log 2}$ or $3 + \dfrac{\log 5}{\log 2}$; 5.3219 **7.** $\dfrac{\log 5}{\log 9}$; 0.7325 **9.** $\dfrac{\log 3 - 7 \log 4}{\log 4}$ or $\dfrac{\log 3}{\log 4} - 7$; -6.2075

11. 11 **13.** $\dfrac{1}{2}$ **15.** $\dfrac{3}{4}$ **17.** $-2, 3$ **19.** 2 **21.** $\dfrac{1}{8}$ **23.** $\dfrac{4 \log 7 + \log 11}{3 \log 7}$ or $\dfrac{1}{3}\left(4 + \dfrac{\log 11}{\log 7}\right)$; 1.7441 **25.** $4, -1$ **27.** $\dfrac{\ln 5}{6}$; 0.2682

29. $9, -9$ **31.** $\dfrac{1}{5}$ **33.** 100 **35.** $\dfrac{192}{127}$ **37.** $\dfrac{2}{3}$ **39.** $\dfrac{-5 + \sqrt{33}}{2}$ **41.** 103 wolves **43.** 210 **45.** 9.9 yr **47.** 1.7 yr

49. 8.8 yr **51.** 24.5 lb **53.** 55.7 in. **55.** 11.9 lb/sq in. **57.** 3.2 mi **59.** 12 weeks **61.** 18 weeks **63.** $-\dfrac{5}{3}$ **65.** $\dfrac{17}{4}$

67. $f^{-1}(x) = \dfrac{x - 2}{5}$ **69.** 2.7 yr **71.** answers may vary

73. 6.93; **75.** -3.68; 0.19 **77.** 1.74 **79.** 0.2

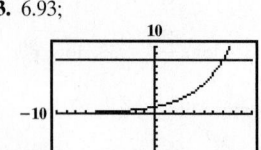 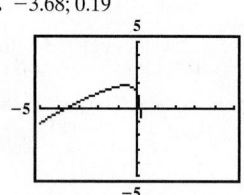 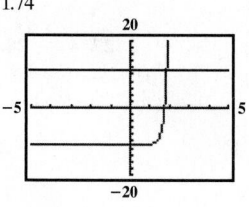 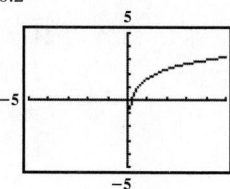

Chapter 12 Vocabulary Check

1. inverse **2.** composition **3.** exponential **4.** symmetric **5.** Natural **6.** Common **7.** vertical; horizontal **8.** logarithmic **9.** Half-life **10.** exponential

Chapter 12 Review

1. $3x - 4$ **3.** $2x^2 - 9x - 5$ **5.** $x^2 + 2x - 1$ **7.** 18 **9.** -2 **11.** one-to-one; $h^{-1} = \{(14, -9), (8, 6), (12, -11), (15, 15)\}$
13. one-to-one;

Rank in Housing Starts for 2014 (Input)	4	3	1	2
US Region (Output)	Northeast	Midwest	South	West

15. a. 3 **b.** 7 **17.** not one-to-one **19.** not one-to-one **21.** $f^{-1}(x) = x + 9$ **23.** $f^{-1}(x) = \dfrac{x - 11}{6}$ **25.** $f^{-1}(x) = \sqrt[3]{x + 5}$

27. $g^{-1}(x) = \dfrac{6x + 7}{12}$ **29.** 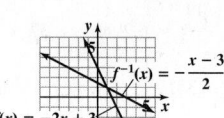 **31.** 3 **33.** $-\dfrac{4}{3}$ **35.** $\dfrac{3}{2}$ **37.** 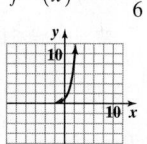 **39.**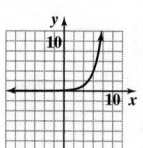

41. $2963.11 **43.** 427,394 **45.** 8 players **47.** $\log_7 49 = 2$ **49.** $\left(\dfrac{1}{2}\right)^{-4} = 16$ **51.** $\dfrac{1}{64}$ **53.** 0 **55.** 5 **57.** 4 **59.** $\dfrac{17}{3}$ **61.** $-1, 4$

63. **65.** $\log_3 32$ **67.** $\log_7 \dfrac{3}{4}$ **69.** $\log_{11} 4$ **71.** $\log_5 \dfrac{x^3}{(x + 1)^2}$ **73.** $3 \log_3 x - \log_3(x + 2)$

75. $\log_2 3 + 2 \log_2 x + \log_2 y - \log_2 z$ **77.** 2.02 **79.** 0.5563 **81.** 0.2231 **83.** 3 **85.** -1 **87.** $\dfrac{e^2}{2}$

89. $\dfrac{e^{-1} + 3}{2}$

91. 1.67 mm **93.** 0.2920 **95.** $1684.66 **97.** $\dfrac{\log 7}{2 \log 3}$; 0.8856 **99.** $\dfrac{\log 6 - \log 3}{2 \log 3}$ or $\dfrac{1}{2}\left(\dfrac{\log 6}{\log 3} - 1\right)$; 0.3155

101. $\dfrac{\log 4 + 5 \log 5}{3 \log 5}$ or $\dfrac{1}{3}\left(\dfrac{\log 4}{\log 5} + 5\right)$; 1.9538 **103.** $\dfrac{\log \frac{1}{2} + \log 5}{\log 5}$ or $-\dfrac{\log 2}{\log 5} + 1$; 0.5693 **105.** $\dfrac{25}{2}$ **107.** $\varnothing$ **109.** $2\sqrt{2}$ **111.** 24.2 yr

113. 38.5 yr **115.** 8.8 yr **117.** 0.69; **119.** -4 **121.** $\dfrac{11}{9}$ **123.** 1 **125.** $-1, 5$ **127.** $e^{-3.2}$ **129.** $2e$

Chapter 12 Getting Ready for the Test

1. B **2.** A **3.** D **4.** C **5.** D **6.** B **7.** D **8.** C **9.** A **10.** B **11.** B **12.** D **13.** B **14.** A **15.** B

Chapter 12 Test

1. $2x^2 - 3x$ **2.** $3 - x$ **3.** 5 **4.** $x - 7$ **5.** $x^2 - 6x - 2$

6. 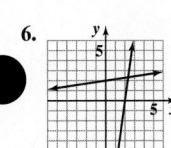   **7.** one-to-one   **8.** not one-to-one   **9.** one-to-one; $f^{-1}(x) = \dfrac{-x + 6}{2}$ **10.** one-to-one; $f^{-1} = \{(0,0),(3,2),(5,-1)\}$

11. not one-to-one   **12.** $\log_3 24$ **13.** $\log_5 \dfrac{x^4}{x+1}$ **14.** $\log_6 2 + \log_6 x - 3\log_6 y$ **15.** -1.53 **16.** 1.0686

17. -1 **18.** $\dfrac{1}{2}\left(\dfrac{\log 4}{\log 3} - 5\right)$; -1.8691 **19.** $\dfrac{1}{9}$ **20.** $\dfrac{1}{2}$ **21.** 22   **22.** $\dfrac{25}{3}$ **23.** $\dfrac{43}{21}$ **24.** -1.0979

25. 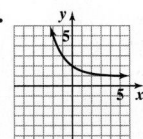   **26.** 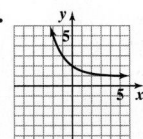   **27.** \$5234.58   **28.** 6 yr   **29.** 6.5%; \$230   **30.** 100,141   **31.** 64,805 prairie dogs
32. 15 yr   **33.** 92%   **34.** 73%

Chapter 12 Cumulative Review

1. a. $\dfrac{64}{25}$ **b.** $\dfrac{1}{20}$ **c.** $\dfrac{5}{4}$; Sec. 1.3, Ex. 4   **3.** 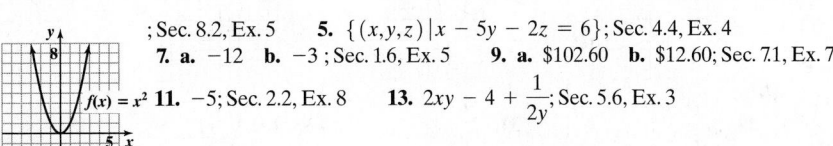   ; Sec. 8.2, Ex. 5   **5.** $\{(x,y,z)\,|\,x - 5y - 2z = 6\}$; Sec. 4.4, Ex. 4
7. a. -12 **b.** -3; Sec. 1.6, Ex. 5   **9. a.** \$102.60   **b.** \$12.60; Sec. 7.1, Ex. 7
11. -5; Sec. 2.2, Ex. 8   **13.** $2xy - 4 + \dfrac{1}{2y}$; Sec. 5.6, Ex. 3

15. $3(m + 2)(m - 10)$; Sec. 6.2, Ex. 9   **17.** $3x - 5$; Sec. 7.3, Ex. 3   **19. a.** 3   **b.** -3 **c.** -5 **d.** not a real number
e. $4x$; Sec. 10.1, Ex. 4   **21. a.** $\sqrt[4]{x^3}$ **b.** $\sqrt[6]{x}$ **c.** $\sqrt[6]{72}$; Sec. 10.2, Ex. 8   **23. a.** $5\sqrt{3} + 3\sqrt{10}$ **b.** $\sqrt{35} + \sqrt{5} - \sqrt{42} - \sqrt{6}$

c. $21x - 7\sqrt{5x} + 15\sqrt{x} - 5\sqrt{5}$ **d.** $49 - 8\sqrt{3}$ **e.** $2x - 25$ **f.** $x + 22 + 10\sqrt{x - 3}$; Sec. 10.4, Ex. 4   **25.** $\dfrac{\sqrt[4]{xy^3}}{3y^2}$; Sec. 10.5, Ex. 3

27. 3; Sec. 10.6, Ex. 4   **29.** $\dfrac{9 + i\sqrt{15}}{6}, \dfrac{9 - i\sqrt{15}}{6}$ or $\dfrac{3}{2} \pm \dfrac{\sqrt{15}}{6}i$; Sec. 11.1, Ex. 8   **31.** $\dfrac{-1 + \sqrt{33}}{4}, \dfrac{-1 - \sqrt{33}}{4}$; Sec. 11.3, Ex. 2

33. $[0, 4]$; Sec. 11.4, Ex. 2   **35.** 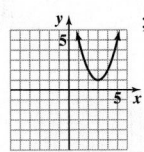   ; Sec. 11.5, Ex. 5   **37.** 63; Sec. 7.6, Ex. 1   **39.** $f^{-1}(x) = x - 3$; Sec. 12.2, Ex. 4

41. a. 2   **b.** -1 **c.** $\dfrac{1}{2}$; Sec. 12.5, Ex. 3

CHAPTER 13 CONIC SECTIONS

Section 13.1

Practice Exercises

1. 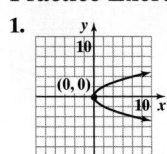   **2.** 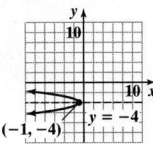   **3.** 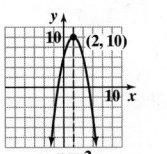   **4.** 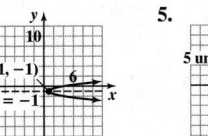   **5.** 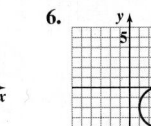   **6.**

7. 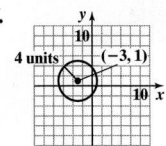   **8.** $(x + 2)^2 + (y + 5)^2 = 81$

Graphing Calculator Explorations 13.1

1. 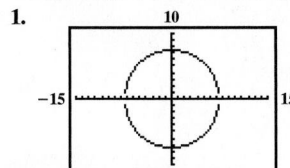   **3.** 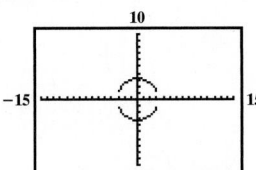   **5.** 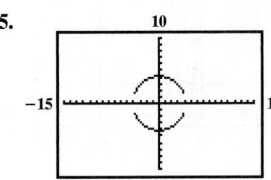   **7.**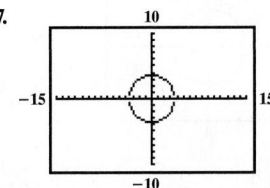

Vocabulary, Readiness & Video Check 13.1

1. conic sections **3.** circle; center **5.** radius **7.** No, their graphs don't pass the vertical line test. **9.** The formula for the standard form of a circle identifies the center and radius, so you just need to substitute these values into this formula and simplify.

Exercise Set 13.1

1. upward **3.** to the left **5.** downward **7.** 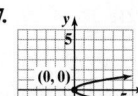 **9.** 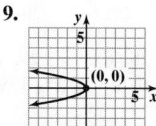 **11.** 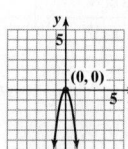 **13.**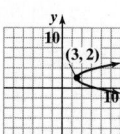

15. 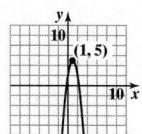 **17.** 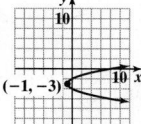 **19.** 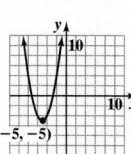 **21.** 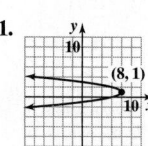 **23.** 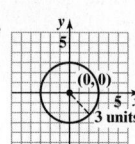 **25.**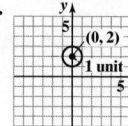

27. 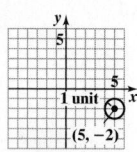 **29.** 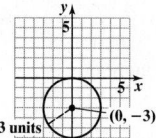 **31.** 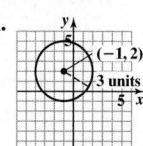 **33.** 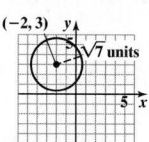 **35.**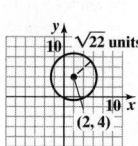

37. 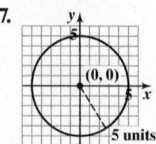 **39.** 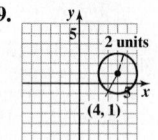 **41.** 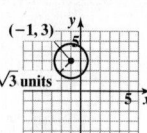 **43.** $(x - 2)^2 + (y - 3)^2 = 36$ **45.** $x^2 + y^2 = 3$
47. $(x + 5)^2 + (y - 4)^2 = 45$

49. 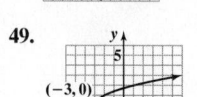 **51.** 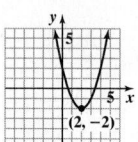 **53.** 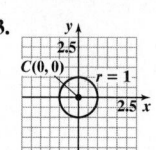 **55.** 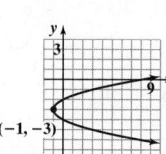 **57.** 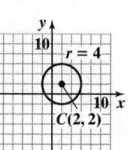 **59.**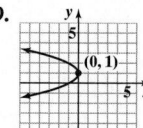

61. 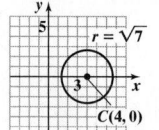 **63.** 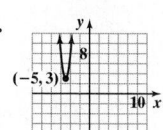 **65.** 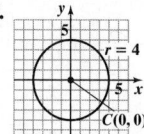 **67.** 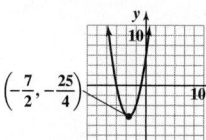 **69.** 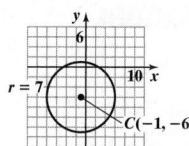 **71.**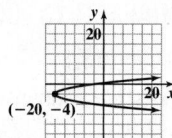

73. 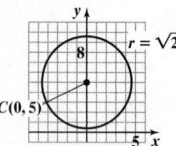 **75.** 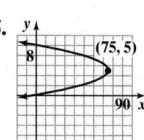 **77.** 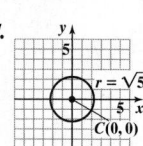 **79.** 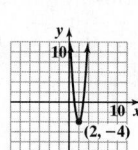 **81.** 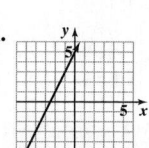 **83.**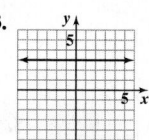

85. $\dfrac{\sqrt{3}}{3}$ **87.** $\dfrac{2\sqrt{42}}{3}$ **89.** The vertex is $(1, -5)$. **91. a.** 16.5 m **b.** 103.67 m **c.** 3.5 m **d.** $(0, 16.5)$ **e.** $x^2 + (y - 16.5)^2 = 16.5^2$
93. a. 125 ft **b.** 14 ft **c.** 139 ft **d.** $(0, 139)$ **e.** $x^2 + (y - 139)^2 = 125^2$ **95.** answers may vary **97.** 20 m
99. 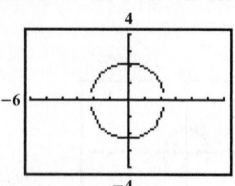 **101.**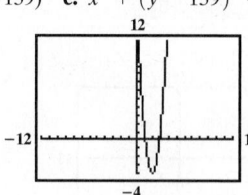

Section 13.2

Practice Exercises

1. 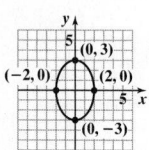 **2.** 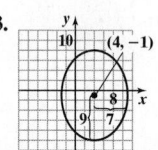 **3.** 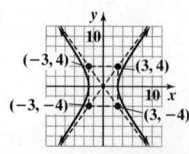 **4.** 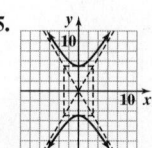 **5.**

Graphing Calculator Explorations 13.2

1. 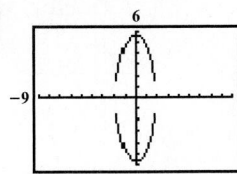 **3.** 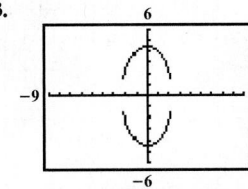 **5.**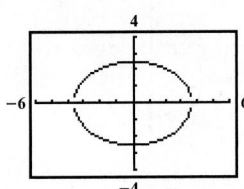

Vocabulary, Readiness & Video Check 13.2

1. hyperbola **3.** focus **5.** hyperbola; $(0,0)$; x; $(a,0)$ and $(-a,0)$ **7.** a and b give us the location of 4 intercepts: $(a,0)$, $(-a,0)$, $(0,b)$, and $(0,-b)$ for $\dfrac{x^2}{a^2} + \dfrac{y^2}{b^2} = 1$ with center $(0,0)$. For Example 2, the values of a and b also give us 4 points of the graph, just not intercepts. Here we move a distance of a units horizontally to the left and right of the center and b units above and below.

Exercise Set 13.2

1. ellipse **3.** hyperbola **5.** hyperbola **7.** 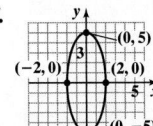 **9.** 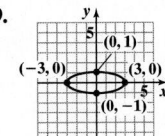 **11.** 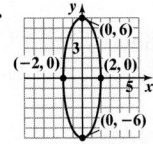 **13.**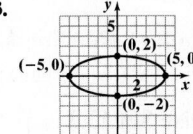

15. 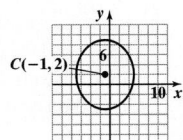 **17.** 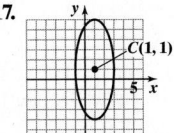 **19.** 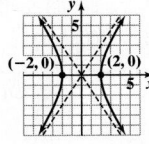 **21.** 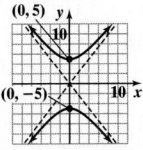 **23.** 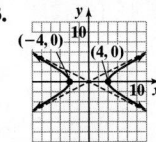 **25.**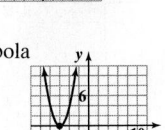

27. 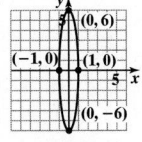 **29.** 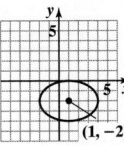 **31.** 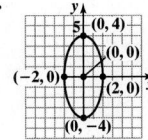 **33.** 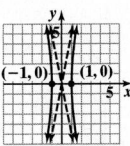 **35.** circle 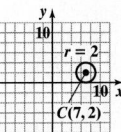 **37.** parabola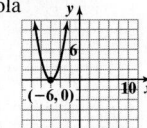

39. hyperbola 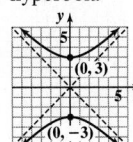 **41.** ellipse 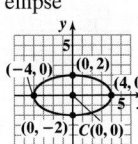 **43.** parabola 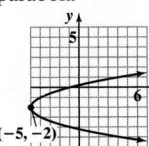 **45.** hyperbola 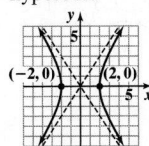 **47.** ellipse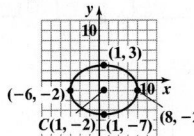

49. circle 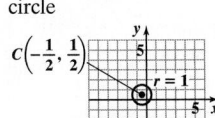 **51.** $-8x^5$ **53.** $-4x^2$ **55.** y-intercepts: 2 units **57.** y-intercepts: 4 units **59.** answers may vary
61. ellipses: C, E, H; circles: B, F; hyperbolas: A, D, G **63.** A: 49, 7; B: 0, 0; C: 9, 3; D: 64, 8; E: 64, 8; F: 0, 0; G: 81, 9; H: 4, 2
65. A: $\dfrac{7}{6}$; B: 0; C: $\dfrac{3}{5}$; D: $\dfrac{8}{5}$; E: $\dfrac{8}{9}$; F: 0; G: $\dfrac{9}{4}$; H: $\dfrac{1}{6}$ **67.** equal to zero **69.** answers may vary
71. $\dfrac{x^2}{1.69 \times 10^{16}} + \dfrac{y^2}{1.5625 \times 10^{16}} = 1$

73. 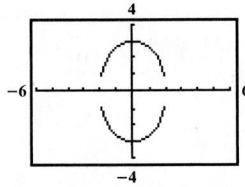 **75.** 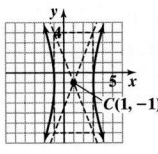 **77.** 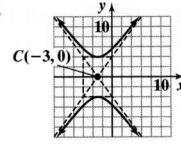 **79.**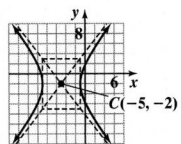

Integrated Review

1. circle 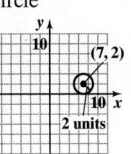 **2.** parabola 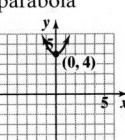 **3.** parabola 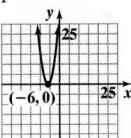 **4.** ellipse 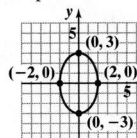 **5.** hyperbola 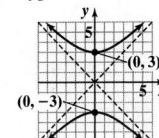 **6.** hyperbola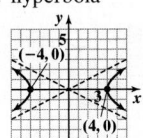

7. ellipse

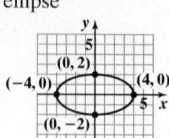

8. circle

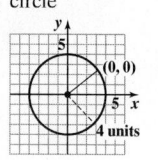

9. parabola

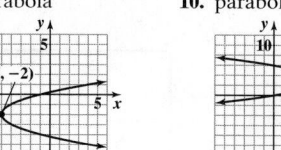

10. parabola

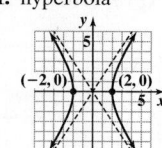

11. hyperbola

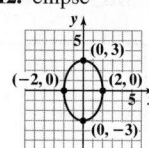

12. ellipse

13. ellipse

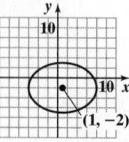

14. hyperbola

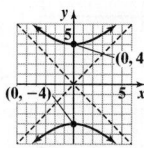

15. circle

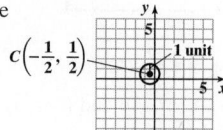

Section 13.3

Practice Exercises

1. $(-4, 3), (0, -1)$ **2.** $(4, -2)$ **3.** $\varnothing$ **4.** $(2, \sqrt{3}), (2, -\sqrt{3}), (-2, \sqrt{3}), (-2, -\sqrt{3})$

Vocabulary, Readiness & Video Check 13.3

1. Solving for y would either introduce tedious fractions (2nd equation) or a square root (1st equation) into the calculations.

Exercise Set 13.3

1. $(3, -4), (-3, 4)$ **3.** $(\sqrt{2}, \sqrt{2}), (-\sqrt{2}, -\sqrt{2})$ **5.** $(4, 0), (0, -2)$ **7.** $(-\sqrt{5}, -2), (-\sqrt{5}, 2), (\sqrt{5}, -2), (\sqrt{5}, 2)$ **9.** $\varnothing$
11. $(1, -2), (3, 6)$ **13.** $(2, 4), (-5, 25)$ **15.** $\varnothing$ **17.** $(1, -3)$ **19.** $(-1, -2), (-1, 2), (1, -2), (1, 2)$ **21.** $(0, -1)$
23. $(-1, 3), (1, 3)$ **25.** $(\sqrt{3}, 0), (-\sqrt{3}, 0)$ **27.** $\varnothing$ **29.** $(-6, 0), (6, 0), (0, -6)$ **31.** $(3, \sqrt{3})$

33. 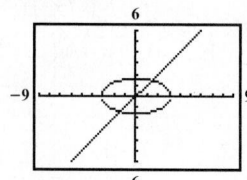   **35.** 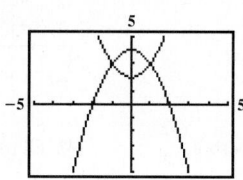   **37.** $(8x - 25)$ in.   **39.** $(4x^2 + 6x + 2)$ m   **41.** answers may vary   **43.** 0, 1, 2, 3, or 4; answers may vary   **45.** 9 and 7; 9 and -7; -9 and 7; -9 and -7 **47.** 15 cm by 19 cm   **49.** 15 thousand compact discs; price: $3.75

51. 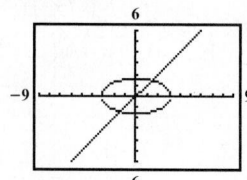   **53.**

Section 13.4

Practice Exercises

1. 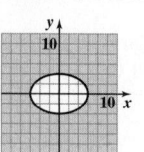   **2.** 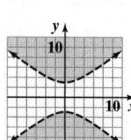   **3.** 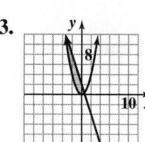   **4.**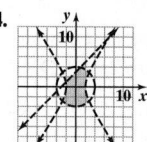

Vocabulary, Readiness & Video Check 13.4

1. For both, we graph the related equation to find the boundary and sketch it as a solid boundary for $\leq$ or $\geq$ and a dashed boundary for $<$ or $>$; also, we choose a test point (or test points) not on the boundary and shade that region if the test point is a solution of the original inequality.

Exercise Set 13.4

1. 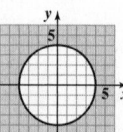   **3.** 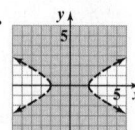   **5.** 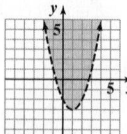   **7.** 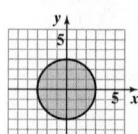   **9.** 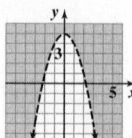   **11.**

13. 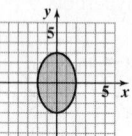   **15.** 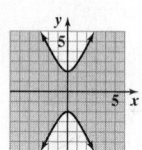   **17.** 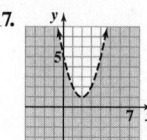   **19.** 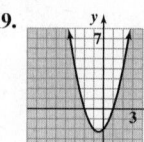   **21.** 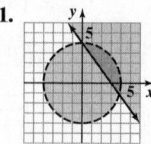   **23.**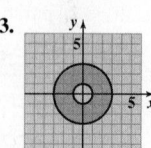

25. 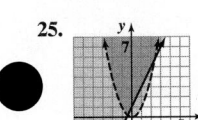 **27.** 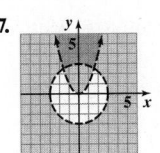 **29.** 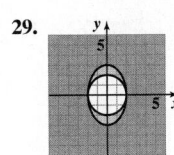 **31.** 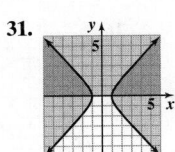 **33.** 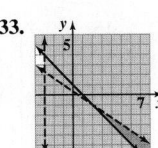 **35.**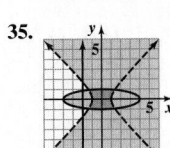

37. not a function **39.** function **41.** 1 **43.** $3a^2 - 2$ **45.** answers may vary **47.**

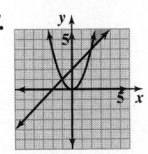

Chapter 13 Vocabulary Check

1. circle; center **2.** nonlinear system of equations **3.** ellipse **4.** radius **5.** hyperbola **6.** conic sections **7.** vertex **8.** diameter

Chapter 13 Review

1. $(x + 4)^2 + (y - 4)^2 = 9$ **3.** $(x + 7)^2 + (y + 9)^2 = 11$ **5.** 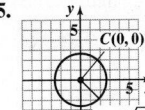 **7.** 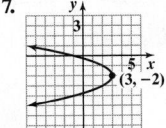 **9.**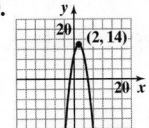

11. 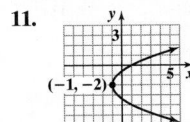 **13.** 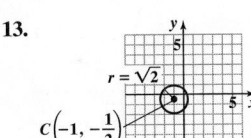 **15.** 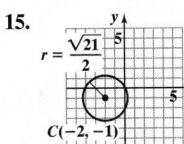 **17.** 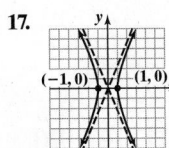 **19.**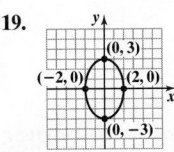

21. 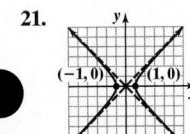 **23.** 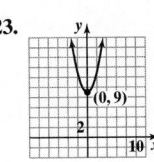 **25.** 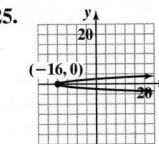 **27.** 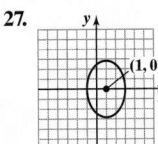 **29.** 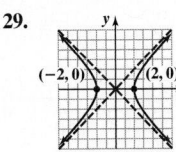 **31.**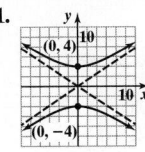

33. $(1, -2), (4, 4)$ **35.** $(-1, 1), (2, 4)$ **37.** $(0, 2), (0, -2)$ **39.** $(1, 4)$ **41.** length: 15 ft; width: 10 ft

43. 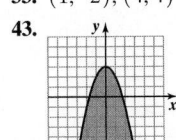 **45.** 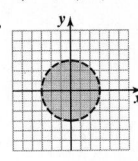 **47.** 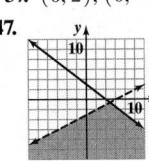 **49.** 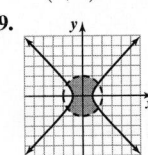 **51.** $(x + 7)^2 + (y - 8)^2 = 25$ **53.**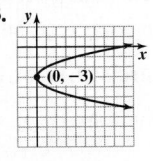

55. 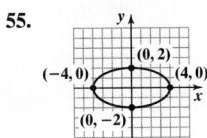 **57.** 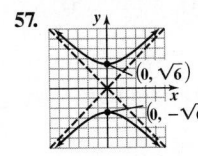 **59.** 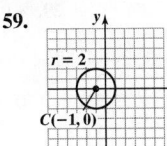 **61.** 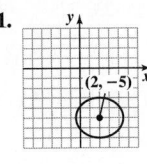 **63.** $(5, 1), (-1, 7)$ **65.**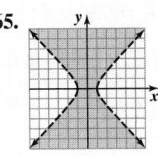

Chapter 13 Getting Ready for the Test

1. C **2.** E **3.** D **4.** A **5.** B **6.** A **7.** D **8.** B **9.** D **10.** C **11.** A **12.** C **13.** C **14.** E **15.** E
16. E **17.** B

Chapter 13 Test

1. 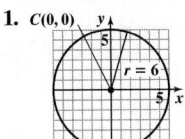 **2.** 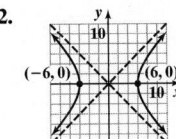 **3.** 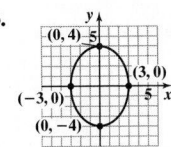 **4.** 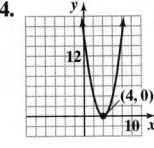 **5.** 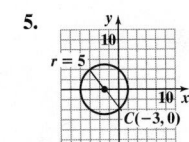 **6.**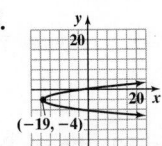

7. 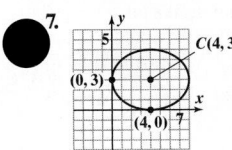 **8.** 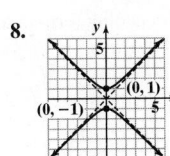 **9.** $(-12, 5), (12, -5)$ **10.** $(-5, -1), (-5, 1), (5, -1), (5, 1)$ **11.** $(6, 12), (1, 2)$
12. $(1, 1), (-1, -1)$

13. 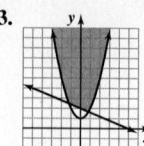 **14.** 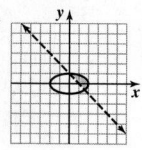 **15.** 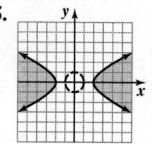 **16.** 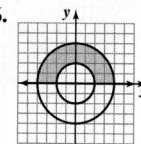 **17.** B **18.** height: 10 ft; width: 30 ft

Chapter 13 Cumulative Review

1. $\varnothing$; Sec. 9.1, Ex. 3 **3.** $\left(-\infty, \frac{13}{5}\right] \cup [4, \infty)$; Sec. 9.1, Ex. 7 **5.** $-2, \frac{4}{5}$; Sec. 9.2, Ex. 2 **7.** $24, -20$; Sec. 9.2, Ex. 3 **9.** $\frac{3}{4}$, 5; Sec. 9.2, Ex. 8

11. $(4, 8)$; Sec. 9.3, Ex. 2 **13.** $\dfrac{xy + 2x^3}{y - 1}$; Sec. 7.7, Ex. 3 **15.** $(-\infty, \infty)$; Sec. 9.3, Ex. 7 **17.** 16; Sec. 5.7, Ex. 4 **19. a.** 1 **b.** -4 **c.** $\dfrac{2}{5}$ **d.** x^2

e. $-3x^5$; Sec. 10.1, Ex. 3 **21. a.** $z - z^{17/3}$ **b.** $x^{2/3} - 3x^{1/3} - 10$; Sec. 10.2, Ex. 5 **23. a.** 2 **b.** $\dfrac{5}{2}\sqrt{x}$ **c.** $14xy^2\sqrt[3]{x}$ **d.** $4a^2b\sqrt[4]{2a}$; Sec. 10.3, Ex. 5

25. a. $\dfrac{5\sqrt{5}}{12}$ **b.** $\dfrac{5\sqrt[3]{7x}}{2}$; Sec. 10.4, Ex. 3 **27.** $\dfrac{\sqrt{21xy}}{3y}$; Sec. 10.5, Ex. 2 **29.** 42; Sec. 10.6, Ex. 1 **31. a.** $-i$ **b.** 1 **c.** -1 **d.** 1; Sec. 10.7, Ex. 6

33. $-1 + \sqrt{5}, -1 - \sqrt{5}$; Sec. 11.1, Ex. 5 **35.** $2 + \sqrt{2}, 2 - \sqrt{2}$; Sec. 11.2, Ex. 3 **37.** $2, -2, i, -i$; Sec. 11.3, Ex. 3 **39.** $[-2, 3)$; Sec. 11.4, Ex. 4

41. 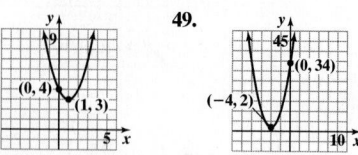 ; Sec. 11.5, Ex. 8 **43.** $(2, -16)$; Sec. 11.6, Ex. 4 **45.** $\sqrt{2} \approx 1.414$; Sec. 10.3, Ex. 6

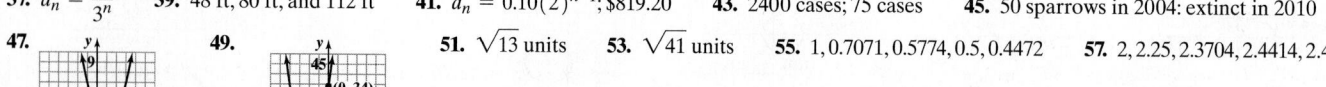

CHAPTER 14 SEQUENCES, SERIES, AND THE BINOMIAL THEOREM

Section 14.1
Practice Exercises

1. 6, 9, 14, 21, 30 **2. a.** $-\dfrac{1}{5}$ **b.** $\dfrac{1}{20}$ **c.** $\dfrac{1}{150}$ **d.** $-\dfrac{1}{95}$ **3. a.** $a_n = 2n - 1$ **b.** $a_n = 3^n$ **c.** $a_n = \dfrac{n}{n + 1}$ **d.** $a_n = -\dfrac{1}{n + 1}$
4. \$2022.40

Vocabulary, Readiness & Video Check 14.1

1. general **3.** infinite **5.** -1 **7.** function; domain; a_n **9.** $a_9 = 0.10(2)^{9-1} = \$25.60$

Exercise Set 14.1

1. 5, 6, 7, 8, 9 **3.** $-1, 1, -1, 1, -1$ **5.** $\dfrac{1}{4}, \dfrac{1}{5}, \dfrac{1}{6}, \dfrac{1}{7}, \dfrac{1}{8}$ **7.** 2, 4, 6, 8, 10 **9.** $-1, -4, -9, -16, -25$ **11.** 2, 4, 8, 16, 32 **13.** 7, 9, 11, 13, 15

15. $-1, 4, -9, 16, -25$ **17.** 75 **19.** 118 **21.** $\dfrac{6}{5}$ **23.** 729 **25.** $\dfrac{4}{7}$ **27.** $\dfrac{1}{8}$ **29.** -95 **31.** $-\dfrac{1}{25}$ **33.** $a_n = 4n - 1$ **35.** $a_n = -2^n$

37. $a_n = \dfrac{1}{3^n}$ **39.** 48 ft, 80 ft, and 112 ft **41.** $a_n = 0.10(2)^{n-1}$; \$819.20 **43.** 2400 cases; 75 cases **45.** 50 sparrows in 2004; extinct in 2010

47. 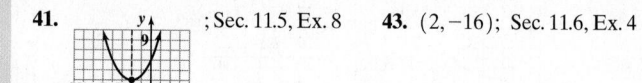 **49.** **51.** $\sqrt{13}$ units **53.** $\sqrt{41}$ units **55.** 1, 0.7071, 0.5774, 0.5, 0.4472 **57.** 2, 2.25, 2.3704, 2.4414, 2.4883

Section 14.2
Practice Exercises

1. 4, 9, 14, 19, 24 **2. a.** $a_n = 5 - 3n$ **b.** -31 **3.** 51 **4.** 47 **5.** $a_n = 54{,}800 + 2200n$; \$61,400 **6.** 8, -24, 72, -216 **7.** $\dfrac{1}{64}$

8. -192 **9.** $a_1 = 3; r = \dfrac{3}{2}$ **10.** 75 units

Vocabulary, Readiness & Video Check 14.2

1. geometric; ratio **3.** first; difference **5.** If there is a common difference between each term and its preceding term in a sequence, it's an arithmetic sequence.

Exercise Set 14.2

1. $4, 6, 8, 10, 12$ **3.** $6, 4, 2, 0, -2$ **5.** $1, 3, 9, 27, 81$ **7.** $48, 24, 12, 6, 3$ **9.** 33 **11.** -875 **13.** -60 **15.** 96 **17.** -28 **19.** 1250
21. 31 **23.** 20 **25.** $a_1 = \dfrac{2}{3}; r = -2$ **27.** answers may vary   **29.** $a_1 = 2; d = 2$ **31.** $a_1 = 5; r = 2$ **33.** $a_1 = \dfrac{1}{2}; r = \dfrac{1}{5}$

35. $a_1 = x; r = 5$ **37.** $a_1 = p; d = 4$ **39.** 19 **41.** $-\dfrac{8}{9}$ **43.** $\dfrac{17}{2}$ **45.** $\dfrac{8}{81}$ **47.** -19 **49.** $a_n = 4n + 50; 130$ seats   **51.** $a_n = 6(3)^{n-1}$

53. $486, 162, 54, 18, 6; a_n = \dfrac{486}{3^{n-1}};$ 6 bounces   **55.** $a_n = 3875 + 125n; \$5375$ **57.** 25 g   **59.** $\dfrac{11}{18}$ **61.** 40 **63.** $\dfrac{907}{495}$

65. $\$11{,}782.40, \$5891.20, \$2945.60, \$1472.80$ **67.** $19.652, 19.618, 19.584, 19.55$ **69.** answers may vary

Section 14.3

Practice Exercises

1. a. $-\dfrac{5}{4}$ **b.** 360 **2. a.** $\displaystyle\sum_{i=1}^{6} 5i$ **b.** $\displaystyle\sum_{i=1}^{4} \left(\dfrac{1}{5}\right)^i$ **3.** $\dfrac{655}{72}$ or $9\dfrac{7}{72}$ **4.** 95 plants

Vocabulary, Readiness & Video Check 14.3

1. infinite   **3.** summation   **5.** partial sum   **7.** sigma/sum, index of summation, beginning value of i, ending value of i, and general term of the sequence

Exercise Set 14.3

1. -2 **3.** 60 **5.** 20 **7.** $\dfrac{73}{168}$ **9.** $\dfrac{11}{36}$ **11.** 60 **13.** 74 **15.** 62 **17.** $\dfrac{241}{35}$ **19.** $\displaystyle\sum_{i=1}^{5}(2i-1)$ **21.** $\displaystyle\sum_{i=1}^{4} 4(3)^{i-1}$

23. $\displaystyle\sum_{i=1}^{6}(-3i+15)$ **25.** $\displaystyle\sum_{i=13}^{4}\dfrac{4}{3^{i-2}}$ **27.** $\displaystyle\sum_{i=1}^{7} i^2$ **29.** -24 **31.** 0 **33.** 82 **35.** -20 **37.** -2 **39.** $1, 2, 3, \ldots, 10; 55$ trees

41. 192 units   **43.** 35 species; 82 species   **45.** 30 opossums; 68 opossums   **47.** 6.25 lb; 93.75 lb   **49.** 16.4 in.; 134.5 in.   **51.** 10 **53.** $\dfrac{10}{27}$

55. 45 **57.** 90 **59. a.** $2 + 6 + 12 + 20 + 30 + 42 + 56$ **b.** $1 + 2 + 3 + 4 + 5 + 6 + 7 + 1 + 4 + 9 + 16 + 25 + 36 + 49$
c. answers may vary   **d.** true; answers may vary

Integrated Review

1. $-2, -1, 0, 1, 2$ **2.** $\dfrac{7}{2}, \dfrac{7}{3}, \dfrac{7}{4}, \dfrac{7}{5}, \dfrac{7}{6}$ **3.** $1, 3, 9, 27, 81$ **4.** $-4, -1, 4, 11, 20$ **5.** 64 **6.** -14 **7.** $\dfrac{1}{40}$ **8.** $-\dfrac{1}{82}$ **9.** $7, 4, 1, -2, -5$

10. $-3, -15, -75, -375, -1875$ **11.** $45, 15, 5, \dfrac{5}{3}, \dfrac{5}{9}$ **12.** $-12, -2, 8, 18, 28$ **13.** 101 **14.** $\dfrac{243}{16}$ **15.** 384 **16.** 185 **17.** -10

18. $\dfrac{1}{2}$ **19.** 50 **20.** 98 **21.** $\dfrac{31}{2}$ **22.** $\dfrac{61}{20}$ **23.** -10 **24.** 5

Section 14.4

Practice Exercises

1. 80 **2.** 1275 **3.** 105 blocks of ice   **4.** $\dfrac{341}{8}$ or $42\dfrac{5}{8}$ **5.** $\$987{,}856$ **6.** $\dfrac{28}{3}$ or $9\dfrac{1}{3}$ **7.** 1764 in.

Vocabulary, Readiness & Video Check 14.4

1. arithmetic   **3.** geometric   **5.** arithmetic   **7.** Use the general term formula from Section 14.2 for the general term of an arithmetic sequence: $a_n = a_1 + (n-1)d$.   **9.** The common ratio r is 3 for this sequence so that $|r| \geq 1$, or $|3| \geq 1; S_\infty$ doesn't exist if $|r| \geq 1$.

Exercise Set 14.4

1. 36 **3.** 484 **5.** 63 **7.** $\dfrac{312}{125}$ **9.** 55 **11.** 16 **13.** 24 **15.** $\dfrac{1}{9}$ **17.** -20 **19.** $\dfrac{16}{9}$ **21.** $\dfrac{4}{9}$ **23.** 185 **25.** $\dfrac{381}{64}$

27. $-\dfrac{33}{4}$ or -8.25 **29.** $-\dfrac{75}{2}$ **31.** $\dfrac{56}{9}$ **33.** $4000, 3950, 3900, 3850, 3800; 3450$ cars; $44{,}700$ cars   **35.** Firm A (Firm A, $\$265{,}000$;
Firm B, $\$254{,}000$)   **37.** $\$39{,}930; \$139{,}230$ **39.** 20 min; 123 min   **41.** 180 ft   **43.** Player A, 45 points; Player B, 75 points   **45.** $\$3050$
47. $\$10{,}737{,}418.23$ **49.** 720 **51.** 3 **53.** $x^2 + 10x + 25$ **55.** $8x^3 - 12x^2 + 6x - 1$ **57.** $\dfrac{8}{10} + \dfrac{8}{100} + \dfrac{8}{1000} + \cdots; \dfrac{8}{9}$
59. answers may vary

Section 14.5

Practice Exercises

1. $p^7 + 7p^6r + 21p^5r^2 + 35p^4r^3 + 35p^3r^4 + 21p^2r^5 + 7pr^6 + r^7$ **2. a.** $\dfrac{1}{7}$ **b.** 840 **c.** 5 **d.** 1
3. $a^9 + 9a^8b + 36a^7b^2 + 84a^6b^3 + 126a^5b^4 + 126a^4b^5 + 84a^3b^6 + 36a^2b^7 + 9ab^8 + b^9$ **4.** $a^3 + 15a^2b + 75ab^2 + 125b^3$
5. $27x^3 - 54x^2y + 36xy^2 - 8y^3$ **6.** $1{,}892{,}352x^5y^6$

Vocabulary, Readiness & Video Check 14.5

1. 1 **3.** 24 **5.** 6 **7.** Pascal's triangle gives you the coefficients of the terms of the expanded binomial; also, the power tells you how many terms the expansion has (1 more than the power on the binomial). **9.** The theorem is in terms of $(a + b)^n$, so if your binomial is of the form $(a - b)^n$, then remember to think of it as $(a + (-b))^n$, so your second term is $-b$.

Exercise Set 14.5

1. $m^3 + 3m^2n + 3mn^2 + n^3$ **3.** $c^5 + 5c^4d + 10c^3d^2 + 10c^2d^3 + 5cd^4 + d^5$ **5.** $y^5 - 5y^4x + 10y^3x^2 - 10y^2x^3 + 5yx^4 - x^5$
7. answers may vary **9.** 8 **11.** 42 **13.** 360 **15.** 56 **17.** $a^7 + 7a^6b + 21a^5b^2 + 35a^4b^3 + 35a^3b^4 + 21a^2b^5 + 7ab^6 + b^7$
19. $a^5 + 10a^4b + 40a^3b^2 + 80a^2b^3 + 80ab^4 + 32b^5$ **21.** $q^9 + 9q^8r + 36q^7r^2 + 84q^6r^3 + 126q^5r^4 + 126q^4r^5 + 84q^3r^6 + 36q^2r^7 + 9qr^8 + r^9$
23. $1024a^5 + 1280a^4b + 640a^3b^2 + 160a^2b^3 + 20ab^4 + b^5$ **25.** $625a^4 - 1000a^3b + 600a^2b^2 - 160ab^3 + 16b^4$ **27.** $8a^3 + 36a^2b + 54ab^2 + 27b^3$
29. $x^5 + 10x^4 + 40x^3 + 80x^2 + 80x + 32$ **31.** $5cd^4$ **33.** d^7 **35.** $-40r^2s^3$ **37.** $6x^2y^2$ **39.** $30a^9b$

41. 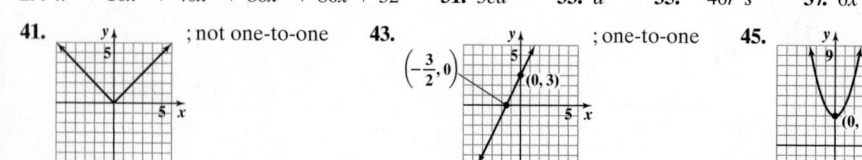 ; not one-to-one **43.** ; one-to-one **45.** ; not one-to-one

47. $x^2\sqrt{x} + 5\sqrt{3}x^2 + 30x\sqrt{x} + 30\sqrt{3}x + 45\sqrt{x} + 9\sqrt{3}$ **49.** 126 **51.** 28 **53.** answers may vary

Chapter 14 Vocabulary Check

1. finite sequence **2.** factorial of n **3.** infinite sequence **4.** geometric sequence; common ratio **5.** series **6.** general term
7. arithmetic sequence; common difference **8.** Pascal's triangle

Chapter 14 Review

1. $-3, -12, -27, -48, -75$ **3.** $\dfrac{1}{100}$ **5.** $a_n = \dfrac{1}{6n}$ **7.** 144 ft, 176 ft, 208 ft **9.** 97,000; 145,500; 218,250; 327,375; 491,063; 491,063 newly infested

acres **11.** $-2, -\dfrac{4}{3}, -\dfrac{8}{9}, -\dfrac{16}{27}, -\dfrac{32}{81}$ **13.** 111 **15.** -83 **17.** $a_1 = 3; d = 5$ **19.** $a_n = \dfrac{3}{10^n}$ **21.** $a_1 = \dfrac{8}{3}, r = \dfrac{3}{2}$ **23.** $a_1 = 7x, r = -2$

25. $8, 6, 4.5, 3.4, 2.5, 1.9$; good **27.** $a_n = 2^{n-1}$; $512; $536,870,912 **29.** $a_n = 150n + 750$; $1650/month **31.** $1 + 3 + 5 + 7 + 9 = 25$

33. $\dfrac{1}{4} - \dfrac{1}{6} + \dfrac{1}{8} = \dfrac{5}{24}$ **35.** $\sum\limits_{i=1}^{6} 3^{i-1}$ **37.** $\sum\limits_{i=1}^{4} \dfrac{1}{4^i}$ **39.** $a_n = 20(2)^n$; n represents the number of 8-hour periods; 1280 yeast **41.** Job A, $48,300;

Job B, $46,600 **43.** -4 **45.** -10 **47.** 150 **49.** 900 **51.** -410 **53.** 936 **55.** 10 **57.** -25 **59.** $30,418; $99,868

61. $58; $553 **63.** 2696 mosquitoes **65.** $\dfrac{5}{9}$ **67.** $x^5 + 5x^4z + 10x^3z^2 + 10x^2z^3 + 5xz^4 + z^5$ **69.** $16x^4 + 32x^3y + 24x^2y^2 + 8xy^3 + y^4$
71. $b^8 + 8b^7c + 28b^6c^2 + 56b^5c^3 + 70b^4c^4 + 56b^3c^5 + 28b^2c^6 + 8bc^7 + c^8$ **73.** $256m^4 - 256m^3n + 96m^2n^2 - 16mn^3 + n^4$ **75.** $35a^4b^3$
77. 130 **79.** 40.5

Chapter 14 Getting Ready for the Test

1. B **2.** C **3.** B **4.** D **5.** A **6.** A **7.** C **8.** A **9.** D **10.** B

Chapter 14 Test

1. $-\dfrac{1}{5}, \dfrac{1}{6}, -\dfrac{1}{7}, \dfrac{1}{8}, -\dfrac{1}{9}$ **2.** 247 **3.** $a_n = \dfrac{2}{5}\left(\dfrac{1}{5}\right)^{n-1}$ or $\dfrac{2}{5^n}$ **4.** $a_n = (-1)^n 9n$ **5.** 155 **6.** -330 **7.** $\dfrac{144}{5}$ **8.** 1 **9.** 10 **10.** -60
11. $a^6 - 6a^5b + 15a^4b^2 - 20a^3b^3 + 15a^2b^4 - 6ab^5 + b^6$ **12.** $32x^5 + 80x^4y + 80x^3y^2 + 40x^2y^3 + 10xy^4 + y^5$ **13.** 925 people; 250 people

initially **14.** $1 + 3 + 5 + 7 + 9 + 11 + 13 + 15$; 64 shrubs **15.** 33.75 cm, 218.75 cm **16.** 320 cm **17.** 304 ft; 1600 ft **18.** $\dfrac{14}{33}$

Chapter 14 Cumulative Review

1. a. -8 **b.** -8 **c.** 9 **d.** -9; Sec. 1.7, Ex. 4 **3.** $-2x - 1$; Sec. 2.1, Ex. 7 **5.** $y = \dfrac{1}{4}x - 3$; Sec. 3.5, Ex. 3 **7.** $-x + 5y = 23$ or $x - 5y = -23$;

Sec. 3.5, Ex. 5 **9.** $x^3 - 4x^2 - 3x + 11 + \dfrac{12}{x + 2}$; Sec. 5.7, Ex. 2 **11. a.** $5\sqrt{2}$ **b.** $2\sqrt[3]{3}$ **c.** $\sqrt{26}$ **d.** $2\sqrt[4]{2}$; Sec. 10.3, Ex. 3 **13.** 10%; Sec. 11.1, Ex. 9

15. 2, 7; Sec. 11.3, Ex. 4 **17.** $\left(-\dfrac{7}{2}, -1\right)$; Sec. 11.4, Ex. 5 **19.** $\dfrac{25}{4}$ ft; $\dfrac{5}{8}$ sec; Sec. 11.6, Ex. 5 **21. a.** 25; 7 **b.** $x^2 + 6x + 9$; $x^2 + 3$; Sec. 12.1, Ex. 2

23. $f^{-1} = \{(1, 0), (7, -2), (-6, 3), (4, 4)\}$; Sec. 12.2, Ex. 3 **25. a.** 4 **b.** $\dfrac{3}{2}$ **c.** 6; Sec. 12.3, Ex. 4 **27. a.** 2 **b.** -1 **c.** 3 **d.** 6; Sec. 12.5, Ex. 5

29. a. $\log_{11} 30$ **b.** $\log_5 6$ **c.** $\log_2(x^2 + 2x)$; Sec. 12.6, Ex. 1 **31.** $2509.30; Sec. 12.7, Ex. 8 **33.** $\dfrac{\log 7}{\log 3} \approx 1.7712$; Sec. 12.8, Ex. 1

35. 18; Sec. 12.8, Ex. 2 **37.** ; Sec. 13.2, Ex. 4 **39.** $(2, \sqrt{2})$; Sec. 13.3, Ex. 2 **41.** ; Sec. 13.4, Ex. 1

$\dfrac{x^2}{16} - \dfrac{y^2}{25} = 1$

$\dfrac{x^2}{9} + \dfrac{y^2}{16} \le 1$

43. 0, 3, 8, 15, 24; Sec. 14.1, Ex. 1 **45.** 72; Sec. 14.2, Ex. 3 **47. a.** $\dfrac{7}{2}$ **b.** 56; Sec. 14.3, Ex. 1 **49.** 465; Sec. 14.4, Ex. 2

APPENDIX A OPERATIONS ON DECIMALS/TABLE OF PERCENT, DECIMAL, AND FRACTION EQUIVALENTS

1. 17.08   **3.** 12.804   **5.** 110.96   **7.** 2.4   **9.** 28.43   **11.** 227.5   **13.** 2.7   **15.** 49.2339   **17.** 80   **19.** 0.07612   **21.** 4.56   **23.** 648.46
25. 767.83   **27.** 12.062   **29.** 61.48   **31.** 7.7   **33.** 863.37   **35.** 22.579   **37.** 363.15   **39.** 7.007

APPENDIX B REVIEW OF ALGEBRA TOPICS

B.1 Practice Exercises

1. -4 **2.** $\frac{5}{6}$ **3.** $-\frac{3}{4}$ **4.** $\frac{5}{4}$ **5.** $-\frac{1}{8}, 2$

Exercise Set B.1

1. $-3, -8$ **3.** -2 **5.** $\frac{1}{4}, -\frac{2}{3}$ **7.** $1, 9$ **9.** 0 **11.** 5 **13.** $\frac{3}{5}, -1$ **15.** no solution   **17.** $\frac{1}{8}$ **19.** 0 **21.** $6, -3$ **23.** all real numbers
25. $\frac{2}{5}, -\frac{1}{2}$ **27.** $\frac{3}{4}, -\frac{1}{2}$ **29.** 29 **31.** -8 **33.** $-\frac{1}{3}, 0$ **35.** $-\frac{7}{8}$ **37.** $\frac{31}{4}$ **39.** 1 **41.** $-7, 4$ **43.** $4, 6$ **45.** $-\frac{1}{2}$ **47. a.** incorrect
b. correct   **c.** correct   **d.** incorrect   **49.** $K = -11$ **51.** $K = 24$ **53.** -4.86 **55.** 1.53

B.2 Practice Exercises

1. a. $3x + 6$ **b.** $6x - 1$ **2.** $4x - 28.8$ **3.** $14, 34, 70$ **4.** $\$450$

Vocabulary & Readiness Check B.2

1. $>$ **3.** $=$ **5.** $31, 32, 33, 34$ **7.** $18, 20, 22$ **9.** $y, y + 1, y + 2$ **11.** $p, p + 1, p + 2, p + 3$

Exercise Set B.2

1. $4y$ **3.** $3z + 3$ **5.** $(65x + 30)$ cents   **7.** $10x + 3$ **9.** $2x + 14$ **11.** -5 **13.** $45, 225, 145$ **15.** approximately 1612.41 million acres
17. 14,940 earthquakes   **19.** 1275 shoppers   **21.** 23%   **23.** 417 employees   **25.** $29°, 35°, 116°$ **27.** 28 m, 36 m, 38 m
29. 18 in., 18 in., 27 in., 36 in.   **31.** $75, 76, 77$ **33.** Fallon's ZIP code is 89406; Fernley's ZIP code is 89408; Gardnerville Ranchos' ZIP code is 89410
35. 1960 million; 2130 million; 2290 million   **37.** nurse: 527 thousand; food service: 422 thousand; customer service: 299 thousand
39. B767-300ER: 207 seats; B737-200: 119 seats; F-100: 87 seats   **41.** $\$430.00$ **43.** $\$496,244$ **45.** 83,328,000   **47.** $40°, 140°$
49. $64°, 32°, 84°$ **51.** square: 18 cm; triangle: 24 cm   **53.** $76, 78, 80$ **55.** Darlington: 61,000; Daytona: 159,000   **57.** New York: 20.6 million;
Delhi: 25.2 million; Tokyo-Yokohama: 37.8 million   **59.** 40.5 ft; 202.5 ft; 240 ft   **61.** incandescent: 1500 bulb hours; fluorescent: 100,000 bulb hours;
halogen: 4000 bulb hours   **63.** Yankees: 84; Blue Jays: 83; Brewers: 82   **65.** Guy's Tower: 469 ft; Queen Mary Hospital: 449 ft; Galter Pavilion: 402 ft

B.3 Practice Exercises

1.

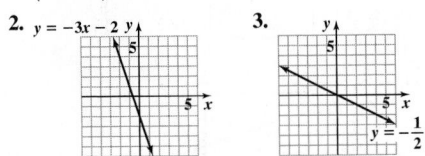

a. Quadrant IV   **b.** y-axis   **c.** Quadrant II   **d.** x-axis   **e.** Quadrant III   **f.** Quadrant I

2. $y = -3x - 2$ **3.**

Exercise Set B.3

1. $(5, 2)$ **3.** $(3, 0)$ **5.** $(-5, -2)$ **7.** $(-1, 0)$ **9.** Quadrant I   **11.** Quadrant II   **13.** Quadrant III
15. y-axis   **17.** Quadrant III   **19.** x-axis   **21.** Quadrant IV   **23.** x-axis   **25.** Quadrant III
27.   **29.**   **31.**   **33.**   **35.** 1 **37.** -4 **39.** $1, 3$ **41.** $g(-1) = -2$

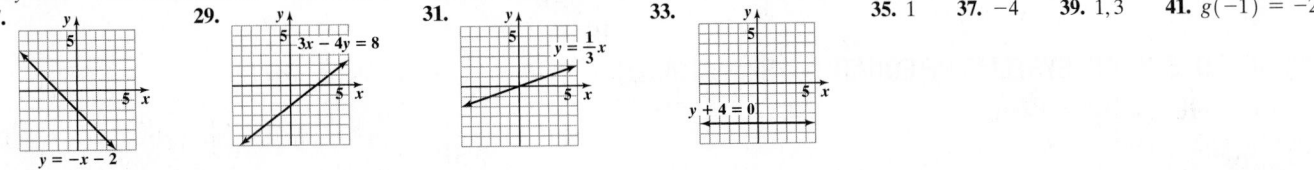

B.4 Practice Exercises

1. a. $3xy(4x - 1)$ **b.** $(7x + 2)(7x - 2)$ **c.** $(5x - 3)(x + 1)$ **d.** $(x^2 + 2)(3 + x)$ **e.** $(2x + 5)^2$ **f.** cannot be factored
2. a. $(4x + y)(16x^2 - 4xy + y^2)$ **b.** $7y^2(x + 3y)(x - 3y)$ **c.** $3(x + 2 + b)(x + 2 - b)$ **d.** $x^2y(xy + 3)(x^2y^2 - 3xy + 9)$
e. $(x + 7 + 9y)(x + 7 - 9y)$

Exercise Set B.4

1. $2y^2 + 2y - 11$ **3.** $x^2 - 7x + 7$ **5.** $25x^2 - 30x + 9$ **7.** $2x^3 - 4x^2 + 5x - 5 + \dfrac{8}{x + 2}$ **9.** $(x - 4 + y)(x - 4 - y)$

11. $x(x - 1)(x^2 + x + 1)$ **13.** $2xy(7x - 1)$ **15.** $4(x + 2)(x - 2)$ **17.** $(3x - 11)(x + 1)$ **19.** $4(x + 3)(x - 1)$ **21.** $(2x + 9)^2$

23. $(2x + 5y)(4x^2 - 10xy + 25y^2)$ **25.** $8x^2(2y - 1)(4y^2 + 2y + 1)$ **27.** $(x + 5 + y)(x^2 + 10x + 25 - xy - 5y + y^2)$ **29.** $(5a - 6)^2$

31. $7x(x - 9)$ **33.** $(b - 6)(a + 7)$ **35.** $(x^2 + 1)(x + 1)(x - 1)$ **37.** $(5x - 11)(2x + 3)$ **39.** $5a^3b(b^2 - 10)$ **41.** prime

43. $10x(x - 10)(x - 11)$ **45.** $a^3b(4b - 3)(16b^2 + 12b + 9)$ **47.** $2(x - 3)(x^2 + 3x + 9)$ **49.** $(3y - 5)(y^4 + 2)$

51. $100(z + 1)(z^2 - z + 1)$ **53.** $(2b - 9)^2$ **55.** $(y - 4)(y - 5)$ **57.** $A = 9 - 4x^2 = (3 + 2x)(3 - 2x)$

B.5 Practice Exercises

1. a. $\dfrac{2n + 1}{n(n - 1)}$ **b.** $-x^2$ **2.** $-\dfrac{y^3}{21(y + 3)}$ **b.** $\dfrac{7x + 2}{x + 2}$ **3. a.** $\dfrac{20p + 3}{5p^4q}$ **b.** $\dfrac{5y^2 + 19y - 12}{(y + 3)(y - 3)}$ **c.** 3   **4.** 1

Exercise Set B.5

1. $\dfrac{1}{2}$ **3.** $\dfrac{1 + 2x}{8}$ **5.** $\dfrac{2(x - 4)}{(x + 2)(x - 1)}$ **7.** 4   **9.** -5 **11.** $\dfrac{2x + 5}{x(x - 3)}$ **13.** -2 **15.** $\dfrac{(a + 3)(a + 1)}{a + 2}$ **17.** $-\dfrac{1}{5}$

19. $\dfrac{4a + 1}{(3a + 1)(3a - 1)}$ **21.** $-1, \dfrac{3}{2}$ **23.** $\dfrac{3}{x + 1}$ **25.** -1 **27. a.** $\dfrac{x}{5} - \dfrac{x}{4} + \dfrac{1}{10}$ **b.** Write each rational expression term so that the

denominator is the LCD, 20.   **c.** $\dfrac{-x + 2}{20}$ **29.** b   **31.** d   **33.** d

APPENDIX C AN INTRODUCTION TO USING A GRAPHING UTILITY

The Viewing Window and Interpreting Window Settings Exercise Set

1. yes   **3.** no   **5.** answers may vary   **7.** answers may vary   **9.** answers may vary

11. Xmin = -12 Ymin = -12 **13.** Xmin = -9 Ymin = -12 **15.** Xmin = -10 Ymin = -25 **17.** Xmin = -10 Ymin = -30

       Xmax = 12   Ymax = 12   Xmax = 9   Ymax = 12   Xmax = 10   Ymax = 25   Xmax = 10   Ymax = 30

       Xscl = 3   Yscl = 3   Xscl = 1   Yscl = 2   Xscl = 2   Yscl = 5   Xscl = 1   Yscl = 3

19. Xmin = -20 Ymin = -30

       Xmax = 30   Ymax = 50

       Xscl = 5   Yscl = 10

Graphing Equations and Square Viewing Window Exercise Set

1. Setting B   **3.** Setting B   **5.** Setting B

7.   **9.**   **11.**   **13.**

15.   **17.**   **19.**   **21.**

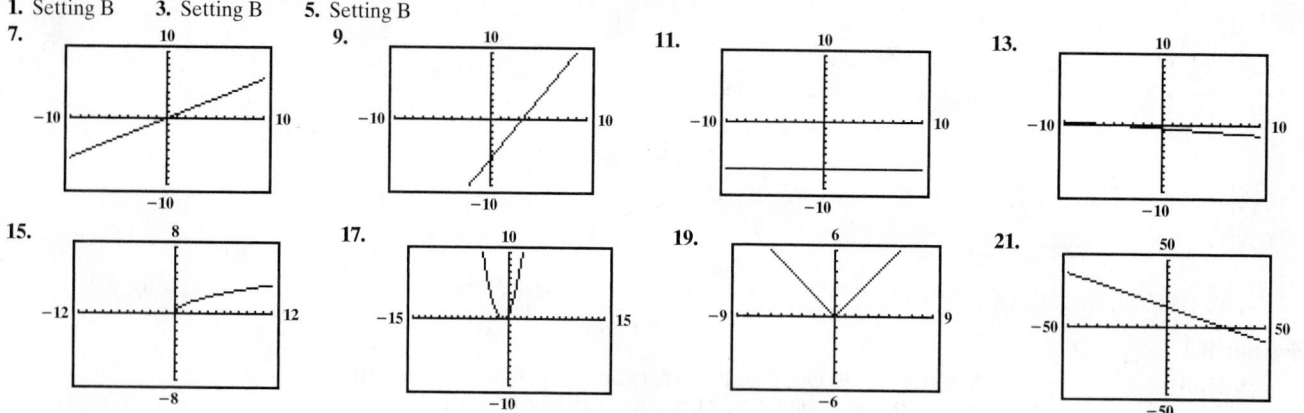

APPENDIX D SOLVING SYSTEMS OF EQUATIONS BY MATRICES

Practice Exercises

1. $(2, -1)$ **2.** $\{ \}$ or $\varnothing$ **3.** $(-1, 1, 2)$

Exercise Set

1. $(2, -1)$ **3.** $(-4, 2)$ **5.** $\{ \}$ or $\varnothing$ **7.** $\{(x, y) \mid 3x - 3y = 9\}$ **9.** $(-2, 5, -2)$ **11.** $(1, -2, 3)$ **13.** $(4, -3)$ **15.** $(2, 1, -1)$

17. $(9, 9)$ **19.** $\{ \}$ or $\varnothing$ **21.** $\{ \}$ or $\varnothing$ **23.** $(1, -4, 3)$ **25.** c

APPENDIX E SOLVING SYSTEMS OF EQUATIONS USING DETERMINANTS

1. 26   **3.** -19 **5.** 0   **7.** $(1, 2)$ **9.** $\{(x, y) \mid 3x + y = 1\}$ **11.** $(9, 9)$ **13.** 8   **15.** 0   **17.** 54   **19.** $(-2, 0, 5)$ **21.** $(6, -2, 4)$

23. 16   **25.** 15   **27.** $\dfrac{13}{6}$ **29.** 0   **31.** 56   **33.** $(-3, -2)$ **35.** $\{ \}$ or $\varnothing$ **37.** $(-2, 3, -1)$ **39.** $(3, 4)$ **41.** $(-2, 1)$

43. $\{(x,y,z)\,|\,x - 2y + z = -3\}$ **45.** $(0, 2, -1)$ **47.** 5   **49.** 0; answers may vary   **51.**

$$\begin{matrix} + & - & + & - \\ - & + & - & + \\ + & - & + & - \\ - & + & - & + \end{matrix}$$

53. -125 **55.** 24

APPENDIX F   MEAN, MEDIAN, AND MODE

1. mean: 29, median: 28, no mode   **3.** mean: 8.1, median: 8.2, mode: 8.2   **5.** mean: 0.6, median: 0.6, mode: 0.2 and 0.6   **7.** mean: 370.9, median: 313.5, no mode   **9.** 1214.8 ft   **11.** 1117 ft   **13.** 6.8   **15.** 6.9   **17.** 85.5   **19.** 73   **21.** 70 and 71   **23.** 9   **25.** 21, 21, 24

APPENDIX G   REVIEW OF ANGLES, LINES, AND SPECIAL TRIANGLES

1. $71°$ **3.** $19.2°$ **5.** $78\frac{3}{4}°$ **7.** $30°$ **9.** $149.8°$ **11.** $100\frac{1}{2}°$ **13.** $m\angle 1 = m\angle 5 = m\angle 7 = 110°; m\angle 2 = m\angle 3 = m\angle 4 = m\angle 6 = 70°$

15. $90°$ **17.** $90°$ **19.** $90°$ **21.** $45°, 90°$ **23.** $73°, 90°$ **25.** $50\frac{1}{4}°, 90°$ **27.** $x = 6$ **29.** $x = 4.5$ **31.** 10   **33.** 12

PRACTICE FINAL EXAM

1. -48 **2.** -81 **3.** $\frac{1}{64}$ **4.** $-\frac{1}{3}$ **5.** $-3x^3 + 5x^2 + 4x + 5$ **6.** $16x^2 - 16x + 4$ **7.** $3x^3 + 22x^2 + 41x + 14$ **8.** $(y - 12)(y + 4)$

9. $3x(3x + 1)(x + 4)$ **10.** $5(6 + x)(6 - x)$ **11.** $(a + b)(3a - 7)$ **12.** $8(y - 2)(y^2 + 2y + 4)$ **13.** $\frac{y^{14}}{x^2}$ **14.** $\frac{25}{7}$ **15.** $(-\infty, -2]$

16. $-7, 1$ **17.** 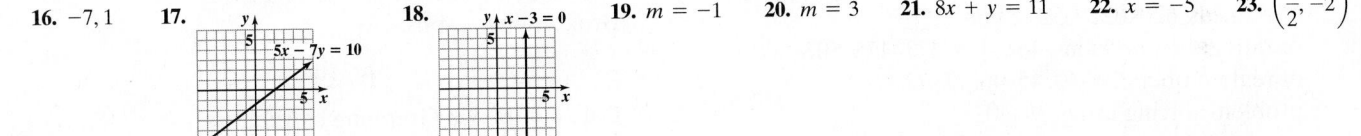   **18.** 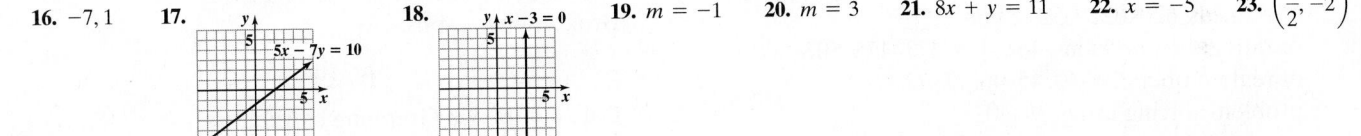   **19.** $m = -1$ **20.** $m = 3$ **21.** $8x + y = 11$ **22.** $x = -5$ **23.** $\left(\frac{1}{2}, -2\right)$

24. no solution; $\{\ \}$ or $\varnothing$ **25.** $9x^2 - 6x + 4 - \dfrac{16}{3x + 2}$ **26. a.** 0   **b.** 0   **c.** 60   **27.** x-intercepts: $(0, 0), (4, 0)$; y-intercept: $(0, 0)$;

domain: $(-\infty, \infty)$; range: $(-\infty, 4]$ **28.** 401, 802   **29.** $2\frac{1}{2}$ hr   **30.** 120 cc   **31.** $\{x \,|\, x$ is a real number, $x \neq -1, x \neq -3\}$ **32.** $\dfrac{19x - 6}{2x + 5}$

33. $\dfrac{2(x + 5)}{x(y + 5)}$ **34.** $\dfrac{3a - 4}{(a - 3)(a + 2)}$ **35.** $\dfrac{5y^2 - 1}{y + 2}$ **36.** $\dfrac{30}{11}$ **37.** -6 **38.** no solution   **39.** 5 or 1   **40.** $6\sqrt{6}$ **41.** 5

42. $\dfrac{8a^{1/3}c^{2/3}}{b^{5/12}}$ **43.** $-x\sqrt{5x}$ **44.** -20 **45.** $1, \frac{2}{3}$ **46.** $\left(\frac{3}{2}, 5\right]$ **47.** $(-\infty, -2) \cup \left(\frac{4}{3}, \infty\right)$ **48.** $\dfrac{3 \pm \sqrt{29}}{2}$ or $\dfrac{3}{2} \pm \dfrac{\sqrt{29}}{2}$ **49.** 2, 3

50. $\left(-\infty, -\frac{3}{2}\right) \cup (5, \infty)$ **51.** 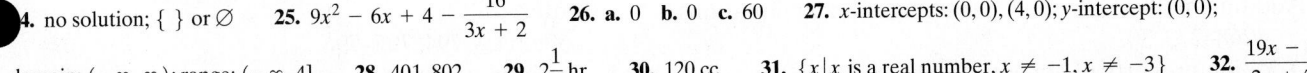   **52.** 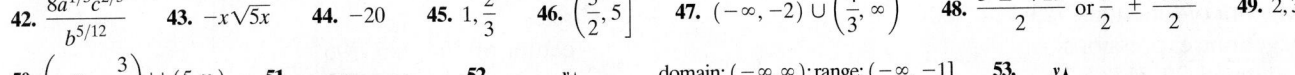 domain: $(-\infty, \infty)$; range: $(-\infty, -1]$ **53.**

54. 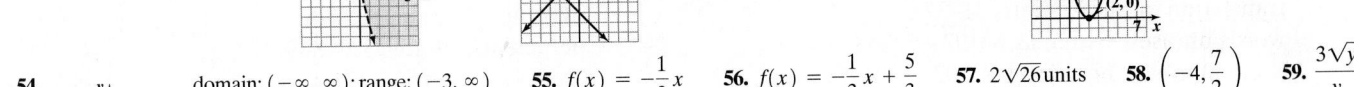 domain: $(-\infty, \infty)$; range: $(-3, \infty)$ **55.** $f(x) = -\frac{1}{2}x$ **56.** $f(x) = -\frac{1}{3}x + \frac{5}{3}$ **57.** $2\sqrt{26}$ units   **58.** $\left(-4, \frac{7}{2}\right)$ **59.** $\dfrac{3\sqrt{y}}{y}$

60. $\dfrac{8 - 6\sqrt{x} + x}{8 - 2x}$ **61.** 16   **62.** 7 ft   **63. a.** 272 ft   **b.** 5.12 sec   **64.** $0 - 2i\sqrt{2}$ **65.** $0 - 3i$ **66.** $7 + 24i$ **67.** $-\frac{3}{2} + \frac{5}{2}i$

68. $(g \circ h)(x) = x^2 - 6x - 2$ **69.** one-to-one; $f^{-1}(x) = \dfrac{-x + 6}{2}$ **70.** $\log_5 \dfrac{x^4}{x + 1}$ **71.** -1 **72.** $\dfrac{1}{2}\left(\dfrac{\log 4}{\log 3} - 5\right)$; -1.8691 **73.** 22

74. $\dfrac{43}{21}$ **75.** $\dfrac{1}{2}$ **76.** 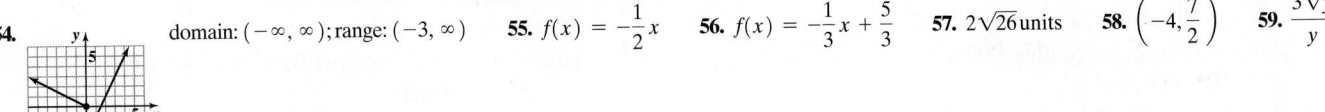   **77.** 64,805 prairie dogs   **78.** 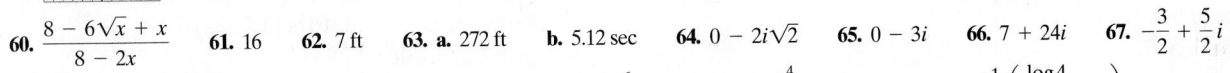   **79.**   **80.**

81. $(-5, -1), (-5, 1), (5, -1), (5, 1)$ **82.** $-\dfrac{1}{5}, \dfrac{1}{6}, -\dfrac{1}{7}, \dfrac{1}{8}, -\dfrac{1}{9}$ **83.** 155   **84.** 1   **85.** 10   **86.** $32x^5 + 80x^4y + 80x^3y^2 + 40x^2y^3 + 10xy^4 + y^5$

Index

A

$a^{-m/n}$, 606–607, 651
$a^{1/2}$, 605, 651
Absolute value, 13–14, 70, 567
Absolute value equations, 567–570, 588–589
Absolute value inequalities, 572–576, 589
Acute angles, 908
Addends, 36–37, 52
Addition
    associative property of, 62–63, 72
    commutative property of, 62, 72
    distributive property of multiplication over, 63–65, 73
    identity for, 65–66, 73
    of complex numbers, 645, 653
    of decimals, 861
    of fractions, 20–22, 70
    of functions, 723–724, 726, 780
    of polynomials, 327–329, 330, 370
    of radicals, 620–622, 652
    of rational expressions, 464–468, 472–475, 507–508
    of real numbers, 36–40, 45–46, 71–72
    problem solving using, 39–40
    words/phrases for, 31
Addition method, for solving systems of linear equations, 268–272, 302
Addition property of equality, 87–90, 91–92, 160
Addition property of inequality, 148–149, 163
Additive inverses, 40–41, 65, 66, 72
Adjacent angles, 909
Algebra of functions, 723–724, 780
Algebraic expressions
    defined, 29, 71
    evaluation of, 29–30, 46–47, 57–58, 71
    simplifying, 79–83, 159, 868–870
    translation of phrases into, 31–32
    words/phrases written as, 83, 92–93, 868–870
All real numbers, notation for, 102
Alternate interior angles, 909
$a^{m/n}$, 605–606, 651
Angles, 48, 908, 909
Applications. *See* Applications index; Problem solving
Approximation
    decimal, 597
    of common logarithms, 767
    of natural logarithms, 769
    of square roots, 597
Area, 117
Arithmetic sequences, 829–832, 842–843, 854, 855
Array of signs, 902
Associative properties of addition/multiplication, 62–63, 72
Asymptotes, 803
$ax^2 + bx + c$, 394–400, 402–406, 437
Axis of symmetry, 698–703, 716

B

Balancing scales, 87
Bar graphs, 172
Bases, exponential, 26, 54, 71, 311, 313
Binomial theorem, 851–852, 855
Binomials
    defined, 322–323, 369
    expansion of, 849–852
    factorials and, 850–851, 855
    factoring, 407–411, 437
    FOIL method for, 341–343, 370
    formula for, 851
    nth terms of, 852
    Pascal's triangle for, 849–850
    squaring, 342–343, 370
Boundary lines, 579–582
Boyle's law, 545
Brackets, in order of operations, 29, 45
Broken-line graphs, 173

C

Calculators. *See* Graphing calculators
Cartesian coordinate system, 878
Center
    of circles, 794, 795–796
    of ellipses, 800, 801
    of hyperbolas, 802, 804
Central tendency, measures of, 906–907
Change of base formula, 771
Circles
    center of, 794, 795–796
    graphing, 794–795, 797, 807, 818
    radius of, 794, 795–796
    standard form of, 794, 807
    writing equations of, 796
Coefficients, 79, 159, 322, 369
Columns, of matrices, 894
Combined variation, problem solving with, 546–548, 553
Combining like terms, 80–81, 159
Common denominators, 464–465, 507. *See also* Least common denominators (LCDs)
Common difference, 829, 854
Common factors, 18–19, 449. *See also* Greatest common factor (GCF)
Common logarithms, 767–769, 782
Common ratio, 832, 855
Common terms, 449
Commutative properties of addition/multiplication, 62, 72
Complementary angles, 48, 908
Completing the square, 663–665, 714
Complex conjugates, 646, 654

Complex fractions
  defined, 499
  ...t common denominators of, 501–502, 865
  simplifying, 499–503, 509
Complex number system, 643
Complex numbers, 643–648, 653–654
Composite functions, 724–726, 780
Composite numbers, 18
Compound inequalities
  compact form of, 561, 562
  defined, 560, 588
  intersection of two sets of, 560, 588
  solving, 152–154, 164, 560–564, 588
  union of two sets of, 563, 588
Compound interest formula, 666
Congruent triangles, 910–911
Conic sections. *See also* Parabolas
  circles, 794–797, 807, 818
  defined, 791
  ellipses, 800–802, 804, 807, 818
  hyperbolas, 802–804, 807, 818–819
Conjugates
  complex, 646, 654
  defined, 628, 652
  rationalizing denominators and numerators using,
    628–629, 630
Consecutive integers
  ...ebraic expressions with, 93, 868
  problem solving for, 110–111, 429–430
Consistent systems of linear equations, 255, 256, 257, 276, 301
Constant of proportionality, 542, 544, 546
Constant of variation, 542, 544, 546
Constants
  defined, 322
  degree of, 323
  terms, 322, 421
Conversions
  measurements, 460–461
  temperature, 117, 119–120
Coordinate plane, 174–175, 241–242, 878
Coplanar lines, 909
Corner points, 583
Corresponding angles, 909
Cramer's rule, 899, 900–901, 903–904
Cross products, 487–488, 508
Cube roots
  defined, 597, 651
  finding, 597–598
  graphing, 600–601
  of perfect cube, 614
  of real numbers, 597, 651
  simplifying, 614–615
Cubes
  ...oring sum or difference of, 410–411
  negative, 598
  perfect, 614, 630

D

Decimal approximations, 597

Decimals
  addition of, 861
  division of, 861
  in linear equations, 100–101
  multiplication of, 54, 861
  rational and irrational numbers written as, 11
  subtraction of, 861
  table of, 863
Denominators. *See also* Least common denominators (LCDs)
  common, 464–465
  defined, 17, 70
  of rational expressions, 464–468, 472–475, 507–508
  rationalizing, 626–629, 652
  unlike, 472–475, 507–508
Dependent equations, 256, 257, 276, 301
Dependent variables, 234
Descending powers, of polynomials, 322
Determinants, solving equations with, 899–904
Diameter, of circles, 794
Difference of squares, 343, 407–409
Difference, common, 829
Direct translation problems, 106–107
Direct variation, problem solving with, 542–544, 553
Discount problems, 131
Discriminants, 674–675
Distance formula, 117, 140, 492, 616–618, 651
Distance problems, 140–142, 492–494, 616–618
Distributive property
  for multiplying polynomials, 335–336
  for parenthesis removal, 81–82, 89, 159
  in combining like terms, 80
  of multiplication over addition, 63–65, 73
Dividends, 56, 358
Division
  by zero, 56
  long, 358–361
  of complex numbers, 646–647, 654
  of decimals, 861
  of fractions, 20, 55–57, 70
  of functions, 723–724, 780
  of polynomials, 357–361, 364–366, 371
  of rational expressions, 457–459, 507
  of real numbers, 55–57, 72
  problem solving using, 58
  symbol for, 56
  synthetic, 364–366, 371
  words/phrases for, 31
Divisors, 56, 360
Domain, 230, 235–236, 244, 445–446, 451

E

Elements
  of matrices, 894
  of sets, 8, 69
Elimination method
  for linear equations, 268–272, 277–280, 303
  for nonlinear equations, 810–811
Ellipses, 800–802, 804, 807, 818
Equal symbol, 8, 30, 70

Equality
  addition property of, 87–90, 91–92, 160
  logarithmic property of, 773, 782
  multiplication property of, 90–92, 160
  words/phrases for, 31
Equations. *See also* Linear equations in one variable; Linear
    equations in two variables; Quadratic equations;
    Rational equations; Systems of linear equations
  absolute value, 567–570, 588–589
  defined, 30, 71, 87
  dependent, 256, 257, 276, 301
  determinants for solving, 899–904
  equivalent, 87, 160
  false, 101
  graphs and graphing utilities for, 187, 202, 891–892
  in point-slope form, 222–223, 244, 518
  independent, 256, 257, 301
  matrices for solving, 894–898
  no solutions to, 101
  nonlinear systems of, 808–811, 819
  of circles, 796
  of exponential functions, 741–742, 773–774
  of inverse functions, 732–734
  of lines, 220–225
  of logarithmic functions, 755–756, 774–775
  ordered pairs as solution to, 177–181, 188–193, 229, 230, 242
  ordered triples as solution to, 276, 302
  percent, 129–130
  radical, 633–636
  slope-intercept form for graphing/writing,
    220–223, 244, 518
  solutions to, 30–31, 71, 87, 97–102, 160, 177–178
  steps to solve, 927–928
  translating sentences into, 32
  variables, solving for, 122
  with rational expressions, 478–482, 508, 885–886
Equivalent equations, 87, 160
Equivalent fractions, 21, 70
Estimation. *See* Approximation
Exam preparation, 5, 916–917, 918, 922
Expanding process for finding determinants, 902
Expansion of binomials, 849–852
Exponential decay, 749–750, 781
Exponential expressions
  bases of, 26, 54, 71, 311, 313
  defined, 311
  evaluating, 26–27, 28, 311–312
  multiplying, 54
  on calculators, 33
  radical, 609
  simplifying, 313, 318–319, 350–351
Exponential functions, 739–745
  defined, 739, 781
  graphing, 739–741, 744–745
  modeling exponential growth and decay, 748–750, 781
  problem solving with, 742–744, 775–776
  solving, 741–742, 773–774
Exponential growth, 748–749, 781
Exponential notation, 26, 311

Exponents
  defined, 26, 71, 311, 369
  negative, 348–350, 370, 502–503
  power rule for, 314–316, 369, 633
  product rule for, 312–314, 369
  quotient rule for, 316–318, 348, 369
  rational, 605–609, 651
  summary of rules for, 350, 607, 651
  zero as, 318, 369
Expressions. *See* Algebraic expressions; Exponential
    expressions; Radical expressions; Rational expressions
Exterior angles, 909
Extraneous solutions, 633

F

Factored form, 379, 381
Factorials, 850–851, 855
Factoring
  binomials, 407–411, 437
  by grouping method, 383–385, 402–406, 436, 437
  defined, 379, 436
  difference of two squares, 407–409
  perfect square trinomials, 398–400, 437
  polynomials, 882–884
  quadratic equations, 418–422, 438
  strategies for, 414–416, 437–438, 882–884
  sum or difference of two cubes, 410–411
  trinomials of form $ax^2 + bx + c$, 394–400, 402–406, 437
  trinomials of form $x^2 + bx + c$, 387–391, 436
Factoring out, 379, 381–383, 391, 398
Factors. *See also* Greatest common factor (GCF)
  common, 18–19, 449
  defined, 17, 70, 379
  zero as, 53
False equations, 101
Finite sequences, 825, 854
Finite series, 837
Focus
  of ellipses, 800
  of hyperbolas, 802
FOIL method, 341–343, 370
Formulas
  area, 117
  binomial, 851
  change of base, 771
  defined, 117, 161
  discount, 131
  distance, 117, 140, 492, 616–618, 651
  for problem solving, 117–123, 161
  interest, 117, 666
  mark-up, 131
  midpoint, 616, 617–618, 652
  perimeter, 117
  Pythagorean theorem, 430, 637, 912
  quadratic, 671–674, 715
  temperature conversion, 117
  variables in, 121–123, 161
  vertex, 709–710
  volume, 117

Fourth roots, 614–615
Fraction bars, 17, 27, 28
Fractions
  addition of, 20–22, 70
  complex, 499–503, 509
  defined, 17, 70
  division of, 20, 55–57, 70
  equivalent, 21, 70
  fundamental principle of, 18
  improper, 22
  in linear equations, 99–100
  in lowest terms, 17, 70
  multiplication of, 19–20, 54, 70, 455
  reciprocal of, 20, 70
  simplified, 17, 70
  subtraction of, 20–22, 70
  table of, 863
  unit, 460, 461
Functions
  addition of, 723–724, 726, 780
  algebra of, 723–724, 780
  composite, 724–726, 780
  defined, 230, 244
  division of, 723–724, 780
  domain of, 230, 235–236, 244, 445–446, 451
  exponential (*See* Exponential functions)
  graphing, 231, 233–234
  horizontal line test for, 730–731, 780
  identifying, 230–231
  inverse, 731–735, 780
  linear, 232, 517–521, 535–536, 552
  linear equations as, 232
  logarithmic (*See* Logarithmic functions)
  multiplication of, 723–724, 780
  nonlinear, 528–530, 536, 553
  notation for, 234–236, 244, 324–325, 369, 518–519, 525–528
  one-to-one, 728–729, 731, 735, 780
  piecewise-defined, 534–535, 553
  radical, 600, 651
  range of, 230, 236, 244
  relations as, 230–231
  subtraction of, 723–724, 780
  vertical line test for, 231–234, 244, 517
Fundamental principle of fractions, 18
Fundamental principle of rational expressions, 447

G

Gaussian Mirror/Lens Formula, 444
General terms, 825–827, 830, 833, 854, 855
Geometric sequences, 832–834, 844–846, 855
Geometry. *See also* Lines; Triangles
  angles, 48, 909
  area, 117
  defined, 908
  plane figures, 908–909
  polygons, 910
  volume, 117

Graphing calculators
  addition of functions on, 726
  circles on, 797
  domain of rational functions on, 451
  ellipses on, 804
  equations containing rational expressions on, 482
  exponential expressions on, 33
  expression evaluation on, 411
  features of, 193
  for linear equation solutions, 102
  for sketching graphs of multiple equations on same axes, 213–214, 530
  for solving equations, 187, 202, 891–892
  interest on, 776–777
  interpreting window settings on, 889–890
  inverse functions on, 735
  linear functions on, 521
  negative numbers on, 58–59
  order of operations on, 33
  patterns on, 225–226
  polynomial addition and subtraction on, 330
  quadratic equations on, 423–424, 667–668, 704
  radical equations on, 638
  real number operations with, 59
  scientific notation on, 354
  square viewing windows on, 891–892
  systems of linear equations on, 258
  TRACE feature, 744–745
  viewing windows of, 889–892
Graphing notation, 148
Graphs and graphing. *See also* Graphing calculators
  bar, 172
  circles, 794–795, 797, 807, 818
  ellipses, 800–802, 807, 818
  exponential functions, 739–741, 744–745
  for systems of linear equations, 254–257
  for systems of linear inequalities, 583–584, 590
  functions, 231, 233–234
  horizontal and vertical lines, 200–201
  hyperbolas, 802–804, 807, 818–819
  inverse functions, 734
  line, 173
  linear equations in two variables, 187–193, 198–200, 879–881
  linear functions, 517–518, 535–536, 552
  linear inequalities, 148, 149, 150, 151–152, 579–582, 589
  logarithmic functions, 756–758
  nonlinear functions, 528–530, 536, 553
  nonlinear inequalities, 813–816, 819
  ordered pairs on, 174–181, 241, 877–879
  paired data, 176–177
  parabolas, 698–703, 706–709, 716, 791–794, 807, 817–818
  piecewise-defined functions, 534–535, 553
  quadratic equations, 423, 698–703, 706–709, 716
  reading, 172–173, 241
  reflecting, 539–540, 553
  relations, 231

slope-intercept form and, 208, 220–221
square and cube root functions, 536–537, 600–601
vertical and horizontal shifting on, 537–539, 553
Greatest common factor (GCF)
defined, 379
factoring out, 381–383, 391, 398, 436
finding, 379–381
Grouping method
defined, 402
factoring by, 383–385, 402–406, 436, 437
Grouping symbols, 27–29

H

Half-life, 750
Half-planes, 579–582
Hooke's law, 543–544
Horizontal axis (*x*-axis), 174, 241, 539
Horizontal line test, 730–731, 780
Horizontal lines
equations of, 224
graphing, 200–201
on coordinate plane, 201
slope of, 209–210, 225, 243, 518
Horizontal shifting, 538–539, 553
Hyperbolas, 802–804, 807, 818–819
Hypotenuse, 430, 912

I

Identities
for addition, 65–66, 73
for multiplication, 65–66, 73
of linear equations, 101, 102
Identity properties, 65–66, 73
Imaginary numbers, 643, 644, 647–648
Imaginary units, 643, 644, 647–648
Improper fractions, 22
Inconsistent systems of linear equations, 255, 256, 257,
    276, 301
Independent equations, 256, 257, 301
Independent variables, 234
Index, 598, 599, 605, 837
Inequalities
absolute value, 572–576, 589
addition property of, 148–149, 163
compound, 152–154, 164, 560–564
linear (*See* Linear inequalities)
multiplication property of, 149–150, 163
nonlinear, 691–695, 715–716, 813–816, 819
on number line, 148
ordered pairs as solution to, 578
polynomial, 691–694, 715
quadratic, 691
rational, 694–695, 715–716
simple, 152
solutions of, 147–148, 929
Inequality symbols, 8–9, 70, 147, 149
Infinite sequences, 825, 845–846, 854
Infinite series, 837

Integers
consecutive, 93, 110–111, 429–430, 868
defined, 10, 69
greatest common factor of list of, 379–380, 436
negative, 10
on number lines, 10–11
positive, 10
quotient of, 11
Intercepts. *See also x*-intercepts; *y*-intercepts
defined, 242
for graphing linear equations, 198–200
identifying, 197–198
Interest problems, 117, 143–144, 666–667
Interior angles, 909
Intersecting lines, 909
Intersection of two sets, 560, 588
Interval notation, 148, 149, 153–154, 561, 562
Inverse functions, 731–735, 780
Inverse variation, problem solving with, 544–545, 553
Inverses
additive, 40–41, 65, 66, 72
multiplicative, 55, 66, 73
Irrational numbers, 11, 69, 597

J

Joint variation, problem solving with, 545–546, 553

L

Least common denominators (LCDs)
defined, 21–22, 466
of complex fractions, 501–502, 865
of rational expressions, 466–468, 472–475, 507
Legs, of right triangles, 430, 912
Like radicals, 620, 652
Like terms, 79–81, 159, 325–326
Line graphs, 173
Linear equations in one variable
decimals in, 100–101
fractions in, 99–100
infinite solutions to, 101
no solutions to, 101
on calculators, 102
solving, 97–102, 160, 177–178, 865–867
writing, 87
Linear equations in two variables
as functions, 232
defined, 187
forms of, 225
graphing, 187–193, 198–200, 879–881
identifying, 187–188
point-slope form of, 222–223, 225, 244
problem solving with, 283–291
slope-intercept form of, 208, 220–223, 225, 244
solutions of, 177–178
standard form of, 187, 225, 242
Linear equations in three variables
problem solving with, 293–294
solving systems of, 276–280, 302–303

Linear functions, 517–521
  graphing, 517–518, 535–536, 552
  linear equations as, 232
  of parallel and perpendicular lines, 520–521
  on graphing calculators, 521
  writing, 518–519, 552
Linear inequalities
  addition property of inequality and, 148–149, 163
  applications of, 154–155
  compound, 152–154, 164
  graphing solutions to, 148, 149, 150, 151–152,
    579–582, 589
  in one variable, 147, 151, 163
  in two variables, 578–582, 589
  multiplication property of inequality and, 149–150, 163
  solving, 147–155, 163–164, 582–584
  systems of, 582–584, 590
Lines
  boundary, 579–582
  coplanar, 909
  equations of, 220–225
  horizontal, 200–201, 209–210, 224, 225, 243, 518
  intersecting, 909
  parallel, 204, 210–212, 225, 243, 520–521, 909, 910
  perpendicular, 210–212, 225, 243, 520–521, 909
  slope of, 205–212, 243–244
  transversals, 909, 910
  vertical, 200–201, 209–210, 223, 225, 243
Logarithmic functions
  change of base formula for, 771
  common, 767–769, 782
  defined, 753, 756, 781, 782
  equality property of, 773, 782
  graphing, 756–758
  natural, 767, 769–770, 782
  notation for, 752–754
  power property of, 762, 764, 782
  problem solving with, 775–776
  product property of, 760–761, 762–764, 782
  properties of, 756, 760–764, 781, 782
  quotient property of, 761–762, 763, 764, 782
  solving, 755–756, 774–775
Logarithmic notation, 752–754
Long division, 358–361
Lowest terms
  of fractions, 17, 70
  of rational expressions, 447

M
Mark-up problems, 131
Mathematical statements, 8, 9–10
Mathematics class
  exam preparation in, 5, 916–917, 918, 922
  in, 5, 922, 924
  homework assignments for, 920–922
  organizational skills for, 2, 918–919, 921
  positive attitude in, 2, 916
  preparation for, 2
  study skills builders for, 915–924
  textbook use in, 3–4, 923–924
  time management in, 6
  tips for success in, 2–6
Matrices
  solving equations with, 894–898
  square, 899
Mean, 906, 907
Measurement conversions, 460–461
Measures of central tendency, 906–907
Median, 906–907
Members, of sets, 8
Midpoint
  formula for, 616, 617–618, 652
  of diameter, 794
Minor, of determinants, 902
Mixed numbers, 22–23
Mixture problems, 133–134, 162
Mode, 906–907
Money problems, 142–143
Monomials, 322–323, 357–358, 369
Multiplication. *See also* Products
  associative property of, 62–63, 72
  commutative property of, 62, 72
  distributive property for polynomials, 335–336
  distributive property over addition, 63–65, 73
  FOIL method of, 341–343, 370
  identity for, 65–66, 73
  of complex numbers, 645–646, 653
  of decimals, 861
  of exponential expressions, 54
  of fractions, 19–20, 54, 70, 455
  of functions, 723–724, 780
  of polynomials, 334–337, 341–345, 370
  of radicals, 623, 652
  of rational expressions, 455–457, 459, 507
  of real numbers, 52–54, 56, 72
  of sum and difference of two terms, 343–344, 370
  problem solving using, 58
  squaring, 342–343, 370
  symbol for, 17
  words/phrases for, 31
Multiplication property of equality, 90–92, 160
Multiplication property of inequality, 149–150, 163
Multiplicative inverses, 55, 66, 73

N
Natural logarithms, 767, 769–770, 782
Natural numbers, 8, 69
Negative exponents, 348–350, 370, 502–503
Negative infinity, 148
Negative integers, 10
Negative numbers
  cube root of, 598
  defined, 12
  on calculators, 58–59
Negative reciprocals, 211, 243
Negative square roots, 596, 598, 650

No solution, notation for, 101
Nonlinear functions, graphing, 528–530, 536, 553
Nonlinear inequalities
   graphing, 813–816, 819
   in one variable, 691–695, 715–716
   systems of, 814–816, 819
Nonlinear systems of equations, 808–811, 819
Notation. *See* Symbols/notation
*n*th roots, 598–599
*n*th terms of binomials, 852
Number lines
   defined, 70
   drawing, 8
   inequalities on, 148
   integers on, 10–11
   real numbers on, 11–12, 36–37
Numbers. *See also* Integers; Ordered pairs; Real
     numbers
   complex, 643–648, 653–654
   composite, 18
   imaginary, 643, 644, 647–648
   in scientific notation, 350–354
   irrational, 11, 69, 597
   mixed, 22–23
   natural, 8, 69
   negative, 12, 58–59, 598
   opposite of, 40–41
   positive, 12
   prime, 17–18
   rational, 11, 69, 445
   signed, 12
   unknown, 106–111, 285–286, 427–428, 490–491,
     870–871
   whole, 8, 10, 69
Numerators, 17, 70, 629–630
Numerical coefficients, 79, 159, 322, 369

O

Obtuse angles, 908
One-to-one functions, 728–729, 731, 735, 780
Opposites, 40–41, 65, 72
Order of operations
   in expression evaluation, 27–29, 45, 71
   on calculators, 33
Order property for real numbers, 13, 70
Ordered pairs
   as equation solutions, 177–181, 188–193, 229, 230, 242
   as inequality solutions, 578
   data represented as, 176–177
   defined, 174
   in system of linear equations, 253
   on graphs, 174–181, 241, 877–879
   plotting, 174, 188–193, 241
Ordered triples
   as equation solutions, 276, 302
   in system of linear equations, 276, 302
Organizational skills, 2, 918–919, 921
Origin, on coordinate plane, 174, 241

P

Paired data, 176–177. *See also* Ordered pairs
Parabolas. *See also* Quadratic equations
   defined, 423, 536
   graphing, 698–703, 706–709, 716, 791–794, 807, 817–818
   standard form of, 791, 807
   vertex of, 698–703, 706–710, 716
Parallel lines
   cut by transversals, 910
   defined, 204, 210, 909
   finding equations of, 520–521
   slope of, 210–212, 225, 243
Parentheses
   distributive property in removal of, 81–82, 89, 159
   for replacement values, 47
   in multiplication of exponential expressions, 54
   in order of operations, 27, 28, 29, 45
Partial sums, 838–839, 842–846, 855
Pascal's triangle, 849–850
Percent problems, 129–133
   discount and mark-up, 131
   increase and decrease, 132–133, 872
   solving equations for, 129–130
   strategies to solve, 162
Percent table, 863, 871
Perfect cubes, 614, 630
Perfect square trinomials, 398–400, 406, 437
Perfect squares, 597, 614
Perimeter, formula for, 117
Perpendicular lines
   defined, 210–211, 909
   finding equations of, 520–521
   slope of, 210–212, 225, 243
Phrases. *See* Words/phrases
Piecewise-defined functions, 534–535, 553
Pixels, 889
Plane figures, 908–909
Point-slope form, 222–225, 244, 518
Polygons, 910
Polynomials. *See also* Binomials; Factoring; Trinomials
   adding, 327–329, 330, 370
   combining like terms of, 325–326
   defined, 322, 369
   degree of, 323–324, 369
   descending powers of, 322
   dividing, 357–361, 364–366, 371
   evaluating, 325
   factoring strategies for, 882–884
   FOIL method for, 341–343, 370
   function notation for, 324–325, 369
   inequalities, 691–694, 715
   monomials, 322–323, 357–358, 369
   multiplying, 334–337, 341–345, 370
   operations on, 881–882
   prime, 390
   quotients of, 445
   simplifying, 325–327
   squaring, 342–343, 370

Polynomials (*continued*)
　subtracting, 328–329, 330, 370
　types of, 322–323
Positive attitude, 2, 916
Positive integers, 10
Positive numbers, 12
Positive square roots, 519, 528, 650
Power property, of logarithmic functions, 762, 764, 782
Power rule, for exponents, 314–316, 369, 633
Power, in exponential expressions, 311
Price problems, 287–288
Prime factorization, 18
Prime numbers, 17–18
Prime polynomials, 390
Principal square roots, 528, 596, 650
Problem solving. *See also* Applications index
　combined variation and, 546–548, 553
　direct variation and, 542–544, 553
　distance problems, 140–142, 492–494, 616–618
　for consecutive integer problems, 110–111
　for direct translation problems, 106–107
　for unknown numbers, 106–111, 285–286, 427–428,
　　490–491, 870–871
　formulas for, 117–123, 161
　general strategy for, 106, 128, 140, 161, 163, 284
　interest problems, 117, 143–144, 666–667
　inverse variation and, 544–545, 553
　joint variation and, 545–546, 553
　mixture problems, 133–134, 162
　money problems, 142–143
　percent problems, 129–133, 162
　proportions for, 488–490, 508–509
　radical equations and, 633–638, 653
　steps for, 870–872
　variation and, 542–548, 553
　with addition, 39–40
　with division, 58
　with exponential functions, 742–744, 775–776
　with inequalities, 154–155
　with linear equations in two variables, 283–291
　with linear equations in three variables, 293–294
　with logarithmic functions, 775–776
　with multiplication, 58
　with point-slope form, 224–225
　with Pythagorean theorem, 430–431, 636–638, 913
　with quadratic equations, 426–431, 438–439, 666–667,
　　675–677, 684–686
　with rational equations, 490–491
　with rational functions, 450–451
　with sequences, 827
　with subtraction, 47
　with systems of linear equations, 283–294, 303–304
　work problems, 491–492
Product rule
　for exponents, 312–314, 369
　for radicals, 612, 614–615
　for square roots, 612
Products. *See also* Multiplication
　cross, 487–488, 508

　defined, 17, 70
　logarithmic property of, 760–761, 762–764, 782
　power rule for, 315, 369
　special, 341–345, 370
　zero involved in, 53, 72
Proportionality, constant of, 542, 544, 546
Proportions
　defined, 486, 508
　for problem solving, 488–490, 508–509
　rational equations and, 487–488
　solving, 486–488
Pure imaginary numbers, 644
Pythagorean theorem
　distance formula and, 616–618
　formula for, 430, 637, 912
　problem solving with, 430–431, 636–638, 913

Q

Quadrants, on coordinate plane, 174
Quadratic equations. *See also* Parabolas
　completing the square to solve, 663–665, 714
　defined, 417, 661
　discriminant and, 674–675
　factoring to solve, 418–422, 438
　graphs of, 423, 698–703, 706–709, 716
　in two variables, 423
　minimum and maximum values, 710–711
　on graphing calculators, 423–424, 667–668, 704
　problem solving with, 426–431, 438–439, 666–667,
　　675–677, 684–686
　quadratic formula to solve, 671–674, 715
　solving, 661–665, 671–674, 681–684, 714–715, 865–867
　square root property to solve, 661–662, 671, 714
　standard form of, 417, 671, 672
　with degree greater than two, 421–422
　x-intercepts of graph of, 423
　zero factor property to solve, 418–420
Quadratic formula, 671–674, 715
Quadratic inequalities, 691
Quotient rule
　for exponents, 316–318, 348, 369
　for radicals, 613, 614, 615–616
　for square roots, 613
Quotients
　logarithmic property of, 761–762, 763, 764, 782
　of integers, 11
　of polynomials, 445
　of real numbers, 55–56, 72
　power rule for, 316, 369
　zero involved in, 56, 72

R

Radical equations
　on graphing calculators, 638
　power rule for, 633
　problem solving and, 633–638, 653
　solving, 633–636
Radical expressions
　defined, 596

rationalizing denominators and numerators of, 626–630, 652–653

simplifying, 609, 612–616, 651–652

Radical functions, 600, 651

Radical sign, 596

Radicals. *See also* Roots; Square roots

adding, 620–622, 652

like, 620, 652

multiplying, 623, 652

product rule for, 612, 614–615

quotient rule for, 613, 614, 615–616

simplifying, 599–600, 609, 612–616, 651–652

subtracting, 620–622, 652

variables in, 599–600, 615–616

Radicands

defined, 596

perfect fourth powers of, 614

perfect square, 597

Radius, of circles, 794, 795–796

Range, 230, 236, 244

Rate of change, 212–213, 243

Rational equations

problem solving with, 490–491, 508–509

proportions and, 487–488

solving number problems modeled by, 490–491

Rational exponents, 605–609, 651

Rational expressions

adding, 464–468, 472–475, 507–508

common denominator of, 464–465, 507

complex, 499–503

defined, 445, 506

dividing, 457–459, 507

equations containing, 478–482, 508, 885–886

evaluating, 445

fundamental principle of, 447

in lowest terms, 447

least common denominators of, 466–468, 472–475, 507

multiplying, 455–457, 459, 507

operations on, 885–886

simplifying, 446–450, 506

subtracting, 464–468, 472–475, 507–508

undefined, 445

unlike denominators of, 472–475, 507–508

writing equivalent forms of, 450, 468–469

Rational functions

defined, 445, 506

domain of, 445–446, 451

problem solving with, 450–451

Rational inequalities, 694–695, 715–716

Rational numbers, 11, 69, 445

Rationalizing denominators, 626–629, 652

Rationalizing numerators, 629–630, 653

Ratios, 486, 508, 832

Real numbers

absolute value of, 13–14, 70, 567

addition of, 36–40, 45–46, 71–72

cube root of, 597, 651

defined, 11, 69

division of, 55–57, 72

multiplication of, 52–54, 56, 72

negative, 12

on calculators, 59

on number lines, 11–12, 36–37

order property for, 13, 70

positive, 12

properties of, 62–66, 72–73

subtraction of, 44–47, 72

Reciprocals. *See also* Multiplicative inverses

defined, 20, 66, 70, 73

finding, 55

inversely proportional, 544

negative, 211, 243

Rectangular coordinate system, 174–175, 241–242, 878

Reflecting graphs, 539–540, 553

Relations

as functions, 230–231

defined, 230, 244

graphs of, 231

Remainder theorem, 366, 371

Right angles, 908

Right triangles, 430, 912–913

Rise, 205

Roots. *See also* Cube roots; Radicals; Square roots

fourth, 614–615

*n*th, 598–599

Rounding. *See* Approximation

Row operations, 894

Row, of matrices, 894

Run, 205

S

Scales, balancing, 87

Scatter diagrams, 176–177

Scientific calculators. *See* Graphing calculators

Scientific notation

converting to standard form, 352–353

defined, 351

on calculators, 354

performing operations with, 353

writing numbers in, 351–352, 371

Sequences. *See also* Series

arithmetic, 829–832, 842–843, 854, 855

defined, 825

Fibonacci, 824, 829

finding general terms of, 826–827

finite, 825, 854

geometric, 832–834, 844–846, 855

infinite, 825, 845–846, 854

notation for, 825, 854

problem solving with, 827

writing terms of, 825–826

Series. *See also* Sequences

defined, 837, 855

finite, 837

infinite, 837

partial sums of, 838–839, 842–846, 855

summation notation for identifying, 837–838

Sets
    defined, 8, 69
    identifying, 10–12
    of integers, 10, 69
    of irrational numbers, 11, 69
    of natural numbers, 8, 69
    of rational numbers, 11, 69
    of real numbers, 11, 69
    of whole numbers, 8, 69
Sigma, 837
Signed numbers, 12
Similar triangles, 489–490, 911–912
Simple inequalities, 152
Simple interest formula, 117, 143, 666
Simplification
    of algebraic expressions, 79–83, 159, 868–870
    of complex fractions, 499–503, 509
    of cube roots, 614–615
    of exponential expressions, 313, 318–319, 350–351
    of fractions, 17, 70
    of negative exponents, 348–350, 502–503
    of polynomials, 325–327
    of radicals, 599–600, 609, 612–616, 651–652
    of rational exponent expressions, 607–608
    of rational expressions, 446–450, 506
    review of, 925–926
Slope
    as rate of change, 212–213, 243
    defined, 205
    of horizontal and vertical lines, 209–210, 225, 243, 518
    of parallel and perpendicular lines, 210–212, 225, 243
    overview, 210
    point-slope form, 222–225, 244, 518
    strategies for finding, 205–212
    undefined, 209, 210
Slope-intercept form
    defined, 208, 220
    of linear equations in two variables, 208, 220–223, 225, 244
    to graph equations, 220–221
    to write equations, 221–223, 244, 518
Solutions
    extraneous, 633
    of absolute value equations, 567–570, 588–589
    of absolute value inequalities, 572–576, 589
    of compound inequalities, 152–154, 164, 560–564, 588
    of equations, 30–31, 71, 87, 97–102, 160, 177–178
    of exponential function equations, 741–742, 773–774
    of inequalities, 147–148, 929
    of linear equations in one variable, 97–102, 160, 177–178, 865–867
    of linear equations in two variables, 177–178
    of linear equations in three variables, 276–280, 302–303
    of linear inequalities, 147–155, 163–164, 582–584
    of logarithmic function equations, 755–756, 774–775
    of nonlinear systems of equations, 808–811, 819
    of quadratic equations, 661–665, 671–674, 681–684, 714–715, 865–867

of radical equations, 633–636
of systems of linear equations, 253, 301
ordered pairs as, 177–181, 188–193, 229, 230, 242, 578
ordered triples as, 276, 302
Special products
    defined, 342
    FOIL method for, 341–343, 370
    squaring binomials, 342–343, 370
    use of, 344–345
Square matrices, 899
Square root property, 661–662, 671, 714
Square roots. *See also* Radicals
    approximating, 597
    defined, 596
    finding, 528, 596–597
    graphing, 536–537, 600–601
    negative, 596, 598, 650
    positive, 528, 619, 650
    product rule for, 612
    quotient rule for, 613
    symbol for, 596
Squares
    completing the square, 663–665, 714
    difference of, 343, 407–409
    factoring difference of, 407–409
    of binomials, 342–343, 370
    perfect, 597, 614
Standard form
    of circles, 794, 807
    of ellipses, 800, 807
    of hyperbolas, 803, 807
    of linear equations, 187, 225, 242, 879
    of parabolas, 791, 807
    of quadratic equations, 417, 671, 672
    of quadratic inequalities, 691
    scientific notation converted to, 352–353
Standard window, in graphing utilities, 193
Straight angles, 908
Study groups, 2, 5
Study skills builders, 915–924
Substitution method
    for linear equations, 261–266, 277, 280, 302
    for nonlinear equations, 808–810
Subtraction
    of complex numbers, 645, 653
    of decimals, 861
    of fractions, 20–22, 70
    of functions, 723–724, 780
    of mixed numbers, 23
    of polynomials, 328–329, 330, 370
    of radicals, 620–622, 652
    of rational expressions, 464–468, 472–475, 507–508
    of real numbers, 44–47, 72
    problem solving using, 47
    words/phrases for, 31, 32
Sum and difference of two terms, multiplying, 343–344, 370
Sum or difference of two cubes, factoring, 410–411
Summation notation, 837–838, 855

Supplementary angles, 48, 908
Symbols/notations. *See also* Scientific notation; Variables; Words/phrases
  additive inverse, 41, 72
  all real numbers, 102
  angles, 909
  division, 56
  equal symbol, 8, 30, 70
  exponential, 26, 311
  for multiplication, 17
  for polynomials, 324–325, 369
  for sequences, 825, 854
  function, 234–236, 244, 324–325, 369, 518–519, 525–528
  graphing, 148
  grouping, 27–29
  in order of operations, 27–29
  inequality symbols, 8–9, 70, 147, 149
  interval, 148, 149, 153–154, 561, 562
  logarithmic, 752–754
  negative infinity, 148
  no solution, 101
  nth root, 598–599
  radical sign, 596
  summation, 837–838, 855
Symmetry, axis of, 698–703, 716
Synthetic division, 364–366, 371
Systems of linear equations, 253–294
  addition method for solving, 268–272, 302
  consistent, 255, 256, 257, 276, 301
  defined, 253
  determinants for solving, 900–902, 903–904
  elimination method for solving, 268–272, 277–280, 303
  graphing to solve, 254–257
  in three variables, 276–280, 302–303
  inconsistent, 255, 256, 257, 276, 301
  nongraph solutions to, 257–258
  on graphing calculators, 258
  problem solving with, 283–294, 303–304
  solutions of, 253, 301
  substitution method for solving, 261–266, 277, 280, 302
Systems of linear inequalities, 582–584, 590
Systems of nonlinear equations, 808–811, 819
Systems of nonlinear inequalities, 814–816, 819

T

Table of values, 179–181
Temperature conversion, 117, 119–120
Terms
  common, 449
  constant, 322, 421
  defined, 79, 322, 369
  degree of, 323–324, 369
  general, 825–827, 830, 833, 854, 855
  greatest common factor of list of, 380–381, 436
  like, 79–81, 159, 325–326
  nth terms of binomials, 852
  of infinite geometric sequences, 845
  of sequences, 825–827
  unlike, 79–80, 159, 325
Test preparation. *See* Exam preparation
Time management, 6
Transversals, 909, 910
Triangles
  congruent, 910–911
  defined, 910
  dimensions of, 430–431
  Pascal's, 849–850
  right, 430, 912–913
  similar, 489–490, 911–912
Trinomials
  defined, 322–323, 369
  of form $ax^2 + bx + c$, 394–400, 402–406, 437
  of form $x^2 + bx + c$, 387–391, 436, 663–665
  perfect square, 398–400, 406, 437
Tutoring services, 3, 5

U

Undefined expressions, 445
Undefined slope, 209, 210
Union of two sets, 563, 588
Unit fractions, 460, 461
Unknown numbers, 106–111, 285–286, 427–428, 490–491, 870–871
Unlike denominators, 472–475, 507–508
Unlike terms, 79–80, 159, 325

V

Variables
  defined, 29, 71
  dependent, 234
  in formulas, 121–123, 161
  independent, 234
  radicals containing, 599–600, 615–616
Variation
  combined, 546–548, 553
  constant of, 542, 544, 546
  direct, 542–544, 553
  inverse, 544–545, 553
  joint, 545–546, 553
  problem solving and, 542–548, 553
Vertex, 698–703, 706–710, 716
Vertex formula, 709–710
Vertical angles, 909
Vertical axis (y-axis), 174, 241, 698, 699, 702
Vertical format, for multiplying polynomials, 337
Vertical line test, 231–234, 244, 517
Vertical lines
  equations of, 223
  graphing, 200–201
  on coordinate plane, 201
  slope of, 209–210, 225, 243
Vertical shifting, 537–538, 553
Viewing windows, 889–892
Volume, 117

W

Whole numbers, 8, 10, 69
Window, in graphing utilities, 193
Words/phrases. *See also* Symbols/notations
    translating into algebraic expressions, 31–32
    writing as algebraic expressions, 83, 92–93, 868–870
Work problems, 491–492

X

x-axis, 174, 241, 539
x-coordinates, 174, 709, 878
x-intercepts
    defined, 197, 242, 880
    finding and plotting, 198–200, 243
    of quadratic equation graphs, 423, 706–707
$x^2 + bx + c$, 387–391, 436, 663–665

Y

y-axis, 174, 241, 698, 699, 702
y-coordinates, 174, 878
y-intercepts
    defined, 197, 242, 880
    finding and plotting, 198–200, 243
    of quadratic equation graphs, 706–707
    slope-intercept form and, 208

Z

Zero
    as exponent, 318, 369
    as identity element for addition, 65
    products involving, 53, 72
    quotients involving, 56, 72
Zero factor property, 418–420, 661

Photo Credits

COVER Sgursolzlu/Fotolia

Chapter 1

Page 1 Rachel Youdelman/Pearson Education Inc.
Page 6 Maciej Noskowski/Veta/Getty Images
Page 10 Axily/Fotolia
Page 16 Astrosystem/Fotolia
Page 29 Speedfighter/Fotolia
Page 40 Sunsinger/Fotolia
Page 42 (top) Em_ek/Fotolia
Page 42 (bottom) Alce/Fotolia
Page 50 (top) Vladimir Kondrachov/Fotolia
Page 50 (bottom) Paylessimages/Fotolia
Page 58 Monkey Business/Fotolia

Chapter 2

Page 78 Umberto Shtanzman/Shutterstock
Page 95 EpicStockMedia/Shutterstock
Page 96 (top) Apdesign/Shutterstock
Page 96 (bottom) Galyna Andrushko/123RF
Page 97 Auremar/123RF
Page 109 Michael Reynolds/Epa european pressphoto agency b.v./Alamy
Page 110 Jason Maehl/Shutterstock
Page 111 EdBockStock/Shutterstock
Page 115 06photo/Shutterstock
Page 116 Ollirg/Shutterstock
Page 118 Michele Cornelius/123RF
Page 126 Hilma Anderson/123RF
Page 127 Elayn Martin Gay
Page 128 (left) Feathercollector/Shutterstock
Page 128 (right) Kwanchaichaiudom/Fotolia
Page 132 Tyler Olson/Shutterstock
Page 133 (top) Photoncatcher36/Fotolia
Page 133 (bottom) Deklofenak/Shutterstock
Page 138 (top) Gillmar/Shutterstock
Page 138 (bottom) Ayzek/Shutterstock
Page 155 Vstock/UpperCut Images/Getty Images
Page 165 Orhan Cam/Shutterstock

Chapter 3

Page 172 Merzzie/Shutterstock
Page 177 Alin Brotea/Shutterstock
Page 180 Dotshock/Shutterstock

Page 183 (top) Snvv/Fotolia
Page 183 (bottom) Artem Kursin/Shutterstock
Page 192 Michaeljung/Fotolia
Page 195 (top) Dmitry Kalinovsky/123RF
Page 195 (bottom left) Shock/Fotolia
Page 195 (bottom right) BlueSkyImages/Fotolia
Page 204 BlueSkyImage/Shutterstock
Page 213 George Burba/123RF
Page 219 (top) Poznyakov/Shutterstock
Page 219 (bottom) AdStock RF/Shutterstock
Page 220 (top) Rscreativeworks/Shutterstock
Page 220 (bottom) Raluca Tudor/Dreamstime LLC
Page 224 GoodMood Photo/Fotolia
Page 228 (left) Marc Xavier/Fotolia
Page 228 (right) ArtWell/Shutterstock
Page 229 Georgiy Pashin/Fotolia
Page 245 Naypong/Fotolia

Chapter 4

Page 252 Janet Foster/Radius Images/Getty Images
Page 268 Tyler Olson/Fotolia
Page 274 Monkey Business Images/Shutterstock
Page 275 Temistocle Lucarelli/Fotolia
Page 287 Yanchenko/Fotolia
Page 295 (top) Urbanhearts/Fotolia
Page 295 (bottom) ElinaManninen/Fotolia
Page 296 Binkski/Shutterstock
Page 298 Michaeljung/Fotolia
Page 299 San Antonio Express-News/Zuma Press/Alamy
Page 300 Anthony Nesmith/CSM/Alamy
Page 307 SF photo/Shutterstock

Chapter 5

Page 310 Victoria/Fotolia
Page 355 (top) Henry Czauderna/Fotolia
Page 355 (bottom) Elenarts/Fotolia
Page 372 Jabiru/Fotolia
Page 373 Byron Moore/Fotolia

Chapter 6

Page 378 WavebreakMediaMicro/Fotolia
Page 413 Kojihirano/Fotolia